The Wildlife Techniques Manual

The Wildlife Techniques Manual | Research

7th edition | volume 1

Edited by NOVA J. SILVY

Published in Affiliation with The Wildlife Society

THE JOHNS HOPKINS UNIVERSITY PRESS | *Baltimore*

© 2012 The Johns Hopkins University Press
All rights reserved. Published 2012
Printed in the United States of America on acid-free paper
9 8 7 6 5 4 3 2 1

The Johns Hopkins University Press
2715 North Charles Street
Baltimore, Maryland 21218-4363
www.press.jhu.edu

Library of Congress Cataloging-in-Publication Data

The wildlife techniques manual / edited by Nova J. Silvy.—7th ed.
    p.   cm.
  Rev. ed. of: Techniques for wildlife investigations and management / edited by Clait E. Braun. 6th ed. Bethesda, Md. : Wildlife Society, 2005.
  Includes bibliographical references and index.
  ISBN-13: 978-1-4214-0159-1 (hardcover : alk. paper)
  ISBN-10: 1-4214-0159-2 (hardcover : alk. paper)
  1. Wildlife management. 2. Wildlife research—Technique.
I. Silvy, Nova J. II. Techniques for wildlife investigations and management.
SK355.R47   2012
639.9—dc23   2011013720

A catalog record for this book is available from the British Library.

Special discounts are available for bulk purchases of this book. For more information, please contact Special Sales at 410-516-6936 or specialsales@press.jhu.edu.

The Johns Hopkins University Press uses environmentally friendly book materials, including recycled text paper that is composed of at least 30 percent post-consumer waste, whenever possible.

CONTENTS

List of Contributors vii
Preface xi
Acknowledgments xiii

Design and Analytical Techniques

1 Research and Experimental Design 1
   EDWARD O. GARTON, JON S. HORNE,
   JOCELYN L. AYCRIGG, AND JOHN T. RATTI

2 Management and Analysis of Wildlife Biology Data 41
   BRET A. COLLIER AND THOMAS W. SCHWERTNER

Capture and Handling Techniques

3 Capturing and Handling Wild Animals 64
   SANFORD D. SCHEMNITZ, GORDON R. BATCHELLER,
   MATTHEW J. LOVALLO, H. BRYANT WHITE, AND
   MICHAEL W. FALL

4 Wildlife Chemical Immobilization 118
   TERRY J. KREEGER

5 Use of Dogs in Wildlife Research and Management 140
   DAVID K. DAHLGREN, R. DWAYNE ELMORE,
   DEBORAH A. SMITH, AIMEE HURT, EDWARD B. ARNETT,
   AND JOHN W. CONNELLY

6 Identifying and Handling Contaminant-Related
   Wildlife Mortality or Morbidity   154
   STEVEN R. SHEFFIELD, JOSEPH P. SULLIVAN,
   AND ELWOOD F. HILL

7 Wildlife Health and Disease: Surveillance,
   Investigation, and Management   181
   MARKUS J. PETERSON AND PAMELA J. FERRO

Identification and Marking Techniques

8 Criteria for Determining Sex and Age of Birds
   and Mammals   207
   EDDIE K. LYONS, MICHAEL A. SCHROEDER,
   AND LESLIE A. ROBB

9 Techniques for Marking Wildlife   230
   NOVA J. SILVY, ROEL R. LOPEZ,
   AND MARKUS J. PETERSON

10 Wildlife Radiotelemetry and Remote
   Monitoring   258
   JOSHUA J. MILLSPAUGH, DYLAN C. KESLER,
   ROLAND W. KAYS, ROBERT A. GITZEN,
   JOHN H. SCHULZ, CHRISTOPHER T. ROTA,
   CATHERINE M. BODINOF, JERROLD L. BELANT,
   AND BARBARA J. KELLER

Measuring Animal Abundance

11 Estimating Animal Abundance   284
   BRIAN L. PIERCE, ROEL R. LOPEZ,
   AND NOVA J. SILVY

12 Use of Remote Cameras in Wildlife
   Ecology   311
   SHAWN L. LOCKE, ISRAEL D. PARKER,
   AND ROEL R. LOPEZ

13 Radar Techniques for Wildlife Research   319
   RONALD P. LARKIN AND ROBERT H. DIEHL

14 Invertebrate Sampling Methods for Use
   in Wildlife Studies   336
   THERESE A. CATANACH

15 Population Analysis in Wildlife Biology   349
   STEPHEN J. DINSMORE AND
   DOUGLAS H. JOHNSON

Measuring Wildlife Habitat

16 Vegetation Sampling and Measurement   381
   KENNETH F. HIGGINS, KURT J. JENKINS,
   GARY K. CLAMBEY, DANIEL W. URESK,
   DAVID E. NAUGLE, ROBERT W. KLAVER,
   JACK E. NORLAND, KENT C. JENSEN,
   AND WILLIAM T. BARKER

17 Modeling Vertebrate Use of Terrestrial
   Resources   410
   LYMAN L. MCDONALD, WALLACE P. ERICKSON,
   MARK S. BOYCE, AND J. RICHARD ALLDREDGE

18 Application of Spatial Technologies in
   Wildlife Biology   429
   THOMAS A. O'NEIL, PETE BETTINGER,
   BRUCE G. MARCOT, WARREN B. COHEN,
   ORIANE TAFT, RICHARD ASH, HOWARD BRUNER,
   CORY LANGHOFF, JENNIFER A. CARLINO,
   VIVIAN HUTCHISON, ROBERT E. KENNEDY,
   AND ZHIQIANG YANG

Research on Individual Animals

19 Animal Behavior   462
   JESSICA R. YOUNG

20 Analysis of Radiotelemetry Data   480
   JOSHUA J. MILLSPAUGH, ROBERT A. GITZEN,
   JERROLD L. BELANT, ROLAND W. KAYS,
   BARBARA J. KELLER, DYLAN C. KESLER,
   CHRISTOPHER T. ROTA, JOHN H. SCHULZ,
   AND CATHERINE M. BODINOF

21 Reproduction and Hormones   502
   JOHN D. HARDER

22 Conservation Genetics and Molecular Ecology
   in Wildlife Management   526
   SARA J. OYLER-MCCANCE AND PAUL L. LEBERG

Literature Cited   547
Index   653

CONTRIBUTORS

J. Richard Alldredge
Department of Statistics
Washington State University
Pullman, WA 99164 USA

Edward B. Arnett
Bat Conservation International
P.O. Box 162603
Austin, TX 78716 USA

Richard Ash
2006 NW Brownly Heights Drive
Corvallis, OR 97330 USA

Jocelyn L. Aycrigg
National Gap Analysis Program
Department of Fish and Wildlife Resources
University of Idaho
Moscow, ID 83844-4408 USA

William T. Barker
3419 Par Street, NE
Fargo, ND 58102 USA

Gordon R. Batcheller
Division of Fish, Wildlife & Marine Resources
New York State
625 Broadway
Albany, NY 12233-4754 USA

viii   CONTRIBUTORS

Jerrold L. Belant
Forest and Wildlife Research
  Center
Mississippi State University
Mississippi State, MS 39762 USA

Pete Bettinger
Warnell School of Forest
  Resources
University of Georgia
Athens, GA 30602 USA

Catherine M. Bodinof
Department of Fisheries and
  Wildlife Sciences
University of Missouri
Columbia, MO 65211 USA

Mark S. Boyce
Department of Biological
  Sciences
University of Alberta
Edmonton, AB T6G 2E9 Canada

Howard Bruner
Western Ecology Division
Dynamac Corporation
200 SW 35th Street
Corvallis, OR 97333 USA

Jennifer A. Carlino
Center for Biological
  Informatics
U.S. Geological Survey
Denver Federal Center
P.O. Box 25046, MS 302
Denver, CO 80225 USA

Therese A. Catanach
Program in Ecology, Evolution,
  and Conservation
University of Illinois
Urbana-Champaign
Urbana, IL 61801 USA

Gary K. Clambey
Department of Biological
  Sciences
North Dakota State University
Fargo, ND 58108-6050 USA

Warren B. Cohen
Pacific Northwest Research Station
U.S. Forest Service
Department of Agriculture
3200 SW Jefferson Way
Corvallis, OR 97331 USA

Bret A. Collier
Institute of Renewable Natural
  Resources
Texas A&M University
College Station, TX 77846-2260
  USA

John W. Connelly
Idaho Department of Fish and
  Game
1345 Barton Road
Pocatello, ID 83204 USA

David K. Dahlgren
Kansas Department of Wildlife
  and Parks
Office: Region 1
1426 U.S. 183 Bypass
Hays, KS 67601-0338 USA

Robert H. Diehl
Department of Biological Sciences
University of Southern Mississippi
118 College Drive [5018]
Hattiesburg, MS 39406-0001 USA

Stephen J. Dinsmore
Department of Natural Resource
  Ecology & Management
Iowa State University
339 Science II
Ames, IA 50011 USA

R. Dwayne Elmore
Department of Natural Resource
  Ecology and Management
008 C Ag Hall
Oklahoma State University
Stillwater, OK 74078-6013 USA

Wallace P. Erikson
Western EcoSystems Technology, Inc.
2003 Central Avenue
Cheyenne, WY 82001 USA

Michael W. Fall
Wildlife Research Center
U.S. Department of Agriculture
4101 LaPorte Avenue
Ft. Collins, CO 80521-2154 USA

Pamela J. Ferro
247-6E Cloverhurst Avenue
Athens, GA 30695 USA

Edward O. Garton
Fish and Wildlife Resources &
  Statistics Departments
University of Idaho
Moscow, ID 83844-1136 USA

Robert A. Gitzen
Department of Fisheries and
  Wildlife Sciences
University of Missouri
Columbia, MO 65211 USA

John D. Harder
Department of Evolution, Ecology
  & Organismal Biology
392 Aronoff Laboratory
318 West 12th Avenue
Ohio State University
Columbus, OH 43210 USA

Kenneth F. Higgins
Box 2140B
Wildlife and Fisheries Sciences
South Dakota State University
Brookings, SD 57007-1696 USA

Elwood F. Hill
P.O. Box 1615
Gardnerville, NV 89410 USA

Jon S. Horne
Department of Fish and Wildlife
  Resources
University of Idaho
Moscow, ID 83844-1136 USA

Aimee Hurt
Working Dogs for Conservation
609 Phillips Street
Missoula, MT 59802 USA

Vivian Hutchison
Center for Biological Informatics
U.S. Geological Survey
Denver Federal Center
P.O. Box 25046, MS 302
Denver, CO 80225 USA

Kurt J. Jenkins
USGS Forest and Rangeland
  Ecosystem Science Center
600 E. Park Avenue
Port Angeles, WA 98362 USA

Kent C. Jensen
Department of Wildlife and
  Fisheries Sciences
SPB 138D, Box 2140B
South Dakota State University
Brookings, SD 57007 USA

Douglas H. Johnson
U.S. Geological Survey
Northern Prairie Wildlife
  Research Center
204 Hodson Hall
University of Minnesota
Saint Paul, MN 55108 USA

Roland W. Kays
New York State Museum
Albany, NY 12230 USA

Barbara J. Keller
Department of Fisheries and
  Wildlife Sciences
University of Missouri
Columbia, MO 65211 USA

Robert E. Kennedy
Department of Forest Ecosystems
  and Society
Oregon State University
321 Richardson Hall
Corvallis, OR 97331 USA

Dylan C. Kesler
Department of Fisheries and
  Wildlife Sciences
University of Missouri
Columbia, MO 65211 USA

Robert W. Klaver
U.S. Geological Survey
Earth Resources Observation and
  Science Center (EROS)
47914 252nd Street
Sioux Falls, SD 57198 USA

Terry J. Kreeger
Veterinary Services Branch
Wyoming Game and Fish
  Department
2362 Highway 34
Wheatland, WY 82201 USA

Cory Langhoff
Northwest Habitat Institute
P.O. Box 855
Corvallis, OR 97339 USA

Ronald P. Larkin
Illinois Natural History Survey
1816 S. Oak Street
Champaign, IL 61820 USA

Paul L. Leberg
P.O. Box 42451
Lafayette, LA 70504 USA

Shawn L. Locke
Department of Wildlife and
  Fisheries Sciences
Texas A&M University
College Station, TX 77843-2258 USA

Roel R. Lopez
Texas A&M Institute of Renewable
  Natural Resources
2632 Broadway, Suite 301 South
San Antonio, TX 78215 USA

Matthew J. Lovallo
Pennsylvania Game Commission
752 Lower George's Valley Road
Spring Mills, PA 16875 USA

Eddie K. Lyons
Harold and Pearl Dripps
  Department of Agricultural
  Sciences
McNeese State University
Lake Charles, LA 70609 USA

Bruce G. Marcot
Pacific Northwest Research Station
U.S. Forest Service
Department of Agriculture
620 SW Main St.
Portland OR 97205 USA

Lyman L. McDonald
Western EcoSystems Technology, Inc.
2003 Central Avenue
Cheyenne, WY 82001 USA

Joshua J. Millspaugh
Department of Fisheries and
  Wildlife Sciences
University of Missouri
Columbia, MO 65211 USA

David E. Naugle
Wildlife Biology Program
FOR 309
32 Campus Drive
University of Montana
Missoula, MT 59812 USA

Jack E. Norland
School of Natural Resource
  Sciences
North Dakota State University
Fargo, ND 58108 USA

Thomas A. O'Neil
Northwest Habitat Institute
P.O. Box 855
Corvallis, OR 97339 USA

Sara J. Oyler-McCance
2150 Centre Ave, Building C
Fort Collins, CO 80526 USA

Israel D. Parker
Department of Wildlife and
  Fisheries Sciences
Texas A&M University
College Station, TX 77843-2258 USA

Markus J. Peterson
Department of Wildlife and
  Fisheries Sciences
Texas A&M University
College Station, TX 77843-2258 USA

Brian L. Pierce
Department of Wildlife and
 Fisheries Sciences
Texas A&M University
College Station, TX 77843-2258 USA

John T. Ratti
P.O. Box 361
New Meadows, ID 83654 USA

Leslie A. Robb
P.O. Box 1077
Bridgeport, WA 98813 USA

Christopher T. Rota
Department of Fisheries and
 Wildlife Sciences
University of Missouri
Columbia, MO 65211 USA

Sanford D. Schemnitz
Department of Fishery and
 Wildlife Sciences
New Mexico State University
Las Cruces, NM 88003 USA

Michael A. Schroeder
Washington Department of Fish
 and Wildlife
P.O. Box 1077
Bridgeport, WA 98813 USA

John H. Schulz
Resource Science Division
Missouri Department of
 Conservation
Columbia, MO 65201 USA

Thomas W. Schwertner
BIO-WEST, Inc.
1063 West 1400 North
Logan, UT 84321-2291 USA

Steven R. Sheffield
Department of Natural Sciences
Bowie State University
Bowie, MD 20715 USA

Nova J. Silvy
Department of Wildlife and
 Fisheries Sciences
Texas A&M University
College Station, TX 77843-2258 USA

Deborah A. Smith
Working Dogs for Conservation
 Foundation
52 Eustis Road
Three Forks, MT 59752 USA

Joseph P. Sullivan
Ardea Consulting
10 1st Street
Woodland, CA 95695 USA

Oriane Taft
Watershed Sciences, Inc.
517 SW 2nd St., Suite 400
Corvallis, OR 97333 USA

Daniel W. Uresk
U.S. Forest Service
Department of Agriculture
231 East St. Joseph Street
Rapid City, SD 57701 USA

H. Bryant White
Furbearer Research Coordinator
Association of Fish and Wildlife
 Agencies
Resource Science Center
Missouri Department of
 Conservation
1110 S. College Avenue
Columbia, MO 65201 USA

Zhiqiang Yang
Oregon State University
321 Richardson Hall
Corvallis, OR 97331 USA

Jessica R. Young
Western State College of Colorado
Gunnison, CO 81231 USA

PREFACE

THIS SEVENTH EDITION of The Wildlife Society's Techniques Manual is unique in several ways. First, it is now a 2-volume set; Volume 1 concerns research techniques, and Volume 2 concerns management techniques. Second, it is more user friendly, because chapter authors have bolded keywords and phrases in each chapter at the request of student reviewers, who wished to locate these words and phrases quickly. Third, the authors have incorporated more information into tables, with accompanying literature citations. Fourth, 7 new chapters have been added, 4 on research techniques and 3 on management techniques. Two revised chapters, "Use of Dogs in Wildlife Research and Management" and "Invertebrate Sampling Methods for Use in Wildlife Studies," have been resurrected from earlier editions of the *Manual*. Fifth, the layout and cover design have been dramatically altered, resulting in a more dynamic format. Lastly, the name of the two-volume set has been changed to reflect the name by which users have long referred to it, *The Wildlife Techniques Manual*. For both volumes, the high standards that have been the hallmark of the *Manual* were maintained and strengthened. The chapter authors are to be commended for their brilliant work.

The organization for this edition has changed considerably with the separation into research and management volumes. In Volume 1, a section on design and analytical techniques describes research design and determination of proper analytic methods (prior to conducting the research). The *Manual* then proceeds to explain the methods and considerations for capturing and handling wild animals during the study. This section is followed by information on identification and marking of captured animals. Finally, Volume 1 addresses measurements of wildlife abundance and habitat, and research on individual animals.

Volume 2 begins with a section on management perspectives, including human considerations. Public outreach is described in a context that encourages engagement prior to initiation of management. An adaptive management approach is described in detail and discussed as a cornerstone of natural resource management. These chapters are followed by a section on managing landscapes and wildlife populations.

The decision to reorganize the material and develop a 2-volume set was made after major university users of the *Manual* were surveyed to determine what chapters they were using in university courses and for what type of courses. Two major use areas were identified from these surveys: (1) courses in wildlife research techniques and (2) courses in wildlife habitat management techniques. Respondents indicated that most wildlife students and professionals would read most, and possibly all, chapters at some point in their education and career and the division between research and management was both practical and logical.

Of the 37 chapters in this edition of the *Manual,* 16 chapters have senior authors that were not participants in prior editions. Overall, 119 individuals (some involved with more than 1 chapter) provided expertise for the 37 chapters. Without the professionalism, persistence, and dedication of all the chapter authors, the volumes would not have been completed. Beyond those who wrote the chapters, special thanks are owed to a number of people. Ruxandra Giura, former program manager for online services (The Wildlife Society [TWS]) located the sixth edition manuscripts, from which revisions could be initiated. Katherine Unger, developmental editor / science writer (TWS) found photographs for the volume covers. Clait E. Braun was especially helpful in more ways than can be recounted. Numerous peer reviewers were engaged for the 37 chapters, and all of them worked diligently and promptly. Students, staff, and fellow faculty at Texas A&M University provided suggestions, reviews, and expertise that improved the quality of the current edition. Texas A&M University's Department of Wildlife and Fisheries Sciences (WFSC) supported this new edition by providing release time for my editing, supplies, and monetary support. Linda Causey, WFSC, worked tireless hours under unimaginable time constraints to bring many of the figures into compliance with publishing standards. Tracy Estabrook (Lubbock, Texas), copyeditor for the *Wildlife Society Bulletin,* and Peter Strupp and copyeditors for Princeton Editorial Associates Inc. (Scottsdale, Arizona), provided excellent editorial support. At the Johns Hopkins University Press, I thank Vincent Burke, executive editor; his assistant, Jennifer Malat; and Julie McCarthy, managing editor, for providing invaluable assistance and encouragement throughout the process. To anyone who assisted with this edition of the *Manual,* but who was not named or implied in one of the groups described above: please accept my apology for the oversight. The wildlife community truly came together to create this edition of the *Manual,* and I am sure I have missed dozens who assisted. Last, but not least, my wife, Valeen, and daughter, Elizabeth, provided support and encouragement whenever I abandoned them to work on the volumes.

Nova J. Silvy, Editor

ACKNOWLEDGMENTS

Chapter 1
We thank J. R. Alldredge, J. H. Bassman, J. Baumgardt, R. A. Black, W. R. Clark, W. C. Conway, F. W. Davis, T. DeMeo, R. A. Fischer, T. K. Fuller, G. D. Hayward, H. R. Jageman, D. J. Johnson, S. L. Johnson, J. A. Kadlec, S. T. Knick, J. A. Manning, L. S. Mills, D. G. Miquelle, D. M. Montgomery, J. M. Peek, K. P. Reese, D. L. Roberts, J. J. Rotella, D. J. Schill, J. M. Scott, W. F. Seybold, D. F. Stauffer, R. K. Steinhorst, K. M. Strickler, A. G. Wells, G. C. White, wildlife students at the University of Idaho, and official reviewers D. H. Johnson and D. F. Stauffer for valuable review comments.

Chapter 2
The authors acknowledge Jonathan R. Bart and William I. Notz, previous authors of this chapter, for their work, as it significantly assisted with this revision

Chapter 3
We are greatly indebted to Julie L. Moore, bibliographer of the Wildlife Bibliographic Services, Las Cruces, New Mexico, who located many helpful obscure references and assisted with expediting interlibrary loan sources; Tyler Rogers, who skillfully assisted with computer typing of the draft, semi-final, and final details of manuscript preparation. We are grateful to Raul Valdez, Head of the Department of Fish, Wildlife and Conservation Ecology, New Mexico State University, Las Cruces, for expediting the completion of this chapter by arranging for manuscript preparation. Michael O'Brien, Nova Scotia Department of Natural Resources, Halifax, Nova Scotia, Canada, provided important information on trap research in Canada. Also we thank Mary M. Schemnitz for her patience, support, and encouragement, which substantially contributed to the completion of this chapter.

Chapter 4
Material in this chapter was adapted from Kreeger and Arnemo (2007): *Handbook of Wildlife Chemical Immobilization*.

Chapter 5

Dr. Simon Thirgood, Macaulay Land Use Institute, Aberdeen, Scotland, United Kingdom, had agreed to be a coauthor for this chapter prior to his tragic death in August 2009. Though he was unable to contribute to this chapter, we acknowledge his interest in this subject and feel a loss at his passing. We also thank the many supporters of this work, including the Jack H. Berryman Institute for Wildlife Damage Management, Logan, Utah; the S. J. and Jessie Quinney Professorship for Wildlife Conflict Management at Utah State University, Logan; Oklahoma State University Extension, Stillwater; Working Dogs for Conservation, Three Forks, Montana; Bat Conservation International, Austin, Texas; and Idaho Fish and Game, Boise. We are grateful to F. Zwickel, who first authored chapters on the use of dogs in wildlife work in The Wildlife Society's *Techniques Manual,* which inspired the reinstatement of this chapter. Last, we acknowledge all the hard working canines in this effort and regret they are only with us for a short time.

Chapter 6

This chapter evolved in part through development of educational tools by the Wildlife Toxicology Working Group of The Wildlife Society, Bethesda, Maryland, to assist field biologists in properly responding to wildlife mortality and/or morbidity events. We thank present and past members of the working group for their assistance. We thank the National Wildlife Health Center in Madison, Wisconsin, and the National Fish and Wildlife Forensic Laboratory in Ashland, Oregon, for helpful discussions on wildlife mortality and morbidity incidents and their experiences and procedures for processing them; J. C. Franson for assistance with a portion of this chapter; K. A. Fagerstone, K. L. Ford, M. I. Goldstein, and M. L. Parker for valuable discussions; and 2 anonymous referees for chapter review and helpful comments.

Chapter 7

We appreciate data and recommendations provided by David Jessup (California Department of Fish and Game, Sacramento), John R. Fisher (Southeastern Cooperative Wildlife Disease Study, Athens, Georgia), Donald S. Davis (Texas A&M University, College Station), and Arturo Angulo, Catherine Barr, Virginia Slutz, and Sonia Lingsweiler (Texas Veterinary Medical Diagnostic Laboratory, College Station). Some sections of this chapter benefited from the analogous chapter in the previous edition written by Thomas J. Roffe and Thierry M. Work (2005; U.S. Fish and Wildlife Service, Honolulu, Hawaii, and U.S. Geological Service National Wildlife Health Center, Madison, Wisconsin, respectively).

Chapter 8

The authors are grateful to N. J. Silvy, editor, for suggestions on the organization of this chapter. E. K. Lyons thanks his wife, J. Lyons, for her patience and direct assistance with myriad computer and scanner issues. Without her help, this chapter would not have been completed. E. K. Lyons also thanks the Department of Agricultural Sciences, McNeese State University, Lake Charles, Louisiana, for allowing the time to work on this chapter. E. K. Lyons is especially indebted to F. C. Lemieux, department head, for his guidance. M. A. Schroeder received financial support from the Washington Department of Fish and Wildlife, Seattle. Last, we appreciate N. J. Silvy, editor, for this opportunity and his guidance, insistence, and patience.

Chapter 9

We acknowledge M. T. Nietfeld and M. W. Barrett for compiling much of the literature on mammal and bird marking techniques (Nietfeld et al. 1994). We have expanded and updated this information and provided a different format for its presentation. We also acknowledge A. L. Hensley, A. D. Lopez, E. K. Lyons, J. S. Wagner, R. E. Walser, and S. W. Whisenant for literature searches and copying of relevant papers; T. M. Johnson for scanning papers and photographs; and M. E. Griffin for proofing an earlier version of the manuscript. To these people, R. E. Bennetts for reviewing an earlier draft of the manuscript, and an anonymous reviewer, we are deeply grateful. Photographs used in this chapter are from the collection in the Department of Wildlife and Fisheries Sciences, Texas A&M University, College Station. Last, we appreciate and respect C. E. Braun, editor, for his help, prodding, and patience for an earlier version of this manuscript.

Chapter 10

We gratefully acknowledge K. Church, M. Fuller, R. Kenward, and M. Samuel, previous authors of this material in the fourth and fifth editions of the *Techniques Manual,* for their contributions to this chapter. We freely used material from previous versions and thank them for their efforts in compiling information.

Chapter 13

We thank F. C. Bellrose, M. M. Horath, and D. B. Quine for radar data and ground-truth; S. J. Franke, D. R. Griffin, E. A. Mueller, and G. W. Swenson, Jr., for insights into radar as a tool; and P. W. Brown for information on waterfowl. T. Adam Kelly contributed material to the chapter.

Chapter 14

Many people have helped directly and indirectly with this chapter. Various collecting trips with entomologists around the world have helped me refine these methods of invertebrate sampling for use in wildlife studies and identify potential strengths and weaknesses of the methods, which may not be apparent when applied to general biodiversity surveys where the techniques are most commonly applied. Aubrey Colvin, Jonathan Cammack, Bernard Smalls, and

Michael Stiller were especially helpful in this regard. Pictures were provided by a number of colleagues, but of particular note is Laura Sands, a photographer who accompanied me on a collecting trip to southern Africa to document collecting methods. I also acknowledge Aubrey Colvin and Nova J. Silvy for providing useful comments on earlier drafts of this chapter.

Chapter 15
W. G. Jobman (U.S. Fish and Wildlife Service, Albuquerque, New Mexico) kindly provided data on whooping crane numbers after 1984. D. R. Anderson, T. A. Bookhout, J. D. Carlson, Jr., W. R. Clark, M. J. Conroy, L. L. Eberhardt, the late G. Caughley, J. D. Nichols, T. L. Shaffer, D. R. Smith, B. S. Bowen, and G. C. White provided comments on earlier drafts of this chapter.

Chapter 16
We thank J. L. Oldemeyer (retired) and the late R. F. Harlow for their earlier help with this chapter. M. A. Rumble, P. J. Happe, A. R. Lewis, R. D. Schilowsky, E. D. Salo, K. A. Sager, F. R. Quamen, S. J. Bandas, and C. P. Lehman assisted with obtaining photographs or literature references and/or proofreading of text and tables. T. L. Symens was responsible for all typing and word processing. We thank N. J. Silvy and anonymous reviewers for editorial suggestions that enhanced the manuscript.

Chapter 17
We thank the authors, John A. Litvaitis, Kimberly Titus, and Eric M. Anderson, of an earlier version of this chapter (Litvaitis et al. 1994), as we freely used material from their chapter. We accept responsibility for any errors in the present edition, but could not have completed the writing without their contribution. We thank John A. Litvaitis, C. E. Braun, and an anonymous reviewer for their suggestions, which improved an earlier version of this chapter (McDonald et al. 2005).

Chapter 18
The lead author thanks Bonneville Power Administration, Portland, Oregon, for support under project 2003-072-00 to redevelop and bring this chapter to fruition in conjunction with our coauthor's expertise. Other sources of support for our coauthors include Office of Science (Biological and Environmental Research), U.S. Department of Energy, Grant ER64360; National Aeronautics and Space Administration's Terrestrial Ecology and Applied Sciences Program; National Park Service Inventory and Monitoring Program; and Northwest Forest Plan Effectiveness Monitoring Program, U.S. Department of Agriculture Forest Service Region 6, Portland, Oregon. We are indebted to Peter Ewins (Arctic Program) and Jeremy Marten of the World Wildlife Fund–Canada, Toronto, Ontario, along with Martyn E. Obbard in Wildlife Research and Development Section of the Ontario Ministry of Natural Resources, Toronto, Ontario, Canada, for allowing us to share their data and website on polar bears. We also thank S. Shin, metadata coordinator, Federal Geographic Data Committee, Denver, Colorado, and Russell Faux, Matthew Boyd, and Brian Kasper of Watershed Sciences, Portland, Oregon, for their expertise on lidar systems, data, and capabilities, and imagery development illustrating the many applications of lidar.

Chapter 19
I acknowledge N. J. Silvy for editorial assistance and S. L. Thode for support and grammar skills. I appreciate the encouragement and support to bridge the fields of animal behavior and wildlife management from my mentors R. D. Howard and C. E. Braun. P. J. French, K. Brown, N. Gauss, and P. J. Muckleroy provided invaluable assistance with revision, format, and research. I thank Western State College of Colorado, Gunnison, for support while this chapter was being prepared.

Chapter 20
We gratefully acknowledge K. Church, M. Fuller, R. Kenward, and M. Samuel, previous authors of this material in the fourth and fifth editions of the *Techniques Manual,* for their contributions to this chapter. We freely used material from previous versions and thank them for their efforts in compiling information.

Chapter 21
Alan Woolf and Roy Kirkpatrick reviewed the original version of this manuscript and made insightful and constructive suggestions that significantly improved the content and balance of this chapter. I thank Dave Dennis for his contributions of photography and graphic illustration, and I am grateful to Donna Harder for her support and assistance in preparation of this chapter.

Chapter 22
We thank Scott Walter and Giridhar Athrey for comments on early drafts of this chapter. Support during the preparation of this chapter was provided to Paul Leberg by the Louisiana Department of Wildlife and Fisheries, Baton Rouge, and by the U.S. Army Basic Research Program's Environmental Quality and Installations Focus Area, as administered by the U.S. Army Engineer Research and Development Center, Vicksburg, Mississippi.

The Wildlife Techniques Manual

1 Research and Experimental Design

EDWARD O. GARTON,
JON S. HORNE,
JOCELYN L. AYCRIGG,
AND JOHN T. RATTI

INTRODUCTION

WILDLIFE MANAGEMENT PROGRAMS must be based on quality scientific investigations that produce objective, relevant information; and quality science is dependent upon carefully designed experiments, estimates, comparisons, and models. This chapter provides an overview of the fundamental concepts of wildlife research and study design and is a revision of Ratti and Garton (1994) and Garton et al. (2005).

Emergence of Rigor in Wildlife Science

Wildlife science is a term the wildlife profession has only recently nurtured. Our profession of wildlife conservation and management was built on natural history observations and conclusions from associations of wildlife population changes with environmental factors, such as weather, habitat loss, or harvest. Thus, we have a long tradition of wildlife management based on laws of association rather than on experimental tests of specific hypotheses (Romesburg 1981).

Although Romesburg (1981, 1989, 1991, 1993) and others (Steidl et al. 2000, Anderson 2001, Anderson et al. 2003, Belovsky et al. 2004) have been critical of wildlife science and its resulting management practices, the wildlife biologist is confronted with tremendous natural variation that might confound the results and conclusions of an investigation. Scientists conducting experiments in cell biology and biochemistry have the ability to control variables associated with an experiment, isolating the key components, and repeating these experiments under the same conditions to confirm their results. They also have the ability to systematically alter the nature or level of specific variables to examine cause and effect.

The wildlife scientist often conducts investigations in natural environments over large geographic areas, making it difficult to control potentially causal factors. Responses, such as density of the species in question, are simultaneously subject to the influences of factors, such as weather, habitat, predators, and competition, that may change spatially and temporally. Thus, rigorous scientific investigation in wildlife science is challenging and requires careful design (Steidl et al. 2000).

Experimental versus Descriptive Research

Most wildlife research prior to 1985 was descriptive. **Experimental research** is the most powerful tool for identifying cause and effect, and it should be used more in wildlife studies. However, descriptive natural history studies, field studies, and carefully designed comparisons based on probability sampling continue to be useful. **Descriptive research** is an essential initial phase of wildlife science and can produce answers to important questions, but it must be expanded to embrace interacting causes and variable results.

Descriptive research often involves broad objectives rather than tests of specific hypotheses. For example, we might have a goal to describe and analyze gray partridge (*Perdix perdix*) breeding ecology. Thus, we might measure characteristics of nesting habitat, clutch size, hatching success, brood use of habitat, food habits of chicks and adult hens, and mortality due to weather events and predators. From this information, we can learn details of gray partridge biology that will help us understand and manage the species. If we observe that 90% of gray partridge nests are in vegetation type "A," 10% in "B," with none in "C" and "D," we are tempted to manage for vegetation type "A" to increase nesting density. However, many alternatives must be investigated. Possibly vegetation type "A" is the best available habitat, but gray partridge experience high nest mortality in this type. Maybe vegetation type "X" is the best habitat for nesting, but it is not available on the study area. What vegetation types do gray partridge in other regions use? How does nest success and predation differ among regions with differing distributions of vegetation types, species of predators present, gray partridge densities, and climatic conditions? These questions show that defining quality nesting habitat is complex.

Combining descriptive studies with other studies published in the scientific literature should provide sufficient information to develop a **research hypothesis** (i.e., theory or conceptual model; Fig. 1.1) that attempts to explain the relationship between vegetation type and nesting success of gray partridge. Such models are general, but can help define specific predictions to be tested to examine validity of the model. These predictions can be stated as hypotheses. We can test hypotheses by gathering more descriptive observations or by conducting an experiment (Fig. 1.1) in which manipulated treatments are compared with controls (no treatment) to measure the change in sign (+ or –) resulting from experimental treatments. Random assignment of plots to treatment and control groups dramatically increases our certainty that measured differences are due to treatment effects rather than some ancillary factor.

Consider again the gray partridge study, and assume we have developed a **theory** (Fig. 1.1) that gray partridge adapted to be most successful at nesting in areas resembling their native habitat in Eurasia with its natural complement of predators, food sources, and vegetation cover. From this theory, we predict that partridge nesting success in grasslands in North America would be highest in undisturbed native prairie resembling native Eurasian gray partridge habitat and least successful in highly modified agricultural monocultures of corn, wheat, etc. We then formulate the hypothesis that gray partridge nesting density and nest success are higher in areas dominated (e.g., >75% of the available landscape) by native prairie than in areas dominated by cultivated fields of corn or wheat. The strongest test of this hypothesis we could perform would involve a **manipulative experiment** (Fig. 1.1), for which we must establish a series of control and experimental study plots. Our **study plots** would be randomly chosen from large blocks of land where agricultural practices have not changed in recent years, which contain the types of agricultural practices common to the region where we want to apply our findings. Some of these study plots (commonly 50%) will be randomly selected to act as control plots and will not change throughout the duration of the study. On the experimental plots (the remaining randomly selected plots in the same region as our control plots), cultivated fields will be planted to native prairie grass to test the validity of our hypothesis and predictions regarding the effect of habitat on gray partridge nesting. This process is difficult, because it requires large blocks of habitat, cooperation from landowners, several years to establish native prairie grass on the experimental plots, and additional years of study to measure the response of gray partridge to vegetative changes. The comparison between control and experimental plots will provide a basis to reject the null hypothesis of no effect, so we can conclude that increasing cover of native prairie grass, which could be within Conservation Reserve Program (CRP) fields in agricultural areas, will increase nesting density and success of gray partridge. If we fail to reject the null hypothesis, we cannot draw a firm conclusion, because the failure to reject might be due to insufficient sample size. If other studies have already shown higher nest success in areas of grass or CRP, then we must move beyond the potentially silly null hypothesis of no effect of grass cover (Johnson 1999, Läärä 2009). Instead we should focus on estimating the magnitude of effects from management efforts directed at gray partridge nesting success, so that we can build predictive models widely applicable to gray partridge management.

Some questions concerning wildlife science are not amenable to experimentation (e.g., effects of weather on populations, or differences in survival rates between gender or age classes). Other potential treatment effects are too expensive or difficult to accomplish. Some treatments may require substantial effort to convince the interested public of the value of applying them in any single treatment area. Finally, the need to evaluate effects of many habitat or population factors simultaneously may preclude experimentation. In these cases, construction of multiple biologically plausible models that seek to explain or predict observable phenomena can be a powerful tool for advancing knowledge (Hilborn and Mangel 1997) when combined with new information theoretic tools designed to identify the most likely explanatory model (Burnham and Anderson 2002). Incorporating modeling into the management process is an effective

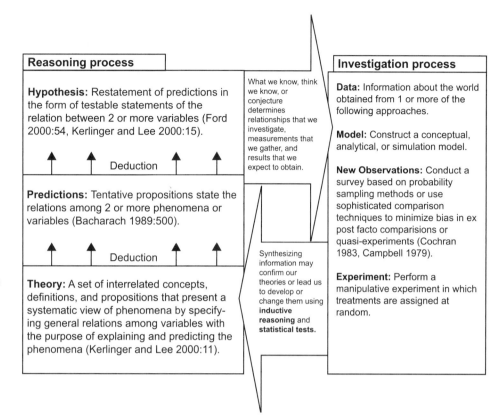

Fig. 1.1. Iterative nature of the scientific method. Data are synthesized inductively to develop theories. These theories form the basis for deductively derived predictions and hypotheses that can be tested empirically by gathering new data with experiments, new observations, or models. *Modified from Ford (2000:6).*

strategy for predicting consequences of management actions while simultaneously learning about key processes affecting wildlife populations and their habitats (Walters 1986). A key requirement for this process to be successful is the need to monitor consequences of management actions through an **adaptive management process** (Walters 1986). This adaptive learning process might be facilitated by application of **Bayesian statistics,** which use additional observations to improve estimates of key relationships assumed prior to the management action (Hilborn and Mangel 1997, Gelman et al. 2003, Bolstad 2007).

Scientific Method

In one of the early papers published on the scientific method in *Science* in 1890 Chamberlin (republished in1965) emphasized the need to examine multiple working hypotheses to explain an observation. Popper (1959, 1968) formalized an approach to testing individual hypotheses, referred to as the **hypothetico-deductive method,** that became the accepted standard in science. The method is a circular process in that previous information is synthesized into a theory; predictions are deduced from the theory; the predictions are stated explicitly in the form of hypotheses; hypotheses are tested through an investigation involving experimentation, observation, models, or a combination of these; the theory is supported, modified, or expanded on the basis of the results of these tests; and the process starts again (Fig. 1.1).

Platt (1964) re-emphasized the importance of **multiple competing hypotheses** and proposed a systematic pattern of inquiry, referred to as **strong inference,** in which the investigator devises alternate hypotheses, develops an experimental design to reject as many hypotheses as possible, conducts the experiment to achieve unambiguous results, and repeats the procedure on the remaining hypotheses. Other major works that provide detailed discussions of the scientific method include Dewey (1938), Bunge (1967), Newton-Smith (1981), Ford (2000), and Gauch (2003).

The most successful applications of the hypothetico-deductive method have been in physics, chemistry, and molecular biology, where experiments can isolate the results from all but a small number of potentially causal factors. The classic methods of natural history observation in wildlife science and other natural sciences have expanded to include experimentation, hypothesis testing, and quantitative modeling. James and McCulloch (1985:1) described this transition for avian biologists: "traditional ornithologists accumulated facts, but did not make generalizations or formulate causal hypotheses . . . modern ornithologists formulate hypotheses, make predictions, check the predictions with new data sets, perform experiments, and do statistical tests." Measuring simultaneous effects of multiple interacting causes (Quinn and Dunham 1983) may be facilitated by application of **information theoretic tools** to models incorporating multiple causes (Burnham and Anderson 2002). In addition

> **BOX 1.1. SYSTEMATIC OUTLINE OF SEQUENTIAL EVENTS IN SCIENTIFIC RESEARCH WITH AN EXAMPLE OF ELK IN THE NORTHERN ROCKY MOUNTAINS**
>
> | | |
> |---|---|
> | 1. Identify the research problem. | What are the influences of environmental factors, such as wildfire and winter severity, on the carrying capacity of elk winter range? |
> | 2. Conduct literature review of relevant topics. | Excellent earlier work by Houston (1982), Merrill and Boyce (1991), DelGiudice (1995), Coughenour and Singer (1996). |
> | 3. Identify broad and basic research objectives. | (a) Determine temporal and spatial differences in food habits that may affect elk nutritional condition during winters of varying severity; (b) examine the relationship between energy intake and mobilization of energy reserves at the population level throughout winter. |
> | 4. Collect preliminary observations and data as necessary. | (a) Winter severity data for 1987–1988, 1988–1989, and 1989–1990 including snow depth; (b) monthly precipitation during 1988 reflecting 100-year drought; (c) wet summers contributed to increases in elk population; (d) substantial winter kill first post-fire winter. |
> | 5. Conduct exploratory data analysis. | (a) Analyze food habits data for 2 different spatial locations pre-fire; (b) estimate energy intake by elk pre-fire. |
> | 6. Formulate a theory (conceptual model or research hypothesis). | Carrying capacity of elk winter range is influenced by wildfire and winter severity. |
> | 7. Formulate predictions from conceptual model as testable hypotheses (Fig. 1.1). | (a) Carrying capacity of elk winter range increases in post-fire areas; (b) carrying capacity of elk winter range decreases with increasing winter severity. |
> | 8. Design research and methodology for each hypothesis. | (a) Collect samples of urine during the same month of each winter to assess nutritional condition of elk from each study area. Only include urine samples from cows and calves. Collect samples in both burned and unburned areas. (b) Construct simulation model to translate individual responses to nutritional condition to population level responses. |
> | 9. Conduct a pilot study to test methodologies and estimate costs and variances. | Pilot study collects urine samples and estimates costs and variances. |

to James and McCulloch (1985), other excellent reviews of scientific approaches applicable to natural systems include Romesburg (1981), Diamond (1986), Eberhardt and Thomas (1991), Murphy and Noon (1991), Sinclair (1991), Hilborn and Mangel (1997), Boitani and Fuller (2000), Williams et al. (2002a), and Morrison et al. (2008).

The first steps in the **scientific method** begin with a clear statement of the research problem (Box 1.1), followed by a careful review of literature on the topic and preliminary observations or data collection. Preliminary data can be combined with published data to conduct an exploratory data analysis (Tukey 1977). Established theory, including principles, concepts and widely accepted models (Pickett et al. 2007), should be combined with creative ideas and potential relationships resulting from the biologist's observations and exploratory data analysis to develop a **conceptual model** (i.e., theoretical framework or general research hypothesis, Andrienko and Andrienko 2006). This conceptual model is essentially a broad theory (Fig. 1.1) that offers explanations and possible solutions, and places the problem in a broader context (Box 1.1). The next step is to develop **predictions** from the conceptual model (i.e., statements that would be true if the conceptual model were true). These predictions are then stated as **multiple testable hypotheses.** Research should be designed to test these hypotheses; ideally experimentation should be used when possible. A **pilot test** at this

10. Estimate required sample sizes and anticipate analysis procedures with assistance from a statistical consultant.	Estimated sample sizes feasible and analysis procedures successful with pilot survey data.
11. Prepare written research proposal that reviews the problem, objectives, hypotheses, methodology, and procedures for data analysis.	Prepare written proposal: Combine steps 1, 3, 6, and 8 to provide background, justification, and methodology for research.
12. Obtain peer review of the research proposal from experts on the research topic and revise if necessary.	Seek out experts in state wildlife agencies as well as authors of papers found during literature review.
13. Perform experiments, collect observational data, or construct a model.	(a) Collected elk urine samples from each winter and each study area; (b) constructed model to simulate energy intake and movements for the elk population.
14. Conduct data analysis.	(a) Non-normally distributed urine sample data analyzed using nonparametric statistics; (b) measured and simulated nutritional conditions compared using urine samples with unpaired t-tests.
15. Evaluate, interpret, and draw conclusions from the data.	Combined use of urine samples and model simulations provided strategic approach for assessing subtle changes in nutritional condition, physical condition, and mortality rates of elk. During winter 1988–1990, snow depth had a pronounced impact on nutritional condition; the most dramatic temporal and spatial effects occurred during the most severe winter in 1989.
16. Speculate on results and formulate new hypotheses.	Carrying capacity of elk winter range influenced more by winter severity than by wildfire.
17. Submit manuscript describing the research for peer-reviewed journal publication, agency publication, and/or presentation at scientific meetings.	Combine steps 9, 11, 12, 13, and 14 to create a well-written and concise manuscript of research findings, which were published in this case as DelGiudice et al. (2001b).
18. Repeat the process with new hypotheses (starting at step 6 or 7).	Repeat process with new hypotheses.

Based on DelGiudice et al. (2001b).

stage is invaluable in testing methodologies and gathering estimates of cost and variances. Included in the design, with the assistance of a statistician, is calculation of the **sample sizes** required to detect the hypothesized effects as well as decisions about how the data will be analyzed. Peers and a statistician should **review** the proposed design before data collection begins. The data are collected using **quality control.** Data **analysis** with appropriate statistical procedures is conducted to test the theory by rejecting fallacious hypotheses, selecting the best models of relationships or differences, obtaining unbiased estimates, or selecting the best alternative. Final conclusions usually result in further speculation, modification of the original conceptual model and hypotheses, and formulation of new hypotheses. The **publication** process is the last, but essential, step, and peer-review comments should be considered carefully before research on new hypotheses is designed.

PHILOSOPHICAL FOUNDATION

Why should wildlife biologists and managers care about the seemingly endless esoteric debates by philosophers of science? One reason is that modern philosophers have reached a perspective on how to gain truth and knowledge that is consistent with the approach of practicing wildlife biologists, managers, and scientists. Modern philosophers assert that classic views of the scientific process are outmoded or inappropriate and propose replacing them with a new integrated approach directly applicable to wildlife science and ecology (Pickett et al. 2007). Their approach is founded on 3 beliefs inherent in **scientific realism** (Boyd 1992). First, the universe is real, and it is possible to gain true knowledge about it (Scheiner 1994). Second, knowledge includes ideas that we posit in theories, but can only sense indirectly (e.g., electrons, plant communities, and carrying capacities). Third, all such theories must ultimately be tested empirically (Scheiner 1994). The goal of wildlife research and experimental design must be to advance our **knowledge** by gathering new information to test and improve our evolving wildlife theory, which consists of a set of interrelated **concepts, definitions,** and **propositions** (i.e., models and confirmed generalizations often referred to as principles).

The scientific method consists of an efficient approach to expanding this evolving knowledge base. This expansion can be accomplished by gathering new observations to obtain unbiased estimates of important characteristics (e.g., age-specific survival rates), testing proposed theories (e.g., harvest and starvation of subadults are compensatory), inferring new patterns or processes (e.g., harvest is additive to cougar mortality in adult elk [*Cervus canadensis;* Polziehn and Strobeck 2002]), and restricting or expanding the domain of inference for models of patterns or processes (e.g., deeper snows decrease winter survival of elk and deer, but thresholds for the effects differ among species). This **integrated approach** estimates strength of contributions (Quinn and Dunham 1983) by multiple simultaneously acting causes (e.g., survival of elk calves depends on date of birth; milk production of cows; quality and quantity of hiding cover; and density of bears, cougars, and wolves) rather than attempting to falsify all but one causal mechanism (Platt 1964).

INITIAL STEPS

Problem Identification

The initial step in most wildlife research is **problem identification** (Box 1.1). Most research is either applied or basic. **Applied research** usually is related to management concerns, political controversy, or public demand. For example, we may study specific populations because the hunting public has demanded greater hunting opportunity or a nongame species decline raises concerns about its long-term survival. Other applied studies may be politically supported due to projected loss of habitat by development or concerns over environmental problems, such as contamination from agricultural or industrial chemicals. **Basic research** seeks to gain knowledge for the sake of knowledge and a more complete understanding of factors that affect behavior, reproduction, density, competition, mortality, habitat use, and population fluctuations. Research on management questions can often be designed so that basic research on underlying principles can be conducted for minimal extra cost as data are gathered to solve the immediate management problem.

Literature Review

Once a research problem has been identified, research should begin with a thorough **literature review,** including collecting published and unpublished management agency data. Searching Google Scholar (http://www.scholar.google.com) and other free online databases provides instant access to titles with links to abstracts and often the full text of published peer-reviewed literature. Membership in The Wildlife Society and other professional organizations (Ecological Society of America, Society for Conservation Biology, American Fisheries Society, etc.) as well as many public libraries provide access to full-text databases of every paper published in societies' refereed journals and monographs. Broadscale Internet searches using Google and other search engines may provide unpublished information of value from public agencies and institutions, but information posted by individuals or unknown organizations should be treated with substantial skepticism. Using a variety of sources for your literature review will ensure that you have compiled the most relevant and recent information pertaining to your objectives.

Biological, Political, and Research Populations

Wildlife professionals work with 3 types of populations that impact study design: biological, political, and research populations. Mayr (1970:424) defined a **biological population** as a group "of potentially interbreeding individuals at a given locality" and **species** as "a reproductively isolated aggregate of interbreeding populations." Thus, a **population** is an aggregation of individuals of the same species that occupies a specific locality at a particular time, and often the boundaries can be described with accuracy. For example, the dusky Canada goose (*Branta canadensis*) population breeds in a relatively small area on the Copper River delta of Alaska and winters in the Willamette Valley near Corvallis, Oregon (Chapman et al. 1969). Between the breeding and wintering grounds of the dusky Canada goose is the more restricted range of the relatively nonmigratory Vancouver Canada goose (Ratti and Timm 1979). Although these 2 populations are contiguous with no physical barriers between their boundaries, they remain reproductively isolated and independent.

For most populations, such as red-winged blackbirds (*Aegolius phoeniceus*), grouping individuals into a hierarchical organization of demes, populations, and metapopulations within the species may require careful consideration of 5 facets (Fig. 1.2, Box 1.2): (1) geographical distribution of individuals, (2) geographical distribution of habitats (resources), (3) correlations in demographic rates (Bjørnstad et al. 1999, Post and Forchhammer 2002, Palsbøll et al. 2006), (4) genetic relationships (Manel et al. 2005), and (5) patterns of movement. Identifying the appropriate level in this hierarchy to

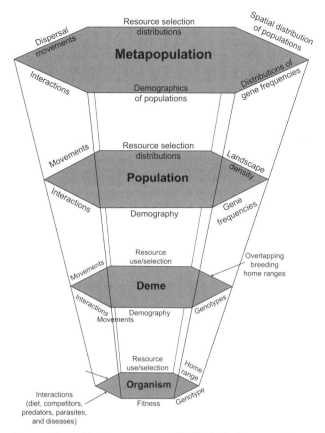

Fig. 1.2. Hierarchical arrangement of individuals from organism to metapopulation, illustrating multiple facets to consider in delineating levels to study: demography, genetics, geographical distribution of individuals, distribution and selection of resources, patterns of movement and interactions (e.g., diet, competitors, predators, parasites, and diseases). Processes operating at 1 level in this ecological hierarchy are influenced by the processes and characteristics at both lower levels (i.e., mechanisms) and higher levels in the ecological hierarchy (i.e., context). *Modified from Pickett et al. (2007:29).*

sample or assign treatments is critical to obtaining precise estimates and performing valid, powerful tests of ideas (i.e., theory), but keep in mind that processes operating at one level are influenced by those occurring at both lower levels in the hierarchy (i.e., mechanisms) and higher levels (i.e., context). Choosing the level in the biological hierarchy to study (Box 1.2) defines the research population or domain (Pickett et al. 2007) to which inferences and conclusions apply.

Beletsky and Orians (1996:152) and refuge biologists studied red-winged blackbirds at Columbia National Wildlife Refuge, Othello, Washington, and demonstrated, with 20 years of banding data, that territorial males and associated females occupying a set of discrete patches of marsh vegetation associated with ponds or streams on the refuge constituted a **deme** (Fig. 1.3). High correlations in demographic rates among demes and genetic similarity due to dispersal among demes make the entire red-winged black-

bird population on the refuge an appropriate biological population for management (Beletsky 1996, Garton 2002:665). Surrounding irrigated farmlands isolate red-winged blackbird populations at refuges from one another to some degree, but populations at refuges throughout the Columbia Basin could be treated as a **metapopulation** within the subspecies (*A. phoeniceus nevadensis;* Fig. 1.3). Another example of biological populations with separate boundaries is the bison (*Bison bison*) populations in Yellowstone National Park in the northwestern United States (Olexa and Gogan 2005). Biological populations for other species may not be so geographically distinct as those for Canada geese, red-winged blackbirds, and Yellowstone bison, in which case the researcher will have to carefully consider from which biological aggregation their samples are selected and to which their findings will apply.

The **political population** has artificial constraints of political boundaries, such as county, state, or international entities. For example, a white-tailed deer (*Odocoileus virginianus*) population in an intensively farmed agricultural region in the Midwest might be closely associated with a river drainage system due to permanent riparian cover and food critical for winter survival. The biological population may extend the entire length of the river drainage, but if the river flows through 2 states, the biological population is often split into 2 political populations that are subjected to different management strategies and harvest regulations. Traditionally, this problem has been common in wildlife management. When biological populations have a political split, it is best to initiate cooperative studies, in which research personnel and funding resources can be pooled to benefit both interested agencies.

Ideally, the **research or statistical population** should conform closely to the biological population, so that inferences can be applied to the chosen biological population. Due to logistical constraints, we often take a sample from this research population (i.e., sample frame; Scheaffer et al. 2005). Thus, sampling methodology is critical, for it provides the only link between samples and the research population. In rare instances, a population may be studied that represents all individuals of a species (e.g., an endangered species with few individuals, such as whooping cranes [*Grus americana*]). Or the research population might represent an entire biological population, such as one of the bison herds in Yellowstone National Park (Olexa and Gogan 2005). However, the research population usually is only a portion of the biological population and a small segment of the species. Carefully specifying a research or statistical population is essential in the planning phase of an investigation and may require thorough investigation of the existing literature on the species to determine breeding biology and dispersal patterns, geographic sampling to identify distribution of individuals and resources, and reviews of the literature on biological aggregations (Mayr 1970, Selander 1971, Stebbins

> **Box 1.2. Hierarchy of spatial population units**
>
> **Deme** — The smallest grouping of individuals approximating random breeding within the constraints of the breeding system, where it is reasonable to estimate birth, death, immigration, and emigration rates. Animals in this grouping are ideally distributed continuously across one patch of homogeneous to heterogeneous habitat, and their movements are restricted to home ranges for breeders during the breeding season. The size of this patch ideally would be related to the dispersal distance of juveniles or perhaps equal an area 20–50 times the size of a female breeding home range (e.g., Fig. 1.3 and Garton [2002] for red-winged blackbirds). Note: for some species demes are not feasible to delineate because of complex mating patterns and movements (e.g., in mallards, *Anas platyrhynchos*, males and females form pair bonds on wintering areas and males follow females to nesting areas the following spring, which may be quite distant from their natal area; Bellrose 1976:236).
>
> **Population** — A collection of demes or individuals at one point in time, typically the breeding season, with strong connections demographically (very high correlations in vital rates), geographically (close proximity), genetically (Manel et al. 2005), and through frequent dispersal. The population occupies a collection of habitat patches (relative to dispersal distance) without large areas of nonhabitat intervening. The area is typically 100 times the size of an average female home range and is not larger than the dispersal distance of 95% of natal dispersers, but it may be much larger if habitat patches are linear in shape and widely dispersed (e.g., all red-winged blackbirds occupying Columbia National Wildlife Refuge during the breeding season might be reasonably treated as a population; Garton 2002; Fig. 1.3). A population is dynamic through time: demes or groups of individuals show correlated fluctuations associated with the effects of broad-scale environmental factors (e.g., weather and fires) or other populations (e.g., competitors, predators, and disease outbreaks).
>
> **Metapopulation** — A collection of populations sufficiently close together that dispersing individuals from source populations occasionally colonize empty habitat resulting from local population extinction (Levins 1969). Populations in a single metapopulation may show low or high correlations in demographic rates, but the low rates of dispersal are sufficient to maintain substantial genetic similarity (e.g., red-winged blackbird populations distributed among the 7 national wildlife refuges along 200 km of the Columbia River in the south-central part of Washington constitute a metapopulation; Garton 2002; Fig. 1.3). Numerous types of metapopulations have been described, from source-sink to nonequilibrium to classic (or Levins) metapopulations (Harrison and Taylor 1997).
>
> **Subspecies** — A collection of populations as well as metapopulations, if present, in a geographic region where very rare dispersals maintain genetic, morphological, and behavioral similarity. However, populations and metapopulations occupy habitat patches that may be separated by large areas of nonhabitat, resulting in substantial demographic independence among populations or metapopulations (Mayr 1982, Garton 2002; Fig. 1.3).
>
> **Species** — The collection of interbreeding populations as well as metapopulations and subspecies, if present, encompassing the entire distribution and geographic range of the populations. The populations may show substantial differences in phenotypes (vegetation association, physiology, and behavior) and genotypes (Garton 2002; Fig. 1.3).
>
> Modified from Garton (2002).

1971, Ratti 1980, Wells and Richmond 1995, Garton 2002, Hanski and Gaggiotti 2004, Cronin 2006).

Conclusions from research are directly applicable only to the research population from which the samples were drawn. However, biologists usually have goals to obtain knowledge and solve problems regarding biological populations and species. The key questions are: (1) Is the sample an unbiased representation of the research population? (2) Is the research population an unbiased representation of the biological population? (3) Is the biological population representative

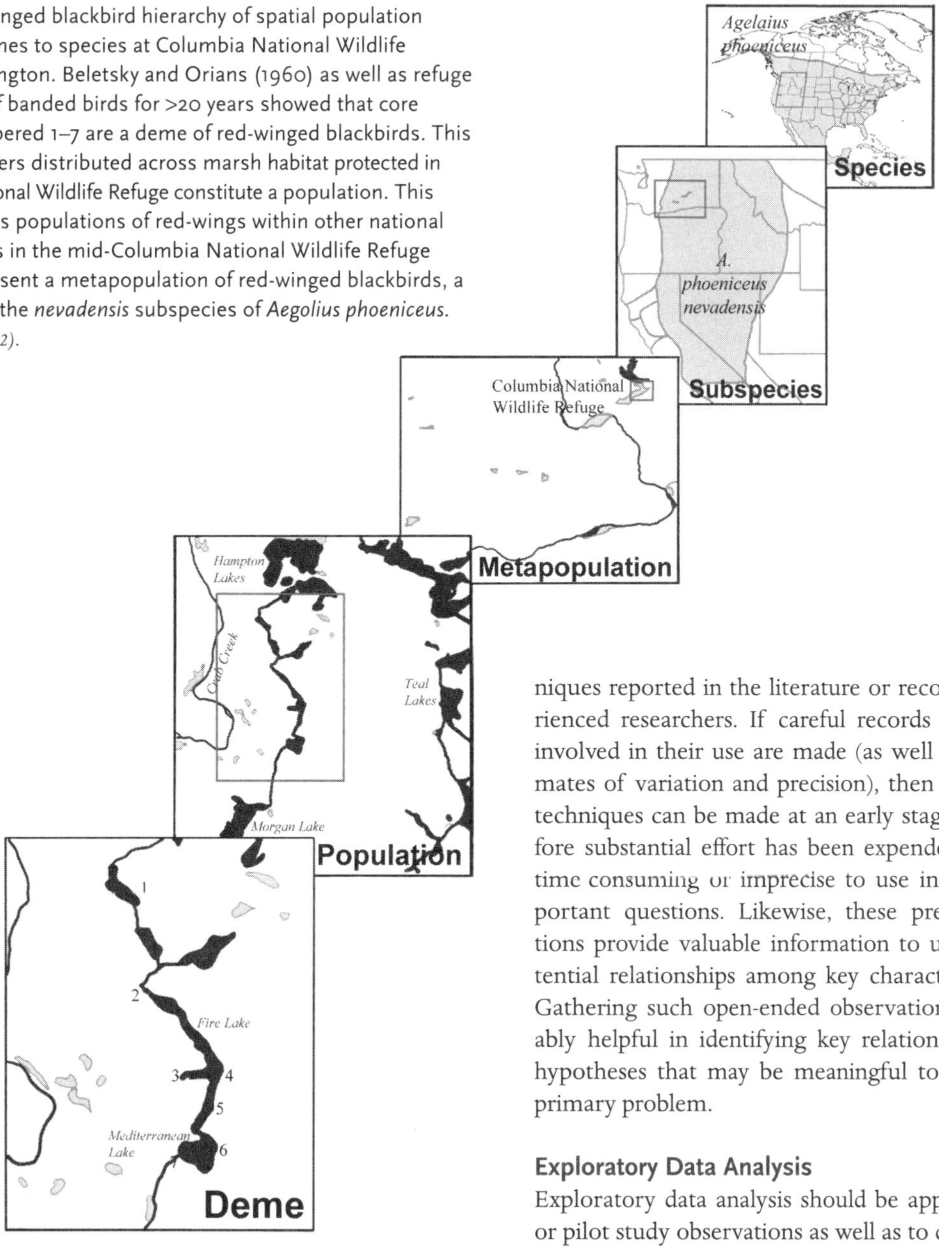

Fig. 1.3. Red-winged blackbird hierarchy of spatial population units from demes to species at Columbia National Wildlife Refuge, Washington. Beletsky and Orians (1960) as well as refuge staff studies of banded birds for >20 years showed that core marshes numbered 1–7 are a deme of red-winged blackbirds. This deme plus others distributed across marsh habitat protected in Columbia National Wildlife Refuge constitute a population. This population plus populations of red-wings within other national wildlife refuges in the mid-Columbia National Wildlife Refuge Complex represent a metapopulation of red-winged blackbirds, a subdivision of the *nevadensis* subspecies of *Aegolius phoeniceus*. *After Garton (2002).*

of the species? Because traits among segments of biological populations (and among populations of a species) often differ, broad conclusions or inferences relative to a research hypothesis should be avoided until several projects from different populations and geographic locations provide similar results. Combining and synthesizing replicate studies across large spatial extents should be a long-term goal, but may require the use of new techniques, such as meta-analysis (Osenberg et al. 1999).

Preliminary Data Collection

Making an effort to gather preliminary observations at this stage can pay great dividends in the end by allowing the researcher to explore a variety of potential research techniques reported in the literature or recommended by experienced researchers. If careful records of time and effort involved in their use are made (as well as preliminary estimates of variation and precision), then optimal choices on techniques can be made at an early stage in the design, before substantial effort has been expended on methods too time consuming or imprecise to use in answering the important questions. Likewise, these preliminary investigations provide valuable information to use in exploring potential relationships among key characteristics of interest. Gathering such open-ended observations also are remarkably helpful in identifying key relationships and alternate hypotheses that may be meaningful to understanding the primary problem.

Exploratory Data Analysis

Exploratory data analysis should be applied to preliminary or pilot study observations as well as to data from the literature or public agencies and institutions (Tukey 1977, James and McCulloch 1985, Andrienko and Andrienko 2006). During this process data are quantitatively analyzed in terms of means, medians, modes, standard deviations, and frequency distributions for important groups, and scatter plots of potential relationships are generated. Exploration of the data should be as complete and biologically meaningful as possible, which may include comparison of data categories (e.g., mean values, proportions, and ratios), multivariate analysis, correlation analysis, and regression. The "basic aim of exploratory data analysis is to look at patterns to see what the data indicate" (James and McCulloch 1985:21). If the research topic has received extensive previous investigation, the exploratory phase might even take the form of a meta-analysis of previous data gathered on the question (Osenberg et al. 1999). This phase often involves extensive discussions with other investigators with field or experimental experience on the topic.

THEORY, MODELS, PREDICTIONS, AND HYPOTHESES

Exploratory data analysis, literature reviews, and perceived associations should lead to the development of a theoretical framework (i.e., conceptual model; Fig. 1.4) of the problem. **Wildlife theories** (Fig. 1.1) are a set or system of interrelated concepts, definitions, assumptions, facts, confirmed generalizations, and propositions (Kerlinger and Lee 2000, Pickett et al. 2007) that present a structured view of wildlife ecology and management by specifying general relations among variables (e.g., waterfowl populations, annual rainfall, abundance of ponds and riparian habitat, and hunter harvest), with the purpose of explaining and predicting the phenomena (e.g., changes in waterfowl abundance; Office of Migratory Bird Management 1999, Ford 2000, Conroy et al. 2005).

We now explore the meaning and value of theory by considering our conceptual model of waterfowl population dynamics (Fig. 1.4), which expresses in a simple way complicated patterns of autumn waterfowl populations being positively influenced by spring breeding population size, number of ponds, and quantity and quality of wetland habitat, and negatively affected by nest predators, whose influence likely interacts with quality and quantity of wetland habitat around ponds. Likewise harvest influences spring population sizes the following year (i.e., $t + 1$ in Fig. 1.4), but the interaction may be complex, with either or both compensatory and additive effects coming into play. Utilizing this theory to understand dynamics of any particular waterfowl population requires stating a domain of interest and inference. For example, Conroy et al. (2005) studied an American black duck (*Anas rubripes*) metapopulation breeding in 3 regions and harvested in 6 regions in Canada and the United States (Box 1.2). Any individual investigation asks important questions and evaluates alternative hypotheses (e.g., models of harvest) in a restricted portion of the entire theory. For example, Conroy et al. (2005) used Bayesian methods to evaluate harvest models for American black ducks for this metapopulation. Often, important variables (e.g., abundance of nest predators) are very difficult to estimate, so their influence must be inferred through changes in nest success and fledging rates resulting from experimental manipulations (e.g., predator removal or manipulation of nesting cover).

Ford (2000:43) identifies 2 parts of a theory, consisting of (1) a working part providing information and a logical basis for making generalizations and (2) a motivational or speculative part that defines a general direction for investigation. Stating our theoretical framework (conceptual model) explicitly requires careful thinking and analysis of accepted generalizations (principles) stated in classic textbooks, reviews, and the published peer-reviewed literature on the topic. **Predictions** or deductive consequences of theory form the basis for hypotheses, which are variously described as assertions subject to verification (Dolby 1982, James and McCulloch 1985; Fig. 1.1) or testable statements derived from or representing various components of theory (Pickett et al. 2007:63; Box 1.3). Normally, the primary research hypothesis is what we initially consider to be the most likely explanation, but if the question has been placed into the proper theoretical framework, several alternate hypotheses are presented as possible explanations for observed facts (Fig. 1.1). Modern hypotheses commonly take the form of quantitative models that explicitly describe the relationships or magnitude of differences (Box 1.3).

We take an important step from descriptive natural history when we formulate conceptual models as research hypotheses. Interpretation of exploratory data analysis, creation of a theoretical framework, deduction of predicted

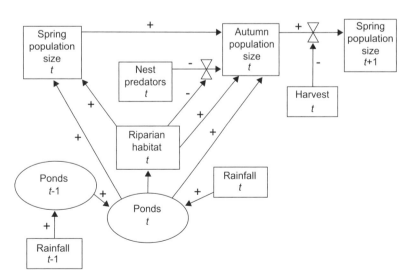

Fig. 1.4. Conceptual model of waterfowl population dynamics.

Box 1.3. Components of Theory

Component	Example
Domain. The scope in space, time, and phenomena addressed by a theory.	An individual waterfowl population or metapopulation in North America (e.g., American black duck, *Anas rubripes*) during 1971–1994.
Assumptions. Conditions needed to build the theory.	Conroy et al. (2005) assumed survival and productivity rates estimated for 3 regions in Canada and harvest rates from 6 regions in Canada and the United States to determine the dynamics of this metapopulation.
Concepts. Labeled regularities in phenomena.	**Harvest** refers to waterfowl shot during a legal hunting season and retrieved by the hunter.
Definitions. Conventions and prescriptions necessary for the theory to work with clarity.	Conroy et al. (2005) defined harvest rate as the probability of harvest based on direct recoveries (hunter reports of banded birds shot or found dead in the hunting season immediately following release; Williams et al. 2002*a*).
Facts. Confirmable records of phenomena.	All data on harvest regulations (season length and bag limit) and hunter numbers for 1971–1994 were obtained from the Canadian Wildlife Service and U.S. Fish and Wildlife Service (Conroy et al. 2005).
Confirmed generalizations. Condensations and abstractions from a body of facts that have been tested or systematically observed.	Harvest rates of male and female waterfowl generally differ, and Conroy et al. (2005) estimated harvest rates for males only to eliminate the need for estimating sex-specific harvest rates.
Laws or principles. Conditional statements of relationship or causation, statements of identity, or statements of process that hold in a domain.	Better wetland habitat conditions positively influence productivity in waterfowl populations (Fig. 1.4).
Models. Conceptual constructs that represent or simplify the structure and interactions in the material world. (**Scientific models** can project consequences of ideas; **statistical models** draw inferences and discriminate among competing ideas based on limited observations).	Conroy et al. (2005) developed statistical models for harvest rates in American black ducks. They found that harvest rates depended on both season length and bag limit, but differed between years and areas during 1971–1994.
Translation. Procedures and concepts needed to move from the abstractions of a theory to the specifics of applications or test or vice versa.	Annual changes in wetland conditions are estimated from aerial strip transect counts of pond densities throughout waterfowl breeding areas in North America (U.S. Fish and Wildlife Service and Canadian Wildlife Service 1987).
Hypotheses. Testable statements derived from or representing various components of theory.	Harvest rates for American black ducks increase with season length and bag limits in an area (tested and confirmed by Conroy et al. 2005).
Framework. Nested causal or logical structure of a theory.	During the fall, groups of American black ducks join with other groups on the same wetlands and other nearby wetlands to form populations that join 3 other populations in Canada during their migration south; they form a metapopulation occupying 6 regions of Canada and the United States (Conroy et al. 2005; Fig. 1.2).

After Pickett et al. (2007:63).

consequences, and formulation of testable hypotheses as alternative models are difficult aspects of science that require creativity and careful reasoning, but they are essential to the future of wildlife science.

OVERVIEW OF STUDY DESIGN

Introduction

Many different study designs are available for answering questions about the biology and management of wildlife species (Eberhardt and Thomas 1991, Morrison et al. 2008; Fig. 1.5). These options differ dramatically in terms of 2 criteria: How certain are the conclusions reached? How widely applicable are the conclusions? No single option is perfect. The biologist must weigh the available options carefully to find the best choice that fits the constraints of time and resources. Here we provide an overview of the most prominent study designs with further explanation in subsequent sections.

Experiments consisting of manipulative trials are underused in wildlife science (Fig. 1.5). **Laboratory experiments,** in which most extraneous factors are controlled, provide the cleanest results with the most certainty, but results generally have only narrow inference to free-ranging wildlife populations, unless they concern basic processes (e.g., disease susceptibility or nutritional biology). **Natural experiments,** in which large-scale perturbations (e.g., wildfires, disease outbreaks, and hurricanes) affect populations and landscapes naturally, provide only weak conclusions because of lack of replication and inability to control extrinsic factors through random assignment of treatments (Diamond 1986, Underwood 1997, Layzer 2008, Diamond and Robinson 2010; Fig. 1.5). **Field experiments,** in which manipulative treatments are applied in the field, combine some of the advantages of laboratory and natural experiments (Hurlbert 1984, Scheiner and Gurevitch 2001; Fig. 1.5). They have singular advantages, because truly replicated field experiments combine both breadth of inference and relatively certain conclusions (Johnson 2002). By assigning treatments to field replicates randomly, we can be certain that conclusions are valid rather than resulting from extrinsic factors beyond our control.

Some questions of importance in wildlife biology and management are not appropriate for experimentation. For example, we may be interested in the effects of weather on a particular animal population, but we cannot manipulate weather at will, in spite of the apparent human impact on its long-term trajectory. In addition, we may be interested in the relative importance of such factors as predation, habitat, and food limitations on population changes (Quinn and Dunham 1983, Mills 2007). In these cases we should formulate primary and alternate hypotheses in the form of models, estimate their maximum likelihood parameters, and test them statistically with likelihood ratios or compare them with information criteria (Hilborn and Mangel 1997, Burnham and Anderson 2002). Case studies consisting of unreplicated natural history descriptions are most useful at early stages in development of the research process (Fig. 1.5). **Pseudoreplicated field studies,** in which replicates are not statistically independent or samples rather than treatments are replicated, are only slightly better than **descriptive natural history studies.** At the other extreme are **replicated field studies,** wherein no manipulation or randomization of treatments occurs, but true replication occurs in a probability sampling framework, and information is gathered to evaluate alternate hypotheses. Conclusions from replicated field studies are broadly applicable, but are less certain than those from replicated field experiments.

Designing good field studies is more difficult than designing good experiments because of the potential for extraneous factors to invalidate tests or comparisons. One key step for both experiments and field studies is designing a sampling procedure to draw observations (experimental units or sample units) from the populations of interest. Only if

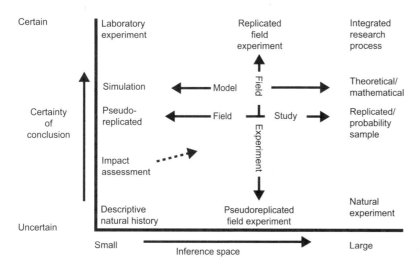

Fig. 1.5. Potential for wildlife study designs to produce conclusions with high certainty (few alternative hypotheses likely) and widespread applicability (a diversity of research populations in which the inferences apply).

this step is done properly can conclusions of the tests be applied to these populations. Survey sampling provides methods that are helpful in designing such sampling procedures (Cochran 1977). These methods are particularly important for field studies, but also are useful in field experiments for drawing experimental units and subsamples (samples within an experimental unit).

Impact assessments are another type of study design, but typically there is no replication, because the impact only occurs at a single site (e.g., an oil spill in a national wildlife refuge). However, they are useful for collecting baseline data as long as the type, time, and place of the impact are known; germane variables can be measured; and spatial and temporal controls exist (Green 1979, Williams et al. 2002a). Frequently, impact assessments are planned (e.g., prescribed fire), which allows for before-and-after measurements. The inference can be improved by monitoring both impact and nonimpact sites at several replicated sites of both types rather than only monitoring impact sites (Anderson 2002b, Williams et al. 2002a).

Models, which are a simplified representation of a system or process, are a versatile way to address a wide range of research questions that emphasizes being certain the conclusions do follow from the estimates, relationships, and assumptions (Fig. 1.5). The inference space of such models spans a continuum from large and general for theoretical or mathematical models (simple differential or difference equations) to smaller, more realistic for simulation models (complex multicausal, multiscale simultaneous differential and/or difference equations; Fig. 1.5). The certainty of conclusions based on models is in part influenced by the measurement of model variables and estimation of model parameters; it can thus be portrayed in predictions of models incorporating both process and estimation uncertainties (Hilborn and Mangel 1997:59). Models provide an important framework from which to begin to understand the processes influencing questions in wildlife science. They can help gauge the influence of one variable on others. For example, models can be used to assess how much juvenile dispersal influences population growth. By holding all other variables that influence population growth constant and then varying juvenile dispersal rates in the model, we can estimate how much the population growth is altered by small or large variation in juvenile dispersal rates. Conducting this type of **sensitivity analysis** makes such models an important tool for wildlife scientists.

In a subsequent section we describe an **integrated research process** that combines many aspects of study design, such as natural history observations, natural experiments, and laboratory experiments (Fig. 1.5). Using this process makes the research inference space large and increases the certainty of research conclusions by combining multiple study components. The complexities of the ecosystems in which wildlife science takes place are best addressed with the integrated research process, because it enables the wildlife scientist to capture more of the natural variability inherent in ecosystems (Clark and Stankey 2006, Morrison et al. 2008).

Once a research option has been chosen for each hypothesis or predictive variable, careful planning of the actual testing process can proceed. We must identify exactly what data will be collected for each hypothesis or predictive variable as well as when, how, how much, and for how long. Furthermore, how will these data be treated statistically? Will the data meet assumptions of the statistical test? Is the sample size adequate? Will the statistical hypothesis provide information directly related to the theory or model? Do biases exist in data collection, research design, or data analysis that might lead to a spurious conclusion? These questions must be considered carefully for each hypothesis before fieldwork begins. Consulting a statistician is important, and the statistician should understand the basic biological problem, the overall objectives, and the research hypotheses.

Peer review (evaluation by independent qualified reviewers) of the proposed research, including both study design and subsequent data collection and analysis, should be obtained from several people with expertise and experience in the research topic. Peer review will usually improve a research design and may disclose serious problems that can be solved during the planning stage. Unfortunately, most peer reviews occur too late for remedial work: after data collection, when the final report or publication manuscript is written.

Laboratory Experiments

Drawing inferences from **laboratory experiments** is easy because of the high level of control, yet this advantage must be weighed against the disadvantages (Table 1.1) in terms of (1) scale—laboratory experiments are restricted to small spatial scales and short time periods, (2) scope—only a restricted set of potential manipulations is possible in the laboratory, (3) realism—the laboratory environment places many unnatural stresses and constraints on animals, and (4) generality—some laboratory results cannot be extrapolated to natural communities. In a continuation of our example, laboratory experiments could be designed to examine whether geese really can select the most nutritious forage when given several alternatives in a cafeteria feeding trial. Diamond (1986) provided examples of the 3 types of experiments (laboratory, natural, and field) and made excellent suggestions for improving each type. Other examples and discussions of experiments are provided by Cook and Campbell (1979), Milliken and Johnson (1984), Kamil (1988), Hairston (1989), Underwood (1997), Tilman et al. (2006), and Chalfoun and Martin (2009).

Laboratory experiments in biology have been most useful for studying basic molecular or biochemical processes common to all organisms of a class. Laboratory experi-

Table 1.1. Strengths and weaknesses of different types of experiments

	Experiment type		
	Laboratory	Field	Natural
Control of independent variables[a]	Highest	Medium	Low
Ease of inference	High	Medium	Low
Potential scale (time and space)	Lowest	Medium	Highest
Scope (range of manipulations)	Lowest	Medium	High
Realism	Low	High	Highest
Generality	Low	Medium	High

Modified from Diamond (1986).

[a] Active regulation and/or site matching.

ments also have provided valuable information on emerging issues, such as wildlife diseases (e.g., Cooke and Berman 2000, Woodhams et al. 2008), efficacy of fertility control (Chambers et al. 1999, Hardy et al. 2006), and interactions between exotic and native species (e.g., Komak and Crossland 2000, Kopp and Jokela 2007).

Identifying one research design as best for all situations is not possible. All options should be considered as possibilities for evaluating hypotheses. Sometimes the best evaluation of a hypothesis involves using a combination of field studies and several types of experiments. For example, field observations by Ratti et al. (1984) indicated that spruce grouse (*Dendragapus canadensis*) fed exclusively on certain trees while ignoring numerous other similar trees of the same species. This observation led to a laboratory experiment with captive birds that tested the hypothesis that trees selected for feeding had higher nutritional content than did trees selected at random (Hohf et al. 1987).

Natural Experiments

Natural experiments are similar to field studies, except that in them we study the effects of uncontrolled treatments, such as wildfires, hurricanes, mass mortality from diseases, agricultural practices, and range expansions by animals or plants (Layzer 2008, Diamond and Robinson 2010). A key problem in evaluating natural experiments is that we cannot assign treatments randomly and therefore cannot be certain that any differences between treated and untreated units are not due to other factors that differed between them before some were "treated." In natural experiments the treatment precedes the hypothesis and most comparisons must be made after the fact. With our Canada goose example, a natural experiment might be to survey farmers in the region to locate pastures that have been fertilized and those that have not been fertilized in recent years. If our observations of feeding geese show more use of pastures that had been fertilized, we have more evidence indicating the birds select more nutritious forage. However, many alternative explanations remain. For example, perhaps those pastures that were fertilized were grazed later in the summer, and geese preferred fields with the shortest grass, where their ability to detect approaching predators is greatest. Many hypotheses of interest to wildlife biologists can be tested only with natural experiments, yet it is difficult to draw inferences from such experiments. The applied nature of wildlife management makes the realism and generality of natural experiments an important advantage, but their applicably to other populations is questionable unless multiple similar natural events are analyzed.

Field Experiments

Field experiments span a range from pseudoreplicated field experiments (Hurlbert 1984), in which no true replication is used (or possible) and conclusions are not certain, to replicated field experiments, for which conclusions are relatively certain (Johnson 2002). **Replicated field experiments** provide conclusions that are broadly applicable to free-ranging wildlife populations. Field experiments offer advantages over natural experiments in terms of certainty of inference and control of confounding factors, but they suffer the disadvantages of restricted scale and lower generality (Table 1.1). Compared to laboratory experiments, field experiments have greater scope and realism. Their main advantage is that we can randomly assign treatments and thereby eliminate fallacious conclusions due to effects of confounding factors. In field experiments, manipulations are conducted, but other factors are not subject to control (e.g., weather). In many situations in wildlife science, field experiments offer the best compromise between the limitations of laboratory and natural experiments (Wiens 1992, Krebs et al. 2001). In our Canada goose example, a subsequent field experiment would be to select random pairs of plots in known foraging areas. One member of each pair would be randomly assigned to be fertilized to learn whether geese select fertilized plots more often than they do the nonfertilized control plots. If they do select fertilized plots more often, a stronger inference about selection of nutritious foods could be made, because random assignment of a large number of plots to fertilization and control groups should have canceled effects of extraneous confounding factors. Interspersion of treatment and control plots (Hurlbert 1984, Johnson 2002) in fields naturally used by geese strengthens our belief that our conclusion would apply in systems where geese typically forage. **Adaptive management** could successfully incorporate field experiments by breaking management zones into replicates that are assigned various treatment levels for comparison to a standard management action (Connelly et al. 2003a). The strong advantages of field experiments are that random assignment of treatments to units interspersed among units to which the conclusions will apply protects against reaching invalid conclusions due to extrinsic factors.

Field Studies

Field studies may appear similar to experiments when they are conducted to test hypotheses, but they differ in that treatments are not assigned at random. For example, in a field study of dietary selection by Canada geese we might randomly select plots where flocks of geese have fed and those where they have not fed to examine whether geese choose areas with vegetation that is more nutritious. If they do, a weak inference would be that geese are choosing nutritious food, but numerous alternative explanations remain untested (e.g., maybe geese preferred hilltop sites, where visibility was good, and coincidentally these also were sites farmers fertilized most heavily to compensate for wind-driven soil erosion from previous years of tillage). Making inferences from field studies is difficult, because we make ex post facto comparisons among groups (Kerlinger 1986). Drawing firm conclusions is difficult, because these groups also differ in many other ways. The important characteristic of a field study is that we have comparison groups (e.g., use versus nonuse plots), but we have no treatments. Well-designed field studies can make important contributions to wildlife science and management (e.g., Paltridge and Southgate 2001), but their limitations must not be overlooked.

Impact Assessment

The most basic form of **impact assessment** compares measurements of wildlife and other characteristics at a site potentially affected by pollution or development to similar measurements at an unaffected reference site (Anderson 2002b; Fig. 1.5). This most simple form of impact assessment provides almost no basis for inference, because the reference site may differ for a multitude of reasons besides absence of the pollution source or development. Green (1979) noted the potential improvement in this design that results from making measurements before and after development at both reference and development sites. The basic before–after/control–impact (**BACI**) design has become standard in impact assessment studies (Anderson 2002b, Morrison et al. 2008) and also has been used in predator removal studies (e.g., Risbey et al. 2000). However, differences from before to after at reference (control) and impacted (treatment) sites are confounded by natural temporal variation and may not be produced by the impact itself (Hurlbert 1984, Underwood 1994, Williams et al. 2002a). In contrast to a well-designed field experiment, neither reference nor impacted sites are chosen randomly over space, and treatments are not assigned randomly. These limitations severely reduce the certainty of conclusions and the application of inferences to other areas. The goal is not to make inferences to all possible sites (Stewart-Oaten et al. 1986) for a power plant, for example, but to the particular power plant site being developed. For larger impact studies in which the goal is to make inferences with more certainty that are applicable to more sites (Fig. 1.5), the basic BACI design must be improved by the addition of replication and randomization (Skalski and Robson 1992, Underwood 1994). Stewart-Oaten et al. (1986) emphasized the value of expanding the BACI design to include temporal replication and noted the advantage of taking samples at irregular time intervals rather than on a fixed schedule. Hurlbert (1984) emphasized that comparing abundances of wildlife from repeated surveys at 1 impact and 1 reference site constitutes pseudoreplication that is only eliminated by having several replicated impact and reference sites. Replicated reference sites with environmental characteristics similar to the impact site are quite possible and highly desirable; however, replicated impact sites are only feasible in large-scale impact studies, typically involving meta-analysis of many single impact site studies.

Modeling

Modeling can be used as a deductive tool to synthesize theoretical understanding together with creative ideas about potential solutions to a problem or question. Creating a quantitative model makes the assumptions, accepted facts, generalizations, and laws or principles explicit for use in making valid and/or testable predictions. Kitching (1983:31) suggested this process of modeling involves 18 steps that correspond exactly to steps in the **scientific method** (Box 1.1; see details below under Modeling). Starfield and Bleloch (1991) describe this process in a straightforward manner with many wildlife examples created in spreadsheets. Clark (2007) presents a very rigorous account of ecological modeling utilizing free statistical and modeling software, such as R (R Development Core Team 2006), and Otto and Day (2007) provide a more mathematical, but very readable treatment of ecological modeling for biologists.

Modeling currently plays an essential role in 2 widely practiced processes of wildlife science: adaptive management and population viability analysis. **Adaptive management** requires building predictive models that summarize what is known or assumed about a management issue to examine alternative management actions. Managers choose one of the alternatives, and monitoring is conducted to: (1) ensure the action was accomplished; (2) evaluate whether the predicted consequences did in fact result; and (3) use feedback of results to improve understanding of the system, its behavior, key parameters, and relationships incorporated into the model. **Population viability analysis** uses models and data for populations to estimate the probability that populations of rare species will persist for specified times into the future (Mills 2007:254). These forecasts are essential to make scientifically defensible decisions concerning the listing or delisting of a species under the Endangered Species Act of 1973 (U.S. Fish and Wildlife Service 1973). Clearly building models such as these is an application of the scientific method that produces knowledge in the form of forecasts, but other applications of modeling strive to increase our general understanding of interrelationships (e.g., long-term impacts and dynamics of wolf, cougar, and coyotes on deer and elk; Garton et al. 1990, Varley and Boyce 2006), which are difficult to manipulate experimentally. Likewise building conceptual and quantitative models acts as a helpful early step in any investigation, because it sharpens our focus on identifying critical relationships and assumptions. It is an essential step in an integrated research process.

Integrated Research Process
The **integrated research process** (Fig. 1.5) builds on a solid base of natural history observations. Field observations and conceptual models should lead to experiments, and the results of natural experiments should lead to field and laboratory experiments. For example, Takekawa and Garton (1984) observed birds feeding heavily on western spruce budworms (*Choristoneura occidentalis*) during a budworm outbreak, which suggested that birds were a major source of budworm mortality. Field experiments were conducted to test this hypothesis by placing netting over trees to exclude birds. Survival of budworms on trees with netting was 3–4 times higher than on the control trees exposed to bird predation (Takekawa and Garton 1984). The level of certainty increases as many predictions from the research hypothesis are supported and alternate hypotheses are rejected in successively more rigorous tests that use replicated research options. After such findings are repeated over broad geographic areas or throughout the range of the species, the research hypothesis may become a principle of wildlife science (Johnson 2002). The integrated research process should be the goal of wildlife science (Clark and Stankey 2006, Morrison et al. 2008).

Outstanding examples of integrated research programs include long-term research on red grouse (*Lagopus lagopus scoticus*) in Scotland (Jenkins et al. 1963, Watson and Moss 1972, Moss et al. 1984, Watson et al. 1994, Kerlin et al. 2007, New et al. 2009), red deer (*Cervus elaphus*) on the Isle of Rhum, Scotland (Lowe 1969, Guinness et al. 1978, Clutton-Brock et al. 1985, Coulson et al. 1997, McLoughlin et al. 2008, Stopher et al. 2008, Owen-Smith 2010), and snowshoe hare (*Lepus americanus*) in North America (Keith 1963, 1974; Windberg and Keith 1976; Keith and Windberg 1978; Keith et al. 1984; Krebs et al. 2001). Research on red grouse and snowshoe hare has focused on hypothesized causes of population cycles, whereas research on red deer has focused on population regulation and density-dependent effects on survival, fecundity, reproductive success, spacing behavior, and emigration. Research on snowshoe hare has evaluated the role of predators (i.e., lynx [*Lynx lynx*] primarily, but other mammals and birds, too) as well as alternate proposed causes of the classic 10-year cycle in snowshoe hare and lynx numbers. For all 3 example species, descriptive studies and field observations formed the groundwork for subsequent research that included a series of innovative field studies and experiments (natural, field, and laboratory).

For example, preliminary studies of red grouse in Scotland (Jenkins et al. 1963) provided information on fundamental population parameters: births, deaths, immigration, and emigration. This information was used to form research hypotheses about causes of population fluctuations. Postulated causes initially included food quality, breeding success, spacing behavior, and genetics (Watson and Moss 1972, Kerlin et al. 2007). Using data from long-term field studies coupled with field and laboratory experiments, Watson and Moss (1972) concluded that quality of spring and summer foods (heather [*Calluna vulgaris*] shoots and flowers) affected egg quality, breeding success (viability of young), and spacing behavior of males and females, but territory size ultimately affected recruitment and population density (but see Bergerud [1988] for a critique of the self-regulation hypothesis and inferences based on red grouse research). Watson et al. (1984*b*) tested these conclusions with innovative field experiments, in which they (1) fertilized fields to assess grouse response to increased nutritional quality of the heather and (2) implanted males with time-release hormones to monitor changes in territory size associated with aggressiveness induced by higher or lower levels of androgens and estrogens (Watson 1967). Additional and more rigorous research rejected hypotheses that nutrition, genetics, and parasitism were causal factors (although Dodson and Hudson [1992] make a counterargument for the role of the parasite *Trichostrongylus tenuis*), and instead focused on emigration as the key factor in population declines (Moss et al. 1984, 1990; Watson et al. 1984*a*; New et al. 2009). These findings led to more research, because the mechanisms underlying density-dependent relationships, including summer and winter emigration, were unclear. Recent research has focused on the hypothesis of kin selection and differential aggression between kin and non-kin to explain cyclic changes in red grouse (Moss and Watson 1991, Watson et al. 1994) and synchronization of cycles across large regions according to weather (Watson et al. 2000, Kerlin et al. 2007). Thus, the integrated research process continues.

EXPERIMENTAL DESIGN

A variety of designs is available for researchers planning an experiment or quasi-experiment. This brief overview of some designs that have seen wide and innovative application to wildlife science should augment information provided in standard courses and references on experimental design (Underwood 1997, Scheiner and Gurevitch 2001, Quinn and Keough 2002, Morrison et al. 2008).

Single-Factor versus Multifactor Designs
Single-factor analyses are the simplest, because they involve only comparisons between 2 or more levels of 1 factor. Evaluating the simultaneous effect of 2 or more independent variables (**multifactor designs**) at once requires

the use of complicated statistical methods, which should be discussed with a statistician. Under many conditions we can test 2 factors at once without expending more effort than would be required to test either of the factors alone. A complicating issue is the potential for interaction among factors (Steel and Torrie 1980). An **interaction** occurs if the effects of one factor on the response variable are not the same at different levels of another factor. For example, if we are interested in the effect of snowmelt date on nest success by arctic-nesting polymorphic snow geese (*Chen caerulescens*), we might discover an interaction between color phase and the onset of spring snow melt. Thus, darker, blue-phase birds would have higher nesting success during early snowmelt years, because they are more cryptically colored once snow has melted and experience less nest predation. During late snowmelt years white-phase birds are more cryptically colored and experience less nest predation. Many observations might be required to clarify possible relationships in these situations.

Dependent Experimental Units

Special designs have been developed to handle many types of dependency in experimental units, where **dependence** means that units tend to be more similar to one another than if we were to pick units at random from the entire population. For example, animals in one group tend to be more similar to one another (e.g., doe–fawn groups of deer have few bucks), and vegetation plots that are spatially proximate tend to be more similar to one another than are plots picked at random from the entire study area. A common design involves pairing. In a **paired design** we match experimental units in pairs that are as similar as possible. The treatment is then applied to one member of each pair at random. If there is a confounding factor, which we succeed in matching in the pairs, this approach will lead to a more powerful test than if pairing is not performed. For example, if we were studying the effects of spring burning on northern bobwhite (*Colinus virginianus*) habitat, we could establish pairs of plots throughout our study area, being careful to place each pair in a homogeneous stand of vegetation. We would then randomly assign one member of each pair to be burned in the spring. The analysis would then examine the differences between the members of a pair and test for a consistent improvement or decline in the burned member of the pair. Pairing would remove the effects of vegetation difference from one part of the study area to another and would result in a more sensitive experiment. If members of pairs are not more similar than members of the general population, the test will be less powerful because of the pairing.

When more than 2 levels of a factor are compared, pairing is referred to as **blocking.** A block is a set of similar experimental units. Treatments are randomly assigned to units in each block, and the effectiveness of blocking can be tested during the analysis. For example, if we expanded our study of burning to include spring and autumn burning as treatments, a block design would be appropriate. Three adjacent plots would be placed in homogeneous vegetation stands, and spring and autumn burning would be applied randomly to 2 of the 3 plots in each block (e.g., set of 3). This powerful design is normally referred to as a **randomized block.**

Another common form of dependency occurs when **repeated measurements** are taken on the same experimental unit through time. This practice is common in wildlife research, wherein the effects of treatments may change over time and must be monitored over a series of years. For example, in our study of spring and autumn burning the effects may be different in the first, second, and third growing seasons after treatment. The plots should be monitored over several years to measure these effects. The measurements are repeated on the same plots, so they are not independent. This repetition must be treated correctly in the analysis by using repeated measures or multivariate analysis of variance (ANOVA; Milliken and Johnson 1984, Johnson and Wichern 1988, Williams et al. 2002*a*). Dependency also is common in count data, especially when animals occur in groups (Eberhardt 1970). This lack of independence is often referred to as **overdispersion.** To properly cope with significant overdispersion the dependency should be modeled. Unless the biologist has extensive training in this topic, close cooperation with a consulting statistician is essential when designing and analyzing experiments involving such complicated designs.

Crossover Experiments

Crossover experiments provide a powerful tool to evaluate treatments that do not produce a long-lasting effect. Selecting pairs of experimental units and randomly assigning one member of each pair to be treated during the first treatment period initiates a crossover experiment. The second member serves as the control during this treatment period. In the second treatment period, the control unit becomes the treatment and the former treatment becomes the control. In this way the effects of any underlying characteristics of experimental units are prevented from influencing the results. This technique is valid only if treatment effects do not persist into the second treatment period.

Consider the following example. Suppose we wanted to test the hypothesis that mowing hay before 4 July decreases ring-necked pheasant (*Phasianus colchicus*) nest success. We could test this idea by dividing our study area into 5 homogeneous hayfield regions and then dividing each region into 2 portions. In one randomly selected portion of each region we could pay farmers not to mow their hay fields until after 4 July (treatments). In the other portion of each region, hay mowing would proceed as in most years, with the first cutting during mid-June; these portions would serve as controls. To monitor nest success, we locate nests by systematic field searches, being sure to search treatment and control areas with identical methodology (e.g., search intensity and seasonal timing). Nest success would be measured with standard techniques. After 1 year, we might measure significantly higher nesting success in the treatment portions (i.e., those areas with delayed hay mowing). However, the number of treatments is small, and we are not able to conclude with confidence whether higher nest success resulted from the treatment or from some undetected, inherent differences in treated portions of each region, such as nest predators. We would implement the crossover experiment by switching in the second year, so the original control portions of the study regions now have mowing delayed until after 4 July (new treatments), and the original treatment portions revert to the standard practice of first cutting in mid-June (new controls). If the portions with late cutting treatments again have higher nest success, we have better evidence that delayed mowing is responsible for higher nest success than we had at the end of the first year (i.e., we have better evidence for a cause-and-effect relationship). If even stronger support for the hypothesis is desired, the crossover experiment might be repeated in the same region and in other farming regions.

Fixed, Random, Mixed, and Nested Effects

One of the most critical decisions we must make in design concerns choosing the population for which we want to make inferences. If only a few levels of a treatment factor are relevant or would occur, we set a limited number of values at which the treatment would be applied, and the factor is termed a **fixed effect** (**Model I**). If we want the conclusion to apply to any level of a treatment factor, we must select the treatment levels as a random sample from the population of potential values, so that a conclusion drawn about the effect of this factor applies across all levels at which it occurs. This design is termed a **random effect** (**Model II**). A **mixed model** (**Model III**) includes both fixed and random effects. In simple 2-factor or multifactor designs all levels of each factor are applied to all levels of other factors, and the design is considered to be a **crossed design.** When this is not possible, the design must use approaches in which one factor is nested in another. A **nested design** can be described as hierarchical, which occurs most commonly where certain levels of one factor only occur in some levels of another factor. For example, a study evaluating the effect of vegetation treatment on bird communities might have 3 plant communities (ecological systems) with treatments of clearcut, burn, partial-cut, and controls. These factors would need to be nested if one of the plant communities was a shrub community where timber harvest does not occur. Decisions about the design of experiments must be reflected correctly in the analysis, as different measures of variance are appropriate for fixed, random, mixed, or nested effects.

Replication

Sample size refers to the number of independent random sample units drawn from the research population. In experiments, sample size is the number of replicates to which a treatment is assigned. For logistical reasons, we may measure numerous **subsamples** closely spaced in a single sample unit. However, we must be careful to distinguish these subsamples from independent random samples. Subsamples are not independent random sample units, because they typically are more similar to one another than are widely spaced samples. Similarly, subsamples in experiments are not true replicates if they cannot be independently assigned to a treatment category. The precision of a statistic is measured by its standard error. **Standard error** is calculated from the variation among the true sample units or replicates and the number of samples. If subsamples are mistakenly treated as true sample units or replicates, sample variance will underestimate the actual amount of variation in the populations; sample size will overestimate true sample size; and we will be overconfident in the precision of the estimate, because its true standard error will be underestimated.

To illustrate this point, suppose we wanted to evaluate the effect of prescribed fire on northern bobwhite habitat in a large valley (1,000 km^2). We might conduct research on a habitat improvement project that involves burning 1 km^2 of grassland and brush (e.g., Wilson and Crawford 1979). We could place 20 permanent plots in the area to be burned and 20 in an adjacent unburned area. Measurements on burned and unburned plots before and after the fire could be compared to examine the effects of fire on bobwhite habitat. However, the 20 plots on the burned area are not really replicates, but are merely subsamples or **pseudoreplicates** (Hurlbert 1984). In fact, we have only one observation, because we have only one fire in a 1-km^2 plot in the 1,000-km^2 valley. What would happen if we were to redesign the study to conduct 20 burns on 20 randomly chosen areas scattered throughout the valley? We would expect to see more variation among these plots than among 20 plots in a single burned area. The fallacy of the first design is obvious. A statistical test would evaluate only whether the burned 1-km^2 area differed from the unburned 1-km^2 area and could lead to false conclusions about effects of burning on bobwhite habitat in this area. A more appropriate design would require randomly selecting 40 sites from throughout the entire valley and randomly assigning 20 to be burned (treatments) and 20 to be control (unburned) sites. Each burned and control site would be sampled with 5 plots to measure bobwhite habitat before and after the treatment, and data would be analyzed by ANOVA; the 40 sites are samples and the 5 plots per site are subsamples. Thus, the 20 sites of each type would be true replicates. Treating the 100 burned and 100 unburned plots as experimental replicates would be an example of **pseudoreplication.** Pseudoreplication is a common problem, and investigators must understand the concept of replication and its importance in ecological research (Hurlbert 1984, Johnson 2002).

Controls

In experimental research, a **control** may be defined as parallel observations used to verify the effects of experimental treatments. Control units are the same as experimental units except they are not treated; they are used to eliminate the effects of confounding factors that could potentially influence conclusions or results. Creative use of controls would improve many wildlife studies. Experimental studies in wildlife that involve repeated measurements through time must include controls because of the importance of weather and other factors that vary with time (Morrison et al. 2008). Without adequate controls, distinguishing treatment effects from other sources of variation is difficult. For example, in the northern bobwhite study, control sites were required to distinguish the effects of burning from those of rainfall and other weather characteristics that affect plant productivity. There might be an increase in grass production in the year following burning because the rainfall was higher that year. Without control sites we cannot tell whether increased grass production resulted from increased rainfall, from burning, or from a combination of both factors. Thus, we cannot evaluate the relative importance of each factor.

Determining Sample Size

One of the more challenging steps prior to starting actual data collection is to set goals for sample size using a prospective **power analysis.** The **power** of any hypothesis test is defined as the probability of rejecting the null hypothesis when, in fact, it is false. Power depends on the magnitude of the effect (e.g., magnitude of difference between treatment and control or a bound on the estimate), variation in the characteristic, significance level (α), and sample size. Zar (1999) provides formulas to calculate power and sample size for hypothesis tests, but a statistician should be consulted for complicated experimental designs and analyses. Many statistical packages (e.g., Statistical Analysis System; SAS Institute 2008) or specialized analysis software (e.g., MARK; White and Burnham 1999) provide capability to generate sample data for analysis to determine in advance how large the sample size should be to detect effects expected.

Effect size (magnitude of effect) is an important factor influencing sample size requirements and the power of a test. However, power and sample size calculations should be based on a biologically meaningful effect size. Identifying a biologically significant effect usually involves expressing the conceptual model as a quantitative model plus value judgments about the importance of a biological response. Estimating power of the test and calculating sample size requirements forces the investigator to evaluate the potential significance of the research prior to beginning fieldwork. Sample size analysis may lead to substantial revision of the goals and objectives of research.

Checklist for Experimental Design

The design of any experiment must be developed carefully or the conclusions reached will be subject to doubt. Four particularly critical elements in the design of a manipulative experiment are (1) specification of the research population, (2) replication with independent units, (3) proper use of controls, and (4) random assignment of treatments to experimental units. An experimental design checklist, such as the one listed in this section, is useful for providing a series of questions to assist in addressing these critical elements. Many of the questions will be helpful with the design of data gathering for studies involving nonexperimental hypothesis testing. Some experimental designs may address several hypotheses simultaneously (e.g., **factorial designs**); in other designs, each hypothesis may require independent experimental testing.

1. **What is the hypothesis to be tested?** The hypothesis developed from the conceptual model must be stated clearly before any experiment can be designed. For example, we could test the hypothesis that nest predation on forest songbirds is higher at sharp edges, such as occur at typical forest clearcuts, than at feathered edges (partial timber removal), such as occur at the boundary of selectively logged areas (Ratti and Reese 1988, Chalfoun et al. 2002, Stephens et al. 2003).

2. **What is the response or dependent variable(s) and how should it be measured?** The **response variable** should be clear from the hypothesis (e.g., nest predation), but selecting the best technique to measure it might be more difficult to determine. We must consider all possible methods and identify one that will simultaneously maximize precision and minimize cost and bias. It is often helpful to contact others who have used the techniques, examine the assumptions of the techniques, and conduct a pilot study to test the potential them. In our example, we might search for naturally occurring nests along forest edges and use a generalized Mayfield estimator of mortality rate (Heisey and Fuller 1985, Jehle et al. 2004, King et al. 2009). This response variable is continuous, and we could apply any of a variety of designs termed **general linear models** (GLM; e.g., ANOVA, linear regression, or analysis of covariance) under a hypothesis testing framework though application of information theoretic methods to these models. Alternately, we could measure the response for each nest as successful (at least one young fledged) or unsuccessful and use appropriate analysis methods, such as chi-squared statistics applied to contingency tables or log-linear models (Fienberg 1970, 1980; Hazler 2004).

3. **What is the independent or treatment variable(s) and what levels of the variable(s) will be tested?** The **indepen-**

dent variable(s) should be clear from the hypothesis (sharp and feathered forest edges in our example), but selecting levels to test will depend on the population for which we want to make inferences. If we want to test the effects of the independent variable at any level, we must select the levels to test at random (random effects [Model II]; Zar 1999). If we are interested in only a few of the levels that our independent variable could take, we use only those levels in our experiment and make inferences only to the levels tested (fixed effects or Model I; Zar 1999). For example, if we wanted to evaluate the effects of forest edges of any type on predation rates, we would select types of forest edges at random from all types that occur and apply a random effects model to analyze the data. In our example we are interested only in the 2 types categorized as sharp and feathered, so a fixed effects model is appropriate. Additionally, our independent variable must be identified and classified clearly or measured precisely. Finally, how can we use controls to expand our understanding? In our example, comparing nest predation in undisturbed forests to predation at the 2 types of edges might be enlightening, and we would analyze the data with fixed effects models. Our final conclusions would not apply to predation rates in all types of forest edges, but only to the 2 types that we compared to undisturbed forest.

An alternative approach to the design would be to treat the independent (treatment) variable as being **continuous** and use regression rather than a classified grouping of treatment categories. Under this design we might specify the treatment would consist of some level of overstory removal on one side of the forest edge, and we would apply regression forms of GLM under either hypothesis testing or information theoretic model evaluations. The response could be measured as the difference in predation rates between the 2 sides of the boundary, which would be predicted from percentage of overstory removed. Here it becomes critical to select **treatment levels** (e.g., percentage of overstory removed) across the full range of forest treatments to which we want to apply our conclusions.

4. **For which population do we want to make inferences?** If the results of the experiment are to be applied to the real world, our experimental units must be drawn from some definable portion of that world, the **research population.** The dependent and independent variables chosen should define the relationship(s) examined and place constraints on the definition of this population. We must also consider the impact of potential **extraneous factors** when selecting the population of interest. If the population is defined so broadly that many extraneous factors affect the results, the variation might be so large that we cannot test the hypothesis (low internal validity). If the population is defined so narrowly that we have essentially a laboratory experiment, application of the results might be severely limited (low generality or external validity).

Reaching the proper balance between internal and external **validity** takes thought and insight. For example, we might want to compare nest predation rates in sharp and feathered forest edges throughout the northern Rocky Mountains, but the logistics and cost would make the study difficult. Thus, we might restrict the study population to one national forest in this region. Next we need to consider the types of forests. We might want to test the hypothesis for the major forest types, but we know the species of birds nesting in these forests and their nest predators differ among forest types. Thus, we may need to restrict our population to one important type of forest to remove extraneous factors that could impact the results if we sampled a large variety of forest types. We need to ask what types of sharp and feathered edges occur to decide which we will sample. Sharp edges are commonly produced by clearcuts, power line rights-of-way, and road rights-of-way. These 3 types differ dramatically in such factors as size, shape, human access, and disturbance after treatment. Additionally, our ability to design a true experiment involving random assignment of treatments is severely limited for all but the clearcuts. Therefore, we might restrict the populations to sharp edges created by clearcuts and feathered edges created by selective harvests.

5. **What is the experimental unit?** What is the smallest unit that is independent of other units, which will allow random assignment of a treatment? This element must be identified correctly or the resulting experiment might not have true replication, but instead represent a case of pseudoreplication (Hurlbert 1984). For example, we might erroneously decide the experimental unit for our nest predation study will be an individual nest. The resulting design might entail selecting 3 areas and randomly assigning them to be clearcut, control, and selectively logged. By intensive searching, we find 20 nests along the edge of each area and monitor them for predation. The resulting data would suggest 20 replicates of each treatment, but, in fact, only a single area was given each treatment. Only 1 area was randomly assigned each treatment, and the 20 nests are subsamples. Thus, pseudoreplication restricts the potential inferences. In effect, we have sampled from populations consisting only of 2 logged areas and 1 unlogged area, and our inferences can be made only for those 3 areas, not to clearcuts, selective cuts, or undisturbed forests in general.

In some situations, **pseudoreplicated designs** are unavoidable, but interpretation of their results is severely restricted, because without replication, confounding factors rather than the treatment could have caused the results. For example, in our nest predation experiment if one of the areas was in the home range of a pair of common ravens

(*Corvus corax*) and the other areas were not, this single confounding factor could affect the results regardless of treatment. A more reliable experiment would require that we identify several areas with potential to be logged, perhaps 15, sufficiently far apart to be independent of one another, and that we randomly assign 5 each to be clearcut, selectively harvested, and controls. We would locate and monitor several nests in each area. The nests in a single area would be correctly treated as subsamples, and their overall success treated as the observation for that area. This approach attempts to remove the effects of confounding factors and to allow development of a conclusion with general application to the populations sampled (i.e., edges created by clearcuts and selective cuts in this habitat type in this region). Including control stands without an edge provides invaluable information for assessing the biological significance of the difference between the 2 types of edges.

6. **Which experimental design is best?** A few of the most widely used designs are described, but we advise consulting texts on experimental design and a statistician before making the final selection (Scheiner and Gurevitch 2001, Quinn and Keough 2002, Morrison et al. 2008). The choice depends primarily on the type of independent and dependent variables (categorical, discrete, or continuous), number of levels of each, ability to block experimental units together, and type of relationship hypothesized (additive or with interactions). For our study of nest predation along 2 types of forest edges, a single-factor design would be appropriate, but Hurlbert's (1984) argument for interspersion of treatments and controls could be incorporated by using a more sophisticated design. For example, 3 adjacent stands in 5 different areas might be randomly assigned to treatment and controls, with areas cast as blocks, resulting in a randomized complete-blocks design (Zar 1999).

7. **How large should the sample size be?** Estimating sample size needed for proper analysis is essential. If the necessary sample size were too costly or difficult to obtain, it would be better to redesign the project or work on a different question that can be answered. Sample size depends on the magnitude of the effect to be detected, variation in the populations, type of relationship that is hypothesized, and desired power for the test. Typically some preliminary data from a pilot test or from the literature are required to estimate variances. These estimates are used in the appropriate formulas available in statistical texts (e.g., Zar 1999) and incorporate a prospective power analysis to ensure that we have a high (80–90%) chance of detecting biologically meaningful differences between the treatment and control categories. Powerful analysis programs like SAS (SAS Institute 2008) provide tools to perform prospective power analysis for complicated designs.

8. **Have you consulted a statistician and received peer review on the design?** Obtaining review by a statistician before the data are gathered is essential. The statistician will not be able to help salvage an inadequate design after a study is completed. Peer review by other biologists having experience with similar studies also could prevent wasted effort if measurements or treatments are proposed that will not work on a large scale in the field. Now is the time to get these comments!

MODELING

"All models are wrong, but some are useful" (Box 1979:2). Rigorously evaluating ideas concerning wildlife habitats and populations by using experimental manipulations may be difficult, because we cannot randomly assign treatments and the high cost of treatments precludes adequate replication in many cases. However, modeling methods provide an alternative route to finding solutions to pressing problems (Starfield and Bleloch 1991, Shenk and Franklin 2001), selecting the best of alternative choices (Holling 1978, Walters 1986, Clemen and Reilly 2001, Conroy and Peterson 2009), determining the relative magnitude of effects from multiple causes acting simultaneously (Wisdom and Mills 1997, Saltelli et al. 2001), and evaluating population viability (Mills 2007:254). A biologist's goal should be to build the simplest model that describes the relationships between causative factors and the effects they produce. It is most likely that a wildlife scientist will select a modeling strategy at the simple, empirical ends of the continua in terms of model complexity (Table 1.2) or in Levins's (1966) terms, sacrifice generality for realism and precision. Long-term monitoring data and extensive measurements of demographic rates and habitat relationships provide the basis for more complex models.

In most cases the goal is to model the responses of wildlife populations or habitats with the smallest number of predictors necessary to make good predictions. Note, this use of the term model corresponds to what Williams et al. (2002*a*:23) refer to as a **scientific model** rather than a statistical model. **Statistical models** are the foundation for all statistical estimation, hypothesis testing, and statistical comparison among competing models through an inductive process based on limited observations (see later sections under Parameter Estimation and Confronting Theories with Data). Scientific models, described in this section, are used deductively to project system dynamics based on a set of ideas expressed as characteristics and relationships estimated inductively from statistical models. We use these 2 types of models cooperatively to help answer important questions about wildlife. Scientific models are commonly referred to as **simulation models,** because they simulate the dynamics of a system described in terms of the assumptions, charac-

Table 1.2. Modeling strategies along gradients of simple to complex for scientific and statistical models

	Gradient	
	Simple	Complex
Scientific models		
Quantification	Conceptual (verbal)	→ Quantitative
Theoretical	General	→ Complex simulation
Relationships	Linear	→ Nonlinear
Variability	Deterministic	→ Stochastic
Time scale	Time-specific	→ Dynamic
Mathematical formulation	Difference equations	→ Differential equations
Number of factors	Single	→ Multifactor
Number of sites	Single site	→ Multisite
Number of species	Single species	→ Multispecies
Statistical models		
Sampling	Simple random	→ Stratified, clustered or multistage
Hypothesis testing	Fixed or random effects	→ Mixed fixed and random effects
Independence of observations	Complete independence	→ Dependence among observations in space, time, or both
Errors	Single term	→ Separate process and observation errors

teristics, relationships, and variability observed. When variability is a key component, they are referred to as **Monte Carlo** scientific models. Kitching (1991) suggested a variation of the following 8 steps to build an ecological model. These steps are directly applicable to building scientific wildlife models.

Steps to Build a Model
Problem Definition
The problem of interest must have been identified earlier as one of the first steps in the scientific method, and the relevant theory, previous observations, conceptual model (Fig. 1.4), predictions, and hypotheses must be stated clearly. Someone proposing to build a model to answer the question must now explain why a numerical or mathematical model is an appropriate way of tackling the problem (Kitching 1991:31). A good example of an appropriate question is: which of the available management options are more likely to recover an endangered species and prevent its extinction? It is important to embrace the modeling approach to this problem as a pragmatic one. "There is no point at all in building an ecological model that is more complex, more complete or more time-consuming than is justified by the terms of reference of the problem to which the model is a response" (Kitching 1991:31). The better the problem(s) is identified, the more useful the model will be.

System Identification
After identifying the problem(s) it is critical to define the **system boundary** and the **level of resolution** to model in the hierarchy of ecological levels (ranging from individual animals with associated spatial extent to population or metapopulation; Fig. 1.2). The biologist must then select a set of components to model (see the examples in Fig. 1.4). One strategy is to pursue a parsimonious approach, making the model as simple as possible, by selecting only critical components essential to describe the system. This approach is used for developing general **theoretical models** (Table 1.2) taking the form of analytical mathematical models. The other extreme is to include all components likely to be involved in the processes of interest. Such models take the form of complex simulation models. The typical route followed in wildlife models is to take the simple empirical approach, and Starfield and Bleloch (1991) recommend tending toward the parsimonious end while including enough complexity to produce realistic predictions. Once the initial set of components is defined to meet the objectives, the nature of their interactions must be defined based on creative thinking and literature as follows: positive, negative, feedback loops, and complex combinations. Creating a simple system diagram is useful for clarifying these relations (e.g., Fig. 1.4) and guiding literature searches.

Model Type Selection
The great variety of model types available (Table 1.2) may seem daunting at first, but the problem definition process described above should guide selection of the appropriate type of model along the continuum from simple to complex, with preference always for the simplest model necessary to meet the needs. Building complex models requires estimating more characteristics with more complex relationships. Fortunately most wildlife problems can be handled with simple, linear models incorporating deterministic effects of a few independent factors at a single or small number of sites. Even forecasts for population viability requiring stochastic models with time lags are easily modeled

with simulations based on estimates obtainable with standard linear regression methods (e.g., Garton et al. 2010).

Mathematical Formulation

Almost all wildlife models are formulated as difference equations because of strong seasonal and annual patterns, which make estimating parameters for **continuous time models** formulated as **differential equations** difficult. Differential equation formulations have been more successful for developing general theoretical models that form the basis for many ecological theories underlying principles of wildlife population ecology (Ginzburg 1986, Turchin 2001, Berryman 2003, Colyvan and Ginzburg 2003), but translating these general models into **stochastic difference equation models** has proven very successful for modeling time series of populations with complex dynamics (e.g., population viability analysis for San Joaquin kit fox [*Vulpes macrotis mutica*] incorporating density dependence and a 2-year lag in rainfall effects on plant productivity; Dennis and Otten 2000).

Computational Method and Program Selection

Simple wildlife models formulated in commonly used general purpose spreadsheet programs can provide remarkable insight into wildlife population dynamics (Starfield and Bleloch 1991). Some specialized software designed for specific purposes, such as population viability analysis, have wide application to projecting persistence of endangered and rare species—for example, RAMAS (Akçakaya 2000b) and VORTEX (Lacy 1993). Programs designed for statistical analysis—for example, SAS (SAS Institute 2008) and R (R Development Core Team 2006)—are equally adaptable to simulating both deterministic and stochastic models as they are to estimating the parameters for these models (Bolker 2008, Garton et al. 2010).

Parameter Estimation

Sampling methods, least squares for GLM, and maximum likelihood methods are all useful for estimating parameters for alternative models. Information-theoretic approaches to evaluating competing models (see further details later under the section Confronting Theories with Data) provide excellent tools to evaluate relative precision of alternative models in predicting responses (Burnham and Anderson 2002). Burnham and Anderson (2002) contend that information theoretic methods, such as using **AIC** to assess the information content of a model, should be applied where we cannot experimentally manipulate causes or predictors. Model averaged parameter estimates are readily calculated within this framework using Akaike weights (Burnham and Anderson 2002:133ff.).

Model Validation

Validation of a model should take at least 2 forms. Comparing the predictions of the model to data that were analyzed to build the model provides a preliminary validation or **verification** (Oreskes et al. 1994) that is always performed as part of constructing the model. Clearly this step is essential to verify the model is performing as the investigator expects. A real test of the **validity** of the model requires comparing output from the model to independent data not used in its construction (Gardner and Urban 2003). The comparison is usually made with standard statistical tools, such as correlation and regression, which may be evaluated from a frequentist perspective by using either hypothesis tests or likelihood measures. Because models using all data possible maximize precision in parameter estimation, approaches, such as jack-knifing, in which each individual observation is predicted from models fit to all the rest of the data are applied (Efron and Tibshirani 1993).

Model Experimentation

Once the previous seven steps are completed, the model is ready to be used to address the original questions that initiated the modeling process. **Experiments** are performed by manipulating key input parameters to assess the response of model output characteristics to anticipated alternative management actions and/or potential environmental trends, changes, or variation. A useful model is an invaluable aid to both research and management, but the **veracity** of any predictions rests firmly on assumptions built into the model structure, the relationships modeled, and the validity of any parameters estimated from field observations. Scheller et al. (2010) provide further details on the approach outlined here, which applies modern software engineering techniques as part of a process to increase the reliability of ecological models. A useful model should be used interactively with ongoing research and management activities, so that modeling exercises help identify critical relationships and parameters that are then investigated in the field by gathering new observations or performing experiments. In the management context this process is **adaptive management:** model predictions guide management actions and continued monitoring provides feedback to validate and improve model assumptions expressed as model parameters and relationships.

SAMPLING

Most information gathered by wildlife biologists is used to meet descriptive rather than experimental objectives, but obtaining precise estimates is equally important for both experiments and descriptive research. Examples include estimates of population size, recruitment, herd composition, annual production of forage species, hunter harvest, and public attitudes. In these efforts biologists attempt to obtain estimates of characteristics that are important for management decisions. We want to obtain the best estimates possible within the constraints of our time and money resources. A large body of statistical literature exists to help; these types of studies are referred to as **surveys,** and the topic is known as **survey sampling** (Cochran 1963, 1983; Scheaffer et al. 2005) or **finite population sampling.**

The research population is typically synonymous with the statistical population, but a powerful approach is to redefine the **statistical population** geographically in terms of units of space or habitat. Defining the statistical population as drainages, forest stands, individual ponds, or square-kilometer blocks often facilitates estimating total numbers of animals and the composition of a population. Sampling smaller units of habitat is more likely to be logistically feasible. Likewise this redefinition of the research (statistical) population makes it feasible to apply the powerful tools for sampling from finite populations.

Sampling also is a critical part of experimental research and the test of formal statistical hypotheses. All field studies and most field experiments require creative sampling designs to reduce variation among observations in the treatment or comparison categories. For example, stratification and clustering can sharpen comparisons, but data collected using these methods require analysis by more complicated designs (e.g., block or split-plot designs; Zar 1999). Choice of specific sampling methods is dependent on the objectives or hypotheses being addressed, the nature of the population being sampled, and many other factors (e.g., species, weather conditions, topography, equipment, personnel, time constraints, and desired sample sizes). A variety of sampling designs is available for biologists to use in wildlife surveys and experimental research (Thompson et al. 1998, Scheaffer et al. 2005, Morrison et al. 2008).

Precision, Bias, and Accuracy

One measure of quality of estimates is their precision. **Precision** refers to the proximity of repeated measurements of the same quantity (Cochran 1963, Krebs 1999, Zar 1999). Precision of an estimate depends on variation in the population and the size of the sample. Indicators of the precision of an estimator are **standard error** and **confidence intervals.** Larger variation in the population leads to lower precision, whereas a larger sample size produces higher precision in the estimator. Another measure of the quality of an estimator is **bias.** Bias describes how far the average value of the estimator is from the true population value. An unbiased estimator centers on the true value for the population. If an estimate is both unbiased and precise, we say that it is **accurate** (defined as an estimator with small mean-squared error; Cochran 1963). Accuracy is the ultimate measure of the quality of an estimate (Fig. 1.6) and refers to the small size of deviations of the estimator from the true population value (Cochran 1963).

Let us illustrate these concepts with a typical population survey. Suppose we were interested in estimating the density of elk on a large winter range. One approach might be to divide the area into a large number of count units of equal size and draw a sample of units to survey from a helicopter. This approach would define the research population in terms of a geographic area rather than in terms of animals. The elements of the target population are count units, and we select a sample of these units using an objective sampling design (a probability sample). Using the helicopter, we search each sampled unit, attempting to count all elk

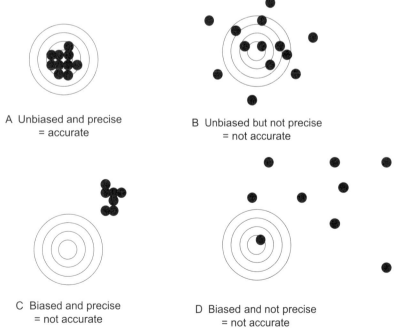

Fig. 1.6. Concepts of bias, precision, and accuracy illustrated with targets and a shot pattern. *Modified from Overton and Davis (1969), White et al. (1982).*

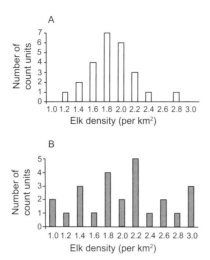

Fig. 1.7. Hypothetical example of elk counts and density estimates in (A) Area 1 and (B) Area 2.

present in it. We divide the number of elk counted in a unit by the size of that unit to obtain a density estimate for each unit (Fig. 1.7A). The histogram suggests little variation in density on this winter range, as most spatial units (80%) have densities between 1.5 and 2.3 elk/km^2. We need a single value that is representative of the entire winter range, and we choose the mean from the sample as the best estimate. The variation from one unit to the next is small; thus, the mean from our sample is a fairly precise estimate. But, suppose we had obtained different results (Fig. 1.7B). Now the variation from one unit to the next is great, and the sample mean is less precise and not as reliable as the previous estimate. Thus, for a given sample size, the former estimate is more precise because of less variation in the population.

Would the mean from the sample in Area A (Fig. 1.7A) be an accurate estimate of the mean density of elk on this winter range? To answer this question, we must evaluate the bias in the estimate. If the winter range was partially forested or had tall brush capable of hiding elk from view, aerial counts in each unit would underestimate the true number of elk present (Samuel et al. 1987). In this example the mean density from the sample would be a biased estimate of elk density on the winter range and, therefore, not highly accurate. If the winter range were a mixture of open brush fields and grasslands, where all animals would be visible, mean density from the sample could be an accurate estimate of elk density on the entire winter range. We strive for accuracy in our estimates by selecting the approach with the least bias and most precision, applying a valid sampling or experimental design, and obtaining a sufficiently large sample size to provide precise estimates.

Evaluating bias in an estimate is difficult and, in the past, has been based on the researcher's biological knowledge and intuition. If the bias is constant, the estimate can be used to make relative comparisons and detect changes (Caughley 1977). Usually it is not constant (Anderson 2001), but its magnitude often can be measured so that a procedure to correct estimates can be developed (Rosenstock et al. 2002, Thompson 2002b). For example, Samuel et al. (1987) measured visibility bias in aerial surveys of elk from helicopters, and Steinhorst and Samuel (1989) developed a procedure to correct aerial surveys for this bias.

Sampling Designs
Simple Random

A **simple random sample** requires that every sample unit in the population has an equal chance of being drawn in the sample and the procedure for selecting units is truly random. This can be accomplished by assigning each member of the population a number and then picking numbers, to identify members to sample, from a table of random numbers or a random number generator on a computer or calculator. For example, suppose that for a special hunt in which a limited number of permits was issued, we wanted to estimate the number of successful hunters. We might decide to contact a sample of permit buyers by telephone after the season to measure their hunting success. A survey design checklist (Box 1.4) helps design such a survey properly. The **population** that we want to make statements about is all persons who obtained a permit. The list of the members of the population is usually called the **sampling frame** (Scheaffer et al. 2005). It is used to draw a random sample from the population. The sampling frame must be developed carefully, or the resulting estimates may be biased. For example, if a portion of the permit buyers did not have telephones and we decided to drop them from the list, the results could be biased if such hunters had different hunting success than did permit buyers with telephones. To draw a random sample for the survey, we could assign a number to each person who purchased a permit and select the numbers to be contacted by using a table of random numbers or a random number generator.

In other types of surveys, obtaining a truly random sample of the population might be difficult. In such instances another method, such as systematic sampling, should be used. When the research population consists of animals that would be difficult to sample randomly, one approach is to change the design. We do this by making small geographic units, such as plots or stands, the **sample units** (or **experimental units,** if we are developing a sampling design for an experimental treatment) and making the measurement on each plot a number or density of animals. Thus, we can take a random sample of spatial units and use it to infer abundance across the entire study area sampled. A valid random sampling procedure must be independent of investigator decisions. For example, an excellent procedure for ran-

Box 1.4. Survey design checklist

Question	Example
1. What is the survey objective?	Estimate the percentage of successful hunters
2. What is the best technique	Telephone survey of permit holders or method?
3. To which population do we make inferences?	Everyone who has a permit for this hunting period
4. What will be the sample unit?	Individual permit holders
5. What is the size of the population to be sampled (N)?	$N = 350$ (for special permit hunt)
6. Which sample design is best?	Simple random sample (Scheaffer et al. 2005).
7. How large should the sample be?	$n = \dfrac{Np(1-p)}{(N-1)B^2/4 + p(1-p)}$ $n = \dfrac{Np(1-p)}{(N-1)(B^2/4) + p(1-p)},$ where: N = population size (350) p = proportion of permit holders who harvested deer (from pilot survey = 0.24) B = bound on the estimate = 0.05 (we want an estimate with $p \pm 0.05$ confidence) Therefore $n = \dfrac{350(0.24)(1-0.24)}{(350-1)(0.05)^2/4 + 0.24(1-0.24)},$ $n = 159$ (i.e., we should contact approximately 160 permit holders
8. Have you contacted a statistician to review design?	Yes!

domly locating plots in a study area would be to use a Landsat image of the study area stored in a Geographic Information System (**GIS**) program, which allows us to select random locations within the boundary of our study area using Universal Transverse Mercator (UTM) coordinates (Fig. 1.8A). The UTM coordinates of these selected plot locations can be entered into a hand-held Global Positioning System (**GPS**) unit that will guide us to the exact location. Random-like methods, referred to as haphazard or representative, have been used in place of truly random designs, but should be avoided, because they are subject to investigator bias. An example of these methods is the technique of facing in a random direction and throwing a pin over the shoulder to determine the center for a vegetation plot. Although this procedure seems random, the odds of a field crew randomly facing away from a dense stand of thorny shrubs, such as multiflora rose (*Rosa multiflora*), and throwing the pin into the middle of such a patch is practically zero. Truly random samples occasionally produce poor estimates by chance due to poor spatial coverage of the area or population of interest (e.g., in an area with a small number of important habitat patches, all patches may be missed by a truly random approach; Hurlbert 1984, Johnson 2002).

Systematic

A **systematic sample** is taken by selecting elements (sampling units) at regular intervals as they are encountered. This method is easier to perform and less subject to investigator errors than simple random sampling. For example, if we wanted to sample birdwatchers leaving a wildlife management area, it would be difficult to draw a truly random sample. However, it would be easy to draw a systematic sample of 10% of the population by sampling every tenth person leaving the area. Systematic sampling also is used extensively in vegetation measurements because of its ease of use in the field. It is almost exclusively used in geographic sampling, because it makes possible evaluation of the spatial pattern of variability (e.g., spatial autocorrelation), which is used for most modern spatial modeling. A valid application requires random placement of the first plot, followed by systematic placement of subsequent plots, usually along a transect or in a grid pattern (Fig. 1.8B). This approach often

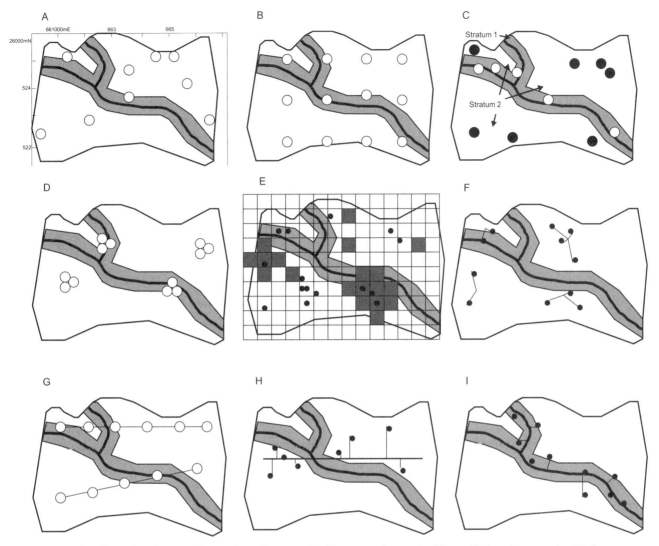

Fig. 1.8. Examples of sampling designs: (A) simple random sample, (B) systematic sample, (C) stratified random sample, (D) cluster sample, (E) adaptive cluster sample. Examples of sampling methods: (F) point sampling, (G) plots along transects, (H) line transect, (I) road sampling.

provides greater information per unit cost than simple random sampling, because the sample is distributed uniformly over the entire population or study area. For random populations (i.e., no serial correlation, cyclic pattern, or long-period trend), systematic samples give estimates with the same variance as simple random samples.

The major danger with systematic samples is they may give biased estimates with **periodic populations** (i.e., with regular or repeating cycles). For example, if we were interested in estimating the number of people using a wildlife management area, we might establish a check station and take a systematic sample of days during the season. This procedure could yield extremely biased results if we chose to take a sample of one-seventh of the days. If the day sampled fell during the workweek, we could obtain different results than if it were during the weekend. Additionally, the estimate of variance would likely be too small, leading us to conclude the estimate was much more precise than it is in reality. In this situation the population sampled obviously is periodic; in other situations the periodicity might be quite subtle. Thus, systematic sampling must be used with caution. The formal procedure is conducted by randomly selecting one of the first k elements to sample and every kth element thereafter. For example, if we wanted to sample 10% of our population, k would equal 10, and we would draw a random number between 1 and 10. Suppose we selected 3; we would then sample the 3rd element and every 10th element thereafter (i.e., 13th, 23rd, 33rd, . . . element). At a check station we might use this strategy to sample 10% of the deer hunters or birdwatchers who came through the station. When locating plots along a transect, we would randomly locate the starting point of the transect and then place plot centers at fixed intervals along the transect, such as every 100 m. Advantages and disadvantages of random and systematic sampling have been reviewed by Thompson et al. (1998), Krebs (1999), and Morrison et al. (2008).

Stratified Random

In many situations, obvious subpopulations exist in the total population. For example, tourists, birdwatchers, and hunters are readily divided into residents and nonresidents. A study area can be divided into habitats. A population of animals can be divided into age or gender groups. If members of these subpopulations are similar in terms of the characteristics we are estimating and the subpopulations themselves differ from one another in the characteristic of interest, a powerful design to use is **stratified random sampling.** Subpopulations are referred to as strata, and we draw a simple random sample of members from each stratum. Stratified random sampling also is useful if we are particularly interested in the estimates for the subpopulations themselves. The strata are chosen so they contain units of identifiably different sample characteristics, usually with lower variance within each stratum.

For example, if the objective of a study of moose (*Alces alces*) is to estimate moose density, we might define strata on the basis of habitats (e.g., bogs and riparian willow [*Salix* spp.] patches, unburned forests, and burned forest). We then draw a simple random sample from each stratum (Fig. 1.8C). If moose density is different among strata, variation in each stratum will be less than the overall variation. Thus, we will obtain a better estimate of moose density for the same or less cost. If strata are not different, stratified estimators may not be as precise as simple random estimators. In some instances the cost of sampling is less for stratified random sampling than for simple random sampling. A final advantage of stratified random sampling is that separate estimates for each stratum (e.g., moose density in willows or in forests) are obtained at no extra cost. The **formal procedure** for stratified random sampling consists of 3 steps: (1) clearly specify the strata (they must be mutually exclusive and exhaustive), (2) classify all sampling units into their stratum, and (3) draw a simple random sample from each stratum. Formulas are available to calculate the sample size and optimal allocation of effort to strata (Krebs 1999, Scheaffer et al. 2005). A pilot survey can be analyzed using ANOVA to learn whether stratification is indicated. If cover types define strata, most GIS software will automatically select random coordinates within cover types, making stratified random samples easy to select.

Cluster Sampling

A **cluster sample** is a simple random sample in which each sample unit is a cluster or collection of observations (Fig. 1.8D). This approach has wide application in wildlife biology, because many birds and mammals occur in groups during all or part of the year. When we draw samples from such populations, we draw clusters of observations (i.e., groups of animals). Likewise, many wildlife user groups (e.g., waterfowl hunters and park visitors) occur in clusters (e.g., boats in wetlands and vehicles along highways). Cluster sampling also is useful where cost or time to travel from one sample unit to the next is prohibitive. This situation is common in surveys of animals and habitat. The formal procedure for cluster sampling consists of 3 steps: (1) specify the appropriate clusters and make a list of all clusters, (2) draw a simple random sample of clusters, and (3) measure all elements of interest in each cluster selected.

Making a formal list of clusters is rarely possible or essential. Instead, we emphasize obtaining a random sample of clusters. If the sample units are animals, which naturally occur in groups, the size of the clusters will vary from group to group, depending on the social behavior of the species. Cluster sampling of habitat is performed by choosing a random sample of locations and then locating multiple plots in a cluster at each location. In this case, the researcher sets the cluster size. The optimal number of plots (**cluster size**) depends on the pattern of variability in habitat. If plots in a cluster tend to be **similar** (i.e., little variability in a cluster), cluster size should be small. If plots in a cluster tend to be **heterogeneous** (high variability within a cluster), it should be large. For other types of cluster samples, such as groups of animals or people in vehicles, cluster size is not under control, but is a characteristic of the population. For example, aerial surveys of elk and deer on winter ranges result in samples of animals in clusters. Estimates of herd composition (e.g., fawn:doe or bull:cow ratios) are readily obtained by treating these data as cluster samples (Bowden et al. 1984).

Adaptive Sampling

Adaptive sampling differs from the methods discussed earlier because the sample size is not set at the start of the sampling effort, but rather depends on the results obtained during sampling. Thompson and Ramsey (1983) pioneered adaptive cluster sampling for gathering information on rare animals and plants, which are often clustered in occurrence. In **adaptive cluster sampling** an initial sample of units is drawn by a random or other standard design, and neighboring units also are sampled for any unit that satisfies a criterion, such as having more than x individuals present (Thompson and Seber 1996, Williams et al. 2002a, Brown 2003, Thompson 2003). The initial sampling unit and neighbors (where sampled) form **neighborhoods** analogous to clusters and are treated as in cluster sampling. The size of clusters does not need to be constant, nor is it known in advance. For spatially clustered animals or plants, the neighborhood consists of adjacent spatial sample units (Fig. 1.8E). Smith et al. (1995a) showed that adaptive cluster sampling would be relatively more efficient than simple random sampling for estimating densities of some species of wintering waterfowl if the right sample unit size and criterion for further sampling in the neighborhood were chosen. The species for which it would be superior show more highly clustered distributions. For other species, conventional sampling designs with fixed sample sizes are superior. Numerous examples of applications of adaptive sampling under conventional sampling designs and estimation methods, as well as applications based on maximum likelihood methods and Bayesian approaches can be found in Thomas et al. (1992), Thompson and Seber (1996), Smith et al. (2003b, 2004), and Noon et al. (2006). Thompson et al. (1998), Williams et al. (2002a), and Morrison et al. (2008) also review the basic concept and provide simple examples.

Sequential Sampling

Sequential sampling differs from the classical statistical approach in that sample size is not fixed in advance (Wald 2004). Instead samples are drawn one at a time, and after each sample is taken the researcher decides whether a conclusion can be reached. Sampling is continued until either the null hypothesis is rejected or the estimate has adequate precision. This type of sampling is applicable to wildlife studies where sampling is performed serially (i.e., the result of including each sample is known before the next sample is drawn; Krebs 1999). The major advantage of this approach is that it usually minimizes sample size, thus saving time and money. After an initial sample of moderately small size is obtained, successive samples are added until the desired precision is met, the null hypothesis can be rejected, or a maximum sample size under a stopping rule has been reached. This approach typically requires 33% the sample size required in a standard design (Krebs 1999:304). For example, if we wanted to survey deer on a winter range to ensure that harvest had not reduced buck abundance below a management guideline of 5% bucks, we would develop a graph (Fig. 1.9) and plot the results of successive samples as shown (Krebs 1999:312). We must choose a level of significance for our test (e.g., $\alpha = 0.10$) and a power for the test $(1 - \beta = 0.90)$ and specify an upper rejection region (>10% bucks), above which we assume the population has not been adversely impacted by buck-only harvests. Once an initial sample of 50 deer has been obtained, sequential groups of deer encountered are added and totals plotted on the graph until the line crosses one of the upper or lower lines or the stopping rule is reached. For example, the lower rejection line is reached at a sample size of 140 (Fig. 1.9). At this point the null hypothesis that bucks constitute >5% of the herd would be rejected, and the conclusion would be there are 5% bucks remaining. An important constraint is the sample must be distributed throughout the entire population, so that a simple random sample of deer groups is obtained. Achieving this sample would be most feasible using aerial surveys from helicopter or fixed-wing aircraft.

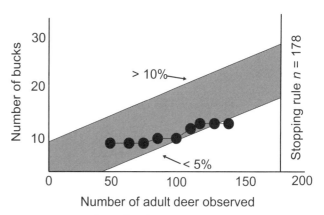

Fig. 1.9. Sequential sampling for percentage of bucks in a deer herd.

Other Sampling Designs

Many other sampling designs are available. For example, **2-stage cluster sampling** involves surveying only a portion of the members of each cluster drawn in the sample. This approach is efficient when clusters are large. Cluster sampling is one version of the more general method referred to as **ratio estimation** (Cochran 1963, Williams et al. 2002a). Related methods are **regression estimation** and **double sampling** (Scheaffer et al. 2005) that have great potential for wide application to wildlife research. The interested reader should consult a standard reference on sampling techniques (Scheaffer et al. 2005) and work with a statistician experienced in survey sampling. Stevens and Olsen (2004) proposed a new, efficient approach that combines the advantages of spatially systematic designs with the proven unbiased nature of random sampling. They described this approach as a **generalized random tessellation stratified (GRTS)** design. GRTS uses a recursive approach that converts a 2-dimensional map into a 1-dimensional one while maintaining spatial closeness in original locations. This conversion allows a valid systematic sample to be drawn that meets the requirements of random sampling while distributing the sample across the entire spatial area. Theobald et al. (2007) have provided free tools (STARMAP Spatial Sampling tools; http://www.stat.colostate.edu/~nsu/starmap/), which make it feasible to apply GRTS to generating spatially balanced probability-based survey designs.

Sampling Methodology

Plots

Plots are widely used to sample habitat characteristics and count animal numbers and sign. **Plots** represent small geographic areas (circular, square, or rectangular) that are the elements of the geographically defined population. The research population size is the number of these geographic areas (plots) that would cover the entire study area. Sufficient time, money, and personnel to study an entire area are usually not available, and a subset of plots is used with the assumption that it is representative of the area. Any of the survey designs (simple random, systematic, stratified random, cluster, etc.; Fig. 1.8) or more complicated designs, such as 2-stage designs, may be applied (Cochran 1963, Williams et al. 2002a). Selecting the best design requires insight into the characteristics and patterns of distribution of species across the landscape. One advantage of using plots is that size of the population is known and totals can be estimated (Seber 1982). Selection of plot size and shape, also an important consideration, has been reviewed by Krebs (1999).

Point Sampling

In **point sampling** a set of points is established throughout the population, and measurements are taken from each sample point (Fig. 1.8F). A common measurement is distance from the point to a member of the population (e.g., plant or calling bird). Examples include point quarter and nearest neighbor methods used widely to estimate the density of trees and shrubs (Mueller-Dombois and Ellenberg 1974), and the variable circular plot or point transect method of estimating songbird density (Reynolds et al. 1980). If observers doing point counts for birds record the distance to each bird detected, as in the variable circular plot approach, transforming distances to areas makes it easy to apply the extensive methods and algorithms developed for line transects referred to as **distance sampling methods** (Buckland et al. 1993, 2001, 2004; Laake et al. 1994). Selection of sample points usually follows a systematic design, but other sample designs can be used, as long as points are spaced sufficiently far apart that few members of the population are sampled more than once. Necessary sample size can be estimated from formulas even if population size is assumed to be large or unknown (Zar 1999).

Transects

A **transect** is a straight line or series of straight line segments placed in the area to be sampled. Transects are used to organize or simplify establishment of a series of sample points or plots and as a sample unit themselves. Transects are widely used to obtain systematic samples of spatially distributed populations (e.g., plants). In these situations, plots along transects are actual sample units (Fig. 1.8G) and should be treated as described for systematic sampling. Plots also can be placed along transects at random intervals. When transects are used as sample units, they are commonly referred to as line transects (Burnham et al. 1980, Williams et al. 2002a). Measurements of perpendicular distance, or sighting distance and angle, to the sampled elements (e.g., flushing animals, groups of animals, carcasses, and snags) are recorded (Fig. 1.8H). These distances are used to estimate the effective width of the area sampled by the transect (Seber 1982; Buckland et al. 1993, 2001, 2004). Each transect is treated as an independent observation, and transects should be nonoverlapping according to established sampling designs (e.g., simple random, systematic, and stratified random). Transects are often easier to establish in rough terrain than are plots, but they must be established carefully with a compass or transit and measuring tape or with a GPS unit. Use of transects is becoming more widespread in aerial survey work because of development of precise navigational systems (Patric et al. 1988, Anthony and Stehn 1994, Marques et al. 2006). The critical assumptions for transect methods for sampling such mobile objects as animals (i.e., 100% detection for objects directly on the line and no movement toward or away from the observer before detection) must be examined carefully before this sampling method is selected (Burnham et al. 1980, Williams et al. 2002a). In certain cases, more sophisticated methods may be used to adjust counts for less-than-perfect detection on the line (Buckland et al. 1993, 2001, 2004; Manly et al. 1996; Quang and Becker 1996; Williams et al. 2002a) or near the points (Kissling and Garton 2006). A strip transect appears similar, but it is really a long, thin plot, because the method assumes all animals or objects in the strip are counted (Krebs 1999).

Road Sampling

Sampling from roads is a widely used method for obtaining observations of species sparsely distributed over large areas or for distributing observations of abundant species over a large geographic area. This sampling method is usually the basis for spotlight surveys of nocturnal species, such as white-tailed deer (Boyd et al. 1986, Collier et al. 2007), black-tailed jackrabbit (*Lepus californicus*; Chapman and Willner 1986), grassland owls (Condon et al. 2005), brood and call counts of upland game birds (Kozicky et al. 1952, Kasprzykowski and Golawski 2009), scent-station surveys (Nottingham et al. 1989, Preuss and Gehring 2007, Mortelliti and Boitani 2008), and the Breeding Bird Survey (Robbins et al. 1986, Sauer et al. 2008). This approach involves drawing a sample from a population defined as that population occupying an area within a distance x of a road (Fig. 1.8I). The distance x is generally unknown and varies with any factor that would affect detection of an animal, such as conspicuousness, density, type of vegetation cover, or background noise for surveys based on aural cues.

Roads rarely provide unbiased estimates for a region, because they are generally placed along ridges or valleys and avoid steep or wet areas. Furthermore, roads modify habitat for many species and may attract some wildlife. For example, during snow periods some bird species will come to roads for grit and spilled grain. Thus, sampling along roads rarely provides a representative sample of habitat (e.g., Hanowski and Niemi 1995) or wildlife populations (Pedrana et al. 2009). Although this bias is well known, it is often ignored in exchange for a method that is cost efficient and easy. As with all indices, every effort should be made to standardize counting conditions along fixed, permanently located routes (Caughley 1977, Sauer et al. 2008); however, this alone does not guarantee reliable counts (Anderson 2001, Thompson 2002b). Sampling along roads can be an efficient approach if it is designed as a random sample from a stratum adjacent to roads that is one element of a stratified random sample of the entire area, including other strata distant from roads (Bate et al. 1999, Langen et al. 2009).

Dependent (Paired) and Independent Observations

If we wish to make population comparisons, **pairing observations** is a powerful tool for detecting differences. If there is a correlation between members of a pair, treating them as dependent or paired observations can improve the power of tests for differences. For example, to compare diets of adult female mountain sheep (*Ovis canadensis*) and lambs, we might treat a ewe with a lamb as a pair and measure the diet of each animal by counting the number of bites of each plant they eat while foraging together. Treating these observations as pairs would sharpen comparisons between age classes, because it would compare animals foraging together and experiencing the same availability of plants. Pairing is a powerful technique in other contexts for which there is dependency between the observations. Pairing should be used only if an association really exists, otherwise the power of comparison will be decreased.

Pairing also can be used to help answer a different question. For example, studies of habitat selection are often made by locating areas used by a species (i.e., nest sites or radio locations) and measuring habitat characteristics at these use sites with sample plots. Available vegetation types are measured from random sample plots throughout the study area (Fig. 1.10A). A comparison of use and random plots can identify characteristics of areas selected by the species. An alternative approach involves pairing use and random plots by selecting a random plot within a certain distance of the use plot (Fig. 1.10B). For analysis, use and random plots are paired (i.e., random plot locations are dependent on use sites). This comparison could produce different results from the unpaired comparison, because it tests for habitat differences in areas used by the species (microhabitat selection). In contrast the unpaired comparison (e.g., independent plots) tests for habitat differences between areas used by the species and typical vegetation types available in the general study area (macrohabitat selection). Choosing a paired or unpaired design will depend on the objectives of the study, but both may be useful when applying a hierarchical approach to studying habitat selection (Wiens 1973, Johnson 1980, Cruz-Angón et al. 2008, Schaefer et al. 2008).

CONFRONTING THEORIES WITH DATA

Confronting theories with data involves evaluation and interpretation, which is a creative phase, similar to hypothesis formulation. The quality of conclusions drawn is dependent on the biologist's past educational and professional experience as well as a willingness to consider standard and less traditional interpretations. One great danger in wildlife science (and other fields) is that researchers often have a conscious or unconscious expectation of results. This **bias** might begin with the development of the overall research objective and carry through to the interpretation phase. This danger is so great that in some fields, such as medicine, experiments are performed with a double-blind approach: neither researcher nor subjects know membership of treatment and nontreatment groups. A scientist must not design research or interpret data in a way that is more likely to support preconceived explanations of biological systems. Biologists who are consciously aware of their own biases and strive to keep an open mind to new ideas are most likely to make revolutionary discoveries.

The objective is to organize, clearly and concisely, the results of data collection, exploratory data analysis, and specific statistical analyses. These results must be transformed from a collection of specific information into a **synthesis** explaining the biological system. Do statistical evaluations support one or more of the theories and hypotheses and clearly reject others? Do the results provide a reasonable ex-

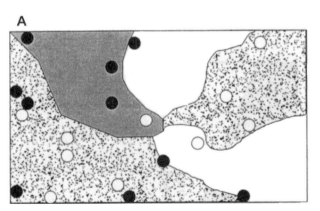
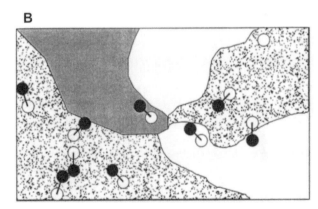

Fig. 1.10. Illustrative examples of (A) use (○) and random plots (●), and (B) use plots paired with random plots.

planation of the biological system? Are there alternative explanations of the data and statistical tests? Are there specific problems with the data that should be identified, such as inadequate sample sizes or unusual variation in specific variables measured? What could have introduced bias into the estimates? Are additional data required? These questions must be considered carefully, and if concerns are identified, they must be noted in reports and publications.

During this phase, the biologist usually reaches some conclusions based on the data and results of statistical evaluations. If the data support the hypothesis, we cannot conclude the theory (model) is true, but only that it has not been rejected (James and McCulloch 1985). The central issue is that we do not prove a research hypothesis or theory to be correct—indeed some would argue that *all* hypotheses are to some degree incorrect. Instead, the **credibility** of the hypothesis increases as more of its predictions are supported and alternative hypotheses are rejected. We can assist other biologists by carefully considering how broadly our conclusions can be generalized to other areas or populations and not allowing our conclusions to go beyond the data. Interpretation of research data must clearly separate conclusions and inferences based on data from speculation. For example, if we demonstrate that droppings from spruce grouse are most abundant under lodgepole pine (*Pinus contorta*) and Engelmann spruce (*Picea engelmannii*), we can conclude that grouse use both tree species for some behaviors, but the type of behavior (e.g., roosting or feeding) is mere speculation without additional data (e.g., observations of feeding activity and crop or fecal analyses). Likewise, **replication** of studies across space and time "provides us greater confidence that certain relationships are general and not specific to the circumstances that prevailed during a single study" (Johnson 2002:930).

Data Collection

Most novice research biologists are anxious to initiate data collection because of the attractiveness of working outdoors and the pleasure derived from observing wildlife-related phenomena. However, the design phase should not be rushed to initiate fieldwork more quickly. Successful research biologists often spend about 40% of their time in design and planning phases, 20% in actual fieldwork, and 40% in data analysis and writing publications. Data collection can be physically difficult and highly repetitious.

All data should be recorded on preprinted data sheets or entered directly into a handheld data logger, computer, or personal digital assistant. This practice ensures that each field person collects exactly the same data, as **consistent** collection of data simplifies analysis. Data sheets should be duplicated after each field day (e.g., computer entry, photocopies, or transcribed) and stored in a separate location from the original data set. Data entered electronically in the field should be downloaded daily and backed up for storage at another location. Transcription of data (including computer data entry) must be followed by careful proofreading, which is greatly facilitated by checking for valid entries by using database queries and spreadsheet scripts. All field personnel should receive careful instructions regarding data collection, and the principal researcher must check periodically to see that each person has similar skills and uses the same methods for observation, measurement, and recording (Kepler and Scott 1981). The principal researcher is responsible for quality control, and the validity of research results depends on the quality of research design and data collection.

Pilot Study

A **pilot study** is a preliminary short-term trial through all phases of a research project. Pilot studies are an important, but often neglected step in the research process. Information can be obtained that will help the researcher avoid potentially disastrous problems during or after the formal research phase. Pilot studies often will disclose hidden costs or identify costs that were over- or underestimated. **Optimal sample allocation** (Scheaffer et al. 2005) incorporates cost estimates to maximize the benefits obtained from limited research budgets. Use of a pilot study should reveal basic logistical problems, such as travel time among study plots being underestimated or expectations for overall sample sizes being infeasible without additional personnel and funding. Statistical procedures for estimating needed sample sizes require variance estimates of variables that will be measured, and these variance estimates are often available only from data gathered in a pilot study. These preliminary data might disclose the variance of the population is so large that obtaining adequate sample sizes will be difficult. It is far better to discover these problems before time, energy, personnel, and critical research dollars are committed to a research project doomed to fail. If the research is part of an ongoing project, or if much research on the topic has been published, costs, methodology, and variance estimates may already be firmly established.

Power Analysis

In descriptive studies, power analysis provides sample size requirements for obtaining an estimate of desired precision and can be calculated after an estimate of population variance is obtained from previous studies or a pilot study. Formulas for sample size are available for standard survey designs (Thompson et al. 1998, Scheaffer et al. 2005) and for typical hypothesis tests (Zar 1999). In studies involving experiments or other types of comparisons, sample size is increased to improve the **power of a hypothesis test** (defined as the probability of detecting a real difference) and to prevent erroneous conclusions. Power analysis for hypothesis tests depends on several factors, including sample size, level of significance (α), variance in the populations, effect size (the true change that occurred), and efficiency of the test or design (Steidl et al. 1997). In contrast to this essential prospective power analysis during the design phase, performing a retrospective power analysis after the data are collected, during the analysis phase, is controversial or contraindicated (Thomas 1996, Steidl et al. 1997). Retrospective power analysis is uninformative unless effect sizes are set independently of the observed effect (Steidl et al. 1997).

To illustrate power of a test, consider the following example. Suppose we were using fawn:doe ratio as an indicator of production for a mule deer (*Odocoileus hemionus*) herd (i.e., the biological population is our research population). We want to know whether the fawn:doe ratio has declined. There are 4 possible outcomes from sampling the herd and testing for a decline in the fawn:doe ratio (i.e., the null hypothesis is there is no change; Table 1.3). We evaluate whether the fawn:doe ratio has declined by comparing the test statistic calculated from our data to a value for this statistic at the chosen level of significance (α). The **level of significance** represents the chance of concluding the ratio changed when in fact it did not. An $\alpha = 0.05$ indicates that we would make this error only 5 times if the population really did not decline and we tested it by drawing a sample 100 times. This error is referred to as a **Type I error.** But, we could make another error. We could conclude the ratio had not changed when in fact it had declined. For the situation where we count 500 deer, we would fail to detect the decline in the fawn:doe ratio 50% of the time (Table 1.3). This type of error is referred to as **Type II error,** and its likelihood is measured by α. When we perform a test, we typically set α low to minimize Type I errors. But, Type II errors might be as important (Alldredge and Ratti 1986, 1992) or even more important than Type I errors. Obviously, we want to detect a change when it occurs; the probability of detecting a change is called the **power of the test.** The power of the test is calculated as the probability of not making a Type II error $(1 - \alpha)$.

We cannot control natural variation in the population or the actual change that occurred, but we can control the other 3 factors (i.e., sample size, efficiency, and significance level). **Parametric tests** (based on a normal distribution, e.g., *t*-tests, *F*-tests, and *Z*-tests) have the highest efficiency for normally distributed populations and for large samples. **Nonparametric tests** (based on distributions other than the normal distribution, e.g., Mann-Whitney, Wilcoxon signed-ranks tests) are superior when sample sizes are small (30) and populations are not normally distributed (Johnson 1995, Cherry 1998). The power of a test declines as the level of significance is made more stringent (decreasing α). In the example (Table 1.3), this problem is critical, because the Type II error (failing to detect declining production) is the more serious error than detecting a declining production when it is actually increasing. It would be preferable to increase α so that power of the test could be increased. In other situations the Type I error will be more serious, and α must be kept low. Increasing sample size increases power of the test. Calculating sample size necessary for a desired level of power is essential to designing a high quality study (Toft and Shea 1983, Forbes 1990, Peterman 1990). However, such calculations should be based on meaningful effect sizes (i.e., one that constitutes a biologically significant result; Reed and Blaustein 1997, Cherry 1998, Johnson 1999).

The importance of **sample size** cannot be overemphasized. Sample size and experimental design are the major factors under the control of the biologist that strongly influence power of the test (i.e., the likelihood of detecting a significant difference when one really occurs). Inadequate sample size usually results from: (1) inadequate consideration of population variance; (2) inability to collect data (e.g., observe a rare species); or (3) insufficient funding, time, or personnel. Often a sample size problem is overlooked initially because of failure to consider sample size reduction throughout the study (i.e., we focus mostly on the initial sample size and not on the final sample size that represents the most important data for consideration of a hypothesis). For example, in a study of mallard (*Anas platyrhynchos*) brood movements almost 10 times as many nests were required to be found as the sample size of broods indicated because of an

Table 1.3. Possible outcomes of a statistical test for declining production in a deer herd. Counts of 500 antlerless deer (adult does and fawns) were obtained each year, and tests of the null hypothesis of no change in the fawn:doe ratio were performed at the 5% level of significance ($\alpha = 0.05$).

	Fawns per 100 does							
	Actual herd value			Count value				
Case	188	1989	Change	1988	1989	Conclusion from test	Result of test	Likelihood of this result (%)
1	60	60	None	61	59	No change	No error	95 $(1-\alpha)$
2	60	60	None	65	50	Declined	Type I error	5 (α)
3	65	50	Declined	65	50	Declined	No error	50 $(1-\alpha)$
4	65	50	Declined	62	57	No change	Type II error	50 (α)

89% sample size reduction from nests located to actual brood data (Rotella and Ratti 1992a, b).

Another common problem is fairly large overall data sets that are not sufficiently similar across years (or seasons) to combine, resulting in annual sample sizes that are too small for analysis. At the beginning of a research project we often set our desired sample size based on combining data collected over several continuous years. However, if the characteristic of interest were different across study years, combining the data would not be valid. For example, in a study of habitat selection by red fox (*Vulpes vulpes*), habitat use might differ between mild and severe winters with heavy snow cover. In this example, combining the data would not be valid, yet the sample size in each year may be too small to detect selection (Alldredge and Ratti 1986, 1992).

Approaches to Data Analysis

At this point, researchers have developed well-planned and biologically meaningful hypotheses; decisions have been made regarding study, experimental, and sampling designs; and empirical data have been collected to shed light on the validity of the hypotheses. Now researchers must decide on a statistical approach. Unfortunately, this decision has become less clear over the past decade (Butcher et al. 2007). General approaches for data analysis include Bayesian versus frequentist paradigms with distinct differences in how probability should be interpreted (Cox 2006). Within the **frequentist** paradigm, one could choose null hypothesis significance testing (**NHST**), point and interval estimation of effect sizes, likelihood-based and information theoretic methods, or some combination of these (Läärä 2009). Unfortunately, the statistical approach that is most familiar and widely used (i.e., NHST) in wildlife science has continued to be criticized (e.g., Yates 1951, Cherry 1998, Johnson 1999, Wade 2000, Fidler et al. 2006, Läärä 2009), causing confusion and frustration for researchers (Butcher et al. 2007). We introduce these various approaches and point out some of the key differences while purposefully not recommending one over another. We think it is more important to expose researchers to the relevant discussions, so they can make an informed selection of the best approach.

Ellison (2004) summarized the main differences between Bayesian and frequentist approaches to statistical inference (also see Ellison 1996, Dennis 1996, Taper and Lele 2004, Hobbs and Hilborn 2006). The first is a difference in what is considered a random outcome. **Frequentist inference** considers the model and the true parameter values to be fixed quantities, whereas the observed data are random outcomes from this process. Thus, frequentists refer to the probability of the data (Y) given a particular hypothesis (H), as defined by the model and parameters: $\text{Prob}(Y|H)$. In contrast, **Bayesian inference** treats both the data and model as random, allowing quantification of the probability of a hypothesis being true given the observed data: $\text{Prob}(H|Y)$.

This distinction brings up the second major difference between these approaches—the definition of ***probability***. Frequentist inference defines probability as the relative frequency of a particular outcome if the process was repeated an infinite number of times. For example, the probability of obtaining a heads with a flip of a coin is the number of times a head turns up divided by the number of flips, where the number of flips is repeated to infinity. Bayesian approaches define probability quite differently: it is the degree of belief in the likelihood of an event occurring.

Finally, the 2 approaches differ in the way **prior knowledge** is incorporated. For Bayesian inference, it is required that prior knowledge is translated into a probability distribution, which is then combined with the sample data to make an inference. Frequentist inference generally uses only the observed data, although prior knowledge can be incorporated by combining likelihoods from previous studies with the likelihood of the observed data (see Hobbs and Hilborn 2006:10). Although the decision of whether to use Bayesian versus frequentist approaches is often made on practical grounds (Lele et al. 2007), we end with a quote from Ellison (2004:517) that we believe is particularly relevant:

> Deciding whether to use Bayesian or frequentist inference demands an understanding of their differing epistemological assumptions. Strong statistical inference demands that ecologists not only confront models with data, but also confront their own assumptions about how the world is structured.

Hypothesis Testing

Significance testing as a statistical approach for confronting hypotheses with empirical data has been the subject of fervent debate in many disciplines (Fidler et al. 2004), including wildlife and ecological science (e.g., Anderson et al. 2000, Eberhardt 2003, Guthery et al. 2005, Lukacs et al. 2007, Steidl 2007, Stephens et al. 2007, Läärä 2009). Nonetheless, it remains a viable option for practicing wildlife researchers (Robinson and Wainer 2002, Butcher et al. 2007). **Hypothesis testing** is rooted in the philosophical idea of falsification, in which an attempt is made to disprove a hypothesis, leaving the alternative to be tentatively accepted (Underwood 1997). Johnson (1999) described the 4 basic steps of statistical hypothesis testing that mirror the approach suggested by Underwood (1997). The researcher develops a hypothesis that reflects his or her ideas about a particular ecological process or the effects of some treatment. The logical opposite of this hypothesis is usually taken as the **null hypothesis**, and data are collected to assess the validity of the null hypothesis. A statistical test of it involves calculating a *P*-value, which is then used to decide the fate of the null hypothesis. Strictly speaking, a ***P*-value** is the probability that if the null hypothesis were true and the test were hypothetically redone, one would observe data at least as extreme as those which were observed. Thus, a study that results in a P-value of 0.05 means that if the null hypothesis were true and the study were repeated 20 times, you would expect only 1 of these 20 studies to produce results at least as different from the null hypothesis as your study. Obviously, the definition is quite cumbersome and likely has led to much confusion, misuse, and misinterpretation of a statistical hypothesis test (Johnson 1999).

To more fully understand the role of hypothesis testing in wildlife science, it is helpful to have some historical perspective. Robinson and Wainer (2002) provide a concise description of hypothesis testing as it was originally intended by the famous statistician R. A. Fisher, who used it to assess potential innovations in agriculture. A few key points from this description are:

1. It is often legitimate to assume a particular innovation would produce no effect, and thus testing a null hypothesis of no effect is not considered trivial.
2. No single test should be the end of the discussion, because there is a chance (depending on the significance level for a particular test) that an effect can be suggested even when there is none, an effect should only be accepted if **repeated studies** continue to provide significant results.
3. Hypothesis testing only makes sense if continued research seeks to identify the **size** and **direction** of the effect.

Given these original intentions, it is not hard to see why so many have been critical of hypothesis testing in wildlife science. Several have argued that it is exceedingly rare to legitimately propose a zero effect or alternatively that some set of parameters are exactly equal (Cherry 1998; Johnson 1999; Anderson et al. 2000, 2001a; Läärä 2009). These point null hypotheses are often deemed silly nulls, because they are almost certain to be false a priori. Additionally, although replication was a cornerstone of Fisher's approach, true replication in wildlife science is not the normal procedure, which instead relies on "single-shot studies" designed to reach conclusions based on a one-time interpretation of a P-value (Robinson and Wainer 2002:265). Although **replication** is an important component of the scientific method regardless of the statistical approach used, because of the definition of a P-value, it is particularly relevant to hypothesis testing. These issues are especially problematic when hypothesis testing is applied to field studies without random assignment of treatments. Many statisticians strongly object to performing hypothesis tests on observational data or recommend alternative approaches for evaluating the data, such as confidence intervals for estimates, information measures for models, or Bayesian confidence measures (Cherry 1998, Johnson 1999, Anderson et al. 2000, Hobbs and Hilborn 2006, Läärä 2009).

Despite these criticisms, most statisticians agree that hypothesis testing can play a valuable, but limited role in data analysis (Cherry 1998, Johnson 1999, Stephens et al. 2007), especially if accompanied by estimates of **effect sizes** and a measurement of the **precision** of these estimates (Robinson and Wainer 2002). One improvement might be for researchers to adopt a trinary decision approach that is likely a more productive route than interpreting results of a hypothesis test (Jones and Tukey 2000). Using this approach, the conclusions of a hypothesis test are either $\mu_1 > \mu_2$, $\mu_2 > \mu_1$, or the direction of the difference is undetermined. Using this language avoids the temptation to accept a null hypothesis that is likely untrue while stressing the need for continued research to determine the direction and magnitude of the effect (Robinson and Wainer 2002).

Information-Theoretic Model Selection

Information-theoretic model selection offers a distinct alternative to hypothesis testing, and the approach has seen widespread growth in wildlife and ecological sciences (Hilborn and Mangel 1997, Burnham and Anderson 2002, Johnson and Omland 2004, Richards 2005). In contrast to hypothesis testing, model selection seeks to identify the hypotheses that are closest to the truth out of a **set of competing ideas** while fully acknowledging that all are wrong or incomplete characterizations of the process. The philosophical basis for this approach is more in line with that of Lakatos (1978:24): "All theories . . . are born refuted and die refuted. But, are they equally good?" He considered it nonsensical to retain only unfalsified hypotheses because of the philosophy that hypotheses may never be truly falsified and, more importantly, science will keep a hypothesis that is known to be wrong if there is not a better one available to take its place. Thus, a hypothesis is falsified only if a better one with greater empirical support is available to replace it. The information-theoretic model selection approach also closely follows Chamberlin's (1890, 1965) view of science by advocating the construction of multiple working hypotheses that are subject to repeated confrontation with empirical data. Those supported by the data tend to be retained, whereas those with little support tend to be dropped from consideration (Burnham and Anderson 2001).

Using the information-theoretic model selection approach, several competing models are suggested to reflect different hypotheses about how a process works or the effects of a particular treatment. An appropriate study is designed to collect empirical data that will be used as the arbiter in a **contest among rival hypotheses.** The metric for deciding among hypotheses is how close each model is to the truth. Due to an explicit link with information theory (hence "information-theoretic"), Kullback-Leibler distance has been promoted as an appropriate measure of the distance each competing model is from the true data-generating model (see Burnham and Anderson 2002:50–54). Several criteria may be used to estimate the relative expected Kullback-Leibler distance (Shibata 1989, Burnham and Anderson 2002), including Takeuchi's information criteria; likelihood cross-validation criteria (Stone 1977); and Akaike's information criteria (**AIC**; Akaike 1973), which are the most common in the wildlife and ecological literature. By focusing on the best explanation for an observed phenomenon, information-theoretic model selection does not rely on a binary decision process characteristic of hypothesis testing, instead allowing models to be differentiated according to the amount of support they receive from the data. Several practical guidelines for using information-theoretic model selection approaches have been published (Anderson et al. 2001a, Anderson and Burnham 2002, Richards 2005). In addition to the comprehensive treatment in Burnham and Anderson (2002), see Guthery et al. (2005) for a more critical review.

Effect Size and Interval Estimation

Most researchers agree that hypothesis testing and model selection are only one component of statistical inference and that estimation of effect sizes and measures of their precision are at least as important (Johnson 1999, Robinson and Wainer 2002, Stephens et al. 2007). Quinn and Dunham (1983:613) suggested: "The objective of biological research typically is to assess the relative contributions of a number of potential causal agents operating simultaneously." If this is the case, then estimation of **effect sizes** is of primary importance to wildlife science and these results should be emphasized in data analysis. Others have echoed this sentiment: "The very basic tools for statistical reasoning on the strength of associations and the sizes of differences and effects are provided by point estimates, their standard errors and associated confidence intervals" (Läärä (2009:152). Reporting effect sizes is not only important for practical interpretation of the focal study, but they also are the critical components for any subsequent meta-analysis (Gurevitch et al. 2001, Hobbs and Hilborn 2006). Läärä (2009) contains several practical recommendations for presenting and interpreting effect sizes that should be especially useful to practicing wildlife professionals.

Regression and General Linear Models

One of the most flexible approaches to identifying predictive and potentially causal relationships between wildlife and environmental or management characteristics involves use of ordinary least squares to estimate parameters of **regression models** or **GLM** (Fig. 1.11). Experimental manipulations that produce different levels of predictor variables are more readily analyzed by ANOVA, regression, or analysis of covariance versions of GLM under a Fisherian philosophy (Fig. 1.11), named after R. A. Fisher, who pioneered a "spirit of reasonable compromise, cautious, but not overly concerned with pathological situations" (Efron 1998:99) in the analysis of experiments. Designing a study to gather data on a variety of potential causal variables rather than manipulating those variables through a designed experiment is an ap-

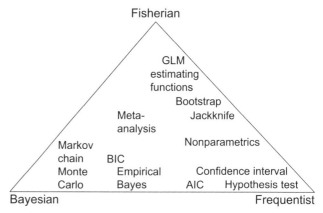

Fig. 1.11. Selecting analysis methods from 3 dominant statistical philosophies. GLM = general linear model, BIC = Bayesian information criteria, AIC = Akaike's information criteria. *Modified from Efron (1998).*

pealing alternative, but yields inferences of much lower certainty (Fig. 1.5). Performing hypothesis tests on such data (e.g., testing point null hypotheses) is easily performed with modern regression programs. However, it may not be justified as an inferential approach and may readily lead into a "fishing expedition" doomed to failure due to high Type I errors. Many statisticians refuse to analyze such data by using hypothesis tests and instead encourage biologists to apply maximum likelihood and information-theoretic model approaches under a modeling perspective, that is, identifying the most parsimonious model with good predictive ability (Milliken and Johnson 1984, Anderson et al. 2000, Burnham and Anderson 2002).

It is essential in designing manipulative or observational studies, if one plans to estimate linear models, to strive to obtain observations throughout the full range of the predictive variables. It is especially important to obtain observations at both low and high values of the predictive variable, because they set limits for the range of values that can be used later for prediction. The values at the ends of this range have the most leverage on slope estimates. If too narrow a range is measured, a significant relationship may not be detected among the variability. However, a relationship may be linear only through a portion of its range, such that beyond a certain level an increasing effect may turn into a negative effect at progressively higher levels. Such situations should be apparent from exploratory data analyses (Anderson 2001, Johnson 2002).

Bayesian Approaches

Bayesian data analyses are described as "practical methods for making inferences from data using probability models for quantities we observe and for quantities about which we wish to learn" (Gelman et al. 2003:3). One of the primary appeals of Bayesian statistics is that after sampling a population and calculating statistics, such as the mean, variance, and confidence interval for the mean, Bayesian analysis allows us to state that we are 95% certain the true mean for the population is within this 95% confidence interval. Johnson (1999) provided an easily understood description of the conceptual differences between the frequentist and Bayesian approaches (Fig. 1.11). A Bayesian analysis requires performing 3 basic steps (Gelman et al. 2003).

1. **Specify a probability distribution for all quantities** (i.e., use prior studies and creative thinking to specify a particular **prior** probability for the parameter[s]). We begin by stating the range of all possible values for the characteristics we are attempting to measure and make our best guess of a probability distribution for a parameter (based on earlier studies and clear thinking) if any values are more likely than others. This step is controversial, because it introduces subjective decisions into the process and has potential for misuse if one's goal is to cook the books to produce a particular result (Dennis 1996). However, well-designed research should gather historical data, so that knowledge is available on the probability distribution of the parameter(s) (Box 1.1) or failing that, it should specify minimum and maximum values with equal chances of intermediate values (i.e., a flat prior).

2. **Use the observed data to calculate a posterior distribution for the parameter of interest as a conditional probability distribution.** This second step in Bayesian analysis follows data collection. We improve our prior guess of the value of the characteristic by combining it with the new data gathered to state conclusively our best posterior guess of the value of the characteristic. This step is performed using **Bayes's rule,** and this Bayesian estimate might be considered as a weighted average estimate based on the sample data and the assumed prior value, where weights are proportional to the precision of the observed and prior values (Gelman et al. 2003:43). As sample size increases, the Bayesian value approaches the maximum likelihood estimate and any influence of the prior probability vanishes. **Markov chain Monte Carlo** methods are used widely for these calculations (Fig. 1.11).

3. **Evaluate the fit of the model and the implications of the resulting posterior distribution.** This step in Bayesian analysis (Gelman et al. 2003:3) consists of "evaluating the fit of the model and the implications of the resulting posterior distribution: does the model fit the data, are the substantive conclusions reasonable, and how sensitive are the results to the modeling assumptions?"

Validating Parametric and Simulation Models

The validation and experimental phases of the modeling process described earlier really constitute an effort to **confront theories with data.** The models express our theoretical understanding of the system, its characteristics, and its processes. Validation and experimentation confront this theory with data, especially when we conduct these activities in an adaptive management framework, where management actions are accompanied by monitoring to simultaneously validate the predictions of the models (our theory or understanding of the system) and probe the behavior of the system (Walters 1986:250). Comparing model predictions to data potentially completes the feedback loop that can be used to improve our understanding, but the natural tendency of managers and biologists is to break the loop by ignoring any inconsistencies detected. This tendency is natural because of the considerable effort expended in developing the models and trade-offs in selecting management actions. Ignoring inconsistencies leads to passive adaptation rather than a probing through experimental management actions. "Conservative, risk-averse decision making creates a particularly difficult situation for learning" (Walters 1986:251), especially when the effects of management are compounded with environmental changes and there are lags inherent in the responses. Where the desired outcome is a harvestable surplus of a game species, the manager and biologist face substantial social, economic, and political pressure to find the "right" answer (see the section Adaptive Management: Connecting Research and Management below). Models are invaluable in efforts to ensure that management and ecological understanding are based on valid estimates and relationships rather than on wishful thinking, but their results are often attacked by the interested public, whose values and "gut instincts" are opposed to model predictions.

SPECULATION AND NEW HYPOTHESES

Rarely does a single research project provide the last word on any problem (Johnson 2002). More commonly, research will generate as many questions as answers. Speculation, based on inconclusive or incomplete evidence, is one of the most important aspects of science. **Speculation** must be identified and should not be confused with conclusions based on data. But, speculation is the fuel for future research. Many facts of nature have been discovered by accident—an unexpected result from some associated research effort. However, most research is directional (i.e., it attempts to support or falsify a theory reached by speculating from facts). **New hypotheses** can be considered a form of speculation, which is verbalized in a more formal fashion than speculation and has a specific testable format. For example, considering spruce grouse, we can formulate a basically untestable hypothesis that spruce grouse have evolved a preference for use of lodgepole pine and Engelmann spruce trees. This statement is too vague and requires historical data that cannot be collected. However, we can hypothesize that spruce grouse use lodgepole pine and Engelmann spruce trees for: (1) feeding or (2) roosting. Testing these hypotheses, we might learn that 80% of the spruce grouse diet is lodgepole pine, even though Engelmann spruce is more abundant. We may then speculate (i.e., hypothesize) that needles from lodgepole pine provide higher nutritional quality than needles from Engelmann spruce.

PUBLICATION

The final step of the scientific method is publication of research. Unfortunately, many research dollars are wasted, because knowledge gained was not published and the information is buried in file cabinets or boxes of data sheets. The **publication process** is the most difficult phase for many biologists. Clear concise scientific writing is difficult, because most biologists have little formal training in or inclination for this activity. Peer review may damage the ego, because we must subject our work to anonymous critiques used by editors to judge whether the manuscript is acceptable for publication.

Agency administrators often do not encourage or reward employees for publishing their work and discourage publication in some instances. Administrators are pressured with calls for immediate answers to management problems; thus, they devalue the long-term benefits of the publication process. Effective administrators recognize that **peer review** and **publication** will: (1) **correct errors** and possibly lead to a better analysis, (2) help authors reach the most **sound conclusions** from their data, (3) make it easier to **defend** controversial policies, (4) help their personnel **grow** as scientists by responding to critical comments and careful consideration of past errors (which may have been overlooked without peer review), and (5) make a **permanent contribution** to wildlife management by placing results in a literature format available to other agencies, researchers, and students.

Publication is essential to science. Peer reviews normally improve the quality of a manuscript, but some research may not be suitable for publication. This observation emphasizes the importance of careful planning, design, data collection, etc. Rarely would any research effort that is properly planned, designed, and executed (including a well-written manuscript) be unpublishable. However, the revision process (i.e., responding to criticisms from the editor and referees) may be painful and frustrating to authors. Overall, the system is necessary to ensure quality publications, and authors should not be discouraged by the necessity to defend their work and revise manuscripts. Research is not complete and does not make a contribution to knowledge and sound management of wildlife resources until results are published in a way that effectively communicates to the scientific community and user groups (e.g., wildlife managers). In addition to publication in peer-reviewed journals, research findings will improve wildlife management immediately if they are communicated in **other forums,** such as professional meetings, workshops, seminars, general technical reports, informational reports, and articles in the popular press.

COMMON PROBLEMS TO AVOID

Procedural Inconsistency

Procedural inconsistency is another common problem that can be prevented with proper research design. Problems of this type occur from seemingly minor variations or alterations in methodology. For example, if a project is dependent on field personnel to accurately identify songs of forest passerine birds, the data set may be biased by identification errors (Cyr 1981). In this situation, the magnitude of the bias will depend on the rate of errors by individuals, differences in the rate of errors among individuals, and relative proportion of data collected by each individual. Research methodology should be defined with great detail, and all individuals collecting data should have similar skills and knowledge of the methods used (Kepler and Scott 1981). If inconsistencies cannot be eliminated through selection and training of field workers, the design must incorporate double sampling or similar procedures to remove inherent biases (Farnsworth et al. 2002). One unfortunate aspect of biases of this type is they are often overlooked (or ignored) as potential problems and are seldom reported in research publications.

Nonuniform Treatments

A common bias stems from **nonuniform treatments.** This problem is illustrated by considering 2 previous research examples. In the discussion of crossover experiments, we described a 2-year study on pheasant nest success, in which mowing on treatment areas was delayed until after 4 July. Assume that in the first year of this study, all treatment areas were mowed between 4 and 7 July, as planned. But, during year 2 of the study, a 3-day rainstorm began on 4 July, and the treatment areas were not cut until 9–12 July. Although this 5-day difference in mowing the treatment areas may seem insignificant, the impact on the results and interpretation of our experiment is really unknown—and may be serious. Thus, the second year of the experiment should be repeated. Because dates of pheasant nesting and plant growth varies from year to year in response to temperature and rainfall patterns, a better way to set the date for the mowing treatment might be based on the cumulated degree-days widely published in farm journals.

In the second example, we want to evaluate effects of sharp and feathered edges on nest success of forest birds. If we had used both clearcuts and road ways as sharp edges, we might have hopelessly confused the treatment results because of differences in the attractiveness of sharp edges near roads, where carrion is an abundant attractant to such generalist predators as ravens. High variability between replicates in nonuniform treatments substantially reduces our power to detect biologically significant effects.

Pseudoreplication

Pseudoreplication occurs when sample or experimental units are **not independent** (i.e., they are really subsamples rather than replicates, but are treated as though they were independent samples or experimental units). This problem is widespread in field ecology (Hurlbert 1984) and should be avoided when possible. Experimental units are independent in manipulative experiments only if we can **randomly assign treatments** to each unit. In field studies, a simple test for pseudoreplication is to ask whether the values for 2 successive observations are more similar than values for 2 observations drawn completely at random from the research population (e.g., Durbin and Watson 1971). If so, the successive observations are probably not true replicates and the research should be **redesigned,** or this lack of independence must be treated correctly in the analysis. This treatment can be done by using cluster sampling, adjusting the degrees of freedom for tests (Porteus 1987, Cressie 1991), or applying Monte Carlo approaches to evaluate test statistics (as is widely done for spatially correlated data; Dale and Fortin 2002).

There must be a direct tie between the sample or experimental unit and the **research population.** If the research population consists of 1 meadow in Yellowstone National Park, then 2 or more samples drawn from that meadow would be replicates. In this example, our inferences or conclusions would apply only to that single meadow. If our research population consisted of all meadows in Yellowstone National Park, then 2 plots in the same meadow would not constitute true replicate samples. Also, repeated sampling of the same radiomarked animal often constitutes a form of pseudoreplication (e.g., if our research population consisted of moose in one ecoregion, repeated observations of habitat use by a single animal would not be true replicates; a similar problem would arise if 2 radiomarked animals were traveling together, so their habitat selection would not be truly independent). The data would have to be summarized into a single value, such as the proportion of observations in a certain habitat, for statistical analysis. This compression would reduce the sample size to the number of radiomarked moose. Treating repeated observations as replicates is strictly justified only when the individual animal is the research population. In this situation, tests for **serial correlation** (Swihart and Slade 1985) should be conducted to ensure that observations are not repeated so frequently they are still pseudoreplicates.

ADAPTIVE MANAGEMENT: CONNECTING RESEARCH AND MANAGEMENT

Wildlife management programs should be developed from the application of scientific knowledge based on research (i.e., we should apply scientific facts and principles resulting from research on specific topics, e.g., population ecology, habitat selection, or behavior). Initially, this practice is a sound one for the development of a new management program. The logic behind formulation of a management program is similar to the formulation of a research hypothesis: both provide opportunity for predictive statements. Our management prediction is that our plan of action will achieve a desired result. However, a major problem with nearly all wildlife management programs throughout the world is the lack of research on the effectiveness of programs (Macnab 1983, Gill 1985). Seldom is the question "does our management lead to the desired result?" addressed in formal, well-designed, long-term research projects. For example, research indicates that spinning-wing decoys make mallard breeding populations more vulnerable to harvest (Szymanski and Afton 2005). A potential long-term management response would be to create more restrictive hunting regulations as the use of spinning-wing decoys increases. The assumption is that if using spinning-wing decoys increases mallard harvest rates, then hunting regulations are needed to ensure mallard populations over the long term do not shrink with increased vulnerability. However, we should consider several important questions. Does increased vulnerability translate to increased harvest? What segments of the mallard populations are most vulnerable to the use of spinning-wing decoys? Will mallards become accustomed to spinning-wing decoys over time and thereby decrease their vulnerability to harvest? These questions and more should be addressed, because imposing more restrictive hunting regulations could backfire if the answers to these questions do not support it.

A second common example is prescribed burning as a management practice to increase deer and elk populations. The effectiveness of this management has not been addressed directly, and most evaluations have only noted increases in browse forage species and changes in animal distributions (Stewart et al. 2002, Van Dyke and Darragh 2007, Long et al. 2008a, b). Increased population levels in response to prescribed burning have not been adequately documented or thoroughly studied (Peek 1989).

A third example is the use of population indices to monitor changes in population levels (e.g., ring-necked pheasant crowing counts, lek counts, track counts, catch-per-unit-effort, and aerial surveys). The primary assumption for use of a population index is the index is directly related to density. Although nearly every wildlife management agency uses trend data from population indices for management decisions, only a few examples of index validation exist (e.g., Rotella and Ratti 1986, Crête and Messier 1987, Marchandeau et al. 2006, Forsyth et al. 2007). Some studies have disclosed that index values are not related to density (Smith et al. 1984, Rotella and Ratti 1986, Nottingham et al. 1989, Rice 2003).

Walters (1986) proposed a systematic solution to these problems, which he called **adaptive management.** It involves a more formal specification of management goals and responses to management actions through the use of **predictive models** (Table 1.2) based on multiple working hypotheses, which can be compared to actual system responses through detailed **monitoring** (Thompson et al. 1998, Sauer and Knutson 2008, Conroy and Peterson 2009). Management actions are treated as experiments, which must be monitored carefully to ascertain whether goals were met and to identify errors in understanding the dynamics of the natural systems being managed. **Actual responses** to management actions are compared to predictions from our models based on current knowledge and assumptions (e.g., adaptive harvest management; Williams and Johnson 1995, Williams et al. 1996, Johnson and Williams 1999, Johnson et al. 2002). **Adaptive resource management** is an interactive process in which learning over time improves management as long as a monitoring program provides feedback to both our understanding of the system and the effects of management (Conroy and Peterson 2009).

Adaptive resources management is a specific case of **structured decision-making,** a process that addresses complexity, uncertainty, multiple objectives, and different perspectives to achieve management objectives (Clemen 1996, Clemen and Reilly 2001). Structured decision-making has multiple steps: problem definition, objectives, alternatives, consequences, trade-offs, uncertainty, risk tolerance, and linked decisions (Conroy et al. 2008). The basic strength of this decision-making approach is that it allows wildlife scientists to make effective decisions more consistently and to provide guidance for working on hard decisions (Clemen 1996, Clemen and Reilly 2001). Wildlife scientists are faced with difficult decisions regarding both the management and conservation of wildlife. For example, how can bison be restored to their former range, which would benefit other threatened prairie species, while also considering the economic and social impacts to cattle ranchers if brucellosis spread from bison to cattle? Both structured decision-making and adaptive resource management are being used increasingly often by wildlife scientists (Conroy et al. 2002, Johnson et al. 2002, Dorazio and Johnson 2003, Regan et al. 2005, Moore and Conroy 2006, McCarthy and Possingham 2007, Martin et al. 2009). Both these approaches differ from **scenario planning** (Kahn 1965, Chermack et al. 2001), practiced in business and other organizations to make flexible long-term plans based on considering multiple assumptions about the future. Such future assumptions are developed from a combination of established facts and multiple plausible forecasts of future changes, especially social changes. Scenario planning by the U.S. National Park and Fish and Wildlife Services in crisis situations, such as British Petroleum's Deepwater Horizon spill of 4.9 million barrels of oil into the Gulf of Mexico in 2010, should provide a foundation for a more measured adaptive management process to restore the damaged wetlands and marine ecosystems.

If wildlife agencies have the responsibility for management of wildlife populations and their habitats, they also have the responsibility to conduct research on the **effectiveness** of management programs. Wildlife agency administrators should strive to develop well-designed, long-term management-research programs as a basic component of annual agency operations.

SUMMARY

Carefully designed wildlife research improves the reliability of knowledge that is the basis of wildlife management. Research biologists must rigorously apply the scientific method and make use of powerful techniques in survey sampling, experimental design, and information theory. Modeling is an effective tool for predicting the consequences of management choices, especially when it is based on carefully designed field studies, long-term monitoring, and management experiments designed to increase understanding. More effort should be dedicated to the design phase of research, including obtaining critiques from other biologists and statisticians and avoiding common problems, such as insufficient sample sizes, procedural inconsistencies, nonuniform treatments, and pseudoreplication. When possible, we must move from observational to experimental studies that provide a more reliable basis for interpretation and conclusions; these studies need to be replicated across space and time. Wildlife biologists have a tremendous responsibility associated with management of animal species experiencing increasing environmental degradation, loss of habitat, and declining populations. We must face these problems armed with knowledge from quality scientific investigations.

2

Management and Analysis of Wildlife Biology Data

BRET A. COLLIER AND
THOMAS W. SCHWERTNER

INTRODUCTION

THE GENERAL FOCUS in this chapter is on outlining the range of options available for the management and analysis of data collected during wildlife biology studies. Topics presented are inherently tied to data analysis, including study design, data collection, storage, and management, as well as graphical and tabular displays best suited for data summary, interpretation, and analysis. The target audience is upper-level undergraduate or masters-level graduate students studying wildlife ecology and management. Statistical theory will only be presented in situations where it would be beneficial for the reader to understand the motivation behind a particular technique (e.g., maximum likelihood estimation) versus attempting to provide in-depth detail regarding all potential assumptions, methods of interpretation, or ways of violating assumptions. In addition, potential approaches that are currently available and are rapidly advancing (e.g., applications of mixed models or Bayesian modeling using Markov Chain Monte Carlo methods) are presented. Also provided are appropriate references and programs for the interested reader to explore further these ideas.

The previous authors (Bart and Notz 2005) of this chapter suggested that before collecting data, visit with a professional statistician. Use the consultation with a professional to serve as a time to discuss the objectives of the research study, outline a general study design, discuss the proposed methods to be implemented, and identify potential problems with initial thoughts on sampling design, as well as to determine what statistical applications to use and how any issues related to sampling will translate to issues with statistical analysis. Remember, no amount of statistical exorcism can save data collected under a poor study design.

NATURE OF KNOWLEDGE

The primary goal of science is to accumulate knowledge on what reliable statements regarding the state of nature can be made. However, defining "knowledge" and "reliable" with respect to science has proven to be a difficult task (Kuhn 1996). **Epistemology** is the branch of philosophy that focuses on determining what constitutes knowledge and how knowledge is acquired. It draws contrasts between the 2 primary methods of knowledge acquisition—**rationalism** and **empiricism.** Rationalists argue that reason alone is sufficient for knowledge accrual (Morrison et al. 2008:4). Conversely, knowledge acquired through experience underlies the empiricist perspective of knowledge acquisition. Scientific research in general and wildlife ecology in particular tends to rely most heavily on empiricism.

Judgment of reliability is another matter. Most wildlife ecologists use the process of **induction** (generalizing from specific results; Guthery 2008) or take the results from a single or group of studies and apply those results widely. Opposed to induction is **deduction,** or generalizing a principle to occurrence of a specific event or events. A good example of deduction might be that all animals are mortal, deer are animals, and thus deer are mortal (Morrison et al. 2008:11). **Retroduction** occurs when deductions are used to determine what conditions might have lead to the observed result (Guthery 2008). Retroduction is common in discussion sections of publications, where scientists try to determine what mechanism lead to the observed results. For example, Dreibelbis et al. (2008) used retroduction from a set of observations (missing egg shell fragments, no evidence of scat or hair) to deduce that a snake was the most likely predator of turkey nests under this set of observations.

In the inductive framework, analyzing data serves to generalize data collected from the samples for use as surrogates for describing the population as a whole (Romesburg 1981, Guthery 2008). Inference is made because there is some biologically, socially, or politically motivated interest in what characterizes a population and what makes it different from other populations, and because knowledge of this difference would be beneficial to management.

REQUIREMENT FOR STATISTICAL INFERENCE

One of the first issues all ecologists are faced with during the initial stage of study development is determining how data of interest will be collected. In the rush to start field work, adequate time and resources may not be assigned to evaluating potential analytical approaches and combining those approaches with sampling designs to address questions of interest. However, appropriate sampling procedures are paramount, because researchers cannot make direct observations of every unique individual in the population (e.g., a census). In practice, the objective of sampling is to generalize population processes based on data collected from a subset of the population of interest (Cochran 1977, Thompson 2002a, Morrison et al. 2008). Although it is nearly always impossible to census populations, by taking a sample ecologists can make inferences about the population in question.

The importance of appropriate sampling cannot be overemphasized when considering data analysis. Predictive models, and hence predictions, will be more accurate and useful when data collection is based on an appropriate sampling design (Harrell 2001). The relationship between analysis and sampling design is clear: population-level inferences require data that represent the population of interest. Consider the situation wherein the sampled population is not equivalent to the target population (e.g., because of the sampling design used, certain elements of the target population have probability zero of being included in the sampled population; Williams et al. 2001). In this case it may be possible to develop a predictive model that seems to provide an accurate representation of population processes. However, the model is unreliable, as it only predicts to a subset of the overall (sampled) population. If inferences are to be made to the target population in this case, then auxiliary data must be collected that shows a consistent relationship between the sampled and target populations (for further discussion see Williams et al. 2002a, Morrison et al. 2008). The framework for all analysis approaches discussed in this chapter hinge on appropriate sampling designs. Therefore, a professional statistician should be consulted before study implementation. Additionally, readers are referred to Chapter 1, This Volume) for a discussion of sampling in ecological studies and to the literature cited in this chapter for books and articles addressing sampling designs and statistical inference.

DATA COLLECTION

What to Collect

To begin a wildlife ecology study, first consider what data to collect to make the study as fruitful as possible. It is imperative to "look before you leap" when deciding on the relevant data. Time spent collecting data that have limited impact on the question of interest should be avoided (e.g., ground litter type when you are estimating canopy bird abundance). Also, data should not be collected that is of different resolution than the question of interest (e.g., hourly precipitation information when interested in estimates of annual elk [*Cervus canadensis*] survival). Data collection is a delicate balancing act requiring care to include potentially useful data for explaining the biological process being studied while omitting irrelevant information. Collect data with future analysis in mind, remembering current time and money limitations.

How to Collect

Without a doubt one of the most important yet unappreciated aspects of wildlife research is the choice of collection instrument. Rarely does data sheet development get discussed in data analysis chapters. Yet data collection protocols and instruments should be given as much consideration as other parts of study development, as they are a critical link in a chain connecting the study subjects with the study results, the failure of which could compromise the reliability of any knowledge drawn from the research.

There are several considerations that students and scientists need to take into account when developing data protocols. As an example, consider a 2-year masters project, during which data collection will occur over several seasons and the study species will undergo several phenological changes. When developing data protocols, determine whether a single data sheet for species-specific data collection across all seasons (e.g., breeding or wintering) will be used, or are separate sheets needed for each season? Will 1 data sheet per individual per day be used, or will individuals (or sample units) be lumped by day onto a single data sheet? Will the same data sheets for each period under study be used? Will data collection occur during a single season over several years, or will data collection be continuous over several seasons, requiring different data types to be collected? Will explanatory data (e.g., vegetative measurements) be collected on the same data sheet as response data (live–dead checks using radiotelemetry), or will multiple sheets be required for each type of data being collected? To assist in answering these questions, a pilot study may be useful to evaluate the appropriateness of the methods used for data collection (transect surveys, point counts, etc.) and to test how the proposed methods for field data collection work in practice.

Although there are keys available to assist with study design (see Thompson et al. 1998), here we provide a short key for data sheet development:

1. Outline study question.
2. Define response variable (e.g., nest survival).
3. Define explanatory and/or descriptive variables that might affect response (e.g., vegetation cover).
4. Define steps for minimizing missing data.
5. Outline data collection approach.
6. Design initial data collection instrument specific to response or explanatory variables.
7. Conduct field test of protocols and data instruments.
8. Evaluate efficiency of data instruments.
9. Repeat steps 2–8 if necessary due to logistical difficulties.
10. Initiate data collection.

There is a wide variety of options available for collecting field data, and not all rely on toting around binders filled with data sheets. Logging data electronically while in the field (Laake et al. 1997) can reduce the additional effort of incorporating those data into electronic storage and can potentially reduce transcription errors. However, whereas computers break and electronic systems fail, batteries never die in a paper notebook. Too often researchers let enthusiasm for the latest electronic gadget drive their selection of study protocols. Data should be collected during field studies using a method the collector is most comfortable with and that accomplishes the study objectives. In the end, the selection of the tool for collection is purely a personal matter, as there is no one right way to collect field data.

When to Collect

Once a question has been defined, a study design determined, and a data instrument selected, thought should next turn to the basics of data collection. Data collection protocols are as important to analysis as choosing the appropriate statistical technique. For example, consider the situation wherein there is interest in evaluating how prescribed fire influences avian survival over the course of an 8-week breeding season. Initially, determine how frequently surveys will be conducted (perhaps through radiotracking). If interested only in basic survival, then locate all the radiotagged individuals at the end of the eighth week, and the number still alive divided by the number radiotagged at the beginning of the study is the estimate of period survival. But, if interested in potential factors that might influence survival over the breeding season, such as determining how nesting activities (nest building, incubation), influence survival (Martin 1995, Dinsmore et al. 2002), data should be collected on when these events occurred and whether the bird survived or died before, during, or after those events. In this situation, tracking should be increased to once per day or more frequently. If there is interest in estimating the impacts of weekly precipitation on survival, then the sampling period would have to be at least weekly and so forth. Ideally, measurement frequency relates back to study objectives, but a rule of thumb is to collect data at the lowest possible measurement frequency, because further aggregation of those data can occur at later times, but pooled data cannot be summarized to a lower level than at which it was collected. Note, however, that measurement frequency also applies to the timing of data collection to coincide with factors that might impact the response of interest. For example, insect abundance has been shown to be important for breeding birds (Rotenberry et al. 1995). If interest is focused on determining how insect abundance affects avian reproductive success, then data on insect abundance should be collected concurrent with reproduction data. Consider the situation where an event that drives availability for capture (e.g., food availability) is more likely to occur immediately following a change to another species phenology (e.g., bat arrival and deposition of guano; Fenolio et al. 2006). Finally, consider conducting research to determine dietary preference of a wildlife species. In this case, diet data must be collected at the same time as data on availability of potential food items to make comparisons between usage and availability (Gray et al. 2007). If events of interest are dependent on other events, then measurement frequency should be tied to the determining event.

Data Management

Data entry should occur as soon as possible after data collection. Scientists often spend weeks or months collecting field data, only to return from the field with a stack of field data and no recollection of what notes, jargon, or mistakes or cross-outs meant for the day in question. Data entry near the day of collection, combined with data proofing, preferably by someone other than the individual who entered it, is beneficial, as most problems with initial data analysis are tied to the data entry process.

There are a number of ways to categorize data. For example, data can be either **qualitative** or **quantitative.** Qualitative data are data that are nonnumeric. Nonnumeric data includes information like study sites or species lists. However, qualitative data can be ordered (e.g., high, medium, or low), and can be used for comparative purposed (grassland, woodland, etc). Quantitative data are numeric and can be broken into 2 more specific categories, **discrete** and **continuous.** Discrete data have a finite number of potential values. For example, the number of nests a sparrow attempts can be 1, 2, or 3 (not 2.65), the number of fox offspring can be 1, 2, or 3 (not 1.45). Continuous data have an infinite number of potential values in any given numerical range. Examples of continuous data include individual weights, lengths, and home range sizes.

Another standard way of categorizing data is as 1 of the following 4 scales of data measurement: **nominal, ordinal, interval,** and **ratio.** Nominal (from the Latin for "name") data consist of data that are differentiated only by name. Data carrying the same name have certain characteristics in common that differentiate them from data with different names. No order or ranking is implied by the names. Examples of nominal data include gender and species. Although nominal data can be assigned a code in which numbers represent the names (0 = male and 1 = female; or 1 = Aves, 2 = Mammalia, 3 = Reptilia, etc.), these codes are simply values that can counted, but not ordered. Ordinal data have a distinct order (ascending or descending), but the difference between the measurement levels is not meaningful. For example, short, medium, and tall are ordinal scale data, in they imply an order, but they provide no insight into the degree of difference between each category. Interval data, in contrast, have constant differences, but no natural zero point. The common example is temperature measured on the Celsius or Fahrenheit scales (0 Fahrenheit does not mean there is no temperature). Note, however, this limitation is a characteristic of the scale on which temperature is measured, not of temperature itself. Temperature measured on the Kelvin scale does, in fact, have a true zero point; but 0 Kelvin does mean there is no temperature. In other words, these data are ratio data. Ratio data are interval data that do have a natural zero point. Other examples include birth weight (e.g., a weight of 0 means nonexistence), height (a 2-m-tall animal is twice as tall as a 1-m-tall animal), and age (a 2-year-old animal is half as old as a 4-year-old animal).

Data can be categorized as quantitative or qualitative while also being nominal, ordinal, interval, or ratio. Thus, sex is both nominal and qualitative, whereas temperature Celsius is both interval and quantitative. It also is worth remembering that data do not always fit neatly into a category. Moreover, the same data may be categorized differently, depending on circumstances (Zar 1999).

What do these ideas have to do with data collection? The structure of data is tied directly to the various approaches used for data analysis. Sooner or later, data need to be formatted for useful storage, retrieval, and analysis. Then data management systems come into play.

Most ecologists will be familiar with the "**flat file**" format for data storage. Flat files are simply files containing all the data collected, usually a single record per line. For simplicity's sake, a **record** is defined as data collected on a single individual during a single survey session. A record also could be a summary of data from multiple surveys, or from multiple individuals, depending on the question you are investigating. Common examples of flat files are files with data arranged in a tabular format, having columns of explanatory data specific to the data record and rows representing each data record. Flat files can be of any type, but are usually electronic files saved as either text or spreadsheet files.

The most common form of a flat file in wildlife biology research is a **spreadsheet** file using a program like Microsoft Excel® or Open Office's Calc®. Spreadsheets are used for several purposes, including data storage, entry, manipulation and summary, statistical analyses, and graphic construction (Morrison et al. 2008:67). Spreadsheets are the most common tool used for entering and storing data, likely due to the widespread availability and inherent simplicity of data entry and manipulation. However, spreadsheets have some drawbacks: they are known to be fraught with considerable mathematical inaccuracies (McCullough and Wilson 1999, 2002, 2005; Almiron et al. 2010) and thus should not be used for intensive data summarization or analyses.

An alternative option for data storage and management is use of a database management system (DBMS). DBMSs control data development, organization, maintenance, and use; they provide users with the ability to extract or retrieve data with the structure of the data still intact (Kroenke 2000). The advantages of using a DBMS are many, but the primary advantage is that it allows for use of data relationships (hence the name relational databases) to define how the system is structured (Codd 1970). Relational structure assists with data integrity, as scientists can minimize the amount of redundant data required. The functionality of a DBMS can be considerable and can be developed to allow for data entry and manipulation, queries to extract information into specific formats for future analysis, summarization, report creation, and linkages relating field-collected biological data to spatial data in Graphical Information System (GIS) or Global Positioning System (GPS) programs. Although the initial learning curve is steep, knowledge of the range of functions of DBMSs for supporting biological data collection would benefit any scientist.

DATA PRESENTATION AND GRAPHICAL ANALYSIS

Tables

The primary function of **tables** is to present numerical data in a format that simplifies the time it takes the reader to evaluate and determine whether specific results or conclusions are supported. In general, tables should present data in an unambiguous manner and should be interpretable without any need to reference the discussion in the accompanying text (Morrison et al. 2008:69). Tables have several uses, including presentation of frequency data on captured or measured individuals, presentation of summary data from statistical analyses, or categorization of data into scientifically relevant sections. Table columns are usually used for separating data into several classes or groups, whereas rows often present the summary data, but this structure is variable, based on the table's function. Table row and column headings should clearly define what the data in each table entry signifies. Table legends should be clear and consistent and should provide the source for the data.

Graphs

An alternative approach used to evaluate and summarize data has historically been the **graph** (Playfair 1786, 1801; Tufte 1983, 2001). Few approaches to visualizing and evaluating data are as powerful as a graph (Chambers et al. 1983), but the use and misuse of graphics is common in wildlife ecology, as there is no general or standardized theory for graph creation (Fienberg 1979), although basic standards for wildlife ecology have been suggested (Collier 2008).

The function of graphs is simple: illustrate the data coherently and at several levels while inducing the viewer to think on the substance of the data rather than on the specifics of the graph construction (Tufte 2001, Collier 2008). However, not all graphs are equally useful or applicable, and is it not necessary to limit the presentation of certain data types to a single type of graph. Rather, any graph design should be considered that will provide the viewer with an unbiased evaluation of the data under question; hence there may be several graphical options for showing the same dataset and no single option is always optimal. But, there are graphs and graphic effects that should be avoided during data evaluation. Collier (2008) and citations provided therein give an overview of the basic graph types used in wildlife ecology.

The purpose of a graph is to display data, but even the basics of graph construction and use can cause frustration for ecologists, as graphs that are difficult to understand or interpret are useless. Consider single parameter graphs, or those graphs showing the relationship between data collected on one specific measurement. Probably the most recognized single parameter graph is the pie chart (not shown). The debate regarding the use and misuse of pie graphs has continued for nearly a century with the current state of though being that pie graphs are inefficient for data interpretation and should not be used (Collier 2008 and references therein). When showing single parameter data, the best choice is the bar graph (Fig. 2.1). **Bar graphs** are one of the most frequently used graphics in ecological science; hence they are more prone to poor usage (Collier 2008). Bar graphs should be used to show relative or absolute frequency data (e.g., counts), never estimates of means and

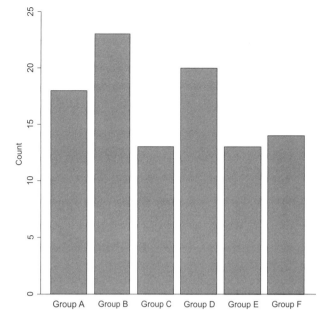

Fig. 2.1. Example of a bar graph showing the counts of individuals in each of 6 groups.

variances. A point graph or a dot plot would better serve the purpose of describing parameter estimates (Figs. 2.2 and 2.3, respectively). When determining whether there is general evidence for relationships between variables, scatter plots are often used to show how a response varies across a variable expected to influence the response (Fig. 2.4).

Historically graphs have been underused in ecological sciences for data analyses. Instead they are used to summarize data and/or results from statistical analyses. However, few tools are as powerful as the graph, and considerable information about the structure and relationships between data can be garnered through appropriate use of a graph (Collier 2008). Consider the example by Anscombe (1973), who used a simple dataset to show how important graphical evaluation of relational data is to appropriate inference. The simple premise was that data analysis ≠ data truth. Anscombe (1973) showed that if ecologists (or anyone) were to rely specifically on the results from a regression analysis to interpret data, then each of the datasets shown had identical summary statistics, regression parameters, residuals, and model fit statistics. However, the data tell a much different story in both form and structure (R Development Core Team 2009; Fig. 2.5). The moral of the story is to use data when deciding on an analysis, as the only error more egregious than looking at the data before analyzing it is not looking at it at all (Harrell 2001).

Good graphs do not force the reader to spend inordinate time trying to understand the data in the graph. Axes should be scaled so they do not change mid-graph (although there are exceptions to this rule, for example, semi-log graphs). Graphs without true zeroes do not need to have intersecting *x*- and *y*-axes (e.g., graphs showing how temperature

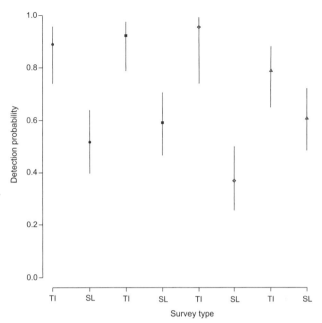

Fig. 2.2. Example of a point graph showing the predicted probability of detection for each of 4 paired thermal-imaging (TI) and spotlight (SL) surveys for white-tailed deer. *From Collier et al. (2007).*

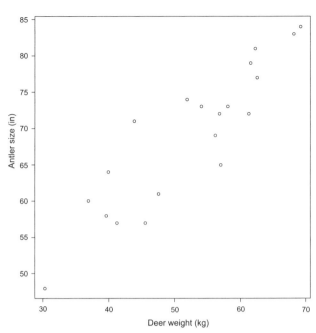

Fig. 2.4. Example of a scatter plot showing a relationship between deer weight and antler size for each individual.

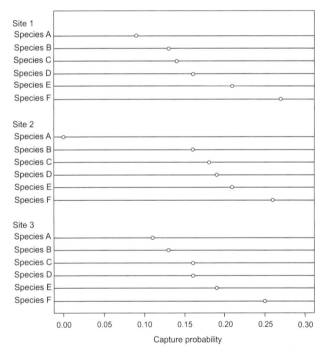

Fig. 2.3. Example of a dot plot showing capture probabilities for each of 6 species at 3 different sites.

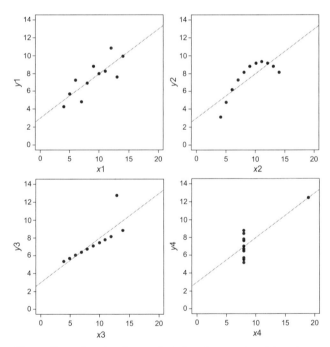

Fig. 2.5. Example using Anscombe's (1973) regression data sets, showing that although the data structure differs, the resultant regression curve (linear line) remains the same. *Data from Anscombe (1973).*

changes by date never have a [0, 0] point). They should not be complicated by **moiré effects** (repetitive use of shading or patterns used to distinguish among categories; Collier 2008). Graph legends should be clear and consistent in both size and font, borders should be crisp and muted, and size should be scaled to the audience and presentation mode. In addition, graph descriptions should be well thought out and detailed, so there is no question as to what the graph is describing: an individual looking at the graph should be able to understand the graph without any other reference material necessary.

Hypothesis and Model Development

The essence of data analysis is to develop a model that accurately and effectively predicts the phenomenon of interest across a wide range of conditions. One would hope that a biological model would exhibit generality, or the model would define and use a set of unique biological conditions that are equally relevant across many different ecological situations and hold true both within and across ecosystem boundaries. However, ecologists by and large have determined the identification of general models in ecological systems is difficult, and models reported are often invalidated by the processes that operate at different spatial, temporal, or organismal scales (Dunham and Beaupre 1998). These limitations do not negate the usefulness of models, as a thoughtful and well-developed model can provide some benefit to wildlife biology and management. The focus in this section is on how scientists should think about modeling wildlife ecology data.

Not all data analysis is tied to a predictive model (e.g., regression model) per se. When appropriate hypotheses are suggested, strong inferences can be made using descriptive statistics (mean, median, mode; Guthery et al. 2005); plots showing the distribution of the data (Loncarich and Krementz 2004); or simple test statistics, such as the t-statistics for comparing means or the chi-square test statistics for testing frequency (count) data (Moore and McCabe 1993, Agresti 1996). Assuming that a nontrivial hypothesis has been suggested (e.g., no silly null hypotheses, where the answer is already known; Cherry 1998; see Chapter 1, This Volume), the objective of statistical modeling is to reduce the chance of finding statistically significant effects that are spurious or finding effects that are statistically significant, but are not a function of the biological process being studied (Anderson et al. 2001a).

Underlying model development is the formulation of hypotheses relative to the question or system of interest. Whether models represent competing scientific hypotheses (Burnham and Anderson 1998, 2002) or competing approaches to data explanation (Guthery et al. 2001) is debatable, but the following discussion should hold for both cases. The research hypotheses should specify the underlying set of predictors to be used in the model (Harrell 2001). Various combinations of those predictors also will come into play during model development, as there are logical arguments for the use of models with additive effects versus those with interactive effects, or of models including polynomial terms of various dimensions (which can allow for curvature in the relationships between response and predictor variables).

Modeling is as much an art in some cases as it is a science. Model selection (Burnham and Anderson 2002) is the process by which the ecologists develops a set of models believed to describe the system of interest and uses the set for data analysis and interpretation. Rarely in wildlife ecology can all potential variables be identified; nor can all the relationships between various predictors be determined. "All models are wrong, but some are useful" (Box 1979:2). In a modeling exercise, an attempt is made to develop models, using the appropriate number of parameters, that provide precise estimates of the parameters of interest for the data at hand. The basic concept is one of **parsimony,** or how much bias will be accepted relative to the amount of variance accepted (Burnham and Anderson 2002). As parameters are added to the models, the model fit gets better (less bias), but the amount of variability also increases as well.

In an extremely simple outline of the vagaries of model development in wildlife ecology, consider a typical masters' study, in which a student is interested in estimating survival of some species of mammal. Assume the student has planned to capture, tag, and track a number of these mammals, at 2 different study sites over 2 years and collect a suite of data considered important in predicting survival.

Using one approach to modeling wildlife data, the student and advisor would develop a detailed list of the potential variables that they hypothesize, based on literature and previous work, have been shown to predict survival. Then after the field work, the student would come back to campus and diligently summarize, look at graphical representations of the data, evaluate relationships via scatter plots or simple t-test and/or contingency table analysis, and use that information to help inform what variables should be used in a predictive model (or models) to predict survival. Using a different approach to modeling wildlife data, the student would sit down with their advisor, develop a detailed list of the potential variables that they hypothesize, based on literature and previous work, have been shown to predict survival. The student then would develop a model set, nested in a global model that includes all factors, relationships, and interactions of interest. At the end of field work, the student would come back to campus and diligently analyze the data collected using only the set of models previously developed. Models would be ranked. Those with lower rankings would have less weight than those with higher rankings, and the formal interpretation of those models would weight the survival estimates based on the evidence garnered from the competing models.

These 2 scenarios represent the extremes of model development, and in truth the optimal scenario lies somewhere in the middle. Some ecologists suggest that model development should be conducted before study implementation, arguing that anything else amounts to failing to think before you act. Others believe a well-informed analysis cannot possibly be based solely on models developed before a study begins, as too many factors can change in the interim.

When developing models, the primary question the ecologist needs to ask is "will this model actually be used?" (Harrell 2001:4; also see Harrell [2001:79–83] for a detailed summary of modeling strategies). For models to be useful, they must be usable: development of an unused model wastes precious resources (and pages in theses and dissertations). In their simplest form useful models should fit the structure of the data in the model in such a way the data are used efficiently and appropriately (Harrell 2001:7). Useful models should then have their mettle tested on datasets independent of the data used to build the predictive model, although this practice is rare in wildlife biology.

Maximum Likelihood Estimation

The workhorse of classical approaches to statistical inference is the likelihood, or the probability of the observed outcome over a particular choice of parameters (Royall 1997). The objective is to find the value of the parameters that maximizes the likelihood (makes it as large as possible) and to use those values (maximum likelihood estimates) as the best guess for the true value of the parameters (Bolker 2008). Simply put, assuming a known statistical distribution (model) for the sample data, these data can be substituted into the known distribution and then that probability viewed as a function of the unknown parameter values one is interested in estimating (Mood et al. 1974, Agresti and Caffo 2000, Agresti et al. 2000, Williams et al. 2002a). The maximum likelihood estimate is thus the parameter value maximizing the probability of the observed data. The binomial $f(x) = \binom{n}{x} p^x q^{n-x}$, where $q = (1 - p)$, is one of the more common distributions used in wildlife sciences and will serve as a simple example of how maximum likelihood estimation works. The basic process for maximum likelihood estimation is to define the likelihood function, take the log of the likelihood function, take the derivative of the likelihood function with respect to the parameter of interest, and set it equal to zero and solve for the parameter of interest. As an example, assume a random variable is a distributed binomial with the above probability density function. Then the likelihood function L is defined as

$$L = p^{\Sigma x_i}(1-p)^{(n-\Sigma x_i)}.$$

The next step in determining the maximum likelihood estimate is to use a little calculus to take the log of L

$$\ln(L(p)) = \Sigma x_i \ln(p) + (n - \Sigma x_i)\ln(1 - p)$$

and then take the derivative of this equation with respect to p,

$$\frac{dL(p)}{dp} = \frac{\Sigma x_i}{p} + \frac{(n - \Sigma x_i)}{1 - p},$$

and set it equal to zero and solve for p:

$$\frac{\Sigma x_i}{p} + \frac{(n - \Sigma x_i)}{1 - p} = 0$$

$$(1 - p)\Sigma x_i - p(n - \Sigma x_i) = 0$$

$$\Sigma x_i - p\Sigma x_i - np + p\Sigma x_i = 0.$$

Canceling out the $p\Sigma x_i$ from the equation leaves

$$\Sigma x_i = np \Leftrightarrow p = \frac{1}{n}\Sigma x_i \Leftrightarrow \frac{\Sigma x_i}{n} = p,$$

which shows, not surprisingly, the value solved for p also is the mean for $\bar{x}$ the data collected. Following the fairly straightforward example put forth in different contexts by Agresti (2007) and Williams et al. (2002a), assume there are 10 individuals on which are recorded successes and failures and of these 10 individuals, 6 are successes and 4 are failures. The likelihood function for the 6 successes would be

$$L(p|y = 6) = \binom{10}{6} p^6 (1-p)^4,$$

using calculus and algebra gives

$$\frac{6}{p} - \frac{4}{(1-p)} = 0,$$

which gives $\hat{p} = 0.6$ or the maximized value for the parameter p based on the data collected.

INFERENCE ON FIELD DATA

Descriptive Statistics

The primary focus of wildlife population studies is the estimation of **parameters** (attributes of the population) for the biological unit of interest. Common estimates of parameters are the sample means, variances, and other measures of central tendency and precision (Cochran 1977). These estimators are common to a wide variety of ecological studies, so only a few important idiosyncrasies will be noted concerning their use and interpretation. There are several potential summary statistics that can be used to describe ecological data. For example, consider the data in Table 2.1, collected on the measurement of interest (centimeters of antler of male white-tailed deer [*Odocoileus virginianus*]). A glance at the table indicates that data have a distribution and there is a need to show the pattern and amount of variation of that distribution.

The first question is: where is the center of the distribution of the data collected? As there are $n = 20$ observations, the arithmetic average (**mean**) of the data can be determined by:

$$\bar{x} = \frac{1}{n}(x_1 + x_2 + \cdots + x_n),$$

which is equal to

$$\bar{x} = \frac{1}{n}\Sigma x_i.$$

However, it is important to note the mean is not resistant to outliers in the dataset. For example, if the data includes one extreme value (e.g., consider adding the number 400 to the response values in Table 2.1), then the mean will be pulled toward 400 ($\bar{x} = 175$ versus $\bar{x} = 185.7$). Thus, another option for evaluating centrality of a data distribution is to look at the **median** M, or the midpoint of the data distribution. Computation of the median is a fairly simple process; first, order the data from lowest to highest. If the total number of data values is odd, then the median is just the center observation located using the formula $(n + 1)/2 = 181.5$. If the total number of data values is even, then the location of the median used the previous formula, but is the average of the 2 center observations in the ordered dataset.

In the above examples the data distribution is based on continuous data. However, sometimes the data of interest can be continuous, nominal, or ordinal. Another measure of central tendency that can be used across a variety of data types is the **mode**. The mode of a data distribution is the value that occurs most frequently in the data. Distributions can be **unimodal**, having a single value that occurs most frequently, or **multimodal**, having several values that occur frequently (also known as a **nonunique** mode). Modes are valuable for summarizing nominal data as well. For example, a sample of trees in plots from the forest, might show that oak is the predominant species; hence, oak would be the mode of the sample.

These types of point estimates are commonly used for summarizing a distribution of data. These estimates provide some insight into the distribution; however, it is often desirable to accompany the point estimate with some measure of variability in the data. For example, when using the mean to describe the data distribution, there also is a need to quantifying the amount of spread in the distribution. One common measure of data spread is the **variance**

$$\sigma^2 = \frac{1}{(n-1)}\left[(x_1 - \bar{x})^2 + (x_2 - \bar{x})^2 + \cdots + (x_n - \bar{x})^2\right] \Leftrightarrow$$

$$\frac{1}{(n-1)}\Sigma(x_i - \bar{x})^2 = 626.6$$

for the example.

Note that due to the squaring, the sum of $(x_i - \bar{x})$ will not always be zero and the unit of measurement is not the same for the mean and the variance. The square root of the variance $\sqrt{\sigma^2}$ is the standard deviation σ (= 25.03 in this example), which is a measure of dispersion about the mean in the appropriate scale (Moore and McCabe 1993).

In practice there is interest in determining the upper and lower bounds for the range of the point estimate. The most common approach is to develop confidence intervals (CIs). CIs are an interval with width estimated from the data, which would be likely to include an unknown population parameter. There is a variety of approaches for estimating CIs for point estimates that are situation and data dependent, but consider the simple example of a population with known mean and variance sampled from a continuous data distribution. Assuming a critical value of 0.05 for a standard normal distribution (Moore and McCabe 1993), the bounds of the CI for a mean can be estimated as

$$\bar{x} - 1.96\,\mathrm{SE}$$

for the lower CI and

$$\bar{x} + 1.96\,\mathrm{SE}$$

for the upper CI, where SE represents the standard error. The assumption when using CIs is if the population of interest were sampled an infinite number of times, for 95% of the samplings their mean would fall between the upper and lower bounds of CI (assuming unbiased sampling) as constructed above (Hoenig and Heisey 2001).

Descriptive statistics have a wide array of uses in wildlife studies. Primarily, descriptive statistics are used to guide future research questions and/or data analyses. Frequently, descriptive statistics are provided in the "results" section of

Table 2.1. Simulated data on antler size (from base to tip of largest antler) of male white-tailed deer with predictor variables representing parasite presence, hoof length, and body weight

Size of antler (cm)	Parasite presence (yes = 1; no = 0)	Hoof length (mm)	Body weight (kg)
122	1	6.56	30.30
183	0	41.31	61.23
196	1	33.13	62.54
147	1	33.67	39.61
145	0	26.73	41.22
152	0	28.25	36.90
155	0	28.42	47.56
201	0	56.91	61.54
165	1	36.46	56.96
175	0	45.40	56.13
163	0	31.84	39.97
185	0	44.82	54.04
188	0	42.61	51.89
145	0	28.93	45.54
180	1	41.46	43.85
206	1	48.70	62.25
183	1	45.41	56.79
185	1	41.85	58.06
211	1	60.04	68.15
213	0	60.22	69.18

manuscripts as either text or graphical summaries (Collier 2008). Graphical summaries are useful to look at the range of the data distribution and perhaps to assess whether data from different groups or classes differs across the range of the data in question. Thus, there are many uses for general descriptive statistics (Guthery 2008). Sometimes there is valuable information to be learned just by looking at the data collected rather than immediately conducting analyses (Anscombe 1973, Harrell 2001).

Descriptive statistics also can include other basic summaries of interest. For example, data collected by ecologists is often represented by discrete counts. Often these counts are stratified according to some response factor of interest (e.g., study area or age), and there is interest in evaluating the associations between these variables. Contingency tables are one common application for evaluating these summary relationships between variables. For example, consider the situation where some plots are treated with herbicide and others are untreated, and species are determined to be present or absence on each plot. Data from this study could be evaluated using a contingency table analysis as shown in Table 2.2.

In this case, the response variable (species presence) has 2 categories. Perhaps there is interest in the odds ratio as another descriptive statistic. Following the example in Agresti (1996:24), for treatment plots the odds of species presence is 0.337 (621/1,840), whereas for the control plots the odds of species presence is 0.065 (193/2,965). Thus, the sample odds ratio is 5.18 (0.337/0.065), showing the odds of species presence in treated locations was approximately 5 times higher than the odds value of the control group. In addition to odds analysis, there are several other contingency table analysis approaches, including looking at differences in proportions, relative risk, and chi-squared or likelihood-ratio tests, all of which work for 2 × 2, 3-way, and higher level contingency tables (see Agresti [1996, 2007] for additional details on categorical data analysis).

Comparative Analyses

Wildlife field studies often focus on comparing responses from different groups and determining whether there is a relationship between there groups. The chi-squared test χ^2 is a statistic evaluating how well the data conform to expectations under the null hypothesis that no association be-

Table 2.2. Cross classification of herbicide treated and control plots on species presence

	Species presence		
Treatment	Yes	No	Total
Herbicide	621	1,840	2,461
Control	193	2,965	3,158

tween the observed and expected data (row and column variables) exist. The chi-squared test statistic is the sum of the squared differences between the observer and expected data divided by the expectation data:

$$\chi^2 = \frac{\Sigma(\text{Observed} - \text{Expected})^2}{\text{Expected}}.$$

Consider the data in Table 2.2 showing the relationship between counts of species being present in treatment and control study plots. A simple chi-squared test on these data provided a test statistics of 406.72 on 1 degree of freedom and a P-value of 0.05; thus, the null hypothesis of no relationship between the row and column data is rejected. Nearly all statistical textbooks cover chi-squared testing in detail, so no further elaboration on the methodology is given here.

As discussed above, scientists classify data based on some set of characteristics that are deemed important. Some of these classifications could include things like vegetation type (woodland or grassland) or areas that have undergone treatments, such as prescribed burning the previous year (yes, no). Additional classifications could be based on characteristics specific to individuals in the population (e.g., age of individuals, sex of individuals, and reproductive status). Often the interest is in determining whether biologically significant differences exist among individuals in different classes. As a simple example, consider the data in the **Length** column shown in Table 2.1. Assume, for the sake of an example, the first 10 records are a simple random sample from population A, and the second 10 are a simple random sample from population B. The hypothesis is there is no difference between the 2 populations. Thus, it would be appropriate to use a 2-sample t-statistic to evaluate whether the means were statistically (and biologically) different:

$$t = \frac{(\bar{x}_1 - \bar{x}_2)}{\sqrt{\left[\frac{(s_1^2)}{n_1} + \frac{(s_2^2)}{n_2}\right]}},$$

where $\bar{x}_i$ is the estimated mean for population i, s_i^2 is the estimated variance for population i, and n_i is the sample size for population i ($i = 1, 2$). Using the data discussed above, the t-statistic is −2.0635 with a CI of −22.06 to 0.0025, which overlaps zero, indicating the null hypothesis, $H_0:\mu_1 = \mu_2$ or the 2 means were different, could not be rejected. There is a variety of t-statistic procedures for equal and unequal variances, 1- and 2-tailed tests that can be found in most statistical texts (e.g., Moore and McCabe 1993).

A corollary to using a simple t-test to evaluate differences between population means also could be to use a one-way analysis of variance (ANOVA; Moore and McCabe 1993, Venables and Ripley 2002). ANOVA is used to evaluate the null hypothesis that all (≥2) population means are equal versus the alternative—they are not equal. Thus, ANOVA could be used to compare the means from several populations and test the statistical hypothesis there is no difference in the means of several populations. In the simple case described in the *t*-test example, use of an ANOVA would provide the same result as the *t*-test, for the *t*-test is the same as an ANOVA comparing only 2 means. One should use a *t*-test if only 2 means are compared and an ANOVA when 3 or more means are compared. Never use a *t*-test to compare more than 2 means. ANOVA should be used cautiously: inferences should not focus on whether differences exist, because those hypotheses are spurious (Johnson 1999), but the magnitudes of those differences are relevant. ANOVA is a widely used statistical procedure both in and outside of wildlife ecology and also can be found in most general statistical texts. Thus no further elaboration on ANOVA approaches is presented here.

Sometimes it is not germane to examine whether 2 variables are the same, but whether they show a positive or negative relationship. For example, interest may be in whether age and antler growth are positively correlated over the life cycle of a white-tailed deer (Koerth and Kroll 2008), or whether mate choice and reproductive success of long-tailed paradise whydah (*Vidua interjecta*) is related to such characteristics as tail feather length (Oakes and Barnard 1994). A statistical correlation provides insight into whether the relationship between 2 variables is positive or negative and the strength of that relationship. Positive correlation occurs when 2 variables change in the same direction; a negative correlation occurs when 2 variables change in opposite directions, such that an increase in one is accompanied by a decline in the other. Correlation is often measured using the correlation coefficient *r*:

$$r = \frac{(n\Sigma xy - [\Sigma x][\Sigma y])}{\sqrt{n(\Sigma x^2) - \Sigma(x)^2} \sqrt{n(\Sigma y^2) - \Sigma(y)^2}}.$$

As seen from the formula, the correlation coefficient's ranges from –1.0 to 1.0 and provides an indication of the strength of the relationship between the variables. The closer to –1.0 or 1.0 the value of *r* is, the stronger is the relationship is between the variables. There is a variety of rules of thumb for interpreting correlation coefficients, but values –0.5 and >0.5 are considered to indicate fairly strong relationships between the variables. Note, however, that correlation coefficients should not be used to detail causation (e.g., the value of *y* is predicted by the value of *x*). Variables can be correlated and not have a predictive relationship. For determining predictive relationships between variables, see the section on linear regression later in the chapter.

It is important to discuss covariance, because correlation and covariance do not provide the same value. Although both measure the extent to which 2 variables vary together, covariance is defined as

$$Cov(x, y) = E[(X - \mu_x) - E(X - \mu_y)],$$

where *Cov*(*x*, *y*) measures the linear relationship between *x* and *y*, but its magnitude does not have meaning, as it depends on the variability in *x* and *y* (Mood et al. 1974). The correlation coefficient is the covariance divided by the product of the standard deviations of *x* and *y*, and it thus provides a better, albeit unitless, measure of the linear relationship between *x* and *y*. Note that when 2 random variables are independent, the covariance is zero, but the inverse does not hold as a covariance of zero does not imply independence (e.g., when *x* and *y* vary in a nonlinear fashion; Mood et al. 1974).

Linear Regression

Ecological variables can be correlated, such as relationships between vegetative communities and soil types, or age and body size, but the 2 variables should not be expected to be entirely dependent on each other. When it is reasonable to assume functional dependence between 2 variables (Zar 1996), linear regression becomes an option ecologists can use to evaluate relationships between the variables. In linear regression the response or dependent variable is assumed to be a function of the predictor variable(s) (also called explanatory, covariate, independent, or descriptive variables; Harrell 2001). Thus, a general interest in linear regression is determining how changes in the response variable are predicated on changes in the predictor variable.

Consider the relationship in the data shown in Table 2.1 and the resulting scatter plot in Figure 2.4. The coupled measurements $(y_1, x_1), (y_2, x_2), \ldots, (y_n, x_n)$ were made on the same experimental unit, leading to the belief these data warrant further examination focusing on the general question: how much does the value of **Y** (column vector of response data) change for a 1 unit change in **X** (column vector of explanatory data)? The typical model form used to evaluate this type of data is a simple linear regression model:

$$Y = \beta_0 + \beta_1(X) + \varepsilon,$$

where β_0 is the model intercept, β_1 is the model slope (for the linear line), and ε is the model residual (error) term. For this model, the values of β_0, β_1, and ε are all unknown quantities, however, β_0 and β_1 are fixed values and can be estimated based on the data in Table 2.1. Thus, the objective is to use the data collected to estimate the regression parameters. By inserting these regression parameters into a linear regression model,

$$\hat{Y} = \hat{\beta}_0 + \hat{\beta}_1 X + \hat{\varepsilon},$$

the mean value of $\hat{Y}$ across a range of values for **X** can be predicted. Estimating the regression parameters is a fairly simple task that requires a bit of calculus and basic algebra. Following the examples outlined in Draper and Smith (1998) and Harrell (2001), the method of least squares is one approach to estimating these regression parameters. For example, given all the underlying assumptions (see below), the sum of squares for the deviation from the true regression line (sum of squares function) is given by

$$C = \sum_{i=1}^{n} \varepsilon_i^2 = \sum_{i=1}^{n}(Y_i - \beta_0 - \beta_1 X_i)^2,$$

Using calculus and following Draper and Smith (1998:23–24), C is differentiated with respect to β_0 and β_1:

$$\frac{\partial C}{\partial B_0} = -2\sum_{i=1}^{n}(Y_i - \beta_0 - \beta_1 X_i) \text{ and } \frac{\partial C}{\partial B_1} = -2\sum_{i=1}^{n} X_i(Y_i - \beta_0 - \beta_1 X_i).$$

Setting the resulting values equal to zero gives

$$\sum_{i=1}^{n}(Y_i - \hat{\beta}_0 - \hat{\beta}_1 X_i) = 0 \text{ and } \sum_{i=1}^{n} X_i(Y_i - \hat{\beta}_0 - \hat{\beta}_1 X_i) = 0,$$

which can be solved for $\hat{\beta}_1$ by using algebra:

$$\hat{\beta}_1 = \frac{(\Sigma X_i Y_i - [(\Sigma X_i)(\Sigma Y_i)]/n)}{\left(\Sigma \frac{X_i^2 - (\Sigma X_i)^2}{n}\right)} = \frac{\Sigma(X_i - \bar{X})(Y_i - \bar{Y})}{\Sigma(X_i - \bar{X})^2}.$$

Knowing $\hat{\beta}_1$ one can solve for $\hat{\beta}_0$:

$$\hat{\beta}_0 = \bar{Y} - \beta_1 \bar{X}.$$

As an example of linear regression, consider the data in Table 2.1 relating the size of white-tailed deer antlers with the deer's body weight. A quick perusal of these data suggests a positive relationship between deer weight and antler size that follows a straight (linear) trend (Fig. 2.4). One way to examine these data is by using linear regression analysis, where $\hat{\beta}_0$ is the intercept along the y-axis and $\hat{\beta}_1$ is the slope of the linear line. The hypothesis being tested supposes the slope of the line is statistically equal to zero (a flat line): therefore, there is no predictable relationship between body weight and antler size. Thus, the alternative hypothesis (H_A: $\hat{\beta}_0 \neq 0$) posits there is such a relationship. The method of least squares seeks to find an appropriate equation for a straight line that minimizes the square of the vertical distances between each data point and the predicted line (Draper and Smith 1998; Fig. 2.6). Using the data outlined in Table 2.1, values for the intercept ($\hat{\beta}_0 = 67.70$) and the slope ($\hat{\beta}_1 = 2.056$) are estimated.

The value for the intercept is an estimate of the mean size of antlers given a body mass of zero. A weight of zero means nonexistence, but it is important to show what can go wrong (this model predicted 67 cm of antler on an individual weighing 0 kg). It is important to recognize that extending the predictive model beyond its data can lead to nonsensical results (Guthery and Bingham 2007). In this case a zero-intercept model likely should have been developed, where the regression line is forced through zero (Draper and Smith 1998).

The sign on the slope parameter is positive, so an increase in weight results in increased in antler size. In addition, given the resulting t-statistic for evaluating the hypothesis $H_A:\hat{\beta}_1 \neq 0$ ($t = 5.498$, $P < 0.05$) there is evidence the slope

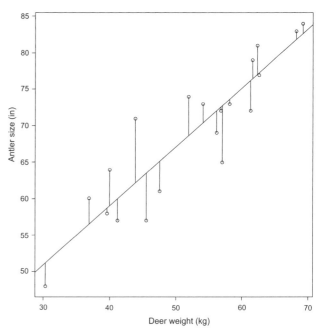

Fig. 2.6. Example of a regression line showing the minimized vertical distance between each data point and the regression line.

is not equal to zero, indicating a relationship between antler size and weight.

General approaches to linear regression and tests of hypotheses regarding slopes have been the workhorses of wildlife ecology studies. Hundreds (if not thousands) of papers and books have been published focused on methods associated with regression techniques, so no further technical details are discussed here.

Multiple Regression

Approaches used for simple linear regression also can be applied to multivariable models (often called multiple or multivariate regressions). Multivariable models are defined as a regression in which ≥2 predictor (x) variables are used in a model having the form

$$\bar{Y} = \hat{\beta}_0 + \hat{\beta}_1 X_1 + \hat{\beta}_2 X_2 + \hat{\varepsilon},$$

where the regression parameter definitions are the same as previously described: the multiple values for $\hat{\beta}_i$ represent the predicted rate of change in the response variable for a 1-unit change in the predictor variable x_i when all other values of x_i are held constant. Note the column vector **X** is now subscripted to have >1 column of data; hence it is now a matrix containing the data collected in the field. The $\hat{\beta}_i$ in a multiple regression are interpreted the same way as in a univariate linear regression: they represent the effect of the predictor variable on the response of interest. Using a multiple regression, the researcher has the ability to construct (often very) complex models that represent hypotheses relating the predictor variables to the response. Consider the data in Table 2.1. Using a multiple regression framework, the response (antler size) can be modeled as an "additive effect" of 2 predictor variables (deer weight, hoof length):

$$\hat{Y} = \hat{\beta}_0 + \hat{\beta}_1 weight_i + \hat{\beta}_2 length_i + \hat{\varepsilon}.$$

Substituting the estimated parameters for the intercept 77.79), slope for the effect weight (1.07), and slope for the effect of length (1.004) into this equation gives

$$\hat{Y} = 79.77 + 1.07\, weight_i + 1.004\, length_i.$$

Using the above formula, the effect of weight on antler size can be predicted, assuming a fixed value for length or vice versa. Biological reasons suggest an additive effect, but there also may be an interactive effect:

$$\hat{Y} = \hat{\beta}_0 + \hat{\beta}_1 weight_i + \hat{\beta}_2 length_i + \hat{\beta}_3 weight_i\, length_i,$$

from which parameters for the main effects (weight and length) and the interactive effect $\hat{\beta}_3$ of weight times length can be estimated. Note, when dealing with an interactive effect, the necessity of modeling and interpreting main effects is unimportant (Venables 1998). The variables interact in some form, so the potential impacts of one predictor by itself on the response of interest is uninteresting, as the data show the interaction exists and is biologically important. Tests of main effects when interactive effects are present are unwise and likely will give spurious answers. The potential range of predictor variable structure is wide, as there are biological reasons to model univariate, multivariate, or perhaps polynomial effects in the regression function (quadratic or cubic models; Guthery and Bingham 2007) to better understand the relationship between predictor variables and biological response. Regression modeling in wildlife sciences is more complex than presented here (for example, non-linear regression models are not discussed). For the interested reader, Draper and Smith (1998), Harrell (2001), and Venables and Ripley (2002) are recommended. Each covers a wide array of regression modeling approaches, including detailed discussion of regression assumptions and specific diagnostics for model checking and validation.

Some Vagaries of Regression

It is important to mention that regression is not an end, but rather a means to an end. There is a wide variety of assumptions used in regression analysis (Draper and Smith 1998, Harrell 2001), many of which can affect the accuracy of the inferences made from regression modeling. Although it is one of the powerful tools in the ecologist's toolkit, if used improperly, it can provide spurious results (Guthery and Bingham 2007). As shown above, in many cases regression can give improper answers even when the model is appropriate, but there is, in theory, a true zero. Zero-intercept regression models, or models that do not estimate the y-intercept (commonly shown in regression equations as $\hat{\beta}_0$) are applicable when the response variable y must be 0 when the predictor variable $x = 0$ (e.g., our example above). However, note that regressions that enforce a zero intercept do not conform to the standard assumptions of minimizing the square of the distances between the regression line and the individual data points used to determine the line; hence statistics like R^2 are inappropriate. Thus, this model requires some additional study regarding the general assumptions of regression-based modeling, which is beyond the scope of this chapter.

What information should be presented when providing results from a modeling exercise in which regression modeling of some sort was used? Often it is not enough to just show or discuss the general form of the model used (e.g., "I used regression analysis to evaluate the effects of insect abundance on avian body mass"), or in the case of competing candidate models, to only show a table with associated model selection statistics (Burnham and Anderson 2002). Obviously a one-size-fits-all approach would be ill advised, but there should be a minimum amount of information regarding regression modeling results provided. First, a complete description of the model(s) used and the logic behind why each model was chosen should be given. This information should include the biological reasoning behind why certain relationships (additive, interactive, polynomial, etc.) were used. Output from the model should include estimates of the model coefficients $\hat{\beta}_i$ and associated standard errors or CIs, or in the case of a set of models, model-averaged estimates of the regression parameters and associated unconditional standard errors. In addition, as reproducibility of research results is the essence of the science method, the range (min, max) of the data used to estimate the regression parameters should be provided, which would ensure inferences are not being made from a model outside the viable range of the data (Harrell 2001, Guthery and Bingham 2007, Guthery 2008). Although often not shown, plots of regression diagnostics (Harrell 2001, Venables and Ripley 2002) can be used as supporting evidence for regression assumptions.

Generalized Linear Models

Data collected in wildlife biology studies often do not conform to the standard normal distribution assumed when implementing ANOVA or linear regression models. Thus, the focus should be on generalized linear models. Generalized linear models aggregate a wide variety of approaches, such as linear regression and ANOVA for continuous responses (as seen above), as well as models for categorical data, including binary responses and count data (logistic regression and log-linear models respectively; see below). Agresti (1996: 72–73) and Venables and Ripley (2002:183) noted that generalized linear models have 3 components:

1. a random component that specifies the distribution of the response variable

2. a systematic component that specifies the explanatory variables to be used to predict the response, and
3. a link function that specifies the relationship between the random and systematic components.

Linear regression and ANOVA models for continuous responses are special cases of generalized linear models, so no time will be spent on them here (see Agresti 1996:73). In short, if a Gaussian (normal) distribution is assumed for the random component and an identity link for the link function, then a generalized linear model is equivalent to using a standard linear regression model.

If the data were nonnormal (binomial, counts, ordinal, etc), generalized linear models can be used to evaluate the effect of predictor variables on the response. Nonnormal data fall in a wide range of categories, but for the sake of this discussion, we consider categorical data and couch the examples as categorical data analysis. Although perhaps not immediately obvious to most wildlife ecologists, categorical data analysis, especially logistic regression, and associated statistical theory underlie a wide range of approaches to population parameter estimation.

Categorical data can exist in several forms. For example, it can be information on the presence or absence of a species at a given location or binomial data on whether the study subject is alive or dead. It might be represented simply as zeroes and ones. Categorical data also can include frequency of occurrence, such as counts of individuals over time (which theoretically have Poisson distributions), or stages of various ecological systems relative to some value of interest (emergent, senescence, etc.; Morrison et al. 2008). Categorical data can be either nominal (order is unimportant) or ordinal (order is important), and models constructed for nominal data can be applied to ordinal data, but not the reverse. Categorical data are often considered as qualitative (rather than quantitative) data, because categorization does not imply quantitative data.

The nature of wildlife research ensures that most information collected by ecologists is structured in a categorical framework. Hence most of these data analyses rely on the application of generalized linear models wherein the response and predictor variables are related via a link function (Agresti 1996). Using a linear regression model (as described previously), the approach is to model the effect of some predictor variables **X** on the response **Y,** where the expected value of **Y** is a linear function of **X** (Hosmer and Lemeshow 2000, Harrell 2001). However, when the response values fall between 0 and 1, as do most parameters of interest in wildlife ecology, other modeling options are needed. The logistic regression (a specific form of generalized linear models) function for binary (0, 1) responses and quantitative predictor variables is

$$logit[\pi(x)] = \log\left(\frac{\pi(x)}{1 - \pi(x)}\right) = \alpha + \beta(x).$$

Logistic regression is used for capture–mark–recapture and survival analysis (wherein the response of interest is binomial: capture or not, alive or dead), resource selection and habitat use evaluations (Boyce and McDonald 1999, Manly et al. 2002), and distribution studies (MacKenzie et al. 2006), among many others. Using this formula, the probability will either increase or decline as a function of the levels of the predictor variable **X.** For example, consider some data of interest in determining whether weight at capture affects fawn survival over some predetermined period (Lomas and Bender 2007, Pojar and Bowden 2004). Using a logistic regression on a set of hypothetical data, parameters for the intercept (0.3) and the slope (0.6) are estimated. The logistic equation for these parameters is

$$\pi(x) = \frac{e^{(\alpha + \beta x)}}{1 + e^{(\alpha + \beta x)}},$$

where $\pi(x)$ is the predicted probability (of survival) based on the β estimate from the regression equation. Using the above parameter estimates and assuming a value for weight (which is the predictor variable) of 3 kg, then estimate the probability of survival is

$$\pi(x) = \frac{e^{0.3 + 0.6(3)}}{1 + \varepsilon^{0.3 + 0.6(3)}} = 0.8909.$$

Using this equation, $\pi(x)$ increases as X increases (Fig. 2.7). Logistic regression also is amenable to the use of several explanatory variables (multiple logistic regression), where both quantitative (e.g., weight or length) and qualitative (e.g., site or species) variables can be used for evaluation (Agresti 1996).

In addition to the analysis of binary data, logistic regression also can be used for evaluation of unordered and ordered multicategory data. These multicategory logit models assume the response (**Y**) values have a multinomial distribution and describe the results as the odds of response in one category instead of a different category (Agresti 1996). For ordered data, the interest is in the probability the response **Y** falls in some category j or below, where the cumulative probabilities reflect the category ordering (see Agresti [1996, 2007] for a detailed discussion of multicategory analysis in a logistic framework).

Many ecological studies have count data as the response of interest, such as number of individuals heard or seen during a survey, number of parasites found on captured individuals, or number of offspring successfully recruited each year. Analysis of these data types is typically accomplished using Poisson regression (Agresti 1996). In this case, the count data are assumed to follow a Poisson distribution, which is unimodal and has a positive parameter that is both the mean and variance of the distribution (Venables and Ripley 2002). Poisson regression models define the log of the mean using the log-link function, although an identity link also can be used (Agresti 2007). For a simple overdispersion pa-

Fig. 2.7. Examples of logistic regression curves showing hypothetical relationship between fawn weight and survival probability. Curve form is shown to be dependent on the sign of the β parameter for weight. (A) Positive β parameter, (B) negative β parameter.

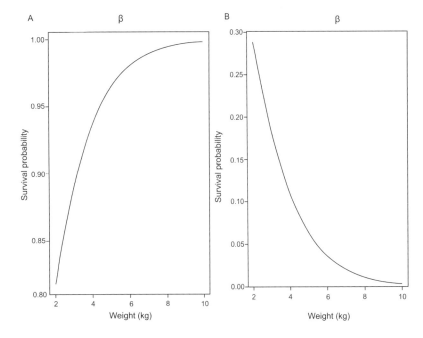

rameter D (which summarizes the amount of dispersion relative to a standard Poisson regression using 1 predictor variable), the Poisson model has the form

$$\log \mu = \alpha + \beta x,$$

where a increase of 1 unit in the predictor variable has a multiplicative effect. For example, exponentiating the above formula (following Agresti 2007) gives

$$\mu = e^{\alpha}(e^{\beta})^x,$$

which shows if $\beta > 0$, then $e^{\beta} > 1$, so the mean of **Y** increases as **X** increases, and if β 0, then the mean of **Y** decreases as **X** increases. Note that count data exhibit greater variability than would be expected if the true underlying distribution were known, a phenomenon called overdispersion. One method for evaluating overdispersion is to use negative binomial regression, wherein an additional parameter is available, allowing the variance to exceed the mean (Cox 1983).

Outside a purely predictive framework, certain approaches to categorical analysis can be used to evaluate patterns of association among categorical variables (Agresti 2007). Log-linear models are useful for the analysis of contingency table that cross-classify individuals into several categorical groups. Interested readers should see books by Agresti (1996, 2007) for details on log-linear modeling approaches and parameter evaluation. These models are useful when ≥2 variables are considered to be response variables, such as how various vegetative characteristics, fire effects (e.g., charring height), food abundance, and avian foraging behaviors are associated under different fire regimes (Pope et al. 2009) or the relationships between wetland size, isolation, and forest cover (Pearce et al. 2007).

Bayesian Analysis

A more recent entrant into the field of wildlife biology is the application of Bayesian methods (Gelman et al. 2003, Lee 2004, Albert 2007, Congdon 2007) to the analysis of biological data (Link and Barker 2009, Kery 2010). Bayesian analysis differs from the classical approach to analysis in several ways; however, in many cases results from either analysis will be approximately equal. Classical approaches as outlined in this chapter view biological data as the observed realization of stochastic systems. **Classical (frequentist) statistics** assume the parameters used to describe data are fixed and unknown and that uncertainty can be evaluated based on long-run assumptions about the distribution of the data based on hypothetical repeated experiments. In contrast, Bayesian approaches estimate the conditional probability distribution of a parameter of interest, given the data and a model, and base statements regarding uncertainty on the posterior distribution of that parameter. Likely, the simplest explanation comes from Bolker (2008:15): "Frequentist statistics assumes there is a 'true' state of the world . . . which gives rise to a distribution of possible experimental outcomes. The Bayesian framework says the experimental outcome (what actually happened) is the truth, while the parameter values of hypotheses have probability distributions."

The foundation for Bayesian inference is Bayes's theorem, which relates the probability of occurrence for one event with the probability of occurrence for a different event. For example, assume 2 events, A and B: what is the probability of event A occurring given that event B has occurred? The theorem states the probability of A occurring, given that B has occurred, is

$$P(A|B) = \frac{P(B|A)P(A)}{P(B)}.$$

But, what does this mean? Here is an example to help explain. When working on Bayesian statistics, it is sometimes easier to think about the conditional probabilities as hypotheses (Lee 1997, Bolker 2008). Adopting that idea here and rewriting the above theorem gives

$$P(H|d) = \frac{P(d|H)P(H)}{P(d)},$$

where H is the probability of the hypothesis occurring and d is the set of data collected. Now assume there are 2 competing hypotheses H_1 and H_2: which is most likely ($H_i|d$), given the data? When using a Bayesian analysis, the assumption is made that $P(H_i)$ is the prior probability of the hypothesis, or the belief in the probability of that hypothesis before field data collection. For now assume the hypotheses are equally likely and thus set the prior probability for both at 0.50. Assume that some data have been collected in the field and based on these data, the probability of d occurring if H_1 is true is 0.15, and the probability of d occurring if H_2 is true is 0.25. Using the above formula, the posterior probability of H_1 occurring given d is

$$P(H_1|d) = \frac{(0.15 \times 0.5)}{[(0.15 \times 0.5)(0.25 \times 0.5)]} = \frac{0.075}{(0.075 + 0.125)}$$

$$= \frac{0.075}{0.20} = 0.375,$$

and the posterior probability of H_2 occurring given d is

$$P(H_2|d) = \frac{(0.25 \times 0.5)}{[(0.25 \times 0.5)(0.15 \times 0.5)]} = \frac{0.125}{(0.125 + 0.075)}$$

$$= \frac{0.125}{0.20} = 0.625.$$

The sum of the posterior probabilities is 1 (0.375 + 0.625), because the value of $P(d)$ (the unconditional probability of observing the data) is actually the sum of the probabilities of observing the data under any of the posited hypotheses (Bolker 2008). Thus, one way to look at $P(d)$ is as

$$\sum_{h=1}^{H_i} P(d|H_i)P(H_i),$$

or the sum of the likelihood of each hypothesis $P(D|H_i)$ times the unconditional probability of that hypothesis occurring $P(H_i)$ (Lee 1997). Interestingly, one of the primary issues historically associated with Bayesian statistics was the difficulties estimating the denominator $P(d)$, because in most common applications this term contains intractable high-dimensional integrals (Gelman et al. 2003). Thus, other approaches are needed so that Bayesian theory can be applied to general problems in wildlife biology. Luckily, such an approach exists in Markov chain Monte Carlo methods (MCMC; Casella and George 1992; Spiegelhalter et al. 2003, 2004). MCMC sampling is a complex topic requiring special software (e.g., WinBUGS; Spiegelhalter et al. 2004) and an intuitive knowledge of how to algebraically structure and code appropriate statistical models beyond the scope of this chapter. For the interested reader, Bolker (2008), Royle and Dorazio (2008), Link and Barker (2010), and Kery (2010) provide a unique perspective on Bayesian analysis for ecological studies.

Survival Analysis

In its most general sense population dynamics is a function of 3 variables: population size, survival of individuals in the population, and recruitment of young into the population. Repeated measures of population size over time may provide insight into the population trajectory, but they do not provide information on what biological processes are driving variability in the population and contributing to changes in its size over time (Williams et al. 2002a). Considering the wide variety of research wildlife ecologists conduct, the topic that occurs most frequently is species demography—more specifically, species survival. Statistical estimation of survival has a rich history in wildlife ecology and has seen a variety of methodological approaches developed for survival estimation (see Williams et al. [2002a] for a review of methods typically used in wildlife; also see Kleinbaum [1996], Hosmer and Lemeshow [1999], and Therneau and Grambsch [2000] for additional approaches).

Survival analysis, or time-to-event modeling, is a set of statistical methods for evaluating how various factors affect the length of time it takes for an event to occur (Kleinbaum 1996, Hosmer and Lemeshow 1999). Events in models can include not only mortality, but also speed of return to a location (e.g., site fidelity), time taken for growth to occur after a habitat treatment, or time needed for incidence of disease to occur or go into remission, or any other experience that may affect an individual (Morrison et al. 2008:182). The primary focus in wildlife research is estimation of survival $\hat{S}$ or apparent survival ϕ for the time frame of interest. The first survival estimate ($\hat{S}$: the probability that an individual survives a specific time period) differs from the second (apparent survival ϕ: the probability the individual survives a specific time period and is available for relocation in the study area of interest). Survival estimates typically come from populations that are studied using radiotelemetry (White and Garrot 1990), for which the probability of relocating and correctly classifying the state (alive or dead) is 1. In contrast, apparent survival estimates typically occur in studies using capture–mark–recapture designs, for which the probabilities for relocation, state assignment, and recovery are less than 1.

Studies of survival in wildlife ecology typically rely on data collected from radiotagged individuals (Winterstein et al.

2001); thus, analysis of those data is the focus of the rest of this section. However, the approaches described here are equally applicable to any time-to-event modeling in which the probability of locating the individual is 1, including analyses of such topics as nest survival (Dinsmore et al. 2002). There is a wealth of literature available for evaluating survival under capture–mark–recapture designs, so these designs will not be covered here in any detail. The interested reader can review Williams et al. (2002a) and Thomson et al. (2009).

There are 3 basic survivorship functions that are used in analysis of time-to-event data (Venables and Ripley 2002). Each function is related, and all rely on the random variable T, which indicates the length of time before a specific event (e.g., death) occurs. Each survivorship function is important, because modeling can be conducted with each function in mind. These 3 survivorship functions are (Venables and Ripley 2002, Morrison et al. 2008:183):

- $S(t) = \Pr(T > t)$ is the most common survivorship function. It describes the probability that an individual survives longer than time T, estimated as the proportion of individuals surviving longer than t.
- $f(t) = 1 - S(t)$ or $f(t) = dF(t)/dt = -dS(t)/dt$. This function is the probability density function for the time until an event occurs, usually referred to as the failure time distribution.
- $h(t) = f(t)/S(t)$. This function is the hazard function and is usually interpreted as the conditional probability of failure rate.

Following the example in Morrison et al. (2008:183), the estimator probably best known to ecologists is the basic Kaplan-Meier product limit estimator (Kaplan and Meier 1958, Pollock et al. 1989):

$$\hat{S}(t) = \prod_{\alpha=1}^{J}\left(1 - \frac{d_i}{r_i}\right),$$

which is the product of J terms for which $a < t$, given that a_j are discrete time points (j) when death occurs, d_j is the number of deaths at time j, and r_j is the number of animals which are at risk at time j. Estimating the probability of surviving from time 0 to a_1 (where a_1 is the interval during which the first death occurs) is

$$\hat{S}(a_1) = 1 - \left(\frac{d_1}{r_1}\right),$$

and the probability of surviving from the time period of the first event a_1 to that of second event a_2 is simply $1 - d_2/r_2$. Therefore, $\hat{S}(a_2)$, or survival until the end of the second time period, is simply the product of these 2 terms:

$$\hat{S}(a_2) = \left(1 - \frac{d_1}{r_1}\right)\left(1 - \frac{d_2}{r_2}\right),$$

which can then be expanded to a_n occasions. A simple graph can then be constructed showing the Kaplan-Meier survival curve, also known as a step function (Fig. 2.8).

These studies focus on whether there are significant (based on biological or statistical criteria) differences in survival between 2 or more groups (e.g., age, sex, and location). A log-rank test is usually used in this approach to evaluate whether survival differs among groups, although several other tests are available (Kleinbaum 1996, Hosmer and Lemeshow 1999).

Another approach gaining favor in wildlife sciences is the use of regression modeling for survival data. Available in many forms, regression modeling of survival data is useful as "biologically plausible models may be easily fit, evaluated, and interpreted" (Hosmer and Lemeshow 1999:1). The Kaplan-Meier approach focused on the probability of surviving past some time t, in other words, the probability of not failing. For regression modeling, the focus shifts to the hazard function: the rate at which failure is expected to occur, given the individual has survived to time t (Hosmer and Lemeshow 1999, Therneau and Grambsch 2000). The hazard function $h(t)$ is related through the abovementioned survivorship functions to $S(t)$: when survival goes up, hazard will decline (Kleinbaum 1996). The Cox proportional hazard model (Cox 1972, Cox and Oakes 1984) is a nonparametric modeling approach to survival analysis in which the hazard of mortality at some time t is a product of 2 statistical quantities: the baseline hazard function $h_0(t)$ and the exponential expression $e\left(\sum_{i=1}^{n}\beta_i x_i\right)$, where the summation is over the n explanatory variables (Kleinbaum 1996, Therneau and Grambsch 2000). The baseline hazard is a function of time

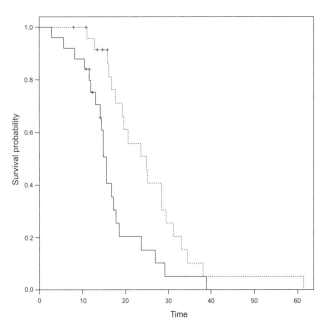

Fig. 2.8. Example of a Kaplan-Meier step function survival curve for 2 sexes (shown as dotted and solid lines).

t, whereas the exponential expression is a function of the explanatory data, but not time. Under this model, measuring the effect of the explanatory data on the hazard function uses the ratios of the βs for individuals with one specific set of predictors relative to individuals with a different set of predictors. The Cox proportional hazard approach to regression modeling of survival data is fairly complex. However, it has been at the forefront of much survival analysis in a wide variety of fields, and new approaches and extensions to the theory are now being developed. For those interested in detailed discussions of this approach, outside the peer-reviewed literature, Kleinbaum (1996) provides a nice initial review with examples; Hosmer and Lemeshow (1999) and Therneau and Grambsch (2000) delve into the mathematical and theoretical details and provide a wide discussion ranging from model development to time-dependent covariate modeling.

A few general remarks about survival analysis in wildlife ecology are in order. Certain criteria must be met for data analysis to be appropriate. Several are presented here, and interested readers should consult Pollock et al. (1989), Tsai et al. (1999), and Williams et al. (2002a) for further information. First, the radiotagged individuals being monitored must be randomly selected from the population of interest. It is assumed the survival times of individuals are independent of one another and the marks (radiotags) do not affect survival. Second, survival should be estimated for a biologically relevant time frame. It makes little sense to estimate daily survival (time step of 1 day) for a study on elephant survival covering 25 years, because most daily survival estimates will be 1.00. The model will be overparameterized (too many parameters for too little data), and the results will be unwieldy. Rather, survival studies should be designed such that at least 1 event (mortality) is expected in each sampling period under study. Third, the concept of censoring (Tsai et al. 1999) also must be addressed. Censoring occurs when the information collected from an individual is incomplete. Hence, censoring occurs when survival times are not known exactly. There are 3 types of censoring —right, middle, and left censoring (Morrison et al. 2008:182), each addressing a time period in which complete information on the individual in question is not available. Potential causes of censoring are various, ranging from radiotag failure or loss to temporary or permanent emigration (Morrison et al. 2008). Careful study design, especially involving the use of pilot studies, can help identify areas where censoring may be more likely and can help investigators decide on the best course of action for reducing the effects of censored observations.

Mixed Effects Models

Mixed effects models are a class of models that allow for the analysis of data displaying some correlation in their structure. Mixed effects models are so named because they incorporate both a fixed effect and a random one (Venables and Ripley 2002, Pinheiro and Bates 2004). Fixed effects are those whose levels in an analysis represent all potential levels of interest in making inferential statements. Random effects are defined as those whose levels in an analysis represent what is expected in a random subset of all plausible levels of the effect. Usually mixed effects models are used when data are collected by using repeated measures on the same sampling units (measurements of individual growth over time, counts of species at specific locations over time, etc.).

There are 3 general model structures: linear mixed-effects models, nonlinear mixed-effects models, and generalized linear mixed-effects models. Linear mixed-effects models are similar to the previously mentioned linear models, except that linear mixed-effects models incorporate additional random effects terms for when data are collected repeatedly on the same experimental unit (Pinheiro and Bates 2000). General linear mixed-effects models typically have the form

$$Y = X\beta + Z\mu + \varepsilon,$$

where μ and ε are ~N(0, A) and ~N(0, C), where A and C represent the variance-covariance matrices for the μ and ε terms (Breslow and Clayton 1993, Pinheiro and Bates 2004).

For some datasets, we are unable to characterize the variation using traditional linear modeling methods; thus, nonlinear mixed-effects models provide an option when the relationship between the response and predictive variables varies nonlinearly (Pinheiro and Bates 2004). Nonlinear mixed-effects models are appropriate when there is variation in the overall mean, as well as within and between individuals on which data are being collected. A good example of the use of nonlinear mixed-effects models comes from tree growth (Gregoire and Schabenberger 1996). Measurements are taken repeatedly on tree bole size over time. Trees grow at different rates and show a nonlinear S-shaped growth curve. Given they reach a stable size at various times, this nonlinear individual growth pattern is easily modeled using nonlinear mixed-effects models (also see Littell et al. 2006).

When the response variable may not follow a normal distribution, rather than use methods transforming the data for use with general linear methods, there is the option to model the data directly with generalized linear mixed-effects models (GLMM). GLMMs are an extension of generalized linear models (see above), in which the response variable is related via a link function to the various predictor variables (Pinheiro and Bates 2004). GLMMs are most useful when the response data are either binary (e.g., 1, 0) or are count data, as these nonnormal data types are easily handled via link functions and with the relationship within the exponential family of distributions (Bolker et al. 2008). Mixed-effects modeling is an extremely complicated and often misused approach (see Bolker et al. [2008], who found that 58% of authors used GLMMs inappropriately); hence caution is suggested when considering using any such model. The literature is rapidly advancing with respect to GLMMs (Pinheiro and Bates 2004), including likelihood estimation, random effects modeling, various estimation methods, and Bayesian approaches (Kery 2010; see Bolker et al. [2008] for an excellent review of GLMM applications in ecology).

Community Analysis

Most wildlife research, and thus most wildlife data analysis, is focused on measuring and comparing attributes of individuals (e.g., mass) or populations of a single species. However, biologists often wish to examine the characteristics of wildlife communities composed of a number of species. Community statistics are useful for this type of work.

Many community ecology studies seek to address the questions: What is the diversity of a community (or communities)? How does it change across time, space, or both? Although there is some disagreement as to the true ecological importance of biodiversity, increased biodiversity is generally viewed as a positive community attribute. From this foundation, biologists can address several practical questions. What will be the effect on biodiversity of a given management activity or environmental perturbation? Which communities should be given highest conservation priority?

Magurran (2004) offers a very thorough treatment of species diversity measures. Krebs (1999) also describes several species diversity measures and their relative merits. Here we discuss some of the more common techniques and their application in wildlife studies.

Species Richness

The most basic approach to assessing diversity is to estimate species richness. Species richness is defined as the number of different species in a given community (McIntosh 1967). Although the concept is straightforward, its application is somewhat more complicated. For the simplest communities, a complete enumeration of species may be achieved. However, in most circumstances, a complete list of species in a community is rarely possible, given limited time, personnel, and budget. Moreover, all species richness measures are strongly influenced by sample size (i.e., the number of individual animals collected from a community). This sensitivity comes about because species are usually not equally abundant in a community: a few species are abundant whereas most are rare. Thus, as sample size increases, there is greater opportunity to capture individuals of the rare species, increasing the observed species richness. This phenomenon makes comparison among communities (or studies) sampled unequally especially difficult, and at the very least, it demands that sample size be reported and thoroughly explained in all species richness studies. Even without these limitation, it is usually impossible to determine when all species have been identified in a community, leaving even the best estimate of species richness uncertain.

Species Enumeration

By far the most common method of estimating species richness in the wildlife literature is a simple enumeration of the species observed, captured, or detected during the study. Inexplicably, enumeration is often used in cases where rather sophisticated analytical methods are used for other aspects of the study. Enumeration provides the minimum number of species present in the community, and thus it will almost always underestimate the true richness of the community. Although the degree to which it underestimates the community richness parameter decreases as sample size increases, Hellmann and Fowler (1999) concluded that it was more negatively biased than other common richness estimates. Although this method may have some value when comparing samples in the same study if sampling effort is equal among sites, it is crude. Most studies could be easily improved employing a more analytically rigorous estimate of species richness.

Richness Indices

Species Indices

Because the number of species in a sample tends to rise as the sample size increases, the first step in developing a useful index of species diversity is to account for sample size. Assume the number of species S in a sample generally increases approximately linearly as a function of log N (where N is the number of individuals in the sample; Hayek and Buzas 1997). Margalef (1957) developed the following simple index of richness d that takes into account sample size:

$$d = \frac{S - 1}{\ln(N)}.$$

Chettri et al. (2005) used Margalef's index to compare species richness of bird communities among various forest vegetation types in the Himalayas of India to assess effects of human disturbance on bird diversity. In South Carolina Metts et al. (2001) assessed species richness of herpetofauna communities at beaver ponds and unimpounded streams using this index, finding that beaver activity tended to reduce species richness.

It is important to remember Margalef's d is an index of species richness, *not* a richness estimate. As such, it cannot be compared with other indices or richness estimates. Moreover, although the use of $\ln(N)$ in the calculation Margalef's index moderates the effect of sample size, the index remains strongly influenced by N. However, its intuitive nature and ease of use make it suitable for some applications, and it is a significant improvement over simple enumeration.

Richness Estimates

Species richness data often are composed of a number of discrete sampling units (e.g., quadrats), in which organisms are (hopefully) completely enumerated. Heltshe and Forrester (1983) stressed that such data are not random samples of individuals, but are instead random samples of spatial units. The species richness estimator chosen depends on whether the data were collected as one large sample or as a number of smaller samples. Although collecting numerous smaller samples is almost always superior to collecting a single large sample, below are a number of richness estimators that address each of these situations in turn.

Data Collected or Pooled as a Single Sample
Rarefaction

Biologists often want to compare species richness among communities by using samples of different sizes. Because species richness increases as a function of sample size, species richness estimated from smaller samples will tend to be underestimated relative to those communities from which a larger number of individuals are collected. Rarefaction (Sanders 1968) is a method of standardizing sample size among all samples to derive a richness estimate comparable among samples. By using the rate at which the species count increases as individuals accumulate in a sample, it is possible to estimate the species richness had sampling been halted at any sample size n smaller than the actual N.

Wildlife studies often result in samples of unequal size, because the number of individuals captured or detected in each unit often varies, despite equal sampling intensity. For example, Kissling and Garton (2008) sampled bird communities in southeastern Alaska. Because the number of bird detections varied among sampling blocks, they standardized species richness to the number of species per 100 detections and compared richness between managed and control blocks.

Although first proposed by Sanders (1968), Hurlbert (1971) and Simberloff (1972) modified the rarefaction algorithm to the correct form used today (Krebs 1999):

$$E(\hat{S}_n) = \sum_{i=1}^{S} \left[1 - \frac{\binom{N - N_i}{n}}{\binom{N}{n}} \right],$$

where $E(\hat{S}_n)$ is the expected number of species in a random sample of n individuals, S is the total number of species in the entire sample, N_i is the number of individual of species i in the entire sample, N is the total number of individuals in the entire sample, n is the sample size for which $E(\hat{S}_n)$ is to be estimated ($n \leq N$), and $\binom{N}{n}$ is the number of combinations of n individuals that can be chosen from a set of N individuals and is defined as

$$\binom{N}{n} = \frac{N!}{n![N-n]!}.$$

Chao 1 Method

Chao (1984) developed a simple estimator of species richness using abundance data. This estimator (the Chao1 method) is especially useful when the dataset is dominated by rare species. It is calculated by using the ratio of species represented by a single individual in the dataset to species represented by exactly 2 individuals:

$$S = S_{obs} + \frac{F_1^2}{2F_2},$$

where S_{obs} is the total number of species observed in the community, F_1^2 is the number of species represented by exactly 1 individual, and F_2 is the number of species represented by exactly 2 individuals.

Data Collected as a Series of Samples
Chao 2 Method

The Chao 1 estimator requires abundance data, at least to the extent of determining which species are represented by exactly 1 or 2 individuals. However, a modification of the Chao 1 method (the Chao 2 method) allows for the estimation of species richness by using presence–absence data, as long as data are collected as a series of samples (quadrats, seine hauls, etc.; Magurran 2004). In this case, the variables of interest are not the number of species represented by 1 or 2 individuals, but the number of species that occur in exactly 1 or exactly 2 samples:

$$S = S_{obs} + \frac{Q_1^2}{2Q_2},$$

where S_{obs} is the total number of species observed in the community, Q_1^2 is the number of species present in exactly 1 sample, and Q_2 is the number of species present in exactly 2 samples.

Jackknife and Bootstrap Estimates

The idea of using a **jackknife** procedure to reduce bias and estimate confidence in a statistic was first introduced by Tukey (1958). Jackknifing refers to the process of estimating a statistic by systematically removing one sample from a dataset of size n and computing the estimate, then returning the sample to the data, removing the next sample and recomputing the estimate, and so on through n iterations. The jackknife procedure has been used to estimate several ecological parameters (Krebs 1999), including species richness.

Heltshe and Forrester (1983) developed a jackknife estimator of species richness called the **first-order jackknife,** because it is based on the number of species that occur in exactly one sample. It is similar to the approach taken by Burnham and Overton (1978) to develop jackknife estimates of population size using mark–recapture data. First-order jackknife species richness $\hat{S}$ is estimated using the equation

$$\hat{S} = y_0 + \left(\frac{n-1}{n} \right) k_1,$$

where n is the total number of samples, y_0 is the total number of species observed across all samples, and k_1 is the number of species occurring in exactly 1 sample.

Smith and van Belle (1984) modified the first-order jackknife procedure by including not only those species occurring in exactly 1 sample, but also those species occurring in exactly 2 samples. Thus, the **second-order jackknife** species richness is estimated as

$$\hat{S} = y_0 + \left[\frac{k_1(2n-3)}{n} - \frac{k_2(n-2)^2}{n(n-1)} \right],$$

where n is the total number of samples, y_0 is the total number of species observed across all samples, k_1 is the number of species occurring in exactly 1 sample, and k_2 is the number of species occurring in exactly 2 samples.

Efron (1979) introduced the concept of the **bootstrap** as an improvement over the jackknife, and Smith and van Belle (1984) extended the bootstrap procedure to the problem of estimating species richness. Like the jackknife, the bootstrap also is derived by resampling the original dataset and is suited to data collected as a series of samples:

$$\hat{S} = S_0 + \sum_{j=1}^{S_0} \left(1 - \frac{Y_{\cdot j}}{n} \right)^n,$$

where n is the total number of samples, S_0 is the total number of species in the original sample, and $Y_{\cdot j}$ is the number of quadrats in which species j is present.

Hellmann and Fowler (1999) showed when n 25% of the community, the second-order jackknife was less biased than the first-order one. However, for larger sample sizes the first-order jackknife was the least biased of the 2, although both showed a tendency to overestimate S. Smith and van Belle (1984) compared the jackknife and bootstrap estimates and determined the jackknife performed better for small sample sizes, but for larger sample sizes the bootstrap was superior.

Species Heterogeneity

Although species richness is an important measure of species diversity, it may fail to fully capture important aspects of the community. For example, suppose 2 communities are composed of 100 individuals representing 5 species. In community A, individuals are evenly distributed among the 5 species, so that each species is represented by 20 individuals. In community B, however, although the same 5 species are present, individuals are unevenly distributed among species, such that 1 species is represented by 80 individuals, whereas each of the remaining 4 species are represented by only 5 individuals each. In this instance, it might be expected the communities would function differently, despite having identical species richness.

The degree to which individuals in a community are evenly distributed among species is referred to as species **heterogeneity.** Although not as intuitive as richness, heterogeneity carries important potential ramifications for diversity and ecosystem function. Several measures of species diversity that incorporate both richness and heterogeneity have been developed. We discuss 2 of the more commonly used ones here.

Shannon–Weiner Function

One of the most popular indices of diversity is the Shannon–Weiner function (Magurran 2004). This function is based on information theory, which attempts to measure the amount of disorder in a system—for example, the amount of disorder (i.e., uncertainty) associated with the information contained in a message. Specifically, the Shannon–Weiner function was developed to measure the amount of uncertainty associated with the next bit of information in a data stream. If the data stream were composed of a series of identical bits, there would be no uncertainty associated with the next bit, and the system would be perfectly ordered. However, as the stream becomes more complex (i.e., disordered or diverse), uncertainty increases.

In the case of wildlife communities, disorder is analogous to species diversity, with more diverse communities having a higher degree of disorder. By way of example, suppose a biologist is sequentially collecting individuals from a given community. Further assume the community is composed of a single species. This monotypic community can be viewed as being highly "ordered": there would be little uncertainty associated with the species identity of the next individual to be captured. That is, if the community were composed of only one species, then the next individual captured would be of that species. However, as the number of species in a community and the degree to which individuals are evenly distributed among these species increases, so does the uncertainty associated with the identity of the next individual and thus with the "disorder" of the system.

The Shannon–Weiner function H' is

$$H' = \sum_{i=1}^{s} p_i \ln(p_i)$$

where s is the total number of species observed in the community, and p_i is the proportion of the total sample belonging the species i.

H' can have a minimum value of 0, indicating a community composed of only one species. Although theoretically without an upper bound, the index rarely exceeds 5 in practice (Washington 1984). However, interpretation of the Shannon–Weiner function is not intuitive, and thus can be problematic. MacArthur (1965) introduced a modified form of the Shannon–Weiner function N_1, calculated as $N_1 = e^{H'}$. This function represents the number of equally common species that would be required to generate the same diversity as H', thus, making interpretation more straightforward.

Note that H' can be calculated using any logarithm base, but comparisons can only be made among values calculated using the same base. Moreover, if a base is used other than e, then the appropriate base must be substituted for e when calculating N_1.

Jobes et al. (2004) used Shannon's index to compare bird community diversity among forest stands subjected to different logging regimes. Lande (1996) noted that although the Shannon–Weiner function is popular, it is prone to serious bias. This bias is most pronounced in communities with a large number of species. He recommended Simpson's index (see below) as a superior alternative.

Simpson's Index

The first measure of biological diversity was developed by Simpson (1949) (Krebs 1999). Simpson's index is a straightforward and intuitive measure of diversity. It is based on the probability that 2 individuals drawn at random will be of the same species. Despite its vintage and simplicity, it is considered by many to be the best measure of species diversity (Lande 1996, Magurran 2004).

As originally formulated, the index was calculated as $D = \Sigma p_i^2$. However, many authors have suggested the complement of Simpson's index $(1 - D)$ as a more intuitive measure. Possible values of this form range from 0 to almost 1, with 0 representing a community with the lowest possible diversity (i.e., containing a single species) and 1 representing a highly diverse community. Because both formulations are commonly used in the literature, it is important to specify which form is used when reporting results.

Simpson's index is one of the most widely used measures of diversity in the wildlife literature. It has been used to characterize species diversity of amphibians (Ward et al. 2008), reptiles (Hampton et al. 2010), birds (Hurteau et al. 2008), and mammals (Phelps and McBee 2009). Magurran (2004) considered Simpson's index to be one of the most meaningful and robust diversity measures. Lande (1996) also found the approach to superior to other methods.

"WHAT SHOULD I USE FOR DATA ANALYSIS?"

Earlier we discussed what the researcher could use for collecting, storing, and managing data from a wildlife study, but the programs available for analyzing data from a wildlife study were not discussed. There are many programs available for conducting statistical analysis of wildlife data (Table 2.3). Each program has advantages and disadvantages (Morrison et al. 2008:67–68); however, in most cases all statistical software listed will conduct standard analyses in a similar

Table 2.3. Commonly used statistical programs for analysis of wildlife biology data

Program name[a]	Source	Primary web link
SPSS	Commercial	www.spss.com
SAS	Commercial	www.sas.com
S-Plus	Commercial	spotfire.tibco.com
R	Open source/freeware	cran.r-project.org
Systat	Commercial	www.systat.com
MatLab	Commercial	www.mathworks.com
Maple	Commercial	www.maplesoft.com
Mathematica	Commercial	www.wolfram.com
WinBUGS (Bayesian)	Freeware	www.mrc-bsu.cam.ac.uk/bugs/
ADMB	Open source/freeware	admb-project.org
OpenBUGS (Bayesian)	Open source/freeware	mathstat.helsinki.fi/openbugs/
Minitab	Commercial	www.minitab.com

[a] ADMB = Automatic Differentiation Model Builder; SAS = Statistical Analysis System; SPSS = Statistical Package of the Social Sciences.

fashion. Although most are commercial and thus have nontrivial costs associated with them, usually student versions are subsidized by universities and licenses can be purchased for a small fraction of the overall program costs. In an effort to look at the range of programs (Table 2.3) used by practicing wildlife scientists, we conducted a survey of volumes 72 and 73 (2008–2009) of the *Journal of Wildlife Management* ($n \approx 384$ articles, excluding editorial articles) and determined which programs were used primarily for wildlife data analysis in these volumes. The most frequently used statistical program was Statistical Analysis System (SAS), used in 28% and 29% of articles in 2008 and 2009, respectively. The statistical programs SPSS (4% and 8%, respectively), R (5% and 5%, respectively), S-Plus (1% and 0%, respectively) and WinBUGS (1% and 1%, respectively) also were used. An "other" category, consisting of programs with a specific, as opposed to general, applicability (e.g., MARK, Presence, Distance, and various genetic programs) was used 20% and 27% of the time, respectively. Surprisingly, 10% and 5%, respectively, of articles reporting analytical results did not provide citations on what programs were used—a serious breach of accepted publication practices.

CONCLUSION

We consider *The Wildlife Techniques Manual* to be a fluid document that is continually being updated. Thus, we have added some techniques to this chapter since its last printing (Bart and Notz 2005). The dictates of brevity and the existence of a multitude of books written on each and every data analysis method available precludes a detailed presentation of a wide variety of statistical procedures in this chapter. We have chosen to leave out classification analysis and stepwise regression, not because of certain peculiarities and shortcomings of these approaches (Rexstad et al. 1988, Harrell 2001, Burnham and Anderson 2002), but because as the manual evolves, its focus should remain on techniques now in common use by students of wildlife ecology. Nonparametric approaches also have been omitted, because when appropriate sampling is used, parametric approaches will usually suffice (Johnson 1995, Stewart-Oaten 1995). We have chosen to include information on approaches to survival analysis, Bayesian analysis, and a significant amount of community diversity statistics, as these topics have become more prevalent in the wildlife literature. For those interested, we have cited the literature extensively, covering a variety of primary literature that should be considered when preparing to develop a wildlife study.

3

Capturing and Handling Wild Animals

SANFORD D. SCHEMNITZ,
GORDON R. BATCHELLER,
MATTHEW J. LOVALLO,
H. BRYANT WHITE, AND
MICHAEL W. FALL

INTRODUCTION

THE ART OF CAPTURING wild animals for food and clothing is as old as human existence on earth. However, in today's world, reasons for catching wild species are more diverse. Millions of wild animals are captured each year as part of damage and disease control programs, population regulation activities, wildlife management efforts, and research studies. Many aspects of animal capture, especially those associated with protected wildlife species, are highly regulated by both state and federal governmental agencies. Animal welfare concerns are important regardless of the reason for capture. In addition, efficiency (the rate at which a device or system catches the intended species) is a critical aspect of wild animal capture systems.

Successful capture programs result from the efforts of experienced wildlife biologists and technicians who have planned, studied, and tested methods prior to starting any new program. State regulations related to animal capture vary widely, and licenses or permits, as well as specialized training, may be required by state wildlife agencies for scientists, managers, and others engaging in animal capture for research, damage management, or fur harvest. Institutional Animal Care and Use Committees, required at universities and research institutions by the Animal Welfare Act (U.S. Department of Agriculture 2002), often question whether scientists capturing animals for research have ensured that pain and distress are minimized by the techniques used. The information in this chapter will assist wildlife management practitioners to identify appropriate equipment and obtain the necessary approvals for its use. Researchers are encouraged to consult Littell (1993) and Gaunt et al. (1997) concerning guidelines and procedures relating to capture and handling **permits.**

Major reviews of bird capture techniques include Canadian Wildlife Service and U.S. Fish and Wildlife Service (1977), Day et al. (1980), Davis (1981), Keyes and Grue (1982), Bloom (1987), Bub (1991), Schemnitz (1994), and Gaunt et al. (1997). Detailed coverage of mammal capture methods include Day et al. (1980), Novak et al. (1987), Schemnitz (1994), Wilson et al. (1996), American Society of Mammalogists (1998), and Proulx (1999a). Mammal capture usually becomes more difficult as animal size increases. Thus, observational techniques and mammalian sign are often more efficient for obtaining both inventory and density information (Jones et al. 1996). Several new techniques to capture mammals ranging in size from small rodents to large carnivores have been developed in recent years. Some of these represent either improved or modified versions of traditional capture methods. Most animals are captured by hand, mechanical devices, remote injection of drugs, or drugs administered orally in baits. The emphasis in this chapter is on methods and equipment other than remotely injected drugs used for capture. Scott (1982), Heyer et al. (1994), Olson et al. (1997), and Simmons (2002) have compiled comprehensive capture references for amphibians.

This chapter is a revision of Schemnitz (2005) and includes additional citations and new methods for the capture and handling of wild animals. Users of this chapter are encouraged to refer to the series on wildlife techniques by Mosby (1960, 1963), Giles (1969), Schemnitz (1980), Bookhout (1994) and Braun (2005). Mammal researchers are encouraged to consult Gannon et al. (2007). They stress the need when live-trapping to provide adequate food, insulation, and avoidance of temperature extremes.

CAPTURING BIRDS

Use of Nets

Dip and Throw Nets

The common **fish dip net** has been used for capture or recapture of radiotagged birds for many years (Table 3.1). Unlike commercial nets, dip nets used to capture wildlife are usually constructed by the investigator. Constructed nets usually have a larger diameter hoop (≥1.5 m) and a longer handle (3–4 m), with mesh size being dependent on the type of animal being captured. Radiotagged birds are first located at night using a "walk in" technique. The bird is located by gradually circling it and then using a flashlight to temporarily blind the bird. A long-handled, large-diameter dip net is then placed over the bird. If several birds roost together (especially a hen with brood), a radiotagged bird can be used to locate a flock, and several other birds also can be trapped. Dark nights with light rain worked best when night lighting birds. This technique can be used on nonradiotagged birds, such as those roosting on roadsides, located on nests, nonflying young on nests or flushed from nests, and birds roosting on water (collected by using boats and long-handled dip nets). The use of dip nets for capturing wildlife is limited only by the investigator's imagination.

Drewien and Clegg (1991) had great success capturing sandhill and whooping cranes (**scientific names** for birds, mammals, reptiles, and amphibians can be found in Appendix 3.1) using a portable generator mounted on an aluminum backpack frame and a 28-volt spotlight mounted on a helmet to locate them (Table 3.2). Cranes were then captured using long-handled (3.0–3.6 m in length) nets, with best success on dark overcast nights when they were roosting in small flocks during summer. Well-trained pointing dogs and 2–3-m-long handled nets have been used to capture nesting and broods of American woodcock (Ammann 1981). Drewien et al. (1999) captured trumpeter swans using

Table 3.1. Dip and throw nets used to capture wildlife

Group/species[a]	Reference
Birds	
American white pelican	Bowman et al. 1994
California gull	Bowman et al. 1994
Common loons	Mitro et al. 2008
Cormorants	Bowman et al. 1994, King et al. 1994
Cranes	Drewien and Clegg 1991
Doves	Morrow et al. 1987, Swanson and Rappole 1994
Eiders	Snow et al. 1990
Greater prairie-chicken	Robel et al. 1970
Greater sage-grouse	Wakkinen et al. 1992
Murrelets	Whitworth et al. 1997
Nightjars	Earlé 1988
Pelagic sea birds	Gill et al. 1970, Bugoni et al. 2008
Swans	Drewien et al. 1999
Mammals	
American beaver	Rosell and Hovde 2001
Jackrabbit	Griffith and Evans 1970
Nutria	Meyer 2006
Amphibians and reptiles	
Aquatic amphibians	Wilson and Maret 2002, Welsh and Lind 2002

[a] Scientific names are given in Appendix 3.1.

night lighting to locate them from a lightweight (180 kg) airboat during severe winter weather. King et al. (1994) successfully captured roosting double-crested cormorants using night lighting from a boat at winter roosts in cypress trees (*Taxodium distichum*; Fig. 3.1). Cormorants were captured with a long-handled net in shallow water. Whitworth et al. (1997) combined the use of dip nets from small boats at sea to capture Xantus murrelets. Mitro et al. (2008) used night lighting to capture adult common loons with chicks. Gill et al. (1970) and Bugoni et al. (2008) described the use of a cast net thrown by hand from a fishing boat to capture scavenging pelagic sea birds attracted by bait thrown into the water.

Bowman et al. (1994) successfully used night lighting to survey, capture, and band island-nesting American white pelicans, double-crested cormorants, and California gulls. Disturbances to birds while night lighting was minimal, and there was no predation by gulls on eggs or chicks. Night lighting was more effective for capturing young than for capturing adults. Snow et al. (1990) night-lighted common eiders during the summer in shoal waters using deep hoop nets 46–61 cm in diameter attached to 3.7–4.3-m-long handles.

Wakkinen et al. (1992) modified night spotlighting techniques by using binoculars in conjunction with a spotlight to locate greater sage-grouse. Binoculars allowed greater detection in 55 of 58 (95%) instances. Capture success increased by >40%.

Throw nets have been used to capture wildlife, but more skill is involved with this technique. These cast-nets are usually used with night lighting to capture birds. **Cast-nets** also

Table 3.2. Night-lighting methods and equipment used to capture wildlife

Group/species[a]	Reference
Birds	
Greater rhea	Martella and Navarro 1992
American white pelican	Bowman et al. 1994
Double-crested cormorant	Bowman et al. 1994, King et al. 1994, 2000
Waterfowl	Glasgow 1957, Lindmeier and Jessen 1961, Cummings and Hewitt 1964, Drewien et al. 1967, Bishop and Barratt 1969, Merendino and Lobpries 1998
Trumpeter swan	Drewien et al. 1999
Common eider	Snow et al. 1990
Ruffed grouse	Huempfener et al. 1975
Greater sage-grouse	Giesen et al. 1982, Wakkinen et al. 1992
Greater prairie-chicken	Labisky 1968
Northern bobwhite	Labisky 1968
Ring-necked pheasant	Drewien et al. 1967, Labisky 1968
Shorebirds	Potts and Sordahl 1979
Sandhill crane	Drewien and Clegg 1991
Whooping crane	Drewien and Clegg 1991
Yellow rail	Robert and Laporte 1997
American woodcock	Rieffenberger and Ferrigno 1970, Shuler et al. 1986
California gull	Bowman et al. 1994
Common nighthawk	Swenson and Swenson 1977
Mammals	
Cottontail rabbit	Drewien et al. 1967, Labisky 1968
Jackrabbit	Griffith and Evans 1970
Muskrat	McCabe and Elison 1986
Mule deer	Steger and Neal 1981

[a] Scientific names are given in Appendix 3.1.

Fig. 3.1. Jon-boat showing positioning of night-lighting equipment (bow rails, lights, converter box, and generator) and personnel. *From King et al. (1994).*

can be used to capture birds on water by using night lighting techniques. Earlé (1988) combined night lighting and a cast-net to capture nightjars (Caprimulgidae) along gravel roads. The 85-cm diameter, circular cast-net had handles to facilitate throwing it

Mist Nets

The number of papers describing the use of mist nets to capture birds or bats are too numerous to include in this chapter. Here we provide the reader with examples of various methods to deploy mist nets and papers that caution the reader on how to use data obtained from this method.

Mist nets continue to be an effective method for sampling bird populations. Ralph and Dunn (2004) summarized and recommended commonly used protocols for monitoring bird populations using mist nets. They discussed a variety of key factors, including annual photography and vegetation assessment at each net site to document vegetation height and density, exact net placement and locations, and type of net used (e.g., net material, mesh size, dimensions, methods used to measure birds, fat scores, and frequency of net checks), thereby allowing comparison of results among independent studies. Length of netting seasons should follow **standardized procedures.** Mist-netting studies should be carefully planned to ensure that sampling design and estimated sample size will allow clearly defined study objectives to be met. Remsen and Good (1996) urged caution in the direct use of mist-net data to estimate relative bird abundance. Corrections should be based on detailed knowledge of the ecology and behavior of the birds involved. Ralph et al. (1993) emphasized the importance of setting nets in locations of similar vegetation density and terrain. Jenni et al. (1996) reported the proportion of birds avoiding mist nets without entering a net shelf depended on the extent of shading and net-shelf height, but not on species, wind speed, or habitat. Dunn et al. (1997) reported that annual capture indices of 13 songbird species based on standardized autumn mist netting were significantly and positively correlated with breeding bird survey data from Michigan and Ontario, Canada. Their results suggested that mist netting could be a useful population monitoring tool. Wang and Finch (2002) noted consistency between the results of mist netting and point counts in assessing land-bird species richness and relative abundance during migration in central New Mexico.

Meyers and Pardieck (1993) developed a lightweight, **low canopy** (1.8–7.3 m) mist-net system using adjustable aluminum telescoping poles. Sims (2004) and Burton (2004) described improvements in net poles and a tool for raising and lowering mist nets. Stokes et al. (2000) perfected a method to deploy mist nests horizontally from a canopy platform in 30-m-tall forests. A connecting wooden bridge can be built between platforms. The nets and net poles were suspended from a support cable and pulled along the cable by a control cord and pulley. This system allowed comparisons of mist net capture rates between forest canopy and understory levels.

Albanese and Piaskowski (1999) perfected an inexpensive ($35.00) **elevated** mist-net apparatus that sampled birds in vegetation strata from ground level to a height of 8.5 m. The equipment consisted of metallic tubs, clothesline cord, and single and double pulleys, and it required only 1 person to operate the system. Bonter et al. (2008) evaluated bird capture success with paired mist nets set at ground level and at elevated heights. They found significantly higher capture rates in nets set at ground level. Meyers (1994a) captured orange-winged parrots by using mist nets in a circular configuration around roost trees. Live parrot decoys were placed within the circle of mist nets and supplemented with playback vocalizations. Catch rate was increased by flushing parrots as the observer rushed toward the nets. Sykes (2006) clustered 3 short mist nets in a triangular array around a heavily baited bird feeder. Observers rushed the feeder, flushing ground-feeding painted bunting into the surrounding mist nets. Wilson and Allan (1996) captured prothonotary warblers and Acadian flycatchers in a forested wetland by placing a mist net in a V-shaped configuration, mounted on a boat. A decoy study mount was placed close to a mist net pole. Barred owls were successfully captured by Elody and Sloan (1984) using 3 mist nets set in an A-shaped configuration with a live barred owl placed in the center as a decoy, along with an outdoor megaphone speaker and cassette tape player broadcasting a recorded call of a barred owl.

Lesage et al. (1997) modified mist net techniques to capture breeding adult and young surf scoters. They placed 2 nets at scoter feeding sites, extending perpendicular from the shore and using copper poles painted black and pushed firmly into the lake bottom. A boat was used to herd the scoters into the net. Capture was successful when nets were placed both above and below the water surface. Breault and Cheng (1990) used **submerged** mist nets to capture eared grebes. They set the nets in waist-deep (1.5 m) water and used 7-g fishing weights attached to the net bottom at 1.5-m intervals to sink the net. Nets were attached to wooden poles. Grebes were driven into the nets by personnel walking or canoeing from behind the birds toward the submerged nets. Avoidance of drowning was achieved by immediate removal of any captured birds from the nets. Bacon and Evrard (1990) successfully captured upland nesting ducks by holding a mist net in a horizontal position over the nest. When the hen flushed, she became entangled in the net mesh. The net was attached between 3-m sections of conduit. Kaiser et al. (1995) placed an array of 3 mist nets floating on rafts to catch marbled murrelets as the birds flew through narrow coastal channels. They used aluminum tubing to support the nets. Nets were set against a forested background to reduce their visibility to approaching murrelets. Pollock and Paxton (2006) devised a technique for capturing birds over deep water by using mist nets suspended between poles kept afloat on compact buoys. Paton et al. (1991) used a large mist net consisting of 5 nets sewn together, elevated by pulleys 45 m into the forest **canopy** (Fig. 3.2) to capture marbled murrelets. Netting sessions were conducted during the main activity periods, 60 minutes before to 60 minutes after sunrise. When not in use, the net was wrapped with a plastic tarp to avoid entanglement with woody debris.

Hilton (1989) used taped fledgling alarm calls along with mist nets near active blue jay nests to successfully capture blue jays. The taped calls were broadcast from a portable tape recorder placed beneath the center of the net. Airola et al. (2006) had more capture success of purple martin with fixed mist nets than with hand-held hoop nets at nest cavity sites. They suggested that a combination of both types of nets might be ideal. They also used purple martin distress calls of captured birds to enhance capture rates. Jones and Cox (2007) efficiently mist netted male Bachman's sparrows during the breeding season by using playback recordings.

Silvy and Robel (1968) placed mist nets at a 45° angle on the **ground** (Fig. 3.3) to intercept greater prairie-chickens walking to booming grounds and found these nets caused fewer behavioral problems with displaying males than did cannon nets. This method also was more efficient for capturing female prairie-chickens. Skinner et al. (1998) combined pointing dogs and mist nets attached to galvanized pipe poles to capture juvenile willow ptarmigan. After the dogs located and pointed the birds, the mist nets were arranged in a V-shaped pattern ahead of the covey. The ptarmigan were then flushed into the nets and captured. Geering (1998) used playback tapes during the breeding season to attract birds to be captured in mist nets. Bull and Cooper (1996) presented 4 new techniques for capturing pileated woodpeckers and Vaux's swifts in roost trees. They camouflaged traps with tree bark or lichens set above the entrance hole. A person on the ground released the trap by pulling a taut line as soon as the bird entered the hole. The lichen-covered trap closed to the side of the hole. Both the bark and the lichen-covered plastic netting were taped to a frame. They also used 2 designs, a mist net on a frame and a mist net suspended between 2 trees (Fig. 3.4) and positioned 3–5 m in front of a nest cavity to capture swifts. Hernandez et al. (2006) tested several capture techniques for Montezuma quail and found a modified (portable) mist net method to be the most successful.

Steenhof et al. (1994) successfully used a tethered great horned owl 1 m behind 2 mist nets to capture American kestrels. Nets were placed 20 m from nest boxes occupied by American kestrels with >5-day-old young. They recommended placement of the nets and a live owl near trees when possible to provide shade and so reduce heat stress on the lure owl. Gard et al. (1989) reported that breeding American kestrels responded less aggressively to taxidermy mounts of great horned owls than to live owls. Rosenfield and Bielefeldt (1993) suggested modifications to Bloom et al. (1992) methods for trap-shy breeding Cooper's hawks. They advised using an elevated great horned owl set, 10–13 m above ground, rather than at or within 0.5 m of the ground, to en-

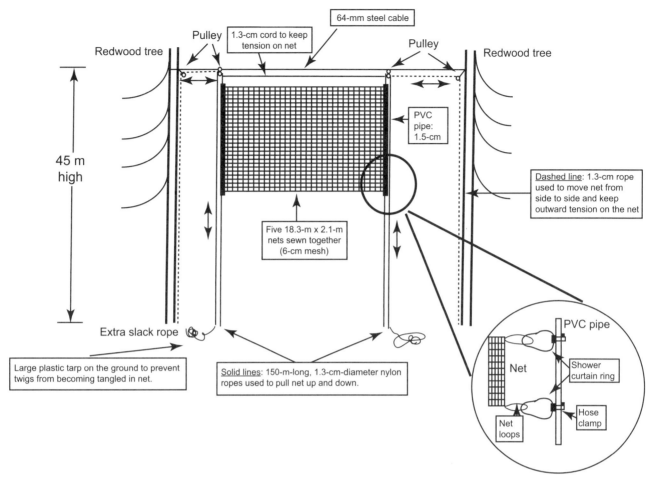

Fig. 3.2. Schematic of mist net used to capture marbled murrelets in the forests of northern California. Branches were on all sides of both trees and were not removed. Diagram not drawn to scale. *From Paton et al. (1991).*

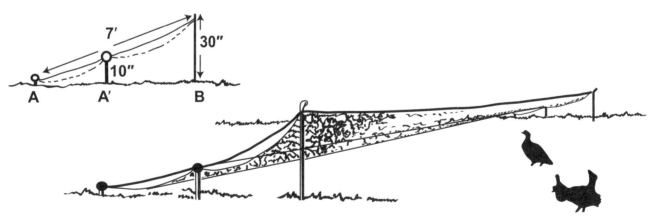

Fig. 3.3. Diagram of erected mist net set at a 45° angle to the ground. The elevated edge of the net should face the path of approaching birds. *From Silvy and Robel (1968).*

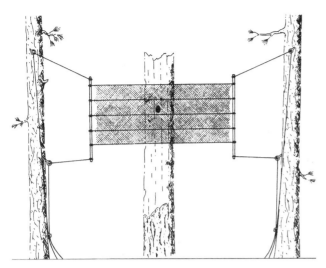

Fig. 3.4. Mist net erected between 2 live trees and positioned in front of a nest cavity. *From Bull and Cooper (1996).*

Table 3.3. Decoys and enticement lures used to capture birds

Group/species[a]	Reference
Waterfowl	
Mallard	Sharp and Lokemoen 1987
Gadwall	Blohm and Ward 1979
Northern pintail	Grand and Fondell 1994, Guyn and Clark 1999
Northern shoveler	Seymour 1974
Blue-winged teal	Garrettson 1998
Canvasback	Anderson et al. 1980
Lesser scaup	Rogers 1964
Barrow's goldeneye	Savard 1985
Galliformes	
Ruffed grouse	Chambers and English 1958, Naidoo 2000
Greater prairie-chicken	Anderson and Hamerstrom 1967, Silvy and Robel 1967
Sharp-tailed grouse	Artmann 1971
Northern bobwhite	Smith et al. 2003c
Ring-necked pheasant	Smith et al. 2003c
Raptors	Berger and Hamerstrom 1962, Bloom 1987, Bloom et al. 1992, Plumpton et al. 1995, Jacobs 1996
Northern goshawk	Meng 1971, McCloskey and Dewey 1999
Cooper's hawk	Rosenfield and Bielefeldt 1993
Red-tailed hawk	Buck and Craft 1995
Northern harrier	Hamerstrom 1963
Crested caracara	Morrison and McGehee 1996
American kestrel	Bryan 1988, Gard et al. 1989, Steenhof et al. 1994
Merlin	Clark 1981
Other birds	
Yellow rail	Robert and Laporte 1997
Virginia rail	Kearns et al. 1998
Sora	Kearns et al. 1998
American woodcock	Norris et al. 1940
Band-tailed pigeon	Drewien et al. 1966
Northern saw-whet owl	Whalen and Watts 1999
Tawny owl	Redpath and Wyllie 1994
Spotted owl	Bull 1987, Johnson and Reynolds 1998
Pileated woodpecker	York et al. 1998
Brown-headed cowbird	Burtt and Giltz 1976
American robin	Dykstra 1968
Loggerhead shrike	Kridelbaugh 1982
Red-winged blackbird	Burtt and Giltz 1970, 1976; Picman 1979
American magpie	Wang and Trost 2000
Regent honeyeater	Geering 1998

[a] Scientific names are given in Appendix 3.1.

hance trapping success. They also advised pre-incubation trapping at or near dawn. Hawks were trapped in mist nets, bow nets, or bal chatris baited with European starlings or ringed turtle doves. Jacobs (1996) reported high trapping success (69% overall) with mist nets set next to a mechanical, mounted great horned owl decoy used to attract red-shouldered, Cooper's, and sharp-shinned hawks (Table 3.3).

Blackshaw (1994) devised a method to secure closed and rolled mist nets that prevented unrolling, tangling, and sagging. She used a 61-cm length of sisal or braided nonslick twine attached to the net and to a long stick placed vertically in the ground near the center of the net. Sykes (1989) used strips of asphalt-saturated, 13.6-kg roofing felt under each tightly furled mist net to prevent accidental capture of birds, small mammals, and large insects, such as beetles, in unattended nets. A chainsaw was used to cut rolls of roofing felt at 22.9-cm intervals.

Dho Gaza Nets

A **dho gaza net** is a large mist net between 2 poles; the net detaches as a bird hits the net and falls to the ground with the bird caught in it. A **fixed dho gaza** has a similar mechanism, but the net does not disconnect from poles; instead it falls in as the whole set. Bierregaard et al. (2008) combined a unique training response that attracted barred owls to a squeaking mouse and then captured them with a dho gaza net. Zuberogoitia et al. (2008) used a combination of a dho gaza and mist net plus an owl lure to capture 13 species of European raptors.

Bloom et al. (1992) evaluated the effectiveness of the dho gaza net baited with a live, tethered great horned owl (Fig. 3.5) as a lure for 11 species of diurnal raptors and 3 species of owls. The technique was most successful when targeting a territorial pair during the reproductive cycle. Playback of audiotaped recordings of great horned owls reduced the time necessary for capture. Net poles should be concealed and the owl lure placed in the shade.

Knittle and Pavelka (1994) simplified attaching a dho gaza net to poles by using fabric hooks and self-adhesive Velcro® as loop fasteners. McCloskey and Dewey (1999) improved success trapping northern goshawks by using a mounted great horned owl decoy that was moved manually while held upright within 1 m of a dho gaza net. The trapper, covered with camouflage netting and holding the

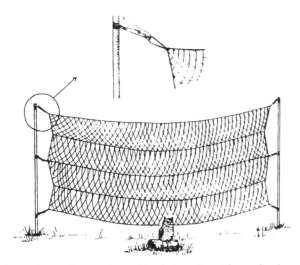

Fig. 3.5. Large dho gaza trap with a tethered great horned owl as an attractant may be used to catch territorial adult raptors. The inset shows a clothespin attachment to a tape tab on a mist net loop. *From Bloom (1987).*

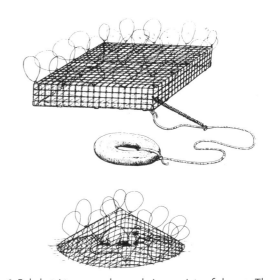

Fig. 3.6. Bal chatri traps can be made in a variety of shapes. The box-shaped bal chatri functions well for accipiters, buteos, and owls, whereas the cone-shaped trap functions best on kestrels and burrowing owls. *From Bloom (1987).*

mounted owl, uttered the 5-note territorial hoot of the great horned owl.

Bal Chatri, Noose Mats, and Halo Traps

A **bal chatri trap** is small wire cage with a rock dove or mouse inside. The cage is covered with monofilament nooses, which twine and trap the raptor's feet. Wang and Trost (2000) caught American magpies with a bal chatri trap baited with a female American magpie and placed under a nest tree. Bierregaard et al. (2008) used a bal chatri noose trap to capture barred owls. Thorstrom (1996) reviewed the methodology used for capturing birds of prey in tropical forests. Baited bal chatri traps (Fig. 3.6) were the most effective and versatile and the simplest to set. He described a modified bal chatri, called an envelope trap, which used as bait the food left behind by a flushed raptor. The bait was enclosed on a semi-flat wire cage with nooses that were tied to the ground. Miranda and Ibanez (2006) successfully used a modified bal chatri trap with horizontal nooses attached to a cage containing a live rabbit to capture Philippine eagles. Crozier and Gawlick (2003) had success using plastic flamingo decoys to attract wading birds. Jacobs and Proudfoot (2002) designed an elevated dho gaza net assembly they used in combination with a great horned owl decoy to capture 5 species of nesting raptors. The owl decoy had a moveable head as described by Jacobs (1996). The net trap was attached to a 2–8-m telescoping pole to allow adjustment to the nest site height and was set within 50 m of the nest tree. Great horned owl vocalizations also were used to attract nesting raptors to the net system.

Smith and Walsh (1981) modified a bal chatri trap for eastern screech owls by placing a 3-mm Plexiglas™ top on a rectangular hardware cloth base. Taped calls were used to

Fig. 3.7. Noose mats may be applied to branches and around burrowing owl nests. *From Bloom (1987).*

attract owls to the mouse-baited trap. Small holes were drilled in the Plexiglas, in which nooses were tied. Blakeman (1990) increased the capture success rate of bal chatri traps by spraying them with flat dark paint. Nylon monofilament used for nooses was soaked for a day in black fabric dye. Both treatments helped camouflage the traps.

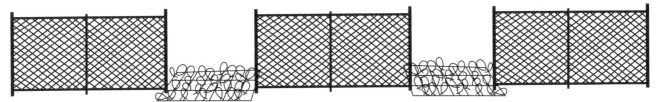

Fig. 3.8. Positioning of lead fences and noose mats to capture wintering shorebirds. *From Mehl et al. (2003).*

Toland (1985) designed a **leather harness** with 15 monofilament slip nooses that he attached to house sparrows to capture trap-wary American kestrels. One end of a monofilament line was attached to a wooden dowel or stick and the other end to the edge of the harness. The wooden weight functioned as a drag when the kestrel attempted to fly away with the harnessed sparrow. Bloom (1987) provided details on the use of a harnessed rock dove for the capture of raptors. Nylon monofilament nooses were tied or cemented to a leather harness that was attached to a rock dove tied on a line to a weight or a nearby shrub.

Noose mat traps are much like bal chatri traps except that monofilament loops are attached to a mat or carpet (Fig. 3.7). McGowan and Simons (2005) used a remote-controlled mechanical decoy to lure territorial adult American oystercatchers for capture in a leg-hold noose mat trap. Paredes et al. (2008) placed a noose carpet attached to a wooden pole on cliff ledges to capture breeding razorbills on the Labrador, Canada, coast. Lightweight noose mats were combined with alternating lead fences by Mehl et al. (2003) to capture wintering shorebirds (Fig. 3.8). Caffrey (2001) was unsuccessful in capturing American crows using a noose carpet. African fish eagles were captured on water by using a floating fish snare vest (Hollamby et al. 2004).

Hilton (1989) described a unique **double halo** nest trap to capture blue jays. The trap consisted of a black metal hanger bent into a "dog-bone shape." Halos at each end had a diameter of 12.5 cm and were connected by a 15-cm wire. Clear nylon, 4–5-kg test monofilament fishing line was tied into nooses similar to those used on bal chatri and other noose traps. Elliptical nooses, 7 × 5 cm, were most successful. The bottom halo was anchored to the branch supporting the nest with 7–8-kg test monofilament tied to a metal washer. The double halo trap was designed to catch a bird by its neck as it arrives or leaves the nest. It was necessary for the bird trapper to remain nearby to prevent strangulation of the bird. The trap was deployed several days after incubation had begun to avoid provoking nest desertion.

Drop Nets

Drop nets (Table 3.4) using **explosive charges** to drop the nets have been deployed to capture wild turkey (Baldwin 1947 and Glazener et al. 1964), band-tailed pigeon (Wooten 1955, Drewien et al. 1966), greater prairie-chicken (Jacobs 1958), shorebirds (Peyton and Shields 1979), and flightless Canada goose (Nastase 1982). Silvy et al. (1990) developed a tension-operated (**nonexplosive**) drop net to capture Attwater's prairie-chicken and king rail (Fig. 3.9). White nets blended into early morning fog and were more efficient at capturing prairie chickens than were dark nets. Bush (2008) developed a similar tension-operated drop net to capture greater sage-grouse. More grouse were captured with gray

Table 3.4. Drop nets used to capture wildlife

Group/species[a]	Reference
Birds	
Attwater's prairie-chicken	Silvy et al. 1990
Canada goose	Nastase 1982
Greater prairie-chicken	Jacobs 1958
Greater sage-grouse	Bush 2008
Wild turkey	Baldwin 1947, Glazener et al. 1964
King rail	Silvy et al. 1990
Band-tailed pigeon	Wooten 1955, Drewien et al. 1966
Shorebirds	Peyton and Shields 1979
Mammals	
White-tailed deer	Ramsey 1968, Conner et al. 1987, DeNicola and Swihart 1997, Lopez et al. 1998
Mule deer	White and Bartmann 1994, D'Eon et al. 2003
Mountain sheep	Fuller 1984, Kock et al. 1987

[a] Scientific names are given in Appendix 3.1.

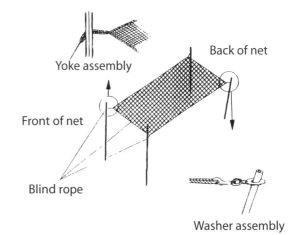

Fig. 3.9. Nonexplosive drop net showing the yoke assembly at the front of the net and the swivel snap-washer assembly for attaching net to back poles. *From Silvy et al. (1990).*

than with black nets. Lockowandt (1993) designed an electromagnetic trigger for drop nets that worked well in cold weather with high winds and ice.

Cannon and Rocket Nets

Cannon and rocket nets (Fig. 3.10) have relative **advantages and disadvantages** with respect to each other. Rocket nets cost more per firing; rocket propellant (charges) cannot be shipped and must be delivered to their place of use, which adds to their cost; and rockets are prone to start fires. Rocket propellant is now solely available through Winn-Star (Marion, IL). Purchasers of rocket propellant should be aware of the type of rockets they are using, as charges used in the old Wildlife Materials (Carbondale, IL) rockets require different changes than do Winn-Star rockets; using the wrong charges can cause the rockets to blow apart. Rockets have the advantage they can be mounted to more readily fire over larger animals (i.e., deer) and the rockets need not be cleaned after firing. Cannons must be cleaned after firing and cannot be mounted above the ground to accommodate larger animals; however, they do not start fires, they are less expensive to fire, no federal permit is required for their use, and charges can be shipped by overnight express companies. Both cannon and rocket net charges must be stored away from buildings and in explosive resistant containers. Also, rocket net charges are prone to explode with age. In recent years, air cannons (i.e., Net Blaster™; Martin Engineering, Neponset, IL) have become available. These cannons are more expensive, but they offer the advantage of not having to use explosives to propel the net. As a result they also cause fewer animal behavioral problems when fired over a given area for several days in succession. Caffrey (2001) captured American crows with camouflaged rocket and cannon nets and a net launcher.

A portable platform for setting rocket nets in **open water** habitats was perfected by Cox and Afton (1994). King et al. (1998) developed a rocket net system consisting of an aluminum box (containing the net) set in 2–4-cm-deep water. Mahan et al. (2002) modified nets and net boxes to enhance the capture of wild turkey. They rotated a 12-m × 12-m net 45° so that it resembled a baseball diamond and attached 3 rockets. One set of drag weights rather than 3 were used.

Table 3.5. Cannon and rocket nets used to capture wildlife

Group/species[a]	Reference
Birds	
American white pelican	King et al. 1998
Waterfowl	Dill and Thornsberry 1950, Turner 1956, Marquardt 1960, Funk and Grieb 1965, Raveling 1966, Moses 1968, Wunz 1984, Zahm et al. 1987, Cox and Afton 1994, Grand and Fondell 1994, Merendino and Lobpries 1998
Great blue heron	King et al. 1998
White ibis	Heath and Frederick 2003
Blue grouse	Lacher and Lacher 1964
Greater sage-grouse	Lacher and Lacher 1964, Giesen et al. 1982
Sharp-tailed grouse	Peterle 1956
Greater prairie-chicken	Silvy and Robel 1968
Ring-necked pheasant	Flock and Applegate 2002
Wild turkey	Austin 1965; Bailey 1976; Wunz 1984, 1987; Davis 1994; Eriksen et al. 1995; Pack et al. 1996; Mahan et al. 2002
Bald eagle	Grubb 1988, 1991
Ruddy turnstone	Thompson and DeLong 1967
Ring-billed gull	Southern 1972
Band-tailed pigeon	Smith 1968, Pederson and Nish 1975, Braun 1976
American crow	Caffrey 2001
Brown-headed cowbird	Arnold and Coon 1972
Mammals	
White-tailed deer	Hawkins et al. 1968, Palmer et al. 1980, Beringer et al. 1996, Cromwell et al. 1999, Haulton et al. 2001
Fallow deer	Nall et al. 1970
Mountain sheep	Jessup et al. 1984
Dall sheep	Heimer et al. 1980

[a] Scientific names are given in Appendix 3.1.

Rocket and cannon nets have been used to trap both birds and mammals (Table 3.5).

Net Guns

Net guns are usually used to capture mammals; however, they also have been employed to capture birds (Table 3.6). Mechlin and Shaiffer (1980) used net guns to capture waterfowl, and O'Gara and Getz (1986) captured golden eagle

Fig 3.10. Photograph of cannon (left) and rocket nets (right) shortly after being fired. Note how the front end of the rocket net comes off the ground, allowing taller animals to be trapped than could be accomplished with a cannon net. *Photo by N. J. Silvy.*

Table 3.6. Net guns used to capture wildlife

Group/species[a]	Reference
Birds	
Waterfowl	Mechlin and Shaiffer 1980
Golden eagle	O'Gara and Getz 1986
Mammals	
Coyote	Barrett et al. 1982, Gese et al. 1987
Moose	Carpenter and Innes 1995
White-tailed deer	Barrett et al. 1982, DeYoung 1988, Potvin and Breton 1988, Ballard et al. 1998, DelGiudice et al. 2001a, Haulton et al. 2001
Mule deer	Barrett et al. 1982, Krausman et al. 1985, White and Bartmann 1994
Caribou	Valkenburg et al. 1983
Pronghorn	Barrett et al. 1982, Firchow et al. 1986
Mountain sheep	Andryk et al. 1983, Krausman et al. 1985, Kock et al. 1987, Jessup et al. 1988
Dall sheep	Barrett et al. 1982

[a] Scientific names are given in Appendix 3.1.

with a net gun. Herring et al. (2008) used a **net gun** to capture nearby (maximum distance, 15 m) wetland birds, whereas Caffrey (2001) was unsuccessful in capturing American crow with one.

Bow Nets

Barclay (2008) developed a technique for nighttime trapping of burrowing owls combining a bow net activated by a solenoid and a live tethered mouse decoy. Jackman et al. (1994) devised a successful radiocontrolled bow net and power snare (Fig. 3.11) to selectively capture bald and golden eagles. The net was completely concealed in loose soil and operated from distances up to 400 m. A recognizable marker was placed just outside the perimeter of the net trap to verify the eagle was in the center of the trap and was feeding with its head down before triggering the trap. Shor (1990a, b) described an easily constructed, simple-to-set bow net that safely caught hawks.

Proudfoot and Jacobs (2001) combined 2-way radios with a conventional home security switch to develop an inexpensive alarm-equipped bow net. The radio alarm eliminated the need to periodically inspect automatic bow nets. The bow net was used to signal the capture of owls, hawks, and loggerhead shrike. Collister and Fisher (1995) tested 4 trap types for capturing loggerhead shrike. They had a higher percentage of trapping successes with a modified Tordoff bow trap. Larkin et al. (2003) perfected an electronic signaling system for prompt removal of an animal from a trap. Herring et al. (2008) developed a solenoid activated **flip trap** for capturing large wetland birds.

Morrison and McGehee (1996) set a **Q-net** (Fuhrman Diversified, Seabrook, TX) similar to a bow net next to a live crested caracara tethered within 100 m of an active nest. The territorial and aggressive resident caracara moved toward the lure bird and was caught in the Q-net when the observer pulled the trigger wire. Modern Q-nets come with a digital radio release that can activate the net from ≤75 m away.

Helinet

Brown (1981) developed the helinet (Fig. 3.12) to capture prairie-chicks and ring-necked pheasant. Lawrence and Silvy (1987) used the helinet to capture and translocate 44 Attwater's prairie-chickens from runways and small areas of prairie habitat adjoining runways of a small airport in Texas. Prairie-chickens were captured by flying over display grounds and flushing an individual bird and then flowing the bird's flight (not pushing the bird) until it landed. After 1 or a few flushes, the bird's primary feathers would become wet, and it could no longer fly and would try to hide in tall grass. The helicopter with a net attached to the struts would then place the net over the hiding bird, and a person riding shotgun in the helicopter would catch the bird by hand from under the net. The passenger door was removed from the helicopter to facilitate capture. Permission had to be obtained from the Federal Aviation Administration prior to attaching anything to a helicopter. This method was the most efficient and cost effective for capturing female prairie-chickens.

SNARES AND NOOSE POLES

Benson and Suryan (1999) described a circular noose (Table 3.7) that allowed safe capture of specific individual black-legged kittiwakes. The leg noose was fitted to the rim of the nest and was remotely triggered. Launay et al. (1999) attached **snares** at 10-cm intervals to a 50-m-long main line at male houbara bustard display areas. They also placed female bustard decoys surrounded by snares at display sites. Nesting females were attracted to dummy eggs made of wood painted to resemble houbara bustard eggs; they were caught with adjacent snares.

Cooper et al. (1995) described a noose trap arrangement used to capture pileated woodpeckers at nest and roost cavities. **Foot nooses** of clear monofilament line were spaced at 1-cm intervals along a main support line, and fence staples were used to secure the line to the tree.

Thorstrom (1996) devised a **noose pole** trap for removing incubating and nestling birds from tree cavities. Young that were out of view in 2-m deep nest cavities were safely extracted. Kramer (1988) designed a noosing apparatus made of wire, plastic straws, and monofilament fishing line that he used to remove nestling bank swallows from their burrows for banding. Thiel (1985) built a similar noosing device to capture adult belted kingfishers as they entered their nesting burrows. Kautz and Seamans (1992) used noose poles to successfully capture rock dove in silos, but not in barns.

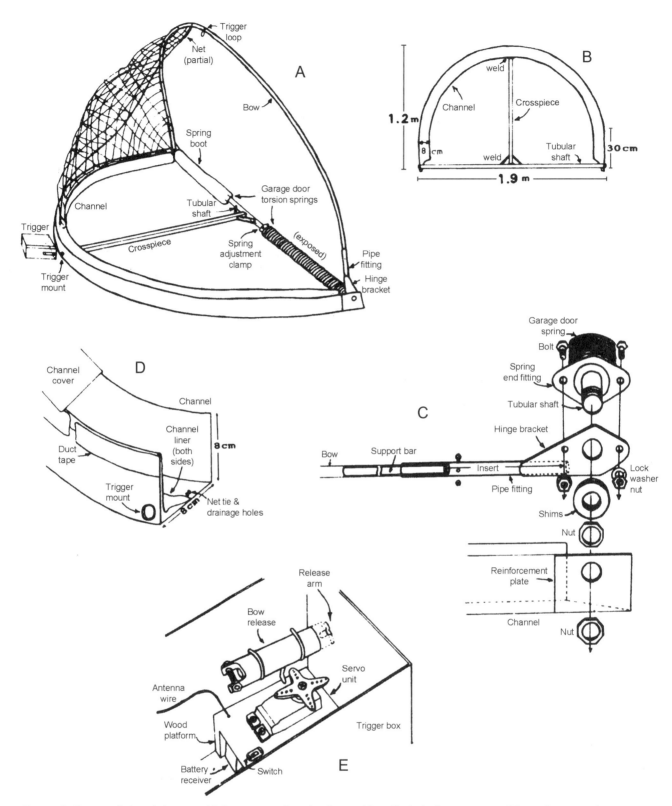

Fig. 3.11. Radiocontrolled eagle bow net. (A) Bow net opening, showing position of principal components; (B) top view, no springs; (C) detail of spring–hinge–bow–channel attachment; (D) cross-section detail of channel at trigger mount; (E) interior detail of trigger box. From Jackman et al. (1994).

Fig. 3.12. Helicopter with helinet attached to the front of its struts. Photo by N. J. Silvy.

Frenzel and Anthony (1982) and Cain and Hodges (1989) described **floating fish snares** with 2 and 4 nooses for capturing bald eagles. Jackman et al. (1993) described a modified floating-fish snare that achieved 40% capture success. They inserted a Styrofoam™ plug in the anterior portion of the fish bait, allowing the tail of the fish to dip more deeply below the water surface. Nooses consisted of 18-kg-test light-green monofilament tied with a slip knot. Two (10–20 cm) nooses were placed in an alternate or lateral position. Sucker (*Catostomus* sp.) or catfish (*Ictalurus* sp.) approximately 40 cm long were used for bait. Fish were anchored and placed in shaded areas during early morning, when the monofilament was less visible to eagles.

McGrady and Grant (1996) designed a radiocontrolled **power snare** similar to that described by Jackman et al. (1994) to capture nesting golden eagles. A nest anchor was used to keep the captured eagle on the nest to avoid injury. Nestlings were isolated in a small chicken-wire cage to avoid fouling the trap snare before firing. A video camera facilitated a clear view of the trap. Territorial golden eagles were caught on the nest efficiently and safely using this design.

Monofilament nooses of 15-kg test line, 5 cm in diameter, were attached to a 1-m-diameter chicken-wire dome and placed over the nest by Ewins and Miller (1993) to capture nesting ospreys. They secured the dome with cords around the base of the nest. Thiel (1985) placed a 20–25-cm monofilament fish-line snare into nest burrows of belted kingfisher. The snare was anchored to a tent stake inserted into the sand bank near the nest burrow entrance.

Winchell and Turman (1992) used a combination of monofilament nooses and wooden dowel rods to capture burrowing owls during the fledging season, when the owls were extremely wary of any change near their burrows or roosts. Several noose rods were placed outside the burrow, and a dowel and weight were inserted beneath the soil surface.

Reynolds and Linkhart (1984) used a telescoping noose pole with an attached 12.5-cm-diameter loop of coated stainless steel line (Zwickel and Bendell 1967) to capture flammulated owl from trees. Scharf (1985) used noose-covered wickets placed around a live male American magpie decoy to capture territorial magpies.

Robertson et al. (2006) used a **pole with a noose** attached to the end to capture common murres in Newfoundland, Canada. Hipfner and Greenwood (2008) used a similar 3-m-long fishing-rod noose pole with an attached monofilament noose to capture common murres in British Columbia, Canada.

Proudfoot (2002) perfected the use of a flexible fiberscope and noose to successfully remove ferruginous pygmy-owl

Table 3.7. Snares and noose poles used to capture birds

Group[a]	Reference
Galliformes	
Greater prairie-chicken	Berger and Hamerstrom 1962
Spruce grouse	Schroeder 1986
Blue grouse	Zwickel and Bendell 1967
Willow ptarmigan	Hoglund 1968
Raptors	Berger and Mueller 1959, Berger and Hamerstrom 1962, Ward and Martin 1968, Jenkins 1979, Dunk 1991
White-tailed kite	Dunk 1991
Rough-legged hawk	Watson 1985
Bald eagle	Frenzel and Anthony 1982; Cain and Hodges 1989; Jackman et al. 1993, 1994
Golden eagle	Jackman et al. 1994, McGrady and Grant 1994, 1996
Osprey	Frenzel and Anthony 1982, Prevost and Baker 1984, Ewins and Miller 1993
Crested caracara	Morrison and McGehee 1996
American kestrel	Wegner 1981, Toland 1985
Prairie falcon	Beauvais et al. 1992
Barn owl	Colvin and Hegdal 1986
Short-eared owl	Kahn and Millsap 1978
Eastern screech-owl	Smith and Walsh 1981
Tropical screech-owl	Thorstrom 1996
Burrowing owl	Barrentine and Ewing 1988, Winchell and Turman 1992
Flammulated owl	Reynolds and Linkhart 1984
Spotted owl	Bull 1987
Other	
Colonial seabirds	Edgar 1968
Double-crested cormorant	Foster and Fitzgerald 1982, Hogan 1985
Black-legged kittiwake	Benson and Suryan 1999
Houbara bustard	Launay et al. 1999
Passerines	
Common nighthawk	McNicholl 1983
Belted kingfisher	Thiel 1985
Pileated woodpecker	Cooper et al. 1995
Loggerhead shrike	Yosef and Lohrer 1992, Collister and Fisher 1995, Doerr et al. 1998
American magpie	Scharf 1985
Bank swallow	Barrentine and Ewing 1988, Kramer 1988
Chipping sparrow	Gartshore 1978

[a] Scientific names are given in Appendix 3.1.

nestlings from oak (*Quercus* spp.) nest cavities without injury. He also suggested using a miniature camera system to assist with nestling removal from cavities.

A live tethered mouse attached to a board surrounded by a monofilament noose lured spotted owls for capture (Johnson and Reynolds 1998). The noose was manually tightened when the owl landed on the mouse. Redpath and Wyllie (1994) captured territorial tawny owls by using a live tethered tawny owl as an attractant in a large modified Chardoneret trap (Fig. 3.13). The territorial owl entered an open lid and lit on a perch that released the trigger, closing the entrance lid.

Drive Nets and Drift Fences

Tomlinson (1963) developed a method for drive-trapping dusky grouse. Clarkson and Gouldie (2003) used a drive net trap to capture moulting harlequin duck. Costanzo et al. (1995) successfully herded large flocks of flightless Canada geese into a moveable catch pen comprised of 6 attached panels (Table 3.8). Each panel was 3.4 m × 1.5 m, made of nylon netting attached to a conduit frame. This trap was inexpensive, portable, and simple to assemble.

Flores and Eddleman (1993) placed **drop-door traps** along 1-m-tall drift fences of 1.8-cm mesh black-plastic bird netting to capture black rail. The netting was stapled to wooden surveyor's stakes. Kearns et al. (1998) combined

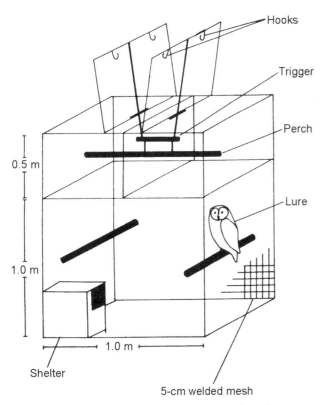

Fig. 3.13. Modified Chardoneret using a captive owl as a lure. Owls flew from an external perch into one of the top compartments, landing on the internal perch and releasing the trigger, which allowed the lid to close. *From Redpath and Wyllie (1994).*

Table 3.8. Drive and drift traps used to capture wildlife

Group/species[a]	Reference
Birds	
Canada goose	Robards 1960, Heyland 1970, Timm and Bromley 1976, Costanzo et al. 1995
Snow goose	Cooch 1953
Wood duck	Tolle and Bookhout 1974
Harlequin duck	Clarkson and Gouldie 2003
Diving ducks	Cowan and Hatter 1952
Blue grouse	Pelren and Crawford 1995
Dusky grouse	Tomlinson 1963
Ruffed grouse	Liscinsky and Bailey 1955, Tomlinson 1963
Greater sage-grouse	Giesen et al. 1982
Greater prairie-chicken	Toepfer et al. 1988, Schroeder and Braun 1991
Lesser prairie-chicken	Haukos et al. 1990
Scaled quail	Schemnitz 1961
Sandhill crane	Logan and Chandler 1987
Clapper rail	Stewart 1951
Black rail	Flores and Eddleman 1993
Virginia rail	Kearns et al. 1998
Sora	Kearns et al. 1998
American coot	Glasgow 1957, Crawford 1977
Shorebirds	Low 1935
American woodcock	Liscinsky and Bailey 1955, Martin and Clark 1964
Mammals	
Snowshoe hare	Keith et al. 1968
White-tailed deer	Stafford et al. 1966, Silvy et al. 1975, DeYoung 1988, Sullivan et al. 1991, Locke et al. 2004
Mule deer	Beasom et al. 1980, Thomas and Novak 1991
Himalayan musk deer	Kattell and Alldredge 1991
Mountain sheep	Kock et al. 1987

[a] Scientific names are given in Appendix 3.1.

2.5-cm-mesh welded-wire cloverleaf traps with ramped funnel entrances and an attached catch box to catch sora and Virginia rails. Drift fences deflected the rails into the traps. Capture rate was increased by using playback of rail vocalizations. The sound system was powered by solar panels. Fuertes et al. (2002) used a modified fish-net trap in the shape of a funnel in pairs with a deflecting drift net in between to capture small rails. They added fruits, vegetables, and cat food as bait. Their traps were easy to transport and place and had a low injury rate. Caudell and Conover (2007) deployed a floating gill net to capture eared grebe in conjunction with a motorboat and a new method (**drive-by netting**).

Haukos et al. (1990) recommended walk-in drift traps (Fig. 3.14) over rocket nets and baited **walk-in traps** for the capture of lesser prairie-chicken in leks in spring. Advantages of the walk-in drift traps included minimal capture stress, no need for observer presence, and the ability to trap the entire lek. Pelren and Crawford (1995) successfully captured blue grouse with walk-in traps that intercepted mov-

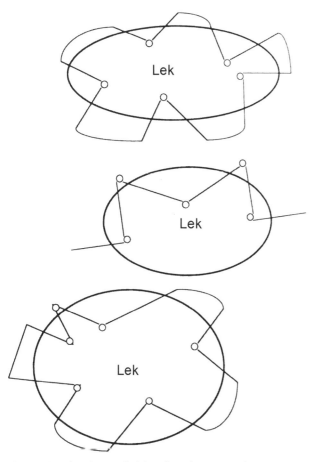

Fig. 3.14. Overhead view of 3 lek walk-in designs used to capture lesser prairie-chickens. *From Haukos et al. (1990).*

ing birds with 60-cm-tall mesh-wire fences. The fences guided the grouse into funnels connected to trap boxes, which were made of plastic netting with fish netting tops to minimize injury to trapped birds.

Nest Traps

Blums et al. (2000) perfected a **multicapture nest box** for cavity-nesting ducks (Table 3.9). This trap featured a swinging false floor, entrance baffle, and counter balance. A scaled down version of this trap can be used to capture smaller cavity-nesting birds. Plice and Balgooyen (1999) designed a remotely operated trap to capture American kestrel by using nest boxes. Kestrels were trapped during prey delivery to nestlings. Cohen and Hayes (1984) perfected a simple device to block the entrance to nest boxes. They used a wooden clothespin or a similarly shaped Plexiglas clothespin attached to a monofilament line. After the bird entered the nest box, the line was pulled, and the entrance was closed. Cohen (1985) used feathers to lure male tree swallows into nest boxes, where they were subsequently captured.

Pribil (1997) developed a clever nest trap for house wrens. The trap consisted of a nest box containing a grass nest with 1 egg (Fig. 3.15). The egg was glued to a lever connected to a spring that closed a door over the entrance hole. The pecking action of the bird pushed the egg down releasing the lever. The lever, attached to a rubber band, pulled a string, which closed the door over the entry hole, thereby

Table 3.9. Nest traps used to capture birds

Trap type/species[a]	Reference
Cavity	
Hooded merganser	Blums et al. 2000
Wood duck	Blums et al. 2000
Acorn woodpecker	Stanback and Koenig 1994
Red-cockaded woodpecker	Jackson and Parris 1991
Pileated woodpecker	Bull and Pedersen 1978
Red-bellied woodpecker	Bull and Pedersen 1978
Tree swallow	Rendell et al. 1989
Bank swallow	Rendell et al. 1989
Nest box	
American kestrel	Plice and Balgooyen 1999
Tree swallow	Lombardo and Kemly 1983, Cohen and Hayes 1984, Cohen 1985, Stutchbury and Robertson 1986
Bluebird	Kibler 1969, Pinkowski 1978
House sparrow	Mock et al. 1999
House wren	Pribil 1997
European starling	DeHaven and Guarino 1969, Lombardo and Kemly 1983
Other passerine birds	Dhondt and van Outryve 1971, Stewart 1971, Yunick 1990
Waterfowl	Harris 1952, Sowls 1955, Addy 1956, Weller 1957, Coulter 1958, Miller 1962, Salyer 1962, Doty and Lee 1974, Zicus 1975, Shaiffer and Krapu 1978, Blums et al. 1983, Zicus 1989, Bacon and Evrard 1990, Dietz et al 1994, Yerkes 1997, Loos and Rohwer 2002
Natural nests	
Pied-billed grebe	Otto 1983
Egrets and herons	Jewell and Bancroft 1991, Mock et al. 1999
White ibis	Frederick 1986
American coot	Crawford 1977
American avocet	Sordahl 1980
Black-necked stilt	Sordahl 1980
Mountain plover	Graul 1979
Snowy plover	Conway and Smith 2000
Wilson's phalarope	Kagarise 1978
Mourning dove	Swank 1952, Stewart 1954, Harris and Morse 1958, Blockstein 1985
White-winged dove	Swanson and Rappole 1994
Raptors	Jacobs and Proudfoot 2002
Osprey	Ewins and Miller 1993
Short-eared owl	Leasure and Holt 1991
Belted kingfisher	Thiel 1985
Passerines	Gartshore 1978
Cliff swallow	Wolinski and Pike 1985
Barn swallow	Wolinski and Pike 1985
Blue jay	Hilton 1989

[a] Scientific names are given in Appendix 3.1.

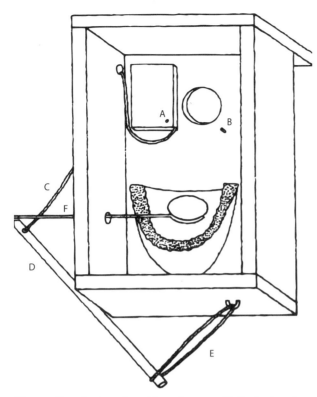

Fig. 3.15. Trapping box viewed from the rear with the back wall removed. A portion of the nest is removed to illustrate the position of the metal lever and the placement of the egg. A = pin around which the wooden door revolves; B = nail protruding from the wall, which keeps the door aligned over the entrance; C = string; D = wooden lever; E = rubber band; F = metal lever. *From Pribil (1997).*

capturing the wren. The wren trapping box should be placed 15–25 m from an active house wren nest. The author had her best trapping success early in the spring breeding season. Stanback and Koenig (1994) developed techniques for capturing acorn woodpecker inside natural cavities. They reached the tree hole with the aid of basic rock-climbing gear and extension ladders. They then cut a triangular door below the cavity entrance, using a folding pruning saw for the main cuts, and held the door in place with nails. The cavity entrance was blocked with a plastic bobber after the bird entered the nest, and the captured bird was then removed.

Dietz et al. (1994) designed an inexpensive **walk-in duck nest trap** with a funnel entrance and lily-pad shape. It was made of welded wire with a top of garden netting. The trap worked most effectively in dense vegetation, where researchers could make a concealed approach to block the entrance. Yerkes (1997) described a portable inexpensive trap for capturing incubating female mallard and redhead ducks that used cylindrical artificial nesting structures. The wire-covered trapdoors at each end of the nesting cylinder were manually triggered with ropes. Loos and Rohwer (2002) found long-handled nets to be more efficient than nest traps for capturing upland nesting ducks. Trapping injuries were far less frequent when long-handled nets were used in comparison to nest traps. Netted females returned to their nest more rapidly than those captured with nest traps. Netting ducks required only 1 trip to the nest, disturbing females less often than with nest traps.

A **self-tripping nest trap** was designed by Frederick (1986) to capture white ibis and other colonial nesting birds. His trap design had the advantage of being suitable for capturing large numbers of birds in a dense nesting site with minimum disturbance where traps were left unattended. A similar automatic trap was developed by Otto (1983) to catch pied-billed grebe. Mock et al. (1999) developed a nest trap that featured a wire door that prevented escape. An electronic-release triggering mechanism allowed the researcher to control the capture at distances ≤200 m. The remote control system was battery operated and inexpensive.

Yunick (1990) suggested blocking the entrance to nest boxes with a broom or rake handle upon approach to prevent escape of an incubating bird. He also described a simple, effective nest box trap of semi-rigid plastic film that hung inside the box entrance. The trap worked on the principle of a hinged flap that could be pushed like a swinging door. The U-shaped film was pinned in place.

Rendell et al. (1989) perfected a manually operated **basket trap,** consisting of a wire skeleton covered with mist netting attached by tape or line. The basket was attached to the end of a lightweight extendable pole and raised to enclose the entrance of a cavity containing a hole-nesting bird, such as a tree or bank swallow. Their trap was simple for 1 person to use, flexible, portable, lightweight, easy to construct, and required few materials.

Robinson et al. (2004) and Friedman et al. (2008) described a simple, inexpensive, and successful nest box trap. Newbrey and Reed (2008) developed an effective nest trap for female yellow-headed blackbirds. Hill and Talent (1990) used a T-shaped spring trap to capture nesting least tern and snowy plover (Fig. 3.16).

Swanson and Rappole (1994) modified a **hoop net trap,** described by Nolan (1961), by attaching mist netting to an aluminum frame from a fishing dip net to capture nesting white-winged doves) in subtropical thorn forest habitat. Conway and Smith (2000) designed a nest trap for snowy plovers. The trap consisted of 1.83-m lengths (2) of 1.25-cm electrical conduit, 16-cm pieces (4) of 1-cm-diameter wooden dowels, and 2 medium-weight strap hinges. The 2 pieces of conduit were bent into equal U shapes and attached to hinges to form the trap frame. Mesh netting was attached to the frame with twine, and black paint was sprayed on the aluminum conduit frame. The trap was anchored and activated with a 50-m-long pull cord by an observer when the incubating bird returned to the nest. The pull cord was at-

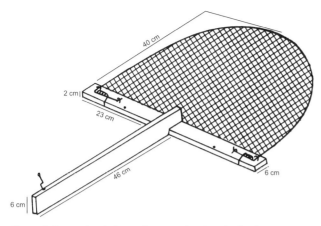

Fig. 3.16. Spring-loaded trap for capturing incubating least terns and snowy plovers. *From Hill and Talent (1990).*

tached to the top piece of conduit. After the bird was caught, the trap was removed to facilitate rapid return of the incubating plover to the nest.

Hines and Custer (1995) collected great blue heron eggs from nests in tall trees by using an extendable net-pole. The device consisted of 4 collapsible 2-m sections with an 11-cm wire loop and an attached 9-cm-deep basket made from nylon stocking material.

Box and Cage Traps

Box and cage traps have been used for years to capture a variety of bird species (Table 3.10). Caffrey (2001) captured American crows and emphasized that crows are extremely wary and difficult to catch. She modified the **Australian crow trap** (Aldous 1936) by adding a drop-door at one end. Bait on trapping days should not be large food items that can be picked up and carried away easily. In all cases, prebaiting and habituating crows to trapping methods were required. Recaptures were infrequent. The Modified Australian crow trap was useful for capturing many species of crop-depredating birds, depending on the size of the entrance (Gadd 1996). Aruch et al. (2003) used a peanut-shaped baited open-door trap with 2 entrances to capture Kalij pheasants in dense Hawaiian forests. Ashley and North (2004) perfected inexpensive automated doors for waterfowl traps, thereby curtailing depredation and escapes. Clark and Plumpton (2005) perfected a simple one-way door design in combination with an artificial burrow to facilitate relocation of western burrowing owls.

Winchell (1999) designed a simplified and efficient push-door wire-mesh trap that readily captured complete broods of burrowing owls. Botelho and Arrowood (1995) constructed a trap for burrowing owls consisting of a 61-cm-long and 10-cm-diameter polyvinyl chloride (PVC) pipe. A hinged one-way Plexiglas door was inserted midway in the PVC pipe, which was placed in the owl burrows. Trapped owls were removed through a hinged door that opened on top of the PVC pipe. Plumpton and Lutz (1992) made multiple captures of burrowing owls by modifying large **Sherman traps** placed in burrow entrances by replacing one end with 2.5-cm wire mesh. They also captured young nestlings by quietly approaching the burrow and grabbing the birds by hand before they retreated completely into the tunnel. Banuelos (1997) advocated using a one-way Plexiglas door trap for burrowing owls. The ease of constructing and setting the trap, potentially high capture rate, and lack of trapping injuries made this simple trap ideal. The one-way door trap captured owls twice as fast as did bal chatri and noose carpet traps.

Harrison et al. (2000) described a trap designed to accommodate tidal water level fluctuations by providing a 1,500-cm^2 floating platform in the trap to curtail mortality from drowning. Mauser and Mensik (1992) constructed a portable **swim-in bait trap** to capture ducks. The trap panels were covered with plastic netting to minimize injuries. A floating catch box allowed trap operation in a variety of water depths. They suggested a loafing platform for birds in the trap.

Wang and Trost (2000) used baited traps with a 50-cm-long funnel entrance with a chicken wire open hoop 20 cm high at the end to catch American magpie. This hoop required the magpie to jump over the hoop to reach the bait.

Buck and Craft (1955) had success catching great horned owl and red-tailed hawk with 2 designs of **walk-in traps.**

Table 3.10. Box and cage traps used to capture birds

Group/species[a]	Reference
Waterfowl	Kutz 1945, Hunt and Dahlka 1953, McCall 1954, Schierbaum and Talmage 1954, Addy 1956, Schierbaum et al. 1959, Mauser and Mensik 1992, Evrard and Bacon 1998, Harrison et al. 2000
Raptors	Ward and Martin 1968, Buck and Craft 1995
Ruffed grouse	Tanner and Bowers 1948, Chambers and English 1958
Sharp-tailed grouse	Hamerstrom and Truax 1938
Greater prairie-chicken	Hamerstrom and Truax 1938
Ring-necked pheasant	Hicks and Leedy 1939, Kutz 1945, Flock and Applegate 2002
Northern bobwhite	Schultz 1950, Smith et al. 1981
Scaled quail	Schemnitz 1961, Smith et al. 1981
Wild turkey	Baldwin 1947, Bailey 1976, Davis 1994
Puffin	Nettleship 1969
Burrowing owl	Martin 1971, Ferguson and Jorgensen 1981, Plumpton and Lutz 1992
Mourning dove	Reeves et al. 1968
Band-tailed pigeon	Drewien et al. 1966, Smith 1968, Braun 1976
Chihuahua raven	Aldous 1936
American magpie	Alsager et al. 1972
House finch	Larsen 1970
House sparrow	Therrien 1996

[a] Scientific names are given in Appendix 3.1.

One type had a welded-wire funnel entrance. The other was activated with a monofilament tripwire that released a trapdoor. Rock doves, domestic chickens, or captive-bred northern bobwhites were enclosed in wire cages and served as live bait. Dieter et al. (2009) evaluated the duck capture success rates of various trap design types. They recommended oval traps.

Decoy Traps and Enticement Lures

Similarly, a **Swedish Goshawk Trap** is a large cage with a trigger mechanism that uses a rock dove in a separate section as bait to trap raptors. Plumpton et al. (1995) successfully used padded and weakened **foothold traps** to capture red-tailed, ferruginous, and Swainson's hawks along roads. Trap springs were weakened by repeatedly hitting them with a hammer. Jaws of size 3 and 3N double-spring foothold traps were padded with 5-mm-thick adhesive-backed foam rubber and then wrapped with cloth friction tape. Traps were baited with a live mouse held in a harness in the form of a 24-gauge steel wire loop. The loop was placed over the head and behind the ears of the mouse. Traps were hidden with a thin covering of sifted soil or snow.

Whalen and Watts (1999) assessed the influence of **audio lures** on capture patterns of northern saw-whet owls. They found a general pattern of decreasing capture frequency with increasing distance from the audio lure. They suggested that capture rates may be maximized by using more lures, each with a small number of nets. Gratto-Trevor (2004) compiled detailed information on procedures to capture shorebirds (Charadriiformes, suborder Charadrii). Play-back distress calls increased shorebird capture rates (Haase 2002).

Various species of upland game birds have been attracted and captured with the use of **recorded calls** (Table 3.11). Breeding male ruffed grouse readily responded to playbacks of recordings of drumming display sounds by approaching to ≤2–9 m of the observer (Naidoo 2000). Playback of recordings of male display sounds near a stuffed decoy could be used to lure ruffed grouse into noosing range for capture. Taped calls and drums of pileated woodpeckers were combined with a mist net by York et al. (1998) to rapidly capture this species with minimum stress to the birds.

Evrard and Bacon (1998) tested 4 duck trap designs. In spring, traps with a live female mallard decoy and traps with a similar decoy and bait were more successful than bait traps without a decoy. Spring trapping was more successful than autumn trapping. Floating bait traps were largely unsuccessful in capturing waterfowl. Conover and Dolbeer (2007) successfully used decoy traps to capture juvenile European starling.

Use of Oral Drugs

O'Hare et al. (2007) provided details on the use of **alpha-chloralose** (A-C) by the U.S. Wildlife Services, Department of Agriculture, to immobilize birds. Bucknall et al. (2006) successfully employed A-C to capture flighted birds affected by an oil spill on the Delaware River. Bergman et al. (2005) described the historical and current use of A-C as an anesthetic to capture or sedate wild turkey.

Stouffer and Caccamise (1991) successfully captured American crow with A-C inserted in fresh chicken eggs. However, McGowan and Caffrey (1994) expressed concern about high mortality of crows captured with A-C. Caccamise and Stouffer (1994) explained the possible cause of mortality and justified the continued use of A-C.

Woronecki et al. (1992) conducted safety, efficacy, and clinical trials required by the U.S. Food and Drug Administration (FDA) to register A-C. They reported the most effective dose to be 30 mg and 60 mg of A-C/kg of body weight for capturing waterfowl and rock dove, respectively. They concluded that A-C was a safe capture drug for these birds. In 1992, the U.S. Wildlife Services was granted approval by the FDA to use A-C nationwide for capturing nuisance waterfowl, American coot, and rock dove (Woronecki and Thomas 1995). Wildlife Services personnel must complete a 12-hour training course and pass a written examination to be certified to use A-C (Belant et al. 1999). The use of A-C 30 days prior to and during the legal waterfowl season for populations that are hunted is prohibited.

Initial use of 60 mg/kg of A-C in field operations yielded a low (6%) capture rate of rock dove. Belant and Seamans (1999) reevaluated doses of A-C used for rock doves and recommended treating corn with 3 mg A-C/corn kernel and 180 mg/kg as an effective dose. Mean time of first effects and mean time to capture at the 180 mg/kg dose rate were significantly less than with lower dosages. Belant and Seamans (1997) also assessed the effectiveness of A-C formulations for immobilizing Canada geese. A-C in tablet form was as effective as A-C in margarine and corn oil in bread baits. Male and female geese responded similarly to A-C immobilization. Seamans and Belant (1999) recommended A-C over DRC-1339 (3-chloro-4-methylbenzenamine hydrochlo-

Table 3.11. Use of tape recordings of calls to attract and expedite capture of game birds

Species[a]	Reference
Ruffed grouse	Healy et al. 1980, Lyons 1981, Naidoo 2000
Blue grouse	Stirling and Bendell 1966
Spruce grouse	MacDonald 1968
Sharp-tailed grouse	Artmann 1971
Greater prairie-chicken	Silvy and Robel 1967
White-tailed ptarmigan	Braun et al. 1973
Chukar partridge	Bohl 1956
Scaled quail	Levy et al. 1966
Gambel's quail	Levy et al. 1966
Montezuma quail	Levy et al. 1966

[a] Scientific names are given in Appendix 3.1.

ride) as a gull population-management chemical, because it was fast acting, humane, and could be used as a nonlethal capture agent.

Scientists at the National Wildlife Research Center (Wildlife Services), Fort Collins, Colorado, have recently developed and tested a tablet form of A-C. These new tablets will be available in 3 sizes, so that combinations of pellets can be used to achieve accurate dose levels for a variety of birds. Tablets should be placed inside bread cube bits for administration to birds. The tablet formulation provides a safer and simpler alternative to the current formulation, which requires mixing a powder prior to use and a syringe for injection of the solution into the bread bait.

Janovsky et al. (2002) tested **tiletamine (zolazepam)**, another oral drug for bird immobilization, at a dosage of 80 mg/kg (applied in powdered form to the surface of fresh meat) on common buzzards in Austria. The deepest anesthesia was produced by fresh-drugged bait administered immediately after preparation. This drug combination had a wide safety margin with little lethal risk of overdosing nontarget birds that might accidentally feed on the bait.

Miscellaneous Capture Methods
Smith et al. (2003c) located radiomarked adult northern bobwhite quail with a brood of young chicks (1–2 days old). They then erected a corral of screen covered panels that surrounded the adult and brood. After flushing the adult, they **hand captured** the chicks in the corral. Thil and Groscolas (2002) caught king penguin by hand and safely immobilized them with tiletamine zolazepam. Kautz and Seamans (1992) described several methods to expedite capture of rock doves. They caught rock doves mainly at night by hand at roost sites in barns and silos by closing the roosting sites with burlap drop window covers to prevent the birds from escaping. They also designed a catch window, consisting of a net bag of 2.5-cm × 2.5-cm mesh nylon gill netting. They developed a stuff sack that allowed placing birds into a burlap bag with 1 hand, a necessity while holding on to a supporting structure. Headlamps with an on-off switch and a rheostat were used to help hand-capture rock doves. Folk et al. (1999) devised a safe and efficient daylight capture technique for whooping cranes. They used a unique capture blind made from a cattle feed trough baited with corn. They grabbed the crane's leg through armholes in the side of the trough while the cranes were feeding on the corn in the trough.

Martella and Navarro (1992) devised a novel method for capturing greater rhea. They blinded the birds using a spotlight at night and captured them using a **boleadoras**, a device consisting of 2 or 3 balls of round stone covered with leather and attached to a long strap of braided leather, 7 mm in diameter and 1-m long. When the bird began to run, the boleadoras was thrown toward the bird's legs. The straps wound around the rhea's legs, causing it to fall and allowing hand capture.

Ostrowski et al. (2001) captured steppe eagle in Saudi Arabia by **vehicle pursuit.** Their method was limited to open habitat, but it was effective on trap-shy individuals. Eagle chases were restricted to a maximum of 15 minutes. Similarly, Ellis et al. (1998) used a **helicopter** to pursue and capture sandhill crane in open habitat.

King et al. (1998) captured American white pelican and great blue heron with modified No. 3 **padded-jaw foothold traps** by replacing both factory coil springs with weaker No. 1.5 coil springs. They also substituted the factory chain with a 20-cm length of aircraft cable and a 30-cm electric shock cord to minimize injury to captured birds. Cormorants also have been captured with padded foothold traps placed in trees with the aid of an 18-m extension ladder. The trap was camouflaged with a flour-water mixture to simulate cormorant guano (King et al. 2000).

CAPTURING MAMMALS

Readers of this chapter are encouraged to review previous major detailed coverage of mammal capture and handling methods. These include Day et al. (1980), Novak et al. (1987), Schemnitz (1994, 2005), Wilson et al. (1996), American Society of Mammalogists (1998), Proulx (1999a), and Feldhamer et al. (2003). Gannon et al. (2007) stressed the need when live trapping to provide adequate food, insulation, and protection from temperature extremes. The newly developed web-based material should be investigated, especially *Best Management Practices for Trapping in the United States,* produced by the Association of Fish and Wildlife Agencies (AFWA 2006a; http://www.fishwildlife.org).

Mammal capture usually becomes more difficult as animal size increases. Thus, observational techniques and mammalian sign are more efficient for obtaining both inventory and density information (Jones et al. 1996). Several new techniques to capture mammals ranging in size from small rodents to large carnivores have been developed in recent years, often for specific research purposes. Some of these represent either improved or modified versions of traditional capture methods. Well-designed commercial traps are available for a variety of species. Biologists and wildlife managers now often use such traps, both for convenience and reliability. Nuisance wildlife control operators and fur trappers use commercial traps almost exclusively. An overwhelming variety of trap types and variations is available from commercial vendors (see Appendix 3.2).

Most animals are captured by hand, mechanical devices, remote injection of drugs, or drugs administered orally in baits. The emphasis in this chapter is on methods and equipment other than remotely injected drugs used for capture (see Chapter 4, This Volume). Powell and Proulx (2003) summarized the importance of mammal trapping ethics, proper handling, and the humane use of various traps for various species.

Use of Nets

Dip Nets

Such mammals as jackrabbits (Griffith and Evans 1970) and skunks are first located with spotlights and then pursued on foot using a flashlight and **dip net.** Dip nets also are used to pull down drugged mammals. Rosell and Hovde (2001) combined a spotlight and the use of nylon mesh landing nets from boats on rivers and on foot on land to catch American beaver. The net, when used in the water, was closed with a drawstring to prevent escape. The netting method resulted in no mortalities, in contrast to 5.3% mortality with snares (McKinstry and Anderson 1998).

Mist and Harp Nets

Kuenzi and Morrison (1998) suggested combining mist net capture with **ultrasonic detection** to identify the presence of bat species. Francis (1989) compared mist nets and 2 designs of **harp traps** for capturing bats (Chiroptera). Large bats (megachiropterans) were captured at similar rates in harp traps and mist nets, but microchiropterans were captured nearly 60 times more frequently in traps. He noted that small bats have teeth with sharp cutting edges and often chewed part of the net around them and escaped. He recommended use of 4-bank harp traps over 2-bank harp traps for capture efficiency. Tidemann and Loughland (1993) devised a trap for capturing large bats. It featured wire cables stretched between rigid uprights. Vertical strings were strung between the cables. Waldien and Hayes (1999) designed a hand-held portable H-net used to capture bats that roosted at night under bridges. The H-net consisted of a mist net attached to PVC pipe and T-couplers. Palmeirim and Rodrigues (1993) described an improved harp trap for bats that was inexpensive and lightweight (4.5 kg) and could be assembled by 1 person in 2 minutes.

Cotterill and Fergusson (1993) described a new trapping device (Fig. 3.17) to capture African free-tailed bats as they left their daylight roosts. They used polythene **plastic sheeting** attached to a rectangular frame of aluminum tubing. Bicycle wheels were attached to each corner of the frame to carry the assembled trap into position below the roost exit. Two people elevated the trap with ropes and pulleys. Bats were caught in a plastic bag and easily removed with a minimum of stress, in contrast to mist nets. Kunz et al. (1996) provided an in-depth review of bat capture methods.

Drop Nets

Drop nets using **explosive charges** have been used to capture white-tailed deer (Ramsey 1968, Conner et al. 1987, and DeNicola and Swihart 1997), mule deer (White and Bartmann 1994, D'Eon et al. 2003), and mountain sheep (Fuller 1984, Kock et al. 1987). Silvy et al. (1990) developed a **nonexplosive** drop net to capture Key deer. Lopez et al. (1998) develop a drop net triggered by a **pull rope** to capture urban deer. Jedrzejewski and Kamler (2004) perfected a modified drop net for capturing ungulates.

Drive Nets and Drift Fences

Silvy et al. (1975) developed a **portable drive net** to capture free-ranging deer. Peterson et al. (2003b) and Locke et al. (2004) described several advantages of a portable drive net for capturing urban white-tailed deer. Okarma and Jedrzejewski (1997) and Musiani and Visalberghi (2001) used fladry to help capture gray wolves. Fladry consists of red flags attached to nylon ropes 60 cm above ground, placed along roads or trails in forested areas. **Beaters,** spaced at 250-m intervals, drove the wolves into nets, where they became entangled and were captured. Drive nets have been widely used to capture large mammals, but they also are useful for trapping small ones. Vernes (1993) devised a drive fence with attached wire-cage traps set parallel to forest edges. Sullivan et al. (1991) compiled data on captures of 430 white-tailed deer using the drive-net technique. The observed capture-related mortality and overall mortality rates were 1.1% and 0.9%, respectively. These rates were lower than those re-

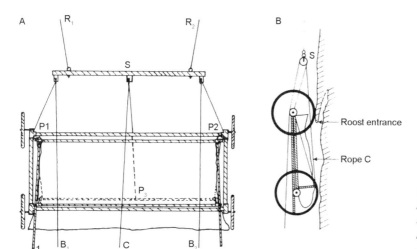

Fig. 3.17. Trap arrangement for catching bats. (A) Assembled trap with ropes and their points of attachment, (B) lateral view of the assembled trap. Aluminum frames are cross-hatched. *From Cotterill and Fergusson (1993).*

ported for other common capture methods. Kattell and Alldredge (1991) used 3–6-m-long, 1.8–2.0-m-high nets to capture Himalayan musk deer in Nepal. After the nets were set, 2 people slowly drove the deer toward the nets, where the animals became entangled. Faulhaber et al. (2005) used **drift fences** to capture Lower Keys marsh rabbits.

Thomas and Novak (1991) described procedures contributing to successful **helicopter** drive-net captures of mule deer. Netting was dyed a dull green or brown color to reduce its visibility. When possible, nets should be placed in or near a drainage bottom, where deer could be herded downhill into the net, which should be concealed by terrain. Net sites providing close hiding cover for observers, which allowed quick access to entangled animals, were essential. Ideal weather conditions consisted of high overcast that reduced glare and net visibility. A steady breeze of 9–18 km/hr blowing downwind from the helicopter toward the deer and net reduced the possibility of animals scenting and avoiding the capture site.

Kelly (1996) captured ringed seals with nets set at breathing holes in the ice. He designed a net that lined a breathing hole and closed below the surface with a weighted triggering device. Three wire hoops were attached to the net to hold it open. He increased seal visitation by cutting holes in the ice.

Cannon and Rocket Nets

Rocket and cannon nets have been used to trap mammals (Table 3.5) for many years. Beringer et al. (1996) noted that if rocket nets are used to capture deer, capture should be limited to ≤3 deer per capture. They advised that handling time be minimized to reduce stress to captured deer. If deer are to be radiotagged, there should be at least 1 person per deer and an extra person to apply the radio collar. Deer should be blindfolded immediately after capture to prevent stress.

Net Guns

Carpenter and Innes (1995) used net guns from helicopters to capture moose with a mortality rate of less than 1%. White and Bartmann (1994) reported that net gunning (Table 3.6) was a more economical, efficient, and safe capture method than drop nets for mule deer fawns. The use of net guns from a helicopter was the most effective method for winter capture of yearling and adult white-tailed deer in non-yarding populations (Ballard et al. 1998). Webb et al. (2008) found the **helicopter** and net gun capture technique for white-tailed deer to be safe compared to other capture techniques.

Snares and Noose Poles

Gray wolves were pursued in Finland with snowmobiles over soft snow 80-cm deep and were captured with a neck hold noose attached to a pole (Kojola et al. 2006). Davis et al. (1996) designed a lightweight noose device attached to ski poles to safely remove mountain lions and bears from trees and cliffs. Grizzly and black bears captured in leg snares exhibited more muscle injury and capture myopathy than did bears captured by helicopter darting or bear drop door traps (Cattet et al. 2008).

Box and Cage Traps

Various box and cage traps are used to capture a large variety of mammals (Table 3.12). Haulton et al. (2001) evaluated 4 methods (Stephenson box traps, Clover traps, rocket nets, and dart guns) to capture deer. They found that smaller deer captured with Clover traps were more susceptible to capture mortality. Anderson and Nielsen (2002) described a modified Stephenson trap to capture deer. It featured lightweight panels that were easily set up and readily movable. They recommended their trap for capturing deer in urban areas. Ballard et al. (1998) used Clover traps and darting from tree stands to capture white-tailed deer. They bolted U-clamps to keep the drop doors on the Clover traps closed to avoid deer escapes and substituted nuts and bolts for welds that broke at sub-zero temperatures.

Table 3.12. Box and cage traps used to capture mammals

Species[a]	Reference
Kangaroo rat	Brock and Kelt 2004, Cooper and Randall 2007
Bushy-tailed woodrat	Lehmkuhi et al. 2006
Dusky-footed woodrat	Innes et al. 2008
Key Largo woodrat	McCleery et al. 2005, 2006
Cotton rat	Sulok et al. 2004, Cameron and Spencer 2008
Deer mouse	Whittaker et al. 1998, Rehmeier et al. 2004, Jung and O'Donovan 2005, Reed et al. 2007
Nine-banded armadillo	Bergman et al. 1999
Snowshoe hare	Aldous 1946, Libby 1957, Cushwa and Burnham 1974, Litvaitis et al. 1985a
Lower Keys marsh rabbit	Faulhaber et al. 2005
Pygmy rabbit	Larrucea and Brussard 2007
Flying squirrel	Carey et al. 1991, Flaherty et al. 2008, Wilson et al. 2008
Red squirrel	Haughland and Larsen 2004, Herbers and Klenner 2007
Gray squirrel	Huggins and Gee 1995, Linders et al. 2004
Fox squirrel	Huggins and Gee 1995; McCleery et al. 2007a, b
Abert's squirrel	Patton et al. 1976, Dodd et al. 2003
Townsend's chipmunk	Carey et al. 1991
Eastern chipmunk	Waldien et al. 2006, Ford and Fahrig 2008
Woodchuck	Trump and Hendrickson 1943, Ludwig and Davis 1975, Maher 2004
California ground squirrel	Horn and Fitch 1946
Pocket gopher	Howard 1952, Sargeant 1966, Baker and Williams 1972, Witmer et al. 1999, Connior and Risch 2009

continued

Table 3.12. continued

Species[a]	Reference
Prairie dog	Dullum et al. 2005, Facka et al. 2008
American beaver	Couch 1942, Hodgdon and Hunt 1953, Collins 1976, Koenen et al. 2005
Mountain beaver	Arjo et al. 2007
Muskrat	Takos 1943, Snead 1950, Stevens 1953, Robicheaux and Linscombe 1978, McCabe and Elison 1986, Lacki et al. 1990
Nutria	Norris 1967, Evans et al. 1971, Palmisano and Dupuie 1975, Linscombe 1976, Robicheaux and Linscombe 1978, Baker and Clarke 1988
Porcupine	Brander 1973, Craig and Keller 1986, Griesemer et al. 1999, Zimmerling 2005
Coyote	Foreyt and Rubenser 1980, Way et al. 2002
Gray fox	AFWA 2006e
Kit fox	Zoellick and Smith 1986
Swift fox	Kamler et al. 2002
Mountain lion	Shuler 1992
Canada Lynx	Mowat et al. 1994
Bobcat	Woolf and Nielson 2002, AFWA 2006b
Black bear	Erickson 1957, Black 1958, Cattet et al. 2008
Brown and grizzly bear	Craighead et al. 1960, Troyer et al. 1962
Raccoon	Robicheaux and Linscombe 1978, Moore and Kennedy 1985, Proulx 1991, Gehrt and Fritzell 1996, AFWA 2006h
American marten	Naylor and Novak 1994, Bull et al. 1996
Virginia opossum	AFWA 2006g
Fisher	Arthur 1988, Frost and Krohn 1994, AFWA 2007b
Striped skunk	Allen and Shapton 1942, AFWA 2009a
Northern river otter	Northcott and Slade 1976; Melquist and Hornocker 1979, 1983; Shirley et al. 1983; Route and Peterson 1988; Serfass et al. 1996; Blundell et al. 1999
Long-tailed weasel	Belant 1992
Short-tailed weasel	Belant 1992
Feral hog	Matschke 1962, Williamson and Pelton 1971, Saunders et al. 1993, Jamison 2002, Mersinger and Silvy 2006
Collared peccary	Neal 1959
Elk	Thompson et al. 1989
White-tailed deer	Bartlett 1938; Ruff 1938; McBeath 1941; Webb 1943; Glazener 1949; Clover 1954, 1956; Hawkins et al. 1967; Sparrowe and Springer 1970; Runge 1972; McCullough 1975; Foreyt and Glazener 1979; Palmer et al. 1980; Rongstad and McCabe 1984; Morgan and Dusek 1992; Naugle et al. 1995; Beringer et al. 1996; Ballard et al. 1998; VerCauteren et al. 1999; DelGiudice et al. 2001a; Haulton et al. 2001; Anderson and Nielsen 2002
Mule deer	Lightfoot and Maw 1963, Roper et al. 1971, D'Eon et al. 2003

[a] Scientific names are given in Appendix 3.1.

Fig. 3.18. Culvert trap for capturing bears. *Photo by the New Mexico Department of Game and Fish.*

Bull et al. (1996) covered wire cage traps with black plastic to protect American marten from rain and snow to reduce the risk of mortality from **hypothermia.** They also placed clumps of wool for insulation in wood boxes to provide warm, dry shelter during winter trapping. Baited culvert traps (Fig. 3.18) have been widely used to capture and transplant nuisance bears (Erickson 1957).

Carey et al. (1991) placed a single-door collapsible wirebox trap 1.5 m above ground in large trees to capture arboreal mammals, such as northern flying squirrels and Townsend's chipmunks. A nest box was inserted behind the trap treadle to minimize stress and hypothermia. Hayes et al. (1994) described a simple and inexpensive modification (Fig. 3.19) of the technique of Carey et al. (1991) to attach live traps to small-diameter trees, 8.5–30.0-cm diameter at breast height, by means of a triangular plywood bracket. The bracket was set tangential to the tree trunk, and 2 aluminum nails were driven through the plywood and into the tree. Nylon twine was tied around the trap and secured to 2 additional nails. Malcolm (1991), Vieira (1998), and Kays (1999) described an **arboreal** mammal box-trap system that could be hoisted to sample arboreal mammal communities. Huggins and Gee (1995) tested 4 cage trap sets for gray and fox squirrels; they found traps set at eye level on a platform attached to tree trunks resulted in the highest rate of capture.

Szaro et al. (1988) assessed the effectiveness of pitfalls and Sherman live traps in measuring small mammal community structure. They found that live traps and pitfalls provided different estimates of species composition and relative abundance. However, live-trapping was significantly more successful than pitfalls in terms of number of new captures per trap night. They recommended the use of both pitfalls and live traps, particularly when shrews (Soricidae), which are not readily caught in live traps, need to be sampled. Slade et al. (1993) advised using a combination of trap types for sampling diverse small mammal faunas.

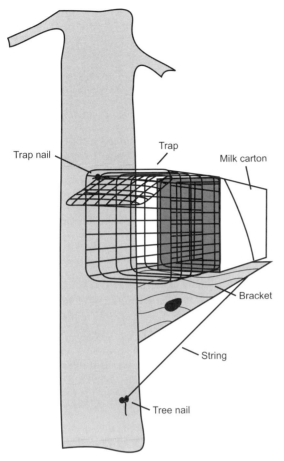

Fig. 3.19. Tomahawk live trap attached to a small-diameter tree by a bracket. *From Hayes et al. (1994).*

Fitzgerald et al. (1999) tested the capture rate of buried and unburied folding Sherman live traps in desert grasslands and desert shrub communities. Traps were set in pairs for 3 consecutive nights. The unburied trap capture rate was significantly greater than that for buried traps. Burying traps may be a cost-effective method of reducing trap fatalities related to temperature fluctuations in desert environments.

Standardization of traps and trapping procedures are needed to adequately sample small-mammal populations. Kirkland and Sheppard (1994) proposed a standard protocol for sampling small-mammal populations with emphasis on shrews. They suggested using Y-shaped arrays of 10 **pitfall traps** (large cans or buckets recessed into the ground) and drift fences. Each arm, which was anchored on a central pitfall, consisted of 3 pitfalls separated by 5-m sections of drift fence. Pitfalls ≥14 cm in diameter and 19-cm deep should be half-filled with water to quickly drown captured animals. They recommended that arrays be operated for 10 consecutive days. This interval totaled 100 trap nights of sampling effort per array per sampling period and allowed easy calculation of relative abundance as the percentage capture success. Handley and Varn (1994) suggested using a small, easily set pitfall array in the form of a triangle with 2.5-cm sides and set in a transect for capturing shrews. Two people set 2 arrays per hour. They used 2-liter, heavy-gauge plastic soft drink bottles with the tops cut off as pitfalls. The plastic bottles were 20-cm deep and 11 cm in diameter. At the center of the array they used a 4-L plastic bottle 18-cm deep and 15 cm in diameter. Pitfalls were arranged with 120° between arms and joined with 1.2-m-long and 30-cm-high drift fence. Tew et al. (1994) tested 2 trap spacings, 24 m and 48 m, using 184 Longworth live traps set in a rectangular grid covering an area of 10 ha. They found the 2 spacings were equally effective in capturing wood mice. They suggested that projects with limited numbers of traps should consider wider trap spacing with an increased trapping period.

A study by Mitchell et al. (1993) in saturated forested wetlands showed that pitfalls in conjunction with drift fences captured significantly greater numbers of small mammals than did isolated pitfall can traps in the same general area. They recommended that different researchers should use the same technique and sampling effort for the same taxa. Moseby and Read (2001) recommended 8 nights of pitfall trapping as the most efficient duration for mammals. Pitfalls should be ≥40 cm deep for small mammals and ≥60 cm for agile species, such as hopping mice.

Hays (1998) devised a new method for live-trapping shrews by inserting small 10-cm Sherman live traps into holes cut in Nalgene plastic jars (25-cm high × 15-cm diameter). The trap entrance was covered with 12-mm wire mesh to exclude mice. Traps were baited with mealworms and cotton batting. Traps were checked daily, and trap mortality was only 1%. Yunger et al. (1992) greatly decreased the mortality of masked shrews (77.5% survival) caught in pitfall traps by providing 7 g of whitefish (*Coregonus* spp.) per pitfall.

Whittaker et al. (1998) evaluated captures of mice in 2 sizes of Sherman live traps. Small Sherman traps captured significantly more white-footed and cotton mice. More rice rats were caught in large Sherman traps. Jorgensen et al. (1994) set paired Sherman and wire-mesh box traps. More rodents were consistently caught in the Sherman traps made of sheet metal. They attributed the capture rate difference to less frequent entry by rodents into wire-mesh traps and a more sensitive treadle in the Sherman traps. In contrast, O'Farrell et al. (1994) experimented with similar sized Sherman and wire-mesh live traps. Captures were significantly greater in mesh traps than in Sherman traps. They surmised that an open trap that can be seen through was preferred to an enclosed box. Their estimates of small mammal density at different sites using wire mesh traps were 15–37% higher than estimates with Sherman traps. They concluded the composition of communities of small mammals might be inaccurately represented based on the type of trap used. McComb et al. (1991) compared capture rates of small mammals and amphibians between pitfall and Museum Special snap traps in mature forests in Oregon.

Fewer small mammal and amphibian species were caught with Museum Special traps than with pitfalls. However, 2 species of salamander were captured only in pitfall traps. Museum Specials baited with peanut butter were more effective than traps baited with meat paste. Pearson and Ruggiero (2003) examined **trap arrangement** in forested areas by comparing transect and grid trapping of small mammals. Transects yielded more total and individual captures and more species than did grid arrangements.

Dizney et al. (2008) evaluated 3 small mammal trap types in the Pacific Northwest. Pitfalls were the most effective trap. Sherman traps significantly outperformed mesh traps. Anthony et al. (2005) compared the effectiveness of Longworth and Sherman live traps. They suggest that using a combination of both traps would ideally sample small mammals with a minimum of bias. Jung and O'Donovan (2005) cautioned the use of Ugglan wire-mesh live traps caused mortality of deer mice, because their upper incisors became entangled in the wire mesh. Kaufman and Kaufman (1989) place wood shelters over Sherman traps at ground squirrel burrows and increased capture success. Waldien et al. (2006) covered Sherman traps with a milk carton sleeve for insulation and used polyfiber batting to provide additional **thermal protection** for captured animals. Umetsu et al. (2006) found pitfalls to be more efficient than Sherman traps for sampling small mammals in the Neotropics. A simplified, easily constructed Tuttle-type collapsible bat trap using PVC tubing was designed by Alvarez (2004). Fuchs et al. (1996) described a technique widely used for catching European rabbits in Scotland that consisted of a buried tip-top galvanized steel box. The earth floor of the trap was covered with wire mesh to prevent escape.

Lambert et al. (2005) detailed an arboreal trapping method for small mammals in tropical forests (Fig. 3.20). Winning and King (2008) perfected a baited pipe trap mounted vertically to a tree to successfully capture squirrel glider in Australia (Fig. 3.21). Waldien et al. (2004) cautioned mammal trappers on the potential mortality of birds captured in Tomahawk™ and Sherman live traps.

Mitchell et al. (1996) reported that use of an **ant insecticide** (Dursban®) did not affect overall capture yield or probability of capture of 12 species of small mammals and that mutilation rates by ants were lower. Gettinger (1990) reported that use of chemical insect repellents increased capture rates.

Yunger and Randa (1999) immersed Sherman live traps for 5 minutes in a 10% bleach solution (sodium hypochlorite) to **decontaminate** them from sin nombre hantavirus. No effect on small mammal capture rate was observed. Cross et al. (1999) tested bleach treatment and found no effect on trap success. Van Horn and Douglass (2000) used a Lysol® disinfectant followed by a fresh water rinse to clean traps. This treatment did not influence subsequent deer mouse capture rates.

Heske (1987) recommended the use of clean live traps to obtain an unbiased demographic sample of small mammals.

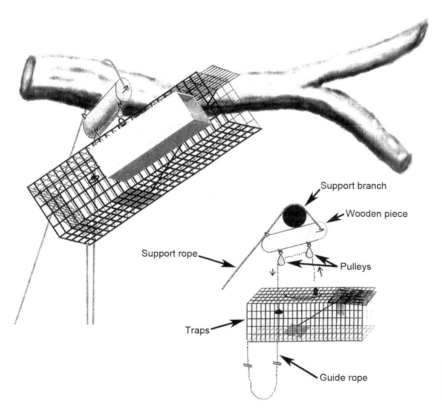

Fig. 3.20. Diagram of the arboreal trapping method used in the southeastern Amazon. *From Lambert et al. (2005).*

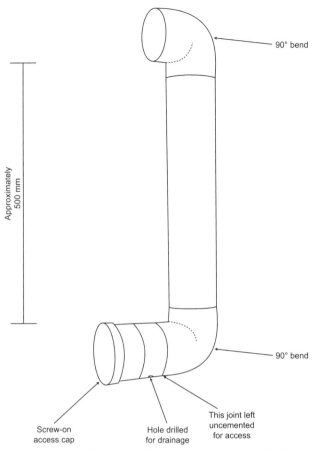

Fig. 3.21. Design of pipe trap. The design uses 90-mm polyvinyl chloride (PVC) pipe and fittings throughout. *From Winning and King (2008).*

Corral Traps

Sweitzer et al. (1997) designed a modified **steel mesh panel trap** for capturing multiple feral hogs with a minimum (5%) of injury. Their traps included a gate entrance with a runway leading to an enlarged corral with a trip line activating a side-hinged squeeze gate. Saunders et al. (1993) suggested attaching fine mesh wire on the inside of trap drop gates to prevent hogs caught inside the trap from gripping the gate with their teeth and lifting it, allowing others to escape. They set traps using a trip wire placed in a back corner of the trap 20 cm above its floor. Jamison (2002) described effective traps for feral hog capture. He emphasized the need for a strong, portable trap the width and length of an average pickup truck bed to facilitate transporting live hogs. Choquenot et al. (1993) used estrous sows as a lure, but no hogs were attracted or captured. West et al. (2009) describe several traps used to capture feral hogs.

Cancino et al. (2002) designed a modified corral trap (Table 3.13) consisting of a 70-ha **enclosure** and an adjacent observation tower. A 4-ha area in the enclosure was irrigated to attract pronghorn. A gate at one end was closed to confine the animals that gradually moved toward the end of the exclosure, attracted by captive pronghorn, mobile feeders, and water, where another gate was closed to confine them. Lee et al. (1998) summarized other pronghorn capture methods. Pérez et al. (1997) perfected a corral trap for capturing Spanish ibex. The trap consisted of a 3-m-high metallic net fence with a 3-m-high net inside. The 2 nets were 1 m apart; salt blocks were used as bait.

Foot Traps and Snares

Since 1997 the Association of Fish and Wildlife Agencies (AFWA), in cooperation with state wildlife agencies and the U.S. Department of Agriculture's Animal and Plant Health Inspection Service, has engaged in a congressionally mandated project evaluating commercial traps for 23 species of North American furbearers in 5 U.S. regions to develop **Best Management Practices (BMP)** for traps and trapping (AFWA

Table 3.13. Corral traps used to capture wildlife

Group/species[a]	Reference
Canvasback	Haramis et al. 1987
Jackrabbit	Henke and Demarais 1990
Collared peccary	Neal 1959
Feral hog	Sweitzer et al. 1997
Deer	Lightfoot and Maw 1963, Hawkins et al. 1967, Rempel and Bertram 1975
Elk	Couey 1949, Mace 1971
Moose	Pimlott and Carberry 1958, LeResche and Lynch 1973
Pronghorn	Spillett and ZoBell 1967, Cancino et al. 2002
Spanish ibex	Pérez et al. 1997

[a] Scientific names are given in Appendix 3.1.

He observed that using soiled traps might cause possible violations of the assumptions of equal catch success of all individuals. He documented that *Microtus* samples were more accurate demographically if all traps were kept clean. Jones et al. (1996) advised cleaning all traps with soap and water after each trapping session to increase consistency in trapping success.

Live trapping **bias** of small mammals varies with gender, age, and species. Results of capture rates to previous trap occupancy depended on gender and age (Gurnell and Little 1992). Wolf and Batzli (2002) reported that white-footed mice were less likely to be captured in live traps that previously held short-tailed shrews. Adult white-footed mice were more likely to be captured in traps previously occupied by conspecific individuals of the opposite gender than in traps previously occupied by the same gender. In contrast, Gurnell and Little (1992) reported no evidence of breeding males or females being attracted to traps containing the odor of the opposite gender. Their studies involved various wood rodents (wood mice, bank voles, and yellow-necked mice).

2006a). Evaluations include performance profiles for commercial traps that include efficiency, selectivity, safety, practicality, and animal welfare, using international standards for humaneness (International Organization for Standardization [ISO] 1999a, b). Numerous documents (cited elsewhere in this chapter) provide data and background information on the AFWA project and are available at the AFWA website, which is continuously updated as new data become available. The technical information and animal welfare information are useful in selecting the most appropriate equipment for particular uses, often help researchers answer the concerns of Institutional Animal Care and Use Committees, help manufacturers design and improve state-of-the-art capture equipment, and help state wildlife agencies maintain healthy wildlife populations using regulated trapping.

Fur trappers, nuisance-wildlife control agents, and researchers have used commercial (see Appendix 3.2 for a list of suppliers) and hand-made traps to capture a variety of mammals, including carnivores, rodents, lagomorphs, and marsupials. These mechanical devices can be divided into 2 broad categories: restraining (live) and killing traps. However, certain trap designs can be included in either category, depending on how they are deployed in the field.

The AFWA documented the performance of foot traps, snares, and other forms of restraining traps in support of the development of BMP (AFWA 2006a). Test traps were selected based on knowledge of commonly used traps, previous research, and input from expert trappers. Data collection, including safety evaluations, was undertaken using procedures specified in ISO Documents 10990-4 and 10990-5 (ISO 1999a, b). Trauma scales used to assess animal welfare performance for restraining traps are presented in ISO Document 10990-5, and BMP research adapted those scales for evaluating injury in captured animals (injury scales ranged from 0 for uninjured animals to 100 for animals found dead in traps). BMP traps are required to consistently yield little to no injury to captured animals (AFWA 2006a), and therefore they are acceptable in many wildlife research applications.

Trap Types

Restraining traps are those designed to capture an animal alive. Three basic types are used to capture mammals. Cage or box traps are manufactured in an array of sizes for small insectivores, rodents, lagomorphs, carnivores, and ungulates. They are constructed of wire or nylon mesh, wood, plastic, or metal. The functional components include the cage box, 1 or 2 self-closing doors, a door lock mechanism, a trigger, and a treadle or trip pan. Foothold traps are commonly used to capture medium-sized mammals, such as wild canids and felids (Fig.3.22). A typical foothold trap has 2 jaws open at 180° when in the set position and closing 90° upon each other when released. Another foothold design includes foot-encapsulating devices, such as the EGG™ trap (Proulx et al. 1993c, Hubert et al. 1996) and Duffer's trap (IAFWA 2000),

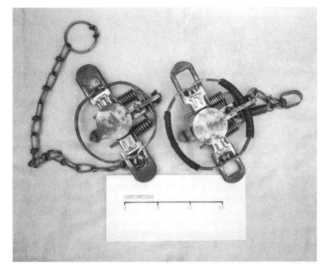

Fig. 3.22. Foothold restraining traps used to capture mammals: Victor No. 1.5 coil spring foothold trap (left) and Victor No. 1.5 Soft-Catch foothold trap with padded jaws (right). *Photo by G. F. Hubert, Jr.*

which have a pull trigger that releases a small striking bar to block an animal's paw as well as a plastic or metal housing that protects the captured limb from torsion or self-inflicted injuries (Fig. 3.23). These traps are species-specific, are considered relatively "dog proof," and are used to capture raccoons and opossums.

Foot snares, such as the Aldrich (Poelker and Hartwell 1973), Åberg™ (Englund 1982), Fremont™ (Skinner and Todd 1990), and Belisle™ (Shivik et al. 2000), are spring-powered cables used to capture and hold medium and large animals by a limb (Fig. 3.24). Modified manual neck snares (McKinstry and Anderson 1998, Pruss et al. 2002) and specialized cable restraints, such as the Collarum™ (Shivik et al. 2000), also can function as restraining traps. The performance of snares as live restraint tools versus killing systems is determined by numerous variables, including set location, snare and lock types, and experience of the trapper (AFWA 2009b).

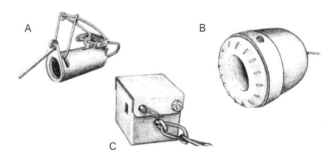

Fig. 3.23. Foot encapsulating traps specifically designed for capturing raccoons (they prevent self-mutilation) and reducing the capture of domestic pets: (A) Lil' Grizz Get'rz, (B) EGG, (C) Duffer's. *Photo courtesy of the Association of Fish and Wildlife Agencies.*

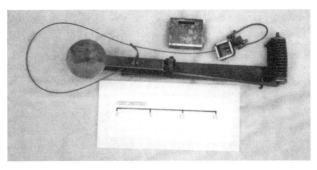

Fig. 3.24. Novak foot snare. *Photo by G. F. Hubert, Jr.*

Killing traps have one or more striking jaws (or a snare noose) activated by one or many springs upon firing by a trigger mechanism. Killing traps come in a variety of sizes, and their method of action varies. Mousetrap-type devices, where one jaw closes 180° on a flat surface, are commonly used to capture commensal and other small rodents. Killing boxes, pincer- and spear-type traps, and certain body-gripping devices are used to capture fossorial rodents and moles. The cage/box and foothold restraining traps also can be used as killing devices by placing them in or near water, so the captured animal is submerged and drowns. This technique is commonly used by fur trappers when harvesting aquatic and semi-aquatic mammals, such as American beaver, mink, muskrat, and northern river otter. Planar traps, in which a spring functions as a killing bar, are used to catch rat-sized rodents and small carnivores (e.g., Mustelidae). Rotating-jaw or body gripping traps have a scissor-like closing action and are used for a variety of mammals ranging in size from tree squirrels to beaver. Finally, manual locking neck and power snares are used to catch and kill medium-sized carnivores, such as foxes, coyotes, and bobcats (Table 3.14).

Trap Research, Performance Standards, and Evaluation

Traps have been and continue to be important and traditional tools for wildlife management and research (Boggess et al. 1990). Nevertheless, the use of these capture devices is not without controversy (Gentile 1987, Andelt et al. 1999). Most concerns are related to **animal welfare.** Consequently, professional wildlife biologists have expressed the need to reduce injury and pain inflicted on animals by trapping (Schmidt and Brunner 1981, Proulx and Barrett 1989). Novak (1987) reviewed traps and trap research related to furbearers. Recent efforts to improve the welfare of animals captured in traps by developing humane trapping standards have met with mixed success. Activities in the United States have primarily focused on the development of BMP for trapping furbearers by using restraining traps under the auspices of the AFWA (AFWA 2006a).

Endeavors through the ISO led to the adoption of 2 international standards—one for methods for testing killing trap systems used on land or underwater (ISO 1999a) and another for methods for testing restraining traps (ISO 1999b). The Canadian General Standards Board first published a national **killing trap standard** in 1984, based on a 180-second time-to-unconsciousness interval (Canadian General Standards Board 1984). Twelve years later this interval was relaxed to 300 seconds for some species (Canadian General Standards Board 1996). However, there are several killing traps currently available that have been shown to kill certain species quicker than the Conibear™ body-gripping series listed as state-of-the-art in 1996. Examples include the C120 Magnum with pitchfork trigger for American marten (Proulx et al. 1989a), the C120 Magnum with pan trigger and the Bionic™ for mink (Proulx et al. 1990, Proulx and Barrett 1991), and the Sauvageau™ 2001-8 for arctic fox (Proulx et al. 1993a).

Numerical scores have often been used to quantify the extent of injury incurred by a trapped animal (e.g., Olsen et al. 1986, 1988; Linhart et al. 1988; Onderka et al. 1990; Phillips et al. 1992; Hubert et al. 1996). Although Linhart and

Table 3.14. Snares and neck collars used to capture mammals

Group/species[a]	Reference
Snowshoe hare	Keith 1965, Brocke 1972, Proulx et al. 1994a
Ground squirrel	Lishak 1976
American beaver	Collins 1976, Mason et al. 1983, Weaver et al. 1985, McKinstry and Anderson 1998, Riedel 1988
Nutria	Evans et al. 1971
Gray wolf	Van Ballenberghe 1984, Schultz et al. 1996
Coyote	Nellis 1968, Guthery and Beasom 1978, Onderka et al. 1990, Phillips et al. 1990b, Skinner and Todd 1990, Phillips 1996, Sacks et al. 1999, Shivik et al. 2000, Pruss et al. 2002
Red fox	Berchielli and Tullar 1980, Novak 1981b, Rowsell et al. 1981, Englund 1982, Proulx and Barrett 1990, Bubela et al. 1998
Gray fox	Berchielli and Tullar 1980
African lion	Frank et al. 2003
Amur (Siberian) tiger	Goodrich et al. 2001
Snow leopard	Jackson et al. 1990
Mountain lion	Pittman et al. 1995, Logan et al. 1999
Canada lynx	Mowat et al. 1994
Black bear	Poelker and Hartwell 1973, Johnson and Pelton 1980b
Raccoon	Berchielli and Tullar 1980
Skunk (Mustelidae)	Novak 1981b
Feral hog	Anderson and Stone 1993
White-tailed deer	Verme 1962, DelGiudice et al. 1990
Mule deer	Ashcraft and Reese 1956
South American Guanaco	Jefferson and Franklin 1986
Pronghorn	Beale 1966

[a] Scientific names are given in Appendix 3.1.

Linscombe (1987) recommended establishment of a standardized numerical system to rank trap-caused injuries, the issue is complicated by the existence of a variety of scoring systems (Proulx 1999b). Engeman et al. (1997) criticized the use of injury scores for judging acceptability of restraining traps. In contrast, Onderka (1999) indicated that numerical scoring reflecting the severity of injuries tended to be consistent and appropriate to assess live-holding devices. The current **international standard** that describes methods for testing restraining traps contains 2 trauma scales (ISO 1999b). One assigns point scores to 34 injury types; the other places these 34 injury types into 4 trauma classes that may be combined to provide an overall measure of animal welfare.

Most recently 2 international agreements, designed to further improve the welfare of trapped animals, have been developed. The United States and the European Union adopted a nonbinding understanding in 1997; the other was signed by Canada, Russia, and the European Union in 1997 and 1998 (Andelt et al. 1999). Since that time, activities in the United States have focused on the development of BMP for trapping furbearers under the auspices of the International Association of Fish and Wildlife Agencies (IAFWA 1997). As part of this project, the best-performing killing traps consider time to death, effectiveness, selectivity, safety, and practicality of field use. Similarly, the best restraining traps will be those based on reduced physical damage to the animal, effectiveness, selectivity, safety, and practicality. The first BMP was completed in 2003 and addresses the use of restraining traps for coyotes in the eastern United States (IAFWA 2003). BMP for all other major furbearer species are under development (IAFWA 1997).

Currently, both the AFWA and the Fur Institute of Canada provide **updated and comprehensive reviews** of traps for use in mammal capture programs (Tables 3.15, 3.16, and 3.17) that comply with BMP standards (AFWA 2009a) or the Agreement on International Humane Trapping Standards (Fur Institute of Canada 2009).

Evaluation and Status of Tranquilizer Trap Devices

Balser (1965) used **tranquilizer trap devices** (TTDs) containing diazepam, a controlled substance not registered for such use by the U.S. Drug Enforcement Administration (Savarie et al. 1993) to reduce injuries to coyotes. Another drug, propiopromazine hydrochloride (PPZH), which acts as a tranquilizer and depresses the central nervous system, was tested on captive coyotes by Savarie and Roberts (1979). Foot injuries to coyotes and other animals caught in foothold traps were reduced substantially when they ingested tranquilizers from tabs attached to trap jaws (Balser 1965).

Linhart et al. (1981) used TTDs containing PPZH to reduce foot and leg injuries to wild coyotes captured in foothold traps. Preliminary data reported by Zemlicka et al. (1997) suggested **significant reduction** in trap related injuries to the feet and legs of 37 gray wolves captured in traps using TTDs containing PPZH. None of 33 nontarget animals captured in traps with TTDs loaded with PPZH succumbed from ingestion of the tranquilizer, and injuries tended to be less severe than among nontarget captures in traps without PPZH TTDs. Sahr and Knowlton (2000) demonstrated that TTDs containing PPZH effectively reduced injuries to limbs of wolves captured in foothold traps, but failed to reduce the severity of tooth injuries. Pruss et al. (2002) evaluated a modified locking neck snare equipped with a diazepam tab for coyotes in an effort to decrease stress, injuries, and unwanted animal captures. This device successfully reduced the incidence of lacerations experienced by captured coyotes without compromising capture efficiency or increasing the capture of nontarget species. Savarie et al. (2004) successfully tested PPZH in a plastic polyethylene pipette reservoir attached to a trap jaw.

The 2 drugs (diazepam and PPZH), used in conjunction with TTDs, are not available for widespread use. Pruss et al. (2002) reported that future use of diazepam in Canada would require a researcher to submit a special request to the Drug Strategy and Controlled Substances Programme, Office of Controlled Substances, Ottawa, Ontario, Canada, and nonresearch use would require the cooperation of a veterinarian. In the United States, diazepam (Valium®) is a Class IV controlled substance (Seal and Kreeger 1987) and has not been authorized as a tranquilizer for traps. Currently, only the U.S. Wildlife Services is authorized to use PPZH in TTDs as part of its wildlife damage-control operations under a special permit issued by the U.S. Food and Drug Administration (T. J. Deliberto, U.S. National Wildlife Research Center, Department of Agriculture, Fort Collins, Colorado, personal communication).

Miscellaneous Capture Methods

Bergman et al. (1999) captured nine-banded armadillo by following a trained **tracking dog** to a burrow. They then placed a 30-cm-high wire fence around the burrow and a cage live trap at the burrow entrance. Godfrey et al. (2000) described a detailed protocol for safe entry into black bear tree dens for capture purposes that minimized risks to biologists and bear mortality.

Karraker (2001) attached a string to hang from the cover board over pitfall traps, allowing small mammals to escape. Perkins and Hunter (2002) reduced small mammal capture by placing wooden sticks in pitfall traps. The rate of amphibian capture was not reduced. Padgett-Flohr and Jennings (2001) perfected a simple and inexpensive small-mammal **safe-house** that is placed in the bottom of pitfall traps (Fig. 3.25). The safe house was constructed of 5-cm-diameter PVC pipe in 12.5-cm lengths and capped at one end. The center of the safe house was one-third filled with 100% cotton batting, and the house was glued to a base of PVC pipe cut in half to a length of 12 cm.

Table 3.15. Live capture devices that meet state-of-the-art animal welfare performance criteria by individual species[a]

Species[b]	Capture method	Trap type
American beaver	Suitcase	Breath Easy™ Live Trap; Hancock™ Live Trap
	Body snare	7×7 weave 0.24 cm (0.94 inch) cable diameter with bent washer lock; 7×7 weave 0.24 cm cable diameter with BMI™ "Slide Free" Lock; 7×7 weave 0.32 cm (0.13 inch) cable diameter with cam lock; 7×7 weave 0.24 cm cable diameter with cam lock; 0.13 cm (1/19 inch) weave 0.24 cm cable diameter with Raymond Thompson™ lock
Bobcat	Foothold	1.5 coiled-spring; 1.5 coiled-spring with padded jaws, 4- coiled 2 coiled-spring; 1.75 coiled-spring; 1.75 coiled-spring with offset, laminated jaws 2 coiled-spring with offset, laminated jaws, 4-coiled; 3 coiled-spring; 3 coiled-spring with laminated jaws; 3 coiled-spring with offset jaws; 3 coiled-spring with offset, laminated jaws; 3 coiled-spring with padded jaws, 4 coiled; 3 double long spring; MJ 600; MB 650-OS with 0.64 cm (0.25 inch) offset jaws
	Foot snare	Bélisle™ Foot Snare No. 6
	Cage	Tomahawk™ 109.5
Coyote	Foothold	1.75 coiled-spring wih offset flat jaws; 1.5 coiled-spring with padded jaws, 4 coiled; 1.75 coiled-spring; 1.75 coiled-spring with forged, offset jaws; 1.75 coiled-spring with offset, laminated jaws; 22 Coyote Cuff™; 2 coiled-spring; 2 coiled-spring with forged, offset jaws; 2 coiled-spring with offset, laminated jaws, 4-coiled; 3 coiled-spring with padded jaws, 4-coiled; 3 Montana Special™ Modified, 2-coiled; MB 650-OS with 0.64 cm (0.25 inch) offset jaws; MJ 600
	Foot snare	Bélisle™ Foot Snare #6
	Neck snare	7×7 weave 0.24 cm cable diameter with Reichart™ washer lock; 7×7 weave 0.24 cm cable diameter with #4 Gregerson™ lock; 7×7 weave 0.24 cm cable diameter with BMI Slide Free lock; 7×19 weave 0.24 cm cable diameter with Reichart washer lock; 7×19 weave 0.24 cm cable diameter with #4 Gregerson lock; 7×19 weave 0.24 cm cable diameter with BMI Slide Free lock; 7×7 weave 0.32 cm cable diameter with Reichart washer lock; 7×7 weave 0.32 cm cable diameter with #4 Gregerson lock; 7×7 weave 0.32 cm cable diameter with BMI Slide Free lock; 7×19 weave 0.32 cm cable diameter with Reichart washer lock; 7×19 weave 0.32 cm cable diameter 7×19 weave 0.32 cm cable diameter with BMI Slide Free lock
Fisher	Foothold	1.5 coiled-spring with padded jaws, 4 coiled
	Cage	Tomahawk 108
Gray fox	Foothold	1.5 coiled-spring with Humane Hold™ pads on jaws; 1.5 coiled-spring with padded and double jaws; 1.5 coiled-spring with padded jaws, 4 coiled; 1.5 coiled-spring with padded jaws and 0.135 spring; 1.75 coiled-spring with offset, laminated jaws; 2 coiled-spring with padded jaws
	Foot snare	Bélisle Foot Snare
	Cage	Tomahawk 108
Nutria	Foothold	1 coiled-spring with padded jaws; 1.5 coiled-spring with padded jaws
Raccoon	Foot-encapsulating	Duffer's™; EGG™; Lil' Grizz Get'rz™
	Foothold	11 double long spring with offset and double jaws; 1.5 coiled-spring with double jaws; 1 coiled-spring; 1.5 coiled-spring with double-jaws and lamination; 1.5 coiled-spring with double-jaws and flat offset; 1.5 coilspring with double-jaws and flat offset, 4-coiled
	Cage	Tomahawk 108
Red Fox	Foothold	1.5 coiled-spring; 1.5 coiled-spring with laminated jaws; 1.5 coiled-spring with padded jaws; 1.5 coiled-spring with padded jaws, 4 coiled; 5 coiled-spring with Humane Hold™ pads; 1.75 coiled-spring; 1.75 coiled-spring with offset laminated jaws; 1.75 coiled-spring with offset wide jaws; 2 coiled-spring with padded jaws; 2 coiled-spring with offset laminated jaws, 4 coiled; 3 coiled-spring with padded jaws, 4 coiled
	Neck snare	7×7 weave 0.24 cm cable diameter with Reichart washer lock; 7×7 weave 0.24 cm cable diameter with #4 Gregerson lock; 7×7 weave 0.24 cm cable diameter with BMI Slide Free lock; 7×19 weave 0.24 cm cable diameter with Reichart washer lock; 7×19 weave 0.24 cm cable diameter with #4 Gregerson lock; 7×19 weave 0.24 cm cable diameter with BMI Slide Free lock; 7×7 weave 0.32 cm cable diameter with Reichart washer lock; 7×7 weave 0.32 cm cable diameter with #4 Gregerson lock; 7×7 weave 0.32 cm cable diameter with BMI Slide Free lock; 7×19 weave 0.32 cm cable diameter with Reichart washer lock; 7×19 weave 0.32 cm cable diameter with #4 Gregerson lock; 7×19 weave 0.32 cm cable diameter with BMI Slide Free lock
	Foot snare	Bélisle Foot Snare No. 6
Northern river otter	Foothold	11 double long spring; 11 double long spring with offset and double jaws; 2 coiled-spring
Striped skunk	Cage	Tomahawk 105.5; Tomahawk 108
Virginia opossum	Foot-encapsulating	EGG
	Foothold	1.5 coiled-spring with double jaws; 1.5 coiled-spring with padded jaws; 1.5 coiled-spring with padded and double jaws; 1.5 coiled-spring with padded jaws, 4-coiled; 1.65 coiled-spring with offset laminated jaws; 1 coiled-spring with padded jaws
	Cage	Tomahawk 108

[a] As listed in *Best Management Practices for Trapping in the United States* species documents (Association of Fish and Wildlife Agencies 2009a,b; http://www.fishwildlife.org/furbearer_resources.html).

[b] Scientific names are given in Appendix 3.1.

Table 3.16. Live capture devices that meet state-of-the-art animal welfare performance criteria by individual species[a]

Species[b]	Capture method	Trap type
Bobcat	Footsnare	Bélisle Footsnare #6
Coyote	Foothold	Bridger #3 equipped with 0.79 cm (0.31-inch) offset, doubled rounded steel jaw laminations 0.48 cm (0.19-inch) on topside of jaw and 0.64 cm (0.25-inch) on underside of jaws), with 4 coiled springs and an anchoring swivel center mounted on a base plate; Oneida Victor #3 Soft Catch equipped with 2 coiled springs
	Footsnare	Bélisle Footsnare #6
Canada lynx	Foothold	Oneida Victor #3 Soft Catch equipped with 2 coiled springs; Oneida Victor #3 ft Soft Catch equipped with 4 coiled springs; Victor #3 equipped with a minimum of 8mm thick, non-offset steel jaws, 4 coiled springs and an anchoring swivel c enter mounted on a base plate
	Footsnare	Bélisle Footsnare #6
Gray wolf	Footsnare	Bélisle Footsnare #8

[a] As certified through Canada's process for implementing the Agreement on International Humane Trapping Standards (Fur Institute of Canada 2009; http://www.fur.ca/index-e/trap_research/index.asp?action=trap_research&page=traps_certified_traps).

[b] Scientific names are given in Appendix 3.1.

Table 3.17. Killing traps that meet state-of-the-art animal welfare performance criteria by individual species[a]

Species[b]	Capture method	Trap type
American	Bodygrip	Bélisle Classic 330; LDL C280; Sauvageau 2001-8; Bélisle Super X 280; LDL C280 beaver Magnum; Sauvageau 2001-11; Bélisle Super X 330; LDL C330; Sauvageau 2001-12; BMI 280 Body Gripper; LDL C330 Magnum; Species-Specific 330 Magnum; BMI 330 Body Gripper; Rudy 280; Species-Specific 440 Dislocator Half Magnum; Bridger 330; Rudy 330; Woodstream Oneida; Victor Conibear 280; Duke 330; Sauvageau 1000-11F; Woodstream Oneida; Victor Conibear 330
Fisher	Bodygrip	Bélisle Super X 120; LDL C220 Magnum; Sauvageau 2001-5; Bélisle Super X 160; Rudy 120 Magnum; Sauvageau 2001-6; Bélisle Super X 220; Rudy 160 Plus; Sauvageau 2001-7; Koro #2 Rudy 220 Plus; Sauvageau 2001-8; LDL C160 Magnum
Canada lynx	Bodygrip	Woodstream Oneida; Victor Conifear 330
American marten	Bodygrip	Bélisle Super X 120; Koro #1; Sauvageau C120 Magnum; Bélisle Super X 160; Northwoods 155; Sauvageau 2001-5; BMI 126 Magnum; Rudy 120 Magnum; Sauvageau 2001-6 Body Gripper; LDL B120 Magnum; Rudy 160 Plus
Muskrat	Bodygrip	Bélisle Super X 120; Duke 120; Sauvageau C120 Magnum; BMI 120; Koro Muskrat; Sauvageau C120; "Reerse Bend"; BMI 120 Magnum; LDL B120 Magnum; Triple M; BMI 126 Magnum; Rudy 120 Magnum; Woodstream Oneida; Victor Conibear 110; Bridger 120; Sauvageau 2001-5; Woodstream Oneida; Victor Conibear 120; Any jaw type trap (body gripping or leghold) set as a submersion set that exerts clamping force on a muskrat and that maintains a muskrat underwater.
Raccoon	Bodygrip	Bélisle Classic 220; Bridger 220; Rudy 160 Plus; Bélisle Super X 160; Duke 160; Rudy 220; Bélisle Super X 220; Duke 220; Rudy 220 Plus; Bélisle Super X 280; LDL C 160; Sauvageau 2001-6; BMI 160 Body Gripper; LDL C 220; Sauvageau 2001-7; BMI 220 Body Gripper; LDL C 220 Magnum; Sauvageau 2001-8; BMI 280′ LDL C 280 Magnum; Species-Specific 220; Dislocator Half Magnum; BMI 280 Magnum; Northwoods 155; Woodstream Oneida Body Gripper; Victor Conibear 160; Bridger 160; Rudy 160; Woodstream Oneida; Victor Conibear 220
Northern river otter	Bodygrip	Bélisle Super X 280; Rudy 280; Woodstream Oneida; Victor Conibear 220; LDL C280 Magnum; Rudy 330; Woodstream Oneida; Victor Conibear 330; Sauvageau 2001-8
Weasel	Snap Trap	Victor Rat Trap

[a] As certified through Canada's process for implementing the Agreement on International Humane Trapping Standards (Fur Institute of Canada 2009; http://www.fur.ca/index-e/trap_research/index.asp?action=trap_research&page=traps_certified_traps).

[b] Scientific names are given in Appendix 3.1.

Scotton and Pletscher (1998) jumped from a hovering **helicopter** to hand capture neonatal Dall sheep. They advocated using smaller, less noisy helicopters to minimize disturbance of ewes and their lambs.

An efficient technique for capturing **swimming deer** (Fig. 3.26) was developed by Boroski and McGlaughlin (1994) for use in lakes and reservoirs. They made a "head bag" from the upper half of a pants leg with a hole for insertion of pipe insulation for flotation. Other materials included a canvas pack cinch, a leather latigo strap, a nylon "piggin" string, and a 1.4-kg weight. A 3-person crew included a boat handler and 2 deer handlers. The piggin string was placed around

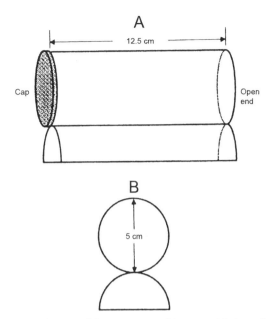

Fig. 3.25. Side (A) and front (B) view of the assembled small-mammal safe-house constructed from 5-cm-diameter polyvinyl chloride (PVC) pipe. *From Padgett-Flohr and Jennings (2001).*

the deer's neck and the head bag was placed over the animal's head to calm the it. The latigo strap was positioned in front of the rear legs. After attachment of a radiocollar to the deer, the restraints and head bag were removed, and the deer previously kept in the water was released and allowed to swim freely. Handling time of captured deer averaged 5.5 minutes.

Ballard et al. (1998) decided that intensive grid ground searching was the most effective method for locating and **hand capturing** neonate white-tailed deer fawns. Franklin and Johnson (1994) hand captured South American guanacos 30–60 minutes after birth, before the neonates could escape by running. Care was taken to avoid separation of the mother from her offspring. Only 5 of 435 captured young guanacos (1.2%) failed to unite or were abandoned by their mothers. They suggested that hand capture and tagging of

Fig. 3.26. Restraint and radiocollar attachment for deer captured while swimming. *From Boroski and McGlaughlin (1994).*

precocial newborns had potential application to a variety of African, Asian, and North American ungulates that live in open habitats.

Lanyon et al. (2006) developed a method for live-capturing dugongs in open water using the **rodeo method,** which involves pursuit of a dugong by boat until it is fatigued, followed by a human catcher jumping off the boat to restrain the dugong. McBride and McBride (2007) successfully, safely, and selectively captured jaguars using trained cat hounds. Omsjoe et al. (2009) used a similar paired-snowmobile pursuit method, entangling a Svalbard reindeer in a net. Capybaras were captured in Venezuela by **lassoing** from horseback (Salas et al. 2004). Corrigan (1998) tested various types of **glue traps** and found them to be largely ineffective for capturing house mice.

Bishop et al. (2007) described the successful use of **vaginal implant transmitters** to aid in the capture of mule deer neonates. Vaginal-implant transmitter modification, including larger holding wings and antennas protruding 1 cm past the vulva, resulted in more successful drops of deer fawns at birth sites (Haskell et al. 2007; Table 3.18).

Benevides et al. (2008) designed a **trap signaling device** with long distance reception (18 km), durability in adverse weather, and light weight, which allowed reduction in the effort required to check traps and quick release of endangered and nontarget species. Nolan et al. (1984) used transmitters for monitoring leg snares set for grizzly bears. Neill et al. (2007) reviewed a Global System for Mobile communication trap alarms attached to padded leg-hold traps that shortened the retention time of capture of Eurasian otters to 22 minutes and reduces trap injuries (Table 3.19).

Use of Attractants

The success of most animal trapping operations depends on a suitable bait or lure to attract animals to traps. Numerous native and commercial foods, artificial and visual lures, agricultural products, and naturally occurring and artificial scents have been used as attractants. Because of the diversity of habitats and species, no universal attractant successfully works for all animals. Consequently, wildlife biologists may need to evaluate several baits or lures before finding

Table 3.18. Use of vaginal implant transmitters for capture of neonates

Species[a]	Reference
Mule deer	Garrott and Bartmann 1984, Johnstone-Yellin et al. 2006, Bishop et al. 2007
White-tailed deer	Bowman and Jacobson 1998, Carstensen et al. 2003, Haskell et al. 2007, Swanson et al. 2008
Elk	Seward et al. 2005, Johnson et al. 2006, Barbknecht et al. 2009

[a] Scientific names are given in Appendix 3.1.

Table 3.19. Systems for signaling successful trap capture

Species[a]	Capture method	Type of signal	Reference
Small Hawaiian carnivores	Tomahawk live trap	Radio transmitter	Benevides et al. 2008
Large mammals	Trap and foot snare	Radio transmitter	Halstead et al. 1995
Mule deer	Clover trap	Telemetry	Hayes 1982
Wild canids	Padded jaw foothold	Electronic	Larkin et al. 2003
Wild canids	Treadle snare	Radio transmitter	Marks 1996
Otter	Padded jaw foothold	Mobile phone technology	Neill et al. 2007
Grizzly bear	Aldrich snare	Radio telemetry	Nolan et al. 1984
Raptors	Bow net	Two-way radio	Proudfoot and Jacobs 2001

[a] Scientific names are given in Appendix 3.1.

those that attract different species in a specific geographical area.

Baits

Prebaiting is generally an important prerequisite to, and baiting an essential part of, any successful trapping program. Carnivores may be attracted to traps by bait made from chunks of meat that is fresh or tainted. For example, holes can be punched in a container of sardines to make a long-lasting attractant. (Bluett 2000) reported that selectivity for certain species, such as raccoons, was enhanced by using **sweet baits,** such as fruit or marshmallows. Saunders and Harris (2000) evaluated bait preferences of captive red fox. Whole mice were the most preferred and horsemeat the least preferred of the 6 animal baits tested. Travaini et al. (2001) simultaneously tested a variety of scented meat baits and 3 ways of delivering these baits to culpeo and Argentine gray foxes in Patagonia. All 4 types of baits used were equally attractive to both species of fox. The percentage of the different types of baits consumed by the 2 species did not differ among bait type, and no differences were detected in visitation rates to the 3 types of bait delivery systems. Andrzejewski and Owadowska (1994) successfully captured bank voles at a significantly greater rate by using conspecific odor foam cube baits rather than food as bait.

Morgan and Dusek (1992) had success capturing white-tailed deer in Clover traps on summer range using **salt blocks** as bait. Alfalfa hay was a successful bait in winter. Naugle et al. (1995) had better deer trapping success using corn rather than salt in summer in agriculture–wetland habitats. Bean and Mason (1995) evaluated the attractiveness of **liquid baits** to white-tailed deer. Apple juice was preferred to cyclamate or saccharin solutions. Volatile apple extract also was an effective lure. Hakim et al. (1996) found the most successful use of liquid bait was in May. They suggested that spring was the best season to attract and capture deer in Virginia. Ballard et al. (1998) reported that white cedar (*Thuja occidentalis*) browse was the best bait for trapping white-tailed deer in winter.

Edalgo and Anderson (2007) evaluated the effects of prebaiting on small-mammal trapping success and concluded that prebaiting was not worthwhile. Barrett et al. (2008) tested various supplements to corn baits and found no increase in deer capture success in Clover traps.

Scents

Fur trappers have used a variety of scents to attract fur-bearing mammals to traps. These lures can be divided into 3 basic categories: gland, food, and curiosity scents. **Gland scents** are made of different parts of animals, such as the reproductive tract and anal glands. Examples of **food scents** include extracts of honey and anise, and fish oil. **Curiosity scents** are typically blends of essential oils, exotic musk, and American beaver and muskrat scent glands. Mason and Blom (1998) listed the common ingredients in lure formulations as well as their sources, methods of preparation, and common uses (Table 3.20).

A variety of scents, including those composed from rotten eggs, decomposed meat, and fish oil, has been used to increase trapping success rates. Other items, such as seal oil, Siberian musk oil, anal glands from foxes and skunks, and mink musk, also are widely used. Clapperton et al. (1994) tested a variety of attractants for feral cats in New Zealand. Catnip (*Nepeta cataria*) and matatabi (*Actinidia polygama*) were the most promising scent lures tried.

Phillips et al. (1990*a*) evaluated seasonal responses of captive coyotes to 9 chemical attractants and tested 26 additional attractants during summer to examine the efficacy of traps, **M-44s** (a tube-like spring-loaded device designed to deliver a lethal dose of sodium cyanide into the mouth of a coyote), and placed baits. Of the 9 attractants tested throughout the year, fatty acid scent (**FAS**) and **W-U lure** (Trimethylammonium decanoate plus sulfides) ranked highest in overall attractiveness. FAS and W-U lure also ranked highest among the 35 attractants tested only during the summer. Kimball et al. (2000) formulated 7 new synthetic coyote attractants by using representative compounds from commercially available attractants with the intention of developing

Table 3.20. Common ingredients in lure formations, methods of preparation, and common applications

Ingredient	Source	Preparation	Use
Muskrat glands/musk	Small glands on either side of vent of males during spring	Fresh ground, preserved, tinctured	Acids in musk are attractive to coyotes
Beaver castor	Large flat glands on each side of vent of both males and females	Fresh ground, preserved, dried, rasped to a powder; tinctured (castorium)	Phenols attractive to coyotes, serve to fix, preserve other ingredients in lures
Beaver sac oil	Long oval-shaped, whitish glands next to the castors	Fresh ground, preserved, oil squeezed from glands	Used alone or mixed with castors and used as a fixative
Mink glands/musk	Glands on either side of vent of males in breeding season	Ground fresh, preserved, tinctured	Contains sulfides, attractive to coyotes
Glands/urine from canids/felids/mustelids	Fox, bobcat, dog, badger, etc.	Ground fresh, preserved, rotted	
Asafetida	Plant	Gum or powdered or tinctured	Contains sulfides, attractive to coyotes
Garlic, onion	Plant	Powders, salts, oils	Contains sulfides, attractive to coyotes
Valerian root	Plant	Powder, oil, extract or salt (i.e., zinc valerate)	Valeric acid, attractive to coyotes
Rue oil	Plant	Oil, 3–5 drops per 0.25 L	Methyl ketones impart a cheesy odor
Skunk musk	Glands on either side of vent in males	Oil, 3–5 drops per 0.25 L used as component, 6–10 drops per 0.25 L as dominant odor	Powerful sulfide (mercaptan) odor odor attractive to coyotes
Orris root	Plant	Powder, oil, tincture, 0.5 tsp of oil/tincture or 0.125 tsp to powder per 0.25 L	Fixative, contains acids attractive to coyotes
Oakmoss	Plant	Resin, tincture, 3–5 drops resin, or 0.25 tsp of tincture per 0.25 L	Fixative
Phenyl acetic acid	Synthetic chemical	Tincture or crystals	Honey-like odor, also found in urines and scent glands
Cilantro oil (coriander leaf oil)	Plant	Oil, 2–4 drops per 0.25 L	Aldehydes attractive to coyotes
Anise oil	Plant	Oil, 3–5 drops per 0.25 L	Licorice odor

Adapted from Mason and Blom (1998).

[a] Scientific names of animals are given in Appendix 3.1.

relatively simple synthetic alternatives. Bioassays with captive coyotes were conducted to compare 9 behavioral responses elicited by the 7 new attractants. Results indicated that each attractant elicited a different behavioral profile. No significant differences among attractants in regard to urinating, sniffing, and licking behaviors were detected, but differences among the attractants existed for rubbing, rolling, scratching, defecating, digging, and pulling behaviors. Saunders and Harris (2000) evaluated 9 chemical attractants for red fox. They reported the strongest preferences were for 2 gustatory additives (sugar and a combination of beef and sugar) and an olfactory attractant (synthetic fermented egg).

Andelt and Woolley (1996) tested the attractiveness of a variety of odors to urban mammals, including cats, dogs, fox squirrels, striped skunks, and raccoons. Deep-fried cornmeal added to bait increased the rate of visitation to scent stations. Harrison (1997) field-tested the attractiveness of 4 scents (Hawbaker's Wildcat 2, synthetic FAS, bobcat urine, and catnip) to wild felids, canids, and Virginia opossum. No differences were noted in visitations to scent stations.

McDaniel et al. (2000) tested scent lures to attract Canada lynx and found beaver castoreum and catnip oil to be most effective.

Fur trappers, especially those who focus on foxes and coyotes, often use **urine** at trap sets to enhance their success. Young and Henke (1999) assessed trap response of cottontail rabbits using wooden cage traps baited with food, block salt and minerals, and urine from nonpregnant female domestic European rabbits. They captured significantly more cottontails in traps baited with rabbit urine.

Plant extractions also may be added to scents. The root of the Asiatic plant asafetida (*Ferula assafoetida*) imparts a strong, persistent odor to scents. The oils from the herbs anise (*Pimpinella anisum*) and valerian (*Valeriana officinalis*) also have been added to scent mixtures.

Scents are used primarily to attract carnivores, but other mammals also are attracted to them. Large rodents, such as beaver and muskrat, can be attracted with scent mixtures containing castoreum from beaver and oil sacs from muskrats. Mason et al. (1993) evaluated salt blocks and several olfactory lures as potential lures for use in attracting white-tailed deer. Such odor stimuli as acorn, apple, and peanut butter significantly enhanced the effectiveness of salt blocks. Mineral blocks were more attractive to deer than salt, molasses, and mineral–molasses blocks; all were scented with apple extract.

Visual Attractants

Visual attractants can enhance trapping success for such species as bobcat that rely heavily on their sense of sight when hunting. Bobcats can be attracted to traps by a piece of fur or feathers suspended 90–120 cm above the wire or string. However, in many states, use of visual attractants by trappers is illegal, because they may attract protected raptors. Knight (1994) and Virchow and Hogeland (1994) described the use of visual attractants in trapping mountain lion and bobcats, respectively.

Species-Specific Traps and Their Performance
American Badger

Limited research in Wyoming indicated that No. 1.5 coil-spring foothold traps with unpadded, laminated, or padded jaws can be used to capture American badgers with only minor injuries (Kern et al. 1994). Also, 78% of 45 badgers captured for a telemetry study in Illinois using Victor™ No. 3 Soft-Catch™ padded foothold traps had no visible injuries (R. E. Warner, University of Illinois, unpublished data). Injuries recorded for the remaining 10 (22%) were minor (e.g., claw loss, mild edema, and small lacerations). No data on the performance of killing traps for badgers are available.

American Beaver

Limited data on restraining traps for beaver are available. Clamshell-type traps, such as the Bailey, Hancock, and Scheffer-Couch, have been used successfully to capture beaver alive for research and management (Couch 1942, Hodgdon and Hunt 1953), but are relatively inefficient, bulky, and expensive. Using Hancock and Bailey traps, Collins (1976) caught >100 beaver with no mortalities. McKinstry and Anderson (1998) reported that 2.38-mm locking snares could be used to efficiently live-capture beaver, but they recorded a mortality rate of 5.3%.

Research in Canada performed under controlled conditions has shown that beaver can be killed in ≤6.1 minutes using standard Conibear 330 and modified (jaws bent inward) Conibear 280 and 330 traps in terrestrial sets (Novak 1981a). Gilbert (1992) reported that Conibear 330 traps with clamping bars rendered 14 beaver unconscious in ≤3 minutes. However, consistent positioning of juvenile beaver in a proper manner was an apparent problem. When captured underwater in locking snares or in drowning sets using No. 3 and No. 4 Victor foothold traps, beaver died in 5.5–10.5 minutes due to CO_2 narcosis or asphyxiation (Novak 1981a, Gilbert and Gofton 1982). Novak (1981a) reported that beaver trapped underwater in modified Conibear 330 traps were killed in 7.0–9.25 minutes. In addition, tests on anesthetized beaver measured the minimum energy forces required to cause death when delivered via a blow to the head, neck, thorax, or chest (Gilbert 1976, Zelin et al. 1983).

An improved, safe beaver live trap was developed by Müller-Schwarze and Haggert (2005). Vantassel (2006) modified the Bailey beaver trap to curtail misfires and increase capture success. McNew et al. (2007) used neck snares to live-capture beavers. Advantages of snares include light weight, low cost, and ease of setting.

BMP for trapping in the United States were based on field studies that captured and evaluated 100 beaver using the Breathe Easy™ Live Trap and the Hancock trap in New Hampshire during 1998–2001 (AFWA 2007a). Both traps met all BMP criteria (Table 3.15). Animal welfare performance was similar for the 2 trap types (cumulative injury score of 13 ISO scale) and efficiency was >92%. Of the 100 beavers captured, there were 2 mortalities: 1 in each trap type.

Snares are the most commonly used trapping technique for capturing beaver by fur trappers in the United States (AFWA 2005). BMP for snare trapping in the United States were based on field studies that captured and evaluated 193 beaver using 6 different snares for live restraint in New Hampshire during 2001–2007 (AFWA 2007a). Cable diameters used were 2.38 mm or 3.17 mm. Cables used during testing were either 7 × 7 multistrand constructions (Fig. 3.27) or 1 × 19 single strand construction (Fig. 3.28). Various locking systems were used, but all locks were either relaxing or positive locking types, no power assisted locks were used (AFWA 2009b). All cable devices tested for live restraint passed BMP criteria for animal welfare (Table 3.15). Efficiency ranged from 58.2% to 91.7%. Of the 193 beaver captured in live restraint cable devices, only 1 mortality occurred.

Bobcat

Relatively few studies have investigated the performance of restraining traps for bobcat. Research in the western United States (Linscombe and Wright 1988, Olsen et al. 1988) and Michigan (Earle et al. 1996) has shown the Victor No. 3 Soft-Catch foothold trap with padded jaws was effective in cap-

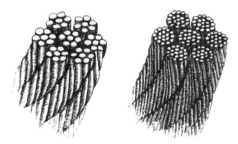

Fig. 3.27. The 7 x 7 multistrand cable has 7 bundles of 7 wires each. The 7 x 19 multistrand cable has 7 bundles of 19 wires each. *Illustration courtesy of the Association of Fish and Wildlife Agencies.*

Fig. 3.28. The 1 x 19 single-strand cable construction consists of 7 wires (twisted right) wrapped by 12 wires (twisted left). *Illustration courtesy of the Association of Fish and Wildlife Agencies.*

turing bobcat with minimal injuries compared to unpadded foothold traps. Modifications to the No. 3 Soft-Catch, such as heavier springs, improved trapping success (Earle et al. 1996). Woolf and Nielson (2002) reported live capture of 96 bobcats in wire cage traps and No. 3 Soft-Catch traps. Trap related injuries were uncommon with both devices and included only minor cuts and bruises. They captured 1.6 bobcats per 100 trap-nights in the cage trap compared with 0.8 per 100 trap-nights using the Soft-Catch trap. Earle et al. (2003) determined the Victor No. 3 Soft-Catch foothold trap with padded jaws was effective in capturing bobcat with minimal injuries compared to unpadded foothold traps.

BMP for trapping bobcats were based on 584 bobcats captured in 16 restraining devices in 16 states during 1998–2006 (AFWA 2006b). All 16 trapping devices evaluated for bobcat met BMP criteria for welfare, efficiency, selectivity, safety, and practicality (Table 3.15). The cage trap had the lowest mean cumulative injury score and the highest efficiency rating. However, animal welfare was acceptable in all trap types tested, and 75% of the traps tested had an efficiency rating for bobcats of >90%.

The most commonly used trap type in the United States for capturing bobcat is the No. 3 coil-spring (IAFWA 1992, AFWA 2005). The standard No. 3 coil-spring trap met all BMP criteria as did the same trap size with modifications, including padded, offset, and laminated jaws and jaws with both offset and lamination. The efficiency of all traps meeting BMP criteria for bobcat ranged from 61% to 100% capture per opportunity. The cage trap was the most efficient, followed by the No. 3 long-spring trap, the No. 1.5 standard coil-spring trap, the No. 2 standard coil-spring trap, and the No. 3 padded coil-spring trap. Trap selectivity for bobcat ranged from 10% to 45%. The No. 3 padded coil-spring trap was the most selective for bobcat, followed by the MJ 600 coil-spring trap, the No. 1.75 offset laminated coil-spring trap, and the No. 3 offset laminated coil-spring trap. No consistent pattern relative to trap type or modifications was apparent for selectivity.

Coyote

More restraining trap research has been conducted on coyotes than on any other North American mammal. Andelt et al. (1999) summarized injury scores and capture rates for 8 coyote traps tested by the Denver Wildlife Research Center. Other investigations of trap performance for coyotes include Linhart et al. (1986, 1988), Linscombe and Wright (1988), Olsen et al. (1988), Onderka et al. (1990), Skinner and Todd (1990), Linhart and Dasch (1992), Phillips et al. (1992, 1996), Gruver et al. (1996), Phillips and Mullis (1996), Hubert et al. (1997), and Shivik et al. (2000). Although Phillips et al. (1996) and Hubert et al. (1997) suggested that laminated traps are likely to be less injurious than standard unpadded foothold traps, the differences in the mean injury scores they observed were not significant. Houben et al. (1993) found no significant difference in mean injury scores assigned to limbs of coyotes captured in modified (heavier springs) No. 3 Soft-Catch padded foothold traps and No. 3 Northwoods™ foothold traps with laminated offset jaws. Padded foothold traps, such as the No. 3 Soft-Catch modified (Gruver et al. 1996) and the No. 3.5 E-Z Grip® (Phillips et al. 1996), have performed best in terms of both animal welfare and efficiency.

Way et al. (2002) tested 4 models of Tomahawk wire cage traps (models 610A, 610B, 610C, and 109) as an alternative capture technique for coyotes in a suburban environment in Massachusetts. These traps proved undesirable for capturing coyotes due to trap expense, time involved in baiting and conditioning coyotes to traps, a high rate of nontarget captures, and difficulty in capturing >1 adult in a social group. On the positive side, those coyotes caught sustained few injuries.

Phillips (1996) tested 3 types of killing neck snares for coyotes. He found that 94% of the coyotes snared by the neck with Kelley locks were dead when snares were checked versus 71% and 68% for the Gregerson and Denver Wildlife Research Center locks, respectively. However, the interval between trap checks was not specified. Phillips et al. (1990b) evaluated 7 types of breakaway snares were for use in coyote control. Maximum tension before breakage for individual snares ranged from 64.5 kg to 221 kg. They indicated that differences in tension loads between coyotes and nontarget species should allow for development of snares that will consistently hold coyotes and release most large nontarget animals.

Phillips and Gruver (1996) evaluated performance of the Paws-I-Trip™ pan tension device on 3 types of foothold traps commonly used to capture coyotes. This device reduced capture of nontarget animals without reducing the effectiveness of the traps for catching coyotes. The mean overall exclusion rates for combined nontarget species in the No. 3 Soft-Catch, Victor 3NM, and No. 4 Newhouse™ foothold traps were 99.1%, 98.1%, and 91%, respectively. Kamler et al. (2002) effectively used modified No. 3 Soft-Catch foothold traps equipped with the Paws-I-Trip device set at 2.15 kg to capture coyotes while excluding swift foxes.

Shivik et al. (2005) compared various coyote trapping devices for efficiency, selectivity, and trap related injuries. Tomahawk cage traps were the least selective and efficient (0%

catch). The Collarum neck restraint, soft catch, and power snare devices had 87–100% catch efficiency. None of the devices used caused major injury.

BMP for capturing coyotes were based on field studies that captured, dispatched, and evaluated 1,285 coyotes using 20 restraining type devices in 19 states during 1998–2005 (AFWA 2006d, e). Sixteen of these devices met or exceeded established BMP criteria for welfare, efficiency, selectivity, safety, and practicality. No coyotes died in any of the trap devices tested, and there were no documented practicability or safety concerns for trappers or nontrappers. Among devices that met BMP established criteria, the nonpowered cable device, Belisle footsnare, offset flat-jaw traps, and offset laminated-jaw traps had lower mean cumulative injury scores than did the standard offset-jaw traps, or offset forged-jaw traps. Also, noteworthy is that 2 regular-jaw traps (No. 1.75 and No. 2 coil-springs) had mean cumulative injury scores lower than standard offset-jaw traps or offset forged-jaw traps (Table 3.15).

The most commonly used trap in the United States for capturing coyotes is the No. 2 coil-spring trap (AFWA 2005). This trap met all established BMP criteria and produced the highest score for the "no injury" category, whereas the 1.75 offset flat-jaw trap had the highest cumulative scores for none, mild, and moderate injuries (99.9%), followed by the No. 3 padded 4-coiled trap (98.1%), the MJ 600 trap (98.0%), and the 1.5 padded, 4-coiled trap (97.9%). All trap devices that meet or exceed BMP standards had ≥83% cumulative injuries in the none, mild, or moderate categories. Trap devices of the No. 3 size typically had the highest efficiency; all had an efficiency of ≥85%. No consistent pattern for selectivity was apparent. However, all traps that meet or exceed BMP criteria had an overall furbearer selectivity of ≥84%.

During BMP studies, nonpowered cable devices and the Belisle No. 6 performed well for restraining coyotes, produced low mean cumulative injury scores (19.3 and 22.7, respectively), and did not result in any mortalities. Of the restraint devices tested, the Belisle No. 6 footsnare and nonpowered cable devices performed well and resulted in either no or mild injuries (AFWA 2006d, e; Table 3.15).

Feral Cat

Wire mesh traps (40 cm × 40 cm × 60 cm) and Victor No. 1.5 Soft-Catch padded jaw foothold traps have been used to trap feral cat in Australia (Molsher 2001). No difference was found in capture efficiency between trap types. Injuries suffered by cats in cage traps were generally minor and usually involved self-inflicted abrasions to the face. Only 1 of 12 cats (8.3%) caught in Soft-Catch traps was more seriously injured. Meek et al. (1995) and Fleming et al. (1998) also used Soft-Catch traps (No. 1.5 and No. 3) to capture feral cat. These researchers reported 100% and 68.6%, respectively, of the cats trapped had no visible trap related injuries or only slight foot or leg edema or both.

Fisher

Fur trappers commonly use cage traps to capture fisher in Massachusetts, but efficiency and animal welfare data for this and other restraining traps are not available. Researchers in Canada have evaluated a variety of killing traps for capturing fisher. Controlled testing on captive animals has shown the Bionic trap cocked to 8 notches consistently killed fisher in 60 seconds (Proulx and Barrett 1993b). The mechanical characteristics of the Sauvageau 2001-8 and modified (stronger springs) Conibear 220 traps surpassed the kill threshold established for fisher, but the standard Conibear 220 and AFK Kania traps did not (Proulx 1990). Double strikes (head and/or neck, and thorax) with a modified Conibear 220 trap equipped with 280-sized springs killed 5 of 6 fisher in an average of 51 seconds (Proulx and Barrett 1993a).

BMP for trapping in the United States were based on field studies that captured and evaluated 74 fishers using both foothold and cage traps in 5 states during 2004–2009 (AFWA 2007b). Two of the devices tested met or exceeded established BMP criteria: the No. 1.5 Soft-Catch foothold trap modified with 4 coil-springs and the Tomahawk 108 cage trap (Table 3.15). Use of the cage trap produced fewer injuries. Efficiency was higher with the cage trap, although efficiency for both traps was >90%. Selectivity was similar among the 2 trap types.

Arctic Fox

Two studies in Canada focused on the Sauvageau 2001-8 (a rotating-jaw killing trap) and the standard Victor No. 1.5 coil-spring foothold trap. Compound testing revealed that 9 arctic foxes caught in the Savageau 2001-8 set in a wire mesh cubby lost consciousness in an average of 74 seconds (Proulx et al. 1993a). During field tests on trap lines in the Northwest Territories, Canada, most arctic foxes captured in the No. 1.5 coil spring trap had only minor injuries when traps were checked daily (Proulx et al. 1994b).

Gray Fox

Berchielli and Tullar (1980) found no difference in trap related injuries of gray fox caught in Victor No. 1.5 coil-spring foothold traps versus those captured with Ezyonem™ leg snares. However, the leg snare was less effective in capturing fox than was the coil-spring foothold trap. Other researchers in the eastern United States have compared the unpadded Victor No. 1.5 coil spring with the padded Victor No. 1.5 Soft-Catch for gray fox. These studies found no difference in capture efficiency between trap types (Tullar 1984, Linscombe and Wright 1988) and a reduction in injuries for foxes captured in padded traps (Tullar 1984, Olsen et al. 1988). Gray fox can be captured in rotating jaw killing traps (e.g., Conibear 220-2) as well as in cage-type restraining traps, but performance data are lacking.

BMP for trapping gray fox were based on 925 foxes that were restrained, dispatched, and evaluated in 13 states during 1998–2003 (AFWA 2006c). Nine of 17 trapping devices evaluated for gray foxes met BMP criteria for welfare, efficiency, selectivity, safety, and practicality (Table 3.15). The No. 1.5 padded coil-spring trap with strengthened coil springs had the lowest mean cumulative injury score, followed by the cage trap and the No. 1 laminated coil-spring trap. The No. 1.5 laminated coil-spring trap, No. 1.5 padded coil-spring trap, and No. 1.65 offset laminated coil-spring trap all had welfare scores slightly higher (5 points) than the BMP criteria. However, all had ≥74% injuries in the lowest 3 classes. In addition, all 3 traps had efficiency ratings of ≥84%. The No. 1.5 padded coil-spring trap and No. 1.65 offset laminated trap both had gray fox selectivity scores higher than the 7 traps that met all criteria. Although the No. 1.5 laminated was not as selective for gray fox, it was selective for furbearers. The most commonly used trap in the United States for capturing gray fox is the No. 1.5 coil-spring (IAFWA 1992, AFWA 2005). This trap met BMP criteria only when modified with padded jaws, padded double jaws, and padded with strengthened coil-springs or with 4 coil-springs.

Efficiency of all traps meeting BMP criteria for gray fox ranged from 41% to 100% capture per opportunity. The cage trap was the most efficient, followed by the No. 1.5 padded 4-coiled coil-spring trap, No. 1.75 offset laminated coil-spring trap, No. 1.5 padded with strengthened coil-springs, and No. 2 padded coil-spring trap. Trap selectivity for gray fox ranged from 16% to 57% for traps meeting BMP criteria. The No. 1.5 with padded and double jaws was the most selective for gray fox, followed by the No. 1.5 padded with strengthened coil-springs, No. 2 padded coil-spring trap, and No. 1.75 offset laminated coil-spring trap.

Kit Fox

Kozlowski et al. (2003) described an enclosure system to live capture denning kit foxes.

Red Fox

The Victor No. 1.5 coil spring is the most common restraining trap used to capture red fox in the United States (IAFWA 1992). Several studies have compared the performance of this trap to the No. 1.5 Soft-Catch foothold trap with padded jaws (Tullar 1984, Linscombe and Wright 1988, Olsen et al. 1988, Kreeger et al. 1990, Kern et al. 1994). The No. 1.5 Soft-Catch proved to be as efficient as its unpadded counterparts, and it caused fewer and less serious injuries to trapped foxes. Kern et al. (1994) also reported that No. 1.5 coil spring traps with laminated or offset jaws were less injurious than those with standard jaws. Some foot snares have been found to be effective restraining traps for foxes under certain conditions (Novak 1981b, Englund 1982). During field tests in southern Ontario, Canada, and powder snow conditions in northern Sweden, the Novak™ and Åberg (Swedish) foot snares virtually eliminated trap related injuries. However, Berchielli and Tullar (1980) reported the Ezyonem foot snare was less effective than the No. 1.5 coil spring foothold traps for capturing foxes, and both devices produced similar trap related injuries. Researchers in Australia found a particular treadle (i.e., foot) snare difficult to set and inefficient; 3 of 71 red foxes they captured using this device had broken legs (Bubela et al. 1998).

Few published data on the performance of killing traps for red fox exist. Limited testing of neck snares indicated that red fox become unconscious ≤6 minutes in power snares, but manual snares may not be suitable killing devices for this species (Rowsell et al. 1981, Proulx and Barrett 1990). Frey et al. (2007) experienced success using neck snares to capture red foxes with very few fatalities.

The development of **BMP** for red fox was based on 654 red foxes captured in 14 devices in 16 states during 1998–2002 (AFWA 2006f). Thirteen of 14 trapping devices evaluated for red fox met BMP criteria for welfare, efficiency, selectivity, safety, and practicality (Table 3.15). The most commonly used trap in the United States is the No. 1.5 coil-spring (IAFWA 1992, AFWA 2005). The Victor No. 1.5 coil-spring was tested and met BMP criteria.

Padded traps with manufacturer-provided integral padding and cable devices had the lowest mean cumulative injury scores. The most efficient devices were the nonpowered cable and Belisle foot snare. Offset laminated and 4-coiled foothold traps followed in efficiency. No consistent pattern was apparent for selectivity, except that none of the 4 most selective devices were padded traps. Efficiency of all traps meeting BMP criteria for red fox ranged from 79% to 100% capture per opportunity. Nonpowered cable devices were the most efficient, followed by the Belisle foot snare, No. 1.75 offset laminated coil-spring trap, No. 3 4-coiled padded coil-spring trap, No. 1.5 4-coiled padded coil-spring trap, and the No. 2 4-coiled offset laminated coil-spring trap. Trap selectivity for red fox ranged from 14% to 34% for traps meeting criteria. The No. 1.75 coil-spring trap with wide offset jaws was the most selective for red foxes, followed by the No. 1.5 coil-spring trap, No. 2 4-coiled offset laminated coil-spring trap, and No. 1.5 laminated coil-spring trap. Selectivity of all furbearers captured in traps tested for red fox ranged from 87% to 94%. The most selective trap was the No. 1.75 coil-spring trap with wide offset jaws, followed by the No. 1.75 coil-spring trap, nonpowered cable device, and No. 1.5 laminated coil-spring trap.

Swift Fox

Baited single door Havahart™ wire cage traps (25.4 cm × 30.5 cm × 81.3 cm) have been successfully used to capture swift fox in Texas (Kamler et al. 2002). The capture rate of swift fox was 48% higher in reverse double sets (which used 2 traps set in opposite directions) than in single sets. No data on trap related injuries were presented.

Gray Wolf

A variety of foothold restraining traps, including the Aldrich™ foot snare, has been evaluated for capturing gray wolf (Van Ballenberghe 1984, Kuehn et al. 1986, Schultz et al. 1996). Van Ballenberghe (1984) reported on trap related injuries to wolves caught in 3 types of long-spring foothold traps and the Aldrich foot snare, but small sample sizes precluded comparison of injuries among trap types. However, suggested methods for reducing injury included shortened chains, center mounting of the chain, and use of tranquilizer tabs. Gray wolf captured in Minnesota using a custom-made No. 14 foothold trap with serrated jaws offset by 0.7 cm had fewer injuries than those caught in No. 4 double long-spring traps (with smooth jaws either not offset or offset by 0.2 cm) and another No. 14 trap with a smaller offset (Kuehn et al. 1986). Schultz et al. (1996) equipped all their wolf traps with drags and checked their sets at least once every 24 hours. They found that 15% of the wolves captured in foothold traps with modified No. 14 Newhouse jaws had moderate to severe injuries. They recommended use of the No. 4 Newhouse trap with modified jaws for capturing wolf pups. Schultz et al. (1996) noted that a pan tension system (Paws-I-Trip) was effective in reducing unwanted captures of other species. No data on the performance of killing traps for wolves are available. Frame and Meir (2007) substantiated that rubber-padded traps minimized capture related injuries to wolves.

Feral Hog

McCann et al. (2004) described various feral pig trap designs (e.g., box and corral) and trapping procedures for island and mainland ecosystems. West et al. (2009) compiled the available data on trapping methods for feral hog.

Jaguar

A safe, selective, and effective procedure for capturing jaguar using trained cat hounds was described in detail by McBride and McBride (2007). Additional orthodox capture methods for jaguar were discussed in detail by Furtado et al. (2008), including leg-hold snares and large cage traps with metal mesh over trap bars to avoid injury.

Canada Lynx

Three restraining traps and 2 killing traps have been evaluated for capturing lynx in Canada. When tested in the Yukon at temperatures ranging from −40° to 0° C, modified Fremont foot snares caused less injury than did the Victor No. 3 Soft-Catch foothold trap with padded jaws (Mowat et al. 1994). Proulx et al. (1995) reported a modified 330 Conibear trap could consistently kill lynx in ≤3 minutes. Breitenmoser (1989) developed a footsnare system to capture lynx and other medium-sized carnivores.

American Marten

The initial research to evaluate performances of killing traps for capturing marten was conducted in Canada using captive animals (Gilbert 1981a, b). Additional comparative testing revealed that standard Conibear 110 and 120 traps could not consistently kill marten in 5 minutes (Novak 1981a, Proulx et al. 1989b). Proulx et al. (1989a) reported 13 of 14 marten caught in the C120 Magnum trap equipped with a pitchfork trigger had an average time to unconsciousness of ≤68 seconds. Field tests in Alberta, Canada, indicated the C120 Magnum placed in elevated box sets was as efficient as foothold traps for harvesting marten (Barrett et al. 1989). During additional field tests in Ontario, Canada, Naylor and Novak (1994) found that wire box traps and the Conibear 120 had similar selectivity, but box traps were less efficient. Novak (1990) experimented with a variety of sets and traps and reported the most efficient and selective set for marten used a killing trap placed in a "trapper's box" on a horizontal pole. Proulx et al. (1994a) designed a snare system that successfully captured snowshoe hare, but allowed snared marten to escape. Their 0.02-gauge stainless steel wire snare was set with a 10.2-cm-diameter loop and equipped with a release device, a 12-gauge high-tensile fence wire shaped into a 5-coil spiral used as a snare anchor.

Fisher et al. (2005) further perfected and tested a snare system to curtail marten mortality and not impact snowshoe hare trapping success. They effectively used 22-gauge brass or 6 strand picture wire.

Mink

Restraining trap research on mink is lacking. Research in Canada under controlled conditions has shown that mink can be killed in terrestrial sets in ≤180 seconds using the C120 Magnum trap with a pan trigger (Proulx et al. 1990, 1993d), the Bionic trap with a 6-cm bait cone (Proulx and Barrett 1991, Proulx et al. 1993d), and the C180 trap with a pan trigger (Novak 1981a). In contrast, the standard Conibear 110 and 120 failed to consistently kill mink in 300 seconds when used on land (Gilbert 1981b, Novak 1981a). Mink died in 240 seconds when captured in drowning sets using foothold traps, but most of them "wet" drown (Gilbert and Gofton 1982). During field tests in Canada, the C120 Magnum with a pan trigger was as efficient for capturing mink as standard foothold traps and the Conibear 120 (Proulx and Barrett 1993a).

Mountain Lion

Logan et al. (1999) used modified foot snares (Schimetz-Aldrich) to trap mountain lion in New Mexico. Most captures (93.3%) resulted in minor or undetectable injuries except for swelling of the capture foot, which ranged from none to >0.2 times normal girth. Mountain lions sustained severe, life-threatening injuries in 2.4% of 209 captures; 4

mountain lions (1.9%) subsequently died. Some problems with mortality of nontarget captures, especially mule deer and oryx, also were encountered.

Muskrat

Lacki et al. (1990) compared the efficiency of 2 cage-type live traps with double doors for capturing muskrat: the Tomahawk was more effective than the Havahart trap. Killing traps for muskrat have been evaluated in Louisiana, New Jersey, and Canada (Palmisano and Dupuie 1975, Linscombe 1976, Penkala 1978, Parker 1983). Tests on anesthetized animals have measured the minimum energy forces required to cause death when delivered via a blow to the head, neck, thorax, and abdomen (Gilbert 1976, Zelin et al. 1983). Novak (1981a) reported that muskrats die in ≤4 minutes if caught in Conibear 110 traps set under water, but standard Conibear 110 and 120 traps failed to consistently kill muskrats in ≤5 minutes when used on land. However, muskrats captured in modified (18-kg springs) Conibear 110 traps set on land died in ≤200 seconds. Controlled experiments have shown that muskrats taken in drowning sets using No. 1.5 long-spring foothold traps died in ≤315 seconds (Novak 1981a), and about half had no injuries (Gilbert and Gofton 1982). Based on a field study in New Jersey using drowning sets, McConnell et al. (1985) reported the Victor No. 1 VG Stoploss with padded jaws caused significantly less damage to limbs of trapped muskrat compared to the unpadded Victor No. 1 VG Stoploss; both traps captured and held muskrat equally well in drowning sets. Conibear 110 traps (standard and modified) set at den entrances were more efficient for capturing muskrat than were a variety of No. 1 size foothold traps placed in similar locations (Penkala 1978). Parker (1983) found that Conibear 110 traps were more humane (i.e., killed a higher percentage of the muskrats caught) and selective for harvesting muskrat than were Victor No. 1 Stoploss and Victor No. 1.5 long-spring footholds.

Nutria

Four field studies, 3 in Louisiana and the other in Great Britain, have evaluated the efficiency of nutria traps. In Great Britain, cage traps set on rafts caught significantly more nutria than traps set on land as well as 50% fewer nontarget animals (Baker and Clarke 1988). Victor No. 1.5 and No. 2 long-spring foothold restraining traps proved more efficient for capturing nutria in Louisiana marshes than were either the Conibear 220 (a killing trap) or the Tomahawk 206 (a cage trap; Palmisano and Dupuie 1975, Linscombe 1976, Robicheaux and Linscombe 1978). The Conibear trap failed to kill about 10% of the nutria caught.

Nolfo and Hammond (2006) used an airboat and a long-handled fishing net to capture nutria in marsh vegetation. Meyer (2006) used a dip net baited with oats to capture nutria when sitting and facing away from the animals. Burke et al. (2008) tested 4 odor lure attractants to enhance capture of nutria with leg-hold traps. All lures increased trapping success, with nutria fur extract being the most effective. Witmer et al. (2008) perfected a multiple-capture box trap for nutria consisting of 2.5-cm PVC tubing with attached welded-mesh wire fencing on sides, top, and bottom. Traps were baited with marsh grass and various vegetable baits (e.g., sweet potatoes, feed corn, and carrots).

BMP for trapping in the United States were based on field studies that captured and evaluated 430 nutria using foothold traps in Louisiana marshes during 1998–2004 (AFWA 2007c). Two devices tested met or exceeded established BMP criteria: the No. 1 Soft-Catch (padded jaw) trap and No. 1.5 Soft-Catch (padded jaw) trap. Animal welfare was similar among traps. Efficiency was >85%, and selectivity >95% for both traps (Table 3.15).

Virginia Opossum

Restraining traps for Virginia opossum have been evaluated on a limited basis, primarily in the eastern United States. Berchielli and Tullar (1980) failed to observe any injuries in 67% of the opossum caught in standard unpadded No. 1.5 coil spring traps, but 20% had fractures. Other reports containing data on restraining trap performance for this species included Turkowski et al. (1984), Linscombe and Wright (1988), and Phillips and Gruver (1996). Hubert et al. (1999) examined injuries of opossums captured in the EGG trap, a foot-encapsulating device, and found severe injuries, such as bone fractures, were limited to animals weighing ≤1.9 kg. Warburton (1982, 1992) examined the performance of several restraining traps for capturing Australian brush-tailed opossum. Hill (1981) noted that certain killing traps appeared to be more efficient for catching Virginia opossum when placed in boxes on the ground rather than above ground level.

BMP for trapping in the United States were based on field studies that captured and evaluated 2,145 Virginia opossums using various restraining trap types. Twenty-two trap types were tested in 20 states during 1998–2001 (AFWA 2006g). BMP criteria were met for 8 of the trap types evaluated, including foothold type traps, a foot-encapsulating trap (EGG), and a wire-mesh cage trap (Tomahawk 108; Table 3.15). Of the foothold trap types that met BMP criteria, all had modifications to the jaws, including padding and/or double-jaws (Fig. 3.29), and offset and lamination. These traps included the Oneida-Victor™ No. 1.5 coil-spring with double jaws, Oneida-Victor No. 1.5 Soft-Catch (with 2 coil-springs and modified with 4 coil-springs), No. 1.5 Soft-Catch with double-jaws, No. 1.65 coil-spring with offset and laminated jaws, and the No. 1 Soft-Catch (padded jaws). Of the traps tested, the Tomahawk 108 cage trap had the lowest mean cumulative injury score (12.5) and was the most selective for opossum (51.9%). Animal welfare (ISO scale) was

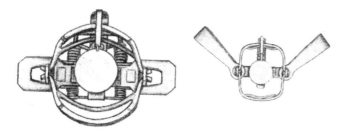
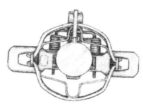

Fig. 3.29. Coil-spring and long-spring traps modified with double jaws. *Illustration courtesy of the Association of Fish and Wildlife Agencies.*

similar among all foothold traps; the EGG trap had cumulative injury scores ranging between 41.1 and 55 points. The efficiency of traps meeting BMP criteria were >87%. The Tomahawk 108 cage trap, EGG trap, No. 1 Soft-Catch, and Bridger No.1.65 offset and laminated jaw trap all had efficiency ratings of 100%.

Porcupine

Single-door cage traps baited with sliced apples and placed at the base of occupied trees have been used successfully to capture porcupine (Griesemer et al. 1999). Traps also have been used to capture porcupines by other researchers (Brander 1973, Craig and Keller 1986). However, injury and efficiency data are lacking for this species. The performance of killing traps for porcupines has not been evaluated.

Pocket Gopher

Witmer et al. (1999) described a variety of killing and cage or box restraining traps for pocket gopher (Geomyidae). They noted that >100 killing trap designs have been developed and tried over the past 140 years, but only a few types remain in common use in North America. Few cage/box restraining-type live traps are available because of a limited market; rectangular box traps of metal construction have been produced by Sherman Traps (Tallahassee, FL) and Don Sprague Sales (Woodburn, OR; Witmer et al. 1999). Sargeant (1966) and Baker and Williams (1972) described cylindrical cage/box restraining traps made of wire mesh and plastic, respectively.

Proulx (1997) evaluated the efficiency of 4 types of killing traps for gophers during the autumn in alfalfa fields. The ConVerT™ box trap was most successful, and was followed, in decreasing success, by the Black Hole™, Guardian™, and Victor Easyset™. Proulx (1999b) tested the experimental pocket-gopher killing trap and found 9 of 9 northern pocket gophers unconscious in ≤78 seconds. He also reported that pocket gophers caught in ConVerT and Sidman killing traps sometimes remained alive if captured in the lower thorax or abdominal regions. Pipas et al. (2000) evaluated the efficiency of 3 types of traps (Cinch [Chinch Trap Company, Hubbard, OR], Macabee [Z. A. Macabee Gopher Trap Company, Los Gatus, CA], and Black Hole Rodent [F. B. N. Plastics, Tulare, CA]) for capturing pocket gophers; they found the Macabee trap to be the most effective.

Raccoon

Numerous studies of restraining traps for raccoons have been conducted. Most research has focused on comparing the capture rate and injuries associated with different trap types. In some instances, injury data from these investigations are difficult to compare, because scoring systems have varied, and several studies reported only injuries to the trapped limb. However, a significant conclusion has been that most serious injuries observed are due to self-mutilation (e.g., Proulx et al. 1993c, Hubert et al. 1996).

Berchielli and Tullar (1980) reported the Blake & Lamb™ No. 1.5 coil spring trap was more efficient for capturing raccoon than the Ezyonem leg snare. They observed self-mutilation in 39% of the raccoons caught in the No. 1.5 coil spring, but were unable to compare injuries between trap types due to the small sample size for the Ezyonem ($n = 2$). However, raccoons caught in the No. 1.5 coil spring had fewer injuries when the traps were covered with sifted soil. Similarly, Novak (1981b) reported a raccoon capture rate of 57% ($n = 113$) for the Novak foot snare compared with 76% ($n = 34$) for the No. 2 coil spring and No. 4 double long-spring traps, both with offset jaws. He noted that 82% of the raccoons caught in the foot snare ($n = 49$), and 50% of those taken in the foothold traps ($n = 22$) had no injuries.

Tullar (1984) was the first researcher to report on the performance of padded foothold traps for raccoons. His data indicated injury scores failed to differ between the unpadded Victor No. 1.5 coil spring and a padded prototype No. 1.5 coil spring. However, 89% ($n = 9$) of the raccoons caught in the padded trap had injury scores ≤15 compared with 50% ($n = 14$) for the unpadded trap. Self-mutilation was observed in 24% ($n = 17$) of the raccoons caught in the unpadded trap.

Most reports published since Tullar (1984) indicate that padded traps failed to preclude self-mutilation behavior and did not significantly reduce injury scores compared to unpadded traps (Olsen et al. 1988, Hubert et al. 1991, Kern et al. 1994). However, Saunders et al. (1988) and Heydon et al. (1993) provided data contrary to this generalization. Padded traps also appeared to be less efficient than unpadded versions for capturing raccoon (Linscombe and Wright 1988, Hubert et al. 1991). Smaller foothold traps seemed to reduce injuries without sacrificing efficiency. The only restraining trap tested to date that has significantly reduced the frequency of self-mutilation and the severity of injuries to trapped raccoon compared with padded and unpadded jaw-type foothold traps is the EGG (Proulx et al. 1993c, Hubert et al. 1996). Based on a field study in Illinois, Hubert et al. (1996) reported the mean total injury score (based on a modified Olsen scale) for raccoon caught in EGG foothold traps was 68 compared to 116 for those trapped with the No. 1 coil spring trap. They reported the EGG trap had a raccoon capture efficiency exceeding that of the unpadded No. 1 coil spring. Proulx (1991) found the raccoon capture efficiency of the EGG was similar to that of cage traps in British Columbia, Canada, but it was less efficient than the Conibear 220 during the latter part of the fur trapping season in Quebec, Canada.

Cage-type restraining traps are commonly used to capture raccoon. Preliminary data contained in a progress report (IAFWA 2000) indicated that 52% ($n = 112$) of the raccoons caught in Tomahawk 108 wire cage traps sustained no injuries. Moore and Kennedy (1985) used Tomahawk and Havahart wire cage traps during a population study and found that capture success was highest in autumn and winter, increased with increasing temperatures, and was negatively correlated with precipitation. Gehrt and Fritzell (1996) reported a gender biased response of raccoons when using Tomahawk cage traps in Texas. Adult males were consistently captured more frequently than were adult females.

Controlled lab tests have been conducted on anesthetized raccoons to measure the minimum energy forces a killing trap must deliver to cause death via a blow to the head and neck (Gilbert 1976, Zelin et al. 1983). Limited data about the effects of clamping force also have been obtained (Zelin et al. 1983). Other research on killing traps conducted in enclosures indicated that raccoon cannot be consistently killed in 5 minutes using standard Conibear 220, 280 (with pan trigger), and 330 traps (Novak 1981a). However, about 60% of the raccoons captured in the Conibear 220 and 280 traps died in 4 minutes. Proulx and Drescher (1994) reported the Savageau 2001-8 and a modified (extra clamping bar) Conibear 280 have the potential to consistently immobilize raccoons and render them irreversibly unconscious in ≤4 minutes, but not in ≤3 minutes. In a separate lab study, the average time to unconsciousness for 4 of 5 immobilized raccoons caught in the BMI 160 (a rotating-jaw trap similar to the Conibear) was 172 ± 16 seconds; the remaining animal was euthanized after 5 minutes (Sabean and Mills 1994). Proulx (1999a) recommended future research should focus on killing systems for raccoon that differ from the rotating-jaw trap type.

The raccoon capture efficiency of the Conibear 220 may be comparable to or better than some restraining traps under certain environmental conditions, but in other instances, it may not (Proulx 1991). Linscombe (1976) reported the Victor No. 2 long spring trap was more efficient than the Conibear 200 for capturing raccoons in brackish marshes. In contrast, Hill (1981) caught a similar number of raccoons per trap night with No. 2 coil spring traps placed in dirt-hole sets and with Conibear 220 traps in boxes placed on the ground.

Kerr et al. (2000) improved trapping success for raccoon by modifying Tomahawk cage traps. They added an extended metal floor that acted as a trip device and wrapped hardware cloth around the back of the trap to reduce missing baits. They also added an elevated bait hook to curtail fire ants. Austin et al. (2004) evaluated EGG and wire cage traps for capturing raccoon. They found that EGG traps (Fig. 3.23) were more effective, especially for capturing males.

Research conducted in support of BMP for trapping in the United States found that No. 1.5 coil-spring foothold traps modified with double jaws reduced self-mutilation and improved animal welfare. Various double-jaw configurations (Fig. 3.29) were tested, and all reduced self-mutilation compared to standard jaw traps. Self-mutilation was reduced to 10% ($n = 128$) when the No. 1.5 coil-spring trap was modified with double jaws compared to a self-mutilation rate of 37.9% ($n = 206$) reported for the No. 1.5 coil-spring trap with standard jaws. Similarly, the No. 11 double long-spring trap modified with double jaws reduced self-mutilation compared to the standard jaw No. 11 ($n = 135$; self-mutilation rate = 27.4%), but only when modified with an offset in the jaws ($n = 35$; self-mutilation rate ≤10%). The efficiency of traps modified with double jaws was similar to that of standard jaw traps.

BMP for trapping in the United States were based on field studies that evaluated 382 raccoons captured in foot encapsulating traps (AFWA 2006h). Three models of foot encapsulating traps were tested during 1998–2004, including the EGG, Duffer's and Lil' Grizz Get'rz (Table 3.15; Fig. 3.23). The foot encapsulating traps passed all BMP criteria. Injury scores ranged from 37.5 to 48.4. Self-mutilation was minimal (2%) due to trap design, which prevents captured animals from accessing the encapsulated foot. Efficiency was higher for these traps types compared to coil-spring and long-spring foothold traps commonly used to capture raccoon. Cage-type restraining traps are frequently used to capture raccoon (AFWA 2005).

Northern River Otter

A variety of restraining traps for the live capture of river otter has been evaluated in Canada and the United States. Capture success with Hancock traps has varied, depending on the season and setting techniques (Northcott and Slade 1976, Melquist and Hornocker 1979, Route and Peterson 1988). In Newfoundland, Canada, Bailey traps proved ineffective (Northcott and Slade 1976). Shirley et al. (1983) reported that a modified Victor No. 11 double long-spring trap was a practical and efficient live trap for otters in Louisiana marsh habitat, but they failed to catch any otters in Tomahawk 208 cage traps. Serfass et al. (1996) compared unpadded Victor No. 11 double long-spring modified (heavier spring added) traps with Victor No. 1.5 Soft-Catch traps with padded jaws for catching otter for relocation. Fewer severe injuries were noted in animals captured with the Soft-Catch trap, but there was no difference in frequency or severity of dental injuries between trap types. More recently, Blundell et al. (1999) compared Hancock and No. 11 Sleepy Creek™ double-jaw foothold traps with long springs for live-capture of northern river otter using blind sets at latrines. They found Hancock traps had slightly lower efficiency, higher escape rate, lower rate of malfunction, and much lower use than the No. 11 Sleepy Creek foothold trap. Otters captured in Hancock traps had significantly more serious injuries to their teeth than animals captured in foothold traps. Although more serious injuries to appendages were observed for animals caught in foothold traps compared with Hancock traps, the difference was not significant. No published research on killing traps for river otter is available.

BMP for trapping in the United States were based on field studies that captured and evaluated 70 river otters using foothold traps. Studies were conducted in 4 states during 2005–2007 (AFWA 2007d). Three foothold traps were tested: No. 2 coil-spring, No. 11 double long-spring, and No. 11 double-jaw double long-spring. All 3 traps met or exceeded established BMP criteria (Table 3.15). The No. 2 coil-spring trap is the most commonly used trap for capturing river otter for fur harvest (AFWA 2005). This trap produced an average cumulative injury score of 45.3, with 81.4% of injuries ranking in the 3 lowest trauma classes (none, mild, and moderate). The efficiency for this trap was 69.9%, and the selectivity for river otter was 25.5%. No published research on killing traps for river otter is available.

Gray and Fox Squirrels

Huggins (1999) presented a detailed review of trapping techniques and equipment for gray and fox squirrels. Based on limited comparative research, cage traps and jaw-type foothold traps were relatively nonselective; rotating-jaw and tunnel-type killing traps were relatively selective for these species. Research needs included welfare and effectiveness testing of killing traps and additional comparative studies of trap types.

Red Squirrel

The Kania 1000, a mouse-type killing trap with a striking bar powered by a coil spring, can reliably cause unconsciousness in red squirrel in ≤90 seconds (Proulx et al. 1993b). When set under conifer branches, it is unlikely the Kania would attract and capture birds (Currie and Robertson 1992). Preliminary field tests showed this trap had the potential to capture red squirrel during the regular harvest season (G. Proulx, Alpha Wildlife Research & Management, unpublished data).

Striped Skunk

The restraining trap research conducted on striped skunk indicated leg injuries of animals caught in unpadded and padded foothold traps were often severe due to the high incidence of self-mutilation (Berchielli and Tullar 1980, Novak 1981b). Novak (1981b) reported that skunk can be captured with few injuries in the Novak foot snare, but this device has a low capture rate and an unacceptable level of efficiency. Numerous pan tension devices have been used on a variety of coyote traps; all have been effective in reducing accidental skunk captures (Turkowski et al. 1984, Phillips and Gruver 1996). The performance of killing traps on striped skunk has not been evaluated.

BMP for trapping in the United States were based on field studies that captured and evaluated 51 striped skunks using cage traps during 2007–2009 (AFWA 2009a). Two models of Tomahawk wire cage traps were tested (models 105.5 and 108), and both met or exceeded established BMP criteria (Table 3.15). These traps were highly effective (capture rate of 100%), and no trap related injuries were reported. Selectivity of traps were 53.8% (model 108) and 67.6% (model 105.5).

Long-Tailed and Short-Tailed Weasels

Research information on traps commonly used for harvesting weasels in North America is not available. During a field study in New Zealand, King (1981) concluded that correctly set Fenn traps killed weasels more humanely than did Gin traps. Typically, North American trapping technique manuals recommend the use of small foothold or rotating-jaw traps as killing traps for these animals.

Belant (1992) tested the efficiency of double-door Havahart, single-door National™, and single-door wooden cage/box traps for capturing long-tailed and short-tailed weasels in New York. Overall success for all 3 types was similar. Trap-related injuries of long-tailed weasel caught in Havahart traps included skin abrasions and broken canines.

Wolverine

Copeland et al. (1995) used a specialized log trap to live-capture wolverine in Idaho. No injuries were noted on individuals captured, but 3 wolverines escaped by chewing holes in the traps. No data are available on the performance of killing traps for wolverine. Copeland et al. (1995) and Lofroth et al. (2008) described and evaluated live-capture techniques for wolverine.

CAPTURING AMPHIBIANS AND REPTILES

Amphibians
Hand Captures

Corn and Bury (1990) described **time-constrained searches** for amphibians and reptiles that were immediately captured by hand. Equal effort was expended in each area searched.

They described another hand collection method for amphibians (surveys of coarse woody debris) and advised searching 30 downed logs per forest stand. Barr and Babbitt (2001) compared 2 techniques for sampling larval stream salamanders. More larvae were captured at high densities using 0.5-m^2 quadrats. Time-constrained sampling for 0.5 hours was more successful at low densities. Pearman et al. (1995) evaluated day and night transects, artificial cover, and plastic washbasins with added leaf litter as sampling methods for amphibians. Significantly more species were found during **nocturnal searches** than with other methods. Parris et al. (1999) compared 3 techniques for sampling amphibians in forests. Nocturnal **stream searches** were the most sensitive and pitfall trapping the least sensitive sampling technique. A minimum of 4 nights of stream searching was recommended to determine the number of amphibian species present at a site. Haan and Desmond (2005) concluded that area-constrained searches for salamanders were superior to pitfall traps, especially during dry periods. Mattfeldt and Campbell-Grant (2007) recommended using both area-contained transects and **leaf litter bags** for improved sampling of stream salamanders.

Dip Nets

Wilson and Maret (2002) reported that **timed dip-net collections** of 5 minutes provided reliable estimates of aquatic amphibian abundance and were superior to **drop box sampling.** Welsh and Lind (2002) sampled amphibians by searching streambed substrates with hardware-cloth catch nets placed downstream and from bank to bank to capture escaping individuals.

Drift Fences with Pitfall and Funnel Traps

Campbell and Christman (1982) developed and described a **standardized** amphibian trapping system. Their system included pitfalls and double-ended funnel traps placed in conjunction with drift fences that diverted moving animals into traps. Data obtained using their technique allowed estimates of species richness and an index of relative abundance of most common terrestrial amphibians and reptiles. Dodd (1991) warned that drift fences used with pitfalls were **biased** in sampling amphibians. Frogs, in particular, readily cross drift fences by climbing over them. Other species burrow under drift fences. Brown (1997) also found that drift fences allowed frogs to escape. She tested pitfall traps and reported that 1% of the individuals placed in pitfall traps escaped.

Scott (1982), Heyer et al. (1994), Olson et al. (1997), and Simmons (2002) have compiled comprehensive capture references for amphibians. Adams and Freedman (1999) evaluated catch **efficiency** of 4 amphibian-sampling methods: pitfall transects, pitfall arrays, quadrat searches, and time-constrained searches in terrestrial habitats. Pitfall arrays sampled the greatest relative abundance and species richness of amphibians. Nadorozny and Barr (1997) designed a **side-flap pail** to capture amphibians that were not readily captured in conventional pitfall traps due to their climbing and jumping ability. This trap design, when used with funnels and drift fencing, was effective for capturing amphibians in terrestrial habitats. Crawford and Kurta (2000) tested capture success of black and white plastic pitfall traps on anurans and masked shrew. Both were caught significantly more often in pitfalls with a black interior than in those with a white one. Adding rims to pitfall traps increased effectiveness by hindering the escape of certain species of salamanders and frogs (Mazerolle 2003). Stevens and Paszkowski (2005) tested 2 pitfall trap designs for sampling boreal anurans. They found that plastic buckets with a polyethylene funnel design were easier to construct and allowed fewer escapes.

Murphy (1993) captured tree frogs with a **modified drift fence** (Fig.3.30) of clear plastic suspended from PVC pipe joined in a T-shaped configuration. Daoust (1991) suggested placing moistened sponges (10 cm × 5 cm × 7 cm) in funnel traps along drift fences to minimize mortality of wood frog from dehydration. Willson (2004) compared **aquatic drift fences** with traditional funnel trapping as a quantitative method for sampling amphibians. Mushet et al. (1997) connected a 200-cm drift fence that directed free-swimming salamanders to the opening of funnel traps. Malone and Laurenco (2004) suggested the use of polystyrene for drift fence

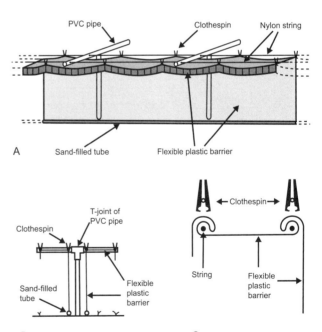

Fig. 3.30. Drift fence for capturing tree frogs as they enter and leave ponds. (A) Front view of the fence. Only a portion of the fence and only one of the plastic barriers are shown. (B) Side view of the fence showing both plastic barriers. (C) Enlarged side view of the fence showing method of attachment of flexible plastic barrier to strings. *From Murphy (1993).*

sampling, because it was economical and easily repaired compared to aluminum or silt fence (silt fence is a woven polypropylene material used to control sediment runoff at construction sites). Rice et al. (2006) combined collapsible minnow traps with PVC pipes attached to a portable drift fence structure to capture various frogs and toads.

Smith and Rettig (1996) sampled amphibian larvae with an **aquatic funnel trap** made of 5-cm-diameter PVC pipe with funnels at each end held in place with a large rubber band. Fronzuto and Verrell (2000) tested the capture efficiency of wire and plastic funnel traps for aquatic salamanders. Plastic funnel traps with a maximum diagonal mesh of 5 mm were superior to 10-mm mesh hardware-cloth wire minnow traps. Mushet et al. (1997) designed a funnel trap for sampling salamanders in wetlands. Casazza et al. (2000) captured aquatic amphibians and reptiles using baited wire-funnel–entrance eel pots with Styrofoam blocks. The blocks allowed the traps to float partly out of the water, avoiding trap mortality from drowning. Richter (1995) used baited aquatic funnel traps made from plastic soda pop bottles attached to a steel rod baited with salmon (Salmonidae) eggs. He captured tadpoles and adult amphibians. Smith and Rettig (1996) increased the catch rate of tadpoles by putting **glow sticks** at night in 3 different funnel trap designs. Jenkins et al. (2002) compared 2 aquatic surveying techniques to sample marbled salamander larvae. Nocturnal visual surveys were less intrusive, less expensive, and more accurate at detecting presence than were the bottle funnel traps described by Richter (1995).

Parris (1999) summarized the **advantages and disadvantages** of various techniques for sampling amphibians in forests and woodlands. Lauck (2004) discussed factors influencing the capture of amphibian larvae in aquatic funnel traps. Willson and Dorcas (2004) verified that funnel traps combined with an aquatic drift fence increased amphibian capture rates. O'Donnell et al. (2007) compared the efficiency of funnel and drift fence trapping, and light touch and destructive sampling of frogs and salamanders in forested seep habitats. Light touch sampling was the most suitable method. Palis et al. (2007) evaluated 2 types of commercially made aquatic funnel traps for capturing ranid frogs and found that both had similar capture rates. They determined that nylon traps were less durable than steel mesh traps. Buech and Egeland (2002) tested 3 types of funnel traps in seasonal forest ponds. Traps with 6-mm mesh captured more wood frog tadpoles than did plastic traps. Traps with 3-mm mesh captured more blue-spotted salamander and spring peepers. Jenkins and McGarigal (2003) tested the catchability of reptiles and amphibians along drift fences using paired funnel and pitfall traps in the northeastern United States. Their results showed funnel traps to be superior to pitfalls in wet or rocky areas. Ghioca and Smith (2007) cautioned against using funnel traps to avoid **biased estimates** of the abundance of larval amphibians. Glow sticks in funnel traps significantly increased capture rates of aquatic amphibians (Grayson and Roe 2007). Willson and Dorcas (2003) found funnel trapping superior to dip-netting for quantitative sampling of stream salamanders.

Pipes

Boughton and Staiger (2000) caught hylid tree frogs in white 3.81-cm-diameter PVC pipe capped at the bottom and hung vertically in hardwood trees, 2 m and 4 m above the ground. The 60-cm-long pipe caught more frogs than did the 30-cm pipe. Moulton (1996) used PVC pipes to capture hylid tree frogs. Bartareau (2004) found that PVC pipes with varied diameters influenced the species and sizes of tree frogs captured in a Florida coastal oak-scrub community. Myers et al. (2007) tallied more captures (81%) of Pacific tree frogs in tree-based than in ground-based pipe refugia. Johnson (2005) designed a novel arboreal pipe trap to capture gray tree frogs using black plastic acylonitrile-butadiene-styrene (ABS) pipe that allowed a constant water depth. Zacharow et al. (2003) sampled 2 species of hylid tree frogs using ground-placed PVC pipes of 3 diameters and identified potential trap biases. The addition of escape ropes to PVC tree pipes used by tree frogs prevented flying squirrel mortality (Borg et al. 2004).

Cover Boards

Trapping methods for herpetofauna are time and labor intensive, and they can result in injury to captured individuals due to physical stress, such as overheating, desiccation, drowning, or predation. **Cover boards** ("boards" placed on the ground under which herpetofauna may hide) avoid these problems. Grant et al. (1992) evaluated cover boards in detail. They recommended that both metal and wood cover boards be used and a wait of at least 2 months after placement before beginning the survey program. They suggested that checks of cover boards be made at different times of day and weather conditions to sample all taxa in residence. They advised that if encounter rates are to be compared among sites, time and weather conditions should be identical.

DeGraaf and Yamasaki (1992) used cover boards to simulate fallen timber to attract and evaluate terrestrial salamander abundance during daylight hours. Their procedure avoided laborious installation of pit traps, as they placed a cluster of 3 boards along transects. They lifted boards 8 times during June–August in a variety of different-aged forest stands. Use of the boards avoided degradation of salamander habitat by turning or breaking existing logs or disrupting forest litter. Hyde and Simons (2001) investigated 4 common sampling techniques to examine variability of salamander catches. They found natural cover transects and artificial cover boards to be the most effective sampling techniques for detecting long-term salamander population trends because of lower sampling variability, good capture success, and ease of use. They associated higher capture rates and lower variability with fewer, but larger plots. **An evaluation** of cover boards for sampling terrestrial salamanders by Houze and Chandler (2002) found that most species were sampled in lower numbers (0.8 salamanders/grid search) than under natural cover (2.3 salamanders/grid search). Temperatures were more variable under cover boards than under natural cover. Carlson and Szuch (2007) found no difference in the use of old and nonweathered cover boards by salamanders. Moore (2005) encountered more red-backed salamanders under native dominant-wood cover boards than under artificial wood cover boards. Luhring and Young (2006) combined a halved PVC pipe with screens at each end attached to a cover board to sample stream-inhabiting salamanders.

Unique Methods

Williams et al. (1981a) used **electroshocking** methods in the Allegheny River, Pennsylvania, to capture hellbender and reported that it was superior to search and seizure, potato rake, and seine herding as a capture method. Soule and Lindberg (1994) used a **peavey** to move large rocks to locate and catch hellbender. The peavey was hooked to the bottom of the rock, which was then manually moved. This technique required a 3-person crew to move rocks and capture the animals. The peavey was much less expensive than electroshocking equipment. Nickerson and Krysko (2003) reviewed a wide array of techniques and their variants used in studying a cryptobranchid salamander and discussed their **advantages and disadvantages.** Electroshocking surveys were strongly discouraged because of the great potential for damaging reproductive success and immune systems, and because they were of questionable effectiveness. Because successful hellbender nesting sites appear to be quite limited, the use of Peavy hooks and crowbars to breakup bedrock or dislodge large cover rocks should be restricted. Currently, **skin-diving** surveys coupled with turning objects is the only method shown to obtain all sizes of gilled larvae and multiple age groups of nongilled and adult hellbenders in brief periods. Foster et al. (2008) compared 3 capture methods for eastern hellbender and found that **rock turning** was most efficient in terms of catch per unit effort. Camp and Lovell (1989) caught blackbelly salamander using a **fishing pole** made from metal coat hangers with barbless hooks baited with earthworms.

Reptiles

To quantify reptile densities, Corn and Bury (1990) used **time-constrained searches** for reptiles that were immediately captured by hand. Equal effort was expended in each area searched. This allowed the calculation of relative densities for each area searched.

Drift Fences with Pitfall and Funnel Traps

Hobbs et al. (1994) tested a variety of pitfall trap designs. A straight line of pit traps with buckets approximately 7 m apart was most effective for sampling reptiles in arid Australia. The use of shade covers reduced heat related mortality. Hobbs and James (1999) reported that foil covers placed inside and at the bottom of buckets reduced pitfall temperature and had minimal influence on trap success. Foil covers were superior to cardboard and plastic. Aboveground covers reduced capture success for mammals, but increased snake captures.

Vogt and Hine (1982) advocated the use of drift fences combined with traps as a practical way to uniformly census reptiles and amphibians. **Aluminum drift fences** (50-cm high) caught more animals per 15 m of fence than did those made of either screening or galvanized metal. A system of 18.9-L traps, 7.6-L traps with funnel rims, and funnel traps was necessary to capture the entire spectrum of amphibians and reptiles in the communities sampled. Funnel traps were more effective for catching lizards than were pit traps, and they also were effective for catching snakes. They recommended at least 4 trapping periods of 3–5 days during April–mid-June.

Moseby and Read (2001) recommended 5 nights of pitfall trapping as the most efficient duration for capturing reptiles. Greenberg et al. (1994) compared sampling effectiveness of pitfalls and single- and double-ended funnel traps used with drift fences. All 3 trap types yielded similar estimates of lizards and frogs, but not snakes. Estimates of relative abundance of large snakes were higher in double-ended funnel traps than in pitfalls or single-ended funnel traps. Captures of snakes were restricted to funnel traps. More surface-active lizards and frogs were captured in pitfalls. They advised that choice of trap type(s) depended on target species and sampling goals. Enge (2001) presented a detailed assessment of the effectiveness of pitfall versus funnel traps. He concluded that salamanders, anurans, lizards, and snakes were captured significantly more often in funnel traps than in pitfall traps. He added that studies that found funnel traps to be less effective than pitfall traps used smaller or poorly constructed or installed funnels. He also reported herpetofaunal mortality rates were generally higher in funnel traps than in pitfall traps. Enge (2001) recommended that traps be checked at least every 3 days to minimize mortality.

Fair and Henke (1997) evaluated the efficiency of capture methods for a low density population of Texas horned lizard. **Road cruising** yielded the highest capture rates, with systematic searches second. Searching resulted in a higher rate of capture than did using pitfall and funnel traps. Sutton et al. (1999) compared pitfalls and drift fences with cover boards for sampling sand skink. They reported that **cover boards** were most efficient in detecting the presence of skinks and were less costly and labor intensive. Allan et al. (2000) developed a successful **habitat trap.** The trap consisted of an artificial replica of a preferred habitat placed on a large sheet of camouflaged plastic. Two people lifted the plastic sheet at all edges once lizards had begun to occupy the artificial habitat, and the animals were trapped.

The artificial habitat consisted of a rock pile or woodpile placed in an excavated shallow pit 15 cm deep covering an area of 1 m².

Doan (1997) captured large lizards by using large (88.5 cm × 31.0 cm × 31.0 cm), collapsible aluminum **Sherman live traps.** Traps were camouflaged with green mosquito netting and fallen branches and leaves. Zani and Vitt (1995) attached a wire-mesh minnow trap over holes in trees, whereas Paterson (1998) used a mesh barrier of bridal veil fabric wrapped around a tree trunk to facilitate hand capture of arboreal lizards.

Gluesenkamp (1995) designed a simple **snake rake** consisting of 120-cm-long, 19-mm-diameter aluminum pipe and 2 pieces of 25-cm-long, 6.5-mm-diameter steel. The 2 pieces were bent 90°, welded together at a 25° angle, and then attached with hose clamps to the end of the aluminum pipe.

Lannom (1962) dangled a barbless dry fly from a support over a buried 1-L glass jar to attract and catch desert lizards. Whitaker (1967) increased his rate of capture of small lizards in pitfall traps by using canned fruit as bait. He also suggested using captive lizards in pitfall traps to attract other curious lizards. Serena (1980) used a **fishing pole** with a line attached to edible palm fruit to attract and capture whiptail lizards. Durden et al. (1995) caught skinks by using crickets (family Gryllidae) threaded onto fishing line attached to a fishing rod. They also baited little Sherman small-mammal traps with crickets tied inside the trap. Small smooth-scaled lizards were captured by Durtsche (1996) using a combination of a pole (fishing pole or collapsible car antenna) with a piece of **sticky pad** fastened to the end. The sticky pad was touched to the back of the lizard, allowing capture. Bauer and Sadleir (1992) used mouse **glue traps** to capture lizards. Corn oil was used to release the animals. Whiting (1998) increased lizard capture success by baiting glue traps with insects and figs. Downes and Borges (1998) captured small lizards with commercial packing tape by creating sticky traps. However, Vargas et al. (2000) cautioned that sticky-trapping of lizards had a higher fatality rate than did capture with a noose or rubber band; sticky-trapping also yielded less reliable gender-biased capture information.

Witz (1996) coated the prongs of a bolt retriever (total length 60 cm) with liquid plastic. This **lizard grabber** grabs the pelvic girdle firmly with minimal chance of escape or injury to the lizard. Strong et al. (1993) caught small fast-moving lizards by chasing them into PVC pipes covered at one end (Fig. 3.31). Brattstrom (1996) used a plastic wastebasket or garbage can as a "skink scooper." When he located a skink, he held the plastic container 15–30 cm away and swept the leaf letter and the skink into the scooper for capture. Sievert et al. (1999) made a "herp scoop" (Fig. 3.32) of pliable plastic for safely capturing herpetofauna from roads at night. They used a flashlight combined with a 1–3-liter clear soft-drink bottle with the bottom removed and a V-shaped

Fig. 3.31. Method for catching lizards by chasing them into tubes placed near a bush. The tubes have one end covered with tape. *From Strong et al. (1993).*

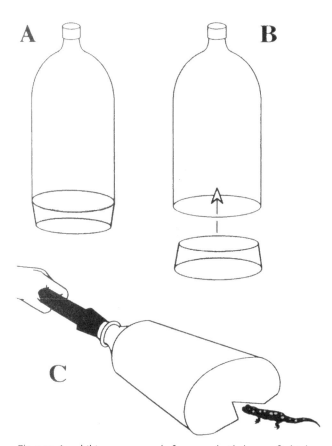

Fig. 3.32. Amphibian scoop made from a polyethylene soft-drink bottle (A) with the base cut off and inverted to act as a lid (B). A V-shaped notch and a flashlight (C) were added to make the scoop more useful. *From Sievert et al. (1999).*

notch cut 3–5 cm wide and 2 cm deep into the bottom lip of the bottle.

Recht (1981) modified a rat trap to block the entrance of burrows of desert and Bolson tortoises to facilitate **hand-capture** as they attempted to re-enter their burrows. Bryan et al. (1991) designed a trap with a spring-loaded arm released by a trigger mechanism activated by a gopher tortoise as it exited its burrow. A net was attached to the trigger to restrain the tortoise.

Graham and Georges (1996) modified collapsible turtle funnel traps by adding PVC pipe as struts to keep the funnels open and in place. They also used a piece of foam as a buoy to expedite trap retrieval. Mansfield et al. (1998) had success capturing spotted turtle in funnel traps by using turtle-shaped decoys of cement poured in plaster-of-Paris casts. Decoys were painted to resemble turtle markings and color. Christiansen and Vandewalle (2000) perfected pitfall traps with wooden flip-top lids along drift fences that were effective in capturing terrestrial turtles (Fig. 3.33). Their traps were more effective in capturing adult terrestrial turtles than were wire box traps or open pitfalls. Feuer (1980) modified the chicken-wire turtle trap described by Iverson (1979) by using oval galvanized hoops with nylon netting. He attached lines to hold the throats of hoop nets in place.

Braid (1974) used a **bal chatri trap** with snares similar in design to a bird trap to capture basking turtles. Unlike bal chatri traps used to catch birds, bait was not necessary. Nooses should be kept upright, and the chicken wire base should be tied to a log. Vogt (1980) used **fyke and trammel nets** to catch aquatic turtles.

Fitch (1992) found that artificial shelters were superior to live traps and random encounters for capturing snakes during a 12-year study. Kjoss and Litvaitis (2001) used black plastic sheets to capture snakes. Their cover sheet method was cheap, limited injuries, required less frequent checks, and was effective in open-canopy habitats. Lutterschmidt and Schaefer (1996) used mist netting with enclosed bait to capture semi-aquatic snakes.

Fritts et al. (1989) successfully captured brown tree snake using **bird odors.** Their funnel traps were baited with chicken and quail manure. Shivik and Clark (1997) found that brown tree snake were attracted to carrion and entered traps baited with dead mice as readily as traps baited with live mice. Engeman (1998) devised a simple method for capturing brown tree snake in trees. He used a branch or stick with a fork at one end that was placed in the middle of the snake, and the stick was then twirled to wind the snake on the stick. The snake would coil around the stick, allowing time to retrieve the stick and snake from the tree for hand capture. Lindberg et al. (2000) tested a variety of **lures** for capturing brown tree snake. They found that visual lures lacking movement were ineffective. Lures combining movement and prey odors were most effective (Shivik 1998). Engeman and Linnell (1998) used modified crawfish traps

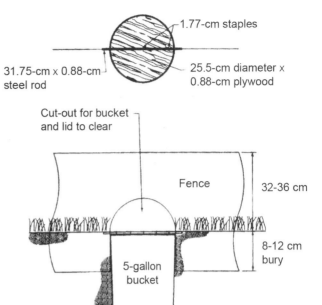

Fig. 3.33. Specifications of flip-top lid on 19-L (5 gallon) bucket set in a drift fences. *From Christiansen and Vandewalle (2000).*

of 10-mm wire mesh with one-way flaps installed at the entrance and baited with a live mouse to capture brown tree snake. Engeman et al. (1999) recommended placing a horizontal bar at the top of chain link fences to facilitate capture of brown tree snake. Captures of these snakes by trapping exceeded those using spotlight searches of fences (Engeman and Vice 2001).

Lizards

Goodman and Peterson (2005) perfected a **pitfall style** trap for lizards consisting of a bucket and a tray of live food (e.g., adult crickets with their hind legs removed or *Tenebrio* larvae). This method was especially effective in rocky habitats. Ferguson and Forstner (2006) perfected a durable and effective predator-exclusion device attached to pitfall traps along a drift fence. An effective, inexpensive tube-trap made of transparent plastic with a one-way door was designed by Khabibullin and Radygina (2005) to sample small terrestrial lizards. Cole (2004) employed a class 1 **laser pointer** to capture arboreal geckos (family Gekkonidae). The geckos chased the laser dot. Estrada-Rodriguez et al. (2004) effectively used a new method, a **water squirting** technique, to hand-capture desert lizards in sand dunes. Horn and Hanula (2006) attached burlap bands on tree trunks to attract and capture various lizards. Lettink (2007) used a double-layered **artificial retreat** made of Onduline™, a lightweight corrugated roofing, in rocky habitat for capturing geckos.

Bennett et al. (2001) described a **noose trap** attached to the side of a tree along with a trigger stick for catching large lizards. Bertram and Cogger (1971) described a noose gun for live lizard captures. The noose gun was made of copper-coated welding wire and used rubber bands to tension the noose and trigger.

Rodda et al. (2005) compared **glueboard** lizard-capture rates with total removal plots on various oceanic islands. Results varied by species, speed, mode of locomotion, and habitat. They concluded that glueboard capture frequencies of arboreal species were less reliable than for terrestrial species. Ribeiro et al. (2006) also indicated that glueboard trapping of lizards provided a useful addition to other sampling methods of neotropical forest lizards. Glor et al. (2000) suggested placing glue traps in shaded areas to avoid heat related mortality in the mainland tropics. Whiting (1998) increased lizard capture success by adding ripe figs and/or live, moving insects as bait to glue traps.

Turtles

Browne and Hecnar (2005) found that capture success for northern map turtle with floating **basking traps** to be superior to baited hoop traps. McKenna (2001) and Gamble (2006; Fig. 3.34) described similar capture results for painted turtles. Robinson and Murphy (1975) perfected a successful net trap for basking softshell turtles. Petokas and Alexander (1979) designed an effective trap for basking turtles made of wood planking and aluminum flashing as a basking platform in a sloping configuration with a chicken-wire bottom and urethane foam. Fratto et al. (2008) evaluated 5 modified hoop net designs. They found that a chimney design was most effective in curtailing turtle bycatch mortality while not reducing catfish catch rates. Barko et al. (2004) found a high mortality of drowned turtles in fyke nets set to capture fish inside the channels of large rivers. They recommended that nets be set several inches above water to avoid turtle mortality. Glorioso and Niemiller (2006) attached a large cork to inexpensive floating, baited, and deep-water crayfish trap nets to successfully catch turtles of various sizes. Sharath and Hegd (2003) designed 2 new traps for sampling black pond turtle. One was a baited **floating pitfall trap;** the other was a baited see-saw board trap. Both were more efficient than a conventional pitfall trap. Fidenci (2005) evaluated the capture efficiency of various traditional turtle-capture methods (e.g., by hand, and using basking and funnel traps) and found his baited wire method to be more effective.

Thomas et al. (2008) tested 3 different baits in **funnel traps** for capturing pond-dwelling turtles. Both canned fish and frozen fish captured more turtles than did canned creamed corn. Kuchling (2003) described a collapsible baited turtle-trap tied to a tree branch that functions in shallow and changing water levels. Kennett (1992) developed a baited **hoop trap** composed of 2 sections, an entry section with funnel entrance to reach the bait, and a holding section from which turtles cannot escape. Plastic floats were placed inside the traps to keep them afloat, thereby allowing trapped

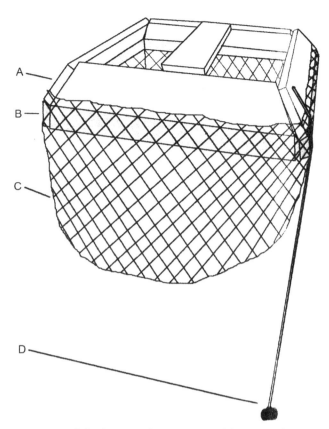

Fig. 3.34. Turtle basking-trap design. A = wood frame, B = foam floats, C = net basket, D = anchor. *From Gamble (2006).*

turtles to breathe. Borden and Langford (2008) caught nesting diamondback terrapin in pitfall traps with self-righting lids attached to drift fences.

Snakes

Dickert (2005) used modified eel **pot traps** with attached Styrofoam floats to capture giant garter snakes. Row and Blouin-Demers (2006a) surrounded snake hibernacula with a perimeter fence and funnel traps for successful snake capture. Mao et al. (2003) designed a new PVC funnel trap with an inverted-T shape and 2 entrances to capture semi-aquatic snakes. Use of live mice in snake traps after rodent suppression enhanced brown tree snake capture rates (Gragg et al. 2007). Keck (1994a) and Winne (2005) both increased aquatic-snake capture success using baited **funnel traps.** Willson et al. (2005) tested escape rates of aquatic snakes and salamanders from various commercially available minnow funnel traps. Plastic and steel minnow traps had the highest retention rates. They recommended plastic traps for sampling small snake species and steel traps for larger species of watersnakes. Camper (2005) warned about potential mortality problems while sampling semi-aquatic snakes in funnel traps due to imported fire ants. Burgdorf et al. (2005) perfected a successful trap design for capturing large terrestrial snakes that consisted of a 4-entrance funnel trap used with perpendicular drift fences and having hinged doors on top to facilitate retrieval of trapped snakes. They suggested frequent trap visits, ant control, and trap placement in shaded areas to curtail snake mortality.

Alligators

Franklin and Hartdegen (1997) sprayed large reptiles in the face with a fine **mist of water** to safely capture American crocodile, American alligator, pythons, and iguanas. Elsey and Trosclair (2004) and Ryberg and Cathey (2004) used baited **box traps** effectively to capture alligators. Chabreck (1965) captured alligators using an airboat at night with a spotlight and a wire **snare** mounted on a stout pole.

Miscellaneous Capture Methods

Lohoefener and Wolfe (1984) designed a **pipe trap** consisting of aluminum window screening, black PVC pipe, and 3 wooden disks. Pipe traps were used with drift fences and were more efficient for capturing salamanders, lizards, and snakes than were pitfall traps. Frogs and toads were more likely to be captured in pitfall traps. A **wire hook** with a blunt end was placed around the tails of lizards by Bedford et al. (1995) to extract the animals from tree and rock crevices. They grasped the lizard by its head with forceps as it emerged from the crevice. Bending the wire at a 90° angle made a handle, and a flashlight was used to help position the wire hook. Enge (1997) recommended silt fencing over aluminum or galvanized drift fencing as inexpensive, easy to install, and durable.

HANDLING CAPTURED ANIMALS

Clark et al. (1992) and Fowler (1995) are excellent sources of information on the restraint and handling of wild animals. Nonchemical handling and physical restraint of captured animals is inexpensive and usually causes lower mortality rates than does retraint involving chemicals (Peterson et al. 2003b).

Birds

Cox and Afton (1998) advised that holding times of **waterfowl** be minimized when large numbers are captured with rocket nets. To minimize subsequent mortality, ducks should be released immediately after they are processed and their plumage is dry. Maechtle (1998) described the Aba (cloak) made from rectangular cotton cloth for restraining **raptors** and other large birds. Wing pockets were stitched, and a strip of elastic tape was sewn onto the back of the cloth to be wrapped around the bird's tarsi. The Aba allows measurements and blood samples to be taken with a minimum of handling. Blood sampling of birds from the brachial and jugular veins did not influence survival, movement, or reproduction (Colwell et al. 1988, Gratto-Trevor et al. 1991, Lanctot 1994). Lecomte et al. (2006) described a successful method of blood sampling of waterfowl embryos.

A 4-pronged pick-up tool was used by Richardson et al. (1998) to remove **red-cockaded woodpecker** nestlings >8 days old from tree cavities. The 4 prongs must be blunted by bending or covered with liquid rubber to avoid injury to the young woodpeckers. Hess et al. (2001) questioned the feasibility of the Richardson et al. (1998) technique because of a high injury rate to red-cockaded woodpecker nestlings.

Cardoza et al. (1995) suggested delaying attempts to capture **wild turkeys** that appear to be wet on arrival at a bait site if a soaking rain had recently occurred. If turkeys become wet from snow or rain during the capture process, they should be allowed to dry in transport boxes before handling to avoid excessive defeathering. Peterson et al. (2003a) developed a modification of the Rio Grande wild-turkey funnel trap to reduce injuries to the birds.

Patterson et al. (1993) facilitated handling of **mourning dove** by designing a modified restraining device similar to one described by DeMaso and Peoples (1993) for **northern bobwhite.** Time of handling and stress and struggling of the captured doves was minimized while leg bands and radio-transmitters were attached.

Ralph (2005) described a body grasp technique that speedily and safely allows removal of birds from **mist nets.** His method allowed an average removal time of 10 seconds per bird. Ponjoan et al. (2008) recommended that handling and restraint of **little bustards** after capture should not exceed 20 minutes to curtail capture myopathy. Abbott et al. (2005) minimized northern bobwhite muscular damage after capture and handling and increased survival by injecting vitamin E and selenium. Rogers et al. (2004) successfully treated cannon-net captured shorebirds in Australia with **capture myopathy** by suspension in a sling.

Mammals

Swann et al. (1997) reviewed the effects of orbital sinus **sampling of blood** on the survival of small mammals and found the results to be variable. White-throated woodrat and deer mouse survival estimates were not adversely affected, but desert pocket mouse and prairie vole survival rates were lower. Douglass et al. (2000) found no difference in handling mortality of 7 species of nonanesthetized wild rodents that were bled versus similar species of rodents that were not bled. They concluded that bleeding in the absence of anesthesia did not affect immediate mortality or subsequent recapture. Parmenter et al. (1998) verified that handling and bleeding procedures for hantavirus had no adverse effect on survival and trap rates of murid rodents (including deer mouse, woodrats, and prairie vole) and cottontail rabbit.

Mills et al. (1995) provided guidelines for personal safety while trapping, handling, and releasing rodents that might be infected with **hantavirus.** Special consideration is essential to provide respiratory protection from aerosolized virus. The use of protective gloves and clothing and suitable disinfectant also is necessary.

Yahner and Mahan (1992) used a polyvinyl Centrap™ cage as a restraining device for **red squirrel.** They used a mesh bag with a cone to minimize mortality from handling shock. Koprowski (2002) safely handled >3,500 squirrels of 7 species with a mortality of 0.01% using a cloth cone and without using an anesthesia, as suggested by Arenz (1997). McCleery et al. (2007a) developed an improved method for handling squirrels and similar-sized mammals.

Frost and Krohn (1994) described the care and handling of **fisher.** Serfass et al. (1996) successfully transported immobilized northern **river otters** in a well-ventilated tube made from 1-m sections of 40-cm-diameter PVC pipe.

Beringer et al. (1996) evaluated the influence of 2 capture methods, rocket nets and Clover traps, on **capture myopathy** in **white-tailed deer.** All deer mortality attributable to capture myopathy was associated with rocket net captures. Mortality attributable to capture myopathy can be reduced by using Clover traps instead of rocket nets when possible. If rocket nets are used, they suggested that capture be limited to ≤3 deer per capture. They advised that handling time be minimized to reduce stress on the animals. Peterson et al. (2003b) found that use of drugs after physical capture of white-tailed deer led to greater mortality than if drugs had not been used.

Byers (1997) described proper precautions for handing young **pronghorn,** including avoidance of handling 6 hours after birth or when coyotes or golden eagles were in sight or known to be within 1 km. Handling time should be brief and avoided during crepuscular hours, when coyotes are active. Byers (1997) concluded that methods he described did not increase mortality risk.

Thompson et al. (2001) concluded that direct release of **mountain sheep** from vehicles was advisable rather than transporting them via helicopter to holding pens. Expenses were less, survival was lower for the sheep kept in holding pens, and no difference was evident in dispersal and group cohesion.

DelGiudice et al. (2005) reviewed major factors influencing margins of safe capture and handling of **white-tailed deer** primarily captured in Clover traps. They stressed the need, when live-trapping, to provide adequate food, insulation, and avoidance of temperature extremes. Powell (2005) studied the blood chemistry effects on black bear captured in Aldrich foot snares and handled in dens. Both met the accepted standards for trap injuries. Forman and Williamson (2005) developed a safe handling device for **small carnivores** captured in a metal box live-trap using a plasterers' float and net bag. Freeman and Lemen (2009) tested various types of leather and recommended deerskin gloves to safely handle various **bat species** while maintaining dexterity. Beasley and Rhodes (2007) evaluated the effects of raccoon tooth removal to determine age and failed to detect any difference in recapture rates between the treated and untreated groups. MacNamara and Blue (2007) designed a portable holding corral system and **TAMER** that allowed physical and safe restraint of wild antelope and goats without the use of immobilizing drugs. The TAMER was constructed with a drop floor and attached electronic weight scale.

Amphibians

Christy (1998) used **elastic straps** and damp gauze attached to a wood base to restrain captured frogs. Rose et al. (2006) restrained captured lizards for measurements in a tray with Velcro strips attached to it. Bourque (2007) used a compression plate and pads to measure frogs without injury. McCallum et al. (2002) made a **frog box** to hold frogs by cutting a round hole in the lid of a Styrofoam ice chest. They then inserted a Styrofoam cup with the bottom removed into the hole, and a second intact cup was inserted inside the first cup to close the hole. The frog box allowed quick collection and secure containment of large numbers of anurans in the field.

Reptiles

King and Duvall (1984) restrained venomous snakes safely in a clear **noose tube** for field and laboratory examination. Quinn and Jones (1974) first developed a snake squeeze box, consisting of a foam rubber pad and Plexiglas, to measure snakes. Hampton and Haertle (2009) modified the snake **squeeze box** described by Cross (2000) and Bergstrom and Larsen (2004) that uses Plexiglas to allow safe dorsal and ventral views. Birkhead et al. (2004) designed "cottonmouth condo," a unique venomous-snake transport device. Penner et al. (2008) followed monkeys habituated to humans in a West Africa forest to efficiently locate and safely capture highly dangerous, venomous rhinoceros vipers. When the monkeys encountered a snake, they gave loud alarm calls, thereby alerting the herpetologists to capture and insert the snake into a custom-made transparent Plexiglas tube with a lockable end. Rivas et al. (1995) described a safe method for handling large nonvenomous snakes, such as anacondas. They placed a **cotton sock** over the snake's head and then wrapped several layers of plastic electrician's tape around the sock. The tape could be removed to release the snake into cloth bags for transport or release. Gregory et al. (1989) developed a portable device made of aluminum tubing to safely restrain **rattlesnakes** in the field. Walczak (1991) safely handled **venomous snakes** by immersing them in a plastic trash barrel partially filled with water. He then placed a clear plastic tube over the snake's head and gently submerged the snake. After the snake entered the tube, its body and the tube end were then grasped firmly with one hand. This method increased handler safety and decreased trauma. Mauldin and Engeman (1999) restrained snakes by using a wire-mesh cable holder. Cross (2000) described a new design for a lightweight squeeze box to allow safe handling of venomous snakes. His squeeze box was made of Plexiglas with a foam rubber lining, sliding doors, and portholes at each end. The squeeze box allowed measurements with a minimum of direct handling of snakes.

Jones and Hayes-Odum (1994) used white PVC pipe with an inside diameter of 0.31 m cut in 3-m lengths to restrain and transport **crocodilians.** Holes of a diameter sufficient for a rope to move freely were drilled at 15-cm intervals in the PVC pipe. One rope was looped around the head and another in front of the hind legs. Pipe diameter and length were chosen to accommodate a variety of alligator sizes.

Tucker (1994) described an easy method to remove **snapping turtle** from Legler™ hoop traps. He grasped the turtle by the tail and the posterior edge of the carapace. The turtle was then upended with the head down. With the animal in a vertical position, it was pressed down over the substrate, forcing the turtle to retract its head. The turtle's hind limbs were held, and it was then removed from the trap. A PVC pipe (10.16 cm in diameter and approximately 60 cm in length) was placed over the heads of snapping turtles for restraining and safe handling by Quinn and Pappas (1997).

Hoefer et al. (2003) placed **ice-cooled** lizards in a petri dish on top of adhesive tape to take measurements. Kwok and Ivanyi (2008) safely extracted venom from helodermatid lizards by using a rubber squeeze bulb. Poulin and Ivanyi (2003) used a locking adjustable hemostat to safely handle **venomous lizards.**

SUMMARY

Many new and innovative capture and handling methods, techniques, and equipment have been described in this chapter, with extensive literature citations for the reader interested in learning more. The coverage of amphibian and reptile capture and handling methods in this chapter is more detailed than was provided in previous editions of the *Wildlife Techniques Manual*. Humane capture and handling techniques continue to be of paramount importance. Tranquilizer trap devices show promise for minimizing injuries to nontarget captures, but unfortunately, they are restricted in their use and availability by the U.S. Food and Drug Administration and a similar agency in Canada. Although complex electronic and mechanized devices have recently been developed to expedite successful and efficient capture, simple variations of existing equipment (e.g., nets) and methods (e.g., the use of live and mounted decoys) continue to be widely described in the literature. The use of different net types and configurations (e.g., bow, cannon, drift, drop, mist, and rocket) continue to be the predominant technique for capturing birds. Mammals are captured primarily with snares and foothold, box, and cage traps. Wild animals may be captured for a variety of purposes, including subsistence, animal damage control, population management, disease control, enhancement of other species, economic benefits, and research. Regardless of the reasons for capture, it is imperative the most humane devices and techniques be used. Finally, all untested capture devices should be evaluated using standardized, scientifically sound protocols that include the documentation of capture-related injuries via whole body necropsies.

APPENDIX 3.1. COMMON NAMES AND SCIENTIFIC NAMES OF ANIMALS MENTIONED IN THE TEXT AND TABLES

The authority for scientific names of North American amphibians, birds, mammals, and reptiles is Banks et al. (1987). The authority for scientific names for non–North American amphibians and reptiles is Sokolov (1988), for non–North American birds is Sibley and Monroe (1990), and for non–North American mammals is Grizimek (1990).

Common name	Scientific name	Common name	Scientific name
Amphibians and reptiles			
Alligator, American	*Alligator mississippiensis*	marbled	*Ambystoma opacum*
Crocodile, American	*Crocodylus acutus*	red-backed	*Plethodon cinereus*
Frog, gray tree	*Hyla versicolor*	Skink, sand	*Neoseps reynoldsi*
Pacific tree	*Pseudacris regilla*	Snake, anaconda	*Eumcetes* spp.
spring peepers	*Pseudacris crucifer*	brown tree	*Boiga irregularis*
tree	*Hyla* spp.	giant garter	*Thamnophis gigas*
wood	*Rana sylvatica*	rattlesnake	*Crotalis* spp.
Hellbender	*Cryptobranchus alleganiensis*	python	*Python* spp.
Iguana	*Iguana* spp.	rhinoceros viper	*Bitis nasicornis*
Lizard, Texas horned	*Phrynosoma cornutum*	Terrapin, diamondback	*Malaclemys terrapin*
whiptail	*Cnemidophorus* spp.	Tortoise, Bolson	*Gopherus flavomarginatus*
Salamander, blackbelly	*Desmognathus quadramaculatus*	desert	*Gopherus agassizii*
blue-spotted	*Ambystoma laterale*	gopher	*Gopherus polyphemus*

continued

Common name	Scientific name	Common name	Scientific name
Turtle, black pond	*Geoclemys hamiltonii*	greater sage-	*Centrocercus urophasianus*
northern map	*Graptemys geographica*	ruffed	*Bonasa umbellus*
snapping	*Chelydra serpentina*	sharp-tailed	*Tympanuchus phasianellus*
spotted	*Clemmys guttata*	spruce	*Falcipennis canadensis*
Birds		Gull, California	*Larus californicus*
Avocet, American	*Recurvirostra americana*	ring-billed gull	*Larus delawarensis*
Blackbird, red-winged	*Agelaius phoeniceus*	Harrier, northern	*Circus cyaneus*
yellow-headed	*Xanthocephalus xanthocephalus*	Hawk, Cooper's	*Accipiter cooperii*
Bluebird	*Sialia* spp.	ferruginous	*Buteo regalis*
Bunting, painted	*Passerina ciris*	northern goshawk	*Accipiter gentilis*
Bustard, houbara	*Chlamydotis undulate*	red-shouldered	*Buteo lineatus*
little	*Tetrax tetrax*	red-tailed	*Buteo jamaicensis*
Buzzard, common	*Buteo buteo*	rough-legged	*Buteo lagopus*
Caracara, crested	*Caracara cheriway*	sharp-shinned	*Accipiter striatus*
Chicken, domestic	*Gallus gallus domesticus*	Swainson's	*Buteo swainsoni*
Coot, American	*Fulica Americana*	Heron, great blue	*Ardea herodias*
Cormorant, double-crested	*Phalacrocorax auritus*	Honeyeater, regent	*Xanthomyza phrygia*
Cowbird, brown-headed	*Molothrus ater*	Ibis, white	*Eudocimus albus*
Crane, sandhill	*Grus canadensis*	Jay, blue	*Cyanocitta cristata*
whooping	*Grus Americana*	Kestrel, American	*Falco sparverius*
Crow, American	*Corvus brachyrhynchos*	Kingfisher, belted	*Ceryle alcyon*
Dove, mourning	*Zenaida macroura*	Kite, white-tailed	*Elanus leucurus*
ringed turtle	*Streptopelia risoria*	Kittiwake, black-legged	*Rissa tridactyla*
rock	*Columba livia*	Loon, common	*Gavia immer*
white-winged	*Zenaida asiatica*	Magpie, American	*Pica hudsonia*
Duck, Barrow's goldeneye	*Bucephala albeola*	Merganser, hooded	*Lophodytes cucullatus*
blue-winged teal	*Anas Discors*	Merlin	*Falco columbarius*
canvasback	*Aythya valisineria)*	Murre, common	*Uria aalge*
gadwall	*Anas strepera*	Murrelet, marbled	*Brachyramphus marmoratus*
harlequin	*Histrionicus histrionicus*	Xantus	*Synthliboramphus hypoleucus*
lesser scaup	*Aythya affinis*	Nighthawk, common	*Chordeiles minor*
mallard	*Anas platyrhynchos*	Nightjars	Family Caprimulgidae
northern pintail	*Anas acuta*	Osprey	*Pandion haliaetus*
northern shoveler	*Anas clypeata)*	Owl, barn	*Tyto alba*
redhead	*Aythya americana*	barred	*Strix varia*
wood	*Aix sponsa*	burrowing	*Athene cunicularia*
Eagle, African fish	*Haliaeetus vocifer*	eastern screech	*Megascops asio*
bald	*Haliaeetus leucocephalus*	flammulated	*Otus flammeolus*
golden	*Aquila chrysaetos*	great horned	*Bubo virginianus*
Philippine	*Pithecophaga jefferyi*	northern saw-whet	*Aegolius acadicus*
steppe	*Aquila nipalensis*	pygmy	*Glaucidium brasilianum*
Eider, common	*Somateria mollissima*	short-eared	*Asio flammeus*
Falcon, prairie	*Falco mexicanus*	spotted	*Strix occidentalis*
Finch, house	*Carpodacus mexicanus*	tawny	*Strix aluco*
Flycather, Acadian	*Empidonax virescens*	tropical screech	*Megascops choliba*
Goose, Canada	*Branta canadensis*	western burrowing	*Athene cunicularia hypugea*
snow	*Chen caerulescens*	Oystercatcher, American	*Haematopus palliatus*
Grebe, eared	*Podiceps nigricollis*	Parrot, orange-winged	*Amazona amazonica*
pied-billed	*Podilymbus podiceps*	Partridge, chukar	*Alectoris chukar*
Grouse, blue	*Dendragapus obscures*	Pelican, American white	*Pelecanus erythrorhynchos*
dusky	*Dendragapus obscurus*	Penquin, king	*Aptenodytes patagonicus*

Common name	Scientific name	Common name	Scientific name
Phalarope, Wilson's	*Phalaropus tricolor*	Wren, house	*Troglodytes aedon*
Pheasant, Kalij	*Lophura leucomelanos*	**Mammals**	
ring-necked	*Phasianus colchicus*	Armadillo, nine-banded	*Dasypus novemcinctus*
Pigeon, band-tailed	*Patagioenas fasciata*	Badger, American	*Taxidea taxus*
Plover, mountain	*Charadrius montanus*	Beaver, American	*Castor canadensis*
snowy	*Charadrius alexandrinus*	Bobcat	*Lynx rufus*
Prairie-chicken, Attwater's	*Tympanuchus cupido*	Bat, African free-tailed	*Tadarida fulminans*
attwateri		Bear, black	*Ursus americanus*
greater	*Tympanuchus cupido*	brown	*Ursus arctos*
lesser	*Tympanuchus pallidicinctus*	grizzly	*Ursus arctos horribilis*
Ptarmigan, white-tailed	*Lagopus leucurus*	Capybara	*Hydrochoerus hydrochaeris*
willow	*Lagopus lagopus*	Caribou	*Rangifer tarandus*
Puffin	*Fratercula* spp.	Cat, feral	*Felis catus*
Purple martin	*Progne subis*	Chipmunk, eastern	*Tamias striatus*
Quail, Gambel's	*Callipepla gambelii*	Townsend's	*Tamias townsendii*
Montezuma	*Cyrtonyx montezumae*	Coyote	*Canis latrans*
northern bobwhite	*Colinus virginianus*	Culpeo	*Pseudalopex culpaeus*
scaled	*Callipepla squamata*	Deer, fallow	*Dama dama*
Rail, black	*Laterallus jamaicensis*	Himalayan musk	*Moschus moschiferus*
clapper	*Rallus longirostris*	Key	*Odocoileus virginianus*
king	*Rallus elegans*)	*clavium*	
sora	*Porzana carolina*	mule	*Odocoileus hemionus*
Virginia	*Rallus limicola*	white-tailed	*Odocoileus virginianus*
yellow	*Coturnicops noveboracensis*	Dog, domestic	*Canis familiaris*
Raven, Chihuahua	*Corvus cryptoleucus*	prairie	*Cynomys* spp.
Razorbill	*Alca torda*	Dugong	*Dugong dugon*
Rhea, greater	*Rhea americana*	Elk	*Cervus canadensis*
Robin, American	*Turdus migratorius*	Fisher	*Martes pennanti*
Scoters, surf	*Melanitta perspicillata*	Fox, Arctic	*Alopex lagopus*
Shrike, loggerhead	*Lanius ludovicianus*	Argentine gray	*Pseudalopex griseus*
Sparrow, Bachman's	*Aimophila aestivalis*	gray	*Urocyon cinereoargenteus*
chipping	*Spizella passerina*	kit	*Vulpes macrotis*
house	*Passer domesticus*	red	*Vulpes vulpes*
Starling, European	*Sturnus vulgaris*	swift	*Vulpes velox*
Stilt, black-necked	*Himantopus mexicanus*	Gopher, northern pocket	*Thomomys talpoides*
Swallows, bank	*Riparia riparia*	pocket	*Geomys breviceps*
barn	*Hirundo rustica*	Guanaco, South American	*Lama guanicoe*
cliff	*Petrochelidon pyrrhonota*	Hare, snowshoe	*Lepus americanus*
tree	*Tachycineta bicolor*	Hog, feral	*Sus scrofa*
Swan, trumpeter	*Cygnus buccinator*	Ibex, Spanish	*Capra pyrenaica*
tundra	*Cygnus columbianus*	Jaguar	*Panthera onca*
Swift, Vaux's	*Chaetura vauxi*	Leopard, snow	*Panthera uncia*
Tern, least	*Sterna antillarum*	Lion, African	*Panthera leo*
Turnstone, ruddy	*Arenaria interpres*	mountain	*Puma concolor*
Turkey, wild	*Meleagris gallopavo*	Lynx, Canada	*Lynx canadensis*
Warbler, prothonotary	*Prothonotaria citrea*	Marten, American	*Martes americana*
Woodcock, American	*Scolopax minor*	Mink	*Mustela vison*
Woodpecker, acorn	*Melanerpes erythrocephalus*	Mouse, cotton	*Peromyscus gossypinus*
pileated	*Drycopus pileatus*	desert pocket	*Chaetodipus penicillatus*
red-bellied	*Melanerpes carolinus*	deer	*Peromyscus maniculatus*
red-cockaded	*Picoides borealis*	hopping	*Notomys* spp.

continued

Common name	Scientific name	Common name	Scientific name
Mouse (continued)		Reindeer, Svalbard	*Rangiver tarandus platyrhynchus*
house	*Mus musculus*		
white-footed	*Peromyscus leucopus*	Seal, ringed	*Phoca hispida*
wood	*Apodemus sylvaticus*	Sheep, mountain	*Ovis canadensis*
yellow-necked	*Apodemus flavicollis*	Dall	*Ovis dalli*
Moose	*Alces alces*	Shrew, masked	*Sorex cinereus*
Mountain beaver	*Aplodontia rufa*	short-tailed	*Blarina brevicauda*
Muskrat	*Ondatra zibethicus*	Skunk, striped	*Mephitis mephitis*
Nutria	*Myocastor coypus*	Squirrel, Abert's	*Sciurus aberti*
Opossum, Australian brush-tailed	*Trichosurus vulpecula*	California ground	*Spermophilus beecheyi*
		fox	*Sciurus niger*
Virginia	*Didelphis virginiana*	gray	*Sciurus carolinensis*
Oryx	*Oryx gazella*	ground	*Spermophilus* spp.
Otter, Eurasian	*Lontra lutra*	northern flying	*Glaucomys sabrinus*
northern river	*Lontra canadensis*	red	*Tamiasciurus hudsonicus*
Peccary, collared	*Tayassu tajacu*	Tiger, Amur (Siberian)	*Panthera tigris altaica*
Porcupine	*Erethizon dorsatum*	Vole, bank	*Clethrionomys glareolus*
Pronghorn	*Antilocapra americana*	prairie	*Microtus ochrogaster*
Rabbit, eastern cottontail	*Sylvilagus floridanus*	Weasel, long-tailed	*Mustela frenata*
European	*Oryctolagus cuniculus*	short-tailed	*Mustela erminea*
Jackrabbit	*Lepus* spp.	Wolf, gray	*Canis lupus*
Lower Keys marsh	*Sylvilagus palustris hefneri*	Wolverine	*Gulo gulo*
pygmy	*Brachylagus idahoensis*	Woodchuck	*Marmota monax*
Raccoon	*Procyon lotor*	Woodrat, bushy-tailed	*Neotoma cinerea*
Rat	*Rattus* spp.	dusky-footed	*Neotoma fuscipes*
cotton	*Sigmodon hispidus*	Key Largo	*Neotoma floridana smalli*
kangaroo	*Dipodomys* spp.	white-throated	*Neotoma albigula*
rice	*Oryzomys palustris*		

APPENDIX 3.2. SOME MANUFACTURERS AND SUPPLIERS OF ANIMAL TRAPS, SNARES, AND RELATED EQUIPMENT

This information is provided for the convenience of readers and offers only a small sampling of the many manufacturers and suppliers of animal traps and related equipment. The authors, their agencies, and The Wildlife Society makes no claim to its accuracy or completeness and neither endorses nor recommends any particular style, brand, manufacturer, or supplier of traps and trapping materials.

Alaska Trap Company
380 Peger Rd.
Fairbanks, AK 99709-4869 USA
Telephone: 907-452-6047

Blue Valley Trap Supply
4174 W Dogwood Rd.
Pickrell, NE 68422 USA
Telephone: 402-673-5935

Butera Manufacturing Industries (BMI)
1068 E 134th St.
Cleveland, OH 44110-2248 USA
Telephone: 216-761-8800

CDR Trap Company
240 Muskingham St.
Freeport, OH 43973 USA
Telephone: 740-658- 4469

J. C. Conners
7522 Mt. Zion Cemetery Rd.
Newcomerstown, OH 43832 USA
Telephone: 740-498-6822

CTM Trapping Equipment
7171 S 1st St.
Hillsdale, IN 47854 USA
Telephone: 765-245-2837

Cumberland's Northwest Trappers Supply
P.O. Box 408
Owatonna, MN 55060 USA
Telephone: 507-451-7607

Duffer's Trap Company
P.O. Box 9
Bern, KS 66408 USA
Telephone: 785-336-3901

Duke Company
P.O. Box 555
West Point, MS 39773 USA
Telephone: 662-494-6767

CAPTURING AND HANDLING WILD ANIMALS 117

The Egg Trap Company
P.O. Box 334
Butte, ND 58723 USA
Telephone: 701-626-7150

Fleming Outdoors
5480 Highway 94
Ramer, AL 36069 USA
Telephone: 800-624-4493

F&T Fur Harvester's Trading Post
10681 Bushey Rd.
Alpena, MI 49707 USA
Telephone: 989-727-8727

Funke Trap Tags & Supplies
2151 Eastman Ave.
State Center, IA 50247 USA
Telephone: 641-483-2597

Halford Hide & Leather Company
2011 39 Ave. NE
Calgary, AB T2E 6R7 Canada
Telephone: 403-283-9197

Hancock Trap Company
P.O. Box 268
Custer, SD 57730-0268 USA
Telephone: 605-673-4128

Kaatz Bros Lures
9986 Wacker Rd.
Savanna, IL 61074 USA
Telephone: 815-273-2344

Kania Industries
63 Centennial Rd.
Nanaimo, BC V9R 6N6 Canada
Telephone: 250-716-1685

Les Entreprises Bélisle
61, Rue Gaston-Dumoulin,
    Bureau 300
Blainville, QC J7C 6B4 Canada
Telephone: 450-433-4242

Les Pieges du Quebec (LPQ)
16125 Demers St.
Hyacinthe, QC J2T 3V4 Canada
Telephone: 450-774-4645

Margo Supplies
P.O. Box 5400
High River, AB T1V 1M5 Canada
Telephone: 403-652-1932

Minnesota Trapline Products
6699 156th Ave. NW
Pennock, MN 56279 USA
Telephone: 320-599-4176

Molnar Outdoor
9191 Leavitt Rd.
Elyria, OH 44035 USA
Telephone: 440-986-3366

Montgomery Fur Company
1539 West 3375 South
Ogden, UT 84401 USA
Telephone: 801-394-4686

National Live Trap Corporation
1416 E Mohawk Dr.
Tomahawk, WI 54487 USA
Telephone: 715-453-2249

Oneida Victor
P.O. Box 32398
Euclid, OH 44132 USA
Telephone: 216-761-9010

PDK Snares
8631 Hirst Rd.
Newark, OH 43055 USA
Telephone: 740-323-4541

Quad Performance Products
Rt. 1, Box 114
Bonnots Mill, MO 65016 USA
Telephone: 573-897-2097

Rally Hess Enterprises
13337 US Highway 169
Hill City, MN 55748 USA
Telephone: 218-697-8113

Rancher's Supply—The Livestock
    Protection Company
P.O. Box 725
Alpine, TX 79831 USA
Telephone: 432-837-3630

R-P Outdoors
505 Polk St., P.O. Box 1170
Mansfield, LA 71052 USA
Telephone: 800-762-2706

Thompson Snares
37637 Nutmeg St.
Anabel, MO 63431 USA
Telephone: 660-699-3782

Rocky Mountain Fur Company
14950 Highway 20/26
Caldwell, ID 83607 USA
Telephone: 208-459-6854

Rudy Traps—LOYS Trapping Supplies
577 Lauzon Ave.
St-Faustin, QC J0T 1J2 Canada
Telephone: 819-688-3387

Sleepy Creek Manufacturing
459 Duckwall Rd.
Berkeley Springs, WV 25411 USA
Telephone: 304-258-9175

The Snare Shop
330 Main, P.O. Box 70
Lidderdale, IA 51452 USA
Telephone: 712-822-5780

Sterling Fur & Tool Company
11268 Frick Rd.
Sterling, OH 44276 USA
Telephone: 330-939-3763

Sullivan's Supply Line
429 Upper Twin
Blue Creek, OH 45616 USA
Telephone: 740-858-4416

Tomahawk Live Trap Company
P.O. Box 323
Tomahawk, WI 54487 USA
Telephone: 800-272-8727

Wildlife Control Products
P.O. Box 115, 107 Packer Dr.
Roberts, WI 54023 USA
Telephone: 715-749-3857

Wildlife Control Supplies
P.O. Box 538
East Granby, CT 06026 USA
Telephone: 877-684-7262

Wildlife-Traps.com
    (Online) SuperStore
P.O. Box 1181
Geneva, FL 32732 USA
Telephone: 407-349-2525

Woodstream Corporation
69 N. Locust St.
Lititz, PA 17543 USA
Telephone: 800-800-1819

4

Wildlife Chemical Immobilization

TERRY J. KREEGER

INTRODUCTION

CHEMICAL IMMOBILIZATION IS the use of drugs to capture or restrain animals. The term **immobilization** describes the actions of such drugs, which can range from tranquilization to paralysis to general anesthesia. Wildlife capture is needed for research, translocation, and public safety, and drugs are just one of many tools the biologist can employ to accomplish this. Animals can be captured using only physical means (e.g., traps), only drugs, or a combination of the 2 (e.g., trap then drug). Every situation is different, requiring all the "capture tools" to be in the toolbox. Using the right drug delivered effectively at the right dose results in an effective and humane capture that not only accomplishes your needs, but also reflects well on your profession.

Reading this chapter will not make you an expert on chemical capture, but it should provide you with a comprehensive overview. Nothing will substitute for expert training coupled with field experience. Many government agencies have staff veterinarians, and there are a few for-profit businesses that conduct excellent training classes. Probably the most important criterion for training is the experience of the instructor. Just because a person is a veterinarian does not mean that he or she has knowledge of injectable immobilants or wildlife biology. Many nonveterinarian biologists have extensive field experience with wildlife captures, and their expertise can be valuable. **Formal training** should consist of 8–12 hours of lecture coupled with equipment handling and, hopefully, actual animal immobilization.

The bulk of this chapter has been taken from the third edition of the *Handbook of Wildlife Chemical Immobilization* (Kreeger and Arnemo 2007). The *Handbook* greatly expands on the information provided below, plus it provides drug recommendations and dosages for >475 species of wildlife supported by >2,400 references. That level of detail obviously cannot be duplicated here, but this resource plus others are included at the end of this volume.

Legal Considerations

Conditions for the use of **drugs** (pharmaceuticals) are established by the **U.S. Food and Drug Administration (FDA).** The FDA verifies the safety and efficacy of drugs as well as ensures manufacturing quality control. Approval by the FDA limits the use of the drug to conditions specified on the label. Only 4 drugs have been specifically approved by the FDA for use on certain wild animals: carfentanil for use on cervids; xylazine for use on elk (*Cervus canadensis*) and fallow (*Dama dama*), mule

(*Odocoileus hemionus*), sika (*Cervus nippon*), and white-tailed deer (*O. virginianus*); yohimbine for use on cervids (deer and elk); and ketamine for use on primates. Any use of these or other drugs on any species not identified on the label is termed **extra label** or **off label.**

However, the **Animal Medicinal Drug Use Clarification Act of 1994** (http://www.fda.gov/AnimalVeterinary/Guidance ComplianceEnforcement/ActsRulesRegulations/ucm 085377.htm) essentially allowed approved animal and human drugs to be used extra label under certain conditions. In general, those conditions are the drug: (1) must be approved by the FDA; (2) used by, or on the lawful written or oral order of, a licensed veterinarian; and (3) used in the context of a valid veterinarian–client–patient relationship. Additionally, if the animal could be consumed by a human, the veterinarian should: (1) establish a substantially extended withdrawal time (the time from the date that a drug was administered to when the animal can safely be consumed by humans); (2) be able to identify the treated animals; and (3) ensure that assigned timeframes for withdrawal are met and no illegal residues occur.

In addition to being **prescription drugs,** some of the drugs used for wildlife immobilization are classified as controlled substances. A **controlled substance** is a drug that is identified in 1 of 5 schedules established by the U.S. Drug Enforcement Administration (DEA). Special regulations govern the recording and storage of these drugs. Regulations regarding drug storage are contained in 21 CFR 1301.75d. The Controlled Substances Act (http://www.justice.gov/dea/pubs/csa.html) requires an individual to have a special DEA registration number to possess controlled substances. Application for this number is made through regional offices of the DEA. If you are unable to determine your regional office, contact the DEA (http://www.deadiversion.usdoj.gov/drugreg/). The following is a brief discussion of the 5 schedules:

> **Schedule I** is reserved for experimental and abused drugs, such as heroin, marijuana, and lysergic acid diethylamide (LSD). No capture drugs are Schedule I.
> **Schedule II (IIN)** includes most of the opioids used for animal immobilization, such as etorphine, fentanyl, sufentanil, and carfentanil and the opioid antagonist diprenorphine.
> **Schedule III (IIIN)** contains ketamine and tiletamine and/or zolazepam.
> **Schedule IV** includes benzodiazepine tranquilizers, such as diazepam and midazolam, and butorphanol.
> **Schedule V** covers small quantities of narcotic drugs included in preparations with non-narcotic active medicinal ingredients. No capture drugs are Schedule V.

Many biologists have obtained a DEA registration number and have been able to procure drugs through veterinary product distributors. Technically, however, even though they are in possession of these drugs, they cannot use them on animals without **veterinary supervision.** This restriction is because all wildlife capture drugs are prescription drugs and must be used by, or on the order of, a licensed veterinarian. Nonveterinarians can legally administer drugs if a valid veterinarian–client–patient relationship is established. That is, the biologist becomes the "client," and the wild animal becomes the "patient." The biologist consults with the veterinarian, who determines whether the dose and application of the drug are appropriate. The veterinarian does not have to be on site during the actual immobilization event, but he or she should be involved in the planning process.

Finally, most states have a **Pharmacy Board** or equivalent. These boards have their own set of regulations that also must be followed and often require licensure. Sometimes, state regulations are more restrictive than FDA or DEA requirements. An excellent and detailed discussion of drug acquisition in Canada can be found in the second edition of *The Chemical Immobilization of Wildlife* (Cattet et al. 2005).

PHARMACOLOGY

General Principles

No perfect capture drug exists. However, the characteristics of an **ideal** injectable **drug** may serve as a guide to the evaluation of currently available drugs. Such **characteristics** include: (1) safe for animals and humans; (2) potent (sufficient dose delivered in a small volume); (3) fast-acting, smooth onset of action; (3) good muscle relaxation; (4) minimal depression of cardiovascular or respiratory systems; (5) capable of being antagonized (reversed); (6) rapid, smooth emergence with minimal side effects; (7) minimum withdrawal time for safe human and/or animal consumption; and (8) low potential for human abuse.

Calculating Drug Doses

Accurate **calculation of drug doses** is critical to reduce the problems associated with under- or overdosing. A dose is the total amount of drug given to an animal. Information required prior to calculating a dose includes: (1) weight of the animal (usually in kg); (2) drug concentration (usually expressed as mg of drug per mL of solvent); and (3) dosage. Dosages are mostly given as mg of drug per kg of animal body weight (mg/kg) and can be found in a variety of references and publications. Armed with these data, you can now calculate the volume (mL) of drug to administer (Box 4.1). If you are using >1 drug, a calculation has to be made for each drug.

Drug Combinations

Capture drugs (i.e., immobilants) are often employed in combination for wildlife immobilization. Effects of immobilants can range from paralysis (the animal is mentally alert, but cannot move) to general anesthesia (unconsciousness).

> **BOX 4.1. FORMULA FOR CALCULATING THE VOLUME (mL) OF DRUG TO ADMINISTER**
>
> The formula is:
>
> $$\text{Volume of drug administered} = \frac{\text{Body weight} \times \text{Dosage}}{\text{Drug concentration}}.$$
>
> For example, consider immobilizing an animal that weighs 80 kg (176 lb) with drug X. The recommended dosage of drug X for this animal is 5 mg/kg. The concentration of drug X is 100 mg/mL. First, calculate the total mg needed for this animal by multiplying the animal's weight (80 kg) by the recommended drug dosage (5 mg/kg):
>
> $$\text{Milligrams of drug X needed} = 80 \text{ kg} \times 5 \text{ mg/kg} = 400 \text{ mg}.$$
>
> Then calculate the volume of drug solution to withdraw from the bottle by dividing the total mg (400 mg) by its concentration (100 mg/mL):
>
> $$\text{Volume (mL) needed} = \frac{400 \text{ mg}}{100 \text{mg/mL}} = 4 \text{ mL of drug solution}.$$
>
> Some points to remember in calculating drug doses: (1) never memorize drug dosages; (2) do not calculate drug doses in your head; (3) double check your math; and (4) check that your answer makes sense. As you gain experience with the drug and the animal, a drug volume that is miscalculated should trigger a mental alarm.

In most cases, the primary drug (e.g., ketamine or carfentanil) is sufficient to immobilize an animal on its own. **Tranquilizers** and/or **sedatives** are often combined with these primary drugs, because they usually improve the overall immobilization process. By themselves, tranquilizers or sedatives only cause sedation, not unconsciousness. Such sedation may be profound to the point that the animal may be safely handled (e.g., deer given only xylazine). However, if sufficiently stimulated, a tranquilized animal can arouse and flee (or attack!). Some **common examples** of immobilants combined with tranquilizers include: ketamine/acepromazine, ketamine/xylazine, ketamine/medetomidine, carfentanil/xylazine, tiletamine/zolazepam, and tiletamine/zolazepam/xylazine.

Advantages of **combining drugs** include: (1) reduction of doses of all drugs (often reducing total cost of drugs); (2) reduction of total drug volume (thus permitting smaller darts to be used); (3) reduction of undesirable side effects (convulsions, muscle rigidity, etc.); (4) decreased induction time; and (5) improved recovery (less stumbling and incoordination). Drug combinations, however, also can exacerbate adverse effects, such as respiratory depression and thermoregulation disruption.

Records

Records are vital for future immobilization events, because they allow review of what worked and what did not. Also, valuable biological data are usually included in the capture record (Box 4.2). **Records** of receipt and use are **required** by federal law for DEA scheduled drugs, but they probably should be kept for all drugs used.

All prescription drugs have an expiration date printed on the label. A drug **expiration date** is the date on which the drug still retains at least 90% of its potency. Although there is good evidence that many drugs remain biologically effective for many years after their expiration date, they should not be used in any circumstance where failure of the drug to work would result in human injury, property damage, or animal harm.

Drug Classes
Neuromuscular Blocking Drugs
The neuromuscular blocking (NMB) drugs were some of the first drugs used for the chemical immobilization of wildlife. They immobilize the animal by inducing **muscle paralysis,** but they have no central nervous system (CNS) effects. Despite their long history of use, NMB drugs are generally inferior to modern drugs. There are 2 major deficiencies of NMB drugs. The first is that NMB drugs have **low therapeutic indices** (ratio of lethal to effective dosages), and dosage errors of only 10% can result in either no effect or death. Overdosing results in diaphragmatic paralysis and death by asphyxia. The second deficiency is that NMBs are virtually devoid of CNS effects because of their inability to cross the

BOX 4.2. SAMPLE ANIMAL CAPTURE FORM

ANIMAL CAPTURE FORM

Date _____ Animal number _____

Name of investigator(s) _____

Species _____ Sex (circle) M   F   UNK

Age _____ mo   yr   (estimated or actual)

Weight _____ lb   kg   (estimated or actual)

Purpose of capture _____

Location of capture _____

Ambient temperature _____ F° C°   Weather conditions _____

Time	Drug	Dose (mg or mL)	Method	Location of injection
____	____	____	____	____
____	____	____	____	____
____	____	____	____	____
____	____	____	____	____
____	____	____	____	____

Time animal immobilized _____ Time animal recovered _____

Vital signs	Time	Temperature	Pulse	Respiration
____	____	____	____	____
____	____	____	____	____
____	____	____	____	____

Condition of animal—indicate: ____ Excellent ____ Good ____ Fair ____ Poor

Injuries or abnormalities noted _____

Sample(s) taken: Time   Type (indicate blood, tissue, tooth, etc.)
                   ____   _____
                   ____   _____

Radiocollar frequency _____   Radio signal checked? _____

Transponder number _____   Transponder checked? _____

Ear tag number(s) and color(s) _____

Other measurements:

   Body length _____ cm   Tail length _____ cm

   Shoulder height _____ cm   Girth _____ cm

Comments:

Table 4.1. Summary of the most common drugs used for wildlife immobilization and their antagonists, if applicable

Drug classification	Drug name	Antagonist
Paralytic	Succinylcholine	None
Tranquilizer or sedative	Acepromazine	None
Tranquilizer or sedative	Diazepam	Flumazenil
Tranquilizer or sedative	Midazolam	Flumazenil
Tranquilizer or sedative	Azaperone	None
Tranquilizer or sedative	Xylazine	Yohimbine, tolazoline, atipamezole
Tranquilizer or sedative	Medetomidine	Atipamezole
Cyclohexane	Ketamine	None
Cyclohexane	Tiletamine (in Telazol®)	None
Opioid	Carfentanil	Naltrexone, naloxone
Opioid	Thiafentanil	Naltrexone, naloxone
Opioid	Sufentanil	Naltrexone, naloxone
Opioid	Butorphanol	Naltrexone, naloxone
Opioid	Etorphine	Naltrexone, naloxone, diprenorphine

blood–brain barrier. Thus, an animal paralyzed with NMB drugs **is conscious,** aware of its surroundings, fully sensory, and can feel pain and experience psychogenic stress yet is physically unable to react. With few exceptions, there are no antagonists for NMBs (Table 4.1).

There are, however, certain definite advantages to a few NMB drugs. They are generally **very fast-acting** (1–3 minutes), and the duration of effect lasts only for a short while (15–30 minutes). **Succinylcholine,** the most commonly used drug of this class, also is fairly safe for humans. Unlike some other drugs, the succinylcholine dose required to immobilize most animals is much lower than the clinically effective dose for humans. Also, animals that have been given only succinylcholine and have died or been euthanized using physical means (i.e., not other drugs) can be **safely eaten** by other animals or, in some countries, by humans. And finally, succinylcholine is extraordinarily **cheap,** perhaps the least expensive immobilizing agent available. Nonetheless, paralytics should be used judiciously, such as in captive environments, where problems can be quickly addressed.

Tranquilizers and Sedatives

Tranquilizers and sedatives (Table 4.1) are used primarily as adjuncts to primary immobilants to hasten and smooth induction and recovery and to reduce the amount of the primary agent required to achieve immobilization. **Tranquilizers** relieve anxiety with minimal sedation; **sedatives** relieve anxiety, making it easier for the animal to rest or sleep. The differences between tranquilizers and sedatives are not terribly important for your needs.

Acepromazine is a phenothiazine antipsychotic drug. Although not used much these days in wildlife immobilization, acepromazine combined with ketamine is a good combination for small to medium (50 kg) mammals. Additionally, acepromazine is readily available from most veterinary clinics. It is not a controlled substance.

Azaperone is a butyrophenone that has been reported to partially counteract opioid respiratory depression in wild animals. Azaperone is enjoying renewed popularity in a drug combination with butorphanol and medetomidine (BAM) discussed later. Azaperone is not a controlled substance.

Benzodiazepine tranquilizers (diazepam, midazolam) are used primarily to **treat convulsions** caused by ketamine. However, they can be combined with ketamine to immobilize small animals (10 kg). They generally are not used to immobilize large mammals (>50 kg), because the volume required is too large to be practical. Benzodiazepine tranquilizers are Schedule IV controlled substances.

The alpha-2 adrenergic agonists (**xylazine, medetomidine**) are potent sedatives that can be completely antagonized. They are usually used as adjuncts with opioids or cyclohexanes to hasten and smooth induction. By themselves, they are capable of heavily sedating animals, particularly ungulates, to the point of relatively safe handling. Immobilization or sedation of highly excited animals using alpha-2 adrenergic agonists alone will be prolonged, if not impossible. Additionally, animals sedated with these agonists generally can be aroused with stimulation and are capable of directed attack. Caution should always be exercised in such animals, even though they appear harmless. These sedatives can cause profound **respiratory depression** and can **alter thermoregulation** (animal overheats or cools).

Cyclohexanes

These drugs are **true anesthetics** and can be used alone to immobilize animals, but they are usually combined with a tranquilizer or sedative. Also termed dissociative anesthetics, these drugs are characterized by producing a **cataleptic state** (a malleable rigidity of the limbs), in which the eyes remain open with intact corneal and light reflexes. There is **no complete antagonist** of the cyclohexanes (Table 4.1), although several drugs appear to antagonize some of their effects.

Ketamine and tiletamine are 2 cyclohexanes in use today. They are probably the most widely used drugs for wildlife anesthesia because of their efficacy and high therapeutic indices. They can **immobilize species** ranging from reptiles to large ungulates. Tiletamine is combined in equal proportions with the benzodiazepine tranquilizer, zolazepam (i.e., Telazol®). Combining these 2 drugs results in few convulsions, good muscle relaxation, and smooth recoveries.

Opioids

Opium is a drug obtained from the juice of the poppy (*Papaver somniferum*) and contains >20 alkaloids. Opioid immobilizing agents are generally congeners of 2 of these alkaloids, morphine and thebaine. The opioids have been used for animal immobilization since the 1960s and are the **most potent drugs** available for this purpose. A major advantage in the use of opioids is the availability of specific antagonists.

Opioids are not general anesthetics. Technically, they are classified as **neuroleptanalgesics** produced by a combination of a tranquilizer or sedative and an opioid analgesic. Animals given opioids often respond to noise, touch, and other stimulation that indicates they are not completely unconscious (a characteristic of general anesthesia). Although most opioids are potent analgesics, they **do not induce surgical anesthesia** and surgical interventions, such as radiotransmitter implantation, should not be done on animals given opioids alone.

The most commonly used opioids are butorphanol, etorphine, thiafentanil, sufentanil, and carfentanil (Table 4.1). **Butorphanol** is 3.5–7.0 times, **etorphine** is 1,000 times, **sufentanil** 4,521 times, and **carfentanil** 9,441 times **more potent than morphine.** Although a weak opioid, butorphanol mixed with azaperone and medetomidine (known as **BAM**) has been used successfully to immobilize a variety of ungulates. The only advantage of BAM is that butorphanol is a Schedule IV controlled substance and does not require the rigorous record keeping of the Schedule II opioids, such as etorphine and carfentanil. The disadvantages of BAM include prolonged induction time (>10 min) and severe respiratory depression.

The potency of **etorphine and carfentanil** is both an advantage and disadvantage. The advantage is the reduced volume of drug required for immobilization makes them the only class of drugs capable of remote capture of large animals. The disadvantage is they are potentially **toxic to humans.** Many people have assumed the human lethal dose of carfentanil, the most potent opioid in use today, is very low. Some have even said that exposure to an almost invisible amount would be fatal. This statement is probably not true. Extra care and concentration, though, is required when working with these drugs. Toxic exposure can be by accidental injection with a syringe or dart; by absorption through the mucous membranes of the mouth, eyes, or nose; or by direct absorption through broken skin. Opioid immobilizing agents should never be used while working alone or without having an antagonist immediately on hand. Although opioids are potentially toxic, keep in mind there have been only 2 recorded human deaths (due to injection with etorphine) and no recorded deaths due to carfentanil or thiafentanil despite >20 years of use and tens of thousands of doses given.

Inhalation Anesthetics

Comprehensive instruction on **inhalation (gas) anesthesia** is beyond the scope of this chapter. Unless you are dealing with the simplest of gas systems, do not attempt to anesthetize animals without hands-on instruction from a veterinarian or an experienced veterinary technician. In the simplest of terms, gas anesthesia is the delivery of vaporized drugs that are breathed directly into the lungs, taken up by the blood, and delivered to the brain, resulting in general anesthesia. Elimination of the drug is mostly by a reversal of this same route. The main advantage of gas anesthesia is the ability to control depth of anesthesia and hasten recoveries.

Inhalation anesthesia sees **limited application** in field immobilizations. However, it can be used effectively for small mammals (Fig. 4.1); some marine mammals, birds, and reptiles; or for maintaining anesthesia in larger animals initially anesthetized with injectable drugs. Its primary use is in zoos and research facilities (West et al. 2007). The most common gas anesthetics used for wildlife are isoflurane and sevoflurane.

The simplest method for delivering inhalation drugs is an **open system.** It can be a jar with ether-soaked cotton balls, in which you place a small mammal until it loses consciousness, or a cone with drug-soaked cotton that is placed over the muzzle of a larger animal. Open systems provide an acceptable means to anesthetize rodents and other small mammals. Cones can be used to maintain anesthesia in animals that have been initially anesthetized in a chamber, but then removed for further handling. Open systems waste a

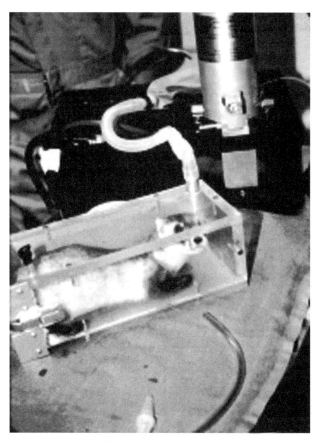

Fig. 4.1. Black-footed ferret (*Mustela nigripes*) in portable sevoflurane anesthesia delivery system.

lot of anesthetic through uncontrolled vaporization. This vaporization also exposes humans to potentially toxic fumes. There is little control of depth of anesthesia with open systems. You must closely observe the depth of anesthesia and remove the animal for a brief period to breathe ambient air when needed.

Rebreathing systems are the best systems for delivering inhalation anesthetics, but are the most expensive and the most complex. Rebreathing systems start with compressed oxygen that passes through a flowmeter to monitor and adjust oxygen flow rate. The oxygen then passes through an adjustable vaporizer, where the anesthetic is vaporized and swept along the hose by the oxygen and delivered to the animal's lungs through a one-way valve. Expired gases are forced by the one-way valve along another tube that passes through a carbon dioxide absorbent canister. Precision vaporizers are gas specific; for example, isoflurane requires an isoflurane vaporizer. Rebreathing systems generally cannot be used for small mammals (2–3 kg) because of their small lung volumes.

You must constantly **monitor the patient** when using gas anesthesia. The biggest mistake novices make with gas anesthesia is to become distracted with other tasks (e.g., blood sampling, measurements, or ultrasound) and ignore the depth of anesthesia. In a very short time, animals can go from an acceptable level of anesthesia to respiratory arrest and death. If you are going to use gas anesthesia, you must have a person dedicated to monitoring the patient.

Antagonists

Some of the more notable pharmacological developments relative to wild animal immobilization have been specific, long-lasting opioid and alpha-2 adrenergic antagonists. The ability to antagonize immobilization and return the animal more quickly to physiological normalcy offers **many advantages,** including: (1) alleviation of problems associated with prolonged recumbency, such as hypothermia; (2) reduced probability of injury or death after recovery due to accident or predation; (3) decreased probability of rejection or interspecific strife due to quicker return to parent, herd, or pack; and (4) decreased personnel and equipment time dedicated to monitoring the recovery process.

In general, opioid and alpha-2 adrenergic antagonists **are safe,** causing adverse effects only at higher doses. It should be remembered that antagonists act on the animal and not on the agonist. Thus, it does not necessarily follow that the more potent the agonist, the greater is the amount of antagonist that needs to be administered. You should select an antagonist that is the most specific for the receptors affected and has the longest biological life in the animal to prevent resedation. Antagonists can be given **intravenously (IV) or intramuscularly (IM),** with IM probably being the easiest to administer while providing the smoothest recovery for the animal.

Opioid antagonists have been in use for >40 years, and their use in combination with potent opioid agonists has made them powerful tools for wildlife capture. Today, they are used extensively to antagonize the effects of such opioid agonists as etorphine, thiafentanil, sufentanil, and carfentanil. **Naloxone and naltrexone** are the most commonly used opioid antagonists; naltrexone is the preferred antagonist, because it has a longer biological half-life.

Alpha-adrenergic antagonists (yohimbine, tolazoline, and atipamezole) are used to antagonize tranquilizers, such as xylazine and medetomidine. For unknown reasons, tolazoline appears to be more effective than yohimbine in some ungulate species. When immobilizing animals with a ketamine–xylazine combination, note that yohimbine or tolazoline antagonizes only the xylazine. These antagonists should not be administered until ≥30 minutes have elapsed since the last ketamine injection. If they are given when ketamine serum concentrations are still high, the xylazine component will be antagonized, resulting in an anesthetic recovery from what is essentially pure ketamine. Such recoveries are characterized by uncontrolled, often violent, body movements and/or severe hyperthermia that can cause injury or death to the animal.

Atipamezole is more potent and more selective than either yohimbine or tolazoline. Only atipamezole should be used to antagonize the effects of medetomidine, because yohimbine or tolazoline may result in incomplete antagonism. A major advantage of using ketamine–medetomidine combinations is the potent medetomidine greatly reduces the amount of ketamine required. Thus, when atipamezole is administered, there is usually little ketamine left in the animal, resulting in quicker, smoother, and more complete recoveries. Atipamezole is generally administered at a dose 5 times higher than medetomidine on a weight-to-weight basis. Atipamezole effectively antagonizes xylazine at a ratio of 1 mg atipamezole for every 10 mg xylazine used.

EQUIPMENT

This section discusses the equipment for delivering drugs to animals and for monitoring the effects of those drugs. There is no single type of system that can be used on all animals at all times. This fact is sometimes difficult to accept, particularly when buying decisions are limited by fiscal constraints. You possibly can get by with only a 13-mm (0.50-caliber) **dart rifle** having variable power settings. The well-equipped professional, however, will have multiple dart guns (pistol, CO_2 and .22 caliber rifles), **pole syringe,** and **blow pipe** (or powered blow pipe). For equipment and suppliers of capture drugs, see Box 4.3.

Syringes and Needles

Hand syringes and needles are the basis for any drug delivery system. Not only are they used to administer drugs di-

> **Box 4.3. Resources (web sites) for chemical immobilization supplies**
>
> Animal Care Equipment and Services, Inc.
> www.animal-care.com
> (Dart guns, darts, blow pipes, other animal capture equipment)
>
> Dan-Inject
> http://www.daninjectdartguns.com/
> (Dart guns, darts, blow pipes)
>
> Palmer Chemical & Equipment Co., Inc.
> www.palmercap-chur.com
> (Dart guns, darts)
>
> Pneu Dart, Inc.
> www.pneudart.com
> (Dart guns, darts, blow pipes)
>
> Telinject USA, Inc.
> www.telinject.com
> (Dart guns, darts, blow pipes)
>
> Wildlife Pharmaceuticals, Inc.
> www.wildpharm.com
> (North American supplier of capture drugs)

rectly to restrained animals, they also are used for measuring and loading immobilization drugs into darts. **Syringes and needles** are required for taking blood samples and administering antibiotics and other drugs. Most syringes and all needles are sterilized and disposable—they are intended to be used once and discarded.

You can never have enough syringes in your kit, because you will consume them rapidly. For example, a syringe used to measure or administer an anesthetic should not be used to administer any antagonist (there could be residual anesthetic in the syringe). Also, any syringe that is used for an intravenous injection in one animal should not be used on another, because it will be contaminated with blood. Syringes are available in an assortment of sizes, but 1 mL, 3 mL, 5–6 mL, and 10–12 mL are the most commonly used.

Likewise, you cannot have too many needles on hand. Needles should be used to withdraw and/or administer only one type of drug. They should not be used on >1 animal and should not be reused for any reason. The basic philosophy here is to avoid cross-contamination of either drugs or animal fluids. Needles also come in a variety of **gauges** (inside diameter measurement) and lengths. The larger the needle gauge, the smaller the inside diameter. For example, a 25-gauge needle is much smaller than a 16-gauge one.

Pole Syringes

Pole syringes are exactly that—a syringe on the end of a pole. They are very useful tools with broad applications, such as administering drugs to trapped or caged animals or safely giving additional drugs to animals not completely immobilized, but approachable. **Pole syringes** are usually limited to administering ≤10 ml of drug, because the animal will usually not hold still long enough to give larger volumes. Homemade pole syringes can be easily and cheaply constructed, but none seem to work as well as the manufactured versions.

Blow Pipes

Blow pipes, or blow guns, are useful devices for delivering small volumes of drugs at short to medium ranges. They operate by propelling a dart through a pipe or tube by rapid expulsion of one's breath, compressed air, or CO_2. Conventional (lung-powered) blow pipes consist of 1- or 2-piece aluminum tubes. Most propel 10-mm-diameter darts having a maximum capacity of 3 mL. Their effective **range is limited** (10 m). Blow pipes are quiet. They usually cause little trauma to the animal, because the dart strikes the animal with little velocity and its method of operation does not cause injury; animals as small as 3 kg (6.6 lb) can be safely treated. Blow pipes are used primarily on captive animals, but they can be used on treed or trapped free-ranging wildlife.

Powered blow pipes consist of a barrel connected to a pistol grip containing a metering device and reservoir. Air is compressed by a foot pump connected by a hose to the pistol grip or by CO_2 cartridges feeding into a reservoir that can be adjusted to increase or decrease the amount of pressure. These devices have a wide effective range of 1–20 m, because the dart flight distance is proportional to the pressure in the reservoir.

Dart Guns

Dart guns **propel darts by** either the gas generated from a .22 caliber blank cartridge, compressed CO_2, or compressed atmospheric air. Effective ranges can be as far as 75 m, but only for larger (>100 kg) animals. Guns can be equipped with a variety of sights; open sights are preferred by those who dart animals from helicopters. Rifle scopes make aiming easier, unless the animal is at close range, where the magnification of the scope makes it difficult to identify where on the animal you are aiming. Also, by closing the opposite eye when aiming through a rifle scope, other animals can easily walk undetected into your shot. Laser or "dot" sights can hardly be seen on very bright days, especially with snow cover.

Pistols are often overlooked because of their perceived limitations. Nonetheless, they can effectively deliver smaller darts ≤25 m, shoot a variety of dart sizes, and are easy to carry. Most pistols are powered by CO_2 (Fig. 4.2). Another overlooked gun is a rifle that compresses air by multiple pumping of a forearm handle. These rifles are lightweight, inexpensive, and versatile. They also have the advantage that no additional propellant is required (i.e., CO_2 cylinders or .22 blanks), so that you never run out of power. The main disadvantage is that compressed air does not develop the propulsive force, and thus the range, comparable to the other systems.

The **best dart rifles** are those that use some kind of metering device to adjust the distance (Fig. 4.2). They are powered by .22 blanks or CO_2 cartridges. Those powered by .22 blanks are much less expensive than CO_2 rifles, but .22 blanks can be frustratingly inconsistent in the distance they propel a dart. The best, and most expensive, rifles are powered by CO_2 and have a metering device that can precisely and consistently adjust dart velocity and thus range. These rifles usually can fire both 11-mm and 13-mm darts by switching barrels. Whatever dart gun is chosen, you should spend time practicing with different dart sizes and power settings.

Darts

Darts can be thought of as **flying syringes**, consisting essentially of a needle, body, plunger, and tailpiece. They differ in the manner in which the plunger is pushed forward to inject the dart's contents and in the materials of construction (Fig. 4.3). Darts discharge their contents by expanding gas from an explosive powder charge, compressed air, butane, or chemical reaction (acid-base). Darts using **explosive charges** expel their contents in 0.001 seconds and thus require large-bore needles to allow the rapid expulsion of liquid. Needle shafts can be smooth or equipped with a variety of barbs or collars to retain the dart in the animal. Smooth-shafted needles are used to deliver the drug and then fall out on their own, eliminating the need to capture the animal to remove the dart. Such darts are commonly used to remotely treat or vaccinate, but not necessarily to capture, animals. If the dart contents are under high pressure, however, smooth-shafted needles can rocket back out of the animal due to the expulsion of the liquid and therefore may not inject any or all of the substance. Either **wire barbs or metal collars** are used to securely retain the dart in the animal. Barbs probably should be used when the dart contains opioids, because barbed darts will stay in the animal, which will allow recovery of the dart as opposed to it falling out somewhere. Most wildlife agencies prefer to recover opioid-containing darts, even when discharged. **Other darts** have been devised that can either simultaneously mark the struck animal and attach a small radiotransmitter for location of the immobilized animal, or take a sample for DNA analysis.

Darts that use a **powder-based injection system** inject their contents explosively. All compressed-air or butane-gas injection systems inject their contents in approximately 0.5–1.0 second. If injection speed is rapid, the underlying tissue

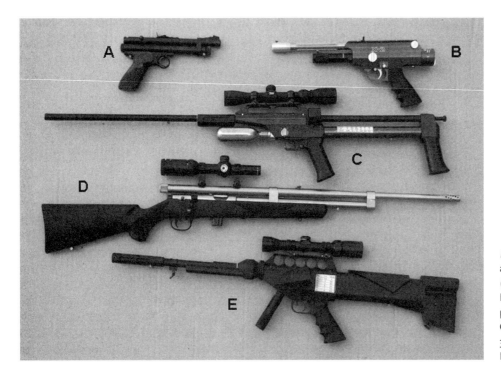

Fig. 4.2. Examples of versatile, adjustable dart projectors: (A) Cap-Chur Short-range Projector, (B) Pneu-Dart X2 CO_2 pistol, (C) Dan-Inject JM Special CO_2 rifle, (D) Pneu-Dart Model 389 .22 caliber rifle, (E) Pneu-Dart X-Caliber CO_2 rifle.

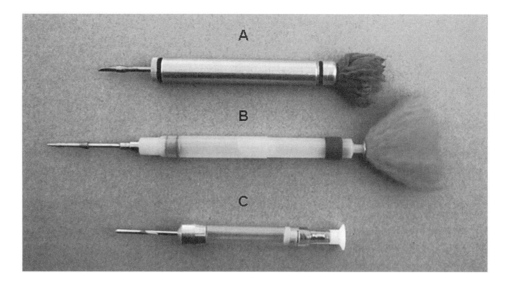

Fig. 4.3. Three common dart types used in North America: (A) Cap-Chur (powder-charge), (B) Dan-Inject (compressed air), (C) Pneu-Dart (powder charge).

will always be injured (e.g., hemorrhage). However, if it is slow, some animals (e.g., carnivores or primates) may have time to remove the dart before all contents have been injected. Lightweight darts may cause less impact when they strike the animal, unless they strike at high velocities, in which case they can cause more harm than heavier, slow-moving darts.

The contents of some darts are pressurized by compressing air or injecting butane into them when they are initially loaded. This type of dart is more prone to leaking or spraying its contents than are darts that do not develop any expulsion pressure until they strike the animal (e.g., powder). Few people use these pressurized darts with opioids because of the risk of leaking. Despite many opinions and unsubstantiated claims, there does not appear to be any difference in induction times among the different types of dart injection methods.

Range Finders

Range finders that utilize the principle of reflected laser beams to accurately measure distance are essential items of equipment when using dart guns. Inaccurate range estimation is probably the single major cause of missed shots.

Range finders are accurate to within ±1 m. Calibrate dart guns with known ranges and different dart sizes before the capture event. That is, set up a target range with measured distances of 10 m, 20 m, 30 m, 40 m, and 50 m (or farther). Fire the darts that you most commonly use (e.g., 1 mL, 2 mL, or 3 mL) at each range. Write down the power settings for each dart size and each range for which the dart is accurately and consistently hitting the target. Prepare a chart from this information and attach it to your dart gun. In the field, all you have to do is quickly get a range from the range finder, look at the chart, and set your gun for the appropriate distance (Fig. 4.4).

Thermometer

Many immobilizing agents disrupt an animal's thermoregulatory capability. Additionally, the physical exertion of being chased or restrained prior to immobilization often results in elevated body temperatures. Either **hyper- or hypothermia** can kill an animal. Thus, monitoring rectal temperatures is important. The glass mercury thermometer is basic equipment, although readings are slow to develop, and the devices are prone to breakage. Inexpensive, electronic digital thermometers sold in human drug stores are rapid and ac-

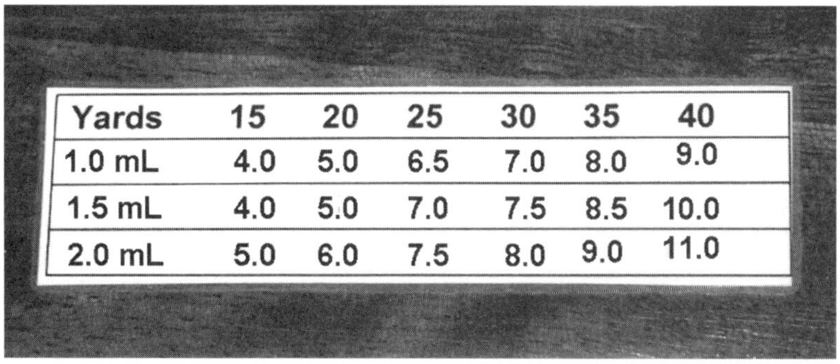

Fig. 4.4. All adjustable dart guns should be calibrated for various dart sizes and ranges. Make a chart of these data and attach it to the gun. This chart, combined with a range finder, will reduce misplaced darts and missed shots.

Yards	15	20	25	30	35	40
1.0 mL	4.0	5.0	6.5	7.0	8.0	9.0
1.5 mL	4.0	5.0	7.0	7.5	8.5	10.0
2.0 mL	5.0	6.0	7.5	8.0	9.0	11.0

curate, but you should always have spare batteries on hand. Other thermometers have a long, flexible temperature probe, allowing deeper probe insertion for large animals (that is essential to obtain accurate temperatures) or enhancing protection of the electronics box by placing it away from the animal.

Pulse Oximeter

Pulse oximeters are electronic devices that measure the percentage of oxygen saturation of hemoglobin in the blood (SpO_2). They provide information on the **respiratory function** of the animal that can be useful, because many immobilizing drugs depress respiration. Oximeters use a clip that can be attached to the tongue or other thin, nonpigmented tissue or a rectal probe to measure SpO_2. Ideally, SpO_2 should be >90%; however, animals anesthetized with potent opioids or alpha-2 adrenergic agonists often have SpO_2 values markedly below this value. The trend of SpO_2 values is more important than any specific value. That is, if the SpO_2 steadily decreases, it can be presumed the animal is in some sort of respiratory crisis.

Vital Signs Monitor

In unusual situations or when immobilizing particularly critical animals, monitoring cardiac function may be required. Additionally, such information may be useful when evaluating a new drug. Portable, rechargeable, **vital signs monitors** can simultaneously display electrocardiogram, blood pressure, temperature, and SpO_2.

CAPTURE PROCEDURE

After reading the above, you should be familiar with the drugs and the equipment used to capture animals. This section now puts this information together. The following "**rules**" have been developed through years of experience, and they should be recalled prior to any capture operation.

1. Most of the effort involved in the handling and treatment of a wild animal is expended in the capture process. Until you get some experience under your belt, you will not believe how much time it takes to locate, get close to, dart, and finally capture a wild animal. Thus, allow plenty of time to both capture the animal and monitor its recovery.

2. Always plan ahead and be prepared for any contingency. Whatever can go wrong, probably will (particularly when someone important is watching!). Most cases of animal loss can be attributed to human error, so think twice and act once. Before you set out to capture an animal, take a few minutes and mentally walk through the process. For each step, try to imagine what equipment is required and what can go wrong. Then make sure you have the appropriate supplies, drugs, etc. to respond to each event that you have visualized.

3. Do not stress about stress. It would be ideal if the capture operation did not stress the animal, but this is unlikely. It is impossible to capture an animal and not stress it. Probably the only way to minimize stress is to capture, process, and release the animal as quickly as possible.

4. When in doubt, dose high! More disastrous capture episodes have occurred when an animal was underdosed than when it was overdosed. An underdosed animal experiences some drug effect, but not enough to become immobilized. In this partially drugged state, it can injure itself, damage property, or escape entirely. Most capture drugs have high therapeutic indices, and high dosages rarely cause life-threatening sequelae. Any medical complications that arise from overdosing can be addressed when you get your hands on the animal. If you are trying to capture a highly stressed animal, increase your dosage by 20–25%. If you are not sure about an animal's weight, estimate it on the heavier side. Remember, the goal of remote capture is 1 dart, 1 animal down!

5. A captured animal becomes a valuable animal—both intrinsically and economically. The loss of any animal is regrettable; the loss of one from an endangered species is tragic. A great deal of personnel and equipment costs are invested in an immobilization; thus, care and treatment of the animal should not be trivialized.

6. Keep records. Good records are essential references for future captures, research, and analyzing disasters (see Box 4.2 for a sample animal capture record).

Considerations Prior to Animal Capture

The species, purpose, and circumstances of the capture must be considered prior to undertaking the drugging of any animal. You should ascertain whether the capture is really necessary and whether the risk of killing the animal justifies the planned gains. If committed to the capture, the following factors should be **considered prior to, during, and after** the process.

1. Species. Drug choice, drug doses, and animal response change across species and may vary within species. Adhering to an inflexible drugging protocol can easily end in disaster. Know the species in terms of body weight range, basic feeding habits, seasonal reproductive and condition cycles, habitat, and response to available drugs. Be prepared to adapt to conditions.

2. Free-ranging versus captive animals. Free-ranging animals usually require higher doses than do captive animals. Free-ranging individuals are usually highly excited and stressed in a capture situation (e.g., chased by a helicopter or trapped in a snare), whereas captive animals tend to be more docile and used to human activity and handling. Effective doses for

free-ranging animals can be twice those for captive individuals of the same species.

3. Age. Young animals (not neonates) require more and older animals usually require less drug per unit body weight than do prime-aged adult animals. Neonates usually require lower dosages. There is a higher risk of complications developing in older animals.

4. Weight. Most current drug dosage literature is based on milligram of drug per kilogram of body weight. Weight estimates accurate to ±20% are easy to make with experience, and doses based on such estimates should be safe. Keeping records of animal weight estimates coupled with actual weights after the animal is captured are useful reference sources. Animal weights may change with seasons or conditions (e.g., winter or drought).

5. Physical condition. A sick, exhausted, injured, or malnourished animal will usually require less drug than will a healthy, well-fed animal. Such compromised animals are high-risk candidates for anesthesia and frequently die after capture, despite everything being done correctly.

6. Pregnancy. Animals in late stage pregnancy may require more drugs for immobilization, but they also may experience more respiratory distress once immobilized, because the large uterus may impinge on the diaphragm. Although anesthetics may depress fetal respiration, there has been no evidence that immobilization during pregnancy results in fetal loss.

7. Psychological condition. As the excitement level of the animal increases, the chances of a successful capture decreases. The calmer the animal, the safer and smoother will be the procedure. An excited animal usually will require a higher drug dose. Failure to consider this phenomenon usually results in underdosing, which leads to even more excitement and increased chances of injury, trauma, hyperthermia, and capture myopathy (CM).

8. Weather. Adverse weather conditions, ambient temperature, and relative humidity must be considered when immobilizing an animal. During extremes of temperature (below −15° C [5° F] or >33° C [91° F]), equipment or facilities should be available to prevent and treat hypo- or hyperthermia. The physiological effects of the chosen drug on an animal's thermoregulation should be understood, so that its response may be anticipated.

9. Hazards. The physical environment must be considered both before and after the capture. A drugged animal cannot choose where it finally becomes immobilized. Water (including water bowls) always presents a drowning hazard. Falls from rocks, ledges, and steep slopes can injure a semi-conscious or ataxic animal. If predation or intraspecific aggression is possible, the animal should be protected or monitored until it recovers.

10. Drugs. Proper selection of the capture drug is critical. The best drug available that will provide the desired result should be selected. Compared to all other factors, drug costs are the least significant. If you cannot afford the proper drug, then the capture is probably unjustified. In general, the best drugs for ungulates are the opioids. An unsubstantiated fear of these drugs should not be an excuse for avoiding them, because less-effective drugs can be dangerous for both animals and humans.

Preparation

The considerations listed above constitute a mental checklist. The following is more of a **physical checklist.**

1. Have everything that you need with you. Before you begin the capture procedure, be sure that you have all drugs, supplies, and equipment that you may need. Fishing tackle boxes usually make good receptacles for all supplies. Vests with multiple pockets, such as a fly fishing or photographer's vest, can be used to carry most items, and they free the hands to carry such things as dart guns and pole syringes.

2. Prepare dart(s) beforehand. Have 1 or more darts loaded before you begin your approach. You will usually expend more darts than you would think possible: darts miss, bounce out, or fail to discharge. Be sure that all loaded darts are safely stored so as to prevent accidental injection. If you are working in freezing weather, be sure to keep the extra darts warm. It is generally best to load darts under controlled conditions, such as inside a heated building, where you can lay everything out and reduce the chance for drug or volume error. When loading multiple darts, do 1 step at a time for all darts to avoid mix-ups.

3. Check the gun before using it. Always inspect your dart gun prior to use to ensure that it is unloaded and the barrel is clean and clear. If you are using any form of electronic sights, be sure they are working (and always carry spare batteries!).

4. Do not load the gun until ready to approach the animal. Until you are actually in a position to approach and dart an animal, it is generally unnecessary to load your dart gun.

5. Approach captive animals quietly and calmly. Even if you are working with captive animals that are restrained in a chute or a trapped wild animal, you should approach it quietly and calmly. If captive animals are used to a routine, such as feeding or cleaning, try to mimic that activity (at the same time of day) to allow a closer approach.

6. Use devices to approach free-ranging animals. Approaching an animal close enough so that you can get a suitable shot with a dart gun can be frustrating. Free-ranging animals, if shot from the ground, are best shot from a blind overlooking a feeding station or some other device that draws the animal into range. Wild animals can often be approached quite closely with a vehicle or on horseback, but you must remain inside the vehicle or on the horse even when taking a shot. If using a vehicle or helicopter to pursue and dart animals, try to limit the length of the chase (5 minutes). Many ungulates have evolved for quick bursts of running only and are physiologically ill-equipped for long-distance pursuits. Such species, if run too hard, will survive the immobilization process only to die several hours or even weeks later due to CM or stress-related diseases.

7. Estimate distance and wind. As stated previously, be familiar with your dart gun and the different settings and distances. Darts are greatly influenced by wind and can drift even at relatively short ranges. If possible, practice under windy conditions to get an appreciation for this factor.

8. Adjust for altitude. A dart gun that is sighted in at one altitude may perform quite differently at another. For example, a gun sighted in at an altitude of 2,000 m will shoot high at 4,000 m and low at sea level (0 m).

Administration Sites

Immobilizing drugs are almost always administered as **IM** injections. The usual injection sites are the large muscle masses of the proximal hind limb and forelimb, with the former being the most commonly used (Fig. 4.5). Areas of large fat deposits also should be avoided, as drug absorption from these sites is slow and unpredictable. For example, bears (*Ursus* spp.) should be injected in the lower regions of the hind limbs or in the shoulder to avoid the fat deposits around the rump.

IV administration is usually reserved for antagonists. IV administration of anesthetics should be done with caution, because the onset of action is often quite rapid and, in some cases, respiratory depression or arrest can occur.

Oral administration is not often used in wildlife capture primarily because of the difficulty in predicting the dose the animal receives. Oral administration is most often used to capture birds, but it has been used to at least partially tranquilize carnivores.

Immobilization

Familiarity with the **signs of immobilization** is essential. You can assess drug effect through changes in behavior, so it is critical to be familiar with the target species. Know what is normal and look for the abnormal. Once the animal is down, you need to assess the depth of anesthesia. Always exercise caution when checking a downed animal. Approach

Fig. 4.5. The preferred areas for intramuscular injection are the large muscles of the hindquarters or the shoulder. In elk and many other ungulates, the area on the rump where the light hair meets the darker hair is a good aiming point.

the animal slowly and quietly; approach dangerous animals from the rear and be sure that you have an escape route. If the animal appears unconscious, check for ear twitch (touch inside of ear), jaw tone (resistance to opening jaws), palpebral reflex (touch eyelashes, animal blinks), and corneal reflex (touch cornea, animal blinks). These reflexes are progressively lost as the animal becomes more anesthetized.

You should always **note the time** when the dart hit the animal. Allow 10–15 minutes (but never >15 minutes) to elapse after the first dart before giving booster doses. If the animal is showing signs of receiving some of the drug, administer 50% of the original dose. If no sign of drug effect is apparent after 10 minutes, then give the animal the same drug(s) and dose(s) again.

If the animal is down and can be handled, but continues to struggle, a low dose of a potent tranquilizer, such as xylazine or medetomidine, often calms the animal enough to allow safer and easier handling.

When an animal is finally down and can be safely handled, **you should:**

1. Position the body. Ruminants should be placed on their sternums; otherwise try to place them on their right sides. Most other animals can be placed on either side or on the sternum.

2. Cover the eyes. Covering the eyes protects them from harmful ultraviolet light from the sun, reduces drying, and prevents dirt and debris from entering them. Coating the eyes with a lubricant further prevents drying. Covering the

eyes also appears to further calm the animal even when it has been effectively immobilized.

3. Hobble the legs. This step is particularly necessary with ungulates to avoid spontaneous kicking that may injure someone. Hobbles also prevent possible escape, should the animal partially or spontaneously recover.

4. Check vital signs. Respiration and temperature are the 2 most important vital signs to monitor. Cardiac impairment due to drugs is rare.

5. Check for wounds, injuries, and general condition. Start from the nose and work toward the tail. Look for blood, swelling, hair loss, and abnormal body configuration (e.g., fractures or luxations).

6. Do not make loud or sharp noises. Animals that have been anesthetized with opioids often spontaneously respond to loud or sharp noises, such as a slammed truck door. The response is usually a kick, but such animals may try to stand. Ear plugs are not recommended, as there will come a time when you will forget to remove them, and most animals cannot dislodge ear plugs on their own.

Urban Wildlife Capture

Many wildlife immobilizations occur in urban settings, when a wild animal wanders into town and poses a hazard not only to itself, but also to people, pets, and property. Probably the most common species that come to town are deer and bears, but moose (*Alces alces*) and mountain lion (*Puma concolor*) are not uncommon. Most of these animals will be quite stressed, because they have been chased by children, dogs, and the police—remember to dose high because of this condition. The above considerations and **preparations** apply in these circumstances **in addition** to the following:

1. Be professional. Urban wildlife capture can be quite stressful because of the public and media scrutinizing your every move. Have a plan before you do anything. Be organized and stay calm, no matter how exciting things might get. Most importantly, assess the situation carefully and try to anticipate any and all problems. Do not just rush to the scene and dart the animal without some consideration of where the animal might go once darted or how you are going to extract it, if it is in a difficult spot.

2. Communicate. Talk over your plan not only with your co-workers, but also with the police and other public officials. If the animal is in a dangerous place and could get hurt (or could look like it was hurt afterward) during the immobilization process, it is wise to communicate your concerns to public officials or the media (but see later). If you apprise everyone that something might go wrong—and it does—then you won't look like an idiot for not having anticipated the outcome. The public does not like to see animals hurt, but generally people are more forgiving when they are forewarned that bad things could happen.

3. Use the right drugs. Your goal in any wildlife capture, and especially urban wildlife capture, is to hit the animal with one dart and immobilize it quickly (10 minutes). Underdosing or using ineffective drugs results in the animal running through town in a partially drugged state, potentially damaging property, itself, or humans. The only effective drugs for immobilizing many ungulates are the opioids, and they should be used despite unfounded fears for public safety. Only the opioids guarantee that once struck, the animal will be immobilized. All bets are off with any other drug combination.

4. Use a spotter. Always have someone accompany the shooter. This person's sole job is to follow the flight of the dart, which can be difficult for the shooter. The spotter should note where the dart hit the animal or, if missed, where the dart landed. Every effort should be made to find missed darts, and a spotter can greatly increase the odds of finding them.

5. Control the public. The police can be of great assistance in accomplishing this task. They understand crowd control, and it also gives them something to do instead of trying to do your job, for which they are untrained.

6. Think about public perception. Even though drugged bears can probably fall dozens of feet out of a tree without getting hurt, it does not appear this way to the public (or the evening news reporter). If the animal is truly in an untenable situation for drugging, it may be wiser to wait the animal out. Many animals will come down a pole or out of a tree when the public leaves and darkness falls. Otherwise, make a show of looking like you are trying to break the animal's fall. Using a tarp is the most common method. A 75-kg bear falling from a 9-m tree generates >415 kg of energy. Four guys holding a tarp are not going to break its fall, but it looks like you did to the public.

7. The media come last. Talk to the media only when the animal is safely in hand. The media might get quite insistent, but you have no obligation to talk to them and waste time better spent catching the animal.

8. Treat the animal respectfully. Treating the downed animal roughly by dragging it on the ground or throwing it into the truck does not set well with the public. Hobble the animal if it is an ungulate and blindfold all animals. Carry them in a tarp or other device.

9. Talk to the public afterward. Once the animal is safely in hand and stable, you might spend a few moments with the public explaining what you did, what age and sex the animal is, and where you are going to take it for release. Although wildlife biologists can take these animals for granted, it should be kept in mind that very few of the public ever gets a chance to see these animals up close. Answer questions; let kids touch the animal; it does not cost anything except some of your time—and your time will be well spent.

Transport

There is much discussion on the appropriate method of **transporting animals** over long distances (e.g., should animals be moved while anesthetized or while awake?). Anesthetized animals may stop breathing or overheat in transit, because they are not continually monitored. However, awake animals may pace continuously, jump, kick, trample one another, overheat, or develop CM. Each species and situation is different. If the species is fairly calm, such as moose, elk, or bighorn sheep (*Ovis canadensis*), then transporting it in an awake, revived condition is best. If the animal is hyperactive, such as deer or pronghorn (*Antilocapra americana*), then moving it while anesthetized (if the animal is stable) might be best. **Tranquilizers,** such as diazepam, might help calm hyperactive species. The males of many ungulate species should have their horns removed or covered with piping or tubes, or they should be transported in individual crates.

Transport trailers should be dark, but not so enclosed as to preclude adequate air circulation. Animals crowded in a closed trailer can quickly overheat, even at subfreezing temperatures. It is more important to allow for good air circulation than it is to have a darkened trailer. Many North American ungulates have been transported safely in unmodified, standard stock or horse trailers.

Carnivores should be crated and shipped separately. They can be transported anesthetized and allowed to recover in the crate. Blocks of ice in a pan can provide needed moisture while preventing spills.

Recovery

An animal recovering from anesthesia should not be left unattended. Ideally, you should remain with the animal until it can walk in a relatively coordinated manner. At the minimum, you should **stay with the animal** until it can at least raise itself to a sternal position. Keep the animal cool or warm (depending on weather conditions), dry, and free from inter- or intraspecific harassment or aggression.

Look around the recovery area for possible hazards, such as sharp rocks, ledges, or water. Either relocate the animal or stay with it through recovery, so as to direct it away from such hazards. When **releasing** animals that have been transported in a trailer, be sure the release area is free of obstacles in the immediate area. Many ungulates simply walk out of the trailer when the door is opened, but some leap out and run in a panic. Enough clear area should be available to allow the animal to quickly orient itself and avoid obstacles. For these same reasons, try to never release animals at night.

Some species may benefit from being released in a temporary enclosure containing food and water. Subsequent release from this site, hours or days later, may result in the group staying together instead of fragmenting as well as their staying closer to the original site.

Euthanasia

Invariably, there will come a time when an animal must be euthanized because it has been critically injured or is terminally ill. If an animal needs to be euthanized, it **should be** done safely and effectively, with some consideration for the dignity of the animal and the sensitivities of the public. Many methods of euthanasia, such as shooting and stunning, are effective and medically acceptable, but are reprehensible to the public. A detailed discussion of euthanasia methods can be found in American Veterinary Medical Association (2001).

For most wildlife, shooting the animal in the head is probably the best method of euthanasia. A "field friendly" method of chemical euthanasia is IV injection of potassium chloride (KCl). A saturated solution can be made by adding about 300 mg of KCl per mL of any kind of water. Shake vigorously and immediately draw into a syringe, as the KCl will quickly settle out of this saturated solution. The solution must be quickly injected (slow, drawn-out administration will not be effective). Administer at a dosage of ≥ 50 mg KCl per kilogram of body weight. The animal should be anesthetized before potassium chloride is administered. With the exception of succinylcholine, an overdose of almost any capture drug will not euthanize the animal.

ANIMAL EMERGENCY MEDICINE

This section is intended to familiarize you with the most common medical emergencies encountered in the chemical capture of wild animals. Many captures are conducted in the field, where monitoring and emergency equipment will be minimal: this section addresses emergencies that arise in such circumstances. Thus, your ability to assess problems will be limited to what you can see, hear, or feel. The conditions listed below are more or less in the order of likelihood of occurrence.

1. **Respiratory depression or arrest.** Respiratory depression or arrest is probably the most common complication encountered in wild animal immobilization. The best advice we can give concerning respiratory arrest is not to panic. You probably have ≥ 5 minutes before irreversible hypoxic brain damage occurs. The tongue, if unpigmented, is a good indicator of oxygenation. If the tongue is pale pink or gray,

Fig. 4.6. Artificial resuscitation can be accomplished on most animals by placing them on their sides and compressing the chest 15–20 times per minute.

you will need to assist ventilation. Ensure the airway is clear of any obstruction. The simplest treatment is artificial resuscitation by laying the animal on its side and pushing down firmly on the chest, 15–20 times per minute (Fig. 4.6). Alternatively, you can: (1) provide supplemental oxygen if you have it; (2) inject 1–2 mg/kg doxapram IV (a respiratory stimulant); (3) try acupuncture by inserting a needle into the upper lip just between and below the nares (Fig. 4.7); or (4) if all else fails, give appropriate antagonists.

2. **Hyperthermia.** Severe hyperthermia (>41° C or 106° F) is a medical emergency, and you must cool the animal immediately. Obtaining a rectal temperature should be one of the first steps taken as soon as the animal can be safely handled. Monitor the temperature throughout the immobilization period. Immersing the body in water (pond, stream, or water tank) is probably the fastest way to cool an animal (Fig. 4.8). Alternatively, you can: (1) spray the entire animal with water, particularly the groin and belly; (2) pack ice or cold water bags on groin and head; (3) douse with isopropyl alcohol (rapid evaporation cools quickly); (4) administer a cold water enema; or (5) give appropriate antagonists.

3. **Hypothermia.** Body temperatures that fall 24° C (75° F) invariably result in death. Because the animal's metabolism is slowed during hypothermia, drug effect is usually prolonged, and recovery will be slow. The only treatment for hypothermia or frostbite is: (1) warming the animal or affected part by placing containers of warm water on it; (2) wrapping it in blankets with some method of external heat applied (heat pad, lights, or hand warmers); or (3) put a small animal inside your coat and warm it with body heat. Regardless of the method used, expect a slow recovery back to temperatures suitable for release of the animal (38° C or 100° F).

4. **Shock.** Shock is a clinical syndrome characterized by ineffective blood perfusion of tissues, resulting in cellular hypoxia. Shock is often seen in animals that have undergone a stressful or strenuous capture or handling. Like CM, there may be little that you can do to treat shock, except prevent it from happening in the first place. Signs of shock include rapid heart rate, low blood pressure (slow capillary refill), pale gums, and perhaps hyperventilation. Administering IV fluids is standard therapy, but few workers have the necessary supplies in the field. Otherwise, consider administering 5 mg/kg dexamethasone IV, give appropriate antagonists quickly, and release the animal.

5. **Bloat.** Bloat develops when gas from normal fermentation accumulates in the rumen of ungulates or from gas-

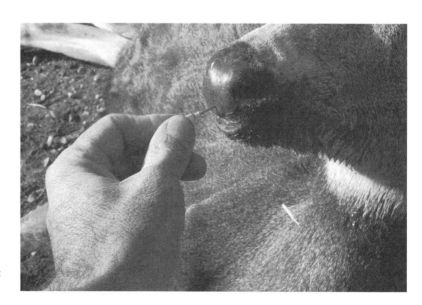

Fig. 4.7. Ungulates and carnivores can be stimulated to breathe by using acupuncture just below the nares. Twirl the needle if a breath is not taken soon after insertion of the needle.

Fig. 4.8. Whole body immersion is probably the quickest and most effective way to cool an overheated animal.

forming bacteria in carnivore stomachs. The rumen or stomach enlarges, compressing the diaphragm and lungs and so impairing respiration. A stomach tube should be standard equipment when immobilizing ruminants, because bloat is a common sequelae to chemical capture. Often, however, you will be through with the procedure and will have antagonized the capture drugs before bloat develops to the point of causing complications. Signs of bloat include increase in abdomen size, labored breathing (rapid and/or shallow), and increased salivation. Treat bloat by positioning the animal on its sternum, passing a stomach tube, or administering the appropriate antagonist.

6. **Vomiting/Aspiration.** Vomiting in and of itself may not be a problem; the aspiration of the vomitus into the respiratory system is. Not only can the animal choke on its vomit and die, the mere aspiration of just a small amount of stomach contents can inoculate the lungs with bacteria, resulting in pneumonia. The pneumonia may not develop for days—long after the animal has been released and often beyond further treatment. Aspiration of large amounts of vomitus has a grim prognosis for the animal, and euthanasia may have to be considered. Prevention by placing the animal on its sternum at all times and clearing any food, vomitus, or other matter from its mouth is about all that can be done. If aspiration is suspected, administer antibiotics and hope for the best.

7. **Capture Myopathy.** CM is a complex condition affecting animals that usually have undergone a particularly stressful or strenuous capture, handling, or transport. It is invariably associated with severe or prolonged physical exertion, but psychological stress is suspected as an important initiator of CM. It occurs predominantly in ungulates. The animal may appear weak, and unable to rise, and its urine may be dark colored. There is no known effective treatment for CM; prevention by not overly exerting animals is the only option.

8. **Seizures/Convulsions.** Most seizures seen in chemical immobilization are due to the use of ketamine, either when used alone or in conjunction with the alpha-2 adrenergic or phenothiazine tranquilizers. Usually seizures do no harm to the animal, but they disrupt handling of the animal and can lead to hyperthermia if left untreated. Seizures accompanying ketamine immobilizations are most common during induction and recovery from anesthesia. In most cases, seizures can be effectively treated by administering 5–10 mg diazepam or midazolam IV slowly or IM.

9. **Wounds.** Antibiotics should always be given to animals that have been darted, particularly those shot with the powder-charged darts. The large-bore dart needles often inoculate surface bacteria, resulting in debilitating abscesses. A combination of procaine penicillin G and benzathine penicillin G provides both fast, high blood concentrations (procaine) plus prolonged therapeutic concentrations (benzathine; 5–7 days). Alternatively, oxytetracycline can be given, but it will not last as long as penicillin. Other wounds should be lavaged with a povidone-iodine solution, 2% chlorhexidine scrub solution, or sterile saline.

HUMAN EMERGENCY MEDICINE

There are many agents used in animal anesthesia that can be harmful to humans. **Accidental exposure** can occur in many ways, but most commonly, drugs are sprayed in the eyes or mouth or injected via a syringe with an unprotected needle. Accidental injection by being hit with a dart or receiving the full dose by some other method is rare. Despite hundreds of thousands of animal captures using drugs, there have been only 2 human deaths due to injection of etorphine. The following are some precautionary steps to take and rules to follow that should decrease the chances of accidental drug exposure.

Preventative Measures

1. Obtain competent training. Safe drug handling and use should not be a self-taught course. Attend courses taught by experienced instructors on the use of capture drugs. Unfortunately, there are several individuals teaching classes about the chemical capture of wildlife, but some of them are woefully inexperienced and misinformed.

2. Be trained in basic first aid and **cardiopulmonary resuscitation (CPR)** techniques. Ideally, everyone involved in the immobilization effort should have this training. Murphy's law dictates that if only one person on the team has such training, that would be the person who needed medical help!

3. Always work in pairs. This condition is absolutely essential when working with drugs that are potentially toxic.

4. Always have appropriate antagonists immediately available. You may not have a second chance to remember the antagonist back in the truck or office. You do not need to have a syringe preloaded with the antagonist; just have the antagonist, syringe, and needle together and close at hand.

5. Wear protective gloves and eye protectors. Drugs can be spilled, sprayed, dripped, dropped, slopped, and leaked in more ways than you can imagine. Do not spray drugs into the air and do not hold loaded syringes in your mouth. Also, do not smoke, eat, drink, rub your eyes or mouth, or work when you have open sores when working with immobilizing drugs.

6. Carefully withdraw drugs from vials. Do not inject excessive air into drug vials; equalize air pressure in vials with a needle before withdrawing the drug. This precaution is particularly important if you work at different altitudes with the same drug vial (vials used at low altitudes will develop high internal pressures at high altitude). Tap the syringe to clear air bubbles only if needed. Use a small-gauge needle (e.g., 21, 22, or 25 gauge) to withdraw the drug, because large-gauge needles will create holes in the rubber stopper from which drug can leak.

7. Avoid using pressurized darts when using potent drugs. Darts whose contents are under pressure by air or butane have their needles capped with a silicone plug or sleeve. When these darts are pressurized, there is a possibility that drug will leak from the sleeve or the needle will fly off. If you must use pressurized darts, place the needle into a test tube or other device that will contain the drug should it leak.

8. Treat syringes and darts as if they were guns. That is, always consider them "loaded" and watch where you point them. Do not carry loaded, unprotected syringes in your pocket; hold them in your hand with a protective cap on them or carry them in a protective case, such as a test tube or cigar case.

9. Know what you are using. Make sure the contents of bottles, tubes, and loaded syringes are marked. If you do not know what a drug is, do not use it!

10. If possible, notify the local emergency care center in advance. Most physicians are ignorant of the drugs used in wildlife anesthesia, and they are unfamiliar with their potency, symptoms, and antagonists. A little communication with the local hospital could save valuable time in an emergency. Either provide the staff, or have on hand, the package inserts of the drugs that you will be using. This information can help the attending physician develop an appropriate treatment. There was a case in which a biologist had injected himself and was refused treatment because of the hospital staff's unfamiliarity with the drug.

11. Be careful with used darts, syringes, and needles. All these items will have residual drugs on them, and many exposures have occurred as a result of careless handling. Store and dispose of used needles and syringes with care. It is very easy to jab a finger when recapping needles—particularly when you are in a hurry or distracted; therefore, used needles **should not be recapped.** Used needles must be placed immediately in a rigid container (sharps disposal container) for disposal. They should never be placed in a plastic trash bag that could permit a needle to penetrate, resulting in the possibility of a puncture wound to someone handling the bag. To comply with state and local waste disposal regulations, the contents of needle collection containers must be encapsulated before they can be placed in a trash dumpster that will go to the local landfill. **Encapsulation** (i.e., formation into a solid matrix) may be accomplished in several ways, such as adding plaster of Paris or melted paraffin to the container.

12. Always clean used darts with care. Wear gloves and goggles when cleaning darts containing potent drugs. You also can submerge reusable darts under water before disassembling, which helps dilute any residual drug.

Rules for Accidental Exposure

The previous discussion was concerned with preventative measures. Below are **rules to follow** when there has been an actual accidental exposure.

1. Do not panic. Stay calm and try to determine how much drug could have been delivered. If there is doubt that a significant amount of drug has been absorbed, you may wish to quietly wait to determine whether any signs of exposure develop. If there are no signs in ≤30 minutes, you can probably assume the amount was clinically insignificant. However, if you know the person has received a significant drug exposure, you want to work fast, but always under control. Panic can obfuscate your thinking processes and costs time. Panic in the exposed person also may cause symptoms, such as lightheadedness, fainting, or hyperventilation, that can be misinterpreted as a drug effect.

2. Tell others of the accident. It is not the time to be embarrassed or feel stupid after you have accidentally injected yourself. Tell someone about the accident immediately!

3. Wash the site. Irrigating the injection or exposure site (particularly mucous membranes) with large volumes of water will greatly reduce further drug absorption. Also, if the drug was sprayed on the skin or clothing, residual drug might be inadvertently picked up and delivered to the mouth or eyes at a much later time. Washing the entire sprayed area can help prevent this secondary exposure.

4. Administer the appropriate antagonist(s). This action applies primarily to opioid exposure, but it also includes exposure to the potent alpha adrenergic agonists, such as medetomidine. Administer antagonists only if you know the person has received a significant amount of drug or is demonstrating any symptoms.

5. Note the time. Despite the harried circumstances surrounding accidental drug exposure, try to remember when the exposure occurred and when treatments were administered. This time could be valuable in assessing the amount of drug absorbed as well as determining an appropriate treatment regimen.

6. Transport the victim to the nearest emergency center. If feasible, have the person walk to the vehicle, but not at a strenuous rate. If the person requires CPR or otherwise cannot be transported, send for help. If there is no one available to send, stay with the patient! Although the exposed person may be conscious when you leave for help, more drug will be absorbed over time, and the person may lapse into a coma, leading to respiratory depression and death.

Specific Emergency Treatments
Opioids: Sufentanil, Carfentanil, Etorphine, and Thiafentanil
Symptoms of opioid toxicity include dizziness, incoordination, lethargy, sedation, nausea, vomiting, pinpoint pupils, slow breathing, collapse, and unconsciousness. If any of these symptoms develop, administer at least 10 mg naloxone, naltrexone, or nalmefene. If you can give the antagonist IV, so much the better, but do not waste time if you cannot locate a vein. If IV is not possible, give IM in the large muscle masses of the shoulder or thigh. Do not administer under the tongue. If there is no improvement in the patient's condition ≤1 minute after giving the antagonist IV (or ≤3–5 minutes for IM administration), repeat the dose. Continue to repeat this dose every 3–5 minutes until central nervous system depression is antagonized. Transport the patient to an emergency center as soon as possible, even if he or she has recovered.

Cyclohexanes: Ketamine and Tiletamine
Symptoms of cyclohexane exposure include disorientation, hallucination, excitement, abnormal behavior, and loss of consciousness. Primary treatment is to transport the patient to an emergency center while monitoring for any respiratory depression. Administer 10 mg of diazepam if patient has convulsions. You can administer diazepam or midazolam IV or IM. If given IV, administer slowly (10–15 seconds). Remember that ketamine and tiletamine are congeners of phencyclidine, also known as PCP or Angel Dust, so the most likely result of cyclohexane injection is abnormal behavior.

Tranquilizers or Sedatives: Xylazine and Medetomidine
Symptoms of alpha-2 adrenergic sedative exposure include decreased respiratory and heart rate, decreased blood pressure, sedation, dizziness, nausea, slurred speech, and unsteady gait. Treatment should support respiration if the patient's respiratory rate falls to 6 breaths per minute or if lips and gums become pale or bluish. If symptoms are severe (respiratory arrest or collapse), administer atipamezole slowly IV and monitor for antagonistic effects. An infusion rate of 5 mg atipamezole over 5 minutes or 100 mg over 20 minutes has been used in healthy adults. For xylazine intoxication only, you also can administer 0.125 mg/kg yohimbine IV; yohimbine can be given IM, but response can take ≤20 minutes.

Death due to overdose of phenothiazine and butyrophenone tranquilizers is rare and highly unlikely, given the doses of these drugs used in wildlife capture. Exposure to these agents would most likely be in conjunction with, and therefore exacerbate the effects of, primary immobilizing agents (opioids and cyclohexanes).

DRUG RECOMMENDATIONS

It is beyond the scope of this chapter to provide all the various drug dosages for all North American species. The reader is referred to Kreeger and Arnemo (2007) for a complete listing of recommended and alternate drug dosages. This section provides an overview of vertebrate immobilization with general suggestions for drugs and dosages.

Fish
Immerse fish in a solution of **tricaine methane sulfonate** (MS-222), 3–10 mg/100 mL water (0.003–0.01%). The longer the fish is immersed in the anesthetic solution, the longer the duration of effect will be. Use higher concentrations for smaller fish and lower concentrations for larger ones. Revive the fish by placing it in clean water. For very large fish, such as sharks, try 20 mg/kg ketamine plus 10 mg/kg xylazine.

Amphibians
Immerse amphibians in a solution of **tricaine methane sulfonate** (MS-222), 2–10 mg/10 mL water (0.02–0.1%). The longer the amphibian is immersed in the anesthetic solution, the longer the duration of effect will be. Always induce immobilization with the lowest concentration possible. Use higher concentrations for smaller amphibians and lower

concentrations for larger ones. Do not completely submerge the animal to prevent drowning. Revive amphibians by washing repeatedly in warm, clean water. Gas anesthesia also works well for amphibians.

Reptiles

There is no entirely satisfactory immobilization regimen for reptiles. Straight **ketamine** can be used for most species. For reptiles weighing between 1 kg and 20 kg, give 25 mg/kg ketamine; for reptiles weighing 20–50 kg, give 12 mg/kg. For large alligators (*Alligator mississippiensis*), give 15 mg/kg ketamine plus 1 mg/kg xylazine. Complete recovery may take days. **Succinylcholine** also can be used on alligators. Inject drugs into the hind legs, the side of the tail just behind the hind legs, or the large jaw muscles.

Birds

Birds should be anesthetized only when necessary, and **gas anesthesia** is preferred. If gas is unavailable, inject drugs into the breast or thigh muscles. For passerine birds weighing between 0.5 and 1.0 kg, give 10 mg/kg ketamine plus 2 mg/kg xylazine. For nonpasserine birds weighing 1 kg, give 15 mg/kg ketamine; for birds weighing 1–5 kg, give 10 mg/kg **ketamine;** for those >5 kg, give 2.5 mg/kg ketamine plus 0.5 mg/kg xylazine. Ducks can be given 50 mg/kg tiletamine-zolazepam (Telazol); geese can be given 20 mg/kg Telazol. Hawks and falcons can be given 30 mg/kg ketamine plus 1.5 mg/kg diazepam IV; eagles, 15 mg/kg Telazol; owls can be given the same dosage as hawks or 10 mg/kg Telazol. If using Telazol, monitor closely for excess salivation to prevent aspiration.

Mammals

If you cannot find a drug dosage for your species, try an initial dosage of 3.0 mg/kg **ketamine plus** 0.1 mg/kg **medetomidine.** This dosage is based on the mean of this combination reported in 56 species. Wait at least 30 minutes after administration of the ketamine-medetomidine to antagonize with 0.5 mg/kg atipamezole.

Tiletamine-zolazepam (Telazol) has been used safely in >200 vertebrate species and is a good drug to try if you have no other suggested drug for a mammal. It is available in lyophilized (freeze-dried) form with 572 mg (286 mg of each drug) per vial. Manufacturer's instructions call for adding 5 mL sterile water to the vial, resulting in an approximate concentration of 100 mg/mL. You can increase this concentration by adding less water. You also can increase the effectiveness of this drug by adding tranquilizers, such as xylazine or medetomidine, instead of water. For example, you can add 2 mL of 100 mg/mL xylazine. One vial of this mixture should immobilize most medium-sized mammals (≤100 kg); 2 vials should work on larger animals (100–250 kg). Alternatively, you can add 1 mL of 100 mg/mL ketamine plus 1 mL of 100 mg/mL xylazine. This combination is good for carnivores, such as bears. The same doses would apply (e.g., 1 vial for ≤100-kg bears).

Ungulates

Cloven-hoofed ruminants are probably the **hardest mammals to immobilize** and are prone to many of the medical problems listed above. The opioids overwhelmingly are the most effective drugs for these animals. For example, **pronghorn,** despite their small size, are almost impossible to effectively immobilize with any drug combination other than the **opioids. Cyclohexane/alpha-adrenergic combinations** have been used to capture ungulates for years, but darted animals often overcome the drugs and rise and stumble away when approached. Two suggestions to decrease this possibility are (1) do not attempt to immobilize excited animals (e.g., chased) and (2) dose high!

Regardless of the drugs used, when you initially get your hands on the downed animal, immediately monitor respiration and temperature and continue to do so. Hobble the front legs to the back legs and place the animal on its sternum, if at all possible. Blindfolding also might help calm ungulates given opioids.

Table 4.2 suggests dosages for various North American ungulates, but there are many other drug combinations and dosages that will work. Most of these suggested dosages are a combination of an **opioid or cyclohexane plus an alpha-adrenergic sedative,** with the exception of the dosage for moose. Moose appear to be very sensitive to sedatives, and

Table 4.2. Recommended immobilizing dosages for ungulates

Species	Immobilizing dosage	Antagonist
Bison (*Bison bison*)	0.005 mg/kg carfentanil plus 0.07 mg/kg xylazine	0.5 mg/kg plus 1 mg/kg tolazoline
Caribou (*Rangifer tarandus*)	2.5 mg/kg ketamine plus 0.25 mg/kg medetomidine	1.25 mg/kg atipamezole
Deer, mule	4.4 mg/kg Telazol® plus 2.2 mg/kg xylazine	2.0 mg/kg tolazoline
Deer, white-tailed	4.4 mg/kg Telazol plus 2.2 mg/kg xylazine	2.0 mg/kg tolazoline
Moose	0.01 mg/kg carfentanil	1.0 mg/kg naltrexone
Mountain goat (*Oreamnos americanus*)	0.035 mg/kg carfentanil	3.5 mg/kg naltrexone
Pronghorn	0.05 mg/kg carfentanil plus 1.0 mg/kg xylazine	5.0 mg/kg naltrexone plus 2.0 mg/kg tolazoline
Sheep, bighorn	0.05 mg/kg carfentanil plus 0.2 mg/kg xylazine	5.0 mg/kg naltrexone plus 2.0 mg/kg tolazoline

Fig. 4.9. Moose immobilized with only opioids (carfentanil, thiafentanil, or etorphine) usually remain sternal with head up. The addition of a tranquilizer increases the probability of pneumonia, because moose have a tendency to roll over and aspirate rumen contents.

it is recommended they be avoided to prevent excess muscle relaxation, which leads to rolling over, regurgitation, aspiration, and development of lethal pneumonia. **Moose** immobilized with just opioids most often remain sternal with their heads up (Fig. 4.9).

Bighorn sheep and mountain goat (*Oreamnos americanus*) are **difficult to chemically immobilize,** and no entirely satisfactory regimen has been found. Regardless of the drug combinations given, sheep and goats often continue to struggle when handled.

Carnivores

Just the opposite of ungulates, carnivores are probably the **easiest mammals to immobilize** with the fewest medical complications. Hyper- and hypothermia are perhaps the most common medical issues. Carnivores can be kept immobilized for hours by administering booster doses. Multiple animals can be safely immobilized and processed at the same time (e.g., wolves, bears, and cubs) if you are organized and have a dedicated person responsible for ensuring that vital signs are routinely checked and all samples and measurements have been taken. Carnivores do not need to be hobbled or blindfolded, but blindfolds can help to keep debris out of the eyes, which usually remain open. Opioids are generally not used on carnivores because of severe respiratory depression.

BEARS

Most bears are initially trapped or are found hibernating and then immobilized. Brown (*Ursus arctos*) and polar (*U. maritimus*) bears are often darted from helicopters. All North American bears can be safely immobilized with 8 mg/kg Telazol. Expect prolonged recoveries (2–4 hours). Ketamine-xylazine combinations have been used for decades on black bears (*Ursus americanus*). Ketamine-medetomidine also is effective, but bears must be constantly monitored for signs of recovery, which can be (heart stoppingly) sudden and complete. Extreme caution should always be used when capturing bears to ensure they are completely immobilized. It may be wise to use a snare to anchor the bear to a tree in the event of sudden arousal. Try to minimize stimulation, such as vocalization of cubs.

CANIDS

Wolves, coyotes (*Canis latrans*), and foxes can all be immobilized with 10 mg/kg Telazol. Many other combinations also are effective. Ketamine-medetomidine with atipamezole antagonism offers the quickest recoveries, if that is important. Medical problems are few. Canids are usually trapped and then immobilized, although wolves can be darted from helicopters.

FELIDS

The felids also are usually trapped or treed before immobilization. Big cats, such as mountain lions, can be immobilized with 5 mg/kg Telazol; give small cats (bobcat [*Lynx rufus*], lynx [*L. canadensis*]) 10 mg/kg Telazol. Ketamine-medetomidine also works well on felids and offers quicker recoveries than does Telazol.

MARINE MAMMALS

Marine mammals (seals, sea lions, and walrus [*Odobenus rosmarus*]) are difficult animals to chemically immobilize and should only be attempted with an experienced biologist present. A major problem is preventing them from returning to the water after being darted or injected. If the animal is small enough to be physically restrained, induction and maintenance with gas anesthesia is preferred. If the immobilization is going to be prolonged, initially immobilize with injectable drugs and then maintain on gas. There are many cyclohexanes/tranquilizer combinations published for a variety of marine mammals, and the reader is again referred to Kreeger and Arnemo (2007) for a complete list. Opioids are generally not used on marine mammals, except for sea otters (*Enhydra lutris*) that are trapped in floating nets and given 0.3 mg/kg fentanyl plus 0.1 mg/kg diazepam. If no published dosage can be found for your species, initially try 2 mg/kg Telazol.

MUSTELIDS

Mustelids are usually trapped and given cyclohexane combinations. Small mustelids can be physically restrained and placed in a gas anesthesia box (Fig. 4.1). Medical complications are few (see Table 4.3 for representative species and suggested dosages).

MISCELLANEOUS SMALL MAMMALS

Most other small mammals can safely and cheaply be anesthetized with ketamine-xylazine combinations or Telazol (Table 4.4). Others can be trapped and placed in a gas anesthesia chamber, but this treatment is more difficult with aquatic mammals, such as beaver (*Castor canadensis*) and muskrat (*Ondatra zibethicus*), because they will hold their breath for long periods of time.

Table 4.3. Recommended immobilizations dosages for mustelids

Species	Immobilization dosage
Ferret (*Mustela nigripes*)	35.0 mg/kg ketamine plus 0.2 mg/kg diazepam
Fisher (*Martes pennanti*)	10.0 mg/kg Telazol®
Marten (*M. americana*)	18.0 mg/kg ketamine plus 2.0 mg/kg xylazine
Mink (*Neovison vison*)	15.0 mg/kg Telazol
Otter (*Lontra canadensis*)	4.0 mg/kg Telazol
Skunk (*Mephitis* spp.)	10.0 mg/kg Telazol
Weasel (*Mustela* spp.)	5.0 mg/kg ketamine plus 0.1 mg/kg medetomidine

Table 4.4. Recommended immobilizations dosages for other small mammals

Species	Immobilization dosage
Badger (*Taxidea taxus*)	4.4 mg/kg Telazol®
Beaver	10.0 mg/kg ketamine plus 1.0 mg/kg xylazine
Marmot (*Marmota* spp.)	80.0 mg/kg ketamine plus 10.0 mg/kg xylazine
Muskrat	50.0 mg/kg ketamine plus 5.0 mg/kg xylazine
Opossum (*Didelphis virginiana*)	10.0 mg/kg ketamine plus 2.0 mg/kg xylazine
Porcupine (*Erethizon dorsatum*)	5.0 mg/kg ketamine plus 2.0 mg/kg xylazine
Rabbit (order Lagomorpha)	30.0 mg/kg ketamine plus 6.0 mg/kg xylazine
Squirrel (family Sciuridae)	10.0 mg/kg Telazol

SUMMARY

Drugs and the equipment used to administer them have become more sophisticated, efficacious, and safe over the past half century. Today, capture drugs should be part of every wildlife management professional's armamentarium. Although capture drugs should be used judiciously, they should always be considered as a primary solution to problems where animal and human safety is uppermost.

5　Use of Dogs in Wildlife Research and Management

DAVID K. DAHLGREN,
R. DWAYNE ELMORE,
DEBORAH A. SMITH,
AIMEE HURT,
EDWARD B. ARNETT, AND
JOHN W. CONNELLY

INTRODUCTION

CONTEMPORARY TECHNIQUES FOR wildlife research and management tend to be relatively expensive and involve an ever expanding technology. In comparison, use of dogs to obtain wildlife data, a relatively old technique, may seem outdated, and a discussion on the use of dogs could appear elementary (Fig. 5.1). However, data collection in the wildlife profession must be field-based if modeling exercises are to represent reality. The use of dogs can provide valuable information that could not otherwise be collected in the face of shrinking budgets and limited personnel, or when more precise estimates are desired and use of dogs is known to surpass other methods. This is especially true given recent advances in techniques and technology, such as use of the Global Positioning Systems (GPS). Though dogs may be used infrequently in North America for wildlife work, European wildlife managers have a long history of using dogs, and their wildlife educational programs require a demonstrated ability to handle dogs (S. Tóth, University of Washington, personal communication). Currently, there is no comprehensive guide for using dogs to aid researchers. Thus, this chapter is intended to provide examples of how dogs can be employed to collect data, provide basic guidance on practices that work, and stimulate thought and discussion of other potential applications. This topic was covered in past editions (Zwickel 1971, 1980) of the *Wildlife Management Techniques Manual,* but was discontinued in later editions. Since the publication of these editions, many additional applications have been devised and are summarized here.

Dogs offer a unique set of skills that otherwise might not be available for collection of wildlife data. The scenting abilities of dogs have been well documented (Johnston 1999, Syrotuck 2000). For instance, dogs can detect scent up to 100 million times better than a human can (Syrotuck 2000) and can detect certain compounds up to 500 parts per trillion (Johnson 1999). Additionally, most dogs offer increased ground coverage with speeds that are up to 4 times faster than a human (Mecozzi and Guthery 2008). These factors illustrate the advantages of using various task-oriented dog breeds for some management and research activities. We have summarized the use of dogs in wildlife management into the following categories: (1) locating wildlife and assessing population status, (2) facilitating specimen and carcass collection, (3) detecting scat, (4) capturing and marking wildlife, (5) studying wildlife behavior, and (6) managing wildlife damage.

Fig. 5.1. "My dog, by the way, thinks I have much to learn about partridges" (Leopold 1970:67). Aldo Leopold with Flick (German shorthaired pointer) at the Riley Game Cooperative, Wisconsin. *Photo courtesy of the Aldo Leopold Foundation (www.aldoleopold.org).*

TYPES OF DOGS

Most experts agree that domestic dogs descended from wolves (*Canis lupus*) rather than other canids (Olsen 1985, Pennisi 2002, Wang and Tedford 2008). However, debate continues concerning timelines and specific selective factors for domestication, as well as the geographic location (Pennisi 2002). Interestingly, Coppinger and Coppinger (2002) describe a theory of natural selection for the evolution of the domestic dog (*Canis lupus familiaris*) that suggests dogs may be the only "self-domesticated" animal.

The American Kennel Club (AKC) currently registers 161 breeds of dog (www.akc.org); however, there are many more breeds throughout the world that are not recognized by AKC. Most individual breeds were developed for specific tasks. Breeds are generally grouped into the following broad categories: sporting, hound, working, terrier, toy, non-sporting, and herding. Hunting breeds (e.g., sporting and hound groups) are most commonly used for wildlife work because of their innate interest in game and other wildlife species, as this interest was the original selected trait. Herding breeds, including protection oriented breeds, also may aid wildlife work, including wildlife damage management. Often breeds in this group (e.g., border collie and German shepherd) exhibit high intelligence, cooperation, and trainability and have been used for tasks unrelated to their herding instincts.

There are too many breeds that could be used in wildlife work to list individually, and there are multiple volumes dedicated to **breed traits** (Fogle 2000, Coile 2005). The proper breed(s) should be chosen for specific tasks (Table 5.1), although, in a given breed or group, individual variation in traits (e.g., drive, intelligence, cooperation, trainability, range, and scenting ability) may be more important than the breed itself. Additionally, certain lines exist within breeds that exhibit specific traits and abilities. For example, English springer spaniels have "show" and "field" lines. The hunting line is commonly referred to as "field bred" English springer spaniels and may be of more value to wildlife work than the show type. Another example is the popularity of the Labrador retriever (labs) as a companion or family dog, and in many cases individuals have been bred irrespective of hunting abilities. Therefore, those interested in wildlife work should obtain labs from proven hunting parentage to ensure the proper traits are present to carry out the desired tasks. Potential for **crossbreeding** among breeds has been suggested in the past (Zwickel 1980); however, the yeoman effort of dedicated breeders to provide consistent heritable traits, combined with the availability of so many potential breeds, behooves the selection of a workable individual from **purebred** lines. This is not to imply that **mixed dogs** are not of use, but their traits will be much less predictable. Moreover, adoption of unwanted or rescue dogs (mixed or purebred) is easy and inexpensive, and may be a reasonable option. For instance, scat and reptile detection work has successfully used mixed breeds and rescue dogs where individuals exhibited specific desirable traits (discussed later). Thorough research into a breed, specific lines, and individual kennels should be undertaken before a dog is obtained for use in wildlife work.

GENERAL INFORMATION ON USE OF DOGS

In most cases, dogs will be used in wildlife work because of their **scenting abilities.** Scent, scenting conditions, and scenting ability of dogs are important when considering their use for field research. Bird scent is thought to be created by rafts of dead skin (continuously shed by birds) that bacteria metabolize, creating residues and secretions of vapor or "scent" (Gutzwiller 1990). However, lipids, fatty acids, and wax produced by the uropygial gland (used in preening) also may be another source of bird scent (Conover 2007). Regardless of scent origin, the ability of scent to be airborne and the impact of environmental conditions on airborne scent are key factors that influence scent-detection ability (Gutzwiller 1990, Conover 2007).

Weather conditions, such as wind, temperature, humidity, and barometric pressure, also play important roles in scenting conditions (Gutzwiller 1990, Shivik 2002, Conover 2007) and should be taken into account when conducting searches with dogs. **Scenting conditions** should be similar

Table 5.1. Dog types and breeds with potential for various wildlife oriented tasks

Dog breed or type[a]	Task								
	Detection of birds	Capture of birds	Harassment of birds	Detection of carcasses	Detection of scat	Livestock protection	Harassment of mammals	Capture of mammals	Detection of reptiles
Pointers									
EP, GSP, GWP, BR	X	X			X			X[b]	X[b]
Setters									
ES, GS, RS	X	X							
Retrievers									
LR,[c] GR,[c] CBR	X		X	X	X				X
Spaniels									
ESS, ECS, FS	X		X	X	X				
Hounds									
BGL, RBH, BTH, WKR, BLTH				X			X	X	
Collies and shepherds									
GSD, BC, AS, AK, KBD		X	X	X	X	X	X	X	X
Other breeds									
GP, AD						X			

[a] This list is not comprehensive but is meant to give an overview and general guidance for the most common breeds. As noted in the text, individual traits vary widely even within breeds and lines and may be the most important factor when considering a dog for specific wildlife tasks. Accordingly, some individuals (mixed or purebred) may work for tasks we have not listed. AD = Akbash dogs; AK = Australian kelpie; AS = Australian shepherd; BC = border collie; BGL = beagle; BLTH = black and tan coonhound; BTH = blue tick coonhound; BR = Brittany; CBR = Chesapeake Bay retriever; ECS = English cocker spaniel; EP = English pointer; ES = English setter; ESS = English springer spaniel; FS = field spaniel; GP = great Pyrenees; GR = golden retriever; GS = Gordon setter; GSD = German shepherd dog; GSP = German shorthaired pointer; GWP = German wirehaired pointer; KBD = Karelian bear dog; LR = Labrador retriever; RBH = red bone coonhound; RS = red setter; WKR = Walker coonhound.

[b] GSP and GWP were originally bred to both point and retrieve feathered game, as well as track and find furred game, though many North American breeding programs have focused more on the bird finding abilities of these and other continental breeds. Therefore, individuals of the continental breeds may vary in their drive for mammals based on past breeding objectives.

[c] Because of the popularity of retriever breeds as companion and family dogs (e.g., Labrador retrievers are the most popular dog in the United States), many lines in these breeds have been bred irrespective of hunting ability. Therefore, individuals used for wildlife work should come from parentage focused on and used for their hunting traits.

between treatment and control groups when collecting data with dogs for experimental practices (Gutzwiller 1990). This requirement necessitates the use of standardized weather parameters as much as possible.

Shivik (2002) tested the ability of dogs to scent humans with scent-adsorptive clothing and found wind variability was negatively correlated with dogs' ability to locate subjects quickly. Gutzwiller (1990) suggested that moderate winds actually enhance scenting conditions, whereas weak or extremely strong winds degrade them. From the authors' personal experience, steady winds between 8 and 40 kph provide optimum scenting conditions, at least for pointing dogs.

In an interesting study, Steen et al. (1996) tested olfaction properties of pointing dogs while they were searching for game. They found that even while exhaling during hunting activities, the dog can maintain a continuous inward air flow for up to 40 seconds or at least 30 respiratory cycles. This ability is due to the **Bernoulli effect,** which results from lower pressure in the mouth cavity than in the nose during inhaling and causes an inward flow of air through the nose (Raphael et al. 2007). This phenomenon only occurs while the dog is running with its head held high and does not occur while it is resting or searching for ground scent (Steen et al. 1996). This phenomenon explains why dogs, and possibly other mammals, can be running and breathing hard (panting) yet continuously scent game.

All dogs are not created equal, and individual dogs differ in their ability to locate subjects (Jenkins et al. 1963, Gutzwiller 1990, Shivik 2002). The differentiating factors between individual dogs can be related to range or ground coverage, scenting ability, and/or age and experience. Thus, during wildlife data collection, individual dogs should be used consistently, and the number of dogs used in a study minimized to reduce bias, much like we minimize human observers (Gutzwiller 1990). Additionally, **individual dog performance** can be variable, even during a given day (Gutzwiller 1990). Therefore, environmental factors that may influence a dog's performance should be taken into account if possible. Physiological factors, such as parasite loads, poor diets, and fatigue, or other negative influences affect a dog's ability to find subjects optimally (Gutzwiller 1990). Furthermore, sociological factors also may influence a dog's performance. Some dogs are more competitive than others and may have ineffective sessions if paired with another dog they feel competitive toward (Gutzwiller 1990). Gutzwiller (1990) provided these standard procedures for **reducing bias** while using dogs in wildlife studies:

1. Use the same dog throughout a study, or balance the use of each of 2 or more dogs in time and space, to avoid observer bias.
2. Ensure dogs are physically fit (before and during searches) and well trained.
3. Search for birds under as similar temperature, wind, precipitation, and barometric conditions as possible, because these factors can affect bird activity, scent, and dog performance.
4. Restrict search to a certain period of the day, because daily cycles in temperature, humidity, and other variables influence scent production and detection. Bird activity and habitat use also vary with time of day.
5. Balance search efforts by using equal numbers of dogs and researchers per unit time and area.

Technology (e.g., **GPS units**) has become essential in wildlife research and management in recent years. The use of GPS has not escaped the realm of working dogs. There are several products currently available for tracking dogs using GPS technology (e.g., RoamEO™, White Bear Technologies, St. Paul, MN; Garmin™ Astro, Garmin International, Olathe, KS). The **Garmin Astro** is specifically designed for hunting dogs. Some researchers have simply attached small GPS units to the dog's collar for tracking purposes (Guthery and Mecozzi 2008), though based on their experience, they encouraged the use of GPS units specifically designed for dogs (G. Mecozzi, Oklahoma State University, personal communication). These units track the handler's path, along with the dogs' path (some units allow multiple dog tracking simultaneously), as well as providing other information about the dog (e.g., speed; distance from handler; direction; distance traveled; and activity, such as pointing or running). This technology may provide increased information concerning biases associated with using dogs and address some of the violated assumption concerns (100% detectability and coverage) of methods, such as the belt transect method (discussed later; Jenkins et al. 1963). Using GPS units designed for dogs can enhance the ability of researchers to collect data for probability detection methodology (e.g., distance sampling; see below; Buckland et al. 1993).

Other considerations when using dogs in wildlife work include **safety** of the subject species and the dog. This aspect of the work must be acknowledged and steps taken to minimize risk. It is especially important in modern research, where Animal Care and Use Committees and wildlife agencies are charged with the task of ensuring minimal harm to wildlife during management and research activities. When using dogs, the predatory instincts of these animals (especially hunting breeds) must not be underestimated. The desire to search out game, or hunt, is merely the first step in a predation event. Appropriate training and cooperation from the dog can keep the pseudo-predation event controlled; however, the intent of the dog remains predatory in nature. In our personal experience, individual dogs vary in their **prey drive** (the desire to chase and/or dispatch game), thus a handler must pay particular attention to each individual's natural drive. A muzzle may be useful for preventing undesired harm, though we could not find an example of its use. Additionally, special care should be given to keep the dog's physical and nutritional condition and its demeanor in order. Dogs can be injured by heat stroke, rattlesnake bites, and porcupine quills (Flake et al. 2010). Again, these concerns should be acknowledged in animal use protocols. Handlers should always maintain an ample water supply and a first aid kit specialized for field dogs.

LOCATING WILDLIFE

Dogs have been widely used for **sampling wildlife populations.** Often this includes counting animals, determining distribution, and/or gathering demographic (e.g., age and sex) information (Table 5.2). These data are then used to project **indices** for a population, such as density or productivity. In many instances dogs can enhance the detection of wildlife or mortalities beyond the ability of an observer alone (Novoa et al. 1996, Homan et al. 2001, Arnett 2006, Dahlgren et al. 2010).

Pointing dogs have been used to estimate densities or indices of the abundance of grouse in several different studies (Jenkins et al. 1963, Thirgood et al. 2000, Amar et al. 2004, Broseth et al. 2005, Dahlgren et al. 2006; Table 5.2). The original method for **estimating grouse density** was developed on red grouse (*Lagopus lagopus scoticus*) in Europe and consisted of using belt transects (Jenkins et al. 1963, Thirgood et al. 2000). In general, the method entails searching an area by working a pointing dog along parallel transects, often spaced approximately 150 m apart. The dog is cast (directed) to either side of the transect line (approx. 75 m), and all birds in the area are assumed to be detected and flushed. However, this assumption is uncertain, because other research indicates that pointing dogs only detected 50% of available radiomarked birds (Stribling and Sisson 1998). Essentially this method is a total (census) strip count that has been validated for consistency, but not accuracy (S. Thirgood, Macaulay Institute and Aberdeen University, UK, personal communication). Additionally, this method does not readily yield error rates for comparison purposes. Interestingly, Broseth and Pedersen (2000) reported detection of willow ptarmigan (*Lagopus lagopus;* known as red grouse in Europe) past 80 m (from the transect line) to be difficult when using dogs, that supports a similar belt transect width reported by Jenkins et al. (1963) and Thirgood et al. (2000).

Use of dogs for **distance sampling** procedures has been suggested to estimate density of birds (Buckland et al. 1993; Table 5.2). Rosenstock et al. (2002) encouraged use of more

Table 5.2. Dogs in wildlife research and management: summary of live animal, nest, and carcass detection[a]

Wildlife species	Dog breed or type[b]	Method	Reference
Study objective: abundance, density, and indices			
Red grouse (*Lagopus lagopus scoticus*)	Pointing dogs (pointers and setters)	Belt transect	Jenkins et al. 1963, Redpath and Thirgood 1999, Thirgood et al. 2000, Park et al. 2001, Thirgood et al. 2002, Amar et al. 2004
Greater sage-grouse (*Centrocercus urophasianus*)	Pointing dogs (German shorthaired pointers)	Belt transect	Dahlgren et al. 2006
Red grouse	Pointing dogs	Line transect distance sampling	Warren and Baines 2007
Ruffed grouse (*Bonasa umbellus*)	Pointing dogs	Belt transect	Berner and Gysel 1969
Sooty grouse (*Dendragapus fuliginosus*)	Pointing dogs	Belt transect	Zwickel 1972
Willow ptarmigan (*Lagopus lagopus*)	Pointing dogs	Line transect distance sampling	Pedersen et al. 2004, Broseth et al. 2005
Northern bobwhite (*Colinus virginianus*)	Pointing dogs	Effective strip width sampling	Guthery and Mecozzi 2008
Desert tortoise (*Gopherus agassizii*)	Herding dogs, Labrador retrievers, and mixed breeds	Systematically searched plots	Cablk and Heaton 2006, Nussear et al. 2008
Ringed seal (*Phoca hispida*)	N/A	Searched likely habitat	Lydersen and Gjertz 1986, Furgal et al. 1996
Study objective: productivity			
Red grouse	Pointing dogs	Belt transects	Redpath 1991, Redpath and Thirgood 1999
Willow ptarmigan	Pointing dogs	Searched entire study area and marked broods	Parker 1985, Schieck and Hannon 1989
Capercaillie (*Tetrao urogallus*)	Pointing dogs	Line transects	Novoa et al. 1996, Storaas et al. 1999
Black grouse (*Tetrao tetrix*)	Pointing dogs	Line transects	Storaas et al. 1999
Greater sage-grouse	Pointing dogs	Line transects and marked broods	Klott and Lindzey 1990, Dahlgren 2009
Columbian sharptailed grouse (*Tympanuchus phasianellus columbianus*)	N/A	Line transects	Klott and Lindzey 1990
Little spotted kiwi (*Apteryx owenii*)	N/A	Searched likely habitat	Colbourne 1992
Study objective: nest searches			
Greater prairie-chicken (*Tympanuchus cupido*)	N/A	Searched likely habitat	Bowen et al. 1976
Willow ptarmigan	Pointing dogs	Searched likely habitat	Schieck and Hannon 1989, Hannon et al. 1993
Capercaillie	Pointing dogs	Line transects	Storaas et al. 1999
Black grouse	Pointing dogs	Line transects	Storaas et al. 1999
Korean pheasant (*Phasianus colchicus karpowi*)	N/A	Searched likely habitat	Wollard et al. 1977
Northern pintail (*Anas acuta*)	N/A	Systematically searched likely habitat	Flint and Grand 1996
Greater golden-plover (*Pluvialis apricaria*)	Pointing dogs	Systematically searched likely habitat	Byrkjedal 1987
Eurasian dotterel (*Charadrius morinellus*)	Pointing dogs	Systematically searched likely habitat	Byrkjedal 1987
Little spotted kiwi	N/A	Searched likely habitat	Colbourne 1992
Yellow rail (*Coturnicops noveboracens*)	German shorthaired pointer	Searched likely habitat	Robert and Laporte 1997
Study objective: capture			
Willow ptarmigan adults and chicks	Pointing dogs	Hand-held nets, by hand, or noose poles	Erikstad and Andersen 1983, Hannon et al. 1990, Broseth and Pedersen 2000
Black grouse broods	Pointing dogs	Large nets dragged over brood or flushed into nets	Caizergues and Ellison 2000, Baines and Richardson 2007
Spruce grouse (*Falcipennis canadensis*)	Pointing dogs	Noose pole	Herzog and Boag 1978
Blue grouse (*Dendragopus* spp.)	Pointing dogs	Noose pole	Zwickel and Bendell 1967, Zwickel 1972
Greater sage-grouse	Pointing dogs (German shorthaired pointers)	By hand and hand-held nets	Connelly et al. 2000, 2003b

continued

Table 5.2. continued

Wildlife species	Dog breed or type[b]	Method	Reference
Aleutian Canada goose (*Branta canadensis leucopareia*)	Border collies	Herded into nets	Shute 1990
Mountain lion (*Puma concolor*)	Hounds	Treed and immobilized	Hornocker 1970
Black bear (*Ursus americanus*)	Hounds	Treed and immobilized	Akenson et al. 2001
Study objective: carcass searches			
Bat and bird fatalities at wind facilities	Labrador retrievers	Systematically searched plots beneath turbines	Arnett 2006

[a] For scat-detection wildlife damage, see the text and MacKay et al. (2008).
[b] N/A = not applicable.

detectability-based density estimates (i.e., distance sampling) in land-bird counting techniques, including the use of dogs while sampling. This method consists of using random or systematic transect lines placed in a specified area and casting the dog as the observer and/or handler walks the transect line (Fig. 5.2). The distance from grouse locations (or dog on point) to the centerline is recorded, as well as number of grouse per flock or cluster. Along with density, program DISTANCE (http://www.ruwpa.st-and.ac.uk/distance/) also calculates probability of detection, an effective strip width (ESW), and error rates. If reliable estimates of density can be obtained, those estimates can be used in a Geographic Information System (GIS) application of Kriging (a group of geostatistical techniques to interpolate the value of a random field) that allows extrapolation of data to obtain a spatial distribution of densities (Warren 2006, Warren and Baines 2007). Distance sampling and subsequent Kriging methods have been applied to red grouse, but further evaluation for additional species is needed. This method likely has application for any gallinaceous (and possibly other species) bird that pointing or flushing breeds commonly detect. However, because pointing dogs generally cover more area than flushing dogs and hold point, likely resulting in more accurate counts and distance measurements, we suggest there is an advantage to using pointing dogs over flushing dogs for distance sampling.

Guthery and Mecozzi (2008) developed a **modification of distance sampling** to obtain northern bobwhite (*Colinus virginianus*) densities using pointing dogs. This method uses a dog's path, recorded by GPS units attached to the dog, as the theoretical centerline for distance sampling. The distance from where a dog establishes a point to the bird(s) is the perpendicular distance, which program DISTANCE uses to estimate an ESW for the transect. The dog's path is then buffered by the ESW on each side to create an area (polygon). Redundancy in a single or multiple dogs' path(s) is then eliminated. Then the number of birds located within the polygon's area yields a density estimate. This method has only been used on northern bobwhites and has not been evaluated for other species. There are some biases with this methodology that should be considered. First, this method does not account for wind direction and assumes equal detectability on either side of the dog despite wind direction. Second, the assumption that all birds are detected in a path, and thus redundancy is wasted effort may not be valid. And third, measuring detection distance based on an established

Fig. 5.2. Example of a transect in a 40.5-ha plot to monitor greater sage-grouse using pointing dogs on Parker Mountain, Utah, 2009. Data were collected using Garmin™ Astro GPS units. Transect line spacing was designed to reduce redundancy in the dog's path and to allow for distance sampling procedures. A problem with this design is that grouse detected at the corners do not have a perpendicular distance to transect line.

point may be a poor assumption, as the distance between the bird and an established point can vary considerably among individual dogs. In the authors' experience, many pointing dogs move well past the location of initial scent detection to approach the bird more closely.

Upland **game bird productivity** has been commonly assessed using dogs to locate hens and their broods (Table 5.2). This method is preferred, as observers without dogs often underestimate brood size (Novoa et al. 1996, Schroeder 1997, Dahlgren et al. 2010). Similarly, Dahlgren et al. (2010) reported that greater sage-grouse (*Centrocercus urophasianus*) chicks were detected more frequently using dogs compared to an observer walking alone. Individual chicks are located by the dog once the general location of the brood is found. Chicks have a tendency to hold tight and let observers pass by without flushing. Once a brood hen is located, a dog can be kept in close proximity to her location and can quickly (10 min) find the vast majority of chicks (Dahlgren et al. 2010).

Habitat use and **breeding characteristics** of various upland game bird species also have been studied with the aid of dogs. Baxter and Wolfe (1968) used dogs to evaluate pheasant (*Phasianus colchicus*) brood cover use in Nebraska. Hines (1986) used dogs to monitor flock characteristics, movement patterns, and home range of sooty grouse (*Dendragapus fuliginosus*) in British Columbia, Canada. Novoa et al. (1996) determined that pointing dog surveys provided better estimates of capercaillie (*Tetrao urogallus*) production than did routes carried out by observers alone. Dogs also have been used to determine breeding status. Hannon and Eason (1995) used dogs to assess the pairing status of male willow ptarmigan by searching their territories for females. Thus, for general distribution and abundance, dogs can greatly increase searcher efficiency and area covered.

Nest searches for gallinaceous and other ground nesting birds have been conducted using the aid of dogs (Table 5.2). Flint and Grand (1996) used dogs to search for northern pintail (*Anas acuta*) nests in Alaska. Byrkjedal (1987) used pointing dogs to locate nests of shorebirds (greater golden-plover [*Pluvialis apricaria*] and Eurasian dotterel [*Charadrius morinellus*]) in Norway. Specific species' reaction to nest disturbance should be considered when using dogs for nest searches. For instance, some species, such as northern pintail (Flint and Grand 1996) and spruce grouse (*Falcipennis canadensis*; Keppie and Herzog 1978), will return to a nest following disturbance by a dog, whereas others, like greater sage-grouse, may be prone to abandonment if flushed from a nest, especially during laying and early incubation (Patterson 1952). Most sage-grouse researchers avoid flushing the hen from the nest because of concerns about observer-induced nest abandonment (Fischer et al. 1993, Sveum et al. 1998, Wik 2002, Chi 2004, Holloran and Anderson 2005, Kaiser 2006, Baxter et al. 2008). Indeed, those who have flushed sage-grouse hens from their nests reported comparatively lower nest success rates (Herman-Brunson 2007, Moynahan et al. 2007). For species whose nest ecology is poorly understood, nest success rates should be carefully monitored for disturbed and undisturbed nests to determine whether the use of dogs is acceptable for that species.

Monitoring wildlife **management actions** is an important strategy for assessing practices and applying adaptive management. Dogs can be used to facilitate monitoring activities. Martin (1970) used dogs to assess greater sage-grouse use of chemically controlled sagebrush (*Artemisia* spp.) areas in Montana. Similarly, Dahlgren et al. (2006) used pointing dogs to monitor greater sage-grouse use of chemically and mechanically treated mountain big sagebrush (*A. tridentata* var. *vaseyana*) in late brood-rearing areas in Utah (Fig. 5.3). Newborn and Foster (2002) used dogs to count red grouse on plots where medicated and nonmedicated grit was applied to evaluate parasite control. Larsen et al. (1994) used dogs to monitor pheasant use of food plots in South Dakota.

Dogs have been used for detecting **species of conservation concern,** especially where other techniques are inefficient at detecting species in low abundance and/or patchy habitats. They have been used to locate a number of endangered species in New Zealand for >100 years (Browne et al. 2006), including the blue duck (*Hymenolaimus malacorhynchos*), kiwi (*Apteryx* spp.), and kakapo (*Strigops habroptila*). Detection dogs are an essential part of little spotted kiwi (*A. owenii*) conservation and data collection in New Zealand because of the difficulty locating nests and young (Colbourne 1992). A German shorthaired pointer successfully found the nests of yellow rail (*Coturnicops noveboracens;* classified as a vulnerable species) which are notoriously difficult to locate, in southern Quebec, Canada (Robert and Laporte 1997). Black rail (*Laterallus jamaicensis*), another difficult bird to locate, have been found using a German shorthaired pointer (R. Elmore, unpublished data). Cablk and Heaton (2006) tested

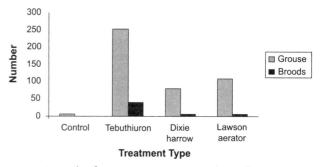

Fig. 5.3. Example of greater sage-grouse use data collected with pointing dogs in 40.5-ha experimental plots on Parker Mountain, Utah, 2003–2004 (see Dahlgren et al. 2006). These data show a preference for tebuthiuron (Spike®; an herbicide) treated plots for both grouse in general and broods specifically. Using dogs allowed the classification of sage-grouse by age and sex, which benefited this project specifically designed to improve late brood-rearing habitat.

the efficacy and reliability of dogs to locate the United States federally listed desert tortoise (*Gopherus agassizii*) above and below ground in the Mojave Desert of the southwestern United States in various climate conditions. They reported that dogs found 90% of the known population and located smaller tortoises than human observers were able to detect. They also suggested using dogs to conduct distance sampling and mark–recapture techniques for this species. Dogs have been used to locate ringed seal (*Phoca hispida*) subnivean structures in the arctic for studies on characteristics of seal predation and territory size (Lydersen and Gjertz 1986, Furgal et al. 1996).

Dogs can be used to locate unknown **grouse leks** (D. Dahlgren and R. Elmore, unpublished data; Fig. 5.4). Although leks are generally easy to locate during the first few hours of the day when the birds are actively displaying, the window of time the birds display is fairly narrow. However, males typically use cover near the lek most of the day. Using pointing dogs that can cover large areas quickly is an ideal method of locating these males. Once a large number of male grouse are flushed, the researcher can mark the location and come back during display periods to do a visual count and determine the precise location of the lek. This method extends survey time in areas with unknown lek locations.

SPECIMEN AND CARCASS COLLECTION

The use of dogs to detect **carcasses** has many wildlife management and research applications involving human–wildlife interactions. Examples include detecting mortality from collisions with manmade structures, poisons, or disease events. The use of dogs around wind farms is proving to be particularly beneficial to determine impacts on avian and bat species. Additionally, dogs can be allies in the search for bird carcasses due to mortality from pesticide use, especially in dense cover (Homan et al. 2001). Finley (1965) used dogs to help locate birds that were affected by use of an insecticide in a Montana forest. Homan et al. (2001) reported that dogs found 92% of house sparrow (*Passer domesticus*) carcasses compared to 45% by human searchers, and dogs provided a greater searching efficiency per unit time. Because scavenging rates may be high in many areas, the ability of dogs to locate carcasses quickly can be beneficial (Homan et al. 2001). Accordingly, dogs have been successfully used to search for lesser prairie chicken (*Tympanuchus pallidicinctus*) fence strike mortalities in Oklahoma (R. Elmore, unpublished data). Dogs also may be used to search for birds dying from natural causes (Jenkins et al. 1963). Dogs have been used to collect ducks with botulism and were able to be trained to select live specimens (Zwickel 1980).

Arnett (2006) used Labrador retrievers to assess the ability of dog–handler teams to recover dead **bats** (and birds) during fatality searches typically performed at **wind energy** facilities to determine fatality rates for birds and bats (Fig.

Fig. 5.4. German shorthaired pointers pointing greater sage-grouse while researchers search for unknown lek (display) sites in the spring in northwestern Utah. Using dogs for this purpose can extend the survey time beyond the grouse display period by locating males using habitat near lek sites. Once males are located, researchers can come back during display periods and determine the exact location of the lek.

5.5). Dogs found 71% of bats used during searcher-efficiency trials at one site and 81% of those at a second site, compared to 42% and 14% for human searchers. Dogs and humans both found a high proportion of trial bats within 10 m of the turbine, usually on open ground (88% and 75%, respectively). During a 6-day fatality search trial at 5 turbines at a wind facility, the dog–handler teams found 45 bat carcasses, of which only 42% ($n = 19$) were found during the same period by humans. In both trials humans found fewer carcasses as vegetation height and density increased, whereas dog–handler teams search efficiency remained high. Arnett (2006) suggested that broad-scale use of dogs to monitor fatalities at wind facilities may be difficult to implement, especially at large facilities, where several trained dogs and handlers would be required. However, dogs could easily be employed to (1) survey smaller facilities (generally those with 20 turbines), particularly when low-visibility habitats prevail; (2) confirm specific questions regarding individual or small numbers of turbines for any facility (e.g., confirm whether bats are killed at nonoperational turbines or meteorological towers); or (3) obtain more precise and accurate estimates of fatality when testing and comparing different approaches to reduce bat fatalities at wind turbines.

Beyond finding carcasses, there are opportunities to use dogs to locate and capture animals. Small dogs have been used to bring fox pups from dens (Zwickel 1980). Small dogs also have been commonly used by the U.S. Department of Agriculture's Wildlife Services to lure coyotes (*Canis latrans*) into gun range for control (M. Conover, Utah State University, personal communication). Further, Johnson (1970) re-

Fig. 5.5. Researcher Ed Arnett, Bat Conservation International, searches for dead bats and birds beneath wind turbines with his Labrador retriever at a facility in south-central Pennsylvania.

ported that raccoons (*Procyon lotor*) captured using dogs provided the most unbiased and representative diet samples, compared to trap-caught animals. Often when nuisance wildlife requires extermination, wildlife agencies will employ dogs (usually hounds) to track individual animals (M. Conover, personal communication). Mecozzi and Guthery (2008) describe characteristics and behaviors of walk-hunters and dogs pursuing northern bobwhite in Oklahoma and Missouri. Shupe et al. (1990) used pointing dogs to facilitate harvest of northern bobwhite in a study on the vulnerability of sex and age of this species. Hardin et al. (2005) used hunters and pointing dogs to spatially analyze northern bobwhite hunting using models that predict daily harvest. By using such models, better understanding and management of quail hunting can be achieved. The use of dogs to harvest wildlife in indigenous Neotropical villages in Nicaragua was a significant predictor of species composition in harvest, and the advantage of using dogs was their ineffectiveness at pursuing species that were vulnerable to overharvest, such as tree dwelling species (e.g., primates), that are much more difficult to detect with dogs (Koster 2008).

SCAT DETECTION

An emerging field in wildlife work is the use of **scat detection** dogs for research and management (Fig. 5.6). An extensive overview of scat detection dogs is found in MacKay et al. (2008) and Hurt and Smith (2009). Based on these 2 main sources and several other recent works, we provide a detailed summary of dogs used for scat detection.

Early uses of scat dogs involved finding scats of such species as black-footed ferrets (*Mustela nigripes;* Dean 1979, Winter 1981); wolves, coyotes, black bears (*Ursus americanus;* P. Paquet, University of Calgary, unpublished data); and lynx (*Lynx Canadensis;* U. Breitenmoser and C. Breitenmoser-Wursten, International Union for Conservation of Nature Species Survival Commission, Cat Specialist Group, unpublished data) to help biologists obtain presence, diet, and other information on populations. In the 1990s, the new benefit of gleaning **DNA information** from scat—such as individual presence and movement, sex ratios, relatedness, habitat selection, and home ranges (Kohn and Wayne 1997) —led to a more formalized and systematic approach to using scat detection dogs (MacKay et al. 2008). In Washington, wildlife researchers and a professional narcotic-detection dog trainer joined forces and applied training techniques similar to narcotic, cadaver, and search-and-rescue disciplines to scat-detection dog methodology (MacKay et al. 2008; Hurt and Smith 2009). Their formally trained dogs were then used to search for scats of bear and other carnivores in the Okanogan National Forest (S. Wasser, B. Davenport, and M. Parker, Okanogan National Forest, Washington, unpublished data). Subsequently, this research team used scat detection dogs to locate brown (*U. arctos*) and black bear scat over a 5,200-km^2 area in Alberta, Canada, and reported that dogs reduced bias in collection methods (Wasser et al.

Fig. 5.6. After alerting her handler to bear scat (upper center of photo) by sitting next to the sample, the scat detection dog now ignores the scat. She chews on her reward toy while her handler prepares to label and collect the sample.

2004). In their study, scat found by dogs helped identify individual bears and sex ratio as well as habitat use patterns, and it provided indices of physiological stress and reproductive activity.

Many dogs have been trained with a similar professional approach to locate scats of **target species** and ignore others both on land and in water (MacKay et al. 2008; Hurt and Smith 2009). In California, Smith et al. (2003a) compared detection and accuracy rates of 4 dogs trained to find scats of San Joaquin kit fox (*Vulpes macrotis mutica*) and demonstrated that dogs could provide large, accurate sample sizes of fox scat for DNA population analyses. They also showed that 1 dog with the lowest detection rate for kit fox scat still equaled the detection rate of 2 experienced humans. Smith et al. (2005) found that scat-detection dog surveys were successful in San Joaquin kit fox population areas that varied in both relative fox density and vegetation type. In addition, other studies using dogs to locate San Joaquin kit fox scat provided current information on status, sex ratio, relatedness, movement patterns, scent marking, and size of home range (Ralls and Smith 2004; Smith et al. 2006a, b).

Long et al. (2007a, b) used scat detection dogs in Vermont to simultaneously locate scat of black bear, fisher (*Martes pennanti*) and bobcat (*L. rufus*). Dogs proved effective for collecting detection–nondetection data on these three species (Long et al. 2007a) and compared to remote cameras and hair snares yielded the highest raw detection rate and probability of detection for each species plus the greatest number of unique detections (Long et al. 2007b). Beckmann (2006) used dogs over multiple years to simultaneously locate scat of black bear, grizzly bear (*U. a. horribilis*), wolf, and mountain lion (*Puma concolor*) in Idaho and Montana. DNA from scat combined with location data identified areas that can support these four species at low densities over time (Beckmann 2006). In New Mexico, a dog trained to find bobcat scats produced approximately 10 times the number of bobcat detections than did remote cameras, hair-snares, and scent stations, suggesting the appropriateness of this method for population monitoring of bobcats (Harrison 2006). Scat detection dogs also have worked well from boats. Rolland et al. (2006) used dogs to locate scats of North Atlantic right whales (*Eubalaena glacialis*) and found rates of scat collection with dogs were >4 times higher than with opportunistic methods. Thus, the use of dogs ensured the required number of samples needed to conduct endocrine, disease, genetic, and biotoxin studies was obtained.

Scat detection dogs also have been used in international research. In Brazil, data from dog-collected scat is helping develop species distribution and landscape models for maned wolf (*Chrysocyon brachyurus*), jaguar (*Panthera onca*), mountain lion, giant anteater (*Myrmecophaga tridactyla*), and giant armadillo (*Priodontes maximus*; C. Vynne, University of Washington, personal communication). Moreover, DeMatteo et al. (2009) reported that a dog could successfully locate scats of the bush dog (*Speothos venaticus*) in dense forest vegetation in Argentina, thereby offering an opportunity to obtain data needed for developing conservation strategies.

In controlled laboratory settings, scat detection dogs have been used for their scent discrimination abilities to match species or individuals from scat samples. This use of dogs can potentially help avoid costly **genetic analysis** and provide reliable information for mark–recapture methods (Smith et al. 2003a, Kerley and Salkina 2007, Wasser et al. 2009). On the species level, dogs showed promise in being able to differentiate scats of grizzly and black bear (A. Hurt, unpublished data); bobcat and sympatric carnivores, such as fox and coyote (Harrison 2006); and San Joaquin kit fox from red fox (*V. vulpes*; Smith et al. 2003a). However, accuracy rates for dogs at this task can be affected by individual dog aptitude and performance, presence of target scats in matching test, and a number of other variables (A. Hurt, unpublished data; Smith et al. 2003a; Harrison 2006). On the individual level, Kerley and Salkina (2007) showed that dogs were 87% accurate at matching individual Amur tigers (*P. tigris altaica*) with their scats. Most recently, Wasser et al. (2009) reported that 3 dogs correctly matched 25 out of 28 scat samples from individual maned wolves, thus demonstrating the potential for dogs to assist researchers even more in obtaining valuable information from scat.

Detection dogs also have been trained to detect guano to locate roosting structures used by bats. Field experiments were first conducted to identify factors that influence the probability of guano detection by scent detection dogs in pinyon–juniper (*Pinus* spp. and *Juniperus* spp., respectively) woodlands in New Mexico (A. Chung-MacCoubrey, U.S. Forest Service Rocky Mountain Research Station, unpublished data). Detection probabilities were higher for larger quantities of scat, and detection probabilities decreased as height of scat increased from ground level to 2 m. There was no effect of dog–handler pair or scat dispersal on detection probabilities, but there were significant interactions among temperature, scat distance from path, and cumulative work time: the probability of detection decreased with increasing values of these variables. Lower detection probabilities also were associated with fatigue (cumulative time worked) and time since scat placement. Researchers in Arizona and New Mexico (A. Chung-MacCoubrey, National Park Service, C. Chambers and L. Mering, Northern Arizona University, and C. Vojta, U.S. Forest Service, unpublished data) then tested the efficiency of dog–handler pairs in locating known roosts of bats in ponderosa pine (*Pinus ponderosa*) snags. Dogs in these trials were trained using a blend of guano from 5 species of bats, and the influence of weather, characteristics of roost trees, roost height, number of bats, and dog–handler pair on the success of identifying roosts was evaluated. The researchers found that dogs located 71% of known roosts and were most successful when roosts had higher numbers of bats and were closer to the ground; there was no difference among dog–handler pairs. Building on previous tests, these researchers also compared the success of dogs finding bags with varying quantities of guano (0 g, 5 g, or 20 g) placed at different heights (2 m or 6 m) throughout 1-ha plots and found that dogs were more likely to find large amounts of guano closer to ground than smaller amounts placed higher. The researchers suggested that detection dogs are likely to be most effective in woodlands (e.g., pinyon–juniper), where bats roost close to ground, or in locating ground (e.g., hidden cave) roosts. In summary, scat detection dogs have and continue to be used for a wide range of species in diverse habitats for multiple research purposes.

CAPTURING AND MARKING WILDLIFE

Capturing and marking of wildlife is often an essential field activity for many research projects. Dogs can provide invaluable service to this end (Table 5.2). Some researchers have suggested using pointing dogs to **capture** grouse chicks approximately ≤4 weeks of age; chicks 2 weeks of age can be picked up by hand in front of the dog, but older chicks should be captured in front of the dog using a long-handled net or noose poles (Hannon et al. 1990; Connelly et al. 2000, 2003b). Border collies were used to capture and relocate the endangered Aleutian Canada goose (*Branta canadensis leucopareia*) to predator-free islands in Alaska (Shute 1990). Dogs in this study were not only more efficient at capturing geese by herding them into nets, but they also spared injury to geese and researchers that had occurred prior to using dogs. However, dogs may increase mortality during capturing activities (Zwickel 1980), and individual dogs vary in their ability to aid in the safe capture of wildlife. This variability is largely related to prey drive and the level of control that handlers have over their dogs.

Large mammalian predators also have been captured and marked by using dogs (hounds; Table 5.2). Akenson et al. (2001) used hounds to estimate black bear density in Oregon. Hornocker (1970) used hounds to capture and mark mountain lions in Idaho for a predation study. Shaw (1989) used his many years of experience to relate the nuances of capturing mountain lions with hounds that could be helpful to others capturing and marking large cats. In contrast Logan et al. (1999) believed that capturing mountain lions with foot-hold snares produced fewer deaths compared to traditional methods of capture—using dogs and immobilization. Germaine et al. (2000) used hounds to monitor presence–absence data for mountain lions in southwestern Arizona. While focusing on likely habitat, hounds were run off horseback, and hound behavior was noted when they detected fresh scent.

STUDIES OF WILDLIFE BEHAVIOR

Studies of **wildlife behavior** can be aided by the use of dogs. Zwickel (1980) reported this area of using dogs in wildlife research may have the most potential for growth. Storaas et al. (1999) used dogs to simulate mammalian predation on capercaillie and black grouse (*Tetrao tetrix*) nests and broods in fragmented habitats. They found that detection distances (behavior of the predator) and reaction-to-predator distances (behavior of the prey) differed between nests and broods when using dogs. Miller et al. (2001) used dogs (accompanied by a handler and alone) to measure disturbance behavior of songbirds (2 grassland species and 1 forest species) and mule deer (*Odocoileus hemionus*) along recreational trails and off trails. They found that wildlife had different responses to different treatment stimuli. Sweeney et al. (1971) used hounds to chase radiomarked white-tailed deer (*O. viginianus*) and monitor escape behavior. They found that individual deer selected different strategies, but most ended up using bodies of water in their tactics. Most deer returned to their home range within a day, and all deer remained in good physical condition. Artificial nest studies are often used in avian research, but they can be compromised by predator behavior of following human scent trails (Donalty and Henke 2001). Donalty and Henke (2001) used a dog when checking artificial nests to test whether the dog's scent could mask human scent and found no difference in treatments and controls.

WILDLIFE DAMAGE MANAGEMENT

Wildlife damage management is an important field due to an extensive wildlife–agricultural interface and expansion of the human–wildlife interface near urban areas. Dogs have been successfully used to target problem animals for capture, harass animals, and protect domestic animals. Because of the protected status of brown bears in North America, Gillin et al. (1997) suggested using animals specifically bred for chasing bear, Laika dogs, for nonlethal management when bear–human conflicts occur. Karelian bear dogs also have been used for this purpose (see www.beardogs.org; C. Hunt, Wind River Bear Institute, unpublished data.). Beckmann et al. (2004) evaluated the effectiveness of **nonlethal deterrents,** including hounds, to manage problem black bears in the Lake Tahoe Basin of the Sierra Nevada range. They found that over time, all nonlethal deterrents were ineffective at keeping nuisance bears from returning to their original home ranges. In contrast, dogs were used to successfully harass nuisance geese in urban settings (Conover and Chasko 1985). Castelli and Sleggs (2000) reported that border collies were successful at controlling nuisance Canada geese (*Branta canadensis*) in New Jersey. Once trained, these dogs can be confined to specific areas by use of electric fences and shock collars (i.e., invisible fences). This has considerable application for golf courses, which have high rates of wildlife conflicts, particularly with Canada geese (M. Conover, personal communication). Dogs also were used to reduce damage caused by white-tailed deer to a tree plantation in Missouri (Beringer et al. 1994). Caley and Ottley

(1995) tested the effectiveness of hunting dogs for removing feral pigs (*Sus scrofa*). They found that effectiveness of dogs decreased with increasing group size of the pigs, and dogs were biased toward catching male pigs. These researchers found that dogs were only effective for removing pigs following control with other methods.

Detection dogs have been used in Guam to locate brown tree snakes (*Boiga irregularis*) in outgoing cargo to prevent the spread of this invasive species (Engeman et al. 1998). However, the dogs only detected 61–64% of known planted brown tree snakes in an efficacy test (Engeman et al. 2002). The handlers' search pattern did not change between detected and undetected cases, but the dog's body language failed to indicate a snake for the latter. This study demonstrated a difference in detection based on indoor or outdoor searches. This discrepancy likely occurred because of variable scenting conditions outside, or possibly because of training issues, as the dogs experienced more distracting stimuli outside.

Livestock protection dogs have been used for thousands of years in Europe and Asia, but have seen less emphasis in the western hemisphere (Green and Woodruff 1980). Smith et al. (2000) reported a document written in 150 B.C. on Roman farm management that described the use of livestock protection dogs. European breeds are the most common, and Great Pyrenees is probably the most popular breed, though many others exist, including the Akbash dog, Anatolian shepherd, Komondor, and sarplaninac (Green and Woodruff 1980, Green et al. 1984). Green and Woodruff (1988) provided an overview of the use of different breeds and their characteristics for livestock protection. Coppinger et al. (1985) reported that mongrel dogs can make effective livestock protectors as well and, historically, have been used for this purpose by the Navajo tribes of the southwestern United States. One common ingredient for a successful protection dog is raising it among the animals that it is to protect. The majority (89%) of livestock producers using dogs feel they are cost effective, and most (80%) individual dogs from the breeds mentioned above become reliable protection dogs (Green et al. 1984). Andelt and Hopper (2000) showed that protection dogs reduced domestic sheep depredation in Colorado and that producers without dogs lost 2.1 and 5.9 times (depending on year) more lambs than did producers with dogs. Savings from protection dogs outweigh the cost in most cases, and some producers saved up to approximately US$14,000 per year with use of dogs (Green et al. 1984).

TRAINING AND HANDLING

Training dogs in general is an art form as much as a science. All dogs used in wildlife research should have basic obedience training (sit, come, stay, heel, etc.) prior to more specialized training for their target species. There are many different modern and traditional techniques for training hunting dogs (e.g., Wolters 1961, Tarrant 1977, Williams 2002). Training can take place at any age, though young dogs (≤2 years) seem most impressionable. Keeping things simple and fun (no pressure) for puppies is important. Each dog matures at different ages, and a good trainer recognizes when to increase the training pressure. The choice to train a dog yourself for specialized work is a personal one. Often a working relationship or bond develops during training that can reward the trainer with a morale boost and a partner with excellent skills, if done properly. If commitment (of time, space, expertise, and desire) is lacking, the training and its incomplete results can often be the most frustrating aspect of using a dog in wildlife work. Another option is to obtain a partially or fully trained dog, though upfront costs are generally higher than those for a puppy. If personal training is an issue, hiring a professional dog trainer is an alternative to get the dog to perform tasks according to specific project needs. However, after ensuring the professional trainer understands these specific needs, the researchers using the dogs need to work with the trainer, so the dog will respond to them as consistently as it does to the trainer.

This requirement presents an inherent problem for wildlife research work. Research projects often rely on graduate students, who may be working on the project for only 2–4 years. These temporary situations may conflict with committed dog ownership (unless the graduate students have their own dogs), because the life span for most working breeds generally ranges between 8 and 12 years. Therefore, it may be preferable for a principal investigator—rather than the university or agency—to have ownership of the animal. Trained dogs can then be used as needed by field personnel. In our experience, university and agency ownership adds considerably to the burden of using dogs in research, as institutional requirements become very restrictive.

Another option is to use **volunteers** from local hunting dog or conservation groups. Such organizations as the North American Versatile Hunting Dog Association or AKC breed clubs have chapters throughout North America and may provide volunteers with highly trained dogs. We caution that dogs and handlers vary greatly, and it may be difficult to ensure quality data collection. Specific protocol must be established and adhered to. However, for general presence–absence studies or animal capture, they can be extremely beneficial. Additionally, in some cases volunteers could be used to determine density or indices to density of animals, such as in an evaluation of habitat treatments (Dahlgren et al. 2006).

Use of modern **electronic collars** (e-collars; the authors have successfully used products from the following companies: Tri-tronics, Tucson, AZ; and Dogtra, Torrance, CA) can be helpful for training purposes (e.g., Dobbs et al. 1993). E-collars are a humane and effective method for dog training if used properly. If used improperly, e-collars are destructive to both the training process and the dog's personality and drive. The most common mistake people make when using e-collars is delivering correction from the e-collar prior to the dog's complete understanding of the command.

A dog must know it disobeyed a command before stimulation from the e-collar can be effective. This understanding from the dog comes from consistent repetitive training sessions prior to e-collar stimulation. Without this understanding, training with an e-collar is completely ineffective. Additionally, dogs vary in how they respond to e-collars. For some dogs, it may not be an effective tool.

Scat detection dogs are motivated by different factors than many dogs trained to find live animals; pointing dogs, for example, have instincts for bird detection that training will hone. In contrast, scat detection dogs are obsessively eager to receive a reward (usually toy-play or food) and are taught to seek a target of no inherent interest by learning that finding that target will result in getting the reward. This noninstinctual reward-based system relates to other differences between scat detection dogs and other wildlife dogs: breed is largely irrelevant and dogs may be taught to detect many targets at once. In all documented studies to date, scat detection dogs have varied by sex, age, and breed (pure and mixed; e.g., Australian cattle dog, Australian shepherd, Belgian malinois, border collie, German shepherd, golden retriever, Labrador retriever; Table 5.1). Of great interest to biologists is that scat detection dogs have the ability to be trained to multiple species, cover large search areas, and locate cryptic or hard-to-find scats of rare and common species. Furthermore, scat detection dogs have been successful in cross-training to detect live animals as well as invasive and rare plants (Cablk and Heaton 2006; Goodwin et al. 2006). In fact, with some live-animal detection, it may be favorable to work a dog with less prey drive and thus less inherent interest in the target (see Calbk and Heaton [2006] for detailed information on training and handling dogs for locating desert tortoise).

Without breed preferences or instinctual interests as a guide, selecting dogs with a specific set of traits is paramount for successfully training a dog for scat detection. Desirable traits, or "drives," in a scat detection dog are similar to those in one selected for narcotic or cadaver detection, but the ideal proportion of these qualities may differ from dogs better suited for other detection disciplines. The relevant drives and characteristics (and ideal drive strength) are pack drive (moderate), play drive (extremely high) or food drive (extremely high), hunt or search drive (extremely high), prey drive (low to moderately high), work ethic (extremely strong), and nerve strength to handle stress and new stimuli (moderate or greater). Scat detection dogs contend with rugged physical environments, like all other wildlife dogs, and so the most versatile and low maintenance dog will be heat tolerant, moderately sized, agile, and fit (Hurt and Smith 2009).

Training is similar to other detection disciplines and consists of: (1) imprinting the target odor so the dog understands that smelling the target results in receiving the reward, (2) training an "alert" (e.g., sit or bark) the dog is to perform upon locating the scent, (3) developing interactive search behavior with the handler, and (4) maintaining fidelity to the target scent. Because individual dogs vary considerably, the selection criteria are strict, and training is intense, not all candidates that begin training end up being used for official work; one dog out of 200–300 will be selected to begin training, and of those selected, 40% become field ready (Hurt and Smith 2009). For additional selection and training information for scat detection, see Smith et al. (2003a), Wasser et al. (2004), and Hurt and Smith (2009).

Arnett (2006) trained Labrador retrievers to locate dead bats by seeding a 10 m × 25 m belt transect with bat carcasses representing different species and in varying stages of decay. The dog was rewarded with a food treat if it performed the task of locating a trial bat, sitting or at least stopping movement when given a whistle command to do so, and leaving the carcass undisturbed. Arnett (2006) chose to begin formal testing based on his perception of the dogs' quickening response to the scent of trial bats, their response to commands, and their ability to consistently find all trial bats during their last few days of training.

SPECIAL CONSIDERATIONS

Reporting research results is a special consideration when using dogs for wildlife studies. Here the researcher must describe the specific attributes of the dogs used and steps taken to control bias and variance among dogs. It is not adequate to simply state that dogs were used in a certain manner to collect data. Specific information needs to be reported, such as breed, temperament (prey drive), range (average and maximum), and ground coverage (average velocity, redundancy in pattern) of the **individual dogs.** Although this requirement may seem a radical departure from traditional research involving humans, it is necessary when using dogs. For human researchers we often have established protocols to attempt to eliminate observer bias. Removing bias is more difficult to achieve with dogs, as they do not completely understand research intent. For example, a dog's range is unique to that animal, as is its velocity of travel. Although the handler can take steps to control for these variations, such as selection of individuals and monitoring range, there are limitations on what can be controlled, and any 2 dogs will behave differently. Therefore, studies using the same methodologies, but with different types of dogs could vary considerably in their results. This variability does not necessarily result in poor studies, but for a researcher attempting to replicate the work, it becomes problematic. However, if specific characteristics of the dog(s) are explicitly described, the reader can more accurately infer results and design future research that is comparable. This practice should be considered true disclosure, just as with any study where potential limitations are described.

We believe **future research needs** for using dogs in wildlife studies could be enhanced by considering the following information. First, the effects of scenting conditions are inherently difficult to deal with and have rarely been modeled. By quantitatively assessing these environmental factors, better data and a greater understanding of scenting conditions can be gathered. Second, with an increase in the use of probability detection techniques, it will be important to know how detection rates may vary by species and individual dogs. Third, we have described various techniques for using dogs to obtain density estimates. Efficacy studies on techniques using dogs have been rare in wildlife research, and there is a need for more work assessing the accuracy of these methods.

SUMMARY

Although dogs have been used in wildlife management and research for many years, there has not been a synthesis of this work, and their use has largely been conducted by trial and error on the part of individuals. We have synthesized the vast array of useful applications of dogs in the field of wildlife management and research. We hope this chapter will not only serve as a reference for field practitioners, but also will stimulate new applications of dogs in our field. Dogs have limitations, and there are special considerations in using them, just as for any technique or tool. Despite their limitations, they provide the wildlife professional with abilities that cannot be otherwise replicated.

The authors' experience and the available literature indicate that dogs are truly an underutilized tool. We hope that professionals find this information useful and consider methods using dogs to better manage wildlife resources. In many instances, the dogs can lend superior skills that lead to better data in the field of wildlife research and management. Advances in techniques from scat detection to GPS technology increase the value of dogs for wildlife work. Consistent and proper training cannot be overemphasized when considering dogs for use in data collection. "Man's best friend" may in fact be a biologist's best asset.

6 Identifying and Handling Contaminant-Related Wildlife Mortality or Morbidity

STEVEN R. SHEFFIELD,
JOSEPH P. SULLIVAN,
AND ELWOOD F. HILL

INTRODUCTION

WILDLIFE BIOLOGISTS HAVE potential for field encounters with wildlife mortality or morbidity incidents as a result of routine monitoring of an area or a call from the general public. **Mortality** is an incidence of death in a population, whereas **morbidity** is an incidence of sickness or ill health. Wildlife mortality or morbidity can be due to natural or accidental causes, disease, or exposure to environmental contaminants. Every year, species of wildlife are subject to exposure to a myriad of different chemical contaminants that make their way into the environment. These **chemical contaminants** include pesticides, metals or metalloids, organics, inorganics, pharmaceuticals, and a wide variety of other compounds in air, soil, sediment, water, plants, and wild and domestic animals. If organisms are exposed to contaminants, there may be no resulting visible effects, suggesting that there were no effects of exposure, or that if there was a negative effect, it was not apparent. However, there may be visible effects from exposure to chemical contaminants, indicating that it caused sickness or was lethal to wildlife species.

Understanding **contaminant impacts** on wildlife includes determining parameters, such as species (or higher taxa) involved, trophic level of the species involved, chemical(s) involved, route(s) of exposure, signs of intoxication, fate and transport (movement) through the environment, environmental compartment (media), and environmental persistence. Not all classes of contaminants pose the same level of risk to all taxonomic groups of animals. For example, mammals may be relatively less sensitive than birds or reptiles to environmental contaminants due to their evolutionarily more advanced **detoxification enzyme systems.** The **physical and chemical properties** of the chemical (e.g., lipid solubility, water solubility, environmental persistence, and volatility), level of toxicity, route(s) of exposure, and **trophic level** (first use) of the animal all affect which taxonomic groups may be more susceptible to different classes of contaminants. The trophic level at which the animal feeds is a major factor in contaminant exposure: animals at higher trophic levels are more susceptible to bioaccumulation and biomagnification of contaminants. **Bioaccumulation** is the process by which chemicals accumulate in an organism at a rate faster than they can be metabolized and excreted. **Biomagnification** is the result of the process of bioaccumulation and biotransfer, by which tissue concentrations of a contaminant increase as it moves up the food chain through 2 or more trophic levels. Herbivorous species may be less susceptible to the effects of contaminant exposure—they are better adapted for detoxifying **foreign chemicals (xenobiotics),** because they routinely encounter **natural plant toxins** (secondary plant compounds) in their diet that require similar detoxification (Vangilder 1983, Ray 1991). Species that feed on soil invertebrates will be more susceptible to exposure to contaminants, such as metals, that remain in soil for long periods of time.

The behavior of sick or intoxicated animals and their location and numbers may be indicative of different classes of contaminants. Once the class of contaminant is identified, characteristics of the class or the specific contaminant will affect what type, where, or how many environmental samples should be collected, and how long after the incident has been discovered the site should be monitored. On discovery of a field mortality or morbidity incident suspected to be caused by **environmental contaminants,** there generally is little time to plan and conduct a research study of the incident. The available time for collecting evidence, such as tissue samples and/or other environmental media (plants, soil, water, sediment, or air) is often a matter of hours to a few days. Chemicals decompose, tissues decay or desiccate, and carcasses are readily scavenged, all of which greatly affect time available for sampling.

The objective of this chapter is to provide guidelines for field biologists to assess wildlife mortality or morbidity incidents and sampling techniques useful in detection and documentation of environmental contaminants impacting wildlife. Previously, there was difficulty finding published procedures for handling wildlife mortality or morbidity incidents. Few specific criteria are available for conclusive diagnosis of wildlife poisoning, other than correlation of effects with chemical residues in critical tissues. Here, we include safe, proper field techniques for collecting, handling, and preserving environmental samples for biological assays or chemical analyses as well as where to look for more information on wildlife mortality or morbidity (Box 6.1). Because time is short and field data and samples are critical, assistance from others with experience in handling contaminant-related issues can be important to gain a full understanding of the entire incident. Careful documentation of the mortality or morbidity incident is necessary, including detailing the appearance of affected individuals, species involved, and likely scenarios leading to the incident.

ENVIRONMENTAL CONTAMINANTS

Human activities have resulted in the pervasive and dynamic nature of contaminants in our environment. Although environmental contamination increased sharply with the rise of the Industrial Revolution in the mid- to late 1800s, the presence of contaminants in the environment has accelerated greatly since the 1940s. For example, pesticide use in the United States increased more than 10-fold, and chemical and mining industries have continued their extensive operations during the post-war economic growth. The United States is the major producer, user, and exporter of pesticides in the world. In 2001, 2.26 billion kg of active ingredient of toxic chemicals were used as pesticides in the United States (Kiely et al. 2004), increasing annually from 2.02 billion kg used in 1994 (Aspelin 1997). The United States also is the major producer, user, and exporter of organic and inorganic chemicals. In 2002, the estimated volume of the top 50 chemicals produced by the chemical industry in the United States was 837 billion kg (Chenier 2003), up from >340 billion kg in 1995 (Chemical and Engineering News 1996). The United States produces or imports about 3,000 different organic chemicals of >454,000 kg each on an annual basis; 43% of these chemicals have no data available on basic toxicity, and only 7% have a full set of basic toxicity data (U.S. Environmental Protection Agency 1998*b*). The United States also is a major world producer of metals and minerals. In 2004, there were 14,478 active mines in the United States (up from 13,925 in 2001), and the annual per capita consumption rate of newly mined metals and minerals reached 21,319 kg in 2004 (National Mining Association 2009), up from 20,870 kg in 2002 (National Mining Association 2004). The high level of production of these industries in the United States resulted in 1.9 billion kg of toxic chemicals released into the air, land, water, and underground in 2007 (Office of Environmental Information 2009*a*); 35,274 chemical and oil spills in 2007 (U.S. Department of Transportation 2009); 1,263 **Superfund hazardous waste sites** (Office of Environmental Information 2009*b*); 40,000 **CERCLIS** (Comprehensive Environmental Response, Compensation, and Liability Act of 1980; http://www.epa.gov/tp/laws/cercla.htm) **hazardous waste sites;** and an estimated 350,000 contaminated commercial or industrial sites and 482,166 leaking underground storage tanks across the country (Office of Solid Waste and Emergency Response 2004). Because of this great potential for chemical contaminants in the environment, it is inevitable that individuals of a variety of different wildlife species will be exposed, and some will become sick and/or die.

The U.S. Environmental Protection Agency reports that state wildlife agencies in the United States annually receive about 3,800 reports of pesticide-related fish, wildlife, and plant incidents (American Society for Testing and Measurement 1997). These reports indicate that pesticide use can pose considerable risk to nontarget species, particularly birds and fish. Most reported incidents are a result of exposure to insecticides and rodenticides, and not herbicides, fungicides, or other pesticides. The greatest number of wildlife mortality or morbidity incidents has been reported for anticholinesterase (anti-ChE; organophosphorous and carbamate) insecticides and anticoagulant rodenticides. **Anticholinesterase insecticides** are chemicals that inhibit the enzyme ChE found in synapses of the nervous system, resulting in overstimulation of muscles. **Anticoagulant rodenticides** are chemicals that

BOX 6.1. RECOMMENDED REFERENCES FOR FISH AND WILDLIFE MORTALITY AND MORBIDITY

Adrian, W. J., editor. 1996. Wildlife forensic field manual. Second edition. Association of Midwest Fish and Game Law Enforcement Officers, Denver, Colorado, USA.

American Society for Testing and Measurement. 1997. Standard guide for fish and wildlife incident monitoring and reporting. Pages 1355–1382 in Biological effects and environmental fate; biotechnology; pesticides. ASTM E 1849-96. American Society for Testing and Measurement, Philadelphia, Pennsylvania, USA.

Briggs, S. A. 1992. Basic guide to pesticides: their characteristics and hazards. Taylor and Francis, Washington, D.C., USA.

Canadian Cooperative Wildlife Health Centre. 1975. Wildlife disease investigation manual. Canadian Cooperative Wildlife Health Centre, University of Saskatchewan, Saskatoon, Canada.

Dierauf, L. A., and F. M. D. Gulland, editors. 2001. CRC handbook of marine mammal medicine. Second edition. CRC Press, Boca Raton, Florida, USA.

Fairbrother, A., L. N. Locke, and G. L. Huff, editors. 1996. Noninfectious diseases of wildlife. Second edition. Iowa State University Press, Ames, USA.

Friend, M., and J. C. Franson, editors. 1999. Field manual of wildlife diseases: general field procedures and diseases of birds. Information and Technology Report 1999–2001, Biological Resources Division, U.S. Geological Survey, Department of the Interior, Washington, D.C., USA.

Hoffman, D. J., B. A. Rattner, G. A. Burton, Jr., and J. Cairns, Jr., editors. 2003. Handbook of ecotoxicology. Second edition. Lewis, Boca Raton, Florida, USA.

Hudson, R. H., R. K. Tucker, and M. A. Haegele. 1984. Handbook of toxicity of pesticides to wildlife. Second edition. Resource Publication 163, U.S. Fish and Wildlife Service, Department of the Interior, Washington, D.C., USA.

Meyer, F. P., and L. A. Barclay, editors. 1990. Field manual for the investigation of fish kills. Resource Publication 177, U.S. Fish and Wildlife Service, Department of the Interior, Washington, D.C., USA.

National Conservation Training Center. 2002. Environmental contaminants: field and laboratory techniques. U.S. Fish and Wildlife Service, Department of the Interior, Shepardstown, West Virginia, USA.

National Wildlife Health Center. 1998. Fish and wildlife mortality incident information workshop. U.S. Geological Survey, Department of the Interior, Madison, Wisconsin, USA.

National Wildlife Health Center. 1999. Wildlife mortality database resource directory. U.S. Geological Survey, Department of the Interior, Madison, Wisconsin, USA.

Smith, G. J. 1987. Pesticide use and toxicology in relation to wildlife: organophosphorus and carbamate compounds. Resource Publication 170, U.S. Fish and Wildlife Service, Department of the Interior, Washington, D.C., USA.

Stroud, R. K., and W. J. Adrian. 1966. Forensic investigational techniques for wildlife law enforcement investigations. Pages 3–18 in A. Fairbrother, L. N. Locke, and G. L. Huff, editors. Noninfectious diseases of wildlife. Second edition. Iowa State University Press, Ames, USA.

Work, T. M. 2000a. Avian necropsy manual for biologists in remote refuges. Hawaii Field Station, National Wildlife Health Center, U.S. Geological Survey, Department of the Interior, Honolulu, USA.

Work, T. M. 2000b. Sea turtle necropsy manual for biologists in remote refuges. Hawaii Field Station, National Wildlife Health Center, U.S. Geological Survey, Department of the Interior, Honolulu, USA.

block the vitamin K cycle, thereby resulting in an inability to produce essential blood clotting factors. There is evidence from field investigations that many pesticides in these 2 categories still on the market today have caused confirmed wildlife mortalities and that avian mortality occurs regularly and frequently in agricultural fields across North America (Mineau 2002).

In Europe, a large investigation of terrestrial wildlife mortality incidents involving pesticides in 18 countries was conducted from 1990 to 1994 (deSnoo et al. 1999). There were high numbers of wildlife mortality incidents in France, The Netherlands, and the United Kingdom, all countries with intensive agricultural programs. Most reported incidents were due to deliberate abuse of pesticides, with few mortality incidents reported for normal agricultural use (deSnoo et al. 1999). Their conclusion that reporting of wildlife mortality incidents was not a reliable tool for obtaining an understanding of the occurrence of wildlife mortality incidents from agricultural pesticide use is most likely valid worldwide.

Deleterious effects of pesticides on wildlife include death from direct exposure and **secondary poisoning** from consuming contaminated prey; reduced survival, growth, and reproductive rates from exposure to sublethal dosages; and habitat reduction through elimination of food resources and refuges. **Sublethal effects** are those that serve to debilitate an exposed organism. In the United States, approximately 3 kg of pesticide/ha are applied to about 160 million ha of land annually (Pimentel et al. 1997). With a large portion of the land area subjected to large quantities of pesticide applications, the impact of pesticides on wildlife could be predicted to be substantial (Pimentel et al. 1992). However, few attempts have been made to estimate the overall magnitude of pesticide effects on wildlife species over a large geographic scale. An existing estimate for effects on birds is substantial (Pimentel et al. 1992), but it does not include such factors as bird losses caused by poisoning of invertebrate prey, eggs or chicks left to die when adults are killed, and those birds suffering **neurological effects** that move from the area to places where they cannot reproduce or survive their exposure(s). The occurrence of these effects has been documented (Pimentel et al. 1992, Hill 1999) following pesticide application, but their importance to a population at the species or regional level has not been quantified.

In addition to pesticides, exposure to other chemicals also can serve as major sources of wildlife mortality and morbidity. These chemicals include metals and metalloids (Fairbrother et al. 1996, Goyer and Clarkson 2001, Hoffman et al. 2003), organic chemicals (Friend and Franson 1999, Bruckner and Warren 2001, Rice et al. 2003), cyanide associated with gold mining (Henny et al. 1994, Eisler et al. 1999), white phosphorus associated with military use (Sparling 2003), pharmaceutical drugs (Friend and Franson 1999, Oaks et al. 2004), and natural plant or animal toxins (Norton 2001, Russell 2001).

The full extent of wildlife mortality from contaminants is difficult to assess, because wildlife species are often secretive, camouflaged, and highly mobile, and they may live in dense habitat. Typical field studies of the effects of pesticides often obtain low estimates of mortality, because carcasses disappear rapidly, well before they can be found and counted. Field studies rarely account for animals that die away from treated areas, and many individuals often hide and die in inconspicuous locations. Studies have demonstrated that only 50% of dead or moribund birds are recovered, even when their location is known (Mineau and Collins 1988). **Carcass searches** are rarely done and, even more rarely, done properly. Most carcasses disappear ≤24–48 hours post-spray, making documentation difficult (Vyas 1999). When known numbers of bird carcasses were placed in identified locations in the field, 62–92% disappeared overnight due to scavengers (Balcomb 1986). Kostecke et al. (2001), using remote cameras, documented heavy scavenging of experimentally placed bird carcasses by mammals (particularly striped skunks, *Mephitis mephitis*) and to a lesser extent by birds. This study demonstrates the potential hazard of secondary poisoning and also the need for careful searches for wildlife mortality or morbidity following pesticide applications.

The full extent of wildlife morbidity from contaminants can be even more difficult to assess, because sick animals may move from the area of exposure or otherwise disappear (e.g., fly from the area or retreat to a burrow), may not always demonstrate visible signs of morbidity, and/or may become more vulnerable to predation or other mortality factors as a result of exposure. Sublethal effects are often subtle, and exposure to chemical contaminants can impact all internal body systems (biochemical, physiological, immunological, etc.) that in turn can reduce the fitness and/or survival of exposed individuals. Many chemical contaminants—including pesticides, pharmaceuticals, and even natural plant chemicals that have the ability to disrupt normal endocrine function in animals—are of particular concern and can have major implications for reproduction in wildlife species (Yamamoto et al. 1996, Gross et al. 2003). Although some sublethal effects can be apparent (e.g., tumors and developmental malformations), many are not, and animals suspected of sublethal poisoning often require close examination and laboratory testing. A formidable problem in identifying and understanding sublethal effects is that baseline data for normal (unexposed) individuals largely are unavailable (Hill 1999).

CLASSES OF CONTAMINANTS

Metals and Metalloids

Metals are natural substances and, in most cases, only become significant contaminants when human activity, such as mining and smelting, releases them from the rocks in which they were deposited (during volcanic activity or subsequent erosion) and relocates them where they can cause environmental problems (anthropogenic enrichment). Metals are nonbiodegradable and, unlike organic compounds, cannot decompose into less harmful components. Detoxification consists of "hiding" active metal ions in a protein (e.g., metallothionein) or depositing them in an insoluble form in intracellular granules for long-term storage or excretion in feces. The term heavy metals generally has been used to refer to metals that are environmental contaminants. However, true heavy metals have a density relative to water >5, which excludes some important contaminants, such as aluminum.

The term **metalloid** is used for elements, such as arsenic and selenium, that are transitional in nature between metals and nonmetals. In the environment, metals and metalloids occur as organic or inorganic complexes, and there are several factors that determine which form is more toxic. For example, inorganic arsenic compounds generally are more toxic to wildlife than are organic arsenic compounds, whereas the opposite is true for mercury and lead.

Essential Metals and Metalloids

All animals require 6 metals (minerals) that are essential **macronutrients,** including sodium, calcium, phosphorus, potassium, magnesium, and sulfur for ionic balance and as integral parts of amino acids, nucleic acids, and structural compounds. All animals also require 12 trace metals or metalloids (minerals) that are essential **micronutrients,** including zinc, copper, manganese, iron, selenium, chromium (Cr^{+3}), nickel, cobalt, molybdenum, iodine, vanadium, and silicon. These trace minerals are essential components of enzymes, enzyme cofactors, and other biochemical structures. Presently, there is insufficient information available to ascertain whether such metals as silver, tin, aluminum, lithium, and boron are essential. All essential metals and metalloids can be toxic to wildlife species if sufficiently concentrated. **Selenium** is an excellent example of an essential metalloid that has caused notable toxicity problems in wildlife species in the United States.

Major environmental sources of selenium are coal-fired and other fossil fuel–burning power plants and mining and smelting operations. Selenium is a naturally occurring component of soils (Ohlendorf 2003). It is an essential micronutrient for wildlife and an integral component of the **glutathione detoxifying enzyme system.** However, there is a fine line between selenium deficiency and selenium **toxicosis** (any disease condition resulting from poisoning). Selenium can become concentrated at relatively high levels from mining activities and agricultural run-off. Although occasionally implicated in mortalities of adult animals, it is more likely to produce sublethal effects, such as developmental abnormalities or embryonic death (Eisler 1985b, Ohlendorf et al. 1986). It has been shown to be highly **teratogenic** (producing malformations) in aquatic birds, causing widespread reproductive failure through decreased egg weight; decreased egg production and hatching success; anemia (decreased numbers of red blood cells or hemoglobin deficiency); and a high incidence of grossly malformed embryos with missing or distorted eyes, beaks, wings, and feet (Eisler 1985b). Excess selenium also causes behavioral modifications, intestinal lesions, and chronic liver damage, and it impacts the immune system (Eisler 1985b). Signs of selenium poisoning include vomiting, lethargy or weakness, diarrhea, increased urination, panting, central nervous system depression, paresis, and prostration, and death can result due to respiratory failure (Osweiler et al. 1985). Selenium is readily bioaccumulated in aquatic and terrestrial food chains, but is not biomagnified through food chains. In the early to mid-1980s at Kesterson National Wildlife Refuge in central California, selenium was the causative agent in numerous cases of waterfowl and wading-bird nesting failure (Ohlendorf et al. 1986). In this case, selenium from irrigation drain water accumulated in the waters of Kesterson, where it caused massive reproductive failure through embryonic mortality and developmental abnormalities of aquatic birds (Ohlendorf et al. 1986). Selenium was deposited in eggs and caused severe developmental abnormalities in chicks. Mean selenium concentrations in livers and kidneys were about 95 ppm dry weight, about 10 times higher than levels in birds from a reference area (Eisler 1985b, Ohlendorf et al. 1986).

Nonessential Metals and Metalloids

Some metals have no known biological function and serve to replace essential metals of like valance in animals. These metals include mercury, lead, cadmium, chromium (Cr^{+6}), and arsenic. All tend to be highly toxic and may exert toxicity by inducing deficiencies of essential metals through competition with them at active sites in biologically significant molecules. Examples include lead replacing calcium in bone and arsenic replacing phosphorus in DNA. Metals and metalloids with no biological function tend to be those of the greatest environmental concern, particularly if they are anthropogenically concentrated in a given area.

MERCURY

Major environmental sources of mercury have been chlorine-alkali (plastics) manufacturing; mining and smelting operations; mercurial seed dressings; mercury-based fungicides; coal-fired power plants; thermometer, battery, and fluorescent bulb manufacture; switches; paints; pulp and paper plants; and dental amalgam (Wiener et al. 2003). The use of mercury in agriculture has been largely curtailed in the United States; sources related to energy production and mining are now of greatest concern. About 25–30% of total atmospheric mercury is anthropogenic (Eisler 1987a). In reducing environments (e.g., sediments), inorganic mercury can be readily **biotransformed** by anaerobic bacteria into methyl mercury, which is extremely toxic.

Mercury deposition since industrial times (mid-1880s) and its subsequent biotransformation to methyl mercury in aquatic systems has created areas where mercury poses a relatively high risk to wildlife, particularly long-lived piscivorous (fish-eating) species (Henny et al. 2002, Wiener et al. 2003). Methyl mercury readily crosses biological membrane barriers, whereas inorganic mercury does not. However, once absorbed, both forms of mercury are highly cytotoxic (toxic to cells), causing histopathological lesions in tissues of the nervous, hepatic, renal, and immune systems (Heinz 1996). The most observable sign of organomercury poisoning is central nervous system dysfunction, leading to respiratory stress and ataxia (lack of muscle coordination). Other common signs of mercury poisoning in wildlife species include anorexia (and resulting emaciation), ataxia, progressive paralysis, tremors or spasms, and loss of sight (Heinz 1996, Wiener et al. 2003).

Mercury is readily bioaccumulated in wildlife and biomagnified through food chains. For birds and mammals that regularly consume fish and other aquatic organisms, total mercury concentrations in prey items should not exceed 100-g/kg fresh weight for birds and 1,100 g/kg for small and medium-sized mammals (Eisler 1987a). In wildlife, concentrations of mercury >1,100 g/kg fresh weight of tissue (liver, kidney, blood, brain, and hair or feathers) should be considered as presumptive evidence of an environmental mercury problem (Eisler 1987a). Although mortality or morbidity from mercury is more of an insidious event involving scattered individuals, a substantial number of mercury-related wildlife mortality or morbidity incidents have been reported. Many of these have involved mortality in grebes (Podicipedidae) in the western United States (Eisler 1987a), common loons (*Gavia immer*) and turkey vultures (*Cathartes aura*) in Canada (Friend and Franson 1999), and reproductive impairment in bald eagles (*Haliaeetus leucocephalus*) in the United States (Friend and Franson 1999).

LEAD

Major environmental sources of lead have been leaded gasoline; paints; pesticides; batteries; mining and smelting operations; metal finishing; petroleum refineries; hunting, fishing, and shooting sports (e.g., trap, skeet, and target shooting); and firearms training activities (Pattee and Pain 2003). Although leaded gasoline, paints, and pesticides are not as prevalent now, lead from these sources continues to persist in the environment. In animals, 10% of dietary lead is absorbed, but >90% of that absorbed is retained in bones. Lead causes anemia and inhibition of the enzyme **δ-aminolevulinic acid dehydratase** and has been demonstrated to cause severe neurotoxic effects in young animals and humans (Pattee and Pain 2003). The exposure and effects of tetraethyl lead, an antiknock agent formerly added to gasoline, have been examined along highways (Grue et al. 1984), and lead shot deposition also has been examined, particularly in wetlands (DiGiulio and Scanlon 1984), around trap and skeet shooting ranges (Stansley and Roscoe 1996), and at firearms training facilities (Lewis et al. 2001).

Lead poisoning is most commonly observed in birds, particularly waterfowl. The first documented report of lead poisoning in waterfowl came from Texas in 1894. Bellrose (1951, 1959) reported widespread waterfowl mortality and illness associated with ingestion of lead shot in the 1950s. In the United States, an estimated 1.6–2.4 million ducks, geese, and swans died annually as a direct result of lead shot ingestion before widespread use of nontoxic shot in the early 1990s (Pattee and Pain 2003). Sanderson and Bellrose (1986) and Beyer et al. (1998) reviewed the problem of lead poisoning in waterfowl. Signs of lead poisoning include gross lesions, impactions of the upper gastrointestinal (GI) tract, submandibular edema (accumulation of fluid), myocardial necrosis (tissue destruction), and biliary discoloration in the liver (Friend and Franson 1999). Field signs include inability or reluctance to fly, weak and/or erratic flight, and poor landings. As the condition worsens, birds become flightless and hold their wings in a characteristic "roof-shaped" position that progresses to wing droop as birds become more moribund (Friend 1987). About 95% of waterfowl diagnosed with lead poisoning had liver lead concentrations of at least 38 ppm dry weight (Friend and Franson 1999).

Ingestion of lead shot by both predatory and scavenging raptors feeding on hunter-killed carcasses has been reported for bald and golden eagles (*Aquila chrysaetos*), red-tailed hawks (*Buteo jamaicensis*), turkey and black (*Coragyps atratus*) vultures, and California condors (*Gymnogyps californianus*; Janssen et al. 1986, Craig et al. 1990). Vultures appear to be highly susceptible to poisoning from ingesting small quantities of lead shot (Eisler 1988b). In addition to lead shot, lead fishing sinkers have contributed to lead-caused mortalities in a number of aquatic bird and mammals, particularly common loons (Pokras and Chafel 1992, Scheuhammer and Norris 1996, Stone and Okoniewski 2001, Sidor et al. 2003). Lead is readily bioaccumulated in wildlife, but it does not appear to be biomagnified in food chains. At least 6 endangered or formerly **endangered species,** including bald eagle, peregrine falcon (*Falco peregrinus*), California condor, brown pelican (*Pelecanus occidentalis*), Mississippi sandhill crane (*Grus canadensis pulla*), and whooping crane (*G. americana*), have been victims of lead poisoning (Friend and Franson 1999).

CADMIUM

Major environmental sources of cadmium include electroplating, zinc and lead mining and smelting, paint and pigments, batteries, plastics, coal-fired power plants, and municipal wastewater and sewage sludge. Cadmium is a known teratogen and affects calcium metabolism, causing excess calcium excretion that negatively impacts both skeletal and cardiovascular systems (Eisler 1985a). In addition, growth retardation, anemia, and testicular damage occur in cadmium-exposed wildlife (Eisler 1985a). Cadmium is readily bioaccumulated, and data suggest that it is biomagnified through food chains (Larison et al. 2000). White-tailed ptarmigan (*Lagopus leucurus*) in Colorado were poisoned by cadmium due to biomagnification (hyperaccumulation) in willow (*Salix* spp.), a primary food plant for these birds (Larison et al. 2000). Cadmium residues in vertebrate kidney or liver that are >10 ppm fresh weight or 2 ppm whole-body fresh weight should be viewed as evidence of probable cadmium toxicity; residues of 200 ppm kidney fresh weight, or >5 ppm whole-body fresh weight are indicative of cadmium poisoning. Wildlife, especially migratory birds that feed on crops growing in fields fertilized with municipal sewage sludge, may be at considerable risk from cadmium toxicity (Eisler 1985a).

CHROMIUM

Major environmental sources of chromium include production of stainless steel (ferrochrome), including electroplating and metal finishing industries; pigments (paint and ink); leather tanning; wood preservatives; coal-fired power plants;

municipal incinerators and publicly owned treatment plants; cement-producing plants; and anticorrosives in cooling systems and boilers. Chromium is most frequently found in the environment in its trivalent (Cr^{+3}) and hexavalent (Cr^{+6}) forms. The biological effects of chromium (Cr^{+6}) may be related to reduction to Cr^{+3} and formation of complexes with intracellular macromolecules that, if it occurs in genetic material, leads to **mutagenesis** (formation of mutations). Chromium (Cr^{+6}) is toxic to embryos; is teratogenic; and causes alterations of blood and serum chemistry, liver and kidney lesions (including acute renal tubular necrosis), and ulcerations in mucous membranes. In wildlife, tissue levels >4.0 mg total chromium/kg dry weight is presumptive evidence of an environmental chromium problem, although the significance of the tissue chromium residues is not known. Chromium is readily bioaccumulated in wildlife, but concentrations usually are highest at the lower trophic levels, and it is not known to be biomagnified in food chains (Eisler 1986a). Wildlife mortality or morbidity as a result of chromium exposure generally is infrequent (Eisler 1986a).

ARSENIC

Major environmental sources of arsenic include copper, zinc, and lead mining and smelting; glass and chemical manufacturing, particularly wood preservatives and arsenic-based herbicides; and coal-fired power plants. There are many different arsenic compounds, and their environmental chemistry is quite complex, but trivalent (As^{+3}) and pentavalent (As^{+5}) forms predominate, and both organic and inorganic forms are common. Arsenic is a teratogen and can traverse placental barriers and produce fetal death and malformations in wildlife (Eisler 1988a). It is highly cytotoxic, affecting mitochondrial enzymes and impairing tissue respiration. Chronic exposure leads to neurotoxicity in peripheral and central nervous systems, liver damage, and peripheral vascular disease (Eisler 1988a). Arsenic is bioaccumulated by wildlife, but is not biomagnified in food chains. Despite its high toxicity, wildlife mortality or morbidity as a result of arsenic exposure generally is infrequent (Eisler 1988a).

Organic and Inorganic Chemicals
Organic Chemicals

Organic chemicals are based on carbon–hydrogen pairs that range from single carbon chains to multiple aromatic rings. Organic chemicals can be released from refineries, oil or gas spills, incinerators, sewage effluent, wood treating, chemical plants, military sites, and other industrial sites. Many pesticides are organic chemicals; pesticides are treated separately —this section pertains only to nonpesticide organic chemicals. Generally, organic chemicals are more hazardous to wildlife than are inorganic chemicals. Some organic chemicals are of concern to wildlife, including organic solvents, ethylene glycol, petroleum products, polychlorinated biphenyls (PCBs), polybrominated diphenyl ethers (PBDEs), and dioxins and furans.

SOLVENTS

Organic solvents generally are refined from petroleum and are used to dissolve, dilute, or disperse other chemicals (including pesticides) that are not soluble in water. They are used widely as degreasers and as constituents of paints, varnishes, lacquers, inks, aerosol sprays, dyes, and adhesives. They also are used as intermediates in chemical synthesis and as fuels and fuel additives. Organic solvents include widely used chemicals, such as chlorinated hydrocarbons (e.g., trichloroethylene, perchloroethylene, methylene chloride, and carbon tetrachloride), aromatic hydrocarbons (e.g., benzene, toluene, xylene, styrene, and ethylbenzene), alcohols (e.g., ethanol and methanol), aldehydes (e.g., formaldehyde), ketones (e.g., acetone), glycols (e.g., ethylene glycol and propylene glycol), glycol ethers, phenols (e.g., phenol and chlorophenol), carbon disulfide, and fuel and fuel additives (e.g., gasoline, methyl tertiary-butyl ether [MTBE], jet fuel, and kerosene). Because of their widespread use, organic solvents are ubiquitous in the environment (Bruckner and Warren 2001). Generally highly lipophilic (having an affinity for lipids), extremely volatile, and of relatively small molecular size and lacking charge, organic solvents are rapidly absorbed across lungs, GI tract, and skin. The most notable negative effect of this group is central nervous system depression (Bruckner and Warren 2001). Other negative effects include carcinogenesis and damage to the hematopoietic system (bone marrow), liver, and kidney (Bruckner and Warren 2001). Organic solvents tend to readily bioaccumulate, but are not known to biomagnify through food chains.

ETHYLENE GLYCOL

A major ingredient in antifreeze and de-icing solutions, **ethylene glycol** is responsible for numerous wildlife deaths in the United States and Canada each year (U.S. Environmental Protection Agency 1998a). It is an oily liquid with a mild odor and a sweet taste that makes it attractive to wildlife. Puddles of antifreeze or brake fluid can accumulate on roads or parking lots, and their color and smell attract many wildlife species. The vast majority of ethylene glycol is released directly into the environment as airport and runway runoff from de-icing activities. An annual release of >26 million kg of ethylene glycol occurs during icing conditions at the 17 busiest airports in the United States (U.S. Environmental Protection Agency 1998a). Ethylene glycol also is used in polyester compounds and as a solvent in the paint and plastics industries, photographic developing solutions, hydraulic brake fluids, and inks.

Wildlife poisoned by ethylene glycol appears intoxicated; signs, including depression, ataxia, and reluctance to move, appear as soon as 2–4 hours following exposure (Stowe et al. 1981). Ethylene glycol metabolizes to oxalic acid and binds to calcium to form calcium oxalate crystals that block renal tubules: death results from acute renal failure (MacNeill and Barnard 1978, Stowe et al. 1981). Kidneys should be collected from carcasses if ethylene glycol poisoning is suspected. Canids and felids are particularly susceptible to ethylene glycol; as little as 4–5 mL/kg is lethal to domestic dogs, and 2–4 mL/kg is lethal to domestic cats (Osweiler et al. 1985). Waterfowl, vultures, and birds of the family Corvidae (jays, crows, ravens, and magpies) occasionally are victims of ethylene glycol poisoning. There is at least one record each of a California condor (Murnane et al. 1995) and a polar bear (*Ursus maritimus;* Amstrup et al. 1989) being lethally poisoned by ethylene glycol.

PETROLEUM PRODUCTS

Petroleum products, including crude oil, diesel, gasoline, and kerosene, are ubiquitous in the environment. Every year, an average of 53 million liters of oil from >10,000 accidental spills flows into fresh and saltwater environments in the United States (Friend and Franson 1999). However, accidental releases account for a small fraction of all oil entering the environment; most oil is introduced through intentional discharges from normal transport and refining operations, industrial and municipal discharges, used lubricant and other waste oil disposal, urban runoff, river runoff, atmospheric deposition, and natural seeps (Eisler 1987b, Jessup and Leighton 1996, Albers 2003). Wildlife exposed to petroleum products can be impacted both externally and internally. Oil contamination of hair and feathers disrupts their normal structure and function, resulting in a loss of insulation and waterproofing (Eisler 1987b). Birds and mammals also can ingest, inhale, and absorb petroleum products when exposed during spill events while preening (grooming) contaminated feathers (hair). In birds, hatching success is reduced when adults are exposed to fuel oil during incubation and transfer oil to their eggs (Jessup and Leighton 1996).

Polycyclic aromatic hydrocarbons (PAHs) contribute heavily to the toxicity of crude petroleum and refined petroleum products, but amounts of these compounds in petroleum products vary widely. PAHs are semi-volatile and occur in the environment from many sources in addition to the petrochemical industry, including natural sources. Low-molecular-weight PAHs cause significant acute toxicity and other adverse effects in wildlife, but are not carcinogenic. However, high-molecular-weight PAHs are usually less acutely toxic, but may be carcinogenic, mutagenic, or teratogenic in a wide variety of wildlife (Eisler 1987b). PAHs, although highly lipid soluble, generally are rapidly metabolized and tend not to bioaccumulate in wildlife; there is little evidence for biomagnification in food chains. PAHs, such as benzo(a) pyrene, naphthalene, anthracene, and styrene, have been investigated for their effects on wildlife (Eisler 1987b). There are no specific regulations regarding the protection of wildlife species from PAHs other than laws governing petroleum products (Eisler 1987b). There is little evidence to indicate that PAHs are likely to produce large numbers of wildlife deaths or sicknesses except when associated with oil spills.

POLYCHLORINATED BIPHENYLS

PCBs were introduced in 1929 for use in dielectric (insulating) fluids. They were used extensively in the electricity generating industry as insulating or cooling agents in transformers and capacitors. Although their manufacture was banned by the U.S. Environmental Protection Agency in 1977, products containing PCBs made prior to that date can still be found. PCBs are still released from hazardous waste sites, illegal or improper disposal of industrial wastes and consumer products, leaks from old electrical transformers, burning of some wastes in incinerators, and aquatic sediments (Eisler 1986c, Eisler and Belisle 1996). The estimated environmental burden of PCBs from these sources is almost 400 million kg (Tanabe 1988, Eisler and Belisle 1996). PCBs bind strongly to organic particles in soil and sediment-forming PCB sinks, where local concentrations can be high. PCBs are transported globally through atmospheric and oceanic processes. There are 209 different PCB **congeners** (forms), but only 100–150 are represented in PCB formulations.

Some PCB congeners are of greater environmental concern than others. In general, PCB congeners with high K_{ow} (a physical characteristic of a chemical correlated with lipid solubility) values and high numbers of substituted chlorines in adjacent positions constitute the greatest environmental threat to wildlife. These congeners include planar PCBs, a group of about 20 PCB congeners that closely resemble dioxins (Eisler and Belisle 1996, Rice et al. 2003). PCBs cause a wide variety of biological effects, including death, developmental abnormalities, reproductive failure, liver damage, tumors, and a wasting syndrome (Eisler and Belisle 1996). Effects on reproduction, endocrine and immune systems, and behavior may have the greatest impacts on wildlife populations. Mink (*Mustela vison*) are one of the most susceptible species to PCBs, and dietary levels as low as 100 g PCB/kg fresh weight cause reproductive failure and death (Aulerich and Ringer 1977, Aulerich et al. 1987). Signs of PCB toxicity in mink include anorexia; bloody stools; disrupted molting patterns; and thickened, elongated, and deformed nails (Aulerich and Ringer 1977). In birds, total PCB levels (g/kg fresh weight) of 3,000 in the diet, 16,000 in the egg, or 54,000 in the brain were associated with PCB poisoning (Eisler 1986c). PCBs have been shown to have severe effects on avian reproduction, mainly decreased productivity and hatching success (embryo mortality), and abnormal breeding behavior (Eisler and Belisle 1996).

PCBs are highly lipid soluble and are readily bioaccumulated in wildlife and biomagnified in both aquatic and terrestrial food chains. Some wildlife species, such as long-lived fish and common snapping turtles (*Chelydra serpentina*), can bioaccumulate and store high levels of PCBs in their tissues, posing a potential hazard for predators, particularly avian piscivores (Eisler 1986c, Eisler and Belisle 1996). Although much of the environmental burden of PCBs is localized, PCBs continue to represent a considerable hazard to exposed wildlife species (Eisler 1986c, Tanabe 1988, Rice et al. 2003). However, continuing impacts of PCBs on wildlife are likely to be related to reproductive impairment and other sublethal effects. Mortality from chronic exposure is unlikely except in sensitive species with high-risk feeding habits (e.g., piscivores; Eisler and Belisle 1996).

POLYBROMINATED DIPHENYL ETHERS

PBDEs are chemicals that have been used extensively over the past several decades as flame retardants in textiles, plastics, furniture, electronic circuitry, and a variety of other products (Alaee and Wenning 2002, Hale et al. 2003). As a consequence of substantial, long-term usage worldwide, PBDEs have been detected in a wide variety of animal tissues as well as in all other environmental media in all ecosystems globally (Alaee and Wenning 2002, Law et al. 2003).

Three commercial PBDE mixtures have been produced and used in the United States and elsewhere: pentabromodiphenyl ether (pentaBDE), octabromodiphenyl ether (octaBDE), and decabromodiphenyl ether (decaBDE). As of 2004, both pentaBDE and octaBDE commercial products were phased out in the United States; however, decaBDE is still used in most locations in this country. As a consequence of continued usage of decaBDE around the world, significant increases in BDE-209 body burdens have been observed in both North American and European birds (Chen and Hale 2010).

Most PBDE congeners are highly persistent in the environment, and tetra- to hexa-bromodiphenyl ethers are the most frequently detected congeners in wildlife species (Law et al. 2003). Consumption of prey species likely is the largest contributor to wildlife exposures to PBDEs; these chemicals are readily bioaccumulated and even biomagnified as predators at higher trophic levels ingest the accumulated concentrations of PBDEs from their prey (Law et al. 2003, Voorspoels et al. 2006, 2007).

The biological effects of exposure to PBDEs have not been well studied to date. The limited data available suggest their adverse effects may be similar to those of PCBs, and include endocrine disruption (thyroid hormones), altered vitamin A and glutathione metabolism, increased oxidative stress, reproductive and developmental impairment, and neurobehavioral effects (McDonald 2002, Darnerud 2003, Fernie et al. 2006). The critical effects of pentaBDEs are those on neurobehavioral development (≥ 0.6 mg/kg body weight) and, at slightly higher doses, thyroid hormone levels in rodents and birds. OctaBDEs cause fetal toxicity or teratogenicity in rats and rabbits (≥ 2 mg/kg body weight). DecaBDEs affect thyroid, liver, and kidney morphology in adult animals (≥ 80 mg/kg body weight; Darnerud 2003).

DIOXINS AND FURANS

Dioxins and furans have no commercial use and are released into the environment as contaminants from combustion; incineration; synthesis of phenoxy herbicides and wood preservatives; and industrial and municipal processes, such as paper manufacturing (Bradbury 1996, Rice et al. 2003). There are approximately 75 different forms of dioxins, with **tetrachlorodibenzo-p-dioxin (2,3,7,8-TCDD)** being the most prevalent in the environment and of most concern to wildlife. There are approximately 135 different forms of furans. Most dioxins and furans are resistant to environmental and biological degradation and, once formed, disperse throughout the atmosphere, soil, and water. Environmental dioxins and furans have resulted in the deaths of many wildlife species and domestic animals (Bradbury 1996, Rice et al. 2003).

Exposure to dioxins and furans can result in a wide range of negative effects, from acute and delayed mortality to teratogenic, histopathological, immunological, and reproductive effects (Rice et al. 2003). Exposure to even minute quantities of 2,3,7,8-TCDD has been shown to result in reproductive failure in mink (Hochstein et al. 1988), wood ducks (*Aix sponsa*; White and Seginak 1994) and ring-necked pheasants (*Phasianus colchicus*; Nosek et al. 1992). Signs of dioxin toxicity include a "wasting syndrome," subcutaneous edema, alterations in lipid metabolism and gluconeogenesis, reproductive effects (teratogenicity or fetal toxicity), decreased immunocompetence, and thymic atrophy (Bradbury 1996). As with PCBs, dioxins and furans are highly lipid soluble, readily bioaccumulated in wildlife, and biomagnified in both aquatic and terrestrial food chains. Wildlife that bioaccumulate and store high levels of dioxins and furans in their tissues pose a potential hazard for predators, particularly avian and mammalian piscivores. It is recommended that 2,3,7,8-TCDD concentrations in water should not exceed 0.01 ppt (parts per trillion) to protect aquatic wildlife species or 10–12 ppt in foods of terrestrial wildlife (Eisler 1986b). Currently, there are no regulations governing dioxins and furans to protect wildlife (Eisler 1986b, Eisler and Belisle 1996).

Inorganic Chemicals

Inorganic chemicals are a diverse group that includes those that do not have carbon and its derivatives as their principal elements. These chemicals include 4 general groups: alkalis and chlorine, industrial gases, inorganic pigments, and industrial inorganic chemicals. Examples of industrial inorganic chemicals include acids; bases; metallic compounds; catalysts; ammonia; and salts derived from sodium, phosphorus, potassium, and sulfur. Inorganic chemicals generally are disposed in hazardous waste streams and do not pose a great threat to wildlife. However, some chemicals are used in processes, such as mining and military activities, and can leak or spill from storage, where they can occur in large volumes in the environment and pose substantial hazards to wildlife. Two inorganic chemicals that pose a particular hazard to wildlife are cyanide and white phosphorus.

CYANIDE

Cyanides are highly toxic chemicals widely used in mining and other industrial processes. Cyanide levels tend to be elevated in the vicinity of metal processing operations, electroplaters, gold-mining facilities, oil refineries, power plants, and solid-waste combustion facilities (Eisler 1991). The most common form of cyanide is hydrocyanic acid, used in electroplating and fumigation. Other chemical forms include sodium cyanide, used in extracting precious metals from raw ore and for predator control (e.g., M-44 ejector device); potassium cyanide; and calcium cyanide. Cyanides are readily absorbed through oral, dermal, and inhalation routes and are distributed throughout the body via the blood. Cyanide is a potent and rapid-acting asphyxiant (reduces tissue oxygen levels), inducing tissue anoxia through inactivation of cytochrome oxidase and thus causing cytotoxic **hypoxia** (lack of oxygen) in the presence of normal hemoglobin oxygenation. Diagnosis of acute lethal cyanide poisoning is difficult, because symptoms are nonspecific and numerous factors modify its toxicity. The most consistent changes in acute cyanide poisoning are inhibition of brain cytochrome oxidase activity and changes in electrical activity in the heart and brain.

Birds, mammals, and other wildlife in the vicinity of gold mining operations are particularly prone to cyanide exposure. Cyanide associated with gold mining activities in Nevada leached into nearby ponds and killed large numbers of migratory birds (Henny et al. 1994) and mammals (Clark and Hothem 1991). In a sampling of Nevada mines, >90 avian species (mainly waterfowl, shorebirds, and passerines), 28 mammalian species (mainly rodents, bats, and lagomorphs), and several reptilian and amphibian species were reported poisoned by cyanide solution ponds between 1986 and 1991 (Henny et al. 1994). For birds and bats, most mortality incidents associated with exposure to cyanide at mining operations are reported in the spring and autumn during migration (Clark 1991, Clark and Hothem 1991, Henny et al. 1994). Eisler et al. (1999) reviewed the specific environmental hazard for wildlife species at gold mining operations.

In addition to mining, cyanide is used in M-44 predator control devices, mostly in the western United States, where mammalian (mainly coyotes [*Canis latrans*]) and avian (mainly golden eagles) predators are subject to cyanide poisoning. From 1986 through 1995, >3,000 cyanide-related mortalities involving about 75 species of birds representing 23 families were reported to the National Wildlife Health Center in Madison, Wisconsin. Waterbirds and passerines (songbirds) were the 2 groups of birds most impacted by cyanide.

WHITE PHOSPHORUS

White phosphorus (P_4) is a highly toxic, incendiary munition extensively used by the military for marking artillery impacts (target practice) and as an obscurant. Areas in and around active (and inactive) military artillery and bombing ranges can concentrate white phosphorus, which can run off into surface waters and move to areas away these ranges. White phosphorus caused the death of an estimated 1,000–2,000 migrating dabbling ducks (*Anas* spp.) and 10–50 swans (*Cygnus buccinator* and *C. columbianus*) per year in the late 1980s and early 1990s at Eagle River Flats, a 1,000-ha estuarine salt marsh at Fort Richardson, Alaska, used for artillery training by the U.S. Army (Racine et al. 1992, Sparling 2003). Signs of white phosphorus poisoning observed in wild waterfowl include lethargy, repeated drinking, and head shaking and rolling with convulsions prior to death (Racine et al. 1992). Although no mortality of predators at Eagle River Flats was found, secondary exposure and poisoning of predators and scavengers, such as bald eagles, herring gulls (*Larus argentatus*), and common ravens (*Corvus corax*), was noted (Roebuck et al. 1994). White phosphorus has been shown to cause significant changes in a wide range of blood parameters in mallards (*Anas platyrhynchos*; Sparling et al. 1998) and mute swans (*Cygnus olor*; Sparling et al. 1999). It also has caused secondary poisoning in American kestrels (*Falco sparverius*; Sparling and Federoff 1997).

Pharmaceuticals

There is a wide diversity of **pharmaceutical drugs,** hormones, and other related organic wastewater contaminants present in waterways of the United States that pose a potential hazard to wildlife. During 1999–2000, a U.S. Geological Survey monitoring effort found 82 of 95 different pharmaceutical drugs tested for in water samples from a network of 139 streams across 30 states (Kolpin et al. 2002). A wide range of residential, industrial, and agricultural drugs and chemicals was found in 80% of all streams tested. Little is known about the potential impact of these drugs and other chemicals on wildlife, particularly the potential interactive effects that may occur from complex mixtures of these and other chemicals in the environment. Numerous wildlife mortality and morbidity incidents occurring from widely used pharmaceutical drugs, such as sodium pentobarbital and diclofenac, provide evidence of the hazard posed by this group of chemicals.

Sodium Pentobarbital

Sodium pentobarbital and related barbiturates are used extensively in veterinary medicine, especially for euthanasia of domestic animals. They result in the deaths of numerous wildlife species across the United States and Canada each year (Friend and Franson 1999). The use of highly concentrated solutions for euthanasia of domestic animals (e.g., cats, dogs, and horses) is routine practice in veterinary medicine. Carcasses that are not incinerated or otherwise properly disposed are subject to scavenging by wildlife and can result in exposure to this family of chemicals. Any wildlife species that scavenges food potentially is at risk from these chemicals. Mortality of wildlife—from bald and golden eagles to grizzly bears (Ursus arctos horribilis) has been reported from landfills and other improper burial sites, where animal carcasses were either left in the open or not disposed of properly. In recent years, the National Wildlife Health Center in Madison, Wisconsin, and the National Fish and Wildlife Forensic Laboratory in Ashland, Oregon, have had verifiable reports of at least 133 eagle deaths resulting from secondary pentobarbital poisoning, most likely only a fraction of the real total.

Diclofenac

Diclofenac is a nonsteroidal anti-inflammatory drug used extensively in veterinary medicine and is administered to livestock and other domestic animals for the relief of pain and arthritis in many countries around the world. Diclofenac was identified as the most likely cause of a mass mortality of 3 species of vultures in Pakistan (Oaks et al. 2004). Vultures consuming dead livestock containing diclofenac were exposed to high levels of the drug in livestock tissues. Necropsies revealed that exposed animals had visceral gout and histopathological lesions, including acute renal tubular necrosis and uric-acid crystal formation in the kidneys and other tissues, that led to acute renal failure and death. Populations of the 3 species of vultures—Oriental white-backed vulture (Gyps bengalensis), long-billed vulture (G. indicus), and slender-billed vulture (G. tenuirostris)—were decimated by as much as 95% in some cases (Oaks et al. 2004). Although this incident occurred in Asia, it clearly demonstrates the potential hazard of pharmaceutical drugs to wildlife.

Pesticides

A **pesticide** is any substance or mixture of substances intended for preventing, destroying, repelling, or mitigating any pest. The term pesticide is a generic name for a variety of agents classified more specifically on the basis of the pattern of use and organism killed. Pesticides include chemicals designed to kill specific groups of organisms, such as insecticides, herbicides, fungicides, rodenticides, miticides, acaricides, larvicides, and molluscicides. They also function as attractants (pheromones), defoliants, desiccants, plant growth regulators, repellents, and fumigants for purposes of reducing the numbers of pest species.

Pesticides are a unique category of environmental contaminants, as they are intentionally released into the environment. Thus, regulations for monitoring pesticide usage and the likelihood of detecting pesticide-related mortality events are enhanced. Insecticides are among the most acutely toxic contaminants in the environment and can produce dramatic mortality and morbidity incidents. **Target species** selectivity of pesticides is not well developed, and nontarget species frequently are affected, because they possess physiological and/or biochemical systems similar to those of the target organisms. Specific classes of pesticides of major concern to wildlife include insecticides; herbicides; fungicides; fumigants; and vertebrate pest-control chemicals, such as rodenticides and avicides.

Insecticides

Most chemical insecticides in use today **are neurotoxicants** and act by poisoning the nervous system of the target organisms. The central nervous system of insects is highly developed and not unlike that of vertebrates. Generally, insecticides are not selective and affect nontarget organisms as readily as target organisms. Target sites and/or mechanism(s) of action may be similar in all species; only the level of exposure (dosage and duration of contact with toxic receptors) influences the intensity of biological effects. Four distinct groups of insecticides—chlorinated hydrocarbons, anti-ChE (organophosphorus and carbamate), synthetic pyrethroids, and other botanicals—are discussed here, as they pose a significant threat to wildlife.

CHLORINATED HYDROCARBONS

Chlorinated hydrocarbon (organochlorine) insecticides are a diverse group belonging to 3 distinct chemical classes: dichlorodiphenylethanes (e.g., dichlorodiphenyltrichloroethane [DDT], dicofol, and methoxychlor), cyclodienes (e.g., heptachlor, dieldrin, and aldrin), and chlorinated benzenes (e.g., lindane; Smith 1991, Blus et al. 1996, Blus 2003). DDT was used extensively in all aspects of agriculture and forestry, in building and structural protection, and in human health situations from the mid-1940s to the early 1970s. The chemical properties of chlorinated hydrocarbon insecticides (e.g., low volatility, chemical stability, lipid solubility, and slow rate of biotransformation and degradation) that made them effective also brought about their demise due to persistence in the environment and bioaccumulation and biomagnification in food chains. Registration for DDT was canceled in the United States and several other countries in 1972, and the cancellation and/or restriction of registration for other chlorinated hydrocarbon insecticides followed. Despite the ban on their use in North America and Europe, chlorinated hydrocarbon insecticides are used extensively in developing countries. They continue to be used in these countries, because they are inexpensive to manufacture, highly effective, relatively safe, few substitutes are available, and the risk–benefit ratio is highly weighted in favor of their continued use for control of insects that devastate crops and human health (Smith 1991, Blus et al. 1996, Blus 2003). As a result of their continued heavy use, they become airborne and are transported globally in the atmosphere, so that deposition occurs on a global basis, particularly at high latitudes (Bidleman et al. 1990).

Definitive studies of both wildlife and laboratory species have demonstrated potent estrogenic and enzyme-inducing properties of chlorinated hydrocarbon insecticides that interfere directly or indirectly with fertility and reproduction in wildlife. In avian species, this interference due to DDT exposure is related to steroid metabolism and the inability of the bird to mobilize calcium to produce sufficiently strong eggshells to withstand incubation (cracking allows bacteria to enter and kill developing embryos; Blus et al. 1996, Blus 2003).

Chlorinated hydrocarbons act as diffuse stimulants or depressants of the central nervous system. Signs of acute toxicity occur within minutes to a few days following exposure, usually in ≤24 hours; may be progressively severe in nature; and can include muscle spasms, seizures, loss of coordination, abnormal walking and/or posture, and excessive salivation (Osweiler et al. 1985). Exposed animals may become comatose and remain so for several hours prior to death or may regain consciousness and fully recover. Pathologic changes associated with acute poisoning by chlorinated hydrocarbons are usually minimal and nonspecific. They include pulmonary congestion, hemorrhages, and edema, particularly in the central nervous system (Osweiler et al. 1985). Chronic exposure to chlorinated hydrocarbons results in alteration of hepatocytes (liver cells; Osweiler et al. 1985).

The highly lipid-soluble nature of chlorinated hydrocarbon insecticides results in crossing of the normally protective **placental and blood-brain barriers** in mammals, leading to direct embryonic/fetal and central nervous system exposure. It also results in these chemicals being sequestered in body tissues (liver, kidney, nervous system, and adipose tissue) having a high lipid content, where the residues either elicit some biological effect or, as in the case of adipose tissue, remain stored and undisturbed until mobilized. Elimination rate and depletion of body storage sites may be enhanced by fasting, resulting in mobilization of adipose tissue and any insecticide contained therein. However, with a high-chlorinated hydrocarbon body burden, there is a possibility of enhanced toxicity from the circulating agent being redistributed to target organs. The most serious effects, such as mortality, reduced reproductive success, population decline, and even extirpation, occurred in birds, particularly raptors, seabirds, and waterbirds in the orders Strigiformes, Falconiformes, Pelecaniformes, Ciconiformes, and Podicipediformes (Blus et al. 1996, Blus 2003).

ANTICHOLINESTERASES

Organophosphorus and carbamate insecticides commonly are grouped together and referred to as anti-ChEs (Mineau 1991, Hill 2003). These insecticides have a common mechanism of action—inhibition of the neurotransmitting enzyme ChE (Baron 1991, Gallo and Lawryk 1991). However, they arise from 2 distinctly different chemical classes: the esters of phosphoric or phosphorothioic acid and those of carbamic acid. Currently, there are some 200 different organophosphorus and about 50 carbamate pesticides (mainly insecticides) on the market, formulated into thousands of different products (Hill 2003). Anti-ChE insecticides are applied mainly on terrestrial landscapes, but they also are used extensively in wetlands and coastal areas for mosquito control. This enzyme is responsible for the destruction and termination of the biological activity of the neurotransmitter acetylcholine. With accumulation of free, unbound acetylcholine at nerve endings of all cholinergic nerves, there is continual stimulation of electrical activity. Following lethal exposure, death results from acute respiratory failure (Hill 2003).

Organophosphorus and carbamate insecticide intoxication has become more complicated in recent years with the recognition of additional and persistent signs of neurotoxicity not previously associated with acute exposure to these chemicals. One condition, an "intermediate syndrome," is a potentially lethal paralytic condition of the neck, limbs, and respiratory muscles. The other condition, in which neuropathic conditions persist indefinitely, is referred to as **organophosphorus-induced delayed neuropathy** (Ecobichon 2001, Hill 2003).

Most widely used anti-ChE insecticides are highly toxic, but relatively short-lived in the environment (usually 2–4 weeks) and are readily metabolized and excreted by birds and mammals (Hill 2003). Carbamates are direct ChE inhibitors that do not require metabolic activation for full potency. Many organophosphorus insecticides are known to become more toxic as a result of metabolism (e.g., diazinon, malathion, and parathion), because the metabolites (the "oxon" form) are more potent inhibitors of ChE (Matsumura 1985, Smith 1987). Thus, there may be some delayed toxicity (and onset of signs) associated with organophosphorus insecticide poisoning. Dietary toxicity experiments have shown that birds that die from carbamate insecticide poisoning do so within a few hours of exposure, but mortality from organophosphorus insecticide poisoning may extend over 5 days (Hill 2003).

Organophosphorus and carbamate insecticides are responsible for more reported wildlife mortality and morbidity incidents than any other category of environmental contaminant. However, only relatively few of these pesticides are responsible for the majority of large-scale incidents of wildlife mortality and morbidity. Birds are highly sensitive to most organophosphorus and carbamate insecticides, and are particularly susceptible to granular formulations. As few as one granule (0.1–5.0 mg/kg) of some anti-ChE insecticides, such as carbofuran, may be lethal in 5 minutes to waterfowl and songbirds (Hill 2003). Extensive records of bird mortality and morbidity from exposure to organophosphorus and carbamate insecticides exist (Smith 1987, Sheffield 1997, Friend and Franson 1999, Mineau et al. 1999). One of the most notable mass mortalities in recent years involved the death of >10,000 Swainson's hawks (*Buteo swainsoni*) on their wintering grounds in the pampas of Argentina in the mid-1990s (Goldstein et al. 1996, 1999). In this case, hawks were poisoned through their consumption of grasshoppers and other prey items in alfalfa fields sprayed with the organophosphorus insecticide monocrotophos. Although mammals generally are less sensitive than birds to organophosphorus and carbamate insecticides, many mammalian mortality incidents also have been reported (Smith 1987). Intensive field research with mammalian exposure to organophosphorus insecticides has documented reproductive and other sublethal effects at environmentally relevant levels (Sheffield and Lochmiller 2001, Sheffield et al. 2001).

Signs of acute exposure to anti-ChE insecticides include lethargy and excessive salivation, lacrimation, urination, and defecation; vomiting may occur along with muscle fasciculation (brief spontaneous contractions of a few muscle fibers) and weakness, dyspnea (difficulty breathing), excessive bronchial secretion, and bradycardia (slowed heart rate; Fairbrother 1996). In severe cases, prostration and convulsions precede death. These signs are useful when sick animals are found on or near an area of recent anti-ChE insecticide application. However, these signs are not uniquely different from poisoning by other neurotoxic chemicals. Inhibition of brain ChE activity by 20% (i.e., activity at 80% of normal) is considered diagnostic of sublethal poisoning, and a dead bird with a >50% reduction in activity generally is diagnostic of anti-ChE poisoning. Activity reductions of 70–95% are commonly reported for birds killed by organophosphorus insecticides (Hill and Fleming 1982, Hill 2003). Conclusive diagnosis depends on biochemical and chemical analyses for brain ChE activity and organophosphorus residues in the carcass (Hill 1999, 2003).

A wide diversity of sublethal effects has been documented to occur following exposure to anti-ChE insecticides, including biochemical, physiological, behavioral, and others that impact survival and fitness of exposed animals (Mineau 1991; Hill 1999, 2003). Many of these effects may be lethal, but also may mask pesticide exposure as the cause of mortality. For example, a group of exposed animals that has become disoriented and less vigilant may become more susceptible to predation or other mortality factors.

Anti-ChE insecticides generally do not bioaccumulate in organisms and do not biomagnify in food chains (Hill 2003). However, prey items, such as arthropods and animal carcasses, can contain sufficiently high levels of these insecticides to cause secondary poisoning in predatory and scavenging birds (particularly raptors) and mammals (Sheffield 1997, Mineau et al. 1999, Shore and Rattner 2001). Bald eagles and red-tailed hawks in British Columbia were found poisoned by consuming unabsorbed pesticide in the stomachs of dead animals up to 6 months following its application (Elliott et al. 1996).

SYNTHETIC PYRETHROIDS

Synthetic pyrethroids are the newest major class of insecticides, entering the market in 1980. By 1982, they accounted for about 30% of worldwide insecticide usage. These synthetics arise from a much older class of botanical insecticides, pyrethrum, that is a mixture of 6 insecticide esters extracted from dried pyrethrum or *Chrysanthemum* flowers (Ray 1991). The increasing demand for pyrethrum has exceeded the limited world production. This led chemists to focus attention on synthesis of new analogs with better stability in light and air, better persistence, more selectivity to target species, and low mammalian and avian toxicity. In addition to extensive agricultural use, synthetic pyrethroids are components of household sprays, flea dips and sprays, and plant sprays for home and greenhouse use. Studies on intact animals have not yielded conclusive, fundamental information concerning the mechanism of action of pyrethroids (Ray 1991, Ecobichon 2001).

Synthetic pyrethroids alter sodium channels in nerve membranes, causing repetitive (sensory and/or motor) neuronal discharge and a prolonged negative after-potential with the effects being similar to those produced by DDT. Other effects noted for synthetic pyrethroids include inhibition of Ca- and Mg-ATPase, the effect of which is to increase intracellular calcium levels accompanied by increased neurotransmitter release and postsynaptic depolarization (Matsumura 1985, Ray 1991).

There have been relatively few reports of wildlife mortality or morbidity as a result of synthetic pyrethroid exposure, and little is known about their sublethal effects on wildlife. The available evidence suggests that synthetic pyrethroids elicit little chronic toxicity to wildlife. In addition, there is little storage or bioaccumulation of pyrethroids, because they are readily biotransformed by the **mixed-function oxidase** system. However, piperonyl butoxide, an inhibitor of **cytochrome P-450s** (an important family of detoxification enzymes) is a synergist added to many synthetic pyrethroid formulations for increased toxicity (10- to 300-fold increase in toxicity; Ray 1991, Ecobichon 2001).

OTHER BOTANICAL INSECTICIDES

Nicotine and rotenone are among the more widely used botanical insecticides. These compounds are natural plant **alkaloids** whose toxic properties have been recognized for hundreds of years (Ray 1991). Nicotine usually is obtained from the dried leaves of *Nicotiana tabacum* and rotenone is derived from the roots of derris (*Derris* spp.; South America) and cubé (*Lonchocarpus* spp.; southeast Asia). Used as an insecticide and piscicide, rotenone is extremely toxic to aquatic vertebrates, particularly fish. Because use of rotenone as a piscicide is so widespread, there is concern about the potential negative effects of rotenone on amphibian (frogs and salamanders) and aquatic reptile (turtles and snakes) species, particularly **neotenic** (attaining reproductive maturity while retaining larval morphology) salamanders that use aquatic respiration (Fontenot et al. 1994). The most frequent signs of rotenone poisoning in wildlife include vomiting, anorexia, dermatitis, irritation of the GI tract, lack of coordination, muscle tremors, and convulsions, with death occurring through respiratory failure (Osweiler et al. 1985).

Herbicides

Herbicides are any compound capable of either killing or severely injuring plants and may be used for elimination of plant growth or killing of plant parts. Many early herbicides contained forms of arsenic and were difficult to handle, highly toxic, relatively nonspecific, or phytotoxic to crops as well as undesirable plants. However, currently used herbicides generally present a much lower hazard to wildlife than those used earlier and are more likely to result in sublethal effects rather than cause wildlife mortality or morbidity (Stevens and Sumner 1991).

Over the past 2 decades, herbicides have represented the most rapidly growing section of the agrochemical pesticide business due in part to (1) monoculture practices, where the risk of weed infestation has increased, because fallowing and crop rotation are no longer standard practice; and (2) mechanization of agricultural practices (planting, cultivating, and harvesting) to counter increased labor costs. The result has been a plethora of chemically diverse compounds rivaling the innovative chemistry of insecticides. The goal of herbicides has been to protect desirable crops and obtain high yields by selectively eliminating unwanted plant species, thereby reducing competition for nutrients, water, and space (Stevens and Sumner 1991).

There are ≥6 broad classes and ≥22 chemical groups of herbicides, including (1) germination inhibitors, such as dinitroanilines (e.g., trifluralin) and chloroacetamides (e.g., alachlor and metolachlor); (2) photosynthesis inhibitors, such as triazines (e.g., atrazine, simazine, and metribuzin); (3) meristem inhibitors, such as sulfonylureas (e.g., chlorsulfuron) and imidazolinones (e.g., imazethapyr and imazapyr); (4) contact action agents, such as bipyridylium (e.g., paraquat and diquat) and arsenicals (e.g., monosodium methanearsonate [MSMA]); (5) auxin growth regulators, such as phenoxy acids (e.g., 2,4-dichlorophenoxyacetic acid [2,4-D]); and (6) foliar grass killers, such as phosphono-amino acids (e.g., glyphosate).

Herbicide classification is based on how and when they are applied. Preplanting herbicides are applied to the soil before a crop is seeded, pre-emergent herbicides are applied to the soil before the usual time of appearance of the unwanted vegetation, and postemergent herbicides are applied to the soil or foliage after germination of the crop and/or weeds.

The chlorphenoxy (e.g., 2,4-D and 2,4,5-trichlorophenoxyacetic acid [2,4,5-T]) and bipyridyl (e.g., paraquat and diquat) herbicides are acutely toxic to wildlife and humans. Paraquat is a contact herbicide and is one of the most specific pulmonary toxicants known. Many countries have banned or severely restricted use of these herbicides (Ecobichon 2001). Another group, the triazines, including atrazine, although considered less acutely toxic, is of concern for wildlife due to their widespread and high volume use. There also is evidence of sublethal effects, such as endocrine disruption and impacts on reproduction and development (Hayes et al. 2002, 2003).

Herbicides show a broad range of persistence (Stevens and Sumner 1991). Some, such as paraquat, may persist for years, whereas others persist for only days or months. Most herbicides occur in either plants or the soil. Because they are not as persistent as organics (e.g., PCBs) or some organochlorine insecticides (e.g., DDT), they tend not to move via the atmosphere to distant locations. However, such herbicides as atrazine and metolachlor are used in high volume throughout the midwestern United States and can result in high atmospheric concentrations and movement. Most herbicides do not bioaccumulate in tissue of any class of animals. Because of the overall limited persistence or tendency to bind to soil particles, there is generally limited movement through the environment. The most frequent signs of herbicide poisoning in wildlife include anorexia; diarrhea; edema; ataxia; inflammation of the GI tract; and congestion of the lungs, liver, and kidneys (Osweiler et al. 1985).

Fungicides

Fungicides are derived from a wide variety of chemicals, ranging from simple inorganic compounds (e.g., sulfur and copper sulfate), through the aryl- and alkyl-mercurial compounds and chlorinated phenols, to metal-containing derivatives of thiocarbamic acid. There are at least 36 different chemical groups of fungicides, a direct result of the great diversity of fungi (Edwards et al. 1991, Ecobichon 2001).

There are 3 general types of fungicides: (1) foliar, applied as liquids or powders to the aerial green parts of plants, producing a protective barrier on the cuticular surface and causing systemic toxicity in developing fungus; (2) soil, applied as liquids, dry powders, or granules, acting either through the vapor phase or by systemic properties; and (3) dressings, applied to seeds prior to planting and to the postharvest crop (cereal grains, tubers, etc.) as liquids or dry powders to prevent fungal infestation of the seed and crop (particularly if it is stored under less-than-optimal conditions of temperature and humidity).

Most fungicides have low acute toxicity to mammals and birds. However, all fungicides are cytotoxic (toxic to cells), and almost all produce positive results in microbial mutagenicity and animal carcinogenicity tests (Ecobichon 2001). Many fungicides also are teratogenic and embryotoxic, and they are endocrine disruptors (Edwards et al. 1991). Fungicide groups of current environmental concern to wildlife include benzimidazoles (e.g., benomyl, carbendazim, and thiabendazole), dithiocarbamates (e.g., maneb, mancozeb, and zineb), aromatics (e.g., chlorothalonil), dinitrophenols (e.g., dinocap), and dicarboximides (e.g., captan and vinclozolin; Ecobichon 2001). Others that were heavily used in the past, but have largely been discontinued in the United States due to the environmental hazard they pose include the organo-mercurials, hexachlorobenzene, pentachlorophenol, captafol, and folpet (Ecobichon 2001). The most frequent signs of fungicide poisoning in wildlife include anorexia and weight loss, lethargy and depression, impaired liver function, and reproductive impairment (Osweiler et al. 1985).

Fumigants

Fumigants are agents used to kill insects, nematodes, weed seeds, and fungi in soil as well as in stored cereal grains, fruits, vegetables, clothes, and other products. They are usually used in enclosed spaces due to high volatility of the compounds. Chemicals used as fumigants include acrylonitrile, carbon disulfide, carbon tetrachloride, ethylene dibromide, chloropicrin, methyl bromide, and ethylene oxide. These chemicals can be liquids that readily vaporize at ambient temperatures, solids that can release a toxic gas on reacting with water or with acid, or gases. They generally are nonselective, highly reactive, and cytotoxic. Fumigants of environmental concern include phosphine (used heavily on grains), ethylene dibromide, and 1,2-dibromo-3-chloropropane; the latter 2 are known animal carcinogens (Ecobichon 2001).

Vertebrate Pest-Control Chemicals

RODENTICIDES

Rodenticides were developed to control pest small mammals (particularly rodents) that cause agricultural damage, carry disease, and are considered by some to be nuisance species. Chemicals used as rodenticides constitute a diverse range of compounds having a variety of mechanisms of action that are partially successful at attaining species selectivity. The design of some rodenticides has taken advantage of unique physiological and biochemical characteristics of rodents. The sites of action are common to most mammals, but the habits of the pest animal are taken into account and/or dosages are defined to minimize effects on nontarget species. Some inorganic compounds have been used as rodenticides, including thallium sulfate, arsenic oxide and other arsenicals, barium carbonate, yellow phosphorus, aluminum phosphide, and zinc phosphide. Several insecticides have been used as rodenticides, including DDT. In addition, some natural plant toxins have been used as rodenticides or to control other mammalian species; they include strychnine, red squill, ricin, and sodium monofluoroacetate (Ray 1991, Eisler 1995). Sodium monofluoroacetate (compound 1080) has been used extensively in prepared baits to control rodents and predators, particularly coyotes. Most mammals are fatally poisoned by 1 mg/kg body weight of sodium monofluoroacetate (Eisler 1995). Domestic sheep have experienced toxic effects from wearing compound 1080-impregnated livestock protection collars (Burns and Connolly 1995).

Currently, anticoagulants are the most significant class of rodenticides in terms of wildlife mortality and morbidity incidents. The basis of efficacy of anticoagulant rodenticides is coumadin (warfarin), which was isolated from spoiled sweet clover (*Melilotus* spp.) and acts as an anticoagulant by antagonizing the actions of vitamin K in the synthesis of clotting factors. Warfarin has been in use since the 1920s, and some rodent populations had developed resistance to it by the 1950s. Since then, the next generation of "super warfarins" has appeared (e.g., brodifacoum, bromadialone, diphenacoum, and diphacinone). These second-generation super warfarins, particularly brodifacoum, have caused a substantial number of wildlife mortality incidents across the United States (Sheffield 1997, Stone et al. 1999). Brodifacoum has been documented to poison nontarget wildlife. Secondary poisoning of raptors (particularly red-tailed hawks and great horned owls [*Bubo virginianus*]) made up 50% of the cases. Gray squirrels (*Sciurus carolinensis*), raccoons (*Procyon lotor*), and white-tailed deer (*Odocoileus virginianus*) were the most frequently poisoned mammals (Stone et al. 1999).

AVICIDES

Avicides were developed to control pest birds, particularly flocking species, such as European starlings (*Sturnus vulgaris*), blackbirds (family Icteridae), and rock pigeons (*Columba livia*), that cause agricultural damage or are considered nuisance species. Several chemicals with avicidal properties have been used, including avitrol, chloralose, endrin, fenthion, methiocarb, and strychnine. Most of these chemicals are no longer registered for avicidal uses. One currently used avicide, DRC-1339, was developed specifically to kill starlings. Although designed to be specific to starlings, there is evidence that it is nonspecific, because it has been shown to pose a hazard to nontarget seed-eating species, such as ring-necked pheasants (Avery et al. 1998).

Because of the great potential for these compounds to kill nontarget vertebrates that may come in contact with them, they were designed to degrade fairly rapidly. Many are unstable and degrade rapidly in water. However, compounds, such as avitrol (Kamrin 1997), some anticoagulants, and compound 1080, require months to decompose in soil. Many of these compounds, such as DRC-1339, are stable in water as well (Kimball and Mishalanie 1994). Soil degradation can last from hours to months, depending on the compound and climatic conditions. Avitrol degrades slowly in sunlight under dry conditions and in flooded soils, but 2.5 cm of rain will wash it away (Betts et al. 1976). Not only is environmental degradation important, but also persistence in the target species. For example, the half-life of bromodiolone in Norway rats (*Rattus norvegicus*) is ≤58 hours (Kamil 1987) allowing for potential exposure of predators and scavengers.

Secondary poisoning has been documented or considered possible for many vertebrate pest-control compounds. Barn owls (*Tyto alba*) are particularly sensitive to DRC-1339, but the residues present in dead birds are usually too low to cause toxicity (Johnston et al. 1999). However, bromadiolone and chlorophacinone have been implicated in secondary poisoning of many predators and scavengers (Berny et al. 1997, McDonald et al. 1998). Avitrol has been shown to be a potential hazard to sharp-shinned hawks (*Accipiter striatus*) and American kestrels (Holler and Schafer 1982).

Vertebrate pest control chemicals include a variety of compounds with a wide range of behavior in different environmental compartments. Anticoagulants and acute toxicants tend to be nonvolatile, whereas fumigants are highly volatile. Fumigants generally are unstable in water, anticoagulants are stable in water, and acute toxicants vary in their water stability. All are fairly stable in dry soil, but fumigants degrade quickly in wet soil. For example, both aluminum and zinc phosphides release highly toxic phosphine gas when in contact with water (Kamrin 1997).

Pest control chemicals vary greatly in mobility in general, and specific media alter their mobility. Anticoagulant rodenticides generally are not mobile in any environmental media, whereas fumigants are mobile in air, but not in water (Kamrin 1997). Acute toxicants are not mobile in air and vary in their mobility in water. Compound 1080 is highly mobile in water, because it is highly soluble. However, because of its high adsorption onto soil particles, it does not penetrate deeply into soil (Irwin et al. 1996). Other acute toxicants, such as avitrol, exhibit moderate water solubility and are not highly mobile in water.

Natural Plant and Animal Toxins

Natural plant and animal toxins are toxic chemicals produced by living organisms, such as bacteria, blue-green algae, fungi, marine invertebrates and fish, vascular plants, and poisonous aquatic and terrestrial animal species. Exposure to certain natural toxins, especially natural plant toxins, may have significant effects on wildlife. There are many different natural plant toxins, also known as secondary plant compounds, that can be highly toxic to wildlife, causing mortality and/or morbidity. Some plant toxins are used as the basis for pesticides (e.g., nicotine, pyrethrum, and rotenone), demonstrating their acute toxicity (Ray 1991). Several factors are involved in exposure of wildlife to natural plant toxins. For example, various parts of the plant (root, stem, leaves, or seeds) often contain different concentrations of a chemical. Plant age, climate, soil, and genetic differences within a plant species also are important factors in variability. Examples of natural plant toxin chemical groups that can be highly toxic are alkaloids, tannins, phenols, lectins, glycosides, and terpenes. Generally, wild herbivorous animals have adapted to avoid or efficiently detoxify endemic toxic plants and are not impacted by exposure to these toxins (Vangilder 1983). However, there have been a number of documented cases of poisoning of wildlife by plant toxins (Ray 1991, Wickstrom 1999, Norton 2001).

Three groups of microscopic organisms—bacteria, algae, and fungi—are capable of producing some of the most deadly toxins known. Probably the most significant natural toxin in terms of wildlife mortality and morbidity is **botulinum toxin** from the bacteria *Clostridium botulinum* (types C and E). Type C botulism causes mortality and morbidity in thousands of waterfowl across the United States and Canada each year, whereas Type E botulism has largely been restricted to causing mortality of fish-eating birds (bald eagles, loons, grebes, and gulls) in the Great Lakes (Friend and Franson 1999, Roffe and Work 2005). The botulinum toxin generally is formed under conditions of low environmental oxygen and is considered to be the most toxic substance known. Waterfowl, especially dabbling ducks, are most susceptible to Type C botulism, but American coots (*Fulica americana*), gulls, and shorebirds (order Charadriiformes) also are commonly killed during an outbreak. In Canada, annual losses of waterfowl in the prairie provinces can reach 100,000–1,000,000 birds (Wickstrom 1999). The neurotoxins produced by *C. botulinum* cause a paralytic effect in birds, which show signs of weakness, dizziness, inability to fly, muscular paralysis, and respiratory distress (Friend and Franson 1999, Roffe and Work 2005).

Blue-green algae (cyanobacteria) blooms commonly occur in fresh and brackish water worldwide. Wildlife that inhabit stagnant, eutrophic water bodies, especially during warm, sunny weather, are most susceptible to algal toxins. Algae in the genera *Nostoc, Oscillatoria, Anabaena,* and *Microcystis* produce hepatotoxic cyclic peptides that disrupt the structure of liver cells, causing massive hemorrhage and necrosis and leading to shock and death within hours. Algae in the genera *Anabaena, Aphanizomenon,* and some *Oscillatoria* produce potent, rapid-acting alkaloid neurotoxins. Anatoxin-a is a potent ChE inhibitor that causes permanent depolarization of postsynaptic membranes and disrupts nerve conduction, leading to muscle tremors, rigidity, paralysis, and death by respiratory failure within minutes. Exposure to this toxin could be confounding to analysis of ChE activity due to organophosphorus or carbamate insecticide exposure.

Aphanitoxins, another group of neurotoxins, act by blocking sodium channels and thus disrupting nerve conduction, leading to muscle tremors, rigidity, paralysis, and death. This group appears to be identical to saxitoxin and neosaxitoxin, the causative agents of paralytic shellfish poisoning in humans. In marine systems, harmful algal blooms produced by phytoplankton containing protozoans (mainly dinoflagellates) together produce some of the most potent toxins known, including domoic acid, brevetoxins, and saxitoxins. These compounds are concentrated in shellfish, are highly neurotoxic, and are commonly lethal to mammals at levels of 1 ?g/kg (ppb) or less. In North America, **harmful algal blooms** have been responsible for the death of wildlife in freshwater and marine systems, including waterfowl, colonial waterbirds and other bird species, wild canids, white-tailed deer, sea turtles, manatees (*Trichechus manatus*), pinnipeds, and whales (Friend and Franson 1999, Wickstrom 1999, Dierauf and Gulland 2001).

Fungi also are known to produce extremely toxic substances collectively known as **mycotoxins** (O'Hara 1996). Generally, wildlife is exposed to mycotoxins through contaminated feed. Although effects on wildlife can be significant, reports of poisoning by mycotoxins are relatively rare, because it is difficult to establish a diagnosis in the field. Aflatoxins, produced by the fungus *Aspergillus flavus* (or *A. parasiticus*), are among the most toxic of the mycotoxins and are common contaminants of corn, peanuts, and other cereal and oil seeds. Wildlife is at risk from eating waste grain, especially during times of restricted access to other feed or forage. The trichothecenes form another group of mycotoxins produced by fungi in the genera *Fusarium, Cephalosporium, Myrothecium,* and *Trichoderma*. These sesquiterpene compounds include T-2 toxin, diacetoxyscirpenol, and vomitoxin and act to inhibit protein synthesis, targeting rapidly dividing cell types in the skin, intestine, and hematopoetic (bone marrow) and lymphoid tissues. These toxins are known to cause anorexia; dermal, oral, and GI necrosis and ulceration; hemorrhage; and impairments of the reproductive and immune systems. Other mycotoxins that may have adverse effects on wildlife include fumonisins, zearalenone, ochratoxin A, ergot alkaloids, and sporidesmin. Although data on the role of mycotoxins in wildlife mortality and morbidity are rare, *Fusarium* (trichothecene) mycotoxins on waste peanuts were implicated in a mass mortality of sandhill cranes involving 9,500 birds in New Mexico and Texas between 1982 and 1987 (Windingstad et al. 1989, Friend and Franson 1999). In this case, the most common sign was an inability to hold the head erect while standing or flying; multiple muscle hemorrhages and submandibular edema were the predominant lesions at necropsy (Windingstad et al. 1989).

CONTAMINANT DIAGNOSTICS

Safety

Personal safety is a primary concern in a wildlife mortality or morbidity incident. Field investigators should not handle carcasses, collect environmental samples, or enter the area of the incident until adequate safety precautions have been taken. If the causative contaminant is known, a Material Safety Data Sheet (MSDS) or other U.S. Occupational Safety and Health Administration (OSHA) safety publication can describe the level of personal protective equipment required. For pesticides, the product label will provide the necessary information. Because some environmental contaminants may produce cancer, reproductive impairment, or birth defects in humans that would not become immediately apparent, the results of not adequately protecting investigators can be severe and long lasting.

Field biologists should take safety precautions when investigating possible wildlife mortality or morbidity incidents stemming from contaminants or disease. Personnel not trained in the use of personal protective equipment (PPE) should not attempt to use such equipment, nor should they enter contaminated sites. If the specific contaminant involved is unknown, only individuals trained to enter hazardous sites should enter until the contaminant is identified and the appropriate PPE determined. Until trained personnel are available, the site should be clearly marked, and access should be carefully controlled if possible. Any individual entering a site where the contaminant is unknown should have completed the 40-hour Health and Safety Training for Hazardous Workers (http://www.ehscompliance.com/safety training.html).

Once the contaminant has been identified, proper protective clothing for the contaminant type should be worn. Impermeable gloves and footwear (generally rubber boots) should always be worn. Some contaminants are readily dissolved in water and can easily penetrate the skin. Therefore, field investigators should keep bare skin protected and should not wade into shallow water. When retrieving carcasses or debilitated animals from water, impermeable gloves and rubber boots should be worn.

Short pants or short-sleeved shirts should not be worn, bare skin should be protected, and dust or fumes should not be inhaled. In wet conditions, waterproof pants may be required. Dust masks or respirators also may be required as well as impermeable clothing (e.g., Tyvek® coveralls or full suit), depending on the situation. In hot or humid weather, this type of equipment can be problematic for the person(s) wearing it, so common sense is needed. Under such conditions, work periods could be shortened to prevent heat stress and a clean, shaded area provided, where workers can remove PPE, cool off, and rehydrate. If clothing becomes contaminated, once the contaminant type has been confirmed, it should be washed or discarded. For some contaminants, washing is not sufficient to allow continued wearing.

If disease, rather than contaminants, is suspected, caution is still required, but the precautions are not as extensive (Roffe and Work 2005). It must be remembered that many contaminants in the environment are toxic to many different taxa, including humans. Further, some wildlife diseases can be transmitted to humans, but diseases generally are more species-specific than contaminants. This species-specificity may provide some support and clues as to whether an incident was mediated by contaminants or infectious disease.

Initial Site Reconnaissance

Three rules govern initiation of any wildlife mortality or morbidity investigation: (1) protect yourself and others involved, (2) obtain the best case history possible, and (3) collect the best specimens possible. Handling and collection of specimens in the field will affect what the laboratory can (and cannot) do with them. When possible, notify a wildlife veterinarian or other trained personnel and wait for their arrival before initiating the incident investigation. If this is not practical prior to starting the incident investigation, an initial reconnaissance of the site can direct the subsequent investigation and save time and money. During the initial reconnaissance, it is critical to assume there will be legal implications of the investigation and the cause may be a highly toxic or contagious agent. Field notes and documentation that begin with the initial stages of the investigation are critically important and impact the entire investigation that follows.

An initial identification of the agent causing the incident should be attempted if it can be accomplished safely. (1) Is there reason to believe contaminants are the source? The approaches to investigating and collecting samples from a disease or contaminant incident differ. (2) Is the incident centralized, and is it downslope, downstream, or downwind from a likely point source? (3) Is the incident on or near agricultural lands? In an agricultural setting, the crops in the area would be a starting point for what pesticides might have been applied. Early identification of the contaminant can dictate the safety precautions needed and direct the types of samples that should be collected and how they should be handled. If the source and cause of the incident are not immediately obvious, the field investigator should err on the side of safety and collect samples in the most inclusive manner possible, given the constraints of time and expertise.

As a starting point to decide whether the cause is a disease or contaminant, consider the species affected. If a single species or group of species is affected, disease is more likely to be the cause. For example, botulism may be indicated if only ducks are found dead or debilitated while other species appear unaffected. However, if many unrelated species are affected, it is more likely that contaminants are responsible. Field biologists should carry an immediate response kit with them at all time. This kit should include protective (e.g., Tyvek) coveralls, respirator or dust mask, plastic or rubber gloves, rubber boots, dark-glass collection bottles or jars, and plastic bags. Filters for respirators are specific for different types of contaminants: if the wrong filter is used, protection can be greatly diminished. For example, filters required to capture particulates are different from those designed to capture vapors or gases. Thus, a variety of respirator filters will be necessary. This kit should be kept in a waterproof container that can be securely closed to prevent contamination.

On initial discovery of the wildlife mortality or morbidity site, the nearest wildlife contaminant or disease expert should be contacted immediately. Experts in these areas may be at a teaching and/or research wildlife hospital or state or federal agency. In the case of pesticides, a county extension agent may be helpful.

Mortality

Personal safety must be the primary consideration before attempting to collect carcasses or samples or spending any time at the site of the incident. If an environmental contaminant is present in sufficient concentrations to kill or debilitate wildlife species, it also may pose a health hazard to the field biologist.

Locating carcasses, especially of small, secretive species, can be difficult. Therefore, finding one or a few carcasses should not preclude the possibility that many additional animals could have been poisoned and either removed by scavengers or moved to another area prior to death. Once dead animals are found, the immediate goals are to prevent further deaths and to identify the cause and source of the environmental contaminant(s) involved. It may not be possible to accomplish the former without first determining the latter. An immediate search of the area for intoxicated or sick animals can help identify the cause by observing their appearance, movements, and behavior. Detailed observations also may provide an opportunity to provide care for their recovery.

In many cases, exposure to environmental contaminants is obvious. Dying birds and mammals observed drinking irrigation runoff water from a field recently sprayed with an organophosphate insecticide more than likely were poisoned by the insecticide. Aquatic birds, mammals, or other wildlife species found dead in a containment pond from a cyanide leaching process most likely died from exposure to cyanide. However, no matter how obvious these causal associations may seem, it is imperative that both carcasses and samples of the apparent source of exposure be chemically analyzed for evidence of environmental toxicants. In other cases, exposure to environmental contaminants is not as obvious. A colonial waterbird rookery with almost complete nesting failure the spring following a severe winter may not be due to the colony being exposed to applications of pesticides in the area, but to exposure of the adults to organic chemicals remobilized in the environment. Chemical remobilization could result from severe scouring of nearby river sediments during heavy winter flows (American Society for Testing and Measurement 1997).

The risk of chemical contaminants to wildlife is dependent on toxicity, concentration, and route of exposure. Acute toxicity of insecticides and vertebrate pest-control chemicals (rodenticides and avicides) to wildlife is high, whereas the acute toxicity of herbicides is low. Exposure routes in wildlife include oral, dermal (including ocular, or through the eyes), and inhalation as well as from maternal sources (deposited in eggs or passed through the placenta). For mammals and birds, the most common route of exposure is oral: contaminants are ingested through the mouth. In addition to consumption of contaminated food items, birds and mammals sprayed directly or exposed to an aerosol suspension of a pesticide would result in oral exposure through preening and grooming behaviors, respectively, that would result in oral ingestion. Secondary poisoning through consumption of contaminated prey items by predatory and scavenging wildlife species is a relatively common occurrence. Mammals and birds also can readily absorb pesticides directly through their feet by standing or perching on a contaminated substrate. This route has been shown to affect red-tailed hawks foraging in orchards during the winter, following applications of organophosphorus insecticide dormant sprays (Hooper et al. 1989). Perching behavior in birds has been exploited by avian pest-control operators, who target perches with toxic chemicals specifically for dermal exposure through the feet. Mortality incidents in birds and mammals through inhalation are difficult to document and relatively uncommon.

Morbidity

Discovering intoxicated or sick (morbid) animals presents the field biologist with a situation in which action has to be taken. Species of wildlife that are intoxicated or sick from exposure to environmental contaminants may be able to fully recover. Depending on the environmental contaminant involved and the concentration, duration, and route of exposure, the negative effects on wildlife may or may not be reversible. However, during a wildlife mortality or morbidity incident, there is the chance that exposed animals have been seriously poisoned and may need to be euthanized (Friend and Franson 1999, Dein et al. 2004).

Treatment or transport of many wildlife species, particularly birds, requires one or more permits. Additionally, a salvage permit is often required to collect dead animals. Before collecting either carcasses or live animals, the necessary permit(s) must be obtained as well as knowledge about how to transport specimens or animals. It also is important to know where the specimens or animals are to be taken, particularly if the animals are still alive. Treatment of intoxicated or sick animals by wildlife rehabilitators requires specific permits. Most veterinarians are not equipped to accept and treat wildlife species, as they do not have the facilities to hold animals apart from their routine domestic patients. Wildlife rehabilitators generally are registered with state wildlife agencies, which can provide a list of wildlife rehabilitators for a given area. Prior to collecting morbid animals, the destination must be identified and appropriate transport containers obtained that will safely hold the animals and provide comfortable conditions for them. Allowing animals to die from improper care during transport is not acceptable. It may be better to humanely euthanize an animal than to subject it to unnecessary stress stemming from inadequate care during transport.

Wildlife species that are intoxicated or otherwise sick from exposure to environmental contaminants invariably demonstrate clinical signs of the poisoning (Table 6.1). Although many clinical signs from exposure to environmental contaminants are somewhat general in nature, the suite of responses exhibited in a given situation can be quite useful for diagnosing the group of contaminants responsible for the intoxication or sickness.

Wildlife Contaminant Investigation

Circumstances involved in a contaminant-related wildlife mortality or morbidity incident and the appearance of exposed wildlife are difficult to distinguish from those caused by disease or natural causes. For example, certain wildlife diseases may resemble wildlife mortality or morbidity caused by contaminants, including botulism, salmonella, trichomoniasis, mycotoxicosis, and duck virus enteritis (American Society for Testing and Measurement 1997). Investigators should rely on a wildlife disease specialist to obtain a definitive diagnosis if disease is suspected (Roffe and Work 2005). Investigations of wildlife mortality or morbidity suspected to be caused by contaminants should proceed as though the cause was unknown. All factors must be checked or eliminated, unless there is solid evidence to support specific conclusions.

If only a few carcasses are involved, external examination is necessary to rule out natural (e.g., predation) or accidental causes. Thus, it is important to be able to differentiate between evidence left by scavenging and true predation. This distinction may not be possible, but it should be attempted. It is possible that predation was successful because the animal was impaired from disease or exposure to an environmental contaminant. Thus, overall condition of the carcass can be important. A wasted or unkempt appearance could be indicative of impairment prior to predation. Large numbers of carcasses are likely related to either disease or environmental contaminants, but they could be the result of an accidental mortality (e.g., bird collisions with communication towers or other man-made structures, or road kills). Therefore, accidents should be considered before assigning the cause to disease or environmental contaminants.

The initial decision as to whether a wildlife mortality or morbidity incident is likely contaminant-related is a process of elimination. If there are no other plausible explanations for the incident, the site should be investigated for contaminants or diseases. Locating and contacting someone with experience in differentiating between disease- and contaminant-related mortality is highly desirable. Thus, it is essential to document the incident with detailed field notes and photographs.

Table 6.1. Overview of clinical signs exhibited by wildlife species by general environmental contaminant group

Clinical sign	Metals	Organic chemicals	Anti-ChE[a] insecticides	Anticoagulant rodenticides
Loss of coordination (ataxia)	X	X	X	
Muscular weakness			X	
Tremors	X			
Convulsions		X	X	X
Lethargy	X	X	X	X
Hyperactivity	X			
Reproductive effects	X	X	X	
Developmental abnormalities	X	X		
Reduced fertility	X	X	X	
Spontaneous abortions	X	X		
Excretory effects		X	X	
Excessive defecation			X	
Bloody feces		X		
Diarrhea			X	
Spasmodic contraction of anal sphincter			X	
Emesis		X	X	
Weight loss or emaciation (anorexia)		X	X	
Excessive thirst			X	
Nasal secretions				
Epistaxis (bleeding from nares)			X	
Salivation	X			
Skin lesions		X		
Imunotoxic response		X		
Depressed ChE[a]			X	
Behavioral effects				
Altered behavior				X
Unkempt appearance		X		
Physiological effects			X	
Hypothermia			X	
Coma			X	X
Paralysis			X	X
Internal bleeding				X
Dyspnea (labored breathing)			X	
Tachypnea (rapid breathing)			X	
Eye or vision problems			X	
Blindness			X	
Contraction of pupils			X	
Dilation of pupils			X	
Ptosis (drooping of eyelids)			X	X
Protrusion of eyes			X	
Lacrimation (excessive tears)			X	
Head and limbs arched back			X	
Piloerection (erection of contour feathers)			X	

[a] ChE = cholinesterase.

The investigator(s) often can obtain a substantial amount of information from the individual(s) reporting the incident, including the extent, whether a field response is necessary, and whether the contaminant(s) may cause more widespread wildlife mortality or morbidity. Important factors in interpretation of the incident scene include location, time and date of incident, species involved, number of dead and/or sick animals, rate of deaths (e.g., did they occur over a short or long period of time?), chance of continuing mortality or morbidity, clinical signs observed, climatic conditions (e.g., precipitation, temperature, and winds) preceding the incident, and any recent change that has occurred in the area. Recent changes in land use, agricultural practices, insect outbreaks, evidence of recent pesticide applications, or other factor in the area of the incident should be noted, as well as other similar incidents in this area and the observations of the person(s) reporting the incident. This information should allow the investigator to decide whether the incident warrants a field investigation. A specific case number should be assigned to each investigation and used on all labels, tags, data sheets, photographs, and other records related to the incident. The investigators must rely on their best professional judgment as to the intensity of the field investigation required and the individuals and agencies to contact.

The investigator's interpretation of the wildlife mortality or morbidity incident scene will affect the type, number, and location of samples taken and the analyses performed. The first few hours after arrival on the scene are most critical, and information should be collected as soon as possible. This consideration is especially important when an incident occurs in association with flowing water in ditches and streams. One reason is that some chemical contaminants, such as most organophosphorus and carbamate insecticides, degrade relatively quickly: chemical and diagnostic signs present at the site (e.g., sick or dying animals and water conditions) may rapidly disappear.

Wildlife mortality or morbidity incidents may be a result of illegal activities, such as a pesticide applied to intentionally kill wildlife, and they have the potential to become legal cases. In any investigation, chain of custody documentation is required to demonstrate that evidence can be accounted for at all times (American Society for Testing and Measurement 1997). Chain of custody is defined as the witnessed, written record of all individuals who have maintained unbroken control over the evidence since acquisition. The chain of custody begins with the collection of an item of evidence and is maintained until its final disposal. Each individual in the chain of custody is responsible for items of evidence, including care, safekeeping, and preservation while under their control. Because it is possible that any item or specimen acquired during the investigation of a wildlife mortality or morbidity incident may have value as evidence, it is important to treat all specimens as evidence and follow chain of custody procedures.

FIELD PROCEDURES

Sample Documentation and Transport

It is critical that samples collected in the field are handled properly to ensure that useable information can be obtained for the best understanding of what may have caused the die-off or adverse effects incident. All samples should be double bagged with a label on the inner bag or placed between the bags. By labeling the inner bag, if the label somehow becomes detached, the outer bag will keep the label with the sample. If adhesive labels are not available, the information can be recorded on notebook paper and included between the bags. Double bagging will help reduce dehydration and protect against loss of a sample, should a bag inadvertently open during shipping or storage. Each specimen should be labeled with sample type, for example, tissue, species, plant, or soil type. The sample location (both overall site name and location in the site), sample date and time, and the sample collector's name must be included. This information is extremely important for subsequent follow-up and interpretation of the sample analysis.

Labels should be written clearly with indelible felt-tip pens or other ink that will not smear when it comes in contact with water. Field biologists commonly use pencils for field notes, because a lead pencil does not smudge when wet. However, when samples are being tracked for possible litigation, pencil is not acceptable, as permanent labeling is required for all sample logs and sample labels. If permanent ink is not available for field records, it is best to make a photocopy of the sample log as soon as possible.

Samples should be placed on ice in the field, as some contaminants can degrade quickly, for example in hours, and tissues or carcasses can deteriorate quickly at warm field temperatures. Once the samples are taken from the field, they should be hard frozen. When multiple specimens are available, some samples should be placed on ice for preliminary pathology analysis while the remaining specimens are frozen. The only exception would be carcasses that may have succumbed to disease. These should be cooled and shipped to a pathologist in ≤ 48 hours of collection (Box 6.2). Samples for contaminant analysis should be transported frozen or on dry ice. It is important that samples not thaw during shipment, because thawing may compromise subsequent contaminant or disease analyses.

Handling

The manner of handling field-collected samples can differ according to the likely contaminant type. Metals generally do not tend to adhere to plastics, nor will storage in plastics interfere with the analysis by the chemist. It is acceptable to use polyethylene bottles for sample shipping and storage—a 1-L bottle is an appropriate size. It is important to acidify water samples to prevent degradation, but only when it is known that metals are the contaminant. Acidification can make other water sample types useless.

Organics, including pesticides, can readily adsorb onto or absorb into plastic, and plasticizers can leach from the container into a water sample, confounding the subsequent chemical analysis. Thus, it is best to use glass when sampling organics, including pesticides. Depending on the type of organic contaminant present, between 40 mL and 2 L should be collected. If freezing without damaging the container is not possible, the sample should be cooled to $4°$ C for storage and shipping. At least 186 g of soil should be collected and frozen. Some pesticides may have a tendency to migrate down through soil. If this is considered likely, a soil core of up to 1 m in depth should be collected.

When animal tissues are collected, great care should be taken to prevent cross-contamination from other samples or sources. Thus, only individuals experienced in dissecting animals for subsequent chemical analysis should do so. If the incident is legally contested and untrained individuals dis-

Box 6.2. Recommended laboratories for fish and wildlife mortality or morbidity incidents

United States

Wildlife Resources Center/Wildlife Pathology Unit
New York State Department of Environmental
    Conservation
108 Game Farm Road
Delmar, NY 12054 USA
Phone: 518-478-2203
http://www.dec.state.ny.gov/animals/6957.html

Southeastern Cooperative Wildlife Disease Study
Wildlife Health Building
College of Veterinary Medicine
University of Georgia
589 D.W. Brooks Drive
Athens, GA 30602 USA
Phone: 706-542-1741; Fax: 706-542-5865
http://www.uga.edu/scwds

National Wildlife Research Center
U.S. Animal and Plant Health Inspection Service (APHIS)
Department of Agriculture
4101 La Porte Avenue
Fort Collins, CO 80521 USA
Phone: 970-266-6000; Fax: 970-266-6032
http://www.aphis.usda.gov/wildlife_damage/nwrc

National Marine Fisheries Service
U.S. National Oceanic and Atmospheric Administration
Department of Commerce
1315 East-West Highway
Silver Spring, MD 20910 USA
Phone: 301-713-2332; Fax: 301-713-0376
http://www.nmfs.noaa.gov/strandings.htm

National Wildlife Health Center
U.S. Geological Survey
Department of the Interior
6006 Schroeder Road
Madison, WI 53711 USA
Phone: 608-270-2400; Fax: 608-270-2415
http://www.nwhc.usgs.gov

Division of Environmental Quality
U.S. Fish and Wildlife Service
Department of the Interior
4401 N. Fairfax Drive, Suite 322
Arlington, VA 22203 USA
Phone: 703-358-2148; Fax: 703-358-1800
http://contaminants.fws.gov

National Fish and Wildlife Forensic Laboratory
U.S. Fish and Wildlife Service
Department of the Interior
1490 East Main Street
Ashland, OR 97520 USA
Phone: 541-482-4191; Fax: 541-482-4989
http://www.lab.fws.gov

National Health and Environmental Effects Research
    Laboratory (NHEERL)
U.S. Environmental Protection Agency
109 TW Alexander Drive
Durham, NC 27709 USA
Phone: 919-541-4577; Fax: 919-541-1831
http://www.epa.gov/nheerl
http://www.epa.gov/ecotox

Canada

Canadian Cooperative Wildlife Health Centre
Veterinary Pathology
Western College of Veterinary Medicine,
University of Saskatchewan
Saskatoon, SK S7N 5B4 Canada
Phone: 306-966-5099, 1-800-567-2033 (Canada);
Fax: 306-966-7439
http://wildlife1.usask.ca/ccwhc2003

Wildlife Toxicology Division
National Wildlife Research Centre
Canadian Wildlife Service
Environment Canada
Carleton University
Ottawa, ON K1A 0H3 Canada
Phone: 819-997-2800, 1-800-668-6767 (Canada);
Fax: 819-953-2225
http://www.cws.ec.gc.ca/nwrc-cnrf/toxic/index_e.cfm

sect the samples, damage can be done to the legal acceptability of the sample analyses. It is best to freeze the samples and allow specialists to perform the dissections.

Tissues can be placed in plastic bags or small glass sample jars that have sterile interiors, and larger samples can be placed in zip-lock bags. Smoky-colored (dark) glass sampling jars should be used for soil, water, and sediment samples, particularly if pesticides or organics are involved. Using dark glass is especially important when handling chemicals that undergo photodegradation. Plastic containers should be avoided for samples that could contain pesticides or organics, as they tend to adsorb onto the plastic. Sampling equipment should be cleaned between processing and collecting samples to prevent cross-contamination. Gloves should be changed between samples or between groups of samples of similar contamination levels to prevent cross-contamination.

Record Keeping

A field log will be useful to make entries regarding each sample collected for analysis. Entries should include the sample identification number, type of sample collected, site name where collected, date, and the name or initials of the sample collector. These entries provide backup identification in case sample labels are damaged or lost, or if confusion ensues over when and where certain samples were taken.

Accurate record keeping is critical for documenting wildlife mortality or morbidity incidents. Detailed incident reports (Appendix 6.1) are essential to identification and confirmation of ecological risks associated with a particular chemical contaminant. Over time, incident reports provide information regarding those chemicals or agricultural practices that are involved most often in wildlife mortality or morbidity incidents. They also identify species that are particularly sensitive to certain chemicals. Incident reports can identify geographic areas or landscape variables most frequently affected by specific chemicals. The more detailed the information that is provided on the field data sheet, the better are the chances that investigators of the incident will be able to understand what happened. The importance of detail in the field data sheet, both to enhance accurate diagnosis and to ensure that appropriate information is provided for forensic purposes, cannot be overemphasized. It is imperative to learn whether the contaminant threat is still present and whether there is a continuing threat of wildlife mortality or morbidity.

Sample Collection

In addition to wildlife tissue samples, which are critical for identifying the cause of the mortality or morbidity incident, other environmental samples are critical. Some contaminants may be metabolized quickly in an animal, and the environmental samples may be the only place where the unaltered contaminant will be found. It also is possible the contaminants were encountered in a location some distance from where the carcasses were discovered. If the exposure occurs off-site, the actual contamination source must be located. Depending on the specific conditions of the situation, soil, water, and vegetation should be sampled. If possible, advice should be sought on the proper sampling techniques for different sample types and for different contaminants.

Environmental samples should be collected from the immediate vicinity of the dead or debilitated animals. Additionally, samples should be collected from areas where the contamination may have moved or have originated. It is possible that dead or debilitated animals are first found in a highly visible location, but that contamination may be greater elsewhere. Those experienced with site and contaminant types can provide advice on number of samples required and how far the samples should be collected from the original site.

Many contaminants act as an emetic when ingested. If vomitus or regurgitated material is found with the specimens, it should be collected. It will often contain high concentrations of the contaminant, possibly higher than in the carcass or GI tract. In acute poisonings, contaminant residues are usually higher in the anterior GI tract than in the post-absorptive tissues.

ANIMAL TISSUES

Animal tissues can be taken directly from necropsies, whereas vomitus, urine, feces, blood, and hair or feathers can be collected at or around the mortality or morbidity site. Collecting samples from a carcass or a group of carcasses becomes more difficult as time elapses. Time since death is a critical factor, as the onset of rigor mortis, decomposition, and scavenging by predators make tissue samples more difficult to obtain. Ideally, whole, fresh carcasses available for sampling tissues would be present at the incident scene, but this frequently is not the case. When the whole carcass cannot be submitted and evidence suggests that specific causes may be involved, tissue samples can be strategically taken and preserved during necropsy (Table 6.2). The best materials for establishing oral exposure to an acute toxicant are in the GI tract (crop and gizzard and/or stomach in birds, stomach in mammals and other wildlife species). Liver tissue, lipid (fat) deposits, and brain tissue generally are considered best for identifying the presence of toxic levels of many of the lipid-soluble contaminants, such as organochlorine insecticides and PCBs, and for trace metals (e.g., lead and mercury). Brain tissue also is important for diagnosing anti-ChE insecticide poisoning, through the measurement of ChE activity. Keratin structures (hair, feathers, or scales) are often used as sublethal samples to detect chronic exposure to heavy metals that may be contributing to an overall decline in fitness of the animals, making them more susceptible to disease or other environmental conditions. Analysis of samples from other environmental media also can assist in establishing routes of exposure and identifying occurrences of exposure to multiple toxic chemicals. In the case of preda-

Table 6.2. Sample selection and preservation from necropsy when whole carcass cannot be submitted and evidence suggests specific causes may be involved

Sample	Suggested test	Preservation	Comment
Lesion	As appropriate	Frozen	Lesion tissues appear abnormal; a portion of each lesion should be saved frozen; fixed tissue important
Lesion[a]	Specimen is fixed, sectioned, and stained for microscopic study	10% formalin	A portion of each lesion should be saved frozen
Liver	Metals and organics	Frozen	Entire liver of birds and small mammals, selected portions from larger species; fixed tissue important
Liver[a]	Specimen is fixed, sectioned, and stained for microscopic study	10% formalin	Specimen portions should be ≥6 mm in thickness
Kidney	Metals	Frozen	Entire kidneys from birds and small mammals, selected portions from larger species; fixed tissue important
Kidney[a]	Specimen is fixed, sectioned, and stained for microscopic study	10% formalin	Specimen portions should be ≥6 mm in thickness
Stomach	OP[b] and carbamate insecticides, plant toxins, mycotoxins, strychnine, cyanide	Frozen	Save entire contents; samples to be checked for cyanide or other toxic gases; must be placed in airtight containers
GI tract[a,b]	Specimen is fixed, sectioned, and stained for microscopic study	10% formalin or Bouin's stain	Small piece of stomach at the ileocecal junction, piece of duodenum (near pancreas), and colon
Brain	Cholinesterase activity, OC[b] insecticide residues, Organomercuric compounds	Frozen	For chemical analysis, sample must be wrapped in clean aluminum foil and placed inside a clean glass bottle; fixed tissue important
Brain, nervous tissue, eye[a]	Specimen is fixed, sectioned, and stained for microscopic study	10% formalin	Divide brain in half (sagittal); place half in formalin, freeze other half
Blood	Lead, cyanide, H_2S, nitrites	Frozen	Samples to be checked for cyanide or other toxic gases must be placed in airtight containers
Gonad[a]	Specimen is fixed, sectioned, and stained for microscopic study	10% formalin or Bouin's stain	Specimen portions should be ≥6 mm in thickness
Lung	Cyanide, H_2S	Frozen	Samples to be checked for cyanide or other toxic gases must be placed in airtight containers
Heart, lung, skeletal muscle, lymph nodes spleen, thymus[a]	Specimen is fixed, sectioned, and stained for microscopic study	10% formalin	Specimen portions should be ≥6 mm in thickness.

Adapted from Friend and Franson (1999).

[a] Histopathological examination (microscopic).

[b] GI = gastrointestinal; OC = organochlorine; OP = organophosphorous.

tors and scavengers, it may be necessary to collect local prey species or scavenged carcasses to examine possible exposure.

If the animals are not dead, but are intoxicated or otherwise sick, nondestructive techniques can be used to collect tissue samples, specifically, blood, hair, feather, or scale, and biopsies or other types of samples, such as foot washes (Fossi and Leonzio 1994). Waste materials, such as urine, feces, and vomitus, can be collected from debilitated animals found at the site by holding them in clean, ventilated containers for a period of time. Fecal and urinary products are useful for analysis of contaminants and can be evaluated for disease as well. For living birds, a foot wash with methanol or isopropyl alcohol and analysis of feathers can be useful for establishing exposure to an aerosolized chemical application, such as a pesticide. Typically, a foot wash must be performed in ≤48–72 hours of exposure to detect the presence of the chemical. After that time, most or the entire chemical will have been absorbed through the skin (Fossi and Leonzio 1994, Friend and Franson 1999, Millam et al. 2000).

If the cause of the incident is unclear or if causes in addition to contaminants are possible, animal carcasses need to be handled in different ways. Freezing animal tissues can cause damage to tissues that make disease identification by a pathologist difficult or impossible. However, failure to freeze tissues for contaminant analysis may allow the contaminant to degrade to the extent they will not appear to be present.

If there are many specimens, some should be frozen (with dry ice) and others kept cool (with ice or refrigeration). Those set aside for disease evaluation should be kept cool and transported to a trained pathologist in ≤48 hours of collection. If transportation will require >48 hours, it is best to freeze all specimens. To prevent contamination, it is best to freeze carcasses that are already deteriorating or have become putrid. None of the carcasses should be dissected prior to sending them to the pathologist.

PLANT TISSUES

Plant residues may be important for identifying how exposure may have occurred and the extent of the contamination. Some contaminants may accumulate in plants via uptake from the roots; however, many others may be present primarily as surface residues. For those contaminants most likely to be deposited on plant surfaces, care must be taken not to dislodge the residues during collection. Contamination during collection of plant samples is of greater concern if surface residues are present.

When collecting plant samples, the plants should be handled as little as possible to prevent dislodging any contaminant residues. If possible, the entire plant, including the roots, should be collected, as the roots may contain the highest residues, making identification of the contaminant more likely. Samples should be collected from both on- and off-site, with areas thought to be least contaminated sampled first. Cross-contamination among samples can be reduced by starting in the least contaminated areas. Consideration should be given to separating animal food items, such as seeds or fruits, from leaves and stems, if it will help with the follow-up investigation. Samples should be frozen as soon as possible and should remain frozen during shipping and storage until contaminant analysis.

SOIL

Soil samples can be useful for measuring the extent and levels of environmental contamination. Samples should be collected from the immediate vicinity of the dead or debilitated animals. Depending on the specifics of the incident, soil should be collected at different distances from the site. Samples should be collected off-site if movement is possible, particularly up or down hill (or up or down wind). Some contaminants have a tendency to move down through soil and may contaminate groundwater. If possible and appropriate, collect soil core samples to a depth of 1 m. Samples should be collected first from areas thought to be least contaminated and then in those areas of highest contamination. It is surprisingly easy to contaminate samples from sampling equipment and even clothing and footwear. All soil samples should be frozen, if possible, at time of collection. If prompt freezing is not possible, the soil should be placed on ice and frozen as soon as practical.

WATER AND SEDIMENT

Water samples are useful for identifying the extent and levels of environmental contamination. Glass containers should be used to sample water, as some contaminants adhere to or absorb into plastics. Samples should be protected from light, and the glass should be brown or wrapped in aluminum foil. Samples should be collected from the immediate incident area (e.g., pond) and up- or downstream. Samples can be collected from nearby surface water as appropriate. However, care should be taken that water samples contain no soil or other debris. Containers should be about half full (to prevent cracking from expansion during freezing), labeled, and placed in a plastic bag. Samples should be frozen immediately if possible, but they can be cooled to 4° C for shipping. During freezing, glass containers should be stored upright. Containers should be shipped upright and kept frozen or cooled to 4° C.

AIR

Air might be the most difficult environmental factor to sample in the field. For soil, water, or vegetation, as long as adequate sample amounts are collected, only portions of the sample are required for subsequent analyses. It is impractical to collect a sample of the air to provide to a chemist for analysis. Since air cannot be taken from the field, contaminants must be extracted from the air or measured during a field visit.

The concentration of a contaminant in the air is measured from a known volume of air sampled in the field. This measurement requires a calibrated air pump or detector. Faulty calibration or leaks will produce inaccurate measurement of the volume sampled and thus inaccurate reporting of the concentrations of contaminants. Monitoring equipment must be checked for air leaks and proper calibration prior to monitoring for contaminants in the field.

Direct measurement of aerial contaminants in the field requires an instrument capable of detecting the presence and concentration of the specific contaminant of concern. If the contaminant of concern is not known before attempting air monitoring, selection of the proper detector will be difficult. Also, some detectors are designed for human health and safety and report only if a contaminant exceeds safe levels for humans. This might not be helpful, as the level harmful to wildlife is often unknown. Other detectors are designed for monitoring organic compounds and might not detect inorganics well, and vice versa.

It also is possible to extract the contaminant from the air and provide the media to a chemist for analysis. Because concentration is based on the volume of air sampled, it is critical the volume be accurately measured and recorded by using a pump calibrated to move a known volume of air during a specific time period (e.g., mL air per sec) for a known period of time. The air being sampled must flow through a filter or liquid capable of extracting the contaminant. For many organic compounds, bubbling air through a solvent like hexane can be an effective sampling procedure. Filters also are available to remove many organic or inorganic compounds. However, the filter must be capable of capturing all the contaminant from the sampled air. The concentration will be underreported if the capacity of a filter is exceeded. Assistance from someone with experience in air quality monitoring will likely be necessary to ensure that measured air concentrations are accurate.

CHEMICAL RESIDUE ANALYSIS

Residue analysis is expensive, and there are many aspects to consider, including detection limits, quality assurance and control, how to read and evaluate a laboratory chemical analysis report, and how to interpret the toxicological data. There are 2 types of detection limits to be considered: instrument and method detection limits. Differences between instrument detection limits are the result of detector sensitivity, the chromatograph system that precedes it, and the injection method. Method detection limits represent the best performance consistently achievable from a method in a particular laboratory with a given set of instruments. Method detection limits are a function of the clean-up and extractive procedure and thus are more closely allied to the chemist's standard operating procedures and technical abilities. Standard operating procedures vary by detector and chemical, based on the relative polarity of the chemical and the environmental media in which it is found.

Interpretation of residue analysis data can be frustrating. Overall, we know little about how body residue levels of environmental contaminants correlate to corresponding effects seen in wildlife species. One excellent source of information on interpretation of residue analysis data is Beyer et al. (1996). Their work is the first major attempt to make sense of residue analysis data as related to accompanying effects.

SUMMARY

A wide variety and substantial volume of chemical contaminants (as well as natural plant and animal toxins) are present in the environment and frequently have been shown to negatively affect wildlife species, causing mortality and/or morbidity. As a result, wildlife mortality or morbidity incidents occur. Thus, there is a strong need for field biologists to be able to adequately identify and handle these incidents. Few biologists receive training in the field of environmental or wildlife toxicology, as this area of interest is relatively specialized. Thus, it is important that field biologists understand and have a source for determining standard operating procedures for successfully handling wildlife mortality or morbidity incidents. This chapter provides wildlife biologists with guidance on understanding wildlife toxicology and describes procedures that should be followed when confronted with a wildlife mortality or morbidity event. It also is important for biologists to have additional sources of information and be aware of the locations of wildlife mortality or morbidity incident databases.

(Appendix 6.1 follows on next page.)

APPENDIX 6.1. SAMPLE WILDLIFE MORTALITY OR MORBIDITY INCIDENT FIELD DATA SHEET

Date: _____

Submitter's name: _____

Submitter's affiliation: _____

Submitter's contact information: _____

Date collected: _____

Method of collection: (found dead, euthanized; if euthanized—technique used)
_____
_____

Incident scene biologist: _____

Incident location: State: _____ County: _____ Latitude/longitude: _____

Specific incident location: _____

Incident area description: (land use, habitat types, etc.)
_____
_____

Environmental factors at incident site: (climatic conditions, description of water bodies, evidence of chemicals, etc.)
_____
_____

Time of onset of incident (date and time): _____ (best estimate)

Species affected: _____

Species that appear unaffected (if known): _____
_____

Age/sex of species affected: _____

Number known dead of each species: _____

Mortality/morbidity ratio: (number of dead ÷ number of sick) _____

Estimated dead (consider scavengers, other removal): _____

Clinical signs: _____

Species at risk: _____

Additional information and observations/comments: _____
_____
_____
_____
_____

Modified from Friend and Franson (1999).

7

Wildlife Health and Disease
Surveillance, Investigation, and Management

MARKUS J. PETERSON
AND PAMELA J. FERRO

INTRODUCTION

WILDLIFE HEALTH AND DISEASE are increasingly important aspects of wildlife conservation, particularly when species at risk of extinction are involved or human health is effected (Daszak et al. 2000; see Appendix 7.1 for definitions of disease-related terminology). For example, Jones et al. (2008) found that 60% of 335 emerging infectious disease events occurring worldwide since 1940 were zoonoses, with the majority of these (72%) originating in wild animals. They also found the proportion of emerging infectious diseases originating in wildlife has increased since the 1940s. Recent concerns over the H5N1 strain of avian influenza and the H1N1 swine flu pandemic, as well as continuing concerns about Lyme disease, rabies, tuberculosis, and West Nile virus, have sensitized people to the role wildlife play in their own well-being. Similarly, wildlife biologists now are well aware of the ramifications of infectious diseases, such as bovine tuberculosis, brucellosis, and chronic wasting disease (CWD) in free-roaming cervid populations and, perhaps more importantly, in public perception (Peterson et al. 2006). Surveillance for infectious agents in wildlife populations, and timely and efficient investigation of wildlife disease outbreaks, are critical to effective management of infectious diseases in wildlife, livestock, and human populations. Despite the importance of diseases to conserving wild species at risk of extinction and the importance of wildlife populations to managing emerging infectious diseases that threaten human and livestock health, wildlife scientists typically are poorly equipped in some ways to participate in effective disease surveillance and management in free-roaming wildlife. The reasons for this situation are largely artifacts of history.

As wildlife science developed as a discipline, host–parasite interactions often were an integral component of wildlife-oriented research (Peterson 2004, 2007). The mammoth *The Grouse in Health and Disease* (Committee of Inquiry on Grouse Disease 1911), among other things, argued that *Trichostrongylus tenuis* (= *T. pergracilis*) was the primary cause of what was known as "the Grouse Disease" in red grouse (*Lagopus lagopus scoticus*) in the British Isles. The principle impetus for this study was to determine whether infectious agents controlled observed variation in grouse abundance across years. This publication undoubtedly stimulated North American game bird researchers not only to attempt similarly massive ecological studies that included disease investigations (e.g., Stoddard 1931, Bump et al. 1947), but also to search for their own version of "the Grouse Disease" (Gross 1925:424, Lack 1954:164) or "the quail disease" (Bass 1939, 1940; Durant and Doll 1941). Aldo

Leopold (1933:325), in his influential *Game Management*, probably increased interest in the infectious agents of wildlife by arguing "the rôle of disease in wild-life conservation has probably been radically underestimated." He also noted that "density fluctuations, such as cycles and irruptions, are almost certainly due to fluctuations in the prevalence of, virulence of, or resistance to [infectious] diseases." Thus, Leopold placed host–parasite interactions on par with other important interspecific relationships, such as predator–prey interactions. He did not offer any empirical or experimental evidence to support his suppositions, however.

By about 1950, many influential wildlife scientists began to assume that infectious agents of free-roaming wildlife were ecologically unimportant, except as almost inanimate extensions of poor habitat conditions or as natural disasters (Trippensee 1948:369–384, Lack 1954:161–169, Taylor 1956: 581–583). In *Wildlife Management*, Gabrielson (1951) did not even mention wildlife diseases or parasitism, suggesting that he believed infectious agents were inconsequential. Similarly, Herman (1969:325) ended his review of how diseases influenced wildlife populations by stating there was only "limited documentation that disease, as an individual factor, can drastically affect population fluctuations," and that "it is imperative that we recognize the dependency of the occurrence of disease in wildlife on habitat conditions." Herman (1963) pointed out elsewhere, however, that few studies have been conducted in such a manner that population-level effects of infectious agents could be documented, even if they occurred. This criticism still largely holds (Peterson 1996, Tompkins et al. 2002). At any rate, perceiving bacterial or viral diseases as simply extensions of poor habitat conditions or as natural disasters, where management could not reasonably be brought to bear (much like hurricanes or volcanic eruptions), led North American wildlife scientists to neglect these important interspecific relationships until relatively recently (Peterson 1991*b*). Most university programs that trained wildlife scientists in North America during this period reflected this neglect in their curriculums.

Conversely, since the early 20th century, those interested in parasite systematics continued to study their favorite taxa in wild hosts. Such efforts tended to emphasize host lists, parasite descriptions, and revisions of taxonomic relationships. Similarly, veterinary pathologists and microbiologists conducted numerous studies of infectious diseases in wildlife designed, at least in part, to determine whether wild species served as reservoir hosts for diseases occurring in livestock or humans. Most of these efforts, although quite useful, lacked an ecological dimension until recently, and thus did not address many issues important to wildlife ecologists and conservationists (Peterson 1991*a*). Research addressing wildlife diseases conducted from a predominately veterinary medical perspective became a subdiscipline of its own that led to the formation of the Wildlife Disease Association in 1952 and publication of the *Journal of Wildlife Diseases* dedicated to this subject (published since 1965).

Although others had previously addressed host–parasite interactions from an ecological perspective, Anderson and May (1978, 1979) and May and Anderson (1978, 1979), in a series of 2-part articles, provided the basic theoretical framework still used by ecologists for evaluating host–parasite interactions. During the past decade, there has been renewed interest in attempting to bridge the disciplinary divides that traditionally separated the fields of wildlife health, livestock health, human health, ecosystem functionality, biodiversity conservation, and wildlife management with a synthetic cross-disciplinary approach, often referred to as conservation medicine (e.g., Aguirre et al. 2002). Others maintain that ecology is the logical synthetic discipline to bridge these perspectives of wildlife diseases (e.g., Hudson et al. 2002, Collinge and Ray 2006, Ostfeld et al. 2008). Regardless, multidisciplinary approaches that include wildlife ecologists and conservationists are required to adequately address emerging infectious diseases, zoonoses, and diseases of importance to wildlife populations.

In light of current public concern about wildlife diseases, such as avian influenza, West Nile virus, bovine tuberculosis, and CWD, many North American wildlife conservation agencies are now firmly ensconced in wildlife disease surveillance and management programs, whether they wish to be or not. For example, CWD surveillance and testing alone accounted for >50% of the Wisconsin Department of Natural Resources' (WDNR) entire budget during 2002–2008 (WDNR 2009:29). Concerns about these expenditures and the effectiveness of WDNR's CWD eradication program led to a legislative audit, completed in November 2006 (State of Wisconsin Legislative Audit Bureau 2006). The auditors found the agency's efforts to eradicate or even control CWD had not been effective. The WDNR is now in the process of completely rethinking its approach to CWD surveillance and management (Garner et al. 2009, WDNR 2009). Without doubt, wildlife biologists and administrators working for the WDNR, at least, are well aware of the importance of wildlife diseases in their state. Comparable management efforts, conundrums, and controversies surrounding wildlife disease surveillance and management also exist in several other states and provinces. Regulatory agencies tasked with conserving wild species can no longer ignore wildlife health and disease. Instead, it is far more productive for wildlife scientists to prepare themselves to work with multidisciplinary teams attempting to address the complex ecological and social issues surrounding diseases in wild species.

It is well beyond the scope of this chapter to even briefly summarize what is known about disease processes in wildlife, the ecology of wildlife macro- or microparasites (see Appendix 7.1 for definitions), or even a small portion of the diseases known to occur in wild animals. Such a discussion would require numerous volumes the size of the one you are reading. Instead, our primary objective is to expedite connecting interested wildlife scientists with needed information on wildlife health and disease in as painless a fashion as possible. Specifically, we begin with a primer on wildlife disease processes that points readers toward key reviews of ecological approaches to host–parasite interactions; specific diseases known to occur in free-roaming wild vertebrates; and regional and national wildlife disease research, diagnostic, and management programs. Next, we summarize aspects of wildlife disease surveillance and evaluation that wildlife biologists might find themselves involved with; these include field observations, appropriate training, and procedures used to collect specimens for laboratory analyses. During this discussion, we point readers toward key reviews of basic laboratory procedures, necropsy, specimen collection, specimen shipment, and other techniques used during surveillance, investigation, and management of wildlife diseases. Finally, we discuss key issues relevant to managing wildlife diseases by presenting 3 case studies that illustrate an array of disease processes, rationales for intervention, and management strategies. Some aspects of this chapter benefited from the chapter on this topic in the previous edition of this manual (Roffe and Work 2005).

WILDLIFE DISEASE

For our purposes, disease refers to an interruption, cessation, or disorder of body functions, systems, or organs. This definition includes toxic, genetic, metabolic, behavioral, neoplastic, and nutritional diseases in addition to those caused by macro- and microparasites (see Appendix 7.1 for definitions). Interactions among these categories occur as well. For example, certain toxicants can cause neoplastic disease in animals, and nutritional deficiencies can render hosts more susceptible to infectious agents.

The variety of taxonomies used by disease specialists to classify and/or describe diseases probably accounts for much of the reason many wildlife ecologists and managers find dealing with wildlife diseases a daunting challenge. For example, veterinarians and physicians often think of diseases based on the organ system involved, which explains the names of many medical specialties, such as cardiology, dermatology, dentistry, neurology, and ophthalmology. Pathologists often describe diseases based on abnormalities found during necropsy and subsequent histopathology (e.g., ulcerative colitis or spongiform encephalopathy). Conversely, microbiologists and parasitologists tend to classify diseases based on the etiological agents responsible for the disease, resulting in such terms as bacterial, mycotic, parasitic, toxic, and viral diseases (this is the approach used in the most comprehensive series of reference books addressing wildlife diseases: Fairbrother et al. 1996, Samuel et al. 2001, Williams and Barker 2001, Atkinson et al. 2007, Thomas et al. 2007). Epidemiologists often categorize infectious diseases based on their mode of transmission among hosts (e.g., direct, indirect, or vectorborne diseases). Since the early 1980s, ecologists typically evaluate infectious agents (e.g., parasites; Appendix 7.1) based primarily on their life history strategies (e.g., macroparasites versus microparasites; Anderson and May 1979). To many wildlife biologists, the most intuitive approach for categorizing wildlife diseases centers on the host species involved. For this reason, many field guides to wildlife diseases that target wildlife biologists and/or the general public classify diseases by host species (diseases of deer, quails, waterfowl, wild turkeys, etc.; for an example of this approach, see Davidson 2006). None of these taxonomies is necessarily superior to the others, and all have merit. What is important for wildlife scientists is they learn to negotiate these taxonomies effectively, so they can fully utilize reference books and the refereed literature devoted to wildlife diseases and communicate effectively with wildlife disease specialists.

Although toxins, such as botulinum toxin and various mycotoxins, and numerous toxic environmental contaminants cause disease in wildlife, we do not address these substances in detail here because Chapter 6 (This Volume) is dedicated to this topic. Similarly, nutritional deficiencies, such as starvation, are probably among the most common diseases of wild animals; however, wildlife nutrition is not covered in this chapter. Instead, this chapter focuses on wildlife diseases caused by infectious agents. We use an ecological perspective that should be more relevant to wildlife scientists than the technical jargon of other disciplines.

Disease Processes, Ecology, and Epidemiology

Epidemiologists attempt to determine the factors that account for patterns of disease occurrence in human or other animal populations by using modeling approaches. Such patterns typically relate to the presence of antagonists, such as microparasites, macroparasites, or toxicants, or the absence of some essential nutritional element, such as a micronutrient or vitamin. Today, the thrust of epidemiology centers on modeling the presence of antagonists in host populations rather than on the absence of essential nutrients. For this reason, epidemiologists perceive disease as a 3-way interaction among susceptible hosts, disease causing agents, and the environment; with the environment acting as a fulcrum influencing whether the host or the disease causing agent predominates (Wobeser 2005; Fig. 7.1A,C).

Traditionally, epidemiologists did not attempt to model how infectious agents might influence human population dynamics, as they assumed there were so many people that even a relatively large mortality event would not alter the trajectory of human population growth. This attitude began to change after the pioneering ecological research of Anderson and May (1978, 1979) and May and Anderson (1978, 1979). If nothing else, their analytical models served notice that if it is reasonable to assume that predators can influence prey population dynamics, it is equally reasonable to suppose that macro- and microparasites have the same potential. Since then, ecologists studying host–parasite interactions have emphasized understanding and modeling how macro- or microparasites influence host population dynamics

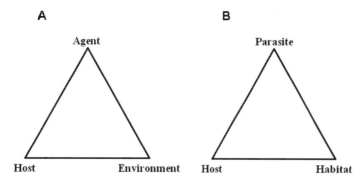

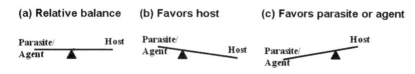

Fig. 7.1. (A) Interactive relationships among a disease agent, host animal, and the environment as perceived by epidemiologists. (B) The interactive relationships among a parasite, host animal, and the habitat of both the host and the parasite as perceived by wildlife disease ecologists. (C) The effect of habitat or environmental factors on the balance between the parasite or disease causing agent and the animal host. This relationship may be (a) relatively balanced, (b) favor the host animal, or (c) favor the parasite or disease causing agent along a continuum. *After Wobeser (2005)*.

(including host population regulation), parasite transmission, parasite population dynamics, patterns of parasitism in space and time, and the importance of parasites to biodiversity conservation (Hudson et al. 2002). It also became clear during this period that some microparasites, such as the human immunodeficiency virus, are associated with negative human population growth in some regions of Earth, and today epidemiologists often account for the influence of infectious agents on human population trends. Most disease ecologists also would replace "environment" in the epidemiological triangle (Fig. 7.1A) with the habitats of both the host and parasite as influenced by climatic and edaphic factors (Fig. 7.1B). This point is discussed later in this subsection.

The reason habitat or environment acts as a fulcrum influencing whether the host or parasite/agent thrives (Fig. 7.1C) is that each host has an inherent ability to resist toxicants and micro- or macroparasitic invasion and colonization, whereas each parasite has an inherent ability to invade and colonize the host. Environmental conditions influence this balance. For example, unusually cold conditions and/or deep, crusted snow could tip the balance toward the etiological agent, whereas mild temperatures and plentiful food supplies should benefit hosts.

Potential hosts resist macro- and microparasitic invasion and colonization by a variety of means. First, not all host species are susceptible to a particular infectious agent. Humans, for example, do not contract canine distemper, a viral disease primarily of canids, but raccoons (*Procyon lotor*), striped skunks (*Mephitis mephitis*), and several other wild species are susceptible to this microparasite. This is an example of genetically based, species-wide resistance and is independent of humoral or cell-mediated immunity (see Appendix 7.1). Hosts also have structural and functional characteristics that aid in resisting disease causing agents attempting to invade

the body. These include a skin thick enough to impede insect vectors (e.g., African elephants; *Loxodonta africana*), intact skin, ciliated epithelium, the sneeze reflex, and intact senses. Finally, host defenses include critically important circulatory agents, such as macrophages, specific antibodies, and cell-mediated immunity. The science of immunology is dedicated to studying specifically acquired resistance to disease, and many recent advances in disease diagnosis, prevention, and control involve acquired immunity.

Each parasite species also has evolved the ability to invade and colonize hosts. Ecologists often find it useful to perceive host–parasite interactions as analogous to the theory of island biogeography (MacArthur and Wilson 1967). Although many hosts (islands) are to varying degrees mobile, distances among susceptible hosts (i.e., host density) and methods of parasite transmission (i.e., access to islands) are critically important to understanding and managing host–parasite interactions. For example, most microparasites causing respiratory diseases are easily transmitted directly to susceptible hosts from infected hosts via sneezing, coughing, or touching susceptible individuals with contaminated skin. Further, hosts that rapidly move among continents may expedite transmission. These factors were the rationale behind the recent public health warnings given to reduce the severity of the H1N1 influenza pandemic in humans. Many microparasites are adapted to survive for a time outside their hosts and thus become environmental contaminants. These microorganisms are transmitted directly to susceptible hosts using the contaminated habitat as well as by contaminated transport hosts and fomites (Appendix 7.1). Several parasitic helminths and protozoans are indirectly transmitted to their definitive hosts, where sexual replication occurs via obligatory intermediate hosts. There also are a number of vectorborne, indirectly transmitted microparasites. For example, the microparasites responsible for malaria (*Plasmodium* spp.) require biting arthropod vectors for transmission; these vectors can easily penetrate the intact skin of most host species and do not require hosts to be in particularly close proximity to one another. Several macro- and microparasite species alter the behavior of their intermediate hosts, rendering them more susceptible to predation by definitive hosts and thus expediting completion of the parasite's lifecycle (Day and Edman 1983, Lafferty and Morris 1996, Thomas and Poulin 1998). Stated differently, these host islands either move more slowly or are more obvious to their predatory definitive hosts. Some species of bacteria have commensal relationships with their hosts when present in certain organ systems, yet are highly virulent when they access other organ systems. Other microparasites, such as *Bacillus anthracis* (the etiological agent responsible for anthrax), multiply at high rates and rapidly kill their hosts (Dragon and Rennie 1995, Gates et al. 2001, Hugh-Jones and de Vos 2002). These bacteria then directly infect and often kill other animals, such as predators and scavengers that rend the carcass and ungulates that consume contaminated plants. Flies that have been in contact with contaminated carcasses also can serve as transport hosts. Moreover, once *B. anthracis* is exposed to oxygen (after scavengers open the carcass), it can form highly resistant spores that persist in suitable soil conditions for decades if not centuries, thus lying in wait for a susceptible host (mobile island), where it can reproduce. These examples represent only a portion of the ways that parasites have evolved to invade and colonize hosts. Regardless, it should be clear that macro- and microparasite transmission among hosts is critically important to individuals attempting to manage wildlife diseases.

Because of the current significance of CWD to North American wildlife conservation agencies, it is worth noting that CWD and other transmissible spongiform encephalopathies (TSEs), such as bovine spongiform encephalopathy (BSE), scrapie in domestic sheep and goats, and Creutzfeldt-Jakob disease (CJD) in humans, do not fit seamlessly within some of the disease taxonomies discussed at the beginning of this section. Disease associated with TSEs is thought to be caused by aberrantly refolded isoforms of normal prion-related protein (Prusiner 1982, 1991) that induce normal prion-related protein in the brain and lymphoid tissue to convert to the abnormal form (Collinge 2001, Williams et al. 2001). The exact mechanism of this transformation is unknown. Pathogenic forms arise sporadically in mammals as the result of genetic mutation or by a susceptible animal ingesting infected tissues or body fluids. Upon accumulation in the brain, the abnormally folded proteins cause spongiform encephalopathy and related clinical disease. Because prions are devoid of nucleic acid and are remarkably resistant to environmental degradation, including cooking, they are in many ways more similar to a unique class of toxicants than to parasites; however, their infectious nature and multiplication in hosts closely resembles the life histories of many microparasites. For this reason and several others, the latter model is more likely to ground successful TSE management than the former.

Although many people are concerned about how CWD influences cervid populations, wildlife viewing, or the future of hunting, it is likely that many others are primarily concerned about livestock or human health. After all, it was only just a few years ago that BSE jumped the species barrier to humans, leading to an inevitably fatal variant of CJD in unusually young patients (Will et al. 1996, Lacey 1998, Collinge 2001). Belay et al. (2001), using epidemiological methods, were unable to tie CWD to a single case of human CJD or its variants, even when these cases occurred in unusually young patients who had consumed venison. Instead, it is likely that humans and cattle are no more susceptible to CWD than to scrapie (Raymond et al. 2000, Kong et al. 2005); no cases of human TSEs have been tied to scrapie since this disease was first described in sheep in the 1730s. This observation certainly does not mean, however, that we can safely rule out transmission of some CWD variant to humans in the future. Regardless, there will be increasing pressure from various directions on North American wildlife conservation agencies regarding how best to address CWD in cervids. Similar pressure also could be brought to bear regarding current or yet unknown emerging infectious diseases.

In light of the complexities of host–parasite interactions outlined above, it is logical to assume that environmental factors can markedly influence a host's ability to resist macro- and microparasites and a parasite's ability to invade and colonize hosts under some circumstances (Fig. 7.1A). When wildlife ecologists consider the triangular relationship among parasites or infectious agents, hosts, and the environment, they might assume the environment is composed of the climatic and edaphic factors present in an area of interest. There is no doubt that weather conditions, which are related to climate, can and do influence wildlife disease occurrence. Similarly, in temperate regions, the occurrence of diseases requiring flying insect vectors is indeed seasonal. Soil conditions also influence disease occurrence. For example, anthrax tends to occur in wild ungulates in the same geographic regions over time, because *B. anthracis* spores survive better in calcareous (alkaline to neutral) soils than in acidic soils (Dragon and Rennie 1995, Hugh-Jones and de Vos 2002). It would be a mistake, however, for conservation biologists and wildlife managers to assume that because they cannot control edaphic and climatic factors, they cannot influence how the environment affects host–parasite interactions.

To epidemiologists and wildlife disease ecologists, studying host–parasite interactions, the environment in this context (Fig. 7.1A) includes the total of all factors impinging on the host and the parasite. Thus, environment includes the habitats of both host and parasite (Fig. 7.1B), not just climatic and edaphic factors influencing these habitats. For example, nutrient availability; movement corridors; and habitat condition, fragmentation, and loss or conversion all form important aspects of the environment in which host–parasite interactions take place. Moreover, nearly every habitat change or conservation practice humans implement has the potential to influence the outcome of host–parasite interactions involving wildlife. Although wildlife managers cannot do much about climate, weather, and soil conditions, they can and do alter habitat conditions, habitat connectivity, land development trends, and wildlife densities. Wildlife scientists' expertise in these areas places them in an excellent position to influence the host and the habitat or environmental portion of the host–parasite–habitat/environment triangle (Fig. 7.1). We discuss this matter further in the section Managing Wildlife Diseases, below.

For wildlife scientists with little or no formal training in wildlife diseases, but who would like to jumpstart their knowledge of wildlife disease processes, we recommend Wobeser's (2005) *Essentials of Disease in Wild Animals* as the point of departure. This textbook is suitable for introductory undergraduate courses on wildlife diseases. Wobeser (2005) does an excellent job of covering the basics of parasite transmission, host defense, effects on infectious agents on host individuals and populations, and the importance of zoonoses. He also addresses disease detection, surveillance, and management. Several excellent publications explore ecological or epidemiological perspectives of infectious agents in wild species (Table 7.1). For primers on host–parasite ecology, we recommend the textbook chapters by Sinclair et al. (2006) and Begon et al. (2006). We suggest reading these chapters in conjunction with Wobeser's (2005) textbook. For those interested in a more detailed perspective on wildlife disease ecology, we recommend Hudson et al. (2002) as an excellent place to begin.

Resources Addressing Specific Diseases of Wildlife

Due to circumstances beyond their control, wildlife biologists often find themselves in the middle of raging controversies concerning what should or should not be done about a specific wildlife disease. For this reason, they require quick access to resources for rapid education. For example, when CWD is first detected in a state or province, field biologists and administrators might want to learn about the disease quickly in preparation for the inevitable challenges they must face. Similarly, because most emerging zoonotic diseases now originate in wild species (Jones et al. 2008), wildlife managers and administrators require rapid access to key information on these diseases, should they occur in their jurisdictions. Although it is certainly possible for wildlife managers and administrators to use electronic search engines, such as the ISI Web of Knowledge (Thompson Reuters) or Google Scholar, to locate hundreds or even thousands of refereed journal articles relevant to a specific wildlife disease, carefully study all these publications, and reach informed conclusions about the disease in question, this may not be the most efficient use of their time. Instead, reviews of specific wildlife diseases (or of the diseases of a host species of interest) written by wildlife disease specialists allow wildlife scientists rapid access to reliable knowledge regarding the most important diseases of wildlife (see Table 7.2 for a list of such reviews).

Wildlife field biologists and administrators needing a quick summary of key facts surrounding a specific wildlife disease probably will want to begin by reading the summary of that disease in 1 or more of the field manuals listed in Table 7.2. Of the 7 field manuals listed, 4 are freely available online. Some state wildlife agencies also publish useful field guides on their departmental websites that were not included in Table 7.2, because we were unable to determine who wrote the guide or when it was written. Nearly all these field guides target a specific state, province, or region of North America. In general, all briefly address important topics for each disease, such as the etiological agent responsible, clinical signs, lesions, host species range, diagnosis, transmission, geographic distribution, significance to wildlife conservation, and public health implications. Typically, authors provide this information in just a few paragraphs to perhaps 2 pages of text. The authors of these manuals specifically targeted wildlife biologists and/or the public as their intended audience, and most of these field guides are replete with high-quality color photographs. Although some wildlife disease specialists might find these field manuals somewhat simplistic, they are useful and inexpensive tools for field biologists, wildlife agency administrators, and the public.

Some wildlife scientists may require a more detailed treatment of a specific disease of wild animals, the diseases of a specific host species, or those of a group of related host species. We are fortunate in there currently is a wealth of such reviews (Table 7.2). Many detailed reviews of wildlife diseases are arranged by the class of etiological agent responsible for the disease (e.g., Fairbrother et al. 1996, Samuel et al. 2001, Williams and Barker 2001, Atkinson et al. 2007, Thomas et al. 2007). Others target the array of diseases known to occur in a single host species (e.g., Davidson et al. 1981) or a related group of host species (e.g., Wobeser 1997; Peterson 2004, 2007), and still others address a single disease in a group of related host species (e.g., Williams et al. 2002c, d; Williams 2005, Sigurdson 2008). Similarly, some detailed reviews attempt to address what is known about all wildlife diseases of interest occurring in a given state or region (e.g., Dieterich 1981, Thorne et al. 1982, Forrester 1992, Forrester and Spalding 2003), whereas others address what is known about specific wildlife diseases worldwide (e.g., Fairbrother et al. 1996, Samuel et al. 2001,

Table 7.1. Selected publications addressing the ecology and epidemiology of host–parasite interactions involving free-roaming wild vertebrate populations

Publication type/Title	Content	Reference
Brief summaries		
Parasites and pathogens. Chapter 11 *in* Wildlife ecology, conservation, and management	An effective primer on host–parasite ecology	Sinclair et al. 2006:179–195
Parasitism and disease. Chapter 12 *in* Ecology: from individuals to ecosystems	An excellent yet brief summary of host–parasite ecology and modeling	Begon et al. 2006:347–380
Book length		
Population dynamics of rabies in wildlife	Ecological and epidemiological approaches to rabies in wild animal populations worldwide	Bacon 1985
Ecology of infectious diseases in natural populations	The first major ecological synthesis of knowledge surrounding host-parasite interactions in unmanaged animal and plant populations, with emphasis on mathematical modeling to explore observed patterns	Grenfell and Dobson 1995
The ecology of wildlife diseases	Detailed ecological treatment of host–parasite interactions such as parasite transmission, influence of parasites on host population dynamics, and the role of parasites in biodiversity conservation	Hudson et al. 2002
Wildlife diseases: landscape epidemiology, spatial distribution and utilization of remote sensing technology	Discusses the use of remote sensing and Geographic Information System technology to track the spread of infectious agents through wildlife populations in a spatially explicit fashion	Majumdar et al. 2005
Disease ecology: community structure and pathogen dynamics	Explores various aspects of community ecology that influence pathogen transmission rates and disease dynamics in a wide variety of study systems	Collinge and Ray 2006
Wildlife and emerging zoonotic diseases: the biology, circumstances and consequences of cross-species transmission	Written from traditional microbiological and epidemiological perspectives but includes a distinctly ecological perspective in several chapters and does an excellent job of addressing emerging zoonotic diseases involving wild vertebrates	Childs et al. 2007
Infectious disease ecology: effects of ecosystems on disease and of disease on ecosystems	Uses a systems approach in an ecological context to attempt integration of the perspectives of various medical, agricultural, and natural resource management disciplines as required to effectively control infectious agents important to human society	Ostfeld et al. 2008

Williams and Barker 2001, Atkinson et al. 2007, Thomas et al. 2007). This breadth of approaches allows readers to choose the format that best addresses their needs, so we included this information in Table 7.2.

Two cautions are in order regarding detailed reviews of wildlife disease. First, the books included in Table 7.2 were approximately 2 years out-of-date when published and have become increasingly dated since then. For most wildlife diseases, this does not matter much. However, in areas where considerable research is ongoing, such as West Nile virus, avian influenza, or CWD, readers will need to update these reviews with current refereed journal articles. Similarly, some newly emerging wildlife diseases, such as white-nose syndrome in bats, were unknown when most of the reviews in Table 7.2 were published. Somewhat less obvious is that authors of older state-level reviews of wildlife diseases often had access to previously unpublished data from their state. For this reason, these publications not only review what was known about wildlife diseases in the state, but also present original data. For example, some authors in Thorne et al. (1982) utilized data on wildlife diseases in Wyoming obtained from the Wyoming State Veterinary Laboratory, Laramie, that appears nowhere else in print. For this reason, if for no other, those interested in wildlife diseases in Alaska, Florida, and Wyoming, for example, should not overlook the reviews by Dieterich (1981), Forrester (1992), and Thorne et al. (1982), respectively, despite the age of these publications.

Sources for Wildlife Disease Expertise

Most wildlife biologists will not be sufficiently interested in wildlife diseases to warrant completing a D.V.M. degree with specialization in the diseases of wildlife, and/or a Ph.D. degree in such fields as veterinary pathology, microbiology, immunology, epidemiology, parasitology, or wildlife disease ecology. Wildlife regulatory biologists, researchers, and administrators, however, may well need access to such exper-

Table 7.2. Selected publications reviewing disease processes in, and/or specific diseases of, free-roaming wild vertebrates[a]

Class/species	Locale[b]	Content	Title	Reference
Birds				
Wide range	Worldwide	Detailed review	Infectious diseases of wild birds	Thomas et al. 2007
Wide range	Worldwide	Detailed review	Parasitic diseases of wild birds	Atkinson et al. 2007
Wide range	North America	Field manual	Field manual of wildlife diseases: general field procedures and diseases of birds[c]	Friend and Franson 1999
Wide range	Florida[d]	Detailed review	Parasites and diseases of wild birds in Florida	Forrester and Spalding 2003
Prairie grouse	Species' range	Detailed review	Parasites and infectious diseases of prairie grouse: should managers be concerned?	Peterson 2004
Quails	Texas[d]	Detailed review	Diseases and parasites of Texas quails	Peterson 2007
Sage grouse	Species' range	Detailed review	Parasites and infectious diseases of greater sage-grouse	Christiansen and Tate 2011
Waterfowl	Worldwide	Detailed review	Diseases of wild waterfowl	Wobeser 1997
Mammals				
Wide range	Worldwide	Detailed review	Infectious diseases of wild mammals	Williams and Barker 2001
Wide range	Worldwide	Detailed review	Parasitic diseases of wild mammals	Samuel et al. 2001
Wide range	Florida[d]	Detailed review	Parasites and diseases of wild mammals in Florida	Forrester 1992
Cervids	North America	Detailed review[e]	Chronic wasting disease of deer and elk: a review with recommendations for management	Williams et al. 2002c
Cervids	North America	Detailed review[e]	Chronic wasting disease: implications and challenges for wildlife managers	Williams et al. 2002d
Cervids	North America	Detailed review[e]	Chronic wasting disease	Williams 2005
Cervids	North America	Detailed review[e]	A prion disease of cervids: chronic wasting disease	Sigurdson 2008
White-tailed deer	Species' range	Detailed review	Diseases and parasites of white-tailed deer	Davidson et al. 1981
Birds and mammals[f]				
Wide range	Southeastern United States	Field manual	Field manual of wildlife diseases in the southeastern states	Davidson 2006
Wide range	Alaska	Field manual	A field guide to common wildlife diseases and parasites in Alaska[c]	Elkin and Zarnke 2001
Wide range[g]	Alaska[d]	Detailed review	Alaskan wildlife diseases	Dieterich 1981
Wide range	British Columbia, Canada	Field manual	Manual of common diseases and parasites of wildlife in Northern British Columbia[c]	Miller et al. 2003
Wide range	Colorado	Field manual	Manual of common wildlife diseases in Colorado	Adrian 1981
Wide range	Ontario, Canada	Field manual	Manual of common parasites, diseases and anomalies of wildlife in Ontario[c]	Fyvie and Addison 1979
Wide range	Saskatchewan, Canada	Field manual	Handbook of diseases of Saskatchewan wildlife	Wobeser 1985
Wide range	Wyoming[d]	Detailed review	Diseases of wildlife in Wyoming	Thorne et al. 1982
Multiple classes				
Wide range	Worldwide	Introductory textbook	Essentials of disease in wild animals[h]	Wobeser 2005
Wide range	Worldwide	Advanced textbook	Disease in wild animals: investigation and management[i]	Wobeser 2007
Wide range	Worldwide	Detailed review	Noninfectious diseases of wildlife	Fairbrother et al. 1996
Wide range	Worldwide	Detailed review	Disease emergence and resurgence: the wildlife–human connection[c]	Friend 2006
Wide range	Worldwide	Detailed review	Wildlife diseases of the Pacific Basin and other countries	Fowler 1981

[a] Does not include publications focused on the diseases of captive wild animals in zoos, aquaria, or rehabilitation centers or the diseases of fishes.

[b] Authors of publications labeled "worldwide" attempted to review what was known about the disease in question regardless of location, but published data undoubtedly reflects national and regional interests.

[c] Available electronically at no cost as a PDF and/or HTML file; URL included in Literature Cited.

[d] Includes considerable material from locales other than the focal location listed.

[e] We found 45 reviews of chronic wasting disease in the refereed literature; here we included only those written specifically for wildlife ecologists and managers (i.e., Williams et al. (2002c, d) and those that address issues important to wildlife scientists in relatively transparent language.

[f] Some of these publications mention reptiles or amphibians in passing, but they essentially deal only with diseases of birds and mammals.

[g] Also addresses the diseases of fish.

[h] Addresses disease processes applicable to a wide array of wild animals, the study of wildlife diseases, and disease management; does not address any specific disease in detail; most examples from birds and mammals.

[i] Does not address any specific disease in detail; most examples from birds and mammals.

tise. Several wildlife health centers employ people with these skills at both the national and regional levels in Canada and the United States (Box 7.1). Some, but not all, of these entities conduct wildlife health and disease research, perform necropsies, offer diagnostic testing, provide training, and participate in wildlife disease management efforts; see each program for its scope and other details.

In Canada, each province has a Canadian Cooperative Wildlife Health Centre or at least a portion of a regional center (Box 7.1). The situation is more complicated in the United States. The Southeast Cooperative Wildlife Disease Study is affiliated with the state wildlife conservation agencies in the southeastern United States (Box 7.1), but there are no comparable regional wildlife health facilities in other regions of the country. Some state wildlife agencies, however, have a long and vibrant history of wildlife health and disease programs (e.g., the California Department of Fish and Game and the Wyoming Game and Fish Department). As of this writing, 22 state wildlife conservation agencies have 1–5 wildlife veterinarians on staff (David Jessup, California Department of Fish and Game, unpublished data; John R. Fisher, Southeastern Cooperative Wildlife Disease Study, unpublished data); some of these agencies also have programs for investigating wildlife diseases, maintain wildlife health laboratories, and/or maintain informative websites that address wildlife diseases relevant to the state. See the websites of each state wildlife conservation agency for details.

Another source of wildlife disease expertise may be found at state-operated veterinary medical diagnostic laboratories. Although the focus of these laboratories is on the diagnosis of diseases occurring in domestic livestock and poultry, some employ staff members with considerable expertise in wildlife diseases and provide expert diagnostic services for wild vertebrates as well. For example, the late Dr. Elizabeth S. Williams, a leading wildlife pathologist, who, among other things, discovered that CWD was a previously unknown TSE (Williams and Young 1980), was employed at the Wyoming State Veterinary Laboratory for many years. Wildlife scientists dealing with disease issues should determine whether someone of Dr. Williams' stature is working at the governmental veterinary medical diagnostic laboratory in their state, and if so, develop a working relationship with this wildlife disease specialist. These laboratories typically maintain websites listing services available and contact information as a place to begin.

We expect that wildlife regulatory agency biologists, researchers, and administrators will increasingly be involved with wildlife disease surveillance and management in the future. Such diseases as avian influenza, bovine tuberculosis, CWD, rabies, white-nose syndrome, and various emerging zoonotic diseases ensure that involvement will occur. After all, wildlife ecologists have the required expertise regarding the life histories of wild vertebrates, habitat management techniques, and wildlife density manipulation. To be effective members of teams conducting disease surveillance or management of wildlife populations, wildlife scientists must become familiar with the basic aspects of disease processes and ecology as well as details of the specific diseases they are working with. Fortunately, there is a wealth of knowledge available regarding disease ecology and epidemiology, disease processes, and specific diseases of wild vertebrates (Tables 7.1 and 7.2). Similarly, wildlife disease specialists are available at some state wildlife conservation agencies and veterinary medical diagnostic laboratories as well as at national and regional wildlife health programs in North America (Box 7.1). Wildlife scientists who make effective use of these resources can quickly become key members of multidisciplinary teams organized to conduct disease surveillance and/or management of wild animal populations.

RESOURCES AND METHODS FOR INVESTIGATING WILDLIFE DISEASES

Field Observations

The wildlife field biologist has a crucial role in any wildlife disease investigation. Most often, he or she is the first investigator on the scene and is more familiar with the habitat and the population(s) involved than anyone else. Field observations are critical, because data providing a detailed and accurate history of the event can greatly narrow the possibilities for diagnosis. Field biologists should collect and record data on key parameters associated with a disease outbreak (Box 7.2). Digital photographs also are useful, but care must be taken to ensure clear depiction of the subject. Investigators also should carefully document photographs, so the important features, including the location, date, species, sex, and lesions, are easily identified.

Environment or Habitat

Establishing potential links between changes in the physical environment and wildlife mortality events is critical. If the disease is occurring where it previously did not occur, changes in habitat or other environmental conditions may be involved (Fig. 7.1). There are many possibilities that should be considered regarding how the interactions among habitat, edaphic factors, and climatic factors influence macro- and microparasitism in individual animals and wildlife populations. These ecological relationships play out with each of the classes of disease parameters discussed here.

Magnitude and Onset

Two of the most difficult issues associated with initiating a wildlife disease investigation include when to assign time of onset, and in the case of mortalities, when to call a mortality event "unusual." Unless the population is closely monitored, and the biologist understands the background mortality in the population, assigning time of onset is problematic. Moreover, most wild species, when ill, seek refuge in hidden

BOX 7.1. EXAMPLES OF NATIONAL AND REGIONAL WILDLIFE HEALTH PROGRAMS IN CANADA AND THE UNITED STATES

Some, but not all, of these entities conduct wildlife health and disease research, offer diagnostic testing, and participate in wildlife disease management efforts; see each program for its scope and other details.

Canada

Canadian Cooperative Wildlife Health Centre
National Headquarters
Western College of Veterinary Medicine
University of Saskatchewan
52 Campus Drive
Saskatoon, SK S7N 5B4 Canada
National information line (Canada): 800-567-2033;
Phone: 306-966-5815, Fax: 306-966-7387
http://www.ccwhc.ca/

Canadian Cooperative Wildlife Health Centre Alberta
Faculty of Veterinary Medicine
University of Calgary
3280 Hospital Drive
Calgary, AB T2N 4Z6 Canada
Phone: 403-210-3824; Fax: 403-210-8121

Canadian Cooperative Wildlife Health Centre Western/Northern
Western College of Veterinary Medicine
University of Saskatchewan
52 Campus Drive
Saskatoon, SK S7N 5B4 Canada
Phone: 306-966-5815; Fax: 306-966-7439

Canadian Cooperative Wildlife Health Centre Ontario/Nunavut
Pathobiology
University of Guelph, Guelph, ON N1G 2W1 Canada
Toll-free (Canada): 866-673-4781; Phone: 519-824-4120
  ext. 54662; Fax: 519-821-7520

Canadian Cooperative Wildlife Health Centre Quebec
Faculté de Médecine Vétérinaire
Université de Montréal
3200 rue Sicotte
Saint-Hyacinthe, QC J2S 2M2 Canada
Phone: 450-773-8521 ext. 8346; Fax: 450-778-8116

Canadian Cooperative Wildlife Health Centre Atlantic
Atlantic Veterinary College
Pathology and Microbiology
University of Prince Edward Island
550 University Avenue
Charlottetown, PEI C1A 4P3 Canada
Phone: 902-628-4314; Fax: 902-566-0871
http://atlantic.ccwhc.ca/

Canadian Cooperative Wildlife Health Centre British Columbia
Centre for Coastal Health
Building 305, Room 406
900 5th Street, Nanaimo, B.C., V9R 5S5 Canada
Phone: 250-740-6366; Fax: 250-740-6366
http://www.centreforcoastalhealth.ca/

United States

Several state wildlife departments, or departments of natural resources, maintain wildlife health laboratories and/or informative websites that address wildlife diseases relevant to the state. See state wildlife agency web sites for details. Some state veterinary medical diagnostic laboratories include staff members with expertise in wildlife diseases and provide expert diagnostic services related to wildlife diseases. See state diagnostic laboratory web sites for details and contacts.

U.S. Geological Survey National Wildlife Health Center
6006 Schroeder Road
Madison, WI 53711-6223 USA
Phone: 608-270-2400; Fax: 608-270-2415
http://www.nwhc.usgs.gov/.

Southeast Cooperative Wildlife Disease Study
University of Georgia
Athens, GA 30602-7393 USA
Phone: 706-542-1741; Fax: 706-542-5865
http://www.uga.edu/scwds/

Wildlife Health Center
University of California–Davis
TB128 (Old Davis Road)
Davis, CA 95616 USA
Phone: 530-752-4167
http://www.vetmed.ucdavis.edu/whc/

The Center for Wildlife Health
274 Ellington Plant Sciences Building
2431 Joe Johnson Drive
University of Tennessee
Knoxville, TN 37996-4563 USA
Phone: 865-974-6173; Fax: 865-974-4714
http://wildlifehealth.tennessee.edu/index.htm

> **BOX 7.2. EXAMPLES OF IMPORTANT PARAMETERS ASSOCIATED WITH A WILDLIFE DISEASE OUTBREAK THAT SHOULD BE ASCERTAINED AND RECORDED**
>
> **Environment/Habitat:**
> - Are there unusual or novel circumstances?
> - Habitat types involved?
> - Habitat conditions?
> - Human land uses?
> - Nearby human land uses?
>
> **Magnitude:**
> - Are both dead and sick animals observed?
> - How many animals are sick or dead?
> - What is the approximate ratio of sick to dead?
> - Anything else unusual?
>
> **Onset:**
> - When did the situation develop?
> - How rapidly did the situation progress?
>
> **Temporal distribution:**
> - How long has the situation occurred?
> - Has this situation been observed in this area before?
>
> **Geographic distribution:**
> - Where did the event occur?
> - How much area is affected by this event?
>
> **Species, age, and sex:**
> - What species are dying or sick compared to what species appear at risk for the disease?
> - What other wild species are present in the area?
> - Do affected animals tend to be of a particular age? Sex?
>
> **Clinical signs:**
> - What physical or behavioral abnormalities are observed?
>
> **Population at risk:**
> - What is the size of the population at risk for the disease?
> - Are livestock in the area? If so, what type and how many?
>
> **Population movement:**
> - Where did the animals come from?
> - Where are the animals going?

areas, thus complicating discovery of carcasses or sick animals. Regular patrols and the cataloging of mortalities by location (using Global Positioning System [GPS] technology), species affected, and estimated stage of decomposition provide valuable data that will assist those attempting to determine whether a mortality event is unusual. Developing knowledge about normal behaviors and movements of wildlife facilitates detection of sick animals. In reality, however, the true onset of a wildlife mortality event often is unknown.

Temporal Distribution
Assessing the temporal scale of the wildlife morbidity or mortality event can provide critical information on its potential cause(s). For example, spills of toxicants or natural toxins may kill large numbers of animals over short periods (days to weeks); this often is the case for avian botulism or organophosphate insecticide poisoning (see Chapter 6). Certain infectious agents, particularly when introduced where animals are concentrated, may kill many animals over several weeks. Mortalities will diminish when predominantly immune animals remain, animals emigrate from the area of disease transmission, the source of the infectious agent is no longer present, or obligatory biological vectors are no longer available. Some diseases cause losses over longer periods (months as opposed to days or weeks), and mortalities can increase or decrease depending on demographic and environmental factors. Additionally, some diseases are only detectable when animals congregate for some reason. For example, acute lead poisoning is most evident when waterfowl are staging for migration in wetlands contaminated with lead shot. Thus, moralities due to lead shot ingestion are most likely to be detected when large numbers of animals are exposed. In contrast, chronic lead poisoning can be less apparent: as animals become increasingly debilitated, they seek refuge where they are difficult to observe, are killed by predators, or disperse. For all these reasons, assessment of the temporal scale of a morbidity or mortality event provides important information that can assist investigators attempting to determine the cause of the event.

Geographic Distribution
Spatial patterns also can assist in determining the cause of a wildlife disease event. Carcasses in one particular area could suggest they were concentrated there by wind or current, or there is a point source for the disease causing agent nearby. The presence of severely decomposed carcasses along with fresher carcasses suggests the mortality event is ongoing.

The location of common foraging and loafing areas as well as daily movement patterns of animals in the affected population should be recorded using a GPS device. Wildlife biologists should pay particular attention to how wildlife morbidity or mortality occurrences move in space, as this information can help investigators map the spread of the disease and determine the appropriate management actions. This is typically done using Geographic Information System (GIS) technology. The coverage containing locations of sick or dead animals is used in conjunction with other coverages, such as soils, land cover, roads, and streams, to help determine why the disease occurred where it did. The use of GPS and GIS technologies has revolutionized the task of evaluating changes in the geographic distribution of wildlife diseases over time and explaining why these changes occurred.

Species, Age, and Sex

Recording the species, age, and sex of affected animals provides valuable indicators as to what may or may not be causing the observed sickness and/or death. For example, starvation tends to affect young or old animals disproportionately. Moreover, some species may be more sensitive to a specific disease than others are. For example, the northern shoveler (*Anas clypeata*) is highly efficient at sifting sediment for invertebrates. Because botulinum toxin is concentrated in invertebrates, northern shovelers typically are among the first animals to be affected during a botulism outbreak. Other etiological agents responsible for wildlife mortality, such as pesticides, are more indiscriminate and affect several species, individuals of various ages, and both sexes. In fact, raptors can die from consuming smaller birds suffering from organophosphate poisoning. Younger animals with less-developed immune systems or different behavior patterns than adults may be more susceptible or receive higher exposure to certain microparasites. They may therefore be overrepresented during an infectious disease outbreak. Other diseases may affect one sex preferentially, because of behavioral or physiological differences. For example, brucellosis, a disease of the reproductive tract in mammals caused by *Brucella* spp., tends to affect females more severely than males and often manifests as spontaneous abortions.

Clinical Signs

Clinical signs of sick animals should be carefully documented and, if possible, captured using a video recorder. Some diseases cause neurological abnormalities, such as the inability to turn right-side-up, tremors, unusual posture of the limbs, abnormal gait or flight, or inability to keep the head upright. Some clinical signs can be pathognomonic for a specific disease. For example, brown pelicans (*Pelecanus occidentalis*) affected by the marine toxin domoic acid exhibit scratching behavior alternating between the left and right legs. Other diseases induce changes in temperament. Rabies, for example, often renders normally wary wild carnivores more docile or aggressive and causes normally nocturnal animals to move about during the day. At times, sick animals will be in unusual locations or apart from others of their species. Finally, the absence of sick animals during a wildlife mortality episode can be just as important a clue as their presence; this fact could help wildlife disease specialists narrow their list of differential diagnoses.

Laboratory Procedures

Although detailed field observations provide a critical starting point for disease investigations, collecting and submitting proper samples in a safe and legal manner to an appropriate diagnostic laboratory are the next logical steps. For many wildlife scientists, once they submit samples to the laboratory for testing, it is almost as if the samples disappear into a black hole, followed at some indeterminate period by the emergence of a report presented in an arcane format and apparently written in a foreign language. Poor communications between laboratory and field personnel can create major problems in the course of a disease investigation and can hinder management strategies designed to address wildlife diseases. Our objective in this section is to help wildlife scientists understand the basic approaches to laboratory testing. Our primary intent is to help wildlife biologists grasp the fundamental advantages, limitations, and sample requirements of each diagnostic discipline. This knowledge should help wildlife scientists communicate more effectively with laboratory staff members and help ensure that field biologists collect the most useful specimens and handle them optimally. Additionally, a basic understanding of the different diagnostic disciplines will help inform sound management decisions based upon laboratory findings.

The primary function of laboratory testing is to identify infectious and noninfectious agents that could cause observed wildlife disease. For this reason, sample collection, handling, and shipping are critically important to laboratory testing. If inappropriate samples are collected, or appropriate samples are handled improperly either in the field or during shipment, test results will be unreliable. Thus, it is critical to know prior to collection and submission what samples are needed and how they should be handled, packaged, and shipped. The best way to ensure these tasks are done correctly is to contact the laboratory prior to sample collection. Box 7.1 includes several programs that have laboratories specializing in wildlife diseases. As discussed earlier, some state wildlife conservation agencies conduct testing for certain wildlife diseases, and most state veterinary medical diagnostic laboratories will accept samples collected from wildlife. If possible, it is best to send the entire carcass to the laboratory as soon as possible following death and allow a veterinary pathologist with expertise in wildlife diseases to conduct the necropsy, collect appropriate samples for testing, and select appropriate testing based on necropsy findings, field observations provided, and experience with wildlife disease diagnostics.

There are times when wildlife biologists might want to collect fecal samples or other materials from live animals and complete some basic laboratory work at their field station. For example, with just a little training, most biologists can do fecal floatations and evaluate samples microscopically for parasite ova. Moreover, some readers might appreciate more details on laboratory procedures than we provide in the brief summaries below. For these reasons, we include citations for some useful reference manuals addressing basic laboratory procedures in Table 7.3.

Pathology

The first step in any investigation involving a dead animal is examination of the carcass. Veterinary pathologists who have experience working with wild animals perform a complete external and internal examination (necropsy) of the carcass. The external exam involves observing and recording external factors, such as the size and weight of the animal, as well as any evident abnormalities, such as emaciation, skin conditions, or lesions. The internal exam involves examination of all internal organs and any abnormalities in color, consistency, appearance, number, distribution, and size are recorded. During this process, samples are collected and preserved for further testing. Tissue samples collected for histopathology are collected in a preservative (typically 10% buffered neutral formalin), although other fixatives, such as formaldehyde or ethanol, can be used.

In the laboratory, fixed samples are embedded in paraffin, and thin sections are cut, placed on microscope slides, stained, and then examined under the microscope by a pathologist. The pathologist looks for abnormalities at the tissue and cellular levels. These results may indicate the need for specialized techniques, such as stains that increase the visual detectability of bacteria, fungi, internal parasites, or viruses. Pathology can provide a wealth of information regarding what did or did not cause the disease, and these results can be used to focus further laboratory and field investigations. However, there are limitations to histopathology, the primary limitation being condition of the tissues. Decomposition, freezing, or improper fixation damages tissues and cells. The more damage caused by these postmortem factors, the less likely it is that pathologist can accurately interpret the effects of the disease. Therefore, submitting the freshest, yet unfrozen, specimens possible is critical to investigating a wildlife disease event.

Microbiology

The term microbiology often is used as a synonym for bacteriology, but it is much broader in application and encompasses several disciplines, including bacteriology, medical mycology, virology, and clinical parasitology. Because of difficulty identifying some microorganisms, the use of the polymerase chain reaction (PCR) has revolutionized their detection and identification, resulting in the emergence of molecular biology in diagnostic microbiology. Microbiology also commonly uses serological approaches. In all these disciplines, tissues, fluids, and/or swabs from tissues taken during necropsy or from live animals are processed in the laboratory to culture bacteria, fungi, or viruses, or for molecular assays with the goal of identifying the microorganism responsible for the morbidity or mortality event. You should submit the freshest possible specimens to the microbiology laboratory during a wildlife disease investigation. In most cases, samples should be frozen prior to submission, but freezing can negatively affect some tests, so contact the laboratory prior to freezing and submitting samples. Regardless of whether fresh or frozen samples are submitted, inclusion of refrigerated or frozen chemical gel ice packs in the package is important to ensure preservation of the samples during transit.

If the submitted sample does not contain the causative agent or the causative agent has degraded, laboratory culture will be negative, but this result does not necessarily im-

Table 7.3. Selected publications addressing basic laboratory procedures relevant to diagnosing diseases in free-roaming vertebrate populations[a]

Title	Content	Reference
Clinical diagnosis and management by laboratory methods	Laboratory procedures for a variety of disciplines, including serology and clinical pathology	Henry 1979
Manual of clinical microbiology	Detailed laboratory procedures for a variety of disciplines, including virology and bacteriology	Murray 2007
Toxic plants of North America	Discussion of native, exotic, and cultivated toxic plants	Burrows and Tyrl 2001
Veterinary clinical parasitology	Diagnostic parasitology for a wide array of vertebrate species	Zajac and Conboy 2006
Veterinary parasitology reference manual	Diagnostic parasitology for a wide array of vertebrate species, including wild animals	Foreyt 2001

[a] Methods in some of these manuals pertain to domestic species but are also directly applicable to free-roaming vertebrates.

ply the microbe was not present initially. Thus, the major drawback to classical culture methods is they are highly dependent on intact, viable microorganisms. However, with the development of such molecular techniques as PCR, some of these limitations can be overcome, because an intact, viable microorganism is not necessary for these techniques, although the condition of the specimen still can be problematic if it is badly degraded. For these reasons, those interpreting microbiology results must consider field events and known sampling limitations as well as the laboratory results. Further, repeated sampling may be needed during an outbreak to confirm the original agent is still responsible for continued morbidity or mortality or to track the pathogen into other populations. This information allows wildlife biologists and conservation agency administrators to manage the situation adaptively for optimum effect.

BACTERIOLOGY/MYCOLOGY

Typically, tissues submitted for bacteriology and mycology are collected using sterile techniques and placed in individual sterile containers, such as plastic specimen collection containers or Whirl-Pak® bags. Alternatively, a sterile swab can be inserted into the tissues and fluids can be aspirated or swabbed and placed in sterile containers. Swabs or tissues should be refrigerated and sent to the laboratory as soon as possible. If shipping in 24–48 hours is not possible, freezing tissues until shipment is acceptable and is preferable to a decomposed sample. Freezing can destroy certain pathogens, however, so one should contact the laboratory where the samples will be submitted prior to freezing samples. Once received by the laboratory, samples are placed in special media that encourages the growth of bacteria or fungi. The media used is based on the ability to selectively grow certain microorganisms and inhibit the growth of others. Once an organism has been cultured, those thought to be significant are identified by using a battery of biochemical tests, special stains, and/or PCR tests.

As stated earlier, the primary limitation for bacteriology is the condition of the specimens with respect to postmortem decomposition and preservation. Opportunistic microorganisms will rapidly colonize a carcass, and the variety of these contaminant microorganisms, as well as the sheer numbers of them, increases with time. Colonization by contaminant organisms can mask the presence of the original pathogen, kill the pathogen, and hinder interpretation of laboratory results. Using poor sampling techniques, or leaving samples unrefrigerated for several days prior to culture, increases the chances for contamination. Thus, sample collection using sterile techniques and immediate shipment of samples to the testing laboratory is critical.

VIROLOGY

As for bacteriology and mycology, virology samples are collected using sterile techniques and transported to the laboratory as soon as possible. Temporary storage in a refrigerator is acceptable, unless >48 hours will pass before shipping the samples; in that case it is best to freeze the samples, preferably at or below –70° C. In the laboratory, samples typically are homogenized in special media designed to inhibit bacterial growth and are placed in cell cultures. Because viruses utilize cells to survive and are highly dependent on host cell machinery to reproduce, alterations to the cell culture —known as cytopathic effects—can be observed visually. Some laboratories utilize the electron microscope, which allows direct visualization of virus particles; however, large amounts of virus are needed to detect viruses by electron microscopy. Additional molecular and immunological tests are used to identify the virus once its presence has been determined. Specialized assays also can be used, such as direct fluorescent antibody techniques using frozen sections or PCR for direct detection of viral nucleic acid.

As with bacteriology and mycology, virology will provide a positive response only if the sample contains virus particles that can be detected. Inactivated, or "dead," viruses cannot be detected by virus isolation methods, such as cell culture, and the result will be negative even if the virus was initially present. False negative results are even more common in virology than in bacteriology because of the more fragile nature of viruses, their greater susceptibility to handling errors, and because many viruses disappear after causing initial organ damage. Thus, like other disciplines, virology is highly dependent on the quality of the samples submitted. When possible, if a viral agent is suspected, collection of specimens from animals in contact with those that succumbed to disease can assist in the identification of the virus.

PARASITOLOGY

Most macroparasites are visible to the naked eye, although microscopy typically is used for their identification. Parasitic protozoans can only be seen using a microscope. Parasitic helminths and arthropods frequently occur in wildlife and may or may not be the cause of sickness or death. If parasites are the causative agent of wildlife morbidity or mortality, they typically will occur in large numbers and will be associated with tissue changes. Knowledge of "normal" parasite intensity is important to interpreting parasitological results. For example, the mere presence of a parasitic helminth in the lungs or small intestine does not necessarily mean the organ was significantly compromised.

As with other diagnostic disciplines, positive results are conclusive, yet negative results do not necessarily mean the animal was not parasitized. False negative results are less common in parasitology than in other diagnostic disciplines, and they are considerably less likely to occur for metazoan (multicellular) parasites. Most macroparasites, such as round worms, tapeworms, fleas, lice, and ticks, are best preserved in 70% ethanol. However, as always, we recommend contacting the laboratory prior to specimen preservation and shipment.

SEROLOGY

When a microparasite infects an animal, the host typically develops an antibody response for that microorganism. The ability to generate an antibody, and the level of antibody production, is host–agent specific. Serology is the technique used to detect and measure levels of specific antibodies in serum.

Serologic assays detect exposure to a specific agent, but they do not detect the presence of disease. For this reason, results of serologic surveys reflect historical exposure to the infectious agent and provide no evidence of disease or its effects on the population. In an individual animal, however, a 4-fold rise in titer of specific antibodies using paired serum samples (one sample taken at initial onset and another one 2–3 weeks later) is considered a reasonable indication the agent of interest is present in the animal and is replicating. The timing of the rise in antibody titer relative to disease and the specificity of the test for antibodies—particularly the type of antibody detected (e.g., immunoglobulin G vs. immunoglobulin M)—are critical to making inferences regarding infectious agents and disease. Whether a specific infectious agent can be implicated as the cause of mortality or morbidity also depends on whether the field investigation and pathology results corroborate the effects that agent would be expected to produce in the host.

Various assays exist to detect antibodies. Some can be used in the field, whereas others are more complicated and must be conducted in specialized laboratories (Table 7.3). Many of these assays were developed for use in domestic livestock or small animals (cats and dogs), but they also are used in wild species, where they have not yet been validated. Despite this drawback, many of these assays provide valuable information regarding exposure to a wide variety of infectious agents. Biologists should interpret serological results with caution, however, and couple their results with additional testing, such as agent identification, when possible.

Clinical Chemistry and Hematology

Clinical chemistry and hematology can assist disease investigators in assessing which organ systems the disease-causing agent is likely to have affected, but they rarely provide a specific diagnosis. Clinical chemistry entails the analysis of serum or plasma for proteins, enzymes, metabolites, and minerals. Damage to, or improper function of, internal organs (e.g., the liver, kidney, pancreas, or muscle) leads to predictable changes in the levels of specific proteins, enzymes, and minerals in the blood. These changes may be an increase or decrease in a particular serum constituent, depending on which organ is affected. However, interpretation is difficult in many wild animals, because normal values for many less commonly studied species are unknown. Veterinary clinicians routinely use clinical chemistry to obtain an idea of how internal organs are functioning and which systems may be involved in the disease process in living animals, but this approach is not useful for dead animals because of changes associated with decomposition.

Hematology entails the analysis of whole blood for quantification and morphological evaluation of cellular components and is only useful in living animals. Blood smears are collected, dried, stained, and examined under a microscope to assess the different types of blood cells present, their relative abundances, and whether blood parasites are observed. Automated hematology analyzers are available for domestic animals; however, normal values for many wild species are unknown. Many veterinary diagnostic laboratories evaluate samples from domestic animals and provide normal values for these species, but they do not have similar data for most wild species. Thus, you should interpret these results with caution and consult appropriate wildlife health specialists as needed (Box 7.1). Alternatively, wildlife biologists can send samples to laboratories with expertise on wild species.

Toxicology

Laboratory assays for chemicals are complex and require expensive equipment. Some chemicals are more resistant to decomposition than others, and some are more likely to be detected in one type of tissue or organ than in others. There also may be additional requirements to procure environmental samples (e.g., water, soil, and food) to trace the source of the chemical in the environment. Because of the expense of testing, analyses for toxins are rarely completed before a thorough field investigation and necropsy suggests that a particular toxin may be involved. There are specific requirements for collecting, handling, packaging, and shipping tissues for toxicological analyses, so biologists should contact the laboratory prior to collecting, processing, or shipping such samples.

Field Procedures

Necropsy and Specimen Collection

As detailed above, wildlife biologists typically begin wildlife disease investigations as they work in the field, and their observations are critical to this process (see the section Field Observations, above). For investigations involving dead animals, necropsies are necessary (Box 7.3). The best approach is to submit fresh, whole carcasses to a diagnostic laboratory employing personnel with expertise and experience working with the species involved. For example, if the disease outbreak involves pinnipeds, it is preferable to submit the specimens to a laboratory with experience working with these marine mammals. Copies of field observations should accompany the specimens. In this scenario, a veterinary pathologist working in a controlled environment will ensure that appropriate samples are collected, histopathology is completed, and suitable samples are submitted for microbiological workup or other laboratory procedures.

> **BOX 7.3. NECROPSY STRATEGY AND CAUTIONS**
>
> 1. Send fresh carcasses to a veterinary diagnostic laboratory with expertise in wildlife diseases if possible (see Box 7.2).
> 2. If step 1 is not feasible, field necropsies should be conducted by a veterinary pathologist or wildlife veterinarian if possible.
> 3. Wildlife biologists should not conduct field necropsies without proper training.
> 4. While performing a necropsy, use caution, as there are numerous zoonotic (infect humans) diseases
> 5. Wear appropriate personal protective equipment when performing a necropsy (e.g., gloves, mask, boots, and coveralls).
> 6. Remember to decontaminate surfaces and equipment and wash your hands thoroughly when finished collecting samples or after handling any specimens, including carcasses.

There are situations, however, when submitting an entire animal is not feasible and field necropsy is required. For example, although it is not difficult to ship a few ducks to a laboratory, it often is impracticable to move a large ungulate or marine mammal from a remote area to the distant laboratory. If possible, a veterinary pathologist or a wildlife veterinarian should conduct field necropsies (Box 7.3). These individuals are well trained and experienced in differentiating pathological from postmortem changes (these differences are not always obvious), and identifying even small lesions or abnormalities requires specific training and experience. They also know what samples should be collected under an array of circumstances and how to do so safely. Because of the risk certain zoonotic diseases pose for those conducting necropsies, such as the pneumonic form of bubonic plague, safety is critically important.

There are times when wildlife biologists must conduct field necropsies or collect samples from potentially infected animals. For several reasons, not the least of which is safety, wildlife biologists should not conduct necropsies or collect disease related samples without appropriate training (Box 7.3). Wildlife conservation agencies that expect their biologists to conduct necropsies or collect samples for disease surveillance (e.g., samples for CWD testing) should provide the required training. In the United States, some state conservation agencies have their own wildlife disease programs and provide training for field biologists if needed. Additionally, some entities listed in Box 7.1 also provide necropsy and sample collection training. For example, the Field Investigation Team of the U.S. Geological Survey National Wildlife Health Center provides training for groups of wildlife biologists employed by federal and state conservation agencies and Indian tribes. Similarly, the Southeast Cooperative Wildlife Disease Study provides training for groups of biologists employed by their member-state wildlife conservation agencies and sometimes biologists employed by other state or federal agencies working with wildlife. Typically, these training sessions last about 2 days. They emphasize classroom lectures and demonstrations outlining necropsy techniques; personal protective equipment; safety; and specimen collection, handling, and submission. They also present summaries of specific diseases relevant to the geographic region and interests of the sponsoring agency. Normally, this training reserves about half a day for participants to conduct necropsies and collect samples under the supervision of wildlife disease specialists.

We have avoided the temptation to provide a simplistic summary of an ideal necropsy here; because of the danger posed by zoonotic diseases, there is no room for a paint-by-numbers approach to necropsy. For those who have appropriate training, we provide a list of excellent field manuals addressing necropsy techniques and sampling procedures for a variety of wild species (Table 7.4). These publications can be used to help refresh your knowledge prior to conducting a necropsy or other disease-related field procedures and to provide insight regarding how to approach necropsies for less-familiar species, such as sea turtles or marine mammals (e.g., Work 2000b, Pugliares et al. 2007). These field manuals are available at no cost online.

Wildlife biologists working in the field are increasingly asked to help collect samples from apparently healthy animals as part of disease surveillance programs. This task may be as simple as collecting fresh fecal material from the environment to more involved procedures, such as capturing animals and collecting blood samples or swabs. For example, it now is common to collect samples for disease testing when animals are captured for translocation to prevent releasing animals carrying certain infectious agents in sites where these diseases do not yet occur. Similarly, wildlife biologists may be asked to collect opportunistic samples for surveillance programs, such as cloacal swabs from freshly killed waterfowl at hunter check stations as part of avian influenza surveillance or samples from hunter-killed deer for CWD testing. Proper training regarding safety, sample col-

Table 7.4. Selected publications addressing wildlife health field procedures (e.g., necropsy) and the investigation, surveillance, and/or management of diseases in free-roaming vertebrate populations

Publication type/Title	Content	Reference
Field manual[a]		
Necropsy of wild animals[b]	Primer on necropsy procedures, safety, sampling, and sample processing for wild carnivores, ungulates, birds, and reptiles	Munson n.d.
Wildlife disease investigation manual	Primer on investigating wildlife diseases; includes wildlife disease contacts in Canada, human safety, and field procedures, such as necropsy, specimen collection, preparation, shipment, and carcass disposal	Canadian Cooperative Wildlife Health Centre 2007
Avian necropsy manual for biologists in remote refuges	Primer on necropsy procedures, safety, sampling, and sample processing for birds	Work 2000*a*
Field manual of wildlife diseases: general field procedures and diseases of birds	Includes primers on field procedures, including avian necropsy, specimen collection, preparation, and shipment as well as wildlife disease management	Friend and Franson 1999
Sea turtle necropsy manual for biologists in remote refuges	Primer on necropsy procedures, safety, sampling, and sample processing for sea turtles	Work 2000*b*
Wild birds and avian influenza: an introduction to applied field research and disease sampling techniques	Field manual sponsored by the United Nations; provides primer on avian influenza and addresses avian capture, handling, sample collection, avian surveys, and monitoring relating to avian influenza in wild birds; most procedures apply equally well to other infectious diseases	Whitworth et al. 2007
Marine mammal necropsy: an introductory guide for stranding responders and field biologists	Detailed and highly informative field guide addressing necropsy in pinnipeds, small cetaceans, and large whales as well as equipment required, safety, sample management, and record keeping	Pugliares et al. 2007
Detailed review		
Conservation medicine: ecological health in practice	Attempts to bridge disciplines that traditionally have addressed wild animal health, human health, ecosystem functionality, and biodiversity conservation; addresses disease monitoring and management	Aguirre et al. 2002
Essentials of disease in wild animals	Introductory textbook addressing disease processes in wild animals, the study of wildlife diseases, and disease management	Wobeser 2005
Disease in wild animals: investigation and management	Textbook offering in-depth treatment of techniques for investigating and managing disease in free-roaming animals	Wobeser 2007
Management of disease in wild mammals	Discusses ecologically grounded approaches to disease management in wild mammal populations; includes topics such of host–parasite interactions, surveillance, epidemiology, planning, modeling, cost-benefit analysis, and biodiversity conservation	Delahay et al. 2009

[a] These field manuals are available electronically at no cost as PDFs; URLs are included in Literature Cited.
[b] Also available in Portuguese, French, and Spanish at no cost from the Wildlife Conservation Society (http://www.wcs.org/resources.aspx).

lection, and sample handling still is required and typically is provided by wildlife disease specialists in the field at the beginning of these projects.

Specimen Selection, Collection, and Handling

We provide this section as a reminder and resource for wildlife biologists who already have received proper training in necropsy and specimen collection techniques and safety procedures. It is important to remember that sample collection and handling is one of the most critical steps in diagnosing disease (Box 7.4). Improper handling of specimens jeopardizes the laboratory testing process and can prevent the correct diagnosis. For example, in the case of histopathology, freezing the tissues causes artifacts, thereby hindering the evaluation of characteristic changes in the tissues that could lead to discovery of the cause of the disease or preventing more targeted testing to narrow down the diagnosis. The collection of specimens from contact animals is highly recommended when possible. Sometimes the etiological agent is no longer detectable in a carcass or severely ill animal, but it may be detectable early in the disease process in a contact animal.

When field observations and necropsy findings suggest that specific categories of etiological agents may be involved, targeted collection of specimens may be warranted. When the field investigations suggest a particular cause of the event, specific specimens should be included in samples collected (Table 7.5). However, when possible, it is best to submit as many different samples as possible, because the cause of the disease event could be more than one agent or could be something unexpected. Regardless, Table 7.5 provides specifics regarding how tissues should be selected, processed, and preserved prior to shipment to the diagnostic laboratory.

> **BOX 7.4. SPECIMEN COLLECTION AND HANDLING POINTERS**
>
> 1. Collect freshest specimens possible.
> 2. Use sharp knife or scalpel for tissue pieces, swab for surfaces and exudates, and sterile needle and syringe for fluids.
> 3. Collect all samples in separate sterile containers except for histopathology.
> 4. Prevent cross-contamination.
> 5. Use sterile sample plastic bags or tubes. Nonadditive blood tubes (red top) also can be used.
> 6. In general, freeze specimens unless they will arrive at the laboratory within 48 hours EXCEPT for histopathology and parasitology samples.
> 7. Allow blood to clot. Centrifuge and remove serum. Freeze serum.
> 8. Label all samples from one specimen with the same unique number. Use indelible ink and labels that stay attached in the freezer.

Specimen Shipment

With increased concerns over bioterrorism and biosafety, there has been an increased emphasis on security for air and ground transportation of all biological specimens. Thus, it is essential that field personnel and receiving laboratories be familiar with U.S. Department of Transportation (DOT) and public health regulations involving shipment of biological samples by ground or air. The Centers for Disease Control (CDC) and DOT regulations generally permit movement of "diagnostic specimens," although certain restrictions on packaging and "select agents" may apply to the shipment. The CDC list of select agents includes pathogens considered to be high risks for bioterrorism or as foreign animal diseases. The Code of Federal Regulations (CFR) is a useful resource that covers DOT shipping requirements for biological material (49 CFR 173; http://www.access.gpo.gov/nara/cfr/waisidx_09/49cfr173_09.html). It includes diagnostic specimens (UN3373), as well as select agent interstate movement and public health issues (42 CFR 73, http://www.access.gpo.gov/nara/cfr/waisidx_09/42cfrv1_09.html). Related information is available on the websites listed in Box 7.5.

Proper packaging of specimens for shipment is critical. If the specimens are improperly packaged, the package may be delayed during shipment. Additionally, samples that need to remain separate may leak and be exposed to one another, thereby resulting in cross-contamination of the specimens. Consult the resources listed in Box 7.5 for more information regarding specimen packaging and shipment. The receiving laboratory also can provide packaging and shipping information, including their preferences. Shipping companies have the final word on the acceptability of a particular shipment; however, knowledge of federal and state transportation regulations can assist in gaining the acceptance of the package and expediting the shipping process. Overnight shipment is the only practical method for perishable materials, even though cold packs are included. Biologists also should consider the time of shipment to ensure the package arrives at the testing laboratory on a workday. With some specimens, collection permits need to be included with the package. In other cases, close coordination with appropriate law enforcement authorities may be necessary to avoid delays, particularly with international shipments. If the case is a legal matter, chain of custody forms also will need to accompany the specimens.

> **BOX 7.5. INFORMATION RELATING TO STATUTES AND REGULATIONS GOVERNING PACKAGING AND SHIPPING OF BIOLOGICAL AND DIAGNOSTIC SPECIMENS**
>
> Official certified training may be required.
>
> **Canada, United States, and Elsewhere**
> International Air Transportation Association
> www.iata.org
>
> **United States**
> Department of Transportation
> www.phmsa.dot.gov/hazmat
>
> Centers for Disease Control
> www.cdc.gov

Table 7.5. Sample selection and preservation from field necropsy when entire carcass cannot be submitted and field observations and necropsy findings suggest specific causes may be involved

Sample	Projected laboratory analysis	Method of preservation	Comment
When microbial infections are expected			
Observed lesions	Microbiology	Frozen	Lesions (abnormal-appearing tissue): a portion of each should be saved frozen and fixed[a]
Heart	Bacteriology	Frozen	Entire heart from birds and small mammals; selected portions from larger animals
Liver	Bacteriology	Frozen	Entire lobe from birds and small mammals; several pieces ≤2 cm^2 or larger in larger animals
Blood or serum	Bacteriology/virology	Frozen	Serum also useful for serology
Spleen	Bacteriology/virology	Frozen	Entire spleen from birds and small mammals; selected portions from larger mammals; fix[a] the remainder
Intestine	Bacteriology/virology	Frozen	Segments from middle or distal (ileum) of small intestine
Brain	Bacteriology/virology	Frozen	If animal exhibited abnormal behavior, save entire head and submit intact to laboratory for removal of brain by laboratory personnel
When toxicants are suspected			
Observed lesions	As appropriate	Frozen	Lesions (abnormal-appearing tissue): a portion of each lesion should be saved frozen; fixed[a] tissue important
Liver	Heavy metals (Pb, Tl)	Frozen	Entire liver from birds and small mammals; selected portions from larger mammals; fixed[a] tissue important
Kidney	Heavy metals (Pb, Hg, Tl, Fe, Cd, Cr)	Frozen	Entire kidneys from birds and small mammals; selected portions from larger mammals; fixed[a] tissue important
Stomach contents	Organophosphates, carbamates, plant poisons, strychnine, cyanide, mycotoxins	Frozen	Save entire contents; samples to be checked for cyanide or H_2S must be placed in airtight containers to prevent loss of these toxic gasses into the air
Brain	Brain cholinesterase, organochloride residues, organomercuric compounds	Frozen	If brain is removed for chemical analysis, it must be wrapped in clean aluminum foil then placed inside a chemically clean glass bottle; fixed[a] tissue important
Lungs	H_2S, cyanide	Frozen	Samples to be checked for cyanide or H_2S must be placed in airtight containers to prevent loss of these toxic gasses into the air
Blood	Lead, cyanide, H_2S, nitrites	Frozen	Samples to be checked for cyanide or H_2S must be placed in airtight containers to prevent loss of these toxic gasses into the air
For microscopic study			
Observed lesions	Histopathology[b]	10% formalin[a]	Lesions (abnormal-appearing tissue): a portion of each lesion should be saved frozen
Liver	Histopathology	10% formalin[a]	Specimen portions should be ≤6 mm in thickness
Kidney	Histopathology	10% formalin[a]	Specimen portions should be ≤6 mm in thickness
Gonad	Histopathology	10% formalin or Bouin's stain[a]	Specimen portions should be ≤6 mm in thickness
Intestinal tract	Histopathology	10% formalin or Bouin's stain[a]	Snippet of stomach at the ileocecal junction, piece of duodenum (near pancreas), and colon
Brain, nervous tissue, eye	Histopathology	10% formalin[a]	Divide brain in half (sagittal); place half in formalin; freeze the other half
Impression, smear	Many laboratory tests	Air-dry on slide	Touching glass microscope slide to cut surface of any organ
Heart, lung, skeletal muscle, lymph node, spleen, thymus	Histopathology	10% formalin[a]	Specimen portions should be ≤6 mm in thickness

After Roffe and Work (2005).

[a] Fixation refers to placing a small sample of tissue in a much larger volume of 10% formalin or similar fixation compound.

[b] Once the fixed samples arrive at the laboratory, they will be sectioned and stained for microscopic study by a veterinary pathologist.

MANAGING WILDLIFE DISEASES

It is likely that wildlife regulatory agency biologists, researchers, and administrators will increasingly participate in multidisciplinary teams conducting wildlife disease research, surveillance, and management. The reasons for this conclusion are twofold. First, wildlife diseases can be important to biodiversity conservation as well as to livestock and human health (Daszak et al. 2000). After all, 60% of 335 emerging infectious disease events occurring since 1940 were zoonoses, and of these, 72% originated in wildlife (Jones et al. 2008). Clearly, the health of wild vertebrates is tightly linked to that of humans and domestic animals. Second, wildlife scientists have knowledge about the life histories of wild vertebrates and skills in wildlife trapping, handling, marking, and monitoring as well as habitat management and density manipulation required for wildlife disease management. For these reasons, regardless of why wildlife disease surveillance and management are implemented, wildlife scientists are bound to be part of the team. Wildlife ecologists, managers, and administrators can use the materials outlined and referenced in this chapter to help them integrate rapidly into effective multidisciplinary teams organized to conduct disease surveillance or management in wild vertebrate populations.

Management strategies for wildlife diseases involve the nexus between management objectives and the ecological relationship among hosts, disease causing agents, and habitat or environment represented in Figure 7.1. Classes of management objectives include (1) preventing an infectious agent from becoming established in an area or population where is does not already occur, (2) controlling an existing disease to tolerable levels, and (3) eradicating an existing disease from a region or larger geographic area. Management manipulation can target the host, the parasite or infectious agent, and/or the habitat or environment. With wildlife, prevention is the preferable alternative, as eradication is extremely unlikely and control often is difficult.

It is beyond the scope of this chapter to provide a detailed treatment of wildlife disease management. Instead, we present 3 case studies that illustrate the sort of ecological interactions discussed in the above section on Disease Processes, Ecology, and Epidemiology and the array of management principles and approaches used to address diseases of free-roaming wild vertebrates. Our objective is to illustrate why certain management approaches are more likely to be effective than others for addressing specific objectives and the particular ecological relationships among parasites and hosts, given the habitats where these interactions play out. Those interested in more comprehensive treatments of wildlife disease management will find the publications on this topic listed in Table 7.4 useful.

Case 1: Necrotic Stomatitis in Elk

Olaus J. Murie, perhaps the foremost elk (*Cervus canadensis*) biologist of the mid-20th century, considered necrotic stomatitis "to be by far the most important elk disease" (Murie 1951:177). Most elk biologists we have spoken with have never heard of necrotic stomatitis or view the disease as an interesting historical anecdote. For these reasons, necrotic stomatitis in elk and its successful control offer an informative window on the interdependent relationships illustrated in Figure 7.1 and on disease management strategies.

When elk densities approach or exceed K-carrying capacity for winter range, elk are forced to browse on coarse twigs and branches, resulting in sharp splinters being embedded between the teeth and gums or elsewhere in the soft tissues of the oral cavity or throat (Murie 1930, 1951:177–188). These wounds then become infected with *Fusobacterium necrophorum* (=*Actinomyces necrophorus*, *Spherophorus necrophorus*, *Bacillus necrophorus*), which produces toxins that kill host tissue, leading to bony necrosis, tooth loss, and reactive proliferation (Murie 1930, 1951; Allred et al. 1944; Leighton 2001). Septicemia also can occur, causing liver and lung necrosis and death. Elk with necrotic stomatitis often exhibit excessive salivation, difficulty manipulating, chewing, and swallowing food (thus the characteristic cheek distended with a bolus of forage), wasting, emaciation, and sometimes death.

Necrotic stomatitis is one of an array of disease syndromes, collectively termed necrobacillosis, caused by *F. necrophorum* (Leighton 2001). This ubiquitous microorganism is part of the normal intestinal and fecal flora of a wide range of mammalian hosts and is an important opportunistic pathogen of humans, domestic animals, and wildlife worldwide. Most commonly, necrobacillosis begins with damage to the epithelium of the feet or mouth that becomes infected with *F. necrophorum* (e.g., foot rot, foot abscesses, calf diphtheria, and necrotic stomatitis). The infection often progresses to the bone, where it causes lysis and reactive proliferation, to the rumen and other areas of the intestinal tract, and then to the liver, lungs, and other internal organs via the bloodstream. Septicemic necrobacillosis typically results in death.

Numerous deaths caused by necrotic stomatitis were documented during the first half of the 20th century in Rocky Mountain elk (*C. c. nelsoni*) on inadequate winter range in Jackson Hole, Wyoming (Murie 1930, 1951:177–188), Yellowstone National Park, and elsewhere in the Greater Yellowstone Area (GYA; Rush 1932, Mills 1936). Deaths due to necrotic stomatitis also were observed in Roosevelt elk (*C. c. roosevelti*) with insufficient winter range in the Olympic Mountains of Washington during this period (Schwarts 1943, Schwarts and Mitchell 1945). Interestingly, providing feed for elk during the winter at the National Elk Refuge near Jackson, Wyoming, did not initially eliminate this disease. During some years, the grass hay being fed contained large numbers of sharp foxtail barley (*Hordeum jubatum*) and, to a lesser degree, cheatgrass (*Bromus tectorum*) seeds and awns (Murie 1930, 1951:177–188; Allred et al. 1944). These seeds and awns became wedged between the teeth and gums or penetrated soft tissues in the mouth and throat, providing a nidus for *F. necrophorum* infection and thus necrotic stomatitis.

Because *F. necrophorum* is part of the normal gut and fecal flora of mammals and occurs worldwide, there was little point in attempting to limit where this ubiquitous microorganism occurred. Instead, the obvious places to focus management efforts were elk winter habitat and elk density. Elk will not eat large volumes of twigs and branches if more suitable winter forage is available, so Murie and other elk biologists of that era working in the GYA and elsewhere used hunters and other methods to reduce elk densities in many herds so they could be supported by winter range during most years. In northwestern Wyoming, managers decided to maintain elk densities above what winter range could support by feeding elk during most winters. To avoid necrotic stomatitis associated with winter feeding, the U.S. Fish and Wildlife Service and Wyoming Game and Fish Department ensured that only high quality hay was fed. Managers also could have made more native forage available for elk during winter by such methods as limiting domestic livestock grazing during the summer on public lands that were important winter ranges for elk. Although the management strategies utilized certainly did not eradicate necrobacillosis, they did control necrotic stomatitis in elk to the point where this disease was no longer considered a problem, let alone the "most important elk disease" (Murie 1951:177), as it previously had been.

Case 2: Rabies in Wild Carnivores

Unlike the bacterium *F. necrophorum*, the rhabdovirus responsible for rabies is far from a commensal organism. Rabies is essentially a disease of mammals, primarily affecting carnivores and bats, and it is most often transmitted through the bite of an infected animal (Rupprecht et al. 2001, Lyles and Rupprecht 2007). The incubation period for rabies can vary from less than a week to several years, but typically is 1–2 months. The length of the incubation period may depend on several factors, such as the proximity of the bite site to the central nervous system, severity of the bite, type and amount of virus introduced into the host, host age, and the immune status of the host (Rupprecht et al. 2001, Lyles and Rupprecht 2007). Once disease develops, rabies is fatal, with death occurring in a few days. Due to the long incubation period, wild carnivores can move about as they normally would for months. This long latency allows them to move from the site where they were exposed before they begin shedding the virus, so they can effectively carry the virus to new locations. Additionally, infected animals often shed the virus several days before becoming clinically ill, which is the rationale for quarantining apparently healthy pet dogs that bite people—these dogs will develop clinical rabies within a few days if they were shedding the virus when they bit the person. Although a broad array of wild mammals contract the rabies virus, wild carnivores as well as vampire bats are more likely to bite other mammals and thus transmit the virus.

Rabies typically occurs at low rates in populations of wild canids, until the density of susceptible hosts exceeds the threshold required for rapid virus transmission and thus triggers a rabies epidemic. For example, Anderson et al. (1981b) found, during a temporally and spatially extensive study of rabies in European red foxes (*Vulpes vulpes*), that rabies epidemics had occurred every 3–5 years since 1939, leading to corresponding cycles in fox numbers. The rabies virus led to time-delayed, density-dependent regulation of fox abundance, with the time of lag determined by how long the fox density was below the threshold level needed for an epidemic. Interestingly, Anderson et al. (1981b) found that cyclic fluctuations in fox abundance were absent in areas with low fox densities. Rabies in wild animals becomes a problem for human society during such epidemics, because the virus is much more likely to spill over into domestic animal and human populations during these periods.

For such diseases as rabies that require the density of susceptible hosts to exceed a threshold before an epidemic will occur, management can profitably be focused on ensuring this threshold is not crossed (Bacon 1985, Rupprecht et al. 2001). For example, in the United States, Wildlife Services could use lethal methods to control coyote (*Canis latrans*), gray fox (*Urocyon cinereoargenteus*), red fox, striped skunk, and/or raccoon densities. Public opposition to the mass slaughter of these charismatic species, however, almost certainly would prevent implementing such an approach over broad areas. Moreover, when these population control measures were attempted, they proved ineffective (Rupprecht et al. 2001).

Instead, public health officials attempting to eliminate or reduce the severity of rabies epidemics in wild carnivores chose a different, more publicly acceptable alternative for reducing the density of susceptible hosts—vaccination of key wildlife species (Cross et al. 2007). Rabies immunization programs have been required by law for domestic dogs and sometimes cats for many decades in North American states and provinces and have proven effective in controlling rabies in these species. The primary objective of these regulations was to protect humans from exposure to the rabies virus. Because pet animals are more likely to contract rabies from wild animals than are humans—and thus expose people to the rabies virus—highly efficacious vaccines were developed for these species. Following the success of vaccination to control rabies in domestic animal populations, researchers developed vaccines that were efficacious when administered orally to key wild species. Simultaneously, others developed attractive baits that could be distributed easily from the air and, when consumed by a target animal, deliver the vaccine (Jojola et al. 2007, Cliquet et al. 2008). Today, public health officials can effectively reduce the density of susceptible wild-animal hosts by immunizing them with highly effective oral rabies vaccines delivered in attractive baits. The vaccine does not need to be delivered to every coyote or fox—just enough of them to prevent the density of susceptible hosts from exceeding the threshold required for an epidemic to occur. This approach now is widely employed to prevent rabies epidemics in foxes and coyotes in North America and elsewhere, whereas oral vaccination for rabies in skunks and raccoons has remained a challenge (Slate et al. 2009). The successes of oral rabies vaccination in wild species not only protects humans and their domestic animals from rabies (the primary objective), but also protects the wild carnivores from this disease, which many people would perceive as a favorable outcome.

Case 3: Brucellosis in Elk

Bovine brucellosis is a disease of the reproductive tract of cattle caused by the bacterium *Brucella abortus*. Cattle, bison (*Bos bison*), and African buffalo (*Syncerus caffer*) can sustain endemic *B. abortus* infections without human intervention (Davis et al. 1990, Madsen and Anderson 1995, Godfroid 2002). The most important effects of bovine brucellosis in both bison and elk include abortion during the last half of gestation and birth of nonviable calves (see reviews by Thorne et al. 1997, Williams et al. 1997, Cheville et al. 1998, Thorne 2001). The primary mode of *B. abortus* transmission in these ruminants is through (1) licking infected fetuses, calves, placentas, or associated vaginal discharges; (2) consuming feed contaminated by these items, (3) consuming contaminated placentas, and (4) licking the genitalia of infected females soon after an abortion or live birth. A wide array of wild ungulates, carnivores, rodents, and lagomorphs; humans; and other mammals, in addition to cattle and other domestic livestock, are susceptible to *B. abortus* (see Davis [1990] for review by host species). It is likely that such mammals as carnivores and rodents acquire bovine brucellosis by consuming *B. abortus* infected tissues, and the disease would be self-limiting in these species if it were eliminated in ruminants. Similarly, Rhyan (2000) maintained that when brucellosis was prevalent in cattle, occasional seropositive wild ungulates, particularly cervids (e.g., elk), were found due to spillover of the infection from cattle. Bison can serve as sources of spillover infection as well. However, elk concentrated in game farms or on winter feed grounds also can maintain the infection without access to either infected cattle or bison (Thorne et al. 1997, Rhyan 2000, Godfroid 2002).

Although bovine brucellosis formerly occurred worldwide in cattle, eradication programs targeting domestic livestock greatly reduced its distribution (Corbel 1997). The U.S. Department of Agriculture was mandated to coordinate the eradication of bovine brucellosis in 1934 (Ragan 2002). Since then, tremendous strides toward this goal have been made. In late 2000, no known *B. abortus*–infected cattle herds were present in the United States, and 48 states were classified as brucellosis free. Thus, reservoirs of *B. abortus*, in the form of infected GYA bison and elk, are considered a clear and present threat to this eradication program. These reservoirs add to the concerns of livestock owners and regulators in the region, as brucellosis vaccine protects only about 65% of cows from abortion.

Brucellosis in elk is a problem of winter feed grounds (Smith 2001, Thorne 2001). Even where free-roaming elk share habitat with infected bison and occupy ranges that overlap to some degree with those of feed-ground elk (Toman et al. 1997), mean seroprevalence is far less than for elk using feed grounds (1.5% of 11,609 free-roaming elk [1931–1998] versus 28.6% of 4,906 feed-ground elk [1930–2000]; Peterson 2003). Where elk are not fed during the winter, do not share range with infected bison or cattle, and do not occupy ranges that overlap with feed-ground elk, mean seroprevalence is essentially zero (0.02% of 5,828, 1946–1999). Clearly, free-roaming elk are unable to maintain brucellosis in the absence of feed grounds. Yet "a single [*B. abortus*-induced] abortion on a crowded elk feedground assures exposure of many elk sharing the feedground" (Thorne 2001: 377), due to the disruption of normal calving behavior caused by high elk densities and resultant abortions occurring on feedlines (Thorne et al. 1997). Because currently available brucellosis vaccines are much less effective at preventing *B. abortus*–induced abortions in elk than in cattle, and vaccine delivery is considerably less certain in elk than in livestock, the prevalence of *B. abortus* in feed-ground elk will not be markedly influenced by vaccination. There is no silver bullet. For all these reasons, if brucellosis management were the top priority of GYA elk management, reducing elk densities to what winter ranges could support while discontinuing winter feeding would lead to an order of magnitude decrease in seroprevalence over time. Further, if brucellosis were eradicated in GYA bison and winter feeding of elk were discontinued, brucellosis would essentially disappear from the region. This approach was used successfully in Elk Island National Park in Alberta, Canada (Tessaro 1986).

Although *B. abortus*–induced abortion probably reduces the annual calf crop of elk herds utilizing feed grounds, such as the National Elk Refuge (Oldemeyer et al. 1993), the increased nutritional plane resulting from winter feeding probably more than compensates for these losses compared to unfed populations, particularly during harsh winters. If it were not for the risk of *B. abortus* transmission to cattle, however small that risk may be, those managing the GYA elk herds probably would be no more concerned about brucellosis than they now are about several other diseases occurring in these herds, such as septicemic pasteurellosis and scabies. Realistically, brucellosis in GYA elk is primarily a sociopolitical problem. The disease is of concern largely because of the risk of transmission to livestock, and the U.S. Department of Agriculture and livestock growers already have expended tremendous energy and resources to eradicate this disease. Simultaneously, there continues to be considerable public pressure to maintain winter feeding of elk on numerous feed grounds in northwestern Wyoming and 2 feed grounds just across the border in Idaho. Most elk hunters, other wildlife enthusiasts, and tourists enjoy the high elk densities provided by winter feeding, so local chambers of commerce and politicians tend to favor continuing this practice. The human–human conflict (Peterson et al. 2010) regarding winter feeding of elk in the GYA and brucellosis has raged for decades. It will not be easily resolved. After all, there currently is no method for greatly reducing the threat of elk as reservoir hosts for *B. abortus* in cattle without reducing elk densities to what winter ranges can support during most years and discontinuing winter feeding, yet many powerful individuals and groups vehemently oppose this strategy.

Précis

Our objective for choosing these 3 case studies of wildlife disease management was to illustrate a few key issues that apply to many diseases of wildlife. The first case study, necrotic stomatitis in elk, exemplifies a disease of concern primarily because of its impact on wildlife. Successful management involved ensuring that elk densities did not exceed what winter ranges could adequately support during most years or feeding high quality hay during the winter. In light of this success, it should come as no surprise that many influential wildlife scientists began to assume that infectious agents of free-roaming wildlife were primarily extensions of poor habitat conditions (e.g., Trippensee 1948:369–384, Lack 1954:161–169, Taylor 1956:581–583).

In contrast, it is difficult to blame rabies, the topic of the second case study, on poor habitat conditions. Although transmission of the rabies virus is dependent on host density, this does not imply that hosts are necessarily anywhere near *K*-carrying capacity during a rabies epidemic. Rabies management in wild carnivores illustrates a case where wildlife disease management was implemented primarily to protect humans from a serious zoonotic disease. The objective was to reduce the density of susceptible hosts below the threshold required for an epidemic rather than to eradicate the disease. Oral rabies vaccine distributed by attractive baits certainly can reach this objective for some carnivore species.

The final case study, brucellosis in elk, demonstrates that sociopolitical controversy can render management of wildlife diseases difficult, even though the technical solution is clear. For example, reducing elk densities in the GYA to what could be supported by winter range during most years and discontinuing winterfeeding would greatly reduce the prevalence of *B. abortus* in elk over time and effectively manage the disease. However, many people do not think that managing brucellosis in elk should take precedence over other objectives for these populations. Specifically, several advocacy groups with considerable political clout maintain that, because herd reduction and eliminating winterfeeding would substantially reduce elk numbers, this approach would be detrimental to the elk hunting industry, wildlife viewing, tourism in general, the local economy, and the lifestyles of many people living in northwestern Wyoming.

We could have chosen any number of other diseases as examples, but space is limited. Those primarily interested in wild birds will find the brief summaries of disease management provided by Friend and Franson (1999) informative. Readers requiring more detailed treatments of wildlife disease surveillance and management should utilize the reviews listed in Table 7.4 that address these topics. Those wanting a primer on wildlife disease management will find the chapter on this topic in Wobeser (2005) useful; readers requiring a more detailed treatment of this topic will find Wobeser (2007) and Delahay et al. (2009) informative.

SUMMARY

Wildlife health and disease are becoming increasingly important aspects of wildlife conservation and management. Reasons for this trend include the threat certain diseases pose for species at risk of extinction, the fact the majority of emerging zoonotic diseases that threaten human health originate in wild animals, and the difficulties such diseases as CWD pose for wildlife conservation agencies. Although most North American universities teaching wildlife science do not address wildlife diseases in the systematic way they do vertebrate taxonomy, population ecology, or habitat management, wildlife scientists have key expertise required by multidisciplinary teams conducting wildlife disease research, surveillance, and management. For example, wildlife biologists possess in-depth knowledge of vertebrate life history; habitat management techniques; wildlife density manipulations; and animal trapping, handling, and marking methodologies, to name just a few key topics.

To become effective members of multidisciplinary teams working with wildlife diseases, however, wildlife scientists and administrators must become familiar with the basic aspects of disease processes and ecology as well as details of the specific diseases they are working with. Wildlife field biologists are likely to be the first investigators on the scene during a disease outbreak and probably will be involved with collecting samples and submitting them to a diagnostic laboratory. For all these reasons, we have provided an overview of wildlife disease processes and ecology, field observations, laboratory procedures, and field procedures. Due to the brevity of this chapter, we included numerous references for excellent resources addressing the ecology and epidemiology of host–parasite interactions, specific diseases of free-roaming vertebrates, laboratory procedures, sources of wildlife disease expertise, field procedures, and disease investigation and management for those interested in more detailed information. Wildlife biologists and administrators who work with wildlife diseases in any capacity should be able to use this chapter and cited resources to quickly become key members of multidisciplinary teams organized to conduct disease research, surveillance, and/or management in wild animal populations.

APPENDIX 7.1. GLOSSARY OF WILDLIFE HEALTH- AND DISEASE-RELATED TERMINOLOGY

Acute: Having a rapid onset, short duration, and relatively severe course, as in acute infection.

Antibody: A large molecular-weight protein produced by host plasma cells (differentiated B lymphocytes) in response to an antigen. It binds specifically to the antigen that elicited its synthesis.

Antigen: Any substance (usually foreign) that induces an immune response in a host animal.

B lymphocyte: White blood cells involved in humoral immunity; B lymphocytes differentiate into plasma cells that produce a specific antibody.

Bacteriology: The branch of science that studies bacteria and bacterial diseases (practitioner: **bacteriologist** or **microbiologist**).

Cell-mediated immunity: Immunity accomplished primarily by T lymphocytes and accessory cells, such as macrophages, rather than by an antibody.

Chronic: Having a prolonged duration (weeks, months, or years), as in chronic infection.

Clinical chemistry: The area of pathology concerned with chemical analysis of body fluids.

Clinical sign: Objective evidence of a disease perceivable to a trained observer.

Cytopathic effects of viruses: Degenerative changes in cells, particularly in cell culture, associated with the multiplication of certain viruses.

Definitive host: The host in protozoan or metazoan life cycles in which the parasite undergoes sexual replication.

Diagnosis: (1) The process of determining the nature of a specific cause of disease and (2) the decision reached during this process.

Disease: Both a broad and a more focused definition are commonly used by wildlife disease specialists:
  1. **Broad:** An interruption, cessation, or disorder of body functions, systems, or organs. This definition includes toxic, genetic, metabolic, behavioral, neoplastic, and nutritional diseases in addition to those caused by macro- and microparasites.
  2. **Focused:** A disturbed or altered condition of an organism either caused by the presence of an antagonist (e.g., macroparasite, microparasite, or toxicant) or the absence of some essential element (e.g., micronutrient or vitamin).

Emerging infectious disease: An infectious disease that has newly appeared in one or more animal populations or that has been known for some time but is rapidly increasing in incidence and/or geographic range.

Endemic [Enzootic]:[a] A disease that occurs with predictable regularity and rate in a given population and area.

Endoparasite: A parasite that lives in host organs, tissues, or cells.

Epidemic [Epizootic]:[a] A disease that is occurring in a time or location where it does not normally occur or at a rate greater than historical norms.

Epidemiology [Epizootiology]:[a] The study of diseases in human or other animal populations, including factors accounting for patterns of disease occurrence (practitioner: **epidemiologist**).

Etiology: The cause of a disease or abnormal condition.

Fomite: An inanimate object that can be contaminated with an infectious agent and become a vehicle for transmission.

Hematology: The medical science that deals with the blood and blood-forming organs (practitioner: **hematologist**).

Histopathology: The branch of pathology concerned with tissue changes characteristic of disease. It employs microscopic examination of tissue to study the manifestations of disease.

Host: An organism that is parasitized.

Humoral immunity: Specific immunity that is mediated by antibodies.

Immune: Having or producing antibodies and/or lymphocytes capable of reacting with a specific antigen on a disease causing agent and thus preventing disease.

Incidence: The number of new cases of a disease occurring during a specified period as a proportion of the number of animals at risk of developing the disease during that time (contrast with prevalence).

Incubation period: The time between the initial exposure of an individual animal to a pathogen and the manifestation of clinical signs caused by that infectious agent.

Indirect transmission: When ≥2 host species are required for the completion of a parasite's life cycle.

Infection: Invasion and replication of a biological agent in a host animal.

Infectious agent: Any biological entity that invades and replicates in a host animal.

Intensity: Number of macro- or microparasites of a specific type per host.

---

[a] Traditionally, the terms including "**dem**" referred to human populations, whereas those including "**zoo**" referred to the identical concept in nonhuman animal populations. Probably because humans are animals, "endemic," "epidemic," and "epidemiology" now are commonly used for both human and nonhuman animal populations.

Intermediate host: Required host for indirect transmission cycles of protozoan and helminthic parasites in which the agent does not undergo sexual replication.

Lesion: A pathological change in host tissue.

Lymphocyte: A white blood cell important to the immune response. B lymphocytes mature to become plasma cells that produce antibodies; T lymphocytes are responsible for cell-mediated immunity.

Macroparasite: A metazoan parasite that grows in or on the definitive host and whose direct multiplication is either absent or occurs at a low rate. It multiplies more slowly than do microparasites by producing infective stages that are released from the host to infect new hosts. Macroparasites tend to induce incomplete and short-duration immunity that is dependent on the number of parasites present, so macroparasites generally occur as chronic, endemic host infections that are more likely to cause morbidity than mortality. Examples include parasitic helminths and arthropods.

Macrophage: A large white blood cell derived in the bone marrow occurring in tissues and the blood stream. Macrophages are important in phagocyctosis, and thence the destruction of foreign particles, and in cell-mediated immunity.

Microbiology: The branch of science that studies microorganisms and their effects on other living species (practitioner: **microbiologist**). This term sometimes is used as a synonym for bacteriology; however, microbiology encompasses a number of disciplines that study microorganisms, including bacteriology, virology, and medical mycology.

Microparasite: A small parasite characterized by short generation times, high rates of direct reproduction in the host (typically in host cells), and a tendency to induce long-lasting immunity to reinfection in surviving hosts. Time infected typically is short relative to the expected lifespan of the host, so microparasitic diseases often occur as epidemics in which the pathogen apparently disappears as susceptible hosts die or become immune, only to reappear when sufficient densities of susceptible hosts are again available in the population. Examples include parasitic bacteria, fungi, protozoa, rickettsia, and viruses.

Morbidity: (1) A diseased state, disability, or poor health due to any cause. (2) Proportion of animals in a group of interest that develop clinical disease caused by a specific etiological agent during a specified time.

Mortality: (1) Dead. (2) Proportion of animals in a group of interest that die during a specified time.

Mycology: The branch of science that studies fungi and in some cases fungal diseases (practitioner: **mycologist**).

Necropsy: Postmortem examination of a nonhuman animal.

Necrosis: Pathological death of one or more cells, or even large masses of tissue, as a result of injury or disease.

Necrotic: Affected by necrosis.

Neoplasia: Pathological process involving uncontrolled cellular proliferation that results in the formation of a tumor or neoplasm.

Neoplasm: Abnormal body tissue that grows by cellular proliferation more rapidly than normal and continues to grow after the stimulus that initiated growth ceases.

Pandemic: An epidemic disease that is widespread across large spatial areas (i.e., continents, or worldwide).

Parasite: Both ecological and systematic (taxonomic) definitions are used:
1. **Ecological:** An organism that at some phase of its lifecycle utilizes a host as habitat, depends on its host nutritionally, and causes the host some degree of harm.
2. **Systematic:** A helminth, arthropod, or protozoan that lives in or on an animal host and depends on this host nutritionally during some phase of its lifecycle.

Parasitology: A branch of biology dealing with parasitic helminths, arthropods, and protozoans and parasitism caused by these organisms in animals (practitioner: **parasitologist**).

Pathogen: A disease causing agent. The term often refers to microparasites that cause disease in hosts (e.g., pathogenic viruses and bacteria).

Pathogenic: Producing disease or pathological changes.

Pathognomonic: Distinctively characteristic or diagnostic of a particular disease.

Pathological: Diseased or relating to disease.

Pathology: The branch of medical science that studies the nature of diseases, especially the structural and functional changes produced by them (practitioner: **pathologist**).

Phagocyctosis: The process of actively engulfing and internalizing particles, such as bacteria or cell fragments, by host cells, such as macrophages, where these particles are then destroyed.

Plasma: The noncellular or liquid portion of blood, which differs from serum in that it contains fibrin and other soluble clotting elements.

Polymerase chain reaction (PCR): A technique in molecular biology used to amplify segments of DNA or RNA by several orders of magnitude, generating thousands to millions of copies of a particular nucleic acid sequence. PCR is used in diagnostic microbiology to rapidly identify viruses and other microorganisms.

Prevalence: The number of animals with a disease-related condition as a proportion of the total number in the group of interest at a specific time (contrast with incidence).

Prion: Particles composed of abnormally folded protein, containing no nucleic acid, that proliferate by inducing normal prion-related protein in the brain and lymphoid tissue to convert to the abnormal form. In mammals pathogenic forms arise sporadically as the result of genetic mutation or by transmission (as by ingestion of infected tissue or other materials); upon accumulation in the brain, they are thought to cause spongiform encephalopathy and related clinical disease.

Reservoir: One or more epidemiologically connected populations of different species in which an infectious agent can be permanently maintained.

Septicemia: The presence of pathogenic bacteria in the bloodstream.

Serology: The branch of medical science dealing with the immunological reactions and properties of blood and sera (practitioner: **serologist or immunologist**).

Serum: The liquid portion of blood remaining after coagulation and clot removal.

Surveillance: Organized collection, collation, and analysis of data related to an infectious agent of interest or a disease occurrence (e.g., chronic wasting disease surveillance programs implemented by wildlife conservation agencies).

T lymphocyte: A white blood cell that is responsible for cell-mediated immunity and acts as helper cells in the humoral immune response.

Toxic: Poisonous.

Toxicant: A toxic agent.

Toxicology: The science that deals with the detection of poisons, including their effects on animals and treatment (practitioner: **toxicologist**).

Toxin: A poisonous substance that is the metabolic product of a living organism. Examples include botulinum toxin and mycotoxins.

Transmissible spongiform encephalopathy (TSE): Any of a group of spongiform encephalopathies thought to be caused and transmitted by prions. Examples of TSE include bovine spongiform encephalopathy (BSE), chronic wasting disease (CWD), Creutzfeldt-Jacob disease (CJD), and scrapie.

Transmission: The process by which an infectious agent or parasite is shed from one host and infects another.

Transport host: An animal that becomes contaminated by an infectious agent and carries it to another host, but in which the agent does not complete a required portion of its lifecycle or multiply.

Vector: An invertebrate that transmits an infectious agent among vertebrates. The infectious agent multiplies or completes some required portion of its lifecycle in the invertebrate.

Virology: The branch of science that studies viruses and viral diseases (practitioner: **virologist**).

Virulence: The ability of an agent to cause disease.

Virulent: Having a marked ability to cause disease or death.

Zoonosis: An infectious disease that can be transmitted naturally between humans and other animals (plural: **zoonoses**).

8

Criteria for Determining Sex and Age of Birds and Mammals

EDDIE K. LYONS,
MICHAEL A. SCHROEDER,
AND LESLIE A. ROBB

INTRODUCTION

ACCURATE CLASSIFICATION of an animal's sex and age is fundamental to wildlife research and management (Leopold 1933, Morris 1972). **Population structure** is partially defined by age and sex ratios, and this information is often used to establish harvest regulations and strategies, assess population trends of the species, and provide an understanding of behavioral ecology. For animals in hand, numerous **physical characteristics** can be measured to obtain sex and age information. For example, body mass in all animals; forearm length in bats (Fig. 8.1); snout–vent length in lizards, frogs, and salamanders; and wing chord or wing notch length in birds are commonly measured. Regardless of the technique used, care needs to be taken to ensure that measurements are standard and results can be replicated (Nisbet et al. 1970). For instance, in birds, **wing chord length** is measured from, and including, the wrist to the tip of the longest primary (Fig. 8.2). However, wing chord can be measured in different ways: (1) nonflattened; (2) flattened, normal camber of wing reduced with gentle pressure; or (3) maximally flattened, normal camber reduced and feathers gently straightened. Wing flattening and feather straightening can add 0.5–5.0% to the nonflattened length; wing drying also can reduce the length (Pyle 1997, Dunning 2008).

In this chapter, we provide an overview of techniques for accurately sexing and aging wildlife. We also emphasize techniques that have **reduced subjectivity,** improved accuracy, and a long history of standardized use. In general, the best techniques for determining accurate estimation of age and sex are those that are versatile and can be used throughout the year with live or dead animals, different body parts, and numerous age categories. In reality, development of particular techniques has often been **affected by time of harvest and/or sampling methodology.** For sexually **dimorphic species** or those exhibiting distinct age-specific differences in appearance, accurate classification can be relatively simple. For example, breeding male birds may have brightly colored plumage and may take part in overt sexual displays, whereas females will have drab colors and be inconspicuous (Fig. 8.3). Accurate classification of an individual's sex and/or age is more complicated for **monomorphic species,** including many species in which young-of-the-year are identifiable, but differentiation among older age classes is difficult. Classification may be especially difficult if only partial information and/or material, such as a wing, jaw, or tooth, are available for evaluation. In such cases difficulties can be exacerbated by the relatively short and/or suboptimal time during which many samples are collected, such as during a hunting season. Therefore, many of the tech-

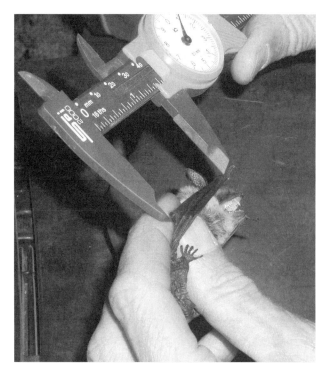

Fig. 8.1. Length of the forearm of bats is the most common measurement taken for them, in addition to mass. The slightly curved forearm of this fringed myotis (*Myotis thysanodes*) is measured as the straight-line distance from the end of the ulna to the base of the thumb, preferably using calipers. *Photo by M. A. Schroeder.*

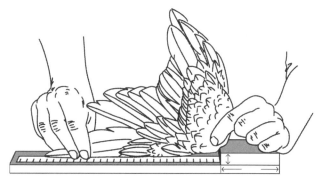

Fig. 8.2. Technique for measuring wing notch length. The measurement is taken from the notch in the wrist to the tip of the flattened primaries. *Modified from Carney (1992); original drawing by A. J. Godin.*

Fig. 8.3. Many species of animals, including lesser prairie-chicken (*Tympanuchus pallidicinctus*) exhibit sex-specific behavior. Male lesser prairie-chickens have a characteristic lek display that is not performed by females.

Fig. 8.4. Behavior is species specific. Males of some species may stand and stretch while urinating, and females may squat. Exceptions are common, as illustrated by this male mule deer (*Odocoileus hemionus*). *Photo by V. Geist; Geist (1981).*

niques discussed in this chapter only are **applicable** for a portion of the year (e.g., the hunting season), and investigators may have to use or develop alternative techniques for different times of the year.

Many species display extensive **variation** in body size and mass associated with subspecies or race, region, season, sex, and age. However, because variation may also exist within each category, there may be substantial overlap in the measurements of specific features. Even though the average male of many species may be heavier than the average female, often there is a range in body mass where the sex could be either. This problem may be exacerbated in monomorphic species, in which the size of young males is similar to that of adult females. **Behavior** for most species varies substantially among sex and age classes. Consequently, behavior can be used to identify sex and age categories of outwardly monomorphic species. Behavioral differences can include calls, songs, visual displays (Fig. 8.3), nest building, clutch incubation, nursing, and urination posture (Fig. 8.4). However, due to the complicated and species-specific nature of behavioral displays (Young 2005), with few exceptions, this chapter focuses on the use of **morphological characteristics** for assessing an individual's sex and/or age.

Molecular techniques that can be used to ascertain sex and age of organisms are growing rapidly in popularity. Information on this complex and dynamic category of research is provided in Chapter 22 (This Volume).

Our objectives in this chapter are to (1) describe general techniques used to classify the sex and age of most vertebrate species, (2) identify techniques and resources used to examine a specific species or group of species, and (3) provide appropriate references for detailed work on specific species. Although we cannot cover every species or group of species, many of the general techniques described here will have application to other groups. If working on a species where sex and/or age identification has not been determined, these general techniques should be some of the first to be considered.

SEX AND AGE CHARACTERISTICS OF BIRDS

Development
Embryonic
Development of embryos can be examined in birds using egg **flotation techniques** (Westerskov 1950, Barth 1953, Hays and LeCroy 1971, Dunn et al. 1979, Nol and Blokpoel 1983, Van Paassen et al. 1984, Alberico 1995) and **candling techniques** (Westerskov 1950, Weller 1956, Young 1988). Some evidence suggests that the age of early-stage clutches may be overestimated, whereas that of late-stage clutches may be underestimated with both egg flotation (Walter and Rusch 1997) and candling (Reiter and Anderson 2008). However, Reiter and Anderson (2008) documented that age estimation with these techniques had minimal effects on estimates of daily survival rates and nest success in Canada geese (*Branta canadensis*).

The degree of **development** of bird embryos prior to hatching varies. Differences between the 2 types of development strategies (i.e., altricial versus precocial) can be observed using the developmental stages in the 14-day incubation period of the altricial mourning dove (*Zenaida macroura*; Muller et al. 1984), the 23-day incubation period of the precocial northern bobwhite (*Colinus virginianus*; Roseberry and Klimstra 1965), and the 26-day incubation period of the precocial wild turkey (*Meleagris gallopavo*; Stoll and Clay 1975). When precocial embryos are approximately two-thirds of the way through their normal incubation period, they are similar to newly hatched altricial birds.

Postnatal
In birds, **altricial young** are sparsely feathered and blind at hatching, and the young remain in the nest until fledging. Hanson and Kossack (1963) provided a photographic guide to aging nestling mourning doves in days. **Precocial young** are covered with down and have open eyes. Age of precocial young can be classified in the field with pattern of down replacement or with measurements of **primaries** and/or their pattern of replacement. This pattern has been well documented for young waterfowl (Bellrose 1980) and various **gallinaceous** birds (Table 8.1). In addition to plumage characteristics, size of young relative to adults as well as flight capability can be used to determine age. These differences have been documented in turkey poults (Williams and Austin 1988), northern bobwhites (Stoddard 1931), ruffed grouse (*Bonasa umbellus*; Bump et al. 1947) and blue grouse (*Dendragapus obscures*; Smith and Buss 1963).

Plumage Characteristics
Birds typically have a natal plumage, followed by a juvenile (or immature) plumage, and then an adult plumage. Although downy natal plumage is easily identifiable (e.g., chukar [*Alectoris chukar*]; Fig. 8.5), juvenile plumage can resemble adult plumage in basic appearance while differing in subtle

Fig. 8.5. Changes in appearance of juvenile chukars with age. Alkon (1982).

Table 8.1. Age (in days) of immature gallinaceous birds and mourning dove when primary wing feathers are replaced during molt. Figures in column A represent age or range in age at which the juvenile feather is molted.

Species	1 A	1 B	2 A	2 B	3 A	3 B	4 A	4 B	5 A	5 B	6 A	6 B	7 A	7 B	8 A	8 B	9 A	9 B	10 A	10 B
Willow ptarmigan	18		25		30		35		40		46		53		65		91[a]		Juvenile not replaced	
Ruffed grouse	14	45	20	49	27	63	35	68	42	77	49	83	61	98	74	119	Juvenile not replaced			
Blue grouse	21–28		28–35		28–35		28–42		35–49		42–56		49–63		63–70		77–91		Juvenile not replaced	
Hungarian partridge	24		32		40		46		52		59		73		87		105		Juvenile not replaced	
Red-legged partridge	29		34		41		49		58		70		86		105		130		Juvenile not replaced	
Ring-necked pheasant	28		35		42		48		56		63		70		82		91		98	112
Northern bobwhite	26–30	54–58	33–37	56–60	40–44	60–64	44–50	70–76	52–58	81–89	58–62	99–107	69–77	120–128	97–105	146–154	Juvenile not replaced			
Coturnix quail	22		23		27		34		34		40		49		Juvenile not replaced					
California quail	29	55	32	62	38	70	46	80	52	90	62	108	72	121	100	141	Juvenile not replaced			
Mourning dove[b]	30.1 ± 5.5		35.5 ± 4.3		39.9 ± 4.9		50.4 ± 8.0		61.2 ± 10.5		74.0 ± 15.8		87.0 ± 9.2		107.5 ± 19.1		123.0 ± 18.9		157.8 ± 36.6	

Data are from willow ptarmigan (*Lagopus lagopus*; Westerkov 1956), ruffed grouse (Bump et al. 1947), blue grouse (Smith and Buss 1963), Hungarian partridge (*Perdix perdix*; McCabe and Hawkins 1946), red-legged partridge (*Alectoris rufa*; Petrides 1951), ring-necked pheasant (*Phasianus colchicus*; Buss 1946), northern bobwhite (Petrides and Nestler 1943), coturnix quail (*Coturnix coturnix*; Wetherbee 1961), California quail (*Callipepla californica*; Raitt 1961), and mourning dove (Morrow et al. 1992).

[a] Bergerud et al. (1963) noted this feather could be identified by its soft quill to 112 days.
[b] The second figure for mourning dove represents 1 standard deviation.

Fig. 8.6. Tail feathers of hatch year (HY) waterfowl may be notched or have a downy plume attached to the tip, whereas those of after hatch-year (AHY) birds are rounded or pointed. *Godin (1960).*

ways, such as **notched tail feathers** (Fig. 8.6), **buffy or worn edges of wing primaries** (Fig. 8.7), and variation in color patterns. Knowledge of **feather type** (Fig. 8.8) and **molt patterns** is extremely important for understanding which feathers offer the best clues to an individual's age and sex. For example, the first juvenile feathers to be replaced by adult feathers in spruce grouse (*Falcipennis canadensis*) are on the upper sides of the breast at about 30 days of age, thus permitting identification of sex (Boag and Schroeder 1992).

Waterfowl

Plumage characteristics can be used to separate most species of waterfowl into **hatch year** (HY) and **after hatch year** (AHY) categories. Species exhibiting subadult plumage may be classified as after second year (ASY). Tail feathers of HY birds typically are notched or have downy plumes attached to tips of the shafts (Fig. 8.6). Most ducks cycle through 2 distinct annual plumages that may differ markedly in color by season and sex. The **nuptial** (also called breeding or alternate) **plumage** is the **definitive plumage** for both sexes of ducks, and most display this plumage for the greater part of the annual cycle. Consequently, wing characteristics of ducks are particularly important for providing an adequate indication of species, sex, and age. Characteristics of sex and age based entirely on wing plumage were described by Carney (1992), and Bellrose (1980) provided descriptions and color photos with helpful clues to sex and age characters of all North American waterfowl. This information is now online at the Northern Prairie Wildlife Research Center (http://www.npwrc.usgs.gov); and an example key is provided in Figure 8.9.

Geese, swans, and whistling ducks have only one plumage per year, which is their definitive plumage. General patterns of plumage in swans and geese are similar in both sexes for all species, and usually can be used only for age (Table 8.2). Sex generally should be verified with **cloacal examination** (Fig. 8.10). Birds are classified as HY (before completion of the prebasic molt) and AHY (after completion of the prebasic molt). All HY swans and geese may have notched tail feathers early in autumn. Only the male AHY mute swan (*Cygnus alar*) has a fleshy knob on its forehead.

Gallinaceous Birds

Most gallinaceous birds retain juvenile primaries 9 and 10 (numbered from P1 [inner] to P10 [outer]; Fig. 8.8) through their first year, and these primaries often differ in appearance from P9 and P10 of adults. Consequently, some gallinaceous birds can be reliably placed into 3 age classes (depending on time of year; Johnsgard 1973; Table 8.3). These

Fig. 8.7. Diagnostic plumage characteristics of adult and juvenile wild turkeys. (A) Tail fans of adult (top) and juvenile (bottom). (B) Outer primaries of juvenile (left) and adult (center and right). Blunt tip of right feather is caused by dragging on ground during strut. *Based on Williams (1961); original report by Petrides (1942).*

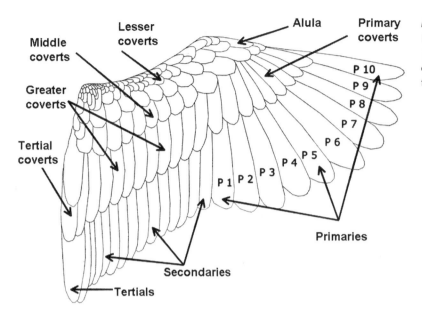

Fig. 8.8. Basic feather types on a typical wing. Primaries are numbered from proximal to distal (P1–P10), and secondaries are numbered from distal to proximal (not individually labeled on figure).

Age and Sex Identification of Green-winged Teal Using Wing Plumage

Wing character	Male Adult	Male Immature	Female Immature	Female Adult
Scapulars	Vermiculated or barred or both		Barred	
Tertials	Uniform gray; tapering to a narrow rounded tip; unfrayed; rarely have narrow light edging; black longitudinal stripe sharply defined along inner edge; may be molting	Small; narrow; rather delicate tips are often badly frayed; usually buff edging		Buff edging including rounded tips; unfrayed; longitudinal stripe often poorly defined along inner edge; may be brown or black; stripe sometimes well defined; may be molting
		Longitudinal stripe usually black and well defined	Longitudinal stripe usually brown and poorly defined along inner edge	
		After molt: Similar to adult male	After molt: Similar to adult female	
Greater tertial coverts	No edging; gray; may be either rounded or pointed; not frayed or faded	Narrow with fine light edging; often faded or frayed to wispy tips		Broadly rounded; usually with wide light edging; not frayed or faded
		After molt: Similar to adult male	After molt: Similar to adult female	
Middle and lesser coverts	Broadly rounded; no edging; gray that matches tertial coverts	Gray with wear around edges; appear ragged; somewhat narrow and trapezoidal; late in year contrast to a variable degree with new (replaced) greater coverts		Broadly rounded; usually with wide light edging but sometimes no edging
Primary coverts	No edging to a trace of light edging on inner web of outer four	Usually with considerable light edging on inner webs of outer four		No edging; or faint, light edging on inner webs of outer four

Fig. 8.9. Example of a key for age and sex criteria for a species of duck (green-winged teal [*Anas carolinensis*]); from the Northern Prairie Wildlife Research Center, http://www.npwrc.usgs.gov/resource/birds/duckplum/index.htm).

Table 8.2. Age characteristics for swans and geese

Species	Age characteristics[a]
Swans	HY birds are usually dull with light gray patches, whereas AHY birds are solid white.
Greater white-fronted goose (*Anser albifrons*)	HY birds have grayish body plumage, yellow legs and bill, and lack a white face patch. AHY birds have white face patch, orange legs, and pink bill (Ely and Dzubin 1994).
Snow goose (*Chen caerulescens*)	HY blue-phase birds may have brownish-gray patches on head, body, legs, and bills. AHY blue-phase birds have slate gray body plumage with white head. HY white-phase birds may have patches of sooty gray on otherwise white plumage and grayish-brown legs and bill. AHY white-phase birds white with black wing tips, red legs, and a pink bill (Mowbray et al. 2000).
Ross' goose (*C. rossii*)	HY birds may have patches of pale gray on otherwise white plumage, and AHY birds are white with black wing tips (Ryder and Alisauskas 1995).
Emperor goose (*C. canagica*)	HY birds may have patches of black-brown on head and neck; their legs and bill are black. AHY birds have a white head and upper neck, yellow legs, and a pink bill (Petersen et al. 1994).
Canada goose	Tail feathers may be notched, breast feathers relatively narrow, and outer primaries more pointed in HY than in AHY birds (Caithamer et al. 1993, Mowbray et al. 2002).
Brant (*B. bernicla*)	HY birds (Atlantic subspecies) have no white on necks until mid-winter; greater and middle wing coverts may be tipped with white. AHY birds have a white crescent on side of neck and the greater and middle coverts are dark brown. HY birds of the black form may have dark plumage with white under-tail coverts and light-gray edging of wing coverts. AHY birds have barred gray and white flanks with dark wing coverts (Reed et al. 1998).

Abbreviated and summarized from Bellrose (1980) and other references listed in the table.

[a] AHY = after hatch year; HY = hatch year.

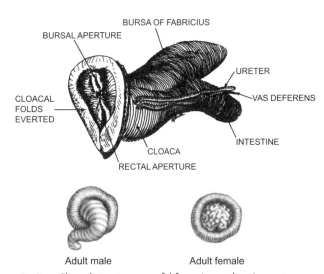

Fig. 8.10. Cloacal structures useful for aging and sexing waterfowl. Lower left shows a male duck cloaca with penis extended and lower right is typical female cloaca. *From Godin (1960).*

classes include hatch year or juvenile (HY), second year or yearling, usually through the prebasic molt in late summer and early autumn (SY), and after second year or adult (ASY). Later in the hunting season and/or following completion of the prebasic molt, SY birds are usually indistinguishable from ASY birds; hence, both are referred to as AHY (after hatch year) birds. In this latter case, only 2 age classes are distinguishable (HY and AHY; Fig. 8.11). Many other species of birds (except for a few with intermediate plumage patterns) can only be differentiated into HY and AHY age classes, or in some cases, no differentiation at all can be made (e.g., after the prebasic molt of mourning doves). Another indicator of age can be the depth of the **bursa of Fabricius,** which decreases with age (Gower 1939, Wight 1956; Fig. 8.10). However, because most gallinaceous birds display some age-specific variation in plumage (Fig. 8.12), measurement of the bursa is usually not necessary.

There can be substantial variation in plumage characteristics associated with region and subspecies. For example, ruffed grouse in southern populations typically have longer tails than those in northern populations (Uhlig 1953, Davis 1969, Servello and Kirkpatrick 1986). Wild turkeys also show regional and subspecific variation (Healy and Nenno 1980). Many juvenile wild turkeys in Florida molt P9, and in some cases P10, in their first autumn (Williams and Austin 1970), in contrast to the normal pattern of gallinaceous birds. The potential variation in appearance and pattern of molt associated with ecological region is not clearly understood, yet this factor may be a problem when samples are drawn from a broad geographic area and/or include multiple subspecies.

Other Birds

There also is substantial information on classification of sex and age in many other species, including American woodcock (Fig. 8.13), shorebirds, pigeons and doves, cranes, rails, and raptors (Table 8.4). In addition to numerous field guides of birds (e.g., Peterson 1998, 2002; Sibley 2000), there are detailed guides for identifying the sex, age, and subspecies of birds (Pyle et al. 1987, Pyle 1997, Dunning 2008). Pyle (1997) and Dunning (2008) provide particularly useful information for evaluating birds in the hand. Additionally, each species in North America has been extensively reviewed in individual species accounts produced by the American Ornithologists' Union, that were placed online in 2003 (Poole and Gill 2003).

Table 8.3. Age and sex characteristics for selected gallinaceous birds. The number of potential age classes is largely dependent on timing of examination relative to completion of prebasic molt. Primaries are numbered from proximal to distal (see Fig. 8.8).[a]

Species	Age characteristics	Sex characteristics
Spruce grouse	Chick age estimated by replacement and growth of primaries (McCourt and Keppie 1975, Quinn and Keppie 1981, Towers 1988). Pointed P9/P10 in HY/SY birds is reliable and easy (Zwickel and Martinsen 1967). P9 (McKinnon 1983) and P1 (Szuba et al. 1987) tend to have smaller shaft diameters in HY/SY birds.	Breast feathers solid black or black tipped with white in males and horizontally barred in females (Ellison 1968, Boag and Schroeder 1992). Rectrices mostly black in males or tipped with light brown and/or white, depending on subspecies and age. Rectrices of females motted black and brown and 1–2 cm shorter for given age category (Zwickel and Martinsen 1967, Boag and Schroeder 1992).
Ruffed grouse	Bursa of Fabricius length may be useful for ascertaining age, but not after January following hatch (Kalla 1991). HY birds tend to have pointed tips and less sheathing on P9/P10 than on P8, but this is less clear with aging (Hale et al. 1954, Dorney and Holzer 1957, Kalla and Dimmick 1995). HY/SY birds have a smaller P9 diameter or ratio of P9:P8 (Davis 1969, Rodgers 1979).	Males have longer ruff feathers on side of neck and 2–3 whitish dots on terminal ends of rump feathers; females have 1 whitish dot on terminal ends of rump feathers (Bump et al. 1947, Hale et al. 1954, Dorney 1966, Davis 1969, Roussel and Ouellet 1975). Starting at about 8 weeks of age, males can usually be distinguished from females by color of the bare patch above the eye; moderate to vivid reddish-orange in males and slight or no pigmentation in females (Palmer 1959). Males have distinct subterminal band on center 2 rectrices; females have indistinct subterminal band; female's tail is about 1 cm shorter for a given age category (Hale et al. 1954, Davis 1969, Rusch et al. 2000).
Blue grouse	Chick age estimated by replacement and growth of primaries (Zwickel and Lance 1966, Schladweiler et al. 1970, Redfield and Zwickel 1976, Zwickel 1992). P9 and P10 are pointed on HY/SY birds and rounded on ASY birds (Van Rossem 1925, Bendell 1955, Smith and Buss 1963, Braun 1971, Hoffman 1985).	Males have cervical apteria edged with white feathers and are 15–25% heavier than females (Caswell 1954, Boag 1965, Bunnell et al. 1977, Zwickel 1992). Males have primaries and rectrices 1–2 cm longer than females (Bendell 1955, Mussehl and Leik 1963, Boag 1965, Braun 1971, Hoffman 1983, Zwickel et al. 1991, Zwickel 1992). Rectrices of males mostly black or black with terminal band of gray, depending on subspecies. Sexual variation appears as early as 6 weeks (Nietfeld and Zwickel 1983).
Sharp-tailed grouse (*Tympanuchus phasianellus*)	P9 and P10 tend to be more pointed and worn in HY/SY than in ASY birds (Hillman and Jackson 1973).	Male crown feathers are dark with buff-colored edge; female crown feathers are barred (Henderson et al. 1967, Connelly et al. 1998). Central 2 rectrices of male are longitudinally striped; comparable feathers in female are horizontally barred (Henderson et al. 1967).
Lesser (*T. pallidicinctus*) and greater prairie-chicken (*T. cupido*)	Chick age estimated by replacement and growth of primaries (Etter 1963), and from descriptive photographs (Baker 1953). P9 and P10 in HY/SY birds tend to be more pointed and worn and have more spotting on their anterior portions (Campbell 1972).	Male undertail coverts are solid with a terminal round spot; crown feathers are dark with a buff-colored edge. Female undertail coverts and crown feathers are barred (Copelin 1963, Henderson et al. 1967, Schroeder and Robb 1993, Hagen and Giesen 2005). Tails of males are solid or lightly barred; those of females are entirely or partially barred (Copelin 1963; Fig. 8.12).
Gunnison (*Centrocerus minimus*) and greater sage-grouse (*C. urophasianus*)	Chick age estimated based on replacement and growth of primaries (Pyrah 1963). The pointed appearance of P9 and P10 in juveniles is distinct (Eng 1955). Primaries tend to be longer in ASY than in HY/SY birds, especially P1, which can differ by about 1.5 cm (Crunden 1963, Schroeder et al. 1999).	Males have black chin, white breast, filoplumes, and white tipped undertail coverts, and are 35–50% smaller for a given age category (Dalke et al. 1963, Schroeder et al. 1999). Male primaries are 1.5–3.5 cm longer and rectrices are 7–10 cm longer for a given age category than for females (Crunden 1963, Schroeder et al. 1999).
White-tailed ptarmigan (*Lagopus leucura*)	Chick age estimated by replacement and growth of primaries (Giesen and Braun 1979). HY/SY birds have dusky brown flecking on P9/P10; this pigmentation is absent in ASY birds (Braun et al. 1993).	Male has prominent eye combs during the breeding season; upper breast, neck, and head feathers are buff and tipped with blackish-gray to dark brown. Female breast feathers are coarsely barred. Sex difficult to distinguish based on plumage during autumn and winter (Braun and Rogers 1967, Braun et al. 1993).
Rock ptarmigan (*L. muta*)	HY/SY birds have more dark pigmentation and less gloss on P9 than on P8; pigmentation tends to be equal or greater on P8 and gloss tends to be equal on ASY birds (Weeden and Watson 1967).	Male has distinct red eye combs and blackish-brown breast during breeding season; female has mostly brown breast. Sex difficult to distinguish based on plumage during autumn and winter (Holder and Montgomerie 1993).

continued

Table 8.3. continued

Species	Age characteristics	Sex characteristics
Willow ptarmigan (*L. lagopus*)	Chick age estimated by replacement and growth of primaries (Bergerud et al. 1963, Parr 1975). HY/SY birds have more dark pigmentation and less gloss on P9 than on P8; in ASY birds pigmentation tends to be equal or greater on P8 and gloss tends to be equal (Bergerud et al. 1963, Weeden and Watson 1967).	Feathers on neck and breast of male are distinctly rufous to chestnut, and eye combs are red during the breeding season. Sex difficult to distinguish during autumn and winter (Hannon et al. 1998). Male has long black rectrices and black central upper tail coverts. Female has shorter and dark brown rectrices and central upper tail coverts (Bergerud et al. 1963).
Wild turkey	In HY/SY birds the central 3 pairs of rectrices are longer than the outer rectrices, P9/P10 tend to be pointed with no bars in distal portions, and the upper secondary covert patch is narrower and duller (Petrides 1942, Williams 1961, Williams and Austin 1970; Fig. 8.7). Spur and beard length increase with age (Kelly 1975), but overlap is large (Steffen et al. 1990). Tarsometatarsus length used with about 75% accuracy (Wakeling et al. 1997).	Skin on side of neck is bare and reddish-pink in male; beard present on older males. Skin on side of neck is lightly feathered and grayish-blue in female; shorter beards are occasionally present (Edminster 1954). Tarsometatarsus measurements are larger in males and have been used to predict sex with about 96% accuracy (Wakeling et al. 1997). Primaries and rectrices are longer in males than in females for a given age category (Wallin 1982).
Montezuma quail (*Cyrtonyx montezumae*)	Greater upper primary coverts edged with buff or buffy bars near base in HY birds, or spotted or barred with white in AHY birds (Johnsgard 1973).	Face and throat of male is marked in bold black and white pattern; face and throat of female is mottled with brown, buff, and white (Leopold 1959).
Northern bobwhite	Chick age estimated based on growth of primaries (Petrides and Nestler 1952). Upper greater primary coverts buffy and tapered in HY birds and gray-brown and rounded in AHY birds. P9/P10 are pointed and dull brown in HY/SY birds and rounded and grayish in ASY birds (Stoddard 1931, Dimmick 1992).	Male has white chin and eye stripe, except masked bobwhite (mostly rufous with black head); female has buffy chin and eye stripe (Dimmick 1992). Base of lower mandible black in males and yellow in females. Middle wing coverts have fine, black, sharply pointed undulations in males, whereas those in females are wide and dull gray (Thomas 1969, Brennan 1999).
Scaled quail (*Callipepla squamata*)	Primary coverts tipped, edged, or mottled with white in HY/SY birds and uniformly gray in ASY birds (Wallmo 1956).	Side of male's face is uniform gray with a brownish ear patch. Side of female's face is dirty gray streaked with black (Wallmo 1956).
Gambel's (*C. gambelii*), and California quail (*C. californica*)	Greater upper primary coverts are mostly buff-tipped and pointed in HY birds and uniformly gray and rounded in AHY birds. P9/P10 are also more pointed and frayed in HY/SY birds (Calkins et al. 1999).	Male has black throat and crest; female has pale or buffy throat and small brown crest (Calkins et al. 1999).
Mountain quail (*Oreortyx pictus*)	HY birds have buff-tipped primary coverts and AHY birds have uniform gray coverts. HY/SY birds also have pointed or frayed P9/P10 (Gutierrez and Delehanty 1999).	Back of neck is gray and plume generally long and black in males. Back of neck is brown and plume is shorter and browner in females (Johnsgard 1975, Brennan and Block 1985, Gutierrez and Delehanty 1999).
Ring-necked pheasant (*Phasianus colchicus*)	Length of P10 may be useful for estimating age of chicks (Etter et al. 1970). Depth of bursa of Fabricius ≤8 or ≤6 mm for AHY males and females, respectively (Johnsgard 1975, Larson and Taber 1980). P1 of ASY birds tend to be longer and thicker than HY/SY birds (Wishart 1969, Greenberg et al. 1972). Spur length and eye-lens weight have not been useful (Stokes 1957, Dahlgren et al. 1965, Gates 1966, Koubek 1993).	Males large and brightly colored throughout with distinct leg spur and longer tail; females mottled brown with no spur and shorter tail (Oates et al. 1985, Rodgers 1985). Day-old males distinguishable from females based on infantile wattle just below eye (Woehler and Gates 1970). Field-dressed males distinct due to their larger sternum (Oates et al. 1985). Bars on male primaries meet rachi at sharp angles except on unbarred tips. Bars on female primaries meet rachi at right angles (Linder et al. 1971).
Chukar	Growth of juveniles described and illustrated in detail (Alkon 1982; Fig. 8.5). P9 <29 mm in HY and ≥29 mm in AHY birds. P9/P10 pointed in HY/SY and rounded in ASY birds (Weaver and Haskell 1968).	Primary measurements generally are greater for males than females (Weaver and Haskell 1968, Cramp and Simmons 1980); sex is difficult to distinguish (Christensen 1996).
Gray partridge (*Perdix perdix*)	P9 covert pointed in HY and rounded in AHY birds. P9/P10 pointed in HY/SY and rounded in ASY birds (Petrides 1942).	Throat and eye stripe buffy-orange for males and buffy for females. Scapulars and median wing coverts lack crossbars in males and have 2–4 crossbars in females (Carroll 1993).

[a] AHY = after hatch year; ASY = after second year; HY = hatch year; P = primary; SY = second year.

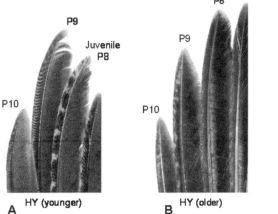
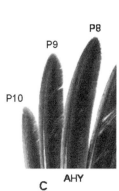

Fig. 8.11. Comparison of HY hatch year (HY; or juvenile) and after hatch year (AHY) female blue grouse wings collected during the autumn harvest. (A) The relatively short juvenile P8 has not yet molted, and P9 and P10 are relatively pointed: the wing is clearly definable as HY. (B) Juvenile P8 has been replaced, and P9 and P10 are both relatively pointed: the wing is from an HY bird. (C) P9 and P10 are relatively rounded, indicating that the bird is AHY. Because the bird has completed its molt, there is no possibility of differentiating between second year and after second-year birds.

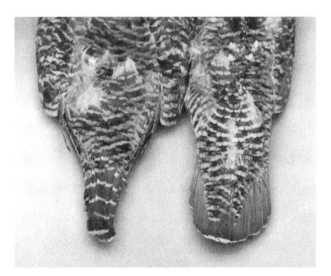

Fig. 8.12. Marking patterns of rectrices of pinnate grouse (*Tympanuchus* spp.) used to ascertain sex. Female (left) has bars across rectrices (upper tail coverts removed), male rectrices lack bars. *Based on Copelin (1963).*

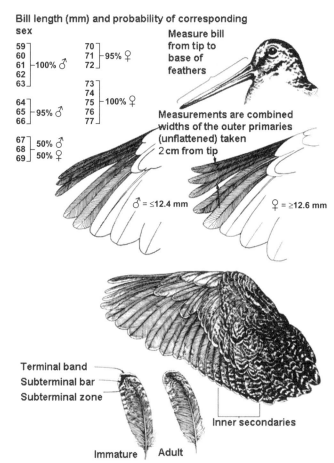

Fig. 8.13. Ascertaining sex and age in American woodcock (*Scolopax minor*). (A) Bill lengths and (B) outer primary widths distinguish sex. (C) Patterns of inner secondary feathers distinguish age class. *Based on Roberts (1988); adapted from Liscinsky (n.d.), Martin (1964).*

Table 8.4. Age and sex characteristics for miscellaneous birds. The number of potential age classes is largely dependent on timing of examination in relation to completion of the prebasic molt. Primaries are numbered from proximal to distal and secondaries from distal to proximal.[a]

Species	Age characteristics	Sex characteritics
American woodcock (*Scolopax minor*)	Depending on time of year, 3 age classes can be recognized based on retention of juvenile secondaries during second year (Sheldon et al. 1958, Martin 1964). Juvenile secondaries have light tips and distinct dark subterminal bars; adult secondaries lack a distinct bar (Petrides 1950a, Martin 1964, Roberts 1988). Coloration of neck, foot, and bill are useful (Shissler et al. 1981).	Females are heavier than males with overlap in 160–190 g range (Owen et al. 1977). Bill length >72 mm, combined width of outer 3 primaries ≥12.6 mm, and wing chord (to tip of P6 or P7) ≥134 mm characterizes female. Measurements <64 mm, ≤12.4 mm, and ≤133 mm, respectively, characterize males (Artmann and Schroeder 1976, Keppie and Whiting 1994). The combination of characteristics minimizes overlap (Fig. 8.13).
Wilson's snipe (*Callinago delicata*)	Juveniles may have a faint black tip on some lesser and median secondary coverts; adults have wide dark terminal shaft line (Dwyer and Dobell 1979). Multivariate analysis with feathers is useful, but there is 20% overlap (McCloskey and Thompson 2000).	Not easily distinguishable by plumage or cloacal characteristics (Fogarty et al. 1977, U.S. Fish and Wildlife Service and Canadian Wildlife Service 1977). Females have shorter outer rectrices and longer bills than do males (Mueller 1999); 10% are unclassifiable by multivariate analysis of skeletal and feather measurements (McCloskey and Thompson 2000).
White-winged dove (*Zenaida asiatica*)	Primary coverts of juveniles have pale tips and primaries may be edged with white or buff (Cottam and Trefethen 1968); juveniles lack black cheek-patch of adults (Schwertner et al. 2002).	Males larger than females with brighter plumage on crown, nape, and hind neck (Cottam and Trefethen 1968). Cloacal examination may be necessary (Brown 1977, Swanson and Rappole 1992).
Mourning dove	Juveniles have white or buffy tipped primary coverts, or buffy edge on P9/P10 (Petrides 1950a; Swank 1955; Wight et al. 1967; Haas and Amend 1976; Cannell 1984; Morrow et al. 1985, 1992). Long breeding season can complicate age classification (Schultz et al. 1995).	Females have tan breast and throat with a brown or brownish-gray crown; males are blue or blue-gray with slightly pink crown (Petrides 1950a, Cannell 1984, Mirarchi and Baskett 1994). Accuracy is not perfect (Menasco and Perry 1978, Schultz et al. 1995).
Band tailed pigeon (*Patagioenas fasciata*)	Juvenile growth has been described in detail (White and Braun 1990). Juveniles have buffy edged primaries, worn outer tips on P9/P10, and no wear on tips on S6 and S7. They retain secondary coverts up to 340 days of age (Silovsky et al. 1968, White and Braun 1978).	Breast and crown dull brown-gray in females and purplish to vinaceous in males (White and Braun 1978, Keppie and Braun 2000). This technique is useful as early as 45 days post hatch.
Sandhill crane (*Grus canadensis*)	Juvenile plumage brownish; the same plumage of adults is grayish (Walkinshaw 1949). Rusty staining can make separation difficult. Forehead of juveniles may be tawny; adults may be pale gray with a red crown (Lewis 1979).	Plumage differences insignificant; males are usually heavier than females (Tacha et al. 1992). Cloacal examination is only 66% accurate (Tacha and Lewis 1978).
Whooping crane (*Grus americana*)	Juveniles have brownish patches or buff-tipped feathers; adults are white with black wing tips and have a red crown (Lewis 1995b).	Sex not distinguishable based on plumage (Walkinshaw 1973), but males tend to be heavier (Lewis 1995b).
Rails	Presence of bursa of Fabricius is used to classify age of clapper rails (*Rallus longirostris*; Adams and Quay 1958); juveniles also have paler bill (Eddleman and Conway 1998). The black throat patch of adult soras (*Porzana carolina*) is absent in immatures (Melvin and Gibbs 1996). Juvenile black rails (*Laterallus jamaicensis*) are slightly duller in plumage than adults (Eddleman et al. 1994). Juvenile rails also tend to have narrower outer rectrices (Pyle 2008).	Male clapper rail are brighter on side and base of bill (Eddleman and Conway 1998). Male sora has lighter-colored bill (Melvin and Gibbs 1996). Male king rail (*R. elegans*) is slightly brighter in coloration (Odom 1977). Male black rail has darker throat (Eddleman et al. 1994). Male yellow rail (*Coturnicops noveboracensis*) has distinct yellow bill during the breeding season (Bookhout 1995). Males are generally heavier than females, although differences can be small.
Purple gallinule (*Porphyrula martinica*) and common moorhen (*Gallinula chloropus*)	Juveniles are brown or grayish with white feathers in throat region; bills and/or frontal shields lack red and yellow of adults (Bannor and Kiviat 2002, West and Hess 2002). Evidence of juvenile age class may persist until spring (Holliman 1977).	Sex is not distinguishable based on plumage, but males are slightly heavier than females in purple gallinule (West and Hess 2002) and ≤100 g heavier in common moorhen (Bannor and Kiviat 2002).

continued

Table 8.4. continued

Species	Age characteristics	Sex characteritics
American coot (*Fulica americana*)	Juveniles are paler than adults with lighter tipped feathers (Brisbin and Mowbray 2002).	Females are smaller than males, but overlap is large (Fredrickson 1968, Eddleman and Knopf 1985).
Raptors	Most raptors have distinct juvenile plumage that is only slightly worn in first autumn (Dunne 1987). Eye color changes with age in accipiters from yellow (juveniles) to red, orange, or brown (adults; Dunne 1987). Bald eagles (*Haliaeetus leucocephalus*) can be differentiated into multiple age categories based on increasing whiteness of the tail and head (McCollough 1989).	Wing chord is often larger for females than males (U.S. Fish and Wildlife Service and Canadian Wildlife Service 1977, Dunne 1987, Pyle 1997). Some raptors are clearly dimorphic in appearance; male northern harrier (*Circus cyaneus*) is gray, whereas the female is brown (MacWhirter and Bildstein 1996); the male American kestrel (*Falco sparverius*) has blue-gray wings, whereas the female's are rusty (Smallwood and Bird 2002). Bald eagles do not differ in plumage coloration (Bortolotti 1984), but females tend to be larger (Buehler 2000).

[a] AHY = after hatch year; ASY = after second year; HY = hatch year; P = primary; S = secondary; SY = second year.

SEX AND AGE CHARACTERISTICS OF MAMMALS

Mammals display much greater variation in size, longevity, productivity, and breeding cycles than do birds. General appearance, such as body size, for many species may differ greatly for most sex and age categories, thus making classification relatively straightforward with general field guides. Many small mammals enter the breeding population in the same year they are born, whereas large mammals can take many years to mature; for example, the house mouse (*Mus musculus*) is sexually mature 5–7 weeks after birth (Bronson 1979), but the gray whale (*Eschrichtius gibbosus*) reaches sexual maturity after at least 8 years (Burt and Grossenheider 1998). These differences add to the complications of assessing mammals, particularly with regard to age. Because field guides (e.g., Hall 1981) are necessarily general in nature, species accounts for individual mammal species produced by the American Society of Mammalogists (first account produced in 1969) may be an essential resource for detailed information (e.g., dentition). These accounts are particularly useful for species receiving little research and management attention. Despite the difficulty of capture and/or collection, current techniques for estimating age of mammals, particularly older mammals, are more effective than comparable techniques for estimating age in birds (Table 8.5).

Development
Embryonic
Fetal development in mammals can be used to estimate age in days, conception date, and/or parturition date (Bookhout 1964). Prenatal development of white-tailed deer (*Odocoileus virginianus*) and mule deer are well described (Armstrong 1950, Hudson and Browman 1959, Salwasser and Holl 1979, Larson and Taber 1980, Hamilton et al. 1985) and may be examined using a portable radiography unit (Ozoga and Verme 1985).

Postnatal
During the period of growth from birth to sexual maturity, **ossification of bones** (Fig. 8.14) and **epiphyseal cartilage in long bones** (Fig. 8.15) are universally present in mammals. Other internal characteristics that are unique to a particular sex may be associated with secondary sex characteristics or directly with reproductive organs, such as with suspensory tuberosities in white-tailed deer and mule deer (Taber 1956; Fig. 8.16). Although suspensory tuberosities are observable in deer 2.5 years old, they are not obvious in deer as young as 1.5 years old. In these cases, the ilio-pectineal eminences can be used to ascertain sex (Edwards et al. 1982; Fig. 8.17). Although internal characteristics are useful, they usually cannot be examined in live animals. Young mammals also differ from adults in numerous ways, such as **body size and weight, horn growth** (Fig. 8.18), **pelt appearance** (Figs. 8.19 and 8.20), **appearance of genitalia** (Figs. 8.21 and 8.22), **and changes in dentition.** Additional examples are provided in Table 8.5.

Eye-Lens Weight
The crystalline eye lens of vertebrates is an indicator of age in mammal species, because it grows without shedding cells (Lord 1959, Sanderson 1961c, Bloemendal 1977). In addition, the accumulation of the insoluble protein tyrosine in the eye lens can be measured and may be useful in determining age of mature animals (Birney et al. 1975). These methods are probably most useful for separating juveniles from adults and are not practical indicators of year class among adults. Eye-lens weight is accurate for many species from small mammals (Dapson and Irland 1972, Birney et al. 1975) to white-tailed deer (Ludwig and Dapson 1977). If properly preserved lens specimens are available, analysis of eye-lens weight can be used to accurately identify younger age classes (Friend 1967, Hearn and Mercer 1988, Koubek 1993, Bruns Stockrahm et al. 1996). An advantage of this technique

Table 8.5. Age and sex characteristics for selected mammals. Appearance of external genitalia is sufficient for classification of sex for most species (and in the case of large ungulates can be used to determine sex from a distance).

Species	Age characteristics	Sex characteristics
White-tailed deer (*Odocoileus virginianus*)	Fawns spotted in summer and smaller with relatively short nose in winter with innominate bone incompletely ossified (Edwards et al. 1982; Fig. 8.14). Tooth eruption and wear (Severinghaus 1949; Fig. 8.24) are used to estimate age, but results are unreliable for older deer (Gilbert and Stolt 1970, DeYoung 1989, Jacobson and Reiner 1989, Gee et al. 2002). Examination of tooth replacement and wear should be used for 3 age classes (fawn, yearling, and adult; Gee et al. 2002), unless reduced accuracy is acceptable. Cementum annuli analysis is effective for older animals (Gilbert 1966, Ransom 1966, Lockard 1972, McCullough and Beier 1986).	With rare exceptions, only males have antlers. First year antlers are usually small and referred to as buttons. Presence of tuberosities on the pelvic girdle distinguishes adult males (≥2.5 years) from females (Taber 1956; Fig. 8.16). Specific differences in the iliopectineal eminence of the pelvic girdle can be used to identify sex in animals 1.5 years old (Edwards et al. 1982; Fig. 8.17).
Mule and black-tailed deer (*O. h. columbianus*)	Fawns are spotted in summer and smaller with a relatively short nose in winter. A general analysis of morphology is complicated by habitat type and/or region (Strickland and Demarais 2000). Pattern of tooth eruption used to estimate age of fawns and yearlings (Rees et al. 1966). For deer >2 years old, tooth wear, eye-lens weight, and molar tooth–ratio techniques are imprecise (Robinette et al. 1957, Connolly et al. 1969a, Erickson et al. 1970, Van Deelen et al. 2000). Counts of cementum annuli from incisors are accurate for older ages (Low and Cowan 1963; Thomas and Bandy 1973, 1975; Hamlin et al. 2000).	With rare exceptions, only males have antlers. Tracks of adult and larger yearling males distinguishable from females by their larger arc width (McCullough 1965). Presence of tuberosities on pelvic girdle distinguishes adult males (≥2.5 years) from females (Taber 1956; Fig. 8.16).
Elk (*Cervus canadensis*)	Head profile and presence/shape of antlers are used to identify calves, yearlings, and adults (≥2 years old; Taber et al. 1982, Smith and McDonald 2002). Head profile is quantifiable with significant variation in rostral length, interorbital width, and ear length for female age classes; yearlings are larger than calves, and adults are larger than yearlings (Smith and McDonald 2002). Yearling males lack brow tines on antlers, whereas antlers of adult males have brow tines and are branched (Taber et al. 1982). Pattern of tooth eruption is used to estimate age through about 3 years (Quimby and Gaab 1957, Peek 1982); accurate estimation of older animals is by cementum annuli analysis (Keiss 1969, Hamlin et al. 2000)	Only males have antlers and upper canines (Greer and Yeager 1967). Antler scars also may be visible following antler drop.
Moose (*Alces alces*)	Calves identifiable by size. Tooth wear is considered for aging (Passmore et al. 1955), but cementum annuli analysis of incisors or molars give valid indication of year class (Sergeant and Pimlott 1959, Wolfe 1969, Gasaway et al. 1978, Haagenrud 1978).	Only males have antlers, and only females have a white vulval patch (Roussel 1975). Differences in sex are detectable by dimension of fecal pellets (MacCracken and Van Ballenberghe 1987).
Caribou (*Rangifer tarandus*)	Calves are identifiable by small size and relatively short head profile (Bergerud 1978). Antlers are usually larger for adults than for yearlings. Tooth eruption pattern is useful for classifying age to about 2 years (Bergerud 1970; Miller 1974a, b, 1982). Cementum annuli analysis is best technique for older animals (McEwan 1963, Bergerud and Russell 1966).	Antlers of males are larger than in females (Miller 1982). Presence of dark vulval patch in females is most consistent characteristic (Bergerud 1978). Mandible length is larger for males than females for a given age category (Bergerud 1964, Miller and McClure 1973).
Muskox (*Ovibus moschatus*)	Calves are small, yearling males are small with straight horns (about 100 mm), yearling females are small (horns about 66 mm), and adults are larger. Tooth emergence is useful for animals ≤6 years old; cementum annuli analysis is more accurate for older animals. Basal depressions of horns in 4-year-old females are maximally developed; bulls are maximally developed by year 6, when horns completely cover their foreheads (Tener 1965).	Horns of yearlings are longer in males than in females (100 versus 66 mm). In 2-year-olds, horns of males tend to be whiter and project straighter from the head (Tener 1965).

continued

Table 8.5. continued

Species	Age characteristics	Sex characteristics
Bison (*Bison bison*)	Cranial fusion is used for 2 age classes (Duffield 1973, Shackleton et al. 1975), horn development for 4 female and 5 male age classes (Fuller 1959, Reynolds et al. 1982), and tooth replacement and wear for 5–7 age classes (Skinner and Kaisen 1947, Fuller 1959, Frison and Reher 1970). Cementum annuli analysis is most reliable for estimating older age classes (Novakowski 1965, Moffitt 1998).	Horns of females are more slender and inwardly curved than those of males (Reynolds et al. 1982). There are numerous differences in horn cores, burrs, and skeletal measurements (Skinner and Kaisen 1947, Duffield 1973).
Wild sheep (*Ovis* spp.)	Lambs are distinguishable by small size. Because horn size increases with age, yearling rams can be classified based on size of curl (Jones et al. 1954). Horn segments are used for older age classes (Geist 1966). Tooth eruption and replacement are used to estimate age to 4 years (Hemming 1969, Lawson and Johnson 1982). Cementum annuli analysis is reliable for older ages (Turner 1977).	Sex is difficult to evaluate for lambs, but males of other age classes have larger horns (Lawson and Johnson 1982). Yearling rams are difficult to differentiate from adult ewes unless scrotum is detected.
Mountain goat (*Oreamnos americanus*)	Kids are distinguishable by size of body and horns less than half ear length in autumn, yearlings have horns about ear length, and adults have longer horns. Replacement of teeth is used to estimate ages ≤3 years, and rings on the horn are used for all ages (Brandborg 1955; Fig. 8.18).	Males stand or stretch while urinating, and females squat. Yearling males may have visible scrotum, and yearling females may have visible vulval patch under tail. Horns of males are generally thicker than those of females, but field interpretation is difficult (Wigal and Coggins 1982).
Pronghorn (*Antilocapra americana*)	Animals with horns longer than ear are usually adult males; maximum horn measurements from 2- and 3-year-old males (Mitchell and Maher 2001). Sequence of tooth eruption, replacement, and wear are used to estimate age (Dow and Wright 1962, Jensen 1998), but cementum annuli analysis of first permanent incisor is used for older age classes (McCutchen 1969, Kerwin and Mitchell 1971).	Horns of females average 42 mm in length and have unsubstantial prongs; horns of yearling males are larger (O'Gara 1969). Adult males have black face to horns and black cheek patch; females have black nose area only (Einarsen 1948, Yoakum 1978).
Collared peccary (*Pecari tajacu*)	Tooth emergence and replacement are used to estimate age to 21.5 months (Kirkpatrick and Sowls 1962). Eye-lens weights of limited value (Richardson 1966).	External dimorphism is limited to genitals. Suspensory tuberosities on pelvic girdle are prominent in males (Lochmiller et al. 1984).
Gray wolf (*Canis lupus*)	Pups are identifiable by small size to 8 months (Carbyn 1987). Tooth eruption, replacement, and size are useful to 26 weeks (Schonberner 1965, Van Ballenberghe and Mech 1975). Fusion of epiphyses of radius and ulna occurs at 12–14 months (Rausch 1967); animal is fully grown at 18 months (Young and Goldman 1944). Cementum annuli analysis of teeth is useful for estimating age of older animals (Goodwin and Ballard 1985, Landon et al. 1998, Gipson et al. 2000). Tooth wear (Landon et al. 1998, Gipson et al. 2000; Fig. 8.25), cranial sutures, and pulp cavity measurements (Landon et al. 1998) have been considered but are less versatile.	Urination posture is used to identify sex (Carbyn 1987). Examination of nipples, penile scar/opening, and testicles are used to identify sex for live wolves or pelts.
Coyote (*Canis latrans*)	Pups are classified by size (Barnum et al. 1979, Bekoff 1982). Permanent canines emerge at 4–5 months and are complete at 8–12 months (Voigt and Berg 1987); width of canine pulp cavity may be useful for estimating age (Root and Payne 1984, Tumlison and McDaniel 1984, Knowlton and Whittemore 2001). Cementum annuli is useful for estimating age >20 months (Linhart and Knowlton 1967, Allen and Kohn 1976, Nellis et al. 1978, Bowen 1982, Root and Payne 1984), particularly for canine teeth (Roberts 1978).	Examination of nipples, penile scar/opening, and testicles are used to identify sex in live animals or pelts (Voigt and Berg 1987). Sagittal crest in males is more developed than in females (Gier 1968, Bekoff 1982).
Fox	Canine teeth replacement is complete at about 1 year (Geiger et al. 1977); roots (Voigt 1987) and pulp cavities (Bradley et al. 1981, Tumlison and McDaniel 1984) are used to estimate age. Cementum annuli analysis is also	Examination of nipples, penile scar/opening, and testicles are used to identify sex in live foxes or pelts (Fritzell 1987). The baculum in males can be detected by palpating.

continued

Table 8.5. continued

Species	Age characteristics	Sex characteristics
	used (Grue and Jensen 1973, 1976; Allen 1974; Johnston et al. 1987), but accuracy decreases with number of annuli (Geiger et al. 1977). Eye-lens weight, baculum, body and skull measurements, and cranial sutures are used, but their reliability is not high (Sullivan and Haugen 1956, Wood 1958, Lord 1961, Geiger et al. 1977, Harris 1978).	
Black (*Ursus americanus*), brown (*U. arctos*), and polar bear (*U. maritimus*)	Eruption of canines used to estimate age to 3–4 years in black bears (Marks and Erickson 1966, Kolenosky and Strathearn 1987) and 2 years in brown bears (Rausch 1969). Cementum annuli analysis is preferred method for estimating age in black bears (Stoneberg and Jonkel 1966, Willey 1974, Carrel 1994, Keay 1995, Costello et al. 2004), brown bears (Craighead et al. 1970), and polar bears (Hensel and Sorensen 1980, Calvert and Ramsay 1998, Medill et al. 2009), but there are occasional errors (Hensel and Sorensen 1980, Kolenosky 1987, Harshyne et al. 1998, Medill et al. 2010). Baculum weight is also used in brown bears (Pearson 1975). A multivariate approach has been used for black bear cubs, including hair length, total length, skull width, and ear length (Bridges et al. 2002).	Males are larger than females but substantially overlap in size (Pearson 1975, Craighead and Mitchell 1982). Lower canines of black bears are used for sex identification (Sauer 1966). Length of mandibular canine alveolus and width of second mandibular molar are also used (Gordon and Morejohn 1975).
Raccoon (*Procyon lotor*)	Bacula of juvenile males is porous at base with cartilaginous tip, <1.2 g in mass and <90 mm in length (Sanderson 1961b, Kaufmann 1982; Fig. 8.21). Uterine horn of juvenile females is translucent and 1–3 mm in diameter with no placental scars (Sanderson 1950); it is opaque and 4–7 mm with placental scars in adults. Tooth eruption is useful to 110 days (Montgomery 1964); disappearance of cranial sutures and closure of epiphyses occurs at about 12 months (Sanderson 1961b, Junge and Hoffmeister 1980). Cementum annuli analysis is used for 4 age classes, including older animals (Grau et al. 1970, Johnson 1970).	Males are slightly larger than females, but overlap makes trait difficult to use. Palpation is used to detect baculum and testes in males (Stuewer 1943, Sanderson 1950, Kramer et al. 1999). Penile scars or nipples can be located on pelts.
American marten (*Martes americana*)	Tooth replacement is useful for estimating age to 18 weeks (Brassard and Bernard 1939). Radiographs of canine pulp cavities permit separation of juveniles from adults (Dix and Strickland 1986b). Cementum annuli analysis is used to estimate age for older animals (Strickland et al. 1982, Archibald and Jessup 1984). Suprafabellar tubercle on femur is used to separate juveniles from adults (Leach et al. 1982), but fusion of the distal femoral epiphysis is not reliable (Dagg et al. 1975). Juvenile males have bacula weighing <0.1 g (Marshall 1951, Brown 1983).	Presence of baculum, preputial orifice on pelt, and larger size of head confirm male and vulva confirms female (Strickland and Douglas 1987). Characteristics of teeth and skull are used to identify sex (Strickland et al. 1982, Brown 1983), but regional variation is large (Nagorsen et al. 1988). Tracks may be useful, although there is overlap (Zalewski 1999).
Northern river and sea otters (*Lontra canadensis, Enhydra lutris*)	Radiographs of teeth (Kuehn and Berg 1983, Melquist and Dronkert 1987) and closure of long bone epiphyses (Hamilton and Eadie 1964) are useful for classifying general age. Cementum annuli analysis is most reliable (Stephenson 1977, Bodkin et al. 1997). Eye-lens weight, baculum and skull characteristics, development of testes, and body size are used with less success (Toweill and Tabor 1982, Melquist and Hornocker 1983).	Relative position of anus and urogenital openings are used to ascertain sex; baculum is detectable with palpation (Thompson 1958).
Wolverine (*Gulo gulo*)	Genitalia and bone fusion are used to separate young-of-the-year from adults (Wright and Rausch 1955, Rausch and Pearson 1972). Body weight, tooth wear, and physiological condition are used to estimate age (Whitman et al. 1986). Best assessment for animals >1 year old are based on cementum annuli analysis (Rausch and Pearson 1972).	Nipples and genitalia (also scars and holes) are used to classify sex of live animals and pelts (Hash 1987). Females weigh 30% less than males (Hall 1981) and have smaller skull condylobasal length (Magoun 1985).

continued

Table 8.5. continued

Species	Age characteristics	Sex characteristics
Fisher (*M. pennanti*)	Suprafabellar tubercle is present on adult femur (Leach et al. 1982). Adults have prominent sagittal crest (Douglas and Strickland 1987); young can be identified with bone epiphyses and pulp cavities (Dagg et al. 1975; Kuehn and Berg 1981; Jenks et al. 1984, 1986; Dix and Strickland 1986*a*). Tooth emergence is useful ≤7 months. Cementum annuli analysis of the first premolar is used for estimating age of adults (Douglas and Strickland 1987, Arthur et al. 1992).	Males are twice as large as females and have larger bones (Leach 1977, Leach and de Kleer 1978). External genitalia or nipples are readily apparent on live animals or pelts. Lower canines of males have root widths >5.64 mm (Parsons et al. 1978) and are longer (Kuehn and Berg 1981, Jenks et al. 1984, Dix and Strickland 1986*a*).
American mink (*Neovison vison*) and other mustelids	Tooth eruption is useful for age ≤3 months in mink (Aulerich and Swindler 1968). Cementum annuli analysis is useful for older animals (Klevezal' and Kleinenberg 1967, Birney and Fleharty 1968). Baculum mass in mink averages 172 mg in juveniles and 398 mg in adults (Lechleitner 1954, Greer 1957, Godin 1960). Head of baculum is distinctly ridged in adult mink (Lechleitner 1954) or expanded in long-tailed weasel (*Mustela frenata*; Wright 1947).	Testes or penis scar identifies male and nipples female (Birney and Fleharty 1966, Eagle and Whitman 1987).
American badger (*Taxidea taxus*)	Techniques used include bone sutures, sagittal crest (Messick 1987), and baculum characteristics (Messick and Hornocker 1981). Cementum annuli analysis is best indicator of adult year classes (Crowe and Strickland 1975, Messick and Hornocker 1981).	Body and skull measurements are useful but overlap (Messick and Hornocker 1981, Messick 1987). Testes, penis, or penis scar is used to classify males and vulva or nipples to classify females (Petrides 1950*b*).
Skunks	Cementum annuli analysis is good estimator of adult year classes (Nicholson and Hill 1981). Other less effective techniques include bone ossification, tooth wear, and eye-lens weight (Allen 1939, Petrides 1950*b*, Mead 1967, Verts 1967, Bailey 1971, Leach et al. 1982).	Testes, penis, or penis scar is used to identify males and vulva or nipples to identify females. Lower canines also may be indicative of sex (Fuller et al. 1984).
Felids	Tooth emergence and replacement are useful for estimating age ≤240 days (Crowe 1975, McCord and Cardoza 1982, Lindzey 1987). Cementum annuli analysis is useful for estimating age in older animals (Crowe 1972, Nellis et al. 1972); it is less successful with cougar. The foramen of the canine tooth closes at 13–18 months in lynx and bobcat (Saunders 1964, Crowe 1972, Johnson et al. 1981*a*). Gum line recession is used to estimate age in older cougars (Laundre et al. 2000); mass, body length, and tail length are used to estimate age in younger ones (Laundre and Hernandez 2002); growth rate may vary by population (Maehr and Moore 1992).	Male genitalia are detectable but less obvious than in other carnivores (McCord and Cardoza 1982, Lindzey 1987, Rolley 1987). Lower canine size is useful for determining sex in bobcat (Friedrich et al. 1983). Body mass differs between male and female cougars, but there is overlap (Lindzey 1987, Laundre and Hernandez 2002).
Pinnipedia	Patterns of tooth eruption and body size are useful for estimating age (Spalding 1966), but cementum annuli analysis of canines is best technique for older animals (Scheffer 1950*b*, Laws 1962, Kenyon and Fiscus 1963, Anas 1970). Eye-lens weight is useful in limited situations (Bauer et al. 1964).	Northern fur seal (*Callorhinus ursinus*), Steller sea lion (*Eumetopias jubatus*), California sea lion (*Zalophus californianus*), northern elephant seal (*Mirounga angustirostris*), walrus (*Odobenus rosmarus*), and gray seal (*Halichoerus grypus*) males are substantially larger than females (King 1983, Riedman 1990). Harp seal (*Phoca groenlandica*) males are only slightly larger than females, but black markings tend to be larger and more distinct. Harbor seal (*P. vitulina*) is an exception, as it is outwardly monomorphic. Canine teeth are larger for males than for females in every age category in northern fur seals (Huber 1994) and for animals >5 months in California sea lion (Lowry and Folk 1990).
Lagomorphs	Epiphyseal grooves on bones are used for age ≤14 months (Hale 1949, Godin 1960, Tiemeier and Plenert 1964, Bothma et al. 1972, Kauhala and Soveri 2001; Fig. 8.15); periosteal layers in mandibles also may be useful (Sullins et al. 1976). Skull length is useful for estimating	Careful examination can reveal the penis (cylindrical organ) or clitoris (flattened posteriorly); young rabbits and hares are difficult to evaluate (Fox and Crary 1972).

continued

Table 8.5. continued

Species	Age characteristics	Sex characteristics
	days after birth (Bray et al. 2002). Eye-lens weight is used to separate juveniles from adults (Lord 1959, Tiemeier and Plenert 1964, Rongstad 1966, Connolly et al. 1969b, Pelton 1970, Keith and Cary 1979, Hearn and Mercer 1988, Kauhala and Soveri 2001).	
Muskrat (*Ondatra zibethicus*)	Pelt primeness varies substantially between adults and juveniles; the underside of the pelt tends to be mottled in adults and broadly patterned in juveniles (Dozier 1942, Kellogg 1946, Applegate and Predmore 1947, Shanks 1948, Godin 1960, Doude Van Troostwijk 1976; Fig. 8.19). Adults have less fluting on first upper molar than do juveniles (Olsen 1959, Proulx and Gilbert 1988), but pelt primeness appears more useful for classifying age (Moses and Boutin 1986). Adults have lower ratio of crown length to total length of first upper molar than do juveniles, but regional variation should be considered (Pankakoski 1980, Erb et al. 1999). Additional characteristics include ossification of the baculum (Elder and Shanks 1962; Fig. 8.21) and zygomatic breadth (Alexander 1951).	Careful examination can reveal the penis or nipples (Dozier 1942, Baumgartner and Bellrose 1943, Schofield 1955, Godin 1960). Sexual dimorphism in teeth is not detectable (Lewis et al. 2002).
American beaver (*Castor canadensis*)	Acceptable accuracy with a small number of age classes can be achieved with radiography of jaws of live or dead animals (Hartman 1992); cementum annuli analysis is useful for additional age classes (Van Nostrand and Stephenson 1964, Larson and Van Nostrand 1968). Evaluation of anal–urogenital opening in females is useful for classifying adults and juveniles (Thompson 1958). Skull characteristics (Buckley and Libby 1955) and tooth-root closure (Van Nostrand and Stephenson 1964) are useful for classifying juveniles and adults of both sexes.	Males are usually larger and heavier than females (Payne 1979). Careful palpation can identify the testes and baculum (Osborn 1955). Color and viscosity of anal gland secretion is a reliable indicator (Schulte et al. 1995).
Tree squirrels	Development of fox (*Sciurus niger*) and eastern gray squirrels (*S. carolinensis*), can be estimated with basic morphology up to 6 weeks (Uhlig 1955). The fur on the lateral rump of adult eastern gray squirrels has a distinct yellowish streak near the base that is absent in juveniles (Barrier and Barkalow 1967); age-specific patterns in tail pelage also are noted (Sharp 1958; Fig. 8.20). Teats are inconspicuous and hidden by hair in juvenile females and large and noticeable in adults (Fig. 8.22A). Male genitalia are larger on adults than on juveniles (Fig. 8.22B). Cementum annuli analysis is useful for estimating age class (Lemnell 1974, Fogl and Mosby 1978). Other techniques include epiphyseal lines in long bones (Petrides 1951, Carson 1961, Nellis 1969), epiphyseal lines in the foot (McCloskey 1977), and eye-lens weight (Beale 1962, Fisher and Perry 1970).	Sex is classified by examination of external genitalia, but skulls are also useful (Nellis 1969).
Woodchuck (*Marmota monax*)	Juveniles weigh 300–450 g by about 15 May and have eye-lens weights averaging 12.3 mg; yearlings have narrow and pointed incisors and eye-lens weights averaging 21.8 mg; adults have broad incisors and eye-lens weights averaging 28.53 mg (Davis 1964).	Careful examination is used to reveal the os penis; testes are often regressed (Kwiecinski 1998).
Virginia opossum (*Didelphis virginiana*)	The pouch is white, shallow, or insignificant in size in juvenile females; it is flabby, fatty, and dark in adults (Petrides 1949). Tooth eruption and emergence are useful characteristics (Lowrance 1949, Petrides 1949, McManus 1974, Tyndale-Biscoe and Mackenzie 1976).	Canines of males are longer and heavier than those of females (Gardner 1982). Males have scrotum and females have pouch (McManus 1974, Gardner 1982).

continued

Table 8.5. continued

Species	Age characteristics	Sex characteristics
Bats	Cartilaginous epiphyseal plates in the fingerbones of juveniles make joints look tapered and less knobby than do joints of adults (Anthony 1988).	External genitalia are visible in males; testes are relatively large when male is in breeding condition (Racey 1988).
Small mammals (insectivores, other rodents)	Eye-lens weight is used (Birney et al. 1975, Gourley and Jannett 1975) with mixed success (Dapson and Irland 1972, Barker et al. 2003); tyrosine content in lens may be more accurate (Dapson and Irland 1972). Tooth eruption (Mitchell and Carsen 1967, Beg and Hoffmann 1977), tail collagen strength (Sherman et al. 1985), adhesion lines in the lower jaw (Millar and Zwickel 1972) and femur (Barker et al. 2003), and cementum annuli analysis (Adams and Watkins 1967, Montgomery et al. 1971) also have been used.	Careful examination of genitals in live animals can be useful with most species. Shape of pelvic girdle can be used when only bones are available (Dunmire 1955).

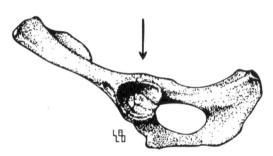

Fig. 8.14. Innominate bone of 1-year-old white-tailed deer. The arrow points at the area of incomplete ossification. *Edwards et al. (1982).*

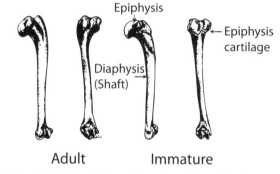

Fig. 8.15. Illustration of the epiphyseal cartilage of the humerus in an immature and adult cottontail rabbits (*Sylvilagus* spp.). *Godin (1960).*

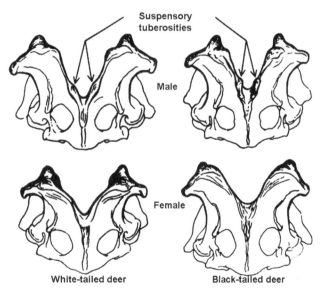

Fig. 8.16. Diagram of pelvic girdle of white-tailed deer and black-tailed deer (*O. h. hemionus*) ≥2.5 years of age, showing suspensory tuberosities for the attachment of the penis ligaments. *Taber (1956).*

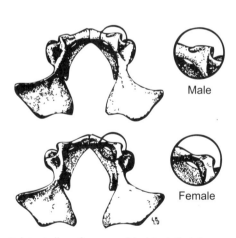

Fig. 8.17. Pelvic girdles of 1.5-year-old white-tailed deer can be classified by gender based on the position of the ilio-pectineal eminences (IPE; insets). The IPE is flattened and on the edge of the acetabular branch of the pubis in females; it is rounded and above the edge of the acetabular branch of the pubis in males. *Edwards et al. (1982).*

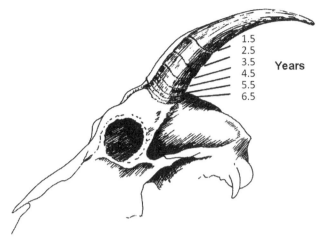

Fig. 8.18. Horns of mountain goats (*Oreamnos americanus*) may have rings that correspond to year class. *From Brandborg (1955).*

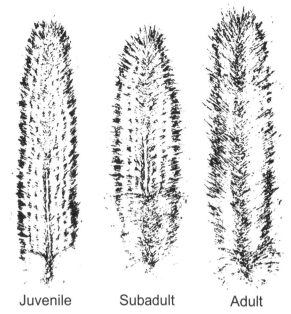

Fig. 8.20. Increased prevalence of short appressed hairs on the ventral surface of a gray squirrel's (*Sciurus carolinensis*) tail alters its age-related appearance. *Godin (1960).*

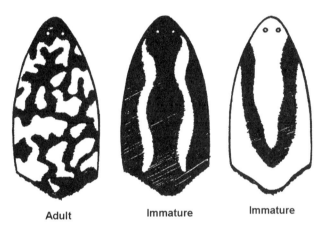

Fig. 8.19. Undersides of muskrat (*Ondatra zibethicus*) pelts have different patterns of light (prime) and dark (unprimed) fur that correspond with general age categories. *Godin (1960).*

is that it is fairly rapid, inexpensive, and does not require an intact skull (McLeod et al. 2006). If fresh specimens are available, this technique is superior to tooth wear procedures and is less costly and time consuming than cementum annuli. However, it is probably not as accurate as cementum annuli analysis for older age classes.

Dentition

The structure, consistent growth patterns, and replacement of teeth are commonly used to classify age and sex of mammals (Fig. 8.23; Table 8.6). General age classes of mammals can be identified by dental characteristics such as **thin root walls, wide-open root tips, ratio of pulp width to tooth width, ratio of dentine to enamel, tooth shape,** and the

timing of tooth emergence (Severinghaus 1949; Jenks et al. 1984; Dix and Strickland 1986a, b; Johnston et al. 1987; Helldin 1997). In white-tailed deer, tooth eruption criteria can be useful for classifying broad age classes (e.g., fawn, yearling, and adult; Fig. 8.24), or as a site-specific comparison technique. However, research has often shown that tooth size and wear can vary by individual, subspecies, region, habitat, diet, and sex (Hesselton and Hesselton 1982, Erb et al. 1999, Van Deelen et al. 2000, Gee et al. 2002). Estimation of age of known-aged deer with tooth emergence and wear techniques has been inaccurate, especially for older age categories (Hamlin et al. 2000, Gee et al. 2002). However, aging deer to older age classes (i.e., ≥3.5 years) based on tooth eruption is still a commonly used and taught technique. There has been substantial effort to use patterns of tooth wear, in addition to emergence and replacement of teeth, to classify older age categories of white-tailed deer and gray wolf (*Canis lupus;* Gipson et al. [2000]; Fig. 8.25). This effort has been accompanied by development of such field techniques as dental impressions (Flyger 1958, Barnes and Longhurst 1960, Clawson and Causey 1995) and reference sets of sex specific mandibles (Thomas and Bandy 1975). Normal variation in tooth wear has been exacerbated by confusion in wear characteristics of teeth necessary to discriminate between age categories (Marchinton et al. 2003). Misinterpretation of these characteristics (3.5-year-old deer incorrectly described in Dimmick and Pelton 1994:193) can result in deer being misclassified (Marchinton et al. 2003).

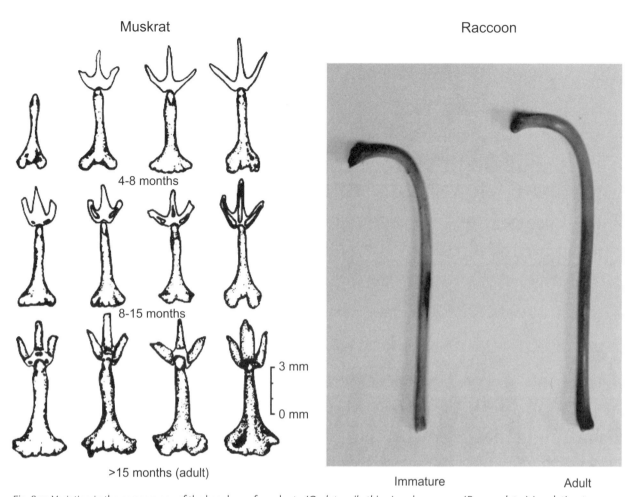

Fig. 8.21. Variation in the appearance of the baculum of muskrats (*Ondatra zibethicus*) and raccoons (*Procyon lotor*) in relation to age. Note that the immature baculum of the raccoon is somewhat porous at both the base and tip. *Illustration of muskrat bacula from Elder and Shanks (1962); photo of raccoon bacula by N. J. Silvy.*

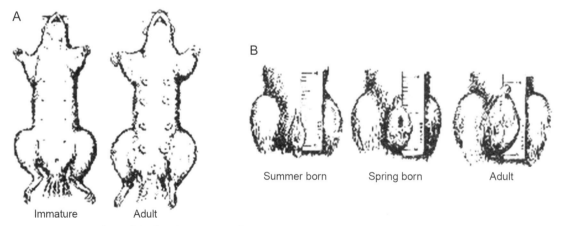

Fig. 8.22. (A) Mastology of the female squirrel. Left: juvenile, with nipple minute and barely discernible. Right: lactating adult, nipples pigmented black with most of hair worn off. (B) Scrotal measurements of the male squirrel. Left: In the summer born animal, the testes are abdominal, and the skin is just beginning to pigment. Center: In the spring born, the testes are large, and the scrotum is pigmented but heavily furred. Right: The adult has shed most of the fur from its scrotum. *After Allen (1943); from Godin (1960).*

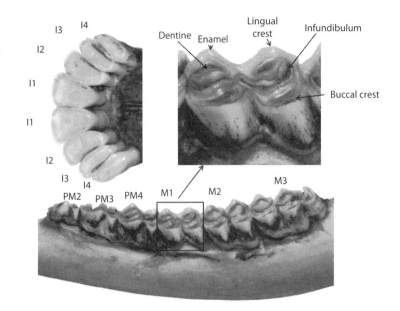

Fig. 8.23. Lateral view of lower left jaw of mule deer (*Odocoileus hemionus*), facing the buccal crest (cheek side). The front of the lower jaw also is shown as well as an enlarged area illustrating the first molar. Teeth are labeled as I (incisor), PM (premolar), and M (molar).

Table 8.6. Approximate age (in months) when permanent molars emerge or incisors, canines, and premolars replace deciduous teeth in the lower jaws of selected North American ungulates

Species	Reference	Incisors			Canines	Premolars				Molars		
		1	2	3	1	2	3	4		1	2	3
White-tailed deer	Severinghaus 1949	<6	<12	<12	<12	<18	~18	~12		2–6	~12	<18
Mule deer	Taber and Dasmann 1958	~12	~12	<18	<24	~24	~24	~24		2–6	6–12	18–24
Elk	Quimby and Gaab 1957	<18	~18	<30	<30	~30	~30	~30		~6	<18	<30
Caribou	Miller 1974b	10–13	12–15	12–15	12–17	22–29	22–29	22–29		<3	10–15	15–24
Pronghorn	Dow and Wright 1962	<15	<27	<39	39–41	<27	<27	<27		<2	<15	<15
Wild sheep	Lawson and Johnson 1982	12–16	24–28	33–36	45–48	24–32	24–30	24–30		1–6	8–16	22–40
Mountain goat	Brandborg 1955	15–16	26–29	38–40	~48	26–29	26–29	26–29		6–10	10–16	15–29

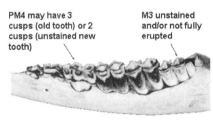

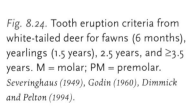

Fig. 8.24. Tooth eruption criteria from white-tailed deer for fawns (6 months), yearlings (1.5 years), 2.5 years, and ≥3.5 years. M = molar; PM = premolar. *Severinghaus (1949), Godin (1960), Dimmick and Pelton (1994).*

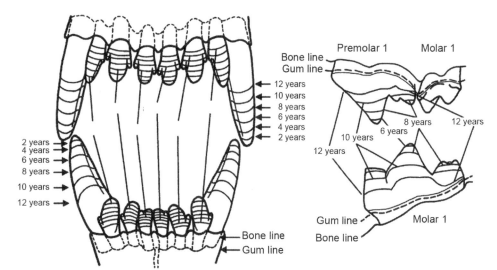

Fig. 8.25. Progressive wear on incisors and canines in 2-year increments for gray wolves. The lines represent averages for a study of known-aged wolves; errors of 1–3 years were observed using this technique. *Gipson et al. (2000)*.

Cementum Annuli

For most species, collection of a tooth for **cementum annuli analysis** is the most accurate method used to estimate age among older age categories (Hamlin et al. 2000). Cementum is deposited annually on the roots of teeth, so the layer closest to the dentine is from the earliest year, and the layer of the current year lies closest to the root. Because sex, physiology, ecological region, and annual variation in weather appear to minimally influence the layers (Allen and Kohn 1976), the cementum of permanent teeth can indicate the number of years following tooth emergence (Klevezal' and Mina 1973; Fig. 8.26). In teeth with distinct layers (e.g., beaver), grinding and polishing a section of the tooth is sufficient for evaluation of age (Van Nostrand and Stephenson 1964). In most situations, however, the tooth must be decalcified, cut into thin histological sections, and stained before evaluation. Techniques also are being expanded and developed to deal with other situations and tooth materials, including archaeological specimens (Lieberman et al. 1990, Beasley et al. 1992). All teeth have layers, but the tooth type used to assess an animal's age varies among species and collecting conditions. Some teeth, such as incisors and premolars, are easier to extract and may be removed from live animals without obvious adverse effects (Nelson 2001, Bleich et al. 2003). Nevertheless, there is some debate about the ethics of tooth removal from live animals, including arguments for (Nelson 2002) and against (Festa-Bianchet et al. 2002) the practice.

There are standard teeth and sections of teeth used for evaluation of cementum annuli. The standard tooth is the first incisor (central) for all ungulates, a lower canine or premolar 1 for most carnivores, and premolar 2 for cougars (*Puma concolor*; Dimmick and Pelton 1994). Premolar 3 or 4 also has been used for American marten (*Martes americana*), the lateral incisor for Canada lynx (*Lynx canadensis*) and bobcats, and an upper canine for bull elk (*Cervus canadensis*).

Standardization minimizes problems associated with differences in eruption time and interpretations of growth layers (Landon et al. 1998). If a nonstandard tooth type is selected for cementum age classification, the tooth must be identified, because differences in eruption time require different interpretations of growth layers. Errors of ≥1 year can result when an unidentified, nonstandard tooth is substituted for the standard. Techniques for tooth removal, mailing, storage, and processing should be selected before initiating research (Bergerud and Russell 1966, Erickson and Seliger 1969, Fancy 1980, Dimmick and Pelton 1994, Harshyne et al. 1998, Nelson 2001).

Use of cementum annuli for age classification appears to be more accurate than tooth wear for older mammals. In an experiment involving 120 known-aged samples from 12 species, there was exact agreement between known and cementum age in 94 individuals; ≤1-year discrepancy in 21 individuals, and >1-year discrepancy in 5 individuals (Dimmick and Pelton 1994). One reason for incorrect age classification using cementum annuli is the presence of double or uneven

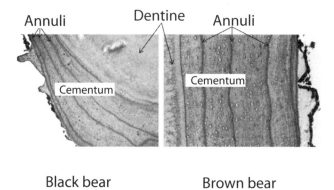

Fig. 8.26. Cementum annuli analysis of 4-year-old black bears and brown bears (*Ursus arctos*). *Photos by G. M. Matson*.

layers of cementum (Kolenosky 1987). This problem can result in errors, particularly the overestimation of age in younger animals and underestimation of age in older ones, such as with polar bears (*Ursus maritimus;* Hensel and Sorensen 1980) and wolves (Landon et al. 1998, Gipson et al. 2000). It is likely that pulp cavities and tooth eruption and replacement are more accurate for ascertaining younger age classes than are cementum annuli; in these cases, use of cementum annuli is unnecessary (Johnston et al. 1987, Jacobson and Reiner 1989, Landon et al. 1998). These characteristics can be examined visually or with radiography (Kuehn and Berg 1981, 1983; Dix and Strickland 1986a, b; Nagorsen et al. 1988; Helldin 1997; Knowlton and Whittemore 2001).

Skeletochronology

Skeletochronology is similar to cementum annuli analysis, but it potentially has a wider array of applications. It is the most commonly used method for evaluating age of amphibians and reptiles. Adhesion lines or annual growth layers in bones can be examined to estimate age. Several studies have addressed this possibility in femur bones of sea turtles with substantial success (Zug et al.1986, 2002; Bjorndal et al. 1988; Klinger and Musick 1992; Klinger et al. 1997; Zug and Glor 1999). Examination of a known age interval following injection with oxytetracycline supported the accuracy of this technique (Coles 1999). However, Eden et al. (2007) suggested that calibration with known-aged individuals was necessary to avoid underestimating age. Adhesion lines in the sectioned femurs of yellow pine chipmunks (*Tamias amoenus*) also appear to accurately indicate age categories (Barker et al. 2003). The technique has been expanded to include toe-clipped samples of amphibians (Parham et al. 1996); a transverse histological section through the midpoint of the toe phalanx appears to be best (avoiding cartilaginous areas near the epiphyses).

SUMMARY

Effective wildlife research and management depends on accurate assessment of sex and age. These assessments often can be conducted using long-established techniques that are relatively simple to perform, including visual examinations of general appearance and/or sex organs. Information also can be gathered through examinations of dentition and/or partial samples, such as wings or teeth. Although some species may appear monomorphic, the vast majority readily can be classified to sex and basic age categories. However, newer techniques are constantly being developed and evaluated, because there often is a need to obtain better estimates of age or sex using limited material. These techniques include improved cementum annuli analysis, skeletochronology, and genetic analysis of small tissue samples. It is likely that these techniques will provide a foundation for evaluation of population demography, establishment of harvest regulations and strategies, and development of protocols to monitor population and ecosystem health.

9

Techniques for Marking Wildlife

NOVA J. SILVY,
ROEL R. LOPEZ, AND
MARKUS J. PETERSON

INTRODUCTION

ALL CAPTIVE-ANIMAL and many field studies involving wildlife require that individuals are marked for future identification. Marked individuals can provide detailed information on population dynamics, movement, behavior, and density estimates. We provide an overview of factors that should be considered before deciding to mark vertebrates (excluding fish) and address factors relevant to the selection of appropriate procedures. Others have addressed these issues previously. Stonehouse (1978) described general marking techniques for animals, and Murray and Fuller (2000) reviewed effects of marking on vertebrates. Marking methods for amphibians, reptiles, birds, and mammals were reviewed by Nietfeld et al. (1994) and Silvy et al. (2005). Methods for marking amphibians and reptiles have been reviewed by Woodbury (1956), Thomas (1977), and Swingland (1978); Ferner (1979) and Donnelly et al. (1994) reviewed marking methods specifically for reptiles and amphibians, respectively. Beausoleil et al. (2004) presented methods to mark amphibians, reptiles, and marine mammals. Spellerberg and Prestt (1978) and Fitch (1987) reviewed methods for marking snakes. Marion and Shamis (1977), the American Ornithologists' Union (AOU; 1988), and Calvo and Furness (1992) reviewed marking methods for birds. The Ornithological Council's 2010 guidelines for the use of wild birds in research can be found online (http://www.nmnh.si.edu/BIRDNET/guide/index.html), and they supersede the AOU guidelines. Bird and Bildstein (2007) reviewed marking methods for raptors. The American Society of Mammalogists (1998) provided general guidelines for marking mammals. Barclay and Bell (1988) gave detailed information for marking bats. Although not covered in this chapter, overviews for marking fish were provided by Wydowsky and Emery (1983) and Parker et al. (1990). Hagler and Jackson (2001) provided an excellent overview of current techniques for marking insects.

Because of the wide diversity among vertebrate species, no single list of approved methods for marking is practical or desirable. The ultimate responsibility for the ethical and scientific validity of methods used rests with the investigator. In general, natural marks have the least adverse effect on individual animals and should be used when possible, whereas invasive techniques have the greatest potential for adverse effects. Moreover, many techniques require capture, recapture, and handling of animals that also might affect their behavior and survival. Separation of these effects from those caused directly by the marking method has yet to be evaluated in most cases.

CONSIDERATIONS PRIOR TO MARKING

Questions to Consider

Before attempting to mark free-ranging wildlife, the following **checklist** of species and situation-dependent questions should be considered.

1. Do the animals need to be marked, or can natural markings be used instead?
2. Do the animals need to be marked as individuals, or can they be marked as a group?
3. Do the animals need to be physically captured prior to marking, or can they be marked without capture?
4. How visible do the marks need to be, and do the animals need to be "recaptured" for the mark to be observed?
5. Will the marking method cause pain and/or decrease the chances of survival of the animal?
6. Will the proposed mark affect the animal's health, reproduction, movement patterns, and/or behavior?
7. How long will the mark be required to last to complete the study, and how durable is the proposed marking method?
8. Will the proposed marking method interfere with other studies?
9. Will the marks promote public concern about the study, and will the marks have to be removed after study completion?
10. Have the appropriate approvals (animal welfare and state and/or federal permits) to mark animals been obtained?

Considerable thought should be given to these questions before the decision to mark wildlife is made. Techniques for marking wildlife fall into 3 main categories: **natural, noninvasive,** and **invasive** marks. If natural marks cannot be used, noninvasive marks are preferable over invasive marks. Although some marking techniques may be unique to a single species, most apply to a wide variety of species. As in Silvy et al. (2005), we present marking information by methods. This approach has eliminated most repetition inherent in presenting this information by animal classes (i.e., amphibians, birds, mammals, and reptiles; Nietfeld et al. 1994). We consolidated general information on proper application of the technique, its retention time and visibility, and any adverse effects of the technique on marked animals (where this information is available). This allows the reader to more easily evaluate and compare individual methods. Additionally, we present these methods in sequence of what we consider most to least preferred (however, this ranking can differ for the different classes of animals). More detailed information, such as species or group, comments, and citations (in chronological order), is presented in tables. Thus, the reader can select an animal class, identify which methods have been used for the species or group of species of interest, and pursue the citations for more detailed information on the method's appropriateness for the specific application.

Marking Permits

Before an animal can be captured and marked, the appropriate local (e.g., animal welfare permits), federal, and/or state or provincial **permits** must be obtained. Wildlife species are regulated within state or provincial boarders by the appropriate wildlife agency. The federal government regulates capture and marking of migratory birds and threatened and endangered species. Authorization to mark migratory birds and threatened and endangered species must be approved by the Bird Banding Laboratory, U.S. Geological Survey, Department of the Interior, Laurel, Maryland 20708-4037, or the Canadian Bird Banding Office, Canadian Wildlife Service, Ottawa, Ontario, Canada KIA OH3. Also, capture, handling, and marking protocols used must be approved by an institutional animal care and use committee (IACUC) to ensure compliance with the Federal Animal Welfare Act and its amendments (U.S. Department of Agriculture 2002) and the Public Health Service Policy on the Humane Care and Use of Laboratory Animals (http://grants.nih.gov/grants/olaw/references/phspol.htm). Prior to the initiation of animal work, an investigator must submit an **Animal Use** Protocol to be reviewed and approved by an IACUC. Additionally, at many institutions, all participants must complete training regarding occupational health and safety issues related to animal work.

Natural Marks

The first questions to be considered when contemplating marking animals are: (1) is marking necessary, (2) can the study be conducted without recognition of individuals or a specific group of animals, and (3) if not, can animals be identified without use of applied marks? Perhaps the ideal method of recognizing individuals is to use their own naturally occurring unique traits, much as we identify other people by their physiognomic traits. Humans may be unable to differentiate individuals in some wildlife species, but there are others whose physical characteristics allow for individual identification using natural markings or distinct morphological characteristics. Many animals exhibit unique coat patterns or can be identified by unique color patterns (Fig. 9.1), scarring, fin or fluke notches, antler configuration, and/or other traits. Natural markings are most efficiently used on individuals with complex patterns, and analysis must be confined to a local population or region (Pennycuick 1978).

Natural markings have been used to identify individual mammals, reptiles, and amphibians more commonly than birds (Table 9.1). Unique plumage or bill patterns can be used as distinguishing features for birds, but such features

Fig. 9.1. Unique spots and stripes on 2 bobcats.

Table 9.1. Natural markings used to identify individual animals

Group/species[a]	Method for identification	Reference
Amphibians and reptiles		
Grass snakes	Ventral pattern	Carlström and Edelstam 1946
Viviparous lizard	Dorsal pattern	Carlström and Edelstam 1946
Slow-worm lizard	Throat pattern	Carlström and Edelstam 1946
Smooth newt	Belly pattern	Hagström 1973
Anoles	Distinctive pattern and tail regeneration	Stamps 1973
Warty newt	Belly pattern	Hagström 1973
Eastern newt	Dorsal spot pattern	Healy 1975
Dusky salamander	Dorsal color pattern	Forester 1977; Tilley 1977, 1980
Snakes	Distinctive characteristic on exuvia	Henley 1981
Snakes	Characteristic of subcaudal scales	Shine et al. 1988
Spotted salamander	Spot pattern	Loafman 1991
Patterned amphibians	Spot and stripe pattern	Doody 1995
Birds		
Bewick's swan	Bill pattern and body features	Scott 1978
Osprey	Head marking pattern	Bretagnolle et al. 1994
Mammals		
Giraffe	Unique coat pattern	Foster 1966
Tiger	Unique coat pattern	Schaller 1967, Karanth 1995, Karanth and Nichols 1998
African lion	Whisker pattern	Pennycuick and Rudnai 1970
Black rhinoceros	Unique ear markings, horn shape and wrinkle pattern	Mukinya 1976
Cetaceans, manatee	Unique color, scars, and fin or fluke notches	Würsig and Würsig 1977, Irvine et al. 1982, Irvine and Scott 1984
Domestic dog	Unique coat pattern	Heussner et al. 1978
African bushbuck	Unique coat pattern	Seydack 1984
Leopard	Pelt characteristics	Seydack 1984
Bobcat	Spot variation	Rolley 1987, Heilbrun et al. 2003
Cheetah	Pelt characteristics	Caro 1994, Kelly 2001
White-tailed deer	Antler, pelt, and body characteristics	Jacobson et al. 1997

[a] Scientific names are given in Appendix 9.1.

are rare in avian populations and may change with molt and/or age. Thus, the potential for natural marking systems in birds is limited, but may have short-term application in conjunction with other markers for some species.

Marking as Individuals or Groups

If a study requires the use of applied marks, do the animals have to be marked as individuals, or can they be marked as groups? Many herd or flock movement and dispersal studies only require that large numbers of individuals be marked in a given area and relocated later. For example, large numbers of white geese could be marked by placing dye in roost ponds and followed by searching for colored geese. Similarly, many mark–recapture or mark–resight studies conducted only to estimate population density do not require that marked individuals be differentiated from others.

Marking without Capture

Capture may stress animals, and marking without capture is preferred where practical. **Remote marking** of animals as individuals or groups has a long history (Table 9.2). Mammals have been marked with paint-tipped arrows (N. J. Silvy, unpublished data) and paint balls (Table 9.2). Animals also have been marked using a manually triggered dye-spraying device, and dyes can be introduced into the animal's food to produce dyed fat, teeth, pelage, and droppings. Self-affixing collars have been developed for several species (Table 9.2). Dye-spraying devices affixed to aircraft have been used to mark large mammals and could be used for marking large numbers of white-colored birds (e.g., white geese, egrets). Dyes also can be placed on eggs and nests, marking the adults as they incubate their eggs (Table 9.2). Subsequent collection or observation of marked animals provides data on dispersal and population dynamics.

Marking after Capture

If animals must be captured, there are numerous marking techniques available. Although the most suitable marking techniques will depend on the needs of the investigator, Barclay and Bell (1988) suggested considering the following factors: duration of study, ability to relocate marked animals, number of animals to be individually identified, and the effect of the mark on the animal. According to Marion and Shamis (1977) and Ferner (1979), an **ideal marking technique** would: (1) involve minimal pain or stress, (2) produce no adverse effects on survival and behavior, (3) permanently mark individuals, (4) be easy to recognize at a distance, (5) be easy to apply, (6) be easy to obtain and/or assemble, and (7) be relatively inexpensive. Additionally, the selected marking technique should not conflict with other studies in the area, and permission to use the technique should be readily obtainable from the appropriate authorities. Most marking techniques do not satisfy all these criteria, and investigators must prioritize prior to mark selection.

Nietfeld et al. (1994) grouped markers into 3 categories relative to retention time: temporary, semi-permanent, and permanent. We prefer 2 groups: permanent and nonpermanent. We define **permanent marks** as those lasting the life of the animal and **nonpermanent** marks as all others. Permanent marks include branding, tattoos, ear notching, toe clipping, and other invasive techniques, although scarring, tearing, and aging may reduce their effectiveness. Nonpermanent marks generally are more visible and can be used with permanent marks to increase visibility of the animal, yet still have the animal marked for life. For example, a white-tailed deer (all scientific names are found in Appendix 9.1) could be given a unique ear tattoo (permanent) as well as a numbered, brightly colored cattle-ear tag (visible). Animal size, however, limits the size of marks that can be applied, but

Table 9.2. Remote marking methods used to mark animals as individuals and in groups

Group/species[a]	Remote marking method	Reference
Birds		
Greater sage-grouse	Aniline dyes in tank attached to spray head and buried in lek	Moffitt 1942
Ruffed grouse	Aluminum and bronze dust in nests found later on shed feathers	Bendell and Fowle 1950
Glaucous-winged gull	Thief detection powder on eggs and nests	Mossman 1960
Nesting terns	Blow dye from bottle using rubber tubing	Moseley and Mueller 1975
Nesting wood duck	Rubber band with color marker in nest box hole	Heusmann et al. 1978
Cattle egret and gull eggs	Rhodamine B dye in oil-based silica gel placed on eggs; adults marked 2–6 months	Paton and Pank 1986, Cavanagh et al. 1992
Roosting red-winged blackbirds	Aerial application of liquid fluorescent pigmented material, visible under UV light in subsequent collections of marked birds	Otis et al. 1986
Wood stork	Pressurized canister with nozzle on pole with control lever	Rodgers 1986
Waterfowl	Fluorescent particles applied to lakes marked waterfowl for 8 weeks	Godfrey et al. 1993
Common tern	Device using refillable bottles filled with dye, remotely controlled	Wendelin et al. 1996
Mammals		
Deer	Treadle-type spray devices	Clover 1954
White-tailed deer	Self-affixing collar	Verme 1962, Siglin 1966, Taylor 1969
Mountain sheep	Manually triggered dye-spraying device and modified Cap-Chur darts	Hansen 1964, Simmons and Phillips 1966, Turner 1982
Moose	Manually triggered dye spraying devices	Taber et al. 1956
Pronghorn	Collar-holder frame over water	Beale 1966
Hares and rabbits	Self-affixing collar	Keith et al. 1968
Dall's sheep	Spraying devices used from aircraft	Simmons 1971
Muskox	Paint-pellet pistols	Jonkel et al. 1975
Elk	Paint-ball guns	Herriges et al. 1989, Herriges et al. 1991
Red squirrel	Remotely applied collars	Mahan et al. 1994

[a] Scientific names are given in Appendix 9.1.

color-coded marks still can enhance recognition. A point to remember when using color-coded marks is that many people are red/green colorblind. Therefore, selection of contrasting colors that can be recognized at a distance by all individuals involved with the project is important.

The use of marks can influence behavior, particularly color marks used on birds, and can increase predation (Kessler 1964, Burley et al. 1982). The combination of stress and mortality associated with capture and the effect of the mark itself could decrease survival more than either capture or marking alone. Thus, it is important to examine whether the necessary data can be obtained without use of marks. If not, researchers must ascertain whether marking animals is likely to result in reliable knowledge that can be used to better manage the population. Further, they should realistically weigh the benefits of this knowledge against the discomfort or harm done to the individual animals. There is no simple checklist that will delineate the most appropriate marking technique(s) for all potential research projects.

NONINVASIVE MARKING TECHNIQUES

Neck Collars

Many different **neck collars** have been designed for field identification of free-ranging animals (Table 9.3). Properly fitted collars (Fig. 9.2) should not restrict feeding, inhibit circulation or breathing, or cause entanglement. Collars may be fixed in size or expandable to allow for growth. Many neck collars are placed too loosely on animals (Fig. 9.2). A loose collar (especially if the collar has the added weight of a radio transmitter) will slip up and down an animal's neck when it lowers and raises its head. This action can cause abrasions and possible open sores that can lead to infection and possibly death. If a collar is extremely loose, the animal may get a foot caught in the collar as it extends its front feet to stand from a bedding position. If a collar is placed too tightly around an animal's neck, the collar may cut off blood circulation that can lead to tissue sloughing, infection, and death. During the rut, the necks of many male ungulates swell,

Table 9.3. Neck collars used on wildlife

Group/species[a]	Material and comment	Reference
Amphibians and reptiles		
American alligator	Vinyl-plastic tape	Chabreck 1965
Birds		
Geese, brant, swans, ducks, and cranes	Plastic collars of flexible vinylite, flexible plastic, rigid acrylic resin, and aluminum with or without letters and numbers with retention up to 11 years on adult geese, but should not be used on goslings because few are retained; icing not a problem with aluminum neckbands, but collared birds may move from breeding areas	Aldrich and Steenis 1955, Helm 1955, Craighead and Stockstad 1956, Idstrom and Lindmeier 1956, Ballou and Martin 1964, Huey 1965, Sherwood 1966, Lensink 1968, MacInnes et al. 1969, Fjetland 1973, Greenwood and Bair 1974, Koerner et al. 1974, Ankney 1975, Chabreck and Schroer 1975, Raveling 1976, Maltby 1977, Craven 1979, Abraham et al. 1983, Zicus et al. 1983, Pirkola and Kalinainen 1984, Hawkins and Simpson 1985, Zicus and Pace 1986, MacInnes and Dunn 1988, Ely 1990, Samuel et al. 1990, Campbell and Becker 1991, Johnson et al. 1995, Castelli, and Trost 1996, Menu et al. 2000, Schmutz and Morse 2000, Wiebe et al. 2000
Game birds	Colored plastic neckbands	Taber and Cowan 1963, Marcstrom et al. 1989
Mammals		
Foxes	Metal collar slit for expansion	Sheldon 1949
Ungulates	Plastic, aluminum, nylon fabrics; polyethylene rope with flags; rubberized machine belting:, and self-adjusting plastic collars for young ungulates	Ealey and Dunnet 1956, Progulske 1957, Fashingbauer 1962, Lightfoot and Maw 1963, Harper and Lightfoot 1966, Knight 1966, Hawkins et al. 1967, Craighead et al. 1969, Hanks 1969, Phillips and Nicholls 1970, Beale and Smith 1973, Brooks 1981, Keister et al. 1988, Hölzenbein 1992
Hares	Leather collar	Hewson 1961
Polar bear	Nylon webbing	Lentfer 1968
African elephant	Rubberized machine belting	Hanks 1969
Feral goats	Galvanized steel chain	Rudge and Joblin 1976
Cetaceans, manatees	Rubberized belt	White et al. 1981
Bats	Spiral bird ring and keychain collar	Moran 1985, Wilkinson 1985
Coyote	Vinyl plastic collar	Gionfriddo and Stoddart 1988

[a] Scientific names are given in Appendix 9.1.

Fig. 9.2. Oversized neck collar (right) that could allow animal to place its leg through collar. Collar should fit snugly around neck just below head (left).

and collars must expand to allow for this swelling. Collars made with nylon elastic will allow expansion of the collar. Collars for fawns may be made entirely of folded nylon elastic with folds stitched together with thread that breaks with the pressure of neck growth and allows the collar to expand with the growing animal (Fig 9.3).

Silvy (1975) developed Boltaron™ (thermal plastic) expandable collars (Fig 9.4) for male white-tailed deer that were 7.4 cm wide and made to fit the neck contours of deer of each gender in each age class. The open ends of the U-shaped collars for female deer were riveted (brass split rivets), and no elastic straps were used (Fig. 9.5). Collars for male deer had elastic straps on the inside that were attached by rivets at the bottom of the "U." Straps passed through brass welding-rod guides embedded in the open ends of the plastic collar permitted expansion and contraction. Because the weight of a radio package was on the elastic straps in the U-shaped collars, the rubber in the elastic straps degraded over time and the collars sagged. This problem was solved by design of a C-shaped collar with ends overlapping at the side of the neck with elastic bands to resist expansion that completely opened the "C." This design allowed the weight of the collar and radio to be supported by the Boltaron and not by the elastic. Once a male's neck returned to normal size after the rut, the Boltaron collar returned to its normal shape and reduced tension on the elastic straps. Collars were of 2 thicknesses (0.2 or 0.3 cm Boltaron) and of 2 colors (black and white). Various colors of scotch-lite reflective tape in the form of numbers, letters, or other symbols were attached to collars for ready identification of deer during both day and night. Radios were mounted (using dental acrylic) on, and antennas were either stainless-steel whips or copper wire embedded in, the Boltaron collar. Stainless-steel whips tended to break due to salt-water etching; this was not a problem with embedded copper wire antennas.

Typically, collars are highly visible, but their longevity depends on the material used, climate, and behavior and gender of the animal involved. Most studies report either no or insignificant adverse effects of neck collars on breeding-related activities, social behavior, and physical damage beyond minor hair or feather wear and irritation. Neck collars on birds (Fig. 9.6), however, have been observed to disrupt pair bonds, lower success in agonistic encounters, contribute to starvation, and increase mortality through severe icing.

Bands

Metal bands (Fig. 9.7) bearing an identification number and return address are the most common method of marking wild birds (Table 9.4). Although states and provinces are required to use their own bands for resident game birds, the U.S. Geological Survey and the Canadian Wildlife Service

Fig. 9.3. Elastic (expandable) radiocollar on white-tailed deer fawn.

Fig. 9.4. Expandable neck collars for male ungulates.

Fig. 9.5. Nonexpandable female-ungulate neck collar with holes for brass-split rivets.

Fig. 9.6. Plastic neck collars on tundra swan. *Photos by D. Watts.*

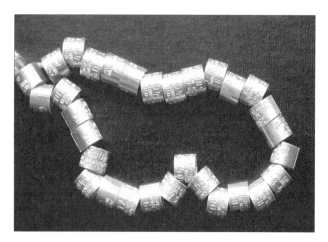

Fig. 9.7. Standard butt-end bands used on the legs of birds.

Colored bands made from plastic or other materials have been used alone or in conjunction with metal bands (Fig. 9.8) to mark individuals of a variety of species (Table 9.4). Colored bands are primarily intended to permit rapid identification of individuals without requiring recapture. Color bands deteriorate relatively quickly and are best for short-term studies. Soft plastic wrap-around bands have the lowest durability and color retention (Anderson 1981), which is somewhat greater in laminated wrap-around bands (Lumsden et al. 1977, Anderson 1981). Retention is higher in wide versus narrow plastic bands. Painted bands are of limited use, because abrasion or paint removal by birds results in rapid marker loss (Childs 1952). Milligan et al. (2003) noted there are errors associated with using colored leg bands to identify wild birds

Arm and Wing Bands

The attachment of bands to the forearms has been the most widely used technique for marking bats and penguins (Table 9.4). **Flipper bands,** made initially of aluminum and more recently from Monel metal and stainless steel, have been used on penguins. Several markers are available for bats, including serially numbered metal bands, color-anodized aluminum bands, numbered and unnumbered colored plas-

issue bands for migratory birds. Aluminum bands are sufficient for marking many species, but are easily damaged by abrasion and corrosion. As a result, Monel™, Incoloy®, stainless steel, and titanium bands sometimes are used for long-lived and marine birds. Koronkiewicz et al. (2005) provided a method to produce aluminum color bands from noncolored aluminum bands.

Table 9.4. Bands used on arms, wings, tails, and legs to mark wildlife

Group/species[a]	Material and comment	Reference
Amphibians and reptiles		
Frogs	Butt-end bird bands on toes	Kaplan 1958
Bullfrog	Plastic waist bands	Emlen 1968
Lizards	Colored metal rings around thigh	Subba Rao and Rajabai 1972
Six-lined racerunners	Colored plastic bands glued to tail	Paulissen 1986
Anurans	Waist bands	Rice and Taylor 1993
Birds		
Passerines, terns, doves, pheasants, grouse, vultures, parakeets, geese, parrots, swallows	Butt-end metal bands	Young 1941; Wandell 1943, 1945; Elmes 1955; Dunbar 1959; MacDonald 1961; Kaczynski and Kiel 1963; Hamerstrom and Mattson 1964; Henckel 1976; Burtt and Tuttle 1983; Hatch and Nisbet 1983a, b; Nisbet and Hatch 1983; Nisbet and Hatch 1985; Bailey et al. 1987; Marcstrom et al. 1989; Meyers 1994b; Powell et al. 2000b; Menu et al. 2001; Davis 2006
Penguins	Flipper bands of aluminum, Teflon, monel™, and stainless steel	Sladen 1952; Penny and Sladen 1966; Cooper and Morant 1981; Sallaberry and Valencia 1985; Boersma and Rebstock 2009, 2010
Waterfowl	Plexiglas™, butt-end bands	Balham and Elder 1953
Doves, waterfowl	Reward bands give higher reporting rates	Bellrose 1955, Tomlinson 1968, Henny and Burnham 1976, Nichols et al. 1991, Reinecke et al. 1992
Raptors	Butt-end and lock-on (can only be removed by eagles) leg bands	Berger and Mueller 1960, Environment Canada 1984, Robson 1986, Young and Kochert 1987, Harmata et al. 2001, Harmata 2002
House sparrow	Colored tape around metal bands	Gullion 1965a
Finches and grouse	Colored anodized and aluminum butt-end	Gullion 1965b, Cohen 1969, Godfrey 1975, Stedman 1990, Hannon and Eason 1995
Small birds	Nylon wing tag fastened with a strap around the humerus	Hewitt and Austin-Smith 1966
Captive birds	Close-ring bands put on nestlings	Cohen 1969, Godfrey 1975
Finches, geese, oystercatchers, loons, cranes, woodpeckers, juncos, owls, ducks, blackbirds, magpies, and American goldfinches	Colored leg bands can affect mate selection, sex ratio of surviving offspring, and longevity	Martin 1963; Ogilvie 1972; Wheeler and Lewis 1972; Reese 1980; Burley 1982, 1985, 1986a, b, 1988; Burley et al. 1982; Goss-Custard et al. 1982; Forsman 1983; Seguin and Cooke 1983; Hoffman 1985; Ratcliffe and Boag 1987; Strong et al. 1987; Hagan and Reed 1988; Cristol et al. 1992; Metz and Weatherhead 1993; Forsman et al. 1996; Houston 1999; Dau et al. 2000; Verner et al. 2000; Watt 2001
Gulls	Butt-end, color bands, and rings	Mills 1972, Kadlec 1975, Spear 1980, Ottaway et al. 1984, Shedden et al. 1985
Raptors, ravens, and American woodcock	Color fabric wrapped around wing	Kochert 1973, Morgenweck and Marshall 1977, Kochert et al. 1983
Ducklings	Florist's wax or plasticine filled	Spencer 1978; Blums et al. 1994, 1999, Amundson and Arnold 2010
Seabirds and sandpipers	Butt-end and color bands; banding tibia rather than tarsus increases longevity and legibility	Anderson 1980, Perdeck and Wassenaar 1981, Zmud 1985, Colclough and Ross 1987, Reed and Oring 1993, Amat 1999, Bart et al. 2001, Breton et al. 2006, Sharpe et al. 2009, Roche et al. 2010
Mammals		
Bats	Bands cause injuries, and neonates need room to grow; best attached to forearm as bands are ineffective if attached to hind leg or pollex; do not band during hibernation, as populations decline	Davis 1963b, Perry and Beckett 1966, Cockrum 1969, Bonaccorso and Smythe 1972, Bateman and Vaughan 1974, Bonaccorso et al. 1976, Bradbury 1977, LaVal et al. 1977, Morrison 1978, Stebbings 1978, Keen and Hitchcock 1980, Hooper 1983, Moran 1985, Phillips 1985, Racey and Swift 1985, Bell et al. 1986, Barclay and Bell 1988
Small rodents	Leg rings	Fullagar and Jewell 1965
African elephant	Plastic tail collar	Viljoen 1986

[a] Scientific names are given in Appendix 9.1.

Fig. 9.8. Butt-end aluminum band (right leg) and colored plastic band (left leg) placed on greater prairie-chicken.

tic bands, and celluloid rings. In bats, injuries caused by bands often result due to motion of the forearms during flight. **Celluloid rings** produce fewer injuries. Bands attached to the bat's back legs are not effective markers due to band loss.

Leg Bands

The **butt-end** or **split-ring** metal band is widely used for most avian species (Table 9.4). Lock-on bands are used on raptors and other birds capable of removing butt-end bands. Rivet bands are used for eagles, which are capable of removing both butt-end and lock-on bands. Closed-ring bands often are used to mark birds raised in captivity.

Bands should fit properly, allowing movement, and young birds may be ringed with the aid of wax or other materials that yield with growth. Morrow et al. (1987) developed equipment to return nestlings to their tree nest following flushing and banding. Birds can mutilate and remove bands, and bands have been lost from nestlings. The main causes of loss of leg bands, however, are abrasion and corrosion from saltwater and feces. Vultures, which excrete down their legs, should not be leg banded, as excrement loading of the band can lead to loss of the leg or foot. Ice build-up on banded passerines in cold climates also may cause impairment of leg movement or leg loss. Colored plastic bands have caused severe leg abrasions (Reed 1953), band constriction has amputated legs (Atherton et al. 1982), and band displacement can cause crippling in web-footed species (Box 9.1). Leg-band loss can lead to inflated mortality estimates and errors in estimations of population size, especially for long-lived species (Nelson et al. 1980).

Nasal Discs and Saddles

Nasal discs and saddles (Fig. 9.9) have been used extensively to mark waterfowl (Table 9.5). **Nasal tags** are generally made from rigid or flexible plastic or nylon, marked with patterns or numbers, and attached by a short nylon or stainless steel pin through the nares. Discs may snag on vegetation and tangle in nets during trapping and probably increase mortality of diving ducks (Table 9.5). **Nasal saddles** that properly fit the size and shape of the bill of particular waterfowl species reduce such hazards. Entanglement in fences and traps has resulted in tag loss and icing on nasal saddles may increase mortality.

Backpacks, Harnesses, and Ponchos

Markers designed to lie on the back have been used frequently to mark upland game birds, waterfowl, and other birds (Table 9.6). **Backpacks** (Fig. 9.10) generally are made from flexible plastics or plastic-coated nylon fabric and are

> **BOX 9.1. SHRINKAGE OF SPIRAL PLASTIC LEG BANDS RESULT IN LEG DAMAGE TO MOURNING DOVES**
>
> Recaptures of mourning doves banded with spiral plastic leg bands revealed these bands were constricting and causing loss or severe damage to the legs (Atherton et al. 1982). Band color and temperature affected band shrinkage. Dark colored bands experienced greater shrinkage than did light colored ones. Higher temperatures caused bands to shrink more than lower temperatures did. Acetone-treated bands fused coils of the band together to help prevent shrinkage. Birds with fleshy legs, such as doves and pigeons, should have their spiral plastic leg bands treated with acetone prior to release of the birds.

Fig. 9.9. Nasal saddle on the bill of a female mallard.

Table 9.5. Nasal discs and saddles used to mark waterfowl

Tag type	Comment	Reference
Nasal disc	Snagged on vegetation and tangled in nets used to trap ducks; tag loss high on geese	Bartonek and Dane 1964, Sherwood 1966
Nasal saddle	Less tangling than nasal discs, but icing may increase mortality; fewer lost when saddles are sized to shape of bill; problems with small ducks due to large size of saddles and shape of duck bill and nares	Sugden and Poston 1968, Doty and Greenwood 1974, Greenwood and Bair 1974, Joyner 1975, Greenwood 1977, Koob 1981, Davey and Fullagar 1985, Lokemoen and Sharp 1985, Evrard 1986, Byers 1987, Pelayo and Clark 2000, Brook and Clark 2002, Regehr and Rodway 2003

Table 9.6. Back packs, harnesses, and ponchos used to mark birds and mammals

Group/mark type/species[a]	Comment	Reference
Birds		
Backpacks with straps		
Gray partridge, grouse, pheasant	Leather retained up to 1 year	Blank and Ash 1956, Gullion et al. 1962, Labisky and Mann 1962, Boag et al. 1973
American coot	Leather retained 1 year	Anderson 1963
Small birds	Cumbersome for small birds	Hester 1963, Furrer 1979
Bald eagles, falcons	Could be seen from long distance	Southern 1964, Kenward et al. 2001
Backpacks glued on back		
Gull chicks	Circular numbered tag to synsacrum	Cuthbert and Southern 1975
Hummingbirds	Glued back tags	Baltosser 1978
Ponchos		
Grouse, partridges, pheasant	Back tag modified into ponchos	Pyrah 1970, Marcstrom et al. 1989
Mammals		
Harnesses		
Collared peccary, deer	Braided rope harness	Bigler 1966

[a] Scientific names are in Appendix 9.1.

attached by a leather or nylon cord **harness** that passes around each wing base. Nylon straps last longer than those of leather. Backpack markers also have been modified into **ponchos**. Back tagging typically is considered too cumbersome for small birds, but a backpack marker that protruded from the bird's back, making it more visible, has been used to mark starling-sized birds. Numbered plastic circles glued to the backs of birds as small as hummingbirds have been used, but they are lost during molt. Rope harnesses have been used to individually mark large mammals (Table 9.6).

Trailing Devices

Trailing devices have been used to study movements of amphibians and reptiles with limited movement (Table 9.7). These devices usually consist of a freewheeling bobbin or spool holding thread or light string attached to an animal's body. In some aquatic situations, lines with floats are attached directly to the animal. **Bobbins** have been glued to an elastic band secured around the animal, or in the case of turtles, attached to the carapace with waterproof tape. To study movements, one end of the line is secured to a stake at the point of capture; as the animal moves, the trailing thread is released along the route of movement. Usefulness of the device depends on the amount of thread the bobbin or spool can hold and the speed and distance moved by the animal. The bulkiness of these devices can interfere with normal movement patterns, and the waistband attachment

Fig. 9.10. Female black grouse with backpack tag.

Table 9.7. Trailing devices applied to amphibians and reptiles to follow movements

Group/species[a]	Material	Comment	Reference
Box turtle	Wooden spool and thread with housing	Attached to carapace with waterproof adhesive tape	Stickel 1950
	Thread trailer and radiotransmitter	Attached to carapace	Lemkau 1970
	35-mm film canisters to hold wooden spool and thread	Attached to caudal end of carapace; avoided interference with mating	Reagan 1974
Green sea turtle	Fiberglass-coated floats attached to 24-m lines; 3-v flashlight bulb powered by batteries attached to float; fiberglass mast topped by orange pennant	No adverse effects reported	Carr et al. 1974
Lizards	Small piece of foil attached to 30-cm light string around lower abdomen	Allowed measurements of subterranean depth of lizards at night, located buried lizards for body temperature readings	Deavers 1972, Judd 1975
Turtles	Low-friction thread-release mechanism	Similar to spincast fishing reels	Scott and Dobie 1980
Northern leopard frog >60 mm long	Glued bobbin to elastic band around waist with stake to mark point of capture with sewing thread tied to it	50 m of thread lasted from 1 hr to 7 days; weighed 8.5 g; shortened jumping ability and hampered swimming and entering crevices; waistband caused skin irritation	Dole 1965, Grubb 1970
Tiger salamander	Sutured numbered plastic float through tail with monofilament line	Line sufficiently long to allow individual to move through the deepest part of lake	Whitford and Massey 1970

[a] Scientific names are given in Appendix 9.1.

can cause skin irritation. These devices have been used to study movement patterns both in terrestrial and aquatic systems, and to determine belowground depth of animals at night.

Nocturnal Tracking Lights

Light sources attached to animals allow them to be visually tracked at night, providing information on movements and foraging behavior. **Chemical** and **radioactive** lights can be used alone or in conjunction with radiotelemetry (Table 9.8). Evidence suggests that use of optical light sources does not increase predation of marked individuals or adversely affect their behavior, although this potential exists. Conversely, marked predators might have less success capturing prey, and a constant light source may cause undue stress in bats.

Cyalume®, a chemical light source, has been used to monitor the activity of wildlife (Table 9.8). The light is obtained by mixing dibutyl phthalate and dimethyl phthalate liquids and sealing the mixture in small, clear spheres that were glued to animals. Varying the proportions of this mixture controls the brightness and duration of light emission. Battery-operated "**pin lights**" and neon lights have been used

Table 9.8. Nocturnal light sources for tracking wildlife

Group/species[a]	Light source	Comment	Reference
Birds			
Black skimmer	Cyalume® or light-emitting diodes	Sealed plastic bulb on back	Clayton et al. 1978
Long-eared owl	Light-emitting diodes	Studied nest behavior	DeLong 1982
Boreal owl	Betalights	On radio antennas	Hayward 1987
Mammals			
Bats	Pin light with battery	Glued to fur	Barbour and Davis 1969
	Cyalume	Glass spheres glued to fur	Buchler 1976, LaVal et al. 1977
	Cyalume in gelatin capsule tag and lightstick tag	Miniature lightsticks provided equal or superior results	Hovorka et al. 1996
Mule deer	Neon light with battery	Neck collars	Carpenter et al. 1977
Europen badger	Betalights	On radiotransmitters	Kruuk 1978
American beaver	Light-emitting diodes	Neck collars	Brooks and Dodge 1978
Rabbits	Betalights	Attached to ear tags	Davey et al. 1980
Wallaby	Light-emitting diodes	Neck collars	Batchelor and McMillan 1980
Rodents	Betalights	Glued on head	Thompson 1982

[a] Scientific names are given in Appendix 9.1.

for nocturnal observations of mammals (Table 9.8). Light intensity or blinking sequence can be varied on neon lights for individual animal identification.

A **light-emitting diode** (LED) and flasher have been used to track wildlife at night (Table 9.8). The device produced consistently timed flashes that could be used for individual identification. A similar system with individually programmable flashes, a light-sensitive flasher, and optional attachment of a radiotransmitter to the same circuit was later developed. Battery size and light source intensity influenced the lifespan and visibility of the marker. Use of binoculars or night vision scopes greatly increased the distance at which these markers could be seen.

Betalights™ are a radioactive light source consisting of phosphor excited by tritium gas in glass capsules. The capsules can be produced in any shape and size with different colors. The useful range varies from about 50 m to 1 km, depending on shape, size, and viewing method. The lifespan of Betalights is about 15–20 years. Acceptable radiation levels should be considered when these light sources are used. Colors at different intensities can be used to increase the number of individuals identifiable. Betalights have been used on crabs (Wolcott 1977), birds, and mammals (Table 9.8). For birds, the most effective location for the Betalight was on a radio antenna away from the bird's body. Betalights did not increase mortality of radiomarked boreal owls, although hunting success could be affected.

Tapes, Streamers, and Bells

Tapes, streamers, and bells have been applied to animals to make them more readily detectable in the natural environment. **Fluorescent tapes** and bells also allow the animal to be detected and located more easily at night. The effect of these methods on animal survival requires further study.

Tapes

Colored tapes have been used to improve band retention and field recognition of birds (Table 9.9). Colored fabric, rip-stop nylon, and reflective tape with or without coded numbers have been used to mark other animals. Highly reflective plastic-tape strips and plastic-covered tape with coded numbers were glued to the heads of bats as temporary individual markers. Colored plastic adhesive tape was used as a durable visual marker on the horns of mountain sheep and as a short-term marker on the quills of porcupines. Labels on colored plastic tape have been used to mark individual eggs in bird nests. The tape label was firmly applied to the egg near the apex, and a different color or color combination was used for each egg laid in a clutch. These markers were not lost prior to hatching.

Streamers

Many types of streamers (Fig. 9.11) and flags made from such materials as fluorescent plastic, polypropylene, polyurethane, hypalon, orthoplast, nylon-coated vinyl, and vinyl

Table 9.9. Tapes, streamers, and bells applied to wildlife for individual or group identification

Group/species[a]	Material	Comment	Reference
Amphibians and reptiles			
American alligator	Flexible chain or plastic strip attached to anchor tag	Beneath skin on side of tail; slow healing	Chabreck 1965
Bullfrog	Nylon waistband painted with black numerals	Recognizable up to 8–12 months with binoculars	Emlen 1968
Iguanas, lizards	Colored Mystik™ cloth tape	Around neck	Minnich and Shoemaker 1970
Green iguana	Bells on fishing line	Around neck	Henderson 1974
Spotted turtle	Adhesive with numbers	On carapace	Ward et al. 1976
Amphibians, lizards	Colored beads	Around neck	Nace and Manders 1982, Fisher and Muth 1989
Skinks	Pressure sensitive tape	Around neck	Zwickel and Allison 1983
Bullfrog	Reflective tape	Cemented to head	Robertson 1984
Birds			
Pheasants	Plastic streamers, tags	Attached to tail feathers, neck	Trippensee 1941, Taber 1949
Stilt, grackle, gull, and heron nestlings	Plasticized PVC[b] tape	Attached to leg	Downing and Marshall 1959, Carrick and Murray 1970, Willsteed and Fetterolf 1986
Wild turkey, blackbirds, gulls, waterfowl, raptors	Leg streamers	Attached on leg through slits in the marker or to bands	Campbell 1960, Fankhauser 1964, Thomas and Marburger 1964, Guarino 1968, Arnold and Coon 1971, Royall et al. 1974, Frentress 1976, Platt 1980, Cline and Clark 1981
Gull eggs	Colored plastic tape	Attached to apex of egg	Hayward 1982

continued

Table 9.9. continued

Group/species[a]	Material	Comment	Reference
Mammals			
Deer and collared peccary	Bells	Used to observe behavior	Jordan 1958, Gruell and Papez 1963, Ellisor and Harwell 1969, Schneegas and Franklin 1972
Gray squirrel	Plasticized PVC[b] tape	Attached around neck with slot and notch system	Downing and Marshall 1959
Ungulates	Colored streamers of plastic, nylon, or nylon-coated fabric (Herculite®, Saflag®, or Annortite®), and plastic ear pennants	Attached to ear, horn, Achilles tendon, or to other marking devices; some reluctance of does to accept tagged fawns, but survival similar to non-tagged fawns	Knowlton et al. 1964, Harper and Lightfoot 1966, Miller and Robertson 1967, Queal and Hlavachick 1968, Downing and McGinnes 1969, Jonkel et al. 1975, Ozoga and Clute 1988, Panagis and Stander 1989
Bats	Reflective plastic tape strips with numbers	Glued to head fur; temporary markers	Williams et al. 1966, Daan 1969
Polar bear	Colored flagging tape	Ear marker	Lentfer 1968
Cetaceans	Streamers and flags	Secured with steel barbs, nylon darts, umbrella anchors, and anchor rivets	Evans et al. 1972, Mitchell and Kozicki 1975, White et al. 1981
Mountain sheep	Colored adhesive tape	On horn	Day 1973
Porcupine	Colored tape or flags	On quills or radiotransmitters	Pigozzi 1988, Griesemer et al. 1999

[a] Scientific names are in Appendix 9.1.

[b] PVC = polyvinyl chloride.

Fig. 9.11. Neck collar and ear streamer on white-tailed deer.

tubing have been used to visibly mark wild animals (Table 9.9). Nylon-coated fabric streamers were retained for several months to years. Different lengths and color codes provided a means of individual identification at a distance. **Streamers** often are attached to plastic or metal tags or collars to increase animal visibility.

Bells

Bells have been used in conjunction with other individual marking methods (e.g., color-coded ear tags and collars) to facilitate locating and monitoring movements of deer, collared peccaries (Fig. 9.12), and green iguanas (Table 9.9). Periods of auditory observation of peccaries provided movement data comparable to those gained from telemetry and allowed activity patterns and habitat use of the animal to be identified. **Bells,** however, could attract predators.

External Color Marks

Dyes, fluorescent pigments, bleaching, inks, and paints have been used as short-term external markers to identify wildlife at a distance (Table 9.10). No adverse physiological effects have been reported for these markers when properly applied

Fig. 9.12. Bell attached to collared peccary that allows investigators to follow herd movements.

Table 9.10. Dyes, paints, stains, pigments, ink, and bleaches used to externally mark wildlife

Group/species[a]	Material	Comment	Reference
Amphibians and reptiles			
Tortoises, turtles, snakes	Colored paint	On carapace of tortoises and on rattles or head of snakes	Woodbury and Hardy 1948, Pough 1966, Bennett et al. 1970, Bayless 1975, Medica et al. 1975, Bennion and Parker 1976, Parker 1976, Brown et al. 1984
Terrapins	Ink	Injected into skin	Burger and Montevecchi 1975, Burger 1976
Frogs, tadpoles	Neutral red, whole-body dye	Some immediate deaths and affected growth	Herreid and Kinney 1966, Guttman and Creasey 1973, Travis 1981
Lizards	Paint, indelible pencil, felt-tipped pen	Lost with shedding; survival same as for toe clipping	Tinkle 1967, 1973; Jenssen 1970; Stebbins and Cohen 1973; Henderson 1974; Vinegar 1975; Fox 1978; Jones and Ferguson 1980; Simon and Bissinger 1983
Salamanders	Fluorescent pigment	Good for short-term studies	Taylor and Deegan 1982, Nishikawa and Service 1988, Ireland 1991
Frogs, toads	Panjet dye	Lasted up to 2 years	Brown 1997
Juvenile frogs	Tetracycline bath	Failed as marker	Hatfield et al. 2001
Birds			
Small birds, ducks, gulls, pheasants, eagles, swifts, terns, geese, swans, blackbirds	Dye	Visibility up to 2 km	Butts 1930, Price 1931, Wadkins 1948, Jones 1950, Winston 1955, Kozlik et al. 1959, Ellis and Ellis 1975, White et al. 1980, Malacarne and Griffa 1987, Underhill and Hofmeyer 1987, Paullin and Kridler 1988, Belant and Seamans 1993
Ruffed grouse, cattle egrets, bird eggs	Printer's ink	Lasted up to 12 months for cattle egret; no harmful effects on eggs	Bendell and Fowle 1950, Boss 1963, Siegfried 1971, Olsen et al. 1982
Mourning doves, northern cardinals	Model airplane paint, spray paint	Preening resulted in feather loss; pair-bond disturbance	Swank 1952, Frankel and Baskett 1963, Goforth and Baskett 1965, Dickson et al. 1982
Mammals			
Squirrels, deer, terrestrial mammals, pinnipeds	Dye (Gentian violet, Biebrich scarlet, picric acid, Nyanzol A, Rhodamine B, Woolite®, clothing and aniline, and human hair dye with peroxide or hair bleach)	Ear tags and toe clipping best for long-term marking	Baumgartner 1940, Fitzwater 1943, Webb 1943, Hansen 1964, Simmons 1971, Day 1973, Brady and Pelton 1976, Bradbury 1977, Gentry 1979, Pitcher 1979, Johnson et al. 1981*b*, Gentry and Holt 1982, Henderson and Johanos 1988, Hurst 1988
African elephants, bovids, bats, antelopes, aquatic mammals	Paint, paint-stick, spray paint	Applied to hide, horn, or pelage; must remain dry for 15–30 minutes	Pienaar et al. 1966, Hanks 1969, Watkins and Schevill 1976, Gentry and Holt 1982, Clausen et al. 1984, Irvine and Scott 1984, McCracken 1984
Seals, small mammals	Fluorescent pigment	Adequate for 2 years for seals and small mammals dusted after trapping; trail followed with UV lamps	Griben et al. 1984, Lemen and Freeman 1985, Boonstra and Craine 1986, Dickman 1988, Mullican 1988, Mikesic and Drickamer 1992, Stapp et al. 1994
Woodrat, rats, pangolin	Capsule containing fluorescent dust	Long-term tracking and trail deposition	Goodyear 1989

[a] Scientific names are in Appendix 9.1.

to mammals. For birds, no obvious behavioral changes were noted other than temporarily increased preening. Certain markings could disrupt pair bonding, however, and altered intraspecific recognition mechanisms in birds may severely alter social interactions (Rohwer 1977).

Dyes

Waterproof dyes should yield an easily recognizable color, resist fading, and be nontoxic, harmless to plumage, capable of use with a wetting agent or solvent to ensure quick penetration and coverage, and fast acting in a cool solution (Patterson 1978). Picric acid, Rhodamine B Extra, and Malachite Green yield strong colors and exhibit good penetration and retention (Handel and Gill 1983). Avian species with light plumage are most effectively marked with dyes. Dipping, brushing (Fig. 9.13), and spraying have been used to apply dyes. To avoid hypothermia in cool weather, dye-marked birds should be thoroughly dried before release.

Bleaching

Bird feathers and mammal furs have been **bleached** and colored using human hair dyes or lighteners mixed with hydrogen peroxide (Table 9.10). Skin and feather damage can occur if tissues are bleached at too high a temperature or for too long. Animals also may be susceptible to hypo- and hyperthermia during the bleaching process.

Fluorescent Pigments

Trapped animals have been dusted with **fluorescent pigments,** so that a fluorescent trail can be traced using ultraviolet (UV) lamps the following night (Table 9.10). The amount of vegetation cover, precipitation, and ambient light influenced trail detection. This technique enables collection of detailed information on home range, movement patterns, and habitat in a few days. To increase the duration of this marker beyond the second night, capsules containing pigments can be attached. A promising marker for aquatic mammals is a paste made from fluorescent pigments, vehicle binder, and solvent. It has visibly marked aquatic mammals for up to 2 years with no adverse behavioral effects or tissue abnormalities. Codit™ white reflective liquid also has been used to mark freshwater animals.

Inks

Ink has been used to mark salamanders, terrapins, turtle eggs, iguanas, lizards, bird eggs, and deer (R. R. Lopez, unpublished data; Table 9.10). On deer, **ink** proved superior to paint for duration and visibility. Marking pens have been used to number eggs in clutches. No harmful effects were observed, but marking pens should be used with discretion until possible toxic effects on embryos are evaluated.

Paints

Liquid and spray **paints** usually are applied to the skin, pelage, horns, or feathers (Table 9.10) and persist for a few weeks to several months. Individuals must be repainted, as paint is lost due to shedding, molting, and grooming. How these marks influence the behavior of species for which colors have seasonal social significance is unknown. Paints should be dry before animals are released.

INVASIVE MARKING TECHNIQUES

Internal Markers

Chemical, particle, and radioactive markers have been injected in or fed to animals to either physically mark individual animals or groups of animals (some chemical markers) or to detect byproducts from marked individuals (fecal markers). These methods require animals to be captured prior to marking.

Chemical Markers

Organic stains placed in the tail-fin cavity or caudal region with a hypodermic needle have been developed as a reasonably permanent marker for amphibians (Table 9.11). During metamorphosis, the mark was reabsorbed with the tail with no ill effects.

Rhodamine B taken orally acts as an internal marker, coloring the gall bladder, gut, feces, urine, and oral and urogenital openings and producing fluorescent banding of feathers in birds (Table 9.11). These bands were most evident in primary and secondary feathers. Rhodamine B may become visible within 24 hours of dosing and can persist for several weeks. Scanning for fluorescence using portable UV lamps allows trapped animals to be examined and released immediately, thus reducing stress. Use of Rhodamine B as a

Fig. 9.13. Colored dye being applied with brush to the white portion of a white-winged dove wing.

Table 9.11. Internal particle and chemical markers used to study wildlife

Group/species[a]	Material	Comment	Reference
Amphibians and reptiles			
Salamanders	2:1 Liquitex® acrylic polymer to distilled water	Injected into lateral, proximal, and caudal regions	Woolley 1973
Salamander larvae	Fine-grained fluorescent pigments mixed as paste	Administered with heated probe; short-term tag	Ireland 1973
Frog and salamander larvae	21:20 ratio of mineral oil to petroleum jelly and stains (Oil Red A and Oil Blue M)	Injected into tail fin cavity with 22-gauge hypodermic needle; no effect on animals	Seale and Boraas 1974
Birds			
Duck and passerine eggs	Food dye	Injected into egg; hatched young marked for few days	Evans 1951, Rotterman and Monnett 1984
Bait consuming birds, raptors	Microtaggants (small, color-coded plastic particles)	Fed in baits	Johns and Thompson 1979, Nietfeld et al. 1994
Bait consumers	Iophenoxic acid and mirex	Iophenoxic acid ineffective	Larson et al. 1981
Waterfowl	Tetracycline	Injected; detected in eggs; egg-laying rate decreased	Haramis et al. 1983, Eadie et al. 1987
Mammals			
Small mammals	Dye in food	To mark fat, teeth, pelage, and feces	New 1958, 1959; Kindel 1960; Nass and Hood 1969
Cottontail rabbit	Dye pellets placed under skin	Observed in urine on snow	Brown 1961
Coyote, rodents, skunks, raccoon, seals, dolphins, whales, bears, white-tailed deer	Tetracycline group	Fed in baits; more intense in mandible and teeth and in young animals	Owen 1961, Yagi et al. 1963, Linhart and Kennelly 1967, Crier 1970, Nelson and Linder 1972, Best 1976, Geraci et al. 1986, Garshelis and Visser 1997, Taylor and Lee 1994, Van Brackle et al. 1994
Collared peccary	Glass beads	Force-fed beads	Sowls and Minnamon 1963
Ground squirrels	Nyanzol A and D fur dyes	Accuracy with field identification	Melchior and Iwen 1965
Snowshoe hare	Picric acid and Rhodamine B	Picric acid worked best	Keith et al. 1968
Nutria	Codit™ white reflective liquid	Fecal tracer; for 30 days	Evans et al. 1971
Nutria	Powered aluminum pigment	Fecal tracer	Evans et al. 1971
Rats, rabbits	Sudan black, orally	Stained fat deposits	Taylor and Quy 1973, Cowan et al. 1984
Rabbits, Virginia opossum	Rhodamine B	Fecal tracer	Evans and Griffith 1973; Morgan 1981; Cowan et al. 1984, 1987
Bait consuming mammals	Fluorescent acetate floss fibers	Fed in bait	Randolph 1973, Johns and Thompson 1979, Cowan et al. 1984
Coyote, gophers, mountain beaver	Rhodamine B	Systemic marker; produces fluorescent banding of claws and hair	Ellenton and Johnston 1975, Johns and Pan 1981, Lindsey 1983
Rats	Quinacrine dehydrochloride	Fluorescent in blood	Johns and Pan 1981
Bait consumers	Microtaggants	Fed in bait	Johns and Thompson 1979
Dogs, foxes	Iophenoxic acid	Fed in bait	Baer et al. 1985, Follmann et al. 1987
Coyote	Chlorinated benzenes	Fed in bait	Johnson et al. 1998

[a] Scientific names are listed in Appendix 9.1.

systemic marker may be limited to certain periods of the year in birds, because banding probably occurs only in actively growing tissue. Rhodamine B has been used to detect bait consumption, estimate densities, and examine movements. Fisher (1999) summarized the literature on Rhodamine B and concluded the long-term effects of a single dose and a short succession of low doses on live animals should be investigated. She recommended Rhodamine WT as an alternative systemic bait marker.

Certain members of the tetracycline family of antibiotics, given orally or intravenously, combine with calcium in bones and teeth of mammals and eggshells of birds to produce a characteristic yellow fluorescence under UV light (Table 9.11). **Tetracyclines** are persistent, quantitative markers that can cross the placental barrier. They have been used to obtain mark–recapture population estimates and to identify the percentage of predators that consumed baits.

Quinacrine dehydrochloride, a fluorescent chemical marker, can be detected in blood with fluorometric and chromatographic analytical techniques (Table 9.11). Iophenoxic acid, an iodine-containing compound, and mirex, an organochlorine pesticide, have been used as blood and tissue markers for bait-consuming birds and mammals. Codit white-reflective liquid and Sudan black also are satisfactory fecal tracers for most mammals.

Particle Markers

Microtaggants, small plastic particles that are coded by colored layers, do not cause bait aversion; remain intact; and, due to their fluorescent and magnetic properties, can be readily recovered from gut or fecal samples (Table 9.11). Fibers of fluorescent acetate floss also have been tested for measuring bait consumption by birds and mammals and individual movements in small mammals. As with microtaggants, floss fibers are quantitative nonpersistent markers. Floss fibers do not affect bait palatability and are more economic than microtaggants. Powdered aluminum placed in baits also has been used as a fecal tracer.

Visible Implant Elastomers

Visible Implant Elastomers (VIEs) use a 2-part silicone-based material that is mixed immediately before use. VIE tags are **injected as a liquid** that soon cures into a pliable, biocompatible solid. The tags are implanted beneath transparent or translucent tissue and remain externally visible. In many amphibians, VIE tags are visible through darkly pigmented skin. VIE tags are widely used for marking an ever-broadening range of reptiles and amphibians. Bailey (2004) evaluated VIE markings using photo identification for terrestrial salamanders. Nauwelaerts et al (2000) used VIE as a method to mark adult anurans. Regester and Woosley (2005) used **Visible Fluorescent Elastomer** to mark salamander egg masses. Campbell et al. (2009) tested the efficacy of VIE and toe-clipping on 4 species of tree frogs in west-central Florida. Of the 840 tree frogs recaptured during a 15-month period, only 1 mark was unreadable. A significantly higher percentage of VIE marks (80%) than toe-clips (55%) remained viable for the duration of the study. On average, toe-clips remained readable for 100 days, and VIE marks remained readable for 112 days.

Radioactive Markers

Radioactive tracers have been used to identify and acquire information on the behavior of amphibians, reptiles, and mammals, but they have received little attention for birds. The 3 primary methods of marking animals with **radioisotopes** are inert implants, external attachments, and metabolizable radionucleotides (Table 9.12). Inert implants are suitable for monitoring specific movements, such as nest visits by birds and small mammals, using a manual or auto-

Table 9.12. Radioisotopes used for marking wildlife

Group/species[a]	Radioactive material	Comment	Reference
Amphibians and reptiles			
Toads, salamanders, snakes	Cobalt	Injected	Karlstrom 1957; Breckenridge and Tester 1961; Barbour et al. 1969a, b; Ashton 1975
Northern fence lizard	Gold	In tubing around waist	O'Brien et al. 1965
Salamanders, turtles, skinks, lizards, snakes	Tantalum	Injected, local ulceration in salamanders	Bennett et al. 1970, Madison and Shoop 1970 Ward et al. 1976, Ferner 1979
Salamander larvae	Sodium	Injected	Shoop 1971
Birds			
Semipalmated plover	Radioactive leg band		Griffin 1952
Ring-necked pheasant	Calcium	Identify chicks from fed hens	McCabe and LePage 1958
Mammals			
Voles	Phosphorus	Injected	Miller 1957
Mammals	Iodine	Injected, capsules on rings, implanted, or fed	Gifford and Griffin 1960, Johanningsmeier and Goodnight 1962
Harvest mouse	Gold	Implanted	Kaye 1960
Small mammals	Cobalt	Implanted or in capsule on rings	Linn and Shillito 1960, Barbour 1963, Schnell 1968
Small mammals	Tantalum	Implanted	Graham and Ambrose 1967, Schnell 1968
Small mammals, Virginia opossum, rabbits, European badger, bobcat, black bear	Zinc	Injected or fed	Nellis et al. 1967, Schnell 1968, Gentry et al. 1971, Pelton and Marcum 1975, Kruuk et al. 1980, Conner 1982
Black bear	Magnesium	Injected	Pelton and Marcum 1975
Small mammals	Sulphur	Passed through mother's milk	Dickman et al. 1983
Rodents	Radionuclides	Mother–offspring relatedness; male reproductive success	Tamarin et al. 1983, Scott and Tan 1985
Raccoon	Cadmium	Injected	Conner and Labisky 1985
Coyote	Several tested	Implanted	Crabtree et al. 1989

[a] Scientific names are in Appendix 9.1.

mated detector (Griffin 1952, Bailey et al. 1973, Linn 1978). Radioactive wires, pins, and capsules containing isotopes have been inserted subcutaneously in small rodents and bats as inert implants. Radioactive material can be attached to external leg bands and forearm tags, or the bands or tags can be made radioactive. Radioactive material can be fed, injected, or implanted into the animal in a metabolizable form. These materials may be incorporated into the tissues of the animal, passed on to offspring, or voided in feces and urine; thus they can be used for many purposes besides tracking (Linn 1978). This approach has been used to estimate population abundances of a number of species.

A major disadvantage of using radioactive markers is the restrictions imposed by state or federal regulations. These tags also can cause illness or death of marked animals, be lost, and can constitute a hazard to other animals, including humans. When selecting a radioactive marker, one should consider availability, type of radiation, energy levels emitted, physical and biological half-life, radiotoxicity, and metabolic characteristics (Pendleton 1956).

Transponders

Passive integrated transponder (**PIT**) **tags** have been developed as permanent markers and have been tested on amphibians, reptiles, birds, and mammals (Table 9.13). The tags consist of an electromagnetic coil and a custom-designed transponder chip that emits a uniquely programmed alphanumeric analog signal when excited by a scanning wand that discharges electromagnetic energy. The PIT-tag reader displays the code and can store this information for later retrieval. PIT tags are implanted subcutaneously (Fig. 9.14) with a special syringe and canula (needle).

No adverse effects of transponders have been observed in animals, but PIT tags are not as permanent as first thought; they can fail and be lost (Box 9.2). The major disadvantage of this system, however, is the reader must be close (a few centimeters) to the animal to record the code, which may necessitate recapturing the animal. Remote readings of transponders can be made (Table 9.13): a reader tube can be inserted into burrows or nesting cavities, or along travel routes, reading the transponder number each time the marked animal passes.

Tattoos

Tattoos provide an efficient means of permanently marking a wide range of species (Table 9.14). Best results are achieved by tattooing lightly pigmented areas free of hair (inside ears [Fig. 9.15], on inner legs or arms, or inside lips) or feathers (under wings). Standard or rotary pliers, electric tattooing pencils, and syringes filled with ink have been used to inject contrasting dye (e.g., green or black; Table 9.14). Small quantities of fluorescent pigments also have been used to make tattoos that are visible only under UV light. Although tattoos generally cause fewer problems (e.g., no added weight, inconspicuous to predators) than other marking techniques, they have the disadvantage of requiring ani-

Table 9.13. Passive integrated transponders (PIT) used to mark wildlife

Group/species[a]	Comment	Reference
Amphibians and reptiles		
Frogs, toads, alligators, snakes, lizards, turtles, sea turtles	Only 1 of 118 PIT tags failed; lasted up to 2 years	Camper and Dixon 1988, Brown 1997
Blunt-nosed leopard lizard	Successfully scanned 250 of 273	Germano and Williams 1993
Pine snake	PIT tags retained on 92%	Elbin and Burger 1994
Neonatal snakes	No effect on growth and movement	Keck 1994b
Rattlesnakes	No effect on growth and movement	Jemison et al. 1995
Desert tortoise	Detected as they entered culverts	Boarman et al. 1998
Great-crested newt larval stage	Up to 2 years	Cummins and Swan 2000
Birds		
Captive birds	Success varied with species and year	Elbin and Burger 1994
Northern bobwhite chicks	PIT tags lost on 5%	Carver et al. 1999
Mammals		
Black-footed ferret	Failed in 6 of 48	Fagerstone and Johns 1987
Sea otter	Successfully scanned 6 of 6	Thomas et al. 1987
Big brown bat	Successfully scanned 17 of 17	Barnard 1989
Mice	Successfully scanned 4 of 4	Rao and Edmondson 1990
Norway rat	Successfully scanned 10 of 10	Ball et al. 1991
Townsend's ground squirrel	No effect on squirrels	Schooley et al. 1993
Captive mammals	Success varied with species and year	Elbin and Burger 1994
Voles	Used to monitor runways	Harper and Batzli 1996
Naked mole rat	Survival not different from toe-clipped	Braude and Ciszek 1998

[a] Scientific names are given in Appendix 9.1.

Fig. 9.14. Implanting a passive integrated transponder tag into a radiomarked fox squirrel.

Fig. 9.15. Numeric characters tattooed inside the ear of a white-tailed deer.

mal recapture for identification. Tattoos are often used with more visible, but less permanent marking methods.

Tags

Tags, as used here, differ from bands in they penetrate some part of the animal's body and generally inflict pain, at least during insertion. With amphibians and reptiles, tags are usually placed through the shell, scutes, fore flipper, scales, tail fin, rattles, or tail (Table 9.15). In birds, tags are generally placed in the patagium of the wing or the webbing of the foot. Tags are typically placed in the ear, webbing of foot, flipper, or dorsal fin of mammals. Tag loss increases with time since tagging and may result from infection, wear, grooming, or fighting. Placing tags bilaterally and using them in conjunction with more permanent markers (e.g., tattoos) minimizes the chance of losing the identity of an animal over a long period. Study duration and required tag visibility are factors that influence tag choice. Many types of tags require recapturing the animal for identification.

Ear

Tags manufactured from metals and plastics (Fig. 9.16) in a variety of shapes, sizes, and colors with identifying numbers stamped into the surface are commonly used for marking mammals (Table 9.15). Tag-closing mechanisms can be interlocking, self-locking, or a rivet design that cannot be easily pried apart once the rivet is flattened. Tags may be self-piercing (Box 9.3) or inserted through a hole pierced with a knife or punch provided with the tagging kit. **Ear tags** usually are placed on the lower, inner region of the ear characterized by heavier cartilage, where the tag is best protected from being torn out. Tags should be loose enough to not interfere with blood circulation; puncture marks should be treated appropriately to prevent infection and ensure healing.

Aluminum, Monel, and plastic tags available for domestic livestock (Fig. 9.17) work well on ungulates. **Fingerling fish tags** have been used in the ears of bats since the 1930s. These tags may not be suitable for large-eared bats or species that exhibit rapid ear movement synchronized with their echolocation emissions, or for medium-sized to large bats due to poor retention. **Delrin® button tags** are satisfactory for marking several species.

> **BOX 9.2. PASSIVE INTEGRATED TRANSPONDERS (PIT) SHOULD NOT BE USED AS SOLE DEVICE TO MARK WILDLIFE**
>
> Recent research using PIT tags to mark fox squirrel had a 17% unsuccessful scan rate 3 months after implantation. Recaptured squirrels also were marked with radio collars. In a separate study on pocket gopher in which PIT tags were the only mark used, only 1 of the original 13 pocket gophers marked was recaptured during 1 year of trapping. Loss of tags, tag breakage, or trap avoidance by previously trapped gophers were possible explanations for the low recapture rate. However, because both the fox squirrels and pocket gophers were tagged in the nape of the neck and both species used areas (holes in trees or burrows in the ground) that rubbed this part of the body, the PIT tags may have been rubbed off or crushed. We recommend that PIT tags not be the sole marking device used to mark wildlife.

Table 9.14. Wildlife marked using tattoo techniques

Group/species[a]	Tattoo location	Comment	Reference
Amphibians and reptiles			
Snakes	Skin	Method was permanent	Woodbury 1956
Frogs	Skin of the venter	Etched grooves with ink	Kaplan 1958
American alligator	Light skin under tail	Legible for several months	Chabreck 1965
Salamander	Subcutaneous	Fluorescent elastomer	Davis and Ovaska 2001
Birds			
European starling	Abdomen	India ink dots using syringe	Ricklefs 1973
Birds of prey	Underside of wing	Captive birds, long lasting	Havelka 1983
Mammals			
Bats	Wing membrane	Slow process	Griffin 1934
Hares, rabbits	Ear	Used Franklin Rotary Tattoo	Thompson and Armour 1954, Keith et al. 1968
Bears	Upper lip, axilla, or groin	Permanent mark	Lentfer 1968, Johnson and Pelton 1980a
Deer fawns	Ear	Permanent mark	Downing and McGinnes 1969
Cottontail rabbit	Ear	Permanent mark	Brady and Pelton 1976
Dolphinids	Fin	Proposed only	White et al. 1981
European badger	Inguinal area	Electrically powered pen	Cheeseman and Harris 1982
Pere David's deer	Ear	Permanent mark	Carnio and Killmar 1983
Beluga whale	Flipper	Unsatisfactory	Geraci et al. 1986
Rats, mice	Ear	Permanent mark	Honma et al. 1986
Marsupial young	Pinnae	Fluorescent pigment	Soderquist and Dickman 1988
Porcupine	Ear	Not necessary with collars	Griesemer et al. 1999
Rodents	Subcutaneous	Chinese ink	Leclercq and Rozenfeld 2001

[a] Scientific names are in Appendix 9.1.

Table 9.15. Tags used to mark wildlife

Group/species[a]	Tag type	Reference
Amphibians and reptiles		
Frogs, toads, snakes	Metal jaw tags	Raney 1940, Stille 1950, Hirth 1966
Frogs, turtles	Bands, rings, or plates fastened through holes in shell	Kaplan 1958, Loncke and Obbard 1977, Graham 1986, Layfield et al. 1988
Alligators	Monel tag to dorsal tail scute	Chabreck 1965
Snakes, turtles	Button in caudal musculature	Pough 1970, Froese and Burghardt 1975
Sea turtles	Monel™ and plastic tags in fore flipper	LeBuff and Beatty 1971, Bacon 1973, Pritchard 1976, Bjorndal 1980, Pritchard 1980, Frazer 1983, Balazs 1985, Eckert and Eckert 1989
Rattlesnakes	Colored discs through rattle	Pendlebury 1972, Stark 1984
Turtles	Titanium disks held by adhesive	Gaymer 1973
Hellbender	Floy® T-tag	Nickerson and Mays 1973
Turtles	Wooden dowel in scute	Davis and Sartor 1975
Snakes	Colored beads on line	Hudnall 1982
Birds		
Waterfowl	Streamers pinned to head	Gullion 1951
Penguins	Flipper bands made of aluminum, Teflon®, monel, and stainless steel	Sladen 1952, Penny and Sladen 1966, Cooper and Morant 1981, Sallaberry and Valencia 1985
American woodcock	Plastic neck tag attached with surgical clip	Westfall and Weeden 1956
Waterfowl, turkey, gulls, cranes, willet, vultures, blackbirds, large passerines, woodpeckers, pigeons, grouse, American coot	Patagial tag using various materials to attach tag through patagium	Anderson 1963; Knowlton et al. 1964; Mudge and Ferns 1978; Tacha 1979; Bartelt and Rusch 1980; Howe 1980; Wallace et al. 1980; Jackson 1982; Seel et al. 1982; Baker 1983; Curtis et al. 1983; Southern and Southern 1983, 1985; Sweeney et al. 1985; Szymczak and Ringelman 1986; Cummings Hart and Hart 1987; Hannon et al. 1990; Seamans et al. 2010
Wood duck, gull chicks, geese, ducklings in eggs	Fingerling fish tags attached to foot web through hole in egg	Grice and Rogers 1965; Alliston 1975; Haramis and Nice 1980; Ryder and Ryder 1981; Seguin and Cooke 1985; Blums et al. 1994, 1999

continued

Table 9.15. continued

Group/species[a]	Tag type	Reference
Mammals		
Bats	Fingerling ear tags	Mohr 1934, Stebbings 1978
Rabbits, squirrels, Steller sea lion, deer, caribou, fox, goats, seals, bears, mice, coyote, elk, porcupine, moose calves, American beaver	Plastic or metal ear tag with or without streamers	Trippensee 1941, Scheffer 1950a, Tyndale-Biscoe 1953, Labisky and Lord 1959, Craighead and Stockstad 1960, Knowlton et al. 1964, Miller 1964, Harper and Lightfoot 1966, Miller and Robertson 1967, Downing and McGinnes 1969, Larsen 1971, Day 1973, Hubert et al. 1976, Rudge and Joblin 1976, Hobbs and Russell 1979, Stirling 1979, Warneke 1979, Johnson and Pelton 1980a, Beasom and Burd 1983, Alt et al. 1985, LeBoulenge-Nguyen and LeBoulenge 1986, Gionfriddo and Stoddart 1988, Ostfeld et al. 1993, Griesemer et al. 1999, Swenson et al. 1999
Fox squirrel	Fingerling toe tags, bands on toes	Linduska 1942, Cooley 1948
Big game	Plastic streamer through slit in ear	Craighead and Stockstad 1960
Hares, nutria, sea otter, seal pups	Tags placed on hind-foot web or rear flipper	Keith et al. 1968, Evans et al. 1971, Johnson 1979a, Miller 1979, Ames et al. 1983, Henderson and Johanos 1988
Cetaceans	Plastic and bolt tags to dorsal fin	Norris and Pryor 1970, Irvine et al. 1982, Tomilin et al. 1983
Whales	Discovery marks and spaghetti tags (stainless steel projectiles) shot from shotgun	Clarke 1971, Evans et al. 1972, Mitchell and Kozicki 1975, Leatherwood et al. 1976, Brown 1978, Irvine and Scott 1984, De La Mare 1985, Miyashita and Rowlett 1985, Kasamatsu et al. 1986

[a] Scientific names are given in Appendix 9.1.

Fig. 9.16. Plastic numbered tags attached to both ears of a collared peccary.

Wing

Wing tags are commonly used on birds (Table 9.15). They are generally made from flexible plastic-coated nylon fabric (Fig. 9.18), and rigid or upholstery plastic and attached through the **patagium** using a stainless steel or nylon pin, pop-rivet, or the marker itself. Durability and colorfastness are functions of material composition and manufacturing (Nesbitt 1979, Young and Kochert 1987), with some materials lasting ≤10 years. Tag loss is generally low the first year (Patterson 1978, Stiehl 1983), but gradually increases in subsequent years (Patterson 1978). Double pinning tags reduced marker loss. Streamers are often used with wing tags to make them visible at a distance. If used, they should be sufficiently large for observational purposes, yet not so large as to hinder flight.

Wing markers often have no consistent effect on birds, although the initial adjustment period ranges from a few days to 2 weeks. Light feather wear and patagium callusing have been commonly noted. Severe abrasion has been observed occasionally with some species, and consistently with falcons. Abnormal replacement of feathers may occur, and flight can be affected. Double pinning greatly reduces feather abrasion and callusing. Reported effects of wing markers on reproductive and social behavior also are variable. For many species, no significant influence on fledging success

> **BOX 9.3. PLACEMENT OF SELF-PIERCING METAL EAR TAGS IS IMPORTANT FOR RETAINING TAGS**
>
> It has been our experience when using self-piercing metal ear tags on white-tailed deer that placement is important for retention of tags. Tags should be placed near the base of the ear, and the metal tag should be flush with the edge of the ear. If space is left between the tag and the edge of the ear, there is greater probability that brush or other foreign objects will become entangled in the tag and rip it from the ear. The tag should not be so tight as to roll the edge of the ear, but it should be flush with that edge. Care also should be taken not to puncture any veins in the ear when applying the tag.

aquatic mammals, and sea birds), **wings** (birds and bats), and **dorsal fins** (cetaceans; Table 9.15). Migration of the tags, injury to the dorsal fin, and covering of the tag with algae were problems associated with dorsal fin tags. For marking fore flippers, Monel tags are more durable than plastic tags, although they may be less visible on marked animals and can exhibit significant rates of loss. Aluminum tags, which wear and corrode easily, are regarded as inferior to stainless steel or Monel tags for species inhabiting seawater.

Self-piercing fingerling-fish tags, Monel tags, plastic and metal ear tags, and Delrin button tags also have been used to mark the hind foot webs of mammals and birds with good retention. **Web tagging** has been used to mark ducklings in pipped eggs—part of the shell and membrane of an egg were removed, a foot extracted, tagged, and replaced, and the hole covered with masking tape. Web tagging did not affect hatching success or survival after nest departure.

Body

Metal and plastic tags have been used to tag the shells of turtles, rattles of snakes, scutes of turtles and alligators, tails of amphibians, and snakes (Table 9.15). With the exception of turtles, other marking methods are typically recommended over body tags.

Jaw

Jaw tags have been used for amphibians and reptiles, but often were lost and caused irritation (Table 9.15). Numbered Monel tags had to be clamped into the corner of the mouth, a technique that has not been widely used and is not recommended.

Branding

Branding provides an inexpensive, permanent, and visible means of marking animals. Hot iron, freeze, chemical, and laser branding have all been used to mark wildlife (Table 9.16). In addition, brand-like marks have been produced by using a special clamp to hold a stencil on either side of the dorsal fin of cetaceans, causing the epithelium under the pressurized area to be exfoliated and replaced by demelanized skin that remained distinct for ≥2 years. This procedure, however, required 4 days for the depigmented tissue to be produced, limiting its value as a field marker.

Hot-Iron Branding

Historically, **hot-iron branding** was used to permanently mark domestic livestock. Hot branding has almost no role in modern wildlife management and is not recommended, because it causes extreme pain and can produce open wounds that become infected. Currently, the only commonly used application of this technique in wildlife involves marking the horns of bovids.

Fig. 9.17. Plastic domestic-livestock ear tag used on white-tailed deer that also has been collared.

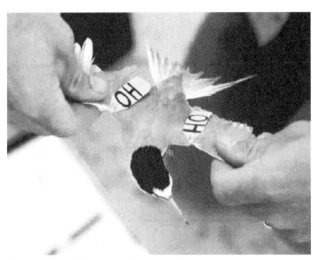

Fig. 9.18. Patagial wing markers on a least tern.

was found when ≥1 adult was marked (Young and Kochert 1987). However, reduced brood size, lengthened mean renesting interval, decreased social status, interference with migration, altered habitat selection, increased mortality, and effects on parental behavior (Brubeck et al. 1981) have been documented. Saunders (1988) contended that patagial tags should not be used on rare, vulnerable, or endangered species unless no other marking technique would work.

Other Appendages

Tags designed for marking ears also have been used to mark foot webs (birds and mammals), interdigital **webbing** of the hind foot (aquatic mammals and birds), **flippers** (sea turtles,

Table 9.16. Wildlife marked using hot-iron, freeze, chemical, and laser branding techniques

Group/species[a]	Brand type	Comment	Reference
Amphibians and reptiles			
Tortoises, snakes, toads, frogs, turtles, anoles, lizards, hellbender	Hot iron	Tortoises and turtles branded on carapace	Woodbury and Hardy 1948, Weary 1969, Clark 1971, Taber et al. 1975
Snakes, sea turtles, frogs, iguanas, salamanders	Freeze	Tailed frogs branded on ventral surface	Lewke and Stroud 1974, Daugherty 1976, Ferner 1979, Bull et al. 1983
Anurans	Chemical	Silver nitrate	Thomas 1975
Turtles, snakes	Laser	Ruby laser	Ferner 1979
American alligator	Freeze	Tail and rear foot pad	Jennings et al. 1991
Birds			
Mallard duckling	Freeze	Branded feather tracts and premaxillae	Greenwood 1975
Mammals			
Mountain sheep, African ungulates, seals, bovids	Hot iron	Branded horns and/or body	Aldous and Craighead 1958, Hanks 1969, Summers and Witthames 1978, Ashton 1978
Livestock, lab animals, pets, white-tailed deer, rodents, squirrels, mongoose, seals, dolphins, bats, American beaver	Freeze	Branded body	Newsom and Sullivan 1968, Farrell et al. 1969, Hadow 1972, Farrell and Johnston 1973, Lazarus and Rowe 1975, Hobbs and Russell 1979, Rood and Nellis 1980, Russell 1981, Irvine et al. 1982, Miller et al. 1983, Pfeifer et al. 1984, Sherwin et al. 2002
Seals	Explosive hot-iron device	Branded body	Homestead et al. 1972
Dolphins	Pressure stencil on dorsal fin	Lasted for at least 2 years	Tomilin et al. 1983

[a] Scientific names are given in Appendix 9.1.

Freeze Branding

Freeze branding, a technique originally developed for livestock, is a more humane marking method than hot-iron branding. Highly conductive branding irons are supercooled, most commonly in a mixture of dry ice and methanol or liquid nitrogen, and placed on a shaved and washed area of the skin. The epidermis is temporarily frozen, destroying the pigment-producing melanocytes in the hair follicles and causing regrowth of white (Fig. 9.19) as opposed to pigmented hair. Freeze branding has been used successfully to mark a variety of wildlife (Table 9.16). Freeze branding, if properly applied, rarely results in infection. However, freezing the skin for too long can cause scab formation or tissue necrosis, resulting in the formation of new cells with intact melanocytes, that creates an indistinct mark. On lightly pigmented animals, however, these can produce a dark mark that can be read at a distance. A disadvantage of freeze branding is the brand cannot be read until after the animal molts its pelage.

Chemical Branding

Anurans have been branded using silver nitrate or a silver nitrate–potassium nitrate mixture. The **silver nitrate** caused a brown mark to form immediately, with the dark mark fading into a light mark in about 2 weeks. The method was recommended for dark-colored amphibians.

Laser Marking

Ruby lasers have been used to mark snakes, but were unsuccessful in marking a turtle (Table 9.16).

Tissue Removal

The effect of most tissue-removal marking methods on survival and fitness is not adequately known and is a topic that should be rigorously investigated (Society for the Study of Amphibians and Reptiles 1987). Alternative marking tech-

Fig. 9.19. Freeze branding mark on hip of Thomson's gazelle.

niques should be used if excessive pain, behavioral changes, or decreased survival is expected.

Feather Imping

Imping (insertion of a colored feather into the clipped shaft of a bird's rectrices or remiges; Fig. 9.20) using a double-pointed needle, cement, or "super glue," and a toothpick has been used to mark birds until molting (Table 9.17). Rectrices typically are used, although remiges are suitable if the replacement feather closely matches the one cut off. Imping is probably less effective than painting feathers.

Feather Clipping

Portions of vanes are clipped in different sizes and shapes from the shaft of several adjacent feathers, creating unique holes in the wings or tail that are used to identify birds (Table 9.17). **Clipping** should be performed so as to not impair flight. This technique is most suitable for gliding species and is of limited value for sedentary species, because the marks cannot be observed on perching birds. Moreover, the number of combinations producing effective marks is limited. Dyed feathers or colored tape attached to natural feathers, attached with wire to the rachis of natural feathers whose vanes have been clipped off, or glued to plumage in unnatural, conspicuous patterns also have been used on birds. All these marks are lost during molt.

Fig. 9.20. During the imping process, a feather of a captured bird (left) is clipped and a feather of contrasting color (right) is attached to it by means of a double-pointed needle.

Fur Removal

The removal of fur in a unique pattern is a nonpermanent humane means of marking mammals (Table 9.17). The marked animal generally is identifiable until the next molt. Hair may be removed with mechanical clippers, chemicals, or heat, thus allowing recognition of individuals at a distance. **Depilatory pastes** have been used to mark numbers on mammals, but can be extremely irritating to the skin of seals. **Hair burning** ("hair branding") produces a sharp, highly visible mark on northern fur seals and does not damage the skin; however, a fire source and a series of irons are required.

Table 9.17. Tissue removal methods used to mark wildlife

Group/species[a]	Type	Comment	Reference
Amphibians and reptiles			
Snakes	Subcaudal scale clipping	Permanent mark (regeneration 4–5 years) scars; marks not lost by tail breakage; marks persisted 4 years; 92% of the time shed skin from clipped racers could be precisely identified	Blanchard and Finster 1933, Carlström and Edelstam 1946, Conant 1948, Woodbury 1956, Weary 1969, Pough 1970, Brown and Parker 1976, Ferner 1979
Turtles	Toe clipping and shell notching	Notches on young turtles may not be permanent	Cagle 1939, Ernst 1971
Frogs, toads, newts, lizards, iguanas, hellbender	Toe clipping	Depending on species, some toe regeneration; should avoid clipping thumbs of toads due to use in amplexus	Martof 1953, Jameson 1957, Efford and Mathias 1969, Briggs and Storm 1970, Brown and Alcala 1970, Minnich and Shoemaker 1970, Hillis and Bellis 1971, Clarke 1972, Dole and Durant 1974, Richards et al. 1975, Daugherty 1976, Jones and Ferguson 1980, Hero 1989, Huey et al. 1990, Dodd 1993, Golay and Durrer 1994, Campbell et al. 2009
Salamanders	Toe clipping	Only successful marking method	Hendrickson 1954, Woodbury 1956, Heatwole 1961, Twitty 1966, Hall and Stafford 1972, Wells and Wells 1976, Davis and Ovaska 2001
Amphibian tadpoles, salamanders	Tail-fin notching	Tadpoles had higher mortality than with staining; salamanders regenerated tail in 1 month	Turner 1960, Orser and Shure 1972, Guttman and Creasey 1973, Ferner 1979

continued

Table 9.17. continued

Group/species[a]	Type	Comment	Reference
American alligator	Toe clip, tail-scute notch, and web punch	Permanent marks	Chabreck 1965, Jennings et al. 1991
Eastern newt	Amputating 1 limb	Not recommended	Healy 1974
Alpine newt	Skin transplantation	95% retention rate after 3 years	Rafinski 1977
Birds			
Large to medium sized	Dyed and painted feathers or colored tape attached to cut feathers	Marking techniques are temporary	Edminster 1938, Kozicky and Weston 1952, Neal 1964, Dickson et al. 1982, Ritchison 1984
Medium and large	Imping	Used double-ended needle or cement	Wright 1939, Hamerstrom 1942, Sowls 1950
Penguins, zoo birds	Web punching	More practical than using leg bands, fighting destroyed marks	Richdale 1951, Reuther 1968
Pheasants, raptors, frigatebird	Feather vane clipping leaving holes in wings or tail	Most suitable for gliding species; reduced breeding success of pheasants	Geis and Elbert 1956, Enderson 1964, Snelling 1970, Gargett 1973, Garnett 1987
Nestling gulls	Grafting the pollex to the skin of the head	Resulted in alula feathers growing from the head region	Coppinger and Wentworth 1966
Mallard	Alula clipping	Did not affect growth rate, behavior, or flight capability	Burger et al. 1970
Nestlings	Toenail and toe clipping	Toenail clipping remained for at least 18 days	Murphy 1981, St. Louis et al. 1989
Mammals			
Bats, seals, nutria, American beaver	Web punching or slits	Distinct after 2 years in fur seals	Aldous 1940, Scheffer 1950a, Davis 1963a
Small mammals, hares, coyote, seal pups	Toe clipping	Best to take only 1 toe per foot	Baumgartner 1940, Dell 1957, Sanderson 1961a, Melchior and Iwen 1965, Ambrose 1972, Andelt and Gipson 1980, Riley and William 1981, Fairley 1982, Gentry and Holt 1982, Pavone and Boonstra 1985, Korn 1987, Wood and Slade 1990
Small mammals	Ear punching or clipping	Some effect on movement and behavior	Blair 1941, Honma et al. 1986, Wood and Slade 1990
Rats, seals	Depilatory paste	Caused extreme skin irritation in seals	Chitty and Shorten 1946, Gentry 1979
Bats	Wing hole punching	White scar lasted 1–5 months	Bonaccorso and Smythe 1972, Bonaccorso et al. 1976, Stebbings 1978
Juvenile bats	Claw clipping	Lasted only a few weeks	Stebbings 1978
Seals	Hair burning	Does not burn skin	Gentry 1979
Seals, European European badger, mice	Fur removal	Lasted until next molt	Gentry 1979, Stewart and Macdonald 1997, Johnson 2001b

[a] Scientific names are given in Appendix 9.1.

Shell Notching

The most commonly used marking technique for turtles is notching the shell (Table 9.17). Marks on turtles may not be permanent. To avoid weakening the shell, marginals at the bridge or junction of the plastron and carapace should not be notched.

Scale Clipping

Scale clipping with scissors or clippers is the most commonly used method for marking snakes (Table 9.17). Pieces should be cut from the subcaudals, leaving "permanent" scars. **Subcaudal cuts** can be numbered on each side, beginning at the proximal end of the tail. No adverse effects have been reported for snakes, but regeneration could be a problem, and clipping is difficult on small or young snakes. **Ventral scales** are larger and easier to clip than subcaudal scales, and scars in this area cannot be lost by tail breakage.

Toenail Clipping

Clipping the toenail rather than toes (Fig. 9.21) is preferable for short-term studies of small mammals and nestling birds (Table 9.17). **Clipped toenails** remained sufficiently blunt at the tip to be distinguished throughout the nestling period, when birds are too young to be banded, although the nails eventually grow back. This method also has been used in bat nursery roosts, but the marks lasted only a few weeks.

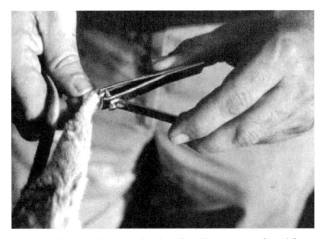

Fig. 9.21. Clipping the toenail rather than the toe is preferred for short-term marking studies of small mammals.

Toe Clipping

Toe clipping is widely used to individually mark anurans, small mammals, small turtles, and lizards (Table 9.17). The nail and first joint of the toe are removed with sterile dissecting scissors. The technique is inexpensive, rapid, and permanent, but at times, clipped toes cannot be distinguished from other causes of toe loss. Kumar (1979) developed a toe-clipping code for identification of ≤9,999 animals using no more than 2 digits clipped per foot. No direct adverse effects of toe clipping were reported for small mammals, and none of the extensive studies documented harmful effects caused by clipping the toes of lizards. **Toe clipping,** however, caused a temporary reduction in capture rates. Toe clipping is not advised for bats, because the toes are essential for roosting and grooming. This technique also has been used for identifying the tracks of marked individuals. Suitable conditions (e.g., snow) are required for track identification. Ecologists generally avoid toe clipping tree frogs and salamanders for long-term studies because of their regenerative capabilities. Although toe-clipping of amphibians and reptiles has disadvantages, it is still the most common marking technique used for anurans.

Ear Punching and Notching

The ears of many small mammals can be marked by punching or clipping them in a variety of coded systems (Table 9.17). Large-eared ungulates, carnivores, and primates have been marked by cutting 1 or 2 notches at preselected coded sites on the margin of the ear, allowing for a number of combinations. **Ear notching** or punching (using a leather punch) for large mammal species permits identification of marked animals at a distance. Notches usually last longer than tags, although they can be distorted by infection, growth, or injury (Ashton 1978). Ear notching is not advisable for mammals that use their ears for orientation and prey location or have valve-like ears that function during deep-sea dives. The ethical implications of these techniques should be considered.

Web Punching

Slits or holes punched into foot webs, flippers, or wing membranes have been used to mark many birds and mammals (Table 9.17). The marks are permanent, but unclean cutting may produce a small scar rather than a hole. **Leather punches** usually produce clean holes. Although some marks on web-footed birds are altered by injury or healing, most marks are identifiable. Some authors reported this method was more practical than leg bands. The major disadvantage of web punching is that birds must be recaptured for the web holes to be read. There are some questions about the ethics of this technique.

Tail Clipping

Notches clipped from a tail fin is a traditional method for marking amphibian tadpoles and some salamanders (Table 9.17). Fin clipping, however, produced higher mortality than did staining techniques. Scutes clipped on the tails of crocodilians have proved useful in long-term studies.

Skin Transplantation

This method involves the removal of skin from one part of the body and transplanting it to another. Although it has been used successfully on amphibians and some birds (Table 9.17), we do not recommend it.

Amputation

Healy (1974) marked post-larval metamorphs of eastern newt by amputating one limb at the middle of the zeugopodium, but few individuals were recaptured (Table 9.17). Newts regenerated the limb, usually within a month. Amputation is not recommended.

USING MUILTIPLE MARKS

It is best to use as few marks as needed for the project objectives. However, there are times when multiple marks may be needed. For behavioral studies, highly visible marks are needed to observe an animal at a distance, especially when they are in a large group, but a radio collar may be needed to first locate the animal (Fig. 9.22). In addition, a more permanent mark, such as a tattoo or PIT tag, may be needed for long-term studies (e.g., survival studies). We recommend the use of as few marks as possible to complete the objectives of a study.

SUMMARY

If there is a need to recognize individual animals, the use of natural markings is the preferred alternative. If this method is not feasible, marking animals without capture is the next

Fig. 9.22. Caribou cow with radiocollar and a highly visible numbered collar attached that allows for individual recognition of the animal if the radiotransmitter fails. *Photo by D. Watts.*

best option. These methods eliminate stress associated with capture. For animals that must be captured prior to marking, noninvasive techniques are preferred, but are not without problems. They can interfere with reproductive behavior (color marks), increase predation risks (color marks), and cause injury or increased mortality (band constriction, icing, entanglement of marks). Noninvasive methods are generally preferred, because the application of many invasive marks causes pain. The advantage of some invasive techniques is they are "permanent." For example, tattoos are probably the most permanent marking method available for many species, but they have the disadvantage of requiring the animal to be in hand (recaptured or found dead) to be identified. The use of PIT tags also offers a relatively permanent marking method (some are lost or become inoperable), but these tags have the same primary disadvantage as tattoos—usually the animals must be recaptured for identification. If animals need to be marked for only a limited time, then permanency of the mark is not a factor. There are both noninvasive (e.g., dyes) and invasive (e.g., toe-nail clipping) marking methods that can be used for short-term studies yet have little effect on the animal. Only use multiple marks when it is absolutely necessary to meet the objectives of the study. The ultimate responsibility regarding which method should be used to mark wildlife for a particular study rests with the investigator, whose choice should be based on the ethical and scientific validity of the method.

APPENDIX 9.1. COMMON AND SCIENTIFIC NAMES OF ANIMALS MENTIONED IN THE TEXT AND TABLES

Authority for scientific names of North American amphibians, birds, mammals, and reptiles is Banks et al. (1987). That for non–North American amphibians and reptiles is Sokolov (1988). Authority for non–North American birds is Sibley and Monroe (1990), and that for non–North American mammals is Grizimek (1990).

Common name	Scientific name	Common name	Scientific name
Amphibians and reptiles		smooth	*Trituris vulgaris*
Alligator, American	*Alligator mississippiensis*	warty	*Trituris cristatus*
Bullfrog	*Rana catesbeiana*	Salamander, dusky	*Desmognathus fuscus*
Frog, northern leopard	*Rana pipiens*	spotted	*Ambystoma maculatum*
Hellbender	*Cryptobranchus alleganiensis*	tiger	*Ambystoma tigrinum*
Iguana, green	*Iguana iguana*	Snake, pine	*Pituophis melanoleucus*
Lizard, blunt-nosed leopard	*Gambelia silus*	Racerunner, six-lined	*Cnemidophorus sexlineatus*
northern fence	*Sceloporus undulatus*	Rattlesnakes	*Crotalus* spp.
slow-worm	*Anguis fragilis*	Terrapin	*Malaclemys terrapin*
viviparous	*Lacerta vivipara*	Tortoise, desert	*Gopherus agassizii*
Newt, alpine	*Trituris alpestris*	Turtle, box	*Terrapene* spp.
eastern	*Notophthalmus viridescens*	green sea	*Chelonia mydas*
great-crested	*Triturus cristatus*	spotted	*Clemmys guttata*

Common name	Scientific name
Birds	
Blackbird, red-winged	*Agelaius phoeniceus*
Brant	*Branta bernicla*
Cardinal, northern	*Cardinalis cardinalis*
Coot, American	*Fulica americana*
Dove, mourning	*Zenaida macroura*
white-winged	*Zenaida asiatica*
Duck, wood	*Aix sponsa*
Eagle, bald	*Haliaeetus leucocephalus*
Egret, cattle	*Bubulcus ibis*
Frigatebird	*Fregata* spp.
Goldfinch, American	*Carduelis tristis*
Grouse, black	*Tetrao tetrix*
ruffed	*Bonasa umbellus*
greater sage-	*Centrocercus urophasianus*
Gull, glaucous-winged	*Larus glaucescens*
Juncos	*Junco* spp.
Mallard	*Anas platyrhynchos*
Osprey	*Pandion haliaetus*
Owl, boreal	*Aegolius funereus*
long-eared	*Asio otus*
Oystercatchers	*Haematopus* spp.
Partridge, gray	*Perdix perdix*
Pheasant, ring-necked	*Phasianus colchicus*
Plover, semipalmated	*Charadrius semipalmatus*
Prairie-chicken, greater	*Tympanuchus cupido*
Quail, northern bobwhite	*Colinus virginianus*
Skimmer, black	*Rynchops niger*
Sparrow, house	*Passer domesticus*
Starling, European	*Sturnus vulgaris*
Stork, wood	*Mycteria americana*
Swan, Bewick's	*Cygnus bewickii*
tundra	*Cygnus columbianus*
Tern, common	*Sterna hirundo*
least	*Sterna antillarum*
Turkey, wild	*Meleagris gallopavo*
Willet	*Catoptrophorus semipalmatus*
Woodcock, American	*Scolopax minor*
Mammals	
Badger, European	*Meles meles*
Beaver, American	*Castor canadensis*
Bobcat	*Lynx rufus*
Bat, big brown	*Eptesicus fuscus*
Bear, black	*Ursus americanus*
polar	*Ursus maritimus*
Bushbuck, African	*Tragelaphus scriptus*
Caribou	*Rangifer tarandus*
Cheetah	*Acinonyx jubatus*
Coyote	*Canis latrans*
Deer, mule	*Odocoileus hemionus*
Pere David's	*Elaphurus davidanus*
white-tailed	*Odocoileus virginianus*
Elephant, African	*Loxodonta africanus*
Elk	*Cervus canadensis*
Ferret, black-footed	*Mustela nigripes*
Gazelle, Thomson's	*Gazella thomsonii*
Giraffe	*Giraffa camelopardalis*
Gopher, pocket	*Geomys breviceps*
Ground squirrel, Townsend's	*Spermophilus townsendii*
Hare, snowshoe	*Lepus americanus*
Leopard	*Panthera pardus*
Lion, African	*Panthera leo*
Manatee	*Trichechus manatus*
Mongoose	*Herpestes* spp.
Moose	*Alces alces*
Mountain beaver	*Aplodontia rufa*
Mouse, harvest	*Reithrodontomys* spp.
Muskox	*Ovibos moschatus*
Nutria	*Myocastor coypus*
Opossum, Virginia	*Didelphis virginiana*
Otter, sea	*Enhydra lutris*
Pangolin	*Manis* spp.
Peccary, collared	*Tayassu tajacu*
Porcupine	*Erethizon dorsatum*
Pronghorn	*Antilocapra americana*
Rabbit, cottontail	*Sylvilagus* spp.
Raccoon	*Procyon lotor*
Rat, naked mole	*Heterocephalus glaber*
Norway	*Rattus norvegicus*
Rhinoceros, black	*Diceros bicomis*
Sea lion, Steller	*Eumetopias jubatus*
Seal, northern fur	*Collorhinus ursinus*
Sheep, mountain	*Ovis canadensis*
Dall's	*Ovis dalli*
Squirrel, fox	*Sciurus niger*
gray	*Sciurus carolinensis*
red	*Tamiasciurus hudsonicus*
Tiger	*Panthera tigria*
Wallaby	*Petrogale* spp.
Whale, beluga	*Delphinapterus leucas*
Woodrat	*Neotoma* spp.

10

Wildlife Radiotelemetry and Remote Monitoring

JOSHUA J. MILLSPAUGH,
DYLAN C. KESLER,
ROLAND W. KAYS,
ROBERT A. GITZEN,
JOHN H. SCHULZ,
CHRISTOPHER T. ROTA,
CATHERINE M. BODINOF,
JERROLD L. BELANT, AND
BARBARA J. KELLER

INTRODUCTION

OVER THE PAST several decades radiotelemetry has been one of the most effective tools in wildlife biology, as the size and cost of radio transmitters has steadily decreased while battery life, transmission range, and types of data that can be collected have increased. Consequently, radiotelemetry is integral to our understanding of wildlife behavior, movement, and demography (Lord et al. 1962, Cochran et al. 1965, White and Garrott 1990, Kenward 2001, Millspaugh and Marzluff 2001, Warnock and Takekawa 2003). Through knowledge of animal locations and status, biologists have the capacity to address critical ecological hypotheses and management questions that might otherwise be impossible to examine. Radiomarked animals can be repeatedly observed or relocated more consistently, systematically, and frequently than animals marked with any other technique. Put another way, radiotelemetry allows scientists to gather information that is not practical or possible through use of other methods.

Even the simplest and most commonly used equipment, **very high frequency** (**VHF**) radiotelemetry, has provided novel information about animal locations and movements in remote locations, in inhospitable habitats, and during inclement weather conditions. Moreover, advances in battery technology, satellite availability, and sensor developments have expanded the types of questions biologists can address and the species for which radiotelemetry studies are feasible. Modern automated systems allow data collection without requiring observers to be present. Current satellite-based systems make it possible to detect and quantify global scale movements that would have been impossible to track from the ground or sea.

To make effective use of radiotelemetry technology, biologists need to know how the technique works and to understand the numerous options for equipment, field methods, and analytical procedures. Biologists are often excited to make use of continually advancing radiotelemetry technology, but each set of equipment comes with its own set of assumptions, strengths, and drawbacks. As we emphasize throughout this chapter, choices among these options must be made in light of each study, to ensure that data are collected at the resolution needed to meet these objectives. For any study, investigators should carefully contrast the advantages and disadvantages of applying radiotelemetry techniques versus using alternative approaches.

This chapter provides an overview of the available technology and its use in the field, along with general considerations to be borne in mind when using this technique. We consider only **Lagrangian** movement data (i.e., sensors attached to an animal, offering potentially continuous movement of known individuals) and not **Eulerian** data types (i.e., fixed camera traps or track plates recording the movements of animals through a space).

STUDY CONSIDERATIONS

At the most basic level, radiotelemetry allows us to locate a radiomarked animal, but the utility of the technology goes far beyond placing location coordinates on a map. Depending on how the technique is applied, it can have wide application in wildlife studies. Location information allows biologists to track animal movements, such as **dispersal** or **migration,** at individual and group levels and to estimate overall space-use patterns (Kernohan et al. 2001). When combined with other technologies, such as the Geographic Information System (GIS), location data allow evaluation of large-scale resource use and selection (McDonald et al. 2006). The same movement data can be used to study intra-specific (e.g., social behavior) or inter-specific (predator–prey) interactions. And the availability of complementary sensors allows for further investigation. For example, activity and mortality sensors can provide additional information about animal activities and demographics (e.g., cause-specific mortality).

Depending on which types of data are needed to meet study objectives, there are specific study designs, equipment, and analytical choices that will be most appropriate. However, there also are general study issues that must be carefully considered in any radiotelemetry investigation, regardless of the specific objectives. In this section, we provide an overview of these critical issues. Where appropriate, we offer more specific guidance relevant to alternative potential radiotelemetry objectives. However, both our general and specific study-design recommendations serve primarily as an introduction to key issues, and we urge readers to consult several sources that offer more detailed guidance about radiotelemetry study design (e.g., White and Garrott 1990; Samuel and Fuller 1994; Garton et al. 2001, 2010; Kenward 2001; Millspaugh and Marzluff 2001).

Although radiotelemetry has wide application to a diversity of biological questions, it is just one of the available tools, and it should be used only after careful consideration of whether it is the most appropriate tool to address the study objectives. When assessing the potential utility of radiotelemetry, it also is important to clarify the scope of the inferences to be made from the results. For what age, gender, and other categories of individual (e.g., social status); time of day or season of year; and what geographic area are the results to be representative? As we discuss below, the scope and type of inferences to be made will affect the number and type of individuals that must be radiomarked, the timing and frequency with which animals are located, and the spatial scale over which animals must be sampled.

Similar to many other wildlife research tools, radiotelemetry has several important assumptions that must be considered during the design phase of a research project (Morrison et al. 2001) and analysis of the resulting data (White and Garrott 1990, Manly et al. 2002). Early considerations in the design phase of a project should address whether the assumptions related to using radiotelemetry will affect resulting information and ultimately the implied management or policy decision (Lyons et al. 2008). Critical assumptions in radiotelemetry studies are ignored too often, which limits the utility of resulting information. Although occasionally it is argued that "some information is better than none," we feel that when important assumptions are ignored, biased data can be worse than no data if they are misleading and lead to faulty management prescriptions. In the following sections, we place important assumptions in the context of general study design issues, and we then discuss 2 critical issues (transmitter effects and location errors) in more detail.

Critical Issues in Designing Radiotelemetry Studies

In most telemetry studies, biologists place radiotransmitters on a subset of animals in the population of interest and assume that data from these animals can be used to support inferences about the entire population. In many studies, the focus is on comparisons of subpopulations in this overall target population (e.g., males versus females), or on comparison of different populations (e.g., those on control versus treatment sites). In all cases, biologists assume that data from the set of observed individuals are representative of patterns in the overall target population(s). There also is a temporal component: biologists may be interested in year-round, seasonal, or shorter-term activity patterns; they also may be interested in diurnal or nocturnal activity. Therefore, we seek to make accurate and defensible inferences about a target population over some period of interest, based on measurements of a subset, or sample, of this population at limited times. As with any sampling situation, this goal is best met by implementing standard sampling-design strategies (see Garton et al. 2010). Proper sampling design also provides a framework for assessing how alternative sampling strategies and amounts of effort (i.e., economic costs) affect the statistical precision of parameter estimates or statistical power to detect differences among subpopulations. In particular, quantitative study design provides a framework for optimizing the number of animals monitored versus the frequency of measurements on each animal, and for determining how to best allocate effort among subpopulations of interest and across subperiods in the temporal period of interest. As discussed below, some realities of wildlife telemetry studies may make it impossible to completely implement a probabilistic sampling design. However, striving to follow proper sampling procedures within biological constraints will maximize study quality. Here we briefly discuss aspects of sampling and study design that are directly pertinent to radiotelemetry studies; see Garton et al. (2010) and other sources for detailed information about wildlife study design in general.

Setting Objectives

The first step in designing a radiotelemetry study is to develop specific study objectives based on a clear idea of why the study is being implemented. This aspect is the most critical of any data collection effort, as all sampling and data analysis decisions must be based on underlying objectives. Fuzzy objectives or even method-based objectives (e.g., "to put transmitters on as many individuals as we can catch") almost guarantee fuzzy results. This step is especially critical given the expense of telemetry studies, the potential for adverse effects on captured individuals, the multitude of data that can be collected simultaneously, and the temptation of biologists to use new equipment. Study objectives should clearly identify the target population (e.g., female northern goshawks (*Accipiter gentilis*) nesting at elevations between 1,000 and 1,500 m in the Grizzly Creek Management Area, WY) and the time period of interest (e.g., between sunrise and 1900 hours during May through July).

In addition to addressing the "who," "where," and "when," objectives obviously should specify and prioritize biological parameters and comparisons of interest. For example, is the focus on absolute home range size for its own sake, on comparative sizes in different treatments or portions of the study area, or is it on simultaneous overlap/dynamic interactions or potentially nonsimultaneous overlap in space use throughout a longer period? Are precise estimates for specific subpopulations (males and females) needed? That telemetry data could be used to estimate home-range size, spatial overlap, migration characteristics, resource selection patterns, survival, and abundance does not mean one's objective should be to collect as much data as possible and figure out which questions are amenable to analysis after data are collected. Such an approach may well produce a large mass of data that is of mediocre quality for addressing *any* of these questions. Although a study may target biological parameters, multiple objectives need to be prioritized to ensure that the study has a high chance of meeting top-priority objectives. Well-designed studies may use available funding efficiently to meet multiple objectives. For example, if both home-range size and survival are high priorities, the study may include an extensive sample of individuals monitored periodically for survival estimation and a subset of these individuals monitored much more intensively to obtain data for estimating space use.

In relation to these specific study objectives, biologists should develop quantitative precision and/or power goals for the study. Before implementing studies, biologists routinely should examine whether available funding will give their study a high chance of success and consider how effort can be allocated to help the study efficiently meet its objectives. **Success,** in this case, is not an ambiguous term but is defined quantitatively: in terms of the desired precision of parameter estimates, the magnitude of differences among subpopulations that we want to have an acceptable power to detect, or the probability that we will be able to discriminate among alternative hypothesized statistical models to identify the model that best describes the true patterns in our target population.

Minimizing Bias When Selecting Animals

Biologists should design a strategy for selecting study animals that maximizes how well the radiomarked animals represent the larger target population of interest. Of all sampling steps discussed in this section, this is one of the most fundamental, but also the one most hindered by factors outside our control, as logistical and biological realities prevent strict application of probabilistic sampling practices. In an ideal world, all animals in the population of interest would have the same probability of being included in the radiomarked sample, and selection of which specific animals to follow would be defined by some random choice based on these probabilities rather than on capture opportunity. Such an approach would minimize bias, or systematic discrepancies between parameter estimates based on the observed subset of animals and true parameter values for the entire target population. It would allow direct statistical inference from the sampled subset to the overall target population. However, wild animals have to be captured before they can be equipped with transmitters. The feasibility of capture depends both on the feasibility of trapping in different vegetation types or locations in the study area (Garton et al. 2001) and on the individual animal's behavioral susceptibility to being captured. If trapping is only feasible in certain vegetation types or localized sites (e.g., constrained flight paths in the case of bats) that are not widely distributed across the area of interest, some portions of the target population may have little chance of being included in the sample. Because of behavioral and personality differences, individuals may vary widely in their likelihood of being captured, and potentially their susceptibility to capture may be directly related to the parameter of interest in the study (e.g., highly elusive or experienced animals may use habitats differently or have different average home-range sizes than animals that are more easily captured). Because of such uncontrollable factors, the validity of our statistical inference from the sampled to the target population is compromised. Instead, we must rely on scientific inference, arguments, and assumptions for why results from the sampled subset are likely to adequately represent patterns in the target population.

In most telemetry studies, the inability to guarantee unbiased inference from the sampled population is a reality that should be recognized and addressed. However, the potential for uncontrollable factors to introduce some bias makes it even more critical to minimize sources of bias that are at least partially controllable. To maximize how representative study animals are of the target population, trapping locations should be distributed in suitable vegetation types across the study area, rather than in just a few locations that are easily accessible and convenient to sample. These locations should be selected objectively rather than by convenience. For example, trapping could be conducted in a systematic sample of 1-km patches across a larger landscape. Rather than collaring the first 50 animals that are trapped regardless of location, it may be necessary to limit how many animals to include from any single trapping location and to try to distribute transmitters among the widely distributed capture locations. Regardless of whether the target population includes both sexes and all age classes or focuses on specific subgroups, differential susceptibility of these different groups to capture should be considered. If a biologist wants an average home-range estimate for an entire population, but juvenile males are much more likely to be captured, collaring animals of different groups in proportion to their frequency of capture will bias results by giving too much weight to estimates based on juvenile males. If it is much more expensive to trap areas away from roads, but animals in those areas are in the target population, then those animals need to have some chance of being included in the sample, even if statistical efficiency is maximized by collaring proportionately more animals in near-road areas. For example, the landscape could be stratified into accessible and less accessible strata. In the absence of information on how the biological parameter of interest varies among near-road versus away-from-road subpopulations and on relative densities and trapability in these strata, the comparative number of animals to target in each stratum could be based on the comparative costs of trapping and monitoring animals and on the sizes of the strata (e.g., Cochran 1977). Such an approach is not perfect (probabilistic sampling of landscape units may not fully ensure representative sampling of the target wildlife population), but it is much better than simply assuming that individuals near and far from roads behave in the same way.

Scheduling Radiotelemetry Data Collection

Biologists should ensure the schedule of data collection fully covers the temporal window of interest. As with selecting animals to be equipped with transmitters, either all portions of this window need to be represented equally, or unequal allocation of effort should be controlled by the survey design (Fieberg 2007a). The biology of the animal is an important component of this issue (e.g., Beyer and Haufler 1994). Practically, this recommendation can be met most easily with a systematic data-collection schedule for each animal across the longer-term period of interest (e.g., season) and across the 24-hour cycle or throughout daylight or nighttime hours, depending on the question and species of interest (Otis and White 1999). A feasible study may combine multiple sampling strategies (e.g., systematic, simple random, and stratified sampling). For example, Fieberg (2007a) simulates an example in which systematic sampling of every nth day was implemented across the study period, and each day was stratified into 2 intervals. If costs or safety concerns produce a need to have lower sampling intensity at night, the relative amount of effort allocated to nighttime versus daytime sampling should be chosen before the study commences, and the relative probability that a nighttime versus a daytime location is included in the sample should be considered during data analysis. For example, if 25% as much sampling effort was devoted to each nighttime hour compared to each daylight hour, each nighttime observation could be weighted proportionately higher than each daytime observation when estimates for the entire 24-hour period are calculated (e.g., Fieberg 2007a).

Estimating the Number of Animals and Locations Needed

We recommend that biologists determine sample sizes and allocation of effort based on quantitative examination in light of specified precision and/or power goals, rather than on broad guesstimates based on the idea that "whatever we can afford has to be good enough." Although conceptually this step is straightforward, in practice it may not be easy to implement in telemetry studies. To examine sample size–precision relationships, we need estimates of the variability in the parameter of interest across the target population or in subpopulations that will be compared in the study, and we may need a preliminary estimate of this parameter. In most studies, individual animals, not locations, should be treated as independent study units (Otis and White 1999). We also need to understand the relationship between the number of locations collected per animal and the uncertainty and potential bias in the estimated value (e.g., home-range size) for each animal. That is, the effect on precision or power of measurement error at the level of individual animals, as well as the variability among animals, affects how to best allocate effort and how to find the optimal balance between measuring as many animals as possible versus measuring each animal as intensively as possible. The costs of capturing and equipping an animal with a transmitter and the average cost per location can be combined with precision–variance relationships to form cost functions for optimizing allocation of sampling effort given different levels of available funding.

Given careful pilot data or adequate information for the species of interest from previous studies, conservative estimates of among-animal variability in the parameter of interest (e.g., home-range size) can be obtained for guiding sample size examinations. In some cases, there are rules of thumb for how intensively each animal must be sampled to obtain acceptably low measurement errors (Otis and White

1999). Seaman et al. (1999) recommended that home-range studies using kernel estimators obtain at least 30–50 locations per animal. However, in determining how to allocate effort between and among individuals, if one is focusing on population-level parameters (e.g., average home-range size), the optimal number of locations to collect per animal may be lower than what is needed to obtain adequate precision in estimates for each animal. This consideration is particularly important if (1) the analytical method remains unbiased with lower sample sizes, (2) there is a trade-off between number of locations per animal and number of animals monitored, and (3) among-animal variability is high compared to the magnitude of measurement error. However, exploring such trade-offs require knowledge of the relationship between number of locations per animal and expected measurement error (e.g., average uncertainty in the estimates of home-range size for each study animal).

Unfortunately, this relationship is difficult to quantify for many types of telemetry investigations, because there is no parametric statistical model for the observation process and thus no equation allowing computation of expected measurement error as a function of number of locations. For example, with modern home-range estimation methods discussed later in this chapter, there is no a priori way to quantify expected measurement error, and only approximate variance estimates may be feasible for individual estimates after data are collected (e.g., bootstrapping approaches). Thus, determining how measurement error decreases as the number of locations increases may require careful investigation of pilot data sets (or those from previous studies of that species in similar landscapes), as well as simulation investigations. For some investigations, such as survival studies, sample sizes can be solved numerically for specific designs (Samuel and Fuller 1994). Given the limited attention this issue has received in the ecological literature, approximate rules of thumb (e.g., collect 50 observations per animal for kernel estimation) probably will continue to be heavily used to determine how much effort to allocate per animal. In this case, effort per animal and expected measurement error can be treated as fixed quantities, and sample-size investigations can focus on how many animals need to be followed to obtain adequate precision, given the estimated among-animal variability.

These issues also require that biologists consider how they intend to analyze data ahead of time. Some methods are more or less data intensive, so knowledge of analytical procedures can help ensure that an adequate amount of data have been collected. In addition, the same strategies involved in allocating effort between and among animals can be used in determining how to allocate effort (in this case, number of animals sampled) among age classes, male versus female, near versus far from road, etc. Such decisions should be based on the relative priority of respective subpopulations (e.g., whether more precise estimates are needed for adult females compared to the rest of the population), how variable the parameter of interest is within and among these subpopulations, and the comparative cost of capturing and tracking members of each subpopulation.

Radiotelemetry Data Management Issues

In all disciplines of natural resource management and ecology, data-gathering programs are increasingly becoming a critical aspect of investigations. Given the amount and complexity of data obtained in radiotelemetry studies, it is especially important for investigators to establish procedures for standardizing how data are obtained and recorded, ensuring the consistency of data collected by multiple observers, promptly converting data into digital formats, identifying discrepancies and removing errors, and archiving data for future use. Many agencies and programs also are required to make data available for examination and use by other scientists, and some journals require public data access as a prerequisite for publication. These mandates underscore the importance of maintaining **metadata,** which document data collection methods, dataset contents, and known errors.

Two issues are especially relevant to radiotelemetry studies. First is the need for increased focus on integrated data-management systems with modern telemetry studies. The volume of traditional animal tracking data usually consisted of a few dozen points collected each day, which can be handled by standard spreadsheet programs. However, Global Positioning System (GPS) tags and sensors can record data at much higher rates, which can create challenges for data management. Fortunately, the field of informatics has provided tools to efficiently handle volumes of data. For example, Cagnacci and Urbano (2008) used open source spatial database tools (PostgreSQL and PostGIS) together with web services modules (R, QGIS, GRASS, MapServer, Ka-Map) to develop a customized system that stores, retrieves, analyzes, and visualizes GPS tracking data from their work with roe deer (*Capreolus capreolus*).

The second issue relates to making data available to other biologists. Regardless of the tracking technology, most animal movement data consist of a time/date stamp and a geographic location. These standard data can then be compared across studies (Ballard et al. 2002) through integrated data banks, which have been successfully used in animal movement studies and in other fields (e.g., DNA sequencing, or natural history collections). For example, **Movebank** provides a museum-based data archive with sophisticated data-rights management tools and basic data editing and visualization functions (www.movebank.org). This type of tool extends the questions that animal tracking studies can address, provides a means for scientists to verify the repeatability of analytical results, and allows researchers to reexamine old data with new analytical techniques.

Addressing Other Common Sources of Bias in Radiotelemetry Studies

The effects of radiotransmitters on animal behavior and the effects of terrain on an observer's ability to accurately locate animals also should be considered. Radiomarked study animals may behave differently because of the weight or irritation caused by radiotransmitters (Garton et al. 2001; see section on transmitter effects below). Animal location information also can be biased if the probability of obtaining a successful location for a study animal and the degree of location error vary systematically with terrain or habitat. For example, animal use of areas on a site where GPS receivers cannot detect satellite transmissions will likely be underrepresented in a dataset based on GPS tracking. The issue becomes particularly important if there is an interaction between habitat use, time of day, and the ability of a biologist to collect observations during those periods. Neglecting to account for these differences could bias study results; we discuss this issue further below.

Effects of Transmitters

Among the most important assumptions needed for making inferences based on radiotelemetry is the assumption that radiotags have no influence on the animal. This assumption is crucial for making proper inferences about parameters of interest and for basic ethical and **animal welfare** considerations. Early radiotelemetry investigators evaluated whether transmitters affected survival, and they worked to ensure that equipment remained attached long enough to obtain useful information (Dwyer 1972, Sayre et al. 1981, Ciofi and Chelazzi 1991, Rappole and Tipton 1991). Biologists concluded there was no effect if a majority of the transmitters remained attached or most of the animals survived the first days after capture and instrumentation (Dwyer 1972, Raim 1978, Rappole and Tipton 1991, Riley and Fistler 1992). At that time, telemetry applications were primarily concerned with basic life-history questions about where animals go and how much area they traverse (Amstrup and Beecham 1976, Cranford 1977, Herzog 1979, Schulz et al. 1983), and as a result more subtle equipment effects might have been missed. More recently, however, quantifying device-induced effects is becoming a greater concern among a wide range professional ecologists, particularly given increased scrutiny of animal care and use protocols (Wilson and McMahon 2006, Casper 2009).

Most previous studies investigating radiotransmitter effects focused on flying animals (e.g., birds or bats) because of concerns about direct effects on flight and indirect ones on behavior and survival. A review of 5 journals from 1972 to 2000 identified only 96 studies that evaluated transmitter effects; 79% of these studies assessed effects on birds, and nearly 50% of all the papers focused on waterfowl and upland game birds (Withey et al. 2001). Whereas earlier studies of flying animals evaluated equipment effects on aerodynamics (Obrecht et al. 1988) or transmitter weight relative to animal size and weight (Caccamise and Hedin 1985, Aldridge and Brigham 1988), recent evaluations have focused on effects on behavior (Hupp et al. 2003, Blomquist and Hunter 2007, Vukovich and Kilgo 2009), nesting or reproduction (Bergmann et al. 1994, Hepp et al. 2002, Kurta and Murray 2002), survival (Larson et al. 2001, Kenow et al. 2003, Reynolds et al. 2004), or other metrics associated with research (Croll et al. 1996, Neudorf and Pitcher 1997, Jones et al. 2002a, Fleskes 2003, Anich et al. 2009). Research biologists also have expanded their efforts to evaluate transmitter effects on a wider variety of animals beyond birds (Mourao and Medri 2002, Weatherhead and Blouin-Demers 2004, Rittenhouse et al. 2005, Martin et al. 2006, Rittenhouse and Semlitsch 2007).

The literature on transmitter effects makes somewhat contradictory conclusions. Dozens of studies conclude that radiotransmitters do not affect a particular aspect of interest for a particular species (e.g., Neudorf and Pitcher 1997; Jones et al. 2002a; Kurta and Murray 2002; Suedkamp et al. 2003; Durnin et al. 2004; Palmer and Wellendorf 2007; Terhune et al. 2007; Davis et al. 2008, 2009; Anich et al. 2009; Vukovich and Kilgo 2009). However, numerous investigations report some effect caused by the transmitter, antenna, and/or attachment device (e.g., Schulz and Ludwig 1985; Pietz et al. 1993; Schulz et al. 1998, 2001, 2005; Bro et al. 1999; Guthery and Lusk 2004; Hamel et al. 2004; Conway and Garcia 2005; Whidden et al. 2007). Other investigations show effects from capture, handling, or sample collection (Beringer et al. 1996, DeNicola and Swihart 1997, Cox and Afton 1998, Abbott et al. 2005).

Collectively, these studies demonstrate the importance of considering properties of the available transmitters and relating them to study objectives and focal species. Transmitter size, shape, and weight are usually a compromise between the biologists' objectives and the manufacturers' constraints on available equipment configurations. For example, if study objectives seek information about long-distance migration movements of a shorebird, researchers should consider a transmitter and attachment method that allows for flight mobility while minimizing weight and frictional drag. Conversely, multi-year projects aimed at learning about seasonal habitat use of large ungulates require transmitters with multi-year batteries. If mark–resight information is one of the objectives, the collars can be constructed of highly visible material, which also might affect subtle social interactions or survival by drawing atypical attention to marked individuals. Although transmitter effects have been less worrisome for larger land animals, these effects can influence the resulting data by causing subtle changes in behavior or socialization among conspecifics (Schulz and Ludwig 1985).

Location Error

All geographic locations estimated with radiotelemetry contain errors. Two major classes of error can be defined for radiotelemetry data: inaccurate geographic locations and missing locations from data sets. The magnitude and distribution of each error type differs among equipment setups. GPS tracking equipment relies on satellite signals, so physical barriers, such as vegetation and terrain, can impede reception and result in variable location accuracy (Rempel et al. 1995, D'Eon et al. 2002). Most users of GPS technology are now reporting the amount of uncertainty in GPS locations at 20–200 m. Missed location fixes are another critical type of error with GPS (Frair et al. 2004, Lewis et al. 2007). The units require a short time to acquire signals from satellites before a geographic location can be estimated, and if the units are not allowed to run through the acquisition process, locations are not recorded. This problem can cause significant bias in animal habitat models if signal acquisition rates are lower in some vegetation types than in others—some researchers have reported rates as low as 13% (Frair et al. 2004). However, recent technological advances have made satellite acquisition less of a problem (Lewis et al. 2007). For VHF systems, location error also is affected by features of the environment, because radio signals can be attenuated or reflected by buildings, vegetation, and moisture. Additionally, movement by the animal, distance to the operator, and differences in equipment can all affect location accuracy in VHF (Heezen and Tester 1967, Springer 1979, Deat et al. 1980, Lee et al. 1985, Saltz and Alkon 1985, Schmutz and White 1990, White and Garrott 1990, Withey et al. 2001) and GPS (D'Eon 2003) systems. Other systems described below, including geolocators, can have location errors in the 100–200 km range (Phillips et al. 2004).

The necessary accuracy of telemetry locations is dictated by study objectives and methods of analysis. In studies of global scale movements, errors of a couple hundred kilometers might be acceptable. However, studies of resource selection, in which the researcher places individual locations on a GIS map to denote vegetation types used by an animal, require much higher accuracy. Several researchers have shown the potential bias that can occur when location error is present in resource selection studies (White and Garrott 1986). Many others have offered options for screening data (Keating 1994) or analytical approaches for reducing these problems (Samuel and Kenow 1992, Frair et al. 2004). The underlying resource conditions also can interplay with location error and can make determining when animals are occupying small patches of habitat problematic (Findholt et al. 1996). When use of some vegetation types may be underrepresented due to lower probability of successfully obtaining a fix in these habitats, analytical methods that adjust for this bias should be used (e.g., Nielson et al. 2009). The choice of analytical method also can affect the importance of location error. Moser and Garton (2007) reported that telemetry location error did not substantially influence fixed kernel home-range estimation, although they acknowledged that measurements for animals with long, linear home ranges might be affected by location error. Many recent movement analyses explicitly model location error and thereby keep it separate from the process model that describes interactions between animals and their environment (Patterson et al. 2008, Schick et al. 2008).

Prior to any radiotelemetry study, biologists should assess their systems and determine whether the magnitude of expected errors will preclude study objectives. Further, published results should report errors (Lee et al. 1985, White and Garrott 1990, Saltz 1994, Moser and Garton 2007). We encourage biologists to read White and Garrott (1990) and other sources (e.g., Lee et al. 1985, Withey et al. 2001) for detailed descriptions of testing and evaluating telemetry errors.

TELEMETRY AND REMOTE TRACKING EQUIPMENT

All too often, biologists push to implement field portions of telemetry studies before thoroughly considering the specific strengths and limitations of their equipment. This lack of foresight and planning can lead to failed projects or biased data that do not accurately represent the behaviors and movements of study animals. To avoid these costly mistakes, investigators should select equipment only after thinking through the natural history of their study species, the region in which investigations will occur, and the budgetary constraints of their project. To facilitate this planning, we review several types of tracking equipment and summarize the primary strengths and weaknesses of each. Technology changes rapidly, however, so biologists also should look for the latest technical specifications from manufacturers before finalizing a study design.

Common Considerations

In any radiotelemetry study, the investigator must consider several elements that are common to every tracking system. Each system includes a battery-powered device that must be attached to study animals. These devices, tags, or radio-markers either send information directly to observers or store data onboard until they are physically recovered by researchers. Systems that transmit information to observers send **electromagnetic radio signals**, which then must be gathered by an antenna and receiver. Simple systems transmit pulses of energy that are then used to detect animal directions and estimate animal locations. More complicated systems transmit geographic coordinates that are generated by animal-mounted computers. In this section, we first review these tag and transmission options and then detail the 3 most prominent tracking technologies: VHF radiotelemetry, satellite tracking systems, and light-level geolocation. We further describe sensors that can be integrated into these systems and present an overview of data-gathering networks.

Several methods have been developed over the past 50 years to attach equipment to study animals. We briefly describe the most popular methods and focus on general issues to be considered when choosing an attachment method. More detailed descriptions of attachment techniques can be found in Kenward (2001) and in the individual citations below.

Mammals are frequently fitted with radiocollars placed around the neck (Fig. 10.1). The size of the collar is adjusted to each individual, so that it is large enough to allow normal growth and movement but sufficiently small to prevent it from slipping over the head and falling off. Radiocollars should distribute weight appropriately and contour the neck as much as possible. Collars can cause skin irritation when ill fitted, especially on ungulates, where they can slide up and down when the animal feeds. Collar sizing can be problematic should the neck size change over the year on adult animals (e.g., ungulates during the rut). When collars are placed on juveniles, biologists sometimes use foam rubber inserts or expandable systems that eventually break down (Strathearn et al. 1984, Jackson et al. 1985). Biodegradable links (e.g., leather) also have been inserted between the ends of collar material to facilitate collar drop-off (Garshelis and McLaughlin 1998). Similarly, necklaces with pendant radiotransmitters often are used on gallinaceous birds, which are suited to heavy front-loading of their large crops (Riley and Fistler 1992, Dobony et al. 2006). The use of necklace transmitters has been a preferred alternative to the poncho attachment technique (Amstrup 1980), which was a method favored by some early investigators because of its ease of application.

Small birds and bats present challenges, because they require lightweight equipment and attachment methods that do not impede flight (Aldridge and Brigham 1988). Early investigators fitted birds and bats with **backpack-style** harnesses that loop around the wings, which seemed like an intuitive approach at the time. However, backpacks proved to be problematic, and the method is now generally discouraged (Foster et al. 1992, Gaunt and Oring 1999). Several alternative and more effective techniques have been developed, and perhaps the most commonly used method for small birds is a design with **harness loops** that wrap around the body and fit over the legs (Rappole and Tipton 1991;

Fig. 10.1. Neck radiocollars for mammals. Fit of collars must be tight enough to not slide over the animal's head, but loose enough to account for seasonal differences in neck sizes (e.g., ungulates during rut) and to avoid restricting the airway or feeding. Collaring juvenile animals that are still growing requires that extra space be left for growth, which is often not possible.
(A) Yellow baboon (*Papio cynocephalus*) with a Global Positioning System collar; (B) coyote (*Canis latrans*) with a very high frequency (VHF) radiocollar; (C) deer mouse (*Peromyscus* spp.) being fitted with a 1-g radiocollar; (D) white-tailed deer fawn (*Odocoileus virginianus*) with a VHF radiocollar. *(A) Photo by C. Markham; (B) photo by D. E. Beyer, Jr.; (C) photo by R. Kays; (D) photo by J. F. Duquette.*

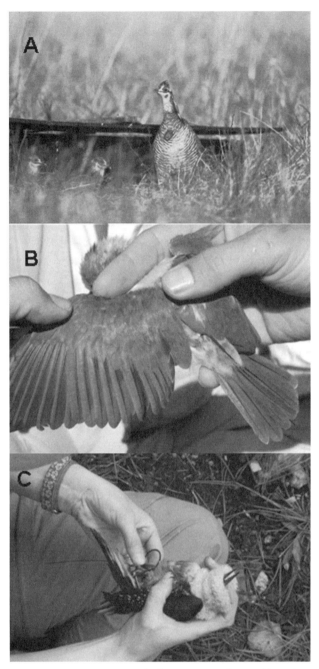

Fig. 10.2. (A) Greater prairie-chicken (*Tympanuchus cupido*) with necklace with pendant very high frequency (VHF) transmitter; (B) Tuamotu kingfisher (*Todiramphus gambieri*) with leg-harness VHF transmitter; (C) black-backed woodpecker (*Picoides arcticus*) with Rappole and Tipton (1991) attachment technique of a VHF transmitter. *(A) Photo by N. Paothong; (C) photo by M. Rumble.*

Fig. 10.2). This basic construction also has been modified to include an extra loop that goes around the breast for wading birds (Haramis and Kearns 2000). A weak link section may be added, which breaks away and allows the transmitter equipment to fall off birds after batteries expire (Karl and Clout 1987, Doerr and Doerr 2002). Breast harnesses made of Teflon® ribbon have been successfully used on larger birds (Kenward 2001, Roshier and Asmus 2009). Considerable experience is needed to determine the proper fit when attaching a harness, because loose equipment can cause the animal to become entangled and tight harnesses may restrict movements and cause skin abrasions.

Transmitters also can be glued directly to skin, feathers, and exoskeletons of many species with cyanoacrylate adhesives and epoxies (Fitzner and Fitzner 1977, Raim 1978, Warnock and Warnock 1993, Spears et al. 2002, Mong and Sandercock 2007; Fig. 10.3). These surgical glues are particularly germane to small birds and bats, because they are temporary and often drop from study subjects during molt. However, cyanoacrylate adhesives produce an exothermic reaction during curing, which can irritate sensitive skin, especially if feathers are trimmed down to the quill. Further problems can result if birds are released before the glue is fully cured, so many researchers also use compounds that accelerate curing. Despite these potential issues, adhesives have been commonly used for bats (Kerth et al. 2001), marine mammals (Yochem et al. 1987), and a diversity of birds (Mong and Sandercock 2007).

Other transmitter mounts used on a range of taxa include subcutaneous anchors (Newman et al. 1999) and coelomic and subcutaneous implants (Schulz et al. 1998; Fig. 10.4). Implanted transmitters require surgery, but they provide an option when externally attached transmitters are not possible (e.g., studies in which transmitters attached with harnesses or adhesives are easily removed, external transmitters affect movements or behavior, or harnesses or adhesives directly cause harm to the study animals). Subcutaneous implants have most often been used on land birds whose flight is affected by externally attached transmitters (Schulz et al. 2001, 2005; Small et al. 2004). Intra-abdominal implants are often used in situations requiring a larger transmitter and battery package or when the skin characteristics of the study animals preclude a subcutaneous transmitter, for example, sea ducks (Hatch et al. 2000, Brown and Luebbert 2002), geese (Hupp et al. 2003), American beavers (*Castor canadensis;* Guynn et al. 1987), otters (Reid et al. 1986, Ralls et al. 2006), and amphibians or reptiles (Wang and Adolph 1995). Implanted transmitter antennas can be configured with an internal or external antenna, depending on the application. Although internal antennas have reduced signal strength (or range), they minimize effects on movement or behavior. Implants with external antennas have signal strengths similar to those of external transmitter attachments; however, infection may be a complication if bacteria enter along the antenna–skin interface, which is an especially important concern with waterbirds. Regardless of the potential difficulties, implanted transmitters have been used on wide variety of animals and have proven to be at least as effective as externally attached transmitters (Korschgen et al. 1996*a, b;* Hatch et al. 2000; Small et al. 2004; Schulz et al. 2005).

Fig. 10.3. Glue attachment technique for transmitters placed on (A) dragonfly (order Odonata), (B) three-toed box turtle (*Terrapene carolina triunguis*), and (C) red bat (*Lasiurus borealis*). *(A) Photo by M. Wikelski; (B) photo by C. Rittenhouse; (C) photo by Northern Research Station, U.S. Forest Service.*

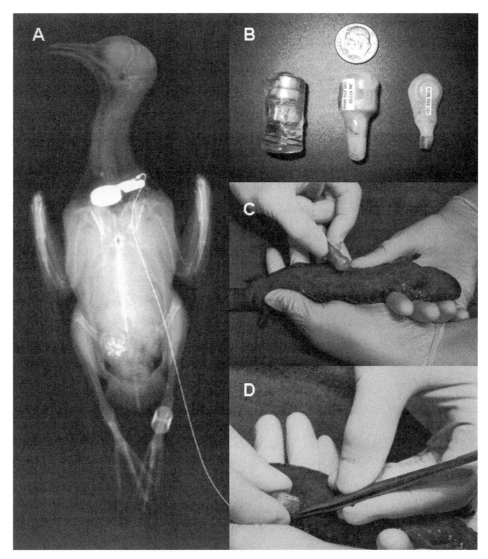

Fig. 10.4. Implant transmitters in (A) mourning dove (*Zenaida macroura*) and (B, C) Ozark hellbenders (*Cryptobranchus alleganiensis bishopi*). *(A) Photo by J. H. Schulz; (B, C) photos by C. Bodinof.*

Surgical implantation of transmitters requires considerable skill, training, specialized equipment, and often some form of professional board certification (Mulcahy 2006, Small et al. 2006). Biologists should not underestimate the level of planning and justification that may be needed to obtain an operational protocol approved and implemented, particularly when study protocols must be approved by an Institutional Animal Care and Use Committee. Investigators will usually require the assistance of a board-certified veterinarian. Some investigators transport study animals to a veterinary clinic equipped with the necessary surgical equipment. In studies where transmitters will be implanted in the field, a specialized portable anesthesia machine is required, along with a customized surgical tool kit suited for the type and size of animal being studied. The importance of conducting practice surgical procedures along with follow-up monitoring and evaluation cannot be overstressed (e.g., effects of handling on nest success or abandonment of young, or how much time is necessary after anesthesia before releasing the animal).

Transmitter weight is a major consideration when developing any radiotelemetry project (Box 10.1). Investigators working with flying or highly mobile animals need to pay special attention to both subtle and overt effects related to equipment weight, placement, and attachment technique. The American Ornithologists Union suggested that the transmitter package should be 5% of the body weight of birds (Gaunt and Oring 1999), and the U.S. Geological Survey Bird Banding Laboratory recommended that transmitters be 3% body weight (USGS-BBL 1999). General guidelines about transmitter mass can prove helpful when selecting a particular transmitter and attachment technique; however, numerous other factors also are critically important. For example, a transmitter located near the center of gravity of a bat or bird is less likely to adversely affect flight dynamics (Caccamise and Hedin 1985, Kenward 2001). In addition, several studies have shown that transmitter weight is of less importance compared to effects related to the attachment technique (Gessaman and Nagy 1988, Gessaman et al. 1991). Also, the 5% recommendation works well for a body mass 70 g (Aldridge and Brigham 1988), but in larger flying animals this rule of thumb results in relatively greater, and potentially unacceptable, reductions in power available for flight (Caccamise and Hedin 1985). We recommend using the lowest weight transmitter possible to accomplish study objectives.

All tags described in this chapter require electricity, and batteries are the most common power source. Batteries remain the heaviest component in most transmission systems, as they are bulky; they are also the only component that regularly wears out. Larger batteries with more capacity translate to longer-lasting equipment and stronger signals. In contrast, smaller and lighter batteries reduce unit sizes enough to be deployed on small species. Several manufactures have integrated solar cells into tags to recharge transmitters or buffer them against rapid battery drain. The most popular power sources for transmitters are lithium and silver oxide batteries. The operational life of a lithium battery can be estimated by dividing the battery capacity (mA-hr) by the current drain; an estimate of life for a silver oxide battery would be about 25% less than that of a lithium battery. Battery life estimates are always approximate, and actual life will vary depending on such factors as transmission rate (i.e., VHF pulses or data streams), shelf-life of battery, GPS search times, temperature, and variation in battery capacity among and within brands.

Overview of Telemetry Systems

Choice of a telemetry system involves careful consideration of study objectives, target species, and logistical and budgetary constraints. In the following sections, we describe 2 major classes of systems: VHF and global (satellite platform terminal transmitter [PTT] location systems, GPS, and geolocators; Table 10.1). VHF radiotelemetry systems are the most commonly used, with observers locating tagged animals from the ground, towers, or aircraft. These systems work well for species with small to medium ranges and for easily accessible study areas. Satellite telemetry equipment and light-level geolocating systems are options if study animals move over great distances or observers cannot access study sites.

VHF Systems

The most commonly used remote tracking equipment is based on animal-mounted transmitters that send signals to observers in the VHF portion of the electromagnetic spectrum (30–300 MHz; Appendix 10.1). Radiotelemetry based on VHF equipment is often the lowest cost and lightest alternative among those described here. In most cases, each animal is tagged with a radio that transmits signals at a unique frequency, so biologists can distinguish locations and movements of particular individuals. Observers with specialized antennas and receiver equipment then detect those signals and estimate animal locations (Appendix 10.2). Many companies specialize in equipment for VHF systems (Appendix 10.3).

VHF Frequency Selection

A range of **frequencies** is available for VHF studies, and selecting which portion of the electromagnetic spectrum to use is among an investigator's first decisions. Only a portion of this spectrum is available for wildlife radiotelemetry, and the available portion changes from country to country. In the United States, frequency allocations are governed by Title 47 of the Code of Federal Regulations (http://wireless.fcc.gov/index.htm?job=rules_and_regulations). The Federal Communications Commission (FCC) provides substantial information about the most recent national and international transmission standards and rules at their web site (http://www.fcc.gov). The FCC allocates frequencies in the range of 40.66–40.70 MHz and 216–220 MHz for "tracking of, and telemetering of scientific data from, ocean buoys and wildlife" (Federal Communications Commission 2009:472). However, other portions of the spectrum are most commonly used for wildlife radiotelemetry, including the 148–152 MHz band, the 162–168 MHz band, and the 170–173 MHz band. Frequency allocations will change from country to country, however, so biologists should research applicable laws and contact equipment manufacturers to determine which frequencies are available.

Researchers should consider several aspects of their study when selecting frequencies. First, constraints on the size of

Box 10.1. Considerations for Transmitter Attachment

- Assume that every transmitter and attachment technique has some effect on marked animals, and realize that subtle transmitter effects may have an insidious influence on the study results beyond obvious overt effects (e.g., rapid mortality or dropped transmitter).
- Although transmitter weight is typically the primary consideration of tag design, other factors are of equal if not greater importance (e.g., attachment technique or antenna configuration, shape, and color).
- A pilot study can reveal attachment technique problems that may not have been problematic for earlier investigators or the manufacturer. Pilot work to evaluate different techniques in the context of the study objectives may reveal trade-offs not apparent in other studies.
- Pilot work is especially critical when choosing an attachment technique. Ease of attachment may not always be the best criteria to evaluate a particular telemetry application, especially with techniques that have considerable subjectivity (e.g., sizing of backpack harnesses or using adhesives). For example, use of adhesives may require that you hold the animal in hand until the adhesive cures and prevent it from becoming entangled in the glue.
- Antenna configuration is critical to obtaining proper signal strength for locating the animal, but it also potentially affects flight mechanics of birds or behavior in mammals during feeding or social interactions with conspecifics. A relatively thin and limp wire antenna is often less obtrusive but breaks easily. A thicker diameter antenna will stand up vertically and provide a stronger signal, but may affect movement and behavior, especially for burrowing animals.
- Subtle changes in shape, color, and location on the animal can affect mate selection and status of the animal in herd situations, potentially producing biased inference and ultimately affecting management recommendations. Observations of a few marked individuals may help assess whether such problems are present.
- There are many useful papers that have evaluated transmitter effects on wildlife. Consider them when planning your study, but carefully note more than weight and transmitter used; also consider whether their evaluation is directly applicable to your situation.
- In addition to published studies, investigators who have attempted or are trying to put transmitters on similar species are an invaluable source of suggestions and information. Online discussion sites are good places to ask questions and search for previous suggestions on the topic (http://community.movebank.org/, TWS-L listserve, Mammal-L listserve, etc.)
- Check with your research institution or agency to determine how to receive approval for animal care and use. Many professional organizations (e.g., American Ornithologists' Union, American Society of Mammalogists) have guidelines for capture, handling, and care of animals and can be a valuable reference for suggestions. Always receive appropriate animal care and use approval for your research.

Table 10.1. Major technological options for tracking individually tagged animals. Weights and costs are estimates: as technology improves, tags get smaller and cost decreases.

Technology[a]	Smallest animal[b] (g)	Smallest tag (g)	Tracking range	Approximate accuracy	Approximate cost per tag (US$)
GPS	400	20	Global	20–200 m	2,000 + optional networking costs
Satellite PTT	100	5	Global	250 m–200 km	3,000 + 75–150/month mandatory data retrieval costs (depending on options)
Satellite GPS/PTT	440	22	Global	10–100 m	3,000 + 75–150/month mandatory data retrieval costs (depending on options)
Light-level geolocation	30	1.5	Global[c]	100–200 km	250
VHF radiotelemetry	4	0.2	0.5–5.0 km	0–500 m[d]	200

[a] GPS = Global Positioning System; PTT = platform terminal transmitter; VHF = very high frequency.
[b] Following the 5% body weight rule (Gaunt and Oring 1999, Murray and Fuller 2000). We recommend the lightest possible transmitter that can meet study objectives.
[c] Tag must be retrieved to download data.
[d] Error with VHF can essentially be zero when homing is used, but it increases when such techniques as triangulation are used to estimate the location of the animal.

the transmitting and receiving equipment may limit the choice of frequencies, because the efficiency of antennas is affected by the ratio of antenna length to radio wavelength. The approximate size of the wavelength λ resulting from a specific frequency is easily calculated as follows:

$$\lambda = \frac{300}{\text{Frequency (in MHz)}},$$

where λ is the wavelength, measured in meters and the constant 300 is derived from the speed at which radio waves travel, which is about 300×10^6 m/sec (Kenward 2001). For example, a 150-MHz frequency has a wavelength of about $300/150 = 2$ m, which affects antenna size (see below). Lower frequencies emit radio waves with longer wavelengths and thus require larger transmitting and receiving antennas than do higher frequencies. Second, the transmission frequency also determines how well signals travel through the environment. Lower frequencies propagate better through water and vegetation; higher frequencies are more subject to attenuation and signal reflectance. However, higher frequencies transmit data more rapidly. These differences are most important for sophisticated systems that transmit coded data (discussed below). Finally, and perhaps most importantly, study area conditions may render some frequencies unusable if other transmissions cause background interference. Interference varies greatly from location to location, so we strongly recommend biologists take the time to scan possible frequencies in their study areas and make efforts to learn of other local tracking projects at an early stage of project design.

Consideration also must be given to the specific frequencies of individual transmitters. Investigators should space radio frequencies far enough apart so that signals do not overlap. Optimally, transmitters should be spaced by at least 10 kHz (e.g., 150.000 MHz and 150.010 MHz; Kenward 2001) and preferably by 25 kHz. This separation protects against frequency drift, which is the tendency of the transmission frequency to vary slightly with time and environmental conditions. If the available spectrum is too narrow, or many transmitters are required, individuals transmitting at the same frequency can be discerned by programming microcontrollers to emit pulses at different rates, or by uniquely coding individual pulses. Another important consideration is the possible conflict of multiple researchers working on different species in the same area where similar frequencies and bandwidths are used.

VHF Transmitters

Transmitters have been developed for use on a great variety of species, ranging from large marine and terrestrial mammals to bats and small passerine birds and insects. Consequently, biologists have many options when selecting which transmitter is most appropriate for their studies. When ordering VHF radiotransmitters, the biologist should be able to provide information about the preferred frequencies and separation between them; pulse rate; pulse width; minimum acceptable radiated power; allowable transmitter antenna length, mass, and operational life; operating temperature range; and how the transmitter will be attached. With basic information about preferred frequencies and separation between them, minimum pulse rate, required range and life of the transmitter, attachment technique, and animal under study, manufacturers (Appendix 10.3) can help researchers select the most appropriate equipment and explain trade-offs. These details highlight the importance of conducting pilot work to ensure that the entire system functions properly and will allow researchers to collect data to meet their objectives.

The basic technology of VHF transmitters has not changed markedly since Cochran and Lord (1963) first published directions for constructing transmitters. The 2 basic transmitter circuits include single-stage and 2-stage designs (Kenward 2001). Generally, single-stage transmitters are lighter and less powerful than 2-stage units, although it depends on the battery. Further developments came in the late 1980s, as manufacturers added lightweight microcontrollers that improved functionality and options available for VHF transmitters (Rodgers 2006). **Microcontrollers** allow precise control of the pulse rate and pulse width and allow the user to program "duty cycles" that switch units off and on during preprogrammed portions of the day or year. However, these advantages come with a trade-off in size. The addition of the microcontroller increases transmitter weight, excluding their use on the smallest species.

Animal-borne VHF tags typically transmit pulsed signals. For example, a 20-msec pulse may be sent once every second. This scheme conserves power and the resulting beeps are easier to hear than a constant signal. The energy of these pulses also can be set by the manufacturer, allowing more powerful pulses that can be detected farther away. However, they also require more energy and thus reduce the lifespan of the equipment. Microcontrollers can further prolong battery life by switching transmitters off during specified times. Pilot studies can be a useful way to determine an optimal trade-off between transmitter range and lifespan in a particular study area.

Transmitter antennas are constructed as whips or loops (Figs. 10.1–10.4). Whip antennas are straight antennas, which are most frequently used in radiotelemetry. The transmission efficiency of a whip antenna depends on the length of the antenna relative to the wavelength. Whip antennas should be at least $\lambda/16$-m or $\lambda/8$-m long, with shorter antennas transmitting radio waves less efficiently. For example, a 150-MHz transmitter should have an antenna at least 12.5 cm ($\lambda/16$ m) in length, but longer antennas will result in more efficient transmissions. The efficiency of whip antennas can be enhanced by adding an additional ground plane antenna, which is an antenna, usually 67% the length of the main whip, oriented at 90° or opposite the main whip antenna (Kenward

2001). Another less frequently used transmitter antenna is the tuned loop, which works well as a collar or an implant. Tuned loops are a popular alternative to the whip antenna for smaller species and social mammals that might damage whip antennas (Fig. 10.1).

VHF Receiving Antennas

Receiving antennas acquire signals from distant transmitters and relay those signals to VHF receivers. Several antenna designs are available for radiotelemetry studies, but as with other equipment, there are trade-offs to consider with each (Fig. 10.5). The first consideration is the purpose of the tracking. Biologists are usually interested either in detecting the presence or absence of a radiomarked animal or in precisely recording its location, and different antennas are better suited to these very different goals. Another consideration to be made is the trade-off between antenna receiving strength and portability in the field. Generally, as the number and length of antenna elements increases, so does signal strength and directionality. However, this increase is accompanied by a resulting decrease in portability in the field. Finally, as mentioned above, each frequency moves through the environment at a specific wavelength. Manufacturers tune antennas to match the frequency ranges of transmitters, so investigators should take care to use the proper antenna for their study.

The strengths and weaknesses of each type of antenna should be considered before committing to a particular type. **Omnidirectional** antennas do not provide directionality but are useful for detecting the presence of a signal. They do, however, relay a stronger signal when closer to the transmitter. Omnidirectional antennas are often designed for easy mounting on vehicles (Fig. 10.5), and they can be used in conjunction with scanning receivers that simultaneously monitor multiple frequencies. This combination allows biologists to quickly scan for signals while traveling. Once a signal is detected, directional antennas can be used to locate the radiomarked animals.

Adcock or "H" antennas are a commonly used directional antenna with 2 elements, which makes them compact and handy for field projects (Fig. 10.5). Some H antennas also have flexible polyvinyl chloride (PVC)-wrapped elements for increased durability under demanding field conditions. The trade-off to increased portability in the field is reduced signal strength and directionality. The H antennas provide the strongest signal when pointing directly at the transmitter, and another, slightly weaker, **peak signal** when pointed 180° from the transmitter. The relative strength of these 2 peak signals may not be clearly discernable, especially if the signal is weak. In these situations, the direction must be determined either through knowledge of the general area used by an animal or through triangulation. Loop antennas are another small and durable antenna that are useful for tracking transmitters with relatively low frequencies (30–40 MHz) that would otherwise require large directional antennas. Like the H antennas, however, they also can give ambiguous directionality.

The Yagi is another frequently used directional antenna. Handheld Yagi antennas often have 3 or 4 elements, and

Fig. 10.5. Antennas for very high frequency radiotracking: (A) omnidirectional; (B) yagi used in aerial tracking; (C) null-peak tracking system with integrated electronic compass, (D) H antennae with portable receiver. *(B) Photo by R. Kays; (C) photo by M. Alleger; (D) photo by J. Millspaugh.*

mounted units can have more. The increased number of elements improves both the power to detect a signal and directional sensitivity. Like the H antenna, peak signals are evident when Yagi antennas are pointed directly toward or away from transmitters. However, the increased directionality of Yagi antennas makes the relative difference in strength much more evident, and there is usually no ambiguity in the direction of the transmitter. The trade-off with increased signal strength and directionality of a Yagi antenna is reduced portability, but most 3-element Yagis have folding elements to mitigate their bulk. One way to overcome this problem is to mount Yagis on the top of a vehicle, which gives an added benefit of height. When conducting telemetry from a vehicle, it is helpful to use an electronic compass to take azimuths (Cox et al. 2002), because manual azimuths based on the approximate angle of the antenna can add an unnecessary source of error.

A commonly used vehicle-mounted antenna setup is a **null-peak system.** Null-peak systems consist of 2 vertical 4–6 element Yagi antennas mounted parallel to each other and connected via an electronic switchbox. A null-peak system can work by increasing the strength and beam width of a signal in the general direction of the transmitter (with approximately 1° accuracy) or by producing a very sharp null when pointed directly at the transmitter (with approx. 0.5° accuracy). Null peak systems operate based on the physical properties of the radio waves, and correct spacing of antennas is crucial to proper operation. Optimal spacing depends on the frequency, and the 2 antennas should be λ m or $\lambda/4$ m apart (Voight and Lotimer 1981).

Antennas also may be mounted on fixed-location towers and can be used both for detecting the presence or absence of study animals or for estimating their locations. If a tower is used to detect animal presence near the center of the study area, an omnidirectional whip antenna can be used (e.g., Castellón and Sieving 2006). Alternatively, a tower with a Yagi antenna can be used if placed at the boundary and pointed toward the study area (Fig. 10.6).

Tower networks also can be used to estimate locations of individual animals. The most commonly used approach is to have rotating Yagi antennas in a network of at least 3 towers and use triangulation to estimate the location of individual animals (Fig. 10.6). Another approach is to use several fixed antennas arrayed in a circle that point away from the tower in different directions (Larkin et al. 1996). The relative amplitude of the pulse at each antenna can then be used to estimate animal direction, and a series of towers can be used to triangulate animal locations. However, this approach may have limited utility for weak transmitters. These types of systems can be operated both manually or automatically but with an increased price associated with the automated systems.

VHF Receivers

Biologists can choose from an impressive array of wildlife telemetry receivers (Fig. 10.7). Although the variety of manufacturers is wide and available features remain large, the basic functions are similar: they acquire a signal through the antenna and process that signal to produce an audio tone (Appendix 10.1). The user needs to take into account several considerations when selecting the appropriate receiving unit. One consideration is the available bandwidth. Many receivers now come with a bandwidth of 4 MHz, which should be suitable for most radiotelemetry applications. However, some less expensive units are useful in only narrow bandwidths, and the user should ensure the available bandwidth is adequate for the number of transmitters that will be tracked at any one time (i.e., if each transmitter is spaced 10 kHz apart, a biologist could get approx. 100 transmitters for each MHz, minus bands to avoid interference). Also, the user needs to ensure that the receiver is set to the correct frequency range: not all receivers work with all frequencies! If transmitters with microcontrollers are used to emit uniquely coded pulses on the same frequency, the user requires a receiver with the capability to decode those pulses.

Fig. 10.6. Towers for very high frequency transmitters. System on left has 2–5 element yagi antennas and is collapsible and portable. System on right is a parallel fixed array of log-periodic antennas mounted on a 42.7-m (140 feet) Rohn 25 tower.
Photograph on left by J. Millspaugh; photograph on right by R. Kays.

Fig. 10.7. (A) Long-billed curlew (*Numenius americanus*) with leg-harness mounted platform terminal transmitter (PTT); (B) Laysan albatross (*Phoebastria immutabilis*) with light-level geolocator mounted on a numbered leg band; (C) short-tailed albatross (*P. albatrus*) with feather-mounted solar PTT. (D) The bird in panel C was marked in Japan and later photographed near the California coast. *(A) Photo by A. Hartman; (B) photo by M. Romano, U.S. Fish and Wildlife Service; (C) photo by R. Suryan; (D) photo by A. Jaramillo.*

Scanning and programming features can improve receiver functionality. Programmable scanners store many frequencies and eliminate the need to repeatedly fine-tune equipment for each study animal. Receivers with scanners rapidly monitor many frequencies and allow researchers to search for many animals simultaneously. Most scanners allow for variable scanning speeds, and different scanning speeds may be important for different applications. For example, faster scanning speeds may be desirable while flying, to avoid missing a pulse, whereas slower scanning speeds may be desirable if scanning from a vehicle or fixed location.

There are increasing options for automatic recording of data from receivers. The most basic option is to simply record the presence or absence of an individual, and that can be done without expensive data-logging receivers. The simplest methods use an audio recorder to record pulse activity. More sophisticated receivers are available with automatic data-logging options. These receivers allow users to scan many frequencies and record information from each frequency. If signal strength is recorded from only 1 omnidirectional antenna, these data can be used to detect the presence and activity of animals. Alternatively, a rotating directional antenna, or multiple fixed antennas, can be used to automatically find the azimuth to a tagged animal (Cochran et al. 1965, Larkin et al. 1996). These automated telemetry receivers have been integrated into complicated systems with tower-mounted antennas and live data flow through wireless networks to obtain continuous telemetry data on animal location, activity, and mortality (Aliaga-Rossel et al. 2006, Crofoot et al. 2008, Lambert et al. 2009).

Background noise and interference can often make it difficult or impossible to hear a signal or discern a strong directionality. Sources of background noise or interference could include static from vehicle engines, power lines, radio towers, or airport communication towers. Some receivers come with built-in noise blocking capability, but they may not sufficiently filter powerful sources of interference. In situations where strong background noise and interference make signal acquisition difficult, noise reduction units may be appropriate. Noise reduction units filter out ambient background noise and deliver a digitally processed signal. A receiver must still be used to acquire a signal, which is delivered to the noise reduction unit via the receiver's headphone jack. The noise reduction unit then digitally filters most noise below about 240 Hz and above about 2,000 Hz, retaining information in the frequency range most commonly used to listen to a telemetry pulse. Noise reduction units substantially reduce background noise, allowing users to detect and follow a signal that may otherwise be difficult to hear. Because the unit is external to the receiver, additional considerations must be made for transportation in the field. The signal also is slightly delayed (1–2 sec) as it is processed, so its utility in detecting signals from airplanes or while rapidly scanning may be limited.

Global Tracking Systems

The ability to effectively track animals beyond localized study sites has been a goal of wildlife biologists for decades, and satellite-based tracking, GPS, and light-level geolocators have made these planet-scale projects a recent reality. The equipment can be constructed with many different configurations that make it most useful for studies of medium-sized and large species. However, the benefits of global-scale tracking systems also are countered by substantial drawbacks. The units can be heavier than VHF transmitters, prices can be an order of magnitude higher, and there are substantial costs associated with data retrieval. Further, the useful life of these systems is usually shorter than that of VHF systems of comparable size, or they may require that study animals be recaptured to retrieve data. Nonetheless, satellite, GPS, and light-level geolocator systems have been widely used, and they have provided examples of spectacular animal movements (e.g., Guilford et al. 2009, Stutchbury et al. 2009). Although GPS tags also have been used on local study sites, these systems are truly global in their potential application and consequently are placed in this category. We now discuss these global tracking systems.

Global Positioning System Technology

GPS tags offer potential for generating high volumes of very accurate location data, regardless of an animal's location on the globe. GPS units use information transmitted from a constellation of satellites to estimate a geographic location. Compared to other wildlife tracking equipment, GPS tags can relay extremely accurate locations, with errors ranging from 20 to 200 m. However, buildings, steep terrain, and thick vegetation block satellite transmissions, so GPS tags historically had limited utility (Bourgoin et al. 2008). Recent technological developments have made the equipment much more robust, and it is now capable of reliably operating under moderate vegetation cover (Holland et al. 2009, Tobler 2009). However, the usefulness of the system for any particular study is limited by the ability of animals to carry the batteries needed for power, so most GPS tags can be programmed with duty cycles that turn the units off and on to record key movement information and avoid repeated data. When combined with **archival logging** units that store geographic information onboard, GPS technology provides a low cost means of gathering accurate location information.

There are 3 options for retrieving GPS data: (1) recovering the tag either by recapturing the animal or by retrieving the tag after it drops off remotely, (2) receiving the data through a transmitted signal from the tag (e.g., remote download), or (3) integrating the GPS component with a communication network, such as Argos PTT or Groupe Spécial Mobile (or Global System for Mobile Communications; GSM™) phone network. The integrated GPS and PTT transmitters have greatly improved the accuracy of global scale satellite tracking, reducing location errors from hundreds of kilometers (typical of Doppler locations) to tens of meters (typical of GPS). Currently, these integrated transmitters weigh approximately twice as much as basic PTT units, and they also cost substantially more.

Satellite PTT Location Systems

Satellite-based receiving equipment may be the only practical way of obtaining transmitter signals from animals that traverse extremely large areas or remote regions with difficult topography (Jouventin and Weimerskirch 1990). The most basic systems use **platform terminal transmitters (PTT)** to send signals, which are received by Argos equipment on 6 U.S. National Oceanic and Atmospheric Administration (NOAA) satellites (Appendix 10.1). The satellites polar-orbit Earth at a relatively low altitude and at a fast speed, and each has a footprint that spans approximately 5,000 km (Argos 2008). Thus there are more opportunities to receive PTT transmissions near Earth's poles and fewer in tropical regions. When the Argos equipment receives transmissions from a PTT unit, it estimates the distance to the transmitter by measuring the change in the broadcast frequency that is caused by a Doppler shift (Appendix 10.1). Four or more measurements are needed to estimate longitude, latitude, the true transmission frequency, and location accuracy.

The PTT transmitters are heavier than VHF equipment, because they require batteries powerful enough to send signals to orbiting satellites (Fig. 10.7). However, developers have employed several technologies to minimize weight and maximize operational lifespan. Some designs include integrated solar cells that recharge onboard batteries. The PTT systems also are manufactured so they transmit signals only during times when programmed to be active. Depending on the objectives of an investigation, duty cycles can turn PTT units on for 1 day out of every 3 or 4 days, or they can make the units operational during specific times when animals are likely to be exposed to the sky. In many ways, the PTT duty cycles are similar to observer sampling regimes for VHF based studies, so investigators should thus invest substantial energy in designing duty cycles that provide unbiased representations of animal locations. Despite solar cell integration and duty cycling, however, PTT systems remain so heavy that their utility is questionable for smaller animals, including 80% of bird species and 65% of mammals (Wikelski et al. 2007).

Data from animal-mounted PTTs are handled by the Argos data collection relay system, which is administered under a joint agreement between the NOAA and the French space agency (Centre National d'Etudes Spatiales). Satellites relay PTT transmission data to French subsidiary firms, Collecte Localisation Satellites (Toulouse, France) and Service Argos (Largo, Maryland; hereafter CLS Argos). Researchers must then pay a subscription fee to the CLS Argos data service to retrieve animal movement reports. Each data report provides PTT location estimates and estimates of location accuracy of >1500 m, 1,500 m, 500 m, and 250 m (Argos 2008).

Data obtained from CLS Argos are often filtered via additional quality assurance or quality control procedure to censor unrealistic locations and further improve accuracy (e.g., CLS Argos does not estimate the accuracy of locations based on 2 or 3 transmissions). Filtering algorithms depend on the natural history of the study species; for example, they remove points that are impossibly distant. Filters also use algorithms to identify unrealistic turn angles and distances between multiple successive points to eliminate erroneous data (Austin et al. 2003, Freitas et al. 2008). Investigators need to have a realistic expectation for the magnitude of location error from PTT systems. Independent tests by wildlife researchers with large study animals equipped with heavier PTT units report errors similar to those described by Argos (White and Sjöberg 2002). However, tests with lightweight PTT systems reported that Argos was unable to estimate many locations and that location error was as large as 439 km (Britten et al. 1999, Soutullo et al. 2007). Accuracy also can be affected by the geometry of the PTT during satellite pass, the number of transmissions received during the 10-minute over-flight, PTT transmitter frequency stability, and power output. Thus, wildlife researchers should expect greater accuracy and more frequent locations from larger study animals fitted with stronger transmitters and from those exposed to the sky more often. Smaller animals and those spending time near human-caused radio interference, below ground, under water, or beneath substantial tree cover are poor candidates for this technology.

Recently, PTT manufacturers have integrated additional location and sensing technology into the PTT units. GPS units are among the new additions, and they have drastically improved the accuracy of satellite-based wildlife tracking, because they reduce location errors from hundreds of kilometers (typical of **Doppler** locations) to tens of meters (typical of GPS technology). Currently, these integrated transmitters weigh approximately twice as much as basic PTT units. Other sensors also have been integrated into PTT systems, which can relay information about altitude, heading, travel speed, temperature, or other ambient information (see sensors below).

Geolocators

Light-level geolocators, or global location sensing (GLS) units, have been refined in recent years for tracking wildlife movements (Burger and Shaffer 2008). The GLS equipment is based on a type of archival tag that must be retrieved from study animals to download data. The GLS tags record time-stamped data about light levels, which are then used to infer the time of sunrise and sunset (Wilson et al. 1992). Light cycles are unique to a particular location on Earth, so these data can then be used to estimate the geographic position of an animal.

As with other tracking equipment, there are benefits and costs associated with GLS tags. The tags weigh 2 g, and they can, therefore, be used on migrant birds. The GLS tags also can last for many years, because both light sensors and logging equipment require little electricity (Burger and Shaffer 2008). Additionally, the tags do not require an antenna, so the compact units can be fitted to the legs and backs of migrant birds. However, estimates of location error are typically 100–200 km (Phillips et al. 2004), and latitude cannot be estimated around the time of the equinox. Nonetheless, techniques for estimating locations are rapidly improving. Some manufacturers of GLS equipment provide software for data processing that attempts to account for cloudiness. Further, some tags are equipped with sea-surface temperature sensors that refine location estimates for marine animals by incorporating known information about ocean conditions (Shaffer et al. 2005). Thus far, the coarse location accuracy of GLS tags, combined with their lightweight and compact characteristics, have made them most useful for migratory birds, marine species, and other animals that traverse very large distances.

Sensors

A host of sensors can be integrated into any animal-borne tag to provide information about the conditions of the animal or its environment. The variety of available sensors increases every year as new technologies are developed to reduce limitations of power and data storage. Simple information can be coded into a broadcast signal in a way that can be decoded by a human observer (e.g., a variable signal pulse rate), whereas more complicated data can be encoded into the signal and streamed live (Cochran et al. 2008, Steiger et al. 2009) or logged into onboard memory that is later downloaded from the tag (Van Oort et al. 2004, Mandel et al. 2008).

The most common sensor relays the activity of an animal by transmitting a constant VHF signal when an animal is resting or a fluctuating signal when it is moving (Theuerkauf and Jedrzejewski 2002). Automated telemetry receivers can be used to obtain a constant and detailed actogram of daily rhythms and to quantify the proportion of the day study subjects are moving (Cochran et al. 1965, Sunquist and Montgomery 1973, Lambert et al. 2009). Tip-switch sensors also are commonly integrated into the animal transmitter to record more specific data on the movement and posture (Janis et al. 1999). Activity sensors are often designed to change the pulse of a transmitter after a certain length of inactivity, typically indicating the death of an animal, and are commonly referred to as **mortality sensors.**

The most detailed activity sensors are 3-axis accelerometers, which measure the fine-scale movement of an animal at a high temporal resolution (e.g., >10 locations/sec). The degree of acceleration can be considered a quantitative degree of activity, which has been found to match energy expenditure in captive animals (Green et al. 2009, Halsey et al. 2009). The patterns in the accelerometer data also may be used to distinguish different behavior types (Watanabe et al. 2005, Shepard et al. 2008), providing the potential for the automated generation of ethograms for completely free-ranging animals (Sakamoto et al. 2009).

Body temperature has long been measured by animal transmitters with sensors placed against the skin or implanted, and a variety of commercial units are available (Dausmann 2005). Although other physiological sensors have been developed, few are commercially available, so that most field biologists have to team up with sensor engineers to develop transmitters that meet their research goals (Cooke et al. 2004).

Heart rate is probably the most sought-after physiological trait, as it can be a direct estimate of metabolism and energy use (Butler 1991, Portugal et al. 2009). Measuring heart rate requires implanted electrodes, which typically connect to implanted data loggers, although streaming data via an external radiotransmitter also is possible (Steiger et al. 2009). Other physiological measurements that have been gathered with specialized sensors are blood flow and chemistry (Ponganis et al. 2007, Wang and Hicks 2008); muscle use (Tobalske et al. 2009); brain activity (Rattenborg et al. 2008, Vyssotski et al. 2009); and stomach temperature, which can be used to indicate feeding (Nevitt et al. 2008)

Sensors worn by animals also can be used to measure aspects of the environment. This feature is common in marine studies, where time–depth recorders have long measured the diving behavior of animals (Austin et al. 2006). This concept has been extended by adding additional sensors to measure water temperature and chemistry to the point where tagged, free-ranging animals are considered to be autonomous samplers of environmental parameters (Charrassin et al. 2002). To date, this concept has not been developed by terrestrial biologists.

The use of animal-borne video and environmental data collection systems (AVEDs) for terrestrial mammals is a recent development, although similar systems have been used for marine mammals for about 20 years (Moll et al. 2007). These systems integrate multiple sensors and collect video, often from the viewpoint of the animal observing its environment, and integrate audio, location, temperature, and acceleration information from other sensors. Researchers have used AVEDs to address numerous hypotheses about animal behavior and foraging tactics (Heithaus et al. 2002, Beringer et al. 2004, Rutz et al. 2007), animal energetics (Williams et al. 2000c, Hays et al. 2007), wildlife damage issues (Grémillet et al. 2006), and inter- and intra-specific interactions (Passaglia et al. 1997, Heithaus and Dill 2002). Terrestrial systems are largely transmission-based (Beringer et al. 2004, Carruthers et al. 2007, Rutz et al. 2007, Millspaugh et al. 2008, Taylor et al. 2008), meaning they transmit video to a receiver that may or may not be portable (Beringer et al. 2004, Rutz et al. 2007). These terrestrial transmission-based systems offer long recording times (>70 hr; Beringer et al. 2004), but they require constant close contact between the receiving station and tagged animal to record transmitted video. Although intermittent data from difficult-to-track species might be useful, battery power is wasted and data are lost if researchers lose contact with a tagged animal because of animal movement or signal attenuation (Millspaugh et al. 2008). Consequently, terrestrial AVED studies have focused on species that are easily tracked or are habituated to humans (Millspaugh et al. 2008). Store-onboard AVEDs overcome these limitations, enabling research on elusive, free-ranging species for which noninvasive behavioral data are often difficult to obtain (Marshall et al. 2007, Moll et al. 2007). Moll et al. (2009) described the first terrestrial, store-onboard AVED developed for large mammals and demonstrated its utility by describing contact rates among white-tailed deer (*Odocoileus virginianus*) in Missouri. The AVED technology is a rapid growing field of research, but it is hampered by the size of equipment its short lifespan.

Data Networks

Many of the methods described in this chapter involve using animal-borne tags to collect detailed data from satellites (i.e., GPS) or other sensors. Typically these data are **stored** (logged) **onboard**, at least initially, and must eventually be retrieved. The simplest retrieval involves recovering the physical tag and downloading data through a cable. However, this method is often impossible or impractical, as most wildlife species are difficult or impossible to recapture. Animal-borne devices can be designed to fall off an animal after the study because of the natural decay of collar material (Garshelis and McLaughlin 1998) or the specifically timed action of a release device (Muller et al. 2009).

Store-on-board strategies are risky, because they depend on the biologists' ability to physically locate a tag at the end of a study. Failed electronics, long-distance dispersal, and Murphy's Law all conspire to make this the option of last resort for most studies. The preferred option is to incrementally retrieve data remotely from the tags as the study is underway. This option not only avoids the risk of complete data loss but also allows researchers to monitor the performance of equipment and use the data during the study to learn more about the species. For example, Zimmermann et al. (2007) downloaded GPS tracks of predators in the field and immediately backtracked these routes to find recent kill sites. Live telemetry data also makes it possible to find dead animals more quickly, so that cause of mortality can be more accurately determined, even in tropical conditions (Aliaga-Rossel et al. 2006).

The most simple remote data-access method involves homing on the location of a tagged animal to within a few hundred meters and downloading the data with a handheld receiver. These "remote download" options do not rely on a more complicated data network, but they do require biologists to repeatedly locate animals and add weight to the transmitter package. Local area networks also can be created to stream telemetry data from a limited area back to a central computer (Crofoot et al. 2008). Finally, commercial satellite and mobile phone networks can be used to obtain data. These options are more expensive than local ones, but they increase the scale and ease of data retrieval. Mobile phone networks, typically GSM networks, have more limited coverage than do satellites, especially in remote areas, but they have the advantage of offering 2-way communication with tags. This capability allows biologists to reprogram the settings on collars during the study. Satellite systems do not offer 2-way communication, but they can provide global coverage.

SUMMARY

Radiotelemetry has become one of the most useful techniques in wildlife ecology and management because of its ability to obtain location and other data remotely, increasingly at a global scale. Radiotracking investigations are most often focused on questions about animal space use and demographics. Radiotelemetry studies can only provide useful information when biologists clearly articulate their objectives, use appropriate study designs with a high likelihood of meeting the study objectives, consider important assumptions, apply appropriate analytical methods, and carefully interpret the results. With all studies, appropriate methods of attaching transmitters to the focal species are critical, as is the assumption that transmitters have no significant effects on study animals. The development of GPS transmitters, light-level geolocators, and satellite telemetry allows investigators to examine movements at continental and global scales, but VHF systems still are appropriate for many studies. Telemetry technology continues to develop at a rapid pace. We encourage biologists to consult the literature, colleagues, and equipment suppliers and manufacturers about equipment capabilities and limitations.

APPENDIX 10.1. HOW TO LOCATE A SIGNAL

There are several ways to locate the source of a signal from an electronically tagged animal. We briefly review the 6 most common approaches.

1. Nondirectional—Presence or Absence

The mere detection of a radio signal by a receiver can be used to register the general location of a tagged animal. Detection presumes the animal is within a certain distance of the receiver, but it does not provide a precise location. Presence or absence systems are typically used with automated receivers that regularly log the presence or absence of an individual at a site.

The area of coverage depends on the range of the transmitter and receiver systems. Typically the systems work for distances ≤2 km ground to ground. Because this method does not locate the animal within this area of detection, this distance also would be the error rate (Breck et al. 2006).

2. Direct Tracking—Location

The simplest way to locate the source of a radio signal is to use a directional antenna and simply walk right up to the broadcasting unit. This approach is not practical for animals that flee or change their behavior when a person approaches, and therefore, it should be reserved for occasions when direct observations are required for other purposes. The method can be particularly useful for tame species or those animals sitting on a nest. This method is required to locate tags that have fallen off an animal or to find tagged animals that have died.

The range of this method is limited to the distance from which a radio signal can be detected. Thus, limitations vary by tag type, antenna size, and terrain, but a reasonable expectation might include ≤2 km for VHF equipment. Presuming the tracker carries a handheld GPS unit, this method can be quite accurate (within 10 m).

3. Triangulation

The most common method of locating VHF radio-tagged animals is through triangulation (Springer 1979, White and Garrott 1990). In this method field-workers use a receiver and antenna to obtain an azimuth from a known location (where they are standing) to the source of the signal (where the animal is located). The intersection of azimuths estimates the location of the animal. Three or more azimuths must be used to also estimate accuracy of the location. This method can be problematic if tagged animals move substantial distances between subsequent detections, because the intersections of those azimuths will not represent the location of the animal. Multiple observers simultaneously recording azimuths from different locations at the same time eliminate this problem.

Azimuths should be obtained relatively close to the target, if possible, because error in the azimuth estimate becomes more critical at larger distances. In addition, azimuths should be obtained from a range of different angles around the animal. Azimuths should be separated by at least 20°, and greater separation improves accuracy substantially. Depending on the relative angle of the azimuths, even small amounts of angular error can cause great location error. Therefore, most observers use 3 or more azimuths to estimate locations and the associated location error. There are a handful of algorithms to triangulate animal locations, and most are available in a variety of software packages (White and Garrott 1990; Appendix 10.3). Most of these algorithms estimate error for each location; these estimates should be used as general guidelines for the relative accuracy of your data and not as a replacement for field-based tests of accuracy (Withey et al. 2001; Figs. 10.A1.1–10.A1.3).

The area over which triangulation methods work is limited by signal detection distances that vary by tag type, antenna size, and terrain. For most typical VHF systems, transmission distances are usually reliably detected between 100 m and 2 km. Listening for signals from aircraft can extend this range up to 10 km. Trail or road networks improve chances of obtaining azimuths. There are a variety of sources of error for triangulation, including moving animals, multipath signal propagation (signal bounce), and error in azimuth estimation. Thus, the accuracy of location estimates also can vary greatly. Studies sometimes publish accuracy estimates from a handful of test transmitters placed in static locations in the field site, and these estimates typically range from 50–500 m.

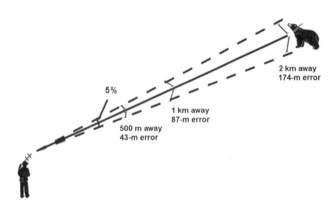

Fig. 10.A1.1. Because azimuths are never perfect, the error of a triangulated location estimate increases with increasing distance between the animal and the observer. In this example, a bearing estimate with 5° error could be 47 m for a target that is 500 m away, but 174 m for one 2 km away.

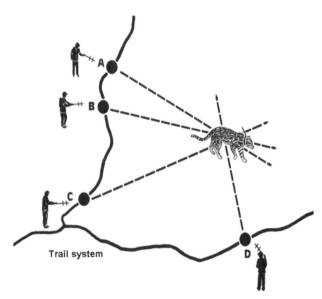

Fig. 10.A1.2. Triangulated location estimates are best with azimuths that differ by ≥20° to reduce the effect of azimuth error on location error. In this example, azimuths from A and B are taken from locations too close together; the addition of azimuths from C and D improve the estimate.

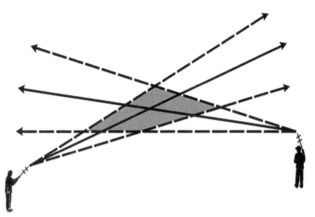

Fig. 10.A1.3. Although location estimates are theoretically possible with 2 azimuths, it is not recommended because of the large error estimate (shaded error polygon) resulting from imperfect azimuth estimation (dashed lines showing bearing error).

4. Global Positioning System Units

GPS units determine location based on the time required for a radio signal to travel from a transmission satellite to an Earth-based receiver. Transmissions originate from a constellation of 24–32 U.S. satellites equipped with precise clocks. Receivers are the small devices attached to study animals. The satellites broadcast precise information about their locations in orbit and the times of their transmissions. Because the transmissions travel at a constant speed, the receiver can precisely estimate the distance between itself and the satellite by determining the time required for the signal to be received. By overlapping these distance estimates from at least 3 satellites, the receiver unit pinpoints its geographic location (Fig. 10.A1.4).

The GPS network has global coverage, but vegetation, water, or geographic features can block satellite signals. Obtaining data from a tag remains an additional challenge that can affect the range over which animals are tracked. Archival units require the tag be retrieved to manually download the data. Various remote download options also are available through satellite, cell phone, or ad-hoc networks.

Handheld GPS units can reach 10-m accuracy by averaging multiple fixes and subcentimeter accuracy by correcting with a base station. Most animal-borne GPS collars are optimized to save battery life by taking only one fix at a time. Thus, 20–200-m accuracy is more typical of wildlife applications (Lewis et al. 2007).

5. Platform Terminal Transmitters

If a transmitter and receiver are in motion relative to each other, there will be an apparent shift in frequency. The satellite based system uses this principal to locate Earth-bound platform terminal transmitters (known as PTTs). The satellite system records the frequency of multiple transmissions and compares them with the true transmitter frequency.

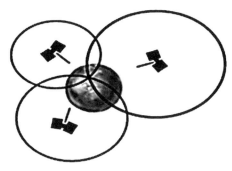

Fig. 10.A1.4. As these 3 Global Positioning System (GPS) satellites orbit Earth, they send out radio signals that are detected by GPS receivers attached to animals. The amount of time it takes these signals to travel to the receiver can be used to estimate the distance, but not direction, and to estimate a circle of potential locations (black circles). The intersection of these circles from the multiple satellites approximates the true location of the GPS receiver.

magnetic interference, such as Europe. In addition, PTT signals can be blocked by vegetation or steep terrain, especially for terrestrial animals. CLS Argos, the commercial company operating the satellite network, provides a quality ranking to help determine the accuracy of each location. Highest quality locations come from the intersection of at least 4 ellipsoids.

6. Global Location Sensing Units

The timing of sunrise and sunset varies in unique and predictable patterns across the surface of the earth, and this variation can be used to estimate a geographic position. GLS units are animal-mounted loggers that record time-stamped daylight records and are then used to estimate geographic position based on light level. Tags can weigh as little as 1.4 g, contain a light meter, clock, data logger, and a small battery, and can archive light levels for many years. They must then be physically retrieved to download data and estimate animal locations and movements. Logged light levels are used to infer the time of sunset and sunrise for each day (Fig. 10.A1.6). The longitude can then be estimated based on the time of local noon (or midnight), and the latitude can be estimated from the total length of the day (combing sunrise and sunset times).

Based on the Doppler principal, with each transmission an ellipsoid of possible locations can be drawn on the surface of Earth. The intersection of 2 ellipsoids from successive transmissions provides 2 possible locations of the animal many hundreds of kilometers apart. One is the true position, the other is known as the mirror location. There are a variety of technical and biological ways to determine the true location. Location estimates with 2 transmissions are more accurate (Fig. 10.A1.5).

Satellites give global coverage, but they may have difficulty detecting signals in regions with high levels of electro-

Light-level geolocation can be used globally, but errors in latitude increase around the solar equinox. Location accuracy is worse than for most other methods, because small errors in day length can lead to large errors in estimated latitude. Errors can vary from 24 km to 1,043 km, but average 200–400 km

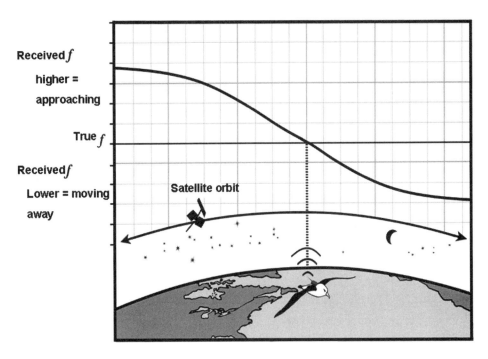

Fig. 10.A1.5. A satellite moving over an earthbound transmitter will detect an apparent shift in the transmission frequency (f) as it moves toward and then away from the source—a phenomenon known as Doppler shift. The exact point of this frequency shift can be used to estimate the location of the earthbound transmitter.

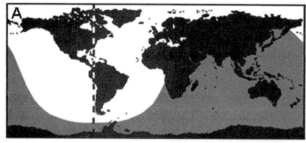
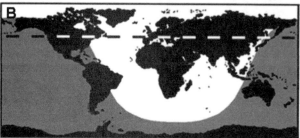
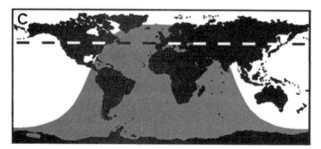

Fig. 10.A1.6. Maps illustrating how the timing of light and dark periods can be used to estimate a location on Earth using light-level geolocation. The time of high noon (A) estimates longitude, whereas the exact times of sunrise (B) and sunset (C) estimate latitude.

APPENDIX 10.2. CONSIDERATIONS FOR VERY HIGH FREQUENCY RADIOTELEMETRY EQUIPMENT

VHF radiotelemetry systems are the most commonly used equipment for tracking wildlife. Researchers and practitioners are presented with a wide array of options that can make the selection of the best materials for a particular study challenging. Each system, and each system component, has associated strengths and weaknesses. When selecting VHF telemetry equipment, biologists should consider the materials in the context of their investigation. In general, researchers should: (1) decide what type of data is needed, (2) decide how accurate animal locations need to be to answer study questions, and then (3) select transmitters and receiver equipment.

Transmitter Considerations

Several types of transmitters are available, and each differs in performance and configuration. In general, smaller species and birds require transmitters with minimized weights, which translates to shorter operational lifespan, lower transmission power, longer pulse intervals, and reduced options. Larger study species can carry much more sophisticated equipment that can last multiple years and can transmit a range of information from much greater distances.

Transmitter Weight

Smaller animals and birds are often limited in their ability to carry VHF equipment, simply because of its weight and bulk. Researchers have previously recommended that transmitters be 5% of body mass for terrestrial animals and 3% for birds that travel long distances. These guidelines preclude the use of VHF for some birds and small mammals, and for other species transmitters must be so light they have reduced lifespans or functionality. Some manufactures make transmitters weighing 0.3 g, but the smallest units have basic functionality, lifespans 1 month, and detection distances 100–200 m.

Transmission Power

Many VHF transmitters can be constructed so they broadcast stronger or weaker signals. Stronger signals originate from more powerful transmissions that use more battery power. Thus, increased transmission strength also translates to larger batteries and heavier transmitters, or to a reduced transmitter lifespan. If study animals are likely to move great distances or if research occurs in heavy vegetation that might block radio signals, researchers should consider the benefits of using more powerful transmission equipment.

Pulse Interval

Transmitters broadcast short pulses of energy that are translated to toned beeps by receiving equipment. Transmitters can be manufactured to send out pulses more or less often.

Pulse Interval Modulation (PIM)

Transmitters can be constructed with additional components that expand their utility greatly when they are used on larger study animals. They can be made to transmit at shorter or longer pulse intervals, depending on activity, physiological conditions, climate, or other data generated by sensors. Perhaps the most common use of PIM is on transmitters with a "mortality mode" that broadcasts transmissions much more frequently after equipment remains motionless for extended durations.

Transmitter Antennas

Transmitters can be constructed with several different antenna configurations. The most compact transmitters include a coiled antenna that is fully encapsulated and is most often used with internally implanted transmitters or those attached to bird leg bands. Loop antennas can be fitted into collars, which makes them useful for mammals or other animals that might bite or pull at an external antenna. Whip antennas are the straight antennas that stand out from transmitters, and they are the most commonly used because they transmit greater distances and provide a more uniform signal.

Antennas and Receivers

Unlike transmitters that must be attached to free-ranging animals, the choice of receiving equipment is often driven more by costs and field conditions. Receiver systems can cost hundreds of U.S. dollars to tens of thousands of dollars. Similarly, receivers can weigh 1 kg, or they can be so heavy they are completely immobile. The basic features of VHF antennas and receivers are presented below.

Hand-Held Antennas

Observers can track study animals on foot when they are using small and light VHF receivers and antennas. The most commonly used hand-held antennas include loop, H, and Yagi antennas. The trade-off among these antennas is one of signal sensitivity and ability to detect transmission direction. In general, larger antennas with more elements are better at detecting a signal, so researchers who are attempting to use triangulation to estimate animal locations should use Yagi antennas with 3 or more elements.

Vehicle-Mounted Antennas

Vehicle-mounted antennas can be used to identify the general location of an animal, so that more precise methods can be subsequently employed, or they can be used to triangulate more exact animal locations. Airplane-mounted Yagi antennas or omnidirectional whip antennas mounted to cars or trucks can be used to detect the presence of study animals from great distances, but observers need to use more precise triangulation or homing techniques to obtain exact locations. When research occurs in areas with reasonably good road access or in flat locations, vehicle-mounted null-peak systems can be used to triangulate study animal locations. Some investigators have constructed real-time triangulation systems that allow field observers to estimate animal locations in real time with null-peak systems that integrate antennas, digital compasses, GPS devices, and a laptop computers running triangulation software.

Antenna Tower Systems

Towers with fixed VHF antennas have been used to study movements of animals across particular landscape features, such as coasts or islands, or animal movements within specific study areas. Tower-mounted antennas are most often fixed in a single direction, so at best, they can be used to make crude location estimates. However, towers are tall, so they can be used to detect signals from great distances, and they are extremely useful for investigations of animal presence and absence.

Receivers

Most VHF receiver systems are relatively small, and nearly all of them can be hand-carried. Further, the sensitivity of receiver systems seems to differ little between the largest and most expensive and the smallest and least costly. Nonetheless, receivers can differ in price by a factor of ten, and there are associated trade-offs. The more expensive systems tend to come with options for attaching automatic data logging equipment, and for use with fixed antenna systems that constantly track animal movements.

APPENDIX 10.3. EQUIPMENT VENDORS AND DISTRIBUTORS FOR RADIOTELEMETRY EQUIPMENT

The source of much of the following information is the Directory of Biotelemetry Equipment Manufacturers (http://www.biotelem.org/manufact.htm).

Advanced Telemetry Systems, United States
http://www.atstrack.com
- VHF transmitters, antennas, receivers, and GPS collars (store onboard with limited remote query)

Andreas Wagener Telemetrieanlagen, Germany
http://www.wagener-telemetrie.de/
- VHF transmitters, antennas, and receivers

AVM Instrument Company, United States
http://www.avminstrument.com
- VHF transmitters, transmitter refurbishment, antennas, and receivers

Ayama Radio Tracking, Spain
http://www.ayama.com/
- VHF transmitters, antennas, and receivers; batteries; and accessories

Biomark, United States
http://www.biomark.com
- Destron and Avid passive integrated transponder tags, implanters, readers, and portable and pass-through antennas

Biotrack, United Kingdom
http://www.biotrack.co.uk
- VHF transmitters, antennas, receivers; GPS collars (remote download via GSM, UHF, or Argos); and geolocators

BlueSky Telemetry, United Kingdom
http://www.blueskytelemetry.com
- VHF transmitters, antennas, and receivers, and wildlife and livestock GPS collars (remote download via GSM or UHF)

British Antarctic Survey, United Kingdom
http://birdtracker.co.uk/
- Light-level geolocators

Data Sciences International, United States
http://www.datasci.com/
- Short-range (8 m) data logging transmitters and repeaters designed to monitor physiological parameters of animals in a lab setting

E-obs Digital Telemetry, Germany
http://www.e-obs.de/
- Lightweight GPS tags with remote data-download capabilities

E-Shepherd Solutions, United Kingdom
http://www.e-shepherd.co.uk
- Solar-powered lightweight GPS data loggers and transmitters (remote download via bluetooth communication)

Environmental Studies, Germany
http://www.environmental-studies.de
- VHF transmitters, GPS collars (remote download via GSM, UHF, Argos or satellite phone), handheld GPS receivers, and collar refurbishment

ENSID Technologies, New Zealand
http://www.ensid.com
- Plastic (food-safe) passive integrated transponder tags, implanters, and scanners designed for fish

Followit, Sweden
http://www.followit.se/wildlife/
- VHF and UHF transmitters, antennas, and receivers, and GPS collars (remote download via UHF, GSM, VHF, and satellite)

Holohil Systems, Canada
http://www.holohil.com
- VHF transmitters only

Hydroacoustic Technology, United States
http://www.htisonar.com
- Acoustic tags, data loggers, and receivers

L.L. Electronics, United States
http://www.radiotracking.com
- VHF transmitters, antennas, receivers, and accessories

Lotek Wireless, Canada
http://www.lotek.com
- VHF transmitters, acoustic transmitters, receivers, hydrophones, GPS collars (remote download via GSM, UHF, Argos, or Iridium), GPS handheld receivers, and geolocators

Merlin Systems, United States
http://www.merlin-systems.com
- VHF transmitters and tranquilizer dart transmitters

Microwave Telemetry, United States
http://www.microwavetelemetry.com/
- Solar and battery powered PTTs, GPS enhanced PTTs (remote download via Argos), UHF antennas, UHF receivers; specializing in avian and fish tags

North Star Science and Technology, United States
http://www.northstarst.com
- Solar, battery powered and GPS enhanced PTTs for avian species, GPS collars (remote download via Globalstar), and Argos PTT locator

The Sexton Company, United States
http://www.thesextonco.com
- Waterproof housing for Telonics VHF receivers

Sigma Eight, Canada
http://www.grant.ca/
- Consumer-programmable VHF transmitters, receivers, and antennas

Sirtrack Limited, New Zealand
http://www.sirtrack.com
- VHF transmitters, antennas, and receivers; GPS enhanced PTTs (remote download via GSM or Argos), and GPS collars (remote download via GSM or Argos)

Sonotronics, United States
http://www.sonotronics.com
- VHF and acoustic transmitters, receivers, and antennas

Telemetry Solutions, United States
http://www.telemetrysolutions.com
- GPS collars (remote download via VHF and hand held receiver) and standalone GPS pods for converting VHF to GPS collars

Telenax, Mexico
http://www.telenax.com
- VHF transmitters, antennas, and receivers

Telonics, United States
  http://www.telonics.com
  - VHF transmitters, antennas, and receivers; GPS enhanced PTTs; GPS collars (remote download via Argos); Argos receivers and antennas

TenXsys, United States
  http://www.tenxsys.com
  - VHF and GPS mammal collars (store on board or Argos optional) and noise and nest monitoring devices

Titley Scientific, Australia
  http://www.titley.com.au
  - VHF transmitters, antennas, and receivers, and GPS collars (store on board only)

Vectronic Aerospace, Germany
  http://www.vectronic-aerospace.com
  - VHF and GPS collars (remote download via UHF, GSM, or Argos)

VEMCO, Canada
  http://www.vemco.com
  - Acoustic transmitters, receivers, hydrophones, and data loggers

Wildlife Tracking Systems, United Kingdom
  http://www.wildlifetracking.co.uk
  - VHF transmitters, antennas, and receivers; specializing in falconry

Wildlife Computers, United States
  http://www.wildlifecomputers.com
  - GPS transmitters (remote download via Argos)

Wildlife Materials, United States
  http://www.wildlifematerials.com
  - VHF transmitters, antennas, receivers, and accessories

Ziboni Tecnofauna, Italy
  http://www.tecnofauna.it
  - VHF transmitters, antennas, receivers, and accessories, and GPS transmitters (remote download via GSM)

ZoHa EcoWorks, Canada
  http://www.zohaecoworks.com
  - Official distributor of wildlife telemetry equipment for Followit (see above) in North America; GPS collars (remote download via UHF, GSM, VHF, and satellite), and VHF antennas and receivers

11

Estimating Animal Abundance

BRIAN L. PIERCE,
ROEL R. LOPEZ, AND
NOVA J. SILVY

INTRODUCTION

CHAPTERS ON **census** methods in The Wildlife Society's "techniques manual" have exploded from 9 pages in the first manual (Wight 1938) to 48 pages in the sixth manual (Lancia et al. 2005). This expansion is testament to the volume of literature produced over the years on this subject, and it has not subsided since the sixth manual. Indeed, the subject has spawned a voluminous literature over the years, including many in-depth books (Caughley 1977; Seber 1982; Caughley and Sinclair 1994; Sutherland 1996; Krebs 1999; Thompson et al. 1998; Buckland et al. 2001, 2004; Borchers et al. 2002; Williams et al. 2002a) on this subject, leading us to ponder how to properly balance coverage of the subject and our intended audience in a limited number of pages.

This chapter differs from those in previous editions (Lancia et al. 1994, 2005) in that we have designed the chapter for use in an undergraduate wildlife techniques class. Our intent is to provide an overview of the basic and most widely used **population estimation techniques.** As pointed out by Lancia et al. (2005), there are several possible approaches to writing a chapter dealing with population estimation that include (1) supplying a detailed treatment that focuses on statistical models and deriving estimators based on these models, (2) providing details on survey protocol design and actually applying different population estimation techniques, or (3) providing the conceptual basis underlying the various estimation methods. Lancia et al. (2005) chose to do the latter. We have chosen the second approach, recognizing, as noted by Lancia et al. (2005), that such an approach has limitations due to the diversity of real-world circumstances and our inability to provide detailed instructions for all possible situations. As such, we do not present all variations of the basic population estimation procedures, but rather provide citations for the relevant literature and computer software where variations of these estimators can be found. However, we believe that a more concise chapter using simple examples will provide a much needed introduction for students, while providing a reference for wildlife biologists and resource managers.

Here we provide an overview of factors that should be considered before choosing a method to estimate population abundance, the pros and cons of using various methods, relevant literature, and available computer software, so the reader may make informed decisions based on their particular needs. For readers with a more quantitative background, literature citations provide access to more detailed coverage of the topics discussed in this chapter.

DEFINITIONS

As terms are used in this chapter, they are defined in relation to population estimation to help the reader understand the material in the chapter. Definitions are based on Overton and Davis (1969), Caughley (1977), Cochran (1977), White et al. (1982), Verner (1985), Caughley and Sinclair (1994), Sokal and Rohlf (1995), Sutherland (1996), Zar (1996), Thompson et al. (1998), Krebs (1999), and Ott and Longnecker (2008).

Population Definitions

Population: A group of animals of the same species occupying a given area (**study area**) at a given time.

Absolute abundance: The number of individuals.

Relative abundance: The number of individuals in a population at one place and/or time period, relative to the number of individuals in a different place and/or time period.

Population density: The number of individuals per unit area.

Relative density: The density in one place and/or time period, relative to the density in another place and/or time period.

Population trend: The change in numbers of individuals over time.

Census: A total count of an animal population.

Census method: The method (e.g., spotlight count) used to obtain data for an estimate of population abundance.

Population estimate: A numerical approximation of total population size.

Population estimator: A mathematical formula used to compute a population estimate calculated from data collected from a sampled animal population.

Closed population: A sampled population in which births, deaths, emigration, and immigration do not occur during the sampling period.

Open population: A sampled population that is not closed.

Population index: A statistic that is assumed to be related to population size.

Detection probability: The probability that an individual animal in a sampled population is detected. Synonyms include **observability, sightability, catchability, detectability,** and **probability of detection.**

Statistical Definitions

Parameter: An attribute (e.g., percentage of females) of a population. If you know the parameters of the population, you do not need statistics.

Statistic: An attribute (e.g., percentage of females) from a sample taken from the population.

Frequency of occurrence: The observed number of occurrences of an attribute relative to total possible number of occurrences of that attribute (e.g., individual was observed on 4 of 5 spotlight counts).

Accuracy: A measure of bias error, or how close a statistic (e.g., a population estimate) taken from a sample is to the population parameter (e.g., actual abundance).

Bias: The difference between an estimate of population abundance and the true population size. However, without knowledge of the true population size, bias is unknown.

Mean estimate: The average of repeated sample population estimates usually taken over a short time period.

Precision: A measure of the variation in estimates obtained from repeated samples. Precision can be measured by (1) **range** (difference between lowest and highest estimates), (2) **variance** (sum of the squared deviations of each n sample measurements from the mean divided by $n - 1$), (3) **standard deviation** (positive square root of the variance), (4) **standard error** (the sample's standard deviation divided by $\sqrt{n}$. It therefore estimates the standard deviation of sampled means based on the population mean), and (5) **confidence interval** (probability that a given estimate will fall within n standard errors of the mean; e.g., a 95% confidence interval would be ±2 standard errors).

Central Limit Theorem: A statistical theorem stating that for large sample sizes (30), the **sampling distribution** of any statistic (e.g., the distribution of means obtained by repeated sampling of the mean from the same population) will be approximately **normally distributed** (form a symmetrical, bell-shaped frequency histogram). Therefore, we can divide the normal curve for the sampling distribution of means into sections represented by n standard deviations above and below the mean. When this is done, 68.26% of the area lies within ±1 standard deviation, and approximately 95% lies within ±2 standard deviations of the mean. Accordingly, a **95% confidence interval** implies a range of values within which 95% of the estimated means would fall. Stated differently, there is a 95% chance the true mean lies within ±2 standard errors of the estimated mean, provided there is no bias in the estimate.

Overton and Davis (1969), in the third edition of *Wildlife Management Techniques Manual*, provided a pictorial presentation (Fig. 11.1) of the relationship between **precision** and **accuracy** that made them easy to visualize. The bull's eye on the rifle target represents the **true population abundance**. If one were to fire 10 shots from a rifle, the 10 bullet strikes would represent the value of each of the 10 individual **population estimates**. The center of the area circumscribed by these 10 shots would then represent where the rifle is firing, on average, or the overall average estimate of population abundance. The distance from the center of all shots fired to the center of the bull's eye represents bias, or the amount of inaccuracy present during those 10 shots.

A Accurate and precise

B Accurate, but not precise

C Precise, but not accurate

D Neither precise nor accurate

Fig. 11.1. An analogy of precision and accuracy when estimating animal abundance or firing a rifle at a target. Note that in target C, shots are biased to lower left, and in target D, shots are biased to right. When estimating animal abundance, we rarely know in which direction our estimation may be biased (either low or high). *Modified from Overton and Davis (1969).*

The spread of the bullet strikes would represent **precision** of the population estimates. **Variance** is used to measure precision; the smaller the spread, the smaller the variance and the better the precision will be. For **perfect precision** and **perfect accuracy,** all 10 shots would strike the bull's eye (Fig. 11.1A). However, one can have **poor precision,** but still maintain overall **mean accuracy** if the center of the area circumscribed by the 10 bullet strikes falls on the bull's eye, thereby giving a **mean estimate** equal to the **true population abundance** (Fig. 11.1B). In the same way, one can have **poor accuracy** with **perfect precision** if all bullet strikes hit in 1 spot biased away from the bull's eye (Fig. 11.1C). The worst-case scenario would be to have **poor accuracy** with **poor precision** (Fig. 11.1D). In the real world of population estimation, one does not ever know where the bull's eye lies; therefore, one can only measure **precision** of the estimates.

In practice, population estimates need to be at least precise to be useful. If estimates can be replicated many times in a short time frame, precision can be increased. And, if an estimator or method has good precision, it might be useful as an indicator of population trend, even if it is not accurate. However, if field conditions change (even during the same field season), precision may not increase (Rakestraw et al. 1998). Furthermore, using trend data to manage wildlife populations can be problematic, as the basic assumption when using trend data is that nothing changes over time except population abundance. So, although precision is easy to compute, in real wildlife populations the true population abundance is never known, and therefore accuracy **cannot** be computed. It can only be implied by the sum of all evidence at hand. As such, if one needs information on population abundance, accuracy is still paramount. Hence the warning precision is no surrogate for accuracy (Lancia et al. 2005).

SURVEY DESIGN

The solution to obtaining a usable estimate of abundance is to choose the right **method** (sampling and/or analysis technique) and to employ proper **survey design or experimental design** (scheme or plan used to obtain samples for abundance or density estimation; see Chapter 1, This Volume). Both must be optimized for the particular circumstance and species to obtain precise (and hopefully accurate) population estimates. Unfortunately, what may work well in some circumstances is useless in others. In addition, there are many combinations of methods and survey designs to choose from, and these can differ by orders of magnitude in their precision and expense. Likewise, there are many opportunities to encounter setbacks and failure. Hence, before any surveying is attempted, the wildlife manager should ask a number of questions:

1. Have I reviewed the relevant literature on the species and/or method?
2. Do I need an estimate of density, or will an index of relative abundance suffice?
3. What methods are available that meet these criteria?
4. What is the extent of the survey area?

5. Are there any limitations on where I can sample?
6. What are the experimental units from which samples will be drawn?
7. How much precision is desired?
8. If comparing areas or time periods, how small a difference must be detected?
9. Given the precision or difference to be detected, how much replication is required?
10. How much replication can I afford?
11. What is the distribution of the species to be surveyed?
12. How will the sample units be distributed?
13. Will sample units be drawn with or without replacement?
14. Do I have the necessary equipment and infrastructure?
15. Do I have sufficient funds to conduct the proposed survey?
16. Is that money better spent on answering another question?
17. Do I have the time required to complete the estimate?
18. Do I have the expertise to collect and analyze the data, or is it available elsewhere?
19. Are there other biologists and biometricians who can provide an independent review?
20. Will I need a pilot study to answer any of the above questions?

Answering the above questions is absolutely necessary, the completion of which should result in a project proposal. Note, the process is iterative and may require several attempts to reach an optimum set of conditions for your particular project. This is typically a good time to contact other biologists and/or biometricians for help, and at the very least to request an independent review of your proposal prior to initiating any work.

Survey Extent

Population estimates typically occur over a defined spatial area, with the estimates representing a specific period of time. As simple as this idea may seem, it is imperative that you define the spatial and temporal extent of the area over which inference is to be made. Answers to these questions will lay the foundation for the statistical analyses ahead and are integral to proper survey design. Integral to this design is an assessment of any nonhabitat and/or nonaccessible areas (private property or dangerous conditions) that may affect species and/or sample distribution.

Experimental Units

Because of the limits of time and costs, a survey of the entire study area of interest is usually not possible. Therefore, an **experimental design** is devised to select a portion of the study area to be sampled (**experimental units**). By definition, experimental units are homogeneous and should be representative of the population or treatment to which inference is to be applied. Experimental units may be time periods, units of space, groups of animals, or an individual animal. It is from **experimental units** that **samples units** are drawn (**replication**). For example, if mice in a cage are given a treatment in diet (e.g., food type A), the cage of animals is the experimental unit, and mice in the cage are sample units. Likewise, if we are comparing abundance among habitat units, the differing habitats are the experimental units, and each survey would be a sample drawn from each of the habitats. In simple surveys, where a population estimate is to be obtained from a single entity with no treatments or controls, there is only one experimental unit.

An **experimental unit** is the smallest entity to which a **treatment** can be randomly assigned (see Chapters 1 and 2, This Volume). If the treatments are manipulative (applied by the experimenter), a **randomization rule** is used to ensure an unbiased assignment of treatments to experimental units. If the treatments are mensurative (categories of time or space; Hurlbert 1984) or organismal (natural categories, such as age class or sex), the randomization rule ensures that experimental units are drawn randomly from each treatment. Thus, proper experimental design helps minimize the effects of uncontrolled variation, allowing you to obtain unbiased estimates of abundance and experimental error (variation among experimental units treated alike).

Sample Units and Sampling Design

Sample units are the entity from which measurements are obtained. Sampling units may be quadrats, transects, or points. Selection of sample units from an experimental unit should be done using a probability sampling scheme, or **sampling design**, where every sample unit has some probability of being selected, and this probability can be accurately determined. Without some type of **randomization rule**, there is no way to avoid discrimination or favoritism in sample unit selection, resulting in **bias** (inaccuracy) and unrepresentative estimates of **variance** (precision) in the estimate of abundance.

Several sampling designs exist to accommodate particular survey conditions (Cochran 1977). The most common sampling design is **simple random sampling**, where sample units are selected randomly to ensure that each sample unit has an equal probability of being selected. You proceed by exhaustively subdividing the experimental unit into sample units, and then you may draw lots, flip a coin, roll dice, or use a random number table to select units to be sampled. During random sampling, sample units may be drawn with replacement (i.e., a sample unit is selected and then placed back into the pool of possible sample units, where it may possibly be drawn again) or without replacement (i.e., sample units may be selected only once). Because sampling without replacement is more precise than sampling with replacement, it is more commonly used in wildlife management (Caughley and Sinclair 1994).

Stratified random sampling is employed when there are implicit differences in sample units that need to be accounted for in the analysis. For instance, differences in habitat quality may produce localized differences in animal density, resulting in increased variance. To reduce variance, the area may be stratified by habitat quality, with sample units selected randomly from each habitat type. For example, a large survey has defined experimental units (areas of homogeneous habitat) by physiognomy (grassland, forest, savanna, desert, etc.). But investigation revealed that controlled burns in each experimental unit created perturbations in the underlying physiognomic matrix (alterations in otherwise homogeneous experimental material). To account for the variability, experimental units are stratified into burned and unburned areas, and sample units are randomly obtained from each stratum.

Systematic sampling (or **systematic random sampling**) is employed to reduce the amount of effort (time or fuel) necessary to navigate among sample units. Systematic sampling typically uses a random start point and the proceeds in an ordered fashion (e.g., a point grid where a sample is collected every 200 m) until the entire area to be covered is sampled. It has the advantage of ensuring thorough coverage of area under investigation, but is susceptible to an array of problems (Cochran 1977), the most pernicious of which is the possible coexistence of an unknown periodic variation in the population being sampled (Krebs 1999). The periodic fluctuation could match the frequency of a systematic sampling design, resulting in a biased estimate with unrepresentative precision (that is unknown to the user).

Several **nonprobabilistic sampling designs** that may be used in error have been described in the literature (Cochran 1977, Krebs 1999), such as accessibility sampling (sampling along trails or roads due to ease of access; later called "convenience sampling" by Anderson [2001]), haphazard sampling (without a plan, as the name implies), or judgmental sampling (selected as "typical" or "representative" on the basis of subjective opinion). Even worse, some sample units may be selected because of the greater opportunity to "see more animals," despite the obvious bias that will result. Regardless of cause or origin, nonprobabilistic sampling designs are likely to yield biased estimates with levels of precision that are not representative of the area of inference, and they should therefore be avoided.

Sampling Intensity and Statistical Power

Sampling intensity is a concept that encompasses desired **precision**, statistical **power**, and the amount of **variability** among the sample units. Determining the sample size required to achieve study objectives is a central question that must be addressed prior to the initiation of work. If the sample obtained is too big, valuable resources will be wasted obtaining excess precision that produces no change in outcome or conclusions. More catastrophic is a sample size that is too small, as the information obtained may be incapable of producing useful results, leading to incorrect conclusions. Sampling intensity also is an ethical consideration. Studies with improper sample size exposes subjects (animals or humans) to risks when little (too many samples) or no (too few samples) gain in useful knowledge is possible. Lenth (2001) observed that for such an important and complex issue, there was an alarming paucity of published literature. Fortunately, most popular statistical packages (R [http://www.r-project.org/], SPSS [http://www.spss.com/], SAS [http://www.sas.com], JMP [http://www.jmp.com/], and Statistica [http://www.statsoft.com/]) have the tools for sample size determination, and there a growing number of resources devoted specifically to the task, including books (Armitage and Colton 2005, Chow et al. 2008, Dattalo 2008, van Belle 2008, Julious 2010), standalone software packages (Thomas and Krebs 1997, Lenth 2001, Faul et al. 2007), and several online calculators (Lenth 2001).

There are 5 interrelated components that influence sample size determinations and the conclusions you might reach from a statistical test in a research project. The logic of statistical inference with respect to these components is often difficult to understand and explain (see Chapter 1, This Volume). Here we clarify the 5 components and describe their interrelationships

1. **Significance level:** The significance level is the odds the observed result is due to chance. This concept includes 2 components that define the types of errors possible in statistical tests. **Type I error** is rejecting the null hypothesis when it is true, and the probability of committing this type of error is controlled by the **alpha level** (α) of the test (frequently $\alpha = 0.05$). **Type II error** is failing to reject the null hypothesis when it is false, and the probability of committing this type of error is controlled by the **beta level** (β) of the test (frequently $\beta = 0.05$). The investigator should adjust the levels of alpha and beta according to experimental needs, being mindful of the potential harm that may result from dogmatically applying "typical" or "established" probability levels.

2. **Power:** Power is the odds that you will observe a treatment effect when it occurs. Defined another way, **power** is the probability of rejecting the null hypothesis when it is false, and it is controlled by adjusting beta (i.e., power = $1 - \beta$). Increased power results in requisite increases in sample size, due to the relationship between power and beta.

3. **Effect size:** Effect size (d^2) is the difference between treatments (e.g., in number of animals seen) relative to the noise in measurements. **Effect size** expresses the magnitude of difference between 2 sample means and therefore is the logical complement to the *P*-values generated from statistical hypothesis tests. Effect size

and the ability to detect it are indirectly related; the smaller the effect, the more difficult it will be to find, therefore requiring a larger sample size. The term "effect size" is sometimes used synonymously with "standardized difference." Effect size can be written as

$$d^2 = \frac{\bar{x}_1 - \bar{x}_2}{s},$$

which scales the difference in population means 1 and 2 ($\bar{x}_1 - \bar{x}_2$) by the standard deviation σ (Cohen 1988, van Belle 2008). Although it is useful to think in these terms, one should recognize the dangers of formulating study objectives exclusively in terms of effect size (Lenth 2001, van Belle 2008). For determining sample sizes, it is important to know the anticipated means and variances under the null and alternative hypotheses for the entities being compared.

4. Variation in the response variable: The sample **variance** (s^2) or **standard deviation** (s) are often used to estimate variability in the parameter of interest (e.g., population mean). The standard deviation is calculated as positive square root of the sample variance:

$$s = \sqrt{s^2},$$

where the variance is

$$s^2 = \frac{\sum_{i=1}^{n}(X_i - \bar{X})^2}{n-1}.$$

where X_i is 1 data point within a sample and $\bar{X}$ is the mean of all data points within the sample. Similar to the requirement to know the anticipated means for the entities being compared, to accurately determine sample size, we also must estimate the **variance or standard deviation** for the entities being compared. They are typically obtained from either the literature or a pilot study.

5. **Sample size:** Sample size (n) is the number of samples required to obtain the desired precision in an estimate or the desired power in a hypothesis test. Larger sample sizes generally lead to parameter estimates with smaller variances, giving you a greater ability to detect a significant difference. Sample size is typically the variable being solved for in the planning stages, but it can be an input variable when you are attempting to estimate power.

For example, to determine the sample size required for comparing 2 populations with equal variance in a 2-tailed hypothesis (Lehr 1992, van Belle 2008):

$$n = \frac{2(z_1 - \alpha/2 + z_1 - \beta)^2}{\left(\frac{\bar{x}_1 - \bar{x}_2}{s}\right)}.$$

When $\alpha = 0.05$ and $\beta = 0.20$ (typical settings for these parameters in wildlife research), the corresponding critical values from a standardized normal probability table (z-values or z-scores) become 1.96 (z-score for α, the probability of committing a Type I error: $z_1 - \alpha/2$) and 0.84 (z-score for β, the probability of committing a Type II error: $z_1 - \beta$), respectively. The z distribution is a normal or Gaussian distribution with a mean of 0 and a standard deviation of 1. Standardized or z-values then represent deviations from the normalized mean in units of standard deviation. The numerator then simplifies to 15.68. Rounded up to 16 and substituted into the equation, it yields a useful rule of thumb for calculating sample size (Lehr 1992, van Belle 2008):

$$n = \frac{16}{d^2},$$

where

$$d^2 = \frac{\bar{x}_1 - \bar{x}_2}{s},$$

the standardized difference, reflects the difference to be detected between treatment means (effect size) divided by the standard deviation.

It is clear, the ideal experimental design would be one that minimizes the probability of Type I and Type II errors while maximizing power, given the particular experimental constraints of time and resources. Likewise, the above example illustrates the 5 components that are necessary for determining sample size and conducting power analysis, are not independent. The usual objectives of a power analysis are to calculate the sample size (5) required to achieve the desired power (2), given effect size (3) and sample variability (4), at a predetermined level of significance (1). In studies with limited resources, the maximum sample size (5) will be known. In these instances, power analysis then becomes necessary to determine whether sufficient power (2) can be achieved with the known sample size (5), for the desired significance values (1), sample means (3), and sample variances (4). The researcher can then evaluate whether the study is worth pursuing. As indicated above, there are many software packages available for calculating sample size and power (Thomas and Krebs 1997, Lenth 2001, Faul et al. 2007, R Development Core Team 2008). Consult the user's manual of the software package you are using to become familiar with these calculations. The goal is to achieve a balance of components that provides the maximum level of power to detect an effect if one exists, given programmatic, logistical, or financial constraints on the other components.

Proposal Generation and Independent Review

We began this section with a list of questions that should be addressed when developing a survey design. By answering these questions, the researcher should have gained sufficient understanding of the task at hand to finalize the process with a research proposal. Although many view the writing of a research proposal as an unnecessary formality, we believe that it is an essential part of wildlife management. The steps required to gather the information necessary to write a research proposal forces the investigator(s) to assess the various parameters that will ultimately determine the success, or failure, of a project. The written proposal then represents the investigator's understanding of the problem at hand, as well as the resources and methods believed to provide the solution, given any limitations. As such, the proposal conveys all the information necessary for an independent review. The independent review provides a critique of the survey design, either confirming a sound design or providing the information necessary to improve on the existing knowledge. Therefore, the independent review serves as either the starting point of a new iterative loop through the whole process or the conclusion of the survey design phase.

METHOD CATEGORIES AND CONSIDERATIONS

Animal survey methods have developed over time, building on established knowledge and growing in sophistication. They can be broadly categorized as census methods or estimates derived from sampling, and they are further subdivided by complete or incomplete detection in samples (Fig. 11.2). Early methods focused on complete **census** of a given population. For animals that were elusive or otherwise difficult to census, methods were developed to census animal **indices.** Indices were typically based on cues or other byproducts of animal activity (fecal pellets, nests, burrows, tracks, calls, scrapes, etc.) that were believed to be proportional to animal abundance or density. At the same time, methods were developed for obtaining trends or abundance estimates from exploited populations. Later, methods capitalized on existing methodology and attempted to **estimate** abundance by obtaining **complete counts from sample areas.** Finally, because it was impossible to ascertain whether a complete count had been obtained (i.e., to prove a negative: "no animals were missed"), newer methods of estimation were developed utilizing **incomplete counts from sample areas.** It is through this general classification (modified from Lancia et al. 2005) that we introduce the basic methodology of estimating animal density and/or abundance (Fig. 11.2).

Considerations

As noted above, the breadth and depth of the subject of abundance estimation for animal populations spans many methods. The combination of method and survey design then, in turn, dictates how samples may be combined to estimate means and variances. Chapters 1 and 2 (This Volume) should be consulted for more in-depth discussions of experimental design and analysis of data. We intend to provide a basic overview of methods available for consideration in each category for assessing animal abundance, providing simple examples from historical methods and references for further investigation. We begin by re-emphasizing 2 factors that must be considered due to their impact on precision and accuracy of methods: distribution of the target species relative to the distribution of samples and detection probability.

Species Distribution

Attempts to manage populations using indices (counts believed to be related to abundance) and complete counts (census) revealed the analytical and practical limitations of these methods. As the size of the area to be surveyed increased, practical limits on available resources were reached, forcing investigators to derive methods for obtaining estimates from samples. Similarly, development of methods for obtaining estimates from samples revealed the importance of sample distribution in relation to species distribution. Resources, and therefore wildlife, are not randomly distributed, which can create bias in estimates of animal abundance. Problems arise when animal distributions are clumped, or when the distribution of samples correlates with the underlying distribution of animals to be sampled. Appropriate survey design is almost always the key component in alleviating this problem, with random sampling or stratified random sampling the most common remedy. Although avoiding the problems resulting from nonrandom distribution of either samples or species is a requisite for obtaining precise and accurate estimates of abundance, defining or describing the underlying distribution of animal abundance is sometimes a necessary objective (Pielou 1974, Cochran 1977, Diggle 1983, Greig-Smith 1983, Ludwig and Reynolds 1988, Ripley 2004). Regardless, we again warn that it is prudent to use probabilistic sampling as the easiest alternative for avoiding unforeseeable problems in obtaining estimates.

Detection Probability

Most animal survey methods do not observe all individuals in the population. Generally, the probability of seeing or trapping an individual animal over a given area is 1. Sampling design and detection probability are major concerns when estimating animal abundance. Usually, one assumes that detection probability is similar across all sampling areas; however, this assumption is not always true, and there may be different detection probabilities for different sampling units. Estimators for these cases take this variation into consideration (Thompson 2002a, Skalski 1994). Lancia et al. (2005) noted that considerable effort in development of abundance estimators has involved ways to estimate detection probability.

Conroy and Nichols (1996), Pollock et al. (2002), and Lancia et al. (2005) noted there are 3 basic approaches used in attempts to deal with variation in detection probability in

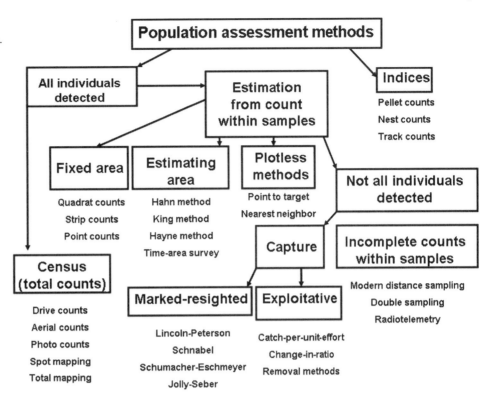

Fig. 11.2. The relationship among population estimation procedures.

surveys. The first is to use **standardized methods** when conducting the surveys. All potential sources of variation in detection probability that are under the control of the investigator should be kept constant (methods, effort, observer experience, weather, etc.).

The second involves use of **covariates** in analyses of survey statistics. If exogenous variables, such as weather conditions or observer identity, account for most of the variation in detection probability, models can be developed for estimating change in population size as a function of the relevant exogenous variables (Overton and Davis1969; Craig et al. 1997; Link and Sauer 1997, 1998, 2002). Lancia et al. (2005) stressed that covariates selected for use in modeling cannot be associated with both detection probability and true abundance. They used the example of vegetation type, because it can affect detection probability, and it also can influence abundance. Therefore, using vegetation type as a covariate affecting detection probability would not be appropriate.

The third approach is to recognize that detection probability is **not constant** over space or time, and that not all exogenous variables can be measured, modeled, or even perceived. This idea leads to implementation of methods that permit direct estimation of detection probability. Of the 3 methods, Lancia et al. (2005) believed this approach was the only one that was scientifically defensible, and they recommended that developers of future surveys and monitoring programs utilize this approach. Despite this recommendation, index use is common in wildlife surveys and monitoring programs throughout the world (e.g., Thompson et al. 1998).

INDICES

Most **indices** collect **frequency** (number of individual animals or animal sign) information along transects, at quadrats, or points. Examples of index methods include the number of animals seen per kilometer of road, the number of animals present per night at a waterhole, fecal pellets per quadrat, and nest or burrow counts per kilometer of transect. Similarly, a **frequency of occurrence index** only collects presence or absence data. A frequency of occurrence index is based on the proportion of sample units (e.g., scent stations) that contain at least 1 animal or animal sign (Scattergood 1954, Caughley 1977, Seber 1982). However, Seber (1982) noted that a population with a highly clumped distribution will yield a lower frequency of occurrence index (proportion of quadrats with at least 1 animal) than a population of similar density with a more uniform or random distribution.

A **density index** can be defined as any measure that correlates with density (Caughley 1977:12). Indices are used most often because of perceived savings in cost, time, and/or labor. Indices differ from population estimation methods in that only relative abundance or relative density can be derived from the indices. Data are usually presented as deer/km, rabbit pellets/m^2, or birds/point. Indices can be used to compare animal numbers between treatment and control areas (e.g., disked with nondisked areas) or to compare the same area over time, based on the assumption that nothing changes except the relative abundance of the animal being studied. The probability of catching, counting, or otherwise detecting an animal in sample units from 2 areas or time periods being compared should be similar. If indices are employed, they should be standardized as to season, time of day, weather conditions, habitat, and observer experience. For multispecies surveys, detection probability will vary with species. Such factors as group status, reproductive cycle, sex and age ratios, and population density also will affect detection probability. For aquatic species, water level, water temperature, and moonlight also affect detection probability.

Data obtained from indices (e.g., relative abundance) are correlated with abundance in some unknown manner. Standardizing methods and using covariates in an analysis can address some sources of variation in index surveys (Lancia et al. 2005). However, as noted by Lancia et al. (2005), other factors that affect detection cannot be handled in these ways or may not even be identified. They recommended caution and skepticism when using and interpreting indices, and they preferred that all indices include an estimate of detection. There are only 2 ways to obtain the relationship of an index to population abundance: (1) estimate detection probability or (2) estimate population abundance and "calibrate" the index (Caughley 1977, Lancia et al. 2005). However, it should be apparent that if an estimate of population abundance can be obtained, there may be no need to do the index survey. We are of the opinion that a calibrated index would only be applicable for the time and place it was done, as conditions typically change over time and for difference areas. Therefore, we will go a step further and recommend that all indices include an estimate of detection probability, and if possible, only population estimation procedures should be used to obtain animal abundance data. Most index surveys can be readily modified to provide information needed in a population estimation procedure. For example, live catch per unit effort (an index) could be easily modified for a mark–recapture population estimation procedure, or deer seen per kilometer could be easily modified for a line-transect population estimation procedure. Regardless, given the advances in sampling methodology, there are relatively few circumstances where index could not be adequately replaced by a more quantitative method.

CENSUS OR TOTAL COUNTS

In few situations are total counts possible. Total counts may be possible for the number of deer in a small paddock or maybe the number of elephants in a small pasture. But for wild populations, it is seldom possible for wildlife managers to obtain a total count of animals in a give study area. As sample area increases, animals are inevitably missed. If it is possible to obtain a total count, then no descriptive statistics are needed nor apply. The data obtained are not a sample, but an enumeration of the whole population (i.e., no variability is present, because you counted them all). Total counts are assumed to be accurate and can be used to calibrate (i.e., estimate probability of detection) extensive field surveys (Lancia et al. 2005). Total counts on small areas can be derived from intensive surveys (Tilton et al. 1987), from a known number of marked individuals, or by other ingenious means (Kuvlesky et al. 1989). Several methods (presented below) have been purported to produce accurate population counts in some circumstances. However, we warn investigators there is always a possibility that an unknown number of animals will be missed (e.g., nonsinging male birds in a spot-mapping survey). When this occurs, there are no means to detect bias or assess the precision of the sample. So, although the methods listed below are in this section on total counts, in most cases, data resulting from these methods should viewed skeptically and probably should be considered indices rather than total counts.

Drive Counts

Drive counts occur over limited areas where "beaters," or herders, drive animals into an enclosure or past counters to count total animals in the study area. The method works best with large, easy to detect animals, such as deer. Drivers remain in sight of one another at all times (to prevent animals from escaping unseen between observers), spaced along a line, and sweep across an area with well-defined boundaries. In the best case scenario, the area would be surrounded by a high fence or water (Tilton et al. 1987). If not, additional observers are placed along the boundaries to count animals that move in or out of the census area. All observers count only those animals that move past them on their right side (this practice eliminates double counting). The census is the sum of the number of animals moving out of the area or back through the line of drivers, minus any moving into the area ahead of the drivers.

McCullough (1979) compared drive counts with population estimates reconstructed from the age of death of individuals in the population. At low densities, drive counts underestimated the true population, and at high densities, they overestimated the true population. Errors could be as large as 20–30%. Thus, drive counts are probably best viewed as an index of population size.

Aerial Photography

Low altitude photography of flocks of birds (or other groups of animals) is often used as a census technique. The entire assemblage of animals is photographed and later counted to give a complete census. However, it is often difficult to ascertain whether all individuals are visible (e.g., some diving ducks may be under water) to be photographed, and errors in counting undoubtedly are made (Bajzak and Piatt 1990).

Spot Mapping or Territorial Mapping

Spot mapping involves plotting locations of individual birds that are seen or heard on a gridded map during repeated visits to a study area. The technique is most suited to birds that regularly sing or call in exclusive territories. Floaters (i.e., nonterritorial birds) and young of the year are usually not surveyed by this technique. The combined data reveal clusters of locations, assumed to represent centers of activity for individual territories during the breeding season. The total number of clusters in the study area equals the number of clusters completely inside the area plus the sum of fractional parts of clusters on the boundaries. Total number of birds is estimated by multiplying the number of clusters by mean number of birds per cluster, which is normally 2, assuming that birds breed in pairs.

Assumptions of the method (Verner 1985, Bibby et al. 2000) are: (1) populations are constant, and birds remain in exclusive spaces or territories during the sampling period; (2) birds in territories produce cues frequently enough to permit repeated location on successive observational visits; (3) estimated proportions of territories along boundaries are accurate; (4) the estimated mean number of birds represented by each cluster is accurate; and (5) observers are skilled, record data accurately, and are consistent. Verner and Milne (1990) provided evidence that spot mapping results should not be considered to be complete counts, as results can vary among observers (Best 1975, O'Conner and Marchant 1981) and map analysts. At best, spot mapping yields an index.

Total Mapping of Bird Territories

This approach is similar to spot mapping, except breeding birds are first trapped and color banded, prior to surveys to delineate territories. This practice facilitates the identification of individuals. Verner (1985:266) believed that, when thoroughly executed, total mapping was probably the most accurate method of estimating population density of breeding birds. He also believed the method should be used as a standard for evaluating the accuracy of other methods of estimating bird density. However, Bibby et al. (2000:42) noted this method only estimates the population of relatively conspicuous birds holding territories, not **floaters** (i.e., nonterritorial birds) or transients. Assumptions are the same as for territorial mapping (described above).

COUNTS ON SAMPLE PLOTS (FIXED AREA)

It may be possible to obtain complete counts of animals on sample units of limited area within some larger area of interest. The sample units must be suitably sized relative to the organism being considered, to ensure that a complete count is obtained. The area being counted is fixed in terms of length and width prior to the start of the survey. Because all individuals are counted, there is no variation associated with the density or number of animals seen on the sample plots (unless counts on each sample plot are replicated). Instead, only geographic (plot to plot) variation is a concern. The mean density from all sample plots is then extrapolated to the entire study area, giving an estimate of average density and/or population abundance for the area of inference. This basic sampling method has been modified to use sample units of various shape (quadrats, strips, plots, etc.) and size, depending on circumstance and target species. We refer the reader to Caughley and Sinclair (1994) for their excellent illustration of the advantages and disadvantages of sampling with replacement versus sampling without replacement, and transects (long, narrow rectangles) versus quadrats (squares). Here we focus on simple estimates derived from sampling units of equal size. We provide examples for strip and point counts. We again warn investigators of the possibility that an unknown number of animals may be missed in some or all sample units, resulting in negatively biased estimates of population size and/or density.

Strip Counts

This method is the one of the most commonly used to measure density. The counting unit is a strip or transect, which is merely a long, narrow rectangle of fixed area. Transects are randomly placed across the grain of the topography and landscape. Transect lines can be traversed on foot or horseback, by truck or boat, or in a helicopter or airplane. The classic strip census uses a preset distance (0.5-strip width) on each side of the transect line, and then only those animals within this predefined distance are counted. Animals observed outside this distance are not counted, and it is assumed that all animals in the strip are counted with certainty. If these assumptions are valid, the population abundance can be estimated using any of the simple population estimators (Cochran 1977, Krebs 1998, Caughley and Sinclair 1994) for samples of equal area, samples of unequal area, or sampling proportional to size. Here we illustrate the calculations of density and abundance from strip counts of equal area, sampled with and without replacement. Density is calculated as the ratio of the sum of counts to the sum of strip areas (see below for variable definitions):

$$D = \frac{\sum x_i}{\sum a}.$$

The density obtained on the sample strips is then multiplied by the size of the study area (area of inference) to obtain populations size:

$$N = DA.$$

By combining the 2 formulas, we obtain the simple strip abundance estimator:

$$N = \frac{A \sum x_i}{2Lwn_s}.$$

The variance is obtained from the strip counts using

$$s_x^2 = \frac{\sum x_i^2 - \frac{(\sum x_i)^2}{n_s}}{n_s - 1}.$$

The strip count standard error is then

$$SE_{\bar{x}} = \sqrt{\frac{s_x^2}{n_s}}.$$

The variance of the population estimate when sampling with replacement (SWR) is

$$s_N^2 = \frac{(n_t)^2}{n_s} s_x^2,$$

where n_t is the total number of samples possible on the area of inference (calculated by A/a). The variance of the population estimate when sampling without replacement (SWOR) is

$$s_N^2 = \frac{(n_t)^2}{n_s} s_{\bar{x}}^2 \left(1 - \frac{n_s}{n_t}\right).$$

Here the added term (finite population correction) reduces the variance of SWOR relative to SWR by 1 minus the proportion of the area sampled (i.e., the number samples taken over the total number of samples possible on the area of inference). The standard error of the population estimate is then

$$SE_N = \sqrt{SE_{\bar{x}}}.$$

Finally, we obtain the 95% confidence interval from

$$95\%CI = N \pm t_{\alpha,df}(SE_N),$$

where N = population abundance
 D = density of animals in strips
 A = area of inference (study area)
 a = area of each strip ($L \times 2w$)
 x_i = number of animals seen on transect i
 w = preset 0.5-strip width (sample area on each side of transect line)
 L = length of transect
 n_s = number of samples (strips)
 n_t = total possible samples (A/a) in study area
 s^2 = sample variance
 t = Student's t for the desired alpha (α) and degrees of freedom ($df = n - 1$)
 SE = standard error
 $\bar{x}$ = mean number of animals seen on all transects
 $95\%CI$ = 95% confidence interval

Example: We wish to estimate the number of grouse on a 2-km² study area. We utilize 5 counting strips, each 100 m in length with a preset sighting distance of 10 m (0.5-strip width). We divide the study area into strips and select 5 to survey using a random number table. We count each strip, flushing a total of 15 grouse (x_i = 4, 3, 3, 2, and 3). The total possible number of samples of this size is 1,000 ($n_t = A/a$). Therefore, the estimated population abundance would be calculated as follows:

$$N = \frac{(2\ \text{km}^2)(15)}{(2)(0.1\ \text{km})(0.01\ \text{km})(5)} = 3,000.$$

The strip count variance ($s^2 = 0.50$) is then used to obtain the strip count standard error ($SE_{\bar{x}} = 0.7071$). The variance of the population estimate when SWR is then

$$s_N^2 = \frac{(1,000)^2}{5}(0.50) = 100,000,$$

and the standard error of the population estimate when SWR is

$$SE_N = \sqrt{100,000} = 316.23.$$

We can then calculate the 95% confidence intervals when SWR as

$$95\%CI = (\pm 2.776)(316.23) = \pm 877.85.$$

The population estimate, $\pm 95\%CI$ when SWR, is 3,000 ± 878 grouse. If we had obtained the counts by SWOR, the population estimate would remain the same, but the variance of the population estimate would change:

$$s_N^2 = \frac{(1,000)^2}{5}(0.50)\left(1 - \frac{5}{1,000}\right) = 99,500.$$

The standard error of the population estimate would become

$$SE_N = \sqrt{99,500} = 315.44,$$

and the resulting 95% confidence interval would be

$$95\%CI = (\pm 2.776)(315.44) = \pm 875.66.$$

The population estimate, $\pm 95\%CI$ when SWOR, is 3,000 ± 876 grouse. The increased precision reveals the additional information obtained from n unique samples using SWOR over the possible redundant information contained in repeated samples gathered using SWR. Regardless, we would report the population estimate as $N \pm SE$ (e.g., 3,000 ± 315.44 grouse), which would allow other investigators to derive confidence intervals of their choice from the data.

Point Counts

Point counts are typically used to estimate bird density. An observer proceeds to a sample point and might, or might not, allow a rest period of specified duration for equilibration of bird activity (Reynolds et al. 1980). The observer then detects (by both sight and sound) birds for a specified count period within a preset distance (radius) from the point. Although it is generally assumed that all birds are detected within the sample radius, this assumption is typically false unless the preset radius is quite small or the target species is quite conspicuous. Therefore, unless complete counts are certain, point counts should be considered as an index to relative density. If the assumption is reasonable, then estimation proceeds similar to strip transect counts (described above), differing only in the form of the equation for the simple population estimate:

$$N = \frac{A \sum x_i}{n \pi r^2},$$

where N = population abundance
A = area of study area
x_i = number of birds seen within a fixed radius r of point i
n = total points sampled
π = pi (ratio of the circumference of a circle to its diameter)
r = preset radial distance

Example: A survey consisting of 10 random points, each with a fixed radius of 50 m, is conducted on a 2-km² study area. Surveyors count 50 birds. The estimated population abundance would be calculated as follows:

$$N = \frac{(2 \text{ km}^2)(50)}{(10)(3.1416)(0.050 \text{ km})^2} = 1{,}273.24.$$

The sample variance (s_x^2), population variance (s_n^2), and population standard error (SE_n) are calculated using the strip count equations for SWOR. We then obtain a point count variance of 0.6667, a population variance of 4,238.13, and population standard error of 65.1. Our calculated 95%CI is then ±147 birds. Therefore, the population estimate (±95%CI) for the study area is approximately 1,910 ± 107 birds. We would report the population estimate $N \pm SE$ (e.g., 1,273 ± 65 birds), which would allow other investigators to derive confidence intervals of their choice from the data.

Sample Units of Unequal Area

Samples units of unequal area require an average density to be calculated from all units sampled, as indicated in the discussion of strip counts (above). The average density (D) is then extrapolated to the survey area using $N = DA$. However, the formulas for SWR and SWOR differ for samples of unequal area (Krebs 1998):

$$_{SWR}s_N^2 = \frac{(n_t)^2}{n_s(n_s - 1)} [\sum x_i^2 + D^2 \sum a_i^2 - 2D\sum(x_i a_i)]$$

$$_{SWOR}s_N^2 = \frac{n_t(n_t - n_s)}{n_s(n_s - 1)} [\sum x_i^2 + D^2 \sum a_i^2 - 2D\sum(x_i a_i)],$$

where x_i = count from sample i
a_i = area of sample i
n_s = number of samples taken
n_t = total number of samples in study area
D = average density from the samples

Example: We wish to estimate the number of grouse on a 2-km² study area. From a total of 784 possible transects, we selected 10 counting strips without replacement. Each strip had a different length, but each was surveyed with a preset sighting distance of 10 m (0.5-strip width) on each side of the centerline. We counted each strip, flushing a total of 50 grouse, with the counts (x) and area (a) of each strip recorded. There are 784 possible transects on the study area. The estimated population abundance would be calculated as follows:

$$D = \frac{50}{0.0255 \text{ km}^2} \; 1{,}960.8$$

and

$$N = (1{,}960.8)(2) = 3{,}922.$$

The variance of the population estimate (SWOR) would be calculated as

$$_{SWOR}s_N^2 = \frac{784(784-1)}{10(10-1)}[256 + (1{,}960.8)^2(0.00006681) -$$

$$(2)(1{,}960.8)(0.1305)] = 7{,}403.4.$$

Using the equations for strip counts, we obtain a population standard error (SE_N) of 86.0. Our calculated 95%CI is then ±229 birds. Therefore, the population estimate (±95%CI) for the study area is approximately 3,922 ± 107 birds. Again, we would report the population estimate $N \pm SE$ (e.g., 3,922 ± 86), which would allow other investigators to derive confidence intervals of their choice from the data.

Sampling with Probability Proportional to Size

Large study areas are seldom homogeneous with respect to resources, species density, or detectability. When this variability occurs, stratification is used to divide the area into units of similar composition. As a result, the units to be sampled are often of unequal size. In these circumstances, one may employ sampling with probability proportional to size (PPS), where the probability of a sample being selected is proportional to the size of the various units being sampled. The PPS method may be used with equal or unequal sized sampling units. Although the PPS method is unbiased and ideally suited for sampling irregular experimental units of differing size, it is limited by design to SWR. Thus, Caughley and Sinclair (1994:202) recommend the method be limited to circumstances where sampling intensity is 15%.

Sampling using the PPS method requires the density to be calculated for each sample, with the average density and variance of the density estimates (s_D^2; calculations are the same as sample variance for strip counts above, except they use the density for each sample rather than the count for each sample) to be calculated from all units sampled. The average density (D) is extrapolated to the survey area using $N = DA$. However, the formula for calculating the variance of the total population differs for PPS estimates (Krebs 1998):

$$_{PPS}s_N^2 = \frac{(A)^2}{n_s} s_D^2,$$

where A = total study area size
n_s = number of samples selected
D = average density from the samples
s_D^2 = variance of sample densities

Example: We wish to estimate the number of grouse on a 2-km² study area consisting of 3 vegetation types. We selected 10 samples using sampling PPS. Each strip was 100 m in length with a preset sighting distance of 10 m (0.5-strip width) on each side of the centerline. We counted each strip, flushing a total of 50 grouse (x_i = 4, 5, 6, 6, 4, 5, 5, 5, 4, 6). Densities for each sample were calculated (d_i in birds/km² = 2,000, 2,500, 3,000, 3,000, 2,000, 2,500, 2,500, 2,500, 2,000, 3,000), yielding an average density of 2,500 grouse/km², with a variance (s^2) of 166,667. The estimated population abundance ($N = DA$) was 5,000 birds. The variance of the total population was calculated as follows:

$$_{PPS}s_N^2 = \frac{(2)^2}{10} 166,667 = 66,667.$$

The standard error of the population estimate (SE_N) was 258, with a 95%CI of ± 687 birds. We would report the population estimate as $N \pm SE$ (e.g., 5,000 ± 258 birds), which would allow other investigators to derive confidence intervals of their choice from the data.

COUNTS ON SAMPLE PLOTS (ESTIMATING AREA)

Considerable attention was given to conducting sample counts prior to 1980. In particular, methodology began to center on methods that would allow an accurate estimate of sample area to be obtained from counts without preset strip widths. The thoughts of the day, summarized by Eberhardt (1968), stated that precision was proportional to the square root of the number of animals seen, and therefore efforts should be focused on methods that would allow all sightings to be used. Sightings were expensive to obtain, particularly when many were discarded for being outside the sample frame. The basic solution had several forms, but each attempted to determine the sample area congruent to the area over which counts were obtained.

The King method (Leopold 1933, Buckland et al. 2001) used the average radial distance to all observed animals to estimate the strip width used in the calculations of animal abundance. Kelker (1945) used perpendicular distances to generate a histogram, and from the histogram subjectively determined the strip width over which all animals were likely detected. Hayne (1949a) developed the first widely recognized line-transect density estimator with a solid mathematical foundation (Buckland et al. 2001), based on the sighting distances and angles to flushed birds. Hahn (1949) used visibility measurements, periodically taken perpendicular to the transect line, to estimate the area over which deer were counted. Density estimates were then based on all detected animals, using average visibility as the estimate of strip width. Robinette et al. (1974) compared the accuracy of these and 6 other early line-transect methods, noting that only the King and Kelker methods showed promise.

We group these methods together based on use of sighting distances to estimate sample area. We refer to this type of distance sampling as **traditional distance sampling.** As **modern distance sampling** has superseded these methods, we provide only the estimators and no examples.

Hahn Method

The Hahn (1949) method is still commonly used to estimate population density. It is very similar to the strip method example provided above, differing only in the use of distances to estimate the strip width defining the sample area. Transects are randomly placed across the grain of the topography and landscape, and they can be traversed on foot, on horseback, or by vehicle. Estimates of maximum visibility are made periodically (e.g., every 200 m) on both sides of the transect, with maximum visibility defined as the maximum distance an observer could see a target animal perpendicular to the transect at each point. The Hahn estimate of population abundance is calculated as

$$N = \frac{A \sum x_i}{2Lv},$$

where N = population abundance
A = area of study area
x_i = number of animals seen on transect i
v = the 0.5-strip width determined by average visibility measurements
L = total length of all transects

King Method

The King method (Leopold 1933, Buckland et al. 2001) used the average radial distance from all observed animals to estimate the 0.5-strip width to be applied in the calculations of density or abundance. Thus, it is similar to the Hahn method:

$$N = \frac{A \sum x_i}{2L\bar{r}},$$

where N = population abundance
A = area of study area
x_i = number of animals seen on transect i
$\bar{r}$ = the 0.5-strip width determined by average sighting radius
L = total length of all transects

Hayne Method

The Hayne (1949a) method was commonly used to estimate population density of flushing birds. The method assumed there was a fixed flushing radius for each bird species and habitat. When an observer walking a transect came within that radial distance, the bird would flush and be spotted. Further, the method assumed the sine of the angle for each observation came from a uniform random distribution ranging from 0 to 1, with an average angle of 32.7° (Hayne 1949a). Later investigators (Robinette et al. 1974, Burnham et al. 1980) determined the mean sighting angles were generally around 40°, with Burnham and Anderson (1976) providing a correction factor for the original Hayne method. The Hayne estimator of density, from Krebs (1998), is

$$D_H = \frac{n}{2L}\left(\frac{1}{n}\sum\frac{1}{r_i}\right)$$

and

$$N = DA,$$

where N = population abundance
D_H = population density
A = area of inference
n = number of animals seen on each transect
r_i = sighting distance to animal i
L = length of transect

The variance associated with this density estimate is calculated as

$$s^2_{D_H} = D_H\left[\frac{s_n^2}{n^2} + \frac{\sum(1/r_i - R)^2}{R^2 n(n-1)}\right],$$

where D_H = population density
n = number of animals seen
s_n^2 = variance of n
r_i = sighting distance to animal i
R = mean of the reciprocals of sighting distances r_i

Time-Area Squirrel Survey

Time-area surveys are a common method used to census tree squirrels (Goodrum 1940:8). They are a point-based example of using distances to estimate the effective sample area of the counts. Sample points are chosen at random, and counters are stationed at each point (base of a tree nearest to the point) before sunrise. Starting at sunrise, counters wearing camouflaged clothes remain quiet and relatively motionless while counting all squirrels that come into view for 30 minutes. The counter determines the distance to each squirrel when first detected using a laser rangefinder. The average distance to all squirrels detected is then used to compute the area over which the squirrels were counted. Under field conditions, the proportion of a circle observed by each counter will vary from point to point. As such, each observer uses a compass to estimate the portion of a circle under surveillance during the count (e.g., 0.75 or 75% sample effort). This estimate is then factored into the estimation equation (mean area observed by each surveyor). Population size is estimated using

$$N = \frac{A\sum x}{n\Delta\pi r},$$

where N = population abundance
A = area of study area
$\sum x_i$ = number of squirrels seen at point i
n = total points sampled
Δ = average effort in terms of portion of circle observed
π = pi (ratio of the circumference of a circle to its diameter)
r = average radial distance to all detections

The simple strip estimator of variance, standard error, and 95%CI can be used with this method.

COUNTS ON SAMPLE PLOTS (PLOTLESS METHODS)

Although methods of fixed area counts were common in both plant and animal sampling, they suffer from boundary effects, where a decision must be made to determine whether to include each target observed on a plot boundary in the sample, and they are time consuming. Plant biologist developed several "plotless" methods to estimate density and abundance that alleviate these problems and are relatively easy to apply, so long as the target species (e.g., bird nests) remains in place or can be measured before they move (Cottam and Curtis 1956). They have sometimes been referred to as distance methods, because they utilize either point-to-target or target-to-nearest-neighbor distances to estimate density and/or the spatial pattern of the target species.

Two general considerations should be weighed when considering use of plotless methods. The first is the execution of the random sampling design often proposed for this method. Random sampling is great in theory, and reviews well in proposals, but it is difficult and time consuming to achieve in the field. There also is an uncanny proportion of "random" points that do not occur in the thick brush, in the deeper portion of the marsh, on the ant bed, or other "random, but inconvenient" places in the field. Further, Pielou (1977) demonstrated that using random points to select random individuals is biased toward isolated individuals. In some circumstances, systematic random sampling is a good compromise, as the starting points are randomly placed, and they provide broad coverage of the area. Regardless, if you utilize random sampling, then establish a map and/or Global Positioning System (GPS)-based navigation system, allow extra time for navigating to the random points, and develop the willpower to place the points objectively where they fall. The second consideration is the distribution of the target species. Although most methods work well when the target species is randomly or uniformly distributed, many have problems when the target species is clumped or severely clumped (Legendre et al. 2004), and this drawback is especially pronounced for the plotless methods (Engeman et al. 1994).

Point-to-Target and Target-to-Nearest-Neighbor Methods

Byth and Ripley (1980) recommend 2 plotless sampling methods for measuring density and an excellent sampling design procedure for obtaining data from both methods simultaneously:

1. Determine sample size (n) for the density estimate.
2. Set out $2n$ points using a systematic random or other probabilistic sampling design.
3. Randomly select half of the $2n$ points, proceed to those points, and measure the distance from the point to the nearest target species (point-to-target or PTT).
4. On the remaining half of the $2n$ points, lay out a circle of radius sufficient to enclose (on average) the 5 nearest targets. Number these individuals and select n at random. From the randomly selected individuals, measure the distance to the nearest target species (target-to-nearest-neighbor or TNN).

The PTT density is estimated by

$$D_{PTT} = \frac{n}{\pi \sum x_i^2},$$

where D = density
n = number of samples
x_i = distance from point i to nearest target

The TNN density is estimated by

$$D_{TNN} = \frac{n}{\pi \sum x_j^2},$$

where D = density
n = number of samples
x_j = distance from target j to nearest neighbor

The variance for both estimates is calculated from the reciprocal of the density,

$$y = \frac{1}{D},$$

with the variance of y calculated as

$$s_y^2 = \frac{y^2}{n}.$$

The standard error of y is then

$$SE_y = \sqrt{\frac{s_y^2}{n}},$$

where D = density from either the PTT or TNN estimator
n = number of samples
y = reciprocal of the density estimate (D)

Example: We wish to estimate the number of active nest on a 2-km^2 study area during the breeding season. We used a map to delineate 20 systematic samples and randomly selected 10 for PTT measurements, reserving the other 10 for TNN measurements. At the PTT locations, we obtained the distances (x_i = 0, 10, 1, 10, 11, 15, 7, 12, 10, 9), with the sum of squared distances (x^2) equal to 921. At the TNN locations, we obtained the distances (x_i = 15, 7, 3, 12, 9, 15, 5, 11, 1, 7), with the sum of squared distances (x^2) equal to 929. As the calculations are the same for each estimator, we illustrate the density estimate from the PTT measurements:

$$D_{PTT} = \frac{10}{(3.14159)(921 \text{ m}^2)} = 0.0035.$$

So we estimate 0.0035 nests/m^2 or 34.56 nests/ha. The variance of the PTT estimate is

$$s_y^2 = \frac{\left(\frac{1}{0.003456}\right)^2}{10} = 8{,}371.8.$$

The standard error of the population estimate (SE_y) is

$$SE_y = \sqrt{\frac{8{,}371.8}{10}} = 28.934.$$

Therefore, the 95%CI for y is

$$95\%CI_y = \pm(2.262)(28.934) = \pm 65.45.$$

The upper and lower bounds on 95%CI are calculated as:

$$\frac{1}{0.003456} + 65.45 = 289.35 + 65.45 = 354.8$$

and

$$\frac{1}{0.003456} + 65.45 = 289.35 - 65.45 = 223.9.$$

We take the reciprocal of the results and multiply by 10,000 to convert to nests per hectare, so

$$\left(\frac{1}{354.8}\right)(10{,}000) = 28.18$$

and

$$\left(\frac{1}{223.9}\right)(10{,}000) = 44.66.$$

Therefore we have a mean of 34.56 nests/ha with 95%CI of 28–45 nests/ha.

Point-Quarter Method

The point-quarter method is a classic for sampling vegetation that dates back to the first land surveys in the United States. Surveyors would locate and describe the 4 trees nearest to each corner of a section (1 square mile) of land. The method was used by Cottam and Curtis (1956) for estimating forest species and continues to be used today. The method has application to animal density estimates as long as the target species (e.g., bird nests) remains in place or can be measured before they move. Using this technique, selected points from a sampling design are located in the field, and the area around the point is precisely divided into 4 (90°) quadrants (either perpendicular to the transect for point-transect sampling, or by compass bearing for random points). The distance from the point to the nearest target within each quadrant is measured, so that 4 distances are obtained at each point. The population density is then calculated as (Pollard 1971, Krebs 1998)

$$D_{PQ} = \frac{4(4n-1)}{\pi \Sigma(x_{ij}^2)},$$

where D = point-quarter estimate of density
n = number of points sampled
x_{ij} = distances from point i to the nearest target in quadrant j

Variance of the density estimate is

$$s_{PQ}^2 = \frac{D_{PQ}}{4n-2}.$$

The standard error of the density estimate is

$$SE_{PQ} = \sqrt{\frac{s_{PQ}^2}{4n}}.$$

The 95%CI can be obtained by

$$95\%CL_{PQ} = \left(\frac{\sqrt{16n-1} \pm 1.96}{\sqrt{\pi \Sigma(x_{ij}^2)}}\right)^2.$$

Example: We wish to estimate the number of active nests on a 2-km² study area during the breeding season. We use a map to delineate a point transect through a patch of forest, with 5 points spaced at 100 m. At the 5 locations, we obtain the distances (x_i = 0, 10, 1, 10, 11, 15, 7, 12, 10, 9, 15, 7, 3, 12, 9, 15, 5, 11, 1, 7), with the sum of squared distances (x_i^2) equal to 1,850. The density estimate is

$$D_{PQ} = \frac{(4)[(4)(5)-1]}{(3.1416)(1850)} = 0.0131$$

with a variance of the density estimate equal to

$$s_{PQ}^2 = \frac{0.01308}{(4)(5)-2} = 0.000727.$$

The standard error of the density estimate is then

$$SE_{PQ} = \sqrt{\frac{0.00072647}{(4)(5)}} = 0.00603,$$

and the lower and upper bounds on the 95%CI are

$$95\%LCL_{PQ} = \left(\frac{\sqrt{(16)(5)-1} - 1.96}{\sqrt{(3.1416)(1,850)}}\right)^2 = 0.00826$$

and

$$95\%LCL_{PQ} = \left(\frac{\sqrt{(16)(5)-1} + 1.96}{\sqrt{(3.1416)(1,850)}}\right)^2 = 0.02025.$$

The above units are in nests per square meter. We multiply by 10,000 to get nests per hectare, so we have a mean of 131 nests/ha, with 95%CI of 83–202 nests/ha.

COUNTS ON SAMPLE PLOTS (DETECTION PROBABILITY)

The preceding methods for estimating population size either reduced the survey area to ensure complete detection or attempted to correct the survey area to allow for unbounded counts with incomplete detection. The strategy was to either standardize or estimate the survey parameters necessary to obtain accurate estimates without direct evaluation of detection probability. The methods that follow use the opposite strategy: to estimate detection probability directly or collect ancillary data necessary to develop models for predicting detection probability.

Double Sampling

Double sampling (Jolly 1969a, b; Eberhardt and Simmons 1987; Pollock and Kendall 1987; Estes and Jameson 1988; Prenzlow and Lovvorn 1996; Anthony et al. 1999; Bart and Earnst 2002) is a modified form of sampling based on ratio estimation, where a large number of samples are obtained using a rapid method, such as point counts, followed by the surveying of a random subsample of those same plots using an intensive method that determines actual density. In the subsampled area, the densities obtained from the intensive method are used to estimate the proportion of animals seen using the rapid method. The relative probability of detection derived from the ratio of the rapid-method results to actual density is then used to correct estimates obtained from the rapid method over the remaining surveyed region.

The estimate of the proportion of animals seen (β) is the ratio of the mean counts (or density estimate) from the rapid method (y) to the mean count (or density estimate) from the intensive method (x):

$$\beta = \frac{y}{x}.$$

We can then use this estimate of the proportion of animals (β) on the subsamples to correct the population estimate (N) using the rapid method on the larger set of samples:

$$N = \frac{A \Sigma y}{na\beta},$$

where A = area of the study area (area of inference)
Σy = sum of counts or density estimates from the rapid method
n = the number of rapid-method samples
a = the area of each rapid-method sample
β = the relative proportion of animals (rapid method verses intensive method)

Jolly (1969a, b) and Pollock and Kendall (1987) presented an estimator for the variance of this estimator.

The **assumptions** of double sampling are the intensive method is accurate and reflects the actual density of the subsamples. Inaccuracy in the intensive method will result in multiplicative bias in the population estimate. For instance, lack of complete detection using the intensive method will create negative bias in the "corrected" population estimator. Similarly, the timing of the counts should coincide, and ideally would be simultaneous, so that both methods sample the same population. Differences in timing will increase variability, and perhaps bias, in the final population estimate.

Double Observer Sampling

Multiple observer methods were developed initially for aerial transect surveys (Caughley 1974, Magnusson et al. 1978, Cook and Jacobson 1979, Grier et al. 1981, Caughley and Grice 1982, Pollock and Kendall 1987, Graham and Bell 1989), but more recently they have been applied to ground point count surveys (e.g., Nichols et al. 2000). These methods can be divided into groups based on use of independent or dependent observers.

Independent Observers

Aerial or surface (ship, car, etc.) transects may be conducted with 2 observers, each collecting observations independently. The animal locations can be annotated on maps by each observer, or precise offset locations (x, y, and time) can be obtained using survey equipment (total station or GPS and offset laser rangefinder), allowing maps to be created post-survey. The mapped data are assigned to categories based on the type of detection: those seen by observer 1, those seen by observer 2, and those seen by both observers as in the equation below. Caughley (1974) demonstrated that data of this sort can be analyzed using the Lincoln–Petersen estimator (see Marked–Resight Methods later in this chapter) to estimate population size in the surveyed area (Grier et al. 1981, Caughley and Grice 1982, Pollock and Kendall 1987):

$$N = \frac{n_1 n_2}{m},$$

where N = population size in the area of inference
n_1 = total number of animals seen by observer 1
n_2 = total number of animals seen by observer 2
m = total number of animals seen by both observers

The method has several assumptions that will affect precision and accuracy:

1. Observations must be independent.
2. Category assignments must be accurate.
3. Targets must have equal detectability.

The **assumption** of independence of sightings between the observers is a strict requirement that may be difficult to achieve. For example, the independence assumption will be violated if the activity of one observer, such as speaking into a tape recorder or writing on a map, alerts the other observer to an animal's presence. Likewise, if separate surveys are conducted (e.g., ground and aerial), different observers should be used to ensure independence. Further, all animals must have equal detection probabilities, but these probabilities can differ between the 2 observers. If some animals differ in detectability (e.g., if males are more conspicuous than females), the resulting heterogeneity will produce negative bias in the Lincoln–Petersen estimator. However, Magnusson et al. (1978) noted the assumption of equal detection probabilities is not critical. Observation locations from each observer must be precise and unambiguous, or categorical assignments will be inaccurate. Similarly, because animal movement may contribute to this problem, surveys of mobile animals should be conducted simultaneously, so that each observer views the same sample population. Immobile targets (nests, middens, lodges, etc.) pose no such problem, and therefore, separate surveys may be made so long as the sample frame remains the same. Chapman (1951) provided a modified estimator with less bias:

$$N = \frac{(n_1 + 1)(n_2 + 1)}{m + 1} - 1,$$

and the variance of N was provided by Seber (1982):

$$s_N^2 = \frac{(n_1 + 1)(n_2 + 1)(n_1 - m)(n_2 - m)}{(m + 1)^2 (m + 2)}.$$

The method also has been used to estimate bird abundance from fixed-radius point counts, using 2 independent observers at each point. The point method requires there be no undetected movement into or out of the fixed radius, and that each observation must be accurately assigned as either inside or outside the fixed radius.

Dependent Observers

Another double observer approach involves 2 observers working in tandem. One is designated as the primary observer, the other as the secondary observer. The primary observer detects animals and reveals all sightings to the secondary observer. The secondary observer then records any additional sightings independently. Animal locations can be annotated on maps by each observer, or precise offset locations (x, y, and time) can be obtained using survey equipment (total station or GPS and offset laser rangefinder), allowing maps to be created post-survey. The mapped data are assigned to categories based on the type of detection: those seen by observer 1 and those additional animals seen by observer 2. Assuming equal detection probabilities for the 2 observers, we can obtain estimation of population size under the 2-sample removal model (Seber 1982, Pollock and Kendall 1987):

$$N = \frac{n_1^2}{n_1 - n_2}.$$

The variance of N is estimated as

$$s_N^2 = \frac{n_1^2 n_2^2 (n_1 + n_2)}{(n_1 - n_2)^4}.$$

The probability of an animal being detected is

$$p = 1 - \left(\frac{n_2}{n_1 + 1}\right)$$

and the variance of the detection probability is

$$s_P^2 = \frac{n_2(n_1 + n_2)}{n_1^3},$$

where N = population size in the area of inference
n_1 = total number of animals seen by observer 1
n_2 = total number of animals seen by observer 2
p = probability of an animal being detected

As with the independent observer approach, heterogeneous detection probabilities will produce negatively biased estimates of population size. Pollock and Kendall (1987) noted this method does not use the number of animals seen by both observers, and it assumes both observers have equal sighting probabilities. Therefore, it may not be as useful as the independent double observer method using the Lincoln–Petersen estimator.

Cook and Jacobson (1979) developed a similar dependent double-observer approach for transect surveys, but it has the 2 observers switch roles midway through the survey to overcome the possible difference in detectability between the observers. This method assumes that swapping roles does not alter the detection probability of the observers, all other assumptions being the same as above. Nichols et al. (2000) suggested applying the Cook and Jacobson (1979) method to estimate bird abundance from fixed-radius point counts, noting the model (DOBSERV; Nichols et al. 2000) permits estimation of observer-specific detection probabilities and bird abundance.

The advantage of the dependent double-observer approach occurs when there are practical or logistical reasons prohibiting the use of the independent double-observer method. The disadvantage is the dependent approach is less efficient than the independent approach, because capture–recapture methods are more efficient than removal methods (Seber 1982:324, Pollock and Kendall 1987:505). Therefore, we agree with Pollock and Kendall (1987) the independent approach using the Lincoln–Petersen estimator is more precise, simpler to understand, and allows the 2 observers to have different sighting probabilities.

Generalizations using program MARK (White and Burnham 1999) or DOBSERV (http://www.mbr-pwrc.usgs.gov/software.html) give the researcher the option to fit generalized Lincoln–Petersen models that allow for detection probability to depend on covariates, such as species, wind speed, and distance. MARK and DOBSERV use Akaike's Information Criterion (AIC; Burnham and Anderson 1998, 2002) to pick the most parsimonious model that explains the data adequately.

Marked Sample

We can use marked animals in a population to estimate detection probabilities. Using this method, some marked animals are released into the population and are therefore available for detection at the time of the survey. Marked and unmarked animals are counted during the survey, and the probability of detection for the marked animals can be estimated as

$$\beta = \frac{m}{n_1}.$$

By rearranging the terms, we get the Lincoln–Petersen estimator:

$$N = \frac{n_2}{\beta} = \frac{n_1 n_2}{m_2},$$

where N = total population size in the surveyed area
n_1 = number of marked animals present in the area at the time of the survey
n_2 = number of animals (both marked and unmarked) seen during the survey
m = number of marked animals seen during the survey
β = probability of detection

In practice, we recommend use the bias-adjusted modification of this estimator provided by Chapman (1951):

$$N = \frac{(n_1 + 1)(n_2 + 1)}{m + 1} - 1.$$

Although this approach is straightforward, the practical aspects require careful consideration. Marked and unmarked individuals must have the same probability of being detected. The mark must be conspicuous, so that no marked animals are erroneously or inadvertently recorded as unmarked. But the mark must not be so obvious that it draws attention to marked animals, making them more visible than unmarked animals. Further, it is necessary to determine how many marked animals are actually present for observation during the survey. The number present, and therefore available for observation, is frequently not equal to the number released. Radiotelemetry is a commonly used approach, as it can be used to determine the number of radiomarked animals in the surveyed area at the time of the survey (e.g., Packard et al. 1985, Samuel et al. 1987) and to verify whether each animal seen is marked. However, it is not necessary to have individually identifiable animals, and batch marks (e.g., collars with no alphanumeric identification code) will suffice, so long as the number of marked animals available for detection can be determined prior to the survey. Similarly, any marked animals seen during the survey that were not known to be present prior to the survey are not included in n_1. They are treated as unmarked in the survey data and included in n_2, but not in m.

Program NOREMARK (White 1996; http://www.cnr.colostate.edu/~gwhite/software.html) provides multiple estimators to determine the number of animals in the study area (Bartmann et al. 1987, White and Garrott 1990, Neal et al. 1993), a simulation capability for anticipating estimator performance, routines for estimating sample sizes, and a simulation for calculating the relative effort to put into marking versus resighting.

Time of Detection

Farnsworth et al. (2002) were the first to recognize that useful information on detection probabilities were available from the times when birds were detected in point count surveys. Their method was a modification of removal methods that used only the time interval when a bird was first detected to estimate detection probabilities. Similar to the development chronology of double observer methods, more recent work (Alldredge et al. 2007a) has extended the approach using a capture–recapture formulation, because capture–recapture methods are generally more efficient than removal methods (Seber 1982:324, Pollock and Kendall 1987: 505). Both approaches capitalize on the common practice of recording data at point counts in temporal intervals, where the number of birds counted is separated into those first observed in the first 3 minutes, those first observed in the next 2 minutes, and those first observed in the final 5 minutes. This procedure was recommended by Ralph et al. (1995) and was originally designed to allow results from 10-minute counts to be comparable with those from studies employing 3- and 5-minute counts.

Using the removal method of Farnsworth et al. (2002), the simplest application of the time of detection approach can be illustrated with just 2 time intervals of equal duration. Suppose that an observer records all birds seen and/or heard in the first 5 minutes and then records any additional birds detected in the second 5 minutes. We can then define x_1 as the number of birds counted in the first time interval and x_2 as the number of new birds (not detected in the first period) detected in the second time interval. The expected values of the random variables x_1 and x_2 are

$$E(x_1) = Np_1$$
$$E(x_2) = N(1-p_1)p_2,$$

where N = total number of birds within the detection radius of the observer
p_1 = detection probability for an individual bird in the first time period
p_2 = detection probability for an individual bird in the second time period

The term $(1-p_1)$ is necessary, because all birds first detected in the second interval must, by definition, have been missed in the first time interval. If we assume the detection probability for the 2 intervals is equal (i.e., $p_1 = p_2 = p$), solving the above equations for p and N produces the moment estimator (Zippin 1958)

$$p = \frac{x_1 - x_2}{x_1}$$

and the population **estimator**

$$N = \frac{x_1^2}{x_1 - x_2} = \frac{x_1}{p}.$$

The estimators can fail if $x_1 \leq x_2$, which is possible when p is small. We present this 2-sample removal estimator to illustrate the approach with the simplest possible situation.

Example: During the first 5 minutes, we observe 20 birds, and during the second 5 minutes, we observe 5 birds that we did not observe during the first 5 minutes. The probability of detection is then

$$p = (20-5)/(20) = 0.75$$

and the population estimate is

$$N = (20)(20)/(20-5) = 20/0.75 = 26.67 = 27 \text{ birds}.$$

In practice, we use >2 intervals, because doing so permits relaxation of the assumption of equal detection for different species. For instance, Farnsworth et al. (2002) present a more general model with 3 count intervals of variable length, allowing for differences in detection probabilities among intervals and heterogeneity of detection among individual birds. These differences are taken into account by assuming there are 2 groups of individuals in an unknown proportion, and that all members of the first group are detected in the first time interval.

Alldredge et al. (2007a) suggested a more efficient approach using a closed population capture–recapture model with k time intervals to account for more sources of variability in the point count data. Their method was specifically designed to account for variation in detection probabilities associated with the singing rate of birds, by modeling both availability and detection bias. They recommended using ≥4 equal intervals to reduce assumptions. For example, the assumption of constant detection rates over time required by the removal model of Farnsworth et al. (2002) is not required in the capture–recapture approach, because it models temporal variation from the full detection history.

Assumptions of the general time-of-detection method (Farnsworth et al. 2002, Alldredge et al. 2007b) model are: (1) there is no change in the population of birds within the detection radius during the point count (i.e., the population is closed and birds do not move into or out of the radius), (2) there is no double counting of individuals, (3) all members of group 1 are detected in the first interval, (4) all members of group 2 that have not yet been detected have a constant per minute probability of being detected, and (5) observers accurately assign birds to within or beyond the radius used for the fixed radius circle. As noted by Alldredge et al (2007b), these restrictions are not trivial, because movement of individuals and difficulties associated with aural detections may result in violation of all assumptions.

Program CAPTURE can produce maximum likelihood estimates for N, as well as the estimated variance of N (Otis et al. 1978, White et al. 1982), for equal time intervals using the method of Farnsworth et al. (2002). **Program MARK** (White and Burnham 1999) can be used to model detection history over k intervals, constant detection probability for all individuals, time effects on detection probability, difference due to previous detection, and unobservable heterogeneity, following the method of Alldredge et al. (2007a, b).

Modern Distance Sampling

Modern distance sampling is a widely used method for estimating size or density of biological populations. It is a comprehensive approach that encompasses study design, data collection, and statistical analyses (Buckland et al. 2001). Modern distance sampling is based on the observation that detection probabilities decrease with increasing distance from the observer (Burnham and Anderson 1984). Distance data are used to estimate the specific shape of the **detection function** relating detection probability to distance for a particular target species and set of conditions. We can define the detection function $g(x)$ as the conditional probability of detecting an animal, given that it is located at some distance (w) from the line. Although the various analyses can be quite sophisticated, the data collected along line transects or points counts for modern distance sampling methods are the same data one would use for traditional distance sampling methods. When properly applied, distance sampling yields estimates of absolute density and detection probability, meeting the requirements for inference put forth by Rosenstock et al. (2002). The history and development of distance sampling is described by Buckland et al. (2001), and extensions to the basic theory are covered in Buckland et al. (2004) and Thomas et al. (2010). An extensive reference archive, covering methodological development and practical applications of modern distance sampling, is available on the Distance Sampling website (http://www.ruwpa.st-and.ac.uk/distancesamplingreferences/).

In distance sampling, counts are assumed to be incomplete. Thus, the proportion of animals present that are actually seen (β) must be estimated, and actual counts must be corrected by these detection probabilities. Perpendicular or radial distance data are used to estimate these detection probabilities. To examine what a detection function looks like, we can plot a **histogram** using the frequency of detections (y axis) grouped into small distance intervals (x axis) from the center of a transect line (distance 0) to the maximum observation distance (w). If our sample size is large, we can approximate the shape of the detection function by drawing a smooth curve through the top of each distance

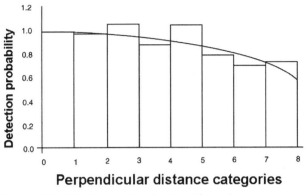

Fig. 11.3. The detection function for the uniform plus one-term detection function for duck nest data. *From Anderson and Pospahala (1970).*

interval in the histogram (Fig. 11.3). In practice, sample sizes are often too small, and this procedure does not work well.

Survey planning, including sampling design and estimates of sample size, can be performed in program DISTANCE. Data collection can be performed using either line transects or points. Analysis of the resulting data typically involves 4 steps: (1) data examination via graphical displays, (2) model fitting using various functions and adjustment terms, (3) model selection using the AIC criteria, and (4) inference under the chosen model. Program DISTANCE allows for the fitting of complex detection functions (half normal, uniform, or hazard rate) using a series of adjustment terms (cosine, simple polynomial, or hermite polynomial). Rather than review these models and associated parameter estimators here, we recommend the excellent book by Buckland et al. (2001). A concise overview of distance sampling and program DISTANCE, including newly available advanced options, can be found in Thomas et al. (2010). Details concerning the actual use of program DISTANCE (Thomas et al. 1998; available at http://www.rupwa.st-and.ac.uk/distance) are contained in the help files provided with the program.

Actual field application of modern distance sampling methods involves many decisions and **considerations** specific to each survey situation. For example, many animals exhibit gregarious behavior and tend to occur in groups. This behavior requires measuring the distance from the line or point to the geometric center of each group and recording the number of animals in present in each cluster. Because groups of animals are easier to detect than individuals, detection bias can occur as a function of group or cluster size. Thus, decisions must be made concerning whether to measure distances to groups or to individuals. The density of groups or clusters along with estimates of cluster size are modeled to improve the precision of estimates of density and population size. Drummer and McDonald (1987) and Otto and Pollock (1990) discussed models for use when detection probability for fixed distance depends on group size.

Another consideration involves grouping of data. Accurate measurement of distances in the field may not be possible; therefore, detections may need to be grouped into distance categories. Even when direct distance measurements are recorded, anomalous patterns may be apparent, such as few objects detected at short distances, heaping of detections at commonly rounded measurements (e.g., 50 m or 100 m), or a relatively large number of detections near the boundary distance. Buckland et al. (2001) recommended truncation of data at distances greater than that at which observations seem likely to be outliers. Further, data may be grouped into a histogram before analysis as a smoothing technique (Buckland et al. 2001). However, exact distance measurements are to be preferred when possible, as they allow the data to be placed into distance intervals during analysis.

Additional problems may arise due to insufficient sample size in terms of observations, transects, or points. The variability between lines and points is an important factor that influences encounter rates (n/k) and detection probability. Failure to obtain a representative sample of the true variability within a population will lead to bias, and too few lines or points will result in lack of precision. The number of lines or points (k) should be 4, and sampling should be probabilistic to adequately represent the area of inference. We also suggest that transect length be selected to provide a minimum of 40 animals detected, and preferably 60–80 (Buckland et al. 2001).

We recommend those planning to conduct a modern distance sampling study consult Buckland et al. (2001) and, if available, published recommendations for specific field situations or species (e.g., Karanth et al. 2002). For instance, Anderson et al. (2001b) used field trials to estimate the abundance of artificial desert tortoise (*Gopherus agassizii*) models to test whether assumptions that underlie distance sampling were met. They found the density estimate of adult tortoise models was relatively unbiased, whereas the estimate for subadult (small) tortoise models was biased low (about 20%). They attributed the bias to failure to detect small tortoises on or near the centerline and presented ideas to better train observers before commencing the survey. And standard distance theory, based on the premise that detection probability is a decreasing function of distance and that nothing else influences detection, can be violated. Breeden et al. (2008) noted the effects of traffic noise on auditory point surveys of urban white-winged doves (*Zenaida asiatica*).

Distances also can be measured to animals (usually land birds) that are counted around a point rather than along a transect. There are advantages and disadvantages associated with use of points rather than line transects. For example, a line transect can yield more data per unit time than can points, particularly when more time is spent traveling between transects or points than actual sampling (Bibby et al. 2000, Rosenstock et al. 2002). Scale also is important, as a typical transect generally covers more spatial area than a typical set of points; thus, the scale of spatial habitat diversity must be commensurate with the scale of transects or points. The main disadvantage with points, according to Bibby et al. (2000:92), is the area surveyed is proportional to the square of the distance from the observer, whereas in transects the area is proportional to lateral distance from the transect line. Thus, density estimates from point data are more susceptible to errors arising from inaccurate distance measurements or from violation of assumptions about detecting animals.

However, points are often preferred to transects in habitats with a variety of small patches of habitat relative to the home range of an animal (Bibby et al. 2000). Likewise, points can be preferred over transects when vegetation or terrain hinders navigation, or when observer movement signals the animals of observer presence. For instance, Reynolds et al. (1980) noted that observers traveling along line transects, in structurally complex vegetation and rough terrain, tended to watch the path of travel, reducing their ability to detect birds. Consequently, they recommended establishing equally spaced observer stations, positioned along a transect of points that could be located randomly. Similarly, Koenen et al. (2002) used point transects to estimate seasonal density and group size of mule deer (*Odocoileus hemionus*) by gender and age class on the Buenos Aires National Wildlife Refuge in southeastern Arizona. The authors believed their survey design balanced the often conflicting objectives of random placement of transects and detecting animals before they moved. Burnham et al. (1980) and Buckland et al. (1993, 2001) provide details for sampling designs of point transects.

The **assumptions** of distance sampling are: (1) points or transects are located randomly with respect to the distribution of animals; (2) all objects at the center of the point or transect are detected with certainty; (3) objects are detected at their initial location prior to any movement in response to the observer; (4) distances are measured accurately (ungrouped data), or objects are counted in the proper distance category (grouped data); and (5) objects are detected independently. Violation of the second assumption is a critical failure and is probably common when conducting bird surveys. This violation will result in negatively biased estimates of density. Similarly, if animals are attracted to the observer, the data are not likely to indicate a problem, resulting in positive bias in the estimate of density.

REMOVAL METHODS

Removal methods of population estimation are old and have been analyzed by numerous investigators. Yet these methods are attractive, because often someone other than the investigator, such as hunters, can collect the removal data. Thus, the investigator may not have to actually capture and mark animals to develop population estimates based on removals, which often makes these methods inexpensive to implement in the field.

Catch-per-Unit-Effort

Catch-per-unit-effort (e.g., catch/day) is based on the premise that as more animals are removed from a population, fewer are available to be "caught," and catch/day **will decline** (Fig. 11.4). Eventually, if all animals are removed, the expected catch will become zero, and the total number of animals removed will equal the initial population size. Because it is generally not desirable (and seldom possible) to remove all individuals in a population, this method involves developing a linear regression of the number of animals removed each day on the cumulative total number of animals removed prior to that day (Leslie and Davis 1939). An advantage of this method is that population estimates can be derived prior to all animals being removed, and they can be used with removals that are a part of routine management activities, such as hunter or fisher harvests. Animals do not have to be physically taken or removed to be "caught." Animals can be trapped, shot, photographed, or seen. If animals are marked (i.e., live-trapping of small mammals), they would be included in the calculation on the day they were trapped and marked, but they would be ignored on subsequent days if re-trapped.

Assumptions for this method include: (1) sampling units are taken at random; (2) the population is closed (e.g., the removal period is kept as short as possible); (3) all individual animals have an equal probability of being caught; and (4) unit of effort is constant, and all the removals are known. Catch-per-unit-effort estimates are not likely to be accurate or precise unless a large proportion of the population is removed (i.e., large enough to cause a decline in catch-per-unit-effort; Krebs 1998, Bishir and Lancia 1996).

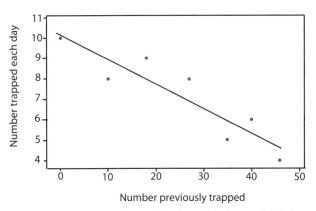

Fig. 11.4. Estimating population abundance by plotting the daily number trapped against the total number previously captured. In this example 10, 8, 9, 8, 5, 6, and 4 mice were trapped on 7 consecutive days. The regression equation for these data is $x = 74.7 - 6.94y$; therefore, when $y = 0$ (all mice are removed), x would equal 75 mice in the population.

The regression equation is not a typical regression, because the catch/day and the cumulative removals depend on the same removals. This lack of independence makes calculation of variances and 95%CI difficult. Bishir and Lancia (1996) have shown that estimates do not follow a normal distribution and, therefore, standard variance equations are not appropriate.

Change-in-Ratio

Kelker (1940) first used this method on selective harvest of male and female deer. Often it is referred to as the **sex-ratio estimator,** because in most cases, sex determines the 2 classes (e.g., male and female deer or pheasants) used with the estimator. However, the method can be used on any 2 classes of animals as long as harvest varies between the classes, such as age classes (e.g., adults and juveniles); species harvested at the same time (e.g., gray [*Sciurus carolinensis*] and fox [*Sciurus niger*] squirrels if one species is selected over the other); if only 1 species is harvested (e.g., deer and cows, where cows are not hunted); or with marked and non-marked populations, where restrictions are placed on harvest of marked animals (e.g., collar-marked deer).

This methods primarily has been used on public hunting areas or on private ranches, where hunts are controlled, and animals taken must come through a check station. In this way, total kill is known. Additional information needed is an estimate of the proportion of each class (e.g., proportion of males and females) in the population just prior to the hunt and an estimate of the proportion of each class after the hunt. **Assumptions** include: (1) the proportion of the classes will change after the hunt due to selective harvest of one class over the other (e.g., more bucks killed than does), (2) observed proportions of the 2 classes are unbiased (a major problem with the estimator to be discussed later), (3) the population is closed, and (4) the number of removals of each class is known. If these assumptions are valid, the population abundance can be estimated using the following equation:

$$N_1 = \frac{[(T)(p_2) - F]}{p_2 - p_1}$$

$$N_2 = N_1 - T,$$

where N_1 = pre-hunt population
T = total kill (all animals harvested regardless of sex class)
F = number of females killed
p_1 = proportion of females in survey before hunt
p_2 = proportion of females in survey after hunt
N_2 = post-hunt population

Example: On a public hunting area, you observed 300 male and 300 female pheasants on a road survey prior to a hunt. During the hunt, 400 male and 25 female pheasants (females are illegal to shoot) are shot and brought through a check station. On a road survey immediately after the hunt, you observe 100 male and 300 female pheasants. The estimated population abundance prior to the hunt would be calculated as

$$N_1 = \frac{[(425)(0.75) - 25]}{0.75 - 0.50} = 1{,}175.$$

Therefore, **post-hunt** population is

$$N_2 = 1{,}175 - 425 = 750.$$

Comment: Note that a small change in p_1 or p_2 will have a great effect on the estimate. As noted for assumption 2 above, observed proportions of the 2 classes should be unbiased. We believe this is a major problem with the estimator, and we discuss this issue in some detail here. Prior to a hunt, animals have probably not been hunted for at least a year, or in the case of released pheasants, not at all. Therefore, sighting probabilities for male and females may be unbiased; however, once hunting centers on 1 sex class, we believe that sex class will have a lower probability of sighting after the hunt, whereas the nonhunted sex will have a higher probability of being sighted. This bias would exist for pheasants or deer and similarly, for different age classes for which larger animals (either trophy deer or larger deer hunted for meat) are harvested more than are young of the year (i.e., fawns). In addition, the probability of sighting different sex and age classes of deer varies by month even if they are not hunted (Downing et al. 1977), thereby giving a bias between pre- and post-hunt observations. We, therefore, do not recommend this method to estimate population abundance, and we know of few people who currently use it. We have presented the method here only because readers may come across this method in the literature and should be aware of its problems.

MARKED–RESIGHT METHODS

Unlike previous editions of the *Techniques* manual, in which this section was usually titled Capture–Recapture or Capture–Mark–Recapture, we prefer the term marked–resight, because animals do not have to be captured to be marked (e.g., they may have natural marks, including DNA, or may be marked remotely with paint-ball guns, etc.; see Chapter 9, This Volume), nor do they need to be recaptured (e.g., they can be observed; photographed; or DNA fingerprinted from hair, feathers, or feces) to determine whether they are marked. In fact, they do not need to be marked at all (we explain this later). There is **only one assumption** for marked–resight methods: the proportion of marked to nonmarked individuals in a sample is the same as it is in the population. All other purported assumptions are just violations of this assumption. We examine this issue more closely later in this section.

We consider marked–resight methods to be the **gold standard** for conducting population estimates. For if done correctly, we believe they produce more accurate and reliable estimates. However, the percentage of marked animals in the population will affect the accuracy of the estimates (Silvy et al. 1977). Silvy et al. (1977) noted that when 50% of the population was marked, more accurate estimates were obtained; however, due to cost of marking animals, they recommended that at least 25% of the population be marked. How does one know when 25% of the animals are marked? When 25% of animals seen on random resight surveys are observed, 25% of the population is marked.

Known Number Alive

Many times when conducting marked–resight studies on small populations (e.g., bobcats [*Lynx rufus*] on small areas), few if any animals are resighted. A common method to estimate abundance is to simply use the number of original captures as an estimate of abundance. **Known-to-be-alive** or **minimum-number-live** estimates are often the most appropriate estimates when conducting these types of studies. These estimates tend to underestimate population density; however, an overestimate of density may lead to inappropriate management action, whereas an underestimate may produce inefficient, but safe management strategies.

Lincoln–Petersen Estimator

A known number of animals in a study is "marked" during a short time period, and then within a few days, a random sample is taken to determine the number marked in the sample. A **rule of thumb** is to use a different method to obtain the sample than was used to mark the animals. For example, do not use a net gun from a helicopter to capture and mark deer and then use a helicopter to obtain the sample, as deer captured and marked may hide from the helicopter, thereby producing a bias in the sample that will cause an overestimation of population abundance. If the assumption given above is valid, an unbiased estimate of population abundance can be obtained using the following equation:

$$N = \frac{Mn}{m},$$

where N = population abundance
M = number marked in study area
n = number of marked and nonmarked animals observed in sample
m = number marked in sample

Example: You mark 100 deer using box traps on a study area, and a week later, you conduct a random road survey and see 50 deer, of which 10 are marked. The estimated population abundance would be calculated as

$$N = \frac{(100)(50)}{10} = 500.$$

Assuming a normal distribution, the 95%*CI* would be approximately ±2 standard errors (*SE*). An estimate for 1 *SE* can be obtained using

$$SE_N = \sqrt{\frac{(m^2n) + (n-m)}{m^3}}.$$

Therefore,

$$SE_N = \sqrt{\frac{[(100)(100)(50)] + (50-10)}{(10)(10)(10)}} = 141.42$$

$$2\,SE = 283;$$

therefore, 95%CI = 217–783 deer. Again, we should replicate the sample several times to obtain a mean estimate, a 95%CI, and conduct a power analysis to determine the sample size needed to detect a desired **effect size.**

Comment: The **only assumption** made in marked–resight methods is the proportion of marked to nonmarked individuals in a sample is the same as it is in the population. If animals lose their marks, then fewer marked animals would be seen than expected, which would cause an overestimation of population abundance. A **rule of thumb** is that any factor (e.g., marked animals leave study area) that causes one to see fewer marked animals than expected will cause an overestimation of population abundance. In contrast, factors (e.g., trap-happy animals) that cause one to see more marked than expected will cause an underestimation of population abundance. There is a premise the Lincoln–Petersen estimator is limited to a closed population. This scenario is best case; however, if the ratio of marked to nonmarked animals leaving a study area is the same as it is in the population, an open population will have **no effect** on the estimate. Similarly, if the same number of nonmarked animals emigrate from and immigrate to the study area, it will have no effect on the estimate. The best way to avoid any problems with population closure is to mark the animals within a short time frame and conduct the resight sample soon thereafter.

Another **misconception** is that animals have to be marked randomly or uniformly throughout the study area. This marking would be ideal; however, if a random resight sample is taken, it does not have to be done, because a random sample should contain the ratio of marked to nonmarked as they are found in the population. To illustrate this idea, we use an extreme example. Say there are 2 identical (e.g., size and vegetation types) islands crossed by a single road with a bridge between them. On the first island, you mark 100 deer, and on the second island you mark none. Later that week you conduct a resight road census over both islands. On the first, you sight 50 deer, of which 10 were marked, and on the second, you sight 50 deer (this result would be expected if the islands were truly identical), of which none were marked. Using the example given above, your estimate for the first island would be 500 deer. Now let us recalculate using all the information from both islands.

Unlike the example above, n now equals 100 (50 seen on each island):

$$N = \frac{(100)(100)}{10} = 1{,}000$$

The 1,000 deer would be expected if the islands were truly identical. We use this illustration to debunk the idea that marked and nonmarked animals must be evenly mixed. What **must be done** is to obtain a random sample across the study area that will give you a true ratio of marked to nonmarked individuals in the population. For large animals, such are deer, that can more easily be trapped and marked along roads, a resight survey using randomly placed infrared motion-sensitive cameras is ideal, especially if neck collars are used to mark the deer. Also, remember the Lincoln–Petersen estimator does not require that animals be individually marked, making this method ideal when photos may not get a good angle of the marked animal. However, additional information (e.g., movements or survival) can be obtained from animals if they are individually marked, and we recommend that you do so.

At the beginning of this section, we made the comment that no animals need be marked to conduct a marked–resight estimate. In south Texas on large ranches, some landowners stock ranches with known numbers of exotic deer. Given this practice, one could use the number of exotic deer as marked animals and all native deer as nonmarked animals to estimate the number of exotic deer and native deer on the ranch, especially if randomly located infrared cameras were used to resight animals. Subtracting the known number of exotic deer from the estimate would give you an estimate of native deer abundance. The **assumption** is that exotic deer and native deer have the same detection probability. Only one's imagination limits the use of marked–resight methods.

In practice, the major problem we find with marked–resight methods is defining the study area. This is not a problem if working on islands or estimating deer abundance within high fences. But it is a real problem when using live traps in a defined grid to mark–resight small mammals. We recommend using the maximum daily movement (i.e., obtained from maximum distance between traps in the grid in which an individual was trapped on consecutive days) of the mammals in question to define the limits outside the grid. For larger animals (e.g., deer), we also recommend using the maximum daily movement (i.e., obtained from maximum distance between daily sightings of marked animals during resight surveys). This distance is then used to expand an area obtained by including all locations of marked animals within a convex polygon using the minimum number of locations to connect all other locations.

Schnabel Estimator

In situations where animals are continually being marked as resight surveys are conducted, there are several ways to analyze the data for a population estimate. A common way is to treat each resight survey as a separate data set (i.e., using the total number marked at the time of the survey) to obtain multiple estimates and then calculate a mean estimate of population abundance using the Lincoln–Petersen estimator. Or, because the number of marked animals in the population affects the estimate, one could use only the data obtained from the final survey to obtain an estimate. If the former is used, then one is giving equal weight to each survey and if only the last survey is used because it has a larger sample size, one is not using all data available. To overcome this problem, Schnabel (1938) developed a method (i.e., weighted average) to use all available data without giving each survey an equal weight. The **assumption** for the Schnabel estimator is the same as for the Lincoln–Petersen estimator; namely, the resight sample has the same ratio of marked to nonmarked animals as is found in the population. If the assumption given above is valid, an unbiased estimate of population abundance can be obtained using the following equation:

$$N = \frac{\sum Mn}{\sum m},$$

where N = population abundance
M = number marked in study area
n = number of marked and nonmarked animals observed in sample
m = number marked in sample

Example: Over a 5-day period, you trapped and marked mice using 100 live traps, with the results shown in Table 11.1. The death of some animals during trapping must be accounted for as noted below. If no animals die, then $A = n$. If animals are found dead in the sample, they must be accounted for (i.e., dead marked animals must then be subtracted from M, and dead nonmarked animals are then not added to M). Using the data from Table 11.1 in the above equation yields

$$N = \frac{1{,}268}{19} = 66.7.$$

If we had run 4 Lincoln–Petersen estimates for days 2–5, our estimates would be 60 mice for day 2, 57 mice for day 3, 78 mice for day 4, and 68 mice for day 5. If we average these estimates, we get 66 mice with a standard error of 4.70 mice. Assuming a normal distribution, we have a 95%CI (about ±2 SE) of 57–75 mice. Even though the mean Lincoln–Petersen estimator (66) and Schnabel estimator (67) are similar, the Schnabel estimator gives greater weight to the last days of trapping when a greater number of mice were marked. Silvy et al. (1977) have shown that accuracy of estimates is greater when more animals are marked; therefore, one should use the Schnabel estimator when there are 1 day of resightings.

Schumacher–Eschmeyer Estimator

The Schumacher–Eschmeyer estimator (Schumacher and Eschmeyer 1943) is a variation of the Schnabel estimator, itself a variation of the Lincoln–Petersen estimator. Like the Schnabel estimator, it uses all available data without giving each survey an equal weight. Using the data from the Schnabel example above, 2 additional columns are calculated (Table 11.1). The **assumption** for the Schumacher–Eschmeyer estimator is the same as for the Lincoln–Petersen and Schnabel estimators: the resight sample has the same ratio of marked to nonmarked animals as is found in the population. If the assumption is valid, an unbiased estimate of population abundance can be obtained using the following equation:

$$N = \frac{\sum nM^2}{\sum mM},$$

where N = population abundance
M = number marked in study area
n = number of marked and nonmarked animals observed in sample
m = number marked in sample

Using data from the Schnabel example above and the last 2 columns of Table 11.1, we obtain

$$N = \frac{42{,}164}{611} = 69.$$

Table 11.1. Hypothetical example of 5 days of trapping and marking mice with data presented in format suitable for estimation of population abundance using the Schnabel and Schumacher–Eschmeyer estimators[a]

Day	Number trapped (n)	Number recaptured (m)	Number alive (A)	Total marked alive prior to date (M)	Mn	nM^2	mM
1	10	0	10	0	0	0	0
2	12	2	11	10	120	1,200	20
3	15	5	15	19	285	5,415	95
4	10	5	9	39	390	15,210	195
5	11	7	11	43	473	20,339	301
Totals		19			1,268	42,164	611

[a] Note that only the first 6 columns are needed for the Schnabel estimator, whereas all 8 columns are needed for the Schumacher–Eschmeyer estimator.

Jolly–Seber Estimator

The Jolly–Seber estimator (Jolly 1965, Seber 1965) is used for open populations and estimates population size, survival rates, and births. A marked–resight experiment is conducted, during which, on ≥3 successive occasions, animals are marked from the population. The identity of marked individuals is recorded on each occasion, unmarked animals are marked, and all animals are released. An estimate of population size is calculated from the simple relationship that population size is equal to the size of the marked population divided by the proportion of animals marked. Estimates can be obtained for each occasion except the first and last. Calculations for the Jolly–Seber estimator are complicated and are best done with available computer programs; therefore, they are not presented here. Estimates of population size, survival rates, and births can be computed directly by **program JOLLY** (Pollock et al. 1990; http://www.mbr-pwrc.usgs.gov/software.html). **Program POPAN-5** (Arnason and Schwarz 1999; http://www.cs.umanitoba.ca/~popan/) is based on a different approach to the Jolly–Seber model (Crosbie and Manly 1985, Schwarz and Arnason 1996). It includes estimation of the total number of individuals that are in the population at any time during the study, and from the program computes an estimate of population size (plus survival rate and recruitment). To achieve this estimate, one must make some **assumptions** about the values of parameters at the beginning and end of the study (Schwarz and Arnason 1996).

Assumptions of the Jolly–Seber estimator are: (1) all individuals have equal probability of capture; (2) every marked animal present in the population has the same probability of survival; (3) marks are not lost or overlooked; (4) all samples are instantaneous, and each release is made immediately after the sample; and (5) every animal in the population is equally likely to emigrate, and all emigration from the population is permanent.

COMPUTER SOFTWARE PACKAGES

Several computer software packages are available that can be used to estimate population abundance using the methods described above plus other methods not covered here. Prior to the use of these packages, however, one must be aware that errors may exist in these programs (e.g., early versions of program CAPTURE). If results obtained from a software program appear unrealistic, compare them to results from a different software package. Also, be aware that input errors also can give unrealistic or erroneous results (i.e., "garbage in is garbage out," and it is not the fault of the software package). Input errors include, but are not limited to (1) data entry, (2) data transfer, (3) column and/or row selection in spreadsheets, and (4) model selection (i.e., assumptions). The best way to test software results is to run a small "known" data set through the software program, where the outcome has been previously determined without the use of software programs. If the result obtained is the same or similar, then the data have been entered correctly and the software program is probably working properly.

Program CAPTURE

Program CAPTURE computes tests to select a model from several possible models and then computes estimates of capture probability and population size for **closed** population marked–resight data. However, some models in CAPTURE do not work with small data sets. For those who want to learn more about Program CAPTURE, references are provided in Box 11.1.

Program MARK

Program MARK (http://warnercnr.colostate.edu/~gwhite/mark/mark.htm) provides parameter estimates from marked animals when they are re-encountered later. Generalizations using program MARK (White and Burnham 1999) or DOBSERV (http://www.mbr-pwrc.usgs.gov/software.html) give the researcher the option to fit generalized Lincoln–Petersen models that allow for probability of detection to depend on covariates, such as species, wind speed, and distance. MARK and DOBSERV use AIC (Burnham and Anderson 1998, 2002) to pick the most parsimonious model that explains the data adequately. Currently, no paper documentation is available for MARK. Electronic documentation can be found at http://warnercnr.colostate.edu/~gwhite/mark/mark.htm. This material can be printed if you want hard copy. A reasonably complete description of MARK can be found in White and Burnham (1999). Other references are given in Box 11.1.

Program DISTANCE

Program DISTANCE provides an analysis of **distance sampling data** to estimate density and abundance of a population. Considerably more detail is provided at the web site (http://www.ruwpa.st-and.ac.uk/distance/), that includes the software and an electronic manual. The methods used by this program are documented in the references listed in Box 11.1.

SUMMARY

Obtaining precise estimates of animal abundance or density in wild populations is difficult, time consuming, and costly. Most techniques have problems related to estimating the probability of capturing or detecting animals during a survey and to taking insufficient and/or nonrandom samples. When using indices, it is assumed the detection probability is constant, but unknown and that over time nothing changes except population abundance. These assumptions may or may not be true, and we caution against use of indi-

> **BOX 11.1. REFERENCES FOR POPULATION ESTIMATION COMPUTER PROGRAMS**
>
> **Program CAPTURE**
>
> Otis, D. L., K. P. Burnham, G. C. White, and D. R. Anderson. 1978. Statistical inference from capture data on closed animal populations. Wildlife Monographs 62.
>
> Rexstad, E., and K. P. Burnham. 1991. Users guide for interactive program CAPTURE. Colorado Cooperative Fish and Wildlife Research Unit, Colorado State University, Fort Collins, USA.
>
> White, G. C., D. R. Anderson, K. P. Burnham, and D. L. Otis. 1982. Capture–recapture and removal methods for sampling closed populations. LA-8787-NERP, Los Alamos National Laboratory, Los Alamos, New Mexico, USA.
>
> **Program MARK**
>
> White, G. C., and K. P. Burnham. 1999. Program MARK: survival estimation from populations of marked animals. Bird Study (Supplement) 46:120–138.
>
> **Program DISTANCE**
>
> Buckland, S. T., D. R. Anderson, K. P. Burnham, and J. L. Laake. 1993. Distance sampling: estimating abundance of biological populations. Chapman and Hall, New York, New York, USA.
>
> Buckland, S. T., D. R. Anderson, K. P. Burnham, J. L. Laake, D. L. Borchers, and L. J. Thomas. 2001. Introduction to distance sampling: estimating abundance of biological populations. Oxford University Press, Oxford, England, UK.
>
> Buckland, S. T., D. R. Anderson, K. P. Burnham, J. L. Laake, D. L. Borchers, and L. J. Thomas, editors. 2004. Advanced distance sampling. Oxford University Press, Oxford, England, UK.
>
> Laake, J. L., S. T. Buckland, D. R. Anderson, and K. P. Burnham. 1994. DISTANCE user's guide V2.1. Colorado Cooperative Fish and Wildlife Research Unit, Colorado State University, Fort Collins, USA.
>
> Thomas, L., S. T. Buckland, K. P. Burnham, D. R. Anderson, J. L. Laake, D. L. Borchers, and S. Strindberg. 2002. Distance sampling. Pages 544–552 in A. H. El-Shaarawi and W. W. Piegorsh, editors. Encyclopedia of environmetrics. Volume 1. John Wiley and Sons, Chichester, England, UK.
>
> Thomas, L., J. L. Laake, S. Strindberg, F. F. C. Marques, S. T. Buckland, D. L. Borchers, D. R. Anderson, K. P. Burnham, S. L. Hedley, and J. H. Pollard. 2002. DISTANCE 4.0., Release 1. Research Unit for Wildlife Population Assessment, University of St. Andrews, Scotland, UK. http://www.ruwpa.st-and.ac.uk/distance/.

ces unless these assumptions can be verified for the comparisons being made. In the case of population estimation, techniques range from complete counts, where sampling concerns dominate, to incomplete counts, where detection concerns dominate. Examples of population estimation procedures include multiple observer, removal, and capture–resight methods.

Before conducting a survey to estimate population abundance, determine what information is needed, for what purpose the information will be used, how precise an estimate is needed, and the time and cost required to conduct the survey. The key to deriving population abundance estimates is to select a method that fits the situation. If necessary, techniques can be adapted to meet a particular need. Generally, a biometrician familiar with population estimation literature should be consulted. However, most biometricians consider a method "better" when it has greater precision than another method, but remember that most of these methods have never been tested for accuracy under field conditions. Great precision does not mean great accuracy.

12

Use of Remote Cameras in Wildlife Ecology

SHAWN L. LOCKE,
ISRAEL D. PARKER, AND
ROEL R. LOPEZ

INTRODUCTION

USE OF CAMERA TECHNOLOGY is deeply rooted in wildlife research and management. It includes remote monitoring, real-time observations, infrared and ultraviolet analysis, and many other technologies (Fig. 12.1). Perhaps the most familiar use of cameras in wildlife research and management is that of remote cameras. **Remote cameras** (commonly known as game cameras, trail cameras, infrared-triggered cameras, or by trade names) have recently become readily available and popular among hunters and other wildlife enthusiasts. However, remote cameras have been used in ecological research for more than 50 years (Kucera and Barrett 1993). Early remote cameras were custom-built by researchers to record various types of wildlife activities (Gysel and Davis 1956, Dodge and Snyder 1960, Cowardin and Ashe 1965). These rudimentary remote cameras were the precursor to a burgeoning industry of commercially available remote cameras with a variety of field applications in wildlife research and management (Kucera and Barrett 1993) that includes identifying nest predators (e.g., Hernandez et al. 1997, Dreibelbis et al. 2008); studying animal activity and behavior (e.g., Foster and Humphrey 1995, Main and Richardson 2002); estimating animal abundance (e.g, Jacobson et al. 1997, Roberts et al. 2006); and monitoring species occurrence, including rare and endangered species (e.g., Karanth and Nichols 1998, Watts et al. 2008, O'Connell et al. 2011).

Camera technology extends beyond solely remote, illustrating the breadth of its application to wildlife research and management. For example, camera equipment orbits Earth, collecting remote sensing data; records habitat and species data in oceans and freshwater bodies; monitors wildlife behavior in real time; reveals the reflective properties of wildlife pelages; monitors zooplankton and invertebrates; and brings nature to citizen scientists via Internet connections. Cameras in wildlife management and research have advantages and disadvantages that researchers should be aware of prior to starting a project (Table 12.1).

The goal of this chapter is to describe different camera systems and their applications in wildlife ecology. Specifically, we discuss aspects of camera equipment, data storage, and use of various camera systems in wildlife research and management. We include the strengths and weakness of different systems and techniques.

EQUIPMENT AND DATA MANAGEMENT

Data collection using camera systems is dependent on the quality of the equipment and ability of the operator(s). The operator must correctly determine the appropriate

Fig. 12.1. Results from the use of cameras in wildlife research and management. (A–C) Infrared-triggered camera photos. (D) Hawk nest monitoring. (E) Using a peep camera to monitor red-cockaded woodpecker nest. (F) Thermal infrared image of Rio Grande wild turkeys foraging. *(A–C) Photos courtesy of I. Parker; (D) photo courtesy of K. Melton; (E, F) photo courtesy of S. Locke.*

use of and need for camera systems, set up the camera systems for optimal data collection, and adequately maintain the equipment. Additionally, the **effectiveness** of camera system equipment is dependent on one or more of the following: (1) battery life, (2) data storage capacity, and (3) picture quality (i.e., resolution). As technology has improved, cameras have become more efficient, more affordable, and easier to use. A brief review of each of these factors is provided here.

In general, **battery longevity** is a product of multiple factors (e.g., temperature, age, number of pictures taken, flash configuration or presence, and battery quality). These are especially salient factors for remote cameras that are left unattended in the field. Jackson et al. (2005) found that battery life for their film-based remote camera systems exceeded 3 months in the harsh winter conditions of snow leopard (*Uncia uncia*) habitats. Battery life that once limited remote cameras has significantly advanced, and current models can remain operational for up to 150 days or ≥1,000 photos using battery-saving technology. Users can opt for higher capacity batteries (e.g., NiMH or lithium), solar chargers, or additional batteries (i.e., external battery packs) to dramatically increase battery life (Brown and Gehrt 2009).

Data storage continues to improve. Relatively inexpensive storage units hold thousands of pictures and videos, with exact numbers depending on image quality and compression (Newbery and Southwell 2009). This capacity is a vast improvement over film-based cameras, which are generally limited to a maximum of 36 images before film replacement (Parker et al. 2008). Additionally, development costs and storage concerns are virtually nonexistent for digital images. Most costs are incurred during start-up in the form of equipment purchases.

Images recorded by cameras can vary in quality and size, depending on equipment specifications, thereby impacting storage capabilities and data collection opportunities from the images. Low resolution (e.g., insufficient pixel count) and videos require less storage space, but researchers must balance resolution requirements with storage capabilities. Image quality is an important consideration for real-time observation (e.g., peep cameras), data transmission via the Internet (e.g., web-based remote cameras), and infrared cameras. Researchers should determine storage and resolution requirements prior to beginning research or management activities.

It also is important to have a data management plan prior to the initiation of a camera project. Some camera companies have **data management** software that is included with purchase. Photo and video management software also can be purchased separately. However one chooses to organize camera data, it should be done such that collection and retrieval of media are quick and easy. When analyzing media from cameras, it is helpful to have a software package with image enhancement tools that allow the user to zoom in or

Table 12.1. Advantages and disadvantages of cameras in general and of remote and thermal infrared cameras specifically

Camera type	Advantages	Disadvantages
General[a]	Declining costs	Dependent on operator skill
	Miniaturization	
	Increased usability	Maintenance and repair
	Increased build quality	Rapid obsolescence
	Uniform data storage	Replacement expensive
	Flexibility	Equipment storage difficult
		Subject to environment
Remote	Invasiveness reduced	Dependent on human placement
	Consistent monitoring	
	Photo/video evidence	May disrupt behavior
	Simultaneous observation	Subject to failure
	Observer bias reduced	Subject to environment
	Declining costs	Maintenance and repair
	Increasing capabilities	Vandalism
	Consistent observation in rough, inclement, or dangerous areas	Limitations of photographic data
	Observe secretive or aggressive animals	
Thermal infrared	Works well in optimal conditions	Cost
		Detection varies among vegetation structure
	Declining costs	
	Detects spectrum outside of human vision	Animal size impacts detectability
	Increasing utility in wildlife disease study and management	Poikilothermic organisms problematic
		Seasonally dependent

[a] Single lens reflex and digital single lens reflex cameras.

adjust brightness, color, and contrast for optimal picture quality and clarity to aid in photograph interpretation.

REMOTE CAMERAS

Remote cameras are often referred to as game cameras, trail cameras, or infrared-triggered cameras and are widely available and affordable. Remote cameras can be categorized into 2 types: active infrared (AIR) and passive infrared (PIR). **Active infrared** or beam-break sensors use an infrared emitter and receiver, creating a beam of infrared light or trip line. When this beam is broken or tripped, the camera takes a picture or video of the intended target area. Currently, AIR camera systems are manufactured by TrailMaster® (Goodson & Associates, Lenexa, KS; Kays and Slauson 2008). **Passive infrared** sensors detect movement and radiation emitted by animals within a field of view (Kays and Slauson 2008). Therefore, when the sensor detects a moving object with a surface temperature different from its surroundings, the object is captured by photo and/or video (Table 12.2). Both AIR and PIR sensors have advantages and disadvantages that should be acknowledged prior to the start of a project (Table 12.3).

Occupancy and Distribution

Documenting the occupancy and distribution of species—particularly rare, endangered, or elusive species—can be difficult (Moruzzi et al. 2002). Traditional methods for documenting species presence include visual surveys, auditory surveys, track counts, scat identification, hair analysis, detection dogs, drive counts, and trapping. Watts et al. (2008) used remote cameras to document the presence and distribution of endangered Florida Key deer (*Odocoileus virginianus clavium*) on outer islands. Watts et al. (2008) suggested that cameras were a practical method for monitoring Key deer on outer islands. Long et al. (2007b) compared remote cameras, hair snares, and detection dogs for detecting black bears (*Ursus americanus*), fishers (*Martes pennanti*), and bobcats (*Lynx rufus*) in Vermont. Detection dogs were the most effective method for detecting the 3 carnivores, with remote cameras less effective than dogs, but more effective than hair snares. Detection dogs were more costly than the other methods on a per site basis. Remote cameras are an effective means for evaluating the presence of wildlife after a treatment, and they can be used to identify the potential for disease transmission or vaccine delivery.

Wildlife-Crossing Structures

Interactions between vehicles and wildlife pose significant problems. Wildlife–vehicle collisions represent significant physical and monetary dangers to wildlife and drivers. Wildlife–vehicle collisions also can be considered take of threatened or endangered species and thus impact road construction projects and development strategies (Lopez et al. 2003). Additionally, roadways can negatively impact wildlife movement patterns, including dispersal, migration, and corridor connectivity (Jackson 2000). One strategy for reducing such problems is construction of **wildlife-crossing structures** (e.g., overpasses, underpasses, or exclusion fencing) to reconnect areas bisected by roadways and provide safe alternative movement corridors for wildlife (Foster and Humphrey 1995, Ng et al. 2004). These structures must fit in a larger mitigation effort that generally includes exclusion fencing, speed limit alterations, and warning signs. Wildlife crossing structures also must undergo rigorous and sustained monitoring over (possibly) many years to ensure proper function (e.g., wildlife acceptance and use; Hardy et al. 2003, Braden et al. 2008). A popular method for monitoring wildlife crossing structures is remote cameras (Ford et al. 2009). Advantages of remote cameras include continuous operation, full coverage of crossing structure, and minimal intrusion by researchers. Given the extended time periods required for appropriate monitoring, remote cameras are often the preferred method for data collection. Disadvantages include risk of vandalism and natural hazards (e.g., flooding; Box 12.1).

Table 12.2. Specifications for commercially available, passive infrared remote cameras[a]

Brand	Capacity (megapixel)	Flash range (m)	Flash type	Video	Expandable memory	Delays	Sensitivity adjustment	Password protection	MSRP (US$)
BuckEye									
Apollo	0.3–3.1	15+	Both	Yes	SD	0.02–120 min	Yes	N/A	595
Orion	0.3–3.1	15+	IR	Yes	SD	0.02–120 min	Yes	N/A	999
Bushnell									
Trophy Cam	3–5	14	IR	Yes	SD	0–1 min	Yes	N/A	260
Trail Scout	2–5	14	White	Yes	SD	0.5 min	No	Yes	326
Trail Scout Pro	3–7	14	IR	Yes	SD	1 min	Yes	Yes	456
Trail Sentry	2–5	14	IR	No	SD	N/A	No	Yes	140
Camtrakker									
MK-8	1.3–3.2	N/A	Both	No	SD	Yes	N/A	No	430
Cuddeback									
Capture	3	15	White	No	SD	0.5–30 min	No	No	200
Capture IR	1.3–5.0	12	IR	No	SD	0.5–30 min	No	No	250
NoFlash	1.3–3.0	12–18	IR	Yes	CF	1–60 min	Yes	Yes	399
ExPert	3.0	18	White	Yes	CF	1–60 min	Yes	Yes	349
ExCite	2.0	12	White	No	CF	1–60 min	Yes	No	249
Leaf River									
DV-5	5.0	N/A	White	Yes	SD	1–90 min	Yes	No	300
IR-5	5.0	N/A	IR	Yes	SD	1–90 min	Yes	No	330
DV-7SS	7.0	N/A	White	Yes	SD	1–90 min	Yes	No	350
IR-7SS	7.0	N/A	IR	Yes	SD	1–90 min	Yes	No	380
Moultrie									
D-40	4.0	14	White	Yes	SD	1–60 min	N/A	No	120
M-45	4.0	15	White	Yes	SD	1–60 min	N/A	No	290
M-65	6.0	15	White	Yes	SD	1–60 min	N/A	Yes	390
I-40	4.0	15	IR	Yes	SD	1–60 min	N/A	No	216
I-45	4.0	15	IR	Yes	SD	1–60 min	N/A	No	290
I-60	6.0	15	IR	Yes	SD	1–60 min	N/A	Yes	320
I-65	6.0	15	IR	Yes	SD	1–60 min	N/A	Yes	390
Recon Outdoors									
Viper	2.1	N/A	IR	Yes	CF	N/A	N/A	No	230
Extreme	3.0	N/A	IR	Yes	CF	30 sec–60 min	Yes	No	350
Extreme	5.0	N/A	IR	Yes	CF	30 sec–60 min	Yes	No	400
Reconyx									
PC90	3.1	11	IR	Yes	CF	0–60 min	Yes	Yes	800
PC85	3.1	18	IR	Yes	CF	0–60 min	Yes	Yes	700
PM75	1.3	15	IR	Yes	CF	0–60 min	Yes	Yes	600
MC65	3.1	15	IR	Yes	CF	0–5 min	Yes	Yes	550
RC60	3.1	11	IR	Yes	CF	0–5 min	Yes	Yes	600
RC55	3.1	18	IR	Yes	CF	0–5 min	Yes	Yes	550
RC45	1.3	15	IR	Yes	CF	0–5 min	Yes	Yes	450
Stealthcam									
Prowler HD	1.3–8.0	12	IR	Yes	SD	1–59 min	No	No	310
Sniper Pro	1.3–8.0	15	White	Yes	SD	1–59 min	No	No	170
Sniper IR	1.3–5.0	9	IR	Yes	SD	1–59 min	Yes	No	230
Rouge IR	1.3–5.0	12	IR	Yes	SD	1–59 min	Yes	No	160
Nomad IR	1.3–5.0	9	IR	Yes	SD	1–59 min	No	No	180
Wildview									
EZ-Cam	1.3	9	White	No	SD	NA	N/A	No	75
Xtreme 2	2.0	9	White	Yes	SD	1–20 min	N/A	No	90
Xtreme 5	5.0	9	White	Yes	SD	1–20 min	N/A	No	150
Infrared	5.0	N/A	IR	Yes	SD	1–20 min	N/A	No	120

[a] CF = compactflash; IR = infrared; MSRP = manufacturer's suggested retail price; N/A = not available; SD = secure digital.

Table 12.3. Comparisons of active infrared and passive infrared camera systems

Feature	Active infrared	Passive infrared
Size	Two larger units (separate from camera)	One smaller unit (housed with camera)
Models	One company	Many companies
Price	Higher	Lower
Ease of use	More complicated	Simpler
Sensitivity	High (but flexible)	Medium (can be flexible)
Detection beam width	Narrow	Narrow or wide
False triggers	Usually fewer	Usually more
Sensitivity in tropical climates	Not affected by temperature	May be lower
Damage by wildlife	Highly susceptible	Lower risk

Adapted from Kays and Slauson (2008).

Disease Transmission and Vaccine Delivery

Issues of wildlife disease transmission and vaccine delivery are important, but difficult to evaluate. Intra- or inter-species disease transmission studies would benefit from increased knowledge of indirect or direct individual contact (e.g., nuzzling, fecal-oral contact, and site visitation). Vaccine delivery studies often provide vaccines to free-ranging species, but they lack direct knowledge of species visitation rates to the baits or individual bait consumption. Filling in these knowledge deficits would aid disease mitigation strategies and vaccine delivery methods, thus lowering costs and increasing effectiveness. Some of these issues can be addressed with the application of remote camera technology. Although remote cameras cannot always provide the clear evidence demonstrating transmission of disease or uptake of vaccine, they can add data critical for inference. For instance, VerCauteren et. al. (2007a, b) provided moment of contact pictures between farmed and wild cervids, demonstrating possible transmission routes for bovine tuberculosis and chronic wasting disease. Garnett et al. (2002) showed badger (*Taxidea taxus*) visits to feed lots, thus providing the possible bovine tuberculosis connection between domestic animals and wildlife species. Several studies (Gortázar et al. 2008, Jennelle et al. 2009) monitored cervid carcasses for possible conduits of bovine tuberculosis and chronic wasting disease transmission from dead wildlife to scavengers.

The delivery of vaccines to wildlife is often complicated by the presence of multiple species in the focal area, vaccine delivery over very large areas (e.g., air drops), and difficulty assessing success of vaccine delivery. Remote cameras are often used to monitor vaccine delivery systems for species visitation. Wolf et al. (2003) and Campbell and Long (2007) placed remote cameras on baits containing vaccines for rabies. Boulanger et al. (2006) used remote cameras to monitor the effectiveness of a new technique to dispense rabies vaccines to raccoons (*Procyon lotor*).

Estimating Abundance

Reliable **population estimates** are vital in the field of wildlife research and management (Jenkins and Marchington 1969) and require cost-effective and accurate methods (Rob-

---

Box 12.1. Pitfalls of Camera Use in Wildlife Research and Management

1. **Security:** To avoid vandalism by humans and damage by wildlife, researchers and managers should ensure that cameras are concealed and securely attached to a solid substrate. Some manufacturers provide additional security options, such as strong boxes and digital security codes.
2. **Invertebrates:** Invertebrates are often attracted to camera housings for shelter, thus exposing researchers and managers to unexpected bites and stings. Invertebrates also can negatively impact camera electronics. Methods of addressing these concerns include sealing openings (e.g., with tape) and using insecticides or repellants.
3. **Environmental conditions:** Moisture intrusion (e.g., hurricanes or high humidity), fire (e.g., prescribed or wild), and sand intrusion (e.g., dust devils) can damage equipment and data. Camera openings should be sealed or equipment removed from the field prior to storms or fires.
4. **Camera placement:** Shadows, movement of vegetation, and sun-facing cameras are often the cause of misfires. Researchers and managers should face cameras in a northern or southern direction and trim problematic vegetation to avoid misfires.
5. **Nontarget species:** To minimize photographs of nontarget species and maximize those of target species, researchers and managers can use exclusion structures (e.g., fencing), species specific baits, nonconsumable baits (e.g., aromatic baits), or repellants. Additionally, researchers and managers can adjust the sensitivity of cameras to better capture the target species.

erts et al. 2006). Traditional methods for estimating abundance include drive counts, strip counts, line transects, removal methods, and mark–recapture techniques (Chapter 11, This Volume). The use of remote cameras for estimating abundance is based on mark–recapture techniques using Lincoln–Petersen estimators (Sweitzer et al. 2000), although there is increasing use of other techniques (Amstrup et al. 2005). Initial and/or subsequent "captures" are conducted via photographs, and individuals may be marked from initial capture or marked based on physical characteristics (e.g., branched antlers, pelage, or other visible markings or features). Remote cameras have been used to estimate abundance for white-tailed deer (*Odocoileus virginianus*; e.g., Jacobson et al. 1997, Koerth et al. 1997, Roberts et al. 2006), bighorn sheep (*Ovis canadensis*; Jaeger et al. 1991), feral hogs (*Sus scrofa*; Sweitzer et al. 2000), bears (*Ursus* spp.; Mace et al. 1994, Matthews et al. 2008), red fox (*Vulpes vulpes*; Sarmento et al. 2009), and felines (Felidae; e.g., Karanth and Nichols 1998, Heilbrun et al. 2006, Jackson et al. 2006, Dillon and Kelly 2007, Larrucea et al. 2007*a*), among other species.

Demographic and geographic closure is often difficult to attain with highly mobile, wide ranging species. Difficulties with closure can be overcome with remote cameras by using short duration surveys; timing surveys to take advantage of animal behavior; or integrating other technologies, such as radiotelemetry, into the survey. Remote camera studies often use baited stations to maximize photographic captures. Baited camera stations (i.e., convenience sampling) may violate the equal catchability assumption, thereby affecting the accuracy and precision of the estimate (White et al. 1982). Watts et al. (2008) suggested that baiting should be avoided when trying to estimate abundance or the time period when baiting was most significant should be excluded from the survey. Larrucea et al. (2007*b*) concluded that due to animal behavior, remote cameras do not always provide unbiased estimates, and camera placement is important to consider to reduce bias.

Compared to other methods of abundance estimation, remote cameras are attractive. Jacobson et al. (1997) concluded that estimates of adult white-tailed deer bucks could be reliably and accurately estimated using remote cameras, and remote cameras may at least provide managers with a minimum population estimate.

Nest Predation

Remote cameras have become a valuable tool for identifying nest predators in many wildlife studies and applications. Nest predation is an extremely influential aspect of nest survival, particularly among ground nesting birds (Rollins and Carroll 2001, Stephens et al. 2005). Traditional methods for identifying nest predators include physical evidence, such as eggshell fragments or animal sign (e.g., hair, scat, or tracks) recovered at the nest site (Major 1991, Larivière 1999). Physical evidence, however, can be subjective and time sensitive, and it also fails to account for multiple predator events or partial predation events (Leimgruber et al. 1994).

Cutler and Swann (1999) suggested that many researchers preferred remote cameras over traditional methods because photographs provided verifiable evidence of predation events, predator identification, and timing of predation. Using remote cameras, Dreibelbis et al. (2008) determined that multiple predator events were common at Rio Grande wild turkey (*Meleagris gallopavo intermedia*) nests. Little research has been conducted to determine the impact of the presence of remote cameras on nests. The increase of human activity around a nest may disrupt normal nesting patterns or attract or deter certain predators. Leimgruber et al. (1994) found that remote cameras had little impact on artificial ground nests. In contrast, Richardson et al. (2009) found that on average, camera equipment reduced nest predation rates, and they provided several recommendations to minimize the potential bias of remote cameras.

Animal Activity

Complex wildlife activity is difficult to observe and is often influenced by the presence of humans. Remote cameras provide sustained monitoring of wildlife behavior that would be impractical for human observers. Researchers and managers use remote cameras to investigate daily and seasonal **wildlife activity patterns** and use of specific resources (e.g., water sources). Larrucea and Brussard (2009) documented activity patterns of pygmy rabbits (*Brachylagus idahoensis*) and found a bimodal daily activity pattern impacted by season. Several studies have evaluated wildlife use of natural and manmade water sources in arid environments by using remote cameras (Morgart et al. 2005, Whiting et al. 2009).

Diet

Wildlife diets are often measured directly via observation or indirectly via scat analysis, prey remains, or animal harvest (i.e., stomach or crop analysis). Remote cameras offer an alternative form of direct observation with the added advantages of being able to monitor multiple sites simultaneously as well as providing photographic evidence that can be scrutinized at a later date. Franzreb and Hanula (1995) evaluated Trailmaster cameras to quantify the diet of nestling red-cockaded woodpeckers (*Picoides borealis*). Using photographs from the cameras, the researchers were able to identify 65% of the arthropods that adults brought to the nestlings.

THERMAL INFRARED CAMERAS

A limiting factor in studying mammals is observing them (Boonstra et al. 1998). Mammals often can be cryptic or nocturnal, making them difficult to see using only human vision. The use of **thermal infrared imagery** devices can aid researchers by converting the invisible infrared spectrum (0.8–14.0 μm) into a visible spectrum. Essentially, these devices convert surface temperatures of objects into an image visible to the human eye. For several decades, researchers have speculated the use of thermal infrared imagery would aid in detecting and observing mammals. Croon et al. (1968) and Graves et al. (1972) were among the first to use aerial thermal infrared imagery to detect large mammals (e.g., white-tailed deer). Both authors noted that thermal infrared imagery had great potential, but was not without significant limitations, such as the difficulty differentiating the thermal signatures of dense vegetation from wildlife.

More recently, thermal cameras have become more accessible; less costly (although cost is still a limiting factor); and smaller, making them highly portable. They have been used primarily to aid in the detection of large mammals, although several studies have evaluated their use for smaller mammals, such as wild boars (*Sus scrofa*), red foxes, European rabbits (*Oryctolagus cuniculu;* Focardi et al. 2001), and bats (Betke et al. 2008, Horn et al. 2008), as well as Rio Grande wild turkeys (Locke et al. 2006).

Thermal infrared cameras are largely thought to assist in detecting more individuals than do standard techniques, thereby improving estimates of density. However, the uses of thermal infrared cameras have expanded beyond density estimation. Infrared cameras have been used as a noninvasive method for detecting diseased mammals. Dunbar and MacCarthy (2006) were able to experimentally detect clinical signs of rabies in raccoons using this technology. Infrared cameras also were used to identify mule deer (*Odocoileus hemionus*) suspected of being infected with foot-and-mouth disease (Dunbar et al. 2009). Researchers have used infrared cameras to better understand thermoregulatory processes via thermal windows in the world's largest terrestrial animal, the African elephant (*Loxodonta africana*).

INNOVATIVE CAMERA TECHNIQUES

Improvements in component miniaturization and capability, storage capacity, build quality, and price have spurred the use of cameras (both still and video) in ecology in a variety of new directions. Cameras are increasingly common in habitat monitoring studies, Internet-based research and outreach, and evaluation of management activities.

Companies are now designing camera (both still and video) systems to answer specific questions. For example, Fuhrman Diversified (Seabrook, TX; R. Fuhrman, Fuhrman Diversified, personal communication) has designed and manufactured 850 video systems for various field, laboratory, educational, interactive, industrial, and scientific applications throughout the world. Rather than using existing cameras systems, many researchers are opting to have custom camera systems designed and manufactured to answer their specific research needs.

Camera monitoring now provides data from a variety of perspectives. Some of these cameras are becoming increasingly interactive and have the ability to disseminate real-time information to classrooms, researchers, and the general public over the Internet, with some providing the ability to tilt, pan, zoom, and otherwise control the cameras (Connolly 2007). State and federal natural resource agencies and nongovernmental conservation organizations provide **live streaming video** and photographs of a variety of wildlife species (e.g., bald eagles [*Haliaeetus leucocephalus*], grizzly bears [*Ursus arctos horribilis*], and barn owls [*Tyto alba*]).

Even as these broad-based initiatives expand the use of cameras beyond traditional wildlife monitoring, the technology continues to evolve and allows researchers to think outside the normal technological paradigm. For example, researchers have mounted cameras on remotely controlled model airplanes (Thome and Thome 2000, Jones 2003), on flexible tubing for burrow and den monitoring (VerCauteren et al. 2002), on blimps (Murden and Risenhower 2000), on floating platforms (Lopez and Silvy 1999), and on satellites (Mehner et al. 2004). Researcher innovations can serve to expand the range of observations, save money, and decrease disturbance of target wildlife species. They also are expanding observation into alternative wavelengths (e.g., infrared or ultraviolet) outside the normal visible spectrum using new types of detectors. For instance, many avian species reflect ultraviolet radiation (Keyser and Hill 1999). Without specialized equipment (i.e., spectrometer), this type of information remained unknown. Alternatives to these technologies have historically required the use of expensive fixed-wing aircraft or helicopters, loud and destructive excavations or intrusions, or the reduction of available data. Limitations inherent to emerging civilian (i.e., nonmilitary) technologies (e.g., relatively short transmission distance for radiocontrolled airplanes or high monetary cost) prevent these techniques from gaining wide use; however, researchers continue to explore these and other methodologies.

Cameras are often used to monitor wildlife when the physical presence of a human would disrupt behavior or prove impractical or dangerous. For instance, **peep cameras** (closed circuit cameras on extendable poles) are commonly used to view the interior of red-cockaded woodpecker nest cavities as the viewer stays safely on the ground (Richardson et al. 1999). These cameras obviate the need to climb the tree, thereby minimizing impacts on bird behavior and exposure of personnel to dangerous conditions. Additionally, cameras have been modified to enter burrows and in some cases are coupled with grappling devices to manipulate objects inside (VerCauteren et al. 2002).

SUMMARY

As cameras and camera equipment become less expensive, better built, increasingly capable, and more user friendly, they are more common and valuable in wildlife research and management. Cameras are a useful tool in wildlife ecology, but their usefulness depends on the quality of the study design and capabilities of the operator. Cameras are appropriate in research where: (1) humans would cause disturbance to wildlife behavior; (2) extended observational periods are required; (3) observation must take place in dangerous, inclement, or remote areas; (4) permanent and verifiable data are needed; or (5) different capabilities from those of the human eye are required. The heterogeneity of ecological research is reflected in its varied uses of cameras and continues to evolve to meet new research challenges.

13

Radar Techniques for Wildlife Research

RONALD P. LARKIN
AND ROBERT H. DIEHL

INTRODUCTION

DURING WORLD WAR II, English ornithologists found the new secret weapon known as RADAR (RAdio Detection And Ranging), while looking for ships and aircraft, was receiving echoes from gannets (*Sula* spp.) and other birds (Lack and Varley 1945:446). They noted that birds gave rise to "several [torpedo boat] scares and at least 1 invasion alarm" and "getting visual confirmation of the source of the echo" was difficult. These pioneers immediately recognized that radar was a powerful tool for monitoring and studying movements of flying animals, providing one could interpret the radar information. These themes reverberate through this introduction to radar as a tool in wildlife conservation and ecology.

Radar is distinct from radiotracking and aerial and satellite remote sensing. Radar operates in a different band of the electromagnetic spectrum and mostly relies on different physical principles. Radiotracking involves placing an active (powered) electronic device on an animal and then locating the signal from that device by direction finding or other means. Remote sensing in wildlife biology usually involves visual or infrared data obtained passively from satellites or high-flying aircraft. Radar directs a high-energy beam, and some of that energy is reflected back from objects, in this case flying animals (Eastwood 1967, Vaughn 1985, Bruderer 1997). Flying animals need no electronics mounted on them, as their bodies reflect the radar beam. Further, the subjects almost certainly do not know they are being observed (Bruderer et al. 1999).

Using radar, wildlife biologists can observe birds and bats flying above vegetation but not in or near it, especially vegetation being moved by the wind. However, small tripod-mounted radars are used routinely by the military to detect moving soldiers on the ground. Radar can observe animals on extended flat surfaces, such as runways or the surface of calm water (Radford et al. 1994). Most large radars have the power and sensitivity to detect birds at great distances when the birds are in the open and can be reached by the radar beam. However, the largest radars are limited in the distance they can observe flying animals, especially those flying at low height above ground level (AGL). Failure to detect low-flying animals usually happens because Earth, and anything bound to it, curves from under the radar beam (Fig. 13.1) or topography prevents a clear view of the animals.

A **radar display** does not reveal which kind of animal produces a radar echo, and without specialized research to relate animals to echoes (ground truth), radar does not directly allow a wildlife scientist or manager to know how many animals

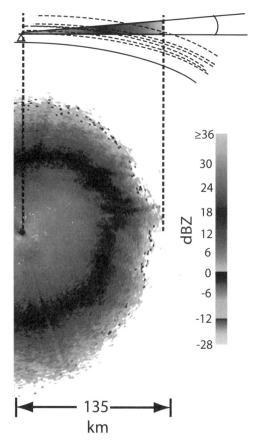

Fig. 13.1. Top: The radar beam detects migrating birds (and almost certainly some insects and bats), then at greater range it passes completely over the layer of animals. Brighter colors represent stronger echoes. Bottom: East half of a map display of a weather surveillance 88 Doppler radar near the middle of the night. Just before the beam passes completely over the biological layer, the lower periphery of the beam encounters only the highest tail of a distribution of birds, causing the radar to register relatively weak echoes (yellow).

are responsible for an echo. These limitations can be frustrating. However, radar allows following animals through the blackest night, inside clouds (Griffin 1973, Larkin and Frase 1988), and occasionally at great distances, contributing to knowledge of animal movements (Alerstam 1996). Further, the technology is becoming more widely available. With minimal computing resources, one can download radar data about every 5–10 minutes from much of the continental United States almost as soon as the radars record the data. The data are free or available at negligible cost. The technology has enormous potential for use and misuse.

Meteorologists, aviation agencies, maritime users, and the military operate radars useful for observing wildlife. Those who want to observe wildlife with radar are encouraged to try it. One should not necessarily believe radar operators, who may have been taught that birds or bats cannot be "detected" with their equipment. For instance, some time ago the director of a sophisticated radar installation looked over the shoulder of an ecologist sitting at a console of the installation's best radar. "Oh, yes," he remarked, looking at the radar echoes filling the large screen, "atmospheric inhomogeneities." Several days later, after the ecologists had shown him "atmospheric inhomogeneities" flapping their wings on radar displays and zooming past the radar considerably faster than the wind, the director quietly admitted that yes, they might be birds. Most radar biologists can relate similar stories.

RADAR 101

This discussion assumes that the radar uses the same antenna to send (transmit) the radar signal and sense (receive) the returning echo and that the radar's operational frequencies are microwaves. The reader is referred to standard texts for further details (Eastwood 1967, Woolcock 1985, Levanon 1988, Skolnik 1990).

How much radar echo a bird or bat produces is determined by a ratio in the **Radar Equation** (Box 13.1). Knowledge of the Radar Equation is useful to understand and be conversant about radar. To help understand the Radar Equation, consider a person shouting across and canyon and listening for an echo. P_r is the loudness of the echo. Technically, this **echo** or "radar return" is back-scattered radiation. Larger values of P_r give more intense dots or brighter colors on a radar display. In simple radar, P_r below the radar's threshold sensitivity is indistinguishable from noise. P_t is roughly the loudness of the person's shout and is usually constant for any given radar. G is the gain, a dimensionless ratio usually stated in decibels (dB, a logarithmic measure calculated as $10\log[\text{ratio}]$). When the person cups a hand behind her ear to hear the faint echo, she experiences an increase in loudness of the sound, or gain. Positive gain occurs only forward (directivity). Large radar antennas can

---

BOX 13.1. THE RADAR EQUATION

$$P_r = \frac{P_t G^2 \lambda^2 \sigma}{(4\pi)^3 R^4},$$

where

P_r = received power (W) from the echo
P_t = radar transmitted power
G = antenna gain, or amplification
λ = wavelength (m) of the radar
σ = radar cross-section (m²)
R = range (straight-line distance; m) to the target.

produce gains in excess of 40 dB (10^4 or 10,000 times). The gain of a microwave antenna is proportional to its frontal area. **Microwaves,** like light (or even the sounds produced by someone shouting across a canyon), can be described in terms of **wavelength (λ),** the distance between successive troughs (or crests) in a traveling wave. Wavelength is as important for radar as transmitter frequency is for radiotracking and color for visual observation. Radar wavelengths include **X-band** (North Atlantic Treaty Organization [NATO] designation I band, about 3 cm), **C-band** (NATO designation G and H, about 5 cm) and **S-band** (NATO designation E and F, about 10 cm). **L-band,** used in aircraft surveillance, is longer than S-band. **K-band** (NATO designation J and K) is shorter than X-band and is becoming quite widely used in such applications as automotive radar. Most radar used with wildlife operates at a single wavelength. One can tell which band radar uses from the size of its waveguide, the metal tubing or "plumbing" that is used in place of wire for conducting microwave energy (Fig. 13.2).

Radar wavelengths are in the same size range as body parts of bats, birds, and even large insects. In this size region, microwaves wrap around objects and otherwise interact with them in a complex fashion. Thus, the amount of echo from even a simple object, such as a sphere, is not exactly proportional to its size (Fig. 13.3). Unfortunately, except for small animals observed at relatively long wavelength and large ones observed at relatively short wavelength, birds and bats mostly lie in the nonmonotonic region (from the middle to the right in Fig 13.3). This implication is profound, as physically larger animals do not necessarily generate stronger echoes. For instance, body parts of size $\lambda/2$ will resonate on radar, producing intense echoes.

Effective **target area (σ)** on radar differs for different wavelengths (Fig 13.3). The smooth left end of the curve represents small insects and becomes linear as it extends down to

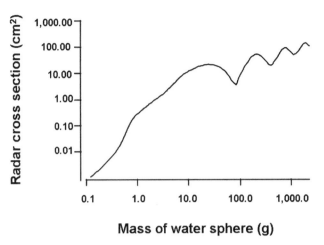

Fig. 13.3. The nonlinear relationship between apparent size of a target on radar (vertical axis) and actual mass (or volume) of the target, at 10-cm wavelength (S-band). *Redrawn from Vaughn (1985).*

include tiny radar **scatterers,** such as raindrops, cloud droplets, and dust. The middle region of the curve, where the amount of radar echo is not directly proportional to target size, represents most bats and birds. The right end of the curve lies near the mass of a goose. Wildlife appears on radar like skin-enclosed water, which is 0.56 as reflective as metal (Eastwood 1967). Engineers conclude that poorly conducting tissues, such as feathers and chitin, are essentially transparent to radar (Edwards and Houghton 1959). Radars cannot measure absolute target area of wildlife with great accuracy, because even in carefully controlled test facilities, moving targets vary by approximately 2 dB or 60% (Dybdal 1987).

Range or slant range (R) is distance along a radar's beam to a target and is part of the acronym RADAR. The speed of light being constant, pulsed radars measure range by timing the delay between transmitting a pulse and receiving an echo off of a target. Because energy spreads in 2 dimensions as it travels from the radar to a flying animal and again in 2 dimensions as it travels back from the animal to the radar, the received echo strength is proportional to the inverse fourth power of range. The maximum range at which an animal of a certain size can be detected is

$$R_{max} = \left[\frac{P_t G^2 \lambda^2 \sigma}{(4\pi)^3 P_{r,min}}\right]^{1/4}.$$

This short introduction to the Radar Equation provides a great deal of information. For instance, consider designing a radar for observing flying animals over a wide area. Most targets will be at a great distance, R, where the returning echo will be weak. Noting the strong nonlinear influence of the inverse fourth power of range, we see that several variables have linear effects and, therefore, cannot make much difference. For instance, from the Radar Equation, a more

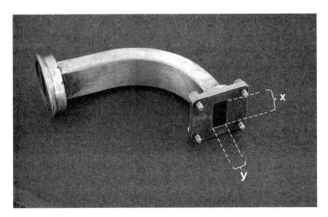

Fig. 13.2. A piece of rectangular waveguide. Microwave energy travels on the inside surface of the tube, which is machined to tight specifications and must be smooth, clean, and dry. Usually 0.9 wavelengths > x > 0.6 wavelengths, and y = 0.5x.

sensitive receiver (able to detect fainter echoes of P_r) will not provide much greater range, nor will size of birds or bats that we want to observe (σ) make much difference. However, the term G, **antenna gain,** in the numerator of the Radar Equation is itself proportional to antenna diameter squared, so that G^2 is proportional to diameter^4. Therefore, increasing antenna diameter increases P_r to the fourth power, which can compensate for R^{-4}. A **large antenna** is necessary to work at long ranges. In addition, if the wildlife biologist wishes to be sure to observe subjects through rain or snow or to minimize confusing echoes from clouds of flying insects, it will help to use longer wavelengths to ensure that interfering echoes are toward the left edge of Figure 13.3. A longer wavelength also will provide more range directly. A large metal antenna for long wavelengths with motors and apparatus to direct it is expensive, comprising up to 40% of the radar's cost. The expense of large radar systems may be daunting, considering most wildlife budgets. In that case, leasing appropriate radar equipment or using already existing radars are alternatives to purchase. Alternatively, wildlife biologists can learn how to obtain data from existing local aviation, weather, maritime, or military radar. Thus, they can invest time in understanding available data rather than building radars and designing the necessary software to gather the data.

ANTENNAS AND SCANNING

A radar antenna's main lobe, or **beam** (Fig. 13.4) points at targets, and its direction, along with range, gives their location. Direction is expressed in polar coordinates familiar to foresters and others (Fig. 13.5). Returning echoes from birds, bats, and other objects take the reverse path, and the antenna concentrates received echoes the same way it concentrates transmitted microwave energy. Large antennas may have beams as narrow as 1°.

Side lobes are weaker concentrations of energy that are, to some extent, symmetrical about the main lobe (Fig. 13.4). All **directional antennas** have side lobes, and they matter. For instance, an X-band radar for field work (Bluestein and Pazmany 2000) has side lobes 22 dB (159×) weaker than the main lobe. Using the R^4 relationship, $10^{2.2/4}$ is a factor of 3.5 and corresponds to a range 30% of that of a bird in the main lobe. So if the antenna is pointing at a bird, another bird of similar size illuminated by a side lobe in a different (deceptive) direction will appear equally prominent on the radar display if it is as close as 0.3R.

Spillover radiation includes any energy that escapes past the edge of the antenna (Fig. 13.4). Like side lobes, spillover radiation produces spurious echoes from the ground, structures, and vegetation (ground clutter) in directions different from the main lobe.

Only a few radar beam shapes are commonly used in wildlife biology (Fig. 13.6). Many radars used in meteorol-

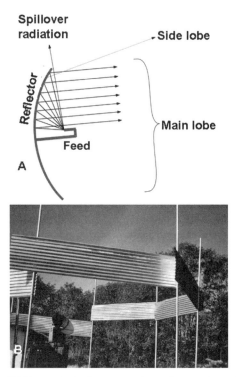

Fig. 13.4. (A) The 3 principal concentrations of energy in a reflector antenna. Microwave energy emerges from the feed, which directs it toward a solid or mesh reflector, in this case a parabolic reflector. (B) A parabolic reflector antenna with a radar fence constructed of 3.04-m (10 ft.) lengths of corrugated sheet metal.

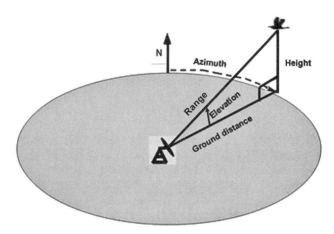

Fig. 13.5. The polar coordinate system used by radars consists of azimuth (angle from north), elevation (angle up from horizontal), and range.

ogy (Fig. 13.6) have a narrow, conical **pencil beam** to provide height information (e.g., on storms). A large radar of this type can operate at great range, but its resolution is coarse even though its beam might be narrow (Wurman et al. 1997). For instance, a 1° beam is 1 km across at a range of approximately 60 km. This type of radar must rotate several times with its antenna at different elevations (a **volume scan**) to achieve coverage of different heights. Smaller, specialized ornithological research radars use spatially precise conical beams to obtain information on animals at specific heights, often even single flocks and individual animals (Larkin 1982, Cooper et al. 1991, Bruderer et al. 1995).

Surveillance radar is any radar designed to scan a wide geographic area repeatedly, usually in the horizontal plane. Some surveillance radars use a beam that is narrow (often 1–2°) in azimuth but broad in elevation (Fig. 13.6); marine radar beams are shaped this way to detect objects on the surface even while the radar platform is pitching and rolling (Williams 1984). These radars provide almost no information on height of flying animals. Animals near the edges of the beam are partially illuminated, and the size of the echo they produce depends on their exact position in the beam. Paths of observed targets flying at high elevation on these radars are distorted (Cohen and Williams 1980). The antennas are usually swiveled (rotated in azimuth) through 360°, but in wildlife biology, they also can be held stationary to monitor bird or bat traffic passing through the beam (Fig. 13.7). Surveillance radars designed to find or follow aircraft

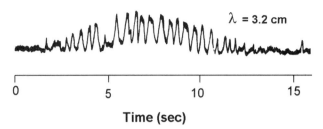

Fig. 13.7. Echo strength from a single flying animal passing through the beam of a stationary vertically pointing pencil-beam radar. Overall echo strength increases as the target comes into the beam, peaks as it flies through the beam's center, and decreases again as it exits. *Adapted from Atlas (1965).*

have even broader beams in elevation to obtain information on fast-moving targets at various heights (Fig. 13.6).

A conical scanning radar antenna may or may not have a **conical beam** and rotates in azimuth while elevation is held constant (Fig. 13.6). A nearly horizontal conical scan generates a **plan position indicator** (**PPI**) display that is projected onto the earth as a map (Fig. 13.1, bottom). This is the display shown in weather forecasts, with stronger targets coded as more intense spots or brighter colors. A PPI scan performed like a windshield wiper (360°) is called a sector scan. At times, a conical beam is held stationary and animals are counted as they fly through the beam (Crawford 1949, Larkin 1982, Korschgen et al. 1984, Smith and Riley 1996). Specially constructed radar that looks vertically and spins rapidly in azimuth provided useful data on insects with simple wing beat patterns (Riley and Reynolds 1979).

Biologists may use conical-beam radars in conjunction with surveillance radars (Cooper et al. 1991). Some radars have a **stacked-beam** arrangement in which several stationary narrow beams are arrayed vertically to provide height information as the array is swiveled in azimuth. **Vertically scanning** radars (Fig. 13.6C) intercept animals crossing the plane of elevation through which the radars scan. Radar antennas are designed to scan sufficiently slowly to receive multiple echoes from each target, reducing the effects of many kinds of noise and clutter, yet sufficiently fast to provide information that is timely. Similarly, **long radar pulses** give stronger echoes, whereas **short pulses** give greater detail in range. In wildlife biology, advantages of rapid data updates and fine spatial detail may outweigh need for detecting weak echoes.

TYPES OF RADAR

A wildlife scientist or manager new to radar should understand what data can be obtained and how the data should be collected to best incorporate the technology into a study design. In this chapter, types of radars, the data they can collect, and their limitations (Richardson 1979, Kelly 2000; Table 13.1) are reviewed.

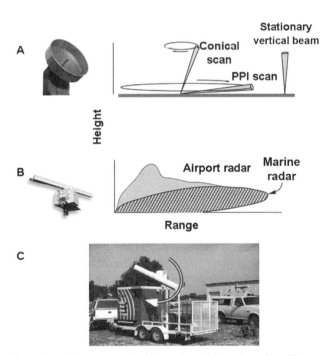

Fig. 13.6. (A) A narrow conical beam (pencil beam) produced by a small feed evenly illuminating the surface of a paraboloid. The antenna swivels in azimuth and tilts in elevation. A short cylindrical cuff partly shields the paraboloid from ground clutter. (B) A view of beams narrow in azimuth but wide in elevation. The hatched pattern is produced by a slotted waveguide antenna typical of marine radars (photo shown), and the additional shaded region at high elevation is typical of airport surveillance radars (not shown). (C) A marine radar modified to perform a vertical scan (arrow).

Table 13.1. Types of radar data collection methods

Radar type[a]	Spatial data collected[b]	Typical maximum range in a terrestrial environment	Spatial	Normal application resolution
S-band marine surveillance radar	X-Y, T, M, track, speed	30 km at 30 kW peak power	Moderate	Measuring track, behavior, and habitat use
X-band marine surveillance radar	X-Y, T, M, track, speed, estimate of target size	5.5 km at 25 kW peak power	High	Measuring track, behavior, and habitat use
X-band stationary beam modified marine radar	Z, D, T, M, track, speed estimate of target size	2.4 km at 25 kW peak power	High	Measuring height and wing beats at one location
X-band conical scan modified marine radar	Z, D, T, M, heading, speed, estimate of target size	2.4 km at 25 kW peak power	High	Measuring height and track at one location
Vertical scan modified marine radar	X-Y, Z, D, T, M, estimate of target size	2.4 km at 25 kW peak power	High	Measuring height and rate of passage across a line
Tracking radar	X-Y, Z, T, M, accurate estimate of target size, heading derived from X-Y and Z	1.5 km at 40 kW peak power	High[c] (X-band) to 80 km at 5 MW (S-band)	Measuring track details, wing beats, rate of climb, and impacts of stimuli
Police or anti-personnel radar	Radial speed and Doppler spectra	<<1 km	N/A[d]	Measuring speed and wing beats
Doppler weather radar (NEXRAD [WSR-88D])	X-Y, T, M, 230 km reflectivity used to estimate aggregate target size, velocity used to derive track		Low, 0.25 km	Measuring track and habitat use; measuring heading if uniform

Adapted from Kays and Slauson (2008).

[a] NEXRAD = Next generation weather radar; WSR-88D = Weather Surveillance Radar 88 Doppler.

[b] D = density (number of birds in a volume of airspace); M = metadata (additional observations, e.g., location or environmental conditions at the time of the observation); T = time observation made; X-Y = coordinates over the ground; Z = height.

[c] Approaching 1 m precision.

[d] N/A = not applicable.

Marine Radar

Marine radars on boats and ocean vessels track other vessels; detect weather; aid in navigating land hazards; and in the fishing industry, spot birds feeding on large schools of fish. Marine radar can be used to record the horizontal tracks of birds as they move through an area, including their size, speed, track, and position. These radars can precisely register the shape of large flocks of birds (Williams et al. 1976) and can be mounted to point upward to measure height, size, and numbers of flying animals passing overhead.

Marine radar has become the most common type used to detect bird targets. These radars are available from marine suppliers, are relatively inexpensive, and most importantly, are reliable and built to operate in the punishing marine environment.

Although marine radars straight "out of the box" can detect vertebrates, some may not have the best characteristics for collecting data on vertebrates. This shortcoming has led radar ornithologists to modify marine radars to better suit their needs. A factory-made **slotted-waveguide antenna** for marine radars has a horizontally narrow (approx. 1–2°) and vertically wide (20–30°) beam for operation on a vessel that is pitching and rolling (Fig. 13.6). Wide beams can be undesirable for bird detection. Greater beam width dissipates the power of radar over a larger volume, reducing the range at which an object can be detected and increasing uncertainty in the object's 3-dimensional location. This limitation can be partly overcome by using a more powerful transmitter or, more effectively, by replacing the slotted antenna with a parabolic dish antenna that concentrates energy into a narrower beam (1–4°). Suitable parabolic antennas with appropriate waveguides can be found as military surplus items.

The other limitation of most marine radar slotted-waveguide antennas, as they come from the manufacturer, is their prominent side lobes. Side lobes are insignificant at sea, where the radar gain is usually set low, because it is unlikely that another ship will be immediately adjacent to the radar. However, on land, side lobes detect the ground, trees, buildings, vehicles, and even people. Stationary objects create echoes that do not move from scan to scan, whereas moving targets create echoes that appear and disappear. When side lobes are reflected, the resulting multipath signal can appear on the radar display at an incorrect range, reducing the useful area of the radar display.

Side lobe echoes can be reduced by careful antenna design; however, the cost of a custom built antenna is prohibitive for most wildlife studies. To reduce both side lobes and ground clutter in a terrestrial environment, radar ornithologists elevate the antenna or shield it with radar absorbing or reflecting materials, such as aluminum, radar absorbing foam, radar fences (Fig. 13.4B), pits or earth berms (Fig. 13.8), or the hard edges of thick vegetation.

Level marine radar directs some of its beam below the horizontal to illuminate the horizon when a ship pitches and rolls in high seas. For land-based radar, this wasted power generates ground clutter, produces multipath echoes, and exacerbates the problems of side lobes. One can rotate the antenna in elevation by attaching a bracket that pushes the front mount of the antenna up by the number of degrees required to place the lower edge of the radar beam parallel to the horizon. However, many marine types of radar are built with rigid metal waveguides that cannot readily be modified. Replacing a portion of the rigid waveguide (Fig. 13.2) with a compatible length of flexible waveguide may help.

Marine radar is a cost effective, reliable solution that has proven to be adaptable to many survey techniques. Its compactness permits flexibility and allows mounting radar on an extensible boom (Cooper and Blaha 2002). It has been modified to be pencil beam radar, conical scanning radar, or vertical scanning radar. The key is to select a scanning technique that fits the data to be collected.

Doppler "Weather" Radar

Doppler radar usually refers to large weather radars, but this name is restrictive, because they also are excellent wildlife radars. Older surveillance weather radars that do not use the Doppler principle have been largely phased out in technically advanced countries, for example, in North America and Europe. Doppler weather radars sharply contrast with modified marine radar for use in ornithology (Gauthreaux and Belser 1998, Koistinen 2000). They have lower spatial resolution, significantly higher power, longer range, and highly sensitive receivers; they are expensive and generally not portable, and, via networking, usually send data rapidly to a central location for display and archiving. Fortunately, data from existing radars can be obtained easily and cheaply.

Current Doppler radars generally produce 3 basic data types: reflectivity, radial velocity, and spectral width. **Reflectivity** is a measure of the amount of energy returned to the radar by a target (P_r), although these radars usually sum the amount of echo caused by multiple targets into a single value. Brighter colors represent more echo (Fig. 13.1), whereas ground clutter produces intense echoes at the center (white and mauve; Fig. 13.1). Reflectivity data are available to 460-km range on U.S. weather radars. The real power of a Doppler unit over non-Doppler radar is the additional data from measuring the Doppler shift produced by targets. **Radial velocity** is a measure of target motion toward or away from the radar (velocity is used synonymously with speed). When a target moves tangentially to the radar, the Doppler shift and radial velocity decrease to zero (Fig. 13.9). **Spectral width** is a measure of variation in radial velocity. Although little used by meteorologists, spectral width is useful for biological targets. Doppler velocity and spectral width data are available to 230-km range on U.S. weather radars.

Data networks exist to deliver meteorological data, including radar data. The primary source of weather radar data in the United States is the national network of **NEXRAD radars** (for NEXt generation weather RADar; Diehl and Larkin 2005), but data also are available from Terminal Doppler Weather Radar in Europe (Dokter et al. 2010), and increasingly from privately owned weather radars. It also might be possible for research and academic institutions to access data from experimental and research radars.

The NEXRAD radar system is composed of WSR-88D radars (Weather Surveillance Radar 88 Doppler), which store

Fig. 13.8. A radar is sited in a gravel pit behind an earth barrier to reduce ground clutter during observations of wildlife flying at low height over a ridge. The operators positioned the trailer-mounted antenna so that the radar's pencil beam can point low over the ridge in the background, while the earth berm 30 m away shields the radar from side lobe reflections off the ridge itself. The aluminum cuff around the antenna further reduces return from side lobes.

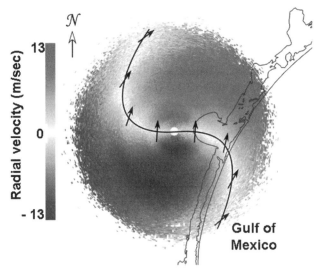

Fig. 13.9. A target moving tangentially to the radar has a Doppler shift and radial velocity of zero. In this typical autumn nighttime Doppler image of migrants from the Corpus Christi, Texas, weather surveillance 88 Doppler radar echoes with negative velocities (blue) approach the radar, positive velocities (red) recede, and a zero velocity line (white) is perpendicular to the direction of travel of migrants at each range on either side of the radar. Birds at farther ranges are higher and are flying more toward the northeast.

user data primarily in 2 formats (Crum et al. 1993). The more comprehensive NEXRAD data format is Level II data, which is polar and fully encodes reflectivity, radial velocity, and spectrum width at their highest spatial and data resolutions. Reflectivity is available in 0.5-dB increments, velocity and spectrum width in 0.5-m/sec increments. A recent upgrade increases spatial resolution of reflectivity 8× and that of velocity and spectrum width 2× (Torres and Curtis 2007). Level III data is a lower resolution version of Level II data and is available in polar or Cartesian formats. Level III reflectivity data are available in either of 2 partially overlapping modes: clear air mode, which emphasizes weaker echoes in 4-dB increments, or precipitation modes for stronger echoes in 5-dB increments. These modes together span much of the radar's full range of reflectivity (−28 dBZ to 75 dBZ, a measure of signal strength used in meteorology).

Several kinds of special artifacts are included with the greater amount of information and wider spatial coverage available from large Doppler radars. The details of Doppler data are beyond the scope of this chapter, but users should be familiar with second-trip echoes, false "wrapped" velocities, malfunctioning automated attempts to unwrap velocities, and other situations discussed by Doviak and Zrnic (1993), Diehl and Larkin (2005), and Rinehart (2010). Radars in the NEXRAD system (WSR-88D radars) presently reject point targets, such as individual soaring turkey vultures (*Cathartes aura*), a widespread, significant low-level hazard to aviation (Defusco 1995).

Airport Surveillance Radar (ASR)

Aircraft are routinely detected with large, often fast-rotating S- or L-band radars with beams that are broad in elevation (Fig. 13.6B). Many older but capable and high-power ASRs are still in service and can detect flying animals, especially larger birds (Nisbet 1963, Gauthreaux 1974, Alfiya 1995). Some newer, lower-power ASRs are specialized for following aircraft equipped with transponders that actively transmit microwave pulses; consequently, they may lack the sensitivity to be of use for studying wildlife beyond the immediate airport environs. **Airport surveillance radars** can be so specialized for large aircraft that they have circuitry to block smaller targets, such as small numbers of birds. Lack of height information also is a problem with these radars; one study astutely circumvented that limitation by comparing data from a sea level ASR radar and another nearby ASR radar placed 1,200 m higher (Williams et al. 1986).

Tracking Radar

Military surplus tracking radars can track a single bird target or flock and map its trajectory (Griffin 1973, Able 1977, Larkin 1978, Bruderer 1994, Bruderer et al. 1995, Liechti and Bruderer 1995). A powerful tracking radar followed single birds at a range of 80 km (Williams et al. 1972). These radars are flexible and can be useful to understand flight dynamics and flocking behavior (Larkin and Szafoni 2008) of birds and have been fitted with telephoto cameras to aid in target identification.

Military radars are designed to be robust, and some fit into artillery shells. However, they are state-of-the-art designs and are often produced in small quantities. Thus, their anticipated reliability might or might not match their actual performance in the field. Capable military radars are expected to appear on the used equipment market in the next decade.

Other Types of Radar

Police (traffic) and military antipersonnel radar use Doppler to detect moving targets in a cluttered environment but provide no information on target range, because they send and receive a fixed-frequency signal continuously rather than in pulsed form. Provided that wind speed and direction are measured accurately, measurement of speeds of birds with police radars can be converted into flight speeds and energetic estimates (Schnell 1965, Schnell and Hellack 1978, Evans and Drickamer 1994, Brigham et al. 1998). **Military antipersonnel radar** appears to function similarly to traffic radar (Martinson 1973). Small stationary radars are used routinely for monitoring traffic flow and, within a few years, most vehicles sold will include tiny (λ shorter than K-band) radars mounted on their front surfaces to serve safety, speed control, and parking functions (Spectrum Planning Team 2001). **Automotive radar** will require much lower price structures and production volumes that are 1,000× greater than any other existing radar application. Consequently, wildlife biologists might use **pinpoint radars** as sensors for camera traps and arrays of radars to investigate such clutter-plagued problems as bird kills at tall structures. **Airborne radar** has been used to monitor insects (Hobbs and Wolf 1996) and can detect birds (Graber and Hassler 1962). Miniature airborne radar might fit into flying "microvehicles" the size of a bird or insect (Fontana et al. 2002).

Harmonic radar is a specialized research technology using a directional transmitter and receiver at different frequencies combined with miniature tags to convert the transmitted frequency to the received frequency (Mascanzoni and Wallin 1986, Riley et al. 1998). A radar transponder requires a larger, active (powered) circuit (French and Priede 1992). Both have potential for following tagged subjects, even terrestrial wildlife, at useful ranges when line-of-sight visibility is possible.

ACQUIRING RADAR DATA

Wildlife researchers and managers work either with existing radar equipment operated by another agency or with dedicated wildlife radar. Whatever the source of data, obtaining information useful for statistical analysis, model development, data visualization, and management is the key to radar study of wildlife as a science rather than as a curiosity.

One can use existing equipment by borrowing a radar, visiting a radar facility, hiring a consulting company to gather radar data on wildlife, or by acquiring already archived data on computer media or from the Internet. Using existing stationary equipment often avoids the cost of acquiring, operating, and maintaining equipment. Frequently, someone is available to help instruct and interpret the data. **Existing radars** can often be directed quickly on a problem, which is an enormous advantage. However, existing radars were not placed with wildlife in mind, and their managers may not be amenable to modifying them or their operation to help a visiting wildlife scientist.

Purchase of a **new or used radar** system or assembly of a radar dedicated to biological studies can be productive. Most work with small- to medium-scale radar has involved enterprising biologists who were unafraid of new tools, new skills, and new collaborators. However, overenthusiastic wildlife professionals place themselves in danger of metamorphosing from wildlife managers or field scientists into low skilled, poorly paid engineers. The purchase cost of a radar that is truly suitable for acquiring the needed data can be too expensive, and months or years can be wasted trying to make cheaper, unsuitable purchased equipment do a job that is beyond its capability. Furthermore, biologists need good engineering advice before attempting to modify or troubleshoot radiofrequency parts, including the antenna, feed, waveguide, magnetron, receiver front end, or **microwave integrated circuit.** Microwave components need to be clean, dry, and adjusted to tight tolerances to operate well. An untrained biologist may do more harm than good. For a short-duration research project or a feasibility study, lease or rental of a radar from a dealer, radar manufacturer, or consulting firm should be considered.

Widespread use of marine radar is a recent development, and a few companies now market **bird radars** built around off-the-shelf hardware. That these products offer ready access to radar technology is an attractive prospect to new users, but there has been no calibration or evaluative comparison of them (Schmaljohann et al. 2008). Prospective consumers wanting to purchase an off-the-shelf radar or contract for the services of a consulting firm to operate a radar should be sufficiently aware of the issues surrounding application of radar to ensure that units perform as needed. In addition, they should look for peer-reviewed publications describing the performance of the units in real-world conditions.

The inside of working radar is not safe. Externally, when working near operating marine radars, one should **stay away** from the motor-driven antenna both for physical safety and as a general precaution around the emitted microwave energy. Most marine radars are not powerful by radar standards (a few tens of kW), and the pulses are so short (about 10^{-7} sec) that there is no known cause for concern for people in the general vicinity, but not close to the antenna. There is seldom need for a user of a large radar to be near the antenna, which may emit up to several megawatts. If one plans to work near a large radar or one with a phased-array antenna, professional advice about safety should be obtained.

The back end of radar, which delivers the data, is at least as important as the front-end antenna and microwave electronics. **Recording radar information** by videography or time-lapse photography has immediate intuitive appeal and can be handy for obtaining images to accompany oral presentations or proposals. However, for monitoring animals, one should avoid photography, preferring direct recording on a computer medium. Signals in the radar exist as voltages (see Box 13.2), and information is lost when the signals are converted into an optical display and subsequently converted back to voltages in a camcorder or camera. More importantly, the deferred labor of quantifying radar data from photographs will quickly become the most expensive part of a radar project and the most tedious. Radars can quickly generate large amounts of data. Infrastructure to clearly label, efficiently quantify, and readily summarize those data is critical and should be accorded the same importance as the data.

One often has access to radar data in more than one form: as easily interpreted images, such as color PPI images;

as tracks drawn across a map; or as lower-level numerical data, such as angles, decibels, and velocities. **Color PPIs and maps** are excellent for making decisions, taking notes, and summarizing, but **numerical data** are superior for quantifying and detecting differences and trends.

If only images of radar data are available, one should try to acquire unembellished versions without thresholds to remove "artifacts," such as biological targets, and without non-data, such as map overlays. Radar data can be registered to soft-copy maps later. Images are usually stored as computer files that code quantities as colors. Reversing the process, colors can be decoded to yield numerical values. Although image files offer a direct route to quantification, they do not prevent one from becoming inundated with vast amounts of information. For instance, in the United States, >150 NEXRAD (WSR-88D) radars each produce reflectivity, radial velocity, and spectrum width data about every 20 seconds. The resulting archive in North Carolina (Crum et al. 1993) is the second-largest unclassified data base after satellite remote sensing data. Access to these data is free (Del Greco and Hall 2003), and tools are available to convert data to a form more familiar to biologists (e.g., shapefile; Del Greco and Ansari 2008).

Radar echoes vary by 7 orders of magnitude or more in strength. To handle that large dynamic range, radar receivers typically generate logarithmic signals. Consequently, radars often display log (P_r) in decibels, which must be converted into linear units prior to averaging or summing (Black and Donaldson 1999). **Old radars** with monochromatic displays may be incapable of displaying echo strengths over a range greater than about 10-fold (Hunt 1973). However, this limitation can be circumvented by clever manipulation of the controls (Gauthreaux 1970; Drake 1981a, b). A radar should be **calibrated** to estimate target size; simple calibration techniques can be used in the field (Atlas 2002).

INTERPRETING RADAR DATA

Enumeration of Flying Animals

Useful data require timely information on antenna position and/or scan pattern. Animal flight takes place in 4 dimensions (3 spatial and 1 time), frequently reported as **direction, speed, height, and time.** Enumeration of flying animals can be accomplished in several ways, depending on the questions asked (Gauthreaux and Belser 1998, 1999; Black and Donaldson 1999). One can convert a radar signal to meaningful measures, such as numbers of animals or biomass. Questions of social behavior or probability of encounter with an animal will require **volumetric densities** (animals per length^3), habitat-related questions will require **areal densities** (animals per length FD) summed over height, and **rate-of-passage** questions will require rates of crossing a line on the earth summed over height (animals per length per time) or rates of passing **through a vertical plane** (animals per length FD per time). Statistical treatments for angular or directional data are available specifically for biology (Batschelet 1981, Zar 1996) and in more depth (Mardia 1972, Fisher 1993).

Vertebrates and Other Sources of Echo

Field workers using radiotracking equipment listen for beeps of radio transmitters against an ever-present background of natural and human-generated noise and gradually become sophisticated at that task. Learning to use radar is no different, and one should not expect to recognize wildlife immediately. **Radar noise** normally refers to intrinsic receiver noise and external radiation, such as celestial background and pulses from other radars using a similar wavelength. Clutter usually refers to spurious received echo from something not regarded as a target.

Close-in **ground clutter** is ubiquitous and can limit a radar's usefulness for detecting nearby animals and those flying at low height. With long-range radars, a special kind of ground clutter occurs when air that is cold, moist, or both lies close to the ground and bends or refracts the radar beam downward. Radar scientists use the term **anomalous propagation** to describe unexpected echoes in conditions of notable refraction. Not uncommonly, layers in the atmosphere can trap part of the radar beam at a certain height above the ground, so that the beam follows Earth's curvature, allowing low objects and flying animals to be detected from great distances (Fig. 13.10).

Clutter reduction is accomplished in several ways. Siting a radar in a shallow pit or depression in the ground or behind an earth berm reduces echoes from the surrounding terrain and can still permit an unobstructed view of the air space (see Bruderer 1971, 1994; Fig. 13.8). Radar also may benefit from radar opaque or radar absorbent material mounted directly on the antenna to "shape" the radar beam (Freeman 1982, Cooper et al. 1991, Kelleher and Hyde 1993, Smith and Riley 1996), or a radar fence (LaGrone et al. 1964, Prickschat 1964, Becker and Sureau 1966, Freeman 1982; Fig. 13.4). Ideally, a radar fence should be of **radar absorbent material (RAM)** to screen the highest point of vegetation, structures, or topography that would otherwise be visible from the tip of the feed of a reflector antenna. A fence of material other than RAM can generate reflections, which also require screening. This screening normally requires completely encircling the radar antenna. Natural vegetation can function as a radar fence in some situations (Seilman et al. 1981, Cooper et al. 1991, Buler and Diehl 2009). Cuffs, shields, and fences do not need to be grounded but should be free of holes and gaps wider than a small fraction of the wavelength. Most biologists rely on a combination of imitation, trial-and-error, and advice from specialists rather than attempt to understand the theoretical basis of such devices.

RAM can reduce radiated power by 95%. Sometimes coated to reduce weathering, RAM is somewhat flexible but

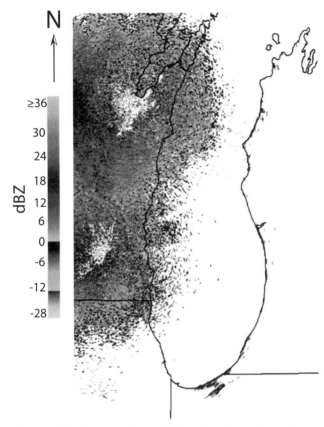

Fig. 13.10. Weather surveillance 88 Doppler radars at Green Bay and west of Milwaukee, Wisconsin, present clear images of perched dunes on the east shore of Lake Michigan 200 km away. These have zero Doppler velocity, confirming that they are on the ground rather than in the air. In a "normal" atmosphere without ducting, the height of the bottom of the radar beam would be >3 km above ground level at such ranges.

ground targets, echoes from animals near them may be suppressed as well. Hills, mountains, and structures can partially or completely obstruct a radar beam, severely limiting some applications (Felix et al. 2008; Fig. 13.11).

Some radars reduce echo with radial velocity near zero by using a **moving target indicator (MTI) filter.** Most targets flying tangential to the radar also will be suppressed, because they, too, have a radial velocity near zero (Drury and Nisbet 1964). Depending on how MTI circuitry is adjusted, echoes from small, slow-moving animals at all azimuths also can be lost. Doppler radars accomplish the same filtering of **stationary echoes** via a filter centered at zero radial velocity (Keeler et al. 1999; Fig. 13.9). At times, notch filters (filters that pass only a narrow range of velocities) can be flexibly and creatively programmed to suit the requirements of an individual project. Bats and birds flying over areas with heavy ground clutter are less likely to be detected, even though filtering eliminates the ground clutter.

Filters and clutter maps are commonly applied to alter the signal from the radar receiver, the data sent to specific displays, or both. **Biological echoes** are frequently the target of filters applied to real-time or published commercial

expensive and can be heavy, especially exposed open-cell foam after a rain. RAM can reduce echoes from unwanted targets when applied in the path of radar side lobes. This technique is used extensively in radar entomology, a field that needs good performance and high radar gain close to the ground (Beerwinkle et al. 1993). One also can install a metal plate extending 20–30 cm forward from the aperture of a marine radar antenna to block the lower part of the radar beam that would otherwise strike the ground or create other unwanted echoes. This shielding must be done carefully or reduction in close-range ground clutter may be accompanied by increase in other forms of clutter. A similar approach has been used to reduce side lobe echoes in a vertical scanning radar, but the shielding was applied to all 4 sides of the radar aperture to prevent side lobes from reaching the ground (T. A. Kelly, DeTect, personal observation).

Because **surveillance radars** usually register echoes from stationary ground targets, such as structures and terrain, they often use computerized clutter maps to subtract known ground clutter from a display. Depending on the size of

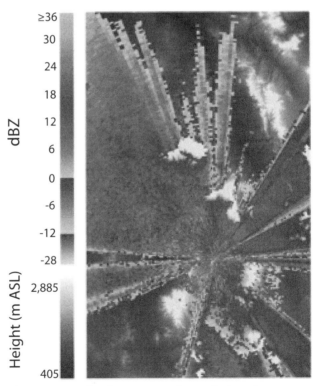

Fig. 13.11. A Doppler plan position indicator display at low elevation angle (0.5°) from a weather surveillance 88 Doppler radar near Tucson, Arizona, shows birds (and probably some insects and bats) during spring migration, except where terrain relief interferes with propagation of the radar beam. Height of terrain above sea level (ASL) is shown in gray-scale and increases from black to white. White areas signify partial or complete blockage of the radar beam.

data from Doppler weather radars. Although displays often show filtered versions of radar data, raw unfiltered data may be stored in digital archives. The wildlife user should prefer unfiltered data, and characteristics of filter settings or clutter maps should be identified, if known, when publishing radar data.

Sources of clutter (in this context, nonwildlife echoes) vary (Fig. 13.12) and can include smoke plumes, vehicles on highway overpasses, wind turbines, trees and ocean waves moving in the wind, and railroad trains. Sometimes clutter echoes show Doppler velocity, defeating MTI filters. Nearby pulsed radars are seldom sources of serious confusion, because they operate at different pulse rates. There is no foolproof way to distinguish birds or bats from other echoes, but an experienced observer can usually do so. Generally, careful language should be used in characterizing the many species that may generate radar echoes at a given time. Although **ground truth** is necessary for radar biology, the following indications also may help.

1. Vertebrates travel at speeds different from the wind. Thus, accurate local wind measurements often permit their identification, if one can measure speed and direction of travel from PPI target motion, Doppler radar (Gauthreaux and Belser 2003), or tracking radar.
2. On many spring and autumn evenings when migrating biological targets orient their bodies similarly, their echoes will be stronger when detected side-on than from the front or rear, producing a characteristic butterfly or dumbbell shape on a PPI image (Riley 1980, Buurma 1995).
3. Weather echoes on Doppler weather radars are characteristically smooth in radial velocity, whereas even dense migrations of birds usually present a somewhat uneven or stippled appearance (Fig. 13.9). Smooth echoes are particularly characteristic of snow and warm front rain. Widespread echoes over land that extend high into the atmosphere are usually weather, not migrating birds. Insect echoes can be either stippled or smooth in appearance.
4. Knowledge of the biology of species active at a certain time of year and time of day can provide excellent evidence of the nature of biological echoes. For instance, many animals begin or cease activity at dawn every day, whereas meteorological phenomena are not tightly synchronized to first light. However, without ground truth, one cannot be certain which flying animal is active at dawn or dusk. For example, evening takeoff of migratory birds and emergence from roosts of local bats happen within minutes of each other (Fullard and Napoleone 2001).
5. Stationary beam and tracking radars provide an opportunity to identify targets by observing wing beats, such as a 2.1/sec modulation characteristic of an insect (Fig. 13.7). Although wing beat frequency by itself does not impart much taxonomic information for vertebrates (Emlen 1974, Vaughn 1974, Williams and Williams 1980, Bruderer 1997, Diehl and Larkin 1998), wing beat signatures can contain fine detail and hold promise for advances in ground truth (Bruderer and Popa-Lisseanu 2005; Box 13.2; Fig. 13.13). The term signature properly applies only to the variation of echo strength over time (Schaefer 1968:53).

Ground Truth

One should acquire sufficient radar data, but equal weight should be given to concomitant field observations that establish the identity and numbers of targets (Fig. 13.14). **Ground truth,** a term borrowed from radar meteorologists, includes visual observations, infrared, sound, and separate small radars (Williams et al. 1981a, Bruderer et al. 1995, Liechti et al. 1995, Gauthreaux and Belser 1998, Larkin et al. 2002, Gauthreaux and Livingston 2006). Ground truth should be simultaneous with radar operations, because daily monitoring of migrating birds on the ground is usually a poor indication of numbers of birds actually migrating overhead (Parslow 1962, Williams et al. 1981b). Although the difficulty of discriminating small birds from insects with radar has been appreciated for a long time (Schaefer 1968), failing to do so remains one of the most common mistakes in design-

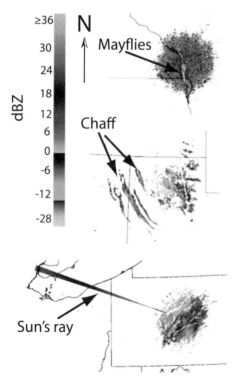

Fig. 13.12. Insects (mayflies, Plecoptera) emerging from the Mississippi River, military chaff over Utah, and a ray from the setting sun over Pennsylvania, as seen on radar images.

Box 13.2. Producing and interpreting an A-scope display

An A-scope display shows time variation of echoes versus range (Fig. 13.13). The radar antenna should be either stationary, with birds and bats flying through the beam, or tracking a bird or bat. On the vertical axis, the radar receiver signal produces a positive logarithmic display. The horizontal axis is the range in km (or delay, corresponding to 150 m/sec). The outgoing radar pulse ("main bang" in radar terminology) and some ground clutter appear on the left (omitted for clarity in Fig. 13.13). No biological targets appear beyond a certain range on the right, largely because of the R^{-4} term in the Radar Equation (see Box 13.1). Stationary ground targets illuminated by side lobes, slowly fluctuating vegetation, and narrow peaks from flying animals that wax and wane as they enter and leave the beam may appear in between. If the beam is pointed along or opposed to the direction of travel of flying targets, the targets will move to the right or left on the A-scope, respectively.

One can construct an A-scope using a suitable oscilloscope (less costly on the used electronics market) and 2 high-frequency cables. One cable feeds the radar "video out" or "rectified video" signal into a vertical "signal in" or "voltage in" connection on the oscilloscope, the other feeds the radar "transmitter trigger" or "pulse out" into the oscilloscope's "trigger in" connection.

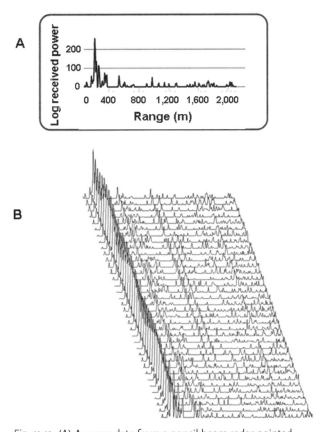

Fig. 13.13. (A) A-scope data from a pencil beam radar pointed across the path of migrating animals. (B) Behavior of the animals over time is revealed in successive traces that descend in 50-msec increments from the top. The outgoing pulse and short-range ground clutter are omitted. Echo strength of the prominent target at slightly over 200-m range varies regularly at 19.4/sec. Field workers noted, "Bird-like target, range a little less than 2 μs." At greater ranges, smaller biological targets wax and then wane over time as they fly through the beam.

ing radar studies of wildlife (Larkin 1991a). **Magnitudes** of radar echoes from insects can be "amazingly high" (Achtemeier 1992:922). Being alert to differences between insect and bird echoes can lead to new insights, such as aerial feeding of insectivorous birds (Puhakka et al. 1986, Russell 1999). **Software algorithms** using wing beat patterns and other signal characteristics can distinguish between bird and insect echoes in many cases (Schmaljohann et al. 2008, Zaugg et al. 2008). Such algorithms require that targets dwell in the radar's beam for a period of time of at least several wing beats, which is not possible when radars scan at high rates as with off-the-shelf marine units.

The wildlife scientist's inclination is to go into the field to find where flying animals congregate and identify and count them, then look on radar to obtain more information about their movements and spatial patterns. However, with roosting birds, feeding flights of waterfowl, and some other wildlife targets, many wildlife practitioners find it more productive to first use radar to **identify places and times** where wildlife seems to be aloft and then go into the field with binoculars (Russell and Gauthreaux 1998, Larkin 2006).

Wildlife as Unwanted Radar Echo

Flying wildlife can be important sources of unwanted echoes for those using radar to observe aircraft and weather. Engineers regard birds as clutter when echoes persist despite use of anticlutter techniques (Edgar et al. 1973). Doppler weather radars used by meteorologists deliver quantitative information, which is often polluted by echoes from flying animals. Wind speeds measured using weather radars may actually be ground speeds of flying birds, warm front showers may be spring passerine migrations, and downbursts may be roosts (Larkin 1991b, Jungbluth et al. 1995, Gauthreaux et al. 1998, Serafin and Wilson 2000). Operational me-

Fig. 13.14. Ground truth as supplied by continuous scanning with binoculars and telescopes. Radar observations have shown that some raptors and storks fly higher than visual observers on the ground can detect.

teorologists are partly aware of the problem and should consult local wildlife experts for guidance. In military radar applications, stealth aircraft, remotely piloted vehicles, and tiny autonomous flying robots are intentionally or unintentionally similar to vertebrates.

APPLICATIONS OF RADAR SYSTEMS TO OBSERVING WILDLIFE

Aviation Safety

Radar has long held the promise of being an effective tool for warning of hazardous birds (Blokpoel 1976) and bats (Williams et al. 1973) aloft and is used for this purpose operationally in several countries. Estimated annual losses from **collisions** of aircraft with wildlife (including "bird strikes") total at least US$500 million (summarized in Dolbeer et al. 2000). A small number of these collisions result in serious damage to aircraft and occasionally the loss of the aircraft and its occupants. Serious damage is more likely in military aircraft, where speeds are higher and aircraft more fragile. The Israeli Air Force estimated that an average of US$30 million per year was saved in that small country as a result of bird migration surveys in which radar was used (Leshem 1995).

Impact forces sustained by an aircraft are proportional to the mass of the bird and the square of the closing speed of the bird and aircraft. Because most aircraft operate at specific cruising speeds and because radar studies have shown that birds fly almost every day and night during the year, the key to avoiding collisions with birds is to fly where there are the fewest and smallest birds. Only radar has the potential to monitor birds over long distances by day and night.

Civilian and military aviation have many similarities concerning the dynamics of **bird strike risk,** but they also differ in some key areas. Military fighter and attack aircraft often train at low height and high speed (>900 km/hr). Civil aircraft operating at low altitudes, such as to conduct wildlife or pipeline surveys and crop dusting, have much lower speeds. All aircraft, military or civilian, experience similar bird strike hazards when departing or arriving.

A radar system that detects birds over large geographic areas is required to warn military aircraft during low altitude training (Buurma 1995). Many low altitude training routes used by the military in the United States pass through the coverage area of several long-range radars, precluding the possibility of using a single radar to warn pilots of bird activity. The U.S. Air Force uses the NEXRAD weather radar network to monitor bird activity in its low altitude airspace using the **Avian Hazard Advisory System** (AHAS, http://www.ahas.com/). A similar network comprised of C-band radars can be used for bird detection in Europe. These systems are often integrated with Geographic Information System (GIS) data (O'Neil et al. 2005) to serve the immediate function of providing near-real-time warning of hazardous birds aloft and the more lasting function of monitoring trends in bird activity. Special rapid-scan radar systems (2–3 sec/scan) designed to eventually provide bird track and height information directly to the cockpit will soon supplement the wide area systems to reduce their intrinsic delay in warning. The best bird strike risk reduction methods use more than one strategy. Scheduling to avoid known high-risk periods and seeking to avoid any remaining birds in real time via data gathered by a radar sensor is an example of combining active and passive risk management.

Human Impacts on Wildlife

Because radar can monitor flying vertebrates at night and beyond human vision, it can provide data helpful for assessing and/or reducing **wildlife collisions** with electrical power distribution lines (Gauthreaux 1985), wind turbines (Kunz et al. 2007, National Research Council, 2007), and tall structures (Larkin and Frase 1988, Gauthreaux 1996). For impact studies, information on height of flying animals is usually essential; consequently, radars providing poor height information, such as marine radars operated horizontally, large weather radars, and other radars with vertically fanned beams, are insufficient. In such work, radar can often provide definitive information (e.g., Desholm et al. 2006) and is often the only technology capable of providing the needed data.

Animal Control and Insect Pests in Agriculture

Birds that flock in large numbers can damage crops, impact farmers economically, and even contribute to food shortages. **Depredating birds** are typically passerines that eat grain (DeGrazio 1989, Elliott 1989), but also include other granivorous species, such as cranes (Gruidae), and even fish-eating species, such as cormorants (Phalacrocoracidae), that can impact commercial fish farms. Bird species that roost near airports also can pose a danger to aircraft on takeoff and landing (Seubert and Meanly 1974).

Radar offers an opportunity for long- and short-term monitoring of flocking species that cause economic hardship. Data collected with radar can be used to assess effectiveness of management techniques and long-term impacts of management on the species. In the United States, blackbirds and similar species (Icteridae and Sturnidae) impact such crops as rice and sunflowers. **Flights** (to and from their communal roosts) are visible on many radar systems, including NEXRAD during portions of the year. They appear in the morning as rings of echoes that radiate from roost locations (Eastwood et al. 1962). On radar, a roost is the center of concentric waves of departing birds (Fig. 13.15). Birds traveling NNW through ESE are fully visible, but echoes of birds traveling NW and SE (tangential to the radar) are suppressed by a Doppler velocity filter (the center 3 velocity ranges are black in the figure). Birds traveling toward the radar also are less evident, because they are flying among urban structures. The velocity of the structures is zero, and echoes from them are suppressed, along with those of the birds flying above and among them. As birds fly into surrounding fields to alight and subsequently feed, the red/orange concentric rings disappear from the radar at the radius of the feeding habitat of these birds on that day (Fig. 13.15).

Monitoring **roost echoes** as birds disperse provides information on how far and to which areas birds from a single roost go to feed. Following application of a new management program, the effect on the species can be evaluated by comparing roost size, extent, and dispersion before and after treatment. If elimination of the roost habitat itself is considered, the roost location can be easily monitored year round to learn whether other species, such as tree swallows (*Tachycineta bicolor*) or purple martins (*Progne subis;* Russell and Gauthreaux 1998) use the same roosting site at other times of the year than do the target birds. Radar provides no information on which roosting species, but requires less effort than driving to a site on a weekly basis to count birds, especially if they are at distant or remote locations (Brugger et al. 1992). Large Doppler radar was instrumental in revealing high-altitude predation by *Tadarida* bats on pest insects of enormous economic importance (Horn and Kunz 2008, McCracken et al. 2008, Westbrook 2008).

Population Monitoring

Radar has potential as a conservation tool in population monitoring. Visual surveys are limited by visibility in daytime and can be biased by the observers' skills and persistence. Radar, in contrast, can detect birds in all light conditions and even through light rain, with the appropriate signal processing.

A good **example** of the application **of marine radar** to population surveys is Burger's (1997) study of marbled murrelets (*Brachyramphus marmoratus*), which demonstrated limitations of audiovisual surveys and strengths of radar (also Cooper et al. 2001). Radar found an influx of murrelets 35–60 minutes before sunrise, which audiovisual surveys failed to detect. Visual surveys indicated a later peak 35 minutes before to 90 minutes after sunrise, when radar showed intensive circling and departure. By making careful radar observations of flight speeds and direction and relating them to visual ground-truth observations, it was possible to distinguish seabirds from other groups of radar targets, such as bats (Hamer et al. 1995). One of the distinct benefits of radar is that it allows remote observations with minimum disturbance to nesting colonies. Marine radar detected at least 4 times as many marbled murrelets as audiovisual surveys in a later study (Cooper and Blaha 2002). However, radar was ineffective in detecting low-flying murrelets in or near the forest canopy or in hilly country; flock sizes could not be estimated with confidence on the radar (Burger 1997).

WSR-88D (NEXRAD) and other radar systems can **detect waterfowl** as birds leave on migration and at stopover and wintering locations when they move from refuges and lakes

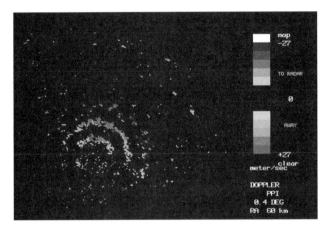

Fig. 13.15. Doppler velocity image of a dawn exodus from a roost of 1.5×10^8 European starlings (*Sturnus vulgaris*) and a few brown-headed cowbirds (*Molothrus ater*). Data were taken with a large research radar then operated by the Illinois State Water Survey, Champaign.

to feed. For overabundant species, such as snow geese (*Chen caerulescens*), radar offers an enhancement to existing tools for monitoring movements and relative abundance (Blokpoel and Richards 1978; Fig. 13.16). However, careful ground truth is required to relate radar reflectivity to actual bird numbers. Monitoring takeoff, passage, and daily feeding flights of waterfowl with surveillance radar offers a productive alternative to more conventional census and survey techniques, but detecting arrival of migrating waterfowl on a stopover or wintering area still offers a great challenge. This challenge is worth examination, as autumn movements of waterfowl are not easily predicted. "In contrast to spring migration, autumn (waterfowl) migration was not strongly correlated with any of the weather variables examined" (Beason 1980:452).

Waterfowl generally migrate later in autumn and earlier in spring than do most other bird species; thus, it is relatively easy to recognize waterfowl activity on a large radar system. However, ground truth may be problematic when many species are aloft, such as migratory takeoffs in a sudden and dramatic burst of activity near sunset. Identifying the specific species active on a given night may depend on the happenstance that an observer was on the ground at the start of migration to make the observation. An observer along the route who recognizes bird calls may be able to deduce that a particular species was involved in a dramatic overnight rise in numbers at a distant location.

Stopover habitat for migrating birds, such as shorebirds (Charadriiformes) and neotropical–temperate migrant passerines, is becoming an important conservation issue (Gauthreaux and Belser 2003). Recent research linking observations of populations of marked birds from breeding and wintering habitats suggests that about 80% of mortality of typical neotropical–temperate passerine migrants is associated with migration rather than breeding or wintering habitats (Sillett and Holmes 2002). Where large numbers of migrants depend on sparse and patchy habitats for food and cover during migratory stops, radar is a powerful tool. Although radar is not helpful for birds on the ground or in vegetation, fortuitously placed surveillance radars can detect them immediately after takeoff (usually at sunset) and can be helpful in locating areas actually used (Gauthreaux and Belser 1998, Bonter et al. 2009, Buler and Diehl 2009). However, careful **ground truth and quantification** of radar echoes is needed before habitats can be evaluated in terms of their value to migrating birds. Radar observation of departures of migratory birds is feasible only within a certain range of the radar. Radar images have immediate visual impact and can convince the public and land managers more effectively than can reports with tables of bird numbers and densities.

THE FUTURE

Forty years ago, scientists were excited about radar as a new tool for studying animal movements and populations. In the late 1990s, this excitement underwent a resurgence fueled by ready accessibility of radar technology. To the extent that wildlife practitioners exercise care with fundamentals, such as ground truth, the coming decade will see the fruits of this resurgence. We should attain better understanding of the **relationship** of animal sizes, taxa, flight patterns, speeds, numbers, and density to radar measurements of the properties of echoes as well as Doppler speed and its variation. This understanding will permit new inferences about wildlife and application of the new knowledge to management. Inferences about types of biological targets aloft will primarily be limited by the amount and quality of groundtruth data available from visual observations, small local radars, and acoustic and infrared sensors.

Networks of Doppler weather radars are revolutionizing atmospheric sciences and will revolutionize knowledge of organisms from pollen and insects to the largest birds, similar to the way satellites revolutionized geography and

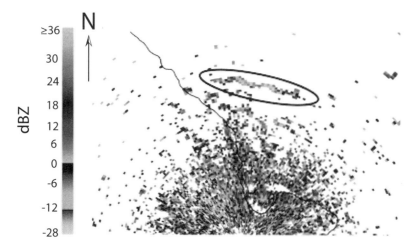

Fig. 13.16. A spatially extended flock of 6 × 10⁴ snow geese departing from Coffeen Lake in west-central Illinois. The birds were counted during an aerial census by the Illinois Natural History Survey and followed on KLSX (Weldon Springs, Missouri, weather surveillance 88 Doppler radar) until after they crossed the Mississippi River (black) into Missouri. The flock (circled) extends 53 km east–west 1050 hours (Central Standard Time) at approximately 100-km range. Other echoes around the radar are caused by ground clutter and probably other biological targets.

earth sciences in the last 20 years of the 20th century. As data from these powerful instruments are fused with data from other sensors, particularly those using GIS technology, opportunities will emerge to compile data useful for land use and other decisions in wildlife management and to put displays of this technology to use directly in the classroom and in nature centers. The technology is advancing rapidly (National Research Council 2002). Soon the NEXRAD system in the United States will add an additional plane of polarization (Mueller and Larkin 1985, Zrnic and Ryzhkov 1998, Bringi and Chandrasekar 2001). This upgrade will improve target discrimination capability, add to the diversity of quantifiable behaviors, and considerably augment the data flowing from the NEXRAD system. On the heels of this upgrade, successors to NEXRAD and other large-scale radars are already under development (Weber et al. 2007, Zrnic et al. 2007, Weadon et al. 2009).

Presently, individual scientists conduct most research in radar ornithology, whereas teams perform research on migrating insect pests. Wildlife scientists may want to change that paradigm to make better use of the new technology.

SUMMARY

Radar has been used for more than half a century to observe flying animals. Its application to wildlife research and management continues to blossom, primarily because of the availability of capable and reliable small radars and copious data from large networks of Doppler radars designed for monitoring weather. Properly placed radar can observe flying animals at different spatial scales without affecting their behavior. However, radar offers little if any information on species identity. Radar is useful for informing aircraft of flying birds and bats; locating roosts; following birds that depredate crops; and monitoring populations, including threatened and overabundant species and waterfowl. Locating areas critical for stopover of migratory birds is an especially useful application. Acquiring radar data necessitates care and accumulation of meaningful numbers. Challenges in using data from radars lie in establishing the identity of radar echoes (ground truth), in recognizing different kinds of artifacts, and especially in coping with large amounts of automatically generated data.

14 Invertebrate Sampling Methods for Use in Wildlife Studies

THERESE A. CATANACH

INTRODUCTION

WILDLIFE BIOLOGISTS have long recognized invertebrates as an integral part of ecosystems both for the services they provide and as prey for many vertebrate species. However, due to disconnect between wildlife biologists and entomologists, projects that would benefit from expertise represented by both have generally not been undertaken by both simultaneously. Invertebrates have many effects on wildlife, most of which are greatly understudied. These **interactions** concern forage, disease vectors, parasites, and (although not specifically a wildlife–invertebrate interaction) indicators of ecosystem health. Studying different interactions requires different methodologies, both in terms of sample procurement and in the processing of samples. This chapter starts with brief descriptions of some invertebrate groups most commonly encountered (in terms of numbers or of those particularly important to wildlife), continues with a short discussion of wildlife–invertebrate interactions, details collecting techniques, and then finishes with information regarding invertebrate processing and curatorial techniques. Although invertebrates as a whole include all animals minus chordates, this chapter concentrates on **arthropods,** with brief mention of some **nonarthropod invertebrates,** as these groups are typically the most important taxa related to wildlife study.

Arthropoda includes a number of distinct groups divided into 4 subphyla: the now extinct Trilobita (trilobites), Chelicerata (including spiders and horseshoe crabs), Crustacea (crustaceans), and Atelocerata (including insects and centipedes). Members of Arthropoda have successfully adapted to virtually all habitats found on Earth and have a variety of life histories. Of them, the insects are the most diverse group. Insect higher level classification and relationships among orders are areas of active research. A new order was described in the past decade, although the status is of it and the splitting of Neuroptera into 3 orders is debated. Triplehorn and Johnson (2005) provide a good starting point for general insect information in North America and include keys to the families in each order, along with brief descriptions of natural history and the number of taxa in North America for each family. For specimens collected outside North America, regional insect guides are often available. On this topic, it is important to mention that use of a key outside its region of coverage can result in **misidentifications.** Below, I provide a brief description of the more important or commonly encountered insect orders along with other invertebrates of interest.

INVERTEBRATE GROUPS OF INTEREST

Orthoptera

This order includes **grasshoppers, katydids, and crickets,** along with more obscure insects. With some exceptions, members are heavy bodied (although not necessarily large, as some taxa are 1 cm long). They are easily recognized by their having well-developed hind legs for jumping. Most Orthoptera are herbivores, and they sometimes occur in large enough numbers to defoliate plants. Orthoptera are commonly consumed by vertebrates, making them common subjects for wildlife food habit studies.

Hemiptera

This order contains insects formally placed in 2 orders, the Homoptera and Heteroptera, and includes many well-known insects, such as **cicadas, assassin bugs, and aphids.** It is difficult to characterize this group because of the huge variation in morphology and size; however, all have piercing or sucking mouthparts. Feeding strategies are varied, with the majority of species feeding on plant juices. Some, such as bedbugs, are blood feeders, whereas others, including the assassin bugs, are predacious. Some members of this order are large enough to attack and consume small vertebrates, including fish and lizards. This group contains a number of vectors both of plant and animal diseases.

Hemiptera are among the most commonly collected orders in both terrestrial and aquatic situations and are often used as bioindicators of ecosystem health, because many have strict habitat requirements. Additionally, many herbivorous Hemiptera are specialized feeders, relying on a single or small number of host species for food. These groups have been especially useful in grassland habitats to measure ecosystem health. Hemiptera also are important in wildlife study as prey items for many insectivorous species. However, not all Hemiptera serve this role, as many are known to sequester plant-derived chemicals, such as cardenolides for defense against predation (Malcolm 1990).

Coleoptera

Beetles make up the insect order with the most species and are one of the most successful groups of animals on the planet. Members exhibit an array of morphological variation, although most members have 2 pairs of wings, with the forewings hardened into a protective structure known as the elytra. Beetles have chewing mouthparts and feed on a variety of items, including plant material, detritus, and other animals. Additionally, adults and immatures are radically different—immatures being worm- or grub-like and often living in protected situations, such as inside logs or underground. The larva are especially important food for a variety of wildlife species. This group also is useful as bioindicators in rainforests (Davis et al. 2001).

Diptera

The order Diptera includes the **flies, mosquitoes, and gnats.** Although some rare members have secondarily lost both their wings and halters (a golf-tee-like structure formed by the modified hind wings and used for balance), the majority of adult members can be easily recognized by their halters and normal forewings. As with Coleoptera, larvae, often known as maggots, are drastically different in appearance and life history from adults. Many species are vectors of diseases affecting wildlife, such as bluetongue, west Nile virus, and avian malaria. In addition to causing disease, some members of this order infest vertebrates either as larva (e.g., botflies) or adults (e.g., flat flies).

Hymenoptera

Hymenoptera include **bees, wasps, ants,** and more obscure groups, such as sawflies. The majority of taxa in this order can be recognized by a highly constricted region between the first and second abdominal segments; however, this character is not present in sawflies and other primitive members. Many Hymenoptera are important pollinators, whereas others are economically important as forest pests or biological control agents. Some groups—in particular, ants—also have been used successfully as indicators of ecosystem health (Englisch et al. 2005, Majer et al. 2007).

Phthiraptera

The order Phthiraptera contains both the **sucking and chewing lice.** These animals are obligate ectoparasites of mammals and birds. Most lice are very host specific, even to the point of being found only on certain body parts; as a result, one individual animal may have many types of lice present. They are not collected by using most invertebrate sampling methods; instead, animals must be caught and lice removed by hand from the specimen.

Lepidoptera

Lepidoptera, **butterflies and moths,** are recognized by having wings covered with scales. These insects are important prey items for many vertebrates, including bats and birds. Bats in particular are common moth predators, and studying predator–prey relationships between these groups has been popular. Additionally, many plants are pollinated by Lepidoptera, and in some situations, these insects are the most common pollinators. Because of their close ties with plant hosts, some groups of butterflies have been useful indicators of ecosystem health. Although these insects are commonly collected using virtually all techniques outlined below, many methods render identification difficult to impossible by a nonexpert, because any liquid killing agent removes the wing scales and matting hairs. For this reason, if Lepidoptera are the target of interest, some type of hand or attractant collecting is required.

Siphonaptera

The order Siphonaptera (fleas) are **obligate ectoparasites** of vertebrates. Unlike lice, they are able to spend extended time off the host. While taking blood meals, they can transmit various diseases, including plague. As with lice, they are not typically collected using standard invertebrate sampling methods, and instead require collection directly from a host or from areas frequented by the host, such as dens or nests.

Acari

Acari includes ticks and mites which are **obligate ectoparasites** of vertebrates. They transmit a number of diseases, many of which affect wildlife species. Mites have a variety of lifestyles, including as predators, parasites, and even herbivores.

WILDLIFE–INVERTEBRATE INTERACTION

Invertebrates play an important role in the **diets** of many wildlife species. In some cases, this role tends to be prominent in a single period of their lives (e.g., upland game birds will consume high numbers of insects as young chicks, but older animals eat insects only occasionally), whereas other species utilize invertebrates over the entire course of their lives. Laboratory and field studies have shown that being deprived of invertebrates during developmental periods was detrimental to greater sage-grouse (*Centrocercus urophasianus*) and other upland game bird chicks (Hill 1985, Johnson and Boyce 1990). Additionally, field studies demonstrated that areas with lower insect abundance had lower chick survival rates (Drut et al. 1994).

Recently, certain invertebrate groups have been used for **ecosystem monitoring.** Aquatic invertebrates have long been known to be closely tied to water quality (Gaufin and Tarzwell 1952). This use is not limited to aquatic systems, as invertebrate indicators have been used in forests, grasslands, and other systems (Davis et al. 2001). See Karr (1999) and Hilty and Merenlender (2000) for discussions on characteristics of groups useful as **biological indicators.** Invertebrates are ideal for ecosystem monitoring, because they are able to survive in small remnant patches where larger animals could not. Many are incredibly specialized, requiring certain microclimates and hosts to survive. Most ecosystem monitoring studies require a high degree of taxonomic detail, as even members of the same genus can react differently to disturbance.

Invertebrates also **vector** a number of wildlife and human **diseases,** including bluetongue, plague, west Nile virus and Lyme disease. Some of these diseases have caused large die-offs in wildlife populations either periodically or for a period of time after the disease is first introduced into a susceptible population. In addition to disease transmission, many invertebrates parasitize vertebrates. Although these infestations typically occur in low densities with little or no effect on the host, there is a possibility of disease transmission. Additionally, under certain conditions, death can occur as a result of being parasitized, especially when hosts are under stress from bad weather, poor nutrition, or extremely high parasite loads (Nelson 1984, Lehmann 1993, Pavel et al. 2008).

COLLECTING METHODS

There are a wide variety of collecting methods for invertebrate sampling. Some of these methods are popular, but each has inherent biases that must be recognized. Briefly described below are a number of (although by no means all) different collecting techniques with a discussion of their strengths and weaknesses.

Aquatic Sampling
Aquatic Net

These **D-shaped nets** typically are made of fine mesh or canvas (Fig. 14.1). The net is dragged along the surface of the water (to collect surface dwelling invertebrates) or is pulled along the bottom (to sample invertebrates living at the bottom; Fig. 14.2). The net contents are then dumped into a white tray (or sorted inside the net itself), and organisms of interest are removed from the muck. This method works well for organisms living near shorelines or in relatively shallow bodies of water. When sampling in fast-moving shallow streams, the net can be placed in a narrowing of the stream and areas upstream disturbed. The disturbance causes invertebrates to be swept downstream into the net. This method is especially useful when sampling invertebrates that cling to rocks or vegetation. Potential drawbacks include difficulty in sampling some areas, such as around submerged roots, and possible difficulty separating certain invertebrates from debris.

Fig. 14.1. Nets for collecting aquatic samples can be used with collecting jar attached or not. If a collecting jar is not used, invertebrates are manually removed from the net. *Photo by M. Pessino.*

INVERTEBRATE SAMPLING METHODS FOR USE IN WILDLIFE STUDIES 339

Fig. 14.2. Example of aquatic sampling technique. Keep in mind that different parts of the water column will support different invertebrate communities. *Photo by M. Pessino.*

Light Traps

Submerged light traps have been used with great success for a variety of aquatic invertebrates, including insects and mites (Hungerford et al. 1955). The trap itself can be purchased from an entomological supplier and then various colors of light can be created using chemiluminescent candles or underwater light-emitting diode (LED) flashlights, which Radwell and Camp (2009) showed to be as effective as candles. **Light color** is important: lighter colors, such as white and yellow, attract more individuals than does blue light. These lights work well in areas where aquatic nets may do poorly, in addition to being more practical for sampling very small invertebrates. However, for the method to be effective, the target invertebrates must be attracted to light.

Terrestrial Sampling

Sweep Net

sweep netting is a common method for collecting terrestrial invertebrates, particularly insects. Nets come in a variety of sizes and mesh types, ranging from canvas for sweeping bushes or dense vegetation to fine mesh for sweeping thin grass or catching flying insects (Fig. 14.3). Nets are swept through the vegetation for some period of time, distance, or number of sweeps (depending on the protocol used) and then either invertebrates are removed from debris or the entire contents of the net are removed and placed in a killing agent or a freezer (Fig. 14.4). This method is often thought to allow for comparisons between study sites, as effort can be standardized. However, under some circumstances, collecting effort will be difficult to standardize, especially when there are differences in vegetation types (i.e., dense versus thin grass) or multiple people are sweeping. If the purpose of the survey is to document biodiversity, sweeping is ideal for collecting many groups of invertebrates, especially those that live near the top or edges of vegetation or spend a rela-

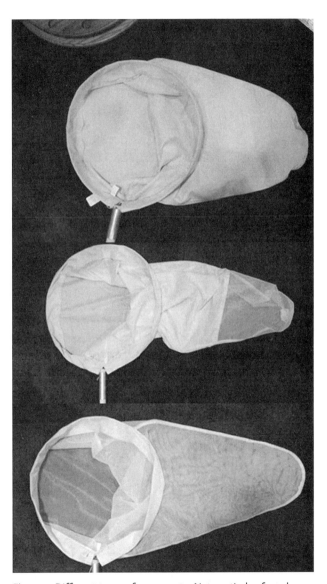

Fig. 14.3. Different types of sweep nets. Nets entirely of mesh (bottom) are best for sampling fragile insects in flight or in thin vegetation, whereas nets of thick cloth (top) are best in situations where vegetation tends to be thick or contain spines that could tear the mesh. Combination nets (center) allow for collecting in both situations.

Fig. 14.4. Example of terrestrial sweeping technique.

tively large amount of time flying. However, invertebrates that primarily live deeper inside vegetation or are ground dwelling will be undersampled. Additionally, different life stages may live in different parts of the plant, making sweeping effective for some stages but not others, such as some nymphal Hemiptera (Schotzko and O'Keeffe 1986). However, a multiyear study also by Schotzko and O'Keeffe (1989) found no significant difference in population (adult plus nymphs) when comparing sweeping, vacuuming, and absolute sampling. Randel et al. (2006) compared sweeping transects with vacuum sampling of square meter quadrats and found that sweeping yielded greater invertebrate biomass and was more likely to result in invertebrates being collected, probably because insects are patchily distributed in the landscape and a random square meter quadrat may fall on a location lacking insects.

Vacuum Sampling

D-vacs or other suction devices have become popular tools for collecting invertebrates, especially in areas with dense vegetation (Fig. 14.5). Methodologies vary, although the more popular techniques include using a ring as a boundary for the vacuumed area (creating a known area) or vacuuming along transects. If the purpose of the study is to conduct a biodiversity survey, it may be of interest to sample stands of a single plant species (to record host associations) or sample a wide variety of species (to record biodiversity of the area as a whole). Suction samples are typically collected into a mesh bag, which will contain not only specimens but also loose vegetation. Separating specimens from vegetation can be done after killing the specimens or by using emergence traps (Fig. 14.6). Emergence traps use light to separate photosensitive insects from the duff. However, as not all insects are photosensitive, some will be missed if this approach is the only one used. Instead, after allowing the sample to sit in an emergence trap for 2–3 hours, the duff should be examined and any additional invertebrates removed. This method works well for small invertebrates; however, Mommertz et al. (1996) showed suction sampling to be unreliable for estimating populations of large bodied insects and spiders. Additionally, Schotzko and O'Keeffe (1986) found vacuum samples overestimated populations compared to absolute sampling, possibly due to invertebrates

Fig. 14.5. Different types of suction devices. The D-vac (left) is commercially available; the other device is a leafblower (right) with a suction option. Both work well, although the D-vac is more powerful (but less portable). *Left photo by L. Sands.*

INVERTEBRATE SAMPLING METHODS FOR USE IN WILDLIFE STUDIES   341

Fig. 14.6. Sorting suction samples by hand. *Photo by L. Sands.*

being sucked in from neighboring vegetation. This phenomenon could be limited by using a tube instead of a ring to mark the sampling boundaries.

Malaise Traps and Flight Intercept Traps

Malaise traps are tent-like mesh structures that passively collect invertebrates (Fig. 14.7). Insects fly into the center or side panels and then crawl upward, eventually reaching a high point, where they then fall into a jar of alcohol. However, heavier bodied invertebrates, such as beetles, tend to fall rather than climb when they hit the panels. **Flight intercept traps** capitalize on this tendency by placing a trough under the panels filled with a killing agent or soapy water. When invertebrates hit the panel, they fall into the trough for easy collection. A single trap can be used both as a malaise trap and a flight intercept trap, allowing for broader taxon sampling. Additionally, these methods can be used for canopy sampling by suspending a malaise trap by a tree branch. This style of trap is most effective where the natural landscape funnels insects into a narrow area, for example, a forest path or streambed. To maximize catch, malaise traps should be placed perpendicular to these openings. **Placement** is very important, as only insects that contact the inner panels have a chance of being caught. Because traps can be left up for a standard period of time, catch effort can easily be standardized. However, because only those insects that contact the inner panels have a chance of being caught, trap placement is crucial. For comparative studies, trap orientation must be identical with regard to natural flyways or attractions (e.g., water sources) to minimize trap biases. In addition to landscape, movement patterns of animals and humans must be taken into account. Large animals may walk through these traps, causing holes or even ruining the trap. Additionally, humans may take the traps, a problem especially common in developing nations. To help limit this loss, I recommend placing a sign with contact information and a statement that the trap is for scientific study.

Pan Traps

Pan traps are simply plastic bowls of various colors with a liquid covering the bottom (Fig. 14.8). **Yellow** attracts the widest variety of invertebrates, although blues, greens, and reds have been used with success to collect a narrower range of invertebrates. Bowls are easily obtainable at a party supply store, with the bright yellow Solo® bowls (http://www.solocup.com/) being the most popular. Typically, pans are placed along a transect or in a grid and then filled to approximately 1–2 cm deep with fluid. **Water** with a drop of soap (to break the surface tension) is the most common fluid used, but many variations exist. The principle is simple: an invertebrate is attracted by the colored bowl and lands inside, where it is unable to get out of the fluid. Because soapy water is not a preservative, bowls should be emptied every day or every other day, depending on temperature. If pans will be run longer than a couple days, various preservatives can be used instead of soapy water, such as soapy saline solutions (Trebicki et al. 2010). Additionally, if bowls will be left for extended periods of time, fill the bowls with more liquid to allow for evaporation. To gather the insects from the bowl, use a fine-mesh aquarium net as a strainer, rinse the catch with water, and then place it in a vial of alcohol for storage. This method has limited applications in

Fig. 14.7. Two types of malaise traps: A 6-m trap placed in a natural corridor leading to a stream (left) and a 2-m trap placed on a ridgeline (right). *Right photo by L. Sands.*

Fig. 14.8. Yellow pan trap with invertebrates ready to collect.

rainy settings or areas prone to flooding, as the bowls can fill and insects or pans wash away. Also, animals or people may disturb the set up. However, it is an easy way to sample for many invertebrates, some of which are difficult to collect using other methods.

Sticky Traps

Sticky traps are widely used for sampling insects in wildlife studies (Fig. 14.9). They are inexpensive and easy to obtain as a card or ribbon coated with an adhesive, or the adhesive itself may be purchased from an entomological supplier and then applied to the surfaces of interest. **Color** can serve as an attractant, and as with pan traps, yellow tends to collect the most invertebrates, although if certain types of invertebrates are the target group, other colors may be preferred (Harman et al. 2007, Blackmer et al. 2008, Wallis and Shaw 2008). Although sticky traps are easy to use and results are easy to compare among locations, invertebrate processing can be a challenge. Specimens can be covered in adhesive, obscuring characters, or stuck in such a way as to render key characters not visible. There are techniques to remove invertebrates, depending on the adhesive used (Murphy 1985). If comparisons among traps are to be made, color, height, orientation, and the times of day traps are active must be held constant, as all these factors can affect catch rates and species composition (Blackmer et al. 2008). Additionally, for some insects, skewed sex ratios were observed using sticky traps, although other trapping methods did not document this observation (Blackmer et al. 2008).

Attractants

Attractants can be used in many collecting techniques. Attractants range from **pheromones** (chemicals created by insects to attract conspecifics), which target a single species to compounds known to attract a wide range of insects (Fig. 14.10). Attractants are often used with some form of adhesive, although some devices using attractants do not require adhesive to collect the invertebrate.

Pheromone traps by definition are targeted (although there is some cross attraction, especially among closely related groups), so they tend to be of limited use in biodiversity studies. However, they are common in agricultural and stored

Fig. 14.9. Sticky trap.

Fig. 14.10. Pheromone trap full of collected moths. *Photo by B. Roble.*

goods settings for monitoring pest species. They also are used to detect introduced pests. For a discussion of various pheromone traps, see Francis et al. (2007), Vitullo et al. (2007), and Reddy et al. (2009).

At the opposite end of the spectrum are attractants that target a wide variety of species. Some attractants have been developed using **chemicals found in flowers,** such as Canada thistle (*Cirsium arvense*), and have been found to attract both pollinators and herbivores (El-Sayed et al. 2008). Attractant traps for invertebrates that feed on vertebrates, such as bedbugs and ticks, utilize **carbon dioxide or heat** (Cançado et al. 2008, Wang et al. 2009). Other types of attractants include **nutrient sources,** such as beer and bread mixtures, fruits, meat, and other food products. Various nutrients have been used with great success to attract Coleoptera, Lepidoptera, Diptera, Hymenoptera, and others (Bharathi et al. 2004, MacGown and Brown 2006, Wang and Bannett 2006).

Light Collecting Methods

Many invertebrates, particularly insects and arachnids, can be collected at light, either because they are attracted to light or to the high concentrations of potential prey items. There are many different types of light, with **black lights and mercury vapor** (MV) being the most common. **Mercury vapor lights** are more expensive and require a generator if an alternating current (AC) power source is not available, but they are thought to attract a wider range of invertebrates and also to be effective over longer distances. When using MV lights with a generator, use self-ballasted bulbs to avoid more equipment. **Black lights** require much less power and can be easily run off of a car battery. They appear to attract fewer types of invertebrates. **LED lights** have recently become available in colors conducive to invertebrate collecting. Currently, these set ups are available through Bioquip Products (Rancho Dominguez, CA), but they are quite expensive. However, they require very little power to operate, making them attractive for field areas difficult to access. Different types of light can be used together to attract a wider variety of invertebrates than is possible with one type of light.

Light traps can either passively collect insects or be actively staffed so that taxa of interest are collected for study. Passive traps can be used to compare locations, if such variables as light type, intensity, and duration are standardized. **Passive collecting** can be achieved in a number of ways, including funneling invertebrates into a jar with a killing agent or running a light trap for a set length of time and then bagging it, killing invertebrates via freezing or an agent, and then gathering the insects (Fig. 14.11). **Actively** staffing light traps is time intensive and is not always conducive to comparative studies due to differences in collecting effort (Fig. 14.12). Recently, pyrethroid insecticides have been applied to **light sheets,** so that invertebrates landing on the sheet are quickly knocked down for easy collecting.

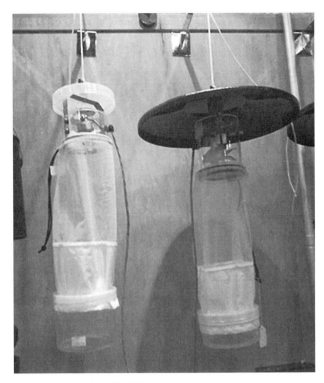

Fig. 14.11. Examples of light traps.

Fig. 14.12. Collecting at a light sheet. *Photo by K. Hill.*

Light collecting methods work well for many nocturnal insects and quite a few diurnal insects. However, to be most efficient, night temperatures must remain high enough for invertebrates to be active. Additionally, in areas with heavy light pollution, light collecting may be less effective. Moon phase is important and should be taken into account when sampling.

Pitfall Traps

Pitfall traps are ideal for collecting invertebrates that dwell at ground level, including beetles, ants, springtails, centipedes, millipedes, and spiders (Santos et al. 2007). Because they collect ground dwelling invertebrates, pitfall traps are ideal for use in studies documenting invertebrate availability for upland game birds or other species unlikely to encounter flying insects. **Pitfalls** can be of various sizes and contain different killing agents, such as propylene glycol. For a discussion of different styles and agents, see Santos et al. (2007), Sasakawa (2007) and Aristophanous (2010).

Lindgren Funnel Trap

Lindgren funnel traps are used primarily for sampling forest dwelling beetles, including a number of introduced and economically important species. Sandoval et al. (2007) used them to examine beetle diversity for artificially created snags, collecting members from 28 families. See Miller and Crowe (2009) for trap description and discussion on the effects of funnel length. These traps can be used with or without an attractant and have a dry killing container (using an insecticide strip) or a wet container (e.g., antifreeze). Miller and Duerr (2008) documented differences in catches between the 2 types of kill methods. Hayes et al. (2008) found that Lindgren funnel traps baited with an aggregation pheromone can be used to assess species presence and relative abundance.

Beat Sheets

Beat sheets are used to sample areas where sweep nets are impractical, such as trees or shrubs. A stick is used to hit vegetation, which knocks invertebrates onto a sheet (Fig. 14.13). Sheets can be angled into a jar, where invertebrates are killed, or the animals can be hand collected. Although this method is idea for heavy bodied invertebrates, such as beetles and spiders, it is ineffective for lighter insects. Wade et al. (2006) compared beat sheets to a number of different methods and found that beat sheet collecting more closely matched counts calculated using an absolute method. As with all methods that are conducted over a short period of time, temporal variation must be taken into account, because invertebrates may be utilizing different vegetation at different times of day.

Fogging

Fogging has become very popular, particularly for sampling in forest canopies. A **pyrethroid pesticide** is vaporized using a commercially available fogger and allowed to rise through the canopy. The best pyrethroids have a strong knock-down component, which quickly causes the invertebrate to fall from the tree onto a sheet. As with beat sheets, the sheet can be angled into a kill jar or the invertebrates hand collected. See Erwin (1989) for a description of fogging techniques. For this method to be effective, the air must be calm; otherwise, the vapor cloud drifts away, and any insects knocked down blow away before hitting the sheet. Early morning (often between 0200 and 0500 hours) appears to be the best time to use this method, at least in tropical and subtropical forests. The equipment required can be costly and difficult to transport to more remote field locations.

Berlese Funnel

This method is ideal for collecting invertebrates living **in soil; leaf litter;** or other structures, such as **nests.** The material harboring invertebrates is placed inside a column with a funnel at the bottom. A light is put on top of the column, causing most invertebrates to fall through the funnel into a kill jar as they attempt to get away from the light.

Other Methods

Other collection methods exist for answering questions about invertebrates in wildlife studies. Many of these questions focus on prey selection, so simply documenting what invertebrates occur in a given area is only part of the answer. Not all insects collected using the sampling methods will actually be available; for example, an upland game bird chick is much more likely to eat ground dwelling invertebrates compared to birds that spend most of their time flying or perched in the upper reaches of shrubs. Likewise, **insect size** is an important consideration, as large insects may be too difficult to handle by young nestlings, or small active invertebrates may not be nutritious enough to make up for the work required to catch them. A few methods have been used to identify what taxa are actually being consumed. When studying nesting birds, it may be possible to use **trip cameras** at nest sites to document what the parents are bringing nestlings to eat. This method is only useful in certain situations, as smaller invertebrates will be difficult or impossible to identify. Other studies rely on **visual observations** to identify prey items. This method is time intensive and often results in a high percentage of unidentified prey items (Scheiffarth 2001). However, when studying a species

Fig. 14.13. Beat sheet used to collect invertebrates from brush that is not practical to sweep.

that consumes larger invertebrates, such as crabs, the percentage of unidentified prey items is much lower (e.g., Stenzel et al. 1976). **Stomach contents, regurgitated pellets, and fecal material** also can be examined for invertebrates (Stenzel et al. 1976, Schneider and Harrington 1981, Scheiffarth 2001). Many invertebrates are not fully digested, so some hard parts, such as mouth parts, are found intact. In some groups, these parts hold enough characters to identify prey items down to the species level.

PROPER CURATORIAL TECHNIQUES

Specimen Handling and Processing

After specimens are collected, they must be **killed, prepared** for long term storage, and then **identified.** Decisions made at this stage have lasting effects on the material and can make or break a project. There are many ways to kill and store invertebrates for scientific study, depending on the collecting technique used and the purpose of the specimens. Popular **killing agents include** freezing, ethanol, ethyl acetate, and cyanide. The merits and drawbacks of each of these agents are discussed in detail below. Once specimens are prepared for storage, there are certain methods for keeping specimens permanently and in such a way that they are useful for continued study or as vouchers. **Voucher specimens** are a necessity for studies on invertebrates. Identification is challenging, and nomenclature is far from stable, so vouchers allow specimens to be reevaluated and if necessary re-identified. The accepted method for long term **preservation** depends on the group of invertebrates and **includes** slide mounting, pinning, pointing, and permanent storage in alcohol. If you are unsure of the best method to preserve your samples (both temporarily and for long term storage), consult an invertebrate taxonomist (preferably one familiar with the specific groups in question) or collection curator. By following the suggestions detailed below or those suggested by a curator familiar with preservation of invertebrates, it is possible for invertebrates to be well preserved and useful for scientific study decades or even centuries after they are collected.

Labeling

Specimens must have **accurate data labels** to be of lasting scientific value. As soon as specimens are collected, a data label must be included in the sample. This label must be written in **indelible ink,** such as an archival quality Pigma® pen. Most permanent markers will run if the label is stored in alcohol, so they are not recommended. If an indelible ink pen is not available, pencil can be used. When in doubt, make a test label and place it in your preserving agent of choice for a few days, then check to make sure the label is still legible. There is nothing worse than discovering that the labels have become unreadable and the sampling has to be repeated. The type of paper used also is important. Typical **label paper** is archival quality paper—thinner, lower quality paper may disintegrate over time. This system works for long term alcohol storage. Some people prefer to use computer generated labels (again, check the ink to ensure that it does not run). A final word of caution about labels for specimens **stored in fluid:** the label should be large enough that it does not move in the vial, otherwise it will damage specimens when the vial is moved.

If specimens are **dried and mounted,** each specimen must have its own label, which is placed on the pin supporting the specimen. Although labels may be handwritten using indelible ink, printing them on label paper is more efficient. Use 4-pt sans serif fonts, and make the label rectangular with as little white space as possible. Typical insect labels for dried specimens are 1 cm x 2 cm with approximately 4 or 5 lines of text per label. If 6 lines of text are needed, multiple labels should be used.

At a minimum the **label should include** the locality data (country; state or province; county or division; an exact location, e.g., mile marker 186 I-40; GPS coordinates; and elevation), collection date (day in Arabic numbers, month in Roman numerals, and 4-digit year), collector (first and middle initial, then last name), collecting method, and a brief description of the habitat (oak forest, prairie, cotton field, etc.). In instances where multiple samples are collected from the same locality but using different treatments, some note of this fact should be included on the label. It may be helpful to create a coding system, so that each collecting event has a **unique identifier** printed on the label. This identifier can reference a collecting notebook that includes additional notes, such as the weather, vegetation composition, and descriptions of treatment procedures.

Specimen Killing and Temporary Preservation Methods

Cyanide was a very popular killing agent, as it is fast acting, and the specimens stay dry (making hairs and bristles easier to observe). However, it is often difficult to obtain and is potentially dangerous, so it has widely been replaced by other agents. It is still in use by researchers with remaining stores of cyanide and also in developing nations, where its purchase is less restricted.

Ethanol has become the killing agent of choice for many entomologists, as it is widely available and inexpensive. The percentage of ethanol is important: 70% is considered to be the lowest that can be used for preservation. However, if specimens are to be used for any type of molecular sequencing, a higher percentage (preferably 95%) is recommended. Additionally, as invertebrates contain a large amount of water in their bodies, the concentration of ethanol decreases over time. This phenomenon is most drastic when large-bodied invertebrates are preserved or a large number of invertebrates are placed in a relatively small amount of ethanol (e.g., a vial is full of specimens with only enough ethanol to cover them). For this reason, ethanol should be **changed out** in a vial approximately 2 days post collection and in cases with large numbers of large invertebrates again a week later. Care must be taken to retain all specimens in the transfer process and to handle specimens carefully, so they are not damaged. Although this procedure may require a lot of time, especially in intense sampling efforts, failure to take these precautions can result in the ethanol concentration dropping below the threshold needed for preservation. When it does, invertebrates will start decomposing, making identification difficult if not impossible. Ethanol is used for permanent specimen storage in many soft bodied groups, such as Trichoptera, and for either permanent or temporary storage in taxa with harder bodies. There are some **potential drawbacks**, however, as killing specimens in ethanol may result in matted hairs or missing appendages (some groups often drop appendages after being submerged in ethanol), making identification more challenging. If ethanol is used for only temporary storage, when the specimens are removed to be mounted for permanent storage, care must be taken during the drying process. Several techniques are used, depending on the taxa in question. Larger insects or those with high levels of sclerotization can be allowed to **air dry** with little chance of harm. However, smaller insects, those with regions that are less sclerotized, and those whose identification requires examination of setal characters need to be specially dried to prevent parts of the body from collapsing or the setae from matting. **Critical point drying, chemical drying** (e.g., using hexamethyldisilazane), or even putting freshly mounted specimens in the freezer have all been used successfully with various groups of invertebrates (Nation 1983, Rumph and Turner 1998, Hochberg and Litvaitis 2000).

Ethyl acetate is another popular killing agent, especially for Lepidoptera. Rather than submerging the specimen in a liquid (which in the case of Lepidoptera would result in scale loss), ethyl acetate is allowed to soak into permeable material, such as a card or plaster, inside an airtight jar. The jar is allowed to charge, and then specimens are placed inside. This method is easy, and specimens typically stay in excellent condition using this method, making identification easier. However, if specimens are to be used for **molecular study**, ethyl acetate should not be used, as it appears to degrade DNA (Reiss et al. 1995, Dillon et al. 1996).

Freezing can be done in a variety of ways. Some of these are suitable for field collecting specimens under certain conditions, whereas others are impractical. Liquid nitrogen has been used to preserve specimens, especially for use in molecular studies. Although this method is often impractical in field situations without much prior planning and extra equipment, it has been done with great success. Another method involves collecting insects as normal and then transporting them to a conventional freezer. This can be done prior to sorting invertebrates from accumulated plant material or after. There are several drawbacks with this method, including possible predation of specimens by spiders or predatory insects in the samples. In addition, when vegetation is not first removed, sorting dead invertebrates from vegetation can be time consuming.

Permanent Preservation and Storage

There are many methods for permanent preservation and storage of invertebrates. Each group has standard methods, so consulting with a curator familiar with the group of interest is wise. In general, minute invertebrates are slide mounted or stored in alcohol, whereas larger invertebrates are either dried (hard bodied invertebrates) and mounted or stored in alcohol (softer bodied invertebrates). No matter what method or methods are chosen, humidity, temperature, and light exposure all must be examined. **Optimal humidity and temperature** depend on the storage method (see Museums and Galleries Commission 1992), although in general, cooler and drier conditions are favored. All specimens should be **stored in the dark** to prevent fading.

Slide Mounting

Minute invertebrates often must be slide mounted for identification, although this process is time consuming and impractical for studies with large amounts of material. For this reason, slide mounting representatives of each taxa and then preserving the rest in ethyl alcohol is recommended. There are many different methods of slide mounting, some of which are designed to be temporary (allowing the specimen to be returned to alcohol after study), whereas others are permanent. Noyes (1982) gives a detailed description for permanent slide mounting using Canada balsam. This protocol can be modified so that specimens are cleared (using chemicals to dissolve nonsclerotized portions of the body) and DNA extracted prior to slide mounting. Depending on mounting technique, slides must be periodically examined to ensure that they have not dried out, and if necessary, more mounting material should be added. If invertebrates are slide mounted, storage methods include slide cabinets, slide boxes, and slide folders, all of which keep specimens in the dark and can be organized to allow specimens to be found quickly.

Fluid Preservation

Larger invertebrates with soft bodies (i.e., Ephemeroptera and many noninsect arthropods) should be permanently preserved in ethyl alcohol. Allowing these taxa to dry out often results in the body collapsing, making identification a challenge or even impossible. Specimens should be placed in standard sized vials with tight fitting lids (preferably screw tops with a Poly-Seal™ lining rather than corks). If the vials are not well sealed, alcohol will evaporate over time, resulting in samples drying out. Even with tight fitting lids, samples should periodically be checked and on occasion the alcohol topped off or replaced. Samples should be kept in the dark and, if possible, kept in a freezer, especially those samples containing specimens for molecular study, as Reiss et al. (1995) documented degradation of DNA if stored at room temperature for long periods of time.

Dry Preservation

Invertebrates with **higher degrees of sclerotization** should be dried, mounted, and then placed in appropriate insect storage containers. There are 2 types of dry mounting: pinned or pointed. Small invertebrates (or large invertebrates with very narrow bodies, e.g., some Mantodea or Phasmatodea) should be pointed. **Pointing** involves gluing a specimen to the tip of a pointed piece of heavy paper. Various types of glues can be used, although simple white glue is among the most common, as it is water soluble (allowing for specimens to be soaked and removed from the point if need be) and dries clear. Shellac is another popular glue of choice, but is harder to obtain. Clear fingernail polish is used for small specimens, but the fumes are irritating to some people. Whichever glue is chosen, the smallest amount possible should be utilized, and care must be taken to only put glue on the part of the specimen that will be touching the point. The point should be bent to mimic the angle created by the upright specimen's thorax, and then the specimen should be carefully placed on the bent part. The insect should be straight in all directions (front to back, side to side, and top to bottom). Heavy bodied specimens might require support until the glue dries. A similar method, **card mounting,** is used in some regions, but pointing is the more widespread method in the United States. **Pinning** invertebrates involves sticking a pin through the right side of the thorax. Insect pins can be bought from a number of distributors and should be used instead of conventional sewing or similar pins. Pins come in a variety of sizes, although the smaller sizes should be avoided, as they tend to bend. Again, care must be taken to ensure that the specimen is straight in all directions. Once the specimen has been pinned, it may be necessary to use paper to support the legs so they are near the body until they dry in position. More compact specimens are less likely to be damaged during handling. Some groups are easier to identify if the wings have been spread. This requires a **spreading board,** wax paper, and additional pins. First the specimen is pinned, but rather than pinning through the right side of the thorax it is pinned in the middle of the thorax. Then the specimen is placed in the grove of the spreading board and the wings carefully opened (use a pin to move the wing at the uppermost vein, the costa) and placed on the sides of the board. The wings are then held down with wax paper tacked down with additional pins. Depending on the size of the specimen, it may take days to weeks before the specimen had dried to the point that it can be removed. Although spread specimens are more attractive than unspread specimens, they require more storage space and are easily broken. It may be more practical to spread 1 side rather than both to cut down on space requirements.

Once invertebrates are mounted, they should be placed inside **airtight boxes.** Insect drawers are available from entomological suppliers, such as Bioquip Products, or can be built, provided the seals are airtight. Insects can either be pinned directly into foam on the drawer bottoms or be put into **insect trays** (thick paper boxes with foam on the bottom; Fig. 14.14). Insect trays tend to be easier to work with, as boxes of insects can be manipulated during processing rather than having to individually remove insects from a drawer. Trays come in a variety of standard sizes, allowing for greater flexibility. Insect drawers are typically stored in **airtight metal cabinets** that keep the specimens in the dark (to prevent fading; Fig. 14.15). Additionally, airtight containers help keep out dermestid beetles or other insects, which can destroy an insect collection quickly. Commercially avail-

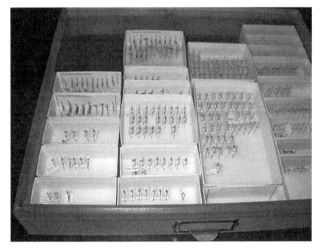

Fig. 14.14. Drawer of labeled insects. Pesticides can be placed inside the drawer or in cabinets to help prevent specimen damage.

Fig. 14.15. Insect drawers are stored in airtight cabinets.

able **repellents,** such as moth balls, should be used to prevent an infestation along with properly closing all drawers and cabinets. If any signs of an infestation are observed, the entire collection should be treated for beetles. Methods of treatment include freezing for several days, heating for several hours, or applying pesticides (see Strang [1992] for lethal temperatures by insect pest).

Specimen Identification

Specimen identification is the most time consuming part of any invertebrate study. Even an hour in the field can result in hundreds (or thousands) of specimens that need to be sorted and identified, if conditions are favorable for insects. To process specimens in the most efficient way possible, first the taxonomic level for identification needs to be set. It is often impractical to identify specimens to the species level due to a lack of available keys and the tendency for many insect species to be delineated based on difficult to interpret characters, such as male genitalia or chaetotaxy (patterns of hairs or bristles). In studies where this level of detail is required, such as some indicator species work, it is recommended that a specialist in the group of interest be consulted. In well-known groups or regions where the invertebrate fauna have been extensively studied, it may be possible to identify specimens to the genus level, but again this may not be practical. The majority of invertebrates sampled for **invertebrate–wildlife research** are identified to the **ordinal or family level.** Placing invertebrates in ordinal groups is typically straightforward, although certain atypical taxa may prove more challenging. However, order may not be detailed enough for some research questions because of the wide range of life histories found in a single order. For example, in the order Hemiptera, some members posses secondary plant chemicals, rendering them unpalatable, whereas others are favored foods of some birds. Additionally, the size of Hemiptera vary from 1 mm to 10 cm in length, making conclusions difficult to draw if size is important to the study question. For this reason, the family level may be a happy medium. There are **keys** readily available to identify insect specimens to family, and keys are available for many other invertebrate groups also. Although some family level classification is in flux, most groups are stable at this level (something that is not true for lower taxonomic levels), which creates less confusion regarding the organisms encompassed by a name.

Most invertebrate keys are created for regional faunas or relatively small taxonomic units. Triplehorn and Johnson (2005) provide keys to North American insect families and orders or families of noninsect arthropods along with brief biological accounts of each family. Keys to invertebrates of specific habitats also are available (Hawking et al. 2009). With the advent of easily accessible Internet access, the numbers of interactive digital keys are rapidly increasing. Interactive keys allow users to start at any place in the key rather than at a predefined starting place (as found in a dichotomous key). This approach allows users to pick features they are comfortable assigning states to, and then the key returns a narrower list of characters to check. Interactive keys are available online, with many including figures and hints to aid identification.

SUMMARY

Invertebrates are important components in wildlife studies that historically have been ignored. This chapter briefly describes important groups of invertebrates, outlines various roles played by invertebrates, discusses collecting techniques, and then finishes with information regarding specimen handling and identification. Although there are hundreds of invertebrate groups, the arthropods (specifically, insects) are the focus of this chapter. Insects play numerous roles in an ecosystem and can be indicators of ecosystem health. Additionally, many wildlife species rely on insects for nutrients during some or all of their life cycle. Not all invertebrates have a positive relationship with wildlife, with many acting as disease vectors. Collecting methods are highly variable, depending on the research question, habitat, and targeted invertebrates. Once specimens are collected, they must be identified and voucher specimens permanently preserved. Incorporating invertebrates into wildlife studies can be time consuming, but they play an important, yet often understudied, role in wildlife ecology.

15

Population Analysis in Wildlife Biology

STEPHEN J. DINSMORE
AND
DOUGLAS H. JOHNSON

INTRODUCTION

WILDLIFE ECOLOGISTS FREQUENTLY ask questions about wildlife populations. How many individuals are there? What are estimates of **vital rates** (birth rate, survival rate) of that population? And, ultimately, is the population increasing or decreasing? A **population,** defined here as a group of organisms of the same species living in a particular space at a particular time (Krebs 1985:157), involves such concepts as birth rate, death rate, sex ratio, and age structure (Cole 1957). These concepts lack meaning for lower levels of biological organization (individuals) or at higher levels (communities or ecosystems). Populations are composed of individual organisms; thus, the study of population dynamics includes the concepts of survival, immigration, and emigration. It is these topics that form the basis for this chapter.

Population analysis also involves the study of **population dynamics:** the changes that occur over time and the causes of those changes. The population could be the number of aphids on a plant, the number of white-tailed deer (*Odocoileus virginianus*) in a woodlot, or the number of snow geese (*Chen caerulescens*) in North America.

Population-level questions are of obvious interest to wildlife managers and scientists. For species that are economic pests, ways of reducing numbers are sought. For game species, managers desire to maintain populations at levels that provide surpluses for harvest. For threatened or endangered species, the goal is to increase their numbers to avoid extinction. To meet any of these objectives, an understanding of the species' population dynamics is the first priority.

The subject of population dynamics includes the number of individuals in a population and the factors that affect that population size: (1) survival of those individuals, (2) their reproduction, and (3) their movements into and out of a population (immigration and emigration). If all factors were known for a population, understanding its dynamics would be straightforward. This is rarely achieved, however, and biologists are forced to make decisions based on an incomplete understanding.

Examples used in this chapter are based on several species, but disproportionate attention is given to a few, especially mallards (*Anas platyrhynchos*), mountain plover (*Charadrius montanus*), and white-tailed deer. This emphasis reflects our own experience and, in some cases, the amount of work that has been done on the species. It is worth recalling Durward Allen's remark that "numbers phenomena tend to be universal. They change only in detail as we shift from fish to fur to fowl" (Allen 1962:36).

A single chapter can only touch lightly on the diversity of techniques used in population analysis. References for further reading are given at appropriate places throughout. For general reference, 3 books stand out. Seber (1982, 1986, 2001; the latter 2 are updates) provided near-encyclopedic coverage of methods used to estimate numbers of wild animals as well as their survival and related parameters. Caughley (1977) described his practical views on population analysis. Williams et al. (2002a) and Amstrup et al. (2006) provide detailed summaries of contemporary population analysis tools, techniques, and practical applications; the serious student is advised to carefully study these texts.

This chapter presents material in several sections. We begin by introducing theoretical models of population growth that form the basis for a discussion of population analyses. We then discuss concepts of parameter estimation and population modeling, population viability analyses and, ultimately, how we make inferences from population data. In each section, we introduce the basic concepts and appropriate formulas, and illustrate them with biological examples. The purpose of showing simple calculations is only to demonstrate them. We acknowledge that today's population analyst will rarely do these calculations by hand and will instead use a computer. The material provides only a general overview of the subject matter, and the reader is encouraged to refer to the specific references in each section for more detailed coverage of that topic.

THEORETICAL MODELS OF POPULATION GROWTH

The growth (or decline) of a population (Box 15.1) can be described by following the number of individuals in the population through time. In the simplest context, the number of individuals (N) in a population at some future time (time $t + 1$) depends on the number of individuals present now (time t) and any gains (births [B] and immigrants [I]) and losses (deaths [D] and emigrants [E]) that occur between times t and $t + 1$. This relationship can be written as a simple difference equation: $N_{t+1} = N_t + B_t + I_t - D_t - E_t$. Although this model is simplistic, it forms an important foundation for building more complex models of population growth. What happens if we place constraints on the relationship between N_t and N_{t+1}? We could consider population growth that is **density independent** (unimpeded population growth) or **density dependent** (population growth depends directly on population density). From this simple equation, we also can discuss concepts of population growth as measured by lambda (λ). The population rate of growth from time t to time $t + 1$ is simply $\lambda = \frac{N_{t+1}}{N_t}$. Is population growth always constant, or does it follow some other pattern?

Population analysis increasingly requires a modeling approach, especially to bridge gaps in knowledge. A **model** is

> **BOX 15.1. THE BEHAVIOR OF SMALL POPULATIONS—THE ALLEE EFFECT**
>
> We have assumed that density dependence will increase mortality rates or decrease birth rates or both as populations become large. Conversely, as populations become small, mortality rates should decline and birth rates increase, according to the model. In reality, small populations may not enjoy favorable demographic rates. Birth rates especially may decline, rather than increase, as populations dwindle. This decline may happen because it is difficult to find mates when the population is small, or because breeding requires social stimulation. Another possibility is illustrated by colonial nesting birds, in which larger colonies provide greater protection from predators and greater reproductive success (Birkhead 1977). The phenomenon of increased mortality rates or decreased birth rates at low population levels is known as the **Allee effect,** after W. C. Allee, who documented numerous situations in which it was manifested (Allee 1931).

an abstraction of a real system that enables us to think more clearly about the real one. Any model must sacrifice at least 1 of 3 desiderata: generality, realism, or precision (Levins 1966). A model may be a complex mathematical exercise, incorporating thousands of variables and equations, and requiring hours of computer time to analyze. Or it may be a simple heuristic concept, such as barn owls (*Tyto alba*) lay larger clutches of eggs in years when food resources are abundant. Which kind of model is appropriate to a scientific or management application depends on the objectives of the model.

This chapter presents a variety of models, from simple, involving only a single parameter, to complex, involving numerous rates and relationships. A **parameter** is something that describes a population (e.g., the annual survival rate of male mallards). We denote such estimates with a "hat" symbol (thus, $\hat{S}$ is an estimate of S) and calculate appropriate measures of uncertainty (typically standard errors). However, we can rarely measure all individuals in the population. Thus, a parameter is an unknown numerical characteristic of the entire population. There is a general trade-off between model complexity and realism. Simple models are relatively easy to interpret, but lack a sense of realism when applied to actual biological situations. Conversely, complex models may be more realistic, but suffer because we seldom have sufficient data to support them.

In addition to varying in complexity, models of populations may vary in other attributes. In **discrete-time models,** events, such as births, occur only at certain times, such as a short breeding season within a year. In **continuous-time models,** events can occur throughout time. Or, as Starfield and Bleloch (1991:9) expressed it, time jumps in a discrete model; time flows in a continuous one. Another distinction is whether random components are included in a model. In a **deterministic model,** parameter values are fixed through time, and the result from the model depends only on the values of the variables. In a **stochastic model,** certain parameters vary randomly; their statistical distributions, rather than exact values, are specified. If the variation of the system is important, stochastic models are usually more suitable than deterministic ones. In the following sections we discuss ways of estimating population change under these and other scenarios.

Populations with Unimpeded Growth

The simplest model assumes the number of animals in a population (N) goes up (or down) by a constant ratio, say λ, during each unit of time (which we will assume is a year). That is, at time $t + 1$, the population size is λ times its value at time t:

$$N_{t+1} = \lambda_t.$$

The population is increasing if λ 1, constant if $\lambda = 1$, and declining if $\lambda > 1$. Sometimes λ is called the **finite rate of population increase.** This formulation is geared toward organisms that reproduce during a short breeding season (discrete growth; **birth-pulse fertility** of Caughley [1977]). If N_0 is population size at some initial year, then repeating the above equation t times gives:

$$N_t = \lambda^t N_0. \quad (1)$$

Consider the natural population of whooping cranes (*Grus americana*), which has been monitored on their wintering ground since 1938. Virtually the entire natural population congregates near Aransas National Wildlife Refuge in Texas. The actual counts (Table 15.1) during 1938–2001 were fitted to equation 1, where year t has been recoded, so that 1938 is year 0 and $\hat{\lambda} = 1.0363$ is an estimate of λ (the hat symbol denotes an estimator of a parameter). Thus, the crane population was growing similarly to a bank deposit with an interest rate of 3.63%, compounded annually (Fig. 15.1). Binkley and Miller (1980, 1988), Boyce and Miller (1985), Boyce (1987), and Nedelman et al. (1987) have presented more detailed analyses of this population.

An alternative expression for population growth has some advantages. If we replace λ by e^r, then $\lambda^t = e^{rt}$. Here e is the base of natural logarithms and r is the **instantaneous rate of increase.** Among the advantages of the exponential formulation (Caughley 1977:52) is the ability to convert easily between time units; for example, if the growth rate of a

Table 15.1. Counts of wintering whooping cranes, Aransas National Wildlife Refuge, Texas, 1938–2001

Year	Adults	Young	Year	Adults	Young
1938	14	4	1970	51	6
1939	15	7	1971	54	5
1940	21	5	1972	46	5
1941	14	2	1973	47	2
1942	15	4	1974	47	2
1943	16	5	1975	49	8
1944	15	3	1976	57	12
1945	18	4	1977	62	10
1946	22	3	1978	68	7
1947	25	6	1979	70	6
1948	27	3	1980	72	6
1949	30	4	1981	71	2
1950	26	5	1982	67	6
1951	20	5	1983	68	7
1952	19	2	1984	71	15
1953	21	3	1985	81	16
1954	21	0	1986	89	21
1955	20	8	1987	109	25
1956	22	2	1988	116	18
1957	22	4	1989	126	20
1958	23	9	1990	133	13
1959	31	2	1991	124	8
1960	30	6	1992	121	15
1961	34	5	1993	127	16
1962	32	0	1994	125	8
1963	26	7	1995	130	28
1964	32	10	1996	144	16
1965	36	8	1997	152	30
1966	38	5	1998	165	18
1967	39	9	1999	171	17
1968	44	6	2000	171	9
1969	48	8	2001	161	15

From Boyce (1987), W. G. Jobman (U.S. Fish and Wildlife Service, personal communication).

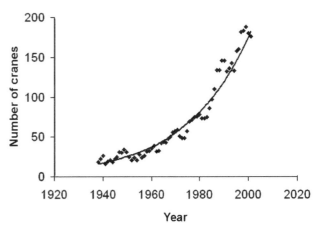

Fig. 15.1. Counts of wintering whooping cranes at Aransas National Wildlife Refuge, Texas, 1938–2001, fitted to an exponential curve.

population per year is 0.10, the growth rate per day is 0.10/365. Also, the time it takes a population to double is $(\log 2)/r = 0.69315/r$ (note: throughout this chapter natural logarithms are used). This formulation is particularly appropriate for continuously growing populations (**birth-flow fertility** of Caughley [1977]), but it works as long as the times (t) when the population is counted represent comparable times in the life cycle, such as the beginning of the breeding season. Then r is $\log(\lambda)$, and equation 1 becomes:

$$N_t = N_0 e^{rt}. \qquad (2)$$

For the whooping crane population, we obtain $\hat{r} = \log(\hat{\lambda}) = \log(1.0363) = 0.03566$, on an annual basis. Thus, the population is growing 3.566% per year and will double every 19.4 years (0.69315/0.03566), should growth continue at this rate.

This model is termed the **exponential growth model,** and it may be realistic when growth is unhindered (i.e., resources are ample and competition is not a factor). Such situations often occur when a species initially invades an optimal habitat, or, as with whooping cranes, when a population rebounds from near extinction and the habitat is not limited. It also can be useful for short-term forecasts (Eberhardt 1987b). This approach is deterministic; that is, no allowance is made for variation caused by randomness or by variables not included in the model. It can be made stochastic (incorporating random events) by considering chance variations in births and deaths (Pielou 1969:13–16).

Estimating r from Counts

The simplest way to estimate the growth rate r in equation 2 from population counts is to take logarithms of both sides, giving:

$$\log(N_t) = \log(N_0) + rt. \qquad (3)$$

Linear regression of $\log(N_t)$ on t for a series of years provides estimates of the regression coefficient (slope, equal to r) and the intercept. These values can be transformed by exponentiating to provide estimates of λ and, if desired, N_0. As an example, consider the whooping crane counts during 1938–2001 (Table 15.1). A linear regression of the logs of the counts (adults plus young) against year (recoded so that 1938 = 0) provides a slope of $\hat{r} = 0.03880$ (the estimated standard error [SE] = 0.00112) and an intercept of 2.7638 ($SE = 0.0408$). Thus, $\hat{N}_0 = e^{2.7638} = 15.9$ ($SE = 0.6471$ by the delta method, a procedure for obtaining estimated standard errors of functions of random variables; Seber 1982:7).

An alternative to linear regression on transformed variables is to use nonlinear regression directly on equation 2. Using the whooping crane data gives $\hat{r} = 0.0421$ ($SE = 0.00123$) and $\hat{N}_0 = 14.0$ ($SE = 0.9129$). These estimates differ from those given previously, because the analytic methods are based on different assumptions. Because errors in estimated population sizes are likely to increase with true population size, the assumption of constant error variance, used in ordinary least squares regression, is more likely to be met with the linearized form of the model represented by equation 3 than by equation 2. For this reason, the linear approach is usually preferred.

Estimating r from Changes in Population Size

The form of the exponential growth model lends itself to another method for estimating r. Consider the ratio of population sizes in successive years. From equation 2, this ratio is

$$\frac{N_t}{N_{t-1}} = \frac{N_0 e^{rt}}{N_0 e^{r(t-1)}} = e^r = \lambda.$$

Thus, the logarithms of the average of these ratios can be used to estimate r. For the whooping crane example, counts for 1938–2001 provided 63 ratios N_t/N_{t-1} that averaged 1.0454 ($SE = 0.0164$). The logarithm of this average gives the estimate $\hat{r} = 0.0444$ ($SE = 0.0157$).

In a comparison of the 3 estimates of whooping crane population growth (not shown here), the fit provided by the linearized model (equation 3) was second best, that of the nonlinear fit (equation 2) was best, and that of the ratios in successive years was worst. Eberhardt (1987b) discussed other estimation methods for this model, including ratio estimators with various weights. He also considered variance estimators. McCullough (1982, 1983) and Van Ballenberghe (1983) present a spirited discussion of the estimation of population growth of a white-tailed deer herd.

Populations with Density-Dependent Growth
Continuous-Time Formulation

We now consider the number of bison (*Bison bison*) on the National Bison Range, Montana, during 1909–1922, when no harvesting occurred (Table 15.2). Fitting the linearized model (equation 3) to the first 10 years of data gives $\hat{N}_0 = 53.45$ ($SE = 1.53$) and $\hat{r} = 0.216$ ($SE = 0.00535$). The observed number of bison at the end of each of those years fits the exponential curve nicely (Fig. 15.2). Projections for 1919–1922, however, are consistently higher than actual numbers (Fig. 15.2). It is conceivable the slowdown in population growth can be attributed to a density-dependent response. The proportional annual change in population ($\hat{R}$) is negatively correlated with population size ($r = -0.77$, $P = 0.001$; here r denotes the correlation coefficient, not the population growth rate, a distinction between usages that should be clear from the context; Fig. 15.3).

It is impossible for any population to continue to grow indefinitely at a constant rate. Most likely, growth will slow as the population becomes large, and some limiting factor exerts an influence. Density dependence is likely to operate (Box 15.2). How can density dependence be included in the model to make it more realistic and useful? In the model of equation 2, the population growth rate per animal,

$$\frac{1}{N_t} \frac{dN_t}{dt} = r,$$

Table 15.2. Counts of bison on the National Bison Range, Montana, 1909–1922

Year	Number at start of year	Number born	Young deaths	Number at end of year
1909	37	11	0	48
1910	51[a]	19	0	70
1911	70	16	1	85
1912	85	19	0	104
1913	104	26	0	130
1914	130	34	0	164
1915	164	32	2	194
1916	194	47	1	240
1917	240	56	1	295
1918	295	73	1	367
1919	367	58	5	420
1920	420	68	9	479
1921	479	82	7	554
1922	554	85	4	635

From Fredin (1984).

[a] Three animals were added to the existing herd.

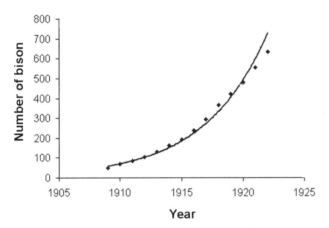

Fig. 15.2. Bison counts on the National Bison Range, Montana, 1909–1922, fitted to an exponential curve.

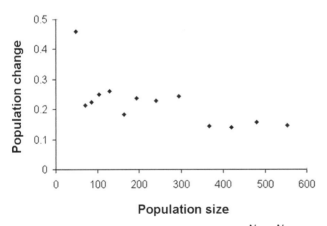

Fig. 15.3. Proportional change in bison population, $\frac{N_{t+1} - N_t}{N_t}$, versus population size, N_t, on the National Bison Range, Montana, 1909–1922.

is constant, regardless of population size. One way to make it depend on population size is to multiply it by a factor that has negligible effect when the population is small, but reduces the growth rate to zero as the population approaches some limit, K (which we might call the carrying capacity). The term $(K - N)/K$ does just that (using this term is the simplest way; there are many others, reviewed by May [1973: 80–81]). It is nearly 1 when N is small, and converges to 0 as N approaches K. If we use this factor, the per capita growth rate becomes

$$\frac{1}{N_t}\frac{dN_t}{dt} = r_m \frac{K - N_t}{K}, \tag{4}$$

where r_m, the maximum rate of growth, replaces r. Equivalently,

$$\frac{dN_t}{dt} = r_m N_t \frac{K - N_t}{K},$$

which can be interpreted (Krebs 1985:213) as rate of increase per unit of time, $\frac{dN_t}{dt}$, equals maximum rate of population growth per capita, r_m, times population size, N_t, times unused opportunity for growth, $\frac{K - N_t}{K}$. Note the factor modifying growth rate depends both on population size (through N) and on the environment (through K). The equation has the solution

$$N_t = \frac{K}{1 + e^{a - r_m t}}, \tag{5}$$

which is known as the **logistic equation.** K is the asymptote or carrying capacity, and r_m is the maximum rate of population growth, the rate that would result if the population was free of constraints caused by the density of the population. The value of the parameter a depends on the time origin used; it measures the size of the population at time 0 relative to the asymptotic (infinite time) population size. Setting $t = 0$ in equation 5, we find $a = \log[(K - N_0)/N_0]$.

The logistic equation has a rich history (Hutchinson 1978, Kingsland 1995) and is a mathematically convenient model that describes growth of a variety of populations. It is not, however, any sort of *law* of population growth. Among the assumptions (e.g., Pielou 1969:30, Poole 1974:63, Krebs 1985:220) are: (1) all individuals, regardless of age, gender, or genotype, are equivalent with respect to survival, reproduction, and susceptibility to crowding; (2) carrying capacity (K) is constant; (3) growth rate of the population responds instantaneously to population size; and (4) the effect of population size on growth rate is linear. Each of these restrictive assumptions can be relaxed, but with a loss of mathematical tractability. Models in which assumptions 1 and 2 are eased are discussed later. Assumption 3 can be modified either with discrete-time formulations or by introducing time lags into the continuous-time model (e.g., May 1973, Krebs 1985:224). One extension that overcomes assumption 4 is the generalized logistic equation (Gilpin and

> **BOX 15.2. HOW DOES DENSITY DEPENDENCE WORK?**
>
> The exact role of density dependence in population regulation has long been a source of controversy (e.g., Krebs 1985:327–347). Density may influence survival or reproduction rates only at extreme densities (e.g., Strong 1986), exact values of which depend on the quantity and quality of habitat available.
>
> Knowlton (1972) presented evidence suggesting a relation between population density and fertility when he compared the average number of uterine swellings per female coyote (*Canis latrans*), an index to fertility, with intensity of efforts used to control coyotes in 7 counties in south Texas. Although sample sizes were limited and levels of coyote control were not randomly assigned to counties, there is a suggestion of an effect of population density on this index of fertility.
>
> | Control effort | County | Sample size | Average number of uterine swellings per female coyote | Average number of uterine swellings per treatment |
> |---|---|---|---|---|
> | Intensive | Zavala | 8 | 8.9 | 7.2 |
> | | Dimmit | 12 | 6.4 | |
> | | Uvalde | 10 | 6.2 | |
> | Moderate | Jim Wells | 21 | 5.3 | 4.5 |
> | | Hildago | 11 | 3.7 | |
> | Light | Jim Hogg | 17 | 4.2 | 3.5 |
> | | Duval | 11 | 2.8 | |

Ayala 1973, Eberhardt 1987b). Pielou (1969:22–30) and others have considered stochastic versions of the logistic model.

Discrete-Time Formulation

The logistic formulation specifically applies to continuously reproducing organisms, although it suffices for populations with discrete breeding seasons if population size is measured at the same time each year, as with the bison example. Otherwise, for animals with discrete breeding seasons, the discrete counterpart of equation 4 is

$$\frac{N_{t+1} - N_t}{N_t} = r_m \frac{K - N_t}{K},$$

with solution

$$N_{t+1} = N_t + r_m\left(1 - \frac{N_t}{K}\right)N_t. \quad (6)$$

This discrete version implicitly has a time delay; the population growth rate at time $t + 1$ depends on population size at time t. In contrast, equation 4 assumes the rate of population change responds instantaneously to changes in the size of the population. Because of the time delay in the discrete version, the behavior of the modeled population depends strikingly on the values of the parameters. The population can smoothly approach the asymptote, approach it in an oscillatory manner, cycle indefinitely, or fluctuate chaotically, depending on the value of r_m (May 1974, May and Oster 1976). That model, along with the realization that a simple deterministic mechanism could produce such a striking array of random-appearing behavior, was one of the early discoveries of what has now become the study of chaos.

Nonlinear least squares can be used directly on equation 5 to estimate the parameters of the logistic equation. Using the bison data for 1909–1922 gave estimates of $\hat{K} = 1,172$ ($SE = 77.4$) for the asymptote, $\hat{r}_m = 0.2479$ ($SE = 0.0078$) for the rate parameter, and $\hat{a} = 3.069$ ($SE = 0.046$) for the origin parameter.

An alternative is to use the discrete form of the logistic model. From equation 6, we have

$$N_{t+1} = N_t + r_m\left(1 - \frac{N_t}{K}\right)N_t = N_t(1 + r_m) + N_t^2\left(\frac{-r_m}{K}\right),$$

and we can perform a regression of N_{t+1} on N_t and N_t^2, excluding an intercept term. The coefficient of N_t will be an estimate of $(1 + r_m)$, and the coefficient of N_t^2 will estimate $-r_m/K$. For the bison example, we obtain $(1 + r_m) = 1.2669$ ($SE = 0.0266$) and $\hat{r}_m = 0.2669$ ($SE = 0.0266$). The estimate of $-r_m/K$ is -0.000238 ($SE = 0.000061$) and $\hat{K} = 0.2669/0.000238 = 1,121.43$ ($SE = 183.24$).

One statistical difficulty with this regression approach is the assumption that explanatory variables, in this case N_t and N_t^2, are measured without error (Walters 1986:136). This is not a problem in the present case, because we believe the bison counts are exact, but the problem arises in most situations. We can illustrate the effect of measurement errors by reanalyzing the bison data, except that we include a small multiplicative error (each count is multiplied by e^z, where z is a normal random deviate with mean zero and standard

Table 15.3. Actual counts of bison on the National Bison Range, Montana, 1909–1922, and counts with multiplicative error

Year	Actual count	Count with error
1909	48	48
1910	70	64
1911	85	84
1912	104	100
1913	130	144
1914	164	178
1915	194	199
1916	240	227
1917	295	292
1918	367	389
1919	420	387
1920	479	496
1921	554	598
1922	635	553

deviation = 0.1; Table 15.3). Results from this analysis give $\hat{r}_m = 0.4295$ ($SE = 0.1138$) and $\hat{K} = 594.88$ ($SE = 72.43$), values far different from estimates obtained using values measured without error (0.2669 and 1,121.43).

Of the 2 estimation techniques applied to the bison data, the nonlinear regression applied to equation 5 gave a better fit (higher r^2 value) than did linear regression of N_{t+1} on N_t and N_t^2. That superiority may not hold in general.

Some Dangers of Detecting Density Dependence

Discovering density dependence in a series of counts of a population is less straightforward than it might appear. First, population size and change in population tend to be negatively correlated, even if the change occurs independently of population size (e.g., Maelzer 1970, St. Amant 1970). Second, any uncertainty in estimating population size tends to add to the appearance of density dependence.

Consider a hypothetical example (Table 15.4), in which we started with $N_0 = 1,000$ animals; $\log(N_0) = \log(1,000) = 6.91$. Adding a random number to the logarithm of the previous population generated the population in each successive year gives

$$\log(N_{+1}) = \log(N_t) + z,$$

where z is a random deviate with mean zero and standard deviation = 0.1. Although z was generated independently of N_t, a negative correlation between the 2 variables was induced; in the example shown (Table 15.4), we have $r = -0.62$ ($P = 0.055$). The reason for this surprising result is that, even in an irregular sequence of numbers, an unusually high value tends to be followed by a decrease (if it was more likely followed by an increase, then it would no longer be an unusually high value), and vice versa (St. Amant 1970). Thus, a negative correlation between population change and previous population size cannot be construed as evidence for density dependence.

If the counts had been made subject to error, the situation is even worse. The appearance of density dependence increases, as the following illustrates. Suppose the population was underestimated in a particular year; this error will make the observed population size in that year more likely to be small than large. Also, it will make the change in observed population size larger than it should be, unless the population is underestimated again the following year. Thus, a smaller-than-expected population size will be associated with a larger-than-expected population change, and a negative correlation will be induced. Consider again the hypothetical example (Table 15.4), except that now the counts were measured with error rather than exactly (call the observed counts O_t):

$$\log(O_t) = \log(N_t) + y,$$

where y is another random deviate, normally distributed with mean zero and standard deviation = 0.2. The correlation between observed population change and observed population size is stronger ($r = -0.73$, $P = 0.017$) than the correlation between true values.

From this observation, we conclude that density dependence should not be inferred from regression analysis on counts of populations, even if they are measured exactly. The same problem arises when performing a regression of $\log(N_{t+1})$ on $\log(N_t)$; regression coefficients 1.0 are expected, even if there is no density dependence (Maelzer 1970). Eberhardt (1970), Slade (1977), Solow (1990), and especially Pollard et al. (1987) presented additional cautions. The converse problem also arises: Gaston and Lawton (1987) found that methods for detecting density dependence from census data consistently failed to do so, even for populations

Table 15.4. Hypothetical example illustrating the appearance of density dependence from annual counts of a population that varies randomly from year to year[a]

Year (t)	$\log(N_t)$	$\Delta_t(N)$	$\log(O_t)$	$\Delta_t(O)$
0	6.91		6.76	
1	7.15	0.24	6.95	0.19
2	7.18	0.03	7.58	0.63
3	7.12	−0.06	7.03	−0.55
4	7.10	−0.02	7.19	0.16
5	7.03	−0.07	7.23	0.04
6	7.06	0.03	7.12	−0.11
7	6.96	−0.10	6.94	−0.18
8	6.87	−0.09	6.89	−0.05
9	7.02	0.15	6.77	−0.12
10	7.12	0.10	7.22	0.45

[a] N_t is actual population size in year t, O_t is observed population size, $\Delta_t = \log\left(\frac{N_t}{N_{t-1}}\right)$ is actual change in population size, and $\Delta_t = \log\left(\frac{O_t}{O_{t-1}}\right)$ is observed change in population size.

known (from independent evidence) to be subject to density-dependent processes.

Immigration and Emigration

Dispersal is a critical process that allows individuals to persist despite degradation of the habitat they currently occupy. Virtually all plant and animal species exhibit dispersal during at least 1 life stage. Caughley (1977:57) defined **dispersal** as the movement an animal makes from its point of origin to the place where it reproduces. He distinguished it from other types of movements, namely, local movement within a home range and migration (back-and-forth movements between discrete locations). Although dispersal is important in population dynamics, it is hard to detect and harder yet to measure. A biologist conducting a population analysis typically ignores dispersal, assumes it to be nonexistent, or hopes that immigration and emigration cancel one another.

Several texts (e.g., Pielou 1969, Poole 1974, Caughley 1977) have discussed models of dispersal, but its estimation has received little attention. Most techniques for detecting or estimating dispersal rely on marking animals and observing where they go or recapturing them. Recently, other tools, such as genetic markers, have been used to address dispersal in animals (see Clobert et al. 2001).

Observation of marked animals has provided most of the evidence of dispersal, including direction, distance, time of occurrence, and length of time between sightings. Fortuitous records, such as a coyote (*Canis latrans*) being trapped a long distance from where it had been marked, are interesting and informative, but reveal little about the dispersal patterns of coyotes in general. For a more complete picture, telemetry studies are needed in which all radio-equipped animals can be followed.

Mark–recapture studies can provide estimates of losses or gains to the population between trapping occasions, although they ordinarily are used to estimate size of a population. With certain designs, losses can be partitioned into deaths and emigration, and gains can be separately estimated as births and immigration (Jackson 1939, Krebs 1985:169, Manly 1985:41–43). Nichols and Pollock (1990) presented a procedure for separately estimating births and immigrants from a robust design mark–recapture study involving primary periods of trapping (well separated in time) and secondary trapping periods (closely spaced in time). Zeng and Brown (1987) proposed a method for distinguishing emigration from death in mark–recapture studies, but it requires recapture of all animals that are still alive and have not dispersed. By comparing estimated survival rates based on local mark–recapture studies (which incorporate probabilities of surviving and returning to the study area) with survival rates from banding studies (which incorporate only survival), one can estimate the return rate (e.g., Anderson and Sterling 1974, Hepp et al. 1987) as 1 minus the probability of permanent dispersal from an area.

Hestbeck et al. (1991) developed models for resighting of individually marked Canada geese (*Branta canadensis*) in the Atlantic Flyway, United States. In 3 years, nearly 29,000 geese were marked and 102,000 resightings were made. The models included survival and resighting probabilities, and the probability of movement from one region to another in successive winters. They found that annual changes in movement probabilities corresponded to variation in the severity of winter. A model incorporating memory and tradition better fit the data, indicating that wintering location of a goose depended not only on where it spent the previous winter, but also on where it had been 2 years before.

Birth and Death Models

We first recognize that population growth is the net result of births and deaths in the population (ignoring emigration and immigration), and refer again to the simple difference equation $N_{t+1} = N_t + B_t + I_t - D_t - E_t$. From this equation, birth rate $\left(b = \dfrac{B_t}{N_t}\right)$ and death rate $\left(d = \dfrac{D_t}{N_t}\right)$ can be defined as simple proportions. In many cases, we can analyze birth and death processes separately, because they may be affected by different environmental variables. Consider again the bison example used earlier. Counts of bison can be divided into young of the year and adult age classes (Table 15.2), and the count of young of the year can be considered the final outcome of the birth process, birth here including not only parturition, but also survival until autumn. The number of deaths of adults also was recorded.

Estimates of annual birth and death rates (defined per individual in the bison population at the start of the year) varied. Birth rates were lower when the population was larger (Fig. 15.4), and death rates increased with population size (Fig. 15.5). If birth and death rates are similar functions of population size, as appears to be true for the bison data, we can work with their difference rather than with individual components. It may be that only one or the other of birth and death rates is density dependent, or the nature of the

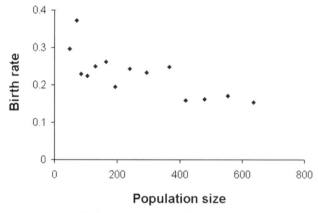

Fig. 15.4. Bison birth rates versus population size at start of year on the National Bison Range, Montana, 1909–1922.

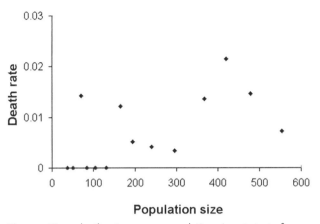

Fig. 15.5. Bison death rates versus population size at start of year on the National Bison Range, Montana, 1909–1922.

relationship differs for the 2 processes. Thus, we should treat the 2 processes separately. Suppose, for example, that instantaneous birth rates (logarithms of finite birth rates) varied with density, say,

$$b = b_0 + b_1 N,$$

but that instantaneous death rates were independent of density:

$$d = d_0.$$

Then the population rate of increase, $r = b - d$, is thus

$$r = (b_0 - d_0) + b_1 N,$$

and it depends on the population size N. That is, population growth is density dependent, but an analysis of population size alone would not indicate which of the 2 processes, birth or death, was density dependent.

Pielou (1969), among others, provided an introduction to stochastic birth and death processes. De Angelis (1976) applied these models to a population of Canada geese.

Estimating Birth Rates

The **fertility** of a population is the number of live births produced over some period of time, generally a year. Because it usually suffices to study the female segment of a population, fertility often is expressed as young females produced per female in the population. A related parameter is **fecundity,** the potential level of reproductive performance of a population, which is ordinarily much greater than the realized reproduction (fertility; the terms fertility and fecundity are not used consistently). A related term, **recruitment,** refers to the addition of new individuals (typically only breeding individuals) to the population through reproduction. To calculate fertility in mammals, for example, we need to know average litter size, average number of litters produced per time interval (year), and the sex ratio at birth (Caughley 1977).

Estimation of fertility rate has received only scattered attention. Typically, rates are based on different criteria for different species groups. For mammals, number of live births is an appropriate measure. For fish, reptiles, and birds, number of eggs laid or number that hatch often is used. It is ordinarily difficult to see newborn young of many species. Thus, fertility is often assessed by the number of young produced that attain a particular life stage or size. In waterfowl studies, some measures of reproduction involve counts of ducklings that are nearly ready to fledge or are members of the autumn (hunted) population. For populations with synchronized, seasonal breeding (birth-pulse fertility), we calculate the number of births for females in a given age **class.** For more-or-less continuous breeders (birth-flow fertility), the number of births for females in a specified age **interval** is appropriate.

Estimation of birth rate, especially by age class, is difficult. Three general approaches are discussed. The first uses age ratios based on direct counts. The second uses mark–recapture methods, and the third involves a potpourri of indirect measures. In addition, change-in-ratio techniques may be applicable in some limited circumstances (e.g., Hanson 1963, Seber 1982:382).

Age Ratios Based on Direct Counts

With whooping cranes or bison, we could count exactly the young produced (and surviving until time of census). More generally, biologists often relate number of young seen to number of adults seen and obtain an index to fertility. For example, ratios of fawns to does may be used for white-tailed deer, number of placental scars in harvested fox squirrels (*Sciurus niger*), or number of successful nests or broods for birds or American crocodiles (*Crocodylus acutus*). For hunted species, the age ratio in the harvest is an index to recruitment that must be adjusted for differential vulnerability of age classes (e.g., Martin et al. 1979). A variety of errors can creep into calculations of fertility. For example, does without fawns may be less conspicuous than those with fawns, squirrels that had borne young may be more likely to be shot (and fertility thus measured) than those that did not, or successful bird nests may be more likely found than unsuccessful ones (Mayfield 1961).

More fundamentally, all components of reproduction should be considered to gain a full understanding of the process. These include age at which animals first breed, incidence of nonbreeding among adults of breeding age, number of breeding cycles per year, size of clutch or litter, and survival to adult stage. For management purposes, a consistent index to reproduction, along with its standard error, will often suffice.

Mark–Recapture Methods

These methods have been described for both **closed populations** (which do not change in number during the period of interest) and **open populations** (which allow births, deaths, emigration, and immigration; Lancia et al. 2005). The primary method for open populations involves the Jolly–Seber model, although the more restrictive Cormack–Jolly–Seber model often is used in practice. This model yields estimates not only of population size, but also of the number of individuals added to the population between trapping occasions (including births and immigrants) and the number removed from the population (including deaths and emigrants). If one can safely assume that no immigration has occurred, the estimated number of additions to the population is a measure of births, although standard errors are usually large. Another approach is to estimate the recruitment rate (f_i) of the population using a reparameterized Jolly–Seber model (Pradel 1996) and program MARK (White and Burnham 1999). Here, f_i is defined as the number of new animals in the population at time i per animal in the population at time $i - 1$. It is calculated using information about population size and apparent survival (ϕ_i; see section below on Estimating Survival Rates) as

$$f_i = \frac{N_{i+1} - N_i \phi_i}{N_i}.$$

Indirect Measures

Reproductive success of a population often is evaluated in terms of some component of reproduction. For example, clutch size and nest success are commonly used as measures in bird studies. Among mammals, characteristics of female reproductive tracts may be used (e.g., Harder and Kirkpatrick 1994). These measures might be perfectly adequate indices to reproduction, but they are only part of the picture. Other factors must be considered, for by focusing on only 1 or 2 of the components of reproduction, we implicitly assume the others vary only slightly, if at all.

Consider how duck productivity might be monitored on a refuge, for example. Counts of breeding pairs are used to estimate the size of the population, and studies of nests give an estimate of nest success rate (Cowardin and Blohm 1992; Box 15.3). These 2 variables may be key components in reproduction, but other variables could have a major, or even a dominant, role (Johnson et al. 1992). Some members of the population might not breed, for example. Some might renest after a nest failure, whereas others do not. Clutch size can differ, and individual eggs may be depredated from nests or not hatch for other reasons. Finally, of the ducklings that hatch from a successful nest, the proportion that ultimately fledges may vary from 0 to 100%. Hence, reliance on only a subset of the components can cause misleading conclusions.

Estimating Survival Rates

In addition to recruitment, survival is an important component of population growth and decline. Here, we first distinguish between **true survival** (the real probability of living) and **apparent survival** (the product of true survival and fidelity). In this context, fidelity refers to an individual's availability for sampling, which naturally implies that it remains faithful to a particular area and does not emigrate. Only if fidelity is 100% will apparent survival equal true survival. Relative to true survival, apparent survival will be biased low when there is permanent emigration. Due largely to the design of the studies we conduct, apparent survival is much easier to estimate than is true survival, because our studies are limited in space, and we may not be able to account for permanent emigrants that tend to bias survival estimates low. Furthermore, we seldom are able to observe the fates of all individuals in a population and use this information as a measure of survival. Increasingly, we rely on estimates of survival from a sample of marked individuals. Thus, there are 5 basic approaches to estimating a survival rate (many involve estimating the complement of survival, the mortality rate, which is equal to 1 − survival rate); each requires a different sort of data.

Observed Survival

In some studies, survival can be observed directly when deaths in the population are known. For captive or other closely watched populations, survival rates can be calculated without difficulty. In other studies, markers attached to animals allow biologists to observe a subset of a population. Radiotelemetry especially affords an opportunity to monitor animals closely and record instances of mortality. Fuller et al. (2005) describe several methods of estimating survival rates from radiomarked animals. The same methods can be used for animals marked in other ways, as long as markers are retained and marked individuals can readily be found. Particularly troublesome are instances in which an animal's signal or marker cannot be located, so the observer is not sure whether the transmitter failed, the animal (along with its transmitter) was destroyed, or the animal left the study area. Also, radio packages (Fuller et al. 2005) or other markers (e.g., Brodsky 1988, Kinkel 1989) may influence behavior and survival. Further, telemetry studies usually have been limited by small samples of animals and relatively short durations.

Ratios of Population Sizes or Indices

If no movement into or out of a population occurs, the mortality between times t and $t + 1$ is the population size at time t minus the number of those that still remain at time $t + 1$. If those survivors can be distinguished from the young that were added to the population, survival can be computed directly. Consider the whooping crane example again. This is a rare example, where all individuals in the population are known, so there were no immigrants. Adults counted in one winter thus represent the survivors of the total population (adults plus young) in the previous winter. From ratios of these counts, survival rates can be com-

Box 15.3. A common bias in estimating nest success

Among birds, one of the most important factors affecting size of a population is the percentage of nests from which young are successfully fledged. Fortunately, it is a parameter that managers often can influence by manipulating habitat or predation rates.

Many studies and monitoring programs of nest success fell victim to a serious bias. Biologists reported nest success rate to be the percentage of the successful nests among the sample of nests they found. This intuitively reasonable procedure is acceptable if all nests can be found at initiation, or if destroyed nests are as likely to be found as successful ones. Many species of birds, however, are secretive when nesting, and biologists are most likely to find nests when tended by an adult. Once destroyed, a nest will be abandoned by the adults and may be difficult to detect by usual nest-searching methods (e.g., Klett et al. 1986). An adult, in contrast, will tend a successful nest from initiation until the young leave. For that reason, it will more likely be found. This disparity in chances of detecting failed and successful nests introduces a major bias into the usual measure of nest success rate. Harold Mayfield (1961) was among the first to recognize this problem, and he proposed a solution. He recommended computing a **daily mortality rate** for nests, based on the number of nests destroyed divided by the total number of days nests were under observation. Subtracting this value from 1.0 gives a daily survival rate, which, when raised to a power equal to the number of days needed for a nest to proceed from initiation to success, gives a much better estimate of the true nest success rate. For example, Dinsmore et al. (2002) studied the nesting success of mountain plovers in Montana. Their apparent nesting success estimate was 0.58. The Mayfield daily survival estimate was 0.9740 (SE = 0.0021). The incubation period is 29 days for this species, so the estimated proportion of successful nests was 0.9740^{29} = 0.47. Johnson (1979b) presented a statistical model and standard errors for the Mayfield procedure. Johnson and Shaffer (1990) outlined situations in which the Mayfield method performed better than the apparent method.

The Mayfield method is clearly superior to estimates of apparent nest success, but it is not always biologically reasonable. In natural settings, daily survival of nests would hardly be expected to be constant in time. Instead, nest survival probably varies as a result of season, habitat, experience of the incubating adult, and other factors. Johnson (1979b), Bart and Robson (1982), and Klett and Johnson (1982) discussed models that relax the assumption of constant nest survival. Additional developments in nest survival modeling include random individual nest effects (Natarajan and McCulloch 1999); daily nest survival across nest stages (e.g., the transition between incubation and nestling stages), where the exact transition date between stages is unknown (Stanley 2000); and the effects of nest checks on the daily nest survival rate following a visit (Rotella et al. 2000). The Bart and Robson (1982) nest survival model is now incorporated into program MARK (Dinsmore et al. 2002) and offers greater flexibility to explore factors affecting daily nest survival rates of birds. Shaffer (2004) discusses how flexible models can be fit using logistic regression.

---

puted. In 1938, for instance, there were 18 birds (14 adults and 4 young; Table 15.1). Of these, 15 were still alive in 1939, giving a survival rate of $15/18 = 0.83$ ($SE = 0.09$, as a binomial variate). Survival rates for other years can be similarly calculated.

It is unusual to have exact counts from which survival rates can be calculated, such as we have for the cranes. Often, however, indices to population size are available; if they are representative of a constant proportion of the population, they can be used equally as well. Consider as an example results from a banding study of female mallards in Minnesota (Table 15.5). In 1968, 338 adult females were banded. Assume that hunters took equal proportions of the banded populations in the 1968 and 1969 hunting seasons (a conclusion supported by a more rigorous analysis in Johnson [1974]; Table 15.5). Thus, the 16 recoveries in 1968 represent the same fraction of the 1968 population that 9 recoveries in 1969 represent of the 1969 population. Based on this logic, we can estimate the survival between hunting seasons to be $9/16 = 0.56$, albeit with a large standard error ($SE = 0.23$, as a ratio of 2 multinomial variates).

Survival also can be estimated if indices do not represent constant fractions of the population, but are known to be in certain proportions. Such methods are based on **catch-effort models.** Suppose it is somehow known that recovery rates of female mallards banded in Minnesota varied during 1968–1970 in the proportions 0.058, 0.056, and 0.100 (Johnson 1974; Table 15.5). The numbers of birds banded in 1968

Table 15.5. Recoveries of female mallards banded as adults in Minnesota, 1968–1970

Year	Number banded	Number recovered		
		1968	1969	1970
1968	338	16	9	5
1969	67		6	5
1970				93	12

From Jonson (1974).

and recovered in 1968, 1969, and 1970 were 16, 9, and 5. The ratios 16/0.058, 9/0.056, and 5/0.100 should represent the same proportion of the population in each of the 3 years. These ratios, 275.86, 160.71, and 50.00, suggest that survival from 1968 to 1969 was 160.71/275.86 = 0.58 (minimum SE = 0.24, assuming that effort is known exactly) and survival from 1969 to 1970 was 50.00/160.71 = 0.31 (minimum SE = 0.17). Catch-effort models are most often applied to fisheries problems, in which fishing effort is well known, or to populations of small mammals, where trapping effort can be calculated.

A similar technique produces estimates of the survival of young animals from ratios of sizes of litters or broods at different ages. For example, Stoudt (1971) computed mortality of canvasback (*Aythya valisineria*) ducklings between young (Class I) and older (Class II) stages to be 1.2 ducklings per brood, based on average brood sizes in those 2 classes. It is necessary, however, to account for the possibility that some litters or broods may have been lost completely. Further, among young waterfowl, it is not uncommon for broods to split into 2 or more groups, or for 2 or more broods to combine into a larger aggregation. These processes can bias estimates of survival rate from brood counts.

Change-in-Ratio Methods

The change-in-ratio technique, usually applied to estimating population size, can be used to estimate the rate of mortality due to exploitation (Paulik and Robson 1969:16, Seber 1982: 380). To do so requires 2 distinguishable types of animals (male and female, or young and adult) and estimates of the fraction of each type in the population before, during, and after harvest. Assumptions required to give good estimates are stringent, however, and should be carefully considered before the method is adopted (Downing 1980:251–252).

Mark–Recapture Methods

These methods are the most widely used approaches for estimating survival. A large class of models exists, each emphasizing particular assumptions or a combination of survival and other parameters.

Consider a study involving J occasions on which animals are captured, marked, and returned to the population. Suppose that all animals are alike in having the same chance of being captured on a particular occasion, call this probability c_i for the ith occasion, and in having the same probability of surviving from occasion i to occasion $i + 1$, say, S_i. Define N_i to be the number of animals in the population on occasion i. Suppose that M_i of these animals had been marked previously. On the ith occasion, n_i animals are captured, of which m_i had been marked already and the remaining u_i had not been marked previously. From these values, we can estimate the population size on occasions 2 through $J - 1$, as well as the number of combined births and immigrants (B_i) between occasions i and $i + 1$ for $i = 2$ through $J - 2$. Of special concern, survival rates S_i, $i = 1$ through $J - 2$, can be estimated. They are

$$\hat{S}_i = \frac{\hat{M}_{i+1}}{\hat{M}_i - m_i + R_i},$$

where

$$\hat{M}_i = m_i + \frac{R_i z_i}{r_i},$$

and R_i is the number of the n_i animals that are released after the ith sampling occasion (normally equal to n_i minus any losses during capture), r_i is the number of the R_i animals released at i that are captured again, and z_i is the number of animals that were captured before i, not captured at i, but captured again later. Estimated standard errors of survival rates are available (e.g., Seber 1982:202, Pollock et al. 1990: 21–22). Methods for estimating N_i, B_i, and c_i are given in Lancia et al. (2005). Cormack (1973) presented a readable justification for the Jolly–Seber model. MARK can be used to estimate parameters of the Cormack–Jolly–Seber model. Seber (1982) described a variety of alternative models. Pollock et al. (1990) discussed mark–recapture methods and developed several new models.

Recently, the Jolly–Seber model has been reparameterized (Pradel 1996) to allow estimation of parameters other than survival, including **seniority** (γ_i, the probability that an animal alive at time i had not entered the population between times $i - 1$ and i, which is useful for estimating recruitment), a recruitment rate into the population (f_i), and the rate of population change (λ). Estimating seniority becomes important because it can be used to estimate the proportion of population change that is due to recruitment (births and immigration) and survival (Franklin 2001). In the terms of the Jolly–Seber model, seniority also can be written as

$$\gamma_i = 1 - \frac{B_i}{N_{i+1}},$$

and λ can be computed as

$$\lambda_i = \frac{S_i}{\gamma_{i+1}}.$$

As an example, consider the data in Table 15.6, derived from a mark–recapture study of meadow voles (*Microtus pennsylvanicus*). There were 6 trapping occasions from late June through December. From these recapture statistics, we can calculate estimates of survival rate from 1 occasion to the next (for the first 4 occasions). We can estimate survival from occasion 1 to occasion 2 from

$$\hat{M}_1 = m_1 + \frac{R_1 z_1}{r_1}$$

$$= 0 + 105\left(\frac{0}{87}\right)$$

$$= 0$$

$$\hat{M}_2 = m_2 + \frac{R_2 z_2}{r_2}$$

$$= 84 + 121\left(\frac{5}{76}\right)$$

$$= 91.96;$$

thus,

$$\hat{S}_1 = \frac{\hat{M}_2}{\hat{M}_1 - m_1 + R_1}$$

$$= \frac{91.96}{0 - 0 + 105}$$

$$= 0.88,$$

and likewise for the remaining values. Pollock et al. (1990: 30) also presented estimates of population size and the number of births.

It should be emphasized that births include all animals added to a population, whether by actual birth or by immigration. Also, the survival rate reflects not only actual survival, but also permanent emigration from the study area; the measure is thus of apparent survival. This method can be used with different kinds of capture on different occasions. Of particular interest is marking animals on the first occasion and using resightings of marked animals on subsequent occasions.

There are other methods of estimating survival rates using mark–recapture data; these approaches have become varied and sophisticated. A class of models called multistate or multistrata models (Arnason 1973, Nichols et al. 1992) at times may be used to estimate survival rates, although they are more appropriate for estimating transition probabilities between specific stages (e.g., the rate at which individuals in a population transition between life stages or age classes or the rate of movement between different sites).

The robust design (Pollock 1982; Kendall and Nichols 1995; Kendall et al. 1995, 1997) incorporates features of both open and closed capture–recapture models and offers several advantages over the traditional Jolly–Seber model. A limitation of the Jolly–Seber model is that it estimates parameters associated with the general population (N_i^0) and not with the subset of the population that is exposed to sampling (N_i; Kendall et al. 1997). Consideration of this sampling issue is necessary, because some individuals may occasionally not be exposed to sampling efforts, effectively making them temporary emigrants. This relationship is clarified by the formula

$$E(N_i) = (1 - \gamma_i)N_i^0,$$

where E represents the individuals not exposed to sampling efforts, γ_i is the probability that an individual is not exposed to capture in period i and is thus a temporary emigrant (note: this is not the same γ_i used earlier to indicate seniority; notation used in mark–recapture models is sometimes confusing!). Furthermore, the relationship between the probability of temporary emigration (γ_i) and the capture probabilities of animals that are exposed to sampling ($p_i^\star$, a pooled capture probability for each sampling period) and those of all individuals in the population (p_i^0) is

$$p_i^0 = (1 - \gamma_i)p_i^\star.$$

The probability of temporary emigration is then estimated as

$$\hat{y}_i = 1 - \frac{\hat{p}_i^0}{\hat{p}_i^\star}.$$

A robust design study includes i primary sampling periods, each with l_i secondary sampling periods. The number of secondary sampling periods in each primary sampling period need not be equal. Closure (no births, deaths, immigration, or emigration) is assumed during the secondary sampling periods within each primary sampling period. The population is "open" to births, deaths, immigration, and emigration in the time interval between primary sampling periods. Information from secondary sampling periods is

Table 15.6. Mark–recapture statistics[a] for a population of meadow voles trapped in 1981, Maryland

Period	Date	n_i	m_i	R_i	r_i	z_i	$\hat{S}_i$	SE
1	27 June–1 July	108	0	105	87	0	0.88	0.039
2	1–5 August	127	84	121	76	5	0.66	0.048
3	29 August–2 September	102	73	101	68	8	0.69	0.049
4	3–7 October	103	73	102	63	3	0.63	0.049
5	31 October–4 November	102	61	100	84	5		
6	4–8 December	149	89	148				

From Pollock et al. (1990:29).

[a] For the ith occasion, n_i animals are captured, of which m_i were already marked; R_i is the number of n_i animals released after the ith sampling occasion; r_i is the number of R_i animals released at i that are captured again; z_i is the number of animals that were captured before i, not captured at i, but captured again later; $\hat{S}_i$ is the estimated survival rate; and SE is the standard error.

used to estimate conditional capture (p_{ij}) and recapture (c_{ij}) probabilities, and population size (N_i). A pooled capture probability ($p_i^\star$) is then estimated for each primary sampling period as

$$p_i^\star = 1 - \prod_{j=1}^{l_i}(1 - p_{ij}),$$

which is the probability that an animal is captured in at least one of the l_i secondary sampling periods in primary sampling period i. The pooled capture probabilities are used to estimate apparent survival and temporary emigration. Temporary emigration is defined by 2 parameters, γ_i'' and γ_i'. Here, γ_i'' is the probability that an animal is a temporary emigrant in period i, given that it was alive and available for sampling in primary sampling period $i - 1$. This probability contrasts with γ_i', which is the probability that an animal that was a temporary emigrant in primary sampling period $i - 1$ remains a temporary emigrant in primary sampling period i. This design allows estimation of apparent survival ($\phi_1, \ldots, \phi_{k-1}$) and population size ($N_1, \ldots, N_k$) in the presence of temporary emigration. By estimating capture probabilities separately for each secondary sampling period, this approach is robust to heterogeneity and trap response in capture probability (Pollock et al. 1990). The advantages of the robust design are many and include the ability to estimate temporary emigration, population size, and apparent survival simultaneously in a single study. Other variations of the robust design permit use of individual covariates (a characteristic unique to each individual, e.g., its body mass or total length) that capture and recapture probabilities (Huggins 1989, 1991).

Methods Based on Tag Recoveries

An important class of models, called tag or band recovery models, is similar to mark–recapture methods, but typically involves recoveries of dead, rather than live, marked animals. Many of these models were developed to use with data from banding programs for game birds, although they also are widely used with other taxa. In banding programs, large numbers of birds are captured each year and banded with individually identifiable bands. Hunters who recover a banded bird are encouraged to report the identification number. The situation is a mark–recapture study with many marking occasions (typically one per year for a series of years), but for an individual bird, only a single recapture is possible. Consider a simple example. Suppose that wood ducks (*Aix sponsa*) are banded for 2 years, just prior to the hunting season of each year (Table 15.7). Define c_1 to be the recovery rate, the probability that a bird is shot and its band reported during the first year. Similarly, c_2 is the probability that a bird, alive at the beginning of the hunting season in year 2, is shot and its band reported. Let S_1 be the probability that a bird survives from the beginning of the first hunt-

Table 15.7. Bandings and recoveries of wood ducks in 1964 and 1965

Year	Number banded	Number recovered	
		1964	1965
1964	1,603	127	44
1965	1,595	62	

From Brownie et al. (1985:22).

ing season to the beginning of the second; this is the survival rate we wish to estimate. If 1,603 birds are banded in year 1, we expect $1,603 \times c_1$ to be shot and reported the first year (Table 15.8); the actual number was 127. From this we calculate an estimate of c_1: $\hat{c}_1 = 127/1,603 = 0.0792$. We also calculate an estimate of c_2: $\hat{c}_2 = 62/1,595 = 0.0389$. Of the 1,603 birds banded the first year, we expect $1,603 \times S_1$ to survive to the beginning of the hunting season in the second year, and a fraction c_2 of them to be shot and reported. The actual number was 44. Thus, $44 = 1,603 \times S_1 \times 0.0389$, or $\hat{S}_1 = 0.7056$.

This procedure of equating observed to expected values works only when the number of parameters to be estimated equals the number of equations, but it does illustrate the principle behind the construction of modern banding models.

Suppose that only adults are banded and released in the program. It may be reasonable to assume that recovery and survival rates vary annually, but do not depend on the year when the bird originally was banded. This is Seber's (1970) model, termed Model 1 in Brownie et al. (1985).

We illustrate this procedure with an example from Brownie et al. (1985:14) involving male wood ducks (Table 15.9). From the summary statistics, we obtain $\hat{c}_1 = 0.0792$, $\hat{c}_2 = 0.0401$, $\hat{c}_3 = 0.0688$, $\hat{S}_1 = 0.6512$, and $\hat{S}_2 = 0.6311$. Calculated standard errors were $SE(\hat{c}_2) = 0.00674$, $SE(\hat{c}_2) = 0.00415$, $Se(\hat{c}_3) = 0.00608$, $SE(\hat{S}_1) = 0.0675$, and $Se(\hat{S}_2) = 0.0647$. An approximate 95% confidence interval is the sample value

Table 15.8. Expected numbers of bandings and recoveries of wood ducks in 1964 and 1965[a]

Year	Number banded	Expected number of recoveries	
		1964	1965
1964	N_1	$N_1 c_1$	$N_1 S_1 c_2$
1965	N_2		$N_2 c_2$

[a] N_1 is the number banded in year 1, N_2 is the number banded in year 2, c_i is the recovery rate in the ith hunting season, and S_i is the probability that a bird survives from the beginning of the ith hunting season to the beginning of the next.

Table 15.9. Banding and recovery data for male wood ducks[a]

Year banded (i)	Number banded	Year of recovery (j)					R_i
		1964	1965	1966	1967	1968	
1964	1,603	127	44	37	40	17	265
1965	1,595		62	76	44	28	210
1966	1,157			82	61	24	167
		$C_j =$	127	106	195	145	69
		$T_j =$	265	348	409	214	69

From Brownie et al. (1985:22).

[a] R_i are row totals, C_j are column totals, and T_j are block totals of the number of recoveries.

minus and plus 1.96 times the standard error. For the first-year recovery rate, for example, we obtain $0.0792 - 1.96 \times 0.00674 = 0.0660$ as a lower limit, and $0.0792 + 1.96 \times 0.00674 = 0.0924$ as an upper limit. Hence, a 95% confidence interval for c_1 is 0.0660–0.0924.

It is possible that more restrictive models fit a particular data set adequately, in which case the relevant parameters may be estimated more precisely. A likely candidate is the model in which survival rates are assumed to be the same each year, but recovery rates vary. This model (Model 2 of Brownie et al. [1985]) often fits data sets well, perhaps because true survival rates do not vary much, and the ability of actual banding data to detect those differences is weak. The most restrictive model assumes that both survival rates and recovery rates are the same each year. This model (Model 0 of Brownie et al. [1985]) might fit small data sets, but is likely actually to be true only in rare circumstances.

Both young and adult birds are often banded in the same program. One cannot safely assume that birds of the 2 age groups have the same survival and recovery rates, so they must be treated differently. Yet, if the young birds survive long enough, they become adults, subject to adult survival and recovery patterns. Several useful models have been developed for this situation. One of the most general is that of Brownie and Robson (1976), termed Model H_1 by Brownie et al. (1985). As before, survival and recovery rates are assumed to vary by year, and young have different survival and recovery rates for their first year only. Estimators are presented in Brownie et al. (1985:60) and are calculated by the recoveries-only model in MARK. An example is presented in Tables 15.10 and 15.11.

More restrictive models can give more precise estimates, if they fit the data adequately. Another reasonable model allows survival rates for both young and adults to vary from year to year, but assumes that rates for the 2 age groups

Table 15.10. Data from a banding study of juvenile and adult male mallards banded preseason in the San Luis Valley, Colorado, 1963–1971

Year banded	Number banded	Year of recovery								
		1963	1964	1965	1966	1967	1968	1969	1970	1971
		Banded as adults								
1963	231	10	13	6	1	1	3	1	2	0
1964	649		58	21	16	15	13	6	1	1
1965	885			54	39	23	18	11	10	6
1966	590				44	21	22	9	9	3
1967	943					55	39	23	11	12
1968	1,077						66	46	29	18
1969	1,250							101	59	30
1970	938								97	22
1971	312									21
		Banded as juveniles								
1963	962	83	35	18	16	6	8	5	3	1
1964	702		103	21	13	11	8	6	6	0
1965	1,132			82	36	26	24	15	18	4
1966	1,201				153	39	22	21	16	8
1967	1,199					109	38	31	15	1
1968	1,155						113	64	29	22
1969	1,131							124	45	22
1970	906								95	25
1971	353									38

Table 15.11. Estimated survival and recovery rates for juvenile and adult male mallards fitted to Model H_1. Estimated standard errors are in parentheses.

Year	Survival rate		Recovery rate	
	Adult	Juvenile	Adult	Juvenile
1963	0.576 (0.113)	0.471 (0.059)	0.0433 (0.0134)	0.0863 (0.0091)
1964	0.636 (0.076)	0.506 (0.070)	0.0856 (0.0092)	0.1467 (0.0134)
1965	0.666 (0.079)	0.589 (0.072)	0.0590 (0.0061)	0.0724 (0.0077)
1966	0.805 (0.098)	0.591 (0.072)	0.0628 (0.0067)	0.1274 (0.0096)
1967	0.650 (0.072)	0.478 (0.061)	0.0520 (0.0050)	0.0909 (0.0083)
1968	0.552 (0.058)	0.652 (0.072)	0.0633 (0.0055)	0.0978 (0.0087)
1969	0.572 (0.066)	0.464 (0.068)	0.0789 (0.0061)	0.1096 (0.0093)
1970	0.542 (0.129)	0.393 (0.113)	0.0888 (0.0080)	0.1049 (0.0102)
1971			0.0673 (0.0142)	0.1076 (0.0165)

fluctuate in parallel. This model, proposed by Johnson (1974), has no closed-form solution and is not included in program BROWNIE (Brownie et al. 1985), but it can be fitted with MARK or with general maximum-likelihood programs. In addition, one can fit models that allow survival rates for 2 age classes to vary from year to year in parallel, but with recovery rates varying independently, or vice versa.

Other restrictions include assuming that survival rates remain the same from year to year (Model H_{02} of Brownie et al. 1985) or that both survival and recovery rates are constant (Model H_{01} of Brownie et al. 1985). Further, the procedure can be generalized to 3 age classes, if birds can be distinguished by age class, and some members from each age class are banded. This situation may pertain to geese, for example.

Two thoughts should be considered when planning a banding program to estimate survival. First, adults need to be included. If only young are banded, little can be estimated from the resulting recovery data unless some strong assumptions are made (e.g., Burnham and Anderson 1979, Anderson et al. 1981a). Second, sample size must be large to obtain meaningful estimates. The program BAND2 (Wilson et al. 1989) estimates required sample sizes for different models. That program should be used, and the handbook by Brownie et al. (1985) carefully reviewed, prior to embarking on a banding program.

Populations with Age-Dependent Birth and Death Rates
Fertility Tables

Both fertility and survival (Box 15.4) are known to vary by age for many species, and considerable effort has been placed in developing models with age-dependent birth and death rates. Consider an age-structured population with a maximum of i age classes, recorded in years. Suppose females of age x produce an average of m_x young females per year. The table giving the number of female offspring per year per

---

BOX 15.4. AGE VARIATION IN SURVIVAL RATES

For vertebrate populations, it is the norm that survival (or, equivalently, mortality) rates vary by age. Typical patterns involve low survival of young animals, higher survival of animals in their prime, and decreasing survival with advancing age. Deviations from this pattern can occur, especially if reproduction imposes an added mortality risk. For the analysis of a population, the difference in survival between young animals and prime-aged animals is usually important. The difference between prime and older years may be less important, especially in exploited populations in which few animals reach advanced ages.

Consider a simple example of the age-specific survival of the mountain plover (Dinsmore et al. 2003). These birds have 2 distinct age classes: juveniles (hatch to first birthday) and adult (>1 year old). A simple model with no age differences produced an estimate of annual survival of $\hat{S} = 0.59$ ($SE = 0.02$). When age effects were considered, the corresponding estimates of annual survival were $\hat{S}_{adult} = 0.68$ ($SE = 0.03$) and $\hat{S}_{juvenile} = 0.46$ ($SE = 0.07$). The model with age effects received far more support than the no-age-effects model using Akaike's Information Criterion for model selection, suggesting that survival differed by age. This result was expected, and the large differences in annual survival by age class are an indication of the importance of modeling age effects in survival.

Table 15.12. Survival and reproduction data for white-tailed deer in central Michigan

Age (yr; x)	Survival rate (s_x)	Fertility rate (m_x)
0	0.58	0.000
1	0.70	0.047
2	0.70	0.503
3	0.70	0.663
4	0.70	0.733
5	0.70	0.743
6	0.70	0.771
>6	0.70	0.644

From Eberhardt (1969).

female of age x is called the **fertility table** (e.g., see Table 15.12 for a population of white-tailed deer in central Michigan). Average fertility rates vary with age and are zero for young of the year, nearly zero for yearlings, and increase up to age 6, after which they decline.

Life Tables

Analogous to the fertility table is the mortality schedule, which describes the pattern of deaths by age class. The probability of a female surviving from the beginning of age class x to the beginning of age class $x + 1$ is defined to be s_x. The survival data (Table 15.12) for the central Michigan deer population (Eberhardt 1969) suggested the survival rate of age classes 1 and above did not differ from one another and the average rate was 0.70.

Consider a cohort of animals (a group born at roughly the same time) that begins with 1,000 individuals at age 0. Thus, there will be $1{,}000 \times s_0$ individuals the next year (at age 1), $1{,}000 \times s_0 \times s_1$ members the following year (at age 2), and so forth. The number of individuals surviving from birth to age class x is termed n_x:

$$n_x = 1{,}000 s_0 s_1 \ldots s_{x-1}.$$

Often, the mortality rate, rather than the survival rate, is expressed:

$$q_x = 1 - s_x.$$

A **life table** gives these and other relevant values. A life table is basically a summary of the survivorship of a population. It also can be used to calculate or estimate mortality rates, by age, under certain assumptions. Life tables were developed for human populations, especially for insurance applications, but also have been applied to wildlife populations. Human life tables generally involve large numbers of individuals for which exact times of death can be ascertained, whereas information for wild animals is typically incomplete. For most animal populations, information is based on a sample; thus, a life table provides

estimates of relevant parameters that are less exact than values for humans.

For many animals, survival and fertility rates differ more sharply by size or life stage than by age. Some life table methods can be used with size classes or stages. Lefkovitch (1965) developed population projection methods for such situations. Usher (1972), Kirkpatrick (1984), Sauer and Slade (1987), and Caswell (2000) provided further details and some applications.

A life table consists of several of the following 6 basic columns:

- x—age, measured in years or some other convenient unit or interval, $(x, x + 1)$;
- n_x—the number of individuals surviving to the beginning of age x from an initial cohort of n_0 members;
- d_x—the number of deaths in the age class $(x, x + 1)$, $d_x = n_x - n_{x+1}$;
- q_x—the mortality rate at age x;
- s_x—the survival rate at age x, $s_x = 1 - q_x$; and
- l_x—the cumulative survival rate from birth until age x,
$$l_x = s_0 \times s_1 \times \ldots \times s_{x-1} = \frac{n_x}{n_0}.$$

The definition of survival rates in the mortality table pertains to the period from the beginning of one age class to the beginning of the next. The fertility table describes reproduction per female in an age class. To use survival and reproductive rates in combination, one must define the age classes similarly in the 2 tables. That is, if reproduction is categorized by number of young produced and surviving to autumn, survival of adults should be assessed from autumn to autumn.

We provide an example of a life table (Table 15.13) and note that examples are given only to illustrate the method, as sample sizes are too small to draw reliable conclusions. Note that d_x can be computed from values of n_x by subtraction, and n_x can be calculated by adding entries in the d_x column from the bottom. Also, q_x is based on d_x and n_x; conversely, the table of n_x for $x > 0$ can be constructed from q_x values. Thus, there

Table 15.13. Example of a life table based on known deaths of 42 gray squirrels (*Sciurus carolinensis*) born in 1954

Age (yr) (x)	Number in population (n_x)	Number of deaths (d_x)	Mortality rate (q_x)	Survival rate (s_x)
0–1	42	22	22/42 = 0.52	20/42 = 0.48
1–2	20	10	10/20 = 0.50	10/20 = 0.50
2–3	10	7	7/10 = 0.70	3/10 = 0.30
3–4	3	2	2/3 = 0.67	1/3 = 0.33
4–5	1	1	1/1 = 1.00	0/1 = 0

From Downing (1980:256).

is only one independent column, and all others can be calculated from any one of them. Depending on the type of data available and the assumptions that can realistically be made, a variety of life tables can be constructed.

Graphs of cohort size or cumulative survivorship (on a logarithmic scale) against age often approximate 1 of 3 characteristic shapes (Fig. 15.6), but possibly with a downward jag reflecting lower survival of newborns (Pearl 1928). **Type I survivorship** curves have low mortality early in life, but higher rates among older individuals. Female elk (*Cervus canadensis*) in the northern Yellowstone herd exemplify this pattern, with the exception of a depressed survival rate of young (Fig. 15.7). **Type II survivorship** curves have mortality rates roughly constant with age, leading to a straight-line relation on a log scale. Adult songbirds are suggested to have similar patterns (Krebs 1985). The **Type III survivorship** curve involves high mortality among young and decreasing mortality as individual's age. Many invertebrates and fish display Type III survivorship; they are vulnerable when they are young and small, but age and growth impart greater security. Siler (1979) and Eberhardt (1985) discuss how survivorship functions might be decomposed into functions representing 3 stages of life: early life, maturity, and senescence.

The Stable Age Distribution

The **age distribution** of a population is the number of individuals of each age class in the population at a particular time. If age-dependent survival and fertility rates remain constant for a fairly long period of time, the proportion of animals in each age class will stabilize. This relationship is true even if the population itself is not constant in size; that is, a population can be expanding or declining and still have constant proportions in each age class. The resulting fractions comprise what is termed the **stable age distribution**. The fraction of the population in age class x will equal C_x:

$$C_x = \frac{e^{-rx}l_x}{\sum_i e^{-ri}l_i}, \quad (7)$$

where r is the growth rate of the population once it attains a stable age distribution, and l_x is the cumulative survival rate from birth until age x.

Suppose that a population was constant in size at N members. The stable age distribution at any time t would contain members in each age class proportional to the survivorship (i.e., $N \times l_0$ of age class 0, $N \times l_1$ of age class 1, and so on). But, if the population has been changing at an annual rate λ, the number of members in age class x at time t would be the number born x years earlier ($N_{t-x}l_0$) times the survivorship of those members (l_x/l_0). Because of population growth, $N_t = N_{t-x}e^{-\lambda x}$. Thus, in year t the fraction of the population in age class x will be $N_{t-x}l_0 \times l_x/l_0 = N_t e^{-\lambda x}l_x$, which is proportional to the numerator of equation 7. The denominator is the sum of such values over all ages, which scales the numbers so they total 1.0.

Alternatively, the size of the age class x relative to that of newborn (Caughley 1977:114) is:

$$\frac{C_x}{C_0} = e^{-rx}l_x.$$

The value of r can be calculated from age-dependent survival and fertility rates according to the following equation:

$$1 = \sum_x e^{-rx}l_x m_x. \quad (8)$$

Equation 8 is the discrete version of **Lotka's equation** (sometimes called Euler's equation; e.g., Mertz 1970, Wilson and Bossert 1971:116, Caughley 1977:107). This equation requires that survival and reproduction schedules remain constant for a long period of time, often an unlikely presumption. It is strictly appropriate only for a birth pulse population, in which births occur instantaneously and the

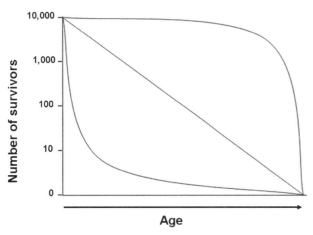

Fig. 15.6. Three characteristic survivorship curves.

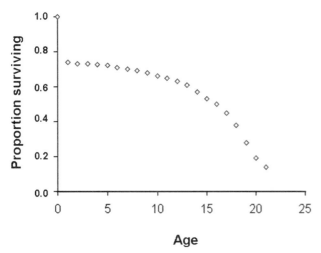

Fig. 15.7. Survivorship curve of female elk in the Northern Yellowstone herd (Houston 1982:55). Age is measured in years.

Table 15.14. Stable age distribution calculated from survival and reproduction data for white-tailed deer in central Michigan

Age (x)	Survival to age (l_x)	Fraction in age (C_x)
0	1.0000	0.3214
1	0.5800	0.1913
2	0.4060	0.1375
3	0.2842	0.0988
4	0.1989	0.0709
5	0.1392	0.0510
6	0.0974	0.0366
7	0.0682	0.0263
8	0.0478	0.0189
9	0.0334	0.0136
>9	$0.58(0.70)^{x-1}$	0.0337

From Eberhardt (1969).

age structure is observed at the same time (Michod and Anderson 1980).

Equation 8 can be solved for r from a schedule of age-dependent cumulative survival rates (l_x) and fertility rates (m_x). For the white-tailed deer data (Table 15.12), the estimated value of r is $\hat{r} = -0.026$. This value suggests, because $e^{-0.026} - 1 = 0.0257$, the deer population was declining (since the rate is negative) at about 2.6% each year.

With this estimate of r, we can now calculate the asymptotic age structure of the deer population from equation 7. The number in age class x is proportional to $e^{-rx}l_x$, which gives the C_x values (Table 15.14). This distribution can be compared to the actual age distribution, if known, to test whether the underlying assumptions are met.

Projecting the Population: Leslie Matrices

If a population has attained a stable age distribution, and we are fortunate enough to know the age-dependent survival and fertility rates, we could learn a lot about the population from studying those rates. Consider an age-structured population (with M age classes) that breeds seasonally and has survival and fertility rates that vary with age, but not annually. Suppose the population is surveyed for several years at the same time each year, say, immediately after the birth season. Let $n_{x,t}$ be the number of individuals of age x in year t. The number of 1-year-olds in year $t + 1$ ($n_{1,t+1}$) will be the number that were born in year t ($n_{0,t}$) times the survival rate of 0-year-olds (s_0):

$$n_{1,t+1} = s_0 n_{0,t}.$$

Similarly, the number of 2-year-olds in year $t + 1$ equals the number of 1-year-olds in the previous year times their survival rate:

$$n_{2,t+1} = s_1 n_{1,t},$$

or, in general,

$$n_{i+1,t+1} = s_i n_{i,t}. \tag{9}$$

Consider next the number of births, which can be allocated according to the different age classes that reproduce. The number of 0-year-olds (births) in year $t + 1$ ($n_{0,t+1}$) represents the number of 1-year-olds in that year ($n_{1,t+1}$) times the fertility rate of 1-year-olds (m_1), plus the number of 2-year-olds in that year ($n_{2,t+1}$) times their fertility rate, and so forth:

$$n_{0,t+1} = m_1 n_{1,t+1} + m_2 n_{2,t+1} + \ldots + m_M n_{M,t+1}.$$

We want to express this number in terms of the population in the previous year and, in light of equation 9, we obtain:

$$n_{0,t+1} = m_1(s_0 n_{0,t}) + m_2(s_1 n_{1,t}) + \ldots + m_M(s_{M-1} n_{M-1,t})$$
$$= (m_1 s_0) n_{0,t} + (m_2 s_1) n_{1,t} + \ldots + (m_M s_{M-1}) n_{M-1,t}$$
$$= g_0 n_{0,t} + g_1 n_{1,t} + \ldots + g_{M-1} n_{M-1,t}, \tag{10}$$

where

$$g_i = m_{i+1} s_i, \text{ for } i = 0, \ldots, M-1. \tag{11}$$

Equation 10 expresses the production of young (number of 0-year-olds) as a linear combination of the number in each age class in the previous year. The values of g_i indicate the number of young that are produced per individual of age i in year t that will be alive in year $t + 1$. We can combine equation 9 for $I = 0, \ldots, M-1$ and equation 10 into a single equation involving matrices:

$$\begin{bmatrix} n_0 \\ n_1 \\ n_2 \\ . \\ . \\ n_M \end{bmatrix}_{t+1} = \begin{bmatrix} g_0 & g_1 & g_2 & \cdots & g_{M-1} & g_M \\ s_0 & 0 & 0 & \cdots & 0 & 0 \\ 0 & s_1 & 0 & \cdots & 0 & 0 \\ . & . & . & \cdots & . & . \\ . & . & . & \cdots & . & . \\ 0 & 0 & 0 & \cdots & s_{M-1} & s_M \end{bmatrix} \times \begin{bmatrix} n_0 \\ n_1 \\ n_2 \\ . \\ . \\ n_M \end{bmatrix}_t$$

or, in matrix notation, $\boldsymbol{n}_{t+1} = L \times \boldsymbol{n}_t$. L is called the **population projection matrix**, or **Leslie matrix**. The term in any particular row and column can be considered to be the contribution of an individual in the age class represented by that *column* in year t to the age class represented by that *row* in year $t + 1$ (Jenkins 1988). Note that we assumed survival and fertility rates were the same each year. This formulation was developed by Bernadelli (1941), Lewis (1942), and Leslie (1945, 1948). Van Groenendael et al. (1988) reviewed the method and applications, and Caswell (2000) gave an excellent overview of the technique. It allows several interesting interpretations. For example, we can project the population from one year to the next or, by repeating the process, k years into future:

$$\boldsymbol{n}_{t+1} = L \times \boldsymbol{n}_t.$$

Thus,

$$n_{t+2} = L \times n_{t+1}$$
$$= L \times L \, n_t,$$

and in general,

$$n_{t+k} = L^k n_t.$$

We also can derive useful properties mathematically from this formulation, including the stable age distribution and population rate of change (e.g., Leslie 1945, Pielou 1969).

The population projection matrix approach has occasionally been misused, generally by using fertility data as g_i values; Wethey (1985) and Jenkins (1988, 1989) presented examples. The parameter g_i is somewhat odd, incorporating both fertility and survival, and measuring the fertility of a cohort of age $i + 1$ times the survival rate from age i to age $i + 1$ (equation 11).

We described this formulation with surveys occurring immediately following the birth season. If counts are conducted at another time, the definition of g_i must be changed to incorporate survival from birth until the time of the census (e.g., Michod and Anderson 1980). In practice, the estimation of g_i is difficult at best (Taylor and Carley 1988).

Age- and Density-Dependent Models

Birth and death models in which the rates depend on both age and density can be constructed (Leslie, 1948, 1959; Williamson 1959; Cooke and Leon 1976; Caswell 2000), but they do not have ready mathematical solutions, and their properties are not well understood. Pennycuick et al. (1968) developed a computer program that allowed elements of a Leslie matrix to be density dependent and also to have time lags. Little is known about how density dependence actually operates, however. More likely to be useful are models that decompose birth and death rates into meaningful components. These components can be related to age, density, or environmental factors, as appropriate.

Estimating Age-Dependent Death Rates

Earlier, we discussed ways of estimating mortality rates not specifically related to age. We now turn to the common problem of estimating mortality rates by age. Several methods are available, based either on following a cohort of animals or on examining the age distribution of a population. The appropriateness of estimators depends on the assumptions they require, how they are met by the population under study, and on how the data are collected.

Estimating Survival by Following a Cohort

Knowing All the Deaths

Suppose we know the complete death history of a cohort of 42 gray squirrels (*Sciurus carolinensis*) born in 1954 (d_x column, Table 15.13). Specifically, we know that 22 died during their first year, 10 died during their second year, 7 died during their third year, 2 died during their fourth year, and 1 died in its fifth year. From this information, we can calculate exact mortality rates by age. If q_i is the probability of dying during year i (which also is during age i), then $q_0 = 22/42 = 0.52$, because 22 of the 42 squirrels died before their first birthday. Likewise, $q_1 = 10/20 = 0.50$, because 10 of the 20 survivors from the first year died during the second. Similarly, $q_2 = 7/10 = 0.70$, $q_3 = 2/3 = 0.67$, and $q_4 = 1/1 = 1.00$. Thus, age-dependent mortality rates can be computed exactly for this known population.

If the 42 animals can be considered a random sample from some larger population, statistical estimation is possible. Because q_0 is the mortality rate of animals in their first year of life, the number of animals expected to die by the beginning of the second year is $42 \, q_0$. That number was actually 22, hence $\hat{q}_0 = 22/42 = 0.52$. Also, the number of animals alive at the beginning of each year is known at that time, and if we assume that individual animals live or die independently of one another, the number dying during year i can be treated as a binomial variate, with n_i representing the number of animals alive at the beginning of the year and rate $= q_i$. The standard error of $\hat{q}_i$ is then estimated by

$$\sqrt{\frac{\hat{q}_i(1-\hat{q}_i)}{N_i}}.$$

In our example, $SE(\hat{q}_0) = 0.52 \times 0.48/42 = 0.077$. Similarly, $SE(\hat{q}_1) = 0.112$, $SE(\hat{q}_2) = 0.145$, $SE(\hat{q}_3) = 0.272$, $SE(\hat{q}_4) = 0$.

Knowing All the Living

Suppose that instead of knowing the age at which the individual squirrels died, we had surveyed the cohort at the beginning of each year. This information forms the basis of the n_x column (Table 15.13). There were 42 squirrels at the beginning of year 0 and 20 at the beginning of year 1. Thus, the survival rate during that year was $20/42 = 0.48$, and the mortality rate was $1 - 0.48 = 0.52$. Mortality rates for the other years also coincide with those calculated from the information on age of death. Likewise, if the sample of animals is representative of a larger population, we can treat the process as binomial and calculate the same estimates of standard errors.

Life tables based on information from following a specific cohort are termed **cohort,** or sometimes **dynamic** or **age-specific life tables.** Unfortunately, only for closely monitored or captive populations do we have situations with such ideal knowledge of the ages at death or of exactly how the size of a particular cohort changes over time.

Following More Than One Cohort

If more than one cohort is followed, an age-specific table can be generated for each of them. Survival rates can then be estimated that vary both by age and by year, although limited sample sizes usually preclude accurate estimates. Alternatively, estimates can be pooled across years to obtain age-dependent estimates (e.g., Downing 1980; Table 15.6) or pooled across ages for year-dependent estimates. Which pooling is more appropriate depends on whether survival rates vary more by age or by year. Loery et al. (1987) presented an example of estimating survival rates by both age and year for black-capped chickadees (*Poecile atricapilla*), based on a long-term mark–recapture study.

Estimating Survival from Age Distributions

Suppose that we do not have complete information from following one or more cohorts through time, but that we have the age composition of a sample of animals from the population at a particular time. That sample must accurately reflect these members either **dying** or **living**. We also require the population to have achieved a stable age distribution and to be constant in size (although the method can be adapted if the population is increasing or decreasing at a known rate). These assumptions are stringent and must be carefully regarded, and the methods that follow work better in theory than in practice.

The Age Distribution of the Living

Suppose that we have the age distribution of a sample from the living members of the population at a particular time in year t. The number of individuals of age x in year t ($n_{x,t}$) is the number that were of age $x-1$ in year $t-1$ ($n_{x-1,t-1}$) times the survival rate for those animals ($s_{x-1,t-1}$):

$$n_{x,t} = n_{x-1,t-1} s_{x-1,t-1},$$

from which we could estimate $s_{x-1,t-1}$ as

$$\hat{s}_{x-1,t-1} = \frac{n_{x,t}}{n_{x-1,t-1}}.$$

This calculation can be done only if we have accurate age distribution data for successive years. By assuming the population is stationary, however, we can obtain estimates from a sample in a single year. Stationarity implies that survival rates (and fertility rates) are constant from year to year ($s_{x,t}$ is independent of t) and that population size and age structure are the same from year to year. That is, $n_{x,t}$ is independent of t. (This is a **critical assumption.**) Hence, we have

$$\hat{s}_{x-1} = \frac{n_x}{n_{x-1}}.$$

Chapman and Robson (1960) recommended adding 1 to the denominator to reduce bias. Life tables formed this way are called **time-specific life tables** and represent a cross-section of ages at a specific time.

A statistical model for these data can be developed. In a sample of n_* animals in a particular year, the number of individuals of age x can be considered a multinomial variate. The probability ϕ_x that an individual will be in age class x is proportional to $n_0 s_0 s_1 \cdots s_{x-1}$. This proportion depends on sampling intensity. These probabilities are estimated by $\hat{\phi}_x = \frac{n_x}{n_*}$. Also, due to the multinomial nature of the data, the estimated (E) probability is

$$E(\hat{\phi}_x) = \phi_x,$$

and the variance of the probability ϕ_x that an individual will be in age class x is

$$\text{Var}(\hat{\phi}_x) = \frac{\phi_x(1-\phi_x)}{n_*},$$

and the covariance between 2 survival rates is

$$\text{Cov}(\hat{\phi}_x, \hat{\phi}_y) = -\frac{\phi_x \phi_y}{n_*}.$$

Survival rates are estimated by the ratio of successive $\hat{\phi}_x$

$$\hat{s}_x = \frac{\hat{\phi}_{x+1}}{\hat{\phi}_x},$$

with standard error estimated from

$$SE^2(\hat{s}_x) = \frac{\phi_{x+1}(\hat{\phi}_x + \hat{\phi}_{x+1})}{n_* \phi_x^3}$$

$$= \frac{\hat{s}_x(1+\hat{s}_x)}{n_* \hat{\phi}_x}.$$

The mortality rate, $\hat{q}_x$, will have the same standard error as the survival rate, $\hat{s}_x$.

We illustrate the procedure with the age distribution of male white-tailed deer on the George Reserve in Michigan, just before the 1956 hunting season (n_x, Table 15.15). These values (n_x) can be used to generate a life table with estimated survival rate (Table 15.15). These estimates appear unrealistic, especially the higher survival rate for younger animals than for older individuals. These aberrancies might in part reflect large sampling errors due to small sample sizes, but more likely they result from the population not being stationary because of year-to-year variation in reproduction (McCullough 1979).

Table 15.15. Life table based on age distribution of male white-tailed deer alive in 1956[a]

Age (x)	n_x	$\hat{d}_x$	$\hat{s}_x$	$SE(\hat{s}_x)$
0	40	17	0.575	0.150
1	23	17	0.261	0.120
2	6	2	0.667	0.430
3	4	3	0.250	0.280
4	1	1	0.000	0.000
>4	0			

Data from McCullough (1979:36).

[a] n_x is the number of animals of age x, $\hat{d}_x$ is the number of deaths of animals of age x, $\hat{s}_x$ is the estimated survival rate, and SE is the standard error.

The Age Distribution of the Dying

Consider the example (Table 15.16) representing ages of white-tailed deer found dead in surveys of carcasses. These data reflect the age distribution of dying members of the population. Suppose the studied population is stationary; the age distribution can then be used as the d_x column of a life table (Table 15.16). From these values, we can estimate the n_x column by adding the d_x entries from the bottom up. The ratio of d_x to $\hat{n}_x$ gives an estimate of q_x, the age-dependent mortality rate (Table 15.16). Note the mortality rate for the last age class will always be 1.0.

Is the Sample of the Living or the Dying?

Surprisingly, the age composition of dead animals may not provide a suitable estimate of the age structure of animals dying (Caughley 1966). For example, if animals are shot unselectively with respect to age, the resulting sample will reflect the population **alive** at the beginning of the collection period and will not reflect the age structure of all animals that died (unless the shooting was the only mortality source). Such data would most appropriately be used in the n_x column of a life table. In contrast, if carcass pick-ups were made of **all** animals that died during a year (Table 15.16), the resulting data would truly reflect mortality and could be used in the d_x column.

The 3 main kinds of data (Seber 1982:401–402) that can be used to construct a time-specific life table are (1) number of animals of each age for a representative sample of live animals, used as the n_x column; (2) number of animals of each age at death for a representative sample of animals killed by an agent independent of age (nonselective collection, natural catastrophe), also used as the n_x column; or (3) number of animals of each age at death for a representative sample of carcasses, used as the d_x column. Biased estimates can arise if younger age classes are less vulnerable to sampling, possibly because when they are alive, they are less detectable than older animals, and when they die, their softer bones do not persist as long. Survival estimates for older age classes are unaffected by this bias (Caughley 1966, Seber 1982:402).

Is the Population Stationary?

We indicated that age distribution data can be used to estimate survival if the population has a stable age distribution and is constant in size (i.e., the population is stationary). The method can be modified if the population is increasing or decreasing at a **known** rate (Caughley 1977:92, Eberhardt 1988). Knowing this rate requires independent information, such as estimates of population trend. The requirement of stable age distributions remains. Survival rates also can be estimated from data for a stable age distribution if appropriate fertility rate are available (Michod and Anderson 1980).

One cannot examine a single age distribution and learn whether the population is stationary or not (Caughley 1966, Seber 1982:403). However, Tait and Bunnell (1980) provided a possible exception if the age at death is known for a large number of animals. A series of age distributions at different times may be used to examine whether a population is stable.

It is tempting to assume a population is stationary and to estimate survival rates from age structure data, concluding from those results the population is stable. Despite warnings to the contrary about this circular argument (Caughley and Birch 1971), the practice has persisted (Lancia and Bishir 1985, Jenkins 1989).

Pooling Ages for Survival Estimation

Because of variation due to small samples, it is often necessary to smooth either the observed age frequencies or the resulting estimators; Caughley (1977:96–97) illustrated the former, and we mention the latter.

If we believe that mortality is constant for ages in a specified interval, pooled estimates of the rate can be obtained. Eberhardt (1985) noted that pooled mortality estimates are biased high if older animals survive at a lower rate than do animals of prime age. For example, it seems reasonable the mortality rate of the deer in Table 15.16 is roughly constant for individuals 1 year of age. A pooled estimator of that adult mortality rate is:

$$\frac{d_1 + d_2 + \ldots}{n_1 + n_2 + \ldots}$$
$$= \frac{(n_1 - n_2) + (n_2 - n_3) + \ldots}{n_1 + n_2 + \ldots}$$
$$= \frac{n_1}{\sum_{j \geq 1} n_j},$$

which in the present example is

$$\frac{18 + 14 + \ldots + 8}{92 + 74 + \ldots + 8} = \frac{92}{383}\ 0.240.$$

Table 15.16. Life table based on age distribution of female white-tailed deer found dead[a]

Age (x)	$\hat{n}_x$	d_x	$\hat{q}_x$
0–1	198	106	0.535
1–2	92	18	0.196
2–3	74	14	0.189
3–4	60	18	0.300
4–5	42	9	0.214
5–6	33	5	0.152
6–7	28	6	0.214
7–8	22	8	0.364
8–9	14	4	0.286
9–10	10	2	0.200
>10	8	8	1.000

Data from Eberhardt (1969:488).

[a] $\hat{n}_x$ is the estimated number of animals of age x, d_x is the number of deaths of animals of age x, and $\hat{q}_x$ is the estimated age-dependent mortality rate.

Average, rather than pooled, estimators also can be formed (Seber 1982:397). More importantly, the unbiased estimator of adult survival rate with smallest variance (Chapman and Robson 1960, Robson and Chapman 1961) is

$$\hat{s}_{CR} = \frac{T}{n + T - 1},$$

where

$$n = \sum_{j \geq 1} n_j$$

and

$$T = \sum_{j \geq 1} j n_j.$$

Its standard error can be estimated from:

$$SE^2(\hat{s}_{CR}) = \hat{s}_{CR}\left(\hat{s}_{CR} - \frac{T-1}{n+T-2}\right).$$

In the example (Table 15.16), we have $n = 92 + 74 + \ldots + 8 = 383$, $T = 1 \times 92 + 2 \times 74 + 3 \times 60 + \ldots + 11 \times 8 = 1{,}365$, so $\hat{s}_{CR} = 1{,}365/(383 + 1{,}365 - 1) = 0.781$, and $\hat{q} = 1 - \hat{s} = 0.219$. Its standard error is calculated from

$$SE^2(\hat{q}) = SE^2(\hat{s}_{CR}) = 0.78134\left(0.78134 - \frac{1{,}365}{383 + 1{,}365 - 2}\right)$$

$$= 0.0000983,$$

so

$$Se = \sqrt{0.0000983} = 0.0099.$$

Comments on Life Tables

Because age composition is typically measured from samples rather than from entire populations, the entries in the life table are estimates, subject to sampling variation. Caughley (1977:95) suggested that life tables based on 150 age determinations were unlikely to be sufficiently accurate for any purpose. Polacheck (1985) found with simulation that analyses based on even larger samples often provided misleading estimates of survival rate.

McCullough (1979:221) analyzed one of the best available sets of data on age structure of white-tailed deer and concluded: "Although numerous attempts have been made to apply life-table methods to the analysis of kill data, . . . most of these methods have not proven to be useful at the practical level." The major problem he identified was meeting the assumption of a stable age distribution, as variable environmental factors had differential effects on different age classes. He suggested that time-specific life tables, although clearly not meeting the assumptions necessary to estimate survival rates, are valuable to the manager of exploited populations, because they show the existence of strong year classes, indicative of good reproduction in a particular year. Seber (1982:393) cautioned that life tables may give an overall picture of a population, but have limited accuracy and should be supported by other methods of estimation.

Jenkins (1989) also cautioned about the limited value of age distribution data, in light of the ease with which they can be obtained, especially by wildlife management agencies that monitor harvests.

Estimating Population Growth from Birth and Death Rates

Knowing the birth and death rates of a population should allow us to examine whether the population was increasing, holding steady, or declining. We can indeed calculate the growth rate of the population from such information. We discuss 2 approaches, a simplified one applicable when birth and death rates are not segregated into many age classes, and the Lotka equation for when they are segregated. The population projection matrix also can be used to measure the growth rate of a population (Leslie 1945, Pielou 1969).

A Direct Method

Consider, following Martin et al. (1979), the female segment of the North American population of mallards. For 1961–1974, the average survival rate for adult females was 0.555, and the average survival rate for immature females was 0.563. Survival was estimated between anniversary dates of 1 September in successive years. Suppose the recruitment rate, measured as young females per adult female on this anniversary date, averaged 1.03. From these survival and fertility rates, we can estimate the average annual change in the population of female mallards. The number of adult females on 1 September of year $t + 1$ represents the adults from the previous year that survived plus the immatures that survived:

$$A_{t+1} = A_t(0.55) + Y_t(0.563),$$

where A_t is the number of adult females in year t and Y_t is the number of young females in year t. We also have from the recruitment rate:

$$Y_t = A_t(1.03),$$

so

$$A_{t+1} = A_t(0.555) + A_t(1.03)(0.563),$$

or

$$A_{t+1} = A_t(1.135).$$

From this equation we can conclude 1 of 3 things: (1) the female segment of the mallard population was growing at a rate of 13.5% per year ($\lambda = 1.135$), (2) the estimates of survival and/or recruitment are wrong, or (3) the model is incorrect. Evidence from annual surveys of mallards during 1961–1974 led Martin et al. (1979) to reject the first possibility, and the simplicity of the model argues against the third. Thus, the authors concluded that certain estimates of survival or recruitment were the problem. In fact, they used this approach as a check on the consistency of their parameter estimates.

Lotka's Equation

If age-dependent schedules of survival (l_x) and fertility (m_x) are available, the growth rate implied by those schedules can be computed from Lotka's formula (equation 8) in an iterative procedure. Caughley (1977:215) presented a short Fortran computer program to perform the necessary calculations. We illustrate use of this equation with the white-tailed deer data (Tables 15.12, 15.14). Using a value of $r = 0$, indicative of a steady population and values of l_x (Table 15.14) and m_x (Table 15.12) in the right-hand-side of equation 8, we obtain (because $e^0 = 1$)

$$1.000 \times 0 + 0.5800 \times 0.047 + 0.4060 \times 0.503 + \ldots + 0.0974 \times 0.771,$$

which sums to 0.89. This value is <1, indicating that $r = 0$ is too high. Using $r = -0.10$ in equation 8 gives 1.44, which also is too large. The value of the sum that we want is 20% of the way between 0.89 and 1.44. We then try a value for r that is 20% of the way between 0 and -0.10, that is, $r = -0.02$. Use of $r = -0.02$ gives 0.97, which is too small, but $r = -0.026$ results in a sum of 0.9995; close enough to stop the iteration. Because $e^{-0.026} = 1 - 0.0257$, this value suggests the deer population was declining at about 2.57% each year.

This approach also could be used with the mallard data of Martin et al. (1979). Age-dependent survival and fertility rates give:

$$l_0 = 1,$$
$$l_x = 0.563(0.555)^{x-1}, x > 0,$$

and

$$m_0 = 0,$$
$$m_x = 1.03, x > 0.$$

Using these values with $r = \log \lambda = \log(1.135) = 0.1266$ in the right-hand side of equation 8 yields 0.99987, negligibly different from 1.0.

Another useful statistic is the **net reproductive rate**, the average number of young produced by an individual during its lifetime:

$$R_0 = \Sigma l_x m_x.$$

Values of $R_0 < 1$ indicate that members of the population are not replacing themselves and the population is declining. Conversely, $R_0 > 1$ denotes an increasing population, and $R_0 = 1$ indicates a stable population.

For the deer population, we obtain $R_0 = 0.89$, indicating a declining population. For the mallard example, we have

$$R_0 = 1 \times 0 + 0.563 \times 1.03 + 0.563 \times 0.555 \times 1.03 + \ldots$$
$$= 0 + 0.580 + 0.322 + 0.179 + 0.099 + 0.055 +$$
$$0.031 + 0.017 + 0.009 + 0.005$$
$$= 1.300,$$

adding through age class 10 (note that terms become small for older age classes, indicating the few old individuals have little effect on the size of the population). This value also suggests an increasing population.

Models with Components of Survival or Birth

The models presented thus far are rather simple, as they really depend only on time. Given the features of a model and the current status of a population, we can predict exactly what will happen at any future time (if the model was correct). This is unrealistic, but simple models have nonetheless proven useful. Their major advantage lies in the way they can be treated mathematically. We now turn to models that are more complex, but are actually often simpler to construct and analyze. The trade-off is that as we gain realism and complexity, we lose the ability to analyze the model mathematically, requiring use of a computer. For that reason, most of these models are simulation models. Some of the most useful population analyses today are based on simulation models.

A population goes up or down during a year, depending on its annual survival and birth rates. The annual survival rate is an overall measure, encompassing the risks encountered by a population, that may vary season to season, day to day, among individuals in the population, and from place to place (Box 15.5). It is often worthwhile to examine survival rate in closer detail, such as by parts of the year. Likewise, fertility rates incorporate a multiplicity of components that may be treated individually. For example, the measure of births for our whooping crane example (Table 15.1) is the number of young recorded in the winter population. This measure reflects the number of adult birds that are paired, the proportion of those that successfully lay eggs, clutch size, proportion of eggs that hatch, survival of young until fledging, and survival from fledging until the winter survey.

By subdividing survival and fertility rates into finer components, we gain several advantages. First, we can consider environmental and other factors, beyond the density of conspecifics, that influence individual components. For example, clutch size of mallards depends primarily on age of the female and date the clutch is initiated. Thus, it can be modeled as a function of those factors. Nest success, however, depends mostly on nesting habitat and predator numbers, and these features can be used to model nest success. The second potential advantage is that we can often obtain better estimates of individual components and the factors that influence them. For example, clutch size can be studied either passively or experimentally, and studies that manipulate clutch size might give good insight into that parameter, but poor information about overall nest success. A third advantage is that we may gain a clearer understanding of the relationships involved in each component by their separate study. This understanding is especially important for man-

Box 15.5. Estimating population trends using life history traits

In many situations, the biologist is ultimately interested in answering the question, "is the population stable, increasing, or decreasing?" To fully answer this question, the trend in the population must be estimated. We can use a wide range of tools to estimate population trends, from simple changes in observed count data to complex models that attempt to explain temporal patterns in populations. The figure illustrates one approach.

Using capture–recapture data, Franklin et al. (2000) built models to explain variation in life history traits (e.g., annual survival) that ultimately affected the rate of population change. They then estimated the annual rate of population change (λ) as a function of annual rates of survival and recruitment, and they further suggested that yearly weather changes best explained annual variation in this species. The study focused on understanding the processes that influence annual population changes in this species and provided a template for how wildlife biologists should model life history traits and incorporate this information into estimates of population change.

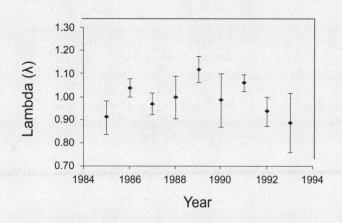

An example of a contemporary approach to modeling population changes is that of Franklin et al. (2000), who estimated trends in annual population counts of the northern spotted owl (*Strix occidentalis caurina*) in California. The annual rate of population change is indicated by λ.

agement applications, in which one or more components may be altered; the effect on the entire system needs to be understood, or else the population response may not be the one desired.

We illustrate this procedure with a model of the production of mallards in the prairie pothole region (area of the northern Great Plains and midgrass and tallgrass prairies containing thousands of shallow wetlands extending from north-central Iowa to central Alberta, Canada; Johnson et al. 1992). Only females are considered, and 2 age classes of breeding females, yearling and older, are identified. Let F_i ($i = 1, 2$) be the number of yearling females and older females, respectively, in the breeding population, and F_0 be the number produced. F_0 can be apportioned according to age class of the adult:

$$F_0 = F_1 R_1 + F_2 R_2,$$

where R_i is the production rate for females in age class i. This value can be further decomposed according to nesting attempts, giving

$$R_i = D_i(Q_{i1} + Q_{i2} + Q_{i3} + Q_{i4} + Q_{i5}),$$

where D_i is the proportion of females of age class i that attempt to breed, and Q_{ij} is the production from the jth nesting attempt of females in age class i. This equation allows a maximum of 5 nesting efforts in a breeding season.

The production from a particular nesting attempt itself involves several factors and can be expressed as

$$Q_{ij} = A_{ij} C_{ij} HEB,$$

where A_{ij} is the probability that a female of age i will make the jth nesting attempt in a breeding season, C_{ij} is the average clutch size of the jth nesting attempt by a female of age i, H is the nest success rate, E is the survival rate of eggs in successful nests, and B is the survival rate of young.

Most parameters are indexed by age of the female and nesting attempt, because age and attempt are known to influence them. Effects of age and attempt on nest success rate (H), survival rate as eggs (E), and survival rate as young (B) have not been demonstrated clearly. The rates of incidence of breeding (D_i) vary most strongly with wetland conditions. Nesting probabilities were formulated to be higher for older females than for yearlings, to be higher when wetland conditions were good, to decline with nesting attempt, and to be lower when nest success is high (because nests are more likely to be destroyed later, when the female is in poorer condition). Clutch size was modeled to decline with nesting attempt and to be 1 egg smaller for yearling females than for older females.

Nest success of mallards is highly variable and is a component amenable to management. It varies according to the condition of the nesting habitat and predator abundance. Egg survival is generally high and in the model does not vary as a function of any environmental variable. Survival of young after hatch is lower, however, and likely depends on weather, predators, and food supplies for ducklings, and possibly on disease; some of these factors may operate in a density-dependent manner.

Johnson et al. (1992) executed the model described above by allowing the parameters to vary about as widely as they seem to do in natural populations. They contrasted results from the model for mallards with results for other species. They concluded that recruitment of mallards was most dependent on wetland conditions and predation. Similar models (e.g., Johnson et al. 1987, Cowardin et al. 1988) have been used to evaluate management options in terms of expected production of mallards anticipated by manipulating various parameters.

Simulation models have been applied to many wildlife species, often to assess the effect of harvest strategies. Useful references on the construction of models for wildlife management include texts by Starfield and Bleloch (1991), Grant et al. (1997), and Williams et al. (2002a).

It is easy to build models—perhaps too easy in this day of ready access to computing power. It is harder to evaluate them (Johnson 2001a). One should compare model results with real data, independent of information used to construct the model and preferably obtained by direct experimentation. If that test is not feasible, a comparison with other models, built on different assumptions, is worthwhile, as is a comparison to analytic solutions.

Competition Models

Let us briefly consider populations not of a single species, but of 2 species that interact. Here we assume they compete for some resource, such as food. If that resource is limited, the habitat will support fewer of species 1 when species 2 is common than when species 2 is rare. Let α ($\alpha > 0$) be the relative impact of an individual of species 2 on the population growth rate of species 1. That is, αN_2 individuals of species 2 have the same effect on species 1 as do N_1 individuals of species 1, or $N_1 = \alpha N_2$ in terms of effect on species 1. Then, generalizing the logistic model of equation 5 and writing $N_1(t)$ and $N_2(t)$ as the sizes of the 2 populations at time t, the per capita growth rate of population 1 is modified, not just by

$$\frac{K_1 - N_1(t)}{K_1},$$

but by

$$\frac{K_1 - N_1(t) - \alpha N_2(t)}{K_1},$$

where $K_1 - \alpha N_2(t)$ can be considered the carrying capacity for species 1, as reduced by the presence of N_2 animals of species 2. From this expression we have

$$\frac{1}{N_1(t)}\frac{dN_1(t)}{dt} = r_1 \frac{K_1 - N_1(t) - \alpha N_2(t)}{K_2},$$

where r_1 and K_1 are the parameters of logistic growth for species 1 in the absence of species 2. Analogously, if β ($\beta > 0$) is the relative effect of an individual of species 1 on the population growth rate of species 2, we obtain

$$\frac{1}{N_2(t)}\frac{dN_2(t)}{dt} = r_2 \frac{K_2 - N_2(t) - \beta N_1(t)}{K_2},$$

where r_2 and K_2 are defined correspondingly. Many of the results can be obtained without recourse to logistic formulation (Maynard Smith 1974:62), but it is a convenience. Parameters α and β are termed the **competition coefficients** of the model developed by Lotka (1925) and Volterra (1926). This system has been the basis of substantial theoretical work in competition (e.g., Levins 1968; MacArthur 1968, 1972; Vandermeer 1972; Pianka 1974; Berryman 1981), but has received little use in wildlife studies, in part because of the inherent difficulty of estimating the relevant parameters.

Mathematical results from the equations above follow from the values of K_1, K_2, α, and β. Only if $K_1/\alpha > K_2$ and $K_2/\beta > K_1$ is it possible for the 2 species to coexist. Basically, then, coexistence is possible only if the growth rate of each species is inhibited more by a member of its own species than by an individual of the other species. Their own density-dependent controls must cause growth to stop before they eliminate the competitor. One way this can happen is if the 2 species do not overlap completely in resource use. Although such results as these are useful theoretically, most actual populations probably do not exhibit similar behavior developed from this simple competition model, but are affected by a variety of other phenomena. For example, patchiness in resources reduces competition by favoring the first species in one kind of habitat and the second species in another kind. Also, because the environment changes with time, the relative competitive abilities of the species may vary.

Predator–Prey Models

A second kind of interaction involves predation. Predator–prey models in a variety of forms have seen extensive use in wildlife studies. Suppose that species 1 serves as a prey for species 2, and the population growth rate of species 1 is inhibited in direct proportion to the number of predators. Then

$$\frac{1}{N_1(t)} \frac{dN_1(t)}{dt} = r_1 - \gamma N_2(t),$$

where the **predation coefficient**, γ, indicates the removal rate of prey per predator. This model includes no inhibitory effects of the population of species 1; that is, in the absence of predators, the prey population would grow exponentially. Also, each predator consumes a number of prey proportional to the abundance of the prey. For predators (species 2), the per capita population growth rate is assumed to be

$$\frac{1}{N_2(t)} \frac{dN_2(t)}{dt} = \delta N_1(t) - d_2,$$

where d_2 is the death rate of predators, assumed to be independent of the population of prey. The coefficient δ represents the *conversion rate* of prey to predators. This model also was developed by Lotka (1925) and Volterra (1926).

The model assumes (e.g., Ricklefs 1979:602): (1) exponential growth by the prey species in the absence of predators (i.e., numbers of prey are limited only by predation), (2) exponential decay by predators in the absence of prey (mortality is independent of the density of predators), and (3) the rate at which prey are consumed is directly proportional to the product of the 2 species' densities (which can be construed as the chance of encountering one another if movements are random). The first 2 assumptions indicate that population growth of each species is controlled by the other species; these assumptions can be relaxed by including a logistic-type self-inhibitory restraint on population growth rates. The third assumption can be replaced by any of a variety of choices (May 1973:81–84, Maynard Smith 1974:25–33).

This model can be analyzed mathematically under the assumption there is no random variation. Depending on the values of the parameters, 2 outcomes of the model are feasible: either the populations of both species reach an equilibrium point and remain there, or populations of both species oscillate over time, with increases in the predator species lagging behind increases in the prey species. Most investigations of actual populations of predators and prey involved invertebrate species in controlled laboratory situations; Tanner (1975) offered an exception that dealt with vertebrates. He concluded that vertebrate predator–prey systems were stable only if the prey species limited its own population or if it had lower (intrinsic) growth rate than the predator species. Populations of snowshoe hares (*Lepus americanus*) and Canada lynx (*Lynx canadensis*), which have roughly equal growth rates, oscillate in a cyclic fashion. Caughley and Krebs (1983) provided a more general view of this issue.

Powell (1979) applied predator–prey modeling to a community involving fisher (*Martes pennanti*) and its primary prey, the porcupine (*Erethizon dorsatum*). He prudently examined 5 variations and extensions of the basic predator–prey model, so the conclusions he drew would be less susceptible to assumptions underlying any single model. He also considered the effects of 2 alternative prey species. Although space does not permit a detailed treatment of the models here, Powell's results suggested the community was stable, but that only small increases in fisher mortality could cause local extinction of that predator.

These general predator–prey models, like other models, are unrealistically simplistic. Nonetheless, they offer useful insight into the general behavior of predator–prey systems, lead to more realistic models, and form the foundation for managing populations for optimal yield. Connelly et al. (2004) discuss situations where the predator is human.

PARAMETER ESTIMATION

In the previous discussion, we introduced theoretical models of population growth under a range of conditions and assumptions. Once a model is chosen for an analysis, it is necessary to estimate the parameters of the model. In the discussion of theoretical models, we illustrated some simple calculations of parameter estimates, but in practice, these calculations are complicated and require the use of a computer for all, but the simplest models. How do we use a computer to estimate the parameters of a particular model?

To estimate a population parameter, say, a survival rate S, we first sample the population of interest and then use statistical procedures to derive the estimate. Parameter estimation has seen considerable recent growth, evolving from simplistic and often inflexible computer programs to more comprehensive ones. Several computer programs are available to aid parameter estimation, and new developments occur regularly. In this section, we outline a general approach to parameter estimation, briefly discuss how to examine whether a model "fits" the data, and introduce the reader to key computer programs used to estimate certain parameters.

Key Steps to Parameter Estimation

Once an underlying theoretical model is chosen for an analysis, there are several steps needed to obtain estimates of parameters. **Modeling** logically precedes the process of parameter estimation, because it can identify important variables. Earlier, we introduced the concept of a population model. Because we often are interested in predicting population responses to one or more variables, we often develop and evaluate competing models to describe a population process, although a single model that included all covariates could be used.

Modeling is an iterative process in which multiple models (a model set) to explain the same phenomenon are compared and a decision is made regarding the appropriateness of one or more of these models to describe the process (Williams et al. [2002a] provided a thorough review). The first and arguably most important step involves thinking about the process and developing a short list of **a priori** biological hypotheses (Lebreton et al. 1992, Anderson and Burnham 2001, Burnham and Anderson 2002). These hypotheses should be based on the biological information available, should address the specific factors of interest, and will form the basis for specific models in the model set. The building of meaningful models should be guided by the desire to answer "Why?" questions about the population of interest, rather than simpler "What?" questions that seek answers without much understanding of the underlying processes. For example, it might be relatively easy to estimate an annual survival rate for a particular species (a "What?" question). It is more difficult to understand factors that might influence annual survival, such as the health or condition of the animal, its age, or the habitat where it resides (the "Why?" questions). Ideally, model development should be guided by these questions, and analyses should seek to understand processes rather than simply describe patterns.

Once a list of hypotheses is formulated, we rely on computers to help us choose one or more of these models for inference and generate parameter estimates. Many models used in population analyses are based on the multinomial distribution and use such procedures as the method of maximum likelihood to estimate the parameters (Williams et al. 2002a), although other distributions and estimation procedures are sometimes used (e.g., a Bayesian approach is especially useful for estimating variance components). Finally, the parameter estimates, whether survival rates or rates of population change, are obtained. Because different models often result in different parameter estimates, key questions are: Which model do I use for inference? Should the estimates come from a single (best) model, or from a number of potential models?

The process of deciding which model or models to use for inference is known as **model selection.** Model selection in itself is a complex process (Franklin et al. 2001, Burnham and Anderson 2002). Model selection methods vary from traditional hypothesis-testing approaches (e.g., likelihood ratio tests) to more complex methods based on information theory (e.g., **Akaike's Information Criterion [AIC];** Akaike 1973). The most popular model selection tool in wildlife science currently is probably AIC (Burnham and Anderson 2002), which optimizes the trade-off between model fit and parsimony. In some cases, a "best" model may be chosen for inference, and parameter estimates from that model are used. However, in many situations, it may be desirable to use parameter estimates that reflect the uncertainty associated with selecting the "best" model. In such cases, parameter estimates may be weighted by the strength of evidence for each model, followed by computation of the parameter estimates (Buckland et al. 1997, Burnham and Anderson 2002). This practice is termed **model averaging** and is used increasingly often to present parameter estimates that incorporate the uncertainty in the model selection process (Box 15.6). Regardless of the approach used, the estimates are reported along with a measure of precision or uncertainty (standard error or confidence interval).

Factors Affecting Parameter Estimates

Biologists should not be satisfied with simply estimating a parameter without gaining some understanding of **why** the parameter varies and what it means. We are particularly interested in answering the question: what affects the parameter? There are several types of factors, called covariates, that can affect a parameter, and these should be incorporated into an analysis if possible.

Some covariates are characteristics that apply at a group level. Gender (male versus female) or geographic location (site A versus site B) are examples of group variables that may influence some parameters, such as survival rate. Some factors can vary in time, such as weather or day of a nesting season. If rainfall is believed to influence survival, then models incorporating some measure of precipitation (e.g., total annual rainfall or daily rainfall) could be used to address this question. Both group and temporal sources of variation can be easily incorporated into a population analysis using MARK and certain other software.

In addition to group covariates, some factors that vary among individuals may influence a parameter of interest, such as survival or reproduction (Franklin 2001, Pollock 2002). The possible types of these variables, called individual covariates, is almost endless, including measures of body condition, size metrics, breeding history, genetic characteristics, and many more. Each of these captures some of the inherent variation among individuals in any population and can be used to model the parameter of interest. MARK handles individual covariates easily. The inclusion of individual covariates allows a more thorough exploration of mechanisms affecting the dynamics of a population (Franklin 2001).

Goodness-of-Fit Tests

Another key step in population analysis involves answering the question: does the model fit the data? To properly answer this question, we must somehow test to see whether the statistical model is appropriate for the data. Generally, this check is done using a **goodness-of-fit test.** There are many goodness-of-fit tests available, each suited for a particular class of models. The general procedure with any of these tests is to compare the data against the expectations under a particular model. For example, a band recovery study

Box 15.6. The process of modeling and parameter estimation

To illustrate the concept of population modeling, we refer to a simple example: modeling the daily nest survival of the mountain plover (Dinsmore et al. 2002). The mountain plover is a ground nesting bird of the Great Plains and has a mating system in which males and females incubate separate nests. The nesting season spans mid-May to early August. Of primary interest was whether male- and female-tended nests had different patterns of daily nest survival. Researchers also were interested in describing the within-season variation in nest survival, expecting that early nests survived better, because they were tended by older, more experienced adults. This a priori knowledge formed the basis for a set of 4 competing models: (1) daily nest survival of male- and female-tended nests differs (S_{gender} model); (2) daily nest survival of male- and female-tended nests differs, and within-year variation follows a linear time trend ($S_{gender+T}$ model); (3) daily nest survival of male- and female-tended nests differs, and within-year variation follows a quadratic time trend ($S_{gender+TT}$ model); and (4) daily nest survival is constant, both within years and between genders (S model).

Each of these competing models was fitted using the nest survival model in program MARK (see table below), including model ranking by Akaike's Information Criterion (AIC), delta AIC, the AIC weight (w_i), number of parameters (K), and deviance for each model (see Burnham and Anderson 2002).

Model	AIC	Delta AIC	w_i	K	Deviance
$S_{gender+TT}$	1,132.50	0.00	0.65	4	1,124.46
$S_{gender+T}$	1,134.96	2.45	0.19	3	1,128.94
S_{gender}	1,136.40	3.90	0.09	2	1,132.39
S	1,137.02	4.52	0.07	1	1,135.02

On the basis of these results, what can we conclude about nest survival? Gender appears to have an effect, because the 3 best models contain this effect, and their AIC weights sum to 0.93. There also appears to be evidence of within-year variation in nest survival, and the form of that variation appears to be quadratic (this model has most of the weight: $w_i = 0.65$). There does not appear to be strong evidence supporting similar nest survival rates by gender or constant rates over the nesting season.

We are also interested in estimates of the daily nest survival rates. What is the probability that a nest will survive a day? Because the 2 best models contain temporal variation, it makes sense that daily nest survival rates vary over time. In the best model ($S_{gender+TT}$), daily nest survival rates begin at about 0.977 at the beginning of the nesting season, slowly decrease to a low of about 0.966 in mid-season, and then gradually climb to peak at about 0.995 at the end of the nesting season. In the model with only gender effects (S_{gender}), the daily nest survival rates are 0.971 for female-tended nests and 0.977 for male-tended nests. The model with neither gender effects nor temporal variation (S) produces an estimate of daily nest survival of 0.975.

How are we to choose which parameter estimate is best? In this example, we would be reasonably safe in basing inference on the best model, although a better approach would be to "average" the daily nest survival rates by allowing each model to influence the estimate proportional to its AIC weight. In this example, the model-averaged estimate of the first daily survival rate is $\hat{S}_1 = 0.974$, which compares to the estimates generated from each model: $S_{gender+TT}(\hat{S}_1 = 0.977)$, $S_{gender+T}(\hat{S}_1 = 0.963)$, $S_{gender}(\hat{S}_1 = 0.971)$, and $S(\hat{S}_1 = 0.975)$. The model-averaged estimate is close to the estimate from the best model, reflecting the influence of its large model weight.

---

would include releases and recoveries of birds in multiple years. The underlying band recovery model (described earlier) can be used to compute the expected number of recoveries each year, given the assumptions of the model. The expected recoveries are compared to the actual recoveries using a chi-square goodness-of-fit test. The model is judged to fit the data adequately if the test result is clearly nonsignificant (e.g., the P value associated with the test statistic is much greater than a critical value of $\alpha = 0.05$); there is lack of fit otherwise. If the model fits, the analysis can proceed. If there is a lack of model fit, the analyst should scrutinize study design and model assumptions, try to uncover the source of the lack of fit, and make adjustments to the data to conform to model assumptions.

Methods for testing model goodness-of-fit are limited to the simpler population models; there are no goodness-of-fit tests for complex models, such as the robust design. Computer programs can compute goodness-of-fit tests for band recovery models (ESTIMATE, Brownie et al. [1985]) and the Jolly–Seber model (RELEASE, Burnham et al. [1987]; U-CARE,

Choquet et al. [2009a]). For other models, either goodness-of-fit tests are lacking, or fit must be assessed using ad hoc methods or by such techniques as bootstrapping.

Computer Programs for Population Analyses

Computer programs to estimate population parameters have become more sophisticated, providing biologists with tools to estimate parameters, such as population size and rates of survival, fecundity, immigration and emigration, and population change. There have been major advances in computer programs to facilitate these complex analyses. Formerly, biologists might have been responsible for learning multiple computer programs (e.g., JOLLY, JOLLYAGE, SURVIV, and others) to complete a population analysis. Many of these programs are now unified in MARK (White and Burnham 1999, White et al. 2001). MARK is a powerful and flexible program that allows most population parameters to be estimated, provided the study is designed well and includes a sufficient sample of marked individuals. This program requires knowledge of basic mathematical and statistical concepts, and the serious population analyst should carefully study the user's guide (Cooch and White 2009).

MARK offers a wide range of modeling features, including (1) the ability to model group (e.g., gender or age classes) effects, (2) the ability to include individual (e.g., a measure of body mass) or time-varying (e.g., daily precipitation) covariates (see Pollock 2002), (3) model selection using AIC (Akaike 1973), (4) the ability to model average parameter estimates across competing models to reflect model-selection uncertainty (see Burnham and Anderson 2002), (5) goodness-of-fit testing for some models, and (6) a Bayesian (Markov-chain Monte Carlo) modeling tool. MARK is available free at http://www.cnr.colostate.edu/~gwhite/mark/mark.htm. The detailed user's guide for MARK also is available from the same link.

MARK is certainly the most comprehensive computer program for population analyses, although other programs, such as POPAN (Arnason and Schwarz 1995, 2002; useful for the analysis of open populations), M-SURGE (Choquet et al. 2005), and E-SURGE (Choquet 2007, Choquet et al. 2009b), are widely used.

POPULATION VIABILITY ANALYSIS

A culmination of our knowledge of population biology is the ability to predict future population viability (Lande 1988, 1993; Noon et al. 1999). The process of making such predictions using available population data is called **population viability analysis** (**PVA;** Boyce 1992). Two general approaches are used to make predictions: (1) the probability is estimated that a population of a specified size will persist for a certain time period (PVA), or (2) the minimum viable population (MVP) needed for a population to persist for a certain time period is estimated (Shaffer 1987). In either case, an underlying population model is used to make predictions about future behavior of the population.

Although appealing to biologists, good PVAs present many challenges (see White 2000). Making meaningful predictions requires a thorough knowledge of the population processes, including detailed information on survival, reproduction, and other facets of population dynamics, all based on studies that are replicated in space and time. Given expense and time requirements, few such studies exist. Most studies are hampered by sparse data, collected for only a few years and at one or a few sites. Should these types of studies form the basis for PVAs? The answer depends on the desired predictions. If only a general idea about population persistence is sought, then predictions based on simpler studies might be useful, but only for planning purposes. If detailed predictions are sought (e.g., the conservation of an endangered species), then only data from long-term, replicated studies should be used. Computer programs, such as RAMAS Metapop 5.0 (Akçakaya and Root 2005) and VORTEX, are available commercially and can perform many of the calculations associated with a PVA.

As with all analyses discussed in this chapter, a thorough understanding of assumptions is necessary to do a PVA. In PVAs, the behavior of the population is influenced by demographic parameters of the population (e.g., survival rate), variation in those parameters (temporal and spatial variations), and individual heterogeneity (e.g., the genetic makeup of an individual). These sources of variation collectively constitute **process variance,** which refers to the real underlying variation in the population growth process. Making predictions about population persistence would be much easier if such analyses were not confounded by another source of variation (**sampling variation**). Sampling variation results from stochasticity (random events) in our measures of these sources of variation and from having only samples, not entire the population, measured.

To realistically model population persistence, all these sources of variation must be included in the population model. In addition, the underlying statistical distribution of demographic parameters (e.g., annual survival rate) across all individuals in the population must be included (White 2000), resulting in increased persistence times for most populations. This omission is common in many PVAs and yields results that are often too pessimistic. Another problem is that estimates of process variation also typically include sampling variation that causes variances to be positively biased and PVA projections to be too pessimistic.

As interest in making long-term population predictions increases, use of PVA is likely to increase. The models themselves are appealing and offer useful predictions necessary for management decisions. However, many PVAs are plagued by too few data and a lack of rigorous estimates of demographic, environmental, and individual variation, thus reducing their real utility to managers and conservationists (Boxes 15.7, 15.8).

INFERENCE

Once a population analysis is complete, we often desire to draw conclusions from the data. The process of formulating hypotheses, testing those hypotheses against data, and then making conclusions based on the study results is a logical scientific process. Typically, we use inductive inference to use the results from localized studies to make broader statements about a larger population of interest. For example, we may estimate the annual survival of mallards from a lo-

Box 15.7. An example of population viability analysis

A population viability analysis (PVA) is used to predict the future behavior of a population, usually related to specific management or conservation goals. As an example, consider the PVA for the California gnatcatcher (*Polioptila californica*), an endangered songbird of the Pacific Coast (Akçakaya and Atwood 1997). Using multiple field studies, the authors gathered information on vital rates (survival, dispersal, fecundity, etc.), habitat use, known range of the species, and characteristics of the habitat patches the species occupied to build a detailed model for predicting future population behavior. The metapopulation model incorporated spatial and stochastic variation and predicted sharp population declines and a high risk of extinction for this species. However, some parameters were poorly understood, leading to variation in estimates of persistence and uncertainty in the interpretation of some results. Modeling results could be used to suggest possible conservation strategies, and the authors suggested that such an exercise may prove useful for evaluating future management activities and conservation measures.

This example provides a template for thinking about PVAs. Emphasis should be placed on estimating demographic parameters well, including multiple sources of population variation in the PVA model, and on a careful interpretation of the results. Models built in this manner can be informative and will provide useful tools for wildlife biologists.

Box 15.8. Metapopulations

Many animals occur in distinct patches of suitable habitat, rather than in a large continuous area. If they move freely among patches, we can treat them as a single population, because the dynamics are likely to be similar in all patches. If there is virtually no movement among patches, the population in each patch should be treated separately, because they can display completely different dynamics. If there is limited movement among patches, the dynamics of the patches may differ, and extinction of a population in a patch can be overcome (rescued) by immigration from another patch. This is the situation in which metapopulation theory applies. A metapopulation is basically a population of populations.

Richard Levins coined the term metapopulation around 1970 (Levins 1970). Interest in the topic has flourished, as exemplified by 2 edited volumes (McCullough 1996, Hanski and Gilpin 1999). Metapopulation dynamics clearly apply to populations that occupy naturally occurring patches of habitat. Moreover, increasing human-induced fragmentation of once-continuous habitats has made metapopulation theory more generally applicable. What used to be large expanses of forest, for example, have been reduced to smaller stands of trees, surrounded by a landscape matrix of habitats unsuitable for many forest-dwelling species. Metapopulation theory has largely replaced island biogeography as a theoretical framework for thinking about fragmentation (Wiens 1996).

Metapopulation theory resides at the intersection of numerous topics of interest to modern wildlife biologists, including landscape ecology, corridors that connect habitat patches, and source–sink population dynamics (Pulliam 1988). Further, it is key to population viability analysis; an important consequence is that, because the dynamics of populations in separate patches may differ, metapopulation structure may enhance the viability of populations (Wiens 1996).

Unfortunately, empirical studies involving metapopulations, and methods to analyze them, lag far behind the theory. Acknowledging that many populations may occur as metapopulations, it is important to consider this topic and recognize that conclusions based on treating a population as continuous may be flawed.

calized study in the Great Plains, but really desire to infer those results to all mallards in the Great Plains. How, then, do we correctly interpret the results of a population analysis?

In a population analysis, emphasis is typically placed on modeling one or more parameters of interest, and different hypotheses about how the population behaves are imbedded in different (competing) models. We then use model selection procedures to come to some conclusion(s) about the system, whether it is that adult and juvenile survival differs or that improved body condition results in greater survival. We seldom use controlled experiments, although such studies offer more reliable results and increase the scope of inference.

Results of a population analysis can be interpreted in several contexts, each directed at answering a different question. How, for example, could we infer that improved body condition resulted in greater survival in male mallards? Ideally, we would conduct a manipulative experiment in which body condition was experimentally controlled (keeping all other factors constant), and then measure the resulting effect(s) on survival. This experiment might allow us to infer causation with some confidence; we could then say that improved body condition does result in greater survival in male mallards. But what happens if we cannot conduct a manipulative experiment? In this example, we would probably attempt to mark a sample of male mallards of known body condition and then estimate survival in the presence of that variable. We could build models with and without a body condition effect on survival and then compare those models to see which receives better support. In this example, we could not infer causation, but would instead demonstrate a correlative relationship between body condition and survival, the strength of which would depend on sample size, study design, and the actual relationship itself.

SUMMARY

This chapter introduces numerous methods for analysis of wildlife population data. From the simplest (but often biased) estimate of observed survival to complex models to estimate survival as it relates to biological processes, today's wildlife biologist has many tools to estimate parameters of interest. We began by introducing theoretical models of population growth, including such concepts as exponential and logistic growth, birth and death processes, and age effects. We considered situations in which population growth also depended on numbers of another species, either a competitor, or a prey or predator. Next, we discussed parameter estimation. Given data and a set of models, how do we generate an estimate of a parameter of interest? We emphasized computer-based modeling of population processes and the need to model complex biological processes as functions of a variety of variables, including environmental factors, age or gender, and variables particular to an individual. We briefly covered methods for predicting the long-term viability of a population and concluded with a section on inference, or how we draw conclusions from data. Each of these components is vital to a proper analysis of population data, and the entire process is intended to contribute to our collective scientific knowledge.

Given the diversity of approaches that can be taken, how is a wildlife biologist to choose? The choice ultimately depends on the specific objectives of the study. An analysis should include those relationships and variables suspected of being most influential on the dynamics of the population under study. Parameters are of interest because they can help describe complex population processes and lead to a better understanding of factors that influence the population. The trade-off between simple and complex models is difficult: simple models are tractable, but may overlook key processes, whereas a complex model may satisfy only its builder.

Wildlife management in essence is based on only 2 primary tools, manipulation of habitat and control of harvest. These activities are effective only if they influence in the desired manner the population dynamics of the target species. To evaluate their actions and know they are doing the right thing, managers must understand those dynamics.

Methods for analyzing population data continue to change rapidly, and they will likely continue to do so into the future. The past decade has seen rapid advances in such topics as model selection, numerical approaches to parameter estimation, the handling of covariates in an analysis, and model averaging, to name a few. But other topics, such as individual heterogeneity, how to handle missing data, and goodness-of-fit testing, still await the development of rigorous approaches. Still other topics, among them Bayesian approaches to capture–recapture modeling and the use of genetic capture–recapture methods, are seeing widespread application to contemporary population analyses. With frequent methodological and technological advances, approaches to population analysis continue to evolve rapidly to make this an exciting and important discipline of wildlife biology.

16

Vegetation Sampling and Measurement

KENNETH F. HIGGINS,
KURT J. JENKINS,
GARY K. CLAMBEY,
DANIEL W. URESK,
DAVID E. NAUGLE,
ROBERT W. KLAVER,
JACK E. NORLAND,
KENT C. JENSEN, AND
WILLIAM T. BARKER

INTRODUCTION

WHAT IS THE UTILITY of vegetation measurements for wildlife managers? In the prairie, savanna, tundra, forest, steppe, and wetland regions of the world, mixtures of plant species provide wildlife with food, cover, and in some circumstances, water; the 3 essential habitat elements necessary to sustain viable wildlife populations. **Habitat** refers to use of a vegetation type by an animal (e.g., deer habitat), and **vegetation type** refers to differences in vegetation stands (e.g., marsh vegetation type versus tall grass prairie vegetation type; Hall et al. 1997). The variety of wildlife using plants ranges from snails and voles to bison (*Bison bison*) and elephants (*Loxodonta* spp.) in uplands and from mosquitoes and ducks to muskrats (*Ondatra zibethicus*) and manatees (*Trichechus manatus*) in wetlands. Through evolutionary processes, some wildlife species are totally dependent on vegetation for all annual life requirements, whereas other species use vegetation only for cover or food. Regardless of the role of vegetation in the sustenance of wildlife, any management or research project that requires evaluation of wildlife and vegetation type relationships on a unit of land will necessitate some form of vegetation measurement.

The term **vegetation** can refer to a single plant or species on a specific site or a community in the landscape. Vegetation may occur naturally or be introduced, and it may be live or dead. Uses of vegetation measurements are many: (1) evaluation of vegetation response to management, (2) estimation of carrying capacity, (3) characterization of cover and habitat components for an endangered species, or (4) long-term monitoring of the general trend of plant vigor or vegetation type condition.

Surveying and measuring **quantity** and **quality** of vegetation in habitats are basic to wildlife research and management. Grassland, shrubland, and woodland vegetation types are comprised of populations in which individual plants are usually too numerous to inventory completely. Consequently, wildlife biologists usually use sampling techniques to make inferences about the total plant population in a given vegetation type.

Vegetation sampling methodologies have evolved in several ecological disciplines (e.g., plant ecology, forestry, and range science) and for a variety of management or research objectives (e.g., estimating forage for ungulates and describing habitat use by passerine birds). Description of every method that has been used to sample vegetation is beyond the scope of this chapter. We describe how to measure **vegetation structure,** which Dansereau (1957) defined as the spatial organization (distribution) of individuals that form a stand. We have organized this chapter into a description of basic methods of vegetation sampling, with examples of how those methods have been applied or modified in wildlife research and management. We assume the investigator has adequate knowledge of the concepts of wildlife ecology, primary habitat requirements of wildlife species under study, and ability to systematically identify the species of wildlife and vascular plants in the geographic area of the investigation.

INITIAL STEPS IN SAMPLING VEGETATION

Development of Objectives

The critical element of any project, whether management or research, is **defining objectives.** Data should not be collected if a project has neither an objective for vegetative measurements nor a defined use for each type of measurement. Collecting vegetation data is time consuming and often difficult, and that time should be used to meet well-defined objectives. It is important to review management or research plans to ensure the information being collected fulfills the objectives and that critical information is not neglected.

Objectives must be specific. They should include what will be sampled, when it will be sampled, and where it will be sampled. Although these factors often are taken for granted, their identification requires the investigator to make a thorough analysis of the biology of the wildlife species to be studied, factors that relate to the study, and management or research needs. Elzinga et al. (1998) elaborate on components of vegetation sampling objectives and provide examples of effective measurement objectives.

General Aspects of Vegetation

After listing the objectives of the study and primary habitat requirements of the wildlife species under study, one then may identify which aspects of the vegetation to sample. Some or all of the following may be important in describing primary wildlife habitat requirements:

1. species composition,
2. vertical and/or horizontal spatial distribution,
3. temporal variation in structure,
4. biomass,
5. overall stand structure, and
6. surrounding environment (landscape structure).

A **reconnaissance survey** of an area is usually sufficient to provide the investigator with an overview of vegetation structure. Reconnaissance can be done on the ground or with aerial photography. In either instance, the objective of a preliminary survey is to decide **whether** to sample, identify **what** to sample, and determine which environmental factors will influence **how** and **when** to sample.

Consider the following example. Suppose the goal of a study is to inventory potential natural nesting sites for wood ducks (*Aix sponsa*). The wood duck nests only in cavities in trees. Because nesting cavities within a reasonable distance of water are a primary habitat requirement of wood ducks, 3 objectives are to

1. quantify the number of wood duck nesting cavities,
2. identify the species of trees containing the cavities, and
3. calculate the age-distribution of trees with cavities.

Assume that a reconnaissance survey has revealed the study area is a riparian system with a permanent stream, riparian vegetation bordering the stream, and farmland bordering the riparian vegetation. Because wood ducks nest in trees, one would not sample the area with crops, but would sample the riparian vegetation. A sample would be designed to randomly select a number of trees for examination. The objectives require identifying the species of trees in which wood duck cavities occur. Because we are interested in estimating the number of potential nest cavities, our sample will need to provide an estimate of tree density, one aspect of horizontal spatial distribution. We are not, however, interested in heights of cavities; thus, vertical distribution will not be of interest. Cavities often are present in older and larger trees and in dead trees, and the age distribution of trees is important. In addition, dead trees are likely to be blown over in windstorms, and we may decide to mark cavity trees and follow them over time to measure the rate of loss. Biomass of trees is not of interest; however, if a mast crop is produced by the trees, we would be interested in biomass of mast, a food item of wood ducks.

Study Site Selection

Study site selection is a critical phase of any vegetation study and is directly related to the objectives. If the objective is to describe vegetation conditions in relation to patterns of animal distribution or abundance, location of vegetation plots may be influenced by locations of animal observations or by wildlife population sampling objectives. If the objective is to describe vegetation conditions of selected habitats, the first issue is to define the targeted sample population and develop an appropriate sampling frame, following sampling principles described by Garton et al. (2005). Elzinga et al. (1998) provide a thorough overview of the step-by-step procedures for vegetation surveys.

A variety of factors influence selection of study sites (e.g., topography, elevation, slope, aspect, soil type, management history, distance to human-caused disturbances, and vegetation). Generally, one is interested in selecting sites that are similar to one another, and care must be taken to select sites so the intersite variation is natural and not affected by some factor not accounted for in the objectives and design of the study. This step may require **mapping** of the project area, so that all vegetation types, their locations, and their sizes are enumerated. The objectives may require that samples be taken in all or in only several sites containing a certain habitat type.

The size of the study site must be sufficiently large, so that vegetation characteristics being measured are not influenced by adjacent habitat types (often called **edge effect**). Edge effect may increase the variation in the sample. Unless such variation is explained by the sampling design (Garton et al. 2005), the results of the sample will be biased with regard to the objectives of the study. For example, if one were to sample browse production in a 100-ha stand of upland willow (*Salix* spp.), one would avoid sampling adjacent to the edge of a vegetation type that offered resting cover for moose (*Alces alces*), because those plants measured in close proximity of resting cover likely would have higher use (and perhaps lower levels of production) than plants measured in the middle of the willow stand.

Visualizing how vegetation sampling plots, or plots along a transect, will appear in field applications can be difficult. Many layout designs are possible (Figs. 16.1–16.3), and the final choice of a design also will depend on the objectives and requirements of the statistical analysis. For those not familiar with statistical principles of sampling, it is important to consult a biometrician before committing project resources to vegetation sampling.

PREPARATIONS AND GETTING STARTED

Leadership

Vegetation sampling is time consuming and demanding (Table 16.1). Good **leadership** is essential to maintaining enthusiasm and quality of data collection. The principal investigator can demonstrate leadership by (1) being enthusiastic and knowledgeable about the study area, research design, equipment, plant identification, and data collection; (2) being organized and efficient during all aspects of vegetation sampling; (3) explaining to other team members how the data will be used to make decisions on resource management; and (4) doing his or her share of the data collection. The principal investigator also should listen to suggestions from team members. They often have ideas that make data collection more efficient. Explaining the entire project, answering questions, and incorporating appropriate sugges-

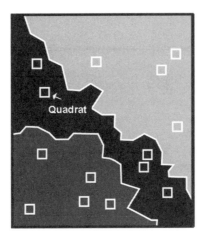
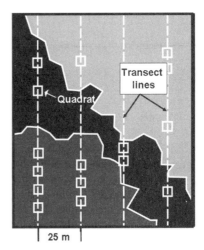

Fig. 16.1. Random (left) and systematic (right) distribution of quadrats with and without use of transect lines on a site with 3 different vegetative cover types.

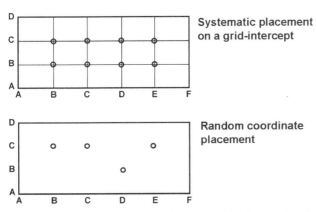

Fig. 16.2. Systematic and random placement of quadrats with grid coordinates.

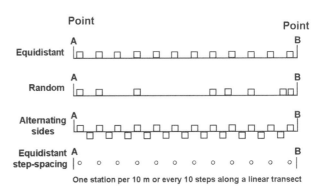

Fig. 16.3. Examples of patterns of quadrat placement along permanent transect lines.

Table 16.1. Representative times to complete a transect or a number of plots for different vegetative sampling techniques in various habitats or for different purposes. The numbers are relative, so they may not apply to a specific project; however, they should help the investigator during initial planning. The times were derived from published literature, personal communication, and personal experience.

Sampling technique	Vegetation type	Estimated time[a]	Number[b]	Reference
Grassland				
30.5-m transect, line intercept method for basal area		1.8–2.5 hr/transect[c]		Johnston 1957
30.5-m transect, point quadrat for basal area		0.5–0.8 hr/transect[c]		Johnston 1957
30.5-m transect, loop method for basal area		0.3–0.4 hr/transect[c]		Johnston 1957
0.30-m² clipped plots	California annual grasses	7 min/plot		Reppert et al. 1962
2.9-m² circular plots, clipped all species	Southeastern U.S.	32 min/plot, one person		Hilmon 1959
Single-point basal hit sampling	Tallgrass prairie	7 hr/3,000–4,000 points/3 persons	~25 ha	Owensby 1973
Foliage density readings (Robel et al. 1970)	Any grassland	8 hr/1,000 readings/2 persons	10 sites	J. M. Callow[d]
Nudds-board foliage density readings (Nudds 1977)	Any grassland	8 hr/100–200 readings/2 persons		L. D. Flake[d]
10-pin point frame (Smith 1959)	Mixed and tallgrass	8 hr/4,000–6,000 points/2 persons		L. L. Manske[d]
10-pin point frame (Smith 1959)	Wet meadow wetland	8 hr/2,000–3,000 points/2 persons		L. L. Manske[d]
Shrubs				
Shrub dimension/production estimates	Boreal forest	2 hr/25 plants/2 persons		Peek 1970
3-m × 5-m clipped plots	Boreal forest	24.7 hr/17 plots[c]		Bobek and Bergstrom 1978
Height × diameter measurements in 3-m × 5-m plots	Boreal forest	2.3 hr/21 plots[c]		Bobek and Bergstrom 1978
Clipped plots	Southern forest	10–50 plots/2 person-days	28–158/site	Harlow 1977
Twig length method to measure browse use	Montane shrub	50 min/50 plants/2 persons	50/site	Jensen and Scotter 1977
30.5-m² plot, weight estimation for twig production	Eastern deciduous forest	1.5 hr/41 plots/2 persons		Shafer 1963
30.5-m² plot, twig count for twig production	Eastern deciduous forest	1.5 hr/39 plots/2 persons		Shafer 1963
30.5-m² plot, clip and weigh for twig production	Eastern deciduous forest	6.5 hr/37 plots/2 persons		Shafer 1963
30.5-m transect, sample every 0.30 m for shrub cover	Chaparral	7 min/transect[c]	4–26/site	Heady et al. 1959
30.5-m line intercept for shrub cover	Chaparral	16 min/transect[c]	9–13/site	Heady et al. 1959
0.1-m × 0.5-m quadrats for shrub cover	Shrub steppe	15–30 min/80 quadrats/2 persons		Hanley 1978
1.2-m × 7.6-m plot for shrub cover mapping	Shrub steppe	12 plots/day/2 persons		Pickford and Stewart 1935
1-m × 5-m quadrats for shrub density	Boreal forest (postburn)	50 quadrats/day[c]		Oldemeyer and Regelin 1980
Trees				
0.1-ha circular plots	Upland forest	10–15/day[c]	5–20/site	Lindsey et al. 1958, James and Shugart 1970
Point-centered quarter method	Upland forest	20–50/day	10–50/site	Lindsey et al. 1958, James and Shugart 1970
Bitterlich variable radius sampling	Upland forest	40–75/day[c]	10–50/site	Lindsey et al. 1958, James and Shugart 1970
Camera on stick to analyze cover from 33-mm slides	Grassland	36 scanned images in 2.5 hours		Bennett et al. 2000
Visual obstruction readings	Grassland	20 stations/transect in 25 minutes		Benkobi et al. 2000
Clipped quadrats	Grassland	6 (0.25-m² circular) in 45 minutes		Benkobi et al. 2000

[a] Estimated time necessary to complete a plot, a practicable number of sample plots, or a transect by 1–3 persons.

[b] Minimum number or range of plots usually necessary to characterize the community's vegetative structure.

[c] One or two persons were used to collect data in specified time.

[d] Personal communication; J. M. Callow is biologist with U.S. Fish and Wildlife Service, Woodworth, North Dakota; L. D. Flake was former biologist with South Dakota Fish and Game Department; and L. L. Manske is with North Dakota State University, Dickinson.

tions will make team members feel they are an integral part of the project (and they are!).

Initial Planning and Preparation

Considerable **office preparation** is required before the team goes into the field to conduct a vegetation study. The development of a **list of supplies** and equipment necessary to complete the task (Table 16.2) is an important first step. Equipment lists will vary, depending on sampling objectives and whether sampling is in grasslands, wetlands, shrublands, or woodlands. These lists should be all-inclusive and should include everything from the number of pencils, color of data sheets and plot markers, and size and shape of the sampling frame to calipers, photometers, seed traps, and field vehicles.

Data Forms

Develop **forms** for recording the field data. Major advances have been made with entry of field data directly into laptop computers at the time of sampling. Remember to electronically download and store information daily and backup digital data to avoid costly loss of information. Alternately, field data forms can be developed to facilitate simple mathematical analysis with conventional calculators or entry onto a personal computer for detailed and complex analysis. In either situation, a set of instruction codes defining what is represented by each number or letter entry should accompany each field data form. Team members must understand the meaning of **zeros** and **blank spaces.** Although a blank space usually means no value was available to measure or no attempt was made to make a measurement, we have found that a hash mark rather than a blank space reduces confusion about whether the blanks were accidental or intentional. We suggest use of different color forms for different sampling tasks to aid organization and recording efficiency. For example, one color might be used for sampling shrub density and another for herbaceous cover when both were measured at a site and required use of 2 different sam-

Table 16.2. Supplies and equipment needed in the field for vegetation sampling

Data forms and notebooks	Camera and film or digital camera
Pencils and ink pens	Hammer, hatchet, and knife
Rulers and tape measures	Transect markers
Plant identification guides	Shovel and hand trowel
Plant press	Global Positioning System unit
Tags and plastic bags	Metal tags and wire
Quadrat frames	Sunscreen lotion
Cover board or poles	Insect repellent
Point frame	Hand gloves
Hand magnifying lens	Backpack on frame
Maps and aerial photos	Compass
Laptop computer or data logger	Batteries and chargers

pling techniques. White paper reflects direct sunlight, and the investigator may want to use colored paper to reduce eyestrain. Waterproof or water-legible paper is more convenient and reliable than regular bond in regions with frequent rainfall or snow.

Preliminary Field Test

It is important and useful to conduct a small-scale preliminary **field test** of a site before initiating full-scale sampling with the entire team. This field test provides the investigator with an opportunity to identify and collect plants for field mounts (Burleson 1975) for technician use, evaluate and test equipment and sampling methods, evaluate and adjust experimental design, and make final estimates of the time required to complete the work. Many research projects and surveys that were designed in the office have been completely abandoned after the first day of fieldwork, because the investigator failed to test the procedures and equipment under field conditions.

Training the Field Crew

An important step to maximize field efficiency is to properly **train** field assistants. Field assistants should have a thorough understanding of the safe and proper use of equipment, be familiar with the plants and study area, understand the correct methods for collecting and recording the data, and thoroughly understand the rationale of the study so that, in the principal investigator's absence, they can make an intelligent and informed decision when an unforeseen situation arises. We have found that several questions and concerns arise during the first week of data collection even when the crew is adequately trained. We suggest that each day end with a short meeting of the entire field crew to answer questions, inspect data forms for completeness and legibility, and discuss problems encountered in collecting data. We suggest that experienced members be teamed with those less experienced and that membership rotate daily if the field crew is divided into smaller teams for collecting data. Daily rotation of field teams increases the number of questions that arise early in the project, and the prompt settlement of problems results in more uniform collection of data and builds better rapport among crew members.

The principal investigator or field team leader is responsible for quality control of the project. We recommend the principal investigator spend at least one day working with each crew member early in the field season. This practice provides the opportunity to discuss the project more fully, provide assistance and guidance in field methodologies, demonstrate enthusiasm about the project, and learn more about the background and interests of the individual crew members. These recommendations all contribute to building a quality field team and improving the quality of data collected. An important point is to make sure all crew members have a personal stake in the quality of data collected. Emphasize creating a sense of ownership in the outcome of the project.

TECHNIQUES FOR SAMPLING VEGETATION

Frequency of Occurrence

Frequency is the proportion of sample units in which a species occurs (Bonham 1989). If, for example, 50 small plots were examined in a study site and bitterbrush (*Purshia tridentata*) occurred in 20 of those plots, the frequency of bitterbrush would be 20/50 × 100, or 40%. Frequency is an easy attribute to estimate, because the plant either occurs in the sample unit or it does not (Fig. 16.4). Frequency is a useful characteristic for describing distribution of plants in a community, and it is a measure that is related to plant density (Mueller-Dombois and Ellenberg 1974). It also is useful for monitoring changes in the plant community over time or comparing different communities (Bonham 1989). If frequency is low (15%), plants have an **aggregated distribution** (occur in clumps) in the community. When frequency is high (>90%), plants are **uniformly distributed.** Most statistical procedures rely on plants being **randomly distributed,** (i.e., having a frequency of 63–86%; Bonham 1989:92). Wild plants generally have an aggregated distribution that is related to the morphological characteristics of the species in the community, the extent and nature of competitive interaction among individuals and species, and environmental patterns (e.g., differences in soil characteristics, fire history, or herbivory; West 1989). Thus, each species may have its own distributional pattern, and the pattern of the plant community may be different from those of component species. Sampling methods to deal with complex plant distribution patterns are not adequately developed (West 1989). In an attempt to identify a best method for sampling complex communities, Etchberger and Krausman (1997) evaluated 5 methods for measuring plant species occurrence in complex desert vegetation communities, where they had a complete census of the vegetation. They found that using a **line-intercept** method, whereby the plants that intercept the line are counted as hits rather than measuring the length of vegetation canopy intercepting the line, most closely estimated the true vegetation census.

Frequency varies with size and shape of the sample unit when compared over time or among communities. Consequently, sample unit size and shape must remain constant, because it is difficult, if not impossible, to compare frequency data among sampling sites when different sizes and shapes of quadrats have been used. The size and shape of the sample unit is a function of whether one is sampling herbaceous vegetation, shrubs, or trees. Cain and Castro (1959:146) recommended **sample unit sizes** for herbaceous vegetation (1–2 m^2), tall herbs and low shrubs (4 m^2), tall shrubs and low trees (10 m^2), and trees (100 m^2)

When the total vegetation of a community is sampled, one size of sample unit will not adequately sample frequency for each form of vegetation. The mean frequency of the several species in a given vegetation form should not be 5% or >95% (Hyder et al. 1965). Nesting plots of different sizes within each other can solve this problem. Preliminary surveys of vegetation may be made using the size recommendations of Cain and Castro (1959). Further refinements of sample unit size may then be made by use of the relationship between density and frequency suggested by Hyder et al. (1965).

Plots may be square, rectangular, or round. Ordinarily, plot boundaries are either marked or measured to size with a ruler or tape measure, or they are defined by the inside dimensions of a frame (Fig 16.4). Frames may be of permanent shape and made of welded steel rod or some other rigid material, or they may be collapsible and made with hinged wood products or jointed polyvinyl chloride (PVC)

Fig. 16.4. Color demarcation and subplot frame attachments (bottom) also are used to provide quick representation of percentage of frame coverage by vegetation.

pipe. Collapsible frames are useful when they enhance efficiency of placement on the ground or travel to remote areas that are inaccessible to vehicle use. We have found that frames with one open end are useful for placing the plot around shrubs or other obstructions in shrubby terrain.

Frequency also may be measured with **points**. A pin (knitting needle or pointed, small-diameter steel rod) is lowered to the ground over herbaceous cover and will either hit or miss a plant part (Fig. 16.5). The percentage of hits gives an estimate of the frequency of a species. A single pin may be used to measure frequency (or cover; Owensby 1973), or commonly, a frame containing several (usually 10) pins is used, and pins may be positioned vertically or at an inclined angle. Spacing of the pins in the frame is dependent on the vegetative type, but it is commonly 4–15 cm (Hays et al. 1981). Although the **point frame** can be placed in random locations, pins are usually spaced systematically. Single point sampling is self-descriptive. Cook and Stubbendieck (1986) provided useful suggestions for making a 10-pin frame. Along a 10-pin frame (Fig. 16.5), the same plant may be intercepted more than once in communities with large or clumped plants. This duplication can result in overestimates of cover for those species (Bonham 1989).

Sample size is a consideration when frequency is estimated. Frequency data have a **binomial distribution,** and confidence limits are wide for small samples. Grieg-Smith (1964:39) recommended that >100 sample units be read to obtain estimates that provide reliable comparisons from one community to another or over time. With a 10-point frame, data from 1,000 (100 frames) points (hits) are usually sufficient to describe grassland vegetation at one location, whereas fewer points (200–500; 20–50 frames) usually will provide data similar to those from a single-point method (Goodall 1952).

Density

Density is the total number of objects (e.g., individual plants or seeds) per unit area. One advantage of the density parameter is that **count data** are straightforward to obtain and interpret, and results obtained from different methods are directly comparable (Gysel and Lyon 1980). A disadvantage of measuring shrub density is that data are tedious to obtain and often are excessively variable. Such variability requires an often prohibitively large sample size for statistical reliability. Density is a useful and often important measurement for evaluation of wildlife habitat for bunchgrasses, annual grasses and forbs, some shrubs, and trees. However, by itself, density is not an adequate descriptor of a plant community, because it does not provide information about how plants are distributed in the community. Combined with frequency, density provides a good description of a plant community. Combined with biomass of individual plants, density may provide estimates of total biomass in a plant community.

Fig. 16.5. An inclined 10-pin frame for frequency estimates.

The definition of an individual plant poses a problem when density is sampled. For perennial grasses and forbs, and shrubs that produce several stems from below ground, definition of an individual plant may be impossible or not sufficiently important to warrant the effort or the potential error. However, individual plant identity may be necessary in studies of plant succession, whereas counting the number of stems, etc., may be sufficient when the objective is to quantify cover or food availability. In these situations, frequency combined with some other measurement, such as cover, may provide more useful descriptions of the plant community. For shrubs, the problem is best resolved either by counting stems at ground level, thereby eliminating the need to define an individual shrub, or by establishing a distance criterion to define individuals arbitrarily. For example, Lyon (1968a) considered stems rooted ≤15 cm of one another to represent a single shrub, whereas those sprouting ≥15 cm apart were counted as separate individuals.

Quadrat Methods

Density can be measured with either **quadrats** or **plotless** methods. Quadrats are frames made of materials with fixed boundaries and are placed over vegetation to demarcate an area in which plants are counted, whereas plotless methods make use of ocular estimates. If quadrats are used, the investigator must distribute quadrats of uniform size representatively throughout each experimental unit and then count each individual in each quadrat. Quadrats require that 3 characteristics be considered (Bonham 1989): (1) distribution of the plants, (2) size and shape of the quadrat, and (3) number of observations needed to obtain adequate estimates of frequency and density.

Sample frames typically are rectangular, square, or circular. **Rectangular plots** have the largest perimeter per unit area and hence the most edge, where decisions must be made about including or excluding a plant. **Circular quadrats** often are more efficient to use than square or rectangular quadrats. Sampling in circular quadrats also is effective for characterizing the vicinity around a point of interest, such as a nest, a den location, or a feeding or resting site. A review of recent wildlife habitat studies reveals frequent use of circular quadrats, that are typically in the range of 0.01–0.1 ha (e.g., Hirst 1975, Pierce and Peek 1984, Ratti et al. 1984, Wiggers and Beasom 1986, DeGraaf and Chadwick 1987, Edge et al. 1987, Bentz and Woodard 1988). For these areas, the radius of a quadrat would range from 5.6 to 17.8 m. Increasing the size of a quadrat generally results in lower variance and reduces the perimeter:area ratio (Bonham 1989). Numerous studies have evaluated quadrat size, and no consistent recommendation has been made about the size to use for herbaceous vegetation, shrubs, or trees.

For herbaceous vegetation, 1-m × 1-m quadrats frequently are used (Bonham 1989). However, in dense vegetation, smaller quadrats, such as 20 cm × 50 cm, may be appropriate (Daubenmire 1959). Eddleman et al. (1964) compared quadrats of 4 sizes and several shapes in alpine vegetation. They recommended against using 100-cm² plots because of high standard deviations and highest frequencies of 50%. Even though plot sizes >400 cm² provided similar estimates of density, they favored 400-cm² rectangular plots, because the chance for counting error was reduced and fewer rectangular plots were required (compared to square plots of the same area) to obtain a 10% standard error of the mean.

Quadrats sufficiently large to contain an average of 4 individuals have been recommended in shrub communities (Curtis and McIntosh 1950, Cottam and Curtis 1956). Although quadrats as small as 1 m² have been used to measure shrub density (Alaback 1982), 4–10-m² plots are more commonly selected (Irwin and Peek 1979). Oldemeyer and Regelin (1980) recommended a 1-m × 5-m quadrat over a 2-m × 5-m quadrat in an Alaskan shrub community, because the smaller one provided nearly the same precision and required only one-half the sampling time as the larger quadrat. Rectangular plots have advantages over square and circular plots in aggregated shrub communities, because they have the greatest chance of overlapping individual clusters of shrubs. A rectangular quadrat 1-m wide of any length may be delineated easily by marking one long side of the rectangle with 2 chaining pins and a chain, and using a meter stick to define the remaining boundaries while one counts shrubs along the strip as the meter stick passes over them.

Quadrats must be quite large when trees are sampled, typically in the range of 0.01–0.1 ha. Curtis (1959) used square quadrats 10 m on a side (0.01 ha) in deciduous and coniferous forests of Wisconsin. Mueller-Dombois and Ellenberg (1974) concluded that forest quadrats typically should be squares of 10 m or 20 m on a side (0.01 ha or 0.04 ha). Quadrats can be positioned with tape measures or other measuring devices and surveyor's pins after sampling points are located. This measurement might require considerable time and effort in dense vegetation or in some types of terrain. To reduce that time, Penfound and Rice (1957) proposed using an elongated 0.0004-ha quadrat established by measuring the width of one's outstretched arms and then, knowing the average pace length, walking the appropriate number of paces along a compass line and recording the trees within reach. It is important to realize that, although this method is faster to implement under natural forest conditions, the area sampled is approximate, and accuracy is sacrificed without careful attention. Further, the advent of **laser rangefinders** now favors use of circular plots as an alternative, relatively quick method of estimating tree density. From any established plot center, the field biologist using a laser rangefinder may quickly estimate distances of trees from plot center. Thus, the biologist may quickly count all trees present within a predetermined fixed radius. The method has the added advantage of minimizing the perimeter:area ratio of the sample plot, but care must be exercised to ensure the laser rangefinder being used is both accurate and precise.

The **number of samples** measured varies from community to community and among the different vegetation forms in a community. Because many species are not randomly distributed, variation normally is quite high, and number of samples required is quite large. To calculate **sample size,** one can use results from the preliminary field test to obtain an estimate of the variance for use in the sample size equation (Garton et al. 2005). Frequently, less common species require a larger number of samples than do the more common species. For example, to obtain a 10% standard error of the mean in a 10-cm × 40-cm plot, Eddleman et al. (1964) concluded that 816 plots would be required for a species with a density of 0.13 (no area units given), whereas 69 plots would be required for a species with a density of 5.6. Oldemeyer and Regelin (1980) reported that 50 1-m × 5-m quadrats produced estimates of shrub density within 2 standard errors of actual (counted) shrub densities. Lyon (1968a), however, reported that >400 1.5-m × 6.1-m quadrat samples would be necessary to obtain an estimate of shrub density within 10% of the true mean 95% of the time. Sample sizes in the hundreds are not an uncommon result. As an alternative, one may plot the running mean density against the number of samples taken (Kershaw 1964). One may stop sampling when the density of the target, or more abundant, species does not significantly change with additional quadrats. Mueller-Dombois and Ellenberg (1974:77) suggested that sampling stop when the running mean of a sample is within 5–10% of a "maximum" sample. Clearly, one must critically evaluate the objectives of a project and the end product of the data when designing a study of plant density. It may not be necessary to have the density (or frequency or cover) estimate be within 5% of the true value; however, it is a waste of time and effort to undersample a community and obtain totally unreliable estimates.

Plotless Methods

Plotless methods of sampling density have been in use since the 1950s. These methods do not use boundaries and are based on the premise that density may be estimated from the mean area occupied per tree; that is, density (trees/m²) = 1/mean area (m²/tree). The challenge then becomes estimating the area occupied per tree from distance measurements that can be obtained in the field.

Cottam and associates (Cottam and Curtis 1949, 1956; Cottam et al. 1953) pioneered research on plotless methods. These methods included the **closest individual method, nearest-neighbor method, random-pairs method,** and **point-centered-quarter method.** Of these, the point-centered-quarter method has been widely used in many vegetation types throughout North America. Using this method, one randomly or systematically selects a number of points in a community and measures the distance to the nearest plant in each of 4 quadrants around the point (Fig. 16.6). Mean area is calculated by squaring the mean distance d between points and individual stems: density = $1/d^2$.

This method may be used to calculate density of all species collectively. Or, density of individual species can be estimated by measuring distances to each species in every quadrant around each point. A reliable estimate of an individual species' density cannot be obtained by using only those distances for an individual of the species that was the closest plant in a sample and the distance was measured to the nearest plant regardless of species. That is, if 25 points were sampled, and distances to 100 plants of several species were measured, the density of all plants can be estimated based on the 100 distances. The density of one of the several species from the sample cannot be estimated reliably, because the distance measured to the plant when it was the closest plant in a particular quadrant may not be the least distance when all plants of that species within the entire circle around the point are considered (e.g., species A in Fig. 16.6).

The point-centered-quarter method has been criticized, because it provides reliable estimates of density only when plants are distributed randomly and not when plants are clumped or uniformly distributed. Studying stands of known density, Oldemeyer and Regelin (1980) concluded this method accurately estimated density of white spruce (*Picea glauca*) saplings, which were more randomly distributed, but underestimated density of paper birch (*Betula papyrifera*) and aspen (*Populus tremuloides*) saplings, which had clumped distributions. The point-centered-quarter method overestimates density in communities with regularly distributed plants (Mueller-Dombois and Ellenberg 1974). This method likely provides reliable estimates of density when total plant density in a community is the only concern. However, Laycock and Batcheler (1975) reported that evaluating composition from the proportion of times each species occurred in the total measurements resulted in biased composition estimates.

Methods have been developed to correct for density estimates in nonrandom plant populations (Morisita 1957, Batcheler 1973). The **angle-order method** (Morisita 1957) measures the distance from the point to the center of the third nearest plant in each quadrant around the point. This method is based on the assumption the area may be divided into several smaller units in which the plants will be distributed randomly or uniformly, even though they are distributed nonrandomly over the larger area. The method was tested on known populations of grasses, forbs, and shrubs (Laycock and Batcheler 1975, Oldemeyer and Regelin 1980) and provided estimates of density that were more accurate than those of the point-centered-quarter method. Oldemeyer and Regelin (1980) reported the method provided density estimates closest to the true density in shrub stands and that its coefficient of variation was lower than other accurate estimators. However, because of the time required, Laycock (1965) recommended against using the angle-order method when density is measured for each species in a community. Bonham (1989) provided a detailed description of the procedures for calculating density and the variance when the angle-order method is used.

The **corrected-point-distance** (Batcheler 1973) is a modification of the point-centered-quarter method that uses measurements to the second and third nearest plants to correct for nonrandomness. That is, from a sample point, one

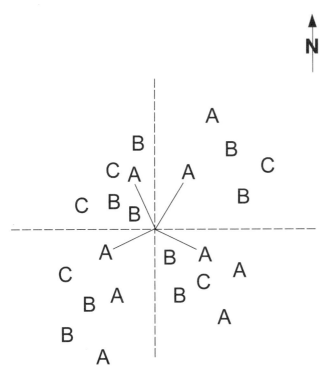

Fig. 16.6. Point-centered quadrat sampling: the point-to-plant distance is measured for the individual of each species nearest the point in each of the 90° quadrants around the point. Species A is being sampled in this figure.

measures the distance to the nearest plant, the distance from that plant to its nearest neighbor, and the distance from the nearest neighbor to its nearest neighbor, exclusive of the first plant measured. In aggregated populations, the distance between the nearest plant and its nearest neighbor is generally less than the distance from the point to the nearest plant. Density is calculated by the equation

$$m = \frac{a}{\pi[\Sigma r_i^2 + (N-1)R^2]},$$

where N is sample size or number of points used, m is density, R is the maximum distance over which a search is made for a plant at any point, a is the number of points at which a plant is found at a distance $\leq R$, and r_i is the ith distance measured. As R decreases, m approaches the true density; however, variance generally will increase, because fewer measurements are included (Bonham 1989). Although this equation is designed for random and nonrandom distributions, densities will be biased in nonrandom populations (Bonham 1989). This problem may be corrected by using a factor based on distances from the nearest plant to its nearest neighbor. Laycock and Batcheler (1975) recommended use of the corrected-point-distance method over other distance methods, because the density estimate was within 12% of the true density, and the method is relatively fast and easy to use.

Engeman et al. (1994) reviewed 24 plotless density estimators and compared the relative biases among methods in relation to simulated differences in plant aggregation patterns. These simulated results verified field results described above, indicating that when plant distributions are clumped, the point-centered-quarter method produces biased results compared to angle-order methods. The authors observe, however, that the added effort of measuring several plants per quarter complicates the method and results in fewer sample points. Generally, for any fixed amount of effort, it is better to sample more independent points than to invest more effort at individual points. Based on this practical consideration and the evaluation of bias under simulated conditions, Engeman et al. (1994) recommend using ordered-distance (Morisita 1957, Pollard 1971) or variable area transect (Parker 1979) sampling methods for estimating density. The ordered-distance estimator involves measuring distance to the third closest tree to sampling points. The variable area transect involves measuring the distance along a fixed-width strip transect (generally, 1–2 m in most field applications) until the gth individual tree is encountered (usually, $g = 3$). The review paper by Engeman et al. (1994) or the original authors cited above should be consulted for the analytical formulas.

The choice of using a plotless method over a quadrat method will depend on the objectives of the study. If the density of 1 or 2 species is required, plotless methods appear to be faster than quadrat methods. If the density of all species in the community is desired, the quadrat method is recommended.

Cover

Cover is defined as the vertical projection of the crown or stem of a plant onto the ground surface. **Canopy cover** serves as a criterion for relative dominance in a community and is of practical importance because of its influence on interception of light or precipitation and on soil temperature (Hanley 1978). It may be used by plant ecologists to describe total vegetation cover, by range managers to define cover of forage for livestock, or by foresters to describe **basal area** of merchantable timber. Cover can be an estimator of biomass when height structure of a community is known. Daubenmire (1959) suggested that canopy cover is the surface area over which a plant has influence; thus, cover provided by seedlings and seed stalks might not be measured, because they have little influence in the ecosystem. Although canopy or crown cover may vary in a season or over years, basal cover is relatively stable. **Basal cover** is a reliable measurement for bunchgrasses, tussocks, and trees. Cover is frequently measured at a height of about 2 cm on bunchgrasses and tussocks (Bonham 1989:98), whereas on single-stemmed trees, it is measured at 1.5 m above ground (Mueller-Dombois and Ellenberg 1974:88). This latter measurement is referred to as **diameter at breast height** (DBH). Basal cover is measured at the ground surface on trees with multiple stems or on trees with buttressed trunks. Cover often is expressed as a percentage, and in a dense or multilayered community, total vegetation cover may exceed 100%. Cover can be measured directly with quadrat-charting (Gibbens and Beck 1988) or pantographic methods (Mueller-Dombois and Ellenberg 1974, Fehmi and Bartolome 2001), an ocular estimation technique (Daubenmire 1959, Mueller-Dombois and Ellenberg 1974), or line-intercept (Canfield 1941) and point-intercept methods (Levy and Madden 1933, Owensby 1973).

Quadrat-Charting Method

This method has its greatest utility in low herbaceous vegetation, where one can stand and look over the vegetation. Cover is mapped to scale on graph paper from a small quadrat (e.g., 1m^2). The idea is to map the crown area or the basal area onto the graph paper. This process may be facilitated by subdividing the larger quadrats into smaller ones. Quadrat charting is useful generally only in long-term studies, when quadrats are permanently marked at each corner and can be exactly relocated for each measurement (Gibbens and Beck 1988). Rather than charting indirectly from what the observer sees on the ground, the observer may use a pantograph (Mueller-Dombois and Ellenberg 1974) or take photographs (Wimbush et al. 1967).

Ocular Estimate Method

Ocular estimates of basal and canopy cover can be obtained with relative ease in grasslands because of their low profile and height. However, the task becomes more difficult in wetland vegetation because of the combination of water depth and plant height, often requiring use of scuba equipment or a ladder.

Cover can be estimated to the nearest percentage point, or to the nearest 5th or 10th percentile; however, most commonly it is estimated according to some form of cover class (Brown 1954, Daubenmire 1959, Braun-Blanquet et al. 1965, Mueller-Dombois and Ellenberg 1974, Floyd and Anderson 1987).

A cover-class scale (Box 16.1) often has been used in grasslands (Daubenmire 1959). Division of class range (%) is facilitated by painting lengths on the frame in different colors. Zero has been used separately as a data integer by some users.

A variety of plot sizes has been used to estimate vegetation cover (Brummer et al. 1994). Daubenmire (1959) recommended using 20-cm × 50-cm quadrats for both shrubs and herbaceous vegetation, because cover is more easily estimated in small quadrats. However, data from a transect, generally having 20–30 quadrats, are summed into one mean for each variable. The transect is the basic unit of sampling. Meter-square frames also have been commonly used to estimate shrub cover. Cook and Bonham (1977) suggested dividing 1-m^2 frames into 5-cm × 5-cm cells, each corresponding to 0.25% cover. One may estimate cover with a gridded quadrat by counting the number of grid cells covered by shrubs and adding the number of obstructed cells to calculate the total percentage. Although ocular estimation is a rapid method of estimating data on basal or canopy cover, there are drawbacks. Ocular estimates are subject to personal bias; thus, estimation error among investigators may add unnecessary variability to the data. Hence, these methods require consistent training and calibration among investigators. Dimensions of plant cover, even on permanently marked plots, also are subject to the influences of precipitation, heat, and sunlight on plant growth. Consequently, care must be exercised in data interpretation, because a reduction in the cover of a species on the same plot in different years may be a result of drought as much as of interspecies competition for the same site.

Line-Intercept Method

The line-intercept method is particularly suited for measuring basal area of bunchgrasses or tussocks and canopy cover of shrubs, particularly in arid or semiarid lands (e.g., sagebrush [*Artemisia* spp.]; see Connelly et al. 2003*b*). The identification of intercept can be quite difficult and prone to error in less clumped forms of vegetation. In this technique, a line or tape measure is placed between 2 stakes, and basal width or canopy width of all plants touching the line or tape is measured, even if only a small part of the plant is in contact with the tape. Cover is expressed as a percentage of the total length of tape intercepted by vertical projections of the canopy. Keeping a tape line taut and straight may be difficult in tall, dense vegetation. Canfield (1941) reported that a minimum of 16 15–30-m transects was necessary to adequately describe shrub vegetation in Arizona rangelands. A 15.2-m transect was adequate in shrub fields with 5–15% shrub cover, whereas a 30.4-m transect was necessary on sites with 5% cover.

The principal advantages of the line-intercept method are the high levels of accuracy and precision that result from direct measurement rather than estimation of vegetation (Cook and Stubbendieck 1986, Connelly et al. 2003*b*). The main limitation of the method is the time required to measure intercepts compared to estimating cover in quadrats. Of the 2 methods, the line-intercept method was more precise, whereas the quadrat method was quicker. Hanley (1978) concluded the line-interception method is preferable to 0.1-m^2 quadrats in scientific research, when precision of the cover estimate may be more important than cost efficiency. Based on comparisons of several techniques, Floyd and Anderson (1987) and Etchberger and Krausman (1997) found the line-intercept method was equal to or better than alternative methods. Wambolt et al. (2006) recently standardized line-intercept methods for estimating shrub cover to ensure that vegetation measures yield reliable results for use in rangeland management.

Point-Intercept Method

Basal and canopy cover also may be measured as the percentage of points whose vertical or angled projections intercept vegetation. The point-intercept method is best suited for estimating cover of herbaceous and low shrub vegetation, but it also has been used to estimate leaf-area index in sagebrush steppe communities (Clark and Seyfried 2001). For relatively large-scale surveys of plant cover, points may

---

BOX 16.1. SCALE OF COVER CLASSES FOR A 2-DM × 5-DM QUADRAT

Data integer	Class range (%)	Midpoint (%)
1	0–5	2.5
2	5–25	15.0
3	25–50	37.5
4	50–75	62.5
5	75–95	85.0
6	95–100	97.5

be defined by putting a V-shaped notch or line in the tip of a boot and using the notch or line as a single point (Evans and Love 1957, Etchberger and Krausman 1997) while walking over a tract of grassland. This method offers rapid assessment or survey of cover, but it may be prone to observer bias and is less repeatable than other point-sampling methods. When more precision is required, generally at a smaller scale of study, points may be defined with a multiple point frame (Levy and Madden 1933, Cook and Stubbendieck 1986) or a single point frame (Owensby 1973). With either method, a single pin is lowered toward the ground. The first strike of any part of the vegetation canopy becomes a canopy cover hit; if it strikes the basal area of a plant, it is a basal hit. Often a pin will miss all vegetation in its line of travel. Percent canopy or basal cover is calculated as the total number of hits divided by the total number of pin placements times 100. The diameter of the pin and the point affect the accuracy of cover estimates. Because a mathematical point does not have a diameter, but the pin point does, cover is generally somewhat overestimated (Winkworth 1955). The point-intercept method is frequently used along transect lines. The user should be aware the line is the sample unit and that it is better to have fewer points per line and more lines than vice versa (Bonham 1989).

Heady et al. (1959) reported that line-intercept and point-intercept procedures produced comparable estimates of shrub cover when ground cover was 3%; however, the point-intercept procedure was quicker and thus preferable. Species with ≤3% cover required extremely large samples with the point-intercept method. Thus, the line-intercept procedure should be used in sparse shrub communities.

Bitterlich Variable Radius Method

The **Bitterlich variable radius method** is a modified point-sampling method developed for use in forestry (Bitterlich 1948, Grosenbaugh 1952) to measure basal area of trees. The method was subsequently modified for use in range habitats to measure canopy cover of shrubs (Cooper 1957). Hyder and Sneva (1960) recommended the method for sampling basal cover of bunchgrasses. Shrubs or trees are viewed with one of several types of sighting devices (**angle gauges**) that delimit a certain sighting angle from randomly located sampling points (Cooper 1957, Mueller-Dombois and Ellenberg 1974:102). The sighting device must be held as nearly horizontal as possible. Shrubs with widths or trees with trunks larger in diameter than a specified angle when seen through the sighting device are reported. To be included in the count, small shrubs or trees must be relatively close to the observer, but larger ones can be farther away and yet exceed the viewing angle. The probabilities of species being sampled are proportional to their size, and the correction factor needed to calculate cover depend on size of the viewing angle. Percentage cover is defined as

$$P = [(n \times W^2)/L^2] \times 25,$$

where P is percentage cover, W is the width of the crosspiece of the sighting device, L is the distance of the crosspiece from the observer's eye, and n is the number of plants counted. Using a sighting device with a width:length ratio of 1:50 gives a viewing angle of $1°10'$, and the count of trees within that angle is numerically equal to the tree basal area in square meters per hectare (Mueller-Dombois and Ellenberg 1974). Generally, a ratio of 1:7.07 is most acceptable for shrub communities (Fisser 1961, Cooper 1963), and the average count per plot is divided by the correction factor 2. Correction factors for different width crosspieces used for sampling shrubs were given by Cooper (1957).

Clear-glass **prisms** have largely replaced wooden sighting sticks as a means of measuring basal areas of trees (Dilworth 1989). The prism is a wedge-shaped piece of glass that refracts light rays to establish the critical angle used to estimate basal area of tree stems. In using the prism, the observer holds the prism immediately over the sample point while viewing tree stems both through the glass and over the top of the prism. Distance of the prism from the viewer's eyes is not a factor, as long as tree stems appear clearly when viewed through the glass. Viewed through the prism, tree trunks appear displaced to one side, due to refraction of light passing through the glass. Basal area is calculated by recording the number of trees whose trunks, when viewed through the prism, appear displaced within the trunkline of the actual tree. The tree is not recorded if the trunk viewed through the prism is completely displaced beyond the trunkline. Trees whose displaced trunklines are even with the actual trunk are counted as a half tree. Prisms are readily available through most forestry equipment suppliers and come in a variety of metric and English "**basal area factors**" that are used to convert stem counts to basal area per hectare or acre, respectively. The stem count per sample point multiplied by the basal area factor gives the total basal area of stems (m^2 or ft^2) per unit of area (ha or acre). Generally, a basal area factor should be chosen that gives a count of 4–8 trees per point (Dilworth 1989).

The utility of the variable radius method for sampling shrub stands is influenced by several factors. The method assumes the plant is round. Thus, the estimate of cover will be overestimated for species or stands with shrub crowns, particularly of irregular shape. Individual shrubs or trees that should be counted, but are shielded from view by another plant may be missed in dense stands. Cooper (1957) reported the method could be used in desert shrub stands when cover was 35% and Fisser (1961) observed that shorter investigators underestimated cover compared to taller investigators. The chief advantage of the variable radius method is that it is quick and requires counts rather than measurements in the field. Several studies have shown this method produced estimates of cover comparable to those of the line-intercept method in shrub fields with 30% shrub cover (Cooper 1957, Kinsinger et al. 1960, Fisser 1961). Kinsinger et al. (1960) reported that readings from only 3–6 variable plots were required to produce the same precision as estimates obtained from 20 30-m long line transects that required considerably more time to measure. Cooper (1963) reasoned this precision, and the lower coefficients of variation, from the variable radius method was because of the larger area covered than that with point or line-transect methods. Kinsinger et al. (1960) concluded that within the stated constraints, the variable radius method was faster and more precise than the line-intercept method, but it could not be used as effectively to study subtle changes in shrub cover.

Tree Canopy Cover

At times, tree canopy cover is an adequate, perhaps even preferred, measure of overstory structure and composition. In this situation, line or point sampling or ocular estimates in plots can be used to estimate canopy cover. Many workers prefer to use a **spherical densiometer** (Lemmon 1957;

Fig. 16.7. A spherical densiometer used to estimate percentage overstory cover in woodlands.

Fig. 16.8. A moosehorn densiometer used to estimate percentage overstory cover in woodlands.

Fig. 16.7) for making these estimates. The spherical densiometer uses a curved, gridded mirror that reflects the overstory at a point and provides estimates of relative amounts of the area covered. Although there are variations (Cook et al. 1995), usually the observer levels the densiometer at about chest height and counts the proportion of quarter cells (etched in the mirror) obscured by the reflected vegetation. Because the mirror is curved, the spherical densiometer measures canopy in a 30–60° angle of view projected upward through the canopy (Cook et al. 1995). Lemmon (1957) concluded that (1) there was no difference in overstory estimates between the spherical densiometer and other instruments used to estimate overstory, (2) variation among replicated measurements increased as overstory cover declined, and (3) reliability was greater when the actual grid count was used rather than broader overstory classes obtained from grouping the counts.

Alternative ocular methods of estimating forest overstory cover include **sighting tubes** (Ganey and Block 1994), **moosehorns** (Cook et al. 1995), and **photographic fisheye lenses** (Chan et al. 1986). The moosehorn (Fig. 16.8) is a sighting tube with a 25-point grid etched in glass on one end. The observer sights through the tube and counts the proportion of dots obscured by overhead vegetation. Because the moosehorn samples a narrower (10°) angle of view than does the densiometer, it produced a truer estimate of the vertical projection of canopy on the ground (Bunnel and Vales 1990, Cook et al. 1995). Although the moosehorn provides the most accurate assessment of vertical projection of overstory canopy, spherical densiometers also may measure biologically relevant influences of tree canopies on an area (i.e., light interception or angular canopy cover; Nuttle 1997). The appropriate measurement tool depends on study objectives and consideration of how tree canopy influences the environmental properties of interest (Nuttle 1997).

Biomass or Standing Crop

One of the best indicators of species importance in a plant community is composition based on dry weight (Daubenmire 1968). Wildlife and land managers frequently require data on **biomass** or **standing crop** rather than on density or cover, because biomass is closely related to forage availability and habitat carrying capacity (Bonham 1989). Here, we use the term biomass to include both live and dead vegetation and synonymously with the term standing crop. Woody biomass and size structure are required to estimate fuel loading, a necessity for formulating fire prescriptions and predicting fire behavior in wildlands. Wildlife managers often are interested in measuring biomass of edible components of browse, such as current annual growth, foliage, or twigs. Total biomass and biomass of edible components may be estimated directly by clipping and weighing or indirectly by dimension analyses or through the use of capacitance meters (Gonzalez et al. 1990).

Clipping Techniques

Plant biomass can be measured directly by removing all vegetation in a sample plot to ground level and measuring its mass immediately (wet mass) or after air- or oven-drying the sample (dry mass). Clipping, drying, and weighing plant material directly is accomplished with minimal variation in results among investigators; however, proper implementation of methods necessary to obtain good data is both labor and time intensive. For consistency, herbage should be clipped at a specific height or location on the plant and may be separated into edible and inedible portions, depending on the objectives. Mean biomass per unit area then may be estimated as the product of mean biomass per plant (e.g., g/plant) and mean density of plants (e.g., plants/m^2). Sample variance may be computed as the variance of a product (Goodman 1960). Data from a site or transect is pooled into a mean. Variances are calculated from across sites or transects from which harvesting was conducted in each quadrat. Because clipping is a destructive sampling method, new plots must be selected in subsequent sampling periods to avoid the effects of previous sampling activities.

In wetlands, biomass samples of macrophytes may be obtained by harvesting all vegetation in a quadrat frame placed above the sediment level (Whigham et al. 1978). Harvesting consists of clipping plants in floating (Tanner and Drumond 1985) or submerged metal rod frames or in an open-ended cylinder or box enclosures (Sefton 1977, Anderson 1978). Water depth also should be measured near the center of each quadrat and recorded. Clipping can be done easily in conventional waders in shallow (1 m) wetlands. However, deeper wetlands (>1 m) may require sampling with specialty gear, such as swimmer's goggles, wetsuits, dredging equipment, modified rakes (Rodusky et al. 2005), or even scuba equipment. Vegetation samples should be dried to a constant weight. Drying temperature is dependent on the purpose of the plant materials; if one is interested only in dry weight, then 80° C for 48 hours may be used. If the plants are to be analyzed for nutritional analysis, lower temperatures (e.g., 60° C for 48 hr) are required to avoid volatilizing nutritional components. If drying and weighing cannot be done onsite, vegetation samples should be frozen or kept at 4° C to stop further respiration activity.

The "**clip-and-weigh**" method also may be used to estimate twig biomass in plots. Clipping all twigs in plots is a highly accurate yet laborious means of measuring browse biomass (Shafer 1963). Several investigations have reported that total browse collection may require 10–120 times as long as estimating browse biomass from dimension analysis or twig count methods (Shafer 1963, Uresk et al. 1977, Bobek and Bergstrom 1978). This time requirement is an important consideration, given high sampling variation inherent in browse estimation.

Ocular Estimations

Herbage biomass also may be ascertained by ocular estimation techniques (Pechanec and Pickford 1937, Ahmed and Bonham 1982, Ahmed et al. 1983, Stohlgren et al. 1998). Requirements of biomass estimation techniques include intensive training of investigators. This may be facilitated by incorporating double sampling procedures into the activity. **Double sampling** requires that ocular biomass estimates be made in each quadrat or for each plant and that a subset of quadrats or plants be clipped and weighed after the estimates are made. Weighing the plants helps the observer develop more accurate ocular estimates. Regression of the estimates and actual weights provides an estimator for the plots or plants for which only estimates were made. Procedures to calculate an adequate ratio of clipped to estimated samples were provided by Ahmed and Bonham (1982), Ahmed et al. (1983), and Reich et al. (1993).

Dimension Analyses

Dimension analysis has been used in forestry for timber attributes and in wildlife and range management for estimating shrub biomass. The technique assumes that plant attributes are related and that one attribute can be predicted from another that is more easily measured (Whittaker 1965). Because clipping, drying, and weighing require so much time, and yet biomass frequently is a critical attribute of a plant community, numerous investigators have developed regression equations of biomass and some more easily measured attribute. Biomass of individual grass plants has been estimated from volume as measured by height and basal diameter (Johnson et al. 1988). Biomass estimates of individual shrubs have been obtained with, as independent variables, measures of basal stem diameter (Telfer 1969b, Brown 1976); maximum plant height (Ohmann et al. 1976); and various crown dimensions, including diameter, area, volume, and height × circumference (Lyon 1968b, Rittenhouse and

Sneva 1977, Uresk et al. 1977, Murray and Jacobson 1982). Common forms of the predictive equations include linear ($y = a + bx$) and power ($y = ab^x$) curves. Traditionally, researchers have linearized the power curve with logarithmic transformations ($\log[y] = \log[a] + x \cdot \log[b]$), but such transformations may introduce bias (Baskerville 1972). There is little reason to transform the nonlinear relationships with nonlinear regression procedures commonly available in statistical software packages. Several independent variables may provide satisfactory estimates of shrub biomass (Oldemeyer 1982), but care must be taken to select those variables that provide the best predictive accuracy and are not correlated.

Generally, one measures stem and crown dimensions from a sample of individual shrubs in the field. The plant material then is clipped, taken to the laboratory, oven-dried, and weighed. A sample of 25 plants per species is usually adequate for calculating predictive equations for total shrub weight (Peek 1970). Care must be taken in the field to adequately sample the full range of plant sizes present, because one may not estimate biomass of plants that fall outside the size range of plants used to develop the regression. We believe more reliable regression equations may be developed if one stratifies the plants in the community into size classes, measures the variance of biomass in each size class, and calculates the number of plants to measure in each size class on the basis of the variance. For example, if the relative variance of the largest size class was 20% and if 25 plants were to be measured for the regression analysis, then 5 plants (0.2×25) would be measured from the largest size class.

Weight–dimension relationships of shrubs vary among sites and years (Oldemeyer 1982), making it necessary to test the influence of site factors on the regression parameters if predictive relationships are to be applied to a broad area. Developing separate predictive equations for each shrub species in each vegetation community of the study area is often necessary. Once satisfactory predictive equations have been developed, biomass can be estimated from data on shrub density and shrub biomass estimates without destroying shrubs. Dimension analysis represents a substantial savings in time and expenditure over traditional clip-and-weigh methods when only one, or at most a few, predictive relationships need to be developed for use for a variety of site conditions. Because the method is nondestructive, plants can be measured annually in the permanent plots.

Dimension analysis has been used to estimate twig and foliage production of individual shrubs in the same manner as described above for total aboveground standing-crop biomass. Production estimates for individual shrubs are obtained by measuring a sample of shrubs in the field; the shrubs then are harvested, and all current annual growth of twigs and foliage is clipped, sorted, and dried. Sampling and analytical considerations are the same as for estimating total shrub biomass.

Lyon (1968b) and Peek (1970) reported that total twig production was related linearly to crown volume and crown area, with the resulting equation explaining >80% of the variation in twig production. Oldemeyer (1982) used multiple regression procedures to estimate twig production as a function of shrub circumference, shrub height, crown length, and number of current annual growth twigs. Despite the high predictive accuracy of the equations, Lyon (1968b) and Peek (1970) warned that production–dimension relationships of shrubs were influenced strongly by site factors, and they varied among species, which necessitated developing unique predictive equations for each shrub species on each distinctive site type. Dimension analysis is a convenient, nondestructive alternative to the traditional clip-and-weigh methods, once predictive equations are developed for a particular site type.

The **twig-count** method (Shafer 1963) for measuring browse biomass is based on the simple conversion of twig counts to browse weight by using an average weight per individual twig. In its basic form, an average browsing diameter of a particular shrub species is calculated from a random sample of 100 browsed twigs. An average weight per twig then is calculated by weighing 50 unbrowsed twigs clipped at the average browsing diameter. Shafer (1963) suggested counting twigs in 9.3-m^2 circular plots. Twig densities then were converted to biomass estimates from a mean twig weight. Irwin and Peek (1979) observed that it was faster and easier to count twigs in 1-m × 1-m or 1-m × 4-m belt transects. Shafer (1963) reported the twig-count method was nearly as accurate as the clip-and-weigh method. The twig-count method also is nondestructive, making it suitable for repeated measurement of permanent plots. Additionally, individual twigs are easily counted and recorded in different height categories, permitting easy assessment of the influence of snow depth and browsing heights on available browse (Potvin and Huot 1983).

A commonly used modification of Shafer's (1963) twig-count method involves development of **weight–diameter** or **weight–length equations** to estimate mean twig weights (Basile and Hutchings 1966, Telfer 1969a, Halls and Harlow 1971). This method is based on the principle that average twig weights may be estimated by regressions of twig diameters or twig lengths. Predictive equations relating twig weight to twig diameter or length may be developed by clipping a number of unbrowsed twigs (50 are recommended), measuring twig length and basal diameter, oven-drying, and weighing to the nearest 0.01 g. Care must be taken to collect the full range of twig sizes from several shrubs and to stratify the sample among lower and upper portions of each shrub (Basile and Hutchings 1966). Because twigs are often elliptical in cross-section, it may be necessary to estimate twig basal diameter as the average of 2 perpendicular measurements. Linear regression produces acceptable predictive equations if the range of twig diameters or lengths is not great (Basile and Hutchings 1966, Halls and Harlow 1971); however, curvilinear regression may be required if twig sizes vary widely (Telfer 1969a). Peek et al. (1971) reported there might be considerable site variation in length–weight and diameter–weight relationships of twigs that would require developing a separate regression equation for each shrub species and each site type under investigation.

Other Attributes
Visual Obstruction

Visual obstruction caused by vegetation may be functionally important to wildlife both as hiding cover (i.e., cover necessary to escape a sense of danger) and as thermal cover (i.e., cover that creates a beneficial thermal environment). The measurement of **horizontal cover** of vegetation has been used extensively by wildlife managers and researchers in assessing wildlife habitat suitability, habitat preference, and impacts of land use practices on wildlife habitats (Griffith and Youtie 1988, Reece et al. 2001, Vermeire and Gillen 2001, Uresk and Juntti 2008). Some measure of horizontal obstruction also has been used by researchers to examine the relative influence of visibility biases associated with wildlife surveys in different vegetation classes. Further, measures of horizontal obstruction have been used reliably in many instances as a relatively rapid surrogate measure to estimate standing crop biomass of grassland vegetation (Harmoney et al. 1997, Volesky et al. 1999, Benkobi et al. 2000, Vermeire and Gillen 2001).

A variety of devices has been used to measure horizontal visual obstruction caused by vegetation. Wight (1939) first proposed use of a **density board,** a 1.83-m tall board, each 30.48-cm mark labeled 1 to 5 (Fig 16.9). Horizontal cover is assessed by placing the board in cover, viewing the board from a distance of 20 m, and adding the numbers unobstructed by vegetation. The method produces an index of horizontal cover that ranges from 0 (no obstruction) to 15 (complete obstruction), but it provides no means of describing the vertical distribution of the obstructing vegetation.

Nudds (1977) devised a **vegetation profile board** that enables the investigator to assess visual obstruction of shrub vegetation in 5 0.5-m vertical intervals above ground. The board is 2.5-m high and 30.48-cm wide and is marked in alternate black and white colors at 0.5-m intervals. Horizontal cover is assessed in each interval by viewing the board from 15 m in a randomly chosen direction. The percentage of each interval concealed by vegetation is recorded as a single-digit score, ranging from 1 to 5, corresponding to 0–20, 21–40, 41–60, 61–80, and 81–100% estimated concealment. Although the vegetation profile board has been widely used, its size, weight, and inconvenience associated with use in remote areas are drawbacks of the technique. The board may, however, be reproduced on thin vinyl or nylon material that is easily rolled and transported in the field; it can be held in place conveniently by a single pole or by a field assistant. Griffith and Youtie (1988) modified the Nudds-type checkerboard into standing and bedded deer silhouettes. Values of the height and percentage of the silhouette blocks covered by vegetation was estimated by eye from the 4 cardinal directions and at 4 0.5-m levels. Haukos et al. (1998) reported sample size, power, and other analytical considerations when using profile boards in wetland plant cover. Naugle et al. (2000) used a profile board to investigate black tern (*Chlidonias niger*) nest site selection in wetland vegetation.

Robel et al. (1970) used a pole-shaped cover board (3 cm × 150 cm) that could be read from a standard distance (4 m) and height (1 m) in any direction (Fig. 16.10). The pole was marked in decimeters, and the height of total visual obstruction was recorded. For example, if the pole was not visible until the fifth decimeter, the reading was 4. Additionally, all vegetation was clipped, dried, and weighed from a 2-dm × 5-dm quadrat next to the pole, and regressions were developed from the average obstruction reading and biomass of 30 transects. The $R^2 = 0.95$ indicated the obstruction reading could be used as a method of estimating biomass in tall grasses to assess prairie-chicken (*Tympanuchus* spp.) habitat. Benkobi et al. (2000) modified the Robel pole with alternating gray and white 2.5-cm rings and clipped vegetation around the pole. Their modified pole greatly improved the precision and accuracy for mid- and short-grass prairie. Sample size for number of pole stations and transects are presented, as are monitoring or sampling protocols for small areas (≤259 ha) to large landscape areas of 1,215–46,560 ha. Uresk and Juntti (2008) further modified the pole with 1.27-cm bands. This modification greatly improved the precision and accuracy of measures of short and sparse vegetation. More importantly, with this modification, critical cover structure can be detected for wildlife before a major change occurs caused by livestock grazing and other factors. Resource guidelines may be developed to meet wildlife ob-

Fig. 16.9. Cover boards used to index or quantify cover or to provide visual records of changes in cover when photographed from the same reference point.

Fig. 16.10. Estimating visual obstruction from a specific height and distance.

jectives to maintain habitat structure. The tool is fast, easy to use, highly accurate, and cost effective. This modified pole provides an assessment of standing crop on grasslands, can be used to monitor livestock grazing, and provides status of vegetation structure for wildlife habitat.

Alternatively, Griffith and Youtie (1988) reported that a 2.5-cm × 200-cm **cover pole,** easily transported in the field, produced measures of horizontal shrub cover indistinguishable from those produced by the vegetation profile board. The cover pole was painted with alternating 0.1-m black and white bands, and 3 red bands divided the board into 0.5-m zones. Visual obstruction in each zone is recorded as the number (1–5) of 0.1-m bands that are ≥25% concealed by vegetation in each 0.5-m level.

Collins and Becker (2001) developed a new point sampling method, the **staff-ball** method (Fig 16.11), to characterize horizontal cover and compared time and precision of use among observers and with 3 other methods (cover pole, profile tube, and checker board). Their results indicate the staff-ball method provided estimates of horizontal cover from 5.1 to 14.3 times faster than other methods and with greater precision, because observers only needed to make yes/no decisions rather than subjective estimates and/or counts. The staff-ball method also can be used in a variety of vegetation types. Staff-ball point cover readings are taken at the point where the ball meets the pole (one side only); the reading consists of determining whether this point is or is not obscured. The ball or balls are positioned at set heights on the pole, depending on vegetation type.

Marlow and Clary (1996) and Dudley et al. (1998) used photography in combination with cover boards to assess vegetation differences. Although the technique enables visual assessment of cover changes through time, it does not provide measurable differences. To ensure comparisons from year to year, photographs must be taken annually from the same point (height, distance, and direction) with similar film,

Fig. 16.11. Estimating horizontal cover using the staff-ball method of Collins and Becker (2001).

date, and time of day. At times, the date is adjusted to phenological characters of specific plant species.

Users of visual estimation techniques to characterize vegetation structure should be aware of the potential for interobserver judgments and associated biases (Schultz et al. 1961). In studies comparing visual estimation data sets to those obtained using instrument measurements, Gotfryd and Hansell (1985) and Block et al. (1987), using univariate and multivariate analyses, found significant differences between observer estimates and measurements for many vegetation variables. Thus, studies that rely solely on visual (ocular) estimation techniques may forfeit accuracy to save on labor and sampling costs.

Herbage Height

Height of herbage is probably the easiest attribute of vegetation to measure in grasslands, but it has received little attention in published literature. Plant height can be estimated with high precision in many grasslands. Plant height correlates well with other structural attributes of herbage important to the management of grasslands. For example, Higgins and Barker (1982) reported that herbage height explained 63% of the foliage density values that were taken concurrently with the use of a modified visual obstruction pole (Robel et al. 1970). Herbage height in grassland habitats has an important role in predator deterrence and prey security. Average plant (stubble) height also can be used to evaluate the impact of livestock grazing on a pasture (reviewed by Clary and Leininger 2000, Turner and Clary 2001).

Herbage height can refer to the tallest portion of a plant or the effective cover height (generally the upper limit of vegetation leafiness), or to the area-height of herbage (where a 30-cm diameter plastic disk is lowered slowly, until it touches the vegetation). Maximum plant height can be measured readily with a calibrated ruler or tape placed next to a plant. Multiple measurements (≥10) usually are expressed as an average height.

Effective plant height usually is measured as the maximum height of leafy cover for grasses and forbs; however, effective plant height of a forb (e.g., alfalfa [*Medicago sativa*]) may be equivalent to its maximum height. Effective herbage height also may be measured by holding a pole or meter stick parallel to the ground and reading the effective height at the point where leafy plant parts touch the horizontal pole in a minimum of 3 places along its length. Bakker et al. (2002) found that effective plant height was associated with savannah sparrow (*Passerculus sandwichensis*) use of grassland habitats in eastern South Dakota.

The height of herbage per unit area can be measured with a disk or plate in combination with a ruler (Higgins and Barker 1982, Gonzalez et al. 1990). Clear or lightly colored plastic allows plant parts to be seen beneath the disk. Maximum area-height measurements are made at the point where the plastic disk is first touched by a plant part. If a weighted disk is used, measurements are made at the lowest point where the disk settles on the vegetation (Bransby et al. 1977, Gonzalez et al. 1990).

Rangeland canopy height also can be measured by counting the number of laser measurements by 1.3-cm height categories and dividing by the total number of laser measurements for a line transect (Ritchie et al. 1992). The laser transmits and receives reflected wavelength signals, and at 4,000 pulses/sec with an aircraft altitude of about 150 m and a speed of 60 m/sec, a vertical measurement is taken at 1.5-cm intervals along the flight line. These data can be obtained with a laser profiler mounted in a fixed-wing aircraft that measures the distance between the aircraft and the defined surface material to be sampled (e.g., vegetation) with this method.

Tree Dimensions and Structural Characteristics

The size and characteristics of individual trees affect the **physiognomic structure** of forested wildlife habitats. In many forested habitats, large trees provide critical structures necessary for nesting, reproduction, or survival. For example, studies of nesting sites of northern owls (*Strix occidentalis caurina*) and California spotted owls (*S. o. occidentalis*) indicate that presence of large old-growth trees or snags is a key characteristic of nesting habitat in western forests (Mills et al. 1993, North et al. 2000). In other situations, a variety of tree sizes, age classes, and structures contributes to habitat complexity and overall diversity of wildlife species inhabiting the forest. Choosing which characteristics of trees to measure depends on study objectives and biological characteristics of the species under study. Morrison et al. (1998:139–167) provides a complete discussion of measuring forest habitat structure. Here we describe a few of the most common measurements.

Three common, interrelated measures of tree size are height, crown volume, and trunk diameter. Height of tall trees may be measured using trigonometric functions based on the horizontal distance of the observer from the tree and angle measurements made to the base and top of the tree (Woodward et al. 2009). Although angle measurements may be made using a standard clinometer, a wide variety of laser rangefinders and hypsometers, available from many forestry outfitters, simplifies the task of measuring tree heights and eliminates the need for subsequent computations. Crown volume may be measured from similar measurements of minimum and maximum canopy heights and canopy diameters measured horizontally (Sturman 1968; reviewed in Morrison et al. [1998]).

Trunk diameter and cross-sectional area are the most common measurements of tree size because of ease of measurement and high correlation with height and volume. Diameter can be measured with a diameter tape (calibrated to give diameter from a measure of circumference) placed around the circumference of a tree trunk or with calipers (Fig. 16.12). By convention, the measurement (DBH) is made 1.4 m above ground level (Spurr 1964) and above the enlarged base of some trees; DBH also is a representative height where measurements can be made consistently and rapidly. Such data often are summarized as numbers of individuals of species per size class per unit of land area. If exact diameter measurements are not needed, a forester's Bilt-

Fig. 16.12. Calipers used to measure tree diameter at breast height.

more stick (Avery 1959) can be used to estimate diameters in size classes.

Trunk cross-sectional area, calculated as $A = \pi r^2$ (r is the radius of the tree at breast height), also is measured at breast height, and the results (commonly identified by the misnomer "basal area") are given as area units of trunk per unit land area. Individual tree areas can be computed from diameter measurements or measured directly with a tape scaled with area equivalent units. Data can be presented as the value just described or as a relative value (percentage of the total contributed by a single species).

In addition to describing dimensions of trees, assessing the presence or frequency of trees with unique structural characteristics also may be important to wildlife managers or biologists in judging habitat suitability or studying wildlife habitat relationships. For example, the presence of trees with broken tops, nesting platforms, or snags of a particular decay class may be important for wildlife with specialized nesting requirements. Until recently, guidelines for assessing the many unique structures of trees were distributed broadly throughout the forestry and wildlife literature. The recent focus of the U.S. Forest Service on describing forest health at the national scale has led to the standardization of many protocols for describing such diverse characteristics of individual trees as live crown ratios, crown dieback, and decay classes of standing dead trees, as well as many other structural characteristics. The biologist concerned with measuring specialized tree characteristics may find it useful to study protocols available on the website of the U.S. Forest Service Inventory and Analysis National Program (http://fia.fs.fed.us/library/field-guides-methods-proc/; e.g., see Bate et al. 2008).

Tree Age

For many wildlife studies, it is sufficient to obtain one or more expressions of tree size, without age, although at times the latter also has value. Age data are beneficial in forest history and dynamics, including predictions of future status. For instance, knowledge of the approximate life span of a tree species aids in assessment of the current tree population age structure and of regeneration success. Past events influencing the forest and its wildlife inhabitants can be revealed by the presence of fire scars or periods of reduced growth.

Some wildlife species have tree-size and age-specific requirements. For example, in longleaf pine (*Pinus palustris*) forests of the southeastern United States, trees >95 years old have been judged important for red-cockaded woodpeckers (*Picoides borealis;* Hooper 1988). Ruffed grouse (*Bonasa umbellus*) in northern forests do best in a mosaic of aspen stands of various ages (Sharp 1963, Dessecker and McAuley 2001).

Age classification of trees is possible, because trunk lateral growth occurs in annular increments related to the seasonality of temperate zone climates (Raven et al. 1986). The increments are especially evident in so-called ring-porous species. Large-pored vascular tissue is formed early in the growing season, followed by small-pored tissue and then termination of growth that year, followed by the onset of obvious spring growth as another growing season begins. Examples of these species include oaks (*Quercus* spp.), ashes (*Fraxinus* spp.), and elms (*Ulmus* spp.). Diffuse-porous angiosperm species, e.g., maples (*Acer* spp.), aspens, and birches, have less apparent growth rings. Conifers, unlike angiosperms, have a somewhat different anatomical structure, yet they, too, typically have easily recognized growth rings. Extra treatments of the wood, such as applying light oil, certain stains, or water; sanding; or shaving with a razor blade can help make growth rings more evident.

Growth rings can be seen on trunk or stump cross-sections. Vegetation sampling in concert with timber harvest or removal of damaged and/or dead trees is an easy way to collect such data. Where destructive sampling is not in order, small cylindrical cores can be collected with a wood increment borer. Cores can be analyzed onsite or stored, for example in soda straws, until they are viewed in the laboratory. They also can be affixed to a grooved board and kept for future reference. Together with classifying age, tree ring analysis can be used to measure growth rate and to date discernable past events, such as fire (resulting in scarred tissue) or varied climatic or competitive regimes (revealed by varied growth ring widths).

Plant Use

Quantification of **plant use** and its effect on the ecosystem are important for estimating the number of herbivores that can use the land without deterioration of the soil base and plant community (Bonham 1989, Clary and Leininger 2000, Turner and Clary 2001). Maintenance of adequate plant and litter cover retards water runoff and reduces erosion. Early methods to evaluate use of range grasses were developed during 1930–1950 (Stoddart 1935, Pechanec 1936, Lommasson and Jensen 1938, Canfield 1944, Roach 1950), and with some modification, they are still used today. Many of the methods of estimating shrub use are modifications of those used for grasses, and we discuss methods for each in the following paragraphs. To avoid confusion, we use the term stems to refer to stems of grasses and twigs to refer to shrubs and saplings.

As with estimation of biomass, plant use may be estimated with ocular methods. These methods require training with ungrazed plants that are clipped to simulate different intensities of grazing. Such estimates vary by individual investigator and may be inconsistent from year to year. Commonly accepted methods of measuring use vary from simply counting used or unused stems, to obtaining "before and after" measures of stem lengths, to regression methods.

The **stem-count method** (Stoddart 1935, Cole 1956) is a minor modification of the range survey method described above, in which used and unused stems are counted rather than estimated. Stems may be counted in plots or along transect lines. Pechanec (1936) observed that stem counts did not compare favorably with other methods for estimating grass use. Stickney (1966) and Jensen and Scotter (1977) reported that proportion of shrub twigs used correlated well with proportions of lengths removed, but the method was insensitive under heavy use. Stickney (1966) observed that virtually all shrub twigs received at least minor browsing at use levels 55% of length for black chokecherry (*Prunus virginiana*) and 60% of length for Saskatoon serviceberry (*Amelanchier alnifolia*). Those wishing to compare among sites that receive >50% use will need to select a method that remains sensitive under a wider range of use.

Use may be estimated by measuring height of grass stems or length of shrub twigs before and after use by herbivores. Relationships are calculated for height/length removed and biomass used (Lommasson and Jensen 1938, Stickney 1966, Jensen and Scotter 1977). For both grasses and shrubs, the relationship is not linear, so curvilinear relationships must be developed. Jensen and Scotter (1977) reported the twig-length method provided a sensitive measure of shrub use across a range of use levels (0–100%). A primary disadvantage of the method is that it requires 2 trips to the field, one prior to and one following the browsing season, yet it provides no estimate of production. Curves must be developed individually for each species, site, and year to accurately estimate use (Bonham 1989).

Browse use also may be estimated with dimension analyses of twigs by predicting the prebrowsing lengths or weights of twigs from diameter–weight or diameter–length relationships (Basile and Hutchings 1966; Telfer 1969a, b; Lyon 1970). Once the diameter–weight or diameter–length equations have been developed, the technique requires 3 additional types of data:

1. an estimate of the percentage of twigs browsed,
2. mean diameters at the point of browsing of a stratified sample of browsed twigs, and
3. mean lengths or weights of the twig parts remaining after browsing.

Prebrowsing weights or lengths of browsed twigs can be estimated from regression equations. Postbrowsing weights of browsed twigs can be measured by clipping and weighing the residual twigs. Alternatively, postbrowsing lengths of browsed twigs can be measured directly. The percentage use can be computed from the formula

$$U = B \times [(P - A)/P] \times 100,$$

where B is the percentage of browsed twigs, P is predicted prebrowsing mean length or weight of browsed twigs, and A is postbrowsing mean length or weight of browsed twigs (Lyon 1970).

As an alternative to the above procedure, several workers have estimated weights of consumed twigs directly by using the diameter at point of browsing in weight–diameter equations (Oldemeyer 1982, Rumble 1987). In that instance, use may be computed as

$$U = [(B \times C)/P] \times 100,$$

where B is the proportion of browsed twigs, P is predicted prebrowsing mean weight of browsed twigs (based on diameter of current annual growth), and C is predicted mean weight of consumed portions of twigs (based on browsing diameter).

Several authors (Jensen and Urness 1981, Provenza and Urness 1981) demonstrated that use estimates obtained from twig diameter measurements are rapid and compare favorably with twig length measurements. Once diameter–weight or diameter–length equations have been developed for a site, the method represents a considerable savings in time over the twig-length method, because all measurements of use can be obtained during a single trip to the field after use has occurred.

Percentage of plants or stems used by herbivores often is used as an estimator of plant use. This method requires a combination of techniques. For grasses, one measures the percentage of biomass removed, using height–weight relationships, and regresses percentage of plants used on biomass removed from a sample of several sites (Roach 1950). Similar regressions can be developed for shrubs with percentage of plants used and results of dimension analysis (Oldemeyer 1982).

Another evaluation technique commonly used to assess levels of plant use at the landscape scale is classification of key browse species into **form and age classes** (Dasmann 1951, Cole 1959, Patton and Hall 1966). In this procedure, ≥25 plants of a key browse species are marked along permanently established survey courses in selected key winter range areas. For each plant in the survey, the observer records:

1. **hedging**—classified as light, moderate, or severe, based on the length and appearance of the previous year's growth below the current leaders;
2. **availability**—classified as available or unavailable, based on shrub height and maximum browsing reach of the principal browsing species; and
3. **age/decadence**—classified as seedling, young, mature, or decadent, based on stem diameter classes (any living plant with ≥25% of the crown dead is classified as decadent).

Hedging, availability, and age class are summarized as percentages of shrubs in each class. The method has the advantage of being quite rapid, allowing for completion of extensive surveys. However, as for all subjective ratings, there is considerable variation among individual examiners in the assignment of form classes.

Keigley (1997) recently proposed a new method of evaluating browse growth form based on explicit definitions of browsing intensity and **plant architecture.** In this procedure, browsing intensity of individual shrub stems is rated as light to moderate or intense, depending on whether the current annual production consistently develops from the previous year's growth (light-to-moderate browsing) or from stem segments >1 year old because the previous year's grow was killed by browsing (intense browsing). At the whole-plant level (including multiple stems that comprise the plant), plant architecture is classified as **uninterrupted growth type** (reflecting light-to-moderate browsing), **arrested type** (reflecting intense browsing), **retrogressed type** (reflecting light-to-moderate changing to intense browsing), or **released type** (reflecting intense changing to light-to-moderate browsing). Explicit definitions are given for each architecture type. Additional details and applications of the method are provided by Keigley et al. (2002a, b).

TECHNIQUES FOR SAMPLING FRUITS

Data on fruit abundance can be quite important when certain species of wildlife are dependent on annual fruit production (DeGange et al. 1989, McShea and Schwede 1993, Wolff 1996, McShea 2000, Suthers et al. 2000). However, few habitat analyses include an inventory of fruit production. An enumeration of the number and size of fruiting plants is often as far as managers go to describe fruit-bearing potential and its value to wildlife. The inconsistent and seasonal fruiting tendencies of plants, coupled with their often sporadic distribution, minimize the usefulness of simple enumeration of plants.

In studies of wildlife food habits, fruits generally are referred to as mast and are divided into 2 categories: hard and soft. Consequently, **mast** can be defined as the fruits and seeds of all plants, both woody and herbaceous, used as food by animals. The importance of fruit as wildlife food is well known; for example, oak mast alone is used by 185 wildlife species and is available for ≤8 months (Van Dersal 1940). Mast is high in food energy, especially carbohydrates and fats (Goodrum et al. 1971).

Soft mast includes fruits with fleshy exteriors, such as berries, drupes, and pomes. **Hard mast,** in contrast, includes fruits with dry or hard exteriors, such as achenes, nuts, samaras, cones, pods, seeds, and capsules. Numerous factors affect fruit production, including age of plant, size of plant, genetics of the individual plant, climate, soil, competition for resources, and previous use by animals (Schupp 1990). Annual variation in yields of wild food plants makes it difficult to estimate fruit production over large land units (Koenig and Knops 1995). Consequently, management practices that provide for the greatest variety of food-producing plants will ensure favorable conditions for the greatest variety of wild animals.

Large or Heavy Fruits of Trees

The sampling design necessary for species with large or heavy fruits depends on whether total mast production or an index of annual mast abundance is desired. Choosing a large number of random points for trap locations may be necessary if the objective is to characterize mast production at the landscape scale. Although this design is costly, it avoids intentional bias and allows statistical inference from the sample to the larger area. Depending on the objectives of the project, one may want to sample only under the canopies of mast-producing trees. This method would be appropriate if one is measuring the production per unit area of mast-producing canopy in the forest or obtaining an annual index of mast production.

Sampling may be random in forests with well-defined stands of trees (Thompson 1962) or stratified by vegetation type, stand age classes, or stand location (edge or interior). Sampling methods have been devised to estimate production by small versus large trees (Minckler and McDermott 1960), to compare production of ≥2 species of oaks (Tryon and Carvell 1962, Koenig and Knops 1995), and to estimate production in mixed oak stands 63–82 years of age (Beck and Olson 1968).

Mast production can be estimated by counts of mast in ground plots (Goodrum et al. 1971), counts of mast on trees (Gysel 1956, Koenig et al. 1994, Koenig and Knops 1995), or use of seed traps (Schupp 1990, Sork et al. 1993, Ostfeld et al. 1996). Counts of mast in plots on the forest floor are generally unreliable estimators of mast production, because mast frequently is taken by wildlife before counts are made; however, such counts, when used with seed traps, may be a good estimator of wildlife use of fallen mast. Total counts of mast on trees may be quite accurate for small trees, but they are difficult and time consuming for large trees. Consequently, many researchers and managers have opted to use relatively rapid indices of mast production rather than more labor-intensive methods. For example, indices based on visual counts have the advantage of being quick and permitting rapid assessment of acorn production. In the most general index, acorn production may be rated on a visual scale from 0 to 4: 0 (no acorns), 1 (a few acorns seen after close scrutiny), 2 (a fair number), 3 (a good crop), and 4 (a bumper crop; Koenig et al. 1994). The obvious disadvantage of such a rating system is its subjectivity. As an example of a more quantitative index, Koenig et al. (1994) and Koenig and Knops (1995) counted as many acorns as they were able on a single tree during a 30-second interval. Although such an index may be limited by the maximum rate at which an observer may count under high acorn abundances, Koenig et al. (1994) found the index was highly correlated with values obtained from acorn traps. Alternatively, Wolff (1996) used visual counts of acorns on 10 randomly selected branches as an index of mast production in oak woodlands of California.

Many kinds of **mast traps** have been used to measure mast fall. Downs and McQuilkin (1944) developed square traps made of hardware cloth on a wood frame. These traps were about 1 m^2 in size, and 2 were placed under each tree. Since that time, several trap designs have been developed, ranging from makeshift types, such as large oil drums; to large fruit baskets; to those made from wood, cardboard, or polyethylene film and particularly designed for catching acorns. Because rodents and other wildlife will eat mast in the traps, early traps used predator guards; however, these deflected mast from the trap, and guards are not recommended. A study of 8 types of traps comparing catching efficiency, durability, and cost (Thompson and McGinnes 1963) revealed 3 types to be most suitable: polyethylene film traps, square wire-cage traps, and paperboard seed traps. The polyethylene conical-shaped seed trap sampled an area of 0.4 m^2 and had an acorn-retention efficiency of 99%. Fifty of these traps can be carried by one person a considerable distance without discomfort. The wire cage trap (Moody et al.1954) sampled 1.0 m^2. With a wire cover, it had an acorn-catching efficiency of 87% and a durability of 10 years. The design was similar to traps used by Downs and McQuilkin (1944). Of the 8 traps compared, the wire cage model was the most expensive to construct. The paperboard seed trap (Klawitter and Stubbs 1961) was a modified version of the pine seed trap (Easley and Chaiken 1951) that has a sampling area of 0.0003 ha (3.2 m^2). The paperboard trap had 96% acorn-retention efficiency and was durable for 2–3 years.

Christisen and Kearby (1984) constructed acorn traps of 8-gauge steel wire formed into a circle 0.73 m in diameter. They attached to the wire clear 4-mil plastic, cut into a semi-circle, forming a cone. Holes punched in the bottom of the cone allowed water to drain. The trap was attached to wooden stakes to hold it off the ground. They concluded the plastic cone was superior to baskets and wire mesh traps, because the soft plastic prevented acorns from bouncing out of the trap, acorn predation was eliminated, and traps were inexpensive and portable. The primary disadvantage was the plastic lasted only 1 year. Sork et al. (1993) and Schroeder and Vangilder (1997) used a similar seed collecting trap made of 6-mil plastic and a trap area of 0.5 m^2. Schupp (1990) studied seedfall from the understory trees in Panama using 1.0-m FD traps constructed of 1.5-mm mesh plastic window screening on 1-m × 1-m frames of 1.25-cm PVC tubing.

Mast production varies considerably among tree species (Sork et al. 1993), among trees of the same species, and among years (Christisen and Kearby 1984, Koenig et al. 1994). Thus, one must design a mast production study with great care. Traps have been placed under trees at a distance of two-thirds the crown radius from the trunk; however, we are not aware that a consistent distance from the trunk is required. Christisen and Kearby (1984) randomly placed 3 traps under each sample tree with the stipulations that no 2 traps were placed in the same direction and that no traps be placed under a side of a tree that lacked canopy. Further, they imagined the canopy as consisting of 2 concentric circles and either placed 2 traps in the inner circle and one in the outer, or vice versa. Traps should be examined at 1–2-week intervals from the time large fruits (e.g., acorns) begin to drop until all have fallen. Fruits removed from traps should be counted and may be placed into categories, such as (1) well developed and sound; (2) well developed, but damaged by birds or squirrels; (3) well developed, but showing insect emergence holes; and (4) imperfectly developed, deformed, or aborted (Downs and McQuilkin 1944, McQuilkin and Musbach 1977, McShea and Schwede 1993).

Gysel (1957) estimated acorn production by multiplying the number of acorns collected per trap and species by 1.1 to compensate for losses by deflection. He then multiplied that value (acorns per unit area of trap) by the average weight of sound acorns and total crown area of the stand to derive an estimate of the weight of acorn production per unit area.

Small or Light Fruits of Trees

Like large mast, smaller seeds and fruits are important wildlife foods used by many small rodents, tree and ground squirrels, and game and nongame birds (Trousdell 1954, Hooven 1958, Yeatman 1960, Abbott 1961, Abbott and Dodge 1961, Asher 1963, Powell 1965, Landers and Johnson 1976, McShea and Schwede 1993, Schroeder and Vangilder 1997, McCracken et al. 1999). Abundance of small or light mast (e.g., pine seeds) varies from year to year, as for all fruiting species. For example, loblolly pine (*Pinus taeda*) seed varied from nearly 0 to as high as 243,000 seeds/ha (Allen and Trousdell 1961). The 2 principal techniques of sampling small or light seed production of trees are placing seed traps in a stand (Lotti and LeGrande 1959, Allen and Trousdell 1961, Graber 1970, McCracken et al. 1999) or counting the number of ripening cones on a tree with binoculars (Wenger 1953). The latter method may be simplified by counting only a portion of the tree (Wenger 1953) or by categorizing the relative abundance of cones on the tree as none, few (1–25 cones), medium (29–90), and heavy (≥100).

Fruits of Shrubs

Soft and hard mast of shrubs often is within reach of a biologist and may be counted (Suthers et al. 2000) or harvested directly from the shrub (Perry et al. 1999). In Georgia, Johnson and Landers (1978) collected all fruits, by species, in 4-m² plots on a monthly basis from April through October. Their small sample of 5 plots per line had such high sampling error they were not able to compare production among the months sampled. Harlow et al. (1980) counted mast on scrub oaks (*Quercus ilicifolia*) in Florida in a series of 0.004-ha circular plots to estimate mast abundance. Total counts of mast were made for each species in each of 20–40 plots in each stand. Stransky and Halls (1980) counted fruits of shrubs and woody vines in 20 1-m² quadrats in 0.6-ha plots in eastern Texas. They dried fresh fruits of each species to obtain an average weight of each fruit and projected the yield per quadrat based on the quadrat counts. Stransky and Halls (1980) further developed regressions between fruit yield and plant height and density to simplify the sampling effort, similar to regressions of browse production. Perry et al. (1999) conducted soft mast surveys for 31 taxa in Arkansas and Oklahoma by counting berries present in 3-m² plots during mid-June, mid-July, and mid-August to coincide with ripening phenology of the major fruit-producing species. To estimate **dry mass production,** they counted and weighed samples of each fruit type and developed wet- to dry-mass conversion factors. They developed species-specific regressions relating seed head volume with dry mass for species with large seed heads containing abundant fruits, so that dry mass was estimated from counts and measurements of seed heads rather than from individual berries. Like most total enumeration methods, estimating total production of berries may be quite time consuming. Consequently, soft mast production may be characterized for extensive surveys on a scale of relative abundance ranging from 0 to 4, in much the same manner as for hard mast (Clark et al. 1994). Further, double sampling methods have been used to calibrate relative abundance indices to actual production by measuring fruit production for a sample of plots on which relative abundance is measured (Noyce and Coy 1989). Biologists must remain aware of the potentially serious variation in relative abundance estimates made by different observers, or among different regions or years.

Fruits of Herbaceous Vegetation

Herbaceous vegetation provides an abundant supply of seeds for wildlife. Sampling seeds of herbaceous species has not been as well developed as for trees, because more plant species are involved, and wildlife that use those seeds generally are less obvious. Sampling for seeds of herbaceous species is a miniature version of sampling for large mast from trees; samples may be taken from the ground, from traps, or directly from the plant. Ripley and Perkins (1965) sampled ground seed supplies (primarily legumes) for northern bobwhite (*Colinus virginianus*) from soil samples. They removed soil cores (7.6-cm diameter × 2.5-cm deep), screened the cores of litter and soil, and counted number of seeds in each core. Eight soil cores were taken at each of 3 points along a transect line, and the 8 samples were combined to project an estimated seed density and weight. Variation among lines was not greater than variation among points; thus, Ripley and Perkins (1965) suggested that random sampling may be as efficient as using lines. They also reported decreased numbers of seeds in the soil cores from autumn to spring, suggesting removal by wildlife. Larger plots and different sampling depths have been used by others. Haugen and Fitch (1955) used 15 30.5-cm × 30.5-cm plots, but they took material only from the soil surface when sampling for lespedeza (*Lespedeza* spp.) and partridge pea senna (*Cassia marilandica*) seeds. Young et al. (1983) used 32-cm × 32-cm open-bottom metal boxes driven 15 cm into soil to estimate abundance of Indian ricegrass (*Oryzopsis hymenoides*) seed in Nevada. They further removed the soil in 2.5-cm depth increments to identify where seed reserves occurred.

Seed traps for herbaceous plant seeds are smaller than those used for tree mast. Traps with fine-screen wire for the bottom and 0.64-cm hardware screen for the top have been used for estimating seed yield for game birds (Davison et al. 1955). Traps of this type eliminate seed predation by wildlife. Others have used traps with adhesives to hold the seeds. A Petri dish containing filter paper sprayed with Tanglefoot® or other nondrying sticky substances was used by Werner (1975), Rabinowitz and Rapp (1980), and Potvin (1988) to sample seed deposition in prairie grasslands. Rabinowitz and Rapp (1980) believed that seed production was underestimated in tallgrass prairie, because leaves closed over the trap, and seeds were intercepted by overhanging leaves. When temperatures dropped below freezing or when traps became covered with snow, they were not effective for catching seed. Huenneke and Graham (1987) used house-construction insulation hangers coated with a smooth surface of adhesive to sample seed rain in grasslands. They observed that height of seedfall affected the proportion of seeds adhering to the trap surface; at 60 cm, only about 3% of the seeds adhered, whereas at 10 cm, 65% adhered to the trap surface. Exposure to light, high temperatures, and dust had little effect on capture rates, but shape and form of seed did affect capture rates.

Seed traps also can be used over water to sample seed production and availability in wetlands. Olinde et al. (1985) constructed 12-cm × 30-cm traps and floated the traps on Styrofoam™ blocks. These blocks were held in place with ropes and stakes driven into the soil, and the blocks could rise and fall with changing water levels.

Laubhan and Fredrickson (1992) and Gray et al. (1999) describe techniques to sample and estimate seed yields in wetland and moist-soil environments. Laubhan and Fredrickson (1992) collected inflorescence measurements and all seeds from inflorescences of 13 common moist-soil plant species in a 25-cm × 25-cm sample frame. Sample stations were randomly placed in distinct vegetation zones or patches in wetland area. They found that seed yield varied widely among plant species. Gray et al. (1999) developed models to predict seed yield per wetland plant species that also required an estimate of plant stem density per species. They used regression calculations of the mean stem density multiplied by mean seed yield per 60 plants per species to provide extrapolated species-specific seed yield data. Metabolizable energy values per seed yields per species can be used to estimate waterfowl carrying capacities per unit area of moist-soil or wetland habitat.

MULTIPLE-SCALE VEGETATION SURVEYS

Vegetation measurement on the ground can be facilitated with several technologies. Ground-based digital imaging systems, such as digital cameras with and without infrared viewing, have been shown to measure vegetation cover, amount of green vegetation, and **leaf area index** in grass and shrub dominated ecosystems (White et al. 2000, Rundquist 2002). Dycam ADC (http://www.dycam.com/agri.html) and Decagon First Cover systems (http://www.decagon.com/) are specifically made for vegetation measurement. Measurement of green vegetation cover is accomplished through automated procedures by classifying pixels in the digital image as green vegetation, bare ground, litter, woody material, or other nongreen material. Software classifies the image as a percentage of each specified material. Species and plant-form coverages are developed by sampling digital images for each component using trained observers and viewer software after field image acquisition. Digital cameras provide rapid field collection with minimally trained personnel and provide an extensive record of field conditions that is easily moved to computers, where further and more detailed analyses can occur.

Wildlife personnel have long desired to have methods to estimate live herbaceous biomass or structure and use at a fine scale over large areas (Olenicki 2001). Ground-based passive sensors or **radiometers** that measure electromagnetic reflectance from vegetation have been used to measure biomass, amount of green cover, and biochemical constituents along with classifying vegetation in grass, shrub, and forest-dominated systems (Van der Meer and de Jong 2001). These sensors measure several areas of the **electromagnetic spectrum** (**multispectral**) and some can measure continuously from the visible to well into the thermal portion (**hyperspectral**). Calibration via ground truthing is often required to accurately relate reflectance to traditional vegetation measurements. Ratios of the amount of energy reflected in different regions or bands of the electromagnetic spectrum are used to develop the relationship between reflectance and traditional vegetation measures. Radiometer readings need to be taken during midday on sunny days, so that incoming electromagnetic radiation is similar for all readings. Differences among ecosystems, plant forms, soils, and changes in vegetation during the year prevent universal calibrations from being developed for ground-based radiometers (Asner 1998). Calibrations using local conditions are needed to ensure the best fit (Moulin et al. 1998). Calibration of radiometers with vegetation measurement techniques that have poor repeatability, such as ocular estimation of cover, will result in poor relationships to vegetation reflectance because of the inherent variability of ocular estimations (Bonham 1989).

Ground-based systems have become lighter, more mobile, and easier to use in the field, enabling operators to take more samples in less time than with traditional field methods. Resolution is in the centimeter to meter range, so large numbers of samples are needed to adequately characterize diverse vegetation types over large areas. These systems also can be automated, so that field personnel do not need to be extensively trained in instrument operation.

Olenicki (2001) proposed that real-time Global Positioning System (GPS) receivers or military precision lightweight GPS receivers can aid relocation of points within 1 m accuracy, making the combination of ground-based radiometers and real-time GPS units ideal for monitoring temporal and spatial changes in vegetation over large areas. As an example, Merrill and Boyce (1991) successfully linked field sampling of herbaceous phytomass in Yellowstone National Park with spectral values taken from **Landsat multispectral scanners** for the same field sites to describe trends in phytomass availability on the northern Yellowstone elk (*Cervus canadensis*) range.

Aerial or satellite based technologies have been used extensively to measure regional vegetation patterns at the largest scale of sampling (Avery and Berlin 1992, Van der Meer and de Jong 2001). Passive sensors ranging from panchromatic (**aerial photographs**) to multispectral to hyperspectral have potential applications to vegetation measurement for wildlife managers. Passive sensors have been used to develop digital land-use coverages that are available from government sources (O'Neil et al. 2005). Most current land use and vegetation type coverages are derived from either aerial photography or **Landsat thematic mapper** multispectral imagery (30-m resolution). Aerial photography provides detailed images with resolutions ranging from a 1 m to 100 m. Aerial photography is limited to expert visual interpretation that requires extensive training and many hours to interpret small numbers of images. In contrast, Landsat thematic mapper multispectral imagery can be processed with an automated classification that increases efficiency. However, the 30-m resolution might not be at the scale that is useful for wildlife managers. Smaller resolution multispectral sensors are available, such as Ikonos (http://www.satimagingcorp.com/gallery-ikonos.html), Quickbird (http://www.digitalglobe.com/index.php/85/QuickBird), Orbview (http://www.glcf.umd.edu/data/orbview/), and special aerial sensors, but efforts at vegetation measurement are project-specific, and full coverage is not available for large-scale areas. Wildlife managers who want vegetation measurements at smaller resolutions will have to initiate and fund projects to acquire such information at increased effort and cost.

Satellite imagery is useful for deriving measures of **net primary productivity** and **vegetation phenology**. Often these measurements have been derived from observations at weather stations several kilometers from the study area. Furthermore, net primary productivity has been generally based on variation in annual (or seasonal) rainfall rather than on more direct measurements of vegetation. Although imagery from the National Oceanic and Atmospheric Administration's Advanced Very High Resolution Radiometer (AVHRR) and the National Aeronautics and Space Administration's Moderate Resolution Imaging Spectroradiometer (MODIS) satellites have coarse spatial resolution (250–1,000 m), they have high temporal coverage (daily) over large geographic areas. Imagery from these (and other) satellites may be used to calculate the **Normalized Difference Vegetation Index** (NDVI), which is related to the density of chlorophyll contained in terrestrial plants (Sellers 1985). NDVI is derived from the red to near-infrared reflectance ratio [NDVI = (NIR − RED)/(NIR + RED), where NIR and RED are the amounts of near-infrared and red light, respectively, reflected by vegetation and recorded by the satellite]. Chlorophyll absorbs red light, and mesophyll leaf structure scatters near infrared light. NDVI ranges between −1 and 1, with negative values corresponding to no vegetation (Asrar et al.1984, Sellers 1985, Myneni et al. 1995). A time-series of NDVI may be used to calculate biologically relevant metrics, such as start of the growing season, rate of green-up, and length of growing season (Reed et al. 1994; Table 16.3). The integral of the seasonal time-series is strongly related to net primary production (Fung et al. 1987, Goward et al. 1987, Running 1990). To remove contamination from clouds and other atmospheric effects, ≥7 NDVI images are generally combined, using their maximum value; still, smoothing of the time-series may be necessary (Reed et al. 1994). NDVI imagery from AVHRR satellites is available globally at 8-km resolution from 1981; for the conterminous United States, imagery is available from 1989 at 1-km resolution. MODIS imagery is available from 2000 at 250 m, 500 m, and 1,000 m resolutions (Pettorelli 2005). Pettorelli (2005) provided a useful review of NDVI imagery for ecological studies.

NDVI imagery and metrics derived from them have proved useful for ecological studies. Rasmussen et al. (2006) compared explanatory power of rainfall and NDVI to predict time-specific conception rate of African elephants (*Loxodonta africana*). They found that NDVI was a more accurate metric than rainfall for the link between ecological variability and demographic parameters, such as mortality, reproduction, and carrying capacity. Sanz et al. (2003) studied the reproductive output of pied flycatchers (*Ficedula hypoleuca*) using NDVI to monitor tree phenology. They found that oak leafing occurred earlier with a concurrent advancement of peak availability of caterpillars. However, the flycatchers did not change their arrival time from Africa, causing a mismatch between timing of peak food supplies and nesting demand. Nesting growth and survival were negatively affected. Pettorelli et al. (2006) studied the yearly variation in mass of roe deer (*Capreolus capreolus*) fawns in 2 regions of France. There was a strong influence of plant productivity, as measured by integrated NDVI, in the region of lower plant productivity, but none in the higher productivity region. This variation demonstrated the need to use these tools in places where there is a strong link between the canopy (that satellites observe) and ground level vegetation (see also Rasmussen et al. 2006).

Table 16.3. Metrics that may be derived from time-series of normalized difference vegetation index (NDVI) imagery and their phenological interpretation

Metric	Phenological interpretation
Temporal NDVI metrics	
Time of onset of greenness	Beginning of measurable photosynthesis
Time of end of greenness	Cessation of measurable photosynthesis
Duration of greenness	Duration of measurable photosynthesis
Time of maximum NDVI	Time of maximum measurable photosynthesis
NDVI-value metrics	
Value of onset of greenness	Level of photosynthetic activity at onset of growing season
Value of end of greenness	Level of photosynthetic activity at end of growing season
Value of maximum NDVI	Maximum measurable level of photosynthetic activity
Range of NDVI	Range of measurable photosynthetic activity
Derived metrics	
Time-integrated NDVI	Net primary photosynthesis
Rate of green-up	Acceleration of photosynthesis
Rate of senescence	Deceleration of photosynthesis
Modality	Periodicity of photosynthetic activity

After Reed et al. (1994).

Although much of the use of aerial or remote sensing methods relates to mapping, recent improvements in spatial accuracy of Geographic Information System (GIS) tools have helped bring science to vegetation measurement and sampling and to detect the lack of vegetation, such as in forest canopy gaps. The importance of forest canopy gaps for location of songbird nests (Fox et al. 2000) was evaluated with data obtained from color-infrared photographs scanned at high resolution and spatially rectified to ground control points. Using aerial stereo photographs and scopes, Fox et al. (2000) created 3-dimensional images of canopy gaps that could be used in Arc/INFO (ESRI, Redlands, California) computer GIS files to aid digital software analyses. Tanaka and Nakashizuka (1997) used similar methodology to analyze long-term (15 years) canopy dynamics of a 25.25-ha mixed deciduous forest in Japan.

Several other methods have been used to detect temporal changes in vegetation structure at 5 scales of resolution that could be transformed to digital data sets, including cameras on sticks (Bennett et al. 2000), tethered balloons or blimps (Mims 1990, Pitt and Glover 1993, Murden and Risenhoover 2000), tower crane with a horizontal jib (Parker et al. 1992), ultralight aircraft (Cohen et al. 1990), fixed-wing aircraft (Everitt and Nixon 1985, Everitt et al. 1991, Ritchie et al. 1992, Blackburn and Milton 1996), and high-altitude remote sensing (Satellite Probatoire d'Observation de la Terre's High Resolution Visible Imaging System panchromatic images and Landsat thematic mapper data; Cohen et al. 1990, Bradshaw and Spies 1992). Reviews of these methods are provided by Everitt et al. (1991), Ritchie et al. (1992), Pitt and Glover (1993), and Blackburn and Milton (1996).

Active sensors are finding more application in vegetation measurement and will have an increasing role in this area. Radar applications are being developed to measure forest canopy and stand characteristics (Waring et al. 1995, Ranson et al. 2001). **Lidar** or laser altimetry is currently being used in canopy measurement for forests and shrubs (Lefsky et al. 2002), and it has been applied to aquatic habitats (Wang and Philpot 2007). The basic measurement by a lidar device is the distance between the aircraft sensor and a target surface that is expressed as the elapsed time between the laser pulse and the time it took to reflect back to the sensor divided by 2. Lidar's capability to characterize 3-dimensional canopies at small resolutions holds much promise for wildlife managers. Lidar technology is being quickly implemented compared to radar, and there are many providers available. Lidar is still limited, in that clouds can interfere with its functions, whereas radar is an all-weather technology. Acoustic (echolocation) methods are active sensors that can be used to sample and map submerged aquatic vegetation and are not hindered by water clarity (Sabol et al. 2002, Warren and Peterson 2007).

Lefsky et al. (2002) discussed the state-of-the-art of applications of lidar remote sensing relative to natural resources. They noted that numerous applications are feasible, but have not yet been explored, making it difficult to predict which applications will be dominant in the future. According to their review, current applications of lidar remote sensing in vegetation and ecological measurements fall into 3 general categories: remote sensing of ground topography; measurement of the 3-dimensional structure of vegetation canopies; and prediction of forest stand attributes, such as aboveground biomass. They also identify efforts of bathymetric lidar systems to measure elevations in shallow bodies of water. According to Lefsky et al. (2002), mapping of topographic features is the largest and fastest growing area of application for lidar remote sensing, mainly for commercial land surveys (Flood and Gutelis 1997), and largely because airborne laser altimetry is more accurate and cost effective than other methods.

Measurements of vegetation canopy and function are of primary interest to wildlife researchers and managers who study forest animal–vegetation relationships. Allometric canopy heights (maximum and mean) and cover or lack of cover (gaps in the canopy) have been computed for temperate (Maclean and Krabill 1986), tropical (Nelson et al. 1997), boreal (Magnussen et al. 1999), and temperate deciduous (Ritchie et al. 1995) forests. Caution must be made if considerable understory vegetation is present under the tree canopy, because it can disrupt exact elevation measures to the ground surface.

Relative to forest-stand structure attributes when species composition was noted, Maclean and Krabill (1986) were able to account for 92% of the variation in timber volume in stands of oak and loblolly pine. Nelson et al. (1997) successfully estimated basal area, volume, and biomass in tropical wet forests. The availability of information and results from lidar devices will increase as technology and analytical skills improve, including satellite lidar devices. Lidar measurements will have application to estimating levels of taxon biodiversity ranging from guilds and communities to specific species (e.g., natural cavities that are natural nesting sites for wood ducks in old-growth forests).

APPLICATIONS OF VEGETATION MEASUREMENT

We have presented methods for measuring plants or plant attributes of different forms of vegetation. We now discuss how some of these methods have been applied to studies of wildlife habitat.

Loft et al. (1987:656) evaluated mule deer (*Odocoileus hemionus*) habitat during 3 growing seasons in California. Their objectives were to "determine the effects of cattle stocking rate on hiding cover structure during the summer grazing season" and to measure levels of herbivory on willows and herbaceous meadow vegetation. Estimates of herbaceous forage production, deer hiding cover, and browse use were made in 0.1-ha cattle exclosures and adjacent sites subjected to moderate and heavy levels of cattle grazing. Herbaceous forage was clipped from 0.1-m^2 plots, oven-dried, and weighed 2–5 times each growing season. Hiding cover in aspen and meadow habitats was estimated at 8 locations around circular plots of 5.65-m radius with a 1-m^2 grid subdivided into 100 cells. A narrower 1.0-m × 0.4-m grid, similar to that described by Nudds (1977), was placed at 2-m intervals along 2 20-m transects, and the grids were read from a distance of 5.65 m in patchy willow habitat, where structure of the shrubs precluded use of the larger grid. The grids were read at 3 0.5-m increments to 1.5 m; the percentage obscured by vegetation from ground level to 1 m was considered hiding cover for fawns, and the percentage obscured from 0.5 m to 1.5 m was considered cover for adult deer. To evaluate the browsing level of willows, Loft et al. (1987) tagged willow branches with ≤24 new shoots and measured the percentage of shoots browsed after cattle were removed from the site.

Litvaitis et al. (1985b:866) studied understory characteristics of snowshoe hare (*Lepus americanus*) habitat in Maine. Their objectives were to "examine hare habitat use and density in 2 areas of Maine with differing forest composition, and determine how those variables were influenced by forest understory characteristics." Snowshoe hare pellets were counted in 105 circular plots of 1-m radius on 7 700-m transects at each of 2 sites at each study area. Vegetation features were measured at each pellet plot. Percentage ground (canopy) cover of softwood, hardwood, herbaceous plants, and moss was estimated in each circular plot by projecting the plant crown to the ground surface. Understory stem density was estimated by counting the number of hardwood and softwood stems ≤7.5 cm DBH and ≥0.5 m tall in 2 15-m × 0.5-m quadrats, beginning at each pellet plot and running perpendicular to the transect. Visual obscurity at each pellet plot was estimated from a distance of 15 m for 3 0.5-m strata 0.50–2.0 m above the plot with profile boards (Nudds 1977). Overstory canopy closure was estimated with a spherical densiometer (Lemmon 1957) at each pellet plot. Correlation coefficients were calculated between each of the vegetation variables and the associated snowshoe hare pellet counts to identify which variables influenced pellet density.

Sedgwick and Knopf (1990:112) studied habitat relationships of cavity-nesting birds in plains cottonwood (*Populus sargentii*) along the South Platte River, Colorado. One of their objectives was to "compare nest sites of cavity-nesting birds with available (random) nesting habitat." Each nest tree was characterized by its species, DBH, height (measured with a clinometer), and the estimated length of dead limbs ≥10 cm diameter. Habitat was characterized in a 0.04-ha circle centered at each nest tree and at 31 random points in the cottonwood-dominated riparian vegetation type. Numbers of snags, trees 23 cm DBH, trees 23–69 cm DBH, and trees >69 cm DBH were counted in each circle to estimate density of the 4 classes. Overstory canopy cover was estimated at 4 points on the perimeter of each circle with a spherical densiometer. Tree basal area was measured in a circle around each tree and random point with a 10-basal-area-factors prism. These data were compared among the species of cavity-nesters, using the cavity to characterize habitat use.

Kirsch et al. (1978) studied habitat characteristics of upland nesting birds, particularly ducks, in North Dakota. One of their objectives was to evaluate the height-density (obstruction) of residual grassland vegetation structure in relationship to success and density of duck nests. Height-density of grassland was measured with a modified version of a visual obstruction pole (Robel et al. 1970); readings of 100% obstruction were taken from a distance of 4 m and an eye-level height of 1 m. Results of their study indicated that higher nest density and success for ducks occurred in residual grassland cover, with the highest average height-density readings at 100% obstruction.

Gilbert and Allwine (1991) studied relationships between small mammals and habitat characteristics of unmanaged Douglas-fir (*Pseudotsuga menziesii*) forests in Oregon. One of their objectives was to identify which environmental factors might be responsible for differences in small mammal communities among young, mature, and old growth Douglas-fir stands. They sampled small mammal abundance and vegetation in 56 young, mature, or old growth stands in 3 locations. At each stand, mammals were sampled in a 6 × 6 pitfall grid or 12 × 12 snap trap grid. In the pitfall grids, 9 points were sampled for vegetation; 16 points were sampled in the snap trap grid. Measurements were made in nested circular plots of 5.6-m and 15-m radius. In the 5.6-m radius plot, cover of logs by decay class, and cover on the ground of bare rock, exposed bare mineral soil, organic litter, moss, and lichen were estimated visually. Cover of foliage to 2-m height and by life form was estimated visually. Number and species of small and medium-sized live trees, snags, and stumps were counted to obtain density. In the larger circular plot, cover of shrubs and trees 2-m height was estimated in 3 canopy layers (midstory, main canopy, and super canopy). Number and species of large live trees and snags were counted. In the larger circle, the presence and type of water and occurrence of rock outcrop and exposed talus were recorded. The number of recent tree-fall mounds with exposed roots and mineral soil was counted. Vegetation components and small mammal numbers were summarized by stand, and data from the 56 stands were analyzed by detrended correspondence analysis (Hill and Gauch 1980) to explore relationships between species abundance and environmental variables.

Hobbs et al. (1982:12) studied carrying capacity of elk in Colorado. Their objectives were to "demonstrate that estimates of nutritional carrying capacity are viable habitat-evaluation procedures and to identify sensitive parameters in the range supply-animal demand algorithm." Estimates of biomass of plants comprising 2% of the elk's diet were necessary to develop the carrying capacity model. They obtained biomass estimates from 32 1-ha stands stratified by vegetation type. In each stand, forbs and grasses were clipped at ground level in 30 0.25-m^2 plots. Ten 2-m^2 plots were sampled for shrubs, and current stem growth was collected between ground level and 2.5-m high. Species were individually separated, dried, and weighed. These data were used to develop biomass estimates for vegetation types in the winter range of elk and were combined with nitrogen concentrations and in vitro dry-matter digestibility to estimate range supply of energy and nitrogen.

Schupp (1990:504) studied seed fall and seedling recruitment of a fruit-producing tree in Panama. Fruits of this tree are eaten by monkeys and birds, and the seeds are eaten by a variety of rodents. Seedlings are eaten by deer and other large browsers. One of the objectives of the study was to examine whether there were "extensive year-to-year differences in viable seed fall, post-dispersal seed predation, seedling emergence, early seedling mortality, and seedling recruitment." Seed fall was monitored with 84 1.0-m² traps constructed of 1.5-mm mesh plastic window screening in 1-m × 1-m frames. Two traps were placed randomly in each of 42 adjacent 20-m × 20-m plots. Traps were not intentionally placed either under or outside the canopy of individual trees, although no traps occurred in large openings. Seeds were counted and removed from traps on a weekly basis. Seedling emergence was studied by scattering a known number of seeds and fruits directly under traps and counting the number of seedlings that emerged. Seedling recruitment was estimated in 3-m × 3-m plots that centered at the seed trap. Newly emerged seedlings were counted twice a year and marked with numbered colored plastic bird bands. The number of seedlings marked in a year was an estimate of that year's seedling emergence. From 58% to 74% of the seedlings marked in the first count of the year were present in the second, indicating moderate mortality of newly emerged seedlings. The total number present at the second count represented the year's seedling recruitment. Predation of individual seeds was measured by gluing a 30-cm piece of nylon fishing line to 576 seeds each year, attaching that line to wire-stake flags, and measuring unnatural changes in position or loss of the seed. Schupp's (1990) experiments showed that removal generally indicated loss to vertebrate seed predators. Among-year variation in viable seed fall, seedling emergence, seedling recruitment, and seedling survival was analyzed with parametric and nonparametric analysis of variance methods. An actuarial life-table method was used to analyze seed predation.

Wildlife ecologists have spent considerable time linking fine-grain vegetation measurements (e.g., quadrats and point-centered-quarters) to wildlife habitat use. In contrast, comparatively little research has been conducted to assess the importance of vegetation characteristics to habitat use at larger scales. As a result, resource managers confronted with conserving ecosystems extrapolate local recommendations to regional levels, because landscape studies are lacking.

Advances in the capabilities of electronic equipment (e.g., computers, video cameras, and GPS units) and increased availability of **landscape scale** data or mapping units for soils, aquatics, vegetation, weather, climate, and land use effects have allowed natural resource researchers and managers to scale up measurements of vegetation made in individual quadrat plots to the landscape level and to combine multiple coverages of other environmental attributes with vegetation data. These advances include digital imaging systems, radiometers, laser altimetry or lidar, and satellite or aerial systems that use active sensors (radar) or passive sensors (panchromatic to hyperspectral imagers). These technologies can be used to measure vegetation at different scales, from high resolution studies in small site-specific areas (measured in cm² or m²) to regional or global assessments. Selection of the correct technology is a function of such factors as cost, time, highest resolution of vegetation to be measured, scale, and availability of technology. In addition, field data can be electronically entered onsite on palm or laptop computers or data loggers (Fig 16.13) and can be identified to specific transects, quadrats, or points using GPS locator Universal Transverse Mercator (UTM) coordinates.

Researchers have used remotely sensed land-cover data to incorporate regional variation in climate and land use into vegetation sampling schemes (Meentemeyer 1989, Bakker et al. 2002). Sampling designs that account for regional variability provide more reliable information to land managers who must deliver habitat programs across large geographic regions. Results from this type of work are being used to direct conservation planning efforts and design na-

Fig. 16.13. Electronic data logger with bar codes referenced to specific attributes of plants or animals.

ture reserves (Askins et al. 1987, Hansson and Angelstam 1991, Pearson 1993).

Scale issues are now widely recognized in wildlife science as a critical concept that influences the way that organisms relate to landscape vegetation patterns. Turner et al. (2001) formally defined **scale** as the spatial or temporal dimension of an object or process. Scale is important, because individual species often perceive the same spatial arrangement of habitats quite differently (Wiens 1989a, Levin 1992). For example, a highly mobile species, such as northern harrier (*Circus cyaneus*), that forages widely may be less sensitive to fine-scale changes in grassland vegetation than a sedentary meadow vole (*Microtus pennsylvanicus*) that uses dense grasslands to escape predation. Although it is easy to acknowledge that scale is an important study component, identifying the "right" scale at which to work remains a challenging issue. A key to selecting appropriate scales is to replace our own human perceptions of scale with a view of how individual wildlife species experience the landscape in space and time (Wiens 1976, Pearson et al. 1996). The concept of **ecological neighborhoods** (Addicott et al. 1987) provides a useful framework for thinking about how space and time components of an organism's behavior may be used to define an appropriate scale for study. However, studies of wildlife habitat at different spatial scales have confirmed there is no single correct scale at which to work: ecologists should identify a suite of appropriate scales at which to analyze their data (Pearson 1993, Sisk et al. 1997, Woodward et al. 2001). Relatively new information-theoretical approaches (Burnham and Anderson 1998) provide statistical methodology for conducting multiscale habitat analyses.

Specific software is needed for display and data analysis to use the new technologies. At the most basic level, workers need to view the imagery. Providers of free viewer products include ESRI (http://www.esri.com/), Erdas (http://www.erdas.com/Homepage.aspx), PCI (http://www.pcigeomatics.com/), ER Mapper (http://www.erdas.com/Homepage.aspx), Leica Geosystems (http://www.leica-geosystems.us/en/index.htm), the University of California at Berkeley, ENVI (http://www.ittvis.com/language/en-US/Company.aspx), and Global Mapper (http://www.globalmapper.com/). Full-featured programs that can view and analyze data range from modestly priced packages, such as Idrisi (http://www.clarklabs.org/), to expensive packages, such as Erdas, ENVI, ER Mapper, PCI, and those from ESRI. Other software packages, such as Adobe Photoshop®, have some use for applications using digital cameras.

Although some techniques are fairly recent in application, by combining a knowledge of the spatial characteristics of tree canopy and canopy gaps with principles of plant ecology, wildlife ecology, and landscape ecology, several inferences can be made regarding the distribution and diversity of wildlife and plant species in habitats across and at edges of geographic ecosystems and gradients of extensive scale.

SUMMARY

Vegetation structure, arrangement, and location are considered the primary components of wildlife conservation and management. Natural resource managers and research biologists use a variety of equipment and techniques to sample and measure vegetation in a multitude of different aquatic and terrestrial plant communities and vegetation types.

Aquatic vegetation assessment is more difficult to accomplish than that for terrestrial vegetation, because it involves floating, submergent, and emergent plant species. In most years, aquatic vegetation assessment is conducted while wading, from a boat, or from aerial photography. In these circumstances, such equipment as quadrat frames must be constructed of materials that will float to facilitate aquatic vegetation sampling and measurement in wetlands.

Terrestrial vegetation assessment is fairly straightforward, but sampling and measurement techniques vary considerably among grassland, shrubland, and woodland vegetation types. For example, rulers and tape measures can be used to measure plant height or canopy coverage (e.g., line intercept) for grass and forb species, whereas prisms, angle gauges, and spherical densiometers are needed to obtain the same measurements on trees. Relative to vegetative food items, the amount of fruit or mast production may be estimated by ocular counts on sample limbs or by collecting falling mast in various traps.

Vegetation sampling and measurement are generally conducted in vegetation patches or field-sized units of a local nature. In contrast, landscape-level assessments of vegetation are usually conducted using satellite, aerial, or video photography coupled with GIS techniques. Recent advances in computer capabilities have enabled managers and researchers to work with larger and more complex data sets to assess vegetation characteristics. These capabilities also enable the integration of other data sets, such as those for animal population demographics, weather, topography, and soils, with the vegetation data for greater in-depth analytical and modeling exercises.

To comprehensively address all possible ways to sample and measure vegetation would require volumes of text and figures. Here we have introduced the reader to as wide an array of vegetation sampling and measuring techniques as possible within the limits of this chapter. We encourage others to explore the literature we have presented and any new literature that will enhance their ability to assess vegetation in a manner that best fits their research or management objectives.

17

Modeling Vertebrate Use of Terrestrial Resources

LYMAN L. MCDONALD,
WALLACE P. ERICKSON,
MARK S. BOYCE, AND
J. RICHARD ALLDREDGE

INTRODUCTION

WE DEFINE HABITAT in its broadest context, including all abiotic and biotic features of the environment. Although other definitions of habitat exist (e.g., Karr 1981, Hall et al. 1997), we consider habitat selection as the association of an animal with these features. It is essential that wildlife management studies identify habitat selection (i.e., vegetation types and foods used) by animals in comparison to those resources in a study area. The availability and use of the environmental components that are necessary for life impact abundance of animals and distribution of their populations in space and time. Although many studies on these topics have been published for common wildlife species, more knowledge is needed about the life requisites of most terrestrial vertebrates. In addition, information about habitat use by a particular species may be needed for a specific region or time period. For example, wildlife biologists have become increasingly involved in assessing effects of human activities, such as urbanization, highway construction, and power line development, on wildlife (Grinder and Krausman 2001). These assessments often require identification of important vegetation patches and food resources in the affected area. As a result, a biologist must collect site-specific information on patterns of vegetation types and food use. But how is such information obtained? What should be considered when a study is designed to identify vegetation types or food use? This chapter provides an outline of the major techniques used to study these issues and some problems likely to be encountered.

Methods for **design and analysis of wildlife studies** have recently been summarized by Morrison et al. (2008) and Williams et al. (2002a). Methods for studying and modeling **resource selection** have been reviewed recently by Alldredge et al. (1998) and Strickland and McDonald (2006), and a general (mathematical and/or statistical) theory for analysis of food and habitat selection studies has been updated by Manly et al. (2002) and Johnson et al. (2006). New methods for analysis of **Global Positioning System (GPS)** locations, collected for **radio-tagged** individuals, have been developed (e.g., Sawyer et al. 2006, 2009).

We have taken different approaches in addressing aspects of habitat and food use to take advantage of other sections of this book. Higgins et al. (Chapter 16, This Volume) discuss vegetation sampling and measurement methods, and Servello et al. (2005) provide additional background for understanding food use patterns. Thus, in this chapter, we focus on conceptual issues of investigating habitat and food use and modeling of resource selection.

Before any study of habitat or food use begins, an understanding of how the results will be used is essential (Chapter 2, This Volume). Is the objective of the study to describe habitat use patterns or food selection for the entire year or during one season that is considered critical? Is there a need to identify limiting factors or simply to document use? Much of the wildlife research to date has been directed at addressing descriptive questions of **how, what, when, and where** (Keppie 1990, Gavin 1991). These investigations have provided a detailed foundation on the natural history of many species. However, understanding why an animal occupies a specific vegetation type (for thermal cover, food abundance, or predator avoidance) or selects a particular forage grass (to maximize energy intake, obtain a specific nutrient, or minimize toxin intake) may reveal much more about the factors that limit a species than simply documenting patterns of use (Gavin 1991, Morrison 2001). Although it may seem obvious, taking the time to **think through a study** and articulate the specific question(s) being addressed is time well spent. One should be able to state concisely to someone else the research question that is being addressed. Results will be only as coherent as the initial conception of the problem (Green 1979, Garton et al. 2005, Morrison et al. 2008). A thorough **review of existing literature** can help investigators develop an understanding of the variability in resource use patterns and avoid the common pitfall of collecting descriptive information simply because it has not been collected in their specific study areas (Hunter 1989). The final question to consider when defining a research project is the application of results. Are the conclusions of the study to be extrapolated from samples collected at 1 area during 1 time period and applied to other regions and time periods? Without consideration of spatial and temporal variations, any extrapolations may mask the effects of spatial and temporal dependencies (Thomas and Taylor 1990, 2006).

MODELING RESOURCE SELECTION

The words use, **selection,** and **preference** have been applied widely and often interchangeably when information on patterns of resource exploitation is presented. Use indicates an association or consumption when habitats or food resources, respectively, are discussed. Selection, however, implies that an animal is choosing among alternative vegetation types or foods that are available in the study area. Use is **selective** if components are exploited disproportional to their availability (Johnson 1980). **Resource availability,** the quantity accessible to the animal or population of animals, is distinguished from **abundance,** which is defined to be the quantity of the resource in a study area. **Preference** for resources is defined as selection independent of availability. Studies to examine preference must allow free access to resources that are provided on an equal basis. Information on preference can be obtained only under special conditions, such as **enclosure experiments** that provide habitat categories in equal abundance, or **cafeteria experiments,** wherein captive animals are presented a variety of foods and allowed to choose among them. Because of the special nature of preference experiments, we focus on developing an understanding of habitat and food selection in study areas rather than on preference.

Definition of the study area influences analyses and fitting of models. In many cases, interest will be in examining how habitat selection changes as habitat units change from one study area to another. When choosing areas, one must consider the distribution of resource units, scale of selection studied, what is truly available to the animals, and labor and budget constraints for collection of data. We use the term **resource units** to indicate habitat units, points in the habitat, or food items.

MODELING OCCUPANCY

A primary application of the methods presented in this chapter is in **monitoring** populations by estimation and mapping of the relative probability that units (grid cells, pixels, etc.) are occupied by a species **as measured by the sampling design.** We emphasize the phrase "as measured by the sampling design," because in survey of units for presence of a species, a unit may be occupied, but the survey protocol fails to detect presence. However, if the probability of detection is high and approximately constant, these modeling methods may be quite useful and economical for estimating and mapping the relative probability that units are occupied, resulting in a powerful monitoring tool, again, as measured by the sampling design. Study protocols requiring multiple independent surveys of units have been developed to obtain **patch occupancy models,** giving clean estimates of the probability of occupancy (MacKenzie et al. 2006). We encourage readers to consider use of these more expensive and time-consuming survey methods when faced with monitoring species distribution where probability of detection varies significantly among vegetation types.

LEVELS OF SELECTION AND EFFECTS OF SCALE

Habitat selection can occur at a variety of levels or scales, with animals selecting habitats according to a **hierarchical** scheme (Johnson 1980). These scales include the biogeographic (e.g., the eastern deciduous forest); home range, or activity points (e.g., a den, nest, or roost site in a home range or selection of particular foods at an activity point; Johnson 1980). Factors that influence selection at each of these scales also vary. For example, climatic extremes may affect the geographic range of a species, whereas vegetation structure may influence home range size and shape, and competition with conspecifics or predation risk (Hebblewhite et al. 2005) may influence territory placement in a home range. The distribution of food and cover is probably most influential in affecting local movements in a home range.

The choice of an appropriate spatial and temporal **scale** of measurement, and consideration of spatial pattern, will directly influence results and their interpretation (Wiens 1981, Otis 1997, Morrison et al. 2008). Although thinking of scale in discrete levels (e.g., time = daily, seasonal, or annual intervals; space = feeding site, home range, or geographic range) is convenient, it is important to recognize that scales of measurement and environmental heterogeneity are continuous (Karr 1983). Choosing the wrong scale of measurement may lead to the interpretation that a species is generalized or specialized in its selection of vegetation types, whereas another scale of measurement might lead to a different interpretation. For example, Wiens (1989b) observed the biogeographic range of Brewer's sparrows (*Spizella breweri*) was associated with shrub-dominated vegetation types. However, at a regional scale (multiple study sites), abundance of sparrows was negatively associated with shrub cover, whereas at a local scale (single study area), shrub abundance and sparrow abundance were not related.

Habitats can be characterized on several spatial scales, for example, from landscape to microenvironment. Wildlife biologists often restrict their studies to a single scale when an examination of several scales may provide great insight to animal–habitat relations (Morris 1984, Sodhi et al. 1999, Apps et al. 2001, Welsh and Lind 2002). For example, the macrohabitat (forest-cover type) and microhabitat (canopy closure and snow depth) components of white-tailed deer (*Odocoileus virginianus*) wintering areas (yards) are important management considerations for this species in the northern portion of its range (Verme 1968). Macrohabitats in this example may describe the interspersion of food and cover selected by deer. Microhabitat features may have a direct effect on thermoregulation (a factor influenced by variation in canopy closure) and energy costs of travel or ability to escape predators (factors influenced by snow depth).

Many investigators are now using tools, such as Geographic Information System (GIS) technology (O'Neil et al. 2005), for discerning land-cover/land-use associations of wildlife species (Pereira and Itami 1991; Hepinstall and Sader 1997; Erickson et al. 1998; Sawyer et al. 2007, 2009). At this **macro scale**, habitat data may be acquired remotely, such as from maps (Mosby 1969), aerial photographs (Avery 1968), or satellite images (Short 1982). Unfortunately, associations between animals and habitat attributes at this level of resolution often are general, because investigators often use GIS data layers that are available, and consequently, many of the predictor covariates are surrogates for the real ecological process that determines distribution and abundance. Care must be taken to ensure that information is gathered at a scale comparable to that at which the research or management question is being addressed (Svancara et al. 2002).

Management Implications of Resource Selection

The **limits of data** should be recognized when applications of research conclusions to manipulations of habitats or populations are considered. Analyses can be helpful in identifying patterns of habitat or food selection. However, biologists should not necessarily conclude biological need from such patterns. For example, suppose that a fictitious species (the blue-nosed yak, *Bos azurostrum*) has demonstrated selection for forests 40–80 years old; that is, yaks are most abundant or spend a disproportionate amount of time in this habitat. Would we be correct to conclude that if all forests 40–80 years old were eliminated from the range of blue-nosed yaks, this species would decline in abundance or go extinct? This conclusion is doubtful. Although we have demonstrated selection, we have not shown how **fitness** (e.g., survival or reproductive success) of yaks varies with different amounts of the selected habitat. We cannot make a biological leap of faith and assume that if we increase the amount of the selected habitat (or food) we will have more yaks.

Van Horne (1983) showed that in certain instances, population density and habitat quality (based on animal fitness) can be inversely correlated. Subordinate individuals (especially juveniles) might become locally abundant in less preferred habitats as a result of avoiding contact with territorial individuals that occupy sites with an abundance of food and cover. As a result, survival and reproductive success in sink habitats are low. Therefore, if our objective is to evaluate the biological importance of a particular habitat, we should consider some type of manipulation experiment in which the amounts of the selected habitat (or food) are varied and fitness is monitored (Van Horne 1983). Although these studies are not always practical, they are essential to demonstrate habitat associations. Such studies may be possible when applied to habitat management programs or large-scale habitat manipulations, such as impounding a river with a dam or logging a forest (Macnab 1983, Sinclair 1991, Williams 1997). Management experiments require considerable planning, because biologists rarely have control over the manipulation. Some habitat or food-based questions may allow the researcher to compare a measure of success among used and less-used resources and evaluate the features that lead to success and, presumably, the basis of selection (e.g., waterfowl nest success and vegetation features that influence concealment).

Sampling Protocols and Study Designs

The researcher must identify the **scale of selection** to study (Johnson 1980), consisting of resolution (grain) and extent (size). The biology of the animal is important (e.g., if the animal being studied is territorial, then selection is commonly studied on a different scale from that used for a nonterritorial animal). For example, an area occupied by one pack of gray wolves (*Canis lupus*) may be unavailable to other packs, and the study area for a new pack should not include the inhabited areas (Mladenoff and Sickley 1998). The study area should be adjusted if animals tend to forage from a central location (e.g., a nest site). Or predictor variables, such as the distance from the nest, should be considered in the analysis.

As a general rule, resource selection studies should consider selection at **1 scale.** One might study selection of home ranges by a wolf pack and selection of locations for hunting in the home range of the pack. Season, gender, age class, behavioral activity, and daily activity pattern of the animal studied often affect resource selection. For example, if radiolocations are used to assess selection of foraging sites, it may be necessary to only record locations during certain hours of the day. If resource selection and/or characteristics of units change across seasons, the study should focus on habitat selection during relatively short periods of time or fit models that allow for the characteristics to change (Schooley 1994, Arthur et al. 1996, Cooper and Millspaugh 1999). Pooling information across times, subpopulations, age classes, or activities may result in erroneous inferences.

Resource selection may be detected and measured by comparing any 2 of the 3 possible sets of resource units (used units, unused units, or units in the study area). On this basis, the following 3 common **sampling protocols** (SP-A, SP-B, and SP-C) have been identified, depending on the 2 sets measured (Manly et al. 2002). In addition, we define a fourth commonly used sampling design (SP-D; Box 17.1).

If all units in a category are sampled, the protocol is a **census.** The same general analysis is conducted whether a census of all units is taken or not. We consider only the most common sampling protocol that arises in practice: **SP-A** (study area units and used units are independently sampled). This protocol also is known as a **use/available design** (Johnson et al. 2006). Manly et al. (2002) discuss the analysis of data falling under the other cases (e.g., food items might be sampled before selection, and unused food items are sampled after selection).

The assumption that units are **randomly sampled** is a strong statement, implying no biases in determining which units are used. It is often difficult to be certain that a unit is used or not used by a species, potentially leading to **false negatives** (i.e., a unit was used [or the species is present], but the use [presence] was not detected). Similarly, **false positives** arise if unused units are identified as used. In general, it is probably best to recognize that classification of used or unused units is determined **as measured by the sampling design.** Special advanced sampling techniques requiring multiple surveys and modeling methods, known as **patch occupancy modeling,** have been developed for cases when the proportion of false negatives varies significantly with vegetation types (MacKenzie et al. 2006).

Three general study designs for evaluating selection have been identified (Thomas and Taylor 1990, 2006). These designs differ on whether selection of units by individual animals can be identified (e.g., locations of radiomarked animals) and whether study areas are defined for the population of animals or for individual animals. Definition of the 3 study designs follows Thomas and Taylor (1990) and Manly et al. (2002).

Design I

Units used by the population of animals are recorded (but use by individual animals is not possible to record). For example, in this design aerial or ground surveys are used to locate animals. Variables that potentially influence selection of units by animals are measured at the locations. For example, vegetation or forage types, food availability, slope, aspect, and density of roads in a plot centered at the locations might be measured at each location and used in a model to predict the relative probability of selection of locations by animals in the population. Maps, aerial photographs, or GIS technology might be used to provide sample or census data on study area plots or pixels. For example, Stinnett and Klebenow (1986) examined cover type selection by California quail (*Callipepla californica*) by classifying flushes observed during ground surveys into cover types. Maps and aerial photography were partitioned into the respective cover types. Erickson et al. (1998) studied habitat selection by moose (*Alces alces*) using SP-A on the Innoko National Wildlife Refuge in Alaska. Aerial line transect surveys were conducted to obtain a sample of locations used by moose during daytime in winter. The assumption was that locations were a random sample of those selected by the population of moose. The study area consisted of river corridors

---

BOX 17.1. SAMPLING PROTOCOLS FOR RESOURCE UNITS

Sampling protocol (SP)	What is sampled?
SP-A	Study area units are randomly sampled, and used units are randomly sampled.
SP-B	Study area units are randomly sampled, and unused units are randomly sampled.
SP-C	Used units are randomly sampled, and unused units are randomly sampled.
SP-D	Available units are randomly sampled and classified as used or unused.

on the refuge. Predictor variables were derived from a GIS for circular buffers centered at the moose locations (used units) and for circular buffers centered at the gird intersections of a systematic sample of points in the river corridors. A **model** called a **resource selection function (RSF)** was fitted to the data to predict the relative probability of selection of a given point (conditional on the specific time period, population density, and sampling protocol).

Design II

In some cases, the study area is defined for a population of animals, but **individual animals are identifiable,** and habitat units selected can be recorded for unique animals. Four examples of this design are provided:

1. A random sample of animals is obtained from the population of interest and is **uniquely identified,** so that a sample of habitat units selected by a given animal can be recorded. Also, a sample of study area units is selected. Predictor variables are measured on units selected by the ith animal and on the sample of study area units. Predictor variables might be measured in the field or from aerial photographs, GIS data, or maps.

2. Pendleton et al. (1998) studied habitat selection by northern goshawks (*Accipiter gentilis*) in southeast Alaska. Goshawks were trapped and **radiomarked,** and their **home ranges** were measured on the assumption that birds captured provided a random sample. A Design II study would involve comparison of proportions of resource types and other variables in home ranges to the same variables measured on similarly sized regions randomly sampled from the entire study area.

3. Roy and Dorrance (1985) compared habitat in coyote (*Canis latrans*) home ranges with the habitat in the entire study area.

4. Prevett et al. (1985) compared food selected by individual snow geese (*Chen caerulescens*) and Canada geese (*Branta canadensis*) with random samples of food from the entire study area.

Design III

In this design, **individuals are uniquely identified** (usually by radiotransmitters) or collected for stomach samples. Data from each animal are analyzed to provide a **RSF for each animal.** Two examples of this design are provided:

1. The animals in a sample are radiomarked, and the **relocations** of an animal provide a sample of resource units selected by that animal. Resource units in an animal's home range also are sampled. Predictor variables are measured on each sampled unit to contrast used units with units in each home range. This type of design was used by Pendleton et al. (1998) to study habitat selection by northern goshawks in southeast Alaska. Habitat variables were measured using a systematic point grid in the minimum convex polygon use area (i.e., estimated home range) of each radiomarked goshawk. A sample of locations selected by the birds was obtained from **radiotracking data.** Habitat selection was evaluated by comparing variables measured at used locations to variables measured on the sample of points from the home ranges.

2. Individual animals might be collected, and stomach analysis performed on each. Predictor variables are measured on prey or food items (e.g., species, color, or size). These data are then compared to measurements from a sample of prey or food items collected in a certain size buffer surrounding the collection site.

Comparison of Designs

Design I has been the most commonly used in the past; however, it has the least specific information. Inferences can be made to resource selection by the **population** of animals, assuming that study design and sampling protocol adequately sample habitat units selected by the population and in the study area.

Designs II and III tend to be preferred, because data are obtained on individual animals and their habitat or food selection. Thus, variation in habitat selection among gender or age classes can be analyzed. However, cost of a resource selection study usually increases when individual animals are captured, marked, and tracked. In Designs II and III, **implicit assumptions** are made that a random sample of animals is obtained. In practice, when trapping or otherwise capturing animals, fulfilling this assumption is difficult. However, every effort should be made to spread the sample of animals over the population. The design then becomes a sequential process: first, selection of a sample of animals and second, selection of samples of used and study area resource units for each animal, followed by measurement of predictor variables on the selected units.

There are several advantages of Designs II and III over Design I. The relocations (used units) of radiomarked animals might be close together in time and space, and hence, they may be **dependent** as opposed to a random sample of units selected by the animals. If data were pooled among animals into a Design I analysis, the mathematical requirements of the analyses might not be satisfied because of the **lack of independence** of locations of used units. Similarly, in food studies, the selection of consecutive prey items might fail to meet the assumption of spatial and temporal independence. When resource selection is analyzed for individual animals in Designs II and III, **all data** are typically analyzed, regardless of whether observations are independent. The next stage, where we consider variation from animal to animal and make inferences to the population of animals, depends only on the assumption the sample of animals was collected by a random procedure, not on the fact that data on individual animals might have lacked independence. The **sample size** is the number of animals in the study (White and Garrott 1990). Thus, inferences rely on random sampling of animals rather than on the assumption the correct statistical model for dependent observations of animals is being used.

Models for RSFs

Resource units usually are defined as individual items of food (with respect to food selection) or blocks of land or points on the landscape (with respect to habitat selection). Each resource unit is characterized by the values it possesses for certain **predictor variables** (also called **independent variables** or **covariates**) $X_1, X_2, \ldots, X_p$, representing such characteristics as size and color of food items, distance from water, and habitat type of habitat units. Three mathematical functions (i.e., models or curves; McDonald et al. 1990, McDonald and Manly 2001) are involved in studies of resource selection. They can be illustrated with a simple **hypothetical example** with one X variable (Fig. 17.1).

The curve labeled **available** in the figure represents the probability distribution function (smooth histogram) of the predictor variable X for the set of units in the study area. For this hypothetical example, it was assumed that X has approximately a normal distribution with mean $\mu = 20$ and variance $\sigma^2 = 2.5$. The function labeled **used** represents the probability distribution function of the predictor variable X on those units selected by the animal. The distribution of X on the used units is approximately a normal distribution with mean $\mu = 22$ and variance $\sigma^2 = 1.9$.

The third curve (Fig. 17.1) is the **RSF**. This function shows **how units must be selected from the study area to produce the distribution of the used set.** If an animal is selecting units from the study area such that a probability of selecting a unit with $X = x$ is proportional to the RSF, this selection will produce the distribution of X shown for the used units. For example, the animal must select units with $X = 25.07$ with about twice the probability of units with $X = 23.68$ to produce the distribution of used units. The RSF provides a way to **rank the relative importance** of different units. Values of the relative probability of selection can be computed for all habitat units and mapped using GIS technology to show areas that are selected with relatively high, medium, or low probability.

Our example (Fig. 17.1) uses approximate normal distributions to illustrate the distributions of X for used and unused units and for used units; it uses a unique formula from Manly (1985:61) to calculate the selection function. However, normal distributions rarely fit all predictor variables that may be of interest in resource selection. A general theory allowing multiple predictor variables, based on exponential models and other special statistical methods has been developed (Manly et al. 2002, Johnson et al. 2006).

The RSF can provide the **relative probability of selection** of different habitat units (food types) among those in the study area. One application of RSFs is to produce **maps** of the relative probability of selection in either 3 dimensions over the study area or as contour lines showing the relative probabilities of selection. Erickson et al. (1998) produced this type of map showing areas with relatively high, medium, and low probabilities of selection by moose on the Innoko National Wildlife Refuge, Alaska, in winter 1994 and 1996. Predictor variables included the percentages of different vegetation types in buffers surrounding used points (as evidenced by sighting a moose group at the point during an aerial survey) and in buffers surrounding a large systematic sample of points on grid lines. A GIS was used to obtain the predictor variables in the buffers.

Assumptions for Estimation of RSFs

There are **6 major assumptions** made when estimating RSFs.

1. The researcher is interested in **ranking habitat units (food units) in a study area** based on the relative probabilities of selection by animals. In other words, an RSF is unique for the study area, and interest is in knowing the relative probabilities that units from the study area are selected as measured by the sampling design. If the shape of the distribution of the predictor variables, Xs, changes for different study areas, the RSF must change to generate the same distribution for used units (Fig. 17.1).

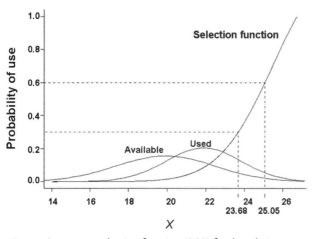

Fig. 17.1. A resource selection function (RSF) for the relative probability of use of resource units with a single variable X. The "available" curve is approximately a normal distribution with a mean of 20 and variance of 2.5. The "used" curve is approximately a normal distribution with a mean of 22 and variance of 1.9. These 2 distributions define the RSF. *Adapted from McDonald and Manly (2001).*

2. **Predictor variables** to be measured on sampled units, $X_1, X_2, \ldots, X_p$, are **correlated** with the **probability of selection** and do not change appreciably during the study period. Given samples from the study area and from the used units and a good set of predictor variables, an approximate model can be constructed to predict the relative probability of selection for that specific location, time, and population.

3. **Measurement errors** for predictor variables, $X_1, X_2, \ldots, X_p$, are relatively **small** in comparison to variation from unit to unit.

4. In **Design I and II** studies, animals in the population have **equal access** to all units in the study area. If this is not the case (e.g., if animals are territorial), then Design III should be used.

5. **Study area units are randomly sampled.** In practice, systematic location of grid lines and selection of units at line intersections, or uniformly spaced units along a linear feature (e.g., a river) can be used to sample habitat units. **Systematic samples** almost always provide better coverage of the landscape than do pure random samples for a given sample size, but see Manly (2009) for guidance.

6. **Selected units are randomly sampled,** or the probability of detection of selected units is approximately constant. If the probability of detection of use of a sampled unit is highly dependent on the vegetation type or other predictor variables, then other, more complex, study designs are required (MacKenzie et al. 2006).

DEFINING STUDY AREAS AND MEASURING HABITAT SELECTION

Selection and use of a particular area or unit of habitat by an animal are the result of **proximate** and/or **ultimate** predictor (independent) variables (Partridge 1978). **Proximate** variables are those features used as cues when an animal evaluates a site (habitat unit). They may include structural features, such as understory cover, canopy height, or slope. Other potential structural features are available from GIS data, such as density of roads in the unit or distance to water. The presence or absence of other animals that may act as competitors or predators also may influence habitat selection. Animals may use such features as cues, but they may not be the same as the variables that have resulted in evolutionary associations between animals and habitat. **Ultimate** variables are those parameters that affect how successful an animal is in a particular habitat. An individual's abilities to reproduce, obtain food, and avoid predators are examples of ultimate variables that influence habitat selection. **Studies of habitat selection usually involve measure of proximate variables and food availability.** However,

with adequate data on ultimate variables, for example, measurements of predator abundance and competition, a more complete understanding of habitat selection can be obtained.

The relationship between a habitat feature being measured and its biological link to the animal often is clear. For example, understory stem density frequently is used as an index of escape cover for small or medium-sized mammals, such as snowshoe hares (*Lepus americanus;* Litvaitis et al. 1985a). However, in other instances, the animal–habitat relationship may be less obvious, such as using the abundance of snags as an index of insect availability for pileated woodpeckers (*Dryocopus pileatus*). In this situation, a structural feature is correlated with a proximate predictor variable. Because it is much easier to inventory and manage snag abundance than insect abundance, one is tempted to use this association when investigating woodpecker–habitat relations. However, the relationship between snags and insects should be verified to ensure that subsequent research conclusions are reliable.

Techniques for Detection of Habitat Selection

Direct and indirect methods have been used to detect wildlife habitat selection. **Direct methods** include observation, capture, and radiotelemetry, whereas **indirect methods** are dependent on some evidence of animal activity in an area or specific site (e.g., bed sites, browsed twigs, feces, nests, or tracks). These measures may be used to detect use of units along systematic transects, in a small-mammal trapping grid, or with other sampling designs appropriate to the animal of interest (Fig. 17.2). Morrison et al. (2008) provide a good introduction to basic sampling procedures in wildlife study design; Hurlbert (1984) and Williams et al. (2002a) provide specific advice on study design.

Direct Methods for Detection of Habitat Selection

Direct observations of animals may allow economical sampling of a large segment of the study area and permit activities to be distinguished within vegetation types (e.g., Biggins and Pitcher 1978, Stinnett and Klebenow 1986, Erickson et al. 1998). Collection of data can be combined with aerial or other survey procedures. Problems to consider are **differential visibility** among vegetation types and the difficulty of recording observations during nocturnal periods.

Advantages of animal capture include being able to examine individuals for age, gender, and other characteristics (e.g., Parren and Capen 1985). Capture procedures can be combined with mark–recapture statistics to estimate abundance. However, differential vulnerability to capture due to age and gender or other factors may bias results, and attractants may cause animals to select vegetation types that are normally not selected.

Radiotelemetry also can be used to measure habitat selection (Nams 1989; Erickson et al. 2001; Sawyer et al. 2007, 2009). Advantages include being able to examine individuals

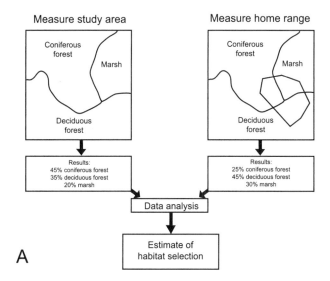
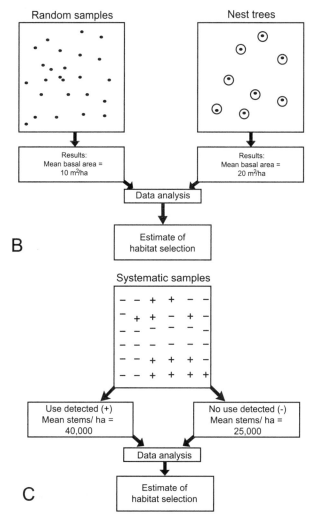

of known age and/or gender and other characteristics for habitat selection. Animals can be located multiple times throughout the day or night and seasons. Habitat selection for important components (e.g., den site selection or roost sites) can be studied. Use of GPS radiotelemetry allows detection of locations of individuals multiple times during a 24-hour period, thus allowing study of resource selection during different periods of the day. However, radiotelemetry is relatively expensive, usually resulting in small sample sizes of animals from the population.

Indirect Methods for Indication of Habitat Selection

Detection of tracks (e.g., Litvaitis et al. 1985a, Thompson et al. 1989) allows one to potentially sample units economically and detect use in a large sample of units in a short time by all segments of the populations. However, the procedure often suffers from lack of good tracking conditions (e.g., uniform snow) and differential visibility of tracks in different vegetation types.

Detection of pellet groups or scat measures selection by all segments of a population (Collins and Urness 1981, Orr and Dodds 1982). On cleared plots, information on seasonal selection is obtained and potentially can be combined with deposition rates to estimate density of animals. However, defecation rates often are unknown or vary with habitat type and activity, and decomposition rates may vary among habitats. Visibility and detection of pellet groups also may vary with vegetation type.

Browsing or feeding may provide evidence of use of habitat units by all segments of a population. Evidence of browsing or feeding may provide additional information on food habitats and can be combined with studies of carrying capacity. Potential biases include competition for the same food by other species, and food species must be present before use of a unit can be documented.

Techniques to Define Study Areas

Delineation of study areas can influence interpretation of habitat selection and subsequent management recommendations. It is often important to model resource selection in **multiple study areas,** because the vegetation types available to the animals will vary, and interest is in how selection changes when vegetation changes. Assessing **availability** of

Fig. 17.2. Representative methods to examine habitat use patterns. (A) Design II and sampling protocol A (SP-A): available habitat is inventoried and compared to the composition of an animal's home range. (B) Design I and SP-A: random samples are compared to characteristics of a sample of sites where use has been detected, such as nests or roost sites. (C) Design I and SP-C: systematic plots (or points) are established and features are compared between sites where use was detected (via captures, tracks, feces, radio relocations, etc.) and sites where use did not occur.

vegetation types in a study area from the **point of view of an animal** is not possible; any effort to estimate availability in a study area is naturally beset with problems (Chesson 1978, Jaenike 1980, Johnson 1980). In studies of habitat selection, biologists often choose administrative units, such as a parks, forests, or refuges, to arbitrarily represent the study area because of the obvious management ramifications. The study area might be much larger than the area occupied, which could produce biased results. If study animals are not radiomarked, they might select a larger area than anticipated, resulting in a biased interpretation of selection.

Issues with the definition of the study area are simplified if the scale of selection is clearly delineated. Johnson (1980) defined **first-order selection** as the selection of physical or geographic range of a species. Few, if any, habitat selection studies are of first-order selection. **Second-order selection** results in the home range of an individual or social group in the physical or geographical range of a species. Second-order selection is of interest in many habitat selection studies and will typically require radiomarking of individuals or social groups (Design II and III studies). **Third-order selection** is specific selection of sites in the home range. Typically, the home range of an individual or social group is calculated (e.g., by the minimum convex polygon method; Williams et al. 2002a), and all units in the home range define the study area. Finally, Johnson (1980) defined **fourth-order selection** as the actual procurement of food items from those available at a feeding site identified by third-order selection. In this case, paired sample data are typically collected: contents of scat or stomach samples are paired against samples of food items collected from the site. It may be advisable to study selection on several scales—for example, third-order selection of local cover types and second-order selection of regional landscapes (Steventon and Major 1982).

Knowing something about the habitat associations of the animal being studied is essential before study area boundaries are delineated for second-order selection studies or for Design I studies where individuals are not radiomarked or otherwise identifiable. For example, including open fields as study units for forest-interior songbirds (e.g., ovenbird [*Seiurus aurocapillus*]) probably would yield 0.0 as the estimated relative probability of selection for open fields. Inclusion of open fields in the analysis would not be detrimental, but the results would be trivial, because we already know that such areas are rarely selected by these species. Inclusion of vegetation types rarely (if ever) selected by members of a population in a study area will not unduly influence the results in an analysis of resource selection that uses estimation of selection ratios or RSFs (Manly et al. 2002). However, delineating study area boundaries should not be so restrictive that potentially important vegetation types are eliminated.

The distribution and size of cover types in a study area also can influence our ability to detect selection patterns. Porter and Church (1987) illustrated this problem by comparing an area where vegetation types were regularly distributed with an area where they were clumped. In areas that had regular or random distributions of cover types, the delineation of the study area had little influence on the models for habitat selection. However, if cover types were aggregated, delineation of study area boundaries substantially influenced the analysis of selection.

The **guidelines** presented below may be helpful when study area boundaries are delineated for second-order selection or Design I studies, but each study is unique.

1. Size of the study area should be substantially larger than the home range of the study species.

2. Numbers of study animals, groups, or social units present on the study area should be, as far as possible, adequate for study.

3. Opportunity should exist for independent locations of animals or independent location of home ranges in the study area (i.e., as close as possible to an unbiased random sample of sites selected by animals or a random sample of home ranges).

4. Study area boundaries should be chosen with consideration of the biology of the animal. Physical barriers, such as rivers or mountain ranges, might make better boundaries than an arbitrary (geopolitical) straight line on a map.

Vegetation types and other landscape features of units in a study area often are measured directly from aerial photographs, maps, satellite images, or data layers in a GIS database. In these situations, we are dealing with known quantities that, although they have measurement error, **do not have sampling error** associated with them. Biologists often have access to vegetation type maps or GIS layers produced for multiple-use planning, such as timber type and plant association maps produced for national forests or private forest industry lands. Although the inventory may cover the area of interest, various approximations and measurement errors are part of these products (e.g., smallest forest stand inventoried is often >1 ha), and an understanding of these limitations is required prior to integration with wildlife data. Partitioning the study area (e.g., home ranges) into discrete grid cells or pixels (e.g., 100 m × 100 m) provides an approach to use data from maps or GIS. Cells are categorized according to the number of captures, observations, or other index of use by study animals. Habitat features then are measured in each cell or a subset of cells (or measured on a data layer of a GIS) with comparisons based on intensity of use or a sample of units in the study area versus a sample of selected cells (Porter and Church 1987, Erickson et al. 1998, Manly et al. 2002, Johnson et al. 2006).

At times, the study area may be large, and the key issue is to randomly locate samples of units throughout the study area. If units are sampled from a GIS, large sample sizes (i.e., ≥5,000) can be easily obtained using computer software.

Sampling errors can essentially be ignored for variables measured with this intensity.

Not only are specific attributes of habitat (amount and size) important, but also **juxtapositions** among habitats, **variability** between habitats, or **spatial pattern** may be influential in affecting habitat suitability (Otis 1997, 1998). One of the most important components of habitat structure is the **spatial heterogeneity** or amount of **edge habitat** present. Spatial heterogeneity not only integrates the absolute values of the vegetation or physiography, but also their variations in space (Wiens 1976). Many birds and mammals rely on >1 vegetation type for feeding, mating, nesting, or denning. A specific vegetation type may have an abundance of one resource, such as food, and not be selected by an animal because it lacks or is distant from sites that provide another necessary resource, such as cover. Additionally, distribution of patches and edge may have ramifications on habitat suitability, such as influencing a predator's ability to stalk or ambush prey. Thus, some measure of vegetation type variation is important. The availability of multiple data layers in a GIS allows easy measurement of some of these important variables (Hepinstall and Sader 1997; Erickson et al. 1998, 2001). For example, the density and/or length of edge between vegetation types, density of roads, or number of contour lines crossed by transects radiating from a used unit might easily be measured in a buffer surrounding a used site or a randomly sampled study unit in a GIS. Many formal methods based on measured variables have been developed to access heterogeneity at a variety of scales using ranges, variances, and coefficients of variation (Williams et al. 2002a, Morrison et al. 2008, Zar 2010), Wiens's heterogeneity index (Wiens 1974, Rotenberry and Wiens 1980), juxtaposition (Heinen and Cross 1983), spatial diversity indices (Thomas et al. 1979, Mead et al. 1981, Heinen and Cross 1983), interspersion index (Baxter and Wolfe 1972), and land surface ruggedness index (Beasom et al. 1983).

Heterogeneity also can be expressed in vertical and horizontal dimensions. Layering of vegetation in plant communities is a common way to express vertical heterogeneity. Techniques, such as those that use a vertical density board, have been used to describe heterogeneity (De Vos and Mosby 1969, Nudds 1977, Noon 1981, Robbins et al. 1989). Biologists working at the landscape level evaluate such parameters as habitat patch dispersion and corridor development (Forman and Godron 1986, Otis 1997). However, the biological interpretation of these characteristics may be less intuitive than that of other habitat features and should be addressed before information is collected.

MEASURING FOOD SELECTION AND AVAILABILITY

Abundance and distribution of food resources are among the major environmental features that influence habitat selection. Because food intake relates to energy needs, reproduction, and ultimately to survival, understanding food selection is a fundamental component of behavioral ecology.

Techniques to Measure Food Availability in Resource Selection Studies

Wildlife **food abundance** can be estimated for an area by measuring the annual production of herbaceous plants, woody stems, fruits, and seeds, or by assessing the abundance of potential prey. Measures of food abundance on study units can serve as predictor variables in models for habitat selection; however, these methods can be time and labor intensive and may only be applicable for use on small-scale resource selection studies. A complication that requires careful consideration is that food **abundance** might not be directly related to **availability**. Availability suggests that a food resource is both accessible and usable (Morrison et al. 1992). Access to food resources can vary with weather or by the presence of predators or competitors. Snow can make forage unavailable to herbivores, or alternatively, snow can alter the vulnerability of prey for carnivores (Halpin and Bissonette 1988, Fuller 1991). In resource selection studies, effects that modify food abundance might be modeled using covariates, such as snow depth or presence of a predator or competitor.

Grasses and Forbs

Clipping and weighing dried samples of aboveground vegetation is the most accurate, but most time consuming, technique for measuring predictor variables for availability of herbaceous plants. Many techniques have been developed to more rapidly estimate vegetative biomass and to avoid destructive sampling. These include the capacity of the grassland vegetation to obstruct vision, as measured by Robel range pole methods (Robel et al. 1970), estimating biomass in small quadrats (Shoop and McIlvain 1963), and estimating percentage cover by species in small sample plots (Daubenmire 1958). Pin intercept methods have been developed using sampling frames containing rows of pins that are pushed through the vegetation. A coarse method often used in rangelands involves sampling the ground cover below a point on the observer's boot. While walking through an area, at regular intervals, the species of plant beneath the tip of the boot is recorded (Owensby 1973, Cook and Stubbendieck 1986).

Browse

Predictive equations have been developed that relate measures of shrub size to forage production (Lyon 1968b, Telfer 1969b, Bobek and Bergstrom 1978) and hence, estimation of the amount of browse in a study unit as predictor variables. Specific equations must be estimated for each species and for individual study sites. The twig count method estimates biomass of browse by calculating the average weight of edible material in a single twig and multiplying that value times the number of twigs (Shafer 1963). A sample of previously browsed twigs is used to estimate the average browsing diameter for each forage species. Mass of browsed twigs is then estimated from a collection of twigs clipped to the size of the average browsed twig. Densities of twigs can be estimated from counts on circular plots (Shafer 1963) or belt transects (Irwin and Peek 1979), and browse biomass is calculated per unit area. Modifications of this technique include development of equations that use unbrowsed twig length or basal diameter to estimate twig mass (Basile and Hutchings 1966, Telfer 1969a).

Fruits and Seeds

Fruits and seeds from low-growing herbs and shrubs can be counted and averaged per plant and summed over an area to estimate biomass in study units. Similarly, hard mast crops, such as acorns, can be collected in funnel traps that sample an area under the canopy (Gysel 1956). Although traps usually prevent animals from taking mast once it has fallen from the tree, information on production can be biased if seeds are consumed before falling to the ground.

DESIGN CONSIDERATIONS AND ANALYSIS

General Considerations

Frequently, investigators have focused on modeling selection of habitat units during specific periods of time of day or behavior: feeding, resting, or rearing young (Stinnett and Klebenow 1986). The most common study design (SP-A) would involve collection of a sample of selected units, such as locations of radiomarked animals, and contrast those units with a sample of units in the study area. In this case, the models yield estimates of the relative probability of selection (i.e., information to the effect that one unit might be selected with twice or 3 times the probability that another unit is selected).

A valid probability sampling procedure (Morrison et al. 2008) is used to subsample a variety of features at each study site, cell, or home range (James and Shugart 1970, Dueser and Shugart 1978, Fridell and Litvaitis 1991). The features selected to describe a sampled unit (e.g., litter depth, understory stem density, canopy closure, distance from roads, aspect, or slope) are assumed to represent or be highly correlated with the variables used by animals to evaluate a site and often include some measurement of food abundance, cover, and structural characteristics (Fig. 17.3).

Measurement of Landscape Variables Using a GIS

A landscape is defined as a mosaic of habitat patches in which a patch of interest is embedded (Dunning et al. 1992). **Landscape variables** include patch size, patch context, and other habitat characteristics (e.g., density of roads, proportion of habitat types, or density of edge between habitat types in a buffer [circle] centered at the site). With development of GIS technologies, many possibilities exist to measure landscape variables (Erickson et al. 1998, Otis 1998)

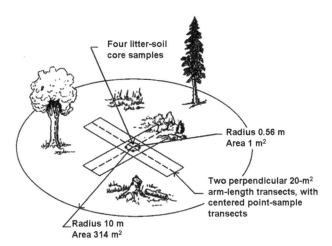

Fig. 17.3. An example of nested plots used to sample ground litter, understory stem density, and overstory composition. Modified from Dueser and Shugart (1978).

that may be important for site selection. Because the actual habitat features influencing selection are not known, measuring several features is appropriate (e.g., Rice et al. 1984). However, as the number of features measured becomes large, the chance of detecting spurious relationships also increases. Therefore, the list of features to be sampled should be limited to those based on biological considerations for the relationships between animals and their habitats (Green 1979, Anderson 2001).

Standard Statistical Analyses

Analytical methods for resource selection studies usually involve comparison of characteristics of **samples or censuses of used units** and **samples of units from the study area.** The first step should involve graphical and descriptive comparisons of the distribution of the predictor variables (also called covariates) that describe each unit for the samples being compared (e.g., used versus study area). Patterns described in these analyses probably will be apparent in any inferential analyses (e.g., hypothesis testing).

Standard statistical procedures (Morrison 2001, Manly 2009, Zar 2010) are appropriate for comparison of used units and study area units or for comparison among other stratifications of the sites or home ranges. Measurements of use also might be partitioned into categories (e.g., rarely used, occasionally used, or frequently used). Comparisons also may be made between portions of home ranges of study animals (activity core versus outside the core). Univariate statistical tests and confidence intervals can be computed on individual variables and multivariate tests used on several variables simultaneously, following standard statistical procedures. In this section, we concentrate on statistical modeling procedures that have been developed specifically for study of resource selection by animals.

Several older hypothesis-testing techniques, such as **chi-square analyses** (Neu et al. 1974, Byers et al. 1984), Fried-

Table 17.1. Characteristics associated with common statistical analysis methods of resource selection (Y = yes, N = no)

Characteristics	Neu et al. (1974)	Johnson (1980)	Friedman (1937)	Compositional Aebischer et al. (1993)	Logistic regression	Log-linear modeling	Discrete choice
Applicable to Design I: individuals not marked and population-level availability	Y	N	N	N	Y	Y	Y
Applicable to Design II: individuals marked and population-level availability	N[a]	Y	Y	Y	Y	Y	Y
Applicable to Design III: individuals marked and availability defined for each animal	N[a]	Y	Y	Y	Y	Y	Y
Temporal independence of relocations assumed	Y	N	N	N	Y/N[b]	Y/N[b]	N
Independence among animals assumed	Y	Y	Y	Y	Y	Y	Y
Categorical covariates allowed (e.g., gender or subgroups)	N	N	N	Y	Y	Y	Y
Continuous covariates allowed (e.g., distance to roads)	N	N	N	N	Y	N	Y
Sample of animals assumed representative of population; inferences are to average selection of larger population from which sample was obtained	Y[c]	Y	Y	Y	Y[d]	Y[d]	Y

Adapted from Erickson et al. (2001).

[a] Neu et al. (1974) method could be applied by pooling data, but this is not recommended.
[b] Independence important if data are pooled across animals, not important when animals are used as the units of replication.
[c] Assumes the indices of use (e.g., observed tracks) are representative of the population of indices.
[d] Inference is to the larger population if animals are the units of replication (i.e., data are not pooled).

man's test (Friedman 1937, Conover 1999) and the similar Quade (1979) method, compositional analysis (Aebischer et al. 1993), and the PREFER method (Johnson 1980), have been extensively reviewed (Alldredge and Ratti 1986, 1992; Alldredge et al. 1998; Table 17.1). However, these methods are being applied less frequently now because of the availability of such techniques as **discrete-choice modeling** (also known as case control or conditional fixed effects; Cooper and Millspaugh 2001, Manly et al. 2002) and **generalized linear modeling** approaches (e.g., log-linear modeling; Heisey 1985, Manly et al. 2002). This shift in emphasis is, in part, due to increased use of GIS technology and because most earlier methods only considered discrete (categorical) resource types (e.g., land cover types). The latter methods can consider multiple continuous and discrete covariates (e.g., log-linear and discrete-choice modeling). These techniques have become more readily available in statistical software computer packages. Many of the most common techniques for analyzing resource selection data have been previously reviewed and summarized (Alldredge et al. 1998, Manly et al. 2002, Strickland and McDonald 2006). White and Garrott (1990) and Erickson et al. (2001) provide in-depth descriptions of many of the approaches for study of resource selection in radiotelemetry studies. Sawyer et al. (2007, 2009) present the **state-of-the-art analysis methods** for development of resource selection models based on GPS radiotelemetry.

Specialized methods have been used for analysis of resource selection data, and their assumptions and applications vary (Table 17.1). We present a summary of the chi-square analysis because of its historical interest and analyze the Neu et al. (1974) data as an example. However, we **emphasize use of RSFs** as the most comprehensive procedure for analysis of resource selection in a specific study area and time period. We show by example that simple selection ratios computed for the Neu et al. (1974) data arise as odds, or relative probability of use of the vegetation types, when a RSF is specialized to the case of a single categorical variable (e.g., vegetation type).

Chi-Square Analyses for Categorical Data

For data based on resource categories (e.g., vegetation types, food types, or categorized continuous predictors), a chi-square test can provide an omnibus answer to the question: **is there evidence of selection or not?** This test is appropriate when individual observations of selected units are considered independent. Chi-square analyses appear most appropriate in Design I studies and are not appropriate when several animals have multiple (dependent) relocations. This approach was first considered by Neu et al. (1974) and later by Byers et al. (1984). Two statistics that have approximate chi-square distributions are most commonly considered. Suppose there are C resource categories, and that some number of animals (or observations of a single animal) have been observed in each category. Further, suppose the null hypothesis of no selection leads to an expected number of observations in each category. The form of test statistic that is most commonly used for this purpose is the chi-square statistic,

$$X_p^2 = \sum_{i=1}^{c}(O_i - E_i)^2/E_i,$$

Table 17.2. Chi-square analysis of 117 observations (based on tracks) of use by moose

Location of tracks	Percentage of area	Use	Expected use	Pearson chi-square
Interior of burned	34.0	25	39.78	5.49
Edge of burned	10.1	22	11.82	8.77
Edge of unburned	10.4	30	12.17	26.13
Interior of unburned	45.5	40	53.24	3.29
Totals	100	117	Chi-square:	43.69
			P-value:	0.001

From Neu et al. (1974).

where O_i is an observed sample frequency, E_i is the expected value of O_i according to the hypothesis being considered, and the summation is over all resource categories.

A caution against small observed and/or expected numbers applies (Zar 2010). Tests and confidence intervals on data with one or more categories with low observed or expected numbers may be suspect. Eighty percent of all cells should have expected values of ≥5; otherwise, the standard chi-square distribution may not be an accurate approximation to the sampling distribution of the statistic.

We provide an example of Design I, SP-A, using the popular data set provided by Neu et al. (1974) of the proportions of 4 vegetation types in a study area containing moose (Table 17.2). These proportions are contrasted to the proportions of observations of moose tracks in each of the vegetation types. The chi-square analysis provides a strong indication that selection is occurring.

MODELING RESOURCE SELECTION

Selection Ratios for Categorical Data

The **selection ratio** for a given resource category is the ratio of the proportion used to the proportion of the category in the study area (Manly et al. 2002). If the **ratio is close to 1**, there is evidently **no selection**. Values <1 indicate selection against that category; large values indicate selection for the category.

To differentiate between the 2 proportions of interest, we represent the true proportion of vegetation type i in the study area by π_i and the true proportion selected by p_i. Let $n_{i,u}$ be the number of animals found in vegetation type i, with total sample size being n_u. Thus, an estimate of p_i is $\hat{p}_i = \frac{n_{i,u}}{n_u}$. The **symbol** w has been used by Manly et al. (2002) for the ratio of proportion selected to proportion available, because it also is known as a **weighting** factor. When π_i is known, $\hat{w}_i = \frac{\hat{p}_i}{\pi_i}$. The selection ratios, $\hat{w}_i$, are often standardized to be between 0 and 1 by using the formula

$$\hat{w}_i = \frac{\hat{w}_i}{\Sigma \hat{w}_{ii}}.$$

Extensive numerical examples, variance calculations, and modifications to these basic formulas are in Manly et al. (2002).

Selection ratios also are applicable to the case of multiple animals with multiple relocations per animal (Erickson et al. 2001, Manly et al. 2002). Making inferences regarding the average selection ratios for a large population of animals can be conducted by averaging individual selection ratios and using the number of animals as the effective sample size, avoiding issues of pseudo-replication (Erickson et al. 2001). This approach assumes the sample of animals is randomly drawn from the population, which may be difficult to approximate.

We now revisit the moose track data (Table 17.2). **Selection ratios** and **standardized selection ratios** are calculated (Table 17.3) with 95% Bonferroni adjusted confidence limits (e.g., Byers et al. 1984, Manly et al. 2002). Based on these confidence limits, one would conclude there is statistically significant selection against the interiors of the burned and unburned areas, with significant selection for the edge of the unburned area. The selection ratio for the edge of the burn, 1.862, indicates selection for this type, but the result is not statistically significant with the conservative Bonferroni procedure, because the confidence interval (0.968–2.755) slightly overlaps the number 1.0. Of course, all conclusions relate to selection as measured by the sampling design implemented by Neu et al. (1974).

The selection intensity (for or against) one resource category can be compared to that of another using tests and confidence intervals for differences in pairs of selection ratios. The number of situations that can arise for the analysis of categorical data is quite large, and the analyses are tedious. In the sections below, we emphasize the **more important case** where multiple continuous and discrete variables are considered for their influences on selection of resources. For example, in the moose track data (Table 17.2), other continuous variables, such as density of roads (km of road/km²) or snow cover, also might affect selection of feeding or resting sites. Manly et al. (2002) provide descriptions of further analyses of resource selection when used units are assigned to categories.

RSF for Categorical Data

Logistic regression is a specialized regression tool for working with multiple continuous and discrete variables (Hosmer

Table 17.3. Statistics for moose data presented in Table 17.2

Location of tracks	Selection ratio	SE[a]	Confidence interval		Standardized ratio
Interior of burned	0.628	0.111	0.350	0.907	0.110
Edge of burned	1.862	0.358	0.968	2.755	0.326
Edge of unburned	2.465	0.388	1.496	3.435	0.432
Interior of unburned	0.751	0.096	0.511	0.992	0.132

[a] *SE* = standard error.

and Lemeshow 1989, Neter et al. 1996). However, logistic regression also can be used to analyze the effect of a single categorical response variable on resource selection (e.g., the effects of the 4 habitat categories [Tables 17.2, 17.3] on selection of locations by moose in Neu et al. [1974]). The technique is presented as a means to analyze selection among categories of a categorical variable and as an introduction to estimation of a RSF when there are multiple variables.

Suppose X is an environmental variable thought to be associated with selection of resources by the animal under study. Details for estimation of the coefficients in the model will be deferred, but for illustration, X could be a simple categorical variable referencing habitat categories. In later examples, the predictor variable X could be a continuous variable, such as distance from features (e.g., water) or density of roads in a unit, used alone or in conjunction with other habitat variables, or it could be a categorical variable other than habitat type. Consider the generic example (Table 17.4) with 4 categories of habitat type: A, B, C, and D. Suppose that category A is chosen as the baseline (the choice is arbitrary), then logistic regression analysis would allow estimation of changes in the **odds** between categories A and B, A and C, and A and D. **Dummy** (or **indicator**) variable X_1 (Table 17.4) indicates a shift from A to B, X_2 from A to C, and X_3 from A to D. There are 4 categories of habitat type; thus, 3 dummy variables are needed, 1 less than the number of categories. Notice the row for category A has every indicator set to zero. Also, the location of the only "1" in any column identifies which habitat that predictor is an indicator for. These are the 2 key features of indicator variable construction. Mechanically, the use of indicator variables leads to a special form of multiple logistic regression with the following model for the RSF or **odds of selection**:

$$odds = \exp(\beta_0 + \beta_1 X_1 + \beta_2 X_2 + \beta X_3) = e^{\beta_0}e^{\beta_1 X_1}e^{\beta_2 X_2}e^{\beta X_3}.$$

In practice, logistic regression will yield numerical values for the coefficients, the βs, in the model. The restricted set of chosen values for the Xs (Table 17.4) lead to the list of possible model expressions and odds ratios. Careful examination of the material (Table 17.4) will aid understanding the use of indicator variables for logistic regression. For example, the odds for category B are obtained by substitution of $X_1 = 1$, $X_2 = 0$, and $X_3 = 0$ into the above equation, and the odds ratio for selection of type B to A is $e^{\beta_0}e^{\beta_1}/e^{\beta_0} = e^{\beta_1}$.

Now assume that X is a continuous variable, such as distance (km) to water. Logistic regression gives a model for the change in the odds (defined as the ratio of probability of use to probability of "not-use") of habitat use according to the following model equation:

$$odds = \exp(\beta_0 + \beta_1 X) = e^{\beta_0}e^{\beta_1 X}.$$

In logistic regression, the impact of changing X to $X + 1$ (e.g., to increase the distance to water by 1 km) is in the ratio of the odds for each value of X:

$$odds\ ratio = \frac{e^{\beta_0}e^{\beta_1(X+1)}}{e^{\beta_0}e^{\beta_1 X}} = e^{\beta_1}.$$

For example, if $\beta_1 = -0.41$, the odds ratio would be $e^{-0.41} = 0.66$, and an increase of 1 km in X (distance to water) is associated with a 33% decrease in the odds of use. If the coefficient were positive for some variable, say, X = density of shrubs, and $\beta_1 = 0.41$, the odds ratio would be $e^{0.41} = 1.5$, and an increase of one unit in shrub density would be associated with a 50% increase in the odds of use.

Estimation of a RSF for Categorical Data Using Logistic Regression

We now reanalyze the moose data (Table 17.2) using logistic regression. We present this analysis to illustrate the relationship between the selection ratios and odds ratios from logistic regression, although logistic regression is generally not used when considering only one categorical variable.

General Theory for Modeling and Estimation of a RSF

We illustrate the general theory for estimation of a RSF for multiple variables by defining 3 dummy variables to model the selection among the 4 vegetation types (Table 17.2). Let $w(X_i) = w(X_{i1}, X_{i2}, X_{i3})$ denote the RSF, that is, the odds or relative probability that a unit with covariate vector $X_i = (X_{i1}, X_{i2}, X_{i3})$ is selected, where the dummy variables are coded (Table 17.5). Given data from SP-A (i.e., a sample of points in the study area and a sample of selected points), we assume the RSF can be modeled by the **exponential function**

$$w(X_i) = e^{\beta_1 X_{i1} + \beta_2 X_{i2} + \beta_3 X_{i3}}.$$

Then conditional on the observed sample sizes, probability of use given X_i is (use $| X_i$) = $e^{\beta_0 + \beta_1 X_{i1} + \beta_2 X_{i2} + \beta_3 X_{i3}}/(1 + e^{\beta_0 + \beta_1 X_{i1} + \beta_2 X_{i2} + \beta_3 X_{i3}})$.

This is the form of a logistic regression function, and we can now examine a simulated sample of available points (coded as $Y = 0$ for the available points [dependent variable] in Table 17.5) and the sample results for the 117 used points (coded as $Y = 1$ for the dependent variable; from Table 17.2). Any **computer software package** for fitting a logistic regression function can be used to obtain estimates of the coefficients and their standard errors. Those estimates can be placed in the exponential function

Table 17.4. Indicator variables for 4 categories, with A chosen as the baseline category

Category	X_1	X_2	X_3	Model	Odds ratio (to A)
A	0	0	0	e^{β_0}	1
B	1	0	0	$e^{\beta_0}e^{\beta_1}$	e^{β_1}
C	0	1	0	$e^{\beta_0}e^{\beta_2}$	e^{β_2}
D	0	0	1	$e^{\beta_0}e^{\beta_3}$	e^{β_3}

Table 17.5. Independent and dependent variables coded for entry into a logistic regression computer software package. The sample of used points is coded $Y = 1$, based on the available sample sizes (see Table 17.2). The simulated sample of 20,000 available points was computed to yield the exact proportions of habitats in the study area (Table 17.2); these points are coded $Y = 0$.

Location of tracks	Depending variable Y	Sample size	Independent variables			
			X_{i0}	X_{i1}	X_{i2}	X_{i3}
Interior of burned	1	25	1	1	0	0
Edge of burned	1	22	1	0	1	0
Edge of unburned	1	30	1	0	0	1
Interior of unburned	1	40	1	0	0	0
Interior of burned	0	6,796	1	1	0	0
Edge of burned	0	2,015	1	0	1	0
Edge of unburned	0	2,073	1	0	0	1
Interior of unburned	0	9,114	1	0	0	0

$$w(X_i) = e^{\beta_1 X_{i1} + \beta_2 X_{i2} + \beta_3 X_{i3}}$$

to obtain the RSF, ignoring the constant β_0. Note the coefficient β_0 is not used in the RSF for a technical reason (the fraction of used points in the sample is unknown). It is important to understand the **RSF** is given by equation

$$w(X_i) = e^{\beta_1 X_{i1} + \beta_2 X_{i2} + \beta_3 X_{i3}}$$

and not by the logistic regression function

$$w_i(X_i) = e^{\beta_0 + \beta_1 X_{i1} + \beta_2 X_{i2} + \beta_3 X_{i3}} / (1 + e^{\beta_0 + \beta_1 X_{i1} + \beta_2 X_{i2} + \beta_3 X_{i3}}).$$

Once the logistic regression equation is fitted, values for the coefficients are estimated, and the RSF, $w(x_i)$ can be evaluated numerically.

Application of the General Theory to Categorical Data

To illustrate the estimation of a RSF for a single categorical variable, we assume the vegetation types in Neu et al. (1974) are in a GIS and that a random or systematic sample of 20,000 points was placed across the area. To that end, we assume that 34% (i.e., 6,796) of 20,000 points were in the interior of the burned habitat, 10.1% (i.e., 2,015) were in the edge of the burned habitat, 10.4% (i.e., 2,073) in the edge of the unburned habitat, and 45.5% (i.e., 9,114) in the interior of the unburned habitat to yield a simulated sample of the available vegetation types based on the reported proportions (Table 17.2). We then coded independent and dependent variables as they would be placed in a standard logistic regression software program (Table 17.5).

Estimates of the coefficients and their standard errors from a logistic regression software package were then obtained (Table 17.6). Upper and lower limits of approximate 95% confidence intervals on the coefficients can be computed from the expressions, $\beta_1 \pm 1.96(SE[\beta_1])$ or equivalently, 95% confidence intervals on the selection ratios can be obtained from $e^{\beta_1 \pm 1.96(SE[\beta_1])}$. We note (Table 17.6) that 2 of the coefficients are large compared to their standard errors (95% confidence intervals will not contain 0.0). Thus, there is significant selection among the vegetation types. We elect to leave all of the vegetation types in the RSF for the categorical variable "vegetation type."

The RSF evaluated for the type "interior of burned" is $w(X_1) = e^{-0.179(1) + 0.907(0)1.188(0)} = e^{-0.179} = 0.836$, where $x_1 = (1,0,0;$ Table 17.5) and the constant b_0 is ignored. The values of the RSF were evaluated for the other vegetation types and standardized so they sum to 1.0 in the third column (Table 17.7). Note the fourth vegetation type, "interior of unburned," is the **reference type** with $X_4 = (0, 0, 0)$. Statistical significance of the other coefficients is judged in comparison to the reference selection ratio, $w(X_4) = e^0 = 1.0$. The coefficient $b_1 = -0.179$ is small compared to its standard error, and one could conclude the "interior of burned" does not add a significant contribution to the RSF compared to the reference level, "interior of unburned." In the terminology of hypothesis testing, there is no significant difference in selection for "interior of burned" and the reference level "interior of unburned." Both coefficients for "edge of burned" and "edge of unburned" are positive and large, indicating there is significant selection for those 2 habitat types in comparison to the reference level "interior of unburned" or "interior of burned."

Table 17.6. Estimated coefficients and their standard errors for the resource selection function, $w(\chi_i) = e^{b_1 X_{i1} + b_2 X_{i2} + b_3 X_{i3}}$, fitted to the moose data (Tables 17.2, 17.5)

Coefficient	Value	SE^a
b_0	−5.4290	0.1581
b_1	−0.1787	0.2549
b_2	0.9073	0.2655
b_3	1.1882	0.2416

From Neu et al. (1974).

[a] SE = standard error.

Table 17.7. Estimated values for the resource selection function (RSF), $w(\chi_i) = e^{b_1 X_{i1} + b_2 X_{i2} + b_3 X_{i3}}$, for each of the habitat types in the moose example (Table 17.2)

Location of tracks	$w(X_i)$	$w(X_i)/\Sigma w(X_i)^a$
Interior of burned	0.836	0.110
Edge of burned	2.478	0.326
Edge of unburned	3.281	0.432
Interior of unburned	1.000	0.132
Totals	7.595	1.000

[a] Values of the RSF standardized to sum to 1.0.

The interpretation of the results (Tables 17.3, 17.7) is the probability of location of moose tracks in the "edge of burned" by the protocol used in the Neu et al. (1974) study was about 2.5 times larger than the probability of location of moose tracks in the "interior of unburned." This interpretation is subject to the specific time, study location, density of the moose population, etc., and specific values for other independent predictor variables (e.g., the depth of snow). Extrapolation of the results to the relative probability of selection of the different vegetation types by moose depends on the additional assumptions that tracks were equally visible in the different vegetation types and that moose selected locations independent of one another (i.e., the animals have free and equal access to the entire study area).

The standardized values (Table 17.7) are the **same** as reported based on the simple selection ratios (Table 17.3). We included the analysis of the Neu et al. (1974) data by use of a computer software package for logistic regression for 2 reasons. First, we wanted to demonstrate the simple selection ratios (Table 17.3) are equivalent to fitting a more general RSF that gives the relative probability (odds) of selection among the vegetation types based on observed data. The interpretation of the values of the selection ratios, $w(X_i)$ (Table 17.7), or the simple selection ratios (Table 17.3) are in reference to each other. For example, the probability of selection for the "edge of burned" by moose was about 2.478, or 2.5 times larger than the probability of selection of the reference habitat "interior of unburned" by moose.

Second, we also wanted to provide a unified procedure for analysis of categorical data that does not depend on tedious use of formulas for the ratios of random variables (Manly et al. 2002). The above procedure using a standard computer program for logistic regression is one of the easiest ways to analyze habitat selection among categories of a single categorical variable.

Multiple Continuous and Discrete Variables

We illustrate the general theory for estimation of a RSF by analysis of resource selection by alder flycatchers (*Empidonax alnorum*) on the Innoko National Wildlife Refuge in Alaska. A complete description of the study is provided in Erickson et al. (2003). Locations of breeding alder flycatchers were gathered by walking transect surveys in the floodplain of the Innoko River during the 2000 and 2001 breeding seasons. Transects were systematically located along the Innoko River perpendicular to the general orientation of the river. The study area for breeding habitat was defined to be the area in the floodplain of the river. This example fits in **Design I,** because breeding areas were sampled for the population of alder flycatchers. **SP-A** fits, because a separate sample of selected locations is contrasted to a sample of locations in the floodplain.

Using **ArcInfo GRID** (http://home.gdal.org/projects/aigrid/aigrid_format.html), the study area polygon (floodplain) was converted into an image with 30-m units (pixels), coincident with **Landsat Thematic Mapper™ imagery** taken on 26 August 1991. Eleven raster layers were coregistered (measured) on each 30-m pixel in the study area and used in subsequent processing. For each pixel, the following 11 **coverages** were considered:

- coverages 1–6: spectral values of Landsat™ bands 1–5 and 7 from 26 August 1991,
- coverage 7: elevation (from U.S. Geological Survey 1:63,360 scale quads),
- coverage 8: slope (derived from elevation),
- coverage 9: aspect (derived from elevation),
- coverage 10: distance from unit to river (rivers defined from Landsat™ imagery), and
- coverage 11: distance from unit to closest lake (lakes defined from Landsat™ imagery).

There were 109 observations of breeding birds (109 "used" points) and a systematic sample of 5,094 pixels in the floodplain (of a total of 616,754 pixels). The values of each of the 11 coverages were recorded for each pixel that intersected a 105-m radius of the center of the pixel containing the observed used point. This collection of pixels formed a buffer surrounding each used point. A similarly sized buffer was defined for each of the pixels in the floodplain sample and for each pixel in the floodplain, so the estimated relative probability of selection could be mapped for the entire study area. Twenty potential predictor variables were defined for the buffer surrounding each pixel:

- variables 1–6: average of Landsat™ bands 1–5 and 7 for all pixels in the buffer,
- variables 7–12: standard deviation of Landsat™ bands 1–5 and 7 for all pixels in the buffer
- variable 13: average elevation of all pixels in the buffer,
- variable 14: average slope of all pixels in the buffer,
- variables 15–18: aspects (the percentage of pixels in the buffer defined as having north, south, east, or west aspect),
- variable 19: average distance to river for all pixels in the buffer, and
- variable 20: average distance to closest lake for all pixels in the buffer.

Twenty variables are too many to realistically fit and evaluate in a RSF using contemporary computer software. In addition, several of the variables were **highly correlated** with one another, and inclusion of ≥2 highly correlated variables (i.e., multicollinearity) leads to unstable coefficients in the fitted models (Neter et al. 1996). Highly correlated variables were eliminated from consideration, resulting in a set of 12 variables (Table 17.8). Because data were collected using SP-A, logistic regression can be used to estimate the coefficients in the RSF:

$$w(X) = e^{\beta_1 X_1 + \cdots + \beta_p X_p}.$$

Again, we emphasize that use of the **logistic regression software is a shortcut** to obtain the coefficients in the exponential model

$$w(X) = e^{\beta_1 X_1 + \cdots + \beta_p X_p}.$$

The function $w(X)$ estimates the relative probability of selection of a point in the landscape given the values of variables X_1 through X_p at that point and coefficients β_1 through β_p for variables in the model. For each selection of p variables among the 12 listed in Table 17.8, the sample of selected locations (with the dependent variable $Y = 1$) and the sample of floodplain locations (with $Y = 0$) were fitted to a logistic regression function using a computer software package. The relative probabilities of selection are calculated by taking the coefficients from the fitted logistic regression and placing them into the formula

$$w(X) = e^{\beta_1 X_1 + \cdots + \beta_p X_p}.$$

rather than using probabilities generated by the computer software. Probabilities from the computer software are generated from the logistic model

$$e^{\beta_1 X_1 + \cdots + \beta_p X_p}/(1 + e^{\beta_1 X_1 + \cdots + \beta_p X_p}),$$

which is the "wrong" model for SP-A (i.e., independent samples of selected units and units in the study area). The logistic regression software is only a convenient shortcut to obtain the maximum likelihood estimates of coefficients in the correct exponential model.

There were 12 variables available for estimating the RSF. All 4,095 ($2^{12} - 1 = 4,095$) possible main-effect models were fitted by logistic regression of the dependent variable Y on the independent predictor variables. Quadratic variables and interaction terms were ignored for this study. These 4,095 models were then ranked according to Bayesian Information Criterion (BIC; Schwarz 1978). The top 5 models (i.e., those with the smallest BICs) from this set were reported as our final best models (Table 17.9). BIC was defined as

$$-2\log(likelihood) + p\log(n),$$

where p is the number of variables in the model, n is the number observed bird locations plus number of locations sampled in the study area, *likelihood* is the value of the logistic likelihood, and log is the natural logarithm.

Relative probabilities of selection were calculated for each pixel in the floodplain based on weights of importance values (Burnham and Anderson 1998) for the top 5 models. For each of the top 5 models, the BIC differences were calculated as

$$\Delta_i = BIC_i - \min(BIC),$$

Table 17.8. Importance values and sign of coefficient based on importance value weighting for variables in the models for alder flycatcher, Innoko National Wildlife Refuge, Alaska

Variable	Variable description[a]	Importance value (sign of coefficient)
River	Average distance from river (m)	0.7 (−)
Lake	Average distance from lake (m)	0
Band 1	Average of band 1 in buffer	0
Band 4	Average of band 4 in buffer	1 (+)
Band 5	Average of band 5 in buffer	0
Std band 1	SD of band 1 in buffer	0.28 (−)
Std band 3	SD of band 3 in buffer	0.89 (+)
Std band 4	SD of band 4 in buffer	0.61 (+)
Std band 7	SD of band 7 in buffer	0.39 (+)
Elevation	Average elevation (m)	1 (−)
Slope	Average slope (deg)	0
Aspect	4 categorical variables: percentage in north, south, east, and west aspect	0

From (Erickson et al. (2003).
[a] SD = standard deviation.

Table 17.9. Top 5 models for alder flycatcher example, Innoko National Wildlife Refuge, Alaska. Rankings based on Bayesian Information Criterion (BIC). Models based on averages and standard errors (SE) of variables in a 105-m radius buffer around each sampled point in the landscape. The floodplain sample was derived from the corridor-wide area.

Model	BIC	Model weight	Variable[a]	Estimate	SE
1	911.36	0.376	River	−0.0008	0.00029
			Band 4	0.0412	0.00701
			Std band 3	0.2278	0.03470
			Std band 4	0.0393	0.01020
			Elevation	−0.1191	0.02260
2	912.35	0.229	Band 4	0.0471	0.00690
			Std band 3	0.2347	0.03420
			Std band 4	0.0517	0.00923
			Elevation	−0.1228	0.02320
3	912.50	0.213	River	−0.0008	0.00029
			Band 4	0.0433	0.00691
			Std band 1	−0.7367	0.19590
			Std band 3	0.2940	0.07320
			Std band 7	0.2039	0.04890
			Elevation	−0.1231	0.02400
4	913.81	0.111	River	−0.0013	0.00030
			Band 4	0.0332	0.00525
			Std band 7	0.1933	0.02680
			Elevation	−0.1018	0.02220
5	914.72	0.070	Band 4	0.0479	0.00687
			Std band 1	−0.9981	0.17920
			Std band 3	0.3839	0.06850
			Std band 7	0.2149	0.04740
			Elevation	−0.1343	0.02470

From Erickson et al. (2003).
[a] See Table 17.8 for descriptions of the variables.

Fig. 17.4. The weighted average of estimated relative probabilities of use by alder flycatchers based on 5 models and weights (see Table 17.9) for the southern section of the floodplain of the Innoko National Wildlife Refuge, Alaska. A general pattern of higher probability of use near the river in the southern section of the study area is apparent. Erickson et al. *(1998).*

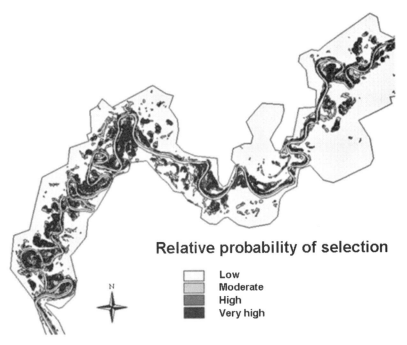

where BIC_i is the BIC for the ith model, and min(BIC) is the minimum BIC value for the 5 models. The BIC weights were calculated as

$$w_i = e^{\frac{1}{2}\Delta_i} / e^{\sum_{i=1}^{5} \exp\left(\frac{1}{2}\Delta_i\right)}.$$

Each model was used to predict the relative probability of selection for a pixel; a **weighted average** of the 5 predictions was then obtained using the BIC weights (Table 17.9, Fig. 17.4). This weighting procedure is similar to the use of Akaike weights (Akaike 1973, Burnham and Anderson 1998). Other model selection criteria, such as Akaike's Information Criterion (Burnham and Anderson 1998), could have been considered.

Although not described in this example, **model verification** and **validation** techniques also are recommended for further evaluation of the fit and robustness of the selected models. Model verification methods, such as Hosmer and Lemeshow's goodness of fit method for logistic regression (Hosmer and Lemeshow 1989), and validation techniques, such as those described in Boyce et al. (2002) and Howlin et al. (2003), should be conducted.

The variables elevation and the average of band 4 were included in each of the top 5 models, the standard deviation of band 3 was in 4 of the 5, and distance to river was included in 3 of the top 5. The probability of selection tends to decrease as elevation and distance to river increases (negative coefficients in the RSFs). The probability of selection tends to increase as the average of band 4 and standard deviation of band 3 increases (positive coefficients in the RSFs). Importance values, following Burnham and Anderson (1998: 326–327), for each of the variables considered, varied (Table 17.8). The variables elevation, band 4, and standard deviation of band 3 had the highest importance values. In this example, interpretation of the mean and standard deviations of the Landsat information is difficult, and other variables (e.g., quadratic effects, interactions, or ratios of bands) might have improved the models. A quality vegetation map was not available; it would have allowed for more direct interpretation of importance of different vegetation types.

SUMMARY

This chapter introduces the study of habitat use and food selection by wildlife in comparison to those resources in a study area to help identify important habitat patches and food resources. Selection is defined to exist during a study period if resources are exploited disproportionately to their availability at a given scale: geographic range of a species or home range of an individual or social group. Selection also can be defined as sites selected within a home range or as a site where food is procured. Generally, resource selection studies should consider selection at >1 scale and should be replicated in time and space.

Common field sampling protocols are introduced, followed by detailed discussion with examples of the most applicable method, SP-A: study area units are randomly sampled and selected units are randomly sampled. The sampling protocol is considered in association with the most common study designs. After presentation of field procedures for recording data on habitat units, food predictor variables, and use by wildlife, we present the estimation of RSFs as a unified theory for study of the relationships among selection of units and predictor variables measured on those units.

These functions allow one to estimate the relative probabilities that habitat units were selected and to rank units according to their value for use by wildlife for the study area, population, and time period.

Modeling of resource selection is introduced for computation of selection ratios among categories of vegetation types or food types. Logistic regression is the basic tool by which all example sets of data in this chapter are analyzed; it is used to analyze data to illustrate the concepts of relative probabilities of selection among vegetation types. Estimation of RSFs using multiple predictor variables is illustrated for estimation of relative probability of selection of breeding sites by alder flycatchers on the floodplain of the Innoko River, Innoko National Wildlife Refuge, Alaska. Information theory and maximum likelihood procedures are used for selection of appropriate predictor variables for the RSFs. Finally, we illustrate the mapping of habitat to identify those areas with the highest relative probability of selection during the study period by alder flycatchers in the Innoko National Wildlife Refuge.

18 Application of Spatial Technologies in Wildlife Biology

THOMAS A. O'NEIL,
PETE BETTINGER,
BRUCE G. MARCOT,
WARREN B. COHEN,
ORIANE TAFT,
RICHARD ASH,
HOWARD BRUNER,
CORY LANGHOFF,
JENNIFER A. CARLINO,
VIVIAN HUTCHISON,
ROBERT E. KENNEDY,
AND ZHIQIANG YANG

VISION STATEMENT

IMAGINE CROSSING GEOGRAPHIC and administrative boundaries to efficiently make policy decisions for regional fish, wildlife, and their habitats. By accessing multiple organizations' data and information, we can begin to successfully manage a region's dynamic natural systems and their inhabiting fish and wildlife. With effort and organization, we can adopt common data-management strategies that incorporate core data elements, data standards, and protocols to enhance information transferability. A data management system connecting numerous entities would create a powerful tool for effective management planning and scientific monitoring (Roger et al. 2007).

INTRODUCTION

Natural systems, including fish, wildlife, and **habitat** (habitat includes water, air, land, and other areas that species occupy), are complex, interrelated, and ever-changing. Data and information about natural systems mirror these properties, and the thing common to nearly all of these datasets is they occur somewhere in space. Currently, ecosystem information is collected across multiple programs and efforts, using many different methods, and it is maintained in many different technical systems. The result is that it is difficult—and in some cases, practically impossible—to assemble the data into ecosystem level views that cross geographic, administrative, and political boundaries. So let us all agree the Information Age has arrived, and that technology plays a large and important role in gathering, compiling, and synthesizing data. The old adage of analyzing wildlife data over "time and space" today implies use of technologies, including their integration into research and monitoring studies as well as evaluation strategies. Thus, resource managers must understand how to use these technologies, especially in regard to evaluating and assessing land at various **scales** (e.g., site, watershed, sub-basin, and basin levels). To assist resource managers with this task, this chapter explores spatial technologies and the applications that are commonly used by wildlife managers to acquire, compile, and interpret data. These include updated spatial application information from O'Neil et al. (2005). This chapter focuses on **Geographic Information System (GIS)** and **Global Positioning System (GPS)** technologies, and on using remotely sensed data, such as imagery from Landsat and, recently, **light detection and ranging (lidar).** This chapter also serves to heighten awareness and understanding of data documentation, data accuracy, and Internet applications.

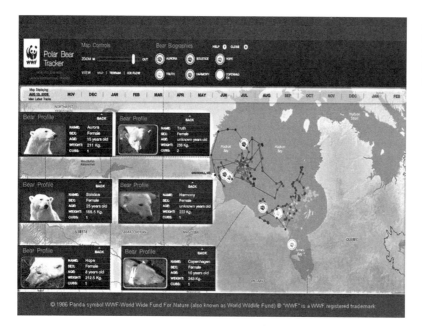

Fig. 18.1. Example of how a Geographic Information System map is combined with data collected by Global Positioning System collars to track individual movements of polar bears (*Ursus maritimus*) in Ontario, Canada.

To start, we direct your attention to Figure 18.1. This figure illustrates how to integrate wildlife species information with data from GIS, GPS, and the Internet. This kind of an approach gives people a dose of science in an easily understandable format. Thus, the primary purpose of this chapter is to get you familiar with the technology and its associated components, so that you will have a better understanding of its use, eventually enabling you to develop your own applications. To help facilitate a better understanding, a series of box inserts is included to briefly touch on key topics, concepts, or applications that might be of interest to users of spatial technologies. Included are such topics as how cartography can be manipulated to produce a great looking map that gives a false impression (Monmonier and H. de Blij 1996; see Box 18.1) and how spatial technologies can support vegetation monitoring at a fine scale. The definitions of key terms for certain technologies also are given in these boxes.

Today's issues and their complexities have a tendency to overwhelm resource managers in a sea of data. Most resource agencies are awash in data, but managers still find themselves with a lack of information. **Spatial technologies** provide tools to incorporate and analyze large datasets in a meaningful manner that produces useful information. Data can be converted or displayed by locations or across a landscape as charts, drawings, symbols, or a map. These technologies provide a way to assess and depict complex relationships among variables, which is useful for incorporating scale and hierarchy concepts into ecosystem-based management approaches (O'Neill 1996, O'Neil et al. 2008) and for helping examine environmental impact (Antunes et al. 2001). Additionally, they allow spatial depictions in 3 dimensions and can display theoretical concepts, such as total diversity of ecosystem functions (O'Neil et al. 2005). The technologies presented here also allow others to see how decisions are made, thus leaving a trail of accountability in the decision-making process to follow. However, as with any analysis and modeling tool, spatial technologies are only as accurate and reliable as the underlying data. Spatial tools on their own cannot improve accuracy, precision, or bias of information.

Spatial technologies should be considered as tools to assist resource managers with mapping. **Maps** are as important to the manager as calculators and vehicles. Using spatial technologies can provide timely information in usable formats for aiding decision-makers, but these tools should not be viewed as making the final decisions per se. Spatial technologies, such as GIS, are frequently described in terms of **hardware** (computers and workstations) and **software** (computer programs); typically, more computing power (speed and memory) in combination with large computing **storage** (disk space) is preferred. Workstations do most of the heavy lifting in handling large and/or complex datasets. Peripherals, such as storage and retrieval systems, and CD-ROM, DVD-RAM, and USB devices are required to effectively transcribe data into and out of systems.

The first spatial technology addressed is GIS, which is a general-purpose technology for handling geographic data in a digital form. **GIS** has the ability to preprocess large amounts of data into a form suitable for analysis and evaluation; support models that perform analysis, calibration, forecasting, and prediction; and postprocessing of results to produce tables, reports, and maps (Goodchild 1993). For a more technical description on what GIS is and steps needed to maintain it, review Koeln et al. (1996) and O'Neil et al. (2005).

Box 18.1. Example of what not to do when depicting information about a species

This example shows what can happen when depicting information about a species without understanding or taking into account the scale or level of information needed to study the problem of interest. The figure is a hypothetical example of an American marten range map depicting various levels of habitat suitability. This map represents a false-positive outcome in that it conveys the impression that we know a great deal about this species and have the ability to rate large expanses of its habitat.

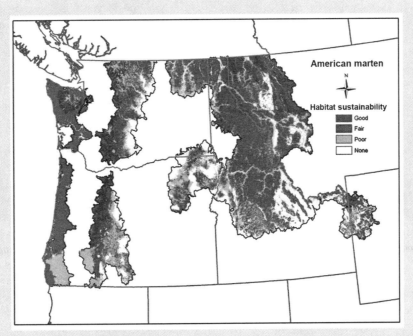

Depiction of the American marten range map in Oregon to demonstrate how a map can appear to be accurate while the information it displays can be misleading. Habitat suitability requires fine scale information.

The fallacy: it is the right question at the wrong scale; maps can simply reflect that we may not know the habitat used by marten in the Oregon and Washington Coast ranges as well as we do the intermountain range. Mapping nonhabitat is confusing, because if it covers large areas in the map, value judgments of good, fair, or poor may just reflect the author's biases or narrow understanding of the information being used to construct the map. The bottom line is that such a map superficially suggests that large areas of a species habitat could be discounted or presumed to be highly unimportant in a conservation strategy, when the reality is that fine-scale elements will dictate what is suitable and what is not. A more appropriate depiction at this coarser scale would be to show just a species range without any qualifiers as to habitat suitability or to show only habitats that are generally associated with a species presence.

There is a strong tendency to want to depict more about a species or a situation than we may really know. We caution users of Geographic Information System technology to be sure that they understand the information being presented. If you are unsure, then ask the following question of someone who does understand: does this make sense?

USING GIS IN THE FIELD OF WILDLIFE

Prior to 1950, vegetation maps were tediously drawn by hand, and wildlife biologists interpreted the maps in terms of habitat for wildlife, typically game animals. The availability of aerial photography and then satellite imagery gave biologists a way to more accurately analyze habitats across broader landscapes. For example, in the early 1970s, Schuerholz (1974) quantified forest edge habitat from aerial photographs, and Cowardin and Myers (1974) identified and classified wetland habitats from remotely sensed images. In the later 1970s and 1980s, as computers became more widely available and more capable, vegetation maps were transcribed into digital images, and habitat analyses became highly automated (e.g., Marcot et al. 1981, Mead et al. 1981, Mayer 1984, Burroughs 1986, Brekke 1988). In practice, the staff of the Chattahoochee-Oconee National Forest in Georgia developed a forest plan (U.S. Forest Service 2004) that reflected a public desire for more wilderness and recreational areas, given the Forest's proximity to the Atlanta metropolitan area. The plan acknowledged that GIS was heavily relied on, using such processes as buffering (e.g., for delineation of scenic corridors) in the development of various land classes. Today, GIS is an indispensable tool for analyzing historical, current, and potential future habitat conditions for wildlife (Fig. 18.2) and for assessing the spatial relationships among landscape features (Box 18.2).

Another case in point is an up-and-coming concept, whereby GIS supports a **habitat accounting and appraisal system** called Combined Habitat Assessment Protocols (CHAP) for evaluating and crediting habitat impacts and mitigation (O'Neil 2010). CHAP is being used in 5 western states for projects that range from evaluating ecosystem restoration of the Los Angeles River to settle a lawsuit for a 25-year wildlife-habitat loss in Oregon for US$133 million. This approach uses site mapping and inventory of individual vegetation types for structural conditions and fine-feature vegetation components or key environmental correlates. The field information is then linked to a peer-reviewed dataset that can tie potential fish and wildlife species to the site's vegetation inventory. Two matrices are generated—species-function and habitat-function matrices—and these values are then combined to generate the overall value of the habitat for fish and wildlife (see Fig. 18.3); the values are determined, stored, and displayed in a GIS geodatabase. GIS also is gaining in popularity for use in evaluating **cumulative effects** of management actions and potential effects of **alternative management decisions** on wildlife habitats, populations, and communities.

GIS and Modeling Wildlife–Habitat Relationships

One of the most common uses of GIS in wildlife management is in analysis of **amounts, patterns,** and **trends** of habitat for individual wildlife species. For example, McComb et al. (2002) used GIS to model potential habitat of northern spotted owls (*Strix occidentalis caurina*) at landscape scales in the Pacific Northwest. Dettmers and Bart (1999) applied GIS to predict forest songbird habitat in southern Ohio. Carroll et al. (1999) used presence data to construct and validate spatial habitat models of fisher (*Martes pennanti*) in central Oregon. Knick and Dyer (1997) used GIS to analyze black-tailed jackrabbit (*Lepus californicus*) habitat in south-

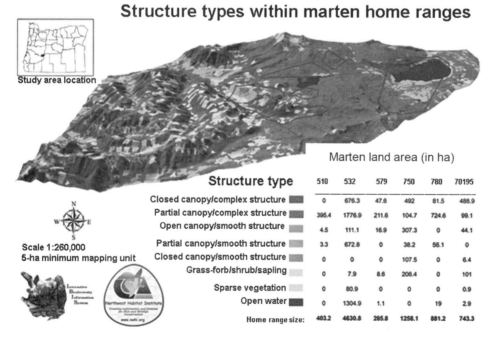

Fig. 18.2. A Geographic Information System allows an integration of data. American marten home ranges are overlaid on structural habitat conditions that are draped over topography. *Image courtesy of Northwest Habitat Institute.*

Box 18.2. What is a spatial relationship?

Spatial relationships may be important for understanding the resources of concern when developing habitat management strategies (Schroeder et al. 1999. Spatial relationships describe the association among landscape features, and they may be characterized by both topological and directional aspects. When developing the topology among landscape features, one is using methods that develop associations among landscape features. Research into spatial relationships has been driven by the work of mathematicians, cognitive scientists, and designers of software for any Geographic Information System (GIS). In a GIS, it may not be sufficient to know just the position of a landscape feature; it also may be necessary to know how the landscape feature relates to other features in the same (or other) databases. For example, it may not be sufficient to just be able to locate a patch of optimal habitat; it also may be important to locate other patches of good or optimal habitat nearby.

Some examples of spatial relationships include:

1. polygons that share a common boundary (e.g., adjacent polygons),
2. polygons of a certain type (e.g., optimal habitat) nearest to other specific polygons (e.g., proposed harvests),
3. polygons that overlap other polygons (e.g., the intersection of soils and timber stands),
4. lines that cross one another (e.g., roads that cross streams),
5. lines that logically flow into one another (e.g., stream networks),
6. lines that are within a certain distance of other landscape features (e.g., roads within a certain distance of streams),
7. points contained in polygons (e.g., bird point sample locations in timber stands), and
8. points that can be seen from certain other points (e.g., as in defining viewsheds).

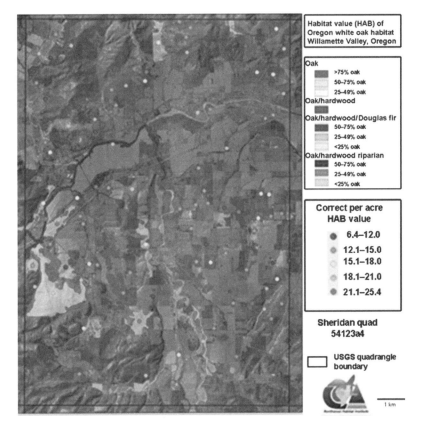

Fig. 18.3. Displaying wildlife habitat values of focal habitats in a landscape.

western Idaho. O'Neil et al. (1995) depicted and mapped all of Oregon's wildlife habitat types using GIS. Following procedures used by O'Neil et al. (1995), Kiilsgaard and Barrett (2000) created the first **wildlife-habitat types map** of the entire Pacific Northwest. Many other examples can be found in the literature. These habitat relationship models use such variables as slope, aspect, and vegetation structure to provide estimates of habitat quality. Variables used are based on factors shown to influence quality of habitat selected by particular species. Research or professional judgment generally provides the knowledge necessary to develop models of habitat relationships. The challenge is to create the appropriate databases necessary to evaluate relationships in a spatial manner. Bettinger (2001) outlined many of the challenges facing integration of wildlife models with remotely sensed imagery and data related to forest structural conditions. Dussault et al. (2001) caution that existing forest cover maps may not be adequate for evaluating wildlife habitat suitability without an examination of the correlation between mapped forest features and **structural conditions** desired by specific wildlife species or species groups.

Other types of models for evaluating habitat quality also have been integrated with GIS. For example, Clevenger et al. (2002) integrated expert-based models to help identify and plan for wildlife habitat corridors, Raphael et al. (2001) integrated Bayesian belief network models of species habitat suitability into GIS analyses, Guisse and Gimblett (1997) evaluated recreation conflicts in state parks by integrating a neural network model with GIS, and Rickel et al. (1998) used a fuzzy logic model in conjunction with GIS to evaluate wildlife habitat quality. The main objective of these approaches has been to develop useful tools for resource managers charged with identifying locations of important areas for wildlife when empirical information is lacking. GIS helps facilitate this process by providing a representation of the spatial features of a landscape in the habitat quality evaluation.

Statistical analyses also have been integrated with GIS processes to evaluate quality of wildlife habitat. For example, Clark et al. (1993a) integrated a multivariate analysis of female black bear (*Ursus americanus*) habitat into a GIS model, and Pereira and Itami (1991) used results of a logistic multiple-regression study of Mt. Graham red squirrel (*Tamiasciurus hudsonicus grahamensis*) in their GIS model of habitat for the species. Software developed from a variety of organizations is increasingly becoming open to integration. As a result, almost any wildlife habitat-quality model that can be described with quantitative relationships can be integrated with GIS.

Some researchers have taken the integration of wildlife habitat relationships and GIS a step further by integrating qualitative wildlife-habitat relationships with forest planning processes. The goal is to allow features of a landscape important for describing habitat to guide development of forest plans. This innovation represents a distinct change in forest planning, because historically, forest plan development has been guided by economic or commodity production objectives, rather than by ecological or social objectives. As a result, **wildlife–habitat relationship** models that include **spatial components** can be incorporated into forest planning processes and guide development of forest plans. Hof and Joyce (1992, 1993) and Hof et al. (1994) were among the first to attempt to integrate wildlife habitat concerns in mathematical forest-planning models. Specific examples of the integration of wildlife–habitat relationships with forest planning processes include elk (*Cervus canadensis*) in Oregon (Bettinger et al. 1997, 1999), birds in the Northwest (Bettinger et al. 2002), birds in the Midwest (Nevo and Garcia 1996), ruffed grouse (*Bonasa umbellus*) habitat in the Midwest (Arthaud and Rose 1996), red-cockaded woodpecker (*Picoides borealis*) habitat in the Southeast (Boston and Bettinger 2001), northern flying squirrel (*Glaucomys sabrinus*) in western Oregon (Calkin et al. 2002), spotted owl in Washington State (Hof and Raphael 1997), and late successional habitat in Oregon (Sessions et al. 2000). **Measures of biodiversity** also have been integrated into forest planning efforts (Kangas and Pukkala 1996). Other habitat-related concerns can be included in forest planning processes (Box 18.3), such as the desire to develop and maintain contiguous core areas of older forest (Öhman and Eriksson 1998, Öhman 2000) and development of connected habitat corridors (Sessions 1991, Williams 1998).

GIS and Modeling Populations

Spatially explicit wildlife population models consider 2 factors of importance to the estimation of populations: the species–habitat relationships and the arrangement of habitat over space and time. Spatial models can assist land managers in their decision-making processes, because landscapes are complex and dynamic. However, many of the **population models** are developed for **one (or a few) species,** and accommodating multiple species across a landscape remains a significant challenge (Turner et al. 1995). Liu et al. (1995) provide an example of the use of a spatially explicit population model in GIS to examine the impact on a nontarget species, Bachman's sparrow (*Aimophila aestivalis*), of a forest plan developed with other goals in mind. The model results allow managers to determine how sparrow population density and distribution may react to planned management activities, such as whether the populations are of a size that meets the minimum management goal for the species, or whether the populations are sensitive to certain projected landscape characteristics. Another example of the prediction of potential population densities is by Mladenoff and Sickley (1998), who used GIS to assess potential population sizes of gray wolf (*Canis lupis*) in the northeastern United States.

GIS has been used as an integral part of **population viability analysis** (PVA; Akçakaya 2000a). Kingston (1995) provides a review of the use of population viability models in a

GIS environment. This analysis uses the vital rates (survival, reproduction, and dispersal) of individuals of a species to calculate population size over time and rate of change. To model PVA in a spatial manner, information about spatial structure (size, shape, and quality) of habitat patches that a particular species might inhabit also is required. This habitat quality information for a landscape can be provided through GIS processes and applications, and it allows land managers and planners to evaluate how management practices may affect the probability of species extinction. Brook et al. (1999) compared 4 PVA processes, including ones that use spatial or mapped representations of populations, and concluded that subtle differences among models can affect results.

Beyond assessing populations and viability, Allen et al. (2001) used GIS gap analysis to model mammal populations and concluded that geographically defining minimum critical areas produced more conservative and defensible maps of species richness than did other methods. Hof and Raphael (1997) developed a geographic model to optimize allocation of northern spotted owl habitat in Washington State. Optimization model parameters included adult survival, fecundity, and occupancy of sites. Some authors have even integrated assessments of population genetics with GIS. For example, Ji and Leberg (2002) evaluated genetic diversity from a regional perspective using GIS. Finally, because intensive ground surveys cannot keep pace with the rate of land use change in some areas of the world, presence–absence models are being developed for use in GIS, in conjunction with remote sensing and other technology, to allow one to map the potential distribution of species at large spatial scales (Osborne et al. 2001, Kilgo et al. 2002).

GIS and Conservation of Wildlife Communities

Another use of GIS in wildlife management is in delineation and conservation of wildlife and ecological **communities.** Delineation of **hot spots** (areas of high species richness or centers of species endemism or rarity; Dobson et al. 1997, Griffin 1999, Ceballos et al. 1998) has become popular for specifying areas with wildlife and plant assemblages and communities needing protection. Mapping species-rich hot spots has been used to delineate potential protected areas or reserves (e.g., Bojorquez-Tapia et al. 1995). Spatial algorithms or processes used to define hot spots usually undergo rigorous evaluation. For example, Araújo and Williams (2001) discovered bias toward marginal populations when delineating complementary hot spots, and NCASI (1996) reported that results of richness hot spots can be highly sensitive to the underlying distribution maps of individual species.

The National Gap Analysis Program (GAP) took delineation of conservation hot spots further by intersecting areas of high species richness with a set of land use allocations, with the goal of identifying areas of high richness that may suffer from lack of protection (Scott et al. 1993). GAP provides an assessment of the extent of representation of native species and communities across a landscape. Those species or **communities not adequately represented** on public lands can be viewed as **gaps** in conservation networks (Pearlstine et al. 2002). The process of identifying gaps has been aided by dividing analysis areas into segments to account for geographic variation and to help cover broad geographic areas (Scott et al. 2001). For example, the Florida GAP project is a geographically extensive analysis. One of the objectives of this project is to provide interested stakeholders with GIS databases related to status of terrestrial vertebrate species and their habitats. Landsat Thematic Mapper satellite imagery is used, as are the National Wetlands Inventory GIS databases, other available databases (e.g., for soils), aerial photography, and on-the-ground surveys (Pearlstine et al. 2002).

GIS also has been used to design potential reserves or protected areas (Fig. 18.4). One of the fundamental issues for biologists and managers is selection or proposal of areas that should be conserved. A variety of techniques has been developed to determine optimal reserve designs, each using GIS databases to guide selection of reserve areas. Wildlife–habitat relationships can be used to delineate areas of special concern, such as the **corridor suitability** GIS database created for Maryland's Green Infrastructure Assessment (Weber and Wolf 2000). For this assessment, GIS databases representing land cover, stream networks, roads, slope classes, aquatic community conditions, and other traits were used to create a database that described the suitability of areas to act as corridors of wildlife movement. This database was then used as input for a model that determined the least-cost pathway between core wildlife management areas.

Other researchers (e.g., McDonnell et al. 2002, Nalle et al. 2002) have devised mathematical algorithms to most efficiently design nature reserve systems to meet biodiversity objectives. In addition, reserve area redundancy (ReVelle et al. 2002), complementarity (Williams et al. 2000a), and representativeness (e.g., Mackey et al. 1988, Powell et al. 2000a, MacNally et al. 2002) have been discussed in the literature and demonstrated through use of GIS processes. Dobson et al. (2001) advocated integrating strategies and objectives to meet multiple needs for people and species.

Efficient identification of potential reserve areas with GIS processes has allowed policy-makers to consider a number of management issues. For example, should reserve areas represent an array of community, productivity, or ecosystem classes, as Stokland (1997) suggested for bird and insect conservation in boreal forest reserves of Norway? Should reserve areas be established mainly for species richness, species rarity, or for other objectives, such as balancing requirements of rare species conservation with a broader biodiversity conservation perspective? Williams et al. (2000a) suggested that more biodiversity could be protected if the few species that attract the most popular support (flagship

> **Box 18.3. An example of integrating habitat relationships into a forest planning process**
>
> We illustrate the potential to integrate habitat relationships into a forest planning process with an example provided by Bettinger and Boston (2008). Wildlife habitat goals can be qualitatively or quantitatively defined. Quantitative goals also can reference spatial information provided by Geographic Information System (GIS) databases, allowing spatial goals to be developed. Spatial goals may include configurations, such as required minimum patch sizes or complementary habitat types, and thus may indicate that, for optimal benefit to a particular species, one type of habitat should be placed next to (or in some proximity of) another. Great gray owls (*Strix nebulosa*), for example, prefer early seral stage forests (clearcuts) for foraging, yet these areas should be adjacent to single-story open-canopy forests containing snags or large trees with broken tops.
>
> One can use spatial relationships to guide the development of a forest plan that seeks to provide the greatest amount of habitat over time while achieving other economic or commodity production goals. Bettinger and Boston (2008) illustrate a planning process where commodity production goals in the Pacific Northwest can influence spotted owl (*Strix occidentalis*) habitat capability levels. The process also was used to constrain scheduled activities such that owl habitat capability levels do not fall much be-
>
>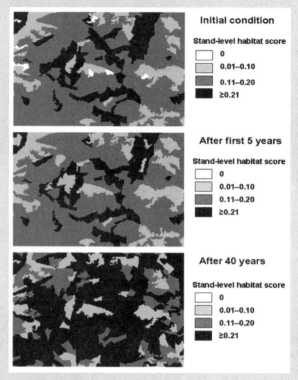
>
> Spatial arrangement of spotted owl habitat on a portion of the landscape.

species) had distributions that covered the broader diversity of organisms across a landscape. Should some level of **redundancy** be built into reserve areas to guard against potential losses from major disturbance events? Finally, should reserve areas be complementary to one another to efficiently and cost-effectively set aside the least amount of land area with highest biological opportunity cost? Each of these questions can be addressed with the appropriate GIS databases and reserve selection processes.

GIS and Wildlife Conservation in a Risk Assessment Context

GIS lends itself well to modeling wildlife populations, habitat relationships, and ecological communities in a spatially explicit risk-assessment framework. Such an approach generally entails overlaying maps of the occurrence of wildlife species or communities and habitats (Glenn and Ripple 2004) with **natural and anthropogenic stressors,** such as disturbance events (e.g., fire or climate change) and patterns of land use, and then evaluating potential changes in conditions that could serve as conservation threats. As an example, Barve et al. (2005) developed a GIS-based procedure for assessing threats to a wildlife sanctuary in India. They mapped current and projected threats from roads and accessibility, human settlements, and livestock density in the sanctuary and devised a composite threat index expressed on a 5-category ordinal scale of intensity. They found that **threat intensity** negatively correlated with tree species diversity and concluded the threat index is a useful protocol that could be applied to other areas.

Integrating remote sensing and GIS tools with risk analysis (evaluating potential threats, their costs or implications, and their probabilities) has long been established in spatial fire risk models. For example, Sturtevant et al. (2009) used GIS with a landscape disturbance and succession model (LANDIS) to estimate fire risk and to help guide management of fire-prone forests in wildland–urban interface areas. Lehmkuhl et al. (2007) combined fire risk and wildlife habi-

low grow-only (no activity) levels, and then to describe the subsequent impact on timber harvest levels. The spatial aspect of the planning process involved the timing and placement of management activities; it included a landscape-level home-range analysis for spotted owl habitat quality. In the latter of these 2 concerns, the forest structure within 2,400 m of each stand was assessed to understand how scheduled activities would affect overall habitat quality. Each stand of trees, in essence, received a habitat score based on the planned activities in other stands within a 2,400-m radius. Stand-level habitat scores were then weighted by the size of each stand, and a landscape-level average was determined. The landscape-level average was then used to control the timing and placement of activities. Although mainly performed using vector GIS data, this process is similar to the moving window approaches commonly used in conjunction with raster GIS data. A few practical constraints also were added to the planning process. These included constraints on minimum harvest age and on maximum clearcut size (48.6 ha).

A heuristic planning technique (i.e., one that determines good, feasible problem solutions that are not necessarily optimal), threshold accepting, was used to develop the forest plan. Threshold accepting is a stochastic neighborhood search process based on accepting, during a search, subsequently lower quality solutions only if they are within some threshold quality of the best solution located during the search. The plans developed were thus informed by the spatial arrangement of scheduled activities and the resulting landscape-level habitat quality (the figure on page 436). When assessing the competing goals, trade-off curves (the figure below) were used to help visualize how the goals interact. The relative effects of policies can then be considered by policy-makers and managers in the assessment of a gain (or decline) in one measure compared to the decline (or gain) of another.

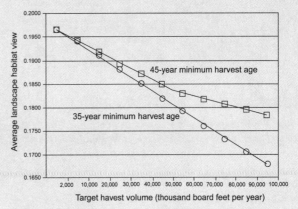

Trade-offs among spotted owl habitat and commodity production goals when the minimum harvest age is altered.

tat models to spatially depict potential benefits for fire hazard reduction and wildlife conservation from alternative fuel-reduction treatments at the landscape scale.

Risk analysis approaches to evaluating wildlife communities typically entail dealing with incomplete information and use of expert judgment. GIS and **remote sensing** information and tools can be used to integrate limited data with expert knowledge and to map uncertainties (Bojorquez-Tapia et al. 1995, Jones and Thornton 2002). Johnson and Gillingham (2004) developed a procedure to evaluate the reliability of predictive species-distribution models when based at least in part on expert opinion. They found that expert-based predictive models can vary according to opinions used and provided a means of mapping such uncertainties of species distributions. The uncertainty maps can provide managers with explicit information on the reliability of expert opinion as the basis for species distribution models (also see Murray et al. 2009). More generally, GIS tools for denoting and representing expert knowledge are being applied in various conservation venues (Bojorquez-Tapia et al. 2003, Gao et al. 2004).

GIS-based tools have been developed for evaluating connectivity of wildlife habitat and for delineating potential wildlife dispersal corridors in the landscape (Clevenger et al. 2002). One such ArcGIS-based tool using a cost distance approach is CorridorDesigner (http://www.corridordesign.org/; Beier et al. 2008). Other approaches to identifying potential wildlife habitat corridors combine GIS models with landscape genetics. Epps et al. (2007) evaluated landscape genetics of desert bighorn sheep (*Ovis canadensis*) in a least-cost GIS model to predict probable movement corridors, evaluate potential connectivity among populations, and determine effects on connectivity from anthropogenic barriers and translocated populations.

Other GIS-based risk assessment tools have been developed to evaluate and help delineate biodiversity hot spots for potential conservation consideration. For example, Hunter et al. (2003) used GIS to merge anthropogenic threats (hous-

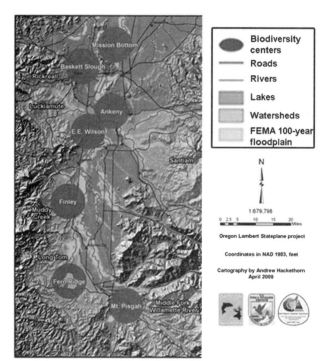

Fig. 18.4. Willamette Valley conservation strategy that links centers of biodiversity (i.e., U.S. Fish and Wildlife Service wildlife refuges, state wildlife management areas, or community conservation areas) with Willamette River 100-year floodplain and 4 watersheds. Land conservation by easements or acquisitions is targeted first in the biodiversity center buffers or in or near the 100-year floodplain, and then in watershed. FEMA = Federal Emergency Management Agency.

ing density) with indicators of biological richness from the Utah GAP to map biodiversity hot spots with high potential threat levels that might require mitigation or conservation, and biodiversity hot spots with low threat levels that could serve as new reserves.

Other examples of using GIS in risk analysis for wildlife conservation and to inform and support natural resource management decisions include the following. Llewellyn et al. (1996) developed a map-based decision-support system to help prioritize sites for restoration on the Mississippi River floodplain; McDonald and McDonald (2003) presented a GIS-based wildlife risk assessment method using resource selection models; and Goparaju et al. (2005) used GIS tools to evaluate effects of forest fragmentation risk on ecological communities.

One promising area of remote sensing and GIS application is community-based, participatory resource inventory and environmental decision-making, empowering local communities with tools they can use to evaluate, economically valuate, and plan for use of their local natural resources, including wildlife. Such GIS applications were presented by Lewis (1995a), Kearns et al. (2003), Guralnick and Neufeld (2005), and Jankowski (2009). These approaches would likely prove fruitful if used with increasing public interest in **citizen science** participatory programs for wildlife observation (Lepczyk 2005), such as Cornell Lab of Ornithology's Project FeederWatch (www.birds.cornell.edu/pfw/).

Another use of GIS is in crafting optimal or satisfactory solutions to delineating potential protected areas across landscapes. Over the years, several GIS-based spatial allocation models have been developed and used, such as SITES (Roberts et al. 2003), Marxan and Marzone (Christensen et al. 2009), Ecospace (Pauly et al. 2000, Christensen et al. 2009), SELES (Fall and Fall 2001), and network flow models of dispersal corridors (Phillips et al. 2008). Such approaches usually entail identifying species richness or key areas of biological diversity, determining existing or desired land-allocation boundaries, and developing a cost-function layer that denotes various aspects of potential wildlife dispersal filters and other real or assumed costs of conservation and land allocations. Results are maps with protected area polygons typically denoting optimal or satisfactory solutions to habitat conservation that meet lowest-cost criteria.

GIS AND THE INTERNET

The rapid development of GIS networks has been concurrent with the ever expanding capabilities of the Internet. GIS technologies have become more complex, growing from isolated workstation-based solutions to fully integrated multi-user enterprise systems that connect many people and organizations to dynamic data and web-based application tools. The Internet has served to disseminate data; support software applications; document the enormous volumes of data being created today; and facilitate professional networking and collaboration among scientists, resource managers, and the public. The Internet is a valuable resource without which any modern GIS user cannot fully function.

When beginning a GIS project, analysts may first need to compile initial geospatial datasets that are pertinent to their location of interest. There is an ever-growing amount of geospatial data being served online via the Internet; one can find, free of charge, the appropriate datasets to support both the analysis and reporting protocols being implemented for any given project. These data are being served by a variety of methods, from simple data clearinghouses, such as the Oregon Geospatial Enterprise office (http://www.oregon.gov/DAS/EISPD/GEO/index.shtml), to complex dynamic representations, such as **Google Earth** or Microsoft's **Virtual Earth.**

There are many sources of data available on the Internet that are collected and maintained by private organizations and are available for a fee (or are free), and public data are available from federal, state, and local governments and various nonprofit sources. A prominent source of data on the Internet in the United States is the Natural Resources Conservation Service Spatial Data Gateway (http://datagateway.nrcs.usda.gov/). Users can select the geographic areas of interest, usually by county and state, and can view the available data, such as imagery, topographic maps, **digital elevation models (DEMs),** soil hydrology, and geographic information. Most, if not all, of the 50 states in the United States operate websites to disseminate GIS data; so they are a great place to start looking. They are usually easily located on the Internet by using a search engine with the state name and "GIS data" as keywords.

Data **clearinghouses** are usually simple websites that offer GIS data files, premade maps, and metadata (data about data) for download just by clicking on the appropriate links. This type of site exemplifies a static concept of data delivery. Files that are available for download from this type of site are rarely or only periodically updated; they represent data that do not regularly change, such as geologic or topographic information. Users simply click on a file to download it and save it to their hard drive. Files are usually served in one or more formats compatible with modern GIS software, such as **Earth Systems Research Institute (ESRI)** shapefiles and coverages and .e00 exchange files. The various incarnations of ESRI geodatabases, such as the Microsoft Access®-based Personal Geodatabase, the standalone file geodatabase or the multi-user geodatabases that run on third-party **relational database management system** platforms, such as SQL or Oracle, are becoming more popular for direct data transfer. Static data delivery requires little server side maintenance or hardware requirements, as long as the data do not change often. Every time there is a change in a dataset, it is necessary to regenerate statistics, metadata, and hyperlinks on the webpage. Users also do not have the option of defining their own areas of interest. Instead, they are forced to download and handle a dataset that is either geographically larger than needed or to piece together multiple smaller files to cover the project area. These types of data are increasingly being served via dynamic websites that provide greater flexibility for users.

Rapidly changing, extremely large datasets that require multiple users to edit them are not suited to distribution by the type of static websites described above. Users of high-resolution aerial imagery, for instance, often require only a portion of an image to cover a site of interest. Downloading an entire set of imagery may be too time consuming and would require extensive clipping or excessive display times. Dynamic data delivery is a model that incorporates server and GIS technologies to reduce the maintenance of the data delivery system by automatically updating linkages to existing datasets while simultaneously reducing data storage redundancy. Instead of constantly having to deal with updating frequently changing datasets or separate copies for multiple users, managers can store, back up, and maintain a single dataset that incorporates input from all users. This system results in a much more robust method of handling large amounts of data, but it is not without high initial costs associated with the hardware, software, and personnel required to run such a system. Dynamic data delivery systems are often supported in an organization's information technology department to facilitate seamless data analysis and dissemination of information while sharing the costs among many departments or agencies. Data managers can link with source data, manipulate changes, and instantly have those changes reflected in the data being served to the public by a single action rather than by having to copy the data to an export file and updating the website code to point to the new file(s). Dynamic mapping applications, such as ESRI's **ArcGIS** Server, also can be set up to automatically incorporate various GIS datasets to provide interactive maps that can be imbedded in websites.

Dynamic web services can be configured to provide this imagery for download or streaming directly to a GIS application. The Oregon Imagery Explorer (http://oregonexplorer.info/imagery/) is a good example of a dynamic web service that allows users to stream or download 0.5-m **National Agriculture Imagery Program** data directly to a GIS application, such as **ArcMap.** The data storage requirements of these datasets alone would require prohibitively large storage servers and/or download times.

Finding where to access some of these data also can be confusing at times. The most popular search engines for quickly locating websites, as determined in a 2006 Nielsen NetRatings survey (http://www.google.com/search?q=2006+Nielsen+NetRatings&rls=com.microsoft:en-us:IE-SearchBox&ie=UTF-8&oe=UTF-8&sourceid=ie7&rlz=1I7SUNA_en), are Google (49.2%), Yahoo (23.9%), and MSN (9.6%). Although these search engines may not directly link to spatial data sources, they often do point to viable data centers or other Internet resources. The Geography Network (http://www.geographynetwork.com/) is a good example of a place to start looking for geographic datasets. Other resources include user groups and help forums. The Society for Conservation GIS (http://www.scgis.org) is an example of an active user group that can help link many organizations in the natural resources field. ESRI operates a support website (http://support.esri.com/) that can be invaluable to GIS users in providing analysis solutions and identifying software bugs. ESRI also hosts many online training opportunities at its training site (http://training.esri.com/gateway/index.cfm) that can help users learn about the basics of GIS analysis and new functionality that is constantly being added to GIS tools. Much of this training is offered free of charge. Another source of online geographic data is the U.S. government's **Geospatial One Stop** website (http://www.geodata.gov). This website contains linkages to many federal, state, and local data sources.

GIS in the International Community

Numerous ongoing efforts are under way to consolidate and enhance spatial information studies at the global level, such as the various United Nations programs to enhance biodiversity and information exchange; the World Wildlife

Fund; and Diversitas, an international program to promote biodiversity. Several governmental and nonprofit agencies have developed forums (websites that allow users to interact with one another in a bulletin board-like setting) to create, analyze, and disseminate spatial information. Such groups as the American Geological Institute, **United Nations Environmental Program's (UNEP) Global and Regional Integrated Data (GRID)** program, and the Global Spatial Data Infrastructure (GSDI) lead the way in connecting scientists and the public to data and information regarding the status of geophysical and biological resources of the planet.

UNEP-GRID is a good example of the coordination of many entities at the global or regional scale. UNEP-GRID is headquartered in Nairobi, Kenya, and provides global leadership in coordinating environmental data management, disseminating GIS decision support tools, and providing GIS training. UNEP-GRID centers around the world coordinate regional data on climate change, vegetation coverage, temperature maps, elevation data, and many other key datasets. In addition to housing regional datasets, each GRID center has additional research responsibilities and interests. For instance, the UNEP-GRID node in Arendal, Norway, is responsible for polar research, and the Sioux Falls, South Dakota, site houses global landcover data collected from U.S. satellites. Currently there are 7 GRID sites around the world, with 5 more being proposed. As more regional centers come on line, there will be further opportunities for every country to participate in the global area of environmental data management.

GSDI is a nonprofit organization dedicated to enhancing and facilitating the development of spatial data to build sustainable social, economic, and environmental systems and fostering informed and responsible use of geographic information. Many groups, including UNEP-GRID, are members of GSDI. Other major players in the global spatial data arena provide similar services to scientists and resource managers around the globe. Another UN group, the World Conservation Monitoring Centre provides a great deal of spatially derived data, such as biodiversity information, species databases, and international environmental agreement protocols.

Other sources of information certainly do exist to address global issues in GIS. ESRI hosts an annual international GIS user's conference in San Diego, California. This conference draws well over 10,000 participants from all over the globe. Attendees can hear presentations by GIS users in many different fields and have access to hundreds of hours of training in GIS technologies. There are often opportunities to receive complimentary or discounted registration passes from ESRI if the applicants are affiliated with educational or nonprofit organizations or are students. There also are projects that address species-specific issues at the global scale. For instance, the Global Owl Project (http://www.globalowlproject.com/) incorporates scientists and resources from many different countries in an effort to conserve and protect world owl populations. Another great source for global spatial information is the **National Biological Information Infrastructure (NBII)** and the **Global Biodiversity Information Facility.** They offer global biodiversity information through a coordinated effort of international scientists. Becoming aware of these data sources will certainly help connect disparate scientific studies to the global context of biodiversity and conservation of wildlife in the future.

USING GLOBAL POSITIONING SYSTEMS

In recording wildlife and related natural resource information, a fundamental component is location. No tool has shaped the nature of location gathering, navigation, and field geographic data collection more than the United States' **Global Positioning System (GPS).** GPS provides accurate 3-dimensional locations (latitude, longitude, and elevation) 24 hours a day to users who have a GPS receiver. Various types of GPS receivers for consumers, mapping professionals, surveyors, and other users requiring high accuracy are available. Manufacturers include Garmin, Trimble Navigation, Topcon, Novatel, and Magellan. A larger list of manufacturers and system integrators can be found in the GPS World Buyer's Guide (http://www.gpsworld.com). As users, we are not required to pay a subscription fee to access GPS signals; thus, GPS is often viewed as "free of charge." In fact, the U.S. government via the taxpayers pays the bill, and current maintenance of the GPS is in the neighborhood of US$750 million per year.

GPS also is known as the **Navstar** system, and it is managed by the U.S. Department of Defense, specifically, the U.S. Air Force. There is no doubt that GPS is primarily a military tool, but its expansion and has created a huge multibillion dollar commercial industry. The Air Force Space Command is required to maintain at least **24 GPS satellites** in orbit, which is considered the baseline constellation; in actuality, the system is currently maintaining ≥30 operational satellites. The system is dynamic, and the GPS constellation and individual satellite status is updated daily (http://www.usno.navy.mil/USNO/time/gps/current-gps-constellation). The system is designed for use anywhere in the world, although the military does maintain control of it and could shut it off or scramble civilian use in any region of the world if they deemed it in the United States' best interests to do so.

As there are other international "GPS" systems in various phases of development, it now becomes good practice to consider the wider term **Global Navigation Satellite System (GNSS).** GNSS is the standard generic term for all satellite navigation systems that provide autonomous geospatial positioning with global coverage. GNSS systems include the U.S. Navstar GPS, the Russian Global Navigation System (GLONASS), the European Galileo system, the Chinese Beidou system, and others. The viable operation and future expansion of all of these systems relies on many political and long-term funding factors. It is undeniable the proven and fully operational U.S. Navstar GPS system is, and will be for some time, the benchmark GNSS system. Because of this system's status and its familiarity, we continue to use the "GPS" acronym throughout this section.

GPS technology and how it works is well documented. More in-depth study of GPS, its inner workings, and highly technical material on signal processing, etc., are available. There are several self-help guides (Letham 1998, Anderson 2002a, Kennedy 2009), if one desires a more detailed understanding. Many manufacturers also are good providers of overviews of the technology. For example, Trimble Navigation's GPS tutorial can be found at their website (www.trimble.com/gps/) and provides a good basic understanding of the technology without overstressing the technical details. Such magazines as GPS World (gpsworld.com) have a great deal of coverage of the GNSS/GPS industries.

GPS satellites transmit signals at specific frequencies in the microwave energy range that GPS receivers can **track and receive.** In simple terms, receivers range (determine their distance to the satellite transmitting the signal) by calculating the signal travel time and rate of speed, which is the speed of light. The signals also provide critical almanac and ephemeris data that tells our receivers exactly where the satellites are positioned in their orbital planes. If the GPS receiver can simultaneously track 3 satellites (4 if elevation is required), it can **triangulate** these ranges and calculate its position (Fig. 18.5, Box 18.4). How accurately the receiver can calculate its position depends on a variety of factors we summarize below.

Fig. 18.5. Three satellites are used for triangulation; the fourth satellite takes another measurement to check the other 3.

BOX 18.4. GLOBAL POSITIONING SYSTEM SATELLITES

Navstar GPS was developed by the U.S. Department of Defense and manufactured by Rockwell International. These 24 satellites are placed in 6 orbital planes 20,200 km (10,900 nautical miles) above Earth. Each plane is inclined 55° relative to the equator. They weigh about 710 kg and are 5.2 m wide with solar panels extended. Their orbital period is about 12 hr, and they pass over one of the ground stations twice a day. The lifespan of these satellites is planned at 7.5 years. Using ≥4 satellites can yield 3-dimensional estimates of location, whereas using 3 satellites can only generate 2-dimensional observations.

How Accurate Is GPS?

As selective availability (the government's intentional dithering of the GPS signal) has been discontinued since May 2000, we will discount its potential error effect (≥100 m) in the unlikely event that selective availability were ever turned back on. For practical discussion, all GPS receivers can typically provide 2-dimensional positioning (i.e., x, y) to within approximately 15.2 m in autonomous mode, that is, without the aid of any other correction augmentation sources. In fact, in autonomous mode, your US$100 recreational hunting receiver and a US$20,000.00 survey-grade rover really are not that far apart. Of course, 15.2-m accuracy is not going to cut it in many applications, including those in natural resource research that demand higher integrity in mapped locations. Thus, it is important to understand that **GPS accuracy** is based on 3 main aspects:

- the **type and quality** of the GPS **receiver** being used,
- the **quality of the signals** being tracked, and
- the **differential correction,** with another GPS receiver (known as a base station) collecting simultaneous GPS positions on a known geodetic survey point.

Types of GPS Equipment and Common GPS Uses

Although the accuracy of a position determined in autonomous (i.e., no correction augmentation) mode is not much different among GPS receivers, the type and quality of a given GPS receiver and antenna will yield significantly different results when accurate correction data are applied (we explain more about corrections in the section on differential GPS below). Thus, the user's receiver—also known as the rover—is the first (but not only) critical bottleneck in the quality of the positions one can expect. It is true that an inexpensive consumer-grade receiver can improve on the presumed 15.2 m inaccuracy of autonomous GPS by means of correction augmentation, but not to a typical accuracy of much better than 3.0–4.6 m, assuming good GPS conditions. If we need better than that, we need to use better receivers!

There are 3 common grades of GPS receivers. The first is consumer grade receivers, which are the least expensive and provide the least accurate positional information. The second is mapping grade receivers, which generally cost between US$1,000 and US$10,000 and are frequently used to develop information for GIS databases. The wide dispersion in cost of mapping grade receivers reflect a variety of accuracy capabilities, depending on make and model. However, many mapping grade applications focus on receivers that can achieve at least 1 m or even 0.3 m horizontal accuracy, which is far superior to what a consumer grade GPS receiver can achieve. As one would expect, these high-accuracy mapping grade systems will be at the higher end of this cost spectrum. Finally, the third grade is survey grade receivers, which are the most expensive and may require more data collection time, yet they can provide centimeter-level positional accuracy. Antennas can be self-contained, as in the case of many consumer and mapping grade receivers, or external, allowing them to be mounted on a pole, a backpack, a truck, or a person.

The choice of GPS receiver generally centers on, foremost, its accuracy capability for the project. Time occupation required (i.e., the efficiency of GPS collection) and cost considerations to achieve this accuracy must be weighed.

Quality of Signal

GPS is a **line of sight technology.** If the GPS receiver cannot "see" (track) the signals from the satellites, then it cannot calculate positions. If it can see them, but the signals are mitigated in their flight, we will not get accurate positions. The quality of these signals is critical to accurate positioning. There are 2 key aspects to signal quality, described below.

Geometric Configuration of the Satellites

If the tracked satellites are spaced apart sufficiently in the sky, the receiver will have better inputs to perform the math required for accurate positioning. If the opposite is true, we get greater dilution of precision (DOP) and our accuracy suffers. The DOP is used to describe the strength of a satellite configuration and its impact on the accuracy of the data being collected. Positional DOP (PDOP) is most commonly used indicator, and it represents the mean of the vertical DOP and the horizontal DOP. A major benefit of many higher quality GPS receivers is their ability to filter or set masks for DOP thresholds, ensuring only low DOP data is used (low DOP implies more accurate postions). Most consumer-grade GPS receivers do not allow the user to set a DOP mask and thus cannot filter out this high dilution of precision.

Satellite Signal Strength and Direct Flight Path

Strong signals from direct observable satellites in view **yield better** positioning **results.** A satellite directly overhead that is tracked by our receiver without being hindered by ground obstructions (trees, buildings, canyon walls, etc.) is an excellent input to the location calculation. Satellite signals can pass through forested canopies, yet they can be blocked by tree trunks or other landscape features. This blockage may lead to either no reception of the signal or reception of a multipathed signal. Other sources of error include atmospheric interference, problems involving the synchronization of clocks (satellite and receiver), and other receiver errors.

In forested conditions, mapping-grade receivers can obtain accuracy levels of ≤3 m after differential correction is applied (Wing 2008, Danskin et al. 2009). Depending on the model, consumer-grade GPS receivers may obtain accuracy levels of ≥5 m (Wing et al. 2005). Positional accuracy seems to increase as slope position increases, is better in the winter time (under leaf-off conditions), and can be improved slightly when the Wide Area Augmentation System (see below) is available. Further, differential correction significantly improves positional accuracy (Wing et al. 2008, Danskin et al. 2009). GPS test sites have been developed in Oregon (Clackamas GPS test network), Pennsylvania (Ridley Creek test network), and Georgia (Whitehall Forest GPS test site) to help study and publish the performance characteristics of different receivers under forest canopy (Box 18.5).

Critical Role of Differential GPS

Differential GPS (DGPS) is **mission critical** to driving our GPS systems to better accuracy beyond what receivers can do alone (i.e., autonomous mode). Simply explained, differential GPS is correcting our rover receiver's answers relative to another GPS receiver that is calculating locations at the same time. We call this second receiver a **base station.** What is so great about this base station? Well, technologically, not a whole lot. Base stations are basically subject to the same kind of **GPS errors** as ordinary receivers. However, they are

---

BOX 18.5. GLOBAL POSITIONING SYSTEM GROUND STATIONS

Known as the control segment, these stations monitor each satellite's functionality and exact position in space. They correct ephemeris errors, such as clock offsets, and transmit corrections to the satellites. There are 5 stations worldwide: Hawaii and Kwajalein in the Pacific Ocean; Diego Garcia in the Indian Ocean; Ascension Island in the Atlantic Ocean; and Colorado Springs, Colorado, which is the master ground station.

placed in excellent GPS conditions, such as on a tall building or tower (above any potential for signal obstructions). What really makes them so useful is they are stationary and placed on known survey locations. So, even when the base station is calculating imprecise GPS autonomous locations, our base station and correction systems can detect it and can correct the same errors that our rovers are seeing! Errors the base station detects, which are common to our rover data, such as atmospheric errors, can be effectively eliminated. DGPS cannot get rid of all errors from our rover, but it gets us closer to perfect data, depending on the quality of receiver and signals we are using, as described in the section above.

Post-Processing versus Real-Time Correction of DGPS

DGPS corrections can be delivered from the base station in 2 main ways, **post-processed and real-time.** Post-processing delivers the correction result after the rover has collected its data from the satellites. This method is adequate for applications where the data is being collected for map production or GIS database updates. The data accuracy is not critical in the field, but may be for the end-product. Real-time DGPS systems allow us to obtain our DGPS corrected accuracy (again, relative to the device type and quality of inputs) on the spot. This speed may be key in mission critical navigation, where autonomous accuracy will not bring us close enough to the desired target. For example, if our navigation map shows us the precise location of a buried well head that we need to excavate, a 15.2-m navigation circle around the exact location is a lot of wasted digging and is not practical! If we instead can use a sub-meter or better mapping-grade receiver with real-time corrections, we can drill into the target area within the standard size of an excavator's bucket.

Post-Processing

Not all GPS receivers support post-processing. To post-process, you will need the following.

- A GPS rover's receiver with internal software that supports collection of raw satellite information.
- Access to correction data collected by a GPS base station, preferably one dedicated to collecting GPS data on a known, verified surveyed position. The U.S. National Geodetic Survey runs a huge network of base stations that provides correction data for post-processing free to the GPS public (http://www.ngs.noaa.gov/CORS/).
- A post-processing software or post-processing service provider that is compatible with the data from your receiver. This would generally be sourced from the manufacturer of your receiver. Users also can use the National Geodetic Survey Online Positioning User Service at their website (http://www.ngs.noaa.gov/OPUS/).

Real-Time DGPS Corrections

Real-time DGPS opens up the ability to obtain your receiver's corrected accuracy right in the field. In the early days of GPS, real-time DGPS stations were more than likely limited to private companies, for example, a large surveying outfit or multimillion dollar mining or drilling operation. However, the real-time DGPS options and infrastructure have been expanding. There are a number of real-time sources, many free of charge, that one can use, depending on receiver capabilities. It should be noted that real-time DGPS depends on acquisition of the corrections by one's receiver hardware (typically by radio or satellite). One needs to determine whether the available real-time source will be blocked by obstructions or natural topography.

Wide Area Augmentation System

Because of GPS utility to fix an airplane's location in real time, the Federal Aviation Agency (FAA) has developed a plan called **Wide Area Augmentation System (WAAS)** that would **extend coverage** for DGPS **for the entire United States.** WAAS is a critical component of the FAA's strategic objective of a seamless satellite navigation system for civil aviation. This system improves the accuracy, availability, and integrity of GPS, thereby improving the system's capacity and safety. Ultimately, WAAS allows GPS to be used as a primary means of navigation from takeoff through Category I precision approach (i.e., very close to the runway, but not zero visibility; Category 3 landings are zero visibility). The ramifications of the FAA to maintain this system go well beyond aviation; because of its design, the system helps ensure that DGPS corrections will be accessible to all who need them. WAAS takes a consumer grade GPS to about 4.6-m accuracy, and with the right piece of mapping grade equipment, you can achieve about 0.6–0.9-m accuracy. Users working in closed canopy conditions will find it quite difficult to consistently obtain and hold the WAAS satellite signal. If you want to know more about WAAS, see the Garmin website (http://www.garmin.com/aboutGPS/waas.html).

U.S. Coast Guard DGPS

The **original** real-time DGPS **source,** the **U.S. Coast Guard** and other international maritime and navigation agencies, have established radiobeacons that transmit corrections to compatible receiver hardware. It is likely you would require a dual-frequency beacon receiver along with your GPS receiver. The radios transmit usually in the 300 kHz range. For information about the Coast Guard DGPS system, go to their website (http://www.navcen.uscg.gov/dgps/Default.htm).

Real-Time Kinematic and Their Networks

As discussed above, WAAS and the Coast Guard beacon DGPS systems have provided a great augmentation infrastructure primarily to our consumer grade and mapping grade receivers. However, WAAS and the Coast Guard beacons cannot directly achieve survey grade accuracy. If you are going to use survey grade GPS, your receiver and its base station will need to be capable of **real-time kinematic (RTK)** to effectively incorporate real-time centimeter positioning. Because the base station requirements in a RTK network require very proximal access to the base station (typically baselines ≤20 km), it has been very common for surveyors and other high precision users to set up their own base stations, transmitting corrections to their RTK rovers on a radio broadcast (e.g., 450–470 MHz).

There is now a push for many areas to fill in RTK networks. These are the RTK correction equivalents to WAAS and the Coast Guard beacons. Because of the short baselines, RTK networks require many more receivers to cover DGPS for a user base. However, we are seeing cooperative infrastructure that is building large RTK networks. The networks also are taking advantage of virtual reference station technology. A virtual reference station can greatly extend the baseline distances allowed to obtain quality RTK corrections to your rover. It also allows for redundant corrections to be applied from multiple base stations in the network to maximize the quality of results. RTK corrections from these networks can be "broadcast" over an Internet connection, making it possible for a user to subscribe to a data cell phone plan and obtain RTK corrections from the cell phone connected to the RTK rover.

A significant by-product of these RTK networks is they eliminate the need for users in the coverage area to set up their own RTK base stations, cutting down significantly on capital expense and operation setup time. Along with survey-grade RTK systems, some of today's sub-meter mapping-grade GPS devices (connected to a compatible data modem or cell phone) can accept RTK corrections from these networks as well. This capability provides a more affordable solution for users that do not need centimeter real-time positioning, but want to navigate to features to within as good as 0.2 m! An example of a sub-meter real-time capable solution is Trimble's 2008 GeoXH® device (http://www.trimble.com/).

GPS Uses in Wildlife Biology

Now that a cursory understanding has been given of what GPS is and how it functions, you may be wondering what its practical applications are in the field for wildlife. Two areas of use are predominant: (1) tracking and recording wildlife movements and (2) inventorying, mapping, and/or surveying wildlife habitats or specific wildlife use areas (Box 18.6). A **GPS tracking collar** can be used to track and record wildlife movements (Fig. 18.1) and provide more accuracy than other tracking systems can (Rempel and Rodgers 1997). Since 1994, there are a number of GPS collars available, and some use the Navstar GPS (e.g., those from Lotek, Telonics, and Televilt). For example, GPS collars have been used to successfully track moose (*Alces alces*; Rodgers et al.

> **BOX 18.6. GLOBAL POSITIONING SYSTEM RECEIVERS**
>
> GPS receivers can be carried by hand or installed in airplanes, boats, cars, or trucks. These receivers detect, decode, and process GPS satellite signals. Typically, hand-held receivers are about the size of a cellular phone or palm computer, and they are getting smaller all the time.

1997), grizzly bears (*Ursus arctos horribilis*; Servheen and Waller 1999), caribou (*Rangifer tarandus*; Dyer 1999), mountain lions (*Puma concolor*; Bleich et al. 2000) and wolves (Merrill and Mech 2000).

These collars now come in different sizes that can be used on small, medium-sized, or large mammals. The weight varies from 100–2,100 g (depending on the collar size), and the collars can store typically up to 10,000 locations non-differentially corrected or 5,000 locations differentially corrected, depending on recording frequency and battery configuration (Box 18.7). They can operate in temperatures ranging from –30° C to 50° C, and the data can be retained in the collar at temperatures ranging from –50° C to 75° C. Collars can be configured to allow periodic data downloads, or all the data can be transferred to a computer when the collar falls off. A source of concern, however, in using GPS collars lies with locating an animal, such as an elk, in a forest of varying density and topography. Rumble and Lindzey (1997) found that nearly 50% of attempted GPS locations failed in stands with >70% overstory canopy cover; in stands with less canopy cover, the percentage of GPS location attempts that failed was lower. Attempts to model the effects suggested a positive linear relationship ($P \leq 0.01$) between percentage of GPS location attempts that failed and the tree density, tree basal area, and index of diameter at breast-height times tree density. Dussault et al. (1999) and Gamo et al. (2000) also noted that **vegetation can block signals** from satellites to GPS radio collars. Therefore, a vegetation dependent bias to telemetry data may occur, which, if quantified, could be accounted for. As this technology evolves, steps to increase GPS efficiency in a forested environment may enhance its usability in these habitat types.

GPS technology also can be used to inventory, map, and monitor marine, fish, and wildlife habitats. For instance, GPS has been used to delineate coral reefs (Field et al. 2000), wildlife habitat types (Kiilsgaard 1999, O'Neil and Barrett 2001) and fish habitats (Martischang 1993, Threloff 1994, Waddle et al. 1997). GPS is first a **navigation tool** (Ander-

Box 18.7. Definition of terms for data capturing standards

Base station—a stationary receiver at a known location that provides the data used in the differential corrections of GPS data acquired by a moving receiver; a rover (field) receiver should be 500 km from the base station when using differential corrections. Depending on the accuracy requirements of your application and the equipment you are using, your baseline (distance between the base station and rover field unit) will be a key consideration in achieving effective corrections. For example, for some survey-grade receivers to achieve centimeter accuracy, they may need to be 10 km from a base station (and collect ample occupation points), whereas for an appropriate mapping-grade receiver to achieve submeter accuracy may allow a more liberal baseline of 150 km or more. Review the manufacturer's specifications for the accuracy requirements of your project.

Datum—a smooth mathematical surface that closely fits the mean sea-level surface, for example, the North American Datum of 1983 (NAD83), a geodetic data system used in satellite navigation systems to translate positions indicated on their products to their real position on Earth.

Elevation mask—a filter that ensures that the rover (field) receiver is using the same set of satellites as the base station. For distances 500 km to the base station, use 15°, for distances 1,000 km, use 20°.

Kinematic mode—the mode in which data points are collected at time intervals that vary, depending on the rate at which you are collecting data. Measurement interval will usually be ≤1 second; these data are stored in the receiver for later downloading and post-processing.

Positional dilution of precision (PDOP)—an indication of the quality of the geometry of the satellite constellation. The lower the PDOP value, the better the receiver can triangulate the satellite positions to provide accurate readings. However, other GPS errors can mitigate accuracy even when a good PDOP is being calculated by the user's receiver; thus, PDOP is just one indicator GPS users should evaluate in the field and in the office. Quality receivers can ignore (filter) satellite readings with unacceptably high PDOP; these receivers provide significant quality control improvements to the user who needs to validate GPS conditions relative to their reported positions.

Signal to noise ratio (SNR)—a measure of a signal's quality at the Global Positioning System (GPS) receiver. The higher the SNR value, the stronger the desired signal is compared with associated noise. A low value would indicate a weaker signal and/or higher levels of noise; for example, a SNR of 6 indicates a signal that should not used.

Spheroid—a spheroid of best fit over the surface of Earth; for example, the Geodetic Reference System 1980 (GRS1980) is a geodetic reference system consisting of a global reference ellipsoid and a gravity field model.

Static mode—the mode in which data points are collected at 1-second fixed intervals; a general guideline is to collect point positions at 1-second intervals. The amount of data collected varies with the type of receiver.

son 2002a) that allows researchers to accurately track their movements and guide themselves to an exact location (e.g., a coral reef) and then record the delineation of the location on a map. These wildlife habitat maps require the developers to interface GPS with a map database that would permit storage of information directly on the map (Box 18.8). Further, the maps need to display the GPS receiver's current position in real time, creating a moving map. By displaying their positions in real time, researchers can be sure of their locations and the location of what they are classifying.

Mobile GIS

Making useful decisions based on geographic information requires more than just accurate positions and GPS. We need GIS technology, but in the past, GIS was relegated to our desktops. We would collect data in the field and return it to our office computers to make decisions. In fact, for much of our field collection, we may have used paper maps and written capture data notes on paper. This process is very inefficient and leads to errors, not to mention extremely slow turnaround of the data back to our field people that

> **BOX 18.8. DESKTOP AND MOBILE VEGETATION MANAGER (VEMA) SOFTWARE RELATIONAL DATABASE**
>
> Resource managers require effective sampling tools to collect, archive, and report the data they use to make their management decisions. Desktop VEMA is a customized Microsoft Access(R) relational database that helps record, calculate, and report vegetation performance based on user-determined performance thresholds. The database was largely designed around a vegetation monitoring protocol developed by a team of agency and academic plant ecologists and expert practitioners to help standardize and automate vegetation monitoring and reporting at mitigation and reference sites. The database enables users to record vegetation data and measure the outcomes against peer-reviewed performance criteria and default or user-determined performance thresholds.
>
> Now, with mobile VEMA, the data collection protocol is handled on a mobile personal digital assistant (PDA). Mobile VEMA was developed to complement desktop VEMA by allowing users to capture vegetation data in the field electronically and to capture the Global Positioning System (GPS) coordinates (decimal degrees) where the data is collected. The field data recorded in mobile VEMA can then be transferred to desktop VEMA using a universal series bus (USB) enabled cradle for the GPS or PDA device, ActiveSync software, and a the embedded Vegetation Table Manager. Automated field data collection helps eliminate many data entry errors, because information is collected and entered into database only once. Although VEMA data can be manipulated and represented in Geographic Information System (GIS) software, as both shapefiles and feature classes in geodatabases, it is not yet a true geodatabase at the outset. We are continuing to strive to increase VEMA's efficiency, GIS compatibility, and field utility. Mobile VEMA is an excellent example of how mobile GIS is increasing field user's efficiency and extending office capabilities into the field.

rely on it. **Mobile GIS** is the extension of GIS from the office into the field. Mobile GIS integrates one or more of the following:

- mobile devices (e.g., personal digital assistants [PDAs]),
- GIS software,
- GPS technology, and
- wireless communications for Internet GIS access.

These technologies allow our field-based users to not only capture GIS features with accurate GPS devices, but also to access existing GIS data (GIS layers, aerial photography, and other necessary supporting data) when and where the decisions have to be made. O'Neil et al. (2007) drove over 80,467 km, visually inventorying and mapping wildlife habitats in the Willamette Valley of Oregon using a GPS device connected to a laptop with GIS data. Such extensive surveys are why mobile GIS is so critical. Now, field professionals could collect positions (from the GPS or other measurement device) and provide detailed database descriptions about the observations. They also can update existing GIS databases directly in the field, virtually eliminating the slow turnaround of paper edits to these databases. They can immediately notify authorities of mission critical events by submitting, from their wireless PDAs, emailed reports or automatic incident updates to the GIS.

ArcPad by ESRI (Redlands, CA), SOLO CE (http://www.timbersys.com/products/handheld-systems/solo-forest), and TerraSync Professional (http://www.trimble.com/terrasync.shtml) are examples of **mobile GIS platforms;** they feature map displays, **interface to GPS** and other measurement systems, and supply data collection and editing capabilities. ArcPad, which is a particularly GIS function- and feature-rich mobile GIS also is customizable for special projects that require particular work flows, data entry and edit forms, and even special tool bars.

USING REMOTELY SENSED DATA

Landsat Imagery

The **Landsat** series of satellites has been in operation since 1972, when Earth Resources Technology Satellite 1, later renamed Landsat 1, was launched. There is a rich historical description of the Landsat program and usage of its data in a variety of published articles (e.g., Cohen and Goward 2004, Williams et al. 2006). The interested reader should refer to those and related articles for a historical perspective. Landsat's contributions to **study of Earth** as a system are legendary, owing largely to its observation characteristics, which have steadily improved with each successive launch. These include global coverage, multiple monthly observations of the same location, excellent radiometric calibration and georegistration, approximately **30-m spatial resolution,** and multispectral sensors that include shortwave–infrared wavelengths.

The latest satellite in the series (number 8), the Landsat Data Continuity Mission (LDCM), is slated for launch in

December 2012. In keeping with the historic character of the Landsat program, the Operational Land Imager (the LDCM sensor) is an improvement over the existing Landsat 5 and 7 Thematic Mapper-class sensors. The existing 2 satellites are being managed for minimum fuel consumption, and hence maximum life span, with the hope their utility can be extended into the LDCM era. If successful, there will be no data gap, and the user community will be extraordinarily fortunate. If either or both Landsat 5 and 7 fail before the launch of LDCM, the sudden lack of new Landsat data will cause a scramble for other current datasets as a replacement.

As the name implies, LDCM is about **data continuity** (http://landsat.usgs.gov/documents/ldcm_factsheet.pdf). Thus, the specifications for both spacecraft and reflectance sensor have qualities at least as good as those of their predecessors. For example, there will be a minimum cross-track swath width of 185 km at the equator; multispectral band spatial resolution of 30 m; panchromatic band spatial resolution of 15 m; about 12 m spatial accuracy; essentially the same 6 spectral reflectance bands and panchromatic band, with 2 additional bands for coastal studies, aerosol detection, and cirrus cloud detection. There is currently no thermal sensor being built for LDCM, but the platform will have a place for it, and current negotiations suggest there will be a thermal sensor on the LDCM satellite, as a separate sensor from Operational Land Imager.

There is one major advancement, or even revolution, in the Landsat program associated with the new data policy. For the first time, all Landsat data acquired by LDCM will be **available free** to all users on a nondiscriminatory basis. Given the anticipated demand for data, to accomplish this goal, a single product recipe will be offered, which includes 30-m (**multispectral**) and 15-m (**panchromatic**) spatial resolutions, Level 1T (terrain corrected) data in GeoTIFF format, Universal Transverse Mercator (UTM) projection (except for Antarctica), World Geodetic System (WGS) 84 (used in satellite navigation systems to translate positions indicated on their products to their real position on Earth) datum, cubic convolution resampling, and web-enabled delivery.

Another extremely important change in the program is the archive of U.S. Geological Survey data from all previous Missile Defense, Thematic Mapper, and Enhanced Thematic Mapper+ sensors. All Thematic Mapper and Enhanced Thematic Mapper+ data in the U.S. archive are now available at no cost in the above standard recipe to everyone (Landsat Science Team 2008). Technical issues with the MSS portion of the archive are delaying the ready release of these data in a timely manner using the standard recipe. It is expected this problem will be solved within the year. Moreover, negotiations with international receiving station countries are expected to lead to a similar data policy for data in their archives.

Beyond LDCM, there is a plan for a National Land Imaging Program; http://www.ostp.gov/pdf/fli_iwg_report_print_ready_low_res.pdf). The National Land Imaging Program is intended to provide a long-term strategy for Landsat-class observations of Earth in perpetuity. At the current time, there is no traction in the program, but the Landsat Science Team is strongly advocating it at appropriate places in the U.S. government. This effort is important, because the road from declaration of a new satellite to actual launch can be long and circuitous.

Landsat Change Detection: 40 Years and Counting

Over the years, a variety of approaches has emerged for using Landsat data to characterize **land cover change** (see review by Coppin et al. [2004]). Most approaches involve 2 dates of imagery, or a series of images at intervals generally >3 years. These interval-based datasets were necessitated largely by the relatively high cost of Landsat data per image and data storage and computation limitations. With the availability of free Landsat data in a highly processed format, inexpensive storage space, and improved computing capabilities, several approaches for characterizing cover change with annual (Kennedy et al. 2007, Schroeder et al. 2007) or near annual (Healey et al. 2006, Powell et al. 2008, Vogelmann et al. 2009, Huang et al. 2010) Landsat time series are emerging.

Landsat Detection of Trends in Disturbance and Recovery

To describe this new time series class of Landsat change detection approaches, we provide a description of the Landsat Detection of Trends in Disturbance and Recovery (LandTrendr) algorithm (Kennedy et al. 2010). The LandTrendr algorithm is based on the concept of **trajectory-based change** detection, first presented by Kennedy et al. (2007).

With LandTrendr, the goal is to assemble an annual stack of images that has been geometrically registered, atmospherically corrected, and radiometrically normalized (Fig. 18.6). Then, for each pixel, the spectral response over time for a given band or spectral index (the pixel's temporal trajectory) is examined, and a set of statistical and knowledge-based rules is imposed to fit segments with consistent trends. This analysis results in a single trajectory being segmented into 1 to n segments, depending on the number of distinct breaks in the trajectory identified by the algorithm. For example, in Figure 18.6, the fitted line for the forested pixel identified as having been clearcut between 1985 and 1986 has 2 segments: 1 for the clearcut and the other for the slower process of vegetation recovery after the disturbance. In contrast, the green fitted line is associated with a forested pixel that has 3 distinct segments: 1 stable segment where spectral response was essentially constant between 1985 and 1991, at which time the forest was thinned (trees were cut to make them less crowded), and a third segment associated with the recovery after thinning.

Temporal trends in pixel trajectories can be noisy (Fig. 18.6), due to sun angle, atmospheric effects, and phenologi-

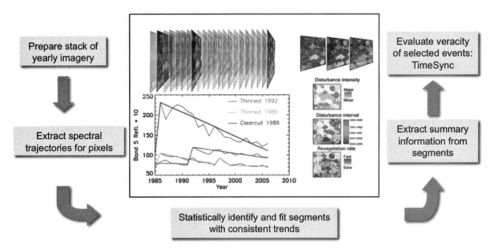

Fig. 18.6. Schematic of how the LandTrendr algorithm works.

cal state differences among images used to create a given stack. For example, the recovery segment for the clearcut pixel response (red fitted line) shows a fine temporal grain instability, whereas the longer-term trend is clearly directional. The statistical fitting rules examine multiple temporal scales of spectral response to best select the break points in the trend (vertices) and thus properly segment the full trajectory.

The map output from LandTrendr is rich and must be summarized for practical utility. The most basic information is maps of disturbance interval and magnitude, in terms of relative or absolute vegetation loss (Fig. 18.7). The vegetation loss is relative if in terms of degree of spectral change, or it is absolute if the spectral band or index has been statistically related to a biophysical variable, such as percentage cover. Most disturbances are of the event type, whereby a sudden change occurs in a 1-year observation period. However, as is usual for recovery, some disturbances also occur over multiple years. This time frame is common for forest insect damage, for example, and leads to several other possible map summaries,

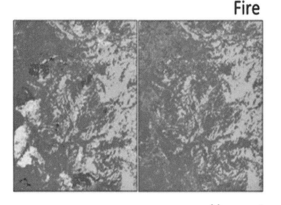

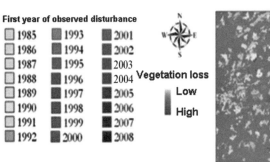

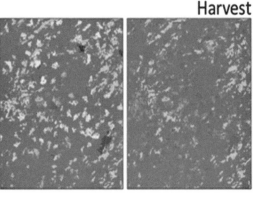

Fig. 18.7. Sample output maps from LandTrendr for areas disturbed by fire and harvest.

such as disturbance and recovery duration and recovery magnitude. Because different types of disturbance (e.g., insects or harvests) have different characteristic durations and spatial patterns, it should be possible to identify different disturbance agents using the LandTrendr algorithms. This possibility is being explored and improved as the algorithm matures (Kennedy et al. 2010).

With LandTrendr and related algorithms that focus on an annual time step, clouds can be a persistent problem in some locations. Moreover, the Landsat 7 Enhanced Thematic Mapper+ instrument developed a problem in 2003 that has led to "wedges" of missing data in every image collected since that time. With the ready availability of free data, it is now possible to operationally composite images in such a way that pixels covered by clouds (and their shadows) and those that contain no data (Landsat 7 from May 2003 to present) are replaced with suitable image data from another relevant time period. Compositing is an important feature of the LandTrendr algorithm.

Creating a set of rules to derive maps of **land and/or vegetation cover** from Landsat data can be expensive and time consuming. But more often than not, it is highly desirable to know not just that an area changed, but also that it changed from one cover type or quality to another. Thus, the typical approach of developing 2 separate land cover maps (1 for each of 2 dates) and comparing them was developed long ago and persists today. Scene-level radiometric normalization procedures (e.g., Canty and Nielsen 2008) were developed to address this issue by facilitating more meaningful application of the same classification rules to 2 dates of imagery, but scene-level normalization procedures are a relatively crude solution for the problems associated with variable sun angle-viewing geometries, atmospheric conditions, and phenological conditions among image dates, which can be pixel specific.

LandTrendr uniquely addresses this issue by performing a pixel-level normalization based on the trajectory itself (Fig. 18.8). Given that annual time series express the inherent noise of Landsat imagery even after scene-level normalization, but that meaningful changes in spectral direction are confidently detectable (Fig. 18.6), the LandTrendr approach uses the fitted trajectory segments to normalize a pixel's spectral trajectory. In other words, the spectral response for all bands and/or indices, over time for a given segment, are expressed as the fitted values for that segment rather than as the original (or scene-level normalized) values. This format allows for a more meaningful use of classification rules developed for a single year of Landsat imagery to be applied to other years of imagery. Furthermore, because we have an annual time series, those rules can be applied to all images in the series, resulting in an annual time series of land and/or vegetation cover maps that express stable class labels for unchanged areas and new labels for changed areas as the times series progresses (Fig. 18.9).

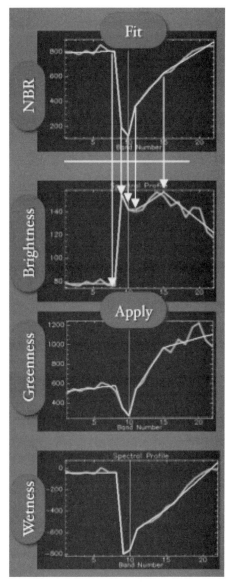

Fig. 18.8. Example of how LandTrendr fits the temporal spectral trend of a given pixel (the magenta line). The fitted line (gray) is based on regression for each segment of a given band or index (in this case the normalized burn ratio [NBR]). The vertices for each segment were derived from the fitting of the NBR, and the same vertices are transferred to the other indices or bands (in this case brightness, greenness, and wetness). Fitting is repeated for each band or index of interest.

Validation of Landsat Time Series Maps

Maps of **forest change** that have an annual time step and are for large areas are **difficult to validate.** As we cannot visit areas today and clearly determine their history, we must rely on extant datasets. Extant datasets are commonly for a single point in time; even for areas visited more than once, it would be the exception rather than the rule they were visited on an annual basis over the period of interest.

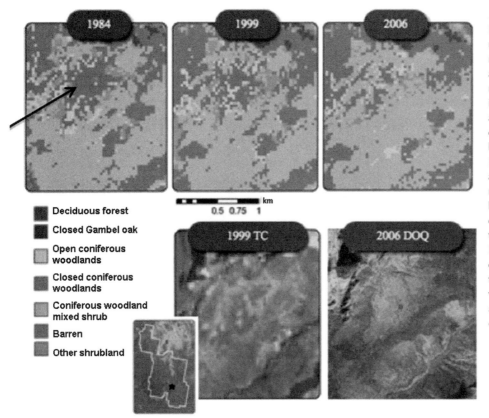

Fig. 18.9. Example of how LandTrendr fitted time series is used with a set of classification rules. Shown are three dates from a time series, where classification rules were applied to fitted imagery. The arrow points to an area that changed from barren to coniferous woodland mixed shrub between 1984 and 2006 in Zion National Park, Utah. Also shown are a 1999 Tasseled Cap transformation (TC) image and a 2006 high-resolution digital orthophoto quadrangle (DOQ) photo. The TC transforms 6 Landsat Thematic Mapper™ data channels to 3 data channels with known characteristics (soil brightness channel, vegetation greenness channel, and moisture content of soil and/ or vegetation [wetness] channel).

Moreover, given that maps are spatially explicit at a 30-m resolution, there are large areas where no extant datasets exist. For Landsat change detection in forested areas, this problem was addressed by Cohen et al. (1998). They found that by observing and interpreting the Landsat images, one can unambiguously detect forest stand replacement disturbances visually in relatively dense forests. This was verified by comparing visual Landsat assessment with **airphoto interpretation** and polygon databases from a federal agency. In the era of time series maps developed over large areas, the problem of validation is exacerbated by the temporal richness and spatial extent of the maps. To address this issue, Cohen et al. (2010) developed a Landsat time-series visualization tool, TimeSync. Similarly, Thomas et al. (2011) have developed a different tool for the same general purpose.

TimeSync derives its name from the concept of syncing automated and human interpretations of Landsat time series. TimeSync has 4 main components a: chip window, trajectory window, Microsoft Access database, and a Google Earth interface. For an area of interest (a 3-pixel × 3-pixel plot), TimeSync displays an image chip series in the chip window and the spectral time trajectory for any Landsat band or vegetation index of interest. For a plot that is spectrally stable, the image chip series will reveal a stable spectral response, and the trajectory in various bands and indices will reveal the normal noise associated with time series, but will show no meaningful breaks in spectral trend (Fig. 18.10). For changed areas, both the chip and trajectory windows will reveal those changes (Fig. 18.11). As the plots are interpreted, the analyst uses the TimeSync interface to enter data into the database that contains the interpretations made on the series of plots under investigation. Because of the dense time series, as with LandTrendr, TimeSync facilitates interpretations of both high and low intensity disturbances and assessments of recovery trajectories.

Google Earth is used to examine a high-resolution **georeferenced image** of plot. This image serves to assist the analyst in describing the plot trajectory, as the date of the image displayed in Google Earth is commonly time stamped. Moreover, the newest version of Google Earth has historical imagery for many locations, facilitating a direct interpretation of change independent of the Landsat imagery.

Using TimeSync, one can design a statistically valid sample that covers the full time period of interest at an annual resolution whenever a Landsat time series was developed for analysis by an automated algorithm. Once the data have been collected with TimeSync, they can be summarized and compared to map output (Fig. 18.12). **Ancillary datasets** from forest harvest records, for example, can still play an important role in the map validation process. Using these data, we can assess the level of consistency between the TimeSync interpretations and the information in the extant databases. When the whole approach is integrated, one has

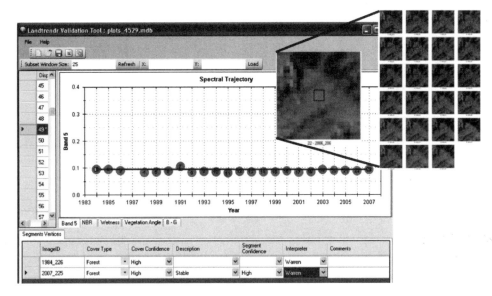

Fig. 18.10. TimeSync visualization, showing a set of image time series chips (upper right) and the spectral trajectory for Landsat band 5 for the 3-pixel × 3-pixel area (plot) at the center of the chips. This plot was unchanged throughout the time series.

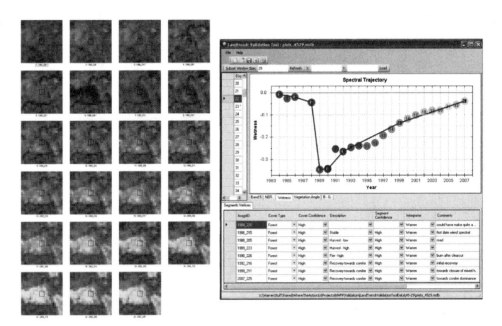

Fig. 18.11. A TimeSync example for a 3-pixel × 3-pixel plot that was disturbed.

a statistically valid assessment of map quality via TimeSync and an independent assessment of TimeSync observation quality. Although it is tempting to call this process an accuracy assessment, in reality, there are errors in all component observations, interpretations, and databases, and it may be best to consider the result a series of agreement analyses. For a further review of how Landsat imagery works and for data sources see O'Neil et al. (2005).

Lidar—Light Detection and Ranging
Characterizing Habitat with Lidar

Lidar (*l*ight *d*etection *a*nd *r*anging) is an optical remote-sensing technology and geospatial mapping tool that allows for the fine-grained yet broad-scale collection of high resolution, accurate terrain surface data that can provide valuable information about the 3-dimensional structure of terrestrial and aquatic ecosystems (Lefsky et al. 2002, Vierling et al. 2008). Although **lidar** technology has been around for >2 decades, recent advances in lidar technology over the past 5 years have revolutionized mapping and have the potential to contribute greatly to our understanding of wildlife–habitat associations. Providing better vertical resolution and sampling density than can be achieved by traditional field assessments, lidar is a powerful tool for investigating questions in wildlife ecology at multiple spatial scales using new, previously unmeasurable, habitat features. Particularly in **remote, rugged, and otherwise inaccessible terrain,** lidar has the potential to become a viable, and in many cases su-

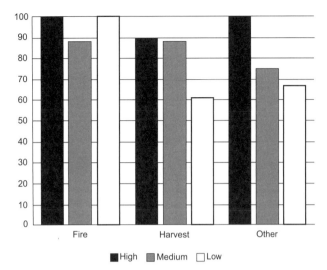

Fig. 18.12. Results of a comparison between LandTrendr output and TimeSync observations for approximately 500 plots. Shown are percentage agreement on disturbance for a variety of TimeSync-observed disturbance agents and intensities.

perior, surrogate for assessing wildlife habitat features traditionally measured in the field.

Lidar data and their applications are young, but are rapidly gaining recognition. Although many researchers have quantified habitat features of relevance to wildlife, few studies have actually used lidar data to aid our understanding of wildlife–habitat relationships, and there are many applications that have yet to be explored. Moreover, the technology is ever-changing, and the possible trajectories of advances make it difficult to predict the dominant and useful applications of lidar to wildlife science in the future. It will be well worth monitoring how the field advances to seize opportunities to better inform wildlife and conservation science.

Despite the recent rapid advances in the ability to achieve high-density lidar data, it is highly important to recognize that **not all lidar datasets are equal**. Differences in **acquisition specifications** (e.g., flight altitude and sensor settings) and **ground truthing** methodologies will result in lidar datasets of differing resolutions and accuracies. Moreover, the data products themselves will determine the utility of the data for different applications. It is essential for any biologist to understand the **strengths and limitations** of a lidar dataset before it is used as a predictor variable from which to draw inferences regarding animal–habitat relationships.

Lidar Technology

There are 3 general classes of lidar instruments that differ in mode and therefore the scale of data acquisition: **ground-based, spaceborne, and airborne lidar.** Ground-based lidar systems can be stationary (mounted on a tripod) or mobile (i.e., vehicle mounted), and collections of dense point clouds from multiple vantage points provide for detailed views of study areas at small spatial scales (meters). Satellite sensors currently in operation include the Geoscience Laser Altimeter System NASA 2009) and the Vegetation Canopy Lidar mission, both administered by NASA. Applications of satellite sensors are typically to answer questions at very large spatial scales, such as deforestation rates on a continental scale (e.g., Potter 1999, DeFries et al. 2006), and coarse measurements of forest structure (Dubayah et al. 1997, Lefsky et al. 2005). Although ground- and satellite-based sensors certainly can be applied to studying very fine or broad scale questions in wildlife biology, airborne sensors have been used to collect the majority of data at spatial scales that biologists are typically most interested in (from the watershed scale down to individual habitat parcels of relevance to wildlife). Among airborne lidar instruments, there are currently 3 primary sensor types with great potential for providing valuable terrain data to inform wildlife biology studies: discrete-return infrared (IR) lidar; full waveform IR lidar; and high-energy, full-waveform green lidar.

Discrete-return IR lidar instruments record the returning laser signal as one or more distinct returns and operate using a laser wavelength of 1,064 nm (the near infrared range of the spectrum) with laser pulses emitted at 10–200 kHz (Haugerud 2009). Discrete returns are simplifications of the continuous waveform into as many as 4 principal peaks corresponding to echos (returns) off of surfaces in the travel path of the laser pulse (Fig. 18.13). Quantity of discrete-return lidar data is measured as pulse density, which typically ranges between 1 and 12 points/m² (0.3–1.0-m spot spacing). Because **infrared light does not penetrate water**,

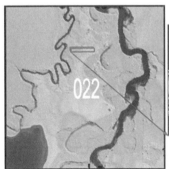
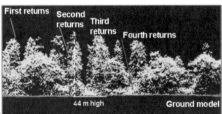

Fig. 18.13. Ten-meter deep cross-section of multistory canopy, showing laser returns (point cloud) of dense vegetation (right) and planar view of the surface derived from classified ground points (bare earth triangulated irregular networks; left). These data were acquired at a pulse rate of 70 kHz, Yamhill River Basin, Oregon. *Lidar data acquired and imagery produced by Watershed Sciences.*

discrete-return IR lidar instruments are principally used to map terrain that is above the water surface at the time of the survey. In contrast to discrete-return lidar instruments, **full waveform sensors** capture the entire return profile, providing highly detailed information and allowing biophysical features, such as canopy structure and aboveground biomass, to be estimated with unprecedented accuracy (e.g., Harding et al. 2001, Hyde et al. 2005).

High-energy, full waveform, green lidar instruments are used to provide bathymetric data at shallow (20 m) water depths. As such, their potential relevance to wildlife studies include characterizing habitat for aquatic wildlife (e.g., amphibians and riverine or coastal marine mammals). Green lidar sensors use information from the entire waveform and operate using a laser wavelength of 532 nm with laser pulses emitted at 0.8–4.0 kHz (Haugerud 2009). Typical pulse densities for full waveform green lidar vary from 0.05 to 0.25 points/m^2 (2–4-m spot spacing), much lower than can be achieved with near IR lidar, yielding data of lower resolution. Green lidar data may be advantageous by providing bathymetric data in shallow streams, where a hydroacoustic survey may not be possible. The prominent available green lidar systems include Compact Hydrographic Airborne Rapid Total Survey (CHARTS; Joint Airborne Lidar Bathymetry Technical Center of Expertise [JALBTCX 2009]; available commercially from Fugro Pelagos), Hawk-Eye (Airborne Hydrography AB 2009), and Experimental Advance Airborne Research Lidar (EAARL; U.S. Geological Survey 2009a). Although not available commercially, the EAARL system's short pulse width (5 kHz) and narrow field of view allows for simultaneous mapping of bare-earth topography and shallow submerged topography, enabling characterization of the littoral zone of coastal and riverine environments. The EAARL system is operated by the U.S. Geological Survey.

Lidar Data

Using rapid pulses of light photons, lidar systems detect and determine the **distance (range) of terrain features** to provide the location of objects in (x, y, z) space. When an emitted laser pulse hits a target surface, portions (most for full waveform) of that laser energy will be returned to the sensor. Given the speed of light, aircraft position and attitude, and time elapsed between laser pulse emission and reception of these returns, the spatial coordinates (latitude, longitude, and particularly, elevation) of reflected laser pulses can be calculated. Reflections off tall or high objects have faster return times than those off the ground surface. The pattern of this information allows discrimination of such features as trees or buildings from the ground surface. Current sensors can resolve laser information fired and returned at rates of up to 200,000 pulses of light per second and as many as 4 returns per pulse. Depending on the flight parameters, the lidar sensor is capable of essentially "painting" the study area features using a multitude of laser pulses that are returned to the sensor.

A lidar system includes a scanning pulse laser scanner and receiver, a GPS device, and an inertial measurement unit (IMU). Throughout the survey, the collection of **real-time GPS** and IMU data enables calculating the (x, y, z) position and exact orientation and attitude (roll, pitch, and yaw) of the lidar sensor with a precise time reference. Combining this information with the time reference encoded for each laser pulse return yields the absolute position of each laser pulse return. Finally, concurrent operation of ground base stations strategically placed to provide geodetic survey support for the mission enables adjustment and correction of the laser positions to result in a dataset of high absolute accuracy.

The spatial data provided by current lidar technology enables the generation of **3-dimensional models** of the surface of **Earth and land cover** that are of amazingly high resolution (Fig. 18.14). Lidar data are far superior to the traditional topographic maps researchers have worked with in the past: current datasets are correct to within a few centimeters of their true absolute elevation in space and to within a few meters horizontally. In addition, compared to traditional survey techniques, lidar has revolutionized the speed with which surface terrain data can be collected over large survey areas, making it relatively inexpensive to collect per unit area.

Raw lidar data are collected as a cloud of (x, y, z) points. The **points** are associated with a myriad of features, including not only the ground, but also manmade structures (buildings, bridges, powerlines, vehicles, etc.) and vegetation. To extract a topographic surface, the data needs to be processed to generate clean **surface models.** Processing includes automated and manual techniques for modeling ground-level and aboveground points and filtering points erroneously classified as ground. A triangular mesh of the points is

Fig. 18.14. Detail of bare earth (top) and highest hit (bottom) lidar-derived digital elevation models (1-m resolution) in the Siskiyou National Forest, Oregon. *Lidar data acquired and imagery produced by Watershed Sciences.*

then output as a grid, or **digital elevation model (DEM)** at resolutions as high as 0.5 m, depending on resulting ground point density and terrain. Hill shading and coloring can be applied to display final maps. The DEM can be generated from only the ground-classified points (bare earth model), or from the first returns (highest hit model). From the data cloud, specialized remote sensing software packages, such as FUSION (McGaughey 2009), are helpful in further characterizing vegetation and generating canopy surface models.

Applications to Wildlife Biology

Two general classes of lidar data products have been useful and provide great promise for studies in wildlife ecology. These include the point cloud of data points, elucidating vegetation structure, and the digital elevation models generated from the ground-classified points, providing **detailed information on surface topography.** Naturally, these 2 classes relate to applications in wildlife ecology in 2 general ways: (1) use of point cloud data and lidar-derived canopy surface models in studies of terrestrial vegetation, primarily forests (Figs. 18.15, 18.16), and (2) use of digital terrain models in studies related to geomorphology and topography, in both terrestrial and aquatic environments (Fig. 18.17).

VEGETATION STRUCTURE

For a great majority of terrestrial wildlife (e.g., birds and mammals), the structure of vegetation (placement, quantity, shape, type, and connectivity) is of great interest for understanding species–habitat associations (Hansen and Rotella 2000; Fig. 18.2). For these studies, of the data provided by lidar instruments, it is the point returns from the vegetation that are of primary interest. Using lidar, **canopy structure** can most simply be measured and defined by canopy height and cover (Lefsky et al. 2002). High agreement between lidar-derived data and field measurements for canopy height, cover, and biomass have been observed in temperate, boreal, and tropical forests a well as desert scrub (e.g., Ritchie et al. 1995; Nelson et al. 1997; Magnussen et al. 1999; Hyde et al. 2005, 2006; Clawges et al. 2007). In addition, compared to other active sensors (e.g., radar) and passive optical sensors (e.g., Landsat and Quickbird 4-band imagery), lidar has been found to be the best single sensor for estimating canopy height and biomass (e.g., Hyde et al. 2006). Additional descriptors of canopy structure, such as the height distribution of outer canopy surfaces quantifying light gaps, also have been made with lidar (Lefsky et al. 1999). The 3-dimensional vertical distribution of all material in the canopy, or foliage-height profiles, can be measured with full waveform IR lidar (Harding et al. 2001). Finally, combining lidar data with ancillary data can provide measurements for a variety of wildlife-relevant vegetation patch characteristics, including stand density, age, and volume; number of snags and downed trees; number of large trees; basal area; understory height and density; ground surface texture; and landscape metrics, such as patch/edge characteristics and the landscape matrix (Dubayah and Drake 2000, Lefsky et al. 2002, Turner et al. 2003, Corona and Fattorini 2008; Fig. 18.16). The Precision Forestry Cooperative of the University of Washington is an active group researching the many forestry applications of remote sensing, including lidar (Precision Forestry Cooperative 2009).

Lidar data has been used as an analytical tool to **predict species distributions** based on prior knowledge of species' habitat preferences. For example, Nelson et al. (2005) used lidar to identify potential habitat for the Delmarva fox squirrel (*Sciurus niger cinereu;* known to prefer tall, dense forests with an open understory) across extensive areas in Delaware to guide survey efforts for this endangered species. In addition, lidar data can be used as an exploratory tool to determine **habitat selection patterns.** Broughton et al. (2006) used lidar data to determine habitat preferences of marsh tits (*Poecile palustris*) by comparing lidar-derived vegetation structure in territories to that in surrounding locations that were not occupied by the species. As determined from the lidar data, birds occupied sites containing mature trees with a subcanopy shrub layer, and they avoided sites containing many small young trees. The relationship between lidar-derived woodland canopy height and structure and songbird (great tit [*Parus major*] and blue tit [*Cyanistes caeruleus*]) breeding success in local territories was investigated to develop a map depicting habitat quality for these species across an entire woodland study forest in England (Hinsley et al.

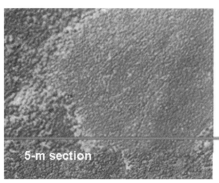 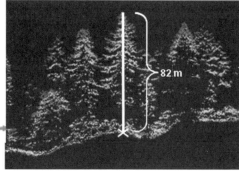

Fig. 18.15. Example of lidar-derived highest-hit digital elevation model of forest habitat (left) with a 5-m cross-section of the lidar point cloud (right). From this data, forest structure can be quantified, and patch characteristics can be inventoried and stratified using data from the multiple laser returns that paint the canopy vegetation. *Lidar data and imagery acquired by Watershed Sciences.*

APPLICATION OF SPATIAL TECHNOLOGIES IN WILDLIFE BIOLOGY 455

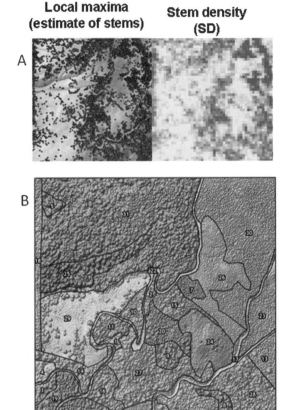

Fig. 18.16. Examples of stem density measurements and forest stand delineations developed from lidar-derived vegetation heights. These can be used as valuable habitat attributes describing components of forest landscape structure of relevance to the habitat associations of wildlife. (A) The tops of individual trees (stems) >2 m in height (local maxima) are identified, enabling the generation of a 30-m resolution stem density grid. (B) In this example, with ancillary Ikonos-2 color infrared data, delineations are based on a number of lidar-derived data outputs, including elevation, terrain aspect, slope, stem number and density, canopy density, vegetation height, and percentage distributions of conifer, deciduous, and nonoverstory vegetation. *Lidar data and imagery acquired by Watershed Sciences.*

in Maryland (including forested wetlands; Goetz et al. 2007) and mixed conifer/aspen forests in South Dakota (Clawges et al. 2008). In particular, Goetz et al. (2007) found that vertical distribution of canopy elements as characterized by lidar was the strongest predictor of species richness; Clawges et al. (2008) found that lidar-derived foliage-height diversity indices were positively correlated with indices of bird species diversity, particularly near the forest floor (5 m from ground). Boelman et al. (2007) used lidar data to determine how an invasive plant species has affected avian species abundance and community composition across a range of Hawaiian submontane forests, finding that total avian abundance and the ratio of native to exotic avifauna were highest in stands with the highest canopy height and cover, but that among biophysically equivalent sites, stands dominated by

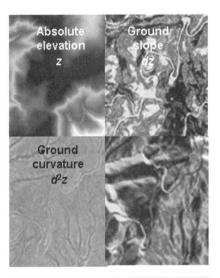

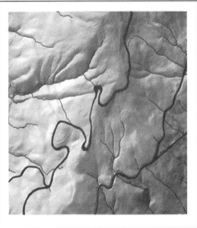

Fig. 18.17. Example of topographic parameters that can be derived from high resolution lidar data and applied to questions of interest to wildlife ecologists. In this example, terrain elevation, slope, and curvature (top) are used to delineate the detailed routing of streams and roads (bottom) for use in hydrologic flow modeling. *Lidar data and imagery acquired by Watershed Sciences.*

2002, Hill et al. 2004). In addition to forested ecosystems, lidar-derived characteristics of vegetation structure have been made to examine wildlife–habitat relationships in low-lying shrubland (Streutker and Glenn 2006), farmed landscapes (e.g., sky lark [*Alauda arvensis*] and other grassland species in English's cropland; Davenport et al. 2000, Mason et al. 2003, Bradbury et al. 2005), and dune vegetation (feral horses on Assateague Island National Seashore, MD; Stoppelaire et al. 2004).

Lidar data also has been used to evaluate wildlife community structure. For example, relationships between lidar-derived measures of vegetation structural diversity and bird species diversity have been found in both deciduous forests

native tree species hosted larger native avian communities than did mixed stands of firetree (*Morella faya*), an invasive evergreen shrub.

TOPOGRAPHY

For characterizing topography, in certain situations, lidar provides a number of advantages over traditional photogrammetric mapping and field methods (e.g., to detect the ground under dense forests where the ground is not visible, and where the time and labor costs would preclude an equivalent volume of data collection). Given that **terrestrial and bathymetric topography** can affect components of landscape structure and ecological processes (e.g., erosion and hydrologic flow; e.g., Fig. 18.17) that impact wildlife, lidar-derived bare earth models (DEMs) also have been used in wildlife studies and hold great promise for further application development. For example, lidar ground models have been used to characterize wetland topography (Irish and Lillycrop 1999); the surface dynamics of beaches and dunes (Krabill et al. 2000); underlying forest topography (Harding and Berghoff 2000); and the structure and elevation of abandoned, paleo, and overflow stream channels (Jones 2006); they also have been used to map river corridors and riparian habitats (see Hilldale et al. 2009, McKean et al. 2009). All these features of terrestrial and aquatic wildlife habitats may be significant determinants of habitat selection, population distributions, and responses to restoration efforts for some species. There are few studies, however, that have directly related a detailed understanding of topography to questions of interest to wildlife biologists. Among these are an effort to understand the impacts of grazing by feral horses on vegetation and dune topography on an island national refuge in Maryland (Stoppelaire et al. 2004) and the use of elevation, among other lidar-derived metrics, as a predictor of bird species richness in temperate forests of Maryland (Goetz et al. 2007).

A number of studies have demonstrated the usefulness and applications of using accurate DEMs. Although many of these researchers used 10–30-m resolution DEMs, it is not difficult to see how their applications would be strengthened by ground surface models of a higher resolution, such as the 1-m DEMs derived from current lidar technology. For example, Jenness (2004) demonstrated a calculation of landscape surface area and topographic roughness relevant to characterizing wildlife habitat using digital elevation data. Goetz (2002) looked at how to use topographic information from DEMs to better resolve different riparian zones of critical importance to fish and wildlife in Idaho and Wyoming. Brandes and Ombalski (2004) modeled **raptor migration pathways** based on knowledge of terrain updrafts calculated using wind direction, terrain slope, and terrain aspect data determined from DEMs in a simulation experiment. The usefulness of DEMs in terrain representation models for ungulates residing in mountainous environments has been repeatedly demonstrated: Wairimu (1997) evaluated the effectiveness of DEMs in locating and mapping wintering areas of white-tailed deer (*Odocoileus virginianus*), Divine et al. (2000) assessed the use of DEMs of varying resolutions to categorize desert bighorn sheep habitat in Mojave Desert mountain ranges, Locke et al. (2005) appraised escape terrain (slopes >60%) using DEMs for desert bighorn sheep to assess the probability of translocation success on public land in Texas, and Wallin and Wells (2006) used DEMs to evaluate predictors of seasonal variation in habitat selection for mountain goat (*Oreamnos americanus*) in the Washington Cascades.

Future Opportunities and Applications

The accuracy and spatial scale of lidar data have already brought it to the forefront of recognition in terms of its utility for certain applications in remote sensing. However, lidar data have only recently become a research tool for wildlife biologists, and we have far to go in terms of tapping into that tool's full research potential (Vierling et al. 2008). The one class of sensors that may truly elevate the utility of lidar, particularly for forest wildlife studies in the future, is full waveform IR. Lefsky et al. (2002) include further information on applications and utility of waveform lidar.

Research applications of lidar are most notably missing in terms of geographic scope, habitat types, and the species groups for which its use has been explored (the majority of studies have only thus far used lidar to characterize habitat and evaluate wildlife–habitat associations in forested systems, primarily for birds and to a lesser degree for mammals, and principally in the United States and England). Some studies have ventured into **evaluating the accuracy and source of errors** in characterizing wetland habitats using lidar (e.g., Hopkinson et al. 2005), and some researchers have evaluated the use of various lidar sensors to map stream channel characteristics for fisheries research (see Bayer and Schei 2009). But the use of lidar to examine the influence of features of aquatic habitats (e.g., rivers and wetlands) that are of relevance to wildlife species distributions, habitat use, movements, survival, and restoration response has yet to be developed. Moreover, applications of ground-based lidar to examine wildlife–habitat associations for **small organisms** (e.g., amphibians and insects) with finer spatial scale habitat requirements is another greatly **untapped area** for lidar applications in wildlife studies, although it has gained recent popularity for urban and corridor mapping.

With future advances in lidar technology and applications, lidar may make its greatest contribution yet to the field of landscape ecology. Many researchers recognize the need to study species–habitat relationships at multiple spatial scales relevant to individuals, populations, and metapopulations to effectively address the pressing global conservation issues of today. For a number of questions of interest to wildlife scientists, lidar data have the potential to revolutionize the scale and dimension of questions posed by landscape ecologists, as well as to refine the resolution and accuracy of detected patterns. In addition, advances made by the use of lidar, particularly when combined with other remote sensing data, are likely to impact large-scale conservation programs and have far-reaching policy and management implications. A prime example includes the Gap Analysis Program led by the U.S. Geological Survey (Scott et al. 1993), in which lidar data are being incorporated into GIS databases to map on a national scale the key habitat classes of importance to native species distributions, areas of high biodiversity, and areas of priority for conservation (Vierling et al. 2008).

Data Availability

At present, discrete-return IR lidar surveys **costs can range widely** (approx. US$1.25–2.50/ha for areas >45,000 ha), depending on data quality, specifications, and size and shape of study area; ground-based lidar data are considerably more costly per unit area than IR lidar data (but is usually purchased only for small areas), and commercially available bathymetric data (e.g., CHARTS) are an order of magnitude more expensive. Although costs for lidar data can be high relative to other remote sensing data, the multi-utility nature of the data has encouraged collaborative partnerships for data sharing and acquisition. For instance, a growing number of states (Iowa, Louisiana, Ohio, Pennsylvania, North Carolina, Texas, and Florida) may soon have discrete-return IR lidar coverage, and others (Oregon and Washington) have formed consortiums of agencies devoted to collecting public lidar data at reduced cost for large, regularly shaped areas of interest (e.g., DOGAMI 2009, PSLC 2009). A visit to the Center for Lidar Information Coordination and Knowledge (CLICK) website (U.S. Geological Survey 2009b) will provide more information on collaborative cost sharing programs for large-scale public acquisition of lidar data.

Accuracy Assessment of Remotely Sensed Data

A variety of devices and techniques, such as **Landsat Imagery** and **Forward Looking Infrared** systems, can be used to record characteristics of Earth's surface from remote positions. However, interpretation of remotely sensed data can introduce error (Janssen and van der Wel 1994). Error in mapping can be generated in several ways: error in thematic classification, both by **omission** and by misclassification (**commission;** Story and Congalton 1986) and error in **cartographic delineation** (location error).

Accuracy assessment of landscape maps generated from remotely sensed data is generally accomplished through field verification of a select subset (samples) of thematic or areal map units. The investigator must identify accuracy assessment objectives as well as the level of error acceptable for accuracy estimates (based on planned uses of the map). To keep the sampling design simple, easy to analyze, and statistically robust, it is important to define the sampling unit and to use a basic probability sampling design (inclusion probabilities are equal and nonzero for all members of the population). Design-based statistical inference can be applied when sampling is of characteristics of a real, explicitly defined population (Stehman 2000). Probability sampling designs can be interpreted as accuracy estimates for the entire population via established statistical estimators that vary according to the particular sampling design (Stehman 1999). Limitations of resources for field verification or site access can constrain a sampling design. Sampling designs that meet the requirements of equal probability sampling are simple random sampling, systematic sampling, stratified systematic unaligned sampling, and one-stage cluster sampling (Stehman and Czaplewski 1998, Stehman 1999).

Investigators initially developed the confusion or error matrix, which permitted calculation of simple test sample ratios (the number of land use classes incorrectly depicted on the map divided by the number of correctly depicted land use classes confirmed by field verification; van Genderen et al. 1978, Fitzpatrick-Lins 1981). Since those efforts, a great variety of error matrix interpretations and new error metrics have been presented in the literature. The most important contributions of recent work for accuracy findings have been the increase in statistical rigor and decrease in confidence intervals (Richards 1996, Stehman 2001).

Variation in size and frequency of thematic cover types necessitates adjustments in sampling intensity that reflect their relative importance. Thus, a cover type with limited occurrence can be sampled with greater frequency, whereas those most common and abundant will be sampled according to statistical parameters. Stehman (2001) reported that sample size required to achieve a standard error of 0.05 for a population estimate reaches a maximum sample size of 100, when population size is ≥10,000 (for populations 10,000, the sample size required to achieve $SE = 0.05$ is a function of $n = N/[0.01N + 1]$, where n is sample size, N is population size, and SE is standard error).

The **error matrix** is composed of orthogonal axis with cover types (Table 18.1) and allows analysis of accuracy and error rate for each cover type. Cover type accuracy is measured by dividing the number of correctly classified sample points for each cover type by total points sampled. **Map accuracy** also can be presented as **user's** (diagonal values divided by row totals for each matrix) and **producer's** (diagonal values divided by column totals for each matrix) values for each cover type, which are the converse of commission and omission error, respectively.

The assessment of map accuracy by field verification could benefit from methods that increase the accuracy of sample

Table 18.1. Error matrix for cover type

Cover class	Cover class					
	A	B	C	D	E	Total
A	2	0	1	0	0	3
B	7	10	3	0	2	22
C	1	0	6	1	0	8
D	0	0	0	9	0	9
E	0	0	0	0	8	8
Totals	10	10	10	10	10	Diagonal total: 35

point capture (Woodcock 1996). This improvement could be accomplished by tagging the sample points with location information (UTM coordinates, or latitude and longitude), which could be targets for the field verification. GPS units could help in quantifying variability encountered in accessing sample points. Further, proximity to each sample point could be quantified and used in the assessment of map accuracy.

The overall objective of performing an accuracy assessment of a map is to provide a quantified measure of how well the map represents reality. If proper procedures are followed in the design, performance, and analysis of sampling, the accuracy assessment results can be used as an integral part of the map.

DATA DOCUMENTATION

Data documentation represents a critical component in the creation of a spatial dataset for wildlife biology and helps complete the dataset. In an age of increased technological abilities and information sharing, a dataset created for a particular effort may be used in many other ways not perceived at the time of development. Thus, detailed documentation enables data users to better understand why the dataset was developed, the process steps taken, when the data was collected, what was collected, and where geographically the collection took place. In addition to assisting a wide variety of data users, a **metadata record** also provides useful information for the data creator, because it includes a detailed **account of how the dataset was developed** before the passage of time interferes with memory. Another use for metadata involves providing information to help the natural resources community avoid data duplication and thus decreasing the cost of data collection. Furthermore, the creation of metadata serves as an institutional memory for an organization (a history of datasets at a given institution). Metadata transcend people and time, allowing a new data manager to continue work with a dataset his/her predecessor may have left behind. Metadata creation retains valuable information about data for internal organizational or external client use and provides a key component in sustaining a biological GIS program in the long term. Finally, delivering metadata records through **clearinghouses** allows users to discover and find data, determine their applicability, and possibly form new research collaborations.

History of Metadata

The metadata concept was formalized at the federal level in 1994 with release of Executive Order 12906 and the Office of Management and Budget's Circular A16 as part of a government-wide effort to reduce duplication of effort when collecting information and to provide a way for federal agencies and taxpayers to access data created with federal funding (Federal Register 1994). The Office of Management and Budget (2002) released a revised circular A16 to reflect technology changes, but kept the core component of establishing a coordinated approach to electronically develop the National Spatial Data Infrastructure (NSDI). As part of the NSDI, release of the Content Standard for Digital Geospatial Metadata (CSDGM) provided a **common set of terms and definitions** needed to document data. All types of **spatial and nonspatial data** can be documented using this standard. Additionally, several profiles of the standard provide users with additional elements for biological, shoreline, or remote sensing data. In 1998, the **Federal Geographic Data Committee** (**FGDC**) approved the Biological Data Profile, an effort led by the National Biological Information Infrastructure. Additional elements include taxonomy, methodology, and analytical tools allowing data managers for wildlife biology to provide more accurate documentation of their data. The Executive Order states that any dataset created with federal funding needs documentation using the CSDGM. Many state and local governments and other organizations that receive federal dollars have adopted the standard. Other standards exist, but many GIS professionals either use or work on data created with federal funds and need a working knowledge of it for their jobs. Crosswalks exist to share metadata among the major standards for nonfederal organizations that choose to use other standards for documentation.

The **International Organization for Standardization** (**ISO**) has released an international geospatial metadata standard as part of an effort by a network of national standards institutes from 145 countries working in partnership with international organizations, governments, industry, business, and consumer representatives (Technical Committee ISO/TC 211 2003.). The CSDGM along with the several other major standards provided many content contributions to the development of ISO 19115. In December 2003, the American National Standards Institute (ANSI) adopted the international standard, ISO 19115, and the ANSI International Committee for Information Technology Standards–

L1 signed an agreement with Canadian General Standards Board Committee on Geomatics to co-develop the **North American Profile,** a regional profile of ISO 19115 Geographic Information—Metadata. Inquiries were made into Mexico's participation early in the profile development process; however, Mexico decided to pursue another option for developing its national profile. In developing the North American Profile, the United States and Canada join many other nations in furthering the Global Spatial Data Infrastructure that extends capabilities for documenting geospatial data to a global scale (Box 18.9). Compared to the CSDGM, the ISO NAP structure uses a Unified Modeling Language or object-oriented structure and supplies a few new elements, such as the addition of language and character sets. The **National Biological Information Infrastructure (NBII)** leads the effort to build a set of biological elements that will extend the ISO NAP standard, so that data managers can enter information pertinent to wildlife biology, such as taxonomy, methodology, and analytical tools. Similar efforts are on-going for topics of concern for other interest groups, such as shoreline information and remote sensing.

Précis of Content Standard for Geospatial Metadata

As the **FGDC** CSDGM standard transitions to ISO NAP, metadata creators and users should be aware of both standards. Many extensive systems of metadata repositories and tools for metadata creation in existence use the CSDGM, and it will take time for the implementation of ISO NAP.

The CSDGM metadata standard is organized into **10 sections** (7 main sections and 3 supporting sections) that provide elements to **answer a series of questions** (Federal Geographic Data Committee 1998, 1999, 2000).

- Who collected and who distributes the data?
- What is the subject, processing, and projection of the data?
- When were the data collected?
- Why were the data collected? (What is the purpose?)
- How were the data collected? How should they be used?
- How much do the data cost?

Although the standard includes many elements, not all require data entry. Those that do are labeled mandatory, and metadata creators may make selections from a series of other elements that directly apply to their data (labeled as mandatory if applicable and optional). Definitions provide clear descriptions about the type of information to include in each field about the dataset.

Standardization of the NAP Content

Aside from the Unified Modeling Language structure of the ISO standard, the NAP is organized into sections, classes, elements/attributes, domains, and code lists (Box 18.10). The ISO NAP includes 16 sections. Classes are a secondary component of the organization of the ISO NAP; they can occur at multiple levels and can contain both subclasses and attributes. Attributes are an important part of the ISO NAP structure, as this is where information is entered by the metadata creator. Attributes can be found in sections, classes, or subclasses (Federal Geographic Data Committee 2009). High-level sections of the ISO NAP include:

- Identification Information,
- Constraint Information,
- Data Quality Information,
- Maintenance Information,
- Spatial Representation,
- Reference System Information,
- Content Information,
- Portrayal Catalog Information,
- Distribution Information,
- Application Schema Information,
- Extent Information,
- Citation Information,
- Date Information,
- Responsible Party Information, and
- Contact Information.

The ISO NAP includes an expanded list of attribute types, such as Boolean (true/false), date, distance, free text, generic name, integer, and URL. Additionally, NAP employs codesets that represent fixed domains. This new and beneficial addition to the metadata standard allows the development of standardized descriptors to enhance search capabil-

---

BOX 18.9. FEDERAL GEOGRAPHIC DATA COMMITTEE METADATA STANDARD

Sections of the standard:
1. Identification[a]
2. Data Quality
3. Spatial Data Organization
4. Spatial Reference
5. Entity and Attribute
6. Distribution
7. Metadata Reference[a]

Supporting sections (reusable):
8. Citation
9. Time Period
10. Contact

[a]Denotes a mandatory section.

> **BOX 18.10. HIERARCHY OF THE INTERNATIONAL ORGANIZATION FOR STANDARDIZATION'S NORTH AMERICAN PROFILE**
>
> **Identification Information (Section)**
> Content Information Online Resource **(Class Name)**
> **(Attributes):** Data filled in here
> + linkage   + name
> + protocol   + description
> + application profile + function

ities. The NAP also introduces a structured hierarchy for the documentation of related levels of data, including dataset series, datasets, features, and data attributes. These categories allow metadata authors to detail metadata content to a certain level (Federal Geographic Data Committee 2009).

New elements contained in the ISO NAP incorporate needs of the international community. Added fields include Dataset Language, Dataset Character Set, Metadata Language, and Metadata Character Set. Other sections of the CSDGM have been expanded in the NAP, such as Data Quality. The ISO NAP adds an area to document geospatial data services, which include such resources as web-mapping, data models, online data catalogs, online data processing, ontologies, thesauri, data hierarchies, and classification systems. Elements related to documenting web services include service type (e.g., OCG Catalog), coupled data resources, and operations the service can perform (Federal Geographic Data Committee 2009).

The NAP deals with keywords in a slightly different way from the CSDGM. At a very high level, the NAP supports Topic Categories as a required element. In addition, an Online Linkage Function Code is built into the NAP to capture the type of link the metadata creator is providing. The NAP also includes a Portrayal Catalog section, in which a metadata creator can provide a citation to standardized (e.g., Anderson Land Use Cover Land Cover Mapping) or internally developed symbologies. An Application Schema provides a method for describing the use of standardized software applications that includes attributes to describe a schema, constraints of the application, software dependencies, and others (Federal Geographic Data Committee 2009).

Software Tools

Software tools provide a way to **create a metadata record** and, in many cases, tools can automatically enter values into elements as users create their data. When selecting a tool, note the pros and cons for your environment. As with any software tool, there are trade-offs. The FGDC provides a review of tools available for creating metadata at its website (http://www.fgdc.gov). The NBII also hosts information about tools that contain the Biological Data Profile option at its website (http://www.nbii.gov). The FGDC website also reviews tools available for creating ISO NAP metadata at its website(http://www.fgdc.gov/metadata/iso-metadata-editor-review).

Distributing and Accessing Metadata

A completed metadata record should be posted on the **Federal Geographic Data Committee clearinghouse** website (http://geodata.gov) and on the NBII Clearinghouse website (http://mercury.ornl.gov/nbii). These clearinghouses provide a single point of entry for discovery of thousands of metadata records. Organizations that want to establish a node in the clearinghouses are provided directions on the websites. Additionally, individual records can be uploaded to the clearinghouse after completing a quality control process. In general, clearinghouses utilize a harvesting method to obtain records. The clearinghouse benefits, because it avoids a hassle with ports of entry, such as Z39.50, being mysteriously unavailable. The organization providing records benefits, because it maintains control of its original metadata records at all times and can easily serve records to multiple clearinghouses to distribute its information more broadly. Clearinghouses provide an opportunity for powerful collaborations to develop, as scientists and other users can discover many types of data.

The websites (http://www.fgdc.gov and http://www.nbii.gov) offer a wide range of tools, training, and information about creating and serving metadata and provide links to a variety of agencies and organizations that specialize in metadata. Visit geodata.gov (http://gos2.geodata.gov/wps/portal/gos) or the NBII Clearinghouse (http://mercury.ornl.gov/nbii) to search for metadata records on a wide variety of topics in geospatial activities or wildlife biology.

SUMMARY

All projects, whether habitat or animal related, occur spatially in wildlife biology and management. Thus, spatial technologies can be used to evaluate research and management efforts. This chapter provides a brief look into using 3 spatial technologies: GIS, GPS, and remotely sensed data (Landsat Imagery and lidar). It also highlights the need to understand data documentation, data accuracy, and Internet applications. Spatial technologies should be considered as tools to assist resource managers with mapping and as a way to merge or incorporate datasets from a variety of sources into one format. Maps can focus discussion by presenting what is known or thought to be known about an area or issue. Additionally, most people readily accept maps, because they are easier to understand at first glance than some tables or figures, and because many people use them to navigate across town or across a country. High resolution data are increasingly becoming more widely available and hold high potential for contributing to questions in wildlife biology, conservation, and management. We have discussed several studies relating habitat use and distributions to vegetative and topographic features derived directly from these data. As the tools and methods for extracting information relevant to wildlife and inherent in these datasets continue to evolve, it is likely that we will see further development on the full potential of GIS, GPS, Landsat imagery, and lidar applications. Spatial technologies rely on computer technologies and currently are expensive to develop and maintain. However, their value outweighs their costs when information is incorporated into products that help managers make wise decisions about natural resources.

19

JESSICA R. YOUNG

Animal Behavior

INTRODUCTION

SINCE THE DAWN of time, the study of animal behavior or **ethology** has been a key component of human ability to utilize and manage other species. The earliest hunter-gatherers had a keen insight into the subtle behavioral traits that both determined the differences in animal groups (now called species) as well as the traits of individuals that were to be their target or prey. Today, members of hunter-gatherer societies, such as the San Bushmen of southern Africa, can tell entire stories from the placement of an animal's single track in the sand. The most skilled hunter or conservationist can likely describe with incredible detail the behaviors of the individual being tracked. A founder of the field of wildlife biology, Aldo Leopold (1949), used animal behavior to illustrate the need for new management paradigms in *A Sand County Almanac*. A species' relationship to its environment and its subsequent management or conservation cannot be interpreted without understanding factors influencing its **social behavior.** Currently, wildlife biologists and managers struggle with understanding animal movements, site fidelity, and social transmission of information in an increasingly fragmented environment. Given the importance of comprehending animal behavior, it is remarkable how few studies combine the fields of wildlife management and animal behavior.

Perhaps one of the most compelling stories for why wildlife managers need to better understand animal behavior comes from comparing studies of red deer (*Cervus elaphus*) movements in Europe to Florida Key deer (*Odocoileus virginianus clavium*) movements in the United States. More than 20 years ago, the Berlin Wall, a fortified barrier of electric fences and barbwire separating Czechoslovakia, East Germany, and West Germany was removed after separating the nations for more than 25 years. The area that was once patrolled by men with machine guns is now one of Europe's largest natural preserves housing red deer. Game managers and biologists began to suspect in the 1990s that something odd was going on with the red deer. After 7 years of tracking deer during the 2000s, biologists on both sides of the border found that deer, all of which have been born since the wall came down, stopped at the border and turned back, creating 2 distinct populations separated by a few hundred meters of contiguous habitat. This amazing **group phenomenon** may be the result of red deer use of traditional trails being a learned trait. Although an occasional male may venture across the invisible line, he always returns to his natal side, and females have not been known to cross, likely a result of females staying with mothers even longer than males do and learning their mother's movement patterns.

Comparing the red deer to Florida Key deer provides an even more intriguing management quandary, as the Florida Key deer not only move across significant barriers, such as fenced highways, they quickly become **adapted** to wildlife underpass tunnels (Braden et al. 2008), and other species easily adapt to wildlife overpasses, which can provide connectivity and prevent genetic isolation in areas with roads (Corlatti et al. 2009).

Despite the importance of understanding animal behavior, its study has not been widely used in the field of conservation or management because of differences in training and study emphases. Sutherland (1998b), in a 1996 survey of the journal *Animal Behaviour,* reported that of 229 papers, none directly related to conservation or management. Martin (1998) reviewed 8 behavioral texts and found that only one had a chapter on the application of behavior studies to management and conservation, while Arcese et al. (1997) found only 2 of 17 animal behavior, ethology, or behavioral biology texts had such chapters. These authors and others have proposed hypotheses about reasons wildlife biology and animal behavior have not become more integrated. One of the major issues may be the unit of study used. **Animal behavioral biologists** tend to focus on an individual's physical or behavioral differences as they apply to survival and reproductive success. **Wildlife biologists** are generally more concerned with study of populations; consequently, as individuals become summarized as numbers in demographic or habitat use models, the importance of behavior of individual animals is often lost (Caro 1998, Martin 1998).

Traditional boundaries between different academic departments reinforce disconnection of animal behavior from management studies (Martin 1998). In the United States and Canada, practitioners in their respective fields are generally in different academic departments in colleges and universities. Although it is fairly common for a behavioral ecologist in a higher education program to include coursework in community and population ecology, wildlife biology programs rarely (5% in the United States; J. R. Young, unpublished data), include a required course in animal behavior or behavioral ecology. Many wildlife biologists continue to believe that most studies in animal behavior have little direct impact on species management. In contrast, conservation biologists are increasingly recognizing the importance of the field of animal behavior to wildlife conservation (Buchholz 2007).

There are notable examples in which wildlife studies and conservation biology studies have thoughtfully integrated behavior into management. Proceedings from wildlife management and conservation symposia (e.g., Weller 1988, Festa-Bianchet and Appollonio 2003), books (Gosling and Sutherland 2000), and book series (Chapman and Hall's [London] Wildlife Ecology and Behaviour) demonstrate an increasing awareness of the importance of studying animal behavior. Although more papers in management journals include aspects of animal behavior, studies generally are focused on **basic descriptions** of foraging behaviors and habitat use and fail to integrate modern developments, such as game theory, optimal foraging, cultural evolution, and the importance of phenotypic plasticity (Sutherland 1998b). Wildlife studies are rarely conducted in a manner that encourages development of **predictive models** of population demography, dispersal, or habitat use based on changes in individual behaviors caused by **anthropogenic activities** or on factors that include variation in individual behavior. Two solutions are needed to correct the disconnect between wildlife biology and animal behavior: (1) researchers and students in each discipline need more exposure to the concepts and ideas of both disciplines, and (2) management teams should include behavioral biologists (Arcese et al. 1997) to incorporate behavioral studies relevant to management or conservation issues.

ANIMAL BEHAVIOR AND WILDLIFE BIOLOGY

Behavior of Whooping Cranes

Wildlife biologists have learned to include insights from animal behavior into management of wildlife populations. For example, the whooping crane (*Grus americana*) has become an international symbol of the challenges of managing and recovering populations of endangered species. This largest of North American cranes was once widely distributed across the north-central United States and southern Canada. During the late 1800s and early 1900s, **overharvesting** by settlers and homesteaders led to large-scale population declines (Allen 1952). By 1939, J. J. Lynch (Aransas National Wildlife Refuge, unpublished data) reported the large size and conspicuous plumage of the whooping crane had made it an easy mark for hunters, and there were only a few scattered pairs left breeding in the wild. By 1939, only 2 populations remained, a migratory population that nested in Canada and wintered in Texas, and a nonmigratory group nesting in Louisiana. Wildlife biologists faced a crisis when, by 1941, only 16 birds remained in the **migratory population** at the Aransas National Wildlife Refuge, Texas (Lewis 1995b) and fewer than a dozen in the nonmigratory population. Today, all whooping cranes are genetic descendants of those 16 individuals. The nonmigratory genetic heritage was lost with the death of Josephine in the New Orleans Zoo (R. C. Drewien, U.S. Fish and Wildlife Service, personal communication). A major challenge for wildlife biologists was to increase the size of the migratory population, because migration routes, nesting locations, and wintering sites were **learned** behaviors. Although in many species, migration appears to be an **innate** behavior (Box 19.1), in whooping cranes, social interactions appear to be key to the establishment of migration routes. In 1975, a field experiment was initiated to re-establish a migratory flock through cross-fostering whooping cranes by using sandhill cranes (*Grus canadensis*) as foster parents (Drewien and Bizeau 1978).

> **BOX 19.1. LEARNED BEHAVIORS**
>
> These are behaviors that are modified by experience and the environment. For example, young whooping cranes learn which types of areas are suitable for foraging by following adults. In contrast, innate behaviors are those performed the same way each time after their initial expression. Innate behaviors are usually "hard-wired" in species' nervous systems. An example of an innate behavior in cranes is performing highly ritualized mating displays, such as head bowing and leaping into the air.

Wildlife biologists cross-fostered whooping cranes by placing them with closely related sandhill crane parents. Whooping cranes generally produce 2 eggs, but raise only 1 chick. Wildlife biologists took advantage of this behavior by removing 1 of the eggs from nests of wild and captive birds and placing them in nests of selected pairs of sandhill cranes at Grays Lake National Wildlife Refuge, Idaho, in 1975 (Drewien and Bizeau 1978).

This experiment initially appeared to be successful, as the foster parents raised whooping crane chicks and successfully taught them feeding habits and the 1,350-km migratory pathway to Bosque Del Apache National Wildlife Refuge, New Mexico (Drewien and Bizeau 1978). However, it became apparent that lack of previous behavioral studies on one key aspect of the social behavior of whooping cranes was critical to understanding the underlying failures of the experiment. Although migratory behavior was a learned behavior, **sexual imprinting** influenced choice of mates. This is a special type of imprinting, in which choice of sexual partner is determined during early development. Exposure in the initial hours after hatching can permanently influence the choice of mates for many species of birds. Unfortunately, no one had studied the extent to which whooping cranes became sexually imprinted on their parents, as there was little opportunity to do so prior to the cross-fostering experiment. Although male whooping cranes raised by foster parents established breeding territories and nests, they did not pair with whooping crane females. However, at least 1 hybrid was produced by a male whooping crane and a female sandhill crane. The last known whooping crane in the experimental population disappeared in 2002.

Wildlife managers and conservation biologists learned from this experience and recently established a new migratory population of whooping cranes that migrate between Wisconsin and Florida following motorized ultralight aircraft. The first autumn migration occurred in 2001. In autumn 2002, 16 young whooping cranes imprinted to ultralight aircraft migrated over 7 states in a 1,900-km journey lasting 49 days (www.operationmigration.org). In 2002, birds from the previous year migrated north without aid of the ultralight aircraft. The success of this experiment, as measured by successful reproduction with members of their own species, will not be known for several years.

The use of information from behavioral studies has been critical to the initial success of the Wisconsin–Florida experiment (Operation Migration) as well as to the releases of >200 fledged juveniles in the Florida nonmigratory population (R.C. Drewien, personal communication). Care has been taken in these experiments to prevent **imprinting** young whooping cranes on humans. In the Wisconsin–Florida experiment, cranes hatched from eggs incubated by captive mothers are taken to Necedah National Wildlife Refuge, Wisconsin, for acclimation to the wild. Human handlers dress in costumes and teach the young cranes critical survival skills, including foraging, predator avoidance, and avoidance of humans (Fig. 19.1). Eventually, these birds are taught to follow their costumed trainer as the trainer operates a motorized ultralight (Fig. 19.2) in an effort to reestablish a migratory population. Efforts for the nonmigratory flock in Florida include teaching young cranes to avoid predators by roosting in water as well as other survival skills. The challenge of teaching learned behaviors to long-lived species that depend on social transmission of behavior is difficult with experimental flocks. The lesson learned from whooping cranes is that prior studies of species' behavior in their natural environment are critical for future management and conservation efforts.

Placement of Wood Duck Nest Boxes

Understanding the mechanisms of social interactions among female wood ducks (*Aix sponsa*) has led to recent manage-

Fig. 19.1. A costumed pilot works to train young whooping cranes by using a puppet head and a loudspeaker playing a soft purring sound that chicks would normally hear from their mothers. *Photo courtesy of Operation Migration.*

Fig. 19.2. Whooping cranes follow an ultralight craft from Wisconsin to Florida, possibly establishing a new migratory population. *Photo courtesy of Operation Migration.*

ment changes. Wood ducks are secondary cavity nesters and inhabit wetlands, including swamps and marshes during the breeding season, and they seek cavities created by other species, such as pileated woodpecker (*Dryocopus pileatus;* Bellrose and Holm 1994). In such cavities, wood ducks may lay a clutch of up to 14 eggs (Semel and Sherman 1992). In a natural setting, wood ducks typically nest solitarily in dispersed and cryptic cavities. Deforestation and wetland habitat loss led ornithologists to predict that wood ducks would be extinct around the turn of the 20th century (Hepp and Bellrose 1995). Several factors have contributed to the existence of the healthy populations observed today. The establishment of the Migratory Bird Act (http://www.fws.gov/laws/lawsdigest/migtrea.html), conservation and creation of wetlands, and use of nest boxes (Fig. 19.3) to replace cavities lost from old growth forest harvesting were major factors leading to wood duck recovery.

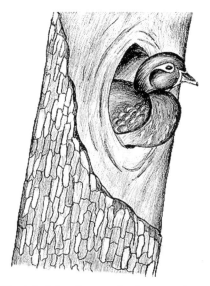

Fig. 19.3. Wood duck female emerging from nest box. *Drawing courtesy of R. W. Henninger.*

Undoubtedly, use of artificial nest boxes has led the overall **recovery** of wood duck populations since the turn of the 20th century (Hepp and Bellrose 1995). Ironically, the mechanism for recovery led to an overall reduction of clutch sizes when artificial boxes were compared to natural cavities. Although establishment of nest boxes helped provide nest cavities for wood ducks, it also increased intra-specific brood **parasitism** (Semel et al. 1990, Semel and Sherman 1993). The most common method of placing wood duck nesting boxes by wildlife managers was to cluster the boxes to make them visible. This clustering changed the solitary nesting practices of wood ducks to semicoloniality. Thus, females had a much higher chance of being observed by conspecifics and, therefore, a higher probability of having their nest parasitized (Semel and Sherman 1986).

Intra-specific brood parasitism occurs when a female lays eggs into the nest of another female (egg dumping); egg dumping is triggered in wood ducks by females observing members of their own species entering or leaving nest sites (Semel et al. 1988). Waterfowl have relatively high rates of nest parasitism, and studies have shown that nest parasitism in wood ducks can exceed 50% (Semel et al.1988, Roy-Nielson et al. 2006). A consequence of intra-specific nest parasitism in natural cavities and nest boxes can be unusually high clutch sizes. Although natural cavity nests of wood ducks have an average clutch size of 9–12 eggs, parasitized nests in nest boxes often contain ≥20 eggs and have lower hatching success (Semel et al. 1988). One possible reason could be that larger clutch sizes have a greater hatching asynchrony, or eggs that hatch at different times, causing fewer chicks to be produced per clutch. Hatching success also may be lower because of aggressive interactions between females when nest parasitism is attempted. At Nauvoo Slough, Illinois, 379 eggs of 76 nesting females were crushed due to female intra-specific aggression (Bellrose and Holm 1994). Recent genetic studies have suggested that rates of intra-specific nest parasitism in some populations using natural cavities are similar to those in populations using nest boxes, and that wildlife managers should focus on identifying other behavioral causes of larger clutch sizes in nest boxes (Roy-Nielson et al. 2006).

Animal behavior biologists continue to team with wildlife managers to find solutions to enhance wood duck populations (Semel and Sherman 2001). Understanding female intra-specific nest-parasitism behavior has led to placement of nest boxes in dispersed and concealed areas in habitats preferred by wood ducks. Although occupation rates may be lower, clutch success should be higher, and the dispersed nesting behavior of the species will be preserved.

Ibex Reintroductions

Understanding natural **behavioral rhythms** is critical, as reintroductions and population augmentations often have a high failure rate initially. One reason for lack of success may be lack of integration of animal behavior into study designs. Unfortunately, despite considerable effort to examine habitat suitability and food availability, rarely do wildlife managers seek to understand how the fundamental behavioral ecology of a species might differ among populations and subspecies or among individuals of different ages and genders.

A classic example of reintroduction efforts not succeeding due to a lack of behavioral knowledge occurred when ibex (*Capra ibex*; Fig. 19.4) from the Tatra Mountains in Czechoslovakia were extirpated during the first half of the 19th century. The Tatra Mountains rise in elevation to 2,600 m and are quite cold in spring, with frequent snowstorms and temperatures below 0° C. Local populations of ibex were adapted for the mountainous environment and bred in midwinter, causing young to be born in late spring (Turcek 1951).

During the beginning of the 20th century, ibex from 3 different populations (from nearby Austria and warmer climates in Asia) and 3 different species (*Capra ibex, C. hircus,* and *C. nubiana*) were **translocated** to the mountains. The 3 species interbred, and their hybrids had different physical and behavioral characteristics than the ancestral stock. The breeding season for the hybrids occurred in late summer, and offspring were born in early spring. Young born in early spring were unable to survive the cold temperatures and storms, so generation after generation failed until they became locally extirpated (Turcek 1951). An understanding of the reproductive behavior of the 3 species would probably have helped provide better management solutions to reestablish ibex to the Tatra Mountains.

SENSORY PERCEPTION

The examples above demonstrate how understanding fundamental aspects of an animal's behavior can lead to better management and conservation decisions. A major challenge for wildlife biologists is to not allow their own sensory limitations to blind them to the importance of sources of infor-

Fig. 19.4. A lack of behavioral knowledge led to an initial failed reintroduction attempt of ibex in the Tatra Mountains in Czechoslovakia. *Photo courtesy of C. Pourre.*

mation for other species. This blindness could lead to **anthropomorphism** of the animal's behavior, or interpretation of the behavior in terms of human contexts, motivations, and biases. When wildlife managers are deciding which behavioral information is important, they should consider that one of the fundamental principles of animal behavior is to learn to comprehend the world through the senses of the animal being studied. Jakob von Uexkull (1864–1944) coined the term **Umwelt** to describe how an animal senses its universe (von Uexkull 1921). He recognized that preconceived ideas about how animals should behave often came from ignorance of how animal senses worked. He challenged us to imagine the world from an animal's perspective by learning what senses were available and active during different stages of their lives.

One example von Uexkull (1921) used that is useful to consider for all who have walked the woods is how a tick's perception of its universe changes depending on environmental stimulus. Ticks have simple sequential behaviors based on which sensory perceptions of their external environments are functioning. While adult ticks are waiting on the end of a piece of vegetation, they are in a form of almost suspended animation until their olfactory senses detect the unique shape of a butyric acid molecule (common in mammalian sweat). Their nervous system reacts by sending signals that allow them to let go of the vegetation. They then quit receiving signals from their butyric acid receptors and can only detect heat. When they detect heat, they burrow toward the sources of greatest heat. Imagine a world from a tick's perspective. One would not see, hear, or feel until one encountered butyric acid. The encounter would turn on a new sensory perspective of the world centered on heat. Other environmental signals would have little meaning.

Bubenik (2007) suggested that a key to understanding moose (*Alces alces*) behavior was to understand the concept of Umwelt. He stated that a basic requirement of understanding why and how moose reacted to different environmental cues was to understand the basic differences between humans and moose in their perception and response to environmental cues. He goes on to further explain that scientists often wonder why an excellent foraging area was not being used by moose or was visited more by one sex or social class than another. His explanation was that such behavior was determined by neurohormonal factors that may be specific to a population, gender, or social class.

Often our lack of understanding of other animals' **sensory perception** and its importance has led to poor management decisions or unintended consequences. Studies contrasting our sensory perception to that of other animals have shown how differently they view the world from us. Animal senses are the mechanisms through which animals find food, mates, and shelter, and they affect timing and processes of migration and hibernation. Most senses, such as hearing, sight, smell, taste, and touch, are familiar to us;

however, sensitivity to the stimuli activating the senses varies greatly. Radar, compasses, and infrared detectors are technologies that allow us to mimic other animals' senses that humans do not possess.

Understanding the importance of animals' sensory mechanisms has become increasingly important in understanding subtle impacts of human activities on the landscape. For example, some forest birds depend on certain light conditions before performing their mating displays (Endler 1997). Further, changes in stream turbidity due to forest management can affect the duration and location of fish mating behaviors (Endler 1997). The major point of understanding sensory perception is that researchers cannot begin to comprehend animal behavior until they let go of their human biases and place themselves in the animal's world.

Hearing

Hearing has a large role in how most mammals interpret their environments. It also has a critical role for many species in individual recognition, mate choice (Howard and Young 1998), prey location (Ryan et al. 1982), predator avoidance, and navigation (Roeder and Treat 1961). One can misunderstand animal behavior, because human hearing is often less acute. Because animals use sound for so many purposes, one often underestimates the extent to which sound pollution interferes with vital functions of wildlife, such as mating activities and foraging behavior (reviewed by Larkin 1996). Of particular concern is the degree to which the sounds produced by commercial, research, and military activities influence marine mammals' abilities to communicate and locate prey (Tyack 2008). There has been increased attention to understand the degree to which anthropogenic activities impact animal populations. For example, whereas Jepson et al. (2003) found that different species of whales (Cetacea) may have beached themselves due to brain hemorrhages caused by strong military sonar, Krausman et al. (2004) found that auditory effects of military activities had little impact on the behavior of an endangered population of Sonoran pronghorn (*Antilocapra americana sonoriensis*).

Mammals and birds hear sounds by detecting pressure waves with use of membranes and **hair cell sensory receptors** (Bradbury and Vehrencamp 1998). The tympanic membrane vibrates in tune to the frequencies of the pressure waves and stimulates sensory cells in the ear that send signals for processing in the brain. Sound waves have no inherent directionality, so one challenge for wildlife is to localize sounds. Most mammals have **pinnae** or external ears to aid in localizing the source of sounds (Bradbury and Vehrencamp 1998). Bats (Chiroptera) and many ungulates, such as mule deer (*O. hemionus*), can rotate their pinnae toward the source of the sound to aid in locating it. Barn owls (*Tyto alba*) have offset ear openings to allow them to localize small running rodents by sound in the complete absence of all light. Their offset ears and facial ruffs allow processing differences in the intensity and timing of the arrival of sound waves of running rodents to help precisely locate them (Proctor and Konishi 1997).

There is significant variation in animal hearing. Elephants (Elephantidae) use low frequency sounds undetectable to humans (20 Hz) for communication across vast expanses of open savannah environments (Heffner and Heffner 1982, Langbauer et al. 1991). Many avian species can hear faint sounds at low frequencies that may warn them of approaching storms or may disrupt their flight when exposed to commercial jet shockwaves (Hagstrum 2000). Bats hear high frequency sounds above our detection abilities (>20,000 kHz) and use these sounds to detect prey as small as mosquitoes. Cave swiftlets (*Aerodramus linchi*) and oilbirds (*Steatornis caripensis*) use relatively high frequencies for **echolocation** to locate entrances to caves and relatively large (≥20 mm) items in their environment in the dark (Griffin and Suthers 1970).

Perhaps one of the most common reasons for animals to produce sounds is for communication. There is a rich body of research examining the different things animals communicate through sound. For example, Gunnison sage-grouse (*Centrocercus minimus*) perform elaborate mating displays (Fig. 19.5), in which their sounds can be heard for ≥1 km. Their vocalizations and mechanical sounds are used to communicate their willingness to mate, warn members of the same gender away from potential mates, and attract females to their mating site (Young 1994). Bull elk (*Cervus canadensis*) produce a variety of calls during rut that serves to defend their territories and attract cows to their location; the calls vary, depending on aggressive context (Feighny et al. 2006). Common ravens (*Corvus corax*) vocalize loudly when finding a large carcass. This behavior seems counterintuitive (why bring competitors for food to the scene?). Careful testing of alternative hypotheses led to the discovery that young ra-

Fig. 19.5. Gunnison sage-grouse perform ritualized displays on their mating grounds. Sounds are produced by both syringeal vocalizations and mechanical release of air from their yellow air sacs. *Photo courtesy of J. D. Sartore.*

vens intruding on territories of older pairs vocalized to draw large groups of conspecifics to the scene to avoid being forced from the food bonanza by the **territory** owners (Heinrich 1988). Begging by young birds is a signal to parents to increase food delivery. This communication between parents and their offspring has been used by such species as brown-headed cowbirds (*Molothrus ater*), in which the young brood parasites produce louder calls than those of their nest mates.

Vision

Although almost all organisms are sensitive to light, vertebrates have developed the ability to capture images from their environment and synthesize information about those images. We can misunderstand the impact of different levels of **visual acuity** unless we understand that humans have only moderate abilities for detecting electromagnetic energy or light. Most other animals have evolved different levels of light sensitivity for finding food, detecting threats, and orienting themselves in their environments (Bradbury and Vehrencamp 1998).

Light is captured in photoreceptor cells that contain **photopigments.** In vertebrates, the cells are called rods and cones and are packed densely in the eyes to form the retina. Rods allow for vision in low light conditions, and cones are responsible for color vision in high light intensity environments. The ability to resolve fine details in the environment depends on the number of receptor cells in a given area, the optical system, and the neurological mechanism for passing the signals to the brain (Bradbury and Vehrencamp 1998).

One of the most obvious ways humans alter visual habitats is the extent to which nights are illuminated. **Photopollution** is the detrimental addition of light into an animal's environment, with nocturnal animals being most at risk. For example, most species of sea turtles (Cheloniidae and Dermochelyidae) are listed as threatened or endangered and will not nest on preferred beaches if the beaches are lit (Witherington 1992). If they do nest on beaches with night lighting, their hatchlings are often at risk as they move toward the light rather than toward the relative safety of the sea. Behavioral research is providing wildlife managers with guidelines for safely lighting beaches without impacting sea turtle behavior (Witherington 1997). Another problem with light pollution occurs from **phototaxis** (light attraction) by birds toward lighted radio towers at night. When weather conditions bring low cloud ceilings or fog, lights on towers cause refraction, creating areas of illumination around the towers. Thousands of migrating birds that have lost their stellar cues are attracted to the lighted area and may die if they collide with the tower or its supports (Avery et al. 1976).

In general, most birds have better distance vision than do mammals and can see 2–3 times farther than humans can (Gill 1995). One of the greatest variations in vision among species is differing abilities to see color. Many mammals, such as hamsters (Cricetinae), Virginia opossum (*Didelphis virginiana*), raccoons (*Procyon lotor*), some monkeys (Primates), and bats, have little to no color vision, whereas birds can sense portions of the color spectrum, such as **ultraviolet** light. For example, Eurasian kestrels (*Falco tinnunculus*) use ultraviolet trails left by small mammals marking their runs with urine to locate prey corridors (Koivula et al. 1999), and female European starlings (*Sturnus vulgaris*) use ultraviolet cues for mate choice (Bennett et al. 1997). Some snakes, such as rattlesnakes (*Crotalis* spp.) and other pit vipers (Viperidae), use wavelengths of light not visible to us at the other end of the visual spectrum and can detect **infrared waves** (Hartline et al. 1978). The ability to sense an infrared wavelength is critical for their success in hunting small mammals emitting such frequencies of light. The position of a tail may send subtle signals to subordinate members of a gray wolf (*Canis lupus*) pack. The same wolves may end a chase of white-tailed deer that wave their tails in a conspicuous flagging behavior, alerting the wolves they have been seen (Fig. 19.6). Such species as Thomson's gazelle (*Gazella thomsonii*) use vigorous "stotting" displays to warn approaching predators they have been seen by their prey (Caro 1986). Stotting occurs when gazelles bound up and down with all 4 legs held stiffly while displaying their white rump patch.

Fig. 19.6. Tail wagging by white-tailed deer provides visual signals to potential prey and predators. Such signals are probably adaptive evolutionary traits, as the prey signal the unprofitability of pursuit to the predator. Thus, both prey and predator can save energy by avoiding energetically expensive escape activities and chases. *Photo courtesy of N. Paothong.*

Olfaction

Smell is poorly understood, and the extent to which anthropogenic activities influence the "smellscape" is only beginning to be recognized. Such species as Pacific salmon (*Oncorhynchus* spp.), an important food item for bears (Ursidae), are significantly impacted by our olfactory pollution of their streams. One mechanism involved in the decline of salmon in the Pacific Northwest appears to be their dependence on olfactory cues for returning to their natal streams (Scholz et al. 1976, Nevitt et al. 1994, Dittman and Quinn 1996).

Olfaction is possible through reception of chemicals and was likely one of the first animal senses to evolve. Most, if not all, wildlife possess some sort of olfactory organ that allows them to detect airborne chemical messages. In general, there is some type of inlet (e.g., mouth or nose) that leads to a chamber carpeted in sensory cells that respond differentially to diverse olfactory chemicals (Bradbury and Vehrencamp 1998). Historically, scientists believed that birds had poor or no sense of smell, due to the relatively small olfactory bulbs in their brains. Currently, we know that most, if not all, birds possess a sense of smell and can detect odors as accurately as mammals can (Clark et al. 1993*b*). Nocturnal birds and carrion eaters, such as vultures, have better senses of smell than do other birds. Turkey vultures (*Cathartes aura*) can detect traces of the chemical ethyl mercaptan that is released from decaying meat (Smith and Paselk 1986). Engineers have taken advantage of the vultures' olfaction abilities by pumping small amounts of ethyl mercaptan into pipelines with breaks and watching where vultures gather (Gill 1995). Leach's storm-petrels (*Oceanodroma leucorhoa*) can detect odors up to 25 km away emitted by krill, small crustacea that occur in groups in the ocean (Clark and Shah 1992). Mammals also have a keen sense of smell, with nocturnal predators having abilities thought to be 10–100-fold as great as those of humans.

There are many advantages of using chemical communication over auditory cues. For example, pheromones in urine are likely to last days or weeks, whereas songs and howls last only until the singer is finished. **Allomones** are chemical signals passed between species, such as between predator and prey. **Pheromones** are chemical messages (organic compounds with a carbon skeleton) passed within the same species. Pheromones are likely used by mammals for mate identification and attraction, territory marking and defense, and as alarms of danger or stress (Bradbury and Vehrencamp 1998).

Sources for pheromones include excretory products, such as urine and feces, or specialized glands on the outside of the animal's body. Some mammals, such as ground squirrels (*Spermophilus* spp.) and goats (family Bovidae), produce pheromones from sebaceous glands associated with their skin and hair follicles. Other mammals, such as mustelids (Mustelidae), have anal glands that produce secretions. Some ungulates, such as pronghorn (*Antilocapra americana*), have pre-orbital glands by their eyes. Most cervids, such as mule deer, produce pheromones from tarsal and metatarsal glands on their legs and tails. They often rub their leg against their head and then rub their forehead on stems and barks to transfer their scent.

Pheromones can be used for courtship, and cervids likely convey information about individual identification and reproductive status, as well as social status (Eisenberg and Kleiman 1972, Johnston 1998). Some birds may use pheromones to elicit sexual responses in males. The ability of mallards (*Anas platyrhynchos*) to mate may be influenced by males detecting pheromones produced by females. Balthazart and Schoffeniels (1979) found that male mallards with their olfactory nerves excised did not exhibit courtship behavior in the presence of females. Animals often also use pheromones for territorial marking. Pheromones in mammalian urine are a common method of territorial marking for large carnivores and other mammals.

Taste or Contact Reception

Although smell entails sensing airborne chemicals, taste depends on contact reception of chemical molecules. We know little about how changes in animals' environments interfere with their sense of taste. Currently, most management activities studying taste are taste aversion studies. **Conditioned taste aversion** is a learned behavior that occurs when negative consequences (getting ill, hurt, shocked, etc.) occur following the consumption of an item. Cowan et al. (2000) published an excellent review of the use of the behavior for reducing predation.

Chemical receptor cells in the mouth are specialized nerve cells that react on contact with different molecules. Taste is shaped by an initial molecule that binds tightly to a specific protein receptor and then undergoes a physical change, causing a neurological signal to be passed to the brain. The intensity of a taste is influenced by the number of cells binding to receptors, types of cells, and density of receptors being activated at a given moment. Chemical receptor cells are some of the shortest-lived cells in an animal's body, only functioning for a few days before they are replaced (Bradbury and Vehrencamp 1998).

The number and kinds of receptors present likely influence taste. Birds have relatively few taste receptors (usually 100), whereas mammals, such as humans, have >10,000 taste buds on their tongues (Gill 1995, Mason and Clark 2000). Both amount and types of receptors are important in the sense of taste.

One way that mammals communicate by taste is through use of a chemosensory organ called the **vomeronasal organ** (Jacobson's organ), a single opening lying between the nasal cavity and the roof of the mouth (Bradbury and Vehrencamp 1998). Some ungulates perform a behavior called **flehmen** (Fig. 19.7) after they contact another individual's urine or secretions, in which they pull back their upper lip

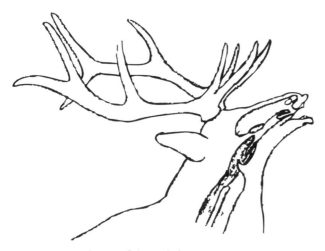

Fig. 19.7. Elk performing flehmen behavior to expose its vomeronasal organ to "taste" the air. *Drawing courtesy of R. W. Henninger (after Bradbury and Vehrencamp 1998).*

to cover their nostrils and raise their head to close off airflow into their epiglottis (Doty 2001). This behavior enables them to draw air into their vomeronasal organ. The receptors in the organ are more structurally similar to taste receptors than those located in the nose for smell. This form of communication is probably important for assessing information about mating status and territoriality. For example, male mountain goats (*Oreamnos americanus*) spend about 4% of their time during rut performing a flehmen behavior after they scent the urine or genitalia of their tended female or nearby females (Mainguy et al. 2008), and scientists speculate that males may be selecting among females in estrus through this sensory mechanism.

Tactile

Tactile sensory systems are commonly used by all wildlife, but are poorly understood and studied. Although the importance of touch in mating rituals, agonistic encounters, social grooming, and other social behaviors is recognized, its role is not well investigated, compared to those of visual and vocal displays. In addition to using touch for communication, it can be a critical sense for navigation in the dark in forests or through subterranean or subnivean (below the snow) environments.

Touch occurs when **mechanoreceptors** are directly stimulated. Although tactile sensory systems are most developed in burrowing mammals, moles (Talpidae), and elephants have tactile sensors in their lips and snout, allowing them to sense their environment through touch. In birds, filoplumes and bristles are specialized feather mechanoreceptors used for sensory functions. Filoplumes associated with flight feathers in the joints of the wings help adjust to minute changes in wind pressures during flight; those associated with general body feathers may provide information about airspeed (Clark and De Cruz 1989, Gill 1995). Flightless birds, such as penguins (Spheniscidae) and ostriches (Struthionidae), are devoid of filoplumes.

The types of mechanoreceptors can vary as well as their location and numbers. Specialized mechanoreceptors occur in the vibrissae (whiskers) of many mammals and in filoplumes and bristles on birds. Touch receptors in the bristles of birds can be seen on flying insectivores, such as common nighthawk (*Chordeiles minor*) and swallows (Hirundinidae), that have bristles around their mouths to sense the lightest touch of a mosquito (Gill 1995). In mammals, vibrissae often occur around the eyes, muzzle, ears, or tail and can provide detailed information to the animal about size, shape, and movement in its environment (Ahl 1986). **Pressure receptors** are often associated with hair in mammals. Social grooming in primates stimulates the receptors and is an important aspect of many primate's social systems. Many small mammals, such as kangaroo rats (*Dipodomys* spp.), use touch for seismic communication in their environment. Kangaroo rats communicate by foot-thumping species-specific and individually recognized patterns (Randall 1997). Blind mole-rats (*Spalax ehrenbergi*) thump their heads against their subterranean tunnel walls to defend territories (Hill 2001).

Several species of animals, including black bears (*Ursus americanus*), shift from using primary sensory modes (e.g., visual) for foraging to using tactile and auditory modes during twilight and night. For example, black bears can have increased foraging success when they cue in on the touch and sounds of salmon splashing as salmon become more active in increased darkness (Klinka and Reimchen 2009).

Other Senses
Barometric Pressure

Many species of birds apparently have some form of mechanical receptors to allow them to assess subtle changes in barometric pressure (Bagg et al. 1950). Many species of songbirds often engage in feeding frenzies prior to low-pressure systems that create winter storms. Birds of the same species are often found in the same altitudes during nocturnal migration, suggesting they can adjust their flights based on pressure differences (Gill 1995).

Magnetic Fields

Another important sense that birds and other species possess is the ability to sense magnetic fields. The exact mechanism behind this sense is still under debate (Walcott et al. 1979, Phillips and Borland 1992); however, there is strong evidence that many avian species can detect weak magnetic fields from Earth (Wiltschko and Wiltschko 2003). There is increasing evidence that some species of birds use a magnetic compass or map, whereas others use a polarity compass that distinguishes pole-ward from the equator rather than north from south; the evidence suggests these birds have receptors in their retinas. Such a system would be quite effective for migration movements from the equator toward the poles and back again (Beason 2005).

Although use of magnetic fields for both orientation and mapping appears to be common in invertebrates (Boles and Lohmann 2003), bacteria (Blakemore 1975), and sea turtles (Lohmann et al. 2001; Fig. 19.8), the extent to which mammalian species can detect magnetic fields is just now being explored. Some evidence suggests that rats are influenced by magnetic fields (Reuss and Oclese 1986).

Sensory Perception Summary

As one learns to better understand the complex and divergent ways in which animals perceive their environments, one can escape from the limitations of one's own sensory perceptions and develop hypotheses relevant to other species' actions based on their perceptions of their environments. One can then better understand how the presence of other species, conspecifics, and changes in species auditory, olfactory, or visual environments might affect management and conservation actions. One also can develop behavioral hypothesis based on individual responses to the perceived environment.

FORMING BEHAVIORAL HYPOTHESES

Our activities and efforts can have unintended consequences on management of a species because of ignorance of how those actions will change the animals' social behavior and social organization (e.g., attempts to control animal populations believed to have an economic impact on our activities). Efforts to control the transmission of bovine tuberculosis by killing European badgers (*Meles meles*), which are hosts of the disease, may actually cause higher transmission rates, as individuals encounter one another at increased rates because of the changing dynamics of their social organization (Swinton et al. 1997). Hunting can have significant effects on population parameters that initially seem unrelated, due to changes in individual behaviors, social systems, or mating systems. For example, killing male mallards after pair-bond initiation can result in lower population reproductive success for yearling female mallards (Lercel et al. 1999). Killing male black bears may result in changes in population sex ratios and an increase in infanticide, lowering reproductive success at the population level (Swenson 2003).

Once wildlife biologists have accepted that animals have different senses and different behavioral responses due to social structure and density, biologists can begin to understand the importance of developing hypotheses about how an animal's behavior can influence its response to efforts to manage or conserve its population. Garton et al. (Chapter 1, This Volume) present a review of critical features for hypothesis testing in wildlife biology and management. However, it is important to understand how formation of hypothesis for testing ideas about behavioral ecology may differ from traditional hypothesis testing for wildlife science.

Tinbergen (1963) recognized that all behavioral hypotheses could be examined using 4 categories: (1) evolution, (2) development/ontogeny, (3) function, and (4) causation. One way to develop hypotheses is to distinguish between **ultimate** (those involving evolution and development) and **proximate** (those involving function and causation) questions. Imagine behavioral biologists coming upon a male mule deer rubbing its antlers back and forth across a small aspen (*Populus tremuloides*) tree or a bush. When forming an evolutionary or ultimate hypothesis, they would ask such questions as: Does the rubbing of its antlers increase its survival or reproductive success? When did rubbing behavior evolve in this species? These questions can turn into testable ultimate hypotheses with specific predictions. Behavioral biologists may form mechanistic or proximate hypotheses about the physiological or neurological mechanisms leading to the rubbing behavior by asking such questions as: Do the male's hormones trigger the rubbing behavior? Are males more likely to rub their antlers during certain seasons? These questions can be turned into testable hypotheses regarding the mechanisms behind the expression of the behavior.

Both ultimate and proximate hypotheses can lead to important insight into the management and conservation of species. For example, if after careful study it was concluded that males who had more rubs on small diameter trees were more likely to mate, subsequent management of aspen tree age classes could influence the distribution and reproductive success of mule deer in that area. If it was found that males only rubbed their antlers during certain seasons, antler rubs would be an indication of population use of habitat on a

Fig. 19.8. Lohmann et al. (2001) outfitted loggerhead sea turtle (*Caretta caretta*) hatchlings with harnesses tethered to an electronic tracking unit that recorded the turtles' positions as they responded to manipulated magnetic fields. The hatchlings processed magnetic information to follow innate migration routes, suggesting they possess a "magnetic map." *Photo courtesy of K. J. Lohmann.*

seasonal basis, promoting management decisions about access and physical disturbance activities during different times of the year. It is important to recognize that although there are 2 major types of behavioral hypotheses, examining one logically leads to examining the other.

DIRECT METHODS FOR OBSERVING BEHAVIOR

Direct observation is generally necessary to gather sufficient independent quantitative data for statistical testing of hypotheses about animal behavior. One of the most important and often overlooked methods for observing and recording behavior is keeping a daily **field journal**, which can lead to formation and testing of hypotheses about animal behavior. A good field journal should always include dates, times, places, weather conditions (wind, clouds, etc.), description of the habitat, description of activities, and distributions of animals in the habitat.

During direct observations, the researcher must consider and implement ways to minimize observer effect on animal behavior, yet still understand and acknowledge that observer presence may cause changes in their subject's behavior. While considering how to mask the observer's presence, one's sensory biases may lead one astray, so that only visual masking with blinds or platforms is considered. Olfactory presence is probably of even greater significance for mammals in general and is especially critical for carnivores.

Visual masking is important, but the observer should acknowledge that animals generally know the observer is present in blinds and towers. Often the observer will arrive well before the animal and leave after the study organism has left. While studying Gunnison sage-grouse in Colorado, Young (1994) often arrived by 0300 hours and spent long, cold hours in a blind until the last bird left the display ground after 0900 hours. Although the visual presence of the observer was somewhat masked, it is uncertain to what extent olfactory presence influenced the birds' breeding behavior. Even when the study organism appears to be habituated or accustomed to the observer's presence, it is difficult to ascertain the extent to which categories of individuals (genders, age classes, etc.) are differentially affected by other sensory indications that an observer is near. Bekoff (2000) provides a more complete review of **observer effects** on study animals.

There are a variety of ways to mark individuals to allow behavioral observation at a distance. The study design may require that marks are observable from a few to hundreds of meters. The common error of overestimating sample sizes in wildlife studies is exacerbated by a tendency for wildlife managers to mark study populations rather than individually mark captured animals, as is more common in behavioral studies. Behavioral ecologists should carefully consider their needs before embarking on individually marking their study organism. One critical question that should be asked is the extent to which retention of the mark is needed. For example, amphibians may rapidly regrow clipped toes, and bands may discolor or be lost from birds. Other considerations should include minimizing stress or possibility of injury to the study animal. Finally, many **marking techniques** may have unintended consequences on the behavior of the study animal or the behavior of individuals interacting with the marked individual. For example, Burley (1988) found that color banding zebra finches (*Poephila guttata*) influenced mate choice. Silvy et al. (Chapter 9, This Volume) provide a more complete review of marking.

After animals are marked, researchers may assign a name for field recognition. Most researchers advocate use of letters or numbers that have no subjective bias associated with them and avoid using personal names. Bekoff (1997) offers a contrasting view that naming an animal increases researchers' respect for their study organism.

STRATEGIES FOR DEFINING BEHAVIORS

Observations

Observing animals leading to increased familiarity is a critical step in defining behaviors for a study organism. Martin and Bateson (1996) suggested that it was vital to get to know the organism and to review the literature before defining terms. There are 2 reasons for informal observations preceding quantitative studies. First, it is generally through informal observations that hypotheses are formed. Second, choosing the appropriate methods to address hypotheses is greatly assisted by a period of observation. Most animal behaviorists would suggest that young researchers or those studying an animal for the first time also should review published literature on how behavioral definitions have been formed and used. However, some researchers would recommend immersing yourself in the animal's environment with no preconceived notions, as was done by renown primate behaviorist, Jane Goodall.

Lehner (1996) distinguished between watching animals and rigorously observing them. Informal observers should understand the difference. A babysitter may watch children, whereas a psychologist observes them. Both are valuable activities, but the psychologist's observations are far more likely to lead to testable hypotheses. Although observations can initially be informal, they should still be able to provide intricate detail about individuals and their social behaviors.

Behavioral Definitions

Behavioral definitions should be sufficiently precise so they can be communicated clearly to other field personnel and researchers, and definitions should avoid **bias**. For example, defining types of movement, such as still, walking, running, or flying, is generally more objective than suggesting the cause of the movement (resting, boredom, fear, migration, etc.). For each term, the researcher should apply a definition that makes the term mutually exclusive from other terms. Walking could be defined as taking 10 steps in a 5-second interval, whereas running could be defined as taking >10 steps in a 5-second interval. Behaviors can either describe structure (posture, movement, etc.), consequences (escape, compete, etc.), or relative position with inter- or intra-specifics (approach, flee). Two warnings are appropriate. Behavioral definitions based on consequences of the behavior and those based on inter- or intra-specific interactions are often larger categories of activities and can be associated with human bias. Definitions based on structure can often generate excessive detail that does not necessarily test general hypotheses well.

Indirect versus Direct Measures of Behavior

It is important to consider types of questions that can be addressed with indirect behavioral observations. For many animals, direct observations are challenging at best, and indirect observations may provide valuable insight into the social behavior of the species. **Indirect observations** of behavior include studying tracks, markings, foraging sites, bite marks, frequency of scars, feces, and even chewing of radio-tracking collars. One exciting method for using indirect behavior has been recording of nocturnal avian migration vocalizations to provide survey measurements (Evans and Mellinger 1999).

TYPES OF SAMPLING

Altmann (1974) provided an excellent paper on sampling methods for behavioral biologists. There are 4 basic sampling methods: (1) ad libitum sampling, (2) focal-animal sampling, (3) scan sampling, and (4) social structure sampling. Each type of sampling method has benefits and costs associated with its use (Table 19.1).

Ad libitum sampling occurs when the observer notes what they see that seems relevant or interesting. Major problems with this type of sampling are that it lacks randomization for meeting the requirement of independence for statistical tests and is likely to bias observers to overestimate conspicuous behaviors. Consider the excitement of watching a coyote (*Canis latrans*) chase a ground squirrel. Although it may be interesting to note the sequence of action and the depredation event, it would not provide a researcher with information on the frequency or duration of predatory events for either species. Ad libitum sampling can be useful when the observer is first observing a species to gather preliminary information for hypotheses formation or to record rare events, but it is rarely useful for actually testing hypotheses.

Focal-animal sampling occurs when one individual or group (when several groups are present) is watched continuously for a set period of time, and all acts which the animal either initiates or has directed toward it are recorded. Generally, the investigator has formed an **ethogram** or a catalog of carefully defined, mutually exclusive behaviors to categorize the behavioral repertoire of the focal animal (Table 19.2). The choice of the focal animal is often randomized to prevent observer bias or is established for experimental design reasons prior to the initiation of the study. One major challenge of focal-animal sampling is the targeted individual may move from sight or leave before the period of time for their sampling is finished. Behavioral biologists form explicit rules about how to proceed with this common occurrence. Although each rule may differ among studies, the critical aspect is they are applied equally by different observers and across time in the same study. For example, while watching a herd of pronghorn, the targeted individual may disappear behind a hill. The observer can either record "time out" for the period it is out of sight or switch observations to a new focal animal. One caution about using this type of sampling occurs if the study animal generally does certain behaviors secretly or with great privacy (e.g., birthing or mating) that would lead to those behaviors being underrepresented. Focal-animal sampling provides good information about rates, durations, sequences, and interactions.

Table 19.1. Types of data obtained and behaviors measured with different sampling methods

Sampling method	State or event sampling	Behavioral measure	Weakness
Ad libitum	Either	Opportunistic measures suitable for field notes or for initial observations prior to hypotheses testing	Overestimates rare behaviors; cannot use statistics on observations, given nonrandomness of sampling
Focal-animal	Either	Provides good information about rates, durations, in animal sequences, and interactions	Difficult to gather data on sequence of interactions among individuals
Scan	State	Provides good estimates of time budgets or percentage of time individuals spend doing different activities in a group	Does not provide duration data and usually is limited to recording a few types of behavioral acts
Sociometric matrix	Event	May provide information about dominance structure and other group social structures	Obtaining random sequences is challenging and rarely done

Modified from Altmann (1974).

Table 19.2. An ethogram for bald eagles (*Haliaeetus leucocephalus*), providing mutually exclusive descriptions of behaviors. This ethogram identifies relative amount of time spent by male and female eagles in foraging, but it does not distinguish between relative frequencies of different types of aggressive behaviors.

Behavior	Description of behavior
Resting (RS)	The eagle is not performing any active behavior, remaining stationary for 10 seconds.
Self-preening (SP)	The eagle is manipulating its own feathers with its beak.
Allo-preening (AP)	The eagle is manipulating feathers of a conspecific.
Foraging (FG)	The eagle is actively attacking or consuming food items.
Nest building (NB)	The eagle is creating a nesting structure or carrying nesting material to a nesting structure.
Nest incubation (NI)	The eagle is sitting in a nest or manipulating an egg in a nest.
Courtship flight (CF)	The eagle is performing aerial flight displays with its mate.
Vocalizing (VC)	The eagle is vocalizing.
Flying (FY)	The eagle is in the air and is not courting a mate.
Walking (WK)	The eagle is hopping or walking on the ground.
Aggression (AG)	The eagle is demonstrating aggressive behaviors toward a conspecific (pecking, erecting feathers, etc.).

It is considered by most biologists to be the type of sampling that will have the highest dividends for testing the original hypothesis and providing information for subsequent studies.

Scan sampling uses a census (total count) of an entire group of animals for a single behavior or a small set of behaviors. For example, while watching a flock of geese (Anatidae), a researcher would scan the group for 15 seconds every 5 minutes to record how many geese in the group have their heads up in an alert manner. Although short sampling periods are important for this methodology, realistically the sampling period will be affected by group size and complexity of behaviors. The important factor is to keep sampling time short and constant across samples. The time it takes will be a function of group size, number of behaviors recorded, and general visibility. Scan sampling is a good method of measuring the distribution of behaviors in a group and also can be used to measure the extent to which behaviors are synchronized in a group. Scan sampling also can help the observer estimate **time budgets** (the percentage of time individuals spend doing different activities). For example, if the researcher wanted to know whether male green-winged teal (*A. carolinensis*) spent more time (and therefore energy) on vigilance behavior than did their mates prior to nesting season, scan samples could be quite useful. Problems with this type of sampling are that it does not provide the ability to estimate the duration of behaviors (unless they are performed for long periods or the sampling is virtually instantaneous) and is generally limited in the number of behaviors that can be investigated.

Social structure sampling examines social structure rather than individual behavior. It is an important type of sampling that can be useful for examining social dynamics between pairs of individuals and among groups (McDonald et al. 2000). Social dynamics are the resultant interaction of an individual's ecology and behavior. Understanding the social dynamics of a species is critical to predicting the consequences of actions taken to manage wildlife.

Historical methods for studying social dynamics include sequence sampling, which allows the investigator to follow a chain of behavioral interactions between individuals or successive animals in a group. It can be challenging to identify the beginning or end of a sequence, and choosing sequences at random is rarely done.

Pairwise interactions between individuals or species that can be measured using sociometric matrix sampling can be useful for identifying social dominance (Lehner 1996) or other types of social behaviors. McDonald et al. (2000) provide an excellent review of current methodologies for examining and organizing questions about social structure.

Each sampling method has strengths and weaknesses. Most behavioral biologists combine more than one methodology to test their research hypotheses. The key to selecting the correct methodology is being certain the methods chosen are best to measure the behavior to be quantified for statistically testing the hypothesis and the methodology has been carefully defined to allow others to repeat the experiments.

TYPES OF BEHAVIORAL MEASUREMENT

Behaviors can be recorded as either events or states. **Events** are behaviors, such as mounting or vocalizing, that occur for a relatively short duration, whereas states are behaviors, such as rutting or foraging, that indicate prolonged activity.

Frequencies are measurements of how often specific behaviors occur during some unit of time (Fig. 19.9). Frequency of events, such as fighting, is often used to measure behaviors for social interactions. In general, frequencies are the most common type of behavior measurement. **Durations** are often used to measure states, such as rutting or migratory behavior. Durations are important measurements in time budget analyses. **Latency,** or the **initiation time,** measures the time interval between when a behavior begins and another behavior ends (Fig. 19.9). For example, a researcher may be interested in the latency to attack after a gray wolf raises the hair on the back of its neck. **Intensity** is an indication of how extreme the behavior expressed is or how loud a vocalization may be. Researchers can associate the presence of certain acts with either high or low intensity to provide more objective measurements. For example, a threat display from a gray wolf may be of higher intensity if it has hair raised, lips curled, and its tail is held in an upright pos-

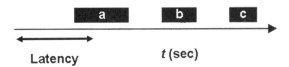

Fig. 19.9. The black bars measure the occurrences of a behavior, such as foraging, by mountain sheep (*Ovis canadensis*) over time. The bars represent 3 occurrences of foraging in time (*t*), and frequency of foraging would be 3/*t*. Latency is the measure of time until the behavior first occurred. The total duration of foraging during the observation period is (*a* + *b* + *c*). The proportion of time spent foraging is (*a* + *b* + *c*)/*t*. *After Risenhoover and Bailey (1985).*

ture. **Bouts** are commonly used in post-field analyses. Bouts reduce behavioral observations into temporal clusters (courting bouts, foraging bouts, singing bouts, etc.) of behavior. A bout is generally recognized as a period of time in which a behavior occurs regularly, but not continuously (Bart et al. 1998). Occasionally, defining bouts is easy, because the behavioral activities occur in discrete clumps of time separated by a standard amount of time. When Gunnison sage-grouse perform mating displays, they have a fairly regular pattern of behaviors followed by a few seconds of rest (Young et al. 1994). However, sage-grouse can be quite variable in how long they perform the displays and how long they rest between bouts of mating displays. Because bouts of displays can vary by individual, absolute definitions of time defining a bout often mask individual variation in a species. There are several techniques that allow observers to define bouts mathematically (Slater and Lester 1982, Sibly et al. 1990, Martin and Bateson 1996).

USEFUL NEW CONCEPTS

Improved technology has made aspects of recording animal behavior easier and allowed us to view previously inaccessible types of behavior. Use of computers, specialized software, and improved camera technology has dramatically aided our ability to record and analyze behavioral data.

Software

Software has been developed for quantifying animal behaviors, social structure, movements, spatial data, and vocalizations. There are numerous programs available that allow a laptop computer or personal digital assistant (PDA) to record and analyze behavioral data. For example, Biobserve (**http://www.biobserve.com/**) provides software for generating ethograms, correlating ultrasound to behavioral observations, recording events and social interactions, and creating *x/y* coordinate systems from videos. The company also provides hardware for using infrared sensor and illumination to record observations of behaviors directly to computers. Many products allowing researchers to capture behavioral data and ethograms work on PDAs and smart phones (e.g., Spectator Go! Professional; http://ptf.com/spectator/ spectator+go+professional/), as well as on more traditional hardware, such as laptops and desktops. Other commercial products include **Forager** (Amber Waves Software, Lancaster, PA), a program designed to simulate current foraging concepts, record data, and produce reports and maps. Many academic laboratories produce software for use on both Windows and Macintosh platforms. For example, JWatcher, developed by researchers at the University of California–Los Angeles and at Macquarie University, Sydney, is a powerful Java program designed to assist in recording focal animal data (http://www.jwatcher.ucla.edu/). The Animal Behavior Society has both a website with current software (http://www.animalbehavior.org/Resources/software.htm) and archives of software, as does an international site from the University of Saskatchewan, Saskatoon, Canada, focusing on applied ethology (http://www.usask.ca/wcvm/herdmed/applied-ethology/links.html#software). Colorado State University (Fort Collins) also has a large collection of software developed for examining mark–recapture data for population estimates and distributions (http://www.cnr.colostate.edu/~gwhite/software.html).

There are several challenges associated with using computers, cell phones, or hand-held PDAs to record observational data. Many researchers have been confronted with the unfortunate reality their electronic information has not been recorded or saved through a variety of failures. Another challenge, not unique to this platform, is that watching behavioral interactions while operating recording computers (or pencil and paper!) causes observers to take their eyes off of their subjects. The advent of voice recognition software will likely replace tape and digital recorders as a solution to this enduring problem (White et al. 2002).

Global Positioning System Telemetry Combined with Activity Sensors

Although Global Positioning System (GPS) collars have revolutionized telemetry (see Chapter 18, This Volume) and provided much more insight into individual use of habitat, their use simply provides an animal's position at a much more frequent rate than what traditional very high frequency (VHF) transmitters provided. A new generation of GPS transmitting collars are equipped with a dual-axis acceleration sensor (Vectronic Aerospace, Berlin, Germany), that can be highly correlated with simple individual behavior categories, such as feeding, resting, and locomotion in ungulates (Löttker et al. 2009) and to levels of activity in brown bears (*Ursus arctos;* Gervas et al. 2006).

Robotics

During the past decade, as technology improves and costs are reduced, some biologists are working with engineers to design robotic technology to address hypotheses about animal behavior. As robotic technology improves, it will provide an opportunity for doing experimentally based behavioral studies that can help explain individual responses to specific visual cues in conspecifics. Robots have been used in the field of animal behavior to understand the evolution of learning and to test specific hypotheses about the mechanisms of behavior (Webb 2000).

Typically, a **robot** is an elecromagnetic machine controlled by a computer program and is designed to mimic actions of humans or animals. Robots can either mimic the appearance of an individual in the species or, in research designed to look at the instinctive behaviors, they may look nothing like the animal itself, but accurately mimic a visual or tactile signal. The general concept of mimicking animal signals or behavioral cues has been applied in audio playback experiments and with mechanical models for decades; however, as robotic technology has become better and less expensive, researchers will increasingly be able to model more complex visual, tactile, and auditory behaviors for focal animal studies in the field.

Currently, a plethora of behavioral studies have used robots to mimic specific behaviors in focal animal studies both in the laboratory and in the field (Michelsen et al. 1992, Patricelli et al. 2002, Göth and Evans 2004b, Martins et al. 2005; Partan et al. 2009). One example of a current use of such technology is illustrated by Gail Patricelli's work (Patricelli and Krakauer 2010) with sage-grouse robots. Patricelli (University of California–Davis, personal communication) is using robotic technology to test hypotheses about male responses to female behavioral cues (Fig. 19.10). Another unique use of the technology that may assist in captive breeding programs is using female robots to collect sperm (T. L. Hicks, Western State College of Colorado, personal communication) from males of species in which traditional collection of sperm would involve invasive methods. Such use of behavioral signals and robotic technology also can lead to an examination of male fertility and inbreeding depression in field populations, in which habitat fragmentation may have led to low genetic variation and possible decline in individual fertility.

Perhaps one of the most promising areas of wildlife and conservation management currently being explored with robotics is using robots to better understand social cueing. **Social cueing** is the mechanism by which individuals in a population make decisions about when to join groups (conspecific attraction) and exhibit social behaviors that can influence choice of habitats for foraging, predator avoidance, and mating interactions. Understanding how individual behaviors influence formation of social groups and habitat choices can be a strong tool for wildlife managers working to promote reintroductions of populations into underutilized habitats or habitats that may have been historically disconnected by fragmentation of core habitats. For example, Ackerman et al. (2006) found that social cueing was increased is some species of waterfowl when decoys were motorized to add wing movements.

Sound Recordings and Sound Arrays

An interesting combination of new technologies involving combining digital audio with video recordings has led to the creation of digital recording arrays, such as acoustic directionality measurement systems (Patricelli et al. 2007) and acoustic location systems (Mennill et al. 2006, Fitzsimmons et al. 2008). These **array systems** measure sounds in the field from animals inside an array of 8–16 microphones. Such technology can not only help explain the directionality and context of different types of vocalizations (alarm calls versus mating calls), but also may allow the possibility of spatially tracking and locating animals on larger spatial scales, helping to clarify their use of the landscape. The use of digital vocal arrays also is leading to current research on the effects of energy development on species dependent on acoustical signaling to attract mates, such as sage-grouse and songbirds in the shrubsteppe habitat in the West. Finally, such systems may become increasingly common for trying to ascertain the presence and location of increasingly rare species as habitat fragmentation and degradation leads to increased extirpation of species. Such information could lead to targeted protection efforts as well as to target population augmentation attempts to maintain small populations.

Cameras

Video can be useful for field studies, because observations can be viewed multiple times, and different sampling methodologies can be used on the same behavioral sample. Video images of field observations also can be used as important training tools to increase interobserver reliability. Video cameras can help observers see details of rapidly performed behaviors and understand the mechanisms of those behaviors. Longer battery life (up to 12 hr) and the advent of compact digital video cameras have made these devices increasingly useful and portable for field studies. Along with changes in size and weight have come improvements in technology. Today, even a relatively inexpensive digital video camera has superior light-gathering ability compared to those commercially available during the past decade, allowing high quality

Fig. 19.10. A female greater sage-grouse robot used to study male responses to females on the lek. *Photo courtesy of A. H. Krakauer.*

records during dusk and dawn. There also are video cameras and optical devices sensitive to infrared light that have infrared light emitters in them for night viewing. Another technology change has been the advent of **fiber optic** cameras that can be placed into nests, dens, and burrows to allow glimpses of behaviors previously unavailable for observation. Inexpensive wireless tilt cameras are increasingly being used to record data as well.

Digital cameras can be used to take pictures of feathers, pelage, skin, or scar patterns on animals, allowing them to be individually identified. Although traditional cameras also can perform this task, the resolution and ability to immediately access and manipulate digital images make them much more useful in the field. Both video and digital cameras can be set to automatically record events in the absence of the investigator (see Chapter 10, This Volume). Although there is some risk of technology failure with such endeavors, the ability to capture rare events of particularly secretive animals or to record depredation events makes automatic recordings a welcome additional field tool.

Internet Access

The advent of increased access to the Internet has enormous potential for allowing exchange of raw data and observations that can enhance research efforts and collaborations. Most peer-reviewed journals no longer publish general observations or common and rare events that can provide insight into behavioral events. The inability to examine other researchers' data in a form that leads to new analyses increases wasteful efforts of repeating studies and hinders population comparisons. Posting of such data on personal websites would be helpful. A second opportunity provided by the Internet is the ability for researchers to share video examples of tools, such as digital ethograms. Stewart et al. (1997) created a multimedia vocal ethogram of European badgers based on digital video. These types of contributions, although not publishable in the traditional sense, enhance communication among scientists and help alleviate disagreements in the published literature based on misunderstandings of basic methodology. Sharing of Internet databases and methodologies led to major advances in DNA research and could do the same in the field of behavioral ecology.

MANAGING HUMAN (ANIMAL) BEHAVIOR

Managing human behavior will be one of the most important wildlife management priorities in the 21st century. Historically, managing human exploitation of resources was the most common form of behavioral management. Currently, human behavior associated with nonconsumptive uses of wildlife is of growing interest and concern. Many states and provinces receive greater economic benefits and associated human impacts from recreational activities, such as wildlife watching, than they do from hunting. In 2006, the 71 million American wildlife watchers (Fig. 19.11) who visited, photographed, and fed wildlife had a reported economic value of US$122.6 billion, including paying US$18.2 billion in state and federal taxes (Leonard 2008). Although federal land-management agencies are beginning to grasp the implications of changing uses of wildlife and their ecosystems, most state and provincial agencies are still focused on hunting and other consumptive uses.

As habitat and wildlife viewing becomes increasingly restricted due to reduced areas of suitable habitat and growing conservation concerns, the impact of human disturbance on animal behavior and fitness will increase (Hockin et al. 1992, Knight and Gutzwiller 1995, Gill and Sutherland 2000, Fernández-Juricic et al. 2005). Animals may avoid areas with high human traffic, noise, or light pollution, or structures (e.g., blinds), resulting in habitat sinks or areas unsuitable for their reproduction and habitation. For example, bald eagles are less likely to use areas with heavy foot and fishing boat traffic, may experience disturbances up to 500 m away, and will delay foraging by several hours if disturbed (Stalmaster and Kaiser 1998). Several studies have demonstrated that at least some species will avoid roads (Mattson et al. 1987, Gill 1996), causing the effects of roads to be greater than the actual area of habitat loss. Sutherland (1998a) developed mathematical models to examine the effects of human disturbance at the species level. Despite the documentation of recreational impacts, recreationists do not perceive their activities as impacting wildlife (Taylor and Knight 2003). A critical role for behavioral biologists and ecologists is to provide information to wildlife managers about the consequences of human disturbance on population parameters, such as density-dependent breeding and mortality.

Increasingly, wildlife managers will need to take proactive steps toward managing the impacts of human behavior on wildlife. Although the formation of "Watchable Wildlife"

Fig. 19.11. The increase in wildlife viewing requires new approaches to managing human behavior to prevent impacts on wildlife. The dancing greater prairie-chicken (*Tympanuchus cupido pinnatus*) on the photography blind was performing a breeding display at a watchable wildlife area at the Nature Conservancy's Dunn Ranch, Eagleville, Missouri.

(http://www.watchablewildlife.org/) sites may help consolidate impacts, it is important that such sites are monitored and evaluated as to their population-level impacts. In the Gunnison Basin, Colorado, the designation of Gunnison sage-grouse as a new species (Young et al. 2000) led to enormous increases in visitation at a designated viewing area. As visitor numbers increased, attendance of males and females at that mating ground declined at a rate greater than at mating grounds not designated for public viewing (J. R. Young, unpublished data). Clear criteria for managing human impacts are needed as well as studies providing wildlife managers with data to make informed decisions. Such incidents as bear attacks and other negative interactions with wildlife are increasing at refuges and national parks. Research, management, and education will be important undertakings as increasing numbers of people seek visitation to such sites with concomitant effects on wildlife and human behaviors.

FUTURE DIRECTIONS

The challenge for future biologists in wildlife management and animal behavior is to examine how studies can better integrate these fields. A study by Huwer et al. (2008) demonstrated the importance of integrating the fields of wildlife biology and animal behavior. They used human imprinted sage-grouse chicks (Fig. 19.12) to test ideas about effects of habitat quality on chick development. Their study was helping resolve important management and conservation questions in sage-grouse management that could not be resolved with traditional radiotracking or observational studies. This study is an excellent example of understanding the concept of Umwelt and using the species instinctive behavior to learn about the foraging ecology of a species. Such insights are leading to further imprinting studies to better understand the effects of grazing on Gunnison sage-grouse. Other areas of investigation for which behavioral studies can pro-

Fig. 19.12. Sherri Huwer uses a field incubator to collect her human-imprinted Gunnison sage-grouse chicks at the end of the field day. *Photo courtesy of J. R. Young.*

Table 19.3. Examples of animal behavior studies that could enhance current efforts to manage and conserve species

Behavior study	Description of study
Allee Effect	The Allee Effect is a decline in individual fitness due to low population density. Recent behavioral studies suggest that mechanisms for reduced fitness at low population densities may be behavioral. An important tool for conservation will be to understand the role of individual behavior in increasing risk of extirpation or extinction in populations of different sizes.
Communication	Detecting sources of communication among conspecifics can enhance understanding of social structure and increase wildlife managers' ability to census populations. Increasingly important is understanding how anthropogenic sources of environmental "noise" in smells, visual disturbance, and sounds influence populations.
Dispersal	Knowing which gender and at what age animals disperse from family groups, distances they travel, as well as the conditions leading to survival success, will be critical for habitat designation in multiple use areas and for reserve design.
Diversity	Research quantifying behavioral diversity among individuals and populations is an important tool for wildlife management. Although many wildlife managers are sensitive to the concept of maintaining morphological and genetic diversity, few consider the importance of managing to maintain variation in behavior. Behavioral flexibility in a population is critical for allowing adaptation to different ecological conditions and changing landscapes.
Foraging behavior	Studies showing the effects of habitat quality on distribution of animals and subsequent changes in behavior in suboptimal habitat patches will be important as increasing human populations intrude into traditional foraging sites of animals. Foraging behavior often tells us much about how to design or conserve species' habitats, as studies of foraging behavior can elucidate how an individual weighs the risks of predation, the quality of food supply, and interactions in and among species.
Learning and cultural	One of the major causes of failure of reintroductions has been that newly released animals fail to know transmission of behavior where to forage, find mates, and hide from predators. Field studies of species investigating the importance of conspecific's roles in the transmission of behavior could be invaluable.
Life history traits	Traits, such as fecundity, age class, and survival, can change due to behavioral responses to exploitation of species. Understanding population responses to exploitation will require an understanding of individual behavior.
Mating systems	A fundamental understanding of how changes in densities of populations and of their preferred habitats affect mating system and facultative sex ratios in species could help managers understand population-level responses of species of concern as well as hunted species. Understanding mating social structure in a population can aid in reducing the myriad of assumptions used to estimate parameters in population viability models.
Migration	Understanding the mechanisms and evolutionary causes of migration will be critical during the next century as human population growth leads to greater habitat degradation, fragmentation, and loss. Knowing why an area is critical for migratory success will allow either preservation or replacement to facilitate traditional movements.
Sensory studies	We are only beginning to understand the unintended consequences of how odor and light pollution impact migrating species. Basic research into animal senses could provide answers to some of the more intractable wildlife management and conservation challenges. A second area requiring more research is learning how to manipulate the distribution of animals through aversion conditioning based on an understanding of their sensory system. Such studies could lead to nonlethal animal control.
Sexual imprinting	Studying the mechanisms behind sexual choice are important steps to successful captive breeding programs, releases, and reintroductions.
Sexual selection	Studies examining the relationship between mating preferences and trait variation among populations could allow predictions of success for population augmentations.
Stress	Changes due to atypical weather, pollution, disturbance, and population density can change hormonal secretions, leading to adaptable changes in behavior. Understanding which factors cause a physiological and behavioral change due to an animal's stress level will be critical for managing human presence and activity in important wildlife habitats.

From Sutherland (1998b), Gosling and Sutherland (2000).

vide new insights into wildlife management include communication, dispersal, diversity, the Allee Effect, and foraging behavior (Table 19.3).

SUMMARY

This chapter introduces methods for studying behavior by wildlife biologists and suggests important future areas of study. Given the global biodiversity crisis, large-scale changes in global weather patterns, and the continued loss of habitat, integrating knowledge of animal behavior with wildlife management and conservation is critical. A fundamental understanding of animal behavior will be necessary to develop robust metapopulation models, understand true effective population sizes, and elucidate consequences of anthropomorphic changes on landscapes. Although conservation often is directed at the ecosystem level, and wildlife management generally deals with populations, all levels of biological inquiry (from genes to communities) are necessary for successful management and recovery of species and their ecosystems. Important first steps to fostering cross-disciplinary approaches include adding classes on animal behavior in wildlife curriculums, holding joint regional and national meetings of disciplinary societies, developing chapters and units on applied animal behavior for courses and programs, and being willing to embrace collaboration across traditional disciplinary boundaries.

20

Analysis of Radiotelemetry Data

JOSHUA J. MILLSPAUGH,
ROBERT A. GITZEN,
JERROLD L. BELANT,
ROLAND W. KAYS,
BARBARA J. KELLER,
DYLAN C. KESLER,
CHRISTOPHER T. ROTA,
JOHN H. SCHULZ, AND
CATHERINE M. BODINOF

INTRODUCTION

THE USE OF RADIOTELEMETRY and other methods of remote sensing provide investigators with data about locations of animals and their fates. These data are used to estimate several individual and population parameters, including resource selection, home ranges, dispersal, migration, and survival; because animals are marked and then resighted with nonmarked cohorts, population estimates can be obtained. Radiotelemetry affords us the ability to investigate features of wild animal populations that would otherwise be impossible (see Chapter 10, This Volume). This chapter provides an overview of the techniques available to analyze radiotelemetry data. The increased use of radiotelemetry in the past 2 decades, ever-growing array of analytical methods, and developments in technology have been accompanied by many publications and books about the technique (Kenward 2001, Millspaugh and Marzluff 2001). However, rapid developments continue on all fronts, making it essential that biologists stay aware of recent advances by investigating the latest literature and corresponding with colleagues and those developing analytical methods and computer software.

GENERAL CONSIDERATIONS

Radiotelemetry is used to address a wide variety of objectives, each with corresponding analytical issues and methods. However, with any radiotelemetry analysis, several general issues are relevant, regardless of the methods being considered. First, choice among analytical methods should be determined by the question of interest and the ecology of the species being investigated. Radiotelemetry is a means to an end, and analyses need to be tailored accordingly. There is no universal best technique for the analysis of radiotelemetry data. Although some analytical approaches need to be avoided, there are often better and worse options. Second, proper sampling design and sufficient sample sizes are vital to determining the strength of study conclusions. No data analysis approach can save a study that was not designed to ensure representative sampling and adequacy of sample sizes. Third, increased sophistication in analytical methods is often a way to make more effective use of data and increase the depth of ecological inference in radiotelemetry studies. Methods that make use of all aspects of telemetry data, particularly the temporal component, often allow more powerful insights than do traditional approaches. Explanatory power will be increased with approaches that directly integrate biology into the analytical model(s). These approaches are clearly the wave of the future in the analysis of radiotelemetry data. Fourth, more than with perhaps any other analytical topic in wildlife ecology, one must be very careful when assessing recommendations in the literature about radiotelemetry data analyses. These recommendations are often based on limited comparison of methods with new techniques. Simulations are valuable for comparing methods, but beware of broad advice based on very limited case studies or simulations, particularly chosen to show a technique at its worst. Additionally, in terms of comparing techniques, real data provide a case study, not a broad perspective, and such case studies do not offer adequate support for sweeping conclusions about the utility of techniques. Fifth, biologists should beware of software variations, because seemingly similar methods can produce remarkably dissimilar results stemming from differences in software. The analysis of radiotelemetry data is often done using standalone computer packages developed for specific analytical techniques, and the utility of these packages is dependent on the programmers' knowledge and skill (Appendix 20.1). Programmers should offer transparency in the coding and assumptions made in their software. Sixth, greater transparency is needed when reporting results from radiotelemetry investigations. We strongly urge biologists to make raw data available (e.g., www.movebank.org or other archive) and report sampling design, sample sizes, user-defined options in analysis, and assumptions in their analysis and interpretation of telemetry data. Not only is this practice essential to meeting basic scientific standards regarding repeatability, but also the quality of inference from radiotelemetry results may depend heavily on methodological details often not reported.

SPACE USE AND MOVEMENTS

There are many motivations for studying animal movement. Nathan et al. (2009) identified 4 fundamental questions about organismal movement: (1) Why does an animal move? (2) How does it move? (3) When and where does it move? (4) What are the ecological and evolutionary consequences of movement? White and Garrott (1990) offered questions of "which," "why," "when," and "where" as motivations for studying movements of wildlife. Building on these basic questions, Nathan et al. (2009) proposed a **Movement Ecology Paradigm** that attempts to unify movement studies of all organisms around a conceptual framework that comprehensively considers the interplay between the organism and the environment. Understanding the proximate causes, mechanisms, and spatiotemporal patterns of movement is critical for understanding the evolutionary causes and consequences of movements, as well as the role of these movements in broader ecological processes (Nathan et al. 2009).

Radiotelemetry is a powerful tool for collecting information about animal locations, rates of movement, and in some cases, general behaviors associated with individual locations.

As a result, radiotelemetry often is essential for addressing the specific scientific and applied motivations of any study or research program examining movements. In this section, we focus on 3 commonly used groups of analyses: (1) estimating general summaries of **space use,** such as home-range or territory size and intensity of use of areas in the home range over a relatively long period (weeks or months); (2) drawing inferences about intra- or inter-specific interactions, such as the amount of overlap between 2 individuals of the same species, and (3) examining more detailed aspects of movement rates and trajectories, including movements in the home range and larger-scale migratory or dispersal movements. Another primary goal of many studies is to relate the intensity of space use to habitat characteristics, a topic that is briefly addressed in the following section on resource selection and more intensively in Chapter 17 (This Volume). Although we focus on these categories individually, one of the most exciting analytical developments over the past decade is mechanistic models that integrate across these artificial categories.

The Utilization Distribution and Home Range Analysis

A major objective of many radiotelemetry studies is to estimate the spatial extent of the area used by an animal and the intensity of use of different portions of this area during a finite period of time. Commonly, studies estimate the size of the **home range,** generally defined as the 2-dimensional area in which the animal conducts its "normal" activities (Burt 1943). Given the ambiguity of "normal," numerous more precise definitions have been proposed, and investigators need to carefully define what "home range" means in the context of their species of interest. For example, a biologist might choose to include paths of excursions to and from outlying locations as part of the home range, but exclude dispersal movements (Kenward 2001). The terms territory and home range are not interchangeable; some territorial species maintain only portions of their home range for "exclusive" or "priority use" (Powell 2000:70). Furthermore, biologists need to carefully consider whether standard definitions of home range make sense for their study situation. Despite some views to the contrary (Gautestad and Mysterud 1993), the home range concept has broad utility: in a defined period of time (White and Garrott 1990) and at a spatial scale of ecological or management interest, there is a finite area in which an animal conducts its activities (Powell 2000), and individuals of many species repeatedly traverse their general area of use. The home range concept may be less useful for species that wander irregularly at large spatial scales or spend large amounts of time at a few discrete locations, perhaps with occasional large scale dispersal or migratory movements. However, estimators discussed below still may be useful for assessing some aspects of space use, such as location and extent of migration corridors.

Kernohan et al. (2001:126) defined home range as "the extent of area with a defined probability of occurrence of an animal during a specified time period." The probabilistic component of this definition clearly identifies the home range as summarizing one aspect of an animal's **utilization distribution** (UD), which estimates the relative intensity or probability of use of areas by an animal (i.e., the relative frequency distribution of locations used by the animal; Jennrich and Turner 1969, Van Winkle 1975). Seen as a probability density function, the UD is the basis for summarizing numerous aspects of space use, such as home range size evaluated at specified probability contours as well as the location and number of **core areas** (portions of the home range with disproportionately high probability of use). The UD also can form the basis for probabilistic estimates of space use overlap between animals (Seidel 1992; see below) and use of different resources (e.g., vegetation types) in the home range (Marzluff et al. 2004). Therefore, methods that estimate a probabilistic UD rather than simply home-range polygon boundaries have broad utility (Kernohan et al. 2001).

Outliers, Home Range Contours, and Core Areas

In studies of space use, investigators need to consider how to handle **outliers:** locations that are comparatively far from most other observations, but that accurately represent forays or excursions outside the area "normally" used by each animal. How outliers are handled depends on the aspect of space use being examined and the analytical technique being used. For the commonly used **minimum convex polygon** (MCP) approach, such observations can cause severe overestimation of home range size. For other approaches, these excursions may not dramatically affect the estimated UD, but the biologist may not be interested in spatial extremes of the UD that have low probability of use. However, the potential for disproportionately high importance of unique resources at these used locations also must be considered (e.g., Powell et al. 2000:75, Fuller et al. 2005). Kenward (2001) and Kenward et al. (2001) summarized methods for statistical identification of individual outliers. The percentage of locations that have disproportionately high effects on the home range size estimate can be evaluated through visual inspection of utilization plots, showing the incremental decrease in estimated size as locations are omitted.

Commonly, investigators focus on inner subsets of the home range, either the minimum area containing a specified percentage of observed locations or the minimum area with the corresponding estimated probability of use (i.e., the minimum area over which the UD probability density function integrates to the specified probability). The latter approach is recommended, because it makes full use of the estimated UD, focusing on probability contours rather than being tied only to a (usually limited) sample of locations (Kernohan et al. 2001:136). The choice of which probability contour to select can be somewhat arbitrary. The 95% contour is a standard summary (White and Garrott 1990, Laver and Kelly 2008). However, this choice should be made not simply based on what other studies have done, but on the biological questions of interest, the underlying biology and patterns of space use by the species of interest, and the management context of the study.

Some authors refer to a core home range as the subset of the home range excluding occasional outlier excursions (Kenward et al. 2001). More frequently, core areas are defined as much smaller subsets of the UD with disproportionately high use compared to the rest of the home range beyond what would be expected from a random distribution. Similar to approaches used for eliminating outlying locations, core areas may be defined based on arbitrary contour levels (e.g., 50%) or graphical tests (see Powell 2000). We encourage biologists to consider whether some arbitrary contour level adequately identifies an area of intense use. In many cases, we argue that such arbitrary designation of core area does not adequately capture the metric of interest. Rather than justifying selection of the 50% contour because others have used it, consider methods that actually directly identify and estimate the area of intense use. For example, Samuel et al. (1985), Samuel and Green (1988) and Seaman and Powell (1996) have examined differential intensity of use in a range to help delineate the core area, regardless of associated probability contours.

Criteria for Comparing Home Range Methods

Before providing an overview of common estimators, we describe important criteria for comparing alternative methods, similar to the framework provided by previous reviews (Kenward 2001, Kernohan et al. 2001). Our criteria partly overlap those of previous reviews, but we emphasize some additional issues that are relevant to current trends in ecological data analysis (Table 20.1).

1. Calculation of a UD. As discussed above, methods that estimate a probabilistic UD allow estimation of home range size based on probability contours and facilitate other UD-based space-use analyses. The use of UDs also is useful in complementary analyses of space use overlap and resource selection.

2. Sample size requirements. Acceptable bias and precision at moderately low sample sizes is a desirable property of home range estimators (e.g., Seaman et al. 1999, Kenward 2001, Kernohan et al. 2001). Although our focus is on minimum sample sizes needed by each method, such technologies such as satellite telemetry now allow automatic collection of hundreds or thousands of locations on some species. The performance of estimators needs to be compared more thoroughly across a broad range of sample sizes, from datasets containing a few dozen observations, as is characteristic

Table 20.1. Comparison of home-range estimators based on selected evaluation criteria[a]

	Home range			Explanatory/mechanistic		
Comparison of estimators	MCP	Cluster	KDE	Model entire set of locations	Sequential movements modeled	Sequential movements as a function of covariates
Calculates utilization distribution	No	No	Yes	Yes	Yes	Yes
Sample-size requirements[b]	No	Yes	Yes	N/A	N/A	N/A
Incorporates temporal dimension	No	No	Yes[c]	No	Yes	Yes
Information theoretic comparison of alternative models	No	No	Yes[c]	Yes	Yes	Yes
Integrates ecological hypotheses into model structure	No	No	No	Yes	No	Yes
Thoroughly evaluated?	Yes	No	Yes	No	No	No

[a] KDE = kernel density estimation; MCP = minimum convex polygon estimation; N/A = not applicable.

[b] Estimated home range extent often stabilizes with 50 data points.

[c] The standard kernel approach does not incorporate the temporal dimension or allow direct statistical ranking of alternative models, but the method has been extended to incorporate the temporal component of locations by using a multivariate kernel estimator and to allow direct ranking of alternative models if bandwidths are selected with likelihood cross-validation (Horne and Garton 2006a, b).

of ground-based very high frequency (VHF) telemetry, to much larger datasets typical of satellite telemetry studies.

3. Use of temporal component or other dimensions of space use. As modern telemetry data analysis progresses, estimators that can make use of the temporal dimension of location data will allow greater insight into animal space-use dynamics (Laver and Kelly 2008, Keating and Cherry 2009). Such estimators also can incorporate other dimensions of space use, such as elevation (Keating and Cherry 2009). Ranking methods higher if they incorporate the temporal dimension is based on the idea the temporal dimension of radiotelemetry data provides important information, rather than being a nuisance element. Past discussion of the temporal component of telemetry studies generally treated autocorrelation as a significant problem, because sequential observations collected too close together in time provide less information about space use than would a set of independent locations (White and Garrott 1990). In some cases, concern about **autocorrelation** leads researchers to discard some hard-earned data (e.g., by removing observations that may have some degree of autocorrelation) or to collect locations too infrequently to define space use by the animal. However, if the study design produces an adequate sample of locations for each animal throughout the temporal frame of interest, autocorrelation is of little concern (e.g., De Solla et al. 1999; Fieberg 2007a, b). As discussed previously in this chapter, an adequate sample generally will be based on a systematic and/or stratified design that ensures samples are collected throughout the daily and seasonal period of interest.

4. Feasibility. Although wildlife ecologists continue to develop and apply increasingly sophisticated statistical tools, for many biologists, the ability to run analyses in standard statistical or Geographic Information System (GIS) software packages still will be an important factor in the choice of an analytical technique. Ideally, the availability of user friendly software would not drive selection of analytical options; in reality, it does. We encourage those developing analytical techniques to also consider the end user.

5. Facilitation of modern explanatory model-based inference. In traditional space use analyses, biologists used a single descriptive statistical model to estimate some property of space use (e.g., home range size) and then treated the resulting estimates as the raw data for subsequent analysis to assess ecological patterns. Modern ecological data analysis attempts to integrate such multistaged inference using models that incorporate alternative hypotheses about both the ecological process of interest and the observation process, along with variability in the data contributed by each process (e.g., Royle and Dorazio 2008). Uncertainty about the best statistical model for a dataset is not ignored, but rather multimodel inference incorporates this model-selection uncertainty. Home range estimators that can fit into this framework may be more powerful than purely descriptive estimators.

6. Extent of evaluation. Frequently, new UD estimators are introduced and promoted based on very limited comparisons with existing estimators and without good understanding of how these estimators compare across a range of distribution types and sample sizes. Estimators that have been evaluated thoroughly support a more informed choice about which method is most appropriate for a particular study.

OVERVIEW OF CURRENT METHODS

In this section, we provide a brief overview of current methods for estimating the UD and home range size. Several formerly commonly discussed estimators, including the bivariate normal and other ellipse approaches, the harmonic mean, Fourier transform, and grid-cell methods, are not discussed here, because we see them as superseded by other available methods. Additionally, currently available forms of these methods and their limitations have been extensively reviewed (White and Garrott 1990, Powell 2000, Kenward 2001, Kernohan et al. 2001, and Fuller et al. 2005). Instead, we discuss methods that currently are widely used or that have been recently developed, including the **MCP** approach, other recently developed polygon or linkage methods that may be of broader utility than the MCP method, kernel density estimation, and nonmechanistic and mechanistic models. The latter 2 approaches share the important advantage of being able to estimate a true probability density function corresponding to the UD.

MCP

The MCP method produces a convex polygon around (1) the least number of locations that will enclose all other locations (Silvy et al. 1979), (2) all estimated animal locations by minimizing the sum of the distances of segments linking the outer locations (Kenward 2001), or (3) the smallest polygon encompassing a specified proportion of estimated animal locations (White and Garrott 1990; Fig. 20.1). The area of this polygon is the MCP estimate of home range size. Although this technique is the oldest home-range estimation approach (Mohr 1947, Hayne 1949b, Powell 2000), it has been and continues to be the most frequently used method (Seaman et al. 1999, Laver and Kelly 2008). Moreover, it still is sometimes recommended over modern methods, such as kernel density estimation (Wauters et al. 2007). However, the continued use and recommendation of MCP as a default method is bizarre, given its severe limitations. For example, Börger et al. (2006:1402) argued the method "should not be used at all for estimating home range size" (see also reviews by White and Garrott [1990], Kernohan et al. [2001], and Fuller et al. [2005]).

Although the MCP method is simple and assumes no underlying statistical distribution, it, "more than any other method, emphasizes only the unstable, boundary properties of a home range and ignores the internal structures" (Powell 2000:80). It is highly dependent on sample size, both because it requires a large sample size to provide adequate estimates (e.g., >100; Harris et al. 1990) and because as the number of samples increases, so does the probability of obtaining an outlying location that has a large effect on the MCP estimate (e.g., Anderson 1982, White and Garrott 1990) that would not heavily influence other methods. Although interior polygons can be formed that eliminate outliers (e.g., by delineating the polygon around the subset of 95% of locations that produces the smallest such polygon), defining such a polygon requires an even larger sample size to be collected, so that a sufficient subsample remains to cal-

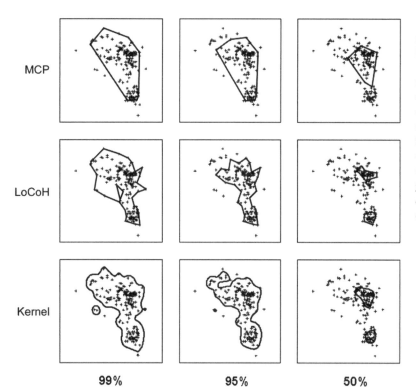

Fig. 20.1. Estimated 99%, 95%, and 50% home-range contours based on 187 radiotelemetry locations of a bull elk in Custer State Park, South Dakota. Results of 3 methods are shown: MCP = minimum convex polygon method; LoCoH = local convex hull method (adaptive form, with parameter *a* set as the maximum distance between observed locations; Getz et al. 2007); Kernel = fixed kernel estimates (normal kernel, plug-in estimation of bandwidth matrix with differential smoothing along *x*- and *y*-axes). *Produced with program R (R Development Core Team 2008) using package adehabitat (Calenge 2006) for MCP and LocoH and package ks (Duong 2008) for fixed kernel.*

culate this interior polygon. The MCP method ignores information provided by points within this outer boundary, and there may be large areas within this boundary that were never used by the animal (Powell 2000, Kenward 2001).

The simplicity of the MCP approach has led to it being recommended as an appropriate estimator for comparisons across studies (Harris et al. 1990). However, its sensitivity to sample sizes, insensitivity to internal use of the home range, and inconsistency in performance depending on the underlying point pattern (Downs and Horner 2008) make this comparability an illusion (Kernohan et al. 2001). In most cases, other modern methods discussed below better address the problems used by some authors as justification for recommending continued use of the MCP method. Problems with comparability among estimates from different studies are not limited to the MCP approach. In home range estimation in general, it is difficult to compare home range estimates from one study to another because of differences in sample sizes used to construct the estimate, choice of techniques (e.g., method of analysis and user defined options), and variations in software. All these factors should be reported and considered when comparing home range estimates across studies, and they argue for the importance of archiving raw data (http://www.movebank.org or other archives) to facilitate robust comparisons.

Other Linkage and Cluster-Based Methods

The MCP approach fits into the general category of methods that use alternative strategies for linking observations based on distance (e.g., between nearest neighbors or from a central point) and forming or merging polygons and/or clusters into one or multiple home-range components (Kenward 2001). More recently, applied approaches in this category can overcome some problems with the MCP method and currently may be the best approaches when there are unusable areas within the home-range outer boundary or other sharp boundaries between used and unused areas, if delineating these sharp boundaries is a priority (Kenward et al. 2001, Getz et al. 2007). One such approach, **incremental cluster analysis,** links locations into clusters that minimize the sum of nearest neighbor distances (Kenward 1987). In contrast to the MCP approach, this method allows multinuclear home ranges with multiple components making up the home range encompassing a specified percentage of locations. Kenward et al. (2001) also incorporate a test for identifying outlier locations that deviate from the distribution of nearest neighbor distances for the location set.

Similarly, local **convex hull** approaches first form local convex polygons for each location, where each polygon encompasses the focal point and either its k nearest neighbors or all neighbors within a specified distance of the point (Getz and Wilmers 2004, Getz et al. 2007; Fig 20.1). Hulls are sorted based on their size or the number of locations they contain; the union of polygons is formed incrementally, starting with the smallest polygons or those with fewest points, until a specified percentage of the locations are contained in the resulting polygon. This polygon is then taken as the $x\%$ contour of the estimated UD (Fig 20.2). An adaptive form of this approach uses variable distances to determine which neighbors to include in each point's convex hull, with shorter distances used for locations in heavily used areas. With local convex hull approaches, the number of neighbors, the distances used, or sum of local distances in the adaptive approach to form local polygons must be specified. Getz et al. (2007) recommends choosing the minimum values for these parameters needed to form polygons that exclude known holes in the UD (e.g., unusable areas, such as lakes for a terrestrial species that does not swim).

These linkage or convex hull methods form clusters and polygons based on linking observed locations (Fig. 20.1). The resulting home ranges therefore will be delimited by observed locations. This is true for both the individual polygons associated with each point or cluster and the resulting overall polygons formed from merging polygons or clusters. As a result, the home range isopleths can follow rigid boundaries and exclude known unusable areas in the home range. However, the property the estimated home range will not contain areas outside the observed set of external locations is not always an advantage: with limited sample sizes, there may be a high probability that nontrivial use occurs outside the outer periphery of locations, particularly if there is clustering of high use at the inner edge of this periphery.

Although the nearest neighbor approach of Kenward et al. (2001) has been evaluated using numerous real datasets, evaluation of local convex hull estimators has mostly focused on a limited set of point patterns with sharp boundar-

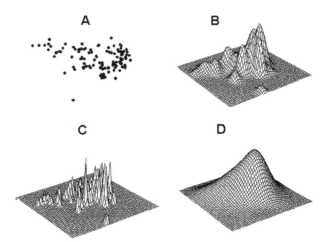

Fig. 20.2. Effect of different bandwidth values on kernel utilization distribution estimation. (A) 146 elk locations. Fixed kernel utilization distributions for these locations based on 3 bandwidth values (h). (B) $h = 500$. (C) $h = 150$. (D) $h = 1,500$. UTM = Universal Transverse Mercator coordinates. *(A) From Millspaugh (1999); kernel plotting routines from Bowman and Azzalini (1997).*

ies and on a location data from a few animals. These estimators need much more extensive examination across a broader range of point patterns and sample sizes. Although with the MCP and these approaches, a series of isopleths can be formed that approximates probability contours of the UD, none of these approaches form a formal probability density function as an estimate of the UD. Resulting isopleths could still be used to modify estimates of space use overlap or resource selection to account for nonrandom space use in the home range, but their use would depend on binning areas into discrete isopleth slices rather than using a continuously defined UD.

Kernel-Based Methods

Kernel density estimation (Silverman 1986; Worton 1987, 1989, 1995) has been widely recommended as the default home-range/UD estimation method (Seaman and Powell 1996, Powell 2000, Kernohan et al. 2001, Gitzen et al. 2006), although this recommendation is challenged frequently (Getz and Wilmers 2004, Hemson et al. 2005, Row and Blouin-Demers 2006b, Downs and Horner 2008). Like other approaches discussed above, the standard kernel method is nonparametric, in that it assumes no specific probability distribution for the underlying UD. Like convex hull methods, the kernel approach is formed by the union of individual components (i.e., kernels) formed for each location; in this case, the overall estimated UD is the sum of individual kernel functions centered at each location (Silverman 1986, Kernohan et al. 2001). This overall probability density function typically is evaluated across a fine mesh grid of points encompassing the observed locations and extending beyond them to allow for nonzero probability of use exterior to the outer periphery of locations. The density value at an evaluation point x (usually a grid point) is given by

$$\hat{f}(x) = \frac{1}{nh^2}\sum_{i=1}^{n} K\left[\frac{\bar{x} - X_i}{h}\right],$$

where X_i contains the coordinates of the ith observed location, n is the number of observed locations, h is the **bandwidth or smoothing parameter** (or vector of (h_x, h_y) to allow differential smoothing along the x- and y-axes), and $K(\cdot)$ is a kernel function. Usually $K(\cdot)$ is a probability density function; most frequently the Gaussian (bivariate normal) kernel has been used, but this choice has the slight disadvantage of allowing locations far from the evaluation point to have a finite if very small contribution to the estimated density. Other kernel functions ensure that points beyond some distance (in relation to the bandwidth) have exactly zero contribution.

The choice of a suitable bandwidth is the most critical, and controversial, factor involved in applying the kernel method (Table 20.2). The bandwidth affects the amount of smoothing applied to the observed locations by controlling how the effect of each location on the estimated density at an evaluation point scales with the distance from the location (Fig. 20.2). Relatively large bandwidths allow locations far from the evaluation point to significantly influence the estimated density function at that point, producing unimodal, smooth estimated UD and large estimates of home range size (Fig. 20.2D). Overly small bandwidths cause the estimated UD to fragment into numerous individual components (Fig. 20.2C). Thus, estimates of home-range size necessarily increase as bandwidth values increase; this phenomenon is not a weakness of the kernel method (Row and Blouin-Demers 2006), but is simply an inherent property. Separate bandwidths for the x- and y-coordinates, or a **bandwidth matrix** that allows smoothing to be oriented along axes rotated from the x- and y-coordinates, are recommended (Wand and Jones 1995, Amstrup et al. 2004), but many home-range software packages do not allow automatic incorporation of such differential smoothing. The optimal bandwidth for a set of locations (i.e., the bandwidth minimizing discrepancy between the estimated and true UDs) is affected by the spatial spread of the data and the overall spatial pattern (e.g., degree of clustering) of observed locations.

In addition to the choice of a general bandwidth selection method, there are options of using either **fixed kernels,** in which the bandwidth stays constant for all locations, or an **adaptive or variable kernel** approach, in which different amounts of smoothing are applied locally, depending on the number of nearby locations. Generally the fixed kernel approach may perform better than the adaptive kernel approach implemented in ecological studies, primarily because the latter produces relatively greater smoothing at the edges of the estimated UD (Seaman and Powell 1996, Seaman et al. 1999). When there are sharp boundaries between used and unused areas, the standard kernel method may overestimate home range size and estimate significant use of unusable areas, regardless of which bandwidth method is used (Getz and Wilmers 2004, Downs and Horner 2008). Thus, territories that have rigid boundaries may be problematic for current kernel estimators. However, the imprecise and ambiguous nature of territorial boundaries (Powell 2000) may mitigate this problem, especially for species with larger ranges (Jetz et al. 2004). Bounded kernels that allow a sharp discontinuity between heavily used and unused areas are possible (e.g., Hazelton and Marshall 2009), but have not been implemented in ecological studies

Nonmechanistic Models

Many noninferential statistical models have been used to analyze movement paths of wildlife. These include random walks, first passage time, multibehavioral models, fractals, and Lévy flights (Hagen et al. 2001). These statistical approaches have been widely reviewed (Turchin 1998, Schick et al. 2008) and applied in the ecological literature. For example, the **fractal index** (D) is a scale-free measure of the tortuosity of the movement path (With 1994). The value of D typically ranges from 1 to 2; a value of 1 implies straight line move-

Table 20.2. Overview of bandwidth selection methods used for kernel density estimation in space-use studies[a]

Method	Description	Strengths	Weaknesses	Selected references
Reference (optimal)	Would minimize discrepancy between true and estimated UDs if the distribution was bivariate normal	Easily calculated; performance has been evaluated thoroughly	Oversmooths greatly, leading to high over-estimates of home range size; not optimal except for unimodal UDs	Silverman 1986, Seaman and Powell 1996
Scaled-reference	Reference bandwidth multiplied by a fixed proportion	Easily calculated; can reduce oversmoothing and outperform unscaled reference bandwidth	Optimal scaling proportion depends on UD shape and must be picked subjectively	Worton 1995
LSCV	Numerical minimization of objective function based on mean integrated squared error of estimated distribution	Bias is low compared to reference method; if it does not fail, it can perform well for UDs consisting mainly of tight clusters of locations	Often fails or identifies inappropriately small bandwidth if there are identical observations or large sample sizes; high sampling variability	Silverman 1986, Seaman et al. 1999, Gitzen and Millspaugh 2003
Likelihood cross-validation	Minimize score function corresponding to the K-L distance between the true and estimated UDs	Outperforms LSCV in terms of integrated squared error at sample sizes 40–60 in terms of K-L distance and sampling variability at all sample sizes and enables multimodel comparisons	Performance versus plug-in or solve-the-equation approaches unknown; size estimates compared to positive bias in home-range size estimates compared to LSCV	Silverman 1986, Horne and Garton 2006a
Plug-in or solve-the-equation	Error function is minimized theoretically, then numerical estimate is iteratively calculated	Always produces a meaningful estimate (unlike LSCV); oversmooths much less than reference bandwith	Oversmooths significantly and may be outperformed by LSCV if the UD has sharp boundaries between used and unused areas or consists mainly of tight clusters of locations	Wand and Jones 1995, Gitzen et al. 2006

[a] K-L = Kullback–Leibler; LSCV = least-squares cross-validation; UD = utilization distribution.

ment and of 2 denotes a random movement path (but see Turchin [1998] for cautions about the use of this index). The primary attraction of these descriptive methods is for linking these movement patterns to patterns of the landscape to determine how species respond to and perceive their environment (With 1994, Nams 2005). For example, With (1994) evaluated how D for 3 species of acridid grasshoppers (Orthoptera) was affected by landscape structure in a grassland mosaic.

However, the linkage between these statistical descriptors and the underlying landscapes has historically been weak and "none of these models has an ability to test for how landscape features actually influence the movement process" (Schick et al. 2008:1344). Instead, new inferential models, such as state-space modeling (Morales et al. 2004, Forester et al. 2007, Eckert et al. 2008, Patterson et al. 2008) and other mechanistic models provide powerful new tools for modeling the movement steps and relating the species' movement patterns to its environment.

MECHANISTIC MODELS

Most home range estimators discussed to this point are models of animal movement based on a statistical description of point-location patterns. Modern approaches to modeling space use and movements include both statistical and conceptual advances over traditional estimators. Traditional home-range analyses have not included the time of locations as an important variable, but only used them to be sure locations were statistically or biologically independent (Swihart and Slade 1985, Doncaster and Macdonald 1997). However, the exact sequence of points in an animal's path contains additional information not exploited by approaches that ignore the temporal aspect, and many models explicitly consider the links between successive locations. Temporal autocorrelation thus is not an issue of concern, but it is directly integrated into the model structure (Patterson et al. 2008, Schick et al. 2008). As discussed earlier in the chapter, location error is present and often significant in all radiotelemetry studies. Furthermore, unless location estimates are made every few seconds, trajectories connecting consecutive locations also will include additional error by presuming a straight line path between points. Modern analytical approaches explicitly incorporate uncertainty in estimated locations and movement paths, with separate components modeling the observational process (to account for location uncertainty) and the biological processes (Schick et al. 2008). The best of these models directly incorporate biological mechanisms relating space use and movement to the underlying landscape, separate biotic and abiotic forcing (e.g., wind or water currents), and recognize that multiple behavioral patterns may be represented in movement data. Rather than basing inference on a single descriptive model, investigators increasingly are using multiple models tied to alternative biological explanations for patterns of space use, using a likelihood framework that allows direct comparison of these alternative models (Burnham and Anderson 2002, Royle and Dorazio 2008).

Thus, there is rapidly increasing focus, both in home-range/UD analyses and in broader analyses of locations and movements, on multimodel inference based on models that directly incorporate both observational and biological uncertainty and that make use of the temporal and behavioral dimensions of radiotelemetry data. Compared to traditional descriptive methods, this approach is well suited for predicting how space use may change under alternative ecological or management scenarios (Moorcroft and Lewis 2006). We use this section to briefly review these approaches and encourage readers to review Lima and Zollner (1996), Morales et al. (2004), Armsworth and Roughgarden (2005), Jonsen et al. (2005), and Schick et al. (2008) for additional background and specific examples.

The key process that most movement models are trying to elucidate is the relationship between a moving animal and its environment. Explanatory and mechanistic models directly incorporate ecological factors hypothesized to affect space use into the UD estimation process, incorporate rules governing movements between sequential locations, or combine both of these approaches to model sequential movements as a function of hypothesized explanatory factors. The first approach does not assess movements per se, but fits explanatory models to observed locations without incorporating the temporal link between sequential observations. For example, Matthiopoulos (2003) used a descriptive kernel of space use, with supplemental information about ecological factors and other factors affecting space use, and therefore can be seen as a link among descriptive and explanatory approaches. Additionally, Horne et al. (2007) estimated a probability density function with a maximum likelihood approach as a function of spatially explicit covariates, such as habitat variables and proximity of potential mates. In both studies, alternative models were compared with an information theoretic approach (Burnham and Anderson 2002); both the probabilistic UD and coefficients for the effects of covariates are estimated simultaneously, thus unifying home range estimation and resource selection analysis into an integrated approach (Horne et al. 2007).

Another category of methods explicitly considers the links between successive locations (Kernohan et al. 2001). Temporal autocorrelation thus is not an issue of concern, but is directly integrated into the model structure. However, these movements are not typically modeled as a function of explanatory variables. This family of approaches is not new. The estimator of Dunn and Gipson (1977) assumes that a stochastic diffusion process determines movements between successive locations, with an underlying UD that is a bivariate normal distribution (White and Garrott 1990). Horne et al. (2007) demonstrate the potential utility of **Brownian bridges** for estimating home range size and other movement characteristics. The Brownian bridge approach models a stochastic movement process (random walk) conditioned on the locations of each pair of sequential location; Horne et al. (2007) build on the Brownian bridge to develop a probability density function for overall space use (i.e., the UD) that also incorporates telemetry uncertainty associated with each recorded location. No assumptions are made about the shape of the UD. Like standard nonmechanistic approaches, the estimation of the UD does not integrate habitat or other ecological factors directly into the estimator.

The final, and most sophisticated, approach to relating animal movement to environment also builds on simple rules determining movements and then scales from individual movements to produce an estimate of the overall probabilistic UD. However, in contrast to the previous approach, here ecological factors are directly incorporated. These methods start with simple statistical descriptions of movement (e.g., diffusion or random walk processes), but then add movement rules as functions of covariates hypothesized to affect movements. Movement rules might include moving toward or away, or changing the movement rate or turning propensity, in response to certain habitat characteristics or social factors (Moorcroft and Lewis 2006). This mechanistic approach allows direct modeling of the primary ecological factors and processes hypothesized to affect space use as well as a prediction of how space use may be affected with changes in these primary factors. Models are fit with maximum likelihood techniques, allowing information theoretic comparison of alternative models and multimodel inference. For example, Moorcroft et al. (1999, 2006) used local landscape data on topography and prey availability to build a model of coyote (*Canis latrans*) movement based on random diffusive movement. They found the prey-driven model was a better fit with telemetry data from the same area. By making sequential movement direction a function of habitat features, Moorcroft and Barnett (2008) demonstrate how a mechanistic model can integrate estimation of the UD and resource selection analyses into a single framework. Christ et al. (2008) perform a similar integration by making the likelihood of a relocation, given the previous location, a function of habitat covariates. Moorcroft and Barnett (2008) essentially used a resource selection function (RSF) embedded in a mechanistic home-range model, but they did not explicitly consider the time series nature of the telemetry data. Models that consider temporal autocorrelation in data, as well as connections with the underlying landscape, are essentially deducing a RSF from movement paths.

The final conceptual advance we review here regarding animal movement modeling is the integration of movement patterns with animal behavior. Movement patterns can be used to infer the behavior of an animal (also known as its state) and then used to better predict its subsequent movement and habitat use. The effect of factors on subsequent movements depends partly on an animal's perceptual range (Mech and Zollner 2002), and this relationship can be directly incorporated into behavioral movement models. When integrated with the other statistical and conceptual advances discussed above, these integrated **movement ecology models** offer the most sophisticated view of animal movement. These models offer the potential for predicting the realistic response of individual animals to changing conditions, be they dynamic environments or management scenarios. Such models will allow us to infer behavior from movement patterns, find relationships between behavior and habitat, and use these relationships to better predict movements.

Although we expect rapid expansion of the models described in this section, several issues must be considered if a biologist is considering using these approaches. First, the ability to develop models that target specific hypotheses about the species of interest is a major advantage, but this specificity also means that models need to be tailored to each species and study situation, if not to every individual animal. Although this approach more accurately reflects the complex reality of animal movement, it also produces a new challenge to make the results of one study more generally relevant, particularly for conservation and management. Second, the mathematical complexity of these approaches is beyond the quantitative skills of many biologists. The examples we describe above include models developed as partial differential equations, hierarchical Bayesian statistics, and state space models, to name a few. The hierarchical aspect of many of these models is functionally useful, because it allows them to factor higher dimensional problems into lower dimensional, conditionally dependent ones (Berliner 1996, Wikle et al. 1998, Clark 2005). However, even quantitatively inclined ecologists may have difficulty implementing such complex models. Finally, their bias and precision, compared to previous estimators, in estimating home range size and parameters related to factors affecting space use have not been evaluated thoroughly. Sample size requirements and evaluation of how frequently data need to be recorded have not been examined in detail (Horne et al. 2007).

SITE FIDELITY, ANIMAL INTERACTIONS, AND SPACE USE OVERLAP

We discuss site fidelity and animal interactions together, because to some extent they both fall under the general umbrella of space use overlap. **Fidelity** is "the tendency of an animal either to return to an area previously occupied or to remain within the same area for an extended period of time" (White and Garrott 1990:133). Fidelity measures components of overlap over time for the same individuals or group, whereas interaction analyses looks at overlap among different animals at the same time or over time. Therefore, some analytical methods used to assess fidelity also are useful for looking at spatiotemporal overlap among individuals of the same or different species.

Analyses of animal interactions frequently are classified as either **static or dynamic** (MacDonald et al. 1980, Doncaster 1990). Static analyses consider spatial interaction and overlap over a long time period, whereas dynamic analyses consider the temporal dimension to look at extent to which animals are in close proximity simultaneously. Techniques used for assessing static animal interactions are equally useful for comparing site fidelity over time. These approaches include comparing changes in one aspect of the UD (e.g., home range centroid) over time, or examining to what degree the overall UD has changed. Distances among centroids or average locations can easily be calculated and compared over time and across different groups of animals; a null hypothesis of no shift in centroids or no differences in the magnitude of shifts for different groups can be compared with multivariate tests (Hotelling's T^2 or a Mann–Whitney method). However, this approach is not sensitive to changes in space use, such as expansions of the area used, that do not change the mean location (White and Garrott 1990).

Broader changes in space use over time or among groups can be assessed by examining whether the sets of locations from different periods or animals come from the same underlying distribution, by assessing overlap among the 2-dimensional home range estimates, or by comparing overall overlap in UDs calculated for different times or animals. A flexible test of whether 2 sets of locations come from the same or different UDs is the multiresponse permutation procedure proposed by Mielke and Berry (1982). This method examines whether the average pairwise distance between locations in a group or time period are similar to the average distances among locations calculated independently of time period. It does not require an estimated UD from a statistical model, but simply uses the observed locations, which may be seen as an advantage of this approach (White and Garrott 1990). However, model-based quantification of overlap allows estimation of the amount of overlap rather than just a test of whether overlap occurred (Fieberg and Kochanny 2005).

Estimates of 2-dimensional overlap are based on the percentage of each home range polygon that overlaps the other polygon in the comparison. If $A_{i,j}$ is the area of animal i's home range that overlaps animal j's home range (with area A_i), then $HR_{i,j} = A_{i,j}/A_i$ is the proportion of animal i's home range that overlaps animal j's home range. For example, $HR_{1,2} = A_{1,2}/A_1$, and $HR_{2,1} = A_{2,1}/A_2$; these 2 indices are not equal unless both animals have the same home range size ($A_1 = A_2$). Although intuitively simple, these indices are appropriate only if use is uniformly random in each home range. If home range use is not uniform, 2-dimensional overlap is a biased estimate of joint space use, because it does not consider each animal's probability of using the overlapping areas (i.e., outer contours of the home range may overlap significantly, even though this overlap occurs where probability of use by each animal is relatively low; Kernohan et al. 2001). Such problems are avoided by overlap indices that consider differential use intensity in each UD, based on joint probabilities of use of the same area. Use of estimated UDs allows direct estimation of the probability that 2 animals have used the same area over the time period for which data were collected. Such UD-based estimates should replace overlap analyses based on 2-dimensional overlap (Millspaugh et al. 2004, Fieberg and Kochanny 2005).

Fieberg and Kochanny (2005) reviewed and compared 3 alternative measures of overlap that consider the complete probability density function for each UD: the **volume of intersection** index, introduced by Seidel (1992); Bhattacharyya's affinity (Bhattacharyya 1943), and an alternative UD overlap index. The volume of intersection index estimates the integrated overlap in density among 2 UDs; Bhattacharyya's affinity index and UD overlap index measure the joint distribution based on the product of 2 UDs. Although these indices correctly rank pairs of UDs based on the true amount of overlap and should be used in preference to 2-dimensional estimators, they are biased. The amount of bias depends on the true amount of overlap. Simulations indicate these indices underestimate overlap when it is high and overestimate overlap when it is moderately low. Bias also is affected by sample size, making comparisons in overlap estimates among studies problematic (Seidel 1992, Fieberg and Kochanny 2005).

Dynamic interaction analyses consider whether 2 animals are using an area at the same time, and thus require near-simultaneous locations for each animal (Minta 1992, Fuller et al. 2005). Traditional dynamic interaction analyses compute indices of co-occurrence. Cole's (1949) coefficient of association compares the proportion of total observations for 2 animals in which the animals occurred in the same location simultaneously. A distance-based index (Kenward 2001) computes whether the average distance between simultaneously taken observations for 2 animals is different than the average pairwise difference among all locations regardless of time; it can indicate avoidance, positive association, or no relationship in simultaneous space use. The utility of dynamic interaction analysis has been facilitated greatly by technology (e.g., Prange et al. 2006), and continued sensor development will help us understand these interactions.

GENERAL MOVEMENT CHARACTERISTICS

In addition to broad properties of space use, investigators often focus on detailed aspects of discrete movements and movement paths between observed locations. As with analyses of UDs, traditional analyses of individual movements have focused on nonmechanistic statistical descriptions of selected movement characteristics. The distance, direction, and rate of movement may be relevant to basic understanding of a species' natural history, may have direct management relevance (e.g., understanding the timing of migration in relation to harvest regulations), or may be compared among groups or habitats to test hypotheses about factors affecting movements. Distance between successive locations at times i and $i + 1$ can be calculated as $d_{i,i+1} = \sqrt{(x_{i+1} - x_i)^2 + (y_{i+1} - y_i)^2}$. White and Garrott (1990), Kernohan et al. (2001), and other sources provide formulas for calculating the direction of travel. Summaries and statistical analyses of movement direction must use methods designed for circular distributions. A summary of such methods is provided by Zar (1996), and comprehensive monographs also are available (e.g., Mardia and Jupp 1999).

Biologists are often interested in quantifying movement characteristics for migratory and dispersal movements, and in comparing these characteristics before and after onset of such larger scale movements. For land managers, understanding where animals go involves both determining the final destination as well as the path or corridor used to get to this destination. Understanding the "which," "what," and "where" aspects usually requires telemetry techniques that enable animals to be located far from their original location as well as a large sample of animals. For example, if the objective is to compare proportions of males versus females that migrate with real—but unknown—proportions of 0.3 and 0.6, respectively, basic power calculations (Zar 1996) indicate that one would need to monitor 33 animals of each sex to have high power (0.8) to detect such a difference at a significance level of 0.10. Any study of long-distance movements, such as migration, needs to develop objective criteria for determining the onset of these movements (i.e., separating these movements from the more normal movements that occur before and after migration or dispersal events). Ideally, these criteria are defined a priori, but in practice, they may have to be defined after exploratory analyses of the data, unless the investigator has a good understanding of what constitutes normal nonmigratory movements in the study population.

As with home range or space use analyses, analysis of specific movement attributes increasingly is done using a multimodel, often hierarchical approach in a likelihood-based framework (Schick et al. 2008). Moreover, the line between analyses of space use versus individual movements is increasingly blurred, as models scale up from properties of individual movements to estimates of an overall UD. Thus, the summary of mechanistic home range or space use models earlier in this chapter provides a starting point for understanding more recently developed approaches to analyzing movement data.

USE OF RADIOTELEMETRY DATA TO STUDY RESOURCE SELECTION

The study of wildlife–habitat relationships has been greatly enriched by use of radiotelemetry. **Resource selection** is a popular and active area of research that addresses the resource choices made by wildlife. Here, we consider resources to be vegetation, objects (e.g., a cavity), or environmental conditions (e.g., wind). In the study of wild animals, resources are tied to spatial features of the environment, making resource selection a spatially explicit issue. Managers often use the results of resource selection studies to guide vegetation management directed at improving conditions for a species, evaluate how species might overlap and compete for resources, or assess how human disturbances affect a species' distribution. There is an implied assumption in resource selection studies of wildlife that when a resource is selected, it promotes fitness of the animal. Radiotelemetry is uniquely able to provide unbiased estimates of resources used by wildlife, provided the important sampling issues discussed above are considered and appropriate analytical procedures are used and interpreted properly. In this section, we provide an overview of the concepts and methods of analyzing wildlife resource selection with radiotelemetry data. McDonald et al. (Chapter 17, This Volume) offer an in-depth, broader discussion of the analytical techniques that assess resource selection and food use and availability using a broad range of techniques.

The study of resource selection comes with its own jargon, and biologists should be aware of basic terms and concepts. Foremost among terminology is **use, availability, selection, and preference.** Use is the quantity of a resource utilized by the consumer in a fixed period (Johnson 1980) and is by default a subset of available ones (Buskirk and Millspaugh 2006). In telemetry studies, use is often defined by telemetry locations, but it also can be defined by a collection of points, such as a UD (Marzluff et al. 2004, Millspaugh et al. 2006) or other metrics (e.g., time, distance traveled, or energy expended; Buskirk and Millspaugh 2006). If resources are selected, they are used disproportionately more than would be expected based on their availability (Johnson 1980). **Availability** is one of the more difficult components to define in resource selection studies, because it is strongly affected by semantics and scale (Buskirk and Millspaugh 2006). Definitions of availability (Johnson 1980, Manly et al. 2002, Fuller et al. 2005) all consider accessibility and the amount of resources. However, such factors as accessibility are nearly impossible to determine in field studies. Instead, biologists consider only the occurrence of a resource in a study area or some finer scale when defining availability. Some recent evaluations and analytical options consider animal movement rates and simultaneous modeling of movements and resources, but most simply consider the maximum area that could have been used as a means of defining resource availability. Preference for a resource reflects the likelihood of that resource being chosen if offered on an equal basis with others (Johnson 1980). Therefore, it is nearly impossible to determine resource preference for free-ranging animals, because resources are not available in equal units and in a random distribution (Marzluff et al. 2001). Thus, resource selection of free-ranging animals is restricted to use and selection, unless resources meet these criteria. **Avoided** refers to use that is disproportionately less than would be expected based on the availability of that resource (Johnson 1980). Based on this standard definition, the animal still may use resources that are "avoided," so the intuitive connotation of the term does not match its definition in this context. "Unused" resources received no use during the study period.

The hierarchical nature of resource selection studies inherently requires those using telemetry to consider the order of selection (Johnson 1980). The selection of areas and resources by animals can be seen as a hierarchical process (Johnson 1980). The importance of **hierarchical selection** should be explicitly considered as part of the research design, because the selection order implicitly identifies the population of interest, the experimental units, and the level of management interpretation (Thomas and Taylor 1990). **First-order selection** considers the selection of the physical or geographic ranges of a species (Johnson 1980). **Second-order selection** evaluates the selection of a home range in the study area or geographic range of the species (i.e., it emphasizes resource selection in the overall study area; Johnson 1980, Thomas and Taylor 1990). **Third-order selection** evaluates the importance of resource components in the animal's home range, emphasizing resource selection patterns in the home range (Thomas and Taylor 1990). Most studies involving radiotelemetry will focus on second- or third-order resource selection.

A second critical element of the study design is the determination of how use and availability data will be collected, which determines which analytical methods are appropriate. The variety of available approaches primarily reflects data reliability, quantity of data recorded at each animal location, and the assumption about data independence among animals and animal locations. Thomas and Taylor (1990) described the 3 study designs for resource selection studies, depending on whether use or availability information is collected for the individual or for the population as whole. Study design I assumes that both use and availability data are collected and summarized at the population level. Study design II records use per animal, but defines availability for the entire population. In contrast, study design III estimates use and availability for each animal. Erickson et al. (2001) added a fourth design that considers recent developments in resource selection studies (Arthur et al. 1996; Cooper and

Millspaugh 1999, 2001). As with the 2 previous designs, design IV summarizes use per animal, but it defines availability for each location rather than assuming that availability is constant throughout the study area or home range. Alldredge et al. (1998) provide a useful discussion of the assumptions of each technique across different designs.

Resource selection analyses most often compare resources used by the animal to resources considered available to the animal or population. Resource availability represents the amount of area of each vegetation type (e.g., meadow or ponderosa pine forest) that is available for use by the population or individual animal. Unfortunately, it is one of the most troubling aspects of resource selection studies that compare use to availability (Cooper and Millspaugh 1999). The biologist's definition of availability might not correspond to what the animal perceives to be available (Marzluff et al. 2001). For example, various factors that may be immeasurable to the biologist might influence what is actually available to an animal (e.g., social hierarchies or predation; Otis 1997, 1998; Mysterud and Ims 1998). Thus, the determination of the area for each vegetation type may not provide an adequate means to measure habitat availability (Johnson 1980, White and Garrott 1990). In addition, definition of the appropriate study area can be problematic and can influence availability (Porter and Church 1987). A common problem in defining availability is the occurrence of widespread vegetation types that are rarely used, possibly indicating this vegetation type is not biologically suitable for the species. Inclusion or exclusion of this vegetation type may have a substantial influence on habitat selection analyses (Johnson 1980), and alternative analyses including and excluding these vegetation types should be presented (Thomas and Taylor 1990). There is limited general guidance on dealing with other problems related to defining availability (Johnson 1980, Porter and Church 1987, Cooper and Millspaugh 1999); appropriate strategies for addressing these problems depend on the species studied.

Resource selection studies may focus on selection of discrete sites or of general habitats. The spatial extent of availability commonly may be based on the entire study area or the area within some specified distance of the focal site, on composite population-wide "home range" in the study area, on individual home ranges, or on multistage approaches combining multiple scales (Buskirk and Millspaugh 2006). Ultimately, the decision of which boundary to use is dependent on what order of selection is being considered, the animal under investigation, and objectives of the study; each potential definition of spatial extent of availability has potential problems. The boundary of a study area may have no biological meaning from the standpoint of the animals being studied, but using the investigator-defined study area boundary may be appropriate in coarse scale evaluations (e.g., Erickson et al. 1998). Porter and Church (1987) demonstrated that when vegetation types are aggregated, the study area measure of availability may provide spurious results. Using composite or individual animal home ranges is problematic, in that a researcher must select an appropriate home range estimator, and previously discussed issues, such as sample size, choice of contour, and the technique used, all affect home range estimates. Placing a buffer around individual locations to define availability helps refine what is considered available to an animal, but it requires the investigator define a biologically meaningful buffer size (Cooper and Millspaugh 1999, 2001). At any of these scales, defining availability may require or be facilitated by GIS layers for resources of interest. However, GIS data layers have limitations that should be carefully considered in resource selection analyses. For example, difficulties in assigning use locations to habitat features and a miscalculation of vegetation type areas or other landscape metrics are possible when map accuracy is low.

Use is usually quantified based on the telemetry points directly or on some summary of those observations via a home range analysis. Although many other measures have been proposed (see Buskirk and Millspaugh 2006), use of these methods is most dominant in the literature. Most studies use time as the currency of use, followed by event sites. However, as technology makes other options more readily available, we expect changes to the interpretation of use and frequency metrics, such as distance traveled.

One important and often overlooked aspect of resource selection relates to the animal's behavior at relocation points. Marzluff et al. (2001) argued that knowing what an animal is doing in a particular resource increases our understanding of why the animal is there. Marzluff et al. (2001:315) developed behavior-specific UDs based on kernel density estimates to evaluate whether resources were used differently for 4 behaviors (foraging, locomotion, perching, and parental care). Using nested discrete-choice models, Cooper and Millspaugh (2001) illustrated how to account for variation in animal behavior in resource selection studies. By comparing RSFs with and without behavior included as a component, they demonstrated potential problems with ignoring the behavioral element.

POPULATION ESTIMATION

Radiomarked animals provide considerable advantages for estimating wildlife population size and density, because information about the locations of animals can be incorporated into the estimates. Initially, especially for territorial pack animals (e.g., gray wolf [*Canis lupus*]; Burch et al. 2005), home range or territory size and percentage overlap of home ranges of radiomarked animals was scaled up to estimate population size based on the number of territories or home ranges the study area likely supported. However, this approach generally underestimates the actual population.

Many wildlife population studies now use data from radiomarked individuals as auxiliary information to refine estimates based on other, independent techniques. There are 4 primary ways that radiomarked animals can be used to augment or improve population estimates (White and Shenk 2001). The first is direct **mark-resighting,** in which animals are marked and later recaptured or observed. The second uses **sightability** models developed from radiomarked animals that serve as correction factors applied to each individual or group of animals observed. The third employs radiomarked animals in conjunction with trapping or **capture grids** that are used to assess demographic closure. The fourth uses radiomarked animals as an auxiliary likelihood with age-at-harvest data in a statistical population reconstruction analysis (Skalski et al. 2005, Broms et al. 2010).

Mark–Resight Estimation

In mark-resight estimation, a sample of animals is initially captured (marked); however, resightings are rarely based on recaptures, but on observations, including direct observations or camera surveys. A major advantage of this technique is that costs of resightings are generally minimal relative to initial capture costs. A disadvantage is that standard mark–recapture closed population estimation models are not appropriate (e.g., Otis et al. 1978), because unmarked animals cannot be marked during resighting occasions. Mark-resighting techniques have been used frequently with ungulates, such as white-tailed deer (*Odocoileus virginianus*; Rice and Harder 1977), mule deer (*O. hemionus*; Bartmann et al. 1987), and bighorn sheep (*Ovis canadensis*; Bartmann et al. 1987, Neal et al. 1993), as well as with carnivores, including black bears (*Ursus americanus*) and grizzly bears (*U. arctos*; Miller et al. 1987, 1997), coyotes (Hein and Andelt 1995), and harbor seals (*Phoca vitulina*; Ries et al. 1998). The technique has broad utility for other species that can be marked and feasibly resighted (e.g., Magle et al. 2007).

There are 2 critical assumptions in mark–resight studies. The first assumption is that radiomarked animals must have the same sightability as unmarked animals. Thus, behavior of marked individuals is comparable to unmarked individuals, and marked animals adequately represent variability in the sightability of unmarked animals. For example, a study area having 3 markedly different understory vegetation types would require marked animals in each of these 3 types. The second assumption is the population is closed, with 2 forms of closure assumed. One form assumes the population has a fixed size with no births, deaths, immigration, or emigration during the study. The second form is geographic closure. Geographic closure is required for most mark–resight models, a notable exception being the joint hypergeometric estimator (White and Shenk 2001). The closure assumption needs to be carefully considered during study design. For example, surveys should be conducted during periods of minimal immigration or emigration. As discussed below, one analytical method allows the closure assumption to be relaxed by modeling immigration and/or emigration (Neal et al. 1993). Statistical tests are available to assess geographic closure of populations (Stanley and Burnham 1999, Stanley and Richards 2003). In studies extending over longer periods, sampling needs to be scheduled in discrete sessions of concentrated surveys (e.g., 1 week/yr), with closure in each session (i.e., secondary occasions, such as 5 days of surveys in a 2-week period), but not necessarily between sessions (primary occasions, such as the survey week each year). However, data from all sessions can be integrated into a single analysis to support more efficient sightability parameters (McClintock et al. 2006). This framework also has been extended to allow estimation of apparent survival and transitions between observable and unobservable states (McClintock and White 2009).

Additional assumptions may be necessary to obtain unbiased population estimates, regarding whether the population is sampled with replacement and regarding heterogeneity in detection probabilities. Individual capture heterogeneity (e.g., heterogeneity in capture probability during the marking phase and in sightability during the resight phase) simply quantifies variation in detection among individuals in a population and has been demonstrated for numerous wildlife species, including black bear (Belant et al. 2005). Furthermore, capture heterogeneity can occur even with individuals and is influenced by behavior, as demonstrated in moose (*Alces alces*; Gasaway et al. 1985). The effect of violating assumptions of closure, sampling with replacement, and capture heterogeneity are dependent on the mark–resight population estimator used (White and Shenk 2001).

Commonly used models in mark–resight population estimation include the joint hypergeometric estimator (JHE; White and Garrott 1990; Neal et al. 1993), an extension of the JHE to accommodate immigration and emigration (Neal et al. 1993), Bowden's estimator (Bowden and Kufeld 1995), and the Minta–Mangel estimator (Minta and Mangel 1989). The JHE is a maximum likelihood estimator that estimates population size using an iterative numerical approach. It assumes the number of individuals in the population and number of marked animals is constant across sampling occasions. However, the probability of observing marked animals is not assumed to be constant across sampling occasions. The extension of the JHE accommodating immigration and emigration incorporates a binomial process (Neal et al. 1993). The likelihood function incorporating immigration and emigration is the product of the probability of an animal in the study area during the sampling occasion (calculated using the binomial distribution) and the joint hypergeometric likelihood (White and Shenk 2001).

Bowden's estimator is grounded in a sampling survey and based on the frequency of observations of marked individuals along with the sum of observations of unmarked individuals (Bowden and Kufeld 1995). This estimator calculates confidence intervals based on the variance of resighting frequencies of marked individuals. An unbiased estimate using this procedure requires identification of ≥90% of marked individuals sighted (White 1996). The Minta–Mangel estimator employs a bootstrap of the population size based on the frequency of marked animal observations (Minta and Mangel 1989). Sighting frequencies of unmarked animals are randomly drawn from the observed sighting frequencies of marked animals, until the total number of captures equals the total number of sightings of unmarked individuals. The number of unmarked animals drawn represents an estimate of the number of unmarked individuals. The population size is then estimated as the number of marked individuals plus the number of unmarked animals sampled. Although this estimator is generally unbiased, confidence interval coverage is not at the expected 95% level, because the number of marked animals observed is fixed as opposed to being a random effect (White 1993). Recently developed estimators using mixed-effects models have expanded analytical options by incorporating random effects into the model heterogeneity when sightability varies among individuals (McClintock et al. 2009a, b).

Choice of model to use for mark–resight population estimation will be determined largely on the validity of model assumptions. Except for the most recently developed estimators, all other estimators described require equal sightability of marked and unmarked animals. All except the Neal et al. (1993) extension to the JHE assume demographic closure (White and Shenk 2001) during each primary sampling occasion. However, remaining assumptions necessary to obtain unbiased estimates vary among the models described. For example, all but the JHE model with immigration and emigration require geographic closure. Neither form of JHE accommodates sampling with replacement or individual capture heterogeneity. In contrast, Bowden's and the Minta–Mangel estimators allow sampling with replacement and individual heterogeneity. Estimators developed by McClintock et al. (2009a, b) are highly flexible in allowing individual heterogeneity, sampling with or without replacement, and estimation without exact knowledge of the number of marked animals currently in the population.

Sightability Models

Sightability models are used frequently in wildlife population estimates, particularly those derived from aerial surveys. Surveys typically involve dividing the study area into sampling units that generally are then selected using a random or stratified random design. Those sampling units selected are then surveyed with the goal of counting every animal in each unit. However, sightability can be influenced by vegetation, topography, group size, animal behavior, and observer (e.g., Floyd et al. 1979, Gasaway et al. 1985, Samuel et al. 1987, Cogan and Diefenbach 1998, Bleich et al. 2001, Walsh et al. 2009). Consequently, detection of animals is imperfect, and radiomarked animals are used to develop sight-ability models that reduce visibility bias stemming from imperfect detection. Sightability models using radiomarked animals have been developed and applied to elk (*Cervus canadensis*; Unsworth et al. 1990, Otten et al. 1993, Walsh et al. 2009), moose (Gasaway et al. 1985, Anderson and Lindzey 1996), mule deer (Ackerman 1988), bighorn sheep (Bodie et al. 1995), mountain goats (*Oreamnos americanus*; Rice et al. 2009), Arabian oryx (*Oryx leucoryx*), and waterfowl (Smith et al. 1995b). Samuel et al. (1992) developed sightability models to correct age and sex bias in surveys of wildlife populations.

To develop sightability models, radiomarked animals are initially located, and a search area is delineated around the locations. An independent observer(s) without knowledge of the locations of marked animals then searches the area. Covariates (e.g., group size or vegetation) considered important are recorded for each marked animal sighted. Animals not sighted are located via telemetry after the search to record the same covariates. A logistic regression model is generally developed to predict the probability of sighting an animal based on the values of covariates in the model. The model is then used to correct sightability bias in the observed counts of animals. Thus, radiotelemetry is used only during the model development phase.

There are 2 important assumptions when using sightability models. The first is that conditions observed during operational surveys must be in the range of conditions that were observed during development of the sightability model. For example, it is inappropriate to use a sightability model that incorporated group sizes up to only 3 individuals for a survey where group sizes of up to 10 individuals were recorded. Also, sightability can be influenced by the type of aircraft, altitude, flight speed, and search pattern or effort; therefore, these factors must be constant between the model development and operational survey phases. The second assumption is there is no measurement error for covariates used to predict sighting probability. For example, under-counting group size during model development will result in a negatively biased population estimate when applied to survey data (Cogan and Diefenbach 1998).

Correction for Density Estimation

In traditional mark–recapture population surveys using a grid of sampling locations (i.e., trapping grids), the naïve density estimate (estimated population size/area of trap grid) is biased. Animals on the periphery of the trapping area spend only a portion of their time in the area, resulting in difficulty in estimating effective trapping area and naïve density estimates that overestimate true density. In a multisite study (e.g., comparison of abundance among different habitat types), comparisons based on absolute abundance estimates will be biased if effective trapping area systematically differs among the site types being compared. Heuristic approaches to estimating the effective trapping area (e.g., nested trapping grids) are ad hoc approaches (Otis et al. 1978, Wilson and Anderson 1985) and are inferior to approaches based on distance sampling (e.g., trapping webs; Anderson et al. 1983).

Radiomarked animals have been used to refine estimates from trapping grids in 3 primary ways. The first is establishing a buffer around the trapping grid, generally one equal to the radius of a circle the size of the average home range during the period or season of inference (Kenward 1987, 2001; Soisalo and Cavalcanti 2006). The second means of correcting bias is to use the proportion of radiomarked animals that spend ≥50% of their time in the trapping grid (White and Shenk 2001). However, this approach would result in a density of zero if no animals spent ≥50% of their time in the area. To account for this problem, the third bias-correction approach is to estimate the proportion of time each radiomarked animal spends in the area trapped. From this estimate, a mean and variance can be calculated that can be applied to the population estimate and variance derived from the trapping. If tissue samples are collected from captured animals or if radiomarked animals are individually identifiable visually, this approach can be similarly applied to camera trap or hair snare grids (Garshelis 1992, Matthews et al. 2008). Populations estimated using the third approach may be biased if animals spending little time in the trapping grid are proportionately less likely to be initially captured and radiocollared.

Use of Radiotelemetry Data in Statistical Population Reconstruction

Statistical population reconstruction is a recent approach used to model age-at-harvest data for many wildlife species (Gove et al. 2002, Skalski et al. 2005, Broms et al. 2010). For many management agencies, age-at-harvest and hunter effort data are commonly used for population estimation, because they are easy to collect, applicable across large spatial extents, and do not require intensive or expensive sampling (Skalski et al. 2005, Millspaugh et al. 2009).

Statistical population reconstruction uses an age-at-harvest likelihood, a reporting likelihood (e.g., incomplete harvest reporting or sampling for sex and/or age composition), and an auxiliary likelihood, such as radiotelemetry, but it could take several forms (Skalski et al. 2005). Auxiliary likelihoods are independent datasets that are used to estimate one or more demographic parameters. For example, use of radiotelemetry data to estimate natural survival could be used as an auxiliary likelihood to refine population estimates (Broms et al. 2010). Alternatively, population estimates obtained from mark–recapture efforts or aerial surveys using radiomarked animals can be used to calibrate estimates obtained using statistical population reconstruction (Skalski et al. 2005). This modeling technique has considerable potential; a thorough description of its development and applications is provided by Skalski et al. (2005).

SURVIVAL

Estimating **survival** of animal populations is an integral part of wildlife ecology and management. Monitoring radiomarked animals provides us knowledge of when animals were marked, the time and position of each consecutive location, and status (e.g., alive, dead, or missing); thus, radiotelemetry is the basis for many survival studies. Use of radiotelemetry allows us to understand cause-specific mortality, rates of survival, and what factors influence rates of survival (e.g., Haines et al. 2005, Bender et al. 2007). For example, radiotelemetry has been used to estimate the effects of humans on survival (Belant 2007, Adams et al. 2008) and predation rates (Vreeland et al. 2004, Knopff et al. 2009), and to compare survival among sex or age cohorts (Chamberlain and Leopold 2001, DelGiudice et al. 2006).

Numerous models have been adapted or developed to estimate survival of radiomarked individuals. The appropriate strategy to use depends on numerous factors, including study objectives, variability in survival rates, occurrence of **immigration** or **emigration,** and whether the fate of all study animals is known. Animals that are missing or animals at the end of a study are considered right-censored, because the death of the animal was either unknown or did not occur during the study; right censoring may occur because of migration from the study area, transmitter failure, or termination of the study. Left-censored individuals entered the study after it started and after some mortality may have already occurred. We provide a brief overview of those techniques most commonly used. More comprehensive treatments are provided by White and Garrott (1990), Kenward (2001), Winterstein et al. (2001), and Murray (2006).

The analysis of survival data from radiomarked animals can be divided into 2 general categories: one that considers estimating a survival rate for a discrete time interval and the other that estimates a continuous survival curve. Interval survival rates assume a constant mortality rate across the study period, so the timing of mortalities does not matter. Instead, the number of days each animal survives in the time interval is important. These interval-based methods are appropriate when the fate of all animals is known, the survival rate is constant throughout the time period, and all animals are radiomarked at the beginning of the interval. However, in many cases, biologists often add new radiomarked animals into the study, animals do not have constant mortality rates throughout an interval, and the fate of some animals may be censored. In such situations, continuous survival models are appropriate.

Heisey–Fuller Method

Originally developed for estimating avian nesting success (Mayfield 1961, 1975), this approach was later adapted for estimating survival of radiomarked wildlife (Trent and Rongstad 1974, Heisey and Fuller 1985) and used extensively to estimate cause-specific mortality. Survival is assumed to be constant in each interval (e.g., day or week). Survival for each time step (e.g., daily survival) is derived from the number of deaths and number of exposure days during each time step (i.e., number of deaths per total radiodays), and the interval survival rate is calculated exponentially using daily survival and the number of time steps during the interval. This method is appropriate when survival is constant for short intervals, but may vary over longer time periods; that is, a study may consist of consecutive intervals with constant survival assumed within, but not among, intervals (Williams et al. 2004). Formulas for variance estimates of survival rates for time steps, as well as confidence intervals for animals located at irregular intervals, have been developed (Bart and Robson 1982).

Kaplan–Meier Estimator

The Kaplan–Meier (product limit) estimator (Kaplan and Meier 1958) was adapted for survival analyses in wildlife radiotelemetry studies during the 1980s (e.g., Krauss et al. 1987, Conroy et al. 1989, Pollock et al. 1989). It differs from the Heisey–Fuller method in the assumption of constant survival is relaxed in time intervals, because mortality events are used to define the time intervals (Murray 2006). For this estimator, the survival function is the probability of an animal in the population surviving for a specified number of time intervals since the study began. The Kaplan–Meier estimator will produce a survival curve showing the probability of survival as a function of time. At the start of the time step, survival is assumed to be 1.0; each mortality in the pool of radiomarked animals is demonstrated graphically by a decrease in survival (Fig. 20.3). The method is attractive, because it can incorporate the addition of radiomarked animals after the beginning of the study and is robust to right censoring.

The generalized form of the Kaplan–Meier estimator is recommended, because it adjusts the conditional probability based on individuals radiomarked after study initiation; however, this form is sensitive to small sample sizes (Pollock et al. 1989). Nelson (1972) and Aalen (1978) developed an alternative estimator that is better in dealing with small sample sizes. White and Garrott (1990) and Skalski et al. (2005) provide details on calculations used by this model. Important attributes of each Kaplan–Meier estimator are that it does not assume constant survival and it allows staggered entry of radiomarked animals (e.g., addition of new radiomarked animals after the initiation of the study). Limitations to the Kaplan–Meier and Heisey–Fuller estimators have been described (Rotella et al. 2004), but these approaches continue to be used and developed. For example, Heisey and Patterson (2006) recently described use of the cumulative incidence function to estimate cause-specific mortality with the Kaplan–Meier estimator.

Hazard Functions

Several multivariate approaches have been developed to address simultaneously the relative contributions and interactions of the multiple factors that can influence survival. Hazard functions estimate the probability of death at a specified point in time conditional on the survival to that time; parameter coefficients are generally estimated using regression techniques. Hazard functions are fit to survival data initially by calculating the cumulative hazard function for the dataset (Klein and Moeschberger 2003, Murray 2006). The Cox proportion hazard model is widely applied in wildlife telemetry studies and uses an exponential regression model containing parameters with constant hazards that are proportional through time. The Cox proportional hazard model does not assume a specific hazard function and is consequently robust, allowing detailed assessment of param-

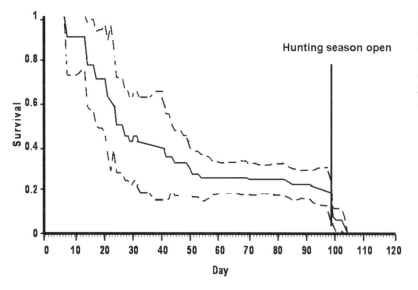

Fig. 20.3. Survival distributions for radiomarked hatching-year (HY) mourning doves (*Zenaida macroura*), calculated using the Kaplan–Meier product limit estimator with staggered entry, during summer and autumn 2006 on the James A. Reed Memorial Wildlife Area, Lee's Summit, Missouri.

eter effects (Hosmer and Lemeshow 1999). Advantages of the Cox proportional hazard model are that variables can be discrete (categorical) or continuous, and the model can be extended to include time-dependent variables (Hosmer and Lemeshow 1999), such as behavior, age, and season. If hazards become discontinuous due to periods when animals are not monitored (e.g., temporary movement from the study area), the Anderson–Gill model can be applied (Anderson and Gill 1982, DelGiudice et al. 2002). Discrete time models also have been applied to radiotelemetry studies to model survival as a nonlinear function of model parameters (White and Burnham 1999, Williams et al. 2002a).

Staggered-Entry and Time Origin

Wildlife studies rarely capture all individuals at or near the same time. In addition, investigators frequently desire to augment the number of radiomarked animals when mortality or other factors (e.g., radiotransmitter failure) reduces samples below desired levels. Situations in which animals are introduced into the marked population over longer time periods are known as staggered entry (Pollock et al. 1989). Using Kaplan–Meier, the addition of animals is accounted for by adjusting the number at risk during respective time intervals (Pollock et al. 1989). The Mayfield method incorporates new animals using the sum of individual exposure days. Regardless of the analytical technique, a critical assumption is that recently marked animals have had the same survival rate since the beginning of the study as animals that were marked at the start.

An important aspect for estimating survival is using the appropriate time origin at which to begin measuring survival. This choice has relevancy at both the individual and population levels. At the individual level, an adjustment period following capture and marking may be necessary to allow the animal to return to what we consider normal behavior. At the population level, the investigator needs to ensure there is an adequate sample size of marked individuals before measuring survival. Measuring survival prematurely using either scenario has considerable potential to bias estimates.

Important Assumptions of All Radiotelemetry Survival Models

Irrespective of the model used for estimating survival from radiomarked animals, the following conditions should be met to minimize error (Bunck 1987, Tsai et al. 1999, Winterstein et al. 2001, Fuller et al. 2005):

1. The sample of radiomarked individuals must be representative of the population being studied. Generally, it is a random sample of individuals in each classification being investigated (e.g., sex or age class).

2. Radiomarked individuals represent independent samples. Animals that are closely associated (e.g., nestlings or pack members) may be subjected to similar mortality factors and thus, provide biased information about mortality rates compared to data obtained from truly independent individuals. In the latter case, the nestlings or pack would serve as the independent sample.

3. Transmitters or tracking equipment should not influence survival. Animals with transmitters should provide an unbiased estimate of the rate of survival for the population of inference; however, this assumption may be violated (e.g., Guthery and Lusk 2004, Steenhof et al. 2006). These first 3 conditions also are required of survival estimates from band recovery and mark–recapture studies (Jolly 1965, Seber 1965, Pollock 1981, Brownie et al. 1985).

4. When the fate of an animal is unknown (i.e., censored), the known survival time is often assumed to be independent of the animal's actual fate. Censored individuals are assumed to be random in the marked population; however, biases likely exist in many wildlife studies (DelGiudice et al. 2002). Appropriate transmitter design, periodically monitoring outside the study area to search for marked animals, and extending study duration to obtain an appropriate number of mortalities have been suggested as means to reduce censoring rates (Garton et al. 2001, Bender et al. 2004).

5. The exact time of death is known. This assumption can be relaxed, however, without substantially affecting survival estimates (Johnson 1979a, Bart and Robson 1982, Heisey and Fuller 1985).

6. Newly marked animals must have the same survival function as previously marked animals.

Additionally, independence of consecutive radio locations is an assumption of the most survival estimators (Winterstein et al. 2001). However, consistency in timing of consecutive locations within and among radiomarked animals is likely of greater importance.

SUMMARY

As with technology, the analysis and application of telemetry data have developed greatly in sophistication and inferential power. Recent methods incorporate modern statistical tools and philosophies, such as hierarchical models and information theoretic approaches, and are increasingly based on mechanistic or process-based models that directly incorporate hypothesized relationships determining animal movements and space use. Although we outlined many general advantages and disadvantages of various analytical procedures, the biological and methodological assumptions appropriate for each analytical technique must be considered and tested prior to collection, analysis, and interpretation of radiotracking data. There are numerous options for all aspects of a radiotelemetry study, from equipment to study design to analyses, and decisions about these aspects must be tailored to the specific study objectives.

APPENDIX 20.1. SOFTWARE TO ANALYZE RADIOTELEMETRY DATA

Home Range Programs

ArcGIS/ArcView extensions
  http://www.esri.com/software/arcgis/
   ABODE extension for ArcGIS
      http://filebox.vt.edu/users/plaver/abode/home.html
   Animal Movements extension for ArcView (future plans for ArcGIS tool)
      http://www.absc.usgs.gov/glba/gistools/
      Requires Spatial Analyst extension for full functionality:
      http://www.esri.com/software/arcgis/extensions/spatialanalyst/index.html
   Convex Hulls for ArcView
      http://www.jennessent.com/arcview/arcview_extensions.htm
   Geospatial Modeling Environment for ArcGIS
      http://www.spatialecology.com/
      Requires program R (http://www.r-project.org/) and StatConn (http://sunsite.univie.ac.at/rcom/)
   Hawth's Analysis Tools for ArcGIS
      http://www.spatialecology.com
   Home Range Extension for ArcView
      http://flash.lakeheadu.ca/~arodgers/hre/
   Home Range Tools for ArcGIS
      http://flash.lakeheadu.ca/~arodgers/hre/
   LoCoH (Local Convex Hull) extension for ArcGIS and ArcView
      http://nature.berkeley.edu/~alyons/locoh/

MATLAB extensions
  http://www.mathworks.com/
   KDE Package (Kernel Density Estimation)
      http://www.ics.uci.edu/~ihler/code/
      See: http://www.mathworks.com/matlabcentral/fileexchange/
      For additional kernel applications in Matlab

MATHMATICA Packages
  http://www.wolfram.com/
   ULYSSES
      http://nhsbig.inhs.uiuc.edu/wes/home_range.html

Program R Packages
  http://www.r-project.org/
   aspace
      http://cran.r-project.org/web/packages/aspace/index.html
   ade4 package
      http://cran.r-project.org/web/packages/ade4/index.html
   adehabitat package
      http://cran.r-project.org/web/packages/adehabitat/index.html
   Kernel Home Ranges
      KDE package: http://sekhon.berkeley.edu/stats/html/density.html
      KS package: http://cran.r-project.org/web/packages/ks/index.html
      splancs package: http://cran.r-project.org/web/packages/splancs/index.html
      spatialkernel package: http://cran.r-project.org/web/packages/spatialkernel/index.html

QUANTUM GIS Plugins (free, open-source GIS)
  http://www.qgis.org/
   HomeRange_plugin
      Requires Program R and ADEhabitat (see above)
      http://www.qgis.org/en/download/plugins.html

Standalone Programs

ANTELOPE
  http://nhsbig.inhs.uiuc.edu/wes/home_range.html
BIOTAS
  http://www.ecostats.com/software/biotas/biotas.htm
CALHOME
  http://nhsbig.inhs.uiuc.edu/wes/home_range.html
DIXON
  http://detritus.inhs.uiuc.edu/wes/home_range.html
HOMER
  http://www.cnr.colostate.edu/~gwhite/software.html
  (see Radiotelemetry Programs)
HomeRange
  http://detritus.inhs.uiuc.edu/wes/home_range.html
Home Range
  http://www.cnrhome.uidaho.edu/fishwild/Garton/tools
Home Ranger
  http://detritus.inhs.uiuc.edu/wes/home_range.html
THE KERNEL
  http://gcmd.nasa.gov/records/ucsb_Kernel.html
KERNELHR
  http://detritus.inhs.uiuc.edu/wes/home_range.html
McPAAL
  http://detritus.inhs.uiuc.edu/wes/home_range.html
RANGES 8
  http://www.anatrack.com/
WILDTRAK
  http://reocities.com/RainForest/3722/index.html

Software for Estimating Demographics
Program R packages
> http://www.r-project.org/
>> **demogR** (age-specific survival and reproduction with life tables)
>> http://cran.r-project.org/web/packages/demogR/index.html
>> **popbio** (stage-specific vital rates using matrix population models)
>> http://cran.r-project.org/web/packages/popbio/index.html
>> **Survival**
>> http://cran.r-project.org/web/packages/survival/index.html

SAS packages
> http://www.sas.com/
>> **PROC LIFEREG**
>> http://support.sas.com/rnd/app/da/new/802ce/stat/chap6/sect3.htm
>> **PROC LIFETEST**
>> http://support.sas.com/documentation/cdl/en/statug/63033/HTML/default/statug_lifetest_sect004.htm
>> **PROC PHREG**
>> http://support.sas.com/rnd/app/da/new/801ce/stat/chap12/sect3.htm

Standalone Programs
BMDP
> http://www.statsol.ie/index.php?pageID=9&productID=6&productContentID=239

CONTRAST
> http://gcmd.nasa.gov/records/USGS_CONTRAST.html

MAYFIELD
> http://www.mbr-pwrc.usgs.gov/software.html

SURVIV
> http://www.cnr.colostate.edu/~gwhite/software.html

Mark–Recapture Programs
Program R packages
> http://www.r-project.org/
>> **mra**
>> http://cran.r-project.org/web/packages/mra/index.html
>> **Rcapture**
>> http://cran.r-project.org/web/packages/Rcapture/index.html

Standalone Programs
BAND2
> http://www.mbr-pwrc.usgs.gov/software.html

BROWNIE
> http://www.mbr-pwrc.usgs.gov/software.html

CAPQUOTA
> http://www.mbr-pwrc.usgs.gov/software.html

CAPTURE
> http://www.mbr-pwrc.usgs.gov/software.html

CENTROID
> http://www.mbr-pwrc.usgs.gov/software.html

CloseTest
> http://www.fort.usgs.gov/Products/Software/clostest/

EAGLES
> http://www.cs.umanitoba.ca/~popan/

ESTIMATE
> http://www.mbr-pwrc.usgs.gov/software.html

GENCAPH1
> http://www.mbr-pwrc.usgs.gov/software.html

JOLLY
> http://www.mbr-pwrc.usgs.gov/software.html

JOLLYAGE
> http://www.mbr-pwrc.usgs.gov/software.html

LOLASURVIV
> http://www.mbr-pwrc.usgs.gov/software.html

MSSURVIV, MSSRVRD, MSSRVRCV, and MSSRVMIS
> http://www.mbr-pwrc.usgs.gov/software.html

MULT
> http://www.mbr-pwrc.usgs.gov/software.html

ORDSURVIV
> http://www.mbr-pwrc.usgs.gov/software.html

POPAN-5 and POPAN-6
> http://www.cs.umanitoba.ca/~popan/

Program MARK
> http://www.cnr.colostate.edu/~gwhite/mark/mark.htm

NOREMARK
> http://www.cnr.colostate.edu/~gwhite/software.html

RDSURVIV
> http://www.mbr-pwrc.usgs.gov/software.html#a

RELEASE
> http://www.mbr-pwrc.usgs.gov/software.html

SMOLT
> http://www.cs.umanitoba.ca/~popan/

SPAS
> http://www.cs.umanitoba.ca/~popan/

SURPH
> http://www.cbr.washington.edu/paramest/surph/index.html

TMSURVIV
> http://www.mbr-pwrc.usgs.gov/software.html

USER
> http://www.cbr.washington.edu/paramest/user/

General Telemetry Programs and Location Estimation
Program R packages
> http://www.r-project.org/

trip (accessing and manipulating spatial data for
animal tracking)
http://cran.r-project.org/web/packages/trip/
index.html

Standalone programs

ARGOS (satellite transmitter monitoring)
http://www.argos-system.org/html/services/
tracking-monitoring_en.html

BIOCHECK
User's guide is in White and Garrott (1990:295–297).
http://www.cnr.colostate.edu/~gwhite/software.html
(see Radiotelemetry Programs)

BIOPLOT
User's guide is in White and Garrott (1990:299).
http://www.cnr.colostate.edu/~gwhite/software.html
(see Radiotelemetry Programs)

FIELDS
http://www.cnr.colostate.edu/~gwhite/software.html
(see Radiotelemetry Programs)

LOAS
http://www.ecostats.com/software/loas/loas.htm

LOCATE III
http://www.locateiii.com/

LOTE
http://www.ecostats.com/software/lote/lote.htm

MAP (used to view locations estimated with TRIANG in AUTOCAD)
http://fwie.fw.vt.edu/wsb/contr.html
AUTOCAD
http://usa.autodesk.com/adsk/servlet/pc/
index?siteID=123112&id=13779270

SPADS (Simplified Plotting and Data Storage)
http://fwie.fw.vt.edu/wsb/contr.html

TRIANG
http://www.cnr.colostate.edu/~gwhite/software.html
(see Radiotelemetry Programs)

Programs for Resources or Habitat Selection Analyses

ArcGIS extensions
http://www.esri.com/software/arcgis/
Marine Geospatial Ecology Tools
http://code.env.duke.edu/projects/mget

SAS packages
http://www.sas.com/
BYCOMP.SAS
http://nhsbig.inhs.uiuc.edu/wes/habitat.html
PROC MDC
http://support.sas.com/documentation/

Program R packages
http://www.r-project.org/
RUF.FIT
http://csde.washington.edu/~handcock/ruf/

EXCEL calculators
http://office.microsoft.com/
Selection Ratios Calculator
http://www.resourceselectionbyanimals.com/
rsba/ProgramListing.aspx

Standalone Programs

Compos Analysis
http://www.smithecology.com/software.htm

Fish Telemetry Analysis Program
http://wildlife.state.co.us/Research/Aquatic/Software/

MacComp
http://detritus.inhs.uiuc.edu/wes/habitat.html

PREFER
http://www.npwrc.usgs.gov/resource/methods/prefer/
index.htm

RANGES 8
http://www.anatrack.com/

RSF Programs by West, Inc.
http://www.resourceselectionbyanimals.com/rsba/
ProgramListing.aspx

RSW (Resource Selection for Windows)
http://www.cnrhome.uidaho.edu/fishwild/Garton/
tools

Movement or Time-Series Analysis Programs

ArcGIS and ArcView extensions
http://www.esri.com/software/arcgis/
Alternate Animal Movement Routes for ArcView
http://www.jennessent.com/arcview/arcview_
extensions.htm
Animal Movements extension for ArcView
(future plans for ArcGIS tool)
http://www.absc.usgs.gov/glba/gistools/
Requires Spatial Analyst extension for full
functionality:
http://www.esri.com/software/arcgis/
extensions/spatialanalyst/index.html
Path with Distances and Azimuths for ArcView
http://www.jennessent.com/arcview/arcview_
extensions.htm
Tracking Analyst extension for ArcGIS
http://www.esri.com/software/arcgis/extensions/
trackinganalyst/index.html

Program R packages
http://www.r-project.org/
tripEstimation
http://cran.r-project.org/web/packages/
tripEstimation/index.html

Standalone Programs

BLOSSOM
http://www.fort.usgs.gov/Products/Software/Blossom/

Fractal
http://nsac.ca/envsci/staff/vnams/Fractal.htm

GIS Programs for Visualizing Locations
ACCUGLOBE (freeware)
http://www.accuglobe.net/
ARCGIS
http://www.esri.com/software/arcgis/
DIVA GIS (freeware)
http://www.diva-gis.org/
CMT software
http://www.cmtinc.com/
ERDAS IMAGINE
http://www.erdas.com/
Geomatica
http://www.pcigeomatics.com/
Geoserver (freeware, open-source)
http://geoserver.org
GRASS (freeware, open-source)
http://grass.itc.it/
MapWindows (freeware, open-source)
http://www.mapwindow.com/
Minerva (freeware, open-source)
http://www.minerva-gis.org/
NRDB (freeware)
http://www.nrdb.co.uk/
OPENJump (freeware, open-source)
http://www.openjump.org/
Orbit GIS
http://www.orbitgis.com/
OSSIM (freeware, open-source)
http://www.ossim.org
QUANTUM GIS (freeware, open-source)
http://www.qgis.org/
SPRING (freeware)
http://www.dpi.inpe.br/spring/english/
TatukGIS
http://www.tatukgis.com/Home/home.aspx
Thuban (freeware)
http://thuban.intevation.org/
TNTmap
http://www.microimages.com/
uDig (freeware, open-source)
http://udig.refractions.net/
UTOOLS (freeware)
http://forsys.cfr.washington.edu/utools_uview.html

Clearinghouses
Animal Behavior Software
http://www.animalbehavior.org/Resources/software.htm
Bill's Wildlife Sites
http://wildlifer.com/wildlifesites/software.html
Colorado State University, Department of Fishery and Wildlife Biology
http://warnercnr.colostate.edu/~gwhite/software.html
CRAN programs for analysis of spatial data
http://cran.r-project.org/web/views/Spatial.html
CRAN programs for ecological and environmental data
http://cran.r-project.org/web/views/Environmetrics.html
Dr. Garton's population links and software download
http://www.cnrhome.uidaho.edu/fishwild/Garton/tools
Evan Cooch's Software Page
http://www.phidot.org/software/
GIS, Remote Sensing, and Telemetry Working Group of The Wildlife Society
http://fwie.fw.vt.edu/tws-gis/wwwsrce.htm
Grant Biotelemetry Software and Data Analysis
http://www.biotelem.org/software.htm
Illinois Natural History Survey
http://nhsbig.inhs.uiuc.edu/
Jenness Enterprises (ESRI extensions and tools)
http://www.jennessent.com/
  ESRI
  http://www.esri.com/about-esri/index.html
Patuxent Wildlife Research Center software archive
http://www.mbr-pwrc.usgs.gov/software.html
Population Analysis Software Group
http://www.cs.umanitoba.ca/~popan/
Software published in *The Wildlife Society Bulletin*
http://fwie.fw.vt.edu/wsb/
USGS Fort Collins Science Center Software
http://www.fort.usgs.gov/Products/Software/
WEST, Inc. Resource Selection Software
http://www.resourceselectionbyanimals.com/rsba/ProgramListing.aspx
Wildlife Ecology Software
http://www.humboldt.edu/~mdj6/585/Wildlife%20Software.htm

21

Reproduction and Hormones

JOHN D. HARDER

INTRODUCTION

MEASURES OF REPRODUCTIVE RATE and natality are central to understanding the biology of populations, and in many cases, they are more readily obtained than other components, such as population size. Estimates of diverse parameters from ovulation rate to neonatal survival provide a basis for calculation of natality and recruitment for a variety of wildlife species. Moreover, during the past 3 decades, hormonal regulation of reproductive cycles of a substantial number of wild species has been clarified, so that it is now possible to assess reproductive status and performance from blood levels of certain hormones. Promising new developments in contraceptive technology provide methods that are safe, effective, and sufficiently practical to offer an alternative, under some conditions, to traditional methods of population control.

This chapter is organized to meet 3 objectives: (1) review reproductive cycles of amphibians, reptiles, birds, and mammals and describe procedures to assess reproductive rates in each of these groups; (2) review basic endocrinology relative to measures of reproduction and stress, and (3) review recent developments in reproductive technology, particularly as they relate to contraception and control of recruitment in wild mammals.

The precipitous worldwide decline in amphibians (Semlitsch 2000, Whitfield et al. 2007, Hayes et al. 2010) and reptiles (Fitzgerald and Painter 2000) increasingly demands the attention of wildlife biologists, and background information and techniques relevant to these groups are highlighted in this chapter. Detailed instructions for some of the most commonly used methods are provided; for others, the reader is referred to primary sources and authoritative reviews (e.g., Zug et al. 2001). Stress is an important issue in wildlife research, first, as a potential factor in natural regulation of populations and second, as a consideration in the welfare of animals held captive for research and conservation. An expanding knowledge of the biology and endocrinology of stress provides measures and standards for meaningful evaluation of this factor in both wild and captive animals. The goal of this chapter is to increase awareness of and appreciation for a full spectrum of reproductive techniques and their potential application to the study and conservation of wild terrestrial vertebrates.

REPRODUCTIVE CYCLES AND MEASURES OF REPRODUCTION

Estimation of **natality** (i.e., number of young produced per unit of population per unit of time) is a fundamental requirement for understanding dynamics of a wild population. In some instances, this information can be obtained indirectly through mark–recapture procedures. However, this approach is difficult; mortality of newborn young is high, and younger age groups may be difficult to trap for marking. Consequently, it is necessary to obtain some measure of **reproductive rate,** such as clutch size in birds or the number of **corpora lutea** (CL) in ovaries (for ovulation rate) or fetuses per female collected during the breeding season. Such estimates of reproductive rate can then be used with information on the sex and age structure of the population to calculate gross natality.

Reproductive rates can be estimated at several points in the reproductive cycle, beginning with courtship and ending with fledging or weaning and dispersal of offspring. The value of estimates made at different points varies with the species or taxon and with the goals of the investigation. For example, observations of singing males or nests in a given area will establish density of reproductive pairs, whereas counts of placental scars in mammals at necropsy provide only a size estimate of previous litters.

Knowledge of the variety of measurements of reproductive performance that can be made throughout the reproductive cycle of a given species provides the investigator with more options for meeting specific study objectives. This knowledge also allows the investigator to consider techniques that might be used at different times of the year to achieve more efficient use of fiscal resources and biological material. Accordingly, the sequential events in the reproductive cycle, from **gametogenesis** and breeding to **fledging** or **weaning,** form an outline for this review of techniques and measurements used in the study of reproduction.

Reproductive Modes and Performance in Females

Amphibians and reptiles exhibit a fascinating array of reproductive patterns compared to the relatively uniform condition among birds and in mammals, and although similar protocols can be used with these taxa to study male reproduction (spermatogenesis), females are more challenging. Fertilization is external in amphibians (Fig. 21.1), but internal in **amniotes** (i.e., reptiles, birds, and mammals). Moreover, all birds and nearly all amphibians are **oviparous** (i.e., eggs are laid and hatch outside the mother), whereas all mammals, except monotremes, are **viviparous** (i.e., embryos develop to term in utero, sustained on nutrients transferred through a placenta). Most reptiles are oviparous, but about one-fifth of all lizards and snakes are viviparous (Blackburn 1993). Even in some oviparous species, shelled eggs are retained and hatch in the uterus. Awareness of these diverse modes and patterns of reproduction is important, because they pose constraints as well as opportunities in planning research. For example, in field studies of frogs, should investigators search for egg masses or sample for tadpoles? With viviparous snakes, can ovulation rate be estimated from counts of CL on the ovary and can pregnancy be diagnosed from plasma progesterone levels, as in mammals? These and similar questions often can be studied effectively through comparative biology, in which knowledge of a process in a well studied taxon (e.g., gestation in mammals) is applied to

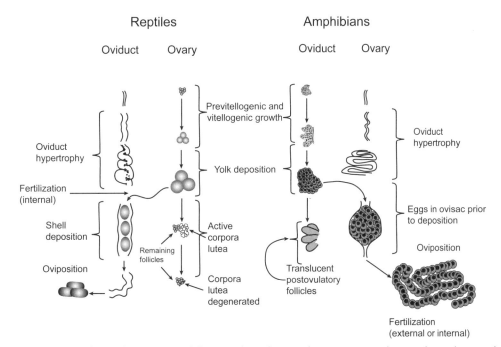

Fig. 21.1. Modes of reproduction in amphibians and reptiles. Fertilization is internal in reptiles and external in amphibians. Postovulatory follicles develop into corpora lutea in most reptiles, but remain empty in most amphibians. In viviparous reptiles, embryos are retained in the uterus without a shell and are nourished via a placenta. *From Zug et al. (2001) with permission from Elsevier.*

the same process in a closely related taxon (e.g., viviparous reptiles).

Three major stages of reproduction in females provide a framework for comparative methodology of reproduction in terrestrial vertebrates: (1) ovarian activation and mating, including seasonal and social activation, follicular development, yolking of eggs, and courtship behavior; (2) ovulation, fertilization, and embryonic development (incubation and gestation); and (3) hatching or birth and parental care of neonates and young.

Ovarian Activation and Mating
Follicular Development

Onset of the breeding season in vertebrates is indicated by an increase in size of the ovary, stemming from an increase in number and size of **tertiary follicles** (i.e., large, yolk-filled [except in mammals] preovulatory follicles). For example in toads (*Bufo* spp.), a small subset of the 30,000–40,000 oocytes in the ovary is responsive to gonadotropin in any given cycle and begins to accumulate yolk. **Vitellogenesis** (i.e., yolking of eggs) is rapid just prior to ovulation, when mature ova reach 10–100 times their original size (Jorgensen 1992). In most oviparous species, the ovary, nearly undetectable in the nonbreeding season, increases to nearly fill the body cavity with yolked eggs, a process wherein ovarian mass increases to represent up to 30% of the total body mass in some taxa. In such species, a **gonadosomatic index** (i.e., ovarian mass divided by body mass) provides a convenient indication of the onset of the breeding season (Licht et al. 1983, Itoh et al. 1990) and, by extrapolation, the proportion of breeding females in a population, whereas body mass alone provides information on the energetic status of individuals (Leary et al. 2008).

Tertiary ovarian follicle counts have been used as measures of reproductive activity in band-tailed pigeons (*Patagioenas fasciata*; March and Sadlier 1970) and mourning doves (*Zenaida macroura*; Guynn and Scanlon 1973). Ankney and MacInnes (1978) were able to distinguish a group of large >20-mm diameter), highly vascularized preovulatory follicles from smaller (10 mm) ones (Fig. 21.2) and thereby estimate potential clutch size in a sample of snow geese (*Chen caerulescens*) shot as they arrived in nesting areas. Because these estimates of ovulation rate are collected at necropsy, they can be analyzed with reference to other data, such as carcass weight, nutrient reserves, and blood hormone concentrations.

Courtship and Estrus

Follicular development and yolking of eggs is accompanied by increased secretion of **estrogen** from large ovarian follicles. **Estrus** is the behavioral state of sexual receptivity associated with elevated estrogen immediately preceding or coincident with ovulation. Elaborate courtship behaviors, which ultimately bring male and female together to fertilize eggs, are obvious and well known in many species of vertebrates. These include calling of male frogs, toads, and passerine birds and visual displays, such as the head-bob and dewlap extension of the male green anole (*Anolis carolinensis*; Crews 1980). In other groups, including salamanders, reptiles, and mammals, social status and readiness to breed may not be communicated in obvious vocal or visual displays, but through **pheromones** (Johnston 2003, Vanden-

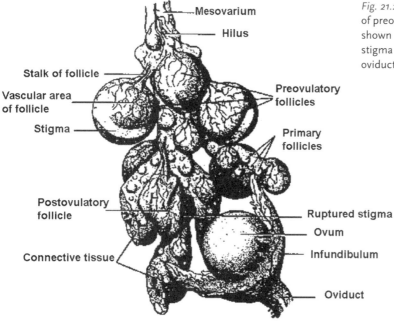

Fig. 21.2. Avian ovary showing a presumptive clutch (group of preovulatory follicles) in a hierarchy of follicles. Also shown is a postovulatory (collapsed) follicle with a ruptured stigma and an ovum entering the infundibulum of the oviduct. *From Nelson (1953) with permission.*

bergh 2006, Harder et al. 2008), which are chemicals released by one individual that elicit a behavioral or physiological response in conspecifics. Pheromones, which may be released in scent marks of urine or feces, or secreted by specialized glands (e.g., cloacal glands in salamanders and skin glands in many snakes; Norris 2007), attract and activate reproductive behavior in the opposite sex and may also signal the loss of attractivity and receptivity in females after mating (Mendonca and Crews 2001). Phermones are widely recognized in sexual signaling (e.g., estrus in females), but in some species, **priming pheromones** also regulate physiological processes. For example, male urine accelerates puberty in female mice (*Mus musculus*), with potential demographic effects (Drickamer 1990), and female opossums (*Monodelphis domestica*) remain reproductively inactive when isolated from males and their scent marks, which contain a nonvolatile estrus-inducing pheromone (Harder et al. 2008).

Wildlife biologists have found that captive breeding programs may be required when prospects for natural reproduction are diminished, as with endangered species, such as the California condor (*Gymnogyps californianus*) and black-footed ferret (*Mustela nigripes*). Zoos holding threatened or endangered species have begun to use more advanced techniques, such as cryopreservation (freezing) of sperm and **in vitro fertilization** in wild species (*Felis nigrips* and *F. margarita*; Herrick et al. 2010) and **embryo transfer** with multiple laparoscopic oocyte retrievals in fishing cats (*Prionailurus viverrinus*), caracals (*Caracal caracal*), and domestic cats (Pope et al. 2006).

Detection of the time of estrus relative to the time of impending ovulation is essential to success of nearly all captive breeding programs. **Artificial insemination** requires placement of sperm into the vagina in a relatively narrow time period within 6–24 hours of ovulation. Proper timing can be accomplished in large species by monitoring follicular development near the time of estrus by laparoscopy or ultrasonography (Ginther 1990) and, in some cases, with predictive hormone profiles (see below; Brown et al. 2004). Hormonal changes that occur throughout the estrous cycle induce changes in the relative proportion of leukocytes and types of epithelial cells in the vaginae of rodents and certain other mammals (Fig. 21.3). A rise in circulating **estradiol** (a type of estrogen) that occurs just prior to estrus stimulates a rapid proliferation and sloughing of keratinized epithelial cells into the lumen of the vagina, a condition indicative of estrus. Because each stage of the cycle is characterized by a particular vaginal cytology, estrous cycles can be monitored, and time of estrus can be predicted (Fig. 21.3, Box 21.1). The technique, first described for domestic rodents (guinea pigs, rats, and mice, [Zarrow et al. 1964]), has been applied

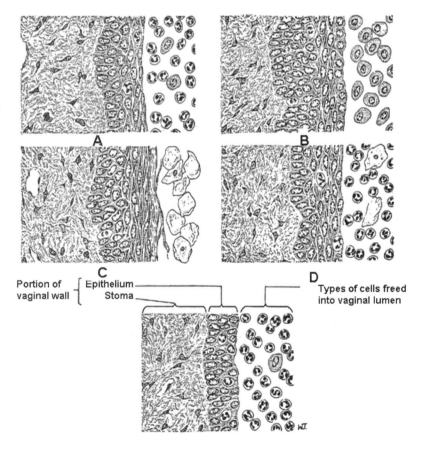

Fig. 21.3. Sections through the vaginal wall of the rat, illustrating changes in the proportions of two types of epithelial cells and leukocytes that are released into the lumen of the vagina during each stage of the estrous cycle. (A) diestrus: a predominance of leucocytes; (B) proestrus: primarily nucleated (basal) epithelial cells; (C) estrus: an abundance of keratinized epithelial cells; (D) metestrus; (E) a female that had been ovariectomized for 6 months (i.e., diestrus). *From Turner and Bagnara (1976) with permission.*

Box 21.1. Characteristic vaginal cytology for each stage of the estrous cycle and procedures for collection and staining of cells from the vagina

Estrous Cycle

The estrous cycle is a sequence of changes in ovarian activity and physiology of the reproductive tract, punctuated with recurring periods of sexual receptivity (estrus) and ovulation. Although estrous cycles of different species vary considerably in length (from a few days to several weeks) and in the timing of cytological changes around estrus, the following description of the 4–5-day estrous cycle of the laboratory rat (adapted from Turner and Bagnara 1976) is reasonably representative of the 4 stages in other species that exhibit cyclic vaginal cytology.

Diestrus is a relatively long stage (60–70 hours) that extends into pregnancy if fertilization occurs or into anestrus during the nonbreeding season of many wild mammals. Corpora lutea (CL) begin to regress, and progesterone levels decline late in this stage. Leukocytes migrate through the thin vaginal mucosa and appear as the predominant cell type in vaginal smears (Fig 21.3A).

Proestrus precedes estrus and lasts for 17–21 hours; it also is known as the follicular phase in species with longer estrous cycles. It is characterized by growth of preovulatory follicles, elevated estrogen levels, and swelling of the uteri. Nucleated epithelial cells dominate the vaginal smears collected at this time (Fig. 21.3B).

Estrus is the period of heat, peak estrogen levels, and ovulation; sexual receptivity is high and limited to this period, which lasts for 9–15 hours. The vaginal epithelium proliferates rapidly, causing the upper layers to exfoliate into the vaginal lumen. Vaginal smears taken at this time are dominated by keratinized (wrinkled) epithelial cells with low numbers of nucleated epithelial cells and leukocytes (Fig. 21.3C).

Metestrus lasts for 10–14 hours in the rat, begins with formation of CL following ovulation, and is characterized by elevated progesterone levels. Metestrus and diestrus are generally known as the luteal phase in species with longer estrous cycles. Large numbers of leukocytes invade the vaginal lumen and often appear clumped around a few keratinized epithelial cells in the vaginal smear.

Vaginal Smear Procedure

1. Appropriate restraint for collection of the smear varies with size and behavior of the animal under study, but many species, ranging from mice to opossums in size, can be handled by grasping the tail. The animal is allowed to stand on the top of a bench or a cage while the tail is raised to expose the vaginal orifice. With this approach, the animal's struggling is reduced and focused on escape, which directs the head (and teeth) away from the handler.
2. To collect cells from the vaginal lumen, a cotton swab is moistened with physiological saline and inserted into the vagina, rotated, and removed. Cells are transferred by rolling the tip of the swab over the surface of a microscope slide. Alternatively, cells can be collected by vaginal lavage, which is the recommended approach for mouse-sized animals. With this procedure, the tip of a fire-polished Pasteur pipet or disposable pipet tip containing a drop of physiological saline is inserted a few millimeters into the vagina. The saline solution is aspirated several times to rinse the vaginal lumen and collect cells. A drop of the aspirated cell suspension is then placed on a microscope slide to dry.
3. The dried vaginal smear may be fixed by gently rinsing with or immersing the slide in 100% ethanol and allowing the smear to dry.
4. The smear is then stained by immersion in a methylene blue solution for 10–15 minutes. After the smear is rinsed gently with distilled water and dried, it is ready for microscopic examination. Reliable monitoring of estrous cycles by vaginal cytology requires that observers be able to recognize 3 types of cells found in the smears (Fig. 21.3A–C).

    Polymorphonuclear leukocyte is a small cell (less than half the size of epithelial cells) with a large, lobed nucleus and little visible cytoplasm. This cell is represented by small, dark-staining, C-shaped nuclei in vaginal smears (Fig. 21.3A).

    Nucleated epithelial cell is a large, rounded cell with a prominent nucleus (basal cell) that appears in greatest numbers during proestrus, but it can be found in smears collected during all stages of the cycle (Fig. 21.3B).

    Keratinized epithelial cell is a large, squamous cell with a wrinkled, "potato chip" appearance. The nucleus is degenerate and is often not visible, even in stained preparations. These cells are present in such large numbers at estrus that a vaginal lavage takes on a milky appearance (Fig. 21.3C).

to diverse species, including pine voles (*Microtus pinetorum;* Kirkpatrick and Valentine 1970), American beaver (*Castor canadensis;* Doboszyńska 1976), domestic dogs and cats (Stabenfeldt and Shille 1977), coyotes (*Canis latrans;* Kennelly and Johns 1976), and Virginia opossums (*Didelphis virginiana;* Jurgelski and Porter 1974).

Ovulation and Embryonic Development
Perhaps the one reproductive parameter of greatest importance to natality and recruitment in all vertebrates is **ovulation rate.** Fortunately, it can be measured directly, or indirectly, with accuracy in most species.

Postovulatory Follicles

After ovulation and release of mature eggs into the oviduct, the collapsed wall of the ovarian follicle in most vertebrates does not form a **corpus luteum** (CL; the acronym is used for both singular and plural [corpora lutea]), as it does in viviparous reptiles and mammals. Instead, **postovulatory follicles** (POF) regress, the process being completed in many avian species within a month after ovulation (Payne 1973). However, in some species, such as the ring-necked pheasant (*Phasianus colchicus*), POF persist for months as small (1–2 mm) pigmented (reddish brown) structures that can be viewed macroscopically (Fig. 21.2). Kabat et al. (1948) reported a high correlation between number of eggs laid and number of POF counted in captive pheasants killed up to 100 days after ovulation, thus providing a potential method for estimating clutch sizes in summer from hens harvested the following autumn. However, Hannon (1981) found that POF in blue grouse (*Dendragapus obscurus*) could not be counted macroscopically beyond approximately 25 days of age, although they could be used to distinguish laying from nonlaying hens shot during the hunting season. Similarly, POF allowed independent observers to correctly distinguish breeding from nonbreeding mallard (*Anas platyrhynchos*) hens 60–90 days after laying and thereby estimate breeding propensity in free-ranging mallard populations (Lindstrom et al. 2006).

Observation of the avian ovary need not be limited to necropsies. **Laparotomy** has been widely used for identifying the sex of monomorphic birds that are captured live in mist nets and other types of traps (Risser 1971). An incision is made on the left side to expose the left testis or ovary, the latter being distinguished by the presence of follicles. The incision is small, and the procedure can be accomplished in the field, providing that appropriate surgical procedures are used (Wingfield and Farner 1976). Although an accurate classification or count of POF is not feasible under such conditions, laparotomy can provide useful information on the stage of follicular development and proportion of birds nearing the egg laying stage.

Clutch Size

Estimation of **clutch size** (number of eggs in a nest) is by far the most popular and practical method of estimating ovulation rate in oviparous vertebrates. However, locating the nest is a challenge, particularly with many species of frogs and salamanders. Even when nests or egg masses are obvious, difficulty may be encountered in developing unbiased sampling procedures that permit projection of estimates to the population level or unit area of habitat. In species whose nests and egg masses can be reliably located, number of nests, their distribution, and survival to time of hatching are recorded, but seldom is the number of eggs per egg mass noted. Enumeration of the tadpole and larval stage is more commonly recorded and used as an index of reproductive rate and hatching.

Clutch size has been most extensively studied and applied to avian population ecology. Most birds lay at most one egg per day immediately prior to incubation. Ovulation precedes laying or deposition of the egg by approximately 26 hours. Thus, close observation of a laying hen provides immediate, real-time data on ovulation rate and an opportunity to study the temporal relationships among courtship behavior, copulation, ovulation, and associated hormone changes.

Avian nests found early in the incubation period provide not only a good estimate of number of eggs laid, but also a basis for estimating egg loss and hatching success. Furthermore, sampling procedures normally provide estimates of nest density that in turn can be used to calculate size of the breeding population of females in monogamous species. In such cases, estimates of reproductive rate and population structure are available, and thus, **gross natality** may be calculated.

Clutch size has been estimated in relatively few amphibians. More often, numbers of nests or egg masses per area or habitat type are recorded. Crouch and Paton (2000) found these data more useful than estimates of the number of calling males as an index to abundance of wood frogs (*Rana sylvatica*). However, even with frogs that lay obvious egg masses, the number of eggs is most often expressed in 10s or 100s and seldom with estimates of mean and variance.

Follicular development (Leyton and Valencia 1992) and clutch size have been estimated for a large number of reptiles. It is possible to visualize even partially calcified eggs in the oviduct with X-ray photography and thereby estimate time of ovulation and future clutch size (Gibbons and Green 1979). However, the short- and long-term effects of radiation exposure on the young remain largely unknown. Vitt (1992) summarized studies of 26 species of squamate reptiles in the Caatinga of Brazil; clutch sizes across species were primarily in the range of 2–15 (maximum of 31) and were positively correlated with body size in some species (Abell 1999, King 2000). Many reptiles, including turtles, lay >1 clutch per season. Congdon and Gibbons (1996) reported clutch size data on nests (>2,500) of 3 species of turtle as part of a long-term multifaceted study of community structure and dynamics in southern Michigan.

Nesting behavior and clutch size have been studied extensively in several species of sea turtles (Cheloniidae). Their nests are concentrated in traditional nesting beaches and thus can be located with relative ease. Nests are the focus of sea turtle biology, because all other phases of the life cycle occur in the open ocean and are largely unavailable for study. Most species of sea turtles require >20 years to reach maturity and have complex multiyear reproductive cycles with evidence of delayed fertilization. Females lay from 2 to 5 clutches during a season. Owens (1995) recommended establishment of colonies of adult turtles held captive in large natural ponds with beaches to study these events in greater detail. This effort could be particularly valuable in clarifying details of male reproductive biology and temporal relationships between insemination and fertilization in a number of sea turtle species. Clutch size in sea turtles usually exceeds 100 (Miller 1997) and shows a positive correlation with body size of the female, at least in green (*Chelonia mydas*) and loggerhead turtles (*Caretta caretta*; Ehrhart 1995). Although largely peripheral to the scope of this chapter, it should be noted that Cheloniidae is one of 19 families of reptiles with species in which sex determination is temperature dependent (Zug et al. 2001). For example, if the average incubation temperature in the nest of many species of turtles exceeds approximately 30° C, nearly all young will be female, whereas the opposite effect is seen in some lizards and alligators (Nelson 2000).

Because the avian egg contains all nutrients and energy required by young from conception through the immediate post-hatching period, egg size and quality are major factors affecting hatching success and survival of avian young. This was demonstrated in studies of yolk and albumen content of eggs of snow geese (Ankney 1980) and the positive relationship between egg size, motor ability, and postnatal weight gain in Australian brush-turkey (*Alectura lathami*) chicks (Göth and Evans 2004a). Vangilder and Peterle (1981) observed a reduction in eggshell thickness and proportion of yolk contained in eggs laid by mallards fed either crude oil or DDE (1,1-dichloro-2,2-bis[p-dichlorodiphenyl]ethylene), and Beckerton and Middleton (1982) demonstrated that increasing protein content (from 7.6% to 20.1%) in isocaloric diets of ruffed grouse (*Bonasa umbellus*) was associated with linear increases ($P < 0.025$) in a series of 9 reproductive parameters, ranging from clutch size and egg weight to chick survival. However, Arnold and Green (2004) cautioned against using allometric regression to characterize the relationship between proportional egg composition and egg size.

Ovarian Function and Ovulation in Mammals

Mammalian eggs are microscopic (10–30 μm), retained in the female reproductive tract, and therefore, cannot be easily counted. Direct enumeration of the number of eggs released in a given estrous cycle requires flushing eggs from the oviduct and/or uterus with isotonic medium and examining the resulting fluid microscopically. This approach is impractical for most wildlife investigations and is seldom used except for embryological studies or intensive studies of reproductive physiology, such as those involving embryo transfer. Instead, ovaries are examined for follicular development or evidence of ovulation, such as the presence of **CL** and related scar structures.

Ovarian analysis requires a basic understanding of anatomy and physiology with respect to the events preceding and following ovulation. The mammalian ovary contains a life-long supply of **primary oocytes** (in prophase of meiosis) at birth, contained in small primary follicles (Fig. 21.4). During each estrous cycle, a small fraction of these follicles begin growing more rapidly and develop a fluid-filled cavity or antrum. These growing follicles secrete increasing amounts of estrogen, which stimulates estrus and, indirectly, ovulation in spontaneous ovulators. Some of these **antral (Graafian) follicles** reach precisely the appropriate stage of preovulatory development to respond to a surge in **luteinizing hormone** (LH) secretion at estrus and ovulate. Most follicles, however, do not reach the preovulatory state, but instead undergo atresia, a degenerative process leading to disassociation of the granulosa layers of the follicle and death

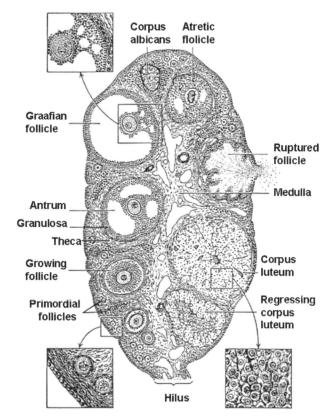

Fig. 21.4. Drawing of the mammalian ovary, illustrating (in clockwise progression) follicular development (from primordial to Graafian follicle), ovulation, development of a corpus luteum (CL), and regression of the CL. The ovum is released with follicular fluid from the ruptured follicle. *From Turner and Bagnara (1976) with permission.*

of the oocyte. Thus, a dynamic balance of follicular development and atresia affects the ovulation rate in any given estrous cycle.

Ovulation results in rupture of the follicle that leaves a **corpus hemorrhagicum** (the ovulation point or blood spot) and initiates immediate luteinization of thecal and granulosa cells of the follicle wall (i.e., they enlarge and sequester lipids). This process results in filling of the cavity and the formation of the CL (Fig. 21.4), a transient endocrine gland that secretes progesterone, a hormone essential for support of pregnancy.

If fertilization or pregnancy fails, CL regresses, and a new estrous cycle begins with growth of preovulatory follicles. However, if conception occurs and embryos implant in the uterus, CL persist and secrete large amounts of **progesterone** throughout a substantial portion of the gestation period. CL of pregnancy become large, often occupying much of the volume of the enlarged ovary (Fig. 21.4). Prior to or coincident with parturition, CL begin to regress, and progesterone secretion declines sharply.

LAPAROSCOPY AND ULTRASONOGRAPHY

Ovarian analysis of wild mammals is usually conducted on material collected during necropsy of carcasses obtained at hunter check stations or other sources, such as traffic accidents. However, with application of fiber optics to surgical instruments, internal examination of live animals has become safer and more convenient through a technique known as **laparoscopy.** The abdominal wall is punctured with a large needle cannula, through which a fiber optic scope (2–10-mm diameter) is inserted. Organs can be manipulated with a probe inserted through a second canula. With this technique, it is possible to observe ovarian follicles and CL as well as uterine swellings. The animal must be anesthetized, but incisions per se are not made, which minimizes surgical trauma and risk of infection. An extensive review of laparoscope methodology is presented in Harrison and Wildt (1980). Although laparoscopy is widely used in medicine and animal science, applications in wildlife research have been limited. However, Nelson and Woolf (1983) observed no complications or mortalities in 20 radiocollared white-tailed deer (*Odocoileus virginianus*) monitored after laparoscopy of ovaries in the field. Laparotomy and laparoscopy both require special equipment and training and are recommended only in situations where (1) multiple observations on the same animals are required, as in pen-based experiments; (2) the animal under study is rare or endangered; or (3) the animals are valuable, and little is known of their basic physiology (e.g., zoo animals).

Echoes of high frequency sound (3–8 megahertz [mHz]), processed by real-time computerized video displays, also can be used to reveal internal morphology. This technique, known as **ultrasound** or **ultrasonography** and widely used in human medicine, is recognized as a practical and reliable approach to monitoring estrous cycles and gestation in livestock and several wild species, including bottlenose dolphins (*Tursiops truncates;* Williamson et al. 1990), red deer (*Cervus elaphus*), black rhinoceros (*Diceros bicornis*), giraffe (*Giraffa camelopardalis*), gaur (*Bos gaurus;* Adams et al. 1991), and several species of rhinos and elephants (Hildebrandt et al. 2006). The procedure is particularly effective in visualizing antral follicles, because the liquid phase absorbs ultrasound and appears black in contrast to surrounding tissue, which emits strong ultrasound echoes. Ultrasonography has been used successfully to count and measure follicles >2 mm in diameter (Fig. 21.5) and to monitor the growth of larger (5–15 mm) individual follicles in cows (Sirois and Fortune 1988). Ultrasonography with this level of resolution has the potential for monitoring ovarian activity in larger carnivores, ungulates, and gravid sea turtles (Kuchling 1999) without use of invasive surgical procedures. More recently, advances in ultra-high frequency (40 MHz) ultrasonography have permitted characterization of structures as small as 70 μm and monitoring of follicular development in mice (Jaiswal et al. 2009).

ENUMERATION OF CL

Follicle counts will not predict ovulation rate in mammals. However, the ovary develops an unambiguous sign of ovulation: the CL, which in most species, grows to occupy most of the volume of the ovary. In large mammals, such as domestic cattle, the CL can be palpated through the rectum with a gloved arm and hand. This procedure, routinely used in the dairy and beef industry, has limited utility in wildlife studies because of size limitations, although it has been used on elk (*Cervus canadensis;* Greer and Hawkins 1967).

CL can be counted and ovulation rate estimated in medium-sized to large mammals through gross (or with the aid of a dissecting microscope) examination of sliced ovaries obtained at necropsy (Box 21.2; Fig. 21.6). This approach has been applied widely to many species, including American beaver (Provost 1962), moose (*Alces alces;* Simkin 1965),

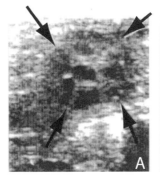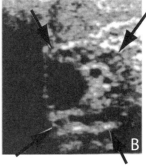

Fig. 21.5. Ultrasound images of ovaries in cattle; arrows mark the periphery of the ovary. (A) Three medium follicles (5–7 mm in diameter) are visible as dark, round objects. (B) Several small (2–3 mm) follicles are visible to the right of a large (12 mm) follicle on this ovary. *From Pierson and Ginther (1988) with permission.*

BOX 21.2. PRESERVATION OF TISSUES AND GROSS OVARIAN ANALYSIS

Preservation of Tissues at Necropsy

Postmortem changes are slowed considerably at low temperatures (e.g., 4° C), and organs may be frozen before subsequent gross examination. However, ice crystals that form in the cytoplasm ruin cells for microscopic study. If histology is planned, organs and tissues must be placed in a fixative solution at necropsy.

Fixatives are used prior to histology to (1) prevent purification, (2) coagulate protein, and (3) protect the tissue against shrinkage and distortion in subsequent procedures. Buffered 10% formalin (1:10 dilution of 40% formaldehyde) is used widely, but it is only one among many options. A histology manual (e.g., Humason 1979) should be consulted for specific recommendations. The volume of fixative should exceed that of the tissue by 5 or 10 times to avoid excessive dilution of the fixative by water diffusing from the tissue.

Gross Examination of Deer Ovaries

1. Ovaries are removed by cutting the mesovarium, the mesentery that suspends the ovary in the body cavity, near the ostium of the oviduct. Ovaries are more easily manipulated if some mesovarium remains with the ovaries, and left and right ovaries can be identified later if extra mesovarium is routinely left on the ovary from one side.
2. After ovaries have been in a fixative, such as formalin, for 36 hours, they will harden sufficiently to withstand slicing. Each is removed from the fixative and rinsed thoroughly in tap water. Caution: formaldehyde is toxic, and so latex gloves should be worn and all work with formalin-fixed material should be done in a fume hood.
3. An ovary can be secured by grasping the mesovarium close to the ovary with curved forceps or hemostat. It is then sliced along the long axis with a scalpel or razor blade, cutting toward the mesovarium and forceps (Fig. 21.6). With practice, horizontal slices of about 2-mm thickness can be cut, stopping just before the mesovarium is reached. In this way, the sliced ovary will stay together like pages of a book, ready for thorough, repeated examination. Hawley (1982) described a razor-blade device used to slice moose ovaries into uniform 1.5-mm sections.
4. Ovaries collected during the breeding season will contain a number of follicles of mixed size and, perhaps, recently ovulated follicles or new corpora lutea (CL) with ovulation points still evident. New CL of pregnancy grow to near full size (7-mm diameter) in the first 2–3 weeks after ovulation and eventually occupy most of the ovarian volume during pregnancy in deer. Thus, they can be "followed" through several slices from one side of the ovary to the other. The sliced surface is solid, cheesy in texture, and creamy white in color. The color varies from yellowish to gray in other species.
5. Far less evident are the small copora albicantia (CA), pigmented scars of the regressing CL of the previous pregnancy (Cheatum 1949). Each slice of ovary must be carefully examined on both sides for these small (1–3-mm diameter) rust-colored structures, which are often compressed into triangular or crescent shapes by surrounding follicles and growing CL. Color is the primary distinguishing characteristic, but it can vary from dark yellow to deep brownish orange.
6. If the ovary is to be saved for further macroscopic or microscopic examination, it should be stored in 70% ethanol to prevent excessive hardening.

red fox (*Vulpes vulpes;* Oleyar and McGinnes 1974), and eastern cottontail rabbits (*Sylvilagus floridanus;* Zepp and Kirkpatrick 1976). **Accessory CL** (i.e., unovulated luteinized follicles) and other structures unrelated to ovulation can, at times, be distinguished by size and appearance.

Counts of CL provide an accurate measure of **ovulation rate,** but are only an index to number of young in utero. This is because CL form during the normal course of each estrous cycle whether or not conception occurs, and even though most animals collected from the wild with active CL will be pregnant, each ovulated follicle forms a CL—whether or not the egg from each follicle is fertilized—and undergoes embryonic development. Thus, if the fertilization rate in a given species is low or if embryonic or fetal losses are high, CL counts will overestimate number of young produced. For example, the Virginia opossum has a high ovulation rate (30 CL/ovary/cycle; Fleming and Harder 1983), but gives birth to only 10–20 young and weans 6–8. Although conception rates in most species are much higher than for opossums, fertilization rates and in utero survival

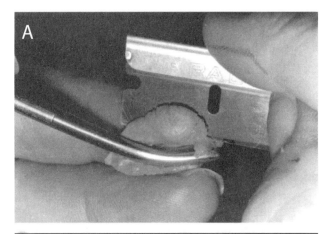
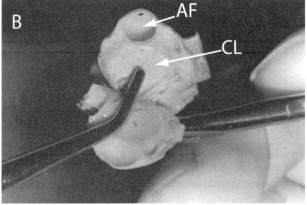

Fig. 21.6. (A) Procedure for slicing a fixed ovary with a razor blade. (B) A view of the sliced ovary, showing an antral follicle (AF) and a corpus luteum (CL). *Photos by D. Dennis.*

vary considerably among species and must be identified for each separately (Brambell 1948).

OVARIAN ANALYSIS OF DEER

All CL leave scar tissue in the ovary as they regress. In most mammals, they are visible only in microscopic examination as whitish bodies of connective tissue known as **corpora albicantia** (CA). However, the large, long-lived CL of pregnancy in cervids regress slowly after parturition and are grossly visible as pigmented CA (sometimes called corpora rubra; Cheatum 1949; Box 21.2). Ovaries of deer are most often collected at hunter check stations in October–December (i.e., often before all does in a population have had an opportunity to ovulate and before fetuses are visible in utero). Therefore, considerable attention has focused on CA as a basis for estimating number of fetuses carried to term in the previous pregnancy (preceding spring).

The value of CA counts in estimation of average litter size in previous pregnancies depends on knowledge of the fertilization rate and longevity of the CA. The **fertilization rate** (fetuses/CL) in deer is high and remarkably constant (85–90%; Roseberry and Klimstra 1970, Barron and Harwell 1973, Woolf and Harder 1979). Unfortunately, the longevity of CA is variable, sometimes remaining grossly visible for more than a year. Identifying the age of CA through histological examination (Mansell 1971) would largely eliminate biases in estimation of previous litter sizes, but this is seldom feasible for management surveys.

Examination of the ovaries of yearling does for CA has great value and utility for estimating the percentage of fawns breeding in a population. With few exceptions, the maximum ovulation rate for fawns that reach puberty and ovulate is 1 ovum per female. Thus, yearlings that are killed during the hunting season or in traffic accidents during July–February will yield reliable CA data, because, for all practical purposes, only 2 possibilities exist: 0 or 1 CA per doe. The proportion of females that conceive in their first year reflects the nutritional status of the herd and varies from 0 (Woolf and Harder 1979) to a high of 77–82% (Nixon 1971, Haugen 1975). This demographic parameter is important, because the 6-month age class is the largest in any population, and therefore, one that has great potential for impact on natality, population growth, and sustainable yield (Harder 1980).

Uterine Analysis in Reptiles and Mammals

Viviparous reptiles and mammals retain developing young in utero and thus, present evidence of reproductive performance equivalent to clutch size in oviparous vertebrates. Enumeration of embryos or fetuses in utero has long been a popular and convenient measure of the reproductive performance of a population. Popular, because the number in utero, especially during the third trimester of gestation, is often a reliable indicator of the number of young that will be born (i.e., **litter size**). Uterine examination for embryos or fetuses is convenient, particularly in later stages of gestation, because fetuses are grossly visible and can be reliably counted by hunters or personnel handling animals killed on highways. The entire uterus with fetuses may be collected and frozen or fixed for later study, or it can be inspected on location. Crown–rump length measurements can be used to estimate age of white-tailed deer (Armstrong 1950) and coyote fetuses (Kennelly et al. 1977) and, by backdating, to estimate breeding dates. The primary sex ratio for a population can be estimated from examination of fetuses, a parameter of practical value in population models and of considerable interest among theoretical and experimental ecologists (Trivers and Willard 1973, Austad and Sunquist 1986, Gosling 1986, Rosenfeld and Roberts 2004).

Fetal counts are most often done at necropsy, but in many instances, this information must be obtained from living animals, such as during investigations of rare and endangered species or animals in zoos. Uterine swellings, indicative of fetuses, and CL of pregnancy can be counted in living animals by laparotomy. This approach has been used on cottontails (Murphy et al. 1973), white-tailed and mule deer (*Odocoileus hemionus*; Zwank 1981), and elk (Follis et al. 1972). Ultrasound readings were used to successfully diagnose pregnancy during the last trimester of gestation in mountain sheep (*Ovis canadensis*; Harper and Cohen 1985) and pronghorn (*Antilocapra americana*; Canon et al. 1997). Several noninvasive methods, based on hormonal and biochemical changes in blood, are available for diagnosing pregnancy.

Birth and Parental Care of Young
Nest Success
The number of young born or hatched per female is the most ecologically important reproductive parameter that can be measured. It is the basis for calculating gross natality and can be obtained with reasonable accuracy for many avian species and some herptiles. If the location of nests and clutch size is known, the number of live young can be obtained by repeated inspection of the nest. This is the basis of the **Mayfield method** for estimating nest success of birds (Dinsmore and Johnson 2005). The eggs of most viviparous reptiles are laid in spring and hatch by late summer, but in a number of turtles (e.g., painted turtles [*Chrysemys picta*]; Breitenbach et al. 1984), emergence of young from the nest is delayed until the following spring, when foraging conditions are more favorable (Fig. 21.7).

Amphibians, particularly frogs, exhibit the greatest diversity of modes of parental care among terrestrial vertebrates. What might be viewed as a typical pattern in north temperate regions (i.e., eggs deposited in vernal pools and hatching into free-foraging tadpoles) is just 1 of 27 different modes recognized by Zug et al. (2001). Seemingly every possible pattern has been observed: from tadpoles emerging from eggs laid on tree leaves and dropping into ponds or streams to eggs that are carried in the dorsal pouch of the female and develop directly into froglets. These diverse patterns present fascinating natural history and opportunities for expanding knowledge of traditional reproductive ecology of a group that is undergoing an alarming decline in diversity and abundance (Semlitsch 2000, Hayes et al. 2010). The challenge is to devise search and sampling procedures that produce estimates useful for population models. The most common approach for estimating number of young at birth in amphibian populations involves use of funnel traps for tadpoles and larvae (Adams et al. 1997) or setting of pitfall traps along a drift fence that encircles a breeding pond (Semlitsch et al.1996).

The nests of many oviparous and viviparous species (some lizards, snakes, and mammals) cannot be located with statistical validity. Thus, the number of young born must be inferred from data collected in utero or, for some frogs, other parts of the body. Ovarian and uterine analyses described in some detail for white-tailed deer could be developed for other viviparous species. More studies of mammals and reptiles are needed to establish the relationship between ovarian scars of CL and the number of young carried and born. Thus, ovaries collected from females shortly after they give birth might provide a measure of litter size. The extent to which uteri of viviparous reptiles might reveal early implantation sites as indicators of potential litter size is largely unknown.

Placental Scars
These scars are pigmented areas of uterine tissue marking sites of previous placental attachment in mammals (Fig. 21.8). Their formation, described by Deno (1937) and Martin et al. (1976), is limited to taxa with deciduous placentae (Wydoski and Davis 1961). Erosion of the uterine endometrium by the embryonic trophoblast and an interdigitation of uterine and chorionic tissue is such that endometrial tissue is torn away when the placenta is expelled at birth (i.e., the placenta is **deciduous;** Vaughan et al. 2000). As the new uterine endometrium grows over this wound, stagnant pools of blood become trapped, and the hemoglobin in the red blood cells is degraded to hemosiderin (an iron-containing pigment) by macrophages. The entrapped hemosiderin remains visible as a placental scar for varying lengths of time.

Species known to develop prominent placental scars belong primarily to the mammalian orders (Insectivora, Chiroptera, Lagomorpha, Rodentia, and Carnivora). However, scarring also has been described in such divergent taxa as elephants (Laws 1967), and it is possible that some squamate reptiles, with mammalian-like chorioallantoic placentae (Blackburn 1993), also develop placental scars. Litter size has been estimated from placental scars in a variety of carnivores, including brown bears (*Ursus arctos*; Hensel et al. 1969), raccoons (*Procyon lotor*; Sanderson 1950), and gray fox (*Urocyon cinereoargenteus*; Oleyar and McGinnes 1974). Placental scars are most useful in mammals that have only 1 or 2 litters per year, such as American beaver (Henry and Bookhout 1969) or gray squirrels (*Sciurus carolinensis*; Nixon et al. 1975). The reliability of the method has been recently confirmed in the

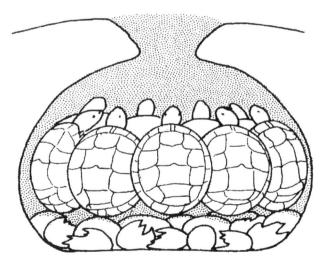

Fig. 21.7. Arrangement of painted turtle hatchlings in an overwintering nest. *From Breitenbach et al. (1984) with permission.*

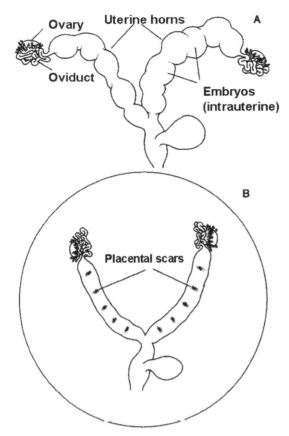

Fig. 21.8. (A) Uterine swellings of a pregnant white-footed mouse (*Peromyscus leucopus*). (B) Postpartum uterine horns compressed between the lid and the inverted base of a petri dish to reveal placental scars. *Drawings by D. Dennis.*

European hare (*Lepus europeus*; Bray et al. 2003). Observations of smaller rodents that have several litters in rapid succession often entail problems in separating scars into sets representing different litters or pregnancies (Rolan and Gier 1967, Martin et al. 1976).

In many species, placental scars can be seen easily in fresh, thawed, or preserved tissue without special treatment. They stand out as darkened spots or bands in the uterine horns. If females are collected soon after parturition, the tissue around the old implantation sites will still be swollen. With increasing time after parturition, however, the scars fade, and additional steps must be taken to clearly see them. A first step, at least with small mammals, is to compress the uterine horns between 2 microscope slides (or with larger uteri, between the nested lid and base of a petri dish; Fig. 21.8). The scars can then be viewed on a dissecting scope, and with backlighting, it is possible to distinguish and count scars in sets. With larger species, it usually is necessary to expose the endometrium of the uterus by cutting longitudinally along the length of each horn with scissors. The scars then appear as darkened bands or discs in the uterine lumen. In some species, placental scars are indistinct, and special clearing or staining procedures are required to make

them more visible (Henry and Bookhout 1969, Humason 1979, Bray et al. 2003). Placental scar count accurately estimated pregnancy rate and litter size in mink (*Mustela vison*) collected up to 3 months postpartum; but with longer intervals, the proportion of females judged to be barren was overestimated, and mean litter size was underestimated (Elmeros and Hammershoj 2006).

Lactation and Nursing

The transition from late gestation to lactation is a critical period in the reproductive cycle of mammals. **Mammary glands** of lactating females grow and fill with milk that can be expressed from the teats of all but the smallest species. If milk cannot be expressed from a female at necropsy, the mammary gland should be sliced open and inspected for pools of milk in the tissue. Nipples of lactating females become swollen and pinkish, and often the fur immediately surrounding the nipple is thin or absent. These and other indications of lactation are only indirect signs that a female is nursing or has recently weaned young. They should be verified, preferably through measurements of females with known reproductive histories, including those nursing young. Some progress has been made in this regard through studies of teat length of yearling white-tailed deer in autumn. Sauer and Severinghaus (1977) concluded that any yearling with 1 of 4 teats 10 mm was without young, and those with 1 teat >15 mm was nursing or had recently weaned a fawn. Both teat length and placental scars merit additional evaluation as indicators of reproductive success in other species, as in fisher (*Martes pennant*; Frost et al. 1999), particularly because data can be collected from carcasses on highways or those brought to hunter check stations.

Because lactating marsupials carry young in their pouch for an extended period (50–60 days in the Virginia opossum), they are attractive subjects for study reproductive ecology (Harder and Fleck 1997, Carusi et al. 2009), particularly of the effect of nutritional status on litter size. For example, Hossler et al. (1994) demonstrated a remarkably close correlation ($R^2 = 0.86$) between a hind-leg fat index and litter size in opossums living near the northern limit of their distribution in New York. Given the reality of substantial neonatal mortality in many wildlife populations, an estimate of the proportion of females in a population that are lactating is more relevant to reproductive success than are parameters measured earlier in the reproductive cycle (e.g., ovulation rate). Coupled with estimates on average litter sizes, lactation indices could substantially improve estimates of net natality (i.e., number of young weaned/female).

Reproductive Patterns and Performance in Males

Reproductive activity in male vertebrates in the breeding season is decidedly noncyclic and less complex than in females, particularly in regard to assessment of reproductive performance. **Spermatogenesis** begins at puberty, and except for periods of seasonal regression, testes normally remain active in production of sperm and secretion of testosterone throughout the life of the individual. Sperm are produced by the millions, well in excess of the few hundred that might be required to fertilize even the largest mass of frog eggs. In most species, **reproductive success** of fertile males is ultimately based on the number of females mated or eggs fertilized. It is often a function of courtship behavior or the outcome of interactions with competing males.

Because testes increase in mass with onset of spermatogenesis, testis weight is most often used to monitor the onset of breeding and as an index of the number of breeding males in populations of herptiles, birds, and mammals. Warm spring weather signals the onset of breeding in the Mexican leaf frog (*Pachymedusa dacnicolor*), when active spermatogenesis is coupled with increased testis weight and plasma androgen levels (Bagnara and Rastogi 1992). Testicular enlargement coincides with ovarian activation in some species of salamanders (Semlitsch 1985), but not all (Marvin 1996). Most reptiles have an **associated** reproductive pattern, in which breeding is associated with elevated androgen secretion and spermatogenesis. However, some squamates, such as the red-sided garter snake (*Thamnophis sirtalis*), have a **dissociated** pattern. In this species, sperm produced during summer are stored by the male for use during the following spring breeding season, when testes are actually regressed and androgen levels are low (Fig. 21.9).

Testis Weight and Counts of Spermatozoa

The **gonads** and reproductive tracts of birds generally remain in a regressed state during the nonbreeding season, but increase markedly during the breeding season. For example, the testis of a mature male white-crowned sparrow (*Zonotrichia leucophrys*) will grow from 10 mg to >600 mg during the breeding season (Wingfield and Farner 1980), and elevated plasma **testosterone** in pintail ducks (*Anus acuta*) coincides high ejaculate quality, as evidenced in number and morphology of sperm (Penfold et al. 2000). Seasonal growth and recrudescence of testes and accessory glands of mammals are much less pronounced and more variable across taxa than in birds. In some, such as tammar wallaby (*Macropus eugenii*) and brush-tailed possum (*Trichosurus vulpecula*), testis weight changes little throughout the year, although prostate weight and testosterone levels show marked seasonal variation (Gilmore 1969, Inns 1982). Most seasonally breeding mammals, however, show noticeable to marked increases in testis size as well as androgen levels (Mirarchi et al. 1977). Weight and volume (estimated by water displacement in a graduated cylinder) can be obtained at necropsy, but 2-dimensional measurements of scrotal testes on live males also provide a valid index of testicular volume and reproductive status. For example, the volume of testes of white-tailed deer increased from a low of 50 cc in June to >150 cc in November (breeding season), during the same time that plasma testosterone increased from basal (near 0) to >3 ng/mL (McMillin et al. 1974).

The presence of **spermatozoa** (sperm) in the testes or **epididymides** provides evidence of reproductive status and is particularly useful for examining age of puberty or onset of seasonal breeding. Sperm counts also have been used to examine seasonal differences in male reproductive activity (Mirarchi et al. 1977), social stress (Sullivan and Scanlon 1976), and exposure to environmental contaminants (Sanders and Kirkpatrick 1975).

An estimate of sperm density or total spermatozoa per testis or epididymis may be obtained by homogenizing a known mass of sliced tissue in a blender or tissue homogenizer with an appropriate culture medium, such as Hank's solution. Triton X-100 (J. T. Baker, Phillipsburg, NJ) can be added (0.01–0.05% by volume) to prevent foaming in the blender (Amann and Lambiase 1969, Sullivan and Scanlon 1976). Alternatively, testes or sections of epididymides of small species can be thoroughly minced with scissors and rinsed repeatedly with a known volume of culture medium to collect spermatozoa. An aliquot of homogenate or rinse is then removed and added to both chambers of a standard hemocytometer. Sperm density relative to tissue mass can then be calculated with the appropriate dilution factors (Box 21.3). A test for the presence of motile spermatozoa involves cutting the tail of the epididymis from a freshly killed specimen and making a smear on a slide for microscopic examination to detect sperm (Kibbe and Kirkpatrick 1971). Morphology and **motility** are commonly used to assess the

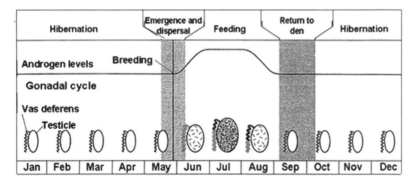

Fig. 21.9. The annual gonadal reproductive cycle of the red-sided garter snake illustrates a dissociated reproductive pattern. Males mate with females in spring, when testes are small and circulating concentrations of androgens are relatively low. Spermatozoa produced during the summer, when testes enlarge and androgen levels are high, are stored in the epididymis until the following spring. *Adapted from Crews and Garstka (1982) with permission.*

> **BOX 21.3. PROCEDURE FOR ESTIMATING DENSITY OF SPERMATOZOA AND OTHER TYPES OF CELLS WITH A HEMOCYTOMETER**
>
> A hemocytometer is a thick glass microscope slide with 2 identical counting chambers of known volume. It was originally developed to provide standard conditions for human blood cell counts, but it has found application in a wide variety of cytological work, including estimation of sperm density for assessment of male fertility. The highly polished floor of each counting chamber is gridded with fine lines spaced at exact intervals as discussed below.
>
> Ridges on the sides of the hemocytometer hold the cover slip exactly 0.1 mm above the floor of the counting chamber. Consequently, cells counted in a given section of the chamber are related to volume, and cell density can be calculated. Counts can be made on fresh, frozen, or fixed material, but fresh material provides the option of rating spermatozoa for motility. In reality, a hemocytometer provides a convenient system of grids that can be readily applied to a wide variety of microscopic sampling problems.
>
> **Procedure**
>
> 1. Spermatozoa may be obtained from a homogenized testis, but if mammalian sperm are to be evaluated for motility, they must be obtained from ejaculated semen or flushed from the tail of the epididymis. Semen or other material should be diluted with an appropriate culture medium, such as Hank's solution rather than saline, particularly if motility is to be assessed. Medium and the hemocytometer should be held on a slide warmer at normal scrotal temperature (e.g., 35° C).
> 2. The source material is diluted with a known volume of medium, and a drop is added to both chambers of the hemocytometer.
> 3. After spermatozoa have settled for 5 minutes, count the number of spermatozoa in sections A, B, C, D, and E of the center grid and multiply the total by 5 to obtain an estimate of the total number in the center grid (1 mm on a side or 0.01 cm²). Alternatively, select one of the 4 corner grids (comprised of 16 squares) at random and count all spermatozoa with heads inside the perimeter lines. If sperm density is high (>40 cells), count within a randomly chosen row of 4 squares and multiply the total by 4 to obtain an estimate of the total in all 16 squares (0.01cm²).
> 4. This procedure should be repeated in a grid in the other counting chamber to permit calculation of mean and standard deviation and an estimate of counting precision.
> 5. The cover slip is held 0.1 mm above the floor of the counting chamber, which, in this example, creates a chamber volume of 1×10^{-4} cm³ or 0.1 µL.
> 6. The concentration or density of sperm in the source material is calculated with the appropriate dilution factors. For example, a 150-mg epididymis is minced and rinsed with 600 µL of medium, and a sample of the resulting cell suspension is placed in a hemocytometer. If 500 spermatozoa are counted in a 0.01 cm² section (0.1 µL) of the hemocytometer; sperm density or concentration (C; i.e., number of spermatozoa per mg of epididymis) would be calculated as: $C = N \times D/0.1 \div S$, where N is number of spermatozoa counted in 0.1 µL of the hemocytometer, D is volume of medium used to rinse or dilute the sample, and S is volume or weight of the original sample. In this case, $C = 500 \times 600/0.1 \div 150 = 20{,}000$ spermatozoa per mg of epididymis.

quality of sperm in human fertility clinics, livestock management, and zoo studies of endangered species (e.g., cheetahs [*Acinonxy jubatus*]; Wildt et al. 1993). However, these measures are made on sperm in semen, which is usually obtained by **electroejaculation,** an approach that is seldom taken in studies of wild animals.

Testicular Biopsy

Counts of spermatozoa are usually made from tissues or fluid collected at necropsy or in ejaculates. However, **needle biopsy** procedures offer an opportunity to obtain testicular tissue and spermatogenic cells from living animals, as reviewd by Berndtson (1977). The animal is anesthetized, and the surface of the scrotum is thoroughly cleansed and disinfected. A testis is punctured with a 19–22-gauge needle, carefully avoiding the epididymis (Sundqvist et al. 1986). Slight pressure applied to the testis or negative pressure in an attached syringe will ensure aspiration of tissue into the needle. The needle is then withdrawn from the testis, and the material contained in the needle is spread onto a microscope slide, dried, and stained prior to examination of the cells. The biopsy smears are scored: low (1–2), when only sertoli cells or **spermatogonia** are present, to high (9–10), when large numbers of mature **spermatids** are counted. This procedure, used in diagnosing human infertility, has been used with ranch mink, in which up to 20% of the males are infertile, as indicated by biopsy scores 7 (Sundqvist et al. 1986). The safety of this procedure has been assessed in mice: spermatogenesis, fertility, and paternity of mature mice were unaffected by their testicular biopsy at 4 weeks of age (Nakane et al. 2005)

ENDOCRINOLGY OF REPRODUCTION AND STRESS

Background

The nervous and endocrine systems act in concert to coordinate physiological processes and behavior. Responses regulated by the endocrine system are generally slower and of longer duration (hours and days) than those provided by the nervous system, which acts on the scale of seconds and minutes. The complex of interrelated glands that compose the endocrine system responds to both internal signals and to external environmental cues by secretion of **hormones** that are carried in the blood stream, usually to other parts of the body, where they have their effect. Environmental stimuli (e.g., photoperiod or social interaction) are relayed through a basal part of the brain, the hypothalamus, which responds by secretion of releasing factors, including **gonadotropin releasing hormone** (GnRH), that control the secretory activity of the **anterior pituitary gland** (Norris 2007). The anterior pituitary gland secretes several peptide hormones, including **follicle stimulating hormone** (FSH) and **LH,** that control gametogenesis and hormone secretion by the gonads The pituitary also secretes **adrenocorticotropic hormone** (ACTH), which stimulates secretion of **corticosteroids** (e.g., cortisol and corticosterone) by the adrenal cortex. Control of circulating levels of hormones is achieved by feedback regulation. For example, when increasing blood levels of testosterone in a male exceed a threshold or set point in the **hypothalamus,** the release of GnRH is inhibited. Thus, less GnRH reaches the anterior pituitary, less LH is released and reaches the testes, and consequently, secretion of testosterone is reduced to appropriate levels.

Hormonal control of reproduction ultimately functions to coordinate development of the reproductive tract, expression of mating behavior, and gametogenesis, so that sperm and eggs are brought together to accomplish fertilization and conception. Each stage in the reproductive cycle is controlled by a sequence of endocrine signals that can be visualized with **hormone profiles** (i.e., plots of concentrations in blood [serum or plasma] over time). Although many hormones are involved, much has been learned about the reproductive processes in wildlife through study of relatively few pituitary hormones and gonadal steroids (Table 21.1). The extent to which endocrine responses to chronic stress (i.e., corticosteroid secretion) affect reproductive processes is a longstanding question in wildlife population ecology. Thus, endocrine measures of stress also are considered in this section.

This introduction provides only a brief outline of the principles and terminology of reproductive endocrinology. Appreciation of information presented in this section will be enhanced with a basic understanding of the anatomic–functional axis consisting of the hypothalamus, pituitary, and gonads (**hypothalamic–pituitary–gonadal axis**) as described in most introductory biology textbooks. More extensive presentations of principles and related primary literature are available in a number of reviews and textbooks, including Austin and Short (1984), Tyndale-Biscoe and Renfree (1987), Bronson (1989), Nelson (2000), Foster and Jackson 2006, and Norris (2007).

Hormones and Reproduction

Hormones and Reproductive Cycles in Females

Peak ovarian mass and numbers of yolked eggs in frogs generally coincide with peak levels of ovarian steroids at the time of amplexsus and oviposition, as in the Mexican leaf frog (Bagnara and Rastogi 1992). Similarly, in female reptiles with associated reproductive patterns, increased ovarian mass and elevated gonadal steroids also coincide with breeding behavior. One of the most thoroughly studied (particularly in behavioral endocrinology) lizards in this group is the green anole, as reviewed by Wade (2005). Unlike most lizards, which ovulate about half of each clutch of eggs from each ovary, the green anole (and other species of *Anolis*) ovulates a single egg, alternately from each ovary, every 14 days. Females are receptive to males during the 7 days preceding ovulation, when the estradiol:progesterone ratio is low (Jones et al. 1983; Table 21.1).

Less is known about factors controlling breeding behavior in dissociated breeders, such as the red-sided garter snake. Activation of courtship behavior in the male of this species does not depend on the presence of gonadal steroids, and for some time, there was little evidence that females were different (Whittier and Tokarz 1992). However, Mendonca and Crews (2001) demonstrated that low levels of circulating estradiol must be present during hibernation if females are to be receptive upon emergence in the spring. Following ovulation, the CL of both oviparous and viviparous lizards secretes progesterone (Norris 2007; Table 21.1).

Owens (1997) synthesized much of the sea turtle literature on gonadotropins and gonadal steroids in a graphic model for endocrine control of reproductive events and behavior in both males and females. Estrogen is highest during follicular growth, rising slightly with each oviposition. Testosterone also is high during migration and the nesting period. Both gonadotropin and progesterone surge during ovulation of each clutch (Table 21.1).

Table 21.1. Hormone data relevant to the assessment of reproductive activity in selected vertebrates. All values are for females, except as noted for males. Hormone concentrations are pg/mL serum or plasma for estrogens and ng/mL serum or plasma for all others, except steroid conjugates in urine (Kasman et al. 1986, Walker et al. 1988, Herrick et al. 2000), which are expressed in µg, ng, or pg per mg of creatinine or in µg/g or ng/g feces. Two concentrations indicate the approximate lows and highs associated with a given reproductive event.

Species	Hormone	Change in concentration	Reproductive event	Reference
Frog (M)	FSH	2–4	Peaks with spermatogenesis	Polzonetti-Magni et al. 1998
	Estradiol	0–8	Peaks with FSH	
	Androgen[a]	0–10	High in recrudescence and breeding	
Bullfrog (M)	LH		Seasonal, highest at amplexus	Licht et al. 1983
	Androgen	2–34	Seasonal elevation 2–4× >T	
	LH	4–32	Surges at the time of ovulation	Licht et al. 1983
Toad	Progesterone	0.1–0.3	Low, highest prior to mating	Itoh et al. 1990
	Estradiol	1,000–3,900	High, peak early in breeding	
Anole lizard (*Anolis carolinensis*)	Estradiol	500–2,100	Lower during sexual receptivity	Jones et al. 1983
Lizard	Progesterone	2–13	Rises from preovulatory to gravid state	Norris 2007
Tortoise (M)	Testosterone	5–25	Low during mating and elevated during spermatogenesis	Kuchling 1999
Sea turtles (from general model)	Estrogen	200	Highest prior to nesting	Owens 1997
	Testosterone	300	Highest during mating	
	Gonadotropin	4	Surge levels at ovulation	
	Progesterone	12	Highest levels at ovulation	
Turkey (*Meleagris gallopavo*)	Prolactin	90–709	Elevated during incubation, low in laying and brooding hens	Wentworth et al. 1983
White-crowned sparrow	Estradiol	35–400	Increases with courtship and egg laying	Wingfield and Farner 1980
Japanese quail (M)	Testosterone	0–5	Increases with day length and onset of breeding season	Follett and Maung 1978
Woodchuck (M) (*Marmota monax*)	Testosterone	0–3	Increased during breeding season	Baldwin et al. 1985
Woodchuck (F)	Progesterone	0–60	Progesterone higher in post-partum than in pregnant females	Concannon et al. 1983
White-tailed deer (F) (*Odocoileus virginianus*)	Progesterone	0–6	Rises with CL formation in early pregnancy	Plotka et al. 1977, Harder and Moorhead 1980
	Estrogens	11–295	Increases near parturition	Harder and Woolf 1976
White-tailed deer (M)	Testosterone	0–3	Increased with hardened antlers and rut	McMillin et al. 1974
Red howler monkey	Progesterone (urinary)	77–212	High in luteal phase of estrous cycle and highest during gestation	Herrick et al. 2000
Little brown bat (*Myotis lucifugus*)	Progesterone	7–136	Increased from early to late pregnancy	Buchanan and Younglai 1986
Indian rhino	Estrone sulfate	47–1	High at estrus and drops rapidly at ovulation	Kasman et al. 1986
Killer whale	Estrone conjugate	0–35	Increases prior to presumptive ovulation	Walker et al. 1988
	Pregnanediol-3-glucuronide	0–100	Elevated during pregnancy	
Tree kangaroo	Progestin[c]	200–900	Rises slowly after ovulation to high levels for 30 days of 59-day cycle	North and Harder 2008
White rhino	Progestin	40,000	Concentrations during gestation higher (100×) than in estrous cycle	Patton et al. 1999
Elk	Progestin	700–2,500	Varies among pregnant females, depending on habitat quality	Creel et al. 2007

[a] CL = corpus luteum; FSH = follicle stimulating hormone; LH = luteinizing hormone; M = male; T = testosterone.

[b] Testosterone and related hormones (e.g., dihydrotestosterone and androstenedione).

[c] Progesterone and its metabolites (e.g., pregnanolone) in feces, expressed in µg/g (elk) or ng/g feces.

The sequence of hormonal signals responsible for the circadian cycle of **ovulation and egg laying** in birds is complex. A major component in the domestic hen is a positive feedback relationship between an increase in progesterone secretion from the ovarian follicle and a surge of LH from the anterior pituitary, which stimulates ovulation (Johnson 2004). Prior to ovulation, elevated estrogen also induces courtship behavior (Table 21.1) and stimulates mobilization of yolk precursors from the liver and their deposition in the egg while it is in the follicle (Van Tienhoven 1983). **Incubation behavior** and **brood patch** formation are stimulated by high **prolactin** levels in association with estradiol and progesterone. Prolactin from the anterior pituitary also stimulates production of "milk" in crop glands of doves and pigeons.

Measurements of circulating hormones are useful in studies of the effects of toxic substances on reproductive performance of birds. Progesterone levels peaked earlier relative to oviposition in mourning doves treated with dietary polychlorinated biphenyls than in controls (Koval et al. 1987). Nonlaying canvasbacks (*Aythya valisineria*) had lower serum concentrations of prolactin and LH than did laying hens. Progesterone levels in laying hens increased during the breeding season, whereas those in the nonlaying birds declined (Bluhm et al. 1983).

The estrous cycle of mammals is a sequence of interrelated physiological events in the hypothalamus, anterior pituitary, ovary, and reproductive tract marked by a period of sexual receptivity (estrus) and ovulation. Preovulatory follicles secrete large amounts of estradiol that stimulates estrus and a preovulatory surge of LH. The LH surge stimulates ovulation and formation of the CL (Figs. 21.4, 21.6). This general pattern appears to hold true for many mammals, as evidenced from data from such widely divergent taxa as tammar wallabies (Harder et al. 1985), rats (Nequin et al. 1979), deer (Plotka et al. 1980), and sheep (Hauger et al. 1977). The rise in estrogen around estrus confirms normal ovarian activity and is potentially valuable as a predictor of time of ovulation in mammals that ovulate spontaneously. This knowledge is essential for artificial insemination and embryo transfer, techniques that are being used with increasing frequency by zoos, endangered species programs, and modern game farms. Jacobson et al. (1989) achieved conception in 75% of 53 trials of artificial insemination in white-tailed deer.

Changes in estrogen levels prior to estrus and ovulation are often small, of short duration, and difficult to detect. In contrast, events following ovulation, namely CL formation and the luteal phase of the estrous cycle or gestation, are of relatively long duration (several days to months in larger species) and are characterized by elevated levels of circulating progesterone (secreted by the CL and/or placenta; Fig. 21.10). Therefore, useful progesterone profiles can be obtained with relatively a low-frequency blood sampling design that is feasible in many wildlife studies. If breeding is highly synchronized, much can be learned from single samples taken from a series of animals killed during the breeding season. With white-tailed deer, this approach revealed a shortened, nonfertile cycle that preceded the first estrus and normal luteal cycle during onset of the breeding season (Harder and Moorhead 1980).

Progesterone profiles are considered the most reliable means of monitoring the 15-week estrous cycles of captive Asian elephants (*Elephus maximus*) and African elephants (*Loxodonta Africana*; Plotka et al. 1988). Until accurate and sensitive progesterone **radioimmunoassay** (**RIA**) was applied to a long series of serum samples from the same cows by Hess et al. (1983), the now well-documented 15-week estrous cycle of the Asian elephant was believed to be only 3 weeks long. More recently, LH and progesterone profiles in Afri-

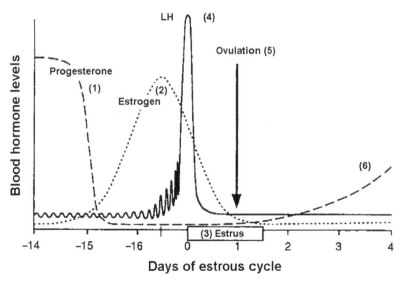

Fig. 21.10. Temporal relationships of estrus and ovulation to circulating levels of progesterone, estrogen, and luteinizing hormone (LH) during the estrous cycle of the ewe, which has preovulatory hormone dynamics similar to many mammalian species with spontaneous ovulation. *From Short (1972) with permission.*

can elephants (Kapustin et al. 1996) and Asian elephants (Brown et al. 2004) have revealed an anovulatory LH surge, followed low progesterone (no ovulation or CL formed) and, predictably, 3 weeks later by a second LH surge that stimulates ovulation. For decades, attempts to breed elephants using artificial insemination have failed, but with refined LH monitoring of the double LH surge and use of an endoscope-guided catheter and transrectal ultrasound to deliver semen into the anterior vagina, artificial insemination was achieved, followed by birth of an Asian elephant calf (Brown 2000, Brown et al. 2004).

The practical value of progesterone data for detecting pregnancy in live animals has been widely recognized, particularly for larger species, in which the gestation period extends well beyond the breeding season. Feces can be collected in the field, and blood can be obtained from trapped or anesthetized animals and assayed later, whereas other techniques (e.g., laparotomy or ultrasonography) require transport of the animal or equipment. Care must be used in selection of drugs and other methods of restraint, because they can affect progesterone levels. Plasma progesterone concentrations in pregnant white-tailed deer are generally >2 ng/mL, whereas those of nonpregnant deer are 1 ng/mL (Abler et al. 1976). The potential errors associated with this generalization (Plotka et al. 1983) notwithstanding, properly validated progesterone assays have permitted accurate (>2% error) diagnosis of pregnancy in white-tailed and mule deer (Wood et al. 1986). Gadsby et al. (1972) reported higher progesterone levels in domestic ewes carrying 2 fetuses than in ewes with a single fetus.

The uterus and placenta secrete numerous nutritive and regulatory proteins during gestation, some of them unique to gestation and therefore useful in identifying pregnant animals. Wood et al. (1986) used a qualitative test for pregnancy-specific protein B (bovine) to identify (4% error) pregnant mule and white-tailed deer. Similar results were reported for this pregnancy test in mountain goats (*Oreamnos americanus;* Houston et al. 1986) and elk (Noyes et al. 1997).

Spermatogenesis and Onset of Breeding in Males

The marked growth of the testis in vertebrates during onset of the breeding season is associated with rapid increases in secretion of FSH and LH from the anterior pituitary gland and testosterone and other androgens from the testes (Fig. 21.11, Table 21.1), but in bullfrogs (*Rana catesbeiana*), androgen levels show greater seasonal variation than does testis size. High androgen levels in spring stimulate breeding behavior in frogs, whereas a peak in FSH and estradiol levels is associated with onset of spermatogenesis (Polzonetti-Magni et al.1998; Table 21.1). In most species, reproductive tract activation (increased size, sperm production, and androgen secretion) is closely associated with breeding behavior. However, some reptiles, such as tortoises (*Gopherus* spp.; Kuchling 1999) and red-sided garter snakes, exhibit dissociated re-

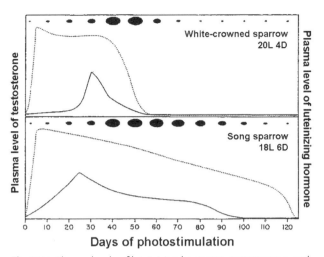

Fig. 21.11. Plasma levels of luteinizing hormone, testosterone, and testis size (solid ovals) in photostimulated male white-crowned sparrows (*Zonotricha leucophrys*) and song sparrows (*Melospiza melodia*). From Wingfield and Farner (1980) with permission.

production, in which testosterone is relatively low during the spring mating season and elevated during spermatogenesis in late summer and early autumn.

In temperate regions of the world, photoperiod is the principle environmental cue for reproductive activation. In tropical regions, where day length is nearly constant, other factors (e.g., onset of the rainy season and increased food availability) stimulate reproduction. Gonad size and singing rate increased in male antbirds (*Hylophylax naevioides*) following addition of crickets to their diet, and even the sight (only) of live crickets increased their singing rate (Hau et al. 2000). Modern techniques permit hormone measurements of small blood samples (50–500 μL), volumes that can be obtained from small wild birds. This ability has facilitated diverse experiments in avian field endocrinology. For example, circulating levels of testosterone in male song sparrows (*Melospiza melodia*) actively defending territories are elevated compared to those with uncontested territorial boundaries (Wingfield 1985). Similar social modulation of testosterone levels in males has been reported in other (e.g., antbird; Wikelski et al. 1999), but not all, avian species (Scriba and Goymann 2010).

In mammals, circulating levels of testosterone and other androgens also increase substantially during the breeding season (Table 21.1), a requisite condition for full spermatogenic activity and breeding behavior, for example, during the **rut** in white-tailed deer (McMillin et al.1974) or during **musth** (an aggressive behavioral state) in Asian elephants (Cooper et al.1990). In species as diverse as the hamster (*Mesocricetes auratus;* Richardson et al. 2004), tammar wallaby (Catling and Sutherland 1980), and human (Roney et al. 2007), elevation of testosterone and, at least in tammars (Paris et al. 2005), semen quality appear to be dependent on or influenced by direct association with females in breeding condition.

Hormone Metabolites in Urine and Feces

For many wild species, even pen-reared individuals, stress associated with restraint and venupuncture often precludes collection of blood samples, so that even minimal information about normal ovarian endocrinology is unavailable. Fortunately, another approach is available. **Steroids** (e.g., progesterone) are metabolized in the liver and excreted as conjugates (primarily sulfates and glucuruonides). Considerable progress has been made in monitoring gestation and estrous cycles of many zoo and wildlife species, including tree kangaroos (*Dendrolagus matschiei*), red howler monkeys (*Alouatta seniculus*), killer whales (*Orcinus orca*), Indian rhinos (*Rhinoceros unicornis*), white rhinos (*Ceratotherium simum*), and elk, through assay of hormone metabolites in their urine and feces (Table 21.1).

Midstream collection is preferred for **urine samples,** and many species, even killer whales, can be trained for this procedure (Walker et al. 1988). It also is possible to collect samples from some arboreal species that urinate from predictable tree-branch locations. This approach was used with individually marked free-ranging red howler monkeys to monitor urinary progesterone concentrations (Table 21.1) and to develop progesterone profiles that characterize temporal features of an apparent 28-day estrous cycle and gestation in this species (Herrick et al. 2000). Individualized feces or urine can be collected from the floors of pens and small enclosures or even from open range. Kirkpatrick et al. (1988) described detailed validation experiments and procedures for estimation of estrone sulfate concentrations in soil soaked with urine by free-roaming feral horses with unique marks. Twelve of 15 mares with estrone sulfate >1.0 µg/mg creatinine later produced foals, whereas no mares with lower concentrations foaled.

Collection of fecal samples from captive and wild animals is convenient, because the source animal can be observed from a distance and the sample collected at extended intervals after defecation. Consequently, hormone fecal analysis is currently the most widely used method for noninvasive (i.e., no restraint or venupuncture) monitoring of hormones in mammals (Safar-Hermann et al. 1987, Wasser et al. 2000, North and Harder 2008). The work of Sam Wasser and colleagues at the University of Washington, Seattle, with scat detection dogs represents some of the most imaginative and promising research to date in wildlife endocrinology. Dogs were trained to detect grizzly bear (*Ursus arctos Linnaeus*) and black bear (*Ursus americanus*) scats over a 5,200-km² study area in Alberta, Canada. DNA analysis of scats indicated sex and identities of individual of grizzly bears, and concentrations of progesterone and cortisol metabolites in fecal samples provided evidence of reproductive activity in females and an index of physiological stress, respectively (Wasser et al. 2004). More recently, Wasser et al. (2009) demonstrated the ability of dogs to identify scats of individual maned wolves (*Chrysocyon brachyurus*), thus reducing or eliminating the need for DNA genotyping of samples.

Control of Reproduction and Wildlife Contraception

The demand for nonlethal control of nuisance wildlife populations has driven research in **contraceptive technology** for >40 years, an effort that has benefited from developments in human medicine and animal science. Contraceptive trials have been conducted on a large number (>100) of mammalian species, mostly ungulates, but also carnivores and rodents. The primary focus of this research has been white-tailed deer and wild horses in North America and elephants in Africa, while Australians have developed a major research program in virally vectored **immunocontraception** aimed at controlling vertebrate pest species (Hardy et al. 2006). Avian species have received less attention in contraceptive research, the notable exception being resident Canada geese (*Branta canadesis*). Bynum et al. (2007) demonstrated a 36% reduction in hatchability of eggs at sites treated with 2,500-ppm nicarbazin bait, a drug that received regulatory approval from the U.S. Environmental Protection Agency (EPA) in 2009 and is commercially available as OvoControl™ G (Innolytics, Rancho Santa Fe, CA) contraceptive bait for Canada geese and OvoControl for pigeons (*Columba livia*; Fagerstone et al. 2010).

Many options are available for interrupting reproductive processes in wildlife, but most fall into 2 categories: hormonal and immunological intervention in females. Males are seldom the subjects of contraceptive programs, because in polygamous species, control of nearly all males in a population would be required to achieve a meaningful reduction in conception rates. Longstanding questions regarding the practicality of wildlife contraception prompted early research on oral application of contraceptive (diethylstilbestrol) to a large, enclosed, but free-ranging population of deer that resulted in a reduction in ovulation rate, but was not as effective as an intramuscular application (Harder and Peterle 1974). Moreover, the relatively low species specificity of bait systems available for oral application of hormones and their analogs entails unknown risks to nontarget species, a disadvantage that has discouraged research in this area.

Subcutaneous implants provide slow, prolonged release of hormone into circulation and thus, have the potential to provide long-term (months to years) contraception. Bell and Peterle (1975) reported early evidence of the effectiveness of this approach in a population of white-tailed deer, and more recently, promising results of long-term fertility control in gray kangaroos (*Macropus fuliginosus*) have been obtained with a related compound, levonorgestrel (Nave et al. 2002). Although remote delivery of contraceptive in a ballistic implant (Jacobsen et al.1995) might eliminate the need for restraining animals for installation of implants, other logistical problems remain. Also, hormones released from implants have the potential to accumulate in tissues of treated animals and present an unknown health risk to wild animals and humans that might consume them (Grandy and Rutberg 2002).

Prostaglandin $F_{2\alpha}$ ($PGF_{2\alpha}$), a hormone secreted by the uterus and other tissues, induces regression of the CL and is an important signal in hormonal control of parturition in mammals. The abortifacient effects of $PGF_{2\alpha}$ have been applied with success in terminating pregnancy in white-tailed deer (Waddell et al. 2001). One important advantage of using $PGF_{2\alpha}$ for wildlife is that it is rapidly metabolized in the lungs and thus, does not accumulate in tissues or present health risks for nontarget species or humans. The principle disadvantage of $PGF_{2\alpha}$ is that its effects are limited to a single reproductive season.

Immunocontraception has received much attention, because it is less physiologically intrusive than hormone treatments. Also, it does not present known heath risks to nontarget species, and animal welfare groups have supported research in this area (Grandy and Rutberg 2002). Immunological methods can involve antibodies to hormones (e.g., GnRH and gonadotropins), but most techniques are directed toward preventing fertilization by interfering with binding of spermatozoa to the zona pellucida, which surrounds the egg. Stimulating production of antibodies against proteins in the **porcine zona pellucida** (PZP) is, by far, the most popular approach. Effective immunocontraceptive technology is currently available for a number of species, including wild horses (Kirkpatrick et al. 1997), white-tailed deer (Naugle et al. 2002), and elephants (Fayrer-Hosken et al. 2000). Animals injected with vaccines, often with dye marking darts, become infertile for periods ranging from 1 to 3 years. Frank et al. (2005) reviewed the biological efficacy of PZP immunocontraception and the timing of booster inoculations in 24 species of ungulates and 5 species of non-ungulates across 10 years of treatment. The collective contraceptive efficacy for 301 individuals in 517 contraceptive intervals was 93.3%, for which no technical problems were identified. The major hurdle in management applications of PZP immunocontraception is one of achieving adequate population control in free-ranging populations, which may be assessed by computer modeling (Mackey et al. 2009). Results of simulation analysis of data from trials on a suburban white-tailed deer population suggest that immunocontraception is useful only on small localized populations with 100 breeding females (Rudolph et al. 2000).

The literature on wildlife contraception has expanded considerably in recent years; several reviews are available, including Kirkpatrick et al. (2002) and Fagerstone et al. (2010). Two **antifertility agents,** an avian reproductive inhibitor containing the active ingredient nicarbazin and an immunocontraceptive vaccine for white-tailed deer, have received approval from the EPA and are commercially available in the United States. OvoControl G for Canada Geese and Ovo Control for pigeons are approved and are available as oral baits. GonaCon™ immunocontraceptive vaccine (http://digitalcommons.unl.edu/usdaaphisfactsheets/49) was registered for injection in female white-tailed deer in September 2009. Several other compounds show promise and are being tested for use in wildlife in the United States, Europe, Australia, and New Zealand (Fagerstone et al. 2010). Fagerstone et al. (2002) concluded that wildlife managers of the future will face 2 challenges: (1) integrating contraceptive methods with traditional approaches to population control, and (2) providing the public with accurate information about the feasibility of fertility control compared to lethal methods for reducing wildlife populations

Endocrinology of Stress

Background and Effects of Stress on Reproduction

Decades of research stimulated by Hans Selye's landmark paper on stress and human health (Selye 1936) have provided both the technical foundation and a conceptual framework for understanding the biology of **stress,** which is defined as a significant disturbance of homeostasis caused by marked or unpredictable environmental change (Wingfield and Raminofsky 1999, Nelson 2000). A **stressor,** in this context, is an environmental change or stimulus, such as pain, fear, cold, blood loss, environmental contaminants, pathogenic microbes, and social tension (Selye 1976). Sudden life-threatening disturbances (e.g., a predator attack) stimulate release of adrenaline from the adrenal medulla and the well-known "fight or flight" response. Acute or short-term stress also activates the hypothalamic–pituitary–adrenal (HPA) axis, wherein release of corticotropin releasing factor (CRF) from the hypothalamus stimulates secretion of ACTH from the anterior pituitary. ACTH, in turn, stimulates secretion of adrenal cortex steroids (**corticosteroids**), including those that regulate glucose metabolism (i.e., glucocorticoids), primarily cortisol and corticosterone (Asterita 1985). This acute response is adaptive, because it allows individuals to maintain essential functions (e.g., by elevation of blood glucose) in the presence of the stressor. However, prolonged, chronic activation of the HPA axis may be associated with pathological conditions, such as gastrointestinal ulcers (Moberg 1985) and, in some cases, reduced reproductive performance.

Laboratory studies have clearly demonstrated the deleterious effect of stress on reproduction in a variety of vertebrates (reviewed by Pottinger 1999, Dobson et al. 2003). For example, social stress and subordination can lead to elevated corticosteroids (Sapolsky 1987) and alteration of reproductive function in some primates (Ziegler et al. 1995). Reduction of gonadotropin secretion has been associated with food restriction in some mammals (Bronson 1989) and overwintering stress in white-throated sparrows (*Zonotrichia albicollis;* Schwabl et al. 1988). Stress of transport and hypoglycemia reduced LH pulse frequency and delayed the LH surge in ewes (Dobson and Smith 2000). Rivier and Rivest (1991) present a model that explains the roles of CRF, ACTH, endorphin, and corticosteroids in modulating the effects of stress on reproductive function.

Christian (1950) and colleagues were quick to recognize the potential relevance of Selye's (1946) observations for population ecology. They postulated that elevated corticosteroid levels interfered with hormonal control of reproduction and induced mortality in high-density populations (Christian 1963, Christian and Davis 1964). Numerous studies have addressed this hypothesis, and it now appears that emigration and other factors, such as nutrition, prevent most natural populations from attaining densities sufficiently high to evoke a response from the HPA axis that would be regulatory at the population level. Moreover, the critical nature of reproduction for the individual has favored evolution of reproductive strategies in some species (e.g., highly seasonal breeders, in which the adrenocortical response to stress is modulated or suppressed; Wingfield and Sapolsky 2003). Experiments with high-density snowshoe hare (*Lepus americanus*; Windberg and Keith 1976, Vaughan and Keith 1981) and deer populations (Seal et al. 1983) support this conclusion. Also, fecal progesterone concentrations in elk declined with increased exposure to wolves (Table 21.1), and progesterone concentrations were a good predictor of calf recruitment in the following year (Creel et al. 2007). However, across populations and years, fecal glucocorticoid concentrations were not related to predator:prey ratios, fecal progesterone levels, or calf production (Creel et al. 2009).

Modulation in responsiveness of the individual to diverse stressors obviously necessitates activation as well as suppression of the HPA axis. Studies that have focused on segments of a population (e.g., sex or social status) have produced some of the clearest evidence for endocrine responses to social stress (McDonald et al. 1981, Sapolsky 1987), which suggests that more attention should be given to stress as a factor in the natural structure and functioning of populations. For example, mating season glucocorticoid levels in adult, but not subadult, Assamese macaques (*Macaca assamensis*) were negatively related with dominance rank and positively related to the amount of aggression received (Ostner et al. 2008). From a meta-analysis of studies involving 7 species of monkeys, Abbott et al. (2003) found that subordinates most often had elevated cortisol levels when they (1) were exposed to high rates of stressors and (2) experienced decreased opportunities for social (including close kin) support.

Indicators and Measures of Stress

A wide range of measures is available for assessment of stress in free-ranging and captive animals. Some are indicators of acute stress, whereas others indicate chronic stress acting over a period of weeks or months. In planning research, it is useful to recognize stress indicators in 1 of 3 categories: (1) **noninvasive**, those that require little or no handling of animals; (2) **invasive;** and (3) **post-mortem.** Noninvasive indicators include behavioral observations, food and water consumption, body weight, respiratory rate, and assay of hormones in saliva, urine, or feces.

Circulating levels of corticosteroids, such as cortisol or corticosterone, provide a direct measure of the endocrine response to acute stress, and with current assay procedures, it is possible to measure hormone concentrations in relatively small (0.05–0.3 mL) plasma samples. Consequently, such measures are increasingly used to assess stress responses in wild animals (e.g., birds during spring migration [Landys-Ciannelli et al. 2002] or care of hatchlings [O'Reilly and Wingfield 2001]) and to evaluate stress and welfare of laboratory animals (Broom and Johnson 1993). The major technical problem with such studies is that blood levels of corticosterone can rise rapidly within 2–3 minutes (Gartner et al. 1980) or less (Roy and Woolf 2001) following initial disturbance of the animal. Moreover, some species show clear diurnal rhythms and even marked hourly fluctuation in plasma corticosteriod levels (Tapp et al. 1984). Corticosteroid responses to stress should be first measured under controlled laboratory conditions (Carruthers and Path 1983) before conducting field trials. In this way, it is possible to assess the potential stress effect of acute environmental changes, such as changes in dove hunting regulations (Roy and Woolf 2001).

Corticosteroids and other hormones are secreted into the blood, continuously metabolized in the liver, and excreted in urine and feces. Because these metabolites accumulate during the hours between defecations, their concentrations in fecal samples represent an average of more variable corticosteroid concentrations in circulation. Fecal glucocorticoid concentrations show promise for assessment of stress in white-tailed deer (Millspaugh et al. 2002), and because the subjects are undisturbed, the confounding effects of stress due to handling and blood collection are avoided. However, procedures for extraction and assay of hormone metabolites in feces can be more complex and interpretation of results more difficult than for hormones in blood. These and other factors to be considered when measuring stress with fecal glucocorticoids are reviewed by von der Ohe and Servheen (2002). Nearly all endocrine and **immune** (e.g., lymphocyte mitogenesis) indicators of stress are considered invasive, because they require restraint of the animal for blood sampling. However, once installed, an indwelling intravenous catheter allows for essentially noninvasive collection of blood samples.

Providing that animals are processed immediately after death, post-mortem examinations or necropsies provide a wealth of information, particularly if glands and organs can be examined for abnormalities (e.g., gastric ulceration) and histopathology. Moreover, all endocrine and immune parameters available in blood samples are available at this time. Increased secretory activity of the adrenal cortex in response to stress is accompanied by an increase in size of the gland and/or the cortico-medullary ratio (Adams and Hane 1972, Han et al. 1998, Terio et al. 2004). Methods for analysis of adrenal gland weight and morphology were reviewed in detail by Harder and Kirkpatrick (1996), and Manser (1992) presented a comprehensive, yet concise and well-organized, guide to assessment of stress in animals. Stress is a complex biological concept, and anyone who hopes to contribute in this field should first read widely (e.g., the reviews of Broom and Johnson [1993], Sapolsky [1998], Balm [1999], and Nelson et al. [2002]); train, or at least consult, with established investigators; and then design their studies and experiments with the greatest of care.

Collection of Samples and Hormone Measurements
Blood Sampling Procedures

Hormone concentrations can be measured in many tissue types and bodily fluids with application of appropriate tissue homogenization and hormone extraction techniques, but blood is by far the most commonly assayed substance. The goal in any sampling procedure should be to collect an adequate volume of blood quickly and efficiently while minimizing stress to the animal. The effect of restraint or anesthetics on hormone levels should be investigated with each species to be studied, and not just in studies of progesterone or adrenal hormones. For example, hormone concentrations in a series of plasma samples collected from bullfrogs recaptured in the field remained relatively constant, but if the frogs were held in a collecting sack, levels of gonadotropin and gonadal steroids began to decline within 2–4 hours, often to basal levels within 20 hours (Licht et al. 1983). Immobilization with succinylcholine chloride elevated circulating progesterone levels in white-tailed deer (Wesson et al. 1979), and certain anesthetics depressed serum concentrations levels of this steroid (Plotka et al. 1983). Apparently, stress from prolonged (15–45 min) restraint can induce significant release of corticosteroids, including progesterone (Plotka et al. 1983).

Blood samples can be collected from quiet animals fitted with an indwelling catheter to investigate the effects of restraint or anesthesia on blood hormone levels. Periodic samples will reveal the natural daily pattern of circulating levels, including episodic changes, which can be distinguished from those that might be related to stress of handling or anesthesia in the same animal. An alternative approach is to collect blood samples immediately upon restraint or sedation of the animal, followed by collection of blood samples at intervals over a period of time in excess of the maximum required for routine collection. If the blood sampling procedure alters secretion of the hormone(s) under study, blood concentrations will change in successive samples relative to that in the initial sample. These experiments will identify the need to standardize timing of blood collection (from time of restraint or anesthesia) in situations where stress to the animal being sampled cannot be avoided. Also, many captive animals, if handled in a consistent manner with gentle or minimal restraint, become conditioned to blood sampling and appear to experience little stress during such procedures. Elephants, for example, can be trained to stand, unrestrained, for blood sampling (Brown et al. 1999).

Peripheral blood (i.e., from the heart or any vein not directly draining the endocrine gland under study) is usually collected from the jugular or a prominent leg or tail vein. The orbital plexus in the corner of the eye may be used to obtain small volumes from mice. The brachial (wing) vein is commonly used in birds, although Arora (1979) concluded the jugular vein was the best source of blood for Japanese quail (*Coturnix japonica*). Blood collection from living animals should be performed only by trained personnel. This is particularly important when blood is to be obtained by cardiac puncture, in which case, the animal must be anesthetized or sedated to alleviate pain and prevent movement of the animal while the needle is in the heart.

If **plasma** is to be separated from blood, the needle and syringe are rinsed with a sterile heparin solution or other anticoagulant. Syringes or tubes containing blood should be placed in crushed ice to cool before the plasma is separated by centrifugation. Alternatively, if **serum** instead of plasma is desired, blood is allowed to clot, and the serum can then be poured carefully from the tube or removed by pipet. Any cellular material remaining with the serum is separated by centrifugation.

Hormone concentrations can be measured in either plasma or serum, but plasma may be preferred, because blood can be chilled and centrifuged immediately after collection. This is important if the hormone under study is temperature sensitive or subject to degradation when it is in contact with blood cells before separation of serum or plasma, as is progesterone in cattle (Vahdat et al. 1984) and muskox (*Ovibos moschatus*; Rowell and Flood 1987). Blood destined for hormone assays, particularly gonadotropins, should be chilled immediately after collection and centrifuged as quickly as possible, with a uniform time interval between collection and centrifugation for all samples (Wiseman et al. 1982). Blood and other biological specimens should be protected from direct sunlight to avoid possible degradation of compounds under study. Plasma and serum also should be stored frozen at −15° C or lower and assayed as soon as practical to avoid degradation of hormones. However, steroid hormone concentrations do not appear to change in plasma and serum stored frozen over periods of 3–8 years.

Timing of collection of blood and other tissue samples for studies of reproductive physiology is important (Figs. 21.9–21.11). In most field studies, the exact stage of egg laying or estrous cycle is not known, and blood samples must be grouped to represent broad categories of reproductive activity (e.g., courtship, incubation, pregnancy, or lactation). The problem of temporal specificity in blood sampling is particularly complex relative to the ovulatory cycle of birds, because many changes occur within a 24–30-hour period.

Collection of Urine and Fecal Samples

Urine and fecal samples can be collected with less effort, if not less time, than blood samples and with little or no disturbance of the animals under study. Moreover, hormone metabolites accumulate in the urine and intestinal contents over time, and thus, their concentrations in urine and feces reflect circulating concentrations over a longer time interval than individual blood samples. The major challenge is identifying the source animal for a given sample, and with urine, this usually involves midstream collection, as described above (Walker et al. 1988, Herrick et al. 2000). Alternatively, when animals are under close observation, individualized urine can be collected from the floors of pens and small enclosures or even from open range (Kirkpatrick et al. 1988).

Feces are generally preferred over urine for hormone analysis. They persist longer in the environment than urine, and more options are available for collection of individualized samples. If captive animals are not individually penned or under prolonged observation, marking material (e.g., "glitter" or dye) can be fed to individuals to mark their feces. Collection of individualized fecal samples in field studies requires direct observation of animals or use of scat-detecting dogs and DNA analysis, as described by Wasser et al. (2004, 2009). Alternatively, feces can be collected (without regard to identity of source animals) over several study areas that differ in some variable of interest (e.g., food availability or predation pressure, as in the study of elk by Creel et al. [2007, 2009]).

Urine and fecal samples are stored in sealed containers at −15° C or lower and assayed as soon as practical to avoid degradation of the hormones. Steroid hormones are highly soluble and stable in ethanol. So, fecal samples may be stored in ethanol and should be, particularly if reliable freezer facilities are not available for storing samples.

Hormone Assays

Because **radioimmunoassay** and the closely related **enzyme-immunoassay (EIA)** are highly specific and sensitive, reliable measurement of hormone concentrations can be obtained from small plasma samples, typically 0.05–0.5 mL; in some cases, as little as 0.02 mL is adequate. Not surprisingly, **RIA** has not only revolutionized the study of reproductive physiology since its first wide application in the early 1970s, but it also has found ready application in field studies and has been a key factor in the development of the emerging field of wildlife endocrinology. In recent years, EIA has been widely applied to the measurement of hormone metabolites in feces (e.g., steroids excreted during the primate menstrual cycle and pregnancy [Wasser et al. 1988], LH in samples from elephants [Graham et al. 2002], and progesterone in feces from elk [Creel et al. 2007]).

RIA uses 2 key reagents: (1) an **antiserum** that selectively binds the hormone under study and (2) a **radiolabeled** form of the hormone (e.g., ^{3}H-progesterone). The antiserum is diluted to where it will bind only about 50% of a fixed amount of labeled hormone, so that addition of unlabeled hormone (from standard solutions or a plasma sample) to the same tubes will displace some of the labeled hormone from the antibody in a dose-related manner. Unbound steroid is adsorbed on charcoal, and the radiolabeled hormone bound to the antibody is decanted and counted in a liquid scintillation spectrometer. Alternatively, the antibody is coated to assay tubes, so that unbound hormone is decanted and the radioactivity adhering to the walls of the tube is counted. EIA is similar in principle; the major difference is the replacement of the radiolabeled hormone with an enzyme-linked hormone, and measurement of light absorbance on a plate reader, instead of radioactivity, is the end point of the assay. Both techniques are highly sensitive and capable of measuring 10 pg/mL plasma (1 pg [picogram] = 1×10^{-12} g; 1 ng [nanogram] = 1×10^{-9} g).

Caution is in order regarding use of RIA or EIA in wildlife research. Underlying seemingly straightforward procedures are complex antigen–antibody interactions that can generate misleading data. Investigators contemplating use of these procedures should train in a laboratory specializing in hormone assays. Most importantly, each laboratory must establish and validate procedures for each hormone in each species under study. Results of validation experiments, designed to demonstrate accuracy, precision, and quality control (Abraham et al. 1977, Jeffcoate 1981) should be published. These procedures are required in manuscripts submitted to many endocrine journals (e.g., *Endocrinology*; http://endo.endojournals.org/misc/itoa.shtml#statistical). Reagents and procedures in commercial RIA or EIA kits most often have been established only for human, rat, or monkey plasma and should not be used for other species unless validated. Improper application of such kits to plasma or serum from other species can lead to highly erroneous and misleading results. Common problems include variable hormone extraction efficiency and nonspecific interference of hormone–antibody binding. For example, high lipid content of blood of laying hens can interfere with steroid extraction as well as the binding and charcoal separation phases of RIA. Gonadotropic hormones, such as **LH**, exhibit species-specific molecular structures that complicate assay validation based on antibodies raised to the gonadotropin of another species. For example, antiserum raised against LH from sheep has been used to measure LH in a wide range of mammalian species. These heterologous assays require rigorous validation and are usually less sensitive than when used to measure LH in sheep.

SUMMARY

The methodologies and techniques used in the study of reproduction in birds and mammals continue to expand more rapidly than for amphibians and reptiles, while the need for research on herptiles is increasingly obvious, particularly as it relates to the worldwide decline in amphibian populations. These taxa exhibit fascinating reproductive patterns that offer opportunities for both basic and applied research. Reproductive processes in many amphibians are particularly sensitive to environmental change, and there is a critical need to understand the mechanisms by which pathogens, endocrine disruptors, and other environmental factors interact to affect recruitment and loss in populations. Methods developed for study of reproduction in birds and mammals (e.g., nest success, CL counts, and ultrasonography) also can be applied to some herptiles, but new methodologies also are needed, particularly for sampling designs and tissue collection techniques relevant to their diverse reproductive patterns.

Since publication of *The Wildlife Techniques Manual* in 1994, significant progress has been made in the application of endocrinology and other reproductive technologies to wildlife conservation, most notably in 3 areas: (1) noninvasive monitoring of reproductive cycles and stress with RIA and EIA of hormone concentrations in feces, (2) assisted reproductive technology, and (3) wildlife contraception. Hormone concentrations are most often monitored by blood sampling in biomedical studies. Indeed, our modern understanding of reproductive physiology is based on serial blood sampling. However, hormone concentrations in saliva, urine, and feces reflect circulating levels, and they can be sampled with little or no disturbance to the animals under study. Of these, feces are most often the material of choice, primarily because they persist over time in the pen or field and can often be individually identified. Moreover, with use of scat detecting dogs and DNA genotyping of samples, the opportunities for research and management in this area seem boundless. Although reproductive technologies, such as artificial insemination and embryo transfer, have limited application to field-based wildlife conservation, they have proven to be invaluable in captive propagation of endangered species, notably, for example, with the first birth, in 2001, of an Asian elephant calf conceived by artificial insemination. Several decades of research in wildlife contraception recently came to fruition with the approval by the EPA of 3 commercial antifertility agents, available in bait for Canada geese and pigeons and in an immunocontraceptive vaccine for white-tailed deer. Wildlife biologists now have at their disposal reproductive techniques and methods for dealing with a wide variety of conservation issues, ranging from the critical need for a better understanding of causes underlying the global decline in amphibians to reproductive suppression of overabundant wildlife and assisted reproduction of critically endangered species held in captivity.

22

Conservation Genetics and Molecular Ecology in Wildlife Management

SARA J. OYLER-MCCANCE
AND PAUL L. LEBERG

INTRODUCTION

PRIOR TO 1980, **genetic techniques** were not typically used in wildlife biology. With recent technological advances, straightforward and rather inexpensive genetic techniques have emerged that can be directly applied to wildlife studies. In this chapter, we discuss molecular genetic techniques and how they can be applied in wildlife biology. This material is intended for wildlife biologists and managers. Geneticists and those interested in detailed descriptions of each technique are referred to Avise (1994) and Hillis et al. (1996). Here, we present a compilation of ideas, techniques, and applications of use to wildlife students and professionals seeking to use molecular genetic techniques.

MOLECULAR GENETIC TECHNIQUES

Nuclear versus Mitochondrial Genomes

All genetic techniques and **molecular markers** described in this chapter examine portions of **DNA** at some scale. Two different genomes are used in genetic studies of animals. The **nuclear genome** is biparentally inherited and is found in the cell nucleus. It is large and not well mapped in most species. The **mitochondrial genome** is housed in the mitochondrion, an organelle involved in cellular metabolism. It is small compared to the nuclear genome and is a circular, maternally inherited molecule that has been well mapped in many species. Nuclear DNA on average evolves slowly, although some portions (e.g., microsatellites) evolve quickly. Mitochondrial DNA (mtDNA) on average evolves more quickly than the nuclear genome and some areas (e.g., control region) evolve very rapidly. These features make mtDNA and some regions of nuclear DNA suitable targets for certain genetic studies (Avise 1994).

Investigating Genetic Variation

Some molecular techniques consider **gene products** (e.g., proteins), and some examine DNA variation at the **nucleotide level** (e.g., DNA sequencing or fragment analysis). In the past, analysis of certain proteins has been easy and economical; however, quantifying variation at the nucleotide level has become a more powerful molecular tool for population genetics and systematics. Some techniques look for differences in actual nucleotide sequence, whereas others infer relatedness based on analysis of fragments and restriction sites.

The advent of the **polymerase chain reaction (PCR)** has revolutionized molecular biology. Essentially, PCR is a reaction in which a region of DNA is targeted and amplified exponentially (Avise 1994, Palumbi 1996). This reaction requires development of unique primers that flank both sides of the targeted region of DNA. Once amplified to large quantities, the targeted region (usually between 100 and 2,000 base pairs) is available for study with a wide variety of molecular techniques. We briefly review several techniques that have been and are currently used in wildlife studies (Table 22.1). More detailed and excellent reviews of these and additional genetic markers available for studying genetic diversity in wildlife populations have been presented elsewhere (Avise 1994, Smith and Wayne 1996, Haig 1998, DeYoung and Honeycutt 2005).

Analysis of Gene Products

Protein electrophoresis is a technique that can be used to examine population subdivision or structure. Proteins are a series of amino acids joined by peptide bonds. Each amino acid has a distinctive side chain, some of which are either positively or negatively charged. Thus, when an electric current is applied, these proteins migrate differentially through a matrix based on their charge, size, and shape. Proteins can then be visualized through histochemical staining or other methods (Murphy et al. 1996). **Mutations** cause changes in the DNA sequences of amino acids forming proteins that, in turn, cause changes in the shape, net charge, and migration rate of proteins. Such changes can be revealed through electrophoresis and provide information showing variability among individuals, populations, or species. Although inexpensive, this technique can examine only a small proportion of the variation present in the DNA that codes for the proteins; differences in proteins are not necessary detected. The subset of proteins typically studied with this approach is called **allozymes**. These proteins, however, may be under selective pressure and may not represent the diversity and divergence present in other genes. Further, the tissue required for this type of analysis typically requires highly invasive or destructive sampling and is logistically difficult to manage in field situations.

Fragment Analysis

Fragment analysis comprises various genetic techniques that explore nucleotide variation indirectly by comparing the size of DNA fragments electrophoretically. Although fragment analysis offers less resolution than does direct DNA sequencing, it is cost effective when examining many individuals and many different loci. Among fragment analysis techniques, some cut DNA in certain areas (e.g., **restriction fragment length polymorphisms [RFLPs]** and **minisatellite fingerprinting**), whereas others amplify many different loci (**amplified fragment length polymorphisms [AFLPs]** and **microsatellites**). With the exception of microsatellites, these techniques produce multiple fragments (bands) per individual (Fig. 22.1). In these cases, individuals are compared by the extent of band sharing among individuals. These markers, with the exception of microsatellites, are considered **dominant**, which refers to the fact they document presence or absence of an allele. **Codominant markers** are those that reveal both alleles at a given locus (i.e., heterozygotes can be distinguished from homozygotes). Thus, they provide much more information and allow for the documentation of heterozygosity and tests of **Hardy–Weinberg Equilibrium** and Mendelian inheritance.

For RFLP analysis, the template DNA is typically a small portion of the nuclear or mitochondrial genome that has been amplified using PCR. RFLPs characterize genetic variation using **restriction endonucleases**, which are enzymes that cut at specific locations in DNA sequences. Restriction enzymes cut at a specific recognition sequence, usually 4–6 base pairs long. The enzyme EcoRI, for example, cuts between G and A when it comes across the sequence GAATTC. Thus, every string of GAATTC in the PCR product will be cut in the same location and will produce many fragments of different sizes. Mutations that cause changes in the cleavage site (e.g., GATTC changed to GATAC) prevent the enzyme from cutting at that location, thereby producing a different series of fragments (different numbers or sizes of fragments). The series of fragments is then compared to examine the similarity of individuals or populations.

Whereas RFLPs look for variation in a single targeted segment of DNA, other fragment-based methods examine

Table 22.1. Applicability of common types of molecular markers for wildlife biologists. The number of Xs indicates the relative applicability (fair, good, very good, and excellent) of each technique to a specific question.

Type of marker[a]	Taxonomic delineations	Regional or subspecific population structure	Genetic diversity and subpopulation structure	Individual identity and paternity/maternity analysis
Allozymes	XXX	XXX	XXX	X
MtDNA sequences	XXXX	XXXX	XX	X
Microsatellites	X	XX	XXXX	XXXX
Minisatellites	X	X	XX	XXXX
AFLPs	X	X	XX	XXX
SNPs	XXX	XX	XX	X

Modified from Mace et al. (1996).

[a] AFLP = amplified fragment length polymorphism; mtDNA = mitochondrial DNA; RFLP = restriction fragment length polymorphism; SNP = single nucleotide polymorphism.

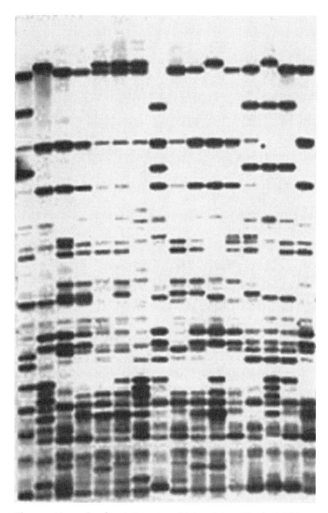

Fig. 22.1. Example of variation at multiple minisatellite loci. This illustration shows variation in and among families of pukeko (*Porphyrio porphyrio*) detected using markers pV47-2 and 3'HVR. *From Lambert et al. (1994).*

variation throughout the genome. **Minisatellites** refer to portions of DNA that have variable numbers of tandem repeats (sometimes called VNTRs); the length of each repeat unit is approximately 20 base pairs long. Typically, genomic DNA is digested into many fragments with restriction enzymes. These fragments are then separated by size using electrophoresis. The number of fragments produced by this process precludes visualization of individual bands, so radioactive or fluorescent probes specific for the minisatellite repeat are used to visualize and compare these sequences (Jeffreys et al. 1988). Because such "repeats" are commonly repeated in the genome, it is not unusual for this technique to produce dozens of bands. Although **DNA fingerprinting** with minisatellites has typically involved analysis with restriction enzymes and labeled probes, PCR-based approaches are becoming more common.

AFLP analysis is another multilocus technique that involves randomly primed loci and requires no a priori knowledge of the target genome (Hill et al. 1996). Analysis of AFLPs involves cutting the genomic DNA with restriction enzymes and ligating short "adapters" of known sequence to the fragment ends. PCR is then used to selectively amplify subsets of these fragments. AFLPs produce a series of hundreds of bands on a gel. Scoring is based on the presence or absence of a particular PCR product. AFLP analysis also is a dominant marker system, but it has the advantage of amplifying several hundred markers using only a few selective PCRs (Mueller and Wolfenbarger 1999, Meudt and Clarke 2007).

Microsatellite analysis, another PCR-based technique, differs from most other fragment analyses, because the attempt is to identify **diploid** (codominant) genotypes for specific loci. Like minisatellites, microsatellites are VNTRs; however, the repeated sequence is short (2–5 base pairs). Mutation rates of these regions are high, and the number of alleles (versions of a particular sequence) per locus in a population also is typically high. Allelic variation is usually in the form of length polymorphism, which can easily be detected on a high-resolution gel. Amplification results in either 1 (homozygote) or 2 (heterozygote) bands (or peaks) per individual (Fig. 22.2). Microsatellite primers are specific to a single locus and are usually specific to a particular species or group of closely related species. Because of this primer specificity, the development of primers for a particular species can be expensive. The advantages of microsatellite analysis include codominance and high levels of polymorphism. Typically, data from several microsatellite loci are used in a particular study.

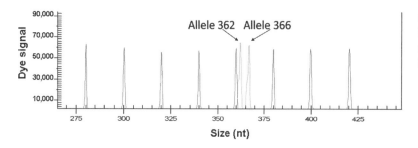

Fig. 22.2. An example of microsatellite data. This locus is heterozygous in this individual, with one allele of 362 base pairs and one allele of 366 base pairs.

DNA Sequence Analysis

Direct **DNA sequencing** (nuclear or mitochondrial) is one of the most widely used techniques today, because it is highly informative and, recently, has become much easier and less expensive to perform. It also is appealing because evolutionary processes can be modeled and integrated into analyses. Further, because the genome is so vast, the amount of information gleaned from sequencing may be quite large. DNA sequencing involves amplifying a target region and then creating a series of labeled (either radioactively or fluorescently) DNA fragments that correspond to each nucleotide (Hillis et al. 1996). The DNA fragments are then separated using electrophoresis and visualized. Recent technological advances have automated the sequencing process using fluorescently labeled DNA fragments (reviewed by Hillis et al. 1996) that are read by a laser and interpreted by computer software.

Single Nucleotide Polymorphisms

Single nucleotide polymorphisms (SNPs) are an emerging class of genetic markers that show great potential for wildlife applications. A SNP is a specific site in a DNA sequence in which a single nucleotide varies, resulting in different alleles (Primmer et al. 2002). Because DNA is comprised of **nucleotides** (A, C, T, or G) strung together to form DNA sequences, each SNP site consists of 4 possible nucleotide variants. Additional variation at a SNP site might include a deletion or insertion of a nucleotide. Once the allele at the SNP site is identified in an individual, it is compared to alleles from other individuals, thereby allowing for the examination of levels of genetic variation or similarity among individuals.

The occurrence of SNPs in the genome is quite common, as they have been documented to occur every 100–300 base pairs in humans (Brown 1999). They occur throughout the genome in both coding and noncoding regions of DNA and their **mode of inheritance** is thought to be well understood, following simple mutation models (Morin et al. 2004). Limitations for this type of marker include difficulty identifying alleles in individuals heterozygous at a particular SNP site and the tediousness and expense of identifying and characterizing SNP sites in non-model organisms (Brumfield et al. 2003, DeYoung and Honeycutt 2005).

Genetic Sampling

For genetic data to be used in a wildlife study, material must be collected from animals in the field. The type of material sampled, sample size, and sampling regime vary according to the questions being asked and the molecular markers being used (reviewed by Baverstock and Moritz 1996). DNA can be extracted from a variety of different tissues, including muscle, heart, liver, blood, skin, hair, feathers, saliva, feces, urine, scales, bone, fins, eggshell membranes, and, potentially, cervid antlers. **DNA extraction techniques** for most tissues are well established and involve the isolation of DNA from proteins and lipids using a digestion with the enzyme proteinase K followed by extraction with organic solvents (Sambrook et al. 1989). Modifications to traditional extraction methods, for example, are needed when using hair or feathers when the DNA is encased in the hardened tissue of the shaft and root (reviewed by Morin and Woodruff 1996). DNA has been successfully extracted and used from museum specimens (Mundy et al. 1997), although these techniques can be highly labor intensive and expensive. When considering what type of tissue to sample, several different factors must be addressed. It must first be decided what quantity and quality of DNA is needed to answer the question of interest. Second, the necessity, feasibility, and logistics of trapping and sampling animals must be examined. Finally, field preservation and sample storage issues should be addressed prior to the beginning of a study.

Some molecular techniques require a reasonable quantity of high-quality DNA (e.g., sequencing large fragments of mtDNA or DNA–DNA hybridization) whereas others (most PCR-based techniques) are much more forgiving. Samples of feathers, hair, feces, and urine may contain small amounts of DNA that may be of low quality (sheared into many fragments), whereas blood, skin, and muscle tissue often yield DNA of high quantity and quality.

The logistics of trapping and sampling wildlife vary greatly, depending on the species of interest. Some species are relatively easy to trap and sample, whereas others are difficult and/or dangerous. **Destructive sampling** refers to instances where the organism is killed during the process of sampling, such as for collection of muscle, heart, liver, or embryo tissue. If an animal is killed (hunting) or found dead (road kill or disease), samples can easily be taken for genetic analysis. **Nondestructive sampling** occurs when a genetic sample can be obtained without sacrificing the animal. Feathers, blood, shell membranes from hatched eggs, skin, hair, feces, and urine can all be collected nondestructively and provide potential sources of DNA for genetic analysis. Genetic samples also can be gathered without having to handle the animal in question (e.g., feathers, hair, feces, and urine); see the section on noninvasive sampling for more details.

In most cases, genetic samples can be stored on ice, refrigerated, dessicated, collected into a preservative buffer, or frozen almost immediately after collection (Table 22.2). When fieldwork occurs in remote areas, sampling certain tissues (e.g., skin or feathers) may be more feasible than such tissues as blood. When working with blood, only a small amount is needed (5 drops) and should be mixed with a preservative, such as ethylenediaminetetraacetic acid (EDTA), or with a blood buffer storage solution, such as Longmire buffer, or stored dry on filter paper. Muscle tissue should be

Table 22.2. Sources of DNA and how samples should be collected

Tissue type	Amount	Quantity	Quality	Preservation method
Blood	5–10 drops	High	Good	EDTA[a] coated tubes
				Lysis buffer (Longmire)
				Filter paper
Muscle	Square 2 cm on a side	High	Good	Buffer
Feather	At least 1	Low	Good	Dry
Eggshell membranes	As much as possible	Depends	Good	Dry
Hair	At least 1	Low	Good	Dry
Scat	Variable	Low	Poor	Ethanol or dry
Teeth	Variable	Low	Depends	Dry
Bone	Variable	Low	Depends	Dry
Buccal swab	Variable	Low	Good	Lysis buffer (Longmire)

[a] EDTA = ethylenediaminetetraacetic acid.

either placed in a preservation buffer or frozen immediately. Contour or wing feathers provide the best source of DNA, but smaller downy feathers can suffice. Feathers from individual birds should be kept in separate bags. Eggshell membranes also can be a good source of DNA, as long as there is vascularization of the membrane. Each membrane should be stored dry in separate bags. Buccal swabs can be collected and stored at room temperature in buffer. Hair, bone, and teeth can be used as a DNA source if they are stored dry. For hair, only the follicle is needed. Scat also can be used, but the quantity and quality of DNA are often low. Scat should be preserved in either liquid ethanol or with silica beads. Detailed protocols for sample collections and descriptions of buffer are available at http://www.absc.usgs.gov/research/genetics/asc_usgs_samplingprotocols.pdf.

TAXONOMY

Species or Subspecies Identification

Taxonomists have been categorizing organisms into hierarchical groups ranging from kingdom and phylum levels to genus and species for hundreds of years. Past classifications have been defined using morphological and behavioral characteristics. Taxonomic delineations derived only from morphological characteristics can be erroneous (Avise 1989, Zink 2004), as they can either fail to recognize distant forms (Avise and Nelson 1989) or they can recognize forms that exhibit little evolutionary differentiation (Laerm et al. 1982). Classifications based on morphology and behavior have been acceptable in the past, yet the use of molecular genetic information can often help resolve discrepancies and refine taxonomic definitions. Although such neutral molecular markers provide important insight into historical and geographic patterns of variation in species, however, using them alone or elevating their significance relative to other forms of evidence, such as morphology or behavior, may, in some situations, mislead conservation efforts.

Although most taxonomic definitions are somewhat arbitrary (subspecies, genera, order), classification at the species level is perceived to be based on real, evolutionary units (Dobzhansky 1970). The debate as to how best to classify organisms into species has been ongoing for >150 years (Darwin 1859, Mayr 1942, Wiley 1978, Cracraft 1983, de Quieroz 1998, Wheeler and Meier 2000). New species concepts are added almost continuously (Hey 2001) to address perceived failures of prior ones, and the debate continues as biologists attempt to place discrete boundaries on a continuous process (Winker et al. 2007). Because the **species definition** is integral to the Endangered Species Act (ESA; U.S. Fish and Wildlife Service 1973) and protection and management of many species, we briefly mention the 2 most commonly used: the **biological species concept** (Dobzhansky 1937) and the **phylogenetic species concept** (Cracraft 1983). The major difference between these 2 species concepts is the biological species concept emphasizes **reproductive isolation,** and the resultant limitation and/or preclusion of **gene flow,** whereas the phylogenetic species concept defines species using the criterion of reciprocal monophyly and typically relies solely on genetic data.

Genetic data can be used to address the species question, regardless of which definition is used. Documenting an absence of gene flow among sympatric populations is one piece of evidence that can be used, along with morphological and behavioral data, to suggest delineation of a species. Constructing **phylogenetic relationships** among individuals to examine whether a monophyletic group exists also can be achieved by comparing DNA sequences.

Until recently, genetic information was difficult and expensive to acquire and, at times, could only be used to resolve differences between distantly related species. Protein electrophoresis (allozymes) became a useful genetic tool to distinguish differences between some species, but it is less useful when delineating the taxonomic relationship among closely related organisms (whether they are species, subspecies, etc.). The advent of PCR and automated sequencing has made it relatively straightforward to collect data at a high resolution in a cost-effective manner from a large number and variety of organisms. Further, sequence data from genes evolving at widely different rates can be gathered, which allows for taxonomic comparisons at immensely different levels (from kingdom/phylum/class to genus/species/subspecies). This ability allows for re-evaluation of taxonomic status using genetic information or for the addition of supplementary data to unresolved taxonomic questions.

There are several molecular techniques with which to assess taxonomic relationships (e.g., DNA–DNA hybridization or protein electrophoresis). Perhaps the most widely used and most applicable to questions in wildlife biology is analysis of the mtDNA sequence, although more and more studies augment such data with multiple loci from the nuclear genome. The **mitochondrial genome** is small (15,000–20,000 base pairs) and contains approximately 37 genes, although the order of these genes is not constant (Avise 1994). It is maternally inherited and does not recombine, as does nuclear DNA. Although comparisons of the gene order of mtDNA have been used in investigations of taxa, direct comparison of sequences has proved to be an effective technique in finer level taxonomic questions (among more closely related species; Avise 1994) that are much more common wildlife management concerns. Mitochondrial DNA is well mapped in many animals (Bibb et al. 1981, Anderson et al. 1982, Roe et al. 1985) and evolves 5–10 times faster than single-copy nuclear genes (Brown et al. 1979, 1982). It also contains a noncoding control region, in which some areas are even more variable (4–5 times more variable than mtDNA as a whole) that can be used to delineate closely related species and populations (Greenberg et al. 1983). Each mtDNA gene evolves at a different rate, allowing for different level comparisons using genes with different mutation rates. Additionally, many studies are moving toward using both mtDNA and nuclear DNA sequences to better resolve taxonomic issues (Barker et al. 2001, Barker 2004).

Once an appropriate **gene** is chosen for the taxonomic issue in question, DNA sequence from that region is obtained, and the relationship among individuals is inferred by comparing the DNA sequences. Metrics, such as the **percentage sequence divergence,** provide some measure of how similar or different the DNA sequences may be. **Genetic distances** or **phylogenetic relationships** (trees) are then estimated using either algorithms (e.g., unweighted pair group method) or optimality criterion (e.g., parsimony or maximum likelihood). These methods are well established and reviewed extensively by Miyamoto and Cracraft (1991) and Swofford et al. (1996). Nucleotide substitution patterns in the mitochondrial control region are quite elaborate, and models that estimate the rate of nucleotide substitutions have been developed (Tamura and Nei 1993, Tamura 1994). Modeling substitution rates circumvents violations of assumptions used by parsimony methods.

Using genetic data to address taxonomic questions becomes important for wildlife management primarily at the species and subspecies level. Wildlife managers are often charged with managing species and subspecies while these definitions are yet unresolved. Further, some subspecies (and even species) are difficult to distinguish in the field without extensive morphological measurements and comparisons with museum type specimens (e.g., Prebles meadow jumping mouse [*Zapus hudsonius preblei*]) or detailed analyses of behavior or song (Southwestern willow flycatcher [*Empidonax traillii extimus*]).

The **ESA** and other national and international environmental programs charge managers with protection of species, **subspecies,** and **distinct population segments** that are deemed threatened or endangered. At times, little is known about the taxonomic status of species or subspecies that are petitioned to be listed as threatened or endangered. This classification also is important for recovery of the species or subspecies, because funding priorities generally are based on taxonomic status (O'Brien and Mayr 1991). **Taxonomic delineations** are often based only on morphological characteristics and could be refined by adding behavioral and genetic characteristics.

The taxonomic status of many different species has recently been re-evaluated using genetic data. For example, sage-grouse (*Centrocercus* spp.) have recently been evaluated using behavioral, morphological, and genetic data, resulting in the recognition of a new species (Box 22.1). Other examples include the Kemp's Ridley sea turtle (*Lepidochelys kempii*), which has been recognized as a separate species qualifying for protection under ESA because of data from a mtDNA study (Bowen et al. 1991). The taxonomic status of right whales (*Eubalaena* spp.), which has historically been based on a single morphological character in the orbital region of the skull, has been redefined as the result of mtDNA data (Rosenbaum et al. 2000). Finally, 2 subspecies of blue grouse (*Dendragapus obscurus obscurus* and *D. o. fuliginosus*) have been elevated to full species as a result of analysis of mtDNA (Barrowclough et al. 2004, Banks et al. 2006).

Hybridization

Defining "hybrid" is as perplexing as is definition of the term species. Classically, **hybridization** and **introgression** are used to describe interbreeding between 2 distinct species. However, because a definitive definition of a species is still nonexistent, "hybridization" is sometimes relaxed to include interbreeding between 2 groups that are genetically different, whereas introgression refers to the movement of genes between 2 genetically differentiated groups (Avise 1994). Hybridization can be positive or negative (Haig 1998). In a positive sense, hybridization events can increase the overall **genetic diversity** of a taxonomic group, it can produce increased **fitness** (hybrid vigor) in some cases, and it can produce progeny that are more adaptable than either parent. However, in some instances, hybrids can have reduced viability and fertility. Further, the effects of **outbreeding depression** (decrease in fitness due to a loss of alleles that are locally adaptive) on a species due to a hybridization event can be quite negative. Because true hybrids are generally not protected by the ESA, hybridization provides interesting challenges for those charged with management and protection of species (O'Brien and Mayr 1991).

Molecular techniques provide an increasingly accurate estimation of taxonomic relationships and history of gene flow (Haig 1998). These techniques are being used to ad-

Box 22.1. Using genetics to help define taxonomic definitions for sage-grouse

Large-scale habitat loss and degradation have resulted in the decline of sage-grouse populations throughout their range (Braun 1998) and have caused an increased concern over their status. Historically, sage-grouse were classified into 2 subspecies: eastern (*Centrocercus urophasianus urophasianus*) and western sage-grouse (*C. u. phaios*), based on plumage and coloration differences in 8 individuals collected from Washington, Oregon, and California (Aldrich 1946). The western sage-grouse presumably occurred in southern British Columbia, central Washington, east-central Oregon, and northeastern California (Aldrich 1946).

Populations in other areas of the range were considered to be eastern sage-grouse. The validity of this taxonomic distinction has been questioned (Johnsgard 1983). Recently, sage-grouse from southwestern Colorado and southeastern Utah were found to be morphologically (Hupp and Braun 1991), behaviorally (Young et al. 1994), and genetically (Kahn et al. 1999, Oyler-McCance et al. 1999) different from sage-grouse throughout the rest of the range. This discovery led to description of a new species, the Gunnison sage-grouse (*C. minimus*; Young et al. 2000; see the figure).

With the validity of the 2 present subspecies in question, Benedict et al. (2003) sequenced a rapidly evolving portion of the control region of the mitochondrial DNA for 16 populations of sage-grouse on both sides of the subspecific boundary. The sequencing results provide no genetic support for the subspecies distinction. The authors suggest that further morphological and behavioral comparisons need to be conducted before overturning the subspecific classifications. This study did, however, identify a population of sage-grouse in the Lyon, Nevada, and Mono, California, areas that was genetically unique from all other sage-grouse populations sampled throughout the species' range (Benedict et al. 2003). This group of sage-grouse is currently being studied morphologically and behaviorally.

Comparison of greater sage-grouse (left) and Gunnison sage-grouse (right).

---

dress questions of hybridization, introgression, and taxonomic status. For example, large canids occupying the southeastern United States have long been classified as the red wolf (*Canis rufus*). Extinction of red wolves in the wild has led to serious conservation efforts to preserve and restore them into the wild. However, mtDNA data and microsatellite data both strongly suggest the red wolf is a hybrid between the gray wolf (*C. lupus*) and coyote (*C. latrans*; Wayne and Jenks 1991, Roy et al. 1994). The hybrid origin of the red wolf has led to debate over its eligibility for protection under the ESA.

Molecular techniques also can be used to identify the maternity and paternity of hybrids. Aldridge et al. (2001) described 2 sage-grouse × sharp-tailed grouse (*Centrocercus urophasianus* × *Tympanuchus phasianellus*) hybrids in Alberta, Canada. Using analysis of mtDNA control region sequence, they demonstrated the mother of each hybrid was a sage-grouse rather than a sharp-tailed grouse. Similarly, hybrids resulting from crosses in both directions of blue (*Balaenoptera musculus*) and fin whales (*B. physalus*) have been documented using both nuclear and mtDNA (Árnason et al. 1991, Spilliaert et al. 1991). The expansion of barred owl (*Strix varia*) into the range of the endangered northern spotted owl (*S. occidentalis caurina*) led to a study investigating the potential for hybridization between the 2 species. Haig et al. (2004) used AFLP data and mtDNA sequence data to confirm instances of hybridization, which has important legal consequences under ESA.

Evolutionary Significant Units and Management Units

Given that genetic analysis can help refine taxonomic relationships, how else can genetic data be used to address management issues? Recently, there has been debate about how to objectively prioritize conservation or management value below the species level. This discussion began with Ryder (1986:9), who defined the term **evolutionary significant unit**

(ESU) as "a subset of the inclusive entity species which possess genetic attributes significant for the present and future generations of the species in question." In an attempt to develop an operational definition more useful to managers, Waples (1991) defined ESUs using 2 criteria. A population or groups of populations had to demonstrate substantial reproductive isolation from other populations of the same species, and at nuclear loci, it had to show significant divergence of allele frequencies. Moritz (1994b:373) further defined ESU as "a population (or set of populations) that is reciprocally monophyletic for mtDNA alleles" and "shows significant divergence of allele frequencies at nuclear loci." Moritz (1994a) defined a second unit called a **Management Unit** as a group with less separation than an ESU, but deserving of specific management attention. This unit was defined to have significant divergence of nuclear or mtDNA allele frequencies, regardless of the phylogenetic differentiation of alleles. Although Moritz's ESUs protect distinct units, allowing for preservation of their long-term genetic variability, his management unit concept allows for shorter term conservation goals. Several other scientists have put forth alternate ideas on the concept of ESUs (Dizon et al. 1992, Avise 1994, Vogler and DeSalle 1994, Crandall et al. 2000, Fraser and Bernatchez 2001). Although definitions of an ESU are as highly debated and diverse (Fraser and Bernatchez 2001) as the species concepts, the ESU is useful if one is aware of the lack of agreement surrounding the best definition. Most genetic studies with applications to management use Moritz's (1994a, b) definitions, because they are well defined when using genetic data. Also, these definitions appear to be among the most well accepted and applied to date. These concepts have been applied to tiger quolls (*Dasyurus maculates*; Firestone et al. 1999; Box 22.2), spotted owls (*Strix occidentalis*; Haig et al. 2001), koalas (*Phascolarctos cinereus*; Houlden et al. 1999), and brown bears (*Ursus arctos*; Waits et al. 2000).

CONSERVATION OF GENETIC DIVERSITY

A focus of conservation genetics is preservation of genetic diversity in and among populations, especially in rare or endangered taxa. Genetic diversity can be estimated using molecular markers or morphological measurements. Although studies have examined the underlying genetic variation and heritability of specific morphological traits in both captive and free-ranging wildlife (Merilä 1997, Kruuk et al. 2000, Réale and Festa-Bianchet 2000), intensive investigations are difficult to implement for many species. Frankham et al. (2002) reviewed the use of **quantitative genetic approaches** in a conservation context to study the effects of multiple genes and environmental variation on such complex traits as morphology and behavior; Ellegren and Sheldon (2008) discuss new approaches useful for studying natural populations. Our primary focus in this chapter is on what molecular markers tell us about the demography and genetics of a population and how that information can be applied to issues in wildlife conservation.

Genetic diversity and genetic variation are often used interchangeably to refer to a dizzying array of population char-

> **BOX 22.2. TAXONOMIC REDEFINITION OF TIGOR QUOLLS IN AUSTRALIA**
>
> Firestone et al. (1999) defined evolutionary significant units (ESUs) and management units (MUs) for tiger quolls, which revised taxonomic classification and management plans for these carnivorous marsupials in Australia. Previously, 2 allopatric subspecies of tiger quoll had been recognized. The smaller subspecies, *Dasyurus maculatus gracilis*, occurs only in northern Australia in northeastern Queensland. The larger subspecies, *D. m. maculates*, occurs in southeastern Australia and Tasmania. Each subspecies has been placed on the International Union for Conservation of Nature list as either endangered or as vulnerable to extinction.
>
> Firestone et al. (1999) used both mitochondrial DNA (mtDNA) sequencing and nuclear microsatellites to survey the genetic relatedness of tiger quolls. Their mtDNA sequencing results show reciprocal monophyly and significant differences in nuclear microsatellite allele frequencies between Tasmanian and all mainland tiger quolls. These results suggest that, even though Tasmanian tiger quolls are recognized as the same subspecies as those in southeastern Australia, they are a separate ESU and that their taxonomic status should be revisited. The 2 subspecies on the mainland do not constitute different ESUs, even though they are considered separate subspecies.
>
> Firestone et al. (1999) suggested that morphological differences between the 2 subspecies may reflect adaptation to climatic differences. Differences in microsatellite allele frequencies and mtDNA haplotypes exist between the 2 subspecies on the mainland, suggesting that they should be considered as distinct MUs. Thus, assessment of genetic data (Firestone et al. 1999) revealed differences between the 2 subspecies at the MU level in Australia, and the classification in Tasmania should be reconsidered to recognize and preserve the unique genetic makeup and evolutionary path of tiger quolls.

acteristics. We use **genetic diversity** to refer to variation in frequencies of alleles at individual genes. It is difficult to quantify total genetic diversity in populations; most studies look at surrogates of this measure based on variation at molecular markers. Four processes are generally thought to influence patterns of genetic diversity: mutation, gene flow, drift, and selection.

Mutation

Normally, mutation does not have a major role in management issues. One exception is the case of exposure of populations to **environmental mutagens.** Animals exposed to radioactive or chemical mutagens might be expected to have more genetic diversity because of an increased number of genetic mutations. This hypothesis was tested using bank voles (*Clethrionomys glareolus*) and barn swallows (*Hirundo rustica*) from the vicinity of the Chernobyl nuclear accident site in Ukraine (Matson et al. 2000). Microsatellite analysis provided evidence of increased mutation rates in the swallows (Ellegren et al. 1997). In the voles, higher levels of mtDNA variation were found near Chernobyl than at reference sites; however, recent work indicates that it is difficult to attribute this increased genetic diversity to increased mutation (Meeks et al. 2007). Other investigations have failed to detect much evidence of increased mutation in wildlife from contaminated sites (Dahl et al. 2001, Stapleton et al. 2001, Berckmoes et al. 2005). Given that mutations are relatively rare events, it is not surprising that it is difficult to detect increased mutation rates in the face of other powerful genetic forces, such as gene flow and drift. Examinations of exposed populations for increased mutation rates will probably expand in coming years, as automated analyses have made it possible to screen large numbers of individuals and genes.

Gene Flow

When organisms disperse to new populations and reproduce, they contribute genetic material to their new populations. This process increases the genetic similarity of populations exchanging individuals. Reductions in gene flow allow populations to diverge through processes of **genetic drift,** the accumulation and spread of different mutations, and selection for local conditions.

Gene flow differs from dispersal as typically measured by studies of animal movement. Radiotelemetry or tagging studies can often provide insight into the proportion of individuals that depart from their natal areas, but they are inadequate for measuring the reproductive contribution of dispersing individuals to their new populations. Gene flow is typically measured through indirect methods using genetic markers (Slatkin 1985a). One of the most common approaches for estimating gene flow involves use of Wright's F_{ST} (1951). One common definition of F_{ST} is the proportion of the total variance in allele frequencies due to differences among populations. An attractive feature of this measure of genetic differentiation is that F_{ST} can be expressed as a function of the **number of migrants per generation (Nm).** Mills and Allendorf (1996) and Whitlock and McCauley (1999) discuss the assumptions necessary to use F_{ST} to estimate Nm and the difficulties of obtaining unbiased estimates of gene flow.

Many estimators of F_{ST} have been developed, and there is a large literature evaluating their merits and performance (reviewed by Neigel 1997, 2002). For other approaches to estimating gene flow, see Slatkin (1985b), Slatkin and Maddison (1990), and Neigel et al. (1991). Recently maximum likelihood and Bayesian approaches have been developed to estimate gene flow and other population parameters (Beerli and Felsenstein 2001; Wilson and Rannala 2003; Kuhner 2006, 2009). These methods can be quite powerful, but demand considerable computational resources.

One problem with most indirect estimates of gene flow is that effects of recent gene flow on gene frequencies are often confounded with **historical gene flow.** If isolation is recent, populations might appear to have high gene flow even if they are completely isolated, because molecular differences have not had time to accumulate (Neigel 1997, 2002). Many estimators of gene flow are based on populations being in an **equilibrium** condition, where population size and number of successful migrants have not changed dramatically for many generations. In cases of population growth or decline, it is assumed the change is constant over time (see Kuhner 2006). These conditions are not typical of many settings in which resource managers wish to estimate gene flow, such as in recently fragmented landscapes. Thus, although often useful in a relative sense, the absolute values of estimates of Nm should be regarded with some caution.

The greater the exchange of individuals between populations, the more that **genetic similarity** of the populations will increase. However, the relationship between gene flow and genetic similarity is not linear (Fig. 22.3); a few successful individuals moving between populations each generation is often sufficient to retard the effects of genetic drift on the similarity of gene frequencies. A consequence of the nonlinear relationship of gene flow and differentiation is that once gene flow is sufficiently high to erase most genetic differences between populations, estimates of F_{ST} approach zero, and it is difficult to estimate the number of migrants per generation (Waples 1998). However, knowledge that gene flow is high enough to minimize F_{ST} should be sufficient for most management decisions—a precise estimate of Nm is not needed.

Sex-Specific Dispersal

Among wildlife species, there is considerable variation in the gender of dispersing individuals. Gender of dispersing individuals is typically beyond the control of wildlife biologists; however, it is important to understand that **breeding**

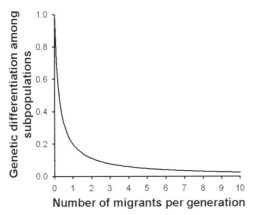

Fig. 22.3. Equilibrium relationship of genetic differentiation among subpopulations (as measured by the statistic F_{ST}) and number of migrants per generation. *Modified from Mills and Allendorf (1996).*

systems and **gender-biased dispersal** are likely to affect estimates of gene flow. Gender-biased dispersal and **social structure** in populations can alter effective population size, influencing rates of loss of genetic diversity (Chesser et al. 1993).

Gender-biased dispersal will have large consequences for data generated by **maternally inherited markers,** such as mtDNA. There is a large set of literature, reviewed by Avise (1994), documenting differences in spatial distribution of biparental nuclear markers (e.g., allozymes and microsatellites) and mtDNA. For example, frequencies of mtDNA genotypes of green turtle (*Chelonia mydas*) differ dramatically among some nesting beaches (Bowen et al. 1992). However, nuclear DNA in this species exhibits much less spatial subdivision than does mtDNA, suggesting that females, but not males, return to their natal beach (Karl et al. 1992).

Given the important contributions of mtDNA studies to understanding of female-specific gene flow, it is clear that markers tracing gene flow in males would be of great value. In mammals, an obvious choice for a paternally inherited marker would be the Y chromosome, which is only passed from males to their sons. Sequences from the Y chromosome have been used to show there was little male-mediated gene flow between a recently colonized population of wolves and more established populations (Sundqvist et al. 2001). Identification of microsatellite loci and other easy-to-assay markers on the Y chromosome has enhanced the ability to characterize gene flow by males in several mammals (Handley and Perrin 2007, Yannic et al. 2008).

Studying male-specific markers in birds is more complicated than in mammals, because birds have **heterogametic (ZW) gender determination,** where the females are heterogametic. Thus, there are no paternally inherited genetic markers similar to the Y chromosome in mammals. Scribner et al. (2001) provide an example of how information from different types of molecular markers, together with theoretical models, can be used to estimate male- and female-specific gene flow in birds. In their study of spectacled eider (*Somateria fisheri*), they used mtDNA, a gender-linked Z-specific microsatellite locus, and biparentally inherited microsatellites to document large differences in sex-specific gene flow. Information about sex-specific movements also can be gained by comparing the spatial genetic structure of adult males and females (Lee et al. 2009).

Population Structure and Fragmentation

Using F_{ST} to quantify **population structure** and gene flow assumes that sampling reflects the underlying population structure; however, choice of sampling locations is often based on other considerations, such as logistics or accessibility. Rather than assessing differences in gene frequencies among sampling locations, it is becoming more common to use **assignment methods** to determine the number of populations in a given sample and to assign individuals to those populations (reviewed in Manel et al. 2005). See Aspi et al. (2009) and Boessenkool et al. (2009) for recent examples of the use of these powerful assignment methods for assessing the structure of wildlife populations. Waples and Gaggiotti (2006) review different methods for identifying population structure based on genetic markers.

If gene flow is limited for many generations by natural barriers to dispersal, populations on opposite sides of the barrier can exhibit striking levels of differentiation. **Fragmentation** of habitat should, given sufficient time, also result in genetic differentiation among recently isolated populations, especially if sizes of populations inhabiting the habitat remnants are small. For example, gene flow was much lower in populations of Sitka deer (*Cervus nippon*) in areas of habitat fragmentation (Goodman et al. 2001). Sometimes even animals capable of long-distance movements, such as migratory songbirds (see Box 22.3) or black bears (*Ursus americanus;* Dixon et al. 2007), experience population subdivision due to habitat fragmentation, suggesting the capability for dispersal might poorly predict actual dispersal.

There is one caveat for most studies of the effects of recent habitat fragmentation on genetic differentiation: they lack temporal control. Although it might be true that recent fragmentation of continuous habitat reduced gene flow, it is often difficult to eliminate the possibility that observed patterns of genetic differentiation are due to events that occurred long before any human activities affected the populations. Analyses of genetic differentiation before and after a fragmentation event are possible, given the ability to isolate **DNA from museum specimens.** For example, Martinez-Cruz and Godoy (2007) found that differentiation among fragmented populations of the Spanish imperial eagle (*Aquila adalberti*) is much greater today than it was in the past, when habitats were presumably more continuous.

> **BOX 22.3. ASSESSING GENETIC DIVERSITY AND STRUCTURE IN 2 ENDANGERED SONGBIRDS**
>
> The golden-cheeked warbler (GCWA; *Dendroica chrysoparia*) and black-capped vireo (BCVI; *Vireo atricapilla*) are migratory songbirds facing a variety of threats to their long-term viabilities (see figure). Both species have restricted breeding ranges (GCWA: central Texas; BCVI: northern Mexico, central Texas, and southern Oklahoma). Within these distributions, the populations have been highly fragmented by habitat loss due to urban, suburban, and agricultural development (Grzybowski 1995, Ladd and Glass 1999). Each species also has been heavily affected by brown-headed cowbird (*Molothrus ater*) nest parasitism.
>
> Microsatellite loci were used to determine whether population fragments had experienced loss of genetic diversity and increased interpopulation differentiation due to bottlenecks and associated genetic drift (see Lindsay et al. [2008] and Barr et al. [2008] for details). Both species had levels of genetic diversity similar to other related songbirds, suggesting that neither had experienced species-wide loss of variation. There also was little evidence that individual population fragments had experienced recent bottlenecks, even though some of them were quite small. However, using both F_{ST} and assignment tests, there was more genetic structure among population fragments than would be expected for songbirds capable of flying hundreds of kilometers during migration. The influence of fragmentation on genetic structure also differed between the species. The greatest genetic differences among GCWA populations were between those separated by agricultural lands; this species depends on mature forests with little edge (Ladd and Glass 1999). The BCVI had greater differences among populations than did the GCWA, but these differences were not strongly influenced by intervening habitat. As a species dependent on early successional shrublands, differentiation may be the result of small numbers of birds colonizing habitats that are only temporarily available. These studies illustrate that habitat fragmentation can restrict gene flow even in species capable of moving great distances and that the habitat requirements of a species will influence its response to fragmentation (Leberg 1991).
>
>
>
> Golden-cheeked warbler (left) and black-capped vireo (right). *Photos by K. Barr*

Gene Flow through Wildlife Translocations

One consequence of **translocation programs** is gene flow (Leberg 1990a). When such translocations might result in the loss of unique genetic characteristics of isolated populations, they should probably be avoided (Moritz 1999). However, when genetic differences have developed through habitat fragmentation resulting from land use, translocations can be used for **genetic restoration.** Many authors have discussed the possibility of using translocations to restore gene flow between populations isolated by habitat loss (Moritz 1999, Tallmon et al. 2004, Bouzat et al. 2009). Based on the relationship between gene flow and F_{ST} (Fig. 22.3), movement of only a few individuals per generation should be adequate to minimize the tendency of isolated populations to genetically differentiate. Mills and Allendorf (1996) provide an excellent review of factors to be considered when designing programs to restore genetic connectivity of populations.

Genetic markers can be used to assess whether translocated individuals reproduced and contributed genetic variation to their recipient populations (Mock et al. 2001, Olsson 2007). For example, Arrendal et al. (2004) found that translocated Eurasian otters (*Lutra lutra*) made genetic contributions to 1 of 2 populations into which they were released. These observations can be useful, because it is often difficult to know whether individuals released in augmentation efforts successfully contributed genetic material to the recipient population.

The success of a translocation program also can be examined by assessing whether patterns of genetic similarity expected from natural dispersal have been disrupted (Latch and Rhodes 2005). For example, Leberg et al. (1994) assessed whether genetic structure of white-tailed deer (*Odocoileus virginianus*) populations in the southeastern United States had been influenced by extensive translocations. They found populations connected by translocations to be more genetically similar than populations that had not had individuals transferred between them. With similar data, Ellsworth et al. (1994a, b) concluded that releases of white-tailed deer had little effect on the genetics of native populations, arguing that most translocated deer had not made genetic contributions to their recipient populations. Additional analysis led Leberg and Ellsworth (1999) to conclude that translocated individuals did contribute to the recovery of recipient populations; however, genetic contributions of the released individuals were restricted to the populations into which they were released. This set of studies illustrates the complexities of understanding the genetic and demographic consequences of translocations when only samples collected after the translocation event are available for analysis.

Considerably more information about translocations could be discerned by obtaining genetic information prior to a translocation (Leberg 1999). Bouzat et al. (2009) provide an excellent example of the insights that can be gained by studying both the genetics and ecology of populations before and after a translocation to restore genetic diversity. They found that although translocations of greater prairie-chickens (*Tympanuchus cupido pinnatus*) restored genetic variation to historic levels and reduced inbreeding depression, the benefits of the effort were limited by habitat conditions.

Drift and Bottlenecks

As a result of chance differences in **reproductive success** and **survival** among individuals with different genotypes, allele frequencies will change from one generation to another. Random change in the frequencies of alleles is referred to as **genetic drift.** The effect of drift on a population is expected to be small when population sizes are large. In large populations, small random changes in allele frequencies will occasionally cause an allele to be lost, but this loss is mitigated by formation of new alleles through mutation. However, when populations are small and isolated from other populations, **gene frequencies** can drift dramatically. A population that is maintained at a small size for several generations has different genetic characteristics than it had prior to the reduction in size. Because of large random changes in allele frequencies, alleles will be lost in a small population faster than they are replaced through mutation, reducing allelic diversity. The average number of genes at which an individual is heterozygous (mean multilocus heterozygosity) also is expected to decrease if a population remains small, because matings between relatives will become unavoidable. Another consequence of drift associated with small population size is increased **genetic differentiation.** Genetic differences, based on neutral molecular markers, between 2 populations will increase rapidly if there is no gene flow between them and at least 1 of them is small enough to experience substantial genetic drift.

When a normally large population goes through a constriction in size, the event is referred to as a **genetic bottleneck.** During bottlenecks, drift is greatly accelerated. Bottlenecks often occur at the establishment of a new population. This type of bottleneck is referred to as a **founder event.** Founder events are often severe bottlenecks, as only a few individuals may establish a population; however, they tend to be of short duration.

The duration and size of the bottleneck have large effects on loss of genetic diversity. During severe bottlenecks of short duration, theory (Nei et al. 1975) and experiments (Spencer et al. 2000) indicate that many alleles will be lost. However, because most alleles are relatively rare in populations, there is no large loss of heterozygosity (Leberg 1992, Spencer et al. 2000). But if the bottleneck is of long duration, relatedness of individuals will increase, along with associated loss of heterozygosity (Nei et al. 1975). Thus, population growth rate can have a large effect on levels of genetic diversity through inbreeding following a reduction in population size.

Detecting Bottlenecks and Drift

A severe reduction in population size will lead to loss of heterozygosity, reduced allelic diversity, and drift of allele frequencies. Because prebottleneck samples are often absent, samples from populations that may have experienced a bottleneck are often compared to populations of the same or related species that are believed to have levels of genetic variation not affected by small population sizes (Leberg 1991, Whitehouse and Harley 2001, Nichols et al. 2001; Box 22.3).

This comparative approach requires the assumption the populations had similar levels of genetic variation prior to the putative bottleneck event (Bouzat 2000). Use of preserved materials provides a more straightforward way to estimate prebottleneck levels of diversity. Matocq and Villablanca (2001) used museum specimens to show that low genetic variation in an endangered species was due to bottlenecks that occurred prior to a known recent reduction in population size. Conversely, museum specimens of greater prairie-chickens provided strong evidence that recent population reductions in Illinois resulted in reduced levels of genetic variation (Bouzat et al. 1998b). Unfortunately, sizes of museum collections from localities of interest are often insufficient to make strong statistical comparisons with contemporary populations. For populations that are likely to be of management concern, it would be appropriate to establish baseline genetic characteristics and preserve DNA samples for monitoring future changes in population size. Schwartz et al. (2007) discuss contributions that DNA from museum archived species have made to conservation biology and advocate for using genetic monitoring to better manage wildlife populations.

The commonly used genetic indices of bottlenecks differ in their sensitivities to population contractions. Loss of allelic diversity is much more sensitive to short population bottlenecks than is heterozygosity (Leberg 1992, Spencer et al. 2000). Not surprisingly, it is easier to detect loss of alleles when using loci with many alleles, such as microsatellites, than with less polymorphic allozyme markers (Spencer et al. 2000). Both simulations and experiments indicate that **temporal change in allele frequencies** also is a much better index of bottleneck severity when drift is estimated with highly polymorphic loci (Richards and Leberg 1996, Luikart et al. 1999, Spencer et al. 2000). Although **allelic richness** is strongly influenced by past size of a population, this parameter also is sensitive to sample size. Thus, when comparing allelic richness among samples, estimates should be adjusted to the smallest sample size of any population used in comparison (Leberg 2002).

Several approaches have been developed to alleviate the need to compare a sample of interest to a reference sample to see whether a population has experienced a loss of genetic variation. These methods are based on expectations of heterozygosity, distributions of frequencies of alleles, or **distributions of allele sizes** for populations that have not experienced bottlenecks (Cornuet and Luikart 1996, Luikart and Cornuet 1998, Luikart et al. 1998, Garza and Williamson 2001). These approaches are dependent on selection of the correct model of mutation used to generate the null distributions. An examination of populations that had experienced known reductions in population size suggests these approaches provide reasonable indices of a population's history of bottlenecks (Luikart and Cornuet 1998, Spencer et al. 2000). However, Larsson et al. (2008) showed these methods where not able to detect a bottleneck of black grouse (*Tetrao tetrix*) that was detectable by comparing contemporary and historical samples. Williamson-Natesan (2005) provides a comparison of these methods and a discussion of their merits.

When considering the effects of bottlenecks on genetic variation, it is critical to realize that not all population reductions will result in measurable losses of **genetic variation**. Population sizes often have to be quite small for several generations to produce a substantial loss of variation. Thus, a 90% reduction in size of a European rabbit (*Oryctolagus cuniculus*) population was insufficient to produce measurable genetic response, because the remnant population was not reduced below approximately 50 individuals and recovered rapidly (Queney et al. 2000). Likewise, experimental populations reduced to 16 individuals for one generation exhibited almost no loss of variation when they rapidly recovered to a large size (Spencer et al. 2000). Gene flow also will make it very difficult to detect the effects of even severe bottlenecks (Busch et al. 2007).

Effective versus Census Population Size

One goal of conservation genetics is to understand how much genetic diversity would be lost from a population reduction or management activity. Genetic diversity is often lost more rapidly than would be predicted from the number of individuals in the population (referred to as the **census population size,** N_c). At times, many individuals in a population are not reproductively active because of age or social constraints, and some individuals are vastly more successful than others in transmitting their genes to the next generation. When individuals differ in their ability to successfully reproduce, genetic diversity will be lost more rapidly than expected on the basis of N. One way of understanding these issues is to estimate the **effective population size,** N_e. N_e is the number of individuals in an ideal population that would lose genetic variation at the same rate as the actual population being studied. An **ideal population** is one where all individuals have an equal chance of producing any progeny making up the next generation. The list of possible factors that can could cause $N_e < N_c$ is large (Crow and Kimura 1970, Hedrick 2000, Leberg 2005). We discuss only those factors likely to have a large effect in wildlife populations, with emphasis on those that might fall under the control of managers.

Temporal variation in population size can have large effects on loss of genetic variation (Crow and Kimura 1970, Vucetich et al. 1997) and may have a strong influence on the effective size of wildlife populations (Frankham 1995*a*). A normally large population that occasionally experiences a large decline in numbers may lose considerable genetic variation during those periods when it is small; this variation is not immediately recovered when the population returns to a large size. Kalinowski and Waples (2002) provide a framework for examining the relationship between N_e and N_c over multiple generations when population size is not stable.

Unequal sex ratios reduce N_e (Wright 1931). If one gender is much more common than the other, members of the more rare gender will disproportionately contribute genes to the next generation. If sex ratios are highly skewed and the rare gender is only represented by a few individuals, then $N_e << N_c$. In species with nonoverlapping generations, highly polygamous mating systems also can result in small estimates of N_e (Nunney 1993).

The **age structure** of a population can complicate efforts to estimate effective population size in wildlife species. Most wildlife populations have overlapping generations; simple formulations of the effects of sex ratio and temporal variation of effective size assume nonoverlapping generations. In some age-structured populations, fairly large numbers of individuals might be too young or too old to reproduce. To make the issue even more confusing, the influences of sex ratio and mating system on N_e are modified by **generation length** in complex ways (Nunney 1993). Methods assuming nonoverlapping generations should be applied with caution when attempting to estimate N_e of wildlife populations (Leberg 2005).

There are several genetic techniques for estimating N_e. One common approach is to quantify genetic changes through time by taking ≥2 temporal samples (Waples 1989, Jorde and Ryman 1995). If insufficient time is available to obtain samples separated by several generations, the genetic characteristics of contemporary populations can be compared to those of museum specimens (Bouzat et al. 1998b, Schwartz et al. 2007, Pertoldi et al. 2008). Recently, there have been a series of powerful methodologies developed to estimate N_e based on this temporal method (Wang and Whitlock 2003, Wang 2005). There are a number of other approaches to estimation of N_e based on genetic data from a single sample (e.g., Tallmon et al. 2008); see Leberg (2005) for a review of issues related to estimating N_e in wildlife populations. In addition to genetic approaches, demographic data can be used to estimate N_e (Harris and Allendorf 1989, Nunney 1993, Engen et al. 2007). When using any of these approaches, it is important to realize that "effective size" can refer to several population genetic parameters and, thus, measure loss of different components of genetic diversity (see Crow and Kimura 1970, Schwartz et al. 1998, Leberg 2005).

Drift and Bottlenecks from Human Activities

Reduced levels of genetic variation have been documented in large numbers of threatened species or populations (Rossiter et al. 2000, Rico et al. 2008, Gebremedhin et al. 2009). Reductions of genetic diversity are often symptomatic of small populations that have become endangered through loss of habitat and other causes. Even in abundant species, individual populations can lose genetic variation when they become isolated in fragments of habitat incapable of supporting large populations (Bouzat 2001, Goodman et al. 2001). Creation of corridors between these fragments or the imposition of gene flow through **translocations** has been suggested as strategies for prevention of loss of diversity in fragmented populations (Hedrick 1995, Mills and Allendorf 1996, Epps et al. 2007).

By definition, **reintroduction programs** create **founder events.** Populations established through releases will often have less genetic diversity than those that are the source of released individuals; this loss is related to the number of individuals released (Stockwell and Leberg 2002, Mock et al. 2004, Sigg 2006). For example, Fitzsimmons et al. (1997) found that populations established with translocated bighorn sheep (*Ovis canadensis*) often had reduced genetic diversity compared to the source of the released individuals. Scribner and Stuwe (1994) found the amount of genetic drift experienced by populations of Alpine ibex (*Capra ibex*) was related to the number and sex ratio of individuals used to establish a population as well as by subsequent population growth. Slow population growth following translocation appears to be responsible for a loss of heterozygosity in a population of elk (*Cervus canadensis*; Williams et al. 2002b).

Not surprisingly, **allele frequencies** of translocated populations often differ from those of their sources (Scribner 1993, Fitzsimmons et al. 1997, Stephen et al. 2005). However, caution should be used when interpreting differences in allele frequencies among translocated populations and their sources. Although differences might be the result of the founder event, they also could have occurred through drift after the translocated population became established (Williams et al. 2000b, Stephen et al. 2005). Vonholdt et al. (2008) provide an example of using genetic markers to monitor the genetic composition and social structure of a reintroduced population of wolves.

Reintroduction strategies that may make sense based on the species' ecology might have the unintended consequence of reducing the **effective population size** of the newly established population (Leberg 1990a). The benefits of such strategies, such as faster initial population growth by releasing more females than males, or of reduced dispersal through release of family groups, should be evaluated in light of their genetic consequences. For example, if it makes sense to release family groups to reduce post-release dispersal, it would be best to release as many groups as possible to avoid inbreeding and loss of genetic variation.

Harvest programs should have little effect on genetic variation, because loss of variation due to drift is small if the population is large. However, harvests can reduce effective population size far below the census population size, creating the potential for rates of drift that might be surprising if only the total population size is considered (Ryman et al. 1981, Laikre and Ryman 1996). For example, harvest regulations and hunter preferences resulting in greater harvests of males can have large effects on the N_e of ungulate populations (Ryman et al. 1981, Coltman 2008). Harris et al. (2002), Coltman (2008), and Allendorf et al. (2008) review the possible effects of game harvests on genetic diversity and effective population size.

Selection

Many genetic markers used by conservation geneticists are thought to be **selectively neutral.** Thus, the specific genotypes associated with these marker systems have little or no effect on the survival or reproduction of individuals. Although this assumption is violated occasionally, most genetic variation examined using many types of markers probably has little consequence for the **fitness** of individuals (Hedrick 2000). Because marker systems are unlikely to be under **direct selection,** they are useful for measuring such phenomena as gene flow, inbreeding, and drift that tend to affect variation throughout the genome and, thus, result in **genetic**

signatures that are detectable with molecular markers. Although the neutrality of molecular markers aids in their usefulness for studying many population processes, it also means the **linkage** of molecular markers and genetic traits of concern to the well-being of individual organisms is at best indirect. The lack of direct concordance often observed between patterns of variation at molecular markers and complex morphological, behavioral, or life history traits has lead to calls for conservation geneticists to more critically evaluate whether molecular data are sufficient for designating conservation priorities (Pearman 2001, Reed and Frankham 2001).

In spite of the general assumption that much of the variation characterized by molecular markers is neutral, there is a large body of work attempting to understand the role of selection in maintaining marker variation in wildlife populations. Initial surveys of natural populations detected higher levels of allozyme variation than expected. This observation generated interest in examining whether individuals that were heterozygous for allozyme loci might have high fitness; such selection would promote high levels of variation (Allendorf and Leary 1986, Reed and Frankham 2003). However, there also have been studies that found no relationship between heterozygosity and traits related to fitness (Britten 1996). In red deer (*Cervus elaphus*), antler growth was actually lower in heterozygotes for some allozymes (Hartl et al. 1995). Furthermore, there is little direct evidence that it is the loci themselves that are producing variation in fitness components. The allozymes might be physically linked, through proximity on chromosomes, to genes producing the effect, or alternatively, high heterozygosity might indicate that an individual's parents were not closely related (Leberg et al. 1990). Understanding relationships between heterozygosity and fitness is being enhanced by examining similar relationships using molecular markers that are probably not under selection. Associations between fitness traits and microsatellite heterozygosity have been detected for several wildlife species (Coulson et al. 1998, Hansson et al. 2001, Höglund et al. 2002). Given that most microsatellite loci do not directly affect phenotype, such associations probably reflect the relatedness of an individual's parents or the physically proximity of assayed microsatellites to other loci affecting the traits of interest. In a recent review of the relationship of heterozygosity to traits related to fitness, Chapman et al. (2009) found that although there are many examples of such correlations, the amount of variance in traits explained by heterozygosity is generally small.

Recently, there has been considerable interest given to examining relationships between **individual viability** and loci in the **major histocompatiblity complex.** These genes are involved in **immune responses,** and there is some evidence that selection maintains variation in populations (Hughes 1991, Hughes and Yeager 1998, Richman et al. 2001). For example, Ditchkoff et al. (2001) found that specific genotypes of the major histocompatibility complex were associated with antler development, body mass, and serum testosterone in white-tailed deer. It is possible that such associations are due to variation in pathogen resistance of different major histocompatibility complex genotypes. Studies also have suggested the major histocompatibility complex might influence mate choice in mammals (Potts et al. 1991, Brown 1998, Penn 2002).

Although examination of correlations between **genotypes** at molecular markers and traits related to individual fitness has been a focus of wildlife genetics, there have been few attempts to apply knowledge in this area directly to management. Any program designed to increase abundance of certain genotypes would be difficult to implement in a natural setting and might be ill advised. Although it has been argued that breeding programs in captive populations should emphasize maintenance of allozyme or major histocompatibility complex diversity because these loci may influence individual survival or fecundity (Wayne et al. 1986, Hughes 1991), **selective breeding schemes** to favor variation at a few molecular markers could result in an increase in the rate of loss of genetic variation at all loci (Hedrick et al. 1986, Vrijenhoek and Leberg 1991, Miller 1995, Lacy 2000). Because there is little understanding of how different genes interact to affect individual well-being, most captive breeding programs advocate maintenance of overall genetic variation and reduction of relatedness. Models also have shown that selection of individuals, on the basis of marker genotype, to be used in reintroduction programs can result in an overall reduction in genetic variation in newly established populations (Haig et al. 1990).

There are promising applications for the use of molecular markers to elucidate variation in fitness traits, and thus improve our understanding of selective pressures faced by wildlife. For example, Slate et al. (2002) used a large number of maps to identify specific genes, referred to as quantitative trait loci for birth weight in wild population of red deer. Ellegren and Sheldon (2008) review using gene mapping and other genomic approaches to understand the genetic basis for variation in fitness traits in wild populations.

Genetic approaches can be used to better understand the implications of **selective harvesting** of wildlife based on hunter preferences or harvest regulations (Allendorf et al. 2008, Coltman 2008). For example, Coltman et al. (2003) used a partly genetically reconstructed pedigree to provide evidence that trophy hunting of bighorn sheep might be selecting for slower horn growth. Because genetic traits are often correlated, selection for one trait, such as horn size, might well affect others, such as body mass or fecundity (Coltman 2008, Sasaki et al. 2009). We are just beginning to understand the effects of selective harvests on wildlife, but at least in some cases, increasing the relative mortality of individuals with larger body sizes—or bigger antlers, tusks, or horns—is likely to reduce not only the sizes of those traits, but also influence other correlated traits that might affect population viability (Allendorf et al. 2008, Coltman 2008, Allendorf and Hard 2009).

Genetic Diversity and Population Viability

Observations of **inbreeding depression** in captive (Lacy et al. 1996) and field populations (Jiménez et al. 1994, Keller et al. 1994, Keller and Waller 2002), and studies of **heterozygosity-fitness relationships** (Reed and Frankham 2003) have led to the realization that loss of genetic variation could affect **population viability** (Gilpin and Soulé 1986, Lacy 1997). Simulation models (Mills and Smouse 1994, Robert et al. 2002) and laboratory studies (Leberg 1990b, Spielman and Frankham 1992, Frankham 1999, Reed and Bryant 2000) have demonstrated decreased population growth and increased extinction rates with loss of genetic variation. Furthermore, observations of wildlife populations that have experienced loss of genetic variation due to bottlenecks also support the conclusion that such losses can affect population productivity (Bouzat et al. 1998a).

Practices that lead to reduced genetic variation, such as establishing populations with only a few individuals or allowing populations to remain small and fragmented, might have serious consequences for population viability (see Chapter 35, Volume II). These concerns about effects of inbreeding on demography occur on a time scale relevant to management activities (e.g., Westemeier et al. 1998, Johnson and Dunn 2006, Ewing et al. 2008). On a longer time scale, managers must be concerned about **loss of allelic variation** that can affect the ability of populations to adapt to new environmental challenges (Allendorf and Leary 1986, Frankham 1995b).

Most conservation geneticists promote maintaining large effective sizes of populations to prevent loss of genetic variation and possible associated reductions in population viability. Recommendations concerning population sizes necessary to prevent adverse genetic consequences vary considerably; there is no general agreement on what appropriate minimum numbers are acceptable for long-term management goals (Gilpin and Soulé 1986, Simberloff 1988, Hedrick and Kalinowski 2000, Reed and Bryant 2000). Most published recommendations of minimum population size are in terms of minimum effective size; the number of breeding age individuals in most populations should be at least 2–4 times as large.

The relationship between loss of genetic diversity and population viability is not as straightforward as the discussion above might suggest. A population with a history of inbreeding might suffer from future inbreeding less than other populations (Fu et al. 1998), and **inbreeding depression** may be influenced by environmental conditions (Bijlsma et al. 1999); however, predicting future inbreeding depression based on population history and environment is difficult (Leberg and Firmin 2008). Furthermore, matings of individuals from genetically differentiated populations, as might occur through translocation, could under some circumstances increase genetic variation in a population while causing a decrease in individual viability (Templeton 1986, Leberg 1993, Edmands 2007). Additionally, other mechanisms besides inbreeding and the loss of genetic variants, such as slow accumulation of mutations with slight deleterious effects, may affect the long-term consequences for small populations (Lande 1995, Jaquiery et al. 2009). Reviews of the mechanisms through which genetic diversity can affect population viability can be found in Soulé (1986), Frankham et al. (2002), Leberg and Firmin (2008), and Chapter 35, Volume 2.

Captive Breeding Programs

When populations decline drastically and only a few individuals remain, biologists often capture some of the remaining individuals in attempts to establish a **captive population.** These animals are bred to expand the captive population, so that individuals can be released into the wild. Because most captive populations are limited in size, they are subject to inbreeding and drift. It has been shown that sound management of the genetic aspects of breeding programs is needed to be successful (Ralls and Ballou 1986, Foose and Ballou 1988, Hedrick and Miller 1992).

Once a captive breeding program has been established, **pedigrees** can be used to avoid matings between close relatives or the over- or underrepresentation of the genes of individual founders in the captive population (Lacy et al. 1995). However, because number of individuals brought into captivity is usually small, inbreeding can be a serious problem if the **founding population** includes related individuals. In such cases, it can be important to consider the genetic identity of animals bred in captivity, so that net genetic variability is maximized and inbreeding is minimized. Molecular genetic techniques have proven to be valuable for inferring relatedness and promise to be useful for examining relatedness of founders (Haig et al. 1994, 1995). Jones et al. (2002b) used microsatellite data to augment wild and captive pedigree information on whooping cranes (*Grus americana*), revealing unknown shared genotype information for founders. Rudnick and Lacy (2008) have shown that improvements in inbreeding avoidance obtained by supplementing information from pedigree relationships with genetic analysis of founder relatedness will be small, unless the founder population included close relatives.

Another issue in captive breeding is **adaptation** to captive conditions. Adaptations like extended reproduction periods and tameness can occur very rapidly and may make it difficult to successfully reintroduce captive bred individuals back into the wild. See Frankham (2008) for a discussion of this problem and possible solutions.

MOLECULAR ECOLOGY

In addition to addressing the traditional concerns of conservation genetics, genetic markers have increasingly provided insight into the ecology of populations. These applications are sometimes referred to as **molecular ecology,** a field that is interwoven with conservation genetics, but that includes applications extending beyond the conservation of genetic variation. As reviewed by Waits and Paetkau (2005), some of these approaches are quite relevant to investigations of wildlife populations.

Noninvasive Sampling

Many wildlife investigators attempt to determine **population size, survival rates,** and **movement patterns** (see Chapters 11 and 20, This Volume). **Mark–recapture methods** are often used to achieve these goals. These approaches usually require capturing individuals and marking them in some way that would allow for their identification if they are ever recaptured. Although these techniques work well for many species, there are others for which this type of study does not. Species that are dangerous and expensive to catch (e.g., bears) and those that are highly elusive (e.g., felids) do not lend themselves to conventional mark and recapture techniques.

Because DNA can be obtained from hair, feathers, shed skin, feces, and urine (Table 22.2), biologists have **noninvasive** ways to obtain genetic information (Waits and Paetkau 2005). Each individual animal has a **unique molecular fingerprint,** so it is possible to use this genetic fingerprint in the same way a biologist might a tag or band in a traditional mark and recapture study. One advantage of this **genetic tag** is that it remains with the individual throughout its lifetime and can even be used to associate the individual with its parents and offspring.

Collecting **noninvasive samples,** such as scat or feathers, often involves searches of locations the organism is expected to use, such as trails, tree rubs, nests, roosts, or den sites (Pearce et al. 1997, Kohn et al. 1999, Piggott et al. 2006). Collection of scat can be aided with the use of specially trained dogs (see Chapter 5, This Volume; Long et al. 2007a, b), some of which can distinguish between the feces of target and nontarget species (Smith et al. 2005). Hair snares have been used to sample a variety of mammals. Hair from bears has been sampled by placing barbed wire around an attractant (Triant et al. 2004, Kendall et al. 2008); a similar approach may work for white-tailed deer (Belant et al. 2007). Hair from felids and other carnivores has been collected from scented hair snares that elicit rubbing behavior (Weaver et al. 2005, Castro-Arellano et al. 2008; Box 22.4). Barbed snares or glue pads for small or medium-sized carnivores have been placed at den entrances (Scheppers et al. 2007) or at the entrance of baited enclosures (Belant 2003, Williams et al. 2009). Snares have even been propelled by a blowgun to collect hair samples from primates (Amendola-Pimenta et al. 2009).

Often **systematic** or **random sampling** is needed for estimating population size or other demographic parameters. Dogs trained for scat detection can be used to sample transects (Smith et al. 2005, Long et al. 2007a, b). Hair snares can be placed along transects, on a sampling grid, or at ran-

---

BOX 22.4. DOCUMENTING THE PRESENCE OF LYNX USING MOLECULAR TECHNIQUES

When Canada lynx (*Lynx canadensis*) populations declined in the contiguous United States, the federal government implemented a survey based partially on DNA approaches. The survey was designed to learn where lynx did or did not occur. Across the potential range of the species south of Canada, transects were established, and hair snares (see figure), designed to snag samples of hair, and attractant were used to collect samples (McDaniel et al. 2000). The technique of Foran et al. (1997) could not be used, because it required amplification of a long fragment of DNA (approx. 900 base pairs) that could not be amplified using degraded DNA from hair samples. Instead, a shorter fragment was used, and sequences of that fragment from hairs were amplified with polymerase chain reactions. Restriction enzymes were then used to create DNA fragments, and hairs of lynx were distinguished from other samples by banding patterns (Mills et al. 2000a).

Baiting a hair snare with catnip.

dom points throughout a study area (Mowat and Paetkau 2002, Castro-Arellano 2008, Williams et al. 2009).

Another advantage of some **noninvasive sampling** is that it does not require handling the organism. Thus, an individual can be sampled repeatedly without influencing the individual's behavior, making it less prone to being sampled. Furthermore, when samples are collected without the aid of attractants, little or no behavior alteration is expected from noninvasive sampling. Behavioral changes following capture and tagging with traditional approaches have the potential for influencing estimates of demographic parameters (see Chapter 11, This Volume).

Estimating Population Size and Survival

For **mark and recapture methods** based on DNA, molecular biologists need to use a genetic marker (or series of markers) that is variable enough so that no 2 individuals will have the same **molecular tag.** Microsatellites are currently the most commonly used marker for this application: each individual's "molecular tag" is based on its genotype for a number of highly polymorphic loci. Using DNA to identify individuals, scientists have been able to estimate population size for a number of species, including humpback whales (*Megaptera novaeangliae;* Palsbøll et al. 1997), fishers (*Martes pennanti*) and American martens (*M. Americana;* Williams et al. 2009), eastern imperial eagles (*Aquila heliaca;* Rudnick et al. 2008), mountain lions (*Puma concolor;* Ernest et al. 2000), lesser horseshoe bats (*Rhinolophus hipposideros;* Puechmaille and Petit 2007), and brush-tailed rock-wallaby (*Petrogale penicillata;* Piggott et al. 2006). Estimates of survival rates, as well as population sizes, have been obtained for Arctic fox (*Alopex lagopus;* Meijer et al. 2008) and grizzly bears (*Ursus arctos horribilis;* Boulanger et al. 2004).

Methods for estimating population size and survival using molecular tags have been reviewed by Lukacs and Burnham (2005*a*). Software has been designed specifically to analyze capture–recapture data based on genetic tags incorporating features like **identification error** (Lukacs and Burnham 2005*b*, Knapp et al. 2009). Other approaches address the possibility of sampling an individual multiple times in the sample collection period and allowing population estimates from only a single period of sample collection (Miller et al. 2005, Petit and Valiere 2006, Puechmaille and Petit 2007). Robinson et al. (2009) found that such models performed better than did multiple occasion capture–recapture estimators. The ability to obtain a population estimate for a single intensive sampling occasion is advantageous when sampling remote study areas that would be difficult to visit multiple times.

Molecular tags have excellent potential for estimating population size (and potentially survival rates) of species that are difficult to trap; however, there are several limitations. The first is the quantity and quality of DNA that is extracted from hair, feathers, feces, and frozen urine. Typically, only small amounts of DNA can be extracted from such samples, and the DNA is often **degraded** (Taberlet et al. 1999; Table 22.2). With low quantity DNA, **contamination** becomes a serious issue, as does a phenomenon known as **allelic dropout** (Taberlet et al. 1999). Allelic dropout occurs when only 1 of 2 alleles of template DNA is amplified by PCR. The consequences are that only 1 allele of a heterozygous genotype is amplified, resulting in incorrect assignment of that individual as a homozygote instead of a heterozygote. Low quality DNA (severed into many short fragments) is undesirable, because it becomes difficult to amplify a microsatellite allele if the template DNA of a certain microsatellite is severed in that region. **Genotyping errors** and **amplification failure** of DNA from collected scat are influenced both by climate and age of the feces (Piggott 2004, Murphy et al. 2007). Such genotyping errors can result in large overestimates of population size (Waits and Leberg 2000). These issues can be addressed by using strict extraction protocols to avoid contamination, adopting repeated PCR amplifications to identify cases of allelic dropout, and using only short microsatellite loci to avoid problems with degraded DNA (Taberlet et al. 1999, Bonin et al. 2004). Alternately, there are statistical approaches to identify genotyping errors (Miller et al. 2002, McKelvey and Schwartz 2004); these approaches could prove useful in reducing costs associated with multiple PCR amplifications and when DNA is limited. Roon et al. (2005*a*) evaluated several approaches to identifying genotyping errors and noted that statistical approaches for filtering errors would provide inadequate resolution, unless genotyping error rates are very low.

The second issue deals with the assumption the method used can **uniquely identify individuals.** For this type of analysis, a sufficient number of highly polymorphic microsatellite loci are needed, so that no 2 individuals will share the same molecular tag. If too few loci are used to identify an individual, it is possible that multiple individuals will have the same molecular tag, resulting in underestimates of population size (Mills et al. 2000*b*, Waits and Leberg 2000). Of course, limited amounts of DNA and increased expense make it undesirable to analyze more loci than necessary to assign unique tags to individuals. There are several approaches for estimating the numbers of loci that should be examined in studies using noninvasive DNA samples (Waits et al. 2001, Hoyle et al. 2005).

Finally, noninvasive samples have the potential to include DNA from multiple individuals. This problem would be most common in cases where several individuals might use a common latrine or leave hair on the same hare snare. For example, Scheppers et al. (2007) found multiple genotypes of badgers at hair snares located at den entrances and along trails; they advocate using DNA from single hairs to identify individuals. This source of bias was examined by Roon et al. (2005*b*), who provide suggestions for addressing the issue of multiple genotypes in a sample.

Tracking Individual Movements

Because individuals can be identified with highly polymorphic markers and sampled through collections of scat or hair, it is possible to obtain information concerning their movements (Kohn and Wayne 1997). Movement data are obtained by "**recapturing**" individuals as a result of multiple collections of their DNA at different locations and times. This method has been applied to a number of mammalian carnivores (Kohn et al. 1999, Ernest et al. 2000, Lucchini et al. 2002). Walker et al. (2008) used DNA from scats to study individual movements and social interactions of the southern hairy-nosed wombat (*Lasiorhinus latifrons*). Smith et al. (2006a) present estimates of movements and home range sizes of kit foxes (*Vulpes macrotis*), based on DNA from scat, and discuss issues associated with using scat as a DNA source for tracking individual movements. Information obtained is often limited by **sampling protocols:** if sampling is confined to roads or paths, an incomplete picture of an individual's use of space will be obtained. Use of specially trained dogs to find scat provides one approach for detecting scat in areas off roads and paths (Smith et al. 2001a). Using DNA from skin samples, Palsbøll et al. (1997) studied long distance migration of individual humpback whales.

At times, it is not necessary to identify "recaptured" individuals to obtain information on movements. If breeding populations differ in genetic composition, it is possible to identify the origin of dispersing or migrating individuals. **Genetic stock identification** allows estimates of the proportion of a sample of individuals that originated from different source populations (Smouse et al. 1990, Xu et al. 1994, Pearce et al. 2000). **Assignment tests** estimate the probability that a specific individual was a member of the different source populations in the sample (Cornuet et al. 1999; Manel et al. 2002, 2005). Variations on these approaches have been used to gain insight into migratory patterns of noctule bats (*Nyctalus noctula;* Petit and Mayer 2000); Wink (2006) reviews the use of DNA markers to study bird migration. Stock identification has proven useful in assigning samples of loggerhead turtles (*Caretta caretta*) collected in foraging areas to their nesting beaches (Bass and Witzell 2000), and in identifying which populations are most affected by incidental captures associated with commercial fisheries (Laurent et al. 1998). Using shed feathers, Rudnick et al. (2008) were able to quantify the degree of natal dispersal and movement in a population of eagles. Gardner-Santana et al. (2009) used assignment tests to examine movements of wild Norway rats (*Rattus norvegicus*) among sites and were able to identify individuals that had dispersed. This approach also has been used to document low amounts of individual dispersal by black bears among habitat fragments (Dixon et al. 2007). In another example, Blanchong et al. (2002) were able to ascertain whether individual white-tailed deer were likely to have been harvested from a specific management unit. These approaches require the genetic composition of possible source populations to be well characterized by a large number of genetic markers and individuals; sampling requirements decrease as genetic differences among populations increase. Although stock identification and assignment tests can be powerful, levels of genetic differentiation in many species, such as northern pintails (*Anas acuta;* Cronin et al. 1996) and double-crested cormorants (*Phalacrocorax auritus;* Green et al. 2006), are sufficiently small to make identification of breeding populations impractical.

Another approach useful for identifying dispersing individuals is **parentage analysis,** a special case of assignment testing (Manel et al. 2005). By determining parent offspring relationships through intensive genetic sampling, it is possible to determine which individuals have dispersed from natal sites (Nutt 2008). Waser et al. (2006) show how estimation of parentage can improve on estimates of natal dispersal rates and distances, even in organisms that can be readily captured and tagged.

Species Identification and Detection

Although accurate individual identification can sometimes be challenging with some noninvasive samples, **species identification** is less problematic (see Foran et al. [1997] for an early example). Wildlife biologists often find **signs** of wildlife, such as feces, tufts of hair, feathers, blood, and even frozen urine, and need to know what species (or individual of a known species) left that sign. This information is particularly important for programs monitoring status of regulated or protected species. DNA extracted from these materials can provide such identification. If a species has **uniquely identifiable populations,** this technique also may be applied to identify which population is the source of a sample.

Species identification can be used to sample for the presence of a rare species, such as in the National Canada Lynx Survey (Box 22.4). Other examples of surveys of the occurrence of a species at a sample site using noninvasive DNA include Dalen et al. (2004), Bidlack et al. (2007), and Ruell and Crooks (2007). A range wide survey of DNA from rabbit droppings was used to determine the current distribution of New England cottontails (*Sylvilagus transitionalis*), a species that has been declining for several decades (Litvaitis et al. 2006).

Other applications of species identification include investigations related to the **illegal harvest** of wildlife (Baker 2008). Cassidy and Gonzales (2005) discuss the need for careful standards when processing samples that might be used in criminal cases. Wasser et al. (2008) discuss the use of assignment tests to help trace poached ivory back to its population of origin.

DNA-based species identification can assist in a variety of other wildlife investigations. Smith et al. (2006b) gained information on habitat use by kit foxes based on the distribution of their scat. Analysis of salvia from bite wounds to sheep was used to determine the predator was a dog rather than a wolf (Sundqvist et al. 2008). Onorato et al. (2006) showed that analysis of scat and hair from the vicinity of ungulate carcasses sites greatly enhanced the ability to determine which predators had visited the site or were involved in the depredation. DNA analysis of tissue remnants has been valuable in identifying which species are involved in bird collisions with aircraft (Dove et al. 2008).

Dietary Analysis

Molecular probes can be used to examine food habits in the absence of recognizable remnants of plant and animal parts, such as hair or seeds (Symondson 2002, Waits and Paetkau 2005, Tollit et al. 2009). Possible sources of dietary information useful for such analyses include stomach contents, mammalian scat, and bird regurgitant. For example, Scribner and Bowman (1998) used microsatellite analysis to distinguish among several species of juvenile waterfowl in stomachs of glaucous gulls (*Larus hyperboreous*). Analysis of scat was used to verify predation by dingos (*Canis familiaris dingo*) on an endangered wombat (*Lasiorhinus krefftii*; Banks et al. 2003). In such analyses, care must be used to select genetic markers with an appropriate level of resolution. If markers only work on a small number of species, some prey will not be identified. However, using approaches that can identify a wide range of species also might detect nondietary items. For example, while attempting to identify large felids from scat, Farrell et al. (2000) detected dipterian DNA that could be the result of flies visiting the feces. Although biases from DNA degradation may result, causing some food types to be over- or underrepresented in molecular analyses of scats, such errors may not be greater than those observed in conventional diet studies (Deagle and Tollit 2007). Clare et al. (2009) give an interesting example of how molecular analyses of bat feces can provide new insights into foraging ecology of species that are difficult to study using traditional analyses of food habits.

Another significant advantage of DNA-based **analysis of scat** is the ability to trace multiple samples back to single individuals. This makes it possible to determine whether individual predators of the same species differ in their **food habits**. For example, Fedriani and Kohn (2001) found that groups of coyotes, and even coyotes within groups, differed in their diets. In a similar study, Prugh et al. (2008) found that spatial and temporal variations in prey availability could explain some of the diet variation among individual coyotes.

Gender Identification

Wildlife biologists studying animals in the field typically need to know the **gender** of individuals to examine differences between males and females. For example, studies of population dynamics often compare survival rates between males and females. In sexually dimorphic wildlife species, it is straightforward to differentiate males from females. However, for some species, it is difficult to accurately assign gender to an individual without invasive procedures. The same problem arises with gender identification from wildlife signs, such as feces, urine, feathers, or hair. Molecular genetic techniques can be used on a variety of different species to assign gender to individuals using only a small sample (e.g., blood, feathers, feces, urine, or hair). Forensic scientists can use **DNA-based identification approaches** when gender of a tissue sample or blood strain might indicate a violation of wildlife harvest regulations (Gilson et al. 1998, Wilson and White 1998, An et al. 2007).

Mammals

Gender can be identified from DNA samples for many groups of mammals, including wombats, rabbits, ungulates, carnivores, seals, primates, and whales (Aasem and Medrano 1990, Griffiths and Tiwari 1993, Reed et al. 1997, Taberlet et al. 1997, Sloane et al. 2000, Wallner et al. 2001, Ensminger and Hoffman 2002, Huber et al. 2002). There are 2 main strategies for detecting gender in mammals using molecular techniques. The first approach is to use PCR to amplify a region specific to the **Y chromosome,** such as the SRY locus, to identify males. If the marker is not detected, the sample is assumed to be from a female. However, because degraded DNA or inhibitory compounds found in some samples can prevent detection of a locus (Kohn and Wayne 1997), it is necessary to have **controls** with other markers to verify there is nothing about the sample that would prevent correct gender identification (Taberlet et al. 1997, Wilson and White 1998). A second approach is to amplify **homologous fragments** of the **X and Y chromosomes,** such as the amelogenin gene (Sullivan et al. 1993, Brinkman and Hundertmark 2009) or zinc finger proteins (Shaw et al. 2003). This approach produces 2 different sized bands, thereby alleviating the need for additional amplification controls. Generally, genetic methods of gender identification have proven to be quite reliable for mammals. However, an approach that works for one set of species might not work for others (Ensminger and Hoffman 2002). Thus, the reliability of any protocol should be verified with samples for which the gender is known. Care also must be taken when using DNA markers from scat to identify the gender of carnivores. Ernest et al. (2000) found that scat from 3 of 4 female mountain lions contained male genotypes. They hypothesized the male genotype might be the result of DNA from male prey, since the SRY marker is not species-specific. This issue can be circumvented in felids by using primers designed for the zinc-finger and amelogenin regions, where deletions in Y-chromosome regions are absent in a wide range of prey species, thus minimizing potential contamination from prey DNA (Pilgrim et al. 2005).

Birds

Gender of birds is typically difficult to assign, as the majority of the world's bird species have males that look identical to females (Griffiths et al. 1998). To address this issue, Griffiths et al. (1998) designed primers around homologous regions in the chromo-helicase-DNA-binding (CHD) gene on **sex chromosomes W and Z** in birds. This technique takes advantage of the fact that chromosomes W and Z evolve at different rates. Homologous regions on sex chromosomes typically are different sizes due to mutations involving insertions and deletions of DNA nucleotides. Their method simultaneously amplifies homologous regions on the W and Z chromosomes followed by a restriction digest that allows for differentiation of males (ZZ – 1 band) and females (ZW – 2 bands) in many species of birds, with the possible exception of Struthioniformes. Ellegren (1996) developed PCR primers for collared flycatchers (*Ficedula albicollis*) in the CHD gene that resulted in gender identification of closely related species without the restriction digest step. Kahn et al. (1998) designed a different set of primers in a more conserved region of the CHD gene that works in most avian species. Bello and Sanchez (1999) further modified this technique to allow for gender identification in ostriches (*Struthio camelus*). This technique has been used to identify gender of many species, including mountain plover (*Charadrius montanus*) using feathers (Dinsmore et al. 2002) and kakapo (*Strigops habroptilus*) from feces (Robertson et al. 1999).

SUMMARY

Molecular genetic techniques represent a relatively new and powerful set of tools that can address both research and management issues in wildlife science. These approaches have shown their utility in wildlife management by helping identify species and appropriate units for conservation. Knowledge gained about the factors affecting distribution and loss of genetic variants has led to refinements in population management, such as maintaining effective population sizes and connectivity between reserves. More recently, the introduction of PCR has allowed noninvasive collection of genetic material from a variety of sources, such as hair, feathers, and feces. Together with the ability to examine highly polymorphic loci and gender-specific markers, noninvasive sampling has allowed genetic assays to contribute to ecological studies of sex ratios, food habits, population size, and mating systems. In this chapter, we provided general theory of population genetics and have identified those techniques and applications currently used in wildlife studies. This body of literature is expanding rapidly, and readers are referred to more detailed accounts of population genetic theory, techniques, and applications. With rapid development of DNA-based technologies, it is likely that currently unforeseen applications of genetic approaches will soon be available to assist wildlife scientists addressing a wide variety of problems.

LITERATURE CITED

Aalen, O. O. 1978. Nonparametric inference for a family of counting processes. Annals of Statistics 6:701–726.

Aasem, E., and J. F. Medrano. 1990. Amplification of the Zfy and Zfx genes for sex identification in humans, cattle, sheep and goats. Bio-Technology 8:1279–1281.

Abbott, C. W., C. B. Dabbert, D. R. Lucia, and R. B. Mitchell. 2005. Does muscular damage during capture and handling handicap radio-marked northern bobwhites? Journal of Wildlife Management 69:664–670.

Abbott, D. H., E. B. Keverne, F. B. Bercovitch, C. A. Shively, S. P. Mendoza, W. Saltzman, C. T. Snowdon, T. E. Ziegler, M. Banjevic, T. Garland, Jr, and R. M. Sapolsky. 2003. Are subordinates always stressed? A comparative analysis of rank differences in cortisol levels among primates. Hormones and Behavior 43:67–82.

Abbott, H. G. 1961. White pine seed consumption by small mammals. Journal of Forestry 59:197–201.

Abbott, H. G., and W. E. Dodge. 1961. Photographic observations of white pine seed destruction by birds and mammals. Journal of Forestry 59:292–294.

Abell, A. J. 1999. Variation in clutch size and offspring size relative to environmental conditions in the lizard (*Sceloporus virgatus*). Journal of Herpetology 33:173–180.

Able, K. P. 1977. The flight behaviour of individual passerine nocturnal migrants: a tracking radar study. Animal Behaviour 25:924–935.

Abler, W. A., D. E. Buckland, R. L. Kirkpatrick, and P. F. Scanlon. 1976. Plasma progestins and puberty in fawns as influenced by energy and protein. Journal of Wildlife Management 40:442–446.

Abraham, G. E., F. S. Manlimos, and R. Gazara. 1977. Radioimmunoassay of steroids. Pages 591–656 in G. E. Abraham, editor. Handbook of radioimmunoassay. Marcel Dekker, New York, New York, USA.

Abraham, K. F., C. D. Ankney, and H. Boyd. 1983. Assortative mating by brant. Auk 100:201–203.

Achtemeier, G. L. 1992. Grasshopper response to rapid vertical displacements within a clear air boundary layer as observed by Doppler radar. Environmental Entomology 21:921–938.

Ackerman, B. B. 1988. Visibility bias of mule deer aerial census procedures in southeast Idaho. Dissertation, University of Idaho, Moscow, USA.

Ackerman, J. T., J. M. Eadie, M. L. Szymanski, J. H. Caswell, M. P. Vrtiska, A. H. Raedeke, J. M. Checkett, A. D. Afton, T. G. Moore, F. D. Caswell, R. A. Walters, D. D. Humburg, and J. L. Yee. 2006. Effectiveness of spinning-wing decoys varies among dabbling duck species and locations. Journal of Wildlife Management 70:799–804.

Adams, D. A., and T. L. Quay. 1958. Ecology of the clapper rail in southeastern North Carolina. Journal of Wildlife Management 22:149–156.

Adams, G. P., E. D. Plotka, C. S. Asa, and O. J. Ginther. 1991. Feasibility of characterizing reproductive events in large nondomestic species by transrectal ultrasonic imaging. Zoo Biology 10:247–259.

Adams, J. D., and B. Freedman. 1999. Comparative catch efficiency of amphibian sampling methods in terrestrial habitats in southern New Brunswick. Canadian Field-Naturalist 113:493–496.

Adams, L., and S. Hane. 1972. Adrenal gland size as an index of adrenocortical secretion rate in the California ground squirrel. Journal of Wildlife Disease 8:19–23.

Adams, L., and S. G. Watkins. 1967. Annuli in tooth cementum indicate age in California ground squirrels. Journal of Wildlife Management 31:836–839.

Adams, L. G., R. O. Stephenson, B. W. Dale, R. T. Ahgook, and D. J. Demma. 2008. Population dynamics and harvest characteristics of wolves in the central Brooks Range, Alaska. Wildlife Monographs 170.

Adams, M .J., K .O. Richter, and W. P. Leonard. 1997. Surveying and monitoring amphibians using aquatic funnel traps. Pages 47–54 in D. A. Olson, W. P. Leonard, and R. B. Bury, editors. Sampling amphibians in lentic habitats: methods and approaches for the Pacific Northwest. Northwest Fauna 4. Society for Northwest Vertebrate Biology, Olympia, Washington, USA.

Addicott, J. F., J. M. Aho, M. F. Antolin, D. K. Padilla, J. S. Richardson, and D. A. Soluk. 1987. Ecological neighborhoods: scaling environmental patterns. Oikos 49:340–346.

Addy, C. E. 1956. Guide to waterfowl banding. U.S. Fish and Wildlife Service, Department of the Interior, Laurel, Maryland, USA.

Adrian, W. J., editor. 1981. Manual of common wildlife diseases in Colorado. Colorado Division of Wildlife, Denver, USA.

Aebischer, N. J., P. A. Robertson, and R. E. Kenward. 1993. Compositional analysis of habitat use from animal radio-tracking data. Ecology 74:1313–1325.

Agresti, A. 1996. An introduction to categorical data analysis. John Wiley and Sons New York, New York, USA.

Agresti, A. 2007. An introduction to categorical data analysis. Second edition. John Wiley and Sons, New York, New York, USA.

Agresti A., J. G. Booth, J. P. Hobert, and B. Caffo. 2000. Random-effects modeling of categorical response data. Sociological Methodology 30:27–80.

Agresti, A., and B. Caffo. 2000. Simple and effective confidence intervals for proportions and differences of proportions result from adding two successes and two failures. American Statistician 54:280–288.

Aguirre, A. A., R. S. Ostfeld, G. M. Tabor, C. House, and M. C. Pearl, editors. 2002. Conservation medicine: ecological health in practice. Oxford University Press, New York, New York, USA.

Ahl, A. S. 1986. The role of vibrissae in behavior: a status review. Veterinary Research Communications 10:245–268.

Ahmed, J., and C. D. Bonham. 1982. Optimum allocation in multivariate double sampling for biomass estimation. Journal of Range Management 35:777–779.

Ahmed, J., C. D. Bonham, and W. A. Laycock. 1983. Comparison of techniques used for adjusting biomass estimates by double sampling. Journal of Range Management 36:217–221.

Airborne Hydrography AB. 2009. Hawk Eye II—general information. http://www.airbornehydro.com/HawkEyeII/hawkeyeII.html.

Airola, D. A., R. Levad., S. Kostka, and D. Garcia. 2006. Using mist nets to capture western purple martins breeding in inaccessible nesting cavities. North American Bird Bander 31:169–174.

Akaike, H. 1973. Information theory and an extension of the maximum likelihood principle. Pages 267–281 in B. N. Petrov and F. Csaki, editors. Second international symposium on information theory. Akademiai Kiado, Budapest, Hungary.

Akçakaya, H. R. 2000a. Population viability analyses with demographically and spatially structured models. Ecological Bulletin 48:23–38.

Akçakaya, H. R. 2000b. RAMAS GIS: linking landscape data with population viability analysis (version 4.0). Applied Biomathematics, Setauket, New York, USA.

Akçakaya, H. R., and J. L. Atwood. 1997. A habitat-based metapopulation model of the California gnatcatcher. Conservation Biology 11:422–434.

Akçakaya, H. R., and W. Root. 2005. RAMAS Metapop: viability analysis for stage-structured metapopulations (version 5.0). Applied Biomathematics. Setauket, New York, USA.

Akenson, J. J., M. G. Henjum, T. L. Wertz, and T. J. Craddock. 2001. Use of dogs and mark–recapture techniques to estimate American black bear density in northeastern Oregon. Ursus 12:203–209.

Alaback, P. B. 1982. Dynamics of understory biomass in Sitka spruce–western hemlock forests of southeast Alaska. Ecology 63:1932–1948.

Alaee, M., and R. Wenning. 2002. The significance of brominated flame retardants in the environment: current understanding, issues and challenges. Chemosphere 46:579–582.

Albanese, G., and V. D. Piaskowski. 1999. An inexpensive elevated mist net apparatus. North American Bird Bander 24:129–134.

Alberico, J. A. R. 1995. Floating eggs to estimate incubation stage does not affect hatchability. Wildlife Society Bulletin 23:212–216.

Albers, P. H. 2003. Petroleum and individual polycyclic aromatic hydrocarbons. Pages 341–371 in D. J. Hoffman, B. A. Rattner, G. A. Burton, Jr., and J. Cairns, Jr., editors. Handbook of ecotoxicology. Second edition. Lewis, Boca Raton, Florida, USA.

Albert, J. 2007. Bayesian computation with R. Springer Science+Business Media, New York, New York, USA.

Aldous, C. M. 1946. Box trap for snowshoe hares and small rodents. Journal of Wildlife Management 10:71–72.

Aldous, M. C., and F. C. Craighead, Jr. 1958. A marking technique for bighorn sheep. Journal of Wildlife Management 22:445–446.

Aldous, S. E. 1936. A cage trap useful in the control of white-necked ravens. Wildlife Research and Management Leaflet BS-27, U.S. Bureau of Biological Survey, Department of Agriculture, Washington, D.C., USA.

Aldous, S. E. 1940. A method of marking beavers. Journal of Wildlife Management 4:145–148.

Aldrich, J. W. 1946. New subspecies of birds from western North America. Proceedings of the Biological Society of Washington 59:129–136.

Aldrich, J. W., and J. H. Steenis. 1955. Neck-banding and other color marking of waterfowl; its merits and shortcomings. Journal of Wildlife Management 19:317–318.

Aldridge, C. L., S. J. Oyler-McCance, and R. M. Brigham. 2001. Occurrence of greater sage-grouse sharp-tailed grouse hybrids in Alberta. Condor 103:657–660.

Aldridge, H. D. J. N., and R. M. Brigham. 1988. Load carrying and maneuverability in an insectivorous bat: a test of the 5% "rule" of radio-telemetry. Journal of Mammalogy 69:379–382.

Alerstam, T. 1996. The geographical scale factor in orientation of migrating birds. Journal of Experimental Biology 199:9–19.

Alexander, M. M. 1951. The aging of muskrats on the Montezuma National Wildlife Refuge. Journal of Wildlife Management 15:175–186.

Alfiya, H. 1995. Surveillance-radar data on nocturnal bird migration over Israel, 1989–1993. Israel Journal of Zoology 41:517–522.

Aliaga-Rossel, E., R. S. Moreno, R. W. Kays, and J. Giacalone. 2006. Ocelot (*Leopardus pardalis*) predation on agouti (*Dasyprocta punctata*). Biotropica 36:691–694.

Alkon, P. U. 1982. Estimating the age of juvenile chukars. Journal of Wildlife Management 46:777–781.

Allan, G. M., C. J. Prelypchan, and P. T. Gregory. 2000. "Habitat trap" for the capture of small- to medium-sized lizards. Herpetological Review 31:160–161.

Alldredge, J. R., and J. T. Ratti. 1986. Comparison of some statistical techniques for analysis of resource selection. Journal of Wildlife Management 50:157–165.

Alldredge, J. R., and J. T. Ratti. 1992. Further comparison of some statistical techniques for analysis of resource selection. Journal of Wildlife Management 56:1–9.

Alldredge, J. R., D. L. Thomas, and L. L. McDonald. 1998. Survey and comparison of methods for study of resource selection. Journal of Agricultural, Biological and Ecological Statistics 3:241–253.

Alldredge, M. W., K. H. Pollock, T. R. Simons, J. A. Collazo, and S. A. Shriner. 2007a. Time-of-detection method for estimating abundance from point-count surveys. Auk. 124:653–664.

Alldredge, M. W., T. R. Simons, K. H. Pollock, and K. Pacifici. 2007b. A field evaluation of the time-of-detection method to estimate population size and density for aural avian point counts. Avian Conservation and Ecology—Écologie et conservation des oiseaux 2:13. http://www.ace-eco.org/vol2/iss2/art13/main.html.

Allee, W. C. 1931. Animal aggregations: a study in general sociology. University of Chicago Press, Chicago, Illinois, USA.

Allen, C. R., L. G. Pearlstine, and W. M. Kitchens. 2001. Modeling viable mammal populations in gap analysis. Biological Conservation 99:135–144.

Allen, D. L. 1939. Winter habits of Michigan skunks. Journal of Wildlife Management 3:212–228.

Allen, D. L. 1943. Michigan fox squirrel management. Publication 100, Game Division, Michigan Department of Conservation, Lansing, USA.

Allen, D. L. 1962. Our wildlife legacy. Revised edition. Funk and Wagnalls, New York, New York, USA.

Allen, D. L., and W. W. Shapton. 1942. An ecological study of winter dens, with special reference to the eastern skunk. Ecology 23:60–68.

Allen, P. H., and K. B. Trousdell. 1961. Loblolly pine seed production in the Virginia–North Carolina Coastal Plain. Journal of Forestry 59:187–190.

Allen, R. P. 1952. The whooping crane. Resource Report 3, National Audubon Society, New York, New York, USA.

Allen, S. H. 1974. Modified techniques for aging red fox using canine teeth. Journal of Wildlife Management 38:152–154.

Allen, S. H., and D. S. C. Kohn. 1976. Assignment of age-classes in coyotes from canine cementum annuli. Journal of Wildlife Management 40:796–797.

Allendorf, F. W., P. R. England, G. Luikart, P. A. Ritchie, and N. Ryman. 2008. Genetic effects of harvest on wild animal populations. Trends in Ecology and Evolution 23:327–337.

Allendorf, F. W., and J. J. Hard. 2009. Human-induced evolution caused by unnatural selection through harvest of wild animals. Proceedings of the National Academy of Sciences of the United States of America 106:9987–9994.

Allendorf, F. W., and R. W. Leary. 1986. Heterozygosity and fitness in natural populations of animals. Pages 51–65 in M. E. Soulé, editor. Conservation biology: the science of scarcity and diversity. Sinauer Associates, Sunderland, Massachusetts, USA.

Alliston, W. G. 1975. Web-tagging ducklings in pipped eggs. Journal of Wildlife Management 39:625–628.

Allred, W. J., R. C. Brown, and O. J. Murie. 1944. Disease kills feed ground elk. Wyoming Wildlife 9:1–8, 27.

Almiron, M. G., B. Lopes, A. L. C. Oliveira, A. C. Medeiros, and A. C. Frery. 2010. On the numerical accuracy of spreadsheets. Journal of Statistical Software 34:4.

Alsager, D. E., J. B. Stenrue, and R. L. Boyles. 1972. Capturing black-billed magpies with circular live traps. Journal of Wildlife Management 36:981–983.

Alt, G. L., C. R. McLaughlin, and K. H. Pollock. 1985. Ear tag loss by black bears in Pennsylvania. Journal of Wildlife Management 49:316–320.

Altmann, J. 1974. Observational study of behavior: sampling methods. Behaviour 49:227–267.

Alvarez, J. A. 2004. An easily constructed Tuttle trap for bats. Wildlife Society Bulletin 32:264–266.

Amann, R. P., and J. T. Lambiase, Jr. 1969. The male rabbit. III. Determination of daily sperm production by means of testicular homogenates. Journal of Animal Science 28:369–374.

Amar, A., B. Arroyo, S. Redpath, and S. Thirgood. 2004. Habitat predicts losses of red grouse to individual hen harriers. Journal of Applied Ecology 41:305–314.

Amat, J. A. 1999. Foot losses of metal banded snowy plovers. Journal of Field Ornithology 70:555–557.

Ambrose, H. W., III. 1972. Effect of habitat familiarity and toe clipping on rate of owl predation in *Microtus pennsylvanicus*. Journal of Mammalogy 53:909–912.

Amendola-Pimenta, M., L. Garcia-Feria, J. C. Serio-Silva, and V. Rico-Gray. 2009. Noninvasive collection of fresh hairs from free-ranging howler monkeys for DNA extraction. American Journal of Primatology 71:359–363.

American Ornithologists' Union. 1988. Report of committee on use of wild birds in research. Auk 105 (Supplement):1A–41A.

American Society for Testing and Measurement. 1997. Standard guide for fish and wildlife incident monitoring and reporting. Pages 1355–1382 *in* Biological effects and environmental fate; biotechnology; pesticides. ASTM E 1849-96, American Society for Testing and Measurement, Philadelphia, Pennsylvania, USA.

American Society of Mammalogists. 1998. Guidelines for the capture, handling and care of mammals as approved by the American Society of Mammalogists. Journal of Mammalogy 79:1416–1431.

American Veterinary Medical Association. 2001. 2000 Report of the AVMA Panel on Euthanasia. Journal of the American Veterinary Medical Association 218(5):669–696.

Ames, J. A., R. A. Hardy, and F. E. Wendell. 1983. Tagging materials and methods for sea otters, *Enhydra lutris*. California Fish and Game 69:243–252.

Ammann, A. 1981. A guide to capturing and banding American woodcock using pointing dogs. Ruffed Grouse Society, Coraopolis, Pennsylvania, USA.

Amstrup, S. C. 1980. A radio-collar for game birds. Journal of Wildlife Management 44:214–217.

Amstrup, S. C., and J. Beecham. 1976. Activity patterns of radio-collared black bears in Idaho. Journal of Wildlife Management 40:340–348.

Amstrup, S. C., G. M. Durner, I. Stirling, and T. L. McDonald. 2005. Allocating harvests among polar bear stocks in the Beaufort Sea. Arctic 58:247–259.

Amstrup, S. C., C. Gardner, K. C. Myers, and F. W. Oehme. 1989. Ethylene glycol (antifreeze) poisoning in a free-ranging polar bear. Veterinary and Human Toxicology 31:317–319.

Amstrup, S. C., T. L. McDonald, and G. M. Durner. 2004. Using satellite radiotelemetry data to delineate and manage wildlife populations. Wildlife Society Bulletin 32:661–679.

Amstrup, S. C., T. L. McDonald, and B. F. J. Manly. 2006. Handbook of capture–recapture analysis. Princeton University Press, Princeton, New Jersey, USA.

Amundson, C. L., and T. W. Arnold. 2010. Effects of radiotransmitters and plasticine bands on mallard duckling survival. Journal of Field Ornithology 81:310–316.

An, J., M. Lee, M.-S. Min, M.-H. Lee, and H. Lee. 2007. A molecular genetic approach for species identification of mammals and sex determination of birds in a forensic case of poaching from South Korea. Forensic Science International 167:59–61.

Anas, R. E. 1970. Accuracy in assigning ages to fur seals. Journal of Wildlife Management 34:844–852.

Andelt, W. F., and P. S. Gipson. 1980. Toe-clipping coyotes for individual identification. Journal of Wildlife Management 44:293–294.

Andelt, W. F., and S. N. Hopper. 2000. Livestock guard dogs reduce predation on domestic sheep in Colorado. Journal of Range Management 53:259–267.

Andelt, W. F., R. L. Phillips, R. H. Schmidt, and R. B. Gill. 1999. Trapping furbearers: an overview of the biological and social issues surrounding a public policy controversy. Wildlife Society Bulletin 27:53–64.

Andelt, W. F., and T. P. Woolley. 1996. Responses of urban mammals to odor attractants and a bait-dispensing device. Wildlife Society Bulletin 24:111–118.

Anderson, A. 1963. Patagial tags for waterfowl. Journal of Wildlife Management 27:284–288.

Anderson, A. 1980. The effects of age and wear on color bands. Journal of Field Ornithology 51:213–219.

Anderson, A. 1981. Making polyvinyl chloride (PVC) colored legbands. Journal of Wildlife Management 45:1067–1068.

Anderson, B. 2002*a*. GPS afloat—GPS navigation made simple. Fernhurst, West Sussex, UK.

Anderson, C. R., Jr., and F. G. Lindzey. 1996. Moose sightability model developed from helicopter surveys. Wildlife Society Bulletin 24:247–259.

Anderson, D. J. 1982. The home range: a new nonparametric estimation technique. Ecology 63:103–112.

Anderson, D. R. 2001. The need to get the basics right in wildlife field studies. Wildlife Society Bulletin 29:1294–1297.

Anderson, D. R., and K. P. Burnham. 2001. Commentary on models in ecology. Bulletin of the Ecological Society of America 82:160–161.

Anderson, D. R., and K. P. Burnham. 2002. Avoiding pitfalls when using information-theroretic methods. Journal of Wildlife Management 66:912–918.

Anderson, D. R., K. P. Burnham, W. R. Gould, and S. Cherry. 2001*a*. Concerns about finding effects that are actually spurious. Wildlife Society Bulletin 29:311–316.

Anderson, D. R., K. P. Burnham, B. C. Lubow, L. Thomas, P. S. Corn, P. A. Medica, and R. W. Marlow. 2001*b*. Field trials of line transect methods applied to estimation of desert tortoise abundance. Journal of Wildlife Management 65:583–597.

Anderson, D. R., K. P. Burnham, and W. L. Thompson. 2000. Null hypothesis testing: problems, prevalence, and an alternative. Journal of Wildlife Management 64:912–923.

Anderson, D. R., K. P. Burnham, G. C. White, and D. L. Otis. 1983. Density estimation of small-mammal populations using a trapping web and distance sampling method. Ecology 64:674–680.

Anderson, D. R., E. G. Cooch, R. J. Gutiérrez, C. J. Krebs, M. S. Lindberg, K. H. Pollock, C. A. Ribic, and T. M. Shenk. 2003. Rigorous science: suggestions on how to raise the bar. Wildlife Society Bulletin 31:296–305.

Anderson, D. R., W. A. Link, D. H. Johnson, and K. P. Burnham. 2001*c*. Suggestions for presenting the results of data analyses. Journal of Wildlife Management 65:373–378.

Anderson, D. R., and R. S. Pospahala. 1970. Correction of bias in belt transects of immotile objects. Journal of Wildlife Management 34:141–146.

Anderson, D. R., and R. T. Sterling. 1974. Population dynamics of molting pintail drakes banded in south-central Saskatchewan. Journal of Wildlife Management 38:266–274.

Anderson, D. R., A. P. Wywialowski, and K. P. Burnham. 1981*a*. Tests of the assumptions underlying life table methods for estimating parameters from cohort data. Ecology 62:1121–1124.

Anderson, M. G. 1978. Distribution and production of sago pondweed (*Potamogeton pectinatus* L.) on a northern prairie marsh. Ecology 59:154–160.

Anderson, M. G., R. D. Sayler, and A. D. Afton. 1980. A decoy trap for diving ducks. Journal of Wildlife Management 44:217–219.

Anderson, P. K., and R. D. Gill. 1982. Cox's regression model for counting processes: a large sample study. Annals of Statistics 10:1100–1120.

Anderson, R. G., and C. K. Nielsen. 2002. Modified Stephenson trap for capturing deer. Wildlife Society Bulletin 30:606–608.

Anderson, R. K., and F. Hamerstrom. 1967. Hen decoys aid in trapping prairie chickens with bownets and noose carpets. Journal of Wildlife Management 31:829–832.

Anderson, R. M., H. C. Jackson, R. M. May, and A. M. Smith. 1981b. Population dynamics of fox rabies in Europe. Nature 289:765–771.

Anderson, R. M., and R. M. May. 1978. Regulation and stability of host–parasite population interactions: I. Regulatory processes. Journal of Animal Ecology 47:219–247.

Anderson, R. M., and R. M. May. 1979. Population biology of infectious diseases: part I. Nature 280:361–367.

Anderson, S., H. L. De Bruijn, A. R. Coulson, I. C. Eperon, F. Sanger, and I. G. Young. 1982. Complete sequence of bovine mitochondrial DNA. Conserved features of the mammalian mitochondrial genome. Journal of Molecular Biology 156:683–717.

Anderson, S. H. 2002b. Managing our wildlife resources. Prentice-Hall, Upper Saddle River, New Jersey, USA.

Anderson, S. J., and C. P. Stone. 1993. Snaring to control feral pigs, *Sus scrofa*, in a remote Hawaiian rain forest. Biological Conservation 63:195–201.

Andrienko, N., and G. Andrienko. 2006. Exploratory analysis of spatial and temporal data: a systematic approach. Springer-Verlag, Berlin, Germany.

Andryk, T. A., L. R. Irby, D. L. Hook, J. McCarthy, and G. Olson. 1983. Comparison of mountain sheep capture techniques: helicopter darting versus net-gunning. Wildlife Society Bulletin 11:184–187.

Andrzejewski, R., and E. Owadowska. 1994. Use of odour bait to catch bank voles. Acta Theriologica 39:221–225.

Anich, N. M., T. J. Benson, and J. C. Bednarz. 2009. Effect of radio transmitters on return rates of Swainson's warblers. Journal of Field Ornithology 80:206–211.

Ankney, C. D. 1975. Neckbands contribute to starvation in female lesser snow geese. Journal of Wildlife Management 39:825–826.

Ankney, C. D. 1980. Egg weight, survival, and growth of lesser snow goose goslings. Journal of Wildlife Management 44:174–182.

Ankney, C. D., and C. D. MacInnes. 1978. Nutrient reserves and reproductive performance of female lesser snow goose. Auk 95:459–471.

Anscombe, F. J. 1973. Graphs in statistical analysis. American Statistician 27:17–21.

Anthony, E. L. P. 1988. Age determination in bats. Pages 47–57 in T. H. Kunz, editor. Ecological and behavioral methods for the study of bats. Smithsonian Institution Press, Washington, D.C., USA.

Anthony, N. M., C. A. Ribic, R. Bautz, and T. Garland, Jr. 2005. Comparative effectiveness of Longworth and Sherman live traps. Wildlife Society Bulletin 33:1018–1026.

Anthony, R. G., M. G. Garrett, and F. B. Isaacs. 1999. Double-survey estimates of bald eagle populations in Oregon. Journal of Wildlife Management 63:794–802.

Anthony, R. M., and R. A. Stehn. 1994. Navigating aerial transects with a laptop computer map. Wildlife Society Bulletin 22:674–676.

Antunes, P., R. Santos, and L. Jordao. 2001. The application of geographic information systems to determine environmental impact significance. Environmental Impact Assessment Review 21:511–535.

Applegate, V. C., and H. B. Predmore, Jr. 1947. Age classes and patterns of primeness in a fall collection of muskrat pelts. Journal of Wildlife Management 11:324–330.

Apps, C. D., B. N. McLellan, T. A. Kinley, and J. P. Flaa. 2001. Scale-dependent habitat selection by mountain caribou, Columbia Mountains, British Columbia. Journal of Wildlife Management 65:65–77.

Araújo, M. B., and P. H. Williams. 2001. The bias of complementarity hotspots toward marginal populations. Conservation Biology 15:1710–1720.

Arcese, P. A., Keller, L. F., and J. R. Cary. 1997. Why hire a behaviorist into a conservation or management team? Pages 48–71 in J. R. Clemmons and R. Buchholz, editors. Behavioral approaches to conservation in the wild. Cambridge University Press, Cambridge, England, UK.

Archibald, W. R., and R. H. Jessup. 1984. Population dynamics of the pine marten (*Martes americana*) in the Yukon Territory. Pages 81–97 in R. Olsen, R. Hastings, and F. Geddes, editors. Northern ecology and resource management memorial essays honoring Don Gill. University of Alberta, Edmonton, Canada.

Arenz, C. L. 1997. Handling fox squirrels: ketamine hydrochloride versus a simple restraint. Wildlife Society Bulletin 25:107–109.

Argos. 2008. User's manual. CLS/Service Argos, Toulouse, France.

Aristophanous, M. 2010. Does your preservative preserve? A comparison of the efficacy of some pitfall trap solutions in preserving the internal reproductive organs of dung beetles. ZooKeys 34:1–16.

Arjo, W. M., R. E. Huenefeld, and D. L. Nolte. 2007. Mountain beaver home ranges, habitat use, and population dynamics in Washington. Canadian Journal of Zoology 85:328–337.

Armitage, P., and P. Colton. 2005. Encyclopedia of biostatistics. John Wiley and Sons, Chichester, England, UK.

Armstrong, R. A. 1950. Fetal development of the northern white-tailed deer (*Odocoileus virginianus borealis* Miller). American Midland Naturalist 43:650–666.

Armsworth, P. R., and J. E. Roughgarden. 2005. The impact of directed versus random movement on population dynamics and biodiversity patterns. American Naturalist 165:449–465.

Arnason, A. N. 1973. The estimation of population size, migration rates, and survival in a stratified population. Research in Population Ecology 15:1–8.

Arnason, A. N., and C. J. Schwarz. 1995. POPAN 4. Enhancements to a system for the analysis of mark–recapture data from an open population. Journal of Applied Statistics 22:785–800.

Arnason, A. N., and C. J. Schwarz. 1999. Using POPAN-5 to analyse banding data. Bird Study 46:S157–S168.

Arnason, A. N., and C. J. Schwarz. 2002. POPAN-6: exploring convergence and estimate properties with SIMULATE. Journal of Applied Statistics 29:649–668.

Árnason, Ú., R. Spilliaert, Á. Palsdottir, and A. Árnason. 1991. Molecular identification of hybrids between the two largest whale species, the blue whale (*Balaenoptera musculus*) and the fin whale (*B. physalus*). Hereditas 115:183–189.

Arnett, E. B. 2006. A preliminary evaluation on the use of dogs to recover bat fatalities at wind energy facilities. Wildlife Society Bulletin 34:1440–1445.

Arnold, K. A., and D. W. Coon. 1971. A technique modification for color-marking birds. Bird-Banding 42:49–50.

Arnold, K. A., and D. W. Coon. 1972. Modifications of the cannon net for use with cowbird studies. Journal of Wildlife Management 36:153–155.

Arnold, T. W., and A. J. Green. 2004. On the allometric relationship between size and composition of avian eggs: a reassessment. Condor 109:705–714.

Arora, K. L. 1979. Blood sampling and intravenous injections in Japanese quail (*Coturnix coturnix japonica*). Laboratory Animal Science 29:114–118.

Arrendal, J., C. W. Walker, A. K. Sundqvist, L. Hellborg, and C. Vila. 2004. Genetic evaluation of an otter translocation program. Conservation Genetics 5:79–88.

Arthaud, G. J., and D. W. Rose. 1996. A methodology for estimating production possibility frontiers for wildlife habitat and timber value at the landscape level. Canadian Journal of Forest Research 26:2191–2200.

Arthur, S. M. 1988. An evaluation of techniques for capturing and radiocollaring fishers. Wildlife Society Bulletin 16:417–421.

Arthur, S. M., R. A. Cross, T. F. Paragi, and W. B. Krohn. 1992. Precision and utility of cementum annuli for estimating ages of fishers. Wildlife Society Bulletin 20:402–405.

Arthur, S. M., B. F. J. Manly, L. L. McDonald, and G. W. Garner. 1996. Assessing habitat selection when availability changes. Ecology 77:215–227.

Artmann, J. W. 1971. Capturing sharp-tailed grouse hens using taped chick distress calls. Journal of Wildlife Management 35:557–559.

Artmann, J. W., and L. D. Schroeder. 1976. A technique for sexing woodcock by wing measurement. Journal of Wildlife Management 40:572–574.

Aruch, S. N. R., T. K. Pratt, and J. P. Vetter. 2003. Methods for capturing and banding Kalij pheasants. North American Bird Bander 28:111–116.

Ashcraft, G., and D. Reese. 1956. An improved device for capturing deer. California Fish and Game 43:193–199.

Asher, W. C. 1963. Squirrels prefer cones from fertilized trees. Research Note SE-3, U.S. Forest Service, Department of Agriculture, Washington, D.C., USA.

Ashley, E. P., and N. R. North. 2004. Automated doors for waterfowl banding traps. Wildlife Society Bulletin 32:273–275.

Ashton, D. G. 1978. Marking zoo animals for identification. Pages 24–34 in B. Stonehouse, editor. Animal marking: recognition marking of animals in research. University Park Press, Baltimore, Maryland, USA.

Ashton, R. E., Jr. 1975. A study of movement, home range, and winter behavior of *Desmognathus fuscus* (Rafinesque). Journal of Herpetology 9:85–91.

Askins, R. A., M. J. Philbrick, and D. S. Sugeno. 1987. Relationship between the regional abundance of forest and the composition of forest bird communities. Biological Conservation 39:129–152.

Asner, G. P. 1998. Biophysical and biochemical sources of variability in canopy reflectance. Remote Sensing of Environment 64:234–253.

Aspelin, A. L. 1997. Pesticide industry sales and usage: 1994 and 1995 market estimates. Biological and Economic Analysis Division, Office of Pesticide Programs, U.S. Environmental Protection Agency, Washington, D.C., USA.

Aspi, J., E. Roininen, J. Kiiskila, M. Ruokonen, I. Kojola, L. Bljudnik, P. Danilov, S. Heikkinen, and E. Pulliainen. 2009. Genetic structure of the northwestern Russian wolf populations and gene flow between Russia and Finland. Conservation Genetics 10:815–826.

Asrar, G., M. Fuchs, E. T. Kanemasu, and J. L. Hatfield. 1984. Estimating absorbed photosynthetic radiation and leaf area index from spectral reflectance in wheat. Agronomic Journal 76:300–306.

Association of Fish and Wildlife Agencies [AFWA]. 2005. Ownership and use of traps by trappers in the United States. Association of Fish and Wildlife Agencies, Washington, D.C., USA.

Association of Fish and Wildlife Agencies [AFWA]. 2006a. Best management practices for trapping in the United States: introduction. Association of Fish and Wildlife Agencies, Washington, D.C., USA.

Association of Fish and Wildlife Agencies [AFWA]. 2006b. Best management practices for trapping bobcat in the United States. Association of Fish and Wildlife Agencies, Washington, D.C., USA.

Association of Fish and Wildlife Agencies [AFWA]. 2006c. Best management practices for trapping coyote in the eastern United States. Association of Fish and Wildlife Agencies, Washington, D.C., USA.

Association of Fish and Wildlife Agencies [AFWA]. 2006d. Best management practices for trapping coyote in the western United States. Association of Fish and Wildlife Agencies, Washington, D.C., USA.

Association of Fish and Wildlife Agencies [AFWA]. 2006e. Best management practices for trapping gray fox in the United States. Association of Fish and Wildlife Agencies, Washington, D.C., USA.

Association of Fish and Wildlife Agencies [AFWA]. 2006f. Best management practices for trapping red fox in the United States. Association of Fish and Wildlife Agencies, Washington, D.C., USA.

Association of Fish and Wildlife Agencies [AFWA]. 2006g. Best management practices for trapping opossum in the United States. Association of Fish and Wildlife Agencies, Washington, D.C., USA.

Association of Fish and Wildlife Agencies [AFWA]. 2006h. Best management practices for trapping raccoon in the United States. Association of Fish and Wildlife Agencies, Washington, D.C., USA.

Association of Fish and Wildlife Agencies [AFWA]. 2007a. Best management practices for trapping beaver in the United States. Association of Fish and Wildlife Agencies, Washington, D.C., USA.

Association of Fish and Wildlife Agencies [AFWA]. 2007b. Best management practices for trapping fisher in the United States. Association of Fish and Wildlife Agencies, Washington, D.C., USA.

Association of Fish and Wildlife Agencies [AFWA]. 2007c. Best management practices for trapping nutria in the United States. Association of Fish and Wildlife Agencies, Washington, D.C., USA.

Association of Fish and Wildlife Agencies [AFWA]. 2007d. Best management practices for trapping river otter in the United States. Association of Fish and Wildlife Agencies, Washington, D.C., USA.

Association of Fish and Wildlife Agencies [AFWA]. 2009a. Best management practices for trapping striped skunk in the United States. Association of Fish and Wildlife Agencies, Washington, D.C., USA.

Association of Fish and Wildlife Agencies [AFWA]. 2009b. Modern snares for capturing mammals: definitions, mechanical attributes and use considerations. Association of Fish and Wildlife Agencies, Washington, D.C., USA.

Asterita, M. F. 1985. The physiology of stress: with special reference to the neuroendocrine system. Human Sciences Press, New York, New York, USA.

Atherton, N. W., M. E. Morrow, A. E. Bivings, IV, and N. J. Silvy. 1982. Shrinkage of spiral plastic leg bands with resulting leg damage to mourning doves. Proceedings of the Annual Conference of the Southeastern Association of Fish and Wildlife Agencies 36:666–670.

Atkinson, C. T., N. J. Thomas, and D. B. Hunter, editors. 2007. Parasitic diseases of wild birds. Wiley-Blackwell Press, Ames, Iowa, USA.

Atlas, D. 2002. Radar calibration: some simple approaches. Bulletin of the American Meteorological Society 83:1313–1316.

Aulerich, R. J., S. J. Bursian, M. G. Evans, J. R. Hochstein, K. A. Koudele, B. A. Olsen, and A. C. Napolitano. 1987. Toxicity of 3, 4, 5, 3', 4', 5'-hexachlorobiphenyl to mink. Archives of Environmental Contamination and Toxicology 16:53–60.

Aulerich, R. J., and R. K. Ringer. 1977. Current status of PCB toxicity to mink, and effect on their reproduction. Archives of Environmental Contamination and Toxicology 6:279–292.

Aulerich, R. J., and D. R. Swindler. 1968. The dentition of mink (*Mustela vison*). Journal of Mammalogy 49:488–494.

Austad, S. N., and M. E. Sunquist. 1986. Sex-ratio manipulation in the common opossum. Nature 324:58–60.

Austin, C. R., and R. V. Short, editors. 1984. Hormonal control of reproduction. Second edition. Cambridge University Press, New York, New York, USA.

Austin, D., W. D. Bowen, J. I. Mcmillan, and S. J. Iverson. 2006. Linking movement, diving, and habitat to foraging success in a large marine predator. Ecology 87:3095–3108.

Austin, D., J. McMillan, and W. Bowen. 2003. A three-stage algorithm for filtering erroneous Argos satellite locations. Marine Mammal Science 19:371–383.

Austin, D. H. 1965. Trapping turkeys in Florida with the cannon net. Proceedings of the Annual Conference of the Southeastern Association of Game and Fish Commissioners 19:16–22.

Austin, J., M. J. Chamberlain, B. D. Leopold, and L. W. Burger, Jr. 2004. An evaluation of EGG™ and wire cage traps for capturing raccoons. Wildlife Society Bulletin 32:351–356.

Avery, M., P. F. Springer, and J. F. Cassel. 1976. The effects of a tall tower on nocturnal bird migration: a portable ceilometer study. Auk 93:281–291.

Avery, M. L., M. J. Kenyon, G. M. Linz, D. L. Bergman, D. G. Decker, and J. S. Humphrey. 1998. Potential risk to ring-necked pheasants from application of toxic bait for blackbird control in South Dakota. Journal of Wildlife Management 62:388–394.

Avery, T. E. 1959. An all-purpose cruiser stick. Journal of Forestry 57:924–925.

Avery, T. E. 1968. Interpretation of aerial photographs. Second edition. Burgess, Minneapolis, Minnesota, USA.

Avery, T. E., and G. L. Berlin. 1992. Fundamentals of remote sensing and airphoto interpretation. Fifth edition. Prentice-Hall, Upper Saddle River, New Jersey, USA.

Avise, J. C. 1989. A role for molecular genetics in the recognition and conservation of endangered species. Trends in Ecology and Evolution 4:279–281.

Avise, J. C. 1994. Molecular markers, natural history, and evolution. Chapman and Hall, New York, New York, USA.

Avise, J. C., and W. S. Nelson. 1989. Molecular genetic relationship of the extinct dusky seaside sparrow. Science 243:646–648.

Bacon, B. R., and J. O. Evrard. 1990. Horizontal mist net for capturing upland nesting ducks. North American Bird Bander 15:18–19.

Bacon, P. J., editor. 1985. Population dynamics of rabies in wildlife. Academic Press, London, England, UK.

Bacon, P. R. 1973. The orientation circle in the beach ascent crawl of the leatherback turtle, *Dermochelys coriacea*, in Trinidad. Herpetologica 29:343–348.

Baer, G. M., J. H. Shaddock, D. J. Hayes, and P. Savarie. 1985. Iophenoxic acid as a serum marker in carnivores. Journal of Wildlife Management 49:49–51.

Bagg, A. M., W. W. H. Gunn, D. S. Miller, J. T. Nichols, W. Smith, and F. P. Wolfarth. 1950. Barometric pressure patterns and spring bird migration. Wilson Bulletin 62:5–19.

Bagnara, J. T., and R. K. Rastogi. 1992. Reproduction in the Mexican leaf frog, (*Pachymedusa dacnicolor*). Pages 98–111 *in* W. C. Hammett, editor. Reproduction in South American

vertebrates. Springer-Verlag, New York, New York, USA.
Bailey, E. E., G. E. Woolfenden, and W. B. Robertson, Jr. 1987. Abrasion and loss of bands from Dry Tortugas sooty terns. Journal of Field Ornithology 58:413–424.
Bailey, G. N. A., I. J. Linn, and P. J. Walker. 1973. Radioactive marking of small mammals. Mammal Review 3:11–23.
Bailey, L. L. 2004. Evaluating elastomer marking and photo identification methods for terrestrial salamanders: marking effects and observer bias. Herpetological Review 35:38–41.
Bailey, R. W. 1976. Live-trapping wild turkeys in North Carolina. North Carolina Wildlife Resources Commission, Raleigh, USA.
Bailey, T. N. 1971. Biology of striped skunks on a southwestern Lake Erie marsh. American Midland Naturalist 85:196–207.
Baines, D., and M. Richardson. 2007. An experimental assessment of the potential effects of human disturbance on black grouse *Tetroa tetrix* in the North Pennines, England. Ibis 149:56–64.
Bajzak, D., and J. F. Piatt. 1990. Computer-aided procedure for counting waterfowl on aerial photographs. Wildlife Society Bulletin 18:125–129.
Baker, C. S. 2008. A truer measure of the market: the molecular ecology of fisheries and wildlife trade. Molecular Ecology 17:3985–3998.
Baker, M. F. 1953. Prairie chickens of Kansas. Miscellaneous Publication 5, Museum of Natural History, University of Kansas, Lawrence, USA.
Baker, R. J., and S. L. Williams. 1972. A live trap for pocket gophers. Journal of Wildlife Management 36:1320–1322.
Baker, S. J., and C. N. Clarke. 1988. Cage trapping coypus (*Myocaster coypus*) on baited rafts. Journal of Applied Ecology 25:41–48.
Baker, W. W. 1983. A non-clamp patagial tag for use on red-cockaded woodpeckers. Pages 110–111 *in* D. A. Wood, editor. Proceedings of the Red-Cockaded Woodpecker Symposium. Florida Game and Freshwater Fish Commission, Tallahasee, USA.
Bakker, K. K., D. E. Naugle, and K. F. Higgins. 2002. Incorporating landscape attributes into models for migratory grassland bird conservation. Conservation Biology 16:1638–1646.
Balazs, G. H. 1985. Retention of flipper tags on hatchling sea turtles. Herpetological Review 16:43–45.
Balcomb, R. 1986. Songbird carcasses disappear rapidly from agricultural fields. Auk 103:817–820.
Baldwin, B. H., B. C. Tennant, T. J. Reimers, R. G. Cowan, and P. W. Concannon. 1985. Circannual changes in serum testosterone concentrations of adult and yearling woodchucks (*Marmota monax*). Biology of Reproduction 32:804–812.
Baldwin, W. P. 1947. Trapping wild turkeys in South Carolina. Journal of Wildlife Management 11:24–36.
Balham, R. W., and W. H. Elder. 1953. Colored leg bands for waterfowl. Journal of Wildlife Management 17:446–449.

Ball, D. J., G. Argentieri, R. Krause, M. Lipinski, R. L. Robinson, R. E. Stoll, and G. E. Visscher. 1991. Evaluation of a microchip implant system used for animal identification in rats. Laboratory Animal Science 41:185–186.
Ballard, W. B., J. C. deVos, J. F. Kamler, and H. A. Whitlaw. 2002. A need for an integrated radiotelemetry database. Wildlife Society Bulletin 30:263–264.
Ballard, W. B., H. A. Whitlaw, D. L. Sabine, R. A. Jenkins, S. J. Young, and G. F. Forbes. 1998. White-tailed deer, *Odocoileus virginianus*, capture techniques in yarding and non-yarding populations in New Brunswick. Canadian Field-Naturalist 112:254–261.
Ballou, R. M., and F. W. Martin. 1964. Rigid plastic collars for marking geese. Journal of Wildlife Management 28:846–847.
Balm, P. H. M., editor. 1999. Stress physiology in animals. CRC Press, Boca Raton, Florida, USA.
Balser, D. S. 1965. Tranquilizer tabs for capturing wild carnivores. Journal of Wildlife Management 29:438–442.
Balthazart, J., and E. Schoffeniels. 1979. Pheromones are involved in the control of sexual behaviour in birds. Naturwissenschaften 66:55–56.
Baltosser, W. H. 1978. New and modified methods for color-marking hummingbirds. Bird-Banding 49:47–49.
Banks, R. C., C. Cicero, J. L. Dunn, A. W. Kratter, P. C. Rasmussen, J. V. Remsen, Jr., J. D. Rising, and D. F. Stotz. 2006. Forty-seventh supplement to the American Ornithologists' Union checklist of North American Birds. Auk 123:926–936.
Banks, R. C., R. M. McDiarmid, and A. L. Gardner. 1987. Checklist of vertebrates of the United States, the U.S. Territories, and Canada. Resource Publication 166, U.S. Fish and Wildlife Service, Department of the Interior, Washington, D.C., USA.
Banks, S. C., A. Horsup, A. N. Wilton, and A. C. Taylor. 2003. Genetic marker investigation of the source and impact of predation on a highly endangered species. Molecular Ecology 12:1663–1667.
Bannor, B. K., and E. Kiviat. 2002. Common moorhen. Account 685 *in* A. Poole and F. Gill, editors. The birds of North America. The Academy of Natural Sciences, Philadelphia, Pennsylvania, and The American Ornithologists' Union, Washington, D.C., USA.
Banuelos, G. 1997. The one-way door trap: an alternative trapping technique for burrowing owls. Journal of Raptor Research 9:122–124.
Barbknecht, A. E., W. S. Fairbanks, J. D. Rogerson, E. J. Maichak, and L. L. Meadows. 2009. Effectiveness of vaginal-implant transmitters for locating elk parturition sites. Journal of Wildlife Management 73:144–148.
Barbour, R. W. 1963. *Microtus*: a simple method of recording time spent in the nest. Science 141:41.
Barbour, R. W., and W. H. Davis. 1969. Bats of America. University of Kentucky Press, Lexington, USA.
Barbour, R. W., J. W. Hardin, J. P. Schafer, and M. J. Harvey. 1969a. Home range, movement,

and activity of thedusky salamander, *Desmognathus fuscus*. Copeia 1969:293–297.
Barbour, R. W., M. J. Harvey, and J. W. Hardin. 1969b. Home range, movement, and activity of the eastern worm snake *Carhophis a. amoenus*. Ecology 50:470–476.
Barclay, J. H. 2008. A technique for nighttime trapping of burrowing owls with a bow net. Journal of Raptor Research 42:142–148.
Barclay, R. M. R., and G. P. Bell. 1988. Marking and observational techniques. Pages 59–76 *in* T. H. Kunz, editor. Ecological and behavioral methods for the study of bats. Smithsonian Institute Press, Washington, D.C., USA.
Barker, F. K. 2004. Monophyly and relationships of wrens (Aves: Troglodytidae): a congruence analysis of heterogeneous mitochondrial and nuclear DNA sequence data. Molecular Phylogenetics and Evolution 31:486–504
Barker, F. K., G. F. Barrowclough, and J. G. Groth. 2001. A phylogenetic hypothesis for passerine birds: taxonomic and biogeographic implications of an analysis of nuclear DNA sequence data. Proceedings of the Royal Society of London B 269:295–308.
Barker, J. M., R. Boonstra, and A. I. Schult-Hostedde. 2003. Age determination in yellow-pine chipmunks (*Tamias amoenus*): a comparison of eye lens masses and bone sections. Canadian Journal of Zoology 81:1774–1779.
Barko, V. A., J. T. Briggler, and D. E. Ostendorf. 2004. Passive fishing techniques: a cause of turtle mortality in the Mississippi River. Journal of Wildlife Management 68:1145–1150.
Barnard, S. M. 1989. The use of microchip implants for identifying big brown bats (*Eptesicus fuscus*). Animal Keepers' Forum 16:50–52.
Barnes, R. D., and W. M. Longhurst. 1960. Techniques for dental impressions, restraining and embedding markers in live-trapped deer. Journal of Wildlife Management 24:224–226.
Barnum, D. A., J. S. Green, J. T. Flinders, and N. L. Gates. 1979. Nutritional levels and growth rates of hand-reared coyote pups. Journal of Mammalogy 60:820–823.
Baron, R. L. 1991. Carbamate insecticides. Pages 1125–1189 *in* W. J. Hayes and E. R. Laws, Jr., editors. Handbook of pesticide toxicology. Volume 3. Classes of pesticides. Academic Press, San Diego, California, USA.
Barr, G. E., and K. J. Babbitt. 2001. A comparison of 2 techniques to sample larval stream salamanders. Wildlife Society Bulletin 29:1238–1242.
Barr, K. R., D. L. Lindsay, G. Athrey, R. F. Lance, T. J. Hayden, S. A. Tweddale, and P. L. Leberg. 2008. Population structure in an endangered songbird: maintenance of genetic differentiation despite high vagility and significant population recovery. Molecular Ecology 17:3628–3639.
Barrentine, C. D., and K. D. Ewing. 1988. A capture technique for burrowing owls. North American Bird Bander 13:107.
Barrett, M. A., S. Morano, G. Delgiudice, and J. Fieberg. 2008. Translating bait preference to capture success of northern white-tailed deer. Journal of Wildlife Management 72:555–560.

Barrett, M. W., J. W. Nolan, and L. D. Roy. 1982. Evaluation of a hand-held net-gun to capture large mammals. Wildlife Society Bulletin 10:108–114.

Barrett, M. W., G. Proulx, D. Hobson, D. Nelson, and J. W. Nolan. 1989. Field evaluation of the C120 magnum trap for marten. Wildlife Society Bulletin 17:299–306.

Barrier, M. J., and F. S. Barkalow, Jr. 1967. A rapid technique for aging gray squirrels in winter pelage. Journal of Wildlife Management 31:715–719.

Barron, J. C., and W. F. Harwell. 1973. Fertilization rates of south Texas deer. Journal of Wildlife Management 37:179–182.

Barrowclough, G. F., J. G. Groth, L. A. Mertz, and R. J. Gutiérrez. 2004. Phylogeographic structure, gene flow and species status in blue grouse (*Dendragapus obscurus*). Molecular Ecology 13:1911–1922.

Bart, J., D. Battaglia, and N. Senner. 2001. Effects of color bands on semipalmated sandpipers banded at hatch. Journal of Field Ornithology 72:521–526.

Bart, J., and S. Earnst. 2002. Double sampling to estimate density and population trends in birds. Auk 119:36–45.

Bart, J., M. A. Fligner, and W. I. Notz. 1998. Sampling and statistical methods for behavioral ecologists. Cambridge University Press, Cambridge, England, UK.

Bart, J., and W. Notz. 1994. Analysis of data. Pages 24–63 *in* T. A. Bookhout, editor. Research and management techniques for wildlife and habitats. The Wildlife Society, Bethesda, Maryland, USA.

Bart, J., and W. Notz. 2005. Analysis of data in wildlife biology. Pages 72–105 *in* C. E. Braun, editor. Research and management techniques for wildlife and habitats. The Wildlife Society, Bethesda, Maryland, USA.

Bart, J., and D. S. Robson. 1982. Estimating survivorship when subjects are visited periodically. Ecology 63:1078–1090.

Bartareau, T. M. 2004. PVC pipe diameter influences the species and sizes of treefrogs captured in a Florida coastal oak scrub community. Herpetological Review 35:150–152.

Bartelt, G. A., and D. H. Rusch. 1980. Comparison of neck bands and patagial tags for marking American coots. Journal of Wildlife Management 44:236–241.

Barth, E. K. 1953. Calculation of egg volume based on loss of weight during incubation. Auk 70:151–159.

Bartlett, I. H. 1938. White-tails—presenting Michigan's deer problem. Michigan Conservation 8(2):4–5, 7, 8(7):8–11.

Bartmann, R. M., G. C. White, L. H. Carpenter, and R. A. Garrott. 1987. Aerial mark–recapture estimates of confined mule deer in pinyon–juniper woodland. Journal of Wildlife Management 51:41–46.

Bartonek, J. C., and C. W. Dane. 1964. Numbered nasal discs for waterfowl. Journal of Wildlife Management 28:688–692.

Barve, N., M. C. Kiran, G. Vanaraj, N. A. Aravind, D. Rao, R. U. Shaanker, K. N. Ganeshaiah, and J. G. Poulsen. 2005. Measuring and mapping threats to a wildlife sanctuary in southern India. Conservation Biology 19:122–130.

Basile, J. V., and S. S. Hutchings. 1966. Twig diameter–length–weight relations of bitterbrush. Journal of Range Management 19:34–38.

Baskerville, G. L. 1972. Use of logarithmic regression in the estimation of plant biomass. Canadian Journal of Forest Research 2:49–53.

Bass, A. L., and W. N. Witzell. 2000. Demographic composition of immature green turtles (*Chelonia mydas*) from the east central Florida coast: evidence from mtDNA markers. Herpetologica 56:357–367.

Bass, C. C. 1939. Observations on the specific cause and the nature of "quail disease" or ulcerative enteritis in quail. Proceedings of the Society for Experimental Biology and Medicine 42:377–380.

Bass, C. C. 1940. Specific cause and nature of ulcerative enteritis of quail. Proceedings of the Society for Experimental Biology and Medicine 46:250–252.

Batcheler, C. L. 1973. Estimating density and dispersion from truncated or unrestricted joint point-distance nearest-neighbor distances. Proceedings of the New Zealand Ecological Society 20:131–147.

Batchelor, T. A., and J. R. McMillan. 1980. A visual marking system for nocturnal animals. Journal of Wildlife Management 44:497–499.

Bate, L. J., E. O. Garton, and M. J. Wisdom. 1999. Estimating snag and large tree densities and distributions on a landscape for wildlife management. General Technical Report PNW-GTR-425, U.S. Forest Service, Department of Agriculture, Washington, D.C., USA.

Bate, L. J., M. J. Wisdom, E. O. Garton, and S. C. Clabough. 2008. Snag PRO: snag and tree sampling and analysis methods for wildlife. General Technical Report PNW-GTR-780, U.S. Forest Service, Department of Agriculture, Portland, Oregon, USA.

Bateman, G. C., and T. A. Vaughan. 1974. Nightly activities of mormoopid bats. Journal of Mammalogy 55:45–65.

Batschelet, E. 1981. Circular statistics in biology. Academic Press, New York, New York, USA.

Bauer, A. M., and R. A. Sadleir. 1992. The use of mouse glue traps to capture lizards. Herpetological Review 23:112–113.

Bauer, R. D., A. M. Johnson, and V. B. Scheffer. 1964. Eye lens weight and age in the fur seal. Journal of Wildlife Management 28:374–376.

Baumgartner, L. L. 1940. Trapping, handling, and marking fox squirrels. Journal of Wildlife Management 4:444–450.

Baumgartner, L. L., and F. C. Bellrose, Jr. 1943. Determination of sex and age in muskrats. Journal of Wildlife Management 7:77–81.

Baverstock, P. R., and C. Moritz. 1996. Project design. Pages 17–28 *in* D. M. Hillis, C. Moritz, and B. K. Mable, editors. Molecular systematics. Sinauer Associates, Sunderland, Massachusetts, USA.

Baxter, R. J., J. T. Flinders, and D. L. Mitchell. 2008. Survival, movements, and reproduction of translocated greater sage-grouse in Strawberry Valley, Utah. Journal of Wildlife Management 72:179–186.

Baxter, W. L., and C. W. Wolfe. 1968. Ecological relationships of wetlands to ring-necked pheasants in Nebraska. Nebraska Game and Parks Commission, White Papers, Conference Presentations, and Manuscripts, University of Nebraska, Lincoln, USA.

Baxter, W. L., and C. W. Wolfe. 1972. The interspersion index as a technique for evaluation of bobwhite quail habitat. Proceedings of the National Bobwhite Quail Symposium 1:158–165.

Bayer, J. M., and J. L. Schei, editors. 2009. PNAMP special publications: remote sensing applications for aquatic resource monitoring. Pacific Northwest Aquatic Monitoring Partnership, Cook, Washington, USA.

Bayless, L. E. 1975. Population parameters for *Chrysemys picta* in a New York pond. American Midland Naturalist 93:168–176.

Beale, D. M. 1962. Growth of the eye lens in relation to age in fox squirrels. Journal of Wildlife Management 26:208–211.

Beale, D. M. 1966. A self-collaring device for pronghorn antelope. Journal of Wildlife Management 30:209–211.

Beale, D. M., and A. D. Smith. 1973. Mortality of pronghorn antelope fawns in western Utah. Journal of Wildlife Management 37:343–352.

Bean, N. J., and J. R. Mason. 1995. Attractiveness of liquid baits containing natural and artificial sweeteners to white-tailed deer. Journal of Wildlife Management 59:610–613.

Beasley, J. C., and O. E. Rhodes, Jr. 2007. Effect of tooth removal on recaptures of raccoons. Journal of Wildlife Management 71:266–270.

Beasley, M. J., W. A. B. Brown, and A. J. Legge. 1992. Incremental banding in dental cementum: methods of preparation of teeth from archaeological sites and for modern comparative specimens. International Journal of Osteoarchaeology 2:37–50.

Beasom, S. L., and J. D. Burd. 1983. Retention and visibility of plastic ear tags on deer. Journal of Wildlife Management 47:1201–1203.

Beasom, S. L., W. Evans, and L. Temple. 1980. The drive net for capturing western big game. Journal of Wildlife Management 44:478–480.

Beasom, S. L., E. P. Wiggers, and J. R. Giardino. 1983. A technique for assessing land surface ruggedness. Journal of Wildlife Management 47:1163–1166.

Beason, R. C. 1980. Orientation of waterfowl migration in the southwestern United States. Journal of Wildlife Management 44:447–455.

Beason, R. C. 2005. Mechanisms of magnetic orientation in birds. Integrative and Comparative Biology 45:565–573

Beausoleil, N. J., D. J. Mellor, and K. J. Stafford. 2004. Methods for marking New Zealand wildlife: amphibians, reptiles and marine mammals. Department of Conservation, Wellington, New Zealand.

Beauvais, G., J. H. Enderson, and A. J. Magro. 1992. Home range, habitat use and behavior of prairie falcons wintering in east-central Colorado. Journal of Raptor Research 26:13–18.

Beck, D. E., and D. F. Olson, Jr. 1968. Seed production in southern Appalachian oak stands. Research Note SE-91, U.S. Forest Service, Department of Agriculture, Washington, D.C., USA.

Becker, J. E., and J.-C. Sureau. 1966. Control of radar site environment by use of fences. IEEE Transactions on Antennas and Propagation 14:768–773.

Beckerton, P. R., and A. L .A. Middleton. 1982. Effects of dietary protein levels on ruffed grouse reproduction. Journal of Wildlife Management 46:569–579.

Beckmann, J. P. 2006. Carnivore conservation and search dogs: the value of a novel, non-invasive technique in the Greater Yellowstone ecosystem. Pages 28–34 in A. Wondrak-Biel, editor. Greater Yellowstone public lands: a century of discovery, hard lessons, and bright prospects. Proceedings of the 8th biennial scientific conference on the Greater Yellowstone Ecosystem. Yellowstone Center for Resources, Yellowstone National Park, Wyoming, USA.

Beckmann, J. P., C. W. Lackey, and J. Berger. 2004. Evaluation of deterrent techniques and dogs to alter behavior of "nuisance" black bears. Wildlife Society Bulletin 32:1141–1146.

Bedford, G. S., K. Christian, and B. Barrette. 1995. A method for catching lizards in trees and rock crevices. Herpetological Review 26:21–22.

Beerli, P., and J. Felsenstein. 2001. Maximum likelihood estimation of a migration matrix and effective population sizes in n subpopulations by using a coalescent approach. Proceedings of the National Academy of Sciences of the United States of America 98:4563–4568.

Beerwinkle, K. R., J. A. Witz, and P. G. Schleider. 1993. An automated, vertical looking, X-band radar system for continuously monitoring aerial insect activity. Transactions of the American Society of Agricultural Engineers 36:965–970.

Beg, M. A., and R. S. Hoffmann. 1977. Age determination and variation in the red-tailed chipmunk, *Eutamias ruficaudus*. Murrelet 58:26–36.

Begon, M., J. L. Harper, and C. R. Townsend. 2006. Ecology: individuals, populations and communities. Fourth edition. Blackwell, Cambridge, Massachusetts, USA.

Begon, M., C. R. Townsend, and J. L. Harper. 2006. Ecology: from individuals to ecosystems. Blackwell, Boston, Massachusetts, USA.

Beier, P., D. R. Majka, and W. D. Spencer. 2008. Forks in the road: choices in procedures for designing wildlife linkages. Conservation Biology 22:836–851.

Bekoff, M. 1982. Coyote. Pages 447–459 in J. A. Chapman and G. A. Feldhamer, editors. Wild mammals of North America. Johns Hopkins University Press, Baltimore, Maryland, USA.

Bekoff, M. 1997. Deep ethology. Pages 35–44 in M. Tobias and K. Solisti, editors. Intimate relationships, embracing the natural world. Beyond Worlds, Hillsboro, Oregon, USA.

Bekoff, M. 2000. Field studies and animal models: the possibility of misleading inferences. Pages 1553–1560 in M. Balls, A.-M. van Zeller, and M. E. Halder, editors. Progress in the reduction, refinement and replacement of animal experimentation. Elsevier Science, Philadelphia, Pennsylvania, USA.

Belant, J. L. 1992. Efficacy of three types of live traps for capturing weasels, *Mustela* spp. Canadian Field-Naturalist 106:394–397.

Belant, J. L. 2003. A hairsnare for forest carnivores. Wildlife Society Bulletin 31:482–485.

Belant, J. L. 2007. Human-caused mortality and population trends of American marten and fisher in a U.S. National Park. Natural Areas Journal 27:155–160.

Belant, J. L., and T. W. Seamans. 1993. Evaluation of dyes and techniques to color-mark incubating herring gulls. Journal of Field Ornithology 64:440–451.

Belant, J. L., and T. W. Seamans. 1997. Comparisons of three formulations of alpha-chloralose for immobilization of Canada geese. Journal of Wildlife Diseases 33:606–610.

Belant, J. L., and T. W. Seamans. 1999. Alpha-chloralose immobilization of rock doves in Ohio. Journal of Wildlife Diseases 35:239–242.

Belant, J. L., T. W. Seamans, and D. Paetkau. 2007. Genetic tagging free-ranging white-tailed deer using hair snares. Ohio Journal of Science 107: 50–56.

Belant, J. L., L. A. Tyson, and T. W. Seamans. 1999. Use of alpha-chloralose by the Wildlife Services program to capture nuisance birds. Wildlife Society Bulletin 27:938–942.

Belant, J. L., J. F. Van Stappen, and D. Paetkau. 2005. American black bear population size and genetic diversity at Apostle Islands National Lakeshore. Ursus 16:85–92.

Belay, E. D., P. Gambetti, L. B. Schonberger, P. Parchi, D. R. Lyon, S. Capellari, J. H. McQuiston, K. Bradley, G. Dowdle, J. M. Crutcher, and C. R. Nichols. 2001. Creutzfeldt–Jakob disease in unusually young patients who consumed venison. Archives of Neurology 58:1673–1678.

Beletsky, L. 1996. The red-winged blackbird: the biology of a strongly polygynous songbird. Academic Press, New York, New York, USA.

Beletsky, L., and L. D. Orians. 1996. Red-winged blackbirds: decision making and reproductive success. University of Chicago Press, Chicago, Illinois, USA.

Bell, G. P., G. A. Bartholomew, and K. A. Nagy. 1986. The roles of energetics, water economy, foraging behavior, and geothermal refugia in the distribution of the bat, *Macrotus californicus*. Journal of Comparative Physiology B 156: 441–450.

Bell, R. L., and T. J. Peterle. 1975. Hormone implants control reproduction in white-tailed deer. Wildlife Society Bulletin 3:152–156.

Bello, N., and A. Sanchez. 1999. The identification of a sex-specific DNA marker in the ostrich using random amplified polymorphic DNA (RAPD) assay. Molecular Ecology 8:667–669.

Bellrose, F. C. 1951. Effects of ingested lead shot upon waterfowl populations. Transactions of the North American Wildlife Conference 16:125–135.

Bellrose, F. C. 1955. A comparison of recoveries from reward and standard bands. Journal of Wildlife Management 19:71–75.

Bellrose, F. C. 1959. Lead poisoning as a mortality factor in waterfowl populations. Illinois Natural History Survey Bulletin 27:235–288.

Bellrose, F. C. 1976. Ducks, geese and swans of North America. Third edition. Stackpole, Harrisburg, Pennsylvania, USA.

Bellrose, F. C. 1980. Ducks, geese and swans of North America. Forth edition. Stackpole, Harrisburg, Pennsylvania, USA.

Bellrose, F. C., and D. J. Holm. 1994. The ecology and management of wood ducks. Wildlife Management Institute, Washington, D.C., and Stackpole, Harrisburg, Pennsylvania, USA.

Belovsky, G. E., D. B. Botkin, T. A. Crowl, K. W. Cummins, J. F. Franklin, M. L. Hunter, Jr., A. Joern, D. B. Lindenmayer, J. A. MacMahon, C. R. Margules, and J. M. Scott. 2004. Ten suggestions to strengthen the science of ecology. BioScience 54:345–351.

Bendell, J. F. 1955. Age, molt and weight characteristics of blue grouse. Condor 57:354–361.

Bendell, J. F. S., and C. D. Fowle. 1950. Some methods for trapping and marking ruffed grouse. Journal of Wildlife Management 14:480–482.

Bender, L. C., L. A. Lomas, and J. Browning. 2007. Condition, survival, and cause-specific mortality of adult female mule deer in north-central New Mexico. Journal of Wildlife Management 71:1118–1124.

Bender, L. C., G. A. Schirato, R. D. Spencer, K. R. McAllister, and B. L. Murphie. 2004. Survival, cause-specific mortality, and harvesting of male black-tailed deer in Washington. Journal of Wildlife Management 68:870–878.

Benedict, N. G., S. J. Oyler-McCance, S. E. Taylor, C. E. Braun, and T. W. Quinn. 2003. Evaluation of the eastern (*Centrocercus urophasianus urophasianus*) and western (*Centrocercus urophasianus phaios*) subspecies of sage-grouse using mitochondrial control-region sequence data. Conservation Genetics 4:301–310.

Benevides, F. L., Jr., H. Hansen, and S. C. Hess. 2008. Design and evaluation of a simple signaling device for live traps. Journal of Wildlife Management 72:1434–1436.

Benkobi, L., D. W. Uresk, G. Schenbeck, and R. M. King. 2000. Protocol for monitoring standing crop in grasslands using visual obstruction. Journal of Range Management 53:627–633.

Bennett, A. T. D., I. C. Cuthill, J. C. Partridge, and K. Lunau. 1997. Ultraviolet plumage colors predict mate preference in starlings. Proceedings of the National Academy of Sciences of the United States of America 94:8618–8621.

Bennett, D., K. Hampson, and V. Yngente. 2001. A noose trap for catching a large arboreal lizard, *Varanus olivaceous*. Herpetological Review 32:167–168.

Bennett, D. H., J. W. Gibbons, and J. C. Franson. 1970. Terrestrial activity in aquatic turtles. Ecology 51:738–740.

Bennett, L. T., T. S. Judd, and M. A. Adams. 2000. Close-range vertical photography for measuring cover changes in perennial grasslands. Journal of Range Management 53:634–641.

Bennion, R. S., and W. S. Parker. 1976. Field observations on courtship and aggressive behavior in desert striped whipsnakes, *Masticophis t. taeniatus*. Herpetologica 32:30–35.

Benson, J., and R. M. Suryan. 1999. A leg-noose for capturing adult kittiwakes at the nest. Journal of Field Ornithology 70:393–399.

Bentz, J. A., and P. M. Woodard. 1988. Vegetation characteristics and bighorn sheep use on burned and unburned areas in Alberta. Wildlife Society Bulletin 16:186–193.

Berchielli, L. T., Jr., and B. F. Tullar, Jr. 1980. Comparison of a leg snare with a standard leg-gripping trap. New York Fish and Game Journal 27:63–71.

Berckmoes, V., J. Scheirs, K. Jordaens, R. Blust, T. Backeljau, and R. Verhagen. 2005. Effects of environmental pollution on microsatellite DNA diversity in wood mouse (*Apodemus sylvaticus*) populations. Environmental Toxicology and Chemistry 24:2898–2907.

Berger, D. D., and F. Hamerstrom. 1962. Protecting a trapping station from raptor predation. Journal of Wildlife Management 26:203–206.

Berger, D. D., and H. C. Mueller. 1959. The bal-chatri: a trap for the birds of prey. Bird-Banding 30:18–26.

Berger, D. D., and H. C. Mueller. 1960. Band retention. Bird-Banding 31:90–91.

Bergerud, A. T. 1964. Relationship of mandible length to sex in Newfoundland caribou. Journal of Wildlife Management 28:54–56.

Bergerud, A. T. 1970. Eruption of permanent premolars and molars for Newfoundland caribou. Journal of Wildlife Management 34:962–963.

Bergerud, A. T. 1978. Caribou. Pages 83–101 *in* J. L. Schmidt and D. L. Gilbert, editors. Big game of North America. Stackpole, Harrisburg, Pennsylvania, USA.

Bergerud, A. T. 1988. Population ecology of grouse. Pages 578–685 *in* A. T. Bergerud and M. W. Gratson, editors. Adaptive strategies and population ecology of northern grouse. Volume II. Theory and synthesis. University of Minnesota Press, Minneapolis, USA.

Bergerud, A. T., S. S. Peters, and R. McGrath. 1963. Determining sex and age of willow ptarmigan in Newfoundland. Journal of Wildlife Management 27:700–711.

Bergerud, A. T., and H. L. Russell. 1966. Extraction of incisors of Newfoundland caribou. Journal of Wildlife Management 30:842–843.

Bergman, D. L., R. D. Bluett, and A. R. Tipton. 1999. An alternative method for capturing armadillos. Southwestern Naturalist 40:414–416.

Bergman, D. L., T. B. Veenendaal, B. F. Wakeling, J. D. Eisemann, and T. W. Seamans. 2005. Current and historical use of alpha-chloralose on wild turkeys. Proceedings of the National Wild Turkey Symposium 9:51–57.

Bergmann, P. J., L. D. Flake, and W. L. Tucker. 1994. Influence of brood rearing on female mallard survival and effects of harness-type transmitters. Journal of Field Ornithology 65:151–159.

Bergstrom, K., and K. Larsen. 2004. Putting the squeeze on venomous snakes: accuracy and precision of length measurements taken with the "squeeze box." Herpetological Review 35:235–238.

Beringer, J., L. P. Hansen, R. A. Heinen, and N. F. Giessman. 1994. Use of dogs to reduce damage by deer to a white pine plantation. Wildlife Society Bulletin 22:627–632.

Beringer, J., L. P. Hansen, W. Wilding, J. Fischer, and S. L. Sherrif. 1996. Factors affecting capture myopathy in white-tailed deer. Journal of Wildlife Management 60:373–380.

Beringer, J., J. J. Millspaugh, J. Sartwell, and R. Woeck. 2004. Real-time video recording of food selection by captive white-tailed deer. Wildlife Society Bulletin 32:648–654.

Berliner, L. M. 1996. Hierarchical Bayesian time series models. Pages 15–22 *in* K. Hanson and R. Silver, editors. Maximum entropy and Bayesian methods. Kluwer Academic, Dordrecht, The Netherlands.

Bernadelli, H. 1941. Population waves. Journal of the Burma Research Society 31:1–18.

Berndtson, W. E. 1977. Methods for quantifying mammalian spermatogenesis: a review. Journal of Animal Science 44:818–833.

Berner, A., and L. W. Gysel. 1969. Habitat analysis and management considerations for ruffed grouse for a multiple use area in Michigan. Journal of Wildlife Management 33:769–778.

Berny, P. J., T. Buronfosse, F. Buronfosse, F. Lamarque, and G. Lorgue. 1997. Field evidence of secondary poisoning of foxes (*Vulpes vulpes*) and buzzards (*Buteo buteo*) by bromadiolone, a 4-year survey. Chemosphere 35:1817–1829.

Berryman, A. A. 1981. Population systems: a general introduction. Plenum Press, New York, New York, USA.

Berryman, A. A. 2003. On principles, laws and theory in population ecology. Oikos 103:695–701.

Bertram, B. P., and H. G. Cogger. 1971. A noosing gun for live captures of small lizards. Copeia 1971:371–373.

Best, L. B. 1975. Interpretational errors in the "mapping method" as a census technique. Auk 92:452–460.

Best, P. B. 1976. Tetracycline marking and the rate of growth layer formation in the teeth of a dolphin *(Lagenorhynchus obscurus)*. South African Journal of Science 72:216–218.

Betke, M., D. E. Hirsh, N. C. Makris, G. F. McCracken, M. Procopio, N. I. Hristov, S. Tang, A. Bagchi, J. Reichard, J. Horn, S. Crampton, C. J. Cleveland, and T. H. Kunz. 2008. Thermal imaging reveals significantly smaller Brazilian free-tailed bat colonies than previously estimated. Journal of Mammalogy 89:18–24.

Bettinger, P. 2001. Challenges and opportunities for linking the modeling of forest vegetation dynamics with landscape planning models. Landscape and Urban Planning 56:107–124.

Bettinger, P., and K. Boston. 2008. Habitat and commodity production trade-offs in Coastal Oregon. Socio-Economic Planning Sciences 42:112–128.

Bettinger, P., K. Boston, and J. Sessions. 1999. Combinatorial optimization of elk habitat effectiveness and timber harvest volume. Environmental Modeling and Assessment 4:143–153.

Bettinger, P., D. Graetz, K. Boston, J. Sessions, and W. Chung. 2002. Eight heuristic planning techniques applied to three increasingly difficult wildlife planning problems. Silva Fennica 36:561–584.

Bettinger, P., J. Sessions, and K. Boston. 1997. Using Tabu search to schedule timber harvests subject to spatial wildlife goals for big game. Ecological Modelling 94:111–123.

Betts, P. M., C. W. Giddings, and J. R. Fleeker. 1976. Degradation of 4-aminopyridine in soil. Journal of Agriculture and Food Chemistry 24:571–574.

Beyer, D. E., Jr., and J. B. Haufler. 1994. Diurnal versus 24-hour sampling of habitat use. Journal of Wildlife Management 58:178–180.

Beyer, W. N., J. C. Franson, L. N. Locke, R. K. Stroud, and L. Sileo. 1998. Retrospective study of the diagnostic criteria in a lead-poisoning survey of waterfowl. Archives of Environmental Contamination and Toxicology 35:506–512.

Beyer, W. N., G. H. Heinz, and A. W. Redmon-Norwood. 1996. Environmental contaminants in wildlife: interpreting tissue concentrations. Lewis, Boca Raton, Florida, USA.

Bharathi, T. E., V. K. R. Sathiyanandam, and P. M. M. David. 2004. Attractiveness of some food baits to the melon fruit fly, *Bactrocera cucurbitae* (Coquillett) (Diptera: Tephritidae). International Journal of Tropical Insect Sciences 24:125–134.

Bhattacharyya, A. 1943. On a measure of divergence between two statistical populations defined by their probability distributions. Bulletin of the Calcutta Mathematical Society 35:99–109.

Bibb, M. J., R. A. Van Etten, C. T. Wright, M. W. Walberg, and D. A. Clayton. 1981. Sequence and gene organization of mouse mitochondrial DNA. Cell 26:167–180.

Bibby, C. J., N. D. Burgess, D. A. Hill, and S. Mustoe. 2000. Bird census techniques. Second edition. Academic Press, London, England, UK.

Bidlack, A. L., S. E. Reed, P. J. Palsboll, and W. M. Getz. 2007. Characterization of a western North American carnivore community using PCR-RFLP of cytochrome b obtained from fecal samples. Conservation Genetics 8:1511–1513.

Bidleman, T. F., G. W. Patton, D. A. Hinckley, M. D. Walla, W. E. Cotham, and B. T. Hargrave. 1990. Chlorinated pesticides and polychlorinated biphenyls in the atmosphere of the Canadian Arctic. Pages 347–372 *in* D. A. Kurtz, editor. Long range transport of pesticides. Lewis, Boca Raton, Florida, USA.

Bierregaard, R. O., Jr., and E. S. Harrold. 2008. Behavioral conditioning and techniques for trapping barred owls (*Strix varia*). Journal of Raptor Research 42:210–214.

Biggins, D. E., and E. J. Pitcher. 1978. Comparative efficiencies of telemetry and visual techniques for studying ungulates, grouse, and raptors on energy development lands in southeastern

Montana. National Wildlife Federation Science Technical Series 4:188–193.

Bigler, W. J. 1966. A marking harness for the collared peccary. Journal of Wildlife Management 30:213–214.

Bijlsma, R., J. Bundgaard, and W. F. Van Putten. 1999. Environmental dependence of inbreeding depression and purging in *Drosophila melanogaster*. Journal of Evolutionary Biology 12:1125–1137.

Binkley, C. S., and R. S. Miller. 1980. Survivorship of the whooping crane, *Grus americana*. Ecology 61:434–437.

Binkley, C. S., and R. S. Miller. 1988. Recovery of the whooping crane *Grus americana*. Biological Conservation 45:11–20.

Bird, D. M., and K. L. Bildstein. 2007. Raptor research and management techniques. Hancock House, Blaine, Washington, USA.

Birkhead, R. D., M. I. Williams, S. M. Boback, and M. P. Greene. 2004. The cottonmouth condo: a novel venomous snake transport device. Herpetological Review 35:153–154.

Birkhead, T. R. 1977. The effect of habitat and density on breeding success in the common guillemot (*Uria aalge*). Journal of Animal Ecology 46:751–764.

Birney, E. C., and E. D. Fleharty. 1966. Age and sex comparisons of wild mink. Transactions of the Kansas Academy of Science 69:139–145.

Birney, E. C., and E. D. Fleharty. 1968. Comparative success in the application of aging techniques to a population of winter-trapped mink. Southwestern Naturalist 13:275–282.

Birney, E. C., R. Jenness, and D. D. Baird. 1975. Eye lens proteins as criteria of age in cotton rats. Journal of Wildlife Management 39:718–728.

Bishir, J. W., and R. A. Lancia. 1996. On catch-effort methods of estimating animal abundance. Biometrics 52:1457–1466.

Bishop, C. J., D. J. Freddy, G. C. White, B. E. Watkins, T. R. Stephenson, and L. L. Wolfe. 2007. Using vaginal implant transmitters to aid in capture of mule deer neonates. Journal of Wildlife Management 71:945–954.

Bishop, R. A., and R. Barratt. 1969. Capturing waterfowl in Iowa by nightlighting. Journal of Wildlife Management 33:956–960.

Bitterlich, W. 1948. Die Winkelzahlprobe. Allgemeine Forst- und Holzwirtschaftliche Zeitung 59:4–5.

Bjorndal, K. A. 1980. Demography of the breeding population of the green turtle, *Chelonia mydas*, at Tortuguero, Costa Rica. Copeia 1980:525–530.

Bjorndal, K. A., A. B. Bolten, R. A. Bennett, E. R. Jacobson, T. J. Wronski, J. J. Valeski, and P. J. Eliazar. 1988. Age and growth in sea turtles: limitations of skeletochronology for demographic studies. Copeia 1:23–30.

Bjørnstad, O. N., N. C. Stenseth, and T. Saitoh. 1999. Synchrony and scaling in dynamics of voles and mice in northern Japan. Ecology 80:622–637.

Black, H. C. 1958. Black bear research in New York. Transactions of the North American Wildlife Conference 23:443–461.

Black, J. E., and N. R. Donaldson. 1999. Comments on "Display of bird movements on the WSR-88D: patterns and quantification." Weather and Forecasting 14:1039–1040.

Blackburn, D. G. 1993. Standardized criteria for the recognition of reproductive modes in squamate reptiles. Herpetologica 49:118–132.

Blackburn, G. A., and E. J. Milton. 1996. Filling the gaps: remote sensing meets woodland ecology. Global Ecology and Biogeography Letters 5:175–191.

Blackmer, J. L., J. A. Byers, and C. Rodriquez-Saona. 2008. Evaluation of color traps for monitoring *Lygus* spp.: design, placement, height, time of day, and non-target effects. Crop Protection 27:171–181.

Blackshaw, S. R. 1994. Tie ups and tie downs: a method for securing rolled nets. North American Bird Bander 19:99.

Blair, W. F. 1941. Techniques for the study of mammal populations. Journal of Mammalogy 22:148–157.

Blakeman, J. A. 1990. Improvement in bal-chatri trap. North American Bird Bander 15:26.

Blakemore, R. 1975. Magnetotactic bacteria. Science 190:377–379.

Blanchard, F. N., and E. B. Finster. 1933. A method of marking living snakes for future recognition, with a discussion of some problems and results. Ecology 14:334–347.

Blanchong, J. A., K. T. Scribner, and S. R. Winterstein. 2002. Assignment of individuals to populations: Bayesian methods and multi-locus genotypes. Journal of Wildlife Management 66:321–329.

Blank, T. H., and J. S. Ash. 1956. Marker for game birds. Journal of Wildlife Management 20:328–330.

Bleich, V. C., C. S. Y. Chun, R. W. Anthes, T. E. Evans, and J. K. Fisher. 2001. Visibility bias and development of a sightability model for tule elk. Alces 37:315–327.

Bleich, V. C., B. M. Pierce, S. G. Torres, and T. Lupo. 2000. Using space age technology to study mountain lion ecology. Outdoor California 3:24–25.

Bleich, V. C., T. R. Stephenson, N. J. Holste, I. C. Snyder, J. P. Marshal, P. W. McGrath, and B. M. Pierce. 2003. Effects of tooth extraction on body condition and reproduction of mule deer. Wildlife Society Bulletin 31:233–236.

Block, W. M., K. A. With, and M. L. Morrison. 1987. On measuring bird habitat: influence of observer variability and sample size. Condor 89:241–251.

Blockstein, D. E. 1985. Active netting to capture nesting mourning doves. North American Bird Bander 10:117–118.

Bloemendal, H. 1977. The vertebrate eye lens. Science 197:127–138.

Blohm, R. J., and P. Ward. 1979. Experience with a decoy trap for male gadwalls. Bird-Banding 50:45–48.

Blokpoel, H. 1976. Bird hazards to aircraft: problems and prevention of bird/aircraft collisions. Clarke, Irwin, and Company, Ottawa, Ontario, Canada.

Blokpoel, H., and W. J. Richards. 1978. Weather and spring migration of snow geese across southern Manitoba. Oikos 30:350–363.

Blomquist, S. M., and M. L. Hunter, Jr. 2007. Externally attached radio-transmitters have limited effects on the antipredator behavior and vagility of *Rana pipiens* and *Rana sylvatica*. Journal of Herpetology 41:430–438.

Bloom, P. H. 1987. Capturing and handling raptors. Pages 99–123 *in* B. A. G. Pendleton, B. A. Millsap, K. W. Cline, and D. M. Bird, editors. Raptor management techniques manual. National Wildlife Federation, Washington, D.C., USA.

Bloom, P. H., J. L. Henckel, E. H. Henckel, J. K. Schmutz, B. Woodbridge, J. R. Bryan, R. L. Anderson, P. J. Detrich, T. L. Maechtle, J. O. McKinley, M. D. McCrary, K. Titus, and P. F. Schempf. 1992. The dho-gaza with great horned owl lure: an analysis of its effectiveness in capturing raptors. Journal of Raptor Research 26:167–178.

Bluestein, H. B., and A. L. Pazmany. 2000. Observations of tornadoes and other convective phenomena with a mobile, 3-mm wavelength, Doppler radar: the spring 1999 field experiment. Bulletin of the American Meteorological Society 81:2939–2952.

Bluett, B. 2000. Trapper education student manual. Division of Wildlife Resources, Illinois Department of Natural Resources, Springfield, USA.

Bluhm, C. K., R. E. Phillips, and W. H. Burke. 1983. Serum levels of luteinizing hormone (LH), prolactin, estradiol, and progesterone in laying and nonlaying canvasback ducks (*Aythya valisineria*). General and Comparative Endocrinology 52:1–16.

Blums, P., J. B. Davis, S. E. Stephens, A. Mednis, and D. M. Richardson. 1999. Evaluation of a plasticine-filled leg band for day-old ducklings. Journal of Wildlife Management 63:656–663.

Blums, P., A. Mednis, and J. D. Nichols. 1994. Retention of web tags and plasticine-filled leg bands applied to day-old ducklings. Journal of Wildlife Management 58:76–81.

Blums, P., V. K. Reders, A. A. Mednis, and J. A. Baumanis. 1983. Automatic drop-door traps for ducks. Journal of Wildlife Management 47:199–203.

Blums, P., C. W. Shaiffer, and L. H. Fredrickson. 2000. Automatic multi-capture nest box trap for cavity-nesting ducks. Wildlife Society Bulletin 28:592–596.

Blundell, G. M., J. W. Kern, R. T. Bowyer, and L. K. Duffy. 1999. Capturing river otters: a comparison of Hancock and leg-hold traps. Wildlife Society Bulletin 27:184–192.

Blus, L. J. 2003. Organochlorine pesticides. Pages 313–339 *in* D. J. Hoffman, B. A. Rattner, G. A. Burton, Jr., and J. Cairns, Jr., editors. Handbook of ecotoxicology. Second edition. Lewis, Boca Raton, Florida, USA.

Blus, L. J., S. N. Wiemeyer, and C. J. Henny. 1996. Organochlorine pesticides. Pages 61–70 *in* A. Fairbrother, L. N. Locke, and G. L. Huff, editors. Noninfectious diseases of wildlife. Second edition. Iowa State University Press, Ames, USA.

Boag, D. A. 1965. Indicators of sex, age, and breeding phenology in blue grouse. Journal of Wildlife Management 29:103–108.

Boag, D. A., and M. A. Schroeder. 1992. Spruce grouse. Account 5 in A. Poole, P. Stettenheim, and F. Gill, editors. The birds of North America. The Academy of Natural Sciences, Philadelphia, Pennsylvania, and The American Ornithologists' Union, Washington, D.C., USA.

Boag, D. A., A. Watson, and R. Parr. 1973. Radio-marking versus back-tabbing red grouse. Journal of Wildlife Management 37:410–412.

Boarman, W. I., M. L. Beigel, G. C. Goodlett, and M. Sazaki. 1998. A passive integrated transponder system for tracking animal movements. Wildlife Society Bulletin 26:886–891.

Bobek, B., and R. Bergstrom. 1978. A rapid method of browse biomass estimation in a forest habitat. Journal of Range Management 31:456–458.

Bodie, W. L., E. O. Garton, E. R. Taylor, and M. McCoy. 1995. A sightability model for bighorn sheep in canyon habitats. Journal of Wildlife Management 59:832–840.

Bodkin, J. L., J. A. Ames, R. J. Jameson, A. M. Johnson, and G. M. Matson. 1997. Estimating age of sea otters with cementum layers in the first premolar. Journal of Wildlife Management 61:967–973.

Boelman, N. T., Asner, G. P., Hart, P. J. and R. E. Martin. 2007. Multi-trophic invasion resistance in Hawaii: bioacoustics, field surveys, and airborne remote sensing. Ecological Applications 17:2137–2144.

Boersma, P. D., and G. A. Rebstock. 2009. Flipper bands do not affect foraging-trip duration of Magellanic penguins. Journal of Field Ornithology 80:408–418.

Boersma, P. D., and G. A. Rebstock. 2010. Effects of double bands on Magellanic penguins. Journal of Field Ornithology 81:195–205.

Boessenkool, S., J. J. Austin, T. H. Worthy, P. Scofield, A. Cooper, P. J. Seddon, and J. M. Waters. 2009. Relict or colonizer? Extinction and range expansion of penguins in southern New Zealand. Proceedings of the Royal Society B 276:815–821.

Boggess, E. K., G. R. Batcheller, R. G. Linscombe, J. W. Greer, M. Novak, S. B. Linhart, D. W. Erickson, A. W. Todd, D. C. Juve, and D. A. Wade. 1990. Traps, trapping, and furbearer management. Technical Review 90-1, The Wildlife Society, Bethesda, Maryland, USA.

Bohl, W. H. 1956. Experiments in locating wild chukar partridge by use of recorded calls. Journal of Wildlife Management 20:83–85.

Boitani, L., and T. K. Fuller, editors. 2000. Research techniques in animal ecology: controversies and consequences. Columbia University Press, New York, New York, USA.

Bojorquez-Tapia, L. A., I. Azuara, P. Balvanera, A. D. Cuaron, L. A. Pena, A. Ramirez, C. Alverez, and M. L. Alquicira. 1995. Predicting and mapping species-rich areas for environmental assessments with limited data. Pages 546–550 in J. A. Bissonette and P. R. Krausman, editors. Integrating people and wildlife for a sustainable future. The Wildlife Society, Bethesda, Maryland, USA.

Bojorquez-Tapia, L. A., L. P. Brower, G. Castilleja, S. Sanchez-Colon, M. Hernandez, W. Calvert, S. Diaz, P. Gomez-Priego, G. Alcantar, and E. D. Melgarejo. 2003. Mapping expert knowledge: redesigning the Monarch Butterfly Biosphere Reserve. Conservation Biology 17:367–379.

Boles, L. C., and K. J. Lohmann. 2003. True navigation and magnetic maps in spiny lobsters. Nature 421:60–63.

Bolker, B. M. 2008. Ecological models and data in R. Princeton University Press, Princeton, New Jersey, USA.

Bolker, B. M., M. E. Brooks, C. J. Clark, S. W. Geange, J. R. Poulsen, M. H. H. Stevens, and J. S. White. 2008. Generalized linear mixed models: a practical guide for ecology and evolution. Trends in Ecology and Evolution 24:127–135.

Bolstad, W. M. 2007. Introduction to Bayesian statistics. Second edition. John Wiley and Sons, Hoboken, New Jersey, USA.

Bonaccorso, F. J., and N. Smythe. 1972. Punch-marking bats: an alternative to banding. Journal of Mammalogy 53:389–390.

Bonaccorso, F. J., N. Smythe, and S. R. Humphrey. 1976. Improved techniques for marking bats. Journal of Mammalogy 57:181–182.

Bonham, C. D. 1989. Measurements for terrestrial vegetation. John Wiley and Sons, New York, New York, USA.

Bonin, A., E. Bellemain, P. B. Eidesen, F. Pompanon, C. Brochmann, and P. Taberlet. 2004. How to track and assess genotyping errors in population genetics studies. Molecular Ecology 13:3261–3273.

Bonter, D. N., E. W. Brooks, and T. M. Donovan. 2008. What are we missing with only ground-level mist nets? Using elevated nets at a stopover site. Journal of Field Ornithology 79:314–320.

Bonter, D. N., S. A. Gauthreaux, Jr., and T. M. Donovan. 2009. Characteristics of important stopover locations for migrating birds: remote sensing with radar in the Great Lakes Basin. Conservation Biology 23:440–448.

Bookhout, T. A. 1964. Prenatal development of snowshoe hares. Journal of Wildlife Management 28:338–345.

Bookhout, T. A. 1995. Yellow rail. Account 139 in A. Poole, P. Stettenheim, and F. Gill, editors. The birds of North America. The Academy of Natural Sciences, Philadelphia, Pennsylvania, and The American Ornithologists' Union, Washington, D.C., USA.

Bookhout, T. A., editor. 1994. Research and management techniques for wildlife and habitats. The Wildlife Society, Bethesda, Maryland, USA.

Boonstra, R., and I. T. M. Craine. 1986. Natal nest location and small mammal tracking with a spool and line technique. Canadian Journal of Zoology 64:1034–1036.

Boonstra, R., C. J. Krebs, and N. C. Stenseth. 1998. Population cycles in small mammals: the problem of explaining the low phase. Ecology 79:1479–1488.

Borchers, D. L., S. T. Buckland, and W. Zucchini. 2002. Estimating animal abundance: closed populations. Statistics for biology and health series, Springer-Verlag, London, England, UK.

Borden, J. A., and G. J. Langford. 2008. A simple pitfall trap for sampling nesting diamondback terrapins. Herpetological Review 39:188–190.

Borg, C. K., S. K. Hoss, L. L. Smith, and L. M. Conner. 2004. A method for preventing flying squirrel mortality in PVC pipe tree frog refugia. Wildlife Society Bulletin 32:1313–1315.

Börger, L., N. Franconi, G. de Michele, A. Gantz, F. Meschi, A. Manica, S. Lovari, and T. Coulson. 2006. Effects of sampling regime on the mean and variance of home range size estimates. Journal of Animal Ecology 75:1393–1405.

Boroski, B. B., and P. L. McGlaughlin. 1994. An efficient technique for capturing swimming deer. California Fish and Game 80:36–42.

Bortolotti, G. R. 1984. Sexual size dimorphism and age-related size variation in bald eagles. Journal of Wildlife Management 48:72–81.

Boss, A. S. 1963. Aging the nests and young of the American coot. Thesis, University of Minnesota, St. Paul, USA.

Boston, K., and P. Bettinger. 2001. Development of spatially feasible forest plans: a comparison of two modeling approaches. Silva Fennica 35:425–435.

Botelho, E. S., and P. C. Arrowood. 1995. A novel, simple, safe and effective trap for burrowing owls and other fossorial animals. Journal of Field Ornithology 66:380–384.

Bothma, J. du P., J. G. Teer, and C. E. Gates. 1972. Growth and age determination of the cotton-tail in south Texas. Journal of Wildlife Management 36:1209–1221.

Boughton, R. G., and J. Staiger. 2000. Use of PVC pipe refugia as a sampling technique for hylid treefrogs. American Midland Naturalist 144:168–177.

Boulanger, J., S. Himmer, and C. Swan. 2004. Monitoring of grizzly bear population trends and demography using DNA mark–recapture methods in the Owikeno Lake area of British Columbia. Canadian Journal of Zoology—Revue Canadienne de Zoologie 82:1267–1277.

Boulanger, J. R., L. L. Bigler, P. D. Curtis, D. H. Lein, and A. J. Lembo. 2006. A polyvinyl chloride bait station for dispensing rabies vaccine to raccoons in suburban landscapes. Wildlife Society Bulletin 34:1206–1211.

Bourgoin, G., M. Garel, D. Dubray, D. Maillard, and J.-M. Gillard. 2008. What determines Global Positioning System fix success when monitoring free-ranging mouflon? European Journal of Wildlife Research 55:603–613.

Bourque, R. M. 2007. An alternative method for restraining frogs. Herpetological Review 38:48–49.

Bouzat, J. L. 2000. The importance of control populations for the identification and management of genetic diversity. Genetica 110:109–115.

Bouzat, J. L. 2001. The population genetic structure of the greater rhea (*Rhea americana*) in an

agricultural landscape. Biological Conservation 99:277–284.
Bouzat, J. L., H. H. Cheng, H. A. Lewin, R. L. Westemeier, J. D. Brawn, and K. N. Paige. 1998a. Genetic evaluation of a demographic bottleneck in the greater prairie chicken. Conservation Biology 12:836–843.
Bouzat, J. L., J. A. Johnson, J. E. Toepfer, S. A. Simpson, T. L. Esker, and R. L. Westemeier. 2009. Beyond the beneficial effects of translocations as an effective tool for the genetic restoration of isolated populations. Conservation Genetics 10:191–201.
Bouzat, J. L., H. A. Lewin, and K. N. Paige. 1998b. The ghost of genetic diversity past: historical DNA analysis of the greater prairie chicken. American Naturalist 152:1–6.
Bowden, D. C., A. E. Anderson, and D. E. Medin. 1984. Sampling plans for mule deer sex and age ratios. Journal of Wildlife Management 48:500–509.
Bowden, D. C., and R. C. Kufeld. 1995. Generalized mark–sight population size estimation applied to Colorado moose. Journal of Wildlife Management 59:840–851.
Bowen, B. W., A. B. Meylen, and J. C. Avise. 1991. Evolutionary distinctiveness of the endangered Kemp's ridley sea turtle. Nature 352:709–711.
Bowen, B. W., A. B. Meylen, J. P. Ross, C. J. Limpus, G. H. Balazs, and J. C. Avise. 1992. Global population structure and natural history of the green turtle (*Chelonia mydas*) in terms of matriarchal phylogeny. Evolution 46:865–881.
Bowen, D. E., R. J. Robel, and P. G. Watt. 1976. Habitat and investigators influence artificial ground nest losses: Kansas. Transactions of the Kansas Academy of Science 79:141–147.
Bowen, W. O. 1982. Determining the age of coyotes, *Canis latrans*, by tooth sections and tooth wear patterns. Canadian Field-Naturalist 96:339–341.
Bowman, A. W., and A. Azzalini. 1997. Applied smoothing techniques for data analysis: the kernel approach with S-Plus illustrations. Clarendon Press, Oxford, England, UK.
Bowman, J. L., and H. A. Jacobson. 1998. An improved vaginal-implant transmitter for locating white-tailed deer birth sites and fawns. Wildlife Society Bulletin 26:295–298.
Bowman, T. D., S. P. Thompson, C. A. Janik, and L. J. Dubuc. 1994. Nightlighting minimizes investigator disturbance in bird colonies. Colonial Waterbirds 17:78–82.
Box, G. E. P. 1979. Robustness in the strategy of scientific model building. Pages 201–236 in R. L. Launer and G. N. Wilkinson, editors. Robustness in statistics. Academic Press, New York, New York, USA.
Boyce, M. S. 1987. Time-series analysis and forecasting of the Aransas/Wood Buffalo whooping crane population. Pages 1–9 in J. C. Lewis, editor. Proceedings of the 1985 crane workshop. Whooping Crane Maintenance Trust, Grand Island, Nebraska, USA.
Boyce, M. S. 1992. Population viability analysis. Annual Review of Ecology and Systematics 23:481–506.

Boyce, M. S., and L. L. McDonald. 1999. Relating populations to habitats using resource selection functions. Trends in Ecology and Evolution 14:268–272.
Boyce, M. S., and R. S. Miller. 1985. Ten-year periodicity in whooping crane census. Auk 102:658–660.
Boyce, M. S., P. R. Vernier, S. E. Nielson, and F. K. A. Schmiegelow. 2002. Evaluating resource selection functions. Ecological Modelling 157:281–300.
Boyd, R. 1992. Constructivism, realism, and philosophical method. Pages 131–198 in J. Earman, editor. Inference, explanation, and other frustrations: essays in the philosophy of science. University of California Press, Berkeley, USA.
Boyd, R. J., A. Y. Cooperrider, P. C. Lent, and J. A. Bailey. 1986. Ungulates. Pages 519–564 in Y. Cooperrider, R. J. Boyd, and H. R. Stuart, editors. Inventory and monitoring of wildlife habitat. Service Center, Bureau of Land Management, Department of the Interior, Denver, Colorado, USA.
Bradbury, J. W. 1977. Lek mating behavior in the hammer-headed bat. Zeitschrift für Tierpsychologie 45:225–255.
Bradbury, J. W., and S. L. Vehrencamp. 1998. Principles of animal communication. Sinauer Associates, Sunderland, Massachusetts, USA.
Bradbury, R. B., R. A. Hill, D. C. Mason, S. A. Hinsley, J. D. Wilson, H. Balzter, G. Q. A. Anderson, M. J. Whittingham, I. J. Davenport, and P. E. Bellamy. 2005. Modeling relationships between birds and vegetation structure using airborne lidar data: a review with case studies from agricultural and woodland environments. Ibis 147:443–452.
Bradbury, S. P. 1996. 2,3,7,8-Tetrachlorodibenzo-p-dioxin. Pages 87–98 in A. Fairbrother, L. N. Locke, and G. L. Huff, editors. Noninfectious diseases of wildlife. Second edition. Iowa State University Press, Ames, USA.
Braden, A. W., R. R. Lopez, C. W. Roberts, N. J. Silvy, C. B. Owen, and P. A. Frank. 2008. Florida Key deer *Odocoileus virginianus clavium* underpass use and movements along a highway corridor. Wildlife Biology 14:155–163.
Bradley, J. A., D. Sescord, and L. Prins. 1981. Age determination in the arctic fox (*Alopex lagopus*). Canadian Journal of Zoology 59:1976–1979.
Bradshaw, G. A., and T. A. Spies. 1992. Characterizing canopy gap structure in forests using wavelet analysis. Journal of Ecology 80:205–215.
Brady, J. R., and M. R. Pelton. 1976. An evaluation of some cottontail rabbit marking techniques. Journal of Tennessee Academy of Science 51:89–90.
Braid, M. R. 1974. A bal-chatri trap for basking turtles. Copeia 1974:539–540.
Brambell, F. W. R. 1948. Prenatal mortality in mammals. Biological Review, Cambridge Philosophical Society 23:370–407.
Brandborg, S. M. 1955. Life history and management of the mountain goat in Idaho. Wildlife Bulletin 2, Idaho Department of Fish and Game, Boise, USA.

Brander, R. B. 1973. Life-history notes on the porcupine in a hardwood-hemlock forest in upper Michigan. Michigan Academician 5:425–433.
Brandes, D., and D. W. Ombalski. 2004. Modeling raptor migration pathways using a fluid-flow analogy. Journal of Raptor Research 38:195–207.
Bransby, D. I., A. G. Matches, and G. F. Krause. 1977. Disk meter for rapid estimation of herbage yield in grazing trials. Agronomy Journal 69:393–396.
Brassard, J. S., and R. Bernard. 1939. Observations on breeding and development of martens, *Martes a. americana* (Ken). Canadian Field-Naturalist 53:15–21.
Brattstrom, B. H. 1996. The "skink scooper": a device for catching leaf litter skinks. Herpetological Review 27:189.
Braude, S., and D. Ciszek. 1998. Survival of naked mole-rats marked by implantable transponders and toe clipping. Journal of Mammalogy 79:360–363.
Braun, C. E. 1971. Determination of blue grouse sex and age from wing characteristics. Game Information Leaflet 86, Colorado Division of Game, Fish and Parks, Fort Collins, USA.
Braun, C. E. 1976. Methods for locating, trapping, and banding band-tailed pigeons in Colorado. Special Report 39, Colorado Division of Wildlife, Fort Collins, USA.
Braun, C. E. 1998. Sage grouse declines in western North America: what are the problems? Proceedings of the Western Association of State Fish and Wildlife Agencies 78:139–156.
Braun, C. E., editor. 2005. Techniques for wildlife investigations and management. The Wildlife Society, Bethesda, Maryland. USA.
Braun, C. E., K. Martin, and L. A. Robb. 1993. White-tailed ptarmigan. Account 68 in A. Poole and F. Gill, editors. The birds of North America. The Academy of Natural Sciences, Philadelphia, Pennsylvania, and The American Ornithologists' Union, Washington, D.C., USA.
Braun, C. E., and G. E. Rogers. 1967. Determination of age and sex of the southern white-tailed ptarmigan. Game Information Leaflet 54, Colorado Division of Game, Fish and Parks, Fort Collins, USA.
Braun, C. E., R. K. Schmidt, Jr., and F. E. Boggers. 1973. Census of Colorado white-tailed ptarmigan with tape recorded calls. Journal of Wildlife Management 37:90–93.
Braun-Blanquet, J., G. D. Fuller, and H. S. Conard. 1965. Plant sociology: the study of plant communities. Second edition. Hafner, London, England, UK.
Bray, Y., S. Champely, and D. Soyez. 2002. Age determination in leverets of European hare *Lepus europaeus* based on body measurements. Wildlife Biology 8:31–39.
Bray, Y., É. Marboutin, R. Péroux, and J. Ferron. 2003. Reliability of stained placental-scar counts in European hares. Wildlife Society Bulletin 31:237–246.
Breault, A. M., and K. M. Cheng. 1990. Use of submerged mist nets to capture diving birds. Journal of Field Ornithology 61:328–330.

Breck, S. W., N. Lance, and J. Bolurassa. 2006. Limitations of receiver/data loggers for monitoring radiocollared animals. Wildlife Society Bulletin 34:111–115.

Breckenridge, W. J., and J. R. Tester. 1961. Growth, local movements and hibernation of the Manitoba toad, *Bufo hemiophrys*. Ecology 42:637–646.

Breeden, J. B., F. Hernández, R. L. Bingham, N. J. Silvy and G. L. Waggerman. 2008. Effects of traffic noise on auditory surveys of urban white-winged doves. Wilson Journal of Ornithology 120:384–389.

Breitenbach, G. L., J. D. Congdon, and R. C. Van Loben Sels. 1984. Winter temperatures of *Chrysemys picta* nests in Michigan: effects on hatchling survival. Herpetologica 40:76–81.

Breitenmoser, U. 1989. A foot-snare for medium sized carnivores. Cat News 11:20.

Brekke, E. B. 1988. Using GIS to determine the effects of CO_2 development on elk calving in south-central Colorado. Technical Note 381, BLM/YA/PT-88/002+6600, U.S. Bureau of Land Management Service Center, Department of the Interior, Denver, Colorado, USA.

Brennan, L. A. 1999. Northern bobwhite. Account 397 *in* A. Poole and F. Gill, editors. The birds of North America. The Academy of Natural Sciences, Philadelphia, Pennsylvania, and The American Ornithologists' Union, Washington, D.C., USA.

Brennan, L. A., and W. M. Block. 1985. Sex determination of mountain quail reconsidered. Journal of Wildlife Management 49:475–476.

Breslow, N. E., and D. G. Clayton. 1993. Approximate inference in generalized linear mixed models. Journal of the American Statistical Association 88:9–25.

Bretagnolle, V., J. C. Thibault, and J. M. Dominici. 1994. Field identification of individual ospreys using head marking pattern. Journal of Wildlife Management 58:175–178.

Breton, A. R., A. W. Diamond, and S. W. Kress. 2006. Surface wear of incoloy and darvic bands on Atlantic puffin adults and chicks. Journal of Field Ornithology 77:111–119.

Bridges, A. S., C. Olfenbuttel, and M. R. Vaughan. 2002. A mixed regression model to estimate neonatal black bear cub age. Wildlife Society Bulletin 30:1253–1258.

Briggs, J. L., and R. M. Storm. 1970. Growth and population structure of the cascade frog, *Rana cascadae* Slater. Herpetologica 26:283–300.

Brigham, R. M., M. B. Fenton, and H. D. N. J. Aldridge. 1998. Flight speed of foraging common nighthawks (*Chordeiles minor*): does the measurement technique matter? American Midland Naturalist 139:325–330.

Bringi, V. N., and V. Chandrasekar. 2001. Polarimetric Doppler weather radar. Cambridge University Press, Cambridge, England, UK.

Brinkman, T. J., and K. Hundertmark. 2009. Sex identification of northern ungulates using low quality and quantity DNA. Conservation Genetics 10:1189–1193.

Brisbin, Jr., I. L., and T. B. Mowbray. 2002. American coot. Account 697 *in* A. Poole and F. Gill, editors. The birds of North America. The Academy of Natural Sciences, Philadelphia, Pennsylvania, and The American Ornithologists' Union, Washington, D.C., USA.

Britten, H. B. 1996. Meta-analyses of the association between multilocus heterozygosity and fitness. Evolution 50:2158–2164.

Britten, M. W., P. L. Kennedy, and S. Ambrose. 1999. Performance and accuracy evaluation of small satellite transmitters. Journal of Wildlife Management 63:1349–1358.

Bro, E., J. Clobert, and F. Reitz. 1999. Effects of radiotransmitters on survival and reproductive success of gray partridge. Journal of Wildlife Management 63:1044–1051.

Brock, R. E., and D. A. Kelt. 2004. Conservation and social structure of Stephens' kangaroo rat: implications from burrow-use behavior. Journal of Mammalogy 85:51–57.

Brocke, R. H. 1972. A live snare for trap-shy snowshoe hares. Journal of Wildlife Management 36:988–991.

Brodsky, L. M. 1988. Ornament size influences mating success in male rock ptarmigan. Animal Behavior 36:662–667.

Broms, K. M., J. R. Skalski, J. J. Millspaugh, C. A. Hagen, and J. H. Schulz. 2010. Using statistical population reconstruction to estimate demographic trends in small game populations. Journal of Wildlife Management 74:310–317.

Bronson, F. H. 1979. The reproductive ecology of the house mouse. Quarterly Review of Biology 54:265–299.

Bronson, F. H. 1989. Mammalian reproductive biology. University of Chicago Press, Chicago, Illinois, USA.

Brook, B. W., J. R. Cannon, R. C. Lacy, C. Mirande, and R. Frankham. 1999. Comparison of the population viability analysis packages GAPPS, INMAT, RAMAS, and VORTEX for the whooping crane (*Grus americana*). Animal Conservation 2:23–31.

Brook, R. W., and R. G. Clark. 2002. Retention and effects of nasal markers and subcutaneously implanted radio transmitters on breeding female lesser scaup. Journal of Field Ornithology 73:206–212.

Brooks, P. M. 1981. Comparative longevity of a plastic and a new machine-belting collar on large African ungulates. South African Journal of Wildlife Research 11:143–145.

Brooks, R. P., and W. E. Dodge. 1978. A night identification collar for beavers. Journal of Wildlife Management 42:448–452.

Broom, D. M., and K. G. Johnson. 1993. Stress and animal welfare. Chapman and Hall, London, England, UK.

Broseth, H., and H. C. Pedersen. 2000. Hunting effort and game vulnerability studies on a small scale: a new technique combining radio-telemetry, GPS, and GIS. Journal of Applied Ecology 37:182–190.

Broseth, H., J. Tufto, H. C. Pedersen, H. Steen, and L. Kastdalen. 2005. Dispersal pattern in a harvested willow ptarmigan population. Journal of Applied Ecology 42:453–459.

Broughton, R. K., S. A. Hinsley, P. E. Bellamy, R. A. Hill, and P. Rothery. 2006. Marsh tit *Poecile palustris* territories in a British broad-leaved wood. Ibis 148:744–752.

Brown, C. S., and J. R. Luebbert. 2002. Arctic research: spectacled eiders, surgery, satellites, and summer solstice. Journal of Avian Medicine and Surgery 16:53–56.

Brown, D. E. 1977. White-winged dove (*Zenaida asiatica*). Pages 247–272 *in* G. C. Sanderson, editor. Management of migratory shore and upland game birds in North America. International Association of Fish and Wildlife Agencies, Washington, D.C., USA.

Brown, D. L. 1981. The helinet: a device for capturing prairie chickens and ring-necked pheasants. Proceedings of the Annual Conference of the Southeastern Association of Fish and Wildlife Agencies 35:92–96.

Brown, D. M. 1954. Methods of surveying and measuring vegetation. Commonwealth Agriculture Bureau, Farnham Royal, Bucks, England, UK.

Brown, J., and S. Gehrt. 2009. The basics of using remote cameras to monitor wildlife. W-21-09, Ohio State University, Columbus, USA.

Brown, J. A. 2003. Designing an efficient adaptive cluster sample. Environmental and Ecological Statistics 10:95–105.

Brown, J. K. 1976. Estimating shrub biomass from basal stem diameters. Canadian Journal of Forest Research 6:153–158.

Brown, J. L. 1998. The new heterozygosity theory of mate choice and the MHC. Genetica 104:215–221.

Brown, J. L. 2000. Reproductive endocrine monitoring of elephants: an essential tool for Assisting captive management. Zoo Biology 19:347–367.

Brown J. L., F. Goritz, N. Pratt-Hawkes, R. Hermes, M. Galloway, L. H. Graham, C. Gray, S. L. Walker, A. Gomez, R. Moreland, S. Murray, D. L. Schmitt, J. Howard, J. Lehnhardt, B. Beck, A. Bellem, R. Montali, and T. B. Hildebrandt. 2004. Successful artificial insemination of an Asian elephant at the National Zoological Park. Zoo Biology 23:45–63.

Brown, J. L., D. L. Schmitt, A. Bellem, L. H. Graham, and J. Lehnhardt. 1999. Hormone secretion in the Asian elephant (*Elephas maximus*): characterization of ovulatory and anovulatory luteinizing hormone surges. Biology of Reproduction 61:1294–1299.

Brown, L. J. 1997. An evaluation of some marking and trapping techniques currently used in the study of anuran population dynamics. Journal of Herpetology 31:410–419.

Brown, L. N. 1961. Excreted dyes used to determine movements of cottontail rabbits. Journal of Wildlife Management 25:199–202.

Brown, M. W. 1983. A morphometric analysis of sexual and age variation in the American marten (*Martes americana*). Thesis, University of Toronto, Toronto, Ontario, Canada.

Brown, S. G. 1978. Whale marking techniques. Pages 71–80 *in* B. Stonehouse, editor. Animal marking: recognition marking of animals in research. MacMillan, London, England, UK.

Brown, T. A. 1999. Genomes. BIOS Scientific, New York, New York, USA.

Brown, W. C., and A. C. Alcala. 1970. Population ecology of the frog, *Rana erythraea*, in southern Negros, Philippines. Copeia 1970:611–622.

Brown, W. M., M. George, Jr., and A. C. Wilson. 1979. Rapid evolution of animal mitochondrial DNA. Proceedings of the National Academy of Sciences of the United States of America 76:1967–1971.

Brown, W. M., E. M. Prager, A. Wang, and A. C. Wilson. 1982. Mitochondrial DNA sequences of primates: tempo and mode of evolution. Journal of Molecular Evolution 18:225–239.

Brown, W. S., V. P. J. Gannon, and D. M. Secoy. 1984. Paint-marking the rattle of rattlesnakes. Herpetological Review 15:75–76.

Brown, W. S., and W. S. Parker. 1976. A ventral scale clipping system for permanently marking snakes (Reptilia, Serpentes). Journal of Herpetology 10:247–249.

Browne, C., K. Stafford, and R. Fordham. 2006. The use of scent-detection dogs. Irish Veterinary Journal 59:97–104.

Browne, C. L., and S. J. Hecnar. 2005. Capture success of northern map turtles (*Graptemys geographica*) and other turtle species in basking vs. baited hoop traps. Herpetological Review 36:145–147.

Brownie, C., D. R. Anderson, K. P. Burnham, and D. S. Robson. 1985. Statistical inference from band recovery data—a handbook. Second edition. Resource Publication 156, U.S. Fish and Wildlife Service, Department of the Interior, Washington, D.C., USA.

Brownie, C., and D. S. Robson. 1976. Models allowing for age-dependent survival rates for band-return data. Biometrics 32:305–323.

Brubeck, M. V., B. C. Thompson, and R. D. Slack. 1981. The effects of trapping, banding, and patagial tagging on the parental behavior of least terns in Texas. Colonial Waterbirds 4:54–60.

Bruckner, J. V., and D. A. Warren. 2001. Toxic effects of solvents and vapors. Pages 869–916 *in* C. D. Klaassen, editor. Casarett and Doull's toxicology: the basic science of poisons. Sixth edition. McGraw-Hill, New York, New York, USA.

Bruderer, B. 1971. Radarbeobachtungen über den Frühlingszug im Schweizerischen Mittelland. Der Ornithologische Beobachter 68:89–158.

Bruderer, B. 1994. Nocturnal bird migration in the Negev (Israel)—a tracking radar study. Ostrich 65:204–212.

Bruderer, B. 1997. The study of bird migration by radar. Part 1. The technical basis. Naturwissenschaften 84:1–8.

Bruderer, B., D. Peter, and T. Steuri. 1999. Behaviour of migrating birds exposed to X-band radar and a bright light beam. Journal of Experimental Biology 202:1015–1022.

Bruderer, B., and A. G. Popa-Lisseanu. 2005. Radar data on wing-beat frequencies and flight speeds of two bat species. Acta Chiropterologica 7:73–82.

Bruderer, B., T. Steuri, and M. Baumgartner. 1995. Short-range high-precision surveillance of nocturnal migration and tracking of single targets. Israel Journal of Zoology 41:207–220.

Brugger, K. E., R. F. Labisky, and D. E. Daneke. 1992. Blackbird roost dynamics at Millers Lake, Louisiana—implications for damage control in rice. Journal of Wildlife Management 56:393–398.

Brumfield, R. T., P. Beerli, D. A. Nickerson, and S. V. Edwards. 2003. The utility of single nucleotide polymorphisms in inferences of population history. Trends in Ecology and Evolution 18:249–256.

Brummer, J. E., J. T. Nichols, R. K. Engel, and K. M. Eskridge. 1994. Efficiency of different quadrat sizes and shapes for sampling standing crop. Journal of Range Management 47:84–89.

Bruns Stockrahm, D. M., B. J. Dickerson, S. L. Adolf, and R. W. Seabloom. 1996. Aging black-tailed prairie dogs by weight of eye lenses. Journal of Mammalogy 77:874–881.

Bryan, J. R. 1988. Radio-controlled bow-net for American kestrels. North American Bird Bander 13:30–32.

Bryan, T. W., E. L. Blankenship, and C. Guyer. 1991. A new method of trapping gopher tortoises (*Gopherus polyphemus*). Herpetological Review 22:19–21.

Bub, H. 1991. Bird trapping and bird banding: a handbook for trapping methods all over the world. Cornell University Press, Ithaca, New York, USA.

Bubela, T. M., R. Bartell, and W. J. Muller. 1998. Factors affecting the trapability of red foxes in Kosciusko National Park. Wildlife Research 25:199–208.

Bubenik, A. 2007. Behavior. Pages 173–221 *in* A. Franzmann and C. Schwartz, editors. Ecology and management of the North American moose. University of Colorado Press, Boulder, USA.

Buchanan G. D., and E.V. Younglai. 1986. Plasma progesterone levels during pregnancy in the little brown bat *Myotis lucifugus* (Vespertilionidae). Biology of Reproduction 34:878–884.

Buchholz, R. 2007. Behavioural biology: an effective and relevant conservation tool. Trends in Ecology and Evolution 22:401–407.

Buchler, E. R. 1976. A chemiluminescent tag for tracking bats and other small nocturnal animals. Journal of Mammalogy 57:173–176.

Buck, J. A., and R. A. Craft. 1995. Two walk-in trap designs for great horned owls and red-tailed hawks. Journal of Field Ornithology 66:133–139.

Buckland, S. T., D. R. Anderson, K. P. Burnham, and J. L. Laake. 1993. Distance sampling: estimating abundance of biological populations. Chapman and Hall, New York, New York, USA.

Buckland, S. T., D. R. Anderson, K. P. Burnham, J. L. Laake, D. L. Borchers, and L. J. Thomas. 2001. Introduction to distance sampling: estimating abundance of biological populations. Oxford University Press, Oxford, England, UK.

Buckland, S. T., D. R. Anderson, K. P. Burnham, J. L. Laake, D. L. Borchers, and L. J Thomas, editors. 2004. Advanced distance sampling. Oxford University Press, Oxford, England, UK.

Buckland, S. T., K. P. Burnham, and N. H. Augustin. 1997. Model selection: an integral part of inference. Biometrics 53:603–618.

Buckley, J. L., and W. L. Libby. 1955. Growth rates and age determination in Alaskan beaver. Transactions of the North American Wildlife Conference 20:495–507.

Bucknall, J., J. Suckow, E. Miller, T. Seamans, C. Stern, and V. L. Trumball. 2006. USDA APHIS Wildlife Services employs cooperative partnerships and use of alternative capture methods to retrieve flighted birds affected by oil on the Delaware River. Proceedings of the International Conference on the Effects of Oil on Wildlife 8:29–36.

Buech, R. R., and L. M. Egeland. 2002. Efficacy of three funnel traps for capturing amphibian larvae in seasonal forest ponds. Herpetological Review 33:182–185.

Buehler, D. A. 2000. Bald eagle. Account 425 *in* A. Poole and F. Gill, editors. The birds of North America. The Academy of Natural Sciences, Philadelphia, Pennsylvania, and The American Ornithologists' Union, Washington, D.C., USA.

Bugoni, L., T. S. Neves, F. V. Peppes, and R. W. Furness. 2008. An effective method for trapping scavenging seabirds at sea. Journal of Field Ornithology 79:308–313.

Buler, J., and R. Diehl. 2009. Quantifying bird density during migratory stopover using weather surveillance radar. IEEE Transactions on Geoscience and Remote Sensing 47:2741–2751.

Bull, E. L. 1987. Capture techniques for owls. Pages 291–294 *in* Biology and conservation of northern forest owls. General Technical Report RM-142, U.S. Forest Service, Department of Agriculture, Washington, D.C., USA.

Bull, E. L., and H. D. Cooper. 1996. New techniques to capture pileated woodpeckers and Vaux's swifts. North American Bird Bander 21:138–142.

Bull, E. L., T. W. Heater, and F. G. Culver. 1996. Live-trapping and immobilizing American martens. Wildlife Society Bulletin 24:555–558.

Bull, E. L., and R. J. Pedersen. 1978. Two methods of trapping adult pileated woodpeckers at their nest cavities. North American Bird Bander 3:95–99.

Bull, E. L., R. Wallace, and D. H. Bennett. 1983. Freeze-branding: a long-term marking technique on long-toed salamanders. Herpetological Review 14:81–82.

Bump, G., R. W. Darrow, F. C. Edminster, and W. F. Crissey. 1947. The ruffed grouse: life history, propagation, and management. New York State Conservation Department, Albany, USA.

Bunck, C. M. 1987. Analysis of survival data from telemetry projects. Journal of Raptor Research 21:132–134.

Bunge, M. 1967. Scientific research. The search for system. Volume I. Springer-Verlag, New York, New York, USA.

Bunnel, F. L., and D. J. Vales. 1990. Comparison of methods for estimating forest overstory cover: differences among techniques. Canadian Journal of Forest Research 20:101–107.

Bunnel, S. D., J. A. Rensel, J. F. Kimball, Jr., and M. L. Wolfe. 1977. Determination of sex and age of dusky blue grouse. Journal of Wildlife Management 41:662–666.

Burch, J. W., L. G. Adams, E. H. Follmann, and E. A. Rexstad. 2005. Evaluation of wolf density estimation from radiotelemetry data. Wildlife Society Bulletin 33:1225–1236.

Burgdorf, S. L., D. C. Rudolph, R. H. Connor, D. Saenz, and R. R. Schaefer. 2005. A successful trap design for capturing large terrestrial snakes. Herpetological Review 36:421–424.

Burger, A. E. 1997. Behavior and numbers of marbled murrelets measured with radar. Journal of Field Ornithology 68:208–223.

Burger, A. E., and S. A. Shaffer. 2008. Application of tracking and data-logging technology in research and conservation of seabirds. Auk 125:253–264.

Burger, G. V., R. J. Greenwood, and R. C. Oldenburg. 1970. Alula removal technique for identifying wings of released waterfowl. Journal of Wildlife Management 34:137–140.

Burger, J. 1976. Temperature relationships in nests of the northern diamondback terrapin, *Malaclemys terrapin terrapin*. Herpetologica 32:412–418.

Burger, J., and W. A. Montevecchi. 1975. Nests site selection in the terrapin, *Malaclemys terrapin*. Copeia 1975:113–119.

Burke, P. W., G. W. Witmer, S. M. Jojola, and D. L. Nolte. 2008. Improving nutria trapping success. Proceedings of the Vertebrate Pest Conference 23:59–62.

Burleson, W. H. 1975. A method of mounting plant specimens in the field. Journal of Range Management 28:240–241.

Burley, N. 1985. Leg-band color and mortality patterns in captive breeding populations of zebra finches. Auk 102:647–651.

Burley, N. 1986a. Comparison of the band-color preferences of two species of estrildid finches. Animal Behaviour 34:1732–1741.

Burley, N. 1986b. Sex-ratio manipulation in color-banded populations of zebra finches. Evolution 40:1191–1206.

Burley, N., G. Krantzberg, and P. Radman. 1982. Influence of colour-banding on the conspecific preferences of zebra finches. Animal Behaviour 30:444–455.

Burley, N. T. 1988. Wild zebra finches have band-colour preferences. Animal Behaviour 36:1235–1237.

Burnham, K. P., and D. R. Anderson. 1976. Mathematical models for nonparametric inferences from line transect data. Biometrics. 32:325–336.

Burnham, K. P., and D. R. Anderson. 1979. The composite dynamic method as evidence for age-specific waterfowl mortality. Journal of Wildlife Management 43:356–366.

Burnham, K. P., and D. R. Anderson. 1984. The need for distance data in transect counts. Journal of Wildlife Management 48:1248–1254.

Burnham, K. P., and D. R. Anderson. 1998. Model selection and inference: a practical information-theoretic approach. Springer-Verlag, New York, New York, USA.

Burnham, K. P., and D. R. Anderson. 2001. Kullback-Leibler information as a basis for strong inference in ecological studies. Wildlife Research 28:111–119.

Burnham, K. P., and D. R. Anderson. 2002. Model selection and multimodel inference: a practical information-theoretic approach. Second edition. Springer, New York, New York, USA.

Burnham, K. P., D. R. Anderson., and J. L. Laake. 1980. Estimation of density from line transect sampling of biological populations. Wildlife Monographs 72.

Burnham, K. P., D. R. Anderson, G. C. White, C. Brownie, and K. H. Pollock. 1987. Design and analysis methods for fish survival experiments based on release-recapture. American Fisheries Society Monograph 5.

Burnham, K. P., and P. S. Overton. 1978. Estimation of population size of a closed population when capture probabilities vary among animals. Biometrika 65:625–633.

Burns, R. J., and G. E. Connolly. 1995. Toxicity of compound 1080 livestock protection collars to sheep. Archives of Environmental Contamination and Toxicology 28:141–144.

Burroughs, P. A. 1986. Principles for geographic information systems for land resource assessment. Oxford University Press, New York, New York, USA.

Burrows, G. E., and R. J. Tyrl. 2001. Toxic plants of North America. Iowa State University Press, Ames, USA.

Burt, W. H. 1943. Territoriality and home range concepts as applied to mammals. Journal of Mammalogy 24:346–352.

Burt, W. H., and R. P. Grossenheider. 1998. A field guide to the mammals: North America north of Mexico. Houghton Mifflin, Boston, Massachusetts, USA.

Burton, K. 2004. A tool for raising and lowering mist nets. North American Bird Bander 29:15.

Burtt, H. E., and M. L. Giltz. 1970. A study of blackbird repeats at a decoy trap. Ohio Journal of Science 70:162–170.

Burtt, H. E., and M. L. Giltz. 1976. Sex differences in the tendency for brown-headed cowbirds and red-winged blackbirds to re-enter a decoy trap. Ohio Journal of Science 76:264–267.

Burtt, E. J., Jr., and R. M. Tuttle. 1983. Effect of timing of banding on reproductive success of tree swallows. Journal of Field Ornithology 54:319–323.

Busch, J. D., P. M. Waser, and J. A. DeWoody. 2007. Recent demographic bottlenecks are not accompanied by a genetic signature in banner-tailed kangaroo rats (*Dipodomys spectabilis*). Molecular Ecology 16:2450–2462.

Bush, K. L. 2008. A pressure-operated drop net for capturing greater sage-grouse. Journal of Field Ornithology 79:64–70.

Buskirk, S. W., and J. J. Millspaugh. 2006. Metrics of use and availability in studies of resource selection. Journal of Wildlife Management 70:358–366.

Buss, I. O. 1946. Wisconsin pheasant populations. Publication 326, A-46, Wisconsin Conservation Department, Madison, USA.

Butcher, J. A., J. E. Groce, C. M. Lituma, M. Constanza Cocimano, Y. Sanchez-Johnson, A. J. Campomizzi, T. L. Pope, K. S. Reyna, and A. C. S. Knipps. 2007. Persistent controversy in statistical approaches in wildlife sciences: a perspective of students. Journal of Wildlife Management 71:2142–2144.

Butler, P. J. 1991. Exercise in birds. Journal of Experimental Biology 160:233–262.

Butts, W. K. 1930. A study of the chickadee and white-breasted nuthatch by means of marked individuals. Part I. Methods of marking birds. Bird-Banding 1:149–168.

Buurma, L. S. 1995. Long-range surveillance radars as indicators of bird numbers aloft. Israel Journal of Zoology 41:221–236.

Byers, C. R., R. K. Steinhorst, and P. R. Krausman. 1984. Clarification of a technique for analysis of utilization-availability data. Journal of Wildlife Management 48:1050–1053.

Byers, J. A. 1997. Mortality risk to young pronghorns from handling. Journal of Mammalogy 78:894–899.

Byers, S. M. 1987. Extent and severity of nasal saddle icing on mallards. Journal of Field Ornithology 58:499–504.

Bynum, K. S., J. D. Eisemann, G. C. Weaver, C. A. Yoder, K. A. Fagerstone, and L. A. Miller. 2007. Nicarbazin OvoControl G bait reduces hatchability of eggs laid by resident Canada geese in Oregon. Journal of Wildlife Management 71:135–143.

Byrkjedal, I. 1987. Antipredator behavior and breeding success in greater golden-plover and Eurasian dotterel. Condor 89:40–47.

Byth, K., and B. D. Ripley. 1980. On sampling spatial patterns by distance methods. Biometrics 36:279–284.

Cablk, M. E., and J. S. Heaton. 2006. Accuracy and reliability of dogs in surveying for desert tortoise (*Gopherus agassizii*). Ecological Applications 16:1926–1935.

Caccamise, D. F., and R. S. Hedin. 1985. An aerodynamic basis for selecting transmitter loads in birds. Wilson Bulletin 97:306–318.

Caccamise, D. F., and P. C. Stouffer. 1994. Risks of using alpha-chloralose to capture crows. Journal of Field Ornithology 65:458–460.

Caffrey, C. 2001. Catching crows. North American Bird Bander 26:137–145.

Cagle, F. R. 1939. A system of marking turtles for future identification. Copeia 1939:170–173.

Cagnacci, F., and F. Urbano. 2008. Managing wildlife: a spatial information system for GPS collars data. Environmental Modelling and Software 23:957–959.

Cain, S. A., and G. M. De O. Castro. 1959. Manual of vegetation analysis. Harper and Brothers, New York, New York, USA.

Cain, S. L., and J. I. Hodges. 1989. A floating-fish snare for capturing bald eagles. Journal of Raptor Research 23:10–13.

Caithamer, D. F., R. J. Gates, J. D. Hardy, and T. C. Tacha. 1993. Field identification of age and sex of interior Canada geese. Wildlife Society Bulletin 21:480–487.

Caizergues, A., and L. N. Ellison. 2000. Age-specific reproductive performance of black grouse *Tetrao tetrix* females. Bird Study 47:344–351.

Calenge, C. 2006. The package adehabitat for the R software: a tool for the analysis of space and habitat use by animals. Ecological Modelling 197:516–519.

Caley, P., and B. Ottley. 1995. The effectiveness of hunting dogs for removing feral pigs (*Sus scrofa*). Wildlife Research 22:147–154.

Calkin, D. E., C. A. Montgomery, N. H. Schumaker, S. Polasky, J. L. Arthur, and D. J. Nalle. 2002. Developing a production possibility set of wildlife species persistence and timber harvest value. Canadian Journal of Forest Research 32:1329–1342.

Calkins, J. D., J. C. Hagelin, and D. F. Lott. 1999. California quail. Account 473 *in* A. Poole and F. Gill, editors. The birds of North America. The Academy of Natural Sciences, Philadelphia, Pennsylvania, and The American Ornithologists' Union, Washington, D.C., USA.

Calvert, W., and M. A. Ramsay. 1998. Evaluation of age determination of polar bears by counts of cementum growth layer groups. Ursus 10:449–453.

Calvo, B., and R. W. Furness. 1992. A review of the use and the effects of marks and devices on birds. Journal of Ringing and Migration 13:129–151.

Cameron, G. N., and S. R. Spencer. 2008. Mechanisms of habitat selection by the hispid cotton rat (*Sigmodon hispidus*). Journal of Mammalogy 89:126–131.

Camp, C. D., and D. G. Lovell. 1989. Fishing for "spring lizards": a technique for collecting blackbelly salamanders. Herpetological Review 20:47.

Campbell, B. H., and E. F. Becker. 1991. Neck collar retention in dusky Canada geese. Journal of Field Ornithology 62:521–527.

Campbell, D. L. 1960. A colored leg strip for marking birds. Journal of Wildlife Management 24:431.

Campbell, H. 1972. A population study of lesser prairie chickens in New Mexico. Journal of Wildlife Management 36:689–699.

Campbell, H. W., and S. P. Christman. 1982. Field techniques for herpetofaunal community analysis. Pages 193–200 *in* N. J. Scott, Jr., editor. Herpetological communities. Symposium of the Society for the Study of Amphibians and Reptiles and the Herpetologist's League. Wildlife Research Report 13, U.S. Fish and Wildlife Service, Department of the Interior, Washington, D.C., USA.

Campbell, T. A., and D. B. Long. 2007. Species-specific visitation and removal of baits for delivery of pharmaceuticals to feral swine. Journal of Wildlife Diseases 43:485–491.

Campbell, T. S., P. Irvin, K. R. Campbell, K. Hoffmann, M. Dykes, A. J. Hardin, and S. A. Johnson. 2009. Evaluation of a new technique for marking anurans. Applied Herpetology 6:247–256.

Camper, J. D. 2005. Observations on problems with using funnel traps to sample semi-aquatic snakes. Herpetological Review 36:288–290.

Camper, J. D., and J. R. Dixon. 1988. Evaluation of a microchip marking system for amphibians and reptiles. Research Publication 7100-159, Texas Parks and Wildlife Department, Austin, USA.

Canadian General Standards Board. 1984. Animal traps—humane, mechanically powered, trigger-activated. Report CAN2-144.1-M84, Canadian General Standards Board, Ottawa, Ontario, Canada.

Canadian General Standards Board. 1996. Animal (mammal) traps—mechanically powered, trigger-activated killing traps for use on land. Report CAN/CGSB-144, 1-96, Canadian General Standards Board, Ottawa, Ontario, Canada.

Canadian Wildlife Service and U.S. Fish and Wildlife Service. 1977. North American bird banding techniques. Volume II. Canadian Wildlife Service, Ottawa, Ontario, Canada, and U.S. Fish and Wildlife Service, Department of the Interior, Washington, D.C., USA.

Cançado, P. H. D., E. M. Piranda, G. M. Mourão, and J. L. H. Faccini. 2008. Spatial distribution and impact of cattle-raising on ticks in the Pantanal region of Brazil by using CO_2 tick trap. Parasitology Research 103:371–377.

Cancino, J., V. Sanchez-Sotomayor, and R. Castellanos. 2002. Alternative capture technique for the peninsular pronghorn. Wildlife Society Bulletin 30:256–258.

Canfield, R. H. 1941. Application of the line interception method in sampling range vegetation. Journal of Forestry 39:388–394.

Canfield, R. H. 1944. Measurement of grazing use by the line interception method. Journal of Forestry 42:192–194.

Cannell, P. F. 1984. A revised age/sex key for mourning doves, with comments on the definition of molt. Journal of Field Ornithology 55:112–114.

Canon, S. K., F. C. Bryant, K. N. Bretslaff, and J. M. Hellman. 1997. Pronghorn pregnancy diagnosis using trans-rectal ultrasound. Wildlife Society Bulletin 25:832–834.

Canty, M. J., and A. A. Nielsen. 2008. Automatic radiometric normalization of multitemporal satellite imagery with the iteratively re-weighted MAD transformation. Remote Sensing of Environment 112:1025–1036.

Carbyn, L. N. 1987. Gray wolf and red wolf. Pages 358–376 *in* M. Novak, J. A. Baker, M. E. Obbard, and B. Malloch, editors. Wild furbearer management and conservation in North America. Ontario Ministry of Natural Resources, Toronto, Canada.

Cardoza, J., B. Eriksen, and H. Kilpatrick. 1995. Procedures and guidelines for handling and transporting wild turkeys. Technical Bulletin 3, National Wild Turkey Federation, Edgefield, South Carolina, USA.

Carey, A. B., B. L. Biswell, and J. W. Witt. 1991. Methods for measuring populations of arboreal rodents. General Technical Report PNW-273, U.S. Forest Service, Department of Agriculture, Washington, D.C., USA.

Carlson, T. A., and E. J. Szuch. 2007. Un-weathered new artificial objects effectively sampled plethodontid salamanders in Michigan. Herpetological Review 38:412–415.

Carlström, D., and C. Edelstam. 1946. Methods of marking reptiles for identification after recapture. Nature 158:748–749.

Carney, S. M. 1992. Species, age and sex identification of ducks using wing plumage. U.S. Fish and Wildlife Service, Department of the Interior, Washington, D.C., USA.

Carnio, J., and L. Killmar. 1983. Identification techniques. Pages 39–52 *in* B. B. Beck and C. Wemmer, editors. The biology and management of an extinct species—Pere David's deer. Noyes, Park Ridge, New Jersey, USA.

Caro, T. M. 1986. The function of stotting: a review of the hypotheses. Animal Behaviour 34:649–662.

Caro, T. M. 1994. Cheetahs of the Serengeti Plains: group living in an asocial species. University of Chicago Press, Chicago, Illinois, USA.

Caro, T. M. 1998. The significance of behavioral ecology for conservation biology. Pages 3–30 *in* T. M. Caro, editor. Behavioral ecology and conservation biology. Oxford University Press, Oxford, England, UK.

Carpenter, L. H., and J. I. Innes. 1995. Helicopter netgunning: a successful moose capture technique. Alces 31:181–184.

Carpenter, L. H., D. W. Reichert, and F. Wolfe, Jr. 1977. Lighted collars to aid night observations of mule deer. Research Notes RN-338, U.S. Forest Service, Department of Agriculture, Washington, D.C., USA.

Carr, A., P. Ross, and S. Carr. 1974. Interesting behavior of the green turtle, *Chelonida mydas*, at a mid-ocean island breeding ground. Copeia 1974:703–706.

Carrel, W. K. 1994. Reproductive history of female black bears from dental cementum. International Conference on Bear Research and Management 9:205–212.

Carrick, R., and D. Murray. 1970. Readable band numbers and "Scotchlite" colour bands for the silver gull. Australian Bird Bander 8:51–56.

Carroll, C., W. J. Zielinski, and R. F. Noss. 1999. Using presence-absence data to build and test spatial habitat models for the fisher in the Klamath Region, U.S.A. Conservation Biology 13:1344–1359.

Carroll, J. P. 1993. Gray partridge. Account 58 *in* A. Poole and F. Gill, editors. The birds of North America. The Academy of Natural Sciences, Philadelphia, Pennsylvania, and The American Ornithologists' Union, Washington, D.C., USA.

Carruthers, A. C., A. L. R. Thomas, and G. K. Taylor. 2007. Automatic aeroelastic devices in the wings of a steppe eagle *Aquila nipalensis*. Journal of Experimental Biology 210:4136–4149.

Carruthers, M., and M. R. C. Path. 1983. Instrumental stress tests. Pages 331–362 *in* H. Selye, editor. Selye's guide to stress research. Volume 2. Scientific and Academic Editions, New York, New York, USA.

Carson, J. D. 1961. Epiphyseal cartilage as an age indicator in fox and gray squirrels. Journal of Wildlife Management 25:90–93.

Carstensen, M., G. D. Delgiudice, and B. A. Sampson. 2003. Using doe behavior and vaginal-implant transmitters to capture neonate white-tailed deer in north-central Minnesota. Wildlife Society Bulletin 31:634–641.

Carusi, L. C., M. I. Farace, M. M. Ribicich, and I. E. G. Villafane. 2009. Reproduction and parasitology of *Didelphis albiventris* (Didelphimorphia) in an agroecosystem landscape in central Argentina. Mammalia 73:89–97.

Carver, A. V., L. W. Burger, Jr., and L. A. Brennan. 1999. Passive integrated transponders and patagial tag markers for northern bobwhite chicks. Journal of Wildlife Management 63:162–166.

Casazza, M. L., G. D. Wylie, and C. J. Gregory. 2000. A funnel trap modification for surface collection of aquatic amphibians and reptiles. Herpetological Review 31:91–92.

Casella, G., and E. I. George. 1992. Explaining the Gibbs sampler. American Statistician 46:167–174.

Casper, R. M. 2009. Guidelines for the instrumentation of wild birds and mammals. Animal Behaviour 78:1477–1483.

Cassidy, B. G., and R. A. Gonzales. 2005. DNA testing in animal forensics. Journal of Wildlife Management 69:1454–1462.

Castelli, P. M., and S. E. Sleggs. 2000. Efficacy of border collies to control nuisance Canada geese. Wildlife Society Bulletin 28:385–392.

Castelli, P. M., and R. E. Trost. 1996. Neck bands reduce survival of Canada geese in New Jersey. Journal of Wildlife Management 60:891–898.

Castellón, T. D., and K. E. Sieving. 2006. An experimental test of matrix permeability and corridor use by an endemic understory bird. Conservation Biology 20:135–145.

Castro-Arellano, I., C. Madrid-Luna, T. E. Lacher, and L. Leon-Paniagua. 2008. Hair-trap efficacy for detecting mammalian carnivores in the tropics. Journal of Wildlife Management 72:1405–1412.

Caswell, E. B. 1954. A method for sexing blue grouse. Journal of Wildlife Management 18:139.

Caswell, H. 2000. Matrix population models: construction, analysis, and interpretation. Sinauer Associates, Sunderland, Massachusetts, USA.

Catling, P. C., and R. L. Sutherland. 1980. Effect of gonadectomy, season, and the presence of female tammar wallabies (*Macropus eugenii*) on concentrations of testosterone, luteinizing hormone and follicle stimulating hormone in the plasma of male tammar wallabies. Journal of Endocrinology 86:25–33.

Cattet, M., J. Boulanger, G. Stenhouse, R. A. Powell, and M. J. Reynold-Hogland. 2008. An evaluation of long-term capture effects in ursids: implications for wildlife welfare and research. Journal of Mammalogy 89:973–990.

Cattet, M, T. Shury, and R. Patenaude. 2005. The chemical immobilization of wildlife. Second edition. Canadian Association of Zoo and Wildlife Veterinarians, Calgary, Alberta, Canada.

Caudell, J. N., and M. R. Conover. 2007. Drive-by netting: a technique for capturing grebes and other diving waterfowl. Human–Wildlife Conflicts 1:49–52.

Caughley, G. 1966. Mortality patterns in mammals. Ecology 47:906–918.

Caughley, G. 1974. Bias in aerial survey. Journal of Wildlife Management 38:921–933.

Caughley, G. 1977. Analysis of vertebrate populations. John Wiley and Sons, New York, New York, USA.

Caughley, G., and L. C. Birch. 1971. Rate of increase. Journal of Wildlife Management 35:658–663.

Caughley, G., and D. Grice. 1982. A correction factor for counting emus from the air and its application to counts in western Australia. Australian Wildlife Research 9:253–259.

Caughley, G., and C. J. Krebs. 1983. Are big mammals simply little mammals writ large? Oecologia 59:7–17.

Caughley, G., and A. R. E. Sinclair. 1994. Wildlife ecology and management. Blackwell Scientific, Boston, Massachusetts, USA.

Cavanagh, P. M., C. R. Griffin, and E. M. Hoopes. 1992. A technique to color-mark incubating gulls. Journal of Field Ornithology 63:264–267.

Ceballos, G., P. Rodriguez, and R. A. Medellin. 1998. Assessing conservation priorities in megadiverse Mexico: mammalian diversity, endemicity, and endangerment. Ecological Applications 8:8–17.

Chabreck, R. H. 1965. Methods of capturing, marking and sexing alligators. Proceedings of the Annual Conference of the Southeastern Association of Game and Fish Commissioners 17:47–50.

Chabreck, R. H., and J. D. Schroer. 1975. Effects of neck-collars on reproduction of snow geese. Bird-Banding 46:346–347.

Chalfoun, A. D., and T. E. Martin. 2009. Habitat structure mediates predation risk for sedentary prey: experimental tests of alternative hypotheses. Journal of Animal Ecology 78:497–503.

Chalfoun, A. D., F. R. Thompson, III, and M. J. Ratnaswamy. 2002. Nest predators and fragmentation: a review and meta-analysis. Conservation Biology 16:306–318.

Chamberlain, M. J., and B. D. Leopold. 2001. Survival and cause-specific mortality of adult coyotes (*Canis latrans*) in central Mississippi. American Midland Naturalist 145:414–418.

Chamberlin, T. C. 1965 (1890). The method of multiple working hypotheses. Science 148:754–759 (reprint of 1890 paper in Science).

Chambers, J. M., W. S. Cleveland, B. Kleiner, and P. A. Tukey. 1983. Graphical methods for data analysis. Wadsworth International Group, Belmont, California, USA.

Chambers, L. K., G. R. Singleton, and L. A. Hinds. 1999. Fertility control of wild mouse populations: the effects of hormonal competence and an imposed level sterility. Wildlife Research 26:579–591.

Chambers, R. E., and P. F. English. 1958. Modifications of ruffed grouse traps. Journal of Wildlife Management 22:200–202.

Chan, S. S., R. W. McCreight, J. D. Walstad, and T. A. Spies. 1986. Evaluating forest vegetative cover with computerized analysis of fisheye photographs. Forest Science 32:1085–1091.

Chao, A. 1984. Non-parametric estimation of the number of classes in a population. Scandinavian Journal of Statistics 11:265–270.

Chapman, D. G. 1951. Some properties of the hypergeometric distribution with applications to zoological censuses. University of California Publication on Statistics 1:131–160.

Chapman, D. G., and D. S. Robson. 1960. The analysis of a catch curve. Biometrics 16:354–368.

Chapman, J. A., C. J. Henny, and H. M. Wight. 1969. The status, population dynamics, and harvest of the dusky Canada goose. Wildlife Monographs 18.

Chapman, J. A., and G. R. Willner. 1986. Lagomorphs. Pages 453–473 in A. Y. Cooperrider, R. J. Boyd, and H. R. Stuart, editors. Inventory and monitoring of wildlife habitat. Service Center, Bureau of Land Management, Department of the Interior, Denver, Colorado, USA.

Chapman, J. R., S. Nakagawa, D. W. Coltman, J. Slate, and B. C. Sheldon. 2009. A quantitative review of heterozygosity-fitness correlations in animal populations. Molecular Ecology 18:2746–2765.

Charrassin, J. B., Y. H. Park, Y. L. Maho, and C. A. Bost. 2002. Penguins as oceanographers unravel hidden mechanisms of marine productivity. Ecology Letters 5:317–319.

Cheatum, E. L. 1949. The use of corpora lutea for determining ovulation incidence and variations in fertility of white-tailed deer. Cornell Veterinarian 39:282–291.

Cheeseman, C. L., and S. Harris. 1982. Methods of marking badgers (*Meles meles*). Journal of Zoology 197:289–292.

Chemical and Engineering News. 1996. Production by the U.S. chemical industry. http://pubs.acs.org/hotartcl/cenear/960624/prod.html.

Chen, D., and R. C. Hale. 2010. A global review of polybrominated diphenyl ether flame retardant contamination in birds. Environment International 36:800–811.

Chenier, P. J. 2003. Survey of industrial chemistry. Third edition. Kluwer, New York, New York, USA.

Chermack, T., S. Lynham, and W. Ruona. 2001. A review of scenario planning literature. Futures Research Quarterly 17:7–31.

Cherry, S. 1998. Statistical tests in publications of The Wildlife Society. Wildlife Society Bulletin 26:947–953.

Chesser, R. K., O. E. Rhodes, Jr., D. W. Sugg, and A. Schnabel. 1993. Effective sizes for subdivided populations. Genetics 135:1221–1232.

Chesson, J. 1978. Measuring preference in selective predation. Ecology 59:211–215.

Chettri, N., D. C. Deb, E. Sharma, and R. Jackson. 2005. The relationship between bird communities and habitat: a study along a trekking corridor in the Sikkim Himalaya. Mountain Research and Development 25:235–243.

Cheville, N. F., D. R. McCullough, and L. R. Paulson. 1998. Brucellosis in the Greater Yellowstone Area. Volume 16. National Research Council, National Academy Press, Washington, D.C., USA.

Chi, R. Y. 2004. Greater sage-grouse reproductive ecology and tebuthiuron manipulation of dense big sagebrush on Parker Mountain. Thesis, Utah State University, Logan, USA.

Childs, J. E., J. S. Machenzie, and J. A. Richt, editors. 2007. Wildlife and emerging zoonotic diseases: the biology, circumstances and consequences of cross-species transmission. Springer, Berlin, Germany.

Childs, H. E., Jr. 1952. Color bands. Western Bird Banding Association News 27:4.

Chitty, D., and M. Shorten. 1946. Techniques for the study of the Norway rat *Rattus norvegicus*. Journal of Mammalogy 27:63–78.

Choquenot, D., R. J. Kilgour, and B. S. Lukins. 1993. An evaluation of feral pig trapping. Wildlife Research 20:15–22.

Choquet, R. 2007. E-SURGE 1.0 user's manual. http://www.cefe.cnrs.fr/BIOM/logiciels.htm.

Choquet, R., J.-D. Lebreton, O. Gimenez, A.-M. Reboulet, and R. Pradel. 2009a. U-CARE: utilities for performing goodness-of-fit tests and manipulating capture–recapture data. Ecography 32:1071–1074.

Choquet, R., A. M. Reboulet, R. Pradel, O. Gimenez, and J.-D. Lebreton. 2005. M-SURGE 1.8 user's manual. http://www.cefe.cnrs.fr/BIOM/logiciels.htm.

Choquet, R., L. Rouan, and R. Pradel. 2009b. Program E-SURGE: a software application for fitting multi-event models. Pages 845–865 in D. L. Thomson, E. G. Cooch, and M. J. Conroy, editors. Modeling demographic processes in marked populations. Series in environmental and ecological statistics. Volume 3. Springer-Verlag, New York, New York, USA.

Chow, S.-C., J. Shao, and H. Wang. 2008. Sample size calculations in clinical research. Chapman and Hall, Boca Raton, Florida, USA.

Christ, A., J. Ver Hoef, and D. L. Zimmerman. 2008. An animal movement model incorporating home range and habitat selection. Environmental and Ecological Statistics 15:27–38.

Christensen, G. C. 1996. Chukar. Account 258 in A. Poole and F. Gill, editors. The birds of North America. The Academy of Natural Sciences, Philadelphia, Pennsylvania, and The American Ornithologists' Union, Washington, D.C., USA.

Christensen, V., Z. Ferdaña, and J. Steenbeek. 2009. Spatial optimization of protected area placement incorporating ecological, social and economical criteria. Ecological Modelling 220:2583–2593.

Christian, J. J. 1950. The adreno-pituitary system and population cycles in mammals. Journal of Mammalogy 31:247–259.

Christian, J. J. 1963. Endocrine adaptive mechanisms and the physiological regulation of population growth. Pages 189–353 in W. Mayer and R. G. VanGelder, editors. Physiological mammology. Volume1. Mammalian populations. Academic Press, New York, New York, USA.

Christian, J. J., and D. E. Davis. 1964. Endocrines, behavior, and population. Social and endocrine factors are integrated in the regulation of growth of mammalian populations. Science 146:1550–1560.

Christiansen, J. L., and T. Vandewalle. 2000. Effectiveness of three trap types in drift fence surveys. Herpetological Review 31:158–160.

Christiansen, T. J., and C. M. Tate. 2011. Parasites and infectious diseases of greater sage-grouse. Pages 113–126 in S. T. Knick and J. W. Connelly, editors. Greater sage-grouse: ecology and conservation of a landscape species and its habitats. Studies in Avian Biology 38. University of California Press, Berkeley, USA.

Christisen, D. M., and W. H. Kearby. 1984. Mast measurement and production in Missouri (with special reference to acorns). Terrestrial Series 13, Missouri Department of Conservation, Columbia, USA.

Christy, M. T. 1998. A portable device for restraining frogs. Herpetological Review 29:90–91.

Ciofi, C., and G. Chelazzi. 1991. Radiotracking of *Coluber viridiflavus* using external transmitters. Journal of Herpetology 25:37–40.

Clapperton, B. K., C. T. Eason, R. J. Weston, A. D. Woolhouse, and D. R. Morgan. 1994. Development and testing of attractants for feral cats, *Felis catus* L. Wildlife Research 21:389–399.

Clare, E. L., E. E. Fraser, H. E. Braid, M. B. Fenton, and P. D. N. Hebert. 2009. Species on the menu of a generalist predator, the eastern red bat (*Lasiurus borealis*): using a molecular approach to detect arthropod prey. Molecular Ecology 18:2532–2542.

Clark, A. C., Jr., and J. B. De Cruz. 1989. Functional interpretation of protruding filoplumes in oscines. Condor 91:962–965.

Clark, D. R., Jr. 1971. Branding as a marking technique for amphibians and reptiles. Copeia 1971:148–151.

Clark, D. R., Jr. 1991. Bats, cyanide, and gold mining. Bats 9:17–18.

Clark, D. R., Jr., and R. L. Hothem. 1991. Mammal mortality at Arizona, California, and Nevada gold mines using cyanide extraction. California Fish and Game 77:61–69.

Clark, H. O., Jr., and D. L. Plumpton. 2005. A simple one-way door design for passive relocation of western burrowing owls. California Fish and Game 91:286–289.

Clark, J. D., D. L. Clapp, K. L. Smith, and B. Ederington. 1994. Black bear habitat use relation to food availability in the interior highlands of Arkansas. International Conference on Bear Research and Management 9:309–318.

Clark, J. D., J. E. Dunn, and K. G. Smith. 1993a. A multivariate model of female black bear habitat use for a geographic information system. Journal of Wildlife Management 57:519–526.

Clark, J. S. 2003. Uncertainty and variability in demography and population growth: a hierarchical approach. Ecology 84:1370–1381.

Clark, J. S. 2005. Why environmental scientists are becoming bayesians. Ecology Letters 8:2–14.

Clark, J. S. 2007. Models for ecological data: an introduction. Princeton University Press, Princeton, New Jersey, USA.

Clark, L., K. V. Avilova, and N. J. Bean. 1993b. Odor thresholds in passerines. Comparative Biochemistry and Physiology 104A:305–312.

Clark, L., and P. S. Shah. 1992. Information content of prey odor plumes: what do foraging Leach's storm petrels know? Pages 421–427 in R. L. Doty and D. Müller-Schwarze, editors. Chemical signals in vertebrates, Volume VI. Plenum, New York, New York, USA.

Clark, P. E., and M. S. Seyfried. 2001. Point sampling for leaf area index in sagebrush steppe communities. Journal of Range Management 54:589–594.

Clark, R. K., A. W. Franzmann, D. A. Jessup, M. D. Kock, N. D. Kock, R. A. Kock, and P. Morkel. 1992. Wildlife restraint series. Revised edition. International Wildlife Veterinary Service, Salinas, California, USA.

Clark, R. N., and G. H. Stankey. 2006. Integrated research in natural resources: the key role of problem framing. USDA General Technical Report PNW-GTR-678, U.S. Department of Agriculture, Washington, D.C., USA.

Clark, W. S. 1981. A modified dho-gaza trap for use at a raptor banding station. Journal of Wildlife Management 45:1043–1044.

Clarke, R. 1971. The possibility of injuring small whales with the standard discovery whale mark. International Whaling Commission Report 21:106–108.

Clarke, R. D. 1972. The effect of toe clipping on survival in Fowler's toad (*Bufo woodhousei fowleri*). Copeia 1972:182–185.

Clarkson, P., and R. I. Gouldie. 2003. Capture techniques and 1993 banding results for molting harlequin ducks in the Strait of Georgia, B.C. Proceedings of the Harlequin Duck Symposium 2:11–14.

Clary, W. P., and W. C. Leininger. 2000. Stubble height as a tool for management of riparian areas. Journal of Range Management 53:562–573.

Clausen, B., P. Hjort, H. Strandgaard, and P. L. Soerensen. 1984. Immobilization and tagging of muskoxen (*Ovibos moschatus*) in Jameson Land, northeastern Greenland. Journal of Wildlife Disease 20:141–145.

Clawges, R., L. A. Vierling, M. Calhoon, and M. P. Toomey. 2007. Use of a ground-based scanning lidar for estimation of biophysical properties of western larch (*Larix occidentalis*). International Journal of Remote Sensing 28:4331–4344.

Clawges, R., K. Vierling, L. Vierling, and E. Rowell. 2008. The use of airborne lidar to assess avian species diversity, density, and occurrence in a pine/aspen forest. Remote Sensing of Environment 112:2064–2073.

Clawson, R. G., and M. K. Causey. 1995. Dental casts for white-tailed deer age estimation. Wildlife Society Bulletin 23:92–94.

Clayton, D. H., C. L. Hartley, and M. Gochfeld. 1978. Two optical tracking devices for nocturnal field studies of birds. Proceedings of the Colonial Waterbird Group 1978:79–83.

Clemen, R. T. 1996. Making hard decisions: an introduction to decision analysis. Duxbury Press, Pacific Grove, California, USA.

Clemen, R. T., and T. Reilly. 2001. Making hard decisions with DecisionTools. Duxbury Press, Pacific Grove, California, USA.

Clevenger, A. P., J. Wierzchowski, B. Chruszcz, and K. Gunson. 2002. GIS-generated, expert-based models for identifying wildlife habitat linkages and planning mitigation passages. Conservation Biology 16:503–514.

Cline, K. W., and W. S. Clark. 1981. Chesapeake Bay bald eagle banding project: 1981 report and five-year summary. Raptor Information Center, National Wildlife Federation, Washington, D.C., USA.

Cliquet, F., J. Barrat, A. L. Guiot, N. Cael, S. Boutrand, J. Maki, and C. L. Schumacher. 2008. Efficacy and bait acceptance of vaccinia vectored rabies glycoprotein vaccine in captive foxes (*Vulpes vulpes*), raccoon dogs (*Nyctereutes procyonoides*) and dogs (*Canis familiaris*). Vaccine 26:4627–4638.

Clobert, J., E. Danchin, A. A. Dhont, and J. D. Nichols. 2001. Dispersal. Oxford University Press, Oxford, England, UK.

Clover, M. R. 1954. A portable deer trap and catchnet. California Fish and Game 40:367–373.

Clover, M. R. 1956. Single-gate deer trap. California Fish and Game 42:199–201.

Clutton-Brock, T. H., M. Major, and F. E. Guinness. 1985. Population regulation in male and female red deer. Journal of Animal Ecology 54:831–846.

Cochran, W. G. 1963. Sampling techniques. Second edition. John Wiley and Sons, New York, New York, USA.

Cochran, W. G. 1977. Sampling techniques. Third edition. John Wiley and Sons, New York, New York, USA.

Cochran, W. G. 1983. Planning and analysis of observational studies. John Wiley and Sons, New York, New York, USA.

Cochran, W. W., M. S. Bowlin, and M. Wikelski. 2008. Wingbeat frequency and flap-pause ratio during natural migratory flight in thrushes. Integrative and Comparative Biology 48:134–151.

Cochran, W. W., and R. D. Lord, Jr. 1963. A radio-tracking system for wild animals. Journal of Wildlife Management 27:9–24.

Cochran, W. W., D. W. Warner, J. R. Tester, and V. B. Kuechle. 1965. Automatic radio-tracking system for monitoring animal movements. BioScience 2:98–100.

Cockrum, E. L. 1969. Migration of the guano bat, *Tadarida brasiliensis*. Miscellaneous Publications, Museum of Natural History, University of Kansas 51:303–336.

Codd, E. F. 1970. A relational model of data for large shared data banks. Communications of the Association for Computing Machinery 13(6):377–387.

Cogan, R. D., and D. R. Diefenbach. 1998. Effect of undercounting and model selection on a sightability-adjustment estimator for elk. Journal of Wildlife Management 62:269–279.

Cohen, B., and T. C. Williams. 1980. Short-range corrections for migrant bird tracks on search radars. Journal of Field Ornithology 51:248–253.

Cohen, J. 1988. Statistical power analysis for the behavioral sciences. Lawrence Erlbaum Associates, Hillsdale, New Jersey, USA.

Cohen, R. 1969. Color-banded house finches. Eastern Bird Banding Association News 32:81–82.

Cohen, R. R. 1985. Capturing breeding male tree swallows with feathers. North American Bird Bander 10:18–21.

Cohen, R. R., and D. J. Hayes. 1984. A simple unattached nest-box trapping device. North American Bird Bander 9:10–11.

Cohen, W. B., M. Fiorella, J. Gray, E. Helmer, and K. Anderson. 1998. An efficient and accurate method for mapping forest clearcuts in the Pacific Northwest using Landsat imagery. Photogrammetric Engineering and Remote Sensing 64:293–300.

Cohen, W. B., and S. N. Goward. 2004. Landsat's role in ecological applications of remote sensing. BioScience 54:535–545.

Cohen, W. B., T. A. Spies, and G. A. Bradshaw. 1990. Semivariograms of digital imagery for analysis of conifer canopy structure. Remote Sensing of Environment 34:167–178.

Cohen, W. B., Z. Yang, and R. E. Kennedy. 2010. Detecting trends in forest disturbance and recovery using yearly Landsat time series: 2. TimeSync—tools for calibration and validation. Remote Sensing of Environment 114:2911–2924.

Coile, D. C. 2005. Encyclopedia of dog breeds. Barron's, Hauppauge, New York, USA.

Colbourne, R. 1992. Little spotted kiwi (*Apteryx owenii*): recruitment and behavior of juveniles on Kapiti Island, New Zealand. Journal of the Royal Society of New Zealand 42:321–328.

Colclough, J. H., and G. J. B. Ross. 1987. Colour band loss in cape gannets. Safring News 16:35–37.

Cole, G. F. 1956. Pronghorn antelope, its range use and food habits in central Montana with special reference to alfalfa. Technical Bulletin 516, Montana Experiment Station, Montana State University, Bozeman, USA.

Cole, G. F. 1959. Key browse survey method. Proceedings of the Western Association of State Fish and Game Commissioners 39:181–185.

Cole, L. C. 1949. The measurement of interspecific association. Ecology 30:411–424.

Cole, L. C. 1957. Sketches of general and comparative demography. Quantitative Biology 22:1–15.

Cole, N. C. 2004. A novel technique for capturing arboreal geckos. Herpetological Review 35:358–359.

Coles, W. C. 1999. Aspects of the biology of sea turtles in the mid-Atlantic Bight. Thesis, College of William and Mary, Williamsburg, Virginia, USA.

Collier, B. A. 2008. Suggestions for basic graph use when reporting wildlife research results. Journal of Wildlife Management 72:1272–1278.

Collier, B. A., S. S. Ditchkoff, J. B. Reglin, and J. M. Smith. 2007. Detection probability and sources of variation in white-tailed deer spotlight surveys. Journal of Wildlife Management 71:272–281.

Collinge, J. 2001. Prion diseases of humans and animals: their causes and molecular basis. Annual Review of Neuroscience 24:519–550.

Collinge, S. K., and C. Ray, editors. 2006. Disease ecology: community structure and pathogen dynamics. Oxford University Press, Oxford, England, UK.

Collins, T. C. 1976. Population characteristics and habitat relationships of beavers, *Castor canadensis*, in northwest Wyoming. Dissertation, University of Wyoming, Laramie, USA.

Collins, W. B., and E. F. Becker. 2001. Estimation of horizontal cover. Journal of Range Management 54:67–70.

Collins, W. B., and P. J. Urness. 1981. Habitat preferences of mule deer as rated by pellet-group distributions. Journal of Wildlife Management 45:969–972.

Collister, D. M., and R. G. Fisher. 1995. Trapping techniques for loggerhead shrikes. Wildlife Society Bulletin 23:88–91.

Coltman, D. W. 2008. Molecular ecological approaches to studying the evolutionary impact of selective harvesting in wildlife. Molecular Ecology 17:221–235.

Coltman, D. W., P. O'Donoghue, J. T. Jorgenson, J. T. Hogg, C. Strobeck, and M. Festa-Bianchet. 2003. Undesirable evolutionary consequences of trophy hunting. Nature 426:655–658.

Colvin, B. A., and P. L. Hegdal. 1986. Techniques for capturing common barn-owls. Journal of Field Ornithology 57:200–207.

Colwell, M. A., C. L. Grotto, L. W. Oring, and A. J. Fivizzani. 1988. Effects of blood sampling on shorebirds: injuries, return rates and clutch desertions. Condor 90:942–945.

Colyvan, M., and L. R. Ginzburg. 2003. Laws of nature and laws of ecology. Oikos 101:649–653.

Committee of Inquiry on Grouse Disease. 1911. The grouse in health and disease. Volume 1–2. Smith, Elder and Company, London, England, UK.

Conant, R. 1948. Regeneration of clipped subcaudal scales in a pilot black snake. Natural History Miscellaneous 13:1–2.

Concannon, P., B. Baldwin, J. Lawless, W. Hornbuckle, and B. Tennant. 1983. Corpora lutea of pregnancy and elevated serum progesterone during pregnancy and postpartum anestrus in wood-chucks (*Marmota monax*). Biology of Reproduction 29:1128–1134.

Condon, A. M., E. L. Kershner, B. L. Sullivan, D. M. Cooper, and D. K. Garcelon. 2005. Spotlight surveys for grassland owls on San Clemente Island, California. Wilson Bulletin 117:177–184.

Congdon, J. D., and J. W. Gibbons. 1996. Structure and dynamics of a turtle community over two decades. Pages 137–159 *in* M. L. Cody and J. A. Smallwood, editors. Long-term studies of vertebrate communities. Academic Press, New York, New York, USA.

Congdon, P. 2007. Bayesian statistical modeling. Second edition. John Wiley and Sons, Chichester, West Sussex, England, UK.

Connelly, J. W., J. H. Gammonley, and J. M. Peek. 2004. Harvest management. Pages 658–690 *in* C. E. Braun, editor. Techniques for wildlife investigations and management. The Wildlife Society, Bethesda, Maryland, USA.

Connelly, J. W., K. P. Reese, E. O. Garton, and M. L. Commons-Kemner. 2003*a*. Response of greater sage-grouse *Centrocerus urophasianus* populations to different levels of exploitation in Idaho, USA. Wildlife Biology 9:335–340.

Connelly, J. W., M. W. Gratson, and K. P. Reese. 1998. Sharp-tailed grouse. Account 354 in A. Poole and F. Gill, editors. The birds of North America. The Academy of Natural Sciences, Philadelphia, Pennsylvania, and The American Ornithologists' Union, Washington, D.C., USA.

Connelly, J. W., K. P. Reese, and M. A. Schroeder. 2003b. Monitoring of greater sage-grouse habitats and populations. Station Bulletin 80, College of Natural Resources Experiment Station, University of Idaho, Moscow, USA.

Connelly, J. W., M. A. Schroeder, A. R. Sands, and C. E. Braun. 2000. Guidelines to manage sage grouse populations and their habitats. Wildlife Society Bulletin 28:967–985.

Conner, M. C. 1982. Determination of bobcat (*Lynx rufus*) and raccoon (*Procyon lotor*) population abundance by radioisotope tagging. Thesis, University of Florida, Gainesville, USA.

Conner, M. C., and R. F. Labisky. 1985. Evaluation of radioisotope tagging for estimating abundance of raccoon populations. Journal of Wildlife Management 49:326–332.

Conner, M. C., E. C. Soutiere, and R. A. Lancia. 1987. Drop-netting deer: costs and incidence of capture myopathy. Wildlife Society Bulletin 15:434–438.

Connior, M. B., and T. S. Risch. 2009. Live trap for pocket gophers. Southwestern Naturalist 54:100–103.

Connolly, C. 2007. Wildlife-spotting robots. Sensor Review 27:282–287.

Connolly, G. E., M. L. Dudzinski, and W. M. Longhurst. 1969a. An improved age-lens weight regression for black-tailed deer and mule deer. Journal of Wildlife Management 33:701–704.

Connolly, G. E., M. L. Dudzinski, and W. M. Longhurst. 1969b. The eye lens as an indicator of age in the black-tailed jack rabbit. Journal of Wildlife Management 33:159–164.

Conover, M. R. 2007. Predator–prey dynamics: the role of olfaction. CRC Press, and Taylor and Francis, New York, New York, USA.

Conover, M. R., and G. G. Chasko. 1985. Nuisance Canada goose problems in the eastern United States. Wildlife Society Bulletin 13:228–233.

Conover, M. R., and R. A. Dolbeer. 2007. Use of decoy traps to protect blueberries from juvenile European starlings. Human–Wildlife Conflicts 1:265–270.

Conover, W. J. 1999. Practical non-parametric statistics. Third edition. John Wiley and Sons, New York, New York, USA.

Conroy, M. J., R. O. Barker, P. W. Dillingham, D. Fletcher, A. M. Gormley, and I. M. Westbrooke. 2008. Application of decision theory to conservation management: recovery of Hector's dolphin. Wildlife Research 35:93–102.

Conroy, M. J., G. R. Costanzo, and D. B. Stotts. 1989. Winter survival of female American black ducks on the Atlantic coast. Journal of Wildlife Management 53:99–109.

Conroy, M. J., C. J. Fonnesbeck, and N. L. Zimpfer. 2005. Modeling regional waterfowl harvest rates using Markov chain Monte Carlo. Journal of Wildlife Management 69:77–90.

Conroy, M. J., M. W. Miller, and J. E. Hines. 2002. Identification and synthetic modeling of factors affecting American black duck populations. Wildlife Monographs 150.

Conroy, M. J., and J. D. Nichols. 1996. Designing a study to assess mammalian diversity. Pages 41–49 in D. E. Wilson, F. R. Cole, J. D. Nichols, R. Rudran, and M. S. Foster, editors. Measuring and monitoring biological diversity: standard methods for mammals. Smithsonian Institution Press, Washington, D.C., USA.

Conroy, M. J., and J. T. Peterson. 2009. Integrating management, research, and monitoring: balancing the 3-legged stool. Pages 2–10 in S. B. Cederbaum, B. C. Faircloth, T. M. Terhune, J. J. Thompson, and J. P. Carroll, editors. Gamebird 2006: quail VI and perdix XII. Warnell School of Forestry and Natural Resources, University of Georgia, Athens, USA.

Conway, C. J., and V. Garcia. 2005. Effects of radiotransmitters on natal recruitment of burrowing owls. Journal of Wildlife Management 69:404–408.

Conway, W. C., and L. M. Smith. 2000. A nest trap for snowy plovers. North American Bird Bander 25:45–47.

Cooch, E., and G. C. White. 2009. Using MARK: a gentle introduction. Seventh edition. http://www.phidot.org/software/mark/docs/book/.

Cooch, G. 1953. Techniques for mass capture of flightless blue and lesser snow geese. Journal of Wildlife Management 17:460–465.

Cook, C. W., and C. D. Bonham. 1977. Techniques for vegetation measurements and analysis for a pre- and post-mining inventory. Range Science Series 28, Colorado State University, Fort Collins, USA.

Cook, C. W., and J. Stubbendieck. 1986. Range research: basic problems and techniques. Society of Range Management, Denver, Colorado, USA.

Cook, J. G., T. W. Stutzman, C. W. Bowers, K. A. Brenner, and L. L. Irwin. 1995. Spherical densiometers produce biased estimates of forest canopy cover. Wildlife Society Bulletin 23:711–717.

Cook, R. D., and J. O. Jacobson. 1979. A design for estimating visibility bias in aerial surveys. Biometrics 35:735–742.

Cook, T. D., and D. T. Campbell. 1979. Quasi-experimentation: design and analysis issues for field settings. Houghton Mifflin, Boston, Massachusetts, USA.

Cooke, B. D., and D. Berman. 2000. Effect of inoculation rate and ambient temperature on the survival time of rabbits, *Oryctolagus cuniculus* (L.), infected with the rabbit haemorrhagic disease virus. Wildlife Research 27:137–142.

Cooke, D., and J. A. Leon. 1976. Stability of population growth determined by 2×2 Leslie matrix with density-dependent elements. Biometrics 32:435–442.

Cooke, S. J., S. G. Hinch, M. Wikelski, R. D. Andrews, L. J. Kuchel, T. G. Wolcott, and P. J. Butler. 2004. Biotelemetry: a mechanistic approach to ecology. Trends in Ecology and Evolution 19:334–343.

Cooley, M. E. 1948. Improved toe-tag for marking fox squirrels. Journal of Wildlife Management 12:213.

Cooper, A. B., and J. J. Millspaugh. 1999. The application of discrete choice models to wildlife resource selection studies. Ecology 80:566–575.

Cooper, A. B., and J. J. Millspaugh. 2001. Accounting for variation in resource availability and animal behavior in resource selection studies. Pages 243–274 in J. J. Millspaugh and J. M. Marzluff, editors. Radio tracking and animal populations. Academic Press, San Diego, California, USA.

Cooper, B. A., and R. J. Blaha. 2002. Comparisons of radar and audio-visual counts of marbled murrelets during inland forest surveys. Wildlife Society Bulletin 30:1182–1194.

Cooper, B. A., R. H. Day, R. J. Ritchie, and C. L. Cranor. 1991. An improved marine radar system for studies of bird migration. Journal of Field Ornithology 62:367–377.

Cooper, B. A., M. G. Raphael, and D. E. Mack. 2001. Radar-based monitoring of marbled murrelets. Condor 103:219–229.

Cooper, C. F. 1957. The variable plot method for estimating shrub density. Journal of Range Management 10:111–115.

Cooper, C. F. 1963. An evaluation of variable plot sampling in shrub and herbaceous vegetation. Ecology 44:565–569.

Cooper, H. D., C. M. Raley, and K. B. Aubry. 1995. A noose trap for capturing pileated woodpeckers. Wildlife Society Bulletin 23:208–211.

Cooper, J., and P. D. Morant. 1981. The design of stainless steel flipper bands for penguins. Ostrich 52:119–123.

Cooper, K. A., J. D. Harder, D. H. Clawson, D. L. Fredrick, G. A. Lodge, H. C. Peachey, T. J. Spellmire, and D. P. Winstel. 1990. Serum testosterone in musth in captive male African and Asian elephants. Zoo Biology 9:297–306.

Cooper, L. S., and J. A. Randall. 2007. Seasonal changes in home ranges of the giant kangaroo rat (*Dipodomys ingens*). Journal of Mammalogy 88:1000–1008.

Copeland, J. P., E. Cesar, J. M. Peek, C. E. Harris, C. D. Long, and D. L. Hunter. 1995. A live trap for wolverines and other forest carnivores. Wildlife Society Bulletin 23:535–538.

Copelin, F. F. 1963. The lesser prairie chicken in Oklahoma. Technical Bulletin 6, Oklahoma Wildlife Conservation Department, Oklahoma City, USA.

Coppin, P., I. Jonckheere, K. Nackaerts, and B. Muys. 2004. Digital change detection methods in ecosystem monitoring: a review. International Journal of Remote Sensing 25:1565–1596.

Coppinger, R., and L. Coppinger. 2002. Dogs: a startling new understanding of canine origin, behavior, and evolution. University of Chicago Press, Chicago, Illinois, USA.

Coppinger, R., C. K. Smith, and L. Miller. 1985. Observations on why mongrels may make effective livestock protecting dogs. Journal of Range Management 38:560–561.

Coppinger, R. P., and B. C. Wentworth. 1966. Identification of experimental birds with the aid of feather autografts. Bird-Banding 37:203–205.

Corbel, M. J. 1997. Brucellosis: an overview. Emerging Infectious Diseases 3:213–221.

Corlatti, L., K. Hacklander, and F. Frey-Roos. 2009. Ability of wildlife overpasses to provide connectivity and prevent genetic isolation. Conservation Biology 23:548–556.

Cormack, R. M. 1973. Common sense estimates from capture–recapture studies. Pages 225–234 in M. S. Bartlett and R. W. Hiorns, editors. The mathematical theory of the dynamics of biological populations. Academic Press, New York, New York, USA.

Corn, P. S., and R. B. Bury. 1990. Sampling methods for terrestrial amphibians and reptiles. General Technical Report PNW-256, U.S. Forest Service, Department of Agriculture, Washington, D.C., USA.

Cornuet, J.-M., and G. Luikart. 1996. Description and power analysis of two tests for detecting recent population bottlenecks from allele frequency data. Genetics 144:2001–2014.

Cornuet, J.-M., S. Piry, G. Luikart, A. Estoup, and M. Solignac. 1999. New methods employing multilocus genotypes to select or exclude populations as origins of individuals. Genetics 153:1989–2000.

Corona, P., and L. Fattorini. 2008. Area-based lidar-assisted estimation of forest standing volume. Canadian Journal of Forest Research 38:2911–2916.

Corrigan, R. M. 1998. The efficacy of glue traps against wild populations of house mice, *Mus domesticus*. Proceedings of the Vertebrate Pest Conference 18:268–275.

Costanzo, G. R., R. A. Williamson, and D. E. Hayes. 1995. An efficient method for capturing flightless geese. Wildlife Society Bulletin 23:201–203.

Costello, C. M., K. H. Inman, D. E. Jones, R. M. Inman, B. C. Thompson, and H. B. Quigley. 2004. Reliability of the cementum annuli technique for estimating age of black bears in New Mexico. Wildlife Society Bulletin 32:169–176.

Cottam, C., and J. B. Trefethen, editors. 1968. Whitewings: the life history, status and management of the white-winged dove. D. Van Nostrand, Princeton, New Jersey, USA.

Cottam, G., and J. T. Curtis. 1949. A method for making rapid surveys of woodlands by means of randomly selected trees. Ecology 30:101–104.

Cottam, G., and J. T. Curtis. 1956. The use of distance measures in phytosociological sampling. Ecology 37:451–460.

Cottam, G., J. T. Curtis, and B. W. Hale. 1953. Some sampling characteristics of a population of randomly dispersed individuals. Ecology 34:741–757.

Cotterill, F. P. D., and R. A. Fergusson. 1993. Capturing free-tailed bats (Chiroptera: Molossidae): the description of a new trapping device. Journal of Zoology (London) 231:645–651.

Couch, L. K. 1942. Trapping and transplanting live beavers. Conservation Bulletin 30, U.S. Fish and Wildlife Service, Department of the Interior, Washington, D.C., USA.

Couey, F. M. 1949. Review and evaluation of big game trapping techniques. Proceedings of the Western Association of State Game and Fish Commissioners 29:110–116.

Coughenour, M. B., and F. J. Singer. 1996. Yellowstone elk population responses to fire—a comparison of landscape carrying capacity and spatial-dynamic ecosystem modeling approaches. Pages 169–179 in J. M. Greenlee, editor. The ecological implications of fire in Greater Yellowstone. International Association of Wildlife Fire, Fairfield, Washington, USA.

Coulson, T., S. Albon, F. Guinness, J. Pemberton, and T. Clutton-Brock. 1997. Population substructure, local density, and calf winter survival in red deer (*Cervus elaphus*). Ecology 78:852–863.

Coulson, T. N., J. M. Pemberton, S. D. Albon, M. Beaumont, T. C. Marshall, J. Slate, F. E. Guinness, and T. H. Clutton-Brock. 1998. Microsatellites reveal heterosis in red deer. Proceedings of the Royal Society of London B 265:489–495.

Coulter, M. W. 1958. A new waterfowl nest trap. Bird-Banding 29:236–241.

Cowan, D. P., J. C. Reynolds, and E. L. Gill. 2000. Reducing predation through conditioned taste aversion. Pages 281–291 in L. M. Gosling and W. J. Sutherland, editors. Behaviour and conservation. Cambridge University Press, Cambridge, England, UK.

Cowan, D. P., J. A. Vaughan, and W. G. Christer. 1987. Bait consumption by the European rabbit in southern England. Journal of Wildlife Management 51:386–392.

Cowan, D. P., J. A. Vaughan, K. J. Prout, and W. G. Christer. 1984. Markers for measuring bait consumption by the European wild rabbit. Journal of Wildlife Management 48:1403–1409.

Cowan, I. M., and J. Hatter. 1952. A trap and technique for the capture of diving waterfowl. Journal of Wildlife Management 16:438–441.

Cowardin, L., and J. Ashe. 1965. An automatic camera device for measuring waterfowl use. Journal of Wildlife Management 29:636–640.

Cowardin, L. M., and R. J. Blohm. 1992. Breeding population inventories and measures of recruitment. Pages 423–445 in B. D. J. Batt, A. D. Afton, M. G. Anderson, C. D. Ankeny, D. H. Johnson, J. A. Kadlec, and G. L. Krapu, editors. Ecology and management of breeding waterfowl. University of Minnesota Press, Minneapolis, USA.

Cowardin, L. M., D. H. Johnson, T. L. Shaffer, and D. W. Sparling. 1988. Application of a simulation model to decisions in mallard management. Technical Report 17, U.S. Fish and Wildlife Service, Department of the Interior, Washington, D.C., USA.

Cowardin, L. M., and V. I. Myers. 1974. Remote sensing for identification and classification of wetland vegetation. Journal of Wildlife Management 38:308–314.

Cox, D. R. 1972. Regression models and life-tables (with discussion). Journal of the Royal Statistical Society B 34:187–220.

Cox, D. R. 1983. Some remarks on overdispersion. Biometrika 70:269–274.

Cox, D. R. 2006. Principles of statistical inference. Cambridge University Press, Cambridge, England, UK.

Cox, D. R., and D. O. Oakes. 1984. Analysis of survival data. Chapman and Hall, London, England, UK.

Cox, R. R., Jr., and A. D. Afton. 1994. Portable platforms for setting rocket nets in open-water habitats. Journal of Field Ornithology 65:551–555.

Cox, R. R., Jr., and A. D. Afton. 1998. Effects of capture and handling on survival of female northern pintails. Journal of Field Ornithology 69:276–287.

Cox, R. R., Jr., J. D. Scalf, B. E. Jamison, and R. S. Lutz. 2002. Using an electronic compass to determine telemetry azimuths. Wildlife Society Bulletin 30:1039–1043.

Crabtree, R. L., F. G. Burton, T. R. Garland, D. A. Cataldo, and W. H. Rickard. 1989. Slow-release radioisotope implants as individual markers for carnivores. Journal of Wildlife Management 53:949–954.

Cracraft, J. 1983. Species concepts and speciation analysis. Pages 159–187 in R. F. Johnston, editor. Current Ornithology. Volume 1. Plenum Press, New York, New York, USA.

Craig, B. A., M. A. Newton, R. A. Garrott, J. E. Reynolds, III, and J. R. Wilcox. 1997. Analysis of aerial survey data on Florida manatee using Markov chain Monte Carlo. Biometrics 53:524–541.

Craig, E. H., and B. L. Keller. 1986. Movements and home range of porcupines, *Erethizon dorsatum*, in Idaho shrub desert. Canadian Field-Naturalist 100:167–173.

Craig, T. H., J. W. Connelly, E. H. Craig, and T. L. Parker. 1990. Lead concentrations in golden and bald eagles. Wilson Bulletin 102:130–133.

Craighead, J. J., F. C. Craighead, Jr., and H. E. McCutchen. 1970. Age determination of grizzly bears from fourth premolar tooth sections. Journal of Wildlife Management 34:353–363.

Craighead, J. J., M. G. Hornocker, M. W. Shoesmith, and R. I. Ellis. 1969. A marking technique for elk. Journal of Wildlife Management 33:906–909.

Craighead, J. J., M. Hornocker, W. Woodgerd, and F. C. Craighead, Jr. 1960. Trapping, immobilizing, and color-marking grizzly bears. Transactions of the North American Wildlife and Natural Resources Conference 25:347–363.

Craighead, J. J., and J. A. Mitchell. 1982. Grizzly bear. Pages 515–556 in J. A. Chapman and G. A. Feldhamer, editors. Wild mammals of North America. Johns Hopkins University Press, Baltimore, Maryland, USA.

Craighead, J. J., and D. S. Stockstad. 1956. A colored neckband for marking birds. Journal of Wildlife Management 20:331–332.

Craighead, J. J., and D. S. Stockstad. 1960. Color marker for big game. Journal of Wildlife Management 24:435–438.

Cramp, S., and K. E. L. Simmons, editors. 1980. The birds of the western Palearctic. Volume 2. Hawks to bustards. Oxford University Press, Oxford, England, UK.

Crandall, K. A., O. R. P. Binida-Emonds, G. M. Mace, and R. K. Wayne. 2000. Considering evolutionary processes in conservation biology. Trends in Ecology and Evolution 17:390–395.

Cranford, J. A. 1977. Home range and habitat utilization by *Neotoma fuscipes* as determined by radiotelemetry. Journal of Mammalogy 58:165–172.

Craven, S. R. 1979. Some problems with Canada goose neckbands. Wildlife Society Bulletin 7:268–273.

Crawford, A. B. 1949. Radar reflections in the lower atmosphere. Proceedings of the Institute of Radio Engineers 37:404–405.

Crawford, E., and A. Kurta. 2000. Color of pitfall affects trapping success for anurans and shrews. Herpetological Review 31:222–224.

Crawford, R. D. 1977. Comparison of trapping methods for American coots. Bird-Banding 48:309–313.

Creel, S., D. Christianson, S. Liley, and J. A. Winnie. 2007. Predation risk affects reproductive physiology and demography of elk. Science 315:960.

Creel, S., J. A. Winnie, and D. Christianson. 2009. Glucocorticoid stress hormones and the effect of predation risk on elk reproduction. Proceedings of the National Academy of Sciences of the United States of America 106:12388–12393.

Cressie, N. A. 1991. Statistics for spatial data. John Wiley and Sons, New York, New York, USA.

Crête, M., and F. Messier. 1987. Evaluation of indices of gray wolf, *Canis lupis*, density in hardwood-conifer forests of southwestern Quebec. Canadian Field-Naturalist 101:147–152.

Crews, D. 1980. Interrelationships among ecological, behavioral, and neuroendocrine processes in the reproductive cycle of *Anolis carolinensis* and other reptiles. Advances in the Study of Behavior 11:1–74.

Crews, D., and W. R. Garstka. 1982. The ecological physiology of a garter snake. Scientific American 247:158–168.

Crier, J. K. 1970. Tetracyclines as a fluorescent marker in bones and teeth of rodents. Journal of Wildlife Management 34:829–834.

Cristol, D. A., C. S. Chiu, S. M. Peckham, and J. F. Stoll. 1992. Color bands do not affect dominance status in captive flocks of wintering dark-eyed juncos. Condor 94:537–539.

Crofoot, M. C., I. C. Gilby, M. C. Wikelski, and R. W. Kays. 2008. The home field advantage: location affects the outcome of asymmetric intergroup contests in *Cebus capucinus*. Proceedings of the National Academy of Sciences of the United States of America 105:577–581.

Croll, D. A., J. K. Jansen, M. E. Goebel, P. L. Boveng, and J. L. Bengtson. 1996. Foraging behavior and reproductive success in chinstrap penguins: the effects of transmitter attachment. Journal of Field Ornithology 67:1–9.

Cromwell, J. A., R. J. Warren, and D. W. Henderson. 1999. Live capture and small-scale relocation of urban deer on Hilton Head Island, South Carolina. Wildlife Society Bulletin 27:1025–1031.

Cronin, M. A. 2006. A proposal to eliminate redundant terminology for intra-species groups. Wildlife Society Bulletin 34:237–241.

Cronin, M. A., J. B. Grand, D. Esler, D. V. Derksen, and K. T. Scribner. 1996. Breeding populations of northern pintails have similar mitochondrial DNA. Canadian Journal of Zoology 74:992–999.

Croon, G., D. McCullough, C. Olson, and L. Queal. 1968. Infrared scanning techniques for big game censusing. Journal of Wildlife Management 32:751–760.

Crosbie, S. F., and B. F. J. Manly. 1985. Parsimonious modeling of capture–mark–recapture studies. Biometrics 41:385–398.

Cross, C. L. 2000. A new design for a lightweight squeeze box for snake field studies. Herpetological Review 31:34.

Cross, M. L., B. M. Buddle, and F. E. Aldwell. 2007. The potential of oral vaccines for disease control in wildlife species. Veterinary Journal 174:472–480.

Cross, T. J., A. W. Reed, B. T. Rutledge, B. R. Laseter, S. A. Maris, and C. N. Doolittle. 1999. A comparison of disinfected and untreated traps for sampling small mammal populations. Journal of the Tennessee Academy of Science 73:35.

Crouch, W. B., and P. W. C. Paton. 2000. Using egg-mass counts to monitor wood frog populations. Wildlife Society Bulletin 28:895–901.

Crow, J. F., and M. Kimura. 1970. An introduction to population genetics theory. Harper and Row, New York, New York, USA.

Crowe, D. M. 1972. The presence of annuli in bobcat tooth cementum layers. Journal of Wildlife Management 36:1330–1332.

Crowe, D. M. 1975. Aspects of aging, growth, and reproduction of bobcats from Wyoming. Journal of Mammalogy 56:177–198.

Crowe, D. M., and M. D. Strickland. 1975. Population structures of some mammalian predators in southeastern Wyoming. Journal of Wildlife Management 39:449–450.

Crozier, G. E., and D. E. Gawlick. 2003. Use of decoys as a research tool in attracting wading birds. Journal of Field Ornithology 74:53–58.

Crum, T. D., R. L. Alberty, and D. W. Burgess. 1993. Recording, archiving, and using WSR-88D data. Bulletin of the American Meteorological Society 74:645–653.

Crunden, C. W. 1963. Age and sex of sage grouse from wings. Journal of Wildlife Management 27:846–849.

Cruz-Angón, A., T. S. Sillett, and P. Greenberg. 2008. An experimental study of habitat selection by birds in a coffee plantation. Ecology 89:921–927.

Cummings, G. E., and O. H. Hewitt. 1964. Capturing waterfowl and marsh birds at night with light and sound. Journal of Wildlife Management 28:120–126.

Cummings, J. L. 1987. Nylon fasteners for attaching leg and wing tags to blackbirds. Journal of Field Ornithology 58:265–269.

Cummins, C. P., and M. J. S. Swan. 2000. Long-term survival and growth of free-living great crested newts (*Triturus cristatus*) pit-tagged at metamorphosis. Herpetological Journal 10:177–182.

Currie, D., and E. Robertson. 1992. Alberta wild fur management study guide. Alberta Forestry, Lands and Wildlife, Edmonton, Canada.

Curtis, J. T. 1959. The vegetation of Wisconsin; an ordination of plant communities. University of Wisconsin Press, Madison, USA.

Curtis, J. T., and R. P. McIntosh. 1950. The interrelations of certain analytic and synthetic phytosociological characters. Ecology 31:434–455.

Curtis, P. D., C. E. Braun, and R. A. Ryder. 1983. Wing markers: visibility, wear, and effects on survival of band-tailed pigeons. Journal of Field Ornithology 54:381–386.

Cushwa, C. T., and K. P. Burnham. 1974. An inexpensive live trap for snowshoe hares. Journal of Wildlife Management 38:939–941.

Cuthbert, F. J., and W. E. Southern. 1975. A method for marking young gulls for individual identification. Bird-Banding 46:252–253.

Cutler, T. L., and D. E. Swann. 1999. Using remote photography in wildlife ecology: a review. Wildlife Society Bulletin 27:571–581.

Cyr, A. 1981. Limitation and variability in hearing ability in censusing birds. Studies in Avian Biology 6:327–333.

Daan, S. 1969. Frequency of displacements as a measure of activity of hibernating bats. Lynx 10:13–18.

Dagg, A. I., D. Leach, and G. Sumner-Smith. 1975. Fusion of the distal femoral epiphysis in male and female marten and fisher. Canadian Journal of Zoology 53:1514–1518.

Dahl, C. R., J. W. Bickham, J. K. Wicklieffe, and T. W. Custer. 2001. Cytochrome b sequences in black-crowned night-herons (*Nycticorax nycticorax*) from heronries exposed to genotoxic contaminants. Ecotoxicology 10:291–297.

Dahlgren, D. K. 2009. Greater sage-grouse ecology, chick survival, and population dynamics, Parker Mountain, Utah. Dissertation, Utah State University, Logan, USA.

Dahlgren, D. K., R. Chi, and T. A. Messmer. 2006. Greater sage-grouse response to sagebrush management in Utah. Wildlife Society Bulletin 34:975–985.

Dahlgren, D. K., T. A. Messmer, E. T. Thacker, and M. R. Guttery. 2010. Evaluation of brood detection techniques: recommendations for estimating greater sage-grouse productivity. Western North American Naturalist 70:233–237.

Dahlgren, R. B., C. M. Twedt, and C. G. Trautman. 1965. Lens weights of ring-necked pheasants. Journal of Wildlife Management 29:212–214.

Dale, M. R., and M. Fortin. 2002. Spatial autocorrelation and statistical tests in ecology. Ecoscience 9:162–167.

Dalen, L., B. Elmhagen, and A. Angerbjorn. 2004. DNA analysis on fox faeces and competition

induced niche shifts. Molecular Ecology 13:2389–2392.
Dalke, P. D., D. B. Pyrah, D. C. Stanton, J. E. Crawford, and E. F. Schlatterer. 1963. Ecology, productivity, and management of sage grouse in Idaho. Journal of Wildlife Management 27:811–841.
Dansereau, P. M. 1957. Biogeography: an ecological perspective. Ronald Press, New York, New York, USA.
Danskin, S., P. Bettinger, T. Jordan, and C. Cieszewski. 2009. A comparison of GPS performance in a southern hardwood forest: exploring low-cost solutions for forestry applications. Southern Journal of Applied Forestry 33:9–16.
Daoust, J.-L. 1991. Coping with dehydration of trapped terrestrial anurans. Herpetological Review 22:95.
Dapson, R. W., and J. M. Irland. 1972. An accurate method of determining age in small mammals. Journal of Mammalogy 53:100–106.
Darnerud, P. O. 2003. Toxic effects of brominated flame retardants in man and in wildlife. Environment International 29:841–853.
Darwin, C. 1859. On the origin of species by means of natural selection. John Murray, London, England, UK.
Dasmann, W. P. 1951. Some deer range survey methods. California Fish and Game 37:43–52.
Daszak, P., A. A. Cunningham, and A. D. Hyatt. 2000. Emerging infectious diseases of wildlife—threats to biodiversity and human health. Science 287:443–449.
Dattalo, P. 2008. Determining sample size: balancing power, precision, and practicality. Oxford University Press, Oxford, England, UK.
Dau, C. P., P. L. Flint, and M. R. Petersen. 2000. Distribution of recoveries of Steller's eiders banded on the Lower Alaska Peninsula, Alaska. Journal of Field Ornithology 71:541–548.
Daubenmire, R. 1958. A canopy-coverage method of vegetational analysis. Northwest Science 33:43–64.
Daubenmire, R. F. 1959. A canopy-coverage method of vegetational analysis. Northwest Science 33:43–64.
Daubenmire, R. F. 1968. Plant communities: a textbook of plant synecology. Harper and Row, New York, New York, USA.
Daugherty, C. H. 1976. Freeze-branding as a technique for marking anurans. Copeia 1976:836–838.
Dausmann, K. H. 2005. Measuring body temperature in the field—evaluation of external vs. implanted transmitters in a small mammal. Journal of Thermal Biology 30:195–202.
Davenport, I. J., R. B. Bradbury, G. Q. A. Anderson, G. R. F. Hayman, J. R. Krebs, D. C. Mason, J. D. Wilson, and N. J. Veck. 2000. Improving bird population models using airborne remote sensing. International Journal of Remote Sensing 21:2705–2717.
Davey, C. C., and P. J. Fullagar. 1985. Nasal saddles for Pacific black duck *Anas superciliosa* and austral teal. Corella 9:123–124.

Davey, C. C., P. J. Fullagar, and C. Kogon. 1980. Marking rabbits for individual identification and a use for betalights. Journal of Wildlife Management 44:494–497.
Davidson, W. R., editor. 2006. Field manual of wildlife diseases in the southeastern states. Third edition. Southeastern Cooperative Wildlife Disease Study, Athens, Georgia, USA.
Davidson, W. R., F. A. Hayes, V. F. Nettles, and F. E. Kellogg, editors. 1981. Diseases and parasites of white-tailed deer. Tall Timbers Research Station, Tallahassee, Florida, USA.
Davis, A. J., J. D. Holloway, H. Huijbregts, J. Krikken, A. H. Kirk-Spriggs, and S. L. Sutton. 2001. Dung beetles as indicators of change in the forests of northern Borneo. Journal of Applied Ecology 38:593–616.
Davis, A. K., N. E. Diggs, R. J. Cooper, and P. P. Marra. 2008. Hematological stress indices reveal no effect of radio-transmitters on wintering hermit thrushes. Journal of Field Ornithology 79:293–297.
Davis, B. E., A. D. Afton, and R. R. Cox, Jr. 2009. Habitat use by female mallard in the lower Mississippi alluvial valley. Journal of Wildlife Management 73:701–709.
Davis, D. E. 1964. Evaluation of characters for determining age of woodchucks. Journal of Wildlife Management 28:9–15.
Davis, D. S. 1990. Brucellosis in wildlife. Pages 321–334 in K. Nielsen, J. R. Nielsen, and J. R. Duncan, editors. Animal brucellosis. CRC Press, Boca Raton, Florida, USA.
Davis, D. S., J. W. Templeton, T. A. Ficht, J. D. Williams, J. D. Kopec, and L. G. Adams. 1990. Brucella abortus in captive bison. I. Serology, bacteriology, pathogenesis, and transmission to cattle. Journal of Wildlife Diseases 26:360–371.
Davis, J. A. 1969. Aging and sexing criteria for Ohio ruffed grouse. Journal of Wildlife Management 33:628–636.
Davis, J. L., C.-L. B. Chetkiewicz, V. C. Bleich, G. Raygorodetsky, B. M. Pierce, J. W. Ostergard, and J. D. Wehausen. 1996. A device to safely remove immobilized mountain lions from trees and cliffs. Wildlife Society Bulletin 24:537–539.
Davis, P. G. 1981. Trapping methods for bird ringers. British Trust of Ornithology, Tring, England, UK.
Davis, R. A. 1963a. Feral coypus in Britain. Proceedings of the Association of Applied Biologists, Great Britain 51:345–348.
Davis, R. D. 1994. A funnel trap for Rio Grande turkey. Proceedings of the Annual Conference of the Southeastern Association of Fish and Wildlife Agencies 48:109–116.
Davis, T. M., and K. Ovaska. 2001. Individual recognition of amphibians: effects of toe clipping and fluorescent tagging on the salamander *Plethodon vehiculum*. Journal of Herpetology 35:217–225.
Davis, W., Jr., and G. Sartor. 1975. A method of observing movements of aquatic turtles. Herpetological Review 6:13–14.
Davis, W. E., Jr. 2006. A long-term bird-banding study in upland tropical rainforest, Paluma Range, northeastern Queensland with notes on breeding. Journal of Field Ornithology 77:91–92.
Davis, W. H. 1963b. Anodizing bat bands. Bat Banding News 4:12–13.
Davison, V. E., L. M. Dickerson, K. Graetz, W. W. Neeley, and L. Roof. 1955. Measuring the yield and availability of game bird foods. Journal of Wildlife Management 19:302–308.
Day, G. I. 1973. Marking devices for big game animals. Arizona Game and Fish Department Research Abstracts 8:1–7.
Day, G. I., S. D. Schemnitz, and R. D. Taber. 1980. Capturing and marking wild animals. Pages 61–88 in S. D. Schemnitz, editor. Wildlife management techniques manual. The Wildlife Society, Washington, D.C., USA.
Day, J. F., and J. D. Edman. 1983. Malaria renders mice susceptible to mosquito feeding when gametocytes are most infective. Journal of Parasitology 69:163–170.
De Angelis, D. L. 1976. Application of stochastic models to a wildlife population. Mathematical Biosciences 31:227–236.
De La Mare, W. K. 1985. Some evidence for mark shedding with discovery whale marks. International Whaling Commission Report 35:477–486.
de Queiroz, K. 1998. The general lineage concept of species, species criteria, and the process of speciation: a conceptual unification and terminological recommendations. Pages 57–75 in D. J. Howard and S. H. Berlocher, editors. Endless forms: species and speciation. Oxford University Press, Oxford, England, UK.
De Solla, S. R., R. Bonduriansky, and R. J. Brooks. 1999. Eliminating autocorrelation reduces biological relevance of home range estimates. Journal of Animal Ecology 68:221–234.
De Vos, A., and H. S. Mosby. 1969. Habitat analysis and evaluation. Pages 135–172 in R. H. Giles, Jr., editor. Wildlife management techniques. The Wildlife Society, Washington, D.C., USA.
Deagle, B. E., and D. J. Tollit. 2007. Quantitative analysis of prey DNA in pinniped faeces: potential to estimate diet composition? Conservation Genetics 8:743–747.
Dean, E. E. 1979. Training of dogs to detect black-footed ferrets. Southwestern Research Institute, Santa Fe, New Mexico, USA.
Deat, A., C. Mauget, R. Mauget, D. Maurel, and A. Sempere. 1980. The automatic, continuous and fixed radio tracking system of the Chize Forest: theoretical and practical analysis. Pages 439–451 in C. J. Amlaner, Jr., and D. W. MacDonald, editors. A handbook on biotelemetry and radio tracking. Pergamon Press, Oxford, England, UK.
Deavers, D. R. 1972. Water and electrolyte metabolism in the arenicolous lizard *Uma notata notata*. Copeia 1972:109–122.
DeFries, R., F. Achard, S. Brown, M. Herold, D. Murdiyarso, B. Schlamadinger, and C. de Souza, Jr. 2006. Reducing greenhouse gas emissions from deforestation in developing countries: considerations for monitoring and measuring. Report of the Global Observation

System (GTOS), Number 46, GOFC-GOLD, Report 26. www.fao.org/gtos/pubs.html.

Defusco, R. P. 1995. Vultures and the U.S. Air Force to share friendly skies. BioScience 45:63.

DeGange, A. R., J. W. Fitzpatrick, J. N. Layne, and G. E. Woolfenden. 1989. Acorn harvesting by Florida scrub jays. Ecology 70:348–356.

DeGraaf, R. M., and N. L. Chadwick. 1987. Forest type, timber size class, and New England breeding birds. Journal of Wildlife Management 51:212–217.

DeGraaf, R. M., and M. Yamasaki. 1992. A non-destructive technique to monitor the relative abundance of terrestrial salamanders. Wildlife Society Bulletin 20:260–264.

DeGrazio, J. W. 1989. Pest birds—an international perspective. Pages 1–8 in R. L. Bruggers and C. H. Elliott, editors. *Quelea* quelea—Africa's bird pest. Oxford University Press, Oxford, England, UK.

DeHaven, R. W., and J. L. Guarino. 1969. A nest-box trap for starlings. Bird-Banding 40:49–50.

Dein, F. J., D. E. Toweill, and K. P. Kenow. 2004. Care and use of wildlife in field research. Pages 185–196 in C. E. Braun, editor. Techniques for wildlife investigations and management. The Wildlife Society, Bethesda, Maryland, USA.

Del Greco, S. A., and S. Ansari. 2008. Radar visualization and data exporter tools to support interoperability and the Global Earth Observation System of Systems (GEOSS), from the National Oceanographic and Atmospheric Administration's (NOAA's) National Climatic Data Center (NCDC). Page 390 in R. W. Babcock, Jr., and R. Walton, editors. ASCE Conference Proceedings. NOAA National Climatic Data Center, Asheville, North Carolina, USA. http://www.ncdc.noaa.gov/oa/radar/jnx/index.php.

Del Greco, S. A., and A. Hall. 2003. NCDC the "one stop shop" for all WSR-88D level II data services. 19th International Conference on Interactive Information Processing Systems (IIPS) for Meteorology, Oceanography, and Hydrology. American Meteorological Society, Long Beach, California, USA.

Delahay, R. J., G. C. Smith, and M. R. Hutchings, editors. 2009. Management of disease in wild mammals. Springer, New York, New York, USA.

DelGiudice, G. D. 1995. Assessing winter nutritional restriction of northern deer with urine in snow: considerations, potential, and limitations. Wildlife Society Bulletin 23:687–693.

DelGiudice, G. D., J. Fieberg, M. R. Riggs, M. Carstensen-Powell, and W. Pan. 2006. A long-term age-specific survival analysis of female white-tailed deer. Journal of Wildlife Management 70:1556–1568.

DelGiudice, G. D., K. E. Kunkel, L. D. Mech, and U. S. Seal. 1990. Minimizing capture-related stress on white-tailed deer with a capture collar. Journal of Wildlife Management 54:299–303.

DelGiudice, G. D., B. A. Mangipane, B. A. Sampson, and C. O. Kochanny. 2001a. Chemical immobilization, body temperature, and post-release mortality of white-tailed deer captured by clover trap and net-gun. Wildlife Society Bulletin 29:1147–1157.

DelGiudice, G. D., R. A. Moen, F. J. Singer, and M. R. Riggs. 2001b. Winter nutritional restriction and simulated body condition of Yellowstone elk and bison before and after the fires of 1988. Wildlife Monographs 147.

DelGiudice, G. D., M. R. Riggs, P. Joly, and W. Pan. 2002. Winter severity, survival, and cause-specific mortality of female white-tailed deer in north-central Minnesota. Journal of Wildlife Management 66:698–717.

DelGiudice, G. D., B. A. Sampson, D. W. Kuehn, M. Carstensen-Powell, and J. F. Fieberg. 2005. Understanding margins of safe capture, chemical mobilization, and handling of free-ranging white-tailed deer. Wildlife Society Bulletin 33:677–687.

Dell, J. 1957. Toe clipping varying hares for track identification. New York Fish and Game Journal 4:61–68.

DeLong, T. R. 1982. Effect of ambient conditions on nocturnal nest behavior in long-eared owls. Thesis, Brigham Young University, Provo, Utah, USA.

DeMaso, S. J., and A. D. Peoples. 1993. A restraining device for handling northern bobwhites. Wildlife Society Bulletin 21:45–46.

DeMatteo, K. E., M. A. Rinas, M. M. Sede, B. Davenport, C. F. Argüelles, K. Lovett, and P. G. Parker. 2009. Detection dogs: an effective technique for bush dog surveys. Journal of Wildlife Management 73:1436–1440.

DeNicola, A. J., and R. K. Swihart. 1997. Capture-induced stress in white-tailed deer. Wildlife Society Bulletin 25:500–503.

Dennis, B. 1996. Discussion: should ecologists become Bayesians? Ecological Applications 6:1095–1103.

Dennis, B., and M. R. M. Otten. 2000. Joint effects of density dependence and rainfall on abundance of San Joaquin kit fox. Journal of Wildlife Management 64:388–400.

Deno, R. A. 1937. Uterine macrophages in the mouse and their relation to involution. American Journal of Anatomy 60:433–471.

D'Eon, R. G., G. Pavan, and P. Lindgren. 2003. A small drop-net versus clover traps for capturing mule deer in southeastern British Columbia. Northwest Science 77:178–181.

D'Eon, R. G., R. Serrouya, G. Smith, and C. O. Kochanny. 2002. GPS radiotelemetry error and bias in mountainous terrain. Wildlife Society Bulletin 30:430–439.

Department of Geology and Mineral Industries (DOGAMI). 2009. LiDAR collection and mapping. http://www.oregongeology.org/sub/projects/olc/default.htm.

Desholm, M., A. D. Fox, P. D. L. Beasley, and L. Kahlert. 2006. Remote techniques for counting and estimating the number of bird-wind turbine collisions at sea: a review. Ibis 148:76–89.

deSnoo, G. R., N. M. I. Scheidegger, and F. M. W. deJong. 1999. Vertebrate wildlife incidents with pesticides: a European survey. Pesticide Science 55:47–54.

Dessecker, D. R., and D. G. McAuley. 2001. Importance of early successional habitat to ruffed grouse and American woodcock. Wildlife Society Bulletin 29:456–465.

Dettmers, R., and J. Bart. 1999. A GIS modeling method applied to predicting forest songbird habitat. Ecological Applications 9:152–163.

Dewey, J. 1938. Scientific method: induction and deduction. Pages 419–441 in J. Dewey, editor. Logic—the theory of inquiry. Henry Holt and Company, New York, New York, USA.

DeYoung, C. A. 1988. Comparison of net-gun and drive-net capture for white-tailed deer. Wildlife Society Bulletin 16:318–320.

DeYoung, C. A. 1989. Aging live white-tailed deer on southern ranges. Journal of Wildlife Management 53:519–523.

DeYoung, R. W., and R. L. Honeycutt. 2005. The molecular toolbox: genetic techniques in wildlife ecology and management. Journal of Wildlife Management 69:1362–1384.

Dhondt, A. A., and E. J. van Outryve. 1971. A simple method for trapping breeding adults in nesting boxes. Bird-Banding 42:119–121.

Diamond, J. 1986. Overview: laboratory experiments, field experiments and natural experiments. Pages 3–22 in J. Diamond and T. J. Case, editors. Community ecology. Harper and Row, New York, New York, USA.

Diamond, J., and J. A. Robinson. 2010. Natural experiments of history. Belknap Press of Harvard University Press, Cambridge, Massachusetts, USA.

Dickert, C. 2005. Giant garter snake surveys at some areas of historic occupation in the Grassland Ecological Area, Merced Co. and Mendota Wildlife Area, Fresno Co., California. California Fish and Game 91:255–269.

Dickman, C. R. 1988. Detection of physical contact interactions among free-living mammals. Journal of Mammalogy 69:865–868.

Dickman, C. R., D. H. King, D. C. C. Happold, and M. J. Howell. 1983. Identification of the filial relationships of free-living small mammals by ^{35}sulfur. Australian Journal of Zoology 31:467–474.

Dickson, J. G., R. N. Conner, and J. H. Williamson. 1982. An evaluation of techniques for marking cardinals. Journal of Field Ornithology 53:420–421.

Diehl, R. H., and R. P. Larkin. 1998. Wing beat frequency of *Catharus* thrushes during nocturnal migration, measured via radio telemetry. Auk 115:591–601.

Diehl, R. H., and R. P. Larkin. 2005. Introduction to the WSR-88D (Nexrad) for ornithological research. Pages 876–888 in C. J. Ralph and T. D. Rich, editors. Bird conservation implementation and integration in the Americas: proceedings of the Third International Partners in Flight Conference 2002. U.S. Forest Service, Department of Agriculture, Albany, California, USA.

Dierauf, L. A., and F. M. D. Gulland, editors. 2001. CRC handbook of marine mammal medicine. Second edition. CRC Press, Boca Raton, Florida, USA.

Dieter, C. D., R. J. Murano, and D. Galster. 2009. Capture and mortality rates of ducks in selected trap types. Journal of Wildlife Management 73:1223–1228.

Dieterich, R. A., editor. 1981. Alaskan wildlife diseases. University of Alaska, Fairbanks, USA.

Dietz, N. J., P. J. Bergmann, and L. D. Flake. 1994. A walk-in trap for nesting ducks. Wildlife Society Bulletin 22:19–22.

Diggle, P. 1983. Statistical analysis of spatial point patterns. Academic Press. New York, New York, USA.

DiGiulio, R. T., and P. F. Scanlon. 1984. Heavy metals in tissues of waterfowl from the Chesapeake Bay, USA. Environmental Pollution A 35:29–48.

Dill, H. H., and W. H. Thornsberry. 1950. A cannon-projected net trap for capturing waterfowl. Journal of Wildlife Management 14:132–137.

Dillon, A., and M. Kelly. 2007. Ocelot *Leopardus pardalis* in Belize: the impact of trap spacing and distance moved on density estimates. Oryx 41:469–477.

Dillon, N., A. D. Austin, and E. Bartowsky. 1996. Comparison of preservation techniques for DNA extraction from hymenopterous insects. Insect Molecular Biology 5:21–24.

Dilworth, J. R. 1989. Log scaling and timber cruising. Oregon State University Book Stores, Corvallis, USA.

Dimmick, R. W. 1992. Northern bobwhite (*Colinus virginianus*). Wildlife resources management manual. Technical Report EL-92-18, Section 4.1.3, Waterways Experiment Station, U.S. Corps of Engineers, Department of the Army, Vicksburg, Mississippi, USA.

Dimmick, R. W., and M. R. Pelton. 1994. Criteria of sex and age. Pages 169–214 *in* T. A. Bookhout, editor. Research and management techniques for wildlife and habitats. The Wildlife Society, Bethesda, Maryland, USA.

Dinsmore, S. J., and D. H. Johnson. 2005. Population analysis. Pages 154–184 *in* C. E. Braun, editor. Techniques for wildlife investigations and management. Sixth edition. The Wildlife Society, Bethesda, Maryland, USA.

Dinsmore, S. J., G. C. White, and F. L. Knopf. 2002. Advanced techniques for modeling avian nest survival. Ecology 83:3476–3488.

Dinsmore, S. J., G. C. White, and F. L. Knopf. 2003. Annual survival and population estimates of mountain plovers in southern Phillips County, Montana. Ecological Applications 13:1013–1026.

Ditchkoff, S. S., R. L. Lochmiller, R. E. Masters, S. R. Hoofer, and R. A. Van Den Bussche. 2001. Major-histocompatibility-complex associated variation in secondary sexual traits of white-tailed deer (*Odocoileus virginianus*): evidence for good-genes advertisement. Evolution 55:616–625.

Dittman, A. H., and T. P. Quinn. 1996. Homing in pacific salmon: mechanisms and ecological basis. Journal of Experimental Biology 199: 83–91.

Divine, D. D., D. W. Ebert, and C. L. Douglas. 2000. Examining desert bighorn habitat using 30-m and 100-m elevation data. Wildlife Society Bulletin 28:986–992.

Dix, L. M., and M. A. Strickland. 1986*a*. Sex and age determination for fisher using radiographs of canine teeth: a critique. Journal of Wildlife Management 50:275–276.

Dix, L. M., and M. A. Strickland. 1986*b*. Use of tooth radiographs to classify martens by sex and age. Wildlife Society Bulletin 14:275–279.

Dixon, J. D., M. K. Oli, M. C. Wooten, T. H. Eason, J. W. McCown, and M. W. Cunningham. 2007. Genetic consequences of habitat fragmentation and loss: the case of the Florida black bear (*Ursus americanus floridanus*). Conservation Genetics 8:455–464.

Dizney, L., P. D. Jones, and L. A. Ruedas. 2008. Efficacy of three types of live traps used for surveying small mammals in the Pacific Northwest. Northwestern Naturalist 89:171–180.

Dizon, A. E., C. Lockyer, W. F. Perrin, D. P. Demaster, and J. Sisson. 1992. Rethinking the stock concept: a phylogeographic approach. Conservation Biology 6:24–36.

Doan, T. M. 1997. A new trap for the live capture of large lizards. Herpetological Review 28:79.

Dobbs, J., P. Dobbs, and A. Woodyard. 1993. Tri-Tronics retriever training. Tri-Tronics, Tucson, Arizona, USA.

Dobony, C. A., B. W. Smith, J. W. Edwards, and T. J. Allen. 2006. Necklace-type transmitter attachment method for ruffed grouse chicks. Page 480–488 *in* S. B. Cederbaum, B. C. Faircloth, T. M. Terhune, J. J. Thompson, and J. P. Carroll, editors. Gamebird 2006: quail VI and perdix XII. Warnell School of Forestry and Natural Resources, Athens, Georgia, USA.

Doboszyńska, T. 1976. A method for collecting and staining vaginal smears from the beaver. Acta Theriologica 21:299–306.

Dobson, A. P., J. P. Rodriguez, and W. M. Roberts. 2001. Synoptic tinkering: integrating strategies for large-scale conservation. Ecological Applications 11:1019–1026.

Dobson, A. P., J. P. Rodriguez, W. M. Roberts, and D. S. Wilcove. 1997. Geographic distribution of endangered species in the United States. Science 275:550–553.

Dobson, H., S. Ghuman, S. Prabhakar, and R. Smith. 2003. A conceptual model of the influence of stress on female reproduction. Reproduction 125:151–163.

Dobson, H., and R. F. Smith. 2000. What is stress, and how does it affect reproduction? Animal Reproduction Science 60–61:743–752.

Dobzhansky, T. G. 1937. Genetics and the origin of species. Columbia University Press, New York, New York, USA.

Dobzhansky, T. G. 1970. Genetics of the evolutionary process. Columbia University Press, New York, New York, USA.

Dodd, C. K., Jr. 1991. Drift fence-associated sampling bias of amphibians at a Florida sandhills temporary pond. Journal of Herpetology 25: 296–301.

Dodd, C. K., Jr. 1993. The effects of toe clipping on sprint performance of the lizard *Cnemidophorus sexlineatus*. Journal of Herpetology 27:209–213.

Dodd, N. L., J. S. States, and S. S. Rosenstock. 2003. Tassel-eared squirrel population, habitat condition, and dietary relationships in north-central Arizona. Journal of Wildlife Management 67:622–633.

Dodge, W., and D. Snyder. 1960. An automatic camera device for recording wildlife activity. Journal of Wildlife Management 24:340–342.

Dodson, A. P., and P. J. Hudson. 1992. Regulation and stability of a free-living host–parasite system: *Trichostrongylus tenuis* in red grouse. II. Population models. Journal of Animal Ecology 61:487–498.

Doerr, E. D., V. A. J. Doerr, and P. B. Stacey. 1998. Two capture methods for black-billed magpies. Western Birds 29:55–58.

Doerr, V. A. J., and E. D. Doerr. 2002. A dissolving leg harness for radio transmitter attachment in treecreepers. Corella 26:19–21.

Dokter, A. M., F. Liechti, H. Stark, L. Delobbe, P. Tabary, and I. Holleman. 2010. Dynamics in bird migration flight altitudes studied by a network of operational weather radars. Journal of the Royal Society Interface doi:10.1098/rsif.2010.0116.

Dolbeer, R. A., S. E. Sright, and E. C. Cleary. 2000. Ranking the hazard level of wildlife species to aviation. Wildlife Society Bulletin 28:372–378.

Dolby, G. R. 1982. The role of statistics in the methodology of the life sciences. Biometrics 38:1069–1083.

Dole, J. W. 1965. Summer movements of adult leopard frogs, *Rana pipiens* Schreber, in northern Michigan. Ecology 46:236–255.

Dole, J. W., and P. Durant. 1974. Movements and seasonal activity of *Atelopus oxyrhynchus* (Anura: Atelopodidae) in a Venezuelan cloud forest. Copeia 1974:230–235.

Donalty, S. M., and S. E. Henke. 2001. Can researchers conceal their scent from predators in artificial nest studies? Wildlife Society Bulletin 29:814–820.

Doncaster, C. P. 1990. Non-parametric estimates of interaction from radio-tracking data. Journal of Theoretical Biology 143:431–443.

Doncaster, C. P., and D. W. Macdonald. 1997. Activity patterns and interactions of red foxes (*Vulpes vulpes*) in Oxford city. Journal of Zoology 241:73–87.

Donnelly, M. A., C. Guyer, J. E. Juterbock, and R. A. Alford. 1994. Techniques for marking amphibians. Pages 277–284 *in* W. R. Heyer, M. A. Donnelly, R. W. McDiarmid, L. C. Hayek, and M. S. Foster, editors. Measuring and monitoring biological diversity. Standard methods for amphibians. Smithsonian Insitution Press, Washington, D.C., USA.

Doody, J. S. 1995. A photographic mark recapture method for patterned amphibians. Herpetological Review 26:19–21.

Dorazio, R. M., and F. A. Johnson. 2003. Bayesian inference and decision theory—a framework for decision making in natural resource management. Ecological Applications 13: 556–563.

Dorney, R. S. 1966. A new method for sexing ruffed grouse in late summer. Journal of Wildlife Management 30:623–625.

Dorney, R. S., and F. V. Holzer. 1957. Spring aging methods for ruffed grouse cocks. Journal of Wildlife Management 21:268–274.

Doty, H. A., and R. J. Greenwood. 1974. Improved

nasal-saddle marker for mallards. Journal of Wildlife Management 38:938–939.

Doty, H. A., and F. B. Lee. 1974. Homing to nest baskets by wild female mallards. Journal of Wildlife Management 38:714–719.

Doty, R. L. 2001. Olfaction. Annual Review of Psychology 52:423–452.

Doude Van Troostwijk, W. J. 1976. Age determination in muskrats, Ondatra zibethicus (L.) in the Netherlands. Lutra 18:33–43.

Douglas, C. W., and M. A. Strickland. 1987. Fisher. Pages 511–529 in M. Novak, J. A. Baker, M. E. Obbard, and B. Malloch, editors. Wild furbearer management and conservation in North America. Ontario Ministry of Natural Resources, Toronto, Canada.

Douglass, R. J., A. J. Kuenzi, T. Wilson, and R. C. Van Horne. 2000. Effects of bleeding nonanesthetized wild rodents on handling mortality and subsequent recapture. Journal of Wildlife Diseases 36:700–704.

Dove, C. J., N. C. Rotzel, M. Heacker, and L. A. Weigt. 2008. Using DNA barcodes to identify bird species involved in birdstrikes. Journal of Wildlife Management 72:1231–1236.

Doviak, R. J., and D. S. Zrnic. 1993. Doppler radar and weather observations. Second edition. Academic Press, San Diego, California, USA.

Dow, S. A., Jr., and P. L. Wright. 1962. Changes in mandibular dentition associated with age in pronghorn antelope. Journal of Wildlife Management 26:1–18.

Downes, S., and P. Borges. 1998. Sticky traps: an effective way to capture small terrestrial lizards. Herpetological Review 29:94–95.

Downing, R. L. 1980. Vital statistics of animal populations. Pages 247–267 in S. D. Schemnitz, editor. Wildlife management techniques manual. The Wildlife Society, Washington, D.C., USA.

Downing, R. L., and C. M. Marshall. 1959. A new plastic tape marker for birds and mammals. Journal of Wildlife Management 23:223–224.

Downing, R. L., and B. S. McGinnes. 1969. Capturing and marking white-tailed deer fawns. Journal of Wildlife Management 33:711–714.

Downing, R. L., E. D. Michael, and R. J. Poux, Jr. 1977. Accuracy of sex and age ratio counts of white-tailed deer. Journal of Wildlife Management 41:709–714.

Downs, A. A., and W. E. McQuilkin. 1944. Seed production of southern Appalachian oaks. Journal of Forestry 42:913–920.

Downs, J. A., and M. W. Horner. 2008. Effects of point pattern shape on home-range estimates. Journal of Wildlife Management 72:1813–1818.

Dozier, H. L. 1942. Identification of sex in live muskrats. Journal of Wildlife Management 6:292–293.

Dragon, D. C., and R. P. Rennie. 1995. The ecology of anthrax spores: tough but not invincible. Canadian Veterinary Journal 36:295–301.

Drake, V. A. 1981a. Quantitative observation and analysis procedures for a manually operated entomological radar. CSIRO Australian Division of Entomology Technical Paper 19:1–41.

Drake, V. A. 1981b. Target density estimation in radar biology. Journal of Theoretical Biology 90:545–571.

Draper, N. R., and H. Smith. 1998. Applied regression analysis. Third edition. John Wiley and Sons, New York, New York, USA.

Dreibelbis, J. Z., K. B. Melton, R. Aguirre, B. A. Collier, J. Hardin, N. J. Silvy, and M. J. Peterson. 2008. Predation of Rio Grande wild turkey nests on the Edwards Plateau, Texas. Wilson Journal of Ornithology 120: 906–910.

Drewien, R. C., and E. G. Bizeau. 1978. Cross-fostering whooping cranes to sandhill crane foster parents. Pages 201–222 in S. A. Temple, editor. Endangered birds: management techniques for preserving threatened species. University of Wisconsin Press, Madison, USA.

Drewien, R. C., and K. R. Clegg. 1991. Capturing whooping cranes and sandhill cranes by night-lighting. Proceedings of the North American Crane Workshop 6:43–49.

Drewien, R. C., K. R. Clegg, and R. E. Shea. 1999. Capturing trumpeter swans by night-lighting. Wildlife Society Bulletin 27:209–215.

Drewien, R. C., H. M. Reeves, P. F. Springer, and T. L. Kuck. 1967. Back-pack unit for capturing waterfowl and upland game by night-lighting. Journal of Wildlife Management 31:778–783.

Drewien, R. C., R. J. Vernimen, S. W. Harris, and C. F. Yocom. 1966. Spring weights of band-tailed pigeons. Journal of Wildlife Management 30:190–192.

Drickamer, L.C. 1990. Urinary chemosignals, reproduction, and population size for house mice (Mus musculus) living in field enclosures. Journal of Chemical Ecology 16:2955–2968.

Drummer, T. D., and L. L. McDonald. 1987. Size bias in line transect sampling. Biometrics 43: 13–21.

Drury, W. H., Jr., and I. C. T. Nisbet. 1964. Radar studies of orientation of songbird migrants in southeastern New England. Bird-Banding 35:69–119.

Drut, M. S., W. H. Pyle, and L. A. Crawford. 1994. Diets and food selection of sage grouse chicks in Oregon. Journal of Range Management 47:90–93.

Dubayah, R. O., J. B. Blair, J. L. Bufton, D. B. Clark, J. JaJa, R. Knox, S. B. Luthcke, S. Prince, and J. Weishampel. 1997. The vegetation canopy lidar mission. Pages 100–112 in Proceedings of land satellite information in the next decade, II: sources and applications. American Society of Photogrammetry and Remote Sensing, Bethesda, Maryland, USA.

Dubayah, R. O., and J. B. Drake. 2000. Lidar remote sensing for forestry. Journal of Forestry 98: 44–46.

Dudley, S. J., C. D. Bonham, S. R. Abt, and J. C. Fischenich. 1998. Comparison of methods for measuring woody riparian vegetation density. Journal of Arid Environments 38:77–86.

Dueser, R. D., and H. H. Shugart, Jr. 1978. Microhabitats in a forest-floor small-mammal fauna. Ecology 59:89–98.

Duffield, L. F. 1973. Aging and sexing the post-cranial skeleton of bison. Plains Anthropologist 18:132–139.

Dullum, J. L. D., K. R. Foresman, and M. R. Matchett. 2005. Efficacy of translocations for restoring populations of black-tailed prairie dogs. Wildlife Society Bulletin 33:842–850.

Dunbar, I. K. 1959. Leg bands in cold climates. Eastern Bird Banding News 22:37.

Dunbar, M. R., S. R. Johnson, J. C. Rhyan, and M. McCollum. 2009. Use of infrared thermography to detect thermo-graphic changes in mule deer (Odocoileus hemionus) experimentally infected with foot-and-mouth disease. Journal of Zoo and Wildlife Medicine 40:296–301.

Dunbar, M. R., and K. MacCarthy. 2006. Use of infrared thermography to detect signs of rabies infection in raccoons (Procyon lotor). Journal of Zoo and Wildlife Medicine 37:518–523.

Dunham, A. E., and S. J. Beaupre. 1998. Ecological experiments: scale, phenomenology, mechanism and the illusion of generality. Pages 27–49 in J. Bernardo and W. Resetarits, editors. Experimental ecology: issues and perspectives. Oxford University Press, New York, New York, USA.

Dunk, J. E. 1991. A selective pole trap for raptors. Wildlife Society Bulletin 19:208–210.

Dunmire, W. W. 1955. Sex dimorphism in the pelvis of rodents. Journal of Mammalogy 36:356–361.

Dunn, E. H., D. J. T. Hussell, and R. J. Adams. 1997. Monitoring songbird population change with autumn mist netting. Journal of Wildlife Management 61:389–396.

Dunn, E. H., D. J. T. Hussell, and R. E. Ricklefs. 1979. The determination of incubation stage in starling eggs. Bird-banding 50:114–120.

Dunn, J. E., and P. S. Gipson. 1977. Analysis of radio telemetry data in studies of home range. Biometrics 33:85–101.

Dunne, P. 1987. Introduction to raptor identification, aging and sexing techniques. Pages 13–21 in B. A. Giron Pendleton, B. A. Millsap, K. W. Cline, and D. M. Bird, editors. Raptor management techniques manual. National Wildlife Federation, Washington, D.C., USA.

Dunning, J. B. 2008. CRC handbook of avian body masses. CRC Press, Boca Raton, Florida, USA.

Dunning, J. B., B. J. Danielson, and H. R. Pulliam. 1992. Ecological processes that affect populations in complex landscapes. Oikos 65:169–175.

Duong, T. 2008. KS: Kernel smoothing: version 1.5.9. Package for program R. http://cran.r-project.org/web/packages/ks/index.html.

Durant, A. J., and E. R. Doll. 1941. Ulcerative enteritis in quail. Research Bulletin 325, Agricultural Experiment Station, University of Missouri, Columbia, USA.

Durbin, J., and G. S. Watson. 1971. Testing for serial correlation in least squares regression. III. Biometrika 58:1–19.

Durden, L. A., E. M. Dotson, and G. N. Vogel. 1995. Two efficient techniques for catching skinks. Herpetological Review 26:137.

Durnin, M. E., R. R. Swaisgood, N. Czekala, and A. Hemin. 2004. Effects of radiocollars on giant panda stress-related behavior and hormones. Journal of Wildlife Management 68:987–992.

Durtsche, R. D. 1996. A capture technique for small smooth-scaled lizards. Herpetological Review 27:12–13.

Dussault, C., R. Courtois, J. Huot, and J. P. Ouellet. 1999. Evaluation of GPS telemetry collar performance for habitat studies in the boreal forest. Wildlife Society Bulletin 27:965–972.

Dussault, C., R. Courtois, J. Huot, and J. P. Ouellet. 2001. The use of forest maps for the description of wildlife habitats: limits and recommendations. Canadian Journal of Forest Research 31:1227–1234.

Dwyer, T. J. 1972. An adjustable radio package for ducks. Bird-Banding 43:282–284.

Dwyer, T. J., and J. V. Dobell. 1979. External determination of age of common snipe. Journal of Wildlife Management 43:754–756.

Dybdal, R. 1987. Radar cross-section measurements. Proceedings of the IEEE 75:498–516.

Dyer, S. J. 1999. Movement and distribution of woodland caribou (*Rangifer tarandus caribou*) in response to industrial development in northeastern Alberta. Thesis, University of Alberta, Edmonton, Canada.

Dykstra, J. N. 1968. A decoy and net for capturing nesting robins. Bird-Banding 39:189–192.

Eadie, J. M., K. M. Cheng, and C. R. Nichols. 1987. Limitations of tetracycline in tracing multiple maternity. Auk 104:330–333.

Eagle, T. C., and J. S. Whitman. 1987. Mink. Pages 615–624 *in* M. Novak, J. A. Baker, M. E. Obbard, and B. Malloch, editors. Wild furbearer management and conservation in North America. Ontario Ministry of Natural Resources, Toronto, Canada.

Ealey, E. H. M., and G. M. Dunnet. 1956. Plastic collars with patterns of reflective tape for marking nocturnal mammals. Australia's Commonwealth Scientific and Industrial Research Organization, Wildlife Research 1:59–62.

Earlé, E. A. 1988. A cast-net for trapping nightjars (and others). Safring News 17:25–28.

Earle, R. D., D. M. Lunning, and V. R. Tuovila. 1996. Assessing injuries to Michigan bobcats held by #3 Soft Catch® traps. Pages 34–35 *in* R. D. Earle, editor. Proceedings of the fourteenth midwest furbearer workshop. Michigan Department of Natural Resources, Lansing, USA.

Earle, R. D., D. M. Lunning, V. R. Tuovila, and J. A. Shivik. 2003. Evaluating injury mitigation and performance of #3 Victor Soft Catch® traps to restrain bobcats. Wildlife Society Bulletin 31:617–629.

Easley, L. T., and L. E. Chaiken. 1951. An expendable seed trap. Journal of Forestry 49:652–653.

Eastwood, E. 1967. Radar ornithology. Methuen, London, England, UK.

Eastwood, E., G. A. Isted, and G. C. Rider. 1962. Radar ring angels and the roosting behavior of starlings. Proceedings of the Royal Society of London B 156:242–267.

Eberhardt, L. L. 1968. A preliminary appraisal of line transects. Journal of Wildlife Management 32:82–88.

Eberhardt, L. L. 1969. Population analysis. Pages 457–495 *in* R. H. Giles, Jr., editor. Wildlife management techniques. The Wildlife Society, Washington, D.C., USA.

Eberhardt, L. L. 1970. Correlation, regression, and density dependence. Ecology 51:306–310.

Eberhardt, L. L. 1985. Assessing the dynamics of wild populations. Journal of Wildlife Management 49:997–1012.

Eberhardt, L. L. 1987*a*. Calibrating population indices by double sampling. Journal of Wildlife Management 51:665–675.

Eberhardt, L. L. 1987*b*. Population projections from simple models. Journal of Applied Ecology 24:103–118.

Eberhardt, L. L. 1988. Using age structure data from changing populations. Journal of Applied Ecology 25:373–378.

Eberhardt, L. L. 2003. What should we do about hypothesis testing? Journal of Wildlife Management 67:241–247.

Eberhardt, L. L., and M. A. Simmons. 1987. Calibrating population indices by double sampling. Journal of Wildlife Management 51:665–675.

Eberhardt, L. L., and J. M. Thomas. 1991. Designing environmental field studies. Ecological Monographs 61:53–73.

Eckert, K. L., and S. A. Eckert. 1989. The application of plastic tags to leatherback sea turtles, *Dermochelys eumochelys corfacea*. Herpetological Review 20:90–91.

Eckert, S. A., J. E. Moore, D. C. Dunn, R. S. van Buiten, K. L. Eckert, and P. Halpin. 2008. Modeling loggerhead turtle movement in the Mediterranean: importance of body size and oceanography. Ecological Applications 18:290–308.

Ecobichon, D. J. 2001. Toxic effects of pesticides. Pages 763–810 *in* C. D. Klaassen, editor. Casarett and Doull's toxicology: the basic science of poisons. Sixth edition. McGraw-Hill, New York, New York, USA.

Edalgo, J., and T. Anderson. 2007. Effects of pre-baiting on small mammal trapping success in a Morrow's honeysuckle-dominated area. Journal of Wildlife Management 71:246–250.

Eddleman, L. E., E. E. Remmenga, and R. T. Ward. 1964. An evaluation of plot methods for alpine vegetation. Bulletin of the Torrey Botanical Club 91:439–450.

Eddleman, W. R., and C. J. Conway. 1998. Clapper rail. Account 340 *in* A. Poole, P. Stenenheim, and F. Gill, editors. The birds of North America. The Academy of Natural Sciences, Philadelphia, Pennsylvania, and The American Ornithologists' Union, Washington, D.C., USA.

Eddleman, W. R., R. E. Flores, and M. L. Legare. 1994. Black rail. Account 123 *in* A. Poole, P. Stettenheim, and F. Gill, editors. The birds of North America. The Academy of Natural Sciences, Philadelphia, Pennsylvania, and The American Ornithologists' Union, Washington, D.C., USA.

Eddleman, W. R., and F. L. Knopf. 1985. Determining age and sex of American coots. Journal of Field Ornithology 56:41–55.

Eden, C. J., H. H. Whiteman, L. Duobinis-Gray, and S. A. Wissinger. 2007. Accuracy assessment of skeletochronology in the Arizona tiger salamander (*Ambystoma tigrinum nebulosum*). Copeia 2:471–477.

Edgar, A. K., E. J. Dodsworth, and M. P. Warden. 1973. The design of a modern surveillance radar. Pages 8–13 *in* Radar—present and future. IEE Conference Publication 105, Institute of Electrical Engineers, London, England, UK.

Edgar, R. L. 1968. Catching colonial seabirds for banding. Bird-Banding 39:41–43.

Edge, W. D., C. L. Marcum, and S. L. Olson-Edge. 1987. Summer habitat selection by elk in western Montana: a multivariate approach. Journal of Wildlife Management 51:844–851.

Edmands, S. 2007. Between a rock and a hard place: evaluating the relative risks of inbreeding and outbreeding for conservation and management. Molecular Ecology 16:463–475.

Edminster, F. C. 1938. The marking of ruffed grouse for field identification. Journal of Wildlife Management 2:55–57.

Edminster, F. C. 1954. American game birds of field and forest. Charles Scribner's Sons, New York, New York, USA.

Edwards, I. R., D. G. Ferry, and W. A. Temple. 1991. Fungicides and related compounds. Pages 1409–1470 *in* W. J. Hayes and E. R. Laws, Jr., editors. Handbook of pesticide toxicology. Volume 2. Classes of pesticides. Academic Press, San Diego, California, USA.

Edwards, J., and E. W. Houghton. 1959. Radar echoing area polar diagrams of birds. Nature 184:1059.

Edwards, J. K., R. L. Marchinton, and G. F. Smith. 1982. Pelvic girdle criteria for sex determination of white-tailed deer. Journal of Wildlife Management 46:544–547.

Efford, I. E., and J. A. Mathias. 1969. A comparison of two salamander populations in Marion Lake, British Columbia. Copeia 1969:723–736.

Efron, B. 1979. Bootstrap methods: another look at the jackknife. Annals of Statistics 7:1–26.

Efron, B. 1998. R. A. Fisher in the 21st century. Statistical Science 13:95–122.

Efron, B., and B. J. Tibshirani. 1993. An introduction to the bootstrap. Chapman and Hall, New York, New York, USA.

Ehrhart, L. M. 1995. A review of sea turtle reproduction. Pages 29–38 *in* K. A. Bjorndal, editor. Biology and conservation of sea turtles. Smithsonian Institution Press, Washington, D.C., USA.

Einarsen, A. S. 1948. The pronghorn antelope and its management. Wildlife Management Institute, Washington, D.C., USA.

Eisenberg, J. F., and D. G. Kleiman. 1972. Olfactory communication in mammals. Annual Review of Ecology and Systematics 3:1–32.

Eisler, R. 1985*a*. Cadmium hazards to fish, wildlife, and invertebrates: a synoptic review. Biological Report 85(1.2), U.S. Fish and Wildlife Service, Department of the Interior, Washington, D.C., USA.

Eisler, R. 1985*b*. Selenium hazards to fish, wildlife, and invertebrates: a synoptic review. Biological Report 85(1.5), U.S. Fish and Wildlife Service,

Department of the Interior, Washington, D.C., USA.
Eisler, R. 1986a. Chromium hazards to fish, wildlife, and invertebrates: a synoptic review. Biological Report 85(1.6), U.S. Fish and Wildlife Service, Department of the Interior, Washington, D.C., USA.
Eisler, R. 1986b. Dioxin hazards to fish, wildlife, and invertebrates: a synoptic review. Biological Report 85(1.8), U.S. Fish and Wildlife Service, Department of the Interior, Washington, D.C., USA.
Eisler, R. 1986c. Polychlorinated biphenyls hazards to fish, wildlife, and invertebrates: a synoptic review. Biological Report 85(1.7), U.S. Fish and Wildlife Service, Department of the Interior, Washington, D.C., USA.
Eisler, R. 1987a. Mercury hazards to fish, wildlife, and invertebrates: a synoptic review. Biological Report 85(1.10), U.S. Fish and Wildlife Service, Department of the Interior, Washington, D.C., USA.
Eisler, R. 1987b. Polycyclic aromatic hydrocarbon hazards to fish, wildlife, and invertebrates: a synoptic review. Biological Report 85(1.11), U.S. Fish and Wildlife Service, Department of the Interior, Washington, D.C., USA.
Eisler, R. 1988a. Arsenic hazards to fish, wildlife, and invertebrates: a synoptic review. Biological Report 85(1.12), U.S. Fish and Wildlife Service, Department of the Interior, Washington, D.C., USA.
Eisler, R. 1988b. Lead hazards to fish, wildlife, and invertebrates: a synoptic review. Biological Report 85(1.14), U.S. Fish and Wildlife Service, Department of the Interior, Washington, D.C., USA.
Eisler, R. 1991. Cyanide hazards to fish, wildlife, and invertebrates: a synoptic review. Biological Report 85(1.23), U.S. Fish and Wildlife Service, Department of the Interior, Washington, D.C., USA.
Eisler, R. 1995. Monosodium fluoroacetate hazards to fish, wildlife, and invertebrates: a synoptic review. Biological Report 85(1.30), U.S. Fish and Wildlife Service, Department of the Interior, Washington, D.C., USA.
Eisler, R., and A. A. Belisle. 1996. Planar PCBs hazards to fish, wildlife, and invertebrates: a synoptic review. Biological Report 85(1.31), U.S. Fish and Wildlife Service, Department of the Interior, Washington, D.C., USA.
Eisler, R., D. R. Clark, Jr., S. N. Wiemeyer, and C. J. Henny. 1999. Sodium cyanide hazards to fish and other wildlife from gold mining operations. Pages 55–67 in J. M. Azcue, editor. Environmental impacts of mining activities: emphasis on mitigation and remedial measures. Environmental Sciences Series. Springer-Verlag, Berlin, Germany.
Elbin, S. B., and J. Burger. 1994. Implantable microchips for individual identification in wild and captive populations. Wildlife Society Bulletin 22:677–683.
Elder, W. H., and C. E. Shanks. 1962. Age changes in tooth wear and morphology of the baculum in muskrats. Journal of Mammalogy 43:144–150.
Elkin, B., and R. L. Zarnke. 2001. A field guide to common wildlife diseases and parasites in Alaska. Alaska Department of Fish and Game, Anchorage, USA. http://www.wc.adfg.state.ak.us/index.cfm?adfg=disease.main.
Ellegren, H. 1996. First gene on the avian W chromosome (CHD) provides a tag for universal sexing of non-ratite birds. Proceedings of the Royal Society of London B 263: 1635–1641.
Ellegren, H., G. Lindgren, C. R. Primmer, and A. P. Moller. 1997. Fitness loss and germline mutations in barn swallows breeding in Chernobyl. Nature 389:593–596.
Ellegren, H., and B. C. Sheldon. 2008. Genetic basis of fitness differences in natural populations. Nature 452:169–175.
Ellenton, J. A., and O. H. Johnston. 1975. Oral biomarkers of calciferous tissues in carnivores. Pages 60–67 in R. E. Chambers, editor. Transactions of Eastern Coyote Workshop. Northeastern Fish and Wildlife Conference, New Haven, Connecticut, USA.
Elliott, C. H. 1989. The pest status of the *Quelea*. Pages 17–34 in R. L. Bruggers and C. H. Elliott, editors. *Quelea quelea*—Africa's bird pest. Oxford University Press, Oxford, England, UK.
Elliott, J. E., K. M. Langelier, P. Mineau, and L. K. Wilson. 1996. Poisoning of bald eagles and red-tailed hawks by carbofuran and fensulfothion in the Fraser Delta of British Columbia, Canada. Journal of Wildlife Diseases 32:486–491.
Ellis, D. H., and C. H. Ellis. 1975. Color marking golden eagles with human hair dyes. Journal of Wildlife Management 39:445–447.
Ellis, D. H., D. Hjertaas, B. W. Johns, and R. P. Urbanek. 1998. Use of a helicopter to capture flighted cranes. Wildlife Society Bulletin 26:103–107.
Ellison, A. M. 2004. Bayesian inference in ecology. Ecology Letters 7:509–520.
Ellison, A. M. 1996. An introduction to Bayesian inference for ecological research and environmental decision-making. Ecological Applications 6:1036–1046.
Ellison, L. N. 1968. Sexing and aging Alaskan spruce grouse by plumage. Journal of Wildlife Management 32:12–16.
Ellisor, J. E., and W. F. Harwell. 1969. Mobility and home range of collared peccary in southern Texas. Journal of Wildlife Management 33:425–427.
Ellsworth, D. L., R. L. Honeycutt, N. J. Silvy, J. W. Bickham, and W. D. Klimstra. 1994a. Historical biogeography and contemporary patterns of mitochondrial DNA variation in white-tailed deer from the southeastern United States. Evolution 48:122–136.
Ellsworth, D. L., R. L. Honeycutt, N. J. Silvy, M. H. Smith, J. W. Bickham, and W. D. Klimstra. 1994b. White-tailed deer restoration to the southeastern United States: evaluating genetic-variation. Journal of Wildlife Management 58:686–697.
Elmeros, M., and M. Hammershoj. 2006. Experimental evaluation of the reliability of placental scar counts in American mink (*Mustela vison*). European Journal of Wildlife Research 52:132–135.
Elmes, R. 1955. Loss of rings. Bird Study 2:153.
Elody, B. I., and N. F. Sloan. 1984. A mist net technique useful for capturing barred owls. North American Bird Bander 9:13–14.
El-Sayed, A. M., J. A. Byers, L. M. Manning, A. Jurgens, V. J. Mitchell, and D. M. Suckling. 2008. Floral scent of Canada thistle and its potential as a generic insect attractant. Journal of Economic Entomology 101:720–727.
Elsey, R. M., and P. L. Trosclair, III. 2004. A new live trap for capturing alligators. Herpetological Review 35:253–255.
Ely, C. R. 1990. Effects of neck bands on the behavior of wintering greater white-fronted geese. Journal of Field Ornithology 61:249–253.
Ely, C. R., and A. X. Dzubin. 1994. Greater white-fronted goose. Account 131 in A. Poole and F. Gill, editors. The birds of North America. The Academy of Natural Sciences, Philadelphia, Pennsylvania, and The American Ornithologists' Union, Washington, D.C., USA.
Elzinga, C. L., D. W. Salzer, and J. W. Willoughby. 1998. Measuring and monitoring plant populations. Technical Reference 1730-1, U.S. Bureau of Land Management, Department of the Interior, Denver, Colorado, USA.
Emlen, S. 1974. Problems in identifying bird species by radar signature analyses: intra-specific variability. Pages 509–524 in Proceedings of a conference on the biological aspects of the bird/aircraft collision problem. Department of Zoology, Clemson University, Clemson, South Carolina, USA.
Emlen, S. T. 1968. A technique for marking anuran amphibians for behavioral studies. Herpetologica 24:172–173.
Enderson, J. H. 1964. A study of the prairie falcon in the central Rocky Mountain region. Auk 81:332–352.
Endler, J. A. 1997. Light, behavior, and conservation of forest dwelling organisms. Pages 329–355 in J. R. Clemmons and R. Buchholz, editors. Behavioral approaches to conservation in the wild. Cambridge University Press, Cambridge, England, UK.
Eng, R. L. 1955. A method for obtaining sage grouse age and sex ratios from wings. Journal of Wildlife Management 19:267–272.
Enge, K. M. 1997. Use of silt fencing and funnel traps for drift fences. Herpetological Review 28:30–31.
Enge, K. M. 2001. The pitfalls of pitfall traps. Journal of Herpetology 35:467–478.
Engeman, R. M. 1998. An easy capture method for brown tree snakes (*Boiga irregularis*). Snake 28:101–102.
Engeman, R. M., H. W. Krupa, and J. Kern. 1997. On the use of injury scores for judging the acceptability of restraining traps. Journal of Wildlife Research 2:124–127.
Engeman, R. M., M. A. Linnell, P. Aguon, A. Manibuson, S. Sayama, and A. Techaira. 1999. Implications of brown tree snake captures from fences. Wildlife Research 26:111–116.
Engeman, R. M., D. V. Rodriquez, M. A. Linnell, and M. E. Pitzler. 1998. A review of the case

histories of the brown tree snake (*Boiga irregularis*) located by detector dogs on Guam. International Biodeterioration and Biodegradation 42:161–165.

Engeman, R. M., R. T. Sugihara, L. F. Pank, and W. E. Dusenberry. 1994. A comparison of plotless density estimators using monte carlo simulation. Ecology 75:1769–1779.

Engeman, R. M., and D. S. Vice. 2001. A direct comparison of trapping and spotlight searches for capturing brown tree snakes on Guam. Pacific Conservation Biology 7:4–8.

Engeman, R. M., D. S. Vice, D. York, and K. S. Gruver. 2002. Sustained evaluation of the effectiveness of detector dogs for locating brown tree snakes in cargo outbound from Guam. International Biodeterioration and Biodegradation 49:101–106.

Engen, S., T. H. Ringsby, B. E. Saether, R. Lande, H. Jensen, M. Lillegard, and H. Ellegren. 2007. Effective size of fluctuating populations with two sexes and overlapping generations. Evolution 61:1873–1885.

Englisch, T., F. M. Steiner, and B. C. Schilick-Steiner. 2005. Fine-scale grassland assemblage analysis in Central Europe: ants tell another story than plants (Hymenoptera: Formicidae; Spermatophyta). Myrmecologische Nachrichten 7:61–67.

Englund, J. 1982. A comparison of injuries to leg-hold trapped and foot-snared red foxes. Journal of Wildlife Management 46:1113–1117.

Ensminger, A. L., and S. M. G. Hoffman. 2002. Sex identification assay useful in great apes is not diagnostic in a range of other primate species. American Journal of Primatology 56:129–134.

Environment Canada. 1984. North American bird banding. Environmental Conservation Service 1:1–3.

Epps, C. W., J. D. Wehausen, V. C. Bleich, S. G. Torres, and J. S. Brashares. 2007. Optimizing dispersal and corridor models using landscape genetics. Journal of Applied Ecology 44:714–724.

Erb, J. D., R. D. Bluett, E. K. Fritzell, and N. F. Payne. 1999. Aging muskrats using molar indices: a regional comparison. Wildlife Society Bulletin 27:628–635.

Erickson, A. W. 1957. Techniques for live-trapping and handling black bears. Transactions of the North American Wildlife Conference 22:520–543.

Erickson, J. A., A. E. Anderson, D. E. Medin, and D. C. Bowden. 1970. Estimating ages of mule deer—an evaluation of technique accuracy. Journal of Wildlife Management 34:523–531.

Erickson, J. A., and W. G. Seliger. 1969. Efficient sectioning of incisors for estimating ages of mule deer. Journal of Wildlife Management 33:384–388.

Erickson, W. P., T. L. McDonald, K. G. Gerow, S. Howlin, and J. W. Kern. 2001. Statistical issues in resource selection studies with radio-marked animals. Pages 211–242 *in* J. J. Millspaugh and J. M. Marzluff, editors. Radio tracking and animal populations. Academic Press, San Diego, California, USA.

Erickson, W. P., T. L. McDonald, and R. Skinner. 1998. Habitat selection using GIS data: a case study. Journal of Agricultural, Biological and Ecological Statistics 3:296–310.

Erickson, W. P., R. Nielson, R. Skinner, B. Skinner, and J. Johnson. 2003. Applications of resource selection modeling using unclassified Landsat TM. Pages 130–140 *in* S. V. Huzurbazar, editor. Resource selection methods and applications. Western EcoSystems Technology, Cheyenne, Wyoming, USA (http://www.west-inc.com).

Eriksen, B., J. Cardoza, J. Pack, and H. Kilpatrick. 1995. Procedures and guidelines for rocket-netting wild turkeys. Technical Bulletin 1, National Wild Turkey Federation, Edgefield, South Carolina, USA.

Erikstad, K. E., and R. Andersen. 1983. The effect of weather on survival, growth rate and feeding time in different sized willow grouse broods. Ornis Scandinavica 14:249–252.

Ernest, H. B., M. C. T. Penedo, B. P. May, M. Syvanen, and W. M. Boyce. 2000. Molecular tracking of mountain lions in the Yosemite Valley region in California: genetic analysis using microsatellites and faecal DNA. Molecular Ecology 9:433–441.

Ernst, C. H. 1971. Population dynamics and activity cycles of *Chrysemys picta* in southeastern Pennsylvania. Journal of Herpetology 5:151–160.

Erwin, T. L. 1989. Canopy arthropod biodiversity: a chronology of sampling techniques and results. Revista Peruana de Entomologia 32:71–77.

Estes, J. A., and R. J. Jameson. 1988. A double-survey estimate for sighting probability of sea otters in California. Journal of Wildlife Management 52:70–76.

Estrada-Rodriguez, J. L., H. Gadsen, S. V. Leyva Pacheco, and H. López Corrujedo. 2004. A new capture technique for the Coahuila fringe-toed lizard and other desert lizards. Herpetological Review 35:244–245.

Etchberger, R. C., and P. R. Krausman. 1997. Evaluation of five methods for measuring desert vegetation. Wildlife Society Bulletin 25:604–609.

Etter, S. L. 1963. Age determination and growth in juvenile greater prairie chickens. Thesis, University of Illinois, Urbana, USA.

Etter, S. L., J. E. Warnock, and G. B. Joselyn. 1970. Modified wing molt criteria for estimating the ages of wild juvenile pheasants. Journal of Wildlife Management 34:620–626.

Evans, C. D. 1951. A method of color marking young waterfowl. Journal of Wildlife Management 15:101–103.

Evans, J., J. O. Ellis, R. D. Nass, and A. L. Ward. 1971. Techniques for capturing and marking nutria. Proceedings of the Annual Conference of the Southeastern Association of Game and Fish Commissioners 25:295–315.

Evans, J., and R. E. Griffith, Jr. 1973. A fluorescent tracer and marker for animal studies. Journal of Wildlife Management 37:73–81.

Evans, R. A., and R. M. Love. 1957. The step-point method of sampling—a practical tool in range research. Journal of Range Management 10:208–212.

Evans, T. R., and L. C. Drickamer. 1994. Flight speeds of birds determined using Doppler radar. Wilson Bulletin 106:154–156.

Evans, W. E., J. D. Hall, A. B. Irvine, and J. S. Leatherwood. 1972. Methods for tagging small cetaceans. Fisheries Bulletin 70:61–65.

Evans, W. R., and D. K. Mellinger. 1999. Monitoring grassland birds in nocturnal migration. Studies in Avian Biology 19:219–229.

Everitt, J. H., D. E. Escobar, R. Villarreal, J. R. Noriega, and M. R. Davis. 1991. Airborne video systems for agricultural assessments. Remote Sensing of Environment 35:231–242.

Everitt, J. H., and P. R. Nixon. 1985. Video imagery: a new remote sensing tool for range management. Journal of Range Management 38:421–424.

Evrard, J. O. 1986. Loss of nasal saddle on mallard. Journal of Field Ornithology 57:170–171.

Evrard, J. O., and B. R. Bacon. 1998. Duck trapping success and mortality using four trap designs. North American Bird Bander 23:110–114.

Ewing, S. R., R. G. Nager, M. A. C. Nicoll, A. Aumjaud, C. G. Jones, and L. F. Keller. 2008. Inbreeding and loss of genetic variation in a reintroduced population of Mauritius kestrel. Conservation Biology 22:395–404.

Ewins, P. J., and M. J. R. Miller. 1993. Noose dome trap for ospreys. North American Bird Bander 18:40.

Facka, A. N., P. L. Ford, and G. W. Roemer. 2008. A novel approach for assessing density and range-wide abundance of prairie dogs. Journal of Mammalogy 89:356–364.

Fagerstone, K. A., M. A. Coffey, P. D. Curtis, R. A. Dolbeer, G. J. Killian, L. A. Miller, and L. M. Wilmont. 2002. Wildlife fertility control. Technical Review 02-2, The Wildlife Society, Bethesda, Maryland, USA.

Fagerstone, K. A., and B. E. Johns. 1987. Transponders as permanent identification markers for domestic ferrets, black-footed ferrets, and other wildlife. Journal of Wildlife Management 51:294–297.

Fagerstone, K. A., L. A. Miller, G. Killian, and C. A. Yoder. 2010. Review of issues concerning the use of reproductive inhibitors, with particular emphasis on resolving human–wildlife conflicts in North America. Integrative Zoology 5:15–30.

Fair, W. S., and S. E. Henke. 1997. Efficacy of capture methods for a low density population of *Phrynosoma cornutum*. Herpetological Review 28:135–137.

Fairbrother, A. 1996. Cholinesterase-inhibiting pesticides. Pages 52–60 *in* A. Fairbrother, L. N. Locke, and G. L. Huff, editors. Non-infectious diseases of wildlife. Second edition. Iowa State University Press, Ames, USA.

Fairbrother, A., L. N. Locke, and G. L. Huff, editors. 1996. Noninfectious diseases of wildlife. Second edition. Iowa State University Press, Ames, USA.

Fairley, J. S. 1982. Short-term effects of ringing and toe-clipping on the recapture of wood mice (*Apodemus sylvaticus*). Journal of Zoology 197:295–297.

Fall, J., and A. Fall. 2001. A domain-specific language for models of landscape dynamics. Ecological Modelling 141:1–18.

Fancy, S. G. 1980. Preparation of mammalian teeth

for age determination by cementum layers: a review. Wildlife Society Bulletin 8:242–248.

Fankhauser, D. 1964. Plastic adhesive tape for color-marking birds. Journal of Wildlife Management 28:594.

Farnsworth, G. L., K. H. Pollock, J. D. Nichols, T. R. Simons, J. E. Hines, and J. R. Sauer. 2002. A removal model for estimating detection probabilities from point-count surveys. Auk 119:414–425.

Farrell, L. E., J. Roman, and M. E. Sunquist. 2000. Dietary separation of sympatric carnivores identified by molecular analysis of scats. Molecular Ecology 9:1583–1590.

Farrell, R. K., and S. D. Johnston. 1973. Identification of laboratory animals: freeze marking. Laboratory Animal Science 23:107–110.

Farrell, R. K., G. A. Laisner, and T. S. Russell. 1969. An international freeze-mark animal identification system. Journal of the American Veterinary Medical Association 154:1561–1572.

Fashingbauer, B. A. 1962. Expanding plastic collar and aluminum collar for deer. Journal of Wildlife Management 26:211–213.

Faul, F., E. Erdfelder, A. G. Lang, and A. Buchner. 2007. G*Power 3: a flexible statistical power analysis program for the social, behavioral, and biomedical sciences. Behavior Research Methods 39:175–191.

Faulhaber, C. A., N. J. Silvy, R. R. Lopez, B. A. Porter, P. A. Frank, and M. J. Peterson. 2005. Use of drift fences to capture Lower Keys marsh rabbits. Wildlife Society Bulletin 33:1160–1163.

Fayrer-Hosken, R. A., D. Grobler, J. J. Van Altena, H. J. Bertschinger, and J. F. Kirkpatrick. 2000. Immunocontraception of African elephants. Nature 407:149.

Federal Communications Commission. 2009. FCC online table of frequency allocations. 47 C.F.R. 2.106. Revised October 15, 2009. http://edocket.access.gpo.gov/cfr_2009/octqtr/pdf/47cfr2.106.pdf.

Federal Geographic Data Committee. 1998. Content standard for digital geospatial metadata (version 2.0), FGDC_STD-001-1998. http://www.fgdc.gov/standards/projects/FGDC-standards-projects/metadata/base-metadata/index_html.

Federal Geographic Data Committee. 1999. Content standard for digital geospatial metadata (version 2.0), FGDC_STD-001-1998, Part 1: biological data profile. Biological Data Working Group, Biological Resources Division, U.S. Geological Survey, Department of the Interior, Washington, D.C., USA. http://www.fgdc.gov/standards/projects/FGDC-standards-projects/metadata/biometadata/biodatap.pdf

Federal Geographic Data Committee. 2000. A guide to writing clearly. Metadata Education Program and the National Metadata Cadre, Reston, Virginia, USA.

Federal Geographic Data Committee. 2009. Preparing for international metadata: North American profile of ISO 19115: geographic information—metadata. Reston, Virginia, USA.

Federal Register. 1994. Coordinating geographic data acquisition and access: the national spatial data infrastructure. Executive Order 12906. 59(71):17671–17674.

Fedriani, J. M., and M. H. Kohn. 2001. Genotyping faeces links individuals to their diet. Ecology Letters 4:477–483.

Fehmi, J. S., and J. W. Bartolome. 2001. A grid-based method for sampling and analyzing spatially ambiguous plants. Journal of Vegetation Science 12:467–472.

Feighny, J. A., K. E. Williamson, and J. A. Clarke. 2006. North American elk bugle vocalizations: male and female bugle call structure and context. Journal of Mammalogy 87:1072–1077.

Feldhamer, G. A., B. C. Thompson, and J. A. Chapman. 2003. Wild mammals of North America: biology, management, and conservation. Second edition. Johns Hopkins University Press, Baltimore, Maryland, USA.

Felix, R. K., Jr., R. H. Diehl, and J. M. Ruth. 2008. Seasonal passerine migratory movements over the arid southwest. Studies in Avian Biology 37:126–137.

Fenolio, D. B., G. O. Graening, B. A. Collier, and J. F. Stout. 2006. Coprophagy in a cave-adapted salamander; the importance of bat guano examined through nutritional and stable isotope analyses. Proceedings of the Royal Society B 273:439–443.

Ferguson, A. W., and M. R. J. Forstner. 2006. A device for excluding predators from pitfall traps. Herpetological Review 37:316–317.

Ferguson, H. L., and P. D. Jorgensen. 1981. An efficient trapping technique for burrowing owls. North American Bird Bander 6:149–150.

Fernández-Juricic E., P. Venier, D. Renison, and D. T. Blumstein. 2005. Sensitivity of wildlife to spatial patterns of recreationist behavior: a critical assessment of minimum approaching distances and buffer areas for grassland birds. Biological Conservation. 125:225–235.

Ferner, J. W. 1979. A review of marking techniques for amphibians and reptiles. Herpetological Circular Number 9, Society for the Study of Amphibian and Reptiles, Marceline, Missouri, USA.

Fernie, K. J., L. Shutt, I. Ritchie, R. Letcher, K. Drouillard, and D. M. Bird. 2006. Changes in the growth, but not the survival, of American kestrels (*Falco sparverius*) exposed to environmentally relevant polybrominated diphenyl ethers. Journal of Toxicology and Environmental Health 69:154–155.

Festa-Bianchet, M., and M. Appollonio, editors. 2003. Animal behavior and wildlife conservation. Island Press, Washington, D.C., USA.

Festa-Bianchet, M., P. Blanchard, J. M. Gaillard, and A. J. M. Hewison. 2002. Tooth extraction is not an acceptable technique to age live ungulates. Wildlife Society Bulletin 30:282–288.

Feuer, R. C. 1980. Underwater traps for aquatic turtles. Herpetological Review 11:107–108.

Fidenci, P. 2005. A new technique for capturing Pacific pond turtles (*Actinemys marmorata*) and a comparison with traditional trapping methods. Herpetological Review 36:266–267.

Fidler, F., M. A. Burgman; G. Cumming, R. Buttrose, and N. Thomason. 2006. Impact of criticism of null-hypothesis significance testing on statistical reporting practices in conservation biology. Conservation Biology 20:1539–1544.

Fidler, F., G. Cumming, M. A. Burgman, and N. Thomason. 2004. Statistical reform in medicine, psychology and ecology. Journal of Socio-Economics 33:615–630.

Fieberg, J. 2007a. Kernel density estimators of home range: smoothing and the autocorrelation red herring. Ecology 88:1059–1066.

Fieberg, J. 2007b. Utilization distribution estimation using weighted kernel density estimators. Journal of Wildlife Management 71:1669–1675.

Fieberg, J., and C. O. Kochanny. 2005. Quantifying home-range overlap: the importance of the utilization distribution. Journal of Wildlife Management 69:1346–1359.

Field, M., P. Chavez, and P. Jokiel. 2000. Interpreting remotely sensed data on coral reefs. PACON 2000 Conference, Honolulu, Hawaii, USA.

Fienberg, S. E. 1970. The analysis of multidimensional contingency tables. Ecology 51:419–433.

Fienberg, S. E. 1979. Graphical methods in statistics. American Statistician 33:165–178.

Fienberg, S. E. 1980. The analysis of cross-classified categorical data. MIT Press, Cambridge, Massachusetts, USA.

Findholt, S. L., B. K. Johnson, L. D. Bryant, and J. W. Thomas. 1996. Corrections for position bias of a LORAN-C radio-telemetry system using DGPS. Northwest Science 70:273–280.

Finley, R. B. 1965. Adverse effects on birds of phosphamidon applied to a Montana forest. Journal of Wildlife Management 29:580–591.

Firchow, K. M., M. R. Vaughan, and W. R. Mytton. 1986. Evaluation of the hand-held net gun for capturing pronghorns. Journal of Wildlife Management 50:320–322.

Firestone, K. B., M. S. Elphinstone, W. B. Sherwin, and B. A. Houlden. 1999. Phylogeographical population structure of tiger quolls *Dasyurus maculatus* (Dasyuridae: Marsupialia), an endangered carnivorous marsupial. Molecular Ecology 8:1613–1625.

Fischer, R. A., A. D. Apa, W. L. Wakkinen, K. P. Reese, and J. W. Connelly. 1993. Nesting-area fidelity of sage grouse in southeastern Idaho. Condor 95:1038–1041.

Fisher, E. W., and A. E. Perry. 1970. Estimating ages of gray squirrels by lens-weights. Journal of Wildlife Management 34:825–828.

Fisher, J. T., C. Twitchell, W. Barney, E. Jenson, and J. Sharpe. 2005. Utilizing behavioral biophysics to mitigate mortality of snared endangered Newfoundland marten. Journal of Wildlife Management 69:1743–1746.

Fisher, M., and A. Muth. 1989. A technique for permanently marking lizards. Herpetological Review 20:45–46.

Fisher, N. I. 1993. Statistical analysis of circular data. Cambridge University Press, Cambridge, England, UK.

Fisher, P. 1999. Review of using Rhodamine B as a marker for wildlife studies. Wildlife Society Bulletin 27:318–329.

Fisser, H. G. 1961. Variable plot, square foot plot, and visual estimate for shrub crown cover measurements. Journal of Range Management 14:202–207.

Fitch, H. S. 1987. Collecting and life-history techniques. Pages 143–164 in R. A. Seigel, J. T. Collins, and S. S. Novak, editors. Snakes: ecology and evolutionary biology. Macmillan, New York, New York, USA.

Fitch, H. S. 1992. Methods of sampling snake populations and their relative success. Herpetological Review 23:17–19.

Fitzgerald, C. S., P. R. Krausman, and M. L. Morrison. 1999. Use of buried and non-buried traps to sample desert rodents. California Fish and Game 85:140–143.

Fitzgerald, L. A., and C. W. Painter. 2000. Rattlesnake commercialization: long-term trends, issues, and implications for conservation. Wildlife Society Bulletin 28:235–253.

Fitzner, R. E., and J. N. Fitzner. 1977. A hot melt glue technique for attaching radiotransmitter tail packages to raptorial birds. North American Bird Bander 2:56–57.

Fitzpatrick-Lins, K. 1981. Comparison of sampling procedures and data analysis for a land-use and land-cover map. Photogrammetric Engineering and Remote Sensing 47:343–351.

Fitzsimmons L. P., J. R. Foote., L. M. Ratcliffe, and D. J. Mennill. 2008. Eavesdropping and communication networks revealed through stereo playback and an acoustic location system. Behavioral Ecology 19:824–829.

Fitzsimmons, N. N., S. W. Buskirk, and M. H. Smith. 1997. Genetic changes in reintroduced Rocky Mountain bighorn sheep populations. Journal of Wildlife Management 61:863–872.

Fitzwater, W. D., Jr. 1943. Color marking of mammals, with special reference to squirrels. Journal of Wildlife Management 7:190–192.

Fjetland, C. A. 1973. Long-term retention of plastic collars on Canada geese. Journal of Wildlife Management 37:176–178.

Flaherty, E. A., W. P. Smith, M. Ben-David, and S. Fyare. 2008. Experimental trials of the northern flying squirrel (*Glaucomys sabrinus*) traversing managed rainforest landscapes: perceptual range and fine-scale movements. Canadian Journal of Zoology 86:1050–1058.

Flake, L. D., J. W. Connelly, T. R. Kirschenmann, and A. J. Lindbloom. 2010. Grouse of plains and mountains—the South Dakota story. South Dakota Department of Game, Fish and Parks, Pierre, USA.

Fleming, M. W., and J. D. Harder. 1983. Luteal and follicular populations in the ovary of the opossum (*Didelphis virginiana*) after ovulation. Journal of Reproduction and Fertility 67:29–34.

Fleming, P. J. S., L. R. Allen, M. J. Berghout, P. D. Meek, P. M. Pavlov, P. Stevens, K. Strong, J. A. Thompson, and P. C. Thomson. 1998. The performance of wild-canid traps in Australia: efficiency, selectivity and trap-related injuries. Wildlife Research 25:327–338.

Fleskes, J. P. 2003. Effects of backpack radiotags on female northern pintails wintering in California. Wildlife Society Bulletin 31:212–219.

Flint, P. L., and J. B. Grand. 1996. Nesting success of northern pintails on the coastal Yukon-Kuskokwim Delta, Alaska. Condor 98:54–60.

Flock, B. E., and R. D. Applegate. 2002. Comparison of trapping methods for ring-necked pheasants in north-central Kansas. North American Bird Bander 27:4–8.

Flood, M., and B. Gutelis. 1997. Commercial implications of topographic terrain mapping using scanning airborne laser radar. Photogrammetric Engineering and Remote Sensing 63:327–366.

Flores, R. E., and W. R. Eddleman. 1993. Nesting biology of the California black rail in southwestern Arizona. Western Birds 24:81–88.

Floyd, D. A., and J. E. Anderson. 1987. A comparison of three methods for estimating plant cover. Journal of Ecology 75:221–228.

Floyd, T. J., L. D. Mech, and M. E. Nelson. 1979. An improved method of censusing deer in deciduous-coniferous forests. Journal of Wildlife Management 43:258–261.

Flyger, V. F. 1958. Tooth impressions as an aid in the determination of age in deer. Journal of Wildlife Management 22:442–443.

Focardi, S., A. M. DeMarinis, M. Rizzotto, and A. Pucci. 2001. Comparative evaluation of thermal infrared imaging and spotlighting to survey wildlife. Wildlife Society Bulletin 29:133–139.

Fogarty, M. J., K. A. Arnold, L. McKibben, L. B. Pospichal, and R. J. Tully. 1977. Common snipe. Pages 189–209 in G. C. Sanderson, editor. Management of migratory shore and upland game birds in North America. International Association of Fish and Wildlife Agencies, Washington, D.C., USA.

Fogle, B. 2000. The new encyclopedia of the dog. Second edition. Dorling Kindersley, New York, New York, USA.

Fogl, J. G., and H. S. Mosby. 1978. Aging gray squirrels by cementum annuli in razor-sectioned teeth. Journal of Wildlife Management 42:444–448.

Folk, M. J., J. A. Schmidt, and S. A. Nesbitt. 1999. A trough-blind for capturing cranes. Journal of Field Ornithology 70:251–256.

Follett, B. K., and S. L. Maung. 1978. Rate of testicular maturation, in relation to gonadotrophin and testosterone levels, in quail exposed to various artificial photoperiods and to natural daylengths. Journal of Endocrinology 78:267–280.

Follis, T. B., W. C. Foote, and J. J. Spillet. 1972. Observation of genitalia in elk by laparotomy. Journal of Wildlife Management 36:171–173.

Follmann, E. H., P. J. Savarie, D. G. Ritter, and G. M. Baer. 1987. Plasma marking of arctic foxes with iophenoxic acid. Journal of Wildlife Disease 23:709–712.

Fontana, R. J., E. A. Richley, A. J. Marzullo, L. C. Beard, R. W. T. Mulloy, and E. J. Knight. 2002. An ultra wideband radar for micro air vehicle applications. Pages 187–191 in Proceedings IEEE conference on ultra wideband systems and technologies, Baltimore, Maryland, USA.

Fontenot, L. W., G. P. Noblet, and S. G. Platt. 1994. Rotenone hazards to amphibians and reptiles. Herpetological Review 25:150–156.

Foose, T. J., and J. D. Ballou. 1988. Population management: theory and practice. International Zoo Yearbook 27:26–41.

Foran, D. R., S. C. Minta, and K. S. Heinemeyer. 1997. DNA-based analysis of hair to identify species and individuals for population research and monitoring. Wildlife Society Bulletin 25:840–847.

Forbes, L. S. 1990. A note on statistical power. Auk 107:438–439.

Ford, A. T., A. P. Clevenger, and A. Bennett. 2009. Comparison of methods of monitoring wildlife crossing-structures on highways. Journal of Wildlife Management 73:1213–1222.

Ford, A. T., and L. Fahrig. 2008. Movement patterns of eastern chipmunks (*Tamias striatus*). Journal of Mammalogy 89:895–903.

Ford, E. D. 2000. Scientific method for ecological research. Cambridge University Press, Cambridge, England, UK.

Forester, D. C. 1977. Comments on the female reproductive cycle and philopatry by *Demognathus ochrophaeus* (Amphibia, Urodela, Plethodontidae). Journal of Herpetology 11:311–316.

Forester, J. D., A. R. Ives, M. G. Turner, D. P. Anderson, D. Fortin, H. L. Beyer, D. W. Smith, and M. S. Boyce. 2007. State-space models link elk movement patterns to landscape characteristics in Yellowstone National Park. Ecological Monographs 77:285–299.

Foreyt, W. J. 2001. Veterinary parasitology reference manual. Fifth edition. Iowa State University Press, Ames, USA.

Foreyt, W. J., and W. C. Glazener. 1979. A modified box trap for capturing feral hogs and white-tailed deer. Southwestern Naturalist 24:377–380.

Foreyt, W. J., and A. Rubenser. 1980. A live trap for multiple capture of coyote pups from dens. Journal of Wildlife Management 44:487–488.

Forman, D. W., and K. Williamson. 2005. An alternative method for handling and marking small carnivores without the need for sedation. Wildlife Society Bulletin 33:313–316.

Forman, R. T. T., and M. Godron. 1986. Landscape ecology. John Wiley and Sons, New York, New York, USA.

Forrester, D. J. 1992. Parasites and diseases of wild mammals in Florida. University of Florida Press, Gainesville, USA.

Forrester, D. J., and M. G. Spalding. 2003. Parasites and diseases of wild birds in Florida. University of Florida Press, Gainesville, USA.

Forsman, E. D. 1983. Methods and materials for locating and studying spotted owls. General Technical Report PNW-162, U.S. Forest Service, Department of Agriculture, Washington, D.C., USA.

Forsman, E. D., A. B. Franklin, F. M. Oliver, and J. P. Ward. 1996. A color band for spotted owls. Journal of Field Ornithology 67:507–510.

Forsyth, D. M., R .J. Barker, G. Morriss, and M. P. Scroggie. 2007. Modeling the relationship between fecal pellet indices and deer density. Journal of Wildlife Management 71:964–970.

Fossi, M. C., and C. Leonzio, editors. 1994. Non-destructive biomarkers in ecotoxicology. Lewis, Boca Raton, Florida, USA.

Foster, C. C., E. D. Forsman, E. C. Meslow, G. S. Miller, J. A. Reid, F. F. Wagner, A. B. Carey, and J. B. Lint. 1992. Survival and reproduction of radio-marked adult spotted owls. Journal of Wildlife Management 56:91–95.

Foster, D. L., and L. M. Jackson. 2006. Puberty in the sheep. Pages 2127–2176 in J. D. Neill, editor. Knobil and Neill's physiology of reproduction. Third edition, Elsevier Science, Amsterdam, The Netherlands.

Foster, J. B. 1966. The giraffe of Nairobi National Park: home range, sex ratios, the herd, and food. East African Wildlife Journal 4:139–148.

Foster, M. L., and S. R. Humphrey. 1995. Use of highway underpasses by Florida panthers and other wildlife. Wildlife Society Bulletin 23:95–100.

Foster, M. S., and L. A. Fitzgerald. 1982. A technique for live-trapping cormorants. Journal of Field Ornithology 53:422–423.

Foster, R. L., A. M. Mcmillan, A. R. Breisch, K. J. Roblee, and D. Schranz. 2008. Analysis and comparison of three capture methods for the eastern hellbender (*Crytobranchus alleganiensis alleganiensis*). Herpetological Review 39:181–186.

Fowler, M. E. 1995. Restraint and handling of wild and domestic animals. Second edition. Iowa State University Press, Ames, USA.

Fowler, M. E., editor. 1981. Wildlife diseases of the Pacific Basin and other countries. Proceedings of the 4th International Conference of the Wildlife Disease Association, Davis, California, USA.

Fox, R. R., and D. D. Crary. 1972. A simple technique for the sexing of newborn rabbits. Laboratory Animal Science 22:556–558.

Fox, S. F. 1978. Natural selection on behavioral phenotypes of the lizard *Uta stansburiana*. Ecology 59:834–847.

Fox, T. J., M. G. Knutson, and R. K. Hines. 2000. Mapping forest canopy gaps using air-photo interpretation and ground surveys. Wildlife Society Bulletin 28:882–889.

Frair, J. L., S. E. Nielson, E. H. Merrill, S. R. Lele, M. S. Boyce, R. H. M. Munro, G. B. Stenhouse, and H. L. Beyer. 2004. Removing GPS collar bias in habitat selection studies. Journal of Applied Ecology 41:201–212.

Frame, P. F., and T. J. Meir. 2007. Field-assessed injury to wolves captured in rubber-padded traps. Journal of Wildlife Management 71:2074–2076.

Francis, A., K. A. Bloem, A. L. Roda, S. L. Lapointe, A. Zhang, and O. Onokpise. 2007. Development of trapping methods with a synthetic sex pheromone of the pink hibiscus mealybug, *Maconellicoccus hirsutus* (Hemiptera: Pseudococcidae). Florida Entomologist 90:440–446.

Francis, C. M. 1989. A comparison of mist nets and two designs of harp traps for capturing bats. Journal of Mammalogy 70:865–870.

Frank, K. M., R. O. Lyda, J. F. Kirkpatrick. 2005. Immunocontraception of captive exotic species—IV. Species differences in response to the porcine zona pellucida vaccine, timing of booster inoculations, and procedural failures. Zoo Biology 24:349–358.

Frank, L., D. Simpson, and R. Woodroffe. 2003. Foot snares: an effective method for capturing African lions. Wildlife Society Bulletin 31:309–314.

Frankel, A. I., and T. S. Baskett. 1963. Color marking disrupts pair bonds of captive mourning doves. Journal of Wildlife Management 27:124–127.

Frankham, R. 1995a. Conservation genetics. Annual Review of Genetics 29:305–327.

Frankham, R. 1995b. Effective population size/adult population size ratios in wildlife: a review. Genetical Research 66:95–107.

Frankham, R. 1999. Resolving conceptual issues in conservation genetics: the roles of laboratory species and meta-analyses. Hereditas 130:195–201.

Frankham, R. 2008. Genetic adaptation to captivity in species conservation programs. Molecular Ecology 17:325–333.

Frankham, R., J. D. Ballou, and D. A. Briscoe. 2002. Introduction to conservation genetics. Cambridge University Press, New York, New York, USA.

Franklin, A. B. 2001. Exploring ecological relationships in survival and estimating rates of population change using program MARK. Pages 350–356 in R. Field, R. J. Warren, H. Okarma, and P. R. Sievert, editors. Wildlife, land, and people: priorities for the 21st century. Proceedings of the Second International Wildlife Management Congress. The Wildlife Society, Bethesda, Maryland, USA.

Franklin, A. B., D. R. Anderson, R. J. Gutierrez, and K. P. Burnham. 2000. Climate, habitat quality, and fitness in northern spotted owl populations in northwestern California. Ecological Monographs 70:539–590.

Franklin, A. B., T. M. Shenk, D. R. Anderson, and K. P. Burnham. 2001. Statistical model selection: an alternative to null hypothesis testing. Pages 75–90 in T. M. Shenk and A. B. Franklin, editors. Modeling in natural resource management: development, interpretation, and application. Island Press, Washington, D.C., USA.

Franklin, C. J., and R. W. Hartdegen. 1997. A safer capture technique for larger reptiles. Herpetological Review 28:197.

Franklin, W. L., and W. E. Johnson. 1994. Hand capture of newborn open-habitat ungulates: the South American guanaco. Wildlife Society Bulletin 22:253–259.

Franzreb, K. E., and J. L. Hanula. 1995. Evaluation of photographic devices to determine nestling diet of the endangered red-cockaded woodpecker. Journal of Field Ornithology 66:253–259.

Fraser, D. J., and L. Bernatchez. 2001. Adaptive evolutionary conservation: towards a unified concept for defining conservation units. Molecular Ecology 10:2741–2752.

Fratto, Z. W., V. A. Barko, P. R. Pitts, S. L. Sheriff, J. T. Briggler, K. P. Sullivan, B. L. Mckeage, and T. R. Johnson. 2008. Evaluation of turtle exclusion and escapement devices for hoopnets. Journal of Wildlife Management 72:1628–1633.

Frazer, N. B. 1983. Survivorship of adult female loggerhead sea turtles, *Caretta caretta*, nesting on Little Cumberland Island, Georgia, USA. Herpetologica 39:436–447.

Frederick, P. C. 1986. A self-tripping trap for use with colonial nesting birds. North American Bird Bander 11:94–95.

Fredin, R. A. 1984. Levels of maximum net productivity in populations of large terrestrial mammals. Pages 381–387 in W. F. Perrin, R. L. Brownell, Jr., and D. P. DeMaster, editors. Special Issue 6. Reports of the International Whaling Commission, Cambridge, England, UK.

Fredrickson, L. H. 1968. Measurements of coots related to sex and age. Journal of Wildlife Management 32:409–411.

Freeman, E. R. 1982. Interference suppression techniques for microwave antennas and transmitters. Artech House, Dedham, Massachusetts, USA.

Freeman, P. W., and C. A. Lemen. 2009. Puncture-resistance of gloves for handling bats. Journal of Wildlife Management 73:1251–1254.

Freitas, C., C. Lydersen, M. Fedak, K. Kovacs. 2008. A simple new algorithm to filter marine mammal Argos locations. Marine Mammal Science 24:315–325.

French, J., and I. G. Priede. 1992. A microwave radar transponder for tracking studies. Pages 41–54 in I. G. Priede, editor. Wildlife telemetry. Ellis Horwood, Sussex, England, UK.

Frentress, C. 1976. "Pop" rivet fasteners for color markers. Inland Bird Banding Association News 47:3–9.

Frenzel, R. W., and R. G. Anthony. 1982. Method for live-capturing bald eagles and osprey over open water. Research Information Bulletin 82-13, U.S. Fish and Wildlife Service, Department of the Interior, Washington, D.C., USA.

Frey, S. N., M. R. Conover, and G. Cook. 2007. Successful use of neck snares to live-capture red foxes. Human–Wildlife Conflicts 1:21–23.

Fridell, R. A., and J. A. Litvaitis. 1991. Influence of resource distribution and abundance on home-range characteristics of southern flying squirrels. Canadian Journal of Zoology 69:2589–2593.

Friedman, M. 1937. The use of ranks to avoid the assumption of normality implicit in the analysis of variance. Journal of the American Statistical Association 32:675–701.

Friedman, S. L., R. L. Brasso, and A. M. Condon. 2008. An improved nest-box trap. Journal of Field Ornithology 79:99–101.

Friedrich, P. D., G. E. Burgoyne, T. M. Cooley, and S. M. Schmidt. 1983. Use of lower canine tooth for determining the sex of bobcats in Michigan. Wildlife Division Report 2960, Michigan Department of Natural Resources, Lansing, USA.

Friend, M. 1967. A review of research concerning eye-lens weight as a criterion of age in animals. New York Fish and Game Journal 14:152–165.

Friend, M. 1987. Field guide to wildlife diseases. Volume 1. General field procedures and diseases of migratory birds. Resource Publication 167, U.S. Fish and Wildlife Service, Department of the Interior, Washington, D.C., USA.

Friend, M. 2006. Disease emergence and resurgence: the wildlife–human connection. U.S. Geological

Survey, Reston, Virginia, USA. http://www.nwhc.usgs.gov/publications/disease_emergence/.

Friend, M., and J. C. Franson, editors. 1999. Field manual of wildlife diseases: general field procedures and diseases of birds. Information and Technology Report 1999–2001, Biological Resources Division, U.S. Geological Survey, Department of the Interior, Washington, D.C., USA.

Frison, G. C., and C. A. Reher. 1970. Age determination of buffalo by teeth eruption and wear. Plains Anthropologist 15:46–50.

Fritts, T. H., N. J. Scott, Jr., and B. E. Smith. 1989. Trapping *Boiga irregularis* on Guam using bird odors. Journal of Herpetology 23:189–192.

Fritzell, E. K. 1987. Gray fox and island gray fox. Pages 408–421 in M. Novak, J. A. Baker, M. E. Obbard, and B. Malloch, editors. Wild furbearer management and conservation in North America. Ontario Ministry of Natural Resources, Toronto, Canada.

Froese, A. D., and G. M. Burghardt. 1975. A dense natural population of the common snapping turtle (*Chelydra s. serpentina*). Herpetologica 31:204–208.

Fronzuto, J., and P. Verrell. 2000. Sampling aquatic salamanders: tests of the efficiency of two funnel traps. Journal of Herpetology 34:146–147.

Frost, H. C., and W. B. Krohn. 1994. Capture, care and handling of fishers (*Martes pennanti*). Technical Bulletin 157, Maine Agriculture and Forest Experiment Station, University of Maine, Orono, USA.

Frost, H. C., E. C. York, W. B. Krohn, K. D. Elowe, T. A. Decker, S. M. Powell, and T. K. Fuller. 1999. An evaluation of parturition indices in fishers. Wildlife Society Bulletin 27:221–230.

Fu, Y. B., G. Namkoong, and J. E. Carlson. 1998. Comparison of breeding strategies for purging inbreeding depression via simulation. Conservation Biology 12:856–864.

Fuchs, R. M. E., W. K. Maclean, C. A. Mackintosh, and I. M. Allan. 1996. The use of tip traps to control rabbit damage in Scotland. Proceedings of the Vertebrate Pest Control Conference 17:199–203.

Fuertes, B., J. Garæia, and J. M. Colino. 2002. Use of fish nets as a method to capture small rails. Journal of Field Ornithology 73:220–223.

Fullagar, P. J., and P. A. Jewell. 1965. Marking small rodents and the difficulties of using leg rings. Journal of Zoology 147:224–228.

Fullard, J. H., and N. Napoleone. 2001. Diel flight periodicity and the evolution of auditory defences in the Macrolepidoptera. Animal Behaviour 62:349–368.

Fuller, A. F. 1984. Drop net capture of bighorn sheep in Arizona. Transactions of the Desert Bighorn Council 28:39–40.

Fuller, M. R., J. J. Millspaugh, K. E. Church, and R. E. Kenward. 2005. Wildlife radiotelemetry. Pages 377–417 in C. E. Braun, editor. Techniques for wildlife investigations and management. The Wildlife Society, Bethesda, Maryland, USA.

Fuller, T. K. 1991. Effect of snow depth on wolf activity and prey selection in north central Minnesota. Canadian Journal of Zoology 69:283–287.

Fuller, T. K., D. P. Hobson, J. R. Gunson, D. B. Schowalter, and D. Heisey. 1984. Sexual dimorphism in mandibular canines of striped skunks. Journal of Wildlife Management 48:1444–1446.

Fuller, W. A. 1959. The horns and teeth as indicators of age in bison. Journal of Wildlife Management 23:342–344.

Fung, I. Y., C. J. Tucker, and K. C. Prentice. 1987. Application of advanced very high resolution radiometer vegetation index to the study of atmosphere-biosphere exchange of CO_2. Journal of Geophysical Research 92:2999–3015.

Funk, H. D., and J. R. Grieb. 1965. Baited cannon-net sampling as an indicator of Canada goose population characteristics. Journal of Wildlife Management 29:253–260.

Fur Institute of Canada. 2009. The agreement on international humane trapping standards (AIHTS). Fur Institute of Canada. http://www.fur.ca/index.php.

Furgal, C. M., S. Innes, and K. M. Kovacs. 1996. Characteristics of ringed seal, *Phoca hispida*, subnivean structures and breeding habitat and their effects on predation. Canadian Journal of Zoology 74:858–874.

Furrer, R. K. 1979. Experiences with a new back-tag for open-nesting passerines. Journal of Wildlife Management 43:245–249.

Furtado, M. M., S. E. Carrillo-Percastegui, A. T. A. Jacomo, G. Powell, L. Silveira, C. Vynne, and R. Sollmann. 2008. Studying jaguars in the wild: past experiences and future perspectives. CAT News: The Jaguar in Brazil 4:41–47.

Fyvie, A., and E. M. Addison. 1979. Manual of common parasites, diseases and anomalies of wildlife in Ontario. Queens Printer, Ontario, Canada. http://www.unbc.ca/nlui/wildlife_diseases/booklet.htm.

Gabrielson, I. N. 1951. Wildlife management. Macmillan, New York, New York, USA.

Gadd, P., Jr. 1996. Use of the modified Australian crow trap for the control of depredating birds in Sonoma County. Proceedings of the Vertebrate Pest Conference 17:103–107.

Gadsby, J. E., R. B. Heap, D. G. Powell, and D. E. Walters. 1972. Diagnosis of pregnancy and of the number of fetuses in sheep from plasma progesterone concentrations. Veterinary Research 90:339–342.

Gallo, M. A., and N. J. Lawryk. 1991. Organic phosphorus pesticides. Pages 917–1123 in W. J. Hayes and E. R. Laws, Jr., editors. Handbook of pesticide toxicity. Volume 3. Classes of pesticides. Academic Press, San Diego, California, USA.

Gamble, T. 2006. The relative efficiency of basking and hoop traps for painted turtles (*Chrysemys picta*). Herpetological Review 37:308–312.

Gamo, R. S., M. A. Rumble, R. Lindzey, and M. Stefanich. 2000 GPS radio collar 3D performance as influenced by forest structure and topography. Pages 464–473 in J. H. Eiler, D. J. Alcorn, and M. R. Neuman., editors. Proceedings of the 15th international symposium on biotelemetry. International Society on Biotelemetry, Juneau, Alaska, USA.

Ganey, J. L., and W. M. Block. 1994. A comparison of two techniques for measuring canopy closure. Western Journal of Applied Forestry 9:21–23.

Gannon, W. L., R. S. Sikes, and The Animal Care and Use Committee of the American Society of Mammalogists. 2007. Guidelines of the American Society of Mammalogists for the use of wild mammals in research. Journal of Mammalogy 88:809–823.

Gao, J., H. Chen, Y. Zhang, and Y. Zha. 2004. Knowledge-based approaches to accurate mapping of mangroves from satellite data. Photogrammetric Engineering and Remote Sensing 70:1241–1248.

Gard, N. W., D. M. Bird, R. Densmore, and M. Hamel. 1989. Responses of breeding American kestrels to live and mounted great horned owls. Journal of Raptor Research 23:99–102.

Gardner, A. L. 1982. Virginia opossum. Pages 3–36 in J. A. Chapman and G. A. Feldharner, editors. Wild mammals of North America. Johns Hopkins University Press, Baltimore, Maryland, USA.

Gardner, R. H., and D. L. Urban. 2003. Model validation and testing: past lessons, present concerns, future prospects. Pages 184–203 in C. D. Canham, J. J. Cole, and W. K. Lauenroth, editors. Models in ecosystem science. Princeton University Press, Princeton, New Jersey, USA.

Gardner-Santana, L. C., D. E. Norris, C. M. Fornadel, E. R. Hinson, S. L. Klein, and G. E. Glass. 2009. Commensal ecology, urban landscapes, and their influence on the genetic characteristics of city-dwelling Norway rats (*Rattus norvegicus*). Molecular Ecology 18:2766–2778.

Gargett, V. 1973. Marking black eagles in the Matopos. Honeyguide 76:26–31.

Garner, D. L., S. Dunwoody, D. Joly, D. O'Brien, M. J. Peterson, and M. J. Pybus. 2009. External panel review of "a plan for managing chronic wasting disease in Wisconsin: the next five years." Wisconsin Department of Natural Resources, Madison, USA. http://dnr.wi.gov/org/land/wildlife/WHEALTH/issues/cwd/doc/External_Review.pdf.

Garnett, B. T., R. J. Delahay, and T. J. Roper. 2002. Use of cattle farm resources by badgers (*Meles meles*) and risk of bovine tuberculosis (*Mycobacterium bovis*) transmission to cattle. Biological Sciences 269:1487–1491.

Garnett, S. 1987. Feather-clipping: a natural technique for short-term recognition of individual birds. Corella 11:30–31.

Garrettson, P. R. 1998. Response of breeding season blue-winged teal to decoy trapping. Prairie Naturalist 30:235–241.

Garrott, R. A., and R. M. Bartmann. 1984. Evaluation of vaginal implants for mule deer. Journal of Wildlife Management 48:646–648.

Garshelis, D. L. 1992. Mark–recapture density estimation for animals with large home ranges. Pages 1098–1111 in D. R. McCullough and R. H. Barrett, editors. Wildlife 2001: populations. Elsevier Applied Science, New York, New York, USA.

Garshelis, D. L., and C. R. McLaughlin. 1998. Review and evaluation of breakaway devices for bear radiocollars. Ursus 10:459–465.

Garshelis, D. L., and L. G. Visser. 1997. Enumerating metapopulations of wild bears with an ingested biomarker. Journal of Wildlife Management 61:466–480.

Gartner, K., D. Buttner, K. Dohler, R. Friedel, J. Lindena, and I. Trautschold. 1980. Stress response of rats to handling and experimental procedures. Laboratory Animal 14:267–274.

Garton, E. O. 2002. Mapping a chimera? Pages 663–666 in J. M. Scott, P. J. Heglund, M. L. Morrison, J. B. Haufler, M. G. Raphael, W. A. Wall, and F. B. Sampson, editors. Predicting species occurrences: issues of accuracy and scale. Island Press, Washington, D.C., USA.

Garton, E. O., J. S. Connelly, J. S. Horne, C. Hagen, A. Moser, and M. Schroeder. 2011. Greater sage-grouse population dynamics and probability of persistence. Pages 293–382 in C. Marti, S. Knick, and J. W. Connelly, editors. Ecology and conservation of greater sage-grouse: a landscape species and its habitats. Studies in avian biology 38. University of California Press, Berkeley, USA.

Garton, E. O., R. L. Crabtree, B. B. Ackerman, and G. Wright. 1990. The potential impact of a reintroduced wolf population on the northern Yellowstone elk herd. Pages 3-59–3-91 in Wolves for Yellowstone? A report to the United States Congress. Volume II: Research and analysis. Yellowstone National Park, National Park Service, Washington, D.C., USA.

Garton, E. O., J. T. Ratti, and J. H. Giudice. 2005. Research and experimental design. Pages 43–71 in C. E. Braun, editor. Techniques for wildlife investigations and management. The Wildlife Society, Bethesda, Maryland, USA.

Garton, E. O., M. J. Wisdom, F. A. Leban, and B. K. Johnson. 2001. Experimental design for radiotelemetry studies. Pages 16–42 in J. J. Millspaugh and J. M. Marzluff, editors. Radio tracking and animal populations. Academic Press, New York, New York, USA.

Gartshore, M. E. 1978. A noose trap for catching nesting birds. North American Bird Bander 3:1–2.

Garza, J. C., and E. G. Williamson. 2001. Detection of reduction in population size using data from microsatellite loci. Molecular Ecology 10:305–318.

Gasaway, W. C., S. D. Dubois, and S. J. Harbo. 1985. Biases in aerial transect surveys for moose during May and June. Journal of Wildlife Management 49:777–784.

Gasaway, W. C., D. B. Harkness, and R. A. Rausch. 1978. Accuracy of moose age determinations from incisor cementum layers. Journal of Wildlife Management 42:558–563.

Gaston, K. J., and J. H. Lawton. 1987. A test of statistical techniques for detecting density dependence in sequential censuses of animal populations. Oecologia 74:404–410.

Gates, C. C., B. Elkin, and D. Dragon. 2001. Anthrax. Pages 396–412 in E. S. Williams and I. K. Barker, editors. Infectious diseases of wild animals. Iowa State University Press, Ames, USA.

Gates, J. M. 1966. Validity of spur appearance as an age criterion in the pheasant. Journal of Wildlife Management 30:81–85.

Gauch, H. G., Jr. 2003. Scientific method in practice. Cambridge University Press, Cambridge, England, UK.

Gaufin, A. R., and C. M. Tarzwell. 1952. Aquatic invertebrates as indicators of stream pollution. Public Health Reports 67:57–64.

Gaunt, A. S., and L. W. Oring, editors. 1999. Guidelines to the use of wild birds in research. Second edition. The Ornithological Council, Washington, D.C., USA.

Gaunt, A. S., L. W. Oring, K. P. Able, D. W. Anderson, L. F. Baptista, J. C. Barlow, and J. C. Wingfield. 1997. Guidelines to the use of wild birds in research. The Ornithological Council, Washington, D.C., USA.

Gautestad, A. O., and I. Mysterud. 1993. Physical and biological mechanisms in animal movement processes. Journal of Applied Ecology 30:523–535.

Gauthreaux, S. A., Jr. 1970. Weather radar quantification of bird migration. BioScience 20:17–20.

Gauthreaux, S. A., Jr. 1974. The detection, quantification, and monitoring of bird movements aloft with airport surveillance radar. Pages 289–307 in Proceedings of a conference on the biological aspects of the bird/aircraft collision problem. Department of Zoology, Clemson University, Clemson, South Carolina, USA.

Gauthreaux, S. A., Jr. 1985. An avian mobile research laboratory. Proceedings of the Hawk Migration Conference 4:339–346.

Gauthreaux, S. A., Jr. 1996. Bird migration: methodologies and major research trajectories (1945–1995). Condor 98:442–453.

Gauthreaux, S. A., Jr., and C. G. Belser. 1998. Displays of bird movements on the WSR-88D: patterns and quantification. Weather and Forecasting 13:453–464.

Gauthreaux, S. A., Jr., and C. G. Belser. 1999. Reply to "comments on 'Display of bird movements on the WSR-88D: patterns and quantification.'" Weather and Forecasting 14:1041–1042.

Gauthreaux, S. A., Jr., and C. G. Belser. 2003. Radar ornithology and biological conservation. Auk 120:266–277.

Gauthreaux, S. A., Jr., and J. W. Livingston. 2006. Monitoring bird migration with a fixed-beam radar and a thermal-imaging camera. Journal of Field Ornithology 77:319–328.

Gauthreaux, S. A., Jr., D. S. Mizrahi, and C. G. Belser. 1998. Bird migration and bias of WSR-88D wind estimates. Weather and Forecasting 13:465–481.

Gavin, T. A. 1991. Why ask "why": the importance of evolutionary biology in wildlife science. Journal of Wildlife Management 55:760–766.

Gaymer, R. 1973. A marking method for giant tortoises and field trials in Aldabra. Journal of Zoology 169:393–401.

Gebremedhin, B., G. F. Ficetola, S. Naderi, H. R. Rezaei, C. Maudet, D. Rioux, G. Luikart, O. Flagstad, W. Thuiller, and P. Taberlet. 2009. Combining genetic and ecological data to assess the conservation status of the endangered Ethiopian walia ibex. Animal Conservation 12:89–100.

Gee, K. L., J. H. Holman, M. K. Causey, A. N. Rossi, and J. B. Armstrong. 2002. Aging white-tailed deer by tooth replacement and wear: a critical evaluation of a time-honored technique. Wildlife Society Bulletin 30:387–393.

Geering, D. J. 1998. Playback tapes as an aid for mist-netting regent honeyeaters. Corella 22:61–63.

Gehrt, S. D., and E. K. Fritzell. 1996. Sex-biased response of raccoons (*Procyon lotor*) to live traps. American Midland Naturalist 135:23–32.

Geiger, G., J. Bromel, and K. H. Habermehl. 1977. Concordance of various methods of determining the age of the red fox (*Vulpes vulpes* L. 1758). Zeitschrift für Jagdwissenschaft 23:57–64.

Geis, A. D., and L. H. Elbert. 1956. Relation of the tail length of cock ring-necked pheasants to harem size. Auk 73:289.

Geist, V. 1966. Validity of horn segment counts in aging bighorn sheep. Journal of Wildlife Management 30:634–635.

Geist, V. 1981. Behavior: adaptive strategies in mule deer. Pages 157–223 in O. C. Wallmo, editor. Mule and black-tailed deer of North America. University of Nebraska Press, Lincoln, USA.

Gelman, A. B., J. B. Carlin, H. S. Stern, and D. B. Rubin. 2003. Bayesian data analysis. Second edition. Chapman and Hall Press, Boca Raton, Florida, USA.

Gentile, J. R. 1987. The evolution of anti-trapping sentiment in the United States: a review and commentary. Wildlife Society Bulletin 15:490–503.

Gentry, R. L. 1979. Adventitious and temporary marks in pinniped studies. Pages 39–43 in L. Hobbs and P. Russell, editors. Report on the pinniped tagging workshop. American Institute of Biological Sciences, Arlington, Virginia, USA.

Gentry, R. L., and J. R. Holt. 1982. Equipment and techniques for handling northern fur seals. Technical Report, Special Scientific Report Fisheries 758, U.S. National Oceanic and Atmospheric Administration, Department of Commerce, Washington, D.C., USA.

Gentry, R. L., M. H. Smith, and R. J. Beyers. 1971. Use of radioactively tagged bait to study movement patterns in small mammals. Annales Zoologici Fennici 8:17–21.

Geraci, J. R., G. J. D. Smith, and T. G. Friesen. 1986. Assessment of marking techniques for beluga whale: final report to World Wildlife Fund Canada. Department of Pathology, University of Guelph, Guelph, Ontario, Canada.

Germaine, S. S., K. D. Bristow, and L. A. Haynes. 2000. Distribution and population status of mountain lions in southwestern Arizona. Southwestern Naturalist 45:333–338.

Germano, D. J., and D. F. Williams. 1993. Field evaluation of using passive integrated transponders (PIT) tags to permanently mark lizards. Herpetological Review 24:54–56.

Gervas V., S. Brunberg, and J. E. Swenson. 2006. An individual-based method to measure animal activity levels: a test on brown bears. Wildlife Society Bulletin 34:1314–1319.

Gese, E. M., O. J. Rongstad, and W. R. Mytton. 1987. Manual and net-gun capture of coyotes from helicopters. Wildlife Society Bulletin 15:444–445.

Gessaman, J. A., and K. A. Nagy. 1988. Transmitter loads affect the flight speed and metabolism of homing pigeons. Condor 90:662–668.

Gessaman, J. A., G. W. Workman, and M. R. Fuller. 1991. Flight performance, energetics and water turn-over of tippler pigeons with a harness and dorsal load. Condor 93:546–554.

Gettinger, R. D. 1990. Effects of chemical insect repellents on small mammal trapping yield. American Midland Naturalist 124:181–184.

Getz, W. M., S. Fortmann-Roe, P. C. Cross, A. J. Lyons, S. J. Ryan, and C. C. Wilmers. 2007. LoCoH: nonparameteric kernel methods for constructing home ranges and utilization distributions. PLoS ONE 2:e207. doi:10.1371/journal.pone.0000207.

Getz, W. M., and C. C. Wilmers. 2004. A local nearest-neighbor convex-hull construction of home ranges and utilization distributions. Ecography 27:489–505.

Ghioca, D. M., and L. M. Smith. 2007. Biases in trapping larval amphibians in playa wetlands. Journal of Wildlife Management 71:991–995.

Gibbens, R. P., and R. F. Beck. 1988. Changes in grass basal area and forb densities over a 64-year period on grassland types of the Jornada Experimental Range. Journal of Range Management 41:186–192.

Gibbons, J. W., and J. L. Green. 1979. X-ray photography: a technique to determine reproductive patterns of freshwater turtles. Herpetological Review 35:86–89.

Gier, H. T. 1968. Coyotes in Kansas. Bulletin 393, Kansas Agricultural Experiment Station, Kansas State University, Manhattan, USA.

Giesen, K. M., and C. E. Braun. 1979. A technique for age determination of juvenile white-tailed ptarmigan. Journal of Wildlife Management 43:508–511.

Giesen, K. M., T. J. Schoenberg, and C. E. Braun. 1982. Methods for trapping sage grouse in Colorado. Wildlife Society Bulletin 10:224–231.

Gifford, C. E., and D. R. Griffin. 1960. Notes on homing and migratory behavior of bats. Ecology 41:378–381.

Gilbert, F. F. 1966. Aging white-tailed deer by annuli in the cementum of the first incisor. Journal of Wildlife Management 30:200–202.

Gilbert, F. F. 1976. Impact energy thresholds for anesthetized raccoons, mink, muskrats, and beavers. Journal of Wildlife Management 40:669–676.

Gilbert, F. F. 1981a. Assessment of furbearer response to trapping devices. Proceedings of the Worldwide Furbearer Conference 3:1599–1611.

Gilbert, F. F. 1981b. Maximizing the humane potential of traps—the Vital and the Conibear 120. Proceedings of the Worldwide Furbearer Conference 3:1630–1646.

Gilbert, F. F. 1992. Aquatic trap testing—Washington State University. Pages 20–21 in Wild fur and the international market place. Fur Institute of Canada, Ottawa, Ontario, Canada.

Gilbert, F. F., and R. Allwine. 1991. Small mammal communities in the Oregon Cascade Range. Pages 257–267 in L. F. Ruggiero, K. B. Aubry, A. B. Carey, and M. H. Huff, technical coordinators. Wildlife and vegetation of unmanaged Douglas-fir forests. General Technical Report PNW-GTR-285, U.S. Forest Service, Department of Agriculture, Washington, D.C., USA.

Gilbert, F. F., and N. Gofton. 1982. Terminal dives in mink, muskrat and beaver. Physiology and Behavior 28:835–840.

Gilbert, F. F., and S. L. Stolt. 1970. Variability in aging Maine white-tailed deer by tooth-wear characteristics. Journal of Wildlife Management 34:532–535.

Giles, R. H., Jr., editor. 1969. Wildlife management techniques. The Wildlife Society, Washington, D.C., USA.

Gill, D. E., W. J. L. Sladen, and C. E. Huntington. 1970. A technique for catching petrels and shearwaters at sea. Bird-Banding 41:111–113.

Gill, F. B. 1995. Ornithology. Second edition. W. H. Freeman and Company, New York, New York, USA.

Gill, J. A. 1996. Habitat choice in pink-footed geese: quantifying the constraints of winter site use. Journal of Applied Ecology 33:884–892.

Gill, J. A., and W. J. Sutherland. 2000. Predicting the consequences of human behavior disturbance from behavioral decisions. Pages 51–64 in L. M. Gosling and W. J. Sutherland, editors. Behaviour and conservation. Cambridge University Press, Cambridge, England, UK.

Gill, R. B. 1985. Wildlife research—an endangered species. Wildlife Society Bulletin 13:580–587.

Gillin, C. M., I Chestin, P. Semchenkov, and J. Claar. 1997. Management of bear-human conflicts using Laika dogs. Bears: Their Biology and Management 9:133–137.

Gilmore, D. P. 1969. Seasonal reproductive periodicity in the male Australian brush-tailed possum (*Trichosurus volpecula*). Journal of Zoology 157:75–98.

Gilpin, M. E., and F. J. Ayala. 1973. Global models of growth and competition. Proceedings of the National Academy of Sciences of the United States of America 70:3590–3593.

Gilpin, M. E., and M. E. Soulé. 1986. Minimum viable populations: processes of species extinctions. Pages 19–34 in M. E. Soulé, editor. Conservation biology: the science of scarcity and diversity. Sinauer Associates, Sunderland, Massachusetts, USA.

Gilson, A., M. Syvanen, K. Levine, and J. Banks. 1998. Deer gender determination by polymerase chain reaction: validation study and application to tissues, bloodstains, and hair forensic samples from California. California Fish and Game 84:159–169.

Ginther, O. J. 1990. Folliculogenesis during the transitional period and early ovulatory season in mares. Journal of Reproduction and Fertility 90:311–320.

Ginzburg, L. R. 1986. Theory of population dynamics: back to principles. Journal of Theoretical Biology 122:385–399.

Gionfriddo, J. P., and L. C. Stoddart. 1988. Comparative recovery rates of marked coyotes. Wildlife Society Bulletin 16:310–311.

Gipson, P. S., W. B. Ballard, R. M. Nowak, and L. D. Mech. 2000. Accuracy and precision of estimating age of gray wolves by tooth wear. Journal of Wildlife Management 64:752–758.

Gitzen, R. A., and J. J. Millspaugh. 2003. Comparison of least-squares cross-validation bandwidth options for kernel home-range estimation. Wildlife Society Bulletin 31:823–831.

Gitzen, R. A., J. J. Millspaugh, and B. J. Kernohan. 2006. Bandwidth selection for fixed kernel analysis of animal range use. Journal of Wildlife Management 70:1334–1344.

Glasgow, L. L. 1957. The night driving of coots for banding on the wintering ground of Louisiana. Bird-Banding 28:153–155.

Glazener, W. C. 1949. Operation deer trap. Texas Game and Fish 7(10):6–7, 17, 19.

Glazener, W. C., A. S. Jackson, and M. L. Cox. 1964. The Texas drop-net turkey trap. Journal of Wildlife Management 28:280–287.

Glenn, E. M., and W. J. Ripple. 2004. On using digital maps to assess wildlife habitat. Wildlife Society Bulletin 32:852–860.

Glor, R. E., T. M. Townsend, M. F. Benard, and A. S. Flecker. 2000. Sampling reptile diversity in the West Indies with mouse glue traps. Herpetological Review 31:88–90.

Glorioso, B. M., and M. L. Niemiller. 2006. Using deep-water crawfish nets to capture aquatic turtles. Herpetological Review 37:185–187.

Gluesenkamp, A. G. 1995. The snake rake: a new tool for collecting reptiles and amphibians. Herpetological Review 26:19.

Godfrey, C. L., K. Needham, M. R. Vaughan, J. Higgins Vashon, D. D. Martin, and G. T. Blank, Jr. 2000. A technique for and risks associated with entering tree dens used by black bears. Wildlife Society Bulletin 28:131–140.

Godfrey, G. A. 1975. Home range characteristics of ruffed grouse broods in Minnesota. Journal of Wildlife Management 39:287–298.

Godfrey, R. D., Jr., A. M. Fedynich, and E. G. Bolen. 1993. Fluorescent particles for marking waterfowl without capture. Wildlife Society Bulletin 21:283–288.

Godfroid, J. 2002. Brucellosis in wildlife. Revue Scientifique et Technique de L'Office International Des Epizooties 21:277–286.

Godin, A. J. 1960. A compilation of diagnostic characteristics used in aging and sexing game birds and mammals. Thesis, University of Massachusetts, Amherst, USA.

Goetz, S., D. Steinberg, R. Dubayah, and B. Blair. 2007. Laser remote sensing of canopy habitat heterogeneity as a predictor of bird species richness in an eastern temperate forest, USA. Remote Sensing of Environment 108:254–263.

Goetz, W. E. 2002. Developing a predictive model for identifying riparian communities at an ecoregion scale in Idaho and Wyoming. Thesis, Utah State University, Logan, USA.

Goforth, W. R., and T. S. Baskett. 1965. Effects of experimental color marking on pairing of captive mourning doves. Journal of Wildlife Management 29:543–553.

Golay, N., and H. Durrer. 1994. Inflammation due to toe clipping in natterjack toads (Bufo calamita). Amphibia-Reptilia 15:81–83.

Goldstein, M. I., T. E. Lacher, B. Woodbridge, M. J. Bechard, S. B. Canavelli, M. E. Zaccagnini, G. P. Cobb, E. J. Scollon, R. Tribolet, and M. J. Hooper. 1999. Monocrotophos-induced mass mortality of Swainson's hawks in Argentina, 1995–96. Ecotoxicology 8:201–214.

Goldstein, M. I., B. Woodbridge, M. E. Zaccagnini, S. B. Canavelli, and A. Lanusse. 1996. An assessment of mortality to Swainson's hawks on wintering grounds in Argentina. Journal of Raptor Research 30:106–107.

Gonzalez, M. A., M. A. Hussey, and B. E. Conrad. 1990. Plant height, disk, and capacitance meters used to estimate bermudagrass herbage mass. Agronomy Journal 82:861–864.

Goodall, D. W. 1952. Some considerations in the use of point quadrats for the analysis of vegetation. Australian Journal of Scientific Research, Series Biological Sciences 5:1–41.

Goodchild, M. F. 1993. The state of GIS for environmental problem solving. Pages 8–15 in M. F. Goodchild, B. O. Parks, and L. T. Steyaert, editors. Environmental Modeling with GIS. Oxford University Press, New York, New York, USA.

Goodman, B. A., and G. N. L. Peterson. 2005. A technique for sampling lizards in rocky habitats. Herpetological Review 36:41–43.

Goodman, L. A. 1960. On the exact variance of products. Journal of the American Statistical Association 55:708–713.

Goodman, S. J., H. B. Tamate, R. A. Wilson, J. Nagata, S. Tatsuzawa, G. M. Swanson, J. M. Pemberton, and D. R. Mccullough. 2001. Bottlenecks, drift and differentiation: the population structure and demographic history of sika deer (Cervus nippon) in the Japanese archipelago. Molecular Ecology 10:1357–1370.

Goodrich, J. M., L. L. Kerley, B. O. Schleyer, D. G. Miquelle, K. S. Quigley, T. N. Smirnov, H. B. Quigley, and M. G. Hornocker. 2001. Capture and chemical anesthesia of Amur (Siberian) tigers. Wildlife Society Bulletin 29:533–542.

Goodrum, P. D. 1940. A population study of the gray squirrel in eastern Texas. Thesis, Agricultural and Mechanical College of Texas, College Station, USA.

Goodrum, P. D., V. H. Reid, and C. E. Boyd. 1971. Acorn yields, characteristics, and management criteria of oaks for wildlife. Journal of Wildlife Management 35:520–532.

Goodwin, E. A., and W. B. Ballard. 1985. Use of tooth cementum for age determination of gray wolves. Journal of Wildlife Management 49:313–316.

Goodwin, K., J. Jacobs, D. Weaver, and R. Engel. 2006. Detecting rare spotted knapweed (Centaurea biebersteinii DC.) plants using trained canines. Department of Land Resources and Environmental Science, Montana State University, Bozeman, USA.

Goodyear, N. C. 1989. Studying fine-scale habitat use in small mammals. Journal of Wildlife Management 53:941–946.

Goparaju, L., A. Tripathi, and C. S. Jha. 2005. Forest fragmentation impacts on phytodiversity—an analysis using remote sensing and GIS. Current Science 88:1264–1274.

Gordon, K. R., and G. V. Morejohn. 1975. Sexing black bear skulls using lower canine and lower molar measurement. Journal of Wildlife Management 39:40–44.

Gortázar, C., M. Torres, J. Vicente, P. Acevedo, M. Reglero, J. de La Fuente, J. Negro, and J. Aznar-Martín. 2008. Bovine tuberculosis in Doñana Biosphere Reserve: the role of wild ungulates as disease reservoirs in the last Iberian lynx strongholds. PLoS ONE 3:1–8.

Gosling, L. M. 1986. Biased sex ratios in stressed animals. American Midland Naturalist 127:893–896.

Gosling, L. M., and W. J. Sutherland. 2000. Behaviour and conservation. Cambridge University Press, Cambridge, England, UK.

Goss-Custard, J. D., S. E. A. Le V. Dit Durell, H. P. Sitters, and R. Swinfen. 1982. Age-structure and survival of a wintering population of oystercatchers. Bird Study 29:83–98.

Gotfryd, A., and R. I. C. Hansell. 1985. The impact of observer bias on multivariate analysis of vegetation structure. Oikos 45:223–234.

Göth A., and C. S. Evans. 2004a. Egg size predicts motor performance and postnatal weight gain of Australian brush-turkey (Alectura lathami) hatchlings. Canadian Journal of Zoology 82:972–979.

Göth A., and C. S. Evans. 2004b. Social responses without early experience: Australian brush-turkey chicks use specific visual cues to aggregate with conspecifics. Journal of Experimental Biology 207:2199–2208.

Gourley, R. S., and F. J. Jannett, Jr. 1975. Pine and montane vole age estimates from eye lens weights. Journal of Wildlife Management 39:550–556.

Gove, N. E., J. R. Skalski, P. Zager, and R. L. Townsend. 2002. Statistical models for population reconstruction using age-at-harvest data. Journal of Wildlife Management 66:310–320.

Goward, S. N., D. Dye, A. Kerber, and V. Kalb. 1987. Comparisons of North and South American biomes from AVHRR observations. Geocarto International 1:27–39.

Gower, W. C. 1939. The use of the bursa of Fabricius as an indication of age in game birds. Transactions of the North American Wildlife Conference 4:426–430.

Goyer, R. A., and T. W. Clarkson. 2001. Toxic effects of metals. Pages 811–867 in C. D. Klaassen, editor. Casarett and Doull's toxicology: the basic science of poisons. Sixth edition. McGraw-Hill, New York, New York, USA.

Graber, R. E. 1970. Natural seed fall in white pine (Pinus strobus L.) stands of varying density. Research Note NE-119, U.S. Forest Service, Department of Agriculture, Washington, D.C., USA.

Graber, R. R., and S. S. Hassler. 1962. The effectiveness of aircraft-type (APS) radar in detecting birds. Wilson Bulletin 74:367–380.

Gragg, J. E., G. H. Rodda, J. A. Savidge, and K. Dean-Bradley. 2007. Response of brown tree snakes to reduction of their rodent prey. Journal of Wildlife Management 71:2311–2317.

Graham, A., and R. Bell. 1989. Investigating observer bias in aerial survey by simultaneous double-counts. Journal of Wildlife Management 53:1009–1016.

Graham, L. H., J. Bolling, G. Miller, N. Pratt-Hawkes, S. Joseph S. 2002. Enzyme-immunoassay for the measurement of luteinizing hormone in the serum of African elephants (Loxodonta africana). Zoo Biology 21:403–408.

Graham, T., and A. Georges. 1996. Struts for collapsible funnel traps. Herpetological Review 27:189–190.

Graham, T. E. 1986. A warning against the use of Petersen disc tags in turtle studies. Herpetological Review 17:42–43.

Graham, W. J., and H. W. Ambrose, III. 1967. A technique for continuously locating small mammals in field enclosures. Journal of Mammalogy 48:639–642.

Grand, J. B., and T. F. Fondell. 1994. Decoy trapping and rocket-netting for northern pintails in spring. Journal of Field Ornithology 65:402–405.

Grandy, J. W., and A. T. Rutberg. 2002. An animal welfare view of wildlife contraception. Pages 1–7 in J. F. Kirkpatrick, B. L. Lasley, W. R. Allen, and C. Doberska, editors. Fertility control in wildlife. Reproduction: Supplement 60.

Grant, B. W., A. D. Tucker, J. E. Lovich, A. M. Mills, P. M. Dixon, and J. W. Gibbons. 1992. The use of coverboards in estimating patterns of reptile and amphibian biodiversity. Pages 379–403 in D. R. McCullough and R. H. Barrett, editors. Wildlife 2001: populations. Elsevier Science, London, England, UK.

Grant, W. E., E. K. Pederson, and S. L. Marin. 1997. Ecology and natural resource management: systems analysis and simulation. John Wiley and Sons, New York, New York, USA.

Gratto-Trevor, C. L. 2004. The North American bander's manual for banding shorebirds. (Charadriiformes, suborder Charadrii). North American Banding Council, Point Reyes Station, California, USA.

Gratto-Trevor, C. L., L. W. Oring, and A. J. Fivizzani. 1991. Effects of blood sampling stress on hormone levels in the semipalmated sandpiper. Journal of Field Ornithology 62:19–27.

Grau, G. A., G. C. Sanderson, and J. P. Rogers. 1970. Age determination of raccoons. Journal of Wildlife Management 34:364–372.

Graul, W. D. 1979. An evaluation of selected techniques for nesting shorebirds. North American Bird Bander 4:19–21.

Graves, H. B., E. D. Bellis, and W. M. Knuth. 1972. Censusing white-tailed deer by airborne thermal infrared imagery. Journal of Wildlife Management 36:875–884.

Gray, M. J., R. M. Kaminski, and G. Weerakkody. 1999. Predicting seed yield of moist-soil plants. Journal of Wildlife Management 63:1261–1268.

Gray, S. S., T. R. Simpson, J. T. Baccus, R. W. Manning, and T. W. Schwertner. 2007. Diets and foraging preference of greater kudu in the Llano Uplift of Texas. Wildlife Biology 13:1–10.

Grayson, K. L., and A. W. Roe. 2007. Glowsticks as effective bait for capturing aquatic amphibians in funnel traps. Herpetological Review 38:168–170.

Green, J. A., L. G. Hasley, R. P. Wilson, and P. B. Frappell. 2009. Estimating energy expenditure of animals using the accelerometry technique: activity, inactivity and comparison with the heart-rate technique. Journal of Experimental Biology 212:471–482.

Green, J. S., and R. A. Woodruff. 1980. Is predator control going to the dogs? Rangelands 2:187–189.

Green, J. S., and R. A. Woodruff. 1988. Breed comparisons and characteristics of use of livestock guarding dogs. Journal of Range Management 41:249–251.

Green, J. S., R. A. Woodruff, and T. T. Tueller. 1984. Livestock-guarding dogs for predator control: costs, benefits, and practicality. Wildlife Society Bulletin 12:44–50.

Green, M. C., J. L. Waits, M. L. Avery, M. E. Tobin, and P. L. Leberg. 2006. Microsatellite variation of double-crested cormorant populations in eastern North America. Journal of Wildlife Management 70:579–583.

Green, R. H. 1979. Sampling design and statistical methods for environmental biologists. John Wiley and Sons, New York, New York, USA.

Greenberg, B. D., J. E. Newbold, and A. Sugino. 1983. Intraspecific nucleotide sequence variability surrounding the origin of replication in human mitochondrial DNA. Gene 21:33–49.

Greenberg, C. H., D. G. Neary, and L. D. Harris. 1994. A comparison of herpetofaunal sampling effectiveness of pitfall, single-ended and double-ended funnel traps used with drift fences. Journal of Herpetology 28:319–324.

Greenberg, R. E., S. L. Etter, and W. L. Anderson. 1972. Evaluation of proximal primary feather criteria for aging wild pheasants. Journal of Wildlife Management 36:700–705.

Greenwood, R. J. 1975. An attempt to freeze-brand mallard ducklings. Bird-Banding 46:204–206.

Greenwood, R. J. 1977. Evaluation of nasal marker for ducks. Journal of Wildlife Management 41:582–585.

Greenwood, R. J., and W. C. Bair. 1974. Ice on waterfowl markers. Wildlife Society Bulletin 2:130–134.

Greer, K. R. 1957. Some osteological characters of known-age ranch minks. Journal of Mammalogy 38:319–330.

Greer, K. R., and W. W. Hawkins, Jr. 1967. Determining pregnancy in elk by rectal palpation. Journal of Wildlife Management 31:145–149.

Greer, K. R., and H. W. Yeager. 1967. Sex and age indications from upper canine teeth of elk (wapiti). Journal of Wildlife Management 31:408–417.

Gregoire, T. G., and O. Schabenberger. 1996. A non-linear mixed effects model to predict cumulative bole volume of standing trees. Journal of Applied Statistics 23:257–271.

Gregory, P. T., G. J. Davies, and J. M. MacCartney. 1989. A portable device for restraining rattlesnakes in the field. Herpetological Review 20:43–44.

Greig-Smith, P. 1983. Quantitative plant ecology. University of California Press, Berkeley, USA.

Grémillet, D., M. R. Enstipp, M. Boudiffa, and H. Liu. 2006. Do cormorants injure fish without eating them? An underwater video study. Marine Biology 148:1081–1087.

Grenfell, B. T., and A. P. Dobson, editors. 1995. Ecology of infectious diseases in natural populations. Cambridge University Press, Cambridge, England, UK.

Griben, M. R., H. R. Johnson, B. B. Gallucci, and V. F. Gallucci. 1984. A new method to mark pinnipeds as applied to the northern fur seal. Journal of Wildlife Management 48:945–949.

Grice, D., and J. P. Rogers. 1965. The wood duck in Massachusetts. Final Report, Federal Aid Project W-19-R, Massachusetts Division of Fish and Game, Amherst, USA.

Grieg-Smith, P. 1964. Quantitative plant ecology. First edition. Plenum Press, New York, New York, USA.

Grier, J. W., J. M. Gerrard, G. D. Hamilton, and P. A. Gray. 1981. Aerial-visibility bias and survey techniques for nesting bald eagles in northwestern Ontario. Journal of Wildlife Management 45:83–92.

Griesemer, S. J., M. O. Hale, U. Roze, and T. K. Fuller. 1999. Capturing and marking adult North American porcupines. Wildlife Society Bulletin 27:310–313.

Griffin, D. R. 1934. Marking bats. Journal of Mammalogy 15:202–207.

Griffin, D. R. 1952. Radioactive tagging of animals under natural conditions. Ecology 33:329–335.

Griffin, D. R. 1973. Oriented bird migration in or between opaque cloud layers. Proceedings of the American Philosophical Society 117:117–141.

Griffin, D. R., and R. A. Suthers. 1970. Sensitivity of echolocation in cave swiftlets. Biological Bulletin 139:495–501.

Griffin, P. C. 1999. Endangered species diversity "hot spots" in Russia and centers of endemism. Biodiversity and Conservation 8:497–511.

Griffith, B., and B. A. Youtie. 1988. Two devices for estimating foliage density and deer hiding cover. Wildlife Society Bulletin 16:206–210.

Griffith, R. E., and J. Evans. 1970. Capturing jackrabbits by night-lighting. Journal of Wildlife Management 34:637–639.

Griffiths, R., M. C. Double, K. Orr, and R. J. G. Dawson. 1998. A DNA test to sex most birds. Molecular Ecology 7:1071–1075.

Griffiths, R., and B. Tiwari. 1993. Primers for the differential amplification of the sex-determining region Y-gene in a range of mammal species. Molecular Ecology 2:405–406.

Grinder, M. I., and P. R. Krausman. 2001. Home range, habitat use, and nocturnal activity of coyotes in an urban environment. Journal of Wildlife Management 65:887–898.

Grizimek, B. 1990. Grizimek's encyclopedia of mammals. McGraw-Hill, New York, New York, USA.

Grosenbaugh, L. R. 1952. Plotless timber estimates, new, fast, easy. Journal of Forestry 50:32–37.

Gross, A. O. 1925. Diseases of the ruffed grouse. Auk 62:423–431.

Gross, T. S., B. S. Arnold, M. S. Sepúlveda, and K. McDonald. 2003. Endocrine disrupting chemicals and endocrine active agents. Pages 1033–1098 in D. J. Hoffman, B. A. Rattner, G. A. Burton, Jr., and J. Cairns, Jr., editors. Handbook of ecotoxicology. Second edition. Lewis, Boca Raton, Florida, USA.

Grubb, J. C. 1970. Orientation in post-reproductive Mexican toads, *Bufo valliceps*. Copeia 1970: 674–680.

Grubb, T. G. 1988. A portable rocket-net system for capturing wildlife. Research Note RM-484, U.S. Forest Service, Department of Agriculture, Washington, D.C., USA.

Grubb, T. G. 1991. Modifications of the portable rocket-net capture system to improve performance. Research Note RM-502, U.S. Forest Service, Department of Agriculture, Washington, D.C., USA.

Grue, C. E., T. J. O'Shea, and D. J. Hoffman. 1984. Lead concentrations and reproduction in highway-nesting barn swallows. Condor 86: 383–389.

Grue, H., and B. Jensen. 1973. Annular structures in canine tooth cementum in red foxes (*Vulpes fulva* L.) of known age. Danish Review of Game Biology 8:1–12.

Grue, H., and B. Jensen. 1976. Annular cementum structures in canine tooth in arctic foxes (*Alopex lagopus* L.) from Greenland and Denmark. Danish Review of Game Biology 10:1–12.

Gruell, G. E., and N. J. Papez. 1963. Movements of mule deer in northeastern Nevada. Journal of Wildlife Management 27:414–422.

Gruver, K. S., R. Phillips, and E. S. Williams. 1996. Leg injuries to coyotes captured in standard and modified Soft Catch™ traps. Proceedings of the Vertebrate Pest Conference 17:91–93.

Grzybowski, J. 1995. Black-capped vireo (*Vireo atricapillus*). Account 181 in A. Poole and F. Gill, editors. The birds of North America. The Academy of Natural Sciences, Philadelphia, and The American Ornithologists' Union, Washington, D.C., USA.

Guarino, J. L. 1968. Evaluation of a colored leg tag for starlings and blackbirds. Bird-Banding 39:6–13.

Guilford, T., J. Meade, J. Willis, R. A. Phillips, D. Boyle, S. Roberts, M. Collett, R. Freeman, and C. M. Perrins. 2009. Migration and stopover in a small pelagic seabird, the Manx shearwater *Puffinus puffinus*: insights from machine learning. Proceedings of the Royal Society B 276:1215–1223.

Guinness, F. E., T. H. Clutton-Brock, and S. D. Albon. 1978. Factors affecting calf mortality in red deer (*Cervus elaphus*). Journal of Animal Ecology 47:817–832.

Guisse, A. W., and H. R. Gimblett. 1997. Assessing and mapping conflicting recreation values in state park settings using neural networks. AI Applications 11:79–89.

Gullion, G. W. 1951. A marker for waterfowl. Journal of Wildlife Management 15:222–223.

Gullion, G. W. 1965a. Another comment on the color-banding of birds. Journal of Wildlife Management 29:401.

Gullion, G. W. 1965b. Improvements in methods for trapping and marking ruffed grouse. Journal of Wildlife Management 29:109–116.

Gullion, G. W., R. L. Eng, and J. J. Kupa. 1962. Three methods for individually marking ruffed grouse. Journal of Wildlife Management 26:404–407.

Guralnick, R. P., and D. Neufeld. 2005. Challenges building online GIS services to support global biodiversity mapping and analysis: lessons from the Mountain and Plains Database and Informatics project. Biodiversity Informatics 2:56–69.

Gurevitch, J., P. S. Curtis, and M. H. Jones. 2001. Meta-analysis in ecology. Advances in Ecological Research 32:199–247.

Gurnell, J., and J. Little. 1992. The influence of trap residual odor on catching woodland rodents. Animal Behaviour 43:623–632.

Guthery, F. S. 2008. A primer on natural resource science. Texas A&M University Press, College Station, USA.

Guthery, F. S., and S. L. Beasom. 1978. Effectiveness and selectivity of neck snares in predator control. Journal of Wildlife Management 42:457–459.

Guthery, F. S., and R. L. Bingham. 2007. A primer on interpreting regression models. Journal of Wildlife Management 71:684–692.

Guthery, F. S., L. A. Brennan, M. J. Peterson, and J. J. Lusk. 2005. Information theory in wildlife science: critique and viewpoint. Journal of Wildlife Management 69:457–465.

Guthery, F. S., and J. J. Lusk. 2004. Radiotelemetry studies: are we radio-handicapping northern bobwhites? Wildlife Society Bulletin 32:194–201.

Guthery, F. S., J. J. Lusk, and M. J. Peterson. 2001. The fall of the null hypothesis: liabilities and opportunities. Journal of Wildlife Management 65:379–384.

Guthery, F. S., and G. E. Mecozzi. 2008. Developing the concept of estimating bobwhite density with pointing dogs. Journal of Wildlife Management 72:1175–1180.

Gutierrez, R. J., and D. J. Delehanty. 1999. Mountain quail. Account 457 in A. Poole and F. Gill, editors. The birds of North America. The Academy of Natural Sciences, Philadelphia, Pennsylvania, and The American Ornithologists' Union, Washington, D.C., USA.

Guttman, S. I., and W. Creasey. 1973. Staining as a technique for marking tadpoles. Journal of Herpetology 7:388.

Gutzwiller, K. J. 1990. Minimizing dog-induced biases in game bird research. Wildlife Society Bulletin 18:351–356.

Guyn, K. L., and R. G. Clark. 1999. Decoy trap bias and effects of marks on reproduction of northern pintails. Journal of Field Ornithology 70:504–513.

Guynn, D. C., J. R. Davis, and A. F. Von Recum. 1987. Pathological potential of intraperitoneal transmitter implants in beavers. Journal of Wildlife Management 51:605–606.

Guynn, D. E., and P. F. Scanlon. 1973. Crop-gland activity in mourning doves during hunting seasons in Virginia. Proceedings of the Annual Conference of the Southeastern Association of Game and Fish Commissioners 27:36–42.

Gysel, L., and E. Davis, Jr. 1956. A simple automatic photographic unit for wildlife research. Journal of Wildlife Management 20:451–453.

Gysel, L. W. 1956. Measurement of acorn crops. Forest Science 2:305–313.

Gysel, L. W. 1957. Acorn production on good, medium, and poor oak sites in southern Michigan. Journal of Forestry 55:570–574.

Gysel, L. W., and L. J. Lyon. 1980. Habitat analysis and evaluation. Pages 305–327 in S. D. Schemnitz, editor. Wildlife management techniques manual. The Wildlife Society, Washington, D.C., USA.

Haagenrud, H. 1978. Layers in secondary dentine of incisors as age criteria in moose (*Alces alces*). Journal of Mammalogy 59:857–858.

Haan, S. S., and M. J. Desmond. 2005. Effectiveness of three capture methods for the terrestrial Sacramento Mountains salamander. Herpetological Review 36:143–145.

Haas, G. H., and S. R. Amend. 1976. Aging immature mourning doves by primary feather molt. Journal of Wildlife Management 40:575–578.

Haase, B. 2002. The use of tape-recorded distress calls to increase shorebird capture rates. Wader Study Group Bulletin 99:58–59.

Hadow, H. H. 1972. Freeze-branding: a permanent marking technique for pigmented mammals. Journal of Wildlife Management 36:645–649.

Hagan, J. M., and J. M. Reed. 1988. Red color bands reduce fledging success in red-cockaded woodpeckers. Auk 105:498–503.

Hagen, C. A., and K. M. Giesen. 2005. Lesser prairie-chicken (*Tympanuchus pallidicinctus*). Account 364 in A. Poole and F. Gill, editors. The birds of North America. The Academy of Natural Sciences, Philadelphia, Pennsylvania, and The American Ornithologists' Union, Washington, D.C., USA.

Hagen, C. A., N. C. Kenkel, D. J. Walker, R. K. Baydack, and C. E. Braun. 2001. Fractal-based spatial analysis of radiotelemetry data. Pages 167–187 in J. J. Millspaugh and J. M. Marzluff, editors. Radio tracking and animal populations. Academic Press, San Diego, California, USA.

Hagler, J. R., and C. G. Jackson. 2001. Methods for marking insects: current techniques and future prospects. Annual Review of Entomology 46:511–543.

Hagström, T. 1973. Identification of newt specimens (*Urodela, Triturus*) by recording the belly pattern and a description of photographic equipment for such registration. British Journal of Herpetology 4:321–326.

Hagstrum, J. T. 2000. Infrasound and the avian navigational map. Journal of Experimental Biology 203:1103–1111.

Hahn, H. C. 1949. A method of censusing deer and its application in the Edwards Plateau of Texas. Final Report for Texas Federal Aid Project 25-R, July 1, 1946 to March 30, 1948. Texas Parks and Wildlife Department, Austin, USA.

Haig, S. M. 1998. Molecular contributions to conservation. Ecology 79:413–425.

Haig, S. M., J. D. Ballou, and N. J. Casna. 1994. Identification of kin structure among Guam rail founders: a comparison of pedigrees and DNA profiles. Molecular Ecology 3:109–119.

Haig, S. M., J. D. Ballou, and N. J. Casna. 1995. Genetic identification of kin in Micronesian kingfishers. Journal of Heredity 86:423–431.

Haig, S. M., J. D. Ballou, and S. R. Derrickson. 1990. Management options for preserving genetic diversity: reintroduction of Guam rails to the wild. Conservation Biology 4:290–300.

Haig, S. M., T. D. Mullins, E. D. Forsman, P. W. Trail, and L. Wennerberg. 2004. Genetic identification of spotted owls, barred owls, and their hybrids: legal implications of hybrid identity. Conservation Biology 18:1347–1357.

Haig, S. M., R. S. Wagner, E. D. Forsman, and T. D. Mullins. 2001. Geographic variation and genetic structure in spotted owls. Conservation Genetics 2:25–40.

Haines, A. M., M. E. Tewes, and L. L. Laack. 2005. Survival and sources of mortality in ocelots. Journal of Wildlife Management 69:255–263.

Hairston, N. G. 1989. Ecological experiments: purpose, design, and execution. Cambridge studies in ecology. Cambridge University Press, New York, New York, USA.

Hakim, S., W. J. McShea, and J. R. Mason. 1996. The attractiveness of a liquid bait to white-tailed deer in the central Appalachian Mountains, Virginia, USA. Journal of Wildlife Diseases 32:395–398.

Hale, J. B. 1949. Aging cottontail rabbits by bone growth. Journal of Wildlife Management 13:216–225.

Hale, J. B., R. F. Wendt, and G. C. Halazon. 1954. Sex and age criteria for Wisconsin ruffed grouse. Technical Wildlife Bulletin 9, Wisconsin Conservation Department, Madison, USA.

Hale, R., M. Alaee, J. B. Manchester-Neevig, H. M. Stapleton, and M.G. Ikonomou. 2003. Polybrominated diphenyl flame retardants in the North American environment. Environment International 29:771–779.

Hall, E. R. 1981. The mammals of North America. Second edition. Volume II. John Wiley and Sons, New York, New York, USA.

Hall, L. S., P. R. Krausman, and M. L. Morrison. 1997. The habitat concept and a plea for standard terminology. Wildlife Society Bulletin 25:1173–1182.

Hall, R. J., and D. P. Stafford. 1972. Studies in the life history of Wehrle's salamander, *Plethondon wehrlei*. Herpetologica 28:300–309.

Halls, L. K., and R. F. Harlow. 1971. Weight–length relations in flowering dogwood twigs. Journal of Range Management 24:236–237.

Halpin, M. A., and J. A. Bissonette. 1988. Influence of snow depth on prey availability and habitat use by red fox. Canadian Journal of Zoology 66:587–592.

Halsey, L.G., J. A. Green, R. P. Wilson, and P. B. Frappell. 2009. Accelerometry to estimate energy expenditure during activity: best practice with data loggers. Physiological and Biochemical Zoology 82:396–404.

Halstead, T. D., K. S. Gruver, R. L. Phillips, and R. E. Johnson. 1995. Using telemetry equipment for monitoring traps and snares. Proceedings, Great Plains Wildlife Damage Control Workshop 12:121–123.

Hamel, N. J., J. K. Parrish, and L. L. Conquest. 2004. Effects of tagging on behavior, provisioning, and reproduction in the common murre (*Uria aalge*), a diving seabird. Auk 121:1161–1171.

Hamer, T. E., B. A. Cooper, and C. J. Ralph. 1995. Use of radar to study the movements of marbled murrelets at inland sites. Northwestern Naturalist 76:73–78.

Hamerstrom, F. 1942. Dominance in winter flocks of chickadees. Wilson Bulletin 54:32–42.

Hamerstrom, F. 1963. The use of great horned owls in catching marsh hawks. Proceedings of the International Ornithological Congress 13:866–869.

Hamerstrom, F. N., Jr., and O. E. Mattson. 1964. A numbered, metal color-band for game birds. Journal of Wildlife Management 28:850–852.

Hamerstrom, F. N., Jr., and M. Truax. 1938. Traps for pinnated and sharp-tailed grouse. Bird-Banding 9:177–183.

Hamilton, R. J., M. L. Tobin, and W. G. Moore. 1985. Aging fetal white-tailed deer. Proceedings of the Annual Conference of the Southeastern Association of Fish and Wildlife Agencies 39:389–395.

Hamilton, W. J., Jr., and W. R. Eadie. 1964. Reproduction in the river otter, *Lutra canadensis*. Journal of Mammalogy 45:242–252.

Hamlin, K. L., D. F. Pac, C. A. Sime, R. M. DeSimone, and G. L. Dusek. 2000. Evaluating the accuracy of ages obtained by two methods for Montana ungulates. Journal of Wildlife Management 64:441–449.

Hampton, P. M., N. B. Ford, and K. Herriman. 2010. Impacts of oil pumps and deer feed plots on amphibian and reptile assemblages in a floodplain. American Midland Naturalist 163:44–53.

Hampton, P. M., and N. E. Haertle. 2009. A new view from a novel squeeze-box design. Herpetological Review 40:44.

Han E. S., T. R. Evans, and J. F. Nelson. 1998. Adrenocortical responsiveness to adrenocorticotropic hormone is enhanced in chronically food-restricted rats. Journal of Nutrition 128:1415–1420.

Handel, C. M., and R. E. Gill, Jr. 1983. Yellow birds stand out in a crowd. North American Bird Bander 8:6–9.

Handley, C. O., Jr., and M. Varn. 1994. The trapline concept applied to pitfall arrays. Pages 285–287 in J. F. Merritt, G. L. Kirkland, Jr., and R. K. Rose, editors. Advances in the biology of shrews. Special Publication 18, Carnegie Museum of Natural History, Pittsburgh, Pennsylvania, USA.

Handley, L. J. L., and N. Perrin. 2007. Advances in our understanding of mammalian sex-biased dispersal. Molecular Ecology 16:1559–1578.

Hanks, J. 1969. Techniques for marking large African mammals. Puku 5:65–86.

Hanley, T. A. 1978. A comparison of the line-interception and quadrat estimation methods of determining shrub canopy coverage. Journal of Range Management 31:60–62.

Hannon, S. J. 1981. Postovulatory follicles as indicators of egg production in blue grouse. Journal of Wildlife Management 45:1045–1047.

Hannon, S. J., and P. Eason. 1995. Color bands, combs, and coverable badges in willow ptarmigan. Animal Behavior 49:53–62.

Hannon, S. J., P. K. Eason, and K. Martin. 1998. Willow ptarmigan. Account 369 in A. Poole and F. Gill, editors. The birds of North America. The Academy of Natural Sciences, Philadelphia, Pennsylvania, and The American Ornithologists' Union, Washington, D.C., USA.

Hannon, S. J., I. Jonsson, and K. Martin. 1990. Patagial tagging of juvenile willow ptarmigan. Wildlife Society Bulletin 18:116–119.

Hannon, S. J., K. Martin, L. Thomas, and J. Schieck. 1993. Investigator disturbance and clutch predation in willow ptarmigan: methods for evaluating impact. Journal of Field Ornithology 64:575–586.

Hanowski, J. M., and G. J. Niemi. 1995. A comparison of on- and off-road bird counts: do you need to go off road to count birds accurately? Journal of Field Ornithology 66:469–483.

Hansen, A., and J. J. Rotella. 2000. Bird responses to forest fragmentation. Pages 201–219 in R. D. Knight, R. W. Smith, W. H. Romme, and S. W. Buskirk, editors. Forest fragmentation in the southern Rockies. University Press of Colorado, Boulder, USA.

Hansen, C. G. 1964. A dye-spraying device for marking desert bighorn sheep. Journal of Wildlife Management 28:584–587.

Hanski, I., and O. E. Gaggiotti. 2004. Metapopulation biology: past, present, and future. Pages 3–22 in I. Hanksi and O. E. Gaggiotti, editors. Ecology, genetics, and evolution of metapopulations. Elsevier Academic Press, Burlington, Massachusetts, USA.

Hanski, I., and M. E. Gilpin, editors. 1999. Metapopulation biology: ecology, genetics, and evolution. Academic Press, San Diego, California, USA.

Hanson, H. C., and C. W. Kossack. 1963. The mourning dove in Illinois. Technical Bulletin 2, Illinois Department of Conservation and the Illinois Natural History Survey, Southern Illinois University Press, Carbondale, USA.

Hanson, W. R. 1963. Calculation of productivity, survival, and abundance of selected vertebrates from sex and age ratios. Wildlife Monographs 9.

Hansson, B., S. Bensch, D. Hasselquist, and M. Akesson. 2001. Microsatellite diversity predicts recruitment of sibling great reed warblers. Proceedings of the Royal Society of London B 268:1287–1291.

Hansson, L., and P. Angelstam. 1991. Landscape ecology as a theoretical basis for nature conservation. Landscape Ecology 5:191–201.

Haramis, G. M., W. G. Alliston, and M. E. Richmond. 1983. Dump nesting in the wood duck traced by tetracycline. Auk 100:729–730.

Haramis, G. M., E. L. Derleth, and D. G. McAuley. 1987. A quick-catch corral trap for wintering canvasbacks. Journal of Field Ornithology 58:198–200.

Haramis, G. M., and G. D. Kearns. 2000. A radio transmitter attachment technique for soras. Journal of Field Ornithology 71:135–139.

Haramis, G. M., and A. D. Nice. 1980. An improved web-tagging technique for waterfowl. Journal of Wildlife Management 44:898–899.

Harder, J. D. 1980. Reproduction of white-tailed deer in the north central United States. Pages 23–35 in R. L. Hine and S. Nehls, editors. White-tailed deer population management in the north central states. North Central Section, The Wildlife Society, Eau Claire, Wisconsin, USA.

Harder, J. D., and D. W. Fleck. 1997. Reproductive ecology of New World marsupials. Pages 173–201 in L. A. Hinds and N. R. Saunders, editors. Recent advances in marsupial biology. University of New South Wales Press, Sydney, Australia.

Harder, J. D., L. A. Hinds, C. A. Horn, and C. H. Tyndale-Biscoe. 1985. Effects of removal in late pregnancy of the corpus luteum, Graafian follicle or ovaries on plasma progesterone, oestradiol, LH, parturition and post-partum oestrus in the tammar wallaby, *Macropus eugenii*. Journal of Reproduction and Fertility 75:449–459.

Harder, J. D., L. M. Jackson, and D. C. Koester. 2008. Behavioral and reproductive responses of female opossums to volatile and nonvolatile components of male suprasternal gland secretion. Hormones and Behavior 54:741–747.

Harder, J. D., and R. L. Kirkpatrick. 1994. Physiological methods in wildlife research. Pages 275–306 in T. H. Bookhout, editor. Research and management techniques for wildlife and habitats. The Wildlife Society, Bethesda, Maryland, USA.

Harder, J. D., and R. L. Kirkpatrick. 1996. Physiological methods in wildlife research. Pages 275–306 in T. A. Bookhout, editor. Research and management techniques for wildlife and habitats. The Wildlife Society, Bethesda, Maryland, USA.

Harder, J. D., and D. L. Moorhead. 1980. Development of corpora lutea and plasma progesterone levels associated with the onset of the breeding season in white-tailed deer (*Odocoileus virginianus*). Biology of Reproduction 22:185–191.

Harder, J. D., and T. J. Peterle. 1974. Effect of diethylstilbestrol on reproductive performance of white-tailed deer. Journal of Wildlife Management 38:183–196.

Harder, J. D., and A. Woolf. 1976. Changes in plasma levels of oestrone and oestradiol during pregnancy and parturition in white-tailed deer. Journal of Reproduction Fertility 47:161–163.

Hardin, J. B., L. A. Brennan, F. Hernandez, E. J. Redeker, and W. P. Kuvlesky, Jr. 2005. Empirical tests of hunter-covey interface models. Journal of Wildlife Management 69:498–514.

Harding, D. J., and G. S. Berghoff. 2000. Fault scarp detection beneath dense vegetation cover: airborne lidar mapping of the Seattle fault zone, Bainbridge Island, Washington State. Proceedings of the American Society for Photogrammetry and Remote Sensing, Annual Conference, Washington, D.C., USA.

Harding, D. J., M. A. Lefsky, G. G. Parker, and J. B. Blair. 2001. Lidar altimeter measurements of canopy structure: methods and validation for closed-canopy, broadleaf forests. Remote Sensing of Environment 76:283–297.

Hardy, A., A. P. Clevenger, M. Huijser, and N. Graham. 2003. An overview of methods and approaches for evaluating the effectiveness of wildlife crossing structures: emphasizing the science in applied science. Pages 319–330 in C. L. Irwin, P. Garrett, and K. McDermott, editors. Proceedings of the International Conference on Ecology and Transportation. Center for Transportation and the Environment, North Carolina State University, Raleigh, USA.

Hardy, C. M., L. A. Hinds, P. J. Kerr, and M. L. Lloyd. 2006. Biological control of vertebrate pests using virally vectored immunocontraception. Journal of Reproductive Immunology 71:102–111.

Harlow, R. F. 1977. A technique for surveying deer forage in the Southeast. Wildlife Society Bulletin 5:185–191.

Harlow, R. F., B. A. Sanders, J. B. Whelan, and L. C. Chappel. 1980. Deer habitat on the Ocala National Forest: improvement through forage management. Southern Journal of Applied Forestry 4:98–102.

Harman, J. A., C. X. Mao, and J. G. Morse. 2007. Selection of colour of sticky trap for monitoring adult bean thrips, *Caliothrips fasciatus* (Thysanoptera: Thripidae). Pest Management Science 63:210–216.

Harmata, A. R. 2002. Encounters of golden eagles banded in the Rocky Mountain West. Journal of Field Ornithology 73:23–32.

Harmata, A. R., M. Restani, G. J. Montopoli, J. R. Zelenak, J. T. Ensign, and P. J. Harmata. 2001. Movements and mortality of ferruginous hawks banded in Montana. Journal of Field Ornithology 72:389–398.

Harmoney, K. R., K. J. Moore, J. R. George, E. C. Brummer, and J. R. Russell. 1997. Determination of pasture biomass using four indirect methods. Agronomy Journal 89:665–672.

Harper, J. A., and W. C. Lightfoot. 1966. Tagging devices for Roosevelt elk and mule deer. Journal of Wildlife Management 30:461–466.

Harper, S. J., and G. O. Batzli. 1996. Monitoring use of runways by voles with passive integrated transponders. Journal of Mammalogy 77:364–369.

Harper, W. L., and R. D. H. Cohen. 1985. Accuracy of Doppler ultrasound in diagnosing pregnancy in bighorn sheep. Journal of Wildlife Management 49:793–796.

Harrell, F. E., Jr. 2001. Regression modeling strategies. Springer Science+Business Media, New York, New York, USA.

Harris, R. B., and F. W. Allendorf. 1989. Genetically effective population size of large mammals: an assessment of estimators. Conservation Biology 3:181–191.

Harris, R. B., W. A. Wall, and F. W. Allendorf. 2002. Genetic consequences of hunting: what do we know and what should we do? Wildlife Society Bulletin. 30:634–643.

Harris, S. 1978. Age determination in the red fox (*Vulpes vulpes*): an evaluation of technique efficiency as applied to a sample of suburban foxes. Journal of the Zoological Society l84: 94–117.

Harris, S., W. J. Cresswell, P. G. Forde, W. J. Trewhella, T. Woolard, and S. Wray. 1990. Home-range analysis using radio-tracking data—a review of problems and techniques particularly as applied to the study of mammals. Mammal Review 20:97–123.

Harris, S. W. 1952. A throw net for capturing waterfowl on the nest. Journal of Wildlife Management 16:515.

Harris, S. W., and M. A. Morse. 1958. The use of mist nets for capturing nesting mourning doves. Journal of Wildlife Management 22: 306–309.

Harrison, M. K., Sr., G. M. Haramis, D. G. Jorde, and D. B. Stotts. 2000. Capturing American black ducks in tidal waters. Journal of Field Ornithology 71:153–158.

Harrison, R. L. 1997. Chemical attractants for Central American felids. Wildlife Society Bulletin 25:93–97.

Harrison, R. L. 2006. A comparison of survey methods for detecting bobcats. Wildlife Society Bulletin 34:548–552.

Harrison, R. M., and D. E. Wildt. 1980. Animal laparoscopy. Williams and Wilkins, Baltimore, Maryland, USA.

Harrison, S., and A. D. Taylor. 1997. Empirical evidence for metapopulation dynamics. Pages 27–42 in I. Hanski and D. Simberloff, editors. Metapopulation biology: ecology, genetics, and evolution. Academic Press, San Diego, California, USA.

Harshyne, W. A., D. R. Diefenbach, G. L. Alt, and G. M. Matson. 1998. Analysis of error from cementum-annuli age estimates of known-age Pennsylvania black bears. Journal of Wildlife Management 62:1281–1291.

Hart, A., and A. D. M. Hart. 1987. Patagial tags for herring gulls: improved durability. Journal of Ringing and Migration 8:19–26.

Hartl, G. B, F. Klein, R. Willing, M. Apollonio, and G. Lang. 1995. Allozymes and the genetics of antler development in red deer (*Cervus elaphus*). Journal of Zoology 237:83–100.

Hartline, P. H., L. Kass, and M. S. Loop. 1978. Merging modalities in the optic tectum: infrared and visual interaction in rattlesnakes. Science 199:1225–1229.

Hartman, G. 1992. Age determination of live beaver by dental x-ray. Wildlife Society Bulletin 20:216–220.

Hash, H. S. 1987. Wolverine. Pages 575–585 in M. Novak, J. A. Bament, M. E. Obbard, and B. Malloch, editors. Wild furbearer management and conservation in North America. Ontario Ministry of Natural Resources, Toronto, Canada.

Haskell, S. P., W. B. Ballard, D. A. Butler, N. M. Tatman, M. C. Wallace, and C. O. Kochanny. 2007. Observations on capturing and aging deer fawns. Journal of Mammalogy 88:1482–1487.

Hatch, J. J., and I. C. T. Nisbet. 1983a. Band wear and band loss in common terns. Journal of Field Ornithology 54:1–16.

Hatch, J. J., and I. C. T. Nisbet. 1983b. Band wear in arctic terns. Journal of Field Ornithology, 54:91.

Hatch, S. A., P. M. Meyers, D. M. Mulcahy, and D. C. Douglas. 2000. Seasonal movements and pelagic habitat use of murres and puffins determined by satellite telemetry. Condor 102:145–154.

Hatfield, J. S., P. F. R. Henry, G. H. Olsen, M. M. Paul, and R. S. Hammerschlag. 2001. Failure of tetracycline as a biomarker in batch-marking juvenile frogs. Journal of Wildlife Diseases 37:318–323.

Hau, M., M. Wikelski, and J. C. Wingfield. 2000. Visual and nutritional food cues fine-tune timing of reproduction in a neotropical rainforest bird. Journal of Experimental Zoology 286:494–504.

Haugen, A. O. 1975. Reproductive performance of white-tailed deer in Iowa. Journal of Mammalogy 56:151–159.

Haugen, A. O., and F. W. Fitch, Jr. 1955. Seasonal availability of certain bush lespedeza and partridge pea seed as determined from ground samples. Journal of Wildlife Management 19:297–301.

Hauger, R. L., F. J. Karsch, and D. L. Foster. 1977. A new concept of the control of the estrous cycle of the ewe based on the temporal relationships between luteinizing hormone, estradiol and progesterone in peripheral serum and evidence that progesterone inhibits tonic LH secretion. Endocrinology 101:807–817.

Haugerud, R. A. 2009. Discussion on remote sensing for aquatic monitoring. Pages 93–100 in J. M. Bayer and J. L. Schei, editors. Remote sensing applications for aquatic resource monitoring. Pacific Northwest Aquatic Monitoring Partnership, Cook, Washington, USA.

Haughland, D. L., and K. W. Larsen. 2004. Ecology of red squirrels across contrasting habitats: relating natal dispersal to habitat. Journal of Mammalogy 85:225–236.

Haukos, D. A., L. M. Smith, and G. S. Broda. 1990. Spring trapping of lesser prairie-chickens. Journal of Field Ornithology 61:20–25.

Haukos, D. A., H. Z. Sun, D. B. Webster, and L. M. Smith. 1998. Sample size, power, and analytical considerations for vertical structure data from profile boards in wetland vegetation. Wetlands 18:203–215.

Haulton, S. M., W. F. Porter, and B. A. Rudolph. 2001. Evaluating 4 methods to capture white-tailed deer. Wildlife Society Bulletin 29:255–264.

Havelka, P. 1983. Registration and marking of captive birds of prey. International Zoological Yearbook 23:125–132.

Hawking, J. H., L. M. Smith, K. Le Busque. 2009. Identification and ecology of Australian freshwater invertebrates. http://www.mdfrc.org.au/bugguide.

Hawkins, L. L., and S. G. Simpson. 1985. Neckband—a handicap in an aggressive encounter between tundra swans. Journal of Field Ornithology 56:182–184.

Hawkins, R. E., D. C. Autry, and W. D. Klimstra. 1967. Comparison of methods used to capture white-tailed deer. Journal of Wildlife Management 31:460–464.

Hawkins, R. E., L. D. Martoglio, and G. G. Montgomery. 1968. Cannon-netting deer. Journal of Wildlife Management 32:191–195.

Hawley, A. W. L. 1982. A simple device for sectioning ovaries. Journal of Wildlife Management 46:247–249.

Hayek, L. C., and M. A. Buzas. 1997. Surveying natural populations. Columbia University Press, New York, New York, USA.

Hayes, C. J., T. E. DeGomez, K. M. Clancy, K. K. Williams, J. D. McMillin, and J. A. Anhold. 2008. Evaluation of funnel traps for characterizing the bark beetle (Coleoptera: Scolytidae) communities in ponderosa pine forests of north-central Arizona. Journal of Economic Entomology 101:1253–1265.

Hayes, J. P., E. G. Horvath, and P. Hounihan. 1994. Securing live traps to small-diameter trees for studies of arboreal mammals. Northwestern Naturalist 75:31–33.

Hayes, R. W. 1982. A telemetry device to monitor big game traps. Journal of Wildlife Management 46:551–555.

Hayes, T. B., A. Collins, M. Lee, M. Mendoza, N. Noriega, A. A. Stuart, and A. Vonk. 2002. Hermaphroditic, demasculinized frogs after exposure to the herbicide atrazine at low ecologically relevant doses. Proceeding of the National Academy of Sciences of the United States of America 99:5476–5480.

Hayes, T. B., P. Falso, S. Gallipeau, and M. Stice. 2010. The cause of global amphibian declines: a developmental endocrinologist's perspective. Journal of Experimental Biology 213:921–933.

Hayes, T. B., H. Haston, M. Tsui, A. Hoang, C. Haeffele, and A. Vonk. 2003. Atrazine-induced hermaphroditism at 0.1 ppb in American leopard frogs (*Rana pipiens*): laboratory and field evidence. Environmental Health Perspectives 111:568–575.

Hayne, D. W. 1949a. An examination of the strip census method for estimating animal populations. Journal of Wildlife Management 13:145–157.

Hayne, D. W. 1949b. Calculation of size of home range. Journal of Mammalogy 30:1–18.

Hays, G. C., G. J. Marshall, and J. A. Seminoff. 2007. Flipper beat frequency and amplitude changes in diving green turtles, *Chelonia mydas*. Marine Biology 150:1003–1009.

Hays, H., and M. LeCroy. 1971. Field criteria for determining incubation stage in eggs of the common tern. Wilson Bulletin 83:425–429.

Hays, R. L., C. Summers, and W. L. Seitz. 1981. Estimating wildlife habitat variables. FWS/OBS-81/47, U.S. Fish and Wildlife Service, Department of the Interior, Washington, D.C., USA.

Hays, W. S. T. 1998. A new method for live-trapping shrews. Acta Theriologica 43:333–335.

Hayward, G. D. 1987. Betalights: an aid in the nocturnal study of owl foraging habitat and behavior. Journal of Raptor Research 21:98–102.

Hayward, J. L., Jr. 1982. A simple egg-marking technique. Journal of Field Ornithology 53:173.

Hazelton, M. L., and J. C. Marshall. 2009. Linear boundary kernels for bivariate density estimation. Statistics and Probability Letters 79:999–1003.

Hazler, K. R. 2004. Mayfield logistic regression: a practical approach for analysis of nest survival. Auk 121:707–716.

Heady, H. F., R. P. Gibbens, and R. W. Powell. 1959. A comparison of the charting, line intercept, and line point methods of sampling shrub types of vegetation. Journal of Range Management 12:180–188.

Healey, S. P., Y. Zhiqiang, W. B. Cohen, and J. Pierce. 2006. Application of two regression-based methods to estimate the effects of partial harvest on forest structure using Landsat data. Remote Sensing of Environment 101:115–126.

Healy, W. M., R. O. Kimmel, D. A. Holdermann, and W. Hunyadi. 1980. Attracting ruffed grouse broods with tape-recorded chick calls. Wildlife Society Bulletin 8:69–71.

Healy, W. M., and E. S. Nenno. 1980. Growth parameters and sex and age criteria for juvenile eastern wild turkeys. Proceedings of the National Wild Turkey Symposium 4:168–185.

Healy, W. R. 1974. Population consequences of alternative life histories in *Notophthalmus v. viridescens*. Copeia 1974:221–229.

Healy, W. R. 1975. Terrestrial activity and home range in efts of *Notophthalmus viridescens*. American Midland Naturalist 93:131–138.

Hearn, B. J., and W. E. Mercer. 1988. Eye-lens weight as an indicator of age in Newfoundland arctic hares. Wildlife Society Bulletin 16:426–429.

Heath, J. A., and P. C. Frederick. 2003. Trapping white ibises with rocket nets and mist nets in the Florida Everglades. Journal of Field Ornithology 74:187–192.

Heatwole, H. 1961. Inhibition of digital regeneration in salamanders and its use in marking individuals for field studies. Ecology 42:593–594.

Hebblewhite, M., C. A. White, C. G. Nietvelt, J. A. McKenzie, T. E. Hurd, J. M. Fryxell, S. E. Bayley, and P. C. Paquet. 2005. Human activity mediates a trophic cascade caused by wolves. Ecology 86:2135–2144.

Hedrick, P. W. 1995. Gene flow and genetic restoration: the Florida panther as a case study. Conservation Biology 9:996–1007.

Hedrick, P. W. 2000. Genetics of populations. Jones and Bartlett, Boston, Massachusetts, USA.

Hedrick, P. W., P. F. Brussard, F. W. Allendorf, J. A. Beardmore, and S. Orzack. 1986. Protein variation, fitness, and captive propagation. Zoo Biology 5:91–99.

Hedrick, P. W., and S. T. Kalinowski. 2000. Inbreeding depression in conservation biology. Annual Review of Ecology and Systematics 31:139–162.

Hedrick, P. W., and P. S. Miller. 1992. Conservation genetics: techniques and fundamentals. Ecological Applications 2:30–46.

Heezen, K. L., and J. R. Tester. 1967. Evaluation of radio-tracking by triangulation with special reference to deer movements. Journal of Wildlife Management 31:124–141.

Heffner, R. S., and H. E. Heffner. 1982. Hearing in the elephant: absolute sensitivity frequency discrimination, and sound localization. Journal of Comparative Physiology 96:926–944.

Heilbrun, R. D., N. J. Silvy, M. J. Peterson, and M. E. Tewes. 2006. Estimating bobcat abundance using automatically triggered cameras. Wildlife Society Bulletin 34:69–73.

Heilbrun, R. D., N. J. Silvy, M. E. Tewes, and M. J. Peterson. 2003. Using automatically triggered cameras to individually identify bobcats. Wildlife Society Bulletin 31:748–755.

Heimer, W. E., S. D. DuBois, and D. G. Kellyhouse. 1980. The usefulness of rocket nets for Dall sheep capture compared with other capture methods. Proceedings of the Biennial Symposium of the Northern Wild Sheep and Goat Council 2:601–613.

Hein, E. W., and W. F. Andelt. 1995. Estimating coyote density from mark–resight surveys. Journal of Wildlife Management 59:164–169.

Heinen, J., and G. H. Cross. 1983. An approach to measure interspersion, juxtaposition, and spatial diversity from cover-type maps. Wildlife Society Bulletin 11:232–237.

Heinrich, B. 1988. Winter foraging at carcasses by three sympatric corvids, with emphasis on recruitment by the raven, *Corvus corax*. Behavioral Ecology and Sociobiology 23:141–156.

Heinz, G. H. 1996. Mercury poisoning in wildlife. Pages 118–127 in A. Fairbrother, L. N. Locke, and G. L. Huff, editors. Noninfectious diseases of wildlife. Second edition. Iowa State University Press, Ames, USA.

Heisey, D. M. 1985. Analyzing selection experiments with log-linear models. Ecology 66:1744–1748.

Heisey, D. M., and T. K. Fuller. 1985. Evaluations of survival and cause-specific mortality rates using telemetry data. Journal of Wildlife Management 49:668–674.

Heisey, D. M., and B. R. Patterson. 2006. A review of methods to estimate cause-specific mortality in presence of competing risks. Journal of Wildlife Management 70:1544–1555.

Heithaus, M. R., and L. M. Dill. 2002. Food availability and tiger shark predation risk influence bottlenose dolphin habitat use. Ecology 83:480–491.

Heithaus, M. R., J. J. McLash, A. Frid, L. M. Dill, and G. J. Marshall. 2002. Novel insights into green sea turtle behaviour using animal-borne video cameras. Journal of Marine Biology Association of U.K. 82:1049–1050.

Helldin, J.-O. 1997. Age determination of Eurasian pine martens by radiographs of teeth in situ. Wildlife Society Bulletin 25:83–88.

Hellmann, J. J., and G. W. Fowler. 1999. Bias, precision, and accuracy of four measures of species richness. Ecological Applications 9:824–834.

Helm, L. G. 1955. Plastic collars for marking geese. Journal of Wildlife Management 19:316–317.

Heltshe, J. F., and N. E. Forrester. 1983. Estimating species richness using the jackknife procedure. Biometrics 39:1–11.

Hemming, J. E. 1969. Cemental deposition, tooth succession, and horn development as criteria of age in Dall sheep. Journal of Wildlife Management 33:552–558.

Hemson, G., P. Johnson, A. South, R. Kenward, R. Ripley, and D. MacDonald. 2005. Are kernels the mustard? Data from global positioning system (GPS) collars suggests problems for kernel home-range analysis with least-squares cross-validation. Journal of Animal Ecology 74:455–463.

Hepinstall, J. A., and S. A. Sader. 1997. Using Bayesian statistics, thematic mapper, and breeding bird survey data to model bird species probability of occurrence in Maine. Photogrammetric Engineering and Remote Sensing 63:1231–1237.

Henckel, R. E. 1976. Turkey vulture banding problem. North American Bird Bander 1:126.

Henderson, F. R., F. W. Brooks, R. E. Wood, and R. B. Dahlgren. 1967. Sexing of prairie grouse by crown feather patterns. Journal of Wildlife Management 31:764–769.

Henderson, J. R., and T. C. Johanos. 1988. Effects of tagging on weaned Hawaiian monk seal pups. Wildlife Society Bulletin 16:312–217.

Henderson, R. W. 1974. Aspects of the ecology of the juvenile common iguana (*Iguana iguana*). Herpetologica 30:327–332.

Hendrickson, J. R. 1954. Ecology and systematics of salamanders of the genus *Batrochoseps*. University of California Publications in Zoology 54:1–46.

Henke, S. E., and S. Demarais. 1990. Capturing jackrabbits by drive corral on grasslands in west Texas. Wildlife Society Bulletin 18:31–33.

Henley, G. B. 1981. A new technique for recognition of snakes. Herpetological Review 12:56.

Henny, C. J., and K. P. Burnham. 1976. A reward band study of mallards to estimate band reporting rates. Journal of Wildlife Management 40:1–14.

Henny, C. J., R. J. Hallock, and E. F. Hill. 1994. Cyanide and migratory birds at gold mines in Nevada, USA. Ecotoxicology 3:45–58.

Henny, C. J., E. F. Hill, D. J. Hoffman, M. G. Spalding, and R. A. Grove. 2002. Nineteenth century mercury: hazard to wading birds and cormorants of the Carson River, Nevada. Ecotoxicology 11:213–231.

Henry, D. B., and T. A. Bookhout. 1969. Productivity of beavers in northeastern Ohio. Journal of Wildlife Management 33:927–932.

Henry, J. B., editor. 1979. Clinical diagnosis and management by laboratory methods. Sixteenth edition. W. B. Saunders, Philadelphia, Pennsylvania, USA.

Hensel, R. J., and F. E. Sorensen, Jr. 1980. Age determination of live polar bears. International Conference on Bear Resource Management 4:93–100.

Hensel, R. J., W. A. Troyer, and A. W. Erickson. 1969. Reproduction in the female brown bear. Journal of Wildlife Management 33:357–365.

Hepp, G. R., and F. C. Bellrose. 1995. Wood duck (*Aix sponsa*). Account 169 *in* A. Poole and F. Gill, editors. The birds of North America. The Academy of Natural Sciences, Philadelphia, and The American Ornithologists' Union, Washington, D.C., USA.

Hepp, G. R., T. H. Folk, and K. M. Hartke. 2002. Effects of subcutaneous transmitters on reproduction, incubation behavior, and annual return rates of female wood ducks. Wildlife Society Bulletin 30:1208–1214.

Hepp, G. R., R. T. Hoppe, and R. A. Kennamer. 1987. Population parameters and philopatry of breeding female wood ducks. Journal of Wildlife Management 51:401–404.

Herbers, J., and W. Klenner. 2007. Effects of logging pattern and intensity on squirrel demography. Journal of Wildlife Management 71:2655–2663.

Herman, C. M. 1963. Disease and infection in the Tetraonidae. Journal of Wildlife Management 27:850–855.

Herman, C. M. 1969. The impact of disease on wildlife populations. BioScience 19:321–325, 330.

Herman-Brunson, K. H. 2007. Nesting and brood-rearing habitat selection of greater sage-grouse and associated survival of hens and broods at the edge of their historic distribution. Thesis, South Dakota State University, Brookings, USA.

Hernandez, F., L. A. Harveson, and C. E. Brewer. 2006. A comparison of trapping techniques for Montezuma quail. Wildlife Society Bulletin 34:1212–1215.

Hernandez, F., D. Rollins, and R. Cantu. 1997. An evaluation of Trailmaster® camera systems for identifying ground-nest predators. Wildlife Society Bulletin 25:848–853.

Hero, J. 1989. A simple code for toe clipping anurans. Herpetological Review 20:66–67.

Herreid, C. F., II, and S. Kinney. 1966. Survival of Alaskan wood frog (*Rana sylvatica*) larvae. Ecology 47:1039–1041.

Herrick, J. R., G. Agoramoorthy, R. Rudran, and J. D. Harder. 2000. Urinary progesterone in free-ranging red howler monkeys (*Alouatta seniculus*): preliminary observations of the estrous cycle and gestation. American Journal of Primatology 51:257–263.

Herrick J. R., M. Campbell, G. Levens, T. Moore, K. Benson, J. D'Agostino, G. West, D. M. Okeson, R. Coke, S. C. Portacio, K. Leiske, C. Kreider, P. J. Polumbo, and W. F. Swanson. 2010. In vitro fertilization and sperm cryopreservation in the black-footed cat (*Felis nigripes*) and sand cat (*Felis margarita*). Biology of Reproduction 82:552–562.

Herriges, J. D., Jr., E. T. Thorne, and S. L. Anderson. 1991. Vaccination to control brucellosis in free-ranging elk on western Wyoming feed grounds. Pages 107–112 *in* R. D. Brown, editor. The biology of deer. Springer-Verlag, New York, New York, USA.

Herriges, J. D., Jr., E. T. Thorne, S. L. Anderson, and H. A. Dawson. 1989. Vaccination of elk in Wyoming with reduced dose of strain 19 *Brucella*: controlled studies and ballistic implant field trials. Proceedings of the Annual Meeting of the United States Animal Health Association 93:640–655.

Herring, G., D. E. Gawlik, and J. M. Beerens. 2008. Evaluating two new methods for capturing large wetland birds. Journal of Field Ornithology 79:102–110.

Herzog, P. W. 1979. Effects of radio-marking on behavior, movements, and survival of spruce grouse. Journal of Wildlife Management 43:316–323.

Herzog, P. W., and D. A. Boag. 1978. Dispersion and mobility in a local population of spruce grouse. Journal of Wildlife Management 42:853–865.

Heske, E. J. 1987. Responses of a population of California voles, *Microtus californicus*, to odor-baited traps. Journal of Mammalogy 68:64–72.

Hess, C. A., P. P. Kelly, R. Costa, and J. H. Carter, III. 2001. Reconsideration of Richardson et al.'s red-cockaded woodpecker nestling removal technique. Wildlife Society Bulletin 29:372–374.

Hess, D. L., A. M. Schmidt, and M. J. Schmidt. 1983. Reproductive cycle of the Asian elephant (*Elephus maximus*) in captivity. Biology of Reproduction 28:767–773.

Hesselton, W. T., and R. M. Hesselton. 1982. White-tailed deer. Pages 878–901 *in* J. A. Chapman and G. A. Feldhamer, editors. Wild mammals of North America. Johns Hopkins University Press, Baltimore, Maryland, USA.

Hestbeck, J. B., J. D. Nichols, and R. A. Malecki. 1991. Estimates of movement and site fidelity using mark–resight data of wintering Canada geese. Ecology 72:523–533.

Hester, A. E. 1963. A plastic wing tag for individual identification of passerine birds. Bird-Banding 34:213–217.

Heusmann, H. W., R. G. Burrell, and R. Bellville. 1978. Automatic short-term color marker for nesting wood ducks. Journal of Wildlife Management 42:429–432.

Heussner, J. C., A. I. Flowers, J. D. Williams, and N. J. Silvy. 1978. Estimating dog and cat populations in an urban area. Animal Regulation Studies 1:203–212.

Hewitt, O. H., and P. J. Austin-Smith. 1966. A simple wing tag for field-marking birds. Journal of Wildlife Management 30:625–627.

Hewson, R. 1961. Collars for marking mountain hares. Journal of Wildlife Management 25:329–331.

Hey, J. 2001. The mind of the species problem. Trends in Ecology and Evolution 16:326–329.

Heydon, C., M. Novak, and H. Rowsell. 1993. Humane attributes of three types of legholding traps. Report. Ontario Ministry of Natural Resources, Toronto, Canada.

Heyer, W. R., M. A. Donnelly, R. W. McDiarmid, L. C. Hayek, and M. S. Foster, editors. 1994. Measuring and monitoring biological diversity. Standard methods for amphibians. Smithsonian Institution Press, Washington, D.C., USA.

Heyland, J. D. 1970. Aircraft-supported Canada goose banding operations in arctic Quebec.

Transactions of the Northeast Section of the Wildlife Society 27:187–198.

Hicks, L. E., and D. L. Leedy. 1939. Techniques for pheasant trapping and population control. Transactions of the North American Wildlife Conference 4:449–461.

Higgins, K. F., and W. T. Barker. 1982. Changes in vegetation structure in seeded nesting cover in the prairie pothole region. Special Scientific Report 242, U.S. Fish and Wildlife Service, Department of the Interior, Washington, D.C., USA.

Hilborn, R., and M. Mangel. 1997. The ecological detective: confronting models with data. Monographs in population biology 28. Princeton University Press, Princeton, New Jersey, USA.

Hildebrandt, T. B., F. Goritz, and R. Hermes. 2006. Ultrasonography: an important tool in captive breeding management in elephants and rhinoceroses. European Journal of Wildlife Research 52:23–27.

Hill, D. A. 1985. The feeding ecology and survival of pheasant chicks on arable farmland. Journal of Applied Ecology 22:645–654.

Hill, E. F. 1999. Wildlife toxicology. Pages 1327–1363 in B. Ballantyne, T. C. Marrs, and T. Syversen, editors. General and applied toxicology. Volume 2. Second edition. MacMillan, London, England, UK.

Hill, E. F. 2003. Wildlife toxicology of organophosphorus and carbamate pesticides. Pages 281–312 in D. J. Hoffman, B. A. Rattner, G. A. Burton, Jr., and J. Cairns, Jr., editors. Handbook of ecotoxicology. Second edition. Lewis, Boca Raton, Florida, USA.

Hill, E. F., and W. J. Fleming. 1982. Anticholinesterase poisoning of birds: field monitoring and diagnosis of acute poisoning. Environmental Toxicology and Chemistry 1:27–38.

Hill, E. P. 1981. Evaluation of improved traps and trapping techniques. Job Progress Report, Federal Aid Project W-44-6, Job IV-B, Alabama Department of Conservation and Natural Resources, Montgomery, USA.

Hill, L. A., and L. G. Talent. 1990. Effects of capture, handling, banding, and radio-marking on breeding least terns and snowy plovers. Journal of Field Ornithology 61:310–319.

Hill, M., H. Witsenboer, M. Zabeau, P. Vos, R. Kesseli, and R. Michelmore. 1996. PCR-based fingerprinting using AFLPs as a tool for studying genetic relationships in Lactuca spp. Theoretical and Applied Genetics 93:1202–1210.

Hill, M. O., and H. G. Gauch. 1980. Detrended correspondence analysis: an improved ordination technique. Vegetation 42:47–58.

Hill, P. S. M. 2001. Vibration and animal communication: a review. American Zoologist 41:1135–1142.

Hill, R. A., S. A. Hinsley, D. L. A. Gaveau, and P. E. Bellamy. 2004. Predicting habitat quality for great tits (Parus major) with airborne laser scanning data. International Journal of Remote Sensing 25:4851–4855.

Hilldale, R. C., J. A. Bountry, and L. A. Piety. 2009. Using bathymetric and bare earth LiDAR in riparian corridors: applications and challenges. Pages 27–34 in J. M. Bayer and J. L. Schei, editors. PNAMP special publication: remote sensing applications for aquatic resource monitoring, Pacific Northwest Aquatic Monitoring Partnership, Cook, Washington, USA.

Hillis, D. M., B. K. Mable, A. Larson, S. K. Davis, and E. A. Zimmer. 1996. Nucleic acids IV: sequencing and cloning. Pages 321–384 in D. M. Hillis, C. Moritz, and B. K. Mable, editors. Molecular systematics. Sinauer Associates, Sunderland, Massachusetts, USA.

Hillis, R. E., and E. D. Bellis. 1971. Some aspects of the ecology of the hellbender, Cryptobranchus alleganiensis alleganiensis. Journal of Herpetology 5:121–126.

Hillman, C. N., and W. W. Jackson. 1973. The sharp-tailed grouse in South Dakota. Technical Bulletin 3, South Dakota Department of Game, Fish and Parks, Pierre, USA.

Hilmon, J. B. 1959. Determination of herbage weight by double-sampling: weight estimate and actual weight. Pages 20–25 in Technique and methods of measuring understory vegetation. Southern and Southeast Forest Experiment Stations, U.S. Forest Service, Department of Agriculture, New Orleans, Louisiana, USA.

Hilton, B., Jr. 1989. Two methods for capturing tree-nesting birds at nests. North American Bird Bander 14:47–48.

Hilty, J., and A. Merenlender. 2000. Faunal indictor taxa selection for monitoring ecosystem health. Biological Conservation 92:185–197.

Hines, J. E. 1986. Social organization, movements, and home ranges of blue grouse in fall and winter. Wilson Bulletin 98:419–432.

Hines, R. K., and T. W. Custer. 1995. Evaluation of an extendable pole-net to collect heron eggs in the canopy of tall trees. Colonial Waterbirds 18:120–122.

Hinsley, S. A., R. A. Hill, D. L. A. Gaveau, and P. E. Bellamy. 2002. Quantifying woodland structure and habitat quality for birds using airborne laser scanning. Functional Ecology 16:851–857.

Hipfner, J. M., and J. L. Greenwood. 2008. Breeding biology of the common murre at Triangle Island, British Columbia, Canada, 2002–2007. Northwestern Naturalist 89:76–84.

Hirst, S. M. 1975. Ungulate–habitat relationships in a South African woodland/savanna ecosystem. Wildlife Monographs 44.

Hirth, H. F. 1966. Weight changes and mortality of three species of snakes during hibernation. Herpetologica 22:8–12.

Hobbs, L., and P. Russell. 1979. Report on the pinniped tagging workshop. American Institute of Biological Sciences, Arlington, Virginia, USA.

Hobbs, N. T., D. L. Baker, J. E. Ellis, D. M. Swift, and R. A. Green. 1982. Energy-and nitrogen-based estimates of elk winter-range carrying capacity. Journal of Wildlife Management 46:12–21.

Hobbs, R. J., and C. D. James. 1999. Influence of shade covers on pitfall trap temperatures and capture success of reptiles and small mammals in arid Australia. Wildlife Research 26:341–349.

Hobbs, R. J., S. R. Morton, P. Masters, and K. R. Jones. 1994. Influence of pit-trap design on sampling of reptiles in arid spinifex grassland. Wildlife Research 21:483–490.

Hobbs, S. E., and W. W. Wolf. 1996. Developments in airborne entomological radar. Journal of Atmospheric and Oceanic Technology 13:58–61.

Hobbs, T. N., and R. Hilborn. 2006. Alternatives to statistical hypothesis testing in ecology: a guide to self teaching. Ecological Applications 16:5–19.

Hochberg, R., and M. K. Litvaitis. 2000. Hexamethyldisilazane for scanning electron microscopy of Gastrotricha. Biotechnic and Histochemistry 75:41–44.

Hochstein, J. R., R. J. Aulerich, and S. J. Bursian. 1988. Acute toxicity of 2,3,7,8-tetrachlorodibenzo-p-dioxin to mink. Archives of Environmental Contamination and Toxicology 17:33–37.

Hockin, D., M. Ounsted, M. Gorman, D. Hill, V. Kellar, and M. A. Barker. 1992. Examination of the effects of disturbance on birds with reference to its importance in ecological assessments. Journal of Environmental Management 36:253–286.

Hodgdon, K. W., and J. H. Hunt. 1953. Beaver management in Maine. Game Division Bulletin 3, Maine Department of Inland Fisheries and Game, Augusta, USA.

Hoefer, A. M., B. A. Goodman, and S. J. Downs. 2003. Two effective and inexpensive methods for restraining small lizards. Herpetological Review 34:223–224.

Hoenig, J. M., and D. M. Heisey. 2001. The abuse of power: the pervasive fallacy of power calculations for data analysis. American Statistician 55:19–24.

Hof, J. G., M. Bevers, L. Joyce, and B. Kent. 1994. An integer programming approach for spatially and temporally optimizing wildlife populations. Forest Science 40:177–191.

Hof, J. G., and L. A. Joyce. 1992. Spatial optimization for wildlife and timber in managed forest ecosystems. Forest Science 38:489–508.

Hof, J. G., and L. A. Joyce. 1993. A mixed integer linear programming approach for spatially optimizing wildlife and timber in managed forest ecosystems. Forest Science 39:816–834.

Hof, J. G., and M. G. Raphael. 1997. Optimization of habitat placement: a case study of the northern spotted owl in the Olympic Peninsula. Ecological Applications 7:1160–1169.

Hoffman, D. J., B. A. Rattner, G. A. Burton, Jr., and J. Cairns, Jr., editors. 2003. Handbook of ecotoxicology. Second edition. Lewis, Boca Raton, Florida, USA.

Hoffman, R. H. 1985. An evaluation of banding sandhill cranes with colored leg bands. North American Bird Bander 10:46–49.

Hoffman, R. W. 1983. Sex classification of juvenile blue grouse from wing characteristics. Journal of Wildlife Management 47:1143–1147.

Hoffman, R. W. 1985. Blue grouse wing analysis: methodology and population inferences. Special Report 60, Colorado Division of Wildlife, Fort Collins, USA.

Hogan, G. G. 1985. Noosing adult cormorants for banding. North American Bird Bander 10:76–77.

Höglund, J., S. B. Piertney, R. V. Alatalo, J. Lindell, A. Lundberg, and P. T. Rintamäki. 2002. Inbreeding depression and male fitness in black grouse. Proceedings of the Royal Society of London B 269:711–715.

Hoglund, N. H. 1968. A method of trapping and marking willow grouse in winter. Viltrevy 5:95–101.

Hohf, R. S., J. T. Ratti, and R. Croteau. 1987. Experimental analysis of winter food selection by spruce grouse. Journal of Wildlife Management 51:159–167.

Holder, K., and R. Montgomerie. 1993. Rock ptarmigan. Account 51 in A. Poole and F. Gill, editors. The birds of North America. The Academy of Natural Sciences, Philadelphia, Pennsylvania, and The American Ornithologists' Union, Washington, D.C., USA.

Hollamby, S., J. Afema-Azikure, W. W. Bowerman, K. N. Cameron, A. Dranzoa, A. R. Gandolf, G. N. Hui, J. B. Kaneene, A. Norris, J. G. Sikarskie, S. D. Fitzgerald, and W. K. Rumbeiha. 2004. Methods for capturing African fish eagles on water. Wildlife Society Bulletin 32:680–684.

Holland, R. A., M. Wikelski, F. Kümmeth, and C. Bosque. 2009. The secret life of oilbirds: new insights into the movement ecology of a unique avian frugivore. PLoS ONE 4:e8264.

Holler, N. R., and E. W. Schafer, Jr. 1982. Potential secondary hazards of Avitrol baits to sharp-shinned hawks and American kestrels. Journal of Wildlife Management 46:457–462.

Holliman, D. C. 1977. Purple gallinule. Pages 105–109 in G. C. Sanderson, editor. Management of migratory shore and upland game birds in North America. International Association of Fish and Wildlife Agencies, Washington, D.C., USA.

Holling, C. S. 1978. Adaptive environmental assessment and management. John Wiley and Sons, Chichester, England, UK.

Holloran, M. J., and S. H. Anderson. 2005. Spatial distribution of greater sage-grouse nests in relatively contiguous sagebrush habitats. Condor 107:742–752.

Hölzenbein, S. 1992. Expandable PVC collar for marking and transmitter support. Journal of Wildlife Management 56:473–476.

Homan, H. J., G. Linz, and D. Peer. 2001. Dogs increase recovery of passerine carcasses in dense vegetation. Wildlife Society Bulletin 29:292–296.

Homestead, R., B. Beck, and D. E. Sergeant. 1972. A portable, instantaneous branding device for permanent identification of wildlife. Journal of Wildlife Management 36:947–949.

Ionma, M., S. Iwaki, A. Kast, and H. Kreuzer. 1986. Experiences with the identification of small rodents. Experimental Animal 35:347–352.

Hooper, J. H. D. 1983. The study of horseshoe bats in Devon caves: a review of progress 1947–1982. Studies in Speleology 4:59–70.

Hooper, M. J., P. J. Detrich, C. P. Weisskopf, and B. W. Wilson. 1989. Organophosphorus insecticide exposure in hawks inhabiting orchards during winter dormant spraying. Bulletin of Environmental Contamination and Toxicology 42:651–659.

Hooper, R. G. 1988. Longleaf pines used for cavities by red-cockaded woodpeckers. Journal of Wildlife Management 52:392–398.

Hooven, E. 1958. Deer mouse and reforestation in the Tillamook burn. Research Note 37, Oregon Forest Lands Research Center, Corvallis, USA.

Hopkinson, C., L. E. Chasmer, G. Sass, I. F. Creed, M. Sitar, W. Kalbfleisch, and P. Treitz. 2005. Vegetation class dependent errors in lidar ground elevation and canopy height estimates in a boreal wetland environment. Canadian Journal of Remote Sensing 31:191–206.

Horn, E. E., and H. S. Fitch. 1946. Trapping the California ground squirrel. Journal of Mammalogy 27:220–224.

Horn, J. W., E. B. Arnett, and T. H. Kunz. 2008. Behavioral responses of bats to operating wind turbines. Journal of Wildlife Management 72:123–132.

Horn, J. W., and T. H. Kunz. 2008. Analyzing NEXRAD Doppler radar images to assess nightly dispersal patterns and population trends in Brazilian free-tailed bats (*Tadarida brasiliensis*). Integrative and Comparative Biology 48:24–39.

Horn, S., and J. L. Hanula. 2006. Burlap bands as a sampling technique for green anoles (*Anolis carolinensis*), and other reptiles commonly found on treeboles. Herpetological Review 37:427–428.

Horne, J. S., and E. O. Garton. 2006a. Likelihood cross-validation versus least squares cross-validation for choosing the smoothing parameter in kernel home-range analysis. Journal of Wildlife Management 70:641–648.

Horne, J. S., and E. O. Garton. 2006b. Selecting the best home range model: an information-theoretic approach. Ecology 87:1146–1152.

Horne, J. S., E. O. Garton, S. M. Krone, and J. S. Lewis. 2007. Analyzing animal movements using Brownian bridges. Ecology 88:2354–2363.

Hornocker, M. G. 1970. An analysis of mountain lion predation upon mule deer and elk in the Idaho primitive area. Wildlife Monographs 21.

Hosmer, D. W., Jr., and S. Lemeshow. 1989. Applied logistic regression. John Wiley and Sons, New York, New York, USA.

Hosmer, D. W., Jr., and S. Lemeshow. 1999. Applied survival analysis. John Wiley and Sons, New York, New York, USA.

Hosmer, D. W., Jr., and S. Lemeshow. 2000. Applied logistic regression. John Wiley and Sons, New York, New York, USA.

Hossler, R. J., J. B. McAninch, and J. D. Harder. 1994. Maternal denning behavior and juvenile survival of opossums in southeastern New York. Journal of Mammalogy 75:60–70.

Houben, J. M., M. Holland, S. W. Jack, and C. R. Boyle. 1993. An evaluation of laminated offset jawed traps for reducing injuries to coyotes. Proceedings, Great Plains Wildlife Damage Control Workshop 11:148–155.

Houlden, B. A., B. H. Costello, D. Sharkey, E. V. Fowler, A. Melzer, W. Ellis, F. Carrick, P. R. Baverstock, and M. S. Elphinstone. 1999. Phylogeographic differentiation in the mitochondrial control region in the koala *Phascolarctos cinereus*. Molecular Ecology 8:999–1011.

Houston, C. S. 1999. Dispersal of great horned owls banded in Saskatchewan and Alberta. Journal of Field Ornithology 70:342–350.

Houston, D. B. 1982. The northern Yellowstone elk: ecology and management. Macmillan, New York, New York, USA.

Houston, D. B., C. T. Robbins, C. A. Ruder, and R. G. Sasser. 1986. Pregnancy detection in mountain goats by assay for pregnancy-specific protein B. Journal of Wildlife Management 50:740–742.

Houze, C. M., Jr., and C. R. Chandler. 2002. Evaluation of coverboards for sampling terrestrial salamanders in south Georgia. Journal of Herpetology 36:75–81.

Hovorka, M. D., C. S. Marks, and E. Muller. 1996. An improved chemiluminescent tag for bats. Wildlife Society Bulletin 24:709–712.

Howard, R. D., and J. R. Young. 1998. Individual variation in male vocal traits and female mating preferences in *Bufo americanus*. Animal Behaviour 55:1165–1179.

Howard, W. E. 1952. A live trap for pocket gophers. Journal of Mammalogy 33:61–65.

Howe, M. A. 1980. Problems with wing tags: evidence of harm to willets. Journal of Field Ornithology 51:72–73.

Howlin, S., W. P. Erickson, and R. M. Nielson. 2003. A proposed validation technique for assessing predictive abilities of resource selection functions. Pages 40–51 in S. V. Huzurbazar, editor. Resource selection methods and applications. Western EcoSystems Technology, Cheyenne, Wyoming, USA (http://www.west-inc.com).

Hoyle, S. D., D. Peel, J. R. Ovenden, D. Broderick, and R. C. Buckworth. 2005. LOCUSEATER and SHADOWBOXER: programs to optimize experimental design and multiplexing strategies for genetic mark–recapture. Molecular Ecology Notes 5:974–976.

Huang, C., S. N. Goward, J. G. Masek, N. Thomas, Z. Zhu, and J. E. Vogelmann. 2010. An automated approach for reconstructing recent forest disturbance history using dense Landsat time series stacks. Remote Sensing of Environment 114:183–198.

Huber, H. R. 1994. A technique for determining sex of northern fur seal pup carcasses. Wildlife Society Bulletin 22:479–483.

Huber, S., U. Bruns, and W. Arnold. 2002. Sex determination of red deer using polymerase chain reaction of DNA from feces. Wildlife Society Bulletin 30:208–212.

Hubert, G. F., Jr., R. D. Bluett, and G. A. Dumonceaux. 1991. Field evaluation of two footholding devices for capturing raccoons in non-drowning water sets. Pages 23–24 in L. Fredrickson and B. Coonrod, editors. Proceedings of the ninth midwest furbearer workshop. South Dakota Department of Game, Fish and Parks, Pierre, USA.

Hubert, G. F., Jr., L. L. Hungerford, and R. D. Bluett. 1997. Injuries to coyotes captured in modified foothold traps. Wildlife Society Bulletin 25:858–863.

Hubert, G. F., Jr., L. L. Hungerford, G. Proulx, R. D. Bluett, and L. Bowman. 1996. Evaluation of two restraining traps to capture raccoons. Wildlife Society Bulletin 24:699–708.

Hubert, G. F., Jr., G. L. Storm, R. L. Phillips, and R. D. Andrews. 1976. Ear tag loss in red foxes. Journal of Wildlife Management 40:164–167.

Hubert, G. F., Jr., K. Wollenberg, L. L. Hungerford, and R. D. Bluett. 1999. Evaluation of injuries to Virginia opossums captured in the EGG™ trap. Wildlife Society Bulletin 27:301–305.

Hudnall, J. A. 1982. New methods for measuring and tagging snakes. Herpetological Review 13:97–98.

Hudson, P., and L. G. Browman. 1959. Embryonic and fetal development of the mule deer. Journal of Wildlife Management 23:295–304.

Hudson, P. J., A. Rizzoli, B. T. Grenfell, H. Heesterbeek, and A. P. Dobson, editors. 2002. The ecology of wildlife diseases. Oxford University Press, Oxford, England, UK.

Huempfner, R. A., S. J. Maxson, G. J. Erickson, and R. J. Shuster. 1975. Recapturing radio-tagged ruffed grouse by nightlighting and snow-burrow netting. Journal of Wildlife Management 39:821–823.

Huenneke, L. F., and C. Graham. 1987. A new sticky trap for monitoring seed rain in grasslands. Journal of Range Management 40:370–372.

Huey, R. B., A. E. Dunham, K. L. Overall, and R. A. Newman. 1990. Variation in locomotor performance in demographically known populations of the lizard *Scleoporus merriami*. Physiological Zoology 63:845–872.

Huey, W. S. 1965. Sight records of color-marked sandhill cranes. Auk 82:640–643.

Huggins, J. G. 1999. Gray and fox squirrel trapping: a review. Pages 117–129 in G. Proulx, editor. Mammal trapping. Alpha Wildlife Research and Management, Sherwood Park, Alberta, Canada.

Huggins, J. G., and K. L. Gee. 1995. Efficiency and selectivity of cage trap sets for gray and fox squirrels. Wildlife Society Bulletin 23:204–207.

Huggins, R. M. 1989. On the statistical analysis of capture–recapture experiments. Biometrika 76:133–140.

Huggins, R. M. 1991. Some practical aspects of a conditional likelihood approach to capture experiments. Biometrics 47:725–732.

Hughes, A. L. 1991. MHC polymorphism and the design of captive breeding programs. Conservation Biology 5:249–251.

Hughes, A. L., and M. Yeager. 1998. Natural selection at major histocompatibility complex loci of vertebrates. Annual Review of Genetics 32:415–435.

Hugh-Jones, M. E., and V. de Vos. 2002. Anthrax and wildlife. Revue Scientifique et Technique de L'Office International des Epizooties 21:359–383.

Humason, G. L. 1979. Animal tissue techniques. Third edition. W. H. Freeman and Company, San Francisco, California, USA.

Hungerford H., P. J. Spangler, and N. A. Walker. 1955. Subaquatic light traps for insects and other animal organisms. Transactions of the Kansas Academy of Science 58:387–407.

Hunt, F. R. 1973. Bird density and the plan position indicator. Publication 63, Committee on Bird Hazards to Aircraft, National Research Council of Canada, Ottawa, Ontario, Canada.

Hunt, G. S., and K. J. Dahlka. 1953. Live trapping of diving ducks. Journal of Wildlife Management 17:92–95.

Hunter, L. M., J. Beal, and T. Dickinson. 2003. Integrating demographic and GAP analysis biodiversity data: useful insight? Human Dimensions of Wildlife 8:145–157.

Hunter, M. L., Jr. 1989. Aardvarks and Arcadia: two principles of wildlife research. Wildlife Society Bulletin 17:350–351.

Hupp, J. W., and C. E. Braun. 1991. Geographic variation among sage grouse in Colorado. Wilson Bulletin 103:255–261.

Hupp, J. W., G. A. Ruhl, J. M. Pearce, D. M. Mulcahy, and M. A. Tomeo. 2003. Effects of implanted radio transmitters with percutaneous antennas on the behavior of Canada geese. Journal of Field Ornithology 74:250–256.

Hurlbert, S. H. 1971. The non-concept of species diversity: a critique and alternative parameters. Ecology 52:577–586.

Hurlbert, S. H. 1984. Pseudoreplication and the design of ecological field experiments. Ecological Monograph 54:187–211.

Hurst, J. L. 1988. A system for the individual recognition of small rodents at a distance, used in free-living and enclosed populations of house mice. Journal of Zoology 215:363–367.

Hurt, A., and D. A. Smith. 2009. Conservation dogs. Pages 175–194 in W. S. Helton, editor. Canine ergonomics: the science of working dogs. CRC Press, Taylor and Francis Group, Boca Raton, Florida, USA.

Hurteau, S. R., T. D. Sisk, W. M. Block, and B. G. Dickson. 2008. Fuel-reduction treatment effects on avian community structure and diversity. Journal of Wildlife Management 72:1168–1174.

Hutchinson, G. E. 1978. An introduction to population ecology. Yale University Press, New Haven, Connecticut, USA.

Huwer, S. L., D. R. Anderson, T. E. Remington, and G. C. White. 2008. Using human-imprinted chicks to evaluate the importance of forbs to sage-grouse. Journal of Wildlife Management 72:1622–1627.

Hwang, Y. T., S. Larivière, and F. Messier. 2007. Local-and landscape-level den selection of striped skunks on the Canadian prairies. Canadian Journal of Zoology 85:33–39.

Hyde, E. J., and T. R. Simons. 2001. Sampling plethodontid salamanders: sources of variability. Journal of Wildlife Management 65:624–632.

Hyde, P., R. Dubayah, B. Peterson, J. B. Blair, M. Hofton, C. Hunsaker, R. Knox, and W. Walker. 2005. Mapping forest structure for wildlife habitat analysis using waveform lidar: validation of montane ecosystems. Remote Sensing of Environment 96:427–437.

Hyde, P., R. Dubayah, W. Walker, J. B. Blair, M. Hofton, and C. Hunsaker. 2006. Mapping forest structure for wildlife habitat analysis using multi-sensor (LiDAR, SAR/InSAR, ETM+, Quickbird) synergy. Remote Sensing of Environment 102:63–73.

Hyder, D. N., R. E. Bement, E. E. Remmenga, and C. Terwilliger, Jr. 1965. Frequency sampling of blue grama range. Journal of Range Management 18:90–94.

Hyder, D. N., and F. A. Sneva. 1960. Bitterlich's plotless method for sampling basal ground cover of bunchgrasses. Journal of Range Management 13:6–9.

Idstrom, J. M., and J. P. Lindmeier. 1956. Some tests of the rubber styrene neck bands for marking waterfowl. Quarterly Program Report of Wildlife Research, Minnesota Department of Conservation 16:134–137.

Innes, R. J., D. H. Vanvuren, and D. A. Kelt. 2008. Characteristics and use of tree houses by dusky-footed woodrats in the northern Sierra Nevada. Northwestern Naturalist 89:109–112.

Inns, R. W. 1982. Seasonal changes in the accessory reproductive system and plasma testosterone levels of the male tammar wallaby, *Macropus eugenii*, in the wild. Journal of Reproduction and Fertility 66:675–680.

International Association of Fish and Wildlife Agencies. 1992. Ownership and use of traps by trappers in the United States in 1992. The Fur Resources Committee of The International Association of Fish and Wildlife Agencies, and The Gallup Organization, Washington, D.C., USA.

International Association of Fish and Wildlife Agencies. 1997. Improving animal welfare in U.S. trapping programs: process recommendations and summaries of existing data. Fur Resources Technical Subcommittee, International Association of Fish and Wildlife Agencies, Washington, D.C., USA.

International Association of Fish and Wildlife Agencies. 2000. Testing restraining traps for the development of best management practices for trapping in the United States. Furbearer Resources Technical Work Group, International Association of Fish and Wildlife Agencies, Washington, D.C. USA.

International Association of Fish and Wildlife Agencies. 2003. Best management practices for trapping coyotes in the eastern United States. International Association of Fish and Wildlife Agencies, Washington, D.C., USA.

International Organization for Standardization. 1999a. Animal (mammal) traps—Part 4. Methods for testing killing-trap systems used on land or underwater. International Standard ISO 10990-4, International Organization for Standardization, Geneva, Switzerland.

International Organization for Standardization. 1999b. Animal (mammal) traps—Part 5. Methods for testing restraining traps. International Standard ISO 10990-5, International Organization for Standardization, Geneva, Switzerland.

Ireland, P. H. 1973. Marking larval salamanders with fluorescent pigments. Southwestern Naturalist 18:252–253.

Ireland, P. H. 1991. A simplified fluorescent marking technique for identification of terrestrial salamanders. Herpetological Review 22:21–22.

Irish, J. L., and W. J. Lillycrop. 1999. Scanning laser mapping of the coastal zone: the SHOALS system. ISPRS Journal of Photogrammetry and Remote Sensing 54:123–129.

Irvine, A. B., and M. D. Scott. 1984. Development and use of marking techniques to study manatees in Florida. Florida Scientist 47:12–26.

Irvine, A. B., R. S. Wells, and M. D. Scott. 1982. An evaluation of techniques for tagging small odontocete cetaceans. Fisheries Bulletin 80: 135–143.

Irwin, K. C., R. T. Podoll, R. I. Starr, and D. J. Elias. 1996. The mobility of [^{14}C] 3-chloro-p-toluidine hydrochloride in a loam soil profile. Environmental Toxicology and Chemistry 15:1671–1675.

Irwin, L. L., and J. M. Peek. 1979. Shrub production and biomass trends following five logging treatments within the cedar-hemlock zone of northern Idaho. Forest Science 25:415–426.

Itoh, M., M. Inoue, and S. Ishii. 1990. Annual cycle of pituitary and plasma gonadotropins and plasma sex steroids in a wild population of the toad, *Bufo japonicus*. General and Comparative Endocrinology 78:242–253.

Iverson, J. B. 1979. Another inexpensive turtle trap. Herpetological Review 10:55.

Jackman, R. E., W. G. Hunt, D. E. Driscoll, and J. M. Jenkins. 1993. A modified floating-fish snare for capture of inland bald eagles. North American Bird Bander 18:98–101.

Jackman, R. E., W. G. Hunt, D. E. Driscoll, and F. J. Lapsansky. 1994. Refinements to selective trapping techniques: a radio-controlled bow net and power snare for bald and golden eagles. Journal of Raptor Research 28:268–273.

Jackson, C. H. N. 1939. The analysis of an animal population. Journal of Animal Ecology 8:238–246.

Jackson, D. H., L. S. Jackson, and W. K. Seitz. 1985. An expandable drop-off transmitter harness for young bobcats. Journal of Wildlife Management 49:46–49.

Jackson, J. A., and S. D. Parris. 1991. A simple, effective net for capturing cavity roosting birds. North American Bird Bander 16:30–31.

Jackson, J. J. 1982. Effect of wing tags on renesting interval in red-winged blackbirds. Journal of Wildlife Management 46:1077–1079.

Jackson, R., G. Ahlborn, and K. B. Shah. 1990. Capture and immobilization of wild snow leopards. International Pedigree Book of Snow Leopards 6:93–102.

Jackson, R., J. Roe, R. Wangchuk, and D. Hunter. 2005. Camera-trapping of snow leopards. Cat News 42:19–21.

Jackson, R. M., J. D. Roe, R. Wangchuk, and O. H. Don. 2006. Estimating snow leopard population abundance using photography and capture–recapture techniques. Wildlife Society Bulletin 34:772–781.

Jackson, S. 2000. Overview of transportation impacts on wildlife movement and populations. Pages 7–20 in T. A. Messmer and B. West, editors. Wildlife and highways: seeking solutions to an ecological and socio-economic dilemma. The Wildlife Society, Bethesda, Maryland, USA.

Jacobs, E. A. 1996. A mechanical owl as a trapping lure for raptors. Journal of Raptor Research 30:31–32.

Jacobs, E. A., and G. A. Proudfoot. 2002. An elevated net assembly to capture nesting raptors. Journal of Raptor Research 36:320–323.

Jacobs, K. F. 1958. A drop-net trapping technique for greater prairie chickens. Proceedings of the Oklahoma Academy of Science 38:154–157.

Jacobsen, N., K. D. A. Jessup, and D. J. Kesler. 1995. Contraception in captive black-tailed deer by remotely delivered norgestomet ballistic implants. Wildlife Society Bulletin 23:718–722.

Jacobson, H. A., H. J. Bearden, and D. B. Whitehouse. 1989. Artificial insemination trials with white-tailed deer. Journal of Wildlife Management 54:224–227.

Jacobson, H. A., J. C. Kroll, R. W. Browning, B. H. Koerth, and M. H. Conway. 1997. Infrared-triggered cameras for censusing white-tailed deer. Wildlife Society Bulletin 25:547–556.

Jacobson, H. A., and R. J. Reiner. 1989. Estimating age of white-tailed deer: tooth wear versus cementum annuli. Proceedings of the Annual Conference of the Southeastern Association of Fish and Wildlife Agencies 43:286–291.

Jaeger, J., J. Wehausen, and V. Bleich. 1991. Evaluation of time-lapse photography to estimate population parameters. Desert Bighorn Council Transactions 35:5–8.

Jaenike, J. 1980. A relativistic measure of variation in preference. Ecology 61:990–991.

Jaiswal, R. S., J. Singh, and G. P. Adams. 2009. High-resolution ultrasound biomicroscopy for monitoring ovarian structures in mice. Reproductive Biology and Endocrinology 7:69. http://www.rbej.com/content/7/1/69.

James, F. C., and C. E. McCulloch. 1985. Data analysis and the design of experiments in ornithology. Pages 1–63 in R. F. Johnston, editor. Current ornithology. Volume 2. Plenum Press, New York, New York, USA.

James, F. C., and H. H. Shugart, Jr. 1970. A quantitative method of habitat description. Audubon Field Notes 24:727–736.

Jameson, D. L. 1957. Population structure and homing responses in the Pacific tree frog. Copeia 1957:221–228.

Jamison, B. 2002. Wild hog trapping. Wildlife Control Technology 9:26–27, 37.

Janis, M. W., J. D. Clark, and C. S. Johnson. 1999. Predicting mountain-lion activity using radio-collars equipped with mercury tip-sensors. Wildlife Society Bulletin 27:19–24.

Jankowski, P. 2009. Towards participatory geographic information systems for community-based environmental decision making. Journal of Environmental Management 90:1966–1971.

Janovsky, M., T. Ruf, and W. Zenker. 2002. Oral administration of tiletamine/zolazepam for immobilization of the common buzzard (*Buteo buteo*). Journal of Raptor Research 36: 188–193.

Janssen, D. L., J. E. Oosterhuis, J. L. Allen, M. P. Anderson, D. G. Kelts, and S. N. Wiemeyer. 1986. Lead poisoning in free-ranging California condors. Journal of the American Veterinary Medical Association 189:1115–1117.

Janssen, L. L. F., and F. J. M. van der Wel. 1994. Accuracy assessment of satellite derived land-cover data: a review. Photogrammetric Engineering and Remote Sensing 60:419–426.

Jaquiery, J., F. Guillaume, and N. Perrin. 2009. Predicting the deleterious effects of mutation load in fragmented populations. Conservation Biology 23:207–218.

Jedrzejewski, W., and J. F. Kamler. 2004. Modified drop-nets for capturing ungulates. Wildlife Society Bulletin 32:1305–1308.

Jeffcoate, S. L. 1981. Efficiency and effectiveness in the endocrine laboratory. Academic Press, New York, New York, USA.

Jefferson, R. T., Jr., and W. L. Franklin. 1986. Behavioral considerations in the live capture of guanacos with spring-activated foot snares. Proceedings of the Iowa Academy of Science 93:48–50.

Jeffreys, A. J., N. J. Royle, V. Wilson, and Z. Wong. 1988. Spontaneous mutation rates to new length alleles at tandem-repetitive hypervariable loci in human DNA. Nature 332:278–281.

Jehle, G., A. A. Yackel Adams, J. A. Savidge, and S. K. Skagen. 2004. Nest survival estimation: a review of alternatives to the Mayfield estimator. Condor 106:472–484.

Jemison, S. C., L. A. Bishop, P. G. May, and T. M. Farrell. 1995. The impact of PIT-tags on growth and movement of the rattlesnake, *Sistrurus miliarius*. Journal of Herpetology 29:129–132.

Jenkins, C. L., and K. McGarigal. 2003. Comparative effectiveness of two techniques for surveying the abundance and diversity of reptiles and amphibians along drift fence arrays. Herpetological Review 34:39–42.

Jenkins, C. L., K. McGarigal, and L. R. Gamble. 2002. A comparison of aquatic surveying techniques used to sample *Ambystoma opacum* larvae. Herpetological Review 33:33–35.

Jenkins, D., A. Watson, and G. R. Miller. 1963. Population studies on red grouse, *Lagopus lagopus scoticus* (Lath.) in north-east Scotland. Journal of Animal Ecology 32:317–376.

Jenkins, J. H., and R. K. Marchington. 1969. Problems in censusing the white-tailed deer. Pages 115–118 in L. K. Halls, editor. White-tailed deer in the southeastern forest habitat: proceedings of a symposium. Southern Forestry Experiment Station, U.S. Forest Service, New Orleans, Louisiana, USA.

Jenkins, M. A. 1979. Tips on constructing monofilament nylon nooses for raptor traps. North American Bird Bander 4:108–109.

Jenkins, S. H. 1988. Use and abuse of demographic models of population growth. Bulletin of the Ecological Society of America 69:201–207.

Jenkins, S. H. 1989. Comments on an inappropriate population model for feral burros. Journal of Mammalogy 70:667–670.

Jenks, J. A., R. T. Bowyer, and A. G. Clark. 1984. Sex and age-class determination for fisher using radiographs of canine teeth. Journal of Wildlife Management 48:626–628.

Jenks, J. A., R. T. Bowyer, and A. G. Clark. 1986. Sex and age determination for fisher using radiographs of canine teeth: a response. Journal of Wildlife Management 50:277–278.

Jennelle, C. S., M. D. Samuel, C. A. Nolden, and E. A. Berkley. 2009. Deer carcass decomposition and potential scavenger exposure to chronic wasting disease. Journal of Wildlife Management 73:655–662.

Jenness, J. S. 2004. Calculating landscape surface area from digital elevation models. Wildlife Society Bulletin 32:829–839.

Jenni, L., M. Leunberger, and F. Rampazzi. 1996. Capture efficiency of mist nets with comments on their role in the assessment of passerine habitat use. Journal of Field Ornithology 67:263–274.

Jennings, M. L., D. N. David, and K. M. Portier. 1991. Effect of marking techniques on growth and survivorship of hatchling alligators. Wildlife Society Bulletin 19:204–207.

Jennrich, R. I., and F. B. Turner. 1969. Measurement of a non-circular home range. Journal of Theoretical Biology 22:227–237.

Jensen, C. H., and G. W. Scotter. 1977. A comparison of twig-length and browsed-twig methods of determining browse utilization. Journal of Range Management 30:64–67.

Jensen, C. H., and P. J. Urness. 1981. Establishing browse utilization from twig diameters. Journal of Range Management 34:113–116.

Jensen, W. 1998. Aging antelope—it's all in the teeth. North Dakota Outdoors 61:16–20.

Jenssen, T. A. 1970. The ethoecology of *Anolis nebulosus* (Sauria, Iguanidae). Journal of Herpetology 4:1–38.

Jepson, P. D., M. Arbelo, R. Deaville, I. A. P. Patterson, P. Castro, J. R. Baker, E. Degollada, H. M. Ross, P. Herráez, A. M. Pocknell, F. Rodríguez, F. E. Howie, A. Espinosa, R. J. Reid, J. R. Jaber, V. Martin, A. A. Cunningham, and A. Fernández. 2003. Was sonar responsible for a spate of whale deaths after an Atlantic military exercise? Nature 425:575–576.

Jessup, D. A., R. K. Clark, R. A. Weaver, and M. D. Kock. 1988. The safety and cost-effectiveness of net-gun capture of desert bighorn sheep (*Ovis canadensis nelsoni*). Journal of Zoo Animal Medicine 19:208–213.

Jessup, D. A., W. E. Clark, and R. C. Mohr. 1984. Capture of bighorn sheep: management recommendations. Administrative Report 24-1, Wildlife Management Branch, California Department of Fish and Game, Sacramento, USA.

Jessup, D. A., and F. A. Leighton. 1996. Oil pollution and petroleum toxicity to wildlife. Pages 141–156 in A. Fairbrother, L. N. Locke, and G. L. Huff, editors. Noninfectious diseases of wildlife. Second edition. Iowa State University Press, Ames, USA.

Jetz, W., C. Carbone, J. Fulford, and J. H. Brown. 2004. The scaling of animal space use. Science 306:266–268.

Jewell, S. D., and J. T. Bancroft. 1991. Effects of nest trapping on nesting success of *Egretta* herons. Journal of Field Ornithology 62:78–82.

Ji, W., and P. Leberg. 2002. A GIS-based approach for assessing the regional conservation status of genetic diversity: an example from the southern Appalachians. Environmental Management 29:531–544.

Jiménez, J. A., K. A. Hughes, G. Alaks, L. Graham, and R. C. Lacy. 1994. An experimental study of inbreeding depression in a natural habitat. Science 266:271–273.

Jobes, A. P., E. Nol, and D. R. Voigt. 2004. Effects of selection cutting on bird communities in contiguous eastern hardwood forests. Journal of Wildlife Management 68:51–60.

Johanningsmeier, A. G., and C. J. Goodnight. 1962. Use of iodine-131 to measure movements of small animals. Science 138:147–148.

Johns, B. E., and H. P. Pan. 1981. Analytical techniques for fluorescent chemicals used as systemic or external wildlife markers. American Society for Testing Materials, Vertebrate Pest Control Management Materials 3:86–93.

Johns, B. E., and R. D. Thompson. 1979. Acute toxicant identification in whole bodies and baits without chemical analysis. Pages 80–88 in E. E. Kenaga, editor. Avian and mammalian wildlife toxicology. Special Technical Publication 693, American Society for Testing Materials, Washington, D.C., USA.

Johnsgard, P. A. 1973. Grouse and quails of North America. University of Nebraska Press, Lincoln, USA.

Johnsgard, P. A. 1975. North American game birds of upland and shoreline. University of Nebraska Press, Lincoln, USA.

Johnsgard, P. A. 1983. The grouse of the world. University of Nebraska Press, Lincoln, USA.

Johnson, A. M. 1979a. Factors contributing to difficulties in the analysis of mark–recapture data. Pages 27–29 in L. Hobbs and P. Russell, editors. Report on the pinniped tagging workshop. Seattle, Washington, USA.

Johnson, A. S. 1970. Biology of the raccoon (*Procyon lotor varius* Nelson and Goldman) in Alabama. Bulletin 402, Alabama Agricultural Experiment Station, Auburn University, Auburn, USA.

Johnson, A. S., and J. L. Landers. 1978. Fruit production in slash pine plantations in Georgia. Journal of Wildlife Management 42:606–613.

Johnson B. K., T. McCoy, C. O. Kochanny, and R. C. Cook. 2006. Evaluation of vaginal implant transmitters in elk (*Cervus elaphus nelsoni*). Journal of Zoo and Wildlife Medicine 37:301–305.

Johnson, C. J., and M. P. Gillingham. 2004. Mapping uncertainty: sensitivity of wildlife habitat ratings to expert opinion. Journal of Applied Ecology 41:1032–1041.

Johnson, C. L., and R. T. Reynolds. 1998. A new trap design for capturing spotted owls. Journal of Raptor Research 32:181–182.

Johnson, D. H. 1974. Estimating survival rates from banding of adult and juvenile birds. Journal of Wildlife Management 38:290–297.

Johnson, D. H. 1979b. Estimating nest success: the Mayfield method and an alternative. Auk 96:651–661.

Johnson, D. H. 1980. The comparison of usage and availability measurements for evaluating resource preference. Ecology 61:65–71.

Johnson, D. H. 1995. Statistical sirens: the allure of nonparametrics. Ecology 76:1998–2000.

Johnson, D. H. 1999. The insignificance of statistical significance testing. Journal of Wildlife Management 63:763–772.

Johnson, D. H. 2001a. Validating and evaluating models. Pages 105–119 in T. M. Shenk and A. B. Franklin, editors. Modeling in natural resource management: development, interpretation, and application. Island Press, Washington, D.C., USA.

Johnson, D. H. 2002. The importance of replication in wildlife research. Journal of Wildlife Management 66:919–932.

Johnson, D. H., J. D. Nichols, and M. D. Schwartz. 1992. Population dynamics of breeding waterfowl. Pages 446–485 in B. D. J. Batt, A. D. Afton, M. G. Anderson, C. D. Ankeny, D. H. Johnson, J. A. Kadlec, and G. L. Krapu, editors. Ecology and management of breeding waterfowl. University of Minnesota Press, Minneapolis, USA.

Johnson, D. H., and T. L. Shaffer. 1990. Estimating nest success: when Mayfield wins. Auk 107:595–600.

Johnson, D. H., D. W. Sparling, and L. M. Cowardin. 1987. A model of the productivity of the mallard duck. Ecological Modelling 38:257–275.

Johnson, F. A., W. L. Kendall, and J. A. Dobovsky. 2002. Conditions and limitations on learning in the adaptive management of mallard harvests. Wildlife Society Bulletin 3:176–185.

Johnson, F. A., and B. K. Williams. 1999. Protocol and practice in the adaptive management of waterfowl harvests. Conservation Ecology 3:8. http://www.ecologyandsociety.org/vol3/iss1/art8/.

Johnson, G. D., and M. S. Boyce. 1990. Feeding trials with insects in the diet of sage grouse chicks. Journal of Wildlife Management 54:89–91.

Johnson, J. A., and P. O. Dunn. 2006. Low genetic variation in the heath hen prior to extinction and implications for the conservation of prairie-chicken populations. Conservation Genetics 7:37–48.

Johnson, J. B., and K. S. Omland. 2004. Model selection in ecology and evolution. Trends in Ecology and Evolution 19:101–108.

Johnson, J. J., L. A. Windberg, C. A. Furcolow, R. M. Engeman, and M. Roetto. 1998. Chlorinated benzenes as physiological markers for coyotes. Journal of Wildlife Management 62:410–421.

Johnson, J. R. 2005. A novel arboreal pipe-trap designed to capture the gray treefrog (*Hyla versicolor*). Herpetological Review 36:274–277.

Johnson, K. G., and M. R. Pelton. 1980a. Marking techniques for black bears. Proceedings of the Annual Conference of the Southeastern Association of Fish and Wildlife Agencies 34:557–562.

Johnson, K. G., and M. R. Pelton. 1980b. Prebaiting and snaring techniques for black bears. Wildlife Society Bulletin 8:46–54.

Johnson, N. F., B. A. Brown, and J. C. Bosomworth. 1981a. Age and sex characteristics of bobcat canines and their use in population assessment. Wildlife Society Bulletin 9:203–206.

Johnson, P. A. 2004. Avian reproduction. Pages 742–764 in W. O. Reece, editor. Dukes' physiology of domestic animals. Comstock and Cornell University Press, Ithaca, New York, USA.

Johnson, P. A., B. W. Johnson, and L. T. Taylor. 1981b. Interisland movement of a young Hawaiian monk seal between Laysan Island and Maro Reef. Elepaio 41:113–114.

Johnson, P. S., C. L. Johnson, and N. E. West. 1988. Estimation of phytomass for ungrazed crested wheatgrass plants using allometric equations. Journal of Range Management 41:421–425.

Johnson, R. A., and D. W. Wichern. 1988. Applied multivariate statistical analysis, Second edition. Prentice-Hall, Englewood Cliffs, New Jersey, USA.

Johnson, S. A., and W. J. Barichivich. 2004. A simple technique for trapping *Siren lacertina*, *Amphiuma means*, and other aquatic vertebrates. Journal of Freshwater Ecology 19:263–268.

Johnson, S. R., J. O. Schieck, and G. F. Searing. 1995. Neck band loss rates for lesser snow geese. Journal of Wildlife Management 59:747–752.

Johnson, W. C. 2001b. A new individual marking technique: positional hair clipping. Southwestern Naturalist 46:126–129.

Johnston, A. 1957. A comparison of the line interception, vertical point quadrat, and loop methods as used in measuring basal area of grassland vegetation. Canadian Journal of Plant Science 37:34–42.

Johnston, D. H., D. G. Joachim, P. Bachmann, K. V. Kardong, R. A. Stewart, L. M. Dix, M. A. Strickland, and I. D. Watt. 1987. Aging furbearers using tooth structure and biomarkers. Pages 228–243 in M. Novak, J. A. Baker, M. E. Obbard, and B. Malloch, editors. Wild furbearer management and conservation in North America. Ontario Ministry of Natural Resources, Toronto, Canada.

Johnston, J. J., D. B. Hurlbut, M. L. Avery, and J. C. Rhyan. 1999. Methods for the diagnosis of acute 3-chloro-p-toluidine hydrochloride poisoning in birds and the estimation of secondary hazards to wildlife. Environmental Toxicology and Chemistry 18:2533–2537.

Johnston, J. M. 1999. Canine detection capabilities: operational implications of recent R & D findings. Institute for Biological Detection Systems, Auburn University, Auburn, Alabama, USA.

Johnston, R. E. 1998. Pheromones, the vomeronasal system, and communication: from hormonal responses to individual recognition. Annals of the NewYork Academy of Sciences 855:333–348.

Johnston, R. E. 2003. Chemical communication in rodents: from pheromones to individual recognition. Journal of Mammalogy 84:1141–1162.

Johnstone-Yellin, T. L., L. A. Shipley, and W. L. Meyers. 2006. Effectiveness of vaginal implant transmitters for locating neonatal mule deer fawns. Wildlife Society Bulletin 34:338–344.

Joint Airborne LiDAR Bathymetry Technical Center of Expertise (JALBTCX). 2009. Compact hydrographic airborne rapid total survey (CHARTS). http://shoals.sam.usace.army.mil/.

Jojola, S. M., S. J. Robinson, and K. C. VerCauteren. 2007. Oral rabies vaccine (ORV) bait uptake by captive striped skunks. Journal of Wildlife Diseases 43:97–106.

Jolly, G. M. 1965. Explicit estimates from capture–recapture data with both death and immigration-stochastic models. Biometrika 52:225–247.

Jolly, G. M. 1969a. Sampling methods for aerial censuses of wildlife populations. East African Agriculture and Forestry Journal 34:46–49.

Jolly, G. M. 1969b. The treatment of errors in aerial counts of wildlife populations. East African Agriculture and Forestry Journal 34:50–56.

Jones, C., W. J. McShea, M. J. Conroy, and T. H. Kunz. 1996. Capturing mammals. Pages 115–155 in D. E. Wilson, F. R. Cole, J. D. Nichols, R. Rudran, and M. S. Foster, editors. Measuring and monitoring biological diversity. Standard methods for mammals. Smithsonian Institution Press, Washington, D.C., USA.

Jones, C. D., and J. A. Cox. 2007. Field procedure for netting Bachman's sparrows. North American Bird Bander 32:114–117.

Jones, D., and L. Hayes-Odum. 1994. A method for the restraint and transport of crocodilians. Herpetological Review 25:14–15.

Jones, F. L., G. Flittner, and R. Gard. 1954. Report on a survey of bighorn sheep and other game in the Santa Rosa Mountains, Riverside County (California). California Department of Fish and Game, Sacramento, USA.

Jones, G. F. 1950. Observations of color-dyed pheasants. Journal of Wildlife Management 14:81–82.

Jones, G. P. J., IV. 2003. The feasibility of using small unmanned aerial vehicles for wildlife research. Thesis, University of Florida, Gainesville, USA.

Jones, I. L., S. Rowe, S. M. Carr, G. Fraser, and P. Taylor. 2002a. Different patterns of parental effort during chick-rearing by female and male thick-billed murres (*Uria lomvia*) at a low-arctic colony. Auk 119:1064–1074.

Jones, J. L. 2006. Side channel mapping and fish habitat suitability analysis using lidar topography and orthophotography. Photogrammetric Engineering and Remote Sensing 72:1202–1206.

Jones, K. E., N. G. Patel, M. A. Levy, A. Storeygard, D. Balk, J. L. Gittleman, and P. Daszak. 2008. Global trends in emerging infectious diseases. Nature 451:990–993.

Jones, K. L., T. C. Glenn, R. C. Lacy, J. R. Pierce, N. Unruh, C. M. Mirande, and F. Chavez-Ramirez. 2002b. Refining the whooping crane studbook by incorporating microsatellite DNA and leg-banding analysis. Conservation Biology 16:789–799.

Jones, L. V., and J. W. Tukey. 2000. A sensible formulation of the significance test. Psychological Methods 5:411–414.

Jones, P. G., and P. K. Thornton. 2002. Spatial modeling of risk in natural resource management. Conservation Ecology 5:27. http://www.consecol.org/vol5/iss2/art27.

Jones, R. E., L. J. Guillette, Jr., C. H. Summers, R. R. Tokarz, and D. Crews. 1983. The relationship among ovarian condition, steroid hormones, and estrous behavior in *Anolis carolinensis*. Journal of Experimental Zoology 227:145–154.

Jones, S. M., and G. W. Ferguson. 1980. The effect of paint marking on mortality in a Texas population of *Sceloporus undulates*. Copeia 1980:850–854.

Jonkel, C. J., D. R. Gray, and B. Hubert. 1975. Immobilizing and marking wild muskoxen in Arctic Canada. Journal of Wildlife Management 39:112–117.

Jonsen, I. D., J. Mills Flemming, and R. A. Myers. 2005. Robust state-space modeling of animal movement data. Ecology 86:2874–2880.

Jordan, P. A. 1958. Marking deer with bells. California Fish and Game 44:183–189.

Jorde, P. E., and N. Ryman. 1995. Temporal allele frequency change and estimation of effective size in populations with overlapping generations. Genetics 139:1077–1090.

Jorgensen, C. B. 1992. Growth and reproduction. Pages 439–466 in M. E. Feder and W. W. Burggren, editors. Environmental physiology of the Amphibia. University of Chicago Press, Chicago, Illinois, USA.

Jorgensen, E. E., S. Demarais, and W. R. Whitworth. 1994. The effect of box-trap design on rodent captures. Southwestern Naturalist 39:291–294.

Jouventin, P., and H. Weimerskirch. 1990. Satellite tracking of wandering albatrosses. Nature 343:746–748.

Joyner, D. E. 1975. Nest parasitism and brood-related behavior of the ruddy duck (*Oxyura jamaicensis rubida*). Thesis, University of Nebraska, Lincoln, USA.

Judd, F. W. 1975. Activity and thermal ecology of the keeled earless lizard, *Holbrookia propinqua*. Herpetologica 31:137–150.

Julious, S. A. 2010. Sample sizes for clinical trials. Taylor and Francis, Boca Raton, Florida, USA.

Jung, T. S., and K. S. O'Donovan. 2005. Mortality of deer mice, *Peromyscus maniculatus*, in wire mesh live-traps: a cautionary note. Canadian Field-Naturalist 119:445–446.

Jungbluth, K., J. Belles, M. Schumacher, and R. Arritt. 1995. Velocity contamination of WSR-88D and wind profiler data due to migrating birds. International Conference on Radar Meteorology 27:666–668.

Junge, R., and D. F. Hoffmeister. 1980. Age determination in raccoons from cranial suture obliteration. Journal of Wildlife Management 44:725–729.

Jurgelski, W., Jr., and M. E. Porter. 1974. The opossum (*Didelphis virginiana* Kerr) as a biomedical model. III. Breeding the opossum in captivity: methods. Laboratory Animal Science 24:412–425.

Kabat, C., I. O. Buss, and R. K. Meyer. 1948. The use of ovulated follicles in determining eggs laid by the ring-necked pheasant. Journal of Wildlife Management 12:399–416.

Kaczynski, C. F., and W. H. Kiel, Jr. 1963. Band loss by nestling mourning doves. Journal of Wildlife Management 27:271–279.

Kadlec, J. A. 1975. Recovery rates and loss of aluminum, titanium, and incoloy bands on herring gulls. Bird-Banding 46:230–235.

Kagarise, C. M. 1978. A simple trap for capturing nesting Wilson's phalaropes. Bird-Banding 49:281–282.

Kahn, H. 1965. On escalation: metaphors and scenarios. Pall Mall Press, London, England, UK.

Kahn, N. W., C. E. Braun, J. R. Young, S. Wood, D. R. Mata, and T. W. Quinn. 1999. Molecular analysis of genetic variation among large and small-bodied sage grouse using mitochondrial control-region sequences. Auk 116:819–824.

Kahn, N. W., J. St. John, and T. W. Quinn. 1998. Chromosome-specific intron size differences in the avian CHD gene provide an efficient method for sex identification in birds. Auk 115:1074–1078.

Kahn, R. H., and B. A. Millsap. 1978. An inexpensive method for capturing short-eared owls. North American Bird Bander 3:54.

Kaiser, G. W., A. E. Derocher, S. Crawford, M. J. Gill, and I. A. Manley. 1995. A capture technique for marbled murrelets in coastal inlets. Journal of Field Ornithology 66:321–333.

Kaiser, R. C. 2006. Recruitment by greater sage-grouse in association with natural gas development in western Wyoming. Thesis, University of Wyoming. Laramic, USA.

Kalinowski, S. T., and R. S. Waples. 2002. Relationship of effective to census size in fluctuating populations. Conservation Biology 16:129–136.

Kalla, P. I. 1991. Studies on the biology of ruffed grouse in the southern Appalachian Mountains. Thesis, University of Tennessee, Knoxville, USA.

Kalla, P. I., and R. W. Dimmick. 1995. Reliability of established aging and sexing methods in ruffed grouse. Proceedings of the Annual Meeting of the Southeastern Association of Fish and Wildlife Agencies 49:580–593.

Kamil, A. C. 1988. Experimental design in ornithology. Pages 313–346 in R. F. Johnston, editor. Current ornithology. Volume 5. Plenum Press, New York, New York, USA.

Kamil, N. 1987. Kinetics of bromadiolone, anticoagulant rodenticide, in the Norway rat (*Rattus norvegicus*). Pharmacological Research Communications 19:767–775.

Kamler, J. F., W. B. Ballard, R. L. Gilliland, P. R. Lemons, II, and K. Motte. 2003. Impacts of coyotes on swift foxes in northwestern Texas. Journal of Wildlife Management 67:317–323.

Kamler, J. F., W. B. Ballard, R. L. Gilliland, and K. Mote. 2002. Improved trapping methods for swift foxes and sympatric coyotes. Wildlife Society Bulletin 30:1262–1266.

Kamrin, M. A. 1997. Pesticide profiles: toxicity, environmental impact, and fate. CRC Press, Boca Raton, Florida, USA.

Kangas, J., and T. Pukkala. 1996. Operationalization of biological diversity as a decision objective in tactical forest planning. Canadian Journal of Forest Research 26:103–111.

Kaplan, E. L., and P. Meier. 1958. Nonparametric estimation from incomplete observations. Journal of the American Statistical Association 53:457–481.

Kaplan, H. M. 1958. Marking and banding frogs and turtles. Herpetologica 14:131–132.

Kapustin, N., J. K. Critser, D. Olson, and P. V. Malven. 1996. Nonluteal estrous cycles of 3-week duration are initiated by anovulatory luteinizing hormone peaks in African elephants. Biology of Reproduction 55:1147–1154.

Karanth, K. U. 1995. Estimating tiger *Panthera tigris* populations from camera-trap data using capture–recapture models. Biological Conservation 71:333–338.

Karanth, K. U., and J. D. Nichols. 1998. Estimation of tiger densities in India using photographic captures and recaptures. Ecology 79:2852–2862.

Karanth, K. U., L. Thomas, and N. S. Kumar. 2002. Field surveys: estimating absolute densities of prey species using line transect sampling. Pages 111–120 in K. U. Karanth and J. D. Nichols, editors. Monitoring tigers and their prey: a manual for wildlife managers, researchers, and conservationists. Centre for Wildlife Studies, Bangalore, India.

Karl, B. J., and M. N. Clout. 1987. Improved radio-transmitter harness with a weak link to prevent snagging. Journal of Field Ornithology 55:73–77.

Karl, S. A., B. W. Bowen, and J. C. Avise. 1992. Global population genetic structure and male-mediated gene flow in the green turtle (*Chelonia mydas*): RFLP analyses of anonymous nuclear loci. Genetics 131:163–173.

Karlstrom, E. L. 1957. The use of Co (60) as a tag for recovering amphibians in the field. Ecology 38:187–195.

Karr, J. R. 1981. Rationale and techniques for sampling avian habitats: introduction. Pages 26–28 in D. E. Capen, editor. The use of muitivariate statistics in studies of wildlife habitat. General Technical Report RM-87, U. S. Forest Service, Department of Agriculture, Washington, D.C., USA.

Karr, J. R. 1983. Commentary. Pages 403–410 in A. H. Brush and G. A. Clark, Jr., editors. Perspectives in ornithology. Cambridge University Press, Cambridge, England, UK.

Karr, J. R. 1999. Defining and measuring river health. Freshwater Biology 41:221–234.

Karraker, N. E. 2001. String theory: reducing mortality of mammals in pitfall traps. Wildlife Society Bulletin 29:1158–1162.

Kasamatsu, F., S. Nishiwaki, and M. Sato. 1986. Results of the test firing of improved .410 streamer marks, February 1985. International Whaling Commission Report 36:201–204.

Kasman, L. H., E. C. Ramsay, and B. L. Lasley. 1986. Urinary steroid evaluations to monitor ovarian function in exotic ungulates. III. Estrone sulfate and pregnanediol-3-glucuronide excretion in the Indian rhinoceros (*Rhinoceros unicornis*). Zoo Biology 5:355–361.

Kasprzykowski, Z., and A. Golawski. 2009. Does the use of playback affect the estimates of numbers of grey partridge *Perdix perdix*? Wildlife Biology 15:123–128.

Kattell, B., and A. W. Alldredge. 1991. Capturing and handling of the Himalayan musk deer. Wildlife Society Bulletin 19:397–399.

Kaufman, D. W., and G. A. Kaufman. 1989. Burrow distribution of the thirteen-lined ground squirrel in grazed mixed-grass prairie: effect of artificial habitat structure. Prairie Naturalist 21:81–83.

Kaufmann, J. H. 1982. Raccoon and allies. Pages 567–585 in J. A. Chapman and G. A. Feldhamer, editors. Wild mammals of North America. Johns Hopkins University Press, Baltimore, Maryland, USA.

Kauhala, K., and T. Soveri. 2001. An evaluation of methods for distinguishing between juvenile and adult mountain hares *Lepus timidus*. Wildlife Biology 7:295–300.

Kautz, J. E., and T. W. Seamans. 1992. Techniques for feral pigeon trapping, tagging, and nest monitoring. North American Bird Bander 17:53–59.

Kaye, S. V. 1960. Gold-198 wires used to study movements of small mammals. Science 131:824.

Kays, R., and K. Slauson. 2008. Remote cameras. Pages 110–140 in R. Long, P. MacKay, W. Zielinski, and J. Ray, editors. Noninvasive survey methods for carnivores. Island Press, Washington, D.C., USA.

Kays, R. W. 1999. A hoistable arboreal mammal trap. Wildlife Society Bulletin 27:298–300.

Kearns, F. R., M. Kelly, and K. A. Tuxen. 2003. Everything happens somewhere: using webGIS as a tool for sustainable natural resource management. Frontiers in Ecology and the Environment 1(10):541–548.

Kearns, G. D., N. B. Kwartin, D. F. Brinker, and G. M. Haramis. 1998. Digital playback and improved trap design enhances capture of migrant soras and Virginia rails. Journal of Field Ornithology 69:466–473.

Keating, K. A. 1994. An alternative index of satellite telemetry location error. Journal of Wildlife Management 58:414–421.

Keating, K. A., and S. Cherry. 2009. Modeling utilization distributions in space and time. Ecology 90:1971–1980.

Keay, J. A. 1995. Accuracy of cementum age assignments for black bears. California Fish and Game 81:113–121.

Keck, M. B. 1994a. A new technique for sampling semi-aquatic snake populations. Herpetological Natural History 2:101–103.

Keck, M. B. 1994b. Test for detrimental effects of pit tags on neonatal snakes. Copeia 1994:226–228.

Keeler, R. J., D. S. Zrnic, and C. L. Frush. 1999. Review of range velocity ambiguity mitigation techniques. International Conference on Radar Meteorology 29:158–163.

Keen, R., and H. B. Hitchcock. 1980. Survival and longevity of the little brown bat (*Myotis lucifugus*) in southeastern Ontario. Journal of Mammalogy 61:1–7.

Keigley, R. B. 1997. A growth form method for describing browse condition. Rangelands 19(3):26–29.

Keigley, R. B., M. R. Frisina, and C. W. Fager. 2002a. Assessing browse trend at the landscape level. Part I. Preliminary steps and field survey. Rangelands 24(3):28–33.

Keigley, R. B., M. R. Frisina, and C. W. Fager. 2002b. Assessing browse trend at the landscape level. Part II. Monitoring. Rangelands 24(3):34–38.

Keiss, R. E. 1969. Comparison of eruption-wear patterns and cementum annuli as age criteria in elk. Journal of Wildlife Management 33:175–180.

Keister, G. P., Jr., C. E. Trainer, and M. J. Willis. 1988. A self-adjusting collar for young ungulates. Wildlife Society Bulletin 16:321–323.

Keith, L. B. 1963. Wildlife's ten-year cycle. University of Wisconsin Press, Madison, USA.

Keith, L. B. 1965. A live snare and a tagging snare for rabbits. Journal of Wildlife Management 29:877–880.

Keith, L. B. 1974. Some features of population dynamics in mammals. Proceedings of the International Congress of Game Biologists 11:17–58.

Keith, L. B., and J. R. Cary. 1979. Eye lens weights from free-living adult snowshoe hares of known age. Journal of Wildlife Management 43:965–969.

Keith, L. B., J. R. Cary, O. J. Rongstad, and M. C. Brittingham. 1984. Demography and ecology of a declining snowshoe hare population. Wildlife Monographs 90.

Keith, L. B., E. C. Meslow, and O. J. Rongstad. 1968. Techniques for snowshoe hare population studies. Journal of Wildlife Management 32:801–812.

Keith, L. B., and L. A. Windberg. 1978. A demographic analysis of the snowshoe hare cycle. Wildlife Monographs 58.

Kelker, G. H. 1940. Estimating deer population by a differential hunting loss in the sexes. Proceedings of the Utah Academy of Sciences, Arts and Letters 17:65–69.

Kelker, G. H. 1945. Measurement and interpretation of forces that determine populations of managed deer. Dissertation, University of Michigan, Ann Arbor, USA.

Kelleher, K. S., and G. Hyde. 1993. Reflector antennas. Pages 17–57 in R. C. Johnson, editor. Antenna Engineering Handbook. Third edition. McGraw-Hill, New York, New York, USA.

Keller, L. F., P. Arcese, J. N. M. Smith, W. M. Hochachka, and S. C. Stearns. 1994. Selection against inbred song sparrows during a natural population bottleneck. Nature 372:356–357.

Keller, L. F., and D. M. Waller. 2002. Inbreeding effects in wild populations. Trends in Ecology and Evolution 17:230–241.

Kellogg, C. E. 1946. Variation in pattern of primeness of muskrat skins. Journal of Wildlife Management 10:38–42.

Kelly, B. P. 1996. Live capture of ringed seals in ice-covered waters. Journal of Wildlife Management 60:678–684.

Kelly, G. 1975. Indices for aging eastern wild turkeys. Proceedings of the National Wild Turkey Symposium 3:205–209.

Kelly, M. J. 2001. Computer-aided photograph matching in studies using individual identification: an example from Serengeti cheetahs. Journal of Mammalogy 82:440–449.

Kelly, T. A. 2000. Radar, remote sensing and risk management. Pages 152–161 in PNAWPPM-III. Proceedings of national avian-wind power planning meeting III. RESOLVE, Washington, D.C., USA.

Kendall, K. C., J. B. Stetz, D. A. Roon, L. P. Waits, J. B. Boulanger, and D. Paetkau. 2008. Grizzly bear density in Glacier National Park, Montana. Journal of Wildlife Management 72:1693–1705.

Kendall, W. L., and J. D. Nichols. 1995. On the use of secondary capture–recapture samples to estimate temporary emigration and breeding proportions. Journal of Applied Statistics 22:751–762.

Kendall, W. L., J. D. Nichols, and J. E. Hines. 1997. Estimating temporary emigration using capture–recapture data with Pollock's robust design. Ecology 78:563–578.

Kendall, W. L., K. H. Pollock, and C. Brownie. 1995. A likelihood-based approach to capture–recapture estimation of demographic parameters under the robust design. Biometrics 51:293–308.

Kennedy, M. 2009. The global positioning system and ArcGIS. Third edition. Taylor and Francis, Boca Raton, Florida, USA.

Kennedy, R. E., W. B. Cohen, and T. A. Schroeder. 2007. Trajectory-based change detection for automated characterization of forest disturbance dynamics. Remote Sensing of Environment 110:370–386

Kennedy, R. E., Z. Yang, and W. B. Cohen. 2010. Detecting trends in forest disturbance and recovery using yearly Landsat time series: 1. LandTrendr—temporal segmentation algorithms. Remote Sensing of Environment 114:2897–2910.

Kennelly, J. J., and B. E. Johns. 1976. The estrous cycle of coyotes. Journal of Wildlife Management 40:272–277.

Kennelly, J. J., B. E. Johns, C. P. Breidenstein, and J. D. Roberts. 1977. Predicting female coyote breeding data from fetal measurements. Journal of Wildlife Management 41:746–750.

Kennett, R. 1992. A new trap design for catching freshwater turtles. Wildlife Research 19:443–445.

Kenow, K. P., M. W. Meyer, F. Fournier, W. H. Karasov, A. Elfessi, and S. Gutreuter. 2003. Effects of subcutaneous transmitter implants on behavior, growth, energetics, and survival of common loon chicks. Journal of Field Ornithology 74:179–186.

Kenward, R. E. 1987. Wildlife radio tagging: equipment, field techniques, and data analysis. Academic Press, San Diego, California, USA.

Kenward, R. E. 2001. A manual for wildlife radio tagging. Academic Press, San Diego, California, USA.

Kenward, R. E., R. T. Clarke, K. H. Hodder, and S. S. Walls. 2001. Density and linkage estimators of home range: nearest-neighbor clustering defines multinuclear cores. Ecology 82:1905–1920.

Kenward, R. E., R. H. Pfeffer, M. A. Al-Bowardi, N. C. Fox, K. E. Riddle, E. A. Bragin, A. Levin, S. S. Walls, and K. H. Hodder. 2001. Setting harness sizes and other marking techniques for a falcon with strong sexual dimorphism. Journal of Field Ornithology 72:244–257.

Kenyon, K. W., and C. H. Fiscus. 1963. Age determination in the Hawaiian monk seal. Journal of Mammalogy 44:280–282.

Kepler, C. B., and J. M. Scott. 1981. Reducing bird count variability by training observers. Studies in Avian Biology 6:366–371.

Keppie, D. M. 1990. To improve graduate student research in wildlife education. Wildlife Society Bulletin 18:453–458.

Keppie, D. M., and C. E. Braun. 2000. Band-tailed pigeon. Account 530 in A. Poole and F. Gill, editors. The birds of North America. The Academy of Natural Sciences, Philadelphia, Pennsylvania, and The American Ornithologists' Union, Washington, D.C., USA.

Keppie, D. M., and P. W. Herzog. 1978. Nest site characteristics and nest success of spruce grouse. Journal of Wildlife Management 42:628–632.

Keppie, D. M., and R. M. Whiting, Jr. 1994. American woodcock. Account 100 in A. Poole and F. Gill, editors. The birds of North America. The Academy of Natural Sciences, Philadelphia, Pennsylvania, and The American Ornithologists' Union, Washington, D.C., USA.

Kerley, L. L., and G. P. Salkina. 2007. Using scent-matching dogs to identify individual Amur tigers from scats. Journal of Wildlife Management 71:1349–1356.

Kerlin, D. H., D. T. Haydon, D. Miller, J. J. Aebischer, A. A. Smith, and S. J. Thirgood. 2007. Spatial synchrony in red grouse population dynamics. Oikos 116:2007–2016.

Kerlinger, F. N. 1986. Foundations of behavioral research. Third edition. Holt, Rinehart and Winston, New York, New York, USA.

Kerlinger, F. N., and H. B. Lee. 2000. Foundations of behavioral research. Fourth edition. Harcourt College, New York, New York, USA.

Kern, J. W., L. L. McDonald, M. D. Strickland, and E. Williams. 1994. Field evaluation and comparison of four foothold traps for terrestrial furbearers in Wyoming. Western Ecosystems Technology, Cheyenne, Wyoming, USA.

Kernohan, B. J., R. A. Gitzen, and J. J. Millspaugh. 2001. Analysis of animal space use and movements. Pages 125–166 in J. J. Millspaugh and J. M. Marzluff, editors. Radio tracking and animal populations. Academic Press, San Diego, California, USA.

Kerr, C. L., S. E. Henke, and R. Tamez. 2000. Cage-trap modifications that enhance the capture success of raccoons. Proceedings of the Vertebrate Pest Conference 19:436–438.

Kershaw, K. A. 1964. Quantitative and dynamic ecology. Edward Arnold, London, England, UK.

Kerth, G., M. Wagner, and B. König. 2001. Roosting together, foraging apart: information transfer about food is unlikely to explain sociality in female Bechstein's bats (*Myotis bechsteinii*). Behavioral Ecology and Sociobiology 50:283–291.

Kerwin, M. L., and G. J. Mitchell. 1971. The validity of the wear-age technique for Alberta pronghorns. Journal of Wildlife Management 35:743–747.

Kery, M. 2010. Introduction to WinBUGS for ecologists—a Bayesian approach to regression, ANOVA, mixed models and related analyses. Academic Press, Burlington, Massachusetts, USA.

Kessler, F. W. 1964. Avian predation on pheasants wearing differently colored plastic markers. Ohio Journal of Science 64:401–402.

Keyes, B. E., and C. E. Grue. 1982. Capturing birds with mist nets: a review. North American Bird Bander 7:2–14.

Keyser, A. J., and G. E. Hill. 1999. Condition-dependent variation in the blue-ultraviolet coloration of a structurally based plumage ornament. Proceedings of the Royal Society of London Bulletin 266:771–777.

Khabibullin, V. F., and M. V. Radygina. 2005. A new trap design to sample small terrestrial lizards. Herpetological Review 36:407.

Kibbe, D. P., and R. L. Kirkpatrick. 1971. Systematic evaluation of late summer breeding in juvenile cottontails, *Sylvilagus floridanus*. Journal of Mammalogy 52:465–467.

Kibler, L. F. 1969. The establishment and maintenance of a bluebird nest-box project. Bird-Banding 40:114–129.

Kiely, T., D. Donaldson, and A. H. Grube. 2004. Pesticides industry sales and usage: 2000 and 2001 market estimates. Biological and Economic Analysis Division, Office of Pesticide Programs, U.S. Environmental Protection Agency, Washington, D.C., USA.

Kiilsgaard, C. 1999. Oregon vegetation: mapping and classification of landscape-level cover types. Final Report, GAP Analysis Program, Biological Resources Division, U.S. Geological Survey, Department of the Interior, Moscow, Idaho, USA.

Kiilsgaard, C., and C. Barrett. 2000. Map—wildlife habitat types of the Pacific Northwest. Northwest Habitat Institute, Corvallis, Oregon, USA.

Kilgo, J. C., D. L. Gartner, B. R. Chapman, J. B. Dunning, Jr., K. E. Franzreb, S. A. Gauthreaux, C. H. Greenberg, D. J. Levey, K. V. Miller, and S. F. Pearson. 2002. A test of an expert-based bird–habitat relationship model in South Carolina. Wildlife Society Bulletin 30:783–793.

Kimball, B. A., J. R. Mason, F. S. Blom, J. J. Johnston, and D. E. Zemlicka. 2000. Development and testing of seven new synthetic coyote attractants. Journal of Agricultural and Food Chemistry 48:1892–1897.

Kimball, B. A., and E. A. Mishalanie. 1994. Stability of 3-chloro-p-toluidine hydrochloride in buffered aqueous solutions. Environmental Science and Technology 28:419–422.

Kindel, F. 1960. Use of dyes to mark ruminant feces. Journal of Wildlife Management 24:429.

King, C. M. 1981. The effects of two types of steel traps upon captured stoats (*Mustela erminea*). Journal of Zoology 195:553–554.

King, D. I., R. B. Chandler, S. Schlossberg, and C. C. Chandler. 2009. Habitat use and nest success of scrub-shrub birds in wildlife and silvicultural openings in western Massachusetts, USA. Forest Ecology and Management 257:421–426.

King, D. T., K. J. Andrews, J. O. King, R. D. Flynt, J. F. Glahn, and J. L. Cummings. 1994. A nightlighting technique for capturing cormorants. Journal of Field Ornithology 65:254–257.

King, D. T., J. D. Paulson, D. J. Leblanc, and K. Bruce. 1998. Two capture techniques for American white pelicans and great blue herons. Colonial Waterbirds 21:258–260.

King, D. T., M. E. Tobin, and M. Bur. 2000. Capture and telemetry techniques for double-crested cormorants (*Phalacrocorax auritus*). Proceedings of the Vertebrate Pest Conference 19:54–57.

King, J. E. 1983. Seals of the world. British Museum, London, England, UK.

King, M. B., and D. Duvall. 1984. Noose tube: a light weight, sturdy, and portable snake restraining apparatus for field and laboratory use. Herpetological Review 15:109.

King, R. B. 2000. Analyzing the relationship between clutch size and female body size in reptiles. Journal of Herpetology 34:148–150.

Kingsland, S. E. 1995. Modeling nature: episodes in the history of population ecology. Second edition. University of Chicago Press, Chicago, Illinois, USA.

Kingston, T. 1995. RAMAS/GIS: linking landscape data with population viability analysis (software review). Conservation Biology 9:966–968.

Kinkel, L. K. 1989. Lasting effects of wing tags on ring-billed gulls. Auk 106:619–624.

Kinsinger, F. E., R. E. Eckert, and P. O. Currie. 1960. A comparison of the line-interception, variable-plot, and loop methods as used to measure shrub-crown cover. Journal of Range Management 13:17–21.

Kirkland, G. L., Jr., and R. K. Sheppard. 1994. Proposed standard protocol for sampling small mammal communities. Pages 277–283 in J. F. Merritt, G. L. Kirkland, Jr., and R. K. Rose, editors. Advances in the biology of shrews. Special Publication 18, Carnegie Museum of Natural History, Pittsburgh, Pennsylvania, USA.

Kirkpatrick, J. F., L. H. Kasman, B. L. Lasley, and J. W. Turner, Jr. 1988. Pregnancy determination in uncaptured feral horses. Journal of Wildlife Management 52:305–308.

Kirkpatrick, J. F., B. L. Lasley, W. R. Allen, and C. Doberska, editors. 2002. Fertility control in wildlife. Proceedings of the Fifth International-Symposium on Fertility Control in Wildlife. August 2001, Skukuza, Kruger National Park, South Africa. Reproduction Supplement 60, Society for Reproduction and Fertility, Cambridge, England, UK.

Kirkpatrick, J. F., J. W. Turner, Jr., I. K. M. Liu, R. A. Fayrer-Hosken, and A. T. Rutberg. 1997. Case studies in wildlife immunocontraception: wild and feral equids and white-tailed deer. Reproduction, Fertility and Development 9:105–110.

Kirkpatrick, M. 1984. Demographic models based on size, not age, for organisms with indeterminate growth. Ecology 65:1874–1884.

Kirkpatrick, R. D., and L. K. Sowls. 1962. Age determination of the collared peccary by the tooth-replacement pattern. Journal of Wildlife Management 26:214–217.

Kirkpatrick, R. L., and G. E. Valentine. 1970. Reproduction in captive pine voles, *Microtus pinetorum*. Journal of Mammalogy 51:779–785.

Kirsch, L. M., H. F. Duebbert, and A. D. Kruse. 1978. Grazing and haying effects on habitats of upland nesting birds. Transactions of the North American Wildlife and Natural Resources Conference 43:486–497.

Kissling, M. L., and E. O. Garton. 2006. Estimating detection probabilities from point count surveys: a combination of distance and double observer sampling. Auk 123:735–752.

Kissling, M. L., and E. O. Garton. 2008. Forested buffer strips and breeding bird communities in southeastern Alaska. Journal of Wildlife Management 72:674–681.

Kitching, R. L. 1983. Systems ecology: an introduction to ecological modelling. University of Queensland Press, St. Lucia, Australia.

Kitching, R. L. 1991. Systems ecology: an introduction to ecological modelling. Second edition. University of Queensland Press, St Lucia, Queensland, Australia.

Kjoss, V. A., and J. A. Litvaitis. 2001. Comparison of 2 methods to sample snake communities in early successional habitats. Wildlife Society Bulletin 29:153–157.

Klawitter, R. A., and J. Stubbs. 1961. A reliable oak seed trap. Journal of Forestry 59:291–292.

Klein, J. P., and M. L. Moeschberger. 2003. Survival analysis. Springer-Verlag, New York, New York, USA.

Kleinbaum, D. G. 1996. Survival analysis. Springer-Verlag, New York, New York, USA.

Klett, A. T., H. F. Duebbert, C. A. Faanes, and K. F. Higgins. 1986. Techniques for studying nest success of ducks in upland habitats in the prairie pothole region. Resource Publication 158, U.S. Fish and Wildlife Service, Department of the Interior, Washington, D.C., USA.

Klett, A. T., and D. H. Johnson. 1982. Variability in nest survival rates and implications to nesting studies. Auk 99:77–87.

Klevezal', G. A., and S. E. Kleinenberg. 1967. Age determination of mammals from annual layers in teeth and bones. USSR Academy of Science, Moscow, Russia.

Klevezal', G. A., and M. V. Mina. 1973. Factors determining the pattern of annual layers in dental tissue and bones of mammals. Zhurnal Obshchei Biologii 34:594–604.

Klinger, R. C., R. H. George, and J. A. Musick. 1997. A bone biopsy technique for determining age and growth in sea turtles. Herpetological Review 28:31–32.

Klinger, R. C., and J. A. Musick. 1992. Annular growth layers in juvenile loggerhead sea turtles (*Caretta caretta*). Bulletin of Marine Science 51:224–230.

Klinka, D. R., and T. E. Reimchen. 2009. Darkness, twilight, and daylight foraging success of bears (*Ursus americanus*) on salmon in coastal British Columbia. Journal of Mammalogy 90:144–149.

Klott, J. H., and F. G. Lindzey. 1990. Brood habitats of sympatric sage-grouse and Columbian sharp-tailed grouse in Wyoming. Journal of Wildlife Management 54:84–88.

Knapp, S. M., B. A. Craig, and L. P. Waits. 2009. Incorporating genotyping rrror into non-invasive DNA-based mark–recapture population estimates. Journal of Wildlife Management 73:598–604.

Knick, S. T., and D. L. Dyer. 1997. Distribution of black-tailed jackrabbit habitat determined by GIS in southwestern Idaho. Journal of Wildlife Management 61:75–85.

Knight, J. E. 1994. Mountain lion. Pages C93–C99 in S. E. Hynstrom, R. M. Timm, and G. E. Larson, editors. Prevention and control of wildlife damage. Cooperative Extension Service, University of Nebraska, Lincoln, USA.

Knight, R. L., and K. J. Gutzwiller. 1995. Wildlife and recreationists: coexistence through management and research. Island Press, Washington, D.C., USA.

Knight, R. R. 1966. Effectiveness of neckbands for marking elk. Journal of Wildlife Management 30:845–846.

Knittle, C. E., and M. A. Pavelka. 1994. Hook and loop tabs for attaching a dho-gaza. Journal of Raptor Research 28:197–198.

Knopff, K. H., A. A. Knopff, M. B. Warren, and M. S. Boyce. 2009. Evaluating global position system telemetry techniques for estimating cougar predation parameters. Journal of Wildlife Management 73:586–597.

Knowlton, F. F. 1972. Preliminary interpretations of coyote population mechanics with some management implications. Journal of Wildlife Management 36:369–382.

Knowlton, F. F., E. D. Michael, and W. C. Glazener. 1964. A marking technique for field recognition of individual turkeys and deer. Journal of Wildlife Management 28:167–170.

Knowlton, F. F., and S. L. Whittemore. 2001. Pulp cavity-tooth width ratios from known-age and wild-caught coyotes determined by radiography. Wildlife Society Bulletin 29:239–244.

Kochert, M. N. 1973. Evaluation of a vinyl wing-marker for raptors. Raptor Research News 7:117–118.

Kochert, M. N., K. Steenhof, and M. Q. Moritsch. 1983. Evaluation of patagial markers for raptors and ravens. Wildlife Society Bulletin 11:271–281.

Kock, M. D., D. A. Jessup, R. K. Clark, C. E. Franti, and R. A. Weaver. 1987. Capture methods in five subspecies of free-ranging bighorn sheep: an evaluation of drop-net, drive-net, chemical immobilization, and the net-gun. Journal of Wildlife Diseases 23:634–640.

Koeln, G. T., L. M. Cowardin, and L. L. Strong. 1996. Geographic information systems. Pages 540–566 in T. A. Booklet, editor. Research and management techniques for wildlife and habitats. The Wildlife Society, Bethesda, Maryland, USA.

Koen, E. L., J. Bowman, C. S. Findlay, and L. Zheng. 2007. Home range and population density of fishers in eastern Ontario. Journal of Wildlife Management 71:1484–1493.

Koenen, K., S. DeStefano, C. Henner, and T. Beroldi. 2005. Capturing beavers in box traps. Wildlife Society Bulletin 33:1153–1159.

Koenen, K. K. G., S. DeStefano, and P. R. Krausman. 2002. Using distance sampling to estimate seasonal densities of desert mule deer in a semidesert grassland. Wildlife Society Bulletin 30:53–63.

Koenig, W. D., and J. Knops. 1995. Why do oaks produce boom-and-bust seed crops? California Agriculture 49(5):7–12.

Koenig, W. D., R. L. Mumme, W. J. Carmen, and M. T. Stanback. 1994. Acorn production by oaks in central coastal California: variation within and among years. Ecology 75:99–109.

Koerner, J. W., T. A. Bookhout, and K. E. Bednarik. 1974. Movements of Canada geese color-marked near southwestern Lake Erie. Journal of Wildlife Management 38:275–289.

Koerth, B. H., and J. C. Kroll. 2008. Juvenile-to-adult antler development in white-tailed deer in south Texas. Journal of Wildlife Management 72:1109–1113.

Koerth, B. H., C. D. McKown, and J. C. Kroll. 1997. Infrared triggered camera versus helicopter counts of white-tailed deer. Wildlife Society Bulletin 25:557–562.

Kohn, M. H., and R. K. Wayne. 1997. Facts from feces revisited. Trends in Ecology and Evolution 12:223–227.

Kohn, M. H., E. C. York, D. A. Kamradt, G. Haugt, R. M. Sauvajot, and R. K. Wayne. 1999. Estimating population size by genotyping faeces. Proceedings of the Royal Society of London B 266:657–663.

Koistinen, J. 2000. Bird migration patterns on weather radars. Physics and Chemistry of the Earth B 25:1185–1193.

Koivula, M., J. Viitala, and E. Korpimäki. 1999. Kestrels prefer scent marks according to species and reproductive status of voles. Ecoscience 6:415–420.

Kojola, I., J. Aspi, A. Hakala, S. Heikkinen, C. Ilmoni, and S. Ronkainen. 2006. Dispersal in an expanding wolf population in Finland. Journal of Mammalogy 87:281–286.

Kolenosky, G. B. 1987. Polar bear. Pages 474–485 in M. Novak, J. A. Baker, M. E. Obbard, and B. Malloch, editors. Wild furbearer management and conservation in North America. Ontario Ministry of Natural Resources, Toronto, Canada.

Kolenosky, G. B., and S. M. Strathearn. 1987. Black bear. Pages 443–454 in M. Novak, J. A. Baker, M. E. Obbard, and B. Malloch, editors. Wild furbearer management and conservation in North America. Ontario Ministry of Natural Resources, Toronto, Canada.

Kolpin, D. W., E. T. Furlong, M. T. Meyer, E. M. Thurman. S. D. Zaugg, L. B. Barber, and H. T. Buxton. 2002. Pharmaceuticals, hormones, and other organic wastewater contaminants in U.S. streams, 1999–2000: a national reconnaissance. Environmental Science and Technology 36:1202–1211.

Komak, S., and M. R. Crossland. 2000. An assessment of the introduced mosquito fish (*Gambusia affinis holbrooki*) as a predator of eggs, hatchlings and tadpoles of native and non-native anurans. Wildlife Research 27:185–189.

Kong, Q. Z., S. H. Huang, W. Q. Zou, D. Vanegas, M. L. Wang, D. Wu, J. Yuan, M. J. Zheng, H. Bai, H. Y. Deng, K. Chen, A. L. Jenny, K. O'Rourke, E. D. Belay, L. B. Schonberger, R. B. Petersen, M. S. Sy, S. G. Chen, and P. Gambetti. 2005. Chronic wasting disease of elk: transmissibility to humans examined by transgenic mouse models. Journal of Neuroscience 25:7944–7949.

Koob, M. D. 1981. Detrimental effects of nasal saddles on male ruddy ducks. Journal of Field Ornithology 52:140–143.

Kopp, K., and J. Jokela. 2007. Resistant invaders can convey benefits to native species. Oikos 116:295–301.

Koprowski, J. L. 2002. Handling tree squirrels with a safe and efficient restraint. Wildlife Society Bulletin 30:101–103.

Korn, H. 1987. Effects of live-trapping and toe-clipping on body weight of European and African rodent species. Oecologia 71:597–600.

Koronkiewicz, T. J., E. H. Paxton, and M. K. Sogge. 2005. A technique to produce aluminum color bands for avian research. Journal of Field Ornithology 76:94–97.

Korschgen, C. E., W. L. Green, W. L. Flock, and E. A. Hibbard. 1984. Use of radar with a stationary antenna to estimate birds in a low-level flight corridor. Journal of Field Ornithology 55:369–375.

Korschgen, C. E., K. P. Kenow, A. Gendron-Fitzpatrick, and W. L. Green. 1996a. Implanting intra-abdominal radiotransmitters with external whip antennas in ducks. Journal of Wildlife Management 60:132–137.

Korschgen, C. E., K. P. Kenow, W. L. Green, M. D. Samuel, and L. Sileo. 1996b. Technique for implanting radio transmitters subcutaneously in day-old ducklings. Journal of Field Ornithology 67:392–397.

Kostecke, R. M., G. M. Linz, and W. J. Bleier. 2001. Survival of avian carcasses and photographic evidence of predators and scavengers. Journal of Field Ornithology 72:439–447.

Koster, J. 2008. The impact of hunting with dogs on wildlife harvests in the Bosawas Reserve, Nicaragua. Environmental Conservation 35:211–220.

Koubek, P. 1993. Eye-lens weight as an indicator of age in captive pheasant chicks (*Phasianus colchicus*). Folia Zoologica 42:237–242.

Koval, P. J., T. J. Peterle, and J. D. Harder. 1987. Effects of polychlorinated biphenyls on mourning dove reproduction and circulating progesterone levels. Bulletin of Environmental Contamination and Toxicology 39:663–670.

Kozicky, E. L., G. O. Henderson, P. G. Homeyer, and E. B. Speaker. 1952. The adequacy of the fall roadside pheasant census in Iowa. Transactions of the North American Wildlife Conference 17:293–305.

Kozicky, E. L., and H. G. Weston, Jr. 1952. A marking technique for ring-necked pheasants. Journal of Wildlife Management 16:223.

Kozlik, F. M., A. W. Miller, and W. C. Rienecker. 1959. Color-marking white geese for determining migration routes. California Fish and Game 45:69–82.

Kozlowski, A. J., T. J. Bennett, E. M. Gese, and W. M. Marjo. 2003. Live capture of denning mammals using an improved box trap enclosure:

kit fox as a test case. Wildlife Society Bulletin 31:630–633.

Krabill, W. B., C. W. Wright, E. B. Frederick, S. S. Manizade, J. K. Yungel, C. F. Martin, J. G. Sonntag, M. Duffy, W. Hulslander, and J. C. Brock. 2000. Airborne laser mapping of Assateague National Seashore Beach. Photogrammetry and Remote Sensing 66:65–71.

Kramer, D. L. 1988. A noose apparatus and its usefulness in capturing nestling bank swallows. North American Bird Bander 13:66–67.

Kramer, M. T., R. J. Warren, M. J. Ratnaswamy, and B. T. Bond. 1999. Determining sexual maturity of raccoons by external versus internal aging criteria. Wildlife Society Bulletin 27:231–234.

Krausman, P. R., L. K. Harris, C. L. Blasch, K. K. G. Koenen, and J. Francene. 2004. Effects of military operations on behavior and hearing of endangered Sonoran pronghorn. Wildlife Monographs 157.

Krausman, P. R., J. J. Hervert, and L. L. Ordway. 1985. Capturing deer and mountain sheep with a net-gun. Wildlife Society Bulletin 13:71–73.

Krauss, G. D., H. B. Graves, and S. M. Zervanos. 1987. Survival of wild and game-farm cock pheasants released in Pennsylvania. Journal of Wildlife Management 51:555–559.

Krebs, C. J. 1985. Ecology: the experimental analysis of distribution and abundance. Third edition. Harper and Row, New York, New York, USA.

Krebs, C. J. 1998. Ecological methodology. Harper and Row, New York, New York, USA.

Krebs, C. J. 1999. Ecological methodology. Second edition. Harper and Row, New York, New York, USA.

Krebs, C. J., S. Boutin, and R. Boonstra, editors. 2001. Ecosystem dynamics of the boreal forest: the Kluane Project. Oxford University Press, Oxford, England, UK.

Kreeger, T. J., and J. M. Arnemo. 2007. Handbook of wildlife chemical immobilization. Third edition. International Wildlife Veterinary Services, Laramie, Wyoming, USA.

Kreeger, T. J., P. J. White, U. S. Seal, and J. R. Tester. 1990. Pathological responses of red foxes to foothold traps. Journal of Wildlife Management 54:147–160.

Kridelbaugh, A. 1982. Improved trapping methods for loggerhead shrikes. North American Bird Bander 7:50–51.

Kroenke, D. M. 2000. Database processing: fundamentals, design and implementation. Seventh edition. Prentice-Hall, Upper Saddle River, New Jersey, USA.

Kruuk, H. 1978. Spatial organization and territorial behavior of the European badger *Meles meles*. Journal of Zoology 184:1–19.

Kruuk, H., M. Gorman, and T. Parrish. 1980. The use of ^{65}Zn for estimating populations of carnivores. Oikos 34:206–208.

Kruuk, L. E. B., T. H. Clutton-Brock, J. Slate, J. M. Pemberton, S. Brotherstone, and F. E. Guinness. 2000. Heritability of fitness in a wild mammal population. Proceedings of the National Academy of Sciences of the United States of America 97:698–703.

Kucera, T. E., and R. H. Barrett. 1993. In my experience: the Trailmaster camera system for detecting wildlife. Wildlife Society Bulletin 21:505–508.

Kuchling, G. 1999. The reproductive biology of the Chelonia. Zoophysiology. Volume 38. Springer-Verlag, New York, New York. USA.

Kuchling, G. 2003. A new underwater trap for catching turtles. Herpetological Review 34:126–128.

Kuehn, D. W., and W. E. Berg. 1981. Use of radiographs to identify age-classes of fisher. Journal of Wildlife Management 45:1009–1010.

Kuehn, D. W., and W. E. Berg. 1983. Use of radiographs to age otters. Wildlife Society Bulletin 11:68–70.

Kuehn, D. W., T. K. Fuller, L. D. Mech, W. J. Paul, S. H. Fritts, and W. E. Berg. 1986. Trap-related injuries to gray wolves in Minnesota. Journal of Wildlife Management 50:90–91.

Kuenzi, A. J., and M. L. Morrison. 1998. Detection of bats by mist-nets and ultrasonic sensors. Wildlife Society Bulletin 26:307–311.

Kuhn, T. S. 1996. The structure of scientific revolutions. Third edition. University of Chicago Press, Chicago, Illinois, USA.

Kuhner, M. K. 2006. LAMARC 2.0: maximum likelihood and Bayesian estimation of population parameters. Bioinformatics 22:768–770.

Kuhner, M. K. 2009. Coalescent genealogy samplers: windows into population history. Trends in Ecology and Evolution 24:86–93.

Kumar, R. K. 1979. Toe-clipping procedure for individual identification of rodents. Laboratory Animal Science 29:679–680.

Kunz T. H., E. B. Arnett, B. M. Cooper, W. P. Erickson, R. P. Larkin, T. Mabee, M. L. Morrison, M. D. Strickland, and J. M. Szewczak. 2007. Assessing impacts of wind-energy development on nocturnally active birds and bats: a guidance document. Journal of Wildlife Management 71:2449–2486.

Kunz, T. H., C. R. Tidemann, and G. R. Richards. 1996. Small volant mammals. Pages 122–143 *in* D. E. Wilson, F. R. Cole, J. D. Nichols, R. Rudran, and M. S. Foster, editors. Measuring and monitoring biological diversity. Standard methods for mammals. Smithsonian Institution Press, Washington, D.C., USA.

Kurta, A., and S. W. Murray. 2002. Philopatry and migration of banded Indiana bats (*Myotis sodalis*) and effects of radio transmitters. Journal of Mammalogy 83:585–589.

Kutz, H. L. 1945. An improved game bird trap. Journal of Wildlife Management 9:35–38.

Kuvlesky, W. P., Jr., B. H. Koerth, and N. J. Silvy. 1989. Problems of estimating northern bobwhite populations at low density. Proceedings of the Annual Conference of the Southeastern Association of Fish and Wildlife Agencies. 43:260–267.

Kwiecinski, G. G. 1998. *Marmota monax*. Number 591 *in* C. E. Rebar, A. V. Lindzey, K. F. Koopman, E. Anderson, and V. Hayssen, editors. Mammalian species. American Society of Mammalogists, Lawrence, Kansas, USA.

Kwok, F. A., and C. Ivanyi. 2008. A minimally evasive method for obtaining venom from helodermatid lizards. Herpetological Review 39:179–181.

Laake, J. L., S. T. Buckland, D. R. Anderson, and K. P. Burnham. 1994. DISTANCE user's guide V2.1. Colorado Cooperative Fish and Wildlife Research Unit, Colorado State University, Fort Collins, USA.

Laake, J. L., J. Calambokidis, S. D. Osmek, and D. J. Rugh. 1997. Probability of detecting harbor porpoise from aerial surveys: estimating g(0). Journal of Wildlife Management 61:63–75.

Läärä, E. 2009. Statistics: reasoning on uncertainty, and the insignificance of testing null. Annales Zoologica Fennici 46:138–157.

Labisky, R. F. 1968. Nightlighting: its use in capturing pheasants, prairie chickens, bobwhites, and cottontails. Biological Notes 62, Illinois Natural History Survey, Urbana, USA.

Labisky, R. F., and R. D. Lord, Jr. 1959. A flexible, plastic eartag for rabbits. Journal of Wildlife Management 23:363–365.

Labisky, R. F., and S. H. Mann. 1962. Backtag markers for pheasants. Journal of Wildlife Management 26:393–399.

Lacey, R. 1998. Bovine spongiform encephalopathy: the fall-out. Reviews in Medical Microbiology 9:119–127.

Lacher, J. R., and D. D. Lacher. 1964. A mobile cannon net trap. Journal of Wildlife Management 28:595–597.

Lack, D. 1954. The natural regulation of animal numbers. Oxford University Press, London, England, UK.

Lack, D., and G. C. Varley. 1945. Detection of birds by radar. Nature 156:446.

Lacki, M. J., W. T. Peneston, and F. D. Vogt. 1990. A comparison of the efficacy of two types of live traps for capturing muskrats, *Ondatra zibethicus*. Canadian Field-Naturalist 104:594–596.

Lacy, R. C. 1993. VORTEX: a computer simulation model for population viability analysis. Wildlife Research 20:45–65.

Lacy, R. C. 1997. Importance of genetic variation to the viability of mammalian populations. Journal of Mammalogy 78:320–335.

Lacy, R. C. 2000. Should we select genetic alleles in our conservation breeding programs? Zoo Biology 19:279–282.

Lacy, R. C., G. Alaks, and A. Walsh. 1996. Hierarchical analysis of inbreeding depression in *Peromyscus polionotus*. Evolution 50:2187–2200.

Lacy, R. C., J. D. Ballou, F. Princée, A. Starfield, and E. A. Thompson. 1995. Pedigree analysis for population management. Pages 57–75 *in* J. D. Ballou, M. Gilpin, and T. J. Foose, editors. Population management for survival and recovery: analytical methods and strategies in small population conservation. Columbia University Press, New York, New York, USA.

Ladd, C., and L. Glass. 1999. Golden-cheeked warbler: *Dendroica chrysoparia*. Account 420 *in* A. Poole, P. Stettenheim, and F. Gill, editors. The birds of North America. The Academy of Natural Sciences, Philadelphia, Pennsylvania, and The American Ornithologists' Union, Washington, D.C., USA.

Laerm, J., J. C. Avise, J. C. Patton, and R. A. Lansman. 1982. Genetic determination of the status of an endangered species of pocket gopher in Georgia. Journal of Wildlife Management 46:513–518.

Lafferty, K. D., and A. K. Morris. 1996. Altered behavior of parasitized killifish increases susceptibility to predation by bird final hosts. Ecology 77:1390–1397.

LaGrone, A. H., A. P. Deam, and G. B. Walker. 1964. Angels, insects, and weather. Radio Science Journal of Research NBS/USNC-URSI 68D:895–901.

Laikre, L., and N. Ryman. 1996. Effects on intraspecific biodiversity from harvesting and enhancing natural populations. Ambio 25:504–509.

Lakatos, I. 1978. Falsification and the methodology of scientific research programmes. Pages 8–101 in J. Worrall and G. Currie, editors. The methodology of scientific research programmes: philosophical papers. Volume 1. Imre Lakatos. Cambridge University Press, Cambridge, England, UK.

Lambert, D. M., C. D. Millar, K. Jack, S. Anderson, and J. L. Craig. 1994. Single- and multilocus DNA fingerprinting of communally breeding pukeko: do copulations or dominance ensure reproductive success? Proceedings of the National Academy of Sciences of the United States of America 91:9641–9645.

Lambert, T. D., R. W. Kays, P. A. Jansen, E. Aliaga-Rossel, and M. Wikelski. 2009. Nocturnal activity by the primarily diurnal Central American agouti (*Dasyprocta punctata*) in relation to environmental conditions, resource abundance and predation risk. Journal of Tropical Ecology 25:211–215.

Lambert, T. D., J. R. Malcolm, and B. L. Zimmerman. 2005. Variation in small mammal species richness by trap height and trap type in southeastern Amazonia. Journal of Mammalogy 86:982–990.

Lancia, R. A., and J. W. Bishir. 1985. Mortality rates of beaver in Newfoundland: a comment. Journal of Wildlife Management 49:879–881.

Lancia, R. A., W. L. Kendall, K. H. Pollock, and J. D. Nichols 2005. Estimating the number of animals in wildlife populations. Pages 106–153 in C. E. Braun, editor. Techniques for wildlife investigations and management. The Wildlife Society, Bethesda, Maryland, USA.

Lancia, R. A., J. D. Nichols, and K. H. Pollock. 1994. Estimating the number of animals in wildlife populations. Pages 215–253 in T. A. Bookhout, editor. Research and management techniques for wildlife and habitats. The Wildlife Society, Bethesda, Maryland, USA

Lanctot, R. B. 1994. Blood sampling in juvenile buff-breasted sandpipers: movement, mass change, and survival. Journal of Field Ornithology 65:534–542.

Lande, R. 1988. Demographic models of the northern spotted owl (*Strix occidentalis caurina*). Oecologia 75:601–607.

Lande, R. 1993. Risks of population extinction from demographic and environmental stochasticity and random catastrophes. American Naturalist 142:911–927.

Lande, R. 1995. Mutation and conservation. Conservation Biology 9:782–791.

Lande, R. 1996. Statistics and partitioning of species diversity, and similarity among communities. Oikos 76:5–13.

Landers, J. L., and A. S. Johnson. 1976. Bobwhite quail food habits. Miscellaneous Publication 4, Tall Timbers Research Station, Tallahassee, Florida, USA.

Landon, D. B., C. A. Waite, R. O. Peterson, and L. D. Mech. 1998. Evaluation of age determination techniques for gray wolves. Journal of Wildlife Management 62:674–682.

Landsat Science Team (C. E. Woodcock, R. Allen, M. Anderson, A. Belward, R.. Bindschadler, W. B. Cohen, F. Gao, S. N. Goward, D. Helder, E. Helmer, R. Nemani, L. Orepoulos, J. Schott, P. Thenkabail, E. Vermote, J. Vogelmann, M. Wulder, and R. Wynne). 2008. Free access to Landsat data. Science 320:1011.

Landys-Ciannelli, M. M., M. Ramenofsky, T. Piersma, J. Jukema, and J. C. Wingfield. 2002. Baseline and stress-induced plasma corticosterone during long-distance migration in the bar-tailed godwit, *Limosa lapponica*. Physiological and Biochemical Zoology 75:101–110.

Langbauer, W. R., Jr., K. B. Payne, R. A. Charif, L. Rapaport, and F. Osborn. 1991. African elephants respond to distant playbacks of low-frequency conspecific calls. Journal of Experimental Biology 157:35–46.

Langen, T. A., K. M. Ogden, and L. L. Schwarting. 2009. Predicting hot spots of herptofauna road mortality along highway networks. Journal of Wildlife Management 73:104–114.

Lannom, J. R., Jr. 1962. A different method of catching the desert lizards, *Callisaurus* and *Uma*. Copeia 1962:437–438.

Lanyon, J. M., R. W. Slade, H. L. Sneath, D. Broderick, J. M. Kirkwood, D. Limpus, C. J. Limpus, and T. Jessop. 2006. A method for capturing dugongs (*Dugong dugon*) in open water. Aquatic Mammals 32:196–201.

Larison, J. R., G. E. Likens, J. W. Fitzpatrick, and J. G. Crock. 2000. Cadmium toxicity among wildlife in the Colorado Rocky Mountains. Nature 406:181–183.

Larivière, S. 1999. Reasons why predators cannot be inferred from nest remains. Condor 101:718–721.

Larkin, R. 1996. Effects of military noise on wildlife: a literature review. Construction Engineering Research Laboratory (CERL) Technical Report, U.S. Army Corps of Engineers, Champaign, Illinois, USA.

Larkin, R. P. 1978. Radar observations of behavior of migrating birds in response to sounds broadcast from the ground. Pages 209–218 in K. Schmidt-Koenig and W. T. Keeton, editors. Animal migration, navigation, and homing. Springer-Verlag, New York, New York, USA.

Larkin, R. P. 1982. Spatial distribution of migrating birds and small-scale atmospheric motion. Pages 28–37 in F. Papi and H.-G. Wallraff, editors. Avian navigation. Springer-Verlag, New York, New York, USA.

Larkin, R. P. 1991a. Flight speeds observed with radar, a correction: slow "birds" are insects. Behavioral Ecology and Sociobiology 29:221–224.

Larkin, R. P. 1991b. Sensitivity of NEXRAD algorithms to echoes from birds and insects. International Conference on Radar Meteorology 25:203–205.

Larkin, R. P. 2006. Locating bird roosts with Doppler radar. Pages 350–356 in R. M. Timm and J. M. O'Brien, editors. Proceedings 22nd vertebrate pest control conference. University of California, Davis, Davis, USA.

Larkin, R. P., W. R. Evans, and R. H. Diehl. 2002. Nocturnal flight calls of dickcissels and Doppler radar echoes over south Texas in spring. Journal of Field Ornithology 73:2–8.

Larkin, R. P., and B. A. Frase. 1988. Circular paths of birds flying near a broadcasting tower in cloud. Journal of Comparative Psychology 102:90–93.

Larkin, R. P., A. Raim, and R. H. Diehl. 1996. Performance of a non-rotating direction finder for automatic radio tracking. Journal of Field Ornithology 67:59–71.

Larkin, R. P., and R. E. Szafoni. 2008. Radar evidence for dispersed groups of migrating vertebrates at night. Integrative and Comparative Biology 48:40–49.

Larkin, R. P., T. R. VanDeelen, R. M. Sabick, T. E. Gosselink, and R. E. Warner. 2003. Electronic signaling for prompt removal of an animal from a trap. Wildlife Society Bulletin 31:391–398.

Larrucea, E. S., and P. F. Brussard. 2007. A method for capturing pygmy rabbits in summer. Journal of Wildlife Management 71:1016–1017.

Larrucea, E. S., and P. F. Brussard. 2009. Diel and seasonal activity patterns of pygmy rabbits (*Brachylagus idahoensis*). Journal of Mammalogy 90:1176–1183.

Larrucea, E. S., P. F. Brussard, M. N. Jaeger, and R. H. Barrett. 2007b. Cameras, coyotes, and the assumption of equal detectability. Journal of Wildlife Management 71:1682–1689.

Larrucea, E. S., G. Serra, M. N. Jaeger, and R. H. Barrett. 2007a. Censusing bobcats using remote cameras. Western North American Naturalist 67:538–548.

Larsen, D. T., P. L. Crookston, and L. D. Flake. 1994. Factors associated with ring-necked pheasant use of winter food plots. Wildlife Society Bulletin 22:620–626.

Larsen, K. H. 1970. A hoop-net trap for passerine birds. Bird-Banding 41:92–96.

Larsen, T. 1971. Capturing, handling, and marking polar bears in Svalbard. Journal of Wildlife Management 35:27–36.

Larson, G. E., P. J. Savarie, and I. Okuno. 1981. Iophenoxic acid and mirex for marking wild, bait-consuming animals. Journal of Wildlife Management 45:1073–1077.

Larson, J. S., and R. D. Taber. 1980. Criteria of sex and age. Pages 143–202 in S. D. Schemnitz, editor. Wildlife techniques manual. The Wildlife Society, Washington, D.C., USA.

Larson, J. S., and F. C. Van Nostrand. 1968. An evaluation of beaver aging techniques. Journal of Wildlife Management 32:99–103.

Larson, M. A., M. E. Clark, and S. R. Winterstein. 2001. Survival of ruffed grouse chicks in northern Michigan. Journal of Wildlife Management 65:880–886.

Larsson, J. K., H. A. H. Jansman, G. Segelbacher, J. Hoglund, and H. P. Koelewijn. 2008. Genetic impoverishment of the last black grouse (*Tetrao tetrix*) population in the Netherlands: detectable only with a reference from the past. Molecular Ecology 17:1897–1904.

Latch, E. K., and O. E. Rhodes. 2005. The effects of gene flow and population isolation on the genetic structure of reintroduced wild turkey populations: are genetic signatures of source populations retained? Conservation Genetics 6:981–997.

Laubhan, M. K., and L. H. Fredrickson. 1992. Estimating seed production of common plants in seasonally flooded wetlands. Journal of Wildlife Management 56:329–337.

Lauck, B. 2004. Using aquatic funnel traps to determine density of amphibian larvae: factors influencing trapping. Herpetological Review 35:248–250.

Launay, F., O. Combreau, S. J. Aspinall, R. A. Loughland, B. Gubin, and F. Karpov. 1999. Trapping of breeding houbara bustard (*Chlamydotis undulata*). Wildlife Society Bulletin 27:603–608.

Laundre, J. W., and L. Hernandez. 2002. Growth curve models and age estimation of young cougars in the northern Great Basin. Journal of Wildlife Management 66:849–858.

Laundre, J. W., and L. Hernandez, D. Streubel, K. Altendorf, and C. Lopez Gonzalez. 2000. Aging mountain lions using gum line recession. Wildlife Society Bulletin 28:963–966.

Laurent, L., P. Casale, M. N. Bradai, B. J. Godley, G. Gerosa, A. C. Broderick, W. Schroth, B. Schierwater, A. M. Levy, D. Freggi, E. M. Abd El-Mawla, D. A. Hadoud, H. E. Gomati, M. Domingo, M. Hadjichristophorou, L. Kornaraky, F. Demirayak, and C. Gautier. 1998. Molecular resolution of marine turtle stock composition in fishery bycatch: a case study in the Mediterranean. Molecular Ecology 7:1529–1542.

LaVal, R. K., R. L. Clawson, M. L. LaVal, and W. Caire. 1977. Foraging behavior and nocturnal activity patterns of Missouri bats, with emphasis on the endangered species *Myotis grisescens* and *Myotis sodalis*. Journal of Mammalogy 58:592–599.

Laver, P. N., and M. J. Kelly. 2008. A critical review of home range studies. Journal of Wildlife Management 72:290–298.

Law, R. J., M. Alaee, C. R. Allchin, J. P. Boon, M. Lebeuf, P. Lepom, and G. A. Stern. 2003. Levels and trends of polybrominated diphenylethers and other brominated flame retardants in wildlife. Environment International 29:757–770.

Lawrence, J. S., and N. J. Silvy. 1987. Movement and mortality of transplanted Attwater's prairie chickens. Journal of the World Pheasant Association 12:57–65.

Laws, R. M. 1962. Age determination of pinnipeds with special reference to growth layers in the teeth. Zeitschrift für Saugetierkunde 27:129–146.

Laws, R. M. 1967. Occurrence of placental scars in the uterus of the African elephant (*Loxodonta africana*). Journal of Reproduction and Fertility 14:445–449.

Lawson, B., and R. Johnson. 1982. Mountain sheep. Pages 1036–1055 *in* J. A. Chapman and G. A. Feldhamer, editors. Wild mammals of North America. Johns Hopkins University Press, Baltimore, Maryland.

Laycock, W. A. 1965. Adaptation of distance measurements for range sampling. Journal of Range Management 18:205–211.

Laycock, W. A., and C. L. Batcheler. 1975. Comparison of distance-measurement techniques for sampling tussock grassland species in New Zealand. Journal of Range Management 28:235–239.

Layfield, J. A., D. A. Galbraith, and R. J. Brooks. 1988. A simple method to mark hatchling turtles. Herpetological Review 19:78–79.

Layzer, J. 2008. Natural experiments: ecosystem-based management and the environment. MIT Press, Cambridge, Massachusetts, USA.

Lazarus, A. B., and F. P. Rowe. 1975. Freeze-marking rodents with a pressurized refrigerant. Mammal Review 5:31–34.

Leach, D. 1977. The descriptive and comparative postcranial osteology of marten (*Martes americana* Turton) and fisher (*Martes pennanti* Erxleben): the appendicular skeleton. Canadian Journal of Zoology 55:199–214.

Leach, D., and V. S. de Kleer. 1978. The descriptive and comparative postcranial osteology of marten (*Martes americana* Turton) and fisher (*Martes pennanti* Erxleben): the axial skeleton. Canadian Journal of Zoology 56:1180–1191.

Leach, D., B. K. Hall, and A. I. Dagg. 1982. Aging marten and fisher by development of the suprafabellar tubercle. Journal of Wildlife Management 46:246–247.

Leary, C. J., A. M. Garcia, R. Knapp, and D. L. Hawkins. 2008. Relationships among steroid hormone levels, vocal effort and body condition in an explosive-breeding toad. Animal Behavior 76:175–185.

Leasure, S. M., and D. W. Holt. 1991. Techniques for locating and capturing nesting short-eared owls (*Asio flammeus*). North American Bird Bander 16:32–33.

Leatherwood, S., D. K. Caldwell, and H. E. Winn. 1976. Whales, dolphins, and porpoises of the western North Atlantic—a guide to their identification. Technical Report, NMFS Special Scientific Report Fisheries 396, U.S. National Oceanic and Atmospheric Administration, Department of Commerce, Washington, D.C., USA.

Leberg, P. L. 1990a. Genetic considerations in the design of introduction programs. Transactions of the North American Wildlife and Natural Resources Conference 55:609–619.

Leberg, P. L. 1990b. Influence of genetic variability on population growth: implications for conservation. Journal of Fish Biology 37:193–195.

Leberg, P. L. 1991. Influence of fragmentation and bottlenecks on genetic divergence of wild turkey populations. Conservation Biology 5:522–530.

Leberg, P. L. 1992. Effects of population bottlenecks on genetic diversity as measured by allozyme electrophoresis. Evolution 46:477–494.

Leberg, P. L. 1993. Strategies for population reintroduction: effects of genetic variability on population growth and size. Conservation Biology 7:194–199.

Leberg, P. L. 1999. Using genetic markers to assess the success of translocation programs. Transactions of the North American Wildlife and Natural Resources Conference 64:174–190.

Leberg, P. L. 2002. Estimating allelic richness: effects of sample size and bottlenecks. Molecular Ecology 11:2445–2449.

Leberg, P. L. 2005. Genetic approaches for estimating the effective size of populations. Journal of Wildlife Management 69:1385–1399.

Leberg, P. L., and D. L. Ellsworth. 1999. Further evaluation of the genetic consequences of translocations on southeastern white-tailed deer populations. Journal of Wildlife Management 63:327–334.

Leberg, P. L., and B. D. Firmin. 2008. Role of inbreeding depression and purging in captive breeding and restoration programmes. Molecular Ecology 17:334–343.

Leberg, P. L., M. H. Smith, and O. E. Rhodes. 1990. The association between heterozygosity and growth of deer fetuses is not explained by effects of the loci examined. Evolution 44:454–458.

Leberg, P. L., P. W. Stangel, H. O. Hillestad, R. L. Marchinton, and M. H. Smith. 1994. Genetic structure of reintroduced wild turkey and white-tailed deer populations. Journal of Wildlife Management 58:698–711.

LeBoulenge-Nguyen, P. Y., and E. LeBoulenge. 1986. New ear-tag for small mammals. Journal of Zoology 209:302–304.

Lebreton, J.-D., K. P. Burnham, J. Clobert, and D. R. Anderson. 1992. Modeling survival and testing hypotheses using marked animals: a unified approach with case studies. Ecological Monographs 62:67–118.

LeBuff, C. R., and R. W. Beatty. 1971. Some aspects of nesting of the loggerhead turtle, *Caretta caretta caretta* (Linne), on the Gulf Coast of Florida. Herpetologica 27:153–156.

Lechleitner, R. R. 1954. Age criteria in mink (*Mustela vison*). Journal of Mammalogy 35:496–503.

Leclercq, G. C., and F. M. Rozenfeld. 2001. A permanent marking method to identify individual small rodents from birth to sexual maturity. Journal of Zoology (London) 254:203–206.

Lecomte, N., G. Gauthier, L. Bernatchez, and J. F. Giroux. 2006. A non-damaging blood sampling technique for waterfowl embryos. Journal of Field Ornithology 77:67–70.

Lee, J. E., G. C. White, R. A. Garrott, R. M. Bartmann, and A. W. Alldredge. 1985. Assessing accuracy of a radiotelemetry system for estimating animal locations. Journal of Wildlife Management 49:658–663.

Lee, J. W., B. S. Jang, D. A. Dawson, T. Burke, and

B. J. Hatchwell. 2009. Fine-scale genetic structure and its consequence in breeding aggregations of a passerine bird. Molecular Ecology 18:2728–2739.

Lee, P. 1997. Bayesian statistics: an introduction. Second edition. Oxford University Press, New York, New York, USA.

Lee, P. 2004. Bayesian statistics: an introduction. Third edition. Oxford University Press, New York, New York, USA.

Lee, R. M., J. D. Yoakum, B. W. O'Gara, T. M. Pojar, and R. A. Ockenfels, editors. 1998. Pronghorn management guides. Eighteenth pronghorn antelope workshop. Arizona Game and Fish Department, Phoenix, USA.

Lefkovitch, L. P. 1965. The study of population growth in organisms grouped by stages. Biometrics 21:1–18.

Lefsky, M. A., W. B. Cohen, S. A. Acker, T. A. Spies, G. G. Parker, and D. Harding. 1999. Lidar remote sensing of the biophysical properties and canopy structure of forests of Douglas-fir and western hemlock. Remote Sensing of Environment 70:339–361.

Lefsky, M. A., W. B. Cohen, G. G. Parker, and D. J. Harding. 2002. Lidar remote sensing for ecosystem studies. BioScience 52:19–30.

Lefsky, M. A., D. J. Harding, M. Keller, W. B. Cohen, C. C. Carabajal, F. Del Born Espirito-Santo, M. O. Hunter, and R. de Oliveira, Jr. 2005. Estimates of forest canopy height and above-ground biomass using ICEsat. Geophysical Research Letters 32:L22S02.

Lehmann, T. 1993. Ectoparasites: direct impacts on host fitness. Parasitology Today 9:8–13.

Lehmkuhl, J. F., M. Kennedy, E. D. Ford, P. H. Singleton, W. L. Gaines, and R. L. Lind. 2007. Seeing the forest for the fuel: integrating ecological values and fuels management. Forest Ecology and Management 246:73–80.

Lehmkuhl, J. F., K. D. Kistler, and J. S. Begley. 2006. Bushy-tailed woodrat abundance in dry forests of eastern Washington. Journal of Mammalogy 87:371–379.

Lehner, P. N. 1996. Handbook of ethological methods. Second edition. Cambridge University Press, New York, New York, USA.

Lehr, R. 1992. Sixteen S-squared over D-squared: a relation for crude sample size estimates. Statistics in Medicine 11:1099–1102.

Leighton, F. A. 2001. *Fusobacterium necrophorum* infection. Pages 493–496 *in* E. S. Williams and I. K. Barker, editors. Infectious diseases of wild mammals. Iowa State University Press, Ames, USA.

Leimgruber, P., W. McShea, and J. Rappole. 1994. Predation on artificial nests in large forest blocks. Journal of Wildlife Management 58:254–260.

Lele, S. R., B. Dennis, and F. Lutscher. 2007. Data cloning: easy maximum likelihood estimation for complex ecological models using Bayesian Markov chain Monte Carlo methods. Ecology Letters 10:551–563.

Lemen, C. A., and P. W. Freeman. 1985. Tracking mammals with fluorescent pigments: a new technique. Journal of Mammalogy 66:134–136.

Lemkau, P. J. 1970. Movements of the box turtle, *Terrapene c. carolina* (Linnaeus), in unfamiliar territory. Copeia 1970:781–783.

Lemmon, P. E. 1957. A new instrument for measuring forest overstory density. Journal of Forestry 55:667–669.

Lemnell, P. A. 1974. Age determination in red squirrels (*Sciurus vulgaris* [L.]). International Congress of Game Biologists 11:573–580.

Lensink, C. J. 1968. Neckbands as an inhibitor of reproduction in black brant. Journal of Wildlife Management 32:418–420.

Lentfer, J. W. 1968. A technique for immobilizing and marking polar bears. Journal of Wildlife Management 32:317–321.

Lenth, R. V. 2001. Some practical guidelines for effective sample size determination. American Statistician 55:187–193.

Leonard, J. 2008. Wildlife watching in the U.S.: the economic impacts on national and state economies in 2006. Report 2006-1, Addendum to the 2006 national survey of fishing, hunting, and wildlife-associated recreation, U.S. Fish and Wildlife Service, Department of the Interior, Washington, D.C., USA.

Leopold, A. 1933. Game management. Charles Scribner's Sons, New York, New York, USA.

Leopold, A. 1949. A Sand County almanac. Oxford University Press, New York, New York, USA.

Leopold, A. 1970. A Sand County almanac: with essays on conservation from Round River. Ballantine Books, New York, New York, USA.

Leopold, A. S. 1959. Wildlife of Mexico: the game birds and mammals. University of California Press, Berkeley, USA.

Lepczyk, C. A. 2005. Integrating published data and citizen science to describe bird diversity across a landscape. Journal of Applied Ecology 42:672–677.

Lercel, B. A., R. M. Kaminski, and R. R. Cox, Jr. 1999. Mate loss in winter affects reproduction of mallards. Journal of Wildlife Management 63:621–629.

LeResche, R. E., and G. M. Lynch. 1973. A trap for free ranging moose. Journal of Wildlife Management 37:87–89.

Lesage, L., J.-P. L. Savard, and A. Reed. 1997. A simple technique to capture breeding adults and broods of surf scoters, *Melanitta perspicillata*. Canadian Field-Naturalist 111:657–659.

Leshem, Y. 1995. Foreword. Proceedings of the international conference on bird migration of the Society for the Protection of Nature in Israel. Israel Journal of Zoology 41:R7–R8.

Leslie, P. H. 1945. On the use of matrices in certain population mathematics. Biometrika 33:183–212.

Leslie, P. H. 1948. Some further notes on the use of matrices in population mathematics. Biometrika 35:213–245.

Leslie, P. H. 1959. The properties of a certain lag type of population growth and the influence of an external random factor on a number of such populations. Physiological Zoology 32:151–159.

Leslie, P. H., and D. H. S. Davis. 1939. An attempt to determine the absolute number of rats on a given area. Journal of Animal Ecology 8:94–113.

Letham, L. 1998. GPS made easy: using global positioning systems in the outdoors. Mountaineers Books, Seattle, Washington, USA.

Lettink, M. 2007. Comparison of two techniques for capturing geckos in rocky habitat. Herpetological Review 38:415–418.

Levanon, N. 1988. Radar principles. John Wiley and Sons, New York, New York, USA.

Levin, S. A. 1992. The problem of pattern and scale in ecology. Ecology 73:1943–1967.

Levins, R. 1966. The strategy of model building in population biology. American Scientist 54:421–431.

Levins, R. 1968. Evolution in changing environments. Princeton University Press, Princeton, New Jersey, USA.

Levins, R. 1969. Some demographic and genetic consequences of environmental heterogeneity for biological control. Bulletin of the Entomological Society of America 15:237–240.

Levins, R. 1970. Extinction. Pages 77–107 *in* M. Gerstenhaber, editor. Some mathematical questions in biology. American Mathematical Society, Providence, Rhode Island, USA.

Levy, E. E., and E. A. Madden. 1933. The point method of pasture analysis. New Zealand Agricultural Journal 46:267–279.

Levy, S. H., J. J Levy, and R. A. Bishop. 1966. Use of tape recorded female quail calls. Journal of Wildlife Management 30:426–428.

Lewis, D. M. 1995*a*. Importance of GIS to community-based management of wildlife: lessons from Zambia. Ecological Applications 5:861–871.

Lewis, E. G. 1942. On the generation and growth of a population. Sankhya 6:93–96.

Lewis, J. C. 1979. Field identification of juvenile sandhill cranes. Journal of Wildlife Management 43:211–214.

Lewis, J. C. 1995*b*. Whooping crane (*Grus americana*). Account 153 *in* A. Poole, P. Stettenheim, and F. Gill, editors. The birds of North America. The Academy of Natural Sciences, Philadelphia, Pennsylvania, and The American Ornithologists' Union, Washington, D.C., USA.

Lewis, J. S., J. L. Rachlow, E. O. Garton, and L. A. Vierling. 2007. Effects of habitat on GPS collar performance: using data screening to reduce location error. Journal of Applied Ecology 44:663–671.

Lewis, L. A., R. J. Poppenga, W. R. Davison, J. R. Fischer, and K. A. Morgan. 2001. Lead toxicosis and trace element levels in wild birds and mammals at a firearms training facility. Archives of Environmental Contamination and Toxicology 41:208–214.

Lewis, P. J., R. Strauss, E. Johnson, and W. C. Conway. 2002. Absence of sexual dimorphism in molar morphology of muskrats. Journal of Wildlife Management 66:1189–1196.

Lewke, R. E., and R. K. Stroud. 1974. Freeze branding as a method of marking snakes. Copeia 1974:997–1000.

Leyton, V., and J. Valencia. 1992. Follicular population dynamics: its relation to clutch and litter size in Chilean *Liolaemus* lizards. Pages 123–134 *in* W. C. Hammett, editor. Reproduction in South American vertebrates. Springer-Verlag, New York, New York, USA.

Libby, W. L. 1957. A better snowshoe hare live trap. Journal of Wildlife Management 21:452.

Licht, P., B. R. McCreery, R. Barnes, and R. Pang. 1983. Seasonal and stress related changes in plasma gonadotropins, sex steroids, and corticosterone in the bullfrog, *Rana catsbeiana*. General and Comparative Endocrinology 50:124–145.

Lieberman, D. E., T. W. Deacon, and R. H. Meadow. 1990. Computer image enhancement and analysis of cementum increments as applied to teeth of *Gazella gazella*. Journal of Archaeological Science 17:519–533.

Liechti, F., and B. Bruderer. 1995. Direction, speed, and composition of nocturnal bird migration in the south of Israel. Israel Journal of Zoology 41:501–515.

Liechti, F., B. Bruderer, and H. Paproth. 1995. Quantification of nocturnal bird migration by moonwatching: comparison with radar and infrared observations. Journal of Field Ornithology 66:457–468.

Lightfoot, W. C., and V. Maw. 1963. Trapping and marking mule deer. Proceedings of the Western Association of State Game and Fish Commissioners 43:138–141.

Lima, S. L., and P. A. Zollner. 1996. Towards a behavioral ecology of ecological landscapes. Trends in Ecology and Evolution 11:131–135.

Lindberg, A. C., J. A. Shivik, and L. Clark. 2000. Mechanical mouse lure for brown tree snakes. Copeia 2000:886–888.

Linder, R. L., R. B. Dahlgren, and C. R. Elliot. 1971. Primary feather pattern as a sex criterion in the pheasant. Journal of Wildlife Management 35:840–843.

Linders, M. J., S. D. West, and W. M. Vander Hagen. 2004. Season variability in the use of space by western gray squirrels in southeastern Washington. Journal of Mammalogy 85:511–516.

Lindmeier, J. P., and R. L. Jessen. 1961. Results of capturing waterfowl in Minnesota by spotlighting. Journal of Wildlife Management 25:430–431.

Lindsay, D. L., K. R. Barr, R. F. Lance, S. A. Tweddale, T. J. Hayden, and P. L. Leberg. 2008. Habitat fragmentation and genetic diversity of an endangered, migratory songbird, the golden-cheeked warbler (*Dendroica chrysoparia*). Molecular Ecology 17:2122–2133.

Lindsey, A. A., J. D. Barton, Jr., and S. R. Miles. 1958. Field efficiencies of forest sampling methods. Ecology 39:428–444.

Lindsey, G. D. 1983. Rhodamine B: a systemic fluorescent marker for studying mountain beavers (*Aplodontia rufa*) and other animals. Northwest Scientist 57:16–21.

Lindstrom, E. B., M. W. Eichholz, and J. M. Eadie. 2006. Postovulatory follicles in mallards: implications for estimates of breeding propensity. Condor 108:925–935.

Linduska, J. P. 1942. A new technique for marking fox squirrels. Journal of Wildlife Management 6:93–94.

Lindzey, F. G. 1987. Mountain lion. Pages 658–668 *in* M. Novak, J. A. Baker, M. E. Obbard, and B. Malloch, editors. Wild furbearer management and conservation in North America. Ontario Ministry of Natural Resources, Toronto, Canada.

Linhart, S. B., F. S. Blom, G. J. Dasch, R. M. Engeman, and G. H. Olsen. 1988. Field evaluation of padded jaw coyote traps: effectiveness and foot injury. Proceedings of the Vertebrate Pest Conference 13:226–229.

Linhart, S. B., and G. J. Dasch. 1992. Improved performance of padded jaw traps for capturing coyotes. Wildlife Society Bulletin 20:63–66.

Linhart, S. B., G. J. Dasch, C. B. Male, and R. M. Engeman. 1986. Efficiency of unpadded and padded steel foothold traps for capturing coyotes. Wildlife Society Bulletin 14:212–218.

Linhart, S. B., G. J. Dasch, and F. J. Turkowski. 1981. The steel leg-hold trap: techniques for reducing foot injury and increasing selectivity. Proceedings of the Worldwide Furbearer Conference 3:1560–1578.

Linhart, S. B., and J. J. Kennelly. 1967. Fluorescent bone labeling of coyotes with demethylchlortetracycline. Journal of Wildlife Management 31:317–321.

Linhart, S. B., and F. F. Knowlton. 1967. Determining age of coyotes by tooth cementum layers. Journal of Wildlife Management 31:362–365.

Linhart, S. B., and R. G. Linscombe. 1987. Test methods for steel foothold traps: criteria and performance standards. Pages 148–158 *in* S. A. Shumake and R. W. Bullard, editors. Vertebrate pest control and management materials. Volume 5. Special Technical Publication 974, American Society for Testing and Materials, Philadelphia, Pennsylvania, USA.

Link, W. A., and R. J. Barker. 2009. Bayesian inference with ecological examples. Academic Press, London, England, UK.

Link, W. A., and R. J. Barker. 2010. Bayesian inference: with ecological applications. Elsevier/Academic, Boston, Massachusetts, USA.

Link, W. A., and J. R. Sauer. 1997. Estimation of population trajectories from count data. Biometrics 53:488–497.

Link, W. A., and J. R. Sauer. 1998. Estimating population change from count data: application to the North American Breeding Bird Survey. Ecological Applications 8:258–268.

Link, W. A., and J. R. Sauer. 2002. A hierarchical analysis of population change with application to cerulean warblers. Ecology 83:2832–2840.

Linn, I. J. 1978. Radioactive techniques for small mammal marking. Pages 177–191 *in* B. Stonehouse, editor. Animal marking: recognition marking of animals in research. MacMillan, London, England, UK.

Linn, I. J., and J. Shillito. 1960. Rings for marking very small mammals. Proceedings of the Zoological Society of London 134:489–495.

Linscombe, R. G. 1976. An evaluation of the no. 2 Victor and 220 Conibear traps in coastal Louisiana. Proceedings of the Annual Conference of the Southeastern Association of Fish and Wildlife Agencies 30:560–568.

Linscombe, R. G., and V. L. Wright. 1988. Efficiency of padded foothold traps for capturing terrestrial furbearers. Wildlife Society Bulletin 16:307–309.

Liscinsky, S. A., and W. J. Bailey, Jr. 1955. A modified shorebird trap for capturing woodcock and grouse. Journal of Wildlife Management 19:405–408.

Lishak, R. S. 1976. A burrow entrance snare for capturing ground squirrels. Journal of Wildlife Management 40:364–365.

Littell, R. 1993. Controlled wildlife. Second edition. Association of Systematic Collections, Washington, D.C., USA.

Littell, R. C., G. A. Milliken, W. W. Stroup, R. D. Wolfinger, and O. Schabenberger. 2006. SAS for mixed models. Second edition. SAS Institute, Cary, North Carolina, USA.

Litvaitis, J. A., J. A. Sherburne, and J. A. Bissonette. 1985*a*. A comparison of methods used to examine snowshoe hare habitat use. Journal of Wildlife Management 49:693–695.

Litvaitis, J. A., J. A. Sherburne, and J. A. Bissonette. 1985*b*. Influence of understory characteristics on snowshoe hare habitat use and density. Journal of Wildlife Management 49:866–873.

Litvaitis, J. A., J. P. Tash, M. K. Litvaitis, M. N. Marchand, A. I. Kovach, and R. Innes. 2006. A range-wide survey to determine the current distribution of New England cottontails. Wildlife Society Bulletin 34:1190–1197.

Litvaitis, J. A., K. Titus, and E. M. Anderson. 1994. Measuring vertebrate use of territorial habitats and foods. Pages 254–274 *in* T. A. Bookhout, editor. Research and management techniques for wildlife and habitats. The Wildlife Society, Bethesda, Maryland, USA.

Liu, J., J. B. Dunning, Jr., and H. R. Pulliam. 1995. Potential effects of a forest management plan on Bachman's sparrows (*Aimophila aestivalis*): linking a spatially explicit model with GIS. Conservation Biology 9:62–75.

Llewellyn, D. W., G. P. Shaffer, N. J. Craig, L. Creasman, D. Pashley, M. Swan, and C. Brown. 1996. A decision-support system for prioritizing restoration sites on the Mississippi River alluvial plain. Conservation Biology 10:1446–1455.

Loafman, P. 1991. Identifying individual spotted salamanders by spot pattern. Herpetological Review 22:91–92.

Lochmiller, R. L., E. C. Hellgren, and W. E. Grant. 1984. Sex and age characteristics of the pelvic girdle in the collared peccary. Journal of Wildlife Management 48:48–52.

Lockard, G. R. 1972. Further studies of dental annuli for aging white-tailed deer. Journal of Wildlife Management 36:46–55.

Locke, S. L., C. E. Brewer, and L. A. Harveson. 2005. Identifying landscapes for desert bighorn sheep translocations in Texas. Texas Journal of Science 57:25–34.

Locke, S. L., M. F. Hess, B. G. Mosley, M. W. Cook, S. Hernandez, I. D. Parker, L. A. Harveson, R. R. Lopez, and N. J. Silvy. 2004. Portable drive-net for capturing urban white-tailed deer. Wildlife Society Bulletin 32:1093–1098.

Locke, S. L., R. R. Lopez, M. J. Peterson, N. J. Silvy, and T. W. Schwertner. 2006. Evaluation of portable infrared cameras for detecting Rio

Grande wild turkeys. Wildlife Society Bulletin 34:839–844.

Lockowandt, S. P. E. 1993. An electromagnetic trigger for drop-nets. Wildlife Society Bulletin 21:140–142.

Loery, G., K. H. Pollock, J. D. Nichols, and J. E. Hines. 1987. Age-specificity of black-capped chickadee survival rates: analysis of capture–recapture data. Ecology 68:1038–1044.

Lofroth, E. C., R. Klafki, J. R. Krebs, and D. Lewis. 2008. Evaluation of live-capture techniques for free-ranging wolverines. Journal of Wildlife Management 72:1253–1261.

Loft, E. R., J. W. Menke, J. G. Kie, and R. C. Bertram. 1987. Influence of cattle stocking rate on the structural profile of deer hiding cover. Journal of Wildlife Management 51:655–664.

Logan, K. A., L. L. Sweanor, J. F. Smith, and M. G. Hornocker. 1999. Capturing pumas with foot-hold snares. Wildlife Society Bulletin 27:201–208.

Logan, T. J., and G. Chandler. 1987. A walk-in trap for sandhill cranes. Pages 221–223 in J. C. Lewis, editor. Proceedings of the 1985 crane workshop. Whooping Crane Maintenance Trust, Grand Island, Nebraska, USA.

Lohmann, K. J., S. D. Cain, S. A. Dodge, and C. M. Lohmann. 2001. Regional magnetic fields as navigational markers for sea turtles. Science 12:364–366.

Lohoefener, R., and J. Wolfe. 1984. A "new" live trap and a comparison with a pit-fall trap. Herpetological Review 15:25–26.

Lokemoen, J. T., and D. E. Sharp. 1985. Assessment of nasal marker materials and designs used on dabbling ducks. Wildlife Society Bulletin 13:53–56.

Lomas, L. A., and L. C. Bender. 2007. Survival and cause-specific mortality of neonatal mule deer fawns, north-central New Mexico. Journal of Wildlife Management 71:884–894.

Lombardo, M. P., and E. Kemly. 1983. A radio-control method for trapping birds in nest boxes. Journal of Field Ornithology 54:194–195.

Lommasson, T., and C. Jensen. 1938. Grass volume tables for determining range utilization. Science 87:444.

Loncarich, F. L., and D. G. Krementz. 2004. External determination of age and sex of the common moorhen. Wildlife Society Bulletin 32:655–660.

Loncke, D. J., and M. E. Obbard. 1977. Tag success, dimensions, clutch size and nesting site fidelity for the snapping turtle, *Chelydra serpentina* (Reptilia, Testudines, Chelydridae) in Algonquin Park, Ontario, Canada. Journal of Herpetology 11:243–244.

Long, R. A., T. M. Donovan, P. Mackay, W. J. Zielinski, and J. S. Buzas. 2007a. Effectiveness of scat detection dogs for detecting forest carnivores. Journal of Wildlife Management 71:2007–2017.

Long, R. A., T. M. Donovan, P. Mackay, W. J. Zielinski, and J. S. Buzas. 2007b. Comparing scat detection dogs, cameras, and hair snares for surveying carnivores. Journal of Wildlife Management 71:2018–2025.

Long, R. A., J. L. Rachlow, and J. G. Kie. 2008a. Effects of season and scale on response of elk and mule deer to habitat manipulation. Journal of Wildlife Management 72:1133–1142.

Long, R. A., J. L. Rachlow, J. G. Kie, and M. Vavra. 2008b. Fuels reduction in a western coniferous forest: effects on quantity and quality of forage for elk. Rangeland Ecology and Mangement 61:302–313.

Loos, E. H., and F. C. Rohwer. 2002. Efficiency of nest traps and long-handled nets for capturing upland nesting ducks. Wildlife Society Bulletin 30:1202–1207.

Lopez, R. R., and N. J. Silvy. 1999. Use of infrared-triggered cameras and monitors in aquatic environments. Proceedings of the Annual Conference of the Southeastern Association of Fish and Wildlife Agencies 53:200–203.

Lopez, R. R., N. J. Silvy, J. D. Sebesta, S. D. Higgs, and M. W. Salazar. 1998. A portable drop net for capturing urban deer. Proceedings of the Annual Conference of the Southeastern Association of Fish and Wildlife Agencies 52:206–209.

Lopez, R. R., M. E. Vieira, N. J. Silvy, P. A. Frank, S. W. Whisenant, and D. A. Jones. 2003. Survival, mortality, and life expectancy of Florida Key deer. Journal of Wildlife Management 67:34–45.

Lord, R. D., Jr. 1959. The lens as an indicator of age in cottontail rabbits. Journal of Wildlife Management 23:358–360.

Lord, R. D., Jr. 1961. The lens as an indicator of age in the gray fox. Journal of Mammalogy 42:109–111.

Lord, R. D., F. C. Bellrose, and W. W. Cochran. 1962. Radiotelemetry of the respiration of a flying duck. Science 137:39–40.

Lotka, A. J. 1925. Elements of physical biology. Williams and Wilkins, Baltimore, Maryland, USA.

Lotti, T., and W. P. LeGrande. 1959. Loblolly pine seed production and seedling crops in the lower coastal plain of South Carolina. Journal of Forestry 57:580–581.

Löttker P., A. Rummel, M. Traube, A. Stache, P. Šustr, J. Müller, and M. Heurich. 2009. New possibilities of observing animal behaviour from distance using activity sensors in GPS-collars—an attempt to calibrate remotely collected activity data with direct behavioral observations in red deer. Wildlife Biology 15:425–434.

Low, S. H. 1935. Methods of trapping shorebirds. Bird-Banding 6:16–22.

Low, W. A., and I. M. Cowan. 1963. Age determination of deer by annular structure of dental cementum. Journal of Wildlife Management 27:466–471.

Lowe, V. P. W. 1969. Population dynamics of red deer (*Cervus elaphus* L.) on Rhum. Journal of Animal Ecology 38:425–457.

Lowrance, E. W. 1949. Variability and growth of the opossum skeleton. Journal of Morphology 85:569–593.

Lowry, M. S., and R. L. Folk. 1990. Sex determination of the California sea lion (*Zalophus californianus californianus*) from canine teeth. Marine Mammal Science 6:25–31.

Lucchini, V., E. Fabbri, F. Marucco, S. Ricci, L. Boitani, and E. Randi. 2002. Noninvasive molecular tracking of colonizing wolf (*Canis lupus*) packs in the western Italian Alps. Molecular Ecology 11:857–868.

Ludwig, J., and D. E. Davis. 1975. An improved woodchuck trap. Journal of Wildlife Management 39:327–329.

Ludwig, J. A., and J. F. Reynolds. 1988. Statistical ecology: a primer on methods and computing. John Wiley and Sons, New York, New York, USA.

Ludwig, J. R., and R. W. Dapson. 1977. Use of insoluble lens proteins to estimate age in white-tailed deer. Journal of Wildlife Management 41:327–329.

Luhring, T. M., and C. A. Young. 2006. Innovative techniques for sampling stream inhabiting salamanders. Herpetological Review 37:181–183.

Luikart, G., F. W. Allendorf, J.-M. Cornuet, and W. B. Sherwin. 1998. Distortion of allele frequency distributions provides a test for recent population bottlenecks. Journal of Heredity 89:238–247.

Luikart, G. and J.-M. Cornuet. 1998. Empirical evaluation of a test for identifying recently bottlenecked populations from allele frequency data. Conservation Biology 12:228–237

Luikart, G., J.-M. Cornuet, and F. W. Allendorf. 1999. Temporal changes in allele frequencies provide estimates of population bottleneck size. Conservation Biology 13:523–530.

Lukacs, P. M., and K. P. Burnham. 2005a. Review of capture–recapture methods applicable to noninvasive genetic sampling. Molecular Ecology 14:3909–3919.

Lukacs, P. M., and K. P. Burnham. 2005b. Estimating population size from DNA-based closed capture–recapture data incorporating genotyping error. Journal of Wildlife Management 69:396–403.

Lukacs, P. M., W. L. Thompson, W. L. Kendall, W. R. Gould, P. F. Doherty, Jr., K. P. Burnham, and D. R. Anderson. 2007. Concerns regarding a call for pluralism of information theory and hypothesis testing. Journal of Applied Ecology 44:456–460.

Lumsden, H. G., V. W. McMullen, and C. L. Hopkinson. 1977. An improvement in fabrication of large plastic leg bands. Journal of Wildlife Management 41:148–149.

Lutterschmidt, W. I., and J. F. Schaefer. 1996. Mist netting: adapting a technique from ornithology for sampling semi-aquatic snake populations. Herpetological Review 27:131–132.

Lydersen, C., and Gjertz, I. 1986. Studies of the ringed seal *Phoca hispida* in its breeding habitat in Kongsfjorden Svalbard Arctic Ocean. Polar Research 4:57–64.

Lyles, D. S., and C. E. Rupprecht. 2007. Rhabdoviridae. Pages 1364–1408 in B. N. Fields and P. M. Howley, editors. Fields virology. Lippincott Williams and Wilkins, Philadelphia, Pennsylvania, USA.

Lyon, L. J. 1968a. An evaluation of density sampling methods in a shrub community. Journal of Range Management 21:16–20.

Lyon, L. J. 1968b. Estimating twig production of serviceberry from crown volumes. Journal of Wildlife Management 32:115–119.

Lyon, L. J. 1970. Length– and weight–diameter relations of serviceberry twigs. Journal of Wildlife Management 34:456–460.

Lyons, J. E., M. C. Runge, H. P. Laskowski, and W. L. Kendall. 2008. Monitoring in the context of structured decision-making and adaptive management. Journal of Wildlife Management 72:1683–1692.

Lyons, K. L. 1981. Use of a chick distress call to capture ruffed grouse hens. Transactions of the Northeastern Fish and Wildlife Conference 38:133–135.

MacArthur, R. H. 1965. Patterns of species diversity. Biological Reviews 40:510–533.

MacArthur, R. H. 1968. The theory of the niche. Pages 159–176 in R. C. Lewontin, editor. Population biology and evolution. Syracuse University Press, Syracuse, New York, USA.

MacArthur, R. H. 1972. Geographical ecology: patterns in the distribution of species. Harper and Row, New York, New York, USA.

MacArthur, R. H., and E. O. Wilson. 1967. The theory of island biogeography. Princeton University Press, Princeton, New Jersey, USA.

MacCracken, J. G., and V. Van Ballenberghe. 1987. Age- and sex-related differences in fecal pellet dimensions of moose. Journal of Wildlife Management 51:360–364.

MacDonald, D. W., F. G. Ball, and N. G. Hough. 1980. The evaluation of home range size and configuration using radio tracking data. Pages 405–424 in C. J. Amlander and D. W. Macdonald, editors. A handbook on biotelemetry and radio tracking. Pergamon Press, Oxford, England, UK.

MacDonald, R. N. 1961. Injury to birds by ice-coated bands. Bird-Banding 32:59.

MacDonald, S. D. 1968. The courtship and territorial behavior of Franklin's race of the spruce grouse. Living Bird 7:5–27.

Mace, G. M., T. B. Smith, M. W. Bruford, and R. K. Wayne. 1996. An overview of the issues. Pages 3–21 in T. B. Smith and R. K. Wayne, editors. Molecular genetic approaches in conservation. Oxford University Press, New York, New York, USA.

Mace, R., S. Minta, T. Manley, and K. Aune. 1994. Estimating grizzly bear population size using camera sightings. Wildlife Society Bulletin 22:74–83.

Mace, R. U. 1971. Trapping and transplanting Roosevelt elk to control damage and establish new populations. Proceedings of the Western Association of State Game and Fish Commissioners 51:464–470.

MacGown, J. A., and R. L. Brown. 2006. Survey of ants (Hymenoptera: Formicidae) of the Tombigbee National Forest in Mississippi. Journal of the Kansas Entomological Society 79:325–340.

MacInnes, C. D., and E. H. Dunn. 1988. Effects of neck bands on Canada geese nesting at the McConnell River. Journal of Field Ornithology 59:239–246.

MacInnes, C. D., J. P. Prevett, and H. A. Edney. 1969. A versatile collar for individual identification of geese. Journal of Wildlife Management 33:330–335.

MacKay, P., D. A. Smith, R. A. Long, and M. Parker. 2008. Scat detection dogs. Pages 183–222 in R. A. Long, P. MacKay, W. J. Zielinski, and J. C. Ray, editors. Noninvasive survey methods for carnivores. Island Press, Washington, D.C., USA.

MacKenzie, D. I., J. D. Nichols, J. A. Royle, K. H. Pollock, L. L. Bailey, and J. E. Hines. 2006. Occupancy estimation and modeling: inferring patterns and dynamics of species occurrence. Academic Press, London, England, UK.

Mackey, B. G., H. A. Nix, M. F. Hutchinson, J. P. Macmahon, and P. M. Fleming. 1988. Assessing representativeness of places for conservation reservation and heritage listing. Environmental Management 12:501–514.

Mackey, R. L., B. R. Page, D. Grobler, and R. Slotow. 2009. Modeling the effectiveness of contraception for controlling introduced populations of elephant in South Africa. African Journal of Ecology 47:747–755.

Maclean, G. A., and W. B. Krabill. 1986. Gross-merchantable timber volume estimation using an airborne LIDAR system. Canadian Journal of Remote Sensing 12:7–18.

Macnab, J. 1983. Wildlife management as scientific experimentation. Wildlife Society Bulletin 11:397–401.

MacNally, R., A. F. Bennett, G. W. Brown, L. F. Lumsden, A. Yen, S. Hinkley, P. Lillywhite, and D. Ward. 2002. How well do ecosystem-based planning units represent different components of biodiversity? Ecological Applications 12:900–912.

MacNamara, M., and A. Blue. 2007. The use of portable corral systems and Tamers, drop-floor chutes for improving and simplifying animal husbandry, veterinary care and management techniques in captive and free-range hoofstock. Animal Keeper's Forum 34:178–180.

MacNeill, A. C., and T. Barnard. 1978. Necropsy results in free-flying and captive Anatidae in British Columbia. Canadian Veterinary Journal 19:17–21.

MacWhirter, R. B., and K. L. Bildstein. 1996. Northern harrier. Account 210 in A. Poole and F. Gill, editors. The birds of North America. The Academy of Natural Sciences, Philadelphia, Pennsylvania, and The American Ornithologists' Union, Washington, D.C., USA.

Madison, D. M., and C. R. Shoop. 1970. Homing behavior, orientation, and home range of salamanders tagged with Tantalum-182. Science 168:1484–1487.

Madsen, M., and E. C. Anderson 1995. Serologic survey of Zimbabwean wildlife for brucellosis. Journal of Zoo and Wildlife Medicine 26:240–245.

Maechtle, T. L. 1998. The Aba: a device for restraining raptors and other large birds. Journal of Field Ornithology 69:66–70.

Maehr, D. S., and C. T. Moore. 1992. Models of mass growth for 3 North American cougar populations. Journal of Wildlife Management 56:700–707.

Maelzer, D. A. 1970. The regression of log N_{n+1} on log N_n as a test of density dependence: an exercise with computer-constructed density-independent populations. Ecology 51:810–822.

Magle, S. B., B. T. McClintock, D. W. Tripp, G. C. White, M. F. Antolin, and K. R. Crooks. 2007. Mark–resight methodology for estimating densities for prairie dogs. Journal of Wildlife Management 71:2067–2073.

Magnussen, S., P. Eggermont, and V. N. LaRiccia. 1999. Recovering tree heights from airborne laser scanner data. Forest Science 45:407–422.

Magnusson, W. E., G. J. Caughley, and G. C. Grigg. 1978. A double-survey estimate of population size from incomplete counts. Journal of Wildlife Management 42:174–176.

Magoun, A. J. 1985. Population characteristics, ecology, and management of wolverines in northwestern Alaska. Dissertation, University of Alaska, Fairbanks, USA.

Magurran, A. E. 2004. Measuring biological diversity. Blackwell Science, Malden, Massachusetts, USA.

Mahan, B. R., D. R. Dufford, N. Emerick, and T. J. Beissel. 2002. Net and net-box modifications for capturing wild turkeys. Wildlife Society Bulletin 30:960–962.

Mahan, C. G., R. H. Yahner, and L. R. Stover. 1994. Development of remote-collaring techniques for red squirrels. Wildlife Society Bulletin 22:270–273.

Maher, C. R. 2004. Intrasexual territoriality in woodchucks (*Marmota monax*). Journal of Mammalogy 85:1087–1094.

Main, M., and L. Richardson. 2002. Response of wildlife to prescribed fire in southwest Florida pine flatwoods. Wildlife Society Bulletin 30:213–221.

Mainguy, J., S. D. Côté, E. Cardinal, and M. Houle. 2008. Mating tactics and mate choice in relation to age and social rank in male mountain goats. Journal of Mammalogy 89:626–635.

Majer, J. D., G. Orabi, and L. Bisevac. 2007. Ants (Hymenoptera: Formicidae) pass the bioindicator scorecard. Myrmecological News 10:69–76.

Major, R. 1991. Identification of nest predators by photography, dummy eggs, and adhesive tape. Auk 108:190–195.

Majumdar, S. K., J. E. Huffman, F. J. Brenner, and A. I. Panah, editors. 2005. Wildlife diseases: landscape epidemiology, spatial distribution and utilization of remote sensing technology. Pennsylvania Academy of Science, Easton, USA.

Malacarne, G., and M. Griffa. 1987. A refinement of Lack's method for swift studies. Sitta 1:175–177.

Malcolm, J. R. 1991. Comparative abundances of Neotropical small mammals by trap height. Journal of Mammalogy 72:188–192.

Malcolm, S. B. 1990. Chemical defense in chewing and sucking insect herbivores: plant-derived cardenolides in the monarch butterfly and oleander aphid. Chemoecology 1:12–21.

Malone, J. H., and D. Laurenco. 2004. The use of polystyrene for drift fence sampling. Herpetological Review 35:142–143.

Maltby, L. S. 1977. Techniques used for the capture, handling and marking of brant in the Canadian high arctic. Program Notes 72, Canadian Wildlife Service, Alberta, Canada.

Mandel, J. T., K. L. Bildstein, G. Bohrer, and D. W. Winkler. 2008. Movement ecology of migration in turkey vultures. Proceedings of the National Academy of Sciences of the United States of America 105:19102–19107.

Manel, S., P. Berthier, and G. Luikart. 2002. Detecting wildlife poaching: identifying the origin of individuals with Bayesian assignment tests and multilocus genotypes. Conservation Biology 16:650–659.

Manel, S., O. E. Gaggiotti, and R. S. Waples. 2005. Assignment methods: matching biological questions with appropriate techniques. Trends in Ecology and Evolution 20:136–142.

Manly, B. F. J. 1985. The statistics of natural selection on animal populations. Chapman and Hall, New York, New York, USA.

Manly, B. F. J. 2009. Statistics for environmental science and management. Second edition. Chapman and Hall and CRC, Boca Raton, Florida, USA.

Manly, B. F. J., L. L. McDonald, and G. W. Garner. 1996. Maximum likelihood estimation for the double-count method with independent observers. Journal of Agricultural, Biological, and Environmental Statistics 1:170–189.

Manly, B. F. J., L. L. McDonald, D. L. Thomas, T. L. McDonald, and W. P. Erickson. 2002. Resource selection by animals: statistical design and analysis for field studies. Second edition. Kluwer Academic, Dordrecht, The Netherlands.

Mansell, W. D. 1971. Accessory corpora lutea in ovaries of white-tailed deer. Journal of Wildlife Management 35:369–374.

Manser, C. E. 1992. The assessment of stress in laboratory animals. Royal Society for the Prevention of Cruelty to Animals, Causeway Horsham, West Sussex, England, UK.

Mansfield, P., E. G. Strauss, and P. J. Auger. 1998. Using decoys to capture spotted turtles (*Clemmys guttata*) in water funnel traps. Herpetological Review 29:157–158.

Mao, J., K. Yen, and G. Norval. 2003. A preliminary test and report on the efficiency of a new funnel trap for semi-aquatic snakes. Herpetological Review 35:350–351.

March, G. L., and R. M. F. S. Sadlier. 1970. Studies on the band-tailed pigeon (*Columba fasciata*) in British Columbia. I. Seasonal changes in gonadal development and crop gland activity. Canadian Journal of Zoology 48:1353–1357.

Marchandeau, S., J. Aubineau, F. Berger, J.-C. Gaudin, A. Roobrouck, E. Corda, and F. Reitz. 2006. Abundance indices: reliability testing is crucial—a field case of wild rabbit *Oryctolagus cuniculus*. Wildlife Biology 12:19–27.

Marchinton, R. L., K. Kammermeyer, and B. Murphy. 2003. Aging white-tailed deer by tooth replacement and wear. Quality Whitetails 10:22–26.

Marcot B., K. Mayer, L. Fox, and R. J. Gutierrez. 1981. Application of remote sensing to wildlife habitat inventory workshop. Wildlife Society Bulletin 9:328.

Marcstrom, V., R. E. Kenward, and M. Karlbom. 1989. Survival of ring-necked pheasants with backpacks, necklaces, and leg bands. Journal of Wildlife Management 53:808–810.

Mardia, K. V. 1972. Statistics of directional data. Academic Press, New York, New York, USA.

Mardia, K. V., and P. E. Jupp. 1999. Statistics of directional data. Second edition. Wiley, New York, New York, USA.

Margalef, R., 1957. La teoria de la informacion en ecologia. Memorias de la Real Academia de Ciencias y Artes de Barcelona 32:373–449.

Marion, W. R., and J. D. Shamis. 1977. An annotated bibliography of bird marking techniques. Bird-Banding 48:42–61.

Marks, C. A. 1996. A radio-telemetry system for monitoring the treadle snare in programs for control of wild canids. Wildlife Research 23:381–386.

Marks, S. A., and A. W. Erickson. 1966. Age determination in the black bear. Journal of Wildlife Management 30:389–410.

Marlow, C. B., and W. P. Clary. 1996. Natural resource monitoring for the Daubenmire disadvantaged. Pages 13–18 in K. E. Evans, compiler. Sharing common ground on western rangelands: proceedings of a livestock/big game symposium. General Technical Report INT-GTR-343, U.S. Forest Service, Department of Agriculture, Washington, D.C., USA.

Marquardt, R. E. 1960. Smokeless powder cannon with lightweight netting for trapping geese. Journal of Wildlife Management 24:425–427.

Marques, T. A., M. Andersen, S. Christensen-Dalsgaard, S. Belikov, A. Boltunov, Ø. Wiig, S. T. Buckland, and J. Aars. 2006. The use of global positioning systems to record distances in a helicopter line-transect survey. Wildlife Society Bulletin 34:759–763.

Marshall, G., M. Bakhtiari, M. Shepard, J. Tweedy, III, D. Rasch, K. Abernathy, B. Joliff, J. C. Carrier, and M. R. Heithaus. 2007. An advanced solid-state animal-borne video and environmental data-logging device ('Crittercam') for marine research. Marine Technology Society Journal 41:31–38.

Marshall, W. H. 1951. An age determination method for the pine marten. Journal of Wildlife Management 15:276–283.

Martella, M. B., and J. L. Navarro. 1992. Capturing and marking greater rheas. Journal of Field Ornithology 63:117–120.

Martin, A. R., V. M. F. DaSilva, and P. R. Rothery. 2006. Does radio tagging affect the survival or reproduction of small cetaceans? A test. Marine Mammal Science 22:17–24.

Martin, D. J. 1971. A trapping technique for burrowing owls. Bird-Banding 42:46.

Martin, F. R. 1963. Colored vinylite bands for waterfowl. Journal of Wildlife Management 27:288–290.

Martin, F. W. 1964. Woodcock age and sex determination from wings. Journal of Wildlife Management 28:287–293.

Martin, F. W., and E. R. Clark. 1964. Summer banding of woodcock, 1962–1964. Administrative Report 43, Migratory Bird Populations Station, U.S. Fish and Wildlife Service, Department of the Interior, Washington, D.C., USA.

Martin, F. W., R. S. Pospahala, and J. D. Nichols. 1979. Assessment and population management of North American migratory birds. Pages 187–239 in J. Cairns, Jr., G. P. Patil, and W. E. Waters, editors. Environmental biomonitoring, assessment, prediction, and management—certain case studies and related quantitative issues. International Cooperative Publishing House, Fairland, Maryland, USA.

Martin, J., M. C. Runge, J. D. Nichols, B. C. Lubow, and W. L. Kendall. 2009. Structured decision making as a conceptual framework to identify thresholds for conservation and management. Ecological Applications 19:1079–1090.

Martin, K. 1998. The role of animal behavior studies in wildlife science and management. Wildlife Society Bulletin 26:911–920.

Martin, K. H., R. A. Stehn, and M. E. Richmond. 1976. Reliability of placental scar counts in the prairie vole. Journal of Wildlife Management 40:264–271.

Martin, N. S. 1970. Sagebrush control related to habitat and sage grouse occurrence. Journal of Wildlife Management 34:313–320.

Martin, P., and P. Bateson. 1996. Measuring behaviour: an introductory guide. Third edition. Cambridge University Press, Cambridge, England, UK.

Martin, T. E. 1995. Avian life history evolution in relation to nest sites, nest predation, and food. Ecological Monographs 65:101–127.

Martinez-Cruz, B., and J. A. Godoy. 2007. Genetic evidence for a recent divergence and subsequent gene flow between Spanish and eastern imperial eagles. BMC Evolutionary Biology 7:170. http://www.biomedcentral.com/1471-2148/7/170.

Martins, E. P., T. J. Ord, and S. W. Davenport. 2005. Combining motions into complex displays: playbacks with a robotic lizard. Behavioral Ecology and Sociobiology 58:351.

Martinson, L. W. 1973. A preliminary investigation of bird classification by Doppler radar. National Aeronautics and Space Administration, Wallops Island, Virginia, USA.

Martischang, M. 1993. A technique for moving existing fish habitat data sets into the spatial environment of a vector geographic information system. Fish Habitat Relationships Technical Bulletin 11, U.S. Forest Service, Department of Agriculture, Washington, D.C., USA.

Martof, B. S. 1953. Territoriality in the green frog, *Rana clamitans*. Ecology 34:165–174.

Marvin, G. A. 1996. Life history and population characteristics of the salamander *Plethodon kentucki*, with a review of *Plethodon* life histories. American Midland Naturalist 136:385–400.

Marzluff, J. M., S. T. Knick, and J. J. Millspaugh. 2001. High-tech behavioral ecology: modeling the distribution of animal activities to better understand wildlife space use and resource selection. Pages 309–380 in J. J. Millspaugh and J. M. Marzluff, editors. Radio tracking and animal populations. Academic Press, San Diego, California, USA.

Marzluff, J. M., J. J. Millspaugh, P. Hurvitz, and

M. A. Handcock. 2004. Relating resources to a probabilistic measure of space use: forest fragments and Steller's jays. Ecology 85:1411–1427.

Mascanzoni, D., and H. Wallin. 1986. The harmonic radar: a new method of tracing insects in the field. Ecological Entomology 11:387–390.

Mason, C. E., E. G. Gluesing, and D. H. Arner. 1983. Evaluation of snares, leg-hold, and Conibear traps for beaver control. Proceedings of the Annual Conference of the Southeastern Association of Fish and Wildlife Agencies 37:201–209.

Mason, D.C., G. Q. A. Anderson, R. B. Bradbury, D. M. Cobby, I. J. Davenport, M. Vandepoll, and J. D. Wilson. 2003. Measurement of habitat predictor variables for organism–habitat models using remote sensing and image segmentation. International Journal of Remote Sensing 24:2515–2532.

Mason, J. R., N. J. Bean, and L. Clark. 1993. Development of chemosensory attractants for white-tailed deer (Odocoileus virginianus). Crop Protection 12:448–452.

Mason, J. R., and S. Blom. 1998. Coyote lure ingredients—what factors determine success? Wildlife Control Technology 5:26–30.

Mason, J. R, and L. Clark. 2000. The chemical senses of birds. Pages 39–56 in G. C. Whittow, editor. Sturkie's avian physiology. Fifth edition. Academic Press, San Diego, California, USA.

Matocq, M. D., and F. X. Villablanca. 2001. Low genetic diversity in an endangered species: recent or historic pattern? Biological Conservation 98:61–68.

Matschke, G. H. 1962. Trapping and handling European wild hogs. Proceedings of the Annual Conference of the Southeastern Association of Game and Fish Commissioners 16:21–24.

Matson, C. W., B. E. Rodgers, R. K. Chesser, and R. J. Baker. 2000. Genetic diversity of Clethrionomys glareolus populations from highly contaminated sites in the Chernobyl region, Ukraine. Environmental Toxicology and Chemistry 19:2130–2135.

Matsumura, F. 1985. Toxicology of insecticides. Second edition. Plenum Press, New York, New York, USA.

Mattfeldt, S. M., and E. H. Campbell-Grant. 2007. Are two methods better than one? Area constrained transects and litter bags for sampling stream salamanders. Herpetological Review 38:43–45.

Matthews, S. M., R. T. Golightly, and J. M. Higley. 2008. Mark–resight population density estimation for American black bears in Hoopa, California. Ursus 19:13–21.

Matthiopoulos, J. 2003. The use of space use by animals as a function of accessibility and preference. Ecological Modelling 159:239–268.

Mattson, D., R. Knight, and B. Blanchard. 1987. The effects of development and primary roads on grizzly bear habitat use in the Yellowstone National Park, Wyoming. International Conference on Bear Research and Management 7:259–273.

Mauldin, R. E., and R. M. Engeman. 1999. A novel snake restraint device. Herpetological Review 30:158.

Mauser, D. M., and J. G. Mensik. 1992. A portable trap for ducks. Wildlife Society Bulletin 20:299–302.

May, R. M. 1973. Stability and complexity in model ecosystems. Princeton University Press, Princeton, New Jersey, USA.

May, R. M. 1974. Biological populations with nonoverlapping generations: stable points, stable cycles, and chaos. Science 186:645–647.

May, R. M., and R. M. Anderson. 1978. Regulation and stability of host–parasite population interactions: II. Destabilizing processes. Journal of Animal Ecology 47:249–267.

May, R. M., and R. M. Anderson. 1979. Population biology of infectious diseases: part II. Nature 280:455–461.

May, R. M., and G. F. Oster. 1976. Bifurcations and dynamic complexity in simple ecological models. American Naturalist 110:573–599.

Mayer, K. E. 1984. A review of selected remote sensing and computer technologies applied to wildlife habitat inventories. California Fish and Game 70:101–112.

Mayfield, H. 1961. Nesting success calculated from exposure. Wilson Bulletin 73:255–261.

Mayfield, H. 1975. Suggestions for calculating nest success. Wilson Bulletin 87:456–466.

Maynard Smith, J. 1974. Models in ecology. Cambridge University Press, New York, New York, USA.

Mayr, E. 1942. Systematics and the origin of species from the viewpoint of a zoologist. Columbia University Press, New York, New York, USA.

Mayr, E. 1970. Populations, species, and evolution: an abridgement of animal species and evolution. Harvard University Press, Cambridge, Massachusetts, USA.

Mayr, E. 1982. Of what use are subspecies? Auk 99:593–595.

Mazerolle, M. J. 2003. Using rims to hinder amphibian escape from pitfall traps. Herpetological Review 34:213–215.

McBeath, D. Y. 1941. Whitetail traps and tags. Michigan Conservationist 10(11):6–7.

McBride, R. T., Jr., and R. T. McBride. 2007. Safe and selective capture technique for jaguars in the Paraguayan Chaco. Southwestern Naturalist 52:570–577.

McCabe, R. A., and A. S. Hawkins. 1946. The Hungarian partridge in Wisconsin. American Midland Naturalist 35:1–75.

McCabe, R. A., and G. A. LePage. 1958. Identifying progeny from pheasant hens given radioactive calcium (Ca^{45}). Journal of Wildlife Management 22:134–141.

McCabe, T. R., and G. Elison. 1986. An efficient live-capture technique for muskrats. Wildlife Society Bulletin 14:282–284.

McCall, J. D. 1954. Portable live trap for ducks, with improved gathering box. Journal of Wildlife Management 18:405–407.

McCallum, M. L., B. A. Wheeler, and S. E. Trauth. 2002. The "Frog Box": a new bulk holding device for anurans. Herpetological Review 33:107–108.

McCann, B. E., K. Ryan, and D. K. Garcelon. 2004. Techniques and approaches for the removal of feral pigs from island and mainland ecosystems. Proceedings of the Vertebrate Pest Conference 21:42–46.

McCarthy, M. A, and H. P. Possingham. 2007. Active adaptive management for conservation. Conservation Biology 21:956–963.

McCleery, R. A., R. R. Lopez, and N. J. Silvy. 2007a. An improved method for handling squirrels and similar-size mammals. Wildlife Biology in Practice 3:39–42.

McCleery, R. A., R. R. Lopez, N. J. Silvy, P. A. Frank, and S. B. Klett. 2006. Population status and habitat selection of the endangered Key Largo woodrat. American Midland Naturalist 155:197–209.

McCleery, R. A., R. R. Lopez, N. J. Silvy, and W. E. Grant. 2005. Effectiveness of supplemental stockings for the endangered Key Largo woodrat. Biological Conservation 124:27–33.

McCleery, R. A., R. R. Lopez, N. J. Silvy, and S. N. Kahlick. 2007b. Habitat use of fox squirrels in an urban environment. Journal of Wildlife Management 71:1149–1157.

McClintock, B. T., and G. C. White. 2009. A less field-intensive robust design for estimating demographic parameters with mark–resight data. Ecology 90:313–320.

McClintock, B. T., G. C. White, M. F. Antolin, and D. W. Tripp. 2009a. Estimating abundance using mark–resight when sampling is with replacement or the number of marked animals is unknown. Biometrics 65:237–246.

McClintock, B. T., G. C. White, and K. P. Burnham. 2006. A robust design mark–resight abundance estimator allowing heterogeneity in resighting probabilities. Journal of Agricultural, Biological, and Ecological Statistics 11:231–248.

McClintock, B. T., G. C. White, K. P. Burnham, and M. A. Pryde. 2009b. A generalized mixed effects model of abundance for mark–resight data when sampling is without replacement. Pages 271–289 in D. L. Thomson, E. G. Cooch, and M. J. Conroy, editors. Modeling demographic processes in marked populations. Springer, New York, New York, USA.

McCloskey, J. T., and S. R. Dewey. 1999. Improving the success of a mounted great horned owl lure for trapping northern goshawks. Journal of Raptor Research 33:168–169.

McCloskey, J. T, and J. E. Thompson. 2000. Aging and sexing common snipe using discriminant analysis. Journal of Wildlife Management 64:960–969.

McCloskey, R. J. 1977. Accuracy of criteria used to determine age of fox squirrels. Proceedings of the Iowa Academy of Science 84:32–34.

McCollough, M. A. 1989. Molting sequence and aging of bald eagles. Wilson Bulletin 101:1–10.

McComb, W. C., R. G. Anthony, and K. McGarigal. 1991. Differential vulnerability of small mammals and amphibians to two trap types and two trap baits in Pacific Northwest forests. Northwest Science 65:109–115.

McComb, W. C., M. T. McGrath, T. A. Spies, and D. Vesely. 2002. Models for mapping potential habitat at landscape scales: an example using northern spotted owls. Forest Science 48:203–216.

McConnell, P. A., D. M. Ferrigno, and D. E. Roscoe. 1985. An evaluation of muskrat trapping systems and new techniques. Job Progress Report, Federal Aid Project W-59-R-7, Job I-C, New Jersey Division of Fish, Game and Wildlife, Trenton, USA.

McCord, C. M., and J. E. Cardoza. 1982. Bobcat and lynx. Pages 728–766 in J. A. Chapman and G. A. Feldhamer, editors. Wild mammals of North America. Johns Hopkins University Press, Baltimore, Maryland, USA.

McCourt, K. H., and D. M. Keppie. 1975. Age determination of juvenile spruce grouse. Journal of Wildlife Management 39:790–794.

McCracken, G. F. 1984. Communal nursing in Mexican free-tailed bat maternity colonies. Science 223:1090–1091.

McCracken, G. F., E. H. Gillam, J. T. Westbrook, Y. Lee, M. L. Jensen, and B. B. Balsley. 2008. Brazilian free-tailed bats (*Tadaraida brasiliensis*: Mollossida, Chiroptera) at high altitude: links to migratory insect populations. Integrative and Comparative Biology 48:107–118.

McCracken, K. E., J. W. Witham, and M. L. Hunter, Jr. 1999. Relationships between seed fall of three tree species and *Peromyscus leucopus* and *Clethrionomys gapperi* during 10 years in an oak–pine forest. Journal of Mammalogy 80:1288–1296.

McCullough, B. D., and B. Wilson. 1999. On the accuracy of statistical procedures in Microsoft Excel 97. Computational Statistics and Data Analysis 31:27–37.

McCullough, B. D., and B. Wilson. 2002. On the accuracy of statistical procedures in Microsoft Excel 2000 and Excel XP. Computational Statistics and Data Analysis 40:713–721.

McCullough, B. D., and B. Wilson. 2005. On the accuracy of statistical procedures in Microsoft Excel 2003. Computational Statistics and Data Analysis 49:1244–1252.

McCullough, D. R. 1965. Sex characteristics of black-tailed deer hooves. Journal of Wildlife Management 29:210–212.

McCullough, D. R. 1975. Modification of the Clover deer trap. California Fish and Game 61:242–244.

McCullough, D. R. 1979. The George Reserve deer herd: population ecology of a k-selected species. University of Michigan Press, Ann Arbor, USA.

McCullough, D. R. 1982. Population growth rate of the George Reserve deer herd. Journal of Wildlife Management 46:1079–1083.

McCullough, D. R. 1983. Rate of increase of white-tailed deer on the George Reserve: a response. Journal of Wildlife Management 47:1248–1250.

McCullough, D. R., editor. 1996. Metapopulations and wildlife conservation. Island Press, Washington, D.C., USA.

McCullough, D. R., and P. Beier. 1986. Upper vs. lower molars for cementum annuli age determination of deer. Journal of Wildlife Management 50:705–706.

McCutchen, H. E. 1969. Age determination of pronghorns by the incisor cementum. Journal of Wildlife Management 33:172–175.

McDaniel, G. W., K. S. McKelvey, J. R. Squires, and L. F. Ruggiero. 2000. Efficacy of lures and hair snares to detect lynx. Wildlife Society Bulletin 28:119–123.

McDonald, D. W., P. D. Stewart, P. Stopka, and N. Yamaguchi. 2000. Measuring the dynamics of mammalian societies: an ecologist's guide to ethological methods. Pages 332–380 in L. Boitani and T. K. Fuller, editors. Research techniques in animal ecology: controversies and consequences. Columbia University Press, New York, New York, USA.

McDonald, I. R., A. K. Lee, A. J. Bradley, and K. A. Than. 1981. Endocrine changes in dasyurid marsupials with differing mortality patterns. General and Comparative Endocrinology 44:292–301.

McDonald, L. L., J. R. Alldredge, M. S. Boyce, and W. P. Erickson. 2005. Measuring availability and vertebrate use of terrestrial habitats and foods. Pages 465–488 in C. E. Braun, editor. Techniques for wildlife investigations and management. The Wildlife Society, Bethesda, Maryland, USA.

McDonald, L. L., and B. F. J. Manly. 2001. Modeling wildlife resource selection: can we do better? Pages 137–145 in T. M. Shenk and A. B. Franklin, editors. Modeling in natural resource management: development, interpretation, and application. Island Press, Washington, D.C., USA.

McDonald, L. L., B. F. J. Manly, and C. M. Rayley. 1990. Analysing foraging and habitat use through selection functions. Studies in Avian Biology 13:325–331.

McDonald, R. A., S. Harris, G. Turnbull, P. Brown, and M. Fletcher. 1998. Anticoagulant rodenticides in stoats (*Mustela erminea*) and weasels (*Mustela nivalis*) in England. Environmental Pollution 103:17–23.

McDonald, T. A. 2002. A perspective on the potential health risks of PBDEs. Chemosphere 46:745–755.

McDonald, T. L., and L. L. McDonald. 2003. A new ecological risk assessment procedure using resource selection models and geographic information systems. Wildlife Society Bulletin 30:1015–1021.

McDonald, T. L., B. F. J. Manly, R. M. Nielson, and L. V. Diller. 2006. Discrete-choice modeling in wildlife studies exemplified by northern spotted owl nighttime habitat selection. Journal of Wildlife Management 70:375–383.

McDonnell, M. D., H. P. Possingham, I. R. Ball, and E. A. Cousins. 2002. Mathematical models for spatially cohesive reserve design. Environmental Modeling and Assessment 7:107–114.

McEwan, E. H. 1963. Seasonal annuli in the cementum of the teeth of barren ground caribou. Canadian Journal of Zoology 41:111–113.

McGaughey, R. J. 2009. FUSION/LDV: software for LIDAR data analysis and visualization. U.S. Forest Service, Department of Agriculture. http://www.fs.fed.us/eng/rsac/fusion/.

McGowan, C. P., and T. R. Simons. 2005. A method for trapping of breeding adult American oystercatchers. Journal of Field Ornithology 76:46–49.

McGowan, K. J., and C. Caffrey. 1994. Does drugging crows for capture cause abnormally high mortality? Journal of Field Ornithology 65:453–457.

McGrady, M. J., and J. R. Grant. 1994. A modified power snare to catch breeding golden eagles (*Aquila chrysaetos*). Journal of Raptor Research 28:61.

McGrady, M. J., and J. R. Grant. 1996. The use of a power snare to capture breeding golden eagles. Journal of Raptor Research 30:28–31.

McIntosh, R. P. 1967. The continuum concept of vegetation. Botanical Review 33:130–187.

McKean, J., D. Isaak, and W. Wright. 2009. Stream and riparian habitat analysis and monitoring with a high-resolution terrestrial-aquatic LiDAR. Pages 7–16 in J. M. Bayer and J. L. Schei, editors. PNAMP special publication: remote sensing applications for aquatic resource monitoring, Pacific Northwest Aquatic Monitoring Partnership, Cook, Washington, USA.

McKelvey, K. S., and M. K. Schwartz. 2004. Genetic errors associated with population estimation using non-invasive molecular tagging: problems and new solutions. Journal of Wildlife Management 68:439–448.

McKenna, K. C. 2001. *Chrysemys* (painted turtle) trapping. Herpetological Review 32:184.

McKinnon, D. T. 1983. Age separation of yearling and adult Franklin's spruce grouse. Journal of Wildlife Management 47:533–535.

McKinstry, M. C., and S. H. Anderson. 1998. Using snares to live-capture beaver, *Castor canadensis*. Canadian Field-Naturalist 112:469–473.

McLeod, S. R., J. P. Druhan, and R. B. Hacker. 2006. Estimating the age of kangaroos using eye lens weight. Wildlife Research 33:25–28.

McLoughlin, P. D., T. Coulson, and T. Clutton-Brock. 2008. Cross-generational effects of habitat and density on life history in red deer. Ecology 89:3317–3326.

McManus, J. J. 1974. *Didelphis virginiana*. Number 40 in S. Anderson, editor. Mammalian species. American Society of Mammalogists, Lawrence, Kansas, USA.

McMillin, J. M., U. S. Seal, K. D. Keenlyne, A. W. Erickson, and J. E. Jones. 1974. Annual testosterone rhythm in the adult white-tailed deer (*Odocoileus virginianus borealis*). Endocrinology 94:1034–1040.

McNew, L. B., C. K. Nielsen, and C. K. Bloomquist. 2007. Use of snares to live-capture beavers. Human–Wildlife Conflicts 1:106–111.

McNicholl, M. K. 1983. Use of a noosing pole to capture common nighthawks. North American Bird Bander 8:104–105.

McQuilkin, R. A., and R. A. Musbach. 1977. Pin oak acorn production on green tree reservoirs in southeastern Missouri. Journal of Wildlife Management 41:218–225.

McShea, W. J. 2000. The influence of acorn crops on annual variation in rodent and bird populations. Ecology 81:228–238.

McShea, W. J., and G. Schwede. 1993. Variable acorn crops: responses of white-tailed deer and other mast consumers. Journal of Mammalogy 74:999–1006.

Mead, R. A. 1967. Age determination in the spotted skunk. Journal of Mammalogy 48:606–616.

Mead, R. A., T. Sharik, and J. T. Heinen. 1981. A computerized spatial analysis system for assessing wildlife habitat from vegetation maps. Canadian Journal of Remote Sensing 7:34–40.

Mech, S. G., and P. A. Zollner. 2002. Using body size to predict perceptual range. Oikos 98:47–52.

Mechlin, L. M., and C. W. Shaiffer. 1980. Net-firing gun for capturing breeding waterfowl. Journal of Wildlife Management 44:895–896.

Mecozzi, G. E., and F. S. Guthery. 2008. Behavior of walk-hunters and pointing dogs during northern bobwhite hunts. Journal of Wildlife Management 72:1399–1404.

Medica, P. A., R. B. Bury, and F. B. Turner. 1975. Growth of the desert tortoise (*Gopherus agassizi*) in Nevada. Copeia 1975:639–643.

Medill S., A. E. Derocher, I. Stirling, and N. Lunn. 2010. Reconstructing the reproductive history of female polar bears using cementum patterns of premolar teeth. Polar Biology 33:115–124.

Medill S., A. E. Derocher, I. Stirling, N. Lunn, and R. A. Moses. 2009. Estimating cementum annuli width in polar bears: identifying sources of variation and error. Journal of Mammalogy 90:1256–1264.

Meek, P. D., D. J. Jenkins, B. Morris, A. J. Ardler, and R. J. Hawksby. 1995. Use of two humane leg-hold traps for catching pest species. Wildlife Research 22:733–739.

Meeks, H. N., J. K. Wickliffe, S. R. Hoofer, R. K. Chesser, B. E. Rodgers, and R. J. Baker. 2007. Mitochondrial control region variation in bank voles (*Clethrionomys glareolus*) is not related to Chernobyl radiation exposure. Environmental Toxicology and Chemistry 26:361–369.

Meentemeyer, V. 1989. Geographical perspectives of space, time, and scale. Landscape Ecology 3:163–173.

Mehl, K. R., K. L. Drake, G. W. Page, P. M. Sanzenbacher, S. M. Haig, and J. E. Thompson. 2003. Capture, breeding, and wintering shorebirds with leg-hold noose-mats. Journal of Field Ornithology 74:401–405.

Mehner, H., M. Cutler, D. Fairbairn, and G. Thompson. 2004. Remote sensing of upland vegetation: the potential of high spatial resolution satellite sensors. Global Ecology and Biogeography 13:359–369.

Meijer, T., K. Noren, P. Hellstrom, L. Dalen, and A. Angerbjorn. 2008. Estimating population parameters in a threatened arctic fox population using molecular tracking and traditional field methods. Animal Conservation 11:330–338.

Melchior, H. R., and F. A. Iwen. 1965. Trapping, restraining, and marking arctic ground squirrels for behavioral observations. Journal of Wildlife Management 29:671–678.

Melquist, W. E., and A. E. Dronkert. 1987. River otter. Pages 627–641 in M. Novak, J. A. Baker, M. E. Obbard, and B. Malloch, editors. Wild furbearer management and conservation in North America. Ontario Ministry of Natural Resources, Toronto, Canada.

Melquist, W. E., and M. G. Hornocker. 1979. Methods and techniques for studying and censusing river otter populations. Technical Report 8, Forestry, Wildlife and Range Experiment Station, University of Idaho, Moscow, USA.

Melquist, W. E., and M. G. Hornocker. 1983. Ecology of river otters in west central Idaho. Wildlife Monographs 83.

Melvin, S. M., and J. P. Gibbs. 1996. Sora. Account 250 in A. Poole, P. Stettenheim, and F. Gill, editors. The birds of North America. The Academy of Natural Sciences, Philadelphia, Pennsylvania, and The American Ornithologists' Union, Washington, D.C., USA.

Menasco, K. A., and H. R. Perry, Jr. 1978. Errors from determining sex of mourning doves by plumage characteristics. Proceedings of the Annual Conference of the Southeastern Association of Fish and Wildlife Agencies 32:224–227.

Mendonca, M. T., and D. Crews. 2001. Control of attractivity and receptivity in female red-sided garter snakes. Hormones and Behavior 40:43–50.

Meng, H. 1971. The Swedish goshawk trap. Journal of Wildlife Management 35:832–43.

Mennill, D. J., J. M. Burt, K. M. Fristrup, and S. L. Vehrencamp. 2006. Accuracy of an acoustic location system for monitoring the position of dueting songbirds in tropical forest. Journal of the Acoustical Society of America 119:2832–2839.

Menu, S., G. Gauthier, and A. Reed. 2001. Survival of juvenile greater snow geese immediately after banding. Journal of Field Ornithology 72:282–290.

Menu, S., J. B. Hestbeck, G. Gauthier, and A. Reed. 2000. Effects of neck bands on survival of greater snow geese. Journal of Wildlife Management 64:544–552.

Merendino, M. T., and D. S. Lobpries. 1998. Use of rocket netting and airboat nightlighting for capturing mottled ducks in Texas. Proceedings of the Annual Conference of the Southeastern Association of Fish and Wildlife Agencies 52:303–308.

Merilä, J. 1997. Expression of genetic variation in body size of the collared flycatcher under different environmental conditions. Evolution 51:526–536.

Merrill, E. H., and M. S. Boyce. 1991. Summer range and elk population dynamics in Yellowstone National Park. Pages 263–273 in R. B. Keiter and M. S. Boyce, editors. The Greater Yellowstone ecosystem: redefining America's wilderness heritage. Yale University Press, New Haven, Connecticut, USA.

Merrill, S. B., and L. D. Mech. 2000. Details of extensive movements by Minnesota wolves (*Canis lupus*). American Midland Naturalist 144:428–433.

Mertz, D. B. 1970. Notes on methods used in life-history studies. Pages 4–17 in J. H. Connell, D. B. Mertz, and W. W. Murdoch, editors. Readings in ecology and ecological genetics. Harper and Row, New York, New York, USA.

Mersinger, R. C., and N. J. Silvy. 2006. Range size and habitat use and dial activity of feral hogs on reclaimed surface-mined lands in east Texas. Human–Wildlife Conflicts 1:161–167.

Messick, J. P. 1987. North American badger. Pages 587–597 in M. Novak, J. A. Baker, M. E. Obbard, and B. Malloch, editors. Wild furbearer management and conservation in North America. Ontario Ministry of Natural Resources, Toronto, Canada.

Messick, J. P., and M. G. Hornocker. 1981. Ecology of the badger in south-western Idaho. Wildlife Monographs 76.

Metts, B. S., J. D. Lanham, and K. R. Russell. 2001. Evaluations of herpetofaunal communities on upland streams and beaver-impounded streams in the Upper Piedmont of South Carolina. American Midland Naturalist 145:54–65.

Metz, K. J., and P. J. Weatherhead. 1993. An experimental test of the contrasting-color hypothesis of red-band effects in red-winged blackbirds. Condor 95:395–400.

Meudt, H. M., and A. C. Clarke. 2007. Almost forgotten or latest practice? AFLP applications, analyses and advances. Trends in Plant Science 12:107–117.

Meyer, J. 2006. Field methods for studying nutria. Wildlife Society Bulletin 34:850–852.

Meyers, J. M. 1994a. Improved capture techniques for psittacines. Wildlife Society Bulletin 22:511–516.

Meyers, J. M. 1994b. Leg bands cause injuries to parakeets and parrots. North American Bird Bander 19:133–136.

Meyers, J. M., and K. L. Pardieck. 1993. Evaluation of three elevated mist-net systems for sampling birds. Journal of Field Ornithology 64:270–277.

Michelsen, A., B. B. Andersen, J. Storm, W. H. Kirchner, and M. Lindauer. 1992. How honeybees perceive communication dances, studied by means of a mechanical model. Behavioral Ecology and Sociobiology 30:143–150.

Michod, R. E., and W. W. Anderson. 1980. On calculating demographic parameters from age frequency data. Ecology 61:265–269.

Mielke, P. W., Jr., and K. J. Berry. 1982. An extended class of permutation techniques for matched pairs. Communications in Statistics: Theory and Methods 11:1197–1207.

Mikesic, D. G., and L. C. Drickamer. 1992. Effects of radiotransmitters and fluorescent powers on activity of wild house mice (*Mus musculus*). Journal of Mammalogy 73:663–667.

Millam, J. R., M. J. Delwiche, C. B. Craig-Veit, J. D. Henderson, and B. W. Wilson. 2000. Noninvasive characterization of the effects of diazinon on pigeons. Bulletin of Environmental Contamination and Toxicology 64:534–541.

Millar, J. S., and F. C. Zwickel. 1972. Determination of age, age structure, and mortality of the pika, *Ochotona princeps* (Richardson). Canadian Journal of Zoology 50:229–232.

Miller, C. R., P. Joyce, and L. P. Waits. 2002. Assessing allelic dropout and genotype reliability using maximum likelihood. Genetics 160:357–366.

Miller, C. R., P. Joyce, and L. P. Waits. 2005. A new method for estimating the size of small populations from genetic mark–recapture data. Molecular Ecology 14:1991–2005.

Miller, D. J. 1979. Sea otter capture and tagging in California. Pages 11–12 in L. Hobbs and P. Russell, editors. Report on the pinniped tagging workshop. Seattle, Washington, USA.

Miller, D. R. 1964. Colored plastic ear markers for beavers. Journal of Wildlife Management 28:859–861.

Miller, D. R., and C. M. Crowe. 2009. Length of multiple-funnel traps affects catches of some bark and wood boring beetles in a slash pine stand in northern Florida. Florida Entomologist 92:506–507.

Miller, D. R., and D. A. Duerr. 2008. Comparison of arboreal beetle catches in wet and dry collection cups with Lindgren multiple funnel traps. Journal of Economic Entomology 101: 107–113.

Miller, D. R., and J. D. Robertson. 1967. Results of tagging caribou at Little Duck Lake, Manitoba. Journal of Wildlife Management 31:150–159.

Miller, D. S., J. Berglund, and M. Jay. 1983. Freeze-mark techniques applied to mammals at the Santa Barbara Zoo. Zoo Biology 2:143–148.

Miller, F. L. 1974a. Age determination of caribou by annulations in dental cementum. Journal of Wildlife Management 38:47–53.

Miller, F. L. 1974b. Biology of the Kaminuriak population of barren ground caribou. Part II. Dentition as an indicator of sex and age; composition and socialization of the population. Report Series 31, Canadian Wildlife Service, Ottawa, Ontario, Canada.

Miller, F. L. 1982. Caribou. Pages 923–959 in J. A. Chapman and G. A. Feldhamer, editors. Wild mammals of North America. Johns Hopkins University Press, Baltimore, Maryland.

Miller, F. L., and R. L. McClure. 1973. Determining age and sex of barren ground caribou from dental variables. Transactions of the Northeastern Section, The Wildlife Society 30:79–100.

Miller, J. D. 1997. Reproduction in sea turtles. Pages 51–82 in P. L. Lutz and J. A. Musick, editors. The biology of sea turtles. CRC Press, New York, New York, USA.

Miller, J. J. R., R. D. Dawson, and H. Schwntje. 2003. Manual of common diseases and parasites of wildlife in Northern British Columbia. University of Northern British Columbia, Prince George, British Columbia, Canada. http://www.unbc.ca/nlui/wildlife_diseases_bc/.

Miller, L. S. 1957. Tracing vole movements by radioactive excretory products. Ecology 38:132–136.

Miller, P. S. 1995. Selective breeding programs for rare alleles: examples from the Przewalski horse and California condor pedigrees. Conservation Biology 9:1262–1273.

Miller, S. D., E. F. Becker, and W. H. Ballard. 1987. Black and brown bear density estimates using modified capture–recapture techniques in Alaska. International Conference on Bear Research and Management 42:471–476.

Miller, S. G., R. L. Knight, and C. K. Miller. 2001. Wildlife responses to pedestrians and dogs. Wildlife Society Bulletin 29:124–132.

Miller, S. G., G. C. Wite, R. A. Sellars, H. V. Reynolds, J. W. Schoen, K. Titus, V. G. Barnes, Jr., R. B. Smith, W. B. Ballard, and C. C. Schwartz. 1997. Brown and black bear density estimation in Alaska using radiotelemetry and replicated mark–resight techniques. Wildlife Monographs 133.

Miller, W. R. 1962. Automatic activating mechanism for waterfowl nest trap. Journal of Wildlife Management 26:402–404.

Milligan, J. L., A. K. Davis, and S. M. Altizer. 2003. Errors associated with using colored leg bands to identify wild birds. Journal of Field Ornithology 74:111–118.

Milliken, G. A., and D. E. Johnson. 1984. Analysis of messy data: designed experiments. Volume I. Van Nostrand Reinhold, New York, New York, USA.

Mills, H. B. 1936. Observations of Yellowstone elk. Journal of Mammalogy 17:250–253.

Mills, J. A. 1972. A difference in band loss from male and female red-billed gulls, *Larus novaehollandiae scopulinus*. Ibis 114:252–255.

Mills, L. S. 2007. Conservation of wildlife populations: demography, genetics, and management. Blackwell, Malden, Massachusetts, USA.

Mills, L. S., and F. W. Allendorf. 1996. The one-migrant-per-generation rule in conservation and management. Conservation Biology 10: 1509–1518.

Mills, L. S., J. J. Citta, K. P. Lair, M. K. Schwartz, and D. A. Tallmon. 2000b. Estimating animal abundance using noninvasive DNA sampling: promise and pitfalls. Ecological Applications 10:283–294.

Mills, L. S., R. J. Fredrickson, and B. B. Moorhead. 1993. Characteristics of old-growth forests associated with northern spotted owls in Olympic National Park. Journal of Wildlife Management 57:315–321.

Mills, L. S., K. L. Pilgrim, M. K. Schwartz, and K. McKelvey. 2000a. Identifying lynx and other North American felids based on mtDNA analysis. Conservation Genetics 1:285–288.

Mills, L. S., and P. E. Smouse. 1994. Demographic consequences of inbreeding in remnant populations. American Naturalist 144:412–431.

Mills, T. M., T. L. Yates, J. E. Childs, R. R. Parmenter, T. G. Ksiasek, P. E. Rollin, and C. J. Peters. 1995. Guidelines for working with rodents potentially infected with hantavirus. Journal of Mammalogy 76:716–722.

Millspaugh, J. J. 1999. Behavioral and physiological responses of elk to human disturbances in the southern Black Hills, South Dakota. Dissertation, University of Washington, Seattle, USA.

Millspaugh, J. J., R. A. Gitzen, B. J. Kernohan, M. A. Larson, and C. L. Clay. 2004. Comparability of three analytical techniques to assess joint space use. Wildlife Society Bulletin 32: 148–157.

Millspaugh, J. J., and J. M. Marzluff. 2001. Radio-tracking and animal populations: past trends and future needs. Pages 383–393 in J. J. Millspaugh and J. M. Marzluff, editors. Radio tracking and animal populations. Academic Press, San Diego, California, USA.

Millspaugh, J. J., R. M. Nielson, L. McDonald, J. M. Marzluff, R. A. Gitzen, C. D. Rittenhouse, M. W. Hubbard, and S. L. Sheriff. 2006. Analysis of resource selection using utilization distributions. Journal of Wildlife Management 70: 384–395.

Millspaugh, J. J., J. Sartwell, R. A. Gitzen, R. J. Moll, and J. Beringer. 2008. A pragmatic view of animal-borne video technology. Trends in Ecology and Evolution 23:294–295.

Millspaugh, J. J., J. R. Skalski, R. L. Townsend, D. R. Diefenbach, M. S. Boyce, L. P. Hansen, and K. Kammermeyer. 2009. An evaluation of sex-age-kill (SAK) model performance. Journal of Wildlife Management 73:442–451.

Millspaugh, J. J., B. E. Washburn, M. A. Milanick, J. Beringer, L. P. Hansen, and T. M. Meyer. 2002. Non-invasive techniques for stress assessment in white-tailed deer. Wildlife Society Bulletin 30:899–907.

Mims, F. M., III. 1990. The amateur scientist: a remote-control camera that catches the wind and captures the landscape. Scientific American 263(October):126–129.

Minckler, L. S., and R. E. McDermott. 1960. Pin oak acorn production and regeneration as affected by stand density, structure and flooding. Research Bulletin 750, Missouri Agricultural Experiment Station, University of Missouri, Columbia, USA.

Mineau, P. 2002. Estimating the probability of bird mortality from pesticide sprays on the basis of the field study record. Environmental Toxicology and Chemistry 21:1497–1506.

Mineau, P., editor. 1991. Cholinesterase-inhibiting insecticides: their impact on wildlife and the environment. Elsevier Scientific, Amsterdam, The Netherlands.

Mineau, P., and B. T. Collins. 1988. Avian mortality in agro-ecosystems—2. Methods of detection. Pages 13–27 in M. P. Greaves, B. D. Smith, and P. W. Greig-Smith, editors. Field methods for the study of environmental effects of pesticides. British Crop Protection Council, Croydon, England, UK.

Mineau, P., M. R. Fletcher, L. C. Glaser, N. J. Thomas, C. Brassard, L. K. Wilson, J. E. Elliott, L. A. Lyon, C. J. Henny, T. Bollinger, and S. L. Porter. 1999. Poisoning of raptors with organophosphorus and carbamate pesticides with emphasis on Canada, U.S. and U.K. Journal of Raptor Research 33:1–37.

Minnich, J. E., and V. H. Shoemaker. 1970. Diet, behavior and water turnover in the desert iguana, *Dipsosaurus dorsalis*. American Midland Naturalist 84:496–509.

Minta, S. C. 1992. Tests of spatial and temporal interaction among animals. Ecological Applications 2:178–188.

Minta, S. C., and M. Mangel. 1989. A simple population estimate based on simulation for capture–recapture and capture–resight data. Ecology 70:1738–1751.

Miranda, H. C., Jr., and J. C. Ibanez. 2006. A modified bal-chatri to capture great Philippine eagles for radio telemetry. Journal of Raptor Research 40:233–235.

Mirarchi, R. E., and T. S. Baskett. 1994. Mourning dove. Account 117 in A. Poole and F. Gill, editors. The birds of North America. The Academy of Natural Sciences, Philadelphia, Pennsylvania, and The American Ornithologists' Union, Washington, D.C., USA.

Mirarchi, R. E., P. F. Scanlon, and R. L. Kirkpatrick. 1977. Annual changes in spermatozoa production and associated organs of white-tailed deer. Journal of Wildlife Management 41: 92–99.

Mitchell, C. D., and C. R. Maher. 2001. Are horn characteristics related to age in male pronghorns? Wildlife Society Bulletin 29:908–916.

Mitchell, E., and V. M. Kozicki. 1975. Prototype visual mark for large whales modified from "Discovery" tag. International Whaling Commission Report 25:236–239.

Mitchell, J. C., S. Y. Erdle, and J. F. Pagels. 1993. Evaluation of capture techniques for amphibian, reptile, and small mammal communities in saturated forested wetlands. Wetlands 13:130–136.

Mitchell, M. S., R. A. Lancia, and E. J. Jones. 1996. Use of insecticide to control destructive activity of ants during trapping of small mammals. Journal of Mammalogy 77:1107–1113.

Mitchell, O. G., and R. A. Carsen. 1967. Tooth eruption in the arctic ground squirrel. Journal of Mammalogy 48:472–474.

Mitro, M. G., D. C. Evers, M. W. Meyer, and W. H. Piper. 2008. Common loon survival rates and mercury in New England and Wisconsin. Journal of Wildlife Management 72:665–673.

Miyamoto, M. M., and J. Cracraft. 1991. Phylogenetic analysis of DNA sequences. Oxford University Press, New York, New York, USA.

Miyashita, T., and R. A. Rowlett. 1985. Test-firing of .410 streamer marks. International Whaling Commission Report 35:305–308.

Mladenoff, D. J., and T. A. Sickley. 1998. Assessing potential gray wolf restoration in the northeastern United States: a spatial prediction of favorable habitat and potential population levels. Journal of Wildlife Management 62:1–10.

Moberg, G. P. 1985. Biological response to stress: key to assessment of animal well-being? Pages 27–49 in G. P. Moberg, editor. Animal stress. Waverly Press, Baltimore, Maryland, USA.

Mock, D. W., P. L. Schwagmeyer, and J. A. Gieg. 1999. A trap design for capturing individual birds at the nest. Journal of Field Ornithology 70:276–282.

Mock, K. E., E. K. Latch, and O. E. Rhodes. 2004. Assessing losses of genetic diversity due to translocation: long-term case histories in Merriam's turkey (*Meleagris gallopavo merriami*). Conservation Genetics 5:631–645.

Mock, K. E., T. C. Theimer, B. F. Wakeling, O.E. Rhodes, D.L. Greenberg, and P. Keim. 2001. Verifying the origins of a reintroduced population of Gould's wild turkey. Journal of Wildlife Management 65:871–879.

Moffitt, J. 1942. Apparatus for marking wild animals with colored dyes. Journal of Wildlife Management 6:312–318.

Moffitt, S. A. 1998. Aging bison by the incremental cementum growth layers in teeth. Journal of Wildlife Management 62:1276–1280.

Mogart, J. R., J. J. Hervert, P. R. Krausman, J. L. Bright, and R. S. Henry. 2005. Sonoran pronghorn use of anthropogenic and natural water sources. Wildlife Society Bulletin 33:51–60.

Mohr, C. E. 1934. Marking bats for later recognition. Proceedings of the Pennsylvania Academy of Sciences 8:26–30.

Mohr, C. O. 1947. Table of equivalent populations of North American small mammals. American Midland Naturalist 37:223–249.

Moll, R. J., J. J. Millspaugh, J. Beringer, J. Sartwell, and Z. He. 2007. A new "view" of ecology and conservation through animal-borne video systems. Trends in Ecology and Evolution 22:660–668.

Moll, R. J., J. J. Millspaugh, J. Beringer, J. Sartwell, Z. He, J. Eggert, and X. Zhao. 2009. A terrestrial animal-borne video system for large mammals. Computers and Electronics in Agriculture 66:133–139.

Molsher, R. L. 2001. Trapping and demographics of feral cats (*Felis catus*) in central New South Wales. Wildlife Research 28:631–636.

Mommertz, S., C. Schauer, N. Kösters, A. Lang, and J. Filser. 1996. A comparison of D-Vac suction, fenced, and unfenced pitfall trap sampling of epigeal arthropods in agroecosystems. Annales Zoologici Fennici 33:117–124.

Mong, T. W., and B. K. Sandercock. 2007. Optimizing radio retention and minimizing radio impacts in a field study of upland sandpipers. Journal of Wildlife Management 71:971–980.

Monmonier, M., and H. J. de Blij. 1996. How to lie with maps. University of Chicago Press, Chicago, Illinois, USA.

Montgomery, G. G. 1964. Tooth eruption in preweaned raccoons. Journal of Wildlife Management 28:582–584.

Montgomery, S. J., D. F. Balph, and D. M. Balph. 1971. Age determination of Uinta ground squirrels by teeth annuli. Southwestern Naturalist 15:400–402.

Mood, A. M., F. A. Graybill, and D. C. Boes. 1974. Introduction to the theory of statistics. Third edition. McGraw-Hill, Boston, Massachusetts, USA.

Moody, R. D., J. O. Collins, and V. H. Reid. 1954. Oak production study underway. Louisiana Conservationist 6(9):6–8.

Moorcroft, P. R., and A. Barnett. 2008. Mechanistic home range models and resource selection analysis: a reconciliation and unification. Ecology 89:1112–1119.

Moorcroft, P. R., and M. A. Lewis. 2006. Mechanistic home range analysis. Princeton University Press, Princeton, New Jersey, USA.

Moorcroft, P. R., M. A. Lewis, and R. L. Crabtree. 1999. Home range analysis using a mechanistic home range model. Ecology 80:1656–1665.

Moorcroft, P. R., M. A. Lewis, and R. L. Crabtree. 2006. Mechanistic home range models capture spatial patterns and dynamics of coyote territories in Yellowstone. Proceedings of the Royal Society B 273:1651–1659.

Moore, C. T., and M. J. Conroy. 2006. Optimal regeneration planning for old-growth forest: addressing scientific uncertainty in endangered species recovery through adaptive management. Forest Science 52:155–172.

Moore, D. S., and G. P. McCabe. 1993. Introduction to the practice of statistics. Second edition. W. H. Freeman and Company, New York, New York, USA.

Moore, D. W., and M. L. Kennedy. 1985. Factors affecting response of raccoons to traps and population size estimation. American Midland Naturalist 114:192–197.

Moore, J.-D. 2005. Use of native dominant wood as a new cover board type for monitoring eastern red-backed salamanders. Herpetological Review 36:268–271.

Morales, J. M., D. T. Haydon, J. Frair, K. E. Holsinger, and J. M. Fryxell. 2004. Extracting more out of relocation data: building movement models as mixtures of random walks. Ecology 85:2436–2455.

Moran, S. 1985. Banding fruit bats. Israel Journal of Zoology 33:91–93.

Morgan, D. R. 1981. Monitoring bait acceptance in brush-tailed possum populations: development of a tracer technique. New Zealand Journal of Forestry Science 11:271–277.

Morgan, J. T., and G. L. Dusek. 1992. Trapping white-tailed deer on summer range. Wildlife Society Bulletin 20:39–41.

Morgart, J. R., J. J. Hervert, P. R. Krausman, J. L. Bright, and R. S. Henry. 2005. Sonoran pronghorn use of anthropogenic and natural water sources. Wildlife Society Bulletin 33: 51–60.

Morgenweck, R. O., and W. H. Marshall. 1977. Wing marker for American woodcock. Bird-Banding 48:224–227.

Morin, P. A., G. Luikart, R. K. Wayne, and the SNP workshop group. 2004. SNPs in ecology, evolution and conservation. Trends in Ecology and Evolution 9:373–375.

Morin, P. A., and D. S. Woodruff. 1996. Noninvasive genotyping for vertebrate conservation. Pages 298–313 in T. B. Smith and R. K. Wayne, editors. Molecular genetic approaches in conservation. Oxford University Press, New York, New York, USA.

Morisita, M. 1957. A new method for the estimation of density by the spacing method applicable to non-randomly distributed populations. Physiological Ecology 7:134–144.

Moritz, C. 1994a. Applications of mitochondrial DNA analysis in conservation: a critical review. Molecular Ecology 3:401–411.

Moritz, C. 1994b. Defining evolutionarily significant units for conservation. Trends in Ecology and Evolution 9:373–375.

Moritz, C. 1999. Conservation units and translocations: strategies for conserving evolutionary processes. Hereditas 130:217–228.

Morris, D. W. 1984. Patterns and scale of habitat use in two temperate-zone small mammal faunas. Candian Journal of Zoology 62:1540–1547.

Morris, P. 1972. A review of mammalian age determination methods. Mammal Review 2:69–104.

Morrison, D. W. 1978. Foraging ecology and energetics of the frugivorous bat *Artibeus jamaicensis*. Ecology 59:716–723.

Morrison, J. L., and S. M. McGehee. 1996. Capture methods for crested caracaras. Journal of Field Ornithology 67:630–636.

Morrison, M. L. 2001. A proposed research emphasis to overcome the limits of wildlife–

habitat relationship studies. Journal of Wildlife Management 65:613–623.

Morrison, M. L., W. M. Block, M. D. Strickland, B. A. Collier, and M. J. Peterson, editors. 2008. Wildlife study design. Second edition. Springer-Verlag, New York, New York, USA.

Morrison, M. L., W. M. Block, M. D. Strickland, and W. L. Kendall. 2001. Wildlife study design. Springer-Verlag, New York, New York, USA.

Morrison, M. L., B. G. Marcot, and R. W. Mannan. 1992. Wildlife–habitat relationships: concepts and applications. University of Wisconsin Press, Madison, USA.

Morrison, M. L., B. G. Marcot, and R. W. Mannan. 1998. Wildlife–habitat relationships: concepts and applications. Second edition. University of Wisconsin Press, Madison, USA.

Morrow, M. E., N. W. Atherton, and N. J. Silvy. 1987. A device for returning nestling birds to their nests. Journal of Wildlife Management 51:202–204.

Morrow, M. E., N. J. Silvy, and W. G. Swank. 1992. Post-juvenile primary feather molt of wild mourning doves in Texas. Proceedings of the Annual Conference of the Southeastern Association of Fish and Wildlife Agencies 46:194–198.

Mortelliti, A., and L. Boitani. 2008. Evaluation of scent station surveys to monitor the distribution of three European carnivore species (*Martes foina, Meles meles, Vulpes vulpes*) in a fragmented landscape. Mammalian Biology 73:287–292.

Moruzzi, T. L., T. K. Fuller, R. M. DeGraaf, R. T. Brooks, and W. J. Li. 2002. Assessing remotely triggered cameras for surveying carnivore distribution. Wildlife Society Bulletin 30:380–386.

Mosby, H. S., editor. 1960. Manual of game investigational techniques. The Wildlife Society, Washington, D.C., USA.

Mosby, H. S., editor. 1963. Wildlife investigational techniques. The Wildlife Society, Washington, D.C., USA.

Mosby, H. S. 1969. Reconnaissance mapping and map use. Pages 119–134 in R. H. Giles, Jr., editor. Wildlife management techniques. The Wildlife Society, Washington, D.C., USA.

Moseby, K. E., and J. L. Read. 2001. Factors affecting pitfall capture rates of small ground vertebrates in arid South Australia, II. Optimum pitfall trapping effort. Wildlife Research 28:61–71.

Moseley, L. J., and H. C. Mueller. 1975. A device for color-marking nesting birds. Bird-Banding 46:341–342.

Moser, B. W., and E. O. Garton. 2007. Effects of telemetry location error on space-use estimates using fixed-kernel density estimator. Journal of Wildlife Management 71:2421–2426.

Moses, E. S. 1968. Experimental techniques for capturing American eiders. Transactions of the Northeast Section of The Wildlife Society 25:89–94.

Moses, R. A., and S. Boutin. 1986. Molar fluting and pelt primeness techniques for distinguishing age classes of muskrats: a reevaluation. Wildlife Society Bulletin 14:403–406.

Moss, R., I. B. Trenholm, A. Watson, and R. Parr. 1990. Parasitism, predation and survival of hen red grouse *Lagopus lagopus scoticus* in spring. Journal of Animal Ecology 59:631–642.

Moss, R., and A. Watson. 1991. Population cycles and kin selection in red grouse *Lagopus lagopus scoticus*. Ibis 133:113–120.

Moss, R., A. Watson, and P. Rothery. 1984. Inherent changes in the body size, viability, and behavior of a fluctuating red grouse (*Lagopus lagopus scoticus*) population. Journal of Animal Ecology 53:171–189.

Mossman, A. S. 1960. A color marking technique. Journal of Wildlife Management 24:104.

Moulin, S., A. Dondeau, and R. Delecolle. 1998. Combining agricultural crop models and satellite observations from field to regional scales. International Journal of Remote Sensing 19:1021–1036.

Moulton, C. A. 1996. The use of PVC pipes to capture hylid frogs. Herpetological Review 27:186–187.

Mourao, G., and I. M. Medri. 2002. A new way of using inexpensive large-scale assembled GPS to monitor giant anteaters in short time intervals. Wildlife Society Bulletin 30:1029–1032.

Mowat, G., and D. Paetkau. 2002. Estimating marten *Martes americana* population size using hair capture and genetic tagging. Wildlife Biology 8:201–209.

Mowat, G., B. G. Slough, and R. Rivard. 1994. A comparison of three live capturing devices for lynx: capture efficiency and injuries. Wildlife Society Bulletin 22:644–650.

Mowbray, T. B., F. Cooke, and B. Ganter. 2000. Snow goose. Account 514 in A. Poole and F. Gill, editors. The birds of North America. The Academy of Natural Sciences, Philadelphia, Pennsylvania, and The American Ornithologists' Union, Washington, D.C., USA.

Mowbray, T. B., C. R. Ely, J. S. Sedinger, and R. E. Trost. 2002. Canada goose. Account 682 in A. Poole and F. Gill, editors. The birds of North America. The Academy of Natural Sciences, Philadelphia, Pennsylvania, and The American Ornithologists' Union, Washington, D.C., USA.

Moynahan, B. J., M. S. Lindbert, J. J. Rotella, and J. W. Thomas. 2007. Factors affecting nest survival of greater sage-grouse in northcentral Montana. Journal of Wildlife Management 71:1773–1783.

Mudge, G. P., and P. N. Ferns. 1978. Durability of patagial tags on herring gulls. Journal of Ringing and Migration 2:42–45.

Mueller, E. A., and R. P. Larkin. 1985. Insects observed using dual-polarization radar. Journal of Atmospheric and Oceanic Technology 2:49–54.

Mueller, H. 1999. Common snipe. Account 417 in A. Poole and F. Gill, editors. The birds of North America. The Academy of Natural Sciences, Philadelphia, Pennsylvania, and The American Ornithologists' Union, Washington, D.C., USA.

Mueller, U. G., and L. L. Wolfenbarger. 1999. AFLP genotyping and fingerprinting. Trends in Ecology and Evolution 14:389–394.

Mueller-Dombois, D., and H. Ellenberg. 1974. Aims and methods of vegetation ecology. John Wiley and Sons, New York, New York, USA.

Mukinya, J. G. 1976. An identification method for black rhinoceros (*Diceros bicomis* Linn. 1758). East African Wildlife Journal 14:335–338.

Mulcahy, D. M. 2006. Are subcutaneous transmitters better than intracoelomic? The relevance of reporting methodology to interpreting results. Wildlife Society Bulletin 34:884–890.

Muller, G., N. Liebsch, and R. P. Wilson. 2009. A new shot at a release mechanism for devices attached to free-living animals. Wildlife Society Bulletin 33:337–342.

Muller, L. I., T. T. Buerger, and R. E. Mirarchi. 1984. Guide for age determination of mourning dove embryos. Circular 272, Alabama Agricultural Experiment Station, Auburn University, Auburn, USA.

Müller-Schwarze, D., and D. P. Haggert. 2005. A better beaver trap—new safety device for live traps. Wildlife Society Bulletin 33:359–362.

Mullican, T. R. 1988. Radio telemetry and fluorescent pigments: a comparison of techniques. Journal of Wildlife Management 52:627–631.

Mundy, N. I., P. Unitt, and D. S. Woodruff. 1997. Skin from feet of museum specimens as a non-destructive source of DNA for avian genotyping. Auk 114:126–129.

Munson, L. No date. Necropsy of wild animals. Wildlife Health Center, School of Veterinary Medicine, University of California–Davis, Davis, USA. http://www.vetmed.ucdavis.edu/whc/pdfs/munsonnecropsy.pdf.

Murden, S. B., and K. L. Risenhoover. 2000. A blimp system to obtain high-resolution, low-altitude photography and videography. Wildlife Society Bulletin 28:958–962.

Murie, O. J. 1930. An epizootic disease of elk. Journal of Mammalogy 11:214–222.

Murie, O. J. 1951. The elk of North America. Stackpole, Harrisburg, Pennsylvania, and Wildlife Management Institute, Washington, D.C., USA.

Murnane, R. D., G. Meerdink, B. A. Rideout, and M. P. Anderson. 1995. Ethylene glycol toxicosis in a captive bred released California condor (*Gymnogyps californianus*). Journal of Zoo and Wildlife Medicine 26:306–310.

Murphy, C. G. 1993. A modified drift fence for capturing treefrogs. Herpetological Review 24:143–145.

Murphy, D. D., and B. D. Noon. 1991. Coping with uncertainty in wildlife biology. Journal of Wildlife Management 55:773–782.

Murphy, M. A., K. C. Kendall, A. Robinson, and L. P. Waits. 2007. The impact of time and field conditions on brown bear (*Ursus arctos*) faecal DNA amplification. Conservation Genetics 8:1219–1224.

Murphy, M. T. 1981. Growth and aging of nestling eastern kingbirds and eastern phoebes. Journal of Field Ornithology 52:309–316.

Murphy, R. W., J. W. Stiles, D. G. Buth, and C. H. Haufler. 1996. Proteins: isozyme electrophoresis. Pages 51–120 *in* D. M. Hillis, C. Moritz, and B. K. Mable, editors. Molecular systematics. Sinauer Associates, Sunderland, Massachusetts, USA.

Murphy, W. F., Jr., P. F. Scanlon, and R. L. Kirkpatrick. 1973. Examination of ovaries in living cottontail rabbits by laparotomy. Proceedings of the Annual Conference of the Southeastern Association of Game and Fish Commissioners 27:343–344.

Murphy, W. L. 1985. Procedures for the removal of insect specimens from sticky-trap material. Annals of the Entomological Society of America 78:881.

Murray, D. L. 2006. On improving telemetry-based survival estimation. Journal of Wildlife Management 70:1530–1543.

Murray, D. L., and M. R. Fuller. 2000. A critical review of the effects of marking on the biology of vertebrates. Pages 15–64 *in* L. Boitani and T. K. Fuller, editors. Research techniques in animal ecology: controversies and consequences. Columbia University Press, New York, New York, USA.

Murray, J. V., A. W. Goldizen, R. A. O'Leary, C. A. McAlpine, H. P. Possingham, and S. L. Choy. 2009. How useful is expert opinion for predicting the distribution of a species within and beyond the region of expertise? A case study using brush-tailed rock-wallabies *Petrogale penicillata*. Journal of Applied Ecology 46:842–851.

Murray, P. R., editor. 2007. Manual of clinical microbiology. Ninth edition. ASM Press, Washington, D.C., USA.

Murray, R. B., and M. Q. Jacobson. 1982. An evaluation of dimension analysis for predicting shrub biomass. Journal of Range Management 35:451–454.

Museums and Galleries Commission. 1992. Standards in the museum care of biological specimens. Museums and Galleries Commission, London, England, UK.

Mushet, D. M., N. H. Euliss, Jr., B. H. Hanson, and S. G. Zodrow. 1997. A funnel trap for sampling salamanders in wetlands. Herpetological Review 28:132–133.

Musiani, M., and E. Visalberghi. 2001. Effectiveness of fladry on wolves in captivity. Wildlife Society Bulletin 29:91–98.

Mussehl, T. W., and T. H. Leik. 1963. Sexing wings of adult blue grouse. Journal of Wildlife Management 27:102–106.

Myers, C. H., L. Eigner, J. A. Harris, R. Hillman, M. D. Johnson, R. Kalinowski, J. J. Muir, M. Reyes, and L. E. Tucci. 2007. A comparison of ground-based and tree-based polyvinyl chloride pipe refugia for capturing *Pseudacris regilla* in northwestern California. Northwestern Naturalist 88:147–154.

Myneni, R. B., F. G. Hall, P. J. Sellers, and A. L. Marshak. 1995. The interpretation of spectral vegetation indexes. IEEE Transactions on Geoscience and Remote Sensing 33:481–486.

Mysterud, A., and R. A. Ims. 1998. Functional responses in habitat use: availability influences relative use in trade-off situations. Ecology 79:1435–1441.

Nace, G. W., and E. K. Manders. 1982. Marking individual amphibians. Journal of Herpetology 16:309–311.

Nadorozny, N. D., and E. D. Barr. 1997. Improving trapping success of amphibians using a side-flap pail-trap. Herpetological Review 28:193–194.

Nagorsen, D. W., J. Forsberg, and G. R. Giannico. 1988. An evaluation of canine radiographs for sexing and aging Pacific Coast martens. Wildlife Society Bulletin 16:421–426.

Naidoo, R. 2000. Response of breeding male ruffed grouse, *Bonasa umbellus*, to playbacks of drumming recordings. Canadian Field-Naturalist 114:320–322.

Nakane, A., Y. Kojima, Y. Hayashi, S. Kurokawa, K. Mizuno, and K. Kohri. 2005. Effect of testicular biopsy in childhood on spermatogenesis, fertility, and paternity in adulthood—mouse model study. Urology 66:682–686.

Nall, R. W., L. S. Philpot, R. D. Smith, and P. W. Sturm. 1970. Use of the cannon-net for capturing fallow deer. Proceedings of the Annual Conference of the Southeastern Association of Game and Fish Commissioners 24:282–291.

Nalle, D. J., J. L. Arthur, and J. Sessions. 2002. Designing compact and contiguous reserve networks with a hybrid heuristic algorithm. Forest Science 48:59–68.

Nams, V. O. 1989. Effects of radiotelemetry error on sample size and bias when testing for habitat selection. Canadian Journal of Zoology 67:1631–1636.

Nams, V. O. 2005. Using animal movement paths to measure response to spatial scale. Oecologia 143:179–188.

Nass, R. D., and G. A. Hood. 1969. Time-specific tracer to indicate bait acceptance by small mammals. Journal of Wildlife Management 33:584–588.

Nastase, A. J. 1982. An inexpensive trap for capturing flightless Canada geese. North American Bird Bander 7:46–48.

Natarajan, R., and C. E. McCulloch. 1999. Modeling heterogeneity in nest survival data. Biometrics 55:553–559.

Nathan, R., W. M. Getz, E. Revilla, M. Holyoak, R. Kadmon, D. Saltz, and P. E. Smouse. 2009. A movement ecology paradigm for unifying organismal movement research. Proceedings of the National Academy of Sciences of the United States of America 105:19052–19059.

Nation, J. L. 1983. A new method using hexamethyldisilazane for preparation of soft insect tissues for scanning electron microscopy. Biotechnic and Histochemistry 58:347–351.

National Aeronautics and Space Administration [NASA]. 2009. Geoscience Laser Altimeter System general information. http://glas.gsfc.nasa.gov.

National Council of the Paper Industry for Air and Stream Improvement [NCASI]. 1996. The National Gap Analysis Program: ecological assumptions and sensitivity to uncertainty. Technical Bulletin 720, National Council of the Paper Industry for Air and Stream Improvement, Research Triangle Park, North Carolina, USA.

National Mining Association. 2004. Mining in the United States: national statistics. http://www.nma.org.

National Mining Association. 2009. Mining in the United States: national statistics. http://www.nma.org.

National Research Council. 2002. Weather radar technology beyond NEXRAD. Committee on Weather Radar Technology Beyond NEXRAD. National Academies Press, Washington, D.C., USA.

National Research Council. 2007. Environmental impacts of wind energy projects. National Academies Press, Washington, D.C., USA.

Naugle, D. E., K. F. Higgins, M. E. Estey, R. R. Johnson, and S. M. Nusser. 2000. Local and landscape-level factors influencing black tern habitat suitability. Journal of Wildlife Management 64:253–260.

Naugle, D. E., B. J. Kernohan, and J. A. Jenks. 1995. Seasonal capture success and bait use of white-tailed deer in an agricultural-wetland complex. Wildlife Society Bulletin 23:198–200.

Naugle, R. E., A. T. Rutberg, H. B. Underwood, J. W. Turner, Jr., and I. K. M. Liu. 2002. Field testing of immunocontraception on white-tailed deer (*Odocoileus virginianus*) on Fire Island National Seashore, New York, USA. Pages 143–153 *in* J. F. Kirkpatrick, B. L. Lasley, W. R. Allen, and C. Doberska, editors. Fertility control in wildlife. Reproduction, Supplement 60.

Nauwelaerts, S., J. Coeck, and P. Aerts. 2000. Visible implant elastomers as a method for marking adult anurans. Histological Review 31:154–155.

Nave, C. D., G. Coulson, A. Poiani, G. Shaw, and M. B. Renfree. 2002. Fertility control in the eastern grey kangaroo using levonorgestrel implants. Journal of Wildlife Management 66:470–477.

Naylor, B. J., and M. Novak. 1994. Catch efficiency and selectivity of various traps and sets used for capturing American martens. Wildlife Society Bulletin 22:489–496.

Neal, A. K., G. C. White, R. B. Gill, D. F. Reed, and J. H. Olterman. 1993. Evaluation of mark–resight model assumptions for estimating mountain sheep numbers. Journal of Wildlife Management 57:436–450.

Neal, B. J. 1959. Techniques of trapping and tagging the collared peccary. Journal of Wildlife Management 23:11–16.

Neal, W. 1964. Extra white feather makes bird important. Inland Bird Banding Association News 36:69–71.

Nedelman, J., J. A. Thompson, and R. J. Taylor. 1987. The statistical demography of whooping cranes. Ecology 68:1401–1411.

Nei, M., T. Maruyama, and R. Chakraborty. 1975. The bottleneck effect and genetic variability in populations. Evolution 29:1–10.

Neigel, J. E. 1997. A comparison of alternative strategies for estimating gene flow from genetic markers. Annual Review of Ecology and Systematics 28:105–128.

Neigel, J. E. 2002. Is FST obsolete? Conservation Genetics 3:167–173.

Neigel, J. E., R. M. Ball, Jr., and J. C. Avise. 1991. Estimation of single generation migration distances from geographic variation in animal mitochondrial DNA. Evolution 45:423–432.

Neill, L. O., A. De Jongh, J. Ozolins, T. De Jong, and J. Rochford. 2007. Minimizing leg-hold trapping trauma for otters with mobile phone technology. Journal of Wildlife Management 71:2776–2780.

Nellis, C. H. 1968. Some methods for capturing coyotes alive. Journal of Wildlife Management 32:402–405.

Nellis, C. H. 1969. Sex and age variation in red squirrel skulls from Missoula County, Montana. Canadian Field-Naturalist 83:324–330.

Nellis, C. H., S. P. Wetmore, and L. B. Keith. 1972. Lynx-prey interactions in central Alberta. Journal of Wildlife Management 36:320–329.

Nellis, C. H., S. P. Wetmore, and L. B. Keith. 1978. Age-related characteristics of coyote canines. Journal of Wildlife Management 42:680–683.

Nellis, D. W., J. H. Jenkins, and A. D. Marshall. 1967. Radioactive zinc as a feces tag in rabbits, foxes, and bobcats. Proceedings of the Annual Conference of the Southeastern Association of Game and Fish Commissioners 21:205–207.

Nelsen, O. E. 1953. Comparative embryology of the vertebrates. McGraw-Hill, New York, New York, USA.

Nelson, M. E. 2001. Tooth extractions from live-captured white-tailed deer. Wildlife Society Bulletin 29:245–247.

Nelson, M. E. 2002. The science, ethics, and philosophy of tooth extractions from live-captured white-tailed deer: a response to Festa-Bianchet et al. Wildlife Society Bulletin 30:284–288.

Nelson, R., G. E. Demas, S. L. Klein, and L. J. Kriegsfeld. 2002. Seasonal patterns of stress, immune function, and disease. Cambridge University Press, Cambridge, England, UK.

Nelson, R., C. Keller, and M. Ratnaswamy. 2005. Locating and estimating the extent of Delmarva fox squirrel habitat using an airborne LiDAR profiler. Remote Sensing of Environment 96:292–301.

Nelson, R., R. Oderwald, and T. G. Gregoire. 1997. Separating the ground and airborne laser sampling phases to estimate tropical forest basal area, volume, and biomass. Remote Sensing of Environment 60:311–326.

Nelson, R. J. 2000. An introduction to behavioral endocrinology. Second edition. Sinauer Associates, Sunderland, Massachusetts, USA.

Nelson, R. L., and R. L. Linder. 1972. Percentage of raccoons and skunks reached by egg baits. Journal of Wildlife Management 36:1327–1329.

Nelson, T. A., and A. Woolf. 1983. Field laparoscopy of female white-tailed deer. Journal of Wildlife Management 47:1213–1216.

Nelson, W. 1972. Theory and applications of hazard plotting for censored failure data. Technometrics 14:946–965.

Nelson, W. A. 1984. Effects of nutrition of animals on their ectoparasites. Journal of Medical Entomology 21:621–635.

Nequin, L. G., J. Alvarez, and N. B. Schwartz. 1979. Measurement of serum steriod and gonadotropin levels and uterine and ovarian variables throughout 4 day and 5 day estrous cycles in the rat. Biology of Reproduction 20:659–670.

Nesbitt, S. A. 1979. An evaluation of four wildlife marking materials. Bird-Banding 50:159.

Neter, J., M. H. Kutner, C. J. Nachtsheim, and W. Wasserman. 1996. Applied linear models. Fourth edition. McGraw-Hill, Boston, Massachusetts, USA.

Nettleship, D. N. 1969. Trapping common puffins. Bird-Banding 40:139–144.

Neu, C. W., C. R. Byers, and J. M. Peek. 1974. A technique for analysis of utilization-availability data. Journal of Wildlife Management 38:541–545.

Neudorf, D. L., and T. E. Pitcher. 1997. Radio transmitters do not affect nestling feeding rates by female hooded warblers. Journal of Field Ornithology 68:64–68.

Nevitt, G. A., A. H. Dittman, T. P. Quinn, and W. J. Moody. 1994. Evidence for a peripheral olfactory memory in imprinted salmon. Proceedings of the National Academy of Sciences of the United States of America 91:4288–4292.

Nevitt, G. A., M. Losekoot, and H. Weimerskirch. 2008. Evidence for olfactory search in wandering albatross, *Diomedea exulans*. Proceedings of the National Academy of Sciences of the United States of America 105:4576–4581.

Nevo, A., and L. Garcia. 1996. Spatial optimization of wildlife habitat. Ecological Modelling 91:271–281.

New, J. G. 1958. Dyes for studying the movements of small mammals. Journal of Mammalogy 39:416–429.

New, J. G. 1959. Additional uses of dyes for studying the movements of small mammals. Journal of Wildlife Management 23:348–351.

New, L. F., J. Matthiopoulos, S. Redpath, and S. T. Buckland. 2009. Fitting models of multiple hypotheses to partial population data: investigating the causes of cycles in red grouse. American Naturalist 174:399–412.

Newbery, K., and C. Southwell. 2009. An automated camera system for remote monitoring in polar environments. Cold Regions Science and Technology 55:47–51.

Newborn, D., and R. Foster. 2002. Control of parasite burden in wild red grouse *Lagopus lagopus scoticus* through the indirect application of anthelmintics. Journal of Applied Ecology 39:909–914

Newbrey, J. L., and W. L. Reed. 2008. An effective nest trap for female yellow-headed blackbirds. Journal of Field Ornithology 79:202–206.

Newman, S. H., J. Y. Takekawa, D. L. Whitworth, and E. E. Burkett. 1999. Subcutaneous anchor attachment increases retention of radio transmitters on seabirds: Xantus' and Marbled murrelets. Journal of Field Ornithology 70:520–534.

Newsom, J. D., and J. S. Sullivan, Jr. 1968. Cryobranding—a marking technique for white-tailed deer. Proceedings of the Annual Conference of the Southeastern Association of Game and Fish Commissioners 22:128–133.

Newton-Smith, W. H. 1981. The rationality of science. Routledge and Kegan Paul, Boston, Massachusetts, USA.

Ng, S. J., J. W. Dole, R. M. Sauvajot, S. P. D. Riley, and T. J. Valone. 2004. Use of highway undercrossings by wildlife in southern California. Biological Conservation 115:499–507.

Nichols, J. D., R. J. Blohn, R. E. Reynolds, R. E. Trost, J. E. Hines, and J. P. Bladen. 1991. Band reporting rates for mallards with reward bands of different dollar values. Journal of Wildlife Management 55:119–126.

Nichols, J. D., J. E. Hines, J. R. Sauer, F. W. Fallon, J. E. Fallon, and P. J. Heglund. 2000. A double-observer approach for estimating detection probability and abundance from point counts. Auk 117:393–408.

Nichols, J. D., and K. H. Pollock. 1990. Estimation of recruitment from immigration versus in situ reproduction using Pollock's robust design. Ecology 71:21–26.

Nichols, J. D., J. R. Sauer, K. H. Pollock, and J. B. Hestbeck. 1992. Estimating transition probabilities for stage-based population matrices using capture–recapture data. Ecology 73:306–312.

Nichols, R. A., M. W. Bruford, and J. J. Groombridge. 2001. Sustaining genetic variation in a small population: evidence from the Mauritius kestrel. Molecular Ecology 10:593–602.

Nicholson, W. S., and E. P. Hill. 1981. A comparison of tooth wear, lens weight, and cementum annuli as indices of age in the gray fox. Pages 355–367 *in* J. A. Chapman and D. Pursely, editors. Worldwide furbearer conference, Frostburg, Maryland, USA.

Nickerson, M. A., and K. L. Krysko. 2003. Surveying for hellbender salamanders, *Cryptobranchus alleganiensis* (Daudin): a review and critique. Applied Herpetology 1:37–44.

Nickerson, M. A., and C. E. Mays. 1973. A study of the Ozark hellbender, *Cryptobronchus alleganiensis bishopi*. Ecology 54:1164–1165.

Nielson, R. M., B. F. J. Manly, L. L. McDonald, H. Sawyer, and T. L. McDonald. 2009. Estimating habitat selection when GPS fix success is less than 100%. Ecology 90:2956–2962.

Nietfeld, M. T., M. W. Barrett, and N. Silvy. 1994. Wildlife marking techniques. Pages 140–168 *in* T. A. Bookhout, editor. Research and management techniques for wildlife and habitats. The Wildlife Society, Bethesda, Maryland, USA.

Nietfeld, M. T., and F. C. Zwickel. 1983. Classification of sex in young blue grouse. Journal of Wildlife Managament 47:1147–1151.

Nisbet, I. C. T. 1963. Quantitative study of migration with 23-centimetre radar. Ibis 105:435–460.

Nisbet, I. C. T., J. Baird, D. V. Howard, and K. S. Anderson. 1970. Statistical comparison on wing lengths measured by four observers. Bird-Banding 41:307–308.

Nisbet, I. C. T., and J. J. Hatch. 1983. Band wear and band loss in roseate terns. Journal of Field Ornithology 54:90.

Nisbet, I. C. T., and J. J. Hatch. 1985. Influence of band size on rates of band loss by common terns. Journal of Field Ornithology 56:178–181.

Nishikawa, K. C., and P. M. Service. 1988. A fluorescent marking technique for individual recognition of terrestrial salamanders. Journal of Herpetology 22:351–353.

Nixon, C. M. 1971. Productivity of white-tailed deer in Ohio. Ohio Journal of Science 71:217–225.

Nixon, C. M., M. W. McClain, and R. W. Donohoe. 1975. Effects of hunting and mast crops on a squirrel population. Journal of Wildlife Management 39:1–25.

Nol, E., and H. Blokpoel. 1983. Incubation period of ring-billed gulls and the egg immersion technique. Wilson Bulletin 95:283–286.

Nolan, J. W., R. H. Russell, and F. Anderka. 1984. Transmitters for monitoring Aldrich snares set for grizzly bears. Journal of Wildlife Management 48:942–985.

Nolan, V., Jr. 1961. A method of netting birds at open nests in trees. Auk 78:643–645.

Nolfo, L. E., and E. E. Hammond. 2006. A novel method for capturing and implanting radio transmitters in nutria. Wildlife Society Bulletin 34:104–109.

Noon, B. R. 1981. Techniques for sampling avian habitats. Pages 42–52 in D. E. Capen, editor. The use of multivariate statistics in studies of wildlife habitat. General Technical Report RM-87, U. S. Forest Service, Department of Agriculture, Washington, D.C., USA.

Noon, B. R., N. M. Ishwar, and K. Vasudevan. 2006. Efficiency of adaptive cluster and random sampling in detecting terrestrial herptofauna in a tropical rainforest. Wildlife Society Bulletin 34:59–68.

Noon, B. R., R. H. Lamberson, M. S. Boyce, and L. L. Irwin. 1999. Population viability analysis: a primer on its principal technical concepts. Pages 87–134 in N. C. Johnson, A. J. Malk, W. T. Sexton, and R. Szaro, editors. Ecological stewardship: a common reference for ecosystem management. Elsevier Science, Oxford, England, UK.

Norris, D. O. 2007. Vertebrate endocrinology. Elsevier Academic Press, New York, New York, USA.

Norris, J. D. 1967. The control of coypus (*Myocastor coypus molina*) by cage trapping. Journal of Applied Ecology 4:167–189.

Norris, K. S., and K. W. Pryor. 1970. A tagging method for small cetaceans. Journal of Mammalogy 51:609–610.

Norris, R. T., J. D. Beule, and A. T. Studholme. 1940. Banding woodcocks on Pennsylvania singing grounds. Journal of Wildlife Management 4:8–14.

North, L. A., and J. D. Harder. 2008. Characterization of the estrous cycle and assessment of reproductive status in Matschie's tree kangaroos (*Dendrolagus matschiei*) with fecal progestin profiles. General and Comparative Endocrinology 156:173–180.

North, M., G. Steger, R. Denton, G. Eberlein, T. Munton, and K. Johnson. 2000. Association of weather and nest-site structure with reproductive success in California spotted owls. Journal of Wildlife Management 64:797–807.

Northcott, T. H., and D. Slade. 1976. A livetrapping technique for river otters. Journal of Wildlife Management 40:163–164.

Norton, S. 2001. Toxic effects of plants. Pages 965–976 in C. D. Klaassen, editor. Casarett and Doull's toxicology: the basic science of poisons. Sixth edition. McGraw-Hill, New York, New York, USA.

Nosek, J. A., S. R. Craven, J. R. Sullivan, S. S. Hurley, and R. E. Peterson. 1992. Toxicity and reproductive effects of 2,3,7,8-tetrachlorodibenzo-p-dioxin in ring-necked pheasants. Journal of Toxicology and Environmental Health 35:187–198.

Nottingham, B. G., Jr., K. G. Johnson, and M. R. Pelton. 1989. Evaluation of scent-station surveys to monitor raccoon density. Wildlife Society Bulletin 17:29–35.

Novak, M. 1981a. Capture tests with underwater snares, leg-hold, Conibear and Mohawk traps. Canadian Trapper April:18–23.

Novak, M. 1981b. The foot-snare and the leg-hold traps: a comparison. Proceedings of the Worldwide Furbearer Conference 3:1671–1685.

Novak, M. 1987. Traps and trap research. Pages 941–969 in M. Novak, J. A. Baker, M. E. Obbard, and B. Malloch, editors. 1987. Wild furbearer management and conservation in North America. Ontario Ministry of Natural Resources, Toronto, Canada.

Novak, M. 1990. Evaluation of LDL, Kania and modified Conibear 120 traps in trapping martens. Progress Report, Wildlife Policy Branch, Ontario Ministry of Natural Resources, Queen's Park, Toronto, Canada.

Novak, M., J. A. Baker, M. E. Obbard, and B. Malloch, editors. 1987. Wild furbearer management and conservation in North America. Ontario Ministry of Natural Resources, Toronto, Canada.

Novakowski, N. S. 1965. Cemental deposition as an age criterion in bison, and the relation of incisor wear, eye lens weight, and dressed bison carcass weight to age. Canadian Journal of Zoology 43:173–178.

Novoa, C., M. Catusse, and L. Ellison. 1996. Capercaillie (*Tetrao urogallus*) summer population census: comparison of counts with pointing dogs and route census. Gibier Faune Sauvage 13:1–11.

Noyce, K. V., and P. L. Coy. 1989. Abundance and productivity of bear food species in different forest types of northcentral Minnesota. International Conference on Bear Research and Management 8:169–181.

Noyes, J. H., R. G. Sasser, B. K. Johnson, L. D. Bryant, and B. Alexander. 1997. Accuracy of pregnancy detection by serum protein (PSPB) in elk. Wildlife Society Bulletin 25:695–698.

Noyes, J. S. 1982. Collecting and preserving chalcid wasps (Hymenoptera:Chalcidoidea). Journal of Natural History 16:315–334.

Nudds, T. D. 1977. Quantifying the vegetative structure of wildlife cover. Wildlife Society Bulletin 5:113–117.

Nunney, L. 1993. The influence of mating system and overlapping generations on effective population size. Evolution 47:1329–1341.

Nussear, K. N., T. C. Esque, J. S. Heaton, M. E. Cablk, K. K. Drake, C. Valentin, J. L. Yee, and P. A. Medica. 2008. Are wildlife detector dogs or people better at finding tortoises? Journal of Herpetological Conservation and Biology 3:103–115.

Nutt, K. J. 2008. A comparison of techniques for assessing dispersal behaviour in gundis: revealing dispersal patterns in the absence of observed dispersal behaviour. Molecular Ecology 17:3541–3556.

Nuttle, T. 1997. Densiometer bias? Are we measuring the forest or the trees? Wildlife Society Bulletin 25:610–611.

Oakes, E. J., and P. Barnard. 1994. Fluctuating asymmetry and mate choice in paradise whydahs, *Vidua paradisaea*: an experimental manipulation. Animal Behavior 48:937–943.

Oaks, J. L., M. Gilbert, M. Z. Virani, R. T. Watson, C. U. Meteyer, B. A. Rideout, H. L. Shivaprasad, S. Ahmed, M. J. I. Chaudhry, M. Arshad, S. Mahmood, A. Ali, and A. A. Khan. 2004. Diclofenac residues as the cause of vulture population decline in Pakistan. Nature 427: 630–633.

Oates, D. W., G. I. Hoilien, and R. M. Lawler. 1985. Sex identification of field-dressed ring-necked pheasants. Wildlife Society Bulletin 13:64–67.

Obrecht, H. H., III, C. J. Pennycuick, and M. R. Fuller. 1988. Wind tunnel experiments to assess the effect of back-mounted radio transmitters on bird body drag. Journal of Experimental Biology 135:263–273.

O'Brien, G. P., H. K. Smith, and J. R. Meyer. 1965. An activity study of a radioisotope-tagged lizard, *Sceloporus undulates hyacinthinus* (Sauria, Iguanidae). Southwestern Naturalist 10:179–187.

O'Brien, S. J., and E. Mayr. 1991. Bureaucratic mischief: recognizing endangered species and subspecies. Science 251:1187–1188.

O'Connell, A. F., J. D. Nichols, and K. U. Karanth. 2011. Camera traps in animal ecology: methods and analyses. Springer, New York, New York, USA.

O'Conner, R. J., and J. H. Marchant. 1981. A field validation of some common bird census techniques. Report from the British Trust for Ornithology to the Nature Conservancy Council, Huntingdon, England, UK.

Odom, R. R. 1977. Sora. Pages 57–65 in G. C. Sanderson, editor. Management of migratory shore and upland game birds in North America. International Association of Fish and Wildlife Agencies, Washington, D.C., USA.

O'Donnell, R. P., T. Quinn, M. P. Hayes., and K. A. Ryding. 2007. Comparison of three methods for surveying amphibians in forested seep habitats in Washington State. Northwest Science 81:274–283.

O'Farrell, M. J., W. A. Clark, F. H. Emmerson, S. M. Juarez, F. R. Kay, T. M. O'Farrell, and T. Y. Goodlett. 1994. Use of a mesh live trap for small mammals: are results from Sherman live traps deceptive? Journal of Mammalogy 75:692–699.

Office of Environmental Information. 2009a. 2007 toxic release inventory (TRI): public data release report. EPA 260-R-09-001, U.S. Environmental Protection Agency, Washington, D.C., USA.

Office of Environmental Information. 2009b. EPA Superfund—final national priorities list sites. U.S. Environmental Protection Agency, Washington, D.C., USA. http://www.epa.gov/superfund/sites/query/queryhtm/npltotal.htm.

Office of Management and Budget. 2002. Coordination of geographic information and related spatial data activities. Circular A-16, Revised. http://www.whitehouse.gov/omb/circulars/a016/a016_rev.html

Office of Migratory Bird Management. 1999. Adaptive harvest management: 1999 duck hunting season. U.S. Fish and Wildlife Service, Department of the Interior, Arlington, Virginia, USA.

Office of Solid Waste and Emergency Response. 2004. Cleaning up the nation's waste sites: markets and technologies. EPA 542-R-04-05, U.S. Environmental Protection Agency, Washington, D.C., USA.

O'Gara, B. W. 1969. Horn casting by female pronghorns. Journal of Mammalogy 50:373–375.

O'Gara, B. W., and D. C. Getz. 1986. Capturing golden eagles using a helicopter and net gun. Wildlife Society Bulletin 14:400–402.

Ogilvie, M. A. 1972. Large numbered leg bands for individual identification of swans. Journal of Wildlife Management 36:1261–1265.

O'Hara, T. M. 1996. Mycotoxins. Pages 24–30 in A. Fairbrother, L. N. Locke, and G. L. Huff, editors. Noninfectious diseases of wildlife. Second edition. Iowa State University Press, Ames, USA.

O'Hare, J. R., J. D. Eisemann, K. A. Fagerstone, L. L. Koch, and T. W. Seamans. 2007. Use of alpha-chloralose by USDA Wildlife Services to immobilize birds. Proceedings of the Wildlife Damage Management Conference 12:103–113.

Ohlendorf, H. M. 2003. Ecotoxicology of selenium. Pages 465–500 in D. J. Hoffman, B. A. Rattner, G. A. Burton, Jr., and J. Cairns, Jr., editors. Handbook of ecotoxicology. Second edition. Lewis, Boca Raton, Florida, USA.

Ohlendorf, H. M., D. J. Hoffman, M. K. Saiki, and T. W. Aldrich. 1986. Embryonic mortality and abnormalities of aquatic birds: apparent impacts of selenium from irrigation drainwater. Science of the Total Environment 52:49–63.

Öhman, K. 2000. Creating continuous areas of old forest in long-term forest planning. Canadian Journal of Forest Research 30:1817–1823.

Öhman, K., and L. O. Eriksson. 1998. The core area concept in forming contiguous areas for long-term forest planning. Canadian Journal of Forest Research 28:1032–1039.

Ohmann, L. F., D. F. Grigal, and R. B. Brander. 1976. Biomass estimation for five shrubs from northeastern Minnesota. Research Paper NC-133, U.S. Forest Service, Department of Agriculture, Washington, D.C., USA.

Okarma, H., and W. Jȩdrzejewski. 1997. Live-trapping wolves with nets. Wildlife Society Bulletin 25:78–82.

Oldemeyer, J. L. 1982. Estimating production of paper birch and utilization by browsers. Canadian Journal of Forest Research 12:52–57.

Oldemeyer, J. L., and W. L. Regelin. 1980. Comparison of 9 methods for estimating density of shrubs and saplings in Alaska. Journal of Wildlife Management 44:662–666.

Oldemeyer, J. L., R. L. Robbins, and B. L. Smith. 1993. Effect of feeding level on elk weights and reproductive success at the National Elk Refuge. Pages 64–67 in R. L. Callas, D. B. Kock, and E. R. Loft, editors. Proceedings of the western states and provinces elk workshop, 15–17 May 1990, Eureka, California, USA.

Olenicki, T. 2001. Ground-based radiometers, real-time GPS receivers, and laser rangefinders—new techniques for estimating vegetation parameters and animal use sites. Intermountain Journal of Sciences 6:384–385.

Olexa, E. M., and P. J. P. Gogan. 2005. Spatial population structure of Yellowstone bison. Journal of Wildlife Management 71:1531–1538.

Oleyar, C. M., and B. S. McGinnes. 1974. Field evaluation of diethylstilbestrol for suppressing reproduction in foxes. Journal of Wildlife Management 38:101–106.

Olinde, M. W., L. S. Perrin, F. Montalbano, III, L. L. Rowse, and M. J. Allen. 1985. Smartweed seed production and availability in southcentral Florida wetlands. Proceedings of the Annual Conference of the Southeastern Association of Fish and Wildlife Agencies 39:459–464.

Olsen, G. H., S. B. Linhart, R. A. Holmes, G. J. Dasch, and C. B. Male. 1986. Injuries to coyotes caught in padded and unpadded steel foothold traps. Wildlife Society Bulletin 14:219–223.

Olsen, G. H., R. G. Linscombe, V. L. Wright, and R. A. Holmes. 1988. Reducing injuries to terrestrial furbearers by using padded foothold traps. Wildlife Society Bulletin 16:303–307.

Olsen, J., T. Billett, and P. Olsen. 1982. A method for reducing illegal removal of eggs from raptor nests. Emu 82:225.

Olsen, P. F. 1959. Dental patterns as age indicators in muskrats. Journal of Wildlife Management 23:228–231.

Olsen, S. J. 1985. Origins of the domestic dog: the fossil record. University of Arizona Press, Tucson, USA.

Olson, D. H., W. P. Leonard, and R. B. Bury, editors. 1997. Sampling amphibians in lentic habitats: methods and approaches for the Pacific Northwest. Northwest Fauna 4. Society for Northwest Vertebrate Biology, Olympia, Washington, USA.

Olsson, O. 2007. Genetic origin and success of reintroduced white storks. Conservation Biology 21:1196–1206.

Omsjoe, E. H., A. Stien, J. Irvine, S. D. Albon, E. Dahl, S. I. Thoresen, E. Rustad, and E. Ropstad. 2009. Evaluating capture stress and its effects on reproductive success in Svalbard reindeer. Canadian Journal of Zoology 87:73–85.

Onderka, D. K. 1999. Pathological examination as an aid for trap selection guidelines: usefulness and limitations. Pages 47–51 in G. Proulx, editor. Mammal trapping. Alpha Wildlife Research and Management, Sherwood Park, Alberta, Canada.

Onderka, D. K., D. L. Skinner, and A. W. Todd. 1990. Injuries to coyotes and other species caused by four models of footholding devices. Wildlife Society Bulletin 18:175–182.

O'Neil, T. A. 2010. Combined Habitat Assessment Protocols (CHAP): a habitat accounting and appraisal aystem. Northwest Habitat Institute, Corvallis, Oregon, USA. http://www.nwhi.org.

O'Neil, T. A., and C. Barrett. 2001. Willamette valley oak and pine habitat conservation project. Final Report, U.S. Bureau of Land Management, Department of the Interior, Eugene, Oregon, USA.

O'Neil, T. A., P. Bettinger, B. G. Marcot, B. W. Luscombe, G. T. Koeln, H. J. Bruner, C. Barrett, J. A. Gaines, and S. Bernatus. 2005. Application of spatial technologies in wildlife biology. Pages 418–447 in C. E. Braun, editor. Techniques for wildlife investigations and management. The Wildlife Society, Bethesda, Maryland, USA.

O'Neil, T. A., C. Langhoff, A. Hackethorn, R. George, C. Russo, C. Kiilsgaard, and C. Barrett. 2007. Willamette Valley white oak, ponderosa pine and riparian habitats inventoried and mapped using NAIP imagery. Northwest Habitat Institute, Corvallis, Oregon, USA.

O'Neil, T. A., C. Langhoff, and A. Johnson. 2008. Mapping at multiple scales using a consistent wildlife habitat classification to improve transportation and conservation planning. Journal of the Transportation Research Board, National Academy of Sciences, Washington, D.C., USA.

O'Neil, T. A., R. J. Steidl, W. D. Edge, and B. Csuti. 1995. Using wildlife communities to improve vegetation classification for conserving biodiversity. Conservation Biology 9:1482–1491.

O'Neill, R. V. 1996. Recent developments in ecological theory: hierarchy and scale. Gap analysis—a landscape approach to biodiversity planning. American Society for Photogrammetry and Remote Sensing, Bethesda, Maryland, USA.

Onorato, D., C. White, P. Zager, and L. P. Waits. 2006. Detection of predator presence at elk mortality sites using mtDNA analysis of hair and scat samples. Wildlife Society Bulletin 34:815–820.

O'Reilly, K. M., and J. C. Wingfield. 2001. Ecological factors underlying the adrenocortical response to capture stress in arctic breeding shorebirds. General and Comparative Endocrinology 124:1–11.

Oreskes, N., K. Shrader-Frechette, and K. Belitz. 1994. Verification, validation, and confirmation of numerical models in the earth sciences. Science 263:641–646.

Orr, C. D., and D. G. Dodds. 1982. Snowshoe hare habitat preferences in Nova Scotia spruce–fir forests. Wildlife Society Bulletin 10:147–150.

Orser, P. N., and D. J. Shure. 1972. Effects of urbanization on the salamander *Desmognathus fuscus fuscus*. Ecology 53:1148–1154.

Osborn, D. J. 1955. Techniques of sexing beaver, *Castor canadensis*. Journal of Mammalogy 36:141–142.

Osborne, P. E., J. C. Alonso, and R. G. Bryant. 2001. Modelling landscape-scale habitat use using GIS and remote sensing: a case study with great bustards. Journal of Applied Ecology 38:458–471.

Osenberg, C. W., O. Sarnelle, S. D. Cooper, and R. D. Holt. 1999. Resolving ecological questions through meta-analysis: goals, metrics, and models. Ecology 80:1105–1117.

Ostfeld, R. S., C. G. Jones, and J. O. Wolff. 1996. Of mice and mast: ecological connections in eastern deciduous forests. BioScience 46:323–330.

Ostfeld, R. S., F. Keesing, and V. Eviner, editors. 2008. Infectious disease ecology: effects of ecosystems on disease and of disease on ecosystems. Princeton University Press, Princeton, New Jersey, USA.

Ostfeld, R. S., M. C. Miller, and J. Schnurr. 1993. Ear tagging increases tick (*Ixodes dammini*) infestation rates of white-footed mice (*Peromyscus leucopus*). Journal of Mammalogy 74:651–655.

Ostner, J., M. Heistermann, and O. Schulke. 2008. Dominance, aggression and physiological stress in wild male Assamese macaques (*Macaca assamensis*). Hormones and Behavior 54:613–619.

Ostrowski, S., E. Fromont, and B.-U. Meyburg. 2001. A capture technique for wintering and migrating steppe eagles in southwestern Saudi Arabia. Wildlife Society Bulletin 29:265–268.

Osweiler, G. D., T. L. Carson, W. B. Buck, and G. A. Van Gelder. 1985. Clinical and diagnostic veterinary toxicology. Third edition. Kendall-Hunt, Dubuque, Iowa, USA.

Otis, D. L. 1997. Analysis of habitat selection studies with multiple patches within cover types. Journal of Wildlife Management 61:1016–1022.

Otis, D. L. 1998. Analysis of the influence of spatial pattern in habitat selection studies. Journal of Agricultural, Biological, and Environmental Statistics 3:254–267.

Otis, D. L., K. P. Burnham, G. C. White, and D. R. Anderson. 1978. Statistical inference from capture data on closed animal populations. Wildlife Monographs 62.

Otis, D. L., C. E. Knittle, and G. M. Linz. 1986. A method for estimating turnover in spring blackbird roosts. Journal of Wildlife Management 50:567–571.

Otis, D. L., and G. C. White. 1999. Autocorrelation of location estimates and the analysis of radio-tracking data. Journal of Wildlife Management 63:1039–1044.

Ott, R. L., and M. T. Longnecker. 2008. An introduction to statistical methods and data analysis. Sixth edition. Duxbury Press, Belmont, California, USA.

Ottaway, J. R., R. Carrick, and M. D. Murray. 1984. Evaluation of leg bands for visual identification of free-living silver gulls. Journal of Field Ornithology 55:287–308.

Otten, M. R., J. B. Haufler, S. R. Winterstein, and L. C. Bender. 1993. An aerial censusing procedure for elk in Michigan. Wildlife Society Bulletin 21:73–80.

Otto, J. E. 1983. An automatic nest trap for pied-billed grebes. North American Bird Bander 8:52–53.

Otto, M. C., and K. H. Pollock. 1990. Size bias in line transect sampling: a field test. Biometrics 46:239–245.

Otto, S. P., and T. Day. 2007. A biologist's guide to mathematical modeling in ecology and evolution. Princeton University Press, Princeton, New Jersey, USA.

Overton, W. S., and D. E. Davis. 1969. Estimating the numbers of animals in wildlife populations. Pages 403–455 in R. H. Giles, Jr., editor. Wildlife management techniques. The Wildlife Society, Washington, D.C., USA.

Owen, L. N. 1961. Fluorescence of tetracyclines in bone tumors, normal bone and teeth. Nature 190:500–502.

Owen, R. B., Jr., J. M. Anerson, J. W. Artmann, E. R. Clark, T. G. Dilworth, L. E. Gregg, F. W. Martin, J. D. Newsom, and S. R. Pursglove. 1977. American woodcock. Pages 149–186 in G. C. Sanderson, editor. Management of migratory shore and upland game birds in North America. International Association of Fish and Wildlife Agencies, Washington, D.C., USA.

Owens, D. W. 1995. The role of reproductive physiology in the conservation of sea turtles. Pages 39–44 in K. A. Bjorndal, editor. Biology and conservation of sea turtles. Smithsonian Institution Press, Washington, D.C., USA.

Owens, D. W. 1997. Hormones in the life history of sea turtles. Pages 315–341 in P. L. Lutz and J. A. Musick, editors. The biology of sea turtles. CRC Press, New York, New York, USA.

Owensby, C. E. 1973. Modified step-point system for botanical composition and basal cover estimates. Journal of Range Management 26:302–303.

Owen-Smith, N., editor. 2010. Dynamics of large herbivore populations in changing environments: towards appropriate models. Wiley-Blackwell Press, Oxford, England, UK.

Oyler-McCance, S. J., N. W. Kahn, K. P. Burnham, C. E. Braun, and T. W. Quinn. 1999. A population genetic comparison of large and small-bodied sage grouse in Colorado using microsatellite and mitochondrial DNA markers. Molecular Ecology 8:1457–1465.

Ozoga, J. J., and R. K. Clute. 1988. Mortality rates of marked and unmarked fawns. Journal of Wildlife Management 52:549–551.

Ozoga, J. J., and L. J. Verme. 1985. Determining fetus age in live white-tailed does by x-ray. Journal of Wildlife Management 49:372–374.

Pack, J. C., C. I. Taylor, D. A. Swanson, and S. A. Warner. 1996. Evaluation of wild turkey trapping techniques in West Virginia. Proceedings of the Annual Conference of the Southeastern Association of Fish and Wildlife Agencies 50:436–441.

Packard, J. M., R. C. Summers, and L. B. Barnes. 1985. Variation of visibility bias during aerial surveys of manatees. Journal of Wildlife Management 49:347–351.

Padgett-Flohr, G., and M. R. Jennings. 2001. An economical safe-house for small mammals in pitfall traps. California Fish and Game 87:72–74.

Palis, J. G., S. M. Adams, and M. J. Pederson. 2007. Evaluation of two types of commercially-made aquatic funnel traps for capturing ranid frogs. Herpetological Review 38:166–167.

Palmeirim, J. M., and L. Rodrigues. 1993. The 2-minute harp trap for bats. Bat Research News 34:60–64.

Palmer, D. T., D. A. Andrews, R. O. Winters, and J. W. Francis. 1980. Removal techniques to control an enclosed deer herd. Wildlife Society Bulletin 8:29–33.

Palmer, W. E., and S. D. Wellendorf. 2007. Effect of radiotransmitters on northern bobwhite annual survival. Journal of Wildlife Management 71:1281–1287.

Palmer, W. L. 1959. Sexing live-trapped juvenile ruffed grouse. Journal of Wildlife Management 23:111–112.

Palmisano, A. W., and H. H. Dupuie. 1975. An evaluation of steel traps for taking fur animals in coastal Louisiana. Proceedings of the Annual Conference of the Southeastern Association of Game and Fish Commissioners 29:342–347.

Palsbøll, P. J., J. Allen, M. Bérube, P. J. Clapham, T. P. Feddersen, P. S. Hammond, R. R. Hudson, H. Jørgensen, S. Katona, A. H. Larsen, F. Larsen, J. Lien, D. K. Mattila, J. Sigurjónsson, R. Sears, T. Smith, R. Sponer, P. Stevick, and N. Øien. 1997. Genetic tagging of humpback whales. Nature 388:767–769.

Palsbøll, P. J., M. Berube, and F. W. Allendorf. 2006. Identification of management units using population genetic data. Trends in Ecology and Evolution 22:11–16.

Paltridge, R., and R. Southgate. 2001. The effect of habitat type and seasonal conditions on fauna in two areas of the Tanami Desert. Wildlife Research 28:247–260.

Palumbi, S. R. 1996. Nucleic acids II: the polymerase chain reaction. Pages 205–248 in D. M. Hillis, C. Moritz, and B. K. Mable, editors. Molecular systematics. Sinauer Associates, Sunderland, Massachusetts, USA.

Panagis, K., and P. E. Stander. 1989. Marking and subsequent movement patterns of springbok lambs in the Etosha National Park, South West Africa/Namibia. Madoqua 16:71–73.

Pankakoski, E. 1980. An improved method for age determination in the muskrat, *Ondatra zibethicus* (L.). Annales Zoologici Fennici 17:113–121.

Paredes, R., I. L. Jones, D. J. Boness, Y. Tremblay, and M. Renner. 2008. Sex-specific differences in diving behavior of two sympatric Alcini species: thick-billed murres and razorbills. Canadian Journal of Zoology 86:610–622.

Parham, J. P., C. K. Dodd, Jr., and G. R. Zug. 1996. Age estimates (skeletochronology) of the Red Hills salamander, *Phaeognathus hubrichti*. Journal of Herpetology 30:401–404.

Paris, D. B. B. P., D. A. Taggart, G. Shaw, P. D. Temple-Smith, and M. B. Renfree. 2005. Changes in semen quality and morphology of the reproductive tract of the male tammar wallaby parallel seasonal breeding activity in the female. Reproduction 130:367–378.

Park, K. J., P. A. Robertson, S. T. Campbell, R. Foster, Z. M. Russell, D. Newborn, and P. J. Hudson. 2001. The role of invertebrates in the diet, growth and survival of red grouse (*Lagopus lagopus scoticus*) chicks. Journal of Zoology (London) 245:137–145.

Parker, G. G., A. P. Smith, and K. P. Hogan. 1992. Access to the upper forest canopy with a large tower crane: sampling the treetops in three dimensions. BioScience 42:664–670.

Parker, G. R. 1983. An evaluation of trap types for harvesting muskrats in New Brunswick. Wildlife Society Bulletin 11:339–343.

Parker, H. 1985. Compensatory reproduction through renesting in willow ptarmigan. Journal of Wildlife Management 49:599–604.

Parker, I., D. Watts, R. McCleery, R. Lopez, N. Silvy, and D. Davis. 2008. Digital versus film-based remote camera systems in the Florida Keys. Wildlife Biology in Practice 4:1–7.

Parker, K. R. 1979. Density estimation by variable area transect. Journal of Wildlife Management 43:484–492.

Parker, N. C., A. E. Giorgi, R. C. Heidinger, D. B. Jester, Jr., E. D. Prince, and G. A. Winans. 1990. Fish-marking techniques. American Fisheries Society, Bethesda, Maryland, USA.

Parker, W. S. 1976. Population estimates, age structure, and denning habits of whipsnakes, *Masticophis t. taeniatus* in a northern Utah *Atriplex–Sarcobatus* community. Herpetologica 32:53–57.

Parmenter, C. A., T. L. Yates, R. R. Parmenter, J. N. Mills, J. E. Childs, M. L. Campbell, J. L. Dunnum, and J. Milner. 1998. Small mammal survival and trapability in mark–recapture monitoring programs for hantavirus. Journal of Wildlife Diseases 34:1–12.

Parr, R. 1975. Aging red grouse chicks by primary molt and development. Journal of Wildlife Management 39:188–190.

Parren, S. G., and D. E. Capen. 1985. Local distribution and coexistence of two species of *Peromyscus* in Vermont. Journal of Mammalogy 66:36–44.

Parris, K. M. 1999. Review: amphibian surveys in forests and woodlands. Contemporary Herpetology 1:1–19.

Parris, K. M., T. W. Norton, and R. B. Cunningham. 1999. A comparison of techniques for sampling amphibians in the forests of southeast Queensland, Australia. Herpetologica 55:271–283.

Parslow, J. L. F. 1962. Immigration of night migrants into southern England in spring 1962. Bird Migration 2:160–175.

Parsons, G. R., M. K. Brown, and G. B. Will. 1978. Determining the sex of fisher from the lower canine teeth. New York Fish and Game Journal 25:42–44.

Partan S. R., C. P. Larco, and M. J. Owens. 2009. Wild tree squirrels respond with multisensory enhancement to conspecific robot alarm behaviour. Animal Behaviour 77:1127–1135.

Partridge, L. 1978. Habitat selection. Pages 351–376 in J. R. Kreb and N. B. Davies, editors. Behavioural ecology: an evolutionary approach. Blackwell Scientific, Oxford, England, UK.

Passaglia, C., F. Dodge, E. Herzog, S. Jackson, and R. Barlow. 1997. Deciphering a neural code for vision. Proceedings of the National Academy of Sciences of the United States of America 94:12649–12654.

Passmore, R. C., R. L. Peterson, and A. T. Cringan. 1955. A study of mandibular tooth-wear as an index to age of moose. Appendix A. Pages 223–238 in R. L. Peterson, editor. North American moose. University of Toronto Press, Ontario, Canada.

Paterson, A. 1998. A new capture technique for arboreal lizards. Herpetological Review 29:159.

Paton, P. W. C., and L. Pank. 1986. A technique to mark incubating birds. Journal of Field Ornithology 57:232–233.

Paton, P. W. C., C. J. Ralph, and J. Seay. 1991. A mist net design for capturing marbled murrelets. North American Bird Bander 16:123–126.

Patric, E. F., T. P. Husband, C. G. McKiel, and W. M. Sullivan. 1988. Potential of LORAN-C for wildlife research along coastal landscapes. Journal of Wildlife Management 52:162–164.

Patricelli, G. L., M. S. Dantzker, and J. W. Bradbury. 2007. Differences in acoustic directionality among vocalizations of the male red-winged blackbird (*Agelaius pheoniceus*) are related to function in communication. Behavioral Ecology and Sociobiology 61:1099–1110.

Patricelli, G. L., and A. H. Krakauer. 2010. Tactical allocation of display effort reduces trade-offs among multiple sexual signals in greater sage-grouse: an experiment with a robotic female. Behavioral Ecology 21:97–106.

Patricelli, G. L., J. A. Uy, G. Walsh, and G. Borgia. 2002. Sexual selection: male displays adjusted to female's response. Nature 415:279–280.

Pattee, O. H., and D. J. Pain. 2003. Lead in the environment. Pages 373–408 in D. J. Hoffman, B. A. Rattner, G. A. Burton, Jr., and J. Cairns, Jr., editors. Handbook of ecotoxicology. Second edition. Lewis, Boca Raton, Florida, USA.

Patterson, I. J. 1978. Tags and other distant-recognition markers for birds. Pages 54–62 in B. Stonehouse, editor. Animal marking: recognition marking of animals in research. MacMillan, London, England, UK.

Patterson, K., C. A. Haskell, and J. H. Schulz. 1993. A modified restraining device for mourning doves. Journal of Field Ornithology 64:413–416.

Patterson, R. L. 1952. The sage grouse in Wyoming. Wyoming Game and Fish Commission. Sage, Denver, Colorado, USA.

Patterson, T. A., L. Thomas, C. Wilcox, O. Ovaskainen, and J. Matthiopoulos. 2008. State-space models of individual animal movement. Trends in Ecology and Evolution 23:87–94.

Patton, D. R., and J. M. Hall. 1966. Evaluating key browse by age and form class. Journal of Wildlife Management 30:476–480.

Patton, D. R., H. G. Hudak, and T. D. Ratcliff. 1976. Trapping, anesthetizing and marking the Abert squirrel. Research Note RM-307, U.S. Forest Service, Department of Agriculture, Washington, D.C., USA.

Patton, M. L., R. R. Swaisgood, N. M. Czekala, A. M. White, G. A. Fetter, J. P. Montagne, R. G. Rieches, and V. A. Lance. 1999. Reproductive cycle length and pregnancy in the southern white rhinoceros (*Ceratotherium simum simum*) as determined by fecal pregnane analysis and observations of mating behavior. Zoo Biology 18:111–127.

Paulik, G. J., and D. S. Robson. 1969. Statistical calculations for change-in-ratio estimators of population parameters. Journal of Wildlife Management 33:1–27.

Paulissen, M. A. 1986. A technique for marking teiid lizards in the field. Herpetological Review 17:16–17.

Paullin, D. G., and E. Kridler. 1988. Spring and fall migration of tundra swans dyed at Malheur National Wildlife Refuge, Oregon. Murrelet 69:1–9.

Pauly, D., V. Christensen, and C. Walters. 2000. Ecopath, ecosim, and ecospace as tools for evaluating ecosystem impact of fisheries. ICES Journal of Marine Science 57:697–706.

Pavel, V., B. Chutny, T. Petruskova, and A. Petrusek. 2008. Blow fly *Trypocalliphora braueri* parasitism on meadow pipit and bluethroat nestlings in central Europe. Journal of Ornithology 149:193–197.

Pavone, L. V., and R. Boonstra. 1985. The effects of toe clipping on the survival of the meadow vole (*Microtus pennsylvanicus*). Canadian Journal of Zoology 63:499–501.

Payne, N. F. 1979. Relationship of pelt size, weight, and age for beaver. Journal of Wildlife Management 43:804–806.

Payne, R. B. 1973. Individual laying histories and the clutch size and numbers of eggs of parasitic cuckoos. Condor 75:414–438.

Pearce, C. M., M. B. Green, and M. R. Baldwin. 2007. Developing habitat models for waterbirds in urban wetlands: a log-linear approach. Urban Ecosystems 10:239–254.

Pearce, J. M., R. L. Fields, and K. T. Scribner. 1997. Nest materials as a source of genetic data for avian ecological studies. Journal of Field Ornithology 68:471–481.

Pearce, J. M., B. J. Pierson, S. L. Talbot, D. V. Derksen, D. Kraege, and K. T. Scribner. 2000. A genetic evaluation of morphology used to identify harvested Canada geese. Journal of Wildlife Management 64:863–874.

Pearl, R. 1928. The rate of living. Alfred A. Knopf, New York, New York, USA.

Pearlstine, L. G., S. E. Smith, L. A. Brandt, C. R. Allen, W. M. Kitchens, and J. Stenberg. 2002. Assessing state-wide biodiversity in the Florida Gap Analysis Project. Journal of Environmental Management 66:127–144.

Pearman, P. B. 2001. Conservation value of independently evolving units: sacred cow or testable hypothesis? Conservation Biology 15:780–783.

Pearman, P. B., A. M. Velasco, and A. Lopez. 1995. Tropical amphibian monitoring: a comparison of methods for detecting inter-site variation in species' composition. Herpetologica 51:325–337.

Pearson, A. M. 1975. The northern interior grizzly bear *Ursus arctos* L. Report Series 34, Canadian Wildlife Service, Ottawa, Ontario, Canada.

Pearson, D. F., and L. F. Ruggiero. 2003. Transects versus grid trapping arrangements for sample-mammal communities. Wildlife Society Bulletin 31:454–459.

Pearson, S. M. 1993. The spatial extent and relative influence of landscape-level factors on wintering bird populations. Landscape Ecology 8:3–18.

Pearson, S. M., M. G. Turner, R. H. Gardner, and R. V. O'Neill. 1996. An organism based perspective of habitat fragmentation. Pages 77–95 in R. C. Szaro, editor. Biodiversity in managed landscapes: theory and practice. Oxford University Press, Covelo, California, USA.

Pechanec, J. F. 1936. Comments on the stem-count method of determining the percentage utilization of range. Ecology 17:329–331.

Pechanec, J. F., and G. D. Pickford. 1937. A weight estimate method for the determination of range or pasture production. Journal of the American Society of Agronomy 29:894–904.

Pedersen, H. C., H. Steen, L. Kastdalen, H. Broseth, R. A. Ims, W. Svendsen, and N. G. Yoccoz. 2004. Weak compensation of harvest despite strong density-dependent growth in willow ptarmigan. Proceedings of the Royal Society of London B 271:381–385.

Pederson, J. C., and D. H. Nish. 1975. The band-tailed pigeon in Utah. Publication 75-1, Division of Wildlife Resources, Utah Department of Natural Resources, Salt Lake City, USA.

Pedrana, J., A. Rodríguez, J. Bustamante, A. Travaini, and J. I. Zañon Martinez. 2009. Failure to estimate reliable sex ratios of guanaco from road-survey data. Canadian Journal of Zoology 87:886–894.

Peek, J. M. 1970. Relation of canopy area and volume to production of three woody species. Ecology 51:1098–1101.

Peek, J. M. 1982. Elk. Pages 851–861 in J. A. Chapman and G. A. Feldhamer, editors. Wild mammals of North America. Johns Hopkins University Press, Baltimore, Maryland, USA.

Peek, J. M. 1989. Another look at burning shrubs in northern Idaho. Pages 157–159 in D. M. Baumgartner, D. W. Breuer, and B. A. Zamora, editors. Proceedings of the symposium on prescribed fire in the intermountain region: forest site preparation and range improvement. Washington State University, Pullman, USA.

Peek, J. M., L. W. Krefting, and J. C. Tappeiner. 1971. Variation in twig diameter–weight relationships in northern Minnesota. Journal of Wildlife Management 35:501–507.

Pelayo, J. T., and R. G. Clark. 2000. Effects of a nasal marker on behavior of breeding female ruddy ducks. Journal of Field Ornithology 71:484–492.

Pelren, E. C., and J. A. Crawford. 1995. A trap for blue grouse. Great Basin Naturalist 55:284–285.

Pelton, M. R. 1970. Effects of freezing on weights of cottontail lenses. Journal of Wildlife Management 34:205–207.

Pelton, M. R., and L. C. Marcum. 1975. The potential use of radioisotopes for determining densities of black bears and other carnivores. Pages 221–236 in R. L. Phillips and C. Jonkel, editors. Proceedings of the 1975 Predator Symposium. Montana Forest and Conservation Experiment Station, University of Montana, Missoula, USA.

Pendlebury, G. B. 1972. Tagging and remote identification of rattlesnakes. Herpetologica 28:349–350.

Pendleton, G. W., K. Titus, R. E. Lowell, E. DeGayner, and C. J. Flatten. 1998. Compositional analysis and GIS for study of habitat selection by goshawks in southeast Alaska. Journal of Agricultural, Biological, and Ecological Statistics 3:280–295.

Pendleton, R. C. 1956. Uses of marking animals in ecological studies: labeling animals with radioisotopes. Ecology 37:686–689.

Penfold, L. M., D. E. Wildt, T. L. Herzog, W. Lynch, L. Ware, S. E. Derrickson, and S. L. Monfort. 2000. Seasonal patterns of LH, testosterone and semen quality in the northern pintail duck (*Anas acuta*). Reproduction Fertility and Development 12:229–235.

Penfound, W. T., and E. L. Rice. 1957. An evaluation of the arms-length rectangle method in forest sampling. Ecology 38:660–661.

Penkala, J. M. 1978. An evaluation of muskrat trapping systems and new techniques. Job Progress Report, Federal Aid Project W-59-R-1, Job I-C, New Jersey Division of Fish, Game and Wildlife, Trenton, USA.

Penn, D. J. 2002. The scent of genetic compatibility: sexual selection and the major histocompatibility complex. Ethology 108:1–21.

Penner, J., C. Fruteau, F. Range, and M. O. Röedel. 2008. Finding a needle in a haystack: new methods of locating and working with rhinoceros vipers (*Bitis rhinoceros*). Herpetological Review 39:310–314.

Pennisi, E. 2002. A shaggy dog story. Science 298:1540–1543.

Penny, R. L., and W. J. L. Sladen. 1966. The use of Teflon for banding penguins. Journal of Wildlife Management 30:847–850.

Pennycuick, C. J. 1978. Identification using natural markings. Pages 147–159 in B. Stonehouse, editor. Animal marking: recognition marking of animals in research. MacMillan, London, England, UK.

Pennycuick, C. J., R. M. Compton, and L. Beckingham. 1968. A computer model for simulating the growth of a population, or of two interacting populations. Journal of Theoretical Biology 18:316–329.

Pennycuick, C. J., and J. Rudnai. 1970. A method of identifying individual lions, *Panthera leo*, with an analysis of the reliability of the identification. Journal of the Zoological Society of London 160:497–508.

Perdeck, A. C., and R. D. Wassenaar. 1981. Tarsus or tibia: where should a bird be ringed? Journal of Ringing and Migration 3:149–157.

Pereira, J. M. C., and R. M. Itami. 1991. GIS-based habitat modeling using logistic multiple regression: a study of the Mt. Graham red squirrel. Photogrammetric Engineering and Remote Sensing 57:1475–1486.

Pérez, J. M., J. E. Granados, I. Ruiz-Martinez, and M. Chirosa. 1997. Capturing Spanish ibexes with corral traps. Wildlife Society Bulletin 25:89–92.

Perkins, D. W., and M. L. Hunter, Jr. 2002. Effects of placing sticks in pitfall traps on amphibian and small mammal capture rates. Herpetological Review 33:282–284.

Perry, A. E., and G. Beckett. 1966. Skeletal damage as a result of band injury in bats. Journal of Mammalogy 47:131–132.

Perry, R. W., R. E. Thill, D. G. Peitz, and P. A. Tappe. 1999. Effects of different silvicultural systems on initial soft mast production. Wildlife Society Bulletin 27:915–923.

Pertoldi, C., S. F. Barker, A. B. Madsen, H. Jorgensen, E. Randi, J. Munoz, H. J. Baagoe, and V. Loeschcke. 2008. Spatio-temporal population genetics of the Danish pine marten (*Martes martes*). Biological Journal of the Linnean Society 93:457–464.

Peterle, T. J. 1956. Trapping techniques and banding returns for Michigan sharp-tailed grouse. Journal of Wildlife Management 20:50–55.

Peterman, R. M. 1990. Statistical power analysis can improve fisheries research and management. Canadian Journal of Fisheries and Aquatic Science 47:2–15.

Petersen, M. R., J. A. Schmutz, and R. F. Rockwell. 1994. Emperor goose. Account 97 in A. Poole and F. Gill, editors. The birds of North America. The Academy of Natural Sciences, Philadelphia, Pennsylvania, and The American Ornithologists' Union, Washington, D.C., USA.

Peterson, M. J. 1991a. Wildlife parasitism, science, and management policy. Journal of Wildlife Management 55:782–789.

Peterson, M. J. 1991b. The Wildlife Society publications are appropriate outlets for the results of host–parasite interaction studies. Wildlife Society Bulletin 19:360–369.

Peterson, M. J. 1996. The endangered Attwater's prairie chicken and an analysis of prairie grouse helminthic endoparasitism. Ecography 19:424–431.

Peterson, M. J. 2003. Infectious agents of concern for the Jackson Hole elk and bison herds: an ecological perspective. Final report, National Elk Refuge and Grand Teton National Park, Department of the Interior, Jackson, Wyoming, USA.

Peterson, M. J. 2004. Parasites and infectious diseases of prairie grouse: should managers be concerned? Wildlife Society Bulletin 32:35–55.

Peterson, M. J. 2007. Diseases and parasites of Texas quails. Pages 89–114 in L. A. Brennan, editor. Texas quails: ecology and management. Texas A&M University Press, College Station, USA.

Peterson, M. N., R. Aguirre, T. A. Lawyer, D. A. Jones, J. N. Schaap, M. J. Peterson, and N. J. Silvy. 2003a. Animal welfare-based modification of the Rio Grande wild turkey funnel trap. Proceedings of the Annual Conference of the Southeastern Association of Fish and Wildlife Agencies 57:208–212.

Peterson, M. N., R. R. Lopez, P. A. Frank, M. J. Peterson, and N. J. Silvy. 2003b. Evaluating capture methods for urban white-tailed deer. Wildlife Society Bulletin 31:1176–1187.

Peterson, M. N., A. G. Mertig, and J. Liu. 2006. Effects of zoonotic disease attributes on public attitudes towards wildlife management. Journal of Wildlife Management 70:1746–1753.

Peterson, M. N., T. R. Peterson, J. L. Birckhead, K. Leong, and M. J. Peterson. 2010. Rearticulating the myth of human–wildlife conflict. Conservation Letters 3:74–82.

Peterson, R. T. 1998. A field guide to western birds: a completely new guide to field marks of all species found in North America west of the 100th meridian and north of Mexico. Houghton Mifflin, Boston, Massachusetts, USA.

Peterson, R. T. 2002. A field guide to the birds of eastern and central North America. Houghton Mifflin, Boston, Massachusetts, USA.

Petit, E., and F. Mayer. 2000. A population genetic analysis of migration: the case of the noctule bat (*Nyctalus noctula*). Molecular Ecology 9:683–690.

Petit, E., and N. Valiere. 2006. Estimating population size with noninvasive capture–mark–recapture data. Conservation Biology 20:1062–1073.

Petokas, P. J., and M. M. Alexander. 1979. A new trap for basking turtles. Herpetological Review 10:90.

Petrides, G. A. 1942. Age determination in American gallinaceous game birds. Transactions of the North American Wildlife Conference 7:308–328.

Petrides, G. A. 1949. Sex and age determination in the opossum. Journal of Mammalogy 30:364–378.

Petrides, G. A. 1950a. Notes on determination of sex and age in the woodcock and mourning dove. Auk 67:357–360.

Petrides, G. A. 1950b. The determination of sex and age ratios in fur animals. American Midland Naturalist 43:355–382.

Petrides, G. A. 1951. Notes on age determination in squirrels. Journal of Mammalogy 32:111–112.

Petrides, G. A., and R. B. Nestler. 1943. Age determination in juvenile bob-white quail. American Midland Naturalist 30:774–782.

Petrides, G. A., and R. B. Nestler. 1952. Further notes on age determination in juvenile bobwhite quails. Journal of Wildlife Management 16:109–110.

Pettorelli, N. 2005. Using the satellite-derived NDVI to assess ecological responses to environmental change. Trends in Ecology and Evolution 20:503–510.

Pettorelli, N., J. M. Gaillard, A. Mysterud, P. Duncan, N. C. Stenseth, D. Delorme, G. Van Laere, C. Toigo, and F. Klein. 2006. Using a proxy of plant productivity (NDVI) to find key periods for animal performance: the case of roe deer. Oikos 112:565–572.

Peyton, L. J., and G. F. Shields. 1979. A drop net for catching shorebirds. North American Bird Bander 4:97–102.

Pfeifer, S., F. H. Wright, and M. Doncarlos. 1984. Freeze-branding beaver tails. Zoo Biology 3:159–162.

Phelps, K. L., and K. McBee. 2009. Ecological characteristics of small mammal communities at a superfund site. American Midland Naturalist 161:57–68.

Phillips, J. B., and S. C. Borland. 1992. Behavioural evidence for use of a light-dependent magnetoreception mechanism by a vertebrate. Nature 359:142–144.

Phillips, R. A., J. R. D. Silk, J. P. Croxall, V. Afanasyev, and D. R. Briggs. 2004. Accuracy of geolocation estimates for flying seabirds. Marine Ecology Progress Series 266:265–272.

Phillips, R. L. 1996. Evaluation of 3 types of snares for capturing coyotes. Wildlife Society Bulletin 24:107–110.

Phillips, R. L., F. S. Blom, G. J. Dasch, and J. W. Guthrie. 1992. Field evaluations of three types of coyote traps. Proceedings of the Vertebrate Pest Conference 15:393–395.

Phillips, R. L., F. S. Blom, and R. M. Engeman. 1990a. Responses of captive coyotes to chemical attractants. Proceedings of the Vertebrate Pest Conference 14:285–290.

Phillips, R. L., F. S. Blom, and R. E. Johnson. 1990b. An evaluation of breakaway snares for use in coyote control. Proceedings of the Vertebrate Pest Conference 14:255–259.

Phillips, R. L., and K. S. Gruver. 1996. Performance of the Paws-I-Trip™ pan tension device on 3 types of traps. Wildlife Society Bulletin 24:119–122.

Phillips, R. L., K. S. Gruver, and E. S. Williams. 1996. Leg injuries to coyotes captured in three types of foothold traps. Wildlife Society Bulletin 24:260–263.

Phillips, R. L., and C. Mullis. 1996. Expanded field testing of the no. 3 Victor Soft Catch trap. Wildlife Society Bulletin 24:128–131.

Phillips, R. L., and T. H. Nicholls. 1970. A collar for marking big game animals. Research Notes NC-I03, U.S. Forest Service, Department of Agriculture, Washington, D.C., USA.

Phillips, S. J., P. Williams, G. Midgley, and A. Archer. 2008. Optimizing dispersal corridors for the *Capte proteaceae* using network flow. Ecological Applications 18:1200–1211.

Phillips, W. R. 1985. The use of bird bands for marking tree-dwelling bats: a preliminary appraisal. Macroderma 1:17–21.

Pianka, E. R. 1974. Evolutionary ecology. Harper and Row, New York, New York, USA.

Pickett, S. T. A., J. Kolasa, and C. G. Jones. 2007. Ecological understanding: the nature of theory and the theory of nature. Second edition. Academic Press, Burlington, Massachusetts, USA.

Pickford, G. D., and G. Stewart. 1935. Coordinate method of mapping low shrubs. Ecology 16:257–261.

Picman, J. 1979. A new technique for trapping red-winged blackbirds. North American Bird Bander 4:56–57.

Pielou, E. C. 1969. An introduction to mathematical ecology. John Wiley and Sons, New York, New York, USA.

Pielou, E. C. 1974. Population and community ecology: principles and methods. Gordon and Breach, New York, New York, USA.

Pielou, E. C. 1977. Mathematical ecology. John Wiley and Sons, New York, New York, USA.

Pienaar, U. D., J. W. Van Niekerk, E. Young, P. Van Wyk, and N. Fairall. 1966. The use of oripavine hydrochlorine (M 99) in the drug immobilization and marking of the wild African elephant (*Loxodonta africana* Blumenbach) in the Kruger National Park. Koedoe 9:108–124.

Pierce, D. J., and J. M. Peek. 1984. Moose habitat use and selection patterns in north-central Idaho. Journal of Wildlife Management 48:1335–1343.

Pierson, R. A., and O. J. Ginther. 1988. Ultrasound imaging of the ovaries and uterus in cattle. Theriogenology 29:21–27.

Pietz, P. J., G. L. Krapu, R. J. Greenwood, and J. T. Lokemoen. 1993. Effects of harness transmitters on behavior and reproduction of wild mallards. Journal of Wildlife Management 57:696–703.

Piggott, M. P. 2004. Effect of sample age and season of collection on the reliability of microsatellite genotyping of faecal DNA. Wildlife Research 31:485–493.

Piggott, M. P., S. C. Banks, N. Stone, C. Banffy, and A. C. Taylor. 2006. Estimating population size of endangered brush-tailed rock-wallaby (*Petrogale penicillata*) colonies using faecal DNA. Molecular Ecology 15:81–91.

Pigozzi, G. 1988. Quill-marking, a method to identify crested porcupines individually. Acta Theriologica 33:138–142.

Pilgrim, K. L., K. S. McKelvey, A. E. Riddle, and M. K. Schwartz. 2005. Felid sex identification based on noninvasive genetic samples. Molecular Ecology Notes 5:60–61.

Pimentel, D., H. Acquay, M. Biltonen, P. Rice, M. Silva, J. Nelson, V. Lipner, S. Giodano, A. Horowitz, and M. D'Amore. 1992. Environmental and economic costs of pesticide use. BioScience 42:750–760.

Pimentel, D., A. Greiner, and T. Bashore. 1997. Economic and environmental costs of pesticide use. Pages 121–150 in J. Rose, editor. Environmental toxicology. Gordon and Breach, London, England, UK.

Pimlott, D. H., and W. J. Carberry. 1958. North American moose transplantations and handling techniques. Journal of Wildlife Management 22:51–62.

Pinheiro, J. C., and D. M. Bates. 2000. Mixed-effects models in S and S-Plus. Springer-Verlag, New York, New York, USA.

Pinheiro, J. C., and D. M. Bates. 2004. Mixed-effects models in S and S-Plus. Springer Science+Business Media, New York, New York, USA.

Pinkowski, B. C. 1978. Habituation of adult eastern bluebirds to a nest-box trap. Bird-Banding 49:125–129.

Pipas, M. J., G. H. Matschke, and G. R. McCann. 2000. Evaluation of the efficiency of three types of traps for capturing pocket gophers. Proceedings of the Vertebrate Pest Conference 19:385–388.

Pirkola, M. K., and P. Kalinainen. 1984. Use of neckbands in studying the movements and ecology of the bean goose *Anser fabalis*. Annales Zoologici Fennici 21:259–263.

Pitcher, K. 1979. Pinniped tagging in Alaska. Pages 3–4 in L. Hobbs and P. Russell, editors. Report on the pinniped tagging workshop. American Institute of Biological Sciences, Arlington, Virginia, USA.

Pitt, D. G., and G. R. Glover. 1993. Large-scale 35-mm aerial photographs for assessment of vegetation-management research plots in eastern Canada. Canadian Journal of Forest Research 23:2159–2169.

Pittman, M. T., B. P. McKinney, and G. Guzman. 1995. Ecology of the mountain lion on Big Bend Ranch State Park in Trans-Pecos, Texas. Proceedings of the Annual Conference of the Southeastern Association of Fish and Wildlife Agencies 49:552–559.

Platt, J. R. 1964. Strong inference. Science 146:347–353.

Platt, S. W. 1980. Longevity of herculite leg jess color markers on the prairie falcon (*Falco mexicanus*). Journal of Field Ornithology 51:281–282.

Playfair, W. 1786. The commercial and political atlas. J. Debrett, London, England, UK.

Playfair, W. 1801. Statistical breviary. T. Bensley, London, England, UK.

Plice, L., and T. G. Balgooyen. 1999. A remotely operated trap for American kestrels using nest boxes. Journal of Field Ornithology 70:158–162.

Plotka, E. D., U. S. Seal, G. C. Schmoller, P. D. Karns, and K. D. Keenlyne. 1977. Reproductive steroids in the white-tailed deer (*Odocoileus virginianus borealis*). Seasonal changes in the female. Biology of Reproduction 16:340–343.

Plotka, E. D., U. S. Seal, L. J. Verme, and J. J. Ozoga. 1980. Reproductive steroids in deer. Luteinizing hormone, estradiol and progesterone around estrus. Biology of Reproduction 22:576–581.

Plotka, E. D., U. S. Seal, L. J. Verme, and J. J. Ozoga. 1983. The adrenal gland in white-tailed deer: a significant source of progesterone. Journal of Wildlife Management 47:38–44.

Plotka, E. D., U. S. Seal, F. R. Zarembka, L. G. Simmons, A. Teare, L. G. Phillips, K. C. Hinshaw, and D. G. Wood. 1988. Ovarian function in the elephant: luteinizing hormone and progesterone cycles in African and Asian elephants. Biology of Reproduction 38:309–314.

Plumpton, D. L., D. I. Downing, D. E. Andersen, and J. M. Lockhart. 1995. A new method of capturing buteonine hawks. Journal of Raptor Research 29:141–143.

Plumpton, D. L., and R. S. Lutz. 1992. Multiple-capture techniques for burrowing owls. Wildlife Society Bulletin 20:426–428.

Poelker, R. J., and H. D. Hartwell. 1973. Black bear of Washington: its biology, natural history and relationship to forest regeneration. Biological Bulletin 14, Washington State Game Department, Olympia, USA.

Pojar, T. M., and D. C. Bowden. 2004. Neonatal mule deer fawn survival in west-central Colorado. Journal of Wildlife Management 68:550–560.

Pokras, M. S., and R. Chafel. 1992. Lead toxicosis from ingested fishing sinkers in adult common loons (*Gavia immer*) in New England. Journal of Zoo and Wildlife Medicine 23:92–97.

Polacheck, T. 1985. The sampling distribution of age-specific survival estimates from an age distribution. Journal of Wildlife Management 49:180–184.

Pollard, E., K. H. Lakhani, and P. Rothery. 1987. The detection of density-dependence from a series of annual censuses. Ecology 68:2046–2055.

Pollard, J. H. 1971. On distance estimators of density in randomly distributed forests. Biometrics 27:991–1002.

Pollock, K. H. 1981. Capture–recapture models allowing for age-dependent survival and capture rates. Biometrics 37:521–529.

Pollock, K. H. 1982. A capture–recapture design robust to unequal probability of capture. Journal of Wildlife Management 46:757–760.

Pollock, K. H. 2002. The use of auxiliary variables in capture–recapture modeling: an overview. Journal of Applied Statistics 29:85–102.

Pollock, K. H., and W. L. Kendall. 1987. Visibility bias in aerial surveys: a review of estimation procedures. Journal of Wildlife Management 51:502–510.

Pollock, K. H., J. D. Nichols, C. Brownie, and J. E. Hines. 1990. Statistical inference for capture–recapture experiments. Wildlife Monographs 107.

Pollock, K. H., J. D. Nichols, T. R. Simons, G. L. Farnsworth, L. L. Bailey, and J. R. Sauer. 2002. Large scale wildlife monitoring studies: statistical methods for design and analysis. Environmetrics 13:105–119.

Pollock, K. H., S. R. Winterstein, C. M. Bunck, and P. D. Curtis. 1989. Survival and analysis in telemetry studies: the staggered entry design. Journal of Wildlife Management 53:7–15.

Pollock, M. G., and E. H. Paxton. 2006. Floating mist nets: a technique for capturing birds in flooded habitat. Journal of Field Ornithology 77:335–338.

Polziehn, R. O., and C. Strobeck. 2002. A phylogenetic comparison of red deer and wapiti using mitochondrial DNA. Molecular Phylogenetics and Evolution 22:342–356.

Polzonetti-Magni, A. M., G. Mosconi, O. Carnevali, K. Yamamoto, Y. Hanaoka, and S. Kikuyama. 1998. Gonadotropins and reproductive function in the anuran amphibian, *Rana esculenta*. Biology of Reproduction 58:88–93.

Ponganis, P. J., T. K. Stockard, J. U. Meir, C. L. Williams, K. V. Ponganis, R. P. van Dam, and R. Howard. 2007. Returning on empty: extreme blood O_2 depletion underlies dive capacity of emperor penguins. Journal of Experimental Biology 210:4279–4285.

Ponjoan, A., G. Bota, E. L. Garcia De La Morena, M. B. Morales, A Woolff, I. Marco, and S. Mañosa. 2008. Adverse effects of capture and handling little bustard. Journal of Wildlife Management 72:315–319.

Poole, A., and F. Gill, editors. 2003. The birds of North America. The Academy of Natural Sciences, Philadelphia, Pennsylvania, and The American Ornithologists' Union, Washington, D.C., USA.

Poole, R. W. 1974. An introduction to quantitative ecology. McGraw-Hill, New York, New York, USA.

Pope, C. E., M. C. Gomez, and B. L. Dresser. 2006. In vitro embryo production and embryo transfer in domestic and non-domestic cats. Theriogenology 66:1518–1524.

Pope, T. L., W. L. Block, and P. Beier. 2009. Prescribed fire effects on wintering, bark-foraging birds in northern Arizona. Journal of Wildlife Management 73:695–700.

Popper, K. R. 1959. The logic of scientific discovery. Hutchinson and Co., London, England, UK.

Popper, K. R. 1968. Conjectures and refutations: the growth of scientific knowledge. Second edition. Harper and Row, New York, New York, USA.

Porter, W. F., and K. E. Church. 1987. Effects of environmental pattern on habitat preference analysis. Journal of Wildlife Management 51:681–685.

Porteus, B. T. 1987. The mutual independence hypothesis for categorical data in complex sampling schemes. Biometrika 74:857–862.

Portugal, S. J., J. A. Green, P. Cassey, P. B. Frappell, and P. J. Butler. 2009. Changing bodies, changing relationships? Heart rate as an indicator of the rate of oxygen consumption throughout the annual cycle of barnacle geese, *Branta leucopsis*. Abstracts of the annual main meeting of the Society of Experimental Biology, 28 June–1 July, Glasgow, England, UK. Comparative Biochemistry and Physiology A: Molecular and Integrative Physiology 153:S102.

Post, E., and M. C. Forchhammer. 2002. Synchronization of animal population dynamics by large-scale climate. Nature 420:168–171.

Potter, C. S. 1999. Terrestrial biomass and the effects of deforestation on the global carbon cycle: results from a model of primary production using satellite observations. BioScience 49:769–778.

Pottinger, T. G. 1999. The impact of stress on animal reproductive activities. Pages 130–163 *in* P. H. M. Balm, editor. Stress physiology in animals. CRC Press, Boca Raton, Florida, USA.

Potts, W. K., C. J. Manning, and E. K. Wakeland. 1991. Mating patterns in seminatural populations of mice influenced by MHC genotype. Nature 352:619–621.

Potts, W. K., and T. A. Sordahl. 1979. The gong method for capturing shorebirds and other ground-roosting species. North American Bird Bander 4:106–107.

Potvin, F., and L. Breton. 1988. Use of a net gun for capturing white-tailed deer, *Odocoileus virginianus*, on Anticosti Island, Quebec. Canadian Field-Naturalist 102:697–700.

Potvin, F., and J. Huot. 1983. Estimating carrying capacity of a white-tailed deer wintering area in Quebec. Journal of Wildlife Management 47:463–475.

Potvin, M. A. 1988. Seed rain on a Nebraska sandhills prairie. Prairie Naturalist 20:81–89.

Pough, H. 1966. Ecological relationships of rattlesnakes in southeastern Arizona with notes on other species. Copeia 1966:676–683.

Pough, H. 1970. A quick method for permanently marking snakes and turtles. Herpetologica 26:428–430.

Poulin, S., and C. S. Ivanyi. 2003. A technique for manual restraint of helodermatid lizards. Herpetological Review 37:194–195.

Powell, G. V. N., J. Barborak, and M. Rodriguez-S. 2000a. Assessing representativeness of protected natural areas in Costa Rica for conserving biodiversity: a preliminary gap analysis. Biological Conservation 93:35–41.

Powell, J. A. 1965. The Florida wild turkey. Technical Bulletin 8, Florida Game and Fresh Water Fish Commission, Tallahassee, USA.

Powell, L. A., M. J. Conroy, J. E. Hines, J. D. Nichols, and D. G. Krementz. 2000b. Simultaneous use of mark–recapture and radiotelemetry to estimate survival, movement, and capture rates. Journal of Wildlife Management 64:302–313.

Powell, R. A. 1979. Fishers, population models, and trapping. Wildlife Society Bulletin 7:149–154.

Powell, R. A. 2000. Animal home ranges and territories and home range estimators. Pages 65–110 in L. Boitani and T. K. Fuller, editors. Research techniques in animal ecology: controversies and consequences. Columbia University Press, New York, New York, USA.

Powell, R. A. 2005. Evaluating welfare of American black bears (*Ursus americanus*) captured in foot snares and in winter dens. Journal of Mammalogy 86:1171–1177.

Powell, R. A., and G. Proulx. 2003. Trapping and marking terrestrial mammals for research: integrating ethics, performance criteria, techniques, and common sense. Ilar Journal 44:259–279.

Powell, S. L., W. B. Cohen, Z. Yang, J. Pierce, and M. Alberti. 2008. Quantification of impervious surface in the Snohomish Water Resources Inventory Area of Western Washington from 1972–2006. Remote Sensing of Environment 112:1895–1908.

Pradel, R. 1996. Utilization of capture–mark–recapture for the study of recruitment and population growth rate. Biometrics 52:703–709.

Prange, S., T. Jordan, C. Hunter, and S. D. Gehrt. 2006. New radiocollars for the detection of proximity among individuals. Wildlife Society Bulletin 34:1333–1344.

Precision Forestry Cooperative. 2009. Precision forestry cooperative. Remote Sensing and Geospatial Analysis Laboratory, University of Washington, Seattle, USA. http://depts.washington.edu/rsgal/.

Prenzlow, D. M., and J. R. Lovvorn. 1996. Evaluation of visibility correction factors for waterfowl surveys in Wyoming. Journal of Wildlife Management 60:286–297.

Preuss, T. S., and T. M. Gehring. 2007. Landscape analysis of bobcat habitat in the northern lower peninsula of Michigan. Journal of Wildlife Management 71:2699–2706.

Prevett, J. P., I. F. Marshall, and V. G. Thomas. 1985. Spring foods of snow and Canada geese at James Bay. Journal of Wildlife Management 49:558–563.

Prevost, Y. A., and J. M. Baker. 1984. A perch snare for catching ospreys. Journal of Wildlife Management 48:991–993.

Pribil, S. 1997. An effective trap for the house wren. North American Bird Bander 22:6–9.

Price, J. B. 1931. An experiment on staining California gulls. Condor 33:123.

Priekschat, F. K. 1964. Screening fences for ground reflection reduction. Microwave Journal 7(8):46–50.

Primmer, C. R., T. Borge, J. Lindell, and G.-P. Sætre. 2002. Single-nucleotide polymorphism characterization in species with limited available sequence information: high nucleotide diversity revealed in the avian genome. Molecular Ecology 11:603–612.

Pritchard, P. C. H. 1976. Post-nesting movements of marine turtles (Cheloniidae and Dermochelyidae) tagged in the Guianas. Copeia 1976:749–754.

Pritchard, P. C. H. 1980. The conservation of sea turtles: practices and problems. American Zoologist 20:609–617.

Proctor, L., and L. M. Konishi. 1997. Representation of sound localization cues in the auditory thalamus of the barn owl. Proceedings of the National Academy of Sciences of the United States of America 94:10421–10425.

Progulske, D. R. 1957. A collar for identification of big game. Journal of Wildlife Management 21:251–252.

Proudfoot, G. A. 2002. Two optic systems assist removal of nestlings from nest cavities. Wildlife Society Bulletin 30:956–959.

Proudfoot, G. A., and E. A. Jacobs. 2001. Bow-net equipped with radio alarm. Wildlife Society Bulletin 29:543–545.

Proulx, G. 1990. Humane trapping program annual report 1989/90. Alberta Research Council, Edmonton, Canada.

Proulx, G. 1991. Humane trapping program annual report 1990/91. Alberta Research Council, Edmonton, Canada.

Proulx, G. 1997. A preliminary evaluation of four types of traps to capture northern pocket gophers, *Thomomys talpoides*. Canadian Field-Naturalist 111:640–643.

Proulx, G. 1999a. A review of current mammal trap technology in North America. Pages 1–46 in G. Proulx, editor. Mammal trapping. Alpha Wildlife Research and Management, Sherwood Park, Alberta, Canada.

Proulx, G. 1999b. Evaluation of the experimental PG trap to effectively kill northern pocket gophers. Pages 89–93 in G. Proulx, editor. Mammal trapping. Alpha Wildlife Research and Management, Sherwood Park, Alberta, Canada.

Proulx, G., and M. W. Barrett. 1989. Animal welfare concerns and wildlife trapping: ethics, standards and commitments. Transactions of the Western Section of The Wildlife Society 25:1–6.

Proulx, G., and M. W. Barrett. 1990. Assessment of power snares to effectively kill red fox. Wildlife Society Bulletin 18:27–30.

Proulx, G., and M. W. Barrett. 1991. Evaluation of the Bionic trap to quickly kill mink (*Mustela vison*) in simulated natural environments. Journal of Wildlife Diseases 27:276–280.

Proulx, G., and M. W. Barrett. 1993a. Evaluation of mechanically improved Conibear 220 traps to quickly kill fisher (*Martes pennanti*) in simulated natural environments. Journal of Wildlife Diseases 29:317–323.

Proulx, G., and M. W. Barrett. 1993b. Evaluation of the Bionic trap to quickly kill fisher (*Martes pennanti*) in simulated natural environments. Journal of Wildlife Diseases 29:310–316.

Proulx, G., M. W. Barrett, and S. R. Cook. 1989a. The C120 magnum: an effective quick-kill trap for marten. Wildlife Society Bulletin 17:294–298.

Proulx, G., M. W. Barrett, and S. R. Cook. 1990. The C120 Magnum with pan trigger: a humane trap for mink (*Mustela vison*). Journal of Wildlife Diseases 26:511–517.

Proulx, G., S. R. Cook, and M. W. Barrett. 1989b. Assessment and preliminary development of the rotating-jaw Conibear 120 trap to effectively kill marten (*Martes americana*). Canadian Journal of Zoology 67:1074–1079.

Proulx, G., and R. K. Drescher. 1994. Assessment of rotating-jaw traps to humanely kill raccoons (*Procyon lotor*). Journal of Wildlife Diseases 30:335–339.

Proulx, G., and F. F. Gilbert. 1988. The molar fluting technique for aging muskrats: a critique. Wildlife Society Bulletin 16:88–89.

Proulx, G., A. J. Kolenosky, M. J. Badry, P. J. Cole, and R. K. Drescher. 1993c. Assessment of the Sauvageau 2001-8 trap to effectively kill arctic fox. Wildlife Society Bulletin 21:132–135.

Proulx, G., A. J. Kolenosky, M. J. Badry, P. J. Cole, and R. K. Drescher. 1994a. A snowshoe hare snare system to minimize capture of marten. Wildlife Society Bulletin 22:639–643.

Proulx, G., A. J. Kolenosky, and P. J. Cole. 1993a. Assessment of the Kania trap to humanely kill red squirrels (*Tamiasciurus hudsonicus*) in enclosures. Journal of Wildlife Diseases 29:324–329.

Proulx, G., A. J. Kolenosky, P. J. Cole, and R. K. Drescher. 1995. A humane killing trap for lynx (*Felis lynx*): the Conibear 330 with clamping bars. Journal of Wildlife Diseases 31:57–61.

Proulx, G., D. K. Onderka, A. J. Kolenosky, P. J. Cole, R. K. Drescher, and M. J. Badry. 1993d. Injuries and behavior of raccoons (*Procyon lotor*) captured in the Soft Catch™ and the EGG™ traps in simulated natural environments. Journal of Wildlife Diseases 29:447–452.

Proulx, G., I. M. Pawlina, D. K. Onderka, M. J. Badry, and K. Seidel. 1994b. Field evaluation of the number 11/2 steel-jawed leghold and the Sauvageau 2001-8 traps to humanely capture arctic fox. Wildlife Society Bulletin 22:179–183.

Proulx, G., I. M. Pawlina, and R. K. Wong. 1993b. Re-evaluation of the C120 Magnum and the Bionic traps to humanely kill mink. Journal of Wildlife Diseases 29:184.

Provenza, F. D., and P. J. Urness. 1981. Diameter–length–weight relations for blackbrush (*Coleogyne ramosissima*) branches. Journal of Range Management 34:215–217.

Provost, E. E. 1962. Morphological characteristics of the beaver ovary. Journal of Wildlife Management 26:272–278.

Prugh, L. R., S. M. Arthur, and C. E. Ritland. 2008. Use of faecal genotyping to determine individual diet. Wildlife Biology 14:318–330.

Prusiner, S. B. 1982. Novel proteinaceous infectious particles cause scrapie. Science 216:136–144.

Prusiner, S. B. 1991. Molecular biology of prion diseases. Science 252:1515–1522.

Pruss, S. D., N. L. Cool, R. J. Hudson, and A. R. Gaboury. 2002. Evaluation of a modified neck

snare to live-capture coyotes. Wildlife Society Bulletin 30:508–516.
Puechmaille, S. J., and E. J. Petit. 2007. Empirical evaluation of non-invasive capture–mark–recapture estimation of population size based on a single sampling session. Journal of Applied Ecology 44:843–852.
Puget Sound LiDAR Consortium (PSLC). 2009. Puget Sound LiDAR Consortium: public-domain high-resolution topography for western Washington. http://pugetsoundlidar.ess.washington.edu/.
Pugliares, K. R., A. Bogomolni, K. M. Touhey, S. M. Herzig, C. T. Harry, and M. J. Moore. 2007. Marine mammal necropsy: an introductory guide for stranding responders and field biologists. Technical Report WHOI-2007-06, Woods Hole Oceanographic Institute, Woods Hole, Maryland, USA. https://darchive.mblwhoilibrary.org/bitstream/handle/1912/1823/WHOI-2007-06.pdf?sequence=3.
Puhakka, T., J. Koistinen, and P. L. Smith. 1986. Doppler radar observation of a sea breeze front. International Conference on Radar Meteorology 23:JP198–JP201.
Pulliam, H. R. 1988. Sources, sinks, and population regulation. American Naturalist 132:652–661.
Pyle, P. 1997. Identification guide to North American birds. Slate Creek Press, Bolinas, California, USA.
Pyle, P. 2008. Identification guide to North American birds. Part 2. Slate Creek Press, Point Reyes Station, California, USA.
Pyle, P., S. N. G. Howell, R. P. Yunick, and D. F. DeSante. 1987. Identification guide to North American passerines. Slate Creek Press, Bolinas, California, USA.
Pyrah, D. 1970. Poncho markers for game birds. Journal of Wildlife Management 34:466–467.
Pyrah, D. B. 1963. Sage grouse investigations. Federal Aid Project W-125-R-2, P-R, Progress Report, Idaho Fish and Game Department, Boise, USA.
Quade, D. 1979. Using weighted rankings in the analysis of complete blocks with additive block effects. Journal of the American Statistical Association 74:680–683.
Quang, P. X., and E. F. Becker. 1996. Combining line transect and double count sampling techniques for aerial surveys. Journal of Agricultural, Biological, and Environmental Statistics 2:230–242.
Queal, L. M., and B. D. Hlavachick. 1968. A modified marking technique for young ungulates. Journal of Wildlife Management 32:628–629.
Queney, G., N. Ferrand, S. Marchandeau, M. Azevedo, F. Mougel, M. Branco, and M. Monnerot. 2000. Absence of a genetic bottleneck in a wild rabbit (*Oryctolagus cuniculus*) population exposed to a severe viral epizootic. Molecular Ecology 9:1253–1264.
Quimby, D. C., and J. E. Gaab. 1957. Mandibular dentition as an age indicator in Rocky Mountain elk. Journal of Wildlife Management 21:435–451.
Quinn, G. P., and M. J. Keough. 2002. Experimental design and data analysis for biologists. Cambridge University Press, Cambridge, England, UK.
Quinn, H., and J. P. Jones. 1974. Squeeze box technique for measuring snakes. Herpetological Review 5:35.
Quinn, H., and T. Pappas. 1997. Restraining and marking method for snapping turtles, *Chelydra serpentina*. Herpetological Review 28:196–197.
Quinn, J. F., and A. E. Dunham. 1983. On hypothesis testing in ecology and evolution. American Naturalist 122:602–617.
Quinn, N. W. S., and D. M. Keppie. 1981. Factors influencing growth of juvenile spruce grouse. Canadian Journal of Zoology 59:1790–1795.
R Development Core Team. 2006. R: a language and environment for statistical computing. R Foundation for Statistical Computing, Vienna, Austria. http://www.r-project.org/.
R Development Core Team. 2008. R: A language and environment for statistical computing. R Foundation for Statistical Computing, Vienna, Austria. http://www.r-project.org.
R Development Core Team. 2009. R: a language and environment for statistical computing. R Foundation for Statistical Computing, Vienna, Austria. http://www.r-project.org.
Rabinowitz, D., and J. K. Rapp. 1980. Seed rain in a North American tall grass prairie. Journal of Applied Ecology 17:793–802.
Racey, P. A. 1988. Reproductive assessment in bats. Pages 31–43 *in* T. H. Kunz, editor. Ecological and behavioral methods for the study of bats. Smithsonian Institution Press, Washington, D.C., USA.
Racey, P. A., and S. M. Swift. 1985. Feeding ecology of *Pipistrellus pipistrellus* (Chiroptera: Vespertilionidae) during pregnancy and lactation. I. Foraging behavior. Journal of Animal Ecology 54:205–215.
Racine, C. H., M. E. Walsh, B. D. Roebuck, C. M. Collins, D. J. Calkins, L. R. Reitsma, P. J. Buchli, and G. Goldfarb. 1992. White phosphorus poisoning of waterfowl in an Alaskan salt marsh. Journal of Wildlife Diseases 28:669–673.
Radford, S. F., R. L. Gran, and R. V. Miller. 1994. Detection of whale wakes with synthetic aperture radar. Marine Technology Society Journal 28:46–52.
Radwell, A. J., and N. B. Camp. 2009. Comparing chemiluminescent and LED light for trapping water mites and aquatic insects. Southeastern Naturalist 8:733–738.
Rafinski, J. N. 1977. Autotransplantation as a method for permanent marking of urodele amphibians (Amphibia, Urodela). Journal of Herpetology 11:241–242.
Ragen, N. V. 2002. The brucellosis eradication program in the United States. Pages 7–15 *in* T. J. Kreeger, editor. Brucellosis in elk and bison in the Greater Yellowstone Area. Wyoming Game and Fish Department, Cheyenne, USA.
Raim, A. 1978. A radio transmitter attachment for small passerine birds. Bird-Banding 49:326–332.
Raitt, R. J., Jr.,. 1961. Plumage development and molts of California quail. Condor 63:294–303.
Rakestraw, D. L., R. J. Stapper, D. B. Fagre, and N. J. Silvy. 1998. A comparison of precision for three deer survey techniques. Proceedings of the Annual Conference of the Southeastern Association of Fish and Wildlife Agencies 52:283–293.
Ralls, K., and J. D. Ballou, editors. 1986. Proceedings of the workshop on genetic management of captive populations. Zoo Biology 5:81–238.
Ralls, K., D. B. Siniff, T. D. Willians, and V. B. Kuechle. 2006. An intraperitoneal radio transmitter for sea otters. Marine Mammal Science 5:376–381.
Ralls, K., and D. A. Smith. 2004. Latrine use by San Joaquin kit foxes (*Vulpes macrotis mutica*) and coyotes (*Canis latrans*). Western North American Naturalist 64:544–547.
Ralph, C. J. 2005. The body-grasp technique: a rapid method of removing birds from mist nets. North American Bird Bander 30:65–70.
Ralph, C. J., and E. H. Dunn, editors. 2004. Monitoring bird populations using mist nets. Studies in Avian Biology 29, Cooper Ornithological Society. University of California Press, Berkeley, USA.
Ralph, C. J., G. R. Geupel, P. Pyle, T. E. Martin, and D. F. DeSante. 1993. Handbook of field methods for monitoring landbirds. General Technical Report PSW-144, U.S. Forest Service, Department of Agriculture, Washington, D.C., USA.
Ralph, C. J., J. R. Sauer, and S. Droege. 1995. Monitoring bird populations by point counts. General Technical Report PSW-GTR-149, U.S. Forest Service, Department of Agriculture, Washington, D.C., USA.
Ramsey, C. W. 1968. A drop-net deer trap. Journal of Wildlife Management 32:187–190.
Randall, J. A. 1997. Species-specific foot drumming in kangaroo rats: *Dipodomys ingens, D. deserti, D. spectabilis*. Animal Behaviour 54:1167–1175.
Randel, C. J., R. B. Aguirre, M. J. Peterson, and N. J. Silvy. 2006. Comparison of 2 techniques for assessing invertebrate availability for wild turkey in Texas. Wildlife Society Bulletin 34:853–855.
Randolph, S. E. 1973. A tracking technique for comparing individual home ranges of small mammals. Journal of Zoology 170:509–520.
Raney, E. C. 1940. Summer movements of the bullfrog, *Rana catesbeiana* Shaw, as determined by the jaw-tag method. American Midland Naturalist 23:733–745.
Ransom, A. B. 1966. Determining age of white-tailed deer from layers in cementum of molars. Journal of Wildlife Management 30:197–199.
Ranson, K. J., G. Sun, R. G. Knox, E. R. Levine, J. F. Weishampel, and S. T. Fifer. 2001. Northern forest ecosystem dynamics using coupled models and remote sensing. Remote Sensing of Environment 75:291–302.
Rao, G. N., and J. Edmondson. 1990. Tissue reaction to an implantable identification device in mice. Toxicological Pathology 18:412–416.
Raphael, L. J., G. J. Borden, and K. S. Harris. 2007. Speech science primer: physiology, acoustics, and perception of speech. Fifth edition. Lippincott Williams and Wilkins, Baltimore, Maryland, USA.

Raphael, M. G., M. J. Wisdom, M. M. Rowland, R. S. Holthausen, B. C. Wales, B. G. Marcot, and T. D. Rich. 2001. Status and trends of habitats of terrestrial vertebrates in relation to land management in the interior Columbia River Basin. Forest Ecology and Management 153:63–87.

Rappole, J. H., and A. R. Tipton. 1991. New harness design for attachment of radio transmitters to small passerines. Journal of Field Ornithology 62:335–337.

Rasmussen, H. B., G. Wittemyer, and I. Douglas-Hamilton. 2006. Predicting time-specific changes in demographic processes using remote-sensing data. Journal of Applied Ecology 43:366–376.

Ratcliffe, L. M., and P. T. Boag. 1987. Effects of colour bands on male competition and sexual attractiveness in zebra finches, *Poephila guttata*. Canadian Journal of Zoology 65:333–338.

Rattenborg, N., B. Voirin, A. L. Vyssotski, R. W. Kays, K. Spoelstra, F. Kuemmeth, W. Heidrich, and M. Wikelski. 2008. Sleeping outside the box: electroencephalographic measures of sleep in sloths inhabiting a rainforest. Biology Letters 4:402–405.

Ratti, J. T. 1980. The classification of avian species and subspecies. American Birds 34:860–866.

Ratti, J. T., and E. O. Garton. 1994. Research and experimental design. Pages 1–23 in T. A. Bookhout, editor. Research and management techniques for wildlife and habitats. The Wildlife Society, Bethesda, Maryland, USA.

Ratti, J. T., D. L. Mackey, and J. R. Alldredge. 1984. Analysis of spruce grouse habitat in north-central Washington. Journal of Wildlife Management 48:1188–1196.

Ratti, J. T., and K. P. Reese. 1988. Preliminary test of the ecological trap hypothesis. Journal of Wildlife Management 52:484–491.

Ratti, J. T., and D. E. Timm. 1979. Migratory behavior of Vancouver Canada geese: recovery rate bias. Pages 208–212 in R. L. Jarvis and J. C. Bartonek, editors. Proceedings, management and biology of Pacific Flyway geese: a symposium. Northwest Section, The Wildlife Society, Portland, Oregon, USA.

Rausch, R. A. 1967. Some aspects of the population ecology of wolves, Alaska. American Zoologist 7:253–265.

Rausch, R. A. 1969. Morphogenesis and age-related structure of permanent canine teeth in the brown bear, *Ursus arctos* L., in arctic Alaska. Zeitschrift für Morphologie der Tiere 66:167–188.

Rausch, R. A., and A. M. Pearson. 1972. Notes on the wolverine in Alaska and the Yukon Territory. Journal of Wildlife Management 36:249–268.

Raveling, D. G. 1966. Factors affecting age ratios of samples of Canada geese caught with cannon-nets. Journal of Wildlife Management 30:682–691.

Raveling, D. G. 1976. Status of giant Canada geese nesting in southeast Manitoba. Journal of Wildlife Management 40:214–226.

Raven, P. H., R. F. Evert, and S. E. Eichhorn. 1986. Biology of plants. Fourth edition. Worth, New York, New York, USA.

Ray, D. E. 1991. Pesticides derived from plants and other organisms. Pages 585–636 in W. J. Hayes and E. R. Laws, Jr., editors. Handbook of pesticide toxicology. Volume 2. Classes of pesticides. Academic Press, San Diego, California, USA.

Raymond, G. J., A. Bossers, L. D. Raymond, K. I. O'Rourke, L. E. McHolland, P. K. Bryant, III, M. W. Miller, E. S. Williams, M. Smits, and B. Caughney. 2000. Evidence of a molecular barrier limiting susceptibility of humans, cattle, and sheep to chronic wasting disease. EMBO Journal 19:4425–4430.

Reagan, D. P. 1974. Habitat selection in the three-toed box turtle, *Terrapene carolina triunguis*. Copeia 1974:512–527.

Réale, D., and M. Festa-Bianchet. 2000. Quantitative genetics of life history traits in a long-lived wild mammal. Heredity 85:593–603.

Recht, M. A. 1981. A burrow-occluding trap for tortoises. Journal of Wildlife Management 45:557–559.

Reddy, G. V. P., Z. T. Cruz, and A. Guerrero. 2009. Development of an efficient pheromone-based trapping method for the banana root borer *Cosmopolites sordidus*. Journal of Chemical Ecology 35:111–117.

Redfield, J. A., and F. C. Zwickel. 1976. Determining the age of young blue grouse: a correction for bias. Journal of Wildlife Management 40:349–351.

Redpath, S. M. 1991. The impact of hen harriers on red grouse breeding success. Journal of Applied Ecology 28:659–671.

Redpath, S. M., and S. J. Thirgood. 1999. Numerical and functional responses in generalist predators: hen harriers and peregrines on Scottish grouse moors. Journal of Animal Ecology 68:879–892.

Redpath, S. M., and I. Wyllie. 1994. Trap for capturing territorial owls. Journal of Raptor Research 28:115–117.

Reece, P. E., J. D. Volesky, and W. H. Schacht. 2001. Cover for wildlife after summer grazing on sandhills rangeland. Journal of Range Management 54:126–131.

Reed, A., D. H. Ward, D. V. Derksen, and J. S. Sedinger. 1998. Brant. Account 337 in A Poole and F. Gill, editors. The birds of North America. The Academy of Natural Sciences, Philadelphia, Pennsylvania, and The American Ornithologists' Union, Washington, D.C., USA.

Reed, A. W., G. L. Kaufman, and B. K. Sandercock. 2007. Demographic response of a grassland rodent to environmental variability. Journal of Mammalogy 88:982–988.

Reed, B. C., J. F. Brown, D. VanderZee, T. R. Loveland, J. W. Merchant, and D. O. Ohlen. 1994. Measuring phenological variability from satellite imagery. Journal of Vegetation Science 5:703–714.

Reed, D. H., and E. H. Bryant. 2000. Experimental tests of minimum viable population size. Animal Conservation 3:7–14.

Reed, D. H., and R. Frankham. 2001. How closely correlated are molecular and quantitative measures of genetic variation? A meta-analysis. Evolution 55:1095–1103.

Reed, D. H., and R. Frankham. 2003. Correlation between fitness and genetic diversity. Conservation Biology 17:230–237.

Reed, D. H., D. J. Tollit, P. M. Thompson, and W. Amos. 1997. Molecular scatology: the use of molecular genetic analysis to assign species, sex and individual identity to seal faeces. Molecular Ecology 6:225–234.

Reed, J. M., and A. R. Blaustein. 1997. Biologically significant population declines and statistical power. Conservation Biology 11:281–282.

Reed, J. M., and L. W. Oring. 1993. Banding is infrequently associated with foot loss in spotted sandpipers. Journal of Field Ornithology 64:145–148.

Reed, P. C. 1953. Danger of leg mutilation from the use of metal color bands. Bird-Banding 24:65–67.

Rees, J. W, R. A. Kainer, and R. W. Davis. 1966. Chronology of mineralization and eruption of mandibular teeth in mule deer. Journal of Wildlife Management 30:629–631.

Reese, K. P. 1980. The retention of colored plastic leg bands by black-billed magpies. North American Bird Bander 5:136–137.

Reeves, H. M., A. E. Geis, and F. C. Kniffin. 1968. Mourning dove capture and banding. Special Scientific Report, Wildlife 117, U.S. Fish and Wildlife Service, Department of the Interior, Washington, D.C., USA.

Regan, H. M., Y. Ben-Haim, B. Langford, W. G. Wilson, P. Lundberg, S. J. Andelman, and M. A. Burgman. 2005. Robust decision making under severe uncertainty for conservation management. Ecological Applications 15:1471–1477.

Regehr, J. M., and M. S. Rodway. 2003. Evaluation of nasal discs and colored leg bands as marker for harlequin ducks. Journal of Field Ornithology 74:129–135.

Regester, K. J., and L. B. Woosley. 2005. Marking salamander egg masses with visible fluorescent elastomer: retention time and effect on embryonic development. American Midland Naturalist 153:52–60.

Rehmeier, R. L., G. A. Kaufman, and D. W. Kaufman. 2004. Long distance movements of the deer mouse. Journal of Mammalogy 85:562–568.

Reich, R. M., C. D. Bonham, and K. K. Remington. 1993. Double sampling revisited. Journal of Range Management 46:88–90.

Reid, D. G., W. E. Melquist, J. D. Woolington, and J. M. Noll. 1986. Reproductive effects of intraperitoneal transmitter implants in river otters. Journal of Wildlife Management 50:92–94.

Reinecke, K. J., C. W. Shaiffer, and D. Delnicki. 1992. Band reporting rates of mallards in the Mississippi alluvial valley. Journal of Wildlife Management 56:526–531.

Reiss, R. A., D. P. Schwert, and A. C. Ashworth. 1995. Field preservation of Coleoptera for molecular genetic analyses. Environmental Entomology 24:716–719.

Reiter, M. E., and D. E. Andersen. 2008. Comparison of the egg flotation and egg candling techniques for estimating incubation day of Canada goose nests. Journal of Field Ornithology 79:429–437.

Rempel, R. D., and R. C. Bertram. 1975. The Stewart modified corral trap. California Fish and Game 61:237–239.

Rempel, R. S., and A. R. Rodgers. 1997. Effects of differential correction on accuracy of a GPS animal location system. Journal of Wildlife Management 61:525–530.

Rempel, R. S., A. R. Rodgers, and K. F. Abraham. 1995. Performance of a GPS animal location system under boreal forest canopy. Journal of Wildlife Management 59:543–551.

Remsen, J. V., Jr., and D. A. Good. 1996. Misuse of data from mist-net captures to assess relative abundance in bird populations. Auk 113:381–398.

Rendell, W. R., B. J. Stutchbury, and R. J. Robertson. 1989. A manual trap for capturing hole-nesting birds. North American Bird Bander 14:109–111.

Reppert, J. N., R. H. Hughes, and D. Duncan. 1962. Herbage yield and its correlation with other plant measurements. Pages 115–121 in Range research methods. Miscellaneous Publication 940, U.S. Forest Service, Department of Agriculture, Washington, D.C., USA.

Reuss, S. P., and J. Oclese. 1986. Magnetic field effects on the rat pineal gland: role of retinal activation by light. Neuroscience Letters 64:97–101.

Reuther, R. T. 1968. Marking animals in zoos. International Zoo Yearbook 8:388–390.

ReVelle, C. S., J. C. Williams, and J. J. Boland. 2002. Counterpart models in facility location science and reserve selection science. Environmental Modeling and Assessment 7:71–80.

Rexstad, E. A., D. D. Miller, C. H. Flather, E. M. Anderson, J. W. Hupp, and D. R. Anderson. 1988. Questionable multivariate statistical inference in wildlife habitat and community studies. Journal of Wildlife Management 52:794–798.

Reynolds, H. W., R. D. Glaholt, and A. W. L. Hawley. 1982. Bison. Pages 972–1007 in J. A. Chapman and G. A. Feldhamer, editors. Wild mammals of North America. Johns Hopkins University Press, Baltimore, Maryland, USA.

Reynolds, J. C., M. J. Short, and R. J. Leigh. 2004. Development of population control strategies for mink (Mustela vison), using floating rafts as monitors and trap sites. Biological Conservation 120:533–543.

Reynolds, R. T., and B. D. Linkhart. 1984. Methods and materials for capturing and monitoring flammulated owls. Great Basin Naturalist 44:49–51.

Reynolds, R. T., J. M. Scott, and R. A. Nussbaum. 1980. A variable circular-plot method for estimating bird numbers. Condor 82:309–313.

Reynolds, R. T., G. C. White, S. M. Joy, and R. W. Mannan. 2004. Effects of radiotransmitters on northern goshawks: do tailmounts lower survival of breeding males? Journal of Wildlife Management 68:25–32.

Rhyan, J. C. 2000. Brucellosis in terrestrial wildlife and marine mammals. Pages 161–184 in C. Brown and C. Bolin, editors. Emerging diseases of animals. ASM Press, Washington, D.C., USA.

Ribeiro, M., A. Júnior., T. A. Gardner, and T. C. S. Avila-Pires. 2006. The effectiveness of glue traps to sample lizards in a tropical rainforest. South American Journal of Herpetology 1:131–137.

Rice, A. N., K. G. Rice, J. H. Waddle, and F. J. Mazzotti. 2006. A portable non-invasive trapping array for sampling amphibians and reptiles. Herpetological Review 37:429–430.

Rice, C. G. 2003. Utility of pheasant call counts and brood counts for monitoring population density and predicting harvest. Western North American Naturalist 63:178–188.

Rice, C. G., K. J. Jenkins, and W. Y. Chang. 2009. A sightability model for mountain goats. Journal of Wildlife Management 73:468–478.

Rice, C. P., P. W. O'Keefe, and T. J. Kubiak. 2003. Sources, pathways, and effects of PCBs, dioxins, and dibenzofurans. Pages 501–573 in D. J. Hoffman, B. A. Rattner, G. A. Burton, Jr., and J. Cairns, Jr., editors. Handbook of ecotoxicology. Second edition. Lewis, Boca Raton, Florida, USA.

Rice, J., B. W. Anderson, and R. D. Ohmart. 1984. Comparison of the importance of different habitat attributes to avian community organization. Journal of Wildlife Management 48:895–911.

Rice, T. M., and D. H. Taylor. 1993. A new method for making waistbands to mark anurans. Herpetological Review 24:141–142.

Rice, W. R., and J. D. Harder. 1977. Application of multiple aerial sampling to a mark–recapture census of white-tailed deer. Journal of Wildlife Management 54:316–322.

Richards, C., and P. L. Leberg. 1996. Temporal changes in allele frequencies and a population's history of severe bottlenecks. Conservation Biology 10:832–839.

Richards, C. M., B. M. Carlson, and S. L. Rogers. 1975. Regeneration of digits and forelimbs in the Kenyan reed frog, Hyperolius viridiflavus ferniquei. Journal of Morphology 146:431–446.

Richards, J. A. 1996. Classifier performance and map accuracy. Remote Sensing of Environment 57:161–166.

Richards, S. A. 2005. Testing ecological theory using the information-theoretic approach: examples and cautionary results. Ecology 86:2805–2814.

Richardson, D. M., J. W. Bradford, B. J. Gentry, and J. L. Hall. 1998. Evaluation of a pick-up tool for removing red-cockaded woodpecker nestlings from cavities. Wildlife Society Bulletin 26:855–858.

Richardson, D. M., J. W. Bradford, P. G. Range, and J. Christensen. 1999. A video probe system to inspect red-cockaded woodpecker cavities. Wildlife Society Bulletin 27:353–356.

Richardson, G. L. 1966. Eye lens weight as an indicator of age in the collared peccary (Pecari tajacu). Thesis, University of Arizona, Tucson, USA.

Richardson, H. N., A. L. A. Nelson, E. I. Ahmed, D. B. Parfitt, R. D. Romeo, and C. L. Sisk. 2004. Female pheromones stimulate release of luteinizing hormone and testosterone without altering GnRH mRNA in adult male Syrian hamsters (Mesocricetus auratus). General and Comparative Endocrinology 138:211–217.

Richardson, T. W., T. Gardali, and S. H. Jenkins. 2009. Review and meta-analysis of camera effects on avian nest success. Journal of Wildlife Management 73:287–293.

Richardson, W. J. 1979. Radar techniques for wildlife studies. National Wildlife Federation, Scientific Technical Series 3:171–179.

Richdale, L. E. 1951. Banding and marking penguins. Bird-Banding 22:47–54.

Richman, A. D., L. G. Herrera, and D. Nash. 2001. MHC class II beta sequence diversity in the deer mouse (Peromyscus maniculatus): implications for models of balancing selection. Molecular Ecology 10:2765–2773.

Richter, K. O. 1995. A simple aquatic funnel trap and its application to wetland amphibian monitoring. Herpetological Review 26:90–91.

Rickel, B. W., B. Anderson, and R. Pope. 1998. Using fuzzy systems, object-oriented programming, and GIS to evaluate wildlife habitat. AI Applications 12:31–40.

Ricklefs, R. E. 1973. Tattooing nestlings for individual recognition. Bird-Banding 44:63.

Ricklefs, R. E. 1979. Ecology. Second edition. Chiron Press, New York, New York, USA.

Rico, Y., C. Lorenzo, F. X. Gonzalez-Cozatl, and E. Espinoza. 2008. Phylogeography and population structure of the endangered Tehuantepec jackrabbit Lepus flavigularis: implications for conservation. Conservation Genetics 9:1467–1477.

Riedel, J. 1988. Snaring as a beaver control technique in South Dakota. General Technical Report RM-154, U.S. Forest Service, Department of Agriculture, Washington, D.C., USA.

Riedman, M. 1990. The pinnipeds: seals, sea lions, and walruses. University of California Press, Berkeley, USA.

Rieffenberger, J. C., and F. Ferrigno. 1970. Woodcock banding on the Cape May Peninsula, New Jersey. Bird Banding 41:1–10.

Ries, E. H., L. R. Hiby, and P. J. H. Reijnders. 1998. Maximum likelihood population size estimation of harbor seals in the Dutch Wadden Sea based on a mark–recapture experiment. Journal of Applied Ecology 35:332–339.

Riley, J., and R. G. William. 1981. A new ear-punch for small rodents. Journal of Institutional Animal Technicians 32:53–55.

Riley, J. R. 1980. Radar as an aid to the study of insect flight. Pages 131–140 in C. J. Amlaner, Jr., and D. W. MacDonald, editors. A handbook on biotelemetry and radio tracking. Pergamon Press, New York, New York, USA.

Riley, J. R., and D. R. Reynolds. 1979. Radar-based studies of the migratory flight of grasshoppers in the middle Niger area of Mali. Proceedings of the Royal Society of London B 204:67–82.

Riley, J. R., P. Valeur, A. D. Smith, D. R. Reynolds, G. M. Poppy, and C. Löfstedt. 1998. Harmonic radar as a means of tracking the pheromone-finding and pheromone-following flight of male moths. Journal of Insect Behavior 11:287–296.

Riley, T. Z., and B. A. Fistler. 1992. Necklace radio transmitter attachment for pheasants. Journal of Iowa Academy of Science 99:65–66.

Rinehart, R. E. 2010. Radar for meteorologists. Fifth edition. Rinehart, Nevada, Missouri, USA.

Ripley, B. D. 2004. Spatial statistics. Wiley-Interscience, Hoboken, New Jersey, USA.

Ripley, T. H., and C. J. Perkins. 1965. Estimating ground supplies of seed available to bobwhites. Journal of Wildlife Management 29:117–121.

Risbey, D. A., M. C. Calver, J. Short, J. S. Bradley, and I. W. Wright. 2000. The impact of cats and foxes on the small vertebrate fauna of Heirisson Prong, Western Australia. A field experiment. Wildlife Research 27:223–235.

Risenhoover, K. L., and J. A. Bailey. 1985. Foraging ecology of mountain sheep: implications for habitat management. Journal of Wildlife Management 49:797–804.

Risser, A. C., Jr. 1971. A technique for performing laparotomy on small birds. Condor 73:376–379.

Ritchie, J. C., J. H. Everitt, D. E. Escobar, T. J. Jackson, and M. R. Davis. 1992. Airborne laser measurements of rangeland canopy cover and distribution. Journal of Range Management 45:189–193.

Ritchie, J. C., K. S. Humes, and M. A. Weltz. 1995. Laser altimeter measurements at Walnut Gulch watershed. Arizona. Journal of Soil and Water Conservation 50:440–442.

Ritchison, G. 1984. A new marking technique for birds. North American Bird Bander 9:8.

Rittenhouse, C. D., J. J. Millspaugh, B. E. Washburn, and M. W. Hubbard. 2005. Effects of radio-transmitters on fecal glucocorticoid metabolite levels of three-toed box turtles in captivity. Wildlife Society Bulletin 33:706–713.

Rittenhouse, L. R., and F. A. Sneva. 1977. A technique for estimating big sagebrush production. Journal of Range Management 30:68–70.

Rittenhouse, T. A. G., and R. D. Semlitsch. 2007. Postbreeding habitat use of wood frogs in a Missouri oak–hickory forest. Journal of Herpetology 41:645–653.

Rivas, J. A., M. D. C. Muñoz, J. Thorbjarnarson, W. Holmstrom, and P. Calle. 1995. A safe method for handling large snakes in the field. Herpetological Review 26:138–139.

Rivier, C., and S. Rivest. 1991. Effects of stress on the activity of the hypothalamic–pituitary–gonadal axis: peripheral and central mechanisms. Biology of Reproduction 45:523–532.

Roach, M. E. 1950. Estimating perennial grass utilization on semidesert cattle ranges by percentage of ungrazed plants. Journal of Range Management 3:182–185.

Robards, F. C. 1960. Construction of a portable goose trap. Journal of Wildlife Management 24:329–331.

Robbins, C. S., D. Bystrak, and P. H. Geissler. 1986. The breeding bird survey: its first 15 years, 1965–1979. Resource Publication 157, U.S. Fish and Wildlife Service, Department of the Interior, Washington, D.C., USA.

Robbins, C. S., D. K. Dawson, and B. A. Dowell. 1989. Habitat area requirements of breeding forest birds of the Middle Atlantic states. Wildlife Monographs 103.

Robel, R. J., J. N. Briggs, A. D. Dayton, and L. C. Hulbert. 1970. Relationships between visual obstruction measurements and weight of grassland vegetation. Journal of Range Management 23:295–297.

Robert, A., D. Couvet, and F. Sarrazin. 2002. Fitness heterogeneity and viability of restored populations. Animal Conservation 5:153–161.

Robert, M., and P. Laporte. 1997. Field techniques for studying breeding yellow rails. Journal of Field Ornithology 68:56–63.

Roberts, C. M., S. Andelman, G. Branch, R. H. Bustamante, J. C. Castilla, J. Dugan, B. S. Halpern, K. D. Lafferty, H. Leslie, and J. Lubchenco. 2003. Ecological criteria for evaluating candidate sites for marine reserves. Ecological Applications (Supplement) 13:S199–S214.

Roberts, C. W., B. L. Pierce, A. W. Braden, R. R. Lopez, N. J. Silvy, P. A. Frank, and D. Ransom. 2006. Comparison of camera and road survey estimates for white-tailed deer. Journal of Wildlife Management 70:263–267.

Roberts, J. D. 1978. Variation in coyote age determination from annuli in different teeth. Journal of Wildlife Management 42:454–456.

Roberts, T. H. 1988. American woodcock (Scolopax minor). Wildlife Resource Management Manual, Technical Report EL-88, U.S. Corps of Engineers, Department of the Army, Vicksburg, Mississippi, USA.

Robertson, B. C., E. O. Minot, and D. M. Lambert. 1999. Molecular sexing of individual kakapo, Strigops haproptilus Aves, from faeces. Molecular Ecology 8:1349–1350.

Robertson, G. J., A. E. Storey, and S. I. Wilhelm. 2006. Local survival rates of common murres breeding in Witless Bay, Newfoundland. Journal of Wildlife Management 70:584–587.

Robertson, J. G. M. 1984. A technique for individually marking frogs in behavioral studies. Herpetological Review 15:56–57.

Robicheaux, B., and G. Linscombe. 1978. Effectiveness of live-traps for capturing furbearers in a Louisiana coastal marsh. Proceedings of the Annual Conference of the Southeastern Association of Fish and Wildlife Agencies 32:208–212.

Robinette, W. L., D. A. Jones, G. E. Rogers, and J. S. Gashwiler. 1957. Notes on tooth development and wear for Rocky Mountain mule deer. Journal of Wildlife Management 21:134–153.

Robinette, W. L., C. M. Loveless, and D. A. Jones. 1974. Field tests of strip census methods. Journal of Wildlife Management 38:81–96.

Robinson, D. H., and H. Wainer. 2002. On the past and future of null hypothesis significance testing. Journal of Wildlife Management 66:263–271.

Robinson, K. M., and G. G. Murphy. 1975. A new method for trapping softshell turtles. Herpetological Review 6:111.

Robinson, S. J., L. P. Waits, and I. D. Martin. 2009. Estimating abundance of American black bears using DNA-based capture–mark–recapture models. Ursus 20:1–11.

Robinson, T. J., L. M. Siefferman, and T. S. Risch. 2004. A quick, inexpensive trap for use with nest boxes. North American Bird Bander 29:116–117.

Robson, D. S., and D. G. Chapman. 1961. Catch curves and mortality rates. Transactions of the American Fisheries Society 90:181–189.

Robson, J. E. 1986. Ring "fit" on blackbreasted snake eagle. Safring News 15:56.

Roche, E. A., T. W. Arnold, J. H. Stucker, and F. J. Cuthbert. 2010. Colored plastic and metal leg bands do not affect survival of piping plover chicks. Journal of Field Ornithology 81:317–324.

Rodda, G. H., K. Dean-Bradley, and T. H. Fritts. 2005. Glueboards for estimating lizard abundance. Herpetological Review 36:252–259.

Rodgers, A. R. 2006. Recent telemetry technology. Pages 82–121 in J. J. Millspaugh and J. M. Marzluff, editors. Radio tracking and animal populations. Academic Press, San Diego, California, USA.

Rodgers, A. R., R. S. Rempel, R. Moen, J. Paczkowski, C. C. Schwartz, E. J. Lawson, and M. J. Gluck. 1997. GPS collars for moose telemetry studies: a workshop. Alces 33:203–209.

Rodgers, J. A., Jr. 1986. A field technique for color-dyeing nestling wading birds without capture. Wildlife Society Bulletin 14:399–400.

Rodgers, R. D. 1979. Ratios of primary calamus diameters for determining age of ruffed grouse. Wildlife Society Bulletin 7:125–127.

Rodgers, R. D. 1985. A field technique for identifying the sex of dressed pheasants. Wildlife Society Bulletin 13:528–533.

Rodusky, A. J., B. Sharfstein, T. L. East, and R. P. Maki. 2005. A comparison of three methods to collect submerged aquatic vegetation in a shallow lake. Environmental Monitoring and Assessment 110:87–97.

Roe, B. A., D.-P. Ma, R. K. Wilson, and J. F.-H. Wong. 1985. The complete nucleotide sequence of the Xenopus laevis mitochondrial genome. Journal of Biological Chemistry 260:9759–9774.

Roebuck, B. D., M. E. Walsh, C. H. Racine, L. R. Reitsma, B. B. Steele, and S. I. Nam. 1994. Predation of ducks poisoned by white phosphorus: exposure and risks to predators. Environmental Toxicology and Chemistry 13:1613–1618.

Roeder, K. D., and A. E. Treat. 1961. The detection and evasion of bats by moths. American Scientist 49:135–148.

Roffe, T. J., and T. M. Work. 2005. Wildlife health and disease investigations. Pages 197–212 in C. E. Braun, editor. Techniques for wildlife investigations and management. The Wildlife Society, Bethesda, Maryland, USA.

Roger, P., T. A. O'Neil, and S. Toshach. 2007. A strategy for managing fish, wildlife, and habitat data. A Draft Report to Northwest Power and Conservation Council, Portland, Oregon, USA.

Rogers, D. I., P. F. Battely, J. Sparrow, A. Koolhass, and C. J. Hassell. 2004. Treatment of capture myopathy in shorebirds: a successful trial in northwestern Australia. Journal of Field Ornithology 75:157–164.

Rogers, J. P. 1964. A decoy trap for male lesser scaups. Journal of Wildlife Management 28:408–410.

Rohwer, S. 1977. Status signaling in Harris sparrows: some experiments in deception. Behaviour 61:107–129.

Rolan, R. G., and H. T. Gier. 1967. Correlation of embryo and placental scar counts of *Peromyscus maniculatus* and *Microtus ochrogaster*. Journal of Mammalogy 48:317–319.

Rolland, R. M., P. K. Hamilton, S. D. Kraus, B. Davenport, R. M. Bower, and S. K. Wasser. 2006. Faecal sampling using detection dogs to study reproduction and health in North Atlantic right whales (*Eubalaena glacialis*). Journal of Cetacean Research and Management 8:121–125.

Rolley, R. E. 1987. Bobcat. Pages 671–681 *in* M. Novak, J. A. Baker, M. E. Obbard, and B. Malloch, editors. Wild furbearer management and conservation in North America. Ontario Ministry of Natural Resources, Toronto, Canada.

Rollins, D., and J. Carroll. 2001. Impacts of predation on northern bobwhite and scaled quail. Wildlife Society Bulletin 29:39–51.

Romesburg, H. C. 1981. Wildlife science: gaining reliable knowledge. Journal of Wildlife Management 45:293–313.

Romesburg, H. C. 1989. More on gaining reliable knowledge. Journal of Wildlife Management 53:1177–1180.

Romesburg, H. C. 1991. On improving natural resources and environmental sciences. Journal of Wildlife Management 55:744–756.

Romesburg, H. C. 1993. On improving natural resources and environmental sciences: a comment. Journal of Wildlife Management 57:182–183.

Roncy, J. R., A. W. Lukaszewski, and A. W. Simmons. 2007. Rapid endocrine responses of young men to social interactions with young women. Hormones and Behavior 52:326–333.

Rongstad, O. J. 1966. A cottontail rabbit lens-growth curve from southern Wisconsin. Journal of Wildlife Management 30:114–121.

Rongstad, O. J., and R. A. McCabe. 1984. Capture techniques. Pages 655–676 *in* L. K. Halls, editor. White-tailed deer: ecology and management. Stackpole, Harrisburg, Pennsylvania, USA.

Rood, J. P., and D. W. Nellis. 1980. Freeze marking mongooses. Journal of Wildlife Management 44:500–502.

Roon, D. A., M. E. Thomas, K. C. Kendall, and L. P. Waits. 2005a. Evaluating mixed samples as a source of error in non-invasive genetic studies using microsatellites. Molecular Ecology 14:195–201.

Roon, D. A., L. P. Waits, and K. C. Kendall. 2005b. A simulation test of the effectiveness of several methods for error-checking non-invasive genetic data. Animal Conservation 8:203–215.

Root, D. A., and N. F. Payne. 1984. Evaluation of techniques for aging gray fox. Journal of Wildlife Management 48:926–933.

Roper, L. A., R. L. Schmidt, and R. B. Gill. 1971. Techniques of trapping and handling mule deer in northern Colorado with notes on using automatic data processing for data analysis. Proceedings of the Western Association of State Game and Fish Commissioners 51:471–477.

Rose, R. J., J. Ng, and J. Melville. 2006. A technique for restraining lizards for field and laboratory measurements. Herpetological Review 37:194–195.

Roseberry, J. L., and W. D. Klimstra. 1965. A guide to age determination of bobwhite quail embryos. Biological Notes 55, Illinois Natural History Survey, Springfield, USA.

Roseberry, J. L., and W. D. Klimstra. 1970. Productivity of white-tailed deer on Crab Orchard National Wildlife Refuge. Journal of Wildlife Management 34:23–28.

Rosell, F., and B. Hovde. 2001. Methods of aquatic and terrestrial netting to capture Eurasian beavers. Wildlife Society Bulletin 29:269–274.

Rosenbaum, H. C., M. G. Egan, P. J. Clapham, R. L. Brownell, Jr., S. Malik, M. W. Brown, B. N. White, P. Walsh, and R. Desalle. 2000. Utility of north Atlantic right whale museum specimens for assessing changes in genetic diversity. Conservation Biology 14:1837–1842.

Rosenfeld, C. S., and R. M. Roberts. 2004. Maternal diet and other factors affecting offspring sex ratio: a review. Biology of Reproduction 71:1063–1070.

Rosenfield, R. N., and J. Bielefeldt. 1993. Trapping techniques for breeding Cooper's hawks: two modifications. Journal of Raptor Research 27:171–172.

Rosenstock, S. S., D. R. Anderson, K. M. Giesen, T. Leukering, and M. F. Carter. 2002. Landbird counting techniques: current practices and an alternative. Auk 119:46–53.

Roshier, D. A., and M. W. Asmus. 2009. Use of satellite telemetry on small-bodied waterfowl in Australia. Marine and Freshwater Research 60:299–305.

Rossiter, S. J., G. Jones, R. D. Ransome, and E. M. Barratt. 2000. Genetic variation and population structure in the endangered greater horseshoe bat *Rhinolophus ferrumequinum*. Molecular Ecology 9:1131–1135.

Rotella, J. J., S. J. Dinsmore, and T. L. Shaffer. 2004. Modeling nest-survival data: a comparison of recently developed methods that can be implemented in MARK and SAS. Animal Biodiversity and Conservation 27:187–205.

Rotella, J. J., and J. T. Ratti. 1986. Test of a critical density index assumption: a case study with gray partridge. Journal of Wildlife Management 50:532–539.

Rotella, J. J., and J. T. Ratti. 1992a. Mallard brood movements and wetland selection in southwestern Manitoba. Journal of Wildlife Management 56:508–515.

Rotella, J. J., and J. T. Ratti. 1992b. Mallard brood survival and wetland habitat conditions in southwestern Manitoba. Journal of Wildlife Management 56:499–507.

Rotella, J. J., M. L. Taper, and A. J. Hansen. 2000. Correcting nesting-success estimates for observer effects: maximum likelihood estimates of daily survival rates with reduced bias. Auk 117:92–109.

Rotenberry, J. T., R. J. Cooper, J. M. Wunderle, and K. G. Smith. 1995. When and how are populations limited? The role of insect outbreaks, fire, and other natural perturbations. Pages 55–84 *in* T. E. Martin and D. E. Finch, editors. Ecology and management of migratory birds: a synthesis and review of critical issues. Oxford University Press, Oxford, England, UK.

Rotenberry, J. T., and J. A. Wiens. 1980. Habitat structure, patchiness, and avian communities in North American steppe vegetation: a multivariate analysis. Ecology 61:1228–1250.

Rotterman, L. M., and C. Monnett. 1984. An embryo-dyeing technique for identification through hatching. Condor 86:79–80.

Roussel, Y. E. 1975. Aerial sexing of anterless moose by white vulval patch. Journal of Wildlife Management 39:450–451.

Roussel, Y. E., and R. Ouellet. 1975. A new criterion for sexing Quebec ruffed grouse. Journal of Wildlife Management 39:443–445.

Route, W. T., and R. O. Peterson. 1988. Distribution and abundance of river otters in Voyageurs National Park, Minnesota. Research/Resource Management Report MWR-10, Midwest Regional Office, National Park Service, Department of the Interior, Omaha, Nebraska, USA.

Row, J. R., and G. Blouin-Demers. 2006a. An effective and durable funnel trap for sampling terrestrial herpetofauna. Herpetological Review 37:183–184.

Row, J. R., and G. Blouin-Demers. 2006b. Kernels are not accurate estimators of home-range size for herpetofauna. Copeia 2006:797–802.

Rowell, J., and P. F. Flood. 1987. Changes in muskox blood progesterone concentration between collection and centrifugation. Journal of Wildlife Management 51:901–903.

Rowsell, H. C., J. Ritcey, and F. Cox. 1981. Assessment of effectiveness of trapping methods in the production of a humane death. Proceedings of the Worldwide Furbearer Conference 3:1647–1670.

Roy, C., and A. Woolf. 2001. Effects of hunting and hunting-hour extension on mourning dove foraging and physiology. Journal of Wildlife Management 65:808–815.

Roy, L. D., and M. J. Dorrance. 1985. Coyote movements, habitat use, and vulnerability in central Alberta. Journal of Wildlife Management 49:307–313.

Roy, M. S., E. Geffen, D. Smith, R. K. Wayne, E. A. Ostrander, and R. K. Wayne. 1994. Patterns of differentiation and hybridization in North American wolflike canids, revealed by analysis of microsatellite loci. Molecular Biology and Evolution 11:553–570.

Roy-Nielsen, C. L., R. J. Gates, and P. G. Parker. 2006. Intraspecific nest parasitism of wood ducks in natural cavities: comparisons with nest boxes. Journal of Wildlife Management 70:835–843.

Royall, R. M. 1997. Statistical evidence: a likelihood paradigm. Chapman and Hall/CRC Press, Boca Raton, Florida, USA.

Royall, W. C., J. L. Guarino, and O. E. Bray. 1974. Effects of color on retention of leg streamers by red-winged blackbirds. Western Bird Bander 49:64–65.

Royle, J. A., and R. M. Dorazio. 2008. Hierarchical modeling and inference in ecology: the analysis of data from populations, metapopulations, and communities. Elsevier, London, England, UK.

Rudge, M. R., and R. J. Joblin. 1976. Comparison of some methods of capturing and marking feral goats (*Capra hircus*). New Zealand Journal of Zoology 3:51–55.

Rudnick, J. A., T. E. Katzner, E. A. Bragin, and J. A. DeWoody. 2008. A non-invasive genetic evaluation of population size, natal philopatry, and roosting behavior of non-breeding eastern imperial eagles (*Aquila heliaca*) in central Asia. Conservation Genetics 9:667–676.

Rudnick, J. A., and R. C. Lacy. 2008. The impact of assumptions about founder relationships on the effectiveness of captive breeding strategies. Conservation Genetics 9:1439–1450.

Rudolph, B. A., W..F. Porter, and H. B. Underwood. 2000. Evaluating immunocontraception for managing suburban white-tailed deer in Irondequoit, New York. Journal of Wildlife Management 64:463–473.

Ruell, E. W., and K. R. Crooks. 2007. Evaluation of noninvasive genetic sampling methods for felid and canid populations. Journal of Wildlife Management 71:1690–1694.

Ruff, F. J. 1938. Trapping deer on the Pisgah National Game Preserve, North Carolina. Journal of Wildlife Management 2:151–161.

Rumble, M., and F. Lindzey. 1997. Effects of forest vegetation and topography on global positioning system collars for elk. Pages 492–501 *in* 1997 American congress on surveying and mapping. American Society for Photogrammetry and Remote Sensing, Annual Convention and Exposition Technical Papers. Volume 4. Resource Technology Institute, Seattle, Washington, USA.

Rumble, M. A. 1987. Using twig diameters to estimate browse utilization on three shrub species in southeastern Montana. Pages 172–175 *in* F. D. Provenza, J. T. Flinders, and E. D. McArthur, editors. Proceedings: symposium on plant–herbivore interactions. General Technical Report INT-222, U.S. Forest Service, Department of Agriculture, Washington, D.C., USA.

Rumph, J. A., and W. J. Turner. 1998. Alternative to critical point drying for soft-bodied insect larvae. Annals of the Entomological Society of America 91:693–699

Rundquist, B. C. 2002. The influence of canopy green vegetation fraction on spectral measurements over native tallgrass prairie. Remote Sensing of Environment 81:129–135.

Runge, W. 1972. An efficient winter live-trapping technique for white-tailed deer. Technical Bulletin 1, Saskatchewan Department Natural Resources, Regina, Canada.

Running, S. W. 1990. Estimating primary productivity by combining remote sensing with ecosystem simulation. Pages 65–86 *in* R. J. Hobbs and H. A. Mooney, editors. Remote sensing of biosphere functioning. Springer-Verlag, New York, New York, USA.

Rupprecht, C. E., K. Stohr, and C. Meredith. 2001. Rabies. Pages 3–36 *in* E. S. Williams and I. K. Barker, editors. Infectious diseases of wild mammals. Iowa State University Press, Ames, USA.

Rusch, D. H., S. DeStefano, M. C. Reynolds, and D. Lauten. 2000. Ruffed grouse. Account 515 *in* A. Poole and F Gill, editors. The birds of North America. The Academy of Natural Sciences, Philadelphia, Pennsylvania, and The American Ornithologists' Union, Washington, D.C., USA.

Rush, W. M. 1932. Northern Yellowstone elk study. Montana Fish and Game Commission, Missoula, USA.

Russell, F. E. 2001. Toxic effects of terrestrial animal venoms and poisons. Pages 945–964 *in* C. D. Klaassen, editor. Casarett and Doull's toxicology: the basic science of poisons. Sixth edition. McGraw-Hill, New York, New York, USA.

Russell, J. K. 1981. Patterned freeze-brands with canned freon. Journal of Wildlife Management 45:1078.

Russell, K. R., and S. A. Gauthreaux, Jr. 1998. Use of weather radar to characterize movements of roosting purple martins. Wildlife Society Bulletin 26:5–16.

Russell, R. W. 1999. Precipitation scrubbing of aerial plankton: inferences from bird behavior. Oecologia 118:381–387.

Rutz, C., L. A. Bluff, A. A. S. Weir, and A. Kacelnik. 2007. Video cameras on wild birds. Science 318:765.

Ryan, M. J., M. D. Tuttle, and A. S. Rand. 1982. Bat predation and sexual advertisement in a neotropical frog. American Naturalist 119:136–139.

Ryberg, W. A., and J. C. Cathey. 2004. A box-trap design to capture alligators in forested wetland habitats. Wildlife Society Bulletin 32:183–187.

Ryder, J. P., and R. T. Alisauskas. 1995. Ross' goose. Account 162 *in* A. Poole and F Gill, editors. The birds of North America. The Academy of Natural Sciences, Philadelphia, Pennsylvania, and The American Ornithologists' Union, Washington, D.C., USA.

Ryder, O. A. 1986. Species conservation and systematics: the dilemma of subspecies. Trends in Ecology and Evolution 1:9–10.

Ryder, P. L., and J. P. Ryder. 1981. Reproductive performance of ring-billed gulls in relation to nest location. Condor 83:57–60.

Ryman, N., R. Baccus, C. Reuterwall, and M. H. Smith. 1981. Effective population size, generation interval, and potential loss of genetic variability in game species under different hunting regimes. Oikos 36:257–266.

Sabean, B., and J. Mills. 1994. Raccoon–6" × 6" body gripping trap study. Report, Nova Scotia Department of Natural Resources, Halifax, Canada.

Sabol, B. M., R. E. Melton, R. Chamberlain, P. Doering, and K. Haunert. 2002. Evaluation of a digital echo sounder system for detection of submersed aquatic vegetation. Estuaries 25:133–141.

Sacks, B. N., K. M. Blejwas, and M. M. Jaeger. 1999. Relative vulnerability of coyotes to removal methods on a northern California ranch. Journal of Wildlife Management 63:939–949.

Safar-Hermann, N., M. N. Ismail, H. S. Choi, E. Mostl, and E. Bamberg. 1987. Pregnancy diagnosis in zoo animals by estrogen determination in feces. Zoo Biology 6:189–193.

Sahr, D. P., and F. F. Knowlton. 2000. Evaluation of tranquilizer trap devices (TTDs) for foothold traps used to capture gray wolves. Wildlife Society Bulletin 28:597–605.

Sakamoto, K. Q., K. Sato, M. Ishizuka, Y. Watanuki, A. Takahashi, F. Daunt, and S. Wanless. 2009. Can ethograms be automatically generated using body acceleration data from free-ranging birds? PLoS ONE 4:e5379.

Salas, V., E. Pannier, C. Galindez-Silva, A. Gols-Ripoll, and E. A. Herrera. 2004. Methods for capturing and marking wild capybaras in Venezuela. Wildlife Society Bulletin 32:202–208.

Sallaberry, A. M., and D. J. Valencia. 1985. Wounds due to flipper bands on penguins. Journal of Field Ornithology 56:275–277.

Saltelli, A., K. Chan, and E. M. Scott. 2001. Sensitivity analysis. John Wiley and Sons, New York, New York, USA.

Saltz, D. 1994. Reporting error measures in radio location by triangulation: a review. Journal of Wildlife Management 58:181–184.

Saltz, D., and P. U. Alkon. 1985. A simple computer-aided method for estimating radio-location error. Journal of Wildlife Management 49:664–668.

Salwasser, H., and S. A. Holl. 1979. Estimating fetus age and breeding and fawning periods in the North Kings River deer herd. California Fish and Game 65:159–165.

Salyer, J. W. 1962. A bow-net trap for ducks. Journal of Wildlife Management 26:219–221.

Sambrook, E., F. Fritsch, and T. Maniatis. 1989. Molecular cloning: a laboratory manual. Cold Springs Harbor Press, Cold Spring Harbor, New York, USA.

Samuel, M. D., and M. R. Fuller. 1994. Wildlife radiotelemetry. Pages 370–418 *in* T. A. Bookhout, editor. Research and management techniques for wildlife and habitats. The Wildlife Society, Bethesda, Maryland, USA.

Samuel, M. D., E. O. Garton, M. W. Schlegel, and R. G. Carson. 1987. Visibility bias during aerial surveys of elk in northcentral Idaho. Journal of Wildlife Management 51:622–630.

Samuel, M. D., and R. E. Green. 1988. A revised test procedure for identifying core areas within the home range. Journal of Animal Ecology 57:1067–1068.

Samuel, M. D., and K. P. Kenow. 1992. Evaluating habitat selection with biotelemetry triangulation error. Journal of Wildlife Management 56:725–734.

Samuel, M. D., D. J. Pierce, and E. O. Garton. 1985. Identifying areas of concentrated use within the home range. Journal of Animal Ecology 54:711–719.

Samuel, M. D., R. K. Steinhorst, E. O. Garton, and J. W. Unsworth. 1992. Estimation of wildlife population ratios incorporating survey design and visibility bias. Journal of Wildlife Management 56:718–725.

Samuel, M. D., N. T. Weiss, D. H. Rusch, S. R. Craven, R. E. Trost, and F. D. Caswell. 1990. Neck-band retention for Canada geese in the Mississippi flyway. Journal of Wildlife Management 54:612–621.

Samuel, W. M., M. J. Pybus, and A. A. Kocan, editors. 2001. Parasitic diseases of wild mammals.

Second edition. Iowa State University Press, Ames, USA.

Sanders, H. L. 1968. Marine benthic diversity: a comparative study. American Naturalist 102:243–282.

Sanders, O. T., and R. L. Kirkpatrick. 1975. Effects of a polychlorinated biphenyl on sleeping times, plasma corticosteroids, and testicular activity of white-footed mice. Environmental Physiology 5:308–313.

Sanderson, G. C. 1950. Methods of measuring productivity in raccoons. Journal of Wildlife Management 14:389–402.

Sanderson, G. C. 1961a. Estimating opossum populations by marking young. Journal of Wildlife Management 25:20–27.

Sanderson, G. C. 1961b. Techniques for determining age of raccoons. Biological Notes 45, Illinois Natural History Survey, Urbana, USA.

Sanderson, G. C. 1961c. The lens as an indicator of age in the raccoon. American Midland Naturalist 65:481–485.

Sanderson, G. C., and F. C. Bellrose. 1986. A review of the problem of lead poisoning in waterfowl. Special Publication 4, Illinois Natural History Survey, Urbana, USA.

Sandoval, S. J., S. P. Cook, F. W. Merickel, and H. L. Osborne. 2007. Diversity of the beetle (Coleoptera) community captured at artificially-created snags of Douglas-fir and grand fir. The Pan-Pacific Entomologist 83:41–49.

Santos, S. A. P., J. E. Cabanas, and J. A. Pereira. 2007. Abundance and diversity of soil arthropods in olive grove ecosystem (Portugal): effect of pitfall trap type. European Journal of Soil Biology 43:77–83.

Sanz, J. J., J. Potti, J. Moreno, S. Merino, and O. Frias. 2003. Climate change and fitness components of a migratory bird breeding in the Mediterranean region. Global Change Biology 9:461–472.

Sapolsky, R. M. 1987. Stress, social status, and reproductive physiology in free-living baboons. Pages 291–322 in D. Crews, editor. Psychobiology of reproductive behavior: an evolutionary perspective. First edition. Prentice-Hall, Englewood Cliffs, New Jersey, USA.

Sapolsky, R. M. 1998. Why zebras don't get ulcers: a guide to stress, stress-related diseases, and coping. W. H. Freeman and Company, New York, New York, USA.

Sargeant, A. B. 1966. A live trap for pocket gophers. Journal of Mammalogy 47:729–73l.

Sarmento, P., J. Cruz, C. Eira, and C. Fonseca. 2009. Evaluation of camera trapping for estimating red fox abundance. Journal of Wildlife Management 73:1207–1212.

SAS Institute. 2008. SAS version 9.2. SAS Institute, Cary, North Carolina, USA.

Sasakawa, K. 2007. Effects of pitfall trap preservatives on specimen condition in carabid beetles. Entomologia Experimentalis et Applicata 125:321–324.

Sasaki, K., S. F. Fox, and D. Duvall. 2009. Rapid evolution in the wild: changes in body size, life-history traits, and behavior in hunted populations of the Japanese Mamushi snake. Conservation Biology 23:93–102.

Sauer, J. R., J. E. Hines, and J. Fallon. 2008. The North American Breeding Bird Survey, results, and analysis 1966–2007. Version 5.15.2008. Patuxent Wildlife Research Center, U.S. Geological Service, Laurel, Maryland, USA.

Sauer, J. R., and M. G. Knutson. 2008. Objectives and metrics for wildlife monitoring. Journal of Wildlife Management 72:1663–1664.

Sauer, J. R., and N. A. Slade. 1987. Size-based demography of vertebrates. Annual Review of Ecology and Systematics 18:71–90.

Sauer, P. R. 1966. Determining sex of black bears from the size of the lower canine tooth. New York Fish and Game Journal 13:140–145.

Sauer, P. R., and C. W. Severinghaus. 1977. Determination and application of fawn reproductive rates from yearling teat length. Transactions of the Northeastern Section, The Wildlife Society 33:133–144.

Saunders, B. P., H. C. Rowsell, and I. W. Hatter. 1988. A better trap, the search continues. Canadian Trapper Winter:3, 16.

Saunders, D. A. 1988. Patagial tags: do benefits outweigh the risks to the animal? Australian Wildlife Research 15:565–569.

Saunders, G., and S. Harris. 2000. Evaluation of attractants and bait preferences of captive red foxes (*Vulpes vulpes*). Wildlife Research 27:237–243.

Saunders, G., B. Kay, and H. Nicol. 1993. Factors affecting bait uptake and trapping success for feral pigs (*Sus scrofa*) in Kosciusko National Park. Wildlife Research 20:653–665.

Saunders, J. K. 1964. Physical characteristics of the Newfoundland lynx. Journal of Mammalogy 45:36–47.

Savard, J.-P. L. 1985. Use of a mirror trap to capture territorial waterfowl. Journal of Field Ornithology 56:177–178.

Savarie, P. J., K. A. Fagerstone, and E. W. Schafer, Jr. 1993. Update on the development of a tranquilizer trap device. Proceedings, Great Plains Wildlife Damage Control Workshop 11:204–208.

Savarie, P. J., and J. D. Roberts. 1979. Evaluation of oral central nervous system depressants in coyotes. Pages 270–277 in J. R. Beck, editor. Vertebrate pest control and management materials. Special Technical Publication 680, American Society for Testing and Materials, Philadelphia, Pennsylvania, USA.

Savarie, P. J., D. S. Vice, L. Bangerter, K. Dustin, W. J. Paul, T. M. Primus, and F. S. Blom. 2004. Operational field evaluation of a plastic bulb reservoir as a tranquilizer trap device for delivering propiopromazine hydrochloride to feral dogs, coyotes and gray wolves. Proceedings of the Vertebrate Pest Conference 21:64–69.

Sawyer, H., M. J. Kauffman, and R. M. Nielson. 2009. Influence of well pad activity on winter habitat selection patterns of mule deer. Journal of Wildlife Management 73:1052–1061.

Sawyer, H., R. M. Nielson, F. G. Lindzey, L. Keith, J. H. Powell, and A. A. Abraham. 2007. Habitat selection of Rocky Mountain elk in a non-forested environment. Journal of Wildlife Management 71:868–874.

Sawyer, H., R. M. Nielson, F. Lindzey, and L. L. McDonald. 2006. Winter habitat selection of mule deer before and during development of a natural gas field. Journal of Wildlife 70:396–403.

Sayre, M. W., T. S. Baskett, and P. Books-Blenden. 1981. Effects of radio-tagging on breeding behavior of mourning doves. Journal of Wildlife Management 45:428–434.

Scattergood, L. W. 1954. Estimating fish and wildlife populations: a survey of methods. Pages 273–285 in O. Kempthorne, T. A. Bancroft, J. W. Gowen, and J. L. Lush, editors. Statistics and mathematics in biology. Iowa State College Press, Ames, USA.

Schaefer, G. W. 1968. Bird recognition by radar: a study in quantitative radar ornithology. Pages 53–85 in R. K. Murton and E. N. Wright, editors. The problems of birds as pests. Academic Press, London, England, UK.

Schaefer, J. A., N. Morellet, D. Pépin, and H. Verheyden. 2008. The spatial scale of habitat selection by red deer. Canadian Journal of Zoology 86:1337–1348.

Schaller, G. B. 1967. Deer and the tiger. University of Chicago, Chicago, Illinois, USA.

Scharf, C. S. 1985. A technique for trapping territorial magpies. North American Bird Bander 10:34–36.

Scheaffer, R. L., W. Mendenhall, III, and R. L. Ott. 2005. Elementary survey sampling. Sixth edition. Duxbury Press, Boston, Massachusetts, USA.

Scheffer, V. B. 1950a. Experiments in the marking of seals and sea lions. Scientific Report Wildlife 4, U.S. Fish and Wildlife Service, Department of the Interior, Washington, D.C., USA.

Scheffer, V. B. 1950b. Growth layers on the teeth of Pinnipedia as an indication of age. Science 112:309–311.

Scheiffarth, G. 2001. The diet of bar-tailed godwits *Limosa lapponica* in the Wadden Sea: combining visual observations and faeces analyses. Ardea 89:481–494.

Scheiner, S. M. 1994. Why ecologists should care about philosophy: a reply to Keddy's reply. Bulletin of the Ecological Society of America 75:50–52.

Scheiner, S. M, and J. Gurevitch. 2001. Design and analysis of ecological experiments. Oxford University Press, New York, New York, USA.

Scheller, R. M., B. R. Sturtevant, E. J. Gustafson, B. C. Ward, and D. J. Mladenoff. 2010. Increasing the reliability of ecological models using modern software engineering techniques. Frontiers in Ecology and the Environment 8:253–260.

Schemnitz, S. D. 1961. Ecology of the scaled quail in the Oklahoma panhandle. Wildlife Monographs 8.

Schemnitz, S. D., editor. 1980. Wildlife management techniques manual. The Wildlife Society, Washington, D.C., USA.

Schemnitz, S. D. 1994. Capturing and handling wild animals. Pages 106–124 in T. A. Bookhout, editor. Research and management techniques for wildlife and habitats. The Wildlife Society, Bethesda, Maryland, USA.

Schemnitz, S. D. 2005. Capturing and handling wild animals. Pages 239–285 in C. E. Braun, editor. Techniques for wildlife investigations and management. The Wildlife Society, Bethesda, Maryland, USA.

Scheppers, T. L. J., A. C. Frantz, M. Schaul, E. Engel, P. Breyne, L. Schley, and T. J. Roper. 2007. Estimating social group size of Eurasian badgers *Meles meles* by genotyping remotely plucked single hairs. Wildlife Biology 13:195–207.

Scheuhammer, A. M., and S. L. Norris. 1996. The ecotoxicology of lead shot and lead fishing weights. Ecotoxicology 5:279–295.

Schick, R. S., S. R. Loarie, F. Colchero, B. D. Best, A. Boustany, D. A. Conde, P. N. Halpin, L. N. Joppa, C. M. McClellan, and J. S. Clark. 2008. Understanding movement data and movement processes: current and emerging directions. Ecology Letters 11:1338–1350.

Schieck, J. O., and S. J. Hannon. 1989. Breeding site fidelity in willow ptarmigan: the influence of previous reproductive success and familiarity with partner and territory. Oecologia 81:465–472.

Schierbaum, D., D. Benson, L. W. DeGraaf, and D. D. Foley. 1959. Waterfowl banding in New York. New York Fish and Game Journal 6:86–102.

Schierbaum, D., and E. Talmage. 1954. A successful diving duck trap. New York Fish and Game Journal 1:116–117.

Schladweiler, P., T. W. Mussehl, and R. J. Greene. 1970. Age determination of juvenile blue grouse by primary development. Journal of Wildlife Management 34:649–652.

Schmaljohann, H., F. Liechti, E. Bächler, T. Steuri, and B. Bruderer. 2008. Quantification of bird migration by radar—a detection probability problem. Ibis 150:342–355.

Schmidt, R. H., and J. G. Brunner. 1981. A professional attitude toward humaneness. Wildlife Society Bulletin 9:289–291.

Schmutz, J. A., and J. A. Morse. 2000. Effects of neck collars and radiotransmitters on survival and reproduction of emperor geese. Journal of Wildlife Management 64:231–237.

Schmutz, J. A., and G. C. White. 1990. Error in telemetry studies: effects of animal movement on triangulation. Journal of Wildlife Management 54:506–510.

Schnabel, Z. E. 1938. The estimation of the total fish population of a lake. American Mathematical Monthly 45:348–352.

Schneegas, E. R., and G. W. Franklin. 1972. The Mineral King deer herd. California Fish and Game 58:133–140.

Schneider, D. C., and B. A. Harrington. 1981. Timing of shorebird migration in relation to prey depletion. Auk 98:801–811.

Schnell, G. D. 1965. Recording the flight-speed of birds by Doppler radar. Living Bird 4:79–87.

Schnell, G. D., and J. J. Hellack. 1978. Flight speeds of brown pelicans, chimney swifts, and other birds. Bird-Banding 49:108–112.

Schnell, J. H. 1968. The limiting effects of natural predation on experimental cotton rat populations. Journal of Wildlife Management 32:698–711.

Schofield, R. D. 1955. Analysis of muskrat age determination methods and their application in Michigan. Journal of Wildlife Management 19:463–466.

Scholz, A., T. Ross, M. Horrall, J. C. Cooper, and A. D. Hasler. 1976. Imprinting to chemical cues: the basis for home stream selection in salmon. Science 192:1247–1249.

Schonberner, V. D. 1965. Beobachtungen zur Fortpflanzungsbiologie des Wolfes, *Canis lupus*. Zeitschrift für Saugetierkunde 30:171–178.

Schooley, R. L. 1994. Annual variation in habitat selection: patterns concealed by pooled data. Journal of Wildlife Management 58:367–374.

Schooley, R. L., B. Van Horne, and K. P. Burnham. 1993. Passive integrated transponders for marking free-ranging Townsend's ground squirrels. Journal of Mammalogy 74:480–484.

Schotzko, D. J., and L. E. O'Keeffe. 1986. Comparison of sweepnet, D-vac, and absolute sampling for *Lygus hesperus* (Heteroptera: Miridae) in lentils. Journal of Economic Entomology 79:224–228.

Schotzko, D. J., and L. E. O'Keeffe. 1989. Comparison of sweep net, D-vac, and absolute sampling, and diel variation of sweep net sampling estimates in lentils for pea aphid (Homoptera: Aphididae), nabids (Hemiptera: Nabidae), lady beetles (Coleoptera: Coccinellidae), and lacewings (Neuroptera: Crysopidae). Journal of Economic Entomology 82:491–506

Schroeder, M. A. 1986. A modified noose pole for capturing grouse. North American Bird Bander 11:42.

Schroeder, M. A. 1997. Unusually high reproductive effort by sage grouse in a fragmented habitat in north-central Washington. Condor 99:933–941.

Schroeder, M. A., and C. E. Braun. 1991. Walk-in traps for capturing greater prairie-chickens on leks. Journal of Field Ornithology 62:378–385.

Schroeder, M. A., and L. A. Robb. 1993. Greater prairie-chicken. Account 36 *in* A. Poole, P. Stettenheim, and F. Gill, editors. The birds of North America. The Academy of Natural Sciences, Philadelphia, Pennsylvania, and The American Ornithologists' Union, Washington, D.C., USA.

Schroeder, M. A., J. R. Young, and C. E. Braun. 1999. Sage grouse. Account 425 *in* A. Poole and F. Gill, editors. The birds of North America. The Academy of Natural Sciences, Philadelphia, Pennsylvania, and The American Ornithologists' Union, Washington, D.C., USA.

Schroeder, R. L., and L. D. Vangilder. 1997. Tests of wildlife habitat models to evaluate oak-mast production. Wildlife Society Bulletin 25:639–646.

Schroeder, T. A., W. B. Cohen, and Z. Yang. 2007. Patterns of forest regrowth following clearcutting in western Oregon as determined from a Landsat time-series. Forest Ecology and Management 243:259–273.

Schuerholz, G. 1974. Quantitative evaluation of edge from aerial photographs. Journal of Wildlife Management 38:913–920.

Schulte, B. A., D. Muller-Schwarze, and L. Sun. 1995. Using anal gland secretion to determine sex in beaver. Journal of Wildlife Management 59:614–618.

Schultz, A. M., R. P. Gibbens, and L. Debano. 1961. Artificial populations for teaching and testing range techniques. Journal of Range Management 14:236–242.

Schultz, J. H., S. L. Sheriff, Z. He, C. E. Braun, R. D. Drobney, R. E. Tomlinson, D. D. Dolton, and R. A. Montogomery. 1995. Accuracy of techniques used to assign mourning dove age and gender. Journal of Wildlife Management 59:759–765.

Schultz, R. N., A. P. Wydeven, and R. A. Megown. 1996. Injury levels with five types of leg-hold traps in Wisconsin. Pages 38–39 *in* R. Earle, editor. Proceedings of the fourteenth midwest furbearer workshop. Michigan Department of Natural Resources, Lansing, USA.

Schultz, V. 1950. A modified Stoddard quail trap. Journal of Wildlife Management 14:243.

Schulz, J. H., A. J. Bermudez, J. L. Tomlinson, J. D. Firman, and Z. He. 1998. Effects of implanted radiotransmitters on captive mourning doves. Journal of Wildlife Management 62:1451–1460.

Schulz, J. H., A. J. Bermudez, J. L. Tomlinson, J. D. Firman, and Z. He. 2001. Comparison of radiotransmitter attachment techniques using captive mourning doves. Wildlife Society Bulletin 29:771–782.

Schulz, J. H., and J. R. Ludwig. 1985. A possible cause of premature loss for deer fawn transmitters. Journal of Mammalogy 66:811–812.

Schulz, J. H., J. Ludwig, and M. Frydendall. 1983. Survival and home range of white-tailed deer fawns in southeastern Minnesota. Minnesota Wildlife Research Quarterly 43:15–23.

Schulz, J. H., J. J. Millspaugh, B. E. Washburn, A. J. Bermudez, J. L. Tomlinson, T. W. Mong, and Z. He. 2005. Physiological effects of radiotransmitters on mourning doves. Wildlife Society Bulletin 33:1092–1100.

Schumacher, F. X., and R. W. Eschmeyer. 1943. The estimate of fish population in lakes or ponds. Journal of the Tennessee Academy of Sciences 18:228–249.

Schupp, E. W. 1990. Annual variation in seedfall, postdispersal predation, and recruitment of a Neotropical tree. Ecology 71:504–515.

Schwabl, H., M. Ramenofsky, I. Schwabl-Benzinger, D. S. Farner, and J. C. Wingfield. 1988. Social status, circulating levels of hormones, and competition for food in winter flocks of the white-throated sparrow. Behaviour 107:107–121.

Schwarts, J. E. 1943. Range conditions and management of the Roosevelt elk on the Olympic Peninsula. U.S. Forest Service, Department of Agriculture, Washington, D.C., USA.

Schwarts, J. E., and G. E. Mitchell. 1945. The Roosevelt elk on the Olympic Peninsula, Washington. Journal of Wildlife Management 9:295–319.

Schwarz, G. 1978. Estimating the dimension of a model. Annals of Statistics 6:461–464.

Schwartz, M. K., G. Luikart, and R. S. Waples. 2007. Genetic monitoring as a promising tool for conservation and management. Trends in Ecology and Evolution 22:25–33.

Schwartz, M. K., D. A. Tallmon, and G. Luikart. 1998. Review of DNA-based census and effective population size estimators. Animal Conservation 1:293–299.

Schwarz, C. J., and A. N. Arnason. 1996. A general methodology for the analysis of capture–recapture experiments in open populations. Biometrics 52:860–873.

Schwertner, T. W., H. A. Mathewson, J. A. Roberson, M. Small, and G. L. Waggerman. 2002. White-winged dove. Account 710 in A. Poole and F. Gill, editors. The birds of North America. The Academy of Natural Sciences, Philadelphia, Pennsylvania, and The American Ornithologists' Union, Washington, D.C., USA.

Scott, A. F., and J. L. Dobie. 1980. An improved design for a thread-trailing device used to study terrestrial movements of turtles. Herpetological Review 11:106–107.

Scott, D. K. 1978. Identification of individual Bewick's swans by bill patterns. Pages 160–168 in B. Stonehouse, editor. Animal marking: recognition marking of animals in research. MacMillan, London, England, UK.

Scott, J. M., F. Davis, B. Csuti, R. Noss, B. Butterfield, C. Groves, H. Anderson, S. Caicco, F. D'erchia, T. C. Edwards, Jr., J. Ulliman, and R. G. Wright. 1993. GAP analysis: a geographic approach to protection of biological diversity. Wildlife Monographs 123.

Scott, J. M., M. Murray, R. G. Wright, B. Csuti, P. Morgan, and R. L. Pressey. 2001. Representation of natural vegetation in protected areas: capturing the geographic range. Biodiversity and Conservation 10:1297–1301.

Scott, M. P., and T. N. Tan. 1985. A radiotracer technique for the determination of male mating success in natural populations. Behavioral Ecology and Sociobiology 17:29–33.

Scott, N. J., editor. 1982. Herpetological communities. Wildlife Research Report 13, U.S. Fish and Wildlife Service, Department of the Interior, Washington, D.C., USA.

Scotton, B. D., and D. H. Pletscher. 1998. Evaluation of a capture technique for neonatal Dall sheep. Wildlife Society Bulletin 26:578–583.

Scriba, M. F., and W. Goymann. 2010. European robins (*Erithacus rubecula*) lack an increase in testosterone during simulated territorial intrusions. Journal of Ornithology 151:607–614.

Scribner, K. T. 1993. Conservation genetics of managed ungulate populations. Acta Theriologica 38 (Supplement 2):89–101.

Scribner, K. T., and T. D. Bowman. 1998. Microsatellites identify depredated waterfowl remains from glaucous gull stomachs. Molecular Ecology 7:1401–1405.

Scribner, K. T., M. R. Petersen, R. L. Fields, S. L. Talbot, J. M. Pearce, and R. K. Chesser. 2001. Sex-biased gene flow in spectacled eiders (Anatidae): inferences from molecular markers with contrasting modes of inheritance. Evolution 55:2105–2115.

Scribner, K. T., and M. Stuwe. 1994. Genetic relationships among alpine ibex (*Capra Ibex*) populations reestablished from a common ancestral source. Biological Conservation 69:137–143.

Seal, U. S., and T. J. Kreeger. 1987. Chemical immobilization of furbearers. Pages 191–215 in M. Novak, J. A. Baker, M. E. Obbard, and B. Malloch, editors. Wild furbearer management and conservation in North America. Ontario Ministry of Natural Resources, Toronto, Canada.

Seal, U. S., L. J. Verme, J. J. Ozoga, and E. D. Plotka. 1983. Metabolic and endocrine responses of white-tailed deer to increasing population density. Journal of Wildlife Management 47:451–462.

Seale, D., and M. Boraas. 1974. A permanent mark for amphibian larvae. Herpetologica 30:160–162.

Seaman, D. E., J. J. Millspaugh, B. J. Kernohan, G. C. Brundige, K. J. Raedeke, and R. A. Gitzen. 1999. Effects of sample size on kernel home range estimates. Journal of Wildlife Management 63:739–747.

Seaman, D. E., and R. A. Powell. 1996. An evaluation of the accuracy of kernel density estimators for home range analysis. Ecology 77:2075–2085.

Seamans, T. W., S. Beckerman, J. Hartmann, J. A. Rader, and B. Blackwell. 2010. Reporting difference for colored patagial tags on ring-billed gulls. Journal of Wildlife Management 74:1926–1930.

Seamans, T. W., and J. L. Belant. 1999. Comparison of DRC-1339 and alpha-chloralose to reduce herring gull populations. Wildlife Society Bulletin 27:729–733.

Seber, G. A. F. 1965. A note on the multiple-recapture census. Biometrika 52:249–259.

Seber, G. A. F. 1970. Estimating time-specific survival and reporting rates for adult birds from band returns. Biometrika 57:313–318.

Seber, G. A. F. 1982. The estimation of animal abundance and related parameters. Second edition. MacMillan, New York, New York, USA.

Seber, G. A. F. 1986. A review of estimating animal abundance. Biometrics 42:267–292.

Seber, G. A. F. 2001. Some new directions in estimating animal population parameters. Journal of Agricultural, Biological, and Environmental Statistics 6:140–151.

Sedgwick, J. A., and F. L. Knopf. 1990. Habitat relationships and nest site characteristics of cavity-nesting birds in cottonwood floodplains. Journal of Wildlife Management 54:112–124.

Seel, D. C., A. G. Thompson, and G. H. Owen. 1982. A wing-tagging system for marking larger passerine birds. Bangor Occasional Papers 14:1–6.

Sefton, D. F. 1977. Productivity and biomass of vascular hydrophytes on the upper Mississippi. Pages 53–61 in C. B. Dewitt and E. Soloway, editors. Wetlands ecology, values and impacts. Proceedings Waubesa Conference on Wetlands. Institute of Environmental Studies, University of Wisconsin, Madison, USA.

Seguin, R. J., and F. Cooke. 1983. Band loss from lesser snow geese. Journal of Wildlife Management 47:1109–1114.

Seguin, R. J., and F. Cooke. 1985. Web tag loss from lesser snow goose. Journal of Wildlife Management 49:420–422.

Seidel, K. S. 1992. Statistical properties and applications of a new measure of joint space use for wildlife. Thesis, University of Washington, Seattle, Washington, USA.

Seilman, M. S., L. A. Sheriff, and T. C. Williams. 1981. Nocturnal migration at Hawk Mountain, Pennsylvania. American Birds 35:906–909.

Selander, R. K. 1971. Systematics and speciation in birds. Pages 57–147 in D. S. Farner and J. R. King, editors. Avian biology. Volume I. Academic Press, New York, New York, USA.

Sellers, P. J. 1985. Canopy reflectance, photosynthesis and transpiration. International Journal of Remote Sensing 6:1335–1375.

Selye, H. 1936. A syndrome produced by diverse nocuous agents. Nature 138:32.

Selye, H. 1946. The general adaptation syndrome and the diseases of adaptation. Journal of Clinical Endocrinology 6:117–230.

Selye, H. 1976. Stress in health and disease. Butterworths, Boston, Massachusetts, USA.

Semel, B., and P. W. Sherman. 1986. Dynamics of nest parasitism in wood ducks. Auk 103:813–816.

Semel, B., and P. W. Sherman. 1992. Use of clutch size to infer brood parasitism in wood ducks. Journal of Wildlife Management 56:495–499.

Semel, B., and P. W. Sherman. 1993. Answering basic questions to address management needs: case studies of wood duck nest box programs. Transactions of the North American Wildlife and Natural Resources Conference 58:537–550.

Semel, B., and P. W. Sherman. 2001. Intraspecific parasitism and nest-site competition in wood ducks. Animal Behaviour. 61:787–803.

Semel, B., P. W. Sherman, and S. M. Byers. 1988. Effects of brood parasitism and nest box placement on wood duck breeding ecology. Condor 90:920–930.

Semel, B., P. W. Sherman, and S. M. Byers. 1990. Nest boxes and brood parasitism in wood ducks: a management dilemma. Pages 163–170 in L. H. Fredrickson, G. V. Burger, S. P. Havera, D. A. Graber, R. E. Kirby, and T. S. Taylor, editors. Proceedings 1988 North American wood duck symposium, St. Louis, Missouri, USA.

Semlitsch, R. D. 1985. Reproductive strategy of a facultatively paedomorphic salamander, *Ambystoma talpoideum*. Oecologia 65:305–313.

Semlitsch, R. D. 2000. Principles for management of aquatic-breeding amphibians. Journal of Wildlife Management 64:615–631.

Semlitsch, R. D., D. E. Scott, J. H. K. Pechmann, and J. W. Gibbons. 1996. Structure and dynamics of an amphibian community, evidence from a 16-year study of a natural pond. Pages 217–248 in M. L. Cody and J. A. Smallwood, editors. Long-term studies of vertebrate communities. Academic Press, New York, New York, USA.

Serafin, R. J., and J. W. Wilson. 2000. Operational weather radar in the United States: progress and opportunity. Bulletin of the American Meteorological Society 81:501–518.

Serena, M. 1980. A new technique for capturing *Cnemidophorus*. Journal of Herpetology 14:91–92.

Serfass, T. L., R. P. Brooks, T. J. Swimley, L. M. Rymon, and A. H. Hayden. 1996. Considerations for capturing, handling, and translocating river otters. Wildlife Society Bulletin 24:25–31.

Sergeant, D. E., and D. H. Pimlott. 1959. Age determination in moose from sectioned incisor teeth. Journal of Wildlife Management 23:315–321.

Servello, F. A., E. C. Hellgren, and S. R. McWilliams. 2005. Techniques for wildlife nutritional ecology. Pages 554–590 in C. E. Braun, editor. Techniques for wildlife investigations and management. The Wildlife Society, Bethesda, Maryland, USA.

Servello, F. A., and R. L. Kirkpatrick. 1986. Sexing ruffed grouse in the Southeast using feather criteria. Wildlife Society Bulletin 14:280–282.

Servheen, C., and J. Waller. 1999. Documenting habitat use and crossing preferences of grizzly bears along highways utilizing GPS technology. Page 3 in The proceedings of the international conference on wildlife ecology and transportation, Missoula, Montana, USA.

Sessions, J. 1991. Solving for habitat connections as a Steiner network problem. Forest Science 38:203–207.

Sessions, J., D. Johnson, J. Ross, and B. Sharer. 2000. The Blodgett Plan, an active management approach to developing mature forest habitat. Journal of Forestry 98(12):29–33.

Seubert, J. L., and B. Meanly. 1974. Relationships of blackbird/starling roosts to bird hazards at airports. Pages 209–219 in Proceedings of a conference on the biological aspects of the bird/aircraft collision problem, Department of Zoology, Clemson University, Clemson, South Carolina, USA.

Severinghaus, C. W. 1949. Tooth development and wear as criteria of age in white-tailed deer. Journal of Wildlife Management 13:195–216.

Seward, N. W., D. S. Maehr, J. W. Gassett, J. J. Cox, and J. L. Larkin. 2005. Field searches vs. vaginal transmitters for locating elk calves. Wildlife Society Bulletin 33:751–755.

Seydack, A. H. W. 1984. Application of a photo-recording device in the census of larger rainforest mammals. South African Journal of Wildlife Research 14:10–14.

Seymour, N. R. 1974. Territorial behavior of wild shovelers at Delta, Manitoba. Wildfowl 25:49–55.

Shackleton, D. M., L. V. Hills, and D. A. Hutton. 1975. Aspects of variation in cranial characters of Plains bison (*Bison bison bison* Linnaeus) from Elk Island National Park, Alberta. Journal of Mammalogy 56:871–887.

Shafer, E. L., Jr. 1963. The twig-count method for measuring hardwood deer browse. Journal of Wildlife Management 27:428–437.

Shaffer, M. L. 1987. Minimum viable populations: coping with uncertainty. Pages 69–86 in M. E. Soulé, editor. Viable populations for conservation. Cambridge University Press, Cambridge, England, UK.

Shaffer, S. A., Y. Tremblay, J. A. Awkerman, R. W. Henry, S. L. H. Teo, D. J. Anderson, D. A. Croll, B. A. Block, and D. P. Costa. 2005. Comparison of light- and SST-based geolocation with satellite telemetry in free-ranging albatrosses. Marine Biology 147:833–843.

Shaffer, T. L. 2004. A unified approach to analyzing nest success. Auk:121:526–540.

Shaiffer, C. W., and G. L. Krapu. 1978. A remote controlled system for capturing nesting waterfowl. Journal of Wildlife Management 42:668–669.

Shanks, C. E. 1948. The pelt-primeness method of aging muskrats. American Midland Naturalist 39:179–187.

Sharath, B. L., and S. N. Hegd. 2003. Two new traps for sampling the black pond turtle (*Melanochelys trijuga*) in the tropical rainforests of the Western Ghats (India). Herpetological Review 34:33–34.

Sharp, D. E., and J. T. Lokemoen. 1987. A decoy trap for breeding-season mallards in North Dakota. Journal of Wildlife Management 51:711–715.

Sharp, W. M. 1958. Aging gray squirrels by use of tail-pelage characteristics. Journal of Wildlife Management 22:29–34.

Sharp, W. M. 1963. The effects of habitat manipulation and forest succession on ruffed grouse. Journal of Wildlife Management 27:664–671.

Sharpe, F., M. Bolton, R. Sheldon, and N. Ratcliffe. 2009. Effects of color banding, radio tagging, and repeated handling on the condition and survival of lapwing chicks and consequences for estimates of breeding productivity. Journal of Field Ornithology 80:101–110.

Shaw, C. N., P. J. Wilson, and B. N. White. 2003. A reliable molecular method of gender determination in mammals. Journal of Mammalogy 84:123–128.

Shaw, H. 1989. Soul among lions: the cougar as peaceful adversary. Johnson, Boulder, Colorado, USA.

Shedden, C. B., P. Monaghan, K. Ensor, and N. B. Metcalfe. 1985. The influence of colour-rings on recovery rates of herring and lesser black-backed gulls. Journal of Ringing and Migration 6:52–54.

Sheffield, S. R. 1997. Owls as biomonitors of environmental contamination. Pages 383–398 in J. R. Duncan, D. H. Johnson, and T. H. Nicholls, editors. Biology and conservation of owls of the northern hemisphere. General Technical Report NC-190, U.S. Forest Service, Department of Agriculture, Washington, D.C., USA.

Sheffield, S. R., and R. L. Lochmiller. 2001. Effects of field exposure to diazinon on small mammals inhabiting a semi-enclosed prairie grassland ecosystem. I. Ecological and reproductive effects. Environmental Toxicology and Chemistry 20:284–296.

Sheffield, S. R., K. Sawicka-Kapusta, J. B. Cohen, and B. A. Rattner. 2001. Rodents and lagomorphs. Pages 215–314 in R. F. Shore and B. A. Rattner, editors. Ecotoxicology of wild mammals. John Wiley and Sons, London, England, UK.

Sheldon, W. G. 1949. A trapping and tagging technique for wild foxes. Journal of Wildlife Management 13:309–311.

Sheldon, W. G., F. Greeley, and J. Kupa. 1958. Aging fall-shot American woodcocks by primary wear. Journal of Wildlife Management 22:310–312.

Shenk, T. M., and A. B. Franklin, editors. 2001. Modeling in natural resource management: development, interpretation, and application. Island Press, Washington, D.C., USA.

Shepard, E. L. C., R. P. Wilson, F. Quintana, A. G. Laich, N. Liebsch, D. A. Albareda, L. G. Hasley, A. Gleiss, D. T. Morgan, A. E. Myers, C. Newman, and D. W. Macdonald. 2008. Identification of animal movement patterns using tri-axial accelerometry. Endangered Species Research 10:47–60.

Sherman, P. W., M. L. Morton, L. M. Hoopes, J. Bochantin, and J. M. Watt. 1985. The use of tail collagen strength to estimate age in Belding's ground squirrels. Journal of Wildlife Management 49:874–879.

Sherwin, R. E., S. Haymond, D. Stricklan, and R. Olsen. 2002. Freeze-branding to permanently mark bats. Wildlife Society Bulletin 30:97–100.

Sherwood, G. A. 1966. Flexible plastic collars compared to nasal discs for marking geese. Journal of Wildlife Management 30:853–855.

Shibata, R. 1989. Statistical aspects of model selection. Pages 215–240 in J. C. Willems, editor. From data to model. Springer-Verlag, New York, New York, USA.

Shine, C., N. Shine, R. Shine, and D. Slip. 1988. Use of subcaudal scale anomalies as an aid in recognizing individual snakes. Herpetological Review 19:79.

Shirley, M. G., R. G. Linscombe, and L. R. Sevin. 1983. A live trapping and handling technique for river otter. Proceedings of the Annual Conference of the Southeastern Association of Fish and Wildlife Agencies 37:182–189.

Shissler, B. P., D. E. Samuel, and D. L. Burkhart. 1981. An aging technique for American woodcock on summer fields. Wildlife Society Bulletin 9:302–305.

Shivik, J. 2002. Odor-adsorptive clothing, environmental factors, and search-dog ability. Wildlife Society Bulletin 30:721–727.

Shivik, J. A. 1998. Brown tree snake response to visual and olfactory cues. Journal of Wildlife Management 62:105–111.

Shivik, J. A., and L. Clark. 1997. Carrion seeking in brown tree snakes: importance of olfactory and visual cues. Journal of Experimental Zoology 279:549–553.

Shivik, J. A., K. S. Gruver, and T. J. DeLiberto. 2000. Preliminary evaluation of new cable restraints to capture coyotes. Wildlife Society Bulletin 28:606–613.

Shivik, J. A., D. J. Martin, M. J. Pipas, J. Turnan, and T. J. Deliberto. 2005. Initial comparison: jaws, cables, and cage-traps to capture coyotes. Wildlife Society Bulletin 33:1375–1383.

Shoop, C. R. 1971. A method for short-term marking of amphibians with 24-sodium. Herpetologica 30:160–162.

Shoop, M. C., and E. H. McIlvain. 1963. The micro-unit forage inventory method. Journal of Range Management 16:172–179.

Shor, W. 1990a. A good bownet design. Hawk Chalk 29:39–46.

Shor, W. 1990b. Installing and using the bownet. Hawk Chalk 29:51–59.

Shore, I. F., and B. A. Rattner, editors. 2001. Ecotoxicology of wild mammals. John Wiley and Sons, London, England, UK.

Short, N. M. 1982. The Landsat tutorial workbook: basics of satellite remote sensing. Reference Publication 1078, U.S. National Aeronautics and Space Administration, Department of Commerce, Washington, D.C., USA.

Short, R. V. 1972. The role of hormones in sex cycles. Pages 43–72 in C. R. Austin and R. V. Short, editors. Hormones in reproduction. Cambridge University Press, New York, New York, USA.

Shuler, J. D. 1992. A cage trap for live-trapping mountain lions. Proceedings of the Vertebrate Pest Conference 15:368–370.

Shuler, J. F., D. E. Samuel, B. P. Shissler, and M. R. Ellingwood. 1986. A modified night-lighting technique for male American woodcock. Journal of Wildlife Management 50:384–387.

Shupe, T. E., F. S. Guthery, and R. L. Bingham. 1990. Vulnerability of bobwhite sex and age classes to harvest. Wildlife Society Bulletin 18:24–26.

Shute, N. 1990. Dogging rare geese to save them. National Wildlife 28:22.

Sibley, C. G., and B. L. Monroe, Jr. 1990. Distribution and taxonomy of birds of the world. Yale University Press, New Haven, Connecticut, USA.

Sibley, D. A. 2000. The Sibley guide to birds. Alfred A. Knopf, New York, New York, USA.

Sibly, R. M., H. M. R. Nott, and D. J. Fletcher. 1990. Splitting behaviour into bouts. Animal Behaviour 39:63–69.

Sidor, I. F., M. A. Pokras, A. R. Major, R. H. Poppenga, K. M. Taylor, and R. M. Miconi. 2003. Mortality of common loons in New England, 1987 to 2000. Journal of Wildlife Diseases 39:306–315.

Siegfried, W. R. 1971. Communal roosting of the cattle egret. Transactions of the Royal Society of South Africa 39:419–443.

Sievert, G. A., P. T. Andreadis, and T. S. Campbell. 1999. A simple device for safely capturing herpetofauna from roads: the "Herp Scoop." Herpetological Review 30:156–157.

Sigg, D. P. 2006. Reduced genetic diversity and significant genetic differentiation after translocation: comparison of the remnant and translocated populations of bridled nailtail wallabies (*Onychogalea fraenata*). Conservation Genetics 7:577–589.

Siglin, R. J. 1966. Marking mule deer with an automatic tagging device. Journal of Wildlife Management 30:631–633.

Sigurdson, C. J. 2008. A prion disease of cervids: chronic wasting disease. Veterinary Research 39:41.

Siler, W. 1979. A competing-risk model for animal mortality. Ecology 60:750–757.

Sillett, T. S., and R. T. Holmes. 2002. Variation in survivorship of a migratory songbird throughout its annual cycle. Journal of Animal Ecology 71:296–308.

Silovsky, G. D., H. M. Wight, L. H. Sisson, T. L. Fox, and S. W. Harris. 1968. Methods for determining age of band-tailed pigeons. Journal of Wildlife Management 32:421–424.

Silverman, B. W. 1986. Density estimation for statistics and data analysis. Chapman and Hall, London, England, UK.

Silvy, N. J. 1975. Population density, movements, and habitat utilization of Key deer (*Odocoileus virginianus clavium*). Dissertation, Southern Illinois University, Carbondale, USA.

Silvy, N. J., J. W. Hardin, and W. D. Klimstra. 1975. Use of a portable net to capture free-ranging deer. Wildlife Society Bulletin 3:27–29.

Silvy, N. J., J. W. Hardin, and W. D. Klimstra. 1977. On the relationship of animals marked to cost and accuracy of Lincoln estimates. Proceedings of the Annual Conference of the Southeastern Association of Fish and Wildlife Agencies. 31:199–203.

Silvy, N. J., R. R. Lopez, and M. J. Peterson. 2005. Wildlife marking techniques. Pages 339–376 in C. E. Braun, editor. Techniques for wildlife investigations and management. The Wildlife Society, Bethesda, Maryland, USA.

Silvy, N. J., M. E. Morrow, E. Shanley, Jr., and R. D. Slack. 1990. An improved drop net for capturing wildlife. Proceedings of the Annual Conference of the Southeastern Association of Fish and Wildlife Agencies 44:374–378.

Silvy, N. J., and R. J. Robel. 1967. Recordings used to help trap booming greater prairie chickens. Journal of Wildlife Management 31:370–373.

Silvy, N. J., and R. J. Robel. 1968. Mist nets and cannon nets compared for capturing prairie chickens on booming grounds. Journal of Wildlife Management 32:175–178.

Silvy, N. J., J. L. Roseberry, and R. A. Lancia. 1979. A computer algorithm for determining home range size using Mohr's minimum home range method. International Wildlife Biotelemetry Conference 2:170–177.

Simberloff, D. 1988. The contribution of population and community biology to conservation science. Annual Review of Ecology and Systematics 19:473–511.

Simberloff, D. S. 1972. Properties of the rarefaction diversity measurement. American Naturalist 106:414–418.

Simkin, D. W. 1965. Reproduction and productivity of moose in northwestern Ontario. Journal of Wildlife Management 29:740–750.

Simmons, J. E. 2002. Herpetological collecting and collections management. Revised edition. Herpetological Circular 13, Society for the Study of Amphibians and Reptiles, Salt Lake City, Utah, USA.

Simmons, N. M. 1971. An inexpensive method of marking large numbers of Dall sheep for movement studies. Transaction of the North American Wild Sheep Conference 1:116–126.

Simmons, N. M., and J. L. Phillips. 1966. Modifications of a dye-spraying device for marking desert bighorn sheep. Journal of Wildlife Management 30:208–209.

Simon, C. A., and B. E. Bissinger. 1983. Paint-marking lizards: does the color affect survivorship? Journal of Herpetology 17:184–186.

Simpson, E. H. 1949. Measurement of diversity. Nature 163:688.

Sims, C. G. 2004. A net pole for the masses. North American Bird Bander 29:178–180.

Sinclair, A. R. E. 1991. Science and the practice of wildlife management. Journal of Wildlife Management 55:767–773.

Sinclair, A. R. E., J. M. Fryxell, and G. Caughley. 2006. Wildlife ecology, conservation, and management. Blackwell, Oxford, England, UK.

Sirois, J., and J. E. Fortune. 1988. Ovarian follicular dynamics during the estrous cycle in heifers monitored by real-time ultrasonography. Biology of Reproduction 39:308–317.

Sisk, T. D., N. M. Haddad, and P. R. Ehrlich. 1997. Bird assemblages in patchy woodlands: modeling the effects of edge and matrix habitats. Ecological Applications 7:1170–1180.

Skalski, J. R. 1994. Estimating wildlife populations based on incomplete area surveys. Wildlife Society Bulletin 22:192–203.

Skalski, J. R., and D. S. Robson. 1992. Techniques for wildlife investigations: design and analysis of capture data. Academic Press, San Diego, California, USA.

Skalski, J. R., K. E. Ryding, and J. J. Millspaugh. 2005. Wildlife demography: analysis of sex, age, and count data. Academic Press. San Diego, California, USA.

Skinner, D. L., and A. W. Todd. 1990. Evaluating efficiency of footholding devices for coyote capture. Wildlife Society Bulletin 18:166–175.

Skinner, M. F, and O. C. Kaisen. 1947. The fossil bison of Alaska and preliminary revision of the genus. American Museum of Natural History Bulletin 89:131–256.

Skinner, W. R., D. P. Snow, and N. F. Payne. 1998. A capture technique for juvenile willow ptarmigan. Wildlife Society Bulletin 26:111–112.

Skolnik, M. I., editor. 1990. Radar handbook. Second edition. McGraw-Hill, New York, New York, USA.

Slade, N. A. 1977. Statistical detection of density dependence from a series of sequential censuses. Ecology 58:1094–1102.

Slade, N. A., M. A. Eifler, N. M. Gruenhagen, and A. L. Davelos. 1993. Differential effectiveness of standard and long Sherman live traps in capturing small mammals. Journal of Mammalogy 74:156–161.

Sladen, W. J. L. 1952. Notes on methods of marking penguins. Ibis 94:541–543.

Slate, D., T. P. Algeo, K. M. Nelson, R. B. Chipman, D. Donovan, J. D. Blanton, M. Niezgoda, and C. E. Rupprecht. 2009. Oral rabies vaccination in North America: opportunities, complexities, and challenges. PLoS Neglected Tropical Diseases 3:e549. doi:510.1371/journal.pntd.0000549.

Slate, J., P. M. Visscher, S. Macgregor, D. Stevens, M. L. Tate, and J. M. Pemberton. 2002. A genome scan for quantitative trait loci in a wild population of red deer (*Cervus elaphus*). Genetics 162:1863–1873.

Slater, P. J. B., and N. P. Lester. 1982. Minimising errors in splitting behaviour into bouts. Behaviour 79:153–161.

Slatkin, M. 1985a. Gene flow in natural populations. Annual Review of Ecology and Systematics 16:393–430.

Slatkin, M. 1985b. Rare alleles as indicators of gene flow. Evolution 39:53–65.

Slatkin, M., and W. P. Maddison. 1990. Detecting isolation by distance using phylogenies of genes. Genetics 126:249–260.

Sloane, M. A., P. Sunnucks, D. Alpers, L. B. Beheregaray, and A. C. Taylor. 2000. Highly reliable genetic identification of individual northern hairy-nosed wombats from single remotely collected hairs: a feasible censusing method. Molecular Ecology 9:1233–1240.

Small, M. F., J. T. Baccus, and F. W. Weckerly. 2006. Are subcutaneous transmitters better than intracoelomic? A response. Wildlife Society Bulletin 34:890–893.

Small, M. F., R. Rosales, J. T. Baccus, F. W. Weckerly, D. N. Phalen, and J. A. Roberson. 2004. A comparison of effects of radiotransmitter attachment techniques on captive white-winged doves. Wildlife Society Bulletin 32:627–637.

Smallwood, J. A., and D. M. Bird. 2002. American kestrel. Account 602 in A. Poole and F. Gill, editors. The birds of North America. The Academy of Natural Sciences, Philadelphia, Pennsylvania, and The American Ornithologists' Union, Washington, D.C., USA.

Smith, A. D., and J. R. Riley. 1996. Signal processing in a novel radar system for monitoring insect migration. Computers and Electronics in Agriculture 15:267–278.

Smith, A. G. 1991. Chlorinated hydrocarbon insecticides. Pages 731–915 in W. J. Hayes and E. R. Laws, Jr., editors. Handbook of pesticide toxicology. Volume 2. Classes of pesticides. Academic Press, San Diego, California, USA.

Smith, B. L. 2001. Winter feeding of elk in western North America. Journal of Wildlife Management 65:173–190.

Smith, B. L., and T. L. McDonald. 2002. Criteria to improve age classification of captive antlerless elk. Wildlife Society Bulletin 30:200–207.

Smith, D. A., K. Ralls, B. Davenport, B. Adams, and J. E. Maldonado. 2001a. Canine assistants for conservationists. Science 291:435.

Smith, D. A., K. Ralls, B. L. Cypher, H. O. Clark, P. A. Kelly, D. F. Williams, and J. E. Maldonado. 2006a. Relative abundance of endangered San Joaquin kit foxes (*Vulpes macrotis mutica*) based on scat-detection dog surveys. Southwestern Naturalist 51:210–219.

Smith, D. A., K. Ralls, B. L. Cypher, and J. E. Maldonado. 2005. Assessment of scat-detection dog surveys to determine kit fox distribution. Wildlife Society Bulletin 33:897–904.

Smith, D. A., K. Ralls, A. Hurt, B. Adams, M. Parker, B. Davenport, M. C. Smith, and J. E. Maldonado. 2003a. Detection and accuracy rates of dogs trained to find scats of San Joaquin kit foxes (*Vulpes macrotis mutica*). Animal Conservation 6:4:339–346.

Smith, D. A., K. Ralls, A. Hurt, B. Adams, M. Parker, and J. E. Maldonado. 2006b. Assessing reliability of microsatellite genotypes from kit fox faecal samples using genetic and GIS analyses. Molecular Ecology 15:387–406.

Smith, D. G., and D. T. Walsh. 1981. A modified bal-chatri trap for capturing screech owls. North American Bird Bander 6:14–15.

Smith, D. R., J. A. Brown, and N. C. H. Lo. 2004. Application of adaptive sampling to biological populations. Pages 77–122 in W. L. Thompson, editor. Sampling rare or elusive species: concepts, designs, and techniques for estimating population parameters. Island Press, Washington, D.C., USA.

Smith, D. R., M. J. Conroy, and D. H. Brakhage. 1995a. Efficiency of adaptive cluster sampling for estimating density of wintering waterfowl. Biometrics 51:777–788.

Smith, D. R., K. J. Reinecke, M. J. Conroy, M. W. Brown, and J. R. Nassar. 1995b. Factors affecting visibility rate of waterfowl surveys in the Mississippi alluvial valley. Journal of Wildlife Management 59:515–527.

Smith, D. R., R. F. Villella, and D. P. Lemarie. 2003b. Applications of adaptive cluster sampling to low-density populations of freshwater mussels. Environmental and Ecological Statistics 10:7–15.

Smith, E. P., and G. van Belle. 1984. Nonparametric estimates of species richness. Biometrics 40:119–129.

Smith, G. J. 1987. Pesticide use and toxicology in relation to wildlife: organophosphorus and carbamate compounds. Resource Publication 170, U.S. Fish and Wildlife Service, Department of the Interior, Washington, D.C., USA.

Smith, G. R., and J. E. Rettig. 1996. Effectiveness of aquatic funnel traps for sampling amphibian larvae. Herpetological Review 27:190–191.

Smith, H. P., F. A. Stormer, and R. G. Godfrey, Jr. 1981. A collapsible quail trap. Research Note RM-400, U.S. Forest Service, Department of Agriculture, Washington, D.C., USA.

Smith, J. G. 1959. Additional modifications of the point frame. Journal of Range Management 12:204–205.

Smith, L. M., I. L. Brisbin, Jr., and G. C. White. 1984. An evaluation of total trapline captures as estimates of furbearer abundance. Journal of Wildlife Management 48:1452–1455.

Smith, M. D., A. D. Hammond, L. W. Burger, Jr., W. E. Palmer, A. V. Carver, and S. D. Wellendorf. 2003c. A technique for capturing northern bobwhite chicks. Wildlife Society Bulletin 31:1054–1060.

Smith, M. E., J. D. C. Linnell, J. Odden, and J. E. Swenson. 2000. Review of methods to reduce livestock depredation: I. Guardian animals. Acta Agriculture Scandinavica 50:279–290.

Smith, N. D., and I. O. Buss. 1963. Age determination and plumage observations of blue grouse. Journal of Wildlife Management 27:566–578.

Smith, S. A., and R. A. Paselk. 1986. Olfactory sensitivity of the turkey vulture (*Cathartes aura*) to three carrion-associated odorants. Auk 103:586–592.

Smith, T. B., and R. K. Wayne. 1996. Molecular genetic approaches in conservation. Oxford University Press, New York, New York, USA.

Smith, W. A. 1968. The band-tailed pigeon in California. California Fish and Game 54:4–16.

Smith, W. K., K. E. Church, J. S. Taylor, D. H. Rusch, and P. S. Gipson. 2001b. Modified decoy trapping of male ring-necked pheasant (*Phasianus colchicus*) and northern bobwhite (*Colinus virginianus*). Game and Wildlife Science 18:581–586.

Smouse, P. E., R. S. Waples, and J. A. Tworek. 1990. A genetic mixture analysis for use with incomplete source population data. Canadian Journal of Fisheries and Aquatic Sciences 47:620–634.

Snead, I. E. 1950. A family type live trap, handling cage, and associated techniques for muskrats. Journal of Wildlife Management 14:67–79.

Snelling, J. C. 1970. Some information obtained from marking large raptors in the Kruger National Park, Republic of South Africa. Ostrich (Supplement) 8:415–427.

Snow, W. D., H. L. Mendall, and W. B. Krohn. 1990. Capturing common eiders by night-lighting in coastal Maine. Journal of Field Ornithology 61:67–72.

Society for the Study of Amphibians and Reptiles. 1987. Guidelines for use of live amphibians and reptiles in field research. Journal of Herpetology 4:1–14.

Soderquist, T. R., and C. R. Dickman. 1988. A technique for marking marsupial pouch young with fluorescent pigment tattoos. Australian Wildlife Research 15:561–563.

Sodhi, N. S., C. A. Paszkowski, and S. Keehn. 1999. Scale-dependent habitat selection by American redstarts in aspen dominated forest fragments. Wilson Bulletin 111:70–75.

Soisalo, M. K., and S. M. C. Cavalcanti. 2006. Estimating the density of a jaguar population in the Brazilian Pantanal using camera-traps and capture–recapture sampling in combination with GPS radio-telemetry. Biological Conservation 129:487–496.

Sokal, R. R., and F. J. Rohlf. 1995. Biometry: the principles and practice of statistics in biological research. W. H. Freeman and Company, New York, New York, USA.

Sokolov, V. E. 1988. Dictionary of animal names: amphibians and reptiles. Russky Yazyk, Moscow, Russia.

Solow, A. R. 1990. Testing for density dependence: a cautionary note. Oecologia 83:47–49.

Sordahl, T. A. 1980. A nest trap for recurvirostrids and other ground-nesting birds. North American Bird Bander 5:1–3.

Sork, V. L., J. Bramble, and O. Sexton. 1993. Ecology of mast-fruiting in three species of North American deciduous oaks. Ecology 74:528–541.

Soulé, M. E. 1986. Conservation biology: the science of scarcity and diversity. Sinauer Associates, Sunderland, Massachusetts, USA.

Soule, N., and A. J. Lindberg. 1994. The use of leverage to facilitate the search for the hellbender. Herpetological Review 25:16.

Southern, L. K., and W. E. Southern. 1983. Responses of ring-billed gulls to cannon-netting and wing-tagging. Journal of Wildlife Management 47:234–237.

Southern, L. K., and W. E. Southern. 1985. Some effects of wing tags on breeding ring-billed gulls. Auk 102:38–42.

Southern, W. E. 1964. Additional observations on winter bald eagle populations: including remarks on biotelemetry techniques and immature plumages. Wilson Bulletin 76:121–137.

Southern, W. E. 1972. Use of cannon-nets in ring-billed gull colonies. Inland Bird Banding News 44:83–93.

Soutullo, A., L. Cadahïa, V. Urios, M. Ferrer, and J. J. Negro. 2007. Accuracy of lightweight satellite telemetry: a case study in the Iberian Peninsula. Journal of Wildlife Management 71:1010–1015.

Sowls, L. K. 1950. Techniques for waterfowl-nesting studies. Transactions of the North American Wildlife Conference 15:478–487.

Sowls, L. K. 1955. Prairie ducks: a study of their behavior, ecology, and management. Wildlife Management Institute, Washington, D.C., and Stackpole, Harrisburg, Pennsylvania, USA.

Sowls, L. K., and P. S. Minnamon. 1963. Glass beads for marking home ranges of mammals. Journal of Wildlife Management 27:299–302.

Spalding, D. J. 1966. Eruption of permanent canine teeth in the northern sea lion. Journal of Mammalogy 47:157–158.

Sparling, D. W. 2003. White phosphorus at Eagle River flats, Alaska: a case history of waterfowl mortality. Pages 767–786 in D. J. Hoffman, B. A. Rattner, G. A. Burton, Jr., and J. Cairns, Jr., editors. Handbook of ecotoxicology. Second edition. Lewis, Boca Raton, Florida, USA.

Sparling, D. W., D. Day, and P. Klein. 1999. Acute toxicity and sublethal effects of white phosphorus in mute swans, *Cygnus olor*. Archives of Environmental Contamination and Toxicology 36:316–322.

Sparling, D. W., and N. E. Federoff. 1997. Secondary poisoning of kestrels by white phosphorus. Ecotoxicology 6:239–247.

Sparling, D. W., S. Vann, and R. A. Grove. 1998. Blood changes in mallards exposed to white phosphorus. Environmental Toxicology and Chemistry 17:2521–2529.

Sparrowe, R. D., and P. F. Springer. 1970. Seasonal activity patterns of white-tailed deer in eastern South Dakota. Journal of Wildlife Management 34:420–431.

Spear, L. 1980. Band loss from the western gull on southeast Farallon Island. Journal of Field Ornithology 51:319–328.

Spears, B. L., W. B. Ballard, M. C. Wallace, R. S. Phillips, D. H. Holdstock, J. H. Brunjest, R. Applegate, P. S. Gipson, M. S. Miller, and T. Barnett. 2002. Retention time of miniature radiotransmitters glued to wild turkey poults. Wildlife Society Bulletin 30:861–867.

Spectrum Planning Team. 2001. A review of automotive radar systems—devices and regulatory frameworks. http://www.atnf.csiro.au/SKA/intmit/autoradar.doc.

Spellerberg, I. P., and I. Prestt. 1978. Marking snakes. Pages 133–141 in B. Stonehouse, editor. Animal marking: recognition marking of animals in research. University Park Press, Baltimore, Maryland, USA.

Spencer, C. C., J. E. Neigel, and P. L. Leberg. 2000. Experimental evaluation of the usefulness of microsatellite DNA for detecting demographic bottlenecks. Molecular Ecology 9:1517–1528.

Spencer, R. 1978. Ringing and related durable methods of marking birds. Pages 45–53 in B. Stonehouse, editor. Animal marking: recognition marking of animals in research. University Park Press, Baltimore, Maryland, USA.

Spiegelhalter, D., A. Thomas, N. Best, and D. Lunn. 2003. WinBUGS user manual, version 1.4. MCR Biostatistics Unit, Cambridge, England, UK.

Spiegelhalter, D., A. Thomas, N. Best, and D. Lunn. 2004. WinBUGS user manual, version 2.10. MRC Biostatistics Unit, Cambridge, England, UK.

Spielman, D., and R. Frankham. 1992. Modeling problems in conservation genetics using captive *Drosophila* populations: improvement of reproductive fitness due to immigration of one individual into small partially inbred populations. Zoo Biology 11:343–351.

Spillett, J. J., and R. S. ZoBell. 1967. Innovations in trapping and handling pronghorn antelopes. Journal of Wildlife Management 31:347–351.

Spilliaert, R., G. Vikingsson, U. Arnason, A. Palsdottir, J. Sigurjonsson, and A. Arnason. 1991. Species hybridization between a female blue whale (*Balaenoptera musculus*) and a male fin whale (*B. physalus*): molecular and morphological documentation. Journal of Heredity 82:269–274.

Springer, J. T. 1979. Some sources of bias and sampling error in radio triangulation. Journal of Wildlife Management 43:926–935.

Spurr, S. H. 1964. Forest ecology. Ronald Press, New York, New York, USA.

St. Amant, J. L. S. 1970. The detection of regulation in animal populations. Ecology 51:823–828.

St. Louis, V. L., J. C. Barlow, and J. P. Sweerts. 1989. Toenail-clipping: a simple technique for marking individual nidicolous chicks. Journal of Field Ornithology 60:211–215.

Stabenfeldt, G. H., and V. M. Shille. 1977. Reproduction in the dog and cat. Pages 499–527 in H. H. Cole and P. T. Cupps, editors. Reproduction in domestic animals. Third edition. Academic Press, New York, New York, USA.

Stafford, S., C. T. Lee, and L. E. Williams, Jr. 1966. Drive trapping white-tailed deer. Proceedings of the Annual Conference of the Southeastern Association of Game and Fish Commissioners 20:63–69.

Stalmaster, M. V., and J. L. Kaiser. 1998. Effects of recreation activity on wintering bald eagles. Wildlife Monographs 137.

Stamps, J. A. 1973. Displays and social organization in female *Anolis aeneus*. Copeia 1973:264–272.

Stanback, M. T., and W. D. Koenig. 1994. Techniques for capturing birds inside natural cavities. Journal of Field Ornithology 65:70–75.

Stanley, T. R. 2000. Modeling and estimation of stage-specific daily survival probabilities of nests. Ecology 81:2048–2053.

Stanley, T. R., and K. P. Burnham. 1999. A closure test for time-specific capture–recapture data. Environmental and Ecological Statistics 6:197–209.

Stanley, T. R., and J. D. Richards. 2003. CloseTest: a program for testing capture–recapture data for closure. Fort Collins Science Center, U.S. Geological Survey, Fort Collins, Colorado, USA.

Stansley, W., and D. E. Roscoe. 1996. The uptake and effects of lead in small mammals and frogs at a trap and skeet range. Archives of Environmental Contamination and Toxicology 30:220–226.

Stapleton, M., P. O. Dunn, J. McCarty, A. Secord, and L. A. Whittingham. 2001. Polychlorinated biphenyl contamination and minisatellite DNA mutation rates of tree swallows. Environmental Toxicology and Chemistry 20:2263–2267.

Stapp, P., J. K. Young, S. Vande Woude, and B. Van Horne. 1994. An evaluation of the pathological effects of fluorescent power on deer mice (*Peromyscus maniculatus*). Journal of Mammalogy 75:704–709.

Starfield, A. M., and A. L. Bleloch. 1991. Building models for conservation and wildlife management. Burgess Press, Edina, Minnesota, USA.

Stark, M. A. 1984. A quick, easy and permanent tagging technique for rattlesnakes. Herpetological Review 15:110.

State of Wisconsin Legislative Audit Bureau. 2006. An evaluation: chronic wasting disease—Department of Natural Resources. Report 06-13, Legislative Audit Bureau, State of Wisconsin, Madison, USA.

Stebbings, R. E. 1978. Marking bats. Pages 81–94 in B. Stonehouse, editor. Animal marking: recognition marking of animals in research. University Park Press, Baltimore, Maryland, USA.

Stebbins, G. L. 1971. Processes of organic evolution. Second edition. Prentice-Hall, Englewood Cliffs, New Jersey, USA.

Stebbins, R. C., and N. W. Cohen. 1973. The effect of parietalectomy on the thyroid and gonads in free-living western fence lizards, *Sceloporus occidentalis*. Copeia 1973:662–672.

Stedman, S. J. 1990. Band opening and removal by house finches. North American Bird Bander 15:136–138.

Steel R. G. D., and J. H. Torrie. 1980. Principles and procedures of statistics: a biometrical approach. Second edition. McGraw-Hill, New York, New York, USA.

Steen, J. B., I. Mohus, T. Kvesetberg, and L. Walloe. 1996. Olfaction in bird dogs during hunting. Acta Physiologica Scandinavica 157:115–119.

Steenhof, K., K. K. Bates, M. R. Fuller, M. N. Kochert, J. O. McKinley, and P. M. Lukacs. 2006. Effects of radiomarking on prairie falcons: attachment failures provide insights about survival. Wildlife Society Bulletin 34:116–126.

Steenhof, K., G. P. Carpenter, and J. C. Bednarz. 1994. Use of mist nets and a live great horned owl to capture breeding American kestrels. Journal of Raptor Research 28:194–196.

Steffen, D. E., C. E. Couvillion, and G. A. Hurst. 1990. Age determination of eastern wild turkey gobblers. Wildlife Society Bulletin 18:119–124.

Steger, G. N., and D. L. Neal. 1981. Night-lighting: a technique to locate and capture fawns. Transactions of the California–Nevada Chapter of The Wildlife Society Annual Conference 17:155–158.

Stehman, S. V. 1999. Basic probability sampling designs for thematic map accuracy assessment. International Journal of Remote Sensing 20: 2423–2441.

Stehman, S. V. 2000. Practical implications of design-based sampling inference for thematic map accuracy assessment. Remote Sensing of Environment 72:35–45.

Stehman, S. V. 2001. Statistical rigor and practical utility in thematic map accuracy assessment. Photogrammetric Engineering and Remote Sensing 67:727–734.

Stehman, S. V., and R. L. Czaplewski. 1998. Design and analysis for thematic map accuracy assessment: fundamental principles. Remote Sensing of Environment 64:331–344.

Steidl, R. J. 2007. Limits of data analysis in scientific inference: reply to Sleep et al. Journal of Wildlife Management 71:2122–2124.

Steidl, R. J., S. DeStefano, and W. Matter. 2000. On increasing the quality, reliability, and rigor of wildlife science. Wildlife Society Bulletin 28:518–521.

Steidl, R. J., J. P. Hayes, and E. Schauber. 1997. Statistical power analysis in wildlife research. Journal of Wildlife Management 61:270–279.

Steiger, S. S., J. P. Kelley, W. W. Cochran, and M. Wikelski. 2009. Low metabolism and inactive lifestyle of a tropical rain forest bird investigated via heart rate telemetry. Physiological and Biochemical Zoology 82:580–589.

Steinhorst, R. K., and M. D. Samuel. 1989. Sightability adjustment methods for aerial surveys of wildlife populations. Biometrics 45:415–425.

Stenzel, L. E., H. R. Huber, and G. W. Page. 1976. Feeding behavior and diet of the long-billed curlew and willet. Wilson Bulletin 88:314–332.

Stephen, C. L., D. G. Whittaker, D. Gillis, L. L. Cox, and O. E. Rhodes. 2005. Genetic consequences of reintroductions: an example from Oregon pronghorn antelope (*Antilocapra americana*). Journal of Wildlife Management 69:1463–1474.

Stephens, P. A., S. W. Buskirk, and C. Martinez del Rio. 2007. Inference in ecology and evolution. Trends in Ecology and Evolution 22:192–197.

Stephens, S., J. Rotella, M. Lindberg, M. Taper, and J. Ringelman. 2005. Duck nest survival in the Missouri Coteau of North Dakota: landscape effects at multiple spatial scales. Ecological Applications 15:2137–2149.

Stephens, S. E., D. N. Koens, J. J. Rotella, and D. W. Willey. 2003. Effects of habitat fragmentation on avian nesting success: a review of the evidence at multiple spatial scales. Biological Conservation 155:101–110.

Stephenson, A. B. 1977. Age determination and morphological variation of Ontario otters. Canadian Journal of Zoology 55:1577–1583.

Stevens, C. E., and C. A. Paszkowski. 2005. A comparison of two pitfall trap designs in sampling boreal anurans. Herpetological Review 34:147–149.

Stevens, D. L., Jr., and A. R. Olsen. 2004. Spatially balanced sampling of natural resources. Journal of the American Statistical Association 99: 262–278.

Stevens, J. T., and D. D. Sumner. 1991. Herbicides. Pages 1317–1408 *in* W. J. Hayes and E. R. Laws, Jr., editors. Handbook of pesticide toxicology. Volume 3. Classes of pesticides. Academic Press, San Diego, California, USA.

Stevens, W. E. 1953. The northwestern muskrat of the Mackenzie delta, Northwest Territories, 1947–48. Wildlife Management Bulletin Series l, Number 8, Canadian Wildlife Service, Ottawa, Ontario, Canada.

Steventon, J. D., and J. T. Major. 1982. Marten use of habitat in a commercially clear-cut forest. Journal of Wildlife Management 46:175–182.

Stewart, K. M., R. T. Bowyer, J. G. Kie, N. J. Cimon, and B. K. Johnson. 2002. Temporospatial distributions of elk, mule deer, and cattle: resource partitioning and competitive displacement. Journal of Mammalogy 83:229–244.

Stewart, P. A. 1954. Combination substratum and automatic trap for nesting mourning doves. Bird-Banding 25:6–8.

Stewart, P. A. 1971. An automatic trap for use on bird nesting boxes. Bird-Banding 42:121–122.

Stewart, P. D., S. A. Ellwood, and D. W. MacDonald. 1997. Video surveillance of wildlife: an introduction from experience with the European badger *Meles meles*. Mammal Review 27:185–209.

Stewart, P. D., and D. W. MacDonald. 1997. Age, sex, and condition as predictors of moult and the efficacy of a novel fur-clip technique for individual marking of the European badger (*Meles meles*). Journal of Zoology 241:543–550.

Stewart, R. E. 1951. Clapper rail populations of the Middle Atlantic states. Transactions of the North American Wildlife Conference 16:421–430.

Stewart-Oaten, A. 1995. Rules and judgments in statistics: three examples. Ecology 76:2001–2009.

Stewart-Oaten, A., W. W. Murdoch, and K. R. Parker. 1986. Environmental impact assessment: . pseudoreplication in time? Ecology 67:929–940.

Stickel, L. F. 1950. Populations and home range relationships of the box turtle, *Terrapene c. carolina* (Linnaeus). Ecological Monographs 20:351–378.

Stickney, P. F. 1966. Browse utilization based on percentage of twig numbers browsed. Journal of Wildlife Management 30:204–206.

Stiehl, R. B. 1983. A new attachment method for patagial tags. Journal of Field Ornithology 54:326–328.

Stille, W. T. 1950. The loss of jaw tags by toads. Natural History Miscellaneous Publications 74, Chicago Natural History Museum, Chicago, Illinois, USA.

Stinnett, D. P., and D. A. Klebenow. 1986. Habitat use of irrigated lands by California quail in Nevada. Journal of Wildlife Management 50:368–372.

Stirling, I. 1979. Tagging Weddell and fur seals and some general comments on long-term marking studies. Pages 13–14 *in* L. Hobbs and P. Russell, editors. Report on the pinniped tagging workshop. American Institute of Biological Sciences, Arlington, Virginia, USA.

Stirling, I., and J. F. Bendell. 1966. Census of blue grouse with recorded calls of a female. Journal of Wildlife Management 30:184–187.

Stockwell, C. A., and P. L. Leberg. 2002. Ecological genetics and the translocation of native fishes: emerging experimental approaches. Western North American Naturalist 62:32–38.

Stoddard, H. L. 1931. The bobwhite quail: its habits, preservation and increase. Charles Scribner's Sons, New York, New York, USA.

Stoddart, L. A. 1935. Range capacity determination. Ecology 16:531–533.

Stohlgren, T. J., K. A. Bull, and Y. Otsuki. 1998. Comparison of rangeland vegetation sampling techniques in the central grasslands. Journal of Range Management 51:164–172.

Stokes, A. E., B. B. Schultz, R. M. DeGraaf, and C. R. Griffin. 2000. Setting mist nets from platforms in the forest canopy. Journal of Field Ornithology 71:57–65.

Stokes, A. W. 1957. Validity of spur length as an age criterion in pheasants. Journal of Wildlife Management 21:248–250.

Stokland, J. N. 1997. Representativeness and efficiency of bird and insect conservation in Norwegian boreal forest reserves. Conservation Biology 11:101–111.

Stoll, R. J., Jr., and D. Clay. 1975. Guide to aging wild turkey embryos. Ohio Fish and Wildlife Report 4, Ohio Department of Natural Resources, Division of Wildlife, Columbus, USA.

Stone, M. 1977. An asymptotic equivalence of choice of model by cross-validation and Akaike's criterion. Journal of the Royal Statistical Society B 39:44–47.

Stone, W. B., and J. C. Okoniewski. 2001. Necropsy findings and environmental contaminants in common loons from New York. Journal of Wildlife Diseases 37:178–184.

Stone, W. B., J. C. Okoniewski, and J. R. Stedelin. 1999. Poisoning of wildlife with anticoagulant rodenticides in New York. Journal of Wildlife Diseases 35:187–193.

Stoneberg, R. P., and C. J. Jonkel. 1966. Age determination of black bears by cementum layers. Journal of Wildlife Management 30:411–414.

Stonehouse, B. 1978. Animal marking: recognition marking of animals in research. University Park Press, Baltimore, Maryland, USA.

Stopher, K. V., J. M. Pemberton, T. H. Clutton-Brock, and T. Coulson. 2008. Individual differences, density dependence, and offspring birth traits in a population of red deer. Proceedings of the Royal Society B 275:2137–2145.

Stoppelaire, G. H., T. W. Gillespie, J. C. Brock, and G. A. Tobin. 2004. Use of remote sensing techniques to determine the effects of grazing on vegetation cover and dune elevation at Assateague Island National Seashore: impact of horses. Environmental Management 34:642–649.

Storaas, T., L. Kastdalen, and P. Wegge. 1999. Detection of forest grouse by mammalian predators: a possible explanation for high brood losses in fragmented landscapes. Wildlife Biology 5:187–192.

Story, M., and R. G. Congalton. 1986. Accuracy assessment: a user's perspective. Photogrammetric Engineering and Remote Sensing 52: 397–399.

Stoudt, J. H. 1971. Ecological factors affecting waterfowl production in the Saskatchewan parklands. Resource Publication 99, U.S. Fish and Wildlife Service, Department of the Interior, Washington, D.C., USA.

Stouffer, P. C., and D. F. Caccamise. 1991. Capturing American crows using alpha-chloralose. Journal of Field Ornithology 62:450–453.

Stowe, C. M., D. M. Barnes, and T. D. Arendt. 1981. Ethylene glycol intoxication in ducks. Avian Diseases 25:538–541.

Strang, T. J. 1992. A review of published temperatures for the control of pest insects in museums. Collection Forum 8:41–67.

Stransky, J. J., and L. K. Halls. 1980. Fruiting of woody plants affected by site preparation and prior land use. Journal of Wildlife Management 44:258–263.

Strathearn, S. M., J. S. Lotimer, G. B. Kolenosky, and W. M. Lintack. 1984. An expanding break-away radio collar for black bear. Journal of Wildlife Management 48:939–942.

Streutker, D. R., and N. F. Glenn. 2006. Lidar measurements of sagebrush steppe vegetation heights. Remote Sensing of Environment 102:135–145.

Stribling, H. L., and D. C. Sisson. 1998. Efficiency of pointing dogs in locating bobwhite quail coveys. Highlights of Agricultural Research 45(3). Auburn University, Auburn, Alabama, USA.

Strickland, B. K., and S. Demarais. 2000. Age and regional differences in antlers and mass of white-tailed deer. Journal of Wildlife Management 64:903–911.

Strickland, M. A., and C. W. Douglas. 1987. Marten. Pages 531–546 in M. Novak, J. A. Baker, M. E. Obbard, and B. Malloch, editors. Wild furbearer management and conservation in North America. Ontario Ministry of Natural Resources, Toronto, Canada.

Strickland, M. A., C. W. Douglas, M. Novak, and N. P. Hunziger. 1982. Marten. Pages 599–612 in J. A. Chapman and G. A. Feldhamer, editors. Wild mammals of North America. Johns Hopkins University Press, Baltimore, Maryland, USA.

Strickland, M. D., and L. L. McDonald. 2006. Introduction to the special section on resource selection. Journal of Wildlife Management 70:321–323.

Strong, D., B. Leatherman, and B. H. Brattstrom. 1993. Two new simple methods for catching small fast lizards. Herpetological Review 24:22–23.

Strong, D. R. 1986. Density vagueness: abiding the variance in the demography of real populations. Pages 257–268 in J. Diamond and T. J. Case, editors. Community ecology. Harper and Row, New York, New York, USA.

Strong, P. I. V., S. A. LaValley, and R. C. Burke, II. 1987. A colored plastic leg band for common loons. Journal of Field Ornithology 58:218–221.

Stuewer, F. W. 1943. Reproduction of raccoons in Michigan. Journal of Wildlife Management 7:60–73.

Sturman, W. A. 1968. Description and analysis of breeding habitats of the chickadees, *Parus atricapillus* and *P. rufescens*. Ecology 49:418–431.

Sturtevant, B. R., B. R. Miranda, J. Yang, H. S. He, E. J. Gustafson, and R. M. Sheller. 2009. Studying fire mitigation strategies in multi-ownership landscapes: balancing the management of fire-dependent ecosystems and fire risk. Ecosystems 12:445–461.

Stutchbury, B. J., and R. J. Robertson. 1986. A simple trap for catching birds in nest boxes. Journal of Field Ornithology 57:64–65.

Stutchbury, B. J. M., S. A Tarof, T. Done, E. Gow, P. M. Kramer, J. Tautin, J. W. Fox, and V. Afanasyev. 2009. Tracking long-distance songbird migration by using geolocators. Science 323:896.

Subba Rao, M. V., and B. S. Rajabai. 1972. Ecological aspects of the agamid lizards *Sitana ponticeriana* and *Calotes nemoricola* in India. Herpetologica 28:285–289.

Suedkamp, K. M., B. E. Washburn, J. J. Millspaugh, M. R. Ryan, and M. Hubbard. 2003. Effects of radiotransmitters on fecal glucocorticoid levels in captive dickcissels. Condor 105:805–810.

Sugden, L. G., and H. J. Poston. 1968. A nasal marker for ducks. Journal of Wildlife Management 32:984–986.

Sullins, G. L., D. O. McKay, and B. J. Verts. 1976. Estimating ages of cottontails by periosteal zonations. Northwest Science 50:17–22.

Sullivan, E. G., and A. O. Haugen. 1956. Age determination of foxes by x-ray of forefeet. Journal of Wildlife Management 20:210–212.

Sullivan, J. A., and P. F. Scanlon. 1976. Effects of grouping and fighting on the reproductive tracts of male white-footed mice (*Peromyscus leucopus*). Research in Population Ecology 17:164–175.

Sullivan, J. B., C. A. DeYoung, S. L. Beasom, J. R. Heffelfinger, S. P. Coughlin, and M. W. Hellickson. 1991. Drive-netting deer: incidence of mortality. Wildlife Society Bulletin 19:393–396.

Sullivan, M., A. Mannucci, P. Kimpton, and P. Gill. 1993. A rapid and quantitative DNA sex test: fluorescence-based PCR analysis of X-Y homologous gene amelogenin. BioTechniques 15:636–641.

Sulok, M., N. A. Slade, and T. J. Doonan. 2004. Effects of supplemental food on movements of cotton rats (*Sigmodon hispidus*) in northeastern Kansas. Journal of Mammalogy 85:1102–1105.

Summers, C. F., and S. R. Witthames. 1978. The value of tagging as a marking technique for seals. Pages 63–70 in B. Stonehouse, editor. Animal marking: recognition marking of animals in research. University Park Press, Baltimore, Maryland, USA.

Sundqvist, A.-K., H. Ellegren, M. Oliver, and C. Vilà. 2001. Y chromosome haplotyping in Scandinavian wolves (*Canis lupus*) based on microsatellite markers. Molecular Ecology 10:1959–1966.

Sundqvist, A.-K., H. Ellegren, and C. Vilà. 2008. Wolf or dog? Genetic identification of predators from saliva collected around bite wounds on prey. Conservation Genetics 9:1275–1279.

Sundqvist, C., A. Lukola, and M. Parvinen. 1986. Testicular aspiration biopsy in evaluation of fertility of mink (*Mustela vison*). Journal of Reproduction and Fertility 77:531–535.

Sunquist, M. E., and G. G. Montgomery. 1973. Activity patterns and rates of movement of two-toed sloths and three-toed sloths (*Choloepus hoffmanni* and *Bradypus infuscatus*). Journal of Mammalogy 54:946–954.

Sutherland, W. J. 1996. Ecological census techniques: a handbook. Cambridge University Press, Cambridge, England, UK.

Sutherland, W. J. 1998a. The effect of local change in habitat quality on populations of migratory species. Journal of Applied Ecology 35:418–421.

Sutherland, W. J. 1998b. The importance of behavioural studies in conservation biology. Animal Behaviour 56:801–809.

Suthers, H. B., J. M. Bickal, and P. G. Rodewald. 2000. Use of successional habitat and fruit resources by songbirds during autumn migration in central New Jersey. Wilson Bulletin 112:249–260.

Sutton, P. E., H. R. Mushinsky, and E. D. McCoy. 1999. Comparing the use of pitfall drift fences and cover boards for sampling the threatened sand skink. (*Neoseps reynoldsi*). Herpetological Review 30:149–151.

Svancara, L. K., E. O. Garton, K.-T. Change, J. M. Scott, P. Zager, and M. Gratson. 2002. The inherent aggravation of aggregation: an example with elk aerial survey data. Journal of Wildlife Management 66:776–787.

Sveum, C. M., W. D. Edge, and J. A. Crawford. 1998. Nesting habitat selection by sage grouse in south-central Washington. Journal of Range Management 51:265–269.

Swank, W. G. 1952. Trapping and marking of adult nesting doves. Journal of Wildlife Management 16:87–90.

Swank, W. G. 1955. Feather molt as an aging technique for mourning doves. Journal of Wildlife Management 19:412–414.

Swann, D. E., A. J. Kuenzi, M. L. Morrison, and S. DeStefano. 1997. Effects of sampling blood on survival of small mammals. Journal of Mammalogy 78:908–913.

Swanson, C. C., J. A. Jenks, C. S. Deperno, R. W. Klaver, R. G. Osborn, and J. A. Tardiff. 2008. Does the use of vaginal-implant transmitters affect neonate survival rate of white-tailed deer *Odocoileus virginianus*? Wildlife Biology 14:272–279.

Swanson, D. A., and J. H. Rappole. 1992. Status of the white-winged dove in southern Texas. Southwestern Naturalist 37:93–97.

Swanson, D. A., and J. H. Rappole. 1994. Capturing nesting white-winged doves in subtropical thornforest habitat in south Texas. Wildlife Society Bulletin 22:500–502.

Sweeney, J. R., R. L. Marchinton, and J. M. Sweeney. 1971. Responses of radio-monitored white-tailed deer chased by hunting dogs. Journal of Wildlife Management 35:707–716.

Sweeney, T. M., J. D. Fraser, and J. S. Coleman. 1985. Further evaluation of marking methods for black and turkey vultures. Journal of Field Ornithology 56:251–257.

Sweitzer, R. A., B. J. Gonzales, I. A. Gardner, D. Van Vuren, J. D. Waithman, and W. M. Boyce. 1997. A modified panel trap and immobilization technique for capturing multiple wild pig. Wildlife Society Bulletin 25:699–705.

Sweitzer, R., D. Van Vuren, I. Gardner, W. Boyce, and J. Waithman. 2000. Estimating sizes of wild pig populations in the north and central coast regions of California. Journal of Wildlife Management 64:531–543.

Swenson, J. E. 2003. Implications of sexually selected infanticide for the hunting of large carnivores. Pages 171–190 in M. Festa-Bianchet and M. Apollonio, editors. Animal behavior and wildlife management. Island Press, Washington, D.C., USA.

Swenson, J. E., and S. Swenson. 1977. Nightlighting as a method for capturing nighthawks and other caprimulgids. Bird-Banding 48:279–280.

Swenson, J. E., K. Wallin, G. Ericsson, G. Cederlund, and F. Sandegren. 1999. Effects of ear-tagging with radiotransmitters on survival of moose calves. Journal of Wildlife Management 63:354–358.

Swihart, R. K., and N. A. Slade. 1985. Testing for independence of observations in animal movements. Ecology 66:1176–1184.

Swingland, I. R. 1978. Marking reptiles. Pages 119–132 in B. Stonehouse, editor. Animal marking: recognition marking of animals in research. University Park Press, Baltimore, Maryland, USA.

Swinton, J., F. Tuyttens, D. W. MacDonald, D. J. Nokes, C. L. Cheeseman, and R. S. Clifton-Hadley. 1997. A comparison of fertility control and lethal control of bovine tuberculosis in badgers: the impact of perturbation induced transmission. Philosophical Transactions of the Royal Society of London B 352:619–631.

Swofford, D., G. Olsen, P. Waddel, and D. M. Hillis. 1996. Phylogenetic inference. Pages 407–514 in D. M. Hillis, C. Moritz, and B. K. Mable, editors. Molecular systematics. Sinauer Associates, Sunderland, Massachusetts, USA.

Sykes, P. W., Jr. 1989. A technique to prevent capturing birds in unattended, furled mist nets. North American Bird Bander 14:45–46.

Sykes, P. W., Jr. 2006. An efficient method of capturing painted buntings and other small granivorous passerines. North American Bird Bander 31:110–115.

Symondson, W. O. C. 2002. Molecular identification of prey in predator diets. Molecular Ecology 11:627–641.

Syrotuck, W. G. 2000. Scent and the scenting dog. Second edition. Arner, Rome, New York, USA.

Szaro, R. C., L. H. Simons, and S. C. Belfit. 1988. Comparative effectiveness of pitfalls and live-traps in measuring small mammal community structure. Pages 282–288 in R. C. Szaro, K. E. Severson, and D. R. Patton, technical coordinators. Management of amphibians, reptiles, and small mammals in North America. General Technical Report RM 166, U.S. Forest Service, Department of Agriculture, Washington, D.C., USA.

Szuba, K. J., J. F. Bendell, and B. J. Naylor. 1987. Age determination of Hudsonian spruce grouse using primary feathers. Wildlife Society Bulletin 15:539–543.

Szymanski, M. L., and A. D. Afton. 2005. Effects of spinning-wing decoys on flock behavior and hunting vulnerability of mallards in Minnesota. Wildlife Society Bulletin 33:993–1001.

Szymczak, M. R., and J. K. Ringelman. 1986. Differential habitat use of patagial-tagged female mallards. Journal of Field Ornithology 57:230–232.

Taber, C. A., R. F. Wilkinson, Jr., and M. S. Topping. 1975. Age and growth of hellbenders in the Niangua River, Missouri. Copeia 1975:633–639.

Taber, R. D. 1949. A new marker for game birds. Journal of Wildlife Management 13:228–231.

Taber, R. D. 1956. Characteristics of the pelvic girdle in relation to sex in black-tailed and white-tailed deer. California Fish and Game 42:15–21.

Taber, R. D., and I. M. Cowan. 1963. Capturing and marking wild animals. Pages 250–283 in H. S. Mosby, editor. Wildlife investigational techniques. Edwards Brothers, Ann Arbor, Michigan, USA.

Taber, R. D., and R. F. Dasmann. 1958. The black-tailed deer of the chaparral: its life history and management in the North Coast Range of California. Game Bulletin 8, California Department of Fish and Game, Sacramento, USA.

Taber, R. D., A. deVos, and M. Altmann. 1956. Two marking devices for large land mammals. Journal of Wildlife Management 20:464–465.

Taber, R. D., K. Raedeke, and D. A. McCaughran. 1982. Population characteristics. Pages 279–300 in J. W. Thomas and D. E. Toweill, editors. Elk of North America: ecology and management. Stackpole, Harrisburg, Pennsylvania, USA

Taberlet, P., J. J. Camarra, S. Griffin, E. Uhres, O. Hanotte, L. P. Waits, C. Duboispaganon, T. Burke, and J. Bouvet. 1997. Noninvasive genetic tracking of the endangered Pyrenean brown bear population. Molecular Ecology 6:869–876.

Taberlet, P., L. P. Waits, and G. Luikart. 1999. Noninvasive genetic sampling: look before you leap. Trends in Ecology and Evolution 14:323–327.

Tacha, T. C., and J. C. Lewis. 1978. Sex determination of sandhill cranes by cloacal examination. Pages 81–83 in J. C. Lewis, editor. Proceedings of the 2nd North American crane workshop. International Crane Foundation, Baraboo, Wisconsin, USA.

Tacha, T. C., S. A. Nesbitt, and P. A. Vohs. 1992. Sandhill crane. Account 31 in A. Poole, P. Stettenheim, and F. Gill, editors. The birds of North America. The Academy of Natural Sciences, Philadelphia, Pennsylvania, and The American Ornithologists' Union, Washington, D.C., USA.

Tait, D. E. N., and R. L. Bunnell. 1980. Estimating rate of increase from age at death. Journal of Wildlife Management 44:296–299.

Takekawa, J. Y., and E. O. Garton. 1984. How much is an evening grosbeak worth? Journal of Forestry 82:426–428.

Takos, M. J. 1943. Trapping and banding muskrats. Journal of Wildlife Management 7:400–407.

Tallmon, D. A., A. Koyuk, G. Luikart, and M. A. Beaumont. 2008. ONeSAMP: a program to estimate effective population size using approximate Bayesian computation. Molecular Ecology Resources 8:299–301.

Tallmon, D. A., G. Luikart, and R. S. Waples. 2004. The alluring simplicity and complex reality of genetic rescue. Trends in Ecology and Evolution 19:489–496.

Tamarin, R. H., M. Sheridan, and C. K. Levy. 1983. Determining matrilineal kinship in natural populations of rodents using radionuclides. Canadian Journal of Zoology 61:271–274.

Tamura, K. 1994. Model selection in the estimation of the number of nucleotide substitutions. Molecular Biology and Evolution 11:154–157.

Tamura, K., and M. Nei. 1993. Estimation of the number of nucleotide substitutions in the control region on mitochondrial DNA in humans and chimpanzees. Molecular Biology and Evolution 10:512–526.

Tanabe, S. 1988. PCB problems in the future: foresight from current knowledge. Environmental Pollution 50:5–28.

Tanaka, H., and T. Nakashizuka. 1997. Fifteen years of canopy dynamics analyzed by aerial photographs in a temperate deciduous forest. Japanese Ecology 78:612–620.

Tanner, G. W., and M. E. Drummond. 1985. A floating quadrat. Journal of Range Management 38:287.

Tanner, J. T. 1975. The stability and the intrinsic growth rates of prey and predator populations. Ecology 56:855–867.

Tanner, W. D., and G. L. Bowers. 1948. A method for trapping ruffed grouse. Journal of Wildlife Management 12:330–331.

Taper, M. L., and S. R. Lele, editors. 2004. The nature of scientific evidence: statistical, philosophical, and empirical considerations. University of Chicago Press, Chicago, Illinois, USA.

Tapp, W. N., J. W. Holaday, and B. H. Natelson. 1984. Ultradian glucocorticoid rhythms in monkeys and rats continue during stress. American Journal of Physiology 247:866–871.

Tarrant, B. 1977. Best way to train your gun dog. Crown, New York, New York, USA.

Taylor, A., and R. L. Knight. 2003. Wildlife responses to recreation and associated visitor perceptions. Ecological Applications 13:951–963.

Taylor, G. K., M. Bacic, R. J. Bomphrey, A. C. Carruthers, J. Gillies, S. M. Walker, and A. L. R. Thomas. 2008. New experimental approaches to the biology of flight control systems. Journal of Experimental Biology 211:258–266.

Taylor, J., and L. Deegan. 1982. A rapid method for mass marking of amphibians. Journal of Herpetology 16:172–173.

Taylor, K. D., and R. J. Quy. 1973. Marking systems for the study of rat movements. Mammal Review 3:30–34.

Taylor, M., and J. S. Carley. 1988. Life table analysis of age structured populations in seasonal environments. Journal of Wildlife Management 52:366–373.

Taylor, M., and J. Lee. 1994. Tetracycline as a biomarker for polar bears. Wildlife Society Bulletin 22:83–89

Taylor, R. H. 1969. Self-attaching collars for marking red deer in New Zealand. Deer 1:404–407.

Taylor, W. P. 1956. The deer of North America. Stackpole, Harrisburg, Pennsylvania, USA.

Technical Committee ISO/TC 211. 2003. ISO19115. International Organization for Standardization, Geneva, Switzerland.

Telfer, E. S. 1969a. Twig weight–diameter relationships for browse species. Journal of Wildlife Management 33:917–921.

Telfer, E. S. 1969b. Weight–diameter relationships for 22 woody plant species. Canadian Journal of Botany 47:1851–1855.

Templeton, A. R. 1986. Coadaptation and outbreeding depression. Pages 105–116 in M. E. Soulé, editor. Conservation biology: the science of scarcity and diversity. Sinauer Associates, Sunderland, Massachusetts, USA.

Tener, J. S. 1965. Musk-oxen in Canada: a biological and taxonomic review. Monograph 2, Canadian Wildlife Service, Ottawa, Ontario, Canada.

Terhune, T. M., D. C. Sisson, J. B. Grand, and H. L. Stribling. 2007. Factors influencing survival of radiotagged and banded northern bobwhites in Georgia. Journal of Wildlife Management 71:1288–1297.

Terio, K. A., L. Marker, and L. Munson. 2004. Evidence for chronic stress in captive but not free-ranging cheetahs (*Acinonyx jubatus*) based on adrenal morphology and function. Journal of Wildlife Diseases 40:259–266.

Tessaro, S. V. 1986. The existing and potential importance of brucellosis and tuberculosis in Canadian wildlife: a review. Canadian Veterinary Journal 27:119–124.

Tew, T. E., I. A. Todd, and D. W. MacDonald. 1994. The effects of trap spacing on population estimation of small mammals. Journal of Zoology (London) 233:340–344.

Theobald, D. M., D. L. Stevens, Jr., D. White, N. S. Urquart, and A. R. Olsen. 2007. Using GIS to generate spatially-balanced random survey designs for natural resource applications. Environmental Management 40:134–146.

Therneau, T. M., and P. M. Grambsch. 2000. Modeling survival data: extending the Cox model. Springer Science+Business Media, New York, New York, USA.

Therrien, J. E. 1996. Testing three cage traps for house sparrow capture. Sialia 18:105–109.

Theuerkauf, J., and W. Jedrzejewski. 2002. Accuracy of radiotelemetry to estimate wolf activity and locations. Journal of Wildlife Management 66:859–864.

Thiel, R. P. 1985. A snare for capturing nesting belted kingfishers. North American Bird Bander 10:2–3.

Thil, M.-A., and R. Groscolas. 2002. Field immobilization of king penguins with tiletamine-zolazepam. Journal of Field Ornithology 73:308–317.

Thirgood, S. J., S. M. Redpath, S. Campbell, and A. Smith. 2002. Do habitat characteristics influence predation on red grouse? Journal of Applied Ecology 39:217–225.

Thirgood, S. J., S. M. Redpath, P. Rothery, and N. J. Aebischer. 2000. Raptor predation and population limitation in red grouse. Journal of Animal Ecology 69:504–516.

Thomas, A., D. J. Spiegelhalter, and W. R. Gilks. 1992. BUGS: a program to perform Bayesian reference using Gibbs sampling. Pages 837–842 in J. M. Bernado, J. O. Bergo, A. P. Dawid, and A. F. M. Smith, editors. Bayesian statistics 4. Clarendon Press, Oxford, England, UK.

Thomas, A. E. 1975. Marking anurans with silver nitrate. Herpetological Review 6:12.

Thomas, D. C., and P. J. Bandy. 1973. Age determination of wild black-tailed deer from dental annulations. Journal of Wildlife Management 37:232–235.

Thomas, D. C., and P. J. Bandy. 1975. Accuracy of dental-wear age estimates of black-tailed deer. Journal of Wildlife Management 39:674–678.

Thomas, D. L., and E. J. Taylor. 1990. Study designs and tests for comparing resource use and availability. Journal of Wildlife Management 54:322–330.

Thomas, D. L., and E. J. Taylor. 2006. Study designs and tests for comparing resource use and availability II. Journal of Wildlife Management 70:324–336.

Thomas, F., and R. Poulin. 1998. Manipulation of a mollusc by a trophically transmitted parasite: convergent evolution or phylogenetic inheritance? Parasitology 116:431–436.

Thomas, J. A., L. H. Cornell, B. E. Joseph, T. D. Williams, and S. Dreischman. 1987. An implanted transponder chip used as a tag for sea otters (*Enhydra lutris*). Marine Mammal Science 3:271–274.

Thomas, J. W., and R. G. Marburger. 1964. Colored leg markers for wild turkeys. Journal of Wildlife Management 28:552–555.

Thomas, J. W., C. Maser, and J. E. Rodeik. 1979. Edges. Pages 48–59 in J. W. Thomas, editor. Wildlife habitat in managed forests—the Blue Mountains of Oregon and Washington. Agriculture Handbook 533, U. S. Forest Service, Department of Agriculture, Washington, D.C., USA.

Thomas, K. P. 1969. Sex determination of bobwhites by wing criteria. Journal of Wildlife Management 33:215–216.

Thomas, L. 1996. Monitoring long-term population change: why are there so many analysis methods? Ecology 77:49–58.

Thomas, L. 1997. Retrospective power analysis. Conservation Biology 11:276–280.

Thomas, L., S. T. Buckland, E. A. Rexstad, J. L. Laake, S. Strindberg, S. L. Hedley, J. R. B. Bishop, T. A. Marques, and K. P. Burnham. 2010. Distance software: design and analysis of distance sampling surveys for estimating population size. Journal of Applied Ecology 47:5–14.

Thomas, L., and C. J. Krebs. 1997. A review of statistical power analysis software. Bulletin of the Ecological Society of America 78:126–138.

Thomas, L., J. L. Laake, J. F. Durry, S. T. Buckland, D. L. Borchers, D. R. Anderson, K. P. Burnham, S. Stringberg, S. L. Hedley, M. L. Burt, F. Marques, J. H. Pollard, and R. M. Fewster. 1998. Program DISTANCE 3.5. Research Unit for Wildlife Population Assessment, University of Saint Andrews, Scotland, UK.

Thomas, N. E., C. Huang, S. N. Goward, S. Powell, K. Rishmawi, K. Schleeweis, and A. Hinds. 2011. Validation of North American forest disturbance dynamics derived from Landsat time series stacks. Remote Sensing of Environment 115:19–32.

Thomas, N. J., D. B. Hunter, and C. T. Atkinson, editors. 2007. Infectious diseases of wild birds. Blackwell, Ames, Iowa, USA.

Thomas, R. A. 1977. Selected bibliography of certain vertebrate techniques. Technical Note 306, U.S. Bureau of Land Management, Department of the Interior, Denver, Colorado, USA.

Thomas, R. B., I. M. Nall, and W. J. House. 2008. Relative efficacy of three different baits for trapping pond-dwelling turtles in east-central Kansas. Herpetological Review 39:186–188.

Thomas, R. B., and B. Novak. 1991. Helicopter drive-netting techniques for mule deer capture on Great Basin ranges. California Fish and Game 77:194–200.

Thome, D. M., and T. M. Thome. 2000. Radio-controlled model airplanes: inexpensive tools for low-level aerial photography. Wildlife Society Bulletin 28:343–346.

Thompson, D. R. 1958. Field techniques for sexing and aging game animals. Special Wildlife Report 1, Wisconsin Conservation Department, Madison, USA.

Thompson, H. V., and C. J. Armour. 1954. Methods of marking wild rabbits. Journal of Wildlife Management 18:411–414.

Thompson, J. R., V. C. Bleich, S. G. Torres, and G. P. Mulcahy. 2001. Translocation techniques for mountain sheep: does the method matter? Southwestern Naturalist 46:87–93.

Thompson, M. C., and R. DeLong. 1967. The use of cannon and rocket-projected nets for trapping shorebirds. Bird-Banding 38:214–218.

Thompson, M. J., R. E. Henderson, T. O. Lemke, and B. A. Sterling. 1989. Evaluation of a collapsible Clover trap for elk. Wildlife Society Bulletin 17:287–290.

Thompson, R. L. 1962. An investigation of some techniques for measuring availability of oak mast and deer browse. Thesis, Virginia Polytechnic Institute and State University, Blacksburg, USA.

Thompson, R. L., and B. S. McGinnes. 1963. A comparison of eight types of mast traps. Journal of Forestry 61:679–680.

Thompson, S. D. 1982. Microhabitat utilization and foraging behavior of bipedal and quadrupedal heteromyid rodents. Ecology 63:1303–1312.

Thompson, S. K. 2002a. Sampling. Second edition. John Wiley and Sons, New York, New York, USA.

Thompson, S. K. 2003. Editorial: special issue on adaptive sampling. Environmental and Ecological Statistics 10:5–6.

Thompson, S. K., and F. L. Ramsey. 1983. Adaptive sampling of populations. Technical Report 82, Department of Statistics, Oregon State University, Corvallis, USA.

Thompson, S. K., and G. A. F. Seber. 1996. Adaptive sampling. John Wiley and Sons, New York, New York, USA.

Thompson, W. L. 2002b. Towards reliable bird surveys: accounting for individuals present but not detected. Auk 119:18–25.

Thompson, W. L., G. C. White, and C. Gowan. 1998. Monitoring vertebrate populations. Academic Press, San Diego, California, USA.

Thomson, D. L., E. G. Cooch, and M. J. Conroy, editors. 2009. Modeling demographic processes in marked populations, environmental and ecological statistics. Volume 3. Springer Science+Business Media, New York, New York, USA.

Thorne, E. T. 2001. Brucellosis. Pages 372–395 in E. S. Williams and I. K. Barker, editors. Infectious disease of wild mammals. Iowa State University Press, Ames, USA.

Thorne, E. T., N. Kingston, W. R. Jolley, and R. C. Bergstrom, editors. 1982. Diseases of wildlife in Wyoming. Second edition. Wyoming Game and Fish Department, Cheyenne, USA.

Thorne, E. T., S. G. Smith, K. Aune, D. Hunter, and T. J. Roffe. 1997. Brucellosis: the disease in elk. Pages 33–44 in E. T. Thorne, M. S. Boyce, P. Nicoletti, and T. J. Kreeger, editors. Brucellosis, bison, elk, and cattle in the Greater Yellowstone area: defining the problem, exploring solutions. Wyoming Game and Fish Department for the Greater Yellowstone Interagency Brucellosis Committee, Cheyenne, USA.

Thorstrom, R. K. 1996. Methods for capturing tropical forest birds of prey. Wildlife Society Bulletin 24:516–520.

Threloff, D. L. 1994. Using a Threloff global positioning system (GPS) to map the distribution of the Cottonball Marsh pupfish. Pages 19–20 in D. A. Hendrickson, editor. Proceedings of the Desert Fishes Council, 1993 annual symposium, Monterrey, Nuevo León, Mexico.

Tidemann, C. R., and R. A. Loughland. 1993. A harp trap for large megachiropterans. Wildlife Research 20:607–611.

Tiemeier, O. W., and M. L. Plenert. 1964. A comparison of three methods for determining the age of black-tailed jackrabbits. Journal of Mammalogy 45:409–416.

Tilley, S. G. 1977. Studies of life histories and reproduction in North American plethodontid salamanders. Pages 1–41 in D. H. Taylor and S. I. Guttman, editors. The reproductive biology of amphibians. Plenum Press, New York, New York, USA.

Tilley, S. G. 1980. Life histories and comparative demography of two salamander populations. Copeia 1980:806–821.

Tilman, D., P. B. Reich, and J. M. H. Knops. 2006. Biodiversity and ecosystem stability in a decade-long grassland experiment. Nature 441:629–632.

Tilton, D. A., J. G. Teer, and N. J. Silvy. 1987. Accuracy of line transect estimation procedures for white-tailed deer densities. Proceedings of the Annual Conference of the Southeastern Association of Fish and Wildlife Agencies 41:424–431.

Timm, D. E., and R. G. Bromley. 1976. Driving Canada geese by helicopter. Wildlife Society Bulletin 4:180–181.

Tinbergen, N. 1963. On aims and methods of ethology. Zeitschrift für tierpsychologie 20:410–433.

Tinkle, D. W. 1967. The life and demography of the side-blotched lizard, Uta stansburiana. Museum Zoological Publication 132, University of Michigan, Ann Arbor, USA.

Tinkle, D. W. 1973. A population analysis of the sagebrush lizard, Sceloporus graciosus in southern Utah. Copeia 1973:284–296.

Tobalske, B. W., T. L. Hedrick, A. A. Biewener, D. R. Warrick, and D. R. Powers. 2009. Effects of flight speed upon muscle activity and metabolic rate in hummingbirds. Abstracts of the annual main meeting of the Society of Experimental Biology, 28 June–1 July, Glasgow, England, UK. Comparative Biochemistry and Physiology A: Molecular and Integrative Physiology 153:S122.

Tobler, M. W. 2009. New GPS technology improves fix success for large mammal collars in dense tropical forests. Journal of Tropical Ecology 25:217–221.

Toepfer, J. E., J. A. Newell, and J. Monarch. 1988. A method for trapping prairie grouse hens on display grounds. Pages 21–31 in A. J. Bjugstad, editor. Prairie chickens on the Sheyenne National Grasslands. General Technical Report RM-159, U.S. Forest Service, Department of Agriculture, Washington, D.C., USA.

Toft, C. A., and P. J. Shea. 1983. Detecting community-wide patterns: estimating power strengthens statistical inference. American Naturalist 122:618–625.

Toland, B. 1985. A trapping technique for trap-wary American kestrels. North American Bird Bander 10:11.

Tolle, D. A., and T. A. Bookhout. 1974. A comparison of two methods for capturing roosting wood ducks. Wildlife Society Bulletin 2:50–55.

Tollit, D. J., A. D. Schulze, A. W. Trites, P. F. Olesiuk, S. J. Crockford, T. S. Gelatt, R. R. Ream, and K. M. Miller. 2009. Development and application of DNA techniques for validating and improving pinniped diet estimates. Ecological Applications 19:889–905.

Toman, T. L., T. Lemke, L. Kuck, B. L. Smith, S. G. Smith, and K. Aune. 1997. Elk in the Greater Yellowstone Area: status and management. Pages 56–64 in E. T. Thorne, M. S. Boyce, P. Nicoletti, and T. J. Kreeger, editors. Brucellosis, bison, elk, and cattle in the Greater Yellowstone area: defining the problem, exploring solutions. Wyoming Game and Fish Department for the Greater Yellowstone Interagency Brucellosis Committee, Cheyenne, USA.

Tomilin, A. G., Y. I. Bliznyuk, and A. V. Zanin. 1983. A new method for marking small cetaceans. International Whaling Commission Report 33:643–645.

Tomlinson, R. E. 1963. A method for drive-trapping dusky grouse. Journal of Wildlife Management 27:563–566.

Tomlinson, R. E. 1968. Reward banding to determine reporting rate of recovered mourning dove bands. Journal of Wildlife Management 32:6–11.

Tompkins, D. M., A. P. Dobson, P. Arneberg, M. E. Begon, I. M. Cattadori, J. V. Greenman, J. A. P. Heesterbeek, P. J. Hudson, D. Newborn, A. Pugliese, A. P. Rizzoli, R. Rosà, F. Rosso, and K. Wilson. 2002. Parasites and host population dynamics. Pages 45–62 in P. J. Hudson, A. Rizzoli, B. T. Grenfell, H. Heesterbeek, and A. P. Dobson, editors. The ecology of wildlife diseases. Oxford University Press, Oxford, England, UK.

Torres, S. M., and C. D. Curtis. 2007. Initial implementation of super-resolution data on the NEXRAD network. 23rd AMS conference on Interactive Information and Processing Systems (IIPS) for meteorology, oceanography, and hydrology. American Meteorological Society, San Antonio, Texas, USA.

Toweill, D. E., and J. E. Tabor. 1982. River otter. Pages 688–703 in J. A. Chapman and G. A. Feldhamer, editors. Wild mammals of North America. Johns Hopkins University Press, Baltimore, Maryland, USA.

Towers, J. 1988. Age determination of juvenile spruce grouse in eastern Canada. Journal of Wildlife Management 52:113–115.

Travaini, A., R. M. Peck, and S. C. Zapata. 2001. Selection of odor attractants and meat delivery methods to control Culpeo foxes (Pseudalopex culpaeus) in Patagonia. Wildlife Society Bulletin 29:1089–1096.

Travis, J. 1981. The effect of staining on the growth of Hyla gratiosa tadpoles. Copeia 1981:193–196.

Trebicki, P., R. M. Harding, B. Rodoni, G. Baxter, and K. S. Powell. 2010. Seasonal activity and abundance of Orosius orientalis (Hemiptera: Cicadellidae) at agricultural sites in southeastern Australia. Journal of Applied Entomology 134:91–97.

Trent, T. T., and O. J. Rongstad. 1974. Home range and survival of cottontail rabbits in southwestern Wisconsin. Journal of Wildlife Management 38:459–472.

Triant, D. A., R. M. Pace, and M. Stine. 2004. Abundance, genetic diversity and conservation of Louisiana black bears (Ursus americanus luteolus) as detected through noninvasive sampling. Conservation Genetics 5:647–659.

Triplehorn, C. A., and N. F. Johnson. 2005. Borror and Delong's introduction to the study of insects. Seventh edition. Thomson Brooks and Cole, Belmont, California, USA.

Trippensee, R. E. 1941. A new type of bird and mammal marker. Journal of Wildlife Management 5:120–124.

Trippensee, R. E. 1948. Wildlife management: upland game and general principles. McGraw-Hill, New York, New York, USA.

Trivers, R. L., and D. E. Willard. 1973. Natural selection of parental ability to vary the sex ratio of offspring. Science 179:90–92.

Trousdell, K. B. 1954. Peak population of seed-eating rodents and shrews occurs 1 year after loblolly stands are cut. Research Note 68, U.S. Forest Service, Department of Agriculture, Washington, D.C., USA.

Troyer, W. A., R. J. Hensel, and K. E. Durley. 1962.

Live-trapping and handling of brown bears. Journal of Wildlife Management 26:330–331.

Trump, R. E., and G. O. Hendrickson. 1943. Methods for trapping and tagging woodchuck. Journal of Wildlife Management 7:420–421.

Tryon, E. H., and K. L. Carvell. 1962. Acorn production and damage. West Virginia University Agricultural Experiment Station Bulletin 466-T, West Virginia University Agricultural Experiment Station, Morgantown, USA.

Tsai, K., K. H. Pollock, and C. Brownie. 1999. Effects of violation of assumptions for survival analysis methods in radiotelemetry studies. Journal of Wildlife Management 63:1369–1375.

Tucker, J. K. 1994. An "easy" method to remove common snapping turtles (*Chelydra serpentina*) from Legler hoop traps. Herpetological Review 25:13.

Tufte, R. 1983. The visual display of quantitative information. Graphics Press, Cheshire, Connecticut, USA.

Tufte, R. 2001. The visual display of quantitative information. Second edition. Graphics Press, Cheshire, Connecticut, USA.

Tukey, J. 1958. Bias and confidence in not quite large samples. Annals of Mathematical Statistics 29:614.

Tukey, J. W. 1977. Exploratory data analysis. Addison-Wesley, Reading, Massachusetts, USA.

Tullar, B. F., Jr. 1984. Evaluation of a padded leg-hold trap for capturing foxes and raccoons. New York Fish and Game Journal 31:97–103.

Tumlison, R., and V. R. McDaniel. 1984. Gray fox age classification by canine tooth pulp cavity radiographs. Journal of Wildlife Management 48:228–230.

Turcek, F. J. 1951. Effect of introductions on two game populations in Czechoslovakia. Journal of Wildlife Management 15:113–114.

Turchin, P. 1998. Quantitative analysis of movement: measuring and modeling population redistribution in animals and plants. Sinauer Associates, Sunderland, Massachusetts, USA.

Turchin, P. 2001. Does population ecology have general laws? Oikos 94:17–26.

Turkowski, F. J., A. R. Armistead, and S. B. Linhart. 1984. Selectivity and effectiveness of pan tension devices for coyote foot-hold traps. Journal of Wildlife Management 48:700–708.

Turner, C. D., and J. T. Bagnara. 1976. General endocrinology. Sixth edition. W. B. Saunders, Philadelphia, Pennsylvania, USA.

Turner, D. L., and W. P. Clary. 2001. Sequential sampling protocol for monitoring pasture utilization using stubble height criteria. Journal of Range Management 54:132–137.

Turner, F. B. 1960. Population structure and dynamics of the western spotted frog, *Rana p. pretiosa* Baird and Girard, in Yellowstone Park, Wyoming. Ecological Monographs 30:251–278.

Turner, J. C. 1977. Cemental annulations as an age criterion in North American sheep. Journal of Wildlife Management 41:211–217.

Turner, J. C. 1982. A modified Cap-Chur dart and dye evaluation for marking desert sheep. Journal of Wildlife Management 46:553–557.

Turner, L. B. 1956. Improved technique in goose trapping with cannon-type net traps. Journal of Wildlife Management 20:201–203.

Turner, M. G., G. J. Arthaud, R. T. Engstrom, S. J. Hejl, J. Liu, S. Loeb, and K. McKelvey. 1995. Usefulness of spatially explicit population models in land management. Ecological Applications 5:12–16.

Turner, M. G., R. H. Gardner, and R. V. O'Neill. 2001. Landscape ecology in theory and practice: pattern and process. Springer-Verlag, New York, New York, USA.

Turner, W., S. Spector, N. Gardiner, M. Fladeland, E. Sterling, and M. Steininger. 2003. Remote sensing for biodiversity science and conservation. Trends in Ecology and Evolution 18:306–314.

Twitty, V. C. 1966. Of scientists and salamanders. W. H. Freeman and Company, San Francisco, California, USA.

Tyack, P. L. 2008. Implications for marine mammals of large-scale changes in the marine acoustic environment. Journal of Mammalogy 89:549–558.

Tyndale-Biscoe, C. H. 1953. A method of marking rabbits for field studies. Journal of Wildlife Management 17:42–45.

Tyndale-Biscoe, C. H., and R. B. Mackenzie. 1976. Reproduction in *Didelphis marsupialis* and *D. albiventris* in Columbia. Journal of Mammalogy 57:249–265.

Tyndale-Biscoe, C. H., and M. B. Renfree. 1987. Reproductive physiology of marsupials. Cambridge University Press, New York, New York, USA.

U.S. Department of Agriculture. 2002. Transportation, sale, and handling of certain animals (Animal Welfare Act). Chapter 54, 7 United States Code 2131-59. http://www.aphis.usda.gov/ac/awa/html.

U.S. Department of Transportation. 2009. National response center statistics: incident by type per year. U.S. Department of Transportation and U.S. Coast Guard, National Response Center, Washington, D.C., USA. http://www.nrc.usge.mil/incident.htm.

U.S. Environmental Protection Agency. 1998*a*. Emergency planning and community right to know. Section 313. Toxic release inventory reporting. Notice of receipt of petition. Federal Register 63:6691–6698.

U.S. Environmental Protection Agency. 1998*b*. U.S. high production volume chemical hazard data availability study. U.S. Environmental Protection Agency, Washington, D.C., USA.

U.S. Fish and Wildlife Service. 1973. Endangered Species Act of 1973 as amended through the 108th Congress. U.S. Fish and Wildlife Service, Department of the Interior, Washington, D.C., USA.

U.S. Fish and Wildlife Service and Canadian Wildlife Service. 1977. North American bird banding manual. Volume II. U.S. Fish and Wildlife Service, Department of the Interior, Washington, D.C., USA.

U.S. Fish and Wildlife Service and Canadian Wildlife Service. 1987. Standard operating procedures for aerial waterfowl breeding ground population and habitat surveys in North America. Office of Migratory Bird Management, Laurel, Maryland, USA.

U.S. Forest Service. 2004. Land and resource management plan, Chattahoochee–Oconee National Forests. Management Bulletin R8-MB 113 A, U.S. Forest Service, Southern Region, Department of Agriculture, Atlanta, Georgia, USA.

U.S. Geological Survey. 2009*a*. Experimental advanced airborne research LiDAR (EAARL). http://coastal.er.usgs.gov/remote-sensing/advancedmethods/eaarl.html.

U.S. Geological Survey. 2009*b*. USGS Center for LiDAR Information Coordination and Knowledge. http://lidar.cr.usgs.gov/.

U.S. Geological Survey–Bird Banding Laboratory [USGS-BBL]. 1999. Memorandum to all banders. http://www.pwrc.usgs.gov/bBL/mtab/mtab83.htm.

Uhlig, H. G. 1953. Weights of ruffed grouse in West Virginia. Journal of Wildlife Management 17:391–392.

Uhlig, H. G. 1955. The determination of age of nestling and sub-adult gray squirrels in West Virginia. Journal of Wildlife Management 19:479–483.

Umetsu, F., L. Naxara, and R. Pardini. 2006. Evaluating the efficiency of pitfall traps for sampling small mammals in the neotropics. Journal of Mammalogy 87:757–765.

Underhill, L., and J. Hofmeyer. 1987. Experience with colour-dyed common terns. Safring News 16:29–30.

Underwood, A. J. 1994. On beyond BACI: sampling designs that might reliably detect environmental disturbances. Ecological Applications 4:3–15.

Underwood, A. J. 1997. Experiments in ecology: their logical design and interpretation using analysis of variance. Cambridge University Press, Cambridge, England, UK.

Unsworth, J. W., L. Kuck, and E. O. Garton. 1990. Elk sightability model validation at the National Bison Range, Montana. Wildlife Society Bulletin 18:113–115.

Uresk, D. W., R. O. Gilbert, and W. H. Rickard. 1977. Sampling big sagebrush for phytomass. Journal of Range Management 30:311–314.

Uresk, D. W., and T. M. Juntti. 2008. Monitoring Idaho fescue grasslands in the Big Horn Mountains, Wyoming, with a modified Robel pole. Western North American Naturalist 68:1–7.

Usher, M. B. 1972. Developments in the Leslie matrix model. Pages 29–60 *in* J. N. R. Jeffers, editor. Mathematical models in ecology. Blackwell Scientific, Oxford, England, UK.

Vahdat, F., B. E. Seguin, H. L. Whitmore, and S. D. Johnston. 1984. Role of blood cells in degradation of progesterone in bovine blood. American Journal of Veterinary Research 45:240–243.

Valkenburg, P., R. D. Boertje, and J. L. Davis. 1983. Effects of darting and netting on caribou in Alaska. Journal of Wildlife Management 47:1233–1237.

Van Ballenberghe, V. 1983. Rate of increase of white-tailed deer on the George Reserve: a re-evaluation. Journal of Wildlife Management 47:1245–1247.

Van Ballenberghe, V. 1984. Injuries to wolves sustained during live-capture. Journal of Wildlife Management 48:1425–1429.

Van Ballenberghe, V., and L. D. Mech. 1975. Weights, growth, and survival of timber wolf pups in Minnesota. Journal of Mammalogy 56:44–63.

van Belle, G. 2008. Statistical rules of thumb. John Wiley and Sons, Hoboken, New Jersey, USA.

Van Brackle, M. D., S. B. Linhart, T. E. Creekmore, V. F. Nettles, and R. L. Marchinton. 1994. Oral biomarking of white-tailed deer with tetracycline. Wildlife Society Bulletin 22:483–488.

Van Deelen, T. R., K. M. Hollis, C. Anchor, and D. R. Etter. 2000. Sex affects age determination and wear of molariform teeth in white-tailed deer. Journal of Wildlife Management 64:1076–1083.

Van der Meer, F. D., and S. M. de Jong. 2001. Imaging spectrometry: basic principles and prospective applications. Kluwer Academic, The Netherlands.

Van Dersal, W. R. 1940. Utilization of oaks by birds and mammals. Journal of Wildlife Management 4:404–428.

Van Dyke, F., and J. A. Darragh. 2007. Response of elk to changes in plant production and nutrition following prescribed burning. Journal of Wildlife Management 71:23–29.

van Genderen, J. L., B. F. Lock, and P. A. Vass. 1978. Remote sensing: statistical testing of thematic map accuracy. Remote Sensing of Environment 7:3–14.

Van Groenendael, J., H. De Kroon, and H. Caswell. 1988. Projection matrices in population biology. Trends in Ecology and Evolution 3:264–269.

Van Horn, R. C., and R. J. Douglass. 2000. Disinfectant effects on capture rates of deer mice (*Peromyscus maniculatus*). American Midland Naturalist 143:257–260.

Van Horne, B. 1983. Density as a misleading indicator of habitat quality. Journal of Wildlife Management 47:893–901.

Van Nostrand, F. C., and A. B. Stephenson. 1964. Age determination for beavers by tooth development. Journal of Wildlife Management 28:430–434.

Van Oort, B. E. H., N. J. C. Tyler, P. V. Storeheier, and K. A. Stokkan. 2004. The performance and validation of a data logger for long-term determination of activity in free-ranging reindeer, *Rangifer tarandus* L. Applied Animal Behaviour Science 89:299–308.

Van Paassen, A. G., D. H. Veldman, and A. J. Beintema. 1984. A simple device for determination of incubation stages in eggs. Wildfowl 35:173–178.

Van Rossem, A. J. 1925. Flight feathers as indicators of age in *Dendragapus*. Ibis (Series 12) 1:417–422.

Van Tienhoven, A. 1983. Reproductive physiology of vertebrates. Second edition. Cornell University Press, Ithaca, New York, USA.

Van Winkle, W. 1975. Comparison of several probabilistic home-range models. Journal of Wildlife Management 39:118–123.

Vandenbergh, J. G. 2006. Pheromones and mammalian reproduction. Pages 2041–2058 in J. D. Neill, editor. Knobil and Neill's physiology of reproduction. Third edition, Elsevier Science, Amsterdam, The Netherlands.

Vandermeer, J. H. 1972. Niche theory. Annual Review of Ecology and Systematics 3:107–132.

Vangilder, L. D. 1983. Reproductive effects of toxic substances on wildlife: an evolutionary view. Pages 250–259 in Czechoslovak–American symposium on toxic effects and reproductive ability in free-living animals. Springer-Verlag, Strblke Pleso, Czechoslovakia.

Vangilder, L. D., and T. J. Peterle. 1981. South Louisiana crude oil or DDE in the diet of mallard hens: effects on egg quality. Bulletin of Environmental Contamination and Toxicology 26:328–336.

Vantassel, F. 2006. The Bailey beaver trap: modifications and sets to improve capture rate. Proceedings of the Vertebrate Pest Conference 22:171–173.

Vargas, G. A., K. L. Krakauer, J. L. Egremy-Hernandez, and M. J. McCoid. 2000. Sticky trapping and lizard survivorship. Herpetological Review 31:23.

Varley, N., and M. S. Boyce. 2006. Adaptive management for reintroductions: updating a wolf recovery model for Yellowstone National Park. Ecological Modelling 193:315–339.

Vaughan, M. R., and L. B. Keith. 1981. Demographic response of experimental snowshoe hare populations to overwinter shortage. Journal of Wildlife Management 45:354–380.

Vaughan, T. A., J. M. Ryan, and N. J. Czaplewski. 2000. Mammalogy. Fourth edition. Saunders College, New York, New York, USA.

Vaughn, C. R. 1974. Intraspecific wingbeat rate variability and species identification using tracking radar. Pages 443–476 in Proceedings of a conference on the biological aspects of the bird/aircraft collision problem. Department of Zoology, Clemson University, Clemson, South Carolina, USA.

Vaughn, C. R. 1985. Birds and insects as radar targets: a review. Proceedings of the IEEE 73:205–227.

Venables, W. N. 1998. Exegeses on linear models. S-Plus user's conference, 8–9 October, 1998. Washington, D.C., USA.

Venables, W. N., and B. D. Ripley. 2002. Modern applied statistics with S. Fourth edition. Springer-Verlag, New York, New York, USA.

VerCauteren, K. C., J. Beringer, and S. E. Hygnstrom. 1999. Use of netted cage traps for capturing white-tailed deer. Pages 155–164 in G. Proulx, editor. Mammal trapping. Alpha Wildlife Research and Management, Sherwood Park, Alberta, Canada.

VerCauteren, K., P. Burke, G. Phillips, J. Fischer, N. Seward, B. Wunder, and M. Lavelle. 2007a. Elk use of wallows and potential chronic wasting disease transmission. Journal of Wildlife Diseases 43:784–788.

VerCauteren, K. C., M. J. Lavelle, N. W. Seward, J. W. Fischer, and G. E. Phillips. 2007b. Fenceline contact between wild and farmed white-tailed deer in Michigan: potential for disease transmission. Journal of Wildlife Management 71:1603–1606.

VerCauteren, K. C., M. J. Pipas, and J. Bourassa. 2002. A camera and hook system for viewing and retrieving rodent carcasses from burrows. Wildlife Society Bulletin 30:1057–1061.

Verme, L. J. 1962. An automatic tagging device for deer. Journal of Wildlife Management 26:387–392.

Verme, L. J. 1968. An index of winter weather severity for northern deer. Journal of Wildlife Management 32:566–574.

Vermeire, L. T., and R. L. Gillen. 2001. Estimating herbage standing crop with visual obstruction in tallgrass prairie. Journal of Range Management 54:57–60.

Verner, J. 1985. Assessment of counting techniques. Current Ornithology 2:247–302.

Verner, J., D. Breese, and K. L. Purcell. 2000. Return rates of banded granivores in relation to band color and number of bands worn. Journal of Field Ornithology 71:117–125.

Verner, J., and K. A. Milne. 1990. Analyst and observer variability in density estimates from spot mapping. Condor 92:313–325.

Vernes, K. 1993. A drive fence for capturing small forest-dwelling macropods. Wildlife Research 20:189–191.

Verts, B. J. 1967. The biology of the striped skunk. University of Illinois Press, Urbana, USA.

Vieira, E. M. 1998. A technique for trapping small mammals in the forest canopy. Mammalia 62:306–310.

Vierling, K. T., L. A. Vierling, W. A. Gould, S. Martinuzzi, and R. M. Clawges. 2008. Lidar: shedding new light on habitat characterization and modeling. Frontiers in Ecology and the Environment 6:90–98.

Viljoen, P. J. 1986. A plastic tail collar for marking wild elephants. South African Journal of Wildlife Research 16:158–159.

Vinegar, M. B. 1975. Life history phenomena in two populations of the lizard *Sceloporous undulatus* in southwestern New Mexico. American Midland Naturalist 93:388–402.

Virchow, D. R., and D. Hogeland. 1994. Bobcat. Pages C35–C43 in S. E. Hynstrom, R. M. Timm, and G. E. Larson, editors. Prevention and control of wildlife damage. Cooperative Extension Service, University of Nebraska, Lincoln, USA.

Vitt, L. J. 1992. Diversity of reproductive strategies among Brazilian lizards and snakes: the significance of lineage and adaptation. Pages 135–149 in W. C. Hammett, editor. Reproduction in South American vertebrates. Springer-Verlag, New York, New York, USA.

Vitullo, J., S. Wang, A. Zhang, C. Mannion, and J. G. Bergh. 2007. Comparison of sex pheromone traps for monitoring pink hibiscus mealybug (Hemiptera: Pseudococcidae). Journal of Economic Entomology 100:405–410.

Vogelmann, J. E., B. Tolk, and Z. Zhu. 2009. Monitoring forest changes in the southwestern United States using multitemporal Landsat data. Remote Sensing of Environment 113: 1739–1748.

Vogler, A. P., and R. DeSalle. 1994. Diagnosing units of conservation management. Conservation Biology 8:354–363.

Vogt, R. C. 1980. New methods for trapping aquatic turtles. Copeia 1980:368–371.

Vogt, R. C., and R. L. Hine. 1982. Evaluation of techniques for assessment of amphibian and reptile populations in Wisconsin. Pages 201–217 in N. J. Scott, Jr., editor. Herpetological communities. Symposium of the Society for the Study of Amphibians and Reptiles. Wildlife Research Report 13, U.S. Fish and Wildlife Service, Department of the Interior, Washington, D.C., USA.

Voight, D. R., and J. S. Lotimer. 1981. Radio tracking terrestrial furbearers: system design, procedures, and data collection. Pages 1151–1188 in J. A. Chapman and D. Pursley, editors. Worldwide furbearer conference. General Technical Report PSW-GTR-157, U.S. Forest Service, Department of the Interior, Washington, D.C., USA.

Voigt, D. R. 1987. Red fox. Pages 379–392 in M. Novak, J. A. Baker, M. E. Obbard, and B. Malloch, editors. Wild furbearers management and conservation in North America. Ontario Ministry of Natural Resources, Toronto, Canada.

Voigt, D. R., and W. E. Berg. 1987. Coyote. Pages 344–357 in M. Novak, J. A. Baker, M. E. Obbard, and B. Malloch, editors. Wild furbearer management and conservation in North America. Ontario Ministry of Natural Resources, Toronto, Canada.

Volesky, J. D., W. D. Schacht, and P. E. Reece. 1999. Leaf area, visual obstruction, and standing crop relationships on Sandhills rangeland. Journal of Range Management 52:494–499.

Volterra, V. 1926. Fluctuations in the abundance of a species considered mathematically. Nature 118:558–560.

von der Ohe, C. G., and C. Servheen. 2002. Measuring stress in mammals using fecal glucocorticoids: opportunities and challenges. Wildlife Society Bulletin 30:1215–1225.

von Uexkull, J. 1921. Umwelt und Innenwelt der Tiere. Springer-Verlag, Berlin, Germany.

Vonholdt, B. M., D. R. Stahler, D. W. Smith, D. A. Earl, J. P. Pollinger, and R. K. Wayne. 2008. The genealogy and genetic viability of reintroduced Yellowstone grey wolves. Molecular Ecology 17:252–274.

Voorspoels, S., A. Covaci, V. L. Jaspers, H. Neels, and P. Schepens. 2007. Biomagnification of PBDEs in three small terrestrial food chains. Environmental Science and Technology 41: 411–416.

Voorspoels, S., A. Covaci, P. Lepom, V. L. Jaspers, and P. Schepens. 2006. Levels and distribution of polybrominated diphenyl ethers in various tissues of birds of prey. Environmental Pollution 144:218–227.

Vreeland, J. K., D. R. Diefenbach, and B. D. Wallingford. 2004. Survival, mortality causes, and habitats of Pennsylvania white-tailed deer fawns. Wildlife Society Bulletin 32:542–553.

Vrijenhoek, R. C., and P. L. Leberg. 1991. Let's not throw the baby out with the bathwater: a comment on management for MHC diversity in captive populations. Conservation Biology 5:252–254.

Vucetich, J. A., T. A. Waite, and L. Nunney. 1997. Fluctuating population size and the ratio of effective to census population size. Evolution 51:2017–2021.

Vukovich, M., and J. C. Kilgo. 2009. Effects of radio transmitters on the behavior of red-headed woodpeckers. Journal of Field Ornithology 80:308–313.

Vyas, N. B. 1999. Factors influencing estimation of pesticide-related wildlife mortality. Toxicology and Industrial Health 15:186–191.

Vyssotski, A. L., G. Dell'Omo, G. Dell'Ariccia, A. N. Abramchuk, A. N. Serkov, A. V. Latanov, A. Loizzo, D. P. Wolfer, and H. P. Lipp. 2009. EEG responses to visual landmarks in flying pigeons. Current Biology 19:1159–1166.

Waddell, R. B., D. A. Osborn, R. J. Warren, J. C. Griffin, and D. J. Kesler. 2001. Prostaglandin $F2_\alpha$-mediated fertility control in captive white-tailed deer. Wildlife Society Bulletin 29:1067–1074.

Waddle, T., K. Bovee, and Z. Bowen. 1997. Two-dimensional habitat modeling in the Yellowstone/Upper Missouri River System. North American Lake Management Society Meeting, Houston, Texas, USA. http://smig.usgs.gov/SMIG/features_0398/habitat.html.

Wade, J. 2005. Current research on the behavioral neuroendocrinology of reptiles. Hormones and Behavior 48:451–460.

Wade, M. R., B. C. G. Scholz, R. J. Lloyd, A. J. Cleary, B. A. Franzmann, and M. P. Zalucki. 2006. Temporal variation in arthropod sampling effectiveness: the case for using the beat sheet method in cotton. Entomologia Experimentalis et Applicata 120:139–153.

Wade, P. R. 2000. Bayesian methods in conservation biology. Conservation Biology 14:1308–1316.

Wadkins, L. A. 1948. Dyeing birds for identification. Journal of Wildlife Management 12:388–391.

Wairimu, S. 1997. Spatial analysis of white-tailed deer wintering habitat in central New York. Dissertation, Cornell University, Ithaca, New York, USA.

Waits, J. L., and P. L. Leberg. 2000. Biases associated with population estimation using molecular tagging. Animal Conservation 3:191–199.

Waits, L. P., G. Luikart, and P. Taberlet. 2001. Estimating the probability of identity among genotypes in natural populations: cautions and guidelines. Molecular Ecology 10:249–256.

Waits, L. P., and D. Paetkau. 2005. Noninvasive genetic sampling tools for wildlife biologists: a review of applications and recommendations for accurate data collection. Journal of Wildlife Management 69:1419–1433.

Waits, L. P., P. Taberlet, J. E. Swenson, F. Sandegren, and R. Franzén. 2000. Nuclear DNA microsatellite analysis of genetic diversity and gene flow in the Scandinavian brown bear (*Ursus arctos*). Molecular Ecology 9:421–431.

Wakeling, B. F., F. E. Phillips, and R. Engel-Wilson. 1997. Age and gender differences in Merriam's turkey tarsometatarsus measurements. Wildlife Society Bulletin 25:706–708.

Wakkinen, W. L., K. P. Reese, J. W. Connelly, and R. A. Fischer. 1992. An improved spotlighting technique for capturing sage grouse. Wildlife Society Bulletin 20:425–426.

Walcott, C., J. L. Gould, and J. L. Kirschvink. 1979. Pigeons have magnets. Science 205:1027–1029.

Walczak, J. T. 1991. A technique for the safe restraint of venomous snakes. Herpetological Review 22:17–18.

Wald, A. 2004. Sequential analysis. Dover, Mineola, New York, USA.

Waldien, D. L., M. M. Cooley, J. Weikel, J. P. Hayes, C. C. Maguire, T. Manning, and T. J. Maier. 2004. Incidental captures of birds in small-mammal traps: a cautionary note for interdisciplinary studies. Wildlife Society Bulletin 32:1260–1268.

Waldien, D. L., and J. P. Hayes. 1999. A technique to capture bats using hand-held mist nets. Wildlife Society Bulletin 27:197–200.

Waldien, D. L., J. P. Hayes, and M. M. P. Huso. 2006. Use of downed wood by Townsend's chipmunks (*Tamias townsendii*) in western Oregon. Journal of Mammalogy 87:454–460.

Walker, F. M., A. C. Taylor, and P. Sunnucks. 2008. Female dispersal and male kinship-based association in southern hairy-nosed wombats (*Lasiorhinus latifrons*). Molecular Ecology 17:1361–1374.

Walker, L. A., L. Cornell, K. Dahl, N. Czekela, C. Dargen, B. Joseph, A. Hsueh, and B. Lasley. 1988. Urinary concentrations of ovarian steroid hormone metabolites and bioactive follicle-stimulating hormone in killer whales (*Orcinus orchus*) during ovarian cycles and pregnancy. Biology of Reproduction 39:1013–1020.

Walkinshaw, L. H. 1949. The sandhill cranes. Cranbrook Institute for Science, Bloomfield Hills, Michigan, USA.

Walkinshaw, L. H. 1973. Cranes of the world. Winchester Press, New York, New York, USA.

Wallace, M. P., P. G. Parker, and S. A. Temple. 1980. An evaluation of patagial markers for cathartid vultures. Journal of Field Ornithology 51:309–314.

Wallin, D. O., and A. G. Wells. 2006. Habitat analysis of mountain goats in the Washington Cascades. Northwestern Naturalist 87: 191.

Wallin, J. A. 1982. Sex determination of Vermont fall-harvested juvenile wild turkeys by the 10th primary. Wildlife Society Bulletin 10:40–43.

Wallis, D. R., and P. W. Shaw. 2008. Evaluation of coloured sticky traps for monitoring beneficial insects in apple orchards. New Zealand Plant Protection 61:328–332.

Wallmo, O. C. 1956. Determination of sex and age of scaled quail. Journal of Wildlife Management 20:154–158.

Wallner, B., S. Huber, and R. Achmann. 2001. Non-invasive PCR sexing of rabbits (*Oryctolagus cuniculus*) and hares (*Lepus europaeus*). Mammalian Biology 66:190–192.

Walsh, D. P., C. F. Page, H. Campa, III, S. R. Winterstein, and D. E. Beyer, Jr. 2009. Incorporating estimates of group size in sightability models for aerial surveys of wildlife populations. Journal of Wildlife Management 73:136–143.

Walter, S. E., and D. H. Rusch. 1997. Accuracy of egg flotation in determining age of Canada goose nests. Wildlife Society Bulletin 25:854–857.

Walters, C. J. 1986. Adaptive management of renewable resources. MacMillan, New York, New York, USA.

Wambolt, C. L., M. R. Frisina, S. J. Knapp, and R. M. Frisina. 2006. Effect of method, site, and taxon on line-intercept estimates of sagebrush cover. Wildlife Society Bulletin 34:440–445.

Wand, M. P., and M. C. Jones. 1995. Kernel smoothing. Chapman and Hall, London, England, UK.

Wandell, W. N. 1943. A multi-marking system for ring-necked pheasants. Journal of Wildlife Management 7:378–382.

Wandell, W. N. 1945. Rapid method for opening and arranging pheasant bands. Journal of Wildlife Management 9:325.

Wang, C., T. Gibb, G. W. Bennett, and S. McKnight. 2009. Bed bug (Heteroptera: Cimicidae) attraction to pitfall traps baited with carbon dioxide, heat, and chemical lure. Journal of Economic Entomology 102:1580–1585.

Wang, C. K., and W. D. Philpot. 2007. Using airborne bathymetric lidar to detect bottom type variation in shallow waters. Remote Sensing of Environment 106:123–135.

Wang, C. T., and G. W. Bennett. 2006. Comparison of cockroach traps and attractants for monitoring German cockroaches (Dictyoptera: Blattellidae). Environmental Entomology 35:765–770.

Wang, J. L. 2005. Estimation of effective population sizes from data on genetic markers. Philosophical Transactions of the Royal Society B 360:1395–1409.

Wang, J. L., and M. C. Whitlock. 2003. Estimating effective population size and migration rates from genetic samples over space and time. Genetics 163:429–446.

Wang, J. P., and S. C. Adolph. 1995. Thermoregulatory consequences of transmitter implant surgery in the lizard *Sceloporus occidentalis*. Journal of Herpetology 29:489–493.

Wang, T., and J. W. Hicks. 2008. Changes in pulmonary blood flow do not affect gas exchange during intermittent ventilation in resting turtles. Journal of Experimental Biology 211:3759–3763.

Wang, X., and R. H. Tedford. 2008. Dogs: their fossil relatives and evolutionary history. Columbia University Press, New York, New York, USA.

Wang, X.-H., and C. H. Trost. 2000. Trapping territorial black-billed magpies. Journal of Field Ornithology 71:730–735.

Wang, Y., and D. M. Finch. 2002. Consistency of mist netting and point counts in assessing landbird species richness and relative abundance during migration. Condor 104:59–72.

Waples, R. S. 1989. A generalized approach for estimating effective population size from temporal changes in allele frequency. Genetics 121:379–391.

Waples, R. S. 1991. Pacific salmon *Oncorhynchus* spp. and the definition of 'species' under the endangered species act. Marine Fisheries Reviews 53(3):11–22.

Waples, R. S. 1998. Separating the wheat from the chaff: patterns of genetic differentiation in high gene flow species. Journal of Heredity 89:438–450.

Waples, R. S., and O. Gaggiotti. 2006. What is a population? An empirical evaluation of some genetic methods for identifying the number of gene pools and their degree of connectivity. Molecular Ecology 15:1419–1439.

Warburton, B. 1982. Evaluation of seven trap models as humane and catch-efficient possum traps. New Zealand Journal of Zoology 9:409–418.

Warburton, B. 1992. Victor foot-hold traps for catching Australian brushtail possums in New Zealand: capture efficiency and injuries. Wildlife Society Bulletin 20:67–73.

Ward, F. P., C. J. Hohmann, J. F. Ulrich, and S. E. Hill. 1976. Seasonal microhabitat selections of spotted turtles (*Clemmys guttata*) in Maryland elucidated by radioisotope tracking. Herpetologica 32:60–64.

Ward, F. P., and D. P. Martin. 1968. An improved cage trap for birds of prey. Bird-Banding 39:18–26.

Ward, R. L., J. T. Anderson, and J. T. Petty. 2008. Effects of road crossings on stream and streamside salamanders. Journal of Wildlife Management 72:760–771.

Waring, R. H., J. B. Way, R. Hunt, Jr., L. Morrisey, K. J. Ranson, J. F. Weishampel, R. Oren, and S. E. Franklin. 1995. Imaging radar for ecosystem studies. BioScience 45:715–723.

Warneke, B. M. 1979. Marking of Australian fur seals, 1966–1977. Pages 7–8 *in* L. Hobbs and P. Russell, editors. Report on the pinniped tagging workshop. National Marine Mammal Laboratory, Seattle, Washington, USA.

Warnock, N, and J. Y. Takekawa. 2003. Use of radiotelemetry in studies of shorebirds' past contributions and future directions. Wader Study Group Bulletin 100:138–150.

Warnock, N., and S. E. Warnock. 1993. Wintering site fidelity and movement patterns of western sandpipers *Calidris mauri* in the San Francisco Bay estuary. Ibis 138:160–167.

Warren, J. D., and B. J. Peterson. 2007. Use of a 600-kHz acoustic Doppler current profiler to measure estuarine bottom type, relative abundance of submerged aquatic vegetation, and eelgrass canopy height. Estuarine, Coastal and Shelf Science 72:53–62.

Warren, P. 2006. Aspects of red grouse *Lagopus lagopus scoticus* population dynamics at a landscape scale in northern England and the implications for grouse moor management. Dissertation, University of Durham, Durham, England, UK.

Warren, P., and D. Baines. 2007. Dispersal distances of juvenile radio-tagged red grouse *Lagopus lagopus scoticus* on moors in northern England. Ibis 149:758–762.

Waser, P. M., J. D. Busch, C. R. McCormick, and J. A. DeWoody. 2006. Parentage analysis detects cryptic precapture dispersal in a philopatric rodent. Molecular Ecology 15:1929–1937.

Washington, H. G. 1984. Diversity, biotic, and similarity indices: a review with special relevance to aquatic ecosystems. Water Research 18:653–694.

Wasser, S. K., W. J. Clark, O. Drori, E. S. Kisamo, C. Mailand, B. Mutayoba, and M. Stephens. 2008. Combating the illegal trade in African elephant ivory with DNA forensics. Conservation Biology 22:1065–1071.

Wasser, S. K., B. Davenport, E. R. Ramage, K. E. Hunt, M. Parker, C. Clarke, and G. Stenhouse. 2004. Scat detection dogs in wildlife research and management: application to grizzly and black bears in the Yellowhead ecosystem, Alberta, Canada. Canadian Journal of Zoology 82:475–492.

Wasser, S. K., K. E. Hunt, J. L. Brown, K. Cooper, C. M. Crockett, U. Bechert, J. J. Millspaugh, S. Larson, and S. L. Monfort. 2000. A generalized fecal glucocorticoid assay for use in a diverse array of nondomestic mammalian and avian species. General and Comparative Endocrinology 120:260–275.

Wasser, S. K., L. Risler, and R. A. Steiner. 1988. Excreted steroids in primate feces over the menstrual cycle and pregnancy. Biology of Reproduction 39:862–872.

Wasser, S. K., H. Smith, L. Madden, N. Marks, and C. Vynne. 2009. Scent-matching dogs determine number of unique individuals from scat. Journal of Wildlife Management 73:1233–1240.

Watanabe, S., M. Izawa, A. Kato, Y. Ropert-Coudert, and Y. Naito. 2005. A new technique for monitoring the detailed behaviour of terrestrial animals: a case study with the domestic cat. Applied Animal Behaviour Science 94:117–131.

Watkins, W. A., and W. E. Schevill. 1976. Underwater paint marking of porpoises. Fisheries Bulletin 74:687–689.

Watson, A. 1967. Social status and population regulation in the red grouse (*Lagopus lagopus scoticus*). Royal Society Population Study Group 2:22–30.

Watson, A., and R. Moss. 1972. A current model of population dynamics in red grouse. Proceedings of the International Ornithological Congress 15:134–149.

Watson, A., R. Moss, and R. Parr. 1984*a*. Effects of food enrichment on numbers and spacing behavior of red grouse. Journal of Animal Ecology 53:663–678.

Watson, A., R. Moss, R. Parr, M. D. Mountford, and P. Rothery. 1994. Kin landownership, differential aggression between kin and non-kin, and population fluctuations in red grouse. Journal of Animal Ecology 63:39–50.

Watson, A., R. Moss, and P. Rothery. 2000. Weather and synchrony in 10-year population cycles of rock ptarmigan and red grouse in Scotland. Ecology 81:2126–2136.

Watson, A., R. Moss, P. Rothery, and R. Parr. 1984b. Demographic causes and predictive models of population fluctuation in red grouse. Journal of Animal Ecology 53:639–662.

Watson, J. W. 1985. Trapping, marking and radio-monitoring rough-legged hawks. North American Bird Bander 10:9–10.

Watt, D. J. 2001. Recapture rate and breeding frequencies of American goldfinches wearing different colored leg bands. Journal of Field Ornithology 72:236–243.

Watts, D. E., I. D. Parker, R. R. Lopez, N. J. Silvy, and D. S. Davis. 2008. Distribution and abundance of endangered Florida Key deer on outer islands. Journal of Wildlife Management 72:360–366.

Wauters, L., D. G. Preatoni, A. Molinari, and G. Tosi. 2007. Radio-tracking squirrels: performance of home rang density and linkage estimators with small range and sample size. Ecological Modelling 202:333–344.

Way, J. G., I. M. Ortega, P. J. Auger, and E. G. Strauss. 2002. Box-trapping eastern coyotes in southeastern Massachusetts. Wildlife Society Bulletin 30:695–702.

Wayne, R. K., L. Forman, A. K. Newman, J. M. Simonson, and S. J. O'Brien. 1986. Genetic markers of zoo populations: morhological and electrophoretic assays. Zoo Biology 5:215–232.

Wayne, R. K., and S. M. Jenks. 1991. Mitochondrial DNA analysis implying extensive hybridization of the endangered red wolf Canis rufus. Nature 351:565–568.

Weadon, M., P. Heinselman, D. Forsythe, W. E. Benner, G. S. Gorok, and J. Kimpel. 2009. Multifunction phased array radar. Bulletin of the American Meteorological Society 90:385–389.

Weary, G. C. 1969. An improved method of marking snakes. Copeia 1969:854–855.

Weatherhead, P. J., and G. Blouin-Demers. 2004. Long-term effects of radiotelemetry on black ratsnakes. Wildlife Society Bulletin 32:900–906.

Weaver, D. K., and J. A. Kadlec. 1970. A method for trapping breeding adult gulls. Bird-Banding 41:28–31.

Weaver, H. R., and W. L. Haskell. 1968. Age and sex determination of the chukar partridge. Journal of Wildlife Management 32:46–50.

Weaver, J. L., P. Wood, D. Paetkau, and L. L. Laack. 2005. Use of scented hair snares to detect ocelots. Wildlife Society Bulletin 33:1384–1391.

Weaver, K. M., D. H. Arner, C. Mason, and J. J. Hartley. 1985. A guide to using snares for beaver capture. Southern Journal of Applied Forestry 9:141–146.

Webb, B. 2000. What does robotics offer animal behaviour? Animal Behaviour 60:545–558.

Webb, S. L., J. S. Lewis, D. G. Hewitt, M. Hellickson, and F. C. Bryant. 2008. Assessing the helicopter and net gun as a capture technique for white-tailed deer. Journal of Wildlife Management 72:310–314.

Webb, W. L. 1943. Trapping and marking white-tailed deer. Journal of Wildlife Management 7:346–348.

Weber M. E., J. Y. N. Cho, J. S. Herd, J. M. Flavin, W. E. Benner, and G. S. Torok. 2007. The next-generation multimission U.S. surveillance radar network. Bulletin of the American Meteorological Society 90:1739–1751.

Weber, T., and J. Wolf. 2000. Maryland's green infrastructure—using landscape assessment tools to identify a regional conservation strategy. Environmental Monitoring and Assessment 63:265–277.

Weeden, R. B., and A. Watson. 1967. Determining the age of rock ptarmigan in Alaska and Scotland. Journal of Wildlife Management 31:825–826.

Wegner, W. A. 1981. A carrion baited noose trap for American kestrels. Journal of Wildlife Management 45:248–250.

Weller, M. W. 1956. A simple field candler for waterfowl eggs. Journal of Wildlife Management 20:111–113.

Weller, M. W. 1957. An automatic nest-trap for waterfowl. Journal of Wildlife Management 21:456–458.

Weller, M. W. 1988. Waterfowl in winter: selected papers from symposium and workshop. University of Minnesota Press, Minneapolis, USA.

Wells, J. V., and M. E. Richmond. 1995. Populations, metapopulations, and species populations: what are they and who should care? Wildlife Society Bulletin 23:458–462.

Wells, K. D., and R. A. Wells. 1976. Patterns of movement in a population of the slimy-salamander, Plethodon glutinosus, with observations on aggregations. Herpetologica 32:156–162.

Welsh, H. H., Jr., and A. J. Lind. 2002. Multiscale habitat relationships of stream amphibians in the Klamath–Siskiyou region of California and Oregon. Journal of Wildlife Management 66:581–602.

Wendelin H., R. Nagel, and P. H. Becker. 1996. A technique to spray dyes on birds. Journal of Field Ornithology 67:442–446.

Wenger, K. F. 1953. The effect of fertilization and injury on the cone and seed production of loblolly pine seed trees. Journal of Forestry 51:570–573.

Wentworth, B. C., J. A. Proudman, H. Opel, J. J. Wineland, N. G. Zimmermann, and A. Lapp. 1983. Endocrine changes in the incubating and brooding turkey hen. Biology of Reproduction 29:87–92.

Werner, P. A. 1975. A seed trap for determining patterns of seed deposition in terrestrial plants. Canadian Journal of Botany 53:810–813.

Wesson, J. A., III, P. F. Scanlon, R. L. Kirkpatrick, H. S. Mosby, and R. L. Butcher. 1979. Influence of chemical immobilization and physical restraint on steroid hormone levels in blood of white-tailed deer. Canadian Journal of Zoology 57:768–776.

West, B. C., A. L. Cooper, and J. B. Armstrong. 2009. Managing wild pigs: a technical guide. Human–Wildlife Interactions. Monograph 1.

West, G., D. Heard, and N. Caulkett, editors. 2007. Zoo animal and wildlife immobilization and anesthesia. Blackwell, Ames, Iowa, USA.

West, N. E. 1989. Spatial pattern—functional interactions in shrub-dominated plant communities. Pages 283–305 in C. M. McKell, editor. The biology and utilization of shrubs. Academic Press, San Diego, California, USA.

West, R. L., and G. K. Hess. 2002. Purple gallinule. Account 626 in A. Poole and F. Gill, editors. The birds of North America. The Academy of Natural Sciences, Philadelphia, Pennsylvania, and The American Ornithologists' Union, Washington, D.C., USA.

Westbrook, J. K. 2008. Noctuid migration in Texas within the nocturnal aerological boundary layer. Integrative and Comparative Biology. 48:99–106.

Westemeier R. L., J. D. Brawn, S. A. Simpson, T. L. Esker, R. W. Jansen, J. W. Walk, E. L. Kershner, J. L. Bouzat, and K. N. Paige. 1998. Tracking the long-term decline and recovery of an isolated population. Science 282:1695–1698.

Westerkov, K. 1950. Methods for determining the age of game bird eggs. Journal of Wildlife Management 14:56–67.

Westerkov, K. 1956. Age determination and dating nesting events in the willow ptarmigan. Journal of Wildlife Management 20:274–279.

Westfall, C. Z., and R. B. Weeden. 1956. Plastic neck markers for woodcock. Journal of Wildlife Management 20:218–219.

Wetherbee, D. 1961. Investigations of the life history of the common coturnix. American Midland Naturalist 65:168–186.

Wethey, D. S. 1985. Catastrophe, extinction, and species diversity: a rocky intertidal example. Ecology 66:445–456.

Whalen, D. M., and B. D. Watts. 1999. The influence of audio-lures on capture patterns of migrant northern saw-whet owls. Journal of Field Ornithology 70:163–168.

Wheeler, Q. D., and R. Meier, editors. 2000. Species concepts and phylogenetic theory: a debate. Columbia University Press, New York, New York, USA.

Wheeler, R. H., and J. C. Lewis. 1972. Trapping techniques for sandhill crane studies in the Platte River Valley. Resource Publication 107, U.S. Fish and Wildlife Service, Department of the Interior, Washington, D.C., USA.

Whidden, S. E., C. T. Williams, A. R. Breton, and C. L. Buck. 2007. Effects of transmitters on the reproductive success of tufted puffins. Journal of Field Ornithology 78:206–212.

Whigham, D. F., J. McCormick, R. E. Good, and R. L. Simpson. 1978. Biomass and primary production in freshwater tidal wetlands of the Middle Atlantic Coast. Pages 3–20 in R. E. Good, D. F. Whigham, and R. L. Simpson, editors. Freshwater wetland ecological processes and management potential. Academic Press, New York, New York, USA.

Whitaker, A. H. 1967. Baiting pitfall traps for small lizards. Herpetologica 23:309–310.

White, D. H., and J. T. Seginak. 1994. Dioxins and furans linked to reproductive impairment in

wood ducks. Journal of Wildlife Management 58:100–106.
White, D. J., A. P. King, and S. D. Duncan. 2002. Voice recognition technology as a tool for behavioral research. Behavior Research Methods, Instruments, and Computers 34:1–5.
White, G. C. 1993. Evaluation of radio tagging marking and sighting estimators of population size using Monte Carlo simulations. Pages 91–103 in J.-D. Lebreton and P. M. North, editors. Marked individuals in the study of bird population. Birkhäuser Verlag, Basel, Switzerland.
White, G. C. 1996. NOREMARK: population estimation from mark–resighting surveys. Wildlife Society Bulletin 24:50–52.
White, G. C. 2000. Population viability analysis: data requirements and essential analyses. Pages 288–331 in L. Boitani and T. K. Fuller, editors. Research techniques in animal ecology: controversies and consequences. Columbia University Press, New York, New York, USA.
White, G. C., D. R. Anderson, K. P. Burnham, and D. L. Otis. 1982. Capture–recapture and removal methods for sampling closed populations. LA-8787-NERP, Los Alamos National Laboratory, Los Alamos, New Mexico, USA.
White, G. C., and R. M. Bartmann. 1994. Drop nets versus helicopter net guns for capturing mule deer fawns. Wildlife Society Bulletin 22:248–252.
White, G. C., and K. P. Burnham. 1999. Program MARK: survival estimation from populations of marked animals. Bird Study (Supplement) 46:120–138.
White, G. C., K. P. Burnham, and D. R. Anderson. 2001. Advanced features of program MARK. Pages 368–377 in R. Field, R. J. Warren, H. Okarma, and P. R. Sievert, editors. Wildlife, land, and people: priorities for the 21st century. Proceedings of the Second International Wildlife Management Congress. The Wildlife Society, Bethesda, Maryland, USA.
White, G. C., and R. A. Garrott. 1986. Effects of biotelemetry triangulation error on detecting habitat selection. Journal of Wildlife Management 50:509–513.
White, G. C., and R. A. Garrott. 1990. Analysis of wildlife radio-tracking data. Academic Press, San Diego, USA.
White, G. C., and T. M. Shenk. 2001. Population estimation with radio-marked animals. Pages 329–350 in J. J. Millspaugh and J. M. Marzluff, editors. Radio tracking and animal populations. Academic Press, San Diego, California, USA.
White, J. A., and C. E. Braun. 1978. Age and sex determination of juvenile band-tailed pigeons. Journal of Wildlife Management 42:564–569.
White, J. A., and C. E. Braun. 1990. Growth of young band-tailed pigeons in captivity. Southwestern Naturalist 35:82–84.
White, M. A., G. P. Asner, R. R. Nemani, J. L. Privette, and S. W. Running. 2000. Measuring fractional cover and leaf area index in arid ecosystems: digital camera, radiation transmittance, and laser altimetry methods. Remote Sensing of Environment 74:45–57.
White, M. J., Jr., J. G. Jennings, W. F. Gandy, and L. H. Cornell. 1981. An evaluation of tagging, marking, and tattooing techniques for small dolphinids. Technical Memorandum 16, U.S. National Oceanic and Atmospheric Administration, Department of Commerce, Washington, D.C., USA.
White, N. A., and M. Sjöberg. 2002. Accuracy of satellite positions from free-ranging grey seals using ARGOS. Polar Biology 25:625–631.
White, S. B., T. A. Bookhout, and E. K. Bollinger. 1980. Use of human hair bleach to mark blackbirds and starlings. Journal of Field Ornithology 51:6–9.
Whitehouse, A. M., and E. H. Harley. 2001. Post-bottleneck genetic diversity of elephant populations in South Africa, revealed using microsatellite analysis. Molecular Ecology 10:2139–2149.
Whitfield, S. M., K. E. Bell, T. Philippi, M. Sasa, F. Bolanos, and G. Chaves. 2007. Amphibian and reptile declines over 35 years at La Selva, Costa Rica. Proceedings of the National Academy of Sciences of the United States of America 104:8352–8356.
Whitford, W. G., and M. Massey. 1970. Responses of a population of *Ambystoma tigrinum* to thermal and oxygen gradients. Herpetologica 26:372–376.
Whiting, J. C., R. T. Bowyer, and J. T. Flinders. 2009. Diel use of water by reintroduced bighorn sheep. Western North American Naturalist 69:407–412.
Whiting, M. J. 1998. Increasing lizard capture success using baited glue traps. Herpetological Review 29:34.
Whitlock, M. C., and D. E. McCauley. 1999. Indirect measures of gene flow and migration: F-ST not equal $1/(4Nm+1)$. Heredity 82:117–125.
Whitman, J. S., W. B. Ballard, and C. L. Gardner. 1986. Home range and habitat use by wolverines in southcentral Alaska. Journal of Wildlife Management 50:460–463.
Whittaker, J. C., G. A. Feldhamer, and E. M. Charles. 1998. Capture of mice, *Peromyscus*, in two sizes of Sherman live traps. Canadian Field-Naturalist 112:527–529.
Whittaker, R. H. 1965. Branch dimensions and estimation of branch production. Ecology 46:365–370.
Whittier, J. M., and R. R. Tokarz. 1992. Physiological regulation of sexual behavior in female reptiles. Pages 24–69 in C. Gans and D. Crews, editors. Biology of the Reptilia. Volume 18. Hormones, brain, and behavior. University of Chicago Press, Chicago, Illinois, USA.
Whitworth, D., S. Newman, T. Mundkur, and P. Harris. 2007. Wild birds and avian influenza: an introduction to applied field research and disease sampling techniques. FAO Animal Production and Health Manual, No. 5., Rome, Italy. ftp://ftp.fao.org/docrep/fao/010/a1521e/a1521e.pdf.
Whitworth, D. L., J. Y. Takekawa, H. R. Carter, and W. R. McIver. 1997. Night-lighting technique for at-sea capture of Xantus' murrelets. Colonial Waterbirds 20:525–531.
Wickstrom, M. 1999. Natural toxins and wildlife. Canadian Cooperative Wildlife Health Centre short course: wildlife toxicology. Canadian Cooperative Wildlife Health Centre, Saskatoon, Saskatchewan, Canada. http://wildlife.usask.ca/english/tox-6.htm.
Wiebe, M. O., J. E. Hines, and G. J. Robertson. 2000. Collar retention of Canada geese and greater white-fronted geese from the western Canadian Arctic. Journal of Field Ornithology 71:531–540.
Wiener, J. G., D. P. Krabbenhoft, G. H. Heinz, and A. M. Scheuhammer. 2003. Exotoxicology of mercury. Pages 373–408 in D. J. Hoffman, B. A. Rattner, G. A. Burton, Jr., and J. Cairns, Jr., editors. Handbook of ecotoxicology. Second edition. Lewis, Boca Raton, Florida, USA.
Wiens, J. A. 1973. Pattern and process in grassland bird communities. Ecological Monographs 43:237–270.
Wiens, J. A. 1974. Habitat heterogeneity and avian community structure in North American grasslands. American Midland Naturalist 91:195–213.
Wiens, J. A. 1976. Population responses to patchy environments. Annual Review of Ecology and Systematics 7:81–120.
Wiens, J. A. 1981. Scale problems in avian censusing. Studies in Avian Biology 6:513–521.
Wiens, J. A. 1989a. Spatial scaling in ecology. Functional Ecology 3:385–397.
Wiens, J. A. 1989b. The ecology of bird communities: foundations and patterns. Volume I. Cambridge University Press, New York, New York, USA.
Wiens, J. A. 1992. Ecological flows across landscape boundaries: a conceptual overview. Pages 217–235 in O. L. Lange, H. A. Mooney, and H. Remmert, editors. Landscape boundaries. Springer-Verlag, New York, New York, USA.
Wiens, J. A. 1996. Wildlife in patchy environments: metapopulations, mosaics, and management. Pages 53–84 in D. R. McCullough, editor. Metapopulations and wildlife conservation. Island Press, Washington, D.C., USA.
Wigal, R. A., and V. L. Coggins. 1982. Mountain goat. Pages 1008–1020 in J. A. Chapman and G. A. Feldhamer, editors. Wild mammals of North America. Johns Hopkins University Press, Baltimore, Maryland, USA.
Wiggers, E. P., and S. L. Beasom. 1986. Characterization of sympatric or adjacent habitats of 2 deer species in west Texas. Journal of Wildlife Management 50:129–134.
Wight, H. M. 1938. Field and laboratory techniques in wildlife management. University of Michigan Press, Ann Arbor, USA.
Wight, H. M. 1939. Field and laboratory techniques in wildlife management. University of Michigan Press, Ann Arbor, USA.
Wight, H. M. 1956. A field technique for bursal inspection of mourning doves. Journal of Wildlife Management 20:94–95.
Wight, H. M., L. H. Blankenship, and R. E. Tomlinson. 1967. Aging mourning doves by outer primary wear. Journal of Wildlife Management 31:832–835.
Wik, P. A. 2002. Ecology of greater sage-grouse in south-central Owyhee County, Idaho. Thesis, University of Idaho, Moscow, USA.

Wikelski, M., M. Hau, and J. C. Wingfield. 1999. Social instability increases plasma testosterone in a year-round territorial neotropical bird. Proceedings of the Royal Society of London B 266:551–556.

Wikelski, M., R. Kays, J. Kasdin, K. Thorup, J. A. Smith, W. W. Cochran, and G. W. Swenson, Jr. 2007. Going wild: what a global small-animal tracking system could do for experimental biologists. Experimental Biology 210:181–186.

Wikle, C. K., L. M. Berliner, and N. Cressie. 1998. Hierarchical Bayesian space–time models. Environmental and Ecological Statistics 5:117–154.

Wildlife Health Centre. 2007. Canadian Cooperative Wildlife Health Centre: wildlife disease investigation manual. Wildlife Damage Management, Internet Center for Canadian Cooperative Wildlife Health Centre: newsletters and publications, University of Nebraska, Lincoln, USA. http://digitalcommons.unl.edu/icwdmccwhcnews/52.

Wildt, D. E., J. L. Brown, M. Bush, M. A. Barone, K. A. Cooper, J. Grisham, and J. G. Howard. 1993. Reproductive status of cheetahs (*Acinonyx jubatus*) in North American zoos: the benefits of physiological surveys for strategic planning. Zoo Biology 12:45–80.

Wiley, E. O. 1978. The evolutionary species concept reconsidered. Systematic Biology 27:17–26.

Wilkinson, G. S. 1985. The social organization of the common vampire bat. I. Pattern and cause of association. Behavioral Ecology and Sociobiology 17:111–121.

Will, R. G., J. W. Ironside, M. Zeidler, S. N. Cousens, K. Estibeiro, A. Alperovitch, S. Poser, M. Pocchiari, A. Hofman, and P. G. Smith. 1996. A new variant of Creutzfeldt–Jakob disease in the UK. Lancet 347:921–925.

Willey, C. H. 1974. Aging black bears from first premolar tooth sections. Journal of Wildlife Management 38:97–100.

Williams, B. K. 1997. Logic and science in wildlife biology. Journal of Wildlife Management 61:1007–1015.

Williams, B. K., and F. A. Johnson. 1995. Adaptive management and the regulation of waterfowl harvests. Wildlife Society Bulletin 23:430–436.

Williams, B. K., F. A. Johnson, and K. Wilkins. 1996. Uncertainty and the adaptive management of waterfowl harvests. Journal of Wildlife Management 60:223–232.

Williams, B. K., J. D. Nichols, and M. J. Conroy. 2002a. Analysis and management of animal populations. Academic Press, San Diego, California, USA.

Williams, B. O. 2002. Bird dog: the instinctive training method. Willow Creek Press, Minocqua, Wisconsin, USA.

Williams, B. W., D. R. Etter, D. W. Linden, K. F. Millenbah, S. R. Winterstein, and K. T. Scribner. 2009. Noninvasive hair sampling and genetic tagging of co-distributed fishers and American martens. Journal of Wildlife Management 73:26–34.

Williams, C. K., R. S. Lutz, and R. D. Applegate. 2004. Winter survival and additive harvest in northern bobwhite coveys in Kansas. Journal of Wildlife Management 68:94–100.

Williams, C. L., T. L. Serfass, R. Cogan, and O. E. Rhodes. 2002b. Microsatellite variation in the reintroduced Pennsylvania elk herd. Molecular Ecology 11:1299–1310.

Williams, D. L., S. Goward, and T. Arvidson. 2006. Landsat: yesterday, today, and tomorrow. Photogrammetric Engineering and Remote Sensing 72:1171–1178.

Williams, E. S. 2005. Chronic wasting disease. Veterinary Pathology 42:530–549.

Williams, E. S., and I. K. Barker, editors. 2001. Infectious diseases of wild mammals. Third edition. Iowa State University Press, Ames, USA.

Williams, E. S., S. L. Cain, and D. S. Davis. 1997. Brucellosis: the disease in bison. Pages 7–19 *in* E. T. Thorne, M. S. Boyce, P. Nicoletti, and T. J. Kreeger, editors. Brucellosis, bison, elk, and cattle in the Greater Yellowstone area: defining the problem, exploring solutions. Wyoming Game and Fish Department for the Greater Yellowstone Interagency Brucellosis Committee, Cheyenne, USA.

Williams, E. S., J. K. Kirkwood, and M. W. Miller. 2001. Chronic wasting disease. Pages 292–301 *in* E. S. Williams and I. K. Barker, editors. Infectious diseases of wild mammals. Iowa State University Press, Ames, USA.

Williams, E. S., M. W. Miller, T. J. Kreeger, R. H. Kahn, and E. T. Thorne. 2002c. Chronic wasting disease of deer and elk: a review with recommendations for management. Journal of Wildlife Management 66:551–563.

Williams, E. S., M. W. Miller, and E. T. Thorne. 2002d. Chronic wasting disease: implications and challenges for wildlife managers. Transactions of the North American Wildlife and Natural Resources Conference 67:87–103.

Williams, E. S., and S. Young. 1980. Chronic wasting disease of captive mule deer: a spongiform encephalopathy. Journal of Wildlife Diseases 16:89–98.

Williams, J. C. 1998. Delineating protected wildlife corridors with multi-objective programming. Environmental Modeling and Assessment 3:77–86.

Williams, L. E., Jr. 1961. Notes on wing molt in the yearling wild turkey. Journal of Wildlife Management 25:439–440.

Williams, L. E., Jr., and D. H. Austin. 1970. Complete post-juvenile (pre-basic) primary molt in Florida turkeys. Journal of Wildlife Management 34:231–233.

Williams, L. E., Jr., and D. H. Austin. 1988. Studies of the wild turkey in Florida. Technical Bulletin 10, Florida Game and Freshwater Fish Commission, Gainesville, USA.

Williams, P. H., N. D. Burgess, and C. Rahbek. 2000a. Flagship species, ecological complementarity and conserving the diversity of mammals and birds in sub-Saharan Africa. Animal Conservation 3:249–260.

Williams, R. D., J. E. Gates, and C. H. Hocutt. 1981a. An evaluation of known and potential sampling techniques for hellbender, *Cryptobranchus alleganiensis*. Journal of Herpetology 15:23–27.

Williams, R. N., O. E. Rhodes, and T. L. Serfass. 2000b. Assessment of genetic variance among source and reintroduced fisher populations. Journal of Mammalogy 81:895–907.

Williams, T. C. 1984. How to use marine radar for bird watching. American Birds 38:982–983.

Williams, T. C., L. C. Ireland, and J. M. Williams. 1973. High altitude flights of the free-tailed bat, *Tadarida brasiliensis*, observed with radar. Journal of Mammalogy 54:807–821.

Williams, T. C., T. J. Klonowski, and P. Berkeley. 1976. Angle of Canada goose V flight formation measured by radar. Auk 93:554–559.

Williams, T. C., J. E. Marsden, T. L. Lloyd-Evans, V. Krauthamer, and H. Krauthamer. 1981b. Spring migration studied by mist netting, ceilometer, and radar. Journal of Field Ornithology 52:177–190.

Williams, T. C., and J. M. Williams. 1980. A Peterson's guide to radar ornithology? American Birds 34:738–741.

Williams, T. C., J. M. Williams, and D. R. Griffin. 1966. The homing ability of the neotropical bat, *Phyllostomus hastatus*, with evidence for visual orientation. Animal Behavior 14:473–486.

Williams, T. C., J. M. Williams, and P. D. Kloeckner. 1986. Airspeed and heading of autumnal migrants over Hawaii. Auk 103:634–635.

Williams, T. C., J. M. Williams, J. M. Teal, and J. W. Kanwisher. 1972. Tracking radar studies of bird migration. Pages 115–128 *in* S. R. Galler, K. Schmidt-Koenig, G. J. Jacobs, and R. E. Belleville, editors. Animal orientation and navigation. U.S. Government Printing Office, Washington, D.C., USA.

Williams, T. M., R. W. Davis, L. A. Fuiman, J. Francis, B. J. Le Boeuf, M. Horning, J. Calambokidis, and D. A. Croll. 2000c. Sink or swim: strategies for cost-efficient diving by marine mammals. Science 288:133–136.

Williamson, M. H. 1959. Some extensions of the use of matrices in population theory. Bulletin of Mathematical Biophysics 21:13–17.

Williamson, M. J., and M. R. Pelton. 1971. New design for a large portable mammal trap. Proceedings of the Annual Conference of the Southeastern Association of Game and Fish Commissioners 25:315–322.

Williamson, P., N. J. Gales, and S. Lister. 1990. Use of real-time B-mode ultrasound for pregnancy diagnosis and measurement of fetal growth rate in captive bottlenose dolphins (*Tursiops truncates*). Journal of Reproduction and Fertility 88:543–548.

Williamson-Natesan, E. G. 2005. Comparison of methods for detecting bottlenecks from microsatellite loci. Conservation Genetics 6:551–562.

Willson, J. D. 2004. A comparison of aquatic drift fences with traditional funnel trapping as a quantitative method for sampling amphibians. Herpetological Review 35:148–152.

Willson, J. D., and M. E. Dorcas. 2003. Quantitative sampling of stream salamanders: comparison of dipnetting and funnel trapping techniques. Herpetological Review 34:128–130.

Willson, J. D., and M. E. Dorcas. 2004. A comparison of aquatic drift fences with traditional funnel trapping as a quantitative method for sampling amphibians. Herpetological Review 35:148–150.

Willson, J. D., C. T. Winne, and L. A. Fedewa. 2005. Unveiling escape and capture rates of aquatic snakes and salamanders (Siren spp. Amphiuma means) in commercial funnel traps. Journal of Freshwater Ecology 20:397–403.

Willsteed, P. M., and P. M. Fetterolf. 1986. A new technique for individually marking gull chicks. Journal of Field Ornithology 57:310–313.

Wilson, D. E., F. R. Cole, J. D. Nichols, R. Rudran, and M. S. Foster. 1996. Measuring and monitoring biological diversity. Standard methods for mammals. Smithsonian Institution Press, Washington, D.C., USA.

Wilson, E. O., and W. H. Bossert. 1971. A primer of population biology. Sinauer Associates, Sunderland, Massachusetts, USA.

Wilson, G. A, and B. Rannala. 2003. Bayesian inference of recent migration rates using multilocus genotypes. Genetics 163:1177–1191.

Wilson, J. A., D. A. Kelt, and D. H. Van Vuren. 2008. Home range and activity of northern flying squirrels (Glaucomys sabrinus) in the Sierra Nevada. Southwestern Naturalist 53: 21–28.

Wilson, J. J., and T. J. Maret. 2002. A comparison of two methods for estimating the abundance of amphibians in aquatic habitats. Herpetological Review 33:108–110.

Wilson, K. R., and D. R. Anderson. 1985. Evaluation of a nested grid approach for estimating density. Journal of Wildlife Management 49:675–678.

Wilson, K. R., J. D. Nichols, and J. E. Hines. 1989. A computer program for sample size computations for banding studies. Technical Report 23, U.S. Fish and Wildlife Service, Department of the Interior, Washington, D.C., USA.

Wilson, M. M., and J. A. Crawford. 1979. Response of bobwhites to controlled burning in south Texas. Wildlife Society Bulletin 7:53–56.

Wilson, P. J., and B. N. White. 1998. Sex identification of elk (Cervus elaphus canadensis), moose (Alces alces), and white-tailed deer (Odocoileus virginianus) using the polymerase chain reaction. Journal of Forensic Sciences 43:477–482.

Wilson, R. P., J. J. Ducamp, W. G. Rees, B. M. Culik, and K. Nickamp. 1992. Estimation of location: global coverage using light intensity. Pages 131–134 in I. G. Priede and S. M. Swift, editors. Wildlife telemetry: remote monitoring and tracking of animals. Ellis Horwood, New York, New York, USA.

Wilson, R. P., and C. R. McMahon. 2006. Measuring devices on wild animals: what constitutes acceptable practice? Frontiers in Ecology and the Environment 4:147–154.

Wilson, R. R., and R. S. Allan. 1996. Mist netting from a boat in forested wetlands. Journal of Field Ornithology 67:82–85.

Wiltschko, R., and W. Wiltschko. 2003. Avian navigation: from historical to modern concepts. Animal Behaviour 65:257–272.

Wimbush, D. J., M. D. Barrow, and A. B. Costin. 1967. Color stereophotography for the measurement of vegetation. Ecology 48:150–152.

Winchell, C. S. 1999. An efficient technique to capture complete broods of burrowing owls. Wildlife Society Bulletin 27:193–196.

Winchell, C. S., and J. W. Turman. 1992. A new trapping technique for burrowing owls—the noose rod. Journal of Field Ornithology 63: 66–70.

Windberg, L. A., and L. B. Keith. 1976. Snowshoe hare population response to artificial high densities. Journal of Mammalogy 57:523–553.

Windingstad, R. M., R. J. Cole, P. E. Nelson, T. J. Roffe, R. R. George, and J. W. Dorner. 1989. Fusarium mycotoxins from peanuts suspected as a cause of sandhill crane mortality. Journal of Wildlife Diseases 25:38–46.

Wing, M. G. 2008. Keeping pace with global positioning system technology in the forest. Journal of Forestry. 106(6):332–338.

Wing, M. G., A. Ecklund, and L. D. Kellogg. 2005. Consumer-grade global positioning system (GPS) accuracy and reliability. Journal of Forestry. 103(4):169–173.

Wing, M. G., A. Ecklund, J. Sessions, and R. Karsky. 2008. Horizontal measurement performance of five mapping-grade global positioning system receiver configurations in several forested settings. Western Journal of Applied Forestry. 23(3):166–171.

Wingfield, J. C. 1985. Short-term changes in plasma levels of hormones during establishment and defense of a breeding territory in male song sparrows, Melospiza melodia. Hormones and Behavior 19:174–187.

Wingfield, J. C., and D. S. Farner. 1976. Avian endocrinology field investigations and methods. Condor 78:570–573.

Wingfield, J. C., and D. S. Farner. 1980. Control of seasonal reproduction in temperate-zone birds. Pages 62–101 in R. J. Reiter and B. K. Follet, editors. Progress in reproductive biology. Volume 5. S. Karger, Basel, Switzerland.

Wingfield, J. C., and M. Raminofsky. 1999. The impact of stress on animal reproductive activities. Pages 130–163 in Stress physiology in animals. P. H. M. Balm, editor. CRC Press, Boca Raton, Florida, USA.

Wingfield, J. C., and R. M. Sapolsky. 2003. Reproduction and resistance to stress: when and how? Journal of Neuroendocrinology 15:711–724.

Wink, M. 2006. Use of DNA markers to study bird migration. Journal of Ornithology 147:234–244.

Winker, K., D. A. Rocque, T. M. Braile, and C. L. Pruett. 2007. Vainly beating the air: species-concept debates need not impede progress in science or conservation. Auk 63:30–44.

Winkworth, R. E. 1955. The use of point quadrats for the analysis of heathland. Australian Journal of Botany 3:68–81.

Winne, C. T. 2005. Increases in capture rates of an aquatic snake (Syminatrix pygaea) using naturally baited minnow traps: evidence for aquatic funnel trapping as a measuring of foraging activity. Herpetological Review 36:411–413.

Winning, G., and J. King. 2008. A new trap design for capturing squirrel gliders and sugar gliders. Australian Mammalogy 29:245–249.

Winston, F. A. 1955. Color marking of waterfowl. Journal of Wildlife Management 19:319.

Winter, W. 1981. Black-footed ferret search dogs. Southwestern Research Institute, Santa Fe, New Mexico, USA.

Winterstein, S. R., K. H. Pollock, and C. M. Bunck. 2001. Analysis of survival data from radio-telemetry studies. Pages 351–380 in J. J. Millspaugh and J. M. Marzluff, editors. Radio tracking and animal populations. Academic Press, San Diego, California, USA.

Wisconsin Department of Natural Resources. 2009. A plan for managing chronic wasting disease in Wisconsin: the next five years [Draft: July 2009]. WM-482-2008, Wisconsin Department of Natural Resources, Madison, USA.

Wisdom, M. J., and L. S. Mills. 1997. Sensitivity analysis to guide population recovery: prairie-chickens as an example. Journal of Wildlife Management 61:302–312.

Wiseman, B. S., D. L. Vincent, P. J. Thomford, N. S. Scheffrahn, G. F. Sargent, and D. J. Kesler. 1982. Changes in porcine, ovine, bovine and equine blood progesterone concentration between collection and centrifugation. Animal Reproductive Science 5:157–165.

Wishart, W. 1969. Age determination of pheasants by measurement of proximal primaries. Journal of Wildlife Management 33:714–717.

With, K. A. 1994. Using fractal analysis to assess how species perceive landscape structure. Landscape Ecology 9:25–36.

Witherington, B. E. 1992. Behavioral responses of nesting sea turtles to artificial lighting. Herpetologica 48:31–39.

Witherington, B. E. 1997. The problem of photo-pollution for sea turtles and other nocturnal animals. Pages 303–328 in J. R. Clemmons and R. Buchholz, editors. Behavioral approaches to conservation in the wild. Cambridge University Press, Cambridge, England, UK.

Withey, J. C., T. D. Bloxton, and J. M. Marzluff. 2001. Effects of tagging and location error in wildlife radiotelemetry studies. Pages 43–75 in J. J. Millspaugh and J. M. Marzluff, editors. Radio tracking and animal populations. Academic Press, San Diego, California, USA.

Witmer, G. W., R. E. Marsh, and G. H. Matschke. 1999. Trapping considerations for the fossorial pocket gopher. Pages 131–139 in G. Proulx, editor. Mammal trapping. Alpha Wildlife Research and Management, Sherwood Park, Alberta, Canada.

Witz, B. W. 1996. A new device for capturing small and medium-sized lizards by hand: the lizard grabber. Herpetological Review 27:130–131.

Wobeser, G. A. 1985. Handbook of diseases of Saskatchewan wildlife. Saskatchewan Parks and Renewable Resources, Regina, Saskatchewan, Canada.

Wobeser, G. A. 1997. Diseases of wild waterfowl. Second edition. Plenum Press, New York, New York, USA.

Wobeser, G. A. 2005. Essentials of disease in wild animals. Blackwell, Ames, Iowa, USA.

Wobeser, G. A. 2007. Disease in wild animals: investigation and management. Second edition. Springer, Berlin, Germany.

Woehler, E. E., and J. M. Gates. 1970. An improved method of sexing ring-necked pheasant chicks. Journal of Wildlife Management 34:228–231.

Wolcott, T. G. 1977. Optical tracking and telemetry for nocturnal field studies. Journal of Wildlife Management 41:309–312.

Wolf, K. N., F. Elvinger, and J. L. Pilcicki. 2003. Infrared-triggered photography and tracking plates to monitor oral rabies vaccine bait contact by raccoons in culverts. Wildlife Society Bulletin 31:387–391.

Wolf, M., and G. O. Batzli. 2002. Relationship of previous trap occupancy to capture of white-footed mice (*Peromyscus leucopus*). Journal of Mammalogy 83:728–733.

Wolfe, M. L. 1969. Age determination in moose from cemental layers of molar teeth. Journal of Wildlife Management 33:428–431.

Wolff, J. O. 1996. Population fluctuations of mast-eating rodents are correlated with production of acorns. Journal of Mammalogy 77:850–856.

Wolinski, R. A., and E. A. Pike. 1985. Hoop-net for the capture of barn and cliff swallows. North American Bird Bander 10:4–5.

Wollard, L. L., R. D. Sparrowe, and G. D. Chambers. 1977. Evaluation of a Korean pheasant introduction in Missouri. Journal of Wildlife Management 41:616–623.

Wolters, R. A. 1961. Gun dog: revolutionary rapid training method. Penguin, New York, New York, USA.

Wood, A. K., R. E. Short, A. E. Darling, G. L. Dusek, R. G. Sasser, and C. A. Ruder. 1986. Serum assays for detecting pregnancy in mule and white-tailed deer. Journal of Wildlife Management 50:684–687.

Wood, J. E. 1958. Age structure and productivity of a gray fox population. Journal of Mamrnalogy 39:74–86.

Wood, M. D., and N. A. Slade. 1990. Comparison of ear-tagging and toe clipping in prairie voles, *Microtus ochrogaster*. Journal of Mammalogy 71:252–255.

Woodbury, A. M. 1956. Uses of marking animals in ecological studies: marking amphibians and reptiles. Ecology 37:670–674.

Woodbury, A. M., and R. Hardy. 1948. Studies of the desert tortoise, *Gopherus agassizii*. Ecological Monographs 18:145–200.

Woodcock, C. E. 1996. On roles and goals for map accuracy assessment: a remote sensing perspective. Pages 535–540 in T. H. Mowrer, R. L. Czaplewski, and R. H. Hamre, editors. Spatial accuracy assessment in natural resources and environmental sciences: second international symposium. General Technical Report RM-GTR-277, U.S. Forest Service, Department of Agriculture, Washington, D.C., USA.

Woodhams, D. C., R. A. Alford, C. J. Briggs, M. Johnson, and L. A. Rollins-Smith. 2008. Life-history trade-offs influence disease in changing climates: strategies of an amphibian pathogen. Ecology 89:1627–1639.

Woodward, A., K. M. Hutten, J. R. Boetsch, S. A. Acker, R. M. Rochefort, M. M. Bivin, and L. L. Kurth. 2009. Forest vegetation monitoring protocol for national parks in the North Coast and Cascades network. U.S. Geological Survey Techniques and Methods 2-A8, U.S. Geological Survey, Corvallis, Oregon, USA. http://pubs.usgs.gov/tm/tm2a8.

Woodward, A. J. W., S. D. Fuhlendorf, D. M. Leslie, Jr., and J. Shackford. 2001. Influence of landscape composition and change on lesser prairie-chicken (*Tympanuchus pallidicinctus*) populations. American Midland Naturalist 145:261–274.

Woolcock, S. C. 1985. Target characteristics. Pages 2(a)1–18 in J. Clarke, editor. Advances in radar techniques. Peter Pergrinus Press, Somerset, England, UK.

Woolf, A., and J. D. Harder. 1979. Population dynamics of a captive white-tailed deer herd with emphasis on reproduction and mortality. Wildlife Monographs 67.

Woolf, A., and C. Nielson. 2002. The bobcat in Illinois. Southern Illinois University, Carbondale, USA.

Woolley, H. P. 1973. Subcutaneous acrylic polymer injections as a marking technique for amphibians. Copeia 1973:340–341.

Wooten, W. A. 1955. A trapping technique for band-tailed pigeons. Journal of Wildlife Management 19:411–412.

Work, T. M. 2000a. Avian necropsy manual for biologists in remote refuges. Hawaii Field Station, National Wildlife Health Center, U.S. Geological Survey, Department of the Interior, Honolulu, USA. http://www.nwhc.usgs.gov/publications/necropsy_manuals/Avian_Necropsy_Manual-English.pdf.

Work, T. M. 2000b. Sea turtle necropsy manual for biologists in remote refuges. Hawaii Field Station, National Wildlife Health Center, U.S. Geological Survey, Department of the Interior, Honolului, USA. http://www.nwhc.usgs.gov/publications/necropsy_manuals/Sea_Turtle_Necropsy_Manual-English.pdf.

Woronecki, P. P., R. A. Dolbeer, T. W. Seamans, and W. R. Lance. 1992. Alpha-chloralose efficacy in capturing nuisance waterfowl and pigeons and current status of FDA registration. Proceedings of the Vertebrate Pest Conference 15:72–78.

Woronecki, P. P., and W. L. Thomas. 1995. Status of alpha-chloralose and other immobilizing/euthanizing chemicals within the animal damage control program. Proceedings of the Eastern Wildlife Damage Control Conference 6:123–127.

Worton B. J. 1987. A review of models of home range for animal movement. Ecological Modelling 38:277–298.

Worton, B. J. 1989. Kernel methods for estimating the utilization distribution in home-range studies. Ecology 70:164–168.

Worton, B. J. 1995. Using Monte Carlo simulation to evaluate kernel-based home range estimators. Journal of Wildlife Management 59:794–800.

Wright, E. G. 1939. Marking birds by imping feathers. Journal of Wildlife Management 3:238–239.

Wright, P. L. 1947. The sexual cycle of the male long-tailed weasel (*Mustela frenata*). Journal of Mammalogy 28:343–352.

Wright, P. L., and R. Rausch. 1955. Reproduction in the wolverine (*Gulo gulo*). Journal of Mammalogy 36:346–355.

Wright, S. 1931. Isolation by distance. Genetics 28:114–138.

Wright, S. 1951. The genetical structure of populations. Annals of Eugenetics 15:323–353.

Wunz, G. A. 1984. Rocket net innovations for capturing wild turkeys and waterfowl. Transactions of the Northeast Section of The Wildlife Society 41:219.

Wunz, G. A. 1987. Rocket-net innovations for capturing wild turkeys. Turkitat 6(2):2–4.

Wurman, J., J. Straka, E. Rasmussen, M. Randall, and A. Zahrai. 1997. Design and deployment of a portable, pencil-beam, pulsed, 3-cm Doppler radar. Journal of Atmospheric and Oceanic Technology 14:1502–1512.

Würsig, B., and M. Würsig. 1977. The photographic determination of group size, composition, and stability of coastal porpoises (*Tursiops truncatus*). Science 198:755–756.

Wydoski, R. S., and D. E. Davis. 1961. The occurrence of placental scars in mammals. Proceedings of the Pennsylvania Academy of Science 35:197–204.

Wydowski, R., and L. Emery. 1983. Tagging and marking. Pages 215–237 in L. A. Nielsen, D. L. Johnson, and S. S. Lampton, editors. Fisheries techniques. American Fisheries Society, Bethesda, Maryland, USA.

Xu, S., C. J. Kobak, and P. E. Smouse. 1994. Constrained least squares estimation of mixed population stock composition from mtDNA haplotype frequency data. Canadian Journal of Fisheries and Aquatic Sciences 51:417–425.

Yagi, T., M. Nishiwaki, and M. Nakajima. 1963. A preliminary study on the method of time marking with lead salt and tetracycline on the teeth of fur seals. Whales Research Institute, Science Report 7:191–195.

Yahner, R. H., and C. G. Mahan. 1992. Use of a laboratory restraining device on wild red squirrels. Wildlife Society Bulletin 20:399–401.

Yamamoto, J. T., R. M. Donohoe, D. M. Fry, M. S. Golub, and J. M. Donald. 1996. Environmental estrogens: implications for reproduction in wildlife. Pages 31–51 in A. Fairbrother, L. N. Locke, and G. L. Huff, editors. Noninfectious diseases of wildlife. Second edition. Iowa State University Press, Ames, USA.

Yannic, G., P. Basset, and J. Hausser. 2008. Phylogeography and recolonization of the Swiss Alps by the Valais shrew (*Sorex antinorii*), inferred with autosomal and sex-specific markers. Molecular Ecology 17:4118–4133.

Yates, F. 1951. The influence of statistical methods for research workers on the development of the science of statistics. Journal of the American Statistical Association 46:19–34.

Yeatman, H. C. 1960. Population studies of seed-eating mammals. Journal of the Tennessee Academy of Science 35:32–48.

Yerkes, T. 1997. A trap for ducks using artificial nesting structures. Journal of Field Ornithology 68:147–149.

Yoakum, J. D. 1978. Pronghorn. Pages 103–121 in J. L. Schmidt and D. L. Gilbert, editors. Big game of North America. Stackpole, Harrisburg, Pennsylvania, USA.

Yochem, P. K., B. S. Stewart, R. L. DeLong, and D. P. DeMaster. 1987. Diel haul-out patterns and site fidelity of harbor seals (*Phoca vitulina richardsi*) on San Miguel Island, California, in autumn. Marine Mammal Science 3:323–332.

York, D. L., J. E. Davis, Jr., J. L. Cummings, and E. A. Wilson. 1998. Pileated woodpecker capture using a mist net and taped call. North American Bird Bander 23:81–82.

Yosef, R., and F. E. Lohrer. 1992. A composite treadle/bal-chatri trap for loggerhead shrikes. Wildlife Society Bulletin 20:116–118.

Young, A. D. 1988. A portable candler for birds' eggs. Journal of Field Ornithology 59:266–268.

Young, J. A., R. A. Evans, and B. A. Roundy. 1983. Quantity and germinability of *Oryzopsis hymenoides* in seed in Lahontan sands. Journal of Range Management 36:82–86.

Young, J. B. 1941. Unusual behavior of a banded cardinal. Wilson Bulletin 53:197–198.

Young, J. G., and S. E. Henke. 1999. Effects of domestic rabbit urine on trap response in cottontail rabbits. Wildlife Society Bulletin 27:306–309.

Young, J. R. 1994. The influence of sexual selection on phenotypic and genetic divergence among sage grouse populations. Dissertation, Purdue University, West Lafayette, Indiana, USA.

Young, J. R. 2005. Animal behavior: its role in wildlife biology. Pages 616–631 *in* C. E. Braun, editor. Techniques for wildlife investigations and management. The Wildlife Society, Bethesda, Maryland, USA.

Young, J. R., C. E. Braun, S. J. Oyler-Mccance, J. W. Hupp, and T. W. Quinn. 2000. A new species of sage-grouse (Phasianidae: *Centrocercus*) from southwestern Colorado. Wilson Bulletin 112:445–453.

Young, J. R., J. W. Hupp, J. W. Bradbury, and C. E. Braun. 1994. Phenotypic divergence of secondary sexual traits among sage grouse, *Centrocercus urophasianus*, populations. Animal Behavior 47:1353–1362.

Young, L. S., and M. N. Kochert. 1987. Marking techniques. National Wildlife Federation, Scientific Technical Series 10:125–156.

Young, S. P., and E. A. Goldman. 1944. The wolves of North America. Part I. Their history, life habits, economic status, and control. American Wildlife Institute, Washington, D.C., USA.

Yunger, J. A., R. Brewer, and R. Snook. 1992. A method for decreasing trap mortality of *Sorex*. Canadian Field-Naturalist 106:249–251.

Yunger, J. A., and L. A. Randa. 1999. Trap decontamination using hypochlorite: effects on trappability of small mammals. Journal of Mammalogy 80:1336–1340.

Yunick, R. P. 1990. Some banding suggestions at nest boxes. North American Bird Bander 15: 146–147.

Zacharow, M., W. J. Barichivich, and C. K. Dodd, Jr. 2003. Using ground-placed PVC pipes to monitor hylid treefrog capture devices. Southeastern Naturalist 2:575–590.

Zahm, G., E. S. Jemisom, and R. E. Kirby. 1987. Behavior and capture of wood ducks in pecan orchards. Journal of Field Ornithology 58:474–479.

Zajac, A. M., and G. A. Conboy. 2006. Veterinary clinical parasitology. Seventh edition. Blackwell, Ames, Iowa, USA.

Zalewski, A. 1999. Identifying sex and individuals of pine marten using snow track measurements. Wildlife Society Bulletin 27:28–31.

Zani, P. A., and L. J. Vitt. 1995. Techniques for capturing arboreal lizards. Herpetological Review 26:136–137.

Zar, J. H. 1996. Biostatistical analysis. Third edition. Prentice-Hall, Englewood Cliffs, New Jersey, USA.

Zar, J. H. 1999. Biostatistical analysis. Fourth edition. Prentice-Hall, Upper Saddle River, New Jersey, USA.

Zar, J. H. 2010. Biostatistical analysis. Fifth edition. Pearson Prentice-Hall, Upper Saddle River, New Jersey, USA.

Zarrow, M. X., J. M. Yochim, and J. L. McCarthy. 1964. Experimental endocrinology: a sourcebook of basic techniques. Academic Press, New York, New York, USA.

Zaugg, S., G. Saporta, E. van Loon, H. Schmaljohann, and F. Liechti. 2008. Automatic identification of bird targets with radar via patterns produced by wing flapping. Journal of the Royal Society Interface 5:1041–1053.

Zelin, S., J. C. Jofriet, K. Percival, and D. J. Abdinoor. 1983. Evaluation of humane traps: momentum thresholds for four furbearers. Journal of Wildlife Management 47:863–868.

Zemlicka, D. E., D. P. Sahr, P. J. Savarie, F. F. Knowlton, F. S. Blom, and J. L. Belant. 1997. Development and registration of a practical tranquilizer trap device (TTD) for foot-hold traps. Proceedings, Great Plains Wildlife Damage Control Workshop 13:42–45.

Zeng, Z., and J. H. Brown. 1987. A method for distinguishing dispersal from death in mark–recapture studies. Journal of Mammalogy 68:656–665.

Zepp, R. L., Jr., and R. L. Kirkpatrick. 1976. Reproduction in cottontails fed diets containing a PCB. Journal of Wildlife Management 40:491–495.

Zicus, M. C. 1975. Capturing nesting Canada geese with mist nets. Bird-Banding 46:168–169.

Zicus, M. C. 1989. Automatic trap for waterfowl using nest boxes. Journal of Field Ornithology 60:109–111.

Zicus, M. C., and R. M. Pace, III. 1986. Neckband retention in Canada geese. Wildlife Society Bulletin 14:388–391.

Zicus, M. C., D. F. Schultz, and J. A. Cooper. 1983. Canada goose mortality from neckband icing. Wildlife Society Bulletin 11:286–290.

Ziegler, T. E., G. Scheffler, and C. T. Snowdon. 1995. The relationship of cortisol levels to social environment and reproductive functioning in female cotton-top tamarins, *Saguinus edipus*. Hormones and Behavior 29:407–424.

Zimmerling, T. N. 2005. The influence of thermal protection on winter den selection by porcupines, *Erethizon dorsatum*, in second-growth conifer forests. Canadian Field-Naturalist 119:159–163.

Zimmermann, B., P. Wabakken, H. Sand, and H. C. L. O. Pedersen. 2007. Wolf movement patterns: a key to estimation of kill rate? Journal of Wildlife Management 71:1177–1182.

Zink, R. M. 2004. The role of subspecies in obscuring avian biological diversity and misleading conservation policy. Proceedings of the Royal Society of London B 271:561–564.

Zippin, C. 1958. The removal method of population estimation. Journal of Wildlife Management 22:82–90.

Zmud, M. E. 1985. Marking of the redshank *Tringa totanus* in the north-western Pricernomorije. The Ring 11:7–15.

Zoellick, B. W., and N. S. Smith. 1986. Capturing desert kit foxes at dens with box traps. Wildlife Society Bulletin 14:284–286.

Zrnic, D. S., J. F. Kimpel, D. E. Forsyth, A. Shapiro, G. Crain, R. Ferek, J. Heimmer, W. Benner, T. J. McNellus, and R. J. Vogt. 2007. Agile-beam phased array radar for weather observations. Bulletin of the American Meteorological Society 88:1753–1766.

Zrnic, D. S., and A. V. Ryzhkov. 1998. Observations of insects and birds with a polarimetric radar. IEEE Transactions on Geoscience and Remote Sensing 36:661–668.

Zuberogoitia, I., J. E. Martínez, J. A. Martínez, J. Zabala, J. F. Calvo, A. Azkona, and I. Pagán. 2008. The dho-gaza and mist net with Eurasian eagle-owl (*Bubo bubo*) lure: effectiveness in capturing thirteen species of European raptors. Journal of Raptor Research 42:48–51.

Zug, G. R., G. H. Balazs, J. A. Wetherall, D. M. Parker, and S. K. K. Murakawa. 2002. Age and growth in Hawaiian green sea turtles (*Chelonia mydas*): skeletochronology. Fishery Bulletin 100:117–127.

Zug, G. R., and R. F. Glor. 1999. Estimates of age and growth in a population of green sea turtles (*Chelonia mydas*) from the Indian River Lagoon system, Florida: a skeletochronological analysis. Canadian Journal of Zoology 76:1497–1506.

Zug, G. R., L. J. Vitt, and J. P. Caldwell. 2001. Herpetology—an introductory biology of amphibians and reptiles. Second edition. Academic Press, New York, New York, USA.

Zug, G. R., A. H. Wynn, and C. Ruckdeschel. 1986. Age determination of loggerhead sea turtles, *Caretta caretta*, by incremental growth marks in the skeleton. Smithsonian Contributions to Zoology 427, Smithsonian Institution, Washington, D.C., USA.

Zwank, P. J. 1981. Effects of field laparotomy on survival and reproduction of mule deer. Journal of Wildlife Management 45:972–975.

Zwickel, F. C. 1971. Use of dogs in wildlife management. Pages 319–324 *in* R. H. Giles, editor. Wildlife management techniques. The Wildlife Society, Washington, D.C., USA.

Zwickel, F. C. 1972. Removal and repopulation of blue grouse in an increasing population. Journal of Wildlife Management 35:1141–1152.

Zwickel, F. C. 1980. Use of dogs in wildlife biology. Pages 531–536 *in* S. D. Schemnitz, editor. Wildlife techniques manual. The Wildlife Society, Washington, D.C., USA.

Zwickel, F. C. 1992. Blue grouse. Account 15 *in* A. Poole, P. Stettenheim, and F. Gill, editors. The birds of North America. The Academy of Natural Sciences, Philadelphia, Pennsylvania, and The American Ornithologists' Union, Washington, D.C., USA.

Zwickel, F. C., and A. Allison. 1983. A back marker for individual identification of small lizards. Herpetological Review 14:82.

Zwickel, F. C., and J. F. Bendell. 1967. A snare for capturing blue grouse. Journal of Wildlife Management 31:202–204.

Zwickel, F. C., M. A. Degner, D. T. McKinnon, and D. A. Boag. 1991. Sexual and subspecific variation in the numbers of rectrices of blue grouse. Canadian Journal of Zoology 69:134–140.

Zwickel, F. C., and A. N. Lance. 1966. Determining the age of young blue grouse. Journal of Wildlife Management 30:712–717.

Zwickel, F. C., and C. F. Martinsen. 1967. Determining age and sex of Franklin spruce grouse by tails alone. Journal of Wildlife Management 31:760–763.

INDEX

Page numbers followed by the letters b, f, and t indicate boxes, figures, and tables, respectively.

abundance
  absolute, 285
  relative, 285
  of resources, 411, 419
  true, 285, 286
abundance estimation, 284–310
  categories of survey methods in, 290–91, 291f
  with census (total counts), 290, 292–93
  with counts on sample plots
    detection probability in, 290–91, 299–304
    estimating area for, 296–97
    in fixed area, 293–96
    plotless methods for, 297–99
  definitions of terms in, 285–86
  dogs used in, 143, 144t
  indices in, 290, 291–92
  marked–resight methods of, 306–9, 493–94
  molecular techniques for, 543
  radiotelemetry in, 492–95
  remote cameras in, 315–16
  removal methods of, 302, 304–6
  software programs in, 309, 310b, 499
  in species richness estimation, 60
  survey design in, 286–90
A-C, 80–81
acarines, description of, 338
accessibility sampling, 288
accuracy
  definition of, 285
  and precision, in abundance estimation, 285–86, 286f
  in sampling, 24–25, 24f

acepromazine, 122
acorns, sampling techniques for, 401–2
ACTH, 516, 521–22
active infrared (AIR) remote cameras, 313, 315t
activity patterns, remote cameras in studies of, 316
activity sensors, 275
acupuncture, for respiratory distress, 133, 133f
acute, definition of, 204
adaptation, to captivity, 541
adaptive cluster sampling, 27f, 28–29
adaptive management, 39–40
  and Bayesian data analyses, 3
  definition of, 39–40
  field experiments in, 14
  modeling in, 15, 23
  monitoring in, 40
adaptive resource management, 40
adhesives. See glue
ad libitum sampling, 473, 473t
adrenaline, 521
adrenocorticotropic hormone (ACTH), 516, 521–22
aerial photography
  census with, 292
  in vegetation sampling, 404, 405–6
aerial remote sensing, vs. radar, 319
aflatoxins, 170
AFLPs, 527–28, 527t
after hatch year (AHY), determination of, 211
after second year (ASY), determination of, 211

AFWA, on capture of mammals, 81, 87–88, 89, 90
age
  birth rates dependent on, 364–68
  in chemical immobilization, 129
  death rates dependent on, 364–68
  in disease investigations, 192
  at harvest, in statistical population reconstruction, 495
  identification of (See age determination)
age class, in fertility estimation, 357
age determination, 207–29
  for birds, 207, 209–13
  for mammals, 207, 218–29
  morphological characteristics in, 208
  physical characteristics in, 207–8
  for trees, 399
age distribution
  definition of, 366
  in population growth models, 366–67
  stable, 366–67, 369–70
  survival rate estimation from, 369–71
age interval, in fertility estimation, 357
age-specific life tables, 368
age structure, and genetic diversity, 538–39
aggregated distribution, of vegetation, 386
agriculture. See also pesticides
  radar applications in, 333, 333f
AHAS, 332
AHY, 211
AIC, 376
airborne lidar, 452–53
airborne radar, 327

air cannons, 72
aircraft
    in airport surveillance radar, 326
    GPS tracking of, 443
    ultralight, migratory behavior taught with, 464, 465f
    wildlife collisions with, 332
air drying, of invertebrate specimens, 346
airport surveillance radar (ASR), 326
air pressure, sensory perception of, 470
AIR remote cameras, 313
air samples, in contaminant investigations, 178–79
Akaike's Information Criterion (AIC), 376
AKC, 141
alarm call recordings, in bird capture, 67
albatross, Laysan, transmitters on, 273f
albatross, short-tailed, transmitters on, 273f
algae
    harmful blooms of, 169–70
    natural toxins produced by, 169–70
alkaloids, in botanical insecticides, 166–67
Allee effect, 350b, 479t
allelic dropout, 543
allelic richness, 538
alligator(s)
    capture methods for, 111
    handling methods for, 113
    invasive marking of, 247t, 249t, 251
alligator, American
    capture methods for, 111
    invasive marking of, 249t, 252t, 254t
    noninvasive marking of, 234t, 241t
allomones, 469
allozymes, 527, 527t
alpha-adrenergic antagonists, 124
alphachloralose (A-C), 80–81
alpha-2 adrenergic agonists, 122, 124
altricial young, 209
American Kennel Club (AKC), 141
American National Standards Institute (ANSI), on metadata, 458–59
amniotes, 503
amphibians. *See also specific species and types*
    age and sex determination for, 229
    capture methods for, 104–7
        cover boards in, 106–7
        by hand, 104–5
        nets in, 65t, 105
        reference sources on, 65
        traps in, 86, 105–6
    chemical immobilization of, 136–37
    clutch size of, 507
    contaminant impacts on, 167
    handling methods for, 112
    marking of
        invasive, 244, 251, 255
        natural, 232t
        noninvasive, 239–40, 240t, 241t
    parental care by, 512
    reproduction of, 503, 503f, 507, 512

    transmitters on, 266
    worldwide decline in, 502
amplification failure, 543
amplified fragment length polymorphisms (AFLPs), 527–28, 527t
amputation, marking with, 255
anaconda, handling methods for, 112
anal glands, 469
analysis of variance (ANOVA), 50–51
anatoxin-a, 169
androgen, 514, 514f, 517t, 519
anesthetics
    inhalation, 123–24, 123f, 137
    true, 122
angle gauges, 392
angle-order method, 389
animal behavior. *See* behavior
animal-borne video and environmental data collection systems (AVEDs), 276
Animal Medicinal Drug Use Clarification Act of 1994, 119
animal tissues
    in contamination investigations, 174–78
    in disease investigations, 192–95, 197, 198b, 199t
    DNA extraction from, 529–30
    marking by removal of, 252–55, 253–54t
animal welfare
    in marking of wildlife, 230, 231
    in radiotelemetry studies, 263
    in trapping of mammals, 88, 89–90, 91t, 92t
Animal Welfare Act, 64
anole
    invasive marking of, 252t
    natural markings of, 232t
anole, green, reproduction of, 504, 516, 517t
anomalous propagation, 328, 329f
ANOVA, 50–51
ANSI, on metadata, 458–59
ant(s)
    description of, 337
    insecticides for, in capture of mammals, 86
antagonists, in chemical immobilization, 122, 122t, 123, 124, 135, 136
antbird, reproduction of, 519
anteater, giant, scat of, detection by dogs, 149
antelope
    handling methods for, 112
    noninvasive marking of, 243t
antennas
    considerations in configuration of, 269b
    efficiency of, 270
    of implant transmitters, 266
    radar, 322–23, 322f, 324–25
    in VHF systems, 270–72, 271f, 272f, 280–81
    whip, 270–71
anthrax, 185
anthropomorphism, and sensory perception, 466
antibiotics, in chemical immobilization, 134

antibodies
    definition of, 204
    in disease investigations, 195
    in immunocontraception, 521
anticholinesterase (anti-ChE) insecticides, 155, 165–66, 173t
anticoagulant rodenticides, 155–56, 168, 173t
antifertility agents, 521
antifreeze, 160
antigens, definition of, 204
antiserum, 524
antler size, analysis of data on, 48–49, 49t, 52–53, 52f
antral (Graafian) follicles, 508, 508f
anurans. *See also* frog(s); toad(s)
    invasive marking of, 246, 252, 252t, 255
    noninvasive marking of, 237t
aphanitoxins, 169–70
aphids, description of, 337
apparent survival, 358, 361
applied research, 6
aquatic drift fences, 105, 106
aquatic ecosystems, invertebrates in monitoring of, 338
aquatic funnel traps, 106
aquatic nets, 338, 338f, 339f
aquatic sampling, of invertebrates, 338–39
arboreal mammals, capture methods for, 84, 85f, 86, 86f
Argos system, 274, 275
armadillo, giant, scat of, detection by dogs, 149
armadillo, nine-banded, capture methods for, 83t, 90
arm bands, 236–38
array systems, sound, 476
arsenic
    in herbicides, 167
    mortality and morbidity from, 157–58, 160
arthropods, 336–48
    classification of, 336–38
    groups of interest, 337–38
artificial insemination, 505, 518, 519
artificial resuscitation, 133, 133f
A-scope display, 331b, 331f
aspen, density of, 389
aspiration pneumonia, 134
ASR, 326
assassin bugs, description of, 337
assays
    hormone, 524
    radioimmunoassays, 518, 524
assignment methods, 535, 544
associated patterns of reproduction, 514, 516
association, laws of, 1
Association of Fish and Wildlife Agencies (AFWA), on capture of mammals, 81, 87–88, 89, 90
assumptions, in theories, 11b
ASY, 211
atipamezole, 124

atmospheric pressure, sensory perception of, 470
atresia, 508–9
attractants
　in capture of mammals, 93–96
　in capture of reptiles, 109
　in collection of invertebrates, 342–43
audio lures, in capture of birds, 80, 80t. *See also* bird call recordings
Australian crow trap, 79
autocorrelation, 483, 487
automobiles. *See* vehicle(s)
AVEDs, 276
Avian Hazard Advisory System (AHAS), 332
avian influenza, H5N1 strain of, 181
avicides, 168–69
avitrol, 168–69
avocet, American, capture methods for, 77t
azaperone, 122

baboon, yellow, radiocollars on, 265f
BACI design, 15
Bacillus anthracis, 185
backpacks
　in marking of wildlife, 238–39, 239f, 239t
　in radiotelemetry, 265
bacteria, natural toxins produced by, 169
bacteriology, 193, 194, 204
badger, American
　age and sex determination for, 222t
　capture methods for, 96
　chemical immobilization of, 139t
　remote cameras in studies of, 315
badger, European
　behavior of, 471, 477
　invasive marking of, 246t, 249t, 254t
　noninvasive marking of, 240t
baits
　in capture of mammals, 94
　in remote camera studies, 316
bal chatri traps, 70, 70f, 109
BAM, 123
band(s), marking of wildlife with, 235–38, 236f, 237t, 238f
band recovery, in survival rate estimation, 362–64
bandwidth, in kernel density estimation, 486, 487t
bandwidth matrices, 486
bar graphs, 45, 45f
barometric pressure, sensory perception of, 470
basal area factors, 392
basal cover, sampling techniques for, 390–93
base stations, GPS, 442–43, 445b
basic research, 6
basket traps, 78
basking traps, 110, 110f
bat(s)
　age and sex determination for, 207, 208f, 224t

capture methods for, 82, 82f
carcasses of, dogs in location of, 145t, 147, 148f, 152
guano of, detection by dogs, 149–50
handling methods for, 112
invasive marking of, 248, 249t, 250t, 252t, 254t, 255
molecular techniques in studies of, 545
noninvasive marking of, 234t, 236–38, 237t, 240, 240t, 241, 242t, 243t
rabies management in, 201
radar in studies of (*See* radar)
sensory perception in, 467, 468
transmitters on, 265, 266
bat, African free-tailed, capture methods for, 82
bat, big brown, invasive marking of, 247t
bat, fringed myotis, age and sex determination for, 208f
bat, lesser horseshoe, population estimation for, 543
bat, little brown, reproduction of, 517t
bat, noctule, molecular techniques in tracking of, 544
bat, red, transmitters on, 267f
batteries
　in remote cameras, 312
　transmitter, 268, 274
Bayesian data analyses, 36–37, 55–56
　in adaptive management, 3
　advantages of, 36–37
　definition of, 36
　limitations of, 56
　vs. other approaches, 34, 36f, 55
　steps in, 37
Bayes's rule (theorem), 37, 55–56
beam shapes, in radar, 322–23, 323f
bear(s)
　capture methods for, 83, 84, 84f
　chemical immobilization of, 130, 138
　invasive marking of, 245t, 249t, 250t
　scat of, detection by dogs, 148–49, 148f
bear, black
　age and sex determination for, 221t, 228f
　behavior of, 470, 471
　capture methods for, 83, 84t, 89t, 90, 145t, 150
　chemical immobilization of, 138
　gene flow in, 535
　habitat models of, 434
　handling methods for, 112
　invasive marking of, 246t
　molecular techniques in tracking of, 544
　population estimation for, 493
　remote cameras in studies of, 313
　reproduction of, 520
　scat of, detection by dogs, 148–49, 520
　sensory perception in, 470
bear, brown
　age and sex determination for, 221t, 228f
　behavior of, 475

capture methods for, 84t
chemical immobilization of, 138
dogs in management of, 150
evolutionary significant units of, 533
reproduction of, 512
scat of, detection by dogs, 148–49
bear, grizzly
　capture methods for, 83, 84t, 93, 94t
　GPS collars on, 444
　population estimation for, 493, 543
　reproduction of, 520
　scat of, detection by dogs, 149, 520
bear, polar
　age and sex determination for, 221t, 229
　chemical immobilization of, 138
　integration of GIS and GPS data on, 430f
　noninvasive marking of, 234t, 242t
beat sheets, 344, 344f
beaver(s)
　capture methods for, 95
　chemical immobilization of, 139, 139t
beaver, American
　age and sex determination for, 223t
　capture methods for, 65t, 82, 84t, 89, 89t, 91t, 96, 96f, 97f
　invasive marking of, 250t, 252t, 254t
　noninvasive marking of, 240t
　reproduction of, 507, 509, 512
　transmitters on, 266
beaver, mountain
　capture methods for, 84t
　invasive marking of, 245t
bedbugs
　collection methods for, 343
　description of, 337
bees, description of, 337
beetles
　collection methods for, 341, 344
　description of, 337
before–after/control–impact (BACI) design, 15
behavior, animal, 462–79
　in age and sex determination, 208, 208f
　anthropomorphism of, 466
　applications in wildlife management for, 462–66
　defining, 472–73
　dogs in studies of, 150
　future of study of, 478–79, 479t
　history of study of, 462
　human behavior and, 477–78, 477f
　hypotheses on, development of, 471–72
　importance of understanding, 462
　marking of wildlife and, 234, 472
　measurement methods for, 473, 474–75, 475f
　and movement patterns, 488–89
　new technologies in, 475–77
　observations of, 472, 473
　in resource selection, 492
　sampling methods for, 473–74, 473t

behavior, animal (*continued*)
  sensory perception in, 466–71
  in taxonomy, 530
  transmitters' effects on, 263
bells, marking of wildlife with, 241–42t, 242, 242f
benzenes, chlorinated, 164–65
benzodiazepine tranquilizers, 122
Berlese funnels, 344
Bernoulli effect, 142
berries, sampling techniques for, 403
Best Management Practices (BMP), in trapping of mammals, 87–88, 90, 96–104
Betalights, 241
bias
  in capture of mammals, 87
  definition of, 31, 285
  in model development, 47
  in radiotelemetry studies, 259, 260–61, 263–64
  in sampling, 24–25, 24f, 30
  ways of avoiding, 31, 38
binoculars, in capture of birds, 65
binomial distribution, 387
bioaccumulation
  definition of, 154
  of metals, 158–60
  of organic chemicals, 160–62
  of pesticides, 166
Biobserve, 475
biodiversity
  in forest planning, 434
  GIS in studies of, 434, 435–36, 437–38, 438f
biogeographic scale, 411, 412
Biological Data Profile, 458
biological echoes, 329–30
biological indicators, invertebrates as, 338
biological populations, 6–9
  criteria for identification of, 7
  definition of, 6
  vs. political populations, 7
biological species concept, 530
biomagnification
  definition of, 154
  of metals, 158–60
  of organic chemicals, 160–62
  of pesticides, 166
biomass
  definition of, 394
  sampling techniques for, 394–96
biopsies, needle, of testes, 515–16
biotransformation, of metals, 158, 159
birch, paper, density of, 389
bird(s). *See also specific species and types*
  abundance estimation for
    with census, 292, 293
    with counts on sample plots, 294–95, 296–97, 302–3
  age and sex determination for, 209–13
    DNA in, 546
    laparotomy in, 507

  physical characteristics in, 207, 208f
  plumage characteristics in, 209–13, 210t, 211f, 212f
  captive, in capture of other birds, 67–71, 69t, 70f, 76, 80
  capture of, 65–81
    nets in, 65–73, 65t, 71t, 72t, 73t, 76, 76t
    night lighting in, 65–66, 66f, 66t
    oral drugs in, 80–81
    reference sources on, 64
    traps in, 73–80
  carcasses of, dogs in location of, 147, 148f
  chemical immobilization of, 137
  contaminant impacts on
    from avicides, 168–69
    from inorganic chemicals, 163
    from metals, 158, 159
    from natural toxins, 169
    from organic chemicals, 161
    from pesticides, 165–66
    types of, 157
  density of, dogs in estimation of, 143–46
  dogs in studies of, 143–46, 144–45t, 147, 148f
  fertility estimation for, 357
  genetic diversity of, 535
  handling of, 111
  hormone measurements for, 523
  marking of
    invasive, 244–55
    natural, 231–32, 232t
    noninvasive, 234–44
    permits for, 231
    remote, 233, 233t
  migratory (*See* migratory birds)
  radar in studies of (*See* radar)
  reproduction of
    clutch size in, 507
    contraception for, 520, 521
    hormones in, 518, 519
    modes of, 503
    nest success in, 512
    ovulation in, 507
    testis weight in, 514
  sea (*See* seabirds)
  sensory perception in, 467–71
  transmitters on
    effects of, 263
    ways of attaching, 265–66, 266f
    weight of, 268
  water (*See* shorebirds; waterfowl)
bird call recordings, in capture techniques, 67, 69–70, 76, 80, 80t
bird eggs. *See* egg(s), bird
bird flu, H5N1 strain of, 181
bird odors
  in capture of reptiles, 109
  detection by dogs, 141
  sources of, 141
bird radar, 327
bird strikes, on aircraft, 332
birth, in reproductive cycle, 512

birth-flow fertility, 352
birth-pulse fertility, 351
birth rates
  age-dependent, 364–68
  individual consideration of components of, 372–74
  methods for estimation of, 356–58
  in population growth estimates, 371–72
bison
  age and sex determination for, 220t
  biological populations of, 7
  brucellosis in, 202
  chemical immobilization of, 137t
  population growth models for
    birth and death rates in, 356–57, 356f, 357f
    continuous-time, 352–53, 353f, 353t
    discrete-time, 354–55, 355t
bitterbrush, frequency of, 386
Bitterlich variable radius method, 392–93
blackbird(s)
  avicides for, 168
  invasive marking of, 249t
  noninvasive marking of, 237t, 241t, 243t
  radar in management of, 333
blackbird, red-winged
  capture methods for, 69t
  hierarchy of spatial population units of, 6–7, 9f
  remote marking of, 233t
blackbird, yellow-headed, capture methods for, 78
black lights, 343
bleach, marking of wildlife with, 242–44, 243t
bloat, emergency medicine for animals with, 133–34
blocking, in experimental design, 17
blood-brain barrier, pesticides and, 165
blood sampling
  DNA in, 529, 530t
  handling techniques in, 111
  for hormones, 522, 523
  procedures for, 111, 523
blow pipes, in chemical immobilization, 124, 125
bluebird, capture methods for, 77t
blue-green algae, 169
B lymphocytes, definition of, 204
BMP, in trapping of mammals, 87–88, 90, 96–104
boar, wild, thermal infrared cameras in studies of, 317
boats
  in capture of birds, 65, 66f, 67
  in capture of mammals, 92–93
  detection of scat by dogs on, 149
bobbins, 239
bobcat
  age and sex determination for, 228
  capture methods for, 84t, 89, 91t, 92t, 96–97
  chemical immobilization of, 138
  invasive marking of, 246t

natural markings of, 232f, 232t
remote cameras in studies of, 313
scat of, detection by dogs, 149
bobwhite, northern
age and sex determination for, 209, 210t, 215t
capture methods for, 66t, 69t, 79t, 81
dogs in studies of, 144t, 145, 148
habitat burning and, 17, 18, 19
handling methods for, 111
invasive marking of, 247t
body mass
in age and sex determination, 207, 208
in chemical immobilization, 129
in radar, 321, 321f
body size
in age and sex determination, 207, 208
of mammals, variation in, 218
body tags, 251
body temperature, sensors for, 275–76
boleadoras, 81
bone(s)
in age and sex determination of mammals, 218, 224f, 229
DNA extraction from, 530, 530t
bootstrap estimator, 61
botanical insecticides, 166–67
botulinum toxin, 169
botulism
dogs in carcass searches related to, 147
investigation of outbreaks, 192
production of toxin causing, 169
boundary effects, in abundance estimation, 297
bouts, in behavioral measurement, 475
bovids
invasive marking of, 252t
noninvasive marking of, 243t
bovine brucellosis, 202–3
bovine spongiform encephalopathy (BSE), 185
bovine tuberculosis, transmission of, 315, 471
Bowden's estimator, 493–94
bow nets, 73, 74f
box traps
capture of birds with, 79–80, 79t
capture of mammals with, 83–84t, 83–87, 88
capture of reptiles with, 111
brain
blood-brain barrier and, 165
in regulation of hormones, 516
branding of wildlife, 251–52, 252t
brant
age and sex determination for, 213t
noninvasive marking of, 234t
breeding seasons. See also reproduction
in fertility estimation, 357
in population growth models, 351, 354
testis weight in, 514
breeding status, dogs in studies of, 146
breeding systems, and gene flow, 534–35

brodifacoum, 168
brood parasitism, intra-specific, 465
brood patches, 518
Brownian bridges, 488
browsing
in habitat selection studies, 417
methods for measurement of, 419–20
brucellosis, in elk, 202–3
brush-turkey, Australian, reproduction of, 508
BSE, 185
buccal swabs, 530, 530t
budworm, western spruce, sources of mortality, 16
buffalo, American, brucellosis in, 202
bullfrog
hormones of, 517t, 519, 523
noninvasive marking of, 237t, 241t
reproduction of, 517t, 519
bunting, painted, capture methods for, 67
burning, prescribed
data collection on effects of, 43
in management of deer and elk populations, 39
bursa of Fabricius, 213
bushbuck, African, natural markings of, 232t
bustard, houbara, capture methods for, 73, 75t
bustard, little, handling methods for, 111
butorphanol, 123
butt-end bands, 238, 238f
butterflies, description of, 337
buzzard, common, capture methods for, 81

CA, 510b, 511
cabinets, insect, 347–48
cadmium, 159
cafeteria experiments, 411
cage traps
capture of birds with, 79–80, 79t
capture of mammals with, 83–84t, 83–87, 88, 96–104
calipers, in vegetation sampling, 398, 398f
calls. See bird call recordings; sound production
cameras. See also photograph(s); remote cameras; video cameras
in behavioral studies, 476–77
Canada, wildlife disease experts in, 189, 190b
candling techniques, 209
canids
capture methods for, 94t
chemical immobilization of, 138
contaminant impacts on, 170
canine distemper, 184
cannon nets
capture of birds with, 72, 72f, 72t
capture of mammals with, 72t, 83
canopy cover
lidar data on, 454, 454f
sampling techniques for, 390–93

canvasback
capture methods for, 69t, 87t
reproduction of, 518
capercaillie, dogs in studies of, 144t, 146, 150
captive breeding, 505, 540, 541
capture
government regulation of, 64
in habitat selection studies, 416
handling after, 111–13
marking after, 233–34
marking without, 233, 233t
methods of, 64–111 (See also specific techniques)
for amphibians, 65, 86, 104–7
for birds, 64, 65–81
efficiency of, 64
for invertebrates, 338–45
for mammals, 64–65, 81–104
reference sources on, 64–65, 81
for reptiles, 107–11
in radiotelemetry studies, 260–61
reasons for, 64, 118
capture grids, in population estimation, 493, 494–95
capture–mark–recapture
in abundance estimation, 306
in birth rate estimation, 357–58
in dispersal studies, 356
molecular techniques in, 542–44
in survival analysis, 56, 57
in survival rate estimation, 360–62
capture myopathy (CM), 111, 112, 134
CAPTURE program, 303, 309, 310b
capture–recapture, in abundance estimation, 306
capybara, capture methods for, 93
car. See vehicle(s)
caracal, reproduction of, 505
caracara, crested, capture methods for, 69t, 73, 75t
carbamate insecticides, 165–66
carbon dioxide, in collection of invertebrates, 343
carcasses
in contaminant investigations, 171–72
collection of tissues from, 176–78
scavenging of, 157, 164
searches for, 157, 171
in disease investigations, 189–92, 195–97
dogs in searches for, 145t, 147–48, 148f, 152
cardinal, northern, noninvasive marking of, 243t
cardiopulmonary resuscitation (CPR), 135
card mounting, of invertebrate specimens, 347
carfentanil, 118, 123, 136
caribou
age and sex determination for, 219t, 227t
capture methods for, 73t
chemical immobilization of, 137t
GPS collars on, 444
invasive marking of, 250t, 256f

carnivores
    chemical immobilization of, 132, 133f, 138
    emergency medicine for, 133f
    invasive marking of, 255
    rabies management in, 201–2, 203
carrying capacity, in population growth
        models, 353
cartography. See maps and mapping
cast-nets, capture of birds with, 65–66
cat, domestic, reproduction of, 505, 507
cat, feral
    capture methods for, 94, 98
    remote cameras in studies of, 316
cat, fishing, reproduction of, 505
cataleptic state, 122
catchability. See detection probability
catch-effort models, 359–60
catch-per-unit-effort, 305, 305f
categorical data, 54
cattle. See cow
causality, challenge of identifying, 1
cavity-nesting birds
    capture methods for, 77, 77t
    vegetation sampling in habitat of, 407
C-band, 321
CDC, on transportation of diagnostic
        specimens, 198
cell-mediated immunity, 184, 204
celluloid rings, 238
cementum annuli, in age and sex determina-
        tion of mammals, 228–29, 228f
censoring, in survival analysis, 58
census, definition of, 285, 413
census methods
    in abundance estimation, 290, 292–93
    definition of, 285, 290
    limitations of, 290
    in resource use studies, 413
census population size, 538–39
Centers for Disease Control (CDC), on
        transportation of diagnostic
        specimens, 198
Central Limit Theorem, 285
cervids
    remote cameras in studies of, 315
    sensory perception in, 469
cetaceans, marking of
    invasive, 250t, 251
    natural, 232t
    noninvasive, 234t, 242t
chain of custody
    in contaminant investigations, 174
    in disease investigations, 198
change-in-ratio method, 305–6, 360
Chao 1 method, 60
Chao 2 method, 60
CHAP, 432
Chardoneret traps, modified, 76, 76f
cheetah, natural markings of, 232t
chemical branding, 251, 252, 252t
chemical communication, 469
chemical contaminants. See contaminants

chemical drying, of invertebrate specimens,
        346
chemical immobilization, 118–39
    animal emergencies during, 132–34
    calculation of drug doses in, 119, 120b,
        136–39
    characteristics of ideal drug for, 119
    classes of drugs in, 120–24, 122t
    definition of, 118
    drug combinations in, 119–20
    equipment for, 124–28, 125b
    formal training in, 118
    guidelines by type of animal, 136–39
    human emergencies during, 134–36
    legal considerations with, 118–19
    preparation and procedures for, 128–32
    record keeping in, 120, 121b
    sites for drug injections, 130, 130f
    in urban areas, 131–32
chemical lights, marking of wildlife with,
        240–41
chemical markers of wildlife, 244–46, 245t
chemical residue analysis, 179
chemistry, clinical
    definition of, 204
    in disease investigations, 195
Chernobyl nuclear accident, 534
chickadee, black-capped, survival rate
        estimation for, 369
chipmunk, eastern, capture methods for, 83t
chipmunk, Townsend's, capture methods for,
        83t, 84
chipmunk, yellow pine, age and sex
        determination for, 229
chi-squared analyses, 50
    of resource selection, 420–22, 422t
chlorinated hydrocarbon (organochlorine)
        insecticides, 164–65
chlorophyll, 405
chromium, 159–60
chronic, definition of, 204
chronic wasting disease (CWD)
    mechanisms of, 185
    scientific attention to, 181, 182
    transmission of, 315
chukar, age and sex determination for, 209,
        209f, 215t
cicadas, description of, 337
circular plots, in vegetation sampling, 387–88
CIs, 24, 36, 49, 285
CJD, 185
CL. See corpora lutea
classical statistics. See frequentist approach
classification analysis, 63
clearinghouses
    metadata, 460
    software, 501
clinical signs
    of contaminant exposure, 172, 173t
    definition of, 204
    of disease, 192
clip-and-weigh method, 394

clipping techniques, for biomass, 394
cloacal structures, in age and sex determina-
        tion, 211, 213f
closed populations
    definition of, 285, 357–58
    estimation of birth rates for, 357–58
closest individual method, 389
clothing, protective, in contaminant
        investigations, 170
clouds, in Landsat imagery, 449
Clover traps, 83
cluster-based methods of home range
        estimation, 483t, 485
cluster sampling, 27f, 28, 29
cluster size, 28
clutch size
    definition of, 507
    measurement of, 507–8
clutter, in radar, 322, 323f, 325, 325f, 328–32
CM, 111, 112, 134
Coast Guard, U.S., GPS use by, 443–44
codominant markers, 527
coexistence of species, models of, 374
cohort(s)
    definition of, 365
    in estimation of survival rates, 368–69
cohort life tables, 368
coleopterans, description of, 337
collars, neck
    marking with, 234–35, 234t, 235f, 236f
    radiocollars, 265, 265f, 444
color
    in collection of invertebrates, 339, 341, 342
    of light, in light traps, 339
    in marking of wildlife, 233–34
        colored bands, 236, 238, 238b, 238f
        colored tape, 241
        external application of, 242–44, 243t
color vision, 468
Combined Habitat Assessment Protocols
        (CHAP), 432
communication
    need for studies on, 479t
    through sensory perception, 467–70
communities
    comparison of species richness among, 60
    diversity of (See species diversity)
    GIS in conservation of, 435–38, 438f
community analysis, 59
community structure, lidar in studies of, 455
comparative analyses, 50–51
competition, interspecific, models of, 374
competition coefficients, 374
compositional analysis, of resource selection,
        421, 421t
compound 1080, 168, 169
computer software. See software programs
concepts, in theories, 11b
conceptual models, in scientific method, 4
condor, California
    captive breeding of, 505
    contaminant impacts on, 159, 161

confidence intervals (CIs), 24, 36, 49, 285
conical beams, 323, 323f
conservation
    of genetic diversity, 533–41
    of wildlife communities
        gaps in, 435
        GIS in, 435–38, 438f
contact reception, 469–70
contaminants, 154–79. *See also specific types*
    classes of, 155, 157–70
    clinical signs of exposure to, 172, 173t
    definition of, 154
    genetic mutations caused by, 534
    history of, 155
    identification of, 155, 171
    impacts of, 154–55, 157
    investigations of, 170–79
        diagnostics in, 170–74
        vs. disease investigations, 170–72, 195
        field procedures in, 174–79
        laboratories recommended for, 174b
        record keeping in, 174, 176, 180f
    prevalence of, 155
    properties of, 154
    references sources on, 156b
    and reproduction, 518
    treatment for exposure to, 172
Content Standard for Digital Geospatial Metadata (CSDGM), 458–59
contingency tables, 50, 50t
continuous data, definition of, 44
continuous-time models
    definition of, 351
    of density-dependent population growth, 352–54
    differential equations in, 23
continuous variables, 20
contraception, wildlife, 520–21
controlled substances
    definition of, 119
    in tranquilizer trap devices, 90
Controlled Substances Act, 119
controls, definition of, 19
convenience sampling, 288
convulsions, during chemical immobilization, 134
coot, American
    age and sex determination for, 218t
    capture methods for, 76t, 77t, 80
    contaminant impacts on, 169
    invasive marking of, 249t
    noninvasive marking of, 239t
core areas, definition of, 482
Cormack–Jolly–Seber estimator, 358, 360
cormorant, capture methods for, 65t, 81
cormorant, double-crested
    capture methods for, 65, 66t, 75t
    molecular techniques in studies of, 544
corpora albicantia (CA), 510b, 511
corpora lutea (CL)
    hormones and, 518
    in mammals, 508, 508f, 509–11, 511f

as measure of reproductive rate, 503, 507
corpus hemorrhagicum, 509
corral traps, 87, 87t
corrected-point-distance method, 389–90
correlation, measurement of, 51
corridors, habitat, GIS in studies of, 435, 437
corridor suitability GIS database, 435
corticosteroids, 516, 521, 522
corticotropin releasing factor (CRF), 521–22
cottonwood, plains, sampling of, 407
cougar, age and sex determination for, 228
courtship, 504–5, 516, 518
covariance, 51, 291
cover, plant
    definition of, 390
    in resource selection studies, 418
    sampling techniques for, 390–93, 404
cover boards
    in capture of amphibians, 106–7
    in capture of reptiles, 107
    in vegetation sampling, 396–97, 396f, 397f
cover change, land, Landsat detection of, 447
cover classes, 391, 391b
cover poles, 397
cow, domestic, reproduction of, 509, 509f
cowbird, brown-headed
    capture methods for, 69t, 72t
    radar in management of, 333f
    sensory perception in, 468
Cox proportional hazard model, 57–58, 496–97
coyote
    age and sex determination for, 220t
    behavior of, 473
    capture methods for, 73t, 84t, 89, 89t, 90, 91t, 92t, 94–95, 97–98
    chemical immobilization of, 138
    contaminant impacts on, 163, 168
    dispersal patterns of, 356
    dogs as lure for, 147
    habitat of, 414
    home range of, 488
    hybridization of, 532
    invasive marking of, 245t, 246t, 250t, 254t
    molecular techniques in studies of, 545
    noninvasive marking of, 234t
    population density and fertility in, 354, 354t
    population estimation for, 493
    rabies management in, 201–2
    radiocollars on, 265f
    reproduction of, 507, 511
    scat of, detection by dogs, 148
CPR, 135
crabs, noninvasive marking of, 241
crane(s)
    capture methods for, 65t
    invasive marking of, 249t
    noninvasive marking of, 234t, 237t
crane, Mississippi sandhill, contaminant impacts on, 159
crane, sandhill
    age and sex determination for, 217t

capture methods for, 65, 66t, 76t, 81
contaminant impacts on, 170
as foster parents for whooping cranes, 463–64
crane, whooping
    age and sex determination for, 217t
    behavior of, 463–64, 464b, 464f, 465f
    capture methods for, 65, 66t, 81
    contaminant impacts on, 159
    genetic diversity of, 541
    population growth models for
        survival rates in, 358–59
        with unimpeded growth, 351–52, 351f, 351t
credibility, of hypotheses, 32
Creutzfeldt–Jakob disease (CJD), 185
CRF, 521–22
crickets, description of, 337
critical point drying, of invertebrate specimens, 346
crocodile, American, capture methods for, 111
crocodilians, handling methods for, 113
crossed design, 18
crossover experiments, 17–18
crow, American, capture methods for, 71, 72, 72t, 73, 79, 80
CSDGM, 458–59
culpeo, capture methods for, 94
culvert traps, 84, 84f
curatorial methods, for invertebrates, 345–48
curiosity scents, in capture of mammals, 94
curlew, long-billed, transmitters on, 273f
CWD. *See* chronic wasting disease
Cyalume, 240
cyanide, 163, 345
cyclodienes, 164–65
cyclohexanes, 122, 122t, 136, 137
cytochrome P-450s, 166
cytopathic effects of viruses, 194, 204

daily mortality rate for nests, 359b
damage management, dogs in, 150–51
dart(s), in chemical immobilization, 126–27, 127f
    accidental exposure of humans to, 134–36
    preparation of, 129
dart guns
    in capture of mammals, 83
    in chemical immobilization, 124, 125–27, 126f
        preparation of, 129
        range finders for, 127, 127f
dart pistols, in chemical immobilization, 124, 126
dart rifles, in chemical immobilization, 124, 125–26, 126f
data
    classifications of, 43–44, 50
    confronting theories with, 31–37
data analysis, 41–63
    Bayesian (*See* Bayesian data analyses)

data analysis (*continued*)
  bias in, 31
  of communities, 59
  comparative, 50–51
  descriptive statistics in, 48–50
  exploratory, as step in research process, 4, 9–10
  frequentist approach to, 34, 55
  generalized linear models in, 53–55
  graphical, 44–47
  linear regression in, 51–52
  mixed effects models in, 58–59
  model development in, 47–48
  multiple regression in, 52–53
  sampling design in, role of, 42
  software programs for, 62, 62t
  of species richness, 59–61
  of survival, 56–58
database management systems (DBMSs), 44
data collection, 42–44
  data entry and proofing after, 32, 43
  dogs used in, 140, 143–50
  identification of data needs in, 42
  methods for, 32, 42–43
  preliminary, as step in research process, 4, 9
  quality control in, 32
  time spent on, 32
  timing and frequency of, 43, 261
data continuity, in Landsat program, 446–47
data distribution, modes of, 49
data documentation, in spatial technologies, 458–60
data loggers, 408, 408f
data management, 43–44
  categories of data in, 43–44
  database management systems in, 44
  flat files in, 44
  in radiotelemetry studies, 262
  with remote cameras, 312–13
data presentation, options for, 44–48
data sharing, in radiotelemetry studies, 262
data sheets
  in contaminant investigations, 176, 180f
  design of, 42–43
  use of, 32, 42–43
  in vegetation sampling, 385
data storage, in remote cameras, 312
datum, geodetic, 445b
DBH, 390, 398, 398f
DBMSs, 44
DDT, 164–65
DEA, U.S., controlled substances under, 119
death (mortality) rates. *See also* mortality
  age-dependent, 364–71
  in estimation of population growth, 371–72
  individual consideration of components of, 372–74
  methods for estimation of, 356–57, 368–71
  survival rates as complement of, 358
deciduous placentae, 512

decision-making, structured, in adaptive management, 40
decoys
  in capture of birds, 67, 69–71, 69t, 73, 80
  and vulnerability to harvest, 39
decoy traps, capture of birds with, 80
deduction, 3f, 42
deer
  abundance estimation of, 29, 29f, 305, 306–7
  capture methods for, 82–83, 87t, 92–93, 93f
  census of, 292
  marking of
    invasive, 249t, 250t
    noninvasive, 239t, 242, 242t, 243t, 244
    remote, 233t
  ovarian analysis of, 510b, 511
  prescribed fires in management of, 39
  reproduction of, 510b, 511, 522
  weight in fawn survival of, 54, 55f
deer, black-tailed, age and sex determination for, 219t
deer, fallow, capture methods for, 72t
deer, Florida Key
  behavior of, 463
  remote cameras in studies of, 313
deer, Himalayan musk, capture methods for, 76t, 83
deer, Key, capture methods for, 82
deer, mule
  abundance estimation of, 304
  age and sex determination for, 208f, 218, 219t, 227f, 227t
  behavior of, 467, 469, 471–72
  capture methods for, 66t, 71t, 73t, 76t, 82, 83, 84t, 89t, 93, 93t, 94t
  chemical immobilization of, 137t
  dogs in studies of, 150
  noninvasive marking of, 240t
  population estimation for, 493, 494
  power analysis of production in, 33, 33t
  reproduction of, 511–12
  sensory perception in, 467, 469
  thermal infrared cameras in studies of, 317
  vegetation sampling in habitat of, 406–7
deer, Pere David's, invasive marking of, 249t
deer, red
  age and sex determination for, 228
  behavior of, 462–63
  genetic diversity of, 540
  integrated research on, 16
  ultrasonography of, 509
deer, Sitka, gene flow in, 535
deer, white-tailed
  age and sex determination for, 218, 219t, 224f, 225, 227f, 227t
  antler size of, 48–49, 49t, 52–53, 52f
  capture methods for, 71t, 72t, 73t, 76t, 82–83, 84t, 89t, 93, 93t, 94t, 96
  chemical immobilization of, 137t
  contaminant impacts on, 168, 170
  dogs in management of, 150

dogs in studies of, 150
  fertility estimation for, 357
  genetic diversity of, 537, 540
  habitat selection by, 412
  handling methods for, 112
  laparoscopy of ovaries of, 509
  lidar in studies of, 456
  marking of
    invasive, 245t, 248f, 250b, 251f, 252t
    natural, 232t
    noninvasive, 235, 235f, 242f
    permanence of, 233
    remote, 233t
  molecular techniques in studies of, 542, 544
  political *vs.* biological populations of, 7
  population estimation for, 493
  population growth models for
    age distribution in, 367, 367t, 369–70, 369t, 370t, 371
    attention given to, 349
    birth and death rates in, 365, 365t, 369–70, 369t, 370t, 372
    growth rate calculation in, 372
  radiocollars on, 265f
  remote cameras in studies of, 316
  reproduction of
    contraception for, 520, 521
    hormones in, 517t, 518, 519
    lactation in, 513
    testis weight in, 514
    uterine analysis of, 509, 511–12
  sensory perception in, 468, 468f
  stress in, 522
definitions, in theories, 11b
definitive hosts, 184–85, 204
definitive plumage, 211
DEM, 454, 456
demes
  criteria for identification of, 6–7, 7f, 8b
  definition of, 8b
density. *See also* population density
  definition of, 387
  of vegetation, 387–90
density boards, 396, 396f
density-dependent population growth, 352–56
  continuous-time model of, 352–54
  dangers of detecting, 355–56, 355t
  definition of, 350
  discrete-time model of, 354–55
  mechanisms of, 354b
density-independent population growth, definition of, 350
density indices, 291–92
dentition
  in age and sex determination of mammals, 218, 225, 227f, 227t, 228–29, 228f
  DNA in, 530, 530t
dependence, in experimental design, 17
dependent double-observer sampling, 300–301
dependent observations, 31
depilatory pastes, 253

descriptive natural history studies, 12
descriptive research, 1–3
descriptive statistics, 48–50
destructive sampling, 529
detectability. *See* detection probability
detection function, 303, 303f
detection probability
    in abundance estimation, 290–91, 299–304
    definition of, 285
deterministic models, definition of, 351
deterrents, nonlethal, dogs as, 150–51
detoxification enzyme systems
    glutathione, 158
    of mammals, 154
DGPS, 442–44
dho gaza nets, 69–70, 70f
diagnosis, definition of, 204
diameter at breast height (DBH), 390, 398, 398f
diazepam, 90
dichlorodiphenylethanes, 164–65
diclofenac, 164
diestrus, 505f, 506b
diet. *See also* food
    design of studies on, 14, 15
    molecular techniques in analysis of, 545
    paired observations of, 31
    remote cameras in studies of, 316
    of vertebrates, invertebrates in, 338, 344–45
differential equations, 23
differential Global Positioning System (DGPS), 442–44
digital elevation model (DEM), 454, 456
digital imaging systems, in vegetation sampling, 404
dilution of precision (DOP), 442
dimension analyses
    for biomass, 394–96
    for plant use, 400
dimorphic species, age and sex determination for, 207
dingo, molecular techniques in studies of, 545
dioxins, 162
diploid genotypes, 528
dip nets
    capture of amphibians with, 105
    capture of birds with, 65, 65t, 66f
    capture of mammals with, 65t, 82
dipterans, description of, 337
direct observation, of animal behavior, 472, 473
direct selection, 539
discrete-choice modeling, of resource selection, 421, 421t
discrete data, definition of, 44
discrete-return infrared (IR) lidar, 452–53, 452f
discrete-time models
    definition of, 351
    of density-dependent population growth, 354–55
disease(s), 181–204. *See also specific diseases*

vs. contaminant exposure, identification of, 171, 172
    definition of, 183, 204
    glossary of, 204–6
    history of research on, 181–82
    impact on populations, 181–82
    invertebrates as vectors of, 338
    investigations of, 189–98
        vs. contaminant investigations, 170–72, 195
        field observations in, 189–92, 191b
        field procedures in, 195–99, 196b, 197t, 198b, 199t
        laboratory procedures in, 192–95, 193t
        safety considerations with, 170, 171, 196
    management of, 200–203
        case studies on, 200–203
        habitat in, 186, 201, 203
        objectives of, 200
    processes of, 183–86
    public concern about, 181, 182
    reference sources on, 186–87, 187t, 188t, 193t, 197t
    resistance to, 184
    sources for experts on, 187–89, 190b
    taxonomies of, 183, 185
    transmission of, 184–85, 315
dispersal
    definition of, 356
    detection and measurement of, 356
    gene flow in, 534–35
    GIS in studies of, 437
    models of, 356
    need for studies on, 479t
    radiotelemetry in studies of, 259, 356, 490
    sex-specific, 534–35
dissociated patterns of reproduction, 514, 514f, 516
DISTANCE program, 145, 303, 309, 310b
distance sampling
    assumptions of, 304
    dogs in, 143–46
    methods of, 30, 296–97
    modern, 296, 303–4, 303f
distance vision, 468
distemper, canine, 184
disturbances
    caused by humans, 477–78
    in landscape, Landsat imagery of, 447–49, 448f, 451f, 452f
diversity. *See* genetic diversity; species diversity
DNA. *See also* genetic(s); molecular ecology
    contamination of, 543
    degraded, 543
    extraction techniques for, 529
    mitochondrial, 526, 527t, 531, 535
    nuclear, 526
    sampling of, 529–30, 530t, 542–44, 542b
    in scat, 148, 529, 542
    sequencing of, 527, 529, 531
DOBSERV program, 301

documentation. *See also* record keeping
    of samples, in contaminant investigations, 174
dog, bush, scat of, detection by dogs, 149
dog, domestic, 140–53
    in behavior studies, 150
    benefits of using, 140
    in capture of birds, 67, 144–45t, 150
    in capture of mammals, 90, 93, 144–45t, 150
    collection of specimens and carcasses by, 147–48, 152
    factors affecting performance of, 141–43, 149–50
    GPS tracking of, 143
    ground coverage abilities of, 140, 142
    history of, 141
    locating wildlife with, 143–47, 144–45t
    in marking of wildlife, 150
    natural markings on, 232t
    prey drive of, 143
    vs. remote cameras, 313
    reporting research results with, 152
    reproduction of, 507
    scat detection by, 148–50, 152, 520, 544
    scenting abilities of, 140–50
    training and handling of, 151–52
    types and breeds of, 141, 142t
    in wildlife damage management, 150–51
dolphin, bottlenose, ultrasonography of, 509
dolphin(s), invasive marking of, 245t, 252t
domain, of theory, 11b
dominant markers, 527
domoic acid, 192
DOP, 442
Doppler radar, 324t, 325–26, 326f, 334–35
dorsal fin tags, 251
DOT, on transport of diagnostic specimens, 198
dot plots, 45, 46f
dotterel, Eurasian, dogs in studies of, 144t
double-blind approach, 31
double halo nest traps, 71
double observer sampling, 300–301
double sampling, 29, 299–300, 394
Douglas-fir, sampling of, 407
dove(s)
    capture methods for, 65t
    noninvasive marking of, 237t, 238b
    reproduction of, 518
dove, mourning
    age and sex determination for, 209, 210t, 217t
    capture methods for, 77t, 79t
    handling methods for, 111
    noninvasive marking of, 238b, 243t
    reproduction of, 504, 518
    survival rates for, 496f
    transmitters on, 267f
dove, rock, capture methods for, 73, 80, 81
dove, white-winged
    abundance estimation of, 304
    age and sex determination for, 217t

dove, white-winged (continued)
    capture methods for, 77t, 78
    noninvasive marking of, 244f
dragonflies, transmitters on, 267f
drawers, insect, 347–48, 347f
DRC-1339, 168, 169
drift fences
    capture of amphibians with, 105–6, 105f
    capture of birds with, 76–77, 76t, 77f
    capture of mammals with, 76–77, 76t, 82–83, 85
    capture of reptiles with, 107–9
drive-by netting, 76
drive counts, 292
drive net traps
    capture of birds with, 76–77, 76t
    capture of mammals with, 76t, 82–83
drop box sampling, of amphibians, 105
drop-door traps, 76
drop nets
    capture of birds with, 71–72, 71f, 71t
    capture of mammals with, 71t, 82
drug(s). *See also* chemical immobilization
    accidental exposure of humans to, 134–36
    in capture of birds, 80–81
    in capture of mammals, 81, 90
    as contaminants, mortality and morbidity from, 163–64
    as controlled substances, 90, 119
    expiration date of, 120
    extra label or off label use of, 119
    FDA approved uses for, 118–19
Drug Enforcement Agency (DEA), U.S., controlled substances under, 119
dry mass, in vegetation sampling, 394, 403
dry preservation, of invertebrate specimens, 347–48, 347f
D-shaped nets, 338, 338f, 339f
duck(s)
    age and sex determination for, 211
    capture methods for, 77, 78, 80
    chemical immobilization of, 137
    contaminant impacts on, 163
    dogs in carcass searches for, 147
    handling methods for, 111
    invasive marking of, 251
    nest success of, 358
    noninvasive marking of, 234t, 237t, 243t
    transmitters on, 266
    vegetation sampling in habitat of, 407
duck, American black, population dynamics of, 10
duck, blue, dogs in studies of, 146
duck, canvasback, survival rates for, 360
duck, harlequin, capture methods for, 76, 76t
duck, mallard
    behavior of, 469, 471
    brood movements in, 33–34
    capture methods for, 69t, 78, 80
    contaminant impacts on, 163
    decoys in vulnerability to harvest, 39
    invasive marking of, 252t, 254t
    nest success of, 374
    noninvasive marking of, 238f
    population growth models for
        attention given to, 349
        growth rate calculation in, 372
        productivity in, 373–74
        survival rates in, 359–60, 360t, 363, 363t, 364t, 371
    reproduction of, 507, 508
    sensory perception in, 469
duck, northern pintail
    capture methods for, 69t
    dogs in studies of, 144t, 146
    molecular techniques in studies of, 544
    reproduction of, 514
duck, redhead, capture methods for, 78
duck, wood
    behavior in nest boxes of, 464–65, 465f
    capture methods for, 76t, 77t
    contaminant impacts on, 162
    invasive marking of, 249t
    remote marking of, 233t
    survival rate estimation for, 362–63, 362t, 363t
    vegetation sampling of habitat of, 382
dugong, capture methods for, 93
duration, in behavioral measurement, 474
D-vacs, 340, 340f
dyes, marking of wildlife with, 233, 242–44, 243t, 244f
dynamic analyses, of animal interactions, 489
dynamic life tables, 368
dynamics, population. *See* population dynamics

EAARL, 453
eagle(s)
    chemical immobilization of, 137
    contaminant impacts on, 164
    molecular techniques in tracking of, 544
    noninvasive marking of, 238, 243t
eagle, African fish, capture methods for, 71
eagle, bald
    behavior of, 474f, 477
    capture methods for, 72t, 73, 74f, 75, 75t
    contaminant impacts on, 159, 163, 166
    noninvasive marking of, 239t
eagle, eastern imperial, population estimation for, 543
eagle, golden
    capture methods for, 72–73, 73t, 74f, 75
    contaminant impacts on, 159
eagle, Philippine, capture methods for, 70
eagle, Spanish imperial, genetic diversity of, 535
eagle, steppe, capture methods for, 81
ear(s), anatomy of, 467
ear notching, 255
ear punching, 255
ear tags, 248, 250b, 250f, 251f
Earth Systems Research Institute (ESRI), 439, 440
echoes, radar, 320–22, 328–30
echolocation, 467
ecological neighborhoods, 409
ecology
    landscape, 456–57
    molecular, 542–46
EcoRI (enzyme), 527
ecosystem(s)
    invertebrates' role in, 336, 338
    invertebrates used in monitoring of, 338
ectoparasites, obligate, description of, 338
edge effects, 383
edge habitat
    in habitat selection studies, 419
    nest predation at, 19, 20–21, 38
    in vegetation sampling, 383
effective herbage height, 398
effective population size, 538–39
effect sizes, 19, 36, 288–89
egg(s), bird
    age and sex determination for, 209
    clutch size of, 507
    hormones in laying of, 518
    invasive marking of, 245t
    noninvasive marking of, 241, 243t, 244
egg(s), turtle, noninvasive marking of, 244
egg dumping, 465
egg flotation techniques, 209
eggshell membranes, DNA extraction from, 530, 530t
egret, cattle
    noninvasive marking of, 243t
    remote marking of, 233t
egret(s), capture methods for, 77t
EIA, 524
eider, common, capture methods for, 65, 66t
eider, spectacled, gene flow in, 535
eider(s), capture methods for, 65t
electrical power lines, wildlife collisions with, 332–33
electroejaculation, 515
electromagnetic radio signals, 264
electronic collars, for dogs, 151–52
electronic data collection, 32, 43, 408, 408f
electrophoresis, protein, 527, 530
electroshocking, in capture of amphibians, 107
elephant(s)
    collection of blood samples from, 523
    reproduction of, 509, 512, 521
    sensory perception in, 467, 470
elephant, African
    noninvasive marking of, 234t, 237t, 243t
    reproduction of, 518–19
    thermal infrared cameras in studies of, 317
    vegetation sampling and, 405
elephant, Asian, reproduction of, 518–19
elevation mask, 445b
elk
    age and sex determination for, 219t, 227t
    brucellosis in, 202–3
    capture methods for, 84t, 87t, 93t

genetic diversity of, 539
habitat of
  in forest planning, 434
  vegetation sampling in, 407
home range estimation for, 484f, 485f
invasive marking of, 250t
necrotic stomatitis in, 200–201, 203
population density of, 24–25, 25f
population estimation for, 494
prescribed fires in management of, 39
remote marking of, 233t
reproduction of
  corpora lutea in, 509
  hormones in, 517t, 519, 520
  stress and, 522
  uterine analysis of, 511–12
sensory perception in, 467, 470f
sites for drug injections on, 130f
steps in research on, 4–5b
survivorship curves for, 366f
elk, Rocky Mountain, necrotic stomatitis in, 200
elk, Roosevelt, necrotic stomatitis in, 200
embryonic development, 507–12
  in birds, 209
  in mammals, 218
embryos, uterine analysis of, 511–12
embryo transfer, 505, 518
emergence traps, 340, 341f
emergency medicine
  for animals, 132–34, 133f
  for humans, 134–36
emerging infectious diseases
  definition of, 204
  with origins in wildlife, 181
emigration
  in population growth models, 356
  in survival estimation, 495
empiricism, 41–42
enclosure experiments, resource preference in, 411
endangered species
  captive breeding of, 505
  conservation of genetic diversity of, 533
  dogs in studies of, 146–47
  lead poisoning in, 159
  permits for marking of, 231
  population viability analysis of, 15
  taxonomic status of, 530, 531
Endangered Species Act of 1973
  hybridization under, 531, 532
  models used for decisions under, 15
  species definition in, 530, 531
endemic diseases, definition of, 204
endocrine system, 516–24. *See also* hormone(s)
  regulation of, 516
  in reproduction, 516–20
  in stress, 502, 516, 521–23
endoparasites, definition of, 204
endorphin, 522
English springer spaniels, 141

enticement lures, in capture of birds, 80
envelope traps, 70
environment
  in disease investigations, 189
  in host–parasite interactions, 184–86, 184f
environmental contaminants. *See* contaminants
environmental mutagens, 534
Environmental Protection Agency, U.S.
  on PCBs, 161
  on pesticide incidents, 155
environmental sensors, 276
enzyme-immunoassay (EIA), 524
epidemic diseases, definition of, 204
epidemiology, 183–86, 204
epididymides, 514
epiphyseal cartilage, in age and sex determination of mammals, 218, 224f
epistemology, 41
epithelial cells, in estrous cycle, 505, 505f, 506b
equilibrium, population, 534
error matrix, in remote sensing, 457, 458t
ESRI, 439, 440
estradiol, 505, 516, 517t, 518, 519
estrogen, in reproduction, 504–7, 516, 517t, 518
estrous cycle, 505–7, 505f, 506b, 518, 518f
estrus, 504–7, 505f, 506b
ESU, 532–33, 533b
ethanol, as killing agent for invertebrates, 345–46
ethograms, 473, 474f, 477
ethology. *See* behavior
ethyl acetate, as killing agent for invertebrates, 346
ethyl alcohol, preservation of invertebrate specimens in, 346–47
ethylene glycol, mortality and morbidity from, 160–61
ethyl mercaptan, 469
etiologies, definition of, 204
etorphine, 123, 136
Eulerian data, 259
Europe
  dogs used in wildlife management in, 140
  pesticide incidents in, 156–57
euthanasia
  methods of, 132
  sodium pentobarbital in, accidental mortality from, 163–64
events, behaviors as, 474
evolutionary significant units (ESU), 532–33, 533b
experiment(s)
  cafeteria, 411
  comparison of types of, 13–14, 14t
  crossover, 17–18
  enclosure, 411
  field, 12–13, 14
  laboratory, 12, 13–14
  manipulative, 2, 3f, 12
  with models, 23
  natural, 12, 14

Experimental Advance Airborne Research Lidar (EAARL), 453
experimental designs, 16–21
  bias in, 31
  challenges of, 1
  checklist for, 19–21
  common problems in, 38–39
  controls in, 19
  philosophical foundation of, 5–6
  replication in, 18–19
  sampling procedures in, 12–13
  types of, 16–18
experimental research, 1–3
  vs. descriptive research, 1–3
  modeling as alternative to, 2–3, 21
experimental units
  definition of, 287
  examples of, 287
  identification of, 20–21, 25–26
  lack of independence in, 39
expiration date, of drugs, 120
explosive charges
  in darts, 126
  drop nets deployed with, 71, 82
exponential growth model, 352
extraneous factors, 20
eye(s)
  in age and sex determination of mammals, 218–25
  anatomy of, 468
  during chemical immobilization, 130–31
eye-lens weight, in age and sex determination of mammals, 218–25

FAA, 443
facts, in theories, 11b
falcon, peregrine, contaminant impacts on, 159
falcon, prairie, capture methods for, 75t
falcon(s), noninvasive marking of, 239t
falsification, of hypotheses, 34, 35
fatty acid scent (FAS), 94
FCC, 268
FDA, 80, 118–19
feathers. *See also* plumage
  clipping of, 253, 254t
  DNA extraction from, 529–30, 530t, 542
  imping of, 253, 253f, 254t
  types of, 212f
fecal samples. *See* scat
fecundity, definition of, 357
Federal Aviation Agency (FAA), 443
Federal Communications Commission (FCC), 268
Federal Geographic Data Committee (FGDC), 458, 459, 459b, 460
feet. *See* foot
felids
  age and sex determination for, 222t
  chemical immobilization of, 138
females
  gene flow in, 535

females (*continued*)
  reproduction by
    contraception for, 520–21
    hormones in, 516–19
    stages of, 504–13
fences, drift. *See* drift fences
ferret, black-footed
  captive breeding of, 505
  chemical immobilization of, 123f, 139t
  invasive marking of, 247t
  scat of, detection by dogs, 148
fertility. *See also* birth rates
  age-dependent, 364–68
  antifertility agents and, 521
  birth-flow, 352
  birth-pulse, 351
  definition of, 357
  estimation of, 357
  pesticides' impact on, 165
  and population density, 354, 354t
fertility tables, 364–65
fertilization
  in vitro, 505
  modes of, 503, 503f
  rates of, 510–11
fetal development. *See* embryonic development
fetuses
  counts of, 511–12
  uterine analysis of, 511–12
FGDC, 458, 459, 459b, 460
fiber optic cameras, 477
fidelity
  as availability for sampling, in survival estimation, 358
  site, methods for analysis of, 489–90
field experiments, 14
  vs. field studies, 12–13, 15
  preliminary, in vegetation sampling, 385
  pros and cons of, 14, 14t
  sampling procedures in, 12–13
field journals, of animal behavior, 472
field observations, in disease investigations, 189–92, 191b
field procedures
  in contaminant investigations, 174–79
  in disease investigations, 195–99, 197t
field studies, 15
  vs. field experiments, 12–13, 15
  sampling procedures in, 12–13
"fight or flight" response, 521
filoplumes, 470
filters
  moving target indicator, 329
  radar, 329–30
  respirator, 171
finch, house, capture methods for, 79t
finch, zebra, behavior of, 472
finch(es), noninvasive marking of, 237t
fingerling fish tags, 248, 251
fingerprinting
  DNA, 528

minisatellite, 527
molecular, 542
finite population sampling, 24
finite rate of population increase, 351
fires
  Landsat imagery of recovery after, 448, 448f
  prescribed, 39, 43
  risk of, GIS in analysis of, 436–37
first-order selection, 418, 491
fish
  chemical immobilization of, 136
  contaminant impacts on, 166–67
  fertility estimation for, 357
fish dip nets, 65, 65t
fisher
  age and sex determination for, 222t
  capture methods for, 84t, 91t, 92t, 98
  chemical immobilization of, 139t
  habitat models of, 432
  handling methods for, 112
  population estimation for, 543
  population growth models for, 375
  remote cameras in studies of, 313
  reproduction of, 513
  scat of, detection by dogs, 149
fisheye lenses, 393
fishing poles
  in capture of amphibians, 107
  in capture of reptiles, 108
fish snares, floating, capture of birds with, 75
fitness
  and genetic diversity, 539, 540, 541
  and habitat selection, 412
  and hybridization, 531
fixed effects, 18, 58
fladry, 82
flat files, 44
fleas, description of, 338
flehmen, 469–70, 470f
flies, description of, 337
flight intercept traps, 341
flipper bands, 236
flipper tags, 251
flip traps, 73
floaters, in abundance estimation, 293
floating pitfall traps, 110
flocking birds, radar in studies of, 333, 333f
flowers, in collection of invertebrates, 343
fluid preservation, of invertebrate specimens, 346–47
fluorescent pigments, marking of wildlife with, 242–44
flycatcher, Acadian, capture methods for, 67
flycatcher, alder, resource selection by, 425–27, 426t, 427f
flycatcher, collared, sex determination for, 546
flycatcher, pied, vegetation sampling and, 405
flying animals, radar in studies of. *See* radar
focal-animal sampling, 473–74, 473t

fogging, in collection of invertebrates, 344
follicles
  antral (Graafian), 508, 508f
  development of, 504, 504f, 508–9, 508f
  postovulatory, 504f, 507
  tertiary, 504
  ultrasonography of, 509, 509f
follicle stimulating hormone (FSH), 516, 517t, 519
fomites, definition of, 204
food abundance, definition of, 419
Food and Drug Administration (FDA), U.S.
  on alphachloralose for capture of birds, 80
  drugs approved by, for wild animals, 118–19
food availability
  definition of, 419
  methods for measurement of, 419–20
food scents, in capture of mammals, 94
food selection
  methods for measurement of, 419–20
  resource units in, 415
  study designs for, 414
foot-and-mouth disease, detection of, 317
foothold traps
  capture of birds with, 80, 81
  capture of mammals with, 87–90, 88f, 96–104, 102f
foot nooses, capture of birds with, 73
foot snares, capture of mammals with, 88, 89f
foot tags, 251
foot washes, 177
Forager program, 475
foraging, need for studies on, 479t
forbs, measurement of availability of, 419
forearm length, in age and sex determination, 207, 208f
forest(s)
  Landsat imagery of change in, 447–50, 448f, 449f, 450f
  nest predation at edges of, 19, 20–21, 38
forest planning, habitat relationships in, 434, 436–37b
founder events, 537, 539
founding population, 541
fourth-order selection, 418
fox(es)
  age and sex determination for, 220–21t
  capture methods for, 89
  chemical immobilization of, 138
  invasive marking of, 245t, 250t
  noninvasive marking of, 234t
  rabies management in, 201–2
  scat of, detection by dogs, 149
fox, arctic
  capture methods for, 89, 98
  population estimation for, 543
fox, Argentine gray, capture methods for, 94
fox, gray
  capture methods for, 84t, 89t, 91t, 98–99
  reproduction of, 512

fox, kit
　capture methods for, 84t, 99
　molecular techniques in studies of, 544–45
fox, red
　capture methods for, 89t, 91t, 94, 95, 99
　habitat selection by, 34
　rabies in, 201
　remote cameras in studies of, 316, 317
　reproduction of, 510
fox, San Joaquin kit, scat of, detection by dogs, 149
fox, swift, capture methods for, 84t, 99
fractal index, 486–87
fragment analysis, 527–28, 528f
frames, in vegetation sampling, 386–88, 386f, 387f, 391
frameworks, definition of, 11b
freeze branding, 251, 252, 252f, 252t
freezing, of invertebrate specimens, 346
frequency of occurrence
　in behavioral measurement, 474
　definition of, 285, 386
　indices of, 291
　in vegetation sampling, 386–87, 386f, 387f
frequency selection, for VHF radiotelemetry, 268–70
frequentist approach, to data analysis, 34, 55
Friedman's test, for analysis of resource selection, 420–21, 421t
frigatebird, invasive marking of, 254t
frog(s)
　capture methods for, 105–6, 105f, 107, 111
　handling methods for, 112
　invasive marking of, 245t, 246, 247t, 249t, 252t, 253t, 255
　noninvasive marking of, 237t, 243t
　reproduction of, 503, 504, 512, 516, 517t
frog, gray tree, capture methods for, 106
frog, Mexican leaf, reproduction of, 514, 516
frog, northern leopard, noninvasive marking of, 240t
frog, Pacific tree, capture methods for, 106
frog, tree, capture methods for, 105, 105f, 106
frog, wood, reproduction of, 507
frog box, 112
fruit(s)
　of herbaceous vegetation, 403–4
　sampling techniques for, 401–4, 407–8, 420
　of shrubs, 402–3
　of trees, 401–2
FSH, 516, 517t, 519
full waveform infrared (IR) lidar, 453, 456
fumigants, 168, 169
fungi, natural toxins produced by, 169, 170
fungicides, 167–68
funnel traps
　capture of amphibians with, 105–6
　capture of invertebrates with, 344
　capture of reptiles with, 107–9, 110–11
furans, 162

fur removal, marking with, 253, 254t
fyke nets, 109, 110

gadwall, capture methods for, 69t
gallinaceous birds, age and sex determination for, 209, 210t, 211–13, 214–15t, 216f
gallinule, purple, age and sex determination for, 217t
game birds
　capture methods for, 80, 80t
　noninvasive marking of, 234t, 235
Gap Analysis Program (GAP), 435, 457
Garmin Astro, 143
gastrointestinal ulcers, 521
gaur, ultrasonography of, 509
gazelle, Thomson's
　invasive marking of, 252f
　sensory perception in, 468
geckos, capture methods for, 109
geese. See goose
gender determination. See sex determination
gene(s). See genetic(s)
gene flow
　definition of, 534
　in genetic diversity, 534–37
　measurement of, 534–35, 535f
　recent vs. historical, 534
　in species definition, 530
　through translocation programs, 536–37
gene products, 526–27
generality
　in laboratory experiments, 13
　in modeling, 350
generalizations, in theories, 10, 11b
generalized linear mixed-effects models (GLMM), 58–59
generalized linear models, 53–55, 421, 421t
generalized random tessellation stratified (GRTS), 29
general linear models (GLM), 19, 36, 36f
general theoretical models, 22
generation length, 539
genetic(s), 526–46
　molecular, 526–30, 527t
　in molecular ecology, 542–46
　population, 435
　sampling techniques in, 529–30, 542–44
　in taxonomy, 530–33
genetic bottlenecks, 537–39
genetic differentiation, 537
genetic distances, 531
genetic diversity, 533–41
　conservation of, 533–41
　definition of, 534
　drift and bottlenecks in, 537–39
　gene flow in, 534–37
　vs. genetic variation, 533–34
　habitat fragmentation and, 535–36, 536b
　human impact on, 539
　hybridization and, 531
　measurement of, 533, 534

　mutation in, 534
　and population viability, 541
　selection in, 539–41
　translocations and, 536–37
genetic drift, 537–39
　definition of, 534, 537
　detection of, 537–38
　from human activities, 539
genetic mutations
　in genetic diversity, 534
　and protein formation, 527
genetic restoration, 536
genetic signatures, 539–40
genetic similarity, 534, 535f
genetic stock identification, 544
genetic tags, 542
genetic variation
　vs. genetic diversity, 533–34
　methods for analysis of, 526–29
genitalia, in age and sex determination of mammals, 218, 226f
genome(s)
　mitochondrial, 526, 531
　nuclear, 526
genotypes, 540
genotyping errors, 543
geographic closure, 493
Geographic Information System (GIS), 432–40
　in conservation efforts, 435–38, 438f
　in disease investigations, 192
　functions of, 430
　in habitat studies, 412, 432–34, 432f, 433f
　integration of GPS and, 430, 430f, 445–46
　in international community, 439–40
　and Internet, 438–40, 445–46
　in measurement of landscape variables, 420
　mobile, 445–46
　in population modeling, 434–35
　in risk assessment, 436–38
　in sampling, 26
　software programs for, 501
　spatial relationships in, 433b
　in vegetation sampling, 405–6, 432
Geography Network, 439
geospatial data, sources of, 438
Geospatial One Stop, 439
gill nets, 76
giraffe
　natural markings of, 232t
　ultrasonography of, 509
GIS. See Geographic Information System
gland scents, 94, 469
GLM, 19, 36, 36f
GLMM, 58–59
Global Biodiversity Information Facility, 440
global location sensing (GLS), 275, 279, 280f
Global Navigation Satellite System (GNSS), 440–41
Global Positioning System (GPS), 440–46
　accuracy of, 278, 441, 442–43

Global Positioning System (*continued*)
applications for, 274, 444–45
in behavioral studies, 475
data management with, 262
data retrieval with, 274, 276, 443
differential, 442–44
in disease investigations, 192
and dogs, 143
ground stations in, 442b
integration of GIS and, 430, 430f, 445–46
location error with, 263, 264, 442–43
mechanism of location estimates by, 278, 279f, 441, 441f, 442
navigation with, 444–45
vs. other telemetry systems, 269t, 274
in radiotelemetry studies, 274, 276, 278, 444–45
receivers in, 440, 441–43, 444b
in sampling designs, 26
terminology of, 445b
in vegetation sampling, 404
Global Spatial Data Infrastructure (GSDI), 440, 459
global tracking systems, 269t, 273–75
glow sticks, 106
GLS, 275, 279, 280f
glue
in dry preservation of invertebrate specimens, 347
transmitters attached with, 266, 267f
glue traps
capture of mammals with, 93
capture of reptiles with, 108, 110
in collection of invertebrates, 342
glutathione detoxifying enzyme system, 158
gnatcatcher, California, population viability analysis of, 379b
gnats, description of, 337
GnRH, 516
GNSS, 440–41
goat, mountain
age and sex determination for, 220t, 225f, 227t
chemical immobilization of, 137t, 138
lidar in studies of, 456
population estimation for, 494
reproduction of, 519
sensory perception in, 470
goat, wild
handling methods for, 112
invasive marking of, 250t
noninvasive marking of, 234t
goat(s), sensory perception in, 469
goldeneye, Barrow's, capture methods for, 69t
goldfinch, American, noninvasive marking of, 237t
gold mining, 163
GonaCon, 521
gonadal steroids, 516
gonadosomatic index, 504
gonadotropin, 516, 517t, 521

gonadotropin releasing hormone (GnRH), 516
goodness-of-fit tests, 376–78
Google Earth, 450
Google Scholar, 6
goose (geese)
age and sex determination for, 211, 213t
behavior of, 474
chemical immobilization of, 137
noninvasive marking of, 232, 234t, 237t, 243t
goose, Aleutian Canada, dogs in capture of, 145t, 150
goose, Canada
age and sex determination for, 209, 213t
capture methods for, 71, 71t, 76, 76t, 80
contraception for, 520, 521
design of studies on diet of, 14, 15
dispersal patterns of, 356
dogs in management of, 150
food selection by, 414
goose, dusky Canada, population of, 6
goose, emperor, age and sex determination for, 213t
goose, greater white-fronted, age and sex determination for, 213t
goose, Ross', age and sex determination for, 213t
goose, snow
age and sex determination for, 213t
capture methods for, 76t
food selection by, 414
nest success of, 17
radar in monitoring of, 334, 334f
reproduction of, 504, 504f, 508
goose, Vancouver Canada, population of, 6
gopher, pocket, capture methods for, 83t, 102
gopher(s), invasive marking of, 245t
goshawk, northern
capture methods for, 69–70, 69t
habitat selection by, 414
GPS. *See* Global Positioning System
Graafian (antral) follicles, 508, 508f
grackle, noninvasive marking of, 241t
graphs, 45–47
design of, 45–47
functions of, 45
misuse of, 45
types of, 45–47, 45f, 46f
grass(es), measurement of availability of, 419
grasshoppers, description of, 337
grassland sampling, time required for, 384t
grebe, eared, capture methods for, 67, 76
grebe, pied-billed, capture methods for, 77t, 78
grebe(s), contaminant impacts on, 159
grid arrangements, traps in, 86
ground-based lidar, 452
ground clutter, in radar, 322, 323f, 325, 325f, 328, 330
ground coverage, by dogs, 140, 142
ground stations, GPS, 442b

ground truth, in radar, 330–31, 331f, 332f
group phenomena, 462
grouse
dogs in capture of, 150
dogs in studies of, 147
invasive marking of, 249t
noninvasive marking of, 237t, 239t
grouse, black
dogs in studies of, 144t, 150
genetic diversity of, 538
noninvasive marking of, 239f
grouse, blue
age and sex determination for, 209, 210t, 214t, 216f
capture methods for, 72t, 75t, 76–77, 76t, 80t
dogs in studies of, 144t
reproduction of, 507
taxonomic status of, 531
grouse, Columbian sharptailed, dogs in studies of, 144t
grouse, dusky, capture methods for, 76, 76t
grouse, eastern sage-, taxonomic status of, 532b
grouse, greater sage-
age and sex determination for, 214t
behavior of, 476, 476f
capture methods for, 65, 65t, 66t, 71–72, 71t, 72t, 76t
dogs in studies of, 144t, 145, 145f, 146, 146f, 147, 147f
invertebrates in diet of, 338
remote marking of, 233t
taxonomic status of, 532f
grouse, Gunnison sage-
age and sex determination for, 214t
behavior of, 467, 467f, 472, 475, 478, 478f
sensory perception in, 467, 467f, 472
taxonomic status of, 532b, 532f
grouse, red
age and sex determination for, 210t
disease in, 181
dogs in studies of, 143, 144t, 145, 146
integrated research on, 16
grouse, ruffed
age and sex determination for, 209, 213, 214t
capture methods for, 66t, 69t, 76t, 79t, 80, 80t
dogs in studies of, 144t
habitat of, in forest planning, 434
noninvasive marking of, 243t
remote marking of, 233t
reproduction of, 508
grouse, sage-
hybridization of, 532
taxonomic status of species of, 531, 532b
grouse, sharp-tailed
age and sex determination for, 214t
capture methods for, 69t, 72t, 79t, 80t
hybridization of, 532
grouse, sooty, dogs in studies of, 144t, 146

grouse, spruce
  age and sex determination for, 211, 214t
  capture methods for, 75t, 80t
  design of studies on diet of, 14
  dogs in studies of, 144t
  use of trees by, 32, 37–38
grouse, western sage-, taxonomic status of, 532b
Grouse Disease, 181
growth. See population growth
growth rings, tree, 399
GRTS, 29
GSDI, 440, 459
guanaco, South American, capture methods for, 89t, 93
gull(s)
  capture methods for, 81
  contaminant impacts on, 169
  marking of
    invasive, 249t, 254t
    noninvasive, 237t, 239t, 241t, 243t
    remote, 233t
gull, California, capture methods for, 65, 65t, 66t
gull, glaucous, molecular techniques in studies of, 545
gull, glaucous-winged, remote marking of, 233t
gull, herring, contaminant impacts on, 163
gull, ring-billed, capture methods for, 72t

habitat
  cluster sampling of, 28
  definition of, 381, 410, 429
  in disease investigations, 189
  in disease management, 186, 201, 203
  essential elements of, 381
  in forest planning, 434, 436–37b
  GIS in studies of, 432–34, 432f, 433f
  in host–parasite interactions, 184, 184f, 186
  lidar in characterization of, 451–52
  in nest success, 2
  spatial relationships in, 433b
habitat accounting and appraisal system, GIS in, 432
habitat corridors, GIS in studies of, 435, 437
habitat edges. See edge habitat
habitat fragmentation, and genetic diversity, 535–36, 536b
habitat patches
  in definition of landscape, 420
  in habitat selection studies, 419
  metapopulations in, 379b
habitat patch occupancy models, 411, 413
habitat selection
  definition of, 410
  lidar in studies of, 454–55
  methods for detection of, 416–17, 417f
  paired observations on, 31, 31f
  resource units in, 415
  and sample sizes, 34
  scales (levels) of, 411–13, 418
  study areas for, 417–19
  study designs for, 412–16, 413b, 420, 420f
habitat traps, capture of reptiles with, 107–8, 109
habitat use, dogs in studies of, 146
Hahn method, 296
hair, DNA extraction from, 529, 530, 530t, 542, 542b
hair burning, 253
hair cell sensory receptors, 467
hair snares, 542–43, 542b, 542f
halo traps, 71
hamster
  hormones in reproduction of, 519
  sensory perception in, 468
hand capturing
  of amphibians, 104–5
  of birds, 81
  of mammals, 93
  of reptiles, 107, 109
handling of samples
  in contaminant investigations, 174–78
  in disease investigations, 192–95, 197, 198b
handling of wildlife. See also capture
  during chemical immobilization, 130–32
  invertebrate specimens, 345–46
  techniques for, 111–13
hantavirus, 111–12
H antennas, 271, 271f
haphazard sampling, 288
hardware, in spatial technologies, 430
Hardy–Weinberg Equilibrium, 527
hare, European, reproduction of, 513
hare, snowshoe
  capture methods for, 76t, 83t, 89t
  integrated research on, 16
  invasive marking of, 245t
  population growth models for, 375
  reproduction of, 522
  vegetation sampling in habitat of, 407
hare(s), marking of
  invasive, 249t, 250t, 254t
  noninvasive, 234t
  remote, 233t
harmonic radar, 327
harnesses
  in marking of wildlife, 239, 239t
  in radiotelemetry, 265–66, 266f
harp traps, capture of mammals with, 82
harrier, northern, capture methods for, 69t
harvest, illegal, detection of, 544
harvest programs. See also hunters
  and genetic diversity, 539, 540–41
  selective, 540–41
hatch year (HY), determination of, 211
hawk(s)
  capture methods for, 73
  chemical immobilization of, 137
  remote cameras in nests of, 312f
hawk, Cooper's, capture methods for, 67, 69, 69t
hawk, ferruginous, capture methods for, 80

hawk, red-shouldered, capture methods for, 69
hawk, red-tailed
  capture methods for, 69t, 79–80
  contaminant impacts on, 159, 166, 168, 172
hawk, rough-legged, capture methods for, 75t
hawk, sharp-shinned
  capture methods for, 69
  contaminant impacts on, 169
hawk, Swainson's
  capture methods for, 80
  contaminant impacts on, 166
Hayne method, 296–97
hazard functions, in survival analysis, 57–58, 496–97
hearing, 467–68
  and behavior, 467–68
  functions of, 467
heart rate, sensors for, 276
heat
  body (See thermoregulation)
  in collection of invertebrates, 343
heavy metals, definition of, 157
Heisey–Fuller method, 495–96
helicopters
  in capture of birds, 73, 75f, 81
  in capture of mammals, 83, 92
  in chemical immobilization of mammals, 138
helinets, 73, 75f
hellbender
  capture methods for, 107
  invasive marking of, 249t, 252t, 253t
hellbender, Ozark, transmitters on, 267f
hematology
  definition of, 204
  in disease investigations, 195
hemipterans
  classification of, 348
  description of, 337
hemocytometers, 515b
herbaceous vegetation, fruits of, 403–4
herbage height, 397–98
herbicides, 167
herbivores
  contaminant impacts on, 154–55
  detoxification of plant toxins by, 155, 169
heron(s)
  capture methods for, 77t
  noninvasive marking of, 241t
heron, great blue, capture methods for, 72t, 79, 81
heterogametic gender determination, 535
high-energy, full-waveform, green lidar, 453
histopathology
  definition of, 204
  in disease investigations, 193
hog, feral
  capture methods for, 84t, 87, 87t, 89t, 100
  dogs in management of, 151
  remote cameras in studies of, 316

home range, 481–89
    contours of, 482
    core areas of, 482
    definitions of, 481–82
    habitat selection in, 411–12
    methods for estimation of size of, 481–89, 483t
    overlaps in, 489–90
    in population size estimation, 492
    software programs for, 498
    vs. territory, 481
    in utilization distribution, 482
honeyeater, Regent, capture methods for, 69t
hooks, in capture of reptiles, 111
hoop net traps, 78–79
hoop traps, 110
horizontal visual obstruction, 396–97, 396f, 397f
hormone(s), 516–25
    in contraception, 520
    excretion of, 520, 522, 524
    methods for measurement of, 518, 519, 520, 523–24
    regulation of secretion of, 516
    in reproduction, 516–20, 517t, 518f
    stress, 502, 516, 521–23
hormone profiles, 516
horns, in age and sex determination, 218, 225f
horse, wild, reproduction of, 520, 521
host(s)
    definition of, 204
    intermediate, 205
    transport, 206
host–parasite interactions. *See also* disease(s)
    ecological view of, 184–86, 184f
    epidemiological view of, 183–86, 184f
    history of research on, 181–82, 183–84
hot-iron branding, 251, 252t
hot spots, 435, 437–38
HPA axis, 521, 522
humans
    emergency medicine for, 134–36
    genetic drift and bottlenecks caused by, 539
    impact of infectious agents on population of, 183–84
    life tables for, 365
    managing behavior of, 477–78, 477f
    scent of, detection by dogs, 142
hummingbirds, noninvasive marking of, 239t
humoral immunity, 184, 204
hunter-gatherers, study of animal behavior by, 462
hunters
    in abundance estimation, 304–6, 495
    animal behavior influenced by, 471
    and genetic diversity, 539, 540–41
    illegal harvest by, 544
    selective harvest by, 540–41
    survey on success of, 25, 26b
    tag recovery by, 362

HY, 211
hybridization, 531–32
hymenopterans, description of, 337
hyperthermia
    during chemical immobilization, 127–28, 133
    emergency medicine for, 133, 134f
hypothalamic–pituitary–adrenal (HPA) axis, 521, 522
hypothalamic–pituitary–gonadal axis, 516
hypothalamus, 516, 518
hypothermia
    in capture of mammals, 84
    during chemical immobilization, 127–28, 133
    emergency medicine for, 133
hypotheses
    credibility of, 32
    in data analysis, 47–48
    definition of, 11b
    development of, 2, 3, 4, 10
    in experimental design checklist, 19
    in modeling, 376
    multiple competing, 3
    multiple testable, 4
    new, as form of speculation, 37–38
    as step in research process, 2, 3, 4, 10–12
    testing of, 2, 3, 4, 34–35
    ultimate *vs.* proximate, 471–72
hypothesis tests, power of, 19, 32–33
hypothetico-deductive method, 3

ibex, Alpine, genetic diversity of, 539
ibex, behavior in reintroductions of, 465–66, 466f
ibex, Spanish, capture methods for, 87, 87t
ibis, white, capture methods for, 72t, 77t, 78
ideal population, 538
identification error, 543
iguana(s)
    capture methods for, 111
    invasive marking of, 252t, 253t
    noninvasive marking of, 241t, 244
iguana, green, noninvasive marking of, 241t, 242
ilio-pectineal eminences (IPE), 218, 224f
immigration
    in population growth models, 356
    in survival estimation, 495
immobilization, chemical. *See* chemical immobilization
immunity. *See also* disease
    cell-mediated, 184, 204
    definition of, 204
    genetics in, 540
    humoral, 184, 204
immunizations. *See* vaccine(s)
immunocontraception, 520, 521
immunology, 184
impact assessments, 13, 15
implant transmitters, 266–68, 267f

imprinting
    human handlers and, 464, 464f, 478, 478f
    sexual, 464, 479t
inbreeding depression, 541
incidence, definition of, 204
incident reports, in contaminant investigations, 176, 180f
incremental cluster analysis, 485
incubation behavior, hormones in, 518
incubation periods, definition of, 204
independent double-observer sampling, 300
independent review. *See* peer review
independent variables, 19–20
indicators, biological, invertebrates as, 338
indirect observation, of animal behavior, 473
indirect transmission, definition of, 204
individual animals
    behavioral biologists' focus on, 463
    molecular techniques in tracking of, 544
    population units of, 6–7, 7f, 8b
    in resource use studies, 414
individual covariates, 376
individual plants, definition of, 387
individual viability, 540
induction, 3f, 42, 379–80
infections, definition of, 204
infectious agents
    definition of, 204
    impact on humans, 183–84
infectious disease. *See* disease(s)
inference(s)
    in population analysis, 379–80
    in radiotelemetry, 259
    statistical (*See* statistical inference)
    strong, 3
influenza, 181
information-theoretic model selection, 35–36
information theoretic tools, 3
infrared cameras. *See* remote cameras; thermal infrared cameras
infrared (IR) lidar
    discrete-return, 452–53, 452f
    full waveform, 453, 456
infrared light, sensory perception of, 468
inhalation anesthetics, 123–24, 123f, 137
inheritance. *See* genetic(s)
initiation time, in behavioral measurement, 474
ink
    for labeling of invertebrates, 345
    marking of wildlife with, 242–44, 243t
innate behaviors
    definition of, 464b
    migration as, 463
inorganic chemicals
    definition of, 162
    human use of, 155
    mortality and morbidity from, 162–63
    as rodenticides, 168
insect(s), 336–48
    classification of, 336–38

collection methods for, 339–45
groups of interest, 337–38
identification of, 348
preservation methods for, 345–48
as radar clutter, 330, 330f, 331
insecticides
anticholinesterase, 155, 165–66
in capture of mammals, 86
clinical signs of exposure to, 173t
mortality and morbidity from, 155, 164–67
as rodenticides, 168
types of, 164–67
insectivores, age and sex determination for, 224t
instantaneous rate of population increase, 351
integrated research process, 6, 13, 16
intensity
in behavioral measurement, 474
of parasites, 204
sampling, 288
threat, 436
interactions, animal. *See also* host–parasite interactions; predator–prey interactions
methods for analysis of, 489–90
vertebrate–invertebrate, 336, 338, 344–45
interactions, in multifactor experimental designs, 17
interbreeding, 531–32
intermediate hosts, definition of, 205
internal markers of wildlife, 244–47
international community
GIS in, 439–40
GPS in, 440–41
International Organization for Standardization (ISO), on metadata, 458–60, 460b
Internet
in behavioral studies, 477
and GIS, 438–40, 445–46
interspecific competition, models of, 374
interval data, definition of, 44
intra-specific brood parasitism, 465
intravenous (IV) administration, in chemical immobilization, 130
introgression, 531
invertebrate(s), 336–48. *See also specific species and types*
classification of, 336–38, 348
collection methods for, 338–45
curatorial methods for, 345–48
definition of, 336
identification of, 345, 348
role in ecosystems, 336, 338
vertebrate interactions with, 336, 338, 344–45
investigation process, in scientific method, 2, 3f
in vitro fertilization, 505
iophenoxic acid, 245–46
IPE, 218, 224f
IR. *See* infrared

island biogeography, 184–85
ISO, on metadata, 458–60, 460b
isoflurane, 123
IV administration, in chemical immobilization, 130
ivory, 544

jackknife estimator, 60–61
jackrabbit, black-tailed, habitat models of, 432–34
jackrabbit(s), capture methods for, 65t, 66t, 82, 87t
jaguar
capture methods for, 93, 100
scat of, detection by dogs, 149
jaw foothold traps, padded, 81
jaw tags, 251
jay, blue, capture methods for, 67, 71, 77t
joint hypergeometric estimator (JHE), 493, 494
JOLLY program, 309
Jolly–Seber estimator, 309, 358, 360, 361
journals, field, of animal behavior, 472
judgmental sampling, 288
jugular vein, 523
junco, noninvasive marking of, 237t
juvenile birds, age and sex determination for, 209–13, 209f, 211f
juvenile dispersal, models of, 13
juvenile mammals, age and sex determination for, 218–29
JWatcher program, 475

kakapo
dogs in studies of, 146
sex determination for, 546
kangaroo, gray, contraception for, 520
kangaroo, tree, reproduction of, 517t, 520
Kaplan-Meier estimator, 57, 57f, 496, 496f, 497
katydids, description of, 337
K-band, 321
keratinized epithelial cells, in estrous cycle, 505f, 506b
kernel density estimation, 483t, 484, 484f, 486, 487t
kestrel, American
capture methods for, 67, 69t, 71, 75t, 77, 77t
contaminant impacts on, 163, 169
kestrel, Eurasian, sensory perception in, 468
ketamine, 122
for birds, 137
emergency treatment for human exposure to, 136
FDA approved use of, 119
for mammals, 137–39
for reptiles, 137
seizures associated with, 134
keys, invertebrate, 348
killing, of invertebrates, 345–46

killing traps
Best Management Practices and standards for, 89–90, 92t, 96–104
capture of mammals with, 89–90, 96–104
kinematic mode, 445b
kingfisher, belted, capture methods for, 73, 75, 75t, 77t
kingfisher, Tuamotu, transmitters on, 266f
King method, 296
kite, white-tailed, capture methods for, 75t
kittiwake, black-legged, capture methods for, 73, 75t
kiwi, dogs in studies of, 146
kiwi, little spotted, dogs in studies of, 144t
knowledge
nature of, 41–42
prior, in frequentist *vs.* Bayesian inference, 34
through reason *vs.* experience, 41–42
known-to-be-alive estimates, 306
koala, evolutionary significant units of, 533
Kriging methods, 145
Kullback-Leibler distance, 35–36

labeling
of invertebrate specimens, 345
of samples, in contaminant investigations, 174
laboratories
recommended, for contaminant investigations, 174b
wildlife disease experts at, 189
laboratory animals, invasive marking of, 252t
laboratory experiments, 12, 13–14
applications for, 13–14
pros and cons of, 13, 14t
laboratory procedures, in disease investigations, 192–95, 193t
Labrador retrievers, 141
lactation, 513
lagomorphs, age and sex determination for, 222–23t
Lagrangian data, 258–59
land cover
Landsat detection of change in, 447
lidar data on, 453, 453f
Landsat Data Continuity Mission (LDCM), 446–47
Landsat imagery, 446–51
access to, 447
algorithms behind, 447–49, 448f, 449f, 450f
of disturbance and recovery trends, 447–49, 448f
of land cover change, 447
validation of, 449–51, 451f
in vegetation sampling, 404, 447–51
landscape(s)
definition of, 420
Landsat imagery of change in, 447–50, 448f, 449f, 450f
spatial relationships in, 433b

landscape ecology, lidar in, 456–57
landscape variables
　definition of, 420
　measurement of, 420
LandTrendr, 447–49, 452f
laparoscopy, 509
laparotomy, 507, 509
laser altimetry, in vegetation sampling, 406
laser branding, 251, 252, 252t
laser pointers, in capture of reptiles, 109
laser rangefinders, 388
lassoing, 93
latency, in behavioral measurement, 474, 475f
laws, in theories, 11b
L-band, 321
LDCM, 446–47
lead, mortality and morbidity from, 159, 191
leadership, in vegetation sampling, 383–85
leaf litter bags, 105
learned behaviors
　definition of, 464b
　migration as, 463
　movement patterns as, 462
LED lights, 241, 343
left censoring, 495
legal considerations, with chemical immobilization, 118–19
leg bands, 236f, 238, 238b
lek sites, dogs' role in location of, 147, 147f
leopard, natural markings of, 232t
leopard, snow
　capture methods for, 89t
　remote cameras in studies of, 312
Leopold, Aldo, 141f, 181–82, 462
lepidopterans, description of, 337
lesions, definition of, 205
Leslie matrices, 367–68
leukocytes, in estrous cycle, 505, 505f, 506b
LH, 508, 516, 517t, 518–19, 519f
lice, description of, 337
lidar, 451–57
　access to, 457
　applications for, 406, 451–52, 454–57
　data provided by, 453–54, 453f
　definition of, 451
　future of, 456–57
　types of, 452–53
life history traits, in estimation of population trends, 373b, 373f
life tables, 365–66
　cohort, 368
　limitations of, 371
　time-specific, 369–70, 371
light, in vision, 468
light detection and ranging. See lidar
light-emitting diode (LED) lights, 241, 343
lighting
　in collection of invertebrates, 339, 340, 343, 343f
　nocturnal
　　in capture of birds, 65–66, 66f, 66t

　　in capture of mammals, 66t
　marking of wildlife with, 240–41, 240t
light-level geolocation systems, 269t, 275, 279
light pollution, 468
light sheets, 343, 343f
light traps
　submerged, 339
　terrestrial, 343, 343f
limb amputation, marking with, 255
Lincoln–Petersen estimator, 300, 301, 306–7, 308
Lindgren funnel traps, 344
linear mixed-effects models, 58
linear regression, 51–52, 52f
line-intercept method, 386, 391, 392
line of sight technology, 442
line transects, 27f, 30
lion, African
　capture methods for, 89t
　natural markings of, 232t
lion, mountain
　capture methods for, 83, 84t, 89t, 96, 100–101, 145t, 150
　chemical immobilization of, 138
　GPS collars on, 444
　population estimation for, 543
　scat of, detection by dogs, 149
　sex determination for, 545
literature review, 4
lithium batteries, transmitter, 268
litter size
　definition of, 511
　estimation of, 512
livestock, domestic
　brucellosis in, 202
　dogs in protection of, 151
　invasive marking of, 252, 252t
lizard(s)
　capture methods for, 107–10, 111
　handling methods for, 112, 113
　marking of
　　invasive, 246t, 247t, 252t, 253t, 255
　　natural, 232t
　　noninvasive, 237t, 240t, 241t, 243t, 244
　reproduction of, 503, 516, 517t
lizard, blunt-nosed leopard, invasive marking of, 247t
lizard, northern fence, invasive marking of, 246t
lizard, Texas horned, capture methods for, 107
lizard, whiptail, capture methods for, 108
lizard grabber, 108
local convex hull method, 484f, 485
locating wildlife, with dogs, 143–47, 144–45t
logistic equation, 353–54
logistic regression, 54, 55f
　applications for, 422
　in resource selection studies, 422–25, 423t, 424t
Longworth traps, 86

loon, common
　capture methods for, 65, 65t
　contaminant impacts on, 159
loon(s), noninvasive marking of, 237t
Lotka's equation, 366–67, 372
lures. See attractants
luteinizing hormone (LH), 508, 516, 517t, 518–19, 519f
lymphocytes, definition of, 205
lynx, Canada
　age and sex determination for, 228
　capture methods for, 84t, 89t, 92t, 100
　chemical immobilization of, 138
　molecular techniques in detection of, 542b, 542f, 544
　population growth models for, 375
　scat of, detection by dogs, 148

macaque, Assamese, stress and reproduction of, 522
macrohabitats, 412
macronutrients, 158
macroparasites, definition of, 205
macrophages, definition of, 205
magnetic fields, sensory perception of, 470–71
magpie, American, capture methods for, 69t, 70, 75, 75t, 79, 79t
magpie(s), noninvasive marking of, 237t
major histocompatibility complex, 540
malaise traps, 341, 341f
malaria, 185
males
　gene flow in, 535
　reproduction of
　　hormones in, 519
　　stages of, 513–16
mammal(s). See also specific species and types
　age and sex determination for, 218–29, 219–24t
　dentition in, 218, 225, 227f, 227t, 228–29, 228f
　DNA in, 545
　eye-lens weight in, 218–25
　physical characteristics in, 207
　capture of, 81–104
　　attractants in, 93–96
　　Best Management Practices and standards for, 87–90, 91t, 92t, 96–104
　　nets in, 65t, 71t, 72, 72t, 73t, 82–83
　　new methods for, development of, 64–65, 81
　　night lighting in, 66t
　　reference sources on, 64, 65, 81
　　techniques by species, 96–104
　　traps in, 76t, 81, 82–90, 96–104
　chemical immobilization of, 137–39, 137t, 139t
　contaminant impacts on, 154, 166, 168
　fertility estimation for, 357
　handling of, 111–12

marking of
    invasive, 244–55
    natural, 232t
    noninvasive, 234–44
    remote, 233, 233t
    remote cameras in studies of, 316–17
    reproduction of
        contraception for, 520–21
        courtship in, 504
        hormones in, 518–19, 518f
        lactation in, 513
        modes of, 503
        ovulation in, 508–11, 508f, 518, 518f
        placental scars in, 512–13
        stress and, 521–22
        testis weight in, 514
        uterine analysis of, 511–12
    sensory perception in, 466–71
    transmitters on, 265, 265f
mammary glands, 513
management. *See* wildlife management
management units (MU), 533, 533b
manatee
    contaminant impacts on, 170
    natural markings of, 232t
    noninvasive marking of, 234t
manipulative experiments, 2, 3f, 12
maps and mapping. *See also* Geographic Information System; Global Positioning System
    importance of, 430
    misuse of data in, 430, 430b
    of resource selection, 415
    spatial technologies in, 430
marine mammals
    chemical immobilization of, 138–39
    effects of sound pollution on, 467
marine radar, 323, 323f, 324–25, 324t
marked–resight methods, 306–9, 493–94
marked sampling, 301–2
marking of wildlife, 230–56
    in abundance estimation, 301–2, 306–9
    behavior affected by, 234, 472
    in behavioral observation, 472
    after capture, 233–34
    considerations before, 231–34
    criteria for choosing method of, 230, 231, 233, 255–56
    in dispersal studies, 356
    dogs used in, 150
    functions of, 230
    as individuals *vs.* groups, 232
    invasive, 231, 244–55
    with multiple marks, 255, 255f
    natural, 231–32, 232t
    noninvasive, 231, 234–44
    permanence of, 233–34, 256
    permits for, 231
    remote, 233, 233t
Markov chain Monte Carlo (MCMC) methods, 37, 56

MARK program, 301, 303, 309, 310b, 378
mark–recapture. *See* capture–mark–recapture
marmots, chemical immobilization of, 139t
marsupials, reproduction of, 513
marten, American
    age and sex determination for, 221t, 228
    capture methods for, 84, 84t, 89, 92t, 100
    chemical immobilization of, 139t
    GIS in studies of, 432f
    population estimation for, 543
    range of, 431b, 431f, 432f
martin, purple
    capture methods for, 67
    radar in studies of, 333
mass. *See* body mass
mast
    definition of, 401
    hard, 401
    sampling techniques for, 401–4, 420
    soft, 401
mast traps, 402
Material Safety Data Sheet (MSDS), 170
maternally inherited markers, 535
mathematical formation, in modeling, 23
mating. *See also* reproduction
    need for studies on, 479t
    sensory perception in, 467, 469
maximum likelihood estimation, 48
Mayfield method, 359b, 497, 512
MCMC methods, 37, 56
MCP approach, 482, 483t, 484–85, 484f
mean, calculation of, 48–49
mean estimate, definition of, 285
mechanistic models of home ranges, 483t, 484, 487–89
mechanoreceptors, 470
medetomidine, 122
    emergency treatment for human exposure to, 136
    for mammals, 137
median, calculation of, 49
mercury, mortality and morbidity from, 158–59
mercury vapor (MV) lights, 343
merganser, hooded, capture methods for, 77t
merlin, capture methods for, 69t
meta-analysis, 9
metadata
    access to, 460
    functions of, 458
    history of, 458–59
    in radiotelemetry studies, 262
    in spatial technologies, 458–59
metal(s)
    clinical signs of exposure to, 173t
    definition of, 157
    essential, 158
    handling of samples of, 174
    human use of, 155
    mortality and morbidity from, 157–60
    nonessential, 158–60

metal bands, marking of wildlife with, 235–36, 236f, 238, 238f
metalloids
    definition of, 157
    essential, 158
    mortality and morbidity from, 157–60
    nonessential, 158–60
metapopulations
    criteria for identification of, 6–7, 7f, 8b, 379b
    definition of, 8b, 379b
    dynamics of, 379b
    origin of term, 379b
    types of, 8b
metestrus, 505f, 506b
M-44s, 94, 163
mice. *See* mouse
microbiology
    definition of, 193, 205
    in disease investigations, 193–95
microcontrollers, in VHF transmitters, 270
microhabitats, 412
micronutrients, 158
microparasites, definition of, 205
microsatellites, 527, 527t, 528, 528f, 543
microtaggants, 246
microwaves, in radar, 320–21, 321f
migrants per generation, number of, 534, 535f
migration
    methods for analysis of, 490
    need for studies on, 479t
    radiotelemetry in studies of, 259
Migratory Bird Act, 465
migratory birds
    learned behaviors in, 463–64
    noninvasive marking of, 236
    radar in monitoring of, 333–34
military aircraft, bird strikes on, 332
military antipersonnel radar, 324t, 326
military surplus tracking radar, 326
minerals
    human use of, 155
    mortality and morbidity from, 158
minimum convex polygon (MCP) approach, 482, 483t, 484–85, 484f
minimum-number-live estimates, 306
minimum viable population (MVP), 378
mining, mortality and morbidity related to, 155, 157, 158, 163
minisatellites, 527, 527t, 528, 528f
mink, American
    age and sex determination for, 222t
    capture methods for, 89, 100
    chemical immobilization of, 139t
    contaminant impacts on, 161, 162
    reproduction of, 513, 516
Minta–Mangel estimator, 493, 494
mist nets
    capture of birds with, 66–69, 68f, 69f
    capture of mammals with, 82
    capture of reptiles with, 109

mist nets (*continued*)
　handling of birds in, 111
mites, description of, 338
mitochondrial DNA (mtDNA), 526, 527t, 531, 535
mitochondrial genome, 526, 531
mitochondrion, 526
mixed effects models, 18, 58–59
model(s) and modeling, 15–16, 21–23. *See also specific types*
　as alternative to experimental research, 2–3, 21
　applications for, 13, 15–16, 21
　complex *vs.* simple, 21, 21t, 22, 350
　definition of, 4, 11b, 13, 350
　deterministic *vs.* stochastic, 351
　development of, 4, 10–12, 10f
　　in data analysis, 47–48
　　steps in, 15, 22–23, 376
　discrete-time *vs.* continuous-time, 351
　goals of, 47, 350
　in information-theoretic model selection, 35–36
　process for selection of, 376–78
　software programs for, 23, 378
　strategies for, 21, 22t
　validation of, 23, 37, 374
model averaging, 376
Model H_1, 363, 364t
modern distance sampling, 296, 303–4, 303f
moiré effects, in graphs, 47
mole(s), sensory perception in, 470
molecular ecology, 542–46
　definition of, 542
　dietary analysis in, 545
　noninvasive sampling in, 542–43
　population size in, 543
　sex determination in, 545–46
　species identification in, 544–45
　tracking movements in, 544
molecular fingerprints, 542
molecular genetics, 526–30, 527t
molecular markers, 526–27, 527t
molecular tags, 543
mole rat, blind, sensory perception in, 470
molt patterns, in age and sex determination, 211, 213
mongoose, invasive marking of, 252t
monkey, red howler, reproduction of, 517t, 520
monkey(s), stress and reproduction of, 522
monomorphic species, age and sex determination for, 207, 208
Monte Carlo methods, 22
moorhen, common, age and sex determination for, 217t
moose
　age and sex determination for, 219t
　capture methods for, 73t, 83, 87t
　chemical immobilization of, 137–38, 137t, 138f
　GPS collars on, 444
　habitat selection by, 413–14
　invasive marking of, 250t
　population density of, 28
　population estimation for, 494
　remote marking of, 233t
　reproduction of, 509
　resource selection by, 415, 422, 422t, 423–25, 424t
　sensory perception in behavior of, 466
moosehorns, 393, 393f
morbidity
　contaminant-related (*See* contaminants)
　definition of, 154, 205
　disease-related (*See* disease)
morphine, 122
morphological characteristics
　in age and sex determination, 208
　in taxonomy, 530
mortality. *See also* death rates
　contaminant-related (*See* contaminants)
　definition of, 154, 205
　disease-related (*See* disease)
mortality sensors, 275
mosquitoes, description of, 337
moths
　collection methods for, 342f
　description of, 337
motility, of sperm, 514–15
mounting, of invertebrate specimens, 345, 346, 347
mouse
　abundance estimation of, 308, 308t
　age at sexual maturity, 218
　capture methods for, 85
　invasive marking of, 247t, 249t, 250t, 254t
　ultrasonography of, 509
mouse, cotton, capture methods for, 85
mouse, deer
　capture methods for, 83t, 86
　handling methods for, 111
　radiocollars on, 265f
mouse, desert pocket, handling methods for, 111
mouse, harvest, invasive marking of, 246t
mouse, house
　capture methods for, 93
　reproduction of, 505
mouse, white-footed
　capture methods for, 85, 87
　reproduction of, 513f
mouse, wood, capture methods for, 85
Movebank, 262
movement, animal
　general characteristics of, 490
　as learned behavior, 462
　methods for analysis of, 481, 490 (*See also* radiotelemetry)
　molecular techniques for tracking, 544
　motivations for studying, 481
movement ecology models, 489
Movement Ecology Paradigm, 481
moving target indicator (MTI) filter, 329
mowing, in nest success, 17–18, 38
MSDS, 170
mtDNA, 526, 527t, 531, 535
MTI filter, 329
MU, 533, 533b
multicapture nest boxes, 77, 77t
multifactor designs, 16–17
multimodal data distribution, 49
multiple regression, 52–53
multiresponse permutation method, 489
murre, common, capture methods for, 75
murrelet, marbled
　capture methods for, 67, 68f
　radar in monitoring of, 333
murrelet, Xantus, capture methods for, 65
murrelet(s), capture methods for, 65t
muscle, DNA extraction from, 529–30, 530t
Museum Special snap traps, 85–86
museum specimens, genetic diversity in, 535, 537–38
muskox
　age and sex determination for, 219t
　measurement of hormones in, 523
　remote marking of, 233t
muskrat
　age and sex determination for, 223t, 225f, 226f
　capture methods for, 66t, 84t, 89, 92t, 95, 101
　chemical immobilization of, 139, 139t
mustelids
　age and sex determination for, 222t
　chemical immobilization of, 139, 139t
　sensory perception in, 469
musth, 519
mutagenesis, 160
mutations. *See* genetic mutations
MV lights, 343
MVP, 378
mycology
　definition of, 205
　in disease investigations, 194
mycotoxins, 170

naloxone, 124
naltrexone, 124
names, in behavioral research, 472
NAP, 459–60, 460b
nasal discs, 238, 239t
nasal saddles, 238, 238f, 239t
natality
　definition of, 502
　methods for estimation of, 502–3, 507, 512
National Agriculture Imagery Program, 439
National Biological Information Infrastructure (NBII), 440, 459, 460
National Land Imaging Program, 447
National Oceanic and Atmospheric Administration (NOAA) satellites, 274

National Resources Conservation Service Spatial Data Gateway, 438–39
National Spatial Data Infrastructure (NSDI), 458
natural experiments, 12, 14, 14t
natural history studies, descriptive, 12
natural markings, on wildlife, 231–32, 232t
natural toxins, 169–70
Navstar system. *See* Global Positioning System
NBII, 440, 459, 460
NDVI, 405, 405t
nearest-neighbor method, 389, 485–86
neck collars
    marking with, 234–35, 234t, 235f, 236f
    radiocollars, 265, 265f
necklace transmitters, 265, 266f
neck snares, capture of mammals with, 88, 89, 89t
necrobacillosis, 200
necropsies
    in contaminant investigations, 176, 177t
    definition of, 205
    in disease investigations, 192–93, 195–97, 196b
    fetal counts in, 511
    formal training for, 196
    gross examination of ovaries in, 509–10, 510b, 511f
    measures of stress in, 522–23
necrosis, definition of, 205
necrotic stomatitis, in elk, 200–201, 203
needle biopsies, of testes, 515–16
needles, in chemical immobilization, 124–25
    accidental injection of humans, 134–36
neighborhoods
    in adaptive cluster sampling, 28–29
    ecological, 409
neon lights, 240–41
neoplasia, definition of, 205
nervous system, and endocrine system, 516
nest(s)
    dogs in searches for, 144t, 146
    eggs in (*See* clutch size; egg(s))
    remote cameras in studies of, 312f, 316, 317, 344
nest boxes, artificial, for wood duck, 464–65, 465f
nested design, 18
nest predation
    at forest edges, 19, 20–21, 38
    remote cameras in studies of, 316
nest success
    bias in estimation of, 359b
    definition of, 512
    habitat in, 2
    as measure of reproductive success, 358
    methods for estimation of, 358, 359b, 512
    mowing in, 17–18, 38
    snowmelt in, 17
nest traps, capture of birds with, 77–79, 77t, 78f

net(s)
    capture of amphibians with, 65t, 105
    capture of birds with, 65–73, 65t, 71t, 72t, 73t, 76, 76t
    capture of invertebrates with, 338–40, 338f, 339f
    capture of mammals with, 65t, 71t, 72, 72t, 73t, 82–83
    capture of reptiles with, 65t, 109, 110
net guns
    capture of birds with, 72–73, 73t
    capture of mammals with, 72, 73t, 83
neuroleptanalgesics, 123
neuromuscular blocking (NMB) drugs, 120–22
neurotoxicants, insecticides as, 164
newt, alpine, invasive marking of, 254t
newt, eastern
    invasive marking of, 254t, 255
    natural marking of, 232t
newt, great-crested, invasive marking of, 247t
newt, smooth, natural markings of, 232t
newt, warty, natural markings of, 232t
newt(s), invasive marking of, 253t
NEXRAD radar system, 325–26, 328, 335
nicotine, as insecticide, 166–67
nighthawk, common
    capture methods for, 66t, 75t
    sensory perception in, 470
nightjars, capture methods for, 65t, 66
night lighting. *See* lighting, nocturnal
nitrogen, liquid, freezing of invertebrates with, 346
NMB drugs, 120–22
NOAA satellites, 274
nocturnal animals
    photopollution and, 468
    thermal infrared cameras in studies of, 316–17
nocturnal searches, for amphibians, 105
nocturnal tracking lights, 240–41, 240t
nominal data, definition of, 44
nondestructive sampling, 529
nonexplosive drop nets, 71, 71f, 82
noninvasive sampling, in molecular ecology, 542–43
nonlinear mixed-effects models, 58
nonmechanistic models of home ranges, 484, 486–87
nonparametric tests, 33, 63
nonprobabilistic sampling designs, 288
nonuniform treatments, 38
noose guns, capture of reptiles with, 109–10
noose mats, capture of birds with, 70–71, 71f
noose poles
    capture of birds with, 73–76, 75t
    capture of mammals with, 83
noose traps, capture of reptiles with, 109
noose tubes, 112
NOREMARK program, 302

Normalized Difference Vegetation Index (NDVI), 405, 405t
North American Profile (NAP), 459–60, 460b
NSDI, 458
nuclear genome, 526
nucleated epithelial cells, in estrous cycle, 505f, 506b
nucleotides, 526
    fragment analysis of, 527–28
    in single nucleotide polymorphisms, 529
nuisance animals
    avicides for, 168–69
    control of reproduction of, 520
    dogs in management of, 150–51
    radar in management of, 333
    rodenticides for, 168
null hypothesis
    definition of, 34
    testing of, 34–35
null-peak antenna systems, 271f, 272
nuptial plumage, 211
nutria
    capture methods for, 65t, 84t, 89t, 91t, 101
    invasive marking of, 245t, 250t, 254t
nutrient sources, in collection of invertebrates, 343

observability. *See* detection probability
observations
    of animal behavior, 472, 473
    dependent, 31
    in disease investigations, 189–92, 191b
    in habitat selection studies, 416
    of invertebrate–vertebrate interactions, 344–45
    paired, 31, 31f
observed survival, 358
observer effects, 472
occupancy of species
    probability of, 411
    remote cameras in studies of, 313
Occupational Safety and Health Administration (OSHA), U.S., on safety in contaminant investigations, 170
ocular estimate method
    for biomass, 394
    for cover, 391, 391b
    for plant use, 399–400
odors. *See* scents
Office of Management and Budget (OMB), on metadata, 458
oilbird, sensory perception in, 467
oil spills, mortality and morbidity from, 161
olfaction, 469, 472. *See also* scents
olfactory pollution, 469
OMB, on metadata, 458
omnidirectional antennas, 271, 271f
oocytes, primary, 508, 508f
open populations
    definition of, 285, 358
    estimation of birth rates for, 358

open systems, for inhalation anesthetics, 123–24
opioids, 122–23, 122t
    antagonists for, 124, 136
    in darts, 126
    emergency treatment for human exposure to, 136
    for mammals, 137–38
    for urban wildlife capture, 131
opossum, brush-tailed, reproduction of, 514
opossum, Virginia
    age and sex determination for, 223t
    capture methods for, 84t, 88, 91t, 101–2, 102f
    chemical immobilization of, 139t
    invasive marking of, 245t, 246t
    reproduction of, 505, 507, 510, 513
    sensory perception in, 468
Oracle, 439
oral drugs
    in capture of birds, 80–81
    in chemical immobilization, 130
ordinal data, definition of, 44
Oregon Imagery Explorer, 439
organ(s), in disease investigations, 195
organic chemicals
    clinical signs of exposure to, 173t
    definition of, 160
    handling of samples of, 174
    human use of, 155
    mortality and morbidity from, 160–62
organochlorine insecticides, 164–65
organo-phosphorus-induced delayed neuropathy, 165
organophosphorus insecticides, 165–66
ornithology, scientific method in, 3
orthopterans, description of, 337
oryx, Arabian, population estimation for, 494
OSHA, U.S., on safety in contaminant investigations, 170
osprey
    capture methods for, 75, 75t, 77t
    natural markings of, 232t
ostrich
    sensory perception in, 470
    sex determination for, 546
otter(s)
    capture methods for, 94t
    transmitters on, 266
otter, Eurasian
    capture methods for, 93
    genetic diversity of, 536
otter, northern river
    age and sex determination for, 221t
    capture methods for, 84t, 89, 91t, 92t, 103–4
    chemical immobilization of, 139t
    handling methods for, 112
otter, sea
    age and sex determination for, 221t
    chemical immobilization of, 138
    invasive marking of, 247t, 250t

outbreeding depression, 531
outliers, in space use, 482
ovaries
    activation of, 504
    gross examination of, 509–11, 510b, 511f
    mammalian, 508–11, 508f, 509f
    reptilian, 512
overdispersion, 17, 55
oviparous species, 503
OvoControl, 520, 521
ovulation, 507–12
    hormones in, 518, 518f
    in mammals, 508–11, 508f, 518, 518f
    measurement of rate of, 507–12
owl(s)
    capture methods for, 67, 73
    chemical immobilization of, 137
    noninvasive marking of, 237t, 240t, 241
owl, barn
    capture methods for, 75t
    contaminant impacts on, 169
    sensory perception in, 467
owl, barred
    capture methods for, 67, 69, 70
    hybridization of, 532
owl, boreal, noninvasive marking of, 240t, 241
owl, burrowing, capture methods for, 73, 75, 75t, 79, 79t
owl, eastern screech, capture methods for, 70, 75t
owl, ferruginous pygmy, capture methods for, 75–76
owl, flammulated, capture methods for, 75, 75t
owl, great gray, habitat of, in forest planning, 436b
owl, great horned
    capture methods for, 79–80
    in capture of other birds, 67–70, 70f
    contaminant impacts on, 168
owl, long-eared, noninvasive marking of, 240t
owl, northern saw-whet, capture methods for, 69t, 80
owl, northern spotted
    estimation of population trends in, 373f
    habitat models of, 432, 435
    hybridization of, 532
owl, short-eared, capture methods for, 75t, 77t
owl, spotted
    capture methods for, 69t, 75t, 76
    evolutionary significant units of, 533
    habitat of, in forest planning, 434, 436–37b, 436f, 437f
owl, tawny, capture methods for, 69t, 76
owl, tropical screech, capture methods for, 75t
owl, western burrowing, capture methods for, 79

oystercatcher, American, capture methods for, 71
oystercatcher, noninvasive marking of, 237t

PAHs, 161
paint, marking of wildlife with, 233, 236, 242–44, 243t
paired design, 17
paired observations, 31, 31f
pandemics, definition of, 205
pangolin, noninvasive marking of, 243t
pan traps, 341–42, 342f
paper, for labeling of invertebrates, 345
parakeets, noninvasive marking of, 237t
paralytic drugs, 120–22
parameters
    definition of, 48, 285, 350
    factors affecting, 376
    methods for estimation of, 23, 48, 375–78, 377b
parametric tests, 33
paraquat, 167
parasites. See also host–parasite interactions
    definition of, 205
parasitism, intra-specific brood, 465
parasitology
    definition of, 205
    in disease investigations, 194–95
parentage analysis, 544
parental care of young, 512–13
parrot, orange-winged, capture methods for, 67
parrot(s), noninvasive marking of, 237t
parsimony, 47
particle markers of wildlife, 244, 245t, 246
partridge, chukar, capture methods for, 80t
partridge, gray
    age and sex determination for, 215t
    habitat in nest success of, 2
    noninvasive marking of, 239t
partridge, Hungarian, age and sex determination for, 210t
partridge, red-legged, age and sex determination for, 210t
passerines
    capture methods for, 77t
    invasive marking of, 249t
    noninvasive marking of, 237t, 238
passive infrared (PIR) remote cameras, 313, 314t, 315t
passive integrated transponder (PIT) tags, 247, 247t, 248b
patch(es). See habitat patches
patch occupancy models, 411, 413
pathogenic, definition of, 205
pathogens, definition of, 205
pathognomonic, definition of, 205
pathological, definition of, 205
pathology, definition of, 205
PBDEs, 162
PCBs, 161–62

PCR, 193, 194, 205, 527, 528
PDOP, 442, 445b
peavey hooks, 107
peccary, collared
   age and sex determination for, 220t
   capture methods for, 84t, 87t
   invasive marking of, 245t, 250f
   noninvasive marking of, 239t, 242, 242f, 242t
pedigrees, 541
peep cameras, 317
peer review
   definition of, 13
   in experimental design, 21, 290
   in publication process, 38
   in research process, 5, 13
   timing of, 13
pelican, American white, capture methods for, 65, 65t, 66t, 72t, 81
pelican, brown
   clinical signs of disease in, 192
   contaminant impacts on, 159
pelt appearance, in age and sex determination, 218, 225f
pelvic girdle, in age and sex determination, 218, 224f
pencil beams, 322–23, 323f
penguin(s)
   invasive marking of, 249t, 254t
   noninvasive marking of, 236, 237t
   sensory perception in, 470
penguin, king, capture methods for, 81
percentage sequence divergence, 531
periodic populations, systematic sampling of, 27
peripheral blood, 523
permits
   capture and handling, 64
   marking, 231
   salvage, 172
personnel protective equipment (PPE), 170
pest(s). *See also* nuisance animals
   invertebrate, collection methods for, 343
   vertebrate, control of reproduction of, 520
pesticides
   in collection of invertebrates, 344
   definition of, 164
   handling of samples of, 174
   history of use, 155, 164
   mortality and morbidity from, 164–69
      dogs in detection of, 147
      prevalence of, 155–57
      research on, 156–57
   as organic chemicals, 160
   types of, 164
petrel, Leach's storm-, sensory perception in, 469
petroleum products, mortality and morbidity from, 161
PGF$_{2\alpha}$, 521
phagocytosis, definition of, 205

phalarope, Wilson's, capture methods for, 77t
pharmaceuticals. *See* chemical immobilization; drug(s)
pharmacy boards, 119
pheasant(s)
   dogs in studies of, 146
   invasive marking of, 254t
   noninvasive marking of, 237t, 239t, 241t, 243t
pheasant, Kalij, capture methods for, 79
pheasant, Korean, dogs in studies of, 144t
pheasant, ring-necked
   age and sex determination for, 210t, 215t
   capture methods for, 66t, 69t, 72t, 73, 79t
   contaminant impacts on, 162, 168
   invasive marking of, 246t
   nest success of, 17–18, 38
   reproduction of, 507, 508f
pheromones
   in collection of invertebrates, 342–43, 342f
   definition of, 342, 469, 505
   in reproduction, 504–5
   sensory perception of, 469
philosophy of science, 5–6
photograph(s)
   aerial
      census with, 292
      in vegetation sampling, 404, 405–6
   in disease investigations, 189
   of radar information, 327
   in vegetation sampling, 397, 404, 405–6
photographic fisheye lenses, 393
photoperiod, in reproduction, 519
photopigments, 468
photopollution, 468
phototaxis, 468
phthirapterans, description of, 337
phylogenetic relationships, 530, 531
phylogenetic species concept, 530
physical characteristics
   in age and sex determination, 207–8
   variation in, 208
pie charts, 45
pig. *See* hog
pigeon(s)
   invasive marking of, 249t
   noninvasive marking of, 238b
   reproduction of, 518
pigeon, band-tailed
   age and sex determination for, 217t
   capture methods for, 69t, 71, 71t, 72t, 79t
   reproduction of, 504
pigeon, rock
   avicides for, 168
   contraception for, 520, 521
pigments, marking of wildlife with, 242–44, 243t
pilot studies
   data collection methods in, 43
   definition of, 32
   purposes of, 4–5, 32
   in radiotelemetry, 269b, 270

PIM, 280
pin lights, 240–41
pinnae, 467
pinning, of invertebrate specimens, 347
pinnipeds
   age and sex determination for, 222t
   contaminant impacts on, 170
   noninvasive marking of, 243t
pinpoint radar, 327
pintail, northern. *See* duck, northern pintail
piperonyl butoxide, 166
pipe traps
   capture of amphibians with, 106
   capture of mammals with, 86, 87f
   capture of reptiles with, 108, 108f, 111
PIR remote cameras, 313, 314t, 315t
pitfall traps
   capture of amphibians with, 105, 106
   capture of invertebrates with, 343–44
   capture of mammals with, 84–86, 90
   capture of reptiles with, 107–9, 109f, 110
   safe-houses inside, 90, 93f
PIT tags, 247, 247t, 248b
pituitary gland, anterior, 516, 518
placental barrier, pesticides and, 165
placental scars, 512–13, 513f
plan position indicator (PPI), 323, 327–28
plant(s). *See also* vegetation
   individual, definition of, 387
plant architecture, 401
plant extractions, in capture of mammals, 95
plant tissues, in contaminant investigations, 178
plant toxins, natural, 169–70
   herbivore detoxification of, 155, 169
plant use, measurement of, 399–401
plasma
   collection of, 523
   definition of, 205
plastic bands, marking of wildlife with, 236–38, 238b, 238f
platform terminal transmitters (PTT) location systems, 274–75, 278–79, 279f
plot(s)
   definition of, 29
   in manipulative experiments, 2
   in sampling methodology, 29–30
   along transects, 27f, 30
plotless methods
   of abundance estimation, 297–99
   of vegetation sampling, 387, 389–90
plover, greater golden-, dogs in studies of, 144t
plover, mountain
   capture methods for, 77t
   daily nest survival of, 377b, 377t
   population growth models for, 349, 359b, 364b
   sex determination for, 546
plover, semipalmated, invasive marking of, 246t

plover, snowy, capture methods for, 77t, 78–79, 79f
plumage characteristics, in age and sex determination, 209–13, 210t, 211f, 212f
pneumonia, aspiration, 134
POF, 504f, 507
point(s), in vegetation sampling, 387, 387f
point-centered-quarter method, 389, 389f
point counts, 294–95
point graphs, 45, 46f
pointing, of invertebrate specimens, 347
point-intercept method, 391–92
point-quarter method, 298–99
point sampling, 27f, 30
point-to-target (PTT) method, 298
poisoning. *See also* contaminants
 secondary, 157
  from avicides, 169
  from inorganic chemicals, 163
  from pesticides, 166
  from rodenticides, 168
 sublethal, 157
Poisson regression, 54–55
polar coordinates, in radar, 322, 322f
pole syringes, in chemical immobilization, 124, 125
police radar, 324t, 326
political populations, definition of, 7
pollution. *See also* contaminants
 light, 468
 olfactory, 469
 overview of sources, 155
 sound, 467
polybrominated diphenyl ethers (PBDEs), 162
polychlorinated biphenyls (PCBs), 161–62
polycyclic aromatic hydrocarbons (PAHs), 161
polymerase chain reaction (PCR), 193, 194, 205, 527, 528
polymorphonuclear leukocyte, 506b
ponchos
 in marking of wildlife, 239, 239t
 in radiotelemetry, 265
POPAN-5 program, 309
population(s)
 biological, 6–9
 closed, 285, 357–58
 definition of, 6, 8b, 285, 349
 equilibrium in, 534
 in experimental design, 20
 founding, 541
 geographic boundaries of, 6
 GIS in studies of, 434–35
 hierarchy of spatial units of, 6–7, 7f, 8b
 ideal, 538
 vs. individual animals, behavior of, 463
 metapopulations, 6–7, 7f, 8b, 379b
 open, 285, 358
 political, 7
 radar in monitoring of, 333–34
 research, 7–9, 20, 24, 39
 reservoir, 206

population analysis, 349–80
 dispersal in, 356
 inference in, 379–80
 modeling approach to, 350
 parameter estimation in, 375–78
 population dynamics in, 349
 of population growth (*See* population growth models)
 of population viability, 15, 378–79, 434–35
 reference sources on, 350
 software programs for, 378
population control, wildlife contraception in, 520–21
population density
 definition of, 285
 in density-dependent population growth, 350, 352–56, 354b
 in density-independent population growth, 350
 in disease management, 201, 203
 dogs in estimation of, 143–46, 144t
 and fertility, 354, 354t
 and habitat quality, 412
 indices of, 291–92
 radiotelemetry in estimation of, 492–95
 relative, 285
 sampling of, 24–25, 25f, 28
population dynamics
 conceptual models of, 10, 10f
 definition of, 349
 host–parasite interactions in, 183–84
 importance of understanding, 349
 major variables in, 56
 in small populations, 350b
population estimates. *See also* abundance estimation
 definition of, 285
 radiotelemetry in, 492–95
population estimators, definition of, 285
population genetics, GIS in studies of, 435
population growth models, 350–75
 birth rates in, 356–58, 364–68, 371–72
 death rates in, 356–57, 364–72
 density-dependent, 350, 352–56, 354b
 density-independent, 350
 dispersal in, 356
 individual components of survival or birth in, 372–74
 interspecific competition in, 374
 predator–prey interactions in, 374–75
 in small populations, 350b
 stable age distribution in, 366–67
 unimpeded, 351–52
population indices
 in abundance estimation, 290, 291–92
 definition of, 285, 290
 dogs in estimation of, 143, 144t
 effectiveness of monitoring changes with, 39
 limitations of, 39, 290
population projection matrices, 367–68
population size. *See also* abundance

 effective *vs.* census, 538–39
 molecular techniques for estimation of, 543
 and population viability, 541
 radiotelemetry in estimation of, 492–95
population structure
 age and sex ratios in, 207
 and gene flow, 535
population trends
 definition of, 285
 life history traits in estimation of, 373b, 373f
population viability, genetic diversity in, 541
population viability analysis (PVA), 378–79
 applications for, 15, 378–79, 379b
 approaches to, 378–79
 GIS in, 434–35
porcine zona pellucida (PZP), 521
porcupine
 capture methods for, 84t, 102
 chemical immobilization of, 139t
 invasive marking of, 249t, 250t
 noninvasive marking of, 241, 242t
 population growth models for, 375
portable drive nets, 82
positional dilution of precision (PDOP), 442, 445b
posterior distribution, in Bayesian data analyses, 37
postnatal development
 of birds, 209
 of mammals, 218
postovulatory follicles (POF), 504f, 507
pot traps, 110
power, statistical, 288–89
power analysis, 19, 21, 32–34, 288–89
power lines, wildlife collisions with, 332–33
power snares, 75, 89
PPE, 170
PPI, 323, 327–28
PPS, 295–96
PPZH, 90
prairie-chicken, Attwater's, capture methods for, 71, 71t, 73
prairie-chicken, greater
 age and sex determination for, 214t
 behavior of, 477f
 capture methods for, 65t, 66t, 67, 68f, 69t, 71, 71t, 72t, 75t, 76t, 79t, 80t
 dogs in studies of, 144t
 genetic diversity of, 537
 noninvasive marking of, 238f
 transmitters on, 266f
prairie-chicken, lesser
 age and sex determination for, 208f, 214t
 capture methods for, 76, 76t, 77f
 dogs in carcass searches for, 147
prairie dog, capture methods for, 84t
precision
 and accuracy, in abundance estimation, 285–86, 286f
 definition of, 24, 285

in modeling, 350
in sampling, 24–25, 24f
precocial young, 209
predation
   dogs in studies of, 150
   marking of wildlife and, 234, 240, 242
   nest (*See* nest predation)
predation coefficient, 375
predator–prey interactions
   models of, 374–75
   sensory perception in, 468
   stress caused by, 522
predictions
   definition of, 10
   as step in research process, 2, 3, 4, 10–12
predictor variables
   definition of, 415
   proximate, 416
   in resource selection functions, 415, 416
   ultimate, 416
PREFER method, 421
pregnancy. *See also* reproduction
   chemical immobilization during, 129
   methods for detection of, 511–13
prescribed burning
   data collection on effects of, 43
   in management of deer and elk populations, 39
preservation methods, for invertebrate specimens, 345–48
pressure receptors, 470
prevalence, definition of, 205
prey. *See* predation; predator–prey interactions
prey drive, of dogs, 143
primary feathers
   in age and sex determination, 210t, 211, 211f
   numbering of, 212f
primary productivity, net, measurement of, 405
primates
   invasive marking of, 255
   reproduction of, 521
   sensory perception in, 468, 470
priming pheromones, 505
principles, in theories, 11b
prions, 185, 205–6
prior knowledge, in frequentist *vs.* Bayesian inference, 34
prisms, clear-glass, 392
probability
   definition of, 34
   of detection (*See* detection probability)
   of selection of resources, 413–16
probability density function, in survival analysis, 57
probability distribution, in Bayesian data analyses, 37
probability proportional to size (PPS), 295–96
problem animals. *See* nuisance animals; pest(s)
problem identification
   in modeling, 22
   in research, 6

procedural inconsistency, in research, 38
process variance, 378
productivity
   dogs in estimation of, 144t, 146
   measurement of net primary, 405
proestrus, 505f, 506b
profile boards, vegetation, 396
progesterone, 509, 516, 517t, 518–19, 520
programs
   management (*See* wildlife management)
   software (*See* software programs)
prolactin, 517t, 518
pronghorn
   age and sex determination for, 220t, 227t
   behavior of, 467, 469, 473
   capture methods for, 73t, 87, 87t, 89t
   chemical immobilization of, 137, 137t
   handling methods for, 112
   remote marking of, 233t
   reproduction of, 512
   sensory perception in, 467, 469
propiopromazine hydrochloride (PPZH), 90
prostaglandin $F_{2\alpha}$ ($PGF_{2\alpha}$), 521
protected areas, GIS in design of, 435–36, 437–38, 438f
protein(s)
   as gene products, 526–27
   structure of, 527
proteinase K, 529
protein electrophoresis, 527, 530
proximate hypotheses, 471–72
proximate predictor variables, 416
pseudoreplication, 18–19, 20–21, 39
ptarmigan, rock, age and sex determination for, 214t
ptarmigan, white-tailed
   age and sex determination for, 214t
   capture methods for, 80t
   contaminant impacts on, 159
ptarmigan, willow
   age and sex determination for, 210t, 215t
   capture methods for, 67, 75t
   dogs in studies of, 143, 144t, 146
PTT location systems, 274–75, 278–79, 279f
PTT method, 298
publication
   challenges of, 38
   as final step in research process, 5, 38
   purposes of, 38
puffin, capture methods for, 79t
pukeko, genetic variation among, 528f
pulse(s), radar, 323
pulse interval modulation (PIM), 280
pulse oximeters, 128
PVA. *See* population viability analysis
P-value, definition of, 34–35
pyrethroids
   in collection of invertebrates, 344
   synthetic, 166
pyrethrum, 166
python, capture methods for, 111
PZP, 521

Q-nets, 73
Quade method, for analysis of resource selection, 420–21
quadrat charting, 390
quadrat methods, of vegetation sampling, 387–88, 390
quail, California
   age and sex determination for, 210t, 215t
   resource use by, 413
quail, coturnix, age and sex determination for, 210t
quail, Gambel's
   age and sex determination for, 215t
   capture methods for, 80t
quail, Japanese
   measurement of hormones of, 523
   reproduction of, 517t
quail, Montezuma
   age and sex determination for, 215t
   capture methods for, 67, 80t
quail, mountain, age and sex determination for, 215t
quail, scaled
   age and sex determination for, 215t
   capture methods for, 76t, 79t, 80t
qualitative data, definition of, 43–44
quality control, in scientific method, 5
quantitative data, definition of, 43–44
quantitative genetic approaches, 533
quinacrine dehydrochloride, 245
quoll, tiger
   evolutionary significant units of, 533, 533b
   taxonomic redefinition of, 533b

rabbit(s)
   chemical immobilization of, 139t
   marking of
     invasive, 245t, 246t, 249t, 250t
     noninvasive, 240t
     remote, 233t
rabbit, cottontail
   age and sex determination for, 224f
   capture methods for, 66t, 95
   handling methods for, 111
   invasive marking of, 245t, 249t
   reproduction of, 510, 511
rabbit, European
   capture methods for, 86, 95
   genetic diversity of, 538
   thermal infrared cameras in studies of, 317
rabbit, Lower Keys marsh, capture methods for, 83, 83t
rabbit, New England cottontail, molecular techniques in detection of, 544
rabbit, pygmy
   capture methods for, 83t
   remote cameras in studies of, 316
rabies
   clinical signs of, 192
   detection of, with thermal infrared cameras, 317

rabies (*continued*)
    management of, 201–2, 203
    vaccines for, 201–2, 203, 315
raccoon
    age and sex determination for, 221t, 226f
    capture methods for, 84t, 88, 88f, 89t, 91t, 92t, 94, 102–3, 148
    contaminant impacts on, 168
    handling methods for, 112
    invasive marking of, 245t, 246t
    remote cameras in studies of, 315, 317
    reproduction of, 512
    sensory perception in, 468
racerunner, six-lined, noninvasive marking of, 237t
radar, 319–35
    acquisition of data from, 327–28
    applications for, 319, 332–35
    body mass of target in, 321, 321f
    display of, 319–20, 320f
    equipment and mechanisms of, 319–23
    future of, 334–35
    interpretation of data from, 328–32
    limitations of, 319–20, 324
    vs. other monitoring techniques, 319
    types of, 323–27, 324t
    in vegetation sampling, 406
radar absorbent material (RAM), 328–29
Radar Equation, 320–22, 320b
radar noise, 328
radial velocity, in Doppler radar, 325–26, 326f
radiation, spillover, 322, 322f
radioactive lights, marking of wildlife with, 240–41
radioactive markers of wildlife, 244, 246–47, 246t
radioimmunoassay (RIA), 518, 524
radiolabeling, 524
radiometers, in vegetation sampling, 404
radiotelemetry, 258–81
    applications for, 258, 259, 480
    assumptions in, 259, 263
    in behavioral studies, 475
    bias in, 259, 260–61, 263–64
    criteria for using, 259, 268, 269t
    data analysis in (*See* radiotelemetry data analysis)
    data collection schedule in, 261
    data management in, 262
    data networks in, 276–77
    data retrieval in, 274, 276–77
    in dispersal studies, 259, 356, 490
    equipment for (*See* radiotelemetry equipment)
    goals of studies with, 480, 481
    GPS in, 274, 276, 278, 444–45
    in habitat selection studies, 416–17
    in home range estimation, 481–89
    locating signals in, 277–79, 278f, 279f
    location error in, 263, 264
    in movement studies, 481–83
    number of animals needed for, 261–62

vs. radar, 319
    in resource selection and use studies, 259, 414, 416–17, 491–92
    selection of animals for, 259, 260–61
    study design with, 259–64
    in survival analysis, 56–57, 358, 495–97
radiotelemetry data analysis, 480–98
    criteria for choosing method of, 480–81
    of general movement characteristics, 490
    of home ranges, 481–89
    of population size and density, 492–95
    of resource selection, 491–92
    software programs in, 481, 498–501
    of survival, 495–97
radiotelemetry equipment, 264–77, 269t
    attachment of, 265–68, 269b
    for global tracking systems, 273–75
    sensors integrated into, 275–76
    vendors and distributors for, 281–83
    for VHF systems, 268–73, 280–81
radio towers, phototaxis toward, 468
radiotransmitters. *See* radiotelemetry; transmitters
rail, black
    capture methods for, 76, 76t
    dogs in studies of, 146–47
rail, clapper, capture methods for, 76t
rail, king, capture methods for, 71, 71t
rail, sora, capture methods for, 76
rail, Virginia, capture methods for, 69t, 76, 76t
rail, yellow
    capture methods for, 66t, 69t
    dogs in studies of, 144t, 146
rail(s), age and sex determination for, 217t
RAM, 328–29
random components, in models, 351
random distribution, of vegetation, 386
random effects, 18, 58
randomization rule, 287
randomized blocks, 17
random-pairs method, 389
random sampling
    in molecular ecology, 542–43
    in plotless methods, 297
    simple, 25–26, 27f, 287
    stratified, 27f, 28, 288
    systematic, 288, 297
range
    of estimates, 285
    of radar, 321–22
rangefinders
    for dart guns, 127, 127f
    in vegetation sampling, 388
raptors
    age and sex determination for, 218t
    capture methods for, 69, 69t, 70, 71, 75t, 77t, 79t, 94t
    handling methods for, 111
    invasive marking of, 245t, 254t
    noninvasive marking of, 237t, 238, 241t
rarefaction, 60

rare species
    conservation of genetic diversity of, 533
    population viability analysis of, 15
rat(s)
    invasive marking of, 245t, 249t, 254t
    noninvasive marking of, 243t
    reproduction of, 505f, 506b
rat, cotton, capture methods for, 83t
rat, kangaroo
    capture methods for, 83t
    sensory perception in, 470
rat, naked mole, invasive marking of, 247t
rat, Norway
    contaminant impacts on, 169
    invasive marking of, 247t
    molecular techniques in tracking of, 544
ratio data, definition of, 44
ratio estimation, 29, 299–300
rationalism, 41–42
rattlesnakes
    handling methods for, 112
    invasive marking of, 247t, 249t
    sensory perception in, 468
raven, Chihuahua, capture methods for, 79t
raven, common
    contaminant impacts on, 163
    sensory perception in, 467–68
raven(s), noninvasive marking of, 237t
razorbill, capture methods for, 71
realism
    of laboratory experiments, 13
    of models, 350
    scientific, 6
real-time kinematic (RTK), 444
reasoning
    inductive *vs.* deductive, 3f, 42
    in rationalism, 41
    in scientific method, 2, 3f
rebreathing systems, 124
receivers
    GPS, 440, 441–43, 444b
    VHF, 272–73, 273f, 281
reconnaissance surveys, of vegetation structure, 382
record(s), in data collection, 44
recorded calls. *See* bird call recordings
record keeping
    in chemical immobilization, 120, 121b
    in contaminant investigations, 174, 176, 180f
    in disease investigations, 189, 191b, 192
recovery
    of animals, from chemical immobilization, 132
    of landscapes, Landsat imagery of, 447–49, 448f
recruitment, definition of, 357
rectangular plots, in vegetation sampling, 387–88
reflectivity, in Doppler radar, 325–26
regression, 29, 36
    assumptions in, 53
    graphs in, 45, 46f

linear, 51–52, 52f
logistic, 54, 55f, 422–25, 423t, 424t
multiple, 52–53
Poisson, 54–55
stepwise, 63
in survival analysis, 57–58
rehabilitation, after exposure to contaminants, 172
reindeer, Svalbard, capture methods for, 93
reintroductions
behavior in, 465–66
and genetic diversity, 539
relative abundance, 285
relative population density, 285
remote cameras, 311–18
active infrared, 313, 315t
applications for, 311, 312f, 313–18, 344
data management with, 312–13
definition of, 311
history of use, 311
innovations in, 317
passive infrared, 313, 314t, 315t
pros and cons of, 313t, 315b
quality of equipment, 311–12
thermal infrared, 316–17
remote marking of wildlife, 233, 233t
remote monitoring. *See* radiotelemetry; remote cameras
remote sensing, 446–58
accuracy assessment of, 457–58
with Landsat, 404, 446–51
with lidar, 406, 451–57
vs. radar, 319
in risk assessment, 437
in vegetation sampling, 405–6
removal methods, of abundance estimation, 302, 304–6
repeated measurements, 17
replicates (sample units). *See* sample units
replication, of studies, 32, 35
reproduction, 502–25
associated *vs.* dissociated patterns of, 514, 514f, 516
effects of stress on, 521–22
female stages of, 503–16
hormones in, 516–20, 517t, 518f
human control of, 520–21
interbreeding in, 531–32
male stages of, 513–16
methods for measurement of, 502–16
modes of, 503–4, 503f
pesticides' impact on, 165
reproductive isolation, in species definition, 530
reproductive rate
methods for measurement of, 503–16
net, 372
reproductive success
of males, 514
methods for measurement of, 358, 514
reptiles. *See also specific species and types*
age and sex determination for, 207, 229

capture methods for, 65t, 107–11
chemical immobilization of, 137
contaminant impacts on, 167
fertility estimation for, 357
handling methods for, 112–13
marking of
invasive, 251, 255
natural, 232t
noninvasive, 239–40, 240t
reproduction of
associated pattern of, 514
clutch size in, 507
courtship in, 504
hormones in, 516, 519
modes in, 503, 503f
nest success in, 512
ovarian analysis of, 512
placental scars in, 512
uterine analysis of, 511–12
transmitters on, 266
worldwide decline in, 502
research, 1–40. *See also* experimental designs
applied *vs.* basic, 6
common problems in, 38–39
experimental *vs.* descriptive, 1–3
exploratory data analysis in, 4, 9–10
hypotheses in, 2, 3, 4, 10–12
integrated approach to, 6, 13, 16
literature review in, 4
modeling in, 2–3
outline of steps in, 3f, 4–5b
philosophical foundation of, 5–6
populations used in, 6–9
predictions in, 2, 3, 4, 10–12
preliminary data collection in, 4, 9
problem identification in, 6
theories in, 2, 3, 4, 10–12
research (statistical) populations, 7–9, 20, 24, 39
research proposals, generation of, 289–90
reserves, GIS in design of, 435–36, 438f
reservoir population, definition of, 206
residue analysis, chemical, 179
resolution, level of, in modeling, 22
resource abundance, definition of, 411
resource availability
definition of, 411, 491, 492
estimation of, 417–18, 491–92
resource avoidance, definition of, 491
resource management, adaptive, 40
resource preference, definition of, 411, 491
resource selection, 410–28
definition of, 411, 491
food availability in studies of, 419–20
hierarchical nature of, 418, 491
management implications of, 412
methods for analysis of, 420–22, 421t
methods for detection of, 416–17
radiotelemetry in studies of, 259, 414, 416–17, 491–92
relative probability of, 413–16
scales (levels) of, 418

study areas for, 417–19
study designs for, 411–16, 413b, 420, 491–92
resource selection functions (RSFs)
applications for, 415, 421
assumptions in estimation of, 415–16
for categorical data, 422–25
definition of, 414
models for, 415, 415f
in study designs, 414
resource selection ratios, 422
resource units, definition of, 411, 415
resource use, 410–28. *See also* food; habitat
applications for studies of, 410, 411
definition of, 411, 491
methods for detection of, 416–17
modeling of, 422–27
radiotelemetry in studies of, 259, 414
scales (levels) of, 411–13
study designs for, 411–16, 413b, 420, 491–92
respirators, in contaminant investigations, 171
respiratory function, during chemical immobilization, 128, 132–33
response variables, 19
restraining traps
Best Management Practices and standards for, 90, 91t, 92t, 96–104
capture of mammals with, 88, 88f, 89f, 96–104
restriction endonucleases, 527
restriction fragment length polymorphisms (RFLPs), 527
retroduction, definition of, 42
RFLPs, 527
rhea, greater, capture methods for, 66t, 81
rhinoceros, black
natural markings of, 232t
ultrasonography of, 509
rhinoceros, Indian, reproduction of, 517t, 520
rhinoceros, ultrasonography of, 509
rhinoceros, white, reproduction of, 517t, 520
Rhodamine B, 244–45
RIA, 518, 524
right censoring, 495
risk assessment, GIS in, 436–38
road(s)
animal behavior around, 477
in capture of reptiles, 107
interactions between wildlife and vehicles on, 313
road sampling, 27f, 30–31
robin, American, capture methods for, 69t
robotics, in behavioral studies, 475–76, 476f
rocket nets
capture of birds with, 72, 72f, 72t
capture of mammals with, 72t, 83
rock turning, in capture of amphibians, 107
rodent(s). *See also specific species and types*
age and sex determination for, 224t
capture methods for, 89
handling methods for, 111–12
invasive marking of, 245t, 246t, 249t, 252t
noninvasive marking of, 237t, 240t

rodenticides, 168
    anticoagulant, 155–56, 168, 173t
    clinical signs of exposure to, 173t
    prevalence of incidents involving, 155
    types of, 168
rodeo method of capture, 93
roosting birds, radar in studies of, 333, 333f
rotenone, 166–67
RSFs. *See* resource selection functions
RTK, 444
ruby lasers, 252

safety considerations
    in contaminant investigations, 170–72
    in disease investigations, 170, 171, 196
    with radar, 327
sage-grouse. *See* grouse
salamander(s)
    capture methods for, 86, 105–7, 111
    invasive marking of, 245t, 246, 246t, 249t, 252t, 253t, 255
    noninvasive marking of, 243t, 244
    reproduction of, 504, 514
salamander, blackbelly, capture methods for, 107
salamander, blue-spotted, capture methods for, 106
salamander, dusky, natural markings of, 232t
salamander, marbled, capture methods for, 106
salamander, red-backed, capture methods for, 107
salamander, spotted, natural markings of, 232t
salamander, tiger, noninvasive marking of, 240t
salmon, Pacific, sensory perception in, 469
salt blocks, in capture of mammals, 94, 95–96
salvage permits, 172
sample allocation, optimal, 32
sample sizes
    calculation of, 5, 19, 21, 288–89
    definition of, 18
    importance of, 33, 288
    in pilot studies, 32
    in power analysis, 33–34
    standardization of, 60
sample units (replicates)
    definition of, 287
    examples of, 287
    lack of independence in, 39
    in random sampling, 25–26
    sampling design for selection of, 287–88
    vs. subsamples, 18–19, 39
    in systematic sampling, 26–28
sampling, 23–31
    accuracy in, 24–25, 24f
    bias in, 24–25, 24f, 30
    design options for, 25–29, 27f, 287–88
    destructive *vs.* nondestructive, 529
    dogs' role in, 143–47, 144–45t
    importance of, 12–13, 42

    methodology of, 29–31
    precision in, 24–25, 24f
    from research populations, 7, 24
sampling distribution, definition of, 285
sampling frame, 25
sampling intensity, 288
sampling variation, 378
sandpipers, noninvasive marking of, 237t
SAS, 62
satellites
    GPS, 440, 441, 441b, 441f, 442
    Landsat, 404, 446–51
    in lidar systems, 452
    vs. radar, remote sensing with, 319
    in radiotelemetry, 274–75
    in vegetation sampling, 404–5
sawflies, description of, 337
S-band, 321, 324t
scale
    definition of, 409
    of laboratory experiments, 13
    in vegetation sampling, 409
scale clipping, 253t, 254
scan sampling, 473t, 474
scat
    detection by dogs, 148–50, 152, 520, 544
    in dietary analysis, 545
    in disease investigations, 193
    DNA extraction from, 148, 529, 530, 530t, 542, 543, 545
    in habitat selection studies, 417
    hormone metabolites in, 520, 522, 524
scatter plots, 45, 46f
scaup, lesser, capture methods for, 69t
scavenging, of contaminated carcasses, 157, 164
scenario planning, 40
scents
    in behavior, 469
    in capture of mammals, 94–96, 95t
    in capture of reptiles, 109
    detection by dogs, 140–50
Schnabel estimator, 307–8
Schumacher–Eschmeyer estimator, 308, 308t
science. *See also* research
    knowledge as goal of, 41
    philosophy of, 5–6
    wildlife, 1
scientific method. *See also* experimental designs; research
    applications in natural systems for, 3–4
    vs. modeling, steps of, 15
    outline of steps in, 3–6, 3f, 4–5b
scientific models, 11b, 21–22
scientific names, lists of, 113–16, 256–57
scientific realism, 6
scientific research. *See* research
scoopers, in capture of reptiles, 108–9, 108f
scope, of laboratory experiments, 13
scoter, surf, capture methods for, 67
scrapie, 185

seabirds
    capture methods for, 65, 65t, 75t
    noninvasive marking of, 237t
seal(s)
    chemical immobilization of, 138–39
    invasive marking of, 245t, 250t, 252t, 254t
    noninvasive marking of, 243t
seal, harbor, population estimation for, 493
seal, ringed
    capture methods for, 83
    dogs in studies of, 144t, 147
sea lion, Stellar, invasive marking of, 250t
sea lion(s), chemical immobilization of, 138–39
search engines, Internet, 439
seasonality. *See also* breeding seasons
    of disease outbreaks, 185
second-order selection, 418, 491
sedatives
    in chemical immobilization, 120, 122, 122t
    emergency treatment for human exposure to, 136
sediment, in contaminant investigations, 178
seeds, sampling techniques for, 402, 403–4, 407–8, 420
seizures, during chemical immobilization, 134
selection, in genetic diversity, 539–41
selection ratios, 422
selective harvesting, 540–41
selenium, 158
self-tripping nest traps, 78
seniority, in estimation of survival rates, 360–61
sensitivity analysis, 13
sensors, in radiotelemetry, 275–76
sensory perception, 466–71
    in direct observations, 472
    importance of understanding, 466–67
    need for studies on, 479t
    types of, 467–71
septicemia, definition of, 206
sequence sampling, behavior, 474
sequential sampling, 29, 29f
serology
    definition of, 206
    in disease investigations, 195
serum
    collection of, 523
    definition of, 206
sevoflurane, 123
sex
    determination of (*See* sex determination)
    in disease investigations, 192
    in dispersal, 534–35
    heterogametic determination of, 535
sex characteristics, secondary, in age and sex determination, 218
sex chromosomes, 535, 545, 546
sex determination, 207–29
    for birds, 207, 209–13, 507, 546
    DNA-based, 545–46

for mammals, 207, 218–29, 545
    morphological characteristics in, 208
    physical characteristics in, 207–8
sex-ratio estimator, 305–6
sex ratios, unequal, 538
sex-specific dispersal, 534–35
sexual imprinting, 464, 479t
sexually dimorphic species, age and sex determination for, 207
sexual selection, 479t
Shannon–Weiner function, 61–62
sheep, age and sex determination for, 220t, 227t
sheep, bighorn
    chemical immobilization of, 137t, 138
    genetic diversity of, 539
    population estimation for, 493, 494
    remote cameras in studies of, 316
sheep, Dall
    capture methods for, 72t, 73t, 92
    remote marking of, 233t
sheep, desert bighorn
    GIS in studies of, 437
    lidar in studies of, 456
sheep, domestic
    reproduction of, 518f, 519
    rodenticides and, 168
sheep, mountain
    behavior of, 475f
    capture methods for, 71t, 72t, 73t, 76t, 82
    handling methods for, 112
    marking of
        invasive, 252t
        noninvasive, 241, 242t
        remote, 233t
    paired observations of diet of, 31
    reproduction of, 512
shellfish, natural toxins in, 169, 170
shell notching, 253t, 254
Sherman traps, 79, 85–86
shock, emergency medicine for animals in, 133
shorebirds
    capture methods for, 66t, 71, 71f, 71t, 76t, 80
    contaminant impacts on, 169
    dogs in studies of, 146
    handling methods for, 111
shoveler, northern
    capture methods for, 69t
    disease in, 192
shrew, masked, capture methods for, 85, 105
shrew, short-tailed, capture methods for, 87
shrew(s), capture methods for, 84, 85
shrike, loggerhead, capture methods for, 69t, 73, 75t
shrubs
    fruits of, 402–3
    sampling of
        in food selection studies, 419–20
        techniques for, 387, 388, 392–93, 395, 400–401
        time required for, 384t

side-flap pails, 105
side lobes, radar, 322, 322f, 324–25
sightability. *See also* detection probability
    in population estimation, 493, 494
sighting tubes, 393
signal to noise ration (SNR), 445b
significance level, 33, 288
silver nitrate, 252
simple random sampling, 25–26, 27f, 287
Simpson's index, 62
simulation models. *See* scientific models
single-factor analyses, 16–17
single nucleotide polymorphisms (SNPs), 527t, 529
siphonapterans, description of, 338
site fidelity, methods for analysis of, 489–90
size. *See* body size
skeletochronology, in age and sex determination of mammals, 229
skimmer, black, noninvasive marking of, 240t
skink
    capture methods for, 107, 108
    invasive marking of, 246t
    noninvasive marking of, 241t
skin transplantation, marking with, 255
skunk(s)
    age and sex determination for, 222t
    capture methods for, 89t
    chemical immobilization of, 139t
    invasive marking of, 245t
skunk, striped, capture methods for, 84t, 91t, 104
slide mounting, of invertebrate specimens, 346
slotted-waveguide antennas, 324
smell. *See* olfaction; scents
smelting, mortality and morbidity related to, 155, 157, 158
smoothing parameter, 486
snake(s)
    capture methods for, 107, 108, 109, 110–11
    handling methods for, 112–13
    marking of
        invasive, 246t, 247t, 249t, 251, 252, 252t, 253t, 254
        natural, 232t
        noninvasive, 243t
    reproduction of, 503
    sensory perception in, 468
snake, brown tree
    capture methods for, 109, 110
    dogs in management of, 151
snake, garter, capture methods for, 110
snake, pine, invasive marking of, 247t
snake, red-sided garter, reproduction of, 514, 514f, 516, 519
snake rake, 108
snares
    cables used for, 96, 96f, 97f
    capture of birds with, 73–75, 75t
    capture of mammals with, 83, 87–90, 89f, 89t, 96–104

capture of reptiles with, 111
list of manufacturers of, 116–17
snipe, Wilson's, age and sex determination for, 217t
snowmelt, in nest success, 17
snowmobiles, in capture of mammals, 83, 93
SNPs, 527t, 529
SNR, 445b
social behavior
    importance of understanding, 462
    robots in studies of, 476
social cueing, 476
social stress, 521, 522
social structure
    and genetic diversity, 535
    sampling of, 473t, 474
Society for Conservation GIS, 439
sodium pentobarbital, accidental mortality from, 163–64
software programs
    for abundance estimation, 309, 310b, 499
    for behavioral studies, 475
    for data analysis, 62, 62t
    for modeling, 23, 378
    for parameter estimation, 375, 378
    for population analysis, 378
    for radar data interpretation, 331
    for radiotelemetry data analysis, 481, 498–501
    for spatial technologies, 430, 460
    for vegetation sampling, 404, 409, 446b
soil
    anthrax in, 185
    samples of, in contaminant investigations, 178
solvents, organic, mortality and morbidity from, 160
sora, capture methods for, 69t, 76t
sound pollution, 467
sound production
    array systems for recording, 476
    by birds, recordings of, in capture techniques, 67, 69–70, 76, 80, 80t
    functions of, 467–68
Southeast Cooperative Wildlife Disease Study, 189, 190b, 196
space use
    definition of, 481
    methods for estimation of, 481–90
    outliers in, 482
    overlaps in, 489–90
sparrow, Bachman's
    capture methods for, 67
    GIS in population models of, 434
sparrow, Brewer's, habitat selection by, 412
sparrow, chipping, capture methods for, 75t
sparrow, house
    capture methods for, 77t, 79t
    dogs in carcass searches for, 147
    noninvasive marking of, 237t
sparrow, song, reproduction of, 519

sparrow, white-crowned, reproduction of, 514, 517t, 519f
sparrow, white-throated, stress in, 521
spatial heterogeneity, in habitat selection studies, 419
spatial relationships, definition of, 433b
spatial technologies, 429–61. *See also specific types*
   applications for, 429, 430
   data documentation in, 458–60
   data standardization in, 458–60
   in habitat mapping, 432–34
   integration of data from multiple, 429, 430
   limitations of, 430
   misuse of, 430, 431b
species
   biological *vs.* phylogenetic concept of, 530
   definition of, 6, 8b, 530, 531
   evolutionary significant units of, 532–33
   molecular techniques for identification of, 544–45
   taxonomic identification of, 530–31
species distribution
   in abundance estimation, 290
   GIS in studies of, 437
   lidar in studies of, 454
   remote cameras in studies of, 313
species diversity
   need for studies on, 479t
   species heterogeneity as measure of, 61–62
   species richness as measure of, 59–61
species enumeration, 59
species heterogeneity, 61–62
   definition of, 61
   methods for estimation of, 61–62
species richness
   comparison among communities, 60
   definition of, 59
   GIS in studies of, 435
   indices of, 59
   lidar in studies of, 455–56
   methods for estimation of, 59–61
specimen collection
   in contaminant investigations, 171
   in disease investigations, 192–97
   dogs used in, 147–48
specimen handling. *See* handling
spectral width, in Doppler radar, 325–26
speculation, role of, 37–38
sperm
   artificial insemination of, 505, 518, 519
   collection by robots, 476
   cryopreservation of, 505
   hormones and, 519
   measurement of, 514–15, 515b
   motility of, 514–15
   production of, 513–14, 519
spermatids, 516
spermatogenesis, 513–14, 519
spermatogonia, 516
spermatozoa, 514–15
spherical densiometer, 393, 393f

spheroid, 445b
spillover radiation, 322, 322f
spot mapping, 293
spreadsheet files, 44
spruce, white, density of, 389
SQL, 439
squirrel(s)
   chemical immobilization of, 139t
   invasive marking of, 245t, 250t, 252t
   noninvasive marking of, 243t
squirrel, Abert's, capture methods for, 83t
squirrel, California ground, capture methods for, 83t
squirrel, Delmarva fox, lidar in studies of, 454
squirrel, eastern gray
   age and sex determination for, 223t, 225f, 226f
   capture methods for, 83t, 84, 104
   contaminant impacts on, 168
   noninvasive marking of, 242t
   population growth models for, 365t, 368
   reproduction of, 512
squirrel, flying, capture methods for, 83t
squirrel, fox
   age and sex determination for, 223t
   capture methods for, 83t, 84, 104
   fertility estimation for, 357
   invasive marking of, 248b, 248f, 250t
squirrel, ground
   capture methods for, 86, 89t
   sensory perception in, 469
squirrel, Mt. Graham red, habitat models of, 434
squirrel, northern flying
   capture methods for, 84
   habitat of, in forest planning, 434
squirrel, red
   capture methods for, 83t, 104
   handling methods for, 112
   remote marking of, 233t
squirrel, Townsend's ground, invasive marking of, 247t
squirrel, tree, abundance estimation of, 297
squirrel glider, capture methods for, 86
stable age distribution, 366–67, 369–70
stacked beams, 323
staff-ball method, 397, 397f
staggered entry, 497
standard deviation
   calculation of, 289
   definition of, 285
standard error
   definition of, 18, 285
   precision and, 24
standing crop. *See* biomass
starling, European
   avicides for, 168
   capture methods for, 77t, 80
   invasive marking of, 249t
   radar in management of, 333f
states, behaviors as, 474
static analyses, of animal interactions, 489

static mode, 445b
statistic(s), definition of, 285
statistical analyses, in research process, 31–32
Statistical Analysis System (SAS), 62
statistical inference
   essential role of, 42
   on field data, 48–62
   frequentist *vs.* Bayesian approaches to, 34
statistical models, 11b, 21
statistical population reconstruction analysis, 493, 495
statistical populations. *See* research populations
statistical power, 288–89
statistical tests, in scientific method, 3, 3f
statisticians, consultation with, 21, 41, 42
stem-count method, 400
Stephenson traps, 83
stepwise regression, 63
sticky pads, in capture of reptiles, 108
sticky traps, in collection of invertebrates, 342, 342f
stilt, black-necked, capture methods for, 77t
stilt, noninvasive marking of, 241t
stochastic difference equation models, 23
stochastic models, definition of, 351
stock identification, genetic, 544
stopover habitat, radar in studies of, 334
storage, of invertebrate specimens, 345–48
stork, wood, remote marking of, 233t
stratified random sampling, 27f, 28, 288
streamers, marking of wildlife with, 241–42, 241–42t, 242f, 250
stream searches, for amphibians, 105
stress
   definition of, 521
   effects on reproduction, 521–22
   endocrinology of, 502, 516, 521–23
   indicators of, 522–23
   measures of, 522–23
   need for studies on, 479t
stressors
   definition of, 521
   in risk assessment, 436
strip counts, 293–94
strip transects, 30
strong inference, 3
study designs, types of, 12–16, 12f
study plots, in manipulative experiments, 2
subcaudal scale clipping, 253t, 254
sublethal effects, of contaminants, 157
subsamples, 18–19, 39
subspecies
   definition of, 8b
   taxonomic identification of, 530–31
succinylcholine, 122, 137
sufentanil, 123, 136
surveillance, in disease investigations, 206
surveillance radar, 323, 323f, 326, 329, 329f
surveys
   definition of, 24
   design checklist for, 25, 26b
survey sampling, 24

survival
    apparent, 358, 361
    observed, 358
    true, 358
survival analysis, 56–58
    criteria for design of, 58
    development of models for, 47
    molecular techniques in, 543
    radiotelemetry in, 56–57, 358, 495–97
    regression in, 57–58
    survivorship functions in, 57
survival rates
    age-dependent, 364–71, 364b
    death rates as complement of, 358
    individual consideration of components of, 372–74
    methods for estimation of, 358–64, 368–71, 495–97, 543
survivorship curves, 366, 366f
    type I, 366
    type II, 366
    type III, 366
survivorship functions, 57, 366
swallow(s)
    noninvasive marking of, 237t
    sensory perception in, 470
swallow, bank, capture methods for, 73, 75t, 77t
swallow, barn
    capture methods for, 77t
    genetic mutation in, 534
swallow, cliff, capture methods for, 77t
swallow, tree
    capture methods for, 77, 77t
    radar in studies of, 333
swan(s)
    age and sex determination for, 211, 213t
    capture methods for, 65t
    contaminant impacts on, 163
    noninvasive marking of, 234t, 243t
swan, Bewick's, natural markings of, 232t
swan, mute
    age and sex determination for, 211
    contaminant impacts on, 163
swan, trumpeter, capture methods for, 65, 66t
swan, tundra, noninvasive marking of, 236f
Swedish goshawk traps, 80
sweep nets, 339–40, 339f, 340f
swift, Vaux's, capture methods for, 67
swift(s), noninvasive marking of, 243t
swiftlet, cave, sensory perception in, 467
swim-in bait traps, 79
swine flu, H1N1 strain of, 181
synthetic pyrethroids, 166
syringes, in chemical immobilization, 124–25
    accidental injection of humans in, 134–36
systematic sampling, 25, 26–28, 27f, 288, 297, 542–43
system boundaries, in modeling, 22

tables, data presentation in, 44–45
tactile sensory systems, 470

tag(s)
    genetic, 542
    marking of wildlife with, 248–51, 249–50t, 250b, 250f, 251f
    molecular, 543
    recovery of, in estimation of survival rates, 362–64
tail clipping, 255
tail feathers, in age and sex determination, 211, 211f
tail fin clipping, 255
TAMER, 112
tapes, marking of wildlife with, 241–42t
target species
    of detection dogs, 149
    of pesticides, 164
target-to-nearest-neighbor (TNN) method, 298
taste, 469–70
taste aversion, conditioned, 469
tattoos, marking of wildlife with, 247–48, 248f, 249t
taxonomic delineations, 531
taxonomy, 530–33
    evolutionary significant units in, 532–33
    genetics in, 530–33
    history of, 530
    hybridization in, 531–32
    at species and subspecies level, 530–31
2,3,7,8-TCDD, 162
teal, blue-winged, capture methods for, 69t
teal, green-winged
    age and sex determination for, 212f
    behavior of, 474
teeth. See dentition
telemetry. See radiotelemetry
teratogens, 158
tern(s)
    noninvasive marking of, 237t, 243t
    remote marking of, 233t
tern, common, remote marking of, 233t
tern, least
    capture methods for, 78, 79f
    invasive marking of, 251f
terrain, lidar data on, 453–54, 453f, 456
terrapin, diamondback
    capture methods for, 110
    noninvasive marking of, 243t, 244
terrestrial resource use. See resource use
terrestrial sampling, of invertebrates, 339–45
territorial mapping, 293
territorial species, home ranges of, 481, 492
territories, vs. home range, 481
testes, 513–16, 519
testosterone, 514, 516, 517t, 519, 519f
tests, power of, 32–33
tetrachlorodibenzo-p-dioxin (2,3,7,8-TCDD), 162
tetracyclines, marking of wildlife with, 245
thebaine, 122
theories, 10–12
    components of, 10, 11b

confronting with data, 31–37
    definition of, 10
    development of, 2, 3, 6, 10, 10f
    testing of, 2, 3, 6
therapeutic indices, definition of, 120
thermal infrared cameras, 313t, 316–17
thermometers, in chemical immobilization, 127–28
thermoregulation
    in capture of mammals, 84
    during chemical immobilization, 127–28, 133
    emergency medicine for problems with, 133, 134f
thiafentanil, 123, 136
third-order selection, 418, 491
threat intensity, GIS in evaluation of, 436
throw nets, capture of birds with, 65–66
ticks
    collection methods for, 343
    description of, 338
    sensory perception in, 466
tiger, Amur
    capture methods for, 89t
    scat of, detection by dogs, 149
tiger, natural markings of, 232t
tiletamine (zolazepam), 81, 122, 136, 137
time
    in modeling, 351 (See also continuous-time models; discrete-time models)
    in radiotelemetry studies, 483
time-area surveys, 297
time budgets, 474
time-constrained searches
    for amphibians, 104–5
    for reptiles, 107
timed dip-net collections, of amphibians, 105
time-of-detection method, 302–3
time origin, in survival analysis, 497
time-specific life tables, 369–70, 371
TimeSync, 450–51, 451f, 452f
time-to-event modeling. See survival analysis
tissues. See animal tissues; plant tissues
tit, blue, lidar in studies of, 454
tit, great, lidar in studies of, 454
tit, marsh, lidar in studies of, 454
T lymphocytes, 206
TNN method, 298
toad(s)
    capture methods for, 111
    invasive marking of, 246t, 247t, 249t, 252t, 253t
    noninvasive marking of, 243t
    reproduction of, 504, 517t
toe clipping, 253–54t, 255
toenail clipping, 254, 255f
tolazoline, 124
Tomahawk traps, 84, 85f, 86
tooth. See dentition
topography, lidar data on, 454, 455f, 456
tortoise(s)
    invasive marking of, 252t

tortoise(s) (*continued*)
  noninvasive marking of, 243t
  reproduction of, 517t, 519
tortoise, Bolson, capture methods for, 109
tortoise, desert
  abundance estimation of, 304
  capture methods for, 109
  dogs in studies of, 144t, 147
  invasive marking of, 247t
total mapping, 293
touch, 470
toxic, definition of, 206
toxicant(s), definition of, 206
toxicology
  definition of, 206
  in disease investigations, 195
toxicosis, definition of, 158
toxins, definition of, 206. *See also* contaminants
tracking lights, nocturnal, 240–41, 240t
tracking radar, 324t, 326
tracks, animal, in habitat selection studies, 417
traditional distance sampling, 296
traffic radar, 326
trailing devices, 239–40, 240t
training
  in chemical immobilization, 118
  in necropsy techniques, 196
  in vegetation sampling, 385–86
trajectory-based change, 447
trammel nets, 109
tranquilizers
  antagonists for, 124
  in chemical immobilization, 120, 122, 122t
  emergency treatment for human exposure to, 136
  during transportation, 132
tranquilizer trap devices (TTDs), 90, 113
transects, 30
  plots along, 27f, 30
  traps along, 86
translation, in theories, 11b
translocation programs, gene flow in, 536–37
transmissible spongiform encephalopathies (TSEs), 185, 206
transmission, definition of, 206
transmitters. *See also* radiotelemetry
  attachment of, 265–68, 269b
  effects on wildlife, 263, 269b
  PTT, 274–75
  VHF, 270–71, 280–81
  weight of, 268, 274, 280
transponders, marking of wildlife with, 247, 247t, 248b
transportation
  of samples
    in contaminant investigations, 174
    in disease investigations, 198, 198b
  of wildlife
    chemical immobilization in, 132
    stress caused by, 521

Transportation, U.S. Department of (DOT), on transport of diagnostic specimens, 198
transport hosts, definition of, 206
trap(s)
  Best Management Practices and standards for, 87–90, 91t, 92t, 96–104
  capture of amphibians with, 86, 105–6
  capture of birds with, 73–80
  capture of invertebrates with, 339–44
  capture of mammals with, 76t, 81, 82–90, 96–104
  capture of reptiles with, 107–11
  list of manufacturers of, 116–17
  signaling devices for, 93, 94t
  in vegetation sampling, 402, 403
trapping grids, in population estimation, 493, 494–95
trays, insect, 347, 347f
treatment levels, 20
tree(s)
  age of, 399
  dimensions of, 398–99, 398f
  fruits of, 401–2
  growth models for, 58
  growth rings of, 399
  sampling of
    techniques for, 388, 389, 392–93, 398–99
    time required for, 384t
  structural characteristics of, 399
triangulation, of radio signals, 277–78, 278f
triazines, 167
tricaine methane sulfonate, 136–37
trichothecenes, 170
trophic level, in contaminant impacts, 154
true survival, 358
TSEs, 185, 206
TTDs, 90, 113
t-test, 50–51
tuberosities, suspensory, 218, 224f
tuned loops, 271
turkey, Rio Grande wild, remote cameras in studies of, 312f, 316, 317
turkey, wild
  age and sex determination for, 209, 211f, 213, 215t
  capture methods for, 71, 71t, 72, 72t, 79t, 80
  handling methods for, 111
  invasive marking of, 249t
  noninvasive marking of, 241t
  reproduction of, 517t
turnstone, ruddy, capture methods for, 72t
turtle(s)
  capture methods for, 109, 109f, 110, 110f
  invasive marking of, 246t, 247t, 249t, 251, 252, 252t, 253t, 254, 255
  noninvasive marking of, 239, 240t, 243t
  reproduction of, 507, 512
turtle, black pond, capture methods for, 110
turtle, box, noninvasive marking of, 240t

turtle, green sea
  genetic diversity of, 535
  noninvasive marking of, 240t
  reproduction of, 508
turtle, Kemp's Ridley sea, taxonomic status of, 531
turtle, loggerhead sea
  molecular techniques in tracking of, 544
  reproduction of, 508
  sensory perception in, 471f
turtle, northern map, capture methods for, 110
turtle, painted
  birth of, 512, 512f
  capture methods for, 110, 110f
turtle, sea
  age and sex determination for, 229
  contaminant impacts on, 170
  invasive marking of, 247t, 249t, 252t
  reproduction of, 507–8, 509, 516, 517t
  sensory perception in, 468, 471
  ultrasonography of, 509
turtle, snapping
  contaminant impacts on, 161–62
  handling methods for, 113
turtle, spotted, noninvasive marking of, 241t
turtle, three-toed box, transmitters on, 267f
turtle eggs, noninvasive marking of, 244
twig-count method, 395, 419
type I errors, 33, 288, 289
type II errors, 33, 288, 289

UD. *See* utilization distribution
Ugglan traps, 86
ulcers, gastrointestinal, 521
ultimate hypotheses, 471–72
ultimate predictor variables, 416
ultralight aircraft, migratory behavior taught with, 464, 465f
ultrasonic detection, in capture of mammals, 82
ultrasonography, of mammal ovaries, 509, 509f
ultraviolet light, sensory perception of, 468
Umwelt, 466, 478
UNEP-GRID, 440
ungulates. *See also* specific species and types
  behavior of, 467, 469–70, 475
  chemical immobilization of
    drugs for, 137–38, 137t
    emergency medicine in, 133, 133f, 134
    transportation during, 132
  invasive marking of, 252t, 255
  noninvasive marking of, 234t, 235f, 236f, 242t
  sensory perception in, 467, 469–70
uniform distribution of vegetation, 386
unimodal data distribution, 49
United Nations Environmental Program's Global and Regional Integrated Data (UNEP-GRID), 440

Universal Transverse Mercator (UTM), 26
urban wildlife, chemical immobilization of, 131–32
urination posture, in sex determination, 208, 208f
urine
 in capture of mammals, 95
 DNA extraction from, 529
 hormone metabolites in, 520, 522, 524
 in reproduction, 505
 sensory perception of, 469–70
uterine analysis, 511–12
utilization distribution (UD)
 definition of, 482
 home range as aspect of, 482
 methods for estimation of, 482, 484–90, 485f
UTM, 26

vaccine(s)
 immunocontraceptive, 521
 rabies, 201–2, 203, 315
 remote cameras in studies of delivery of, 315
vacuum sampling, 340–41, 340f, 341f
vagina, 505, 505f, 506b
vaginal implant transmitters, 93, 93t
vaginal smear procedure, 506b
validation, model, 23, 37
variables
 challenge of controlling, 1
 independent (treatment), 19–20
 predictor, 415
 response (dependent), 19
variance
 in abundance estimation, 285, 286, 290–91
 analysis of (See analysis of variance)
 calculation of, 49, 289
 definition of, 285
vectors, definition of, 206
vegetation
 definition of, 381
 types of, 381
vegetation manager (VEMA) software, 446b
vegetation phenology, 405
vegetation sampling, 381–409
 applications of, 381, 406–9
 equipment for, 385, 385t
 GIS in, 405–6, 432
 goals of, 382–83
 initial steps in, 382–83
 Landsat imagery in, 404, 447–51
 lidar in, 406, 454–56
 preparations for, 383–86
 software programs for, 404, 409, 446b
 study site selection in, 382–83, 383f
 techniques of, 386–401
  for biomass, 394–96
  for cover, 390–93, 404
  for density, 387–90
  in food selection studies, 419–20

 for frequency of occurrence, 386–87, 386f, 387f
 for fruit, 401–4, 420
 for herbage height, 397–98
 multiple-scale, 404–6
 for plant use, 399–401
 for visual obstruction, 396–97
 time required for, 383, 384t
vegetation structure
 definition of, 381
 lidar data on, 454–56, 454f, 455f
 reconnaissance surveys of, 382
vehicle(s)
 in capture of birds, 81
 interactions between wildlife and, 313
 radar in, 326–27
VEMA software, 446b
ventral scale clipping, 254
verification, model, 23
vertebrate(s). See also specific species and types
 invertebrate interactions with, 336, 338, 344–45
 pest control chemicals aimed at, 168–69
vertically scanning radar, 323
very high frequency (VHF) radiotelemetry, 268–73
 antennas in, 270–72, 271f, 272f, 280–81
 applications for, 258
 frequency selection for, 268–70
 location error in, 264
 vs. other systems, 268, 269t
 receivers in, 272–73, 273f, 281
 transmitters in, 270–71, 280–81
veterinarians, in chemical immobilization, 119
VHF. See very high frequency
viability
 individual, 540
 population (See population viability)
vibrissae, 470
video cameras
 in behavioral studies, 476–77
 live streaming, 317
VIEs, 246
viper, pit, sensory perception in, 468
viper, rhinoceros, handling methods for, 112
vireo, black-capped, genetic diversity of, 536b, 536f
virology
 definition of, 206
 in disease investigations, 194
virulence, definition of, 206
virulent, definition of, 206
viruses
 cytopathic effects of, 194, 204
 in disease investigations, 194
visible implant elastomers (VIEs), 246
vision, 468
visual acuity, 468
visual attractants
 in capture of mammals, 96
 in capture of reptiles, 109

visual observations, of invertebrate–vertebrate interactions, 344–45
visual obstruction, in vegetation sampling, 396–97, 396f, 397f
vital rates. See birth rates; survival rates
vital signs monitoring, in chemical immobilization, 128, 131
vitellogenesis, 504
viviparous species, 503
VNTRs, 528
vocalization. See sound production
vole, bank
 capture methods for, 94
 genetic mutation in, 534
vole, meadow, survival rate estimation for, 361, 361t
vole, pine, reproduction of, 507
vole, prairie, handling methods for, 111
vole(s), invasive marking of, 246t, 247t
volume of intersection index, 490
vomeronasal organ, 469–70, 470f
vomiting, during chemical immobilization, 134
voucher specimens, 345
vulture(s)
 contaminant impacts on, 164
 invasive marking of, 249t
 noninvasive marking of, 237t, 238
 sensory perception in, 469
vulture, black, contaminant impacts on, 159
vulture, long-billed, contaminant impacts on, 164
vulture, Oriental white-backed, contaminant impacts on, 164
vulture, slender-billed, contaminant impacts on, 164
vulture, turkey
 contaminant impacts on, 159
 sensory perception in, 469

WAAS, 443–44
walk-in duck nest traps, 78
walk-in traps, capture of birds with, 76–77, 77f, 78, 79–80
wallaby, brush-tailed rock-, population estimation for, 543
wallaby, noninvasive marking of, 240t
wallaby, tammar, reproduction of, 514, 519
walrus, chemical immobilization of, 138–39
warbler, golden-cheeked, genetic diversity of, 536b, 536f
warbler(s), capture methods for, 67
warfarin, 168
wasps, description of, 337
waterfowl
 age and sex determination for, 211, 211f, 213f
 capture methods for, 66t, 69t, 72, 72t, 73t, 77t, 79t, 80
 contaminant impacts on, 159, 163, 169, 170
 handling methods for, 111

waterfowl (*continued*)
    marking of
        invasive, 245t, 249t
        noninvasive, 237t, 238, 241t
        remote, 233t
    nest parasitism, 465
    population dynamics of, 10, 10f
    population estimation for, 494
    radar in monitoring of, 333–34
water quality, invertebrates in monitoring of, 338
water samples, in contaminant investigations, 178
water squirting, in capture of reptiles, 109, 111
wavelengths, radar, 321–22
W chromosomes, 546
weasel
    capture methods for, 92t
    chemical immobilization of, 139t
weasel, long-tailed, capture methods for, 84t, 104
weasel, short-tailed, capture methods for, 84t, 104
weather conditions
    in chemical immobilization, 129
    in disease outbreaks, 185
    and dogs' scenting ability, 141–42
weather radar, 323, 325–26, 334–35
web punching, 255
web tagging, 251
weight, body. *See* body mass
weight–diameter equations, 395–96
weight–length equations, 395–96
welfare. *See* animal welfare
wetlands
    capture methods for birds in, 73
    sampling techniques for biomass in, 394
wet mass, in vegetation sampling, 394
whale(s)
    contaminant impacts on, 170
    effects of sound pollution on, 467
    invasive marking of, 245t, 250t
whale, beluga, invasive marking of, 249t
whale, blue, hybridization of, 532
whale, fin, hybridization of, 532
whale, gray, age at sexual maturity, 218
whale, humpback, molecular techniques in studies of, 543, 544
whale, killer, reproduction of, 517t, 520
whale, North Atlantic right, scat of, detection by dogs, 149
whale, right, taxonomic status of, 531
whip antennas, 270–71
whiskers, 470
white phosphorus, 163
Wide Area Augmentation System (WAAS), 443–44
wildlife conservation. *See* conservation
wildlife-crossing structures, 313

wildlife damage management, dogs in, 150–51
wildlife–habitat relationship models, 432–34
wildlife–habitat types map, 434
wildlife management
    adaptive (*See* adaptive management)
    animal behavior in, 462–66
    dogs in monitoring of, 146
    effectiveness of, need for research on, 39, 40
    GIS in, 432–38
    human behavior management in, 477–78
    laws of association in, 1
    resource selection in, 412
    rise of scientific rigor in, 1
Wildlife Materials rockets, 72
wildlife science
    emergence of rigor in, 1
    origins of term, 1
Wildlife Services, U.S.
    drugs in capture of birds by, 80
    drugs in capture of mammals by, 90
Wildlife Society, The, 6
wildlife viewing, 477–78, 477f
willet, invasive marking of, 249t
wind, and dogs' scenting ability, 142
wind farms, dogs in study of effects of, 147, 148f
wing(s), in age and sex determination, 207, 208f, 211
wing bands, 236–38
wing beat frequency, 330, 331f
wing cord length, in age and sex determination, 207, 208f
wing tags, 250–51, 251f
Winn-Star rockets, 72
wolf, gray
    age and sex determination for, 220t, 225, 228f, 229
    behavior of, 468, 474–75
    capture methods for, 82, 83, 89t, 90, 92t, 100
    chemical immobilization of, 138
    dogs descended from, 141
    GIS in population models of, 434
    GPS collars on, 444
    habitat selection by, 413
    hybridization of, 532
    scat of, detection by dogs, 148, 149
    sensory perception in, 468
wolf, maned, scat of, detection by dogs, 149, 520
wolf, red, taxonomic status of, 532
wolverine
    age and sex determination for, 221t
    capture methods for, 104
wombat, southern hairy-nosed, molecular techniques for tracking, 544
woodchuck
    age and sex determination for, 223t
    capture methods for, 83t
    reproduction of, 517t

woodcock, American
    age and sex determination for, 213, 216f, 217t
    capture methods for, 65, 66t, 69t, 76t
    invasive marking of, 249t
    noninvasive marking of, 237t
woodpecker(s)
    invasive marking of, 249t
    noninvasive marking of, 237t
woodpecker, acorn, capture methods for, 77t, 78
woodpecker, black-backed, transmitters on, 266f
woodpecker, pileated
    capture methods for, 67, 69t, 73, 75t, 77t, 80
    habitat selection by, 416
woodpecker, red-bellied, capture methods for, 77t
woodpecker, red-cockaded
    capture methods for, 77t
    habitat of, in forest planning, 434
    handling methods for, 111
    remote cameras in nests of, 312f, 316, 317
woodrat, bushy-tailed, capture methods for, 83t
woodrat, dusky-footed, capture methods for, 83t
woodrat, Key Largo, capture methods for, 83t
woodrat, white-throated, handling methods for, 111
woodrat(s), noninvasive marking of, 243t
wounds, in chemical immobilization, 134
wren, house, capture methods for, 77–78, 77t, 78f
writing, scientific, difficulty of, 38
W-U lure, 94

X-band, 321, 324t
X chromosomes, 545
xenobiotics, 155
xylazine, 122
    emergency treatment for human exposure to, 136
    FDA approved use of, 118–19

Yagi antennas, 271–72, 271f, 272f
Y chromosomes, 535, 545
yellow (color), in collection of invertebrates, 341, 342f
yohimbine, 119, 124
young
    altricial, 209
    parental care of, 512–13
    precocial, 209

Z chromosomes, 546
zolazepam. *See* tiletamine
zoonoses
    definition of, 206
    prevalence of, 181, 200